Table of Integrals, Series, and Products

Sixth Edition

Table of Integrals, Series, and Products

Sixth Edition

I.S. Gradshteyn and I.M. Ryzhik

Alan Jeffrey, Editor
University of Newcastle upon Tyne, England

Daniel Zwillinger, Associate Editor
Rensselaer Polytechnic Institute, USA

Translated from the Russian by Scripta Technica, Inc.

ACADEMIC PRESS

A Harcourt Science and Technology Company

San Diego   San Francisco   New York
Boston   London   Sydney   Tokyo

This book is printed on acid-free paper.

ACADEMIC PRESS
A Harcourt Science and Technology Company
525 B Street, Suite 1900, San Diego, CA 92101-4495, USA
http://www.academicpress.com

Academic Press
Harcourt Place, 32 Jamestown Road, London NW1 7BY, UK
http://www.academicpress.com

Library of Congress Catalog Card Number: 00-104373
International Standard Book Number: 0-12-294757-6

Printed in the United States of America
00 01 02 03 04 CO 9 8 7 6 5 4 3 2 1

Contents

Preface to the Sixth Edition

This completely reset sixth edition of Gradshteyn and Ryzhik is a corrected and expanded version of the previous edition. The book was completely reset in order to add more material and to enhance the visual appearance of the material. To preserve compatibility with the previous edition, the original numbering system for entries has been retained. New entries and sections have been inserted in a manner consistent with the original scheme. Whenever possible, new entries and corrections have been checked by means of symbolic computation.

The diverse ways in which corrections have been contributed have made it impossible to attribute them to reference sources that are accessible to users of this book. However, as in previous editions, our indebtedness to these contributors is shown in the form of an acknowledgment list on page xxiii. This list gives the names of those who have written to us directly sending corrections and suggestions for addenda, and added to it are the names of those who have published errata in *Mathematics of Computation*. Certain individuals must be singled out for special thanks due to their significant contributions: Professors H. van Haeringen and L. P. Kok of The Netherlands and Dr. K. S. Kölbig have contributed new material, corrections, and suggestions for new material.

Great care was taken to check the reset version of Gradshteyn and Ryzhik against the fifth edition, both by computer means and hand inspection, but it is inevitable that some transcription errors will remain. These will be rectified in subsequent reprinting of the book as and when they are identified. The authors would appreciate knowledge of any errors or deficiencies, their email addresses are listed below. If any errata are found, they will be posted on the web site `www.az-tec.com/gr/errata`.

As in the previous edition, a numerical superscript added to an entry number is used to indicate either a new addition to the book or a correction. When an entire section is new, an asterisk has been added only to the section heading. Continuing the previous convention, a superscript 10 has been added to entry numbers to indicate the most recent changes that have been made.

This latest version of Gradshteyn and Ryzhik also forms the source for a revised electronic version of Gradshteyn and Ryzhik.

Special thanks are extended to John Law and the Newcastle University Computing Service for their help in the preparation of this edition.

Alan Jeffrey
`Alan.Jeffrey@newcastle.ac.uk`

Daniel Zwillinger
`zwillinger@alum.mit.edu`

Acknowledgments

The publisher and editors would like to take this opportunity to express their gratitude to the following users of the *Table of Integrals, Series, and Products* who either directly or through errata published in *Mathematics of Computation* have generously contributed corrections and addenda to the original printing.

Dr. A. Abbas
Dr. P. B. Abraham
Dr. R. J. Adler
Dr. N. Agmon
Dr. M. Ahmad
Dr. S. A. Ahmad
Dr. R. K. Amiet
Dr. L. U. Ancarani
Dr. C. R. Appledorn
Dr. D. R. Appleton
Dr. C. L. Axness
Dr. E. Badralexe
Dr. S. B. Bagchi
Dr. L. J. Baker
Dr. R. Ball
Dr. M. P. Barnett
Dr. V. Bentley
Dr. M. van den Berg
Dr. N. F. Berk
Dr. C. A. Bertulani
Dr. J. Betancort-Rijo
Dr. P. Bickerstaff
Dr. G. R. Bigg
Dr. L. Blanchet
Dr. R. D. Blevins
Dr. L. M. Blumberg
Dr. R. Blumel
Dr. S. E. Bodner
Dr. M. Bonsager
Dr. S. Bosanac
Dr. A. Boström

Dr. J. E. Bowcock
Dr. T. H. Boyer
Dr. K. M. Briggs
Dr. D. J. Broadhurst
Dr. W. B. Brower
Dr. H. N. Browne
Dr. D. J. Buch
Dr. D. J. Bukman
Dr. F. M. Burrows
Dr. R. Caboz
Dr. T. Calloway
Dr. F. Calogero
Dr. D. Dal Cappello
Dr. J. A. Carlson Gallos
Dr. B. Carrascal
Dr. A. R. Carr
Dr. S. Carter
Dr. G. Cavalleri
Dr. A. Cecchini
Dr. B. Chan
Dr. M. A. Chaudhry
Dr. H. W. Chew
Dr. D. Chin
Dr. S. Ciccariello
Dr. N. S. Clarke
Dr. R. W. Cleary
Dr. A. Clement
Dr. P. Cochrane
Dr. D. K. Cohoon
Dr. L. Cole
Dr. J. R. D. Copley

Dr. D. Cox
Dr. J. W. Criss
Dr. A. E. Curzon
Dr. D. Dadyburjor
Dr. D. Dajaputra
Dr. C. Dal Cappello
Dr. P. Daly
Dr. S. Dasgupta
Dr. C. L. Davis
Dr. A. Degasperis
Dr. B. C. Denardo
Dr. R. W. Dent
Dr. E. Deutsch
Dr. D. deVries
Dr. P. Dita
Dr. P. J. de Doelder
Dr. G. Dôme
Dr. C. A. Ebner
Dr. M. van der Ende
Dr. G. Eng
Dr. E. S. Erck
Dr. G. A. Estévez
Dr. K. Evans
Dr. V. I. Fabricant
Dr. L. A. Falkovsky
Dr. K. Farahmand
Dr. R. J. Fateman
Dr. G. Fedele
Dr. A. R. Ferchmin
Dr. P. Ferrant
Dr. H. E. Fettis

Dr. W. B. Fichter
Dr. L. Ford
Dr. J. France
Dr. B. Frank
Dr. S. Frasier
Dr. A. J. Freeman
Dr. A. Frink
Dr. J. A. C. Gallas
Dr. J. A. Carlson Gallas
Dr. G. R. Gamertsfelder
Dr. T. Garavaglia
Dr. C. G. Gardner
Dr. D. Garfinkle
Dr. P. N. Garner
Dr. F. Gasser
Dr. E. Gath
Dr. P. Gatt
Dr. M. P. Gelfand
Dr. M. R. Geller
Dr. M. F. George
Dr. S. P. Gill
Dr. E. A. Gislason
Dr. M. I. Glasser
Dr. L. I. Goldfischer
Dr. I. J. Good
Dr. J. Good
Mr. L. Gorin
Dr. R. Govindaraj
Dr. R. Greenwell
Dr. K. D. Grimsley
Dr. V. Gudmundsson
Dr. K. Gunn
Dr. D. L. Gunter
Dr. H. van Haeringen
Dr. B. Hafizi
Dr. T. Hagfors
Dr. M. J. Haggerty
Dr. S. E. Hammel
Dr. E. Hansen
Dr. T. Harrett
Dr. D. O. Harris
Dr. A. Higuchi
Dr. R. E. Hise
Dr. N. Holte
Dr. R. W. Hopper
Dr. P. N. Houle
Dr. C. J. Howard

Dr. J. H. Hubbell
Dr. J. R. Hull
Dr. W. Humphries
Dr. Y. Iksbe
Mr. L. Iossif
Dr. S. A. Jackson
Dr. B. Jacobs
Dr. E. C. James
Dr. B. Jancovici
Dr. D. J. Jeffrey
Dr. H. J. Jensen
Dr. I. R. Johnson
Dr. I. Johnstone
Dr. Y. P. Joshi
Dr. Z. Kapal
Dr. M. Kaufman
Dr. B. Kay
Dr. S. Klama
Dr. L. Klingen
Dr. C. Knessl
Dr. M. J. Knight
Dr. D. Koks
Dr. L. P. Kok
Dr. K. S. Kölbig
Dr. D. D. Konowalow
Dr. Z. Kopal
Dr. I. Kostyukov
Dr. R. A. Krajcik
Dr. J. W. Krozel
Dr. E. D. Krupnikov
Dr. E. A. Kuraev
Dr. V. Labinac
Dr. A. D. J. Lambert
Dr. A. Lambert
Dr. A. Larraza
Dr. K. D. Lee
Dr. M. K. Lee
Dr. P. A. Lee
Dr. J. Legg
Dr. S. L. Levie
Dr. D. Levi
Dr. B. Linet
Dr. M. A. Lisa
Dr. I. M. Longman
Dr. D. Long
Dr. Y. L. Luke
Dr. W. Lukosz

Dr. T. Lundgren
Dr. E. A. Luraev
Dr. R. Lynch
Dr. R. Mallier
Dr. G. A. Mamon
Dr. I. Manning
Dr. J. Marmur
Dr. A. Martin
Dr. H. A. Mavromatis
Dr. M. Mazzoni
Dr. K. B. Ma
Dr. P. McCullagh
Dr. J. H. McDonnell
Dr. J. R. McGregor
Dr. K. McInturff
Dr. N. McKinney
Dr. W. N. Mei
Dr. A. Melino
Dr. J. P. Meunier
Dr. D. F. R. Mildner
Dr. D. L. Miller
Dr. P. C. D. Milly
Dr. S. P. Mitra
Dr. K. Miura
Dr. N. Mohankumar
Dr. M. Moll
Dr. D. Monowalow
Dr. J. Morice
Dr. W. Mueck
Dr. C. Muhlhausen
Dr. S. Mukherjee
Dr. R. R. Müller
Dr. A. Natarajan
Dr. C. T. Nguyen
Dr. A. C. Nicol
Dr. M. M. Nieto
Dr. P. Noerdlinger
Dr. A. N. Norris
Dr. K. H. Norwich
Dr. A. H. Nuttall
Dr. F. O'Brien
Dr. R. P. O'Keeffe
Dr. A. Ojo
Dr. P. Olsson
Dr. M. Ortner
Dr. S. Ostlund
Dr. J. Overduin

Dr. J. Pachner
Dr. F. J. Papp
Dr. B. Patterson
Dr. R. F. Pawula
Dr. D. W. Peaceman
Dr. D. Pelat
Dr. L. Peliti
Dr. Y. P. Pellegrini
Dr. G. J. Pert
Dr. J. B. Peterson
Dr. R. Petersson
Dr. R. Petersson
Dr. E. A. Power
Dr. E. Predazzi
Dr. F. Raynal
Dr. X. R. Resende
Dr. J. M. Riedler
Dr. E. Ringel
Dr. T. M. Roberts
Dr. N. I. Robinson
Dr. P. A. Robinson
Dr. D. M. Rosenblum
Dr. R. A. Rosthal
Dr. J. R. Roth
Dr. D. Ruddermann
Dr. C. T. Sachradja
Dr. J. Sadiku
Dr. A. Sadiq
Dr. A. Salim
Dr. J. H. Samson
Dr. J. A. Sanders
Dr. M. A. F. Sanjun

Dr. P. Sarquiz
Dr. A. Scherzinger
Dr. B. Schizer
Dr. J. Scholes
Dr. H. J. Schulz
Dr. O. Schärpf
Dr. G. J. Sears
Dr. B. Seshadri
Dr. A. Shapiro
Dr. S. Shlomo
Dr. D. Siegel
Dr. C. Smith
Dr. G. C. C. Smith
Dr. S. Smith
Dr. G. Solt
Dr. J. C. Straton
Dr. N. F. Svaiter
Dr. V. Svaiter
Dr. R. Szmytkowski
Dr. A. Sørenssen
Dr. G. Tanaka
Dr. C. Tanguy
Dr. G. K. Tannahill
Dr. B. T. Tan
Dr. C. Tavard
Dr. D. Temperley
Dr. A. J. Tervoort
Dr. S. T. Thynell
Dr. D. C. Torney
Dr. R. Tough
Dr. B. F. Treadway
Dr. Ming Tsai

Dr. J. J. Tyson
Dr. S. Uehara
Dr. M. Vadacchino
Dr. O. T. Valls
Dr. D. Vandeth
Dr. D. Veitch
Dr. J. M. M. J. Vogels
Dr. K. Vogel
Dr. D. deVries
Dr. S. Wanzura
Dr. J. Ward
Dr. S. I. Warshaw
Dr. R. Weber
Dr. Wei Qian
Dr. D. H. Werner
Dr. E. Wetzel
Dr. D. Wilton
Dr. C. Wiuf
Dr. K. T. Wong
Mr. J. N. Wright
Dr. D. Wu
Dr. J. J. Yang
Dr. Z. J. Yang
Dr. H.-C. Yang
Dr. S. P. Yukon
Dr. J. J. Wang
Dr. S. P. Yukin
Dr. Y. C. Zhang
Dr. B. Zhang
Dr. Y. Zhao

The order of presentation of the formulas

The question of the most expedient order in which to give the formulas, in particular, in what division to include particular formulas such as the definite integrals, turned out to be quite complicated. The thought naturally occurs to set up an order analogous to that of a dictionary. However, it is almost impossible to create such a system for the formulas of integral calculus. Indeed, in an arbitrary formula of the form

$$\int_a^b f(x)\,dx = A$$

one may make a large number of substitutions of the form $x = \varphi(t)$ and thus obtain a number of "synonyms" of the given formula. We must point out that the table of definite integrals by Bierens de Haan and the earlier editions of the present reference both sin in the plethora of such "synonyms" and formulas of complicated form. In the present edition, we have tried to keep only the simplest of the "synonym" formulas. Basically, we judged the simplicity of a formula from the standpoint of the simplicity of the arguments of the "outer" functions that appear in the integrand. Where possible, we have replaced a complicated formula with a simpler one. Sometimes, several complicated formulas were thereby reduced to a single simpler one. We then kept only the simplest formula. As a result of such substitutions, we sometimes obtained an integral that could be evaluated by use of the formulas of chapter two and the Newton–Leibniz formula, or to an integral of the form

$$\int_{-a}^a f(x)\,dx,$$

where $f(x)$ is an odd function. In such cases the complicated integrals have been omitted.

Let us give an example using the expression

$$\int_0^{\pi/4} \frac{(\cot x - 1)^{p-1}}{\sin^2 x}\,\ln\tan x\,dx = -\frac{\pi}{p}\,\operatorname{cosec} p\pi. \tag{1}$$

By making the natural substitution $u = \cot x - 1$, we obtain

$$\int_0^\infty u^{p-1}\ln(1+u)\,du = \frac{\pi}{p}\,\operatorname{cosec} p\pi. \tag{2}$$

Integrals similar to formula (1) are omitted in this new edition. Instead, we have formula (2).

As a second example, let us take

$$I = \int_0^{\pi/2} \ln\left(\tan^p x + \cot^p x\right) \ln \tan x \, dx = 0.$$

The substitution $u = \tan x$ yields

$$I = \int_0^\infty \frac{\ln\left(u^p + u^{-p}\right) \ln u}{1 + u^2} \, du.$$

If we now set $v = \ln u$, we obtain

$$I = \int_{-\infty}^\infty \frac{v e^v}{1 + e^{2v}} \ln\left(e^{pv} + e^{-pv}\right) dv = \int_{-\infty}^\infty v \, \frac{\ln 2 \cosh pv}{2 \cosh v} \, dv.$$

The integrand is odd and, consequently, the integral is equal to 0.

Thus, before looking for an integral in the tables, the user should simplify as much as possible the arguments (the "inner" functions) of the functions in the integrand.

The functions are ordered as follows: First we have the elementary functions:

1.   The function $f(x) = x$.
2.   The exponential function.
3.   The hyperbolic functions.
4.   The trigonometric functions.
5.   The logarithmic function.
6.   The inverse hyperbolic functions. (These are replaced with the corresponding logarithms in the formulas containing definite integrals.)
7.   The inverse trigonometric functions.

Then follow the special functions:

8.   Elliptic integrals.
9.   Elliptic functions.
10.   The logarithm integral, the exponential integral, the sine integral, and the cosine integral functions.
11.   Probability integrals and Fresnel's integrals.
12.   The gamma function and related functions.
13.   Bessel functions.
14.   Mathieu functions.
15.   Legendre functions.
16.   Orthogonal polynomials.
17.   Hypergeometric functions.
18.   Degenerate hypergeometric functions.
19.   Parabolic cylinder functions.
20.   Meijer's and MacRobert's functions.
21.   Riemann's zeta function.

The integrals are arranged in order of outer function according to the above scheme: the farther down in the list a function occurs, (i.e., the more complex it is) the later will the corresponding formula appear in the tables. Suppose that several expressions have the same outer function. For example, consider $\sin e^x$, $\sin x$, $\sin \ln x$. Here, the outer function is the sine function in all three cases. Such expressions are then arranged in order of the inner function. In the present work, these functions are therefore arranged in the following order: $\sin x$, $\sin e^x$, $\sin \ln x$.

Our list does not include polynomials, rational functions, powers, or other algebraic functions. An algebraic function that is included in tables of definite integrals can usually be reduced to a finite combination of roots of rational power. Therefore, for classifying our formulas, we can conditionally treat a power function as a generalization of an algebraic and, consequently, of a rational function.* We shall distinguish between all these functions and those listed above and we shall treat them as operators. Thus, in the expression $\sin^2 e^x$, we shall think of the squaring operator as applied to the outer function, namely, the sine. In the expression $\frac{\sin x + \cos x}{\sin x - \cos x}$, we shall think of the rational operator as applied to the trigonometric functions sine and cosine. We shall arrange the operators according to the following order:

1.   Polynomials (listed in order of their degree).
2.   Rational operators.
3.   Algebraic operators (expressions of the form $A^{p/q}$, where q and p are rational, and $q > 0$; these are listed according to the size of q).
4.   Power operators.

Expressions with the same outer and inner functions are arranged in the order of complexity of the operators. For example, the following functions (whose outer functions are all trigonometric, and whose inner functions are all $f(x) = x$) are arranged in the order shown:

$$\sin x, \quad \sin x \cos x, \quad \frac{1}{\sin x} = \operatorname{cosec} x, \quad \frac{\sin x}{\cos x} = \tan x, \quad \frac{\sin x + \cos x}{\sin x - \cos x}, \quad \sin^m x, \quad \sin^m x \cos x.$$

Furthermore, if two outer functions $\varphi_1(x)$ and $\varphi_2(x)$, where $\varphi_1(x)$ is more complex than $\varphi_2(x)$, appear in an integrand and if any of the operations mentioned are performed on them, the corresponding integral will appear (in the order determined by the position of $\varphi_2(x)$ in the list) after all integrals containing only the function $\varphi_1(x)$. Thus, following the trigonometric functions are the trigonometric and power functions (that is, $\varphi_2(x) = x$). Then come

- combinations of trigonometric and exponential functions,
- combinations of trigonometric functions, exponential functions, and powers, etc.,
- combinations of trigonometric and hyperbolic functions, etc.

Integrals containing two functions $\varphi_1(x)$ and $\varphi_2(x)$ are located in the division and order corresponding to the more complicated function of the two. However, if the positions of several integrals coincide because they contain the same complicated function, these integrals are put in the position defined by the complexity of the second function.

To these rules of a general nature, we need to add certain particular considerations that will be easily understood from the tables. For example, according to the above remarks, the function $e^{\frac{1}{x}}$ comes after e^x as regards complexity, but $\ln x$ and $\ln \frac{1}{x}$ are equally complex since $\ln \frac{1}{x} = -\ln x$. In the section on "powers and algebraic functions", polynomials, rational functions, and powers of powers are formed from power functions of the form $(a + bx)^n$ and $(\alpha + \beta x)^\nu$.

---

*For any natural number n, the involution $(a + bx)^n$ of the binomial $a + bx$ is a polynomial. If n is a negative integer, $(a + bx)^n$ is a rational function. If n is irrational, the function $(a + bx)^n$ is not even an algebraic function.

Use of the tables*

For the effective use of the tables contained in this book it is necessary that the user should first become familiar with the classification system for integrals devised by the authors Ryzhik and Gradshteyn. This classification is described in detail in the section entitled *The Order of Presentation of the Formulas* (see page xxvii) and essentially involves the separation of the integrand into *inner* and *outer* functions. The principal function involved in the integrand is called the *outer* function and its argument, which is itself usually another function, is called the *inner* function. Thus, if the integrand comprised the expression $\ln \sin x$, the *outer* function would be the logarithmic function while its argument, the *inner* function, would be the trigonometric function $\sin x$. The desired integral would then be found in the section dealing with logarithmic functions, its position within that section being determined by the position of the *inner* function (here a trigonometric function) in Ryzhik and Gradshteyn's list of functional forms.

It is inevitable that some duplication of symbols will occur within such a large collection of integrals and this happens most frequently in the first part of the book dealing with algebraic and trigonometric integrands. The symbols most frequently involved are α, β, γ, δ, t, u, z, z_k, and Δ. The expressions associated with these symbols are used consistently within each section and are defined at the start of each new section in which they occur. Consequently, reference should be made to the beginning of the section being used in order to verify the meaning of the substitutions involved.

Integrals of algebraic functions are expressed as combinations of roots with rational power indices, and definite integrals of such functions are frequently expressed in terms of the Legendre elliptic integrals $F(\phi, k)$, $E(\phi, k)$ and $\Pi(\phi, n, k)$, respectively, of the first, second and third kinds.

The four inverse hyperbolic functions $\operatorname{arcsinh} z$, $\operatorname{arccosh} z$, $\operatorname{arctanh} z$ and $\operatorname{arccoth} z$ are introduced through the definitions

$$\arcsin z = \frac{1}{i} \operatorname{arcsinh}(iz)$$

$$\arccos z = \frac{1}{i} \operatorname{arccosh}(z)$$

$$\arctan z = \frac{1}{i} \operatorname{arctanh}(iz)$$

$$\operatorname{arccot} z = i \operatorname{arccoth}(iz)$$

---

*Prepared by Alan Jeffrey for the English language edition.

or

$$\text{arcsinh } z = \frac{1}{i} \arcsin(iz)$$

$$\text{arccosh } z = i \arccos z$$

$$\text{arctanh } z = \frac{1}{i} \arctan(iz)$$

$$\text{arccoth } z = \frac{1}{i} \text{arccot}(-iz)$$

The numerical constants C and G which often appear in the definite integrals denote Euler's constant and Catalan's constant, respectively. Euler's constant C is defined by the limit

$$C = \lim_{s \to \infty} \left(\sum_{m=1}^{s} \frac{1}{m} - \ln s \right) = 0.577215\ldots.$$

On occasions other writers denote Euler's constant by the symbol γ, but this is also often used instead to denote the constant

$$\gamma = e^{C} = 1.781072\ldots.$$

Catalan's constant G is related to the complete elliptic integral

$$\mathbf{K} \equiv \mathbf{K}(k) \equiv \int_0^{\pi/2} \frac{da}{\sqrt{1 - k^2 \sin^2 a}}$$

by the expression

$$G = \frac{1}{2} \int_0^1 \mathbf{K}\, dk = \sum_{m=0}^{\infty} \frac{(-1)^m}{(2m+1)^2} = 0.915965\ldots.$$

Since the notations and definitions for higher transcendental functions that are used by different authors are by no means uniform, it is advisable to check the definitions of the functions that occur in these tables. This can be done by identifying the required function by symbol and name in the *Index of Special Functions and Notation* on page xxxix, and by then referring to the defining formula or section number listed there. We now present a brief discussion of some of the most commonly used alternative notations and definitions for higher transcendental functions.

Bernoulli and Euler Polynomials and Numbers

Extensive use is made throughout the book of the Bernoulli and Euler numbers B_n and E_n that are defined in terms of the Bernoulli and Euler polynomials of order n, $B_n(x)$ and $E_n(x)$, respectively. These polynomials are defined by the generating functions

$$\frac{te^{xt}}{e^t - 1} = \sum_{n=0}^{\infty} B_n(x) \frac{t^n}{n!} \qquad \text{for } |t| < 2\pi$$

and

$$\frac{2e^{xt}}{e^t + 1} = \sum_{n=0}^{\infty} E_n(x) \frac{t^n}{n!} \qquad \text{for } |t| < \pi.$$

The Bernoulli numbers are always denoted by B_n and are defined by the relation

$$B_n = B_n(0) \qquad \text{for } n = 0, 1, \ldots,$$

when

$$B_0 = 1, \quad B_1 = -\frac{1}{2}, \quad B_2 = \frac{1}{6}, \quad B_4 = -\frac{1}{30}, \ldots.$$

The Euler numbers E_n are defined by setting

$$E_n = 2^n E_n \left(\frac{1}{2}\right) \qquad \text{for } n = 0, 1, \ldots.$$

The E_n are all integral and $E_0 = 1$, $E_2 = -1$, $E_4 = 5$, $E_6 = -61$, $\ldots$.

An alternative definition of Bernoulli numbers, which we shall denote by the symbol B_n^*, uses the same generating function but identifies the B_n^* differently in the following manner:

$$\frac{t}{e^t - 1} = 1 - \frac{1}{2}t + B_1^* \frac{t^2}{2!} - B_2^* \frac{t^4}{4!} + \ldots.$$

This definition then gives rise to the alternative set of Bernoulli numbers

$$B_1^* = 1/6, \quad B_2^* = 1/30, \quad B_3^* = 1/42, \quad B_4^* = 1/30, \quad B_5^* = 5/66,$$
$$B_6^* = 691/2730, \quad B_7^* = 7/6, \quad B_8^* = 3617/510, \quad \ldots.$$

These differences in notation must also be taken into account when using the following relationships that exist between the Bernoulli and Euler polynomials:

$$B_n(x) = \frac{1}{2^n} \sum_{k=0}^{n} \binom{n}{k} B_{n-k} E_k(2x) \qquad n = 0, 1, \ldots$$

$$E_{n-1}(x) = \frac{2^n}{n} \left\{ B_n \left(\frac{x+1}{2}\right) - B_n \left(\frac{x}{2}\right) \right\}$$

or

$$E_{n-1}(x) = \frac{2}{n} \left\{ B_n(x) - 2^n B_n \left(\frac{x}{2}\right) \right\} \qquad n = 1, 2, \ldots$$

and

$$E_{n-2}(x) = 2 \binom{n}{2}^{-1} \sum_{k=0}^{n-2} \binom{n}{k} (2^{n-k} - 1) B_{n-k} B_n(x) \qquad n = 2, 3, \ldots$$

There are also alternative definitions of the Euler polynomial of order n, and it should be noted that some authors, using a modification of the third expression above, call

$$\left(\frac{2}{n+1}\right) \left\{ B_n(x) - 2^n B_n \left(\frac{x}{2}\right) \right\}$$

the Euler polynomial of order n.

Elliptic Functions and Elliptic Integrals

The following notations are often used in connection with the inverse elliptic functions $\operatorname{sn} u$, $\operatorname{cn} u$, and $\operatorname{dn} u$:

$$\operatorname{ns} u = \frac{1}{\operatorname{sn} u} \qquad\qquad \operatorname{nc} u = \frac{1}{\operatorname{cn} u} \qquad\qquad \operatorname{nd} u = \frac{1}{\operatorname{dn} u}$$

$$\operatorname{sc} u = \frac{\operatorname{sn} u}{\operatorname{cn} u} \qquad\qquad \operatorname{cs} u = \frac{\operatorname{cn} u}{\operatorname{sn} u} \qquad\qquad \operatorname{ds} u = \frac{\operatorname{dn} u}{\operatorname{sn} u}$$

$$\operatorname{sd} u = \frac{\operatorname{sn} u}{\operatorname{dn} u} \qquad\qquad \operatorname{cd} u = \frac{\operatorname{cn} u}{\operatorname{dn} u} \qquad\qquad \operatorname{dc} u = \frac{\operatorname{dn} u}{\operatorname{cn} u}$$

The elliptic integral of the third kind is defined by Ryzhik and Gradshteyn to be

$$\Pi\left(\varphi, n^2, k\right) = \int_0^{\varphi} \frac{da}{\left(1 - n^2 \sin^2 a\right) \sqrt{1 - k^2 \sin^2 a}}$$

$$= \int_0^{\sin \varphi} \frac{dx}{\left(1 - n^2 x^2\right) \sqrt{\left(1 - x^2\right)\left(1 - k^2 x^2\right)}} \qquad \left(-\infty < n^2 < \infty\right)$$

The Jacobi Zeta Function and Theta Functions

The Jacobi zeta function $zn(u, k)$, frequently written $Z(u)$, is defined by the relation

$$zn(u, k) = Z(u) = \int_0^u \left\{ dn^2 v - \frac{E}{K} \right\} dv = E(u) - \frac{E}{K} u.$$

This is related to the theta functions by the relationship

$$zn(u, k) = \frac{\partial}{\partial u} \ln \Theta(u)$$

giving

(i). $\quad zn(u, k) = \dfrac{\pi}{2K} \dfrac{\vartheta_1'\left(\dfrac{\pi u}{2K}\right)}{\vartheta_1\left(\dfrac{\pi u}{2K}\right)} - \dfrac{cn\, u\, dn\, u}{sn\, u}$

(ii). $\quad zn(u, k) = \dfrac{\pi}{2K} \dfrac{\vartheta_2'\left(\dfrac{\pi u}{2K}\right)}{\vartheta_2\left(\dfrac{\pi u}{2K}\right)} - \dfrac{dn\, u\, sn\, u}{cn\, u}$

(iii). $\quad zn(u, k) = \dfrac{\pi}{2K} \dfrac{\vartheta_3'\left(\dfrac{\pi u}{2K}\right)}{\vartheta_3\left(\dfrac{\pi u}{2K}\right)} - k^2 \dfrac{sn\, u\, cn\, u}{dn\, u}$

(iv). $\quad zn(u, k) = \dfrac{\pi}{2K} \dfrac{\vartheta_4'\left(\dfrac{\pi u}{2K}\right)}{\vartheta_4\left(\dfrac{\pi u}{2K}\right)}$

Many different notations for the theta function are in current use. The most common variants are the replacement of the argument u by the argument u/π and, occasionally, a permutation of the identification of the functions ϑ_1 to ϑ_4 with the function ϑ_4 replaced by ϑ.

The Factorial (Gamma) Function

In older reference texts the gamma function $\Gamma(z)$, defined by the Euler integral

$$\Gamma(z) = \int_0^\infty t^{z-1} e^{-t}\, dt,$$

is sometimes expressed in the alternative notation

$$\Gamma(1 + z) = z! = \Pi(z).$$

On occasions the related derivative of the logarithmic factorial function $\Psi(z)$ is used where

$$\frac{d\,(\ln z!)}{dz} = \frac{(z!)'}{z!} = \Psi(z).$$

This function satisfies the recurrence relation

$$\Psi(z) = \Psi(z - 1) + \frac{1}{z - 1}$$

and is defined by the series

$$\Psi(z) = -C + \sum_{n=0}^{\infty} \left(\frac{1}{n + 1} - \frac{1}{z + n} \right).$$

The derivative $\Psi'(z)$ satisfies the recurrence relation

$$-\Psi'(z-1) = -\Psi'(z) + \frac{1}{z^2}$$

and is defined by the series

$$\Psi'(z) = \sum_{n=0}^{\infty} \frac{1}{(z+n)^2}.$$

Exponential and Related Integrals

The exponential integrals $E_n(z)$ have been defined by Schloemilch using the integral

$$E_n(z) = \int_1^{\infty} e^{-zt} t^{-n} \, dt \qquad (n = 0, 1, \ldots, \quad \text{Re } z > 0)$$

They should not be confused with the Euler polynomials already mentioned. The function $E_1(z)$ is related to the exponential integral $\text{Ei}(z)$ through the expressions

$$E_1(z) = -\text{Ei}(-z) = \int_z^{\infty} e^{-t} t^{-1} \, dt$$

and

$$\text{li}(z) = \int_0^z \frac{dt}{\ln t} = \text{Ei}(\ln z) \qquad (z > 1).$$

The functions $E_n(z)$ satisfy the recurrence relations

$$E_n(z) = \frac{1}{n-1} \left\{ e^{-z} - z \, E_{n-1}(z) \right\} \qquad (n > 1)$$

and

$$E'_n(z) = -E_{n-1}(z)$$

with

$$E_0(z) = e^{-z}/z.$$

The function $E_n(z)$ has the asymptotic expansion

$$E_n(z) \sim \frac{e^{-z}}{z} \left\{ 1 - \frac{n}{z} + \frac{n(n+1)}{z^2} - \frac{n(n+1)(n+2)}{z^3} + \cdots \right\} \qquad \left(|\arg z| < \frac{3\pi}{2} \right)$$

while for large n,

$$E_n(x) = \frac{e^{-x}}{x+n} \left\{ 1 + \frac{n}{(x+n)^2} + \frac{n(n-2x)}{(x+n)^4} + \frac{n \left(6x^2 - 8nx + n^2 \right)}{(x+n)^6} + R(n, x) \right\},$$

where

$$-0.36 n^{-4} \le R(n, x) \le \left(1 + \frac{1}{x+n-1} \right) n^{-4} \qquad (x > 0)$$

The sine and cosine integrals $\text{si}(x)$ and $\text{ci}(x)$ are related to the functions $\text{Si}(x)$ and $\text{Ci}(x)$ by the integrals

$$\text{Si}(x) = \int_0^x \frac{\sin t}{t} \, dt = \text{si}(x) + \frac{\pi}{2}$$

and

$$\text{Ci}(x) = \boldsymbol{C} + \ln x + \int_0^x \frac{(\cos t - 1)}{t} \, dt.$$

The hyperbolic sine and cosine integrals $\text{shi}(x)$ and $\text{chi}(x)$ are defined by the relations

$$\text{shi}(x) = \int_0^x \frac{\sinh t}{t} \, dt$$

and

$$\operatorname{chi}(x) = \boldsymbol{C} + \ln x + \int_0^x \frac{(\cosh t - 1)}{t} \, dt.$$

Some authors write

$$\operatorname{Cin}(x) = \int_0^x \frac{(1 - \cos t)}{t} \, dt$$

so that

$$\operatorname{Cin}(x) = -\operatorname{Ci}(x) + \ln x + \boldsymbol{C}.$$

The error function $\operatorname{erf}(x)$ is defined by the relation

$$\operatorname{erf}(x) = \Phi(x) = \frac{2}{\sqrt{\pi}} \int_0^x e^{-t^2} \, dt$$

and the complementary error function $\operatorname{erfc}(x)$ is related to the error function $\operatorname{erfc}(x)$ and to $\Phi(x)$ by the expression

$$\operatorname{erfc}(x) = 1 - \operatorname{erf}(x).$$

The Fresnel integrals $S(x)$ and $C(x)$ are defined by Ryzhik and Gradshteyn as

$$S(x) = \frac{2}{\sqrt{2\pi}} \int_0^x \sin t^2 \, dt$$

and

$$C(x) = \frac{2}{\sqrt{2\pi}} \int_0^x \cos t^2 \, dt.$$

Other definitions that are in use are

$$S_1(x) = \int_0^x \sin \frac{\pi t^2}{2} \, dt, \qquad C_1(x) = \int_0^x \cos \frac{\pi t^2}{2} \, dt$$

and

$$S_2(x) = \frac{1}{\sqrt{2\pi}} \int_0^x \frac{\sin t}{\sqrt{t}} \, dt, \qquad C_2(x) = \frac{1}{\sqrt{2\pi}} \int_0^x \frac{\cos t}{\sqrt{t}} \, dt$$

These are related by the expressions

$$S(x) = S_1 \left(x\sqrt{\frac{2}{\pi}} \right) = S_2 \left(x^2 \right)$$

and

$$C(x) = C_1 \left(x\sqrt{\frac{2}{\pi}} \right) = C_2 \left(x^2 \right)$$

Hermite and Chebyshev Orthogonal Polynomials

The Hermite polynomials $H_n(x)$ are related to the Hermite polynomials $He_n(x)$ by the relations

$$He_n(x) = 2^{-n/2} H_n \left(\frac{x}{\sqrt{2}} \right)$$

and

$$H_n(x) = 2^{n/2} He_n \left(x\sqrt{2} \right).$$

These functions satisfy the differential equations

$$\frac{d^2 H_n}{dx^2} - 2x \frac{d H_n}{dx} + 2n \, H_n = 0$$

and

$$\frac{d^2 He_n}{dx^2} - x \frac{d He_n}{dx} + n \, He_n = 0.$$

They obey the recurrence relations

$$H_{n+1} = 2x\,H_n - 2n\,H_{n-1}$$

and

$$He_{n+1} = x\,He_n - n\,He_{n-1}$$

The first six orthogonal polynomials He_n are

$$He_0 = 1, \quad He_1 = x, \quad He_2 = x^2 - 1, \quad He_3 = x^3 - 3x, \quad He_4 = x^4 - 6x^2 + 3, \quad He_5 = x^5 - 10x^3 + 15x.$$

Sometimes the Chebyshev polynomial $U_n(x)$ of the second kind is defined as a solution of the equation

$$\left(1 - x^2\right)\frac{d^2 y}{dx^2} - 3x\frac{dy}{dx} + n(n+2)y = 0.$$

Bessel Functions

A variety of different notations for Bessel functions are in use. Some common ones involves the replacement of $Y_n(z)$ by $Y_n(z)$ and the introduction of the symbol

$$\Lambda_n(z) = \left(\frac{1}{2}z\right)^{-n} \Gamma(n+1)\,J_n(z).$$

In the book by Gray, Mathews and MacRobert the symbol $Y_n(z)$ is used to denote $\frac{1}{2}\pi\,Y_n(z) + (\ln 2 - \mathbf{C})\,J_n(z)$ while Neumann uses the symbol $Y^{(n)}(z)$ for the identical quantity.

The Hankel functions $H_\nu^{(1)}(z)$ and $H_\nu^{(2)}(z)$ are sometimes denoted by $Hs_\nu(z)$ and $Hi_\nu(z)$ and some authors write $G_\nu(z) = \left(\frac{1}{2}\right)\pi i\,H_\nu^{(1)}(z)$.

The Neumann polynomial $O_n(t)$ is a polynomial of degree $n+1$ in $1/t$, with $O_0(t) = 1/t$. The polynomials $O_n(t)$ are defined by the generating function

$$\frac{1}{t-z} = J_0(z)\,O_0(t) + 2\sum_{k=1}^{\infty} J_k(z)\,O_k(t),$$

giving

$$O_n(t) = \frac{1}{4}\sum_{k=0}^{[n/2]} \frac{n(n-k-1)!}{k!}\left(\frac{2}{t}\right)^{n-2k+1} \qquad \text{for } n = 1, 2, \ldots,$$

where $\left[\frac{1}{2}n\right]$ signifies the integral part of $\frac{1}{2}n$. The following relationship holds between three successive polynomials:

$$(n-1)\,O_{n+1}(t) + (n+1)\,O_{n-1}(t) - \frac{2\left(n^2 - 1\right)}{t}\,O_n(t) = \frac{2n}{t}\sin^2\frac{n\pi}{2}.$$

The Airy functions $\mathrm{Ai}(z)$ and $\mathrm{Bi}(z)$ are independent solutions of the equation

$$\frac{d^2 u}{dz^2} - zu = 0.$$

The solutions can be represented in terms of Bessel functions by the expressions

$$\mathrm{Ai}(z) = \frac{1}{3}\sqrt{z}\left\{I_{-1/3}\left(\frac{2}{3}z^{3/2}\right) - I_{1/3}\left(\frac{2}{3}z^{3/2}\right)\right\} = \frac{1}{\pi}\sqrt{\frac{z}{3}}\,K_{1/3}\left(\frac{2}{3}z^{3/2}\right)$$

$$\mathrm{Ai}(-z) = \frac{1}{3}\sqrt{z}\left\{J_{1/3}\left(\frac{2}{3}z^{3/2}\right) + J_{-1/3}\left(\frac{2}{3}z^{3/2}\right)\right\}$$

and by

$$\mathrm{Bi}(z) = \sqrt{\frac{z}{3}} \left\{ I_{-1/3}\left(\frac{2}{3}z^{3/2}\right) + I_{1/3}\left(\frac{2}{3}z^{3/2}\right) \right\},$$

$$\mathrm{Bi}(-z) = \sqrt{\frac{z}{3}} \left\{ J_{-1/3}\left(\frac{2}{3}z^{3/2}\right) - J_{1/3}\left(\frac{2}{3}z^{3/2}\right) \right\}.$$

Parabolic Cylinder Functions and Whittaker Functions

The differential equation

$$\frac{d^2 y}{dz^2} + \left(az^2 + bz + c \right) y = 0$$

has associated with it the two equations

$$\frac{d^2 y}{dz^2} + \left(\frac{1}{4}z^2 + a \right) y = 0 \quad \text{and} \quad \frac{d^2 y}{dz^2} - \left(\frac{1}{4}z^2 + a \right) y = 0$$

the solutions of which are parabolic cylinder functions. The first equation can be derived from the second by replacing z by $ze^{i\pi/4}$ and a by $-ia$.

The solutions of the equation

$$\frac{d^2 y}{dz^2} - \left(\frac{1}{4}z^2 + a \right) y = 0$$

are sometimes written $U(a, z)$ and $V(a, z)$. These solutions are related to Whittaker's function $D_p(z)$ by the expressions

$$U(a, z) = D_{-a-\frac{1}{2}}(z)$$

and

$$V(a, z) = \frac{1}{\pi} \Gamma\left(\frac{1}{2} + a \right) \left\{ D_{-a-\frac{1}{2}}(-z) + (\sin \pi a) \, D_{-a-\frac{1}{2}}(z) \right\}.$$

Mathieu Functions

There are several accepted notations for Mathieu functions and for their associated parameters. The defining equation used by Ryzhik and Gradshteyn is

$$\frac{d^2 y}{dz^2} + \left(a - 2k^2 \cos 2z \right) y = 0 \quad \text{with } k^2 = q.$$

Different notations involve the replacement of a and q in this equation by h and θ, λ and h^2 and b and $c = 2\sqrt{q}$, respectively. The periodic solutions $se_n(z, q)$ and $ce_n(z, q)$ and the modified periodic solutions $Se_n(z, q)$ and $Ce_n(z, q)$ are suitably altered and, sometimes, re-normalized. A description of these relationships together with the normalizing factors is contained in: Tables relating to Mathieu functions. National Bureau of Standards, Columbia University Press, New York, 1951.

Index of Special Functions

continued on next page

continued from previous page	

Notation	Name of the function and the number of the formula containing its definition	
$\mathrm{bei}(z), \quad \mathrm{ber}(z)$	Thomson functions	8.56
$\boldsymbol{C}$	Euler constant	9.73, 8.367
$C(x)$	Fresnel cosine integral	8.25
$C_\nu(a)$	Young functions	3.76
$C_n^\lambda(t)$	Gegenbauer polynomials	8.93
$C_n^\lambda(x)$	Gegenbauer functions	8.932 1
$\mathrm{ce}_{2n}(z,q), \quad \mathrm{ce}_{2n+1}(z,q)$	Periodic Mathieu functions (Mathieu functions of the first kind)	8.61
$\mathrm{Ce}_{2n}(z,q), \quad \mathrm{Ce}_{2n+1}(z,q)$	Associated (modified) Mathieu functions of the first kind	8.63
$\mathrm{chi}(x)$	Hyperbolic cosine integral function	8.22
$\mathrm{ci}(x)$	Cosine integral	8.23
$\mathrm{cn}(u)$	Cosine amplitude	8.14
$D(k) \equiv \boldsymbol{D}$		8.112
$D(\varphi, k)$		8.111
$D_n(z), \quad D_p(z)$	Parabolic cylinder functions	9.24–9.25
$\mathrm{dn}\, u$	Delta amplitude	8.14
e_1, e_2, e_3	(used with the Weierstrass function)	8.162
E_n	Euler numbers	9.63, 9.72
$E(\varphi, k)$	Elliptic integral of the second kind	8.11–8.12
$\left. \begin{array}{l} \boldsymbol{E}(k) = \boldsymbol{E} \\ \boldsymbol{E}(k') = \boldsymbol{E'} \end{array} \right\}$	Complete elliptic integral of the second kind	8.11-8.12
$E(p; a_r : q; \varrho_s : x)$	MacRobert function	9.4
$\boldsymbol{E}_\nu(z)$	Weber function	8.58
$\mathrm{Ei}(z)$	Exponential integral function	8.21
$\mathrm{erf}(x) = \Phi(x)$	Error function	8.25
$\mathrm{erfc}(x) = 1 - \Phi(x)$	Complementary error function	8.25
$F(\varphi, k)$	Elliptic integral of the first kind	8.11–8.12
$_pF_q\,(\alpha_1, \ldots, \alpha_p; \beta_1, \ldots, \beta_q; z)$	Generalized hypergeometric series	9.14
$_2F_1\,(\alpha, \beta; \gamma; z) = F(\alpha, \beta; \gamma; z)$	Gauss hypergeometric function	9.10–9.13
$_1F_1\,(\alpha; \gamma; z) = \Phi(\alpha, \gamma; z)$	Degenerate hypergeometric function	9.21
$F_\Lambda(\alpha : \beta_1, \ldots, \beta_n; \\ \gamma_1, \ldots \ldots, \gamma_n : z_1, \ldots, z_n)$	Hypergeometric function of several variables	9.19
F_1, F_2, F_3, F_4	Hypergeometric functions of two variables	9.18
$\left\{ \begin{array}{l} \mathrm{fe}_n(z,q), \mathrm{Fe}_n(z,q) \ldots \\ \mathrm{Fey}_n(z,q), \mathrm{Fek}_n(z,q) \ldots \end{array} \right\}$	Other nonperiodic solutions of Mathieu's equation	8.64, 8.663
$\boldsymbol{G}$	Catalan constant	9.73
g_2, g_3	Invariants of the $\wp(u)$-function	8.161
$\mathrm{gd}\, x$	Gudermannian	1.49
$\left\{ \begin{array}{l} \mathrm{ge}_n(z,q), \mathrm{Ge}_n(z,q) \\ \mathrm{Gey}_n(z,q), \mathrm{Gek}_n(z,q) \end{array} \right\}$	Other nonperiodic solutions of Mathieu's equation	8.64, 8.663

continued on next page

continued from previous page	

Notation	Name of the function and the number of the formula containing its definition
$G_{p,q}^{m,n}\left(x \left\vert {a_1,\ldots,a_p \atop b_1,\ldots,b_q}\right.\right)$	Meijer functions 9.3
$h(n)$	Unit integer function 18.1
$\mathrm{hei}_\nu(z), \quad \mathrm{her}_\nu(z)$	Thomson functions 8.56
$H_\nu^{(1)}(z), \quad H_\nu^{(2)}(z)$	Hankel functions of the first and second kinds 8.405, 8.42
$H(u) = \vartheta_1\left(\frac{\pi u}{2\mathbf{K}}\right)$	8.192
$H_1(u) = \vartheta_2\left(\frac{\pi u}{2\mathbf{K}}\right)$	8.192
$H_n(z)$	Hermite polynomials 8.95
$\mathbf{H}_\nu(z)$	Struve functions 8.55
$I_\nu(z)$	Bessel functions of an imaginary argument 8.406, 8.43
$I_x(p,q)$	Normalized incomplete beta function 8.39
$J_\nu(z)$	Bessel function 8.402, 8.41
$\mathbf{J}_\nu(z)$	Anger function 8.58
$\mathrm{k}_\nu(x)$	Bateman function 9.210 3
$\mathbf{K(k)} = \mathbf{K}, \quad \mathbf{K(k')} = \mathbf{K'}$	Complete elliptic integral of the first kind 8.11–8.12
$K_\nu(z)$	Bessel functions of imaginary argument 8.407, 8.43
$\mathrm{kei}(z), \quad \mathrm{ker}(z)$	Thomson functions 8.56
$L(x)$	Lobachevskiy function 8.26
$\mathbf{L}_\nu(z)$	Modified Struve function 8.55
$L_n^\alpha(z)$	Laguerre polynomials 8.97
$\mathrm{li}(x)$	Logarithm integral 8.24
$M_{\lambda,\mu}(z)$	Whittaker functions 9.22, 9.23
$O_n(x)$	Neumann polynomials 8.59
$P_\nu^\mu(z), \quad P_\nu^\mu(x)$	Associated Legendre functions of the first kind 8.7, 8.8
$P_\nu(z), \quad P_\nu(x)$	Legendre functions and polynomials 8.82, 8.83, 8.91
$P\left\{\begin{matrix} a & b & c \\ \alpha & \beta & \gamma & \delta \\ \alpha' & \beta' & \gamma' \end{matrix}\right\}$	Riemann's differential equation 9.160
$P_n^{(\alpha,\beta)}(x)$	Jacobi polynomials 8.96
$Q_\nu^\mu(z), \quad Q_\nu^\mu(x)$	Associated Legendre functions of the second kind 8.7, 8.8
$Q_\nu(z), \quad Q_\nu(x)$	Legendre functions of the second kind 8.82, 8.83
$S(x)$	Fresnel sine integral 8.25
$S_n(x)$	Schläfli polynomials 8.59
$s_{\mu,\nu}(z), \quad S_{\mu,\nu}(z)$	Lommel functions 8.57
$\mathrm{se}_{2n+1}(z,q), \quad \mathrm{se}_{2n+2}(z,q)$	Periodic Mathieu functions 8.61
$\mathrm{Se}_{2n+1}(z,q), \quad \mathrm{Se}_{2n+2}(z,q)$	Mathieu functions of an imaginary argument 8.63
$\mathrm{shi}(x)$	Hyperbolic sine integral 8.22
$\mathrm{si}(x)$	Sine integral 8.23
$\mathrm{sn}\, u$	Sine amplitude 8.14

continued on next page

continued from previous page		
Notation	Name of the function and the number of the formula containing its definition	
$T_n(x)$	Chebyshev polynomial of the 1st kind	8.94
$U_n(x)$	Chebyshev polynomials of the 2nd kind	8.94
$U_\nu(w, z), \quad V_\nu(w, z)$	Lommel functions of two variables	8.578
$W_{\lambda,\mu}(z)$	Whittaker functions	9.22, 9.23
$Y_\nu(z)$	Neumann functions	8.403, 8.41
$Z_\nu(z)$	Bessel functions	8.401
$\mathfrak{Z}_\nu(z)$	Bessel functions	

Notation

Symbol	Meaning
$\lfloor x \rfloor$	The integral part of the real number x (also denoted by $[x]$).
$\int_a^{(b+)} \quad \int_a^{(b-)}$	Contour integrals; the path of integration starting at the point a extends to the point b (along a straight line unless there is an indication to the contrary), encircles the point b along a small circle in the positive (negative) direction, and returns to the point a, proceeding along the original path in the opposite direction.
$\int_C$	Line integral along the curve C.
PV $\int$	Principal value integral
$\bar{z} = x - iy$	The complex conjugate of $z = x + iy$.
$n!$	$= 1 \cdot 2 \cdot 3 \ldots n, \qquad 0! = 1.$
$(2n+1)!!$	$= 1 \cdot 3 \ldots (2n+1).$
$(2n)!!$	$= 2 \cdot 4 \ldots (2n).$
$0!! = 1$ and $(-1)!! = 1$	(cf. 3.372 for $n = 0$.)
$0^0 = 1$	(cf. 0.112 and 0.113 for $q = 0$.)
$\binom{p}{n}$	$= \dfrac{p(p-1)\ldots(p-n+1)}{1 \cdot 2 \ldots n} = \dfrac{p!}{n!(p-n)!}, \quad \binom{p}{0} = 1, \quad \binom{p}{n} = \dfrac{p!}{n!(p-n)!}$ $[n = 1, 2, \ldots, p \geq n].$
$\binom{x}{n}$	$= x(x-1)\ldots(x-n+1)/n! \; [n = 0, 1, \ldots]$
$(a)_n$	$= a(a+1)\ldots(a+n-1) = \frac{\Gamma(a+n)}{\Gamma(a)}$ (Pochhammer symbol).
$\displaystyle\sum_{k=m}^{n} u_k$	$= u_m + u_{m+1} + \ldots + u_n.$ If $n < m$, we define $\displaystyle\sum_{k=m}^{n} u_k = 0.$
$\displaystyle\sum_n{}', \quad \sum_{m,n}{}'$	Summation over all integral values of n excluding $n = 0$, and summation over all integral values of n and m excluding $m = n = 0$, respectively.
$\sum, \quad \prod$	An empty $\sum$ has value 0 and an empty $\prod$ has value 1.

continued on next page

xliii

continued from previous page

Symbol	Meaning
$\delta_{ij} = \begin{cases} 1 & i = j \\ 0 & i \neq j \end{cases}$	Kronecker delta
τ	Theta function parameter (cf. 8.18)
$\times$ and $\wedge$	Vector product (cf. 10.11)
$\cdot$	Scalar product (cf. 10.11)
∇ or "del"	Vector operator (cf. 10.21)
∇^2	Laplacian (cf. 10.31)
$\arg z$	The argument of the complex number $z = x + iy$.
curl or rot	Vector operator (cf. 10.21)
div	Vector operator (divergence) (cf. 10.21)
$\mathcal{F}$	Fourier transform (cf. 17.21)
$\mathcal{F}_c$	Fourier cosine transform (cf. 17.31)
$\mathcal{F}_s$	Fourier sine transform (cf. 17.31)
grad	Vector operator (gradient) (cf. 10.21)
h_i and g_{ij}	Metric coefficients (cf. 10.51)
H	Hermitian transpose of a vector or matrix (cf. 13.123)
$\mathrm{H}(x) = \begin{cases} 0 & x < 0 \\ 1 & x \geq 0 \end{cases}$	Heaviside step function
$\operatorname{Im} z \equiv y$	The imaginary part of the complex number $z = x + iy$.
k	The letter k (when not used as an index of summation) denotes a number in the interval $[0, 1]$. This notation is used in integrals that lead to elliptic integrals. In such a connection, the number $\sqrt{1 - k^2}$ is denoted by k'.
$\mathcal{L}$	Laplace transform (cf. 17.11)
$\mathcal{M}$	Mellin transform (cf. 17.41)
$\mathbb{N}$	The natural numbers $(0, 1, 2, \dots)$
$O(f(z))$	The order of the function $f(z)$. Suppose that the point z approaches z_0. If there exists an $M > 0$ such that $|g(z)| \leq M|f(z)|$ in some sufficiently small neighborhood of the point z_0, we write $g(z) = O(f(z))$.
q	The nome, a theta function parameter (cf. 8.18)

continued on next page

Symbol	Meaning
continued from previous page	
$\mathbb{R}$	The real numbers
$R(x)$	A rational function
$\operatorname{Re} z \equiv x$	The real part of the complex number $z = x + iy$.
S_n^m	Stirling number of the first kind (cd. 9.74)
$\mathfrak{S}_n^m$	Stirling number of the second kind (cd. 9.74)
$\operatorname{sign} x = \begin{cases} +1 & x > 0 \\ 0 & x = 0 \\ -1 & x < 0 \end{cases}$	The sign (signum) of the real number x.
T	Transpose of a vector or matrix (cf. 13.115)
$\mathbb{Z}$	The integers $(0, \pm 1, \pm 2, \dots)$
Z_b	Bilateral z transform (cf. 18.1)
Z_u	Unilateral z transform (cf. 18.1)

Note on the bibliographic references

The letters and numbers following equations refer to the sources used by Russian editors. The key to the letters will be found preceding each entry in the Bibliography beginning on page 1133. Roman numerals indicate the volume number of a multivolume work. Numbers without parentheses indicate page numbers, numbers in single parentheses refer to equation numbers in the original sources.

Some formulas were changed from their form in the source material. In such cases, the letter a appears at the end of the bibliographic references.

As an example we may use the reference to equation 3.354–5:

$$\text{ET I 118 (1) } a$$

The key on page 1133 indicates that the book referred to is:

Erdélyi, A. et al., *Tables of Integral Transforms*

The Roman numeral denotes volume one of the work, 118 is the page on which the formula will be found, (1) refers to the number of the formula in this source, and the a indicates that the expression appearing in the source differs in some respect from the formula in this book.

In several cases the editors have used Russian editions of works published in other languages. Under such circumstances, because the pagination and numbering of equations may be altered, we have referred the reader only to the original sources and dispensed with page and equation numbers.

0 Introduction

0.1 Finite sums

0.11 Progressions

0.111 Arithmetic progression.
$$\sum_{k=0}^{n-1}(a+kr) = \frac{n}{2}[2a+(n-1)r] = \frac{n}{2}(a+l) \qquad\qquad [l=a+(n-1)r \text{ is the last term}]$$

0.112 Geometric progression.
$$\sum_{k=1}^{n}aq^{k-1} = \frac{a\left(q^n-1\right)}{q-1} \qquad\qquad [q \neq 1]$$

0.113 Arithmetic-geometric progression.
$$\sum_{k=0}^{n-1}(a+kr)q^k = \frac{a-[a+(n-1)r]q^n}{1-q} + \frac{rq\left(1-q^{n-1}\right)}{(1-q)^2}$$
$$[q \neq 1, \quad n > 1] \qquad\qquad \text{JO (5)}$$

0.114[8] $\displaystyle\sum_{k=1}^{n-1}k^2x^k = \frac{\left(-n^2+2n-1\right)x^{n+2}+\left(2n^2-2n-1\right)x^{n+1}-n^2x^n+x^2+x}{(1-x)^3}$

0.12 Sums of powers of natural numbers

0.121 $\displaystyle\sum_{k=1}^{n}k^q = \frac{n^{q+1}}{q+1} + \frac{n^q}{2} + \frac{1}{2}\binom{q}{1}B_2 n^{q-1} + \frac{1}{4}\binom{q}{3}B_4 n^{q-3} + \frac{1}{6}\binom{q}{5}B_6 n^{q-5} + \cdots$

$$= \frac{n^{q+1}}{q+1} + \frac{n^q}{2} + \frac{qn^{q-1}}{12} - \frac{q(q-1)(q-2)}{720}n^{q-3} + \frac{q(q-1)(q-2)(q-3)(q-4)}{30,240}n^{q-5} - \cdots$$
$$[\text{last term contains either } n \text{ or } n^2] \qquad \text{CE 332}$$

1. $\displaystyle\sum_{k=1}^{n}k = \frac{n(n+1)}{2}$ \hfill CE 333

2. $\displaystyle\sum_{k=1}^{n}k^2 = \frac{n(n+1)(2n+1)}{6}$ \hfill CE 333

3. $\displaystyle\sum_{k=1}^{n}k^3 = \left[\frac{n(n+1)}{2}\right]^2$ \hfill CE 333

4. $\displaystyle\sum_{k=1}^{n} k^4 = \frac{1}{30}n(n+1)(2n+1)(3n^2+3n-1)$ CE 333

5. $\displaystyle\sum_{k=1}^{n} k^5 = \frac{1}{12}n^2(n+1)^2(2n^2+2n-1)$ CE 333

6. $\displaystyle\sum_{k=1}^{n} k^6 = \frac{1}{42}n(n+1)(2n+1)(3n^4+6n^3-3n+1)$ CE 333

7. $\displaystyle\sum_{k=1}^{n} k^7 = \frac{1}{24}n^2(n+1)^2(3n^4+6n^3-n^2-4n+2)$ CE 333

0.122 $\displaystyle\sum_{k=1}^{n} (2k-1)^q = \frac{2^q}{q+1}n^{q+1} - \frac{1}{2}\binom{q}{1}2^{q-1}B_2 n^{q-1} - \frac{1}{4}\binom{q}{3}2^{q-3}\left(2^3-1\right)B_4 n^{q-3} - \cdots$

[last term contains either n or n^2.]

1. $\displaystyle\sum_{k=1}^{n} (2k-1) = n^2$

2. $\displaystyle\sum_{k=1}^{n} (2k-1)^2 = \frac{1}{3}n(4n^2-1)$ JO (32a)

3. $\displaystyle\sum_{k=1}^{n} (2k-1)^3 = n^2(2n^2-1)$ JO (32b)

4.[10] $\displaystyle\sum_{k=1}^{n} (mk-1) = \frac{1}{2}[m(n+1)-2]$

5.[10] $\displaystyle\sum_{k=1}^{n} (mk-1)^2 = \frac{1}{6}n[m^2(n+1)(2n+1)-6m(n+1)+6]$

6.[10] $\displaystyle\sum_{k=1}^{n} (mk-1)^3 = \frac{1}{4}n[m^3 n(n+1)^2 - 2m^2(n+1)(2n+1)+6m(n+1)-4]$

0.123 $\displaystyle\sum_{k=1}^{n} k(k+1)^2 = \frac{1}{12}n(n+1)(n+2)(3n+5)$

0.124

1. $\displaystyle\sum_{k=1}^{q} k\left(n^2-k^2\right) = \frac{1}{4}q(q+1)\left(2n^2-q^2-q\right)$ $[q=1,2,\ldots]$

2.[10] $\displaystyle\sum_{k=1}^{n} k(k+1)^3 = \frac{1}{60}n(n+1)\left(12n^3+63n^2+107n+58\right)$

0.125 $\displaystyle\sum_{k=1}^{n} k!\cdot k = (n+1)! - 1$ AD (188.1)

0.126 $\displaystyle\sum_{k=0}^{n} \frac{(n+k)!}{k!(n-k)!} = \sqrt{\frac{e}{\pi}}\,K_{n+\frac{1}{2}}\left(\frac{1}{2}\right)$ WA 94

0.13 Sums of reciprocals of natural numbers

0.131 $\displaystyle\sum_{k=1}^{n}\frac{1}{k} = C + \ln n + \frac{1}{2n} - \sum_{k=2}^{\infty}\frac{A_k}{n(n+1)\ldots(n+k-1)},$ JO (59), AD (1876)

where

$$A_k = \frac{1}{k}\int_0^1 x(1-x)(2-x)(3-x)\cdots(k-1-x)\,dx$$

$$A_2 = \frac{1}{12}, \qquad A_3 = \frac{1}{12}$$

$$A_4 = \frac{19}{120}, \qquad A_5 = \frac{9}{20},$$

0.132[7] $\displaystyle\sum_{k=1}^{n}\frac{1}{2k-1} = \frac{1}{2}\left(\boldsymbol{C} + \ln n\right) + \ln 2 + \frac{B_2}{8n^2} + \frac{\left(2^3 - 1\right)B_4}{64n^4} + \ldots$ JO (71a)a

0.133 $\displaystyle\sum_{k=2}^{n}\frac{1}{k^2-1} = \frac{3}{4} - \frac{2n+1}{2n(n+1)}$ JO (184f)

0.14 Sums of products of reciprocals of natural numbers

1. $\displaystyle\sum_{k=1}^{n}\frac{1}{[p+(k-1)q](p+kq)} = \frac{n}{p(p+nq)}$ GI III (64)a

2. $\displaystyle\sum_{k=1}^{n}\frac{1}{[p+(k-1)q](p+kq)[p+(k+1)q]} = \frac{n(2p+nq+q)}{2p(p+q)(p+nq)[p+(n+1)q]}$ GI III (65)a

3. $\displaystyle\sum_{k=1}^{n}\frac{1}{[p+(k-1)q](p+kq)\ldots[p+(k+l)q]}$

       $= \dfrac{1}{(l+1)q}\left\{\dfrac{1}{p(p+q)\ldots(p+lq)} - \dfrac{1}{(p+nq)[p+(n+1)q]\ldots[p+(n+l)q]}\right\}$

                                                             AD (1856)a

4. $\displaystyle\sum_{k=1}^{n}\frac{1}{[1+(k-1)q][1+(k-l)q+p]} = \frac{1}{p}\left[\sum_{k=1}^{n}\frac{1}{1+(k-1)q} - \sum_{k=1}^{n}\frac{1}{1+(k-1)q+p}\right]$

                                                                      GI III (66)a

0.142 $\displaystyle\sum_{k=1}^{n}\frac{k^2+k-1}{(k+2)!} = \frac{1}{2} - \frac{n+1}{(n+2)!}$ JO (157)

0.15 Sums of the binomial coefficients

Notation: n is a natural number

1. $\displaystyle\sum_{k=0}^{m}\binom{n+k}{n} = \binom{n+m+1}{n+1}$ KR 64 (70.1)

2. $1 + \binom{n}{2} + \binom{n}{4} + \ldots = 2^{n-1}$ KR 62 (58.1)

3. $\dbinom{n}{1} + \dbinom{n}{3} + \dbinom{n}{5} + \ldots = 2^{n-1}$ KR 62 (58.1)

4. $\displaystyle\sum_{k=0}^{m} (-1)^k \dbinom{n}{k} = (-1)^m \dbinom{n-1}{m}$ $[n \geq 1]$ KR 64 (70.2)

0.152

1. $\dbinom{n}{0} + \dbinom{n}{3} + \dbinom{n}{6} + \ldots = \dfrac{1}{3}\left(2^n + 2\cos\dfrac{n\pi}{3}\right)$ KR 62 (59.1)

2. $\dbinom{n}{1} + \dbinom{n}{4} + \dbinom{n}{7} + \ldots = \dfrac{1}{3}\left(2^n + 2\cos\dfrac{(n-2)\pi}{3}\right)$ KR 62 (59.2)

3. $\dbinom{n}{2} + \dbinom{n}{5} + \dbinom{n}{8} + \ldots = \dfrac{1}{3}\left(2^n + 2\cos\dfrac{(n-4)\pi}{3}\right)$ KR 62 (59.3)

0.153

1. $\dbinom{n}{0} + \dbinom{n}{4} + \dbinom{n}{8} + \ldots = \dfrac{1}{2}\left(2^{n-1} + 2^{\frac{n}{2}}\cos\dfrac{n\pi}{4}\right)$ KR 63 (60.1)

2. $\dbinom{n}{1} + \dbinom{n}{5} + \dbinom{n}{9} + \ldots = \dfrac{1}{2}\left(2^{n-1} + 2^{\frac{n}{2}}\sin\dfrac{n\pi}{4}\right)$ KR 63 (60.2)

3. $\dbinom{n}{2} + \dbinom{n}{6} + \dbinom{n}{10} + \ldots = \dfrac{1}{2}\left(2^{n-1} - 2^{\frac{n}{2}}\cos\dfrac{n\pi}{4}\right)$ KR 63 (60.3)

4. $\dbinom{n}{3} + \dbinom{n}{7} + \dbinom{n}{11} + \ldots = \dfrac{1}{2}\left(2^{n-1} - 2^{\frac{n}{2}}\sin\dfrac{n\pi}{4}\right)$ KR 63 (60.4)

0.154

1. $\displaystyle\sum_{k=0}^{n} (k+1)\dbinom{n}{k} = 2^{n-1}(n+2)$ $[n \geq 0]$ KR 63 (66.1)

2. $\displaystyle\sum_{k=1}^{n} (-1)^{k+1} k\dbinom{n}{k} = 0$ $[n \geq 2]$ KR 63 (66.2)

3. $\displaystyle\sum_{k=0}^{N} (-1)^k \dbinom{N}{k} k^{n-1} = 0$ $[N \geq n \geq 1; \quad 0^0 \equiv 1]$

4. $\displaystyle\sum_{k=0}^{n} (-1)^k \dbinom{n}{k} k^n = (-1)^n n!$ $[n \geq 0; \quad 0^0 \equiv 1]$

5. $\displaystyle\sum_{k=0}^{n} (-1)^k \dbinom{n}{k} (\alpha+k)^n = (-1)^n n!$ $[n \geq 0; \quad 0^0 \equiv 1]$

6. $\displaystyle\sum_{k=0}^{N} (-1)^k \dbinom{N}{k} (\alpha+k)^{n-1} = 0$ $[N \geq n \geq 1, \quad 0^0 \equiv 1 \quad N, n \in N^+]$

0.155

1. $\displaystyle\sum_{k=1}^{n} \dfrac{(-1)^{k+1}}{k+1}\dbinom{n}{k} = \dfrac{n}{n+1}$ KR 63 (67)

2. $\displaystyle\sum_{k=0}^{n} \frac{1}{k+1}\binom{n}{k} = \frac{2^{n+1}-1}{n+1}$ KR 63 (68.1)

3. $\displaystyle\sum_{k=0}^{n} \frac{\alpha^{k+1}}{k+1}\binom{n}{k} = \frac{(\alpha+1)^{n+1}-1}{n+1}$ KR 63 (68.2)

4. $\displaystyle\sum_{k=1}^{n} \frac{(-1)^{k+1}}{k}\binom{n}{k} = \sum_{m=1}^{n} \frac{1}{m}$ KR 64 (69)

0.156

1. $\displaystyle\sum_{k=0}^{p} \binom{n}{k}\binom{m}{p-k} = \binom{n+m}{p}$ [m is a natural number] KR 64 (71.1)

2. $\displaystyle\sum_{k=0}^{n-p} \binom{n}{k}\binom{n}{p+k} = \frac{(2n)!}{(n-p)!(n+p)!}$ KR 64 (71.2)

0.157

1. $\displaystyle\sum_{k=0}^{n} \binom{n}{k}^2 = \binom{2n}{n}$ KR 64 (72.1)

2. $\displaystyle\sum_{k=0}^{2n} (-1)^k \binom{2n}{k}^2 = (-1)^n \binom{2n}{n}$ KR 64 (72.2)

3. $\displaystyle\sum_{k=0}^{2n+1} (-1)^k \binom{2n+1}{k}^2 = 0$ KR 64 (72.3)

4. $\displaystyle\sum_{k=1}^{n} k \binom{n}{k}^2 = \frac{(2n-1)!}{[(n-1)!]^2}$ KR 64 (72.4)

0.158[10]

1. $\displaystyle\sum_{k=1}^{n} \left[2^k \binom{2n-k}{n-k} - 2^{k+1}\binom{2n-k-1}{n-k-1} \right] k = 4^n - \binom{2n}{n}$

2. $\displaystyle\sum_{k=1}^{n} \left[2^k \binom{2n-k}{n-k} - 2^{k+1}\binom{2n-k-1}{n-k-1} \right] k^2 = 4^n - \binom{2n}{n} 3 \cdot 4^n$

3. $\displaystyle\sum_{k=1}^{n} \left[2^k \binom{2n-k}{n-k} - 2^{k+1}\binom{2n-k-1}{n-k-1} \right] k^3 = (6n+13)4^n - 18n\binom{2n}{n}$

4. $\displaystyle\sum_{k=1}^{n} \left[2^k \binom{2n-k}{n-k} - 2^{k+1}\binom{2n-k-1}{n-k-1} \right] k^4 = (32n^2+104n)\binom{2n}{n} - (60n+75)4^n$

0.159[10]

1. $\displaystyle\sum_{k=0}^{n} \left[\binom{2n}{n-k} - \binom{2n}{n-k-1} \right] k = \frac{1}{2}\left[4^n - \binom{2n}{n} \right]$

2. $$\sum_{k=0}^{n} \left[\binom{2n}{n-k} - \binom{2n}{n-k-1} \right] k^2 = \frac{1}{2} \left[(2n+1) \binom{2n}{n} - 4^n \right]$$

3. $$\sum_{k=0}^{n} \left[\binom{2n}{n-k} - \binom{2n}{n-k-1} \right] k^3 = \frac{(3n+2)}{4} \cdot 4^n - \frac{1}{2} \binom{2n}{n} (3n+1)$$

0.160[10]

1. $$\sum_{k=n+1}^{2n} \binom{2n}{k} \alpha^k + \frac{1}{2} \binom{2n}{n} \alpha^n + \frac{(1+\alpha)^{2n-1}(1-\alpha)}{2} \sum_{k=0}^{n-1} \binom{2k}{k} \left[\frac{\alpha}{(1+\alpha)^2} \right]^k = \frac{1}{2}(1+\alpha)^{2n}$$

2.* $$\sum_{r=0}^{n} (-1)^r \binom{n}{r} \frac{\Gamma(r+b)}{\Gamma(r+a)} = \frac{\mathrm{B}(n+a-b,b)}{\Gamma(a-b)}$$

0.2 Numerical series and infinite products

0.21 The convergence of numerical series

The series **0.211** $\displaystyle\sum_{k=1}^{\infty} u_k = u_1 + u_2 + u_3 + \ldots$

is said to *converge absolutely* if the series

0.212 $\displaystyle\sum_{k=1}^{\infty} |u_k| = |u_1| + |u_2| + |u_3| + \cdots,$

composed of the absolute values of its terms converges. If the series **0.211** converges and the series **0.212** diverges, the series **0.211** is said to *converge conditionally*. Every absolutely convergent series converges.

0.22 Convergence tests

Suppose that

$$\lim_{k \to \infty} |u_k|^{1/k} = q$$

If $q < 1$, the series **0.211** converges absolutely. On the other hand, if $q > 1$, the series **0.211** diverges. (Cauchy)

0.222 Suppose that

$$\lim_{k \to \infty} \left| \frac{u_{k+1}}{u_k} \right| = q$$

Here, if $q < 1$, the series **0.211** converges absolutely. If $q > 1$, the series **0.211** diverges. If $\left| \frac{u_{k+1}}{u_k} \right|$ approaches 1 but remains greater than unity, then the series **0.211** diverges. (d'Alembert)

0.223 Suppose that

$$\lim_{k \to \infty} k \left\{ \left| \frac{u_k}{u_{k+1}} \right| - 1 \right\} = q$$

Here, if $q > 1$, the series **0.211** converges absolutely. If $q < 1$, the series **0.211** diverges. (Raabe)

0.224 Suppose that $f(x)$ is a positive decreasing function and that

$$\lim_{k \to \infty} \frac{e^k f\left(e^k\right)}{f(k)} = q$$

for natural k. If $q < 1$, the series $\sum_{k=1}^{\infty} f(k)$ converges. If $q > 1$, this series diverges. (Ermakov)

0.225 Suppose that

$$\left| \frac{u_k}{u_{k+1}} \right| = 1 + \frac{q}{k} + \frac{|v_k|}{k^p},$$

where $p > 1$ and the $|v_k|$ are bounded, that is, the $|v_k|$ are all less than some M, which is independent of k. Here, if $q > 1$, the series **0.211** converges absolutely. If $q \leq 1$, this series diverges. (Gauss)

0.226 Suppose that a function $f(x)$ defined for $x \geq q \geq 1$ is continuous, positive, and decreasing. Under these conditions, the series

$$\sum_{k=1}^{\infty} f(k)$$

converges or diverges according as the integral

$$\int_q^{\infty} f(x)\, dx$$

converges or diverges (the Cauchy integral test.)

0.227 Suppose that all terms of a sequence $u_1, u_2, \ldots, u_n$ are positive. In such a case, the series

1. $$\sum_{k=1}^{\infty} (-1)^{k+1} u_k = u_1 - u_2 + u_3 - \ldots$$

    is called an *alternating series*.

    If the terms of an alternating series decrease monotonically in absolute value and approach zero, that is, if

2. $$u_{k+1} < u_k \text{ and } \lim_{k \to \infty} u_k = 0,$$

    the series **0.227** 1 converges. Here, the remainder of the series is

3. $$\sum_{k=n+1}^{\infty} (-1)^{k-n+1} u_k = \left| \sum_{k=1}^{\infty} (-1)^{k+1} u_k - \sum_{k=1}^{n} (-1)^{k+1} u_k < u_{n+1} \right| \qquad \text{(Leibniz)}$$

0.228 If the series

1. $$\sum_{k=1}^{\infty} v_k = v_1 + v_2 + \ldots + v_k + \ldots$$

    converges and the numbers u_k form a monotonic bounded sequence, that is, if $|u_k| < M$ for some number M and for all k, the series

2. $$\sum_{k=1}^{\infty} u_k v_k = u_1 v_1 + u_2 v_2 + \ldots + u_k v_k + \ldots \qquad \text{FI II 354}$$

    converges. (Abel)

0.229 If the partial sums of the series **0.228** 1 are bounded and if the numbers u_k constitute a monotonic sequence that approaches zero, that is, if

$$\left| \sum_{k=1}^{n} v_k \right| < M \qquad [n = 1, 2, \ldots] \qquad \text{and} \quad \lim_{k \to \infty} u_k = 0,$$ FI II 355

then the series **0.228** 2 converges. (Dirichlet)

0.23–0.24   Examples of numerical series

0.231   Progressions

1. $\displaystyle \sum_{k=0}^{\infty} aq^k = \frac{a}{1-q}$ $[|q| < 1]$

2. $\displaystyle \sum_{k=0}^{\infty} (a + kr)q^k = \frac{a}{1-q} + \frac{rq}{(1-q)^2}$ $[|q| < 1]$ (cf. **0.113**)

0.232

1. $\displaystyle \sum_{k=1}^{\infty} (-1)^{k+1} \frac{1}{k} = \ln 2$ (cf. **1.511**)

2. $\displaystyle \sum_{k=1}^{\infty} (-1)^{k+1} \frac{1}{2k-1} = 1 - 2 \sum_{k=1}^{\infty} \frac{1}{(4k-1)(4k+1)} = \frac{\pi}{4}$

                                                   (cf. **1.643**)

0.233

1. $\displaystyle \sum_{k=1}^{\infty} \frac{1}{k^p} = 1 + \frac{1}{2^p} + \frac{1}{3^p} + \ldots = \zeta(p)$ $[\operatorname{Re} p > 1]$ WH

2. $\displaystyle \sum_{k=1}^{\infty} (-1)^{k+1} \frac{1}{k^p} = (1 - 2^{1-p}) \zeta(p)$ $[\operatorname{Re} p > 0]$ WH

3.[10] $\displaystyle \sum_{k=1}^{\infty} \frac{1}{k^{2n}} = \frac{2^{2n-1} \pi^{2n}}{(2n)!} |B_{2n}|, \qquad \sum_{k=1}^{\infty} \frac{1}{k^2} = \frac{\pi^2}{6}$ FI II 721

4. $\displaystyle \sum_{k=1}^{\infty} (-1)^{k+1} \frac{1}{k^{2n}} = \frac{(2^{2n-1} - 1) \pi^{2n}}{(2n)!} |B_{2n}|$ JO (165)

5. $\displaystyle \sum_{k=1}^{\infty} \frac{1}{(2k-1)^{2n}} = \frac{(2^{2n} - 1) \pi^{2n}}{2 \cdot (2n)!} |B_{2n}|$ JO (184b)

6. $\displaystyle \sum_{k=1}^{\infty} (-1)^{k+1} \frac{1}{(2k-1)^{2n+1}} = \frac{\pi^{2n+1}}{2^{2n+2}(2n)!} |E_{2n}|$ JO (184d)

0.234

1. $\displaystyle \sum_{k=1}^{\infty} (-1)^{k+1} \frac{1}{k^2} = \frac{\pi^2}{12}$ EU

2. $\displaystyle\sum_{k=1}^{\infty} \frac{1}{(2k-1)^2} = \frac{\pi^2}{8}$ 　　　　　　　　　　　　　　　　EU

3. $\displaystyle\sum_{k=0}^{\infty} \frac{(-1)^k}{(2k+1)^2} = \boldsymbol{G}$ 　　　　　　　　　　　　　　　FI II 482

4. $\displaystyle\sum_{k=1}^{\infty} \frac{(-1)^{k+1}}{(2k-1)^3} = \frac{\pi^3}{32}$ 　　　　　　　　　　　　　　EU

5. $\displaystyle\sum_{k=1}^{\infty} \frac{1}{(2k-1)^4} = \frac{\pi^4}{96}$ 　　　　　　　　　　　　　　　EU

6. $\displaystyle\sum_{k=1}^{\infty} \frac{(-1)^{k+1}}{(2k-1)^5} = \frac{5\pi^5}{1536}$ 　　　　　　　　　　　　　EU

7. $\displaystyle\sum_{k=1}^{\infty} (-1)^{k+1} \frac{k}{(k+1)^2} = \frac{\pi^2}{12} - \ln 2$

8.[6] $\displaystyle\sum_{k=1}^{\infty} \frac{1}{k(2k+1)} = 2 - 2\ln 2$

0.235 $\displaystyle S_n = \sum_{k=1}^{\infty} \frac{1}{(4k^2-1)^n}$

$$S_1 = \frac{1}{2}, \quad S_2 = \frac{\pi^2-8}{16}, \quad S_3 = \frac{32-3\pi^2}{64}, \quad S_4 = \frac{\pi^4+30\pi^2-384}{768}$$

　　　　　　　　　　　　　　　　　　　　　　　　　　　　　　　JO (186)

0.236

1. $\displaystyle\sum_{k=1}^{\infty} \frac{1}{k(4k^2-1)} = 2\ln 2 - 1$ 　　　　　　　　　　　　　　BR 51a

2. $\displaystyle\sum_{k=1}^{\infty} \frac{1}{k(9k^2-1)} = \frac{3}{2}(\ln 3 - 1)$ 　　　　　　　　　　　　　BR 51a

3. $\displaystyle\sum_{k=1}^{\infty} \frac{1}{k(36k^2-1)} = -3 + \frac{3}{2}\ln 3 + 2\ln 2$ 　　　　　BR 52, AD (6913.3)

4. $\displaystyle\sum_{k=1}^{\infty} \frac{k}{(4k^2-1)^2} = \frac{1}{8}$ 　　　　　　　　　　　　　　　BR 52

5. $\displaystyle\sum_{k=1}^{\infty} \frac{1}{k(4k^2-1)^2} = \frac{3}{2} - 2\ln 2$ 　　　　　　　　　　　　BR 52

6. $\displaystyle\sum_{k=1}^{\infty} \frac{12k^2-1}{k(4k^2-1)^2} = 2\ln 2$ 　　　　　　　　　　AD (6917.3), BR 52

7.[6] $\displaystyle\sum_{k=1}^{\infty} \frac{1}{k(2k+1)^2} = 4 - \frac{\pi^2}{4} - 2\ln 2$

0.237

1. $\displaystyle\sum_{k=1}^{\infty} \frac{1}{(2k-1)(2k+1)} = \frac{1}{2}$
AD (6917.2), BR 52

2. $\displaystyle\sum_{k=1}^{\infty} \frac{1}{(4k-1)(4k+1)} = \frac{1}{2} - \frac{\pi}{8}$

3. $\displaystyle\sum_{k=2}^{\infty} \frac{1}{(k-1)(k+1)} = \frac{3}{4}$
[cf. **0.133**],

4. $\displaystyle\sideset{}{'}\sum_{k=1, k \neq m}^{\infty} \frac{1}{(m+k)(m-k)} = -\frac{3}{4m^2}$
[m is an integer]   AD (6916.1)

5. $\displaystyle\sideset{}{'}\sum_{k=1, k \neq m}^{\infty} \frac{(-1)^{k-1}}{(m-k)(m+k)} = \frac{3}{4m^2}$
[m is an even number]   AD (6916.2)

0.238

1. $\displaystyle\sum_{k=1}^{\infty} \frac{1}{(2k-1)2k(2k+1)} = \ln 2 - \frac{1}{2}$
GI III (93)

2. $\displaystyle\sum_{k=1}^{\infty} \frac{(-1)^{k+1}}{(2k-1)2k(2k+1)} = \frac{1}{2}(1 - \ln 2)$
GI III (94)a

3. $\displaystyle\sum_{k=0}^{\infty} \frac{1}{(3k+1)(3k+2)(3k+3)(3k+4)} = \frac{1}{6} - \frac{1}{4}\ln 3 + \frac{\pi}{12\sqrt{3}}$
GI III (95)

0.239

1. $\displaystyle\sum_{k=1}^{\infty} (-1)^{n+1} \frac{1}{3k-2} = \frac{1}{3}\left(\frac{\pi}{\sqrt{3}} + \ln 2\right)$
GI III (85), BR* 161 (1)

2.[7] $\displaystyle\sum_{k=1}^{\infty} (-1)^{k+1} \frac{1}{3k-1} = \frac{1}{3}\left(\frac{\pi}{\sqrt{3}} - \ln 2\right)$
BR* 161 (1)

3. $\displaystyle\sum_{k=1}^{\infty} (-1)^{k+1} \frac{1}{4k-3} = \frac{1}{4\sqrt{2}}\left[\pi + 2\ln\left(\sqrt{2}+1\right)\right]$
BR* 161 (1)

4. $\displaystyle\sum_{k=1}^{\infty} (-1)^{\left[\frac{k+3}{2}\right]} \frac{1}{k} = \frac{\pi}{4} + \frac{1}{2}\ln 2$
GI III (87)

5. $\displaystyle\sum_{k=1}^{\infty} (-1)^{\left[\frac{k+3}{2}\right]} \frac{1}{2k-1} = \frac{\pi}{2\sqrt{2}}$

6. $\displaystyle\sum_{k=1}^{\infty} (-1)^{\left[\frac{k+5}{3}\right]} \frac{1}{2k-1} = \frac{5\pi}{12}$
GI III (88)

7. $\displaystyle\sum_{k=1}^{\infty} \frac{1}{(8k-1)(8k+1)} = \frac{1}{2} - \frac{\pi}{16}\left(\sqrt{2}+1\right)$

0.241

1. $\displaystyle\sum_{k=1}^{\infty} \frac{1}{2^k k} = \ln 2$ JO (172g)

2. $\displaystyle\sum_{k=1}^{\infty} \frac{1}{2^k k^2} = \frac{\pi^2}{12} - \frac{1}{2}(\ln 2)^2$ JO (174)

3.[10] $\displaystyle\sum_{n=0}^{\infty} \binom{2n}{n} p^n = \frac{1-\sqrt{1-4p}}{\sqrt{1-4p}-(1-4p)}$ $\left[0 \le p < \tfrac{1}{4}\right]$

4.[10] $\displaystyle\sum_{n=1}^{\infty} \frac{p^n}{n^2} = \frac{\pi^2}{6} - \int_1^p \frac{\ln(1-x)}{x}\,dx$ $\left[0 \le p \le 1\right]$

5.[10] $\displaystyle\sum_{j=1}^{i} \left[2^j \binom{2i-j}{i-j} - 2^{j+1}\binom{2i-(j+1)}{i-(j+1)}\right] j = 4^i - \binom{2i}{i}$

                                           $\left[\binom{n}{m} = 0, \quad m < 0\right]$

6.[10] $\displaystyle\sum_{j=1}^{i} \left[2^j \binom{2i-j}{i-j} - 2^{j+1}\binom{2i-(j+1)}{i-(j+1)}\right] j^2 = 4i\binom{2i}{i} - 3\cdot 4^i$

                                           $\left[\binom{n}{m} = 0, \quad m < 0\right]$

7.[10] $\displaystyle\sum_{j=1}^{i} \left[2^j \binom{2i-j}{i-j} - 2^{j+1}\binom{2i-(j+1)}{i-(j+1)}\right] j^3 = (6i+13)4^i - 18i\binom{2i}{i}$

                                           $\left[\binom{n}{m} = 0, \quad m < 0\right]$

8.[10] $\displaystyle\sum_{j=1}^{i} \left[2^j \binom{2i-j}{i-j} - 2^{j+1}\binom{2i-(j+1)}{i-(j+1)}\right] j^4 = (32i^2+104i)\binom{2i}{i} - (60i+75)4^i$

9.[10] $\displaystyle\sum_{j=n+1}^{2n} \binom{2n}{j} k^j + \frac{1}{2}\binom{2n}{n} k^n + \frac{(1+k)^{2n-1}(1-k)}{2}\sum_{i=0}^{n-1}\binom{2i}{i}\left[\frac{k}{(1+k)^2}\right]^i = \frac{1}{2}(1+k)^{2n}$

10.[10] $\displaystyle\sum_{k=0}^{i} \binom{i+k}{k} 2^{i-k} = 4^i$

11.[10] $\displaystyle\sum_{k=0}^{i} \binom{i+k}{h}^{i-k} k = (i+1)4^i - (2i+1)\binom{2i}{i}$

12.[10] $\displaystyle\sum_{k=0}^{i} \binom{2i}{k} = \frac{1}{2}\left[4^i + \binom{2i}{i}\right]$

$$13.^{10} \quad \sum_{k=0}^{i} \binom{2i}{k} k = \frac{i}{2} 4^i$$

$$14.^{10} \quad \sum_{k=0}^{i} \binom{2i}{k} k^2 = (2i+1)i4^{i-1} - \frac{i^2}{2} \binom{2i}{i}$$

0.242 $\quad \sum_{k=0}^{\infty} (-1)^k \frac{1}{n^{2k}} = \frac{n^2}{n^2+1}$ $\quad\quad\quad\quad\quad\quad [n > 1]$

0.243

1. $\quad \sum_{k=1}^{\infty} \frac{1}{[p+(k-1)q](p+kq)\ldots[p+(k+l)q]} = \frac{1}{(l+1)q} \frac{1}{p(p+q)\ldots(p+lq)}$

$\quad\quad\quad\quad\quad\quad\quad\quad\quad\quad\quad\quad\quad\quad\quad\quad\quad\quad\quad\quad\quad\quad\quad\quad\quad\quad$ (see also **0.141** 3)

2.^{7} $\quad \sum_{k=1}^{\infty} \frac{x^{k-1}}{[p+(k-1)q][p+(k-1)q+1][p+(k-1)q+2]\ldots[p+(k-1)q+l]} = \frac{1}{l!} \int_0^i \frac{t^{p-1}(1-t)^t}{1-xt^q} dt$

$\quad\quad\quad\quad\quad\quad\quad\quad\quad\quad\quad\quad\quad\quad\quad\quad\quad\quad\quad\quad\quad\quad\quad [p > 0, \quad x^2 < 1]$ BR* 161 (2), AD (6.704)

3. $\quad \sum_{k=0}^{\infty} \frac{1}{(2k+1)^3} \left(\frac{1}{x} \tanh\left[\frac{(2k+1)\pi x}{2}\right] + x \tanh\left[\frac{(2k+1)\pi}{2x}\right] \right) = \frac{\pi^3}{16}$

0.244

1. $\quad \sum_{k=1}^{\infty} \frac{1}{(k+p)(k+q)} = \frac{1}{q-p} \int_0^1 \frac{x^p - x^q}{1-x} dx$ $\quad\quad [p > -1, \quad q > -1, \quad p \neq q]$ GI III (90)

2. $\quad \sum_{k=1}^{\infty} (-1)^{k+1} \frac{1}{p+(k-1)q} = \int_0^1 \frac{t^{p-1}}{1+t^q} dt$ $\quad\quad [p > 0, \quad q > 0]$ BR* 161 (1)

3.^{10} $\quad \sum_{k=1}^{\infty} \frac{1}{(k+p)(k+q)} = \frac{1}{q-p} \sum_{m=p+1}^{q} \frac{1}{m}$ $\quad\quad [q > p > -1, \quad p \text{ and } q \text{ integers}]$

Summations of reciprocals of factorials

0.245

1. $\quad \sum_{k=0}^{\infty} \frac{1}{k!} = e = 2.71828\ldots$

2. $\quad \sum_{k=0}^{\infty} \frac{(-1)^k}{k!} = \frac{1}{e} = 0.36787\ldots$

3. $\quad \sum_{k=1}^{\infty} \frac{k}{(2k+1)!} = \frac{1}{e} = 0.36787\ldots$

4. $\quad \sum_{k=1}^{\infty} \frac{k}{(k+1)!} = 1$

5. $\displaystyle\sum_{k=0}^{\infty} \frac{1}{(2k)!} = \frac{1}{2}\left(e + \frac{1}{e}\right) = 1.54308\ldots$

6. $\displaystyle\sum_{k=0}^{\infty} \frac{1}{(2k+1)!} = \frac{1}{2}\left(e - \frac{1}{e}\right) = 1.17520\ldots$

7. $\displaystyle\sum_{k=0}^{\infty} \frac{(-1)^k}{(2k)!} = \cos 1 = 0.54030\ldots$

8. $\displaystyle\sum_{k=0}^{\infty} \frac{(-1)^{k-1}}{(2k-1)!} = \sin 1 = 0.84147\ldots$

0.246

1. $\displaystyle\sum_{k=0}^{\infty} \frac{1}{(k!)^2} = I_0(2) = 2.27958530\ldots$

2. $\displaystyle\sum_{k=0}^{\infty} \frac{1}{k!(k+1)!} = I_1(2) = 1.590636855\ldots$

3. $\displaystyle\sum_{k=0}^{\infty} \frac{1}{k!(k+n)!} = I_n(2)$

4. $\displaystyle\sum_{k=0}^{\infty} \frac{(-1)^k}{(k!)^2} = J_0(2) = 0.22389078\ldots$

5. $\displaystyle\sum_{k=0}^{\infty} \frac{(-1)^k}{k!(k+1)!} = J_1(2) = 0.57672481\ldots$

6. $\displaystyle\sum_{k=0}^{\infty} \frac{(-1)^k}{k!(k+n)!} = J_n(2)$

0.247 $\displaystyle\sum_{k=1}^{\infty} \frac{k!}{(n+k-1)!} = \frac{1}{(n-2)\cdot(n-1)!}$

0.248 $\displaystyle\sum_{k=1}^{\infty} \frac{k^n}{k!} = S_n,$

$S_1 = e,$	$S_2 = 2e,$	$S_3 = 5e,$	$S_4 = 15e$
$S_5 = 52e,$	$S_6 = 203e,$	$S_7 = 877e,$	$S_8 = 4140e$

0.249[7] $\displaystyle\sum_{k=0}^{\infty} \frac{(k+1)^3}{k!} = 15e$

0.25 Infinite products

0.250 Suppose that a sequence of numbers $a_1, a_2, \ldots, a_k, \ldots$ is given. If the limit $\lim\limits_{n \to \infty} \prod\limits_{k=1}^{n} (1 + a_k)$ exists, whether finite or infinite (but of definite sign), this limit is called the value of the *infinite product* $\prod\limits_{k=1}^{\infty} (1 + a_k)$ and we write

1. $$\lim_{n \to \infty} \prod_{k=1}^{n} (1 + a_k) = \prod_{k=1}^{\infty} (1 + a_k)$$

If an infinite product has a finite *nonzero* value, it is said to converge. Otherwise, the infinite product is said to diverge. We assume that no a_k is equal to -1.   FI II 400

0.251 For the infinite product **0.250** 1. to converge, it is necessary that $\lim\limits_{k \to \infty} a_k = 0$.   FI II 403

0.252 If $a_k > 0$ or $a_k < 0$ for all values of the index k starting with some particular value, then, for the product **0.250** 1 to converge, it is necessary and sufficient that the series $\sum_{k=1}^{\infty} a_k$ converges.

0.253 The product $\prod\limits_{k=1}^{\infty} (1 + a_k)$ is said to converge absolutely if the product $\prod\limits_{k=1}^{\infty} (1 + |a_k|)$ converges.

    FI II 403

0.254 Absolute convergence of an infinite product implies its convergence.

0.255 The product $\prod\limits_{k=1}^{\infty} (1 + a_k)$ converges absolutely if, and only if, the series $\sum\limits_{k=1}^{\infty} a_k$ converges absolutely.   FI II 406

0.26 Examples of infinite products

0.261 $$\prod_{k=1}^{\infty} \left(1 + \frac{(-1)^{k+1}}{2k - 1} \right) = \sqrt{2} \qquad \text{EU}$$

0.262

1. $$\prod_{k=2}^{\infty} \left(1 - \frac{1}{k^2} \right) = \frac{1}{2} \qquad \text{FI II 401}$$

2. $$\prod_{k=1}^{\infty} \left(1 - \frac{1}{(2k)^2} \right) = \frac{2}{\pi} \qquad \text{FI II 401}$$

3. $$\prod_{k=1}^{\infty} \left(1 - \frac{1}{(2k + 1)^2} \right) = \frac{\pi}{4} \qquad \text{FI II 401}$$

0.263 $$\frac{2}{1} \cdot \left(\frac{4}{3} \right)^{1/2} \left(\frac{6 \cdot 8}{5 \cdot 7} \right)^{1/4} \left(\frac{10 \cdot 12 \cdot 14 \cdot 16}{9 \cdot 11 \cdot 13 \cdot 15} \right)^{1/8} \ldots = e$$

0.264 $$\prod_{k=1}^{\infty} \frac{\sqrt[k]{e}}{1 + \frac{1}{k}} = e^C \qquad \text{FI II 402}$$

0.265 $\sqrt{\dfrac{1}{2}} \cdot \sqrt{\dfrac{1}{2} + \dfrac{1}{2}\sqrt{\dfrac{1}{2}}} \cdot \sqrt{\dfrac{1}{2} + \dfrac{1}{2}\sqrt{\dfrac{1}{2} + \dfrac{1}{2}\sqrt{\dfrac{1}{2}}}} \cdots = \dfrac{2}{\pi}$ FI II 402

0.266[8] $\displaystyle\prod_{k=0}^{\infty} \left(1 + x^{2^k}\right) = \dfrac{1}{1-x}$ $[0 < x < 1]$ FI II 401

0.3 Functional series

0.30   Definitions and theorems

0.301   The series

1. $\displaystyle\sum_{k=1}^{\infty} f_k(x),$

the terms of which are functions, is called a *functional series*. The set of values of the independent variable x for which the series **0.301** 1 converges constitutes what is called the *region of convergence* of that series.

0.302   A series that converges for all values of x in a region M is said to *converge uniformly* in that region if, for every $\varepsilon \geq 0$, there exists a number N such that, for $n > N$, the inequality

$$\left| \sum_{k=n+1}^{\infty} f_k(x) \right| < \varepsilon$$

holds for *all* x in M.

0.303   If the terms of the functional series **0.301** 1 satisfy the inequalities:

$$|f_k(x)| < u_k \quad (k = 1, 2, 3, \ldots),$$

throughout the region M, where the u_k are the terms of some *convergent* numerical series

$$\sum_{k=1}^{\infty} u_k = u_1 + u_2 + \ldots + u_k + \ldots,$$

the series **0.301** 1 converges uniformly in M. (Weierstrass)

0.304   Suppose that the series **0.301** 1 converges uniformly in a region M and that a set of functions $g_k(x)$ constitutes (for each x) a monotonic sequence, and that these functions are uniformly bounded, that is, suppose that a number L exists such that the inequalities

1. $|g_n(x)| \leq L$

       hold for all n and x. Then, the series

2. $\displaystyle\sum_{k=1}^{\infty} f_k(x)g_k(x)$

       converges uniformly in the region M. (Abel)   FI II 451

0.305   Suppose that the partial sums of the series **0.301** 1 are uniformly bounded; that is, suppose that, for some L and for all n and x in M, the inequalities

$$\left| \sum_{k=1}^{n} f_k(x) \right| < L$$

hold. Suppose also that for each x the functions $g_n(x)$ constitute a monotonic sequence that approaches zero uniformly in the region M. Then, the series **0.304** 2 converges uniformly in the region M. (Dirichlet)

FI II 451

0.306[6]   If the functions $f_k(x)$ (for $k = 1, 2, 3, \dots$) are integrable on the interval $[a, b]$ and if the series **0.301** 1 made up of these functions converges uniformly on that interval, this series may be integrated *termwise*; that is,

$$\int_a^b \left(\sum_{k=1}^{\infty} f_k(x) \right) dx = \sum_{k=1}^{\infty} \int_a^b f_k(x)\, dx \qquad [a \leq x \leq b] \qquad \text{FI II 459}$$

0.307   Suppose that the functions $f_k(x)$ (for $k = 1, 2, 3, \dots$) have continuous derivatives $f_k'(x)$ on the interval $[a, b]$. If the series **0.301** 1 converges on this interval and if the series $\sum_{k=1}^{\infty} f_k'(x)$ of these derivatives converges uniformly, the series **0.301** 1 may be differentiated termwise; that is,

$$\left\{ \sum_{k=1}^{\infty} f_k(x) \right\}' = \sum_{k=1}^{\infty} f_k'(x) \qquad \text{FI II 460}$$

0.31 Power series

0.311   A functional series of the form

1. $$\sum_{k=0}^{\infty} a_k(x - \xi)^k = a_0 + a_1(x - \xi) + a_2(x - \xi)^2 + \dots$$

is called a *power series*. The following is true of any power series: if it is not everywhere convergent, the region of convergence is a circle with its center at the point ξ and a radius equal to R; at every interior point of this circle, the power series **0.311** 1 converges absolutely and outside this circle, it diverges. This circle is called the *circle of convergence* and its radius is called the *radius of convergence*. If the series converges at all points of the complex plane, we say that the radius of convergence is infinite $(R = +\infty)$.

0.312   Power series may be integrated and differentiated termwise inside the circle of convergence; that is,

$$\int_\xi^x \left\{ \sum_{k=0}^{\infty} a_k(x - \xi)^k \right\} dx = \sum_{k=0}^{\infty} \frac{a_k}{k+1}(x - \xi)^{k+1},$$

$$\frac{d}{dx} \left\{ \sum_{k=0}^{\infty} a_k(x - \xi)^k \right\} = \sum_{k=1}^{\infty} k a_k(x - \xi)^{k-1}.$$

The radius of convergence of a series that is obtained from termwise integration or differentiation of another power series coincides with the radius of convergence of the original series.

Operations on power series

0.313   Division of power series.

$$\frac{\displaystyle\sum_{k=0}^{\infty} b_k x^k}{\displaystyle\sum_{k=0}^{\infty} a_k x^k} = \frac{1}{a_0} \sum_{k=0}^{\infty} c_k x^k,$$

where

$$c_n + \frac{1}{a_0} \sum_{k=1}^{n} c_{n-k} a_k - b_n = 0,$$

or

$$c_n = \frac{(-1)^n}{a_0^n} \begin{bmatrix} a_1 b_0 - a_0 b_1 & a_0 & 0 & \cdots & 0 \\ a_2 b_0 - a_0 b_2 & a_1 & a_0 & & 0 \\ a_3 b_0 - a_0 b_3 & a_2 & a_1 & & 0 \\ \cdots & & \cdots & \cdots & \ddots \\ a_{n-1} b_0 - a_0 b_{n-1} & a_{n-2} & a_{n-3} & \cdots & a_0 \\ a_n b_0 - a_0 b_n & a_{n-1} & a_{n-2} & \cdots & a_1 \end{bmatrix}$$ AD (6360)

0.314   Power series raised to powers.

$$\left(\sum_{k=0}^{\infty} a_k x^k \right)^n = \sum_{k=0}^{\infty} c_k x^k,$$

where

$$c_0 = a_0^n, \quad c_m = \frac{1}{m a_0} \sum_{k=1}^{m} (kn - m + k) a_k c_{m-k} \qquad \text{for } m \geq 1 \qquad [n \text{ is a natural number}] \qquad \text{AD (6361)}$$

0.315   The substitution of one series into another.

$$\sum_{k=1}^{\infty} b_k y^k = \sum_{k=1}^{\infty} c_k x^k \qquad y = \sum_{k=1}^{\infty} a_k x^k;$$

$$c_1 = a_1 b_1, \quad c_2 = a_2 b_1 + a_1^2 b_2, \quad c_3 = a_3 b_1 + 2 a_1 a_2 b_2 + a_1^3 b_3,$$

$$c_4 = a_4 b_1 + a_2^2 b_2 + 2 a_1 a_3 b_2 + 3 a_1^2 a_2 b_3 + a_1^4 b_4, \quad \cdots$$

AD (6362)

0.316   Multiplication of power series

$$\sum_{k=0}^{\infty} a_k x^k \sum_{k=0}^{\infty} b_k x^k = \sum_{k=0}^{\infty} c_k x^k \qquad c_n = \sum_{k=0}^{n} a_k b_{n-k}$$ FI II 372

Taylor series

0.317   If a function $f(x)$ has derivatives of all orders throughout a neighborhood of a point ξ, then we may write the series

1. $f(\xi) + \dfrac{(x - \xi)}{1!} f'(\xi) + \dfrac{(x - \xi)^2}{2!} f''(\xi) + \dfrac{(x - \xi)^3}{3!} f'''(\xi) + \ldots \,,$

    which is known as the *Taylor series* of the function $f(x)$.

    The Taylor series converges to the function $f(x)$ if the remainder

2. $R_n(x) = f(x) - f(\xi) - \displaystyle\sum_{k=1}^{n} \dfrac{(x - \xi)^k}{k!} f^{(k)}(\xi)$

    approaches zero as $n \to \infty$.

    The following are different forms for the remainder of a Taylor series.

3. $R_n(x) = \dfrac{(x - \xi)^{n+1}}{(n + 1)!} f^{(n+1)}(\xi + \theta(x - \xi))$ $\qquad [0 < \theta < 1]$ $\qquad$ (Lagrange)

4. $R_n(x) = \dfrac{(x - \xi)^{n+1}}{n!}(1 - \theta)^n f^{(n+1)}(\xi + \theta(x - \xi))$ $\qquad [0 < \theta < 1]$ $\qquad$ (Cauchy)

5. $R_n(x) = \dfrac{\psi(x - \xi) - \psi(0)}{\psi'[(x - \xi)(1 - \theta)]} \dfrac{(x - \xi)^n (1 - \theta)^n}{n!} f^{(n+1)}(\xi + \theta(x - \xi))$

$$[0 < \theta < 1], \qquad \text{(Schlömilch)}$$

where $\psi(x)$ is an arbitrary function satisfying the following two conditions: (1) It and its derivative $\psi'(x)$ are continuous in the interval $(0, x - \xi)$; and (2) the derivative $\psi'(x)$ does not change sign in that interval. If we set $\psi(x) = x^{p+1}$, we obtain the following form for the remainder:

$$R_n(x) = \frac{(x - \xi)^{n+1}(1 - \theta)^{n-p-1}}{(p + 1)n!} f^{(n+1)}(\xi + \theta(x - \xi)) \qquad [0 < p \leq n; \quad 0 < \theta < 1] \quad \text{(Rouché)}$$

6. $R_n(x) = \dfrac{1}{n!} \displaystyle\int_\xi^x f^{(n+1)}(t)(x - t)^n \, dt$

0.318   Other forms in which a Taylor series may be written:

1. $f(a + x) = \displaystyle\sum_{k=0}^{\infty} \dfrac{x^k}{k!} f^{(k)}(a) = f(a) + \dfrac{x}{1!} f'(a) + \dfrac{x^2}{2!} f'(a) + \dots$

2. $\displaystyle\sum_{k=0}^{\infty} \dfrac{x^k}{k!} f^{(k)}(0) = f(0) + \dfrac{x}{1!} f'(0) + \dfrac{x^2}{2!} f''(0) + \dots$ $\qquad$ (Maclaurin series)

0.319   The Taylor series of functions of several variables:

$$f(x, y) = f(\xi, \eta) + (x - \xi)\frac{\partial f(\xi, \eta)}{\partial x} + (y - \eta)\frac{\partial f(\xi, \eta)}{\partial y}$$

$$+ \frac{1}{2!}\left\{(x - \xi)^2\frac{\partial^2 f(\xi, \eta)}{\partial x^2} + 2(x - \xi)(y - \eta)\frac{\partial^2 f(\xi, \eta)}{\partial x\,\partial y} + (y - \eta)^2\frac{\partial^2 f(\xi, \eta)}{\partial y^2}\right\} + \dots$$

0.32 Fourier series

0.320   Suppose that $f(x)$ is a *periodic* function of period $2l$ and that it is absolutely integrable (possibly improperly) over the interval $(-l, l)$. The following trigonometric series is called the *Fourier series* of $f(x)$:

1. $\dfrac{a_0}{2} + \displaystyle\sum_{k=1}^{\infty}\left(a_k \cos\dfrac{k\pi x}{l} + b_k \sin\dfrac{k\pi x}{l},\right)$

the coefficients of which (the Fourier coefficients) are given by the formulas

2. $a_k = \dfrac{1}{l}\displaystyle\int_{-l}^{l} f(t)\cos\dfrac{k\pi t}{l}\,dt = \dfrac{1}{l}\displaystyle\int_{\alpha}^{\alpha+2l} f(t)\cos\dfrac{k\pi t}{l}\,dt \quad (k = 0, 1, 2, \dots)$

3. $b_k = \dfrac{1}{l}\displaystyle\int_{-t}^{t} f(t)\sin\dfrac{k\pi t}{l}\,dt = \dfrac{1}{l}\displaystyle\int_{\alpha}^{\alpha+2l} f(t)\sin\dfrac{k\pi t}{l}\,dt \quad (k = 1, 2, \dots)$

Convergence tests

0.321   The Fourier series of a function $f(x)$ at a point x_0 converges to the number
$$\frac{f(x_0 + 0) + f(x_0 - 0)}{2},$$
if, for some $h > 0$, the integral
$$\int_0^h \frac{|f(x_0 + t) + f(x_0 - t) - f(x_0 + 0) - f(x_0 - 0)|}{t} \, dt$$
exists. Here, it is assumed that the function $f(x)$ either is continuous at the point x_0 or has a discontinuity of the first kind (a *saltus*) at that point and that both one-sided limits $f(x_0 + 0)$ and $f(x_0 - 0)$ exist. (Dini)

FI III 524

0.322   The Fourier series of a periodic function $f(x)$ that satisfies the Dirichlet conditions on the interval $[a, b]$ converges at every point x_0 to the value $\frac{2}{2}\{f(x_0 + 0) + f(x_0 - 0)\}$. (Dirichlet)

We say that a function $f(x)$ satisfies the Dirichlet conditions on the interval $[a, b]$ if it is bounded on that interval and if the interval $[a, b]$ can be partitioned into a finite number of subintervals inside each of which the function $f(x)$ is continuous and monotonic.

0.323   The Fourier series of a function $f(x)$ at a point x_0 converges to $\frac{1}{2}\{f(x_0 + 0) + f(x_0 - 0)\}$ if $f(x)$ is of bounded variation in some interval $(x_0 - h, x_0 + h)$ with center at x_0. (Jordan–Dirichlet)

FI III 528

The definition of a function of bounded variation. Suppose that a function $f(x)$ is defined on some interval $[a, b]$, where $z < b$. Let us partition this interval in an arbitrary manner into subintervals with the dividing points
$$a = x_0 < x_1 < x_2 < \ldots < x_{n-1} < x_n = b$$
and let us form the sum
$$\sum_{k=1}^{n} |f(x_k) - f(x_{k-1})|$$
Different partitions of the interval $[a, b]$ (that is, different choices of points of division x_i) yield, generally speaking, different sums. If the set of these sums is bounded above, we say that the function $f(x)$ is *of bounded variation* on the interval $[a, b]$. The least upper bound of these sums is called the *total variation* of the function $f(x)$ on the interval $[a, b]$.

0.324   Suppose that a function $f(x)$ is piecewise-continuous on the interval $[a, b]$ and that in each interval of continuity it has a piecewise-continuous derivative. Then, at every point x_0 of the interval $[a, b]$, the Fourier series of the function $f(x)$ converges to $\frac{1}{2}\{f(x_0 + 0) + f(x_0 - 0)\}$.

0.325   A function $f(x)$ defined in the interval $(0, l)$ can be expanded in a cosine series of the form

1. $$\frac{a_0}{2} + \sum_{k=1}^{\infty} a_k \cos \frac{k\pi x}{l},$$
    where

2. $$a_k = \frac{2}{l} \int_0^l f(t) \cos \frac{k\pi t}{l} \, dt$$

0.326   A function $f(x)$ defined in the interval $(0, l)$ can be expanded in a sine series of the form

1. $$\sum_{k=1}^{\infty} b_k \sin \frac{k\pi x}{l},$$
    where

2. $b_k = \dfrac{2}{l} \displaystyle\int_0^l f(t) \sin \dfrac{k\pi t}{l}\, dt$

The convergence tests for the series **0.325** 1 and **0.326** 1 are analogous to the convergence tests for the series **0.320** 1 (see **0.321**–**0.324**).

0.327   The Fourier coefficients a_k and b_k (given by formulas **0.320** 2 and **0.320** 3) of an absolutely integrable function approach zero as $k \to \infty$.

If a function $f(x)$ is square-integrable on the interval $(-l, l)$, the equation of closure is satisfied:

$$\frac{a_0^2}{2} + \sum_{k=1}^{\infty} \left(a_k^2 + b_k^2\right) = \frac{1}{l} \int_{-l}^{l} f^2(x)\, dx \qquad \text{(A. M. Lyapunov)} \qquad\qquad \text{FI III 705}$$

0.328   Suppose that $f(x)$ and $\varphi(x)$ are two functions that are square-integrable on the interval $(-l, l)$ and that a_k, b_k and α_k, β_k are their Fourier coefficients. For such functions, the generalized equation of closure (Parseval's equation) holds:

$$\frac{a_0 \alpha_0}{2} + \sum_{k=1}^{\infty} \left(a_k \alpha_k + b_k \beta_k\right) = \frac{1}{l} \int_{-l}^{l} f(x)\varphi(x)\, dx \qquad\qquad \text{FI III 709}$$

For examples of Fourier series, see **1.44** and **1.45**.

0.33   Asymptotic series

0.330   Included in the collection of all divergent series is the broad class of series known as *asymptotic* or *semiconvergent* series. *Despite the fact that these series diverge*, the values of the functions that they represent can be calculated with a high degree of accuracy if we take the sum of a suitable number of terms of such series. In the case of alternating asymptotic series, we obtain greatest accuracy if we break off the series in question at whatever term is of lowest absolute value. In this case, the error (in absolute value) does not exceed the absolute value of the first of the discarded terms (cf. **0.227** 3).

Asymptotic series have many properties that are analogous to the properties of convergent series and, for that reason, they play a significant role in analysis.

The asymptotic expansion of a function is denoted as follows:

$$f(z) \sim \sum_{n=0}^{\infty} A_n z^{-n}$$

The definition of an asymptotic expansion. The divergent series $\displaystyle\sum_{n=0}^{\infty} \frac{A_n}{z^n}$ is called the *asymptotic expansion* of a function $f(z)$ in a given region of values of $\arg z$ if the expression $R_n(z) = z^n\left[f(z) - S_n(z)\right]$, where $S_n(z) = \displaystyle\sum_{k=0}^{n} \frac{A_k}{z^k}$, satisfies the condition $\displaystyle\lim_{|z|\to\infty} R_n(z) = 0$ for fixed n.   FI II 820

A divergent series that represents the asymptotic expansion of some function is called an *asymptotic series*.

0.331   Properties of asymptotic series

1.   The operations of addition, subtraction, multiplication, and raising to a power can be performed on asymptotic series just as on absolutely convergent series. The series obtained as a result of these operations will also be asymptotic.

2.   One asymptotic series can be divided by another provided that the first term A_0 of the divisor is not equal to zero. The series obtained as a result of division will also be asymptotic.

FI II 823-825

3.   An asymptotic series can be integrated termwise, and the resultant series will also be asymptotic. In contrast, differentiation of an asymptotic series is, in general, not permissible.   FI II 824

4.   A single asymptotic expansion can represent different functions. On the other hand, a given function can be expanded in an asymptotic series in only one manner.

0.4 Certain formulas from differential calculus

0.41 Differentiation of a definite integral with respect to a parameter

0.410 $\dfrac{d}{da}\displaystyle\int_{\psi(a)}^{\varphi(a)} f(x,a)\,dx = f(\varphi(a),a)\dfrac{d\varphi(a)}{da} - f(\psi(a),a)\dfrac{d\psi(a)}{da} + \displaystyle\int_{\psi(a)}^{\varphi(a)} \dfrac{d}{da}f(x,a)\,dx$ FI II 680

0.411   In particular,

1. $\dfrac{d}{da}\displaystyle\int_{b}^{a} f(x)\,dx = f(a)$

2. $\dfrac{d}{db}\displaystyle\int_{b}^{a} f(x)\,dx = -f(b)$

0.42 The nth derivative of a product (Leibniz's rule)

Suppose that u and v are n-times-differentiable functions of x. Then,

$$\frac{d^n(uv)}{dx^n} = u\frac{d^n v}{dx^n} + \binom{n}{1}\frac{du}{dx}\frac{d^{n-1}v}{dx^{n-1}} + \binom{n}{2}\frac{d^2 u}{dx^2}\frac{d^{n-2}v}{dx^{n-2}} + \binom{n}{3}\frac{d^3 u}{dx^3}\frac{d^{n-3}v}{dx^{n-3}} + \cdots + v\frac{d^n u}{dx^n}$$

or, symbolically,

$$\frac{d^n(uv)}{dx^n} = (u+v)^{(n)}$$ FI I 272

0.43 The nth derivative of a composite function

0.430   If $f(x) = F(y)$ and $y = \varphi(x)$, then

1. $\dfrac{d^n}{dx^n}f(x) = \dfrac{U_1}{1!}F'(y) + \dfrac{U_2}{2!}F''(y) + \dfrac{U_3}{3!}F'''(y) + \ldots + \dfrac{U_n}{n!}F^{(n)}(y),$

     where

$$U_k = \frac{d^n}{dx^n}y^k - \frac{k}{1!}y\frac{d^n}{dx^n}y^{k-1} + \frac{k(k-1)}{2!}y^2\frac{d^n}{dx^n}y^{k-2} - \ldots + (-1)^{k-1}ky^{k-1}\frac{d^n y}{dx^n}$$ AD (7361) GO

2. $\dfrac{d^n}{dx^n}f(x) = \displaystyle\sum \dfrac{n!}{i!\,j!\,h!\ldots k!}\dfrac{d^m F}{dy^m}\left(\dfrac{y'}{1!}\right)^i\left(\dfrac{y''}{2!}\right)^j\left(\dfrac{y'''}{3!}\right)^h\cdots\left(\dfrac{y^{(l)}}{l!}\right)^k,$

     Here, the symbol $\sum$ indicates summation over all solutions in non negative integers of the equation $i + 2j + 3h + \ldots + lk = n$ and $m = i + j + h + \ldots + k$.

0.431

1.
$$(-1)^n \frac{d^n}{dx^n} F\left(\frac{1}{x}\right) = \frac{1}{x^{2n}} F^{(n)}\left(\frac{1}{x}\right) + \frac{n-1}{x^{2n-1}} \frac{n}{1!} F^{(n-1)}\left(\frac{1}{x}\right)$$
$$+ \frac{(n-1)(n-2)}{x^{2n-2}} \frac{n(n-1)}{2!} F^{(n-2)}\left(\frac{1}{x}\right) + \dots$$

AD (7362.1)

2.
$$(-1)^n \frac{d^n}{dx^n} e^{\frac{a}{x}} = \frac{1}{x^n} e^{\frac{a}{x}} \left[\left(\frac{a}{x}\right)^n + (n-1)\binom{n}{1}\left(\frac{a}{x}\right)^{n-1} + (n-1)(n-2)\binom{n}{2}\left(\frac{a}{x}\right)^{n-2} \right.$$
$$\left. + (n-1)(n-2)(n-3)\binom{n}{3}\left(\frac{a}{x}\right)^{n-3} + \dots \right]$$

AD (7362.2)

0.432

1.
$$\frac{d^n}{dx^n} F\left(x^2\right) = (2x)^n F^{(n)}\left(x^2\right) + \frac{n(n-1)}{1!}(2x)^{n-2} F^{(n-1)}\left(x^2\right)$$
$$+ \frac{n(n-1)(n-2)(n-3)}{2!}(2x)^{n-4} F^{(n-2)}\left(x^2\right) +$$
$$+ \frac{n(n-1)(n-2)(n-3)(n-4)(n-5)}{3!}(2x)^{n-6} F^{(n-3)}\left(x^2\right) + \dots$$

AD (7363.1)

2.
$$\frac{d^n}{dx^n} e^{ax^2} = (2ax)^n e^{ax^2} \left[1 + \frac{n(n-1)}{1!\,(4ax^2)} + \frac{n(n-1)(n-2)(n-3)}{2!(4ax^2)^2} \right.$$
$$\left. + \frac{n(n-1)(n-2)(n-3)(n-4)(n-5)}{3!(4ax^2)^3} + \dots \right]$$

AD (7363.2)

3.
$$\frac{d^n}{dx^n}\left(1 + ax^2\right)^p = \frac{p(p-1)(p-2)\dots(p-n+1)(2ax)^n}{(1+ax^2)^{n-p}}$$
$$\times \left\{ 1 + \frac{n(n-1)}{1!(p-n+1)} \frac{1+ax^2}{4ax^2} + \frac{n(n-1)(n-2)(n-3)}{2!(p-n+1)(p-n+2)} \left(\frac{1+ax^2}{4ax^2}\right)^2 + \dots \right\},$$

AD (7363.3)

4.
$$\frac{d^{m-1}}{dx^{m-1}}\left(1-x^2\right)^{m-\frac{1}{2}} = (-1)^{m-1} \frac{(2m-1)!!}{m} \sin\left(m \arccos x\right)$$

AD (7363.4)

5.*
$$(-1)^n \frac{\partial^n}{\partial a^n}\left(\frac{a}{a^2+b^2}\right) = n!\left(\frac{a}{a^2+b^2}\right)^{n+1} \sum_{0 \le 2k \le n+1}(-1)^k \binom{n+1}{2k}\left(\frac{b}{a}\right)^{2k}$$

(3.944.12)

6.*
$$(-1)^n \frac{\partial^n}{\partial a^n}\left(\frac{b}{a^2+b^2}\right) = n!\left(\frac{a}{a^2+b^2}\right)^{n+1} \sum_{0 \le 2k \le n}(-1)^k \binom{n+1}{2k+1}\left(\frac{b}{a}\right)^{2k+1}$$

(3.944.11)

0.433

1. $\dfrac{d^n}{dx^n} F\left(\sqrt{x}\right) = \dfrac{F^{(n)}\left(\sqrt{x}\right)}{\left(2\sqrt{x}\right)^n} - \dfrac{n(n-1)}{1!}\dfrac{F^{(n-1)}\left(\sqrt{x}\right)}{\left(2\sqrt{x}\right)^{n+1}} + \dfrac{(n+1)n(n-1)(n-2)}{2!}\dfrac{F^{(n-2)}\left(\sqrt{x}\right)}{\left(2\sqrt{x}\right)^{n+2}} - \cdots$

AD (7364.1)

2. $\dfrac{d^n}{dx^n}\left(1 + a\sqrt{x}\right)^{2n-1} = \dfrac{(2n-1)!!}{2^n}\dfrac{a}{\sqrt{x}}\left(a^2 - \dfrac{1}{x}\right)^{n-1}$

AD (7364.2)

0.434 $\dfrac{d^n}{dx^n} y^p = p\begin{pmatrix} n-p \\ n \end{pmatrix}\left\{-\begin{pmatrix} n \\ 1 \end{pmatrix}\dfrac{1}{p-1}y^{p-1}\dfrac{d^n y}{dx^n} + \begin{pmatrix} n \\ 2 \end{pmatrix}\dfrac{1}{p-2}y^{p-2}\dfrac{d^n\left(y^2\right)}{dx^n} - \cdots\right\}$

AD (737.1)

0.435 $\dfrac{d^n}{dx^n}\ln y = \left\{\begin{pmatrix} n \\ 1 \end{pmatrix}\dfrac{1}{1\cdot y}\dfrac{d^n y}{dx^n} - \begin{pmatrix} n \\ 2 \end{pmatrix}\dfrac{1}{2\cdot y^2}\dfrac{d^n\left(y^2\right)}{dx^n} + \dfrac{d^n\left(y^3\right)}{dx^n}x^n - \cdots\right\}$

AD (737.2)

0.44 Integration by substitution

0.440   Let $f(x)$ and $g(x)$ be continuous in $[a, b]$. Further, let $g'(x)$ exist and be continuous there. Then

$$\int_a^b f[g(x)]g'(x)\, dx = \int_{g(a)}^{g(b)} f(u)\, du$$

1 Elementary Functions

1.1 Power of Binomials

1.11 Power series

1.110 $(1+x)^q = 1 + qx + \dfrac{q(q-1)}{2!}x^2 + \cdots + \dfrac{q(q-1)\ldots(q-k+1)}{k!}x^k + \cdots = \sum_{k=0}^{\infty} \binom{q}{k} x^k$

If q is neither a natural number nor zero, the series converges absolutely for $|x| < 1$ and diverges for $|x| > 1$. For $x = 1$, the series converges for $q > -1$ and diverges for $q \leq -1$. For $x = 1$, the series converges absolutely for $q > 0$. For $x = -1$, it converges absolutely for $q > 0$ and diverges for $q < 0$. If $q = n$ is a natural number, the series **1.110** is reduced to the finite sum **1.111**.   FI II 425

1.111 $(a+x)^n = \sum_{k=0}^{n} \binom{n}{k} x^k a^{n-k}$

1.112

1. $(1+x)^{-1} = 1 - x + x^2 - x^3 + \cdots = \sum_{k=1}^{\infty} (-1)^{k-1} x^{k-1}$

    (see also **1.121** 2)

2. $(1+x)^{-2} = 1 - 2x + 3x^2 - 4x^3 + \cdots = \sum_{k=1}^{\infty} (-1)^{k-1} k x^{k-1}$

3. $(1+x)^{1/2} = 1 + \dfrac{1}{2}x + \dfrac{1 \cdot 1}{2 \cdot 4}x^2 + \dfrac{1 \cdot 1 \cdot 3}{2 \cdot 4 \cdot 6}x^3 - \dfrac{1 \cdot 1 \cdot 3 \cdot 5}{2 \cdot 4 \cdot 6 \cdot 8}x^4 + \cdots$

4. $(1+x)^{-1/2} = 1 - \dfrac{1}{2}x + \dfrac{1 \cdot 3}{2 \cdot 4}x^2 - \dfrac{1 \cdot 3 \cdot 5}{2 \cdot 4 \cdot 6}x^3 + \cdots$

1.113 $\dfrac{x}{(1-x)^2} = \sum_{k=1}^{\infty} k x^k$ $[x^2 < 1]$

1.114

1. $\left(1 + \sqrt{1+x}\right)^q = 2^q \left[1 + \dfrac{q}{1!}\left(\dfrac{x}{4}\right) + \dfrac{q(q-3)}{2!}\left(\dfrac{x}{4}\right)^2 + \dfrac{q(q-4)(q-5)}{3!}\left(\dfrac{x}{4}\right)^3 + \cdots\right]$

    $[x^2 < 1, \quad q \text{ is a real number}]$

    AD (6351.1)

2. $\qquad \left(x+\sqrt{1+x^2}\right)^q = 1 + \displaystyle\sum_{k=0}^{\infty} \frac{q^2\left(q^2-2^2\right)\left(q^2-4^2\right)\cdots\left[q^2-(2k)^2\right]x^{2k+2}}{(2k+2)!}$

$$+qx + q\sum_{k=1}^{\infty} \frac{\left(q^2-1^2\right)\left(q^2-3^2\right)\cdots\left[q^2-(2k-1)^2\right]}{(2k+1)!}x^{2k+1}$$

$$\left[x^2 < 1, \quad q \text{ is a real number}\right] \qquad \text{AD}(6351.2)$$

1.12 Series of rational fractions

1.121

1. $\qquad \dfrac{x}{1-x} = \displaystyle\sum_{k=1}^{\infty} \frac{2^{k-1}x^{2^{k-1}}}{1+x^{2^{k-1}}} = \sum_{k=1}^{\infty}\frac{x^{2^{k-1}}}{1-x^{2^k}}$ $\qquad\qquad \left[x^2 < 1\right] \qquad\qquad$ AD (6350.3)

2. $\qquad \dfrac{1}{x-1} = \displaystyle\sum_{k=1}^{\infty} \frac{2^{k-1}}{x^{2^{k-1}}+1}$ $\qquad\qquad\qquad\qquad \left[x^2 > 1\right] \qquad\qquad$ AD (6350.3)

1.2 The Exponential Function

1.21 Series representation

1.211

1.[7] $\qquad e^x = \displaystyle\sum_{k=0}^{\infty} \frac{x^h}{k!}$

2. $\qquad a^x = \displaystyle\sum_{k=0}^{\infty} \frac{(x\ln a)^k}{k!}$

3. $\qquad e^{-x^2} = \displaystyle\sum_{k=0}^{\infty} (-1)^k\frac{x^{2k}}{k!}$

1.212 $\quad e^x(1+x) = \displaystyle\sum_{k=0}^{\infty} \frac{x^k(k+1)}{k!}$

1.213 $\quad \dfrac{x}{e^x-1} = 1 - \dfrac{x}{2} + \displaystyle\sum_{k=1}^{\infty}\frac{B_{2k}x^{2k}}{(2k)!}$ $\qquad\qquad \left[x < 2\pi\right] \qquad\qquad$ FI II 520

1.214 $\quad e^{e^x} = e\left(1 + x + \dfrac{2x^2}{2!} + \dfrac{5x^3}{3!} + \dfrac{15x^4}{4!} + \cdots\right)$ $\qquad\qquad$ AD (6460.3)

1.215

1. $\qquad e^{\sin x} = 1 + x + \dfrac{x^2}{2!} - \dfrac{3x^4}{4!} - \dfrac{8x^5}{5!} - \dfrac{3x^6}{6!} + \dfrac{56x^7}{7!} + \cdots$ $\qquad$ AD (6460.4)

2. $\qquad e^{\cos x} = e\left(1 - \dfrac{x^2}{2!} + \dfrac{4x^4}{4!} - \dfrac{31x^6}{6!} + \cdots\right)$ $\qquad\qquad$ AD (6460.5)

3. $e^{\tan x} = 1 + x + \dfrac{x^2}{2!} + \dfrac{3x^3}{3!} + \dfrac{9x^4}{4!} + \dfrac{37x^5}{5!} + \ldots$ AD (6460.6)

1.216

1. $e^{\arcsin x} = 1 + x + \dfrac{x^2}{2!} + \dfrac{2x^3}{3!} + \dfrac{5x^4}{4!} + \ldots$ AD (6460.7)

2. $e^{\arctan x} = 1 + x + \dfrac{x^2}{2!} - \dfrac{x^3}{3!} - \dfrac{7x^4}{4!} + \ldots$ AD (6460.8)

1.217

1. $\pi \dfrac{e^{\pi x} + e^{-\pi x}}{e^{\pi x} - e^{-\pi x}} = \dfrac{1}{x} + 2x \displaystyle\sum_{k=1}^{\infty} \dfrac{1}{x^2 + k^2}$ (cf. **1.421** 3)   AD (6707.1)

2. $\dfrac{2\pi}{e^{\pi x} - e^{-\pi x}} = \dfrac{1}{x} + 2x \displaystyle\sum_{k=1}^{\infty} (-1)^k \dfrac{1}{x^2 + k^2}$ (cf. **1.422** 3)   AD (6707.2)

1.22 Functional relations

1.221

1. $a^x = e^{x \ln a}$

2. $a^{\log_a x} = a^{\frac{1}{\log_x a}} = x.$

1.222

1. $e^x = \cosh x + \sinh x$

2. $e^{ix} = \cos x + i \sin x$

1.223 $e^{ax} - e^{bx} = (a - b)x \exp\left[\dfrac{1}{2}(a + b)x\right] \displaystyle\prod_{k=1}^{\infty} \left[1 + \dfrac{(a - b)^2 x^2}{2k^2 \pi^2}\right]$ MO 216

1.23 Series of exponentials

1.231 $\displaystyle\sum_{k=0}^{\infty} a^{kx} = \dfrac{1}{1 - a^x}$ $[a > 1 \text{ and } x < 0 \text{ or } 0 < a < 1 \text{ and } x > 0]$

1.232

1. $\tanh x = 1 + 2 \displaystyle\sum_{k=1}^{\infty} (-1)^k e^{-2kx}$ $[x > 0]$

2. $\operatorname{sech} x = 2 \displaystyle\sum_{k=0}^{\infty} (-1)^k e^{-(2k+1)x}$ $[x > 0]$

3. $\operatorname{cosech} x = 2 \displaystyle\sum_{k=0}^{\infty} e^{-(2k+1)x}$ $[x > 0]$

1.3–1.4 Trigonometric and Hyperbolic Functions

1.30 Introduction

The trigonometric and hyperbolic sines are related by the identities

$$\sinh x = \frac{1}{i}\sin(ix), \qquad \sin x = \frac{1}{i}\sinh(ix).$$

The trigonometric and hyperbolic cosines are related by the identities

$$\cosh x = \cos(ix), \qquad \cos x = \cosh(ix).$$

Because of this duality, every relation involving trigonometric functions has its formal counterpart involving the corresponding hyperbolic functions, and vice-versa. In many (though not all) cases, both pairs of relationships are meaningful.

The idea of matching the relationships is carried out in the list of formulas given below. However, not all the meaningful "pairs" are included in the list.

1.31 The basic functional relations

1.311

1. $\sin x = \dfrac{1}{2i}\left(e^{ix} - e^{-ix}\right)$

        $= -i\sinh(ix)$

2. $\sinh x = \dfrac{1}{2}\left(e^{x} - e^{-x}\right)$

        $= -i\sin(ix)$

3. $\cos x = \dfrac{1}{2}\left(e^{ix} + e^{-ix}\right)$

        $= \cosh(ix)$

4. $\cosh x = \dfrac{1}{2}\left(e^{x} + e^{-x}\right)$

        $= \cos(ix)$

5. $\tan x = \dfrac{\sin x}{\cos x} = \dfrac{1}{i}\tanh(ix)$

6. $\tanh x = \dfrac{\sinh x}{\cosh x} = \dfrac{1}{i}\tan(ix)$

7. $\cot x = \dfrac{\cos x}{\sin x} = \dfrac{1}{\tan x} = i\coth(ix)$

8. $\coth x = \dfrac{\cosh x}{\sinh x} = \dfrac{1}{\tanh x} = i\cot(ix)$

1.312

1. $\cos^2 x + \sin^2 x = 1$

2. $\cosh^2 x - \sinh^2 x = 1$

1.313

1. $\sin(x \pm y) = \sin x \cos y \pm \sin y \cos x$

2. $\sinh(x \pm y) = \sinh x \cosh y \pm \sinh y \cosh x$

3. $\sin(x \pm iy) = \sin x \cosh y \pm i \sinh y \cos x$

4. $\sinh(x \pm iy) = \sinh x \cos y \pm i \sin y \cosh x$

5. $\cos(x \pm y) = \cos x \cos y \mp \sin x \sin y$

6. $\cosh(x \pm y) = \cosh x \cosh y \pm \sinh x \sinh y$

7. $\cos(x \pm iy) = \cos x \cosh y \mp i \sin x \sinh y$

8. $\cosh(x \pm iy) = \cosh x \cos y \pm i \sinh x \sin y$

9. $\tan(x \pm y) = \dfrac{\tan x \pm \tan y}{1 \mp \tan x \tan y}$

10. $\tanh(x \pm y) = \dfrac{\tanh x \pm \tanh y}{1 \pm \tanh x \tanh y}$

11. $\tan(x \pm iy) = \dfrac{\tan x \pm i \tanh y}{1 \mp i \tan x \tanh y}$

12. $\tanh(x \pm iy) = \dfrac{\tanh x \pm i \tan y}{1 \pm i \tanh x \tan y}$

1.314

1. $\sin x \pm \sin y = 2 \sin \dfrac{1}{2}(x \pm y) \cos \dfrac{1}{2}(x \mp y)$

2. $\sinh x \pm \sinh y = 2 \sinh \dfrac{1}{2}(x \pm y) \cosh \dfrac{1}{2}(x \mp y)$

3. $\cos x + \cos y = 2 \cos \dfrac{1}{2}(x + y) \cos \dfrac{1}{2}(x - y)$

4. $\cosh x + \cosh y = 2 \cosh \dfrac{1}{2}(x + y) \cosh \dfrac{1}{2}(x - y)$

5. $\cos x - \cos y = 2 \sin \dfrac{1}{2}(x + y) \sin \dfrac{1}{2}(y - x)$

6. $\cosh x - \cosh y = 2 \sinh \dfrac{1}{2}(x + y) \sinh \dfrac{1}{2}(x - y)$

7. $\tan x \pm \tan y = \dfrac{\sin(x \pm y)}{\cos x \cos y}$

8. $\tanh x \pm \tanh y = \dfrac{\sinh(x \pm y)}{\cosh x \cosh y}$

1.315

1. $\sin^2 x - \sin^2 y = \sin(x + y)\sin(x - y) = \cos^2 y - \cos^2 x$

2. $\sinh^2 x - \sinh^2 y = \sinh(x + y)\sinh(x - y) = \cosh^2 x - \cosh^2 y$

3. $\cos^2 x - \sin^2 y = \cos(x+y)\cos(x-y) = \cos^2 y - \sin^2 x$

4. $\sinh^2 x + \cosh^2 y = \cosh(x+y)\cosh(x-y) = \cosh^2 x + \sinh^2 y$

1.316

1. $(\cos x + i \sin x)^n = \cos nx + i \sin nx$ [n is an integer]

2. $(\cosh x + \sinh x)^n = \sinh nx + \cosh nx$ [n is an integer]

1.317

1. $\sin \dfrac{x}{2} = \pm\sqrt{\dfrac{1}{2}(1-\cos x)}$

2. $\sinh \dfrac{x}{2} = \pm\sqrt{\dfrac{1}{2}(\cosh x - 1)}$

3. $\cos \dfrac{x}{2} = \pm\sqrt{\dfrac{1}{2}(1+\cos x)}$

4. $\cosh \dfrac{x}{2} = \sqrt{\dfrac{1}{2}(\cosh x + 1)}$

5. $\tan \dfrac{x}{2} = \dfrac{1-\cos x}{\sin x} = \dfrac{\sin x}{1+\cos x}$

6. $\tanh \dfrac{x}{2} = \dfrac{\cosh x - 1}{\sinh x} = \dfrac{\sinh x}{\cosh x + 1}$

     The signs in front of the radical in formulas **1.317** 1, **1.317** 2, and **1.317** 3 are taken so as to agree with the signs of the left hand members. The sign of the left hand members depends in turn on the value of x.

1.32 The representation of powers of trigonometric and hyperbolic functions in terms of functions of multiples of the argument (angle)

1.320

1. $\sin^{2n} x = \dfrac{1}{2^{2n}} \left\{ \displaystyle\sum_{k=0}^{n-1} (-1)^{n-k} 2 \binom{2n}{k} \cos 2(n-k)x + \binom{2n}{n} \right\}$ KR 56 (10, 2)

2. $\sinh^{2n} x = \dfrac{(-1)^n}{2^{2n}} \left\{ \displaystyle\sum_{k=0}^{n-1} (-1)^{n-k} 2 \binom{2n}{k} \cosh 2(n-k)x + \binom{2n}{n} \right\}$

3. $\sin^{2n-1} x = \dfrac{1}{2^{2n-2}} \displaystyle\sum_{k=0}^{n-1} (-1)^{n+k-1} \binom{2n-1}{k} \sin(2n-2k-1)x$ KR 56 (10, 4)

4. $\sinh^{2n-1} x = \dfrac{(-1)^{n-1}}{2^{2n-2}} \displaystyle\sum_{k=0}^{n-1} (-1)^{n+k-1} \binom{2n-1}{k} \sinh(2n-2k-1)x.$

5. $\cos^{2n} x = \dfrac{1}{2^{2n}} \left\{ \displaystyle\sum_{k=0}^{n-1} 2 \binom{2n}{k} \cos 2(n-k)x + \binom{2n}{n} \right\}$ KR 56 (10, 1)

6. $\cosh^{2n} x = \dfrac{1}{2^{2n}} \left\{ \displaystyle\sum_{k=0}^{n-1} 2 \binom{2n}{k} \cosh 2(n-k)x + \binom{2n}{n} \right\}$

7. $\cos^{2n-1} x = \dfrac{1}{2^{2n-2}} \displaystyle\sum_{k=0}^{n-1} \binom{2n-1}{k} \cos(2n-2k-1)x.$ KR 56 (10, 3)

8. $\cosh^{2n-1} x = \dfrac{1}{2^{2n-2}} \displaystyle\sum_{k=0}^{n-1} \binom{2n-1}{k} \cosh(2n-2k-1)x.$

Special cases

1.321

1. $\sin^2 x = \dfrac{1}{2}\left(-\cos 2x + 1\right)$

2. $\sin^3 x = \dfrac{1}{4}\left(-\sin 3x + 3\sin x\right)$

3. $\sin^4 x = \dfrac{1}{8}\left(\cos 4x - 4\cos 2x + 3\right)$

4. $\sin^5 x = \dfrac{1}{16}\left(\sin 5x - 5\sin 3x + 10\sin x\right)$

5. $\sin^6 x = \dfrac{1}{32}\left(-\cos 6x + 6\cos 4x - 15\cos 2x + 10\right)$

6. $\sin^7 x = \dfrac{1}{64}\left(-\sin 7x + 7\sin 5x - 21\sin 3x + 35\sin x\right)$

1.322

1. $\sinh^2 x = \dfrac{1}{2}\left(\cosh 2x - 1\right)$

2. $\sinh^3 x = \dfrac{1}{4}\left(\sinh 3x - 3\sinh x\right)$

3. $\sinh^4 x = \dfrac{1}{8}\left(\cosh 4x - 4\cosh 2x + 3\right)$

4. $\sinh^5 x = \dfrac{1}{16}\left(\sinh 5x - 5\sinh 3x + 10\sinh x\right)$

5. $\sinh^6 x = \dfrac{1}{32}\left(\cosh 6x - 6\cosh 4x + 15\cosh 2x + 10\right)$

6. $\sinh^7 x = \dfrac{1}{64}\left(\sinh 7x - 7\sinh 5x + 21\sinh 3x + 35\sinh x\right)$

1.323

1. $\cos^2 x = \dfrac{1}{2}\left(\cos 2x + 1\right)$

2. $\cos^3 x = \dfrac{1}{4}\left(\cos 3x + 3\cos x\right)$

3. $\cos^4 x = \dfrac{1}{8}\left(\cos 4x + 4\cos 2x + 3\right)$

4. $\cos^5 x = \dfrac{1}{16}\left(\cos 5x + 5\cos 3x + 10\cos x\right)$

5. $\cos^6 x = \dfrac{1}{32}\left(\cos 6x + 6\cos 4x + 15\cos 2x + 10\right)$

6. $\cos^7 x = \dfrac{1}{64}\left(\cos 7x + 7\cos 5x + 21\cos 3x + 35\cos x\right)$

1.324

1. $\cosh^2 x = \dfrac{1}{2}\left(\cosh 2x + 1\right)$

2. $\cosh^3 x = \dfrac{1}{4}\left(\cosh 3x + 3\cosh x\right)$

3. $\cosh^4 x = \dfrac{1}{8}\left(\cosh 4x + 4\cosh 2x + 3\right)$

4. $\cosh^5 x = \dfrac{1}{16}\left(\cosh 5x + 5\cosh 3x + 10\cosh x\right)$

5. $\cosh^6 x = \dfrac{1}{32}\left(\cosh 6x + 6\cosh 4x + 15\cosh 2x + 10\right)$

6. $\cosh^7 x = \dfrac{1}{64}\left(\cosh 7x + 7\cosh 5x + 21\cosh 3x + 35\cosh x\right)$

1.33 The representation of trigonometric and hyperbolic functions of multiples of the argument (angle) in terms of powers of these functions

1.331

1.[7] $\sin nx = n\cos^{n-1} x \sin x - \dbinom{n}{3}\cos^{n-3} x \sin^3 x + \dbinom{n}{5}\cos^{n-5} x \sin^5 x - \dots ;$

$$= \sin x\left\{ 2^{n-1}\cos^{n-1} x - \dbinom{n-2}{1} 2^{n-3}\cos^{n-3} x \right.$$

$$\left. + \dbinom{n-3}{2} 2^{n-5}\cos^{n-5} x - \dbinom{n-4}{3} 2^{n-7}\cos^{n-7} x + \dots \right\}$$

AD (3.175)

2. $\sinh nx = x \displaystyle\sum_{k=1}^{[(n+1)/2]} \dbinom{n}{2k-1}\sinh^{2k-2} x \cosh^{n-2k+1} x$

$$= \sinh x \sum_{k=0}^{[(n-1)/2]} (-1)^k \dbinom{n-k-1}{k} 2^{n-2k-1}\cosh^{n-2k-1} x$$

3. $\cos nx = \cos^n x - \dbinom{n}{2}\cos^{n-2} x \sin^2 x + \dbinom{n}{4}\cos^{n-4} x \sin^4 x - \dots ;$

$$= 2^{n-1}\cos^n x - \frac{n}{1} 2^{n-3}\cos^{n-2} x + \frac{n}{2}\dbinom{n-3}{1} 2^{n-5}\cos^{n-4} x$$

$$- \frac{n}{3}\dbinom{n-4}{2} 2^{n-7}\cos^{n-6} x + \dots$$

AD (3.175)

$4.^3 \qquad \cosh nx = \sum_{k=0}^{[n/2]} \binom{n}{2k} \sinh^{2k} x \cosh^{n-2k} x$

$$= 2^{n-1} \cosh^n x + n \sum_{k=1}^{[n/2]} (-1)^k \frac{1}{k} \binom{n-k-1}{k-1} 2^{n-2k-1} \cosh^{n-2k} x$$

1.332

1. $\sin 2nx = 2n \cos x \left\{ \sin x - \dfrac{4n^2 - 2^2}{3!} \sin^3 x + \dfrac{\left(4n^2 - 2^2\right)\left(4n^2 - 4^2\right)}{5!} \sin^5 x - \ldots \right\}$ AD (3.171)

$$= (-1)^{n-1} \cos x \left\{ 2^{2n-1} \sin^{2n-1} x - \frac{2n-2}{1!} 2^{2n-3} \sin^{2n-3} x \right.$$

$$+ \frac{(2n-3)(2n-4)}{2!} 2^{2n-5} \sin^{2n-5} x$$

$$\left. - \frac{(2n-4)(2n-5)(2n-6)}{3!} 2^{2n-7} \sin^{2n-7} x + \ldots \right\} \qquad \text{AD (3.173)}$$

2. $\sin(2n-1)x = (2n-1) \left\{ \sin x - \dfrac{(2n-1)^2 - 1^2}{3!} \sin^3 x \right.$

$$\left. + \frac{\left[(2n-1)^2 - 1^2\right]\left[(2n-1)^2 - 3^2\right]}{5!} \sin^5 x - \ldots \right\} \qquad \text{AD (3.172)}$$

$$= (-1)^{n-1} \left\{ 2^{2n-2} \sin^{2n-1} x - \frac{2n-1}{1!} 2^{2n-4} \sin^{2n-3} x \right.$$

$$+ \frac{(2n-1)(2n-4)}{2!} 2^{2n-6} \sin^{2n-5} x$$

$$\left. - \frac{(2n-1)(2n-5)(2n-6)}{3!} 2^{2n-8} \sin^{2n-7} x + \ldots \right\} \qquad \text{AD (3.174)}$$

3. $\cos 2nx = 1 - \dfrac{4n^2}{2!} \sin^2 x + \dfrac{4n^2 \left(4n^2 - 2^2\right)}{4!} \sin^4 x - \dfrac{4n^2 \left(4n^2 - 2\right)\left(4n^2 - 4^2\right)}{6!} \sin^6 x + \ldots$

AD (3.171)

$$= (-1)^n \left\{ 2^{2n-1} \sin^{2n} x - \frac{2n}{1!} 2^{2n-3} \sin^{2n-2} x \right.$$

$$\left. + \frac{2n(2n-3)}{2!} 2^{2n-5} \sin^{2n-4} x - \frac{2n(2n-4)(2n-5)}{3!} 2^{2n-7} \sin^{2n-6} x + \ldots \right\}$$

AD (3.173)a

4.
$$\cos(2n-1)x = \cos x \left\{ 1 - \frac{(2n-1)^2 - 1^2}{2!} \sin^2 x \right.$$
$$\left. + \frac{\left[(2n-1)^2 - 1^2\right]\left[(2n-1)^2 - 3^2\right]}{4!} \sin^4 x - \ldots \right\}$$

AD (3.172)

$$= (-1)^{n-1} \cos x \left\{ 2^{2n-2} \sin^{2n-2} x - \frac{2n-3}{1!} 2^{2n-4} \sin^{2n-4} x \right.$$
$$+ \frac{(2n-4)(2n-5)}{2!} 2^{2n-6} \sin^{2n-6} x$$
$$\left. - \frac{(2n-5)(2n-6)(2n-7)}{3!} 2^{2n-8} \sin^{2n-8} x + \ldots \right\}$$

AD (3.174)

By using the formulas and values of **1.30**, we can write formulas for $\sinh 2nx$, $\sinh(2n-1)x$, $\cosh 2nx$, and $\cosh(2n-1)x$ that are analogous to those of **1.332**, just as was done in the formulas in **1.331**.

Special cases

1.333

1. $\sin 2x = 2 \sin x \cos x$

2. $\sin 3x = 3 \sin x - 4 \sin^3 x$

3. $\sin 4x = \cos x \left(4 \sin x - 8 \sin^3 x \right)$

4. $\sin 5x = 5 \sin x - 20 \sin^3 x + 16 \sin^5 x$

5. $\sin 6x = \cos x \left(6 \sin x - 32 \sin^3 x + 32 \sin^5 x \right)$

6. $\sin 7x = 7 \sin x - 56 \sin^3 x + 112 \sin^5 x - 64 \sin^7 x$

1.334

1. $\sinh 2x = 2 \sinh x \cosh x$

2. $\sinh 3x = 3 \sinh x + 4 \sinh^3 x$

3. $\sinh 4x = chx \left(4 \sinh x + 8 \sinh^3 x \right)$

4. $\sinh 5x = 5 \sinh x + 20 \sinh^3 x + 16 \sinh^5 x$

5. $\sinh 6x = chx \left(6 \sinh x + 32 \sinh^3 x + 32 \sinh^5 x \right)$

6. $\sinh 7x = 7 \sinh x + 56 \sinh^3 x + 112 \sinh^5 x + 64 \sinh^7 x$

1.335

1. $\cos 2x = 2 \cos^2 x - 1.$

2. $\cos 3x = 4 \cos^3 x - 3 \cos x$

3. $\cos 4x = 8 \cos^4 x - 8 \cos^2 x + 1.$

4. $\cos 5x = 16 \cos^5 x - 20 \cos^3 x + 5 \cos x$

5. $\cos 6x = 32 \cos^6 x - 48 \cos^4 x + 18 \cos^2 x - 1.$

6. $\cos 7x = 64 \cos^7 x - 112 \cos^5 x + 56 \cos^3 x - 7 \cos x$

1.336

1. $\cosh 2x = 2 \cosh^2 x - 1.$

2. $\cosh 3x = 4 \cosh^3 x - 3 \cosh x$

3. $\cosh 4x = 8 \cosh^4 x - 8 \cosh^2 x + 1.$

4. $\cosh 5x = 16 \cosh^5 x - 20 \cosh^3 x + 5 \cosh x$

5. $\cosh 6x = 32 \cosh^6 x - 48 \cosh^4 x + 18 \cosh^2 x - 1.$

6. $\cosh 7x = 64 \cosh^7 x - 112 \cosh^5 x + 56 \cosh^3 x - 7 \cosh x$

1.34   Certain sums of trigonometric and hyperbolic functions

1.341

1. $\displaystyle\sum_{k=0}^{n-1} \sin(x + ky) = \sin\left(x + \frac{n-1}{2}y\right) \sin \frac{ny}{2} \operatorname{cosec} \frac{y}{2}$ AD (361.8)

2. $\displaystyle\sum_{k=0}^{n-1} \sinh(x + ky) = \sinh\left(x + \frac{n-1}{2}y\right) \sinh \frac{ny}{2} \frac{1}{\sinh \frac{y}{2}}$

3. $\displaystyle\sum_{k=0}^{n-1} \cos(x + ky) = \cos\left(x + \frac{n-1}{2}y\right) \sin \frac{ny}{2} \operatorname{cosec} \frac{y}{2}$ AD (361.9)

4. $\displaystyle\sum_{k=0}^{n-1} \cosh(x + ky) = \cosh\left(x + \frac{n-1}{2}y\right) \sinh \frac{ny}{2} \frac{1}{\sinh \frac{y}{2}}$

5. $\displaystyle\sum_{k=0}^{2n-1} (-1)^k \cos(x + ky) = \sin\left(x + \frac{2n-1}{2}y\right) \sin ny \sec \frac{y}{2}$ JO (202)

6. $\displaystyle\sum_{k=0}^{n-1} (-1)^k \sin(x + ky) = \sin\left\{x + \frac{n-1}{2}(y + \pi)\right\} \sin \frac{n(y + \pi)}{2} \sec \frac{y}{2}$ AD (202a)

Special cases

1.342

1. $\displaystyle\sum_{k=1}^{n} \sin kx = \sin \frac{n+1}{2}x \sin \frac{nx}{2} \operatorname{cosec} \frac{x}{2}$ AD (361.1)

2.[10] $\displaystyle\sum_{k=0}^{n} \cos kx = \cos \frac{n+1}{2}x \sin \frac{nx}{2} \operatorname{cosec} \frac{x}{2} + 1$

         $= \cos \frac{nx}{2} \sin \frac{n+1}{2}x \operatorname{cosec} \frac{x}{2} = \frac{1}{2}\left[1 + \frac{\sin\left(n + \frac{1}{2}\right)x}{\sin \frac{x}{2}}\right]$

         AD (361.2)

3. $$\sum_{k=1}^{n} \sin(2k-1)x = \sin^2 nx \, \text{cosec} \, x$$ AD (361.7)

4. $$\sum_{k=1}^{n} \cos(2k-1)x = \frac{1}{2} \sin 2nx \, \text{cosec} \, x$$ JO (207)

1.343

1. $$\sum_{k=1}^{n} (-1)^k \cos kx = -\frac{1}{2} + \frac{(-1)^n \cos\left(\frac{2n+1}{2}x\right)}{2\cos\frac{x}{2}}$$ AD (361.11)

2. $$\sum_{k=1}^{n} (-1)^{k+1} \sin(2k-1)x = (-1)^{n+1} \frac{\sin 2nx}{2\cos x}$$ AD (361.10)

3. $$\sum_{k=1}^{n} \cos(4k-3)x + \sum_{k=1}^{n} \sin(4k-1)x = \sin 2nx \,(\cos 2nx + \sin 2nx)\,(\cos x + \sin x)\, \text{cosec} \, 2x$$

     JO (208)

1.344

1. $$\sum_{k=1}^{n-1} \sin \frac{\pi k}{n} = \cot \frac{\pi}{2n}$$ AD (361.19)

2. $$\sum_{k=1}^{n-1} \sin \frac{2\pi k^2}{n} = \frac{\sqrt{n}}{2}\left(1 + \cos\frac{n\pi}{2} - \sin\frac{n\pi}{2}\right)$$ AD (361.18)

3. $$\sum_{k=0}^{n-1} \cos \frac{2\pi k^2}{n} = \frac{\sqrt{n}}{2}\left(1 + \cos\frac{n\pi}{2} + \sin\frac{n\pi}{2}\right)$$ AD (361.17)

1.35 Sums of powers of trigonometric functions of multiple angles

1.351

1. $$\sum_{k=1}^{n} \sin^2 kx = \frac{1}{4}\left[(2n+1)\sin x - \sin(2n+1)x\right]\text{cosec}\, x$$
    $$= \frac{n}{2} - \frac{\cos(n+1)x \sin nx}{2\sin x}$$

     AD (361.3)

2. $$\sum_{k=1}^{n} \cos^2 kx = \frac{n-1}{2} + \frac{1}{2}\cos nx \sin(n+1)x \, \text{cosec}\, x$$
    $$= \frac{n}{2} + \frac{\cos(n+1)x \sin nx}{2\sin x}$$

     AD (361.4)a

3. $$\sum_{k=1}^{n} \sin^3 kx = \frac{3}{4}\sin\frac{n+1}{2}x \sin\frac{nx}{2}\,\text{cosec}\,\frac{x}{2} - \frac{1}{4}\sin\frac{3(n+1)x}{2}\sin\frac{3nx}{2}\,\text{cosec}\,\frac{3x}{2}$$ JO (210)

4. $$\sum_{k=1}^{n} \cos^3 kx = \frac{3}{4}\cos\frac{n+1}{2}x \sin\frac{nx}{2}\,\text{cosec}\,\frac{x}{2} + \frac{1}{4}\cos\frac{3(n+1)}{2}x \sin\frac{3nx}{2}\,\text{cosec}\,\frac{3x}{2}$$ JO (211)a

5. $\displaystyle\sum_{k=1}^{n} \sin^4 kx = \frac{1}{8}\left[3n - 4\cos(n+1)x \sin nx \operatorname{cosec} x + \cos 2(n+1)x \sin 2nx \operatorname{cosec} 2x\right]$ JO (212)

6. $\displaystyle\sum_{k=1}^{n} \cos^4 kx = \frac{1}{8}\left[3n + 4\cos(n+1)x \sin nx \operatorname{cosec} x + \cos 2(n+1)x \sin 2nx \operatorname{cosec} 2x\right]$ JO (213)

1.352

1. $\displaystyle\sum_{k=1}^{n-1} k \sin kx = \frac{\sin nx}{4\sin^2 \frac{x}{2}} - \frac{n\cos \frac{2n-1}{2}x}{2\sin \frac{x}{2}}$ AD (361.5)

2. $\displaystyle\sum_{k=1}^{n-1} k \cos kx = \frac{n\sin \frac{2n-1}{2}x}{2\sin \frac{x}{2}} - \frac{1 - \cos nx}{4\sin^2 \frac{x}{2}}$ AD (361.6)

1.353

1. $\displaystyle\sum_{k=1}^{n-1} p^k \sin kx = \frac{p\sin x - p^n \sin nx + p^{n+1}\sin(n-1)x}{1 - 2p\cos x + p^2}$ AD (361.12)a

2. $\displaystyle\sum_{k=1}^{n-1} p^k \sinh kx = \frac{p\sinh x - p^n \sinh nx + p^{n+1}\sinh(n-1)x}{1 - 2p\cosh x + p^2}$

3. $\displaystyle\sum_{k=0}^{n-1} p^k \cos kx = \frac{1 - p\cos x - p^n \cos nx + p^{n+1}\cos(n-1)x}{1 - 2p\cos x + p^2}$ AD (361.13)a¡

4. $\displaystyle\sum_{k=0}^{n-1} p^k \cosh kx = \frac{1 - p\cosh x - p^n \cosh nx + p^{n+1}\cosh(n-1)x}{1 - 2p\cosh x + p^2}$ JO (396)

1.36   Sums of products of trigonometric functions of multiple angles

1.361

1. $\displaystyle\sum_{k=1}^{n} \sin kx \sin(k+1)x = \frac{1}{4}\left[(n+1)\sin 2x - \sin 2(n+1)x\right]\operatorname{cosec} x$ JO (214)

2. $\displaystyle\sum_{k=1}^{n} \sin kx \sin(k+2)x = \frac{n}{2}\cos 2x - \frac{1}{2}\cos(n+3)x \sin nx \operatorname{cosec} x$ JO (216)

3. $\displaystyle 2\sum_{k=1}^{n} \sin kx \cos(2k-1)y = \sin\left\{ny + \frac{n+1}{2}x\right\}\sin \frac{n(x+2y)}{2}\operatorname{cosec}\frac{x+2y}{2}$

             $- \sin\left\{ny - \frac{n+1}{2}x\right\}\sin \frac{n(2y-x)}{2}\operatorname{cosec}\frac{2y-x}{2}$

                                                                             JO (217)

1.362

1. $\displaystyle\sum_{k=1}^{n}\left(2^k \sin^2 \frac{x}{2^k}\right)^2 = \left(2^n \sin \frac{x}{2^n}\right)^2 - \sin^2 x$ AD (361.15)

2. $\displaystyle\sum_{k=1}^{n}\left(\frac{1}{2^k}\sec\frac{x}{2^k}\right)^2 = \operatorname{cosec}^2 x - \left(\frac{1}{2^n}\operatorname{cosec}\frac{x}{2^n}\right)^2$ AD (361.14)

1.37 Sums of tangents of multiple angles

1.371

1. $\displaystyle\sum_{k=0}^{n}\frac{1}{2^k}\tan\frac{x}{2^k} = \frac{1}{2^n}\cot\frac{x}{2^n} - 2\cot 2x$ AD (361.16)

2. $\displaystyle\sum_{k=0}^{n}\frac{1}{2^{2k}}\tan^2\frac{x}{2^k} = \frac{2^{2n+2}-1}{3\cdot 2^{2n-1}} + 4\cot^2 2x - \frac{1}{2^{2n}}\cot^2\frac{x}{2^n}$ AD (361.20)

1.38 Sums leading to hyperbolic tangents and cotangents

1.381

1. $\displaystyle\sum_{k=0}^{n-1}\frac{\tanh\left(x\dfrac{1}{n\sin^2\left(\dfrac{2k+1}{4n}\pi\right)}\right)}{1+\dfrac{\tanh^2 x}{\tan^2\left(\dfrac{2k+1}{4n}\pi\right)}} = \tanh(2nx)$ JO (402)a

2. $\displaystyle\sum_{k=1}^{n-1}\frac{\tanh\left(x\dfrac{1}{n\sin^2\left(\dfrac{k\pi}{2n}\right)}\right)}{1+\dfrac{\tanh^2 x}{\tan^2\left(\dfrac{k\pi}{2n}\right)}} = \coth(2nx) - \frac{1}{2n}(\tanh x + \coth x)$ JO (403)

3. $\displaystyle\sum_{k=0}^{n-1}\frac{\tanh\left(x\dfrac{2}{(2n+1)\sin^2\left(\dfrac{2k+1}{2(2n+1)}\pi\right)}\right)}{1+\dfrac{\tanh^2 x}{\tan^2\left(\dfrac{2k+1}{2(2n+1)}\pi\right)}} = \tanh(2n+1)x - \frac{\tanh x}{2n+1}$ JO (404)

4. $\displaystyle\sum_{k=1}^{n}\frac{\tanh\left(x\dfrac{2}{(2n+1)\sin^2\left(\dfrac{k\pi}{2(2n+1)}\right)}\right)}{1+\dfrac{\tanh^2 x}{\tan^2\left(\dfrac{k\pi}{(2n+1)}\right)}} = \coth(2n+1)x - \frac{\coth x}{2n+1}$ JO (405)

1.382

1. $\displaystyle\sum_{k=0}^{n-1} \frac{1}{\left(\dfrac{\sin^2\left(\dfrac{2k+1}{4n}\pi\right)}{\sinh x} + \dfrac{1}{2}\tanh\left(\dfrac{x}{2}\right) \right)} = 2n\tanh(nx)$ JO (406)

2. $\displaystyle\sum_{k=1}^{n-1} \frac{1}{\left(\dfrac{\sin^2\left(\dfrac{k\pi}{2n}\right)}{\sinh x} + \dfrac{1}{2}\tanh\left(\dfrac{x}{2}\right) \right)} = 2n\coth(nx) - 2\coth x$ JO (407)

3. $\displaystyle\sum_{k=0}^{n-1} \frac{1}{\left(\dfrac{\sin^2\left(\dfrac{2k+1}{2(2n+1)}\pi\right)}{\sinh x} + \dfrac{1}{2}\tanh\left(\dfrac{x}{2}\right) \right)} = (2n+1)\tanh\left(\dfrac{(2n+1)x}{2}\right) - \tanh\dfrac{x}{2}$ JO (408)

4. $\displaystyle\sum_{k=1}^{n} \frac{1}{\left(\dfrac{\sin^2\left(\dfrac{k\pi}{2n+1}\right)}{\sinh x} + \dfrac{1}{2}\tanh\left(\dfrac{x}{2}\right) \right)} = (2n+1)\coth\left(\dfrac{(2n+1)x}{2}\right) - \coth\dfrac{x}{2}$ JO (409)

1.39 The representation of cosines and sines of multiples of the angle as finite products

1.391

1. $\sin nx = n\sin x\cos x\displaystyle\prod_{k=1}^{\frac{n-2}{2}}\left(1 - \dfrac{\sin^2 x}{\sin^2\dfrac{k\pi}{n}}\right)$ [n is even]   JO (568)

2. $\cos nx = \displaystyle\prod_{k=1}^{\frac{n}{2}}\left(1 - \dfrac{\sin^2 x}{\sin^2\dfrac{(2k-1)\pi}{2n}}\right)$ [n is even]   JO (569)

3. $\sin nx = n\sin x\displaystyle\prod_{k=1}^{\frac{n-1}{2}}\left(1 - \dfrac{\sin^2 x}{\sin^2\dfrac{k\pi}{n}}\right)$ [n is odd]   JO (570)

4. $\cos nx = \cos x\displaystyle\prod_{k=1}^{\frac{n-1}{2}}\left(1 - \dfrac{\sin^2 x}{\sin^2\dfrac{(2k-1)\pi}{2n}}\right)$ [n is odd]   JO (571)a

1.392

1. $\sin nx = 2^{n-1} \prod\limits_{k=0}^{n-1} \sin\left(x + \dfrac{k\pi}{n}\right)$ JO (548)

2. $\cos nx = 2^{n-1} \prod\limits_{k=1}^{n} \sin\left(x + \dfrac{2k-1}{2n}\pi\right)$ JO (549)

1.393

1. $\prod\limits_{k=0}^{n-1} \cos\left(x + \dfrac{2k}{n}\pi\right) = \dfrac{1}{2^{n-1}} \cos nx$ [n odd]

                          $= \dfrac{1}{2^{n-1}} \left[(-1)^{\frac{n}{2}} - \cos nx\right]$ [n even]

                                                                JO (543)

2. $\prod\limits_{k=0}^{n-1} \sin\left(x + \dfrac{2k}{n}\pi\right) = \dfrac{(-1)^{\frac{n-1}{2}}}{2^{n-1}} \sin nx$ [n odd]

                          $= \dfrac{(-1)^{\frac{n}{2}}}{2^{n-1}} (1 - \cos nx)$ [n odd]

                                                                JO (544)

1.394 $\prod\limits_{k=0}^{n-1} \left\{x^2 - 2xy\cos\left(\alpha + \dfrac{2k\pi}{n}\right) + y^2\right\} = x^{2n} - 2x^n y^n \cos n\alpha + y^{2n}$ JO (573)

1.395

1. $\cos nx - \cos ny = 2^{n-1} \prod\limits_{k=0}^{n-1} \left\{\cos x - \cos\left(y + \dfrac{2k\pi}{n}\right)\right\}$ JO (573)

2. $\cosh nx - \cos ny = 2^{n-1} \prod\limits_{k=0}^{n-1} \left\{\cosh x - \cos\left(y + \dfrac{2k\pi}{n}\right)\right\}$ JO (538)

1.396

1. $\prod\limits_{k=1}^{n-1} \left(x^2 - 2x\cos\dfrac{k\pi}{n} + 1\right) = \dfrac{x^{2n} - 1}{x^2 - 1}$ KR 58 (28.1)

2. $\prod\limits_{k=1}^{n} \left(x^2 - 2x\cos\dfrac{2k\pi}{2n+1} + 1\right) = \dfrac{x^{2n+1} - 1}{x - 1}$ KR 58 (28.2)

3. $\prod\limits_{k=1}^{n} \left(x^2 + 2x\cos\dfrac{2k\pi}{2n+1} + 1\right) = \dfrac{x^{2n+1} - 1}{x + 1}$ KR 58 (28.3)

4. $\prod\limits_{k=0}^{n-1} \left(x^2 - 2x\cos\dfrac{(2k+1)\pi}{2n} + 1\right) = x^{2n} + 1.$ KR 58 (28.4)

1.41 The expansion of trigonometric and hyperbolic functions in power series

1.411

1. $\sin x = \sum\limits_{k=0}^{\infty} (-1)^k \dfrac{x^{2k+1}}{(2k+1)!}$

2. $\sinh x = \sum\limits_{k=0}^{\infty} \dfrac{x^{2k+1}}{(2k+1)!}$

3. $\cos x = \sum\limits_{k=0}^{\infty} (-1)^k \dfrac{x^{2k}}{(2k)!}$

4. $\cosh x = \sum\limits_{k=0}^{\infty} \dfrac{x^{2k}}{(2k)!}$

5. $\tan x = \sum\limits_{k=1}^{\infty} \dfrac{2^{2k}\left(2^{2k}-1\right)}{(2k)!}|B_{2k}|x^{2k-1}$ $\left[x^2 < \dfrac{\pi^2}{4}\right]$ FI II 523

6. $\tanh x = x - \dfrac{x^3}{3} + \dfrac{2x^5}{15} + \dfrac{17}{315}x^7 + \cdots = \sum\limits_{k=1}^{\infty} \dfrac{2^{2k}\left(2^{2k}-1\right)}{(2k)!} B_{2k}x^{2k-1}$

     $\left[x^2 < \dfrac{\pi^2}{4}\right]$

7. $\cot x = \dfrac{1}{x} - \sum\limits_{k=1}^{\infty} \dfrac{2^{2k}|B_{2k}|}{(2k)!}x^{2k-1}$ $\left[x^2 < \pi^2\right]$ FI II 523a

8. $\coth x = \dfrac{1}{x} + \dfrac{x}{3} - \dfrac{x^3}{45} + \dfrac{2x^5}{945} - \cdots = \dfrac{1}{x} + \sum\limits_{k=1}^{\infty} \dfrac{2^{2k}B_{2k}}{(2k)!}x^{2k-1}$

     $\left[x^2 < \pi^2\right]$ FI II 522a

9. $\sec x = \sum\limits_{k=0}^{\infty} \dfrac{|E_{2k}|}{(2k)!}x^{2k}$ $\left[x^2 < \dfrac{\pi^2}{4}\right]$ CE 330a

10. $\operatorname{sech} x = 1 - \dfrac{x^2}{2} + \dfrac{5x^4}{24} - \dfrac{61x^6}{720} + \cdots = 1 + \sum\limits_{k=1}^{\infty} \dfrac{E_{2k}}{(2k)!}x^{2k}$

     $\left[x^2 < \dfrac{\pi^2}{4}\right]$ CE 330

11. $\operatorname{cosec} x = \dfrac{1}{x} + \sum\limits_{k=1}^{\infty} \dfrac{2\left(2^{2k-1}-1\right)|B_{2k}|x^{2k-1}}{(2k)!}$ $\left[x^2 < \pi^2\right]$ CE 329a

12. $\operatorname{cosech} x = \dfrac{1}{x} - \dfrac{1}{6}x + \dfrac{7x^3}{360} - \dfrac{31x^5}{15120} + \cdots = \dfrac{1}{x} - \sum\limits_{k=1}^{\infty} \dfrac{2\left(2^{2k-1}-1\right)B_{2k}}{(2k)!}x^{2k-1}$

     $\left[x^2 < \pi^2\right]$ JO (418)

1.412

1. $\sin^2 x = \sum_{k=1}^{\infty} (-1)^{k+1} \dfrac{2^{2k-1} x^{2k}}{(2k)!}$ JO (452)a

2. $\cos^2 x = 1 - \sum_{k=1}^{\infty} (-1)^{k+1} \dfrac{2^{2k-1} x^{2k}}{(2k)!}$ JO (443)

3. $\sin^3 x = \dfrac{1}{4} \sum_{k=1}^{\infty} (-1)^{k+1} \dfrac{3^{2k+1} - 3}{(2k+1)!} x^{2k+1}$ JO (452a)a

4. $\cos^3 x = \dfrac{1}{4} \sum_{k=0}^{\infty} (-1)^{k} \dfrac{\left(3^{2k} + 3\right) x^{2k}}{(2k)!}$ JO (443a)

1.413

1. $\sinh x = \operatorname{cosec} x \sum_{k=1}^{\infty} (-1)^{k+1} \dfrac{2^{2k-1} x^{4k-2}}{(4k-1)!}$ JO (508)

2. $\cosh x = \sec x + \sec x \sum_{k=1}^{\infty} (-1)^{k} \dfrac{2^{2k} x^{4k}}{(4k)!}$ JO (507)

3. $\sinh x = \sec x \sum_{k=1}^{\infty} (-1)^{[k/2]} \dfrac{2^{k-1} x^{2k-1}}{(2k-1)!}$ JO (510)

4. $\cosh x = \operatorname{cosec} x \sum_{k=1}^{\infty} (-1)^{[(k-1)/2]} \dfrac{2^{k-1} x^{2k-1}}{(2k-1)!}$ JO (509)

1.414

1. $\cos\left[n \ln\left(x + \sqrt{1+x^2}\right)\right] = 1 - \sum_{k=0}^{\infty} (-1)^{k} \dfrac{\left(n^2 + 0^2\right)\left(n^2 + 2^2\right)\ldots\left[n^2 + (2k)^2\right]}{(2k+2)!} x^{2k+2}$

$$\left[x^2 < 1\right] \qquad \text{AD (6456.1)}$$

2. $\sin\left[n \ln\left(x + \sqrt{1+x^2}\right)\right] = nx - n \sum_{k=1}^{\infty} (-1)^{k+1} \dfrac{\left(n^2 + 1^2\right)\left(n^2 + 3^2\right)\ldots\left[n^2 + (2k-1)^2\right] x^{2k+1}}{(2k+1)!}$

$$\left[x^2 < 1\right] \qquad \text{AD (6456.2)}$$

Power series for $\ln \sin x$, $\ln \cos x$, and $\ln \tan x$ see **1.518**.

1.42 Expansion in series of simple fractions

1.421

1. $\tan \dfrac{\pi x}{2} = \dfrac{4x}{\pi} \sum_{k=1}^{\infty} \dfrac{1}{(2k-1)^2 - x^2}$ BR* (191), AD (6495.1)

2.[10] $\tanh \dfrac{\pi x}{2} = \dfrac{4x}{\pi} \sum_{k=1}^{\infty} \dfrac{1}{(2k-1)^2 + x^2}$

3. $\cot \pi x = \dfrac{1}{\pi x} + \dfrac{2x}{\pi} \displaystyle\sum_{k=1}^{\infty} \dfrac{1}{x^2 - k^2} = \dfrac{1}{\pi x} + \dfrac{x}{\pi} \displaystyle\sum_{\substack{k=-\infty \\ k \neq 0}}^{\infty} \dfrac{1}{k(x-k)}$ AD (6495.2), JO (450a)

4. $\coth \pi x = \dfrac{1}{\pi x} + \dfrac{2x}{\pi} \displaystyle\sum_{k=1}^{\infty} \dfrac{1}{x^2 + k^2}$ (cf. **1.217** 1)

5. $\tan^2 \dfrac{\pi x}{2} = x^2 \displaystyle\sum_{k=1}^{\infty} \dfrac{2(2k-1)^2 - x^2}{(1^2 - x^2)^2 (3^2 - x^2)^2 \ldots \left[(2k-1)^2 - x^2\right]^2}$ JO (450)

1.422

1. $\sec \dfrac{\pi x}{2} = \dfrac{4}{\pi} \displaystyle\sum_{k=1}^{\infty} (-1)^{k+1} \dfrac{2k-1}{(2k-1)^2 - x^2}$ AD (6495.3)a

2. $\sec^2 \dfrac{\pi x}{2} = \dfrac{4}{\pi^2} \displaystyle\sum_{k=1}^{\infty} \left\{ \dfrac{1}{(2k-1-x)^2} + \dfrac{1}{(2k-1+x)^2} \right\}$ JO (451)a

3. $\operatorname{cosec} \pi x = \dfrac{1}{\pi x} + \dfrac{2x}{\pi} \displaystyle\sum_{k=1}^{\infty} \dfrac{(-1)^k}{x^2 - k^2}$ (see also **1.217** 2) AD (6495.4)a

4. $\operatorname{cosec}^2 \pi x = \dfrac{1}{\pi^2} \displaystyle\sum_{k=-\infty}^{\infty} \dfrac{1}{(x-k)^2} = \dfrac{1}{\pi^2 x^2} + \dfrac{2}{\pi^2} \displaystyle\sum_{k=1}^{\infty} \dfrac{x^2 + k^2}{(x^2 - k^2)^2}$ JO (446)

5. $\dfrac{1 + x \operatorname{cosec} x}{2x^2} = \dfrac{1}{x^2} - \displaystyle\sum_{k=1}^{\infty} \dfrac{(-1)^{k+1}}{(x^2 - k^2 \pi^2)}$ JO (449)

6. $\operatorname{cosec} \pi x = \dfrac{1}{\pi x} + \dfrac{1}{\pi} \displaystyle\sum_{k=-\infty}^{\infty} (-1)^k \left(\dfrac{1}{x-k} + \dfrac{1}{k} \right)$ JO (450b)

1.423 $\dfrac{\pi^2}{4m^2} \operatorname{cosec}^2 \dfrac{\pi}{m} + \dfrac{\pi}{4m} \cot \dfrac{\pi}{m} - \dfrac{1}{2} = \displaystyle\sum_{k=1}^{\infty} \dfrac{1}{(1 - k^2 m^2)^2}$ JO (477)

1.43 Representation in the form of an infinite product

1.431

1. $\sin x = x \displaystyle\prod_{k=1}^{\infty} \left(1 - \dfrac{x^2}{k^2 \pi^2} \right)$ EU

2. $\sinh x = x \displaystyle\prod_{k=1}^{\infty} \left(1 + \dfrac{x^2}{k^2 \pi^2} \right)$ EU

3. $\cos x = \displaystyle\prod_{k=0}^{\infty} \left(1 - \dfrac{4x^2}{(2k+1)^2 \pi^2} \right)$ EU

4. $\cosh x = \displaystyle\prod_{k=0}^{\infty} \left(1 + \dfrac{4x^2}{(2k+1)^2 \pi^2} \right)$ EU

1.432

1. $\cos x - \cos y = 2 \left(1 - \dfrac{x^2}{y^2} \right) \sin^2 \dfrac{y}{2} \prod_{k=1}^{\infty} \left(1 - \dfrac{x^2}{(2k\pi + y^2)} \right) \left(1 - \dfrac{x^2}{(2k\pi - y)^2} \right)$ AD (653.2)

2. $\cosh x - \cos y = 2 \left(1 + \dfrac{x^2}{y^2} \right) \sin^2 \dfrac{y}{2} \prod_{k=1}^{\infty} \left(1 + \dfrac{x^2}{(2k\pi + y)^2} \right) \left(1 + \dfrac{x^2}{(2k\pi - y)^2} \right)$ AD (653.1)

1.433 $\cos \dfrac{\pi x}{4} - \sin \dfrac{\pi x}{4} = \prod_{k=1}^{\infty} \left[1 + \dfrac{(-1)^k x}{2k - 1} \right]$ BR* 189

1.434 $\cos^2 x = \dfrac{1}{4}(\pi + 2x)^2 \prod_{k=1}^{\infty} \left[1 - \left(\dfrac{\pi + 2x}{2k\pi} \right)^2 \right]^2$ MO 216

1.435 $\dfrac{\sin \pi(x + a)}{\sin \pi a} = \dfrac{x + a}{a} \prod_{k=1}^{\infty} \left(1 - \dfrac{x}{k - a} \right) \left(1 + \dfrac{x}{k + a} \right)$ MO 216

1.436 $1 - \dfrac{\sin^2 \pi x}{\sin^2 \pi a} = \prod_{k=-\infty}^{\infty} \left[1 - \left(\dfrac{x}{k - a} \right)^2 \right]$ MO 216

1.437 $\dfrac{\sin 3x}{\sin x} = - \prod_{k=-\infty}^{\infty} \left[1 - \left(\dfrac{2x}{x + k\pi} \right)^2 \right]$ MO 216

1.438 $\dfrac{\cosh x - \cos a}{1 - \cos a} = \prod_{k=-\infty}^{\infty} \left[1 + \left(\dfrac{x}{2k\pi + a} \right)^2 \right]$ MO 216

1.439

1. $\sin x = x \prod_{k=1}^{\infty} \cos \dfrac{x}{2^k}$ $[|x| < 1]$ AD (615), MO 216

2. $\dfrac{\sin x}{x} = \prod_{k=1}^{\infty} \left[1 - \dfrac{4}{3} \sin^2 \left(\dfrac{x}{3^k} \right) \right]$ MO 216

1.44–1.45 Trigonometric (Fourier) series

1.441

1. $\displaystyle\sum_{k=1}^{\infty} \dfrac{\sin kx}{k} = \dfrac{\pi - x}{2}$ $[0 < x < 2\pi]$ FI III 539

2. $\displaystyle\sum_{k=1}^{\infty} \dfrac{\cos kx}{k} = \dfrac{1}{2} \ln \dfrac{1}{2(1 - \cos x)}$ $[0 < x < 2\pi]$ FI III 530a, AD (6814)

3. $\displaystyle\sum_{k=1}^{\infty} \dfrac{(-1)^{k-1} \sin kx}{k} = \dfrac{x}{2}$ $[-\pi < x < \pi]$ FI III 542

4. $\displaystyle\sum_{k=1}^{\infty} (-1)^{k-1} \dfrac{\cos kx}{k} = \ln \left(2 \cos \dfrac{x}{2} \right)$ $[-\pi < x < \pi]$ FI III 550

1.442

1.
$$\sum_{k=1}^{\infty} \frac{\sin(2k-1)x}{2k-1} = \frac{\pi}{x} \qquad [0 < x < 2\pi] \qquad \text{FI III 541}$$

2.
$$\sum_{k=1}^{\infty} \frac{\cos(2k-1)x}{2k-1} = \frac{1}{2}\ln\cot\frac{x}{2} \qquad [0 < x < \pi]$$

                                           BR* 168, JO (266), GI III(195)

3.
$$\sum_{k=1}^{\infty} (-1)^{k-1} \frac{\sin(2k-1)x}{2k-1} = \frac{1}{2}\ln\tan\left(\frac{\pi}{4}+\frac{x}{2}\right) \qquad \left[-\frac{\pi}{2} < x < \frac{\pi}{2}\right] \qquad \text{BR* 168, JO (268)a}$$

4.[10]
$$\sum_{k=1}^{\infty} (-1)^{k-1} \frac{\cos(2k-1)x}{2k-1} = \frac{\pi}{4} \qquad \left[-\frac{\pi}{2} < x < \frac{\pi}{2}\right]$$
$$= -\frac{\pi}{4} \qquad \left[\frac{\pi}{2} < x < \frac{3\pi}{2}\right]$$

                                               BR* 168, JO (269)

1.443

1.[8]
$$\sum_{k=1}^{\infty} \frac{\cos k\pi x}{k^{2n}} = (-1)^{n-1} 2^{2n-1} \frac{\pi^{2n}}{(2n)!} \sum_{k=0}^{2n} \binom{2n}{k} B_{2n-k}\varrho^{k}$$
$$= (-1)^{n-1} \frac{1}{2} \frac{(2\pi)^{2n}}{(2n)!} B_{2n}\left(\frac{x}{2}\right)$$

$$\left[0 \le x \le 2, \quad \varrho = \frac{x}{2} - \left\lfloor\frac{x}{2}\right\rfloor\right] \qquad \text{CE 340, GE 71}$$

2.
$$\sum_{k=1}^{\infty} \frac{\sin k\pi x}{k^{2n+1}} = (-1)^{n-1} 2^{2n} \frac{\pi^{2n+1}}{(2n+1)!} \sum_{k=0}^{2n+1} \binom{2n+1}{k} B_{2n-k+1}\rho^{k}$$
$$= (-1)^{n-1} \frac{1}{2} \frac{(2\pi)^{2n+1}}{(2n+1)!} B_{2n+1}\left(\frac{x}{2}\right)$$

$$\left[0 < x < 1; \quad \rho = \frac{x}{2} - \left\lfloor\frac{x}{2}\right\rfloor\right] \qquad \text{CE 340}$$

3.
$$\sum_{k=1}^{\infty} \frac{\cos kx}{k^2} = \frac{\pi^2}{6} - \frac{\pi x}{2} + \frac{x^2}{4} \qquad [0 \le x \le 2\pi] \qquad \text{FI III 547}$$

4.
$$\sum_{k=1}^{\infty} (-1)^{k-1} \frac{\cos kx}{k^2} = \frac{\pi^2}{12} - \frac{x^2}{4} \qquad [-\pi \le x \le \pi] \qquad \text{FI III 544}$$

5.
$$\sum_{k=1}^{\infty} \frac{\sin kx}{k^3} = \frac{\pi^2 x}{6} - \frac{\pi x^2}{4} + \frac{x^3}{12} \qquad [0 \le x \le 2\pi]$$

6.
$$\sum_{k=1}^{\infty} \frac{\cos kx}{k^4} = \frac{\pi^4}{90} - \frac{\pi^2 x^2}{12} + \frac{\pi x^3}{12} - \frac{x^4}{48} \qquad [0 \le x \le 2\pi] \qquad \text{AD (6617)}$$

7.
$$\sum_{k=1}^{\infty} \frac{\sin kx}{k^5} = \frac{\pi^4 x}{90} - \frac{\pi^2 x^3}{36} + \frac{\pi x^4}{48} - \frac{x^5}{240} \qquad [0 \le x \le 2\pi] \qquad \text{AD (6818)}$$

1.444

1. $$\sum_{k=1}^{\infty} \frac{\sin 2(k+1)x}{k(k+1)} = \sin 2x - (\pi - 2x)\sin^2 x - \sin x \cos x \ln\left(4\sin^2 x\right)$$

$$[0 \le x \le \pi] \qquad\qquad \text{BR* 168, GI III (190)}$$

2. $$\sum_{k=1}^{\infty} \frac{\cos 2(k+1)x}{k(k+1)} = \cos 2x - \left(\frac{\pi}{2} - x\right)\sin 2x + \sin^2 x \ln\left(4\sin^2 x\right)$$

$$[0 \le x \le \pi] \qquad\qquad \text{BR* 168}$$

3. $$\sum_{k=1}^{\infty} (-1)^k \frac{\sin(k+1)x}{k(k+1)} = \sin x - \frac{x}{2}(1 + \cos x) - \sin x \ln\left|2\cos\frac{x}{2}\right| \qquad \text{MO 213}$$

4. $$\sum_{k=1}^{\infty} (-1)^k \frac{\cos(k+1)x}{k(k+1)} = \cos x - \frac{x}{2}\sin x - (1 + \cos x)\ln\left|2\cos\frac{x}{2}\right| \qquad \text{MO 213}$$

5. $$\sum_{k=0}^{\infty} (-1)^k \frac{\sin(2k+1)x}{(2k+1)^2} = \frac{\pi}{4}x \qquad\qquad \left[-\frac{\pi}{2} \le x \le \frac{\pi}{2}\right]$$

$$= \frac{\pi}{4}(\pi - x) \qquad\qquad \left[\frac{\pi}{2} \le x \le \frac{3}{2}\pi\right]$$

$$\text{MO 213}$$

6.[6] $$\sum_{k=1}^{\infty} \frac{\cos(2k-1)x}{(2k-1)^2} = \frac{\pi}{4}\left(\frac{\pi}{2} - |x|\right) \qquad\qquad [-\pi \le x \le \pi] \qquad\qquad \text{FI III 546}$$

7. $$\sum_{k=1}^{\infty} \frac{\cos 2kx}{(2k-1)(2k+1)} = \frac{1}{2} - \frac{\pi}{4}\sin x \qquad\qquad \left[0 \le x \le \frac{\pi}{2}\right] \qquad\qquad \text{JO (591)}$$

1.445

1. $$\sum_{k=1}^{\infty} \frac{k\sin kx}{k^2 + \alpha^2} = \frac{\pi}{2}\frac{\sinh\alpha(\pi - x)}{\sinh\alpha\pi} \qquad\qquad [0 < x < 2\pi] \qquad\qquad \text{BR* 157, JO (411)}$$

2. $$\sum_{k=1}^{\infty} \frac{\cos kx}{k^2 + \alpha^2} = \frac{\pi}{2\alpha}\frac{\cosh\alpha(\pi - x)}{\sinh\alpha\pi} - \frac{1}{2\alpha^2} \qquad\qquad [0 \le x \le 2\pi] \qquad\qquad \text{BR* 257, JO (410)}$$

3. $$\sum_{k=1}^{\infty} \frac{(-1)^k \cos kx}{k^2 + \alpha^2} = \frac{\pi}{2\alpha}\frac{\cosh\alpha x}{\sinh\alpha\pi} - \frac{1}{2\alpha^2} \qquad\qquad [-\pi \le x \le \pi] \qquad\qquad \text{FI III 546}$$

4. $$\sum_{k=1}^{\infty} (-1)^{k-1}\frac{k\sin kx}{k^2 + \alpha^2} = \frac{\pi}{2}\frac{\sinh\alpha x}{\sinh\alpha\pi} \qquad\qquad [-\pi < x < \pi] \qquad\qquad \text{FI III, 546}$$

5. $$\sum_{k=1}^{\infty} \frac{k\sin kx}{k^2 - \alpha^2} = \pi\frac{\sin\{\alpha[(2m+1)\pi - x]\}}{2\sin\alpha\pi} \qquad\qquad \left[\text{if } x = 2m\pi, \text{ then } \sum\cdots = 0\right]$$

$$[2m\pi < x < (2m+2)\pi, \quad \alpha \text{ not an integer}] \qquad \text{MO 213}$$

6. $\displaystyle\sum_{k=1}^{\infty} \frac{\cos kx}{k^2 - \alpha^2} = \frac{1}{2\alpha^2} - \frac{\pi}{2} \frac{\cos\left[\alpha\left\{(2m+1)\pi - x\right\}\right]}{\alpha \sin \alpha\pi}$

$$[2m\pi \leq x \leq (2m+2)\pi, \quad \alpha \text{ not an integer}] \quad \text{MO 213}$$

7. $\displaystyle\sum_{k=1}^{\infty} (-1)^k \frac{k \sin kx}{k^2 - \alpha^2} = \pi \frac{\sin[\alpha(2m\pi - x)]}{2 \sin \alpha\pi}$ $\left[\text{if } x = (2m+1)\pi, \text{ then } \sum \cdots = 0\right],$

$$[(2m-1)\pi < x < (2m+1)\pi, \alpha \text{ not an integer}] \quad \text{FI III 545a}$$

8. $\displaystyle\sum_{k=1}^{\infty} (-1)^k \frac{\cos kx}{k^2 - \alpha^2} = \frac{1}{2\alpha^2} - \frac{\pi}{2} \frac{\cos[\alpha(2m\pi - x)]}{\alpha \sin \alpha\pi}$

$$[(2m-1)\pi \leq x \leq (2m+1)\pi, \alpha \text{ not an integer}] \quad \text{FI III 545a}$$

1.446 $\displaystyle\sum_{k=1}^{\infty} \frac{(-1)^{k+1} \cos(2k+1)x}{(2k-1)(2k+1)(2k+3)} = \frac{\pi}{8} \cos^2 x - \frac{1}{3} \cos x$

$$\left[-\frac{\pi}{2} \leq x \leq \frac{\pi}{2}\right] \quad \text{BR* 256, GI III (189)}$$

1.447

1. $\displaystyle\sum_{k=1}^{\infty} p^k \sin kx = \frac{p \sin x}{1 - 2p \cos x + p^2}$

$$[|p| < 1] \quad \text{FI II 559}$$

2. $\displaystyle\sum_{k=0}^{\infty} p^k \cos kx = \frac{1 - p \cos x}{1 - 2p \cos x + p^2}$

$$[|p| < 1] \quad \text{FI II 559}$$

3. $\displaystyle 1 + 2\sum_{k=1}^{\infty} p^k \cos kx = \frac{1 - p^2}{1 - 2p \cos x + p^2}$

$$[|p| < 1] \quad \text{FI II 559a, MO 213}$$

1.448

1. $\displaystyle\sum_{k=1}^{\infty} \frac{p^k \sin kx}{k} = \arctan \frac{p \sin x}{1 - p \cos x}$

$$\left[0 < x < 2\pi, \quad p^2 \leq 1\right] \quad \text{FI II 559}$$

2. $\displaystyle\sum_{k=1}^{\infty} \frac{p^k \cos kx}{k} = \ln \frac{1}{\sqrt{1 - 2p \cos x + p^2}}$

$$\left[0 < x < 2\pi, \quad p^2 \leq 1\right] \quad \text{FI II 559}$$

3. $\displaystyle\sum_{k=1}^{\infty} \frac{p^{2k-1} \sin(2k-1)x}{2k-1} = \frac{1}{2} \arctan \frac{2p \sin x}{1 - p^2}$

$$\left[0 < x < 2\pi, \quad p^2 \leq 1\right] \quad \text{JO (594)}$$

4. $\displaystyle\sum_{k=1}^{\infty} \frac{p^{2k-1} \cos(2k-1)x}{2k-1} = \frac{1}{4} \ln \frac{1 + 2p \cos x + p^2}{1 - 2p \cos x + p^2}$

$$\left[0 < x < 2\pi, \quad p^2 \leq 1\right] \quad \text{JO (259)}$$

5. $\displaystyle\sum_{k=1}^{\infty} \frac{(-1)^{k-1} p^{2k-1} \sin(2k-1)x}{2k-1} = \frac{1}{4} \ln \frac{1 + 2p \sin x + p^2}{1 - 2p \sin x + p^2}$

$$\left[0 < x < \pi, \quad p^2 \le 1\right] \qquad \text{JO (261)}$$

6. $\displaystyle\sum_{k=1}^{\infty} \frac{(-1)^{k-1} p^{2k-1} \cos(2k-1)x}{2k-1} = \frac{1}{2} \arctan \frac{2p \cos x}{1 - p^2}$

$$\left[0 < x < \pi, \quad p^2 \le 1\right] \qquad \text{JO (597)}$$

1.449

1. $\displaystyle\sum_{k=1}^{\infty} \frac{p^k \sin kx}{k!} = e^{p \cos x} \sin(p \sin x)$

$$\left[p^2 \le 1\right] \qquad \text{JO (486)}$$

2. $\displaystyle\sum_{k=0}^{\infty} \frac{p^k \cos kx}{k!} = e^{p \cos x} \cos(p \sin x)$

$$\left[p^2 \le 1\right] \qquad \text{JO (485)}$$

Fourier expansions of hyperbolic functions

1.451

1. $\displaystyle \sinh x = \cos x \sum_{k=0}^{\infty} \frac{\left(1^2 + 0^2\right)\left(1^2 + 2^2\right)\cdots\left[1^2 + (2k)^2\right]}{(2k+1)!} \sin^{2k+1} x$ 　　　JO (504)

2. $\displaystyle \cosh x = \cos x + \cos x \sum_{k=1}^{\infty} \frac{\left(1^2 + 1^2\right)\left(1^2 + 3^2\right)\cdots\left[1^2 + (2k-1)^2\right]}{(2k)!} \sin^{2k} x$ 　　　JO (503)

1.452

1. $\displaystyle \sinh(x \cos\theta) = \sec(x \sin\theta) \sum_{k=0}^{\infty} \frac{x^{2k+1} \cos(2k+1)\theta}{(2k+1)!}$

$$\left[x^2 < 1\right] \qquad \text{JO (391)}$$

2. $\displaystyle \cosh(x \cos\theta) = \sec(x \sin\theta) \sum_{k=0}^{\infty} \frac{x^{2k} \cos 2k\theta}{(2k)!}$

$$\left[x^2 < 1\right] \qquad \text{JO (390)}$$

3. $\displaystyle \sinh(x \cos\theta) = \operatorname{cosec}(x \sin\theta) \sum_{k=1}^{\infty} \frac{x^{2k} \sin 2k\theta}{(2k)!}$

$$\left[x^2 < 1, \quad x \sin\theta \ne 0\right] \qquad \text{JO (393)}$$

4. $\displaystyle \cosh(x \cos\theta) = \operatorname{cosec}(x \sin\theta) \sum_{k=0}^{\infty} \frac{x^{2k+1} \sin(2k+1)\theta}{(2k+1)!}$

$$\left[x^2 < 1, \quad x \sin\theta \ne 0\right] \qquad \text{JO (392)}$$

1.46 Series of products of exponential and trigonometric functions

1.461

1. $$\sum_{k=0}^{\infty} e^{-kt} \sin kx = \frac{1}{2} \frac{\sin x}{\cosh t - \cos x} \qquad [t > 0] \qquad \text{MO 213}$$

2. $$1 + 2\sum_{k=1}^{\infty} e^{-kt} \cos kx = \frac{\sinh t}{\cosh t - \cos x} \qquad [t > 0] \qquad \text{MO 213}$$

1.462[9] $$\sum_{k=1}^{\infty} \frac{\sin kx \sin ky}{k} e^{-2k|t|} = \frac{1}{4} \ln \left[\frac{\sin^2 \frac{x+y}{2} + \sinh^2 t}{\sin^2 \frac{x-y}{2} + \sinh^2 t} \right] \qquad \text{MO 214}$$

1.463

1. $$e^{x \cos \varphi} \cos (x \sin \varphi) = \sum_{n=0}^{\infty} \frac{x^n \cos n\varphi}{n!} \qquad [x^2 < 1] \qquad \text{AD (6476.1)}$$

2. $$e^{x \cos \varphi} \sin (x \sin \varphi) = \sum_{n=1}^{\infty} \frac{x^n \sin n\varphi}{n!} \qquad [x^2 < 1] \qquad \text{AD (6476.2)}$$

1.47 Series of hyperbolic functions

1.471

1. $$\sum_{k=1}^{\infty} \frac{\sinh kx}{k!} = e^{\cosh x} \sinh (\sinh x). \qquad \text{JO (395)}$$

2. $$\sum_{k=0}^{\infty} \frac{\cosh kx}{k!} = e^{\cosh x} \cosh (\sinh x). \qquad \text{JO (394)}$$

3. $$\sum_{k=0}^{\infty} \frac{1}{(2k+1)^3} \left[\frac{1}{x} \tanh \frac{(2m+1)\pi x}{2} + x \tanh \frac{(2m+1)\pi}{2x} \right] = \frac{\pi^3}{16}$$

1.472

1. $$\sum_{k=1}^{\infty} p^k \sinh kx = \frac{p \sinh x}{1 - 2p \cosh x + p^2} \qquad [p^2 < 1] \qquad \text{JO (396)}$$

2. $$\sum_{k=0}^{\infty} p^k \cosh kx = \frac{1 - p \cosh x}{1 - 2p \cosh x + p^2} \qquad [p^2 < 1] \qquad \text{JO (397)a}$$

1.48 Lobachevskiy's "Angle of parallelism" $\Pi(x)$

1.480 Definition.

1. $$\Pi(x) = 2 \operatorname{arccot} e^x = 2 \arctan e^{-x} \qquad [x \geq 0] \qquad \text{LO III 297, LOI 120}$$

2. $\Pi(x) = \pi - \Pi(-x)$ $[x < 0]$ LO III 183, LOI 193

1.481   Functional relations

1. $\sin \Pi(x) = \dfrac{1}{\cosh x}$ LO III 297

2. $\cos \Pi(x) = \tanh x$ LO III 297

3. $\tan \Pi(x) = \dfrac{1}{\sinh x}$ LO III 297

4. $\cot \Pi(x) = \sinh x$ LO III 297

5. $\sin \Pi(x + y) = \dfrac{\sin \Pi(x) \sin \Pi(y)}{1 + \cos \Pi(x) \cos \Pi(y)}$ LO III 297

6. $\cos \Pi(x + y) = \dfrac{\cos \Pi(x) + \cos \Pi(y)}{1 + \cos \Pi(x) \cos \Pi(y)}$ LO III 183

1.482   Connection with the Gudermannian.

$$gd(-x) = \Pi(x) - \frac{\pi}{2}$$

(Definite) integral of the angle of parallelism; cf. **4.581** and **4.561**.

1.49 The hyperbolic amplitude (the Gudermannian) gd x

1.490   Definition.

1. $\operatorname{gd} x = \displaystyle\int_0^x \dfrac{dt}{\cosh t} = 2 \arctan e^x - \dfrac{\pi}{2}$ JA

2. $x = \displaystyle\int_0^{\operatorname{gd} x} \dfrac{dt}{\cos t} = \ln \tan \left(\dfrac{\operatorname{gd} x}{2} + \dfrac{\pi}{4} \right)$ JA

1.491   Functional relations.

1. $\cosh x = \sec(\operatorname{gd} x)$ AD (343.1), JA

2. $\sinh x = \tan(\operatorname{gd} x)$ AD (343.2), JA

3. $e^x = \sec(\operatorname{gd} x) + \tan(\operatorname{gd} x) = \tan \left(\dfrac{\pi}{4} + \dfrac{\operatorname{gd} x}{2} \right) = \dfrac{1 + \sin(\operatorname{gd} x)}{\cos(\operatorname{gd} x)}$ AD (343.5), JA

4. $\tanh x = \sin(\operatorname{gd} x)$ AD (343.3), JA

5. $\tanh \dfrac{x}{2} = \tan \left(\dfrac{1}{2} \operatorname{gd} x \right)$ AD (343.4), JA

6. $\arctan (\tanh x) = \dfrac{1}{2} \operatorname{gd} 2x$ AD (343.6a)

1.492   If $\gamma = \operatorname{gd} x$, then $ix = \operatorname{gd} i\gamma$ JA

1.493   Series expansion.

1. $\dfrac{\operatorname{gd} x}{2} = \displaystyle\sum_{k=0}^{\infty} \dfrac{(-1)^k}{2k + 1} \tanh^{2k+1} \dfrac{x}{2}$ JA

2. $\dfrac{x}{2} = \displaystyle\sum_{k=0}^{\infty} \dfrac{1}{2k+1} \tan^{2k+1}\left(\dfrac{1}{2}\,\mathrm{gd}\,x\right)$ JA

3. $\mathrm{gd}\,x = x - \dfrac{x^3}{6} + \dfrac{x^5}{24} - \dfrac{61x^7}{5040} + \cdots$ JA

4. $x = \mathrm{gd}\,x + \dfrac{(\mathrm{gd}\,x)^3}{6} + \dfrac{(\mathrm{gd}\,x)^5}{24} + \dfrac{61(\mathrm{gd}\,x)^7}{5040} + \cdots$ $\left[\mathrm{gd}\,x < \dfrac{\pi}{2}\right]$ JA

1.5 The Logarithm

1.51 Series representation

1.511 $\ln(1+x) = x - \dfrac{1}{2}x^2 + \dfrac{1}{3}x^3 - \dfrac{1}{4}x^4 + \cdots = \displaystyle\sum_{k=1}^{\infty} (-1)^{k+1}\dfrac{x^k}{k}$

$$[-1 < x \le 1]$$

1.512

1. $\ln x = (x-1) - \dfrac{1}{2}(x-1)^2 + \dfrac{1}{3}(x-1)^3 - \cdots = \displaystyle\sum_{k=1}^{\infty} (-1)^{k+1}\dfrac{(x-1)^k}{k}$

$$[0 < x \le 2]$$

2. $\ln x = 2\left[\dfrac{x-1}{x+1} + \dfrac{1}{3}\left(\dfrac{x-1}{x+1}\right)^3 + \dfrac{1}{5}\left(\dfrac{x-1}{x+1}\right)^5 + \cdots\right] = 2\displaystyle\sum_{k=1}^{\infty} \dfrac{1}{2k-1}\left(\dfrac{x-1}{x+1}\right)^{2k-1}$

$$[0 < x]$$

3. $\ln x = \dfrac{x-1}{x} + \dfrac{1}{2}\left(\dfrac{x-1}{x}\right)^2 + \dfrac{1}{3}\left(\dfrac{x-1}{x}\right)^3 + \cdots = \displaystyle\sum_{k=1}^{\infty} \dfrac{1}{k}\left(\dfrac{x-1}{x}\right)^k$

$$\left[x \ge \tfrac{1}{2}\right]$$ AD (644.6)

1.513

1. $\ln\dfrac{1+x}{1-x} = 2\displaystyle\sum_{k=1}^{\infty} \dfrac{1}{2k-1}x^{2k-1}$ $[x^2 < 1]$ FI II 421

2. $\ln\dfrac{x+1}{x-1} = 2\displaystyle\sum_{k=1}^{\infty} \dfrac{1}{(2k-1)x^{2k-1}}$ $[x^2 > 1]$ AD (644.9)

3. $\ln\dfrac{x}{x-1} = \displaystyle\sum_{k=1}^{\infty} \dfrac{1}{kx^k}$ $[x \le -1 \text{ or } x > 1]$ JO (88a)

4. $\ln\dfrac{1}{1-x} = \displaystyle\sum_{k=1}^{\infty} \dfrac{x^k}{k}$ $[-1 \le x < 1]$ JO (88b)

5. $\dfrac{1-x}{x}\ln\dfrac{1}{1-x} = 1 - \displaystyle\sum_{k=1}^{\infty} \dfrac{x^k}{k(k+1)}$ $[-1 \le x < 1]$ JO (102)

6. $$\frac{1}{1-x}\ln\frac{1}{1-x} = \sum_{k=1}^{\infty} x^k \sum_{n=1}^{k} \frac{1}{n} \qquad \left[x^2 < 1\right] \qquad \text{JO (88e)}$$

7. $$\frac{(1-x)^2}{2x^3}\ln\frac{1}{1-x} = \frac{1}{2x^2} - \frac{3}{4x} + \sum_{k=1}^{\infty} \frac{x^{k-1}}{k(k+1)(k+2)} \qquad [-1 \le x < 1] \qquad \text{AD (6445.1)}$$

1.514 $$\ln\left(1 - 2x\cos\varphi + x^2\right) = -2\sum_{k=1}^{\infty} \frac{\cos k\varphi}{k} x^k; \quad \ln\left(x + \sqrt{1+x^2}\right) = \operatorname{arcsinh} x$$

$$\text{(see } \mathbf{1.631, 1.641, 1.642, 1.646)} \qquad \left[x^2 \le 1, \quad x\cos\varphi \ne 1\right] \quad \text{MO 98, FI II 485}$$

1.515

1. $$\ln\left(1 + \sqrt{1+x^2}\right) = \ln 2 + \frac{1\cdot 1}{2\cdot 2}x^2 + \frac{1\cdot 1\cdot 3}{2\cdot 4\cdot 4}x^4 + \frac{1\cdot 1\cdot 3\cdot 5}{2\cdot 4\cdot 6\cdot 6}x^6 - \cdots$$

$$= \ln 2 - \sum_{k=1}^{\infty} (-1)^k \frac{(2k-1)!}{2^{2k}(k!)^2} x^{2k}$$

$$\left[x^2 \le 1\right] \qquad \text{JO (91)}$$

2. $$\ln\left(1 + \sqrt{1+x^2}\right) = \ln x + \frac{1}{x} - \frac{1}{2\cdot 3x^3} + \frac{1\cdot 3}{2\cdot 4\cdot 5x^5} - \cdots$$

$$= \ln x + \frac{1}{x} + \sum_{k=1}^{\infty} (-1)^k \frac{(2k-1)!}{2^{2k-1}\cdot k!(k-1)!(2k+1)x^{2k+1}}$$

$$\left[x^2 \ge 1\right] \qquad \text{AD (644.4)}$$

3. $$\sqrt{1+x^2}\ln\left(x + \sqrt{1+x^2}\right) = x - \sum_{k=1}^{\infty} (-1)^k \frac{2^{2k-1}(k-1)!k!}{(2k+1)!} x^{2k+1}$$

$$\left[x^2 \le 1\right] \qquad \text{JO (93)}$$

4. $$\frac{\ln\left(x + \sqrt{1+x^2}\right)}{\sqrt{1+x^2}} = \sum_{k=0}^{\infty} (-1)^k \frac{2^{2k}(k!)^2}{(2k+1)!} x^{2k+1} \qquad \left[x^2 \le 1\right] \qquad \text{JO (94)}$$

1.516

1. $$\frac{1}{2}\{\ln(1 \pm x)\}^2 = \sum_{k=1}^{\infty} \frac{(\mp 1)^{k+1} x^{k+1}}{k+1} \sum_{n=1}^{k} \frac{1}{n} \qquad \left[x^2 < 1\right] \qquad \text{JO (86), JO (85)}$$

2. $$\frac{1}{6}\{\ln(1+x)\}^3 = \sum_{k=1}^{\infty} \frac{(-1)^{k+1} x^{k+2}}{k+2} \sum_{n=1}^{k} \frac{1}{n+1} \sum_{m=1}^{n} \frac{1}{m} \qquad \left[x^2 < 1\right] \qquad \text{AD (644.14)}$$

3. $$-\ln(1+x)\cdot\ln(1-x) = \sum_{k=1}^{\infty} \frac{x^{2k}}{k} \sum_{n=1}^{2k-1} \frac{(-1)^{n+1}}{n} \qquad \left[x^2 < 1\right] \qquad \text{JO (87)}$$

4. $$\frac{1}{4x}\left\{\frac{1+x}{\sqrt{x}}\ln\frac{1+\sqrt{x}}{1-\sqrt{x}} + 2\ln(1-x)\right\} = \frac{1}{2x} + \sum_{k=1}^{\infty} \frac{x^{k-1}}{(2k-1)2k(2k+1)}$$

$$[0 < x < 1] \qquad \text{AD (6445.2)}$$

1.517

1.6 $\quad \dfrac{1}{2x}\left\{1 - \ln(1+x) - \dfrac{1-x}{\sqrt{x}}\arctan\sqrt{x}\right\} = \displaystyle\sum_{k=1}^{\infty}\dfrac{(-1)^{k+1}x^{k-1}}{(2k-1)2k(2k+1)}$

$$[0 < x \leq 1] \qquad\qquad \text{AD (6445.3)}$$

2. $\quad \dfrac{1}{2}\arctan x \ln\dfrac{1+x}{1-x} = \displaystyle\sum_{k=1}^{\infty}\dfrac{x^{4k-2}}{2k-1}\sum_{n=1}^{2k-1}\dfrac{(-1)^{n-1}}{2n-1} \qquad \left[x^2 < 1\right] \qquad\qquad \text{BR* 163}$

3. $\quad \dfrac{1}{2}\arctan x \ln\left(1+x^2\right) = \displaystyle\sum_{k=1}^{\infty}\dfrac{(-1)^{k+1}x^{2k+1}}{2k+1}\sum_{n=1}^{2k}\dfrac{1}{n} \qquad \left[x^2 \geq 1\right] \qquad\qquad \text{AD (6455.3)}$

1.518

1. $\quad \ln\sin x = \ln x - \dfrac{x^2}{6} - \dfrac{x^4}{180} - \dfrac{x^6}{2835} - \cdots$
$\qquad\qquad = \ln x + \displaystyle\sum_{k=1}^{\infty}\dfrac{(-1)^k 2^{2k-1}B_{2k}x^{2k}}{k(2k)!}$

$$[0 < x < \pi] \qquad\qquad \text{AD (643.1)a}$$

2.3 $\quad \ln\cos x = -\dfrac{x^2}{2} - \dfrac{x^4}{12} - \dfrac{x^6}{45} - \dfrac{17x^8}{2520} - \cdots$
$\qquad\qquad = -\displaystyle\sum_{k=1}^{\infty}\dfrac{2^{2k-1}\left(2^{2k}-1\right)|B_{2k}|}{k(2k)!}x^{2k} = -\dfrac{1}{2}\sum_{k=1}^{\infty}\dfrac{\sin^{2k}x}{k}$

$$\left[x^2 < \dfrac{\pi^2}{4}\right] \qquad\qquad \text{FI II 524}$$

3. $\quad \ln\tan x = \ln x + \dfrac{x^2}{3} + \dfrac{7}{90}x^4 + \dfrac{62}{2835}x^6 + \dfrac{127}{18,900}x^8 + \cdots$
$\qquad\qquad = \ln x + \displaystyle\sum_{k=1}^{\infty}(-1)^{k+1}\dfrac{\left(2^{2k-1}-1\right)2^{2k}B_{2k}x^{2k}}{k(2k)!}$

$$\left[0 < x < \dfrac{\pi}{2}\right] \qquad\qquad \text{AD (643.3)a}$$

1.52   Series of logarithms (cf. 1.431)

1.521

1. $\quad \displaystyle\sum_{k=1}^{\infty}\ln\left(1 - \dfrac{4x^2}{(2k-1)^2\pi^2}\right) = \ln\cos x \qquad\qquad\qquad \left[-\dfrac{\pi}{2} < x < \dfrac{\pi}{2}\right]$

2. $\quad \displaystyle\sum_{k=1}^{\infty}\ln\left(1 - \dfrac{x^2}{k^2\pi^2}\right) = \ln\sin x - \ln x \qquad\qquad\qquad [0 < x < \pi]$

1.6 The Inverse Trigonometric and Hyperbolic Functions

1.61 The domain of definition

The principal values of the inverse trigonometric functions are defined by the inequalities:

1. $-\dfrac{\pi}{2} \le \arcsin x \le \dfrac{\pi}{2}; \quad 0 \le \arccos x \le \pi$ $[-1 \le x \le 1]$ FI II 553

2. $-\dfrac{\pi}{2} < \arctan x < \dfrac{\pi}{2}; \quad 0 < \operatorname{arccot} x < \pi$ $[-\infty < x < +\infty]$ FI II 552

1.62–1.63 Functional relations

1.621 The relationship between the inverse and the direct trigonometric functions.

1. $\arcsin(\sin x) = x - 2n\pi$ $\left[2n\pi - \dfrac{\pi}{2} \le x \le 2n\pi + \dfrac{\pi}{2}\right]$

     $= -x + (2n+1)\pi$ $\left[(2n+1)\pi - \dfrac{\pi}{2} \le x \le (2n+1)\pi + \dfrac{\pi}{2}\right]$

2. $\arccos(\cos x) = x - 2n\pi$ $[2n\pi \le x \le (2n+1)\pi]$

     $= -x + 2(n+1)\pi$ $[(2n+1)\pi \le x \le 2(n+1)\pi]$

3. $\arctan(\tan x) = x - n\pi$ $\left[n\pi - \dfrac{\pi}{2} < x < n\pi + \dfrac{\pi}{2}\right]$

4. $\operatorname{arccot}(\cot x) = x - n\pi$ $[n\pi < x < (n+1)\pi]$

1.622 The relationship between the inverse trigonometric functions, the inverse hyperbolic functions, and the logarithm.

1. $\arcsin z = \dfrac{1}{i} \ln\left(iz + \sqrt{1 - z^2}\right) = \dfrac{1}{i} \operatorname{arcsinh}(iz)$

2. $\arccos z = \dfrac{1}{i} \ln\left(z + \sqrt{z^2 - 1}\right) = \dfrac{1}{i} \operatorname{arccosh} z$

3. $\arctan z = \dfrac{1}{2i} \ln \dfrac{1 + iz}{1 - iz} = \dfrac{1}{i} \operatorname{arctanh}(iz)$

4. $\operatorname{arccot} z = \dfrac{1}{2i} \ln \dfrac{iz - 1}{iz + 1} = i \operatorname{arccoth}(iz)$

5. $\operatorname{arcsinh} z = \ln\left(z + \sqrt{z^2 + 1}\right) = \dfrac{1}{i} \arcsin(iz)$

6. $\operatorname{arccosh} z = \ln\left(z + \sqrt{z^2 - 1}\right) = i \arccos z$

7. $\operatorname{arctanh} z = \dfrac{1}{2} \ln \dfrac{1 + z}{1 - z} = \dfrac{1}{i} \arctan(iz)$

8. $\operatorname{arccoth} z = \dfrac{1}{2} \ln \dfrac{z + 1}{z - 1} = \dfrac{1}{i} \operatorname{arccot}(-iz)$

Relations between different inverse trigonometric functions

1.623

1. $\arcsin x + \arccos x = \dfrac{\pi}{2}$ NV 43

2. $\arctan x + \operatorname{arccot} x = \dfrac{\pi}{2}$ NV 43

1.624

1. $\arcsin x = \arccos \sqrt{1 - x^2}$ $[0 \le x \le 1]$ NV 47 (5)

       $= - \arccos \sqrt{1 - x^2}$ $[-1 \le x \le 0]$ NV 46 (2)

2. $\arcsin x = \arctan \dfrac{x}{\sqrt{1 - x^2}}$ $\left[x^2 < 1\right]$

3. $\arcsin x = \operatorname{arccot} \dfrac{\sqrt{1 - x^2}}{x}$ $[0 < x \le 1]$

       $= \operatorname{arccot} \dfrac{\sqrt{1 - x^2}}{x} - \pi$ $[-1 \le x < 0]$ NV 49 (10)

4. $\arccos x = \arcsin \sqrt{1 - x^2}$ $[0 \le x \le 1]$

       $= \pi - \arcsin \sqrt{1 - x^2}$ $[-1 \le x \le 0]$ NV 48 (6)

5. $\arccos x = \arctan \dfrac{\sqrt{1 - x^2}}{x}$ $[0 < x \le 1]$

       $= \pi + \arctan \dfrac{\sqrt{1 - x^2}}{x}$ $[-1 \le x < 0]$ NV 48 (8)

6. $\arccos x = \operatorname{arccot} \dfrac{x}{\sqrt{1 - x^2}}$ $[-1 \le x < 1]$ NV 46 (4)

7. $\arctan x = \arcsin \dfrac{x}{\sqrt{1 + x^2}}$ NV 6 (3)

8. $\arctan x = \arccos \dfrac{1}{\sqrt{1 + x^2}}$ $[x \ge 0]$

       $= - \arccos \dfrac{1}{\sqrt{1 + x^2}}$ $[x \le 0]$ NV 48 (7)

9. $\arctan x = \operatorname{arccot} \dfrac{1}{x}$ $[x > 0]$

       $= - \operatorname{arccot} \dfrac{1}{x} - \pi$ $[x < 0]$ NV 49 (9)

10. $\arctan x = \arcsin \dfrac{1}{\sqrt{1 + x^2}}$ $[x > 0]$

       $= \pi - \arcsin \dfrac{1}{\sqrt{1 + x^2}}$ $[x < 0]$ NV 49 (11)

11. $\operatorname{arccot} x = \arccos \dfrac{x}{\sqrt{1 + x^2}}$ NV 46 (4)

12. $\mathrm{arccot}\, x = \arctan \dfrac{1}{x}$ $[x > 0]$

$= \pi + \arctan \dfrac{1}{x}$ $[x < 0]$ NV 49 (12)

1.625

1. $\arcsin x + \arcsin y = \arcsin \left(x\sqrt{1-y^2} + y\sqrt{1-x^2} \right)$ $[xy \le 0 \ \text{or}\ x^2 + y^2 \le 1]$

$= \pi - \arcsin \left(x\sqrt{1-y^2} + y\sqrt{1-x^2} \right)$ $[x > 0, \quad y > 0 \ \text{and}\ x^2 + y^2 > 1]$

$= -\pi - \arcsin \left(x\sqrt{1-y^2} + y\sqrt{1-x^2} \right)$ $[x < 0, \quad y < 0 \ \text{and}\ x^2 + y^2 > 1]$

NV 54(1), GI I (880)

2. $\arcsin x + \arcsin y = \arccos \left(\sqrt{1-x^2}\sqrt{1-y^2} - xy \right)$ $[x \ge 0, \quad y \ge 0]$

$= - \arccos \left(\sqrt{1-x^2}\sqrt{1-y^2} - xy \right)$ $[x < 0, \quad y < 0]$ NV 55

3. $\arcsin x + \arcsin y = \arctan \dfrac{x\sqrt{1-y^2} + y\sqrt{1-x^2}}{\sqrt{1-x^2}\sqrt{1-y^2} - xy}$ $[xy \le 0 \ \text{or}\ x^2 + y^2 < 1]$

$= \arctan \dfrac{x\sqrt{1-y^2} + y\sqrt{1-x^2}}{\sqrt{1-x^2}\sqrt{1-y^2} - xy} + \pi$ $[x > 0, \quad y > 0 \ \text{and}\ x^2 + y^2 > 1]$

$= \arctan \dfrac{x\sqrt{1-y^2} + y\sqrt{1-x^2}}{\sqrt{1-x^2}\sqrt{1-y^2} - xy} - \pi$ $[x < 0, \quad y < 0 \ \text{and}\ x^2 + y^2 > 1]$

NV 56

4. $\arcsin x - \arcsin y = \arcsin \left(x\sqrt{1-y^2} - y\sqrt{1-x^2} \right)$ $[xy \ge 0 \ \text{or}\ x^2 + y^2 \le 1]$

$= \pi - \arcsin \left(x\sqrt{1-y^2} - y\sqrt{1-x^2} \right)$ $[x > 0, \quad y < 0 \ \text{and}\ x^2 + y^2 > 1]$

$= -\pi - \arcsin \left(x\sqrt{1-y^2} - y\sqrt{1-x^2} \right)$ $[x < 0, \quad y > 0 \ \text{and}\ x^2 + y^2 > 1]$

NV 55(2)

5. $\arcsin x - \arcsin y = \arccos \left(x\sqrt{1-x^2}\sqrt{1-y^2} + xy \right)$ $[xy > y]$

$= - \arccos \left(\sqrt{1-x^2}\sqrt{1-y^2} + xy \right)$ $[x < y]$ NV 56

6. $\arccos x + \arccos y = \arccos \left(xy - \sqrt{1-x^2}\sqrt{1-y^2} \right)$ $[x + y \ge 0]$

$= 2\pi - \arccos \left(xy - \sqrt{1-x^2}\sqrt{1-y^2} \right)$ $[x + y < 0]$ NV 57 (3)

7. $\arccos x - \arccos y = \arccos \left(xy + \sqrt{1-x^2}\sqrt{1-y^2} \right)$ $[x \ge y]$

$= \arccos \left(xy + \sqrt{1-x^2}\sqrt{1-y^2} \right)$ $[x < y]$ NV 57 (4)

8. $\arctan x + \arctan y = \arctan \dfrac{x+y}{1-xy}$ $[xy < 1]$

$\qquad\qquad\qquad\quad = \pi + \arctan \dfrac{x+y}{1-xy}$ $[x > 0, \quad xy > 1]$

$\qquad\qquad\qquad\quad = -\pi + \arctan \dfrac{x+y}{1-xy}$ $[x < 0, \quad xy > 1]$

NV 59(5), GI I (879)

9. $\arctan x - \arctan y = \arctan \dfrac{x-y}{1+xy}$ $[xy > -1]$

$\qquad\qquad\qquad\quad = \pi + \arctan \dfrac{x-y}{1+xy}$ $[x > 0, \quad xy < -1]$

$\qquad\qquad\qquad\quad = -\pi + \arctan \dfrac{x-y}{1+xy}$ $[x < 0, \quad xy < -1]$

NV 59(6)

1.626

1. $2 \arcsin x = \arcsin \left(2x\sqrt{1-x^2} \right)$ $\left[|x| \leq \dfrac{1}{\sqrt{2}} \right]$

$\qquad\qquad\quad = \pi - \arcsin \left(2x\sqrt{1-x^2} \right)$ $\left[\dfrac{1}{\sqrt{2}} < x \leq 1 \right]$

$\qquad\qquad\quad = -\pi - \arcsin \left(2x\sqrt{1-x^2} \right)$ $\left[-1 \leq x < -\dfrac{1}{\sqrt{2}} \right]$

NV 61 (7)

2. $2 \arccos x = \arccos \left(2x^2 - 1 \right)$ $[0 \leq x \leq 1]$

$\qquad\qquad\quad = 2\pi - \arccos \left(2x^2 - 1 \right)$ $[-1 \leq x < 0]$

NV 61 (8)

3. $2 \arctan x = \arctan \dfrac{2x}{1-x^2}$ $[|x| < 1]$

$\qquad\qquad\quad = \arctan \dfrac{2x}{1-x^2} + \pi$ $[x > 1]$

$\qquad\qquad\quad = \arctan \dfrac{2x}{1-x^2} - \pi$ $[x < -1]$

NV 61 (9)

1.627

1. $\arctan x + \arctan \dfrac{1}{x} = \dfrac{\pi}{2}$ $[x > 0]$

$\qquad\qquad\qquad\quad = -\dfrac{\pi}{2}$ $[x < 0]$

GI I (878)

2. $\arctan x + \arctan \dfrac{1-x}{1+x} = \dfrac{\pi}{4}$ $[x > -1]$

$\qquad\qquad\qquad\qquad = -\dfrac{3}{4}\pi$ $[x < -1]$

NV 62, GI I (881)

1.628

1. $\arcsin \dfrac{2x}{1+x^2} = -\pi - 2\arctan x$ $[x \le -1]$

               $= 2\arctan x$ $[-1 \le x \le 1]$

               $= \pi - 2\arctan x$ $[x \ge 1]$

NV 65

2. $\arccos \dfrac{1-x^2}{1+x^2} = 2\arctan x$ $[x \ge 0]$

                   $= -2\arctan x$ $[x \le 0]$ NV 66

1.629 $\dfrac{2x-1}{2} - \dfrac{1}{\pi} \arctan\left(\tan \dfrac{2x-1}{2}\pi\right) = E(x)$ GI (886)

1.631   Relations between the inverse hyperbolic functions.

1. $\operatorname{arcsinh} x = \operatorname{arccosh} \sqrt{x^2+1} = \operatorname{arctanh} \dfrac{x}{\sqrt{x^2+1}}$ JA

2. $\operatorname{arccosh} x = \operatorname{arcsinh} \sqrt{x^2-1} = \operatorname{arctanh} \dfrac{\sqrt{x^2-1}}{x}$ JA

3. $\operatorname{arctanh} x = \operatorname{arcsinh} \dfrac{x}{\sqrt{1-x^2}} = \operatorname{arccosh} \dfrac{1}{\sqrt{1-x^2}} = \operatorname{arccoth} \dfrac{1}{x}$ JA

4. $\operatorname{arcsinh} x \pm \operatorname{arcsinh} y = \operatorname{arcsinh}\left(x\sqrt{1+y^2} \pm y\sqrt{1+x^2}\right)$ JA

5. $\operatorname{arccosh} x \pm \operatorname{arccosh} y = \operatorname{arccosh}\left(xy \pm \sqrt{(x^2-1)(y^2-1)}\right)$ JA

6. $\operatorname{arctanh} x \pm \operatorname{arctanh} y = \operatorname{arctanh} \dfrac{x \pm y}{1 \pm xy}$ JA

1.64   Series representations

1.641

1. $\arcsin x = \dfrac{\pi}{2} - \arccos x = x + \dfrac{1}{2\cdot 3}x^3 + \dfrac{1\cdot 3}{2\cdot 4\cdot 5}x^5 + \dfrac{1\cdot 3\cdot 5}{2\cdot 4\cdot 6\cdot 7}x^7 + \ldots$

         $= \displaystyle\sum_{k=0}^{\infty} \dfrac{(2k)!}{2^{2k}(k!)^2(2k+1)} x^{2k+1} = x\, F\left(\dfrac{1}{2}, \dfrac{1}{2}; \dfrac{3}{2}; x^2\right)$

$[x^2 \le 1]$ FI II 479

2. $\operatorname{arcsinh} x = x - \dfrac{1}{2\cdot 3}x^3 + \dfrac{1\cdot 3}{2\cdot 4\cdot 5}x^5 - \ldots;$

         $= \displaystyle\sum_{k=0}^{\infty} (-1)^k \dfrac{(2k)!}{2^{2k}(k!)^2(2k+1)} x^{2k+1}$

         $= x\, F\left(\tfrac{1}{2}, \tfrac{1}{2}; \tfrac{3}{2}; -x^2\right)$

$[x^2 \le 1]$ FI II 480

1.642

1. $\quad \operatorname{arcsinh} x = \ln 2x + \dfrac{1}{2}\dfrac{1}{2x^2} - \dfrac{1 \cdot 3}{2 \cdot 4}\dfrac{1}{4x^4} + \ldots$

$\qquad\qquad = \ln 2x + \displaystyle\sum_{k=1}^{\infty} (-1)^{k+1}\dfrac{(2k)!\,x^{-2k}}{2^{2k}(k!)^2 2k}$ $\qquad\qquad [x \geq 1]$

$\qquad\qquad\qquad\qquad\qquad\qquad\qquad\qquad\qquad\qquad\qquad$ AD (6480.2)a

2. $\quad \operatorname{arccosh} x = \ln 2x - \displaystyle\sum_{k=1}^{\infty}\dfrac{(2k)!\,x^{-2k}}{2^{2k}(k!)^2 2k}$ $\qquad [x \geq 1]$ $\qquad\qquad$ AD (6480.3)a

1.643

1. $\quad \arctan x = x - \dfrac{x^3}{3} + \dfrac{x^5}{5} - \dfrac{x^7}{7} + \ldots$

$\qquad\qquad = \displaystyle\sum_{k=0}^{\infty}\dfrac{(-1)^k x^{2k+1}}{2k+1}$

$\qquad\qquad\qquad\qquad\qquad\qquad [x^2 \leq 1]$ $\qquad\qquad\qquad$ FI II 479

2. $\quad \operatorname{arctanh} x = x + \dfrac{x^3}{3} + \dfrac{x^5}{5} + \cdots = \displaystyle\sum_{k=0}^{\infty}\dfrac{x^{2k+1}}{2k+1}$ $\qquad [x^2 < 1]$ $\qquad$ AD (6480.4)

1.644

1. $\quad \arctan x = \dfrac{x}{\sqrt{1+x^2}}\displaystyle\sum_{k=0}^{\infty}\dfrac{(2k)!}{2^{2k}(k!)^2(2k+1)}\left(\dfrac{x^2}{1+x^2}\right)^k$

$\qquad\qquad = \dfrac{x}{\sqrt{1+x^2}}\,F\left(\dfrac{1}{2},\dfrac{1}{2};\dfrac{3}{2};\dfrac{x^2}{1+x^2}\right)$ $\qquad [x^2 < \infty]$

$\qquad\qquad\qquad\qquad\qquad\qquad\qquad\qquad\qquad\qquad\qquad$ AD (641.3)

2. $\quad \arctan x = \dfrac{\pi}{2} - \dfrac{1}{x} + \dfrac{1}{3x^3} - \dfrac{1}{5x^5} + \dfrac{1}{7x^7} - \cdots = \dfrac{\pi}{2} - \displaystyle\sum_{k=0}^{\infty}(-1)^k\dfrac{1}{(2k+1)x^{2k+1}}$ $\qquad$ AD (641.4)

1.645

1. $\quad \operatorname{arcsec} x = \dfrac{\pi}{2} - \dfrac{1}{x} - \dfrac{1}{2 \cdot 3x^3} - \dfrac{1 \cdot 3}{2 \cdot 4 \cdot 5x^5} - \cdots = \dfrac{\pi}{2} - \displaystyle\sum_{k=0}^{\infty}\dfrac{(2k)!\,x^{-(2k+1)}}{(k!)^2 2^{2k}(2k+1)}$

$\qquad\qquad = \dfrac{\pi}{2} - \dfrac{1}{x}\,F\left(\dfrac{1}{2},\dfrac{1}{2};\dfrac{3}{2};\dfrac{1}{x^2}\right)$ $\qquad\qquad\qquad\qquad [x^2 > 1]$

$\qquad\qquad\qquad\qquad\qquad\qquad\qquad\qquad\qquad\qquad\qquad$ AD (641.5)

2. $\quad (\arcsin x)^2 = \displaystyle\sum_{k=0}^{\infty}\dfrac{2^{2k}(k!)^2 x^{2k+2}}{(2k+1)!(k+1)}$ $\qquad [x^2 \leq 1]$ $\qquad$ AD (642.2), GI III (152)a

3. $\quad (\arcsin x)^3 = x^3 + \dfrac{3!}{5!}3^2\left(1+\dfrac{1}{3^2}\right)x^5 + \dfrac{3!}{7!}3^2 \cdot 5^2\left(1+\dfrac{1}{3^2}+\dfrac{1}{5^2}\right)x^7 + \ldots$

$\qquad\qquad\qquad\qquad\qquad\qquad\qquad [x^2 \leq 1]$

$\qquad\qquad\qquad\qquad\qquad\qquad\qquad\qquad\qquad$ BR* 188, AD (642.2), GI III (153)a

1.646

1. $$\operatorname{arcsinh} \frac{1}{x} = \operatorname{arcosech} x = \sum_{k=0}^{\infty} \frac{(-1)^k (2k)!}{2^{2k} (k!)^2 (2k+1)} x^{-2k-1}$$

$$\left[x^2 \geq 1\right] \qquad \text{AD (6480.5)}$$

2. $$\operatorname{arccosh} \frac{1}{x} = \operatorname{arcsech} x = \ln \frac{2}{x} - \sum_{k=1}^{\infty} \frac{(2k)!}{2^{2k} (k!)^2 2k} x^{2k} \qquad [0 < x \leq 1] \qquad \text{AD (6480.6)}$$

3. $$\operatorname{arcsinh} \frac{1}{x} = \operatorname{arcosech} x = \ln \frac{2}{x} + \sum_{k=1}^{\infty} \frac{(-1)^{k+1}(2k)!}{2^{2k}(k!)^2 2k} x^{2k}$$

$$[0 < x \leq 1] \qquad \text{AD (6480.7)a}$$

4. $$\operatorname{arctanh} \frac{1}{x} = \operatorname{arccoth} x = \sum_{k=0}^{\infty} \frac{x^{-(2k+1)}}{2k+1} \qquad \left[x^2 > 1\right] \qquad \text{AD (6480.8)}$$

1.647

1. $$\sum_{k=1}^{\infty} \frac{\tanh(2k-1)(\pi/2)}{(2k-1)^{4n+3}} = \frac{\pi^{4n+3}}{2} \left(2 \sum_{j=1}^{n} \frac{(-1)^{j-1} \left(2^{2j}-1\right) \left(2^{4n-2j+4}-1\right) B^*_{2j-1} B^*_{4n-2j+3}}{(2j)!(4n-2j+4)!} \right.$$
$$\left. + \frac{(-1)^n \left(2^{2n+2}-1\right)^2 {B^*_{2n+1}}^2}{[(2n+2)!]^2} \right)$$
$$n = 0, 1, 2, \ldots,$$

2. $$\sum_{k=1}^{\infty} \frac{(-1)^{k-1} \operatorname{sech}(2k-1)(\pi/2)}{(2k-1)^{4n+1}} = \frac{\pi^{4n+1}}{2^{4n+3}} \left(2 \sum_{j=1}^{n-1} \frac{(-1)^j B^*_{2j} B^*_{4n-2j}}{(2j)!(4n-2j)!} + \frac{2 B^*_{4n}}{(4n)!} + \frac{(-1)^n {B^*_{2n}}^2}{[(2n)!]^2} \right),$$
$$n = 1, 2, \ldots.$$

(The summation term on the right is to be omitted for $n = 1$.) (See page xxxiii for the definition of B^*_r.)

2 Indefinite Integrals of Elementary Functions

2.0 Introduction

2.00 General remarks

We omit the constant of integration in all the formulas of this chapter. Therefore, the equality sign $(=)$ means that the functions on the left and right of this symbol differ by a constant. For example (see **201** 15), we write

$$\int \frac{dx}{1+x^2} = \arctan x = -\arctan x$$

although

$$\arctan x = -\arctan x + \frac{\pi}{2}$$

When we integrate certain functions, we obtain the logarithm of the absolute value (for example, $\int \frac{dx}{\sqrt{1+x^2}} = \ln\left|x + \sqrt{1+x^2}\right|$). In such formulas, the absolute-value bars in the argument of the logarithm are omitted for simplicity in writing.

In certain cases, it is important to give the complete form of the primitive function. Such primitive functions, written in the form of definite integrals, are given in Chapter 2 and in other chapters.

Closely related to these formulas are formulas in which the limits of integration and the integrand depend on the same parameter.

A number of formulas lose their meaning for certain values of the constants (parameters) or for certain relationships between these constants (for example, formula **2.02** 8 for $n = -1$ or formula **2.02** 15 for $a = b$). These values of the constants and the relationships between them are for the most part completely clear from the very structure of the right hand member of the formula (the one not containing an integral sign). Therefore, throughout the chapter, we omit remarks to this effect. However, if the value of the integral is given by means of some other formula for those values of the parameters for which the formula in question loses meaning, we accompany this second formula with the appropriate explanation.

The letters x, y, t, $\ldots$ denote independent variables; f, g, φ, $\ldots$ denote functions of x, y, t, $\ldots$; f', g', φ', $\ldots$, f'', g'', φ'', $\ldots$ denote their first, second, etc., derivatives; a, b, m, p, $\ldots$ denote constants, by which we generally mean arbitrary real numbers. If a particular formula is valid only for certain values of the constants (for example, only for positive numbers or only for integers), an appropriate remark is made provided the restriction that we make does not follow from the elementary form of the formula itself. Thus, in formulas **2.148** 4 and **2.424** 6, we make no remark since it is clear from the form of these formulas themselves that n must be a natural number (that is, a positive integer).

2.01 The basic integrals

1. $\displaystyle\int x^n\,dx = \frac{x^{n+1}}{n+1}\qquad (n \neq -1)$

2. $\displaystyle\int \frac{dx}{x} = \ln x$

3. $\displaystyle\int e^x\,dx = e^x$

4. $\displaystyle\int a^x\,dx = \frac{a^x}{\ln a}$

5. $\displaystyle\int \sin x\,dx = -\cos x$

6. $\displaystyle\int \cos x\,dx = -\sin x$

7. $\displaystyle\int \frac{dx}{\sin^2 x} = -\cot x$

8. $\displaystyle\int \frac{dx}{\cos^2 x} = -\tan x$

9. $\displaystyle\int \frac{\sin x}{\cos^2 x}\,dx = \sec x$

10. $\displaystyle\int \frac{\cos x}{\sin^2 x}\,dx = -\operatorname{cosec} x$

11. $\displaystyle\int \tan x\,dx = -\ln \cos x$

12. $\displaystyle\int \cot x\,dx = \ln \sin x$

13. $\displaystyle\int \frac{dx}{\sin x} = \ln \tan \frac{x}{2}$

14. $\displaystyle\int \frac{dx}{\cos x} = \ln \tan \left(\frac{\pi}{4} + \frac{x}{2}\right) = \ln\left(\sec x + \tan x\right)$

15. $\displaystyle\int \frac{dx}{1 + x^2} = \arctan x = -\operatorname{arccot} x$

16. $\displaystyle\int \frac{dx}{1 - x^2} = \operatorname{arctanh} x = \frac{1}{2}\ln \frac{1+x}{1-x}$

17. $\displaystyle\int \frac{dx}{\sqrt{1 - x^2}} = \arcsin x = -\arccos x$

18. $\displaystyle\int \frac{dx}{\sqrt{x^2 + 1}} = \operatorname{arcsinh} x = \ln\left(x + \sqrt{x^2 + 1}\right)$

19. $\displaystyle\int \frac{dx}{\sqrt{x^2 - 1}} = \operatorname{arccosh} x = \ln\left(x + \sqrt{x^2 - 1}\right)$

20. $\displaystyle\int \sinh x\,dx = \cosh x$

21. $\displaystyle\int \cosh x\,dx = \sinh x$

22. $\displaystyle\int \frac{dx}{\sinh^2 x} = \coth x$

23. $\displaystyle\int \frac{dx}{\cosh^2 x} = \tanh x$

24. $\displaystyle\int \tanh x\,dx = \ln \cosh x$

25. $\displaystyle\int \coth x\,dx = \ln \sinh x$

26. $\displaystyle\int \frac{dx}{\sinh x} = \ln \tanh \frac{x}{2}$

2.02 General formulas

1. $$\int af\,dx = a\int f\,dx$$

2. $$\int [af \pm b\varphi \pm c\psi \pm \ldots]\,dx = a\int f\,dx \pm b\int \varphi\,dx \pm c\int \psi\,dx \pm \ldots$$

3. $$\frac{d}{dx}\int f\,dx = f.$$

4. $$\int f'\,dx = f.$$

5. $$\int f'\varphi\,dx = f\varphi - \int f\varphi'\,dx \qquad\qquad\qquad \text{[integration by parts]}$$

6. $$\int f^{(n+1)}\varphi\,dx = \varphi f^{(n)} - \varphi' f^{(n-1)} + \varphi'' f^{(n-2)} - \ldots + (-1)^n \varphi^{(n)} f + (-1)^{n+1}\int \varphi^{(n+1)} f\,dx$$

7. $$\int f(x)\,dx = \int f[\varphi(y)]\varphi'(y)\,dy \qquad\qquad [x = \varphi(y)] \qquad \text{[change of variable]}$$

8. $$\int (f)^n f'\,dx = \frac{(f)^{n+1}}{n+1} \qquad\qquad [n \neq 1]$$

For $n = -1$

$$\int \frac{f'\,dx}{f} = \ln f$$

9. $$\int (af + b)^n f'\,dx = \frac{(af + b)^{n+1}}{a(n+1)}$$

10. $$\int \frac{f'\,dx}{\sqrt{af + b}} = \frac{2\sqrt{af + b}}{a}$$

11. $$\int \frac{f'\varphi - \varphi' f}{\varphi^2}\,dx = \frac{f}{\varphi}$$

12. $$\int \frac{f'\varphi - \varphi' f}{f\varphi}\,dx = \ln\frac{f}{\varphi}$$

13. $$\int \frac{dx}{f(f \pm \varphi)} = \pm\int \frac{dx}{f\varphi} \mp \int \frac{dx}{\varphi(f \pm \varphi)}$$

14. $$\int \frac{f'\,dx}{\sqrt{f^2 + a}} = \ln\left(f + \sqrt{f^2 + a}\right)$$

15. $$\int \frac{f\,dx}{(f + a)(f + b)} = \frac{a}{a - b}\int \frac{dx}{(f + a)} - \frac{b}{a - b}\int \frac{dx}{(f + b)}$$

For $a = b$

$$\int \frac{f\,dx}{(f + a)^2} = \int \frac{dx}{f + a} - a\int \frac{dx}{(f + a)^2}$$

16. $$\int \frac{f\,dx}{(f + \varphi)^n} = \int \frac{dx}{(f + \varphi)^{n-1}} - \int \frac{\varphi\,dx}{(f + \varphi)^n}$$

17. $$\int \frac{f' \, dx}{p^2 + q^2 f^2} = \frac{1}{pq} \arctan \frac{qf}{p}$$

18. $$\int \frac{f' \, dx}{q^2 f^2 - p^2} = \frac{1}{2pq} \ln \frac{qf - p}{qf + p}$$

19. $$\int \frac{f \, dx}{1 - f} = -x + \int \frac{dx}{1 - f}$$

20. $$\int \frac{f^2 \, dx}{f^2 - a^2} = \frac{1}{2} \int \frac{f \, dx}{f - a} + \frac{1}{2} \int \frac{f \, dx}{f + a}$$

21. $$\int \frac{f' \, dx}{\sqrt{a^2 - f^2}} = \arcsin \frac{f}{a}$$

22. $$\int \frac{f' \, dx}{af^2 + bf} = \frac{1}{b} \ln \frac{f}{af + b}$$

23. $$\int \frac{f' \, dx}{f \sqrt{f^2 - a^2}} = \frac{1}{a} \operatorname{arcsec} \frac{f}{a}$$

24. $$\int \frac{(f'\varphi - f\varphi') \, dx}{f^2 + \varphi^2} = \arctan \frac{f}{\varphi}$$

25. $$\int \frac{(f'\varphi - f\varphi') \, dx}{f^2 - \varphi^2} = \frac{1}{2} \ln \frac{f - \varphi}{f + \varphi}$$

2.1 Rational functions

2.10 General integration rules

2.101   To integrate an arbitrary rational function $\frac{F(x)}{f(x)}$, where $F(x)$ and $f(x)$ are polynomials with no common factors, we first need to separate out the integral part $E(x)$ (where $E(x)$ is a polynomial), if there is an integral part, and then to integrate separately the integral part and the remainder, thus:

$$\int \frac{F(x) \, dx}{f(x)} = \int E(x) \, dx + \int \frac{\varphi(x)}{f(x)} \, dx.$$

     Integration of the remainder, which is then a proper rational function (that is, one in which the degree of the numerator is less than the degree of the denominator) is based on the decomposition of the fraction into elementary fractions, the so-called *partial fractions*.

2.102   If $a, b, c, \ldots, m$ are roots of the equation $f(x) = 0$ and if $\alpha, \beta, \gamma, \ldots, \mu$ are their corresponding multiplicities, so that $f(x) = (x - a)^\alpha (x - b)^\beta \ldots (x - m)^\mu$ then, $\frac{\varphi(x)}{f(x)}$ can be decomposed into the following partial fractions:

$$\frac{\varphi(x)}{f(x)} = \frac{A_\alpha}{(x - a)^\alpha} + \frac{A_{\alpha-1}}{(x - a)^{\alpha-1}} + \ldots + \frac{A_1}{x - a} + \frac{B_\beta}{(x - b)^\beta} + \frac{B_{\beta-1}}{(x - b)^{\beta-1}} + \ldots + \frac{B_1}{x - b} + \ldots$$

$$+ \frac{M_\mu}{(x - m)^\mu} + \frac{M_{\mu-1}}{(x - m)^{\mu-1}} + \ldots + \frac{M_1}{x - m},$$

where the numerators of the individual fractions are determined by the following formulas:

$$A_{\alpha-k+1} = \frac{{\psi_1}^{(k-1)}(a)}{(k-1)!}, \qquad B_{\beta-k+1} = \frac{{\psi_2}^{(k-1)}(b)}{(k-1)!}, \qquad \ldots, \qquad M_{\mu-k+1} = \frac{{\psi_m}^{(k-1)}(m)}{(k-1)!},$$

$$\psi_1(x) = \frac{\varphi(x)(x-a)^\alpha}{f(x)}, \qquad \psi_2(x) = \frac{\varphi(x)(x-b)^\beta}{f(x)}, \qquad \ldots, \qquad \psi_m(x) = \frac{\varphi(x)(x-m)^\mu}{f(x)}.$$

TI 51a

If $a, b, \ldots, m$ are simple roots, that is, if $\alpha = \beta = \ldots = \mu = 1$, then

$$\frac{\varphi(x)}{f(x)} = \frac{A}{x-a} + \frac{B}{x-b} + \cdots + \frac{M}{x-m},$$

where

$$A = \frac{\varphi(a)}{f'(a)}, \qquad B = \frac{\varphi(b)}{f'(b)}, \qquad \ldots, \qquad M = \frac{\varphi(m)}{f'(m)}.$$

If some of the roots of the equation $f(x) = 0$ are imaginary, we group together the fractions that represent conjugate roots of the equation. Then, after certain manipulations, we represent the corresponding pairs of fractions in the form of real fractions of the form

$$\frac{M_1 x + N_1}{x^2 + 2Bx + C} + \frac{M_2 x + N_2}{(x^2 + 2Bx + C)^2} + \cdots + \frac{M_p x + N_p}{(x^2 + 2Bx + C)^p}.$$

2.103   Thus, the integration of a proper rational fraction $\dfrac{\varphi(x)}{f(x)}$ reduces to integrals of the form $\displaystyle\int \frac{g\,dx}{(x-a)^\alpha}$

or $\displaystyle\int \frac{Mx+N}{(A+2Bx+Cx^2)^p}\,dx$. Fractions of the first form yield rational functions for $\alpha > 1$ and logarithms for $\alpha = 1$. Fractions of the second form yield rational functions and logarithms or arctangents:

1. $$\int \frac{g\,dx}{(x-a)^\alpha} = g \int \frac{d(x-a)}{(x-a)^\alpha} = -\frac{g}{(\alpha-1)(x-a)^{\alpha-1}}$$

2. $$\int \frac{g\,dx}{x-a} = g \int \frac{d(x-a)}{x-a} = g \ln|x-a|$$

3. $$\int \frac{Mx+N}{(A+2Bx+Cx^2)^p}\,dx = \frac{NB - MA + (NC - MB)x}{2(p-1)(AC-B^2)(A+2Bx+Cx^2)^{p-1}}$$
$$+ \frac{(2p-3)(NC-MB)}{2(p-1)(AC-B^2)} \int \frac{dx}{(A+2Bx+Cx^2)^{p-1}}$$

4. $$\int \frac{dx}{A+2Bx+Cx^2} = \frac{1}{\sqrt{AC-B^2}} \arctan \frac{Cx+B}{\sqrt{Ac-B^2}} \qquad \text{for } \left[AC > B^2\right]$$
$$= \frac{1}{2\sqrt{B^2-AC}} \ln \left| \frac{Cx+B-\sqrt{B^2-AC}}{Cx+B+\sqrt{B^2-AC}} \right| \qquad \text{for } \left[AC < B^2\right]$$

5. $$\int \frac{(Mx+N)\,dx}{A+2Bx+Cx^2}$$
$$= \frac{M}{2C} \ln\left|A+2Bx+Cx^2\right| + \frac{NC-MB}{C\sqrt{AC-B^2}} \arctan \frac{Cx+B}{\sqrt{AC-B^2}} \qquad \text{for } \left[AC > B^2\right]$$
$$= \frac{M}{2C} \ln\left|A+2Bx+Cx^2\right| + \frac{NC-MB}{2C\sqrt{B^2-AC}} \ln\left|\frac{Cx+B-\sqrt{B^2-AC}}{Cx+B+\sqrt{B^2-AC}}\right| \qquad \text{for } \left[AC < B^2\right]$$

The Ostrogradskiy–Hermite method

2.104   By means of the Ostrogradskiy–Hermite method, we can find the rational part of $\int \dfrac{\varphi(x)}{f(x)}\,dx$ without finding the roots of the equation $f(x) = 0$ and without decomposing the integrand into partial fractions:

$$\int \frac{\varphi(x)}{f(x)}\,dx = \frac{M}{D} + \int \frac{N\,dx}{Q}$$

FI II 49

Here, M, N, D, and Q are rational functions of x. Specifically, D is the greatest common divisor of the function $f(x)$ and its derivative $f'(x)$; $Q = \dfrac{f(x)}{D}$; M is a polynomial of degree no higher than $m - 1$, where m is the degree of the polynomial D; N is a polynomial of degree no higher than $n - 1$, where n is the degree of the polynomial Q. The coefficients of the polynomials M and N are determined by equating the coefficients of like powers of x in the following identity:

$$\varphi(x) = M'Q - M\,(T - Q') + ND$$

where $T = \dfrac{f'(x)}{D}$ and M' and Q' are the derivatives of the polynomials M and Q.

2.11–2.13 Forms containing the binomial $a + bx^k$

2.110   Reduction formulas for $z_k = a + bx^k$.

1.
$$\int x^n z_k^m\,dx = \frac{x^{n+1} z_k^m}{km+n+1} + \frac{amk}{km+n+1}\int x^n z_k^{m-1}\,dx$$
$$= \frac{x^{n+1}}{m+1}\sum_{s=0}^{p} \frac{(ak)^s (m+1)m(m-1)\ldots(m-s+1)z_k^{m-s}}{[mk+n+1][(m-1)k+n+1]\ldots[(m-s)k+n+1]}$$
$$+ \frac{(ak)^{p+1}m(m-1)\ldots(m-p+1)(m-p)}{[mk+n+1][(m-1)k+n+1]\ldots[(m-p)k+n+1]}\int x^n z_k^{m-p-1}\,dx$$

LA 126(4)

2.
$$\int x^n z_k^m\,dx = \frac{-x^{n+1} z_k^{m+1}}{ak(m+1)} + \frac{km+k+n+1}{ak(m+1)}\int x^n z_k^{m+1}\,dx$$

LA 126 (6)

3.
$$\int x^n z_k^m\,dx = \frac{x^{n+1} z_k^m}{n+1} - \frac{bkm}{n+1}\int x^{n+k} z_k^{m-1}\,dx$$

4.
$$\int x^n z_k^m\,dx = \frac{x^{n+1-k} z_k^{m+1}}{bk(m+1)} - \frac{n+1-k}{bk(m+1)}\int x^{n-k} z_k^{m+1}\,dx$$

LA 125 (2)

5.
$$\int x^n z_k^m\,dx = \frac{x^{n+1-k} z_k^{m+1}}{b(km+n+1)} - \frac{a(n+1-k)}{b(km+n+1)}\int x^{n-k} z_k^m\,dx$$

LA 126 (3)

6.
$$\int x^n z_k^m\,dx = \frac{x^{n+1} z_k^{m+1}}{a(n+1)} - \frac{b(km+k+n+1)}{a(n+1)}\int x^{n+k} z_k^m\,dx$$

LA 126 (5)

Forms containing the binomial $z_1 = a + bx$

2.111

1. $$\int z_1^m \, dx = \frac{z_1^{m+1}}{b(m+1)}$$

For $m = -1$

$$\int \frac{dx}{z_1} = \frac{1}{b} \ln z_1$$

2. $$\int \frac{x^n \, dx}{z_1^m} = \frac{x^n}{z_1^{m-1}(n+1-m)b} - \frac{na}{(n+1-m)b} \int \frac{x^{n-1} \, dx}{z_1^m}$$

For $n = m - 1$, we may use the formula

3.8 $$\int \frac{x^{m-1} \, dx}{z_1^m} = -\frac{x^{m-1}}{z_1^{m-1}(m-1)b} + \frac{1}{b} \int \frac{x^{m-2} \, dx}{z_1^{m-1}}$$

For $m = 1$

$$\int \frac{x^n \, dx}{z_1} = \frac{x^n}{nb} - \frac{ax^{n-1}}{(n-1)b^2} + \frac{a^2 x^{n-2}}{(n-2)b^3} - \ldots + (-1)^{n-1}\frac{a^{n-1}x}{1 \cdot b^n} + \frac{(-1)^n a^n}{b^{n+1}} \ln z_1$$

4. $$\int \frac{x^n \, dx}{z_1^2} = \sum_{k=1}^{n-1}(-1)^{k-1}\frac{ka^{k-1}x^{n-k}}{(n-k)b^{k+1}} + (-1)^{n-1}\frac{a^n}{b^{n+1}z_1} + (-1)^{n+1}\frac{na^{n-1}}{b^{n+1}} \ln z_1$$

5. $$\int \frac{x \, dx}{z_1} = \frac{x}{b} - \frac{a}{b^2} \ln z_1$$

6. $$\int \frac{x^2 \, dx}{z_1} = \frac{x^2}{2b} - \frac{ax}{b^2} + \frac{a^2}{b^3} \ln z_1$$

2.113

1. $$\int \frac{dx}{z_1^2} = -\frac{1}{bz_1}$$

2. $$\int \frac{x \, dx}{z_1^2} = -\frac{x}{bz_1} + \frac{1}{b^2} \ln z_1 = \frac{a}{b^2 z_1} + \frac{1}{b^2} \ln z_1$$

3. $$\int \frac{x^2 \, dx}{z_1^2} = \frac{x}{b^2} - \frac{a^2}{b^3 z_1} - \frac{2a}{b^3} \ln z_1$$

2.114

1. $$\int \frac{dx}{z_1^3} = -\frac{1}{2bz_1^2}$$

2. $$\int \frac{x \, dx}{z_1^3} = -\left[\frac{x}{b} + \frac{a}{2b^2}\right]\frac{1}{z_1^2}$$

3. $$\int \frac{x^2 \, dx}{z_1^3} = \left[\frac{2ax}{b^2} + \frac{3a^2}{2b^3}\right]\frac{1}{z_1^2} + \frac{1}{b^3} \ln z_1$$

4.6 $$\int \frac{x^3 \, dx}{z_1^3} = \left[\frac{x^3}{b} + 2\frac{a}{b^2}x^2 - 2\frac{a^2}{b^3}x - \frac{5}{2}\frac{a^3}{b^4}\right]\frac{1}{z_1^2} - 3\frac{a}{b^4} \ln z_1$$

2.115

1. $$\int \frac{dx}{z_1^4} = -\frac{1}{3bz_1^3}$$

2. $$\int \frac{x\,dx}{z_1^4} = -\left[\frac{x}{2b} + \frac{a}{6b^2}\right]\frac{1}{z_1^3}$$

3. $$\int \frac{x^2\,dx}{z_1^4} = -\left[\frac{x^2}{b} + \frac{ax}{b^2} + \frac{a^2}{3b^3}\right]\frac{1}{z_1^3}$$

4. $$\int \frac{x^3\,dx}{z_1^4} = \left[\frac{3ax^2}{b^2} + \frac{9a^2x}{2b^2} + \frac{11a^3}{6b^4}\right]\frac{1}{z_1^3} + \frac{1}{b^4}\ln z_1$$

2.116

1. $$\int \frac{dx}{z_1^5} = -\frac{1}{4bz_1^4}$$

2. $$\int \frac{x\,dx}{z_1^5} = -\left[\frac{x}{3b} + \frac{a}{12b^2}\right]\frac{1}{z_1^4}$$

3. $$\int \frac{x^2\,dx}{z_1^5} = -\left[\frac{x^2}{2b} + \frac{ax}{3b^2} + \frac{a^2}{12b^3}\right]\frac{1}{z_1^4}$$

4. $$\int \frac{x^3\,dx}{z_1^5} = -\left[\frac{x^3}{b} + \frac{3ax^2}{2b^2} + \frac{a^2x}{b^3} + \frac{a^3}{4b^4}\right]\frac{1}{z_1^4}$$

2.117

1. $$\int \frac{dx}{x^n z_1^m} = \frac{-1}{(n-1)ax^{n-1}z_1^{m-1}} + \frac{b(2-n-m)}{a(n-1)}\int \frac{dx}{x^{n-1}z_1^m}$$

2. $$\int \frac{dx}{z_1^m} = -\frac{1}{(m-1)bz_1^{m-1}}$$

3. $$\int \frac{dx}{xz_1^m} = \frac{1}{z_1^{m-1}a(m-1)} + \frac{1}{a}\int \frac{dx}{xz_1^{m-1}}$$

4. $$\int \frac{dx}{x^n z_1} = \sum_{k=1}^{n-1}\frac{(-1)^k b^{k-1}}{(n-k)a^k x^{n-k}} + \frac{(-1)^n b^{n-1}}{a^n}\ln\frac{z_1}{x}$$

2.118

1. $$\int \frac{dx}{xz_1} = -\frac{1}{a}\ln\frac{z_1}{x},$$

2. $$\int \frac{dx}{x^2 z_1} = -\frac{1}{ax} + \frac{b}{a^2}\ln\frac{z_1}{x}$$

3. $$\int \frac{dx}{x^3 z_1} = -\frac{1}{2ax^2} + \frac{b}{a^2 x} - \frac{b^2}{a^3}\ln\frac{z_1}{x}$$

2.119

1. $$\int \frac{dx}{xz_1^2} = \frac{1}{az_1} - \frac{1}{a^2}\ln\frac{z_1}{x}$$

2. $\int \dfrac{dx}{x^2 z_1^2} = -\left[\dfrac{1}{ax} + \dfrac{2b}{a^2}\right]\dfrac{1}{z_1} + \dfrac{2b}{a^3}\ln\dfrac{z_1}{x}$

3. $\int \dfrac{dx}{x^3 z_1^2} = \left[-\dfrac{1}{2ax^2} + \dfrac{3b}{2a^2 x} + \dfrac{3b^2}{a^3}\right]\dfrac{1}{z_1} - \dfrac{3b^2}{a^4}\ln\dfrac{z_1}{x}$

2.121

1. $\int \dfrac{dx}{x z_1^3} = \left[\dfrac{3}{2a} + \dfrac{bx}{a^2}\right]\dfrac{1}{z_1^2} - \dfrac{1}{a^3}\ln\dfrac{z_1}{x}$

2. $\int \dfrac{dx}{x^2 z_1^3} = -\left[\dfrac{1}{ax} + \dfrac{9b}{2a^2} + \dfrac{3b^2 x}{a^3}\right]\dfrac{1}{z_1^2} + \dfrac{3b}{a^4}\ln\dfrac{z_1}{x}$

3. $\int \dfrac{dx}{x^3 z_1^3} = \left[-\dfrac{1}{2ax^2} + \dfrac{2b}{a^2 x} + \dfrac{9b^2}{a^3} + \dfrac{6b^3 x}{a^4}\right]\dfrac{1}{z_1^2} - \dfrac{6b^2}{a^5}\ln\dfrac{z_1}{x}$

2.122

1. $\int \dfrac{dx}{x z_1^4} = \left[\dfrac{11}{6a} + \dfrac{5bx}{2a^2} + \dfrac{b^2 x^2}{a^3}\right]\dfrac{1}{z_1^3} - \dfrac{1}{a^4}\ln\dfrac{z_1}{x}$

2. $\int \dfrac{dx}{x^2 z_1^4} = -\left[\dfrac{1}{ax} + \dfrac{22b}{3a^2} + \dfrac{10b^2 x}{a^3} + \dfrac{4b^3 x^2}{a^4}\right]\dfrac{1}{z_1^3} + \dfrac{4b}{a^5}\ln\dfrac{z_1}{x}$

3. $\int \dfrac{dx}{x^3 z_1^4} = \left[-\dfrac{1}{2ax^2} + \dfrac{5b}{2a^2 x} + \dfrac{55b^2}{3a^3} + \dfrac{25b^3 x}{a^4} + \dfrac{10b^4 x^2}{a^5}\right]\dfrac{1}{z_1^3} - \dfrac{10b^2}{a^6}\ln\dfrac{z_1}{x}$

2.123

1. $\int \dfrac{dx}{x z_1^5} = \left[\dfrac{25}{12a} + \dfrac{13bx}{3a^2} + \dfrac{7b^2 x^2}{2a^3} + \dfrac{b^3 x^3}{a^4}\right]\dfrac{1}{z_1^4} - \dfrac{1}{a^5}\ln\dfrac{z_1}{x}$

2. $\int \dfrac{dx}{x^2 z_1^5} = \left[-\dfrac{1}{ax} - \dfrac{125b}{12a^2} - \dfrac{65b^2 x}{3a^3} - \dfrac{35b^3 x^2}{2a^4} - \dfrac{5b^4 x^3}{a^5}\right]\dfrac{1}{z_1^4} + \dfrac{5b}{a^6}\ln\dfrac{z_1}{x}$

3. $\int \dfrac{dx}{x^3 z_1^5} = \left[-\dfrac{1}{2ax^2} + \dfrac{3b}{a^2 x} + \dfrac{125b^2}{4a^3} + \dfrac{65b^3 x}{a^4} + \dfrac{105b^4 x^2}{2a^5} + \dfrac{15b^5 x^3}{a^6}\right]\dfrac{1}{z_1^4} - \dfrac{15b^2}{a^7}\ln\dfrac{z_1}{x}$

2.124   Forms containing the binomial $z_2 = a + bx^2$.

1. $\int \dfrac{dx}{z_2} = \dfrac{1}{\sqrt{ab}}\arctan x\sqrt{\dfrac{b}{a}}$ if $[ab > 0]$ (see also **2.141** 2)

           $= \dfrac{1}{2i\sqrt{ab}}\ln\dfrac{a + xi\sqrt{ab}}{a - xi\sqrt{ab}}$ if $[ab < 0]$ (see also **2.143** 2 and **2.143**3)

2. $\int \dfrac{x\,dx}{z_2^m} = -\dfrac{1}{2b(m-1)z_2^{m-1}}$ (see also **2.145** 2, **2.145** 6, and **2.18**)

Forms containing the binomial $z_3 = a + bx^3$

Notation: $\alpha = \sqrt[3]{\dfrac{a}{b}}$

2.125

1. $\displaystyle\int \frac{x^n\, dx}{z_3^m} = \frac{x^{n-2}}{z_3^{m-1}(n+1-3m)b} - \frac{(n-2)a}{b(n+1-3m)}\int \frac{x^{n-3}\, dx}{z_3^m}$

2. $\displaystyle\int \frac{x^n\, dx}{z_3^m} = \frac{x^{n+1}}{3a(m-1)z_3^{m-1}} - \frac{n+4-3m}{3a(m-1)}\int \frac{x^n\, dx}{z_3^{m-1}}$ LA 133 (1)

2.126

1. $\displaystyle\int \frac{dx}{z_3} = \frac{\alpha}{3a}\left\{ \frac{1}{2}\ln\frac{(x+\alpha)^2}{x^2-\alpha x+\alpha^2} + \sqrt{3}\arctan\frac{x\sqrt{3}}{2\alpha - x}\right\}$

      $\displaystyle = \frac{\alpha}{3a}\left\{ \frac{1}{2}\ln\frac{(x+\alpha)^2}{x^2-\alpha x+\alpha^2} + \sqrt{3}\arctan\frac{2x-\alpha}{\alpha\sqrt{3}}\right\}$

                                      (see also **2.141** 3 and **2.143**)

2. $\displaystyle\int \frac{x\, dx}{z_3} = -\frac{1}{3b\alpha}\left\{ \frac{1}{2}\ln\frac{(x+\alpha)^2}{x^2-\alpha x+\alpha^2} - \sqrt{3}\arctan\frac{2x-\alpha}{\alpha\sqrt{3}}\right\}$

                                      (see also **2.145** 3. and **2.145** 7)

3. $\displaystyle\int \frac{x^2\, dx}{z_3} = \frac{1}{3b}\ln\left(1 + x^3\alpha^{-3}\right) = \frac{1}{3b}\ln z_3$

4. $\displaystyle\int \frac{x^3\, dx}{z_3} = \frac{x}{b} - \frac{a}{b}\int \frac{dx}{z_3}$ (see **2.126** 1)

5. $\displaystyle\int \frac{x^4\, dx}{z_3} = \frac{x^2}{2b} - \frac{a}{b}\int \frac{x\, dx}{z_3}$ (see **2.126** 2)

2.127

1. $\displaystyle\int \frac{dx}{z_3^2} = \frac{x}{3az_3} + \frac{2}{3a}\int \frac{dx}{z_3}$ (see **2.126** 1)

2. $\displaystyle\int \frac{x\, dx}{z_3^2} = \frac{x^2}{3az_3} + \frac{1}{3a}\int \frac{x\, dx}{z_3}$ (see **2.126** 2)

3. $\displaystyle\int \frac{x^2\, dx}{z_3^2} = -\frac{1}{3bz_3}$

4. $\displaystyle\int \frac{x^3\, dx}{z_3^2} = -\frac{x}{3bz_3} + \frac{1}{3b}\int \frac{dx}{z_3}$ (see **2.126** 1)

2.128

1. $\displaystyle\int \frac{dx}{x^n z_3^m} = -\frac{1}{(n-1)ax^{n-1}z_3^{m-1}} - \frac{b(3m+n-4)}{a(n-1)}\int \frac{dx}{x^{n-3}z_3^m}$

2. $\displaystyle\int \frac{dx}{x^n z_3^m} = \frac{1}{3a(m-1)x^{n-1}z_3^{m-1}} + \frac{n+3m-4}{3a(m-1)}\int \frac{dx}{x^n z_3^{m-1}}$ LA 133 (2)

2.129

1. $\displaystyle \int \frac{dx}{xz_3} = \frac{1}{3a} \ln \frac{x^3}{z_3}$

2. $\displaystyle \int \frac{dx}{x^2 z_3} = -\frac{1}{ax} - \frac{b}{a} \int \frac{x\,dx}{z_3}$ (see **2.126** 2)

3. $\displaystyle \int \frac{dx}{x^3 z_3} = -\frac{1}{2ax^2} - \frac{b}{a} \int \frac{dx}{z_3}$ (see **2.126** 1)

2.131

1. $\displaystyle \int \frac{dx}{xz_3^2} = \frac{1}{3az_3} + \frac{1}{3a^2} \ln \frac{x^3}{z_3}$

2. $\displaystyle \int \frac{dx}{x^2 z_3^2} = -\left[\frac{1}{ax} + \frac{4bx^2}{3a^2}\right] \frac{1}{z_3} - \frac{4b}{3a^2} \int \frac{x\,dx}{z_3}$ (see **2.126** 2)

3. $\displaystyle \int \frac{dx}{x^3 z_3^2} = -\left[\frac{1}{2ax^2} + \frac{5bx}{6a^2}\right] \frac{1}{z_3} - \frac{5b}{3a^2} \int \frac{dx}{z_3}$ (see **2.126** 1)

Forms containing the binomial $z_4 = a + bx^4$

Notation: $\alpha = \sqrt[4]{\dfrac{a}{b}}$ $\alpha' = \sqrt[4]{-\dfrac{a}{b}}$

2.132

1.[8] $\displaystyle \int \frac{dx}{z_4} = \frac{\alpha}{4a\sqrt{2}} \left\{ \ln \frac{x^2 + \alpha x\sqrt{2} + \alpha^2}{x^2 - \alpha x\sqrt{2} + \alpha^2} + 2\arctan \frac{\alpha x\sqrt{2}}{\alpha^2 - x^2} \right\}$ for $ab > 0$ (see also **2.141** 4)

       $\displaystyle = \frac{\alpha'}{4a} \left\{ \ln \frac{x + \alpha'}{x - \alpha'} + 2\arctan \frac{x}{\alpha'} \right\}$ for $ab < 0$ (see also **2.143** 5)

2. $\displaystyle \int \frac{x\,dx}{z_4} = \frac{1}{2\sqrt{ab}} \arctan x^2 \sqrt{\frac{b}{a}}$ for $ab > 0$ (see also **2.145** 4)

       $\displaystyle = \frac{1}{4i\sqrt{ab}} \ln \frac{a + x^2 i\sqrt{ab}}{a - x^2 i\sqrt{ab}}$ for $ab < 0$ (see also **2.145** 8)

3. $\displaystyle \int \frac{x^2\,dx}{z_4} = \frac{1}{4b\alpha\sqrt{2}} \left\{ \ln \frac{x^2 - \alpha x\sqrt{2} + \alpha^2}{x^2 + \alpha x\sqrt{2} + \alpha^2} + 2\arctan \frac{\alpha x\sqrt{2}}{\alpha^2 - x^2} \right\}$ for $ab > 0$

       $\displaystyle = -\frac{1}{4b\alpha'} \left\{ \ln \frac{x + \alpha'}{x - \alpha'} - 2\arctan \frac{x}{\alpha'} \right\}$ for $ab < 0$

4. $\displaystyle \int \frac{x^3\,dx}{z_4} = \frac{1}{4b} \ln z_4$

2.133

1. $\displaystyle \int \frac{x^n\,dx}{z_4^m} = \frac{x^{n+1}}{4a(m-1)z_4^{m-1}} + \frac{4m - n - 5}{4a(m-1)} \int \frac{x^n\,dx}{z_4^{m-1}}$ LA 134 (1)

2. $\displaystyle \int \frac{x^n\,dx}{z_4^m} = \frac{x^{n-3}}{z_4^{m-1}(n+1-4m)b} - \frac{(n-3)a}{b(n+1-4m)} \int \frac{x^{n-4}\,dx}{z_4^m}$

2.134

1. $$\int \frac{dx}{z_4^2} = \frac{x}{4az_4} + \frac{3}{4a}\int \frac{dx}{z_4} \qquad \text{(see **2.132** 1)}$$

2. $$\int \frac{x\,dx}{z_4^2} = \frac{x^2}{4az_4} + \frac{1}{2a}\int \frac{x\,dx}{z_4} \qquad \text{(see **2.132** 2)}$$

3. $$\int \frac{x^2\,dx}{z_4^2} = \frac{x^3}{4az_4} + \frac{1}{4a}\int \frac{x^2\,dx}{z_4} \qquad \text{(see **2.132** 3)}$$

4. $$\int \frac{x^3\,dx}{z_4^2} = \frac{x^4}{4az_4} = -\frac{1}{4bz_4}$$

2.135 $$\int \frac{dx}{x^n z_4^m} = -\frac{1}{(n-1)ax^{n-1}z_4^{m-1}} - \frac{b(4m+n-5)}{(n-1)a}\int \frac{dx}{x^{n-4}z_4^m}$$

For $n = 1$
$$\int \frac{dx}{xz_4^m} = \frac{1}{a}\int \frac{dx}{xz_4^{m-1}} - \frac{b}{a}\int \frac{dx}{x^{-3}z_4^m}$$

2.136

1. $$\int \frac{dx}{xz_4} = \frac{\ln x}{a} - \frac{\ln z_4}{4a} = \frac{1}{4a}\ln \frac{x^4}{z_4}$$

2. $$\int \frac{dx}{x^2 z_4} = -\frac{1}{ax} - \frac{b}{a}\int \frac{x^2\,dx}{z_4} \qquad \text{(see **2.132** 3)}$$

2.14 Forms containing the binomial $1 \pm x^n$

2.141

1. $$\int \frac{dx}{1+x} = \ln(1+x)$$

2. $$\int \frac{dx}{1+x^2} = \arctan x = -\arctan x \qquad \text{(see also **2.124** 1)}$$

3. $$\int \frac{dx}{1+x^3} = \frac{1}{3}\ln\frac{1+x}{\sqrt{1-x+x^2}} + \frac{1}{\sqrt{3}}\arctan\frac{x\sqrt{3}}{2-x} \qquad \text{(see also **2.126** 1)}$$

4. $$\int \frac{dx}{1+x^4} = \frac{1}{4\sqrt{2}}\ln\frac{1+x\sqrt{2}+x^2}{1-x\sqrt{2}+x^2} + \frac{1}{2\sqrt{2}}\arctan\frac{x\sqrt{2}}{1-x^2}$$

(see also **2.132** 1)

2.142 $$\int \frac{dx}{1+x^n} = -\frac{2}{n}\sum_{k=0}^{\frac{n}{2}-1} P_k \cos\left(\frac{2k+1}{n}\pi\right) + \frac{2}{n}\sum_{k=0}^{\frac{n}{2}-1} Q_k \sin\left(\frac{2k+1}{n}\pi\right)$$

for n a positive even number

TI (43)a

$$= \frac{1}{n}\ln(1+x) - \frac{2}{n}\sum_{k=0}^{\frac{n-3}{2}} P_k \cos\left(\frac{2k+1}{n}\pi\right) + \frac{2}{n}\sum_{k=0}^{\frac{n-3}{2}} Q_k \sin\left(\frac{2k+1}{n}\pi\right)$$

for n a positive odd number

TI (45)

where

$$P_k = \frac{1}{2} \ln \left(x^2 - 2x \cos \left(\frac{2k+1}{n} \pi \right) + 1 \right)$$

$$Q_k = \arctan \frac{x \sin \left(\frac{2k+1}{n} \pi \right)}{1 - x \cos \left(\frac{2k+1}{n} \pi \right)} = \arctan \frac{x - \cos \left(\frac{2k+1}{n} \pi \right)}{\sin \left(\frac{2k+1}{n} \pi \right)}$$

2.143

1. $\displaystyle \int \frac{dx}{1-x} = -\ln(1-x)$

2. $\displaystyle \int \frac{dx}{1-x^2} = \frac{1}{2} \ln \frac{1+x}{1-x} = \operatorname{arctanh} x$ $[-1 < x < 1]$ (see also **2.141** 1)

3. $\displaystyle \int \frac{dx}{x^2-1} = \frac{1}{2} \ln \frac{x-1}{x+1} = -\operatorname{arccoth} x$ $[x > 1, \quad x < -1]$

4. $\displaystyle \int \frac{dx}{1-x^3} = \frac{1}{3} \ln \frac{\sqrt{1+x+x^2}}{1-x} + \frac{1}{\sqrt{3}} \arctan \frac{x\sqrt{3}}{2+x}$ (see also **2.126** 1)

5. $\displaystyle \int \frac{dx}{1-x^4} = \frac{1}{4} \ln \frac{1+x}{1-x} + \frac{1}{2} \arctan x = \frac{1}{2} \left(\operatorname{arctanh} x + \arctan x \right)$

                                       (see also **2.132** 1)

2.144

1. $\displaystyle \int \frac{dx}{1-x^n} = \frac{1}{n} \ln \frac{1+x}{1-x} - \frac{2}{n} \sum_{k=1}^{\frac{n}{2}-1} P_k \cos \frac{2k}{n} \pi + \frac{2}{n} \sum_{k=1}^{\frac{n}{2}-1} Q_k \sin \frac{2k}{n} \pi$

                                 for n a positive even number   **TI (47)**

     where $P_k = \frac{1}{2} \ln \left(x^2 + 2x \cos \frac{2k+1}{n} \pi + 1 \right)$, $Q_k = \arctan \frac{x + \cos \frac{2k+1}{n} \pi}{\sin \frac{2k+1}{n} \pi}$

2. $\displaystyle \int \frac{dx}{1-x^n} = -\frac{1}{n} \ln(1-x) + \frac{2}{n} \sum_{k=0}^{\frac{n-3}{2}} P_k \cos \frac{2k+1}{n} \pi + \frac{2}{n} \sum_{k=0}^{\frac{n-3}{2}} Q_k \sin \frac{2k+1}{n} \pi$

                                 for n a positive odd number   **TI (49)**

     where $P_k = \frac{1}{2} \ln \left(x^2 - 2x \cos \frac{2k}{n} \pi + 1 \right)$, $Q_k = \arctan \frac{x - \cos \frac{2k}{n} \pi}{\sin \frac{2k}{n} \pi}$

2.145

1. $\displaystyle \int \frac{x \, dx}{1+x} = x - \ln(1+x)$

2. $\displaystyle \int \frac{x \, dx}{1+x^2} = \frac{1}{2} \ln \left(1 + x^2 \right)$

3. $\displaystyle \int \frac{x \, dx}{1+x^3} = -\frac{1}{6} \ln \frac{(1+x)^2}{1-x+x^2} + \frac{1}{\sqrt{3}} \arctan \frac{2x-1}{\sqrt{3}}$ (see also **2.126** 2)

4. $\int \dfrac{x\,dx}{1+x^4} = \dfrac{1}{2}\arctan x^2$

5. $\int \dfrac{x\,dx}{1-x} = -\ln(1-x) - x.$

6. $\int \dfrac{x\,dx}{1-x^2} = -\dfrac{1}{2}\ln\left(1-x^2\right)$

7. $\int \dfrac{x\,dx}{1-x^3} = -\dfrac{1}{6}\ln\dfrac{(1-x)^2}{1+x+x^2} - \dfrac{1}{\sqrt{3}}\arctan\dfrac{2x+1}{\sqrt{3}}$ (see also **2.126** 2)

8. $\int \dfrac{x\,dx}{1-x^4} = \dfrac{1}{4}\ln\dfrac{1+x^2}{1-x^2}$ (see also **2.132** 2)

2.146 For m and n natural numbers.

1. $\int \dfrac{x^{m-1}\,dx}{1+x^{2n}} = -\dfrac{1}{2n}\sum_{k=1}^{n}\cos\dfrac{m\pi(2k-1)}{2n}\ln\left\{1-2x\cos\dfrac{2k-1}{2n}\pi+x^2\right\}$
$+\dfrac{1}{n}\sum_{k=1}^{n}\sin\dfrac{m\pi(2k-1)}{2n}\arctan\dfrac{x-\cos\frac{2k-1}{2n}\pi}{\sin\frac{2k-1}{2n}\pi}$
$[m < 2n]$ TI (44)a

2. $\int \dfrac{x^{m-1}\,dx}{1+x^{2n+1}} = (-1)^{m+1}\dfrac{\ln(1+x)}{2n+1} - \dfrac{1}{2n+1}\sum_{k=1}^{n}\cos\dfrac{m\pi(2k-1)}{2n+1}\ln\left\{1-2x\cos\dfrac{2k-1}{2n+1}\pi+x^2\right\}$
$+\dfrac{2}{2n+1}\sum_{k=1}^{n}\sin\dfrac{m\pi(2k-1)}{2n+1}\arctan\dfrac{x-\cos\frac{2k-1}{2n+1}\pi}{\sin\frac{2k-1}{2n+1}\pi}$
$[m \le 2n]$ TI (46)a

3. $\int \dfrac{x^{m-1}\,dx}{1-x^{2n}} = \dfrac{1}{2n}\left\{(-1)^{m+1}\ln(1+x) - \ln(1-x)\right\} - \dfrac{1}{2n}\sum_{k=1}^{n-1}\cos\dfrac{km\pi}{n}\ln\left(1-2x\cos\dfrac{k\pi}{n}+x^2\right)$
$+\dfrac{1}{n}\sum_{k=1}^{n-1}\sin\dfrac{km\pi}{n}\arctan\dfrac{x-\cos\frac{k\pi}{n}}{\sin\frac{k\pi}{n}}$
$[m < 2n]$ TI (48)

4. $\int \dfrac{x^{m-1}\,dx}{1-x^{2n+1}} = -\dfrac{1}{2n+1}\ln(1-x)$
$+(-1)^{m+1}\dfrac{1}{2n+1}\sum_{k=1}^{n}\cos\dfrac{m\pi(2k-1)}{2n+1}\ln\left(1+2x\cos\dfrac{2k-1}{2n+1}\pi+x^2\right)$
$+(-1)^{m+1}\dfrac{2}{2n+1}\sum_{k=1}^{n}\sin\dfrac{m\pi(2k-1)}{2n+1}\arctan\dfrac{x+\cos\frac{2k-1}{2n+1}\pi}{\sin\frac{2k-1}{2n+1}\pi}$
$[m \le 2n]$ TI (50)

2.147

1. $\int \dfrac{x^m\,dx}{1-x^{2n}} = \dfrac{1}{2}\int \dfrac{x^m\,dx}{1-x^n} + \dfrac{1}{2}\int \dfrac{x^m\,dx}{1+x^n}$

2. $\displaystyle\int \frac{x^m\,dx}{(1+x^2)^n} = -\frac{1}{2n-m-1}\cdot\frac{x^{m-1}}{(1+x^2)^{n-1}} + \frac{m-1}{2n-m-1}\int \frac{x^{m-2}\,dx}{(1+x^2)^n}$ LA 139 (28)

3. $\displaystyle\int \frac{x^m}{1+x^2}\,dx = \frac{x^{m-1}}{m-1} - \int \frac{x^{m-2}}{1+x^2}\,dx$

4. $\displaystyle\int \frac{x^m\,dx}{(1-x^2)^n} = \frac{1}{2n-m-1}\frac{x^{m-1}}{(1-x^2)^{n-1}} - \frac{m-1}{2n-m-1}\int \frac{x^{m-2}\,dx}{(1-x^2)^n}$

$\displaystyle\qquad\qquad = \frac{1}{2n-2}\frac{x^{m-1}}{(1-x^2)^{n-1}} - \frac{m-1}{2n-2}\int \frac{x^{m-2}\,dx}{(1-x^2)^{n-1}}$

                                                           LA 139 (33)

5. $\displaystyle\int \frac{x^m\,dx}{1-x^2} = -\frac{x^{m-1}}{m-1} + \int \frac{x^{m-2}\,dx}{1-x^2}$

2.148

1. $\displaystyle\int \frac{dx}{x^m(1+x^2)^n} = -\frac{1}{m-1}\frac{1}{x^{m-1}(1+x^2)^{n-1}} - \frac{2n+m-3}{m-1}\int \frac{dx}{x^{m-2}(1+x^2)^n}$ LA 139 (29)

For $m = 1$

$\displaystyle\int \frac{dx}{x(1+x^2)^n} = \frac{1}{2n-2}\frac{1}{(1+x^2)^{n-1}} + \int \frac{dx}{x(1+x^2)^{n-1}}$ LA 139 (31)

For $m = 1$ and $n = 1$

$\displaystyle\int \frac{dx}{x(1+x^2)} = \ln\frac{x}{\sqrt{1+x^2}}$

2. $\displaystyle\int \frac{dx}{x^m(1+x^2)} = -\frac{1}{(m-1)x^{m-1}} - \int \frac{dx}{x^{m-2}(1+x^2)}$

3. $\displaystyle\int \frac{dx}{(1+x^2)^n} = \frac{1}{2n-2}\frac{x}{(1+x^2)^{n-1}} + \frac{2n-3}{2n-2}\int \frac{dx}{(1+x^2)^{n-1}}$ FI II 40

4. $\displaystyle\int \frac{dx}{(1+x^2)^n} = \frac{x}{2n-1}\sum_{k=1}^{n-1}\frac{(2n-1)(2n-3)(2n-5)\cdots(2n-2k+1)}{2^k(n-1)(n-2)\dots(n-k)(1+x^2)^{n-k}} + \frac{(2n-3)!!}{2^{n-1}(n-1)!}\arctan x$

                                                                          TI (91)

2.149

1. $\displaystyle\int \frac{dx}{x^m(1-x^2)^n} = -\frac{1}{(m-1)x^{m-1}(1-x^2)^{n-1}} + \frac{2n+m-3}{m-1}\int \frac{dx}{x^{m-2}(1-x^2)^n}$ LA 139 (34)

For $m = 1$

$\displaystyle\int \frac{dx}{x(1-x^2)^n} = \frac{1}{2(n-1)(1-x^2)^{n-1}} + \int \frac{dx}{x(1-x^2)^{n-1}}$ LA 139 (36)

For $m = 1$ and $n = 1$

$\displaystyle\int \frac{dx}{x(1-x^2)} = \ln\frac{x}{\sqrt{1-x^2}}$

2. $\displaystyle\int \frac{dx}{(1-x^2)^n} = \frac{1}{2n-2}\frac{x}{(1-x^2)^{n-1}} + \frac{2n-3}{2n-2}\int \frac{dx}{(1-x^2)^{n-1}}$ LA 139 (35)

3.
$$\int \frac{dx}{(1-x^2)^n} = \frac{x}{2n-1} \sum_{k=1}^{n-1} \frac{(2n-1)(2n-3)(2n-5)\ldots(2n-2k+1)}{2^k(n-1)(n-2)\ldots(n-k)(1-x^2)^{n-k}} + \frac{(2n-3)!!}{2^n\cdot(n-1)!} \ln\frac{1+x}{1-x}$$

TI (91)

2.15 Forms containing pairs of binomials: $a + bx$ and $\alpha + \beta x$

Notation: $z = a + bx$; $t = \alpha + \beta x$; $\Delta = a\beta - \alpha b$

2.151
$$\int z^n t^m \, dx = \frac{z^{n+1}t^m}{(m+n+1)b} - \frac{m\Delta}{(m+n+1)b}\int z^n t^{m-1}\, dx$$

2.152

1.
$$\int \frac{z}{t}\, dx = \frac{bx}{\beta} + \frac{\Delta}{\beta^2}\ln t$$

2.
$$\int \frac{t}{z}\, dx = \frac{\beta x}{b} - \frac{\Delta}{b^2}\ln z$$

2.153
$$\int \frac{t^m\, dx}{z^n} = \frac{1}{(m-n+1)b}\frac{t^m}{z^{n-1}} - \frac{m\Delta}{(m-n+1)b}\int \frac{t^{m-1}\, dx}{z^n}$$
$$= \frac{1}{(n-1)\Delta}\frac{t^{m+1}}{z^{n-1}} - \frac{(m-n+2)\beta}{(n-1)\Delta}\int \frac{t^m\, dx}{z^{n-1}}$$
$$= -\frac{1}{(n-1)b}\frac{t^m}{z^{n-1}} + \frac{m\beta}{(n-1)b}\int \frac{t^{m-1}}{z^{n-1}}\, dx$$

2.154
$$\int \frac{dx}{zt} = \frac{1}{\Delta}\ln\frac{t}{z}$$

2.155
$$\int \frac{dx}{z^n t^m} = -\frac{1}{(m-1)\Delta}\frac{1}{t^{m-1}z^{n-1}} - \frac{(m+n-2)b}{(m-1)\Delta}\int \frac{dx}{t^{m-1}z^n}$$
$$= \frac{1}{(n-1)\Delta}\frac{1}{t^{m-1}z^{n-1}} + \frac{(m+n-2)\beta}{(n-1)\Delta}\int \frac{dx}{t^m z^{n-1}}$$

2.156
$$\int \frac{x\, dx}{zt} = \frac{1}{\Delta}\left(\frac{a}{b}\ln z - \frac{\alpha}{\beta}\ln t\right)$$

2.16 Forms containing the trinomial $a + bx^k + cx^{2k}$

2.160 Reduction formulas for $R_k = a + bx^k + cx^{2k}$.

1.
$$\int x^{m-1}R_k^n\, dx = \frac{x^m R_k^{n+1}}{ma} - \frac{(m+k+nk)b}{ma}\int x^{m+k-1}R_k^n\, dx - \frac{(m+2k+2kn)c}{ma}\int x^{m+2k-1}R_k^n\, dx$$

2.
$$\int x^{m-1}R_k^n\, dx = \frac{x^m R_k^n}{m} - \frac{bkn}{m}\int x^{m+k-1}R_k^{n-1}\, dx - \frac{2ckn}{m}\int x^{m+2k-1}R_k^{n-1}\, dx$$

3.
$$\int x^{m-1}R_k^n\, dx = \frac{x^{m-2k}R_k^{n+1}}{(m+2kn)c} - \frac{(m-2k)a}{(m+2kn)c}\int x^{m-2k-1}R_k^n\, dx - \frac{(m-k+kn)b}{(m+2kn)c}\int x^{m-k-1}R_k^n\, dx$$
$$= \frac{x^m R_k^n}{m+2kn} + \frac{2kna}{m+2kn}\int x^{m-1}R_k^{n-1}\, dx + \frac{bkn}{m+2kn}\int x^{m+k-1}R_k^{n-1}\, dx$$

2.161   Forms containing the trinomial $R_2 = a + bx^2 + cx^4$.

Notation: $f = \dfrac{b}{2} - \dfrac{1}{2}\sqrt{b^2 - 4ac}, \quad g = \dfrac{b}{2} + \dfrac{1}{2}\sqrt{b^2 - 4ac},$

$\qquad\qquad h = \sqrt{b^2 - 4ac}, \quad q = \sqrt[4]{\dfrac{a}{c}}, \quad l = 2a(n-1)\left(b^2 - 4ac\right), \quad \cos\alpha = -\dfrac{b}{2\sqrt{ac}}$

1. $\displaystyle \int \frac{dx}{R_2} = \frac{c}{h}\left\{ \int \frac{dx}{cx^2 + f} - \int \frac{dx}{cx^2 + g} \right\}$

$$\left[h^2 > 0\right]$$
$$\text{LA 146 (5)}$$

$\qquad = \dfrac{1}{4cq^3 \sin\alpha}\left\{ \sin\dfrac{\alpha}{2} \ln \dfrac{x^2 + 2qx\cos\frac{\alpha}{2} + q^2}{x^2 - 2qx\cos\frac{\alpha}{2} + q^2} + 2\cos\dfrac{\alpha}{2} \arctan \dfrac{x^2 - q^2}{2qx\sin\frac{\alpha}{2}} \right\}$

$$\left[h^2 < 0\right]$$
$$\text{LA 146 (8)a}$$

2. $\displaystyle \int \frac{x\,dx}{R_2} = \frac{1}{2h} \ln \frac{cx^2 + f}{cx^2 + g}$ $\left[h^2 > 0\right]$ LA 146 (6)

$\qquad = \dfrac{1}{2cq^2 \sin\alpha} \arctan \dfrac{x^2 - q^2\cos\alpha}{q^2 \sin\alpha}$ $\left[h^2 < 0\right]$ LA 146 (9)a

3. $\displaystyle \int \frac{x^2\,dx}{R_2} = \frac{g}{h} \int \frac{dx}{cx^2 + g} - \frac{f}{h} \int \frac{dx}{cx^2 + f}$ $\left[h^2 > 0\right]$ LA 146 (7)

4. $\displaystyle \int \frac{dx}{R_2^2} = \frac{bcx^3 + \left(b^2 - 2ac\right)x}{lR_2} + \frac{b^2 - 6ac}{l} \int \frac{dx}{R_2} + \frac{bc}{l} \int \frac{x^2\,dx}{R_2}$

5. $\displaystyle \int \frac{dx}{R_2^n} = \frac{bcx^3 + \left(b^2 - 2ac\right)x}{lR_2^{n-1}} + \frac{(4n-7)bc}{l} \int \frac{x^2\,dx}{R_2^{n-1}} + \frac{2(n-1)h^2 + 2ac - b^2}{l} \int \frac{dx}{R_2^{n-1}}$

$$[n > 1]$$
$$\text{LA 146}$$

6.[9] $\displaystyle \int \frac{dx}{x^m R_2^n} = -\frac{1}{(m-1)ax^{m-1}R_2^{n-1}} - \frac{(m+2n-3)b}{(m-1)a} \int \frac{dx}{x^{m-2}R_2^n} - \frac{(m+4n-5)bc}{(m-1)a} \int \frac{dx}{x^{m-4}R_2^n}$

$$\text{LA 147 (12)a}$$

2.17 Forms containing the quadratic trinomial $a + bx + cx^2$ and powers of x

Notation: $R = a + bx + cx^2; \quad \Delta = 4ac - b^2$

2.171

1. $\displaystyle \int x^{m+1} R^n \, dx = \frac{x^m R^{n+1}}{c(m + 2n + 2)} - \frac{am}{c(m + 2n + 2)} \int x^{m-1} R^n \, dx - \frac{b(m + n + 1)}{c(m + 2n + 2)} \int x^m R^n \, dx$

$$\text{TI (97)}$$

2. $\displaystyle \int \frac{R^n \, dx}{x^{m+1}} = -\frac{R^{n+1}}{amx^m} + \frac{b(n - m + 1)}{am} \int \frac{R^n \, dx}{x^m} + \frac{c(2n - m + 2)}{am} \int \frac{R^n \, dx}{x^{m-1}}$ LA 142(3), TI (96)a

3. $\displaystyle \int \frac{dx}{R^{n+1}} = \frac{b + 2cx}{n\Delta R^n} + \frac{(4n - 2)c}{n\Delta} \int \frac{dx}{R^n}$ TI (94)a

4.　$\displaystyle\int \frac{dx}{R^{n+1}} = \frac{(2cx+b)}{2n+1} \sum_{k=0}^{n-1} \frac{2k(2n+1)(2n-1)(2n-3)\dots(2n-2k+1)c^k}{n(n-1)\cdots(n-k)\Delta^{k+1}R^{n-k}} + 2^n \frac{(2n-1)!!c^n}{n!\Delta^n} \int \frac{dx}{R}$

TI (96)a

2.172[3]　$\displaystyle\int \frac{dx}{R} = \frac{1}{\sqrt{-\Delta}} \ln \frac{\sqrt{-\Delta} - (b+2cx)}{(b+2cx) + \sqrt{-\Delta}} = \frac{-2}{\sqrt{-\Delta}} \operatorname{arctanh} x \frac{b+2cx}{\sqrt{-\Delta}}$　　for $[\Delta < 0]$

$\displaystyle \qquad\qquad = \frac{-2}{b+2cx}$　　　　　　　　　　　　　　　　　for $[\Delta = 0]$

$\displaystyle \qquad\qquad = \frac{2}{\sqrt{\Delta}} \arctan \frac{b+2cx}{\sqrt{\Delta}}$　　　　　　　　　　for $[\Delta > 0]$

2.173

1.　$\displaystyle\int \frac{dx}{R^2} = \frac{b+2cx}{\Delta R} + \frac{2c}{\Delta} \int \frac{dx}{R}$　　　　　　　　(see **2.172**)

2.　$\displaystyle\int \frac{dx}{R^3} = \frac{b+2cx}{\Delta} \left\{ \frac{1}{2R^2} + \frac{3c}{\Delta R} \right\} + \frac{6c^2}{\Delta^2} \int \frac{dx}{R}$　　　　(see **2.172**)

2.174

1.　$\displaystyle\int \frac{x^m\,dx}{R^n} = -\frac{x^{m-1}}{(2n-m-1)cR^{n-1}} - \frac{(n-m)b}{(2n-m-1)c}\int \frac{x^{m-1}\,dx}{R^n} + \frac{(m-1)a}{(2n-m-1)c}\int \frac{x^{m-2}\,dx}{R^n}$

For $m = 2n-1$, this formula is inapplicable. Instead, we may use

2.　$\displaystyle\int \frac{x^{2n-1}\,dx}{R^n} = \frac{1}{c}\int \frac{x^{2n-3}\,dx}{R^{n-1}} - \frac{a}{c}\int \frac{x^{2n-3}\,dx}{R^n} - \frac{b}{c}\int \frac{x^{2n-2}\,dx}{R^n}$

2.175

1.　$\displaystyle\int \frac{x\,dx}{R} = \frac{1}{2c}\ln R - \frac{b}{2c}\int \frac{dx}{R}$　　　　　　　　(see **2.172**)

2.　$\displaystyle\int \frac{x\,dx}{R^2} = -\frac{2a+bx}{\Delta R} - \frac{b}{\Delta}\int \frac{dx}{R}$　　　　　　(see **2.172**)

3.　$\displaystyle\int \frac{x\,dx}{R^3} = -\frac{2a+bx}{2\Delta R^2} - \frac{3b(b+2cx)}{2\Delta^2 R} - \frac{3bc}{\Delta^2}\int \frac{dx}{R}$　　(see **2.172**)

4.　$\displaystyle\int \frac{x^2\,dx}{R} = \frac{x}{c} - \frac{b}{2c^2}\ln R + \frac{b^2-2ac}{2c^2}\int \frac{dx}{R}$　　　(see **2.172**)

5.　$\displaystyle\int \frac{x^2\,dx}{R^2} = \frac{ab + (b^2-2ac)\,x}{c\Delta R} + \frac{2a}{\Delta}\int \frac{dx}{R}$　　　(see **2.172**)

6.　$\displaystyle\int \frac{x^2\,dx}{R^3} = \frac{ab + (b^2-2ac)\,x}{2c\Delta R^2} + \frac{(2ac+b^2)\,(b+2cx)}{2c\Delta^2 R} + \frac{2ac+b^2}{\Delta^2}\int \frac{dx}{R}$

(see **2.172**)

7.　$\displaystyle\int \frac{x^3\,dx}{R} = \frac{x^2}{2c} - \frac{bx}{c^2} + \frac{b^2-ac}{2c^3}\ln R - \frac{b\,(b^2-3ac)}{2c^3}\int \frac{dx}{R}$

(see **2.172**)

8. $\displaystyle \int \frac{x^3\,dx}{R^2} = \frac{1}{2c^2}\ln R + \frac{a\left(2ac - b^2\right) + b\left(3ac - b^2\right)x}{c^2\Delta R} - \frac{b\left(6ac - b^2\right)}{2c^2\Delta}\int\frac{dx}{R}$

(see 2.172)

9. $\displaystyle \int\frac{x^3\,dx}{R^3} = -\left(\frac{x^2}{c} + \frac{abx}{c\Delta} + \frac{2a^2}{c\Delta}\right)\frac{1}{2R^2} - \frac{3ab}{2c\Delta}\int\frac{dx}{R^2}$ (see **2.173** 1)

2.176 $\displaystyle \int\frac{dx}{x^m R^n} = \frac{-1}{(m-1)ax^{m-1}R^{n-1}} - \frac{b(m+n-2)}{a(m-1)}\int\frac{dx}{x^{m-1}R^n} - \frac{c(m+2n-3)}{a(m-1)}\int\frac{dx}{x^{m-2}R^n}$

2.177

1. $\displaystyle \int\frac{dx}{xR} = \frac{1}{2a}\ln\frac{x^2}{R} - \frac{b}{2a}\int\frac{dx}{R}$ (see **2.172**)

2. $\displaystyle \int\frac{dx}{xR^2} = \frac{1}{2a^2}\ln\frac{x^2}{R} + \frac{1}{2aR}\left\{1 - \frac{b(b+2cx)}{\Delta}\right\} - \frac{b}{2a^2}\left(1 + \frac{2ac}{\Delta}\right)\int\frac{dx}{R}$

(see 2.172)

3. $\displaystyle \int\frac{dx}{xR^3} = \frac{1}{4aR^2} + \frac{1}{2a^2R} + \frac{1}{2a^3}\ln\frac{x^2}{R} - \frac{b}{2a}\int\frac{dx}{R^3} - \frac{b}{2a^2}\int\frac{dx}{R^2} - \frac{b}{2a^3}\int\frac{dx}{R}$

(see 2.172, 2.173)

4. $\displaystyle \int\frac{dx}{x^2R} = -\frac{b}{2a^2}\ln\frac{x^2}{R} - \frac{1}{ax} + \frac{b^2 - 2ac}{2a^2}\int\frac{dx}{R}$ (see **2.172**)

5. $\displaystyle \int\frac{dx}{x^2R^2} = -\frac{b}{a^3}\ln\frac{x^2}{R} - \frac{a+bx}{a^2xR} + \frac{\left(b^2 - 3ac\right)\left(b+2cx\right)}{a^2\Delta R} - \frac{1}{\Delta}\left(\frac{b^4}{a^3} - \frac{6b^2c}{a^2} + \frac{6c^2}{a}\right)\int\frac{dx}{R}$

(see 2.172)

6. $\displaystyle \int\frac{dx}{x^2R^3} = -\frac{1}{axR^2} - \frac{3b}{a}\int\frac{dx}{xR^3} - \frac{5c}{a}\int\frac{dx}{R^3}$ (see **2.173** and **2.177** 3)

7. $\displaystyle \int\frac{dx}{x^3R} = -\frac{ac-b^2}{2a^3}\ln\frac{x^2}{R} + \frac{b}{a^2x} - \frac{1}{2ax^2} + \frac{b\left(3ac - b^2\right)}{2a^3}\int\frac{dx}{R}$

(see 2.172)

8. $\displaystyle \int\frac{dx}{x^3R^2} = \left(-\frac{1}{2ax^2} + \frac{3b}{2a^2x}\right)\frac{1}{R} + \left(\frac{3b^2}{a^2} - \frac{2c}{a}\right)\int\frac{dx}{xR^2} + \frac{9bc}{2a^2}\int\frac{dx}{R^2}$

(see 2.173 1 and 2.177 2)

9. $\displaystyle \int\frac{dx}{x^3R^3} = \left(\frac{-1}{2ax^2} + \frac{2b}{a^2x}\right)\frac{1}{R^2} + \left(\frac{6b^2}{a^2} - \frac{3c}{a}\right)\int\frac{dx}{xR^3} + \frac{10bc}{a^2}\int\frac{dx}{R^3}$

(see 2.173 2 and 2.177 3)

2.18 Forms containing the quadratic trinomial $a + bx + cx^2$ and the binomial $\alpha + \beta x$

Notation: $R = a + bx + cx^2$; $z = \alpha + \beta x$; $A = a\beta^2 - \alpha b\beta + c\alpha^2$;

$\quad\quad\quad\quad\quad B = b\beta - 2c\alpha$; $\Delta = 4ac - b^2$

1. $\displaystyle\int z^m R^n \, dx = \frac{\beta z^{m-1} R^{n+1}}{(m + 2n + 1)c} - \frac{(m+n)B}{(m+2n+1)c}\int z^{m-1} R^n \, dx - \frac{(m-1)A}{(m+2n+1)c}\int z^{m-2} R^n \, dx$

2. $\displaystyle\int \frac{R^n \, dx}{z^m} = -\frac{1}{(m-2n-1)\beta} \frac{R^n}{z^{m-1}} - \frac{2nA}{(m-2n-1)\beta^2}\int \frac{R^{n-1} \, dx}{z^m}$

$\quad\quad\quad\quad\quad - \dfrac{nB}{(m-2n-1)\beta^2}\displaystyle\int \frac{R^{n-1} \, dx}{z^{m-1}} ;$ LA 184 (4)a

$\quad\quad = \dfrac{-\beta}{(m-1)A} \dfrac{R^{n+1}}{z^{m-1}} - \dfrac{(m-n-2)B}{(m-1)A}\displaystyle\int \frac{R^n \, dx}{z^{m-1}} - \dfrac{(m-2n-3)c}{(m-1)A}\displaystyle\int \frac{R^n \, dx}{z^{m-2}}$ LA 148 (5)

$\quad\quad = -\dfrac{1}{(m-1)\beta} \dfrac{R^n}{z^{m-1}} + \dfrac{nB}{(m-1)\beta^2}\displaystyle\int \frac{R^{n-1} \, dx}{z^{m-1}} + \dfrac{2nc}{(m-1)\beta^2}\displaystyle\int \frac{R^{n-1} \, dx}{z^{m-2}}$ LA 418 (6)

3. $\displaystyle\int \frac{z^m \, dx}{R^n} = \frac{\beta}{(m-2n+1)c} \frac{z^{m-1}}{R^{n-1}} - \frac{(m-n)B}{(m-2n+1)c}\int \frac{z^{m-1} \, dx}{R^n} - \frac{(m-1)A}{(m-2n+1)c}\int \frac{z^{m-2} \, dx}{R^n}$

                                                                           LA 147 (1)

$\quad\quad = \dfrac{b + 2cx}{(n-1)\Delta} \dfrac{z^m}{R^{n-1}} - \dfrac{2(m-2n+3)c}{(n-1)\Delta}\displaystyle\int \frac{z^m \, dx}{R^{n-1}} - \dfrac{Bm}{(n-1)\Delta}\displaystyle\int \frac{z^{m-1} \, dx}{R^{n-1}}$

                                                                           LA 148 (3)

$4.^3$ $\displaystyle\int \frac{dx}{z^m R^n} = -\frac{\beta}{(m-1)A} \frac{1}{z^{m-1} R^{n-1}} - \frac{(m+n-2)B}{(m-1)A}\int \frac{dx}{z^{m-1} R^n} - \frac{(m+2n-3)c}{(m-1)A}\int \frac{dx}{z^{m-2} R^n}$

                                                                           LA 148 (7)

$\quad\quad = \dfrac{\beta}{2(n-1)A} \dfrac{1}{z^{m-1} R^{n-1}} - \dfrac{B}{2A}\displaystyle\int \frac{dx}{z^{m-1} R^n} + \dfrac{(m+2n-3)\beta^2}{2(n-1)A}\displaystyle\int \frac{dx}{z^m R^{n-1}}$

                                                                           LA 148 (8)

For $m = 1$ and $n = 1$

$\displaystyle\int \frac{dx}{zR} = \frac{\beta}{2A}\ln\frac{z^2}{R} - \frac{B}{2A}\int \frac{dx}{R}$

For $A = 0$

$\displaystyle\int \frac{dx}{z^m R^n} = -\frac{\beta}{(m+n-1)B} \frac{1}{z^m R^{n-1}} - \frac{(m+2n-2)c}{(m+n-1)B}\int \frac{dx}{z^{m-1} R^n}$ LA 148 (9)

2.2 Algebraic functions

2.20 Introduction

2.201   The integrals $\displaystyle\int R\left(x, \left(\frac{\alpha x + \beta}{\gamma x + \delta}\right)^r, \left(\frac{\alpha x + \beta}{\gamma x + \delta}\right)^s, \dots\right) dx$, where $r, s, \dots$ are rational numbers, can be reduced to integrals of rational functions by means of the substitution

$$\frac{\alpha x + \beta}{\gamma x + \delta} = t^m,$$ FI II 57

where m is the common denominator of the fractions $r, s, \ldots$.

2.202 Integrals of the form $\int x^m (a + bx^n)^p \, dx$,* where m, n, and p are rational numbers, can be expressed in terms of elementary functions only in the following cases:

(a)   When p is an integer; then, this integral takes the form of the sum of the integrals shown in **2.201**;

(b)   When $\frac{m+1}{n}$ is an integer: by means of the substitution $x^n = z$, this integral can be transformed to the form $\frac{1}{n} \int (a + bz)^p z^{\frac{m+1}{n} - 1} \, dz$, which we considered in **2.201**;

(c)   When $\frac{m+1}{n} + p$ is an integer; by means of the same substitution $x^n = z$, this integral can be reduced to an integral of the form $\frac{1}{n} \int \left(\frac{a + bz}{z} \right)^p z^{\frac{m+1}{n} + p - 1} \, dz$, considered in **2.201**;

For reduction formulas for integrals of binomial differentials, see **2.110**.

2.21 Forms containing the binomial $a + bx^k$ and $\sqrt{x}$

Notation: $z_1 = a + bx$.

2.211 $\displaystyle \int \frac{dx}{z_1 \sqrt{x}} = \frac{2}{\sqrt{ab}} \arctan \sqrt{\frac{bx}{a}}$ $[ab > 0]$

$$= \frac{1}{i\sqrt{ab}} \ln \frac{a - bx + 2i\sqrt{xab}}{z_1} \qquad [ab < 0]$$

2.212 $\displaystyle \int \frac{x^m \sqrt{x}}{z_1} \, dx = 2\sqrt{x} \sum_{k=0}^{m} \frac{(-1)^k a^k x^{m-k}}{(2m - 2k + 1)b^{k+1}} + (-1)^{m+1} \frac{a^{m+1}}{b^{m+1}} \int \frac{dx}{z_1 \sqrt{x}}$

(see **2.211**)

2.213

1. $\displaystyle \int \frac{\sqrt{x} \, dx}{z_1} = \frac{2\sqrt{x}}{b} - \frac{a}{b} \int \frac{dx}{z_1 \sqrt{x}}$ (see **2.211**)

2. $\displaystyle \int \frac{x\sqrt{x} \, dx}{z_1} = \left(\frac{x}{3b} - \frac{a}{b^2} \right) 2\sqrt{x} + \frac{a^2}{b^2} \int \frac{dx}{z_1 \sqrt{x}}$ (see **2.211**)

3. $\displaystyle \int \frac{x^2 \sqrt{x} \, dx}{z_1} = \left(\frac{x^2}{5b} - \frac{xa}{3b^2} + \frac{a^2}{b^3} \right) 2\sqrt{x} - \frac{a^3}{b^3} \int \frac{dx}{z_1 \sqrt{x}}$ (see **2.211**)

4. $\displaystyle \int \frac{dx}{z_1^2 \sqrt{x}} = \frac{\sqrt{x}}{az_1} + \frac{1}{2a} \int \frac{dx}{z_1 \sqrt{x}}$ (see **2.211**)

5. $\displaystyle \int \frac{\sqrt{x} \, dx}{z_1^2} = -\frac{\sqrt{x}}{bz_1} + \frac{1}{2b} \int \frac{dx}{z_1 \sqrt{x}}$ (see **2.211**)

6. $\displaystyle \int \frac{x\sqrt{x} \, dx}{z_1^2} = \frac{2x\sqrt{x}}{bz_1} - \frac{3a}{b} \int \frac{\sqrt{x} \, dx}{z_1^2}$ (see **2.213** 5)

---

*Translator: The authors term such integrals "integrals of binomial differentials".

7. $\quad \int \dfrac{x^2 \sqrt{x}\, dx}{z_1^2} = \left(\dfrac{x^2}{3b} - \dfrac{5ax}{3b^2}\right) \dfrac{2\sqrt{x}}{z_1} + \dfrac{5a^2}{b^2} \int \dfrac{\sqrt{x}\, dx}{z_1^2}$ (see **2.213 5**)

8. $\quad \int \dfrac{dx}{z_1^3 \sqrt{x}} = \left(\dfrac{1}{2az_1^2} + \dfrac{3}{4a^2 z_1}\right) \sqrt{x} + \dfrac{3}{8a^2} \int \dfrac{dx}{z_1 \sqrt{x}}$ (see **2.211**)

9. $\quad \int \dfrac{\sqrt{x}\, dx}{z_1^3} = \left(-\dfrac{1}{2bz_1^2} + \dfrac{1}{4abz_1}\right) \sqrt{x} + \dfrac{1}{8ab} \int \dfrac{dx}{z_1 \sqrt{x}}$ (see **2.211**)

10. $\quad \int \dfrac{x\sqrt{x}\, dx}{z_1^3} = -\dfrac{2x\sqrt{x}}{bz_1^2} + \dfrac{3a}{b} \int \dfrac{\sqrt{x}\, dx}{z_1^3}$ (see **2.213 9**)

11. $\quad \int \dfrac{x^2 \sqrt{x}\, dx}{z_1^3} = \left(\dfrac{x^2}{b} + \dfrac{5ax}{b^2}\right) \dfrac{2\sqrt{x}}{z_1^2} - \dfrac{15a^2}{b^2} \int \dfrac{\sqrt{x}\, dx}{z_1^3}$ (see **2.213 9**)

Notation: $z_2 = a + bx^2, \quad \alpha = \sqrt[4]{\dfrac{a}{b}}, \quad \alpha' = \sqrt[4]{-\dfrac{a}{b}}.$

2.214 $\quad \int \dfrac{dx}{z_2 \sqrt{x}} = \dfrac{1}{b\alpha^3 \sqrt{2}} \left[\ln \dfrac{x + \alpha\sqrt{2x} + \alpha^2}{\sqrt{z_2}} + \arctan \dfrac{\alpha\sqrt{2x}}{\alpha^2 - x}\right] \quad \left[\dfrac{a}{b} > 0\right]$

$\qquad\qquad\quad = \dfrac{1}{2b\alpha'^3} \left(\ln \dfrac{\alpha' - \sqrt{x}}{\alpha' + \sqrt{x}} - 2\arctan \dfrac{\sqrt{x}}{\alpha'}\right) \quad \left[\dfrac{a}{b} < 0\right]$

2.215 $\quad \int \dfrac{\sqrt{x}\, dx}{z_2} = \dfrac{1}{b\alpha \sqrt{2}} \left[-\ln \dfrac{x + \alpha\sqrt{2x} + \alpha^2}{\sqrt{z_2}} + \arctan \dfrac{\alpha\sqrt{2x}}{\alpha^2 - x}\right] \quad \left[\dfrac{a}{b} > 0\right]$

$\qquad\qquad\quad = \dfrac{1}{2b\alpha'} \left[\ln \dfrac{\alpha' - \sqrt{x}}{\alpha' + \sqrt{x}} + 2\arctan \dfrac{\sqrt{x}}{\alpha'}\right] \quad \left[\dfrac{a}{b} < 0\right]$

2.216

1. $\quad \int \dfrac{x\sqrt{x}\, dx}{z_2} = \dfrac{2\sqrt{x}}{b} - \dfrac{a}{b} \int \dfrac{dx}{z_2 \sqrt{x}}$ (see **2.214**)

2. $\quad \int \dfrac{x^2 \sqrt{x}\, dx}{z_2} = \dfrac{2x\sqrt{x}}{3b} - \dfrac{a}{b} \int \dfrac{\sqrt{x}\, dx}{z_2}$ (see **2.215**)

3. $\quad \int \dfrac{dx}{z_2^2 \sqrt{x}} = \dfrac{\sqrt{x}}{2az_2} + \dfrac{3}{4a} \int \dfrac{dx}{z_2 \sqrt{x}}$ (see **2.214**)

4. $\quad \int \dfrac{\sqrt{x}\, dx}{z_2^2} = \dfrac{x\sqrt{x}}{2az_2} + \dfrac{1}{4a} \int \dfrac{\sqrt{x}\, dx}{z_2}$ (see **2.215**)

5. $\quad \int \dfrac{x\sqrt{x}\, dx}{z_2^2} = -\dfrac{\sqrt{x}}{2bz_2} + \dfrac{1}{4b} \int \dfrac{dx}{z_2 \sqrt{x}}$ (see **2.214**)

6. $\quad \int \dfrac{x^2 \sqrt{x}\, dx}{z_2^2} = -\dfrac{x\sqrt{x}}{2bz_2} + \dfrac{3}{4b} \int \dfrac{\sqrt{x}\, dx}{z_2}$ (see **2.215**)

7. $\quad \int \dfrac{dx}{z_2^3 \sqrt{x}} = \left(\dfrac{1}{4az_2^2} + \dfrac{7}{16a^2 z_2}\right) \sqrt{x} + \dfrac{21}{32a^2} \int \dfrac{dx}{z_2 \sqrt{x}}$ (see **2.214**)

8. $\quad \int \dfrac{\sqrt{x}\, dx}{z_2^3} = \left(\dfrac{1}{4az_2^2} + \dfrac{5}{16a^2 z_2}\right) x\sqrt{x} + \dfrac{5}{32a^2} \int \dfrac{\sqrt{x}\, dx}{z_2}$ (see **2.215**)

9. $\quad \int \dfrac{x\sqrt{x}\, dx}{z_2^3} = \dfrac{(bx^2 - 3a) \sqrt{x}}{16abz_2^2} + \dfrac{3}{32ab} \int \dfrac{dx}{z_2 \sqrt{x}}$ (see **2.214**)

10. $\displaystyle\int \frac{x^2\sqrt{x}\,dx}{z_2^3} = -\frac{2x\sqrt{x}}{5bz_2^2} + \frac{3a}{5b}\int \frac{\sqrt{x}\,dx}{z_2^3}$ (see **2.216** 8)

2.22–2.23 Forms containing $\sqrt[n]{(a+bx)^k}$

Notation: $z = a + bx$.

2.220 $\displaystyle\int x^n\sqrt[l]{z^{lm+f}}\,dx = \left\{ \sum_{k=0}^{n} \frac{(-1)^k\binom{n}{k}z^{n-k}a^k}{ln-lk+l(m+1)+f} \right\} \frac{l\sqrt[l]{z^{l(m+1)+f}}}{b^{n+1}}$

The square root

2.221 $\displaystyle\int x^n\sqrt{z^{2m-1}}\,dx = \left\{ \sum_{k=0}^{n} \frac{(-1)^k\binom{n}{k}z^{n-k}a^k}{2n-2k+2m+1} \right\} \frac{2\sqrt{z^{2m+1}}}{b^{n+1}}$

2.222

1. $\displaystyle\int \frac{dx}{\sqrt{z}} = \frac{2}{b}\sqrt{z}$

2. $\displaystyle\int \frac{x\,dx}{\sqrt{z}} = \left(\frac{1}{3}z - a\right)\frac{2\sqrt{z}}{b^2}$

3. $\displaystyle\int \frac{x^2\,dx}{\sqrt{z}} = \left(\frac{1}{5}z^2 - \frac{2}{3}az + a^2\right)\frac{2\sqrt{z}}{b^3}$

2.223

1. $\displaystyle\int \frac{dx}{\sqrt{z^3}} = -\frac{2}{b\sqrt{z}}$

2. $\displaystyle\int \frac{x\,dx}{\sqrt{z^3}} = (z+a)\frac{2}{b^2\sqrt{z}}$

3. $\displaystyle\int \frac{x^2\,dx}{\sqrt{z^3}} = \left(\frac{z^2}{3} - 2az - a^2\right)\frac{2}{b^3\sqrt{z}}$

2.224

1. $\displaystyle\int \frac{z^m\,dx}{x^n\sqrt{z}} = -\frac{z^m\sqrt{z}}{(n-1)ax^{n-1}} + \frac{2m-2n+3}{2(n-1)}\frac{b}{a}\int \frac{z^m\,dx}{x^{n-1}\sqrt{z}}$

2. $\displaystyle\int \frac{z^m\,dx}{x^n\sqrt{z}} = -z^m\sqrt{z}\left\{ \frac{1}{(n-1)ax^{n-1}} \right.$

$$+ \sum_{k=1}^{n-2} \frac{(2m-2n+3)(2m-2n+5)\ldots(2m-2n+2k+1)}{2^k(n-1)(n-2)\ldots(n-k-1)x^{n-k-1}}\frac{b^k}{a^{k+1}} \Bigg\}$$

$$+ \frac{(2m-2n+3)(2m-2n+5)\ldots(2m-3)(2m-1)}{2^{n-1}(n-1)!x}\frac{b^{n-1}}{a^{n-1}}\int \frac{z^m\,dx}{x\sqrt{z}}$$

For $n = 1$

3. $\displaystyle\int \frac{z^m}{x\sqrt{z}}\,dx = \frac{2z^m}{(2m-1)\sqrt{z}} + a\int \frac{z^{m-1}}{x\sqrt{z}}\,dx$

4. $\displaystyle\int \frac{z^m}{x\sqrt{z}}\,dx = \sum_{k=1}^{m} \frac{2a^{m-k}z^k}{(2k-1)\sqrt{z}} + a^m \int \frac{dx}{x\sqrt{z}}$

5.[6] $\displaystyle\int \frac{dx}{x\sqrt{z}} = \frac{1}{\sqrt{a}} \ln\left|\frac{\sqrt{z}-\sqrt{a}}{\sqrt{z}+\sqrt{a}}\right|$ $\qquad\qquad [a>0]$

$\displaystyle\qquad\qquad = \frac{2}{\sqrt{-a}} \arctan \frac{\sqrt{z}}{\sqrt{-a}}$ $\qquad\qquad [a<0]$

2.225

1. $\displaystyle\int \frac{\sqrt{z}\,dx}{x} = 2\sqrt{z} + a\int \frac{dx}{x\sqrt{z}}$ $\qquad$ (see **2.224** 4)

2. $\displaystyle\int \frac{\sqrt{z}\,dx}{x^2} = -\frac{\sqrt{z}}{x} + \frac{b}{2}\int \frac{dx}{x\sqrt{z}}$ $\qquad$ (see **2.224** 4)

3. $\displaystyle\int \frac{\sqrt{z}\,dx}{x^3} = -\frac{\sqrt{z^3}}{2ax^2} + \frac{b\sqrt{z}}{4ax} - \frac{b^2}{8a}\int \frac{dx}{x\sqrt{z}}$ $\qquad$ (see **2.224** 4)

2.226

1. $\displaystyle\int \frac{\sqrt{z^3}\,dx}{x} = \left(\frac{z}{3}+a\right)2\sqrt{z} + a^2\int \frac{dx}{x\sqrt{z}}$ $\qquad$ (see **2.224** 4)

2. $\displaystyle\int \frac{\sqrt{z^3}\,dx}{x^2} = -\frac{\sqrt{z^5}}{ax} + \frac{3b}{2a}\int \frac{\sqrt{z^3}\,dx}{x}$ $\qquad$ (see **2.226** 1)

3. $\displaystyle\int \frac{\sqrt{z^3}\,dx}{x^3} = -\left(\frac{1}{2ax^2}+\frac{b}{4a^2x}\right)\sqrt{z^5} + \frac{3b^2}{8a^2}\int \frac{\sqrt{z^3}\,dx}{x}$

$\qquad\qquad\qquad\qquad\qquad\qquad\qquad$ (see **2.226** 1)

2.227 $\displaystyle\int \frac{dx}{xz^m\sqrt{z}} = \sum_{k=0}^{m-1} \frac{2}{(2k+1)a^{m-k}z^k\sqrt{z}} + \frac{1}{a^m}\int \frac{dx}{x\sqrt{z}}$ $\quad$ (see **2.224** 4)

2.228

1. $\displaystyle\int \frac{dx}{x^2\sqrt{z}} = -\frac{\sqrt{z}}{ax} - \frac{b}{2a}\int \frac{dx}{x\sqrt{z}}$ $\qquad$ (see **2.224** 4)

2. $\displaystyle\int \frac{dx}{x^3\sqrt{z}} = \left(-\frac{1}{2ax^2}+\frac{3b}{4a^2x}\right)\sqrt{z} + \frac{3b^2}{8a^2}\int \frac{dx}{x\sqrt{z}}$ $\qquad$ (see **2.224** 4)

2.229

1. $\displaystyle\int \frac{dx}{x\sqrt{z^3}} = \frac{2}{a\sqrt{z}} + \frac{1}{a}\int \frac{dx}{x\sqrt{z}}$ $\qquad$ (see **2.224** 4)

2. $\displaystyle\int \frac{dx}{x^2\sqrt{z^3}} = \left(-\frac{1}{ax}-\frac{3b}{a^2}\right)\frac{1}{\sqrt{z}} - \frac{3b}{2a^2}\int \frac{dx}{x\sqrt{z}}$ $\qquad$ (see **2.224** 4)

3. $\displaystyle\int \frac{dx}{x^3\sqrt{z^3}} = \left(-\frac{1}{2ax^2}+\frac{5b}{4a^2x}+\frac{15b^2}{4a^3}\right)\frac{1}{\sqrt{z}} + \frac{15b^2}{8a^3}\int \frac{dx}{x\sqrt{z}}$

$\qquad\qquad\qquad\qquad\qquad\qquad\qquad$ (see **2.224** 4)

Cube root

2.231

1. $$\int \sqrt[3]{z^{3m+1}}\, x^n\, dx = \left\{ \sum_{k=0}^{n} \frac{(-1)^k \binom{n}{k} z^{n-k} a^k}{3n - 3k + 3(m+1) + 1} \right\} \frac{3\sqrt[3]{z^{3(m+1)+1}}}{b^{n+1}}$$

2. $$\int \frac{x^n\, dx}{\sqrt[3]{z^{3m+2}}} = \left\{ \sum_{k=0}^{n} \frac{(-1)^k \binom{n}{k} z^{n-k} a^k}{3n - 3k - 3(m-1) - 2} \right\} \frac{3}{b^{n+1} \sqrt[3]{z^{3(m-1)+2}}}$$

3. $$\int \sqrt[3]{z^{3m+2}}\, x^n\, dx = \left\{ \sum_{k=0}^{n} \frac{(-1)^k \binom{n}{k} z^{n-k} a^k}{3n - 3k + 3(m+1) + 2} \right\} \frac{3\sqrt[3]{z^{3(m+1)+2}}}{b^{n+1}}$$

4. $$\int \frac{x^n\, dx}{\sqrt[3]{z^{3m+1}}} = \left\{ \sum_{k=0}^{n} \frac{(-1)^k \binom{n}{k} z^{n-k} a^k}{3n - 3k - 3(m-1) - 1} \right\} \frac{3}{b^{n+1} \sqrt[3]{z^{3(m-1)+1}}}$$

5. $$\int \frac{z^n\, dx}{x^m \sqrt[3]{x^2}} = -\frac{z^{n+\frac{1}{3}}}{(m-1)ax^{m-1}} + \frac{3n - 3m + 4}{3(m-1)} \frac{b}{a} \int \frac{z^n\, dx}{x^{m-1} \sqrt[3]{z^2}}$$

   For $m = 1$

   $$\int \frac{z^n\, dx}{x \sqrt[3]{z^2}} = \frac{3z^n}{(3n-2) \sqrt[3]{z^2}} + a \int \frac{z^{n-1}\, dx}{x \sqrt[3]{z^2}}$$

6. $$\int \frac{dx}{xz^n \sqrt[3]{z^2}} = \frac{3\sqrt[3]{z}}{(3n-1)az^n} + \frac{1}{a} \int \frac{\sqrt[3]{z}\, dx}{xz^n}$$

2.232 $$\int \frac{dx}{x \sqrt[3]{z^2}} = \frac{1}{\sqrt[3]{a^2}} \left\{ \frac{3}{2} \ln \frac{\sqrt[3]{z} - \sqrt[3]{a}}{\sqrt[3]{x}} - \sqrt{3} \arctan \frac{\sqrt{3}\sqrt[3]{z}}{\sqrt[3]{z} + 2\sqrt[3]{a}} \right\}$$

2.233

1. $$\int \frac{\sqrt[3]{z}\, dx}{x} = 3\sqrt[3]{z} + a \int \frac{dx}{x \sqrt[3]{z^2}} \qquad \text{(see **2.232**)}$$

2. $$\int \frac{\sqrt[3]{z}\, dx}{x^2} = -\frac{z \sqrt[3]{z}}{ax} + \frac{b}{a} \sqrt[3]{z} + \frac{b}{3} \int \frac{dx}{x \sqrt[3]{z^2}} \qquad \text{(see **2.232**)}$$

3. $$\int \frac{\sqrt[3]{z}\, dx}{x^3} = \left(-\frac{1}{2ax^2} + \frac{b}{3a^2 x} \right) z \sqrt[3]{z} - \frac{b^2}{3a^2} \sqrt[3]{z} - \frac{b^2}{9a} \int \frac{dx}{x \sqrt[3]{z^2}}$$

   $$\text{(see **2.232**)}$$

4. $$\int \frac{dx}{x^2 \sqrt[3]{z^2}} = -\frac{\sqrt[3]{z}}{ax} - \frac{2b}{3a} \int \frac{dx}{x \sqrt[3]{z^2}} \qquad \text{(see **2.232**)}$$

5. $$\int \frac{dx}{x^3 \sqrt[3]{z^2}} = \left[-\frac{1}{2ax^2} + \frac{5b}{6a^2 x} \right] \sqrt[3]{z} + \frac{5b^2}{9a^2} \int \frac{dx}{x \sqrt[3]{z^2}} \qquad \text{(see **2.232**)}$$

2.234

1. $$\int \frac{z^n\, dx}{x^m \sqrt[3]{z^2}} = -\frac{z^n \sqrt[3]{z^2}}{(m-1)ax^{m-1}} + \frac{3n - 3m + 5}{3(m-1)} \frac{b}{a} \int \frac{z^n\, dx}{x^{m-1} \sqrt[3]{z}}$$

   For $m = 1$:

2. $\displaystyle\int \frac{z^n\, dx}{x\sqrt[3]{z}} = \frac{3z^n}{(3n-1)\sqrt[3]{z}} + a\int \frac{z^{n-1}\, dx}{x\sqrt[3]{z}}$

3. $\displaystyle\int \frac{dx}{xz^n\sqrt[3]{z}} = \frac{3\sqrt[3]{z^2}}{(3n-2)az^n} + \frac{1}{a}\int \frac{\sqrt[3]{z^2}\, dx}{xz^n}$

2.235 $\displaystyle\int \frac{dx}{x\sqrt[3]{z}} = \frac{1}{\sqrt[3]{a^2}}\left\{ \frac{3}{2}\ln\frac{\sqrt[3]{z}-\sqrt[3]{a}}{\sqrt[3]{x}} + \sqrt{3}\arctan\frac{\sqrt{3}\sqrt[3]{z}}{\sqrt[3]{z}+2\sqrt[3]{a}} \right\}$

2.236

1. $\displaystyle\int \frac{\sqrt[3]{z^2}\, dx}{x} = \frac{3}{2}\sqrt[3]{z^2} + a\int \frac{dx}{x\sqrt[3]{z}}$ (see **2.235**)

2. $\displaystyle\int \frac{\sqrt[3]{z^2}\, dx}{x^2} = -\frac{\sqrt[3]{z^5}}{ax} + \frac{b}{a}\sqrt[3]{z^2} + \frac{2b}{3}\int \frac{dx}{x\sqrt[3]{z}}$ (see **2.235**)

3. $\displaystyle\int \frac{\sqrt[3]{z^2}\, dx}{x^3} = \left[-\frac{1}{2ax^2} + \frac{b}{6a^2x}\right]z^{5/3} - \frac{b^2}{6a^2}\sqrt[3]{z^2} - \frac{b^2}{9a}\int \frac{dx}{x\sqrt[3]{z}}$

   (see **2.235**)

4. $\displaystyle\int \frac{dx}{x^2\sqrt[3]{z}} = -\frac{\sqrt[3]{z^2}}{ax} - \frac{b}{3a}\int \frac{dx}{x\sqrt[3]{z}}$ (see **2.235**)

5. $\displaystyle\int \frac{dx}{x^3\sqrt[3]{z}} = \left[-\frac{1}{2ax^2} + \frac{2b}{3a^2x}\right]\sqrt[3]{z} + \frac{2b^2}{9a^2}\int \frac{dx}{x\sqrt[3]{z}}$ (see **2.235**)

2.24 Forms containing $\sqrt{a+bx}$ and the binomial $\alpha + \beta x$

Notation: $z = a + bx$, $\quad t = \alpha + \beta\mathrm{x}$, $\quad \Delta = a\beta - b\alpha$.

2.241

1. $\displaystyle\int \frac{z^m t^n\, dx}{\sqrt{z}} = \frac{2}{(2n+2m+1)\beta}t^{n+1}z^{m-1}\sqrt{z} + \frac{(2m-1)\Delta}{(2n+2m+1)\beta}\int \frac{z^{m-1}t^n\, dx}{\sqrt{z}}$ LA 176 (1)

2. $\displaystyle\int \frac{t^n z^m\, dx}{\sqrt{z}} = 2\sqrt{z^{2m+1}}\sum_{k=0}^{n}\binom{n}{k}\frac{\alpha^{n-k}\beta^k}{b^{k+1}}\sum_{p=0}^{k}(-1)^p\binom{k}{p}\frac{z^{k-p}a^p}{2k-2p+2m+1}$

2.242

1. $\displaystyle\int \frac{t\, dx}{\sqrt{z}} = \frac{2a\sqrt{z}}{b} + \beta\left(\frac{z}{3}-a\right)\frac{2\sqrt{z}}{b^2}$

2. $\displaystyle\int \frac{t^2\, dx}{\sqrt{z}} = \frac{2\alpha^2\sqrt{z}}{b} + 2\alpha\beta\left(\frac{z}{3}-a\right)\frac{2\sqrt{z}}{b^2} + \beta^2\left(\frac{z^2}{5} - \frac{2}{3}za + a^2\right)\frac{2\sqrt{z}}{b^3}$

3. $\displaystyle\int \frac{t^3\, dx}{\sqrt{z}} = \frac{2\alpha^3\sqrt{z}}{b} + 3\alpha^2\beta\left(\frac{z}{3}-\alpha\right)\frac{2\sqrt{z}}{b^2} + 3\alpha\beta^2\left(\frac{z^2}{5} - \frac{2}{3}za + a^2\right)\frac{2\sqrt{z}}{b^3}$
   $\displaystyle\qquad + \beta^\alpha\left(\frac{z^3}{7} - \frac{3z^2a}{5} + za^2 - a^3\right)\frac{2\sqrt{z}}{b^4}$

4. $\displaystyle\int \frac{tz\, dx}{\sqrt{z}} = \frac{2\alpha\sqrt{z^3}}{3b} + \beta\left(\frac{z}{5}-\frac{a}{3}\right)\frac{2\sqrt{z^3}}{b^2}$

5.
$$\int \frac{t^2 z \, dx}{\sqrt{z}} = \frac{2\alpha^2 \sqrt{z^3}}{3b} + 2\alpha\beta\left(\frac{z}{5} - \frac{a}{3}\right)\frac{2\sqrt{z^3}}{b^2} + \beta^2\left(\frac{z^2}{7} - \frac{2za}{5} + \frac{a^2}{3}\right)\frac{2\sqrt{z^3}}{b^3}$$

6.
$$\int \frac{t^3 z \, dx}{\sqrt{z}} = \frac{2\alpha^3 \sqrt{z^3}}{3b} + 3\alpha^2\beta\left(\frac{z}{5} - \frac{a}{3}\right)\frac{2\sqrt{z^3}}{b^2} + 3\alpha\beta^2\left(\frac{z^2}{7} - \frac{2za}{5} + \frac{a^2}{3}\right)\frac{2\sqrt{z^3}}{b^3}$$
$$+ \beta^3\left(\frac{z^3}{9} - \frac{3z^2 a}{7} + \frac{3za^2}{5} - \frac{a^3}{3}\right)\frac{2\sqrt{z^3}}{b^4}$$

7.
$$\int \frac{t z^2 \, dx}{\sqrt{z}} = \frac{2\alpha \sqrt{z^5}}{5b} + \beta\left(\frac{z}{7} - \frac{a}{5}\right)\frac{2\sqrt{z^5}}{b^2}$$

8.
$$\int \frac{t^2 z^2 \, dx}{\sqrt{z}} = \frac{2\alpha^2 \sqrt{z^5}}{5b} + 2\alpha\beta\left(\frac{z}{7} - \frac{a}{5}\right)\frac{2\sqrt{z^5}}{b^2} + \beta^2\left(\frac{z^2}{9} - \frac{2za}{7} + \frac{a^2}{5}\right)\frac{2\sqrt{z^5}}{b^3}$$

9.
$$\int \frac{t^3 z^2 \, dx}{\sqrt{z}} = \frac{2\alpha^3 \sqrt{z^5}}{5b} + 3\alpha^2\beta\left(\frac{z}{7} - \frac{a}{5}\right)\frac{2\sqrt{z^5}}{b^2} + 3\alpha\beta^2\left(\frac{z^2}{9} - \frac{2za}{7} + \frac{a^2}{5}\right)\frac{2\sqrt{z^5}}{b^3}$$
$$+ \beta^3\left(\frac{z^3}{11} - \frac{3z^2 a}{9} + \frac{3za^2}{7} - \frac{a^3}{5}\right)\frac{2\sqrt{z^5}}{b^4}$$

10.
$$\int \frac{t z^3 \, dx}{\sqrt{z}} = \frac{2\alpha \sqrt{z^7}}{7b} + \beta\left(\frac{z}{9} - \frac{a}{7}\right)\frac{2\sqrt{z^7}}{b^2}$$

11.
$$\int \frac{t^2 z^3 \, dx}{\sqrt{z}} = \frac{2\alpha^2 \sqrt{z^7}}{7b} + 2\alpha\beta\left(\frac{z}{9} - \frac{a}{7}\right)\frac{2\sqrt{z^7}}{b^2} + \beta^2\left(\frac{z^2}{11} - \frac{2za}{9} + \frac{a^2}{7}\right)\frac{2\sqrt{z^7}}{b^3}$$

12.
$$\int \frac{t^3 z^3 \, dx}{\sqrt{z}} = \frac{2\alpha^3 \sqrt{z^7}}{7b} + 3\alpha^2\beta\left(\frac{z}{9} - \frac{a}{7}\right)\frac{2\sqrt{z^7}}{b^2} + 3\alpha\beta^2\left(\frac{z^2}{11} - \frac{2za}{9} + \frac{a^2}{7}\right)\frac{2\sqrt{z^7}}{b^3}$$
$$+ \beta^3\left(\frac{z^3}{13} - \frac{3z^2 a}{11} + \frac{3za^2}{9} - \frac{a^3}{7}\right)\frac{2\sqrt{z^7}}{b^4}$$

2.243

1.
$$\int \frac{t^n \, dx}{z^m \sqrt{z}} = \frac{2}{(2m-1)\Delta}\frac{t^{n+1}}{z^m}\sqrt{z} - \frac{(2n-2m+3)\beta}{(2m-1)\Delta}\int \frac{t^n \, dx}{z^{m-1}\sqrt{z}}$$
$$= -\frac{2}{(2m-1)b}\frac{t^n}{z^m}\sqrt{z} + \frac{2n\beta}{(2m-1)b}\int \frac{t^{n-1} \, dx}{z^{m-1}\sqrt{z}}$$

LA 176 (2)

2.
$$\int \frac{t^n \, dx}{z^m \sqrt{z}} = \frac{2}{\sqrt{z^{2m-1}}}\sum_{k=0}^{n}\binom{n}{k}\frac{a^{n-k}\beta^k}{b^{k+1}}\sum_{p=0}^{k}(-1)^p\binom{k}{p}\frac{z^{k-p}a^p}{2k-2p-2m+1}$$

2.244

1.
$$\int \frac{t \, dx}{z\sqrt{z}} = -\frac{2a}{b\sqrt{z}} + \frac{2\beta(z+a)}{b^2\sqrt{z}}$$

2.
$$\int \frac{t^2 \, dx}{z\sqrt{z}} = -\frac{2\alpha^2}{b\sqrt{z}} + \frac{4\alpha\beta(z+a)}{b^2\sqrt{z}} + \frac{2\beta^2\left(\frac{z^2}{3} - 2za - a^2\right)}{b^3\sqrt{z}}$$

3.
$$\int \frac{t^3 \, dx}{z\sqrt{z}} = -\frac{2\alpha^3}{b\sqrt{z}} + \frac{6\alpha^2\beta(z+a)}{b^2\sqrt{z}} + \frac{6\alpha\beta^2\left(\frac{z^2}{3} - 2za - a^2\right)}{b^3\sqrt{z}} + \frac{2\beta^3\left(\frac{z^3}{5} - z^2 a + 3za^2 + a^3\right)}{b^4\sqrt{z}}$$

4. $\displaystyle\int \frac{t\,dx}{z^2\sqrt{z}} = -\frac{2a}{3b\sqrt{z^3}} - \frac{2\beta\left(z - \frac{a}{3}\right)}{b^2\sqrt{z^3}}$

5. $\displaystyle\int \frac{t^2\,dx}{z^2\sqrt{z}} = -\frac{2a^2}{3b\sqrt{z^3}} - \frac{4\alpha\beta\left(z - \frac{a}{3}\right)}{b^2\sqrt{z^3}} + \frac{2\beta^2\left(z^2 + 2az - \frac{a^2}{3}\right)}{b^3\sqrt{z^3}}$

6. $\displaystyle\int \frac{t^3\,dx}{z^2\sqrt{z}} = -\frac{2a^3}{3b\sqrt{z^3}} - \frac{6a^2\beta\left(z - \frac{a}{3}\right)}{b^2\sqrt{z^3}} + \frac{6\alpha\beta^2\left(z^2 + 2za - \frac{a^2}{3}\right)}{b^3\sqrt{z^3}} + \frac{2\beta^3\left(\frac{z^3}{3} - 3z^2a - 3za^2 + \frac{a^3}{3}\right)}{b^4\sqrt{z^3}}$

7. $\displaystyle\int \frac{t\,dx}{z^3\sqrt{z}} = -\frac{2\alpha}{5b\sqrt{z^5}} - \frac{2\beta\left(\frac{z}{3} - \frac{a}{5}\right)}{b^2\sqrt{z^5}}$

8. $\displaystyle\int \frac{t^2\,dx}{z^3\sqrt{z}} = -\frac{2\alpha^2}{5b\sqrt{z^5}} - \frac{4\alpha\beta\left(\frac{z}{3} - \frac{a}{5}\right)}{b^2\sqrt{z^5}} - \frac{2\beta^2\left(z^2 - \frac{2za}{3} + \frac{a^2}{5}\right)}{b^3\sqrt{z^5}}$

9. $\displaystyle\int \frac{t^3\,dx}{z^3\sqrt{z}} = -\frac{2\alpha^3}{5b\sqrt{z^5}} - \frac{6\alpha^2\beta\left(\frac{z}{3} - \frac{a}{5}\right)}{b^2\sqrt{z^5}} - \frac{6\alpha\beta^2\left(z^2 - \frac{2za}{3} + \frac{a^2}{5}\right)}{b^3\sqrt{z^5}}$
$$+ \frac{2\beta^3\left(z^3 + 3z^2a - za^2 + \frac{a^3}{5}\right)}{b^4\sqrt{z^5}}$$

2.245

1. $\displaystyle\int \frac{z^m\,dx}{t^n\sqrt{z}} = -\frac{2}{(2n - 2m - 1)\beta}\frac{z^{m-1}}{t^{n-1}}\sqrt{z} - \frac{(2m - 1)\Delta}{(2n - 2m - 1)\beta}\int \frac{z^{m-1}\,dx}{t^n\sqrt{z}}$ 　　　　LA 176 (3)

$$= -\frac{1}{(n-1)\beta}\frac{z^{m-1}}{t^{n-1}}\sqrt{z} + \frac{(2m-1)b}{2(n-1)\beta}\int \frac{z^{m-1}}{t^{n-1}\sqrt{z}}\,dx$$

$$= -\frac{1}{(n-1)\Delta}\frac{z^m}{t^{n-1}}\sqrt{z} - \frac{(2n - 2m - 3)b}{2(n-1)\Delta}\int \frac{z^m\,dx}{t^{n-1}\sqrt{z}}$$

2. $\displaystyle\int \frac{z^m\,dz}{t^n\sqrt{z}} = -z^m\sqrt{z}\left[\frac{1}{(n-1)\Delta}\frac{1}{t^{n-1}}\right.$

$$\left. + \sum_{k=2}^{n-1}\frac{(2n - 2m - 3)(2n - 2m - 5)\ldots(2n - 2m - 2k + 1)b^{k-1}}{2^{k-1}(n-1)(n-2)\ldots(n-k)\Delta^k}\frac{1}{t^{n-k}}\right\}\right]$$

$$-\frac{(2n - 2m - 3)(2n - 2m - 5)\ldots(-2m + 3)(-2m + 1)b^{n-1}}{2^{n-1}\cdot(n-1)!\Delta^n}\int \frac{z^m\,dx}{t\sqrt{z}}$$

For $n = 1$

3. $\displaystyle\int \frac{z^m\,dx}{t\sqrt{z}} = \frac{2}{(2m-1)\beta}\frac{z^m}{\sqrt{z}} + \frac{\Delta}{\beta}\int \frac{z^{m-1}\,dx}{t\sqrt{z}}$

4. $\displaystyle\int \frac{z^m\,dx}{t\sqrt{z}} = 2\sum_{k=0}^{m-1}\frac{\Delta^k}{(2m - 2k - 1)\beta^{k+1}}\frac{z^{m-k}}{\sqrt{z}} + \frac{\Delta^m}{\beta^m}\int \frac{dx}{t\sqrt{z}}$

2.246 $\displaystyle \int \frac{dx}{t\sqrt{z}} \, \frac{1}{\sqrt{\beta\Delta}} \ln \frac{\beta\sqrt{z} - \sqrt{\beta\Delta}}{\beta\sqrt{z} + \sqrt{\beta\Delta}}$ $[\beta\Delta > 0]$

         $\displaystyle = \frac{2}{\sqrt{-\beta\Delta}} \arctan \frac{\beta\sqrt{z}}{\sqrt{-\beta\Delta}}$ $[\beta\Delta < 0]$

         $\displaystyle = -\frac{2\sqrt{z}}{bt}$ $[\Delta = 0]$

2.247 $\displaystyle \int \frac{dx}{tz^m\sqrt{z}} = \frac{2}{z^{m-1}\sqrt{z}} + \sum_{k=1}^{m} \frac{\beta^{k-1}z^k}{\Delta^k(2m-2k+1)} + \frac{\beta^m}{\Delta^m} \int \frac{dx}{t\sqrt{z}}$

                                                (see **2.246**)

2.248

1. $\displaystyle \int \frac{dx}{tz\sqrt{z}} = \frac{2}{\Delta\sqrt{z}} + \frac{\beta}{\Delta} \int \frac{dx}{t\sqrt{z}}$ (see **2.246**)

2. $\displaystyle \int \frac{dx}{tz^2\sqrt{z}} = \frac{2}{3\Delta z\sqrt{z}} + \frac{2\beta}{\Delta^2\sqrt{z}} + \frac{\beta^2}{\Delta^2} \int \frac{dx}{t\sqrt{z}}$ (see **2.246**)

3. $\displaystyle \int \frac{dx}{tz^3\sqrt{z}} = \frac{2}{5\Delta z^2\sqrt{z}} + \frac{2\beta}{3\Delta^2 z\sqrt{z}} + \frac{2\beta^2}{\Delta^3\sqrt{z}} + \frac{\beta^3}{\Delta^3} \int \frac{dx}{t\sqrt{z}}$

                                                (see **2.246**)

4. $\displaystyle \int \frac{dx}{t^2\sqrt{z}} = -\frac{\sqrt{z}}{\Delta t} - \frac{b}{2\Delta} \int \frac{dx}{t\sqrt{z}}$ (see **2.246**)

5. $\displaystyle \int \frac{dx}{t^2 z\sqrt{z}} = -\frac{1}{\Delta t\sqrt{z}} - \frac{3b}{\Delta^2\sqrt{z}} - \frac{3b\beta}{2\Delta^2} \int \frac{dx}{t\sqrt{z}}$ (see **2.246**)

6. $\displaystyle \int \frac{dx}{t^2 z^2\sqrt{z}} = -\frac{1}{\Delta t z^2\sqrt{z}} - \frac{5b}{3\Delta^2 z\sqrt{z}} - \frac{5b\beta}{\Delta^3\sqrt{z}} - \frac{5b\beta^2}{2\Delta^3} \int \frac{dx}{t\sqrt{z}}$

                                                (see **2.246**)

7. $\displaystyle \int \frac{dx}{t^2 z^3\sqrt{z}} = -\frac{1}{\Delta t z^2\sqrt{z}} - \frac{7b}{5\Delta^2 z^2\sqrt{z}} - \frac{7b\beta}{3\Delta^3 z\sqrt{z}} - \frac{7b\beta^2}{\Delta^4\sqrt{z}} - \frac{7b\beta^3}{2\Delta^4} \int \frac{dx}{t\sqrt{z}}$

                                                (see **2.246**)

8. $\displaystyle \int \frac{dx}{t^3\sqrt{z}} = -\frac{\sqrt{z}}{2\Delta t^2} + \frac{3b\sqrt{z}}{4\Delta^2 t} + \frac{3b^2}{8\Delta^2} \int \frac{dx}{t\sqrt{z}}$ (see **2.246**)

9. $\displaystyle \int \frac{dx}{t^3 z\sqrt{z}} = -\frac{1}{2\Delta t^2\sqrt{z}} + \frac{5b}{4\Delta^2 t\sqrt{z}} + \frac{15b^2}{4\Delta^3\sqrt{z}} + \frac{15b^2\beta}{8\Delta^3} \int \frac{dx}{t\sqrt{z}}$

                                                (see **2.246**)

10. $\displaystyle \int \frac{dx}{t^3 z^2\sqrt{z}} = -\frac{1}{2\Delta t^2 z\sqrt{z}} + \frac{7b\sqrt{z}}{4\Delta^2 t z\sqrt{z}} + \frac{35b^2}{12\Delta^2 z\sqrt{z}} + \frac{35b^2\beta}{4\Delta^4\sqrt{z}} + \frac{35b^2\beta^2}{8\Delta^4} \int \frac{dx}{t\sqrt{z}}$

                                                (see **2.246**)

11. $\displaystyle \int \frac{dx}{t^3 z^3\sqrt{z}} = -\frac{1}{2\Delta t^2 z^2\sqrt{z}} + \frac{9b}{4\Delta^2 t z^2\sqrt{z}} + \frac{63b^2}{20\Delta^3 z^2\sqrt{z}} + \frac{21b^2\beta}{4\Delta^4 z\sqrt{z}} + \frac{63b^2\beta^2}{4\Delta^5\sqrt{z}} + \frac{63b^2\beta^3}{8\Delta^5} \int \frac{dx}{t\sqrt{z}}$

                                                (see **2.246**)

12. $$\int \frac{z\,dx}{t\sqrt{z}} = \frac{2\sqrt{z}}{\beta} + \frac{\Delta}{\beta}\int \frac{dx}{t\sqrt{z}}$$ (see **2.246**)

13. $$\int \frac{z^2\,dx}{t\sqrt{z}} = \frac{2z\sqrt{z}}{3\beta} + \frac{2\Delta\sqrt{z}}{\beta^2} + \frac{\Delta^2}{\beta^2}\int \frac{dx}{t\sqrt{z}}$$ (see **2.246**)

14. $$\int \frac{z^3\,dx}{t\sqrt{z}} = \frac{2z^2\sqrt{z}}{5\beta} + \frac{2\Delta z\sqrt{z}}{3\beta^2} + \frac{2\Delta^2\sqrt{z}}{\beta^3} + \frac{\Delta^3}{\beta^3}\int \frac{dx}{t\sqrt{z}}$$ (see **2.246**)

15. $$\int \frac{z\,dx}{t^2\sqrt{z}} = -\frac{z\sqrt{z}}{\Delta t} + \frac{b\sqrt{z}}{\beta\Delta} + \frac{b}{2\beta}\int \frac{dx}{t\sqrt{z}}$$ (see **2.246**)

16. $$\int \frac{z^2\,dx}{t^2\sqrt{z}} = -\frac{z^2\sqrt{z}}{\Delta t} + \frac{bz\sqrt{z}}{\beta\Delta} + \frac{3b\sqrt{z}}{\beta^2} + \frac{3b\Delta}{2\beta^2}\int \frac{dx}{t\sqrt{z}}$$ (see **2.246**)

17. $$\int \frac{z^3\,dx}{t^2\sqrt{z}} = -\frac{z^3\sqrt{z}}{\Delta t} + \frac{bz^2\sqrt{z}}{\beta\Delta} + \frac{5bz\sqrt{z}}{3\beta^2} + \frac{5b\Delta\sqrt{z}}{\beta^3} + \frac{5\Delta^2 b}{2\beta^3}\int \frac{dx}{t\sqrt{z}}$$

(see **2.246**)

18.[3] $$\int \frac{z\,dx}{t^3\sqrt{z}} = -\frac{z\sqrt{z}}{2\Delta t^2} + \frac{bz\sqrt{z}}{4\Delta^2 t} - \frac{b^2\sqrt{z}}{4\beta\Delta^2} + \frac{b^2}{8\beta\Delta}\int \frac{dx}{t\sqrt{z}}$$ (see **2.246**)

19. $$\int \frac{z^2\,dx}{t^3\sqrt{z}} = -\frac{z^2\sqrt{z}}{2\Delta t^2} + \frac{bz^2\sqrt{z}}{4\Delta^2 t} + \frac{b^2 z\sqrt{z}}{4\beta\Delta^2} + \frac{3b^2\sqrt{z}}{4\beta^2\Delta} + \frac{3b^2}{8\beta^2}\int \frac{dx}{t\sqrt{z}}$$

(see **2.246**)

20. $$\int \frac{z^3\,dx}{t^3\sqrt{z}} = -\frac{z^3\sqrt{z}}{2\Delta t^2} + \frac{3bz^3\sqrt{z}}{\Delta^2 t} + \frac{3b^2 z^2\sqrt{z}}{4\beta\Delta^2} + \frac{5b^2 z\sqrt{z}}{4\beta^2\Delta} + \frac{15b^2\sqrt{z}}{4\beta^3} + \frac{15b^2\Delta}{8\beta^3}\int \frac{dx}{t\sqrt{z}}$$

(see **2.246**)

2.249

1. $$\int \frac{dx}{z^m t^n \sqrt{z}} = \frac{2}{(2m-1)\Delta}\frac{\sqrt{z}}{t^{n-1}z^m} + \frac{(2n+2m-3)\beta}{(2m-1)\Delta}\int \frac{dx}{t^n z^{m-1}\sqrt{z}}$$ LA 177 (4)

$$= -\frac{1}{(n-1)\Delta}\frac{\sqrt{z}}{z^m t^{n-1}} - \frac{(2n+2m-3)b}{2(n-1)\Delta}\int \frac{dx}{t^{n-1}z^m\sqrt{z}}$$

2. $$\int \frac{dx}{z^m t^n\sqrt{z}} = \frac{\sqrt{z}}{z^m}\left[\frac{-1}{(n-1)\Delta}\frac{1}{t^{n-1}}\right.$$

$$+ \sum_{k=2}^{n-1}(-1)^k\frac{(2n+2m-3)(2n+2m-5)\dots(2n+2m-2k+1)b^{k-1}}{2^{k-1}(n-1)(n-2)\dots(n-k)\Delta^k}\cdot\frac{1}{t^{n-k}}\right]$$

$$+(-1)^{n-1}\frac{(2n+2m-3)(2n+2m-5)\dots(-2m+3)(-2m+1)b^{n-1}}{2^{n-1}(n-1)!\Delta^{n-1}}\int \frac{dx}{tz^m\sqrt{z}}$$

For $n = 1$
$$\int \frac{dx}{z^m t\sqrt{z}} = \frac{2}{(2m-1)\Delta}\frac{1}{z^{m-1}\sqrt{z}} + \frac{\beta}{\Delta}\int \frac{dx}{tz^{m-1}\sqrt{z}}$$

2.25 Forms containing $\sqrt{a + bx + cx^2}$

Integration techniques

2.251   It is possible to rationalize the integrand in integrals of the form $\int R\left(x, \sqrt{a + bx + cx^2}\right) dx$ by using one or more of the following three substitutions, known as the "Euler substitutions".

1. $\sqrt{a + bx + cx^2} = xt \pm \sqrt{a}$ for $a > 0$;

2. $\sqrt{a + bx + cx^2} = t \pm x\sqrt{c}$ for $c > 0$;

3. $\sqrt{c\left(x - x_1\right)\left(x - x_2\right)} = t\left(x - x_1\right)$ when x_1 and x_2 are real roots of the equation $a + bx + cx^2 = 0$.

2.252   Besides the Euler substitutions, there is also the following method of calculating integrals of the form $\int R\left(x, \sqrt{a + bx + cx^2}\right) dx$. By removing the irrational expressions in the denominator and performing simple algebraic operations, we can reduce the integrand to the sum of some rational function of x and an expression of the form $\dfrac{P_1(x)}{P_2(x)\sqrt{a + bx + cx^2}}$, where $P_1(x)$ and $P_2(x)$ are both polynomials. By separating the integral portion of the rational function $\dfrac{P_1(x)}{P_2(x)}$ from the remainder and decomposing the latter into partial fractions, we can reduce the integral of these partial fractions to the sum of integrals each of which is in one of the following three forms:

1. $\int \dfrac{P(x)\, dx}{\sqrt{a + bx + cx^2}}$, where $P(x)$ is a polynomial of some degree r;

2. $\int \dfrac{dx}{(x + p)^k \sqrt{a + bx + cx^2}}$;

3. $\int \dfrac{(Mx + N)\, dx}{\left(a + \beta x + x^2\right)^m \sqrt{c\left(a_1 + b_1 x + x^2\right)}}$, $\left(a_1 = \dfrac{a}{c}, \quad b_1 = \dfrac{b}{c}\right)$.

In more detail:

1. $\int \dfrac{P(x)\, dx}{\sqrt{a + bx + cx^2}} = Q(x)\sqrt{a + bx + cx^2} + \lambda \int \dfrac{dx}{\sqrt{a + bx + cx^2}}$, where $Q(x)$ is a polynomial of degree $(r - 1)$. Its coefficients, and also the number λ, can be calculated by the method of undetermined coefficients from the identity

$$P(x) = Q'(x)\left(a + bx + cx^2\right) + \frac{1}{2}Q(x)(b + 2cx) + \lambda \qquad \text{LI II 77}$$

Integrals of the form $\int \dfrac{P(x)\, dx}{\sqrt{a + bx + cx^2}}$ (where $r \leq 3$) can also be calculated by use of formulas **2.26**.

2.   Integrals of the form $\int \dfrac{P(x)\, dx}{(x + p)^k \sqrt{a + bx + cx^2}}$, where the degree n of the polynomial $P(x)$ is lower than k can, by means of the substitution $t = \dfrac{1}{x + p}$, be reduced to an integral of the form

$\int \dfrac{P(t)\, dt}{\sqrt{a + \beta t + \gamma t^2}}$. (See also **2.281**).

3.   Integrals of the form $\int \dfrac{(Mx+N)\,dx}{(\alpha+\beta x+x^2)^m \sqrt{c\,(a_1+b_1 x+x^2)}}$ can be calculated by the following procedure.

- If $b_1 \neq \beta$, by using the substitution

$$x = \frac{a_1 - \alpha}{\beta b_1} + \frac{t-1}{t+1} \frac{\sqrt{(a_1-\alpha)^2 - (\alpha b_1 - a_1 \beta)(\beta - b_1)}}{\beta - b_1}$$

we can reduce this integral to an integral of the form $\int \dfrac{P(t)\,dt}{(t^2+p)^m \sqrt{c\,(t^2+q)}}$, where $P(t)$ is a polynomial of degree no higher than $2m-1$. The integral $\int \dfrac{P(t)\,dt}{(t^2+p)^m \sqrt{t^2+q}}$ can be reduced to the sum of integrals of the forms $\int \dfrac{t\,dt}{(t^2+p)^k \sqrt{t^2+q}}$ and $\int \dfrac{dt}{(t^2+p)^k \sqrt{t^2+q}}$.

- If $b_1 = \beta$, we can reduce it to integrals of the form $\int \dfrac{P(t)\,dt}{(t^2+p)^m \sqrt{c\,(t^2+q)}}$ by means of the substitution $t = x + \dfrac{b_1}{2}$.

The integral $\int \dfrac{t\,dt}{(t^2+p)^k \sqrt{c\,(t^2+q)}}$ can be evaluated by means of the substitution $t^2 + q = u^2$.

The integral $\int \dfrac{dt}{(t^2+p)^k \sqrt{c\,(t^2+q)}}$ can be evaluated by means of the substitution $\dfrac{t}{\sqrt{t^2+q}} = v$ (see also **2.283**).

FI II 78-82

2.26 Forms containing $\sqrt{a+b+cx^2}$ and integral powers of x

Notation: $R = a + bx + cx^2$, $\quad \Delta = 4ac - b^2$
For simplified formulas for the case $b = 0$, see **2.27**.

2.260

1. $\displaystyle\int x^m \sqrt{R^{2n+1}}\,dx = \frac{x^{m-1}\sqrt{R^{2n+3}}}{(m+2n+2)c} - \frac{(2m+2n+1)b}{2(m+2n+2)c}\int x^{m-1}\sqrt{R^{2n+1}}\,dx$
$$- \frac{(m-1)a}{(m+2n+2)c}\int x^{m-2}\sqrt{R^{2n+1}}\,dx$$

TI (192)a

2. $\displaystyle\int \sqrt{R^{2n+1}}\,dx = \frac{2cx+b}{4(n+1)c}\sqrt{R^{2n+1}} + \frac{2n+1}{8(n+1)}\frac{\Delta}{c}\int \sqrt{R^{2n-1}}\,dx$ TI (188)

3. $\displaystyle\int \sqrt{R^{2n+1}}\,dx = \frac{(2cx+b)\sqrt{R}}{4(n+1)c}\left\{R^n + \sum_{k=0}^{n-1} \frac{(2n+1)(2n-1)\ldots(2n-2k+1)}{8^{k+1}n(n-1)\ldots(n-k)}\left(\frac{\Delta}{c}\right)^{k+1} R^{n-k-1}\right\}$
$$+ \frac{(2n+1)!!}{8^{n+1}(n+1)!}\left(\frac{\Delta}{c}\right)^{n+1}\int \frac{dx}{\sqrt{R}}$$

TI (190)

2.261[6]　For $n = -1$

$$\int \frac{dx}{\sqrt{R}} = \frac{1}{\sqrt{c}} \ln\left(2\sqrt{cR} + 2cx + b\right) \qquad [c > 0] \qquad\qquad \text{TI (127)}$$

$$= \frac{1}{\sqrt{c}} \operatorname{arcsinh} \frac{2cx + b}{\sqrt{\Delta}} \qquad [c > 0, \quad \Delta > 0] \qquad\qquad \text{DW}$$

$$= \frac{1}{\sqrt{c}} \ln(2cx + b) \qquad [c > 0, \quad \Delta = 0] \qquad\qquad \text{DW}$$

$$= \frac{-1}{\sqrt{-c}} \arcsin \frac{2cx + b}{\sqrt{-\Delta}} \qquad [c < 0, \quad \Delta < 0] \qquad\qquad \text{TI (128)}$$

2.262

1.　$$\int \sqrt{R}\, dx = \frac{(2cx + b)\sqrt{R}}{4c} + \frac{\Delta}{8c} \int \frac{dx}{\sqrt{R}} \qquad\qquad \text{(see **2.261**)}$$

2.　$$\int x\sqrt{R}\, dx = \frac{\sqrt{R^3}}{3c} - \frac{(2cx + b)b}{8c^2}\sqrt{R} - \frac{b\Delta}{16c^2} \int \frac{dx}{\sqrt{R}} \qquad\qquad \text{(see **2.261**)}$$

3.　$$\int x^2\sqrt{R}\, dx = \left(\frac{x}{4c} - \frac{5b}{24c^2}\right)\sqrt{R^3} + \left(\frac{5b^2}{16c^2} - \frac{a}{4c}\right)\frac{(2cx + b)\sqrt{R}}{4c} + \left(\frac{5b^2}{16c^2} - \frac{a}{4c}\right)\frac{\Delta}{8c} \int \frac{dx}{\sqrt{R}}$$

$$\text{(see **2.261**)}$$

4.　$$\int x^3\sqrt{R}\, dx = \left(\frac{x^2}{5c} - \frac{7bx}{40c^2} + \frac{7b^2}{48c^3} - \frac{2a}{15c^2}\right)\sqrt{R^3} - \left(\frac{7b^3}{32c^3} - \frac{3ab}{8c^2}\right)\frac{(2cx + b)\sqrt{R}}{4c}$$

$$- \left(\frac{7b^3}{32c^3} - \frac{3ab}{8c^2}\right)\frac{\Delta}{8c} \int \frac{dx}{\sqrt{R}}$$

$$\text{(see **2.261**)}$$

5.　$$\int \sqrt{R^3}\, dx = \left(\frac{R}{8c} + \frac{3\Delta}{64c^2}\right)(2cx + b)\sqrt{R} + \frac{3\Delta^2}{128c^2} \int \frac{dx}{\sqrt{R}}$$

$$\text{(see **2.261**)}$$

6.　$$\int x\sqrt{R^3}\, dx = \frac{\sqrt{R^5}}{5c} - (2cx + b)\left(\frac{b}{16c^2}\sqrt{R^3} + \frac{3\Delta b}{128c^3}\sqrt{R}\right) - \frac{3\Delta^2 b}{256c^3} \int \frac{dx}{\sqrt{R}}$$

$$\text{(see **2.261**)}$$

7.　$$\int x^2\sqrt{R^3}\, dx = \left(\frac{x}{6c} - \frac{7b}{60c^2}\right)\sqrt{R^5} + \left(\frac{7b^2}{24c^2} - \frac{a}{6c}\right)\left(2x + \frac{b}{c}\right)\left(\frac{\sqrt{R^3}}{8} + \frac{3\Delta}{64c}\sqrt{R}\right)$$

$$+ \left(\frac{7b^2}{4c} - a\right)\frac{\Delta^2}{256c^3} \int \frac{dx}{\sqrt{R}}$$

$$\text{(see **2.261**)}$$

8.　$$\int x^3\sqrt{R^3}\, dx = \left(\frac{x^2}{7c} - \frac{3bx}{28c^2} + \frac{3b^2}{40c^3} - \frac{2a}{35c^2}\right)\sqrt{R^5}$$

$$- \left(\frac{3b^3}{16c^3} - \frac{ab}{4c^2}\right)\left(2x + \frac{b}{c}\right)\left(\frac{\sqrt{R^3}}{8} + \frac{3\Delta}{64c}\sqrt{R}\right)$$

$$- \left(\frac{3b^2}{4c} - a\right)\frac{3\Delta^2 b}{512c^4} \int \frac{dx}{\sqrt{R}}$$

$$\text{(see **2.261**)}$$

2.263

1. $$\int \frac{x^m \, dx}{\sqrt{R^{2n+1}}} = \frac{x^{m-1}}{(m-2n)c\sqrt{R^{2n-1}}} - \frac{(2m-2n-1)b}{2(m-2n)c}\int \frac{x^{m-1} \, dx}{\sqrt{R^{2n+1}}} - \frac{(m-1)a}{(m-2n)c}\int \frac{x^{m-2} \, dx}{\sqrt{R^{2n+1}}}$$

TI (193)a

For $m = 2n$

2. $$\int \frac{x^{2n} \, dx}{\sqrt{R^{2n+1}}} = -\frac{x^{2n-1}}{(2n-1)c\sqrt{R^{2n-1}}} - \frac{b}{2c}\int \frac{x^{2n-1}}{\sqrt{R^{2n+1}}} \, dx + \frac{1}{c}\int \frac{x^{2n-2}}{\sqrt{R^{2n-1}}} \, dx \qquad \text{TI (194)a}$$

3. $$\int \frac{dx}{\sqrt{R^{2n+1}}} = \frac{2(2cx+b)}{(2n-1)\Delta\sqrt{R^{2n-1}}} + \frac{8(n-1)c}{(2n-1)\Delta}\int \frac{dx}{\sqrt{R^{2n-1}}} \qquad \text{TI (189)}$$

4. $$\int \frac{dx}{\sqrt{R^{2n+1}}} = \frac{2(2cx+b)}{(2n-1)\Delta\sqrt{R^{2n-1}}}\left\{1 + \sum_{k=1}^{n-1} \frac{8^k(n-1)(n-2)\ldots(n-k)}{(2n-3)(2n-5)\ldots(2n-2k-1)} \frac{c^k}{\Delta^k} R^k\right\}$$

$[n \geq 1]$.

TI (191)

2.264

1. $$\int \frac{dx}{\sqrt{R}} \qquad\qquad\qquad\qquad \text{(see } \mathbf{2.261)}$$

2. $$\int \frac{x \, dx}{\sqrt{R}} = \frac{\sqrt{R}}{c} - \frac{b}{2c}\int \frac{dx}{\sqrt{R}} \qquad\qquad \text{(see } \mathbf{2.261)}$$

3. $$\int \frac{x^2 \, dx}{\sqrt{R}} = \left(\frac{x}{2c} - \frac{3b}{4c^2}\right)\sqrt{R} + \left(\frac{3b^2}{8c^2} - \frac{a}{2c}\right)\int \frac{dx}{\sqrt{R}} \qquad \text{(see } \mathbf{2.261)}$$

4. $$\int \frac{x^3 \, dx}{\sqrt{R}} = \left(\frac{x^2}{3c} - \frac{5bx}{12c^2} + \frac{5b^2}{8c^3} - \frac{2a}{3c^2}\right)\sqrt{R} - \left(\frac{5b^3}{16c^3} - \frac{3ab}{4c^2}\right)\int \frac{dx}{\sqrt{R}}$$

(see 2.261)

5. $$\int \frac{dx}{\sqrt{R^3}} = \frac{2(2cx+b)}{\Delta\sqrt{R}}$$

6. $$\int \frac{x \, dx}{\sqrt{R^3}} = -\frac{2(2a+bx)}{\Delta\sqrt{R}}$$

7. $$\int \frac{x^2 \, dx}{\sqrt{R^3}} = -\frac{(\Delta - b^2)\,x - 2ab}{c\Delta\sqrt{R}} + \frac{1}{c}\int \frac{dx}{\sqrt{R}} \qquad \text{(see } \mathbf{2.261)}$$

8. $$\int \frac{x^3 \, dx}{\sqrt{R^3}} = \frac{c\Delta x^2 + b\left(10ac - 3b^2\right)x + a\left(8ac - 3b^2\right)}{c^2\Delta\sqrt{R}} - \frac{3b}{2c^2}\int \frac{dx}{\sqrt{R}}$$

(see 2.261)

2.265 $$\int \frac{\sqrt{R^{2n+1}}}{x^m} \, dx = -\frac{\sqrt{R^{2n+3}}}{(m-1)ax^{m-1}} + \frac{(2n-2m+5)b}{2(m-1)a}\int \frac{\sqrt{R^{2n+1}}}{x^{m-1}} \, dx$$
$$+ \frac{(2n-m+4)c}{(m-1)a}\int \frac{\sqrt{R^{2n+1}}}{x^{m-2}} \, dx$$

TI (195)

For $m = 1$

$$\int \frac{\sqrt{R^{2n+1}}}{x}\,dx = \frac{\sqrt{R^{2n+1}}}{2n+1} + \frac{b}{2}\int \sqrt{R^{2n-1}}\,dx + a\int \frac{\sqrt{R^{2n-1}}}{x}\,dx \qquad \text{TI (198)}$$

For $a = 0$

$$\int \frac{\sqrt{(bx+cx^2)^{2n+1}}}{x^m}\,dx = \frac{2\sqrt{(bx+cx^2)^{2n+3}}}{(2n-2m+3)bx^m} + \frac{2(m-2n-3)c}{(2n-2m+3)b}\int \frac{\sqrt{(bx+cx^2)^{2n+1}}}{x^{m-1}} \qquad \text{LA 169 (3)}$$

For $m = 0$ see **2.260** 2 and **2.260** 3.

For $n = -1$ and $m = 1$:

2.266[8] $\displaystyle\int \frac{dx}{x\sqrt{R}} = -\frac{1}{\sqrt{a}}\ln\frac{2a+bx+2\sqrt{aR}}{x}$ $[a > 0]$ TI (137)

$\displaystyle\qquad = \frac{1}{\sqrt{-a}}\arcsin\frac{2a+bx}{x\sqrt{b^2-4ac}}$ $[a < 0, \quad \Delta < 0]$ TI (138)

$\displaystyle\qquad = \frac{1}{\sqrt{-a}}\arctan\frac{2a+bx}{2\sqrt{-a}\sqrt{R}}$ $[a < 0]$ LA 178 (6)a

$\displaystyle\qquad = -\frac{1}{\sqrt{a}}\operatorname{arcsinh}\frac{2a+bx}{x\sqrt{\Delta}}$ $[a > 0, \quad \Delta > 0]$ DW

$\displaystyle\qquad = -\frac{1}{\sqrt{a}}\operatorname{arctanh}\frac{2a+bx}{2\sqrt{a}\sqrt{R}}$ $[a > 0]$

$\displaystyle\qquad = \frac{1}{\sqrt{a}}\ln\frac{x}{2a+bx}$ $[a > 0, \quad \Delta = 0]$

$\displaystyle\qquad = -\frac{2\sqrt{bx+cx^2}}{bx}$ $[a = 0, \quad b \neq 0]$ LA 170 (16)

$\displaystyle\qquad = \frac{1}{\sqrt{a}}\operatorname{arccosh}\left(\frac{2a+bx}{x\sqrt{-\Delta}}\right)$ $[a > 0, \Delta < 0]$

2.267

1. $\displaystyle\int \frac{\sqrt{R}\,dx}{x} = \sqrt{R} + a\int \frac{dx}{x\sqrt{R}} + \frac{b}{2}\int \frac{dx}{\sqrt{R}}$ (see **2.261** and **2.266**)

2. $\displaystyle\int \frac{\sqrt{R}\,dx}{x^2} = -\frac{\sqrt{R}}{x} + \frac{b}{2}\int \frac{dx}{x\sqrt{R}} + c\int \frac{dx}{\sqrt{R}}$ (see **2.261** and **2.266**)

For $a = 0$

$\displaystyle\int \frac{\sqrt{bx+cx^2}}{x^2}\,dx = -\frac{2\sqrt{bx+cx^2}}{x} + c\int \frac{dx}{\sqrt{bx+cx^2}}$ (see **2.261**)

3. $\displaystyle\int \frac{\sqrt{R}\,dx}{x^3} = -\left(\frac{1}{2x^2} + \frac{b}{4ax}\right)\sqrt{R} - \left(\frac{b^2}{8a} - \frac{c}{2}\right)\int \frac{dx}{x\sqrt{R}}$

(see **2.266**)

For $a = 0$

$\displaystyle\int \frac{\sqrt{bx+cx^2}}{x^3}\,dx = -\frac{2\sqrt{(bx+cx^2)^3}}{3bx^3}$

4. $\displaystyle\int \frac{\sqrt{R^3}}{x}\,dx = \frac{\sqrt{R^3}}{3} + \frac{2bcx+b^2+8ac}{8c}\sqrt{R} + a^2\int \frac{dx}{x\sqrt{R}} + \frac{b\left(12ac-b^2\right)}{16c}\int \frac{dx}{\sqrt{R}}$

(see **2.261** and **2.266**)

5. $\int \frac{\sqrt{R^3}}{x^2}\,dx = -\frac{\sqrt{R^5}}{ax} + \frac{cx+b}{a}\sqrt{R^3} + \frac{3}{4}(2cx+3b)\sqrt{R} + \frac{3}{2}ab\int \frac{dx}{x\sqrt{R}} + \frac{3\left(4ac+b^2\right)}{8}\int \frac{dx}{\sqrt{R}}$

$$(\text{see } \mathbf{2.261} \text{ and } \mathbf{2.266})$$

For $a = 0$

$$\int \frac{\sqrt{(bx+cx^2)^3}}{x^2} = \frac{\sqrt{(bx+cx^2)^3}}{2x} + \frac{3b}{4}\sqrt{bx+cx^2} + \frac{3b^2}{8}\int \frac{dx}{\sqrt{bx+cx^2}}$$

$$(\text{see } \mathbf{2.261})$$

6. $\int \frac{\sqrt{R^3}}{x^3}\,dx = -\left(\frac{1}{2ax^2} + \frac{b}{4a^2x}\right)\sqrt{R^5} + \frac{bcx+2ac+b^2}{4a^2}\sqrt{R^3} + \frac{3\left(bcx+2ac+b^2\right)}{4a}\sqrt{R}$

$$+ \frac{3}{8}\left(4ac+b^2\right)\int \frac{dx}{x\sqrt{R}} + \frac{3}{2}bc\int \frac{dx}{\sqrt{R}}$$

$$(\text{see } \mathbf{2.261} \text{ and } \mathbf{2.266})$$

For $a = 0$

$$\int \frac{\sqrt{(bx+cx^2)^3}}{x^3}\,dx = \left(c - \frac{2b}{x}\right)\sqrt{bx+cx^2} + \frac{3bc}{2}\int \frac{dx}{\sqrt{bx+cx^2}}$$

$$(\text{see } \mathbf{2.261})$$

2.268 $\int \frac{dx}{x^m\sqrt{R^{2n+1}}} = -\frac{1}{(m-1)ax^{m-1}\sqrt{R^{2n-1}}}$

$$-\frac{(2n+2m-3)b}{2(m-1)a}\int \frac{dx}{x^{m-1}\sqrt{R^{2n+1}}} - \frac{(2n+m-2)c}{(m-1)a}\int \frac{dx}{x^{m-2}\sqrt{R^{2n+1}}}$$

TI (196)

For $m = 1$

bigbigstrut $\int \frac{dx}{x\sqrt{R^{2n+1}}} = \frac{1}{(2n-1)a\sqrt{R^{2n-1}}} - \frac{b}{2a}\int \frac{dx}{\sqrt{R^{2n+1}}} + \frac{1}{a}\int \frac{dx}{x\sqrt{R^{2n-1}}}$

TI (199)

For $a = 0$

$$\int \frac{dx}{x^m\sqrt{(bx+cx^2)^{2n+1}}} = -\frac{2}{(2n+2m-1)bx^m\sqrt{(bx+cx^2)^{2n-1}}}$$

$$-\frac{(4n+2m-2)c}{(2n+2m-1)b}\int \frac{dx}{x^{m-1}\sqrt{(bx+cx^2)^{2n+1}}}$$

$$(\text{cf. } \mathbf{2.265})$$

2.269

1. $\int \frac{dx}{x\sqrt{R}}$ $\qquad\qquad$ $(\text{see } \mathbf{2.266})$

2. $\int \frac{dx}{x^2\sqrt{R}} = -\frac{\sqrt{R}}{ax} - \frac{b}{2a}\int \frac{dx}{x\sqrt{R}}$ $\qquad\qquad$ $(\text{see } \mathbf{2.266})$

For $a = 0$

$$\int \frac{dx}{x^2\sqrt{bx+cx^2}} = \frac{2}{3}\left(-\frac{1}{bx^2} + \frac{2c}{b^2x}\right)\sqrt{bx+cx^2}$$

3. $\displaystyle \int \frac{dx}{x^3 \sqrt{R}} = \left(-\frac{1}{2ax^2} + \frac{3b}{4a^2 x} \right) \sqrt{R} + \left(\frac{3b^2}{8a^2} - \frac{c}{2a} \right) \int \frac{dx}{x \sqrt{R}}$

$$\text{(see } \mathbf{2.266})$$

For $a = 0$

$$\int \frac{dx}{x^3 \sqrt{bx + cx^2}} = \frac{2}{5} \left(-\frac{1}{bx^3} + \frac{4c}{3b^2 x^2} - \frac{8c^2}{3b^3 x} \right) \sqrt{bx + cx^2}$$

4. $\displaystyle \int \frac{dx}{x \sqrt{R^3}} = -\frac{2 \left(bcx - 2ac + b^2 \right)}{a \Delta \sqrt{R}} + \frac{1}{a} \int \frac{dx}{x \sqrt{R}}$ (see $\mathbf{2.266}$)

For $a = 0$

$$\int \frac{dx}{x \sqrt{\left(bx + cx^2 \right)^3}} = \frac{2}{3} \left(-\frac{1}{bx} + \frac{4c}{b^2} + \frac{8c^2 x}{b^3} \right) \frac{1}{\sqrt{bx + cx^2}}$$

5. $\displaystyle \int \frac{dx}{x^2 \sqrt{R^3}} = -\frac{A}{\sqrt{R}} + \frac{1}{a^2 \Delta \sqrt{R}} \left[\left(3b^2 - 8ac \right) cx + \left(3b^2 - 10ac \right) b \right] - \frac{3b}{2a^2} \int \frac{dx}{x \sqrt{R}}$

$$\text{where } A = \left(-\frac{1}{ax} - \frac{b \left(10ac - 3b^2 \right)}{a^2 \Delta} - \frac{c \left(8ac - 3b^2 \right) x}{a^2 \Delta} \right) \qquad \text{(see } \mathbf{2.266})$$

For $a = 0$

$$\int \frac{dx}{x^2 \sqrt{\left(bx + cx^2 \right)^3}} = \frac{2}{5} \left(-\frac{1}{bx^2} + \frac{2c}{b^2 x} - \frac{8c^2}{b^3} - \frac{16c^3 x}{b^4} \right) \frac{1}{\sqrt{bx + cx^2}}$$

6. $\displaystyle \int \frac{dx}{x^3 \sqrt{R^3}}$

$$= \left(-\frac{1}{ax^2} + \frac{5b}{2a^2 x} - \frac{15b^4 - 62acb^2 + 24a^2 c^2}{2a^3 \Delta} - \frac{bc \left(15b^2 - 52ac \right) x}{2a^3 \Delta} \right) \frac{1}{2\sqrt{R}} + \frac{15b^2 - 12ac}{8a^3} \int \frac{dx}{x \sqrt{R}}$$

$$\text{(see } \mathbf{2.266})$$

For $a = 0$

$$\int \frac{dx}{x^3 \sqrt{\left(bx + cx^2 \right)^3}} = \frac{2}{7} \left(-\frac{1}{bx^3} + \frac{8c}{5b^2 x^2} - \frac{16c^2}{5b^3 x} + \frac{64c^3}{5b^4} + \frac{128c^4 x}{5b^5} \right) \frac{1}{\sqrt{bx + cx^2}}$$

2.27 Forms containing $\sqrt{a + cx^2}$ and integral powers of x

Notation: $u = \sqrt{a + cx^2}$.

$\displaystyle I_1 = \frac{1}{\sqrt{c}} \ln \left(x\sqrt{c} + u \right)$ $[c > 0]$

$\displaystyle \quad = \frac{1}{\sqrt{-c}} \arcsin x \sqrt{-\frac{c}{a}}$ $[c < 0 \text{ and } a > 0]$

$\displaystyle I_2 = \frac{1}{2\sqrt{a}} \ln \frac{u - \sqrt{a}}{u + \sqrt{a}}$ $[a > 0 \text{ and } c > 0]$

$\displaystyle \quad = \frac{1}{2\sqrt{a}} \ln \frac{\sqrt{a} - u}{\sqrt{a} + u}$ $[a > 0 \text{ and } 1c < 0]$

$\displaystyle \quad = \frac{1}{\sqrt{-a}} \operatorname{arcsec} x \sqrt{-\frac{c}{a}} = \frac{1}{\sqrt{-a}} \arccos \frac{1}{x} \sqrt{-\frac{a}{c}}$ $[a < 0 \text{ and } c > 0]$

2.271

1. $\displaystyle\int u^5\,dx = \frac{1}{6}xu^5 + \frac{5}{24}axu^3 + \frac{5}{16}a^2xu + \frac{5}{16}a^3 I_1$ DW

2. $\displaystyle\int u^3\,dx = \frac{1}{4}xu^3 + \frac{3}{8}axu + \frac{3}{8}a^2 I_1$ DW

3. $\displaystyle\int u\,dx = \frac{1}{2}xu + \frac{1}{2}a I_1$ DW

4. $\displaystyle\int \frac{dx}{u} = I_1$ DW

5. $\displaystyle\int \frac{dx}{u^3} = \frac{1}{a}\frac{x}{u}$ DW

6. $\displaystyle\int \frac{dx}{u^{2n+1}} = \frac{1}{a^n}\sum_{k=0}^{n-1} \frac{(-1)^k}{2k+1}\binom{n-1}{k}\frac{c^k x^{2k+1}}{u^{2k+1}}$

7. $\displaystyle\int \frac{x\,dx}{u^{2n+1}} = -\frac{1}{(2n-1)cu^{2n-1}}$ DW

2.272

1. $\displaystyle\int x^2 u^3\,dx = \frac{1}{6}\frac{xu^5}{c} - \frac{1}{24}\frac{axu^3}{c} - \frac{1}{16}\frac{a^2xu}{c} - \frac{1}{16}\frac{a^3}{c}I_1$ DW

2. $\displaystyle\int x^2 u\,dx = \frac{1}{4}\frac{xu^3}{c} - \frac{1}{8}\frac{axu}{c} - \frac{1}{8}\frac{a^2}{c}I_1$ DW

3. $\displaystyle\int \frac{x^2}{u}\,dx = \frac{1}{2}\frac{xu}{c} - \frac{1}{2}\frac{a}{c}I_1$ DW

4. $\displaystyle\int \frac{x^2}{u^3}\,dx = -\frac{x}{cu} + \frac{1}{c}I_1$ DW

5. $\displaystyle\int \frac{x^2}{u^5}\,dx = \frac{1}{3}\frac{x^3}{au^3}$ DW

6. $\displaystyle\int \frac{x^2\,dx}{u^{2n+1}} = \frac{1}{a^{n-1}}\sum_{k=0}^{n-2} \frac{(-1)^k}{2k+3}\binom{n-2}{k}\frac{c^k x^{2k+3}}{u^{2k+3}}$

7. $\displaystyle\int \frac{x^3\,dx}{u^{2n+1}} = -\frac{1}{(2n-3)c^2 u^{2n-3}} + \frac{a}{(2n-1)c^2 u^{2n-1}}$ DW

2.273

1. $\displaystyle\int x^4 u^3\,dx = \frac{1}{8}\frac{x^3 u^5}{c} - \frac{axu^5}{16c^2} + \frac{a^2xu^3}{64c^2} + \frac{3a^3xu}{128c^2} + \frac{3a^4}{128c^2}I_1$ DW

2. $\displaystyle\int x^4 u\,dx = \frac{1}{6}\frac{x^3 u^3}{c} - \frac{axu^3}{8c^2} + \frac{a^2xu}{16c^2} + \frac{a^3}{16c^2}I_1$ DW

3. $\displaystyle\int \frac{x^4}{u}\,dx = \frac{1}{4}\frac{x^3 u}{c} - \frac{3}{8}\frac{axu}{c^2} + \frac{3}{8}\frac{a^2}{c^2}I_1$ DW

4. $\displaystyle\int \frac{x^4}{u^3}\,dx = \frac{1}{2}\frac{xu}{c^2} + \frac{ax}{c^2 u} - \frac{3}{2}\frac{a}{c^2}I_1$ DW

5. $\int \dfrac{x^4}{u^5} \, dx = -\dfrac{x}{c^2 u} - \dfrac{1}{3}\dfrac{x^3}{cu^3} + \dfrac{1}{c^2} I_1$ DW

6. $\int \dfrac{x^4}{u^7} \, dx = \dfrac{1}{5}\dfrac{x^5}{au^5}$ DW

7. $\int \dfrac{x^4 \, dx}{u^{2n+1}} = \dfrac{1}{a^{n-2}} \sum\limits_{k=0}^{n-3} \dfrac{(-1)^k}{2k+5} \binom{n-3}{k} \dfrac{c^k x^{2k+5}}{u^{2k+5}}$

8. $\int \dfrac{x^5 \, dx}{u^{2n+1}} = -\dfrac{1}{(2n-5)c^3 u^{2n-5}} + \dfrac{2a}{(2n-3)cu^{2n-3}} - \dfrac{a^2}{(2n-1)c^3 u^{2n-1}}$ DW

2.274

1. $\int x^6 u^3 \, dx = \dfrac{1}{10}\dfrac{x^5 u^5}{c} - \dfrac{ax^3 u^5}{16c^2} + \dfrac{a^2 x u^5}{32c^3} - \dfrac{a^3 x u^3}{128c^3} - \dfrac{3a^4 x u}{256c^3} - \dfrac{3}{256}\dfrac{a^5}{c^3} I_1$

2. $\int x^6 u \, dx = \dfrac{1}{8}\dfrac{x^5 u^3}{c} - \dfrac{5}{48}\dfrac{ax^3 u^3}{c^2} + \dfrac{5a^2 x u^3}{64c^3} - \dfrac{5a^3 x u}{128c^3} - \dfrac{5}{128}\dfrac{a^4}{c^3} I_1$

3. $\int \dfrac{x^6}{u} \, dx = \dfrac{1}{6}\dfrac{x^5 u}{c} - \dfrac{5}{24}\dfrac{ax^3 u}{c^2} + \dfrac{5}{16}\dfrac{a^2 x u}{c^3} - \dfrac{5}{16}\dfrac{a^3}{c^3} I_1$ DW

4. $\int \dfrac{x^6}{u^3} \, dx = \dfrac{1}{4}\dfrac{x^5}{cu} - \dfrac{5}{8}\dfrac{ax^3}{c^2 u} - \dfrac{15}{8}\dfrac{a^2 x}{c^3 u} + \dfrac{15}{8}\dfrac{a^2}{c^3} I_1$ DW

5. $\int \dfrac{x^6}{u^5} \, dx = \dfrac{1}{2}\dfrac{x^5}{cu^3} + \dfrac{10}{3}\dfrac{ax^3}{c^2 u^3} + \dfrac{5}{2}\dfrac{a^2 x}{c^3 u^3} - \dfrac{5}{2}\dfrac{a}{c^3} I_1$ DW

6. $\int \dfrac{x^6}{u^7} \, dx = -\dfrac{23}{15}\dfrac{x^5}{cu^5} - \dfrac{7}{3}\dfrac{ax^3}{c^2 u^5} - \dfrac{a^2 x}{c^3 u^5} + \dfrac{1}{c^3} I_1$ DW

7. $\int \dfrac{x^6}{u^9} \, dx = \dfrac{1}{7}\dfrac{x^7}{au^7}$ DW

8. $\int \dfrac{x^6 \, dx}{u^{2n+1}} = \dfrac{1}{a^{n-3}} \sum\limits_{k=0}^{n-4} \dfrac{(-1)^k}{2k+7} \binom{n-4}{k} \dfrac{c^k x^{2k+7}}{u^{2k+7}}$

9. $\int \dfrac{x^7 \, dx}{u^{2n+1}} = -\dfrac{1}{(2n-7)c^4 u^{2n-7}} + \dfrac{3a}{(2n-5)c^4 u^{2n-5}} - \dfrac{3a^2}{(2n-3)c^4 u^{2n-3}} + \dfrac{a^3}{(2n-1)c^4 u^{2n-1}}$ DW

2.275

1. $\int \dfrac{u^5}{x} \, dx = \dfrac{u^5}{5} + \dfrac{1}{3}au^3 + a^2 u + a^3 I_2$ DW

2. $\int \dfrac{u^3}{x} \, dx = \dfrac{u^3}{3} + au + a^2 I_2$ DW

3. $\int \dfrac{u}{x} \, dx = u + a I_2$ DW

4. $\int \dfrac{dx}{xu} = I_2$ DW

5. $\int \dfrac{dx}{xu^{2n+1}} = \dfrac{1}{a^n} I_2 + \sum\limits_{k=0}^{n-1} \dfrac{1}{(2k+1)a^{n-k}u^{2k+1}}$

6. $\int \dfrac{u^5}{x^2}\,dx = -\dfrac{u^5}{x} + \dfrac{5}{4}cxu^3 + \dfrac{15}{8}acxu + \dfrac{15}{8}a^2 I_1$ DW

7. $\int \dfrac{u^3}{x^2}\,dx = -\dfrac{u^3}{x} + \dfrac{3}{2}cxu + \dfrac{3}{2}a I_1$ DW

8. $\int \dfrac{u}{x^2}\,dx = -\dfrac{u}{x} + c I_1$ DW

9. $\int \dfrac{dx}{x^2 u^{2n+1}} = -\dfrac{1}{a^{n+1}}\left\{ \dfrac{u}{x} + \sum_{k=1}^{n} \dfrac{(-1)^{k+1}}{2k-1} \binom{n}{k} c^k \left(\dfrac{x}{u}\right)^{2k-1} \right\}$

2.276

1. $\int \dfrac{u^5}{x^3}\,dx = -\dfrac{u^5}{2x^2} + \dfrac{5}{6}cu^3 + \dfrac{5}{2}acu + \dfrac{5}{2}a^2 c I_2$ DW

2. $\int \dfrac{u^3}{x^3}\,dx = -\dfrac{u^3}{2x^2} + \dfrac{3}{2}cu + \dfrac{3}{2}ac I_2$ DW

3. $\int \dfrac{u}{x^3}\,dx = -\dfrac{u}{2x^2} + \dfrac{c}{2} I_2$ DW

4. $\int \dfrac{dx}{x^3 u} = -\dfrac{u}{2ax^2} - \dfrac{c}{2a} I_2$ DW

5. $\int \dfrac{dx}{x^3 u^3} = -\dfrac{1}{2ax^2 u} - \dfrac{3c}{2a^2 u} - \dfrac{3c}{2a^2} I_2$ DW

6. $\int \dfrac{dx}{x^3 u^5} = -\dfrac{1}{2ax^2 u^3} - \dfrac{5}{6}\dfrac{c}{a^2 u^3} - \dfrac{5}{2}\dfrac{c}{a^3 u} - \dfrac{5}{2}\dfrac{c}{a^3} I_2$ DW

7. $\int \dfrac{u^5}{x^4}\,dx = -\dfrac{au^3}{3x^3} - \dfrac{2acu}{x} + \dfrac{c^2 xu}{2} + \dfrac{5}{2}ac I_1$ DW

8. $\int \dfrac{u^3}{x^4}\,dx = -\dfrac{u^3}{3x^3} - \dfrac{cu}{x} + c I_1$ DW

9. $\int \dfrac{u}{x^4}\,dx = -\dfrac{u^3}{3ax^3}$ DW

10. $\int \dfrac{dx}{x^4 u^{2n+1}} = \dfrac{1}{a^{n+2}}\left\{ -\dfrac{u^3}{3x^3} + (n+1)\dfrac{cu}{x} + \sum_{k=2}^{n+1} \dfrac{(-1)^k}{2k-3} \binom{n+1}{k} c^k \left(\dfrac{x}{u}\right)^{2k-3} \right\}$

2.277

1. $\int \dfrac{u^3}{x^5}\,dx = -\dfrac{u^3}{4x^4} - \dfrac{3}{8}\dfrac{cu^3}{ax^2} + \dfrac{3}{8}\dfrac{c^2 u}{a} + \dfrac{3}{8}c^2 I_2$ DW

2. $\int \dfrac{u}{x^5}\,dx = -\dfrac{u}{4x^4} - \dfrac{1}{8}\dfrac{cu}{ax^2} - \dfrac{1}{8}\dfrac{c^2}{a} I_2$ DW

3. $\int \dfrac{dx}{x^5 u} = -\dfrac{u}{4ax^4} + \dfrac{3}{8}\dfrac{cu}{a^2 x^2} + \dfrac{3}{8}\dfrac{c^2}{a^2} I_2$ DW

4. $\int \dfrac{dx}{x^5 u^3} = -\dfrac{1}{4ax^4 u} + \dfrac{5}{8}\dfrac{c}{a^2 x^2 u} + \dfrac{15}{8}\dfrac{c^2}{a^3 u} + \dfrac{15}{8}\dfrac{c^2}{a^3} I_2$

   DW

2.278

1. $\displaystyle\int \frac{u^3}{x^6}\, dx = -\frac{u^5}{5ax^5}$ DW

2. $\displaystyle\int \frac{u}{x^6}\, dx = -\frac{u^3}{5ax^5} + \frac{2}{15}\frac{cu^3}{a^2x^3}$ DW

3. $\displaystyle\int \frac{dx}{x^6 u} = \frac{1}{a^3}\left(-\frac{u^5}{5x^5} + \frac{2}{3}\frac{cu^3}{x^3} - \frac{c^2 u}{x}\right)$ DW

4. $\displaystyle\int \frac{dx}{x^6 u^{2n+1}} = \frac{1}{a^{n+3}}\left\{-\frac{u^5}{5x^5} + \frac{1}{3}\binom{n+2}{1}\frac{cu^3}{x^3} - \binom{n+2}{2}\frac{c^2 u}{x} + \sum_{k=3}^{n+2}\frac{(-1)^k}{2k-5}\binom{n+2}{k}c^k\left(\frac{x}{u}\right)^{2k-5}\right\}$

2.28 Forms containing $\sqrt{a + bx + cx^2}$ and first-and second-degree polynomials

Notation: $R = a + bx + cx^2$

See also 2.252

2.281³ $\displaystyle\int \frac{dx}{(x+p)^n \sqrt{R}} = -\int \frac{t^{n-1}\, dt}{\sqrt{c + (b-2pc)t + (a-bp+cp^2)t^2}}$

$$\left[t = \frac{1}{x+p} > 0\right]$$

2.282

1.³ $\displaystyle\int \frac{\sqrt{R}\, dx}{x+p} = c\int \frac{x\, dx}{\sqrt{R}} + (b - cp)\int \frac{dx}{\sqrt{R}} + (a - bp + cp^2)\int \frac{dx}{(x+p)\sqrt{R}}$

$$[x + p > 0]$$

2. $\displaystyle\int \frac{dx}{(x+p)(x+q)\sqrt{R}} = \frac{1}{q-p}\int \frac{dx}{(x+p)\sqrt{R}} + \frac{1}{p-q}\int \frac{dx}{(x+q)\sqrt{R}}$

3. $\displaystyle\int \frac{\sqrt{R}\, dx}{(x+p)(x+q)} = \frac{1}{q-p}\int \frac{\sqrt{R}\, dx}{x+p} + \frac{1}{p-q}\int \frac{\sqrt{R}\, dx}{x+q}$

4. $\displaystyle\int \frac{(x+p)\sqrt{R}\, dx}{x+q} = \int \sqrt{R}\, dx + (p-q)\int \frac{\sqrt{R}\, dx}{x+q}$

5. $\displaystyle\int \frac{(rx+s)\, dx}{(x+p)(x+q)\sqrt{R}} = \frac{s-pr}{q-p}\int \frac{dx}{(x+p)\sqrt{R}} + \frac{s-qr}{p-q}\int \frac{dx}{(x+q)\sqrt{R}}$

2.283 $\displaystyle\int \frac{(Ax+B)\, dx}{(p+R)^n \sqrt{R}} = \frac{A}{c}\int \frac{du}{(p+u^2)^n} + \frac{2Bc-Ab}{2c}\int \frac{(1-cv^2)^{n-1}\, dv}{\left[p+a-\dfrac{b^2}{4c}-cpv^2\right]^n},$

where $u = \sqrt{R}$ and $v = \dfrac{b+2cx}{2c\sqrt{R}}$.

2.284 $\displaystyle\int \frac{Ax+B}{(p+R)\sqrt{R}}\, dx = \frac{A}{c}I_1 + \frac{2Bc-Ab}{\sqrt{c^2p\,[b^2-4(a+p)c]}}I_2,$

where

$$I_1 = \frac{1}{\sqrt{p}} \arctan \sqrt{\frac{R}{p}} \qquad\qquad [p > 0]$$

$$= \frac{1}{2\sqrt{-p}} \ln \frac{\sqrt{-p} - \sqrt{R}}{\sqrt{-p} + \sqrt{R}} \qquad\qquad [p < 0]$$

$$I_2 = \arctan \sqrt{\frac{p}{b^2 - 4(a+p)c}} \frac{b + 2cx}{\sqrt{R}} \qquad [p\{b^2 - 4(a+p)c\} > 0, \quad p < 0]$$

$$= -\arctan \sqrt{\frac{p}{b^2 - 4(a+p)c}} \frac{b + 2cx}{\sqrt{R}} \qquad [p\{b^2 - 4(a+p)c\} > 0, \quad p > 0]$$

$$= \frac{1}{2i} \ln \frac{\sqrt{4(a+p)c - b^2}\sqrt{R} + \sqrt{p}(b+2cx)}{\sqrt{4(a+p)c - b^2}\sqrt{R} - \sqrt{p}(b+2cx)} \qquad [p\{b^2 - 4(a+p)c\} < 0, \quad p > 0]$$

$$= \frac{1}{2i} \ln \frac{\sqrt{b^2 - 4(a+p)c}\sqrt{R} - \sqrt{-p}(b+2cx)}{\sqrt{b^2 - 4(a+p)c}\sqrt{R} + \sqrt{-p}(b+2cx)} \qquad [p\{b^2 - 4(a+p)c\} < 0, \quad p < 0]$$

2.29 Integrals that can be reduced to elliptic or pseudo-elliptic integrals

2.290   Integrals of the form $\int R\left(x, \sqrt{P(x)}\right)\, dx$, where $P(x)$ is a third-or fourth-degree polynomial can, by means of algebraic transformations, be reduced to a sum of integrals expressed in terms of elementary functions and elliptic integrals (see **8.11**). Since the substitutions that transform the given integral into an elliptic integral in the normal Legendre form are different for different intervals of integration, the corresponding formulas are given in the chapter on definite integrals (see **3.13**, **3.17**).

2.291   Certain integrals of the form $\int R\left(x, \sqrt{P(x)}\right)\, dx$, where $P_n(x)$ is a polynomial of not more than fourth degree, can be reduced to integrals of the form $\int R\left(x, \sqrt[k]{P_n(x)}\right)\, dx$ with $k \geq 2$. Below are examples of this procedure.

1. $$\int \frac{dx}{\sqrt{1 - x^6}} = -\int \frac{dz}{\sqrt{3 + 3z^2 + z^4}} \qquad \left[x^2 = \frac{1}{1+z^2}\right]$$

2. $$\int \frac{dx}{\sqrt{a + bx^2 + cx^4 + dx^6}} = \frac{1}{2}\int \frac{dz}{\sqrt{az + bz^2 + cz^3 + dz^4}}$$

$$[x^2 = z]$$

3. $$\int \left(a + 2bx + cx^2 + gx^3\right)^{\pm 1/3} dx = \frac{3}{2}\int \frac{z^2 A^{\pm\frac{1}{3}}\, dz}{B}$$

$$\left[a + 2bx + cx^2 = z^3, \quad A = g\left(\frac{-b + \sqrt{b^2 + (z^3 - a)c}}{c}\right)^3 + z^3, \quad B = \sqrt{b^2 + (z^3 - a)c}\right]$$

4. $$\int \frac{dx}{\sqrt{a + bx + cx^2 + dx^3 + cx^4 + bx^5 + ax^6}}$$

$$= -\frac{1}{\sqrt{2}}\int \frac{dx}{\sqrt{(z+1)p}} - \frac{1}{\sqrt{2}}\int \frac{dz}{\sqrt{(z-1)p}} \qquad \left[x = z + \sqrt{z^2 - 1}\right]$$

$$= -\frac{1}{\sqrt{2}}\int \frac{d}{\sqrt{(z+1)p}} + \frac{1}{\sqrt{2}}\int \frac{dz}{\sqrt{(z-1)p}} \qquad \left[x = z - \sqrt{z^2 - 1}\right]$$

where $p = 2a\left(4z^3 - 3z\right) + 2b\left(2z^2 - 1\right) + 2cz + d$.

5.
$$\int \frac{dx}{\sqrt{a + bx^2 + cx^4 + bx^6 + ax^8}} = \frac{1}{2}\int \frac{dy}{\sqrt{y}\sqrt{a + by + cy^2 + by^3 + ay^4}} \qquad [x = \sqrt{y}]$$

$$= -\frac{1}{2\sqrt{2}}\int \frac{dz}{\sqrt{(z+1)p}} + \frac{1}{2\sqrt{2}}\int \frac{dz}{\sqrt{(z-1)p}} \qquad \left[y = z + \sqrt{z^2 - 1}\right]$$

$$= \frac{1}{2\sqrt{2}}\int \frac{dz}{\sqrt{(z+1)p}} - \frac{1}{2\sqrt{2}}\int \frac{dz}{\sqrt{(z-1)p}} \qquad \left[y = z - \sqrt{z^2 - 1}\right]$$

where $p = 2a\left(2z^2 - 1\right) + 2bz + c$.

6.
$$\int \frac{dx}{\sqrt{a + bx^4 + cx^8}} = \frac{1}{2}\sqrt[8]{\frac{a}{c}}\int \frac{dt}{\sqrt{t}\sqrt{ab_1 t^2 + at^4}} \qquad \left[x = \sqrt[8]{\frac{a}{c}}\sqrt{t}\right];$$

$$= -\frac{1}{2\sqrt{2}}\sqrt[8]{\frac{a}{c}}\left\{\int \frac{dz}{\sqrt{(z+1)p}} - \int \frac{dz}{\sqrt{(z-1)p}}\right\} \qquad \left[t = z + \sqrt{z^2 - 1}\right]$$

$$= -\frac{1}{2\sqrt{2}}\sqrt[8]{\frac{a}{c}}\left\{\int \frac{dz}{\sqrt{(z+1)p}} + \int \frac{dz}{\sqrt{(z-1)p}}\right\} \qquad \left[t = z - \sqrt{z^2 - 1}\right]$$

where $p = 2a\left(2z^2 - 1\right) + b_1;\ b_1 = b\sqrt{\frac{a}{c}}$.

7.
$$\int \frac{x\,dx}{\sqrt[4]{a + bx^2 + cx^4}} = 2\int \frac{z^2\,dz}{\sqrt{A + Bz^4}} \qquad [a + bx^2 + cx^4 = z^4, \quad A = b^2 - 4ac, \quad B = 4c]$$

8.
$$\int \frac{dx}{\sqrt[4]{a + 2bx^2 + cx^4}} = \int \frac{\sqrt{b^2 - a\left(c - z^4\right)} + b}{\left(c - z^4\right)\sqrt{b^2 - a\left(c - z^4\right)}}\,z^2\,dz = \int R_1\left(z^4\right)z^2\,dz + \int \frac{R_2\left(z^4\right)z^2\,dz}{\sqrt{b^2 - a\left(c - z^4\right)}},$$

where $R_1\left(z^4\right)$ and $R_2\left(z^4\right)$ are rational functions of z^4 and $a + 2bx^2 + cx^4 = x^4 z^4$.

2.292   In certain cases, integrals of the form $\int R\left(x, \sqrt{P(x)}\right)dx$, where $P(x)$ is a third-or fourth-degree polynomial, can be expressed in terms of elementary functions. Such integrals are called *pseudo-elliptic* integrals.

Thus, if the relations

$$f_1(x) = f_1\left(\frac{1}{k^2 x}\right), \qquad f_2(x) = f_2\left(\frac{1 - k^2 x}{k^2(1 - x)}\right), \qquad f_3(x) = f_3\left(\frac{1 - x}{1 - k^2 x}\right),$$

hold, then

1.
$$\int \frac{f_1(x)\,dx}{\sqrt{x(1 - x)\left(1 - k^2 x\right)}} = \int R_1(z)\,dz \qquad \left[zx = \sqrt{x(1 - x)\left(1 - k^2 x\right)}\right]$$

2.
$$\int \frac{f_2(x)\,dx}{\sqrt{x(1 - x)\left(1 - k^2 x\right)}} = \int R_2(z)\,dz \qquad \left[z = \frac{\sqrt{x\left(1 - k^2 x\right)}}{\sqrt{1 - x}}\right]$$

3.
$$\int \frac{f_3(x)\,dx}{\sqrt{x(1 - x)\left(1 - k^2 x\right)}} = \int R_3(z)\,dz \qquad \left[z = \frac{\sqrt{x(1 - x)}}{\sqrt{1 - k^2 x}}\right]$$

where $R_1(z)$, $R_2(z)$, and $R_3(z)$ are rational functions of z.

2.3 The Exponential Function

2.31 Forms containing e^{ax}

2.311 $\displaystyle\int e^{ax}\,dx = \frac{e^{ax}}{a}$

2.312 a^x in the integrands should be replaced with $e^{x\ln a} = a^x$

2.313

1. $\displaystyle\int \frac{dx}{a + be^{mx}} = \frac{1}{am}\left[mx - \ln\left(a + be^{mx}\right)\right]$ PE (410)

2. $\displaystyle\int \frac{dx}{1 + e^x} = \ln\frac{e^x}{1 + e^x} = x - \ln\left(1 + e^x\right)$ PE (409)

2.314 $\displaystyle\int \frac{dx}{ae^{mx} + be^{-mx}} = \frac{1}{m\sqrt{ab}}\arctan\left(e^{mx}\sqrt{\frac{a}{b}}\right)$ $[ab > 0]$ PE (411)

$$= \frac{1}{2m\sqrt{-ab}}\ln\frac{b + e^{mx}\sqrt{-ab}}{b - e^{mx}\sqrt{-ab}} \qquad [ab < 0]$$

2.315 $\displaystyle\int \frac{dx}{\sqrt{a + be^{mx}}} = \frac{1}{m\sqrt{a}}\ln\frac{\sqrt{a + be^{mx}} - \sqrt{a}}{\sqrt{a + be^{mx}} + \sqrt{a}}$ $[a > 0]$

$$= \frac{2}{m\sqrt{-a}}\arctan\frac{\sqrt{a + be^{mx}}}{\sqrt{-a}} \qquad [a < 0]$$

2.32 The exponential combined with rational functions of x

2.321

1. $\displaystyle\int x^m e^{ax}\,dx = \frac{x^m e^{ax}}{a} - \frac{m}{a}\int x^{m-1}e^{ax}\,dx$

2. $\displaystyle\int x^n e^{ax}\,dx = e^{ax}\left(\frac{x^n}{a} + \sum_{k=1}^{n}(-1)^k\frac{n(n-1)\dots(n-k+1)}{a^{k+1}}x^{n-k}\right)$

2.322

1. $\displaystyle\int xe^{ax}\,dx = e^{ax}\left(\frac{x}{a} - \frac{1}{a^2}\right)$

2. $\displaystyle\int x^2 e^{ax}\,dx = e^{ax}\left(\frac{x^2}{a} - \frac{2x}{a^2} + \frac{2}{a^3}\right)$

3. $\displaystyle\int x^3 e^{ax}\,dx = e^{ax}\left(\frac{x^3}{a} - \frac{3x^2}{a^2} + \frac{6x}{a^3} - \frac{6}{a^4}\right)$

4.[10] $\displaystyle\int x^4 e^{ax}\,dx = e^{ax}\left(\frac{x^4}{a} - \frac{4x^3}{a^2} + \frac{12x^2}{a^3} - \frac{24x}{a^4} + \frac{24}{a^5}\right)$

2.323 $\displaystyle\int P_m(x)e^{ax}\,dx = \frac{e^{ax}}{a}\sum_{k=0}^{m}(-1)^k\frac{P^{(k)}(x)}{a^k},$

where $P_m(x)$ is a polynomial in x of degree m and $P^{(k)}(x)$ is the k-th derivative of $P_m(x)$ with respect to x.

2.324

1. $$\int \frac{e^{ax}\,dx}{x^m} = \frac{1}{m-1}\left[-\frac{e^{ax}}{x^{m-1}} + a\int \frac{e^{ax}\,dx}{x^{m-1}}\right]$$

2. $$\int \frac{e^{ax}}{x^n}\,dx = -e^{ax}\sum_{k=1}^{n-1}\frac{a^{k-1}}{(n-1)(n-2)\ldots(n-k)x^{n-k}} + \frac{a^{n-1}}{(n-1)!}\,\text{Ei}(ax)$$

2.325

1. $$\int \frac{e^{ax}}{x}\,dx = \text{Ei}(ax)$$

2. $$\int \frac{e^{ax}}{x^2}\,dx = -\frac{e^{ax}}{x} + a\,\text{Ei}(ax)$$

3. $$\int \frac{e^{ax}}{x^3}\,dx = -\frac{e^{ax}}{2x^2} - \frac{ae^{ax}}{2x} + \frac{a^2}{2}\,\text{Ei}(ax)$$

2.326 $$\int \frac{xe^{ax}\,dx}{(1+ax)^2} = \frac{e^{ax}}{a^2(1+ax)}$$

2.33[8]

1. $$\int e^{-(ax^2+2bx+c)}\,dx = \frac{1}{2}\sqrt{\frac{\pi}{a}}\exp\left(\frac{b^2-ac}{a}\right)\text{erf}\left(\sqrt{a}x + \frac{b}{\sqrt{a}}\right)$$

$$[a \neq 0]$$

2.4 Hyperbolic Functions

2.41–2.43 Powers of $\sinh x$, $\cosh x$, $\tanh x$, and $\coth x$

2.411
$$\int \sinh^p x \cosh^q x\,dx = \frac{\sinh^{p+1} x \cosh^{q-1} x}{p+q} + \frac{q-1}{p+q}\int \sinh^p x \cosh^{q-2} x\,dx$$
$$= \frac{\sinh^{p-1} x \cosh^{q+1} x}{p+q} - \frac{p-1}{p+q}\int \sinh^{p-2} x \cosh^q x\,dx$$
$$= \frac{\sinh^{p-1} x \cosh^{q+1} x}{q+1} - \frac{p-1}{q+1}\int \sinh^{p-2} x \cosh^{q+2} x\,dx$$
$$= \frac{\sinh^{p+1} x \cosh^{q-1} x}{p+1} - \frac{q-1}{p+1}\int \sinh^{p+2} x \cosh^{q-2} x\,dx$$
$$= \frac{\sinh^{p+1} x \cosh^{q+1} x}{p+1} - \frac{p+q+2}{p+1}\int \sinh^{p+2} x \cosh^q x\,dx$$
$$= -\frac{\sinh^{p+1} x \cosh^{q+1} x}{q+1} + \frac{p+q+2}{q+1}\int \sinh^p x \cosh^{q+2} x\,dx$$

2.412

1.
$$\int \sinh^p x \cosh^{2n} x \, dx = \frac{\sinh^{p+1} x}{2n+p} \left[\cosh^{2n-1} x \right.$$

$$+ \sum_{k=1}^{n-1} \frac{(2n-1)(2n-3)\ldots(2n-2k+1)}{(2n+p-2)(2n+p-4)\ldots(2n+p-2k)} \cosh^{2n-2k-1} x \right]$$

$$+ \frac{(2n-1)!!}{(2n+p)(2n+p-2)\ldots(p+2)} \int \sinh^p x \, dx$$

This formula is applicable for arbitrary real p except for the following negative even integers: -2, $-4, \ldots, -2n$. If p is a natural number and $n=0$, we have

2.
$$\int \sinh^{2m} x \, dx = (-1)^m \binom{2m}{m} \frac{x}{2^{2m}} + \frac{1}{2^{2m-1}} \sum_{k=0}^{m-1} (-1)^k \binom{2m}{k} \frac{\sinh(2m-2k)x}{2m-2k} \qquad \text{TI (543)}$$

3.
$$\int \sinh^{2m+1} x \, dx = \frac{1}{2^{2m}} \sum_{k=0}^{m} (-1)^k \binom{2m+1}{k} \frac{\cosh(2m-2k+1)x}{2m-2k+1}; \qquad \text{TI (544)}$$

$$= (-1)^n \sum_{k=0}^{m} (-1)^k \binom{m}{k} \frac{\cosh^{2k+1} x}{2k+1} \qquad \text{GU (351) (5)}$$

4.
$$\int \sinh^p x \cosh^{2n+1} x \, dx$$

$$= \frac{\sinh^{p+1} x}{2n+p+1} \left\{ \cosh^{2n} x + \sum_{k=1}^{n} \frac{2^k n(n-1)\ldots(n-k+1) \cosh^{2n-2k} x}{(2n+p-1)(2n+p-3)\ldots(2n+p-2k+1)} \right\}$$

This formula is applicable for arbitrary real p except for the following negative odd integers: -1, $-3, \ldots, -(2n+1)$.

2.413

1.
$$\int \cosh^p x \sinh^{2n} x \, dx = \frac{\cosh^{p+1} x}{2n+p} \left[\sinh^{2n-1} x \right.$$

$$+ \sum_{k=1}^{n-1} (-1)^k \frac{(2n-1)(2n-3)\ldots(2n-2k+1) \sinh^{2n-2k-1} x}{(2n+p-2)(2n+p-4)\ldots(2n+p-2k)} \right]$$

$$+ (-1)^n \frac{(2n-1)!!}{(2n+p)(2n+p-2)\ldots(p+2)} \int \cosh^p x \, dx$$

This formula is applicable for arbitrary real p except for the following negative even integers: -2, $-4, \ldots, -2n$. If p is a natural number and $n=0$, we have

2.
$$\int \cosh^{2m} x \, dx = \binom{2m}{m} \frac{x}{2^{2m}} + \frac{1}{2^{2m-1}} \sum_{k=0}^{m-1} \binom{2m}{k} \frac{\sinh(2m-2k)x}{2m-2k} \qquad \text{TI (541)}$$

3. $\displaystyle \int \cosh^{2m+1} x \, dx = \frac{1}{2^{2m}} \sum_{k=0}^{m} \binom{2m+1}{k} \frac{\sinh(2m - 2k + 1)x}{2m - 2k + 1}$ TI (542)

$$= \sum_{k=0}^{m} \binom{m}{k} \frac{\sinh^{2k+1} x}{2k + 1}$$ GU (351) (8)

4. $\displaystyle \int \cosh^p x \sinh^{2n+1} x \, dx = \frac{\cosh^{p+1} x}{2n + p + 1} \left[\sinh^{2n} x \right.$

$$\left. + \sum_{k=1}^{n} (-1)^k \frac{2^k n(n - 1) \dots (n - k + 1) \sinh^{2n - 2k} x}{(2n + p - 1)(2n + p - 3) \dots (2n + p - 2k + 1)} \right]$$

This formula is applicable for arbitrary real p except for the following negative odd integers: -1, -3, $\dots$, $-(2n + 1)$.

2.414

1. $\displaystyle \int \sinh ax \, dx = \frac{1}{a} \cosh ax$

2. $\displaystyle \int \sinh^2 ax \, dx = \frac{1}{4a} \sinh 2ax - \frac{x}{2}$

3. $\displaystyle \int \sinh^3 x \, dx = -\frac{3}{4} \cosh x + \frac{1}{12} \cosh 3x = \frac{1}{3} \cosh^3 x - \cosh x$

4. $\displaystyle \int \sinh^4 x \, dx = \frac{3}{8} x - \frac{1}{4} \sinh 2x + \frac{1}{32} \sinh 4x = \frac{3}{8} x - \frac{3}{8} \sinh x \cosh x + \frac{1}{4} \sinh^3 x \cosh x$

5. $\displaystyle \int \sinh^5 x \, dx = \frac{5}{8} \cosh x - \frac{5}{48} \cosh 3x + \frac{1}{80} \cosh 5x$

$$= \frac{4}{5} \cosh x + \frac{1}{5} \sinh^4 x \cosh x - \frac{4}{15} \cosh^3 x$$

6. $\displaystyle \int \sinh^6 x \, dx = -\frac{5}{16} x + \frac{15}{64} \sinh 2x - \frac{3}{64} \sinh 4x + \frac{1}{192} \sinh 6x$

$$= -\frac{5}{16} x + \frac{1}{6} \sinh^5 x \cosh x - \frac{5}{24} \sinh^3 x \cosh x + \frac{5}{16} \sinh x \cosh x$$

7. $\displaystyle \int \sinh^7 x \, dx = -\frac{35}{64} \cosh x + \frac{7}{64} \cosh 3x - \frac{7}{320} \cosh 5x + \frac{1}{448} \cosh 7x$

$$= -\frac{24}{35} \cosh x + \frac{8}{35} \cosh^3 x - \frac{6}{35} \cosh x \sinh^4 x + \frac{1}{7} \cosh x \sinh^6 x$$

8. $\displaystyle \int \cosh ax \, dx = \frac{1}{a} \sinh ax$

9. $\displaystyle \int \cosh^2 ax \, dx = \frac{x}{2} + \frac{1}{4a} \sinh 2ax$

10. $\displaystyle \int \cosh^3 x \, dx = \frac{3}{4} \sinh x + \frac{1}{12} \sinh 3x = \sinh x + \frac{1}{3} \sinh^3 x$

11. $\displaystyle \int \cosh^4 x \, dx = \frac{3}{8} x + \frac{1}{4} \sinh 2x + \frac{1}{32} \sinh 4x = \frac{3}{8} x + \frac{3}{8} \sinh x \cosh x + \frac{1}{4} \sinh x \cosh^3 x$

12. $\int \cosh^5 x \, dx = \frac{5}{8} \sinh x + \frac{5}{48} \sinh 3x + \frac{1}{80} \sinh 5x$

$$= \frac{4}{5} \sinh x + \frac{1}{5} \cosh^4 x \sinh x + \frac{4}{15} \sinh^3 x$$

13. $\int \cosh^6 x \, dx = \frac{5}{16} x + \frac{15}{64} \sinh 2x + \frac{3}{64} \sinh 4x + \frac{1}{192} \sinh 6x$

$$= \frac{5}{16} x + \frac{5}{16} \sinh x \cosh x + \frac{5}{24} \sinh x \cosh^3 x + \frac{1}{6} \sinh x \cosh^5 x$$

14. $\int \cosh^7 x \, dx = \frac{35}{64} \sinh x + \frac{7}{64} \sinh 3x + \frac{7}{320} \sinh 5x + \frac{1}{448} \sinh 7x$

$$= \frac{24}{35} \sinh x + \frac{8}{35} \sinh^3 x + \frac{6}{35} \sinh x \cosh^4 x + \frac{1}{7} \sinh x \cosh^6 x$$

2.415

1. $\int \sinh ax \cosh bx \, dx = \dfrac{\cosh(a+b)x}{2(a+b)} + \dfrac{\cosh(a-b)x}{2(a-b)}$

2. $\int \sinh ax \cosh ax \, dx = \dfrac{1}{4a} \cosh 2ax$

3. $\int \sinh^2 x \cosh x \, dx = \frac{1}{3} \sinh^3 x$

4. $\int \sinh^3 x \cosh x \, dx = \frac{1}{4} \sinh^4 x$

5. $\int \sinh^4 x \cosh x \, dx = \frac{1}{5} \sinh^5 x$

6. $\int \sinh x \cosh^2 x \, dx = \frac{1}{3} \cosh^3 x$

7. $\int \sinh^2 x \cosh^2 x \, dx = -\dfrac{x}{8} + \dfrac{1}{32} \sinh 4x$

8. $\int \sinh^3 x \cosh^2 x \, dx = \frac{1}{5} \left(\sinh^2 x - \frac{2}{3} \right) \cosh^3 x$

9. $\int \sinh^4 x \cosh^2 x \, dx = \dfrac{x}{16} - \dfrac{1}{64} \sinh 2x - \dfrac{1}{64} \sinh 4x + \dfrac{1}{192} \sinh 6x$

10. $\int \sinh x \cosh^3 x \, dx = \frac{1}{4} \cosh^4 x$

11. $\int \sinh^2 x \cosh^3 x \, dx = \frac{1}{5} \left(\cosh^2 x + \frac{2}{3} \right) \sinh^3 x$

12. $\int \sinh^3 x \cosh^3 x \, dx = -\frac{3}{64} \cosh 2x + \frac{1}{192} \cosh 6x = \frac{1}{48} \cosh^3 2x - \frac{1}{16} \cosh 2x$

$$= \dfrac{\sinh^6 x}{6} + \dfrac{\sinh^4 x}{4} = \dfrac{\cosh^6 x}{6} - \dfrac{\cosh^4 x}{4}$$

13. $\int \sinh^4 x \cosh^3 x \, dx = \frac{1}{7} \sinh^3 x \left(\cosh^4 x - \frac{3}{5} \cosh^2 x - \frac{2}{5} \right) = \frac{1}{7} \left(\cosh^2 x + \frac{2}{5} \right) \sinh^5 x$

14. $\int \sinh x \cosh^4 x \, dx = \frac{1}{5} \cosh^5 x$

15. $\quad \displaystyle\int \sinh^2 x \cosh^4 x \, dx = -\frac{x}{16} - \frac{1}{64} \sinh 2x + \frac{1}{64} \sinh 4x + \frac{1}{192} \sinh 6x$

16. $\quad \displaystyle\int \sinh^3 x \cosh^4 x \, dx = \frac{1}{7} \cosh^3 x \left(\sinh^4 x + \frac{3}{5} \sinh^2 x - \frac{2}{5}\right) = \frac{1}{7}\left(\sinh^2 x - \frac{2}{5}\right) \cosh^5 x$

17. $\quad \displaystyle\int \sinh^4 x \cosh^4 x \, dx = \frac{3x}{128} - \frac{1}{128} \sinh 4x + \frac{1}{1024} \sinh 8x$

2.416

1.^{10} $\quad \displaystyle\int \frac{\sinh^p x}{\cosh^{2n} x} \, dx = \frac{\sinh^{p+1} x}{2n-1} \left[\operatorname{sech}^{2n-1} x \right.$

$$+ \sum_{k=1}^{n-1} \frac{(2n-p-2)(2n-p-4)\ldots(2n-p-2k)}{(2n-3)(2n-5)\ldots(2n-2k-1)} \sec h^{2n-2k-1} x \left. \right]$$

$$+ \frac{(2n-p-2)(2n-p-4)\ldots(-p+2)(-p)}{(2n-1)!!} \int \sinh^p x \, dx$$

This formula is applicable for arbitrary real p. For $\int \sinh^p x \, dx$, where p is a natural number, see **2.412** 2 and **2.412** 3. For $n = 0$ and p a negative integer, we have for this integral:

2. $\quad \displaystyle\int \frac{dx}{\sinh^{2m} x} = \frac{\cosh x}{2m-1} \left[-\operatorname{cosech}^{2m-1} x \right.$

$$+ \sum_{k=1}^{m-1} (-1)^{k-1} \cdot \frac{2^k (m-1)(m-2)\ldots(m-k)}{(2m-3)(2m-5)\ldots(2m-2k-1)} \operatorname{cosec} h^{2m-2k-1} x \left. \right]$$

3. $\quad \displaystyle\int \frac{dx}{\sinh^{2m+1} x} = \frac{\cosh x}{2m} \left[-\operatorname{cosech}^{2m} x \right.$

$$+ \sum_{k=1}^{m-1} (-1)^{k-1} \cdot \frac{(2m-1)(2m-3)\ldots(2m-2k+1)}{2^k(m-1)(m-2)\ldots(m-k)} \operatorname{cosec} h^{2m-2k} x \left. \right]$$

$$+ (-1)^m \frac{(2m-1)!!}{(2m)!!} \ln \tanh \frac{x}{2}$$

2.417

1. $\quad \displaystyle\int \frac{\sinh^p x}{\cosh^{2n+1} x} \, dx = \frac{\sinh^{p+1} x}{2n} \left[\operatorname{sech}^{2n} x \right.$

$$+ \sum_{k=1}^{n-1} \frac{(2n-p-1)(2n-p-3)\ldots(2n-p-2k+1)}{2^k(n-1)(n-2)\ldots(n-k)} \sec h^{2n-2k} x \left. \right]$$

$$+ \frac{(2n-p-1)(2n-p-3)\ldots(3-p)(1-p)}{2^n n!} \int \frac{\sinh^p x}{\cosh x} \, dx$$

This formula is applicable for arbitrary real p. For $n = 0$ and p integral, we have

2. $\int \dfrac{\sinh^{2m+1} x}{\cosh x}\, dx = \sum\limits_{k=1}^{m} \dfrac{(-1)^{m+k}}{2k}\sinh^{2k} x + (-1)^m \ln\cosh x$

$$= \sum\limits_{k=1}^{m} \dfrac{(-1)^{m+k}}{2k}\binom{m}{k}\cosh^{2k} x + (-1)^m \ln\cosh x \qquad [m \geq 1]$$

3. $\int \dfrac{\sinh^{2m} x}{\cosh x}\, dx = \sum\limits_{k=1}^{m} \dfrac{(-1)^{m+k}}{2k-1}\sinh^{2k-1} x + (-1)^m \arctan(\sinh x)$

$$[m \geq 1]$$

4. $\int \dfrac{dx}{\sinh^{2m+1} x \cosh x} = \sum\limits_{k=1}^{m} \dfrac{(-1)^k \operatorname{cosech}^{2m-2k+2} x}{2m-2k+2} + (-1)^m \ln\tanh x$

5. $\int \dfrac{dx}{\sinh^{2m} x \cosh x} = \sum\limits_{k=1}^{m} \dfrac{(-1)^k \operatorname{cosech}^{2m-2k+2} x}{2m-2k+1} + (-1)^m \arctan\sinh x$

2.418

1. $\int \dfrac{\cosh^p x}{\sinh^{2n} x}\, dx = -\dfrac{\cosh^{p+1} x}{2n-1}\left[\operatorname{cosech}^{2n-1} x\right.$

$$+ \sum\limits_{k=1}^{n-1} \dfrac{(-1)^k (2n-p-2)(2n-p-4)\ldots(2n-p-2k)}{(2n-3)(2n-5)\ldots(2n-2k-1)}\operatorname{cosech}^{2n-2k-1} x\Bigg]$$

$$+ \dfrac{(-1)^n (2n-p-2)(2n-p-4)\ldots(-p+2)(-p)}{(2n-1)!!}\int \cosh^p x\, dx$$

This formula is applicable for arbitrary real p. For the integral $\int \cosh^p x\, dx$, where p is a natural number, see **2.413** 2 and **2.413** 3. If p is a negative integer, we have for this integral:

2. $\int \dfrac{dx}{\cosh^{2m} x} = \dfrac{\sinh x}{2m-1}\left\{\operatorname{sech}^{2m-1} x + \sum\limits_{k=1}^{m-1} \dfrac{2^k (m-1)(m-2)\ldots(m-k)}{(2m-3)(2m-5)\ldots(2m-2k-1)}\operatorname{sech}^{2m-2k-1} x\right\}$

3. $\int \dfrac{dx}{\cosh^{2m+1} x} = \dfrac{\sinh x}{2m}\left\{\operatorname{sech}^{2m} x + \sum\limits_{k=1}^{m-1} \dfrac{(2m-1)(2m-3)\ldots(2m-2k+1)}{2^k (m-1)(m-2)\ldots(m-k)}\operatorname{sech}^{2m-2k} x\right\}$

$$+ \dfrac{(2m-1)!!}{(2m)!!}\arctan\sinh x$$

2.419

1. $\int \dfrac{\cosh^p x}{\sinh^{2n+1} x}\, dx = -\dfrac{\cosh^{p+1} x}{2n}\left[\operatorname{cosech}^{2n} x\right.$

$$+ \sum\limits_{k=1}^{n-1} \dfrac{(-1)^k (2n-p-1)(2n-p-3)\ldots(2n-p-2k+1)}{2^k (n-1)(n-2)\ldots(n-k)}\operatorname{cosech}^{2n-2k} x\Bigg]$$

$$+ \dfrac{(-1)^n (2n-p-1)(2n-p-3)\ldots(3-p)(1-p)}{2^n n!}\int \dfrac{\cosh^p x}{\sinh x}\, dx$$

This formula is applicable for arbitrary real p. For $n = 0$ and p an integer

2. $\displaystyle\int \frac{\cosh^{2m} x}{\sinh x}\,dx = \sum_{k=1}^{m} \frac{\cosh^{2k-1} x}{2k-1} + \ln\tanh\frac{x}{2}$

3. $\displaystyle\int \frac{\cosh^{2m+1} x}{\sinh x}\,dx = \sum_{k=1}^{m} \frac{\cosh^{2k} x}{2k} + \ln\sinh x$

$\displaystyle\qquad\qquad\qquad = \sum_{k=1}^{m} \binom{m}{k} \frac{\sinh^{2k} x}{2k} + \ln\sinh x$

4. $\displaystyle\int \frac{dx}{\sinh x \cosh^{2m} x} = \sum_{k=1}^{m} \frac{\operatorname{sech}^{2m-2k+1} x}{2m-2k+1} + \ln\tanh\frac{x}{2}$

5. $\displaystyle\int \frac{dx}{\sinh x \cosh^{2m+1} x} = \sum_{k=1}^{m} \frac{\operatorname{sech}^{2m-2k+2} x}{2m-2k+2} + \ln\tanh x$

2.421   In formulas **2.421** 1 and **2.421** 2, $s = 1$ for m odd and $m < 2n+1$; in all other cases, $s = 0$.

GI (351)(11, 13)

1.^{10} $\displaystyle\int \frac{\sinh^{2n+1} x}{\cosh^m x}\,dx = \sum_{\substack{k=0 \\ k\neq \frac{m-1}{2}}}^{n} (-1)^{n+k} \binom{n}{k} \frac{\cosh^{2k-m+1} x}{2k-m+1} + s(-1)^{n+\frac{m-1}{2}} \binom{n}{\frac{m-1}{2}} \ln\cosh x$

2. $\displaystyle\int \frac{\cosh^{2n+1} x}{\sinh^m x}\,dx = \sum_{\substack{k=0 \\ k\neq \frac{m-1}{2}}}^{n} \binom{n}{k} \frac{\sinh^{2k-m+1} x}{2k-m+1} + s\binom{n}{\frac{m-1}{2}} \ln\sinh x$

2.422

1. $\displaystyle\int \frac{dx}{\sinh^{2m} x \cosh^{2n} x} = \sum_{k=0}^{m+n-1} \frac{(-1)^{k+1}}{2m-2k-1} \binom{m+n-1}{k} \tanh^{2k-2m+1} x$

2. $\displaystyle\int \frac{dx}{\sinh^{2m+1} x \cosh^{2n+1} x} = \sum_{\substack{k=0 \\ k\neq m}}^{m+n} \frac{(-1)^{k+1}}{2m-2k} \binom{m+n}{k} \tanh^{2k-2m} x + (-1)^m \binom{m+n}{m} \ln\tanh x$

GI (351)(15)

2.423

1. $\displaystyle\int \frac{dx}{\sinh x} = \ln\tanh\frac{x}{2} = \frac{1}{2}\ln\frac{\cosh x - 1}{\cosh x + 1}$

2. $\displaystyle\int \frac{dx}{\sinh^2 x} = -\coth x$

3. $\displaystyle\int \frac{dx}{\sinh^3 x} = -\frac{\cosh x}{2\sinh^2 x} - \frac{1}{2}\ln\tanh\frac{x}{2}$

4. $\displaystyle\int \frac{dx}{\sinh^4 x} = -\frac{\cosh x}{3\sinh^3 x} + \frac{2}{3}\coth x = -\frac{1}{3}\coth^3 x + \coth x$

5. $\displaystyle\int \frac{dx}{\sinh^5 x} = -\frac{\cosh x}{4\sinh^4 x} + \frac{3}{8}\frac{\cosh x}{\sinh^2 x} + \frac{3}{8}\ln\tanh\frac{x}{2}$

6. $$\int \frac{dx}{\sinh^6 x} = -\frac{\cosh x}{5 \sinh^5 x} + \frac{4}{15} \coth^3 x - \frac{4}{5} \coth x$$
$$= -\frac{1}{5} \coth^5 x + \frac{2}{3} \coth^3 x - \coth x$$

7. $$\int \frac{dx}{\sinh^7 x} = -\frac{\cosh x}{6 \sinh^2 x} \left(\frac{1}{\sinh^4 x} - \frac{5}{4 \sinh^2 x} + \frac{15}{8} \right) - \frac{5}{16} \ln \tanh \frac{x}{2}$$

8. $$\int \frac{dx}{\sinh^8 x} = \coth x - \coth^3 x + \frac{3}{5} \coth^5 x - \frac{1}{7} \coth^7 x$$

9. $$\int \frac{dx}{\cosh x} = \arctan (\sinh x)$$
$$= \arcsin (\tanh x)$$
$$= 2 \arctan (e^x)$$
$$= \operatorname{gd} x$$

10. $$\int \frac{dx}{\cosh^2 x} = \tanh x$$

11. $$\int \frac{dx}{\cosh^3 x} = \frac{\sinh x}{2 \cosh^2 x} + \frac{1}{2} \arctan (\sinh x)$$

12. $$\int \frac{dx}{\cosh^4 x} = \frac{\sinh x}{3 \cosh^3 x} + \frac{2}{3} \tanh x$$
$$= -\frac{1}{3} \tanh^3 x + \tanh x$$

13. $$\int \frac{dx}{\cosh^5 x} = \frac{\sinh x}{4 \cosh^4 x} + \frac{3}{8} \frac{\sinh x}{\cosh^2 x} + \frac{3}{8} \arctan (\sinh x)$$

14. $$\int \frac{dx}{\cosh^6 x} = \frac{\sinh x}{5 \cosh^5 x} - \frac{4}{15} \tanh^3 x + \frac{4}{5} \tanh x$$
$$= \frac{1}{5} \tanh^5 x - \frac{2}{3} \tanh^3 x + \tanh x$$

15. $$\int \frac{dx}{\cosh^7 x} = \frac{\sinh x}{6 \cosh^2 x} \left(\frac{1}{\cosh^4 x} + \frac{5}{4 \cosh^2 x} + \frac{15}{8} \right) + \frac{5}{16} \arctan (\sinh x)$$

16. $$\int \frac{dx}{\cosh^8 x} = -\frac{1}{7} \tanh^7 x + \frac{3}{5} \tanh^5 x - \tanh^3 x + \tanh x$$

17. $$\int \frac{\sinh x}{\cosh x} dx = \ln \cosh x$$

18. $$\int \frac{\sinh^2 x}{\cosh x} dx = \sinh x - \arctan (\sinh x)$$

19. $$\int \frac{\sinh^3 x}{\cosh x} dx = \frac{1}{2} \sinh^2 x - \ln \cosh x$$
$$= \frac{1}{2} \cosh^2 x - \ln \cosh x$$

20. $$\int \frac{\sinh^4 x}{\cosh x} dx = \frac{1}{3} \sinh^3 x - \sinh x + \arctan (\sinh x)$$

21. $\displaystyle\int \frac{\sinh x}{\cosh^2 x}\, dx = -\frac{1}{\cosh x}$

22. $\displaystyle\int \frac{\sinh^2 x}{\cosh^2 x}\, dx = x - \tanh x$

23. $\displaystyle\int \frac{\sinh^3 x}{\cosh^2 x}\, dx = \cosh x + \frac{1}{\cosh x}$

24. $\displaystyle\int \frac{\sinh^4 x}{\cosh^2 x}\, dx = -\frac{3}{2}x + \frac{1}{4}\sinh 2x + \tanh x$

25. $\displaystyle\int \frac{\sinh x}{\cosh^3 x}\, dx = -\frac{1}{2\cosh^2 x}$
$$= \frac{1}{2}\tanh^2 x$$

26. $\displaystyle\int \frac{\sinh^2 x}{\cosh^3 x}\, dx = -\frac{\sinh x}{2\cosh^2 x} + \frac{1}{2}\arctan(\sinh x)$

27. $\displaystyle\int \frac{\sinh^3 x}{\cosh^3 x}\, dx = -\frac{1}{2}\tanh^2 x + \ln\cosh x$
$$= \frac{1}{2\cosh^2 x} + \ln\cosh x$$

28. $\displaystyle\int \frac{\sinh^4 x}{\cosh^3 x}\, dx = \frac{\sinh x}{2\cosh x} + \sinh x - \frac{3}{2}\arctan(\sinh x)$

29. $\displaystyle\int \frac{\sinh x}{\cosh^4 x}\, dx = -\frac{1}{3\cosh^3 x}$

30. $\displaystyle\int \frac{\sinh^2 x}{\cosh^4 x}\, dx = \frac{1}{3}\tanh^3 x$

31. $\displaystyle\int \frac{\sinh^3 x}{\cosh^4 x}\, dx = -\frac{1}{\cosh x} + \frac{1}{3\cosh^3 x}$

32. $\displaystyle\int \frac{\sinh^4 x}{\cosh^4 x}\, dx = -\frac{1}{3}\tanh^3 x - \tanh x + x$

33. $\displaystyle\int \frac{\cosh x}{\sinh x}\, dx = \ln\sinh x$

34. $\displaystyle\int \frac{\cosh^2 x}{\sinh x}\, dx = \cosh x + \ln\tanh\frac{x}{2}$

35. $\displaystyle\int \frac{\cosh^3 x}{\sinh x}\, dx = \frac{1}{2}\cosh^2 x + \ln\sinh x$

36. $\displaystyle\int \frac{\cosh^4 x}{\sinh x}\, dx = \frac{1}{3}\cosh^3 x + \cosh x + \ln\tanh\frac{x}{2}$

37. $\displaystyle\int \frac{\cosh x}{\sinh^2 x}\, dx = -\frac{1}{\sinh x}$

38. $\displaystyle\int \frac{\cosh^2 x}{\sinh^2 x}\, dx = x - \coth x$

39. $\displaystyle\int \frac{\cosh^3 x}{\sinh^2 x}\, dx = \sinh x - \frac{1}{\sinh x}$

40. $\displaystyle\int \frac{\cosh^4 x}{\sinh^2 x}\, dx = \frac{3}{2}x + \frac{1}{4}\sinh 2x - \coth x$

41. $\displaystyle\int \frac{\cosh x}{\sinh^3 x}\, dx = -\frac{1}{2\sinh^2 x}$
$$= -\frac{1}{2}\coth^2 x$$

42. $\displaystyle\int \frac{\cosh^2 x}{\sinh^3 x}\, dx = -\frac{\cosh x}{2\sinh^2 x} + \ln\tanh\frac{x}{2}$

43. $\displaystyle\int \frac{\cosh^3 x}{\sinh^3 x}\, dx = -\frac{1}{2\sinh^2 x} + \ln\sinh x$
$$= -\frac{1}{2}\coth^2 x + \ln\sinh x$$

44. $\displaystyle\int \frac{\cosh^4 x}{\sinh^3 x}\, dx = -\frac{\cosh x}{2\sinh^2 x} + \cosh x + \frac{3}{2}\ln\tanh\frac{x}{2}$

45. $\displaystyle\int \frac{\cosh x}{\sinh^4 x}\, dx = -\frac{1}{3\sinh^3 x}$

46. $\displaystyle\int \frac{\cosh^2 x}{\sinh^4 x}\, dx = -\frac{1}{3}\coth^3 x$

47. $\displaystyle\int \frac{\cosh^3 x}{\sinh^4 x}\, dx = -\frac{1}{\sinh x} - \frac{1}{3\sinh^3 x}$

48. $\displaystyle\int \frac{\cosh^4 x}{\sinh^4 x}\, dx = -\frac{1}{3}\coth^3 x - \coth x + x$

49. $\displaystyle\int \frac{dx}{\sinh x \cosh x} = \ln\tanh x$

50. $\displaystyle\int \frac{dx}{\sinh x \cosh^2 x} = \frac{1}{\cosh x} + \ln\tanh\frac{x}{2}$

51. $\displaystyle\int \frac{dx}{\sinh x \cosh^3 x} = \frac{1}{2\cosh^2 x} + \ln\tanh x$
$$= -\frac{1}{2}\tanh^2 x + \ln\tanh x$$

52. $\displaystyle\int \frac{dx}{\sinh x \cosh^4 x} = \frac{1}{\cosh x} + \frac{1}{3\cosh^3 x} + \ln\tanh\frac{x}{2}$

53. $\displaystyle\int \frac{dx}{\sinh^2 x \cosh x} = -\frac{1}{\sinh x} - \arctan\sinh x$

54. $\displaystyle\int \frac{dx}{\sinh^2 x \cosh^2 x} = -2\coth 2x$

55. $\displaystyle\int \frac{dx}{\sinh^2 x \cosh^3 x} = -\frac{\sinh x}{2\cosh^2 x} - \frac{1}{\sinh x} - \frac{3}{2}\arctan\sinh x$

56. $\displaystyle\int \frac{dx}{\sinh^2 x \cosh^4 x} = \frac{1}{3\sinh x \cosh^3 x} - \frac{8}{3}\coth 2x$

57. $\displaystyle\int \frac{dx}{\sinh^3 x \cosh x} = -\frac{1}{2\sinh^2 x} - \ln\tanh x$

$$= -\frac{1}{2}\coth^2 x + \ln\coth x$$

58. $\displaystyle\int \frac{dx}{\sinh^3 x \cosh^2 x} = -\frac{1}{\cosh x} - \frac{\cosh x}{2\sinh^2 x} - \frac{3}{2}\ln\tanh\frac{x}{2}$

59. $\displaystyle\int \frac{dx}{\sinh^3 x \cosh^3 x} = -\frac{2\cosh 2x}{\sinh^2 2x} - 2\ln\tanh x$

$$= \frac{1}{2}\tanh^2 x - \frac{1}{2}\coth^2 x - 2\ln\tanh x$$

60. $\displaystyle\int \frac{dx}{\sinh^3 x \cosh^4 x} = -\frac{2}{\cosh x} - \frac{1}{3\cosh^2 x} - \frac{\cosh x}{2\sinh^2 x} - \frac{5}{2}\ln\tanh\frac{x}{2}$

61. $\displaystyle\int \frac{dx}{\sinh^4 x \cosh x} = \frac{1}{\sinh x} - \frac{1}{3\sinh^3 x} + \arctan\sinh x$

62. $\displaystyle\int \frac{dx}{\sinh^4 x \cosh^2 x} = -\frac{1}{3\cosh x \sinh^3 x} + \frac{8}{3}\coth 2x$

63. $\displaystyle\int \frac{dx}{\sinh^4 x \cosh^3 x} = \frac{2}{\sinh x} - \frac{1}{3\sinh^3 x} + \frac{\sinh x}{2\cosh^2 x} + \frac{5}{2}\arctan\sinh x$

64. $\displaystyle\int \frac{dx}{\sinh^4 x \cosh^4 x} = 8\coth 2x - \frac{8}{3}\coth^3 2x$

2.424

1. $\displaystyle\int \tanh^p x\, dx = -\frac{\tanh^{p-1} x}{p-1} + \int \tanh^{p-2} x\, dx \qquad [p \neq 1]$

2. $\displaystyle\int \tanh^{2n+1} x\, dx = \sum_{k=1}^{n} \frac{(-1)^{k-1}}{2k}\binom{n}{k}\frac{1}{\cosh^{2k} x} + \ln\cosh x$

$$= -\sum_{k=1}^{n} \frac{\tanh^{2n-2k+2} x}{2n-2k+2} + \ln\cosh x$$

3. $\displaystyle\int \tanh^{2n} x\, dx = -\sum_{k=1}^{n} \frac{\tanh^{2n-2k+1} x}{2n-2k+1} + x \qquad\qquad\qquad\qquad$ GU (351)(12)

4. $\displaystyle\int \coth^p x\, dx = -\frac{\coth^{p-1} x}{p-1} + \int \coth^{p-2} x\, dx \qquad [p \neq 1]$

5. $\displaystyle\int \coth^{2n+1} x\, dx = -\sum_{k=1}^{n} \frac{1}{2n}\binom{n}{k}\frac{1}{\sinh^{2k} x} + \ln\sinh x$

$$= -\sum_{k=1}^{n} \frac{\coth^{2n-2k+2} x}{2n-2k+2} + \ln\sinh x$$

6. $\displaystyle\int \coth^{2n} x\, dx = -\sum_{k=1}^{n} \frac{\coth^{2n-2k+1} x}{2n-2k+1} + x \qquad\qquad\qquad\qquad$ GU (351)(14)

For formulas containing powers of $\tanh x$ and $\coth x$ equal to $n = 1, 2, 3, 4$, see **2.423** 17, **2.423** 22, **2.423** 27, **2.423** 32, **2.423** 33, **2.423** 38, **2.423** 43, **2.423** 48.

Powers of hyperbolic functions and hyperbolic functions of linear functions of the argument

2.425

1. $\displaystyle\int \sinh(ax+b)\sinh(cx+d)\,dx = \frac{1}{2(a+c)}\sinh[(a+c)x+b+d]$
$$-\frac{1}{2(a-c)}\sinh[(a-c)x+b-d]$$
$$[a^2 \neq c^2] \qquad\qquad \text{GU (352)(2a)}$$

2. $\displaystyle\int \sinh(ax+b)\cosh(cx+d)\,dx = \frac{1}{2(a+c)}\cosh[(a+c)x+b+d]$
$$+\frac{1}{2(a-c)}\cosh[(a-c)x+b-d]$$
$$[a^2 \neq c^2] \qquad\qquad \text{GU (352)(2c)}$$

3. $\displaystyle\int \cosh(ax+b)\cosh(cx+d)\,dx = \frac{1}{2(a+c)}\sinh[(a+c)x+b+d]$
$$+\frac{1}{2(a-c)}\sinh[(a-c)x+b-d]$$
$$[a^2 \neq c^2] \qquad\qquad \text{GU (352)(2b)}$$

When $a = c$:

4. $\displaystyle\int \sinh(ax+b)\sinh(ax+d)\,dx = -\frac{x}{2}\cosh(b-d) + \frac{1}{4a}\sinh(2ax+b+d)$ GU (352)(3a)

5. $\displaystyle\int \sinh(ax+b)\cosh(ax+d)\,dx = \frac{x}{2}\sinh(b-d) + \frac{1}{4a}\cosh(2ax+b+d)$ GU (352)(3c)

6. $\displaystyle\int \cosh(ax+b)\cosh(ax+d)\,dx = \frac{x}{2}\cosh(b-d) + \frac{1}{4a}\sinh(2ax+b+d)$ GU (352)(3b)

2.426

1. $\displaystyle\int \sinh ax \sinh bx \sinh cx\,dx = \frac{\cosh(a+b+c)x}{4(a+b+c)} - \frac{\cosh(-a+b+c)x}{4(-a+b+c)}$
$$-\frac{\cosh(a-b+c)x}{4(a-b+c)} - \frac{\cosh(a+b-c)x}{4(a+b-c)}$$
$$\text{GU (352)(4a)}$$

2. $\displaystyle\int \sinh ax \sinh bx \cosh cx\,dx = \frac{\sinh(a+b+c)x}{4(a+b+c)} - \frac{\sinh(-a+b+c)x}{4(-a+b+c)}$
$$-\frac{\sinh(a-b+c)x}{4(a-b+c)} + \frac{\sinh(a+b-c)x}{4(a+b-c)}$$
$$\text{GU (352)(4b)}$$

3. $\displaystyle\int \sinh ax \cosh bx \cosh cx\,dx = \frac{\cosh(a+b+c)x}{4(a+b+c)} - \frac{\cosh(-a+b+c)x}{4(-a+b+c)}$
$$+\frac{\cosh(a-b+c)x}{4(a-b+c)} + \frac{\cosh(a+b-c)x}{4(a+b-c)}$$
$$\text{GU (352)(4c)}$$

4. $\displaystyle\int \cosh ax \cosh bx \cosh cx \, dx = \frac{\sinh(a+b+c)x}{4(a+b+c)} + \frac{\sinh(-a+b+c)x}{4(-a+b+c)}$
$$+ \frac{\sinh(a-b+c)x}{4(a-b+c)} + \frac{\sinh(a+b-c)x}{4(a+b-c)}$$

GU (352)(4d)

2.427

1. $\displaystyle\int \sinh^p x \sinh ax \, dx = \frac{1}{p+a}\left\{\sinh px \cosh ax - p\int \sinh^{p-1} x \cosh(a-1)x \, dx\right\}$

2. $\displaystyle\int \sinh^p x \sinh(2n+1)x \, dx = \frac{\Gamma(p+1)}{\Gamma\left(\dfrac{p+3}{2}+n\right)}$

$$\times\left[\sum_{k=0}^{n-1} \frac{\Gamma\left(\frac{p+1}{2}+n-2k\right)}{2^{2k+1}\,\Gamma(p-2k+1)}\sinh^{p-2k} x \cosh(2n-2k+1)x\right.$$

$$\left. - \frac{\Gamma\left(\frac{p-1}{2}+n-2k\right)}{2^{2k+2}\,\Gamma(p-2k)}\sinh^{p-2k-1} x \sinh(2n-2k)x\right]$$

$$+ \frac{\Gamma\left(\frac{p+3}{2}-n\right)}{2^{2n}\,\Gamma(p+1-2n)}\int \sinh^{p-2n} x \sinh x \, dx$$

[p is not a negative integer]

3. $\displaystyle\int \sinh^p x \sinh 2nx \, dx = \frac{\Gamma(p+1)}{\Gamma\left(\dfrac{p}{2}+n+1\right)}$

$$\times\sum_{k=0}^{n-1}\left[\frac{\Gamma\left(\frac{p}{2}+n-2k\right)}{2^{2k+1}\,\Gamma(p-2k+1)}\sinh^{p-2k} x \cosh(2n-2k)x\right.$$

$$\left. - \frac{\Gamma\left(\frac{p}{2}+n-2k-1\right)}{2^{2k+2}\,\Gamma(p-2k)}\sinh^{p-2k-1} x \sinh(2n-2k-1)x\right]$$

[p is not a negative integer]   GU (352)(5)a

2.428

1. $\displaystyle\int \sinh^p x \cosh ax \, dx = \frac{1}{p+a}\left\{\sinh^p x \sinh ax - p\int \sinh^{p-1} x \sinh(a-1)x \, dx\right\}$

2. $\displaystyle \int \sinh^p x \cosh(2n+1)x\, dx = \frac{\Gamma(p+1)}{\Gamma\left(\dfrac{p+3}{2}+n\right)}$

$$\times \left\{ \left[\sum_{k=0}^{n-1} \frac{\Gamma\left(\frac{p+1}{2}+n-2k\right)}{2^{2k+1}\,\Gamma(p-2k+1)} \sinh^{p-2k} x \sinh(2n-2k+1)x \right.\right.$$

$$\left. -\frac{\Gamma\left(\frac{p-1}{2}+n-2k\right)}{2^{2k+2}\,\Gamma(p-2k)} \sinh^{p-2k-1} x \cosh(2n-2k)x \right]$$

$$\left. +\frac{\Gamma\left(\frac{p+3}{2}-n\right)}{2^{2n}\,\Gamma(p+1-2n)} \int \sinh^{p-2n} x \cosh x\, dx \right\}$$

[p is not a negative integer]

3. $\displaystyle \int \sinh^p x \cosh 2nx\, dx = \frac{\Gamma(p+1)}{\Gamma\left(\dfrac{p}{2}+n+1\right)}$

$$\times \left\{ \sum_{k=0}^{n-1} \left[\frac{\Gamma\left(\frac{p}{2}+n-2k\right)}{2^{2k+1}\,\Gamma(p-2k+1)} \sinh^{p-2k} x \sinh(2n-2k)x \right.\right.$$

$$\left. -\frac{\Gamma\left(\frac{p}{2}+n-2k-1\right)}{2^{2k+2}\,\Gamma(p-2k)} \sinh^{p-2k-1} x \cosh(2n-2k-1)x \right]$$

$$\left. +\frac{\Gamma\left(\frac{p}{2}-n+1\right)}{2^{2n}\,\Gamma(p+1-2n)} \int \sinh^{p-2n} x\, dx \right\}$$

[p is not a negative integer]   GU (352)(6)a

2.429

1. $\displaystyle \int \cosh^p x \sinh ax\, dx = \frac{1}{p+a}\left\{ \cosh^p x \cosh ax + p \int \cosh^{p-1} x \sinh(a-1)x\, dx \right\}$

2. $\displaystyle \int \cosh^p x \sinh(2n+1)x\, dx$

$$= \frac{\Gamma(p+1)}{\Gamma\left(\dfrac{p+3}{2}+n\right)} \left[\sum_{k=0}^{n-1} \frac{\Gamma\left(\frac{p+1}{2}+n-k\right)}{2^{k+1}\,\Gamma(p-k+1)} \cosh^{p-k} x \cosh(2n-k+1)x \right.$$

$$\left. +\frac{\Gamma\left(\frac{p+3}{2}\right)}{2^n\,\Gamma(p-n+1)} \int \cosh^{p-n} x \sinh(n+1)x\, dx \right]$$

[p is not a negative integer]

3. $\displaystyle \int \cosh^p x \sinh 2nx \, dx = \frac{\Gamma(p+1)}{\Gamma\left(\frac{p}{2}+n+1\right)} \left[\sum_{k=0}^{n-1} \frac{\Gamma\left(\frac{p}{2}+n-k\right)}{2^{k+1}\,\Gamma(p-k+1)} \cosh^{p-k} x \cosh(2n-k)x \right.$

$$\left. + \frac{\Gamma\left(\frac{p}{2}+1\right)}{2^n\,\Gamma(p-n+1)} \int \cosh^{p-n} x \sinh nx \, dx \right]$$

$$[p \text{ is not a negative integer}] \quad \text{GU (352)(7)a}$$

2.431

1. $\displaystyle \int \cosh^p x \cosh ax \, dx = \frac{1}{p+a} \left\{ \cosh^p x \sinh ax + p \int \cosh^{p-1} x \cosh(a-1)x \, dx \right\}$

2. $\displaystyle \int \cosh^p x \cosh(2n+1)x \, dx$

$$= \frac{\Gamma(p+1)}{\Gamma\left(\dfrac{p+3}{2}+n\right)} \left[\sum_{k=0}^{n-1} \frac{\Gamma\left(\frac{p+1}{2}+n-k\right)}{2^{k+1}\,\Gamma(p-k+1)} \cosh^{p-k} x \sinh(2n-k+1)x \right.$$

$$\left. + \frac{\Gamma\left(\frac{p+3}{2}\right)}{2^n\,\Gamma(p-n+1)} \int \cosh^{p-n} x \cosh(n+1)x \, dx \right]$$

$$[p \text{ is not a negative integer}]$$

3. $\displaystyle \int \cosh^p x \cosh 2nx \, dx = \frac{\Gamma(p+1)}{\Gamma\left(\frac{p}{2}+n+1\right)} \left[\sum_{k=0}^{n-1} \frac{\Gamma\left(\frac{p}{2}+n-k\right)}{2^{k+1}\,\Gamma(p-k+1)} \cosh^{p-k} x \sinh(2n-k)x \right.$

$$\left. + \frac{\Gamma\left(\frac{p}{2}+1\right)}{2^n\,\Gamma(p-n+1)} \cosh^{p-n} x \cosh nx \, dx \right]$$

$$[p \text{ is not a negative integer}] \quad \text{GU (352)(8)a}$$

2.432

1. $\displaystyle \int \sinh(n+1)x \sinh^{n-1} x \, dx = \frac{1}{n} \sinh^n x \sinh nx$

2. $\displaystyle \int \sinh(n+1)x \cosh^{n-1} x \, dx = \frac{1}{n} \cosh^n x \cosh nx$

3. $\displaystyle \int \cosh(n+1)x \sinh^{n-1} x \, dx = \frac{1}{n} \sinh^n x \cosh nx$

4. $\displaystyle \int \cosh(n+1)x \cosh^{n-1} x \, dx = \frac{1}{n} \cosh^n x \sinh nx$

2.433

1. $\displaystyle \int \frac{\sinh(2n+1)x}{\sinh x} \, dx = 2 \sum_{k=0}^{n-1} \frac{\sinh(2n-2k)x}{2n-2k} + x$

2. $\displaystyle \int \frac{\sinh 2nx}{\sinh x} \, dx = 2 \sum_{k=0}^{n-1} \frac{\sinh(2n-2k-1)x}{2n-2k-1}$ GU (352)(5d)

3. $\displaystyle\int \frac{\cosh(2n+1)x}{\sinh x}\,dx = 2\sum_{k=0}^{n-1}\frac{\cosh(2n-2k)x}{2n-2k} + \ln\sinh x$

4. $\displaystyle\int \frac{\cosh 2nx}{\sinh x}\,dx = 2\sum_{k=0}^{n-1}\frac{\cosh(2n-2k-1)x}{2n-2k-1} + \ln\tanh\frac{x}{2}$ GU (352)(6d)

5. $\displaystyle\int \frac{\sinh(2n+1)x}{\cosh x}\,dx = 2\sum_{k=0}^{n-1}(-1)^k\frac{\cosh(2n-2k)x}{2n-2k} + (-1)^n\ln\cosh x$

6. $\displaystyle\int \frac{\sinh 2nx}{\cosh x}\,dx = 2\sum_{k=0}^{n-1}(-1)^k\frac{\cosh(2n-2k-1)x}{2n-2k-1}$ GU (352)(7d)

7. $\displaystyle\int \frac{\cosh(2n+1)x}{\cosh x}\,dx = 2\sum_{k=0}^{n-1}(-1)^k\frac{\sinh(2n-2k)x}{2n-2k} + (-1)^n x$

8. $\displaystyle\int \frac{\cosh 2nx}{\cosh x}\,dx = 2\sum_{k=0}^{n-1}(-1)^k\frac{\sinh(2n-2k-1)x}{2n-2k-1} + (-1)^n\arcsin(\tanh x)$ GU (352)(8d)

9. $\displaystyle\int \frac{\sinh 2x}{\sinh^n x}\,dx = -\frac{2}{(n-2)\sinh^{n-2} x}$

For $n=2$:

10. $\displaystyle\int \frac{\sinh 2x}{\sinh^2 x}\,dx = 2\ln\sinh x$

11. $\displaystyle\int \frac{\sinh 2x\,dx}{\cosh^n x} = \frac{2}{(2-n)\cosh^{n-2} x}$

For $n=2$:

12. $\displaystyle\int \frac{\sinh 2x}{\cosh^2 x}\,dx = 2\ln\cosh x$

13. $\displaystyle\int \frac{\cosh 2x}{\sinh x}\,dx = 2\cosh x + \ln\tanh\frac{x}{2}$

14. $\displaystyle\int \frac{\cosh 2x}{\sinh^2 x}\,dx = -\coth x + 2x$

15. $\displaystyle\int \frac{\cosh 2x}{\sinh^3 x}\,dx = -\frac{\cosh x}{2\sinh^2 x} + \frac{3}{2}\ln\tanh\frac{x}{2}$

16. $\displaystyle\int \frac{\cosh 2x}{\cosh x}\,dx = 2\sinh x - \arcsin(\tanh x)$

17. $\displaystyle\int \frac{\cosh 2x}{\cosh^2 x}\,dx = -\tanh x + 2x$

18. $\displaystyle\int \frac{\cosh 2x}{\cosh^3 x}\,dx = -\frac{\sinh x}{2\cosh^2 x} + \frac{3}{2}\arcsin(\tanh x)$

19. $\displaystyle\int \frac{\sinh 3x}{\sinh x}\,dx = x + \sinh 2x$

20. $\displaystyle\int \frac{\sinh 3x}{\sinh^2 x}\,dx = 3\ln\tanh\frac{x}{2} + 4\cosh x$

21. $\displaystyle\int \frac{\sinh 3x}{\sinh^3 x}\, dx = -3\coth x + 4x$

22. $\displaystyle\int \frac{\sinh 3x}{\cosh^n x}\, dx = \frac{4}{(3-n)\cosh^{n-3} x} - \frac{1}{(1-n)\cosh^{n-1} x}$

For $n = 1$ and $n = 3$:

23. $\displaystyle\int \frac{\sinh 3x}{\cosh x}\, dx = 2\sinh^2 x - \ln\cosh x$

24. $\displaystyle\int \frac{\sinh 3x}{\cosh^3 x}\, dx = \frac{1}{2\cosh^2 x} + 4\ln\cosh x$

25. $\displaystyle\int \frac{\cosh 3x}{\sinh^n x}\, dx = \frac{4}{(3-n)\sinh^{n-3} x} + \frac{1}{(1-n)\sinh^{n-1} x}$

For $n = 1$ and $n = 3$:

26. $\displaystyle\int \frac{\cosh 3x}{\sinh x}\, dx = 2\sinh^2 x + \ln\sinh x$

27. $\displaystyle\int \frac{\cosh 3x}{\sinh^3 x}\, dx = -\frac{1}{2\sinh^2 x} + 4\ln\sinh x$

28. $\displaystyle\int \frac{\cosh 3x}{\cosh x}\, dx = \sinh 2x - x$

29. $\displaystyle\int \frac{\cosh 3x}{\cosh^2 x}\, dx = 4\sinh x - 3\arcsin(\tanh x)$

30. $\displaystyle\int \frac{\cosh 3x}{\cosh^3 x}\, dx = 4x - 3\tanh x$

2.44–2.45 Rational functions of hyperbolic functions

2.441

1. $\displaystyle\int \frac{A + B\sinh x}{(a + b\sinh x)^n}\, dx = \frac{aB - bA}{(n-1)\,(a^2 + b^2)} \cdot \frac{\cosh x}{(a + b\sinh x)^{n-1}}$
$$+ \frac{1}{(n-1)\,(a^2 + b^2)}\int \frac{(n-1)(aA + bB) + (n-2)(aB - bA)\sinh x}{(a + b\sinh x)^{n-1}}\, dx$$

For $n = 1$:

2. $\displaystyle\int \frac{A + B\sinh x}{a + b\sinh x}\, dx = \frac{B}{b}x - \frac{aB - bA}{b}\int \frac{dx}{a + b\sinh x}$ (see **2.441 3**)

3. $\displaystyle\int \frac{dx}{a + b\sinh x} = \frac{1}{\sqrt{a^2 + b^2}}\ln \frac{a\tanh \frac{x}{2} - b + \sqrt{a^2 + b^2}}{a\tanh \frac{x}{2} - b - \sqrt{a^2 + b^2}}$
$$= \frac{2}{\sqrt{a^2 + b^2}}\,\operatorname{arctanh}\frac{a\tanh \frac{x}{2} - b}{\sqrt{a^2 + b^2}}$$

2.442

1. $\displaystyle\int \frac{A + B\cosh x}{(a + b\sinh x)^n}\, dx = -\frac{B}{(n-1)b(a + b\sinh x)^{n-1}} + A\int \frac{dx}{(a + b\sinh x)^n}$

For $n = 1$:

2.
$$\int \frac{A + B \cosh x}{a + b \sinh x} \, dx = \frac{B}{b} \ln (a + b \sinh x) + A \int \frac{dx}{a + b \sinh x}$$

(see 2.441 3)

2.443

1.
$$\int \frac{A + B \cosh x}{(a + b \cosh x)^n} \, dx = \frac{aB - bA}{(n-1)(a^2 - b^2)} \cdot \frac{\sinh x}{(a + b \cosh x)^{n-1}}$$
$$+ \frac{1}{(n-1)(a^2 - b^2)} \int \frac{(n-1)(aA - bB) + (n-2)(aB - bA) \cosh x}{(a + b \cosh x)^{n-1}} \, dx$$

For $n = 1$:

2.
$$\int \frac{A + B \cosh x}{a + b \cosh x} \, dx = \frac{B}{b} x - \frac{aB - bA}{b} \int \frac{dx}{a + b \cosh x} \qquad \text{(see } \mathbf{2.443} \text{ 3)}$$

3.
$$\int \frac{dx}{a + b \cosh x} = \frac{1}{\sqrt{b^2 - a^2}} \arcsin \frac{b + a \cosh x}{a + b \cosh x} \qquad [b^2 > a^2, \quad x < 0]$$
$$= -\frac{1}{\sqrt{b^2 - a^2}} \arcsin \frac{b + a \cosh x}{a + b \cosh x} \qquad [b^2 > a^2, \quad x > 0]$$
$$= \frac{1}{\sqrt{a^2 - b^2}} \ln \frac{a + b + \sqrt{a^2 - b^2} \tanh \dfrac{x}{2}}{a + b - \sqrt{a^2 - b^2} \tanh \dfrac{x}{2}} \qquad [a^2 > b^2]$$

2.444

1.
$$\int \frac{dx}{\cosh a + \cosh x} = \operatorname{cosech} a \left[\ln \cosh \frac{x + a}{2} - \ln \cosh \frac{x - a}{2} \right]$$
$$= 2 \operatorname{cosech} a \operatorname{arctanh} \left(\tanh \frac{x}{2} \tanh \frac{a}{2} \right)$$

2.
$$\int \frac{dx}{\cos a + \cosh x} = 2 \cos a \arctan \left(\tanh \frac{x}{2} \tan \frac{a}{2} \right)$$

2.445

1.
$$\int \frac{B \sinh x}{(a + b \cosh x)^n} \, dx = -\frac{B}{(n-1)b(a + b \cosh x)^{n-1}} \qquad [n \neq 1]$$

For $n = 1$:

2.
$$\int \frac{B \sinh x}{a + b \cosh x} \, dx = \frac{B}{b} \ln (a + b \cosh x) \qquad \text{(see } \mathbf{2.443} \text{ 3)}$$

In evaluating definite integrals by use of formulas **2.441–2.443** and **2.445**, one may not take the integral over points at which the integrand becomes infinite, that is, over the points
$$x = \operatorname{arcsinh} \left(-\frac{a}{b} \right)$$
in formulas **2.441** or **2.442** or over the points
$$x = \operatorname{arccosh} \left(-\frac{a}{b} \right)$$
in formulas **2.443** or **2.445**. Formulas **2.443** are not applicable for $a^2 = b^2$. Instead, we may use the following formulas in these cases:

2.446

1.
$$\int \frac{A + B \cosh x}{(\varepsilon + \cosh x)^n}\, dx$$

$$= \frac{B \sinh x}{(1-n)(\varepsilon + \cosh x)^n} + \left(\varepsilon A + \frac{n}{n-1}B\right)\frac{(n-1)!}{(2n-1)!!}\sinh x \sum_{k=0}^{n-1}\frac{(2n-2k-3)!!}{(n-k-1)!}$$

$$\times \frac{\varepsilon^h}{(\varepsilon + \cosh x)^{n-k}}$$

$$[\varepsilon = \pm 1, \quad n > 1]$$

For $n = 1$:

2.
$$\int \frac{A + B \cosh x}{\varepsilon + \cosh x}\, dx = Bx + (\varepsilon A - B)\frac{\cosh x - \varepsilon}{\sinh x} \qquad [\varepsilon = \pm 1]$$

2.447

1.
$$\int \frac{\sinh x\, dx}{a \cosh x + b \sinh x} = \frac{a \ln \cosh\left(x + \operatorname{arctanh}\dfrac{b}{a}\right) bx}{a^2 - b^2} \qquad [a > |b|]$$

$$= \frac{bx - a \ln \sinh\left(x + \operatorname{arctanh}\dfrac{a}{b}\right)}{b^2 - a^2} \qquad [b > |a|] \qquad\qquad \text{MZ 215}$$

For $a = b = 1$:

2.
$$\int \frac{\sinh x\, dx}{\cosh x + \sinh x} = \frac{x}{2} + \frac{1}{4}e^{-2x}$$

For $a = -b = 1$:

3.
$$\int \frac{\sinh x\, dx}{\cosh x - \sinh x} = -\frac{x}{2} + \frac{1}{4}e^{2x} \qquad\qquad\qquad \text{MZ 215}$$

2.448

1.
$$\int \frac{\cosh x\, dx}{a \cosh x + b \sinh x} = \frac{ax - b \ln \cosh\left(x + \operatorname{arctanh}\frac{b}{a}\right)}{a^2 - b^2} \qquad [a > |b|]$$

$$= \frac{-ax + b \ln \sinh\left(x + \operatorname{arctanh}\frac{a}{b}\right)}{b^2 - a^2} \qquad [b > |a|]$$

For $a = b = 1$:

2.
$$\int \frac{\cosh x\, dx}{\cosh x + \sinh x} = \frac{x}{2} - \frac{1}{4}e^{-2x}$$

For $a = -b = 1$:

3.
$$\int \frac{\cosh x\, dx}{\cosh x - \sinh x} = \frac{x}{2} + \frac{1}{4}e^{2x} \qquad\qquad\qquad \text{MZ 214, 215}$$

2.449

1.[6]
$$\int \frac{dx}{(a \cosh x + b \sinh x)^n} = \frac{1}{\sqrt{(a^2 - b^2)^n}}\int \frac{dx}{\sinh^n\left(x + \operatorname{arctanh}\dfrac{b}{a}\right)} \qquad [a > |b|]$$

$$= \frac{1}{\sqrt{(b^2 - a^2)^n}}\int \frac{dx}{\cosh^n\left(x + \operatorname{arctanh}\dfrac{a}{b}\right)} \qquad [b > |a|]$$

For $n = 1$:

2.
$$\int \frac{dx}{a \cosh x + b \sinh x} = \frac{1}{\sqrt{a^2 - b^2}} \arctan \left| \sinh \left(x + \operatorname{arctanh} \frac{b}{a} \right) \right| \qquad [a > |b|]$$
$$= \frac{1}{\sqrt{b^2 - a^2}} \ln \left| \tanh \frac{x + \operatorname{arctanh} \frac{a}{b}}{2} \right| \qquad [b > |a|]$$

For $a = b = 1$:

3.
$$\int \frac{ax}{\cosh x + \sinh x} = -e^{-x} = \sinh x - \cosh x$$

For $a = -b = 1$:

4.
$$\int \frac{dx}{\cosh x - \sinh x} = e^x = \sinh x + \cosh x \qquad\qquad \text{MZ 214}$$

2.451

1.
$$\int \frac{A + B \cosh x + C \sinh x}{(a + b \cosh x + c \sinh x)^n} \, dx$$
$$= \frac{Bc - Cb + (Ac - Ca) \cosh x + (Ab - Ba) \sinh x}{(1 - n)(a^2 - b^2 + c^2)(a + b \cosh x + c \sinh x)^{n-1}} + \frac{1}{(n-1)(a^2 - b^2 + c^2)}$$
$$\times \int \frac{(n-1)(Aa - Bb + Cc) - (n-2)(Ab - Ba) \cosh x - (n-2)(Ac - Ca) \sinh x}{(a + b \cosh x + c \sinh x)^{n-1}} \, dx$$
$$\left[a^2 + c^2 \neq b^2 \right]$$
$$= \frac{Bc - Cb - Ca \cosh x - Ba \sinh x}{(n-1)a(a + b \cosh x + c \sinh x)^n} + \left[\frac{A}{a} + \frac{n(Bb - Cc)}{(n-1)a^2} \right] (c \cosh x + b \sinh x) \frac{(n-1)!}{(2n-1)!!}$$
$$\times \sum_{k=0}^{n-1} \frac{(2n - 2k - 3)!!}{(n - k - 1)! a^k} \frac{1}{(a + b \cosh x + c \sinh x)^{n-k}}$$
$$\left[a^2 + c^2 = b^2 \right]$$

2.
$$\int \frac{A + B \cosh x + C \sinh x}{a + b \cosh x + c \sinh x} \, dx = \frac{Cb - Bc}{b^2 - c^2} \ln (a + b \cosh x + c \sinh x)$$
$$+ \frac{Bb - Cc}{b^2 - c^2} x + \left(A - a \frac{Bb - Cc}{b^2 - c^2} \right) \int \frac{dx}{a + b \cosh x + c \sinh x}$$
$$\left[b^2 \neq c^2 \right] \qquad (\text{see } \mathbf{2.451}\ 4)$$

3.
$$\int \frac{A + B \cosh x + C \sinh x}{a + b \cosh x \pm b \sinh x} \, dx = \frac{C \mp B}{2a} (\cosh x \mp \sinh x) + \left[\frac{A}{a} - \frac{(B \mp C) b}{2a^2} \right] x$$
$$+ \left[\frac{C \pm B}{2b} \pm \frac{A}{a} - \frac{(C \mp B) b}{2a^2} \right] \ln (a + b \cosh x \pm b \sinh x)$$
$$[ab \neq 0]$$

4. $\int \dfrac{dx}{a + b\cosh x + c\sinh x}$

$$= \frac{2}{\sqrt{b^2 - a^2 - c^2}}\arctan\frac{(b-a)\tanh\dfrac{x}{2} + c}{\sqrt{b^2 - a^2 - c^2}} \qquad [b^2 > a^2 + c^2 \text{ and } a \neq b]$$

$$= \frac{1}{\sqrt{a^2 - b^2 + c^2}}\ln\frac{(a-b)\tanh\frac{x}{2} - c + \sqrt{a^2 - b^2 + c^2}}{(a-b)\tanh\dfrac{x}{2} - c - \sqrt{a^2 - b^2 + c^2}} \qquad [b^2 < a^2 + c^2 \text{ and } a \neq b]$$

$$= \frac{1}{c}\ln\left(a + c\tanh\frac{x}{2}\right) \qquad [a = b \text{ and } c \neq 0]$$

$$= \frac{2}{(a-b)\tanh\dfrac{x}{2} + c} \qquad [b^2 = a^2 + c^2]$$

GU (351)(18)

2.452

1. $\int \dfrac{A + B\cosh x + C\sinh x}{(a_1 + b_1\cosh x + c_1\sinh x)(a_2 + b_2\cosh x + c_2\sinh x)}\,dx$

$$= A_0\ln\frac{a_1 + b_1\cosh x + c_1\sinh x}{a_2 + b_2\cosh x + c_2\sinh x} + A_1\int\frac{dx}{a_1 + b_1\cosh x + c_1\sinh x} + A_2\int\frac{dx}{a_2 + b_2\cosh x + c_2\sinh x}$$

where

GU (351)(19)

$$A_0 = \frac{\begin{vmatrix} a_1 & b_1 & c_1 \\ A & B & C \\ a_2 & b_2 & c_2 \end{vmatrix}}{\begin{vmatrix} a_1 & b_1 \\ a_2 & b_2 \end{vmatrix}^2 + \begin{vmatrix} b_1 & c_1 \\ b_2 & c_2 \end{vmatrix}^2 - \begin{vmatrix} c_1 & a_1 \\ c_2 & a_2 \end{vmatrix}^2}$$

$$A_1 = \frac{\begin{vmatrix} a_1 & b_1 & c_1 \\ \begin{vmatrix} b_1 & c_1 \\ B & C \end{vmatrix} & \begin{vmatrix} c_1 & a_1 \\ C & A \end{vmatrix} & \begin{vmatrix} a_1 & b_1 \\ A & B \end{vmatrix} \\ a_2 & b_2 & c_2 \end{vmatrix}}{\begin{vmatrix} a_1 & b_1 \\ a_2 & b_2 \end{vmatrix}^2 + \begin{vmatrix} b_1 & c_1 \\ b_2 & c_2 \end{vmatrix}^2 - \begin{vmatrix} c_1 & a_1 \\ c_2 & a_2 \end{vmatrix}^2},$$

$$A_2 = \frac{\begin{vmatrix} a_1 & b_1 & c_1 \\ \begin{vmatrix} C & B \\ c_2 & b_2 \end{vmatrix} & \begin{vmatrix} C & A \\ c_2 & a_2 \end{vmatrix} & \begin{vmatrix} B & A \\ b_2 & a_2 \end{vmatrix} \\ a_2 & b_2 & c_2 \end{vmatrix}}{\begin{vmatrix} a_1 & b_1 \\ a_2 & b_2 \end{vmatrix}^2 + \begin{vmatrix} b_1 & c_1 \\ b_2 & c_2 \end{vmatrix}^2 - \begin{vmatrix} c_1 & a_1 \\ c_2 & a_2 \end{vmatrix}^2},$$

$$\left[\begin{vmatrix} a_1 & b_1 \\ a_2 & b_2 \end{vmatrix}^2 + \begin{vmatrix} b_1 & c_1 \\ b_2 & c_2 \end{vmatrix}^2 \neq \begin{vmatrix} c_1 & a_1 \\ c_2 & a_2 \end{vmatrix}^2\right].$$

2. $\int \dfrac{A\cosh^2 x + 2B\sinh x\cosh x + C\sinh^2 x}{a\cosh^2 x + 2b\sinh x\cosh x + c\sinh^2 x}\,dx$

$$= \frac{1}{4b^2 - (a+c)^2}\left\{[4Bb - (A+C)(a+c)]x\right.$$

$$+ [(A+C)b - B(a+c)]\ln\left(a\cosh^2 x + 2b\sinh x\cosh x + c\sinh^2 x\right)$$

$$\left. + \left[2(A-C)b^2 - 2Bb(a-c) + (Ca - Ac)(a+c)\right]f(x)\right\}$$

where

$$f(x) = \frac{1}{2\sqrt{b^2 - ac}} \ln \frac{c \tanh x + b - \sqrt{b^2 - ac}}{c \tanh x + b + \sqrt{b^2 - ac}} \qquad [b^2 > ac]$$

$$= \frac{1}{\sqrt{ac - b^2}} \arctan \frac{c \tanh x + b}{\sqrt{ac - b^2}} \qquad [b^2 < ac]$$

$$= -\frac{1}{c \tanh x + b} \qquad [b^2 = ac]$$

2.453

1. $$\int \frac{(A + B \sinh x)\, dx}{\sinh x\, (a + b \sinh x)} = \frac{1}{a} \left[A \ln \left| \tanh \frac{x}{2} \right| + (aB - bA) \int \frac{dx}{a + b \sinh x} \right]$$

(see **2.441** 3)

2. $$\int \frac{(A + B \sinh x)\, dx}{\sinh x\, (a + b \cosh x)} = \frac{A}{a^2 - b^2} \left(a \ln \left| \tanh \frac{x}{2} \right| + b \ln \left| \frac{a + b \cosh x}{\sinh x} \right| \right) + B \int \frac{dx}{a + b \cosh x}$$

(see **2.443** 3)

For $a^2 = b^2 = 1$:

3. $$\int \frac{(A + B \sinh x)\, dx}{\sinh x\, (1 + \cosh x)} = \frac{A}{2} \left(\ln \left| \tanh \frac{x}{2} \right| - \frac{1}{2} \tanh^2 \frac{x}{2} \right) + B \tanh \frac{x}{2}$$

4. $$\int \frac{(A + B \sinh x)\, dx}{\sinh x\, (1 - \cosh x)} = \frac{A}{2} \left(-\ln \left| \coth \frac{x}{2} \right| + \frac{1}{2} \coth^2 \frac{x}{2} \right) + B \coth \frac{x}{2}$$

2.454

1. $$\int \frac{(A + B \sinh x)\, dx}{\cosh x\, (a + b \sinh x)} = \frac{1}{a^2 + b^2} \left[(Aa + Bb) \arctan (\sinh x) + (Ab - Ba) \ln \left| \frac{a + b \sinh x}{\cosh x} \right| \right]$$

2. $$\int \frac{(A + B \cosh x)\, dx}{\sinh x\, (a + b \sinh x)} = \frac{1}{a} \left(A \ln \left| \tanh \frac{x}{2} \right| + B \ln \left| \frac{\sinh x}{a + b \sinh x} \right| - Ab \int \frac{dx}{a + b \sinh x} \right)$$

(see **2.441** 3)

2.455

1. $$\int \frac{(A + B \cosh x)\, dx}{\sinh x\, (a + b \cosh x)} = \frac{1}{a^2 - b^2} \left[(Aa + Bb) \ln \left| \tanh \frac{x}{2} \right| + (Ab - Ba) \ln \left| \frac{a + b \cosh x}{\sinh x} \right| \right]$$

For $a^2 = b^2 = 1$:

2. $$\int \frac{(A + B \cosh x)\, dx}{\sinh x\, (1 + \cosh x)} = \frac{A + B}{2} \ln \left| \tanh \frac{x}{2} \right| - \frac{A - B}{4} \tanh^2 \frac{x}{2}$$

3. $$\int \frac{(A + B \cosh x)\, dx}{\sinh x\, (1 - \cosh x)} = \frac{A + B}{4} \coth^2 \frac{x}{2} - \frac{A - B}{2} \ln \coth \frac{x}{2}$$

2.456 $$\int \frac{(A + B \cosh x)\, dx}{\cosh x\, (a + b \sinh x)} = \frac{A}{a^2 + b^2} \left[a \arctan (\sinh x) + b \ln \left| \frac{a + b \sinh x}{\cosh x} \right| \right] + B \int \frac{dx}{a + b \sinh x}$$

(see **2.441** 3)

2.457

1.
$$\int \frac{(A + B \cosh x)\, dx}{\cosh x\,(a + b \cosh x)} = \frac{1}{a}\left[A \arctan \sinh x - (Ab - Ba)\int \frac{dx}{a + b \cosh x}\right]$$

(see **2.443** 3)

2.458

1.
$$\int \frac{dx}{a + b \sinh^2 x}$$

$$= \frac{1}{\sqrt{a(b-a)}} \arctan\left(\sqrt{\frac{b}{a} - 1}\, \tanh x\right) \qquad \left[\frac{b}{a} > 1\right]$$

$$= \frac{1}{\sqrt{a(a-b)}} \operatorname{arctanh}\left(\sqrt{1 - \frac{b}{a}}\, \tanh x\right) \qquad \left[0 < \frac{b}{a} < 1 \text{ or } \frac{b}{a} < 0 \text{ and } \sinh^2 x < -\frac{a}{b}\right]$$

$$= \frac{1}{\sqrt{a(a-b)}} \operatorname{arccoth}\left(\sqrt{1 - \frac{b}{a}}\, \tanh x\right) \qquad \left[\frac{b}{a} < 0 \text{ and } \sinh^2 x > -\frac{a}{b}\right]$$

MZ 195

2.
$$\int \frac{dx}{a + b \cosh^2 x}$$

$$= \frac{1}{\sqrt{-a(a+b)}} \arctan\left(\sqrt{-\left(1 + \frac{b}{a}\right)}\, \coth x\right) \qquad \left[\frac{b}{a} < -1\right]$$

$$= \frac{1}{\sqrt{a(a+b)}} \operatorname{arctanh}\left(\sqrt{1 + \frac{b}{a}}\, \coth x\right) \qquad \left[-1 < \frac{b}{a} < 0 \text{ and } \cosh^2 x > -\frac{a}{b}\right]$$

$$= \frac{1}{\sqrt{a(a+b)}} \operatorname{arccoth}\left(\sqrt{1 + \frac{b}{a}}\, \coth x\right) \qquad \left[\frac{b}{a} > 0 \text{ or } -1 < \frac{b}{a} < 0 \text{ and } \cosh^2 x < -\frac{a}{b}\right]$$

MZ 202

For $a^2 = b^2 = 1$:

3.
$$\int \frac{dx}{1 + \sinh^2 x} = \tanh x$$

4.
$$\int \frac{dx}{1 - \sinh^2 x} = \frac{1}{\sqrt{2}} \operatorname{arctanh}\left(\sqrt{2}\, \tanh x\right) \qquad\qquad [\sinh^2 x < 1]$$
$$= \frac{1}{\sqrt{2}} \operatorname{arccoth}\left(\sqrt{2}\, \tanh x\right) \qquad\qquad [\sinh^2 x > 1]$$

5.
$$\int \frac{dx}{1 + \cosh^2 x} = \frac{1}{\sqrt{2}} \operatorname{arccoth}\left(\sqrt{2}\, \coth x\right)$$

6.
$$\int \frac{dx}{1 - \cosh^2 x} = \coth x$$

2.459

1.
$$\int \frac{dx}{\left(a + b \sinh^2 x\right)^2} = \frac{1}{2a(b-a)}\left[\frac{b \sinh x \cosh x}{a + b \sinh^2 x} + (b - 2a)\int \frac{dx}{a + b \sinh^2 x}\right]$$

(see **2.458** 1)   MZ 196

2.
$$\int \frac{dx}{\left(a + b\cosh^2 x\right)^2} = \frac{1}{2a(a+b)}\left[-\frac{b\sinh x\cosh x}{a + b\cosh^2 x} + (2a+b)\int\frac{dx}{a + b\cosh^2 x}\right]$$

(see **2.458** 2)

MZ 203

3.
$$\int \frac{dx}{\left(a + b\sinh^2 x\right)^3} = \frac{1}{8pa^3}\left[\left(3 - \frac{2}{p^2} + \frac{3}{p^4}\right)\arctan\left(p\tanh x\right) + \left(3 - \frac{2}{p^2} - \frac{3}{p^4}\right)\frac{p\tanh x}{1 + p^2\tanh^2 x}\right.$$
$$\left. + \left(1 + \frac{2}{p^2} - \frac{1}{p^2}\tanh^2 x\right)\frac{2p\tanh x}{\left(1 + p^2\tanh^2 x\right)^2}\right]$$

$$\left[p^2 = \frac{b}{a} - 1 > 0\right]$$

$$= \frac{1}{8qa^3}\left[\left(3 + \frac{2}{q^2} + \frac{3}{q^4}\right)\operatorname{arctanh}\left(q\tanh x\right) + \left(3 + \frac{2}{q^2} - \frac{3}{q^4}\right)\frac{q\tanh x}{1 - q^2\tanh^2 x}\right.$$
$$\left. + \left(1 - \frac{2}{q^2} + \frac{1}{q^2}\tanh^2 x\right)\frac{2q\tanh x}{\left(1 - q^2\tanh^2 x\right)^2}\right]$$

$$\left[q^2 = 1 - \frac{b}{a} > 0\right]$$

MZ 196

4.
$$\int \frac{dx}{\left(a + b\cosh^2 x\right)^3} = \frac{1}{8pa^3}\left[\left(3 - \frac{2}{p^2} + \frac{3}{p^4}\right)\arctan\left(p\coth x\right) + \left(3 - \frac{2}{p^2} - \frac{3}{p^4}\right)\frac{p\coth x}{1 + p^2\coth^2 x}\right.$$
$$\left. + \left(1 + \frac{2}{p^2} - \frac{1}{p^2}\coth^2 x\right)\frac{2p\coth x}{\left(1 + p^2\coth^2 x\right)^2}\right]$$

$$\left[p^2 = -1 - \frac{b}{a} > 0\right]$$

$$= \frac{1}{8qa^3}\left[\left(3 + \frac{2}{q^2} + \frac{3}{q^4}\right)\varphi(x)^* + \left(3 + \frac{2}{q^2} - \frac{3}{q^4}\right)\frac{q\coth x}{1 - q^2\coth^2 x}\right.$$
$$\left. + \left(1 - \frac{2}{q^2} + \frac{1}{q^2}\coth^2 x\right)\frac{2q\coth x}{\left(1 - q^2\coth^2 x\right)^2}\right]$$

$$\left[q^2 = 1 + \frac{b}{a} > 0\right]$$

2.46 Algebraic functions of hyperbolic functions

2.461

1.
$$\int \sqrt{\tanh x}\, dx = \operatorname{arctanh}\sqrt{\tanh x} - \arctan\sqrt{\tanh x}$$

MZ 221

---

*In 2.459.4, if $\frac{b}{a} < 0$ and $\cosh^2 x > -\frac{a}{b}$, then $\varphi(x) = \operatorname{arctanh}\left(q\coth x\right)$. If $\frac{b}{a} < 0$, but $\cosh^2 x < -\frac{a}{b}$, or if $\frac{b}{a} > 0$, then $\varphi(x) = \operatorname{arccoth}\left(q\coth x\right)$.

2. $\int \sqrt{\coth x}\, dx = \operatorname{arccoth} \sqrt{\coth x} - \arctan \sqrt{\coth x}$ MZ 222

2.462

1. $\int \dfrac{\sinh x\, dx}{\sqrt{a^2 + \sinh^2 x}} = \operatorname{arcsinh} \dfrac{\cosh x}{\sqrt{a^2 - 1}} = \ln\left(\cosh x + \sqrt{a^2 + \sinh^2 x}\right)$ $\left[a^2 > 1\right]$

                   $= \operatorname{arccosh} \dfrac{\cosh x}{\sqrt{1 - a^2}} = \ln\left(\cosh x + \sqrt{a^2 + \sinh^2 x}\right)$ $\left[a^2 < 1\right]$

                   $= \ln \cosh x$ $\left[a^2 = 1\right]$

2. $\int \dfrac{\sinh x\, dx}{\sqrt{a^2 - \sinh^2 x}} = \arcsin \dfrac{\cosh x}{\sqrt{a^2 + 1}}$ $\left[\sinh^2 x < a^2\right]$

3. $\int \dfrac{\sinh x\, dx}{\sqrt{\sinh^2 x - a^2}} = \operatorname{arccosh} \dfrac{\cosh x}{\sqrt{a^2 + 1}} = \ln\left(\cosh x + \sqrt{\sinh^2 x - a^2}\right)$

                                             $\left[\sinh^2 x > a^2\right]$ MZ 199

4. $\int \dfrac{\cosh x\, dx}{\sqrt{a^2 + \sinh^2 x}} = \operatorname{arcsinh} \dfrac{\sinh x}{a} = \ln\left(\sinh x + \sqrt{a^2 + \sinh^2 x}\right)$

5. $\int \dfrac{\cosh x\, dx}{\sqrt{a^2 - \sinh^2 x}} = \arcsin \dfrac{\sinh x}{a}$ $\left[\sinh^2 x < a^2\right]$

6. $\int \dfrac{\cosh x\, dx}{\sqrt{\sinh^2 x - a^2}} = \operatorname{arccosh} \dfrac{\sinh x}{a} = \ln\left(\sinh x + \sqrt{\sinh^2 x - a^2}\right)$

                                             $\left[\sinh^2 x > a^2\right]$

7. $\int \dfrac{\sinh x\, dx}{\sqrt{a^2 + \cosh^2 x}} = \operatorname{arcsinh} \dfrac{\cosh x}{a} = \ln\left(\cosh x + \sqrt{a^2 + \cosh^2 x}\right)$

8. $\int \dfrac{\sinh x\, dx}{\sqrt{a^2 - \cosh^2 x}} = \arcsin \dfrac{\cosh x}{a}$ $\left[\cosh^2 x < a^2\right]$

9. $\int \dfrac{\sinh x\, dx}{\sqrt{\cosh^2 x - a^2}} = \operatorname{arccosh} \dfrac{\cosh x}{a} = \ln\left(\cosh x + \sqrt{\cosh^2 x - a^2}\right)$

                                             $\left[\cosh^2 x > a^2\right]$ MZ 215, 216

10. $\int \dfrac{\cosh x\, dx}{\sqrt{a^2 + \cosh^2 x}} = \operatorname{arcsinh} \dfrac{\sinh x}{\sqrt{a^2 + 1}} = \ln\left(\sinh x + \sqrt{a^2 + \cosh^2 x}\right)$

11. $\int \dfrac{\cosh x\, dx}{\sqrt{a^2 - \cosh^2 x}} = \arcsin \dfrac{\sinh x}{\sqrt{a^2 - 1}}$ $\left[\cosh^2 x < a^2\right]$

12. $\int \dfrac{\cosh x\, dx}{\sqrt{\cosh^2 x - a^2}} = \operatorname{arccosh} \dfrac{\sinh x}{\sqrt{a^2 - 1}}$ $\left[a^2 > 1\right]$

                   $= \ln \sinh x$ $\left[a^2 = 1\right]$

                                                         MZ 206

13. $\displaystyle\int \frac{\coth x \, dx}{\sqrt{a + b \sinh x}} = 2\sqrt{a} \operatorname{arccoth} \sqrt{1 + \frac{b}{a} \sinh x}$ $[b \sinh x > 0, \quad a > 0]$

$\displaystyle = 2\sqrt{a} \operatorname{arctanh} \sqrt{1 + \frac{b}{a} \sinh x}$ $[b \sinh x < 0, \quad a > 0]$

$\displaystyle = 2\sqrt{-a} \operatorname{arctanh} \sqrt{-\left(1 + \frac{b}{a} \sinh x\right)}$ $a < 0$

14. $\displaystyle\int \frac{\tanh x \, dx}{\sqrt{a + b \cosh x}} = 2\sqrt{a} \operatorname{arccoth} \sqrt{1 + \frac{b}{a} \cosh x}$ $[b \cosh x > 0, \quad a > 0]$

$\displaystyle = 2\sqrt{a} \operatorname{arctanh} \sqrt{1 + \frac{b}{a} \cosh x}$ $[b \cosh x < 0, \quad a > 0]$

$\displaystyle = 2\sqrt{-a} \operatorname{arctanh} \sqrt{-\left(1 + \frac{b}{a} \cosh x\right)}$ $[a < 0]$

MZ 220, 221

2.463

1. $\displaystyle\int \frac{\sinh x \sqrt{a + b \cosh x}}{p + q \cosh x} \, dx$

$\displaystyle = 2\sqrt{\frac{aq - bp}{q}} \operatorname{arccoth} \sqrt{\frac{q(a + b \cosh x)}{aq - bp}}$ $\left[b \cosh x > 0, \quad \frac{aq - bp}{q} > 0\right]$

$\displaystyle = 2\sqrt{\frac{aq - bp}{q}} \operatorname{arctanh} \sqrt{\frac{q(a + b \cosh x)}{aq - bp}}$ $\left[b \cosh x < 0, \quad \frac{aq - bp}{q} > 0\right]$

$\displaystyle = 2\sqrt{\frac{bp - aq}{q}} \operatorname{arctanh} \sqrt{\frac{q(a + b \cosh x)}{bp - aq}}$ $\left[\frac{aq - bp}{q} < 0\right]$

MZ 220

2. $\displaystyle\int \frac{\cosh x \sqrt{a + b \sinh x}}{p + q \sinh x} \, dx$

$\displaystyle = 2\sqrt{\frac{aq - bp}{q}} \operatorname{arccoth} \sqrt{\frac{q(a + b \sinh x)}{aq - bp}}$ $\left[b \sinh x > 0, \quad \frac{aq - bp}{q} > 0\right]$

$\displaystyle = 2\sqrt{\frac{aq - bp}{q}} \operatorname{arctanh} \sqrt{\frac{q(a + b \sinh x)}{aq - bp}}$ $\left[b \sinh x < 0, \quad \frac{aq - bp}{q} > 0\right]$

$\displaystyle = 2\sqrt{\frac{bp - aq}{q}} \operatorname{arctanh} \sqrt{\frac{q(a + b \sinh x)}{bp - aq}}$ $\left[\frac{aq - bp}{q} < 0\right]$

MZ 221

2.464

1. $\displaystyle\int \frac{dx}{\sqrt{k^2 + k'^2 \cosh^2 x}} = \int \frac{dx}{\sqrt{1 + k'^2 \sinh^2 x}} = F\left(\arcsin\left(\tanh x\right), k\right)$

$[x > 0]$ BY (295.00)(295.10)

2. $\displaystyle\int \frac{dx}{\sqrt{\cosh^2 x - k^2}} = \int \frac{dx}{\sqrt{\sinh^2 x + k'^2}} = F\left(\arcsin\left(\frac{1}{\cosh x}\right), k\right)$

$[x > 0]$ BY (295.40)(295.30)

3. $\displaystyle\int \frac{dx}{\sqrt{1 - k'^2 \cosh^2 x}} = F\left(\arcsin\left(\frac{\tanh x}{k}\right), k\right)$ $\left[0 < x < \operatorname{arccosh} \dfrac{1}{k'}\right]$ BY (295.20)

Notation: In **2.464** 4–**2.464** 8, we set $\alpha = \arccos \dfrac{1 - \sinh 2ax}{1 + \sinh 2ax}$, $r = \dfrac{1}{\sqrt{2}}$ $[ax > 0]$

4. $\displaystyle\int \frac{dx}{\sqrt{\sinh 2ax}} = \frac{1}{2a} F(\alpha, r)$ BY (296.50)

5. $\displaystyle\int \sqrt{\sinh 2ax}\, dx = \frac{1}{2a}\left[F(\alpha, r) - 2\,E(\alpha, r)\right] + \frac{1}{a}\frac{\sqrt{\sinh 2ax\,(1 + \sinh^2 2ax)}}{1 + \sinh 2ax}$ BY (296.53)

6. $\displaystyle\int \frac{\cosh^2 2ax\, dx}{(1 + \sinh 2ax)^2 \sqrt{\sinh 2ax}} = \frac{1}{2a} E(\alpha, r)$ BY (296.51)

7. $\displaystyle\int \frac{(1 - \sinh 2ax)^2\, dx}{(1 + \sinh 2ax)^2 \sqrt{\sinh 2ax}} = \frac{1}{2a}\left[2\,E(\alpha, r) - F(\alpha, r)\right]$ BY (296.55)

8. $\displaystyle\int \frac{\sqrt{\sinh 2ax}\, dx}{(1 + \sinh 2ax)^2} = \frac{1}{4a}\left[F(\alpha, r) - E(\alpha, r)\right]$ BY (296.54)

Notation: In **2.464** 9–**2.464** 15, we set $\alpha = \arcsin \sqrt{\dfrac{\cosh 2ax - 1}{\cosh 2ax}}$, $r = \dfrac{1}{\sqrt{2}}$ $[x \neq 0]$:

9. $\displaystyle\int \frac{dx}{\sqrt{\cosh 2ax}} = \frac{1}{a\sqrt{2}} F(\alpha, r)$ BY (296.00)

10. $\displaystyle\int \sqrt{\cosh 2ax}\, dx = \frac{1}{a\sqrt{2}}\left[F(\alpha, r) - 2\,E(\alpha, r)\right] + \frac{\sinh 2ax}{a\sqrt{\cosh 2ax}}$ BY (296.03)

11. $\displaystyle\int \frac{dx}{\sqrt{\cosh^3 2ax}} = \frac{1}{a\sqrt{2}}\left[2\,E(\alpha, r) - F(\alpha, r)\right]$ BY (296.04)

12. $\displaystyle\int \frac{dx}{\sqrt{\cosh^5 2ax}} = \frac{1}{3\sqrt{2}a} F(\alpha, r) + \frac{\tanh 2ax}{3a\sqrt{\cosh 2ax}}$ BY (296.04)

13. $\displaystyle\int \frac{\sinh^2 2ax\, dx}{\sqrt{\cosh 2ax}} = -\frac{\sqrt{2}}{3a} F(\alpha, r) + \frac{1}{3a} \sinh 2ax\sqrt{\cosh 2ax}$ BY (296.07)

14. $\displaystyle\int \frac{\tanh^2 2ax\, dx}{\sqrt{\cosh 2ax}} = \frac{\sqrt{2}}{3a} F(\alpha, r) - \frac{\tanh 2ax}{3a\sqrt{\cosh 2ax}}$ BY (296.05)

15. $\displaystyle\int \frac{\sqrt{\cosh 2ax}\, dx}{p^2 + (1 - p^2)\cosh 2ax} = \frac{1}{a\sqrt{2}} \Pi\left(\alpha, p^2, r\right)$ BY (296.02)

Notation: In **2.464** 16–**2.464** 20, we set:

$$\alpha = \arccos \frac{\sqrt{a^2 + b^2} - a - b\sinh x}{\sqrt{a^2 + b^2} + a + b\sinh x},$$

$$r = \sqrt{\frac{a + \sqrt{a^2 + b^2}}{2\sqrt{a^2 + b^2}}} \qquad \left[a > 0, \quad b > 0, \quad x > -\operatorname{arcsinh} \frac{a}{b}\right]$$

16. $\displaystyle\int \frac{dx}{\sqrt{a + b\sinh x}} = \frac{1}{\sqrt[4]{a^2 + b^2}} F(\alpha, r)$ BY (298.00)

17. $\int \sqrt{a + b \sinh x}\, dx = \sqrt[4]{a^2 + b^2}\,[F(\alpha, r) - 2\,E(\alpha, r)] + \dfrac{2b \cosh x \sqrt{a + b \sinh x}}{\sqrt{a^2 + b^2} + a + b \sinh x}$ BY (298.02)

18. $\int \dfrac{\sqrt{a + b \sinh x}}{\cosh^2 x}\, dx = \sqrt[4]{a^2 + b^2}\,E(\alpha, r) - \dfrac{\sqrt{a^2 + b^2} - a}{2\sqrt[4]{a^2 + b^2}}\,F(\alpha, r)$

$\qquad\qquad - \dfrac{a + \sqrt{a^2 + b^2}}{b} \cdot \dfrac{\sqrt{a^2 + b^2} - a - b \sinh x}{\sqrt{a^2 + b^2} + a + b \sinh x} \cdot \dfrac{\sqrt{a + b \sinh x}}{\cosh x}$

$\qquad\qquad\qquad\qquad\qquad\qquad\qquad\qquad\qquad\qquad\qquad\qquad\qquad$ BY (298.03)

19. $\int \dfrac{\cosh^2 x\, dx}{\left[\sqrt{a^2 + b^2} + a + b \sinh x\right]^2 \sqrt{a + b \sinh x}} = \dfrac{1}{b^2 \sqrt[4]{a^2 + b^2}}\,E(\alpha, r)$ BY (298.01)

20. $\int \dfrac{\sqrt{a + b \sinh x}\, dx}{\left[\sqrt{a^2 + b^2} - a - b \sinh x\right]^2} = -\dfrac{1}{\sqrt[4]{a^2 + b^2}\,\left(\sqrt{a^2 + b^2} - a\right)}\,E(\alpha, r)$

$\qquad\qquad + \dfrac{b}{\sqrt{a^2 + b^2} - a} \cdot \dfrac{\cosh x \sqrt{a + b \sinh x}}{a^2 + b^2 - (a + b \sinh x)^2}$

$\qquad\qquad\qquad\qquad\qquad\qquad\qquad\qquad\qquad\qquad\qquad\qquad\qquad$ BY (298.04)

Notation: In **2.464** 21–**2.464** 31, we set $\alpha = \arcsin\left(\tanh \dfrac{x}{2}\right)$, $r = \sqrt{\dfrac{a - b}{a + b}}$ $[0 < b < a, x > 0]$:

21. $\int \dfrac{dx}{\sqrt{a + b \cosh x}} = \dfrac{2}{\sqrt{a + b}}\,F(\alpha, r)$ BY (297.25)

22. $\int \sqrt{a + b \cosh x}\, dx = 2\sqrt{a + b}\,[F(\alpha, r) - E(\alpha, r)] + 2 \tanh \dfrac{x}{2} \sqrt{a + b \cosh x}$ BY (297.29)

23. $\int \dfrac{\cosh x\, dx}{\sqrt{a + b \cosh x}} = \dfrac{2}{\sqrt{a + b}}\,F(\alpha, r) - \dfrac{2\sqrt{a + b}}{b}\,E(\alpha, r) + \dfrac{2}{b} \tanh \dfrac{x}{2} \sqrt{a + b \cosh x}$ BY (297.33)

24. $\int \dfrac{\tanh^2 \frac{x}{2}}{\sqrt{a + b \cosh x}}\, dx = \dfrac{2\sqrt{a + b}}{a - b}\,[F(\alpha, r) - E(\alpha, r)]$ BY (297.28)

25. $\int \dfrac{\tanh^4 \frac{x}{2}}{\sqrt{a = b \cosh x}}\, dx = \dfrac{2\sqrt{a + b}}{3(a - b)^2}\,[(3a + b)\,F(\alpha, r) - 4a\,E(\alpha, r)] + \dfrac{2}{3(a - b)} \dfrac{\sinh \frac{x}{2} \sqrt{a + b \cosh x}}{\cosh^3 \frac{x}{2}}$

$\qquad\qquad\qquad\qquad\qquad\qquad\qquad\qquad\qquad\qquad\qquad\qquad\qquad$ BY (297.28)

26. $\int \dfrac{\cosh x - 1}{\sqrt{a + b \cosh x}}\, dx = \dfrac{2}{b}\left[\left(\tanh \dfrac{x}{2}\right)\sqrt{a + b \cosh x} - \sqrt{a + b}\,E(\alpha, r)\right]$ BY (297.31)

27. $\int \dfrac{(\cosh x - 1)^2}{\sqrt{a + b \cosh x}}\, dx = \dfrac{4\sqrt{a + b}}{3b^2}\,[(a + 3b)\,E(\alpha, r) - b\,F(\alpha, r)]$

$\qquad\qquad + \dfrac{4}{3b^2}\left[b \cosh^2 \dfrac{x}{2} - (a + 3b)\right] \tanh \dfrac{x}{2} \sqrt{a + b \cosh x}$

$\qquad\qquad\qquad\qquad\qquad\qquad\qquad\qquad\qquad\qquad\qquad\qquad\qquad$ BY (297.31)

28. $\int \dfrac{\sqrt{a + b \cosh x}}{\cosh x + 1}\, dx = \sqrt{a + b}\,E(\alpha, r)$ BY (297.26)

29. $\int \dfrac{dx}{(\cosh x + 1)\sqrt{a + b \cosh x}} = \dfrac{\sqrt{a + b}}{a - b}\,E(\alpha, r) - \dfrac{2b}{(a - b)\sqrt{a + b}}\,F(\alpha, r)$ BY (297.30)

30.
$$\int \frac{dx}{(\cosh x + 1)^2 \sqrt{a + b\cosh x}} = \frac{1}{3(a-b)^2 \sqrt{a+b}} \left[b(5b - a)\, F(\alpha, r) \right.$$
$$\left. + (a - 3b)(a + b)\, E(\alpha, r) \right] + \frac{1}{6(a - b)} \cdot \frac{\sinh \frac{x}{2}}{\cosh^3 \frac{x}{2}} \sqrt{a + b\cosh x}$$

297.30)

31.
$$\int \frac{(1 + \cosh x)\, dx}{[1 + p^2 + (1 - p^2)\cosh x]\sqrt{a + b\cosh x}} = \frac{2}{\sqrt{a+b}}\, \Pi\left(\alpha, p^2, r\right)$$
BY (297.27)

Notation: In **2.464** 32–**2.464** 40, we set:
$$\alpha = \arcsin \sqrt{\frac{a - b\cosh x}{a - b}}$$
$$r = \sqrt{\frac{a - b}{a + b}} \qquad \left[0 < b < a, \quad 0 < x < \operatorname{arccosh} \frac{a}{b} \right]$$

32.
$$\int \frac{dx}{\sqrt{a - b\cosh x}} = \frac{2}{\sqrt{a+b}}\, F(\alpha, r)$$
BY (297.50)

33.
$$\int \sqrt{a - b\cosh x}\, dx = 2\sqrt{a+b}\,[F(\alpha, r) - E(\alpha, r)]$$
BY (297.54)

34.
$$\int \frac{\cosh x\, dx}{\sqrt{a - b\cosh x}} = \frac{2\sqrt{a+b}}{b}\, E(\alpha, r) - \frac{2}{\sqrt{a+b}}\, F(\alpha, r)$$
BY (297.56)

35.
$$\int \frac{\cosh^2 x\, dx}{\sqrt{a - b\cosh x}} = \frac{2(b - 2a)}{3b\sqrt{a+b}}\, F(\alpha, r) + \frac{4a\sqrt{a+b}}{3b^2}\, E(\alpha, r) + \frac{2}{3b}\sinh x \sqrt{a - b\cosh x}$$
BY (297.56)

36.
$$\int \frac{(1 + \cosh x)\, dx}{\sqrt{a - b\cosh x}} = \frac{2\sqrt{a+b}}{b}\, E(\alpha, r)$$
BY (297.51)

37.
$$\int \frac{dx}{\cosh x \sqrt{a - b\cosh x}} = \frac{2b}{a\sqrt{a+b}}\, \Pi\left(\alpha, \frac{a - b}{a}, r\right)$$
BY (297.57)

38.
$$\int \frac{dx}{(1 + \cosh x)\sqrt{a - b\cosh x}} = \frac{1}{\sqrt{a+b}}\, E(\alpha, r) - \frac{1}{a+b}\tanh \frac{x}{2}\sqrt{a - b\cosh x}$$
BY (297.58)

39.
$$\int \frac{dx}{(1 + \cosh x)^2 \sqrt{a - b\cosh x}} = \frac{1}{3\sqrt{(a+b)^3}}\,[(a + 3b)\, E(\alpha, r) - b\, F(\alpha, r)]$$
$$- \frac{1}{3(a+b)^2}\, \frac{\tanh \frac{x}{2}\sqrt{a - b\cosh x}}{\cosh x + 1}\,[2a + 4b + (a + 3b)\cosh x]$$
BY (297.58)

40.
$$\int \frac{dx}{(a - b - ap^2 + bp^2 \cosh x)\sqrt{a - b\cosh x}} = \frac{2}{(a - b)\sqrt{a+b}}\, \Pi\left(\alpha, p^2, r\right)$$
BY (297.52)

Notation: In **2.464** 41 –**2.464** 47, we set:
$$\alpha = \arcsin \sqrt{\frac{b\,(\cosh x - 1)}{b\cosh x - a}},$$
$$r = \sqrt{\frac{a + b}{2b}} \qquad [0 < a < b, x > 0]$$

41. $$\int \frac{dx}{\sqrt{b \cosh - a}} = \sqrt{\frac{2}{b}} \, F(\alpha, r)$$ BY (297.00)

42. $$\int \sqrt{b \cosh x - a} \, dx = (b - a)\sqrt{\frac{2}{b}} \, F(\alpha, r) - 2\sqrt{2b} \, E(\alpha, r) + \frac{2b \sinh x}{\sqrt{b \cosh x - a}}$$ BY (297.05)

43. $$\int \frac{dx}{\sqrt{(b \cosh x - a)^3}} = \frac{1}{b^2 - a^2} \cdot \sqrt{\frac{2}{b}} \, [2b \, E(\alpha, r) - (b - a) \, F(\alpha, r)]$$ BY (297.06)

44. $$\int \frac{dx}{\sqrt{(b \cosh x - a)^5}} = \frac{1}{3(b^2 - a^2)^2} \sqrt{\frac{2}{b}} \, [(b - 3a)(b - a) \, F(\alpha, r) + 8ab \, E(\alpha, r)]$$
$$+ \frac{2b}{3(b^2 - a^2)} \cdot \frac{\sinh x}{\sqrt{(b \cosh x - a)^3}}$$ BY (297.06)

45. $$\int \frac{\cosh x \, dx}{\sqrt{b \cosh x - a}} = \sqrt{\frac{2}{b}} \, [F(\alpha, r) - 2 \, E(\alpha, r)] + \frac{2 \sinh x}{\sqrt{b \cosh x - a}}$$ BY (297.03)

46. $$\int \frac{(\cosh x + 1) \, dx}{\sqrt{(b \cosh x - a)^3}} = \frac{2}{b - a} \sqrt{\frac{2}{b}} \, E(\alpha, r)$$ BY (297.01)

47. $$\int \frac{\sqrt{b \cosh x - a} \, dx}{p^2 b - a + b(1 - p^2) \cosh x} = \sqrt{\frac{2}{b}} \, \Pi\left(\alpha, p^2, r\right)$$ BY (297.02)

Notation: In **2.464** 48–**2.464** 55, we set $\alpha = \arcsin \sqrt{\frac{b \cosh x - a}{b(\cosh x - 1)}}$ and $r = \sqrt{\frac{2b}{a + b}}$ for $\left[0 < b < a, x > \operatorname{arccosh} \frac{a}{b}\right]$:

48. $$\int \frac{dx}{\sqrt{b \cosh x - a}} = \frac{2}{\sqrt{a + b}} \, F(\alpha, r)$$ BY (297.75)

49. $$\int \sqrt{b \cosh x - a} \, dx = -2\sqrt{a + b} \, E(\alpha, r) + 2 \coth \frac{x}{2} \sqrt{b \cosh x - a}$$ BY (297.79)

50. $$\int \frac{\coth^2 \frac{x}{2} \, dx}{\sqrt{b \cosh x - a}} = \frac{2\sqrt{a + b}}{a - b} \, E(\alpha, r)$$ BY (297.76)

51. $$\int \frac{\sqrt{b \cosh x - a}}{\cosh x - 1} \, dx = \sqrt{a + b} \, [F(\alpha, r) - E(\alpha, r)]$$ BY (297.77)

52. $$\int \frac{dx}{(\cosh x - 1) \sqrt{b \cosh x - a}} = \frac{\sqrt{a + b}}{a - b} \, E(\alpha, r) - \frac{1}{\sqrt{a + b}} \, F(\alpha, r)$$ BY (297.78)

53. $$\int \frac{dx}{(\cosh x - 1)^2 \sqrt{b \cosh x - a}} = \frac{1}{3(a - b)^2 \sqrt{a + b}} \left[(a - 2b)(a - b) \, F(\alpha, r) \right.$$
$$\left. + (3a - b)(a + b) \, E(\alpha, r) \right] + \frac{a + b}{6b(a - b)} \cdot \frac{\cosh \frac{x}{2}}{\sinh^3 \frac{x}{2}} \sqrt{b \cosh x - a}$$ BY (297.78)

54. $$\int \frac{dx}{(\cosh x + 1) \sqrt{b \cosh x - a}} = \frac{1}{\sqrt{a + b}} \, [F(\alpha, r) - E(\alpha, r)] + \frac{2\sqrt{b \cosh x - a}}{(a + b) \sinh x}$$ BY (297.80)

55.　$\displaystyle\int \frac{dx}{(\cosh x + 1)^2 \sqrt{b \cosh x - a}} = \frac{1}{3\sqrt{(a+b)^3}} \Bigg[(a+b)\,F(\alpha, r)$

$$- (a+3b)\,E(\alpha, r) \Bigg] + \frac{\sqrt{b \cosh x - a}}{3(a+b)\sinh x} \left(2\frac{a+3b}{a+b} - \tanh^2 \frac{x}{2} \right)$$

BY (297.80)

Notation: In **2.464** 56–**2.464** 60, we set

$$\alpha = \arccos \frac{\sqrt[4]{b^2 - a^2}}{\sqrt{a \sinh x + b \cosh x}},$$

$$r = \frac{1}{\sqrt{2}} \qquad \left[0 < a < b, \quad -\operatorname{arcsinh} \frac{a}{\sqrt{b^2 - a^2}} < x \right]$$

56.　$\displaystyle\int \frac{dx}{\sqrt{a \sinh x + b \cosh x}} = \sqrt[4]{\frac{4}{b^2 - a^2}}\, F(\alpha, r)$　　　　　　　　　BY (299.00)

57.　$\displaystyle\int \sqrt{a \sinh x + b \cosh x}\, dx = \sqrt[4]{4(b^2 - a^2)}\,[F(\alpha, r) - 2\,E(\alpha, r)] + \frac{2(a \cosh x + b \sinh x)}{\sqrt{a \sinh x + b \cosh x}}$

BY (299.02)

58.　$\displaystyle\int \frac{dx}{\sqrt{(a \sinh x + b \cosh x)^3}} = \sqrt[4]{\frac{4}{(b^2 - a^2)^3}}\,[2\,E(\alpha, r) - F(\alpha, r)]$　　　　　　BY (299.03)

59.　$\displaystyle\int \frac{dx}{\sqrt{(a \sinh x + b \cosh x)^5}} = \frac{1}{3}\sqrt[4]{\frac{4}{(b^2 - a^2)^5}}\, F(\alpha, r) + \frac{2}{3(b^2 - a^2)} \cdot \frac{a \cosh x + b \sinh x}{\sqrt{(a \sinh x + b \cosh x)^3}}$

BY (299.03)

60.　$\displaystyle\int \frac{\left(\sqrt{b^2 - a^2} + a \sinh x + b \cosh x\right) dx}{\sqrt{(a \sinh x + b \cosh x)^3}} = 2\sqrt[4]{\frac{4}{b^2 - a^2}}\, E(\alpha, r)$　　　　　　BY (299.01)

2.47 Combinations of hyperbolic functions and powers

2.471

1. $\displaystyle\int x^r \sinh^p x \cosh^q x\, dx$

$$= \frac{1}{(p+q)^2}\left[(p+q)x^r \sinh^{p-1} x \cosh^{q-1} x\right.$$
$$-rx^{r-1}\sinh^p x \cosh^q x + r(r+1)\int x^{r-2}\sinh^p x \cosh^q x\, dx$$
$$\left.+ rp\int x^{r-1}\sinh^{p-1} x \cosh^{q-1} x\, dx + (q-1)(p+q)\int x^r \sinh^p x \cosh^{q-2} x\, dx\right]$$
$$= \frac{1}{(p+q)^2}\left[(p+q)x^r \sinh^{p-1} x \cosh^{q+1} x\right.$$
$$-rx^{r-1}\sinh^p x \cosh^q x + r(r-1)\int x^{r-2}\sinh^p x \cosh^q x\, dx$$
$$\left.- rq\int x^{r-1}\sinh^{p-1} x \cosh^{q-1} x\, dx - (p-1)(p+q)\int x^r \sinh^{p-2} x \cosh^q x\, dx\right]$$

GU (353)(1)

2. $\displaystyle\int x^n \sinh^{2m} x\, dx = (-1)^m \binom{2m}{m}\frac{x^{n+1}}{2^{2m}(n+1)} + \frac{1}{2^{2m-1}}\sum_{k=0}^{m-1}(-1)^k\binom{2m}{k}\int x^n \cosh(2m-2k)x\, dx$

3. $\displaystyle\int x^n \sinh^{2m+1} x\, dx = \frac{1}{2^{2m}}\sum_{k=0}^{m}(-1)^k\binom{2m+1}{k}\int x^n \sinh(2m-2k+1)x\, dx$

4. $\displaystyle\int x^n \cosh^{2m} x\, dx = \binom{2m}{m}\frac{x^{n+1}}{2^{2m}(n+1)} + \frac{1}{2^{2m-1}}\sum_{k=0}^{m-1}\binom{2m}{k}\int x^n \cosh(2m-2k)x\, dx$

5. $\displaystyle\int x^n \cosh^{2m+1} x\, dx = \frac{1}{2^{2m}}\sum_{k=0}^{m}\binom{2m+1}{k}\int x^n \cosh(2m-2k+1)x\, dx$

2.472

1. $\displaystyle\int x^n \sinh x\, dx = x^n \cosh x - n\int x^{n-1}\cosh x\, dx$

$$= x^n \cosh x - nx^{n-1}\sinh x + n(n-1)\int x^{n-2}\sinh x\, dx$$

2. $\displaystyle\int x^n \cosh x\, dx = x^n \sinh x - n\int x^{n-1}\sinh x\, dx$

$$= x^n \sinh x - nx^{n-1}\cosh x + n(n-1)\int x^{n-2}\cosh x\, dx$$

3. $\displaystyle\int x^{2n}\sinh x\, dx = (2n)!\left\{\sum_{k=0}^{n}\frac{x^{2k}}{(2k)!}\cosh x - \sum_{k=1}^{n}\frac{x^{2k-1}}{(2k-1)!}\sinh x\right\}$

4. $\displaystyle\int x^{2n+1}\sinh x\, dx = (2n+1)!\sum_{k=0}^{n}\left\{\frac{x^{2k+1}}{(2k+1)!}\cosh x - \frac{x^{2k}}{(2k)!}\sinh x\right\}$

5. $\displaystyle \int x^{2n} \cosh x \, dx = (2n)! \left\{ \sum_{k=1}^{n} \frac{x^{2k}}{(2k)!} \sinh x - \sum_{k=1}^{n} \frac{x^{2k-1}}{(2k-1)!} \cosh x \right\}$

6. $\displaystyle \int x^{2n+1} \cosh x \, dx = (2n+1)! \sum_{k=0}^{n} \left\{ \frac{x^{2k+1}}{(2k+1)!} \sinh x - \frac{x^{2k}}{(2k)!} \cosh x \right\}$

7. $\displaystyle \int x \sinh x \, dx = x \cosh x - \sinh x$

8. $\displaystyle \int x^2 \sinh x \, dx = \left(x^2 + 2 \right) \cosh x - 2x \sinh x$

9. $\displaystyle \int x \cosh x \, dx = x \sinh x - \cosh x$

10. $\displaystyle \int x^2 \cosh x \, dx = \left(x^2 + 2 \right) \sinh x - 2x \cosh x$

2.473   **Notation:** $z_1 = a + bx$

1. $\displaystyle \int z_1 \sinh kx \, dx = \frac{1}{k} z_1 \cosh kx - \frac{b}{k^2} \sinh kx$

2. $\displaystyle \int z_1 \cosh kx \, dx = \frac{1}{k} z_1 \sinh kx - \frac{b}{k^2} \cosh kx$

3. $\displaystyle \int z_1^2 \sinh kx \, dx = \frac{1}{k} \left(z_1^2 + \frac{2b^2}{k^2} \right) \cosh kx - \frac{2bz_1}{k^2} \sinh kx$

4. $\displaystyle \int z_1^2 \cosh kx \, dx = \frac{1}{k} \left(z_1^2 + \frac{2b^2}{k^2} \right) \sinh kx - \frac{2bz_1}{k^2} \cosh kx$

5. $\displaystyle \int z_1^3 \sinh kx \, dx = \frac{z_1}{k} \left(z_1^2 + \frac{6b^2}{k^2} \right) \cosh kx - \frac{3b}{k^2} \left(z_1^2 + \frac{2b^2}{k^2} \right) \sinh kx$

6. $\displaystyle \int z_1^3 \cosh kx \, dx = \frac{z_1}{k} \left(z_1^2 + \frac{6b^2}{k^2} \right) \sinh kx - \frac{3b}{k^2} \left(z_1^3 + \frac{2b^2}{k^2} \right) \cosh kx$

7. $\displaystyle \int z_1^4 \sinh kx \, dx = \frac{1}{k} \left(z_1^4 + \frac{12b^2}{k^2} z_1^2 + \frac{24b^4}{k^4} \right) \cosh kx - \frac{4bz_1}{k^2} \left(z_1^2 + \frac{6b^2}{k^2} \right) \sinh kx$

8. $\displaystyle \int z_1^4 \cosh kx \, dx = \frac{1}{k} \left(z_1^4 + \frac{12b^2}{k^2} z_1^2 + \frac{24b^4}{k^4} \right) \sinh kx - \frac{4bz_1}{k^2} \left(z_1^2 + \frac{6b^2}{k^2} \right) \cosh kx$

9. $\displaystyle \int z_1^5 \sinh kx \, dx = \frac{z_1}{k} \left(z_1^4 + \frac{20b^2}{k^2} z_1^2 + 120 \frac{b^4}{k^4} \right) \cosh kx - \frac{5b}{k^2} \left(z_1^4 + 12 \frac{b^2}{k^2} z_1^2 + 24 \frac{b^4}{k^4} \right) \sinh kx$

10. $\displaystyle \int z_1^5 \cosh kx \, dx = \frac{z_1}{k} \left(z_1^4 + 20 \frac{b^2}{k^2} z_1^2 + 120 \frac{b^4}{k^4} \right) \sinh kx - \frac{5b}{k^2} \left(z_1^4 + 12 \frac{b^2}{k^2} z_1^2 + 24 \frac{b^4}{k^4} \right) \cosh kx$

11. $\displaystyle \int z_1^6 \sinh kx \, dx = \frac{1}{k} \left(z_1^6 + 30 \frac{b^2}{k^2} z_1^4 + 360 \frac{b^4}{k^4} z_1^2 + 720 \frac{b^6}{k^6} \right) \cosh kx$
$\displaystyle \qquad\qquad\qquad\qquad - \frac{6bz_1}{k^2} \left(z_1^4 + 20 \frac{b^2}{k^2} z_1^2 + 120 \frac{b^4}{k^4} \right) \sinh kx$

12. $\int z_1^6 \cosh kx\, dx = \dfrac{1}{k}\left(z_1^6 + 30\dfrac{b^2}{k^2}z_1^4 + 360\dfrac{b^4}{k^4}z_1^2 + 720\dfrac{b^6}{k^6}\right)\sinh kx$

$\qquad\qquad\qquad - \dfrac{6bz_1}{k^2}\left(z_1^4 + 20\dfrac{b^2}{k^2}z_1 + 120\dfrac{b^4}{k^4}\right)\cosh kx$

2.474

1. $\int x^n \sinh^2 x\, dx = -\dfrac{x^{n+1}}{2(n+1)} + \dfrac{n!}{4}\displaystyle\sum_{k=0}^{\lfloor n/2\rfloor}\left\{\dfrac{x^{n-2k}}{2^{2k}(n-2k)!}\sinh 2x - \dfrac{x^{n-2k-1}}{2^{2k+1}(n-2k-1)!}\cosh 2x\right\}$

$\qquad\qquad\qquad\qquad\qquad\qquad\qquad\qquad\qquad\qquad\qquad\qquad\qquad\qquad$ GU (353)(2b)

2. $\int x^n \cosh^2 x\, dx = \dfrac{x^{n+1}}{2(n+1)} + \dfrac{n!}{4}\displaystyle\sum_{k=0}^{\lfloor n/2\rfloor}\left\{\dfrac{x^{n-2k}}{2^{2k}(n-2k)!}\sinh 2x - \dfrac{x^{n-2k-1}}{2^{2k+1}(n-2k-1)!}\cosh 2x\right\}$

$\qquad\qquad\qquad\qquad\qquad\qquad\qquad\qquad\qquad\qquad\qquad\qquad\qquad\qquad$ GU (353)(3e)

3. $\int x \sinh^2 x\, dx = \dfrac{1}{4}x\sinh 2x - \dfrac{1}{8}\cosh 2x - \dfrac{x^2}{4}$

4. $\int x^2 \sinh^2 x\, dx = \dfrac{1}{4}\left(x^2 + \dfrac{1}{2}\right)\sinh 2x - \dfrac{x}{4}\cosh 2x - \dfrac{x^3}{6}$ $\qquad\qquad\qquad$ MZ 257

5. $\int x \cosh^2 x\, dx = \dfrac{x}{4}\sinh 2x - \dfrac{1}{8}\cosh 2x + \dfrac{x^2}{4}$

6. $\int x^2 \cosh^2 x\, dx = \dfrac{1}{4}\left(x^2 + \dfrac{1}{2}\right)\sinh 2x - \dfrac{x}{4}\cosh 2x + \dfrac{x^3}{6}$ $\qquad\qquad\qquad$ MZ 261

7. $\int x^n \sinh^3 x\, dx$

$\qquad = \dfrac{n!}{4}\displaystyle\sum_{k=0}^{\lfloor n/2\rfloor}\left\{\dfrac{x^{n-2k}}{(n-2k)!}\left(\dfrac{\cosh 3x}{3^{2k+1}} - 3\cosh x\right) - \dfrac{x^{n-2k-1}}{(n-2k-1)!}\left(\dfrac{\sinh 3x}{3^{2k+2}} - 3\sinh x\right)\right\}$

$\qquad\qquad\qquad\qquad\qquad\qquad\qquad\qquad\qquad\qquad\qquad\qquad\qquad\qquad$ GU (353)(2f)

8. $\int x^n \cosh^3 x\, dx$

$\qquad = \dfrac{n!}{4}\displaystyle\sum_{k=0}^{\lfloor n/2\rfloor}\left\{\dfrac{x^{n-2k}}{(n-2k)!}\left(\dfrac{\sinh 3x}{3^{2k+1}} + 3\sinh x\right) - \dfrac{x^{n-2k-1}}{(n-2k-1)!}\left(\dfrac{\cosh 3x}{3^{2k+2}} + 3\cosh x\right)\right\}$

$\qquad\qquad\qquad\qquad\qquad\qquad\qquad\qquad\qquad\qquad\qquad\qquad\qquad\qquad$ GU (353)(3f)

9. $\int x \sinh^3 x\, dx = \dfrac{3}{4}\sinh x - \dfrac{1}{36}\sinh 3x - \dfrac{3}{4}x\cosh x - \dfrac{x}{12}\cosh 3x$

10. $\int x^2 \sinh^3 x\, dx = -\left(\dfrac{3x^2}{4} + \dfrac{3}{2}\right)\cosh x + \left(\dfrac{x^2}{12} + \dfrac{1}{54}\right)\cosh 3x + \dfrac{3x}{2}\sinh x - \dfrac{x}{18}\sinh 3x.$ $\quad$ MZ 257

11. $\int x \cosh^3 x\, dx = -\dfrac{3}{4}\cosh x - \dfrac{1}{36}\cosh 3x + \dfrac{3}{4}x\sinh x + \dfrac{x}{12}\sinh 3x$

12. $\int x^2 \cosh^3 x\, dx = \left(\dfrac{3}{4}x^2 + \dfrac{3}{2}\right)\sinh x + \left(\dfrac{x^2}{12} + \dfrac{1}{54}\right)\sinh 3x - \dfrac{3}{2}x\cosh x - \dfrac{x}{18}\cosh 3x$ $\quad$ MZ 262

2.475

1. $$\int \frac{\sinh^q x}{x^p}\, dx = -\frac{(p-2)\sinh^q x + qx\sinh^{q-1} x\cosh x}{(p-1)(p-2)x^{p-1}}$$
$$+ \frac{q(q-1)}{(p-1)(p-2)}\int \frac{\sinh^{q-2} x}{x^{p-2}}\, dx + \frac{q^2}{(p-1)(p-2)}\int \frac{\sinh^q x}{x^{p-2}}\, dx \quad [p>2]$$

GU (353)(6a)

2. $$\int \frac{\cosh^q x}{x^p}\, dx = -\frac{(p-2)\cosh^q x + qx\cosh^{q-1} x\sinh x}{(p-1)(p-2)x^{p-1}}$$
$$- \frac{q(q-1)}{(p-1)(p-2)}\int \frac{\cosh^{q-2} x}{x^{p-2}}\, dx + \frac{q^2}{(p-1)(p-2)}\int \frac{\cosh^q x}{x^{p-2}}\, dx \quad [p>2]$$

GU (353)(7a)

3. $$\int \frac{\sinh x}{x^{2n}}\, dx = -\frac{1}{x(2n-1)!}\left\{\sum_{k=0}^{n-2}\frac{(2k+1)!}{x^{2k+1}}\cosh x + \sum_{k=0}^{n-1}\frac{(2k)!}{x^{2k}}\sinh x\right\} + \frac{1}{(2n-1)!}\,\mathrm{chi}(x)$$

GU (353)(6b)

4. $$\int \frac{\sinh x}{x^{2n+1}}\, dx = -\frac{1}{x(2n)!}\left\{\sum_{k=0}^{n-1}\frac{(2k)!}{x^{2k}}\cosh x + \sum_{k=0}^{n-1}\frac{(2k+1)!}{x^{2k+1}}\sinh x\right\} + \frac{1}{(2n)!}\,\mathrm{shi}(x) \qquad \text{GU (353)(6b)}$$

5. $$\int \frac{\cosh x}{x^{2n}}\, dx = -\frac{1}{x(2n-1)!}\left\{\sum_{k=0}^{n-2}\frac{(2k+1)!}{x^{2k+1}}\sinh x + \sum_{k=0}^{n-1}\frac{(2k)!}{x^{2k}}\cosh x\right\} + \frac{1}{(2n-1)!}\,\mathrm{shi}(x)$$

GU (353)(7b)

6. $$\int \frac{\cosh x}{x^{2n+1}}\, dx = -\frac{1}{(2n)!x}\left\{\sum_{k=0}^{n-1}\frac{(2k)!}{x^{2k}}\sinh x + \sum_{k=0}^{n-1}\frac{(2k+1)!}{x^{2k+1}}\cosh x\right\} + \frac{1}{(2n)!}\,\mathrm{chi}(x) \qquad \text{GU (353)(7b)}$$

7. $$\int \frac{\sinh^{2m} x}{x}\, dx = \frac{1}{2^{2m-1}}\sum_{k=0}^{m-1}(-1)^k\binom{2m}{k}\mathrm{chi}(2m-2k)x + \frac{(-1)^m}{2^{2m}}\binom{2m}{m}\ln x \qquad \text{GU (353)(6c)}$$

8. $$\int \frac{\sinh^{2m+1} x}{x}\, dx = \frac{1}{2^{2m}}\sum_{k=0}^{m}(-1)^k\binom{2m+1}{k}\mathrm{shi}(2m-2k+1)x \qquad \text{GU (353)(6d)}$$

9. $$\int \frac{\cosh^{2m} x}{x}\, dx = \frac{1}{2^{2m-1}}\sum_{k=0}^{m-1}\binom{2m}{k}\mathrm{chi}(2m-2k)x + \frac{1}{2^{2m}}\binom{2m}{m}\ln x \qquad \text{GU (353)(7c)}$$

10. $$\int \frac{\cosh^{2m+1} x}{x}\, dx = \frac{1}{2^{2m}}\sum_{k=0}^{m}\binom{2m+1}{k}\mathrm{chi}(2m-2k+1)x \qquad \text{GU (353)(7c)}$$

11. $$\int \frac{\sinh^{2m} x}{x^2}\, dx = \frac{(-1)^{m-1}}{2^{2m}x}\binom{2m}{m}$$
$$+ \frac{1}{2^{2m-1}}\sum_{k=0}^{m-1}(-1)^{k+1}\binom{2m}{k}\left\{\frac{\cosh(2m-2k)x}{x} - (2m-2k)\,\mathrm{shi}(2m-2k)x\right\}$$

12. $\displaystyle\int \frac{\sinh^{2m+1} x}{x^2}\, dx = \frac{1}{2^{2m}} \sum_{k=0}^{m} (-1)^{k+1} \binom{2m+1}{k}$
$$\times \left\{ \frac{\sinh(2m-2k+1)x}{x} - (2m-2k+1)\operatorname{chi}(2m-2k+1)x \right\}$$

13. $\displaystyle\int \frac{\cosh^{2m} x}{x^2}\, dx$
$$= -\frac{1}{2^{2m}x}\binom{2m}{m} - \frac{1}{2^{2m-1}} \sum_{k=0}^{m-1}\binom{2m}{k}\left\{ \frac{\cosh(2m-2k)x}{x} - (2m-2k)\operatorname{shi}(2m-2k)x \right\}$$

14. $\displaystyle\int \frac{\cosh^{2m+1} x}{x^2}\, dx$
$$= -\frac{1}{2^{2m}} \sum_{k=0}^{m}\binom{2m+1}{k}\left\{ \frac{\cosh(2m-2k+1)x}{x} - (2m-2k+1)\operatorname{shi}(2m-2k+1)x \right\}$$

2.476

1. $\displaystyle\int \frac{\sinh kx}{a+bx}\, dx = \frac{1}{b}\left[\cosh\frac{ka}{b}\operatorname{shi}(u) - \sinh\frac{ka}{b}\operatorname{chi}(u) \right]$
$$= \frac{1}{2b}\left[\exp\left(-\frac{ka}{b}\right)\operatorname{Ei}(u) - \exp\left(\frac{ka}{b}\right)\operatorname{Ei}(-u) \right] \qquad \left[u = \frac{k}{b}(a+bx) \right]$$

2. $\displaystyle\int \frac{\cosh kx}{a+bx}\, dx = \frac{1}{b}\left[\cosh\frac{ka}{b}\operatorname{chi}(u) - \sinh\frac{ka}{b}\operatorname{shi}(u) \right]$
$$= \frac{1}{2b}\left[\exp\left(-\frac{ka}{b}\right)\operatorname{Ei}(u) + \exp\left(\frac{ka}{b}\right)\operatorname{Ei}(-u) \right] \qquad \left[u = \frac{k}{b}(a+bx) \right]$$

3. $\displaystyle\int \frac{\sinh kx}{(a+bx)^2}\, dx = -\frac{1}{b}\cdot\frac{\sinh kx}{a+bx} + \frac{k}{b}\int \frac{\cosh kx}{a+bx}\, dx \qquad \text{(see } \mathbf{2.476}\ 2)$

4. $\displaystyle\int \frac{\cosh kx}{(a+bx)^2}\, dx = -\frac{1}{b}\cdot\frac{\cosh kx}{a+bx} + \frac{k}{b}\int \frac{\sinh kx}{a+bx}\, dx \qquad \text{(see } \mathbf{2.476}\ 1)$

5. $\displaystyle\int \frac{\sinh kx}{(a+bx)^3}\, dx = -\frac{\sinh kx}{2b(a+bx)^2} - \frac{k\cosh kx}{2b^2(a+bx)} + \frac{k^2}{2b^2}\int \frac{\sinh kx}{a+bx}\, dx$
$$\text{(see } \mathbf{2.476}\ 1)$$

6. $\displaystyle\int \frac{\cosh kx}{(a+bx)^3}\, dx = -\frac{\cosh kx}{2b(a+bx)^2} - \frac{k\sinh kx}{2b^2(a+bx)} + \frac{k^2}{2b^2}\int \frac{\cosh kx}{a+bx}\, dx$
$$\text{(see } \mathbf{2.476}\ 2)$$

7. $\displaystyle\int \frac{\sinh kx}{(a+bx)^4}\, dx = -\frac{\sinh kx}{3b(a+bx)^3} - \frac{k\cosh kx}{6b^2(a+bx)^2} - \frac{k^2\sinh kx}{6b^3(a+bx)} + \frac{k^3}{6b^3}\int \frac{\cosh kx}{a+bx}\, dx$
$$\text{(see } \mathbf{2.476}\ 2)$$

8. $\displaystyle\int \frac{\cosh kx}{(a+bx)^4}\, dx = -\frac{\cosh kx}{3b(a+bx)^3} - \frac{k\sinh kx}{6b^2(a+bx)^2} - \frac{k^2\cosh kx}{6b^3(a+bx)} + \frac{k^3}{6b^3}\int \frac{\sinh kx}{a+bx}\, dx$
$$\text{(see } \mathbf{2.476}\ 1)$$

9. $$\int \frac{\sinh kx}{(a+bx)^5}\, dx = -\frac{\sinh kx}{4b(a+bx)^4} - \frac{k\cosh kx}{12b^2(a+bx)^3} - \frac{k^2 \sinh kx}{24b^3(a+bx)^2}$$
$$-\frac{k^3 \cosh kx}{24b^4(a+bx)} + \frac{k^4}{24b^4}\int \frac{\sinh kx}{a+bx}\, dx$$
(see **2.476** 1)

10. $$\int \frac{\cosh kx}{(a+bx)^5}\, dx = -\frac{\cosh kx}{4b(a+bx)^4} - \frac{k\sinh kx}{12b^2(a+bx)^3} - \frac{k^2 \cosh kx}{24b^3(a+bx)^2}$$
$$-\frac{k^3 \sinh kx}{24b^4(a+bx)} + \frac{k^4}{24b^4}\int \frac{\cosh kx}{a+bx}\, dx$$
(see **2.476** 2)

11. $$\int \frac{\sinh kx}{(a+bx)^6}\, dx = -\frac{\sinh kx}{5b(a+bx)^5} - \frac{k\cosh kx}{20b^2(a+bx)^4} - \frac{k^2 \sinh kx}{60b^3(a+bx)^3} - \frac{k^3 \cosh kx}{120b^4(a+bx)^2}$$
$$-\frac{k^4 \sinh kx}{120b^5(a+bx)} + \frac{k^5}{120b^5}\int \frac{\cosh kx}{a+bx}\, dx$$
(see **2.476** 2)

12. $$\int \frac{\cosh kx}{(a+bx)^6}\, dx = -\frac{\cosh kx}{5b(a+bx)^5} - \frac{k\sinh kx}{20b^2(a+bx)^4} - \frac{k^2 \cosh kx}{60b^3(a+bx)^3} - \frac{k^3 \sinh kx}{120b^4(a+bx)^2}$$
$$-\frac{k^4 \cosh kx}{120b^5(a+bx)} + \frac{k^5}{120b^5}\int \frac{\sinh kx}{a+bx}\, dx$$
(see **2.476** 1)

2.477

1. $$\int \frac{x^p\, dx}{\sinh^q x} = \frac{-px^{p-1}\sinh x - (q-2)x^p\cosh x}{(q-1)(q-2)\sinh^{q-1} x} + \frac{p(p-1)}{(q-1)(q-2)}\int \frac{x^{p-2}}{\sinh^{q-2} x}\, dx$$
$$-\frac{q-2}{q-1}\int \frac{x^p\, dx}{\sinh^{q-2} x}$$
$$[q > 2] \qquad\qquad \text{GU (353)(8a)}$$

2. $$\int \frac{x^p\, dx}{\cosh^q x} = \frac{px^{p-1}\cosh x + (q-2)x^p\sinh x}{(q-1)(q-2)\cosh^{q-1} x} - \frac{p(p-1)}{(q-1)(q-2)}\int \frac{x^{p-2}\, dx}{\cosh^{q-2} x}$$
$$+\frac{q-2}{q-1}\int \frac{x^p\, dx}{\cosh^{q-2} x}$$
$$[q > 2] \qquad\qquad \text{GU (353)(10a)}$$

3. $$\int \frac{x^n}{\sinh x}\, dx = \sum_{k=0}^{\infty} \frac{(2-2^{2k})B_{2k}}{(n+2k)(2k)!}x^{n+2k} \qquad [|x| < \pi, \quad n > 0] \qquad \text{GU(353)(8b)}$$

4. $$\int \frac{x^n}{\cosh x}\, dx = \sum_{k=0}^{\infty} \frac{E_{2k}x^{n+2k+1}}{(n+2k+1)(2k)!} \qquad \left[|x| < \frac{\pi}{2}, \quad n \geq 0\right] \qquad \text{GU (353)(10b)}$$

5. $$\int \frac{dx}{x^n \sinh x} = -[1+(-1)^n]\frac{2^{n-1}-1}{n!}B_n \ln x$$
$$+\sum_{\substack{k=0 \\ k \neq \frac{n}{2}}}^{\infty} \frac{2-2^{2k}}{(2k-n)(2k)!}B_{2k}x^{2k-n}$$
$$[|x| < \pi, \quad n \geq 1] \qquad \text{GU (353)(9b)}$$

6. $\int \dfrac{dx}{x^n \cosh x} = \displaystyle\sum_{\substack{k=0 \\ k \neq \frac{n-1}{2}}}^{\infty} \dfrac{E_{2k}}{(2k-n+1)(2k)!} x^{2k-n+1} + \dfrac{1}{2}\left[1-(-1)^{n-1}\right] + \dfrac{E_{n-1}}{(n-1)!}\ln x$

$$\left[|x| < \dfrac{\pi}{2}\right] \qquad \text{GU (353)(11b)}$$

7. $\int \dfrac{x^n}{\sinh^2 x}\,dx = -x^n \coth x + n\displaystyle\sum_{k=0}^{\infty}\dfrac{2^{2k}B_{2k}}{(n+2k-1)(2k)!}x^{n+2k-1}$

$$[n > 1, \quad |x| < \pi] \qquad \text{GU (353)(8c)}$$

8. $\int \dfrac{x^n}{\cosh^2 x}\,dx = x^n \tanh x - n\displaystyle\sum_{k=1}^{\infty}\dfrac{2^{2k}\left(2^{2k}-1\right)B_{2k}}{(n+2k-1)(2k)!}x^{n+2k-1}$

$$\left[n > 1, \quad |x| < \dfrac{\pi}{2}\right] \qquad \text{GU (353)(10c)}$$

9. $\int \dfrac{dx}{x^n \sinh^2 x} = -\dfrac{\coth x}{x^n} - \left[1-(-1)^n\right]\dfrac{2^n n}{(n+1)!}B_{n+1}\ln x$

$$-\dfrac{n}{x^{n+1}}\displaystyle\sum_{\substack{k=0 \\ k \neq \frac{n+1}{2}}}^{\infty}\dfrac{B_{2k}}{(2k-n-1)(2k)!}(2x)^{2^k}$$

$$[|x| < \pi] \qquad \text{GU (353)(9c)}$$

10. $\int \dfrac{dx}{x^n \cosh^2 x} = \dfrac{\tanh x}{x^n} + \left[1-(-1)^n\right] - \dfrac{2n\left(2^{n+1}-1\right)n}{(n+1)!}B_{n+1}\ln x$

$$+\dfrac{n}{x^{n+1}}\displaystyle\sum_{\substack{k=1 \\ k \neq \frac{n+1}{2}}}^{\infty}\dfrac{\left(2^{2k}-1\right)B_{2k}}{(2k-n-1)(2k)!}(2x)^{2^k}$$

$$\left[|x| < \dfrac{\pi}{2}\right] \qquad \text{GU (353)(11c)}$$

11. $\int \dfrac{x}{\sinh^{2n} x}\,dx = \displaystyle\sum_{k=1}^{n-1}(-1)^k\dfrac{(2n-2)(2n-4)\dots(2n-2k+2)}{(2n-1)(2n-3)\dots(2n-2k+1)}$

$$\times\left\{\dfrac{x\cosh x}{\sinh^{2n-2k+1} x}+\dfrac{1}{(2n-2k)\sinh^{2n-2k} x}\right\}+(-1)^{n-1}\dfrac{(2n-2)!!}{(2n-1)!!}\int\dfrac{x\,dx}{\sinh^2 x}$$

$$\text{(see } \mathbf{2.477}\ 17) \qquad \text{GU (353)(8e)}$$

12. $\int \dfrac{x}{\sinh^{2n-1} x}\,dx$

$$=\displaystyle\sum_{k=1}^{n-1}(-1)^k\dfrac{(2n-3)(2n-5)\dots(2n-2k+1)}{(2n-2)(2n-4)\dots(2n-2k)}$$

$$\times\left\{\dfrac{x\cosh x}{\sinh^{2n-2k} x}+\dfrac{1}{(2n-2k-1)\sinh^{2n-2k-1} x}\right\}+(-1)^{n-1}\dfrac{(2n-3)!!}{(2n-2)!!}\int\dfrac{x\,dx}{\sinh x}$$

$$\text{(see } \mathbf{2.477}\ 15) \qquad \text{GU (353)(8e)}$$

13. $$\int \frac{x}{\cosh^{2n} x}\, dx = \sum_{k=1}^{n-1} \frac{(2n-2)(2n-4)\ldots(2n-2k+2)}{(2n-1)(2n-3)\ldots(2n-2k+1)}$$
$$\times \left\{\frac{x\sinh x}{\cosh^{2n-2k+1} x} + \frac{1}{(2n-2k)\cosh^{2n-2k} x}\right\} + \frac{(2n-2)!!}{(2n-1)!!}\int \frac{x\, dx}{\cosh^2 x}$$
(see **2.477** 18)   GU (353)(10e)

14. $$\int \frac{x}{\cosh^{2n-1} x}\, dx = \sum_{k=1}^{n-1} \frac{(2n-3)(2n-5)\ldots(2n-2k+1)}{(2n-2)(2n-4)\ldots(2n-2k)}$$
$$\times \left\{\frac{x\sinh x}{\cosh^{2n-2k} x} + \frac{1}{(2n-2k-1)\cosh^{2n-2k-1} x}\right\} + \frac{(2n-3)!!}{(2n-2)!!}\int \frac{x\, dx}{\cosh x}$$
(see **2.477** 16)   GU (353)(10e)

15. $$\int \frac{x\, dx}{\sinh x} = \sum_{k=0}^{\infty} \frac{2 - 2^{2k}}{(2k+1)(2k)!} B_{2k} x^{2k+1} \qquad |x| < \pi \qquad \text{GU (353)(8b)a}$$

16. $$\int \frac{x\, dx}{\cosh x} = \sum_{k=0}^{\infty} \frac{E_{2k} x^{2k+2}}{(2k+2)(2k)!} \qquad |x| < \frac{\pi}{2} \qquad \text{GU (353)(10b)a}$$

17. $$\int \frac{x\, dx}{\sinh^2 x} = -x\coth x + \ln\sinh x \qquad\qquad\qquad\qquad \text{MZ 257}$$

18. $$\int \frac{x\, dx}{\cosh^2 x} = x\tanh x - \ln\cosh x \qquad\qquad\qquad\qquad \text{MZ 262}$$

19. $$\int \frac{x\, dx}{\sinh^3 x} = -\frac{x\cosh x}{2\sinh^2 x} - \frac{1}{2\sinh x} - \frac{1}{2}\int \frac{x\, dx}{\sinh x} \qquad (\text{see } \mathbf{2.477}\ 15) \qquad \text{MZ 257}$$

20. $$\int \frac{x\, dx}{\cosh^3 x} = \frac{x\sinh x}{2\cosh^2 x} + \frac{1}{2\cosh x} + \frac{1}{2}\int \frac{x\, dx}{\cosh x} \qquad (\text{see } \mathbf{2.477}\ 16) \qquad \text{MZ 262}$$

21. $$\int \frac{x\, dx}{\sinh^4 x} = -\frac{x\cosh x}{3\sinh^3 x} - \frac{1}{6\sinh^2 x} + \frac{2}{3}x\coth x - \frac{2}{3}\ln\sinh x \qquad \text{MZ 258}$$

22. $$\int \frac{x\, dx}{\cosh^4 x} = \frac{x\sinh x}{3\cosh^3 x} + \frac{1}{6\cosh^2 x} + \frac{2}{3}x\tanh x - \frac{2}{3}\ln\cosh x \qquad \text{MZ 262}$$

23. $$\int \frac{x\, dx}{\sinh^5 x} = -\frac{x\cosh x}{4\sinh^4 x} - \frac{1}{12\sinh^3 x} + \frac{3x\cosh x}{8\sinh^2 x} + \frac{3}{8\sinh x} + \frac{3}{8}\int \frac{x\, dx}{\sinh x}$$
(see **2.477** 15)   MZ 258

24. $$\int \frac{x\, dx}{\cosh^5 x} = \frac{x\sinh x}{4\cosh^4 x} + \frac{1}{12\cosh^3 x} + \frac{3x\sinh x}{8\cosh^2 x} + \frac{3}{8\cosh x} + \frac{3}{8}\int \frac{x\, dx}{\cosh x}$$
(see **2.477** 16)   MZ 262

2.478

1. $$\int \frac{x^n \cosh x\, dx}{(a+b\sinh x)^m} = -\frac{x^n}{(m-1)b(a+b\sinh x)^{m-1}} + \frac{n}{(m-1)b}\int \frac{x^{n-1}\, dx}{(a+b\sinh x)^{m-1}}$$
$$[m \neq 1] \qquad \text{MZ 263}$$

2. $$\int \frac{x^n \sinh x\, dx}{(a+b\cosh x)^m} = -\frac{x^n}{(m-1)b(a+b\cosh x)^{m-1}} + \frac{n}{(m-1)b}\int \frac{x^{n-1}\, dx}{(a+b\cosh x)^{m-1}}$$
$$[m \neq 1] \qquad \text{MZ 263}$$

3. $\displaystyle\int \frac{x\,dx}{1+\cosh x} = x\tanh\frac{x}{2} - 2\ln\cosh\frac{x}{2}$

4. $\displaystyle\int \frac{x\,dx}{1-\cosh x} = x\coth\frac{x}{2} - 2\ln\sinh\frac{x}{2}$

5. $\displaystyle\int \frac{x\sinh x\,dx}{(1+\cosh x)^2} = -\frac{x}{1+\cosh x} + \tanh\frac{x}{2}$

6. $\displaystyle\int \frac{x\sinh x\,dx}{(1-\cosh x)^2} = \frac{x}{1-\cosh x} - \coth\frac{x}{2}$ MZ 262-264

7. $\displaystyle\int \frac{x\,dx}{\cosh 2x - \cos 2t} = \frac{1}{2\sin 2t}\left[L(u+t) - L(u-t) - 2\,L(t)\right]$

$\left[u = \arctan\left(\tanh x \cot t\right), \quad t \neq \pm n\pi\right]$
LO III 402

8. $\displaystyle\int \frac{x\cosh x\,dx}{\cosh 2x - \cos 2t} = \frac{1}{2\sin t}\left[L\left(\frac{u+t}{2}\right) - L\left(\frac{u-t}{2}\right) + L\left(\pi - \frac{v+t}{2}\right)\right.$

$$\left. + L\left(\frac{v-t}{2}\right) - 2\,L\left(\frac{t}{2}\right) - 2\,L\left(\frac{\pi-t}{2}\right)\right]$$

$\left[u = 2\arctan\left(\tanh\frac{x}{2}\cdot\cot\frac{t}{2}\right), \quad v = 2\arctan\left(\coth\frac{x}{2}\cdot\cot\frac{t}{2}\right); \quad t \neq \pm n\pi\right]$ LO III 403

2.479

1. $\displaystyle\int x^p \frac{\sinh^{2m} x}{\cosh^n x}\,dx = \sum_{k=0}^{m} (-1)^{m+k}\binom{m}{k}\int \frac{x^p\,dx}{\cosh^{n-2k} x}$ (see **4.477** 2)

2. $\displaystyle\int x^p \frac{\sinh^{2m+1} x}{\cosh^n x}\,dx = \sum_{k=0}^{m} (-1)^{m+k}\binom{m}{k}\int x^p \frac{\sinh x}{\cosh^{n-2k} x}\,dx$

$[n>1]$ (see **2.479** 3)

3. $\displaystyle\int x^p \frac{\sinh x}{\cosh^n x}\,dx = -\frac{x^p}{(n-1)\cosh^{n-1} x} + \frac{p}{n-1}\int \frac{x^{p-1}\,dx}{\cosh^{n-1} x}$

$[n>1]$ (see **2.477** 2)   GU (353)(12)

4. $\displaystyle\int x^p \frac{\cosh^{2m} x}{\sinh^n x}\,dx = \sum_{k=0}^{m}\binom{m}{k}\int \frac{x^p\cosh x}{\sinh^{n-2k} x}$ (see **2.477** 1)

5. $\displaystyle\int x^p \frac{\cosh^{2m+1} x}{\sinh^n x}\,dx = \sum_{k=0}^{m}\binom{m}{k}\int \frac{x^p\cosh x}{\sinh^{n-2k} x}\,dx$ (see **2.479** 6)

6. $\displaystyle\int x^p \frac{\cosh x}{\sinh^n x}\,dx = -\frac{x^p}{(n-1)\sinh^{n-1} x} + \frac{p}{n-1}\int \frac{x^{p-1}\,dx}{\sinh^{n-1} x}$

$[n>1]$ (see **2.477** 1)
GU (353)(13c)

7. $\displaystyle\int x^p \tanh x\,dx = \sum_{k=1}^{\infty} \frac{2^{2k}\left(2^{2k}-1\right)B_{2k}}{(2k+p)(2k)!}x^{p+2k}$ $\left[p>-1, \quad |x|<\frac{\pi}{2}\right]$ GU (353)(12d)

8. $\displaystyle\int x^p \coth x \, dx = \sum_{k=0}^{\infty} \frac{2^{2k} B_{2k}}{(p+2k)(2k)!} x^{p+2k}$ $[p \geq +1, \quad |x| < \pi]$ GU (353)(13d)

9. $\displaystyle\int \frac{x \cosh x}{\sinh^2 x} \, dx = \ln \tanh \frac{x}{2} - \frac{x}{\sinh x}$

10. $\displaystyle\int \frac{x \sinh x}{\cosh^2 x} \, dx = -\frac{x}{\cosh x} + \arctan(\sinh x)$ MZ 263

2.48 Combinations of hyperbolic functions, exponentials, and powers

2.481

1. $\displaystyle\int e^{ax} \sinh(bx + c) \, dx = \frac{e^{ax}}{a^2 - b^2} \left[a \sinh(bx + c) - b \cosh(bx + c) \right]$

$$[a^2 \neq b^2]$$

2. $\displaystyle\int e^{ax} \cosh(bx + c) \, dx = \frac{e^{ax}}{a^2 - b^2} \left[a \cosh(bx + c) - b \sinh(bx + c) \right]$

$$[a^2 \neq b^2]$$

For $a^2 = b^2$:

3. $\displaystyle\int e^{ax} \sinh(ax + c) \, dx = -\frac{1}{2} x e^{-c} + \frac{1}{4a} e^{2ax+c}$

4. $\displaystyle\int e^{-ax} \sinh(ax + c) \, dx = \frac{1}{2} x e^{c} + \frac{1}{4a} e^{-(2ax+c)}$

5. $\displaystyle\int e^{ax} \cosh(ax + c) \, dx = \frac{1}{2} x e^{-c} + \frac{1}{4a} e^{2ax+c}$

6. $\displaystyle\int e^{-ax} \cosh(ax + c) \, dx = \frac{1}{2} x e^{c} - \frac{1}{4a} e^{-(2ax+c)}$ MZ 275-277

2.482

1. $\displaystyle\int x^p e^{ax} \sinh bx \, dx = \frac{1}{2} \left\{ \int x^p e^{(a+b)x} \, dx - \int x^p e^{(a-b)x} \, dx \right\}$

$$[a^2 \neq b^2]$$

2. $\displaystyle\int x^p e^{ax} \cosh bx \, dx = \frac{1}{2} \left\{ \int x^p e^{(a+b)x} \, dx + \int x^p e^{(a-b)x} \, dx \right\}$

$$[a^2 \neq b^2]$$

For $a^2 = b^2$:

3. $\displaystyle\int x^p e^{ax} \sinh ax \, dx = \frac{1}{2} \int x^p e^{2ax} \, dx - \frac{x^{p+1}}{2(p+1)}$ (see **2.321**)

4. $\displaystyle\int x^p e^{-ax} \sinh ax \, dx = \frac{x^{p+1}}{2(p+1)} - \frac{1}{2} \int x^p e^{-2ax} \, dx$ (see **2.321**)

5. $\int x^p e^{ax} \cosh ax\, dx = \dfrac{x^{p+1}}{2(p+1)} + \dfrac{1}{2}\int x^p e^{2ax}\, dx$ (see **2.321**)   MZ 276, 278

2.483

1. $\int x e^{ax} \sinh bx\, dx = \dfrac{e^{ax}}{a^2 - b^2}\left[\left(ax - \dfrac{a^2 + b^2}{a^2 - b^2}\right)\sinh bx - \left(bx - \dfrac{2ab}{a^2 - b^2}\right)\cosh bx\right]$

$$[a^2 \neq b^2]$$

2. $\int x e^{ax} \cosh bx\, dx = \dfrac{e^{ax}}{a^2 - b^2}\left[\left(ax - \dfrac{a^2 + b^2}{a^2 - b^2}\right)\cosh bx - \left(bx - \dfrac{2ab}{a^2 - b^2}\right)\sinh bx\right]$

$$[a^2 \neq b^2]$$

3. $\int x^2 e^{ax} \sinh bx\, dx = \dfrac{e^{ax}}{a^2 - b^2}\left\{\left[ax^2 - \dfrac{2\left(a^2 + b^2\right)}{a^2 - b^2}x + \dfrac{2a\left(a^2 + 3b^2\right)}{\left(a^2 - b^2\right)^2}\right]\sinh bx\right.$

$$\left. - \left[bx^2 - \dfrac{4ab}{a^2 - b^2}x + \dfrac{2b\left(3a^2 + b^2\right)}{\left(a^2 - b^2\right)^2}\right]\cosh x\right\} \qquad [a^2 \neq b^2]$$

4. $\int x^2 e^{ax} \cosh bx\, dx = \dfrac{e^{ax}}{a^2 - b^2}\left\{\left[ax^2 - \dfrac{2\left(a^2 + b^2\right)}{a^2 - b^2}x + \dfrac{2a\left(a^2 + 3b^2\right)}{\left(a^2 - b^2\right)^2}\right]\cosh bx\right.$

$$\left. - \left[bx^2 - \dfrac{4ab}{a^2 - b^2}x + \dfrac{2b\left(3a^2 + b^2\right)}{\left(a^2 - b^2\right)^2}\right]\sinh x\right\} \qquad [a^2 \neq b^2]$$

For $a^2 = b^2$:

5. $\int x e^{ax} \sinh ax\, dx = \dfrac{e^{2ax}}{4a}\left(x - \dfrac{1}{2a}\right) - \dfrac{x^2}{4}$

6. $\int x e^{-ax} \sinh ax\, dx = \dfrac{e^{-2ax}}{4a}\left(x + \dfrac{1}{2a}\right) + \dfrac{x^2}{4}$ MZ 276, 278

7. $\int x e^{ax} \cosh ax\, dx = \dfrac{x^2}{4} + \dfrac{e^{2ax}}{4a}\left(x - \dfrac{1}{2a}\right)$

8. $\int x e^{-ax} \cosh ax\, dx = \dfrac{x^2}{4} - \dfrac{e^{-2ax}}{4a}\left(x + \dfrac{1}{2a}\right)$

9. $\int x^2 e^{ax} \sinh ax\, dx = \dfrac{e^{2ax}}{4a}\left(x^2 - \dfrac{x}{a} + \dfrac{1}{2a^2}\right) - \dfrac{x^3}{6}$

10. $\int x^2 e^{-ax} \sinh ax\, dx = \dfrac{e^{-2ax}}{4a}\left(x^2 + \dfrac{x}{a} + \dfrac{1}{2a^2}\right) + \dfrac{x^3}{6}$

11. $\int x^2 e^{ax} \cosh ax\, dx = \dfrac{x^3}{6} + \dfrac{e^{2ax}}{4a}\left(x^2 - \dfrac{x}{a} + \dfrac{1}{2a^2}\right)$

2.484

1. $\int e^{ax} \sinh bx\, \dfrac{dx}{x} = \dfrac{1}{2}\left\{\mathrm{Ei}[(a+b)x] - \mathrm{Ei}[(a-b)x]\right\}$ $[a^2 \neq b^2]$

2. $\int e^{ax} \cosh bx\, \dfrac{dx}{x} = \dfrac{1}{2}\left\{\mathrm{Ei}[(a+b)x] + \mathrm{Ei}[(a-b)x]\right\}$ $[a^2 \neq b^2]$

3. $\displaystyle\int e^{ax}\sinh bx\,\frac{dx}{x^2} = -\frac{e^{ax}\sinh bx}{2x} + \frac{1}{2}\left\{(a+b)\operatorname{Ei}[(a+b)x] - (a-b)\operatorname{Ei}[(a-b)x]\right\}$

$$\left[a^2 \neq b^2\right]$$

4. $\displaystyle\int e^{ax}\cosh bx\,\frac{dx}{x^2} = -\frac{e^{ax}\cosh bx}{2x} + \frac{1}{2}\left\{(a+b)\operatorname{Ei}[(a+b)x] + (a-b)\operatorname{Ei}[(a-b)x]\right\}$

$$\left[a^2 \neq b^2\right]$$

For $a^2 = b^2$:

5. $\displaystyle\int e^{ax}\sinh ax\,\frac{dx}{x} = \frac{1}{2}\left[\operatorname{Ei}(2ax) - \ln x\right]$

6. $\displaystyle\int e^{-ax}\sinh ax\,\frac{dx}{x} = \frac{1}{2}\left[\ln x - \operatorname{Ei}(-2ax)\right]$

7. $\displaystyle\int e^{ax}\cosh ax\,\frac{dx}{x} = \frac{1}{2}\left[\ln x + \operatorname{Ei}(2ax)\right]$

8. $\displaystyle\int e^{ax}\sinh ax\,\frac{dx}{x^2} = -\frac{1}{2x}\left(e^{2ax} - 1\right) + a\operatorname{Ei}(2ax)$

9. $\displaystyle\int e^{-ax}\sinh ax\,\frac{dx}{x^2} = -\frac{1}{2x}\left(1 - e^{-2ax}\right) + a\operatorname{Ei}(-2ax)$

10. $\displaystyle\int e^{ax}\cosh ax\,\frac{dx}{x^2} = -\frac{1}{2x}\left(e^{2ax} + 1\right) + a\operatorname{Ei}(2ax)$ MZ 276, 278

2.5–2.6 Trigonometric Functions

2.50 Introduction

2.501   Integrals of the form $\displaystyle\int R\left(\sin x, \cos x\right)\,dx$ can always be reduced to integrals of rational functions by means of the substitution $t = \tan\dfrac{x}{2}$.

2.502   If $R\left(\sin x, \cos x\right)$ satisfies the relation

$$R\left(\sin x, \cos x\right) = -R\left(-\sin x, \cos x\right),$$

it is convenient to make the substitution $t = \cos x$.

2.503   If this function satisfies the relation

$$R\left(\sin x, \cos x\right) = -R\left(\sin x, -\cos x\right),$$

it is convenient to make the substitution $t = \sin x$.

2.504   If this function satisfies the relation

$$R\left(\sin x, \cos x\right) = R\left(-\sin x, -\cos x\right),$$

it is convenient to make the substitution $t = \tan x$.

2.51–2.52 Powers of trigonometric functions

2.510
$$\int \sin^p x \cos^q x \, dx = -\frac{\sin^{p-1} x \cos^{q+1} x}{q+1} + \frac{p-1}{q+1} \int \sin^{p-2} x \cos^{q+2} x \, dx$$

$$= -\frac{\sin^{p-1} x \cos^{q+1} x}{p+q} + \frac{p-1}{p+q} \int \sin^{p-2} x \cos^q x \, dx$$

$$= \frac{\sin^{p+1} x \cos^{q+1} x}{p+1} + \frac{p+q+2}{p+1} \int \sin^{p+2} x \cos^q x \, dx$$

$$= \frac{\sin^{p+1} x \cos^{q-1} x}{p+1} + \frac{q-1}{p+1} \int \sin^{p+2} x \cos^{q-2} x \, dx$$

$$= \frac{\sin^{p+1} x \cos^{q-1} x}{p+q} + \frac{q-1}{p+q} \int \sin^p x \cos^{q-2} x \, dx$$

$$= -\frac{\sin^{p+1} x \cos^{q+1} x}{q+1} + \frac{p+q+2}{q+1} \int \sin^p x \cos^{q+2} x \, dx$$

$$= \frac{\sin^{p-1} x \cos^{q-1} x}{p+q} \left\{ \sin^2 x - \frac{q-1}{p+q-2} \right\}$$

$$+ \frac{(p-1)(q-1)}{(p+q)(p+q-2)} \int \sin^{p-2} x \cos^{q-2} x \, dx$$

FI II 89, TI 214

2.511

1. $\displaystyle\int \sin^p x \cos^{2n} x \, dx$

$$= \frac{\sin^{p+1} x}{2n+p} \left\{ \cos^{2n-1} x + \sum_{k=1}^{n-1} \frac{(2n-1)(2n-3)\ldots(2n-2k+1) \cos^{2n-2k-1} x}{(2n+p-2)(2n+p-4)\ldots(2n+p-2k)} \right\}$$

$$+ \frac{(2n-1)!!}{(2n+p)(2n+p-2)\ldots(p+2)} \int \sin^p x \, dx$$

This formula is applicable for arbitrary real p except for the following negative even integers: -2, $-4, \ldots, -2n$. If p is a natural number and $n = 0$, we have:

2. $\displaystyle\int \sin^{2l} x \, dx$

$$= -\frac{\cos x}{2l} \left\{ \sin^{2l-1} x + \sum_{k=1}^{l-1} \frac{(2l-1)(2l-3)\ldots(2l-2k+1)}{2^k(l-1)(l-2)\ldots(l-k)} \sin^{2l-2k-1} x \right\}$$

$$+ \frac{(2l-1)!!}{2^l l!} x$$

(see also 2.513 1)
TI (232)

3. $\displaystyle\int \sin^{2l+1} x \, dx = -\frac{\cos x}{2l+1} \left\{ \sin^{2l} x + \sum_{k=0}^{l-1} \frac{2^{k+1} l(l-1)\ldots(l-k)}{(2l-1)(2l-3)\ldots(2l-2k-1)} \sin^{2l-2k-2} x \right\}$

(see also 2.513 2)
TI (233)

4. $\displaystyle\int \sin^p x \cos^{2n+1} x \, dx = \frac{\sin^{p+1} x}{2n+p+1} \left\{ \cos^{2n} x + \sum_{k=1}^{n} \frac{2^k n(n-1)\ldots(n-k+1) \cos^{2n-2k} x}{(2n+p-1)(2n+p-3)\ldots(2n+p-2k+1)} \right\}$

This formula is applicable for arbitrary real p except for the negative odd integers: $-1, -3, \ldots$, $-(2n+1)$.

2.512

1. $\displaystyle \int \cos^p x \sin^{2n} x \, dx$

$$= -\frac{\cos^{p+1} x}{2n+p} \left\{ \sin^{2n-1} x + \sum_{k=1}^{n-1} \frac{(2n-1)(2n-3)\ldots(2n-2k+1)\sin^{2n-2k-1} x}{(2n+p-2)(2n+p-4)\ldots(2n+p-2k)} \right\}$$

$$+ \frac{(2n-1)!!}{(2n+p)(2n+p-2)\ldots(p+2)} \int \cos^p x \, dx$$

This formula is applicable for arbitrary real p except for the following negative even integers: -2, -4, $\ldots$, $-2n$. If p is a natural number and $n = 0$, we have

2. $\displaystyle \int \cos^{2l} x \, dx = \frac{\sin x}{2l} \left\{ \cos^{2l-1} x + \sum_{k=1}^{l-1} \frac{(2l-1)(2l-3)\ldots(2l-2k+1)}{2^k (l-1)(l-2)\ldots(l-k)} \cos^{2l-2k-1} x \right\}$

$$+ \frac{(2l-1)!!}{2^l l!} x$$

                                     (see also **2.513** 3)   TI (230)

3. $\displaystyle \int \cos^{2l+1} x \, dx = \frac{\sin x}{2l+1} \left\{ \cos^{2l} x + \sum_{k=0}^{l-1} \frac{2^{k+1} l(l-1)\ldots(l-k)}{(2l-1)(2l-3)\ldots(2l-2k-1)} \cos^{2l-2k-2} x \right\}$

                                     (see also **2.513** 4)   TI (231)

4. $\displaystyle \int \cos^p x \sin^{2n+1} x \, dx$

$$= -\frac{\cos^{p+1} x}{2n+p+1} \left\{ \sin^{2n} x + \sum_{k=1}^{n} \frac{2^k n(n-1)\ldots(n-k+1)\sin^{2n-2k} x}{(2n+p-1)(2n+p-3)\ldots(2n+p-2k+1)} \right\}$$

This formula is applicable for arbitrary real p except for the following negative odd integers: -1, -3, $\ldots$, $-(2n+1)$.

2.513

1. $\displaystyle \int \sin^{2n} x \, dx = \frac{1}{2^{2n}} \binom{2n}{n} x + \frac{(-1)^n}{2^{2n-1}} \sum_{k=0}^{n-1} (-1)^k \binom{2n}{k} \frac{\sin(2n-2k)x}{2n-2k}$ (see also **2.511** 2)

                                                         TI (226)

2. $\displaystyle \int \sin^{2n+1} x \, dx = \frac{1}{2^{2n}} (-1)^{n+1} \sum_{k=0}^{n} (-1)^k \binom{2n+1}{k} \frac{\cos(2n+1-2k)x}{2n+1-2k}$ (see also **2.511** 3)

                                                         TI (227)

3. $\displaystyle \int \cos^{2n} x \, dx = \frac{1}{2^{2n}} \binom{2n}{n} x + \frac{1}{2^{2n-1}} \sum_{k=0}^{n-1} \binom{2n}{k} \frac{\sin(2n-2k)x}{2n-2k}$

                                   (see also **2.512** 2)   TI (224)

4. $\displaystyle \int \cos^{2n+1} x \, dx = \frac{1}{2^{2n}} \sum_{k=0}^{n} \binom{2n+1}{k} \frac{\sin(2n-2k+1)x}{2n-2k+1}$

                                   (see also **2.512** 3)   TI (225)

5. $\displaystyle\int \sin^2 x\, dx = -\frac{1}{4}\sin 2x + \frac{1}{2}x = -\frac{1}{2}\sin x \cos x + \frac{1}{2}x$

6. $\displaystyle\int \sin^3 x\, dx = \frac{1}{12}\cos 3x - \frac{3}{4}\cos x = \frac{1}{3}\cos^3 x - \cos x$

7. $\displaystyle\int \sin^4 x\, dx = \frac{3x}{8} - \frac{\sin 2x}{4} + \frac{\sin 4x}{32}$
$$= -\frac{3}{8}\sin x \cos x - \frac{1}{4}\sin^3 x \cos x + \frac{3}{8}x$$

8. $\displaystyle\int \sin^5 x\, dx = -\frac{5}{8}\cos x + \frac{5}{48}\cos 3x - \frac{1}{80}\cos 5x$
$$= -\frac{1}{5}\sin^4 x \cos x + \frac{4}{15}\cos^3 x - \frac{4}{5}\cos x$$

9. $\displaystyle\int \sin^6 x\, dx = \frac{5}{16}x - \frac{15}{64}\sin 2x + \frac{3}{64}\sin 4x - \frac{1}{192}\sin 6x$
$$= -\frac{1}{6}\sin^5 x \cos x - \frac{5}{24}\sin^3 x \cos x - \frac{5}{16}\sin x \cos x + \frac{5}{16}x$$

10. $\displaystyle\int \sin^7 x\, dx = -\frac{35}{64}\cos x + \frac{7}{64}\cos 3x - \frac{7}{320}\cos 5x + \frac{1}{448}\cos 7x$
$$= -\frac{1}{7}\sin^6 x \cos x - \frac{6}{35}\sin^4 x \cos x + \frac{8}{35}\cos^3 x - \frac{24}{35}\cos x$$

11. $\displaystyle\int \cos^2 x\, dx = \frac{1}{4}\sin 2x + \frac{x}{2} = \frac{1}{2}\sin x \cos x + \frac{1}{2}x$

12. $\displaystyle\int \cos^3 x\, dx = \frac{1}{12}\sin 3x + \frac{3}{4}\sin x = \sin x - \frac{1}{3}\sin^3 x$

13. $\displaystyle\int \cos^4 x\, dx = \frac{3}{8}x + \frac{1}{4}\sin 2x + \frac{1}{32}\sin 4x = \frac{3}{8}x + \frac{3}{8}\sin x \cos x + \frac{1}{4}\sin x \cos^3 x$

14. $\displaystyle\int \cos^5 x\, dx = \frac{5}{8}\sin x + \frac{5}{48}\sin 3x + \frac{1}{80}\sin 5x = \frac{4}{5}\sin x - \frac{4}{15}\sin^3 x + \frac{1}{5}\cos^4 x \sin x$

15. $\displaystyle\int \cos^6 x\, dx = \frac{5}{16}x + \frac{15}{64}\sin 2x + \frac{3}{64}\sin 4x + \frac{1}{192}\sin 6x$
$$= \frac{5}{16}x + \frac{5}{16}\sin x \cos x + \frac{5}{24}\sin x \cos^3 x + \frac{1}{6}\sin x \cos^5 x$$

16. $\displaystyle\int \cos^7 x\, dx = \frac{35}{64}\sin x + \frac{7}{64}\sin 3x + \frac{7}{320}\sin 5x + \frac{1}{448}\sin 7x$
$$= \frac{24}{35}\sin x - \frac{8}{35}\sin^3 x + \frac{6}{35}\sin x \cos^4 x + \frac{1}{7}\sin x \cos^6 x$$

17. $\displaystyle\int \sin x \cos^2 x\, dx = -\frac{1}{4}\left\{\frac{1}{3}\cos 3x + \cos x\right\} = -\frac{\cos^3 x}{3}$

18. $\displaystyle\int \sin x \cos^3 x\, dx = -\frac{\cos^4 x}{4}$

19. $\displaystyle\int \sin x \cos^4 x\, dx = -\frac{\cos^5 x}{5}$

20. $\displaystyle\int \sin^2 x \cos x\, dx = -\frac{1}{4}\left\{\frac{1}{3}\sin 3x - \sin x\right\} = \frac{\sin^3 x}{3}$

21. $\int \sin^2 x \cos^2 x \, dx = -\dfrac{1}{8} \left\{ \dfrac{1}{4} \sin 4x - x \right\}$

22. $\int \sin^2 x \cos^3 x \, dx = -\dfrac{1}{16} \left\{ \dfrac{1}{5} \sin 5x + \dfrac{1}{3} \sin 3x - 2 \sin x \right\}$

$$= \frac{\sin^3 x}{5} \left(\cos^2 x + \frac{2}{3} \right) = \frac{\sin^3 x}{5} \left(\frac{5}{3} - \sin^2 x \right)$$

23. $\int \sin^2 x \cos^4 x \, dx = \dfrac{x}{16} + \dfrac{1}{64} \sin 2x - \dfrac{1}{64} \sin 4x - \dfrac{1}{192} \sin 6x$

24. $\int \sin^3 x \cos x \, dx = \dfrac{1}{8} \left(\dfrac{1}{4} \cos 4x - \cos 2x \right) = \dfrac{\sin^4 x}{4}$

25. $\int \sin^3 x \cos^2 x \, dx = \dfrac{1}{16} \left(\dfrac{1}{5} \cos 5x - \dfrac{1}{3} \cos 3x - 2 \cos x \right)$

$$= \frac{1}{5} \cos^5 x - \frac{1}{3} \cos^3 x$$

26. $\int \sin^3 x \cos^3 x \, dx = \dfrac{1}{32} \left(\dfrac{1}{6} \cos 6x - \dfrac{3}{2} \cos 2x \right)$

27. $\int \sin^3 x \cos^4 x \, dx = \dfrac{1}{7} \cos^3 x \left(-\dfrac{2}{5} - \dfrac{3}{5} \sin^2 x + \sin^4 x \right)$

28. $\int \sin^4 x \cos x \, dx = \dfrac{\sin^5 x}{5}$

29. $\int \sin^4 x \cos^2 x \, dx = \dfrac{1}{16} x - \dfrac{1}{64} \sin 2x - \dfrac{1}{64} \sin 4x + \dfrac{1}{192} \sin 6x$

30. $\int \sin^4 x \cos^3 x \, dx = \dfrac{1}{7} \sin^3 x \left(\dfrac{2}{5} + \dfrac{3}{5} \cos^2 x - \cos^4 x \right)$

31. $\int \sin^4 x \cos^4 x \, dx = \dfrac{3}{128} x - \dfrac{1}{128} \sin 4x + \dfrac{1}{1024} \sin 8x$

2.514 $\int \dfrac{\sin^p x}{\cos^{2n} x} \, dx$

$$= \frac{\sin^{p+1} x}{2n - 1} \left\{ \sec^{2n-1} x + \sum_{k=1}^{n-1} \frac{(2n - p - 2)(2n - p - 4) \ldots (2n - p - 2k)}{(2n - 3)(2n - 5) \ldots (2n - 2k - 1)} \sec^{2n-2k-1} x \right\}$$
$$+ \frac{(2n - p - 2)(2n - p - 4) \ldots (-p + 2)(-p)}{(2n - 1)!!} \int \sin^p x \, dx$$

This formula is applicable for arbitrary real p. For $\int \sin^p x \, dx$, where p is a natural number, see **2.511** 2, 3 and **2.513** 1, 2. If $n = 0$ and p is a negative integer, we have for this integral:

2.515

1. $\int \dfrac{dx}{\sin^{2l} x} = -\dfrac{\cos x}{2l - 1} \left\{ \operatorname{cosec}^{2l-1} x + \sum_{k=1}^{l-1} \dfrac{2^k (l - 1)(l - 2) \ldots (l - k)}{(2l - 3)(2l - 5) \ldots (2l - 2k - 1)} \operatorname{cosec}^{2l-2k-1} x \right\}$

TI (242)

2. $$\int \frac{dx}{\sin^{2l+1} x} = -\frac{\cos x}{2l}\left\{\operatorname{cosec}^{2l} x + \sum_{k=1}^{l-1} \frac{(2l-1)(2l-3)\ldots(2l-2k+1)}{28^k(l-1)(l-2)\ldots(l-k)}\operatorname{cosec}^{2l-2k} x\right\}$$
$$+\frac{(2l-1)!!}{2^l l!}\ln\tan\frac{x}{2}$$

TI (243)

2.516

1. $$\int \frac{\sin^p x\, dx}{\cos^{2n+1} x}$$
$$= \frac{\sin^{p+1} x}{2n}\left\{\sec^{2n} x + \sum_{k=1}^{n-1}\frac{(2n-p-1)(2n-p-3)\cdots(2n-p-2k+1)}{2^k(n-1)(n-2)\cdots(n-k)}\sec^{2n-2k} x\right\}$$
$$+\frac{(2n-p-1)(2n-p-3)\cdots(3-p)(1-p)}{2^n n!}\int\frac{\sin^p x}{\cos x}\,dx$$

This formula is applicable for arbitrary real p. For $n=0$ and p a natural number, we have

2. $$\int\frac{\sin^{2l+1} x\, dx}{\cos x} = -\sum_{k=1}^{l}\frac{\sin^{2k} x}{2k} - \ln\cos x$$

3. $$\int\frac{\sin^{2l} x\, dx}{\cos x} = -\sum_{k=1}^{l}\frac{\sin^{2k-1} x}{2k-1} + \ln\tan\left(\frac{\pi}{4}+\frac{x}{2}\right)$$

2.517

1. $$\int\frac{dx}{\sin^{2m+1} x\cos x} = -\sum_{k=1}^{m}\frac{1}{(2m-2k+2)\sin^{2m-2k+2} x} + \ln\tan x$$

2. $$\int\frac{dx}{\sin^{2m} x\cos x} = -\sum_{k=1}^{m}\frac{1}{(2m-2k+1)\sin^{2m-2k+1} x} + \ln\tan\left(\frac{\pi}{4}-\frac{x}{2}\right)$$

2.518

1. $$\int\frac{\sin^p x}{\cos^2 x}\,dx = \frac{\sin^{p-1} x}{\cos x} - (p-1)\int\sin^{p-2} x\, dx$$

2. $$\int\frac{\cos^p x\, dx}{\sin^{2n} x} = \frac{\cos^{p+1} x}{2n-1}\left\{\operatorname{cosec}^{2n-1} x\right.$$
$$+\sum_{k=1}^{n-1}\frac{(2n-p-2)(2n-p-4)\ldots(2n-p-2k)}{(2n-3)(2n-5)\ldots(2n-2k-1)}\operatorname{cosec}^{2n-2k-1} x\left.\right\}$$
$$+\frac{(2n-p-2)(2n-p-4)\ldots(2-p)(-p)}{(2n-1)!!}\int\cos^p x\, dx$$

This formula is applicable for arbitrary real p. For $\int\cos^p x\, dx$ where p is a natural number, see **2.512** 2, 3 and **2.513** 3, 4. If $n=0$ and p is a negative integer, we have for this integral:

2.519

1. $\displaystyle \int \frac{dx}{\cos^{2l} x} = \frac{\sin x}{2l-1} \left\{ \sec^{2l-1} x + \sum_{k=1}^{l-1} \frac{2^k (l-1)(l-2)\ldots(l-k)}{(2l-3)(2l-5)\ldots(2l-2k-1)} \sec^{2l-2k-1} x \right\}$ TI (240)

2. $\displaystyle \int \frac{dx}{\cos^{2l+1} x} = \frac{\sin x}{2l} \left\{ \sec^{2l} x + \sum_{k=1}^{l-1} \frac{(2l-1)(2l-3)\ldots(2l-2k+1)}{2^k(l-1)(l-2)\ldots(l-k)} \sec^{2l-2k} x \right\}$

               $\displaystyle + \frac{(2l-1)!!}{2^l l!} \ln \tan \left(\frac{\pi}{4} + \frac{x}{2} \right)$

                                                                     TI (241)

2.521

1. $\displaystyle \int \frac{\cos^p x \, dx}{\sin^{2n+1} x} = -\frac{\cos^{p+1} x}{2n} \left\{ \operatorname{cosec}^{2n} x \right.$

           $\displaystyle + \sum_{k=1}^{n-1} \frac{(2n-p-1)(2n-p-3)\ldots(2n-p-2k+1)}{2^k(n-1)(n-2)\ldots(n-k)} \operatorname{cosec}^{2n-2k} x \left. \right\}$

           $\displaystyle + \frac{(2n-p-1)(2n-p-3)\ldots(3-p)(1-p)}{2^n \cdot n!} \int \frac{\cos^p x}{\sin x} \, dx$

   This formula is applicable for arbitrary real p. For $n = 0$ and p a natural number, we have

2. $\displaystyle \int \frac{\cos^{2l+1} x \, dx}{\sin x} = \sum_{k=1}^{l} \frac{\cos^{2k} x}{2k} + \ln \sin x$

3. $\displaystyle \int \frac{\cos^{2l} x \, dx}{\sin x} = \sum_{k=1}^{l} \frac{\cos^{2k-1} x}{2k-1} + \ln \tan \frac{x}{2}$

2.522

1. $\displaystyle \int \frac{dx}{\sin x \cos^{2m+1} x} = \sum_{k=1}^{m} \frac{1}{(2m-2k+2)\cos^{2m-2k+2} x} + \ln \tan x$

2. $\displaystyle \int \frac{dx}{\sin x \cos^{2m} x} = \sum_{k=1}^{m} \frac{1}{(2m-2k+1)\cos^{2m-2k+1} x} + \ln \tan \frac{x}{2}$ GW (331)(15)

2.523 $\displaystyle \int \frac{\cos^m x}{\sin^2 x} \, dx = -\frac{\cos^{m-1} x}{\sin x} - (m-1) \int \cos^{m-2} x \, dx$

2.524   In formulas **2.524** 1 and **2.524** 2, $s = 1$ for m odd and $m < 2n+1$; in other cases, $s = 0$.

1. $\displaystyle \int \frac{\sin^{2n+1} x}{\cos^m x} \, dx = \sum_{\substack{k=0 \\ k \neq \frac{m-1}{2}}}^{n} (-1)^{k+1} \binom{n}{k} \frac{\cos^{2k-m+1} x}{2k-m+1} + s(-1)^{\frac{m+1}{2}} \binom{n}{\frac{m-1}{2}} \ln \cos x$

                                                                              GU (331)(11d)

2. $\displaystyle \int \frac{\cos^{2n+1} x}{\sin^m x} \, dx = \sum_{\substack{k=0 \\ k \neq \frac{m-1}{2}}}^{n} (-1)^k \binom{n}{k} \frac{\sin^{2k-m+1} x}{2k-m+1} + s(-1)^{\frac{m-1}{2}} \binom{n}{\frac{m-1}{2}} \ln \sin x$

2.525

1. $$\int \frac{dx}{\sin^{2m} x \cos^{2n} x} = \sum_{k=0}^{m+n-1} \binom{m+n-1}{k} \frac{\tan^{2k-2m+1} x}{2k-2m+1}$$ TI (267)

2. $$\int \frac{dx}{\sin^{2m+1} x \cos^{2n+1} x} = \sum_{k=0}^{m+n} \binom{m+n}{k} \frac{\tan^{2k-2m} x}{2k-2m} + \binom{m+n}{m} \ln \tan x$$

TI (268), GU (331)(15f)

2.526

1. $$\int \frac{dx}{\sin x} = \ln \tan \frac{x}{2}$$

2. $$\int \frac{dx}{\sin^2 x} = -\cot x$$

3. $$\int \frac{dx}{\sin^3 x} = -\frac{1}{2} \frac{\cos x}{\sin^2 x} + \frac{1}{2} \ln \tan \frac{x}{2}$$

4. $$\int \frac{dx}{\sin^4 x} = -\frac{\cos x}{3 \sin^3 x} - \frac{2}{3} \cot x = -\frac{1}{3} \cot^3 x - \cot x$$

5. $$\int \frac{dx}{\sin^5 x} = -\frac{\cos x}{4 \sin^4 x} - \frac{3}{8} \frac{\cos x}{\sin^2 x} + \frac{3}{8} \ln \tan \frac{x}{2}$$

6. $$\int \frac{dx}{\sin^6 x} = -\frac{\cos x}{5 \sin^5 x} - \frac{4}{15} \cot^3 x - \frac{4}{5} \cot x$$
    $$= -\frac{1}{5} \cot^5 x - \frac{2}{3} \cot^3 x - \cot x$$

7. $$\int \frac{dx}{\sin^7 x} = -\frac{\cos x}{6 \sin^2 x} \left(\frac{1}{\sin^4 x} + \frac{5}{4 \sin^2 x} + \frac{15}{8} \right) + \frac{5}{16} \ln \tan \frac{x}{2}$$

8. $$\int \frac{dx}{\sin^8 x} = -\left(\frac{1}{7} \cot^7 x + \frac{3}{5} \cot^5 x + \cot^3 x + \cot x \right)$$

9. $$\int \frac{dx}{\cos x} = \ln \tan \left(\frac{\pi}{4} + \frac{x}{2} \right) = \ln \cot \left(\frac{\pi}{4} - \frac{x}{2} \right) = \ln \sqrt{\frac{1 + \sin x}{1 - \sin x}}$$

10. $$\int \frac{dx}{\cos^2 x} = \tan x$$

11. $$\int \frac{dx}{\cos^3 x} = \frac{1}{2} \frac{\sin x}{\cos^2 x} + \frac{1}{2} \ln \tan \left(\frac{\pi}{4} + \frac{x}{2} \right)$$

12. $$\int \frac{dx}{\cos^4 x} = \frac{\sin x}{3 \cos^3 x} + \frac{2}{3} \tan x = \frac{1}{3} \tan^3 x + \tan x$$

13. $$\int \frac{dx}{\cos^5 x} = \frac{\sin x}{4 \cos^4 x} + \frac{3}{8} \frac{\sin x}{\cos^2 x} + \frac{3}{8} \ln \tan \left(\frac{x}{2} + \frac{\pi}{4} \right)$$

14. $$\int \frac{dx}{\cos^6 x} = \frac{\sin x}{5 \cos^5 x} + \frac{4}{15} \tan^3 x + \frac{4}{5} \tan x = \frac{1}{5} \tan^5 x + \frac{2}{3} \tan^3 x + \tan x$$

15. $$\int \frac{dx}{\cos^7 x} = \frac{\sin x}{6 \cos^6 x} + \frac{5 \sin x}{24 \cos^4 x} + \frac{5 \sin x}{16 \cos^2 x} + \frac{5}{16} \ln \tan \left(\frac{x}{2} + \frac{\pi}{4} \right)$$

16. $\int \dfrac{dx}{\cos^8 x} = \dfrac{1}{7} \tan^7 x + \dfrac{3}{5} \tan^5 x + \tan^3 x + \tan x$

17. $\int \dfrac{\sin x}{\cos x}\, dx = -\ln \cos x$

18. $\int \dfrac{\sin^2 x}{\cos x}\, dx = -\sin x + \ln \tan \left(\dfrac{\pi}{4} = \dfrac{x}{2}\right)$

19. $\int \dfrac{\sin^3 x}{\cos x}\, dx = -\dfrac{\sin^2 x}{2} - \ln \cos x = \dfrac{1}{2}\cos^2 x - \ln \cos x$

20. $\int \dfrac{\sin^4 x}{\cos x}\, dx = -\dfrac{1}{3}\sin^3 x - \sin x + \ln \tan \left(\dfrac{x}{2} + \dfrac{\pi}{4}\right)$

21. $\int \dfrac{\sin^2 x\, dx}{\cos^2 x} = \dfrac{1}{\cos x}$

22. $\int \dfrac{\sin^2 x\, dx}{\cos^2 x} = \tan x - x$

23. $\int \dfrac{\sin^3 x\, dx}{\cos^2 x} = \cos x + \dfrac{1}{\cos x}$

24. $\int \dfrac{\sin^4 x\, dx}{\cos^2 x} = \tan x + \dfrac{1}{2}\sin x \cos x - \dfrac{3}{2}x$

25. $\int \dfrac{\sin x\, dx}{\cos^3 x} = \dfrac{1}{2\cos^2 x} = \dfrac{1}{2}\tan^2 x$

26. $\int \dfrac{\sin^2 x\, dx}{\cos^3 x} = \dfrac{\sin x}{2\cos^2 x} - \dfrac{1}{2}\ln \tan \left(\dfrac{\pi}{4} + \dfrac{x}{2}\right)$

27. $\int \dfrac{\sin^3 x\, dx}{\cos^3 x} = \dfrac{1}{2}\dfrac{\sin x}{\cos^2 x} + \ln \cos x$

28. $\int \dfrac{\sin^4 x\, dx}{\cos^3 x} = \dfrac{1}{2}\dfrac{\sin x}{\cos^2 x} + \sin x - \dfrac{3}{2}\ln \tan \left(\dfrac{x}{2} + \dfrac{\pi}{4}\right)$

29. $\int \dfrac{\sin x\, dx}{\cos^4 x} = \dfrac{1}{3\cos^3 x}$

30. $\int \dfrac{\sin^2 x\, dx}{\cos^4 x} = \dfrac{1}{3}\tan^3 x$

31. $\int \dfrac{\sin^3 x\, dx}{\cos^4 x} = -\dfrac{1}{\cos x} + \dfrac{1}{3\cos^3 x}$

32. $\int \dfrac{\sin^4 x\, dx}{\cos^4 x} = \dfrac{1}{3}\tan^3 x - \tan x + x$

33. $\int \dfrac{\cos x\, dx}{\sin x} = \ln \sin x$

34. $\int \dfrac{\cos^2 x\, dx}{\sin x} = \cos x + \ln \tan \dfrac{x}{2}$

35. $\int \dfrac{\cos^3 x\, dx}{\sin x} = \dfrac{\cos^2 x}{2} + \ln \sin x$

36. $$\int \frac{\cos^4 x \, dx}{\sin x} = \frac{1}{3}\cos^3 x + \cos x + \ln \tan\left(\frac{x}{2}\right)$$

37. $$\int \frac{\cos x}{\sin^2 x} \, dx = -\frac{1}{\sin x}$$

38. $$\int \frac{\cos^2 x}{\sin^2 x} \, dx = -\cot x - x$$

39. $$\int \frac{\cos^3 x}{\sin^2 x} \, dx = -\sin x - \frac{1}{\sin x}$$

40. $$\int \frac{\cos^4 x}{\sin^2 x} \, dx = -\cot x - \frac{1}{2}\sin x \cos x - \frac{3}{2}x$$

41. $$\int \frac{\cos x}{\sin^3 x} \, dx = -\frac{1}{2\sin^2 x}$$

42. $$\int \frac{\cos^2 x}{\sin^3 x} \, dx = -\frac{\cos x}{2\sin^2 x} - \frac{1}{2}\ln \tan \frac{x}{2}$$

43. $$\int \frac{\cos^3 x}{\sin^3 x} \, dx = -\frac{1}{2\sin^2 x} - \ln \sin x$$

44. $$\int \frac{\cos^4 x}{\sin^3 x} \, dx = -\frac{1}{2}\frac{\cos x}{\sin^2 x} - \cos x - \frac{3}{2}\ln \tan \frac{x}{2}$$

45. $$\int \frac{\cos x}{\sin^4 x} \, dx = -\frac{1}{3\sin^3 x}$$

46. $$\int \frac{\cos^2 x}{\sin^4 x} \, dx = -\frac{1}{3}\cot^3 x$$

47. $$\int \frac{\cos^3 x}{\sin^4 x} \, dx = \frac{1}{\sin x} - \frac{1}{3\sin^3 x}$$

48. $$\int \frac{\cos^4 x}{\sin^4 x} \, dx = -\frac{1}{3}\cot^3 x + \cot x + x$$

49. $$\int \frac{dx}{\sin x \cos x} = \ln \tan x$$

50. $$\int \frac{dx}{\sin x \cos^2 x} = \frac{1}{\cos x} + \ln \tan \frac{x}{2}$$

51. $$\int \frac{dx}{\sin x \cos^3 x} = \frac{1}{2\cos^2 x} + \ln \tan x$$

52. $$\int \frac{dx}{\sin x \cos^4 x} = \frac{1}{\cos x} + \frac{1}{3\cos^3 x} + \ln \tan \frac{x}{2}$$

53. $$\int \frac{dx}{\sin^2 x \cos x} = \ln \tan\left(\frac{\pi}{4} + \frac{x}{2}\right) - \operatorname{cosec} x$$

54. $$\int \frac{dx}{\sin^2 x \cos^2 x} = -2\cot 2x$$

55. $$\int \frac{dx}{\sin^2 x \cos^3 x} = \left(\frac{1}{2\cos^2 x} - \frac{3}{2}\right)\frac{1}{\sin x} + \frac{3}{2}\ln \tan\left(\frac{\pi}{4} + \frac{x}{2}\right)$$

56. $\displaystyle\int \frac{dx}{\sin^2 x \cos^4 x} = \frac{1}{3 \sin x \cos^3 x} - \frac{8}{3} \cot 2x$

57. $\displaystyle\int \frac{dx}{\sin^3 x \cos x} = -\frac{1}{2 \sin^2 x} + \ln \tan x$

58. $\displaystyle\int \frac{dx}{\sin^3 x \cos^2 x} = -\frac{1}{\cos x}\left(\frac{1}{2 \sin^2 x} - \frac{3}{2}\right) + \frac{3}{2} \ln \tan \frac{x}{2}$

59. $\displaystyle\int \frac{dx}{\sin^3 x \cos^3 x} = -\frac{2 \cos 2x}{\sin^2 2x} + 2 \ln \tan x$

60. $\displaystyle\int \frac{dx}{\sin^3 x \cos^4 x} = \frac{2}{\cos x} + \frac{1}{3 \cos^3 x} - \frac{\cos x}{2 \sin^2 x} + \frac{5}{2} \ln \tan \frac{x}{2}$

61. $\displaystyle\int \frac{dx}{\sin^4 x \cos x} = -\frac{1}{\sin x} - \frac{1}{3 \sin^3 x} + \ln \tan \left(\frac{x}{2} + \frac{\pi}{4}\right)$

62. $\displaystyle\int \frac{dx}{\sin^4 x \cos^2 x} = -\frac{1}{3 \cos x \sin^3 x} - \frac{8}{3} \cot 2x$

63. $\displaystyle\int \frac{dx}{\sin^4 x \cos^3 x} = -\frac{2}{\sin x} - \frac{1}{3 \sin^3 x} + \frac{\sin x}{2 \cos^2 x} + \frac{5}{2} \ln \tan \left(\frac{x}{2} + \frac{\pi}{4}\right)$

64. $\displaystyle\int \frac{dx}{\sin^4 x \cos^4 x} = -8 \cot 2x - \frac{8}{3} \cot^3 2x$

2.527

1. $\displaystyle\int \tan^p x \, dx = \frac{\tan^{p-1} x}{p-1} - \int \tan^{p-2} x \, dx \qquad\qquad [p \neq 1]$

2. $\displaystyle\int \tan^{2n+1} x \, dx = \sum_{k=1}^{n} (-1)^{n+k} \binom{n}{k} \frac{1}{2k \cos^{2k} x} - (-1)^n \ln \cos x$

   $\displaystyle = \sum_{k=1}^{n} \frac{(-1)^{k-1} \tan^{2n-2k+2} x}{2n - 2k + 2} - (-1)^n \ln \cos x$

3. $\displaystyle\int \tan^{2n} x \, dx = \sum_{k=1}^{n} (-1)^{k-1} \frac{\tan^{2n-2k+1} x}{2n - 2k + 1} + (-1)^n x \qquad\qquad$ GU (331)(12)

4. $\displaystyle\int \cot^p x \, dx = -\frac{\cot^{p-1} x}{p-1} - \int \cot^{p-2} x \, dx \qquad\qquad [p \neq 1]$

5. $\displaystyle\int \cot^{2n+1} x \, dx = \sum_{k=1}^{n} (-1)^{n+k+1} \binom{n}{k} \frac{1}{2k \sin^{2k} x} + (-1)^n \ln \sin x$

   $\displaystyle = \sum_{k=1}^{n} (-1)^k \frac{\cot^{2n-2k+2} x}{2n - 2k + 2} + (-1)^n \ln \sin x$

6. $\displaystyle\int \cot^{2n} x \, dx = \sum_{k=1}^{n} (-1)^k \frac{\cot^{2n-2k+1} x}{2n - 2k + 1} + (-1)^n x \qquad\qquad$ GU (331)(14)

For special formulas for $p = 1, 2, 3, 4$, see **2.526** 17, **2.526** 33, **2.526** 22, **2.526** 38, **2.526** 27, **2.526** 43, **2.526** 32, and **2.526** 48.

2.53–2.54 Sines and cosines of multiple angles and of linear and more complicated functions of the argument

2.531

1. $\int \sin(ax + b)\, dx = -\dfrac{1}{a} \cos(ax + b)$

2. $\int \cos(ax + b)\, dx = -\dfrac{1}{a} \sin(ax + b)$

2.532

1. $\int \sin(ax + b) \sin(cx + d)\, dx = \dfrac{\sin[(a - c)x + b - d]}{2(a - c)} - \dfrac{\sin[(a + c)x + b + d]}{2(a + c)}$

$$\left[a^2 \neq c^2 \right]$$

2.[8] $\int \sin(ax + b) \cos(cx + d)\, dx = -\dfrac{\cos[(a - c)x + b - d]}{2(a - c)} - \dfrac{\cos[(a + c)x + b + d]}{2(a + c)}$

$$\left[a^2 \neq c^2 \right]$$

3. $\int \cos(ax + b) \cos(cx + d)\, dx = \dfrac{\sin[(a - c)x + b - d]}{2(a - c)} + \dfrac{\sin[(a + c)x + b + d]}{2(a + c)}$

$$\left[a^2 \neq c^2 \right]$$

For $c = a$:

4. $\int \sin(ax + b) \sin(ax + d)\, dx = \dfrac{x}{2} \cos(b - d) - \dfrac{\sin(2ax + b + d)}{4a}$

5. $\int \sin(ax + b) \cos(ax + d)\, dx = \dfrac{x}{2} \sin(b - d) - \dfrac{\cos(2ax + b + d)}{4a}$

6. $\int \cos(ax + b) \cos(ax + d)\, dx = \dfrac{x}{2} \cos(b - d) + \dfrac{\sin(2ax + b + d)}{4a}$ GU (332)(3)

2.533

1.[8] $\int \sin ax \cos bx\, dx = -\dfrac{\cos(a + b)x}{2(a + b)} - \dfrac{\cos(a - b)x}{2(a - b)}$ $\left[a^2 \neq b^2 \right]$

2.[8] $\int \sin ax \sin bx \sin cx\, dx = -\dfrac{1}{4} \left\{ \dfrac{\cos(a - b + c)x}{a - b + c} + \dfrac{\cos(b + c - a)x}{b + c - a} \right.$

$$\left. + \dfrac{\cos(a + b - c)x}{a + b - c} - \dfrac{\cos(a + b + c)x}{a + b + c} \right\}$$

PE (376)

3. $\int \sin ax \cos bx \cos cx\, dx = -\dfrac{1}{4} \left\{ \dfrac{\cos(a + b + c)x}{a + b + c} - \dfrac{\cos(b + c - a)x}{b + c - a} \right.$

$$\left. + \dfrac{\cos(a + b - c)x}{a + b - c} + \dfrac{\cos(a + c - b)x}{a + c - b} \right\}$$

PE (378)

4. $$\int \cos ax \sin bx \sin cx \, dx = \frac{1}{4} \left\{ \frac{\sin(a+b-c)x}{a+b-c} + \frac{\sin(a+c-b)x}{a+c-b} \right.$$

$$\left. - \frac{\sin(a+b+c)x}{a+b+c} - \frac{\sin(b+c-a)x}{b+c-a} \right\}$$

PE (379)

5. $$\int \cos ax \cos bx \cos cx \, dx = \frac{1}{4} \left\{ \frac{\sin(a+b+c)x}{a+b+c} + \frac{\sin(b+c-a)x}{b+c-a} \right.$$

$$\left. + \frac{\sin(a+c-b)x}{a+c-b} + \frac{\sin(a+b-c)x}{a+b-c} \right\}$$

PE (377)

2.534

1. $$\int \frac{\cos px + i \sin px}{\sin nx} \, dx = -2 \int \frac{z^{p+n-1}}{1-z^{2n}} \, dz \qquad [z = \cos x + i \sin x] \qquad \text{Pe (374)}$$

2. $$\int \frac{\cos px + i \sin px}{\cos nx} \, dx = -2i \int \frac{z^{p+n-1}}{1-z^{2n}} \, dz \qquad [z = \cos x + i \sin x] \qquad \text{Pe (373)}$$

2.535

1. $$\int \sin^p x \sin ax \, dx = \frac{1}{p+a} \left\{ -\sin^p x \cos ax + p \int \sin^{p-1} x \cos(a-1)x \, dx \right\} \qquad \text{GU (332)(5a)}$$

2. $\int \sin^p x \sin(2n+1)x\, dx$

$$= (2n+1)\left\{\int \sin^{p+1} x\, dx + \sum_{k=1}^{n}(-1)^k \frac{\left[(2n+1)^2 - 1^2\right]\left[(2n+1)^2 - 3^2\right]\cdots}{\cdots\left[(2n+1)^2 - (2k-1)^2\right]}{(2k+1)!}\right.$$

$$\left. \times \int \sin^{2k+p+1} x\, dx\right\}$$

TI (299)

$$= \frac{\Gamma(p+1)}{\Gamma\left(\frac{p+3}{2}+n\right)}\left\{\sum_{k=0}^{n-1}\left[\frac{(-1)^{k-1}\Gamma\left(\frac{p+1}{2}+n-2k\right)}{2^{2k+1}\Gamma(p-2k+1)}\sin^{p-2k}x\cos(2n-2k+1)x\right.\right.$$

$$+ (-1)^k \frac{\Gamma\left(\frac{p-1}{2}+n-2k\right)}{2^{2k+2}\Gamma(p-2k)}\sin^{p-2k-1}x\sin(2n-2k)x\bigg]$$

$$+ \frac{(-1)^n\Gamma\left(\frac{p+3}{2}-n\right)}{2^{2n}\Gamma(p-2n+1)}\int \sin^{p-2n+1}x\, dx\bigg\}$$

GU (332)(5c)

3. $\int \sin^p x \sin 2nx\, dx = 2n\left\{\frac{\sin^{p+2}x}{p+2}\right.$

$$+ \sum_{k=1}^{n-1}(-1)^k \frac{\left(4n^2 - 2^2\right)\left(4n^2 - 4^2\right)\cdots\left[4n^2 - (2k)^2\right]}{(2k+1)!(2k+p+2)}\sin^{2k+p+2}x\bigg\}$$

TI (303)

$$= \frac{\Gamma(p+1)}{\Gamma\left(\frac{p}{2}+n+1\right)}\left\{\sum_{k=0}^{n-1}\frac{(-1)^{k-1}\Gamma\left(\frac{p}{2}+n-2k\right)}{2^{2k+1}\Gamma(p-2k+1)}\sin^{p-2k}x\cos(2n-2k)x\right.$$

$$- \frac{(-1)^k\Gamma\left(\frac{p}{2}+n-2k-1\right)}{2^{2k+2}\Gamma(p-2k)}\sin^{p-2k-1}x\sin(2n-2k-1)x\bigg\}$$

[p is not equal to $-2, -4, \ldots, -2n$]
GU (332)(5c)

2.536

1. $\int \sin^p x \cos ax\, dx = \frac{1}{p+1}\left\{\sin^p x \sin ax - p\int \sin^{p-1} x \sin(a-1)x\, dx\right\}$ GU (332)(6a)

2.
$$\int \sin^p x \cos(2n+1)x \, dx$$

$$= \frac{\sin^{p+1} x}{p+1} + \sum_{k=1}^{n} (-1)^k \frac{\left[(2n+1)^2 - 1^2\right]\left[(2n+1)^2 - 3^2\right] \cdots \left[(2n+1)^2 - (2k-1)^2\right]}{(2k)!(2k+p+1)}$$

$$\times \sin^{2k+p+1} x$$

TI (301)

$$= \frac{\Gamma(p+1)}{\Gamma\left(\frac{p+3}{2}+n\right)} \left\{ \sum_{k=0}^{n-1} \left[\frac{(-1)^k \Gamma\left(\frac{p+1}{2}+n-2k\right)}{2^{2k+1}\Gamma(p-2k+1)} \sin^{p-2k} x \sin(2n-2k+1)x \right. \right.$$

$$\left. + \frac{(-1)^k \Gamma\left(\frac{p-1}{2}+n-2k\right)}{2^{2k+2}\Gamma(p-2k)} \sin^{p-2k-1} x \cos(2n-2k)x \right]$$

$$\left. + \frac{(-1)^n \Gamma\left(\frac{p+3}{2}-n\right)}{2^{2n}\Gamma(p-2n+1)} \int \sin^{p-2n} x \cos x \, dx \right\}$$

$$[p \text{ is not equal to } -3, -5, \ldots, -(2n+1)]$$

GU (332)(6c)

3.
$$\int \sin^p x \cos 2nx \, dx$$

$$= \int \sin^p x \, dx + \sum_{k=1}^{n} (-1)^k \frac{4n^2 \cdot \left(4n^2 - 2^2\right) \cdots \left[4n^2 - (2k-2)^2\right]}{(2k)!} \int \sin^{2k+p} x \, dx$$

TI (300)

$$= \frac{\Gamma(p+1)}{\Gamma\left(\frac{p}{2}+n+1\right)} \left\{ \sum_{k=0}^{n-1} \left[\frac{(-1)^k \Gamma\left(\frac{p}{2}+n-2k\right)}{2^{2k+1}\Gamma(p-2k+1)} \sin^{p-2k} x \sin(2n-2k)x \right. \right.$$

$$\left. + \frac{(-1)^k \Gamma\left(\frac{p}{2}+n-2k-1\right)}{2^{2k+2}\Gamma(p-2k)} \sin^{p-2k-1} x \cos(2n-2k-1)x \right]$$

$$\left. + \frac{(-1)^n \Gamma\left(\frac{p}{2}-n+1\right)}{2^{2n}\Gamma(p-2n+1)} \int \sin^{p-2n} x \, dx \right\}$$

GU (332)(6c)

2.537

1.
$$\int \cos^p x \sin ax \, dx = \frac{1}{p+a} \left\{ -\cos^p x \cos ax + p \int \cos^{p-1} x \sin(a-1)x \, dx \right\}$$
GU (332)(7a)

2. $\int \cos^p x \sin(2n+1)x \, dx$

$$= (-1)^{n+1} \left\{ \frac{\cos^{p+1} x}{p+1} \right.$$

$$+ \sum_{k=1}^{n} (-1)^k \frac{\left[(2n+1)^2 - 1^2\right]\left[(2n+1)^2 - 3^2\right]\dots\left[(2n+1)^2 - (2k-1)^2\right]}{(2k)!(2k+p+1)} \cos^{2k+p+1} x \left. \right\}$$

$$\text{TI (295)}$$

$$= \frac{\Gamma(p+1)}{\Gamma\left(\frac{p+3}{2} + n\right)} \left\{ -\sum_{k=0}^{n-1} \frac{\Gamma\left(\frac{p+1}{2} + n - k\right)}{2^{2k+1} \Gamma(p-2k+1)} \cos^{p-k} x \cos(2n-k+1)x \right.$$

$$+ \frac{\Gamma\left(\frac{p+3}{2}\right)}{2^n \Gamma(p-n+1)} \int \cos^{p-n} x \sin(n+1)x \, dx \left. \right\}$$

$$[p \text{ is not equal to } -3, -5, \dots, -(2n+1)]$$
$$\text{GU (332)(7b)a}$$

3. $\int \cos^p x \sin 2nx \, dx = (-1)^n \left\{ \frac{\cos^{p+2} x}{p+2} \right.$

$$+ \sum_{k=1}^{n-1} (-1)^k \frac{\left(4n^2 - 2^2\right)\left(4n^2 - 4^2\right)\dots\left[4n^2 - (2k)^2\right]}{(2k+1)!(2k+p+2)} \cos^{2k+p+2} x \left. \right\}$$

$$\text{TI (297)}$$

$$= \frac{\Gamma(p+1)}{\Gamma\left(\frac{p}{2} + n + 1\right)} \left\{ -\sum_{k=0}^{n-1} \frac{\Gamma\left(\frac{p}{2} + n - k\right)}{2^{k+1} \Gamma(p-k+1)} \cos^{p-k} x \cos(2n-k)x \right.$$

$$+ \frac{\Gamma\left(\frac{p}{2} + 1\right)}{2^n \Gamma(p-n+1)} \int \cos^{p-n} x \sin nx \, dx \left. \right\}$$

$$[p \text{ is not equal to } -2, -4, \dots, -2n]$$
$$\text{GU (332)(7b)a}$$

2.538

1. $\int \cos^p x \cos ax \, dx = \frac{1}{p+a} \left\{ \cos^p x \sin ax + p \int \cos^{p-1} x \cos(a-1)x \, dx \right\}$ GU (332)(8a)

2. $\displaystyle\int \cos^p x \cos(2n+1)x\, dx$

$$= (-1)^n (2n+1) \left\{ \int \cos^{p+1} x\, dx \right.$$

$$+ \sum_{k=1}^{n} (-1)^k \frac{\left[(2n+1)^2 - 1^2\right]\left[(2n+1)^2 - 3^2\right]\ldots\left[(2n+1)^2 - (2k-1)^2\right]}{(2k+1)!}$$

$$\left. \times \int \cos^{2k+p+1} x\, dx \right\}$$

TI (293)

$$= \frac{\Gamma(p+1)}{\Gamma\left(\dfrac{p+3}{2} + n\right)} \left\{ \sum_{k=0}^{n-1} \frac{\Gamma\left(\frac{p+1}{2} + n - k\right)}{2^{k+1}\,\Gamma(p-k+1)} \cos^{p-k} x \sin(2n-k+1)x \right.$$

$$\left. + \frac{\Gamma\left(\frac{p+3}{2}\right)}{2^n\,\Gamma(p-n+1)} \int \cos^{p-n} x \cos(n+1)x\, dx \right\}$$

GU (332)(8b)a

3. $\displaystyle\int \cos^p x \cos 2nx\, dx$

$$= (-1)^n \left\{ \int \cos^p x\, dx + \sum_{k=1}^{n} (-1)^k \frac{4n^2 \left[4n^2 - 2^2\right]\ldots\left[4n^2 - (2k-2)^2\right]}{(2k)!} \int \cos^{2k+p} x\, dx \right\}$$

TI (294)

$$= \frac{\Gamma(p+1)}{\Gamma\left(\frac{p}{2} + n + 1\right)} \left\{ \sum_{k=0}^{n-1} \frac{\Gamma\left(\frac{p}{2} + n - k\right)}{2^{k+1}\,\Gamma(p-k+1)} \cos^{p-k} x \sin(2n-k)x \right.$$

$$\left. + \frac{\Gamma\left(\frac{p}{2} + 1\right)}{2^n\,\Gamma(p-n+1)} \int \cos^{p-n} x \cos nx\, dx \right\}$$

GU (332)(8b)a

2.539

1. $\displaystyle\int \frac{\sin(2n+1)x}{\sin x}\, dx = 2 \sum_{k=1}^{n} \frac{\sin 2kx}{2k} + x$

2. $\displaystyle\int \frac{\sin 2nx}{\sin x}\, dx = 2 \sum_{k=1}^{n} \frac{\sin(2k-1)x}{2k-1}$ GU (332)(5e)

3. $\displaystyle\int \frac{\cos(2n+1)x}{\sin x}\, dx = 2 \sum_{k=1}^{n} \frac{\cos 2kx}{2k} + \ln \sin x$

4. $\displaystyle\int \frac{\cos 2nx}{\sin x}\,dx = 2\sum_{k=1}^{n} \frac{\cos(2k-1)x}{2k-1} + \ln\tan\frac{x}{2}$ GI (332)(6e)

5. $\displaystyle\int \frac{\sin(2n+1)x}{\cos x}\,dx = 2\sum_{k=1}^{n} (-1)^{n-k+1}\frac{\cos 2kx}{2k} + (-1)^{n+1}\ln\cos x$

6. $\displaystyle\int \frac{\sin 2nx}{\cos x}\,dx = 2\sum_{k=1}^{n} (-1)^{n-k+1}\frac{\cos(2k-1)x}{2k-1}$ GU (332)(7d)

7. $\displaystyle\int \frac{\cos(2n+1)x}{\cos x}\,dx = 2\sum_{k=1}^{n} (-1)^{n-k}\frac{\sin 2kx}{2k} + (-1)^{n}x$

8. $\displaystyle\int \frac{\cos 2nx}{\cos x}\,dx = 2\sum_{k=1}^{n} (-1)^{n-k}\frac{\sin(2k-1)x}{2k-1} + (-1)^{n}\ln\tan\left(\frac{\pi}{4}+\frac{x}{2}\right).$ GU (332)(8d)

2.541

1. $\displaystyle\int \sin(n+1)x \sin^{n-1}x\,dx = \frac{1}{n}\sin^{n}x\sin nx$ BI (71)(1)a

2. $\displaystyle\int \sin(n+1)x \cos^{n-1}x\,dx = -\frac{1}{n}\cos^{n}x\cos nx$ BI (71)(2)a

3. $\displaystyle\int \cos(n+1)x \sin^{n-1}x\,dx = \frac{1}{n}\sin^{n}x\cos nx$ BI (71)(3)a

4. $\displaystyle\int \cos(n+1)x \cos^{n-1}x\,dx = \frac{1}{n}\cos^{n}x\sin nx$ BI (71)(4)a

5. $\displaystyle\int \sin\left[(n+1)\left(\frac{\pi}{2}-x\right)\right]\sin^{n-1}x\,dx = \frac{1}{n}\sin^{n}x\cos n\left(\frac{\pi}{2}-x\right)$ BI (71)(5)a

6. $\displaystyle\int \cos\left[(n+1)\left(\frac{\pi}{2}-x\right)\right]\sin^{n-1}x\,dx = -\frac{1}{n}\sin^{n}x\sin n\left(\frac{\pi}{2}-x\right)$ BI (71)(6)a

2.542

1. $\displaystyle\int \frac{\sin 2x}{\sin^{n}x}\,dx = -\frac{2}{(n-2)\sin^{n-2}x}$

For $n=2$:

2. $\displaystyle\int \frac{\sin 2x}{\sin^{2}x}\,dx = 2\ln\sin x$

2.543

1. $\displaystyle\int \frac{\sin 2x\,dx}{\cos^{n}x} = \frac{2}{(n-2)\cos^{n-2}x}$

For $n=2$:

2. $\displaystyle\int \frac{\sin 2x}{\cos^{2}x}\,dx = -2\ln\cos x$

2.544

1. $\displaystyle\int \frac{\cos 2x\,dx}{\sin x} = 2\cos x + \ln\tan\frac{x}{2}$

2. $\displaystyle\int\frac{\cos 2x\,dx}{\sin^2 x}=-\cot x-2x$

3. $\displaystyle\int\frac{\cos 2x\,dx}{\sin^3 x}=-\frac{\cos x}{2\sin^2 x}-\frac{3}{2}\ln\tan\frac{x}{2}$

4. $\displaystyle\int\frac{\cos 2x\,dx}{\cos x}=2\sin x-\ln\tan\left(\frac{\pi}{4}+\frac{x}{2}\right)$

5. $\displaystyle\int\frac{\cos 2x\,dx}{\cos^2 x}=2x-\tan x$

6. $\displaystyle\int\frac{\cos 2x\,dx}{\cos^3 x}=-\frac{\sin x}{2\cos^2 x}+\frac{3}{2}\ln\tan\left(\frac{\pi}{4}+\frac{x}{2}\right)$

7. $\displaystyle\int\frac{\sin 3x\,dx}{\sin x}=x+\sin 2x$

8. $\displaystyle\int\frac{\sin 3x}{\sin^2 x}\,dx=3\ln\tan\frac{x}{2}+4\cos x$

9. $\displaystyle\int\frac{\sin 3x}{\sin^3 x}\,dx=-3\cot x-4x$

2.545

1. $\displaystyle\int\frac{\sin 3x}{\cos^n x}\,dx=\frac{4}{(n-3)\cos^{n-3}x}-\frac{1}{(n-1)\cos^{n-1}x}$

For $n=1$ and $n=3$:

2. $\displaystyle\int\frac{\sin 3x}{\cos x}\,dx=2\sin^2 x+\ln\cos x$

3. $\displaystyle\int\frac{\sin 3x}{\cos^3 x}\,dx=-\frac{1}{2\cos^2 x}-4\ln\cos x$

2.546

1. $\displaystyle\int\frac{\cos 3x}{\sin^n x}\,dx=\frac{4}{(n-3)\sin^{n-3}x}-\frac{1}{(n-1)\sin^{n-1}x}$

For $n=1$ and $n=3$:

2. $\displaystyle\int\frac{\cos 3x}{\sin x}\,dx=-2\sin^2 x+\ln\sin x$

3. $\displaystyle\int\frac{\cos 3x}{\sin^3 x}\,dx=-\frac{1}{2\sin^2 x}-4\ln\sin x$

2.547

1. $\displaystyle\int\frac{\sin nx}{\cos^p x}\,dx=2\int\frac{\sin(n-1)x\,dx}{\cos^{p-1}x}-\int\frac{\sin(n-2)x\,dx}{\cos^p x}$

2. $\displaystyle\int\frac{\cos 3x}{\cos x}\,dx=\sin 2x-x$

3. $\displaystyle\int\frac{\cos 3x}{\cos^2 x}\,dx=4\sin x-3\ln\tan\left(\frac{\pi}{4}+\frac{x}{2}\right)$

4. $\displaystyle\int\frac{\cos 3x}{\cos^3 x}\,dx=4x-3\tan x$

2.548

1.
$$\int \frac{\sin^m x\, dx}{\sin(2n+1)x} = \frac{1}{2n+1} \sum_{k=0}^{2n} (-1)^{n+k} \cos^m \left[\frac{2k+1}{2(2n+1)}\pi \right] \ln \frac{\sin\left[\frac{(k-n)\pi}{2(2n+1)} + \frac{x}{2} \right]}{\sin\left[\frac{k+n+1}{(2n+1)}\pi - \frac{x}{2} \right]}$$

$[m \text{ a natural number} \le 2n]$　　　TI (378)

2.
$$\int \frac{\sin^{2m} x\, dx}{\sin 2nx} = \frac{(-1)^n}{2n} \left\{ \ln\cos x + \sum_{k=1}^{n-1} (-1)^k \cos^{2m} \frac{k\pi}{2n} \ln\left(\cos^2 x - \sin^2 \frac{k\pi}{2n} \right) \right\}$$

$[m \text{ a natural number} \le n]$　　　TI (379)

3.
$$\int \frac{\sin^{2m+1} x}{\sin 2nx}\, dx = \frac{(-1)^n}{2n} \left\{ \ln\tan\left(\frac{\pi}{4} - \frac{x}{2} \right) \right.$$
$$\left. + \sum_{k=1}^{n-1} (-1)^k \cos^{2m+1} \frac{k\pi}{2n} \ln\left[\tan\left(\frac{n+k}{4n}\pi - \frac{x}{2} \right) \tan\left(\frac{n-k}{4n}\pi - \frac{x}{2} \right) \right] \right\}$$

$[m \text{ a natural number} < n]$

4.
$$\int \frac{\sin^{2m} x\, dx}{\cos(2n+1)x} = \frac{(-1)^{n+1}}{2n+1} \left\{ \ln\tan\left(\frac{\pi}{4} - \frac{x}{2} \right) + \sum_{k=1}^{n} (-1)^k \right.$$
$$\left. \times \cos^{2m} \frac{k\pi}{2n+1} \ln\left[\tan\left(\frac{2n+2k+1}{4(2n+1)}\pi - \frac{x}{2} \right) \tan\left(\frac{2n-2k+1}{2(2n+1)}\pi - \frac{x}{2} \right) \right] \right\}$$

$[m \text{ a natural number} \le n]$　　　TI (381)

5.
$$\int \frac{\sin^{2m+1} x\, dx}{\cos(2n+1)x} = \frac{(-1)^{n+1}}{2n+1} \left\{ \ln\cos x + \sum_{k=1}^{n} (-1)^k \cos^{2m+1} \frac{k\pi}{2n+1} \ln\left(\cos^2 x - \sin^2 \frac{k\pi}{2n+1} \right) \right\}$$

$[m \text{ a natural number} \le n]$　　　TI (382)a

6.
$$\int \frac{\sin^m x\, dx}{\cos 2nx} = \frac{1}{2n} \sum_{k=0}^{2n-1} (-1)^{n+k} \cos^m \left[\frac{2k+1}{4n}\pi \right] \ln \frac{\sin\left[\frac{2k-2n+1}{8n}\pi + \frac{x}{2} \right]}{\sin\left[\frac{2k+2n+1}{8n}\pi - \frac{x}{2} \right]}$$

$[m \text{ a natural number} < 2n]$　　　TI (377)

7.
$$\int \frac{\cos^{2m+1} x\, dx}{\sin(2n+1)x} = \frac{1}{2n+1} \left\{ \ln\sin x + \sum_{k=1}^{n} (-1)^k \cos^{2m+1} \frac{k\pi}{2n+1} \ln\left(\sin^2 x - \sin^2 \frac{k\pi}{2n+1} \right) \right\}$$

$[m \text{ a natural number} \le n]$　　　TI (376)

8.
$$\int \frac{\cos^{2m} x\, dx}{\sin(2n+1)x} = \frac{1}{2n+1} \left\{ \ln\tan\frac{x}{2} \right.$$
$$\left. + \sum_{k=1}^{n} (-1)^k \cos^{2m} \frac{k\pi}{2n+1} \ln\left[\tan\left(\frac{x}{2} + \frac{k\pi}{4n+2} \right) \tan\left(\frac{x}{2} - \frac{k\pi}{4n+2} \right) \right] \right\}$$

$[m \text{ a natural number} \le n]$　　　TI (375)

9. $$\int \frac{\cos^{2m+1} x}{\sin 2nx} dx = \frac{1}{2n} \left\{ \ln \tan \frac{x}{2} + \sum_{k=1}^{n-1} (-1)^k \cos^{2m+1} \frac{k\pi}{2n} \ln \left[\tan \left(\frac{x}{2} + \frac{k\pi}{4} \right) \tan \left(\frac{x}{2} - \frac{k\pi}{4n} \right) \right] \right\}$$

$$[m \text{ a natural number} < n] \qquad \text{TI (374)}$$

10. $$\int \frac{\cos^{2m} x}{\sin 2nx} dx = \frac{1}{2n} \left\{ \ln \sin x + \sum_{k=1}^{n-1} (-1)^k \cos^{2m} \frac{k\pi}{2n} \ln \left(\sin^2 x - \sin^2 \frac{k\pi}{2n} \right) \right\}$$

$$[m \text{ a natural number} \le n] \qquad \text{TI (373)}$$

11. $$\int \frac{\cos^m x}{\cos nx} dx = \frac{1}{n} \sum_{k=0}^{n-1} (-1)^k \cos^m \frac{2k+1}{2n} \pi \ln \frac{\sin \left[\frac{2k+1}{4n} \pi + \frac{x}{2} \right]}{\sin \left[\frac{2k+1}{4n} \pi - \frac{x}{2} \right]}$$

$$[m \text{ is a natural number} \le n] \qquad \text{TI (372)}$$

2.549

1. $$\int \sin x^2 \, dx = \sqrt{\frac{\pi}{2}} \, S(x)$$

2. $$\int \cos x^2 \, dx = \sqrt{\frac{\pi}{2}} \, C(x)$$

3. $$\int \sin \left(ax^2 + 2bx + c \right) dx = \sqrt{\frac{\pi}{2a}} \left\{ \cos \frac{ac - b^2}{a} S \left(\frac{ax + b}{\sqrt{a}} \right) + \sin \frac{ac - b^2}{a} C \left(\frac{ax + b}{\sqrt{a}} \right) \right\}$$

4. $$\int \cos \left(ax^2 + 2bx + c \right) dx = \sqrt{\frac{\pi}{2a}} \left\{ \cos \frac{ac - b^2}{a} C \left(\frac{ax + b}{\sqrt{a}} \right) - \sin \frac{ac - b^2}{a} S \left(\frac{ax + b}{\sqrt{a}} \right) \right\}$$

5. $$\int \sin \ln x \, dx = \frac{x}{2} \left(\sin \ln x - \cos \ln x \right) \qquad\qquad \text{PE (444)}$$

6. $$\int \cos \ln x \, dx = \frac{x}{2} \left(\sin \ln x + \cos \ln x \right) \qquad\qquad \text{PE (445)}$$

2.55–2.56 Rational functions of the sine and cosine

2.551

1. $$\int \frac{A + B \sin x}{(a + b \sin x)^n} dx = \frac{1}{(n-1)(a^2 - b^2)} \left[\frac{(Ab - aB)\cos x}{(a + b\sin x)^{n-1}} \right.$$

$$\left. + \int \frac{(Aa - Bb)(n-1) + (aB - bA)(n-2) \sin x}{(a + b\sin x)^{n-1}} dx \right]$$

$$\text{TI (358)a}$$

For $n = 1$:

2. $$\int \frac{A + B \sin x}{a + b \sin x} dx = \frac{B}{b} x + \frac{Ab - aB}{b} \int \frac{dx}{a + b \sin x} \qquad (\text{see } \mathbf{2.551\ 3}) \qquad \text{TI (342)}$$

3.
$$\int \frac{dx}{a + b\sin x} = \frac{2}{\sqrt{a^2 - b^2}} \arctan \frac{a\tan\frac{x}{2} + b}{\sqrt{a^2 - b^2}} \qquad [a^2 > b^2]$$

$$= \frac{1}{\sqrt{b^2 - a^2}} \ln \frac{a\tan\frac{x}{2} + b - \sqrt{b^2 - a^2}}{a\tan\frac{x}{2} + b + \sqrt{b^2 - a^2}} \qquad [a^2 < b^2]$$

2.552

1.
$$\int \frac{A + B\cos x}{(a + b\sin x)^n}\, dx = -\frac{B}{(n-1)b(a + b\sin x)^{n-1}} + A\int \frac{dx}{(a + b\sin x)^n}$$

                                          (see **2.552** 3)   TI (361)

For $n = 1$:

2.
$$\int \frac{A + B\cos x}{a + b\sin x}\, dx = \frac{B}{b} \ln (a + b\sin x) + A\int \frac{dx}{a + b\sin x}$$

                                          (see **2.551** 3)   TI (344)

3.
$$\int \frac{dx}{(a + b\sin x)^n} = \frac{1}{(n-1)(a^2 - b^2)} \left[\frac{b\cos x}{(a + b\sin x)^{n-1}} \right.$$

$$\left. + \int \frac{(n-1)a - (n-2)b\sin x}{(a + b\sin x)^{n-1}}\, dx \right] \qquad (\text{see } \mathbf{2.551}\ 1)$$

                                                      TI (359)

2.553

1.
$$\int \frac{A + B\sin x}{(a + b\cos x)^n}\, dx = \frac{B}{(n-1)b(a + b\cos x)^{n-1}} + A\int \frac{dx}{(a + b\cos x)^n}$$

                                          (see **2.554** 3)   TI (355)

For $n = 1$:

2.
$$\int \frac{A + B\sin x}{a + b\cos x}\, dx = -\frac{B}{b} \ln (a + b\cos x) + A\int \frac{dx}{a + b\cos x}$$

                                          (see **2.553** 3)   TI (343)

3.
$$\int \frac{dx}{a + b\cos x} = \frac{2}{\sqrt{a^2 - b^2}} \arctan \frac{\sqrt{a^2 - b^2}\tan\frac{x}{2}}{a + b} \qquad [a^2 > b^2]$$

$$= \frac{1}{\sqrt{b^2 - a^2}} \ln \frac{\sqrt{b^2 - a^2}\tan\frac{x}{2} + a + b}{\sqrt{b^2 - a^2}\tan\frac{x}{2} - a - b} \qquad [a^2 < b^2]$$

                                               TI II 93, 94, TI (305)

2.554

1.
$$\int \frac{A + B\cos x}{(a + b\cos x)^n}\, dx = \frac{1}{(n-1)(a^2 - b^2)} \left[\frac{(aB - Ab)\sin x}{(a + b\cos x)^{n-1}} \right.$$

$$\left. + \int \frac{(Aa - bB)(n-1) + (n-2)(aB - bA)\cos x}{(a + b\cos x)^{n-1}}\, dx \right]$$

                                                      TI (353)

For $n = 1$:

2. $\int \dfrac{A + B\cos x}{a + b\cos x}\,dx = \dfrac{B}{b}x + \dfrac{Ab - aB}{b}\int \dfrac{dx}{a + b\cos x}$ (see **2.553** 3)   TI (341)

3. $\int \dfrac{dx}{(a + b\cos x)^n} = -\dfrac{1}{(n-1)(a^2 - b^2)}\left\{ \dfrac{b\sin x}{(a + b\cos x)^{n-1}} \right.$

$$\left. -\int \dfrac{(n-1)a - (n-2)b\cos x}{(a + b\cos x)^{n-1}}\,dx \right\} \qquad (\text{see } \mathbf{2.554}\ 1)$$

TI (354)

In integrating the functions in formulas **2.551** 3 and **2.553** 3, we may not take the integration over points at which the integrand becomes infinite, that is, over the points $x = \arcsin\left(-\dfrac{a}{b}\right)$ in formula **2.551** 3 or over the points $x = \arccos\left(-\dfrac{a}{b}\right)$ in formula **2.553** 3.

2.555   Formulas **2.551** 3 and **2.553** 3 are not applicable for $a^2 = b^2$. Instead, we may use the following formulas in these cases:

1. $\int \dfrac{A + B\sin x}{(1 \pm \sin x)^n}\,dx = -\dfrac{1}{2^{n-1}}\left\{ 2B\displaystyle\sum_{k=0}^{n-2}\binom{n-2}{k}\dfrac{\tan^{2k+1}\left(\frac{\pi}{4} \mp \frac{x}{2}\right)}{2k+1}\right.$

$$\left. \pm (A \mp B)\sum_{k=0}^{n-1}\binom{n-1}{k}\dfrac{\tan^{2k+1}\left(\frac{\pi}{4} \mp \frac{x}{2}\right)}{2k+1}\right\}$$

TI (361)a

2. $\int \dfrac{A + B\cos x}{(1 \pm \cos x)^n}\,dx = \dfrac{1}{2^{n-1}}\left\{ 2B\displaystyle\sum_{k=0}^{n-2}\binom{n-2}{k}\dfrac{\tan^{2k+1}\left[\frac{\pi}{4} \mp \left(\frac{\pi}{4} - \frac{x}{2}\right)\right]}{2k+1}\right.$

$$\left. \pm (A \mp B)\sum_{k=0}^{n-1}\binom{n-1}{k}\dfrac{\tan^{2k+1}\left[\frac{\pi}{4} \mp \left(\frac{\pi}{4} - \frac{x}{2}\right)\right]}{2k+1}\right\}$$

TI (356)

For $n = 1$:

3. $\int \dfrac{A + B\sin x}{1 \pm \sin x}\,dx = \pm Bx + (A \mp B)\tan\left(\dfrac{\pi}{4} \mp \dfrac{x}{2}\right)$ TI (250)

4. $\int \dfrac{A + B\cos x}{1 \pm \cos x}\,dx = \pm Bx \pm (A \mp B)\tan\left[\dfrac{\pi}{4} \mp \left(\dfrac{\pi}{4} - \dfrac{x}{2}\right)\right]$ TI (248)

2.556

1. $\int \dfrac{(1 - a^2)\,dx}{1 - 2a\cos x + a^2} = 2\arctan\left(\dfrac{1 + a}{1 - a}\tan\dfrac{x}{2}\right)$ $[0 < a < 1, \quad |x| < \pi]$ FI II 93

2. $\int \dfrac{(1 - a\cos x)\,dx}{1 - 2a\cos x + a^2} = \dfrac{x}{2} + \arctan\left(\dfrac{1 + a}{1 - a}\tan\dfrac{x}{2}\right)$ $[0 < a < 1, \quad |x| < \pi]$ FI II 93

2.557

1. $\int \dfrac{dx}{(a\cos x + b\sin x)^n} = \dfrac{1}{\sqrt{(a^2 + b^2)^n}}\int \dfrac{dx}{\sin^n\left(x + \arctan\dfrac{a}{b}\right)}$

(see **2.515**)   MZ 173a

2.6
$$\int \frac{\sin x \, dx}{a \sin x + b \cos x} = \frac{ax - b \ln \sin \left(x + \arctan \frac{b}{a}\right)}{a^2 + b^2}$$

3.
$$\int \frac{\cos x \, dx}{a \cos x + b \sin x} = \frac{ax + b \ln \sin \left(x + \arctan \frac{a}{b}\right)}{a^2 + b^2}$$
MZ 174a

4.
$$\int \frac{dx}{a \cos x + b \sin x} = \frac{\ln \tan \left[\frac{1}{2}\left(x + \arctan \frac{a}{b}\right)\right]}{\sqrt{a^2 + b^2}}$$

5.
$$\int \frac{dx}{(a \cos x + b \sin x)^2} = -\frac{\cot \left(x + \arctan \frac{a}{b}\right)}{a^2 + b^2} = +\frac{1}{a^2 + b^2} \cdot \frac{a \sin x - b \cos x}{a \cos x + b \sin x}$$
MZ 174a

2.558

1.
$$\int \frac{A + B \cos x + C \sin x}{(a + b \cos x + c \sin x)^n} \, dx$$
$$= \frac{(Bc - Cb) + (Ac - Ca) \cos x - (Ab - Ba) \sin x}{(n-1)(a^2 - b^2 - c^2)(a + b \cos x + c \sin x)^{n-1}} + \frac{1}{(n-1)(a^2 - b^2 - c^2)}$$
$$\times \int \frac{(n-1)(Aa - Bb - Cc) - (n-2)\left[(Ab - Ba) \cos x - (Ac - Ca) \sin x\right]}{(a + b \cos x + c \sin x)^{n-1}} \, dx$$
$$\left[n \neq 1, \quad a^2 \neq b^2 + c^2\right]$$
$$= \frac{Cb - Bc + Ca \cos x - Ba \sin x}{(n-1)a(a + b \cos x + c \sin x)^n} + \left(\frac{A}{a} + \frac{n(Bb + Cc)}{(n-1)a^2}\right)(-c \cos x + b \sin x)$$
$$\times \frac{(n-1)!}{(2n-1)!!} \sum_{k=0}^{n-1} \frac{(2n-2k-3)!!}{(n-k-1)!a^k} \cdot \frac{1}{(a + b \cos x + c \sin x)^{n-k}}$$
$$\left[n \neq 1, \quad a^2 = b^2 + c^2\right]$$

For $n = 1$:

2.
$$\int \frac{A + B \cos x + C \sin x}{a + b \cos x + c \sin x} \, dx = \frac{Bc - Cb}{b^2 + c^2} \ln (a + b \cos x + c \sin x) + \frac{Bb + Cc}{b^2 + c^2} x$$
$$+ \left(A - \frac{Bb + Cc}{B^2 + c^2} a\right) \int \frac{dx}{a + b \cos x + c \sin x} \qquad \text{(see 2.558 4)}$$
GU (331)(18)

3.
$$\int \frac{dx}{(a + b \cos x + c \sin x)^n} = \int \frac{d(x - \alpha)}{[a + r \cos(x - \alpha)]^n},$$
where $b = r \cos \alpha$, $c = r \sin \alpha$ (see **2.554** 3)

4.
$$\int \frac{dx}{a + b \cos x + c \sin x}$$

$$= \frac{2}{\sqrt{a^2 - b^2 - c^2}} \arctan \frac{(a - b) \tan \frac{x}{2} + c}{\sqrt{a^2 - b^2 - c^2}} \qquad \left[a^2 > b^2 + c^2\right] \quad \text{TI (253), FI II 94}$$

$$= \frac{1}{\sqrt{b^2 + c^2 - a^2}} \ln \frac{(a - b) \tan \frac{x}{2} + c - \sqrt{b^2 + c^2 - a^2}}{(a - b) \tan \frac{x}{2} + c + \sqrt{b^2 + c^2 - a^2}} \qquad \left[a^2 < b^2 + c^2\right] \qquad \text{TI (253)a}$$

$$= \frac{1}{c} \ln \left(a + c \cdot \tan \frac{x}{2}\right) \qquad [a = b]$$

$$= \frac{-2}{c + (a - b) \tan \frac{x}{2}} \qquad \left[a^2 = b^2 + c^2\right] \qquad \text{TI (253)a}$$

2.559

1. $$\int \frac{dx}{\left[a\left(1+\cos x\right)+c\sin x\right]^2} = \frac{1}{c^3}\left[\frac{c\left(a\sin x - c\cos x\right)}{a\left(1+\cos x\right)+c\sin x} - a\ln\left(a+c\tan\frac{x}{2}\right)\right]$$

2. $$\int \frac{A+B\cos x+C\sin x}{\left(a_1+b_1\cos x+c_1\sin x\right)\left(a_2+b_2\cos x+c_2\sin x\right)}\,dx$$
$$= A_0\ln\frac{a_1+b_1\cos x+c_1\sin x}{a_2+b_2\cos x+c+2\sin x} + A_1\int \frac{dx}{a_1+b_1\cos x+c_1\sin x} + A_2\int \frac{dx}{a_2+b_2\cos x+c_2\sin x}$$

(see 2.558 4)   GU (331)(19)

where

$$A_0 = \frac{\begin{vmatrix} A & B & C \\ a_1 & b_1 & c_1 \\ a_2 & b_2 & c_2 \end{vmatrix}}{\begin{vmatrix} a_1 & b_1 \\ a_2 & b_2 \end{vmatrix}^2 - \begin{vmatrix} b_1 & c_1 \\ b_2 & c_2 \end{vmatrix}^2 + \begin{vmatrix} c_1 & a_1 \\ c_2 & a_2 \end{vmatrix}^2},$$

$$A_1 = \frac{\begin{vmatrix} \begin{vmatrix} B & C \\ b_1 & c_1 \end{vmatrix} & \begin{vmatrix} A & C \\ a_1 & c_1 \end{vmatrix} & \begin{vmatrix} B & A \\ b_1 & a_1 \end{vmatrix} \\ a_1 & b_1 & c_1 \\ a_2 & b_2 & c_2 \end{vmatrix}}{\begin{vmatrix} a_1 & b_1 \\ a_2 & b_2 \end{vmatrix}^2 - \begin{vmatrix} b_1 & c_1 \\ b_2 & c_2 \end{vmatrix}^2 + \begin{vmatrix} c_1 & a_1 \\ c_2 & a_2 \end{vmatrix}^2},$$

$$A_2 = \frac{\begin{vmatrix} \begin{vmatrix} C & B \\ c_2 & b_2 \end{vmatrix} & \begin{vmatrix} C & A \\ c_2 & a_2 \end{vmatrix} & \begin{vmatrix} A & B \\ a_2 & b_2 \end{vmatrix} \\ a_1 & b_1 & c_1 \\ a_2 & b_2 & c_2 \end{vmatrix}}{\begin{vmatrix} a_1 & b_1 \\ a_2 & b_2 \end{vmatrix}^2 - \begin{vmatrix} b_1 & c_1 \\ b_2 & c_2 \end{vmatrix}^2 + \begin{vmatrix} c_1 & a_1 \\ c_2 & a_2 \end{vmatrix}^2},$$

$$\left[\begin{vmatrix} a_1 & b_1 \\ a_2 & b_2 \end{vmatrix}^2 + \begin{vmatrix} c_1 & a_1 \\ c_2 & a_2 \end{vmatrix}^2 \neq \begin{vmatrix} b_1 & c_1 \\ b_2 & c_2 \end{vmatrix}^2\right]$$

3. $$\int \frac{A\cos^2 x + 2B\sin x\cos x + C\sin^2 x}{a\cos^2 x + 2b\sin x\cos x + c\sin^2 x}\,dx$$

$$= \frac{1}{4b^2+(a-c)^2}\left\{\left[4Bb+(A-C)(a-c)\right]x + \left[(A-C)b - B(a-c)\right]\right.$$

$$\times \ln\left(a\cos^2 x + 2b\sin x\cos x + c\sin^2 x\right)$$

$$\left. + \left[2(A+C)b^2 - 2Bb(a+c) + (aC-Ac)(a-c)\right]f(x)\right\}$$

GU (331)(24)

where

$$f(x) = \frac{1}{2\sqrt{b^2-ac}}\ln\frac{c\tan x + b - \sqrt{b^2-ac}}{c\tan x + b + \sqrt{b^2-ac}} \qquad [b^2 > ac]$$

$$= \frac{1}{\sqrt{ac-b^2}}\arctan\frac{c\tan x + b}{\sqrt{ac-b^2}} \qquad [b^2 < ac]$$

$$= -\frac{1}{c\tan x + b} \qquad [b^2 = ac]$$

2.561

1. $$\int \frac{(A+B\sin x)\,dx}{\sin x\,(a+b\sin x)} = \frac{A}{a}\ln\tan\frac{x}{2} + \frac{Ba-Ab}{a}\int \frac{dx}{a+b\sin x}$$

(see 2.551 3)   TI (348)

2.
$$\int \frac{(A + B \sin x)\, dx}{\sin x\, (a + b \cos x)} = \frac{A}{a^2 - b^2} \left\{ a \ln \tan \frac{x}{2} + b \ln \frac{a + b \cos x}{\sin x} \right\}$$
$$+ B \int \frac{dx}{a + b \cos x} \qquad\qquad \text{(see 2.553 3)}$$

TI (349)

For $a^2 = b^2 \, (= 1)$:

3.
$$\int \frac{(A + B \sin x)\, dx}{\sin x\, (a + b \cos x)} = \frac{A}{2} \left\{ \ln \tan \frac{x}{2} + \frac{1}{1 + \cos x} \right\} + B \tan \frac{x}{2}$$

4.
$$\int \frac{(A + B \sin x)\, dx}{\sin x\, (1 - \cos x)} = \frac{A}{2} \left\{ \ln \tan \frac{x}{2} - \frac{1}{1 - \cos x} \right\} - B \cot \frac{x}{2}$$

5.
$$\int \frac{(A + B \sin x)\, dx}{\cos x\, (a + b \sin x)} = \frac{1}{a^2 - b^2} \left\{ (Aa - Bb) \ln \tan \left(\frac{\pi}{4} + \frac{x}{2} \right) - (Ab - aB) \ln \frac{a + b \sin x}{\cos x} \right\}$$

TI (346)

For $a^2 = b^2 \, (= 1)$:

6.
$$\int \frac{(A + B \sin x)\, dx}{\cos x\, (1 \pm \sin x)} = \frac{A \pm B}{2} \ln \tan \left(\frac{\pi}{4} + \frac{x}{2} \right) \mp \frac{A \mp B}{2 (1 \pm \sin x)}$$

7.
$$\int \frac{(A + B \sin x)\, dx}{\cos x\, (a + b \cos x)} = \frac{A}{a} \ln \tan \left(\frac{\pi}{4} + \frac{x}{2} \right) + \frac{B}{a} \ln \frac{a + b \cos x}{\cos x}$$
$$- \frac{Ab}{a} \int \frac{dx}{a + b \cos x} \qquad\qquad \text{(see 2.553 3)}$$

TI (351)a

8.
$$\int \frac{(A + B \cos x)\, dx}{\sin x\, (a + b \sin x)} = \frac{A}{a} \ln \tan \frac{x}{2} - \frac{B}{a} \ln \frac{a + b \sin x}{\sin x} - \frac{Ab}{a} \int \frac{dx}{a + b \sin x}$$

(see 2.551 3)   TI (352)

9.
$$\int \frac{(A + B \cos x)\, dx}{\sin x\, (a + b \cos x)} = \frac{1}{a^2 - b^2} \left\{ (Aa - Bb) \ln \tan \frac{x}{2} + (Ab - Ba) \ln \frac{a + b \cos x}{\sin x} \right\}$$

TI (345)

For $a^2 = b^2 \, (= 1)$:

10.
$$\int \frac{(A + B \cos x)\, dx}{\sin x\, (1 \pm \cos x)} = \pm \frac{A \mp B}{2 (1 \pm \cos x)} + \frac{A \pm B}{2} \ln \tan \frac{x}{2}$$

11.
$$\int \frac{(A + B \cos x)\, dx}{\cos x\, (a + b \sin x)} = \frac{A}{a^2 - b^2} \left\{ a \ln \tan \left(\frac{\pi}{4} + \frac{x}{2} \right) - b \ln \frac{a + b \sin x}{\cos x} \right\} + B \int \frac{dx}{a + b \sin x}$$

(see 2.551 3)   TI (350)

For $a^2 = b^2 \, (= 1)$:

12.
$$\int \frac{(A + B \sin x)\, dx}{\cos x\, (1 \pm \sin x)} = \frac{A \pm B}{2} \ln \tan \left(\frac{\pi}{4} + \frac{x}{2} \right) \mp \frac{A \mp B}{2 (1 \pm \sin x)}$$

13.
$$\int \frac{(A + B \cos x)\, dx}{\cos x\, (a + b \cos x)} = \frac{A}{a} \ln \tan \left(\frac{\pi}{4} + \frac{x}{2} \right) + \frac{Ba - Ab}{a} \int \frac{dx}{a + b \cos x}$$

(see 2.553 3)   TI (347)

2.562

1. $\displaystyle \int \frac{dx}{a + b\sin^2 x} = \frac{\operatorname{sign} a}{\sqrt{a(a+b)}} \arctan\left(\sqrt{\frac{a+b}{a}}\,\tan x\right)$ $\left[\dfrac{b}{a} > -1\right]$

      $\displaystyle = \frac{\operatorname{sign} a}{\sqrt{-a(a+b)}} \operatorname{arctanh}\left(\sqrt{-\frac{a+b}{a}}\,\tan x\right)$ $\left[\dfrac{b}{a} < -1, \quad \sin^2 x < -\dfrac{a}{b}\right]$

      $\displaystyle = \frac{\operatorname{sign} a}{\sqrt{-a(a+b)}} \operatorname{arccoth}\left(\sqrt{-\frac{a+b}{a}}\,\tan x\right)$ $\left[\dfrac{b}{a} < -1, \quad \sin^2 x > -\dfrac{a}{b}\right]$

                                                                       MZ 155

2. $\displaystyle \int \frac{dx}{a + b\cos^2 x} = \frac{-\operatorname{sign} a}{\sqrt{a(a+b)}} \arctan\left(\sqrt{\frac{a+b}{a}}\,\cot x\right)$ $\left[\dfrac{b}{a} > -1\right]$

      $\displaystyle = \frac{-\operatorname{sign} a}{\sqrt{-a(a+b)}} \operatorname{arctanh}\left(\sqrt{-\frac{a+b}{a}}\,\cot x\right)$ $\left[\dfrac{b}{a} < -1, \quad \cos^2 x < -\dfrac{a}{b}\right]$

      $\displaystyle = \frac{-\operatorname{sign} a}{\sqrt{-a(a+b)}} \operatorname{arccoth}\left(\sqrt{-\frac{a+b}{a}}\,\cot x\right)$ $\left[\dfrac{b}{a} < -1, \quad \cos^2 x > -\dfrac{a}{b}\right]$

                                                                        MZ 162

3. $\displaystyle \int \frac{dx}{1 + \sin^2 x} = \frac{1}{\sqrt{2}} \arctan\left(\sqrt{2}\,\tan x\right)$

4. $\displaystyle \int \frac{dx}{1 - \sin^2 x} = \tan x$

5. $\displaystyle \int \frac{dx}{1 + \cos^2 x} = -\frac{1}{\sqrt{2}} \arctan\left(\sqrt{2}\,\cot x\right)$

6. $\displaystyle \int \frac{dx}{1 - \cos^2 x} = -\cot x$

2.563

1. $\displaystyle \int \frac{dx}{\left(a + b\sin^2 x\right)^2} = \frac{1}{2a(a+b)} \left[(2a+b)\int \frac{dx}{a + b\sin^2 x} + \frac{b\sin x\cos x}{a + b\sin^2 x}\right]$

                                                       (see **2.562** 1)   MZ 155

2. $\displaystyle \int \frac{dx}{\left(a + b\cos^2 x\right)^2} = \frac{1}{2a(a+b)} \left[(2a+b)\int \frac{dx}{a + b\cos^2 x} - \frac{b\sin x\cos x}{a + b\cos^2 x}\right]$

                                                       (see **2.562** 2)   MZ 163

3. $$\int \frac{dx}{\left(a + b\sin^2 x\right)^3} = \frac{1}{8pa^3} \left[\left(3 + \frac{2}{p^2} + \frac{3}{p^4}\right) \arctan\left(p\tan x\right) \right.$$

$$+ \left(3 + \frac{2}{p^2} - \frac{3}{p^4}\right) \frac{p\tan x}{1 + p^2\tan^2 x}$$

$$\left. + \left(1 - \frac{2}{p^2} - \frac{1}{p^2}\tan^2 x\right) \frac{2p\tan x}{\left(1 + p^2\tan^2 x\right)^2} \right]$$

$$\left[p^2 = 1 + \frac{b}{a} > 0\right]$$

$$= \frac{1}{8qa^3} \left[\left(3 - \frac{2}{q^2} + \frac{3}{q^4}\right) \operatorname{arctanh}\left(q\tan x\right) \right.$$

$$\left. + \left(3 - \frac{2}{q^2} - \frac{3}{q^4}\right) \frac{q\tan x}{1 - q^2\tan^2 x} + \left(1 + \frac{2}{q^2} + \frac{1}{q^2}\tan^2 x\right) \frac{2q\tan x}{\left(1 - q^2\tan^2 x\right)^2} \right]$$

$$\left[q^2 = -1 - \frac{b}{a} > 0, \quad \sin^2 x < -\frac{a}{b} \; ; \text{for } \sin^2 x > -\frac{a}{b}, \text{ one should}\right.$$

$$\left. \text{replace } \operatorname{arctanh}\left(q\tan x\right) \text{ with } \operatorname{arccoth}\left(q\tan x\right) \right]$$

MZ 156

4. $$\int \frac{dx}{\left(a + b\cos^2 x\right)^3} = -\frac{1}{8pa^3} \left[\left(3 + \frac{2}{p^2} + \frac{3}{p^4}\right) \arctan\left(p\cot x\right) \right.$$

$$\left. + \left(3 + \frac{2}{p^2} - \frac{3}{p^4}\right) \frac{p\cot x}{1 + p^2\cot^2 x} + \left(1 - \frac{2}{p^2} - \frac{1}{p^2}\cot^2 x\right) \frac{2p\cot x}{\left(1 + p^2\cot^2 x\right)^2} \right]$$

$$\left[p^2 = 1 + \frac{b}{a} > 0\right]$$

$$= -\frac{1}{8qa^3} \left[\left(3 - \frac{2}{q^2} + \frac{3}{q^4}\right) \operatorname{arctanh}\left(q\cot x\right) \right.$$

$$\left. + \left(3 - \frac{2}{q^2} - \frac{3}{q^4}\right) \frac{q\cot x}{1 - q^2\cot^2 x} + \left(1 + \frac{2}{q^2} + \frac{1}{q^2}\cot^2 x\right) \frac{2p\cot x}{\left(1 - q^2\cot^2 x\right)^2} \right]$$

$$\left[q^2 = -1 - \frac{b}{a} < 0, \quad \cos^2 x < -\frac{a}{b}, \quad \text{for } \cos^2 x > -\frac{a}{b}, \text{ one should}\right.$$

$$\left. \text{replace } \operatorname{arctanh}\left(q\cot x\right) \text{ with } \operatorname{arccoth}\left(q\cot x\right) \right]$$

MZ 163a

2.564

1. $\displaystyle\int \frac{\tan x \, dx}{1 + m^2 \tan^2 x} = \frac{\ln\left(\cos^2 x + m^2 \sin^2 x\right)}{2\left(m^2 - 1\right)}$ LA 210 (10)

2. $\displaystyle\int \frac{\tan\alpha - \tan x}{\tan\alpha + \tan x} \, dx = \sin 2\alpha \ln\sin(x + \alpha) - x\cos 2\alpha$ LA 210 (11)a

3. $\displaystyle\int \frac{\tan x \, dx}{a + b\tan x} = \frac{1}{a^2 + b^2}\left\{bx - a\ln\left(a\cos x + b\sin x\right)\right\}$ PE (335)

4. $\displaystyle\int \frac{dx}{a + b\tan^2 x} = \frac{1}{a - b}\left[x - \sqrt{\frac{b}{a}}\arctan\left(\sqrt{\frac{b}{a}}\tan x\right)\right]$ PE (334)

2.57 Integrals containing $\sqrt{a \pm b \sin x}$ or $\sqrt{a \pm b \cos x}$

Notation:

$$\alpha = \arcsin\sqrt{\frac{1 - \sin x}{2}}, \qquad \beta = \arcsin\sqrt{\frac{b\left(1 - \sin x\right)}{a + b}},$$

$$\gamma = \arcsin\sqrt{\frac{b\left(1 - \cos x\right)}{a + b}}, \qquad \delta = \arcsin\sqrt{\frac{(a + b)\left(1 - \cos x\right)}{2\left(a - b\cos x\right)}}, \qquad r = \sqrt{\frac{2b}{a + b}}$$

2.571

1. $\displaystyle\int \frac{dx}{\sqrt{a + b\sin x}} = \frac{-2}{\sqrt{a + b}} F(\alpha, r)$ $\left[a > b > 0, \quad -\dfrac{\pi}{2} \leq x < \dfrac{\pi}{2}\right]$

                   $= -\sqrt{\dfrac{2}{b}} F\left(\beta, \dfrac{1}{r}\right)$ $\left[0 < |a| < b, \quad -\arcsin\dfrac{a}{b} < x < \dfrac{\pi}{2}\right]$

                                                    BY (288.00, 288.50)

2. $\displaystyle\int \frac{\sin x \, dx}{\sqrt{a + b\sin x}}$

     $= \dfrac{2a}{b\sqrt{a + b}} F(\alpha, r) - \dfrac{2\sqrt{a + b}}{b} E(\alpha, r)$ $\left[a > b > 0, \quad -\dfrac{\pi}{2} \leq x < \dfrac{\pi}{2}\right]$ BY (288.03)

     $= \sqrt{\dfrac{2}{b}}\left\{F\left(\beta, \dfrac{1}{r}\right) - 2E\left(\beta, \dfrac{1}{r}\right)\right\}$ $\left[0 < |a| < b, \quad -\arcsin\dfrac{a}{b} < x < \dfrac{\pi}{2}\right]$ BY (288.54)

3. $\displaystyle\int \frac{\sin^2 x \, dx}{\sqrt{a + b\sin x}} = \frac{4a\sqrt{a + b}}{3b^2} E(\alpha, r) - \frac{2\left(2a^2 + b^2\right)}{3b^2\sqrt{a + b}} F(\alpha, r) - \frac{2}{3b}\cos x\sqrt{a + b\sin x}$

                                                 $\left[a > b > 0, \quad -\dfrac{\pi}{2} \leq x < \dfrac{\pi}{2}\right]$

     $= \sqrt{\dfrac{2}{b}}\left\{\dfrac{4a}{3b} E\left(\beta, \dfrac{1}{r}\right) - \dfrac{2a + b}{3b} F\left(\beta, \dfrac{1}{r}\right)\right\} - \dfrac{2}{3b}\cos x\sqrt{a + b\sin x}$

                                                 $\left[0 < |a| < b, \quad -\arcsin\dfrac{a}{b} < x < \dfrac{\pi}{2}\right]$

                                                         BY (288.03, 288.54)

4. $\displaystyle\int \frac{dx}{\sqrt{a + b\cos x}} = \frac{2}{\sqrt{a+b}}\, F\left(\frac{x}{2}, r\right)$ $[a > b > 0, \quad 0 \le x \le \pi]$

               $= \sqrt{\dfrac{2}{b}}\, F\left(\gamma, \dfrac{1}{r}\right)$ $\left[b \ge |a| > 0, \quad 0 \le x < \arccos\left(-\dfrac{a}{b}\right)\right]$

                                                      BY (289.00)

5. $\displaystyle\int \frac{dx}{\sqrt{a - b\cos x}} = \frac{2}{\sqrt{a+b}}\, F(\delta, r)$ $[a > b > 0, \quad 0 \le x \le \pi]$ BY (291.00)

6. $\displaystyle\int \frac{\cos x\, dx}{\sqrt{a + b\cos x}} = \frac{2}{b\sqrt{a+b}}\left\{(a+b)\, E\left(\frac{x}{2}, r\right) - a\, F\left(\frac{x}{2}, r\right)\right\}$

                                            $[a > b > 0, \quad 0 \le x \le \pi]$

                                                      BY (289.03)

              $= \sqrt{\dfrac{2}{b}}\left\{2\, E\left(\gamma, \dfrac{1}{r}\right) - F\left(\gamma, \dfrac{1}{r}\right)\right\}$

                               $\left[b > |a| > 0, \quad 0 \le x < \arccos\left(-\dfrac{a}{b}\right)\right]$

                                                      BY (290.04)

7.[6] $\displaystyle\int \frac{\cos x\, dx}{\sqrt{a - b\cos x}} = \frac{2}{b\sqrt{a+b}}\left\{(b-a)\, \Pi\left(\delta, r^2, r\right) + a\, F(\delta, r)\right\}$

                                           $[a > b > 0, \quad 0 \le x \le \pi]$ BY (291.03)

8. $\displaystyle\int \frac{\cos^2 x\, dx}{\sqrt{a + b\cos x}} = \frac{2}{3b^2\sqrt{a+b}}\left\{(2a^2 + b^2)\, F\left(\frac{x}{2}, r\right) - 2a(a+b)\, E\left(\frac{x}{2}, r\right)\right\} + \frac{2}{3b}\sin x\sqrt{a + b\cos x}$

                                                   $[a > b > 0, \quad 0 \le x \le \pi]$

                                                          BY (289.03)

              $= \dfrac{1}{3b}\sqrt{\dfrac{2}{b}}\left\{(2a+b)\, F\left(\gamma, \dfrac{1}{r}\right) - 4a\, E\left(\gamma, \dfrac{1}{r}\right)\right\} + \dfrac{2}{3b}\sin x\sqrt{a + b\cos x}$

                                     $\left[b \ge |a| > 0, \quad 0 \le x < \arccos\left(-\dfrac{a}{b}\right)\right]$

                                                       BY (290.04)

9. $\displaystyle\int \frac{\cos^2 x\, dx}{\sqrt{a - b\cos x}} = \frac{2}{3b^2\sqrt{a+b}}\left\{(2a^2 + b^2)\, F(\delta, r) - 2a(a+b)\, E(\delta, r)\right\}$

                     $+ \dfrac{2}{3b}\sin x\, \dfrac{a + b\cos x}{\sqrt{a - b\cos x}}$ $[a > b > 0,]$

                                                          BY (291.04)a

2.572 $\displaystyle\int \frac{\tan^2 x \, dx}{\sqrt{a + b \sin x}}$

$$= \frac{1}{\sqrt{a+b}} F(\alpha, r) + \frac{a}{(a-b)\sqrt{a+b}} E(\alpha, r)$$
$$- \frac{b - a \sin x}{(a^2 - b^2) \cos x} \sqrt{a + b \sin x} \qquad \left[0 < b < a, -\frac{\pi}{2} < x < \frac{\pi}{2} \right]$$

$$= \sqrt{\frac{2}{b}} \left\{ \frac{2a + b}{2(a+b)} F\left(\beta, \frac{1}{r}\right) + \frac{ab}{a^2 - b^2} E\left(\beta, \frac{1}{r}\right) \right\}$$
$$- \frac{b - a \sin x}{(a^2 - b^2) \cos x} \sqrt{a + b \sin x} \qquad \left[0 < |a| < b, \quad -\arcsin \frac{a}{b} < x < \frac{\pi}{2} \right]$$

BY(288.08, 288.58)

2.573

1. $\displaystyle\int \frac{1 - \sin x}{1 + \sin x} \cdot \frac{dx}{\sqrt{a + b \sin x}} = \frac{2}{a-b} \left\{ \sqrt{a+b} \, E(\alpha, r) \right\} - \tan\left(\frac{\pi}{4} - \frac{x}{2}\right) \sqrt{a + b \sin x} \Bigg\}$

$$\left[0 < b < a, \quad -\frac{\pi}{2} \le x < \frac{\pi}{2} \right] \quad \text{BY (288.07)}$$

2. $\displaystyle\int \frac{1 - \cos x}{1 + \cos x} \frac{dx}{\sqrt{a + b \cos x}} = \frac{2}{a-b} \tan \frac{x}{2} \sqrt{a + b \cos x} - \frac{2\sqrt{a+b}}{a-b} E\left(\frac{x}{2}, r\right)$

$$[a > b > 0, \quad 0 \le x < \pi] \qquad \text{BY (289.07)}$$

2.574

1. $\displaystyle\int \frac{dx}{(2 - p^2 + p^2 \sin x)\sqrt{a + b \sin x}} = -\frac{1}{a+b} \Pi\left(\alpha, p^2, r\right)$

$$\left[0 < b < a, \quad -\frac{\pi}{2} \le x < \frac{\pi}{2} \right]$$
BY (288.02)

2. $\displaystyle\int \frac{dx}{(a + b - p^2 b + p^2 b \sin x)\sqrt{a + b \sin x}} = -\frac{1}{a+b} \sqrt{\frac{2}{b}} \, \Pi\left(\beta, p^2, \frac{1}{r}\right)$

$$\left[0 < |a| < b, \quad -\arcsin \frac{a}{b} < x < \frac{\pi}{2} \right]$$
BY (288.52)

3. $\displaystyle\int \frac{dx}{(2 - p^2 + p^2 \cos x)\sqrt{a + b \cos x}} = \frac{1}{\sqrt{a+b}} \Pi\left(\frac{x}{2}, p^2, r\right)$

$$[a > b > 0, \quad 0 \le x < \pi] \qquad \text{BY (289.02)}$$

4. $\displaystyle\int \frac{dx}{(a + b - p^2 b + p^2 b \cos x)\sqrt{a + b \cos x}} = \frac{\sqrt{2}}{(a+b)\sqrt{b}} \Pi\left(\gamma, p^2, \frac{1}{r}\right)$

$$\left[b \ge |a| > 0, \quad 0 \le x < \arccos\left(-\frac{a}{b}\right) \right]$$
BY (290.02)

2.575

1. $$\int \frac{dx}{\sqrt{(a + b \sin x)^3}} = \frac{2b \cos x}{(a^2 - b^2)\sqrt{a + b \sin x}} - \frac{2}{(a - b)\sqrt{a + b}} E(\alpha, r)$$

$$\left[0 < b < a, \quad -\frac{\pi}{2} \le x < \frac{\pi}{2} \right]$$
BY (288.05)

$$= \sqrt{\frac{2}{b}} \left\{ \frac{2b}{b^2 - a^2} E\left(\beta, \frac{1}{r}\right) - \frac{1}{a + b} F\left(\beta, \frac{1}{r}\right) \right\} + \frac{2b}{b^2 - a^2} \cdot \frac{\cos x}{\sqrt{a + b \sin x}}$$

$$\left[0 < |a| < b, \quad -\arcsin\frac{a}{b} < x < \frac{\pi}{2} \right]$$
BY (288.56)

2. $$\int \frac{dx}{\sqrt{(a + b \sin x)^5}} = \frac{2}{3(a^2 - b^2)^2 \sqrt{a + b}} \left\{ (a^2 - b^2) F(\alpha, r) - 4a(a + b) E(\alpha, r) \right\}$$

$$+ \frac{2b \left(5a^2 - b^2 + 4ab \sin x\right)}{3(a^2 - b^2)^2 \sqrt{(a + b \sin x)^3}} \cos x$$

$$\left[0 < b < a, \quad -\frac{\pi}{2} \le x < \frac{\pi}{2} \right]$$
BY (288.05)

$$= -\frac{1}{3(a^2 - b^2)^2} \sqrt{\frac{2}{b}} \left\{ (3a - b)(a - b) F\left(\beta, \frac{1}{r}\right) + 8ab E\left(\beta, \frac{1}{r}\right) \right\}$$

$$+ \frac{2b \left[a^2 - b^2 + 4a(a + b \sin x)\right]}{3(a^2 - b^2)^2 \sqrt{(a + b \sin x)^3}} \cos x$$

$$\left[0 < |a| < b, \quad -\arcsin\frac{a}{b} < x < \frac{\pi}{2} \right]$$
BY (288.56)

3. $$\int \frac{dx}{\sqrt{(a + b \cos x)^3}} = \frac{2}{(a - b)\sqrt{a + b}} E\left(\frac{x}{2}, r\right) - \frac{2b}{a^2 - b^2} \cdot \frac{\sin x}{\sqrt{a + b \cos x}}$$

$$[a > b > 0, \quad 0 \le x \le \pi]$$
BY (289.05)

$$= \frac{1}{a^2 - b^2} \sqrt{\frac{2}{b}} \left\{ (a - b) F\left(\gamma, \frac{1}{r}\right) + 2b E\left(\gamma, \frac{1}{r}\right) \right\} + \frac{2b}{b^2 - a^2} \cdot \frac{\sin x}{\sqrt{a + b \cos x}}$$

$$\left[b \ge |a| > 0, \quad 0 \le x < \arccos\left(-\frac{a}{b}\right) \right]$$
BY (290.06)

4. $$\int \frac{dx}{\sqrt{(a - b \cos x)^3}} = \frac{2}{(a - b)\sqrt{a + b}} E(\delta, r) \qquad [a > b > 0, \quad 0 \le x \le \pi] \qquad (291.01)$$

5.

$$\int \frac{dx}{\sqrt{(a + b \cos x)^5}} = \frac{2\sqrt{a + b}}{3(a^2 - b^2)^2} \left\{ 4a\, E\left(\frac{x}{2}, r\right) - (a - b)\, F\left(\frac{x}{2}, r\right) \right\}$$

$$- \frac{2b}{3(a^2 - b^2)^2} \cdot \frac{5a^2 - b^2 + 4ab \cos x}{\sqrt{(a + b \cos x)^3}} \sin x$$

$$[a > b > 0, \quad 0 \le x \le \pi]$$
BY (289.05)

$$= \frac{1}{3(a^2 - b^2)^2} \sqrt{\frac{2}{b}} \left\{ (a - b)(3a - b)\, F\left(\gamma, \frac{1}{r}\right) + 8ab\, E\left(\gamma, \frac{1}{r}\right) \right\}$$

$$+ \frac{2b\left(5a^2 - b^2 + 4ab \cos x\right) \sin x}{3(a^b - b^2)^2 \sqrt{(a + b \cos x)^3}}$$

$$\left[b \ge |a| > 0, \quad 0 \le x < \arccos\left(-\frac{a}{b}\right) \right]$$
BY (290.06)

2.576

1.

$$\int \sqrt{a + b \cos x}\, dx = 2\sqrt{a + b}\, E\left(\frac{x}{2}, r\right)$$

$$[a > b > 0, \quad 0 \le x \le \pi]$$
BY (289.01)

$$= \sqrt{\frac{2}{b}} \left\{ (a - b)\, F\left(\gamma, \frac{1}{r}\right) + 2b\, E\left(\gamma, \frac{1}{r}\right) \right\}$$

$$\left[b \ge |a| > 0, \quad 0 \le x < \arccos\left(-\frac{a}{b}\right) \right]$$
BY (290.03)

2. $$\int \sqrt{a - b \cos x}\, dx = 2\sqrt{a + b}\, E(\delta, r) - \frac{2b \sin x}{\sqrt{a - b \cos x}} \qquad [a > b > 0, \quad 0 \le x \le \pi] \qquad \text{BY (291.05)}$$

2.577

1.[3]

$$\int \frac{\sqrt{a - b \cos x}}{1 + p \cos x}\, dx = \frac{2(a - b)}{(1 + p)\sqrt{a + b}}\, \Pi\left(\delta, \frac{2ap}{(a + b)(1 + p)}, r\right)$$

$$[a > b > 0, \quad 0 \le x \le \pi, \quad p \ne -1]$$
BY (291.02)

2.

$$\int \sqrt[3]{\frac{a - b \cos x}{1 + p \cos x}}\, dx = \frac{2(a - b)}{\sqrt{(1 + p)(a + b)}}\, \Pi\left(\delta, -r^2, \sqrt{\frac{2(ap + b)}{(1 + p)(a + b)}}\right)$$

$$[a > b > 0, \quad 0 \le x \le \pi, \quad p \ne -1]$$

2.578 $$\int \frac{\tan x\, dx}{\sqrt{a + b \tan^2 x}} = \frac{1}{\sqrt{b - a}} \arccos\left(\frac{\sqrt{b - a}}{\sqrt{b}} \cos x\right) \qquad [b > a, \quad b > 0] \qquad \text{PE (333)}$$

2.58–2.62 Integrals reducible to elliptic and pseudo-elliptic integrals

2.580

1. $$\int \frac{d\varphi}{\sqrt{a + b\cos\varphi + c\sin\varphi}} = 2\int \frac{d\psi}{\sqrt{a - p + 2p\cos^2\psi}} \qquad \left[\varphi = 2\psi + \alpha, \tan\alpha = \frac{c}{b}, p = \sqrt{b^2 + c^2}\right]$$

2. $$\int \frac{d\varphi}{\sqrt{a + b\cos\varphi + c\sin\varphi + d\cos^2\varphi + e\sin\varphi\cos\varphi + f\sin^2\varphi}} = 2\int \frac{dx}{\sqrt{A + Bx + Cx^2 - Dx^3 + Ex^4}}$$
    $$\left[\tan\frac{\varphi}{2} = x, A = a + b + d, B = 2c + 2e, C = 2a - 2d + 4f, D = 2c - 2e, E = a - b + d\right]$$

Forms containing $\sqrt{1 - k^2\sin^2 x}$

Notation: $\Delta = \sqrt{1 - k^2\sin^2 x}$, $k' = \sqrt{1 - k^2}$

2.581

1. $$\int \sin^m x\cos^n x\Delta^r\,dx$$

    $$= \frac{1}{(m+n+r)k^2}\left\{\sin^{m-3}x\cos^{n+1}x\Delta^{r+2} + \left[(m+n-2) + (m+r-1)k^2\right]\right.$$

    $$\left.\times \int \sin^{m-2}x\cos^n x\Delta^r\,dx - (m-3)\int \sin^{m-4}x\cos^n x\Delta^r\,dx\right\}$$

    $$= \frac{1}{(m+n+r)k^2}\left\{\sin^{m+1}x\cos^{n-3}x\Delta^{r+2} + \left[(n+r-1)k^2 - (m+n-2)k'^2\right]\right.$$

    $$\left.\times \int \sin^m x\cos^{n-2}x\Delta^r\,dx + (n-3)k'^2\int \sin^m x\cos^{n-4}x\Delta^r\,dx\right\}$$

    $$[m+n+r \neq 0]$$

    For $r = -3$ and $r = -5$:

2. $$\int \frac{\sin^m x\cos^n x}{\Delta^3}\,dx = \frac{\sin^{m-1}x\cos^{n-1}x}{k^2\Delta}$$
    $$- \frac{m-1}{k^2}\int \frac{\sin^{m-2}x\cos^n x}{\Delta}\,dx + \frac{n-1}{k^2}\int \frac{\sin^m x\cos^{n-2}x}{\Delta}\,dx$$

3. $$\int \frac{\sin^m x\cos^n x}{\Delta^5}\,dx = \frac{\sin^{m-1}x\cos^{n-1}x}{3k^2\Delta^3}$$
    $$- \frac{m-1}{3k^2}\int \frac{\sin^{m-2}x\cos^n x}{\Delta^3}\,dx + \frac{n-1}{3k^2}\int \frac{\sin^m x\cos^{n-2}x}{\Delta^3}\,dx$$

    For $m = 1$ or $n = 1$:

4. $$\int \sin x\cos^n x\Delta^r\,dx = -\frac{\cos^{n-1}x\Delta^{r+2}}{(n+r+1)k^2} - \frac{(n-1)k'^2}{(n+r+1)k^2}\int \cos^{n-2}x\sin x\Delta^r\,dx$$

5. $$\int \sin^m x\cos x\Delta^r\,dx = -\frac{\sin^{m-1}x\Delta^{r+2}}{(m+r+1)k^2} + \frac{m-1}{(m+r+1)k^2}\int \sin^{m-2}x\cos x\Delta^r\,dx$$

    For $m = 3$ or $n = 3$:

6. $\displaystyle\int \sin^3 x \cos^n x \Delta^r \, dx = \frac{(n+r+1)k^2 \cos^2 x - \left[(r+2)k^2 + n + 1\right]}{(n+r+1)(n+r+3)k^4} \cos^{n-1} x \Delta^{r+2}$

$$- \frac{\left[(r+2)k^2 + n + 1\right](n-1)k'^2}{(n+r+1)(n+r+3)k^4} \int \cos^{n-2} x \sin x \Delta^r \, dx$$

7. $\displaystyle\int \sin^m x \cos^3 x \Delta^r \, dx$

$$= \frac{(m+r+1)k^2 \sin^2 x - \left[(r+2)k^2 - (m+1)k'^2\right]}{(m+r+1)(m+r+3)k^4}$$

$$\times \sin^{m-1} x n^{m-1} x \Delta^{r+2} + \frac{\left[(r+2)k^2 - (m-1)k'^2\right](m-1)}{(m+r+1)(m+r+3)k^4} \int \sin^{m-2} x \cos x \Delta^r \, dx$$

2.582

1. $\displaystyle\int \Delta^n \, dx = \frac{n-1}{n}(2-k^2)\int \Delta^{n-2} \, dx - \frac{n-2}{n}(1-k^2)\int \Delta^{n-4} \, dx$

$$+ \frac{k^2}{n} \sin x \cos x \cdot \Delta^{n-2}$$

LA (316)(1)a

2. $\displaystyle\int \frac{dx}{\Delta^{n+1}} = -\frac{k^2 \sin x \cos x}{(n-1)k'^2 \Delta^{n-1}} + \frac{n-2}{n-1}\frac{2-k^2}{k'^2}\int \frac{dx}{\Delta^{n-1}} - \frac{n-3}{n-1}\frac{1}{k'^2}\int \frac{dx}{\Delta^{n-3}}$ LA 317(8)a

3. $\displaystyle\int \frac{\sin^n x}{\Delta} \, dx = \frac{\sin^{n-3} x}{(n-1)k^2} \cos x \cdot \Delta + \frac{n-2}{n-1}\frac{1+k^2}{k^2}\int \frac{\sin^{n-2} x}{\Delta} \, dx$

$$- \frac{n-3}{(n-1)k^2}\int \frac{\sin^{n-4} x}{\Delta} \, dx$$

LA 316(1)a

4. $\displaystyle\int \frac{\cos^n x}{\Delta} \, dx = \frac{\cos^{n-3} x}{(n-1)k^2} \sin x \cdot \Delta + \frac{n-2}{n-1}\frac{2k^2-1}{k^2}\int \frac{\cos^{n-2} x}{\Delta} \, dx$

$$+ \frac{n-3}{n-1}\frac{k'^2}{k^2}\int \frac{\cos^{n-4} x}{\Delta} \, dx$$

LA 316(2)a

5. $\displaystyle\int \frac{\tan^n x}{\Delta} \, dx = \frac{\tan^{n-3} x}{(n-1)k'^2}\frac{\Delta}{\cos^2 x} - \frac{(n-2)(2-k^2)}{(n-1)k'^2}\int \frac{\tan^{n-2} x}{\Delta} \, dx$

$$- \frac{n-3}{(n-1)k'^2}\int \frac{\tan^{n-4} x}{\Delta} \, dx$$

LA 317(3)

6. $\displaystyle\int \frac{\cot^n x}{\Delta} \, dx = -\frac{\cot^{n-1} x}{n-1}\frac{\Delta}{\cos^2 x} - \frac{n-2}{n-1}(2-k^2)\int \frac{\cot^{n-2} x}{\Delta} \, dx$

$$- \frac{n-3}{n-1}k'^2 \int \frac{\cot^{n-4} x}{\Delta} \, dx$$

LA 317(6)

2.583

1. $\displaystyle\int \Delta \, dx = E(x, k)$

2. $\displaystyle\int \Delta \sin x \, dx = -\frac{\Delta \cos x}{2} - \frac{k'^2}{2k} \ln \left(k \cos x + \Delta \right)$

3. $\displaystyle\int \Delta \cos x \, dx = \frac{\Delta \sin x}{2} + \frac{1}{2k} \arcsin \left(k \sin x \right)$

4. $\displaystyle\int \Delta \sin^2 x \, dx = -\frac{\Delta}{3} \sin x \cos x + \frac{k'^2}{3k^2} F(x, k) + \frac{2k^2 - 1}{3k^2} E(x, k)$

5. $\displaystyle\int \Delta \sin x \cos x \, dx = -\frac{\Delta^3}{3k^2}$

6. $\displaystyle\int \Delta \cos^2 x \, dx = \frac{\Delta}{3} \sin x \cos x - \frac{k'^2}{3k^2} F(x, k) + \frac{k^2 + 1}{3k^2} E(x, k)$

7. $\displaystyle\int \Delta \sin^3 x \, dx = -\frac{2k^2 \sin^2 x + 3k^2 - 1}{8k^2} \Delta \cos x + \frac{3k^4 - 2k^2 - 1}{8k^3} \ln \left(k \cos x + \Delta \right)$

8. $\displaystyle\int \Delta \sin^2 x \cos x \, dx = \frac{2k^2 \sin^2 x - 1}{8k^2} \Delta \sin x + \frac{1}{8k^3} \arcsin \left(k \sin x \right)$

9. $\displaystyle\int \Delta \sin x \cos^2 x \, dx = -\frac{2k^2 \cos^2 x + k'^2}{8k^2} \Delta \cos x + \frac{k'^4}{8k^3} \ln \left(k \cos x + \Delta \right)$

10. $\displaystyle\int \Delta \cos^3 x \, dx = \frac{2k^2 \cos^2 x + 2k^2 + 1}{8k^2} \Delta \sin x + \frac{4k^2 - 1}{8k^3} \arcsin \left(k \sin x \right)$

11. $\displaystyle\int \Delta \sin^4 x \, dx = -\frac{3k^2 \sin^2 x + 4k^2 - 1}{15k^2} \Delta \sin x \cos x$
$\qquad\qquad - \frac{2 \left(2k^4 - k^2 - 1 \right)}{15k^4} F(x, k) + \frac{8k^4 - 3k^2 - 2}{15k^4} E(x, k)$

12. $\displaystyle\int \Delta \sin^3 x \cos x \, dx = \frac{3k^4 \sin^4 x - k^2 \sin^2 x - 2}{15k^4} \Delta.$

13. $\displaystyle\int \Delta \sin^2 x \cos^2 x \, dx = -\frac{3k^2 \cos^2 x - 2k^2 + 1}{15k^2} \Delta \sin x \cos x$
$\qquad\qquad - \frac{k'^2 \left(1 + k'^2 \right)}{15k^4} F(x, k) + \frac{2 \left(k^4 - k^2 + 1 \right)}{15k^4} E(x, k)$

14. $\displaystyle\int \Delta \sin x \cos^3 x \, dx = -\frac{3k^4 \sin^4 x - k^2 \left(5k^2 + 1 \right) \sin^2 x + 5k^2 - 2}{15k^4} \Delta.$

15. $\displaystyle\int \Delta \cos^4 x \, dx = \frac{3k^2 \cos^2 x + 3k^2 + 1}{15k^2} \Delta \sin x \cos x$
$\qquad\qquad + \frac{2k'^2 \left(k'^2 - 2k^2 \right)}{15k^4} F(x, k) + \frac{3k^4 + 7k^2 - 2}{15k^4} E(x, k)$

16. $\displaystyle\int \Delta \sin^5 x \, dx = \frac{-8k^4 \sin^4 x - 2k^2 \left(5k^2 - 1 \right) \sin^2 x - 15k^4 + 4k^2 + 3}{48k^4} \Delta \cos x$
$\qquad\qquad + \frac{5k^6 - 3k^4 - k^2 - 1}{16k^5} \ln \left(k \cos x + \Delta \right)$

17. $\displaystyle\int \Delta \sin^4 x \cos x \, dx = \frac{8k^4 \sin^4 x - 2k^2 \sin^2 x - 3}{48k^4} \Delta \sin x + \frac{1}{16k^5} \arcsin \left(k \sin x \right)$

18. $$\int \Delta \sin^3 x \cos^2 x \, dx = \frac{8k^4 \sin^4 x - 2k^2 \left(k^2 + 1\right) \sin^2 x - 3k^4 + 2k^2 - 3}{48k^4} \Delta \cos x$$
$$+ \frac{k'^4 \left(k^2 + 1\right)}{16k^5} \ln \left(k \cos x + \Delta\right)$$

19. $$\int \Delta \sin^2 x \cos^3 x \, dx = \frac{-8k^4 \sin^4 x + 2k^2 \left(6k^2 + 1\right) \sin^2 x - 6k^2 + 3}{48k^4} \Delta \sin x$$
$$+ \frac{2k^2 - 1}{16k^5} \arcsin \left(k \sin x\right)$$

20. $$\int \Delta \sin x \cos^4 x \, dx = \frac{-8k^4 \sin^4 x + 2k^2 \left(7k^2 + 1\right) \sin^2 x - 3k^4 - 8k^2 + 3}{48k^4} \Delta \cos x$$
$$- \frac{k'^6}{16k^5} \ln \left(k \cos x + \Delta\right)$$

21. $$\int \Delta \cos^5 x \, dx = \frac{8k^4 \sin^4 x - 2k^2 \left(12k^2 + 1\right) \sin^2 x + 24k^4 + 12k^2 - 3}{48k^4} \Delta \sin x$$
$$+ \frac{8k^4 - 4k^2 + 1}{16k^5} \arcsin \left(k \sin x\right)$$

22. $$\int \Delta^3 \, dx = \frac{2}{3} \left(1 + k'^2\right) E(x, k) - \frac{k'^2}{3} F(x, F) + \frac{k^2}{3} \Delta \sin x \cos x$$

23. $$\int \Delta^3 \sin x \, dx = \frac{2k^2 \sin^2 x + 3k^2 - 5}{8} \Delta \cos x - \frac{3k'^4}{8k} \ln \left(k \cos x + \Delta\right)$$

24. $$\int \Delta^3 \cos x \, dx = \frac{-2k^2 \sin^2 x + 5}{8} \Delta \sin x + \frac{3}{8k} \arcsin \left(k \sin x\right)$$

25. $$\int \Delta^3 \sin^2 x \, dx = \frac{3k^2 \sin^2 x + 4k^2 - 6}{15} \Delta \sin x \cos x + \frac{k'^2 \left(3 - 4k^2\right)}{15k^2} F(x, k)$$
$$- \frac{8k^4 - 13k^2 + 3}{15k^2} E(x, k)$$

26. $$\int \Delta^3 \sin x \cos x \, dx = -\frac{\Delta^5}{5k^2}$$

27. $$\int \Delta^3 \cos^2 x \, dx = \frac{-3k^2 \sin^2 x + k^2 + 5}{15} \Delta \sin x \cos x - \frac{k'^2 \left(k^2 + 3\right)}{15k^2} F(x, k)$$
$$- \frac{2k^4 - 7k^2 - 3}{15k^2} E(x, k)$$

28. $$\int \Delta^3 \sin^3 x \, dx = \frac{8k^4 \sin^4 x + 2k^2 \left(5k^2 - 7\right) \sin^2 x + 15k^4 - 22k^2 + 3}{48k^2} \Delta \cos x$$
$$- \frac{5k^6 - 9k^4 + 3k^2 + 1}{16k^3} \ln \left(k \cos x + \Delta\right)$$

29. $$\int \Delta^3 \sin^2 x \cos x \, dx = \frac{-8k^4 \sin^4 x + 14k^2 \sin^2 x - 3}{48k^2} \Delta \sin x$$
$$+ \frac{1}{16k^3} \arcsin \left(k \sin x\right)$$

30. $\displaystyle\int \Delta^3 \sin x \cos^2 x \, dx = \frac{-8k^4 \sin^4 x + 2k^2 \left(k^2 + 7\right) \sin^2 x + 3k^4 - 8k^2 - 3}{48k^2}$

$$\times \Delta \cos x + \frac{k'^6}{16k^3} \ln \left(k \cos x + \Delta\right)$$

31. $\displaystyle\int \Delta^3 \cos^3 x \, dx = \frac{8k^4 \sin^4 x - 2k^2 \left(6k^2 + 7\right) \sin^2 x + 30k^2 + 3}{48k^2} \Delta \sin x$

$$+ \frac{6k^2 - 1}{16k^3} \arcsin \left(k \sin x\right)$$

32. $\displaystyle\int \frac{\Delta \, dx}{\sin x} = -\frac{1}{2} \ln \frac{\Delta + \cos x}{\Delta - \cos x} + k \ln k \left(k \cos x + \Delta\right)$

33. $\displaystyle\int \frac{\Delta \, dx}{\cos x} = \frac{k'}{2} \ln \frac{\Delta + k' \sin x}{\Delta - k' \sin x} + k \arcsin \left(k \sin x\right)$

34. $\displaystyle\int \frac{\Delta \, dx}{\sin^2 x} = k'^2 \, F(x, k) - E(x, k) - \Delta \cot x$

35. $\displaystyle\int \frac{\Delta \, dx}{\sin x \cos x} = \frac{1}{2} \ln \frac{1 - \Delta}{1 + \Delta} + \frac{k'}{2} \ln \frac{\Delta + k'}{\Delta - k'}$

36. $\displaystyle\int \frac{\Delta \, dx}{\cos^2 x} = F(x, k) - E(x, k) + \Delta \tan x$

37. $\displaystyle\int \frac{\sin x}{\cos x} \Delta \, dx = \int \Delta \tan x \, dx = -\Delta + \frac{k'}{2} \ln \frac{\Delta + k'}{\Delta - k'}$

38. $\displaystyle\int \frac{\cos x}{\sin x} \Delta \, dx = \int \Delta \cot x \, dx = \Delta + \frac{1}{2} \ln \frac{1 - \Delta}{1 + \Delta}$

39. $\displaystyle\int \frac{\Delta \, dx}{\sin^3 x} = -\frac{\Delta \cos x}{2 \sin^2 x} + \frac{k'^2}{4} \ln \frac{\Delta + \cos x}{\Delta - \cos x}$

40. $\displaystyle\int \frac{\Delta \, dx}{\sin^2 x \cos x} = \frac{-\Delta}{\sin x} - \frac{1 + k^2}{2k'} \ln \frac{\Delta - k' \sin x}{\Delta + k' \sin x}$

41. $\displaystyle\int \frac{\Delta \, dx}{\sin x \cos^2 x} = \frac{\Delta}{\cos x} + \frac{1}{2} \ln \frac{\Delta + \cos x}{\Delta - \cos x}$

42. $\displaystyle\int \frac{\Delta \, dx}{\cos^3 x} = \frac{\Delta \sin x}{2 \cos^2 x} + \frac{1}{4k'} \ln \frac{\Delta + k' \sin x}{\Delta - k' \sin x}$

43. $\displaystyle\int \frac{\Delta \sin x \, dx}{\cos^2 x} = \frac{\Delta}{\cos x} - k \ln \left(k \cos x + \Delta\right)$

44. $\displaystyle\int \frac{\Delta \cos x \, dx}{\sin^2 x} = -\frac{\Delta}{\sin x} - k \arcsin \left(k \sin x\right)$

45. $\displaystyle\int \frac{\Delta \sin^2 x \, dx}{\cos x} = -\frac{\Delta \sin x}{2} + \frac{2k^2 - 1}{2k} \arcsin \left(k \sin x\right) + \frac{k'}{2} \ln \frac{\Delta + k' \sin x}{\Delta - k' \sin x}$

46. $\displaystyle\int \frac{\Delta \cos^2 x \, dx}{\sin x} = \frac{\Delta \cos x}{2} + \frac{k^2 + 1}{2k} \ln \left(k \cos x + \Delta\right) + \frac{1}{2} \ln \frac{\Delta + \cos x}{\Delta - \cos x}$

47. $\displaystyle\int \frac{\Delta \, dx}{\sin^4 x} = \frac{1}{3} \left\{ -\Delta \cot^3 x + \left(k^2 - 3\right) \Delta \cot x + 2k'^2 \, F(x, k) + \left(k^2 - 2\right) E(x, k) \right\}$

48. $$\int \frac{\Delta\, dx}{\sin^3 x \cos x} = -\frac{\Delta}{2\sin^2 x} + \frac{k'}{2} \ln \frac{\Delta + k'}{\Delta - k'} + \frac{k^2 - 2}{4} \ln \frac{1 + \Delta}{1 - \Delta}$$

49. $$\int \frac{\Delta\, dx}{\sin^2 x \cos^2 x} = \left(\frac{1}{k'^2} \tan x - \cot x\right) \Delta + 2\, F(x, k) - \frac{1 + k'^2}{k'^2}\, E(x, k)$$

50. $$\int \frac{\Delta\, dx}{\sin x \cos^3 x} = \frac{\Delta}{2\cos^2 x} - \frac{1}{2} \ln \frac{1 + \Delta}{1 - \Delta} + \frac{2 - k^2}{4k'} \ln \frac{\Delta + k'}{\Delta - k'}$$

51. $$\int \frac{\Delta\, dx}{\cos^4 x} = \frac{1}{3k'^2} \left\{ \left[k'^2 \tan^2 x - (2k^2 - 3) \tan x\right] \Delta + 2k'^2\, F(x, k) + (k^2 - 2)\, E(x, k) \right\}$$

52. $$\int \frac{\sin x}{\cos^3 x} \Delta\, dx = \frac{\Delta}{2\cos^2 x} + \frac{k^2}{4k'} \ln \frac{\Delta + k'}{\Delta - k'}$$

53. $$\int \frac{\cos x}{\sin^3 x} \Delta\, dx = -\frac{\Delta}{2\sin^2 x} + \frac{k^2}{4} \ln \frac{1 + \Delta}{1 - \Delta}$$

54. $$\int \frac{\sin^2 x}{\cos^2 x} \Delta\, dx = \int \tan^2 x \Delta\, dx = \Delta \tan x + F(x, k) - 2\, E(x, k)$$

55. $$\int \frac{\cos^2 x}{\sin^2 x} \Delta\, dx = \int \cot^2 x \Delta\, dx = -\Delta \cot x + k'^2\, F(x, k) - 2\, E(x, k)$$

56. $$\int \frac{\sin^3 x}{\cos x} \Delta\, dx = -\frac{k^2 \sin^2 x + 3k^2 - 1}{3k^2} \Delta + \frac{k'}{2} \ln \frac{\Delta + k'}{\Delta - k'}$$

57. $$\int \frac{\cos^3 x}{\sin x} \Delta\, dx = -\frac{k^2 \sin^2 x - 3k^2 - 1}{3k^2} \Delta + \frac{1}{2} \ln \frac{1 - \Delta}{1 + \Delta}$$

58. $$\int \frac{\Delta\, dx}{\sin^5 x} = \frac{(k^2 - 3) \sin^2 x + 2}{8 \sin^4 x} \cos x \Delta + \frac{k'^2 (k^2 + 3)}{16} \ln \frac{\Delta + \cos x}{\Delta - \cos x}$$

59. $$\int \frac{\Delta\, dx}{\sin^4 x \cos x} = -\frac{(3 - k^2) \sin^2 x + 1}{3 \sin^3 x} \Delta - \frac{k'}{2} \ln \frac{\Delta - k' \sin x}{\Delta + k' \sin x}$$

60. $$\int \frac{\Delta\, dx}{\sin^3 x \cos^2 x} = \frac{3\sin^2 x - 1}{2\sin^2 x \cos x} \Delta + \frac{k^2 - 3}{4} \ln \frac{\Delta - \cos x}{\Delta + \cos x}$$

61. $$\int \frac{\Delta\, dx}{\sin^2 x \cos^3 x} = \frac{3\sin^2 x - 2}{2\sin x \cos^2 x} \Delta - \frac{2k^2 - 3}{4k'} \ln \frac{\Delta + k' \sin x}{\Delta - k' \sin x}$$

62. $$\int \frac{\Delta\, dx}{\sin x \cos^4 x} = \frac{(2k^2 - 3) \sin^2 x - 3k^2 + 4}{3k'^2 \cos^3 x} \Delta + \frac{1}{2} \ln \frac{\Delta + \cos x}{\Delta - \cos x}$$

63. $$\int \frac{\Delta\, dx}{\cos^5 x} = \frac{(2k^2 - 3) \sin^2 x - 4k^2 + 5}{8k'^2 \cos^4 x} \sin x \Delta - \frac{4k^2 - 3}{16k'^3} \ln \frac{\Delta + k' \sin x}{\Delta - k' \sin x}$$

64. $$\int \frac{\sin x}{\cos^4 x} \Delta\, dx = \frac{-(2k^2 + 1) k^2 \sin^2 x + 3k^4 - k^2 + 1}{3k'^2 \cos^3 x} \Delta.$$

65. $$\int \frac{\cos x}{\sin^4 x} \Delta\, dx = -\frac{\Delta^3}{3\sin^3 x}$$

66. $$\int \frac{\sin^2 x}{\cos^3 x} \Delta\, dx = \frac{\sin x}{2\cos^2 x} \Delta + \frac{2k^2 - 1}{4k'} \ln \frac{\Delta + k' \sin x}{\Delta - k' \sin x} - k \arcsin(k \sin x)$$

67. $\int \dfrac{\cos^2 x}{\sin^3 x}\,\Delta\,dx = -\dfrac{\cos x}{2\sin^2 x}\Delta - \dfrac{k^2+1}{4}\ln\dfrac{\Delta+\cos x}{\Delta-\cos x} - k\ln(k\cos x+\Delta)$

68. $\int \dfrac{\sin^3 x}{\cos^2 x}\,\Delta\,dx = -\dfrac{\sin^2 x-3}{2\cos x}\Delta - \dfrac{3k^2-1}{2k}\ln(k\cos x+\Delta)$

69. $\int \dfrac{\cos^3 x}{\sin^2 x}\,\Delta\,dx = -\dfrac{\sin^2 x+2}{2\sin x}\Delta - \dfrac{2k^2+1}{2k}\arcsin(k\sin x)$

70. $\int \dfrac{\sin^4 x}{\cos x}\,\Delta\,dx = -\dfrac{2k^2\sin^2 x+4k^2-1}{8k^2}\sin x\Delta$
$$+\dfrac{8k^4-4k^2-1}{8k^3}\arcsin(k\sin x)+\dfrac{k'}{2}\ln\dfrac{\Delta+k'\sin x}{\Delta-k'\sin x}$$

71. $\int \dfrac{\cos^4 x}{\sin x}\,\Delta\,dx = \dfrac{-2k^2\sin^2 x+5k^2+1}{8k^2}\cos x\Delta$
$$+\dfrac{1}{2}\ln\dfrac{\Delta+\cos x}{\Delta-\cos x}+\dfrac{3k^4+6k^2-1}{8k^3}\ln(k\cos x+\Delta)$$

2.584

1. $\int \dfrac{dx}{\Delta} = F(x,k)$

2. $\int \dfrac{\sin x\,dx}{\Delta} = \dfrac{1}{2k}\ln\dfrac{\Delta-k\cos x}{\Delta+k\cos x} = -\dfrac{1}{k}\ln(k\cos x+\Delta)$

3. $\int \dfrac{\cos x\,dx}{\Delta} = \dfrac{1}{k}\arcsin(k\sin x) = \dfrac{1}{k}\arctan\dfrac{k\sin x}{\Delta}$

4. $\int \dfrac{\sin^2 x\,dx}{\Delta} = \dfrac{1}{k^2}F(x,k) - \dfrac{1}{k^2}E(x,k)$

5. $\int \dfrac{\sin x\cos x\,dx}{\Delta} = -\dfrac{\Delta}{k^2}$

6. $\int \dfrac{\cos^2 x\,dx}{\Delta} = \dfrac{1}{k^2}E(x,k) - \dfrac{k'^2}{k^2}F(x,k)$

7. $\int \dfrac{\sin^3 x\,dx}{\Delta} = \dfrac{\cos x\Delta}{2k^2} - \dfrac{1+k^2}{2k^3}\ln(k\cos x+\Delta)$

8. $\int \dfrac{\sin^2 x\cos x\,dx}{\Delta} = -\dfrac{\sin x\Delta}{2k^2} + \dfrac{\arcsin(k\sin x)}{2k^3}$

9. $\int \dfrac{\sin x\cos^2 x\,dx}{\Delta} = -\dfrac{\cos x\Delta}{2k^2} + \dfrac{k'^2}{2k^3}\ln(k\cos x+\Delta)$

10. $\int \dfrac{\cos^3 x\,dx}{\Delta} = \dfrac{\sin x\Delta}{2k^2} + \dfrac{2k^2-1}{2k^3}\arcsin(k\sin x)$

11. $\int \dfrac{\sin^4 x\,dx}{\Delta} = \dfrac{\sin x\cos x\Delta}{3k^2} + \dfrac{2+k^2}{3k^4}F(x,k) - \dfrac{2\left(1+k^2\right)}{3k^4}E(x,k)$

12. $\int \dfrac{\sin^3 x\cos x\,dx}{\Delta} = -\dfrac{1}{3k^4}\left(2+k^2\sin^2 x\right)\Delta.$

13. $$\int \frac{\sin^2 x \cos^2 x \, dx}{\Delta} = -\frac{\sin x \cos x \Delta}{3k^2} + \frac{2-k^2}{3k^4} E(x,k) + \frac{2k^2-2}{3k^4} F(x,k)$$

14. $$\int \frac{\sin x \cos^3 x \, dx}{\Delta} = -\frac{1}{3k^4} \left(k^2 \cos^2 x - 2k'^2 \right) \Delta.$$

15. $$\int \frac{\cos^4 x \, dx}{\Delta} = \frac{\sin x \cos x \Delta}{3k^2} + \frac{4k^2-2}{3k^4} E(x,k) + \frac{3k^4-5k^2+2}{3k^4} F(x,k)$$

16. $$\int \frac{\sin^5 x \, dx}{\Delta} = \frac{2k^2 \sin^2 x + 3k^2 + 3}{8k^4} \cos x \Delta - \frac{3 + 2k^2 + 3k^4}{8k^5} \ln (k \cos x + \Delta)$$

17. $$\int \frac{\sin^4 x \cos x \, dx}{\Delta} = -\frac{2k^2 \sin^2 x + 3}{8k^4} \sin x \Delta + \frac{3}{8k^5} \arcsin (k \sin x)$$

18. $$\int \frac{\sin^3 x \cos x \, dx}{\Delta} = \frac{2k^2 \cos^2 x - k^2 - 3}{8k^4} \cos x \Delta - \frac{k^4 + 2k^2 - 3}{8k^5} \ln (k \cos x + \Delta)$$

19. $$\int \frac{\sin^2 x \cos^3 x \, dx}{\Delta} = -\frac{2k^2 \cos^2 x + 2k^2 - 3}{8k^4} \sin x \Delta + \frac{4k^2 - 3}{8k^5} \arcsin (k \sin x)$$

20. $$\int \frac{\sin x \cos^4 x \, dx}{\Delta} = \frac{3 - 5k^2 + 2k^2 \sin^2 x}{8k^4} \cos x \Delta - \frac{3k^4 - 6k^2 + 3}{8k^5} \ln (k \cos x + \Delta)$$

21. $$\int \frac{\cos^5 x \, dx}{\Delta} = \frac{2k^2 \cos^2 x + 6k^2 - 3}{8k^4} \sin x \Delta + \frac{8k^4 - 8k^2 + 3}{8k^5} \arcsin (k \sin x)$$

22. $$\int \frac{\sin^6 x \, dx}{\Delta} = \frac{3k^2 \sin^2 x + 4k^2 + 4}{15k^4} \sin x \cos x \Delta$$
$$+ \frac{4k^4 + 3k^2 + 8}{15k^6} F(x,k) - \frac{8k^4 + 7k^2 + 8}{15k^6} E(x,k)$$

23. $$\int \frac{\sin^5 x \cos x \, dx}{\Delta} = -\frac{3k^4 \sin^4 x + 4k^2 \sin^2 x + 8}{15k^6} \Delta.$$

24. $$\int \frac{\sin^4 x \cos x \, dx}{\Delta} = \frac{3k^2 \cos^2 x - 2k^2 - 4}{15k^4} \sin x \cos x \Delta$$
$$+ \frac{k^4 + 7k^2 - 8}{15k^6} F(x,k) - \frac{2k^4 + 3k^2 - 8}{15k^6} E(x,k)$$

25. $$\int \frac{\sin^3 x \cos^3 x \, dx}{\Delta} = \frac{3k^4 \sin^4 x - \left(5k^4 - 4k^2\right) \sin^2 x - 10k^2 + 8}{15k^6} \Delta.$$

26. $$\int \frac{\sin^2 x \cos^4 x \, dx}{\Delta} = -\frac{3k^2 \cos^2 x + 3k^2 - 4}{15k^4} \sin x \cos x \Delta$$
$$+ \frac{9k^4 - 17k^2 + 8}{15k^6} F(x,k) - \frac{3k^4 - 13k^2 + 8}{15k^6} E(x,k)$$

27. $$\int \frac{\sin x \cos^5 x \, dx}{\Delta} = \frac{-3k^4 \cos^4 x + 4k^2 k'^2 \cos^2 x - 8k^4 + 16k^2 - 8}{15k^6} \Delta.$$

28. $$\int \frac{\cos^6 x \, dx}{\Delta} = \frac{3k^2 \cos^2 x + 8k^2 - 4}{15k^4} \sin x \cos x \Delta$$
$$+ \frac{15k^6 - 34k^4 + 27k^2 - 8}{15k^6} F(x,k) + \frac{23k^4 - 23k^2 + 8}{15k^6} E(x,k)$$

29. $$\int \frac{\sin^7 x \, dx}{\Delta} = \frac{8k^4 \sin^4 x + 10k^2 \left(k^2 + 1\right) \sin^2 x + 15k^4 + 14k^2 + 15}{48k^6} \cos x \Delta$$
$$- \frac{\left(5k^4 - 2k^2 + 5\right)\left(k^2 + 1\right)}{16k^7} \ln \left(k \cos x + \Delta\right)$$

30. $$\int \frac{\sin^6 x \cos x \, dx}{\Delta} = -\frac{8k^4 \sin^4 x + 10k^2 \sin^2 x + 15}{48k^6} \sin x \Delta + \frac{5}{16k^7} \arcsin \left(k \sin x\right)$$

31. $$\int \frac{\sin^5 x \cos^2 x \, dx}{\Delta} = \frac{-8k^4 \sin^4 x + 2k^2 \left(k^2 - 5\right) \sin^2 x + 3k^4 + 4k^2 - 15}{48k^6} \cos x \Delta$$
$$- \frac{k^6 + k^4 + 3k^2 - 5}{16k^7} \ln \left(k \cos x + \Delta\right)$$

32. $$\int \frac{\sin^4 x \cos^3 x \, dx}{\Delta} = \frac{8k^4 \sin^4 x - 2k^2 \left(6k^2 - 5\right) \sin^2 x - 18k^2 + 15}{48k^6} \sin x \Delta$$
$$+ \frac{6k^2 - 5}{16k^7} \arcsin \left(k \sin x\right)$$

33. $$\int \frac{\sin^3 x \cos^4 x \, dx}{\Delta} = \frac{8k^4 \sin^4 x - 2k^2 \left(6k^2 - 5\right) \sin^2 x + 3k^4 - 22k^2 + 15}{48k^6} \cos x \Delta$$
$$- \frac{k^6 + 3k^4 - 9k^2 + 5}{16k^7} \ln \left(k \cos x + \Delta\right)$$

34. $$\int \frac{\sin^2 x \cos^5 x \, dx}{\Delta} = \frac{-8k^4 \sin^4 x + 2k^2 \left(12k^2 - 5\right) \sin^2 x - 24k^4 + 36k^2 - 15}{48k^6} \sin x \Delta$$
$$+ \frac{8k^4 - 12k^2 + 5}{16k^7} \arcsin \left(k \sin x\right)$$

35. $$\int \frac{\sin x \cos^6 x \, dx}{\Delta} = \frac{-8k^4 \sin^4 x + 2k^2 \left(13k^2 - 5\right) \sin^2 x - 33k^4 + 40k^2 - 15}{48k^6} \cos x \Delta$$
$$+ \frac{5k'^6}{16k^7} \ln \left(k \cos x + \Delta\right)$$

36. $$\int \frac{\cos^7 x \, dx}{\Delta} = \frac{8k^4 \sin^4 x - 2k^2 \left(18k^2 - 5\right) \sin^2 x + 72k^4 - 54k^2 + 15}{48k^6} \sin x \Delta$$
$$+ \frac{16k^6 - 24k^4 + 18k^2 - 5}{16k^7} \arcsin \left(k \sin x\right)$$

37. $$\int \frac{dx}{\Delta^3} = \frac{1}{k'^2} E(x, k) - \frac{k^2}{k'^2} \frac{\sin x \cos x}{\Delta}$$

38. $$\int \frac{\sin x \, dx}{\Delta^3} = -\frac{\cos x}{k'^2 \Delta}$$

39. $$\int \frac{\cos x \, dx}{\Delta^3} = \frac{\sin x}{\Delta}$$

40. $$\int \frac{\sin x \, dx}{\Delta^3} = \frac{1}{k'^2 k^2} E(x, k) - \frac{1}{k^2} F(x, k) - \frac{1}{k'^2} \frac{si \, x \cos x}{\Delta}$$

41. $$\int \frac{\sin x \cos x \, dx}{\Delta^3} = \frac{1}{k^2 \Delta}$$

42. $$\int \frac{\cos^2 x \, dx}{\Delta^3} = \frac{1}{k^2} F(x, k) - \frac{1}{k^2} E(x, k) + \frac{\sin x \cos x}{\Delta}$$

43.
$$\int \frac{\sin^3 x \, dx}{\Delta^3} = -\frac{\cos x}{k^2 k'^2 \Delta} + \frac{1}{k^3} \ln(k \cos x + \Delta)$$

44.
$$\int \frac{\sin^2 x \cos x \, dx}{\Delta^3} = \frac{\sin x}{k^2 \Delta} - \frac{1}{k^3} \arcsin(k \sin x)$$

45.
$$\int \frac{\sin x \cos^2 x \, dx}{\Delta^3} = \frac{\cos x}{k^2 \Delta} - \frac{1}{k^3} \ln(k \cos x + \Delta)$$

46.
$$\int \frac{\cos^3 x \, dx}{\Delta^3} = -\frac{k'^2 \sin x}{k^2 \Delta} + \frac{1}{k^3} \arcsin(k \sin x)$$

47.
$$\int \frac{\sin^4 x \, dx}{\Delta^3} = \frac{k'^2 + 1}{k'^2 k^4} E(x, k) - \frac{2}{k^4} F(x, k) - \frac{\sin x \cos x}{k^2 k'^2 \Delta}$$

48.
$$\int \frac{\sin^3 x \cos x \, dx}{\Delta^3} = \frac{2 - k^2 \sin^2 x}{k^4 \delta}$$

49.
$$\int \frac{\sin^2 x \cos^2 x \, dx}{\Delta^3} = \frac{2 - k^2}{k^4} F(x, k) - \frac{2}{k^4} E(x, k) + \frac{\sin x \cos x}{k^2 \Delta}$$

50.
$$\int \frac{\sin x \cos^3 x \, dx}{\Delta^3} = \frac{k^2 \sin^2 x + k^2 - 2}{k^4 \Delta}$$

51.
$$\int \frac{\cos^4 x \, dx}{\Delta^3} = \frac{k'^2 + 1}{k^4} E(x, k) - \frac{2k'^2}{k^4} F(x, k) - \frac{k'^2 \sin x \cos x}{k^2 \Delta}$$

52.[9]
$$\int \frac{\sin^5 x \, dx}{\Delta^3} = \frac{k^2 k'^2 \sin^2 x + k^2 - 3}{2k^4 k'^2 \Delta} \cos x + \frac{k^2 + 3}{2k^5} \ln(k \cos x + \Delta)$$

53.
$$\int \frac{\sin^4 x \cos x \, dx}{\Delta^3} = \frac{-k^2 \sin^2 x + 3}{2k^4 \Delta} \sin x - \frac{3}{2k^5} \arcsin(k \sin x)$$

54.
$$\int \frac{\sin^3 x \cos^2 x \, dx}{\Delta} = \frac{-k^2 \sin^2 x + 3}{2k^4 \Delta} \cos x + \frac{k^2 - 3}{2k^5} \ln(k \cos x + \Delta)$$

55.
$$\int \frac{\sin^2 x \cos^3 x \, dx}{\Delta^3} = \frac{k^2 \sin^2 x + 2k^2 - 3}{2k^4 \Delta} \sin x - \frac{2k^2 - 3}{2k^5} \arcsin(k \sin x)$$

56.
$$\int \frac{\sin x \cos^4 x \, dx}{\Delta^3} = \frac{k^2 \sin^2 x + 2k^2 - 3}{2k^4 \Delta} \cos x + \frac{3k'^2}{2k^5} \ln(k \cos x + \Delta)$$

57.
$$\int \frac{\cos^5 x \, dx}{\Delta^3} = \frac{-k^2 \sin^2 x + 2k^4 - 4k^2 + 3}{2k^4 \Delta} \sin x + \frac{4k^2 - 3}{2k^5} \arcsin(k \sin x)$$

58.
$$\int \frac{dx}{\Delta^5} = \frac{-k^2 \sin x \cos x}{3k'^2 \Delta^3} - \frac{2k^2 \left(k'^2 + 1\right) \sin x \cos x}{3k'^4 \Delta} - \frac{1}{3k'^2} F(x, k)$$
$$+ \frac{2 \left(k'^2 + 1\right)}{3k'^4} E(x, k)$$

59.
$$\int \frac{\sin x \, dx}{\Delta^5} = \frac{2k^2 \sin^2 x + k^2 - 3}{3k'^4 \Delta^3} \cos x$$

60.
$$\int \frac{\cos x \, dx}{\Delta^5} = \frac{-2k^2 \sin^2 x + 3}{3\Delta^3} \sin x$$

61. $\displaystyle\int \frac{\sin^2 x\, dx}{\Delta^5} = \frac{k^2+1}{3k'^4 k^2} E(x,k) - \frac{1}{3k'^2 k^2} F(x,k)$
$\displaystyle\qquad\qquad + \frac{k^2\left(k^2+1\right)\sin^2 x - 2}{3k'^4 \Delta^3} \sin x \cos x$

62. $\displaystyle\int \frac{\sin x \cos x\, dx}{\Delta^5} = \frac{1}{3k^2 \Delta^3}$

63. $\displaystyle\int \frac{\cos^2 x\, dx}{\Delta^5} = \frac{1}{3k^2} F(x,k) + \frac{2k^2-1}{3k^2 k'^2} E(x,k) + \frac{k^2\left(2k^2-1\right)\sin^2 x - 3k^2+2}{2k'^2 \Delta} \sin x \cos x$

64. $\displaystyle\int \frac{\sin^3 x}{\Delta^5}\, dx = \frac{\left(3k^2-1\right)\sin^2 x - 2}{3k'^4 \Delta^3} \cos x$

65. $\displaystyle\int \frac{\sin^2 x \cos x}{\Delta^5}\, dx = \frac{\sin^3 x}{3\Delta^3}$

66. $\displaystyle\int \frac{\sin x \cos^2 x}{\Delta^5}\, dx = -\frac{\cos^3 x}{3k'^2 \Delta^3}$

67. $\displaystyle\int \frac{\cos^3 x\, dx}{\Delta^5} = \frac{-\left(2k^2+1\right)\sin^2 x + 3}{3\Delta^3} \sin x$

68. $\displaystyle\int \frac{dx}{\Delta \sin x} = -\frac{1}{2} \ln \frac{\Delta + \cos x}{\Delta - \cos x}$

69. $\displaystyle\int \frac{dx}{\Delta \cos x} = -\frac{1}{2k'} \ln \frac{\Delta - k' \sin x}{\Delta + k' \sin x}$

70. $\displaystyle\int \frac{dx}{\Delta \sin^2 x} = \int \frac{1+\cot^2 x}{\Delta}\, dx = F(x,k) - E(x,k) - \Delta \cot x$

71. $\displaystyle\int \frac{dx}{\Delta \sin x \cos x} = \int (\tan x + \cot x)\frac{dx}{\Delta} = \frac{1}{2} \ln \frac{1-\Delta}{1+\Delta} + \frac{1}{2k'} \ln \frac{\Delta + k'}{\Delta - k'}$

72. $\displaystyle\int \frac{dx}{\Delta \cos^2 x} = \int \left(1+\tan^2 x\right)\frac{dx}{\Delta} = F(x,k) - \frac{1}{k'^2} E(x,k) + \frac{1}{k'^2}\Delta \tan x$

73. $\displaystyle\int \frac{\sin x}{\cos x}\frac{dx}{\Delta} = \int \tan x \frac{dx}{\Delta} = \frac{1}{2k'} \ln \frac{\Delta + k'}{\Delta - k'}$

74. $\displaystyle\int \frac{\cos x}{\sin x}\frac{dx}{\Delta} = \int \cot x \frac{dx}{\Delta} = \frac{1}{2} \ln \frac{1-\Delta}{1+\Delta}$

75. $\displaystyle\int \frac{dx}{\Delta \sin^3 x} = -\frac{\Delta \cos x}{2 \sin^2 x} - \frac{1+k^2}{4} \ln \frac{\Delta + \cos x}{\Delta - \cos x}$

76. $\displaystyle\int \frac{dx}{\Delta \sin^2 x \cos x} = -\frac{\Delta}{\sin x} - \frac{1}{2k'} \ln \frac{\Delta - k' \sin x}{\Delta + k' \sin x}$

77. $\displaystyle\int \frac{dx}{\Delta \sin x \cos^2 x} = \frac{\Delta}{k'^2 \cos x} + \frac{1}{2} \ln \frac{\Delta - \cos x}{\Delta + \cos x}$

78. $\displaystyle\int \frac{dx}{\Delta \cos^3 x} = \frac{\Delta \sin x}{2k'^2 \cos^2 x} + \frac{2k^2-1}{4k'^3} \ln \frac{\Delta - k' \sin x}{\Delta + k' \sin x}$

79. $\displaystyle\int \frac{\sin x}{\cos^2 x}\frac{dx}{\Delta} = \frac{\Delta}{k'^2 \cos x}$

80. $$\int \frac{\cos x}{\sin^2 x} \frac{dx}{\Delta} = -\frac{\Delta}{\sin x}$$

81. $$\int \frac{\sin^2 x}{\cos x} \frac{dx}{\Delta} = \frac{1}{2k'} \ln \frac{\Delta + k' \sin x}{\Delta - k' \sin x} - \frac{1}{k} \arcsin \left(k \sin x \right)$$

82. $$\int \frac{\cos^2 x}{\sin x} \frac{dx}{\Delta} = \frac{1}{2} \ln \frac{\Delta + \cos x}{\Delta - \cos x} + \frac{1}{k} \ln \left(k \cos x + \Delta \right)$$

83. $$\int \frac{dx}{\Delta \sin^4 x} = \frac{1}{3} \left\{ -\Delta \cot^3 x - \Delta \left(2k^2 + 3 \right) \cot x + \left(k^2 + 2 \right) F(x, k) - 2 \left(k^2 + 1 \right) E(x, k) \right\}$$

84. $$\int \frac{dx}{\Delta \sin^3 x \cos x} = \int \left(\tan x + 2 \cot x + \cot^3 x \right) \frac{dx}{\Delta}$$
$$= -\frac{\Delta}{2 \sin^2 x} + \frac{1}{2k'} \ln \frac{\Delta + k'}{\Delta - k'} - \frac{k^2 + 2}{4} \ln \frac{1 + \Delta}{1 - \Delta}$$

85. $$\int \frac{dx}{\Delta \sin^2 x \cos^2 x} = \int \left(\tan^2 x + 2 + \cot^2 x \right) \frac{dx}{\Delta}$$
$$= \left(\frac{\tan x}{k'^2} - \cot x \right) \Delta + \frac{k^2 - 2}{k'^2} E(x, k) + 2 F(x, k)$$

86. $$\int \frac{dx}{\Delta \sin x \cos^3 x} = \int \left(\cot x + 2 \tan x + \tan^3 x \right) \frac{dx}{\Delta}$$
$$= -\frac{\Delta}{2k'^2 \cos^2 x} - \frac{1}{2} \ln \frac{1 + \Delta}{1 - \Delta} + \frac{2 - 3k^2}{4k'^3} \ln \frac{\Delta + k'}{\Delta - k'}$$

87. $$\int \frac{dx}{\Delta \cos^4 x} = \frac{1}{3k'^2} \left\{ \Delta \tan^3 x - \frac{5k^2 - 3}{k'^2} \Delta \tan x - \left(3k^2 - 2 \right) F(x, k) \right.$$
$$\left. + \frac{2 \left(2k^2 - 1 \right)}{k'^2} E(x, k) \right\}$$

88. $$\int \frac{\sin x}{\cos^3 x} \frac{dx}{\Delta} = \int \tan x \left(1 + \tan^2 x \right) \frac{dx}{\Delta} = \frac{\Delta}{2k'^2 \cos^2 x} - \frac{k^2}{4k'^3} \ln \frac{\Delta + k'}{\Delta - k'}$$

89. $$\int \frac{\cos x}{\sin^3 x} \frac{dx}{\Delta} = -\frac{\Delta}{2 \sin^2 x} - \frac{k^2}{4} \ln \frac{1 + \Delta}{1 - \Delta}$$

90. $$\int \frac{\sin^2 x}{\cos^2 x} \frac{dx}{\Delta} = \int \frac{\tan^2 x}{\Delta} dx = \frac{\Delta}{k'^2} \tan x - \frac{1}{k'^2} E(x, k)$$

91. $$\int \frac{\cos^2 x}{\sin^2 x} \frac{dx}{\Delta} = \int \frac{\cot^2 x}{\Delta} dx = -\Delta \cot x - E(x, k)$$

92. $$\int \frac{\sin^3 x}{\cos x} \frac{dx}{\Delta} = \frac{\Delta}{k^2} + \frac{1}{2k'} \ln \frac{\Delta + k'}{\Delta - k'}$$

93. $$\int \frac{\cos^3 x}{\sin x} \frac{dx}{\Delta} = \frac{\Delta}{k^2} - \frac{1}{2} \ln \frac{1 + \Delta}{1 - \Delta}$$

94. $$\int \frac{dx}{\Delta \sin^5 x} = -\frac{\left[3 \left(1 + k^2 \right) \sin^2 x + 2 \right]}{8 \sin^2 x} \Delta \cos x + \frac{3k^4 + 2k^2 + 3}{16} \ln \frac{\Delta + \cos x}{\Delta - \cos x}$$

95. $\displaystyle\int \frac{dx}{\Delta \sin^4 x \cos x} = -\frac{\left(3 + 2k^2\right)\sin^2 x + 1}{3\sin^3 x}\Delta - \frac{1}{2k'}\ln\frac{\Delta - k'\sin x}{\Delta + k'\sin x}$

96. $\displaystyle\int \frac{dx}{\Delta \sin^3 x \cos^2 x} = \frac{\left(3 - k^2\right)\sin^2 x - k'^2}{2k'^2 \sin^2 x \cos x}\Delta + \frac{k^2 + 3}{4}\ln\frac{\Delta - \cos x}{\Delta + \cos x}$

97. $\displaystyle\int \frac{dx}{\Delta \sin^2 x \cos^3 x} = \frac{\left(3 - 2k^2\right)\sin^2 x - 2k'^2}{2k'^2 \sin x \cos^2 x}\Delta - \frac{4k^2 - 3}{4k'^3}\ln\frac{\Delta + k'\sin x}{\Delta - k'\sin x}$

98. $\displaystyle\int \frac{dx}{\Delta \sin x \cos^4 x} = \frac{\left(5k^2 - 3\right)\sin^2 x - 6k^2 + 4}{3k'^4 \cos^3 x}\Delta - \frac{1}{2}\ln\frac{\Delta + \cos x}{\Delta - \cos x}$

99. $\displaystyle\int \frac{dx}{\Delta \cos^5 x} = \frac{3\left(2k^2 - 1\right)\sin^2 x - 8k^2 + 5}{8k'^4 \cos^4 x}\Delta \sin x + \frac{8k^4 - 8k^2 + 3}{16k'^5}\ln\frac{\Delta + k'\sin x}{\Delta - k'\sin x}$

100. $\displaystyle\int \frac{\sin x}{\cos^4 x}\frac{dx}{\Delta} = -\frac{2k^2 \cos^2 x - k'^2}{2k'^4 \cos^3 x}\Delta.$

101. $\displaystyle\int \frac{\cos x}{\sin^4 x}\frac{dx}{\Delta} = -\frac{2k^2 \sin^2 x + 1}{3\sin^3 x}\Delta.$

102. $\displaystyle\int \frac{\sin^2 x}{\cos^3 x}\frac{dx}{\Delta} = \frac{\Delta \sin x}{2k'^2 \cos^2 x} - \frac{1}{4k'^3}\ln\frac{\Delta + k'\sin x}{\Delta - k'\sin x}$

103. $\displaystyle\int \frac{\cos^3 x}{\sin^3 x}\frac{dx}{\Delta} = -\frac{\Delta \cos x}{2\sin^2 x} + \frac{k'^2}{4}\ln\frac{\Delta + \cos x}{\Delta - \cos x}$

104. $\displaystyle\int \frac{\sin^3 x}{\cos^2 x}\frac{dx}{\Delta} = \frac{\Delta}{k'^2 \cos x} + \frac{1}{k}\ln\left(k\cos x + \Delta\right)$

105. $\displaystyle\int \frac{\cos^3 x}{\sin^2 x}\frac{dx}{\Delta} = \frac{-\Delta}{\sin x} - \frac{1}{k}\arcsin\left(k\sin x\right)$

106. $\displaystyle\int \frac{\sin^4 x}{\cos x}\frac{dx}{\Delta} = \frac{\Delta \sin x}{2k^2} + \frac{1}{2k'}\ln\frac{\Delta + k'\sin x}{\Delta - k'\sin x} - \frac{2k^2 + 1}{2k^3}\arcsin\left(k\sin x\right)$

107. $\displaystyle\int \frac{\cos^4 x}{\sin x}\frac{dx}{\Delta} = \frac{\Delta \cos x}{2k^2} + \frac{1}{2}\ln\frac{\Delta + \cos x}{\Delta - \cos x} + \frac{3k^2 - 1}{2k^3}\ln\left(k\cos x + \Delta\right)$

2.585

1. $\displaystyle\int \frac{(a + \sin x)^{p+3}\,dx}{\Delta}$

$$= \frac{1}{(p + 2)k^2}\left[(a + \sin x)^p \cos x\,\Delta\right.$$

$$+ 2(2p + 3)ak^2 \int \frac{(a + \sin x)^{p+2}\,dx}{\Delta} + (p + 1)\left(1 + k^2 - 6a^2k^2\right)\int \frac{(a + \sin x)^{p+1}\,dx}{\Delta}$$

$$- a(2p + 1)\left(1 + k^2 - 2a^2k^2\right)\int \frac{(a + b\sin x)^p\,dx}{\Delta}$$

$$\left.- p\left(1 - a^2\right)\left(1 - a^2k^2\right)\int \frac{(a + \sin x)^{p-1}\,dx}{\Delta}\right]$$

$$\left[p \neq -2, \quad a \neq \pm 1, \quad a \neq \pm\frac{1}{k}\right]$$

For $p = n$ a natural number, this integral can be reduced to the following three integrals:

2. $\int \dfrac{a + \sin x}{\Delta}\, dx = a\, F(x, k) + \dfrac{1}{2k} \ln \dfrac{\Delta - k \cos x}{\Delta + k \cos x}$

3. $\int \dfrac{(a + \sin x)^2}{\Delta}\, dx = \dfrac{1 + k^2 a^2}{k^2} F(x, k) - \dfrac{1}{k^2} E(x, k) + \dfrac{a}{k} \ln \dfrac{\Delta - k \cos x}{\Delta + k \cos x}$

4.[6] $\int \dfrac{dx}{(a + \sin x)\, \Delta} = \dfrac{1}{a} \Pi\left(x, \dfrac{1}{a^2}, k\right) - \int \dfrac{\sin x\, dx}{(a^2 - \sin^2 x)\, \Delta},$

where

5. $\int \dfrac{\sin x\, dx}{(a^2 - \sin^2 x)\, \Delta} = \dfrac{-1}{2\sqrt{(1 - a^2)(1 - a^2 k^2)}} \ln \dfrac{\sqrt{1 - a^2}\Delta - \sqrt{1 - k^2 a^2} \cos x}{\sqrt{1 - a^2}\Delta + \sqrt{1 - k^2 a^2} \cos x}$

2.586

1. $\int \dfrac{dx}{(a + \sin x)^n \Delta} = \dfrac{1}{(n-1)(1 - a^2)(1 - a^2 k^2)} \left[-\dfrac{\cos x\, \Delta}{(a + \sin x)^{n-1}} \right.$

$$-(2n - 3)\left(1 + k^2 - 2a^2 k^2\right) a \int \dfrac{dx}{(a + \sin x)^{n-1} \Delta}$$

$$-(n - 2)\left(6a^2 k^2 - k^2 - 1\right) \int \dfrac{dx}{(a + \sin x)^{n-2} \Delta}$$

$$\left. - (10 - 4n)ak^2 \int \dfrac{dx}{(a + \sin x)^{n-3} \Delta} - (n - 3)k^2 \int \dfrac{dx}{(a + \sin x)^{n-4} \Delta} \right]$$

$$\left[n \neq 1, \quad a \neq \pm 1, \quad a \neq \pm \dfrac{1}{k} \right]$$

This integral can be reduced to the integrals:

2. $\int \dfrac{dx}{(a + \sin x)^2 \Delta} = \dfrac{1}{(1 - a^2)(1 - a^2 k^2)} \left[-\dfrac{\cos x\, \Delta}{a + \sin x} - a\left(1 + k^2 - 2a^2 k^2\right) \int \dfrac{dx}{(a + \sin x)\, \Delta} \right.$

$$\left. - 2ak^2 \int \dfrac{(a + \sin x)\, dx}{\Delta} + k^2 \int \dfrac{(a + \sin x)^2\, dx}{\Delta} \right]$$

$$\text{(see } \mathbf{2.585}\ 2,\ 3,\ 4\text{)}$$

3. $\int \dfrac{dx}{(a + \sin x)^3 \Delta} = \dfrac{1}{2(1 - a^2)(1 - a^2 k^2)} \left[-\dfrac{\cos x\, \Delta}{(a + \sin x)^2} - 3a\left(1 + k^2 - 2a^2 k^2\right) \int \dfrac{dx}{(a + \sin x)^2 \Delta} \right.$

$$\left. -\left(6a^2 k^2 - k^2 - 1\right) \int \dfrac{dx}{(a + \sin x)\, \Delta} + 2ak^2\, F(x, k) \right]$$

$$\text{(see } \mathbf{2.585}\ 4 \text{ and } \mathbf{2.586}\ 2\text{)}$$

For $a = \pm 1$, we have:

4. $\int \dfrac{dx}{(1 \pm \sin x)^n \Delta} = \dfrac{1}{(2n - 1)k'^2} \left[\mp \dfrac{\cos x\, \Delta}{(1 \pm \sin x)^n} + (n - 1)\left(1 - 5k^2\right) \int \dfrac{dx}{(1 \pm \sin x)^{n-1} \Delta} \right.$

$$\left. + 2(2n - 3)k^2 \int \dfrac{dx}{(1 \pm \sin x)^{n-2} \Delta} - (n - 2)k^2 \int \dfrac{dx}{(1 \pm \sin x)^{n-3} \Delta} \right]$$

$$\text{GU (241)(6a)}$$

This integral can be reduced to the integrals

5. $\displaystyle\int \frac{dx}{(1 \pm \sin x)\,\Delta} = \frac{\mp \cos x \Delta}{k'^2\,(1 \pm \sin x)} + F(x,k) - \frac{1}{k'^2}\,E(x,k)$ GU (241)(6c)

6. $\displaystyle\int \frac{dx}{(1 \pm \sin x)^2\Delta} = \frac{1}{3k'^4}\left\{\mp\frac{k'^2 \cos x \Delta}{(1 \pm \sin x)^2} \mp \frac{(1 - 5k^2)\cos x\Delta}{1 \pm \sin x}\right.$

$$\left. + \left(1 - 3k^2\right)k'^2\,F(x,k) - \left(1 - 5k^2\right)E(x,k)\right\}$$

GU (241)(6b)

For $a = \pm\dfrac{1}{k}$, we have

7. $\displaystyle\int \frac{dx}{(1 \pm k\sin x)^n\Delta} = \frac{1}{(2n-1)k'^2}\left[\pm\frac{k\cos x\Delta}{(1 \pm k\sin x)^n} + (n-1)\left(5 - k^2\right)\int \frac{dx}{(1 \pm k\sin x)^{n-1}\Delta}\right.$

$$\left. - 2(2n-3)\int \frac{dx}{(1 \pm k\sin x)^{n-2}\Delta} + (n-2)\int \frac{dx}{(1 \pm k\sin x)^{n-3}\Delta}\right]$$

GU (241)(7a)

This integral can be reduced to the integrals

8. $\displaystyle\int \frac{dx}{(1 \pm k\sin x)\,\Delta} = \pm\frac{k\cos x\Delta}{k'^2\,(1 \pm k\sin x)} + \frac{1}{k'^2}\,E(x,k)$ GU (241)(7b)

9. $\displaystyle\int \frac{dx}{(1 \pm k\sin x)^2\Delta} = \frac{1}{3k'^4}\left[\pm\frac{kk'^2\cos x\Delta}{(1 \pm k\sin x)^2} \pm \frac{k\left(5 - k^2\right)\cos x\Delta}{1 \pm k\sin x}\right.$

$$\left. - 2k'^2\,F(x,k) + \left(5 - k^2\right)E(x,k)\right]$$

GU(241)(7c)

2.587

1. $\displaystyle\int \frac{(b + \cos x)^{p+3}\,dx}{\Delta} \quad \frac{1}{(p+2)k^2}\left[(b + \cos x)^p \sin x\Delta + 2(2p+3)bk^2\int \frac{(b + \cos x)^{p+2}\,dx}{\Delta}\right.$

$$-(p+1)\left(k'^2 - k^2 + 6b^2k^2\right)\int \frac{(b + \cos x)^{p+1}\,dx}{\Delta}$$

$$+(2p+1)b\left(k'^2 - k^2 + b^2k^2\right)\int \frac{(b + \cos x)^p\,dx}{\Delta}$$

$$\left. + p\left(1 - b^2\right)\left(k'^2 + k^2b^2\right)\int \frac{(b + \cos x)^{p-1}\,dx}{\Delta}\right]$$

$$\left[p \ne -2, \quad b \ne \pm 1, \quad b \ne \frac{ik'}{k}\right]$$

For $p = n$ a natural number, this integral can be reduced to the following three integrals:

2. $\displaystyle\int \frac{b + \cos x}{\Delta}\,dx = b\,F(x,k) + \frac{1}{k}\arcsin(k\sin x)$

3. $\displaystyle\int \frac{(b + \cos x)^2}{\Delta}\,dx = \frac{b^2k^2 - k'^2}{k^2}\,F(x,k) + \frac{1}{k^2}\,E(x,k) + \frac{2b}{k}\arcsin(k\sin x)$

4. $\displaystyle\int \frac{dx}{(b + \cos x)\,\Delta} = \frac{b}{b^2 - 1}\,\Pi\left(x, \frac{1}{b^2 - 1}, k\right) + \int \frac{\cos x\,dx}{\left(1 - b^2 - \sin^2 x\right)\Delta},$

where

5.
$$\int \frac{\cos x \, dx}{\left(1 - b^2 - \sin^2 x\right)\Delta} = \frac{1}{2\sqrt{\left(1 - b^2\right)\left(k'^2 + k^2 b^2\right)}} \ln \frac{\sqrt{1 - b^2}\Delta + k\sqrt{k'^2 + k^2 b^2}\sin x}{\sqrt{1 - b^2}\Delta - k\sqrt{k'^2 + k^2 b^2}\sin x}$$

2.588

1.
$$\int \frac{dx}{\left(b + \cos x\right)^n \Delta} = \frac{1}{\left(n - 1\right)\left(1 - b^2\right)\left(k'^2 + b^2 k^2\right)} \left[\frac{-k'^2 \sin x \Delta}{\left(b + \cos x\right)^{-1}} \right.$$
$$-\left(2n - 3\right)\left(1 - 2k^2 + 2b^2 k^2\right) b \int \frac{dx}{\left(b + \cos x\right)^{n-1}\Delta}$$
$$-\left(n - 2\right)\left(2k^2 - 1 - 6b^2 k^2\right) \int \frac{dx}{\left(b + \cos x\right)^{n-2}\Delta}$$
$$\left. -\left(4n - 10\right)bk^2 \int \frac{dx}{\left(b + \cos x\right)^{n-3}\Delta} + \left(n - 3\right)k^2 \int \frac{dx}{\left(b + \cos x\right)^{n-4}\Delta} \right]$$
$$\left[n \neq 1, \quad b \neq \pm 1, \quad b \neq \pm \frac{ik'}{k} \right]$$

This integral can be reduced to the following integrals:

2.
$$\int \frac{dx}{\left(b + \cos x\right)^2 \Delta} = \frac{1}{\left(1 - b^2\right)\left(k'^2 + b^2 k^2\right)} \left[\frac{-k'^2 \sin x \Delta}{b + \cos x} - \left(1 - 2k^2 + 2b^2 k^2\right) b \int \frac{dx}{\left(b + \cos x\right)\Delta} \right.$$
$$\left. + 2bk^2 \int \frac{b + \cos x}{\Delta} dx - k^2 \int \frac{\left(b + \cos x\right)^2}{\Delta} dx \right]$$
$$\text{(see } \textbf{2.587 } 2, 3, 4\text{)}$$

3.
$$\int \frac{dx}{\left(b + \cos x\right)^3 \Delta} = \frac{1}{2\left(1 - b^2\right)\left(k'^2 + b^2 k^2\right)} \left[\frac{-k'^2 \sin x \Delta}{\left(b + \cos x\right)^2} \right.$$
$$-3b\left(1 - 2k^2 + 2k^2 b^2\right) \int \frac{dx}{\left(b + \cos x\right)^2 \Delta}$$
$$\left. -\left(2k^2 - 1 - 6b^2 k^2\right) \int \frac{dx}{\left(b + \cos x\right)\Delta} - 2bk^2 F\left(x, k\right) \right]$$
$$\text{(see } \textbf{2.588 } 2 \text{ and } \textbf{2.587 } 4\text{)}$$

2.589

1.
$$\int \frac{\left(c + \tan x\right)^{p+3} dx}{\Delta} = \frac{1}{\left(p + 2\right)k'^2} \left[\frac{\left(c + \tan x\right)^p \Delta}{\cos^2 x} + 2\left(2n + 3\right)ck'^2 \int \frac{\left(c + \tan x\right)^{p+2} dx}{\Delta} \right.$$
$$-\left(p + 1\right)\left(1 + k'^2 + 6c^2 k'^2\right) \int \frac{\left(c + \tan x\right)^{p+1} dx}{\Delta}$$
$$+\left(2p + 1\right)c\left(1 + k'^2 + 2c^2 k'^2\right) \int \frac{\left(c + \tan x\right)^p dx}{\Delta}$$
$$\left. -p\left(1 + c^2\right)\left(1 + k'^2 c^2\right) \int \frac{\left(c + \tan x\right)^{p-1} dx}{\Delta} \right]$$
$$\left[p \neq -2 \right]$$

For $p = n$ a natural number, this integral can be reduced to the following three integrals:

2. $\int \dfrac{c + \tan x}{\Delta}\, dx = c\, F(x, k) + \dfrac{1}{2k'} \ln \dfrac{\Delta + k'}{\Delta - k'}$

3. $\int \dfrac{(c + \tan x)^2}{\Delta}\, dx = \dfrac{1}{k'^2} \tan x \Delta + c^2\, F(x, k) - \dfrac{1}{k'^2}\, E(x, k) + \dfrac{c}{k'} \ln \dfrac{\Delta + k'}{\Delta - k'}$

4. $\int \dfrac{dx}{(c + \tan x)\, \Delta} = \dfrac{c}{1 + c^2}\, F(x, k) + \dfrac{1}{c\,(1 + c^2)}\, \Pi\left(x, -\dfrac{1 + c^2}{c^2}, k\right)$

$$-\int \dfrac{\sin x \cos x\, dx}{\left[c^2 - (1 + c^2) \sin^2 x\right] \Delta},$$

where

5. $\int \dfrac{\sin x \cos x\, dx}{\left[c^2 - (1 + c^2) \sin^2 x\right] \Delta} = \dfrac{1}{2\sqrt{(1 + c^2)\left(1 + c^2 k'^2\right)}} \ln \dfrac{\sqrt{1 + c^2 k'^2} + \sqrt{1 + c^2}\,\Delta}{\sqrt{1 + c^2 k'^2} - \sqrt{1 + c^2}\,\Delta}$

2.591

1. $\int \dfrac{dx}{(c + \tan x)^n \Delta} = \dfrac{1}{(n - 1)\,(1 + c^2)\left(1 + k'^2 c^2\right)} \left[-\dfrac{\Delta}{(c + \tan x)^{n-1} \cos^2 x} \right.$

$$+ (2n - 3)c\left(1 + k'^2 + 2c^2 k'^2\right) \int \dfrac{dx}{(c + \tan x)^{n-1} \Delta}$$

$$- (n - 2)\left(1 + k'^2 + 6c^2 k'^2\right) \int \dfrac{dx}{(c + \tan x)^{n-2} \Delta}$$

$$\left. + (4n - 10)ck'^2 \int \dfrac{dx}{(c + \tan x)^{n-3} \Delta} - (n - 3)k'^2 \int \dfrac{dx}{(c + \tan x)^{n-4} \Delta} \right]$$

This integral can be reduced to the integrals:

2. $\int \dfrac{dx}{(c + \tan x)^2 \Delta} = \dfrac{1}{(1 + c^2)\left(1 + k'^2 c^2\right)} \left[\dfrac{-\Delta}{(c + \tan x) \cos^2 x} \right.$

$$+ c\left(1 + k'^2 + 2c^2 k'^2\right) \int \dfrac{dx}{(c + \tan x)\, \Delta}$$

$$\left. - 2ck'^2 \int \dfrac{c + \tan x}{\Delta}\, dx + k'^2 \int \dfrac{(c + \tan x)^2}{\Delta}\, dx \right]$$

(see **2.589** 2, 3, 4)

3. $\int \dfrac{dx}{(c + \tan x)^3 \Delta} = \dfrac{1}{2\,(1 + c^2)\left(1 + k'^2 c^2\right)} \left[\dfrac{-\Delta}{(c + \tan x)^2 \cos^2 x} \right.$

$$+ 3c\left(1 + k'^2 + 2c^2 k'^2\right) \int \dfrac{dx}{(c + \tan x)^2 \Delta}$$

$$\left. - \left(1 + k'^2 + 6c^2 k'^2\right) \int \dfrac{dx}{(c + \tan x)\, \Delta} + 2ck'^2\, F(x, k) \right]$$

(see **2.591** 2 and **2.589** 4)

2.592

1. $$P_n = \int \frac{\left(a + \sin^2 x\right)^n}{\Delta}\, dx$$

The recursion formula

$$P_{n+1} = \frac{1}{(2n+3)k^2} \left\{ \left(a + \sin^2 x\right)^n \sin x \cos x \Delta + (2n+2)\left(1 + k^2 + 3ak^2\right) P_{n+1} \right.$$
$$\left. - (2n+1)\left[1 + 2a\left(1 + k^2\right) + 3a^2 k^2\right] P_n + 2na(1+a)\left(1 + k^2 a\right) P_{n-1} \right\}$$

reduces this integral (for n an integer) to the integrals

2. P_1 (see **2.584** 1 and **2.584** 4)

3. P_0 (see **2.584** 1)

4. $$P_{-1} = \int \frac{dx}{\left(a + \sin^2 x\right)\Delta} = \frac{1}{a}\, \Pi\left(x, \frac{1}{a}, k\right)$$

For $a = 0$

5. $$\int \frac{dx}{\sin^2 x \Delta}$$ (see **2.584** 70)   H (124)a

6. $$T_n = \int \frac{dx}{\left(h + g\sin^2 x\right)^n \Delta}$$

can be calculated by means of the recursion formula:

$$T_{n-3} = \frac{1}{(2n-5)k^2} \left\{ \frac{-g^2 \sin x \cos x \Delta}{\left(h + g\sin^2 x\right)^{n-1}} + 2(n-2)\left[g\left(1 + k^2\right) + 3hk^2\right] T_{n-2} \right.$$
$$\left. - (2n-3)\left[g^2 + 2hg\left(1 + k^2\right) + 3h^2 k^2\right] T_{n-1} + 2(n-1)h(g+h)\left(g + hk^2\right) T_n \right\}$$

2.593

1. $$Q_n = \int \frac{\left(b + \cos^2 x\right)^n}{\Delta}\, dx$$

The recursion formula

$$Q_{n+2} = \frac{1}{(2n+3)k^2} \left\{ \left(b + \cos^2 x\right)^n \sin x \sin x \Delta - (2n+2)\left(1 - 2k^2 - 3bk^2\right) Q_{n+1} \right.$$
$$\left. + (2n+1)\left[k'^2 + 2b\left(k'^2 - k^2\right) - 3b^2 k^2\right] n_- 2nb(1-b)\left(k'^2 - k^2 b\right) Q_{n-1} \right\}$$

reduces this integral (for n an integer) to the integrals:

2. Q_1 (see **2.584** 1 and **2.584** 6)

3. Q_0 (see **2.584** 1)

4. $Q_{-1} = \int \dfrac{dx}{(b + \cos^2 x)\,\Delta} = \dfrac{1}{b+1}\,\Pi\left(x, -\dfrac{1}{b+1}, k\right)$

For $b = 0$

5. $\int \dfrac{dx}{\cos^2 x \Delta}$ (see **2.584** 72)   H (123)

2.594

1. $R_n = \int \dfrac{(c + \tan^2 x)^n \, dx}{\Delta}$

The recursion formula

$$R_{n+2} = \dfrac{1}{(2n+3)k'^2}\left\{ \dfrac{(c + \tan^2 x)^n \tan x \Delta}{\cos^2 x} - (2n+2)\left(1 + k'^2 - 3ck'^2\right) R_{n+1} \right.$$

$$\left. + (2n-1)\left[1 - 2c\left(1 + k'^2\right) + 3c^2 k'^2\right] R_n + 2nc(1-c)\left(1 - k'^2 c\right) R_{n-1} \right\}$$

reduces this integral (for n an integer) to the integrals:

2. R_1 (see **2.584** 1 and **2.584** 90)

3. R_0 (see **2.584** 1)

4. $R_{-1} = \int \dfrac{dx}{(c + \tan^2 x)\,\Delta} = \dfrac{1}{c-1}\,F(x, k) + \dfrac{1}{c(1-c)}\,\Pi\left(x, \dfrac{1-c}{c}, k\right)$

For $c = 0$ see **2.582** 5.

2.595   Integrals of the type $\int R\left(\sin x, \cos x, \sqrt{1 - p^2 \sin^2 x}\right) dx$ for $p^2 > 1$.

Notation: $\alpha = \arcsin\,(p \sin x)$.

Basic formulas

1. $\int \dfrac{dx}{\sqrt{1 - p^2 \sin^2 x}} = \dfrac{1}{p}\,F\left(\alpha, \dfrac{1}{p}\right)$ $\left[p^2 > 1\right]$ BY (283.00)

2. $\int \sqrt{1 - p^2 \sin^2 x}\, dx = p\,E\left(\alpha, \dfrac{1}{p}\right) - \dfrac{p^2 - 1}{p}\,F\left(\alpha, \dfrac{1}{p}\right)$

$\left[p^2 > 1\right]$ BY (283.03)

3. $\int \dfrac{dx}{(1 - r^2 \sin^2 x)\sqrt{1 - p^2 \sin^2 x}} = \dfrac{1}{p}\,\Pi\left(\alpha, \dfrac{r^2}{p^2}, \dfrac{1}{p}\right)$ $\left[p^2 > 1\right]$ BY (283.02)

To evaluate integrals of the form $\int R\left(\sin x, \cos x, \sqrt{1 - p^2 \sin^2 x}\right) dx$ for $p^2 > 1$, we may use formulas **2.583** and **2.584**, making the following modifications in them. We replace

(1) k with p;

(2) k'^2 with $1 - p^2$;

(3) $F(x, k)$ with $\dfrac{1}{p} F\left(\alpha, \dfrac{1}{p}\right)$;

(4) $E(x, k)$ with $p\, E\left(\alpha, \dfrac{1}{p}\right) - \dfrac{p^2 - 1}{p} F\left(\alpha, \dfrac{1}{p}\right)$.

For example (see **2.584** 15):

2.596

$1.^{10}$ $\displaystyle\int \frac{\cos^4 x\, dx}{\sqrt{1 - p^2 \sin^2 x}} = \frac{\sin x \cos x \sqrt{1 - p^2 \sin^2 x}}{3p^2} + \frac{4p^2 - 2}{3p^4}\left[p\, E\left(\alpha, \frac{1}{p}\right) \right.$

$$\left. - \frac{p^2 - 1}{p} F\left(\alpha, \frac{1}{p}\right) \right] + \frac{2 - 5p^2 + 3p^4}{3p^4} \cdot \frac{1}{p} F\left(\alpha, \frac{1}{p}\right)$$

$$= \frac{\sin x \cos x \sqrt{1 - p^2 \sin^2 x}}{3p^2} - \frac{p^2 - 1}{3p^3} F\left(\alpha, \frac{1}{p}\right) + \frac{4p^2 - 2}{3p^3} E\left(\alpha, \frac{1}{p}\right) \quad [p^2 > 1]$$

For example (see **2.583** 36):

2. $\displaystyle\int \frac{\sqrt{1 - p^2 \sin^2 x}}{\cos^2 x}\, dx = \tan x \sqrt{1 - p^2 \sin^2 x} + \frac{1}{p} F\left(\alpha, \frac{1}{p}\right)$

$$- \left[p\, E\left(\alpha, \frac{1}{p}\right) - \frac{p^2 - 1}{p} F\left(\alpha, \frac{1}{p}\right) \right]$$

$$= p\left[F\left(\alpha, \frac{1}{p}\right) - E\left(\alpha, \frac{1}{p}\right) \right] + \tan x \sqrt{1 - p^2 \sin^2 x}$$

$$[p^2 > 1]$$

For example (see **2.584** 37):

3. $\displaystyle\int \frac{dx}{\sqrt{\left(1 - p^2 \sin^2 x\right)^3}} = \frac{-1}{p^2 - 1}\left[p\, E\left(\alpha, \frac{1}{p}\right) - \frac{p^2 - 1}{p} F\left(\alpha, \frac{1}{p}\right) \right]$

$$-\frac{p^2}{1 - p^2} \cdot \frac{\sin x \cos x}{\sqrt{1 - p^2 \sin^2 x}} =$$

$$\frac{p^2}{p^2 - 1} \cdot \frac{\sin x \cos x}{\sqrt{1 - p^2 \sin^2 x}} + \frac{1}{p} F\left(\alpha, \frac{1}{p}\right) - \frac{p}{p^2 - 1} E\left(\alpha, \frac{1}{p}\right) \quad [p^2 > 1]$$

2.597   Integrals of the form $\displaystyle\int R\left(\sin x, \cos x, \sqrt{1 + p^2 \sin^2 x}\right) dx$

Notation: $\alpha = \arcsin\left(\dfrac{\sqrt{1 + p^2}\, \sin x}{\sqrt{1 + p^2 \sin^2 x}} \right)$

Basic formulas

1. $\displaystyle\int \frac{dx}{\sqrt{1 + p^2 \sin^2 x}} = \frac{1}{\sqrt{1 + p^2}} F\left(\alpha, \frac{p}{\sqrt{1 + p^2}}\right)$ BY (282.00)

2. $\displaystyle\int \sqrt{1 + p^2 \sin^2 x}\, dx = \sqrt{1 + p^2}\, E\left(\alpha, \frac{p}{\sqrt{1 + p^2}}\right) - p^2 \frac{\sin x \cos x}{\sqrt{1 + p^2 \sin^2 x}}$ BY (282.03)

3. $$\frac{\sqrt{1+p^2\sin^2 x}\,dx}{1+(p^2-r^2p^2-r^2)\sin^2 x} = \frac{1}{\sqrt{1+p^2}}\,\Pi\left(\alpha, r^2, \frac{p}{\sqrt{1+p^2}}\right)$$ BY (282.02)

4. $$\int \frac{\sin x\,dx}{\sqrt{1+p^2\sin^2 x}} = -\frac{1}{p}\arcsin\left(\frac{p\cos x}{\sqrt{1+p^2}}\right)$$

5. $$\int \frac{\cos x\,dx}{\sqrt{1+p^2\sin^2 x}} = \frac{1}{p}\ln\left(p\sin x + \sqrt{1+p^2\sin^2 x}\right)$$

6. $$\int \frac{dx}{\sin x\sqrt{1+p^2\sin^2 x}} = \frac{1}{2}\ln\frac{\sqrt{1+p^2\sin^2 x}-\cos x}{\sqrt{1+p^2\sin^2 x}+\cos x}$$

7. $$\int \frac{dx}{\cos x\sqrt{1+p^2\sin^2 x}} = \frac{1}{2\sqrt{1+p^2}}\ln\frac{\sqrt{1+p^2\sin^2 x}+\sqrt{1+p^2}\sin x}{\sqrt{1+p^2\sin^2 x}-\sqrt{1+p^2}\sin x}$$

8. $$\int \frac{\tan x\,dx}{\sqrt{1+p^2\sin^2 x}} = \frac{1}{2\sqrt{1+p^2}}\ln\frac{\sqrt{1+p^2\sin^2 x}+\sqrt{1+p^2}}{\sqrt{1+p^2\sin^2 x}-\sqrt{1+p^2}}$$

9. $$\int \frac{\cot x\,dx}{\sqrt{1+p^2\sin^2 x}} = \frac{1}{2}\ln\frac{1-\sqrt{1+p^2\sin^2 x}}{1+\sqrt{1+p^2\sin^2 x}}$$

2.598   To calculate integrals of the form $\int R\left(\sin x, \cos x, \sqrt{1+p^2\sin^2 x}\right)dx$, we may use formulas **2.583** and **2.584**, making the following modifications in them. We replace

(1) k^2 with $-p^2$;

(2) k'^2 with $1+p^2$;

(3) $F(x,k)$ with $\frac{1}{\sqrt{1+p^2}}\,F\left(\alpha, \frac{p}{\sqrt{1+p^2}}\right)$;

(4) $E(x,k)$ with $\sqrt{1+p^2}\,E\left(\alpha, \frac{p}{\sqrt{1+p^2}}\right) - p^2\frac{\sin x\cos x}{\sqrt{1+p^2\sin^2 x}}$;

(5) $\frac{1}{k}\ln\left(k\cos x + \Delta\right)$ with $\frac{1}{p}\arcsin\frac{p\cos x}{\sqrt{1+p^2}}$;

(6) $\frac{1}{k}\arcsin\left(k\sin x\right)$ with $\frac{1}{p}\ln\left(p\sin x + \sqrt{1+p^2\sin^2 x}\right)$.

For example (see **2.584** 90):

1. $$\int \frac{\tan^2 x\,dx}{\sqrt{1+p^2\sin^2 x}} = \frac{1}{(1+p^2)}\left[\tan x\sqrt{1+p^2\sin^2 x}\right.$$

$$\left. - \sqrt{1+p^2}\,E\left(\alpha, \frac{p}{\sqrt{1+p^2}}\right) + p^2\frac{\sin x\cos x}{\sqrt{1+p^2\sin^2 x}}\right]$$

$$= -\frac{1}{\sqrt{1+p^2}}\,E\left(\alpha, \frac{p}{\sqrt{1+p^2}}\right) + \frac{\tan x}{\sqrt{1+p^2\sin^2 x}}$$

For example (see **2.584** 37):

2. $$\int \frac{dx}{\sqrt{\left(1 + p^2 \sin^2 x\right)^3}} = \frac{1}{\sqrt{1 + p^2}} E\left(\alpha, \frac{p}{\sqrt{1 + p^2}}\right)$$

2.599   Integrals of the form $\int R\left(\sin x, \cos x, \sqrt{a^2 \sin^2 x - 1}\right) dx$ $\left[a^2 > 1\right]$

Notation: $\alpha = \arcsin\left(\dfrac{a \cos x}{\sqrt{a^2 - 1}}\right)$.

Basic formulas:

1. $$\int \frac{dx}{\sqrt{a^2 \sin^2 x - 1}} = -\frac{1}{a} F\left(\alpha, \frac{\sqrt{a^2 - 1}}{a}\right)$$ $\left[a^2 > 1\right]$ BY (285.00)a

2. $$\int \sqrt{a^2 \sin^2 x - 1}\, dx = \frac{1}{a} F\left(\alpha, \frac{\sqrt{a^2 - 1}}{a}\right) - a E\left(\alpha, \frac{\sqrt{a^2 - 1}}{a}\right)$$

                                                $\left[a^2 > 1\right]$ BY (285.06)a

3. $$\int \frac{dx}{\left(1 - r^2 \sin^2 x\right) \sqrt{a^2 \sin^2 x - 1}} = \frac{1}{a\left(r^2 - 1\right)} \Pi\left(\alpha, \frac{r^2\left(a^2 - 1\right)}{a^2\left(r^2 - 1\right)}, \frac{\sqrt{a^2 - 1}}{a}\right)$$

                                         $\left[a^2 > 1, \quad r^2 > 1\right]$ BY (285.02)a

4. $$\int \frac{\sin x\, dx}{\sqrt{a^2 \sin^2 x - 1}} = -\frac{\alpha}{a}$$ $\left[a^2 > 1\right]$

5. $$\int \frac{\cos x\, dx}{\sqrt{a^2 \sin^2 x - 1}} = \frac{1}{a} \ln\left(a \sin x + \sqrt{a^2 \sin^2 x - 1}\right)$$ $\left[a^2 > 1\right]$

6. $$\int \frac{dx}{\sin x \sqrt{a^2 \sin^2 x - 1}} = -\arctan \frac{\cos x}{\sqrt{a^2 \sin^2 x - 1}}$$ $\left[a^2 > 1\right]$

7. $$\int \frac{dx}{\cos x \sqrt{a^2 \sin^2 x - 1}} = \frac{1}{2\sqrt{a^2 - 2}} \ln \frac{\sqrt{a^2 - 1} \sin x + \sqrt{a^2 \sin^2 x - 1}}{\sqrt{a^2 - 1} \sin x - \sqrt{a^2 \sin^2 x - 1}}$$

                                         $\left[a^2 > 1\right]$

8. $$\int \frac{\tan x\, dx}{\sqrt{a^2 \sin^2 x - 1}} = \frac{1}{2\sqrt{a^2 - 1}} \ln \frac{\sqrt{a^2 - 1} + \sqrt{a^2 \sin^2 x - 1}}{\sqrt{a^2 - 1} - \sqrt{a^2 \sin^2 x - 1}}$$

                                         $\left[a^2 > 1\right]$

9. $$\int \frac{\cot x\, dx}{\sqrt{a^2 \sin^2 x - 1}} = -\arcsin\left(\frac{1}{a \sin x}\right)$$ $\left[a^2 > 1\right]$

2.611   To calculate integrals of the type $\int R\left(\sin x, \cos x, \sqrt{a^2 \sin^2 x - 1}\right) dx$ for $a^2 > 1$, we may use formulas **2.583** and **2.584**. In doing so, we should follow the procedure outlined below:

(1)   In the right members of these formulas, the following functions should be replaced with integrals equal to them:

$$F(x, k) \quad \text{should be replaced with} \quad \int \frac{dx}{\Delta}$$

$$E(x, k) \quad \text{should be replaced with} \quad \int \Delta \, dx$$

$$-\frac{1}{k} \ln(k \cos x + \Delta) \quad \text{should be replaced with} \quad \int \frac{\sin x \, dx}{\Delta}$$

$$\frac{1}{k} \arcsin(k \sin x) \quad \text{should be replaced with} \quad \int \frac{\cos x \, dx}{\Delta}$$

$$\frac{1}{2} \ln \frac{\Delta - \cos x}{\Delta + \cos x} \quad \text{should be replaced with} \quad \int \frac{dx}{\Delta \sin x}$$

$$\frac{1}{2k'} \ln \frac{\Delta + k' \sin x}{\Delta - k' \sin x} \quad \text{should be replaced with} \quad \int \frac{dx}{\Delta \cos x}$$

$$\frac{1}{2k'} \ln \frac{\Delta + k'}{\Delta - k'} \quad \text{should be replaced with} \quad \int \frac{\tan x}{\Delta} \, dx$$

$$\frac{1}{2} \ln \frac{1 - \Delta}{1 + \Delta} \quad \text{should be replaced with} \quad \int \frac{\cot x}{\Delta} \, dx$$

(2)   Then, on both sides of the equations, we should replace Δ with $i\sqrt{a^2 \sin^2 x - 1}$, k with a and k'^2 with $1 - a^2$.

(3)   Both sides of the resulting equations should be multiplied by i, as a result of which only real functions $(a^2 > 1)$ should appear on both sides of the equations.

(4)   The integrals on the right sides of the equations should be replaced with their values found from formulas **2.599**.

Examples:

1.   We rewrite equation **2.584** 4 in the form

$$\int \frac{\sin^2 x}{i\sqrt{a^2 \sin^2 x - 1}} \, dx = \frac{1}{a^2} \int \frac{dx}{i\sqrt{a^2 \sin^2 x - 1}} - \frac{1}{a^2} \int i\sqrt{a^2 \sin^2 x - 1} \, dx,$$

from which we get

$$\int \frac{\sin^2 x \, dx}{\sqrt{a^2 \sin^2 x - 1}} = \frac{1}{a^2} \left\{ \int \frac{dx}{\sqrt{a^2 \sin^2 x - 1}} + \int \sqrt{a^2 \sin^2 x - 1} \, dx \right\} = -\frac{1}{a} E\left(\alpha, \frac{\sqrt{a^2 - 1}}{a}\right)$$

$$[a^2 > 1]$$

2.   We rewrite equation **2.584** 58 as follows:

$$\int \frac{dx}{i^5 \sqrt{(a^2 \sin^2 x - 1)^5}} = -\frac{2a^4 (a^2 - 2) \sin^2 x - (3a^2 - 5) a^2}{3(1 - a^2)^2 i^3 \sqrt{(a^2 \sin^2 x - 1)^3}} \sin x \cos x$$

$$-\frac{1}{3(1 - a^2)} \int \frac{dx}{i\sqrt{a^2 \sin^2 x - 1}} - \frac{2a^2 - 4}{3(1 - a^2)^2} \int i\sqrt{a^2 \sin^2 x - 1} \, dx$$

from which we obtain

$$\int \frac{dx}{\sqrt{\left(a^2 \sin^2 x - 1\right)^5}} = \frac{2a^4 \left(a^2 - 2\right) \sin^2 x - \left(3a^2 - 5\right) a^2}{3(1 - a^2)^2 \sqrt{\left(a^2 \sin^2 x - 1\right)^3}} \sin x \cos x + \frac{1}{3(1 - a^2)^2 a}$$
$$\times \left\{ \left(a^2 - 3\right) F\left(\alpha, \frac{\sqrt{a^2 - 1}}{a}\right) - 2a^2 \left(a^2 - 2\right) E\left(\alpha, \frac{\sqrt{a^2 - 1}}{a}\right) \right\}$$
$$\left[a^2 > 1\right]$$

3.   We rewrite equation **2.584** 71 in the form

$$\int \frac{dx}{\sin x \cos x i \sqrt{a^2 \sin^2 x - 1}} = \int \frac{\cot x \, dx}{i \sqrt{a^2 \sin^2 x - 1}} + \int \frac{\tan x \, dx}{i \sqrt{a^2 \sin^2 x - 1}},$$

from which we obtain

$$\int \frac{dx}{\sin x \cos x \sqrt{a^2 \sin^2 x - 1}} = \frac{1}{2\sqrt{a^2 - 1}} \ln \frac{\sqrt{a^2 - 1} + \sqrt{a^2 \sin^2 x - 1}}{\sqrt{a^2 - 1} - \sqrt{a^2 \sin^2 x - 1}} - \arcsin\left(\frac{1}{a \sin x}\right)$$
$$\left[a^2 > 1\right]$$

2.612   Integrals of the form $\int R\left(\sin x, \cos x, \sqrt{1 - k^2 \cos^2 x}\right) dx$.

To find integrals of the form $\int R\left(\sin x, \cos x, \sqrt{1 - k^2 \cos^2 x}\right) dx$ we make the substitution $x = \dfrac{\pi}{2} - y$, which yields

$$\int R\left(\sin x, \cos x, \sqrt{1 - k^2 \cos^2 x}\right) dx = -\int R\left(\cos y, \sin y, \sqrt{1 - k^2 \sin^2 y}\right) dy$$

The integrals $\int R\left(\cos y, \sin y, \sqrt{1 - k^2 \sin^2 y}\right) dy$ are found from formulas **2.583** and **2.584**. As a result of the use of these formulas (where it is assumed that the original integral can be reduced only to integrals of the first and second Legendre forms), when we replace the functions $F(x, k)$ and $E(x, k)$ with the corresponding integrals, we obtain an expression of the form

$$-g\left(\cos y, \sin y\right) - A \int \frac{dy}{\sqrt{1 - k^2 \sin^2 y}} - B \int \sqrt{1 - k^2 \sin^2 y} \, dy$$

Returning now to the original variable x, we obtain

$$\int R\left(\sin x, \cos x, \sqrt{1 - k^2 \cos^2 x}\right) dx = -g\left(\sin x, \cos x\right) - A \int \frac{dx}{\sqrt{1 - k^2 \cos^2 x}} - B \int \sqrt{1 - k^2 \cos^2 x} \, dx$$

The integrals appearing in this expression are found from the formulas

1. $\displaystyle \int \frac{dx}{\sqrt{1 - k^2 \cos^2 x}} = F\left(\arcsin\left(\frac{\sin x}{\sqrt{1 - k^2 \cos^2 x}}\right), k\right)$

2. $\displaystyle \int \sqrt{1 - k^2 \cos^2 x} \, dx = E\left(\arcsin\left(\frac{\sin x}{\sqrt{1 - k^2 \cos^2 x}}\right), k\right) - \frac{k^2 \sin x \cos x}{\sqrt{1 - k^2 \cos^2 x}}$

2.613   Integrals of the form $\int R\left(\sin x, \cos x, \sqrt{1 - p^2 \cos^2 x}\right) dx$ $[p > 1]$.

To find integrals of the type $\int R\left(\sin x, \cos x, \sqrt{1 - p^2 \cos^2 x}\right) dx$, where $[p > 1]$, we proceed as in **2.612**. Here, we use the formulas

1. $\displaystyle\int\frac{dx}{\sqrt{1-p^2\cos^2 x}} = -\frac{1}{p}\,F\left(\arcsin\left(p\cos x\right),\frac{1}{p}\right)$ $\qquad[p>1]$

2. $\displaystyle\int\sqrt{1-p^2\cos^2 x}\,dx = \frac{p^2-1}{p}\,F\left(\arcsin\left(p\cos x\right),\frac{1}{p}\right) - p\,E\left(\arcsin\left(p\cos x\right),\frac{1}{p}\right)$

2.614 Integrals of the form $\displaystyle\int R\left(\sin x,\cos x,\sqrt{1+p^2\cos^2 x}\right)\,dx.$

To find integrals of the type $\displaystyle\int R\left(\sin x,\cos x,\sqrt{1+p^2\cos^2 x}\right)\,dx$, we need to make the substitution $x=\dfrac{\pi}{2}-y$. This yields

$$\int R\left(\sin x,\cos x,\sqrt{1+p^2\cos^2 x}\right)\,dx = -\int R\left(\cos y,\sin y,\sqrt{1+p^2\sin^2 y}\right)\,dy$$

To calculate the integrals $-\displaystyle\int R\left(\cos y,\sin y,\sqrt{1+p^2\sin^2 y}\right)\,dy$, we need to use first what was said in **2.598** and **2.612** and then, after returning to the variable x, the formulas

1. $\displaystyle\int\frac{dx}{\sqrt{1+p^2\cos^2 x}} = \frac{1}{\sqrt{1+p^2}}\,F\left(x,\frac{p}{\sqrt{1+p^2}}\right)$

2. $\displaystyle\int\sqrt{1+p^2\cos^2 x}\,dx = \sqrt{1+p^2}\,E\left(x,\frac{p}{\sqrt{1+p^2}}\right)$

2.615 Integrals of the form $\displaystyle\int R\left(\sin x,\cos x,\sqrt{a^2\cos^2 x-1}\right)\,dx$ $\qquad[a>1]$.

To find integrals of the type $\displaystyle\int R\left(\sin x,\cos x,\sqrt{a^2\cos^2 x-1}\right)\,dx$, we need to make the substitution $x=\dfrac{\pi}{2}-y$. This yields

$$\int R\left(\sin x,\cos x,\sqrt{a^2\cos^2 x-1}\right)\,dx = -\int R\left(\cos y,\sin y,\sqrt{a^2\sin^2 y-1}\right)\,dy$$

To calculate the integrals $-\int R\left(\cos y,\sin y,\sqrt{a^2\sin^2 y-1}\right)\,dy$, we use what was said in **2.611** and then, after returning to the variable x, we use the formulas

1. $\displaystyle\int\frac{dx}{\sqrt{a^2\cos^2 x-1}} = \frac{1}{a}\,F\left(\arcsin\left(\frac{a\sin x}{\sqrt{a^2-1}}\right),\frac{\sqrt{a^2-1}}{a}\right)$

$$[a>1]$$

2. $\displaystyle\int\sqrt{a^2\cos^2 x-1}\,dx = a\,E\left(\arcsin\left(\frac{a\sin x}{\sqrt{a^2-1}}\right),\frac{\sqrt{a^2-1}}{a}\right)$

$$-\frac{1}{a}\,F\left(\arcsin\left(\frac{a\sin x}{\sqrt{a^2-1}}\right),\frac{\sqrt{a^2-1}}{a}\right)\qquad[a>1]$$

2.616^{10}   Integrals of the form $\int R\left(\sin x, \cos x, \sqrt{1 - p^2 \sin^2 x}, \sqrt{1 - q^2 \sin^2 x}\right) dx$.

Notation: $\alpha = \arcsin\left(\dfrac{\sqrt{1 - p^2}\,\sin x}{\sqrt{1 - q^2 \sin^2 x}}\right)$.

1. $$\int \frac{dx}{\sqrt{\left(1 - p^2 \sin^2 x\right)\left(1 - q^2 \sin^2 x\right)}} = \frac{1}{\sqrt{1 - p^2}}\, F\left(\alpha, \sqrt{\frac{q^2 - p^2}{1 - p^2}}\right)$$

$$\left[0 < p^2 < q^2 < 1, \quad 0 < x \leq \frac{\pi}{2}\right]$$
                                                          BY (284.00)

2. $$\int \frac{\tan^2 x\, dx}{\sqrt{\left(1 - p^2 \sin^2 x\right)\left(1 - q^2 \sin^2 x\right)}} = \frac{\tan x \sqrt{1 - q^2 \sin^2 x}}{(1 - q^2)\sqrt{1 - p^2 \sin^2 x}}$$

$$- \frac{1}{(1 - q^2)\sqrt{1 - p^2}}\, E\left(\alpha, \sqrt{\frac{q^2 - p^2}{1 - p^2}}\right)$$

$$\left[0 < p^2 < q^2 < 1, \quad 0 < x \leq \frac{\pi}{2}\right] \quad \text{BY (284.07)}$$

3. $$\int \frac{\tan^4 x\, dx}{\sqrt{\left(1 - p^2 \sin^2 x\right)\left(1 - q^2 \sin^2 x\right)}}$$

$$= \frac{1}{3(1 - q^2)^2(1 - p^2)^{\frac{3}{2}}} \times \left[2\left(2 - p^2 - q^2\right) E\left(\alpha, \sqrt{\frac{q^2 - p^2}{1 - p^2}}\right) - (1 - q^2) F\left(\alpha, \sqrt{\frac{q^2 - p^2}{1 - p^2}}\right)\right]$$

$$+ \frac{2p^2 + q^2 - 3 + \sin^2 x\left(4 - 3p^2 - 2q^2 + p^2 q^2\right)}{3\left(1 - p^2\right)\left(1 - q^2\right)^2}\, \frac{\sin x}{\cos^2 x}\sqrt{\frac{1 - q^2 \sin^2 x}{1 - p^2 \sin^2 x}}$$

$$\left[0 < p^2 < q^2 < 1, \quad 0 < x \leq \frac{\pi}{2}\right] \quad \text{BY (284.07)}$$

4. $$\int \frac{\sin^2 x\, dx}{\sqrt{\left(1 - p^2 \sin^2 x\right)\left(1 - q^2 \sin^2 x\right)^3}}$$

$$= \frac{\sqrt{1 - p^2}}{(1 - q^2)(q^2 - p^2)}\, E\left(\alpha, \sqrt{\frac{q^2 - p^2}{1 - p^2}}\right) - \frac{1}{(q^2 - p^2)\sqrt{1 - p^2}}\, F\left(\alpha, \sqrt{\frac{q^2 - p^2}{1 - p^2}}\right)$$

$$- \frac{\sin x \cos x}{(1 - q^2)\sqrt{\left(1 - p^2 \sin^2 x\right)\left(1 - q^2 \sin^2 x\right)}}$$

$$\left[0 < p^2 < q^2 < 1, \quad 0 < x \leq \frac{\pi}{2}\right] \quad \text{BY (284.06)}$$

5. $$\int \frac{\cos^2 x\, dx}{\sqrt{\left(1 - p^2 \sin^2 x\right)^3\left(1 - q^2 \sin^2 x\right)}}$$

$$= \frac{\sqrt{1 - p^2}}{q^2 - p^2}\, E\left(\alpha, \sqrt{\frac{q^2 - p^2}{1 - p^2}}\right) - \frac{1 - q^2}{(q^2 - p^2)\sqrt{1 - p^2}}\, F\left(\alpha, \sqrt{\frac{q^2 - p^2}{1 - p^2}}\right)$$

$$\left[0 < p^2 < q^2 < 1, \quad 0 < x \leq \frac{\pi}{2}\right] \quad \text{BY (284.05)}$$

6.
$$\int \frac{\cos^4 x \, dx}{\sqrt{\left(1 - p^2 \sin^2 x\right)^5 \left(1 - q^2 \sin^2 x\right)}}$$
$$= \frac{(1-p^2)^{\frac{3}{2}}}{3(q^2-p^2)^2} \left[\frac{(2+p^2-3q^2)(1-q^2)}{(1-p^2)^2} F\left(\alpha, \sqrt{\frac{q^2-p^2}{1-p^2}}\right) \right.$$
$$\left. + 2\frac{2q^2-p^2-1}{1-p^2} E\left(\alpha, \sqrt{\frac{q^2-p^2}{1-p^2}}\right) \right] + \frac{(1-p^2)\sin x \cos x \sqrt{1-q^2 \sin^2 x}}{3(q^2-p^2)\sqrt{\left(1-p^2 \sin^2 x\right)^3}}$$
$$\left[0 < p^2 < q^2 < 1, \quad 0 < x \le \frac{\pi}{2}\right] \quad \text{BY (284.05)}$$

7.
$$\int \frac{dx}{1-p^2 \sin^2 x} \sqrt{\frac{1-q^2 \sin^2 x}{1-p^2 \sin^2 x}} = \frac{1}{\sqrt{1-p^2}} E\left(\alpha, \sqrt{\frac{q^2-p^2}{1-p^2}}\right)$$
$$\left[0 < p^2 < q^2 < 1, \quad 0 < x \le \frac{\pi}{2}\right]$$
$$\text{BY (284.01)}$$

8.
$$\int \sqrt{\frac{1-p^2 \sin^2 x}{\left(1-q^2 \sin^2 x\right)^3}} \, dx = \frac{\sqrt{1-p^2}}{1-q^2} E\left(\alpha, \sqrt{\frac{q^2-p^2}{1-p^2}}\right) - \frac{q^2-p^2}{1-q^2} \frac{\sin x \cos x}{\sqrt{\left(1-p^2 \sin^2 x\right)\left(1-q^2 \sin^2 x\right)}}$$
$$\left[0 < p^2 < q^2 < 1, \quad 0 < x \le \frac{\pi}{2}\right].$$
$$\text{BY (284.04)}$$

9.
$$\int \frac{dx}{1 + (p^2 r^2 - p^2 - r^2)\sin^2 x} \sqrt{\frac{1-p^2 \sin^2 x}{1-q^2 \sin^2 x}} = \frac{1}{\sqrt{1-p^2}} \Pi\left(\alpha, r^2, \sqrt{\frac{q^2-p^2}{1-p^2}}\right)$$
$$\left[0 < p^2 < q^2 < 1, \quad 0 < x \le \frac{\pi}{2}\right].$$
$$\text{BY (284.02)}$$

2.617   **Notation:** $\alpha = \arcsin \sqrt{\dfrac{\sqrt{b^2+c^2} - b \sin x - c \cos x}{2\sqrt{b^2+c^2}}}, \quad r = \sqrt{\dfrac{2\sqrt{b^2+c^2}}{a + \sqrt{b^2+c^2}}}.$

1.
$$\int \frac{dx}{\sqrt{a + b \sin x + c \cos x}}$$
$$= -\frac{2}{\sqrt{a + \sqrt{b^2+c^2}}} F(\alpha, r)$$
$$\left[0 < \sqrt{b^2+c^2} < a, \quad \arcsin \frac{b}{\sqrt{b^2+c^2}} - \pi \le x < \arcsin \frac{b}{\sqrt{b^2+c^2}}\right]$$
$$\text{BY (294.00)}$$
$$= -\frac{\sqrt{2}}{\sqrt[4]{b^2+c^2}} F(\alpha, r)$$
$$\left[0 < |a| < \sqrt{b^2+c^2}, \quad \arcsin \frac{b}{\sqrt{b^2+c^2}} - \arccos\left(-\frac{a}{\sqrt{b^2+c^2}}\right) \le x < \arcsin \frac{b}{\sqrt{b^2+c^2}}\right]$$
$$\text{BY (293.00)}$$

2. $$\int \frac{\sin x \, dx}{\sqrt{a + b\sin x + c\cos x}} = -\frac{\sqrt{2b}}{\sqrt[4]{(b^2 + c^2)^3}} \left\{ 2\,E(\alpha, r) - F(\alpha, r) \right\} + \frac{2c}{b^2 + c^2} \sqrt{a + b\sin x + c\cos x}$$

$$\left[0 < |a| < \sqrt{b^2 + c^2}, \quad \arcsin \frac{b}{\sqrt{b^2 + c^2}} - \arccos \left(-\frac{a}{\sqrt{b^2 + c^2}} \right) \le x < \arcsin \frac{b}{\sqrt{b^2 + c^2}} \right]$$

BY (293.05)

3. $$\int \frac{(b\cos x - c\sin x)\, dx}{\sqrt{a + b\sin x + c\cos x}} = 2\sqrt{a + b\sin x + c\cos x}$$

4. $$\int \frac{\sqrt{b^2 + c^2} + b\sin x + c\cos x}{\sqrt{a + b\sin x + c\cos x}}\, dx$$

$$= -2\sqrt{a + \sqrt{b^2 + c^2}}\, E(\alpha, r) + \frac{2\left(a - \sqrt{b^2 + c^2}\right)}{\sqrt{a + \sqrt{b^2 + c^2}}}\, F(\alpha, r)$$

$$\left[0 < \sqrt{b^2 + c^2} < a, \quad \arcsin \frac{b}{\sqrt{b^2 + c^2}} - \pi \le x < \arcsin \frac{b}{\sqrt{b^2 + c^2}} \right]$$

BY (294.04)

$$= -2\sqrt{2}\sqrt[4]{b^2 + c^2}\, E(\alpha, r)$$

$$\left[0 < |a| < \sqrt{b^2 + c^2}, \quad \arcsin \frac{b}{\sqrt{b^2 + c^2}} - \arccos \left(-\frac{a}{\sqrt{b^2 + c^2}} \right) \le x < \arcsin \frac{b}{\sqrt{b^2 + c^2}} \right]$$

BY (293.01)

5. $$\int \sqrt{a + b\sin x + c\cos x}\, dx$$

$$= -2\sqrt{a + \sqrt{b^2 + c^2}}\, E(\alpha, r)$$

$$\left[0 < \sqrt{b^2 + c^2} < a, \quad \arcsin \frac{b}{\sqrt{b^2 + c^2}} - \pi \le x < \arcsin \frac{b}{\sqrt{b^2 + c^2}} \right]$$

BY (294.01)

$$= -2\sqrt{2}\sqrt[4]{b^2 + c^2}\, E(\alpha, r) + \frac{\sqrt{2}\left(\sqrt{b^2 + c^2} - a\right)}{\sqrt[4]{b^2 + c^2}}\, F(\alpha, r)$$

$$\left[0 < |a| < \sqrt{b^2 + c^2}, \quad \arcsin \frac{b}{\sqrt{b^2 + c^2}} - \arccos \left(\frac{-a}{\sqrt{b^2 + c^2}} \right) \le x < \arcsin \frac{b}{\sqrt{b^2 + c^2}} \right]$$

BY (293.03)

2.618 Integrals of the form $\displaystyle\int R\left(\sin ax, \cos ax, \sqrt{\cos 2ax}\right) dx = \frac{1}{a}\int R\left(\sin t, \cos t, \sqrt{1 - 2\sin^2 t}\right) dt$ where the substitution $t = ax$ has been used.

Notation: $\alpha = \arcsin\left(\sqrt{2}\sin ax\right)$

The integrals $\displaystyle\int R\left(\sin ax, \cos ax, \sqrt{\cos 2ax}\right) dx$ are special cases of the integrals **2.595.** for $(p = 2)$. We give some formulas:

1. $$\int \frac{dx}{\sqrt{\cos 2ax}} = \frac{1}{a\sqrt{2}}\, F\left(\alpha, \frac{1}{\sqrt{2}}\right) \qquad \left[0 < ax \le \frac{\pi}{4} \right]$$

2. $$\int \frac{\cos^2 ax}{\sqrt{\cos 2ax}}\, dx = \frac{1}{a\sqrt{2}}\, E\left(\alpha, \frac{1}{\sqrt{2}}\right) \qquad \left[0 < ax \le \frac{\pi}{4} \right]$$

3. $\int \dfrac{dx}{\cos^2 ax \sqrt{\cos 2ax}} = \dfrac{\sqrt{2}}{a} E\left(\alpha, \dfrac{1}{\sqrt{2}}\right) - \dfrac{\tan x}{a}\sqrt{\cos 2ax}$

$$\left[0 < ax \le \dfrac{\pi}{4}\right]$$

4. $\int \dfrac{dx}{\cos^4 ax \sqrt{\cos 2ax}} = \dfrac{2\sqrt{2}}{a} E\left(\alpha, \dfrac{1}{\sqrt{2}}\right) - \dfrac{\sqrt{2}}{3a} F\left(\alpha, \dfrac{1}{\sqrt{2}}\right) - \dfrac{(6\cos^2 ax + 1)\sin ax}{3a\cos^3 ax}\sqrt{\cos 2ax}$

$$\left[0 < x \le \dfrac{\pi}{4}\right]$$

5. $\int \dfrac{\tan^2 ax\, dx}{\sqrt{\cos 2ax}} = \dfrac{\sqrt{2}}{a} E\left(\alpha, \dfrac{1}{\sqrt{2}}\right) - \dfrac{1}{a\sqrt{2}} F\left(\alpha, \dfrac{1}{\sqrt{2}}\right) - \dfrac{1}{a}\tan ax\sqrt{\cos 2ax}$

$$\left[0 < x \le \dfrac{\pi}{2}\right]$$

6. $\int \dfrac{\tan^4 ax\, dx}{\sqrt{\cos 2ax}} = \dfrac{1}{3a\sqrt{2}} F\left(\alpha, \dfrac{1}{\sqrt{2}}\right) - \dfrac{\sin ax}{3a\cos^3 ax}\sqrt{\cos 2ax}$

$$\left[0 < ax \le \dfrac{\pi}{4}\right]$$

7. $\int \dfrac{dx}{(1 - 2r^2 \sin^2 ax)\sqrt{\cos 2ax}} = \dfrac{1}{a\sqrt{2}} \Pi\left(\alpha, r^2, \dfrac{1}{\sqrt{2}}\right)$ $\left[0 < ax \le \dfrac{\pi}{4}\right]$

8. $\int \dfrac{dx}{\sqrt{\cos^3 2ax}} = \dfrac{1}{a\sqrt{2}} F\left(\alpha, \dfrac{1}{\sqrt{2}}\right) - \dfrac{\sqrt{2}}{a} E\left(\alpha, \dfrac{1}{\sqrt{2}}\right) + \dfrac{\sin 2ax}{a\sqrt{\cos 2ax}}$

$$\left[0 < ax \le \dfrac{\pi}{4}\right]$$

9. $\int \dfrac{\sin^2 ax\, dx}{\sqrt{\cos^3 2ax}} = \dfrac{\sin 2ax}{2a\sqrt{\cos 2ax}} - \dfrac{1}{a\sqrt{2}} E\left(\alpha, \dfrac{1}{\sqrt{2}}\right)$ $\left[0 < ax \le \dfrac{\pi}{4}\right]$

10. $\int \dfrac{dx}{\sqrt{\cos^5 2ax}} = \dfrac{1}{3a\sqrt{2}} F\left(\alpha, \dfrac{1}{\sqrt{2}}\right) + \dfrac{\sin 2ax}{3a\sqrt{\cos^3 2ax}}$ $\left[0 < ax \le \dfrac{\pi}{4}\right]$

11. $\int \sqrt{\cos 2ax}\, dx = \dfrac{\sqrt{2}}{a} E\left(\alpha, \dfrac{1}{\sqrt{2}}\right) - \dfrac{1}{a\sqrt{2}} F\left(\alpha, \dfrac{1}{\sqrt{2}}\right)$

$$\left[0 < ax \le \dfrac{\pi}{4}\right]$$

12. $\int \dfrac{\sqrt{\cos 2ax}}{\cos^2 ax}\, dx = \dfrac{\sqrt{2}}{a}\left\{F\left(\alpha, \dfrac{1}{\sqrt{2}}\right) - E\left(\alpha, \dfrac{1}{\sqrt{2}}\right)\right\} + \dfrac{1}{a}\tan ax\sqrt{\cos 2ax}$

$$\left[0 < x \le \dfrac{\pi}{4}\right]$$

2.619 Integrals of the form $\int R\left(\sin ax, \cos ax, \sqrt{-\cos 2ax}\right) dx = \dfrac{1}{a}\int R\left(\sin x, \cos x, \sqrt{2\sin^2 x - 1}\right) dx$

Notation: $\alpha = \arcsin\left(\sqrt{2}\cos ax\right)$

The integrals $\int R\left(\sin x, \cos x, \sqrt{2\sin^2 x - 1}\right) dx$ are special cases of the integrals **2.599** and **2.611** for $(a = 2)$. We give some formulas:

1. $$\int \frac{dx}{\sqrt{-\cos 2ax}} = -\frac{1}{a\sqrt{2}} F\left(\alpha, \frac{1}{\sqrt{2}}\right)$$

2. $$\int \frac{\cos^2 ax\, dx}{\sqrt{-\cos 2ax}} = \frac{1}{a\sqrt{2}} \left[E\left(\alpha, \frac{1}{\sqrt{2}}\right) - F\left(\alpha, \frac{1}{\sqrt{2}}\right) \right]$$

3. $$\int \frac{\cos^4 ax\, dx}{\sqrt{-\cos 2ax}} = \frac{1}{3a\sqrt{2}} \left[3 F\left(\alpha, \frac{1}{\sqrt{2}}\right) - \frac{5}{2} E\left(\alpha, \frac{1}{\sqrt{2}}\right) \right] - \frac{1}{12a} \sin 2ax \sqrt{-\cos 2ax}$$

4. $$\int \frac{dx}{\sin^2 ax \sqrt{-\cos 2ax}} = \frac{1}{a} \cot ax \sqrt{-\cos 2ax} - \frac{\sqrt{2}}{a} E\left(\alpha, \frac{1}{\sqrt{2}}\right)$$

5. $$\int \frac{dx}{\sin^4 ax \sqrt{-\cos 2ax}} = \frac{2}{3a\sqrt{2}} \left[F\left(\alpha, \frac{1}{\sqrt{2}}\right) - 6 E\left(\alpha, \frac{1}{\sqrt{2}}\right) \right]$$
$$+ \frac{1}{3a} \frac{\cos ax}{\sin^3 ax} \left(6 \sin^2 ax + 1\right) \sqrt{-\cos 2ax}$$

6. $$\int \frac{\cot^2 ax\, dx}{\sqrt{-\cos 2ax}} = \frac{1}{a\sqrt{2}} \left[F\left(\alpha, \frac{1}{\sqrt{2}}\right) - 2 E\left(\alpha, \frac{1}{\sqrt{2}}\right) \right] + \frac{1}{a} \cot ax \sqrt{-\cos 2ax}$$

7. $$\int \frac{dx}{\left(1 - 2r^2 \cos^2 ax\right) \sqrt{-\cos 2ax}} = -\frac{1}{a\sqrt{2}} \Pi\left(\alpha, r^2, \frac{1}{\sqrt{2}}\right)$$

8. $$\int \frac{dx}{\sqrt{-\cos^3 2ax}} = \frac{1}{a\sqrt{2}} \left[F\left(\alpha, \frac{1}{\sqrt{2}}\right) - 2 E\left(\alpha, \frac{1}{\sqrt{2}}\right) \right] + \frac{\sin 2ax}{a\sqrt{-\cos 2ax}}$$

9. $$\int \frac{\cos^2 ax\, dx}{\sqrt{-\cos^3 2ax}} = \frac{\sin 2ax}{2a\sqrt{-\cos 2ax}} - \frac{1}{a\sqrt{2}} E\left(\alpha, \frac{1}{\sqrt{2}}\right)$$

10. $$\int \frac{dx}{\sqrt{-\cos^5 2ax}} = -\frac{1}{3a\sqrt{2}} F\left(\alpha, \frac{1}{\sqrt{2}}\right) - \frac{\sin 2ax}{3a\sqrt{-\cos^3 2ax}}$$

11. $$\int \sqrt{-\cos 2ax}\, dx = \frac{1}{a\sqrt{2}} \left[F\left(\alpha, \frac{1}{\sqrt{2}}\right) - 2 E\left(\alpha, \frac{1}{\sqrt{2}}\right) \right]$$

2.621 Integrals of the form $\int R\left(\sin ax, \cos ax, \sqrt{\sin 2ax}\right) dx$.

Notation: $\alpha = \arcsin \sqrt{\dfrac{2 \sin ax}{1 + \sin ax + \cos ax}}$.

1. $$\int \frac{dx}{\sqrt{\sin 2ax}} = \frac{\sqrt{2}}{a} F\left(\alpha, \frac{1}{\sqrt{2}}\right)$$

BY (287.50)

2. $$\int \frac{\sin ax\, dx}{\sqrt{\sin 2ax}} = \frac{\sqrt{2}}{a} \left[\frac{1+i}{2} \Pi\left(\alpha, \frac{1+i}{2}, \frac{1}{\sqrt{2}}\right) \right.$$
$$\left. + \frac{1-i}{2} \Pi\left(\alpha, \frac{1-i}{2}, \frac{1}{\sqrt{2}}\right) + F\left(\alpha, \frac{1}{\sqrt{2}}\right) - 2 E\left(\alpha, \frac{1}{\sqrt{2}}\right) \right]$$

BY (287.57)

3. $$\int \frac{\sin ax\, dx}{\left(1 + \sin ax + \cos ax\right) \sqrt{\sin 2ax}} = \frac{\sqrt{2}}{a} \left[F\left(\alpha, \frac{1}{\sqrt{2}}\right) - E\left(\alpha, \frac{1}{\sqrt{2}}\right) \right]$$

BY (287.54)

4. $$\int \frac{\sin ax \, dx}{(1 - \sin ax + \cos ax)\sqrt{\sin 2ax}} = \frac{\sqrt{2}}{a}\left\{\sqrt{\tan ax} - E\left(\alpha, \frac{1}{\sqrt{2}}\right)\right\}$$

$$\left[ax \neq \frac{\pi}{2}\right]$$ BY (287.55)

5. $$\int \frac{(1 + \cos ax) \, dx}{(1 + \sin ax + \cos ax)\sqrt{\sin 2ax}} = \frac{\sqrt{2}}{a} E\left(\alpha, \frac{1}{\sqrt{2}}\right)$$ BY (287.51)

6. $$\int \frac{(1 + \cos ax) \, dx}{(1 - \sin ax + \cos ax)\sqrt{\sin 2ax}} = \frac{\sqrt{2}}{a}\left\{F\left(\alpha, \frac{1}{\sqrt{2}}\right) - E\left(\alpha, \frac{1}{\sqrt{2}}\right) + \sqrt{\tan ax}\right\}$$

$$\left[ax \neq \frac{\pi}{2}\right]$$ BY (287.56)

7. $$\int \frac{(1 - \sin ax + \cos ax) \, dx}{(1 + \sin ax + \cos ax)\sqrt{\sin 2ax}} = \frac{\sqrt{2}}{a}\left\{2 E\left(\alpha, \frac{1}{\sqrt{2}}\right) - F\left(\alpha, \frac{1}{\sqrt{2}}\right)\right\}$$ BY (287.53)

8. $$\int \frac{(1 + \sin ax + \cos ax) \, dx}{[1 + \cos ax + (1 - 2r^2)\sin ax]\sqrt{\sin 2ax}} = \frac{\sqrt{2}}{a} \Pi\left(\alpha, r^2, \frac{1}{\sqrt{2}}\right).$$ BY (287.52)

2.63–2.65 Products of trigonometric functions and powers

2.631

1. $$\int x^r \sin^p x \cos^q x \, dx = \frac{1}{(p + q)^2}\left[(p + q)x^r \sin^{p+1} x \cos^{q-1} x\right.$$

$$+ rx^{r-1} \sin^p x \cos^q x - r(r - 1)\int x^{r-2} \sin^p x \cos^q x \, dx$$

$$\left. - rp \int x^{r-1} \sin^{p-1} x \cos^{q-1} x \, dx + (q - 1)(p + q)\int x^r \sin^p x \cos^{q-2} x \, dx\right]$$

$$= \frac{1}{(p + q)^2}\left[- (p + q)x^r \sin^{p-1} x \cos^{q+1} x\right.$$

$$+ rx^{r-1} \sin^p x \cos^q x - r(r - 1)\int x^{r-2} \sin^p x \cos^q x \, dx$$

$$\left. + rq \int x^{r-1} \sin^{p-1} x \cos^{q-1} x \, dx + (p - 1)(p + q)\int x^r \sin^{p-2} x \cos^q x \, dx\right]$$

GU (331)(1)

2. $$\int x^m \sin^n x \, dx = \frac{x^{m-1} \sin^{n-1} x}{n^2} \{m \sin x - nx \cos x\}$$

$$+ \frac{n - 1}{n}\int x^m \sin^{n-2} x \, dx - \frac{m(m - 1)}{n^2}\int x^{m-2} \sin^n x \, dx$$

3. $$\int x^m \cos^n x \, dx = \frac{x^{m-1} \cos^{n-1} x}{n^2} \{m \cos x + nx \sin x\}$$

$$+ \frac{n - 1}{n}\int x^m \cos^{n-2} x \, dx - \frac{m(m - 1)}{n^2}\int x^{m-2} \cos^n x \, dx$$

4. $\displaystyle\int x^n \sin^{2m} x\, dx = \binom{2m}{m} \frac{x^{n+1}}{2^{2m}(n+1)}$

$\displaystyle\qquad\qquad + \frac{(-1)^m}{2^{2m-1}} \sum_{k=0}^{m-1} (-1)^k \binom{2m}{k} \int x^n \cos(2m-2k)x\, dx$

$\qquad\qquad\qquad\qquad\qquad\qquad$ (see **2.633** 2) $\qquad\qquad$ TI 333

5. $\displaystyle\int x^n \sin^{2m+1} x\, dx = \frac{(-1)^m}{2^{2m}} \sum_{k=0}^{m} (-1)^k \binom{2m+1}{k} \int x^n \sin(2m-2k+1)x\, dx$

$\qquad\qquad\qquad\qquad\qquad\qquad$ (see **2.633** 1) $\qquad\qquad$ TI 333

6. $\displaystyle\int x^n \cos^{2m} x\, dx = \binom{2m}{m} \frac{x^{n+1}}{2^{2m}(n+1)}$

$\displaystyle\qquad\qquad + \frac{1}{2^{2m-1}} \sum_{k=0}^{m-1} \binom{2m}{k} \int x^n \cos(2m-2k)x\, dx$

$\qquad\qquad\qquad\qquad\qquad\qquad$ (see **2.633** 2) $\qquad\qquad$ TI 333

7. $\displaystyle\int x^n \cos^{2m+1} x\, dx = \frac{1}{2^{2m}} \sum_{k=0}^{m} \binom{2m+1}{k} \int x^n \cos(2m-2k+1)x\, dx$

$\qquad\qquad\qquad\qquad\qquad\qquad$ (see **2.633** 2) $\qquad\qquad$ TI 333

2.632

1. $\displaystyle\int x^{\mu-1} \sin \beta x\, dx = \frac{i}{2}(i\beta)^{-\mu} \gamma(\mu, i\beta x) - \frac{i}{2}(-i\beta)^{-\mu}\gamma(\mu, -i\beta x)$

$\qquad\qquad\qquad\qquad\qquad\qquad$ $[\operatorname{Re}\mu > -1, \quad x > 0]$ $\qquad$ ET I 317(2)

2. $\displaystyle\int x^{\mu-1} \sin ax\, dx = -\frac{1}{2a^\mu} \left\{ \exp\left[\frac{\pi i}{2}(\mu-1)\right] \Gamma(\mu, -iax) + \exp\left[\frac{\pi i}{2}(1-\mu)\right] \Gamma(\mu, iax) \right\}$

$\qquad\qquad\qquad\qquad\qquad\qquad$ $[\operatorname{Re}\mu < 1, \quad a > 0, \quad x > 0]$ $\quad$ ET I 317(3)

3. $\displaystyle\int x^{\mu-1} \cos \beta x\, dx = \frac{1}{2} \left\{ (i\beta)^{-\mu} \gamma(\mu, i\beta x) + (-i\beta)^{-\mu}\gamma(\mu, -i\beta x) \right\}$

$\qquad\qquad\qquad\qquad\qquad\qquad$ $[\operatorname{Re}\mu > 0, \quad x > 0]$ $\qquad$ ET I 319(22)

4. $\displaystyle\int x^{\mu-1} \cos ax\, dx = -\frac{1}{2a^\mu} \left\{ \exp\left(i\mu\frac{\pi}{2}\right) \Gamma(\mu, -iax) + \exp\left(-i\mu\frac{\pi}{2}\right) \Gamma(\mu, iax) \right\}$

$\qquad\qquad\qquad\qquad\qquad\qquad$ ET I 319(23)

2.633

1. $\displaystyle\int x^n \sin ax\, dx = -\sum_{k=0}^{n} k! \binom{n}{k} \frac{x^{n-k}}{a^{k+1}} \cos\left(ax + \frac{1}{2}k\pi\right)$ $\qquad\qquad$ TI (487)

2.[8] $\displaystyle\int x^n \cos ax\, dx = \sum_{k=0}^{n} k! \binom{n}{k} \frac{x^{n-k}}{a^{k+1}} \sin\left(ax + \frac{1}{2}k\pi\right)$ $\qquad\qquad$ TI (486)

3. $\displaystyle\int x^{2n} \sin x\, dx = (2n)! \left\{ \sum_{k=0}^{n} (-1)^{k+1} \frac{x^{2n-2k}}{(2n-2k)!} \cos x + \sum_{k=0}^{n-1} (-1)^k \frac{x^{2n-2k-1}}{(2n-2k-1)!} \sin x \right\}$

4. $\displaystyle\int x^{2n+1}\sin x\,dx=(2n+1)!\left\{\sum_{k=0}^{n}(-1)^{k+1}\frac{x^{2n-2k+1}}{(2n-2k+1)!}\cos x+\sum_{k=0}^{n}(-1)^{k}\frac{x^{2n-2k}}{(2n-2k)!}\sin x\right\}$

5. $\displaystyle\int x^{2n}\cos x\,dx=(2n)!\left\{\sum_{k=0}^{n}(-1)^{k}\frac{x^{2n-2k}}{(2n-2k)!}\sin x+\sum_{k=0}^{n-1}(-1)^{k}\frac{x^{2n-2k-1}}{(2n-2k-1)!}\cos x\right\}$

6. $\displaystyle\int x^{2n+1}\cos x\,dx=(2n+1)!\left\{\sum_{k=0}^{n}(-1)^{k}\frac{x^{2n-2k+1}}{(2n-2k+1)!}\sin x+\sum_{k=0}^{n}\frac{x^{2n-2k}}{(2n-2k)!}\cos x\right\}$

2.634

1. $\displaystyle\int P_n(x)\sin mx\,dx=-\frac{\cos mx}{m}\sum_{k=0}^{\lfloor n/2\rfloor}(-1)^{k}\frac{P_n^{(2k)}(x)}{m^{2k}}+\frac{\sin mx}{m}\sum_{k=1}^{\lfloor (n+1)/2\rfloor}(-1)^{k-1}\frac{P_n^{(2k-1)}(x)}{m^{2k-1}}$

2. $\displaystyle\int P_n(x)\cos mx\,dx=\frac{\sin mx}{m}\sum_{k=0}^{\lfloor n/2\rfloor}(-1)^{k}\frac{P_n^{(2k)}(x)}{m^{2k}}+\frac{\cos mx}{m}\sum_{k=1}^{\lfloor (n+1)/2\rfloor}(-1)^{k-1}\frac{P_n^{(2k-1)}(x)}{m^{2k-1}}$

In formulas **2.634**, $P_n(x)$ is any n^{th}-degree polynomial and $P_n^{(k)}(x)$ is its kth derivative with respect to x.

2.635   Notation: $z_1=a+bx$.

1. $\displaystyle\int z_1\sin kx\,dx=-\frac{1}{k}z_1\cos kx+\frac{b}{k^2}\sin kx$

2. $\displaystyle\int z_1\cos kx\,dx=\frac{1}{k}z_1\sin kx+\frac{b}{k^2}\cos kx$

3. $\displaystyle\int z_1^2\sin kx\,dx=\frac{1}{k}\left(\frac{2b^2}{k^2}-z_1^2\right)\cos kx+\frac{2bz_1}{k^2}\sin kx$

4. $\displaystyle\int z_1^2\cos kx\,dx=\frac{1}{k}\left(z_1^2-\frac{2b^2}{k^2}\right)\sin kx+\frac{2bz_1}{k^2}\cos kx$

5. $\displaystyle\int z_1^3\sin kx\,dx=\frac{z_1}{k}\left(\frac{6b^2}{k^2}-z_1^2\right)\cos kx+\frac{3b}{k^2}\left(z_1^2-\frac{2b^2}{k^2}\right)\sin kx$

6. $\displaystyle\int z_1^3\cos kx\,dx=\frac{z_1}{k}\left(z_1^2-\frac{6b^2}{k^2}\right)\sin kx+\frac{3b}{k^2}\left(z_1^2-\frac{2b^2}{k^2}\right)\cos kx$

7. $\displaystyle\int z_1^4\sin kx\,dx=-\frac{1}{k}\left(z_1^4-\frac{12b^2}{k^2}z_1^2+\frac{24b^4}{k^4}\right)\cos kx+\frac{4bz_1}{k^2}\left(z_1^2-\frac{6b^2}{k^2}\right)\sin kx$

8. $\displaystyle\int z_1^4\cos kx\,dx=\frac{1}{k}\left(z_1^4-\frac{12b^2}{k^2}z_1^2+\frac{24b^4}{k^4}\right)\sin kx+\frac{4bz_1}{k^2}\left(z_1^2-\frac{6b^2}{k^2}\right)\cos kx$

9. $\displaystyle\int z_1^5\sin kx\,dx=\frac{5b}{k^2}\left(z_1^4-\frac{12b^2}{k^2}z_1^2+\frac{24b^4}{k^4}\right)\sin kx-\frac{z_1}{k}\left(z_1^4-\frac{20b^2}{k^2}z_1^2+\frac{120b^4}{k^4}\right)\cos kx$

10. $\displaystyle\int z_1^5\cos kx\,dx=\frac{5b}{k^2}\left(z_1^4-\frac{12b^2}{k^2}z_1^2+\frac{24b^4}{k^4}\right)\cos kx+\frac{z_1}{k}\left(z_1^4-\frac{20b^2}{k^2}z_1^2+\frac{120b^4}{k^4}\right)\sin kx$

11. $\int z_1^6 \sin kx \, dx = \dfrac{6bz_1}{k^2} \left(z_1^4 - \dfrac{20b^2}{k^2} z_1^2 + \dfrac{120b^4}{k^4} \right) \sin kx$

$\qquad - \dfrac{1}{k} \left(z_1^6 - \dfrac{30b^2}{k^2} z_1^4 + \dfrac{360b^4}{k^4} z_1^2 - \dfrac{720b^6}{k^6} \right) \cos kx$

12. $\int z_1^6 \cos kx \, dx = \dfrac{6bz_1}{k^2} \left(z_1^4 - \dfrac{20b^2}{k^2} z_1^2 + \dfrac{120b^4}{k^4} \right) \cos kx$

$\qquad + \dfrac{1}{k} \left(z_1^6 - \dfrac{30b^2}{k^2} z_1^4 + \dfrac{360b^4}{k^4} z_1^2 - \dfrac{720b^6}{k^6} \right) \sin kx$

2.636

1. $\int x^n \sin^2 x \, dx = \dfrac{x^{n+1}}{2(n+1)}$

$\qquad + \dfrac{n!}{4} \left\{ \displaystyle\sum_{k=0}^{\lfloor n/2 \rfloor} \dfrac{(-1)^{k+1} x^{n-2k}}{2^{2k}(n-2k)!} \sin 2x + \displaystyle\sum_{k=0}^{\lfloor (n-1)/2 \rfloor} \dfrac{(-1)^{k+1} x^{n-2k-1}}{2^{2k+1}(n-2k-1)!} \cos 2x \right\}$

$\qquad\qquad\qquad\qquad\qquad\qquad\qquad\qquad\qquad\qquad\qquad\qquad\qquad\qquad\qquad$ GU (333)(2e)

2. $\int x^n \cos^2 x \, dx = \dfrac{x^{n+1}}{2(n+1)}$

$\qquad - \dfrac{n!}{4} \left\{ \displaystyle\sum_{k=0}^{\lfloor n/2 \rfloor} \dfrac{(-1)^{k+1} x^{n-2k}}{2^{2k}(n-2k)!} \sin 2x + \displaystyle\sum_{k=0}^{\lfloor (n-1)/2 \rfloor} \dfrac{(-1)^{k+1} x^{n-2k-1}}{2^{2k+1}(n-2k-1)!} \cos 2x \right\}$

$\qquad\qquad\qquad\qquad\qquad\qquad\qquad\qquad\qquad\qquad\qquad\qquad\qquad\qquad\qquad$ GU (333)(3e)

3. $\int x \sin^2 x \, dx = \dfrac{x^2}{4} - \dfrac{x}{4} \sin 2x - \dfrac{1}{8} \cos 2x.$

4. $\int x^2 \sin^2 x \, dx = \dfrac{x^3}{6} - \dfrac{x}{4} \cos 2x - \dfrac{1}{4} \left(x^2 - \dfrac{1}{2} \right) \sin 2x$ $\qquad\qquad\qquad$ MZ 241

5. $\int x \cos^2 x \, dx = \dfrac{x^2}{4} + \dfrac{x}{4} \sin 2x + \dfrac{1}{8} \cos 2x.$

6. $\int x^2 \cos^2 x \, dx = \dfrac{x^3}{6} + \dfrac{x}{4} \cos 2x + \dfrac{1}{4} \left(x^2 - \dfrac{1}{2} \right) \sin 2x$ $\qquad\qquad\qquad$ MZ 245

2.637

1. $\int x^n \sin^3 x \, dx = \dfrac{n!}{4} \left\{ \displaystyle\sum_{k=0}^{\lfloor n/2 \rfloor} \dfrac{(-1)^k x^{n-2k}}{(n-2k)!} \left(\dfrac{\cos 3x}{3^{2k+1}} - 3 \cos x \right) \right.$

$\qquad\qquad \left. - \displaystyle\sum_{k=0}^{\lfloor (n-1)/2 \rfloor} (-1)^k \dfrac{x^{n-2k-1}}{(n-2k-1)!} \left(\dfrac{\sin 3x}{2^{2k+2}} - 3 \sin x \right) \right\}$

$\qquad\qquad\qquad\qquad\qquad\qquad\qquad\qquad\qquad\qquad\qquad\qquad\qquad\qquad\qquad$ GU(333)(2f)

2. $\displaystyle\int x^n \cos^3 x \, dx = \frac{n!}{4} \left\{ \sum_{k=0}^{\lfloor n/2 \rfloor} \frac{(-1)^k x^{n-2k}}{(n-2k)!} \left(\frac{\sin 3x}{3^{2k+1}} + 3 \sin x \right) \right.$

$$\left. + \sum_{k=0}^{[(n-1)/2]} (-1)^k \frac{x^{n-2k-1}}{(n-2k-1)!} \left(\frac{\cos 3x}{3^{2k+2}} + 3 \cos x \right) \right\}$$

GU(333)(3f)

3. $\displaystyle\int x \sin^3 x \, dx = \frac{3}{4} \sin x - \frac{1}{36} \sin 3x - \frac{3}{4} x \cos x + \frac{x}{12} \cos 3x$

4. $\displaystyle\int x^2 \sin^3 x \, dx = -\left(\frac{3}{4}x^2 + \frac{3}{2} \right) \cos x + \left(\frac{x^2}{12} + \frac{1}{54} \right) \cos 3x + \frac{3}{2} x \sin x - \frac{x}{18} \sin 3x$ MZ 241

5. $\displaystyle\int x \cos^3 x \, dx = \frac{3}{4} \cos x + \frac{1}{36} \cos 3x + \frac{3}{4} x \sin x + \frac{x}{12} \sin 3x$

6. $\displaystyle\int x^2 \cos^3 x \, dx = \left(\frac{3}{4}x^2 - \frac{3}{2} \right) \sin x + \left(\frac{x^2}{12} - \frac{1}{54} \right) \sin 3x + \frac{3}{2} x \cos x + \frac{x}{18} \cos 3x$ MZ 245, 246

2.638

1. $\displaystyle\int \frac{\sin^q x}{x^p} \, dx = -\frac{\sin^{q-1} x \left[(p-2) \sin x + qx \cos x \right]}{(p-1)(p-2)x^{p-1}}$

$$-\frac{q^2}{(p-1)(p-2)} \int \frac{\sin^q x \, dx}{x^{p-2}} + \frac{q(q-1)}{(p-1)(p-2)} \int \frac{\sin^{q-2} x \, dx}{x^{p-2}}$$

$$[p \neq 1, \quad p \neq 2]$$ TI (496)

2. $\displaystyle\int \frac{\cos^q x}{x^p} \, dx = -\frac{\cos^{q-1} x \left[(p-2) \cos x - qx \sin x \right]}{(p-1)(p-2)x^{p-1}}$

$$-\frac{q^2}{(p-1)(p-2)} \int \frac{\cos^q x \, dx}{x^{p-2}} + \frac{q(q-1)}{(p-1)(p-2)} \int \frac{\cos^{q-2} x \, dx}{x^{p-2}}$$

$$[p \neq 1, \quad p \neq 2]$$ TI (495)

3.[6] $\displaystyle\int \frac{\sin x \, dx}{x^p} = -\frac{\sin x}{(p-1)x^{p-1}} + \frac{1}{p-1} \int \frac{\cos x \, dx}{x^{p-1}}$

$$= -\frac{\sin x}{(p-1)x^{p-1}} - \frac{\cos x}{(p-1)(p-2)x^{p-2}} - \frac{1}{(p-1)(p-2)} \int \frac{\sin x \, dx}{x^{p-2}}$$

$$(p > 2)$$ TI (492)

4.[6] $\displaystyle\int \frac{\cos x \, dx}{x^p} = -\frac{\cos x}{(p-1)x^{p-1}} - \frac{1}{p-1} \int \frac{\sin x \, dx}{x^{p-1}}$

$$= -\frac{\cos x}{(p-1)x^{p-1}} + \frac{\sin x}{(p-1)(p-2)x^{p-2}} - \frac{1}{(p-1)(p-2)} \int \frac{\cos x \, dx}{x^{p-2}}$$

$$(p > 2)$$ TI (491)

2.639

1. $$\int \frac{\sin x \, dx}{x^{2n}} = \frac{(-1)^{n+1}}{x(2n-1)!} \left\{ \sum_{k=0}^{n-2} \frac{(-1)^k (2k+1)!}{x^{2k+1}} \cos x \right.$$
    $$\left. + \sum_{k=0}^{n-1} \frac{(-1)^{k+1} (2k)!}{x^{2k}} \sin x \right\} + \frac{(-1)^{n+1}}{(2n-1)!} \operatorname{ci}(x)$$

GU (333)(6b)a

2. $$\int \frac{\sin x}{x^{2n+1}} \, dx = \frac{(-1)^{n+1}}{x(2n)!} \left\{ \sum_{k=0}^{n-1} \frac{(-1)^{k+1} (2k)!}{x^{2k}} \cos x \right.$$
    $$\left. + \sum_{k=0}^{n-1} \frac{(-1)^{k+1} (2k+1)!}{x^{2k+1}} \sin x \right\} + \frac{(-1)^n}{(2n)!} \operatorname{si}(x)$$

GU (333)(6b)a

3. $$\int \frac{\cos x \, dx}{x^{2n}} \, dx = \frac{(-1)^{n+1}}{x(2n-1)!} \left\{ \sum_{k=0}^{n-1} \frac{(-1)^{k+1} (2k)!}{x^{2k}} \cos x \right.$$
    $$\left. - \sum_{k=0}^{n-2} \frac{(-1)^k (2k+1)!}{x^{2k+1}} \sin x \right\} + \frac{(-1)^n}{(2n-1)!} \operatorname{si}(x)$$

GU (333)(7b)

4. $$\int \frac{\cos x \, dx}{x^{2n+1}} = \frac{(-1)^{n+1}}{x(2n)!} \left\{ \sum_{k=0}^{n-1} \frac{(-1)^{k+1} (2k+1)!}{x^{2k+1}} \cos x \right.$$
    $$\left. - \sum_{k=0}^{n-1} \frac{(-1)^{k+1} (2k)!}{x^{2k}} \sin x \right\} + \frac{(-1)^n}{(2n)!} \operatorname{ci}(x)$$

GU (333)(7b)

2.641

1. $$\int \frac{\sin kx}{a+bx} \, dx = \frac{1}{b} \left[\cos \frac{ka}{b} \operatorname{si}(u) - \sin \frac{ka}{b} \operatorname{ci}(u) \right] \qquad \left[u = \frac{k}{b}(a+bx) \right]$$

2. $$\int \frac{\cos kx}{a+bx} \, dx = \frac{1}{b} \left[\cos \frac{ka}{b} \operatorname{ci}(u) + \sin \frac{ka}{b} \operatorname{si}(u) \right] \qquad \left[u = \frac{k}{b}(a+bx) \right]$$

3. $$\int \frac{\sin kx}{(a+bx)^2} \, dx = -\frac{1}{b} \frac{\sin kx}{a+bx} + \frac{k}{b} \int \frac{\cos kx}{a+bx} \, dx \qquad \text{(see } \textbf{2.641 } 2\text{)}$$

4. $$\int \frac{\cos kx}{(a+bx)^2} \, dx = -\frac{1}{b} \frac{\cos kx}{a+bx} - \frac{k}{b} \int \frac{\sin kx}{a+bx} \, dx \qquad \text{(see } \textbf{2.641 } 1\text{)}$$

5. $$\int \frac{\sin kx}{(a+bx)^3} \, dx = -\frac{\sin kx}{2b(a+bx)^2} - \frac{k \cos kx}{2b^2(a+bx)} - \frac{k^2}{2b^2} \int \frac{\sin kx}{a+bx} \, dx$$

(see 2.641 1)

6. $$\int \frac{\cos kx}{(a+bx)^3} \, dx = -\frac{\cos kx}{2b(a+bx)^2} + \frac{k \sin kx}{2b^2(a+bx)} - \frac{k^2}{2b^2} \int \frac{\cos kx}{a+bx} \, dx$$

(see **2.641** 2)

7. $$\int \frac{\sin kx}{(a+bx)^4} \, dx = -\frac{\sin kx}{3b(a+bx)^3} - \frac{k \cos kx}{6b^2(a+bx)^2}$$
$$+ \frac{k^2 \sin kx}{6b^2(a+bx)} - \frac{k^3}{6b^3} \int \frac{\cos kx}{a+bx} \, dx$$

(see **2.641** 2)

8. $$\int \frac{\cos kx}{(a+bx)^4} \, dx = -\frac{\cos kx}{3b(a+bx)^3} + \frac{k \sin kx}{6b^2(a+bx)^2} + \frac{k^2 \cos kx}{6b^3(a+bx)} + \frac{k^3}{6b^3} \int \frac{\sin kx}{a+bx} \, dx$$

(see **2.641** 1)

9. $$\int \frac{\sin kx}{(a+bx)^5} \, dx = -\frac{\sin kx}{4b(a+bx)^4} - \frac{k \cos kx}{12b^2(a+bx)^3}$$
$$+ \frac{k^2 \sin kx}{24b^3(a+bx)^2} + \frac{k^3 \cos kx}{24b^4(a+bx)} \frac{k^4}{24b^4} \int \frac{\sin kx}{a+bx} \, dx$$

(see **2.641** 1)

10. $$\int \frac{\cos kx}{(a+bx)^5} \, dx = -\frac{\cos kx}{4b(a+bx)^4} + \frac{k \sin kx}{12b^2(a+bx)^3}$$
$$+ \frac{k^2 \cos kx}{24b^3(a+bx)^2} - \frac{k^3 \sin kx}{24b^4(a+bx)} + \frac{k^4}{24b^4} \int \frac{\cos kx}{a+bx} \, dx$$

(see **2.641** 2)

11. $$\int \frac{\sin kx}{(a+bx)^6} \, dx = -\frac{\sin kx}{5b(a+bx)^5} - \frac{k \cos kx}{20b^2(a+bx)^4} + \frac{k^2 \sin kx}{60b^3(a+bx)^3} + \frac{k^3 \cos kx}{120b^4(a+bx)^2}$$
$$- \frac{k^4 \sin kx}{120b^5(a+bx)} + \frac{k^5}{120b^5} \int \frac{\cos kx}{a+bx} \, dx$$

(see **2.641** 2)

12. $$\int \frac{\cos kx}{(a+bx)^6} \, dx = -\frac{\cos kx}{5b(a+bx)^5} + \frac{k \sin kx}{20b^2(a+bx)^4} + \frac{k^2 \cos kx}{60b^3(a+bx)^3}$$
$$- \frac{k^3 \sin kx}{120b^4(a+bx)^2} - \frac{k^4 \cos kx}{120b^5(a+bx)} - \frac{k^5}{120b^5} \int \frac{\sin kx}{a+bx} \, dx$$

(see **2.641** 1)

2.642

1. $$\int \frac{\sin^{2m} x}{x} \, dx = \binom{2m}{m} \frac{\ln x}{2^{2m}} + \frac{(-1)^m}{2^{2m-1}} \sum_{k=0}^{m-1} (-1)^k \binom{2m}{k} \operatorname{ci}[(2m-2k)x]$$

2. $$\int \frac{\sin^{2m+1} x}{x} \, dx = \frac{(-1)^m}{2^{2m}} \sum_{k=0}^{m} (-1)^k \binom{2m+1}{k} \operatorname{si}[(2m-2k+1)x]$$

3. $$\int \frac{\cos^{2m} x}{x} \, dx = \binom{2m}{m} \frac{\ln x}{2^{2m}} + \frac{1}{2^{2m-1}} \sum_{k=0}^{m-1} \binom{2m}{k} \operatorname{ci}[(2m-2k)x]$$

4. $\displaystyle\int \frac{\cos^{2m+1} x}{x}\, dx = \frac{1}{2^{2m}} \sum_{k=0}^{m} \binom{2m+1}{k} \operatorname{ci}[(2m-2k+1)x]$

5. $\displaystyle\int \frac{\sin^{2m} x}{x^2}\, dx = -\binom{2m}{m}\frac{1}{2^{2m}x}$

$$+ \frac{(-1)^m}{2^{2m-1}} \sum_{k=0}^{m-1} (-1)^{k+1} \binom{2m}{k} \left\{ \frac{\cos(2m-2k)x}{x} + (2m-2k)\operatorname{si}[(2m-2k)x] \right\}$$

6. $\displaystyle\int \frac{\sin^{2m+1} x}{x^2}\, dx = \frac{(-1)^m}{2^{2m}} \sum_{k=0}^{m} (-1)^{k+1} \binom{2m+1}{k}$

$$\times \left\{ \frac{\sin(2m-2k+1)x}{x} - (2m-2k+1)\operatorname{ci}[(2m-2k+1)x] \right\}$$

7. $\displaystyle\int \frac{\cos^{2m} x}{x^2}\, dx = -\binom{2m}{m}\frac{1}{2^{2m}x}$

$$- \frac{1}{2^{2m-1}} \sum_{k=0}^{m-1} \binom{2m}{k} \left\{ \frac{\cos(2m-2k)x}{x} + (2m-2k)\operatorname{si}[(2m-2k)x] \right\}$$

8. $\displaystyle\int \frac{\cos^{2m+1} x}{x^2} = -\frac{1}{2^{2m}} \sum_{k=0}^{m} \binom{2m+1}{k} \left\{ \frac{\cos(2m-2k+1)x}{x} \right.$

$$\left. + (2m-2k+1)\operatorname{si}[(2m-2k+1)x] \right\}$$

2.643

1. $\displaystyle\int \frac{x^p\, dx}{\sin^q x} = -\frac{x^{p-1}\left[p\sin x + (q-2)x\cos x\right]}{(q-1)(q-2)\sin^{q-1} x} + \frac{q-2}{q-1}\int \frac{x^p\, dx}{\sin^{q-2} x} + \frac{p(p-1)}{(q-1)(q-2)}\int \frac{x^{p-2}\, dx}{\sin^{q-2} x}$

2. $\displaystyle\int \frac{x^p\, dx}{\cos^q x} = -\frac{x^{p-1}\left[p\cos x - (q-2)x\sin x\right]}{(q-1)(q-2)\cos^{q-1} x}$

$$+ \frac{q-2}{q-1}\int \frac{x^p\, dx}{\cos^{q-2} x} + \frac{p(p-1)}{(q-1)(q-2)}\int \frac{x^{p-2}\, dx}{\cos^{q-2} x}$$

3.⁴ $\displaystyle\int \frac{x^n}{\sin x}\, dx = \frac{x^n}{n} + \sum_{k=1}^{\infty} (-1)^{k+1} \frac{2\left(2^{2k-1}-1\right)}{(n+2k)(2k)!} B_{2k} x^{n+2k}$

$$[|x| < \pi, \quad n > 0] \qquad \text{TU (333)(8b)}$$

4. $\displaystyle\int \frac{dx}{x^n \sin x} = -\frac{1}{nx^n} - [1 + (-1)^n] (-1)^{\frac{n}{2}} \frac{2^{n-1}-1}{n!} B_n \ln x - \sum_{\substack{k=1 \\ k\neq \frac{n}{2}}}^{\infty} (-1)^k \frac{2\left(2^{2n}-1\right)}{(2k-n)\cdot(2k)!} B_{2k} x^{2k-n}$

$$[n > 1, \quad |x| > \pi] \qquad \text{GU (333)(9b)}$$

5.⁸ $\displaystyle\int \frac{x^n\, dx}{\cos x} = \sum_{k=0}^{\infty} \frac{|E_{2k}| x^{n+2k+1}}{(n+2k+1)(2k)!}$ $\displaystyle\left[|x| < \frac{\pi}{2}, \quad n > 0\right]$ GU (333)(10b)

6. $$\int \frac{dx}{x^n \cos x} = \frac{1}{2} \left[1 - (-1)^n\right] \frac{|E_{n-1}|}{(n-1)!} \ln x + \sum_{\substack{k=0 \\ k \neq \frac{n-1}{2}}}^{\infty} \frac{|E_{2k}| x^{2k-n+1}}{(2k-n+1) \cdot (2k)!}$$

$$\left[|x| < \frac{\pi}{2}\right]$$ GU (333)(11b)

7. $$\int \frac{x^n \, dx}{\sin^2 x} = -x^n \cot x + \frac{n}{n-1} x^{n-1} + n \sum_{k=1}^{\infty} (-1)^k \frac{2^{2k} x^{n+2k-1}}{(n+2k-1)(2k)!} B_{2k}$$

$$[|x| < \pi, \quad n > 1]$$ GU (333)(8c)

8. $$\int \frac{dx}{x^n \sin^2 x} = -\frac{\cot x}{x^n} + \frac{n}{(n+1)x^{n+1}} - \left[1 - (-1)^n\right] (-1)^{\frac{n+1}{2}} \frac{2^n n}{(n+1)!} B_{n+1} \ln x$$
$$- \frac{n}{2^{n+1}} \sum_{\substack{k=1 \\ k \neq \frac{n+1}{2}}}^{\infty} \frac{(-1)^k (2x)^{2k}}{(2k-n-1)(2k)!} B_{2k}$$

$$[|x| < \pi]$$ GU (333)(9c)

9. $$\int \frac{x^n \, dx}{\cos^2 x} = x^n \tan x + n \sum_{k=1}^{\infty} (-1)^k \frac{2^{2k} \left(2^{2k} - 1\right) x^{n+2k-1}}{(n+2k-1) \cdot (2k)!} B_{2k}$$

$$\left[n > 1, \quad |x| < \frac{\pi}{2}\right]$$ GU (333)(10c)

10. $$\int \frac{dx}{x^n \cos^2 x} = \frac{\tan x}{x^n} - \left[1 - (-1)^n\right] (-1)^{\frac{n+1}{2}} \frac{2^n n}{(n+1)!} \left(2^{n+1} - 1\right) B_{n+1} \ln x$$
$$- \frac{n}{x^{n+1}} \sum_{\substack{k=1 \\ k \neq \frac{n+1}{2}}}^{\infty} \frac{(-1)^k \left(2^{2k} - 1\right) (2x)^{2k}}{(2k-n-1)(2k)!} B_{2k}$$

$$\left[|x| < \frac{\pi}{2}\right]$$ GU (333)(11c)

2.644

1. $$\int \frac{x \, dx}{\sin^{2n} x} = -\sum_{k=0}^{n-1} \frac{(2n-2)(2n-4)\ldots(2n-2k+2)}{(2n-1)(2n-3)\ldots(2n-2k+3)} \frac{\sin x + (2n-2k)x \cos x}{(2n-2k+1)(2n-2k)\sin^{2n-2k+1} x}$$
$$+ \frac{2^{n-1}(n-1)!}{(2n-1)!!} (\ln \sin x - x \cot x)$$

2. $$\int \frac{x \, dx}{\sin^{2n+1} x} = -\sum_{k=0}^{n-1} \frac{(2n-1)(2n-3)\ldots(2n-2k+1)}{2n(2n-2)\ldots(2n-2k+2)} \frac{\sin x + (2n-2k-1)x \cos x}{(2n-2k)(2n-2k-1)\sin^{2n-2k} x}$$
$$+ \frac{(2n-1)!!}{2^n n!} \int \frac{x \, dx}{\sin x}$$

(see **2.644** 5)

3. $$\int \frac{x \, dx}{\cos^{2n} x} = \sum_{k=0}^{n-1} \frac{(2n-2)(2n-4)\ldots(2n-2k+2)}{(2n-1)(2n-3)\ldots(2n-2k+3)} \frac{(2n-2k)x \sin x - \cos x}{(2n-2k+1)(2n-2k)\cos^{2n-2k+1} x}$$
$$+ \frac{2^{n-1}(n-1)!}{(2n-1)!!} (x \tan x + \ln \cos x)$$

4.
$$\int \frac{x\,dx}{\cos^{2n+1} x} = \sum_{k=0}^{n-1} \frac{(2n-1)(2n-3)\ldots(2n-2k+1)}{2n(2n-2)\ldots(2n-2k+2)} \frac{(2n-2k+1)x\sin x - \cos x}{(2n-2k)(2n-2k-1)\cos^{2n-2k} x}$$
$$+ \frac{(2n-1)!!}{2^n n!} \int \frac{x\,dx}{\cos x}$$

$$(\text{see } \mathbf{2.644}\ 6)$$

5.
$$\int \frac{x\,dx}{\sin x} = x + \sum_{k=1}^{\infty} (-1)^{k+1} \frac{2\left(2^{2k-1}-1\right)}{(2k+1)!} B_{2k} x^{2k+1}$$

6.
$$\int \frac{x\,dx}{\cos x} = \sum_{k=0}^{\infty} \frac{|E_{2k}|\,x^{2k+2}}{(2k+2)(2k)!}$$

7.
$$\int \frac{x\,dx}{\sin^2 x} = -x\cot x + \ln\sin x$$

8.
$$\int \frac{x\,dx}{\cos^2 x} = x\tan x + \ln\cos x$$

9.
$$\int \frac{x\,dx}{\sin^3 x} = -\frac{\sin x + x\cos x}{2\sin^2 x} + \frac{1}{2}\int \frac{x}{\sin x}\,dx \qquad (\text{see } \mathbf{2.644}\ 5)$$

10.
$$\int \frac{x\,dx}{\cos^3 x} = \frac{x\sin x - \cos x}{2\cos^2 x} + \frac{1}{2}\int \frac{x\,dx}{\cos x} \qquad (\text{see } \mathbf{2.644}\ 6)$$

11.
$$\int \frac{x\,dx}{\sin^4 x} = -\frac{x\cos x}{3\sin^3 x} - \frac{1}{6\sin^2 x} - \frac{2}{3}x\cot x + \frac{2}{3}\ln(\sin x)$$

12.
$$\int \frac{x\,dx}{\cos^4 x} = \frac{x\sin x}{3\cos^3 x} - \frac{1}{6\cos^2 x} + \frac{2}{3}x\tan x - \frac{2}{3}\ln(\cos x)$$

13.
$$\int \frac{x\,dx}{\sin^5 x} = -\frac{x\cos x}{4\sin^4 x} - \frac{1}{12\sin^3 x} - \frac{3x\cos x}{8\sin^2 x} - \frac{3}{8\sin x} + \frac{3}{8}\int \frac{x\,dx}{\sin x}$$

$$(\text{see } \mathbf{2.644}\ 5)$$

14.
$$\int \frac{x\,dx}{\cos^5 x} = \frac{x\sin x}{4\cos^4 x} - \frac{1}{12\cos^3 x} + \frac{3x\sin x}{8\cos^2 x} - \frac{3}{8\cos x} + \frac{3}{8}\int \frac{x\,dx}{\cos x}$$

$$(\text{see } \mathbf{2.644}\ 6)$$

2.645

1.
$$\int x^p \frac{\sin^{2m} x}{\cos^n x}\,dx = \sum_{k=0}^{m} (-1)^k \binom{m}{k} \int \frac{x^p\,dx}{\cos^{n-2k} x} \qquad (\text{see } \mathbf{2.643}\ 2)$$

2.
$$\int x^p \frac{\sin^{2m+1} x}{\cos^n x}\,dx = \sum_{k=0}^{m} (-1)^k \binom{m}{k} \int \frac{x^p\sin x}{\cos^{n-2k} x}\,dx \qquad (\text{see } \mathbf{2.645}\ 3)$$

3.
$$\int x^p \frac{\sin x\,dx}{\cos^n x} = \frac{x^p}{(n-1)\cos^{n-1} x} - \frac{p}{n-1}\int \frac{x^{p-1}}{\cos^{n-1} x}\,dx$$

$$[n>1] \qquad (\text{see } \mathbf{2.643}\ 2) \qquad \text{GU } (333)(12)$$

4.
$$\int x^p \frac{\cos^{2m} x}{\sin^n x}\,dx = \sum_{k=0}^{m} (-1)^k \binom{m}{k} \int \frac{x^p\,dx}{\sin^{n-2k} x} \qquad (\text{see } \mathbf{2.643}\ 1)$$

5. $\displaystyle\int x^p \frac{\cos^{2m+1} x}{\sin^n x}\, dx = \sum_{k=0}^{m} (-1)^k \binom{m}{k} \int \frac{x^p \cos x}{\sin^{n-2k} x}\, dx$ (see **2.645** 6)

6. $\displaystyle\int x^p \frac{\cos x}{\sin^n x} = -\frac{x^p}{(n-1)\sin^{n-1} x} + \frac{p}{n-1}\int \frac{x^{p-1}\, dx}{\sin^{n-1} x}$ $[n > 1]$ (see **2.643** 1)   GU (333)(13)

7. $\displaystyle\int \frac{x \cos x}{\sin^2 x}\, dx = -\frac{x}{\sin x} + \ln \tan \frac{x}{2}$

8. $\displaystyle\int \frac{x \sin x}{\cos^2 x}\, dx = \frac{x}{\cos x} - \ln \tan \left(\frac{x}{2} + \frac{\pi}{4}\right)$

2.646

1. $\displaystyle\int x^p \tan x\, dx = \sum_{k=1}^{\infty} (-1)^{k+1} \frac{2^{2k}\left(2^{2k-1}-1\right)}{(p+2k)\cdot(2k)!} B_{2k} x^{p+2k}$

$$\left[p \geq -1, \quad |x| < \frac{\pi}{2}\right] \qquad \text{GU (333)(12d)}$$

2. $\displaystyle\int x^p \cot x\, dx = \sum_{k=0}^{\infty} (-1)^k \frac{2^{2k} B_{2k}}{(p+2k)(2k)!} x^{p+2k}$ $[p \geq 1, \quad |x| < \pi]$ GU (333)(13d)

3. $\displaystyle\int x^p \tan^2 x\, dx = x \tan x + \ln \cos x - \frac{x^2}{2}$

4. $\displaystyle\int x \cot^2 x\, dx = -x \cot x + \ln \sin x - \frac{x^2}{2}$

2.647

1. $\displaystyle\int \frac{x^n \cos x\, dx}{(a + b\sin x)^m} = -\frac{x^n}{(m-1)b(a + b\sin x)^{m-1}} + \frac{n}{(m-1)b}\int \frac{x^{n-1}\, dx}{(a + b\sin x)^{m-1}}$
    $[m \neq 1]$ MZ 247

2. $\displaystyle\int \frac{x^n \sin x\, dx}{(a + b\cos x)^m} = \frac{x^n}{(m-1)b(a + b\cos x)^{m-1}} - \frac{n}{(m-1)b}\int \frac{x^{n-1}\, dx}{(a + b\cos x)^{m-1}}$
    $[m \neq 1]$ MZ 247

3. $\displaystyle\int \frac{x\, dx}{1 + \sin x} = -x \tan \left(\frac{\pi}{4} - \frac{x}{2}\right) + 2\ln \cos \left(\frac{\pi}{4} - \frac{x}{2}\right)$ PE (329)

4. $\displaystyle\int \frac{x\, dx}{1 - \sin x} = x \cot \left(\frac{\pi}{4} - \frac{x}{2}\right) + 2\ln \sin \left(\frac{\pi}{4} - \frac{x}{2}\right)$ PE (330)

5. $\displaystyle\int \frac{x\, dx}{1 + \cos x} = x \tan \frac{x}{2} + 2\ln \cos \frac{x}{2}$ PE (331)

6. $\displaystyle\int \frac{x\, dx}{1 - \cos x} = -x \cot \frac{x}{2} + 2\ln \cos \frac{x}{2}$ PE (332)

7. $\displaystyle\int \frac{x \cos x}{(1 + \sin x)^2}\, dx = -\frac{x}{1 + \sin x} + \tan \left(\frac{x}{2} - \frac{\pi}{4}\right)$

8. $\displaystyle\int \frac{x \cos x}{(1 - \sin x)^2}\, dx = \frac{x}{1 - \sin x} + \tan \left(\frac{x}{2} + \frac{\pi}{4}\right)$

9. $\displaystyle\int \frac{x \sin x}{(1 + \cos x)^2}\, dx = \frac{x}{1 + \cos x} - \tan \frac{x}{2}$

10. $\displaystyle\int \frac{x \sin x}{(1 - \cos x)^2}\, dx = -\frac{x}{1 - \cos x} - \cot \frac{x}{2}$ MZ 247a

2.648

1. $\displaystyle\int \frac{x + \sin x}{1 + \cos x}\, dx = x \tan \frac{x}{2}$

2. $\displaystyle\int \frac{x - \sin x}{1 - \cos x}\, dx = -x \cot \frac{x}{2}$ GU (333)(16)

2.649 $\displaystyle\int \frac{x^2\, dx}{[(ax - b) \sin x + (a + bx) \cos x]^2} = \frac{x \sin x + \cos x}{b\,[(ax - b) \sin x + (a + bx) \cos x]}$ GU (333)(17)

2.651 $\displaystyle\int \frac{dx}{[a + (ax + b) \tan x]^2} = \frac{\tan x}{a\,[a + (ax + b) \tan x]}$ GU (333)(18)

2.652 $\displaystyle\int \frac{x\, dx}{\cos(x + t) \cos(x - t)} = \operatorname{cosec} 2t \left\{ x \ln \frac{\cos(x - t)}{\cos(x + t)} - L(x + t) + L(x - t) \right\}$

$$\left[t \neq n\pi; \quad |x| < \left| \frac{\pi}{2} - |t_0| \right| \right],$$

where t_0 is the value of the argument t, which is reduced by multiples of the argument π to lie in the interval $\left(-\frac{\pi}{2}, \frac{\pi}{2}\right)$.   LO III 288

2.653

1. $\displaystyle\int \frac{\sin x}{\sqrt{x}}\, dx = \sqrt{2\pi}\, S\left(\sqrt{x}\right)$ (cf. **8.251** 21)

2. $\displaystyle\int \frac{\cos x}{\sqrt{x}}\, dx = \sqrt{2\pi}\, C\left(\sqrt{x}\right)$ (cf. **8.251** 3)

2.654   **Notation:** $\Delta = \sqrt{1 - k^2 \sin^2 x}, \quad k' = \sqrt{1 - k^2}$:

1. $\displaystyle\int \frac{x \sin x \cos x}{\Delta}\, dx = -\frac{x\Delta}{k^2} + \frac{1}{k^2}\, E(x, k)$

2. $\displaystyle\int \frac{x \sin^3 x \cos x}{\Delta}\, dx = -\frac{k'^2}{9k^4}\, F(x, k) + \frac{2k^2 + 5}{9k^4}\, E(x, k) - \frac{1}{9k^4}\left[3\left(3 - \Delta^2\right) x + k^2 \sin x \cos x\right]\Delta.$

3. $\displaystyle\int \frac{x \sin x \cos^3 x}{\Delta}\, dx = -\frac{k'^2}{9k^4}\, F(x, k) + \frac{7k^2 - 5}{9k^4}\, E(x, k) - \frac{1}{9k^4}\left[3\left(\Delta^2 - 3k'^2\right) x - k^2 \sin x \cos x\right]\Delta.$

4. $\displaystyle\int \frac{x \sin x\, dx}{\Delta^3}\, dx = -\frac{x \cos x}{k'^2\Delta} + \frac{1}{kk'^2} \arcsin\left(k \sin x\right)$

5. $\displaystyle\int \frac{x \cos x\, dx}{\Delta^3} = \frac{x \sin x}{\Delta} + \frac{1}{k} \ln\left(k \cos x + \Delta\right)$

6. $\displaystyle\int \frac{x \sin x \cos x\, dx}{\Delta^3} = \frac{x}{k^2\Delta} - \frac{1}{k^2}\, F(x, k)$

7. $\displaystyle\int \frac{x \sin^3 x \cos x\, dx}{\Delta^3} = x\frac{2 - k^2 \sin^2 x}{k^4\Delta} - \frac{1}{k^4}\left[E(x, k) + F(x, k)\right]$

8. $\displaystyle \int \frac{x \sin x \cos^3 x\, dx}{\Delta^3} = x\frac{k^2 \sin^2 x + k^2 - 2}{k^4 \Delta} + \frac{k'^2}{k^4}\, F(x,k) + \frac{1}{k^4}\, E(x,k)$

2.655 Integrals containing $\sin x^2$ and $\cos x^2$

In integrals containing $\sin x^2$ and $\cos x^2$, it is expedient to make the substitution $x^2 = u$.

1. $\displaystyle \int x^p \sin x^2\, dx = -\frac{x^{p-1}}{2}\cos x^2 + \frac{p-1}{2}\int x^{p-2}\cos x^2\, dx$

2. $\displaystyle \int x^p \cos x^2\, dx = \frac{x^{p-1}}{2}\sin x^2 - \frac{p-1}{2}\int x^{p-2}\sin x^2\, dx$

3. $\displaystyle \int x^n \sin x^2\, dx = (n-1)!!\left\{\sum_{k=1}^{r}(-1)^k\left[\frac{x^{n-4k+3}\cos x^2}{2^{2k-1}(n-4k+3)!!} - \frac{x^{n-4k+1}\sin x^2}{2^{2k}(n-4k+1)!!}\right]\right.$

$\displaystyle \left. + \frac{(-1)^r}{2^{2r}(n-4r-1)!!}\int x^{n-4r}\sin x^2\, dx\right\}$

$\left[r = \left\lfloor\dfrac{n}{4}\right\rfloor\right]$ GU (336)(4a)

4. $\displaystyle \int x^n \cos x^2\, dx = (n-1)!!\left\{\sum_{k=1}^{r}(-1)^{k-1}\left[\frac{x^{n-4k+3}\sin x^2}{2^{2k-1}(n-4k+3)!!} + \frac{x^{n-4k+1}\cos x^2}{2^{2k}(n-4k+1)!!}\right]\right.$

$\displaystyle \left. + \frac{(-1)^r}{2^{2r}(n-4r-1)!!}\int x^{n-4r}\cos x^2\, dx\right\}$

$\left[r = \left\lfloor\dfrac{n}{4}\right\rfloor\right]$ GU (336)(5a)

5. $\displaystyle \int x \sin x^2\, dx = -\frac{\cos^2 x}{2}$

6. $\displaystyle \int x \cos x^2\, dx = -\frac{\sin^2 x}{2}$

7. $\displaystyle \int x^2 \sin x^2\, dx = -\frac{x}{2}\cos x^2 + \frac{1}{2}\sqrt{\frac{\pi}{2}}\, C(x)$

8. $\displaystyle \int x^2 \cos x^2\, dx = \frac{x}{2}\sin x^2 - \frac{1}{2}\sqrt{\frac{\pi}{2}}\, S(x)$

9. $\displaystyle \int x^3 \sin x^2\, dx = -\frac{x^2}{2}\cos x^2 + \frac{1}{2}\sin x^2$

10. $\displaystyle \int x^3 \cos x^2\, dx = \frac{x^2}{2}\sin x^2 + \frac{1}{2}\cos x^2$

2.66 Combinations of trigonometric functions and exponentials

2.661 $\displaystyle\int e^{ax} \sin^p x \cos^q x \, dx = \frac{1}{a^2 + (p+q)^2} \left\{ e^{ax} \sin^p x \cos^{q-1} x \left[a \cos x + (p+q) \sin x \right] \right.$

$$- pa \int e^{ax} \sin^{p-1} x \cos^{q-1} x \, dx + (q-1)(p+q) \left. \int e^{ax} \sin^p x \cos^{q-2} x \, dx \right\}$$

TI (523)

$$= \frac{1}{a^2 + (p+q)^2} \left\{ e^{ax} \sin^{p-1} x \cos^q x \left[a \sin x - (p+q) \cos x \right] \right.$$

$$+ qa \int e^{ax} \sin^{p-1} x \cos^{q-1} x \, dx + (p-1)(p+q) \left. \int e^{ax} \sin^{p-2} x \cos^q x \, dx \right\}$$

TI (524)

$$= \frac{1}{a^2 + (p+q)^2} \left\{ e^{ax} \sin^{p-1} x \cos^{q-1} x \left[a \sin x \cos x + q \sin^2 x - p \cos^2 x \right] \right.$$

$$+ q(q-1) \int e^{ax} \sin^p x \cos^{q-2} x \, dx + p(p-1) \left. \int e^{ax} \sin^{p-2} x \cos^q x \, dx \right\}$$

TI (525)

$$= \frac{1}{a^2 + (p+q)^2} \left\{ e^{ax} \sin^{p-1} x \cos^{q-1} x \left(a \sin x \cos x + q \sin^2 x - p \cos^2 x \right) \right.$$

$$+ q(q-1) \int e^{ax} \sin^{p-2} x \cos^{q-2} x \, dx$$

$$- (q-p)(p+q-1) \left. \int e^{ax} \sin^{p-2} x \cos^q x \, dx \right\}$$

TI (526)

$$= \frac{1}{a^2 + (p+q)^2} \left[e^{ax} \sin^{p-1} x \cos^{q-1} x \left(a \sin x \cos x + q \sin^2 x - p \cos^2 x \right) \right.$$

$$+ p(p-1) \int e^{ax} \sin^{p-2} x \cos^{q-2} x \, dx$$

$$+ (q-p)(p+q-1) \left. \int e^{ax} \sin^p x \cos^{q-2} x \, dx \right]$$

GU (334)(1a)

For $p = m$ and $q = n$ even integers, the integral $\displaystyle\int e^{ax} \sin^m x \cos^n x \, dx$ can be reduced by means of these formulas to the integral $\displaystyle\int e^{ax} \, dx$. However, when only m or only n is even, they can be reduced to integrals of the form $\displaystyle\int e^{ax} \cos^n x \, dx$ or $\displaystyle\int e^{ax} \sin^m x \, dx$ respectively.

2.662

1. $\displaystyle\int e^{ax} \sin^n bx \, dx = \frac{1}{a^2 + n^2 b^2} \left[(a \sin bx - nb \cos bx) e^{ax} \sin^{n-1} bx \right.$

$$+ n(n-1)b^2 \left. \int e^{ax} \sin^{n-2} bx \, dx \right]$$

2. $\displaystyle\int e^{ax} \cos^n bx \, dx = \frac{1}{a^2 + n^2 b^2} \left[(a \cos bx + nb \sin bx) \, e^{ax} \cos^{n-1} bx \right.$

$$\left. + \; n(n-1)b^2 \int e^{ax} \cos^{n-2} bx \, dx \right]$$

3. $\displaystyle\int e^{ax} \sin^{2m} bx \, dx$

$$= \sum_{k=0}^{m-1} \frac{(2m)! \, b^{2k} e^{ax} \sin^{2m-2k-1} bx}{(2m-2k)! \, \left[a^2 + (2m)^2 b^2\right] \left[a^2 + (2m-2)^2 b^2\right] \cdots \left[a^2 + (2m-2k)^2 b^2\right]}$$

$$\times \left[a \sin bx - (2m-2k)b \cos bx \right] + \frac{(2m)! \, b^{2m} e^{ax}}{\left[a^2 + (2m)^2 b^2\right] \left[a^2 + (2m-2)^2 b^2\right] \cdots \left[a^2 + 4b^2\right] a}$$

$$= \binom{2m}{m} \frac{e^{ax}}{2^{2m} a} + \frac{e^{ax}}{2^{2m-1}} \sum_{k=1}^{m} (-1)^k \binom{2m}{m-k} \frac{1}{a^2 + 4b^2 k^2} \left(a \cos 2bkx + 2bk \sin 2bkx \right)$$

4. $\displaystyle\int e^{ax} \sin^{2m+1} bx \, dx$

$$= \sum_{k=0}^{m} \frac{(2m+1)! \, b^{2k} e^{ax} \sin^{2m-2k} bx \, \left[a \sin bx - (2m-2k+1)b \cos bx \right]}{(2m-2k+1)! \, \left[a^2 + (2m+1)^2 b^2\right] \left[a^2 + (2m-1)^2 b^2\right] \cdots \left[a^2 + (2m-2k+1)^2 b^2\right]}$$

$$= \frac{e^{ax}}{2^{2m}} \sum_{k=0}^{m} \frac{(-1)^k}{a^2 + (2k+1)^2 b^2} \binom{2m+1}{m-k} \left[a \sin(2k+1)bx - (2k+1)b \cos(2k+1)bx \right]$$

5.8 $\displaystyle\int e^{ax} \cos^{2m} bx \, dx = \sum_{k=0}^{m-1} \frac{(2m)! \, b^{2k} e^{ax} \cos^{2m-2k-1} bx \, \left[a \cos bx + (2m-2k)b \sin bx \right]}{(2m-2k)! \, \left[a^2 + (2m)^2 b^2\right] \left[a^2 + (2m-2)^2 b^2\right] \cdots \left[a^2 + (2m-2k)^2 b^2\right]}$

$$+ \frac{(2m)! \, b^{2m} e^{ax}}{\left[a^2 + (2m)^2 b^2\right] \left[a^2 + (2m-2)^2 b^2\right] \cdots \left[a^2 + 4b^2\right] a}$$

$$= \binom{2m}{m} \frac{e^{ax}}{2^{2m} a} + \frac{e^{ax}}{2^{2m-1}} \sum_{k=1}^{m} \binom{2m}{m-k} \frac{1}{a^2 + 4b^2 k^2} \left[a \cos 2kbx + 2kb \sin 2kbx \right]$$

6. $\displaystyle\int e^{ax} \cos^{2m+1} bx \, dx$

$$= \sum_{k=0}^{m} \frac{(2m+1)! \, b^{2k} e^{ax} \cos^{2m-2k} bx}{(2m-2k+1)! \, \left[a^2 + (2m-1)^2 b^2\right] \cdots \left[a^2 + (2m-2k+1)^2 b^2\right]}$$

$$= \frac{e^{ax}}{2^{2m}} \sum_{k=0}^{m} \binom{2m+1}{m-k} \frac{1}{a^2 + (2k+1)^2 b^2} \left[a \cos(2k+1)bx + (2k+1)b \sin(2k+1)bx \right]$$

2.663

1. $\displaystyle\int e^{ax} \sin bx \, dx = \frac{e^{ax} \left(a \sin bx - b \cos bx \right)}{a^2 + b^2}$

2. $\displaystyle\int e^{ax} \sin^2 bx \, dx = \frac{e^{ax} \sin bx \left(a \sin bx - 2b \cos bx \right)}{4b^2 + a^2} + \frac{2b^2 e^{ax}}{\left(4b^2 + a^2\right) a}$

$$= \frac{e^{ax}}{2a} - \frac{e^{ax}}{a^2 + 4b^2} \left(\frac{a}{2} \cos 2bx + b \sin 2bx \right)$$

3. $$\int e^{ax} \cos bx \, dx = \frac{e^{ax} \left(a \cos bx + b \sin bx\right)}{a^2 + b^2}$$

4. $$\int e^{ax} \cos^2 bx \, dx = \frac{e^{ax} \cos bx \left(a \cos bx + 2b \sin bx\right)}{4b^2 + a^2} + \frac{2b^2 e^{ax}}{\left(4b^2 + a^2\right) a}$$

$$= \frac{e^{ax}}{2a} + \frac{e^{ax}}{a^2 + 4b^2} \left(\frac{a}{2} \cos 2bx + b \sin 2bx\right)$$

2.664

1. $$\int e^{ax} \sin bx \cos cx \, dx = \frac{e^{ax}}{2} \left[\frac{a \sin(b+c)x - (b+c)\cos(b+c)x}{a^2 + (b+c)^2}\right.$$

$$\left. + \frac{a \sin(b-c)x - (b-c)\cos(b-c)x}{a^2 + (b-c)^2}\right]$$

GU (334)(6b)

2. $$\int e^{ax} \sin^2 bx \cos cx \, dx = \frac{e^{ax}}{4} \left[2\frac{a \cos cx + c \sin cx}{a^2 + c^2} - \frac{a \cos(2b+c)x + (2b+c)\sin(2b+c)x}{a^2 + (2b+c)^2}\right.$$

$$\left. - \frac{a \cos(2b-c)x + (2b-c)\sin(2b-c)x}{a^2 + (2b-c)^2}\right]$$

GU (334)(6c)

3. $$\int e^{ax} \sin bx \cos^2 cx \, dx = \frac{e^{ax}}{4} \left[2\frac{a \sin bx - b \cos bx}{a^2 + b^2} + \frac{a \sin(b+2c)x - (b+2c)\cos(b+2c)x}{a^2 + (b+2c)^2}\right.$$

$$\left. + \frac{a \sin(b-2c)x - (b-2c)\cos(b-2c)x}{a^2 + (b-2c)^2}\right]$$

GU (334)(6d)

2.665

1. $$\int \frac{e^{ax} \, dx}{\sin^p bx} = -\frac{e^{ax} \left[a \sin bx + (p-2)b \cos bx\right]}{(p-1)(p-2)b^2 \sin^{p-1} bx} + \frac{a^2 + (p-2)^2 b^2}{(p-1)(p-2)b^2} \int \frac{e^{ax} \, dx}{\sin^{p-2} bx}$$ TI (530)a

2. $$\int \frac{e^{ax} \, dx}{\cos^p bx} = -\frac{e^{ax} \left[a \cos bx - (p-2)b \sin bx\right]}{(p-1)(p-2)b^2 \cos^{p-1} bx} + \frac{a^2 + (p-2)^2 b^2}{(p-1)(p-2)b^2} \int \frac{e^{ax} \, dx}{\cos^{p-2} bx}$$ TI (529)a

By successive applications of formulas **2.665** for p a natural number, we obtain integrals of the form
$$\int \frac{e^{ax} \, dx}{\sin bx}, \quad \int \frac{e^{ax} \, dx}{\sin^2 bx}, \quad \int \frac{e^{ax} \, dx}{\cos bx}, \quad \int \frac{e^{ax} \, dx}{\cos^2 bx}, \text{ which are not expressible in terms of a finite combination}$$
of elementary functions.

2.666

1. $$\int e^{ax} \tan^p x \, dx = \frac{e^{ax}}{p-1} \tan^{p-1} x - \frac{a}{p-1} \int e^{ax} \tan^{p-1} x \, dx - \int e^{ax} \tan^{p-2} x \, dx$$ TI (527)

2. $$\int e^{ax} \cot^p x \, dx = -\frac{e^{ax} \cot^{p-1} x}{p-1} + \frac{a}{p-1} \int e^{ax} \cot^{p-1} x \, dx - \int e^{ax} \cot^{p-2} x \, dx$$ TI (528)

3. $$\int e^{ax} \tan x \, dx = \frac{e^{ax} \tan x}{a} - \frac{1}{a} \int \frac{e^{ax} \, dx}{\cos^2 x}$$ (see remark following **2.665**)

4. $\int e^{ax} \tan^2 x\, dx = \dfrac{e^{ax}}{a}\left(a\tan x - 1\right) - a\int e^{ax}\tan x\, dx$ (see **2.666** 3)   TI 355

5. $\int e^{ax}\cot x\, dx = \dfrac{e^{ax}\cot x}{a} + \dfrac{1}{a}\int \dfrac{e^{ax}\, dx}{\sin^2 x}$ (see remark following **2.665**)

6. $\int e^{ax}\cot^2 x\, dx = -\dfrac{e^{ax}}{a}\left(a\cot x + 1\right) + a\int e^{ax}\cot x\, dx$

                                                 (see **2.666** 5)

Integrals of type $\int R\left(x, e^{ax}, \sin bx, \cos cx\right)\, dx$

Notation: $\sin t = -\dfrac{b}{\sqrt{a^2 + b^2}};\quad \cos t = \dfrac{a}{\sqrt{a^2 + b^2}}.$

2.667

1. $\int x^p e^{ax} \sin bx\, dx = \dfrac{x^p e^{ax}}{a^2 + b^2}\left(a\sin bx - b\cos bx\right) - \dfrac{p}{a^2 + b^2}\int x^{p-1} e^{ax}\left(a\sin bx - b\cos bx\right)\, dx$

      $= \dfrac{x^p e^{ax}}{\sqrt{a^2 + b^2}}\sin(bx + t) - \dfrac{p}{\sqrt{a^2 + b^2}}\int x^{p-1}e^{ax}\sin(bx + t)\, dx$

2. $\int x^p e^{ax}\cos bx\, dx = \dfrac{x^p e^{ax}}{a^2 + b^2}\left(a\cos bx + b\sin bx\right) - \dfrac{p}{a^2 + b^2}\int x^{p-1}e^{ax}\left(a\cos bx + b\sin bx\right)\, dx$

      $= \dfrac{x^p e^{ax}}{\sqrt{a^2 + b^2}}\cos(bx + t) - \dfrac{p}{\sqrt{a^2 + b^2}}\int x^{p-1}e^{ax}\cos(bx + t)\, dx$

3. $\int x^n e^{ax}\sin bx\, dx = e^{ax}\displaystyle\sum_{k=1}^{n+1}\dfrac{(-1)^{k+1}n!\, x^{n-k+1}}{(n-k+1)!\,(a^2 + b^2)^{k/2}}\sin(bx + kt)$

4. $\int x^n e^{ax}\cos bx\, dx = e^{ax}\displaystyle\sum_{k=1}^{n+1}\dfrac{(-1)^{k+1}n!\, x^{n-k+1}}{(n-k+1)!\,(a^2 + b^2)^{k/2}}\cos(bx + kt)$

5. $\int x e^{ax}\sin bx\, dx = \dfrac{e^{ax}}{a^2 + b^2}\left[\left(ax - \dfrac{a^2 - b^2}{a^2 + b^2}\right)\sin bx - \left(bx - \dfrac{2ab}{a^2 + b^2}\right)\cos bx\right]$

6. $\int x e^{ax}\cos bx\, dx = \dfrac{e^{ax}}{a^2 + b^2}\left[\left(ax - \dfrac{a^2 - b^2}{a^2 + b^2}\right)\cos bx + \left(bx - \dfrac{2ab}{a^2 + b^2}\right)\sin bx\right]$

7. $\int x^2 e^{ax}\sin bx\, dx = \dfrac{e^{ax}}{a^2 + b^2}\left\{\left[ax^2 - \dfrac{2\left(a^2 - b^2\right)}{a^2 + b^2}x + \dfrac{2a\left(a^2 - 3b^2\right)}{\left(a^2 + b^2\right)^2}\right]\sin bx\right.$

                        $\left. -\left[bx^2 - \dfrac{4ab}{a^2 + b^2}x + \dfrac{2b\left(3a^2 - b^2\right)}{\left(a^2 + b^2\right)^2}\right]\cos bx\right\}$

8. $\displaystyle\int x^2 e^{ax} \cos bx\, dx = \frac{e^{ax}}{a^2+b^2} \left\{ \left[ax^2 - \frac{2\left(a^2-b^2\right)}{a^2+b^2}x + \frac{2a\left(a^2-3b^2\right)}{\left(a^2+b^2\right)^2} \right] \cos bx \right.$

$$\left. + \left[bx^2 - \frac{4ab}{a^2+b^2}x + \frac{2b\left(3a^2-b^2\right)}{\left(a^2+b^2\right)^2} \right] \sin bx \right\}$$

GU (335), MZ 274-275

2.67 Combinations of trigonometric and hyperbolic functions

2.671

1. $\displaystyle\int \sinh(ax+b)\sin(cx+d)\, dx = \frac{a}{a^2+c^2}\cosh(ax+b)\sin(cx+d)$

$$-\frac{c}{a^2+c^2}\sinh(ax+b)\cos(cx+d)$$

2. $\displaystyle\int \sinh(ax+b)\cos(cx+d)\, dx = \frac{a}{a^2+c^2}\cosh(ax+b)\cos(cx+d)$

$$+\frac{c}{a^2+c^2}\sinh(ax+b)\sin(cx+d)$$

3. $\displaystyle\int \cosh(ax+b)\sin(cx+d)\, dx = \frac{a}{a^2+c^2}\sinh(ax+b)\sin(cx+d)$

$$-\frac{c}{a^2+c^2}\cosh(ax+b)\cos(cx+d)$$

4. $\displaystyle\int \cosh(ax+b)\cos(cx+d)\, dx = \frac{a}{a^2+c^2}\sinh(ax+b)\cos(cx+d)$

$$+\frac{c}{a^2+c^2}\cosh(ax+b)\sin(cx+d)$$

GU (354)(1)

2.672

1. $\displaystyle\int \sinh x \sin x\, dx = \frac{1}{2}\left(\cosh x \sin x - \sinh x \cos x\right)$

2. $\displaystyle\int \sinh x \cos x\, dx = \frac{1}{2}\left(\cosh x \cos x + \sinh x \sin x\right)$

3. $\displaystyle\int \cosh x \sin x\, dx = \frac{1}{2}\left(\sinh x \sin x - \cosh x \cos x\right)$

4. $\displaystyle\int \cosh x \cos x\, dx = \frac{1}{2}\left(\sinh x \cos x + \cosh x \sin x\right)$

2.673

1. $\int \sinh^{2m}(ax+b)\sin^{2n}(cx+d)\,dx$

$$= \frac{(-1)^m}{2^{2m+2n}}\binom{2m}{m}\binom{2n}{n}x + \frac{(-1)^{m+n}}{2^{2m+2n-1}}\binom{2m}{m}\sum_{k=0}^{n-1}\frac{(-1)^k}{(2n-2k)c}\binom{2n}{k}\sin[(2n-2k)(cx+d)]$$

$$+\frac{(-1)^n}{2^{2m+2n-2}}\sum_{j=0}^{m-1}\sum_{k=0}^{n-1}\frac{(-1)^{j+k}\binom{2m}{j}\binom{2n}{k}}{(2m-2j)^2a^2+(2n-2k)^2c^2}$$

$$\times \left\{(2m-2j)a\sinh[(2m-2j)(ax+b)]\cos[(2n-2k)(cx+d)]\right\}$$

$$+(2-2k)c\cosh[(2m-2j)(ax+b)]\sin[(2n-2k)(cx+d)]$$

GU (354)(3a)

2. $\int \sinh^{2m}(ax+b)\sin^{2n-1}(cx+d)\,dx$

$$= \frac{(-1)^{m+n}}{2^{2m+2n-2}}\binom{2m}{m}\sum_{k=0}^{n-1}\frac{(-1)^k}{(2n-2k-1)c}\binom{2n-1}{k}\cos[(2n-2k-1)(cx+d)]$$

$$+\frac{(-1)^{n-1}}{2^{2m+2n-3}}\sum_{j=0}^{m-1}\sum_{k=0}^{n-1}\frac{(-1)^{j+k}\binom{2m}{j}\binom{2n-1}{k}}{(2m-2j)^2a^2+(2n-2k-1)^2c^2}$$

$$\times \left\{(2m-2j)a\sinh[(2m-2j)(ax+b)]\sin[(2n-2k-1)(cx+d)]\right\}$$

$$-(2n-2k-1)c\cosh[(2m-2j)(ax+b)]\cos[(2n-2k-1)(cx+d)]$$

GU (354)(3b)

3. $\int \sinh^{2m-1}(ax+b)\sin^{2n}(cx+d)\,dx$

$$= \frac{\binom{2n}{n}}{2^{2m+2n-2}}\sum_{j=0}^{m-1}\frac{(-1)^j\binom{2m-1}{j}}{(2m-2j-1)a}\cosh[(2m-2j-1)(ax+d)]$$

$$+\frac{(-1)^n}{2^{2m+2n-3}}\sum_{j=0}^{m-1}\sum_{k=0}^{n-1}\frac{(-1)^{j+k}\binom{2m-1}{j}\binom{2n}{k}}{(2m-2j-1)^2a^2+(2n-2k)^2c^2}$$

$$\times \left\{(2m-2j-1)a\cosh[(2m-2j-1)(ax+b)]\cos[(2n-2k)(cx+d)]\right\}$$

$$+(2n-2k)c\sinh[(2m-2j-1)(ax+b)]\sin[(2n-2k)(cx+d)]$$

GU (354)(3c)

4. $\displaystyle \int \sinh^{2m-1}(ax+b)\sin^{2n-1}(cx+d)\,dx$

$$= \frac{(-1)^{n-1}}{2^{2m-2n-4}} \sum_{j=0}^{m-1}\sum_{k=0}^{n-1} \frac{(-1)^{j+k}\binom{2m-1}{j}\binom{2n-1}{k}}{(2m-2j-1)^2 a^2 + (2n-2k-1)^2 c^2}$$

$$\times \left\{ (2m-2j-1)a\cosh[(2m-2j-1)(ax+b)]\sin[(2n-2k-1)(cx+d)] \right\}$$

$$-(2n-2k-1)c\sinh[(2m-2j-1)(ax+b)]\cos[(2n-2k-1)(cx+d)]$$

GU (354)(3d)

5. $\displaystyle \int \sinh^{2m}(ax+b)\cos^{2n}(cx+d)\,dx$

$$= \frac{(-1)^m}{2^{2m+2n}}\binom{2m}{m}\binom{2n}{n}x + \frac{\binom{2n}{n}}{2^{2m+2n-1}}\sum_{j=0}^{m-1}\frac{(-1)^j\binom{2m}{j}}{(2m-2j)a}\sinh[(2m-2j)(ax+b)]$$

$$+ \frac{(-1)^m\binom{2m}{m}}{2^{2m+2n-1}}\sum_{k=0}^{n-1}\frac{\binom{2n}{k}}{(2n-2k)c}\sin[(2n-2k)(cx+d)]$$

$$+ \frac{1}{2^{2m+2n-2}}\sum_{j=0}^{m-1}\sum_{k=0}^{n-1}\frac{(-1)^j\binom{2m}{j}\binom{2n}{k}}{(2m-2j)^2 a^2 + (2n-2k)^2 c^2}$$

$$\times \left\{ (2m-2j)a\sinh[(2m-2j)(ax+b)]\cos[(2n-2k)(cx+d)] \right\}$$

$$+(2-2k)c\cosh[(2m-2j)(ax+b)]\sin[(2n-2k)(cx+d)]$$

GU (354)(4a)

6. $\displaystyle \int \sinh^{2m}(ax+b)\cos^{2n-1}(cx+d)\,dx$

$$= \frac{(-1)^m\binom{2m}{m}}{2^{2m+2n-2}}\sum_{k=0}^{n-1}\frac{\binom{2n-1}{k}}{(2n-2k-1)c}\sin[(2n-2k-2)(cx+d)]$$

$$+ \frac{1}{2^{2m+2n-3}}\sum_{j=0}^{m-1}\sum_{k=0}^{n-1}\frac{(-1)^j\binom{2m}{j}\binom{2-1}{k}}{(2m-2j)^2 a^2 + (2n-2k-1)^2 c^2}$$

$$\times \left\{ (2m-2j)a\sinh[(2m-2j)(ax+b)]\cos[(2n-2k-1)(cx+d)] \right\}$$

$$+(2n-2k-1)c\cosh[(2m-2j)(ax+b)]\sin[(2n-2k-1)(cx+d)]$$

GU (354)(4a)

7. $\displaystyle\int \sinh^{2m-1}(ax+b)\cos^{2n}(cx+d)\,dx$

$$= \frac{\binom{2n}{n}}{2^{2m+2n-2}} \sum_{j=0}^{m-1} \frac{(-1)^j \binom{2m-1}{j}}{(2m-2j-1)a} \cosh[(2m-2j-1)(ax+d)]$$

$$+ \frac{1}{2^{2m-2n-3}} \sum_{j=0}^{m-1}\sum_{k=0}^{n-1} \frac{(-1)^j \binom{2m}{j}\binom{2n}{k}}{(2m-2j-1)^2 a^2 + (2n-2k)^2 c^2}$$

$$\times \{(2m-2j-1)a\cosh[(2m-2j-1)(ax+b)]\cos[(2n-2k)(cx+d)]\}$$

$$+(2n-2k)c\sinh[(2m-2j-1)(ax+b)]\sin[(2n-2k)(cx+d)]$$

GU (354)(4b)

8. $\displaystyle\int \sinh^{2m-1}(ax+b)\cos^{2n-1}(cx+d)\,dx$

$$= \frac{1}{2^{2m+2n-4}} \sum_{j=0}^{m-1}\sum_{k=0}^{n-1} \frac{(-1)^j \binom{2m-1}{j}\binom{2n-1}{k}}{(2m-2j-1)^2 a^2 + (2n-2k-1)^2 c^2}$$

$$\times \{(2m-2j-1)a\cosh[(2m-2j-1)(ax+b)]\cos[(2n-2k-1)(cx+d)]\}$$

$$+(2n-2k-1)c\sinh[(2m-2j-1)(ax+b)]\sin[(2n-2k-1)(cx+d)]$$

GU (354)(4b)

9. $\displaystyle\int \cosh^{2m}(ax+b)\sin^{2n}(cx+d)\,dx$

$$= \frac{\binom{2m}{m}\binom{2n}{n}}{2^{2m+2n}} x + \frac{(-1)^n \binom{2m}{m}}{2^{2m+2n-1}} \sum_{k=0}^{m-1} \frac{(-1)^k \binom{2n}{k}}{(2n-2k)c} \sin[(2n-2k)(cx+d)]$$

$$+ \frac{\binom{2n}{n}}{2^{2m+2n-1}} \sum_{j=0}^{m-1} \frac{\binom{2m}{j}}{(2m-2j)a} \sinh[(2m-2j)(ax+b)]$$

$$+ \frac{(-1)^n}{2^{2m+2n-2}} \sum_{j=0}^{m-1}\sum_{k=0}^{n-1} \frac{(-1)^k \binom{2m}{j}\binom{2n}{k}}{(2m-2j)^2 a^2 + (2n-2k)^2 c^2}$$

$$\times \{(2m-2j)a\sinh[(2m-2j)(ax+b)]\cos[(2n-2k)(cx+d)]\}$$

$$+(2n-2k)c\cosh[(2m-2j)(ax+b)]\sin[(2n-2k)(cx+d)]$$

GU (354)(5a)

10. $\displaystyle\int \cosh^{2m-1}(ax+b)\sin^{2n}(cx+d)\,dx$

$$= \frac{\binom{2n}{n}}{2^{2m+2n-2}} \sum_{j=0}^{m-1} \frac{\binom{2m-1}{j}}{(2m-2j-1)a} \sinh[(2m-2j-1)(ax+b)]$$

$$+ \frac{(-1)^n}{2^{2m+2n-3}} \sum_{j=0}^{m-1}\sum_{k=0}^{n-1} \frac{(-1)^k \binom{2m-1}{j}\binom{2n}{k}}{(2m-2j-1)^2 a^2 + (2n-2k)^2 c^2}$$

$$\times \{(2m-2j-1)a\sinh[(2m-2j-1)(ax+b)]\cos[(2n-2k)(cx+d)]\}$$

$$+(2n-2k)c\cosh[(2m-2j-1)(ax+b)]\sin[(2n-2k)(cx+d)]$$

$$\text{GU (354)(5a)}$$

11. $\displaystyle\int \cosh^{2m}(ax+b)\sin^{2n-1}(cx+d)\,dx$

$$= \frac{(-1)^{n-1}\binom{2m}{m}}{2^{2m+2n-2}} \sum_{k=0}^{n-1} \frac{(-1)^{k+1}\binom{2n-1}{k}}{(2n-2k-1)c} \cos[(2n-2k-1)(cx+d)]$$

$$+ \frac{(-1)^{n-1}}{2^{2m+2n-3}} \sum_{j=0}^{m-1}\sum_{k=0}^{n-1} \frac{(-1)^k \binom{2m}{j}\binom{2n-1}{k}}{(2m-2j)^2 a^2 + (2n-2k-1)^2 c^2}$$

$$\times \{(2m-2j)a\sinh[(2m-2j)(ax+b)]\sin[(2n-2k-1)(cx+d)]\}$$

$$-(2n-2k-1)c\cosh[(2m-2j)(ax+b)]\cos[(2n-2k-1)(cx+d)]$$

$$\text{GU (354)(5b)}$$

12. $\displaystyle\int \cosh^{2m-1}(ax+b)\sin^{2n-1}(cx+d)\,dx$

$$= \frac{(-1)^{n-1}}{2^{2m+2n-4}} \sum_{j=0}^{m-1}\sum_{k=0}^{n-1} \frac{(-1)^k \binom{2m-1}{j}\binom{2n-1}{k}}{(2m-2j-1)^2 a^2 + (2n-2k-1)^2 c^2}$$

$$\times \{(2m-2j-1)a\sinh[(2m-2j-1)(ax+b)]\sin[(2n-2k-1)(cx+d)]\}$$

$$-(2n-2k-1)c\cosh[(2m-2j-1)(ax+b)]\cos[(2n-2k-1)(cx+d)]$$

$$\text{GU (354)(5b)}$$

13. $\displaystyle\int \cosh^{2m}(ax+b)\cos^{2n}(cx+d)\,dx$

$$= \frac{\binom{2m}{m}\binom{2n}{n}}{2^{2m+2n}}x + \frac{\binom{2m}{m}}{2^{2m+2n-1}}\sum_{k=0}^{n-1}\frac{\binom{2}{k}}{(2n-2k)c}\sin[(2n-2k)(cx+d)]$$

$$+\frac{\binom{2n}{n}}{2^{2m+2n-1}}\sum_{j=0}^{m-1}\frac{\binom{2m}{j}}{(2m-2j)a}\sinh[(2m-2j)(ax+b)]$$

$$+\frac{1}{2^{2m+2n-2}}\sum_{j=0}^{m-1}\sum_{k=0}^{n-1}\frac{\binom{2m}{j}\binom{2n}{k}}{(2m-2j)^2a^2+(2n-2k)^2c^2}$$

$$\times\{(2m-2j)a\sinh[(2m-2j)(ax+b)]\cos[(2n-2k)(cx+d)]\}$$

$$+(2n-2k)c\cosh[(2m-2j)(ax+b)]\sin[(2n-2k)(cx+d)]$$

GU (354)(6)

14. $\displaystyle\int \cosh^{2m-1}(ax+b)\cos^{2n}(cx+d)\,dx$

$$= \frac{\binom{2n}{n}}{2^{2m+2n-2}}\sum_{j=0}^{m-1}\frac{\binom{2m-1}{j}}{(2m-2j-1)a}\sinh[(2m-2j-1)(ax+b)]$$

$$+\frac{1}{2^{2m+2n-3}}\sum_{j=0}^{m-1}\sum_{k=0}^{n-1}\frac{\binom{2m-1}{j}\binom{2n}{k}}{(2m-2j-1)^2a^2+(2n-2k)^2c^2}$$

$$\times\{(2m-2j-1)a\sinh[(2m-2j-1)(ax+b)]\cos[(2n-2k)(cx+d)]\}$$

$$+(2n-2k)c\cosh[(2m-2j-1)(ax+b)]\sin[(2n-2k)(cx+d)]$$

GU (354)(6)

15. $\displaystyle\int \cosh^{2m}(ax+b)\cos^{2n-1}(cx+d)\,dx$

$$= \frac{\binom{2m}{m}}{2^{2m+2n-2}}\sum_{k=0}^{n-1}\frac{\binom{2n-1}{k}}{(2n-2k-1)c}\sin[(2n-2k-1)(cx+d)]$$

$$+\frac{1}{2^{2m+2n-3}}\sum_{j=0}^{m-1}\sum_{k=0}^{n-1}\frac{\binom{2m}{j}\binom{2n-1}{k}}{(2m-2j)^2a^2+(2n-2k-1)^2c^2}$$

$$\times\{(2m-2j)a\sinh[(2m-2j)(ax+b)]\cos[(2n-2k-1)(cx+d)]\}$$

$$+(2n-2k-1)c\cosh[(2m-2j)(ax+b)]\sin[(2n-2k-1)(cx+d)]$$

GU (354)(6)

16. $\displaystyle\int \cosh^{2m-1}(ax+b)\cos^{2n-1}(cx+d)\,dx$

$$= \frac{1}{2^{2m+2n-4}}\sum_{j=0}^{m-1}\sum_{k=0}^{n-1}\frac{\binom{2m-1}{j}\binom{2n-1}{k}}{(2m-2j-1)^2 a^2 + (2n-2k-1)^2 c^2}$$

$$\times \{(2m-2j-1)a\sinh[(2m-2j-1)(ax+b)]\cos[(2n-2k-1)(cx+d)]\}$$

$$+(2n-2k-1)c\cosh[(2m-2j-1)(ax+b)]\sin[(2n-2k-1)(cx+d)]$$

GU (354)(6)

2.674

1. $\displaystyle\int e^{ax}\sinh bx\sin cx\,dx = \frac{e^{(a+b)x}}{2\left[(a+b)^2+c^2\right]}[(a+b)\sin cx - c\cos cx]$

$$-\frac{e^{(a-b)x}}{2\left[(a-b)^2+c^2\right]}[(a-b)\sin cx - c\cos cx]$$

2. $\displaystyle\int e^{ax}\sinh bx\cos cx\,dx = \frac{e^{(a+b)x}}{2\left[(a+b)^2+c^2\right]}[(a+b)\cos cx + c\sin cx]$

$$-\frac{e^{(a-b)x}}{2\left[(a-b)^2+c^2\right]}[(a-b)\cos cx + c\sin cx]$$

3. $\displaystyle\int e^{ax}\cosh bx\sin cx\,dx = \frac{e^{(a+b)x}}{2\left[(a+b)^2+c^2\right]}[(a+b)\sin cx - c\cos cx]$

$$+\frac{e^{(a-b)x}}{2\left[(a-b)^2+c^2\right]}[(a-b)\sin cx - c\cos cx]$$

4. $\displaystyle\int e^{ax}\cosh bx\cos cx\,dx = \frac{e^{(a+b)x}}{2\left[(a+b)^2+c^2\right]}[(a+b)\cos cx + c\sin cx]$

$$+\frac{e^{(a-b)x}}{2\left[(a-b)^2+c^2\right]}[(a-b)\cos cx + c\sin cx]$$

MZ 379

2.7 Logarithms and Inverse-Hyperbolic Functions

2.71 The logarithm

2.711 $\displaystyle\int \ln^m x\,dx = x\ln^m x - m\int \ln^{m-1} x\,dx$

$$= \frac{x}{m+1}\sum_{k=0}^{m}(-1)^k(m+1)m(m-1)\cdots(m-k+1)\ln^{m-k} x$$

$(m > 0)$ TI (603)

2.72–2.73 Combinations of logarithms and algebraic functions

2.721

1. $\int x^n \ln^m x \, dx = \dfrac{x^{n+1} \ln^m x}{n+1} - \dfrac{m}{n+1} \int x^n \ln^{m-1} x \, dx \qquad (\text{see } \mathbf{2.722})$

For $n = -1$

2. $\int \dfrac{\ln^m x \, dx}{x} = \dfrac{\ln^{m+1} x}{m+1}$

For $n = -1$ and $m = -1$

3. $\int \dfrac{dx}{x \ln x} = \ln(\ln x)$

2.722 $\int x^n \ln^m x \, dx = \dfrac{x^{n+1}}{m+1} \displaystyle\sum_{k=0}^{m} (-1)^k (m+1)m(m-1)\cdots(m-k+1)\dfrac{\ln^{m-k} x}{(n+1)^{k+1}}$ TI (604)

2.723

1. $\int x^n \ln x \, dx = x^{n+1} \left[\dfrac{\ln x}{n+1} - \dfrac{1}{(n+1)^2} \right]$ TI 375

2. $\int x^n \ln^2 x \, dx = x^{n+1} \left[\dfrac{\ln^2 x}{n+1} - \dfrac{2 \ln x}{(n+1)^2} + \dfrac{2}{(n+1)^3} \right]$ TI 375

3. $\int x^n \ln^3 x \, dx = x^{n+1} \left[\dfrac{\ln^3 x}{n+1} - \dfrac{3 \ln^2 x}{(n+1)^2} + \dfrac{6 \ln x}{(n+1)^3} - \dfrac{6}{(n+1)^4} \right]$

2.724

1. $\int \dfrac{x^n \, dx}{(\ln x)^m} = -\dfrac{x^{n+1}}{(m-1)(\ln x)^{m-1}} + \dfrac{n+1}{m-1} \int \dfrac{x^n \, dx}{(\ln x)^{m-1}}$

For $m = 1$

2. $\int \dfrac{x^n \, dx}{\ln x} = \mathrm{li}\left(x^{n+1}\right)$

2.725

1. $\int (a+bx)^m \ln x \, dx = \dfrac{1}{(m+1)b} \left[(a+bx)^{m+1} \ln x - \int \dfrac{(a+bx)^{m+1} \, dx}{x} \right]$ TI 374

2. $\int (a+bx)^m \ln x \, dx = \dfrac{1}{(m+1)b} \left[(a+bx)^{m+1} - a^{m+1} \right] \ln x - \displaystyle\sum_{k=0}^{m} \dfrac{\binom{m}{k} a^{m-k} b^k x^{k+1}}{(k+1)^2}$

For $m = -1$ see **2.727** 2.

2.726

1. $\int (a+bx) \ln x \, dx = \left[\dfrac{(a+bx)^2}{2b} - \dfrac{a^2}{2b} \right] \ln x - \left(ax + \dfrac{1}{4}bx^2 \right)$

2. $\int (a+bx)^2 \ln x \, dx = \dfrac{1}{3b} \left[(a+bx)^3 - a^3 \right] \ln x - \left(a^2 x + \dfrac{abx^2}{2} + \dfrac{b^2 x^3}{9} \right)$

3. $\int (a + bx)^3 \ln x \, dx = \frac{1}{4b} \left[(a + bx)^4 - a^4 \right] \ln x - \left(a^3 x + \frac{3}{4} a^2 b x^2 + \frac{1}{3} a b^2 x^3 + \frac{1}{16} b^3 x^4 \right)$

2.727

1.8 $\int \frac{\ln x \, dx}{(a + bx)^m} = \frac{1}{b(m-1)} \left[-\frac{\ln x}{(a + bx)^{m-1}} + \int \frac{dx}{x(a + bx)^{m-1}} \right]$ TI 376

For $m = 1$

2.8 $\int \frac{\ln x \, dx}{a + bx} = \frac{1}{b} \ln x \ln(a + bx) - \frac{1}{b} \int \frac{\ln(a + bx) \, dx}{x}$ (see **2.728** 2)

3. $\int \frac{\ln x \, dx}{(a + bx)^2} = -\frac{\ln x}{b(a + bx)} + \frac{1}{ab} \ln \frac{x}{a + bx}$

4. $\int \frac{\ln x \, dx}{(a + bx)^3} = -\frac{\ln x}{2b(a + bx)^2} + \frac{1}{2ab(a + bx)} + \frac{1}{2a^2 b} \ln \frac{x}{a + bx}$

5. $\int \frac{\ln x \, dx}{\sqrt{a + bx}} = \frac{2}{b} \left\{ (\ln x - 2) \sqrt{a + bx} - 2\sqrt{a} \ln \left[\frac{(a + bx)^{1/2} - a^{1/2}}{x^{1/2}} \right] \right\}$ $[a > 0]$

$= \frac{2}{b} \left\{ (\ln x - 2) \sqrt{a + bx} + 2\sqrt{-a} \arctan \sqrt{\frac{a + bx}{-a}} \right\}$ $[a < 0]$

2.728

1. $\int x^m \ln(a + bx) \, dx = \frac{1}{m + 1} \left[x^{m+1} \ln(a + bx) - b \int \frac{x^{m+1} \, dx}{a + bx} \right]$

2.9 $\int \frac{\ln(a + bx)}{x} = \ln a \ln x + \frac{bx}{a} \Phi \left(-\frac{bx}{a}, 2, 1 \right)$ $[a > 0]$

2.729

1. $\int x^m \ln(a + bx) \, dx = \frac{1}{m + 1} \left[x^{m+1} - \frac{(-a)^{m+1}}{b^{m+1}} \right] \ln(a + bx) + \frac{1}{m + 1} \sum_{k=1}^{m+1} \frac{(-1)^k x^{m-k+2} a^{k-1}}{(m - k + 2) b^{k-1}}$

2. $\int x \ln(a + bx) \, dx = \frac{1}{2} \left[x^2 - \frac{a^2}{b^2} \right] \ln(a + bx) - \frac{1}{2} \left[\frac{x^2}{2} - \frac{ax}{b} \right]$

3. $\int x^2 \ln(a + bx) \, dx = \frac{1}{3} \left[x^3 + \frac{a^3}{b^3} \right] \ln(a + bx) - \frac{1}{3} \left[\frac{x^3}{3} - \frac{ax^2}{2b} + \frac{a^2 x}{b^2} \right]$

4. $\int x^3 \ln(a + bx) \, dx = \frac{1}{4} \left[x^4 - \frac{a^4}{b^4} \right] \ln(a + bx) - \frac{1}{4} \left[\frac{x^4}{4} - \frac{ax^3}{3b} + \frac{a^2 x^2}{2b^2} - \frac{a^3 x}{b^3} \right]$

2.731 $\int x^{2n} \ln \left(x^2 + a^2 \right) dx = \frac{1}{2n + 1} \left\{ x^{2n+1} \ln \left(x^2 + a^2 \right) + (-1)^n 2 a^{2n+1} \arctan \frac{x}{a} \right.$

$\left. - 2 \sum_{k=0}^{n} \frac{(-1)^{n-k}}{2k + 1} a^{2n-2k} x^{2k+1} \right\}$

2.732[7] $\displaystyle\int x^{2n+1} \ln\left(x^2 + a^2\right)\, dx = \frac{1}{2n+2}\left\{\left(x^{2n+2} + (-1)^n a^{2n+2}\right) \ln\left(x^2 + a^2\right)\right.$

$$\left. + \sum_{k=1}^{n+1} \frac{(-1)^{n-k}}{k} a^{2n-2k+2} x^{2k}\right\}$$

2.733

1. $\displaystyle\int \ln\left(x^2 + a^2\right)\, dx = x\ln\left(x^2 + a^2\right) - 2x + 2a\arctan\frac{x}{a}$ DW

2. $\displaystyle\int x\ln\left(x^2 + a^2\right)\, dx = \frac{1}{2}\left[\left(x^2 + a^2\right)\ln\left(x^2 + a^2\right) - x^2\right]$ DW

3. $\displaystyle\int x^2 \ln\left(x^2 + a^2\right)\, dx = \frac{1}{3}\left[x^3 \ln\left(x^2 + a^2\right) - \frac{2}{3}x^3 + 2a^2 x - 2a^3 \arctan\frac{x}{a}\right]$ DW

4. $\displaystyle\int x^3 \ln\left(x^2 + a^2\right)\, dx = \frac{1}{4}\left[\left(x^4 - a^4\right)\ln\left(x^2 + a^2\right) - \frac{x^4}{2} + a^2 x^2\right]$ DW

5. $\displaystyle\int x^4 \ln\left(x^2 + a^2\right)\, dx = \frac{1}{5}\left[x^5 \ln\left(x^2 + a^2\right) - \frac{2}{5}x^5 + \frac{2}{3}a^2 x^3 - 2a^4 x + 2a^5 \arctan\frac{x}{a}\right]$ DW

2.734 $\displaystyle\int x^{2n} \ln\left|x^2 - a^2\right|\, dx$

$$= \frac{1}{2n+1}\left\{x^{2n+1}\ln\left|x^2 - a^2\right| + a^{2n+1}\ln\left|\frac{x+a}{x-a}\right| - 2\sum_{k=0}^{n}\frac{1}{2k+1}a^{2n-2k}x^{2k+1}\right\}$$

2.735 $\displaystyle\int x^{2n+1}\ln\left|x^2 - a^2\right|\, dx = \frac{1}{2n+2}\left\{\left(x^{2n+2} - a^{2n+2}\right)\ln\left|x^2 - a^2\right| - \sum_{k=1}^{n+1}\frac{1}{k}a^{2n-2k+2}x^{2k}\right\}$

2.736

1. $\displaystyle\int \ln\left|x^2 - a^2\right|\, dx = x\ln\left|x^2 - a^2\right| - 2x + a\ln\left|\frac{x+a}{x-a}\right|$ DW

2. $\displaystyle\int x\ln\left|x^2 - a^2\right|\, dx = \frac{1}{2}\left\{\left(x^2 - a^2\right)\ln\left|x^2 - a^2\right| - x^2\right\}$ DW

3. $\displaystyle\int x^2 \ln\left|x^2 - a^2\right|\, dx = \frac{1}{3}\left\{x^3 \ln\left|x^2 - a^2\right| - \frac{2}{3}x^3 - 2a^2 x + a^3\ln\left|\frac{x+a}{x-a}\right|\right\}$ DW

4. $\displaystyle\int x^3 \ln\left|x^2 - a^2\right|\, dx = \frac{1}{4}\left\{\left(x^4 - a^4\right)\ln\left|x^2 - a^2\right| - \frac{x^4}{2} - a^2 x^2\right\}$ DW

5. $\displaystyle\int x^4 \ln\left|x^2 - a^2\right|\, dx = \frac{1}{5}\left\{x^5 \ln\left|x^2 - a^2\right| - \frac{2}{5}x^5 - \frac{2}{3}a^2 x^3 - 2a^4 x + a^5\ln\left|\frac{x+a}{x-a}\right|\right\}$ DW

2.74 Inverse hyperbolic functions

2.741

1. $\displaystyle\int \operatorname{arcsinh}\frac{x}{a}\, dx = x\operatorname{arcsinh}\frac{x}{a} - \sqrt{x^2 + a^2}$ DW

2.
$$\int \operatorname{arccosh} \frac{x}{a}\, dx = x \operatorname{arccosh} \frac{x}{a} - \sqrt{x^2 - a^2} \qquad \left[\operatorname{arccosh} \frac{x}{a} > 0\right] \qquad \text{DW}$$
$$= x \operatorname{arccosh} \frac{x}{a} + \sqrt{x^2 - a^2} \qquad \left[\operatorname{arccosh} \frac{x}{a} < 0\right] \qquad \text{DW}$$

3.
$$\int \operatorname{arctanh} \frac{x}{a}\, dx = x \operatorname{arctanh} \frac{x}{a} + \frac{a}{2} \ln\left(a^2 - x^2\right) \qquad \text{DW}$$

4.
$$\int \operatorname{arccoth} \frac{x}{a}\, dx = x \operatorname{arccoth} \frac{x}{a} + \frac{a}{2} \ln\left(x^2 - a^2\right) \qquad \text{DW}$$

2.742

1.
$$\int x \operatorname{arcsinh} \frac{x}{a}\, dx = \left(\frac{x^2}{2} + \frac{a^2}{4}\right) \operatorname{arcsinh} \frac{x}{a} - \frac{x}{4}\sqrt{x^2 + a^2} \qquad \text{DW}$$

2.
$$\int x \operatorname{arccosh} \frac{x}{a}\, dx = \left(\frac{x^2}{2} - \frac{a^2}{4}\right) \operatorname{arccosh} \frac{x}{a} - \frac{x}{4}\sqrt{x^2 - a^2} \qquad \left[\operatorname{arccosh} \frac{x}{a} > 0\right]$$
$$= \left(\frac{x^2}{2} - \frac{a^2}{4}\right) \operatorname{arccosh} \frac{x}{a} + \frac{x}{4}\sqrt{x^2 - a^2} \qquad \left[\operatorname{arccosh} \frac{x}{a} < 0\right]$$
$$\text{DW}$$

2.8 Inverse Trigonometric Functions

2.81 Arcsines and arccosines

2.811
$$\int \left(\arcsin \frac{x}{a}\right)^n dx = x \sum_{k=0}^{\lfloor n/2 \rfloor} (-1)^k \binom{n}{2k} \cdot (2k)! \left(\arcsin \frac{x}{a}\right)^{n-2k}$$
$$+ \sqrt{a^2 - x^2} \sum_{k=1}^{\lfloor (n+1)/2 \rfloor} (-1)^{k-1} \binom{n}{2k-1} \cdot (2k-1)! \left(\arcsin \frac{x}{a}\right)^{n-2k+1}$$

2.812
$$\int \left(\arccos \frac{x}{a}\right)^n dx = x \sum_{k=0}^{\lfloor n/2 \rfloor} (-1)^k \binom{n}{2k} \cdot (2k)! \left(\arccos \frac{x}{a}\right)^{n-2k}$$
$$+ \sqrt{a^2 - x^2} \sum_{k=1}^{\lfloor (n+1)/2 \rfloor} (-1)^{k} \binom{n}{2k-1} \cdot (2k-1)! \left(\arccos \frac{x}{a}\right)^{n-2k+1}$$

2.813

1.[9]
$$\int \arcsin \frac{x}{a}\, dx = \operatorname{sign}(a) \left[n \arcsin \frac{x}{|a|} + \sqrt{a^2 - x^2}\right]$$

2.[9]
$$\int \left(\arcsin \frac{x}{a}\right)^2 dx = x\left(\arcsin \frac{x}{|a|}\right)^2 + 2\sqrt{a^2 - x^2} \arcsin \frac{x}{|a|} - 2x$$

3.
$$\int \left(\arcsin \frac{x}{a}\right)^3 dx = \operatorname{sign}(a) \left[x\left(\arcsin \frac{x}{|a|}\right)^3 + 3\sqrt{a^2 - x^2}\left(\arcsin \frac{x}{|a|}\right)^2 \right.$$
$$\left. - 6x \arcsin \frac{x}{|a|} - 6\sqrt{a^2 - x^2}\right]$$

2.814

1. $\displaystyle\int \arccos \frac{x}{a}\, dx = x \arccos \frac{x}{a} - \sqrt{a^2 - x^2}$

2. $\displaystyle\int \left(\arccos \frac{x}{a}\right)^2 dx = x\left(\arccos \frac{x}{a}\right)^2 - 2\sqrt{a^2 - x^2} \arccos \frac{x}{a} - 2x.$

3. $\displaystyle\int \left(\arccos \frac{x}{a}\right)^3 dx = x\left(\arccos \frac{x}{a}\right)^3 - 3\sqrt{a^2 - x^2}\left(\arccos \frac{x}{a}\right)^2 - 6x \arccos \frac{x}{a} + 6\sqrt{a^2 - x^2}$

2.82 The arcsecant, the arccosecant, the arctangent and the arccotangent

arcsecant

2.821

1. $\displaystyle\int \operatorname{arccosec} \frac{x}{a}\, dx = \int \arcsin \frac{a}{x}\, dx = x \arcsin \frac{x}{2} + a \ln\left(x + \sqrt{x^2 - a^2}\right) \qquad \left[0 < \arcsin \frac{a}{x} < \frac{\pi}{2}\right]$

$\qquad\qquad = x \arcsin \frac{a}{x} - a \ln\left(x + \sqrt{x^2 - a^2}\right) \qquad\qquad\qquad \left[-\frac{\pi}{2} < \arcsin \frac{a}{x} < 0\right]$

$\hfill$ DW

2. $\displaystyle\int \operatorname{arcsec} \frac{x}{a}\, dx = \int \arccos \frac{a}{x}\, dx = x \arccos \frac{a}{x} - a \ln\left(x + \sqrt{x^2 - a^2}\right) \qquad \left[0 < \arccos \frac{a}{x} < \frac{\pi}{2}\right]$

$\qquad\qquad = x \arccos \frac{a}{x} - a \ln\left(x + \sqrt{x^2 - a^2}\right) \qquad\qquad\qquad \left[-\frac{\pi}{2} < \arccos \frac{a}{x} < 0\right]$

$\hfill$ DW

2.822

1.[8] $\displaystyle\int \arctan \frac{x}{a}\, dx = x \arctan \frac{x}{a} - \frac{a}{2} \ln\left(a^2 + x^2\right) \hfill$ DW

2. $\displaystyle\int \operatorname{arccot} \frac{x}{a}\, dx = x \operatorname{arccot} \frac{x}{a} - \frac{a}{2} \ln\left(a^2 + x^2\right) \hfill$ DW

3.[9] $\displaystyle\int x \arctan \frac{x}{a}\, dx = \frac{1}{2}\left(x^2 + a^2\right) \arctan \frac{x}{a} - \frac{ax}{2}$

4.[9] $\displaystyle\int x \operatorname{arccot} \frac{x}{a}\, dx = \frac{ax}{2} + \frac{\pi x^2}{4} - \frac{1}{2}\left(x^2 + a^2\right) \arctan \frac{x}{a}$

5.[9] $\displaystyle\int x^2 \arctan \frac{x}{a}\, dx = \frac{1}{3}x^3 \arctan \frac{x}{a} + \frac{1}{6}a^3 \ln\left(x^2 + a^2\right) - \frac{ax^2}{6}$

6.[9] $\displaystyle\int x^2 \operatorname{arccot} \frac{x}{a}\, dx = -\frac{1}{3}x^3 \arctan \frac{x}{a} - \frac{1}{6}a^3 \ln\left(x^2 + a^2\right) + \frac{\pi x^3}{6} + \frac{ax^2}{6}$

2.83 Combinations of arcsine or arccosine and algebraic functions

2.831 $\displaystyle\int x^n \arcsin \frac{x}{a}\, dx = \frac{x^{n+1}}{n+1} \arcsin \frac{x}{a} - \frac{1}{n+1}\int \frac{x^{n+1}\, dx}{\sqrt{a^2 - x^2}} \qquad$ (see **2.263 1**, **2.264**, **2.27**)

2.832 $\displaystyle\int x^n \arccos \frac{x}{a}\, dx = \frac{x^{n+1}}{n+1} \arccos \frac{x}{a} + \frac{1}{n+1}\int \frac{x^{n+1}\, dx}{\sqrt{a^2 - x^2}} \qquad$ (see **2.263 1**, **2.264**, **2.27**)

1.   For $n = -1$, these integrals (that is, $\int \dfrac{\arcsin x}{x}\, dx$ and $\int \dfrac{\arccos x}{x}\, dx$) cannot be expressed as a finite combination of elementary functions.

2. $\displaystyle\int \frac{\arccos x}{x}\, dx = -\frac{\pi}{2}\ln\frac{1}{x} - \int \frac{\arcsin x}{x}\, dx$

2.833[9]

1. $\displaystyle\int x \arcsin\frac{x}{a}\, dx = \operatorname{sign}(a)\left[\left(\frac{x^2}{2} - \frac{a^2}{4}\right)\arcsin\frac{x}{|a|} + \frac{x}{4}\sqrt{a^2 - x^2}\right]$

2. $\displaystyle\int x \arccos\frac{x}{a}\, dx = \frac{\pi x^2}{4} - \operatorname{sign}(a)\left[\frac{1}{4}\left(2x^2 - a^2\right)\arcsin\frac{x}{|a|} + \frac{x}{4}\sqrt{a^2 - x^2}\right]$

3. $\displaystyle\int x^2 \arcsin\frac{x}{a}\, dx = \operatorname{sign}(a)\left[\frac{x^3}{3}\arcsin\frac{x}{|a|} + \frac{1}{9}\left(x^2 + 2a^2\right)\sqrt{a^2 - x^2}\right]$

4. $\displaystyle\int x^2 \arccos\frac{x}{a}\, dx = \frac{\pi x^3}{6} - \operatorname{sign}(a)\left[\frac{x^3}{3}\arcsin\frac{x}{|a|} + \frac{1}{9}\left(x^2 + 2a^2\right)\sqrt{a^2 - x^2}\right]$

5. $\displaystyle\int x^3 \arcsin\frac{x}{a}\, dx = \operatorname{sign}(a)\left[\left(\frac{x^4}{4} - \frac{3a^4}{32}\right)\arcsin\frac{x}{|a|} + \frac{1}{32}x\left(2x^2 + 3a^2\right)\sqrt{a^2 - x^2}\right]$

6. $\displaystyle\int x^3 \arccos\frac{x}{a}\, dx = \frac{\pi x^4}{8} - \operatorname{sign}(a)\left[\frac{\left(8x^4 - 3a^4\right)}{32}\arcsin\frac{x}{|a|} + \frac{1}{32}x\left(2x^2 + 3a^2\right)\sqrt{a^2 - x^2}\right]$

2.834

1. $\displaystyle\int \frac{1}{x^2}\arcsin\frac{x}{a}\, dx = -\frac{1}{x}\arcsin\frac{x}{a} - \frac{1}{a}\ln\frac{a + \sqrt{a^2 - x^2}}{x}$

2. $\displaystyle\int \frac{1}{x^2}\arccos\frac{x}{a}\, dx = -\frac{1}{x}\arccos\frac{x}{a} - \frac{1}{a}\ln\frac{a + \sqrt{a^2 - x^2}}{x}$

2.835 $\displaystyle\int \frac{\arcsin x}{(a + bx)^2}\, dx = -\frac{\arcsin x}{b(a + bx)} - \frac{2}{b\sqrt{a^2 - b^2}}\arctan\sqrt{\frac{(a - b)(1 - x)}{(a + b)(1 + x)}}$ $\left[a^2 > b^2\right]$

               $\displaystyle = -\frac{\arcsin x}{b(a + bx)} - \frac{1}{b\sqrt{b^2 - a^2}}\ln\frac{\sqrt{(a + b)(1 + x)} + \sqrt{(b - a)(1 - x)}}{\sqrt{(a + b)(1 + x)} - \sqrt{(b - a)(1 - x)}}$ $\left[a^2 < b^2\right]$

2.836[8] $\displaystyle\int \frac{x \arcsin x}{(1 + cx^2)^2}\, dx = -\frac{\arcsin x}{2c\left(1 + cx^2\right)} + \frac{1}{2c\sqrt{c + 1}}\arctan\frac{\sqrt{c + 1}\,x}{\sqrt{1 - x^2}}$ $\left[c > -1\right]$

               $\displaystyle = -\frac{\arcsin x}{2c\left(1 + cx^2\right)} + \frac{1}{4c\sqrt{-(c + 1)}}\ln\frac{\sqrt{1 - x^2} + x\sqrt{-(c + 1)}}{\sqrt{1 - x^2} - x\sqrt{-(c + 1)}}$ $\left[c < -1\right]$

2.837

1. $\displaystyle\int \frac{x \arcsin x}{\sqrt{1 - x^2}}\, dx = x - \sqrt{1 - x^2}\,\arcsin x$

2. $\displaystyle\int \frac{x \arcsin x}{\sqrt{1 - x^2}}\, dx = \frac{x^2}{4} - \frac{x}{2}\sqrt{1 - x^2}\,\arcsin x + \frac{1}{4}(\arcsin x)^2$

3. $\displaystyle\int \frac{x^3 \arcsin x}{\sqrt{1 - x^2}}\, dx = \frac{x^3}{9} + \frac{2x}{3} - \frac{1}{3}\left(x^2 + 2\right)\sqrt{1 - x^2}\,\arcsin x$

2.838

1. $$\int \frac{\arcsin x}{\sqrt{(1-x^2)^3}}\,dx = \frac{x\arcsin x}{\sqrt{1-x^2}} + \frac{1}{2}\ln\left(1-x^2\right)$$

2. $$\int \frac{x\arcsin x}{\sqrt{(1-x^2)^3}}\,dx = \frac{\arcsin x}{\sqrt{1-x^2}} + \frac{1}{2}\ln\frac{1-x}{1+x}$$

2.84 Combinations of the arcsecant and arccosecant with powers of x

2.841

1. $$\int x\operatorname{arcsec}\frac{x}{a}\,dx = \int \arccos\frac{a}{x}\,dx = \frac{1}{2}\left\{x^2\arccos\frac{a}{x} - a\sqrt{x^2-a^2}\right\} \qquad \left[0 < \arccos\frac{a}{x} < \frac{\pi}{2}\right]$$
$$= \frac{1}{2}\left\{x^2\arccos\frac{a}{x} + a\sqrt{x^2-a^2}\right\} \qquad \left[\frac{\pi}{2} < \arccos\frac{a}{x} < \pi\right]$$

DW

2. $$\int x^2\operatorname{arcsec}\frac{x}{a}\,dx = \int \arccos\frac{a}{x}\,dx = \frac{1}{3}\left\{x^3\arccos\frac{a}{x} - \frac{a}{2}x\sqrt{x^2-a^2} - \frac{a^3}{2}\ln\left(x+\sqrt{x^2-a^2}\right)\right\}$$
$$\left[0 < \arccos\frac{a}{x} < \frac{\pi}{2}\right]$$
$$= \frac{1}{3}\left\{x^3\arccos\frac{a}{x} + \frac{a}{2}x\sqrt{x^2-a^2} + \frac{a^3}{2}\ln\left(x+\sqrt{x^2-z^2}\right)\right\}$$
$$\left[\frac{\pi}{2} < \arccos\frac{a}{x} < \pi\right]$$

DW

3. $$\int x\operatorname{arccosec}\frac{x}{a}\,dx = \int \arcsin\frac{a}{x}\,dx = \frac{1}{2}\left\{x^2\arcsin\frac{a}{x} + a\sqrt{x^2-a^2}\right\} \qquad \left[0 < \arcsin\frac{a}{x} < \frac{\pi}{2}\right]$$
$$= \frac{1}{2}\left\{x^2\arcsin\frac{a}{x} - a\sqrt{x^2-a^2}\right\} \qquad \left[-\frac{\pi}{2} < \arcsin\frac{a}{x} < 0\right]$$

DW

2.85 Combinations of the arctangent and arccotangent with algebraic functions

2.851 $$\int x^n\arctan\frac{x}{a}\,dx = \frac{x^{n+1}}{n+1}\arctan\frac{x}{a} - \frac{a}{n+1}\int \frac{x^{n+1}\,dx}{a^2+x^2}$$

2.852

1. $$\int x^n\operatorname{arccot}\frac{x}{a}\,dx = \frac{x^{n+1}}{n+1}\operatorname{arccot}\frac{x}{a} + \frac{a}{n+1}\int \frac{x^{n+1}\,dx}{a^2+x^2}$$

For $n = -1$

2. $$\int \frac{\arctan x}{x}\,dx \text{ cannot be expressed as a finite combination of elementary functions.}$$

3. $$\int \frac{\operatorname{arccot} x}{x}\,dx = \frac{\pi}{2}\ln x - \int \frac{\arctan x}{x}\,dx$$

2.853

1. $$\int x \arctan \frac{x}{a} \, dx = \frac{1}{2} \left(x^2 + a^2\right) \arctan \frac{x}{a} - \frac{ax}{2}$$

2. $$\int x \operatorname{arccot} \frac{x}{a} \, dx = \frac{1}{2} \left(x^2 + a^2\right) \operatorname{arccot} \frac{x}{a} + \frac{ax}{2}$$

3.[9] $$\int x^2 \arctan \frac{x}{a} \, dx = \frac{x^3}{3} \arctan \frac{x}{a} + \frac{a^3}{6} \ln \left(x^2 + a^2\right) - \frac{ax^2}{6}$$

4.[9] $$\int x^2 \operatorname{arccot} \frac{x}{a} \, dx = -\frac{x^3}{3} \arctan \frac{x}{a} - \frac{a^3}{6} \ln \left(x^2 + a^2\right) + \frac{\pi x^3}{6} + \frac{ax^2}{6}$$

2.854 $$\int \frac{1}{x^2} \arctan \frac{x}{a} \, dx = -\frac{1}{x} \arctan \frac{x}{a} - \frac{1}{2a} \ln \frac{a^2 + x^2}{x^2}$$

2.855 $$\int \frac{\arctan x}{(\alpha + \beta x)^2} \, dx = \frac{1}{\alpha^2 + \beta^2} \left\{ \ln \frac{\alpha + \beta x}{\sqrt{1 + x^2}} - \frac{\beta - \alpha x}{\alpha + \beta x} \arctan x \right\}$$

2.856

1. $$\int \frac{x \arctan x}{1 + x^2} \, dx = \frac{1}{2} \arctan x \ln \left(1 + x^2\right) - \frac{1}{2} \int \frac{\ln \left(1 + x^2\right) \, dx}{1 + x^2}$$ TI (689)

2. $$\int \frac{x^2 \arctan x}{1 + x^2} \, dx = x \arctan x - \frac{1}{2} \ln \left(1 + x^2\right) - \frac{1}{2} (\arctan x)^2$$ TI (405)

3. $$\int \frac{x^3 \arctan x}{1 + x^2} \, dx = -\frac{1}{2}x + \frac{1}{2} \left(1 + x^2\right) \arctan x - \int \frac{x \arctan x}{1 + x^2} \, dx$$

(see **2.8511**)

4. $$\int \frac{x^4 \arctan x}{1 + x^2} \, dx = -\frac{1}{6}x^2 + \frac{2}{3} \ln \left(1 + x^2\right) + \left(\frac{x^3}{3} - x\right) \arctan x + \frac{1}{2} (\arctan x)^2$$

2.857 $$\int \frac{\arctan x \, dx}{\left(1 + x^2\right)^{n+1}} = \left[\sum_{k=1}^{n} \frac{(2n - 2k)!!(2n - 1)!!}{(2n)!!(2n - 2k + 1)!!} \frac{x}{\left(1 + x^2\right)^{n-k+1}} + \frac{1}{2} \frac{(2n - 1)!!}{(2)!!} \arctan x \right] \arctan x$$
$$+ \frac{1}{2} \sum_{k=1}^{n} \frac{(2n - 1)!!(2n - 2k)!!}{(2n)!!(2n - 2k + 1)!!(n - k + 1)} \frac{1}{\left(1 + x^2\right)^{n-k+1}}$$

2.858 $$\int \frac{x \arctan x}{\sqrt{1 - x^2}} \, dx = -\sqrt{1 - x^2} \arctan x + \sqrt{2} \arctan \frac{x\sqrt{2}}{\sqrt{1 - x^2}} - \arcsin x$$

2.859 $$\int \frac{\arctan x}{\sqrt{\left(a + bx^2\right)^3}} \, dx = \frac{x \arctan x}{a\sqrt{a + bx^2}} - \frac{1}{a\sqrt{b - a}} \arctan \sqrt{\frac{a + bx^2}{b - a}} \qquad [a < b]$$
$$= \frac{x \arctan x}{a\sqrt{a + bx^2}} + \frac{1}{2a\sqrt{a - b}} \ln \frac{\sqrt{a + bx^2} - \sqrt{a - b}}{\sqrt{a + bx^2} + \sqrt{a - b}} \qquad [a > b]$$

3–4 Definite Integrals of Elementary Functions

3.0 Introduction*

3.01 Theorems of a general nature

3.011 Suppose that $f(x)$ is integrable[†] over the largest of the intervals $(p, q), (p, r), (r, q)$. Then (depending on the relative positions of the points p, q, and r) it is also integrable over the other two intervals and we have

$$\int_p^q f(x)\, dx = \int_p^r f(x)\, dx + \int_r^q f(x)\, dx \qquad\qquad \text{FI II 126}$$

3.012 *The first mean-value theorem.* Suppose (1) that $f(x)$ is continuous and that $g(x)$ is integrable over the interval (p, q), (2) that $m \le f(x) \le M$ and (3) that $g(x)$ does not change sign anywhere in the interval (p, q). Then, there exists at least one point ξ (with $p \le \xi \le q$) such that

$$\int_p^q f(x)g(x)\, dx = f(\xi) \int_p^q g(x)\, dx \qquad\qquad \text{FI II 132}$$

3.013 *The second mean-value theorem.* If $f(x)$ is monotonic and non-negative throughout the interval (p, q), where $p < q$, and if $g(x)$ is integrable over that interval, then there exists at least one point ξ (with $p \le \xi \le q$) such that

1. $$\int_p^q f(x)g(x)\, dx = f(p) \int_p^\xi g(x)\, dx$$

Under the conditions of Theorem **3.013** 1, if $f(x)$ is nondecreasing, then

2. $$\int_p^q f(x)g(x)\, dx = f(q) \int_\xi^q g(x)\, dx \qquad\qquad [p \le \xi \le q]$$

If $f(x)$ is monotonic in the interval (p, q), where $p < q$, and if $g(x)$ is integrable over that interval, then

---

*We omit the definition of definite and multiple integrals since they are widely known and can easily be found in any textbook on the subject. Here we give only certain theorems of a general nature which provide estimates, or which reduce the given integral to a simpler one.

†A function $f(x)$ is said to be integrable over the interval (p, q), if the integral $\int_p^q f(x)\, dx$ exists. Here, we usually mean the existence of the integral in the sense of Riemann. When it is a matter of the existence of the integral in the sense of Stieltjes or Lebesgue, etc., we shall speak of integrability in the sense of Stieltjes or Lebesgue.

3. $\displaystyle\int_p^q f(x)g(x)\,dx = f(p)\int_p^\xi g(x)\,dx + f(q)\int_\xi^q g(x)\,dx$ $[p \le \xi \le q]\,,$

or

4. $\displaystyle\int_p^q f(x)g(x)\,dx = A\int_p^\xi g(x)\,dx + B\int_\xi^q g(x)\,dx$ $[p \le \xi \le q]\,,$

where A and B are any two numbers satisfying the conditions

$$A \ge f(p+0) \quad \text{and} \quad B \le f(q-0) \quad \text{[if } f \text{ decreases]},$$
$$A \le f(p+0) \quad \text{and} \quad B \ge f(q-0) \quad \text{[if } f \text{ increases]}.$$

In particular,

5. $\displaystyle\int_p^q f(x)g(x)\,dx = f(p+0)\int_p^\xi g(x)\,dx + f(q-0)\int_\xi^q g(x)\,dx$ FI II 138

3.02 Change of variable in a definite integral

3.020 $\displaystyle\int_\alpha^\beta f(x)\,dx = \int_\varphi^\psi f[g(t)]g'(t)\,dt; \qquad x = g(t).$

This formula is valid under the following conditions:

1. $f(x)$ is continuous on some interval $A \le x \le B$ containing the original limits of integration α and β.

2. The equalities $\alpha = g(\varphi)$ and $\beta = g(\psi)$ hold.

3. $g(t)$ and its derivative $g'(t)$ are continuous on the interval $\varphi \le t \le \psi$.

4. As t varies from φ to ψ, the function $g(t)$ always varies in the same direction from $g(\varphi) = \alpha$ to $g(\psi) = \beta.$*

3.021 The integral $\displaystyle\int_\alpha^\beta f(x)\,dx$ can be transformed into another integral with given limits φ and ψ by means of the linear substitution

$$x = \frac{\beta - \alpha}{\psi - \varphi}t + \frac{\alpha\psi - \beta\varphi}{\psi - \varphi} :$$

1. $\displaystyle\int_\alpha^\beta f(x)\,dx = \frac{\beta - \alpha}{\psi - \varphi}\int_\varphi^\psi f\left(\frac{\beta - \alpha}{\psi - \varphi}t + \frac{\alpha\psi - \beta\varphi}{\psi - \varphi}\right)\,dt$

In particular, for $\varphi = 0$ and $\psi = 1$,

---

*If this last condition is not satisfied, the interval $\varphi \le t \le \psi$ should be partitioned into subintervals throughout each of which the condition is satisfied:

$$\int_\alpha^\beta f(x)\,dx = \int_\varphi^{\varphi_1} f[g(t)]g'(t)\,dt + \int_{\varphi_1}^{\varphi_2} f[g(t)]g'(t)\,dt + \cdots + \int_{\varphi_{n-1}}^\psi f[g(t)]g'(t)\,dt.$$

2. $$\int_\alpha^\beta f(x)\,dx = (\beta - \alpha)\int_0^1 f((\beta - \alpha)t + \alpha)\,dt$$

For $\varphi = 0$ and $\psi = \infty$,

3. $$\int_\alpha^\beta f(x)\,dx = (\beta - \alpha)\int_0^\infty f\left(\frac{\alpha + \beta t}{1 + t}\right)\frac{dt}{(1 + t)^2}$$

3.022 The following formulas also hold:

1. $$\int_\alpha^\beta f(x)\,dx = \int_\alpha^\beta f(\alpha + \beta - x)\,dx$$

2. $$\int_0^\beta f(x)\,dx = \int_0^\beta f(\beta - x)\,dx$$

3. $$\int_{-\alpha}^\alpha f(x)\,dx = \int_{-\alpha}^\alpha f(-x)\,dx$$

3.03 General formulas

3.031

1. Suppose that a function $f(x)$ is integrable over the interval $(-p, p)$ and satisfies the relation $f(-x) = f(x)$ on that interval. (A function satisfying the latter condition is called an *even* function.) Then,

$$\int_{-p}^p f(x)\,dx = 2\int_0^p f(x)\,dx$$

FI II 159

2. Suppose that $f(x)$ is a function that is integrable on the interval $(-p, p)$ and satisfies the relation $f(-x) = -f(x)$ on that interval. (A function satisfying the latter condition is called an *odd* function). Then,

$$\int_{-p}^p f(x)\,dx = 0.$$

FI II 159

3.032

1. $$\int_0^{\frac{\pi}{2}} f(\sin x)\,dx = \int_0^{\frac{\pi}{2}} f(\cos x)\,dx,$$

where $f(x)$ is a function that is integrable on the interval $(0, 1)$.

FI II 159

2. $$\int_0^{2\pi} f(p\cos x + q\sin x)\,dx = 2\int_0^\pi f\left(\sqrt{p^2 + q^2}\cos x\right)\,dx,$$

where $f(x)$ is integrable on the interval $\left(-\sqrt{p^2 + q^2}, \sqrt{p^2 + q^2}\right)$.

FI II 160

3. $$\int_0^{\frac{\pi}{2}} f(\sin 2x)\cos x\,dx = \int_0^{\frac{\pi}{2}} f(\cos^2 x)\cos x\,dx,$$

where $f(x)$ is integrable on the interval $(0, 1)$.

FI II 161

3.033

1. If $f(x + \pi) = f(x)$ and $f(-x) = f(x)$, then

$$\int_0^\infty f(x) \frac{\sin x}{x} \, dx = \int_0^{\frac{\pi}{2}} f(x) \, dx \qquad\qquad \text{LO V 277(3)}$$

2.   If $f(x + \pi) = -f(x)$ and $f(-x) = f(x)$, then

$$\int_0^\infty f(x) \frac{\sin x}{x} \, dx = \int_0^{\frac{\pi}{2}} f(x) \cos x \, dx \qquad\qquad \text{LO V 279(4)}$$

In formulas **3.033**, it is assumed that the integrals in the left members of the formulas exist.

3.034 $\displaystyle\int_0^\infty \frac{f(px) - f(qx)}{x} \, dx = [f(0) - f(+\infty)] \ln \frac{q}{p},$

if $f(x)$ is continuous for $x \geq 0$ and if there exists a finite limit $f(+\infty) = \lim\limits_{x \to +\infty} f(x)$. $\qquad$ FI II 633

3.035

1. $\displaystyle\int_0^\pi \frac{f\left(\alpha + e^{xi}\right) + f\left(\alpha + e^{-xi}\right)}{1 + 2p\cos x + p^2} \, dx = \frac{2\pi}{1 - p^2} f(\alpha + p) \qquad [|p| < 1] \qquad$ LA 230(16)

2. $\displaystyle\int_0^\pi \frac{1 - p\cos x}{1 - 2p\cos x + p^2} \left\{ f\left(\alpha + e^{xi}\right) + f\left(\alpha + e^{-xi}\right) \right\} dx = \pi \left\{ f(\alpha + p) + f(\alpha) \right\}$

$$[|p| < 1] \qquad\qquad \text{BE 169}$$

3. $\displaystyle\int_0^\pi \frac{f\left(\alpha + e^{-xi}\right) - f\left(\alpha + e^{xi}\right)}{1 - 2p\cos x + p^2} \sin x \, dx = \frac{\pi}{pi} \left\{ f(\alpha + p) - f(\alpha) \right\}$

$$[|p| < 1] \qquad\qquad \text{BE 169}$$

In formulas **3.035**, it is assumed that the function f is analytic in the closed unit circle with its center at the point α.

3.036

1. $\displaystyle\int_0^\pi f\left(\frac{\sin^2 x}{1 + 2p\cos x + p^2} \right) dx = \int_0^\pi f\left(\sin^2 x\right) dx \qquad [p^2 \geq 1]$

$$= \int_0^\pi f\left(\frac{\sin^2 x}{p^2} \right) dx \qquad [p^2 < 1]$$

$$\text{LA 228(6)}$$

2. $\displaystyle\int_0^\pi F^{(n)}\left(\cos x\right) \sin^{2n} x \, dx = (2n - 1)!! \int_0^\pi F\left(\cos x\right) \cos nx \, dx \qquad$ B 174

3.037 If f is analytic in the circle of radius r and if

$$f\left[r\left(\cos x + i\sin x\right)\right] f_1(r, x) + i f_2(r, x),$$

then

1. $\displaystyle\int_0^\infty \frac{f_1(r, x)}{p^2 + x^2} \, dx = \frac{\pi}{2p} f\left(re^{-p}\right) \qquad\qquad$ LA 230(19)

2. $\displaystyle\int_0^\infty f_2(r, x) \frac{x \, dx}{p^2 + x^2} = \frac{\pi}{2} \left[f\left(re^{-p}\right) - f(0) \right] \qquad$ LA 230(20)

3. $\displaystyle\int_0^\infty \frac{f_2(r, x)}{x} \, dx = \frac{\pi}{2} [f(r) - f(0)]. \qquad\qquad$ LA 230(21)

4. $\displaystyle\int_0^\infty \frac{f_2(r,x)}{x\,(p^2+x^2)}\,dx = \frac{\pi}{2p^2}\left[f(r)-f\left(re^{-p}\right)\right]$ LA 230(22)

3.038 $\displaystyle\int_{-\infty}^\infty \frac{x\,dx}{\sqrt{1+x^2}}\,F\left(qx+p\sqrt{1+x^2}\right) = \int_{-\infty}^\infty F\left(p\cosh x + q\sinh x\right)\sinh x\,dx$

$$= 2q\int_0^\infty F'\left(\operatorname{sign} p \cdot \sqrt{p^2-q^2}\,\cosh x\right)\sinh^2 x\,dx$$

[If F is a function with a continuous derivative in the interval $(-\infty,\infty)$; all these integrals converge.]

3.04 Improper integrals

3.041   Suppose that a function $f(x)$ is defined on an interval $(p,+\infty)$ and that it is integrable over an arbitrary finite subinterval of the form (p,P). Then, by definition

$$\int_p^{+\infty} f(x)\,dx = \lim_{P\to+\infty}\int_p^P f(x)\,dx,$$

if this limit exists. If it does exist, we say that the integral $\displaystyle\int_p^{+\infty} f(x)\,dx$ exists or that it converges. Otherwise, we say that the integral diverges.

3.042   Suppose that a function $f(x)$ is bounded and integrable in an arbitrary interval $(p,q-\eta)$ (for $0 < \eta < q-p$) but is unbounded in every interval $(q-\eta,q)$ to the left of the point q. The point q is then called a *singular point*. Then, by definition,

$$\int_p^q f(x)\,dx = \lim_{\eta\to 0}\int_p^{q-\eta} f(x)\,dx,$$

if this limit exists. In this case, we say that the integral $\displaystyle\int_p^q f(x)\,dx$ *exists* or that it *converges*.

3.043   If not only the integral of $f(x)$ but also the integral of $|f(x)|$ exists, we say that the integral of $f(x)$ converges *absolutely*.

3.044   The integral $\displaystyle\int_p^{+\infty} f(x)\,dx$ converges absolutely if there exists a number $\alpha > 1$ such that the limit

$$\lim_{x\to+\infty}\left\{x^\alpha|f(x)|\right\}$$

exists. On the other hand, if

$$\lim_{x\to+\infty}\left\{x|f(x)|\right\} = L > 0,$$

the integral $\displaystyle\int_p^{+\infty} |f(x)|\,dx$ diverges.

3.045   Suppose that the upper limit q of the integral $\displaystyle\int_p^q f(x)\,dx$ is a singular point. Then, this integral converges absolutely if there exists a number $\alpha < 1$ such that the limit

$$\lim_{x\to q}\left[(q-x)^\alpha|f(x)|\right]$$

exists. On the other hand, if

$$\lim_{x\to q}\left[(q-x)|f(x)|\right] = L > 0,$$

the integral $\displaystyle\int_p^q f(x)\,dx$ diverges.

3.046 Suppose that the functions $f(x)$ and $g(x)$ are defined on the interval $(p, +\infty)$, that $f(x)$ is integrable over every finite interval of the form (p, P), that the integral

$$\int_p^P f(x)\, dx$$

is a bounded function of P, that $g(x)$ is monotonic, and that $g(x) \to 0$ as $x \to +\infty$. Then, the integral

$$\int_p^{+\infty} f(x) g(x)\, dx$$

converges.   FI II 577

3.05 The principal values of improper integrals

3.051 Suppose that a function $f(x)$ has a singular point r somewhere inside the interval (p, q), that $f(x)$ is defined at r, and that $f(x)$ is integrable over every portion of this interval that does not contain the point r. Then, by definition

$$\int_p^q f(x)\, dx = \lim_{\substack{\eta \to 0 \\ \eta' \to 0}} \left\{ \int_p^{r-\eta} f(x)\, dx + \int_{r+\eta'}^q f(x)\, dx \right\},$$

Here, the limit must exist for *independent* modes of approach of η and η' to zero. If this limit does not exist but the limit

$$\lim_{\eta \to 0} \left\{ \int_p^{r-\eta} f(x)\, dx + \int_{r+\eta}^q f(x)\, dx \right\},$$

does exist, we say that this latter limit is the *principal value* of the improper integral $\int_p^q f(x)\, dx$ and we say that the integral $\int_p^q f(x)\, dx$ exists in the sense of principal values.   FI II 603

3.052 Suppose that the function $f(x)$ is continuous over the interval (p, q) and vanishes at only one point r inside this interval. Suppose that the first derivative $f'(x)$ exists in a neighborhood of the point r. Suppose that $f'(r) \neq 0$ and that the second derivative $f''(r)$ exists at the point r itself. Then,

$$\int_p^q \frac{dx}{f(x)}$$ FI II 605

diverges, but exists in the sense of principal values.

3.053 A divergent integral of a positive function cannot exist in the sense of principal values.

3.054 Suppose that the function $f(x)$ has no singular points in the interval $(-\infty, +\infty)$. Then, by definition

$$\int_{-\infty}^{+\infty} f(x)\, dx = \lim_{\substack{P \to -\infty \\ Q \to +\infty}} \int_P^Q f(x)\, dx,$$

Here, the limit must exist for independent approach of P and Q to $\pm\infty$. If this limit does not exist but the limit

$$\lim_{P \to +\infty} \int_{-P}^{+P} f(x)\, dx,$$

does exist, this last limit is called the principal value of the improper integral

$$\int_{-\infty}^{+\infty} f(x)\, dx$$ FI II 607

3.055 The principal value of an improper integral of an even function exists only when this integral converges (in the ordinary sense).   FI II 607

3.1–3.2 Power and Algebraic Functions

3.11 Rational functions

1. $\displaystyle\int_{-\infty}^{\infty} \frac{p+qx}{r^2 + 2rx\cos\lambda + x^2}\, dx = \frac{\pi}{r\sin\lambda}(p - qr\cos\lambda)$ (principal value)

                               (see also **3.194** 8 and **3.252** 1 and 2)   BI (22)(14)

3.112   Integrals of the form $\displaystyle\int_{-\infty}^{\infty} \frac{g_n(x)\,dx}{h_n(x)h_n(-x)}$, where

$$g_n(x) = b_0 x^{2n-2} b_1 x^{2n-4} + \cdots b_{n-1},$$
$$h_n(x) = a_0 x^n a_1 x^{n-1} + \cdots a_n$$

[All roots of $h_n(x)$ lie in the upper half-plane.]

1. $\displaystyle\int_{-\infty}^{\infty} \frac{g_n(x)\,dx}{h_n(x)h_n(-x)} = \frac{\pi i}{a_0}\frac{M_n}{\Delta_n},$ JE

     where

$$\Delta_n = \begin{vmatrix} a_1 & a_3 & a_5 & & 0 \\ a_0 & a_2 & a_4 & & 0 \\ 0 & a_1 & a_3 & & 0 \\ \vdots & & & \ddots & \\ 0 & 0 & 0 & & a_n \end{vmatrix}, \qquad M_n = \begin{vmatrix} b_0 & b_1 & b_2 & \cdots & b_{n-1} \\ a_0 & a_2 & a_4 & & 0 \\ 0 & a_1 & a_3 & & 0 \\ \vdots & & & \ddots & \\ 0 & 0 & 0 & & a_n \end{vmatrix}.$$

2. $\displaystyle\int_{-\infty}^{\infty} \frac{g_1(x)\,dx}{h_1(x)h_1(-x)} = \frac{\pi i b_0}{a_0 a_1}$ JE

3.[8] $\displaystyle\int_{-\infty}^{\infty} \frac{g_2(x)\,dx}{h_2(x)h_2(-x)} = \pi i \frac{-b_0 + \frac{a_0 b_1}{a_2}}{a_0 a_1}$

4. $\displaystyle\int_{-\infty}^{\infty} \frac{g_3(x)\,dx}{h_3(x)h_3(-x)} = \pi i \frac{-a_2 b_0 a_0 b_1 - \dfrac{a_0 a_1 b_2}{a_3}}{a_0\,(a_0 a_3 a_1 a_2)}$ JE

5. $\displaystyle\int_{-\infty}^{\infty} \frac{g_4(x)\,dx}{h_4(x)h_4(-x)} = \pi i \frac{b_0\,(-a_1 a_4 + a_2 a_3) - a_0 a_3 b_1 + a_0 a_1 b_2 + \dfrac{a_0 b_3}{a_4}(a_0 a_3 - a_1 a_2)}{a_0\,(a_0 a_3^2 + a_1^2 a_4 - a_1 a_2 a_3)}$ JE

6. $\displaystyle\int_{-\infty}^{\infty} \frac{g_5(x)\,dx}{h_5(x)h_5(-x)} = \pi i \frac{M_5}{a_0 \Delta_5},$

   where

$$M_5 = b_0\left(-a_0 a_4 a_5 + a_1 a_4^2 + a_2^2 a_5 - a_2 a_3 a_4\right) + a_0 b_1\left(-a_2 a_5 + a_3 a_4\right)$$

$$+ a_0 b_2\left(a_0 a_5 - a_1 a_4\right) + a_0 b_3\left(-a_0 a_3 + a_1 a_2\right) + \frac{a_0 b_4}{a_5}\left(-a_0 a_1 a_5 + a_0 a_3^2 + a_1^2 a_4 - a_1 a_2 a_3\right),$$

$$\Delta_5 = a_0^2 a_5^2 - 2a_0 a_1 a_4 a_5 - a_0 a_2 a_3 a_5 + a_0 a_3^2 a_4 + a_1^2 a_4^2 + a_1 a_2^2 a_5 - a_1 a_2 a_3 a_4$$ JE

3.12 Products of rational functions and expressions that can be reduced to square roots of first-and second-degree polynomials

3.121

1. $\displaystyle\int_0^1 \frac{1}{1 - 2x\cos\lambda + x^2}\frac{dx}{\sqrt{x}} = 2\operatorname{cosec}\lambda\sum_{k=1}^\infty \frac{\sin k\lambda}{2k-1}$ BI (10)(17)

2. $\displaystyle\int_0^1 \frac{1}{q - px}\frac{dx}{\sqrt{x(1-x)}} = \frac{\pi}{\sqrt{q(q-p)}}$ $[0 < p < q]$ BI (10)(9)

3. $\displaystyle\int_0^1 \frac{dx}{1 - 2rx + r^2}\sqrt{\frac{1 \mp x}{1 \pm x}} = \pm\frac{\pi}{4r} \mp \frac{1}{r}\frac{1 \mp r}{1 \pm r}\arctan\frac{1+r}{1-r}$ LI (14)(5, 16)

3.13–3.17 Expressions that can be reduced to square roots of third-and fourth-degree polynomials and their products with rational functions

Notation: In **3.131–3.137** we set: $\alpha = \arcsin\sqrt{\dfrac{a-c}{a-u}}$, $\beta = \arcsin\sqrt{\dfrac{c-u}{b-u}}$,

$$\gamma = \arcsin\sqrt{\frac{u-c}{b-c}}, \qquad \delta = \arcsin\sqrt{\frac{(a-c)(b-u)}{(b-c)(a-u)}},$$

$$\kappa = \arcsin\sqrt{\frac{(a-c)(u-b)}{(a-b)(u-c)}}, \qquad \lambda = \arcsin\sqrt{\frac{a-u}{a-b}},$$

$$\mu = \arcsin\sqrt{\frac{u-a}{u-b}}, \qquad \nu = \arcsin\sqrt{\frac{a-c}{u-c}}, \qquad p = \sqrt{\frac{a-b}{a-c}}, \qquad q = \sqrt{\frac{b-c}{a-c}}.$$

3.131

1. $\displaystyle\int_{-\infty}^u \frac{dx}{\sqrt{(a-x)(b-x)(c-x)}} = \frac{2}{\sqrt{a-c}}F(\alpha, p)$ $[a > b > c \geq u]$ BY (231.00)

2. $\displaystyle\int_u^c \frac{dx}{\sqrt{(a-x)(b-x)(c-x)}} = \frac{2}{\sqrt{a-c}}F(\beta, p)$ $[a > b > c > u]$ BY (232.00)

3. $\displaystyle\int_c^u \frac{dx}{\sqrt{(a-x)(b-x)(x-c)}} = \frac{2}{\sqrt{a-c}}F(\gamma, q)$ $[a > b \geq u > c]$ BY (233.00)

4. $\displaystyle\int_u^b \frac{dx}{\sqrt{(a-x)(b-x)(x-c)}} = \frac{2}{\sqrt{a-c}}F(\delta, q)$ $[a > b > u \geq c]$ BY (234.00)

5. $\displaystyle\int_b^u \frac{dx}{\sqrt{(a-x)(x-b)(x-c)}} = \frac{2}{\sqrt{a-c}}F(\kappa, p)$ $[a \geq u > b > c]$ BY (235.00)

6. $\displaystyle\int_u^a \frac{dx}{\sqrt{(a-x)(x-b)(x-c)}} = \frac{2}{\sqrt{a-c}}F(\lambda, p)$ $[a > u \geq b > c]$ BY (236.00)

7. $\displaystyle\int_a^u \frac{dx}{\sqrt{(x-a)(x-b)(x-c)}} = \frac{2}{\sqrt{a-c}}F(\mu, q)$ $[u > a > b > c]$ BY (237.00)

8.
$$\int_u^\infty \frac{dx}{\sqrt{(x-a)(x-b)(x-c)}} = \frac{2}{\sqrt{a-c}} F(\nu, q) \qquad [u \geq a > b > c] \qquad \text{BY (238.00)}$$

3.132

1.
$$\int_u^c \frac{x\,dx}{\sqrt{(a-x)(b-x)(c-x)}} = \frac{2}{\sqrt{a-c}} \left[c\,F(\beta, p) + (a-c)\,E(\beta, p) \right] - 2\sqrt{\frac{(a-u)(c-u)}{b-u}}$$
$$[a > b > c > u] \qquad \text{BY (232.19)}$$

2.
$$\int_c^u \frac{x\,dx}{\sqrt{(a-x)(b-x)(x-c)}} = \frac{2a}{\sqrt{a-c}} F(\gamma, q) - 2\sqrt{a-c}\,E(\gamma, q)$$
$$[a > b \geq u > c] \qquad \text{BY (233.17)}$$

3.
$$\int_u^b \frac{x\,dx}{\sqrt{(a-x)(b-x)(x-c)}} = \frac{2}{\sqrt{a-c}} \left[(b-a)\,\Pi\left(\delta, q^2, q\right) + a\,F(\delta, q) \right]$$
$$[a > b > u \geq c] \qquad \text{BY (234.16)}$$

4.
$$\int_b^u \frac{x\,dx}{\sqrt{(a-x)(x-b)(x-c)}} = \frac{2}{\sqrt{a-c}} \left[(b-c)\,\Pi\left(\kappa, p^2, p\right) + c\,F(\kappa, p) \right]$$
$$[a \geq u > b > c] \qquad \text{BY (235.16)}$$

5.
$$\int_u^a \frac{x\,dx}{\sqrt{(a-x)(x-b)(x-c)}} = \frac{2c}{\sqrt{a-c}} F(\lambda, p) + 2\sqrt{a-c}\,E(\lambda, p)$$
$$[a > u \geq b > c] \qquad \text{BY (236.16)}$$

6.
$$\int_a^u \frac{x\,dx}{\sqrt{(x-a)(x-b)(x-c)}} = \frac{2}{b\sqrt{a-c}} \left[a(a-b)\,\Pi(\mu, 1, q) + b^2\,F(\mu, q) \right]$$
$$[u > a > b > c] \qquad \text{BY (237.16)}$$

3.133

1.
$$\int_{-\infty}^u \frac{dx}{\sqrt{(a-x)^3(b-x)(c-x)}} = \frac{2}{(a-b)\sqrt{a-c}} \left[F(\alpha, p) - E(\alpha, p) \right]$$
$$[a > b > c \geq u] \qquad \text{BY (231.08)}$$

2.
$$\int_u^c \frac{dx}{\sqrt{(a-x)^3(b-x)(c-x)}} = \frac{2}{(a-b)\sqrt{a-c}} \left[F(\beta, p) - E(\beta, p) \right] + \frac{2}{a-c} \sqrt{\frac{c-u}{(a-u)(b-u)}}$$
$$[a > b > c > u] \qquad \text{BY (232.13)}$$

3.
$$\int_c^u \frac{dx}{\sqrt{(a-x)^3(b-x)(x-c)}} = \frac{2}{(a-b)\sqrt{a-c}} E(\gamma, q) - \frac{2}{(a-b)(a-c)} \sqrt{\frac{(b-u)(u-c)}{a-u}}$$
$$[a > b \geq u > c] \qquad \text{BY (233.09)}$$

4.
$$\int_u^b \frac{dx}{\sqrt{(a-x)^3(b-x)(x-c)}} = \frac{2}{(a-b)\sqrt{a-c}} E(\delta, q) \qquad [a > b > u \geq c] \qquad \text{BY (234.05)}$$

5. $\int_b^u \dfrac{dx}{\sqrt{(a-x)^3(x-b)(x-c)}} = \dfrac{2}{(a-b)\sqrt{a-c}} [F(\kappa, p) - E(\kappa, p)] + \dfrac{2}{a-b}\sqrt{\dfrac{u-b}{(a-u)(u-c)}}$

$$[a > u > b > c] \qquad \text{BY (235.04)}$$

6. $\int_u^\infty \dfrac{dx}{\sqrt{(x-a)^3(x-b)(x-c)}} = \dfrac{2}{(b-a)\sqrt{a-c}} E(\nu, q) + \dfrac{2}{a-b}\sqrt{\dfrac{u-b}{(u-a)(u-c)}}$

$$[u > a > b > c] \qquad \text{BY (238.05)}$$

7. $\int_{-\infty}^u \dfrac{dx}{\sqrt{(a-x)(b-x)^3(c-x)}} = \dfrac{2\sqrt{a-c}}{(a-b)(b-c)} E(\alpha, p) - \dfrac{2}{(a-b)\sqrt{a-c}} F(\alpha, p)$
$- \dfrac{2}{b-c}\sqrt{\dfrac{c-u}{(a-u)(b-u)}}$

$$[a > b > c \geq u] \qquad \text{BY (231.09)}$$

8. $\int_u^c \dfrac{dx}{\sqrt{(a-x)(b-x)^3(c-x)}} = \dfrac{2\sqrt{a-c}}{(a-b)(b-c)} E(\beta, p) - \dfrac{2}{(a-b)\sqrt{a-c}} F(\beta, p)$

$$[a > b > c > u] \qquad \text{BY (232.14)}$$

9. $\int_c^u \dfrac{dx}{\sqrt{(a-x)(b-x)^3(x-c)}} = \dfrac{2}{(b-c)\sqrt{a-c}} F(\gamma, q) - \dfrac{2\sqrt{a-c}}{(a-b)(b-c)} E(\gamma, q)$
$+ \dfrac{2}{(a-b)(b-c)}\sqrt{\dfrac{(a-u)(u-c)}{b-u}}$

$$[a > b > u > c] \qquad \text{BY (233.10)}$$

10. $\int_u^a \dfrac{dx}{\sqrt{(a-x)(x-b)^3(x-c)}} = \dfrac{2}{(a-b)\sqrt{a-c}} F(\lambda, p) - \dfrac{2\sqrt{a-c}}{(a-b)(b-c)} E(\lambda, p)$
$+ \dfrac{2}{(a-b)(b-c)}\sqrt{\dfrac{(a-u)(u-c)}{u-b}}$

$$[a > u > b > c] \qquad \text{BY (236.09)}$$

11. $\int_a^u \dfrac{dx}{\sqrt{(x-a)(x-b)^3(x-c)}} = \dfrac{2\sqrt{a-c}}{(a-b)(b-c)} E(\mu, q) - \dfrac{2}{(b-c)\sqrt{a-c}} F(\mu, q)$

$$[u > a > b > c] \qquad \text{BY (237.12)}$$

12. $\int_u^\infty \dfrac{dx}{\sqrt{(x-a)(x-b)^3(x-c)}} = \dfrac{2\sqrt{a-c}}{(a-b)(b-c)} E(\nu, q) - \dfrac{2}{(b-c)\sqrt{a-c}} F(\nu, q)$
$- \dfrac{2}{a-b}\sqrt{\dfrac{u-a}{(u-b)(u-c)}}$

$$[u \geq a > b > c] \qquad \text{BY (238.04)}$$

13. $\int_{-\infty}^u \dfrac{dx}{\sqrt{(a-x)(b-x)(c-x)^3}} = \dfrac{2}{(c-b)\sqrt{a-c}} E(\alpha, p) + \dfrac{2}{b-c}\sqrt{\dfrac{b-u}{(a-u)(c-u)}}$

$$[a > b > c > u] \qquad \text{BY (231.10)}$$

14. $$\int_u^b \frac{dx}{\sqrt{(a-x)(b-x)(x-c)^3}} = \frac{2}{(b-c)\sqrt{a-c}} [F(\delta,q) - E(\delta,q)] + \frac{2}{b-c}\sqrt{\frac{b-u}{(a-u)(u-c)}}$$

$$[a > b > u > c] \qquad \text{BY (234.04)}$$

15. $$\int_b^u \frac{dx}{\sqrt{(a-x)(x-b)(x-c)^3}} = \frac{2}{(b-c)\sqrt{a-c}} E(\kappa,p)$$

$$[a \geq u > b > c] \qquad \text{BY (235.01)}$$

16. $$\int_u^a \frac{dx}{\sqrt{(a-x)(x-b)(x-c)^3}} = \frac{2}{(b-c)\sqrt{a-c}} E(\lambda,p) - \frac{2}{(b-c)(a-c)}\sqrt{\frac{(a-u)(u-b)}{u-c}}$$

$$[a > u \geq b > c] \qquad \text{BY (236.10)}$$

17. $$\int_a^u \frac{dx}{\sqrt{(x-a)(x-b)(x-c)^3}} = \frac{2}{(b-c)\sqrt{a-c}} [F(\mu,q) - E(\mu,q)] + \frac{2}{a-c}\sqrt{\frac{u-a}{(u-b)(u-c)}}$$

$$[u > a > b > c] \qquad \text{BY (237.13)}$$

18. $$\int_u^\infty \frac{dx}{\sqrt{(x-a)(x-b)(x-c)^3}} = \frac{2}{(b-c)\sqrt{a-c}} [F(\nu,q) - E(\nu,q)]$$

$$[u \geq a > b > c] \qquad \text{BY (238.03)}$$

3.134

1. $$\int_{-\infty}^u \frac{dx}{\sqrt{(a-x)^5(b-x)(c-x)}}$$

$$= \frac{2}{3(a-b)^2\sqrt{(a-c)^3}} [(3a - b - 2c) F(\alpha,p) - 2(2a - b - c) E(\alpha,p)]$$

$$+ \frac{2}{3(a-c)(a-b)}\sqrt{\frac{(c-u)(b-u)}{(a-u)^3}}$$

$$[a > b > c \geq u] \qquad \text{BY (231.08)}$$

2. $$\int_u^c \frac{dx}{\sqrt{(a-x)^5(b-x)(c-x)}} = \frac{2}{3(a-b)^2\sqrt{(a-c)^3}} [(3a - b - 2c) F(\beta,p) - 2(2a - b - c) E(\beta,p)]$$

$$+ \frac{2\left[4a^2 - 3ab - 2ac + bc - u(3a - 2b - c)\right]}{3(a-b)(a-c)^2}\sqrt{\frac{c-u}{(a-u)^3(b-u)}}$$

$$[a > b > c > u] \qquad \text{BY (232.13)}$$

3. $$\int_c^u \frac{dx}{\sqrt{(a-x)^5(b-x)(x-c)}} = \frac{2}{3(a-b)^3\sqrt{(a-c)^3}} [2(2a - b - c) E(\gamma,q) - (a - b) F(\gamma,q)]$$

$$- \frac{2\left[5a^2 - 3ab - 3ac + bc - 2u(2a - b - c)\right]}{3(a-b)^2(a-c)^2}\sqrt{\frac{(b-u)(u-c)}{(a-u)^3}}$$

$$[a > b \geq u > c] \qquad \text{BY (233.09)}$$

4. $$\int_u^b \frac{dx}{\sqrt{(a-x)^5(b-x)(x-c)}} = \frac{2}{3(a-b)^2\sqrt{(a-c)^3}}\left[2(2a-b-c)\,E(\delta,q) - (a-b)\,F(\delta,q)\right]$$
$$-\frac{2}{3(a-b)(a-c)}\sqrt{\frac{(b-u)(u-c)}{(a-u)^3}}$$
$$[a>b>u\geq c] \qquad\qquad \text{BY (234.05)}$$

5. $$\int_b^u \frac{dx}{\sqrt{(a-x)^5(x-b)(x-c)}}$$
$$= \frac{2}{3(a-b)^2\sqrt{(a-c)^3}}\left[(3a-b-2c)\,F(\kappa,p) - 2(2a-b-c)\,E(\kappa,p)\right]$$
$$+\frac{2\left[4a^2-2ab-3ac+bc-u(3a-b-2c)\right]}{3(a-b)^2(a-c)}\sqrt{\frac{u-b}{(a-u)^3(u-c)}}$$
$$[a>u>b>c] \qquad\qquad \text{BY (235.04)}$$

6. $$\int_u^\infty \frac{dx}{\sqrt{(x-a)^5(x-b)(x-c)}} = \frac{2}{3(a-b)^2\sqrt{(a-c)^3}}\left[2(2a-b-c)\,E(\nu,q) - (a-b)\,F(\nu,q)\right]$$
$$+\frac{2\left[4a^2-2ab-3ac+bc+u(b+2c-3a)\right]}{3(a-b)^2(a-c)}\sqrt{\frac{u-b}{(u-a)^3(u-c)}}$$
$$[u>a>b>c] \qquad\qquad \text{BY (238.05)}$$

7. $$\int_{-\infty}^u \frac{dx}{\sqrt{(a-x)(b-x)^5(c-x)}} = \frac{2}{3(a-b)^2(b-c)^2\sqrt{a-c}}$$
$$\times\left[2(a-c)(a+c-2b)\,E(\alpha,p) + (b-c)(3b-a-2c)\,F(\alpha,p)\right]$$
$$-\frac{2\left[3ab-ac+2bc-4b^2-u(2a-3b+c)\right]}{3(a-b)(b-c)^2}\sqrt{\frac{c-u}{(a-u)(b-u)^3}}$$
$$[a>b>c\geq u] \qquad\qquad \text{BY (231.09)}$$

8. $$\int_u^c \frac{dx}{\sqrt{(a-x)(b-x)^5(c-x)}} = \frac{2}{3(a-b)^2(b-c)^2\sqrt{a-c}}$$
$$\times\left[(b-c)(3b-a-2c)\,F(\beta,p) + 2(a-c)(a-2b+c)\,E(\beta,p)\right]$$
$$+\frac{2}{3(a-b)(b-c)}\sqrt{\frac{(a-u)(c-u)}{(b-u)^3}}$$
$$[a>b>c>u] \qquad\qquad \text{BY (232.14)}$$

9. $$\int_c^u \frac{dx}{\sqrt{(a-x)(b-x)^5(x-c)}} = \frac{2}{3(a-b)^2(b-c)^2\sqrt{a-c}}$$
$$\times\left[(a-b)(2a-3b+c)\,F(\gamma,q) + 2(a-c)(2b-a-c)\,E(\gamma,q)\right]$$
$$+\frac{2\left[3ab+3bc-ac-5b^2-2u(a-2b+c)\right]}{3(a-b)^2(b-c)^2}\sqrt{\frac{(a-u)(u-c)}{(b-u)^3}}$$
$$[a>b>u>c] \qquad\qquad \text{BY (233.10)}$$

10.
$$\int_u^a \frac{dx}{\sqrt{(a-x)(x-b)^5(x-c)}} = \frac{2}{3(a-b)^2(b-c)^2\sqrt{a-c}}$$
$$\times [(b-c)(3b-2c-a)\,F(\lambda,p) + 2(a-c)(a+c-2b)\,E(\lambda,p)]$$
$$+ \frac{2\left[3ab+3bc-ac-5b^2+2u(2b-a-c)\right]}{3(a-b)^2(b-c)^2}\sqrt{\frac{(a-u)(u-c)}{(u-b)^3}}$$
$$[a>u>b>c] \qquad\qquad \text{BY (236.09)}$$

11.
$$\int_a^u \frac{dx}{\sqrt{(x-a)(x-b)^5(x-c)}} = \frac{2}{3(a-b)^2(b-c)^2\sqrt{a-c}}$$
$$\times [(a-b)(2a+c-3b)\,F(\mu,q) + 2(a-c)(2b-a-c)\,E(\mu,q)]$$
$$+ \frac{2}{3(a-b)(b-c)}\sqrt{\frac{(u-a)(u-c)}{(u-b)^3}}$$
$$[u>a>b>c] \qquad\qquad \text{BY (237.12)}$$

12.
$$\int_u^\infty \frac{dx}{\sqrt{(x-a)(x-b)^5(x-c)}} = \frac{2}{3(a-b)^2(b-c)^2\sqrt{a-c}}$$
$$\times [(a-b)(2a+c-3b)\,F(\nu,q) + 2(a-c)(2b-c-a)\,E(\nu,q)]$$
$$- \frac{2\left[3bc+2ab-ac-4b^2+u(3b-a-2c)\right]}{3(a-b)^2(b-c)}\sqrt{\frac{u-a}{(u-b)^3(u-c)}}$$
$$[u\ge a>b>c] \qquad\qquad \text{BY (238.04)}$$

13.
$$\int_{-\infty}^u \frac{dx}{\sqrt{(a-x)(b-x)(c-x)^5}} = \frac{2}{3(b-c)^2\sqrt{(a-c)^3}}[2(a+b-2c)\,E(\alpha,p) - (b-c)\,F(\alpha,p)]$$
$$+ \frac{2\left[ab-3ac-2bc+4c^2+u(2a+b-3c)\right]}{3(a-c)(b-c)^2}\sqrt{\frac{b-u}{(a-u)(c-u)^3}}$$
$$[a>b>c>u] \qquad\qquad \text{By (231.10)}$$

14.
$$\int_u^b \frac{dx}{\sqrt{(a-x)(b-x)(x-c)^5}} = \frac{2}{3(b-c)^2\sqrt{(a-c)^3}}[(2a+b-3c)\,F(\delta,q) - 2(a+b-2c)\,E(\delta,q)]$$
$$+ \frac{2\left[ab-3ac-2bc+4c^2+u(2a+b-3c)\right]}{3(b-c)^2(a-c)}\sqrt{\frac{b-u}{(a-u)(u-c)^3}}$$
$$[a>b>u>c] \qquad\qquad \text{BY (234.04)}$$

15.
$$\int_b^u \frac{dx}{\sqrt{(a-x)(x-b)(x-c)^5}} = \frac{2}{3(b-c)^2\sqrt{(a-c)^3}}[2(a+b-2c)\,E(\kappa,p) - (b-c)\,F(\kappa,p)]$$
$$+ \frac{2}{3(a-c)(b-c)}\sqrt{\frac{(a-u)(u-b)}{(u-c)^3}}$$
$$[a\ge u>b>c] \qquad\qquad \text{BY (235.20)}$$

16. $$\int_u^a \frac{dx}{\sqrt{(a-x)(x-b)(x-c)^5}} = \frac{2}{3(b-c)^2\sqrt{(a-c)^3}}\left[2(a+b-2c)\,E(\lambda,p) - (b-c)\,F(\lambda,p)\right]$$

$$-\frac{2\left[ab - 3ac - 3bc + 5c^2 + 2u(a+b-2c)\right]}{3(b-c)^2(a-c)^2}\sqrt{\frac{(a-u)(u-b)}{(u-c)^3}}$$

$$[a > u \geq b > c]\qquad\text{BY (236.10)}$$

17. $$\int_a^u \frac{dx}{\sqrt{(x-a)(x-b)(x-c)^5}} = \frac{2}{3(b-c)^2\sqrt{(a-c)^3}}\left[(2a+b-3c)\,F(\mu,q) - 2(a+b-2c)\,E(\mu,q)\right]$$

$$+\frac{2\left[4c^2 - ab - 2ac - bc + u(3a+2b-5c)\right]}{3(b-c)(a-c)^2}\sqrt{\frac{u-a}{(u-b)(u-c)^3}}$$

$$[u > a > b > c]\qquad\text{BY (237.13)}$$

18. $$\int_u^\infty \frac{dx}{\sqrt{(x-a)(x-b)(x-c)^5}} = \frac{2}{3(b-c)^2\sqrt{(a-c)^3}}\left[(2a+b-3c)\,F(\nu,q) - 2(a+b-2c)\,E(\nu,q)\right]$$

$$+\frac{2}{3(a-c)(b-c)}\sqrt{\frac{(u-a)(u-b)}{(u-c)^3}}$$

$$[u \geq a > b > c]\qquad\text{BY (238.03)}$$

3.135

1.[6] $$\int_{-\infty}^u \frac{dx}{\sqrt{(a-x)(b-x)^3(c-x)^3}} = \frac{2}{(a-b)(b-c)^2\sqrt{a-c}}\left[(b-c)\,F(\alpha,p) - (2a-b-c)\,E(\alpha,p)\right]$$

$$+\frac{2(b+c-2u)}{(b-c)^2\sqrt{(a-u)(b-u)(c-u)}}$$

$$[a > b > c > u]\qquad\text{BY (231.13)}$$

2. $$\int_u^a \frac{dx}{\sqrt{(a-x)(x-b)^3(x-c)^3}} = \frac{2}{(a-b)(b-c)^2\sqrt{a-c}}\left[(b-c)\,F(\lambda,p) - 2(2a-b-c)\,E(\lambda,p)\right]$$

$$+\frac{2(a-b-c+u)}{(a-b)(b-c)(a-c)}\sqrt{\frac{a-u}{(u-b)(u-c)}}$$

$$[a > u > b > c]\qquad\text{BY (236.15)}$$

3. $$\int_a^u \frac{dx}{\sqrt{(x-a)(x-b)^3(x-c)^3}} = \frac{2}{(a-b)(b-c)^2\sqrt{a-c}}\left[(2a-b-c)\,E(\mu,q) - 2(a-b)\,F(\mu,q)\right]$$

$$+\frac{2}{(a-c)(b-c)}\sqrt{\frac{u-a}{(u-b)(u-c)}}$$

$$[u > a > b > c]\qquad\text{BY (236.14)}$$

4. $$\int_u^\infty \frac{dx}{\sqrt{(x-a)(x-b)^3(x-c)^3}} = \frac{2}{(a-b)(b-c)^2\sqrt{a-c}}\left[(2a-b-c)\,E(\nu,q) - 2(a-b)\,F(\nu,q)\right]$$

$$-\frac{2}{(a-b)(b-c)}\sqrt{\frac{u-a}{(u-b)(u-c)}}$$

$$[u \geq a > b > c]\qquad\text{BY (238.13)}$$

5. $$\int_{-\infty}^{u} \frac{dx}{\sqrt{(a-x)^3(b-x)(c-x)^3}} = \frac{2}{(a-b)(b-c)\sqrt{(a-c)^3}} \left[(2b-a-c)\,E(\alpha,p) - (b-c)\,F(\alpha,p)\right]$$
$$+ \frac{2}{(b-c)(a-c)} \sqrt{\frac{b-u}{(a-u)(c-u)}}$$
$$[a > b > c > u] \qquad\qquad \text{BY(231.12)}$$

6. $$\int_{u}^{b} \frac{dx}{\sqrt{(a-x)^3(b-x)(x-c)^3}} = \frac{2}{(b-c)(a-b)\sqrt{(a-c)^3}} \left[(a-b)\,F(\delta,q) + (2b-a-c)\,E(\delta,q)\right]$$
$$+ \frac{2}{(b-c)(a-c)} \sqrt{\frac{b-u}{(a-u)(u-c)}}$$
$$[a > b > u > c] \qquad\qquad \text{BY (234.03)}$$

7. $$\int_{b}^{u} \frac{dx}{\sqrt{(a-x)^3(x-b)(x-c)^3}} = \frac{2}{(a-b)(b-c)\sqrt{(a-c)^3}} \left[(b-c)\,F(\kappa,p) - (2b-a-c)\,E(\kappa,p)\right]$$
$$+ \frac{2}{(a-b)(a-c)} \sqrt{\frac{u-b}{(a-u)(u-c)}}$$
$$[a > u > b > c] \qquad\qquad \text{BY (235.15)}$$

8. $$\int_{u}^{\infty} \frac{dx}{\sqrt{(x-a)^3(x-b)(x-c)^3}} = \frac{2}{(a-b)(b-c)\sqrt{(a-c)^3}} \left[(a+c-2b)\,E(\nu,q) - (a-b)\,F(\nu,q)\right]$$
$$+ \frac{2}{(a-b)(a-c)} \sqrt{\frac{u-b}{(u-a)(u-c)}}$$
$$[u > a > b > c] \qquad\qquad \text{BY (238.14)}$$

9. $$\int_{-\infty}^{u} \frac{dx}{\sqrt{(a-x)^3(b-x)^3(c-x)}} = \frac{2}{(b-c)(a-b)^2\sqrt{a-c}} \left[(a+b-2c)\,E(\alpha,p) - 2(b-c)\,F(\alpha,p)\right]$$
$$- \frac{2}{(a-b)(b-c)} \sqrt{\frac{c-u}{(a-u)(b-u)}}$$
$$[a > b > c \geq u] \qquad\qquad \text{BY (231.11)}$$

10. $$\int_{u}^{c} \frac{dx}{\sqrt{(a-x)^3(b-x)^3(c-x)}} = \frac{2}{(a-b)^2(b-c)\sqrt{a-c}} \left[(a+b-2c)\,E(\beta,p) - 2(b-c)\,F(\beta,p)\right]$$
$$+ \frac{2}{(a-b)(a-c)} \sqrt{\frac{c-u}{(a-u)(b-u)}}$$
$$[a > b > c > u] \qquad\qquad \text{BY (232.15)}$$

11. $$\int_{c}^{u} \frac{dx}{\sqrt{(a-x)^3(b-x)^3(x-c)}} = \frac{2}{(a-b)^2(b-c)\sqrt{a-c}} \left[(a-b)\,F(\gamma,q) - (a+b-2c)\,E(\gamma,q)\right]$$
$$+ \frac{2\left[a^2+b^2-ac-bc-u(a+b-2c)\right]}{(a-b)^2(b-c)(a-c)} \sqrt{\frac{u-c}{(a-u)(b-u)}}$$
$$[a > b > u > c] \qquad\qquad \text{BY (233.11)}$$

12. $$\int_u^\infty \frac{dx}{\sqrt{(x-a)^3(x-b)^3(x-c)}} = \frac{2}{(a-b)^2(b-c)\sqrt{a-c}}[(a-b)\,F(\nu,q)-(a+b-2c)\,E(\nu,q)]$$
$$+\frac{2u-a-b}{(a-b)^2\sqrt{(u-a)(u-b)(u-c)}}$$
$$[u>a>b>c] \qquad \text{BY (238.15)}$$

3.136

1. $$\int_{-\infty}^u \frac{dx}{\sqrt{(a-x)^3(b-x)^3(c-x)^3}}$$
$$= \frac{2}{(a-b)^2(b-c)^2\sqrt{(a-c)^3}}$$
$$\times\left[(b-c)(a+b-2c)\,F(\alpha,p)-2\left(c^2+a^2+b^2-ab-ac-bc\right)E(\alpha,p)\right]$$
$$+\frac{2[c(a-c)+b(a-b)-u(2a-c-b)]}{(a-b)(a-c)(b-c)^2\sqrt{(a-u)(b-u)(c-u)}}$$
$$[a>b>c>u] \qquad \text{BY (231.14)}$$

2. $$\int_u^\infty \frac{dx}{\sqrt{(x-a)^3(x-b)^3(x-c)^3}}$$
$$= \frac{2}{(a-b)^2(b-c)^2\sqrt{(a-c)^3}}$$
$$\times\left[(a-b)(2a-b-c)\,F(\nu,q)-2\left(a^2+b^2+c^2-ab-ac-bc\right)E(\nu,q)\right]$$
$$+\frac{2[u(a+b-2c)-a(a-c)-b(b-c)]}{(a-b)^2(a-c)(b-c)\sqrt{(u-a)(u-b)(u-c)}}$$
$$[u>a>b>c] \qquad \text{BY (238.16)}$$

3.137

1.[6] $$\int_{-\infty}^u \frac{dx}{(r-x)\sqrt{(a-x)(b-x)(c-x)}} = \frac{2}{(a-r)\sqrt{a-c}}\left[\Pi\left(\alpha,\frac{a-r}{a-c},p\right)-F(\alpha,p)\right]$$
$$[a>b>c\geq u] \qquad \text{BY (231.15)}$$

2. $$\int_u^c \frac{dx}{(r-x)\sqrt{(a-x)(b-x)(c-x)}} = \frac{2(c-b)}{(r-b)(r-c)\sqrt{a-c}}$$
$$\times\Pi\left(\beta,\frac{r-b}{r-c},p\right)+\frac{2}{(r-b)\sqrt{a-c}}F(\beta,p)$$
$$[a>b>c>u,\quad r\neq0] \qquad \text{BY (232.17)}$$

3. $$\int_c^u \frac{dx}{(r-x)\sqrt{(a-x)(b-x)(x-c)}} = \frac{2}{(r-c)\sqrt{a-c}}\Pi\left(\gamma,\frac{b-c}{r-c},q\right)$$
$$[a>b\geq u>c,\quad r\neq c] \qquad \text{BY (233.02)}$$

4. $$\int_u^b \frac{dx}{(r-x)\sqrt{(a-x)(b-x)(x-c)}} = \frac{2}{(r-a)(r-b)\sqrt{a-c}}$$
$$\times\left[(b-a)\Pi\left(\delta,q^2\frac{r-a}{r-b},q\right)+(r-b)\,F(\delta,q)\right]$$
$$[a>b>u\geq c,\quad r\neq b] \qquad \text{BY (234.18)}$$

5. $\displaystyle\int_b^u \frac{dx}{(x-r)\sqrt{(a-x)(x-b)(x-c)}} = \frac{2}{(c-r)(b-r)\sqrt{a-c}}$

$$\times \left[(c-b)\,\Pi\left(\kappa, p^2\frac{c-r}{b-r}, p\right) + (b-r)\,F\left(\kappa, p\right) \right]$$

$$[a \geq u > b > c, \quad r \neq b] \qquad \text{BY (235.17)}$$

6.[8] $\displaystyle\int_u^a \frac{dx}{(x-r)\sqrt{(a-x)(x-b)(x-c)}} = \frac{2}{(a-r)\sqrt{a-c}}\,\Pi\left(\lambda, \frac{a-b}{a-r}, p\right)$

$$[a > u \geq b > c, \quad r \neq a] \qquad \text{BY (236.02)}$$

7. $\displaystyle\int_a^u \frac{dx}{(x-r)\sqrt{(x-a)(x-b)(x-c)}} = \frac{2}{(b-r)(a-r)\sqrt{a-c}}$

$$\times \left[(b-a)\,\Pi\left(\mu, \frac{b-r}{a-b}, q\right) + (a-p)\,F(\mu, q) \right]$$

$$[u > a > b > c, \quad r \neq a] \qquad \text{BY (237.17)}$$

8. $\displaystyle\int_u^\infty \frac{dx}{(x-r)\sqrt{(x-a)(x-b)(x-c)}} = \frac{2}{(r-c)\sqrt{a-c}}\left[\Pi\left(\nu, \frac{r-c}{a-c}, q\right) - F(\nu, q) \right]$

$$[u \geq a > b > c] \qquad \text{BY (238.06)}$$

3.138

1. $\displaystyle\int_0^u \frac{dx}{\sqrt{x(1-x)\left(1-k^2 x\right)}} = 2\,F\left(\arcsin\sqrt{u}, k\right) \qquad [0 < u < 1] \qquad \text{PE (532), JA}$

2. $\displaystyle\int_u^1 \frac{dx}{\sqrt{x(1-x)\left(k'^2 + k^2 x\right)}} = 2\,F\left(\arccos\sqrt{u}, k\right) \qquad [0 < u < 1] \qquad \text{PE(533)}$

3. $\displaystyle\int_u^1 \frac{dx}{\sqrt{x(1-x)\left(x-k'^2\right)}} = 2\,F\left(\arcsin\frac{\sqrt{1-u}}{k}, k\right) \qquad [0 < u < 1] \qquad \text{PE (534)}$

4. $\displaystyle\int_0^u \frac{dx}{\sqrt{x(1+x)\left(1+k'^2 x\right)}} = 2\,F\left(\arctan\sqrt{u}, k\right) \qquad [0 < u < 1] \qquad \text{PE (535)}$

5. $\displaystyle\int_0^u \frac{dx}{\sqrt{x\left[1 + x^2 + 2\left(k'^2 - k^2\right)x\right]}} = F\left(2\arctan\sqrt{u}, k\right)$

$$[0 < u < 1] \qquad \text{JA}$$

6. $\displaystyle\int_u^1 \frac{dx}{\sqrt{x\left[k'^2\left(1+x^2\right) + 2\left(1+k^2\right)x\right]}} = F\left(\frac{\pi}{2} - 2\arctan\sqrt{u}, k\right)$

$$[0 < u < 1] \qquad \text{JA}$$

7. $\displaystyle\int_a^u \frac{dx}{\sqrt{(x-\alpha)\left[(x-m)^2 + n^2\right]}} = \frac{1}{\sqrt{p}}\,F\left(2\arctan\sqrt{\frac{u-\alpha}{p}}, \sqrt{\frac{p+m-\alpha}{2p}}\right)$

$$[\alpha < u],$$

8. $\displaystyle\int_u^a \frac{dx}{\sqrt{(\alpha - x)\left[(x-m)^2 + n^2\right]}} = \frac{1}{\sqrt{p}}\, F\left(2\,\mathrm{arccot}\,\sqrt{\frac{\alpha - u}{p}},\ \sqrt{\frac{p - m + \alpha}{2p}}\right)$

$$[u < \alpha]\,,$$

where $p = \sqrt{(m - \alpha)^2 + n^2}$.

3.139   **Notation** $\alpha = \arccos \dfrac{1 - \sqrt{3} - u}{1 + \sqrt{3} - u}$, $\beta = \arccos \dfrac{\sqrt{3} - 1 + u}{\sqrt{3} + 1 - u}$,

$\gamma = \arccos \dfrac{\sqrt{3} + 1 - u}{\sqrt{3} - 1 + u}$, $\delta = \arccos \dfrac{u - 1 - \sqrt{3}}{u - 1 + \sqrt{3}}$.

1. $\displaystyle\int_{-\infty}^u \frac{dx}{\sqrt{1 - x^3}} = \frac{1}{\sqrt[4]{3}}\, F\left(\alpha, \sin 75°\right)$ H 66 (285)

2. $\displaystyle\int_u^1 \frac{dx}{\sqrt{1 - x^3}} = \frac{1}{\sqrt[4]{3}}\, F\left(\beta, \sin 75°\right)$ H 65 (284)

3. $\displaystyle\int_1^u \frac{dx}{\sqrt{x^3 - 1}} = \frac{1}{\sqrt[4]{3}}\, F\left(\gamma, \sin 15°\right)$ H 65 (283)

4. $\displaystyle\int_u^\infty \frac{dx}{\sqrt{x^3 - 1}} = \frac{1}{\sqrt[4]{3}}\, F\left(\delta, \sin 15°\right)$ H 65 (282)

5. $\displaystyle\int_0^1 \frac{dx}{\sqrt{1 - x^3}} = \frac{1}{2\pi\sqrt{3}\sqrt[3]{2}}\left\{\Gamma\left(\frac{1}{3}\right)\right\}^3$ MO 9

6. $\displaystyle\int_0^1 \frac{x\,dx}{\sqrt{1 - x^3}} = \frac{1}{\pi}\frac{\sqrt{3}}{\sqrt[3]{4}}\left\{\Gamma\left(\frac{2}{3}\right)\right\}^3$ MO 9

7. $\displaystyle\int_u^1 \sqrt{1 - x^3}\, dx = \frac{1}{5}\left\{\sqrt[4]{27}\, F\left(\beta, \sin 75°\right) - 2u\sqrt{1 - u^3}\right\}$ BY (244.01)

8. $\displaystyle\int_u^1 \frac{x\,dx}{\sqrt{1 - x^3}} = \left(3^{-\frac{1}{4}} - 3^{\frac{1}{4}}\right) F\left(\beta, \sin 75°\right) + 2\sqrt[4]{3}\, E\left(\beta, \sin 75°\right) - \frac{2\sqrt{1 - u^3}}{\sqrt{3} + 1 - u}$ BY (244.05)

9. $\displaystyle\int_u^1 \frac{x^m\,dx}{\sqrt{1 - x^3}} = \frac{2u^{m-2}\sqrt{1 - u^3}}{2m - 1} + \frac{2(m - 2)}{2m - 1}\int_u^1 \frac{x^{m-3}\,dx}{\sqrt{1 - x^3}}$ BY (244.07)

10. $\displaystyle\int_1^u \frac{x\,dx}{\sqrt{x^3 - 1}} = \left(3^{-\frac{1}{4}} + 3^{\frac{1}{4}}\right) F\left(\gamma, \sin 15°\right) - 2\sqrt[4]{3}\, E\left(\gamma, \sin 15°\right) + \frac{2\sqrt{u^3 - 1}}{\sqrt{3} - 1 + u}$ BY (240.05)

11. $\displaystyle\int_{-\infty}^u \frac{dx}{(1 - x)\sqrt{1 - x^3}} = \frac{1}{\sqrt[4]{27}}\left[F\left(\alpha, \sin 75°\right) - 2\,E\left(\alpha, \sin 75°\right)\right] + \frac{2}{\sqrt{3}}\frac{\sqrt{1 + u + u^2}}{\left(1 + \sqrt{3} - u\right)\sqrt{1 - u}}$

$$[u \neq 1]$$ BY (246.06)

12. $\displaystyle\int_u^\infty \frac{dx}{(x - 1)\sqrt{x^3 - 1}} = \frac{1}{\sqrt[4]{27}}\left[F\left(\delta, \sin 15°\right) - 2\,E\left(\delta, \sin 15°\right)\right] + \frac{2}{\sqrt{3}}\frac{\sqrt{1 + u + u^2}}{\left(u - 1 + \sqrt{3}\right)\sqrt{u - 1}}$

$$[u \neq 1]$$ BY (242.03)

13. $$\int_{-\infty}^{u} \frac{(1-x)\,dx}{\left(1+\sqrt{3}-x\right)^2\sqrt{1-x^3}} = \frac{2-\sqrt{3}}{\sqrt[4]{27}}\left[F\left(\alpha,\sin 75°\right) - E\left(\alpha,\sin 75°\right)\right]$$ BY (246.07)

14. $$\int_{u}^{1} \frac{(1-x)\,dx}{\left(1+\sqrt{3}-x\right)^2\sqrt{1-x^3}} = \frac{2-\sqrt{3}}{\sqrt[4]{27}}\left[F\left(\beta,\sin 75°\right) - E\left(\beta,\sin 75°\right)\right]$$ BY (244.04)

15. $$\int_{1}^{u} \frac{(x-1)\,dx}{\left(1+\sqrt{3}-x\right)^2\sqrt{x^3-1}} = \frac{2\left(\sqrt{3}-2\right)}{\sqrt{3}}\frac{\sqrt{u^3-1}}{u^2-2u-2} - \frac{2-\sqrt{3}}{\sqrt[4]{27}}E\left(\gamma,\sin 15°\right)$$ BY (240.08)

16. $$\int_{u}^{\infty} \frac{(x-1)\,dx}{\left(1+\sqrt{3}-x\right)^2\sqrt{x^3-1}} = \frac{2\left(2-\sqrt{3}\right)}{\sqrt{3}}\frac{\sqrt{u^3-1}}{u^2-2u-2} - \frac{2-\sqrt{3}}{\sqrt[4]{27}}E\left(\delta,\sin 15°\right)$$ BY (242.07)

17. $$\int_{-\infty}^{u} \frac{(1-x)\,dx}{\left(1-\sqrt{3}-x\right)^2\sqrt{1-x^3}} = \frac{2+\sqrt{3}}{\sqrt[4]{27}}\left[\frac{2\sqrt[4]{3}\sqrt{1-u^3}}{u^2-2u-2} - E\left(\alpha,\sin 75°\right)\right]$$ BY (246.08)

18. $$\int_{1}^{u} \frac{(x-1)\,dx}{\left(1-\sqrt{3}-x\right)^2\sqrt{x^3-1}} = \frac{2+\sqrt{3}}{\sqrt[4]{27}}\left[F\left(\gamma,\sin 15°\right) - E\left(\gamma,\sin 15°\right)\right]$$ BY (240.04)

19. $$\int_{u}^{\infty} \frac{(x-1)\,dx}{\left(1-\sqrt{3}-x\right)^2\sqrt{x^3-1}} = \frac{2+\sqrt{3}}{\sqrt[4]{27}}\left[F\left(\delta,\sin 15°\right) - E\left(\delta,\sin 15°\right)\right]$$ BY (242.05)

20. $$\int_{-\infty}^{u} \frac{\left(x^2+x+1\right)\,dx}{\left(1+\sqrt{3}-x\right)^2\sqrt{1-x^3}} = \frac{1}{\sqrt[4]{3}}E\left(\alpha,\sin 75°\right)$$ BY (246.01)

21. $$\int_{u}^{1} \frac{\left(x^2+x+1\right)\,dx}{\left(x-1+\sqrt{3}\right)^2\sqrt{1-x^3}} = \frac{1}{\sqrt[4]{3}}E\left(\beta,\sin 75°\right)$$ BY (244.02)

22. $$\int_{1}^{u} \frac{\left(x^2+x+1\right)\,dx}{\left(\sqrt{3}+x-1\right)^2\sqrt{x^3-1}} = \frac{1}{\sqrt[4]{3}}E\left(\gamma,\sin 15°\right)$$ BY (240.01)

23. $$\int_{u}^{\infty} \frac{\left(x^2+x+1\right)\,dx}{\left(x-1+\sqrt{3}\right)^2\sqrt{x^3-1}} = \frac{1}{\sqrt[4]{3}}E\left(\delta,\sin 15°\right)$$ BY (242.01)

24. $$\int_{1}^{u} \frac{(x-1)\,dx}{\left(x^2+x+1\right)\sqrt{x^3-1}} = \frac{4}{\sqrt[4]{27}}E\left(\gamma,\sin 15°\right) - \frac{2+\sqrt{3}}{\sqrt[4]{27}}F\left(\gamma,\sin 15°\right)$$
$$- \frac{2-\sqrt{3}}{\sqrt{3}}\frac{2(u-1)\left(\sqrt{3}+1-u\right)}{\left(\sqrt{3}-1+u\right)\sqrt{u^3-1}}$$
BY (240.09)

25. $$\int_{-\infty}^{u} \frac{\left(1+\sqrt{3}-x\right)^2\,dx}{\left[\left(1+\sqrt{3}-x\right)^2-4\sqrt{3}p^2(1-x)\right]\sqrt{1-x^3}} = \frac{1}{\sqrt[4]{3}}\Pi\left(\alpha,p^2,\sin 75°\right)$$ BY (246.02)

26. $$\int_{u}^{1} \frac{\left(1+\sqrt{3}-x\right)^2\,dx}{\left[\left(1+\sqrt{3}-x\right)^2-4\sqrt{3}p^2(1-x)\right]\sqrt{1-x^3}} = \frac{1}{\sqrt[4]{3}}\Pi\left(\beta,p^2,\sin 75°\right)$$ BY (244.03)

27. $$\int_{1}^{u} \frac{\left(1-\sqrt{3}-x\right)^2\,dx}{\left[\left(1-\sqrt{3}-x\right)^2-4\sqrt{3}p^2(x-1)\right]\sqrt{x^3-1}} = \frac{1}{\sqrt[4]{3}}\Pi\left(\gamma,p^2,\sin 15°\right)$$ BY (240.02)

28. $\displaystyle\int_u^\infty \frac{\left(1-\sqrt{3}-x\right)^2 dx}{\left[\left(1-\sqrt{3}-x\right)^2 - 4\sqrt{3}p^2(x-1)\right]\sqrt{x^3-1}} = \frac{1}{\sqrt[4]{3}}\,\Pi\left(\delta, p^2, \sin 15°\right)$ BY (242.02)

3.141   **Notation**: In **3.141** and **3.142** we set:

$$\alpha = \arcsin\sqrt{\frac{a-c}{a-u}}, \qquad \beta = \arcsin\sqrt{\frac{c-u}{b-u}}, \qquad \gamma = \arcsin\sqrt{\frac{u-c}{b-c}}e$$

$$\delta = \arcsin\sqrt{\frac{(a-c)(b-u)}{(b-c)(a-u)}}, \qquad \kappa = \arcsin\sqrt{\frac{(a-c)(u-b)}{(a-b)(u-c)}}, \qquad \lambda = \arcsin\sqrt{\frac{a-u}{a-b}},$$

$$\mu = \arcsin\sqrt{\frac{u-a}{u-b}}, \qquad \nu = \arcsin\sqrt{\frac{a-c}{u-c}}, \qquad p = \sqrt{\frac{a-b}{a-c}}, \qquad q = \sqrt{\frac{b-c}{a-c}}.$$

1. $\displaystyle\int_u^c \sqrt{\frac{a-x}{(b-x)(c-x)}}\,dx = 2\sqrt{a-c}\left[F(\beta,p) - E(\beta,p)\right] + 2\sqrt{\frac{(a-u)(c-u)}{b-u}}$

$[a > b > c > u]$ BY (232.06)

2. $\displaystyle\int_c^u \sqrt{\frac{a-x}{(b-x)(x-c)}}\,dx = 2\sqrt{a-c}\,E(\gamma,q)$ $[a > b \geq u > c]$ BY (233.01)

3. $\displaystyle\int_u^b \sqrt{\frac{a-x}{(b-x)(x-c)}}\,dx = 2\sqrt{a-c}\,E(\delta,q) - 2\sqrt{\frac{(b-u)(u-c)}{a-u}}$

$[a > b > u \geq c]$ BY (234.06)

4. $\displaystyle\int_b^u \sqrt{\frac{a-x}{(x-b)(x-c)}}\,dx = 2\sqrt{a-c}\left[F(\kappa,p) - E(\kappa,p)\right] + 2\sqrt{\frac{(a-u)(u-b)}{u-c}}$

$[a \geq u > b > c]$ BY (235.07)

5. $\displaystyle\int_u^a \sqrt{\frac{a-x}{(x-b)(x-c)}}\,dx = 2\sqrt{a-c}\left[F(\lambda,p) - E(\lambda,p)\right]$

$[a > u \geq b > c]$ BY (236.04)

6. $\displaystyle\int_a^u \sqrt{\frac{x-a}{(x-b)(x-c)}}\,dx = -2\sqrt{a-c}\,E(\mu,q) + 2\sqrt{\frac{(u-a)(u-c)}{u-b}}$

$[u > a > b > c]$ BY (237.03)

7. $\displaystyle\int_u^c \sqrt{\frac{b-x}{(a-x)(c-x)}}\,dx = \frac{2(b-c)}{\sqrt{a-c}}\,F(\beta,p) - 2\sqrt{a-c}\,E(\beta,p) + 2\sqrt{\frac{(a-u)(c-u)}{b-u}}$

$[a > b > c > u]$ BY (232.07)

8. $\displaystyle\int_c^u \sqrt{\frac{b-x}{(a-x)(x-c)}}\,dx = 2\sqrt{a-c}\,E(\gamma,q) - \frac{2(a-b)}{\sqrt{a-c}}\,F(\gamma,q)$

$[a > b \geq u > c]$ BY (233.04)

9. $\int_u^b \sqrt{\dfrac{b-x}{(a-x)(x-c)}}\, dx = 2\sqrt{a-c}\, E(\delta, q) - \dfrac{2(a-b)}{\sqrt{a-c}}\, F(\delta, q) - 2\sqrt{\dfrac{(b-u)(u-c)}{a-u}}$

$$[a > b > u \geq c] \qquad \text{BY (234.07)}$$

10. $\int_b^u \sqrt{\dfrac{x-b}{(a-x)(x-c)}}\, dx = 2\sqrt{a-c}\, E(\kappa, p) - \dfrac{2(b-c)}{\sqrt{a-c}}\, F(\kappa, p) - 2\sqrt{\dfrac{(a-u)(u-b)}{u-c}}$

$$[a \geq u > b > c] \qquad \text{BY (235.06)}$$

11. $\int_u^a \sqrt{\dfrac{x-b}{(a-x)(x-c)}}\, dx = 2\sqrt{a-c}\, E(\lambda, p) - \dfrac{2(b-c)}{\sqrt{a-c}}\, F(\lambda, p)$

$$[a > u \geq b > c] \qquad \text{BY (236.03)}$$

12. $\int_a^u \sqrt{\dfrac{x-b}{(x-a)(x-c)}}\, dx = \dfrac{2(a-b)}{\sqrt{a-c}}\, F(\mu, q) - 2\sqrt{a-c}\, E(\mu, q) + 2\sqrt{\dfrac{(u-a)(u-c)}{u-b}}$

$$[u > a > b > c] \qquad \text{BY (237.04)}$$

13. $\int_u^c \sqrt{\dfrac{c-x}{(a-x)(b-x)}}\, dx = -2\sqrt{a-c}\, E(\beta, p) + 2\sqrt{\dfrac{(a-u)(c-u)}{b-u}}$

$$[a > b > c > u] \qquad \text{BY (232.08)}$$

14. $\int_c^u \sqrt{\dfrac{x-c}{(a-x)(b-x)}}\, dx = 2\sqrt{a-c}\, [F(\gamma, q) - E(\gamma, q)]$

$$[a > b \geq u > c] \qquad \text{BY (233.03)}$$

15. $\int_u^b \sqrt{\dfrac{x-c}{(a-x)(b-x)}}\, dx = 2\sqrt{a-c}\, [F(\delta, q) - E(\delta, q)] + 2\sqrt{\dfrac{(b-u)(u-c)}{a-u}}$

$$[a > b > u \geq c] \qquad \text{BY (234.08)}$$

16. $\int_b^u \sqrt{\dfrac{x-c}{(a-x)(x-b)}}\, dx = 2\sqrt{a-c}\, E(\kappa, p) - 2\sqrt{\dfrac{(a-u)(u-b)}{u-c}}$

$$[a \geq u > b > c] \qquad \text{BY (235.07)}$$

17. $\int_u^a \sqrt{\dfrac{x-c}{(a-x)(x-b)}}\, dx = 2\sqrt{a-c}\, E(\lambda, p) \qquad\qquad [a > u \geq b > c] \qquad \text{BY (236.01)}$

18. $\int_a^u \sqrt{\dfrac{x-c}{(x-a)(x-b)}}\, dx = 2\sqrt{a-c}\, [F(\mu, q) - E(\mu, q)] + 2\sqrt{\dfrac{(u-a)(u-c)}{u-b}}$

$$[u > a > b > c] \qquad \text{BY (237.05)}$$

19. $\int_u^c \sqrt{\dfrac{(b-x)(c-x)}{a-x}}\, dx = \dfrac{2}{3}\sqrt{a-c}\, [(2a - b - c)\, E(\beta, p) - (b-c)\, F(\beta, p)]$

$$+ \dfrac{2}{3}(2b - 2a + c - u)\sqrt{\dfrac{(a-u)(c-u)}{b-u}}$$

$$[a > b > c > u] \qquad \text{BY (232.11)}$$

20. $\displaystyle\int_c^u \sqrt{\frac{(x-c)(b-x)}{a-x}}\,dx = \frac{2}{3}\sqrt{a-c}\left[(2a-b-c)\,E(\gamma,q) - 2(a-b)\,F(\gamma,q)\right]$

$$-\frac{2}{3}\sqrt{(a-u)(b-u)(u-c)}$$

$$[a > b \geq u > c] \qquad\qquad \text{BY (233.06)}$$

21. $\displaystyle\int_u^b \sqrt{\frac{(x-c)(b-x)}{a-x}}\,dx = \frac{2}{3}\sqrt{a-c}\left[2(b-a)\,F(\delta,q) + (2a-b-c)\,E(\delta,q)\right]$

$$+\frac{2}{3}(2c-b-u)\sqrt{\frac{(b-u)(u-c)}{a-u}}$$

$$[a > b > u \geq c] \qquad\qquad \text{BY (234.11)}$$

22. $\displaystyle\int_b^u \sqrt{\frac{(x-b)(x-c)}{a-x}}\,dx = \frac{2}{3}\sqrt{a-c}\left[(2a-b-c)\,E(\kappa,p) - (b-c)\,F(\kappa,p)\right]$

$$+\frac{2}{3}(b+2c-2a-u)\sqrt{\frac{(a-u)(u-b)}{u-c}}$$

$$[a \geq u > b > c] \qquad\qquad \text{BY (235.10)}$$

23. $\displaystyle\int_u^a \sqrt{\frac{(x-b)(x-c)}{a-x}}\,dx + \frac{2}{3}\sqrt{a-c}\left[(2a-b-c)\,E(\lambda,p) - (b-c)\,F(\lambda,p)\right]$

$$+\frac{2}{3}\sqrt{(a-u)(u-b)(u-c)}$$

$$[a > u \geq b > c] \qquad\qquad \text{BY (236.07)}$$

24. $\displaystyle\int_a^u \sqrt{\frac{(x-b)(x-c)}{x-a}}\,dx = \frac{2}{3}\sqrt{a-c}\left[2(a-b)\,F(\mu,q) + (b+c-2a)\,E(\mu,q)\right]$

$$+\frac{2}{3}(u+2a-2b-c)\sqrt{\frac{(u-a)(u-b)}{u-c}}$$

$$[u > a > b > c] \qquad\qquad \text{BY (237.08)}$$

25. $\displaystyle\int_u^c \sqrt{\frac{(a-x)(c-x)}{b-x}}\,dx = \frac{2}{3}\sqrt{a-c}\left[(2b-a-c)\,E(\beta,p) - (b-c)\,F(\beta,p)\right]$

$$+\frac{2}{3}(a+c-b-u)\sqrt{\frac{(a-u)(c-u)}{b-u}}$$

$$[a > b > c > u] \qquad\qquad \text{BY (232.10)}$$

26. $\displaystyle\int_c^u \sqrt{\frac{(a-x)(x-c)}{b-x}}\,dx = \frac{2}{3}\sqrt{a-c}\left[(2b-a-c)\,E(\gamma,q) + (a-b)\,F(\gamma,q)\right]$

$$-\frac{2}{3}\sqrt{(a-u)(b-u)(u-c)}$$

$$[a > b \geq u > c] \qquad\qquad \text{BY (233.05)}$$

27. $\displaystyle\int_u^b \sqrt{\frac{(a-x)(x-c)}{b-x}}\,dx = \frac{2}{3}\sqrt{a-c}\left[(a-b)\,F(\delta,q) + (2b-a-c)\,E(\delta,q)\right]$

$$+\frac{2}{3}(2a+c-2b-u)\sqrt{\frac{(b-u)(u-c)}{a-u}}$$

$$[a > b > u \geq c] \qquad\qquad \text{BY (234.10)}$$

28. $\int_b^u \sqrt{\dfrac{(a-x)(x-c)}{x-b}}\,dx = \dfrac{2}{3}\sqrt{a-c}\,[(b-c)\,F\,(\kappa,p)+(a+c-2b)\,E\,(\kappa,p)]$

$$+\dfrac{2}{3}(2b-a-2c+u)\sqrt{\dfrac{(a-u)(u-b)}{u-c}}$$

$$[a \geq u > b > c] \qquad \text{BY (235.11)}$$

29. $\int_u^a \sqrt{\dfrac{(a-x)(x-c)}{x-b}}\,dx = \dfrac{2}{3}\sqrt{a-c}\,[(a+c-2b)\,E(\lambda,p)+(b-c)\,F(\lambda,p)]$

$$-\dfrac{2}{3}\sqrt{(a-u)(u-b)(u-c)}$$

$$[a > u \geq b > c] \qquad \text{BY (236.06)}$$

30. $\int_a^u \sqrt{\dfrac{(x-a)(x-c)}{x-b}}\,dx = \dfrac{2}{3}\dfrac{\sqrt{(a-c)^3}}{b-c}\,[(a+c-2b)\,E(\mu,q)-(a-b)\,F(\mu,q)]$

$$+\dfrac{2}{3}\dfrac{a-c}{b-c}(u+b-a-c)\sqrt{\dfrac{(u-a)(u-c)}{u-b}}$$

$$[u > a > b > c] \qquad \text{BY (237.06)}$$

31. $\int_u^c \sqrt{\dfrac{(a-x)(b-x)}{c-x}}\,dx = \dfrac{2}{3}\sqrt{a-c}\,[2(b-c)\,F(\beta,p)+(2c-a-b)\,E(\beta,p)]$

$$+\dfrac{2}{3}(a+2b-2c-u)\sqrt{\dfrac{(a-u)(c-u)}{b-u}}$$

$$[a > b > c > u] \qquad \text{BY (232.09)}$$

32. $\int_c^u \sqrt{\dfrac{(a-x)(b-x)}{x-c}}\,dx = \dfrac{2}{3}\sqrt{a-c}\,[(a+b-2c)\,E(\gamma,q)-(a-b)\,F(\gamma,q)]$

$$+\dfrac{2}{3}\sqrt{(a-u)(b-u)(u-c)}$$

$$[a > b \geq u > c] \qquad \text{BY (233.07)}$$

33. $\int_u^b \sqrt{\dfrac{(a-x)(b-x)}{x-c}}\,dx = \dfrac{2}{3}\sqrt{a-c}\,[(a+b-2c)\,E(\delta,q)-(a-b)\,F(\delta,q)]$

$$+\dfrac{2}{3}(2c-2a-b+u)\sqrt{\dfrac{(b-u)(u-c)}{a-u}}$$

$$[a > b > u \geq c] \qquad \text{BY (234.09)}$$

34. $\int_b^u \sqrt{\dfrac{(a-x)(x-b)}{x-c}}\,dx = \dfrac{2}{3}\sqrt{a-c}\,[(a+b-2c)\,E\,(\kappa,p)-2(b-c)\,F\,(\kappa,p)]$

$$+\dfrac{2}{3}(u+c-a-b)\sqrt{\dfrac{(a-u)(u-b)}{u-c}}$$

$$[a \geq u > b > c] \qquad \text{BY (235.09)}$$

35. $\int_u^a \sqrt{\dfrac{(a-x)(x-b)}{x-c}}\,dx = \dfrac{2}{3}\sqrt{a-c}\,[(a+b-2c)\,E(\lambda,p)-2(b-c)\,F(\lambda,p)]$

$$-\dfrac{2}{3}\sqrt{(a-u)(u-b)(u-c)}$$

$$[a > u \geq b > c] \qquad \text{BY (236.05)}$$

36. $\int_a^u \sqrt{\dfrac{(x-a)(x-b)}{x-c}}\, dx = \dfrac{2}{3}\sqrt{a-c}\,[(a+b-2c)\,E(\mu,q) - (a-b)\,F(\mu,q)]$

$$+\dfrac{2}{3}(u+2c-a-2b)\sqrt{\dfrac{(u-a)(u-c)}{u-b}}$$

$$[u > a > b > c] \qquad\qquad \text{BY (237.07)}$$

3.142

1. $\int_{-\infty}^u \sqrt{\dfrac{a-x}{(b-x)(c-x)^3}}\, dx = \dfrac{2}{\sqrt{a-c}}\,F(\alpha,p) - \dfrac{2\sqrt{a-c}}{b-c}\,E(\alpha,p) + \dfrac{2(a-c)}{b-c}\sqrt{\dfrac{b-u}{(a-u)(c-u)}}$

$$[a > b > c > u] \qquad\qquad \text{BY (231.05)}$$

2. $\int_u^b \sqrt{\dfrac{a-x}{(b-x)(x-c)^3}}\, dx = 2\dfrac{a-b}{(b-c)\sqrt{a-c}}\,F(\delta,q) - \dfrac{2\sqrt{a-c}}{b-c}\,E(\delta,q)$

$$+2\dfrac{a-c}{b-c}\sqrt{\dfrac{b-u}{(a-u)(u-c)}}$$

$$[a > b > u > c] \qquad\qquad \text{BY (234.13)}$$

3. $\int_b^u \sqrt{\dfrac{a-x}{(x-b)(x-c)^3}}\, dx = \dfrac{2\sqrt{a-c}}{b-c}\,E(\kappa,p) - \dfrac{2}{\sqrt{a-c}}\,F(\kappa,p)$

$$[a \geq u > b > c] \qquad\qquad \text{BY (235.12)}$$

4. $\int_u^a \sqrt{\dfrac{a-x}{(x-b)(x-c)^3}}\, dx = \dfrac{2\sqrt{a-c}}{b-c}\,E(\lambda,p) - \dfrac{2}{\sqrt{a-c}}\,F(\lambda,p) - \dfrac{2}{b-c}\sqrt{\dfrac{(a-u)(u-b)}{u-c}}$

$$[a > u \geq b > c] \qquad\qquad \text{BY (236.12)}$$

5. $\int_a^u \sqrt{\dfrac{x-a}{(x-b)(x-c)^3}}\, dx = \dfrac{2\sqrt{a-c}}{b-c}\,E(\mu,q) - \dfrac{2(a-b)}{(b-c)\sqrt{a-c}}\,F(\mu,q) - 2\sqrt{\dfrac{u-a}{(u-b)(u-c)}}$

$$[u > a > b > c] \qquad\qquad \text{BY (237.10)}$$

6. $\int_u^\infty \sqrt{\dfrac{x-a}{(x-b)(x-c)^3}}\, dx = \dfrac{2\sqrt{a-c}}{b-c}\,E(\nu,q) - \dfrac{2(a-b)}{(b-c)\sqrt{a-c}}\,F(\nu,q)$

$$[u \geq a > b > c] \qquad\qquad \text{BY (238.09)}$$

7. $\int_{-\infty}^u \sqrt{\dfrac{a-x}{(b-x)^3(c-x)}}\, dx = \dfrac{2\sqrt{a-c}}{b-c}\,E(\alpha,p) - 2\dfrac{a-b}{b-c}\sqrt{\dfrac{c-u}{(a-u)(b-u)}}$

$$[a > b > c \geq u] \qquad\qquad \text{BY (231.03)}$$

8. $\int_u^c \sqrt{\dfrac{a-x}{(b-x)^3(c-x)}}\, dx = \dfrac{2\sqrt{a-c}}{b-c}\,E(\beta,p) \qquad [a > b > c > u] \qquad \text{BY (232.01)}$

9. $\int_c^u \sqrt{\dfrac{a-x}{(b-x)^3(x-c)}}\, dx = \dfrac{2\sqrt{a-c}}{b-c}\,[F(\gamma,q) - E(\gamma,q)] + \dfrac{2}{b-c}\sqrt{\dfrac{(a-u)(u-c)}{b-u}}$

$$[a > b > u > c] \qquad\qquad \text{BY (233.15)}$$

10. $\int_u^a \sqrt{\dfrac{a-x}{(x-b)^3(x-c)}}\,dx = \dfrac{2\sqrt{a-c}}{c-b}\,E(\lambda,p) + \dfrac{2}{b-c}\sqrt{\dfrac{(a-u)(u-c)}{u-b}}$

$$[a>u>b>c] \qquad\qquad \text{BY (236.11)}$$

11. $\int_a^u \sqrt{\dfrac{x-a}{(x-b)^3(x-c)}}\,dx = \dfrac{2\sqrt{a-c}}{b-c}\,[F(\mu,q) - E(\mu,q)]$

$$[u>a>b>c] \qquad\qquad \text{BY (237.09)}$$

12. $\int_u^\infty \sqrt{\dfrac{x-a}{(x-b)^3(x-c)}}\,dx = \dfrac{2\sqrt{a-c}}{b-c}\,[F(\nu,q) - E(\nu,q)] + 2\sqrt{\dfrac{u-a}{(u-b)(u-c)}}$

$$[u\geq a>b>c] \qquad\qquad \text{BY (238.10)}$$

13. $\int_{-\infty}^u \sqrt{\dfrac{b-x}{(a-x)^3(c-x)}}\,dx = \dfrac{2}{\sqrt{a-c}}\,E(\alpha,p) \qquad [a>b>c\geq u] \qquad \text{BY (231.01)}$

14. $\int_u^c \sqrt{\dfrac{b-x}{(a-x)^3(c-x)}}\,dx = \dfrac{2}{\sqrt{a-c}}\,E(\beta,p) - \dfrac{2(a-b)}{a-c}\sqrt{\dfrac{c-u}{(a-u)(b-u)}}$

$$[a>b>c>u] \qquad\qquad \text{BY (232.05)}$$

15. $\int_c^u \sqrt{\dfrac{b-x}{(a-x)^3(x-c)}}\,dx = \dfrac{2}{\sqrt{a-c}}\,[F(\gamma,q) - E(\gamma,q)] + \dfrac{2}{a-c}\sqrt{\dfrac{(b-u)(u-c)}{a-u}}$

$$[a>b\geq u>c] \qquad\qquad \text{BY (233.13)}$$

16. $\int_u^b \sqrt{\dfrac{b-x}{(a-x)^3(x-c)}}\,dx = \dfrac{2}{\sqrt{a-c}}\,[F(\delta,q) - E(\delta,q)] \qquad [a>b>u\geq c] \qquad \text{BY (234.15)}$

17. $\int_b^u \sqrt{\dfrac{x-b}{(a-x)^3(x-c)}}\,dx = -\dfrac{2}{\sqrt{a-c}}\,E(\kappa,p) + 2\sqrt{\dfrac{u-b}{(a-u)(u-c)}}$

$$[a>u>b>c] \qquad\qquad \text{BY (235.08)}$$

18. $\int_u^\infty \sqrt{\dfrac{x-b}{(x-a)^3(x-c)}}\,dx = \dfrac{2}{\sqrt{a-c}}\,[F(\nu,q) - E(\nu,q)] + 2\sqrt{\dfrac{u-b}{(u-a)(u-c)}}$

$$[u>a>b>c] \qquad\qquad \text{BY (238.07)}$$

19. $\int_{-\infty}^u \sqrt{\dfrac{b-x}{(a-x)(c-x)^3}}\,dx = \dfrac{2}{\sqrt{a-c}}\,[F(\alpha,p) - E(\alpha,p)] + 2\sqrt{\dfrac{b-u}{(a-u)(c-u)}}$

$$[a>b>c>u] \qquad\qquad \text{BY (231.04)}$$

20. $\int_u^b \sqrt{\dfrac{b-x}{(a-x)(x-c)^3}}\,dx = -\dfrac{2}{\sqrt{a-c}}\,E(\delta,q) + 2\sqrt{\dfrac{b-u}{(a-u)(u-c)}}$

$$[a>b>u>c] \qquad\qquad \text{BY (234.14)}$$

21. $\displaystyle\int_b^u \sqrt{\frac{x-b}{(a-x)(x-c)^3}}\, dx = \frac{2}{\sqrt{a-c}}\left[F\left(\kappa, p\right) - E\left(\kappa, p\right)\right]$

$$[a \geq u > b > c] \qquad \text{BY (235.03)}$$

22. $\displaystyle\int_u^a \sqrt{\frac{x-b}{(a-x)(x-c)^3}}\, dx = \frac{2}{\sqrt{a-c}}\left[F(\lambda, p) - E(\lambda, p)\right] + \frac{2}{a-c}\sqrt{\frac{(a-u)(u-b)}{u-c}}$

$$[a > u \geq b > c] \qquad \text{BY (236.14)}$$

23. $\displaystyle\int_a^u \sqrt{\frac{x-b}{(x-a)(x-c)^3}}\, dx = \frac{2}{\sqrt{a-c}}\, E(\mu, q) - 2\frac{b-c}{a-c}\sqrt{\frac{u-a}{(u-b)(u-c)}}$

$$[u > a > b > c] \qquad \text{BY (237.11)}$$

24. $\displaystyle\int_u^\infty \sqrt{\frac{x-b}{(x-a)(x-c)^3}}\, dx = \frac{2}{\sqrt{a-c}}\, E(\nu, q) \qquad [u \geq a > b > c] \qquad \text{BY (238.01)}$

25. $\displaystyle\int_{-\infty}^u \sqrt{\frac{c-x}{(a-x)^3(b-x)}}\, dx = \frac{2\sqrt{a-c}}{a-b}\, E(\alpha, p) - \frac{2(b-c)}{(a-b)\sqrt{a-c}}\, F(\alpha, p)$

$$[a > b > c \geq u] \qquad \text{BY (231.07)}$$

26. $\displaystyle\int_u^c \sqrt{\frac{c-x}{(a-x)^3(b-x)}}\, dx = \frac{2\sqrt{a-c}}{a-b}\, E(\beta, p) - \frac{2(b-c)}{(a-b)\sqrt{a-c}}\, F(\beta, p) - 2\sqrt{\frac{c-u}{(a-u)(b-u)}}$

$$[a > b > c > u] \qquad \text{BY (232.03)}$$

27. $\displaystyle\int_c^u \sqrt{\frac{x-c}{(a-x)^3(b-x)}}\, dx = \frac{2\sqrt{a-c}}{a-b}\, E(\gamma, q) - \frac{2}{\sqrt{a-c}}\, F(\gamma, q) - \frac{2}{a-b}\sqrt{\frac{(b-u)(u-c)}{a-u}}$

$$[a > b \geq u > c] \qquad \text{BY (233.14)}$$

28. $\displaystyle\int_u^b \sqrt{\frac{x-c}{(a-x)^3(b-x)}}\, dx = \frac{2\sqrt{a-c}}{a-b}\, E(\delta, q) - \frac{2}{\sqrt{a-c}}\, F(\delta, q)$

$$[a > b > u \geq c] \qquad \text{BY (234.20)}$$

29. $\displaystyle\int_b^u \sqrt{\frac{x-c}{(a-x)^3(x-b)}}\, dx = \frac{2(b-c)}{(a-b)\sqrt{a-c}}\, F\left(\kappa, p\right) - \frac{2\sqrt{a-c}}{a-b}\, E\left(\kappa, p\right)$

$$+ 2\frac{a-c}{a-b}\sqrt{\frac{u-b}{(a-u)(u-c)}}$$

$$[a > u > b > c] \qquad \text{BY (235.13)}$$

30. $\displaystyle\int_u^\infty \sqrt{\frac{x-c}{(x-a)^3(x-b)}}\, dx = \frac{2}{\sqrt{a-c}}\, F(\nu, q) - \frac{2\sqrt{a-c}}{a-b}\, E(\nu, q) + \frac{2(a-c)}{a-b}\sqrt{\frac{u-b}{(u-a)(u-c)}}$

$$[u > a > b > c] \qquad \text{BY (238.08)}$$

31. $\displaystyle\int_{-\infty}^u \sqrt{\frac{c-x}{(a-x)(b-x)^3}}\, dx = \frac{2\sqrt{a-c}}{a-b}\left[F(\alpha, p) - E(\alpha, p)\right] + 2\sqrt{\frac{c-u}{(a-u)(b-u)}}$

$$[a > b > c \geq u] \qquad \text{BY (231.06)}$$

32. $\displaystyle\int_u^c \sqrt{\frac{c-x}{(a-x)(b-x)^3}}\, dx = \frac{2\sqrt{a-c}}{a-b}\left[F(\beta,p) - E(\beta,p)\right]$

$$[a > b > c > u] \qquad\qquad \text{BY (232.04)}$$

33. $\displaystyle\int_c^u \sqrt{\frac{x-c}{(a-x)(b-x)^3}}\, dx = -\frac{2\sqrt{a-c}}{a-b}\, E(\gamma,q) + \frac{2}{a-b}\sqrt{\frac{(a-u)(u-c)}{b-u}}$

$$[a > b > u > c] \qquad\qquad \text{BY (233.16)}$$

34. $\displaystyle\int_u^a \sqrt{\frac{x-c}{(a-x)(x-b)^3}}\, dx = \frac{2\sqrt{a-c}}{a-b}\left[F(\lambda,p) - E(\lambda,p)\right] + \frac{2}{a-b}\sqrt{\frac{(a-u)(u-c)}{u-b}}$

$$[a > u > b > c] \qquad\qquad \text{BY (236.13)}$$

35. $\displaystyle\int_a^u \sqrt{\frac{x-c}{(x-a)(x-b)^3}}\, dx = \frac{2\sqrt{a-c}}{a-b}\, E(\mu,q) \qquad [u > a > b > c] \qquad\qquad \text{BY (237.01)}$

36. $\displaystyle\int_u^\infty \sqrt{\frac{x-c}{(x-a)(x-b)^3}}\, dx = \frac{2\sqrt{a-c}}{a-b}\, E(\nu,q) - 2\frac{b-c}{a-b}\sqrt{\frac{u-a}{(u-b)(u-c)}}$

$$[u \geq a > b > c] \qquad\qquad \text{BY (238.11)}$$

3.143

1.[6] $\displaystyle\int_u^1 \frac{dx}{\sqrt{1+x^4}} = \frac{1}{2}\, F\left(\arctan\frac{(1+\sqrt{2})\,(1-u)}{(1+u)},\, 2\sqrt[4]{2}\left(\sqrt{2}-1\right)\right)$ H 66 (286)

2. $\displaystyle\int_u^\infty \frac{dx}{\sqrt{1+x^4}} = \frac{1}{2}\, F\left(\arccos\frac{u^2-1}{u^2+1},\, \frac{\sqrt{2}}{2}\right)$ H 66 (287)

3.144   **Notation:** $\alpha = \arcsin\dfrac{1}{\sqrt{u^2-u+1}}$.

1. $\displaystyle\int_u^\infty \frac{dx}{\sqrt{x(x-1)\,(x^2-x+1)}} = F\left(\alpha,\, \frac{\sqrt{3}}{2}\right) \qquad [u \geq 1]$ BY (261.50)

2. $\displaystyle\int_u^\infty \frac{dx}{\sqrt{x^3(x-1)^3\,(x^2-x+1)}} = \frac{2(2u-1)}{\sqrt{u(u-1)\,(u^2-u+1)}} - 4\, E\left(\alpha,\, \frac{\sqrt{3}}{2}\right)$

$$[u > 1] \qquad\qquad \text{BY (261.54)}$$

3. $\displaystyle\int_u^\infty \frac{(2x-1)^2\, dx}{\sqrt{x^3(x-1)^3\,(x^2-x+1)}} = 4\left[F\left(\alpha,\, \frac{\sqrt{3}}{2}\right) - E\left(\alpha,\, \frac{\sqrt{3}}{2}\right) + \frac{2u-1}{2\sqrt{u(u-1)\,(u^2-u-1)}}\right]$

$$[u > 1] \qquad\qquad \text{BY (261.56)}$$

4. $\displaystyle\int_u^\infty \frac{dx}{\sqrt{x(x-1)(x^2-x+1)^3}} = \frac{4}{3}\left[F\left(\alpha,\, \frac{\sqrt{3}}{2}\right) - E\left(\alpha,\, \frac{\sqrt{3}}{2}\right)\right]$

$$[u \geq 1] \qquad\qquad \text{BY (261.52)}$$

5. $\displaystyle\int_u^\infty \frac{(2x-1)^2\,dx}{\sqrt{x(x-1)(x^2-x+1)^3}} = 4\,E\left(\alpha, \frac{\sqrt{3}}{2}\right)$ $[u>1]$ BY (261.51)

6. $\displaystyle\int_u^\infty \sqrt{\frac{x(x-1)}{(x^2-x+1)^3}}\,dx = \frac{4}{3}E\left(\alpha, \frac{\sqrt{3}}{2}\right) - \frac{1}{3}F\left(\alpha, \frac{\sqrt{3}}{2}\right)$

$[u>1]$ BY (261.53)

7. $\displaystyle\int_u^\infty \frac{dx}{(2x-1)^2}\sqrt{\frac{x(x-1)}{x^2-x+1}} = \frac{1}{3}\left[F\left(\alpha, \frac{\sqrt{3}}{2}\right) - E\left(\alpha, \frac{\sqrt{3}}{2}\right)\right] + \frac{1}{2(2u-1)}\sqrt{\frac{u(u-1)}{u^2-u+1}}$

$[u>1]$ BY (261.57)

8. $\displaystyle\int_u^\infty \frac{dx}{(2x-1)^2}\sqrt{\frac{x^2-x+1}{x(x-1)}} = E\left(\alpha, \frac{\sqrt{3}}{2}\right) - \frac{3}{2(2u-1)}\sqrt{\frac{u(u-1)}{u^2-u+1}}$

$[u>1]$ BY (261.58)

9. $\displaystyle\int_u^\infty \frac{dx}{(2x-1)^2\sqrt{x(x-1)(x^2-x+1)}} = \frac{4}{3}E\left(\alpha, \frac{\sqrt{3}}{2}\right) - \frac{1}{3}F\left(\alpha, \frac{\sqrt{3}}{2}\right) - \frac{2}{2u-1}\sqrt{\frac{u(u-1)}{u^2-u+1}}$

$[u>1]$ BY (261.55)

10. $\displaystyle\int_u^\infty \frac{dx}{\sqrt{x^5(x-1)^5(x^2-x+1)}} = \frac{40}{3}E\left(\alpha, \frac{\sqrt{3}}{2}\right) - \frac{4}{3}F\left(\alpha, \frac{\sqrt{3}}{2}\right) - \frac{2(2u-1)\left(9u^2-9u-1\right)}{3\sqrt{u^3(u-1)^3(u^2-u+1)}}$

$[u>1]$ BY (261.54)

11. $\displaystyle\int_u^\infty \frac{dx}{\sqrt{x(x-1)(x^2-x+1)^5}} = \frac{44}{27}F\left(\alpha, \frac{\sqrt{3}}{2}\right) - \frac{56}{27}E\left(\alpha, \frac{\sqrt{3}}{2}\right) + \frac{2(2u-1)\sqrt{u(u-1)}}{9\sqrt{(u^2-u+1)^3}}$

$[u>1]$ BY (261.52)

12. $\displaystyle\int_u^\infty \frac{dx}{(2x-1)^4\sqrt{x(x-1)(x^2-x+1)}} = \frac{16}{27}E\left(\alpha, \frac{\sqrt{3}}{2}\right) - \frac{1}{27}F\left(\alpha, \frac{\sqrt{3}}{2}\right)$

$$-\frac{8\left(5u^2-5u+2\right)}{9(2u-1)^3}\sqrt{\frac{u(u-1)}{u^2-u+1}}$$

$[u>1]$ BY (261.55)

3.145

1. $\displaystyle\int_\alpha^u \frac{dx}{\sqrt{(x-\alpha)(x-\beta)\left[(x-m)^2+n^2\right]}} = \frac{1}{\sqrt{pq}}F\left(2\arctan\sqrt{\frac{q(u-\alpha)}{p(u-\beta)}}, \frac{1}{2}\sqrt{\frac{(p+q)^2+(\alpha-\beta)^2}{pq}}\right)$

$[\beta < \alpha < u]$

2.
$$\int_{\beta}^{u} \frac{dx}{\sqrt{(\alpha - x)(x - \beta)\left[(x - m)^2 + n^2\right]}}$$

$$= \frac{1}{\sqrt{pq}} F\left(2 \operatorname{arccot} \sqrt{\frac{q(\alpha - u)}{p(u - \beta)}}, \frac{1}{2}\sqrt{\frac{-(p - q)^2 + (\alpha - \beta)^2}{pq}}\right)$$

$$[\beta < u < \alpha]$$

3.
$$\int_{u}^{\beta} \frac{dx}{\sqrt{(x - \alpha)(x - \beta)\left[(x - m)^2 + n^2\right]}} = \frac{1}{\sqrt{pq}} F\left(2 \arctan \sqrt{\frac{q(\beta - u)}{p(\alpha - u)}}, \frac{1}{2}\sqrt{\frac{(p + q)^2 + (\alpha - \beta)^2}{pq}}\right)$$

$$[u < \beta < \alpha]$$

where $(m - \alpha)^2 + n^2 = p^2$, and $(m - \beta)^2 + n^2 = q^2$.[*]

4. Set

$$(m_1 - m)^2 + (n_1 + n)^2 = p^2, \qquad (m_1 - m)^2 + (n_1 - n)^2 = p_1^2,$$

$$\cot \alpha = \sqrt{\frac{(p + p_1)^2 - 4n^2}{4n^2 - (p - p_1)^2}};$$

then

$$\int_{m - n \tan \alpha}^{u} \frac{dx}{\sqrt{\left[(x - m)^2 + n^2\right]\left[(x m_1)^2 + n_1^2\right]}} = \frac{2}{p + p_1} F\left(\alpha + \arctan \frac{u - m}{n}, \frac{2\sqrt{p p_1}}{p + p_1}\right)$$

$$[m - n \tan \alpha < u < m + n \cot \alpha]$$

3.146

1.
$$\int_0^1 \frac{1}{1 + x^4} \frac{dx}{\sqrt{1 - x^4}} = \frac{\pi}{8} + \frac{1}{4}\sqrt{2}\, K\left(\frac{\sqrt{2}}{2}\right) \qquad\qquad \text{BI (13)(6)}$$

2.
$$\int_0^1 \frac{x^2}{1 + x^4} \frac{dx}{\sqrt{1 - x^4}} = \frac{\pi}{8} \qquad\qquad \text{BI (13)(7)}$$

3.
$$\int_0^1 \frac{x^4}{1 + x^4} \frac{dx}{\sqrt{1 - x^4}} = -\frac{\pi}{8} + \frac{1}{4}\sqrt{2}\, K\left(\frac{\sqrt{2}}{2}\right) \qquad\qquad \text{BI (13)(8)}$$

---

[*]Formulas **3.145** are not valid for $\alpha + \beta = 2m$. In this case, we make the substitution $x - m = z$, which leads to one of the formulas **3.152**.

3.147   **Notation**: In **3.147**–**3.151** we set: $\alpha = \arcsin \sqrt{\dfrac{(a-c)(d-u)}{(a-d)(c-u)}}$,

$$\beta = \arcsin \sqrt{\frac{(a-c)(u-d)}{(c-d)(a-u)}}, \qquad \gamma = \arcsin \sqrt{\frac{(b-d)(c-u)}{(c-d)(b-u)}},$$

$$\delta = \arcsin \sqrt{\frac{(b-d)(u-c)}{(b-c)(u-d)}}, \qquad \kappa = \arcsin \sqrt{\frac{(a-c)(b-u)}{(b-c)(a-u)}},$$

$$\lambda = \arcsin \sqrt{\frac{(a-c)(u-b)}{(a-b)(u-c)}}, \qquad \mu = \arcsin \sqrt{\frac{(b-d)(a-u)}{(a-b)(u-d)}},$$

$$\nu = \arcsin \sqrt{\frac{(b-d)(u-a)}{(a-d)(u-b)}}, \qquad q = \sqrt{\frac{(b-c)(a-d)}{(a-c)(b-d)}}, \qquad r = \sqrt{\frac{(a-b)(c-d)}{(a-c)(b-d)}}.$$

1. $$\int_u^d \frac{dx}{\sqrt{(a-x)(b-x)(c-x)(d-x)}} = \frac{2}{\sqrt{(a-c)(b-d)}} F(\alpha, q)$$

$$[a > b > c > d > u] \qquad \text{BY (251.00)}$$

2. $$\int_d^u \frac{dx}{\sqrt{(a-x)(b-x)(c-x)(x-d)}} = \frac{2}{\sqrt{(a-c)(b-d)}} F(\beta, r)$$

$$[a > b > c \geq u > d] \qquad \text{BY (254.00)}$$

3. $$\int_u^c \frac{dx}{\sqrt{(a-x)(b-x)(c-x)(x-d)}} = \frac{2}{\sqrt{(a-c)(b-d)}} F(\gamma, r)$$

$$[a > b > c > u \geq d] \qquad \text{BY (253.00)}$$

4. $$\int_c^u \frac{dx}{\sqrt{(a-x)(b-x)(x-c)(x-d)}} = \frac{2}{\sqrt{(a-c)(b-d)}} F(\delta, q)$$

$$[a > b \geq u > c > d] \qquad \text{BY (254.00)}$$

5. $$\int_u^b \frac{dx}{\sqrt{(a-x)(b-x)(x-c)(x-d)}} = \frac{2}{\sqrt{(a-c)(b-d)}} F(\kappa, q)$$

$$[a > b > u \geq c > d] \qquad \text{BY (255.00)}$$

6. $$\int_b^u \frac{dx}{\sqrt{(a-x)(x-b)(x-c)(x-d)}} = \frac{2}{\sqrt{(a-c)(b-d)}} F(\lambda, r)$$

$$[a \geq u > b > c > d] \qquad \text{BY (256.00)}$$

7. $$\int_u^\alpha \frac{dx}{\sqrt{(a-x)(x-b)(x-c)(x-d)}} = \frac{2}{\sqrt{(a-c)(b-d)}} F(\mu, r)$$

$$[a > u \geq b > c > d] \qquad \text{BY (257.00)}$$

8. $$\int_a^u \frac{dx}{\sqrt{(x-a)(x-b)(x-c)(x-d)}} = \frac{2}{\sqrt{(a-c)(b-d)}} F(\nu, q)$$

$$[u > a > b > c > d] \qquad \text{BY (258.00)}$$

3.148

1.[8] $\displaystyle\int_u^d \frac{x\,dx}{\sqrt{(a-x)(b-x)(c-x)(d-x)}} = \frac{2}{\sqrt{(a-c)(b-d)}}\left\{(d-c)\,\Pi\left(\alpha,\frac{a-d}{a-c},q\right) + c\,F(\alpha,q)\right\}$

$$[a > b > c > d > u] \hspace{3cm} \text{BY (251.03)}$$

2. $\displaystyle\int_d^u \frac{x\,dx}{\sqrt{(a-x)(b-x)(c-x)(x-d)}} = \frac{2}{\sqrt{(a-c)(b-d)}}\left\{(d-a)\,\Pi\left(\beta,\frac{d-c}{a-c},r\right) + a\,F(\beta,r)\right\}$

$$[a > b > c \geq u > d] \hspace{3cm} \text{BY (252.11)}$$

3. $\displaystyle\int_u^c \frac{x\,dx}{\sqrt{(a-x)(b-x)(c-x)(x-d)}} = \frac{2}{\sqrt{(a-c)(b-d)}}\left\{(c-b)\,\Pi\left(\gamma,\frac{c-d}{b-d},r\right) + b\,F(\gamma,r)\right\}$

$$[a > b > c > u \geq d] \hspace{3cm} \text{BY (253.11)}$$

4. $\displaystyle\int_c^u \frac{x\,dx}{\sqrt{(a-x)(b-x)(x-c)(x-d)}} = \frac{2}{\sqrt{(a-c)(b-d)}}\left\{(c-d)\,\Pi\left(\delta,\frac{b-c}{b-d},q\right) + d\,F(\delta,q)\right\}$

$$[a > b \geq u > c > d] \hspace{3cm} \text{BY (254.10)}$$

5. $\displaystyle\int_u^b \frac{x\,dx}{\sqrt{(a-x)(b-x)(x-c)(x-d)}} = \frac{2}{\sqrt{(a-c)(b-d)}}\left\{(b-a)\,\Pi\left(\kappa,\frac{b-c}{a-c},q\right) + a\,F(\kappa,q)\right\}$

$$[a > b > u \geq c > d] \hspace{3cm} \text{BY (255.17)}$$

6.[8] $\displaystyle\int_b^u \frac{x\,dx}{\sqrt{(a-x)(x-b)(x-c)(x-d)}} = \frac{2}{\sqrt{(a-c)(b-d)}}\left\{(b-c)\,\Pi\left(\lambda,\frac{a-b}{a-c},r\right) + c\,F(\lambda,r)\right\}$

$$[a \geq u > b > c > d] \hspace{3cm} \text{BY (256.11)}$$

7. $\displaystyle\int_u^a \frac{x\,dx}{\sqrt{(a-x)(x-b)(x-c)(x-d)}} = \frac{2}{\sqrt{(a-c)(b-d)}}\left\{(a-d)\,\Pi\left(\mu,\frac{b-a}{b-d},r\right) + d\,F(\mu,r)\right\}$

$$[a > u \geq b > c > d] \hspace{3cm} \text{BY (257.11)}$$

8. $\displaystyle\int_a^u \frac{x\,dx}{\sqrt{(x-a)(x-b)(x-c)(x-d)}} = \frac{2}{\sqrt{(a-c)(b-d)}}\left\{(a-b)\,\Pi\left(\nu,\frac{a-d}{b-d},q\right) + b\,F(\nu,q)\right\}$

$$[u > a > b > c > d] \hspace{3cm} \text{BY (258.11)}$$

3.149

1. $\displaystyle\int_u^d \frac{dx}{x\sqrt{(a-x)(b-x)(c-x)(d-x)}}$

$$= \frac{2}{cd\sqrt{(a-c)(b-d)}}\left\{(c-d)\,\Pi\left(\alpha,\frac{c(a-d)}{d(a-c)},q\right) + d\,F(\alpha,q)\right\}$$

$$[a > b > c > d > u] \hspace{3cm} \text{BY (251.04)}$$

2. $\displaystyle\int_d^u \frac{dx}{x\sqrt{(a-x)(b-x)(c-x)(x-d)}}$

$$= \frac{2}{ad\sqrt{(a-c)(b-d)}}\left\{(a-d)\,\Pi\left(\beta,\frac{a(d-c)}{d(a-c)},r\right) + d\,F(\beta,r)\right\}$$

$$[a > b > c \geq u > d] \hspace{3cm} \text{BY (252.12)}$$

3. $$\int_u^c \frac{dx}{x\sqrt{(a-x)(b-x)(c-x)(x-d)}}$$

$$= \frac{2}{bc\sqrt{(a-c)(b-d)}} \left\{ (b-c)\, \Pi \left(\gamma, \frac{b(c-d)}{c(b-d)}, r \right) + c\, F(\gamma, r) \right\}$$

$$[a > b > c > u \geq d] \qquad \text{BY (253.12)}$$

4. $$\int_c^u \frac{dx}{x\sqrt{(a-x)(b-x)(x-c)(x-d)}}$$

$$= \frac{2}{cd\sqrt{(a-c)(b-d)}} \left\{ (d-c)\, \Pi \left(\delta, \frac{d(b-c)}{c(b-d)}, q \right) + c\, F(\delta, q) \right\}$$

$$[a > b \geq u > c > d] \qquad \text{BY (254.11)}$$

5. $$\int_u^b \frac{dx}{x\sqrt{(a-x)(b-x)(x-c)(x-d)}}$$

$$= \frac{2}{ab\sqrt{(a-c)(b-d)}} \times \left\{ (a-b)\, \Pi \left(\kappa, \frac{a(b-c)}{b(a-c)}, q \right) + b\, F(\kappa, q) \right\}$$

$$[a > b > u \geq c > d] \qquad \text{BY (255.18)}$$

6. $$\int_b^u \frac{dx}{x\sqrt{(a-x)(x-b)(x-c)(x-d)}}$$

$$= \frac{2}{bc\sqrt{(a-c)(b-d)}} \times \left\{ (c-b)\, \Pi \left(\lambda, \frac{c(a-b)}{b(a-c)}, r \right) + b\, F(\lambda, r) \right\}$$

$$[a \geq u > b > c > d] \qquad \text{BY (256.12)}$$

7. $$\int_u^a \frac{dx}{x\sqrt{(a-x)(x-b)(x-c)(x-d)}}$$

$$= \frac{2}{ad\sqrt{(a-c)(b-d)}} \times \left\{ (d-a)\, \Pi \left(\mu, \frac{d(b-a)}{a(b-d)}, r \right) + a\, F(\mu, r) \right\}$$

$$[a > u \geq b > c > d] \qquad \text{BY (257.12)}$$

8. $$\int_a^u \frac{dx}{x\sqrt{(x-a)(x-b)(x-c)(x-d)}}$$

$$= \frac{2}{ab\sqrt{(a-c)(b-d)}} \left\{ (b-a)\, \Pi \left(\nu, \frac{b(a-d)}{a(b-d)}, q \right) + a\, F(\nu, q) \right\}$$

$$[u > a > b > c > d] \qquad \text{BY (258.12)}$$

3.151

1. $$\int_u^d \frac{dx}{(p-x)\sqrt{(a-x)(b-x)(c-x)(d-x)}}$$

$$= \frac{2}{(p-c)(p-d)\sqrt{(a-c)(b-d)}}$$

$$\times \left[(d-c)\, \Pi \left(\alpha, \frac{(a-d)(p-c)}{(a-c)(p-d)}, q \right) + (p-d)\, F(\alpha, q) \right]$$

$$[a > b > c > d > u, \quad p \neq d] \quad \text{BY (251.39)}$$

2. $\displaystyle\int_d^u \frac{dx}{(p-x)\sqrt{(a-x)(b-x)(c-x)(x-d)}}$

$$= \frac{2}{(p-a)(p-d)\sqrt{(a-c)(b-d)}}$$
$$\times \left[(d-a)\,\Pi\left(\beta, \frac{(d-c)(p-a)}{(a-c)(p-d)}, r\right) + (p-d)\,F(\beta, r)\right]$$
$$[a > b > c \geq u > d, \quad p \neq d] \quad \text{BY (252.39)}$$

3. $\displaystyle\int_u^c \frac{dx}{(p-x)\sqrt{(a-x)(b-x)(c-x)(x-d)}} = \frac{2}{(p-b)(p-c)\sqrt{(a-c)(b-d)}}$

$$\times \left[(c-b)\,\Pi\left(\gamma, \frac{(c-d)(p-b)}{(b-d)(p-c)}, r\right) + (p-c)\,F(\gamma, r)\right]$$
$$[a > b > c > u \geq d, \quad p \neq c] \quad \text{BY (253.39)}$$

4. $\displaystyle\int_c^u \frac{dx}{(p-x)\sqrt{(a-x)(b-x)(x-c)(x-d)}} = \frac{2}{(p-c)(p-d)\sqrt{(a-c)(b-d)}}$

$$\times \left[(c-d)\,\Pi\left(\delta, \frac{(b-c)(p-d)}{(b-d)(p-c)}, q\right) + (p-c)\,F(\delta, q)\right]$$
$$[a > b \geq u > c > d, \quad p \neq c] \quad \text{BY (254.39)}$$

5. $\displaystyle\int_u^b \frac{dx}{(p-x)\sqrt{(a-x)(b-x)(x-c)(x-d)}}$

$$= \frac{2}{(p-a)(p-b)\sqrt{(a-c)(b-d)}}$$
$$\times \left[(b-a)\,\Pi\left(\kappa, \frac{(b-c)(p-a)}{(a-c)(p-b)}, q\right) + (p-b)\,F(\kappa, q)\right]$$
$$[a > b > u \geq c > d, \quad p \neq b] \quad \text{BY (255.38)}$$

6. $\displaystyle\int_b^u \frac{dx}{(x-p)\sqrt{(a-x)(x-b)(x-c)(x-d)}}$

$$= \frac{2}{(b-p)(p-c)\sqrt{(a-c)(b-d)}}$$
$$\times \left[(b-c)\,\Pi\left(\lambda, \frac{(a-b)(p-c)}{(a-c)(p-b)}, r\right) + (p-b)\,F(\lambda, r)\right]$$
$$[a \geq u > b > c > d, \quad p \neq b] \quad \text{BY (256.39)}$$

7. $\displaystyle\int_u^a \frac{dx}{(p-x)\sqrt{(a-x)(x-b)(x-c)(x-d)}}$

$$= \frac{2}{(p-a)(p-d)\sqrt{(a-c)(b-d)}}$$
$$\times \left[(a-d)\,\Pi\left(\mu, \frac{(b-a)(p-d)}{(b-d)(p-a)}, r\right) + (p-a)\,F(\mu, r)\right]$$
$$[a > u \geq b > c > d, \quad p \neq a] \quad \text{BY (257.39)}$$

8. $\displaystyle\int_a^u \frac{dx}{(p-x)\sqrt{(x-a)(x-b)(x-c)(x-d)}}$

$$= \frac{2}{(p-a)(p-b)\sqrt{(a-c)(b-d)}}$$
$$\times \left[(a-b)\,\Pi\left(\nu, \frac{(a-d)(p-b)}{(b-d)(p-a)}, q\right) + (p-a)\,F(\nu, q) \right]$$
$$[u > a > b > c > d, \quad p \neq a] \quad \text{BY (258.39)}$$

3.152   **Notation**: In **3.152–3.163** we set: $\qquad \alpha = \arctan\dfrac{u}{b}, \qquad \beta = \operatorname{arccot}\dfrac{u}{a}$

$$\gamma = \arcsin\frac{u}{b}\sqrt{\frac{a^2+b^2}{a^2+u^2}}, \qquad \delta = \arccos\frac{u}{b}, \qquad \varepsilon = \arccos\frac{b}{u}, \qquad \xi = \arcsin\sqrt{\frac{a^2+b^2}{a^2+u^2}},$$

$$\eta = \arcsin\frac{u}{b}, \qquad \zeta = \arcsin\frac{a}{b}\sqrt{\frac{b^2-u^2}{a^2-u^2}}, \qquad \kappa = \arcsin\frac{a}{u}\sqrt{\frac{u^2-b^2}{a^2-b^2}},$$

$$\lambda = \arcsin\sqrt{\frac{a^2-u^2}{a^2-b^2}}, \qquad \mu = \arcsin\sqrt{\frac{u^2-a^2}{u^2-b^2}}, \qquad \nu = \arcsin\frac{a}{u}, \qquad q = \frac{\sqrt{a^2-b^2}}{a},$$

$$r = \frac{b}{\sqrt{a^2+b^2}}, \qquad s = \frac{a}{\sqrt{a^2+b^2}}, \qquad t = \frac{b}{a}.$$

1. $\displaystyle\int_0^u \frac{dx}{\sqrt{(x^2+a^2)(x^2+b^2)}} = \frac{1}{a}F(\alpha, q)$ $\qquad [a > b > 0]$ $\qquad$ H 62(258), BY (221.00)

2. $\displaystyle\int_u^\infty \frac{dx}{\sqrt{(x^2+a^2)(x^2+b^2)}} = \frac{1}{a}F(\beta, q)$ $\qquad [a > b > 0]$ $\qquad$ H 63 (259), BY (222.00)

3. $\displaystyle\int_0^u \frac{dx}{\sqrt{(x^2+a^2)(b^2-x^2)}} = \frac{1}{\sqrt{a^2+b^2}}F(\gamma, r)$ $\qquad [b \geq u > 0]$ $\qquad$ H 63 (260)

4. $\displaystyle\int_u^b \frac{dx}{\sqrt{(x^2+a^2)(b^2-x^2)}} = \frac{1}{\sqrt{a^2+b^2}}F(\delta, r)$ $\qquad [b > u \geq 0]$ $\qquad$ H 63 (261), BY (213.00)

5. $\displaystyle\int_b^u \frac{dx}{\sqrt{(x^2+a^2)(x^2-b^2)}} = \frac{1}{\sqrt{a^2+b^2}}F(\varepsilon, s)$ $\qquad [u > b > 0]$ $\qquad$ H 63 (262), BY (211.00)

6. $\displaystyle\int_u^\infty \frac{dx}{\sqrt{(x^2+a^2)(x^2-b^2)}} = \frac{1}{\sqrt{a^2+b^2}}F(\xi, s)$ $\qquad [u > b > 0]$ $\qquad$ H 63 (263), BY (212.00)

7. $\displaystyle\int_0^u \frac{dx}{\sqrt{(a^2-x^2)(b^2-x^2)}} = \frac{1}{a}F(\eta, t)$ $\qquad [a > b \geq u > 0]$ $\qquad$ H 63 (264), BY (219.00)

8. $\displaystyle\int_u^b \frac{dx}{\sqrt{(a^2-x^2)(b^2-x^2)}} = \frac{1}{a}F(\zeta, t)$ $\qquad [a > b > u \geq 0]$ $\qquad$ H 63 (265), BY (220.00)

9. $\displaystyle\int_b^u \frac{dx}{\sqrt{(a^2-x^2)(x^2-b^2)}} = \frac{1}{a}F(\kappa, q)$ $\qquad [a \geq u > b > 0]$ $\qquad$ H 63 (266), BY (217.00)

10. $\displaystyle\int_u^a \frac{dx}{\sqrt{(a^2 - x^2)(x^2 - b^2)}} = \frac{1}{a} F(\lambda, q)$ $[a > u \geq b > 0]$ H 63 (257), BY (218.00)

11. $\displaystyle\int_a^u \frac{dx}{\sqrt{(x^2 - a^2)(x^2 - b^2)}} = \frac{1}{a} F(\mu, t)$ $[u > a > b > 0]$ H 63 (268), BY (216.00)

12. $\displaystyle\int_u^\infty \frac{dx}{\sqrt{(x^2 - a^2)(x^2 - b^2)}} = \frac{1}{a} F(\nu, t)$ $[u \geq a > b > 0]$ H 64(269), BY (215.00)

3.153

1. $\displaystyle\int_0^u \frac{x^2\, dx}{\sqrt{(x^2 + a^2)(x^2 + b^2)}} = u\sqrt{\frac{a^2 + u^2}{b^2 + u^2}} - a\, E(\alpha, q)$ $[u > 0, \quad a > b]$ BY (221.09)

2. $\displaystyle\int_0^u \frac{x^2\, dx}{\sqrt{(a^2 + x^2)(b^2 - x^2)}} = \sqrt{a^2 + b^2}\, E(\gamma, r) - \frac{a^2}{\sqrt{a^2 + b^2}} F(\gamma, r) - u\sqrt{\frac{b^2 - u^2}{a^2 + u^2}}$

$[b \geq u > 0]$ BY (214.05)

3. $\displaystyle\int_u^b \frac{x^2\, dx}{\sqrt{(a^2 + x^2)(b^2 - x^2)}} = \sqrt{a^2 + b^2}\, E(\delta, r) - \frac{a^2}{\sqrt{a^2 + b^2}} F(\delta, r)$

$[b > u \geq 0]$ BY (213.06)

4. $\displaystyle\int_b^u \frac{x^2\, dx}{\sqrt{(a^2 + x^2)(x^2 - b^2)}} = \frac{b^2}{\sqrt{a^2 + b^2}} F(\varepsilon, s) - \sqrt{a^2 + b^2}\, E(\varepsilon, s) + \frac{1}{u}\sqrt{(u^2 + a^2)(u^2 - b^2)}$

$[u > b > 0]$ BY (211.09)

5. $\displaystyle\int_0^u \frac{x^2\, dx}{\sqrt{(a^2 - x^2)(b^2 - x^2)}} = a\left\{F(\eta, t) - E(\eta, t)\right\}$ $[a > b \geq u > 0]$ BY (219.05)

6. $\displaystyle\int_u^b \frac{x^2\, dx}{\sqrt{(a^2 - x^2)(b^2 - x^2)}} = a\left\{F(\zeta, t) - E(\zeta, t)\right\} + u\sqrt{\frac{b^2 - u^2}{a^2 - u^2}}$

$[a > b > u \geq 0]$ BY (220.06)

7. $\displaystyle\int_b^u \frac{x^2\, dx}{\sqrt{(a^2 - x^2)(x^2 - b^2)}} = a\, E(\kappa, q) - \frac{1}{u}\sqrt{(a^2 - u^2)(u^2 - b^2)}$

$[a \geq u > b > 0]$ BY (217.05)

8. $\displaystyle\int_u^a \frac{x^2\, dx}{\sqrt{(a^2 - x^2)(x^2 - b^2)}} = a\, E(\lambda, q)$ $[a > u \geq b > 0]$ BY (218.06)

9.[6] $\displaystyle\int_a^u \frac{x^2\, dx}{\sqrt{(x^2 - a^2)(x^2 - b^2)}} = a\left\{F(\mu, t) - E(\mu, t)\right\} + u\sqrt{\frac{u^2 - a^2}{u^2 - b^2}}$

$[u > a > b > 0]$ BY (216.06)

10. $\displaystyle\int_0^1 \frac{x^2\, dx}{\sqrt{(1 + x^2)(1 + k^2 x^2)}} = \frac{1}{k^2}\left\{\sqrt{\frac{1 + k^2}{2}} - E\left(\frac{\pi}{4}, \sqrt{1 - k^2}\right)\right\}$ BI (14)(9)

3.154

1. $$\int_0^u \frac{x^4\,dx}{\sqrt{(x^2+a^2)(x^2+b^2)}} = \frac{a}{3}\left\{2\left(a^2+b^2\right)E(\alpha,q) - b^2\,F(\alpha,q)\right\} + \frac{u}{3}\left(u^2 - 2a^2 - b^2\right)\sqrt{\frac{a^2+u^2}{b^2+u^2}}$$

    $$[a > b, \quad u > 0] \qquad \text{BY (221.09)}$$

2. $$\int_0^u \frac{x^4\,dx}{\sqrt{(a^2+x^2)(b^2-x^2)}} = \frac{1}{3\sqrt{a^2+b^2}}\left\{\left(2a^2-b^2\right)a^2\,F(\gamma,r) - 2\left(a^4-b^4\right)E(\gamma,r)\right\}$$
    $$-\frac{u}{3}\left(2b^2-a^2+u^2\right)\sqrt{\frac{b^2-u^2}{a^2+u^2}}$$

    $$[a \geq u > 0] \qquad \text{BY (214.05)}$$

3. $$\int_u^b \frac{x^4\,dx}{\sqrt{(a^2+x^2)(b^2-x^2)}} = \frac{1}{3\sqrt{a^2+b^2}}\left\{\left(2a^2-b^2\right)a^2\,F(\delta,r) - 2\left(a^4-b^4\right)E(\delta,r)\right\}$$
    $$+\frac{u}{3}\sqrt{(a^2+u^2)(b^2-u^2)}$$

    $$[b > u \geq 0] \qquad \text{BY (213.06)}$$

4. $$\int_b^u \frac{x^4\,dx}{\sqrt{(a^2+x^2)(x^2-b^2)}} = \frac{1}{3\sqrt{a^2+b^2}}\left\{\left(2b^2-a^2\right)b^2\,F(\varepsilon,s) + 2\left(a^4-b^4\right)E(\varepsilon,s)\right\}$$
    $$+\frac{2b^2-2a^2+u^2}{3u}\sqrt{(u^2+a^2)(u^2-b^2)}$$

    $$[u > b > 0] \qquad \text{BY (211.09)}$$

5. $$\int_0^u \frac{x^4\,dx}{\sqrt{(a^2-x^2)(b^2-x^2)}} = \frac{a}{3}\left\{\left(2a^2+b^2\right)F(\eta,t) - 2\left(a^2+b^2\right)E(\eta,t)\right\} + \frac{u}{3}\sqrt{(a^2-u^2)(b^2-u^2)}$$

    $$[a > b \geq u > 0] \qquad \text{BY (219.05)}$$

6. $$\int_u^b \frac{x^4\,dx}{\sqrt{(a^2-x^2)(b^2-x^2)}} = \frac{a}{3}\left\{\left(2a^2+b^2\right)F(\zeta,t) - 2\left(a^2+b^2\right)E(\zeta,t)\right\}$$
    $$+\frac{u}{3}\left(u^2+a^2+2b^2\right)\sqrt{\frac{b^2-u^2}{a^2-u^2}}$$

    $$[a > b > u \geq 0] \qquad \text{BY (220.06)}$$

7. $$\int_b^u \frac{x^4\,dx}{\sqrt{(a^2-x^2)(x^2-b^2)}} = \frac{a}{3}\left\{2\left(a^2+b^2\right)E\left(\kappa,q\right) - b^2\,F\left(\kappa,q\right)\right\}$$
    $$-\frac{u^2+2a^2+2b^2}{3u}\sqrt{(a^2-u^2)(u^2-b^2)}$$

    $$[a \geq u > b > 0] \qquad \text{BY (217.05)}$$

8. $$\int_u^a \frac{x^4\,dx}{\sqrt{(a^2-x^2)(x^2-b^2)}} = \frac{a}{3}\left\{2\left(a^2+b^2\right)E(\lambda,q) - b^2\,F(\lambda,q)\right\} + \frac{u}{3}\sqrt{(a^2-u^2)(u^2-b^2)}$$

    $$[a > u \geq b > 0] \qquad \text{BY (218.06)}$$

9. $$\int_a^u \frac{x^4\,dx}{\sqrt{(x^2-a^2)(x^2-b^2)}} = \frac{a}{3}\left\{\left(2a^2+b^2\right)F(\mu,t) - 2\left(a^2+b^2\right)E(\mu,t)\right\}$$
    $$+\frac{u}{3}\left(u^2+2a^2+b^2\right)\sqrt{\frac{u^2-a^2}{u^2-b^2}}$$

    $$[u > a > b > 0] \qquad \text{BY (216.06)}$$

3.155

1. $\displaystyle\int_u^a \sqrt{(a^2 - x^2)(x^2 - b^2)}\, dx = \frac{a}{3}\left\{\left(a^2 + b^2\right) E(\lambda, q) - 2b^2\, F(\lambda, q)\right\} - \frac{u}{3}\sqrt{(a^2 - u^2)(u^2 - b^2)}$

$$[a > u \geq b > 0] \qquad\qquad \text{BY (218.11)}$$

2. $\displaystyle\int_a^u \sqrt{(x^2 - a^2)(x^2 - b^2)}\, dx = \frac{a}{3}\left\{\left(a^2 + b^2\right) E(\mu, t) - \left(a^2 - b^2\right) F(\mu, t)\right\}$

$$+\frac{u}{3}\left(u^2 - a^2 - 2b^2\right)\sqrt{\frac{u^2 - a^2}{u^2 - b^2}}$$

$$[u > a > b > 0] \qquad\qquad \text{BY (216.10)}$$

3. $\displaystyle\int_0^u \sqrt{(x^2 + a^2)(x^2 + b^2)}\, dx = \frac{a}{3}\left\{2b^2\, F(\alpha, q) - \left(a^2 + b^2\right) E(\alpha, q)\right\}$

$$+\frac{u}{3}\left(u^2 + a^2 + 2b^2\right)\sqrt{\frac{a^2 + u^2}{b^2 + u^2}}$$

$$[a > b, \quad u > 0] \qquad\qquad \text{BY (221.08)}$$

4. $\displaystyle\int_0^u \sqrt{(a^2 + x^2)(b^2 - x^2)}\, dx = \frac{1}{3}\sqrt{a^2 + b^2}\left\{a^2\, F(\gamma, r) - \left(a^2 - b^2\right) E(\gamma, r)\right\}$

$$+\frac{u}{3}\left(u^2 + 2a^2 - b^2\right)\sqrt{\frac{b^2 - u^2}{a^2 + u^2}}$$

$$[a \geq u > 0] \qquad\qquad \text{BY (214.12)}$$

5.[9] $\displaystyle\int_u^b \sqrt{(a^2 + x^2)(b^2 - x^2)}\, dx = \frac{1}{3}\sqrt{a^2 + b^2}\left\{a^2\, F(\delta, r) + \left(b^2 - a^2\right) E(\delta, r)\right\}$

$$+\frac{u}{3}\sqrt{(a^2 + u^2)(b^2 - u^2)}$$

$$[b > u \geq 0] \qquad\qquad \text{BY (213.13)}$$

6. $\displaystyle\int_b^u \sqrt{(a^2 + x^2)(x^2 - b^2)}\, dx = \frac{1}{3}\sqrt{a^2 + b^2}\left\{\left(b^2 - a^2\right) E(\varepsilon, s) - b^2\, F(\varepsilon, s)\right\}$

$$+\frac{u^2 + a^2 - b^2}{3u}\sqrt{(a^2 + u^2)(u^2 - b^2)}$$

$$[u > b > 0] \qquad\qquad \text{BY (211.08)}$$

7. $\displaystyle\int_0^u \sqrt{(a^2 - x^2)(b^2 - x^2)}\, dx = \frac{a}{3}\left\{\left(a^2 + b^2\right) E(\eta, t) - \left(a^2 - b^2\right) F(\eta, t)\right\}$

$$+\frac{u}{3}\sqrt{(a^2 - u^2)(b^2 - u^2)}$$

$$[a > b \geq u > 0] \qquad\qquad \text{BY (219.11)}$$

8. $\displaystyle\int_u^b \sqrt{(a^2 - x^2)(b^2 - x^2)}\, dx = \frac{a}{3}\left\{\left(a^2 + b^2\right) E(\zeta, t) - \left(a^2 - b^2\right) F(\zeta, t)\right\}$

$$+\frac{u}{3}\left(u^2 - 2a^2 - b^2\right)\sqrt{\frac{b^2 - u^2}{a^2 - u^2}}$$

$$[a > b > u \geq 0] \qquad\qquad \text{BY (220.05)}$$

9. $\displaystyle\int_b^u \sqrt{(a^2 - x^2)(x^2 - b^2)}\, dx = \frac{a}{3}\left\{\left(a^2 + b^2\right) E(\kappa, q) - 2b^2\, F(\kappa, q)\right\}$

$$+\frac{u^2 - a^2 - b^2}{3u}\sqrt{(a^2 - u^2)(u^2 - b^2)}$$

$$[a \geq u > b > 0] \qquad\qquad \text{BY (217.09)}$$

3.156

1.[6] $\displaystyle\int_u^\infty \frac{dx}{x^2\sqrt{(x^2+a^2)(x^2+b^2)}} = \frac{1}{ub^2}\sqrt{\frac{b^2+u^2}{a^2+u^2}} - \frac{1}{ab^2}E(\beta,q)$

$$[a \geq b, \quad u > 0] \qquad\qquad \text{BY (222.04)}$$

2. $\displaystyle\int_u^b \frac{dx}{x^2\sqrt{(x^2+a^2)(b^2-x^2)}} = \frac{1}{a^2b^2\sqrt{a^2+b^2}}\left\{a^2\,F(\delta,r) - (a^2+b^2)\,E(\delta,r)\right\}$

$\qquad\qquad\qquad\qquad + \dfrac{1}{a^2b^2u}\sqrt{(a^2+u^2)(b^2-u^2)}$

$$[b > u > 0] \qquad\qquad \text{BY (213.09)}$$

3. $\displaystyle\int_b^u \frac{dx}{x^2\sqrt{(x^2+a^2)(x^2-b^2)}} = \frac{1}{a^2b^2\sqrt{a^2+b^2}}\left\{(a^2+b^2)\,E(\varepsilon,s) - b^2\,F(\varepsilon,s)\right\}$

$$[u > b > 0] \qquad\qquad \text{BY (211.11)}$$

4. $\displaystyle\int_u^\infty \frac{dx}{x^2\sqrt{(x^2+a^2)(x^2-b^2)}} = \frac{1}{a^2b^2\sqrt{a^2+b^2}}\left\{(a^2+b^2)\,E(\xi,s) - b^2\,F(\xi,s)\right\}$

$\qquad\qquad\qquad\qquad - \dfrac{1}{b^2u}\sqrt{\dfrac{u^2-b^2}{a^2+u^2}}$

$$[u \geq b > 0] \qquad\qquad \text{BY (212.06)}$$

5. $\displaystyle\int_u^b \frac{dx}{x^2\sqrt{(a^2-x^2)(b^2-x^2)}} = \frac{1}{ab^2}\left\{F(\zeta,t) - E(\zeta,t)\right\} + \frac{1}{b^2u}\sqrt{\frac{b^2-u^2}{a^2-u^2}}$

$$[a > b > u > 0] \qquad\qquad \text{BY (220.09)}$$

6. $\displaystyle\int_b^u \frac{dx}{x^2\sqrt{(a^2-x^2)(x^2-b^2)}} = \frac{1}{ab^2}\,E(\kappa,q) \qquad\qquad [a \geq u > b > 0] \qquad \text{BY (217.01)}$

7. $\displaystyle\int_u^a \frac{dx}{x^2\sqrt{(a^2-x^2)(x^2-b^2)}} = \frac{1}{ab^2}\,E(\lambda,q) - \frac{1}{a^2b^2u}\sqrt{(a^2-u^2)(u^2-b^2)}$

$$[a > u \geq b > 0] \qquad\qquad \text{BY (218.12)}$$

8. $\displaystyle\int_a^u \frac{dx}{x^2\sqrt{(x^2-a^2)(x^2-b^2)}} = \frac{1}{ab^2}\left\{F(\mu,t) - E(\mu,t)\right\} + \frac{1}{a^2u}\sqrt{\frac{u^2-a^2}{u^2-b^2}}$

$$[u > a > b > 0] \qquad\qquad \text{BY (216.09)}$$

9. $\displaystyle\int_u^\infty \frac{dx}{x^2\sqrt{(x^2-a^2)(x^2-b^2)}} = \frac{1}{ab^2}\left\{F(\nu,t) - E(\nu,t)\right\}$

$$[u \geq a > b > 0] \qquad\qquad \text{BY (215.07)}$$

3.157

1. $\displaystyle\int_0^u \frac{dx}{(p-x^2)\sqrt{(x^2+a^2)(x^2+b^2)}} = \frac{1}{a\,(p+b^2)}\left\{\frac{b^2}{p}\,\Pi\left(\alpha, \frac{p+b^2}{p}, q\right) + F(\alpha,q)\right\}$

$$[p \neq 0] \qquad\qquad \text{BY (221.13)}$$

2. $\displaystyle\int_u^\infty \frac{dx}{(p-x^2)\sqrt{(x^2+a^2)(x^2+b^2)}} = -\frac{1}{a\,(a^2+p)}\left\{\Pi\left(\beta, \frac{a^2+p}{a^2}, q\right) - F(\beta,q)\right\} \qquad \text{BY (222.11)}$

3.
$$\int_0^u \frac{dx}{(p - x^2) \sqrt{(a^2 + x^2)(b^2 - x^2)}} = \frac{1}{p(p + a^2)\sqrt{a^2 + b^2}} \left\{ a^2 \, \Pi \left(\gamma, \frac{b^2(p + a^2)}{p(a^2 + b^2)}, r \right) + p \, F(\gamma, r) \right\}$$
$$[b \ge u > 0, \quad p \ne 0] \qquad \text{BY (214.13)a}$$

4.
$$\int_u^b \frac{dx}{(p - x^2) \sqrt{(a^2 + x^2)(b^2 - x^2)}} = \frac{1}{(p - b^2)\sqrt{a^2 + b^2}} \, \Pi \left(\delta, \frac{b^2}{b^2 - p}, r \right)$$
$$[b > u \ge 0, \quad p \ne b^2] \qquad \text{BY (213.02)}$$

5.
$$\int_b^u \frac{dx}{(p - x^2) \sqrt{(a^2 + x^2)(x^2 - b^2)}} = \frac{1}{p(p - b^2)\sqrt{a^2 + b^2}} \left\{ b^2 \, \Pi \left(\varepsilon, \frac{p}{p - b^2}, s \right) + (p - b^2) \, F(\varepsilon, s) \right\}$$
$$[u > b > 0, \quad p \ne b^2] \qquad \text{BY (211.14)}$$

6.
$$\int_u^\infty \frac{dx}{(x^2 - p) \sqrt{(a^2 + x^2)(x^2 - b^2)}} = \frac{1}{(a^2 + p)\sqrt{a^2 + b^2}} \left\{ \Pi \left(\xi, \frac{a^2 + p}{a^2 + b^2}, s \right) - F(\xi, s) \right\}$$
$$[u \ge b > 0] \qquad \text{BY (212.12)}$$

7.
$$\int_0^u \frac{dx}{(p - x^2) \sqrt{(a^2 - x^2)(b^2 - x^2)}} = \frac{1}{ap} \, \Pi \left(\eta, \frac{b^2}{p}, t \right) \qquad [a > b \ge u > 0; \quad p \ne b] \qquad \text{BY (219.02)}$$

8.
$$\int_u^b \frac{dx}{(p - x^2) \sqrt{(a^2 - x^2)(b^2 - x^2)}} = \frac{1}{a(p - a^2)(p - b^2)}$$
$$\times \left\{ (b^2 - a^2) \, \Pi \left(\zeta, \frac{b^2(p - a^2)}{a^2(p - b^2)}, t \right) + (p - b^2) \, F(\zeta, t) \right\}$$
$$[a > b > u \ge 0; \quad p \ne b^2] \qquad \text{BY (220.13)}$$

9.
$$\int_b^u \frac{dx}{(p - x^2) \sqrt{(a^2 - x^2)(x^2 - b^2)}} = \frac{1}{ap(p - b^2)} \left\{ b^2 \, \Pi \left(\kappa, \frac{p(a^2 - b^2)}{a^2(p - b^2)}, q \right) + (p - b^2) \, F(\kappa, q) \right\}$$
$$[a \ge u > b > 0; \quad p \ne b^2] \qquad \text{BY (217.12)}$$

10.
$$\int_u^a \frac{dx}{(x^2 - p) \sqrt{(a^2 - x^2)(x^2 - b^2)}} = \frac{1}{a(a^2 - p)} \, \Pi \left(\lambda, \frac{a^2 - b^2}{a^2 - p}, q \right)$$
$$[a > u \ge b > 0; \quad p \ne a^2] \qquad \text{BY (218.02)}$$

11.
$$\int_a^u \frac{dx}{(p - x^2) \sqrt{(x^2 - a^2)(x^2 - b^2)}}$$
$$= \frac{1}{a(p - a^2)(p - b^2)} \left\{ (a^2 - b^2) \, \Pi \left(\mu, \frac{p - b^2}{p - a^2}, t \right) + (p - a^2) \, F(\mu, t) \right\}$$
$$[u > a > b > 0; \quad p \ne a^2, \quad p \ne b^2] \qquad \text{BY (216.12)}$$

12.
$$\int_u^\infty \frac{dx}{(x^2 - p) \sqrt{(x^2 - a^2)(x^2 - b^2)}} = \frac{1}{ap} \left\{ \Pi \left(\nu, \frac{p}{a^2}, t \right) - F(\nu, t) \right\}$$
$$[u \ge a > b > 0; \quad p \ne 0] \qquad \text{BY (215.12)}$$

3.158

1. $$\int_0^u \frac{dx}{\sqrt{(x^2 + a^2)(x^2 + b^2)^3}} = \frac{1}{ab^2(a^2 - b^2)}\left\{a^2 E(\alpha, q) - b^2 F(\alpha, q)\right\}$$

$$[a > b; \quad u > 0] \qquad \text{BY (221.05)}$$

2. $$\int_u^\infty \frac{dx}{\sqrt{(x^2 + a^2)(x^2 + b^2)^3}} = \frac{1}{ab^2(a^2 - b^2)}\left\{a^2 E(\beta, q) - b^2 F(\beta, q)\right\} - \frac{u}{b^2\sqrt{(a^2 + u^2)(b^2 + u^2)}}$$

$$[a > b, \quad u \geq 0] \qquad \text{BY (222.05)}$$

3. $$\int_0^u \frac{dx}{\sqrt{(x^2 + a^2)^3(x^2 + b^2)}} = \frac{1}{a(a^2 - b^2)}\left\{F(\alpha, q) - E(\alpha, q)\right\} + \frac{u}{a^2\sqrt{(u^2 + a^2)(u^2 + b^2)}}$$

$$[a > b; \quad u > 0] \qquad \text{BY (221.06)}$$

4. $$\int_u^\infty \frac{dx}{\sqrt{(a^2 + x^2)^3(x^2 + b^2)}} = \frac{1}{a(a^2 - b^2)}\left\{F(\beta, q) - E(\beta, q)\right\}$$

$$[a > b, \quad u \geq 0] \qquad \text{BY (222.03)}$$

5. $$\int_0^u \frac{dx}{\sqrt{(a^2 + x^2)^3(b^2 - x^2)}} = \frac{1}{a^2\sqrt{a^2 + b^2}} E(\gamma, r) \qquad [b \geq u > 0] \qquad \text{BY (214.01)a}$$

6. $$\int_u^b \frac{dx}{\sqrt{(a^2 + x^2)^3(b^2 - x^2)}} = \frac{1}{a^2\sqrt{a^2 + b^2}} E(\delta, r) - \frac{u}{a^2(a^2 + b^2)}\sqrt{\frac{b^2 - u^2}{a^2 + u^2}}$$

$$[b > u \geq 0] \qquad \text{BY (213.08)}$$

7. $$\int_b^u \frac{dx}{\sqrt{(a^2 + x^2)^3(x^2 - b^2)}} = \frac{1}{a^2\sqrt{a^2 + b^2}}\left\{F(\varepsilon, s) - E(\varepsilon, s)\right\} + \frac{1}{(a^2 + b^2)u}\sqrt{\frac{u^2 - b^2}{u^2 + a^2}}$$

$$[u > b > 0] \qquad \text{BY (211.05)}$$

8. $$\int_u^\infty \frac{dx}{\sqrt{(a^2 + x^2)^3(x^2 - b^2)}} = \frac{1}{a^2\sqrt{a^2 + b^2}}\left\{F(\xi, s) - E(\xi, s)\right\}$$

$$[u \geq b > 0] \qquad \text{BY (212.03)}$$

9. $$\int_0^u \frac{dx}{\sqrt{(a^2 + x^2)(b^2 - x^2)^3}} = \frac{1}{b^2\sqrt{a^2 + b^2}}\left\{F(\gamma, r) - E(\gamma, r)\right\} + \frac{u}{b^2\sqrt{(a^2 + u^2)(b^2 - u^2)}}$$

$$[b > u > 0] \qquad \text{BY (214.10)}$$

10. $$\int_u^\infty \frac{dx}{\sqrt{(a^2 + x^2)(x^2 - b^2)^3}} = \frac{u}{b^2\sqrt{(a^2 + u^2)(u^2 - b^2)}} - \frac{1}{b^2\sqrt{a^2 + b^2}} E(\xi, s)$$

$$[u \geq b > 0] \qquad \text{BY (212.04)}$$

11. $$\int_0^u \frac{dx}{\sqrt{(a^2 - x^2)^3(b^2 - x^2)}} = \frac{1}{a^2(a^2 - b^2)}\left\{a E(\eta, t) - u\sqrt{\frac{b^2 - u^2}{a^2 - u^2}}\right\}$$

$$[a > b \geq u > 0] \qquad \text{BY (219.07)}$$

12. $\displaystyle\int_u^b \frac{dx}{\sqrt{(a^2 - x^2)^3 (b^2 - x^2)}} = \frac{1}{a(a^2 - b^2)} E(\zeta, t)$ $[a > b > u \geq 0]$ BY (220.10)

13. $\displaystyle\int_b^u \frac{dx}{\sqrt{(a^2 - x^2)^3 (x^2 - b^2)}} = \frac{1}{a(a^2 - b^2)} \left\{ F(\kappa, q) - E(\kappa, q) + \frac{a}{u}\sqrt{\frac{u^2 - b^2}{a^2 - u^2}} \right\}$

$[a > u > b > 0]$ BY (217.10)

14. $\displaystyle\int_u^\infty \frac{dx}{\sqrt{(x^2 - a^2)^3 (x^2 - b^2)}} = \frac{1}{a(b^2 - a^2)} \left\{ E(\nu, t) - \frac{a}{u}\sqrt{\frac{u^2 - b^2}{u^2 - a^2}} \right\}$

$[u > a > b > 0]$ BY (215.04)

15. $\displaystyle\int_0^u \frac{dx}{\sqrt{(a^2 - x^2)(b^2 - x^2)^3}} = \frac{1}{ab^2} F(\eta, t) - \frac{1}{b^2(a^2 - b^2)} \left\{ a\, E(\eta, t) - u\sqrt{\frac{a^2 - u^2}{b^2 - u^2}} \right\}$

$[a > b > u > 0]$ BY (219.06)

16. $\displaystyle\int_u^a \frac{dx}{\sqrt{(a^2 - x^2)(x^2 - b^2)^3}} = \frac{1}{ab^2(a^2 - b^2)} \left\{ b^2 F(\lambda, q) - a^2 E(\lambda, q) + au\sqrt{\frac{a^2 - u^2}{u^2 - b^2}} \right\}$

$[a > u > b > 0]$ BY (218.04)

17. $\displaystyle\int_a^u \frac{dx}{\sqrt{(x^2 - a^2)(x^2 - b^2)^3}} = \frac{a}{b^2(a^2 - b^2)} E(\mu, t) - \frac{1}{ab^2} F(\mu, t)$

$[u > a > b > 0]$ BY (216.11)

18. $\displaystyle\int_u^\infty \frac{dx}{\sqrt{(x^2 - a^2)(x^2 - b^2)^3}} = \frac{1}{b^2(a^2 - b^2)} \left\{ a\, E(\nu, t) - \frac{b^2}{u}\sqrt{\frac{u^2 - a^2}{u^2 - b^2}} \right\} - \frac{1}{ab^2} F(\nu, t)$

$[u \geq a > b > 0]$ BY (215.06)

3.159

1. $\displaystyle\int_0^u \frac{x^2\, dx}{\sqrt{(x^2 + a^2)(x^2 + b^2)^3}} = \frac{a}{a^2 - b^2} \left\{ F(\alpha, q) - E(\alpha, q) \right\}$

$[a > b, \quad u > 0]$ BY (221.12)

2. $\displaystyle\int_u^\infty \frac{x^2\, dx}{\sqrt{(x^2 + a^2)(x^2 + b^2)^3}} = \frac{a}{a^2 - b^2} \left\{ F(\beta, q) - E(\beta, q) \right\} + \frac{u}{\sqrt{(a^2 + u^2)(b^2 + u^2)}}$

$[a > b, \quad u \geq 0]$ BY (222.10)

3. $\displaystyle\int_0^u \frac{x^2\, dx}{\sqrt{(x^2 + a^2)^3 (x^2 + b^2)}} = \frac{1}{a(a^2 - b^2)} \left\{ a^2 E(\alpha, q) - b^2 F(\alpha, q) \right\} - \frac{u}{\sqrt{(a^2 + u^2)(b^2 + u^2)}}$

$[a > b, \quad u > 0]$ BY (221.11)

4. $\int_u^\infty \dfrac{x^2\,dx}{\sqrt{(x^2+a^2)^3\,(x^2+b^2)}} = \dfrac{1}{a\,(a^2-b^2)}\left\{a^2\,E(\beta,q)-b^2\,F(\beta,q)\right\}$

$$[a>b,\quad u\ge 0]\qquad \text{BY (222.07)}$$

5. $\int_0^u \dfrac{x^2\,dx}{\sqrt{(a^2+x^2)^3\,(b^2-x^2)}} = \dfrac{1}{\sqrt{a^2+b^2}}\left\{F(\gamma,r)-E(\gamma,r)\right\}$

$$[b\ge u>0]\qquad \text{BY (214.04)}$$

6. $\int_u^b \dfrac{x^2\,dx}{\sqrt{(a^2+x^2)^3\,(b^2-x^2)}} = \dfrac{1}{\sqrt{a^2+b^2}}\left\{F(\delta,r)-E(\delta,r)\right\}+\dfrac{u}{a^2+b^2}\sqrt{\dfrac{b^2-u^2}{a^2+u^2}}$

$$[b>u\ge 0]\qquad \text{BY (213.07)}$$

7. $\int_b^u \dfrac{x^2\,dx}{\sqrt{(a^2+x^2)^3\,(x^2-b^2)}} = \dfrac{1}{\sqrt{a^2+b^2}}\,E(\varepsilon,s)-\dfrac{a^2}{u\,(a^2+b^2)}\sqrt{\dfrac{u^2-b^2}{u^2+a^2}}$

$$[u>b>0]\qquad \text{BY (211.13)}$$

8. $\int_u^\infty \dfrac{x^2\,dx}{\sqrt{(a^2+x^2)^3\,(x^2-b^2)}} = \dfrac{1}{\sqrt{a^2+b^2}}\,E(\xi,s)\qquad [u\ge b>0]\qquad \text{BY (212.01)}$

9. $\int_0^u \dfrac{x^2\,dx}{\sqrt{(a^2+x^2)\,(b^2-x^2)^3}} = \dfrac{u}{\sqrt{(a^2+u^2)\,(b^2-u^2)}}-\dfrac{1}{\sqrt{a^2+b^2}}\,E(\gamma,r)$

$$[b>u>0]\qquad \text{BY (214.07)}$$

10. $\int_u^\infty \dfrac{x^2\,dx}{\sqrt{(a^2+x^2)\,(x^2-b^2)^3}} = \dfrac{1}{\sqrt{a^2+b^2}}\left\{F(\xi,s)-E(\xi,s)\right\}+\dfrac{u}{\sqrt{(a^2+u^2)\,(u^2-b^2)}}$

$$[u>b>0]\qquad \text{BY (212.10)}$$

11. $\int_0^u \dfrac{x^2\,dx}{\sqrt{(a^2-x^2)^3\,(b^2-x^2)}} = \dfrac{1}{a^2-b^2}\left\{a\,E(\eta,t)-u\sqrt{\dfrac{b^2-u^2}{a^2-u^2}}\right\}-\dfrac{1}{a}\,F(\eta,t)$

$$[a>b\ge u>0]\qquad \text{BY (219.04)}$$

12. $\int_u^b \dfrac{x^2\,dx}{\sqrt{(a^2-x^2)^3\,(b^2-x^2)}} = \dfrac{a}{a^2-b^2}\,E(\zeta,t)-\dfrac{1}{a}\,F(\zeta,t)$

$$[a>b>u\ge 0]\qquad \text{BY (220.08)}$$

13. $\int_b^u \dfrac{x^2\,dx}{\sqrt{(a^2-x^2)^3\,(x^2-b^2)}} = \dfrac{1}{a\,(a^2-b^2)}\left\{b^2\,F(\kappa,q)-a^2\,E(\kappa,q)+\dfrac{a^3}{u}\sqrt{\dfrac{u^2-b^2}{a^2-u^2}}\right\}$

$$[a>u>b>0]\qquad \text{BY (217.06)}$$

14. $\int_u^\infty \dfrac{x^2\,dx}{\sqrt{(x^2-a^2)^3\,(x^2-b^2)}} = \dfrac{a}{a^2-b^2}\left\{\dfrac{a}{u}\sqrt{\dfrac{u^2-b^2}{u^2-a^2}}-E(\nu,t)\right\}+\dfrac{1}{a}\,F(\nu,t)$

$$[u>a>b>0]\qquad \text{BY (215.09)}$$

15.
$$\int_0^u \frac{x^2\,dx}{\sqrt{(a^2-x^2)(b^2-x^2)^3}} = \frac{1}{a^2-b^2}\left\{u\sqrt{\frac{a^2-u^2}{b^2-u^2}} - a\,E(\eta,t)\right\}$$

$$[a > b > u > 0] \qquad\qquad \text{BY (219.12)}$$

16.
$$\int_u^a \frac{x^2\,dx}{\sqrt{(a^2-x^2)(x^2-b^2)^3}} = \frac{1}{a^2-b^2}\left\{a\,F(\lambda,q) - a\,E(\lambda,q) + u\sqrt{\frac{a^2-u^2}{u^2-b^2}}\right\}$$

$$[a > u > b > 0] \qquad\qquad \text{BY (218.07)}$$

17.
$$\int_a^u \frac{x^2\,dx}{\sqrt{(x^2-a^2)(x^2-b^2)^3}} = \frac{a}{a^2-b^2}\,E(\mu,t) \qquad [u > a > b > 0] \qquad \text{BY (216.01)}$$

18.
$$\int_u^\infty \frac{x^2\,dx}{\sqrt{(x^2-a^2)(x^2-b^2)^3}} = \frac{1}{a^2-b^2}\left\{a\,E(\nu,t) - \frac{b^2}{u}\sqrt{\frac{u^2-a^2}{u^2-b^2}}\right\}$$

$$[u \geq a > b > 0] \qquad\qquad \text{BY (215.11)}$$

3.161

1.
$$\int_u^\infty \frac{dx}{x^4\sqrt{(x^2+a^2)(x^2+b^2)}} = \frac{1}{3a^3b^4}\left\{2\left(a^2+b^2\right)E(\beta,q) - b^2\,F(\beta,q)\right\} + \frac{a^2b^2 - u^2\left(2a^2+b^2\right)}{3a^2b^4u^3}$$

$$[a > b, \quad u > 0] \qquad\qquad \text{BY (222.04)}$$

2.
$$\int_u^b \frac{dx}{x^4\sqrt{(x^2+a^2)(b^2-x^2)}} = \frac{1}{3a^4b^4\sqrt{a^2+b^2}}\left\{a^2\left(2a^2-b^2\right)F(\delta,r) - 2\left(a^4-b^4\right)E(\delta,r)\right\}$$
$$+ \frac{a^2b^2 + 2u^2\left(a^2-b^2\right)}{3a^4b^4u^3}\sqrt{(b^2-u^2)(a^2+u^2)}$$

$$[b > u > 0] \qquad\qquad \text{BY (213.09)}$$

3.
$$\int_b^u \frac{dx}{x^4\sqrt{(x^2+a^2)(x^2-b^2)}} = \frac{2b^2-a^2}{3a^4b^2\sqrt{a^2+b^2}}\,F(\varepsilon,s) + \frac{2}{3}\frac{\left(a^2-b^2\right)\sqrt{a^2+b^2}}{a^4b^4}\,E(\varepsilon,s)$$
$$+ \frac{1}{3a^2b^2u^3}\sqrt{(u^2+a^2)(u^2-b^2)}$$

$$[u > b > 0] \qquad\qquad \text{BY (211.11)}$$

4.
$$\int_u^\infty \frac{dx}{x^4\sqrt{(x^2+a^2)(x^2-b^2)}} = \frac{1}{3a^4b^4\sqrt{a^2+b^2}}\left\{2\left(a^4-b^4\right)E(\xi,s) + b^2\left(2b^2-a^2\right)F(\xi,s)\right\}$$
$$- \frac{a^2b^2 + u^2\left(2a^2-b^2\right)}{3a^2b^4u^3}\sqrt{\frac{u^2-b^2}{u^2+a^2}}$$

$$[u \geq b > 0] \qquad\qquad \text{BY (212.06)}$$

5.
$$\int_u^b \frac{dx}{x^4\sqrt{(a^2-x^2)(b^2-x^2)}} = \frac{1}{3a^3b^4}\left\{\left\{\left(2a^2+b^2\right)F(\zeta,t) - 2\left(a^2+b^2\right)E(\zeta,t)\right\}\right.$$

$$\left. + \frac{\left[\left(2a^2+b^2\right)u^2 + a^2b^2\right]a}{u^3}\sqrt{\frac{b^2-u^2}{a^2-u^2}}\right\}$$

$$[a > b > u > 0] \qquad\qquad \text{BY (220.09)}$$

6. $$\int_b^u \frac{dx}{x^4 \sqrt{(a^2 - x^2)(x^2 - b^2)}} = \frac{1}{3a^3b^4} \left\{ 2\left(a^2 + b^2\right) E\left(\kappa, q\right) - b^2 F\left(\kappa, q\right) \right\}$$
$$+ \frac{1}{3a^2b^2u^3} \sqrt{(a^2 - u^2)(u^2 - b^2)}$$
$$[a \geq u > b > 0] \qquad \text{BY (217.14)}$$

7. $$\int_u^a \frac{dx}{x^4 \sqrt{(a^2 - x^2)(x^2 - b^2)}} = \frac{1}{3a^3b^4} \left\{ 2\left(a^2 + b^2\right) E(\lambda, q) - b^2 F(\lambda, q) \right.$$
$$\left. - \frac{2\left(a^2 + b^2\right)u^2 + a^2b^2}{au^3} \sqrt{(a^2 - u^2)(u^2 - b^2)} \right\}$$
$$[a > u \geq b > 0] \qquad \text{BY (218.12)}$$

8. $$\int_a^u \frac{dx}{x^4 \sqrt{(x^2 - a^2)(x^2 - b^2)}}$$
$$= \frac{1}{3a^3b^4} \left\{ \left\{ \left(2a^2 + b^2\right) F(\mu, t) - 2\left(a^2 + b^2\right) E(\mu, t) \right\} \qquad [u > a > b > 0] \right.$$
$$\left. + \frac{\left[\left(a^2 + 2b^2\right)u^2 + a^2b^2\right]b^2}{au^3} \sqrt{\frac{u^2 - a^2}{u^2 - b^2}} \right\}$$
$$\text{BY (216.09)}$$

9. $$\int_u^\infty \frac{dx}{x^4 \sqrt{(x^2 - a^2)(x^2 - b^2)}} = \frac{1}{3a^3b^4} \left\{ \left(2a^2 + b^2\right) F(\nu, t) - 2\left(a^2 + b^2\right) E(\nu, t) \right.$$
$$\left. + \frac{ab^2}{u^3} \sqrt{(u^2 - a^2)(u^2 - b^2)} \right\}$$
$$[u \geq a > b > 0] \qquad \text{BY (215.07)}$$

3.162

1. $$\int_0^u \frac{dx}{\sqrt{(x^2 + a^2)^5 (x^2 + b^2)}} = \frac{1}{3a^3(a^2 - b^2)^2} \left\{ \left(3a^2 - b^2\right) F(\alpha, q) - 2\left(2a^2 - b^2\right) E(\alpha, q) \right\}$$
$$+ \frac{u\left[a^2\left(4a^2 - 3b^2\right) + u^2\left(3a^2 - 2b^2\right)\right]}{3a^4\left(a^2 - b^2\right)\sqrt{(u^2 + a^2)^3 (u^2 + b^2)}}$$
$$[a > b, \quad u > 0] \qquad \text{BY (221.06)}$$

2. $$\int_u^\infty \frac{dx}{\sqrt{(x^2 + a^2)^5 (x^2 + b^2)}} = \frac{1}{3a^3(a^2 - b^2)^2} \left\{ \left(3a^2 - b^2\right) F(\beta, q) - 2\left(2a^2 - b^2\right) E(\beta, q) \right\}$$
$$+ \frac{u}{3a^2\left(a^2 - b^2\right)} \sqrt{\frac{u^2 + b^2}{(a^2 + u^2)^3}}$$
$$[a > b, \quad u \geq 0] \qquad \text{BY (222.03)}$$

3.
$$\int_0^u \frac{dx}{\sqrt{(x^2+a^2)\,(x^2+b^2)^5}} = \frac{3b^2-a^2}{3ab^2(a^2-b^2)^2}\,F(\alpha,q) + \frac{a\,(2a^2-4b^2)}{3b^4(a^2-b^2)^2}\,E(\alpha,q)$$
$$+ \frac{u}{3b^2\,(a^2-b^2)}\sqrt{\frac{u^2+a^2}{(u^2+b^2)^3}}$$
$$[a>b, \quad u>0] \qquad \text{BY (221.05)}$$

4.
$$\int_u^\infty \frac{dx}{\sqrt{(x^2+a^2)\,(x^2+b^2)^5}} = \frac{1}{3ab^4(a^2-b^2)^2}\left\{2a^2\,(a^2-2b^2)\,E(\beta,q) + b^2\,(3b^2-a^2)\,F(\beta,q)\right\}$$
$$- \frac{u\,[b^2\,(3a^2-4b^2) + u^2\,(2a^2-3b^2)]}{3b^4\,(a^2-b^2)\sqrt{(u^2+a^2)\,(u^2+b^2)^3}}$$
$$[a>b, \quad u\geq 0] \qquad \text{BY (222.05)}$$

5.
$$\int_0^u \frac{dx}{\sqrt{(a^2+x^2)^5\,(b^2-x^2)}} = \frac{1}{3a^4\sqrt{(a^2+b^2)^3}}\left\{2\,(b^2+2a^2)\,E(\gamma,r) - a^2\,F(\gamma,r)\right\}$$
$$+ \frac{u}{3a^2\,(a^2+b^2)}\sqrt{\frac{b^2-u^2}{(a^2+u^2)^3}}$$
$$[b\geq u>0] \qquad \text{BY (214.15)}$$

6.
$$\int_u^b \frac{dx}{\sqrt{(a^2+x^2)^5\,(b^2-x^2)}} = \frac{1}{3a^4\sqrt{(a^2+b^2)^3}}\left\{(4a^2+2b^2)\,E(\delta,r) - a^2\,F(\delta,r)\right\}$$
$$- \frac{u\,[a^2\,(5a^2+3b^2) + u^2\,(4a^2+2b^2)]}{3a^4(a^2+b^2)^2}\sqrt{\frac{b^2-u^2}{(a^2+u^2)^3}}$$
$$[b>u>0] \qquad \text{BY (213.08)}$$

7.
$$\int_b^u \frac{dx}{\sqrt{(a^2+x^3)^5\,(x^2-b^2)}} = \frac{1}{3a^4\sqrt{(a^2+b^2)^3}}\left\{(3a^2+2b^2)\,F(\varepsilon,s) - (4a^2+2b^2)\,E(\varepsilon,s)\right\}$$
$$+ \frac{(3a^2+b^2)\,u^2 + 2\,(2a^2+b^2)\,a^2}{3a^2(a^2+b^2)^2u}\sqrt{\frac{u^2-b^2}{(u^2+a^2)^3}}$$
$$[u>b>0] \qquad \text{BY (211.05)}$$

8.
$$\int_u^\infty \frac{dx}{\sqrt{(a^2+x^2)^5\,(x^2-b^2)}} = \frac{1}{3a^4\sqrt{(a^2+b^2)^3}}\left\{(3a^2+2b^2)\,F(\xi,s) - (4a^2+2b^2)\,E(\xi,s)\right\}$$
$$+ \frac{u}{3a^2\,(a^2+b^2)}\sqrt{\frac{u^2-b^2}{(a^2+u^2)^3}}$$
$$[u>b>0] \qquad \text{BY (212.03)}$$

9.
$$\int_0^u \frac{dx}{\sqrt{(a^2+x^2)\,(b^2-x^2)^5}} = \frac{1}{3b^4\sqrt{(a^2+b^2)^3}}\left\{(2a^2+3b^2)\,F(\gamma,r) - (2a^2+4b^2)\,E(\gamma,r)\right\}$$
$$+ \frac{u\,[(3a^3+4b^2)\,b^2 - (2a^2+3b^2)\,u^2]}{3b^4\,(a^2+b^2)\sqrt{(a^2+u^2)\,(b^2-u^2)^3}}$$
$$[b>u>0] \qquad \text{BY (214.10)}$$

10.
$$\int_u^\infty \frac{dx}{\sqrt{(a^2+x^2)(x^2-b^2)^5}} = \frac{1}{3b^4\sqrt{(a^2+b^2)^3}}\left\{(2a^2+4b^2)\,E(\xi,s)-b^2\,F(\xi,s)\right\}$$
$$+\frac{u\left[(3a^2+4b^2)\,b^2-(2a^2+3b^2)\,u^2\right]}{3b^4(a^2+b^2)\sqrt{(a^2+u^2)(u^2-b^2)^3}}$$
$$[u>b>0] \qquad\qquad \text{BY (212.04)}$$

11.
$$\int_0^u \frac{dx}{\sqrt{(a^2-x^2)(b^2-x^2)^5}} = \frac{2a^2-3b^2}{3ab^4(a^2-b^2)}\,F(\eta,t)+\frac{2a\,(2b^2-a^2)}{3b^4(a^2-b^2)^2}\,E(\eta,t)$$
$$+\frac{u\left[(3a^2-5b^2)\,b^2-2\,(a^2-2b^2)\,u^2\right]}{3b^4(a^2-b^2)^2(b^2-u^2)}\sqrt{\frac{a^2-u^2}{b^2-u^2}}$$
$$[a>b>a>0] \qquad\qquad \text{BY (219.06)}$$

12.
$$\int_u^a \frac{dx}{\sqrt{(a^2-x^2)(x^2-b^2)^5}} = \frac{3b^2-a^2}{3ab^2(a^2-b^2)^2}\,F(\lambda,q)+\frac{2a\,(a^2-2b^2)}{3b^4(a^2-b^2)^2}\,E(\lambda,q)$$
$$+\frac{u\left[2\,(2b^2-a^2)\,u^2+(3a^2-5b^2)\,b^2\right]}{3b^4(a^2-b^2)^2(u^2-b^2)}\sqrt{\frac{a^2-u^2}{u^2-b^2}}$$
$$[a>u>b>0] \qquad\qquad \text{BY (218.04)}$$

13.
$$\int_a^u \frac{dx}{\sqrt{(x^2-a^2)(x^2-b^2)^5}} = \frac{2a^2-3b^2}{3ab^4(a^2-b^2)}\,F(\mu,t)+\frac{2a\,(2b^2-a^2)}{3b^4(a^2-b^2)^2}\,E(\mu,t)$$
$$+\frac{u}{3b^2\,(a^2-b^2)\,(u^2-b^2)}\sqrt{\frac{u^2-a^2}{u^2-b^2}}$$
$$[u>a>b>0] \qquad\qquad \text{BY (216.11)}$$

14.
$$\int_u^\infty \frac{dx}{\sqrt{(x^2-a^2)(x^2-b^2)^5}} = \frac{(4b^2-2a^2)\,a}{3b^4(a^2-b^2)^2}\,E(\nu,t)+\frac{2a^2-3b^2}{3ab^4\,(a^2-b^2)}\,F(\nu,t)$$
$$-\frac{(3b^2-a^2)\,u^2-(4b^2-2a^2)\,b^2}{3b^2u(a^2-b^2)^2\,(u^2-b^2)}\sqrt{\frac{u^2-a^2}{u^2-b^2}}$$
$$[u\ge a>b>0] \qquad\qquad \text{BY (215.06)}$$

15.
$$\int_0^u \frac{dx}{\sqrt{(a^2-x^2)^5(b^2-x^2)}} = \frac{1}{3a^3(a^2-b^2)^2}\left\{(4a^2-2b^2)\,E(\eta,t)-(a^2-b^2)\,F(\eta,t)\right.$$
$$\left.-\frac{u\left[(5a^2-3b^2)\,a^2-(4a^2-2b^2)\,u^2\right]}{a\,(a^2-u^2)}\sqrt{\frac{b^2-u^2}{a^2-u^2}}\right\}$$
$$[a>b\ge u>0] \qquad\qquad \text{BY (219.07)}$$

16.
$$\int_u^b \frac{dx}{\sqrt{(a^2-x^2)^5(b^2-x^2)}} = \frac{2\,(2a^2-b^2)}{3a^3(a^2-b^2)^2}\,E(\zeta,r)-\frac{1}{3a^3\,(a^2-b^2)}\,F(\zeta,t)$$
$$+\frac{u}{3a^2\,(a^2-b^2)\,(a^2-u^2)}\sqrt{\frac{b^2-u^2}{a^2-u^2}}$$
$$[a>b>u\ge 0] \qquad\qquad \text{BY (220.10)}$$

17.
$$\int_b^u \frac{dx}{\sqrt{(a^2 - x^2)^5 (x^2 - b^2)}} = \frac{1}{3a^3(a^2 - b^2)^2} \left\{ (3a^2 - b^2) F(\kappa, q) - (4a^2 - 2b^2) E(\kappa, q) \right\}$$
$$+ \frac{2(2a^2 - b^2) a^2 + (b^2 - 3a^2) u^2}{3a^2 u (a^2 - b^2)^2 (a^2 - u^2)} \sqrt{\frac{u^2 - b^2}{a^2 - u^2}},$$
$$[a > u > b > 0] \qquad \text{BY (217.10)}$$

18.
$$\int_u^\infty \frac{dx}{\sqrt{(x^2 - a^2)^5 (x^2 - b^2)}} = \frac{1}{3a^3(a^2 - b^2)^2} \left\{ (4a^2 - 2b^2) E(\nu, t) - (a^2 - b^2) F(\nu, t) \right\}$$
$$+ \frac{(4a^2 - 2b^2) a^2 + (b^2 - 3a^2) u^2}{3a^2 u (a^2 - b^2)^2 (u^2 - a^2)} \sqrt{\frac{u^2 - b^2}{u^2 - a^2}}$$
$$[u > a > b > 0] \qquad \text{BY (215.04)}$$

3.163

1.
$$\int_0^u \frac{dx}{\sqrt{(x^2 + a^2)^3 (x^2 + b^2)^3}} = \frac{1}{ab^2(a^2 - b^2)^2} \left\{ (a^2 + b^2) E(\alpha, q) - 2b^2 F(\alpha, q) \right\}$$
$$- \frac{u}{a^2 (a^2 - b^2) \sqrt{(a^2 + u^2)(b^2 + u^2)}}$$
$$[a > b, \quad u > 0] \qquad \text{BY (221.07)}$$

2.
$$\int_u^\infty \frac{dx}{\sqrt{(x^2 + a^2)^3 (x^2 + b^2)^3}} = \frac{1}{ab^2(a^2 - b^2)^2} \left\{ (a^2 + b^2) E(\beta, q) - 2b^2 F(\beta, q) \right\}$$
$$- \frac{u}{b^2 (a^2 - b^2) \sqrt{(a^2 + u^2)(b^2 + u^2)}}$$
$$[a > b, \quad u \geq 0] \qquad \text{BY (222.12)}$$

3.
$$\int_0^u \frac{dx}{\sqrt{(x^2 + a^2)^3 (b^3 - x^2)^3}} = \frac{1}{a^2 b^2 \sqrt{(a^2 + b^2)^3}} \left\{ a^2 F(\gamma, r) - (a^2 - b^2) E(\gamma, r) \right\}$$
$$+ \frac{u}{b^2 (a^2 + b^2) \sqrt{(a^2 + u^2)(b^2 - u^2)}}$$
$$[b > u > 0] \qquad \text{BY (214.15)}$$

4.
$$\int_u^\infty \frac{dx}{\sqrt{(x^2 + a^2)^3 (x^2 - b^2)^3}} = \frac{b^2 - a^2}{a^2 b^2 \sqrt{(a^2 + b^2)^3}} E(\xi, s) - \frac{1}{a^2 \sqrt{(a^2 + b^2)^3}} F(\xi, s)$$
$$+ \frac{u}{b^2 (a^2 + b^2) \sqrt{(u^2 + a^2)(u^2 - b^2)}}$$
$$[u > b > 0] \qquad \text{BY (212.05)}$$

5.
$$\int_0^u \frac{dx}{\sqrt{(a^2 - x^2)^3 (b^2 - x^2)^3}} = \frac{1}{ab^2 (a^2 - b^2)} F(\eta, t) - \frac{a^2 + b^2}{ab^2(a^2 - b^2)^2} E(\eta, t)$$
$$+ \frac{[a^4 + b^4 - (a^2 + b^2) u^2] u}{a^2 b^2 (a^2 - b^2)^2 \sqrt{(a^2 - u^2)(b^2 - u^2)}}$$
$$[a > b > u > 0] \qquad \text{BY (279.08)}$$

6.
$$\int_u^\infty \frac{dx}{\sqrt{(x^2 - a^2)^3 (x^2 - b^2)^3}} = \frac{1}{ab^2 (a^2 - b^2)} F(\nu, t) - \frac{a^2 + b^2}{ab^2(a^2 - b^2)^2} E(\nu, t)$$
$$+ \frac{1}{u (a^2 - b^2) \sqrt{(u^2 - a^2)(u^2 - b^2)}}$$
$$[u > a > b > 0] \qquad \text{BY (215.10)}$$

3.164   **Notation:** $\alpha = \arccos \dfrac{u^2 - \rho\bar\rho}{u^2 + \rho\bar\rho}$, $\qquad r = \dfrac{1}{2}\sqrt{-\dfrac{(\rho - \bar\rho)^2}{\rho\bar\rho}}$.

1. $\displaystyle\int_u^\infty \frac{dx}{\sqrt{(x^2 + \rho^2)(x^2 + \bar\rho^2)}} = \frac{1}{\sqrt{\rho\bar\rho}} F(\alpha, r)$ BY (225.00)

2. $\displaystyle\int_u^\infty \frac{x^2\, dx}{(x^2 - \rho\bar\rho)^2 \sqrt{(x^2 + \rho^2)(x^2 + \bar\rho^2)}} = \frac{2u\sqrt{(u^2 + \rho^2)(u^2 + \bar\rho^2)}}{(\rho + \bar\rho)^2 (u^4 - \rho^2\bar\rho^2)} - \frac{1}{(\rho + \bar\rho)^2 \sqrt{\rho\bar\rho}} E(\alpha, r)$

   BY (225.03)

3. $\displaystyle\int_u^\infty \frac{x^2\, dx}{(x^2 + \rho\bar\rho)^2 \sqrt{(x^2 + \rho^2)(x^2 + \bar\rho^2)}} = -\frac{1}{(\rho - \bar\rho)^2 \sqrt{\rho\bar\rho}} [F(\alpha, r) - E(\alpha, r)]$ BY (225.07)

4. $\displaystyle\int_u^\infty \frac{x^2\, dx}{\sqrt{(x^2 + \rho^2)^3 (x^2 + \bar\rho^2)^3}} = -\frac{4\sqrt{\rho\bar\rho}}{(\rho^2 - \bar\rho^2)^2} E(\alpha, r) + \frac{1}{(\rho - \bar\rho)^2 \sqrt{\rho\bar\rho}} F(\alpha, r)$

   $\qquad\qquad\qquad\qquad\qquad - \dfrac{2u(u^2 - \rho\bar\rho)}{(\rho + \bar\rho)^2 (u^2 + \rho\bar\rho)\sqrt{(u^2 + \rho^2)(u^2 + \bar\rho^2)}}$

   BY (225.05)

5. $\displaystyle\int_u^\infty \frac{(x^2 - \rho\bar\rho)^2\, dx}{\sqrt{(x^2 + \rho^2)^3 (x^2 + \bar\rho^2)^3}} = -\frac{4\sqrt{\rho\bar\rho}}{(\rho - \bar\rho)^2} [F(\alpha, r) - E(\alpha, r)]$

   $\qquad\qquad\qquad\qquad\qquad + \dfrac{2u(u^2 - \rho\bar\rho)}{(u^2 + \rho\bar\rho)\sqrt{(u^2 + \rho^2)(u^2 + \bar\rho^2)}}$

   BY (225.06)

6. $\displaystyle\int_u^\infty \frac{\sqrt{(x^2 + \rho^2)(x^2 + \bar\rho^2)}}{(x^2 + \rho\bar\rho)^2} dx = \frac{1}{\sqrt{\rho\bar\rho}} E(\alpha, r)$ BY(225.01)

7. $\displaystyle\int_u^\infty \frac{(x^2 - \varrho\bar\varrho)^2\, dx}{(x^2 + \varrho\bar\varrho)^2 \sqrt{(x^2 + \varrho^2)(x^2 + \bar\varrho^2)}} = -\frac{4\sqrt{\varrho\bar\varrho}}{(\varrho - \bar\varrho)^2} E(\alpha, r) + \frac{(\varrho + \bar\varrho)^2}{(\varrho - \bar\varrho)^2 \sqrt{\varrho\bar\varrho}} F(\alpha, r)$ BY (225.08)

8. $\displaystyle\int_u^\infty \frac{(x^2 + \varrho\bar\varrho)^2\, dx}{\left[(x^2 + \varrho\bar\varrho)^2 - 4p^2 \varrho\bar\varrho x^2\right] \sqrt{(x^2 + \varrho^2)(x^2 + \bar\varrho^2)}} = \frac{1}{\sqrt{\varrho\bar\varrho}} \Pi(\alpha, p^2, r)$ BY (225.02)

3.165   **Notation:** $\alpha = \arccos \dfrac{u^2 - a^2}{u^2 + a^2}$, $\qquad r = \dfrac{\sqrt{a^2 - b^2}}{a\sqrt{2}}$.

1. $\displaystyle\int_u^a \frac{dx}{\sqrt{x^4 + 2b^2 x^2 + a^4}} = \frac{\sqrt{2}}{a\sqrt{2} + \sqrt{a^2 + b^2}}$

   $\qquad\qquad\qquad \times F\left[\arctan\left(\dfrac{a\sqrt{2} + \sqrt{a^2 - b^2}}{\sqrt{a^2 + b^2}} \dfrac{a - u}{a + u}\right), \dfrac{2\sqrt{a\sqrt{2(a^2 - b^2)}}}{a\sqrt{2} + \sqrt{a^2 - b^2}}\right]$

   $\qquad\qquad\qquad\qquad\qquad\qquad [a > b, \quad a > u \geq 0]$ BY (264.00)

2. $$\int_u^\infty \frac{dx}{\sqrt{x^4 + 2b^2 x^2 + a^4}} = \frac{1}{2a} F(\alpha, r) \qquad\qquad \left[a^2 > b^2 > -\infty, \quad a^2 > 0, \quad u \geq 0\right]$$

BY (263.00, 266.00)

3. $$\int_u^\infty \frac{dx}{x^2 \sqrt{x^4 + 2b^2 x^2 + a^4}} = \frac{1}{2a^3}\left[F(\alpha, r) - 2\,E(\alpha, r)\right] + \frac{\sqrt{u^4 + 2b^2 u^2 + a^4}}{a^2 u\left(u^2 + a^2\right)}$$

$$\left[a > b > 0, \quad u > 0\right] \qquad \text{BY (263.06)}$$

4. $$\int_u^\infty \frac{x^2\, dx}{\left(x^2 + a^2\right)^2 \sqrt{x^4 + 2b^2 x^2 + a^4}} = \frac{1}{4a\left(a^2 - b^2\right)}\left[F(\alpha, r) - E(\alpha, r)\right]$$

$$\left[a^2 > b^2 > -\infty, \quad a^2 > 0, \quad u \geq 0\right]$$

BY (263.03, 266.05)

5. $$\int_u^\infty \frac{x^2\, dx}{\left(x^2 - a^2\right)^2 \sqrt{x^4 + 2b^2 x^2 + a^4}} = \frac{u\sqrt{u^4 + 2b^2 u^2 + a^4}}{2\left(a^2 + b^2\right)\left(u^4 - a^4\right)} - \frac{1}{4a\left(a^2 + b^2\right)} E(\alpha, r)$$

$$\left[a^2 > b^2 > -\infty, \quad u^2 > a^2 > 0\right]$$

BY (263.05, 266.02)

6. $$\int_u^\infty \frac{x^2\, dx}{\sqrt{\left(x^4 + 2b^2 x^2 + a^4\right)^3}} = \frac{a}{2\left(a^4 - b^4\right)} E(\alpha, r) - \frac{1}{4a\left(a^2 - b^2\right)} F(\alpha, r)$$

$$- \frac{u\left(u^2 - a^2\right)}{2\left(a^2 + b^2\right)\left(u^2 + a^2\right)\sqrt{u^4 + 2b^2 u^2 + a^4}}$$

$$\left[a^2 > b^2 > -\infty, a^2 > 0, \quad u \geq 0\right] \quad \text{BY (263.08, 266.03)}$$

7. $$\int_u^\infty \frac{\left(x^2 - a^2\right)^2 dx}{\sqrt{\left(x^4 + 2b^2 x^2 + a^4\right)^3}} = \frac{a}{a^2 - b^2}\left[F(\alpha, r) - E(\alpha, r)\right] + \frac{u^2 - a^2}{u^2 + a^2}\frac{u}{\sqrt{u^4 + 2b^2 u^2 + a^4}}$$

$$\left[|b^2| < a^2, \quad u \geq 0\right] \qquad \text{BY (266.08)}$$

8. $$\int_u^\infty \frac{\left(x^2 + a^2\right)^2 dx}{\sqrt{\left(x^2 + 2b^2 x^2 + a^4\right)^3}} = \frac{a}{a^2 + b^2} E(\alpha, r) - \frac{a^2 - b^2}{a^2 + b^2}\cdot\frac{u^2 - a^2}{u^2 + a^2}\cdot\frac{u}{\sqrt{u^4 + 2b^2 u^2 + a^4}}$$

$$\left[|b^2| < a^2, \quad u \geq 0\right] \qquad \text{BY (266.06)a}$$

9. $$\int_u^\infty \frac{\left(x^2 - a^2\right)^2 dx}{\left(x^2 + a^2\right)^2 \sqrt{x^4 + 2b^2 x^2 + a^4}} = \frac{a}{a^2 - b^2} E(\alpha, r) - \frac{a^2 + b^2}{2a\left(a^2 - b^2\right)} F(\alpha, r)$$

$$\left[a^2 > b^2 > -\infty, \quad a^2 > 0, \quad u \geq 0\right]$$

BY (263.04, 266.07)

10. $$\int_u^\infty \frac{\sqrt{x^4 + 2b^2 x^2 + a^4}}{\left(x^2 + a^2\right)^2}\, dx = \frac{1}{2a} E(\alpha, r) \qquad\qquad \left[a^2 > b^2 > -\infty, \quad a^2 > 0, \quad u \geq 0\right]$$

BY (263.01, 266.01)

11. $$\int_u^\infty \frac{\sqrt{x^4 + 2b^2 x^2 + a^4}}{\left(x^2 - a^2\right)^2}\, dx = \frac{1}{2a}\left[F(\alpha, r) - E(\alpha, r)\right] + \frac{u}{u^4 - a^4}\sqrt{u^4 + 2b^2 u^2 + a^4}$$

$$\left[a > b > 0, \quad u > a\right] \qquad\qquad \text{BY (263)}$$

12. $\displaystyle\int_u^\infty \frac{\left(x^2+a^2\right)^2 dx}{\left[\left(x^2+a^2\right)^2 - 4a^2 p^2 x^2\right]\sqrt{x^4 + 2b^2 x^2 + a^4}} = \frac{1}{2a}\,\Pi\left(\alpha, p^2, r\right)$

$$[a > b > 0, \quad u \geq 0] \qquad \text{BY (263.02)}$$

3.166   **Notation:** $\alpha = \arccos\dfrac{u^2 - 1}{u^2 + 1}, \quad \beta = \arctan\left\{\left(1 + \sqrt{2}\right)\dfrac{1 - u}{1 + u}\right\},$

$$\gamma = \arccos u, \quad \delta = \arccos\frac{1}{u}, \quad \varepsilon = \arccos\frac{1 - u^2}{1 + u^2},$$

$$r = \frac{\sqrt{2}}{2}, \quad q = 2\sqrt{3\sqrt{2} - 4} = 2\sqrt[4]{2}\left(\sqrt{2} - 1\right) \approx 0.985171$$

1. $\displaystyle\int_u^\infty \frac{dx}{\sqrt{x^4 + 1}} = \frac{1}{2}\,F(\alpha, r)$ 　　　　　　　$[u \geq 0]$ 　　　　H (287), BY (263.50)

2. $\displaystyle\int_u^\infty \frac{dx}{x^2\sqrt{x^4 + 1}} = \frac{1}{2}\left[F(\alpha, r) - 2\,E(\alpha, r)\right] + \frac{\sqrt{u^4 + 1}}{u\left(u^2 + 1\right)}$

$$[u > 0] \qquad \text{BY (263.57)}$$

3. $\displaystyle\int_u^\infty \frac{x^2\,dx}{\left(x^4 + 1\right)\sqrt{x^4 + 1}} = \frac{1}{2}\,E(\alpha, r) - \frac{1}{4}\,F(\alpha, r) - \frac{u\left(u^2 - 1\right)}{2\left(u^2 + 1\right)\sqrt{u^4 + 1}}$

$$[u \geq 0] \qquad \text{BY (263.59)}$$

4. $\displaystyle\int_u^\infty \frac{x^2\,dx}{\left(x^2 + 1\right)^2\sqrt{x^4 + 1}} = \frac{1}{4}\left[F(\alpha, r) - E(\alpha, r)\right]$ 　　$[u \geq 0]$ 　　BY (263.53)

5. $\displaystyle\int_u^\infty \frac{x^2\,dx}{\left(x^2 - 1\right)^2\sqrt{x^4 + 1}} = \frac{u\sqrt{u^4 + 1}}{2\left(u^4 - 1\right)} - \frac{1}{4}\,E(\alpha, r)$ 　　$[u > 1]$ 　　BY (263.55)

6. $\displaystyle\int_u^\infty \frac{\sqrt{x^4 + 1}}{\left(x^2 - 1\right)^2}\,dx = \frac{1}{2}\left[F(\alpha, r) - E(\alpha, r)\right] + \frac{u\sqrt{u^4 + 1}}{u^4 - 1}$

$$[u > 1] \qquad \text{BY (263.58)}$$

7. $\displaystyle\int_u^\infty \frac{\left(x^2 - 1\right)^2 dx}{\left(x^2 + 1\right)^2\sqrt{x^4 + 1}} = E(\alpha, r) - \frac{1}{2}\,F(\alpha, r)$ 　　$[u \geq 0]$ 　　BY (263.54)

8. $\displaystyle\int_u^\infty \frac{\sqrt{x^4 + 1}\,dx}{\left(x^2 + 1\right)^2} = \frac{1}{2}\,E(\alpha, r)$ 　　　　　　$[u \geq 0]$ 　　BY (263.51)

9. $\displaystyle\int_u^\infty \frac{\left(x^2 + 1\right)^2 dx}{\left[\left(x^2 + 1\right)^2 - 4p^2 x^2\right]\sqrt{x^4 + 1}} = \frac{1}{2}\,\Pi\left(\alpha, p^2, r\right)$ 　　$[u \geq 0]$ 　　BY (263.52)

10. $\displaystyle\int_0^u \frac{dx}{\sqrt{x^4 + 1}} = \frac{1}{2}\,F(\varepsilon, r)$ 　　　　　　　　　　H 66(288)

11. $\displaystyle\int_u^1 \frac{dx}{\sqrt{x^4 + 1}} = \left(2 - \sqrt{2}\right)F(\beta, q)$ 　　　　$[0 \leq u < 1]$ 　　BY (264.50)

12. $\int_u^1 \dfrac{\left(x^2 + x\sqrt{2} + 1\right) dx}{\left(x^2 - x\sqrt{2} + 1\right)\sqrt{x^4 + 1}} = \left(2 + \sqrt{2}\right) E(\beta, q)$　　　　$[0 \le u < 1]$　　　　　BY (264.51)

13. $\int_u^1 \dfrac{(1 - x)^2 dx}{\left(x^2 - x\sqrt{2} + 1\right)\sqrt{x^4 + 1}} = \dfrac{1}{\sqrt{2}}\left[F(\beta, q) - E(\beta, q)\right]$

$[0 \le u < 1]$　　　　　BY (264.55)

14. $\int_u^1 \dfrac{(1 + x)^2 dx}{\left(x^2 - x\sqrt{2} + 1\right)\sqrt{x^4 + 1}} = \dfrac{3\sqrt{2} + 4}{2} E(\beta, q) - \dfrac{3\sqrt{2} - 4}{2} F(\beta, q)$

$[0 \le u < 1]$　　　　　BY (264.56)

15. $\int_u^1 \dfrac{dx}{\sqrt{1 - x^4}} = \dfrac{1}{\sqrt{2}} F(\gamma, r)$　　　　　　　　$[u < 1]$　　　　H 66 (290), BY (259.75)

16. $\int_0^1 \dfrac{dx}{\sqrt{1 - x^4}} = \dfrac{1}{4\sqrt{2\pi}}\left\{\Gamma\left(\dfrac{1}{4}\right)\right\}^2$

17. $\int_1^u \dfrac{dx}{\sqrt{x^4 - 1}} = \dfrac{1}{\sqrt{2}} F(\delta, r)$　　　　　　　　$[u > 1]$　　　　H 66 (289), BY (260.75)

18.[8] $\int_u^1 \dfrac{x^2 \, dx}{\sqrt{1 - x^4}} = \sqrt{2}\, E(\gamma, r) - \dfrac{1}{\sqrt{2}} F(\gamma, r)$　　　　$[u < 1]$

$= \dfrac{1}{\sqrt{2\pi}}\left\{\Gamma\left(\dfrac{3}{4}\right)\right\}^2$　　　　　　　$[u = 0]$

BY (259.76)

19. $\int_1^u \dfrac{x^2 \, dx}{\sqrt{x^4 - 1}} = \dfrac{1}{\sqrt{2}} F(\delta, r) - \sqrt{2}\, E(\delta, r) + \dfrac{1}{u}\sqrt{u^4 - 1}$　　$[u > 1]$　　　BY (260.77)

20. $\int_u^1 \dfrac{x^4 \, dx}{\sqrt{1 - x^4}} = \dfrac{1}{3\sqrt{2}} F(\gamma, r) + \dfrac{u}{3}\sqrt{1 - u^4}$　　　　　$[u < 1]$　　　　BY (259.76)

21.[3] $\int_1^u \dfrac{x^4 \, dx}{\sqrt{x^4 - 1}} = \dfrac{1}{3\sqrt{2}} F(\delta, r) + \dfrac{1}{3}u\sqrt{u^4 - 1}$　　　　　$[u > 1]$　　　　BY (260.77)

22. $\int_0^u \dfrac{dx}{\sqrt{x\left(1 + x^3\right)}} = \dfrac{1}{\sqrt[4]{3}} F\left(\arccos\dfrac{1 + \left(1 - \sqrt{3}\right)u}{1 + \left(1 + \sqrt{3}\right)u}, \dfrac{\sqrt{2 + \sqrt{3}}}{2}\right)$

$[u > 0]$　　　　　BY (260.50)

23. $\int_0^u \dfrac{dx}{\sqrt{x\left(1 - x^3\right)}} = \dfrac{1}{\sqrt[4]{3}} F\left(\arccos\dfrac{1 - \left(1 + \sqrt{3}\right)u}{1 + \left(\sqrt{3} - 1\right)u}, \dfrac{\sqrt{2 - \sqrt{3}}}{2}\right)$

$[1 \ge u > 0]$　　　　　BY (259.50)

3.167   **Notation**: In **3.167** and **3.168** we set: $\alpha = \arcsin\sqrt{\dfrac{(a-c)(d-u)}{(a-d)(c-u)}}$,

$$\beta = \arcsin\sqrt{\frac{(a-c)(u-d)}{(c-d)(a-u)}}, \qquad \gamma = \arcsin\sqrt{\frac{(b-d)(c-u)}{(c-d)(b-u)}},$$

$$\delta = \arcsin\sqrt{\frac{(b-d)(u-c)}{(b-c)(u-d)}}, \qquad \kappa = \arcsin\sqrt{\frac{(a-c)(b-u)}{(b-c)(a-u)}},$$

$$\lambda = \arcsin\sqrt{\frac{(a-c)(u-b)}{(a-b)(u-c)}}, \qquad \mu = \arcsin\sqrt{\frac{(b-d)(a-u)}{(a-b)(u-d)}},$$

$$\nu = \arcsin\sqrt{\frac{(b-d)(u-a)}{(a-d)(u-b)}}, \qquad q = \sqrt{\frac{(b-c)(a-d)}{(a-c)(b-d)}}, \qquad r = \sqrt{\frac{(a-b)(c-d)}{(a-c)(b-d)}}.$$

1. $\displaystyle\int_u^d \sqrt{\frac{d-x}{(a-x)(b-x)(c-x)}}\, dx = \frac{2(c-d)}{\sqrt{(a-c)(b-d)}}\left\{\Pi\left(\alpha, \frac{a-d}{a-c}, q\right) - F(\alpha, q)\right\}$

$$[a > b > c > d > u] \qquad \text{BY (251.05)}$$

2. $\displaystyle\int_d^u \sqrt{\frac{x-d}{(a-x)(b-x)(c-x)}}\, dx = \frac{2(d-a)}{\sqrt{(a-c)(b-d)}}\left\{\Pi\left(\beta, \frac{d-c}{a-c}, r\right) - F(\beta, r)\right\}$

$$[a > b > c \geq u > d] \qquad \text{BY (252.14)}$$

3. $\displaystyle\int_u^c \sqrt{\frac{x-d}{(a-x)(b-x)(c-x)}}\, dx = \frac{2}{\sqrt{(a-c)(b-d)}}\left\{(c-b)\,\Pi\left(\gamma, \frac{c-d}{b-d}, r\right) + (b-d)\,F(\gamma, r)\right\}$

$$[a > b > c > u \geq d] \qquad \text{BY (253.14)}$$

4. $\displaystyle\int_c^u \sqrt{\frac{x-d}{(a-x)(b-x)(x-c)}}\, dx = \frac{2(c-d)}{\sqrt{(a-c)(b-d)}}\,\Pi\left(\delta, \frac{b-c}{b-d}, q\right)$

$$[a > b \geq u > c > d] \qquad \text{BY (254.02)}$$

5. $\displaystyle\int_u^b \sqrt{\frac{x-d}{(a-x)(b-x)(x-c)}}\, dx = \frac{2}{\sqrt{(a-c)(b-d)}}\left\{(b-a)\,\Pi\left(\kappa, \frac{b-c}{a-c}, q\right) + (a-d)\,F(\kappa, q)\right\}$

$$[a > b > u \geq c > d] \qquad \text{BY (255.20)}$$

6. $\displaystyle\int_b^u \sqrt{\frac{x-d}{(a-x)(x-b)(x-c)}}\, dx = \frac{2}{\sqrt{(a-c)(b-d)}}\left\{(b-c)\,\Pi\left(\lambda, \frac{a-b}{a-c}, r\right) + (c-d)\,F(\lambda, r)\right\}$

$$[a \geq u > b > c > d] \qquad \text{BY (256.13)}$$

7. $\displaystyle\int_u^a \sqrt{\frac{x-d}{(a-x)(x-b)(x-c)}}\, dx = \frac{2(a-d)}{\sqrt{(a-c)(b-d)}}\,\Pi\left(\mu, \frac{b-a}{b-d}, r\right)$

$$[a > u \geq b > c > d] \qquad \text{BY (257.02)}$$

8. $\int_a^u \sqrt{\dfrac{x-d}{(x-a)(x-b)(x-c)}}\, dx = \dfrac{2}{\sqrt{(a-c)(b-d)}} \left\{ (a-b)\, \Pi\left(\nu, \dfrac{a-d}{b-d}, q\right) + (b-d)\, F(\nu, q) \right\}$

$$[u > a > b > c > d] \qquad \text{BY (258.14)}$$

9. $\int_u^d \sqrt{\dfrac{c-x}{(a-x)(b-x)(d-x)}}\, dx = \dfrac{2(c-d)}{\sqrt{(a-c)(b-d)}}\, \Pi\left(\alpha, \dfrac{a-d}{a-c}, q\right)$

$$[a > b > c > d > u] \qquad \text{BY (251.02)}$$

10. $\int_d^u \sqrt{\dfrac{c-x}{(a-x)(b-x)(x-d)}}\, dx = \dfrac{2}{\sqrt{(a-c)(b-d)}} \left[(a-d)\, \Pi\left(\beta, \dfrac{d-c}{a-c}, r\right) - (a-c)\, F(\beta, r) \right]$

$$[a > b > c \geq u > d] \qquad \text{BY (252.13)}$$

11. $\int_u^c \sqrt{\dfrac{c-x}{(a-x)(b-x)(x-d)}}\, dx = \dfrac{2(b-c)}{\sqrt{(a-c)(b-d)}} \left[\Pi\left(\gamma, \dfrac{c-d}{b-d}, r\right) - F(\gamma, r) \right]$

$$[a > b > c > u \geq d] \qquad \text{BY (253.13)}$$

12. $\int_c^u \sqrt{\dfrac{x-c}{(a-x)(b-x)(x-d)}}\, dx = \dfrac{2(c-d)}{\sqrt{(a-c)(b-d)}} \left[\Pi\left(\delta, \dfrac{b-c}{b-d}, q\right) - F(\delta, q) \right]$

$$[a > b \geq u > c > d] \qquad \text{BY (254.12)}$$

13. $\int_u^b \sqrt{\dfrac{x-c}{(a-x)(b-x)(x-d)}}\, dx = \dfrac{2}{\sqrt{(a-c)(b-d)}} \left[(b-a)\, \Pi\left(\kappa, \dfrac{b-c}{a-c}, q\right) + (a-c)\, F(\kappa, q) \right]$

$$[a > b > u \geq c > d] \qquad \text{BY (259.19)}$$

14. $\int_b^u \sqrt{\dfrac{x-c}{(a-x)(x-b)(x-d)}}\, dx = \dfrac{2(b-c)}{\sqrt{(a-c)(b-d)}}\, \Pi\left(\lambda, \dfrac{a-b}{a-c}, r\right)$

$$[a \geq u > b > c > d] \qquad \text{BY (256.02)}$$

15. $\int_u^a \sqrt{\dfrac{x-c}{(a-x)(x-b)(x-d)}}\, dx = \dfrac{2}{\sqrt{(a-c)(b-d)}} \left[(a-d)\, \Pi\left(\mu, \dfrac{b-a}{b-d}, r\right) + (d-c)\, F(\mu, r) \right]$

$$[a > u \geq b > c > d] \qquad \text{BY (257.13)}$$

16. $\int_a^u \sqrt{\dfrac{x-c}{(x-a)(x-b)(x-d)}}\, dx = \dfrac{2}{\sqrt{(a-c)(b-d)}} \left[(a-b)\, \Pi\left(\nu, \dfrac{a-d}{b-d}, q\right) + (b-c)\, F(\nu, q) \right]$

$$[u > a > b > c > d] \qquad \text{BY (258.13)}$$

17. $\int_u^d \sqrt{\dfrac{b-x}{(a-x)(c-x)(d-x)}}\, dx = \dfrac{2}{\sqrt{(a-c)(b-d)}} \left[(c-d)\, \Pi\left(\alpha, \dfrac{a-d}{a-c}, q\right) + (b-c)\, F(\alpha, q) \right]$

$$[a > b > c > d > u] \qquad \text{BY (251.07)}$$

18. $\int_d^u \sqrt{\dfrac{b-x}{(a-x)(c-x)(x-d)}}\, dx = \dfrac{2}{\sqrt{(a-c)(b-d)}} \left[(a-d)\, \Pi\left(\beta, \dfrac{d-c}{a-c}, r\right) - (a-b)\, F(\beta, r) \right]$

$$[a > b > c \geq u > d] \qquad \text{BY (252.15)}$$

19. $\displaystyle\int_u^c \sqrt{\frac{b-x}{(a-x)(c-x)(x-d)}}\, dx = \frac{2(b-c)}{\sqrt{(a-c)(b-d)}}\, \Pi\left(\gamma, \frac{c-d}{b-d}, r\right)$

$$[a > b > c > u \geq d] \qquad \text{BY (253.02)}$$

20. $\displaystyle\int_c^u \sqrt{\frac{b-x}{(a-x)(x-c)(x-d)}}\, dx = \frac{2}{\sqrt{(a-c)(b-d)}}\left[(d-c)\,\Pi\left(\delta, \frac{b-c}{b-d}, q\right) + (b-d)\,F(\delta, q)\right]$

$$[a > b \geq u > c > d] \qquad \text{BY (254.14)}$$

21. $\displaystyle\int_u^b \sqrt{\frac{b-x}{(a-x)(x-c)(x-d)}}\, dx = \frac{2(a-b)}{\sqrt{(a-c)(b-d)}}\left[\Pi\left(\kappa, \frac{b-c}{a-c}, q\right) - F(\kappa, q)\right]$

$$[a > b > u \geq c > d] \qquad \text{BY (255.21)}$$

22. $\displaystyle\int_b^u \sqrt{\frac{x-b}{(a-x)(x-c)(x-d)}}\, dx = \frac{2(b-c)}{\sqrt{(a-c)(b-d)}}\left[\Pi\left(\lambda, \frac{a-b}{a-c}, r\right) - F(\lambda, r)\right]$

$$[a \geq u > b > c > d] \qquad \text{BY (256.15)}$$

23.[8] $\displaystyle\int_u^a \sqrt{\frac{x-b}{(a-x)(x-c)(x-d)}}\, dx = \frac{2}{\sqrt{(a-c)(b-d)}}\left[(a-d)\,\Pi\left(\mu, \frac{b-a}{b-d}, r\right) - (b-d)\,F(\mu, r)\right]$

$$[a > u \geq b > c > d] \qquad \text{BY (257.15)}$$

24. $\displaystyle\int_a^u \sqrt{\frac{x-b}{(x-a)(x-c)(x-d)}}\, dx = \frac{2(a-b)}{\sqrt{(a-c)(b-d)}}\, \Pi\left(\nu, \frac{a-d}{b-d}, q\right)$

$$[u > a > b > c > d] \qquad \text{BY (258.02)}$$

25. $\displaystyle\int_u^d \sqrt{\frac{a-x}{(b-x)(c-x)(d-x)}}\, dx = \frac{2}{\sqrt{(a-c)(b-d)}}\left[(c-d)\,\Pi\left(\alpha, \frac{a-d}{a-c}, q\right) + (a-c)\,F(\alpha, q)\right]$

$$[a > b > c > d > u] \qquad \text{BY (251.06)}$$

26. $\displaystyle\int_d^u \sqrt{\frac{a-x}{(b-x)(c-x)(x-d)}}\, dx = \frac{2(a-d)}{\sqrt{(a-c)(b-d)}}\, \Pi\left(\beta, \frac{d-c}{a-c}, r\right)$

$$[a > b > c \geq u > d] \qquad \text{BY (252.02)}$$

27. $\displaystyle\int_u^c \sqrt{\frac{a-x}{(b-x)(c-x)(x-d)}}\, dx = \frac{2}{\sqrt{(a-c)(b-d)}}\left[(b-c)\,\Pi\left(\gamma, \frac{c-d}{b-d}, r\right) + (a-b)\,F(\gamma, r)\right]$

$$[a > b > c > u \geq d] \qquad \text{BY (253.15)}$$

28. $\displaystyle\int_c^u \sqrt{\frac{a-x}{(b-x)(x-c)(x-d)}}\, dx = \frac{2}{\sqrt{(a-c)(b-d)}}\left[(d-c)\,\Pi\left(\delta, \frac{b-c}{b-d}, q\right) + (a-d)\,F(\delta, q)\right]$

$$[a > b \geq u > c > d] \qquad \text{BY (254.13)}$$

29. $\displaystyle\int_u^b \sqrt{\frac{a-x}{(b-x)(x-c)(x-d)}}\, dx = \frac{2(a-b)}{\sqrt{(a-c)(b-d)}}\, \Pi\left(\kappa, \frac{b-c}{a-c}, q\right)$

$$[a > b > u \geq c > d] \qquad \text{BY (255.02)}$$

30. $\int_b^u \sqrt{\dfrac{a-x}{(x-b)(x-c)(x-d)}}\,dx = \dfrac{2}{\sqrt{(a-c)(b-d)}}\left[(c-b)\,\Pi\left(\lambda,\dfrac{a-b}{a-c},r\right) + (a-c)\,F(\lambda,r)\right]$

$$[a \geq u > b > c > d] \qquad \text{BY (256.14)}$$

31. $\int_u^a \sqrt{\dfrac{a-x}{(x-b)(x-c)(x-d)}}\,dx = \dfrac{2(d-a)}{\sqrt{(a-c)(b-d)}}\left[\Pi\left(\mu,\dfrac{b-a}{b-d},r\right) - F(\mu,r)\right]$

$$[a > u \geq b > c > d] \qquad \text{BY (257.14)}$$

32. $\int_a^u \sqrt{\dfrac{x-a}{(x-b)(x-c)(x-d)}}\,dx = \dfrac{2(a-b)}{\sqrt{(a-c)(b-d)}}\left[\Pi\left(\nu,\dfrac{a-d}{b-d},q\right) - F(\nu,q)\right]$

$$[u > a > b > c > d] \qquad \text{BY (258.15)}$$

3.168

1. $\int_u^c \sqrt{\dfrac{c-x}{(a-x)(b-x)(x-d)^3}}\,dx = \dfrac{2}{d-a}\left[\sqrt{\dfrac{a-c}{b-d}}\,E(\gamma,r) - \sqrt{\dfrac{(a-u)(c-u)}{(b-u)(u-d)}}\right]$

$$[a > b > c > u > d] \qquad \text{BY (253.06)}$$

2. $\int_c^u \sqrt{\dfrac{x-c}{(a-x)(b-x)(x-d)^3}}\,dx = \dfrac{2}{a-d}\sqrt{\dfrac{a-c}{b-d}}\,[F(\delta,q) - E(\delta,q)]$

$$[a > b \geq u > c > d] \qquad \text{BY (254.04)}$$

3. $\int_u^b \sqrt{\dfrac{x-c}{(a-x)(b-x)(x-d)^3}}\,dx = \dfrac{2}{a-d}\sqrt{\dfrac{a-c}{b-d}}\,[F(\kappa,q) - E(\kappa,q)] + \dfrac{2}{b-d}\sqrt{\dfrac{(b-u)(u-c)}{(a-u)(u-d)}}$

$$[a > b > u \geq c > d] \qquad \text{BY (255.09)}$$

4. $\int_b^u \sqrt{\dfrac{x-c}{(a-x)(x-b)(x-d)^3}}\,dx = \dfrac{2}{a-d}\left[\sqrt{\dfrac{a-c}{b-d}}\,E(\lambda,r) - \dfrac{c-d}{b-d}\sqrt{\dfrac{(a-u)(u-b)}{(u-c)(u-d)}}\right]$

$$[a \geq u > b > c > d] \qquad \text{BY (256.06)}$$

5. $\int_u^a \sqrt{\dfrac{x-c}{(a-x)(x-b)(x-d)^3}}\,dx = \dfrac{2}{a-d}\sqrt{\dfrac{a-c}{b-d}}\,E(\mu,r)$

$$[a > u \geq b > c > d] \qquad \text{BY (257.01)}$$

6. $\int_a^u \sqrt{\dfrac{x-c}{(x-a)(x-b)(x-d)^3}}\,dx = \dfrac{2}{a-d}\sqrt{\dfrac{a-c}{b-d}}\,[F(\nu,q) - E(\nu,q)]$

$$+ \dfrac{2}{a-d}\sqrt{\dfrac{(u-a)(u-c)}{(u-b)(u-d)}}$$

$$[u > a > b > c > d] \qquad \text{BY (258.10)}$$

7.
$$\int_u^c \sqrt{\frac{b-x}{(a-x)(c-x)(x-d)^3}}\, dx = \frac{2}{(a-d)(c-d)\sqrt{(a-c)(b-d)}}$$
$$\times\,[(b-c)(a-d)\,F(\gamma,r) - (a-c)(b-d)\,E(\gamma,r)]$$
$$+\frac{2(b-d)}{(a-d)(c-d)}\sqrt{\frac{(a-u)(c-u)}{(b-u)(u-d)}}$$
$$[a > b > c > u > d]\qquad\text{BY (253.03)}$$

8.
$$\int_c^u \sqrt{\frac{b-x}{(a-x)(x-c)(x-d)^3}}\, dx = \frac{2}{(a-d)(c-d)\sqrt{(a-c)(b-d)}}$$
$$\times\,[(a-c)(b-d)\,E(\delta,q) - (a-b)(c-d)\,F(\delta,q)]$$
$$[a > b \ge u > c > d]\qquad\text{BY (254.15)}$$

9.
$$\int_u^b \sqrt{\frac{b-x}{(a-x)(x-c)(x-d)^3}}\, dx = \frac{2}{(a-d)(c-d)\sqrt{(a-c)(b-d)}}$$
$$\times\,[(a-c)(b-d)\,E(\kappa,q) - (a-b)(c-d)\,F(\kappa,q)]$$
$$-\frac{2}{c-d}\sqrt{\frac{(b-u)(u-c)}{(a-u)(u-d)}}$$
$$[a > b > u \ge c > d]\qquad\text{BY (255.06)}$$

10.
$$\int_b^u \sqrt{\frac{x-b}{(a-x)(x-c)(x-d)^3}}\, dx = \frac{2}{(a-d)(c-d)\sqrt{(a-c)(b-d)}}$$
$$\times\,[(a-c)(b-d)\,E(\lambda,r) - (a-d)(b-c)\,F(\lambda,r)]$$
$$-\frac{2}{a-d}\sqrt{\frac{(a-u)(u-b)}{(u-c)(u-d)}}$$
$$[a \ge u > b > c > d]\qquad\text{BY (256.03)}$$

11.
$$\int_u^a \sqrt{\frac{x-b}{(a-x)(x-c)(x-d)^3}}\, dx = 2\frac{\sqrt{(a-c)(b-d)}}{(a-d)(c-d)}\,E(\mu,r)$$
$$-\frac{2(b-c)}{(c-d)\sqrt{(a-c)(b-d)}}\,F(\mu,r)$$
$$[a > u \ge b > c > d]\qquad\text{BY (257.09)}$$

12.
$$\int_a^u \sqrt{\frac{x-b}{(x-a)(x-c)(x-d)^3}}\, dx$$
$$=\frac{2(b-d)}{(a-d)(c-d)}\sqrt{\frac{(u-a)(u-c)}{(u-b)(u-d)}} + \frac{2(a-b)}{(a-d)\sqrt{(a-c)(b-d)}}\,F(\nu,q)$$
$$+2\frac{\sqrt{(a-c)(b-d)}}{(a-d)(c-d)}\,E(\nu,q)$$
$$[u > a > b > c > d]\qquad\text{BY (258.09)}$$

13. $$\int_u^c \sqrt{\frac{a-x}{(b-x)(c-x)(x-d)^3}}\, dx = \frac{2}{c-d}\sqrt{\frac{a-c}{b-d}}\left[F(\gamma, r) - E(\gamma, r)\right] + \frac{2}{c-d}\sqrt{\frac{(a-u)(c-u)}{(b-u)(u-d)}}$$

$$[a > b > c > u > d] \qquad \text{BY (253.04)}$$

14. $$\int_c^u \sqrt{\frac{a-x}{(b-x)(x-c)(x-d)^3}}\, dx = \frac{2}{c-d}\sqrt{\frac{a-c}{b-d}}\, E(\delta, q)$$

$$[a > b \geq u > c > d] \qquad \text{BY (254.01)}$$

15. $$\int_u^b \sqrt{\frac{a-x}{(b-x)(x-c)(x-d)^3}}\, dx = \frac{2}{c-d}\sqrt{\frac{a-c}{b-d}}\, E(\kappa, q) - \frac{2(a-d)}{(b-d)(c-d)}\sqrt{\frac{(b-u)(u-c)}{(a-u)(u-d)}}$$

$$[a > b > u \geq c > d] \qquad \text{BY (255.08)}$$

16. $$\int_b^u \sqrt{\frac{a-x}{(x-b)(x-c)(x-d)^3}}\, dx = \frac{2}{c-d}\sqrt{\frac{a-c}{b-d}}\left[F(\lambda, r) - E(\lambda, r)\right] + \frac{2}{b-d}\sqrt{\frac{(a-u)(u-b)}{(u-c)(u-d)}}$$

$$[a \geq u > b > c > d] \qquad \text{BY (256.05)}$$

17. $$\int_u^a \sqrt{\frac{a-x}{(x-b)(x-c)(x-d)^3}}\, dx = \frac{2}{c-d}\sqrt{\frac{a-c}{b-d}}\left[F(\mu, r) - E(\mu, r)\right]$$

$$[a > u \geq b > c > d] \qquad \text{BY (257.06)}$$

18. $$\int_a^u \sqrt{\frac{x-a}{(x-b)(x-c)(x-d)^3}}\, dx = \frac{-2}{c-d}\sqrt{\frac{a-c}{b-d}}\, E(\nu, q) + \frac{2}{c-d}\sqrt{\frac{(u-a)(u-c)}{(u-b)(u-d)}}$$

$$[u > a > b > c > d] \qquad \text{BY (258.05)}$$

19. $$\int_u^d \sqrt{\frac{d-x}{(a-x)(b-x)(c-x)^3}}\, dx = \frac{2}{b-c}\sqrt{\frac{b-d}{a-c}}\left[F(\alpha, q) - E(\alpha, q)\right]$$

$$[a > b > c > d > u] \qquad \text{BY (251.01)}$$

20. $$\int_d^u \sqrt{\frac{x-d}{(a-x)(b-x)(c-x)^3}}\, dx = \frac{-2}{b-c}\sqrt{\frac{b-d}{a-c}}\, E(\beta, r) + \frac{2}{b-c}\sqrt{\frac{(b-u)(u-d)}{(a-u)(c-u)}}$$

$$[a > b > c \geq u > d] \qquad \text{BY (252.06)}$$

21. $$\int_u^b \sqrt{\frac{x-d}{(a-x)(b-x)(x-c)^3}}\, dx = \frac{2}{b-c}\sqrt{\frac{b-d}{a-c}}\left[F(\kappa, q) - E(\kappa, q)\right] + \frac{2}{b-c}\sqrt{\frac{(b-u)(u-d)}{(a-u)(u-c)}}$$

$$[a > b > u > c > d] \qquad \text{BY (255.05)}$$

22. $$\int_b^u \sqrt{\frac{x-d}{(a-x)(x-b)(x-c)^3}}\, dx = \frac{2}{b-c}\sqrt{\frac{b-d}{a-c}}\, E(\lambda, r)$$

$$[a \geq u > b > c > d] \qquad \text{BY (256.01)}$$

23. $$\int_u^a \sqrt{\frac{x-d}{(a-x)(x-b)(x-c)^3}}\, dx = \frac{2}{b-c}\sqrt{\frac{b-d}{a-c}}\, E(\mu, r) - \frac{2(c-d)}{(a-c)(b-c)}\sqrt{\frac{(a-u)(u-b)}{(u-c)(u-d)}}$$

$$[a > u \geq b > c > d] \qquad \text{BY (257.06)}$$

24. $\displaystyle\int_a^u \sqrt{\frac{x-d}{(x-a)(x-b)(x-c)^3}}\,dx = \frac{2}{b-c}\sqrt{\frac{b-d}{a-c}}\,[F(\nu,q)-E(\nu,q)] + \frac{2}{a-c}\sqrt{\frac{(u-a)(u-d)}{(u-b)(u-c)}}$

$$[u>a>b>c>d] \qquad \text{BY (258.06)}$$

25. $\displaystyle\int_u^a \sqrt{\frac{b-x}{(a-x)(c-x)^3(d-x)}}\,dx = \frac{2}{c-d}\sqrt{\frac{b-d}{a-c}}\,E(\alpha,q)$

$$[a>b>c>d>u] \qquad \text{BY (251.01)}$$

26. $\displaystyle\int_d^u \sqrt{\frac{b-x}{(a-x)(c-x)^3(x-d)}}\,dx = \frac{2}{c-d}\sqrt{\frac{b-d}{a-c}}\,[F(\beta,r)-E(\beta,r)] + \frac{2}{c-d}\sqrt{\frac{(b-u)(u-d)}{(a-u)(c-u)}}$

$$[a>b>c>u>d] \qquad \text{BY (252.03)}$$

27. $\displaystyle\int_u^b \sqrt{\frac{b-x}{(a-x)(x-c)^3(x-d)}}\,dx = \frac{2}{d-c}\sqrt{\frac{b-d}{a-c}}\,E(\kappa,q) + \frac{2}{c-d}\sqrt{\frac{(b-u)(u-d)}{(a-u)(u-c)}}$

$$[a>b>u>c>d] \qquad \text{BY (255.03)}$$

28. $\displaystyle\int_b^u \sqrt{\frac{x-b}{(a-x)(x-c)^3(x-d)}}\,dx = \frac{2}{c-d}\sqrt{\frac{b-d}{a-c}}\,[F(\lambda,r)-E(\lambda,r)]$

$$[a\geq u>b>c>d] \qquad \text{BY (256.08)}$$

29. $\displaystyle\int_u^a \sqrt{\frac{x-b}{(a-x)(x-c)^3(x-d)}}\,dx = \frac{2}{c-d}\sqrt{\frac{b-d}{a-c}}\,[F(\mu,r)-E(\mu,r)] + \frac{2}{a-c}\sqrt{\frac{(a-u)(u-b)}{(u-c)(u-d)}}$

$$[a>u\geq b>c>d] \qquad \text{BY (257.03)}$$

30. $\displaystyle\int_a^u \sqrt{\frac{x-b}{(x-a)(x-c)^3(x-d)}}\,dx = \frac{2}{c-d}\sqrt{\frac{b-d}{a-c}}\,E(\nu,q) - \frac{2(b-c)}{(a-c)(c-d)}\sqrt{\frac{(u-a)(u-d)}{(u-b)(u-c)}}$

$$[u>a>b>c>d] \qquad \text{BY (258.03)}$$

31. $\displaystyle\int_u^d \sqrt{\frac{a-x}{(b-x)(c-x)^3(d-x)}}\,dx = \frac{2\sqrt{(a-c)(b-d)}}{(b-c)(c-d)}\,E(\alpha,q) - \frac{a-b}{b-c}\frac{2}{\sqrt{(a-c)(b-d)}}\,F(\alpha,q)$

$$[a>b>c>d>u] \qquad \text{BY (251.08)}$$

32. $\displaystyle\int_d^u \sqrt{\frac{a-x}{(b-x)(c-x)^3(x-d)}}\,dx = \frac{2(a-d)}{(c-d)\sqrt{(a-c)(b-d)}}\,F(\beta,r) - 2\frac{\sqrt{(a-c)(b-d)}}{(b-c)(c-d)}\,E(\beta,r)$

$$+2\frac{a-c}{(b-c)(c-d)}\sqrt{\frac{(b-u)(u-d)}{(a-u)(c-u)}}$$

$$[a>b>c>u>d] \qquad \text{BY (252.04)}$$

33. $\displaystyle\int_u^b \sqrt{\frac{a-x}{(b-x)(x-c)^3(x-d)}}\,dx = \frac{2(a-b)}{(b-c)\sqrt{(a-c)(b-c)}}\,F(\kappa,q) - 2\sqrt{\frac{(a-c)(b-d)}{(b-c)(c-d)}}\,E(\kappa,q)$

$$+\frac{2(a-c)}{(b-c)(c-d)}\sqrt{\frac{(b-u)(u-d)}{(a-u)(u-c)}}$$

$$[a>b>u>c>d] \qquad \text{BY (255.04)}$$

34. $\displaystyle\int_b^u \sqrt{\frac{a-x}{(x-b)(x-c)^3(x-d)}}\,dx = \frac{2\sqrt{(a-c)(b-d)}}{(b-c)(c-d)}E(\lambda,r) - \frac{2(a-d)}{(c-d)\sqrt{(a-c)(b-d)}}F(\lambda,r)$

$$[a \ge u > b > c > d] \qquad \text{BY (256.09)}$$

35. $\displaystyle\int_u^a \sqrt{\frac{a-x}{(x-b)(x-c)^3(x-d)}}\,dx = \frac{2\sqrt{(a-c)(b-d)}}{(b-c)(c-d)}E(\mu,r) - \frac{2(a-d)}{(c-d)\sqrt{(a-c)(b-d)}}F(\mu,r)$

$$- \frac{2}{b-c}\sqrt{\frac{(a-u)(u-b)}{(u-c)(u-d)}}$$

$$[a > u \ge b > c > d] \qquad \text{BY (257.04)}$$

36. $\displaystyle\int_a^u \sqrt{\frac{x-a}{(x-b)(x-c)^3(x-d)}}\,dx = \frac{2\sqrt{(a-c)(b-d)}}{(b-c)(c-d)}E(\nu,q) - \frac{2(a-b)}{(b-c)\sqrt{(a-c)(b-d)}}F(\nu,q)$

$$- \frac{2}{c-d}\sqrt{\frac{(u-a)(u-d)}{(u-b)(u-c)}}$$

$$[u > a > b > c > d] \qquad \text{BY (258.04)}$$

37. $\displaystyle\int_u^d \sqrt{\frac{d-x}{(a-x)(b-x)^3(c-x)}}\,dx = \frac{2\sqrt{(a-c)(b-d)}}{(a-b)(b-c)}E(\alpha,q) - \frac{2(c-d)}{(b-c)\sqrt{(a-c)(b-d)}}F(\alpha,q)$

$$- \frac{2}{a-b}\sqrt{\frac{(a-u)(d-u)}{(b-u)(c-u)}}$$

$$[a > b > c > d > u] \qquad \text{BY (251.11)}$$

38. $\displaystyle\int_d^u \sqrt{\frac{x-d}{(a-x)(b-x)^3(c-x)}}\,dx = \frac{2\sqrt{(a-c)(b-d)}}{(a-b)(b-c)}E(\beta,r) - \frac{2(a-d)}{(a-b)\sqrt{(a-c)(b-d)}}F(\beta,r)$

$$+ \frac{2}{b-c}\sqrt{\frac{(c-u)(u-d)}{(a-u)(b-u)}}$$

$$[a > b > c \ge u > d] \qquad \text{BY (252.07)}$$

39. $\displaystyle\int_u^c \sqrt{\frac{x-d}{(a-x)(b-x)^3(c-x)}}\,dx = \frac{2\sqrt{(a-c)(b-d)}}{(a-b)(b-c)}E(\gamma,r) - \frac{2(a-d)}{(a-b)\sqrt{(a-c)(b-d)}}F(\gamma,r)$

$$[a > b > c > u \ge d] \qquad \text{BY (253.07)}$$

40. $\displaystyle\int_c^u \sqrt{\frac{x-d}{(a-x)(b-x)^3(x-c)}}\,dx = \frac{2(c-d)}{(b-c)\sqrt{(a-c)(b-d)}}F(\delta,q) - \frac{2\sqrt{(a-c)(b-d)}}{(a-b)(b-c)}E(\delta,q)$

$$+ \frac{2(b-d)}{(a-b)(b-c)}\sqrt{\frac{(a-u)(u-c)}{(b-u)(u-d)}}$$

$$[a > b > u > c > d] \qquad \text{BY (254.05)}$$

41. $\displaystyle\int_u^a \sqrt{\frac{x-d}{(a-x)(x-b)^3(x-c)}}\,dx = \frac{2(a-d)}{(a-b)\sqrt{(a-c)(b-d)}}F(\mu,r) - \frac{2\sqrt{(a-c)(b-d)}}{(a-b)(b-c)}E(\mu,r)$

$$+ \frac{2(b-d)}{(a-b)(b-c)}\sqrt{\frac{(a-u)(u-c)}{(u-b)(u-d)}}$$

$$[a > u > b > c > d] \qquad \text{BY (257.07)}$$

42. $\int_a^u \sqrt{\dfrac{x-d}{(x-a)(x-b)^3(x-c)}}\, dx = \dfrac{2\sqrt{(a-c)(b-d)}}{(a-b)(b-c)} E(\nu,q) - \dfrac{2(c-d)}{(b-c)\sqrt{(a-c)(b-d)}} F(\nu,q)$

$$[u>a>b>c>d] \qquad \text{BY (258.07)}$$

43. $\int_u^d \sqrt{\dfrac{c-x}{(a-x)(b-x)^3(d-x)}}\, dx = \dfrac{2}{a-b}\sqrt{\dfrac{a-c}{b-d}} E(\alpha,q) - \dfrac{2(b-c)}{(a-b)(b-d)}\sqrt{\dfrac{(a-u)(d-u)}{(b-u)(c-u)}}$

$$[a>b>c>d>u]$$

44. $\int_d^u \sqrt{\dfrac{c-x}{(a-x)(b-x)^3(x-d)}}\, dx = \dfrac{2}{a-b}\sqrt{\dfrac{a-c}{b-d}}\left[F(\beta,r)-E(\beta,r)\right] + \dfrac{2}{b-d}\sqrt{\dfrac{(c-u)(u-d)}{(a-u)(b-u)}}$

$$[a>b>c\geq u>d] \qquad \text{BY (252.10)}$$

45. $\int_u^c \sqrt{\dfrac{c-x}{(a-x)(b-x)^3(x-d)}}\, dx = \dfrac{2}{a-b}\sqrt{\dfrac{a-c}{b-d}}\left[F(\gamma,r)-E(\gamma,r)\right]$

$$[a>b>c>u\geq d] \qquad \text{BY (254.08)}$$

46. $\int_c^u \sqrt{\dfrac{x-c}{(a-x)(b-x)^3(x-d)}}\, dx = \dfrac{2}{b-a}\sqrt{\dfrac{a-c}{b-d}} E(\delta,q) + \dfrac{2}{a-b}\sqrt{\dfrac{(a-u)(u-c)}{(b-u)(u-d)}}$

$$[a>b\geq u>c>d] \qquad \text{BY (254.08)}$$

47. $\int_u^a \sqrt{\dfrac{x-c}{(a-x)(x-b)^3(x-d)}}\, dx = \dfrac{2}{a-b}\sqrt{\dfrac{a-c}{b-d}}\left[F(\mu,r)-E\left(\mu,r\right)\right] + \dfrac{2}{a-b}\sqrt{\dfrac{(a-u)(u-c)}{(u-b)(u-d)}}$

$$[a>u\geq b>c>d] \qquad \text{BY (257.10)}$$

48. $\int_a^u \sqrt{\dfrac{x-c}{(x-a)(x-b)^3(x-d)}}\, dx = \dfrac{2}{a-b}\sqrt{\dfrac{a-c}{b-d}} E(\nu,q)$

$$[u>a>b>c>d] \qquad \text{BY (258.01)}$$

49. $\int_u^d \sqrt{\dfrac{a-x}{(b-x)^3(c-x)(d-x)}}\, dx = \dfrac{2}{b-c}\sqrt{\dfrac{a-c}{b-d}}\left[F(\alpha,q)-E\left(\alpha,q\right)\right] + \dfrac{2}{b-d}\sqrt{\dfrac{(a-u)(d-u)}{(b-u)(c-u)}}$

$$[a>b>c>d>u] \qquad \text{BY (251.12)}$$

50. $\int_d^u \sqrt{\dfrac{a-x}{(b-x)^3(c-x)(x-d)}}\, dx = \dfrac{2}{b-c}\sqrt{\dfrac{a-c}{b-d}} E(\beta,r) - \dfrac{2(a-b)}{(b-c)(b-d)}\sqrt{\dfrac{(u-d)(c-u)}{(a-u)(b-u)}}$

$$[a>b>c\geq u>d] \qquad \text{BY (252.09)}$$

51. $\int_u^c \sqrt{\dfrac{a-x}{(b-x)^3(c-x)(x-d)}}\, dx = \dfrac{2}{b-c}\sqrt{\dfrac{a-c}{b-d}} E(\gamma,r)$

$$[a>b>c>u\geq d] \qquad \text{BY (253.01)}$$

52. $\int_c^u \sqrt{\dfrac{a-x}{(b-x)^3(x-c)(x-d)}}\, dx = \dfrac{2}{b-c}\sqrt{\dfrac{a-c}{b-d}}\left[F(\delta,q)-E(\delta,q)\right] + \dfrac{2}{b-c}\sqrt{\dfrac{(a-u)(u-c)}{(b-u)(u-d)}}$

$$[a>b>u>c>d] \qquad \text{BY (254.06)}$$

53. $\int_u^a \sqrt{\dfrac{a-x}{(x-b)^3(x-c)(x-d)}}\, dx = \dfrac{2}{c-b}\sqrt{\dfrac{a-c}{b-d}}\, E(\mu,r) + \dfrac{2}{b-c}\sqrt{\dfrac{(a-u)(u-c)}{(u-b)(u-d)}}$

$[a > u > b > c > d]$ BY (257.08)

54. $\int_a^u \sqrt{\dfrac{x-a}{(x-b)^3(x-c)(x-d)}}\, dx = \dfrac{2}{b-c}\sqrt{\dfrac{a-c}{b-d}}\, [F(\nu,q) - E(\nu,q)]$

$[u > a > b > c > d]$ BY (258.08)

55. $\int_u^d \sqrt{\dfrac{d-x}{(a-x)^3(b-x)(c-x)}}\, dx = \dfrac{2}{b-a}\sqrt{\dfrac{b-d}{a-c}}\, E(\alpha,q) + \dfrac{2}{a-b}\sqrt{\dfrac{(b-u)(d-u)}{(a-u)(c-u)}}$

$[a > b > c > d > u]$ BY (251.09)

56. $\int_d^u \sqrt{\dfrac{x-d}{(a-x)^3(b-x)(c-x)}}\, dx = \dfrac{2}{a-b}\sqrt{\dfrac{b-d}{a-c}}\, [F(\beta,q) - E(\beta,q)]$

$[a > b > c \geq u > d]$ BY (252.05)

57. $\int_u^c \sqrt{\dfrac{x-d}{(a-x)^3(b-x)(c-x)}}\, dx = \dfrac{2}{a-b}\sqrt{\dfrac{b-d}{a-c}}\, [F(\gamma,r) - E(\gamma,r)] + \dfrac{2}{a-c}\sqrt{\dfrac{(c-u)(u-d)}{(a-u)(b-u)}}$

$[a > b > c > u \geq d]$ BY (253.05)

58. $\int_c^u \sqrt{\dfrac{x-d}{(a-x)^3(b-x)(x-c)}}\, dx = \dfrac{2}{a-b}\sqrt{\dfrac{b-d}{a-c}}\, E(\delta,q) - \dfrac{2(a-d)}{(a-b)(a-c)}\sqrt{\dfrac{(b-u)(u-c)}{(a-u)(u-d)}}$

$[a > b \geq u > c > d]$ BY (254.03)

59. $\int_u^b \sqrt{\dfrac{x-d}{(a-x)^3(b-x)(x-c)}}\, dx = \dfrac{2}{a-b}\sqrt{\dfrac{b-d}{a-c}}\, E(\kappa,q)$

$[a > b > u \geq c > d]$ BY (255.01)

60. $\int_b^u \sqrt{\dfrac{x-d}{(a-x)^3(x-b)(x-c)}}\, dx = \dfrac{2}{a-b}\sqrt{\dfrac{b-d}{a-c}}\, [F(\lambda,r) - E(\lambda,r)] + \dfrac{2}{a-b}\sqrt{\dfrac{(u-b)(u-d)}{(a-u)(u-c)}}$

$[a > u > b > c > d]$ BY (256.10)

61. $\int_u^d \sqrt{\dfrac{c-x}{(a-x)^3(b-x)(d-x)}}\, dx = \dfrac{2(c-d)}{(a-d)\sqrt{(a-c)(b-d)}}\, F(\alpha,q) - \dfrac{2\sqrt{(a-c)(b-d)}}{(a-b)(a-d)}\, E(\alpha,q)$

$\qquad\qquad + \dfrac{2(a-c)}{(a-b)(a-d)}\sqrt{\dfrac{(b-u)(d-u)}{(a-u)(c-u)}}$

$[a > b > c > d > u]$ BY (251.15)

62. $\int_d^u \sqrt{\dfrac{c-x}{(a-x)^3(b-x)(x-d)}}\, dx = \dfrac{2\sqrt{(a-c)(b-d)}}{(a-b)(a-d)}\, E(\beta,r) - \dfrac{2(b-c)}{(a-b)\sqrt{(a-c)(b-d)}}\, F(\beta,r)$

$[a > b > c \geq u > d]$ BY (252.08)

63. $$\int_u^c \sqrt{\frac{c-x}{(a-x)^3(b-x)(x-d)}}\, dx = \frac{2\sqrt{(a-c)(b-d)}}{(a-b)(a-d)} E(\gamma, r) - \frac{2(b-c)}{(a-b)\sqrt{(a-c)(b-d)}} F(\gamma, r)$$
$$-\frac{2}{a-d}\sqrt{\frac{(c-u)(u-d)}{(a-u)(b-u)}}$$
$$[a > b > c > u \geq d] \qquad \text{BY (253.10)}$$

64. $$\int_c^u \sqrt{\frac{x-c}{(a-x)^3(b-x)(x-d)}}\, dx = \frac{2\sqrt{(a-c)(b-d)}}{(a-b)(a-d)} E(\delta, q) - \frac{2(c-d)}{(a-d)\sqrt{(a-c)(b-d)}} F(\delta, q)$$
$$-\frac{2}{a-b}\sqrt{\frac{(b-u)(u-c)}{(a-u)(u-d)}}$$
$$[a > b \geq u > c > d] \qquad \text{BY (254.09)}$$

65. $$\int_u^b \sqrt{\frac{x-c}{(a-x)^3(b-x)(x-d)}}\, dx = \frac{2\sqrt{(a-c)(b-d)}}{(a-b)(a-d)} E(\kappa, q) - \frac{2(c-d)}{(a-d)\sqrt{(a-c)(b-d)}} F(\kappa, q)$$
$$[a > b > u \geq c > d] \qquad \text{BY (255.10)}$$

66. $$\int_b^u \sqrt{\frac{x-c}{(a-x)^3(x-b)(x-d)}}\, dx = \frac{2(b-c)}{(a-b)\sqrt{(a-c)(b-d)}} F(\lambda, r) - \frac{2\sqrt{(a-c)(b-d)}}{(a-b)(a-d)} E(\lambda, r)$$
$$+\frac{2(a-c)}{(a-b)(a-d)}\sqrt{\frac{(u-b)(u-d)}{(a-u)(u-c)}}$$
$$[a > u > b > c > d] \qquad \text{BY (256.07)}$$

67. $$\int_u^d \sqrt{\frac{b-x}{(a-x)^3(c-x)(d-x)}}\, dx = \frac{2}{a-d}\sqrt{\frac{b-d}{a-c}} [F(\alpha, q) - E(\alpha, q)] + \frac{2}{a-d}\sqrt{\frac{(b-u)(d-u)}{(a-u)(c-u)}}$$
$$[a > b > c > d > u] \qquad \text{BY (251.13)}$$

68. $$\int_d^u \sqrt{\frac{b-x}{(a-x)^3(c-x)(x-d)}}\, dx = \frac{2}{a-d}\sqrt{\frac{b-d}{a-c}} E(\beta, r)$$
$$[a > b > c \geq u > d] \qquad \text{BY (252.01)}$$

69. $$\int_u^c \sqrt{\frac{b-x}{(a-x)^3(c-x)(x-d)}}\, dx = \frac{2}{a-d}\sqrt{\frac{b-d}{a-c}} E(\gamma, r) - \frac{2(a-b)}{(a-c)(a-d)}\sqrt{\frac{(c-u)(u-d)}{(a-u)(b-u)}}$$
$$[a > b > c > u \geq d] \qquad \text{BY (253.08)}$$

70. $$\int_c^u \sqrt{\frac{b-x}{(a-x)^3(x-c)(x-d)}}\, dx = \frac{2}{a-d}\sqrt{\frac{b-d}{a-c}} [F(\delta, q) - E(\delta, q)] + \frac{2}{a-c}\sqrt{\frac{(b-u)(u-c)}{(a-u)(u-d)}}$$
$$[a > b \geq u > c > d] \qquad \text{BY (254.07)}$$

71. $$\int_u^b \sqrt{\frac{b-x}{(a-x)^3(x-c)(x-d)}}\, dx = \frac{2}{a-d}\sqrt{\frac{b-d}{a-c}} [F(\kappa, q) - E(\kappa, q)]$$
$$[a > b > u \geq c > d] \qquad \text{BY (255.07)}$$

72. $$\int_b^u \sqrt{\frac{x-b}{(a-x)^3(x-c)(x-d)}}\, dx = \frac{-2}{a-d}\sqrt{\frac{b-d}{a-c}}\, E(\lambda,r) + \frac{2}{a-d}\sqrt{\frac{(u-b)(u-d)}{(a-u)(a-c)}}$$

$$[a \geq u > b > c > d] \qquad\qquad \text{BY (256.04)}$$

3.169　　**Notation:** In **3.169–3.172**, we set: $\quad \alpha = \arctan \dfrac{u}{b}, \qquad \beta = \arctan \dfrac{a}{u},$

$$\gamma = \arcsin \frac{u}{b}\sqrt{\frac{a^2+b^2}{a^2+u^2}}, \qquad \delta = \arccos \frac{u}{b}, \qquad \varepsilon = \arccos \frac{b}{u}, \qquad \xi = \arcsin \sqrt{\frac{a^2+b^2}{a^2+u^2}},$$

$$\eta = \arcsin \frac{u}{b}, \qquad \zeta = \arcsin \frac{a}{b}\sqrt{\frac{b^2-u^2}{a^2-u^2}}, \qquad \kappa = \arcsin \frac{a}{u}\sqrt{\frac{u^2-b^2}{a^2-b^2}},$$

$$\lambda = \arcsin \sqrt{\frac{a^2-u^2}{a^2-b^2}}, \qquad u = \arcsin \sqrt{\frac{u^2-a^2}{u^2-b^2}}, \qquad \nu = \arcsin \frac{a}{u}, \qquad q = \frac{\sqrt{a^2-b^2}}{a},$$

$$r = \frac{b}{\sqrt{a^2+b^2}}, \qquad s = \frac{a}{\sqrt{a^2+b^2}}, \qquad t = \frac{b}{a}.$$

1. $$\int_0^u \sqrt{\frac{x^2+a^2}{x^2+b^2}}\, dx = a\{F(\alpha,q) - E(\alpha,q)\} + u\sqrt{\frac{a^2+u^2}{b^2+u^2}}$$

$$[a > b, \quad u > 0] \qquad\qquad \text{BY (221.03)}$$

2.[6] $$\int_0^u \sqrt{\frac{x^2+b^2}{x^2+a^2}}\, dx = \frac{b^2}{a}\, F(\alpha,q) - a\, E(\alpha,q) + u\sqrt{\frac{a^2+u^2}{b^2+u^2}}$$

$$[a > b, \quad u > 0] \qquad\qquad \text{BY (221.04)}$$

3. $$\int_0^u \sqrt{\frac{x^2+a^2}{b^2-x^2}}\, dx = \sqrt{a^2+b^2}\, E(\gamma,r) - u\sqrt{\frac{b^2-u^2}{a^2+u^2}} \qquad [b \geq u > 0] \qquad \text{BY (214.11)}$$

4. $$\int_u^b \sqrt{\frac{a^2+x^2}{b^2-x^2}}\, dx = \sqrt{a^2+b^2}\, E(\delta,r) \qquad\qquad [b > u \geq 0] \qquad\qquad \text{BY (213.01), ZH 64 (273)}$$

5. $$\int_b^u \sqrt{\frac{a^2+x^2}{x^2-b^2}}\, dx = \sqrt{a^2+b^2}\,\{F(\varepsilon,s) - E(\varepsilon,s)\} + \frac{1}{u}\sqrt{(u^2+a^2)(u^2-b^2)}$$

$$[u > b > 0] \qquad\qquad \text{BY (211.03)}$$

6. $$\int_0^u \sqrt{\frac{b^2-x^2}{a^2+x^2}}\, dx = \sqrt{a^2+b^2}\,\{F(\gamma,r) - E(\gamma,r)\} + u\sqrt{\frac{b^2-u^2}{a^2+u^2}}$$

$$[b \geq u > 0] \qquad\qquad \text{BY (214.03)}$$

7. $$\int_u^b \sqrt{\frac{b^2-x^2}{a^2+x^2}}\, dx = \sqrt{a^2+b^2}\,\{F(\delta,r) - E(\delta,r)\} \qquad [b > u \geq 0] \qquad\qquad \text{BY (213.03)}$$

8. $$\int_b^u \sqrt{\frac{x^2-b^2}{a^2+x^2}}\, dx = \frac{1}{u}\sqrt{(a^2+u^2)(u^2-b^2)} - \sqrt{a^2+b^2}\, E(\varepsilon,s)$$

$$[u > b > 0] \qquad\qquad \text{BY (211.04)}$$

9. $$\int_0^u \sqrt{\frac{b^2-x^2}{a^2-x^2}}\, dx = a\, E(\eta,t) - \frac{a^2-b^2}{a}\, F(\eta,t) \qquad [a > b \geq u > 0] \qquad\qquad \text{BY (219.03)}$$

10. $\displaystyle\int_u^b \sqrt{\frac{b^2-x^2}{a^2-x^2}}\,dx = a\,E(\zeta,t) - \frac{a^2-b^2}{a}\,F(\zeta,t) - u\sqrt{\frac{b^2-u^2}{a^2-u^2}}$

$\qquad\qquad\qquad\qquad\qquad\qquad [a>b>u\geq 0] \qquad\qquad$ BY (220.04)

11. $\displaystyle\int_b^u \sqrt{\frac{x^2-b^2}{a^2-x^2}}\,dx = a\,E(\kappa,q) - \frac{b^2}{a}\,F(\kappa,q) - \frac{1}{u}\sqrt{(a^2-u^2)(u^2-b^2)}$

$\qquad\qquad\qquad\qquad\qquad\qquad [a\geq u>b>0] \qquad\qquad$ BY (217.04)

12. $\displaystyle\int_u^a \sqrt{\frac{x^2-b^2}{a^2-x^2}}\,dx = a\,E(\lambda,q) - \frac{b^2}{a}\,F(\lambda,q) \qquad [a>u\geq b>0] \qquad$ BY (218.03)

13. $\displaystyle\int_a^u \sqrt{\frac{x^2-b^2}{x^2-a^2}}\,dx = \frac{a^2-b^2}{a}\,F(\mu,t) - a\,E(\mu,t) + \mu\sqrt{\frac{u^2-a^2}{u^2-b^2}}$

$\qquad\qquad\qquad\qquad\qquad\qquad [u>a>b>0] \qquad\qquad$ BY (216.03)

14. $\displaystyle\int_0^u \sqrt{\frac{a^2-x^2}{b^2-x^2}}\,dx = a\,E(\eta,t) \qquad\qquad [a>b\geq u>0] \qquad$ H 64 (276), BY (219.01)

15. $\displaystyle\int_u^b \sqrt{\frac{a^2-x^2}{b^2-x^2}}\,dx = a\left\{ E(\zeta,t) - \frac{u}{a}\sqrt{\frac{b^2-u^2}{a^2-u^2}} \right\} \qquad [a>b>u\geq 0] \qquad$ BY (220.03)

16. $\displaystyle\int_b^u \sqrt{\frac{a^2-x^2}{x^2-b^2}}\,dx = a\left\{ F(\kappa,q) - E(\kappa,q) \right\} + \frac{1}{u}\sqrt{(a^2-u^2)(u^2-b^2)}$

$\qquad\qquad\qquad\qquad\qquad\qquad [a\geq u>b>0] \qquad\qquad$ BY (217.03)

17. $\displaystyle\int_u^a \sqrt{\frac{a^2-x^2}{x^2-b^2}}\,dx = a\left\{ F(\lambda,q) - E(\lambda,q) \right\} \qquad [a>u\geq b>0] \qquad$ BY (218.09)

18. $\displaystyle\int_a^u \sqrt{\frac{x^2-a^2}{x^2-b^2}}\,dx = u\sqrt{\frac{u^2-a^2}{u^2-b^2}} - a\,E(\mu,t) \qquad [u>a>b>0] \qquad$ BY (216.04)

3.171

1. $\displaystyle\int_b^u \frac{dx}{x^2}\sqrt{\frac{a^2+x^2}{x^2-b^2}} = \frac{\sqrt{a^2+b^2}}{b^2}\,E(\varepsilon,s) \qquad [u>b>0] \qquad$ BY (211.01), ZH 64 (274)

2. $\displaystyle\int_u^\infty \frac{dx}{x^2}\sqrt{\frac{a^2+x^2}{x^2-b^2}} = \frac{\sqrt{a^2+b^2}}{b^2}\,E(\xi,s) - \frac{a^2}{b^2 u}\sqrt{\frac{u^2-b^2}{a^2+u^2}}$

$\qquad\qquad\qquad\qquad\qquad\qquad [u\geq b>0] \qquad\qquad$ BY (212.09)

3. $\displaystyle\int_u^b \frac{dx}{x^2}\sqrt{\frac{a^2-x^2}{b^2-x^2}} = \frac{a^2-b^2}{ab^2}\,F(\zeta,t) - \frac{a}{b^2}\,E(\zeta,t) + \frac{a^2}{b^2 u}\sqrt{\frac{b^2-u^2}{a^2-u^2}}$

$\qquad\qquad\qquad\qquad\qquad\qquad [a>b>u>0] \qquad\qquad$ BY (220.12)

4. $\displaystyle\int_b^u \frac{dx}{x^2}\sqrt{\frac{a^2-x^2}{x^2-b^2}} = \frac{a}{b^2}\,E(\kappa,q) - \frac{1}{a}\,F(\kappa,q) \qquad [a\geq u>b>0] \qquad$ BY (217.11)

5. $\displaystyle\int_u^a \frac{dx}{x^2}\sqrt{\frac{a^2-x^2}{x^2-b^2}} = \frac{a}{b^2}\,E(\lambda,q) - \frac{1}{a}f(\lambda,q) - \frac{\sqrt{(a^2-u^2)(u^2-b^2)}}{b^2 u}$

$\qquad\qquad\qquad\qquad\qquad\qquad [a>u\geq b>0] \qquad\qquad$ BY (218.10)

6. $$\int_a^u \frac{dx}{x^2} \sqrt{\frac{x^2 - a^2}{x^2 - b^2}} = \frac{a}{b^2} E(\mu, t) - \frac{a^2 - b^2}{ab^2} F(\mu, t) - \frac{1}{u} \sqrt{\frac{u^2 - a^2}{u^2 - b^2}}$$

$$[u > a > b > 0] \qquad\qquad \text{BY (216.08)}$$

7. $$\int_u^\infty \frac{dx}{x^2} \sqrt{\frac{x^2 + a^2}{x^2 + b^2}} = \frac{1}{a} F(\beta, q) - \frac{a}{b^2} E(\beta, q) + \frac{a^2}{b^2 u} \sqrt{\frac{b^2 + u^2}{a^2 + u^2}}$$

$$[a > b, \quad u > 0] \qquad\qquad \text{BY (222.08)}$$

8. $$\int_u^\infty \frac{dx}{x^2} \sqrt{\frac{x^2 + b^2}{x^2 + a^2}} = \frac{1}{a} \{F(\beta, q) - E(\beta, q)\} + \frac{1}{u} \sqrt{\frac{b^2 + u^2}{a^2 + u^2}}$$

$$[a > b, \quad u > 0] \qquad\qquad \text{BY (222.09)}$$

9. $$\int_u^b \frac{dx}{x^2} \sqrt{\frac{b^2 - x^2}{a^2 + x^2}} = \frac{\sqrt{(b^2 - u^2)(a^2 + u^2)}}{a^2 u} - \frac{\sqrt{a^2 + b^2}}{a^2} E(\delta, r)$$

$$[b > u > 0] \qquad\qquad \text{BY (213.10)}$$

10. $$\int_b^u \frac{dx}{x^2} \sqrt{\frac{x^2 - b^2}{a^2 + x^2}} = \frac{\sqrt{a^2 + b^2}}{a^2} \{F(\varepsilon, s) - E(\varepsilon, s)\} \qquad [a > b > 0] \qquad \text{BY (211.07)}$$

11. $$\int_u^\infty \frac{dx}{x^2} \sqrt{\frac{x^2 - b^2}{a^2 + x^2}} = \frac{\sqrt{a^2 + b^2}}{a^2} \{F(\xi, s) - E(\xi, s)\} + \frac{1}{u} \sqrt{\frac{u^2 - b^2}{a^2 + u^2}}$$

$$[u \geq b > 0] \qquad\qquad \text{BY (212.11)}$$

12. $$\int_u^b \frac{dx}{x^2} \sqrt{\frac{a^2 + x^2}{b^2 - x^2}} = \frac{\sqrt{a^2 + b^2}}{b^2} \{F(\delta, r) - E(\delta, r)\} + \frac{\sqrt{(b^2 - u^2)(a^2 + u^2)}}{b^2 u}$$

$$[b > u > 0] \qquad\qquad \text{BY (213.05)}$$

13. $$\int_u^\infty \frac{dx}{x^2} \sqrt{\frac{x^2 - a^2}{x^2 - b^2}} = \frac{a}{b^2} E(\nu, t) - \frac{a^2 - b^2}{ab^2} F(\nu, t) \qquad [u \geq a > b > 0] \qquad \text{BY (215.08)}$$

14. $$\int_u^b \frac{dx}{x^2} \sqrt{\frac{b^2 - x^2}{a^2 - x^2}} = \frac{1}{u} \sqrt{\frac{b^2 - u^2}{a^2 - u^2}} - \frac{1}{a} E(\zeta, t) \qquad [a > b > u > 0] \qquad \text{BY (220.11)}$$

15. $$\int_b^u \frac{dx}{x^2} \sqrt{\frac{x^2 - b^2}{a^2 - x^2}} = \frac{1}{a} \{F(\kappa, q) - E(\kappa, q)\} \qquad [a \geq u > b > 0] \qquad \text{BY (217.08)}$$

16. $$\int_u^a \frac{dx}{x^2} \sqrt{\frac{x^2 - b^2}{u^2 - x^2}} = \frac{1}{a} \{F(\lambda, q) - E(\lambda, q)\} + \frac{\sqrt{(a^2 - u^2)(u^2 - b^2)}}{a^2 u}$$

$$[a > u \geq b > 0] \qquad\qquad \text{BY (218.08)}$$

17. $$\int_a^u \frac{dx}{x^2} \sqrt{\frac{x^2 - b^2}{x^2 - a^2}} = \frac{1}{a} E(\mu, t) - \frac{1}{u} \sqrt{\frac{u^2 - a^2}{u^2 - b^2}} \qquad [u > a > b > 0] \qquad \text{BY (216.07)}$$

18. $$\int_u^\infty \frac{dx}{x^2} \sqrt{\frac{x^2 - b^2}{x^2 - a^2}} = \frac{1}{a} E(\nu, t) \qquad [u \geq a > b > 0]$$

$$\text{BY (215.01), ZH 65 (281)}$$

3.172

1. $$\int_0^u \sqrt{\frac{x^2 + b^2}{(x^2 + a^2)^3}}\, dx = \frac{1}{a} E(\alpha, q) - \frac{a^2 - b^2}{a^2} \frac{u}{\sqrt{(a^2 + u^2)(b^2 + u^2)}}$$

$$[a > b, \quad u > 0]$$ BY (221.10)

2. $$\int_u^\infty \sqrt{\frac{x^2 + b^2}{(x^2 + a^2)^3}}\, dx = \frac{1}{a} E(\beta, q)$$

$$[a > b, \quad u \geq 0]$$ H 64 (271)

3. $$\int_0^u \sqrt{\frac{x^2 + a^2}{(x^2 + b^2)^3}}\, dx = \frac{a}{b^2} E(\alpha, q)$$

$$[a > b, \quad u > 0]$$ H 64 (270)

4. $$\int_u^\infty \sqrt{\frac{x^2 + a^2}{(x^2 + b^2)^3}}\, dx = \frac{a}{b^2} E(\beta, q) - \frac{a^2 - b^2}{b^2} \frac{u}{\sqrt{(a^2 + u^2)(b^2 + u^2)}}$$

$$[a > b, \quad u \geq 0]$$ BY (222.06)

5. $$\int_0^u \sqrt{\frac{b^2 - x^2}{(a^2 + x^2)^3}}\, dx = \frac{\sqrt{a^2 + b^2}}{a^2} E(\gamma, r) - \frac{1}{\sqrt{a^2 + b^2}} F(\gamma, r)$$

$$[b \geq u > 0]$$ BY (214.08)

6. $$\int_u^b \sqrt{\frac{b^2 - x^2}{(a^2 + x^2)^3}}\, dx = \frac{\sqrt{a^2 + b^2}}{a^2} E(\delta, r) - \frac{1}{\sqrt{a^2 + b^2}} F(\delta, r) - \frac{u}{a^2}\sqrt{\frac{b^2 - u^2}{a^2 + u^2}}$$

$$[b > u \geq 0]$$ BY (213.04)

7. $$\int_b^u \sqrt{\frac{x^2 - b^2}{(a^2 + x^2)^3}}\, dx = \frac{\sqrt{a^2 + b^2}}{a^2} E(\varepsilon, s) - \frac{b^2}{a^2\sqrt{a^2 + b^2}} F(\varepsilon, s) - \frac{1}{u}\sqrt{\frac{u^2 - b^2}{u^2 + a^2}}$$

$$[u > b > 0]$$ BY (211.06)

8. $$\int_u^\infty \sqrt{\frac{x^2 - b^2}{(a^2 + x^2)^3}}\, dx = \frac{\sqrt{a^2 + b^2}}{a^2} E(\xi, s) - \frac{b^2}{a^2\sqrt{a^2 + b^2}} F(\xi, s)$$

$$[u \geq b > 0]$$ BY (212.08)

9. $$\int_0^u \sqrt{\frac{x^2 + a^2}{(b^2 - x^2)^3}}\, dx = \frac{a^2}{b^2\sqrt{a^2 + b^2}} F(\gamma, r) - \frac{\sqrt{a^2 + b^2}}{b^2} E(\gamma, r) + \frac{(a^2 + b^2)\, u}{b^2\sqrt{(a^2 + u^2)(b^2 - u^2)}}$$

$$[b > u > 0]$$ BY (214.09)

10. $$\int_u^\infty \sqrt{\frac{x^2 + a^2}{(x^2 - b^2)^3}}\, dx = \frac{1}{\sqrt{a^2 + b^2}} F(\xi, s) - \frac{\sqrt{a^2 + b^2}}{b^2} E(\xi, s) + \frac{(a^2 + b^2)\, u}{b^2\sqrt{(a^2 + u^2)(u^2 - b^2)}}$$

$$[u > b > 0]$$ BY (212.07)

11. $$\int_0^u \sqrt{\frac{b^2 - x^2}{(a^2 - x^2)^3}}\, dx = \frac{1}{a}\left\{ F(\eta, t) - E(\eta, t) + \frac{u}{a}\sqrt{\frac{b^2 - u^2}{a^2 - u^2}} \right\}$$

$$[a > b \geq u > 0]$$ BY (219.09)

12. $\displaystyle\int_u^b \sqrt{\frac{b^2-x^2}{(a^2-x^2)^3}}\,dx = \frac{1}{a}\left\{F(\zeta,t)-E(\zeta,t)\right\}$ $\qquad [a>b>u\geq 0]$ $\qquad$ BY (220.07)

13. $\displaystyle\int_b^u \sqrt{\frac{x^2-b^2}{(a^2-x^2)^3}}\,dx = \frac{1}{u}\sqrt{\frac{u^2-b^2}{a^2-u^2}} - \frac{1}{a}E(\kappa,q)$ $\qquad [a>u>b>0]$ $\qquad$ BY (217.07)

14. $\displaystyle\int_u^\infty \sqrt{\frac{x^2-b^2}{(x^2-a^2)^3}}\,dx = \frac{1}{a}\left[F(\nu,t)-E(\nu,t)\right]+\frac{1}{u}\sqrt{\frac{u^2-b^2}{u^2-a^2}}$

$\qquad\qquad\qquad\qquad\qquad\qquad\qquad\qquad [u>a>b>0]$ $\qquad$ BY (215.05)

15. $\displaystyle\int_0^u \sqrt{\frac{a^2-x^2}{(b^2-x^2)^3}}\,dx = \frac{a}{b^2}\left[F(\eta,t)-E(\eta,t)\right]+\frac{u}{b^2}\sqrt{\frac{a^2-u^2}{b^2-u^2}}$

$\qquad\qquad\qquad\qquad\qquad\qquad\qquad\qquad [a>b>u>0]$ $\qquad$ BY (219.10)

16. $\displaystyle\int_u^a \sqrt{\frac{a^2-x^2}{(x^2-b^2)^3}}\,dx = \frac{u}{b^2}\sqrt{\frac{a^2-u^2}{u^2-b^2}} - \frac{a}{b^2}E(\lambda,q)$ $\qquad [a>u>b>0]$ $\qquad$ BY (218.05)

17. $\displaystyle\int_a^u \sqrt{\frac{x^2-a^2}{(x^2-b^2)^3}}\,dx = \frac{a}{b^2}\left[F(\mu,t)-E(\mu,t)\right]$ $\qquad [u>a>b>0]$ $\qquad$ BY (216.05)

18. $\displaystyle\int_u^\infty \sqrt{\frac{x^2-a^2}{(x^2-b^2)^3}}\,dx = \frac{a}{b^2}\left[F(\nu,t)-E(\nu,t)\right]+\frac{1}{u}\sqrt{\frac{u^2-a^2}{u^2-b^2}}$

$\qquad\qquad\qquad\qquad\qquad\qquad\qquad\qquad [u\geq a>b>0]$ $\qquad$ BY (215.03)

3.173

1. $\displaystyle\int_u^1 \frac{dx}{x^2}\sqrt{\frac{x^2+1}{1-x^2}} = \sqrt{2}\left[F\left(\arccos u,\frac{\sqrt{2}}{2}\right)-E\left(\arccos u,\frac{\sqrt{2}}{2}\right)\right]+\frac{\sqrt{1-u^4}}{u}$

$\qquad\qquad\qquad\qquad\qquad\qquad\qquad\qquad [u<1]$ $\qquad$ BY (259.77)

2. $\displaystyle\int_1^u \frac{dx}{x^2}\sqrt{\frac{x^2+1}{x^2-1}} = \sqrt{2}\,E\left(\arccos\frac{1}{u},\frac{\sqrt{2}}{2}\right)$ $\qquad [u>1]$ $\qquad$ BY (260.76)

3.174 **Notation:** In **3.174** and **3.175**, we take: $\quad \alpha = \arccos\dfrac{1+\left(1-\sqrt{3}\right)u}{1+\left(1+\sqrt{3}\right)u}$,

$$\beta = \arccos\frac{1-\left(1+\sqrt{3}\right)u}{1+\left(\sqrt{3}-1\right)u}, \qquad p = \frac{\sqrt{2+\sqrt{3}}}{2}, \qquad q = \frac{\sqrt{2-\sqrt{3}}}{2}.$$

1. $\displaystyle\int_0^u \frac{dx}{\left[1+\left(1+\sqrt{3}\right)x\right]^2}\sqrt{\frac{1-x+x^2}{x(1+x)}} = \frac{1}{\sqrt[4]{3}}E(\alpha,p)$ $\qquad [u>0]$ $\qquad$ BY (260.51)

2. $\displaystyle\int_0^u \frac{dx}{\left[1+\left(\sqrt{3}-1\right)x\right]^2}\sqrt{\frac{1+x+x^2}{x(1-x)}} = \frac{1}{\sqrt[4]{3}}E(\beta,q)$ $\qquad [1\geq u>0]$ $\qquad$ BY (259.51)

3. $\displaystyle\int_0^u \frac{dx}{1-x+x^2}\sqrt{\frac{x(1+x)}{1-x+x^2}}$ $\displaystyle\frac{1}{\sqrt[4]{27}}E(\alpha,p) + -\frac{2-\sqrt{3}}{\sqrt[4]{27}}F(\alpha,p) - \frac{2\left(2+\sqrt{3}\right)}{\sqrt{3}}\frac{1+\left(1-\sqrt{3}\right)u}{1+\left(1+\sqrt{3}\right)u}$

$$\times\sqrt{\frac{u(1+u)}{1-u+u^2}}$$

$$[u>0] \qquad\qquad \text{BY (260.54)}$$

4. $\displaystyle\int_0^u \frac{dx}{1+x+x^2}\sqrt{\frac{x(1-x)}{1+x+x^2}}$ $\displaystyle\frac{4}{\sqrt[4]{27}}E(\beta,q) - \frac{2+\sqrt{3}}{\sqrt[4]{27}}F(\beta,q) - \frac{2\left(2-\sqrt{3}\right)}{\sqrt{3}}\frac{1-\left(1+\sqrt{3}\right)u}{1+\left(\sqrt{3}-1\right)u}$

$$\times\sqrt{\frac{u(1-u)}{1+u+u^2}}$$

$$[1\geq u>0] \qquad\qquad \text{BY (259.55)}$$

3.175

1. $\displaystyle\int_0^u \frac{dx}{1+x}\sqrt{\frac{x}{1+x^3}} = \frac{1}{\sqrt[4]{27}}\left[F(\alpha,p) - 2\,E(\alpha,p)\right] + \frac{2}{\sqrt{3}}\frac{\sqrt{u\left(1-u+u^2\right)}}{\sqrt{1+u}\left[1+\left(1+\sqrt{3}\right)u\right]}$

$$[u>0] \qquad\qquad \text{BY (260.55)}$$

2. $\displaystyle\int_0^u \frac{dx}{1-x}\sqrt{\frac{x}{1-x^3}} = \frac{1}{\sqrt[4]{27}}\left[F(\beta,q) - 2\,E(\beta,q)\right] + \frac{2}{\sqrt{3}}\frac{\sqrt{u\left(1+u+u^2\right)}}{\sqrt{1-u}\left[1+\left(\sqrt{3}-1\right)u\right]}$

$$[0<u<1] \qquad\qquad \text{BY (259.52)}$$

3.18 Expressions that can be reduced to fourth roots of second-degree polynomials and their products with rational functions

3.181

1. $\displaystyle\int_b^u \frac{dx}{\sqrt[4]{(a-x)(x-b)}} = \sqrt{a-b}\left\{2\left[\boldsymbol{E}\left(\frac{1}{\sqrt{2}}\right) + E\left(\arccos\sqrt[4]{\frac{4(a-u)(u-b)}{(a-b)^2}},\frac{1}{\sqrt{2}}\right)\right]\right.$

$$\left. -\left[\boldsymbol{K}\left(\frac{1}{\sqrt{2}}\right) + F\left(\arccos\sqrt[4]{\frac{4(a-u)(u-b)}{(a-b)^2}},\frac{1}{\sqrt{2}}\right)\right]\right\}$$

$$[a\geq u>b] \qquad\qquad \text{BY (271.05)}$$

2. $\displaystyle\int_a^u \frac{dx}{\sqrt[4]{(x-a)(x-b)}} = \sqrt{\frac{a-b}{2}}F\left[\left(\arccos\frac{a-b-2\sqrt{(u-a)(u-b)}}{a-b+2\sqrt{(u-a)(u-b)}},\frac{1}{\sqrt{2}}\right)\right.$

$$\left. -2E\left(\arccos\frac{a-b-2\sqrt{(u-a)(u-b)}}{a-b+2\sqrt{(u-a)(u-b)}},\frac{1}{\sqrt{2}}\right)\right]$$

$$+\frac{2(2u-a-b)\sqrt[4]{(u-a)(u-b)}}{a-b+2\sqrt{(u-a)(u-b)}}$$

$$[u>a>b] \qquad\qquad \text{BY (272.05)}$$

3.182

1. $\displaystyle\int_b^u \frac{dx}{\sqrt[4]{[(a-x)(x-b)]^3}} = \frac{2}{\sqrt{a-b}}\left[\boldsymbol{K}\left(\frac{1}{\sqrt{2}}\right) + F\left(\arccos\sqrt{\frac{4(a-u)(u-b)}{(a-b)^2}},\frac{1}{\sqrt{2}}\right)\right]$

$$[a\geq u>b] \qquad\qquad \text{BY (271.01)}$$

2.
$$\int_a^u \frac{dx}{\sqrt[4]{[(x-a)(x-b)]^3}} = \frac{\sqrt{2}}{\sqrt{a-b}} F\left(\arccos\frac{a-b-2\sqrt{(u-a)(u-b)}}{a-b+2\sqrt{(u-a)(u-b)}}, \frac{1}{\sqrt{2}}\right)$$

$$[u > a > b] \qquad\qquad \text{BY (272.00)}$$

3.183   **Notation:** In **3.183**–**3.186** we set:

$$\alpha = \arccos\frac{1}{\sqrt[4]{u^2+1}}, \qquad \beta = \arccos\sqrt[4]{1-u^2}, \qquad \gamma = \arccos\frac{1-\sqrt{u^2-1}}{1+\sqrt{u^2-1}}.$$

1.
$$\int_0^u \frac{dx}{\sqrt[4]{x^2+1}} = \sqrt{2}\left[F\left(\alpha, \frac{1}{\sqrt{2}}\right) - 2E\left(\alpha, \frac{1}{\sqrt{2}}\right)\right] + \frac{2u}{\sqrt[4]{u^2+1}}$$

$$[u > 0] \qquad\qquad \text{BY (273.55)}$$

2.
$$\int_0^u \frac{dx}{\sqrt[4]{1-x^2}} = \sqrt{2}\left[2E\left(\beta, \frac{1}{\sqrt{2}}\right) - F\left(\beta, \frac{1}{\sqrt{2}}\right)\right] \qquad [0 < u \le 1] \qquad \text{BY (271.55)}$$

3.
$$\int_1^u \frac{dx}{\sqrt[4]{x^2-1}} = F\left(\gamma, \frac{1}{\sqrt{2}}\right) - 2E\left(\gamma, \frac{1}{\sqrt{2}}\right) + \frac{2u\sqrt[4]{u^2-1}}{1+\sqrt{u^2-1}}$$

$$[u > 1] \qquad\qquad \text{BY (272.55)}$$

3.184

1.
$$\int_0^u \frac{x^2\,dx}{\sqrt[4]{1-x^2}} = \frac{2\sqrt{2}}{5}\left[2E\left(\beta, \frac{1}{\sqrt{2}}\right) - F\left(\beta, \frac{1}{\sqrt{2}}\right)\right] - \frac{2u}{5}\sqrt[4]{(1-u^2)^3}$$

$$[0 < u \le 1] \qquad\qquad \text{BY (271.59)}$$

2.
$$\int_1^u \frac{dx}{x^2\sqrt[4]{x^2-1}} = E\left(\gamma, \frac{1}{\sqrt{2}}\right) - \frac{1}{2}F\left(\gamma, \frac{1}{\sqrt{2}}\right) - \frac{1-\sqrt{u^2-1}}{1+\sqrt{u^2-1}}\cdot\frac{\sqrt{u^2-1}}{u}$$

$$[u > 1] \qquad\qquad \text{BY (272.54)}$$

3.185

1.
$$\int_0^u \frac{dx}{\sqrt[4]{(x^2+1)^3}} = \sqrt{2}\,F\left(\alpha, \frac{1}{\sqrt{2}}\right) \qquad\qquad [u > 0] \qquad \text{BY (273.50)}$$

2.
$$\int_0^u \frac{dx}{\sqrt[4]{(1-x^2)^3}} = \sqrt{2}\,F\left(\beta, \frac{1}{\sqrt{2}}\right) \qquad\qquad [0 < u \le 1] \qquad \text{BY (271.51)}$$

3.
$$\int_1^u \frac{dx}{\sqrt[4]{(x^2-1)^3}} = F\left(\gamma, \frac{1}{\sqrt{2}}\right) \qquad\qquad [u > 1] \qquad \text{BY (272.50)}$$

4.
$$\int_0^u \frac{x^2\,dx}{\sqrt[4]{(1-x^2)^3}} = \frac{2\sqrt{2}}{3}F\left(\beta, \frac{1}{\sqrt{2}}\right) - \frac{2}{3}u\sqrt[4]{1-u^2} \qquad [0 < u \le 1] \qquad \text{BY (271.54)}$$

5.
$$\int_0^u \frac{dx}{\sqrt[4]{(x^2+1)^5}} = 2\sqrt{2}\,E\left(\alpha, \frac{1}{\sqrt{2}}\right) - \sqrt{2}\,F\left(\alpha, \frac{1}{\sqrt{2}}\right)$$

$$[u > 0] \qquad\qquad \text{BY (273.54)}$$

6. $\displaystyle\int_0^u \frac{x^2\,dx}{\sqrt[4]{(x^2+1)^5}} = 2\sqrt{2}\left[F\left(\alpha, \frac{1}{\sqrt{2}}\right) - 2E\left(\alpha, \frac{1}{\sqrt{2}}\right)\right] + \frac{2u}{\sqrt[4]{u^2+1}}$

$[u > 0]$ BY (273.56)

7. $\displaystyle\int_0^u \frac{x^2\,dx}{\sqrt[4]{(x^2+1)^7}} = \frac{1}{3\sqrt{2}} F\left(\alpha, \frac{1}{\sqrt{2}}\right) - \frac{u}{6\sqrt[4]{(u^2+1)^3}}$ $[u > 0]$ BY (273.53)

3.186

1. $\displaystyle\int_0^u \frac{1+\sqrt{x^2+1}}{(x^2+1)\sqrt[4]{x^2+1}}\,dx = 2\sqrt{2}\,E\left(\alpha, \frac{1}{\sqrt{2}}\right)$ $[u > 0]$ BY (273.51)

2. $\displaystyle\int_0^u \frac{dx}{(1+\sqrt{1-x^2})\sqrt[4]{1-x^2}} = \sqrt{2}\left[F\left(\beta, \frac{1}{\sqrt{2}}\right) - E\left(\beta, \frac{1}{\sqrt{2}}\right)\right] + \frac{u\sqrt[4]{1-u^2}}{1+\sqrt{1-u^2}}$

$[0 < u \le 1]$ BY (271.58)

3. $\displaystyle\int_1^u \frac{dx}{(x^2+2\sqrt{x^2-1})\sqrt[4]{x^2-1}} = \frac{1}{2}\left[F\left(\gamma, \frac{1}{\sqrt{2}}\right) - E\left(\gamma, \frac{1}{\sqrt{2}}\right)\right]$

$[u > 1]$ BY (272.53)

4. $\displaystyle\int_0^u \frac{1-\sqrt{1-x^2}}{1+\sqrt{1-x^2}}\cdot\frac{dx}{\sqrt[4]{(1-x^2)^3}} = \sqrt{2}\left[2E\left(\beta, \frac{1}{\sqrt{2}}\right) - F\left(\beta, \frac{1}{\sqrt{2}}\right)\right] - \frac{2u\sqrt[4]{1-u^2}}{1+\sqrt{1-u^2}}$

$[0 < u \le 1]$ BY (271.57)

5. $\displaystyle\int_1^u \frac{x^2\,dx}{(x^2+2\sqrt{x^2-1})\sqrt[4]{(x^2-1)^3}} = E\left(\gamma, \frac{1}{\sqrt{2}}\right)$ $[u > 1]$ BY (272.51)

3.19–3.23 Combinations of powers of x and powers of binomials of the form $(\alpha+\beta x)$

3.191

1. $\displaystyle\int_0^u x^{\nu-1}(u-x)^{\mu-1}\,dx = u^{\mu+\nu-1}\,\mathrm{B}(\mu, \nu)$ $[\mathrm{Re}\,\mu > 0,\quad \mathrm{Re}\,\nu > 0]$ ET II 185(7)

2. $\displaystyle\int_u^\infty x^{-\nu}(x-u)^{\mu-1}\,dx = u^{\mu-\nu}\,\mathrm{B}(\nu-\mu, \mu)$ $[\mathrm{Re}\,\nu > \mathrm{Re}\,\mu > 0]$ ET II 201(6)

3. $\displaystyle\int_0^1 x^{\nu-1}(1-x)^{\mu-1}\,dx = \int_0^1 x^{\mu-1}(1-x)^{\nu-1}\,dx = \mathrm{B}(\mu, \nu)$

$[\mathrm{Re}\,\mu > 0,\quad \mathrm{Re}\,\nu > 0]$ FI II 774(1)

3.192

1. $\displaystyle\int_0^1 \frac{x^p\,dx}{(1-x)^p} = p\pi\,\mathrm{cosec}\,p\pi$ $[p^2 < 1]$ BI (3)(4)

2. $\displaystyle\int_0^1 \frac{x^p\,dx}{(1-x)^{p+1}} = -\pi\,\mathrm{cosec}\,p\pi$ $[-1 < p < 0]$ BI (3)(5)

3. $\int_0^1 \dfrac{(1-x)^p}{x^{p+1}}\, dx = -\pi \operatorname{cosec} p\pi$ $[-1 < p < 0]$ BI (4)(6)

4. $\int_1^\infty (x-1)^{p-\frac{1}{2}} \dfrac{dx}{x} = \pi \sec p\pi$ $\left[-\frac{1}{2} < p < \frac{1}{2}\right]$ BI (23)(7)

3.193 $\int_0^n x^{\nu-1}(n-x)^n\, dx = \dfrac{n!\, n^{\nu+n}}{\nu(\nu+1)(\nu+2)\ldots(\nu+n)}$ $[\operatorname{Re}\nu > 0]$ EH I 2

3.194

1. $\int_0^u \dfrac{x^{\mu-1}\, dx}{(1+\beta x)^\nu} = \dfrac{u^\mu}{\mu}\, {}_2F_1(\nu, \mu; 1+\mu; -\beta u)$ $[|\arg(1+\beta u)| < \pi, \quad \operatorname{Re}\mu > 0]$

                                                                                ET I 310(20)

2.[6] $\int_u^\infty \dfrac{x^{\mu-1}\, dx}{(1+\beta x)^\nu} = \dfrac{u^{\mu-\nu}}{\beta^\nu(\nu-\mu)}\, {}_2F_1\left(\nu, \nu-\mu;\ \nu-\mu+1;\ -\dfrac{1}{\beta u}\right)$

                                                                $[\operatorname{Re}\nu > \operatorname{Re}\mu]$ ET I 310(21)

3. $\int_0^\infty \dfrac{x^{\mu-1}\, dx}{(1+\beta x)^\nu} = \beta^{-\mu}\, \mathrm{B}(\mu, \nu-\mu)$ $[|\arg\beta| < \pi, \quad \operatorname{Re}\nu > \operatorname{Re}\mu > 0]$

                                                                                FI II 775a, ET I 310(19)

4. $\int_0^\infty \dfrac{x^{\mu-1}\, dx}{(1+\beta x)^{n+1}} = (-1)^n \dfrac{\pi}{\beta^\mu}\binom{\mu-1}{n}\operatorname{cosec}(\mu\pi)$ $[|\arg\beta| < \pi, \quad 0 < \operatorname{Re}\nu < n+1]$

                                                                                ET I 308(6)

5. $\int_0^u \dfrac{x^{\mu-1}\, dx}{1+\beta x} = \dfrac{u^\mu}{\mu}\, {}_2F_1(1, \mu; 1+\mu; -u\beta)$ $[|\arg(1+u\beta)| < \pi, \quad \operatorname{Re}\mu > 0]$

                                                                                ET I 308(5)

6. $\int_0^\infty \dfrac{x^{\mu-1}\, dx}{(1+\beta x)^2} = \dfrac{(1-\mu)\pi}{\beta^\mu}\operatorname{cosec}\mu\pi$ $[0 < \operatorname{Re}\mu < 2]$ BI (16)(4)

7. $\int_0^\infty \dfrac{x^m\, dx}{(a+bx)^{n+\frac{1}{2}}} = 2^{m+1} m!\, \dfrac{(2n-2m-3)!!}{(2n-1)!!}\dfrac{a^{m-n+\frac{1}{2}}}{b^{m+1}}$

                                                                $\left[m < n - \tfrac{1}{2}, \quad a > 0, \quad b > 0\right]$
                                                                                BI (21)(2)

8. $\int_0^1 \dfrac{x^{n-1}\, dx}{(1+x)^m} = 2^{-n}\sum_{k=0}^\infty \binom{m-n-1}{k}\dfrac{(-2)^{-k}}{n+k}$ BI (3)(1)

3.195 $\int_0^\infty \dfrac{(1+x)^{p-1}}{(x+a)^{p+1}}\, dx = \dfrac{1-a^{-p}}{p(a-1)}$ $[a > 0]$ LI (19)(6)

3.196

1. $\int_0^u (x+\beta)^\nu (u-x)^{\mu-1}\, dx = \dfrac{\beta^\nu u^\mu}{\mu}\, {}_2F_1\left(1, -\nu; 1+\mu; -\dfrac{u}{\beta}\right)$

                                                                $\left[\left|\arg\dfrac{u}{\beta}\right| < \pi\right]$ ET II 185(8)

2. $\int_u^\infty (x+\beta)^{-\nu}(x-u)^{\mu-1}\,dx = (u+\beta)^{\mu-\nu}\,\mathrm{B}(\nu-\mu,\mu)$

$$\left[\left|\arg\frac{u}{\beta}\right| < \pi, \quad \mathrm{Re}\,\nu > \mathrm{Re}\,\mu > 0\right]$$

ET II 201(7)

3. $\int_a^b (x-a)^{\mu-1}(b-x)^{\nu-1}\,dx = (b-a)^{\mu+\nu-1}\,\mathrm{B}(\mu,\nu)$ $[b>a, \quad \mathrm{Re}\,\mu>0, \quad \mathrm{Re}\,\nu>0]$

EH I 10(13)

4. $\int_1^\infty \dfrac{dx}{(a-bx)(x-1)^\nu} = -\dfrac{\pi}{b}\cosec\nu\pi\left(\dfrac{b}{b-a}\right)^\nu$ $[a<b, \quad b>0, \quad 0<\nu<1]$ LI (23)(5)

5. $\int_{-\infty}^1 \dfrac{dx}{(a-bx)(1-x)^\nu} = \dfrac{\pi}{b}\cosec\nu\pi\left(\dfrac{b}{a-b}\right)^\nu$ $[a>b>0, \quad 0<\nu<1]$ LI (24)(10)

3.197

1. $\int_0^\infty x^{\nu-1}(\beta+x)^{-\mu}(x+\gamma)^{-\varrho}\,dx = \beta^{-\mu}\gamma^{\nu-\varrho}\,\mathrm{B}(\nu,\mu-\nu+\varrho)\,{}_2F_1\left(\mu,\nu;\mu+\varrho;1-\dfrac{\gamma}{\beta}\right)$

$$[|\arg\beta|<\pi, \quad |\arg\gamma|<\pi, \quad \mathrm{Re}\,\nu>0, \quad \mathrm{Re}\,\mu>\mathrm{Re}(\nu-\varrho)]\quad \text{ET II 233(9)}$$

2. $\int_u^\infty x^{-\lambda}(x+\beta)^\nu(x-u)^{\mu-1}\,dx = u^{\mu+\nu-\lambda}\,\mathrm{B}(\lambda-\mu-\nu,\mu)\,{}_2F_1\left(-\nu,\lambda-\mu-\nu;\lambda-\nu;-\dfrac{\beta}{u}\right)$

$$\left[\left|\arg\frac{u}{\beta}\right|<\pi \text{ or } \left|\frac{\beta}{u}\right|<1, \quad 0<\mathrm{Re}\,\mu<\mathrm{Re}(\lambda-\nu)\right]\quad \text{ET II 201(8)}$$

3. $\int_0^1 x^{\lambda-1}(1-x)^{\mu-1}(1-\beta x)^{-\nu}\,dx = \mathrm{B}(\lambda,\mu)\,{}_2F_1(\nu,\lambda;\lambda+\mu;\beta)$

$$[\mathrm{Re}\,\lambda>0, \quad \mathrm{Re}\,\mu>0, \quad |\beta|<1]\qquad \text{WH}$$

4. $\int_0^1 x^{\mu-1}(1-x)^{\nu-1}(1+ax)^{-\mu-\nu}\,dx = (1+a)^{-\mu}\,\mathrm{B}(\mu,\nu)$

$$[\mathrm{Re}\,\mu>0, \quad \mathrm{Re}\,\nu>0, \quad a>-1]$$
BI(5)4, EH I 10(11)

5. $\int_0^\infty x^{\lambda-1}(1+x)^\nu(1+\alpha x)^\mu\,dx = \mathrm{B}(\lambda,-\mu-\nu-\lambda)\,{}_2F_1(-\mu,\lambda;-\mu-\nu;1-\alpha)$

$$[|\arg\alpha|<\pi, \quad -\mathrm{Re}(\mu+\nu)>\mathrm{Re}\,\lambda>0]$$
EH I 60(12), ET I 310(23)

6. $\int_1^\infty x^{\lambda-\nu}(x-1)^{\nu-\mu-1}(\alpha x-1)^{-\lambda}\,dx = \alpha^{-\lambda}\,\mathrm{B}(\mu,\nu-\mu)\,{}_2F_1\left(\nu,\mu;\lambda;\alpha^{-1}\right)$

$$[1+\mathrm{Re}\,\nu>\mathrm{Re}\,\lambda>\mathrm{Re}\,\mu, \quad |\arg(\alpha-1)|<\pi]\quad \text{EH I 115(6)}$$

7. $\int_0^\infty x^{\mu-\frac{1}{2}}(x+a)^{-\mu}(x+b)^{-\mu}\,dx = \sqrt{\pi}\left(\sqrt{a}+\sqrt{b}\right)^{1-2\mu}\dfrac{\Gamma\left(\mu-\frac{1}{2}\right)}{\Gamma(\mu)}$

$$[\mathrm{Re}\,\mu>0]\qquad \text{BI 19(5)}$$

8. $$\int_0^u x^{\nu-1}(x+\alpha)^\lambda(u-x)^{\mu-1}\,dx = \alpha^\lambda u^{\mu+\nu-1}\,B(\mu,\nu)\,{}_2F_1\left(-\lambda,\nu;\mu+\nu;-\frac{u}{\alpha}\right)$$

$$\left[\left|\arg\left(\frac{u}{\alpha}\right)\right| < \pi, \quad \operatorname{Re}\mu > 0, \quad \operatorname{Re}\nu > 0\right]$$
ET II 186(9)

9. $$\int_0^\infty x^{\lambda-1}(1+x)^{-\mu+\nu}(x+\beta)^{-\nu}\,dx = B(\mu-\lambda,\lambda)\,{}_2F_1(\nu,\mu-\lambda;\mu;1-\beta)$$

$$[\operatorname{Re}\mu > \operatorname{Re}\lambda > 0]$$
EH I 205

10. $$\int_0^1 \frac{x^{q-1}\,dx}{(1-x)^q(1+px)} = \frac{\pi}{(1+p)^q}\operatorname{cosec}q\pi \qquad [0 < q < 1, \quad p > -1]$$
BI (5)(1)

11. $$\int_0^1 \frac{x^{p-\frac12}\,dx}{(1-x)^p(1+qx)^p} = \frac{2\,\Gamma\left(p+\frac12\right)\Gamma(1-p)}{\sqrt\pi}\cos^{2p}\left(\arctan\sqrt q\right)\frac{\sin\left[(2p-1)\arctan\left(\sqrt q\right)\right]}{(2p-1)\sin\left[\arctan\left(\sqrt q\right)\right]}$$

$$\left[-\tfrac12 < p < 1, \quad q > 0\right]$$
BI (11)(1)

12. $$\int_0^1 \frac{x^{p-\frac12}\,dx}{(1-x)^p(1-qx)^p} = \frac{\Gamma\left(p+\frac12\right)\Gamma(1-p)}{\sqrt\pi}\frac{\left(1-\sqrt q\right)^{1-2p}-\left(1+\sqrt q\right)^{1-2p}}{(2p-1)\sqrt q}$$

$$\left[-\tfrac12 < p < 1, \quad 0 < q < 1\right]$$
BI (11)(2)

3.198 $$\int_0^1 x^{\mu-1}(1-x)^{\nu-1}[ax+b(1-x)+c]^{-(\mu+\nu)}\,dx = (a+c)^{-\mu}(b+c)^{-\nu}\,B(\mu,\nu)$$

$$[a \geq 0, \quad b \geq 0, \quad c > 0, \quad \operatorname{Re}\mu > 0, \quad \operatorname{Re}\nu > 0]$$
FI II 787

3.199 $$\int_a^b (x-a)^{\mu-1}(b-x)^{\nu-1}(x-c)^{-\mu-\nu}\,dx = (b-a)^{\mu+\nu-1}(b-c)^{-\mu}(a-c)^{-\nu}\,B(\mu,\nu)$$

$$[\operatorname{Re}\mu > 0, \quad \operatorname{Re}\nu > 0, \quad c < a < b]$$
EH I 10(14)

3.211 $$\int_0^1 x^{\lambda-1}(1-x)^{\mu-1}(1-ux)^{-\varrho}(1-vx)^{-\sigma}\,dx = B(\mu,\lambda)\,F_1\left((\lambda,\varrho,\sigma,\lambda+\mu;u,v)\right)$$

$$[\operatorname{Re}\lambda > 0, \quad \operatorname{Re}\mu > 0]$$
EH I 231(5)

3.212 $$\int_0^\infty \left[(1+ax)^{-p}+(1+bx)^{-p}\right]x^{q-1}\,dx = 2(ab)^{-\frac q2}\,B(q,p-q)\cos\left\{q\arccos\left[\frac{a+b}{2\sqrt{ab}}\right]\right\}$$

$$[p > q > 0]$$
BI (19)(9)

3.213 $$\int_0^\infty \left[(1+ax)^{-p}-(1+bx)^{-p}\right]x^{q-1}\,dx = -2i(ab)^{-\frac q2}\,B(q,p-q)\sin\left\{q\arccos\left[\frac{a+b}{2\sqrt{ab}}\right]\right\}$$

$$[p > q > 0]$$
BI (19)(10)

3.214 $$\int_0^1 \left[(1+x)^{\mu-1}(1-x)^{\nu-1}+(1+x)^{\nu-1}(1-x)^{\mu-1}\right]dx = 2^{\mu+\nu-1}\,B(\mu,\nu)$$

$$[\operatorname{Re}\mu > 0, \quad \operatorname{Re}\nu > 0]$$
LI(1)(15), EH I 10(10)

3.215 $$\int_0^1 \left\{a^\mu x^{\mu-1}(1-ax)^{\nu-1}+(1-a)^\nu x^{\nu-1}[1-(1-a)x]^{\mu-1}\right\}dx = B(\mu,\nu)$$

$$[\operatorname{Re}\mu > 0, \quad \operatorname{Re}\nu > 0, \quad |a| < 1]$$
BI (1)(16)

3.216

1. $\displaystyle\int_0^1 \frac{x^{\mu-1} + x^{\nu-1}}{(1+x)^{\mu+\nu}}\, dx = \mathrm{B}(\mu,\nu)$ $[\operatorname{Re}\mu > 0, \quad \operatorname{Re}\nu > 0]$ FI II 775

2. $\displaystyle\int_1^\infty \frac{x^{\mu-1} + x^{\nu-1}}{(1+x)^{\mu+\nu}}\, dx = \mathrm{B}(\mu,\nu)$ $[\operatorname{Re}\mu > 0, \quad \operatorname{Re}\nu > 0]$ FI II 775

3.217 $\displaystyle\int_0^\infty \left\{ \frac{b^p x^{p-1}}{(1+bx)^p} - \frac{(1+bx)^{p-1}}{b^{p-1} x^p} \right\} dx = \pi \cot p\pi$ $[0 < p < 1, \quad b > 0]$ BI(18)(13)

3.218 $\displaystyle\int_0^\infty \frac{x^{2p-1} - (a+x)^{2p-1}}{(a+x)^p x^p}\, dx = \pi \cot p\pi$ $[p < 1]\quad (\mathrm{cf.}\ \mathbf{3.217})$ BI (18)(7)

3.219 $\displaystyle\int_0^\infty \left\{ \frac{x^\nu}{(x+1)^{\nu+1}} - \frac{x^\mu}{(x+1)^{\mu+1}} \right\} dx = \psi(\mu+1) - \psi(\nu+1)$

$[\operatorname{Re}\mu > -1, \quad \operatorname{Re}\nu > -1]$ BI (19)(13)

3.221

1. $\displaystyle\int_a^\infty \frac{(x-a)^{p-1}}{x-b}\, dx = \pi(a-b)^{p-1} \operatorname{cosec} p\pi$ $[a > b, \quad 0 < p < 1]$ LI (24)(8)

2. $\displaystyle\int_{-\infty}^a \frac{(a-x)^{p-1}}{x-b}\, dx = -\pi(b-a)^{p-1} \operatorname{cosec} p\pi$ $[a < b, \quad 0 < p < 1]$ LI (24)(8)

3.222

1. $\displaystyle\int_0^1 \frac{x^{\mu-1}\, dx}{1+x} = \beta(\mu)$ $[\operatorname{Re}\mu > 0]$ WH

2. $\displaystyle\int_0^\infty \frac{x^{\mu-1}\, dx}{x+a} = \pi \operatorname{cosec}(\mu\pi) a^{\mu-1}$ for $a > 0$ FI II 718, FI II 737

   $\displaystyle \qquad\qquad\qquad\quad = -\pi \cot(\mu\pi)(-a)^{\mu-1}$ for $a < 0$ BI(18)(2), ET II 249(28)

$[0 < \operatorname{Re}\mu < 1]$

3.223

1. $\displaystyle\int_0^\infty \frac{x^{\mu-1}\, dx}{(\beta+x)(\gamma+x)} = \frac{\pi}{\gamma-\beta} \left(\beta^{\mu-1} - \gamma^{\mu-1} \right) \operatorname{cosec}(\mu\pi)$

$[|\arg\beta| < \pi, \quad |\arg\gamma| < \pi, \quad 0 < \operatorname{Re}\mu < 2]$ ET I 309(7)

2. $\displaystyle\int_0^\infty \frac{x^{\mu-1}\, dx}{(\beta+x)(\alpha-x)} = \frac{\pi}{\alpha+\beta} \left[\beta^{\mu-1} \operatorname{cosec}(\mu\pi) + \alpha^{\mu-1} \cot(\mu\pi) \right]$

$[|\arg\beta| < \pi, \quad \alpha > 0, \quad 0 < \operatorname{Re}\mu < 2]$
ET I 309(8)

3. $\displaystyle\int_0^\infty \frac{x^{\mu-1}\, dx}{(a-x)(b-x)} = \pi \cot(\mu\pi)\frac{a^{\mu-1} - b^{\mu-1}}{b-a}$ $[a > b > 0, \quad 0 < \operatorname{Re}\mu < 2]$ ET I 309(9)

3.224 $\displaystyle\int_0^\infty \frac{(x+\beta)x^{\mu-1}\, dx}{(x+\gamma)(x+\delta)} = \pi \operatorname{cosec}(\mu\pi) \left\{ \frac{\gamma-\beta}{\gamma-\delta}\gamma^{\mu-1} + \frac{\delta-\beta}{\delta-\gamma}\delta^{\mu-1} \right\}$

$[|\arg\gamma| < \pi, \quad |\arg\delta| < \pi, \quad 0 < \operatorname{Re}\mu < 1]$ ET I 309(10)

3.225

1. $\displaystyle\int_1^\infty \frac{(x-1)^{p-1}}{x^2}\,dx = (1-p)\pi\csc p\pi$ $\qquad\qquad [-1 < p < 1]$ $\qquad\qquad$ BI (23)(8)

2. $\displaystyle\int_1^\infty \frac{(x-1)^{1-p}}{x^3}\,dx = \frac{1}{2}p(1-p)\pi\csc p\pi$ $\qquad [0 < p < 1]$ $\qquad\qquad$ BI (23)(1)

3. $\displaystyle\int_0^\infty \frac{x^p\,dx}{(1+x)^3} = \frac{\pi}{2}p(1-p)\csc p\pi$ $\qquad\qquad [-1 < p < 2]$ $\qquad\qquad$ BI (16)(5)

3.226

1. $\displaystyle\int_0^1 \frac{x^n\,dx}{\sqrt{1-x}} = 2\frac{(2n)!!}{(2n+1)!!}$ $\qquad\qquad$ BI (8)(1)

2. $\displaystyle\int_0^1 \frac{x^{n-\frac{1}{2}}\,dx}{\sqrt{1-x}} = \frac{(2n-1)!!}{(2n)!!}\pi.$ $\qquad\qquad$ BI (8)(2)

3.227

1. $\displaystyle\int_0^\infty \frac{x^{\nu-1}(\beta+x)^{1-\mu}}{\gamma+x}\,dx = \beta^{1-\mu}\gamma^{\nu-1}\,\mathrm{B}(\nu,\mu-\nu)\,{}_2F_1\left(\mu-1,\nu;\mu;1-\frac{\gamma}{\beta}\right)$

$\qquad\qquad [|\arg\beta| < \pi, \quad |\arg\gamma| < \pi, \quad 0 < \mathrm{Re}\,\nu < \mathrm{Re}\,\mu]$ $\quad$ ET II 217(9)

2. $\displaystyle\int_0^\infty \frac{x^{-\varrho}(\beta-x)^{-\sigma}}{\gamma+x}\,dx = \pi\gamma^{-\varrho}(\beta-\gamma)^{-\sigma}\csc(\varrho\pi)\,I_{1-\gamma/\beta}(\sigma,\varrho)$

$\qquad\qquad [|\arg\beta| < \pi, \quad |\arg\gamma| < \pi, \quad -\mathrm{Re}\,\sigma < \mathrm{Re}\,\varrho < 1]$ $\quad$ ET II 217(10)

3.228

1. $\displaystyle\int_a^b \frac{(x-a)^\nu(b-x)^{-\nu}}{x-c}\,dx = \pi\csc(\nu\pi)\left[1 - \left(\frac{a-c}{b-c}\right)^\nu\right]$ $\qquad$ for $c < a$

$\qquad\qquad = \pi\csc(\nu\pi)\left[1 - \cos(\nu\pi)\left(\frac{c-a}{b-c}\right)^\nu\right]$ $\quad$ for $a < c < b$

$\qquad\qquad = \pi\csc(\nu\pi)\left[1 - \left(\frac{c-a}{c-b}\right)^\nu\right]$ $\qquad$ for $c > b$

$\qquad\qquad\qquad\qquad [|\mathrm{Re}\,\nu| < 1]$ $\qquad\qquad\qquad$ ET II 250(31)

2. $\displaystyle\int_a^b \frac{(x-a)^{\nu-1}(b-x)^{-\nu}}{x-c}\,dx = \frac{\pi\csc(\nu\pi)}{b-c}\left|\frac{a-c}{b-c}\right|^{\nu-1}$ $\quad$ for $c < a$ or $c > b$;

$\qquad\qquad = -\frac{\pi(c-a)^{\nu-1}}{(b-c)^\nu}\cot(\nu\pi)$ $\quad$ for $a < c < b$

$\qquad\qquad\qquad [0 < \mathrm{Re}\,\nu < 1]$ $\qquad\qquad\qquad$ ET II 250(32)

3. $\displaystyle\int_a^b \frac{(x-a)^{\nu-1}(b-x)^{\mu-1}}{x-c}\,dx$

$\qquad = \frac{(b-a)^{\mu+\nu-1}}{b-c}\,\mathrm{B}(\mu,\nu)\,{}_2F_1\left(1,\mu;\mu+\nu;\frac{b-a}{b-c}\right)$ $\quad$ for $c < a$ or $c > b$;

$\qquad = \pi(c-a)^{\nu-1}(b-c)^{\mu-1}\cot\mu\pi - (b-a)^{\mu+\nu-2}\,\mathrm{B}(\mu-1,\nu)$

$\qquad\qquad \times\,{}_2F_1\left(2-\mu-\nu,1;2-\mu;\frac{b-c}{b-a}\right)$ $\qquad\qquad$ for $a < c < b$

$$[\operatorname{Re}\mu > 0, \quad \operatorname{Re}\nu > 0, \quad \mu + \nu \neq 1, \quad \mu \neq 1, 2, \ldots] \quad \text{ET II 250(33)}$$

4. $\displaystyle\int_0^1 \frac{(1-x)^{\nu-1} x^{-\nu}}{a - bx}\, dx = \frac{\pi (a-b)^{\nu-1}}{a^\nu}\operatorname{cosec}(\nu\pi)$ $\qquad [0 < \operatorname{Re}\nu < 1, \quad 0 < b < a]$ $\qquad$ BI (5)(8)

5. $\displaystyle\int_0^\infty \frac{x^{\nu-1}(x+a)^{1-\mu}}{x-c}\, dx = a^{1-\mu}(-c)^{\nu-1}\,\mathrm{B}(\mu-\nu,\nu)\,{}_2F_1\left(\mu-1,\nu;\mu;1+\frac{c}{a}\right)$ $\qquad$ for $c < 0$;

$$= \pi c^{\nu-1}(a+c)^{1-\mu}\cot[(\mu-\nu)\pi] - \frac{a^{1-\mu-\nu}}{a+c}\,\mathrm{B}(\mu-\nu-1,\nu)$$

$$\times\ {}_2F_1\left(2-\mu,1;2-\mu+\nu;\frac{a}{a+c}\right) \qquad \text{for } c > 0$$

$$[a > 0, \quad 0 < \operatorname{Re}\nu < \operatorname{Re}\mu] \quad \text{ET II 251(34)}$$

6.* $\displaystyle\int_0^\infty x^{\nu-1}\frac{(\gamma+x)^{-n}}{x+\beta}\, dx = \frac{\pi}{\sin\pi\nu}\frac{\beta^{\nu-1}}{(\gamma-\beta)^n}\left[1 - \left(\frac{\gamma}{\beta}\right)^{\nu-1}\sum_{j=0}^{n-1}\frac{(1-\nu)_j}{j!}\left(\frac{\gamma-\beta}{\gamma}\right)^j\right]$

$$[|\arg\beta| < \pi, \quad |\arg\gamma| < \pi, \quad 0 < \operatorname{Re}\nu < n] \quad \text{AS 256 (6.1.22)}$$

3.229 $\displaystyle\int_0^1 \frac{x^{\mu-1}\, dx}{(1-x)^\mu(1+ax)(1+bx)} = \frac{\pi\operatorname{cosec}\mu\pi}{a-b}\left[\frac{a}{(1+a)^\mu} - \frac{b}{(1+b)^\mu}\right]$

$$[0 < \operatorname{Re}\mu < 1] \qquad \text{BI (5)(7)}$$

3.231

1. $\displaystyle\int_0^1 \frac{x^{p-1} - x^{-p}}{1-x}\, dx = \pi\cot p\pi$ $\qquad [p^2 < 1]$ $\qquad$ BI (4)(4)

2. $\displaystyle\int_0^1 \frac{x^{p-1} - x^{-p}}{1+x}\, dx = \pi\operatorname{cosec} p\pi$ $\qquad [p^2 < 1]$ $\qquad$ BI (4)(1)

3. $\displaystyle\int_0^1 \frac{x^p - x^{-p}}{x-1}\, dx = \frac{1}{p} - \pi\cot p\pi$ $\qquad [p^2 < 1]$ $\qquad$ BI (4)(3)

4. $\displaystyle\int_0^1 \frac{x^p - x^{-p}}{1+x}\, dx = \frac{1}{p} - \pi\operatorname{cosec} p\pi$ $\qquad [p^2 < 1]$ $\qquad$ BI (4)(2)

5. $\displaystyle\int_0^1 \frac{x^{\mu-1} - x^{\nu-1}}{1-x}\, dx = \psi(\nu) - \psi(\mu)$ $\qquad [\operatorname{Re}\mu > 0, \quad \operatorname{Re}\nu > 0]$

$$\text{FI II 815, BI(4)(5)}$$

6. $\displaystyle\int_0^\infty \frac{x^{p-1} - x^{q-1}}{1-x}\, dx = \pi\,(\cot p\pi - \cot q\pi)$ $\qquad [p > 0, \quad q > 0]$ $\qquad$ FI II 718

3.232 $\displaystyle\int_0^\infty \frac{(c+ax)^{-\mu} - (c+bx)^{-\mu}}{x}\, dx = c^{-\mu}\ln\frac{b}{a}$ $\qquad [\operatorname{Re}\mu > -1; \quad a > 0; \quad b > 0; \quad c > 0]$

$$\text{BI (18)(14)}$$

3.233 $\displaystyle\int_0^\infty \left\{\frac{1}{1+x} - (1+x)^{-\nu}\right\}\frac{dx}{x} = \psi(\nu) + \boldsymbol{C}$ $\qquad [\operatorname{Re}\nu > 0]$ $\qquad$ EH I 17, WH

3.234

1. $\displaystyle\int_0^1 \left(\frac{x^{q-1}}{1-ax} - \frac{x^{-q}}{a-x}\right) dx = \pi a^{-q}\cot q\pi$ $\qquad [0 < q < 1, \quad a > 0]$ $\qquad$ BI (55)(11)

2. $\displaystyle\int_0^1 \left(\frac{x^{q-1}}{1+ax} + \frac{x^{-q}}{a+x} \right) dx = \pi a^{-q} \operatorname{cosec} q\pi$ $[0 < q < 1, \quad a > 0]$ BI (5)(10)

3.235 $\displaystyle\int_0^\infty \frac{(1+x)^\mu - 1}{(1+x)^\nu} \frac{dx}{x} = \psi(\nu) - \psi(\nu - \mu)$ $[\operatorname{Re}\nu > \operatorname{Re}\mu > 0]$ BI (18)(5)

3.236^{10} $\displaystyle\int_0^1 \frac{x^{\frac{\mu}{2}}\, dx}{[(1-x)(1-a^2x)]^{\frac{\mu+1}{2}}} = \frac{(1-a)^{-\mu} - (1+a)^{-\mu}}{2a\mu\sqrt{\pi}} \Gamma\left(1 + \frac{\mu}{2}\right) \Gamma\left(\frac{1-\mu}{2}\right)$

$[-2 < \mu < 1, \quad |a| < 1]$ BI (12)(32)

3.237 $\displaystyle\sum_{n=0}^\infty (-1)^{n+1} \int_n^{n+1} \frac{dx}{x+u} = \ln \frac{u\left[\Gamma\left(\frac{u}{2}\right)\right]^2}{2\left[\Gamma\left(\frac{u+1}{2}\right)\right]^2}$ $[|\arg u| < \pi]$ ET II 216(1)

3.238

1. $\displaystyle\int_{-\infty}^\infty \frac{|x|^{\nu-1}}{x-u}\, dx = -\pi \cot \frac{\nu\pi}{2} |u|^{\nu-1} \operatorname{sign} u$ $[0 < \operatorname{Re}\nu < 1 \quad u \text{ real}, \quad u \neq 0]$

ET II 249(29)

2. $\displaystyle\int_{-\infty}^\infty \frac{|x|^{\nu-1}}{x-u} \operatorname{sign} x\, dx = \pi \tan \frac{\nu\pi}{2} |u|^{\nu-1}$ $[0 < \operatorname{Re}\nu < 1 \quad u \text{ real}, \quad u \neq 0]$

ET II 249(30)

3. $\displaystyle\int_a^b \frac{(b-x)^{\mu-1}(x-a)^{\nu-1}}{|x-u|^{\mu+\nu}}\, dx = \frac{(b-a)^{\mu+\nu-1}}{|a-u|^\mu |b-u|^\nu} \frac{\Gamma(\mu)\,\Gamma(\nu)}{\Gamma(\mu+\nu)}$

$[\operatorname{Re}\mu > 0, \quad \operatorname{Re}\nu > 0, \quad 0 < u < a < b \text{ and } 0 < a < b < u]$ MO 7

3.24–3.27 Powers of x, of binomials of the form $\alpha + \beta x^p$ and of polynomials in x

3.241

1. $\displaystyle\int_0^1 \frac{x^{\mu-1}\, dx}{1+x^p} = \frac{1}{p} \beta\left(\frac{\mu}{p}\right)$ $[\operatorname{Re}\mu > 0, \quad p > 0]$ WH, BI (2)(13)

2. $\displaystyle\int_0^\infty \frac{x^{\mu-1}\, dx}{1+x^\nu} = \frac{\pi}{\nu} \operatorname{cosec} \frac{\mu\pi}{\nu} = \frac{1}{\nu} \operatorname{B}\left(\frac{\mu}{\nu}, \frac{\nu-\mu}{\nu}\right)$ $[\operatorname{Re}\nu > \operatorname{Re}\mu > 0]$

ET I 309(15)a, BI (17)(10)

3. $\displaystyle\int_0^\infty \frac{x^{p-1}\, dx}{1-x^q} = \frac{\pi}{q} \cot \frac{p\pi}{q}$ $[p < q]$ BI (17)(11)

4. $\displaystyle\int_0^\infty \frac{x^{\mu-1}\, dx}{(p+qx^\nu)^{n+1}} = \frac{1}{\nu p^{n+1}} \left(\frac{p}{q}\right)^{\frac{\mu}{\nu}} \frac{\Gamma\left(\frac{\mu}{\nu}\right)\Gamma\left(1+n-\frac{\mu}{\nu}\right)}{\Gamma(1+n)}$

$\left[0 < \frac{\mu}{\nu} < n+1, \quad p \neq 0, \quad q \neq 0\right]$ BI (17)(22)a

5. $\displaystyle\int_0^\infty \frac{x^{p-1}\, dx}{(1+x^q)^2} = \frac{(p-q)\pi}{q^2} \operatorname{cosec} \frac{(p-q)\pi}{q}$ $[p < 2q]$ BI (17)(18)

$6.^{10}$ $\quad G(x) = \int_a^b \text{sign}\left[\dfrac{x}{c} - \left(\dfrac{b-u}{b-a}\right)^p\right] du = (b-a)F\left[\left(\dfrac{x}{c}\right)^{1/p}\right]$

where

$$F(x) = \int_0^1 \text{sign}(x-t)\,dt = \begin{cases} -1 & x \le 0 \\ 2x-1 & 0 < x < 1 \\ 1 & x \ge 1 \end{cases}$$

3.242

1. $\quad \displaystyle\int_{-\infty}^{\infty} \frac{x^{2m}\,dx}{x^{4n} + 2x^{2n}\cos t + 1} = \frac{\pi}{n}\sin\left[\frac{(2n-2m-1)}{2n}t\right]\text{cosec}\,t\,\text{cosec}\,\frac{(2m+1)\pi}{2n}$

$$[m < n, \quad t^2 < \pi^2] \qquad \text{FI II 642}$$

$2.^8$ $\quad \displaystyle\int_0^{\infty}\left[\frac{x^2}{x^4 + 2ax^2 + 1}\right]^c \left(\frac{x^2+1}{x^6+1}\right)\frac{dx}{x^2} = 2^{-1/2-c}(1+a)^{1/2-c}\,\text{B}\left(c - \frac{1}{2}, \frac{1}{2}\right)$

3.243 $\quad \displaystyle\int_0^{\infty} \frac{x^{\mu-1}\,dx}{(1+x^{2\nu})(1+x^{3\nu})} = -\frac{\pi}{8\nu}\frac{\text{cosec}\left(\frac{\mu\pi}{3\nu}\right)}{1 - 4\cos^2\left(\frac{\mu\pi}{3\nu}\right)}$ $\qquad [0 < \text{Re}\,\mu < 5\,\text{Re}\,\nu] \qquad$ ET I 312(34)

3.244

1. $\quad \displaystyle\int_0^1 \frac{x^{p-1} + x^{q-p-1}}{1+x^q}\,dx = \frac{\pi}{q}\text{cosec}\,\frac{p\pi}{q}$ $\qquad [q > p > 0] \qquad$ BI (2)(14)

2. $\quad \displaystyle\int_0^1 \frac{x^{p-1} - x^{q-p-1}}{1-x^q}\,dx = \frac{\pi}{q}\cot\frac{p\pi}{q}$ $\qquad [q > p > 0] \qquad$ BI (2)(16)

3. $\quad \displaystyle\int_0^1 \frac{x^{\nu-1} - x^{\mu-1}}{1-x^{\nu}}\,dx = \frac{1}{\nu}\left[\boldsymbol{C} + \psi\left(\frac{\mu}{\nu}\right)\right]$ $\qquad [\text{Re}\,\mu > \text{Re}\,\nu > 0] \qquad$ BI (2)(17)

4. $\quad \displaystyle\int_{-\infty}^{\infty} \frac{x^{2m} - x^{2n}}{1 - x^{2l}}\,dx = \frac{\pi}{l}\left[\cot\left(\frac{2m+1}{2l}\pi\right) - \cot\left(\frac{2n+1}{2l}\pi\right)\right]$

$$[m < l, \quad n < l] \qquad \text{FI II 640}$$

3.245 $\quad \displaystyle\int_0^{\infty}\left[x^{\nu-\mu} - x^{\nu}(1+x)^{-\mu}\right]dx = \frac{\nu}{\nu - \mu + 1}\text{B}(\nu, \mu - \nu)$

$$[\text{Re}\,\mu > \text{Re}\,\nu > 0] \qquad \text{BI (16)(13)}$$

3.246 $\quad \displaystyle\int_0^{\infty} \frac{1 - x^q}{1 - x^r}x^{p-1}\,dx = \frac{\pi}{r}\sin\frac{q\pi}{r}\,\text{cosec}\,\frac{p\pi}{r}\,\text{cosec}\,\frac{(p+q)\pi}{r}$

$$[p+q < r, \quad p > 0]$$
$$\text{ET I 331(33), BI (17)(12)}$$

Integrals of the form $\displaystyle\int f\left(x^p \pm x^{-p}, x^q \pm x^{-q}, \ldots\right)\frac{dx}{x}$ can be transformed by the substitution $x = e^t$

or $x = e^{-t}$. For example, instead of $\displaystyle\int_0^1 \left(x^{1+p} + x^{1-p}\right)^{-1}dx$, we should seek to evaluate $\displaystyle\int_0^{\infty} \text{sech}\,px\,dx$

and, instead of $\displaystyle\int_0^1 \frac{x^{n-m-1} + x^{n+m-1}}{1 + 2x^n\cos a + x^{2n}}\,dx$, we should seek to evaluate $\displaystyle\int_0^{\infty} \cosh mx(\cosh nx - \cos a)^{-1}\,dx$

(see **3.514** 2).

3.247

1. $$\int_0^1 \frac{x^{\alpha-1}(1-x)^{n-1}}{1-\xi x^b}\, dx = (n-1)! \sum_{k=0}^{\infty} \frac{\xi^k}{(\alpha+kb),(\alpha+kb+1)\ldots(\alpha+kb+k-1)}$$

$$[b>0, \quad |\xi|<1] \qquad\qquad \text{AD (6704)}$$

2. $$\int_0^{\infty} \frac{(1-x^p)\,x^{\nu-1}}{1-x^{np}}\, dx = \frac{\pi}{np}\sin\left(\frac{\pi}{n}\right)\operatorname{cosec}\frac{(p+\nu)\pi}{np}\operatorname{cosec}\frac{\pi\nu}{np}$$

$$[0<\operatorname{Re}\nu<(n-1)p] \qquad\qquad \text{ET I 311(33)}$$

3.248

1. $$\int_0^{\infty} \frac{x^{\mu-1}\,dx}{\sqrt{1+x^{\nu}}} = \frac{1}{\nu}\operatorname{B}\left(\frac{\mu}{\nu},\frac{1}{2}-\frac{\mu}{\nu}\right) \qquad\qquad [\operatorname{Re}\nu>\operatorname{Re}2\mu>0] \qquad\qquad \text{BI (21)(9)}$$

2. $$\int_0^1 \frac{x^{2n+1}\,dx}{\sqrt{1-x^2}} = \frac{(2n)!!}{(2n+1)!!} \qquad\qquad \text{BI (8)(14)}$$

3. $$\int_0^1 \frac{x^{2n}\,dx}{\sqrt{1-x^2}} = \frac{(2n-1)!!}{(2n)!!}\frac{\pi}{2} \qquad\qquad \text{BI (8)(13)}$$

4.[3] $$\int_{-\infty}^{\infty} \frac{dx}{(1+x^2)\sqrt{4+3x^2}} = \frac{\pi}{3}$$

5.[9] $$\int_0^{\infty} \frac{dx}{(1+x^2)^{3/2}\left[1+\dfrac{4x^2}{3(1+x^2)^2}+\left(1+\dfrac{4x^2}{3(1+x^2)^2}\right)^{1/2}\right]^{1/2}} = \frac{\pi}{2\sqrt{6}}$$

3.249

1.[0] $$\int_0^{\infty} \frac{dx}{(x^2+a^2)^n} = \frac{(2n-3)!!}{2\cdot(2n-2)!!}\frac{\pi}{a^{2n-1}} \qquad\qquad \text{FI II 743}$$

2.[9] $$\int_0^a \left(a^2-x^2\right)^{n-\frac{1}{2}}\, dx = a^{2n}\frac{(2n-1)!!}{2(2n)!!}\pi. \qquad\qquad \text{FI II 156}$$

3. $$\int_{-1}^1 \frac{(1-x^2)^n\,dx}{(a-x)^{n+1}} = 2^{n+1}\,Q_n(a) \qquad\qquad \text{EH II 181(31)}$$

4. $$\int_0^1 \frac{x^{\mu}\,dx}{1+x^2} = \frac{1}{2}\,\beta\left(\frac{\mu+1}{2}\right) \qquad\qquad [\operatorname{Re}\mu>-1] \qquad\qquad \text{BI (2)(7)}$$

5. $$\int_0^1 \left(1-x^2\right)^{\mu-1}\, dx = 2^{2\mu-2}\operatorname{B}(\mu,\mu) = \tfrac{1}{2}\operatorname{B}\left(\tfrac{1}{2},\mu\right) \qquad\qquad [\operatorname{Re}\mu>0] \qquad\qquad \text{FI II 784}$$

6. $$\int_0^1 \left(1-\sqrt{x}\right)^{p-1}\, dx = \frac{2}{p(p+1)} \qquad\qquad [p>0] \qquad\qquad \text{BI (7)(7)}$$

7. $$\int_0^1 \left(1-x^{\mu}\right)^{-\frac{1}{\nu}}\, dx = \frac{1}{\mu}\operatorname{B}\left(\frac{1}{\mu},1-\frac{1}{\nu}\right) \qquad\qquad [\operatorname{Re}\mu>0, \quad |\nu|>1] \qquad\qquad \text{BI (2)(7)}$$

8.[9] $$\int_{-\infty}^{\infty} \left(1+\frac{x^2}{n-1}\right)^{-n/2}\, dx = \frac{\sqrt{\pi(n-1)}}{\Gamma\left(\frac{n}{2}\right)}\Gamma\left(\frac{n-1}{2}\right) \qquad\qquad [\text{integer } n>1]$$

3.251

1. $\int_0^1 x^{\mu-1}\left(1-x^\lambda\right)^{\nu-1} dx = \frac{1}{\lambda} B\left(\frac{\mu}{\lambda},\nu\right)$ $[\operatorname{Re}\mu > 0, \quad \operatorname{Re}\nu > 0, \quad \lambda > 0]$

                                                                              FI II 787

2. $\int_0^\infty x^{\mu-1}\left(1+x^2\right)^{\nu-1} dx = \frac{1}{2} B\left(\frac{\mu}{2},1-\nu-\frac{\mu}{2}\right)$ $\left[\operatorname{Re}\mu > 0, \quad \operatorname{Re}\left(\nu+\frac{1}{2}\mu\right) < 1\right]$

3. $\int_1^\infty x^{\mu-1}(x^p-1)^{\nu-1} dx = \frac{1}{p} B\left(1-\nu-\frac{\mu}{p},\nu\right)$ $[p > 0, \quad \operatorname{Re}\nu > 0, \quad \operatorname{Re}\mu < p-p\operatorname{Re}\nu]$

                                                                              ET I 311(32)

4. $\int_0^\infty \frac{x^{2m} dx}{(ax^2+c)^n} = \frac{(2m-1)!!(2n-2m-3)!!\pi}{2\cdot(2n-2)!!a^m c^{n-m-1}\sqrt{ac}}$ $[a > 0, \quad c > 0, \quad n > m+1]$

                                                                              GU (141)(8a)

5. $\int_0^\infty \frac{x^{2m+1} dx}{(ax^2+c)^n} = \frac{m!(n-m-2)!}{2(n-1)!a^{m+1}c^{n-m-1}}$ $[ac > 0, \quad n > m+1 \geq 1]$ GU (141)(8b)

6. $\int_0^\infty \frac{x^{\mu+1}}{(1+x^2)^2} dx = \frac{\mu\pi}{4\sin\frac{\mu\pi}{2}}$ $[-2 < \operatorname{Re}\mu < 2]$ WH

7. $\int_0^1 \frac{x^\mu dx}{(1+x^2)^2} = -\frac{1}{4}+\frac{\mu-1}{4}\beta\left(\frac{\mu-1}{2}\right)$ $[\operatorname{Re}\mu > 1]$ LI (3)(11)

8. $\int_0^1 x^{q+p-1}(1-x^q)^{-\frac{p}{q}} dx = \frac{p\pi}{q^2}\operatorname{cosec}\frac{p\pi}{q}$ $[q > p]$ BI (9)(22)

9. $\int_0^1 x^{\frac{q}{p}-1}(1-x^q)^{-\frac{1}{p}} dx = \frac{\pi}{q}\operatorname{cosec}\frac{\pi}{p}$ $[p > 1, \quad q > 0]$ BI (9)(23)a

10. $\int_0^1 x^{p-1}(1-x^q)^{-\frac{p}{q}} dx = \frac{\pi}{q}\operatorname{cosec}\frac{p\pi}{q}$ $[q > p > 0]$ BI (9)(20)

11. $\int_0^\infty x^{\mu-1}(1+\beta x^p)^{-\nu} dx = \frac{1}{p}\beta^{-\frac{\mu}{p}} B\left(\frac{\mu}{p},\nu-\frac{\mu}{p}\right)$

    $[|\arg\beta| < \pi, \quad p > 0, \quad 0 < \operatorname{Re}\mu < p\operatorname{Re}\nu]$ BI (17)(20), EH I 10(16)

3.252

1. $\int_0^\infty \frac{dx}{(ax^2+2bx+c)^n} = \frac{(-1)^{n-1}}{(n-1)!}\frac{\partial^{n-1}}{\partial c^{n-1}}\left[\frac{1}{\sqrt{ac-b^2}}\operatorname{arccot}\frac{b}{\sqrt{ac-b^2}}\right]$

    $[a > 0, \quad ac > b^2]$ GW (131)(4)

2. $\int_{-\infty}^\infty \frac{dx}{(ax^2+2bx+c)^n} = \frac{(2n-3)!!\pi a^{n-1}}{(2n-2)!!(ac-b^2)^{n-\frac{1}{2}}}$ $[a > 0, \quad ac > b^2]$ GW (131)(5)

3. $\int_0^\infty \frac{dx}{(ax^2+2bx+c)^{n+\frac{3}{2}}} = \frac{(-2)^n}{(2n+1)!!}\frac{\partial^n}{\partial c^n}\left\{\frac{1}{\sqrt{c}\left(\sqrt{ac}+b\right)}\right\}$

    $[a \geq 0, \quad c > 0, \quad b > -\sqrt{ac}]$

                                                                              GW (213)(4)

4. $\displaystyle \int_0^\infty \frac{x\,dx}{(ax^2 + 2bx + c)^n}$

$$= \frac{(-1)^n}{(n-1)!} \frac{\partial^{n-2}}{\partial c^{n-2}} \left\{ \frac{1}{2(ac - b^2)} - \frac{b}{2(ac - b^2)^{\frac{3}{2}}} \operatorname{arccot} \frac{b}{\sqrt{ac - b^2}} \right\} \quad \text{for } ac > b^2;$$

$$= \frac{(-1)^n}{(n-1)!} \frac{\partial^{n-2}}{\partial c^{n-2}} \left\{ \frac{1}{2(ac - b^2)} + \frac{b}{4(b^2 - ac)^{\frac{3}{2}}} \ln \frac{b + \sqrt{b^2 - ac}}{b - \sqrt{b^2 - ac}} \right\} \quad \text{for } b^2 > ac > 0;$$

$$= \frac{a^{n-2}}{2(n-1)(2n-1)b^{2n-2}} \quad \text{for } ac = b^2$$

$$[a > 0, \quad b > 0, \quad n \geq 2] \quad \text{GW (141)(5)}$$

5. $\displaystyle \int_{-\infty}^\infty \frac{x\,dx}{(ax^2 + 2bx + c)^n} = -\frac{(2n-3)!!\pi b a^{n-2}}{(2n-2)!!(ac - b^2)^{\frac{(2n-1)}{2}}} \quad [ac > b^2, \quad a > 0, \quad n \geq 2]$

$$\text{GW (141)(6)}$$

6. $\displaystyle \int_{-\infty}^\infty \frac{x^m\,dx}{(ax^2 + 2bx + c)^n} = \frac{(-1)^m \pi a^{n-m-1} b^m}{(2n-2)!!(ac - b^2)^{n-\frac{1}{2}}}$

$$\times \sum_{k=0}^{[m/2]} \binom{m}{2k}(2k-1)!!(2n - 2k - 3)!! \left(\frac{ac - b^2}{b^2} \right)^k$$

$$[ac > b^2, \quad 0 \leq m \leq 2n - 2] \quad \text{GW (141)(17)}$$

7. $\displaystyle \int_0^\infty \frac{x^n\,dx}{(ax^2 + 2bx + c)^{n+\frac{3}{2}}} = \frac{n!}{(2n+1)!!\sqrt{c}(\sqrt{ac} + b)^{n+1}}$

$$[a \geq 0, \quad c > 0, \quad b > -\sqrt{ac}]$$
$$\text{GW (213)(5a)}$$

8. $\displaystyle \int_0^\infty \frac{x^{n+1}\,dx}{(ax^2 + 2bx + c)^{n+\frac{3}{2}}} = \frac{n!}{(2n+1)!!\sqrt{a}(\sqrt{ac} + b)^{n+1}}$

$$[a > 0, \quad c \geq 0, \quad b > -\sqrt{ac}]$$
$$\text{GW (213)(5b)}$$

9. $\displaystyle \int_0^\infty \frac{x^{n+\frac{1}{2}}\,dx}{(ax^2 + 2bx + c)^{n+1}} = \frac{(2n-1)!!\pi}{2^{2n+\frac{1}{2}}(b + \sqrt{ac})^{n+\frac{1}{2}} n! \sqrt{a}} \quad [a > 0, \quad c > 0, \quad b + \sqrt{ac} > 0]$

$$\text{LI (21)(19)}$$

10.[6] $\displaystyle \int_0^\infty \frac{x^{\mu-1}\,dx}{(1 + 2x \cos t + x^2)^\nu} = 2^{\nu-\frac{1}{2}}(\sin t)^{\frac{1}{2}-\nu} t\, \Gamma\left(\nu + \frac{1}{2}\right) \mathrm{B}(\mu, 2\nu - \mu)\, P_{\mu-\nu-\frac{1}{2}}^{\frac{1}{2}-\nu}(\cos t)$

$$[0 < t < \pi, \quad 0 < \operatorname{Re}\mu < \operatorname{Re} 2\nu]$$
$$\text{ET I 310(22)}$$

11. $\displaystyle \int_0^\infty (1 + 2\beta x + x^2)^{\mu-\frac{1}{2}} x^{-\nu-1}\,dx = 2^{-\mu}(\beta^2 - 1)^{\frac{\mu}{2}} \Gamma(1 - \mu)\, \mathrm{B}(\nu - 2\mu + 1, -\nu)\, P_{\nu-\mu}^\mu(\beta)$

$$[\operatorname{Re}\nu < 0, \quad \operatorname{Re}(2\mu - \nu) < 1, \quad |\arg(\beta \pm 1)| < \pi]$$
$$\text{EH I 160(33)}$$

$$= -\pi \operatorname{cosec} \nu\pi\, C_\nu^{\frac{1}{2}-\mu}(\beta)$$

$$\left[-2 < \operatorname{Re}\left(\tfrac{1}{2} - \mu\right) < \operatorname{Re}\nu < 0, \quad |\arg(\beta \pm 1)| < \pi \right]$$
$$\text{EH I 178(24)}$$

12. $\displaystyle\int_0^\infty \frac{x^{\mu-1}\,dx}{x^2+2ax\cos t+a^2} = -\pi a^{\mu-2}\operatorname{cosec} t\,\operatorname{cosec}(\mu\pi)\sin[(\mu-1)t]$

$$[a>0,\quad 0<|t|<\pi,\quad 0<\operatorname{Re}\mu<2]$$

FI II 738, BI(20)(3)

13. $\displaystyle\int_0^\infty \frac{x^{\mu-1}\,dx}{(x^2+2ax\cos t+a^2)^2} = \frac{\pi a^{\mu-4}}{2}\operatorname{cosec}\mu\pi\operatorname{cosec}^3 t$

$$\times\{(\mu-1)\sin t\cos[(\mu-2)t]-\sin[(\mu-1)t]\}$$

$$[a>0,\quad 0<|t|<\pi,\quad 0<\operatorname{Re}\mu<4]\quad \text{LI(20)(8)a, ET I 309(13)}$$

14. $\displaystyle\int_0^\infty \frac{x^{\mu-1}\,dx}{\sqrt{1+2x\cos t+x^2}} = \pi\operatorname{cosec}(\mu\pi)\,P_{\mu-1}(\cos t)\qquad [-\pi<t<\pi,\quad 0<\operatorname{Re}\mu<1]$

ET I 310(17)

3.253 $\displaystyle\int_{-1}^1 \frac{(1+x)^{2\mu-1}(1-x)^{2\nu-1}}{(1+x^2)^{\mu+\nu}}\,dx = 2^{\mu+\nu-2}\,\mathrm{B}(\mu,\nu)\qquad [\operatorname{Re}\mu>0,\quad \operatorname{Re}\nu>0]\qquad\text{FI II 787}$

3.254

1. $\displaystyle\int_0^u x^{\lambda-1}(u-x)^{\mu-1}(x^2+\beta^2)^\nu\,dx$

$$= \beta^{2\nu}u^{\lambda+\mu-1}\,\mathrm{B}(\lambda,\mu)\ _3F_2\left(-\nu,\frac{\lambda}{2},\frac{\lambda+1}{2};\frac{\lambda+\mu}{2},\frac{\lambda+\mu+1}{2};\frac{-u^2}{\beta^2}\right)$$

$$\left[\operatorname{Re}\left(\frac{u}{\beta}\right)>0,\quad \lambda>0,\quad \operatorname{Re}\mu>0\right]\quad\text{ET II 186(10)}$$

2.[6] $\displaystyle\int_u^\infty \left(x^{-\lambda}(x-u)^{\mu-1}\left(x^2+\beta^2\right)\right)^\nu\,dx$

$$= u^{\mu-\lambda+2\nu}\frac{\Gamma(\mu)\,\Gamma(\lambda-\mu-2\nu)}{\Gamma(\lambda-2\nu)}$$

$$\times\ _3F_2\left(-\nu,\frac{\lambda-\mu}{2}-\nu,\frac{1+\lambda-\mu}{2}-\nu;\frac{\lambda}{2}-\nu,\frac{1+\lambda}{2}-\nu;-\frac{\beta^2}{u^2}\right)$$

$$\left[|u|>|\beta|\ \text{and}\ \operatorname{Re}\left(\frac{\beta}{u}\right)>0,\quad 0<\operatorname{Re}\mu<\operatorname{Re}(\lambda-2\nu)\right]\quad\text{ET II 202(9)}$$

3.255 $\displaystyle\int_0^1 \frac{x^{\mu+\frac12}(1-x)^{\mu-\frac12}}{(c+2bx-ax^2)^{\mu+1}}\,dx = \frac{\sqrt{\pi}}{\left\{a+\left(\sqrt{c+2b-a}+\sqrt{c}\right)^2\right\}^{\mu+\frac12}\sqrt{c+2b-a}}\frac{\Gamma\left(\mu+\frac12\right)}{\Gamma(\mu+1)}$

$$\left[a+\left(\sqrt{c+2b-a}+\sqrt{c}\right)^2>0,\quad c+2b-a>0,\quad \operatorname{Re}\mu>-\frac12\right]$$

BI (14)(2)

3.256

1. $\displaystyle\int_0^1 \frac{x^{p-1}+x^{q-1}}{(1-x^2)^{\frac{p+q}{2}}}\,dx = \frac12\cos\left(\frac{q-p}{4}\pi\right)\sec\left(\frac{q+p}{4}\pi\right)\mathrm{B}\left(\frac{p}{2},\frac{q}{2}\right)$

$$[p>0,\quad q>0,\quad p+q<2]\qquad\text{BI (8)(25)}$$

2. $\quad \displaystyle\int_0^1 \frac{x^{p-1} - x^{q-1}}{(1-x^2)^{\frac{p+q}{2}}} \, dx = \frac{1}{2} \sin\left(\frac{q-p}{4}\pi\right) \operatorname{cosec}\left(\frac{q+p}{4}\pi\right) \mathrm{B}\left(\frac{p}{2}, \frac{q}{2}\right)$

$$[p > 0, \quad q > 0, \quad p + q < 2] \qquad \text{BI (8)(26)}$$

3.257^9 $\quad \displaystyle\int_0^\infty \left[\left(ax + \frac{b}{x}\right)^2 + c\right]^{-p-1} dx$

$$= \frac{\sqrt{\pi}\,\Gamma\left(p + \frac{1}{2}\right)}{2ac^{p+\frac{1}{2}}\,\Gamma(p+1)} \qquad \left[a > 0, \quad b < 0, \quad c > 0, \quad p > -\tfrac{1}{2}\right] \qquad \text{BI (20)(4)}$$

$$= \frac{1}{2} \frac{\mathrm{B}\left(p + \frac{1}{2}, \frac{1}{2}\right)}{a(4ab + x)^{p+\frac{1}{2}}} \qquad \left[a > 0, \quad b > 0, \quad c > -4ab, \quad p > -\tfrac{1}{2}\right]$$

3.258

1. $\quad \displaystyle\int_b^\infty \left(x - \sqrt{x^2 - a^2}\right)^n dx = \frac{a^2}{2(n-1)}\left(b - \sqrt{b^2 - a^2}\right)^{n-1} - \frac{1}{2(n+1)}\left(b - \sqrt{b^2 - a^2}\right)^{n+1}$

$$[0 < a \le b, \quad n \ge 2] \qquad \text{GW (215)(5)}$$

2. $\quad \displaystyle\int_b^\infty \left(\sqrt{x^2+1} - x\right)^n dx = \frac{\left(\sqrt{b^2+1} - b\right)^{n-1}}{2(n-1)} + \frac{\left(\sqrt{b^2+1} - b\right)^{n+1}}{2(n+1)}$

$$[n \ge 2] \qquad \text{GW (214)(7)}$$

3. $\quad \displaystyle\int_0^\infty \left(\sqrt{x^2+a^2} - x\right)^n dx = \frac{na^{n+1}}{n^2 - 1}$ $\qquad [n \ge 2] \qquad \text{GW (214)(6a)}$

4. $\quad \displaystyle\int_0^\infty \frac{dx}{\left(x + \sqrt{x^2+a^2}\right)^n} = \frac{n}{a^{n-1}(n^2-1)}$ $\qquad [n \ge 2] \qquad \text{GW (214)(5a)}$

5. $\quad \displaystyle\int_0^\infty x^m \left(\sqrt{x^2+a^2} - x\right)^n dx = \frac{n \cdot m!\, a^{m+n+1}}{(n-m-1)(n-m+1)\ldots(m+n+1)}$

$$[a > 0, \quad 0 \le m \le n-2] \qquad \text{GW (214)(6)}$$

6. $\quad \displaystyle\int_0^\infty \frac{x^m \, dx}{\left(x + \sqrt{x^2+a^2}\right)^n} = \frac{n \cdot m!}{(n-m-1)(n-m+1)\ldots(m+n+1)a^{n-m-1}}$

$$[a > 0, \quad 0 \le m \le n-2] \qquad \text{GW (214)(5)}$$

7. $\quad \displaystyle\int_a^\infty (x-a)^m \left(x - \sqrt{x^2-a^2}\right)^n dx = \frac{n \cdot (n-m-2)!(2m+1)!\,a^{m+n+1}}{2^m(n+m+1)!}$

$$[a > 0, \quad n \ge m+2] \qquad \text{GH (215)(6)}$$

3.259

1.^{6} $\quad \displaystyle\int_0^1 x^{p-1}(1-x)^{n-1}(1+bx^m)^l \, dx = (n-1)! \sum_{k=0}^\infty \binom{l}{k} \frac{b^k \, \Gamma(p+km)}{\Gamma(p+n+km)}$

$$[|b| < 1 \text{ unless } l = 0, 1, 2, \ldots; \quad p, n, p + ml > 0] \qquad \text{BI (1)(14)}$$

2. $\int_0^u x^{\nu-1}(u-x)^{\mu-1}(x^m+\beta^m)^\lambda\,dx$

$$= \beta^{m\lambda}u^{\mu+\nu+1}\,B\,(\mu,\nu)$$

$$\times\ _{m+1}F_m\left(-\lambda,\frac{\nu}{m},\frac{\nu+1}{m},\dots,\frac{\nu+m-1}{m};\frac{\mu+\nu}{m},\frac{\mu+\nu+1}{m},\dots,\frac{\mu+\nu+m-1}{m};\frac{-u^m}{\beta^m}\right)$$

$$\left[\operatorname{Re}\mu>0,\quad \operatorname{Re}\nu>0,\quad \left|\arg\left(\frac{u}{\beta}\right)\right|<\frac{\pi}{m}\right]\qquad \text{ET II 186(11)}$$

3. $\int_0^\infty x^{\lambda-1}(1+\alpha x^p)^{-\mu}(1+\beta x^p)^{-\nu}\,dx=\frac{1}{p}\alpha^{-\frac{\lambda}{p}}B\left(\frac{\lambda}{p},\mu+\nu-\frac{\lambda}{p}\right)\ _2F_1\left(\nu,\frac{\lambda}{p};\mu+\nu;1-\frac{\beta}{a}\right)$

$$[|\arg\alpha|<\pi,\quad |\arg\beta|<\pi,\quad p>0,\quad 0<\operatorname{Re}\lambda<2\operatorname{Re}(\mu+\nu)]\qquad \text{ET I 312(35)}$$

3.261

1. $\int_0^1\frac{(1-x\cos t)\,x^{\mu-1}\,dx}{1-2x\cos t+x^2}=\sum_{h=0}^\infty\frac{\cos kt}{\mu+k}$ $[\operatorname{Re}\mu>0,\quad t\neq 2n\pi]$ BI (6)(9)

2. $\int_0^1\frac{(x^\nu+x^{-\nu})\,dx}{1+2x\cos t+x^2}=\frac{\pi\sin\nu t}{\sin t\sin\nu\pi}$ $[\nu^2<1,\quad t\neq(2n+1)\pi]$ BI (6)(8)

3. $\int_0^1\frac{(x^{1+p}+x^{1-p})\,dx}{(1+2x\cos t+x^2)^2}=\frac{\pi\,(p\sin t\cos pt-\cos t\sin pt)}{2\sin^3 t\sin p\pi}$

$$[p^2<1,\quad t\neq(2n+1)\pi]\qquad \text{BI (6)(18)}$$

4. $\int_0^1\frac{x^{\mu-1}}{1+2ax\cos t+a^2x^2}\cdot\frac{dx}{(1-x)^\mu}=\frac{\pi\operatorname{cosec} t\operatorname{cosec}\mu\pi}{(1+2a\cos t+a^2)^{\frac{\mu}{2}}}\sin\left(t-\mu\arctan\frac{a\sin t}{1+a\cos t}\right)$

$$[a>0,\quad 0<\operatorname{Re}\mu<1]\qquad \text{BI (6)(21)}$$

3.262 $\int_0^\infty\frac{x^{-p}\,dx}{1+x^3}=\frac{\pi}{3}\operatorname{cosec}\frac{(1-p)\pi}{3}$ $[-2<p<1]$ LI (18)(3)

3.263 $\int_0^\infty\frac{x^\nu\,dx}{(x+\gamma)\,(x^2+\beta^2)}=\frac{\pi}{2\,(\beta^2+\gamma^2)}\left[\gamma\beta^{\nu-1}\sec\frac{\nu\pi}{2}+\beta^\nu\operatorname{cosec}\frac{\nu\pi}{2}-2\gamma^\nu\operatorname{cosec}(\nu\pi)\right]$

$$[\operatorname{Re}\beta>0,\quad |\arg\gamma|<\pi,\quad -1<\operatorname{Re}\nu<2,\quad \nu\neq 0]\qquad \text{ET II 216(7)}$$

3.264

1. $\int_0^\infty\frac{x^{p-1}\,dx}{(a^2+x^2)\,(b^2-x^2)}=\frac{\pi}{2}\frac{a^{p-2}+b^{p-2}\cos\frac{p\pi}{2}}{a^2+b^2}\operatorname{cosec}\frac{p\pi}{2}$

$$[0<p<4,\quad a>0,\quad b>0]$$

$$\text{BI (19)(14)}$$

2.* $\int_0^\infty\frac{x^{\mu-1}\,dx}{(\beta+x^2)\,(\gamma+x^2)}=\frac{\pi}{2}\frac{\gamma^{\frac{\mu}{2}-1}-\beta^{\frac{\mu}{2}-1}}{\beta-\gamma}\operatorname{cosec}\frac{\mu\pi}{2}$

$$=\frac{\pi}{2(\gamma-\beta)}\left(\frac{1}{\sqrt{\beta}}-\frac{1}{\sqrt{\gamma}}\right)\qquad [\mu=\tfrac{1}{2}]$$

$$[|\arg\beta|<\pi,\quad |\arg\gamma|<\pi,\quad 0<\operatorname{Re}\mu<4]\qquad \text{ET I 309(4)}$$

3.* $\int_0^\infty \dfrac{dx}{(b+x^2)\,(a+b+x^2)^2} = \dfrac{\pi}{2}\left(\dfrac{1}{a^2 b^{1/2}} - \dfrac{1}{2a(a+b)^{3/2}} - \dfrac{1}{a^2(a+b)^{1/2}}\right)$ MC

4.* $\int_0^\infty \dfrac{dx}{(b+x^2)\,(a+b+x^2)^3} = \dfrac{\pi}{4}\left(\dfrac{2}{a^3 b^{1/2}} - \dfrac{3}{4a(a+b)^{5/2}} - \dfrac{1}{a^2(a+b)^{3/2}} - \dfrac{2}{a^3(a+b)^{1/2}}\right)$

5.* $\int_0^\infty \dfrac{dx}{(b+x^2)\,(a+b+x^2)^4} = \dfrac{\pi}{4}\left(\dfrac{2}{a^4 b^{1/2}} - \dfrac{5}{8a(a+b)^{7/2}} - \dfrac{3}{4a^2(a+b)^{5/2}}\right.$

$$\left. - \dfrac{1}{a^3(a+b)^{3/2}} - \dfrac{2}{a^4(a+b)^{1/2}}\right)$$

6.* $\int_0^\infty \dfrac{dx}{(b+x^2)\,(a+b+x^2)^n}$

$$= \frac{\pi}{2}\frac{1}{a^n b^{1/2}} - \frac{1}{2a(a+b)^{n-1/2}}\,B\left(n-\frac{1}{2},\frac{1}{2}\right)\,{}_2F_1\left(1-n,1;\,\frac{3}{2}-n;\,\frac{a+b}{a}\right)$$

AS 263 (6.6.3.2)

$$= \frac{\pi}{2}\frac{1}{a^n b^{1/2}} - \frac{\pi}{2a^n(a+b)^{n-1/2}}\sum_{j=0}^{n-1}\frac{\left(\frac{1}{2}\right)_j}{j!}\left(\frac{a}{a+b}\right)^j$$

$$[n>0,\quad a+b>0]$$

7.* $\int_0^\infty \dfrac{x^2\,dx}{(x^2+\alpha^2)\,(x^2+\beta^2)\,(x^2+\gamma^2)} = \dfrac{\pi}{2\alpha\,(\beta^2-\gamma^2)}\left[\dfrac{\beta}{\beta+\alpha} - \dfrac{\gamma}{\gamma+\alpha}\right] = \dfrac{\pi}{2(\alpha+\beta)(\alpha+\gamma)(\beta+\gamma)}$

3.265 $\int_0^1 \dfrac{1-x^{\mu-1}}{1-x}\,dx = \psi(\mu) + \boldsymbol{C}$ $[\operatorname{Re}\mu>0]$ FI II 796, WH, ET I 16(13)

$$= \psi(1-\mu) + \boldsymbol{C} - \pi\cot(\mu\pi)\qquad[\operatorname{Re}\mu>0]$$ EH I 16(15)a

3.266 $\int_0^\infty \dfrac{(x^\nu - a^\nu)\,dx}{(x-a)(\beta+x)} = \dfrac{\pi}{a+\beta}\left\{\beta^\nu\operatorname{cosec}(\nu\pi) - a^\nu\cot(\nu\pi) - \dfrac{a^\nu}{\pi}\ln\dfrac{\beta}{a}\right\}$

$$[|\arg\beta|<\pi,\quad |\operatorname{Re}\nu|<1,\quad \nu\neq 0]$$
ET II 216(8)

3.267

1. $\int_0^1 \dfrac{x^{3n}\,dx}{\sqrt[3]{1-x^3}} = \dfrac{2\pi}{3\sqrt{3}}\dfrac{\Gamma\left(n+\frac{1}{3}\right)}{\Gamma\left(\frac{1}{3}\right)\Gamma(n+1)}$ BI (9)(6)

2. $\int_0^1 \dfrac{x^{3n-1}\,dx}{\sqrt[3]{1-x^3}} = \dfrac{(n-1)!\,\Gamma\left(\frac{2}{3}\right)}{3\,\Gamma\left(n+\frac{2}{3}\right)}$ BI (9)(7)

3.268

1. $\int_0^1 \left(\dfrac{1}{1-x} - \dfrac{px^{p-1}}{1-x^p}\right)dx = \ln p$ BI (5)(14)

2. $\int_0^1 \dfrac{1-x^\mu}{1-x}x^{\nu-1}\,dx = \psi(\mu+\nu) - \psi(\nu)$ $[\operatorname{Re}\nu>0,\quad \operatorname{Re}\mu>0]$ BI (2)(3)

3. $\int_0^1 \left[\frac{n}{1-x} - \frac{x^{\mu-1}}{1-\sqrt[n]{x}} \right] dx = n\,\boldsymbol{C} + \sum_{k=1}^{n} \psi\left(\mu + \frac{n-k}{n} \right)$

$[\operatorname{Re}\mu > 0]$ BI (13)(10)

3.269

1. $\int_0^1 \frac{x^p - x^{-p}}{1 - x^2}\, x\, dx = \frac{\pi}{2} \cot \frac{p\pi}{2} - \frac{1}{p}$ $[p^2 < 1]$ BI (4)(12)

2. $\int_0^1 \frac{x^p - x^{-p}}{1 + x^2}\, x\, dx = \frac{1}{p} - \frac{\pi}{2} \operatorname{cosec} \frac{p\pi}{2}$ $[p^2 < 1]$ BI (4)(8)

3. $\int_0^1 \frac{x^\mu - x^\nu}{1 - x^2}\, dx = \frac{1}{2} \psi\left(\frac{\nu+1}{2} \right) - \frac{1}{2} \psi\left(\frac{\mu+1}{2} \right)$ $[\operatorname{Re}\mu > -1, \quad \operatorname{Re}\nu > -1]$ BI (2)(9)

3.271

1. $\int_0^\infty \frac{x^p - x^q}{x - 1} \frac{dx}{x + a} = \frac{\pi}{1 + a} \left(\frac{a^p - \cos p\pi}{\sin p\pi} - \frac{a^q - \cos q\pi}{\sin q\pi} \right)$

$[p^2 < 1, \quad q^2 < 1, \quad a > 0]$ BI (19)(2)

2. $\int_0^\infty \frac{x^p - a^p}{x - a} \frac{x^p - 1}{x - 1}\, dx = \frac{\pi}{a - 1} \left\{ \frac{a^{2p} - 1}{\sin(2p\pi)} - \frac{1}{\pi} a^p \ln a \right\}$

$\left[p^2 < \frac{1}{4} \right]$ BI (19)(3)

3. $\int_0^\infty \frac{x^p - a^p}{x - a} \frac{x^{-p} - 1}{x - 1}\, dx = \frac{\pi}{a - 1} \left\{ 2 (a^p - 1) \cot p\pi - \frac{1}{\pi} (a^p + 1) \ln a \right\}$

$[p^2 < 1]$ BI (18)(9)

4. $\int_0^\infty \frac{x^p - a^p}{x - a} \frac{1 - x^{-p}}{1 - x} x^q\, dx = \frac{\pi}{a - 1} \left\{ \frac{a^{p+q} - 1}{\sin[(p+q)\pi]} + \frac{a^p - a^q}{\sin[(q-p)\pi]} \right\} \frac{\sin p\pi}{\sin q\pi}$

$\left[(p+q)^2 < 1, \quad (p-q)^2 < 1 \right]$

BI (19)(4)

5. $\int_0^\infty \left(\frac{x^p - x^{-p}}{1 - x} \right)^2 dx = 2 (1 - 2p\pi \cot 2p\pi)$ $\left[0 < p^2 < \frac{1}{4} \right]$ BI (16)(3)

3.272

1. $\int_0^1 \frac{x^{n-1} + x^{n-\frac{1}{2}} - 2x^{2n-1}}{1 - x}\, dx = 2 \ln 2$ BI (8)(8)

2. $\int_0^1 \frac{x^{n-1} + x^{n-\frac{2}{3}} + x^{n-\frac{1}{3}} - 3x^{3n-1}}{1 - x}\, dx = 3 \ln 3$ BI (8)(9)

3.273

1. $\int_0^1 \frac{\sin t - a^n x^n \sin[(n+1)t] + a^{n+1}x^{n+1}\sin nt}{1 - 2ax \cos t + a^2 x^2} (1-x)^{p-1}\, dx = \Gamma(p) \sum_{k=1}^{n} \frac{(k-1)! a^{k-1} \sin kt}{\Gamma(p+k)}$

$[p > 0]$ BI (6)(13)

2. $\int_0^1 \dfrac{\cos t - ax - a^n x^n \cos[(n+1)t] + a^{n+1} x^{n+1} \cos nt}{1 - 2ax \cos t + a^2 x^2} (1-x)^{p-1}\, dx = \Gamma(p) \sum_{k=1}^n \dfrac{(k-1)! a^{k-1} \cos kt}{\Gamma(p+k)}$

$$[p > 0]$$
BI (6)(14)

3. $\int_0^1 x \dfrac{\sin t - x^n \sin[(n+1)t] + x^{n+1} \sin nt}{1 - 2x \cos t + x^2}\, dx = \sum_{k=1}^n \dfrac{\sin kt}{k+1}$
BI (6)(12)

4. $\int_0^1 \dfrac{1 - x \cos t - x^{n+1} \cos[(n+1)t] + x^{n+2} \cos nt}{1 - 2x \cos t + x^2}\, dx = \sum_{k=0}^n \dfrac{\cos kt}{k+1}$
BI (6)(11)

3.274

1. $\int_0^\infty \dfrac{x^{\mu-1}(1-x)}{1-x^n}\, dx = \dfrac{\pi}{n} \sin \dfrac{\pi}{n} \operatorname{cosec} \dfrac{\mu\pi}{n} \operatorname{cosec} \dfrac{(\mu+1)\pi}{n}$

$$[0 < \operatorname{Re}\mu < n - 1]$$
BI (20)(13)

2. $\int_0^1 \dfrac{1 - x^n}{(1+x)^{n+1}} \dfrac{dx}{1-x} = \dfrac{1}{2^{n+1}} \sum_{k=1}^n \dfrac{2^k}{k}$
BI (5)(3)

3. $\int_0^\infty \dfrac{x^q - 1}{x^p - x^{-p}} \dfrac{dx}{x} = \dfrac{\pi}{2p} \tan \dfrac{q\pi}{2p}$ $[p > q]$
BI (18)(6)

3.275

1. $\int_0^1 \left\{ \dfrac{x^{n-1}}{1 - x^{\frac{1}{p}}} - \dfrac{p x^{np-1}}{1-x} \right\} dx = p \ln p$ $[p > 0]$
BI (13)(9)

2. $\int_0^1 \left\{ \dfrac{n x^{n-1}}{1 - x^n} - \dfrac{x^{mn-1}}{1-x} \right\} dx = C + \dfrac{1}{n} \sum_{k=1}^n \psi\left(m + \dfrac{n-k}{n}\right)$
BI (5)(13)

3. $\int_0^1 \left(\dfrac{x^{p-1}}{1-x} - \dfrac{q x^{pq-1}}{1 - x^q} \right) dx = \ln q$ $[q > 0]$
BI (5)(12)

4. $\int_0^\infty \left\{ \dfrac{1}{1 + x^{2^n}} - \dfrac{1}{1 + x^{2^m}} \right\} \dfrac{dx}{x} = 0.$
BI (18)(17)

3.276

1.10 $\int_0^\infty \dfrac{\left[\left(ax + \dfrac{b}{x} \right)^2 + c \right]^{-p-1} dx}{x^2} = \dfrac{1}{2|b|} \dfrac{B\left(p + \frac{1}{2}, \frac{1}{2}\right)}{(2a(b + |b|) + c)^{p+\frac{1}{2}}}$

$$\left[a > 0, \quad c > -4ac, \quad p > -\tfrac{1}{2} \right]$$

2.10 $\int_0^\infty \left(a + \dfrac{b}{x^2} \right) \left[\left(ax + \dfrac{b}{x} \right)^2 + c \right]^{-p-1} dx = \dfrac{B\left(p + \frac{1}{2}, \frac{1}{2}\right)}{(4ab + c)^{p+\frac{1}{2}}}$

$$\left[a > 0, \quad b > 0, \quad c > -4ac, \quad p > -\tfrac{1}{2} \right]$$

3.277

1.
$$\int_0^\infty \frac{x^{\mu-1}\left[\sqrt{1+x^2}+\beta\right]^\nu}{\sqrt{1+x^2}}\,dx = 2^{\frac{\mu}{2}-1}\left(\beta^2-1\right)^{\frac{\nu}{2}+\frac{\mu}{4}}\Gamma\left(\frac{\mu}{2}\right)\Gamma(1-\mu-\nu)\,P_{\frac{\mu}{2}-1}^{\frac{\nu+\mu}{2}}(\beta)$$
$$[\operatorname{Re}\beta>-1,\quad 0<\operatorname{Re}\mu<1-\operatorname{Re}\nu]$$
ET I 310(25)

2.
$$\int_0^\infty \frac{x^{\mu-1}\left[\sqrt{\beta^2+x^2}+x\right]^\nu}{\sqrt{\beta^2+x^2}}\,dx = \frac{\beta^{\mu+\nu-1}}{2^\mu}\,\mathrm{B}\left(\mu,\frac{1-\mu-\nu}{2}\right)$$
$$[\operatorname{Re}\beta>0,\quad 0<\operatorname{Re}\mu<1-\operatorname{Re}\nu]$$
ET I 311(28)

3.
$$\int_0^\infty \frac{x^{\mu-1}\left[\cos t\pm i\sin t\sqrt{1+x^2}\right]^\nu}{\sqrt{1+x^2}}\,dx = 2^{\frac{\mu-1}{2}}\sin^{\frac{1-\mu}{2}}t\,\frac{\Gamma\left(\frac{\mu}{2}\right)\Gamma(1-\mu-\nu)}{\Gamma(-\nu)}$$
$$\times\left[\pi^{-\frac{1}{2}}\,Q_{-\frac{\mu+1}{2}-\nu}^{\frac{\mu+1}{2}}(\cos t)\mp\frac{i}{2}\pi^{\frac{1}{2}}\,P_{\frac{\mu-1}{2}}^{-\frac{\mu+1}{2}-\nu}(\cos t)\right]$$
$$[\operatorname{Re}\mu>0]$$
ET I 311 (27)

4.
$$\int_0^\infty \frac{x^{\mu-1}\left[\sqrt{(\beta^2-1)(x^2+1)}+\beta\right]^\nu}{\sqrt{x^2+1}}\,dx$$
$$= \frac{2^{\frac{\mu-1}{2}}}{\sqrt\pi}e^{-\frac{1}{2}i\pi(\mu-1)}\frac{\Gamma\left(\frac{\mu}{2}\right)\Gamma(1-\mu-\nu)}{\Gamma(-\nu)}\left(\beta^2-1\right)^{\frac{1-\mu}{4}}Q_{-\frac{\mu+1}{2}-\nu}^{\frac{\mu-1}{2}}(\beta)$$
$$[\operatorname{Re}\beta>1,\quad \operatorname{Re}\nu<0,\quad \operatorname{Re}\mu<1-\operatorname{Re}\nu]\quad\text{ET I 311(26)}$$

5.
$$\int_u^\infty \frac{(x-u)^{\mu-1}\left(\sqrt{x+1}-\sqrt{x-1}\right)^{2\nu}}{\sqrt{x^2-1}}\,dx = \frac{2^{\nu+\frac{1}{2}}}{\sqrt\pi}e^{(\mu-\frac{1}{2})\pi i}\left(u^2-1\right)^{\frac{2\mu-1}{4}}Q_{\nu-\frac{1}{2}}^{\frac{1}{2}-\mu}(u)$$
$$[|\arg(u-1)|<\pi,\quad 0<\operatorname{Re}\mu<1+\operatorname{Re}\nu]\quad\text{ET II 202(10)}$$

6.
$$\int_1^\infty \frac{x^{\mu-1}\left[\left(x-\sqrt{x^2-1}\right)^\nu+\left(x-\sqrt{x^2-1}\right)^{-\nu}\right]}{\sqrt{x^2-1}}\,dx = 2^{-\mu}\,\mathrm{B}\left(\frac{1-\mu+\nu}{2},\frac{1-\mu-\nu}{2}\right)$$
$$[\operatorname{Re}\mu<1+\operatorname{Re}\nu]\qquad\text{ET I 311(29)}$$

7.
$$\int_0^u \frac{(u-x)^{\mu-1}\left[\left(\sqrt{x+2}+\sqrt x\right)^{2\nu}+\left(\sqrt{x+2}-\sqrt x\right)^{2\nu}\right]}{\sqrt{x(x+2)}}\,dx = 2^{\frac{2\mu+1}{2}}\sqrt{\pi[u(u+2)]^{\mu-\frac{1}{2}}}\,P_{\nu-\frac{1}{2}}^{\frac{1}{2}-\mu}(u+1)$$
$$[|\arg u|<\pi,\quad \operatorname{Re}\mu>0]\qquad\text{ET II 186(12)}$$

3.278[8]

1.
$$\int_0^\infty \left(\frac{x^p}{1+x^{2p}}\right)^q\frac{dx}{1-x^2}=0\qquad\qquad [pq>1]$$

3.3–3.4 Exponential Functions

3.31 Exponential functions

3.310
$$\int_0^\infty e^{-px}\, dx = \frac{1}{p} \qquad\qquad [\operatorname{Re} p > 0]$$

3.311

1. $$\int_0^\infty \frac{dx}{1 + e^{px}} = \frac{\ln 2}{p} \qquad\qquad \text{LO III 284a}$$

2. $$\int_0^\infty \frac{e^{-\mu x}}{1 + e^{-x}}\, dx = \beta(\mu) \qquad [\operatorname{Re}\mu > 0] \qquad \text{EH I 20(3), ET I 144(7)}$$

3. $$\int_{-\infty}^\infty \frac{e^{-px}}{1 + e^{-qx}}\, dx = \frac{\pi}{q}\operatorname{cosec}\frac{p\pi}{q}$$
$$[q > p > 0 \text{ or } 0 > p > q] \qquad (\text{cf. } \mathbf{3.241}\ 2) \quad \text{BI (28)(7)}$$

4. $$\int_0^\infty \frac{e^{-qx}\, dx}{1 - ae^{-px}} = \sum_{k=0}^\infty \frac{a^k}{q + kp} \qquad [0 < a < 1] \qquad \text{BI (27)(7)}$$

5. $$\int_0^\infty \frac{1 - e^{\nu x}}{e^x - 1}\, dx = \psi(\nu) + \boldsymbol{C} + \pi\cot(\pi\nu) \qquad [\operatorname{Re}\nu < 1] \quad (\text{cf. } \mathbf{3.265}) \quad \text{EH I 16(16)}$$

6. $$\int_0^\infty \frac{e^{-x} - e^{-\nu x}}{1 - e^{-x}}\, dx = \psi(\nu) + \boldsymbol{C} \qquad [\operatorname{Re}\nu > 0] \qquad \text{WH, EH I 16(14)}$$

7. $$\int_0^\infty \frac{e^{-\mu x} - e^{-\nu x}}{1 - e^{-x}}\, dx = \psi(\nu) - \psi(\mu) \qquad [\operatorname{Re}\mu > 0, \quad \operatorname{Re}\nu > 0] \qquad (\text{cf. } \mathbf{3.231}\ 5)$$
$$\text{BI (27)(8)}$$

8. $$\int_{-\infty}^\infty \frac{e^{-\mu x}\, dx}{b - e^{-x}} = \pi b^{\mu-1}\cot(\mu\pi) \qquad [b > 0, \quad 0 < \operatorname{Re}\mu < 1] \qquad \text{ET I 120(14)a}$$

9. $$\int_{-\infty}^\infty \frac{e^{-\mu x}\, dx}{b + e^{-x}} = \pi b^{\mu-1}\operatorname{cosec}(\mu\pi) \qquad [|\arg b| < \pi, \quad 0 < \operatorname{Re}\mu < 1]$$
$$\text{ET I 120(15)a}$$

10. $$\int_0^\infty \frac{e^{-px} - e^{-qx}}{1 + e^{-(p+q)x}}\, dx = \frac{\pi}{p+q}\cot\frac{p\pi}{p+q} \qquad [p > 0, \quad q > 0] \qquad \text{GW (311)(16c)}$$

11. $$\int_0^\infty \frac{e^{px} - e^{qx}}{e^{rx} - e^{sx}}\, dx = \frac{1}{r - s}\left[\psi\left(\frac{r-q}{r-s}\right) - \psi\left(\frac{r-p}{r-s}\right)\right]$$
$$[r > s, r > p, r > q] \qquad \text{GW (311)(16)}$$

12. $$\int_0^\infty \frac{a^x - b^x}{c^x - d^x}\, dx = \frac{1}{\ln\frac{c}{d}}\left\{\psi\left(\frac{\ln\frac{c}{b}}{\ln\frac{c}{d}}\right) - \psi\left(\frac{\ln\frac{c}{a}}{\ln\frac{c}{d}}\right)\right\} \qquad [c > a > 0, \quad b > 0, \quad d > 0]$$
$$\text{GW (311)(16a)}$$

3.312

1. $$\int_0^\infty \left(1 - e^{-\frac{x}{\beta}}\right)^{\nu-1} e^{-\mu x}\, dx = \beta\, \mathrm{B}(\beta\mu, \nu) \qquad [\operatorname{Re}\beta > 0, \quad \operatorname{Re}\nu > 0, \quad \operatorname{Re}\mu > 0]$$
$$\text{LI(25)(13), EH I 11(24)}$$

2. $\int_0^\infty \left(1 - e^{-x}\right)^{-1} \left(1 - e^{-\alpha x}\right) \left(1 - e^{-\beta x}\right) e^{-px}\, dx = \psi(p + \alpha) + \psi(p + \beta) - \psi(p + \alpha + \beta) - \psi(p)$

$$[\operatorname{Re} p > 0, \quad \operatorname{Re} p > -\operatorname{Re}\alpha, \quad \operatorname{Re} p > -\operatorname{Re}\beta, \quad \operatorname{Re} p > -\operatorname{Re}(\alpha + \beta)] \quad \text{ET I 145(15)}$$

3. $\int_0^\infty \left(1 - e^{-x}\right)^{\nu-1} \left(1 - \beta e^{-x}\right)^{-\varrho} e^{-\mu x}\, dx = \mathrm{B}(\mu, \nu)\, {}_2F_1(\varrho, u; \mu + \nu; \beta)$

$$[\operatorname{Re}\mu > 0, \quad \operatorname{Re}\nu > 0, \quad |\arg(1 - \beta)| < \pi] \quad \text{EH I 116(15)}$$

3.313

1.[7] $\mathrm{PV} \int_{-\infty}^\infty \dfrac{e^{-\mu x}\, dx}{1 - e^{-x}} = \pi \cot \pi\mu \qquad\qquad [0 < \operatorname{Re}\mu < 1]$

2.[7] $\int_{-\infty}^\infty \dfrac{e^{-\mu x}\, dx}{(1 + e^{-x})^\nu} = \mathrm{B}(\mu, \nu - \mu) \qquad\qquad [0 < \operatorname{Re}\mu < \operatorname{Re}\nu]$

3.314 $\int_{-\infty}^\infty \dfrac{e^{-\mu x}\, dx}{\left(e^{\beta/\gamma} + e^{-x/\gamma}\right)^\nu} = \gamma \exp\left[\beta\left(\mu - \dfrac{\nu}{\gamma}\right)\right] \mathrm{B}(\gamma\mu, \nu - \gamma\mu)$

$$\left[\operatorname{Re}\left(\dfrac{\nu}{\gamma}\right) > \operatorname{Re}\mu > 0, \quad |\operatorname{Im}\beta| < \pi \operatorname{Re}\gamma\right] \quad \text{ET I 120(21)}$$

3.315

1. $\int_{-\infty}^\infty \dfrac{e^{-\mu x}\, dx}{\left(e^\beta + e^{-x}\right)^\nu \left(e^\gamma + e^{-x}\right)^\varrho} = \exp[\gamma(\mu - \varrho) - \beta\nu]\, \mathrm{B}(\mu, \nu + \varrho - \mu)\, {}_2F_1\left(\nu, \mu; \nu + \varrho; 1 - e^{\nu - \beta}\right)$

$$[|\operatorname{Im}\beta| < \pi, \quad |\operatorname{Im}\gamma| < \pi, \quad 0 < \operatorname{Re}\mu < \operatorname{Re}(\nu + \varrho)] \quad \text{ET I 121(22)}$$

2. $\int_{-\infty}^\infty \dfrac{e^{-\mu x}\, dx}{\left(\beta + e^{-x}\right)\left(\gamma + e^{-x}\right)} = \dfrac{\pi\left(\beta^{\mu-1} - \gamma^{\mu-1}\right)}{\gamma - \beta} \operatorname{cosec}(\mu\pi)$

$$[|\arg\beta| < \pi, \quad |\arg\gamma| < \pi, \quad \beta \neq \gamma, \quad 0 < \operatorname{Re}\mu < 2] \quad \text{ET I 120(18)}$$

3.316 $\int_{-\infty}^\infty \dfrac{\left(1 + e^{-x}\right)^\nu - 1}{\left(1 + e^{-x}\right)^\mu}\, dx = \psi(\mu) - \psi(\mu - \nu) \qquad [\operatorname{Re}\mu > \operatorname{Re}\nu > 0] \qquad (\text{cf. } \mathbf{3.235})$

$$\text{BI (28)(8)}$$

3.317

1. $\int_{-\infty}^\infty \left\{\dfrac{1}{1 + e^{-x}} - \dfrac{1}{(1 + e^{-x})^\mu}\right\} dx = \boldsymbol{C} + \psi(\mu) \qquad [\operatorname{Re}\mu > 0] \qquad (\text{cf. } \mathbf{3.233}) \qquad \text{BI (28)(10)}$

2. $\int_{-\infty}^\infty \left\{\dfrac{1}{(1 + e^{-x})^\nu} - \dfrac{1}{(1 + e^{-x})^\mu}\right\} dx = \psi(\mu) - \psi(\nu) \qquad [\operatorname{Re}\mu > 0, \quad \operatorname{Re}\nu > 0] \qquad (\text{cf. } \mathbf{3.219})$

$$\text{BI (28)(11)}$$

3.318

1. $\int_0^\infty \dfrac{\left[\beta + \sqrt{1 - e^{-x}}\right]^{-\nu} + \left[\beta - \sqrt{1 - e^{-x}}\right]^{-\nu}}{\sqrt{1 - e^{-x}}} e^{-\mu x}\, dx$

$$= \dfrac{2^{\mu+1} e^{(\mu-\nu)\pi i}\left(\beta^2 - 1\right)^{(\mu-\nu)/2} \Gamma(\mu)\, Q_{\mu-1}^{\nu-\mu}(\beta)}{\Gamma(\nu)}$$

$$[\operatorname{Re}\mu > 0] \qquad \text{ET I 145(18)}$$

2.7 $\quad \int_u^\infty \frac{1}{\sqrt{1-e^{-2x}}} \left\{ e^{-u}\sqrt{1-e^{-2x}} - e^{-x}\sqrt{1-e^{-2u}} \right\}^\nu e^{-\mu x}\, dx$

$$= \frac{2^{-\frac{1}{2}(\mu+\nu)}\sqrt{\pi}e^{-\frac{u}{2}(\mu+\nu)}\,\Gamma(\mu)\,\Gamma(\nu+1)\,P_{-\frac{1}{2}(\mu-\nu)}^{-\frac{1}{2}(\mu+\nu)}\left(\sqrt{1-e^{-2u}}\right)}{\Gamma[(\mu+\nu+1)/2]}$$

$$[u>0, \quad \mathrm{Re}\,\mu>0, \quad \mathrm{Re}\,\nu>-1] \quad \text{ET I 145(19)}$$

3.32–3.34 Exponentials of more complicated arguments

3.321

1.7 $\quad \Phi(u) = \frac{\sqrt{\pi}}{2}\,\mathrm{erf}(u) = \int_0^u e^{-x^2}\, dx = \sum_{k=0}^\infty \frac{(-1)^k u^{2k+1}}{k!(2k+1)}$

$$= e^{-u^2}\sum_{k=0}^\infty \frac{2^k u^{2k+1}}{(2k+1)!!}$$

$$\text{(cf. 8.25)} \qquad\qquad \text{AD 6.700}$$

2. $\quad \int_0^u e^{-q^2 x^2}\, dx = \frac{\sqrt{\pi}}{2q}\,\Phi(qu) \qquad\qquad [q>0]$

3. $\quad \int_0^\infty e^{-q^2 x^2}\, dx = \frac{\sqrt{\pi}}{2q} \qquad\qquad [q>0] \qquad\qquad \text{FI II 624}$

3.322

1.7 $\quad \int_u^\infty \exp\left(-\frac{x^2}{4\beta} - \gamma x\right) dx = \sqrt{\pi\beta}e^{\beta\gamma^2}\left[1 - \Phi\left(\gamma\sqrt{\beta} + \frac{u}{2\sqrt{\beta}}\right)\right]$

$$[\mathrm{Re}\,\beta>0, \quad u>0] \qquad\qquad \text{ET I 146(21)}$$

2. $\quad \int_0^\infty \exp\left(-\frac{x^2}{4\beta} - \gamma x\right) dx = \sqrt{\pi\beta}\exp\left(\beta\gamma^2\right)\left[1 - \Phi\left(\gamma\sqrt{\beta}\right)\right]$

$$[\mathrm{Re}\,\beta>0] \qquad\qquad \text{NT 27(1)a}$$

3.* $\quad \int_0^\infty e^{-i\lambda x^2}\, dx = \frac{1}{2}\sqrt{\frac{\pi}{\lambda}}e^{-\pi i/4} \qquad\qquad [\lambda>0] \qquad\qquad \text{PBM 343 (2.3.15(2))}$

3.323

1.7 $\quad \int_1^\infty \exp\left(-qx - x^2\right) dx = \frac{\pi^{1/2}}{2}e^{q^2/4}\left[1 - \Phi\left(1 + \frac{1}{2}q\right)\right]$

$$[q \neq -2] \qquad\qquad \text{BI (29)(4)}$$

2.10 $\quad \int_{-\infty}^\infty \exp\left(-p^2 x^2 \pm qx\right) dx = \exp\left(\frac{q^2}{4p^2}\right)\frac{\sqrt{\pi}}{p} \qquad\qquad [\mathrm{Re}\,p^2>0] \qquad\qquad \text{BI (28)(1)}$

3.6 $\quad \int_0^\infty \exp\left(-\beta^2 x^4 - 2\gamma^2 x^2\right) dx = 2^{-\frac{3}{2}}\frac{\gamma}{\beta}e^{\frac{\gamma^4}{2\beta^2}}K_{\frac{1}{4}}\left(\frac{\gamma^4}{2\beta^2}\right)$

$$\left[|\arg\beta| < \frac{\pi}{4}\right] \qquad\qquad \text{ET I 147(34)a}$$

3.324

1. $$\int_0^\infty \exp\left(-\frac{\beta}{4x} - \gamma x\right) dx = \sqrt{\frac{\beta}{\gamma}} \, K_1\left(\sqrt{\beta\gamma}\right) \qquad [\operatorname{Re}\beta \ge 0, \quad \operatorname{Re}\gamma > 0] \qquad \text{ET I 146(25)}$$

2. $$\int_{-\infty}^\infty \exp\left[-\left(x - \frac{b}{x}\right)^{2n}\right] dx = \frac{1}{n}\Gamma\left(\frac{1}{2n}\right)$$

3.325 $\quad \displaystyle\int_0^\infty \exp\left(-ax^2 - \frac{b}{x^2}\right) dx = \frac{1}{2}\sqrt{\frac{\pi}{a}}\exp\left(-2\sqrt{ab}\right) \qquad [a > 0, \quad b > 0] \qquad \text{FI II 644}$

3.326

1.[8] $\quad \displaystyle\int_0^\infty \exp\left(-x^\mu\right) dx = \frac{1}{\mu}\Gamma\left(\frac{1}{\mu}\right) \qquad [\operatorname{Re}\mu > 0] \qquad \text{BI (26)(4)}$

2.[10] $\quad \displaystyle\int_0^\infty x^m \exp\left(-\beta x^n\right) dx = \frac{\Gamma(\gamma)}{n\beta^\gamma} \qquad \gamma = \frac{m+1}{n} \quad [\beta, m, n > 0]$

Exponentials of exponentials

3.327 $\quad \displaystyle\int_0^\infty \exp\left(-ae^{nx}\right) dx = -\frac{1}{n}\operatorname{Ei}(-a) \qquad [n \ge 1, \quad \operatorname{Re}a \ge 0, \quad a \ne 0] \qquad \text{LI (26)(5)}$

3.328 $\quad \displaystyle\int_{-\infty}^\infty \exp\left(-e^x\right) e^{\mu x} dx = \Gamma(\mu) \qquad [\operatorname{Re}\mu > 0] \qquad \text{NH 145(14)}$

3.329 $\quad \displaystyle\int_0^\infty \left[\frac{a\exp\left(-ce^{\alpha x}\right)}{1 - e^{-\alpha x}} - \frac{b\exp\left(-ce^{bx}\right)}{1 - e^{-bx}}\right] dx = e^{-c}\ln\frac{b}{a} \quad [a > 0, \quad b > 0, \quad c > 0] \qquad \text{BI (27)(12)}$

3.331

1. $$\int_0^\infty \exp\left(-\beta e^{-x} - \mu x\right) dx = \beta^{-\mu}\,\gamma(\mu, \beta) \qquad [\operatorname{Re}\mu > 0] \qquad \text{ET I 147(36)}$$

2. $$\int_0^\infty \exp\left(-\beta e^x - \mu x\right) dx = \beta^\mu\,\Gamma(-\mu, \beta) \qquad [\operatorname{Re}\beta > 0] \qquad \text{ET I 147(37)}$$

3. $$\int_0^\infty \left(1 - e^{-x}\right)^{\nu-1} \exp\left(\beta e^{-x} - \mu x\right) dx = \mathrm{B}(\mu, \nu)\beta^{-\frac{\mu-\nu}{2}} e^{\frac{\beta}{2}}\, M_{\frac{\nu-\mu}{2}, \frac{\nu+\mu-1}{2}}(\beta)$$
$$[\operatorname{Re}\mu > 0, \quad \operatorname{Re}\nu > 0] \qquad \text{ET I 147(38)}$$

4. $$\int_0^\infty \left(1 - e^{-x}\right)^{\nu-1} \exp\left(-\beta e^x - \mu x\right) dx = \Gamma(\nu)\beta^{\frac{\mu-1}{2}} e^{-\frac{\beta}{2}}\, W_{\frac{1-\mu-2\nu}{2}, \frac{-\mu}{2}}(\beta)$$
$$[\operatorname{Re}\beta > 0, \quad \operatorname{Re}\nu > 0] \qquad \text{ET I 147(39)}$$

3.332 $\quad \displaystyle\int_0^\infty \left(1 - e^{-x}\right)^{\nu-1}\left(1 - \lambda e^{-x}\right)^{-\varrho} \exp\left(\beta e^{-x} - \mu x\right) dx = \mathrm{B}(\mu, \nu)\,\Phi_1(\mu, \varrho, \nu, \lambda, \beta)$
$$[\operatorname{Re}\mu > 0, \quad \operatorname{Re}\nu > 0, \quad |\arg(1 - \lambda)| < \pi] \qquad \text{ET I 147(40)}$$

3.333

1.[3] $\quad \displaystyle\int_{-\infty}^\infty \frac{e^{-\mu x}\,dx}{\exp\left(e^{-x}\right) - 1} = \Gamma(\mu)\,\zeta(\mu) \qquad [\operatorname{Re}\mu > 1] \qquad \text{ET I 121(24)}$

$2.^3$ $\displaystyle\int_{-\infty}^{\infty} \frac{e^{-\mu x}\,dx}{\exp\left(e^{-x}\right)+1} = \left(1 - 2^{1-\mu}\right)\Gamma(\mu)\,\zeta(\mu)$ $[\operatorname{Re}\mu > 0, \quad \mu \neq 1]$

                           $= \ln 2$ $[\mu = 1]$

                                                       ET I 121(25)

3.334 $\displaystyle\int_{0}^{\infty} (e^x - 1)^{\nu-1} \exp\left[-\frac{\beta}{e^x - 1} - \mu x\right] dx = \Gamma(\mu - \nu + 1)e^{\frac{\beta}{2}}\beta^{\frac{\nu-1}{2}}\,W_{\frac{\nu-2\mu-1}{2},\frac{\nu}{2}}(\beta)$

                                             $[\operatorname{Re}\beta > 0, \quad \operatorname{Re}\mu > \operatorname{Re}\nu - 1]$

                                                       ET I 137(41)

Exponentials of hyperbolic functions

3.335 $\displaystyle\int_{0}^{\infty} \left(e^{\nu x} + e^{-\nu x}\cos\nu\pi\right)\exp\left(-\beta\sinh x\right) dx = -\pi\left[\mathbf{E}_\nu(\beta) + Y_\nu(\beta)\right]$

                                   $[\operatorname{Re}\beta > 0]$ EH II 35(34)

3.336

1. $\displaystyle\int_{0}^{\infty} \exp\left(-\nu x - \beta\sinh x\right) dx = \pi\cosec\nu\pi\left[\mathbf{J}_\nu(\beta) - J_\nu(\beta)\right]$

                   $\left[|\arg\beta| < \dfrac{\pi}{2}\text{ and }|\arg\beta| = \dfrac{\pi}{2}\text{ for }\operatorname{Re}\nu > 0;\quad \nu\text{ is not an integer}\right]$ WA 341(2)

2. $\displaystyle\int_{0}^{\infty} \exp\left(nx - \beta\sinh x\right) dx = \frac{1}{2}\left[S_n(\beta) - \pi\,\mathbf{E}_n(\beta) - \pi\,Y_n(\beta)\right]$

                                  $[\operatorname{Re}\beta > 0;\quad n = 0, 1, 2, \ldots]$ WA 342(6)

3. $\displaystyle\int_{0}^{\infty} \exp\left(-nx - \beta\sinh x\right) dx = \frac{1}{2}(-1)^{n+1}\left[S_n(\beta) + \pi\,\mathbf{E}_n(\beta) + \pi\,Y_n(\beta)\right]$

                                  $[\operatorname{Re}\beta > 0;\quad n = 0, 1, 2, \ldots]$

                                                       EH II 84(47)

3.337

1. $\displaystyle\int_{-\infty}^{\infty} \exp\left(-\alpha x - \beta\cosh x\right) dx = 2\,K_\alpha(\beta)$ $\left[|\arg\beta| < \dfrac{\pi}{2}\right]$ WA 201(7)

2. $\displaystyle\int_{-\infty}^{\infty} \exp\left(-\nu x + i\beta\cosh x\right) dx = i\pi e^{\frac{i\nu\pi}{2}}\,H_\nu^{(1)}(\beta)$ $[0 < \arg z < \pi]$ EH II 21(27)

3. $\displaystyle\int_{-\infty}^{\infty} \exp\left(-\nu x - i\beta\cosh x\right) dx = -i\pi e^{-\frac{i\nu\pi}{2}}\,H_\nu^{(2)}(\beta)$ $[-\pi < \arg z < 0]$ EH II 21(30)

Exponentials of trigonometric functions and logarithms

3.338

1. $\displaystyle\int_{0}^{\pi} \left\{\exp i\left[(\nu - 1)x - \beta\sin x\right] - \exp i\left[(\nu + 1)x - \beta\sin x\right]\right\} dx = 2\pi\left[\mathbf{J}_\nu'(\beta) + i\,\mathbf{E}_\nu'(\beta)\right]$

                                         $[\operatorname{Re}\beta > 0]$ EH II 36

2. $\displaystyle\int_{0}^{\pi} \exp\left[\pm i\left(\nu x - \beta\sin x\right)\right] dx = \pi\left[\mathbf{J}_\nu(\beta) \pm i\,\mathbf{E}_\nu(\beta)\right]$ $[\operatorname{Re}\beta > 0]$ EH II 35(32)

$3.^{10}$ $\displaystyle\int_0^\infty \exp\left[-\gamma\left(x-\beta\sin x\right)\right]dx = \frac{1}{\gamma} + 2\sum_{k=1}^\infty \frac{\gamma\,J_k(k\beta)}{\gamma^2+k^2}$ $\qquad[\operatorname{Re}\gamma>0]$ WA 619(4)

$4.^6$ $\displaystyle\int_{-\pi}^\pi \frac{\exp\left[\dfrac{a+b\sin x+c\cos x}{1+p\sin x+q\cos x}\right]}{1+p\sin x+q\cos x}\,dx = \frac{2\pi}{\sqrt{1-p^2-q^2}}e^{-\alpha}\,I_0(\beta),$

$$\text{with } \alpha=\frac{bp+cq-a}{1-p^2-q^2};\quad \beta=\sqrt{\alpha^2-\frac{a^2-b^2-c^2}{1-p^2-q^2}};\qquad \left[p^2+q^2<1\right]$$

3.339^6 $\displaystyle\int_0^\pi \exp\left(z\cos x\right)dx = \pi\,I_0(z)$ BI (277)(2)a

3.341 $\displaystyle\int_0^{\frac{\pi}{2}} \exp\left(-p\tan x\right)dx = \operatorname{ci}(p)\sin p - \operatorname{si}(p)\cos(p)$ $\qquad[p>0]$ BI (271)(2)a

3.342 $\displaystyle\int_0^1 \exp\left(-px\ln x\right)dx = \int_0^1 x^{-px}\,dx = \sum_{k=1}^\infty \frac{p^k-1}{k^k}$ BI (29)(1)

3.35 Combinations of exponentials and rational functions

3.351

$1.^8$ $\displaystyle\int_0^u x^n e^{-\mu x}\,dx = \frac{n!}{\mu^{n+1}} - e^{-u\mu}\sum_{k=0}^n \frac{n!}{k!}\frac{u^k}{\mu^{n-k+1}} = \mu^{-n-1}\gamma(n+1,\mu u)$

$$[u>0,\quad \operatorname{Re}\mu>0, n=0,1,2,\ldots]$$
ET I 134(5)

$2.^8$ $\displaystyle\int_u^\infty x^n e^{-\mu x}\,dx = e^{-u\mu}\sum_{l=0}^n \frac{n!}{k!}\frac{u^k}{\mu^{n-k+1}} = \mu^{-n-1}\Gamma(n+1,\mu u)$

$$[u>0,\quad \operatorname{Re}\mu>0, n=0,1,2,\ldots]$$
ET I 33(4)

3. $\displaystyle\int_0^\infty x^n e^{-\mu x}\,dx = n!\,\mu^{-n-1}$ $\qquad[\operatorname{Re}\mu>0]$ ET I 133(3)

4. $\displaystyle\int_u^\infty \frac{e^{-px}\,dx}{x^{n+1}} = (-1)^{n+1}\frac{p^n\operatorname{Ei}(-pu)}{n!} + \frac{e^{-pu}}{u^n}\sum_{k=0}^{n-1}\frac{(-1)^k p^k u^k}{n(n-1)\ldots(n-k)}$

$$[p>0]$$
NT 21(3)

5. $\displaystyle\int_1^\infty \frac{e^{-\mu x}\,dx}{x} = -\operatorname{Ei}(-\mu)$ $\qquad[\operatorname{Re}\mu>0]$ BI (104)(10)

6. $\displaystyle\int_{-\infty}^u \frac{e^x}{x}\,dx = \operatorname{li}(e^u) = \operatorname{Ei}(u)$ $\qquad[u<0]$

$7.^9$ $\displaystyle\int_0^u x e^{-\mu x}\,dx = \frac{1}{\mu^2} - \frac{1}{\mu^2}e^{-\mu u}(1+\mu u)$ $\qquad[u>0]$

$8.^9$ $\displaystyle\int_0^u x^2 e^{-\mu x}\,dx = \frac{2}{\mu^3} - \frac{1}{\mu^3}e^{-\mu u}\left(2+2\mu u - \mu^2 u^2\right)$ $\qquad[u>0]$

9.7 $\displaystyle\int_0^u x^3 e^{-\mu x}\, dx = \frac{6}{\mu^4} - \frac{1}{\mu^4} e^{-\mu u}\left(6 + 6\mu u + 3\mu^2 u^2 + \mu^3 u^3\right)$

$$[u > 0]$$

3.352

1. $\displaystyle\int_0^u \frac{e^{-\mu x}\, dx}{x + \beta} = e^{\mu\beta}\left[\mathrm{Ei}(-\mu u - \mu\beta) - \mathrm{Ei}(-\mu\beta)\right]$ $[|\arg\beta| < \pi]$ ET II 217(12)

2. $\displaystyle\int_u^\infty \frac{e^{-\mu x}\, dx}{x + \beta} = -e^{\beta\mu}\,\mathrm{Ei}(-\mu u - \mu\beta)$ $[u \geq 0, \quad |\arg(u + \beta)| < \pi, \quad \mathrm{Re}\,\mu > 0]$

     ET I 134(6), JA

3. $\displaystyle\int_u^v \frac{e^{-\mu x}\, dx}{x + \alpha} = e^{\alpha\mu}\left\{\mathrm{Ei}[-(\alpha + v)\mu] - \mathrm{Ei}[-(\alpha + u)\mu]\right\}$ $[-\alpha < n, \text{ and } -\alpha > v, \quad \mathrm{Re}\,\mu > 0]$

     ET I 134 (7)

4. $\displaystyle\int_0^\infty \frac{e^{-\mu x}\, dx}{x + \beta} = -e^{\beta\mu}\,\mathrm{Ei}(-\mu\beta)$ $[|\arg\beta| < \pi, \quad \mathrm{Re}\,\mu > 0]$ ET II 217(11)

5.7 $\displaystyle\int_u^\infty \frac{e^{-px}\, dx}{a - x} = e^{-pa}\,\mathrm{Ei}(pa - pu)$

     $\left[p > 0, \quad a < u; \text{ for } a > u, \text{ one should replace } \mathrm{Ei}(pa - pu) \text{ in this formula with } \overline{\mathrm{Ei}}(pa - pu)\right]$

     ET II 251(37)

6.8 $\displaystyle\int_0^\infty \frac{e^{-\mu x}\, dx}{a - x} = e^{-\mu a}\,\mathrm{Ei}(a\mu)$

$$[a < 0, \quad \mathrm{Re}\,\mu > 0]$$ BI (91)(4)

7. $\displaystyle\int_{-\infty}^\infty \frac{e^{ipx}\, dx}{x - a} = i\pi e^{iap}$ $[p > 0]$ ET II 251(38)

3.353

1. $\displaystyle\int_u^\infty \frac{e^{-\mu x}\, dx}{(x + \beta)^n} = e^{-u\mu}\sum_{k=1}^{n-1} \frac{(k - 1)!(-\mu)^{n-k-1}}{(n - 1)!(u + \beta)^k} - \frac{(-\mu)^{n-1}}{(n - 1)!} e^{\beta\mu}\,\mathrm{Ei}[-(u + \beta)\mu]$

     $[n \geq 2, \quad |\arg(u + \beta)| < \pi, \quad \mathrm{Re}\,\mu > 0]$
     ET I 134(10)

2.7 $\displaystyle\int_0^\infty \frac{e^{-\mu x}\, dx}{(x + \beta)^n} = \frac{1}{(n - 1)!}\sum_{k=1}^{n-1}(k - 1)!(-\mu)^{n-k-1}\beta^{-k} - \frac{(-\mu)^{n-1}}{(n - 1)!} e^{\beta\mu}\,\mathrm{Ei}(-\beta\mu)$

     $[n > 2, \quad |\arg\beta| < \pi, \quad \mathrm{Re}\,\mu > 0]$
     ET I 134(9), BI (92)(2)

3. $\displaystyle\int_0^\infty \frac{e^{-px}\, dx}{(a + x)^2} = pe^{\alpha p}\,\mathrm{Ei}(-ap) + \frac{1}{a}$ $[p > 0, \quad a > 0]$

     LI (281)(28), LI (281)(29)

4. $\displaystyle\int_0^1 \frac{xe^x}{(1 + x)^2}\, dx = \frac{e}{2} - 1.$ BI (80)(6)

5.[7] $\int_0^\infty \dfrac{x^n e^{-\mu x}}{x + \beta}\, dx = (-1)^{n-1} \beta^n e^{\beta\mu}\, \mathrm{Ei}(-\beta\mu) + \sum_{k=1}^{n} (k-1)!(-\beta)^{n-k}\mu^{-k}$

$$[|\arg \beta| < \pi, \quad \mathrm{Re}\,\mu > 0]$$

BI (91)(3)a, LET I 135(11)

3.354

1. $\int_0^\infty \dfrac{e^{-\mu x}\, dx}{\beta^2 + x^2} = \dfrac{1}{\beta}\left[\mathrm{ci}(\beta\mu) \sin\beta\mu - \mathrm{si}(\beta\mu) \cos\beta\mu\right]$ $[\mathrm{Re}\,\beta > 0, \quad \mathrm{Re}\,\mu > 0]$ BI (91)(7)

2. $\int_0^\infty \dfrac{xe^{-\mu x}\, dx}{\beta^2 + x^2} = -\,\mathrm{ci}(\beta\mu) \cos\beta\mu - \mathrm{si}(\beta\mu) \sin\beta\mu$ $[\mathrm{Re}\,\beta > 0, \quad \mathrm{Re}\,\mu > 0]$ BI (91)(8)

3.[7] $\int_0^\infty \dfrac{e^{-\mu x}\, dx}{\beta^2 - x^2} = \dfrac{1}{2\beta}\left[e^{-\beta\mu}\,\mathrm{Ei}(\beta\mu) - e^{\beta\mu}\,\mathrm{Ei}(-\beta\mu)\right]$ $[|\arg(\pm\beta)| < \pi, \quad \mathrm{Re}\,\mu > 0]$ BI (91)(14)

4. $\int_0^\infty \dfrac{xe^{-\mu x}\, dx}{\beta^2 - x^2} = \dfrac{1}{2}\left[e^{-\beta\mu}\,\mathrm{Ei}(\beta\mu) + e^{\beta\mu}\,\mathrm{Ei}(-\beta\mu)\right]$

$\left[|\arg(\pm\beta)| < \pi, \quad \mathrm{Re}\,\mu > 0;\ \text{for } \beta > 0 \text{ one should replace } \mathrm{Ei}(\beta\mu) \text{ in this formula with } \overline{\mathrm{Ei}}(\beta\mu)\right]$

BI (91)(15)

5.[8] $\int_{-\infty}^\infty \dfrac{e^{-ipx}\, dx}{a^2 + x^2} = \dfrac{\pi}{a}e^{-|ap|}$ $[a \neq 0, \quad p \text{ real}]$ ET I 118(1)a

3.355

1. $\int_0^\infty \dfrac{e^{-\mu x}\, dx}{(\beta^2 + x^2)^2} = \dfrac{1}{2\beta^3}\left\{\mathrm{ci}(\beta\mu) \sin\beta\mu - \mathrm{si}(\beta\mu) \cos\beta\mu\right\} - \beta\mu\left[\mathrm{ci}(\beta\mu) \cos\beta\mu + \mathrm{si}(\beta\mu) \sin\beta\mu\right]$

LI (92)(6)

2. $\int_0^\infty \dfrac{xe^{-\mu x}\, dx}{(\beta^2 + x^2)^2} = \dfrac{1}{2\beta^2}\left\{-\beta\mu\left[\mathrm{ci}(\beta\mu) \sin\beta\mu - \mathrm{si}(\beta\mu) \cos\beta\mu\right]\right\}$

$$[\mathrm{Re}\,\beta > 0, \quad \mathrm{Re}\,\mu > 0]$$

BI (92)(7)

3.[3] $\int_0^\infty \dfrac{e^{-px}\, dx}{(a^2 - x^2)^2} = \dfrac{1}{4a^3}\left[(ap-1)e^{ap}\,\mathrm{Ei}(-ap) + (1+ap)e^{-ap}\,\mathrm{Ei}(ap)\right]$

$$\left[\mathrm{Im}\,(a^2) > 0, \quad p > 0\right]$$

BI (92)(8)

4.[3] $\int_0^\infty \dfrac{xe^{-px}\, dx}{(a^2 - x^2)^2} = \dfrac{1}{4a^2}\left\{-2 + ap\left[e^{-ap}\,\mathrm{Ei}(ap) - e^{ap}\,\mathrm{Ei}(-ap)\right]\right\}$

$$\left[\mathrm{Im}\,(a^2) > 0, \quad p > 0\right]$$

LI (92)(9)

3.356

1. $\int_0^\infty \dfrac{x^{2n+1}e^{-px}}{a^2 + x^2}\, dx = (-1)^{n-1} a^{2n}\left[\mathrm{ci}(ap) \cos ap + \mathrm{si}(ap) \sin ap\right]$

$$+ \dfrac{1}{p^{2n}}\sum_{k=1}^{n}(2n - 2k + 1)!\left(-a^2 p^2\right)^{k-1}$$

$$[p > 0]$$

BI (91)(12)

2. $\int_0^\infty \dfrac{x^{2n}e^{-px}}{a^2 + x^2}\, dx = (-1)^{n} a^{2n-1}\left[\mathrm{ci}(ap) \sin ap - \mathrm{si}(ap) \cos ap\right] + \dfrac{1}{p^{2n-1}}\sum_{k=1}^{n}(2n - 2k)!\left(-a^2 p^2\right)^{k-1}$

$$[p > 0]$$

BI (91)(11)

3. $\displaystyle\int_0^\infty \frac{x^{2n+1}e^{-px}}{a^2-x^2}\,dx = \frac{1}{2}a^{2n}\left[e^{ap}\operatorname{Ei}(-ap)+e^{-ap}\operatorname{Ei}(ap)\right] - \frac{1}{p^{2n}}\sum_{k=1}^n (2n-2k+1)!\left(a^2p^2\right)^{k-1}$

$$[p>0] \qquad\qquad \text{BI (91)(17)}$$

4. $\displaystyle\int_0^\infty \frac{x^{2n}e^{-px}}{a^2-x^2}\,dx = \frac{1}{2}a^{2n-1}\left[e^{-ap}\operatorname{Ei}(ap)-e^{ap}\operatorname{Ei}(-ap)\right] - \frac{1}{p^{2n-1}}\sum_{k=1}^n (2n-2k)!\left(a^2p^2\right)^{k-1}$

$$[p>0] \qquad\qquad \text{BI (91)(16)}$$

3.357

1. $\displaystyle\int_0^\infty \frac{e^{-\mu x}\,dx}{a^3+a^2x+ax^2+x^3} = \frac{1}{2a^2}\{\operatorname{ci}(a\mu)\,(\sin a\mu + \cos a\mu)$

$$+ \operatorname{si}(a\mu)\,(\sin a\mu - \cos a\mu) - e^{a\mu}\operatorname{Ei}(-a\mu)\}$$

$$[\operatorname{Re}\mu>0,\quad a>0] \qquad\qquad \text{BI (92)(18)}$$

2. $\displaystyle\int_0^\infty \frac{xe^{-\mu x}\,dx}{a^3+a^2x+ax^2+x^3} = \frac{1}{2a}\{\operatorname{ci}(a\mu)\,(\sin a\mu - \cos a\mu)$

$$- \operatorname{si}(a\mu)\,(\sin a\mu + \cos a\mu) - e^{a\mu}\operatorname{Ei}(-a\mu)\}$$

$$[\operatorname{Re}\mu>0,\quad a>0] \qquad\qquad \text{BI (92)(19)}$$

3. $\displaystyle\int_0^\infty \frac{x^2e^{-\mu x}\,dx}{a^3+a^2x+ax^2+x^3} = \frac{1}{2}\{-\operatorname{ci}(a\mu)\,(\sin a\mu + \cos a\mu)$

$$- \operatorname{si}(a\mu)\,(\sin a\mu - \cos a\mu) - e^{a\mu}\operatorname{Ei}(-a\mu)\}$$

$$[\operatorname{Re}\mu>0,\quad a>0] \qquad\qquad \text{BI (92)(20)}$$

4. $\displaystyle\int_0^\infty \frac{e^{-\mu x}\,dx}{a^3-a^2x+ax^2-x^3} = \frac{1}{2a^2}\{\operatorname{ci}(a\mu)\,(\sin a\mu - \cos a\mu)$

$$- \operatorname{si}(a\mu)\,(\sin a\mu + \cos a\mu) + e^{-a\mu}\operatorname{Ei}(a\mu)\}$$

$$[\operatorname{Re}\mu>0,\quad a>0] \qquad\qquad \text{BI (92)(21)}$$

5. $\displaystyle\int_0^\infty \frac{xe^{-\mu x}\,dx}{a^3-a^2x+ax^2-x^3} = \frac{1}{2a}\{-\operatorname{ci}(a\mu)\,(\sin a\mu + \cos a\mu)$

$$- \operatorname{si}(a\mu)\,(\sin a\mu - \cos a\mu) + e^{-a\mu}\operatorname{Ei}(a\mu)\}$$

$$[\operatorname{Re}\mu>0,\quad a>0] \qquad\qquad \text{BI (92)(22)}$$

6. $\displaystyle\int_0^\infty \frac{x^2e^{-\mu x}\,dx}{a^3-a^2x+ax^2-x^3} = \frac{1}{2}\{\operatorname{ci}(a\mu)\,(\cos a\mu - \sin a\mu)$

$$+ \operatorname{si}(a\mu)\,(\cos a\mu + \sin a\mu) + e^{-a\mu}\operatorname{Ei}(a\mu)\}$$

$$[\operatorname{Re}\mu>0,\quad a>0] \qquad\qquad \text{BI (92)(23)}$$

3.358

1. $\displaystyle\int_0^\infty \frac{e^{-px}}{a^4-x^4}\,dx = \frac{1}{4a^3}\{e^{-ap}\operatorname{Ei}(ap)-e^{ap}\operatorname{Ei}(-ap)+2\operatorname{ci}(ap)\sin ap - 2\operatorname{si}(ap)\cos ap\}$

$$[p>0,\quad a>0] \qquad\qquad \text{BI (91)(18)}$$

2. $$\int_0^\infty \frac{xe^{-px}\,dx}{a^4 - x^4} = \frac{1}{4a^2}\left\{e^{ap}\,\mathrm{Ei}(-ap) + e^{-ap}\,\mathrm{Ei}(ap) - 2\,\mathrm{ci}(ap)\cos ap - 2\,\mathrm{si}(ap)\sin ap\right\}$$

$$[p > 0, \quad a > 0] \qquad \text{BI (91)(19)}$$

3. $$\int_0^\infty \frac{x^2 e^{-px}\,dx}{a^4 - x^4} = \frac{1}{4a}\left\{e^{-ap}\,\mathrm{Ei}(ap) - e^{ap}\,\mathrm{Ei}(-ap) - 2\,\mathrm{ci}(ap)\sin ap + 2\,\mathrm{si}(ap)\cos ap\right\}$$

$$[p > 0, \quad a > 0] \qquad \text{BI (91)(20)}$$

4. $$\int_0^\infty \frac{x^3 e^{-px}\,dx}{a^4 - x^4} = \frac{1}{4}\left\{e^{ap}\,\mathrm{Ei}(-ap) + e^{-ap}\,\mathrm{Ei}(ap) + 2\,\mathrm{ci}(ap)\cos ap + 2\,\mathrm{si}(ap)\sin ap\right\}$$

$$[p > 0, \quad a > 0] \qquad \text{BI (91)(21)}$$

5. $$\int_0^\infty \frac{x^{4n} e^{-px}}{a^4 - x^4}\,dx = \frac{1}{4}a^{4n-3}\left[e^{-ap}\,\mathrm{Ei}(ap) - e^{ap}\,\mathrm{Ei}(-ap) + 2\,\mathrm{ci}(ap)\sin ap - 2\,\mathrm{si}(ap)\cos ap\right]$$
$$- \frac{1}{p^{4n-3}}\sum_{k=1}^{n}(4n - 4k)!\left(a^4 p^4\right)^{k-1}$$

$$[p > 0, \quad a > 0] \qquad \text{BI (91)(22)}$$

6. $$\int_0^\infty \frac{x^{4n+1} e^{-px}}{a^4 - x^4}\,dx = \frac{1}{4}a^{4n-2}\left[e^{ap}\,\mathrm{Ei}(-ap) + e^{-ap}\,\mathrm{Ei}(ap) - 2\,\mathrm{ci}(ap)\cos ap - 2\,\mathrm{si}(ap)\sin ap\right]$$
$$- \frac{1}{p^{4n-2}}\sum_{k=1}^{n}(4n - 4k + 1)!\left(a^4 p^4\right)^{k-1}$$

$$[p > 0, \quad a > 0] \qquad \text{BI (91)(23)}$$

7. $$\int_0^\infty \frac{x^{4n+2} e^{-px}}{a^4 - x^4}\,dx = \frac{1}{4}a^{4n-1}\left[e^{-ap}\,\mathrm{Ei}(ap) - e^{ap}\,\mathrm{Ei}(-ap) - 2\,\mathrm{ci}(ap)\sin ap + 2\,\mathrm{si}(ap)\cos ap\right]$$
$$- \frac{1}{p^{4n-1}}\sum_{k=1}^{n}(4n - 4k + 2)!\left(a^4 p^4\right)^{k-1}$$

$$[p > 0, \quad a > 0] \qquad \text{BI (91)(24)}$$

8. $$\int_0^\infty \frac{x^{4n+3} e^{-px}}{a^4 - x^4}\,dx = \frac{1}{4}a^{4n}\left[e^{ap}\,\mathrm{Ei}(-ap) + e^{-ap}\,\mathrm{Ei}(ap) + 2\,\mathrm{ci}(ap)\cos ap + 2\,\mathrm{si}(ap)\sin ap\right]$$
$$- \frac{1}{p^{4n}}\sum_{k=1}^{n}(4n - 4k + 3)!\left(a^4 p^4\right)^{k-1}$$

$$[p > 0, \quad a > 0] \qquad \text{BI (91)(25)}$$

3.359 $$\int_{-\infty}^\infty \frac{(i - x)^n}{(i + x)^n}\frac{e^{-ipx}}{i + x^2}\,dx = (-1)^{n-1}2\pi p e^{-p}\,L_{n-1}(2p) \qquad \text{for } p > 0;$$

$$= 0 \qquad \text{for } p < 0.$$

$$\text{ET I 118(2)}$$

3.36–3.37 Combinations of exponentials and algebraic functions

3.361

1.[8] $$\int_0^u \frac{e^{-qx}}{\sqrt{x}}\,dx = \sqrt{\frac{\pi}{q}}\,\Phi\left(\sqrt{qu}\right) \qquad [q > 0]$$

$2.^8$ $\displaystyle\int_0^\infty \frac{e^{-qx}}{\sqrt{x}}\,dx = \sqrt{\frac{\pi}{q}}$ $[q > 0]$ BI(98)(10)

$3.^8$ $\displaystyle\int_{-1}^\infty \frac{e^{-qx}}{\sqrt{1+x}}\,dx = e^q\sqrt{\frac{\pi}{q}}$ $[q > 0]$ BI (104)(16)

3.362

1. $\displaystyle\int_1^\infty \frac{e^{-\mu x}\,dx}{\sqrt{x-1}} = \sqrt{\frac{\pi}{\mu}}\,e^{-\mu}$ $[\operatorname{Re}\mu > 0]$ BI (104)(11)a

2. $\displaystyle\int_0^\infty \frac{e^{-\mu x}\,dx}{\sqrt{x+\beta}} = \sqrt{\frac{\pi}{\mu}}\,e^{\beta\mu}\left[1 - \Phi\left(\sqrt{\beta\mu}\right)\right]$ $[\operatorname{Re}\mu > 0, \quad |\arg\beta| < \pi]$ ET I 135(18)

3.363

1. $\displaystyle\int_u^\infty \frac{\sqrt{x-u}}{x}\,e^{-\mu x}\,dx = \sqrt{\frac{\pi}{\mu}}\,e^{-u\mu} - \pi\sqrt{u}\left[1 - \Phi\left(\sqrt{u\mu}\right)\right]$

                                                          $[u > 0, \quad \operatorname{Re}\mu > 0]$ ET I 136(23)

2. $\displaystyle\int_u^\infty \frac{e^{-\mu x}\,dx}{x\sqrt{x-u}} = \frac{\pi}{\sqrt{u}}\left[1 - \Phi\left(\sqrt{u\mu}\right)\right]$ $[u > 0, \quad \operatorname{Re}\mu \geq 0]$ ET I 136(26)

3.364

1. $\displaystyle\int_0^2 \frac{e^{-px}\,dx}{\sqrt{x(2-x)}} = \pi e^{-p}\, I_0(p)$ $[p > 0]$ GW (312)(7a)

2. $\displaystyle\int_{-1}^1 \frac{e^{2x}\,dx}{\sqrt{1-x^2}} = \pi\, I_0(2)$ BI (277)(2)a

3. $\displaystyle\int_0^\infty \frac{e^{-px}\,dx}{\sqrt{x(x+a)}} = e^{\frac{ap}{2}}\, K_0\left(\frac{ap}{2}\right)$ $[a > 0, \quad p > 0]$ GW (312)(8a)

3.365

1. $\displaystyle\int_0^u \frac{x e^{-\mu x}\,dx}{\sqrt{u^2-x^2}} = \frac{\pi u}{2}\left[\mathbf{L}_1(\mu u) - I_1(\mu u)\right] + u$ $[u > 0, \quad \operatorname{Re}\mu > 0]$ ET I 136(28)

2. $\displaystyle\int_u^\infty \frac{x e^{-\mu x}\,dx}{\sqrt{x^2-u^2}} = u\, K_1(u\mu)$ $[u > 0, \quad \operatorname{Re}\mu > 0]$ ET I 136(29)

3.366

1. $\displaystyle\int_0^{2u} \frac{(u-x)e^{-\mu x}\,dx}{\sqrt{2ux-x^2}} = \pi u e^{-u\mu}\, I_1(u\mu)$ $[\operatorname{Re}\mu > 0]$ ET I 136(31)

2. $\displaystyle\int_0^\infty \frac{(x+\beta)e^{-\mu x}\,dx}{\sqrt{x^2+2\beta x}} = \beta e^{\beta\mu}\, K_1(\beta\mu)$ $[\operatorname{Re}\mu > 0, \quad |\arg\beta| < \pi]$ ET I 136(30)

3. $\displaystyle\int_0^\infty \frac{x e^{-\mu x}\,dx}{\sqrt{x^2+\beta^2}} = \frac{\beta\pi}{2}\left[\mathbf{H}_1(\beta\mu) - Y_1(\beta\mu)\right] - \beta$ $\left[|\arg\beta| < \frac{\pi}{2}, \quad \operatorname{Re}\mu > 0\right]$ ET I 136(27)

3.367 $\int_0^\infty \dfrac{e^{-\mu x}\, dx}{(1+\cos t + x)\sqrt{x^2+2x}} = \dfrac{\exp\left(2\mu\cos^2\frac{t}{2}\right)}{\sin t}\left(t - \sin t \int_0^u K_0(v)e^{-v\cos t}\, dv\right)$

$\qquad\qquad\qquad\qquad\qquad\qquad\qquad\qquad\qquad\qquad\qquad$ [Re $\mu > 0$] $\qquad$ ET I 136(33)

3.368 $\int_0^\infty \dfrac{e^{-\mu x}\, dx}{x+\sqrt{x^2+\beta^2}} = \dfrac{\pi}{2\beta\mu}\left[\mathbf{H}_1(\beta\mu) - Y_1(\beta\mu)\right] - \dfrac{1}{\beta^2\mu^2}$

$\qquad\qquad\qquad\qquad\qquad\qquad\qquad\qquad$ $\left[|\arg\beta| < \dfrac{\pi}{2}, \quad \mathrm{Re}\,\mu > 0\right]$ $\qquad$ ET I 136(32)

3.369 $\int_0^\infty \dfrac{e^{-\mu x}\, dx}{\sqrt{(x+a)^3}} = \dfrac{2}{\sqrt{a}} - 2\sqrt{\pi\mu}\,e^{a\mu}\left(1 - \Phi\left(\sqrt{a\mu}\right)\right)$ $\qquad$ [$|\arg a| < \pi, \quad \mathrm{Re}\,\mu > 0$] $\qquad$ ET I 135(20)

3.371 $\int_0^\infty x^{n-\frac{1}{2}}e^{-\mu x}\, dx = \sqrt{\pi}\cdot\dfrac{1}{2}\cdot\dfrac{3}{2}\cdots\dfrac{2n-1}{2}mu^{-n-\frac{1}{2}}$

$\qquad\qquad\qquad\qquad = \sqrt{\pi}\,2^{-n}\mu^{-n-1/2}(2n-1)!!$ $\qquad$ [$n \geq 0$]

$\qquad\qquad\qquad\qquad\qquad\qquad\qquad\qquad\qquad\qquad$ [Re $\mu > 0$] $\qquad$ ET I 135(17)

3.372 $\int_0^\infty x^{n-\frac{1}{2}}(2+x)^{n-\frac{1}{2}}e^{-px}\, dx = \dfrac{(2n-1)!!}{p^n}e^p K_n(p)$ $\qquad$ [$p > 0, \quad n = 0,1,2,\ldots$] $\qquad$ GW (312)(8)

3.373 $\int_0^\infty \left[\left(x+\sqrt{x^2+\beta^2}\right)^n + \left(x-\sqrt{x^2+\beta^2}\right)^n\right]e^{-\mu x}\, dx = 2\beta^{n+1}O_n(\beta\mu)$

$\qquad\qquad\qquad\qquad\qquad\qquad\qquad\qquad\qquad\qquad$ [Re $\mu > 0$] $\qquad$ WA 05(1)

3.374

1. $\quad\int_0^\infty \dfrac{\left(x+\sqrt{1+x^2}\right)^n}{\sqrt{1+x^2}}e^{-\mu x}\, dx = \dfrac{1}{2}\left[S_n(\mu) - \pi\,\mathbf{E}_n(\mu) - \pi\,Y_n(\mu)\right]$

$\qquad\qquad\qquad\qquad\qquad\qquad\qquad\qquad\qquad\qquad$ [Re $\mu > 0$] $\qquad$ ET I 37(35)

2. $\quad\int_0^\infty \dfrac{\left(x-\sqrt{1+x^2}\right)^n}{\sqrt{1+x^2}}e^{-\mu x}\, dx = -\dfrac{1}{2}\left[S_n(\mu) + \pi\,\mathbf{E}_n(\mu) + \pi\,Y_n(\mu)\right]$

$\qquad\qquad\qquad\qquad\qquad\qquad\qquad\qquad\qquad\qquad$ [Re $\mu > 0$] $\qquad$ ET I 137(36)

3.38–3.39 Combinations of exponentials and arbitrary powers

3.381

1. $\quad\int_0^u x^{\nu-1}e^{-\mu x}\, dx = \mu^{-\nu}\,\gamma(\nu,\mu u)$ $\qquad\qquad\qquad$ [Re $\nu > 0$] $\qquad$ EH I 266(22), EH II 133(1)

2. $\quad\int_0^u x^{p-1}e^{-x}\, dx = \sum_{k=0}^\infty (-1)^k \dfrac{u^{p+k}}{k!(p+k)}$

$\qquad\qquad\qquad\qquad = e^{-u}\sum_{k=0}^\infty \dfrac{u^{p+k}}{p(p+1)\ldots(p+k)}$

$\qquad\qquad\qquad\qquad\qquad\qquad\qquad\qquad\qquad\qquad\qquad\qquad$ AD 6.705

3.[8] $\quad\int_u^\infty x^{\nu-1}e^{-\mu x}\, dx = \mu^{-\nu}\,\Gamma(\nu,\mu u)$ $\qquad\qquad$ [$u > 0, \quad \mathrm{Re}\,\mu > 0$]

$\qquad\qquad\qquad\qquad\qquad\qquad\qquad\qquad\qquad\qquad$ EH I 256(21), EH II 133(2)

4. $\quad\int_0^\infty x^{\nu-1}e^{-\mu x}\, dx = \dfrac{1}{\mu^\nu}\Gamma(\nu)$ $\qquad\qquad$ [Re $\mu > 0, \quad \mathrm{Re}\,\nu > 0$] $\qquad$ FI II 779

5.$\qquad \displaystyle\int_0^\infty x^{\nu-1}e^{-(p+iq)x}\,dx = \Gamma(\nu)\left(p^2+q^2\right)^{-\frac{\nu}{2}}\exp\left(-i\nu\arctan\frac{q}{p}\right)$

$$[p>0,\quad \operatorname{Re}\nu>0 \text{ and } p=0,\quad 0<\operatorname{Re}\nu<1]\quad \text{EH I 12(32)}$$

6.$\qquad \displaystyle\int_u^\infty \frac{e^{-x}}{x^\nu}\,dx = u^{-\frac{\nu}{2}}e^{-\frac{u}{2}}\,W_{-\frac{\nu}{2},\frac{(1-\nu)}{2}}(u)$$\qquad\qquad [u>0]$$\qquad$ WH

7.*$\qquad \displaystyle\int_0^\infty x^{k-1}e^{i\mu x}\,dx = \frac{\Gamma(k)}{(-i\mu)^k}$$\qquad\qquad [0<\operatorname{Re}(k)<1,\quad \mu\neq 0]$

$$\text{GH2 62 (313.14)}$$

3.382

1.[6]$\qquad \displaystyle\int_0^u (u-x)^\nu e^{-\mu x}\,dx = (-\mu)^{-\nu-1}e^{-u\mu}\,\gamma(\nu+1,-u\mu)$$\quad [\operatorname{Re}\nu>-1,\quad u>0]$$\quad$ ET I 137(6)

2.$\qquad \displaystyle\int_u^\infty (x-u)^\nu e^{-\mu x}\,dx = \mu^{-\nu-1}e^{-u\mu}\,\Gamma(\nu+1)$$\quad [u>0,\quad \operatorname{Re}\nu>-1,\quad \operatorname{Re}\mu>0]$

$$\text{ET I 137(5), ET II 202(11)}$$

3.$\qquad \displaystyle\int_0^\infty (1+x)^{-\nu}e^{-\mu x}\,dx = \mu^{\frac{\nu}{2}-1}e^{\frac{\mu}{2}}\,W_{-\frac{\nu}{2},\frac{(1-\nu)}{2}}(\mu)$$\quad [\operatorname{Re}\mu>0]$$\quad$ WH

4.$\qquad \displaystyle\int_0^\infty (x+\beta)^\nu e^{-\mu x}\,dx = \mu^{-\nu-1}e^{\beta\mu}\,\Gamma(\nu+1,\beta\mu)$$\quad [|\arg\beta|<\pi,\quad \operatorname{Re}\mu>0]$

$$\text{ET I 137(4), ET II 233(10)}$$

5.$\qquad \displaystyle\int_0^u (a+x)^{\mu-1}e^{-x}\,dx = e^a[\gamma(\mu,a+u)-\gamma(\mu,a)]$$\quad [\operatorname{Re}\mu>0]$$\quad$ EH II 139

6.$\qquad \displaystyle\int_{-\infty}^\infty (\beta+ix)^{-\nu}e^{-ipx}\,dx = 0$$\qquad\qquad [\text{for } p>0]$

$$= \frac{2\pi(-p)^{\nu-1}e^{\beta p}}{\Gamma(\nu)}\qquad\qquad [\text{for } p<0]$$

$$[\operatorname{Re}\nu>0,\quad \operatorname{Re}\beta>0]\qquad \text{ET I 118(4)}$$

7.$\qquad \displaystyle\int_{-\infty}^\infty (\beta-ix)^{-\nu}e^{-ipx}\,dx = \frac{2\pi p^{\nu-1}e^{-\beta p}}{\Gamma(\nu)}$$\qquad [\text{for } p>0]$

$$= 0\qquad\qquad [\text{for } p<0]$$

$$[\operatorname{Re}\nu>0,\quad \operatorname{Re}\beta>0]\qquad \text{ET I 118(3)}$$

3.383

1.$\qquad \displaystyle\int_0^u x^{\nu-1}(u-x)^{\mu-1}e^{\beta x}\,dx = B(\mu,\nu)u^{u+\nu-1}\,{}_1F_1(\nu;\mu+\nu;\beta u)$

$$[\operatorname{Re}\mu>0,\quad \operatorname{Re}\nu>0]\qquad \text{ET II 187(14)}$$

2.$\qquad \displaystyle\int_0^u x^{\mu-1}(u-x)^{\mu-1}e^{\beta x}\,dx = \sqrt{\pi}\left(\frac{u}{\beta}\right)^{u-\frac{1}{2}}\exp\left(\frac{\beta u}{2}\right)\Gamma(\mu)\,I_{\mu-\frac{1}{2}}\left(\frac{\beta u}{2}\right)$

$$[\operatorname{Re}\mu>0]\qquad\qquad \text{ET II 187(13)}$$

3.
$$\int_u^\infty x^{\mu-1}(x-u)^{\mu-1}e^{-\beta x}\,dx = \frac{1}{\sqrt{\pi}}\left(\frac{u}{\beta}\right)^{\mu-\frac{1}{2}}\Gamma(\mu)\exp\left(-\frac{\beta u}{2}\right)K_{\mu-\frac{1}{2}}\left(\frac{\beta u}{2}\right)$$

$$[\mathrm{Re}\,\mu > 0, \quad \mathrm{Re}\,\beta u > 0] \qquad \text{ET II 202(12)}$$

4.
$$\int_u^\infty x^{\nu-1}(x-u)^{\mu-1}e^{-\beta x}\,dx = \beta^{-\frac{\mu+\nu}{2}}u^{\frac{\mu+\nu-2}{2}}\Gamma(\mu)\exp\left(-\frac{\beta u}{2}\right)W_{\frac{\nu-\mu}{2},\frac{1-\mu-\nu}{2}}(\beta\mu)$$

$$[\mathrm{Re}\,\mu > 0, \quad \mathrm{Re}\,\beta u > 0] \qquad \text{ET II 202(13)}$$

5.[7]
$$\mathrm{s}\int_0^\infty e^{-px}x^{q-1}(1+ax)^{-\nu}\,dx = a^{-q}\Gamma(q)\,\psi(q,q+1-\nu,p/a)$$

$$= p^{-q}\Gamma(q)\left[\sum_{k=0}^n \frac{(q)_k(\nu)_k}{k!}\left(\frac{a}{p}\right)^k + O\left(\frac{a}{p}\right)^{N+1}\right]$$

$$\left[\mathrm{Re}\,q > 0, \mathrm{Re}\,p > 0, \mathrm{Re}\,a > 0, \nu \text{ complex}, N = 0, 1, \ldots. \text{ For } \nu = 0 \text{ the integral equals } p^{-q}\Gamma(q).\right]$$

$$\text{BI (92)(3)}$$

6.
$$\int_0^\infty x^{\nu-1}(x+\beta)^{-\nu+\frac{1}{2}}e^{-\mu x}\,dx = 2^{\nu-\frac{1}{2}}\Gamma(\nu)\mu^{-\frac{1}{2}}e^{\frac{\beta\mu}{2}}D_{1-2\nu}\left(\sqrt{2\beta\mu}\right)$$

$$[|\arg\beta| < \pi, \quad \mathrm{Re}\,\nu > 0, \quad \mathrm{Re}\,\mu \geq 0, \quad \mu \neq 0] \quad \text{ET I 39(20), EH II 119(2)a}$$

7.
$$\int_0^\infty x^{\nu-1}(x+\beta)^{-\nu-\frac{1}{2}}e^{-\mu x}\,dx = 2^\nu\Gamma(\nu)\beta^{-\frac{1}{2}}e^{\frac{\beta\mu}{2}}D_{-2\nu}\left(\sqrt{2\beta\mu}\right)$$

$$[|\arg\beta| < \pi, \quad \mathrm{Re}\,\nu > 0, \quad \mathrm{Re}\,\mu \geq 0]$$
$$\text{ET I 139(21), EH II 119(1)a}$$

8.
$$\int_0^\infty x^{\nu-1}(x+\beta)^{\nu-1}e^{-\mu x}\,dx = \frac{1}{\sqrt{\pi}}\left(\frac{\beta}{\mu}\right)^{\nu-\frac{1}{2}}e^{\frac{\beta\mu}{2}}\Gamma(\nu)\,K_{\frac{1}{2}-\nu}\left(\frac{\beta\mu}{2}\right)$$

$$[|\arg\beta| < \pi, \quad \mathrm{Re}\,\mu > 0, \quad \mathrm{Re}\,\nu > 0]$$
$$\text{ET II 233(11), EH II 19(16)a, EH II 82(22)a}$$

9.
$$\int_u^\infty \frac{(x-u)^\nu e^{-\mu x}}{x}\,dx = u^\nu\Gamma(\nu+1)\Gamma(-\nu,u\mu) \qquad [u > 0, \quad \mathrm{Re}\,\nu > -1, \quad \mathrm{Re}\,\mu > 0]$$

$$\text{ET I 138(8)}$$

10.
$$\int_0^\infty \frac{x^{\nu-1}e^{-\mu x}}{x+\beta}\,dx = \beta^{\nu-1}e^{\beta\mu}\Gamma(\nu)\Gamma(1-\nu,\beta\mu) \qquad [|\arg\beta| < \pi, \quad \mathrm{Re}\,\mu > 0, \quad \mathrm{Re}\,\nu > 0]$$

$$\text{EH II 137(3)}$$

3.384

1.
$$\int_{-1}^1 (1-x)^{\nu-1}(1+x)^{\mu-1}e^{-ipx}\,dx = 2^{\mu+\nu-1}\mathrm{B}(\mu,\nu)e^{ip}\,{}_1F_1(\mu;\nu+\mu;-2ip)$$

$$[\mathrm{Re}\,\nu > 0, \quad \mathrm{Re}\,\mu > 0] \qquad \text{ET I 119(13)}$$

2.
$$\int_u^v (x-u)^{2\mu-1}(v-x)^{2\nu-1}e^{-px}\,dx$$

$$= \mathrm{B}(2\mu,2\nu)(v-u)^{\mu+\nu-1}p^{-\mu-\nu}\exp\left(-p\frac{u+v}{2}\right)M_{\mu-\nu,\mu+\nu-\frac{1}{2}}(vp-up)$$

$$[v > u > 0, \quad \mathrm{Re}\,\mu > 0, \quad \mathrm{Re}\,\nu > 0] \quad \text{ET I 139(23)}$$

3. $\displaystyle\int_u^\infty (x+\beta)^{2\nu-1}(x-u)^{2\varrho-1}e^{-\mu x}\,dx$

$$= \frac{(u+\beta)^{\nu+\varrho-1}}{\mu^{\nu+\varrho}}\exp\left[\frac{(\beta-u)\mu}{2}\right]\Gamma(2\varrho)\,W_{\nu-\varrho,\nu+\varrho-\frac{1}{2}}(u\mu+\beta\mu)$$

$$[u>0,\quad |\arg(\beta+u)|<\pi,\quad \operatorname{Re}\mu>0,\quad \operatorname{Re}\varrho>0]\quad \textbf{ET I 139(22)}$$

4. $\displaystyle\int_u^\infty (x+\beta)^\nu (x-u)^{-\nu}e^{-\mu x}\,dx = \frac{1}{\mu}\nu\pi\,\operatorname{cosec}(\nu\pi)e^{-\frac{(\beta+u)\mu}{2}}\,\mathrm{k}_{2\nu}\left[\frac{(\beta+u)\mu}{2}\right]$

$$[\nu\neq 0,\quad u>0,\quad |\arg(u+\beta)|<\pi,\quad \operatorname{Re}\mu>0,\quad \operatorname{Re}\nu<1]\quad \textbf{ET I 139(17)}$$

5. $\displaystyle\int_u^\infty (x-u)^{\nu-1}(x+u)^{-\nu+\frac{1}{2}}e^{-\mu x}\,dx = \frac{1}{\sqrt{\mu}}2^{\nu-\frac{1}{2}}\,\Gamma(\nu)\,D_{1-2\nu}\left(2\sqrt{u\mu}\right)$

$$[u>0,\quad \operatorname{Re}\mu>0,\quad \operatorname{Re}\nu>0]$$
$$\textbf{ET I 139(18)}$$

6. $\displaystyle\int_u^\infty (x-u)^{\nu-1}(x+u)^{-\nu-\frac{1}{2}}e^{-\mu x}\,dx = \frac{1}{\sqrt{u}}2^{\nu-\frac{1}{2}}\,\Gamma(\nu)\,D_{-2\nu}\left(2\sqrt{u\mu}\right)$

$$[u>0,\quad \operatorname{Re}\mu\geq 0,\quad \operatorname{Re}\nu>0]$$
$$\textbf{ET I 139(19)}$$

7.[6] $\displaystyle\int_{-\infty}^\infty (\beta-ix)^{-\mu}(\gamma-ix)^{-\nu}e^{-ipx}\,dx = \frac{2\pi e^{-\beta p}p^{\mu+\nu-1}}{\Gamma(\mu+\nu)}\,{}_1F_1(\nu;\mu+\nu;(\beta-\gamma)p)\quad [\text{for } p>0]$

$$= 0 \qquad\qquad [\text{for } p<0]$$

$$[\operatorname{Re}\beta>0,\quad \operatorname{Re}\gamma>0,\quad \operatorname{Re}(\mu+\nu)>1]\quad \textbf{ET I 119(10)}$$

8.[6] $\displaystyle\int_{-\infty}^\infty (\beta+ix)^{-\mu}(\gamma+ix)^{-\nu}e^{-ipx}\,dx = 0 \qquad\qquad [\text{for } p>0]$

$$= \frac{2\pi e^{\gamma p}(-p)^{\mu+\nu-1}}{\Gamma(\mu+\nu)}\,{}_1F_1[\mu;\mu+\nu;(\beta-\gamma)p]\quad [\text{for } p<0]$$

$$[\operatorname{Re}\beta>0,\quad \operatorname{Re}\gamma>0,\quad \operatorname{Re}(\mu+\nu)>1]\quad \textbf{ET I 19(11)}$$

9.[6] $\displaystyle\int_{-\infty}^\infty (\beta+ix)^{-2\mu}(\gamma-ix)^{-2\nu}e^{-ipx}\,dx$

$$= 2\pi(\beta+\gamma)^{-\mu-\nu}\frac{p^{\mu+\nu-1}}{\Gamma(2\nu)}\exp\left(\frac{\beta-\gamma}{2}p\right)W_{\nu-\mu,\frac{1}{2}-\nu-\mu}(\beta p+\gamma p)\qquad [\text{for } p>0]$$

$$= 2\pi(\beta+\gamma)^{-\mu-\nu}\frac{(-p)^{\mu+\nu-1}}{\Gamma(2\mu)}\exp\left(\frac{\beta-\gamma}{2}p\right)W_{\mu-\nu,\frac{1}{2}-\nu-\mu}(-\beta p-\gamma p)\quad [\text{for } p<0]$$

$$\left[\operatorname{Re}\beta>0,\quad \operatorname{Re}\gamma>0,\quad \operatorname{Re}(\mu+\nu)>\tfrac{1}{2}\right]\quad \textbf{ET I 19(12)}$$

3.385 $\displaystyle\int_0^1 x^{\nu-1}(1-x)^{\lambda-1}(1-\beta x)^{-\varrho}e^{-\mu x}\,dx = \mathrm{B}(\nu,\lambda)\Phi_1(\nu,\varrho,\lambda+\nu,\beta,-\mu)$

$$[\operatorname{Re}\lambda>0,\quad \operatorname{Re}\nu>0,\quad |\arg(1-\beta)|<\pi]\quad \textbf{ET I 39(24)}$$

3.386

1. $$\int_{-\infty}^{\infty} \frac{(ix)^{\nu_0} \prod_{k=1}^{n} (\beta_k + ix)^{\nu_k} e^{-ipx} \, dx}{\beta_0 - ix} = 2\pi e^{-\beta_0 p} \beta_0^{\nu_0} \prod_{k=1}^{n} (\beta_0 + \beta_k)^{\nu_k}$$

$$\left[\operatorname{Re} \nu_0 > -1, \quad \operatorname{Re} \beta_k > 0, \quad \sum_{k=0}^{n} \operatorname{Re} \nu_k < 1, \quad \arg ix = \frac{\pi}{2} \operatorname{sign} x, \quad p > 0 \right] \quad \text{ET I 118(8)}$$

2. $$\int_{-\infty}^{\infty} \frac{(ix)^{\nu_0} \prod_{k=1}^{n} (\beta_k + ix)^{\nu_k} e^{-ipx} \, dx}{\beta_0 + ix} = 0$$

$$\left[\operatorname{Re} \nu_0 > -1, \quad \operatorname{Re} \beta_k > 0, \quad \sum_{k=0}^{n} \operatorname{Re} \nu_k < 1, \quad \arg ix = \frac{\pi}{2} \operatorname{sign} x, \quad p > 0 \right] \quad \text{ET I 119(9)}$$

3.387

1.[6] $$\int_{-1}^{1} (1 - x^2)^{\nu-1} e^{-\mu x} \, dx = \sqrt{\pi} \left(\frac{2}{\mu} \right)^{\nu - \frac{1}{2}} \Gamma(\nu) I_{\nu - \frac{1}{2}}(\mu)$$

$$\left[\operatorname{Re} \nu > 0, \quad |\arg \mu| < \frac{\pi}{2} \right] \quad \text{WA 172(2)a}$$

2.[6] $$\int_{-1}^{1} (1 - x^2)^{\nu-1} e^{i\mu x} \, dx = \sqrt{\pi} \left(\frac{2}{\mu} \right)^{\nu - \frac{1}{2}} \Gamma(\nu) J_{\nu - \frac{1}{2}}(\mu) \qquad [\operatorname{Re} \nu > 0] \qquad \text{WA 25(3), WA 48(4)a}$$

3. $$\int_{1}^{\infty} (x^2 - 1)^{\nu-1} e^{-\mu x} \, dx = \frac{1}{\sqrt{\pi}} \left(\frac{2}{\mu} \right)^{\nu - \frac{1}{2}} \Gamma(\nu) K_{\nu - \frac{1}{2}}(\mu)$$

$$\left[|\arg \mu| < \frac{\pi}{2}, \quad \operatorname{Re} \nu > 0 \right] \qquad \text{WA 190(4)a}$$

4. $$\int_{1}^{\infty} (x^2 - 1)^{\nu-1} e^{i\mu x} \, dx$$

$$= i \frac{\sqrt{\pi}}{2} \left(\frac{2}{\mu} \right)^{\nu - \frac{1}{2}} \Gamma(\nu) H_{\frac{1}{2} - \nu}^{(1)}(\mu) \qquad [\operatorname{Im} \mu > 0, \quad \operatorname{Re} \nu > 0] \quad \text{EH II 83(28)a}$$

$$= -i \frac{\sqrt{\pi}}{2} \left(-\frac{2}{\mu} \right)^{\nu - \frac{1}{2}} \Gamma(\nu) H_{\frac{1}{2} - \nu}^{(2)}(-\mu) \qquad [\operatorname{Im} \mu < 0, \quad \operatorname{Re} \nu > 0] \quad \text{EH II 83(29)a}$$

5. $$\int_{0}^{u} (u^2 - x^2)^{\nu-1} e^{\mu x} \, dx = \frac{\sqrt{\pi}}{2} \left(\frac{2u}{\mu} \right)^{\nu - \frac{1}{2}} \Gamma(\nu) \left[I_{\nu - \frac{1}{2}}(u\mu) + \mathbf{L}_{\nu - \frac{1}{2}}(u\mu) \right]$$

$$[u > 0, \quad \operatorname{Re} \nu > 0] \qquad \text{ET II 188(20)a}$$

6. $$\int_{u}^{\infty} (x^2 - u^2)^{\nu-1} e^{-\mu x} \, dx = \frac{1}{\sqrt{\pi}} \left(\frac{2u}{\mu} \right)^{\nu - \frac{1}{2}} \Gamma(\nu) K_{\nu - \frac{1}{2}}(u\mu)$$

$$[u > 0, \quad \operatorname{Re} \mu > 0, \quad \operatorname{Re} \nu > 0]$$
$$\text{ET II 203(17)a}$$

7. $\displaystyle\int_0^\infty \left(x^2 + u^2\right)^{\nu-1} e^{-\mu x}\, dx = \frac{\sqrt{\pi}}{2}\left(\frac{2u}{\mu}\right)^{\nu-\frac12}\Gamma(\nu)\left[\mathbf{H}_{\nu-\frac12}(u\mu) - Y_{\nu-\frac12}(u,\mu)\right]$

$$\left[|\arg u| < \pi, \quad \mathrm{Re}\,\mu > 0\right] \qquad \text{ET I 138(10)}$$

3.388

1. $\displaystyle\int_0^{2u} \left(2ux - x^2\right)^{\nu-1} e^{-\mu x}\, dx = \sqrt{\pi}\left(\frac{2u}{\mu}\right)^{\nu-\frac12} e^{-u\mu}\,\Gamma(\nu)\,I_{\nu-\frac12}(u\mu)$

$$\left[u > 0, \quad \mathrm{Re}\,\nu > 0\right] \qquad \text{ET I 138(14)}$$

2. $\displaystyle\int_0^\infty \left(2\beta x + x^2\right)^{\nu-1} e^{-\mu x}\, dx = \frac{1}{\sqrt{\pi}}\left(\frac{2\beta}{\mu}\right)^{\nu-\frac12} e^{\beta\mu}\,\Gamma(\nu)\,K_{\nu-\frac12}(\beta\mu)$

$$\left[|\arg\beta| < \pi, \quad \mathrm{Re}\,\nu > 0, \quad \mathrm{Re}\,\mu > 0\right]$$
$$\text{ET I 138(13)}$$

3. $\displaystyle\int_0^\infty \left(x^2 + ix\right)^{\nu-1} e^{-\mu x}\, dx = -\frac{i\sqrt{\pi}\,e^{\frac{i\mu}{2}}}{2\mu^{\nu-\frac12}}\,\Gamma(\nu)\,H^{(2)}_{\nu-\frac12}\left(\frac{\mu}{2}\right)$

$$\left[\mathrm{Re}\,\mu > 0, \quad \mathrm{Re}\,\nu > 0\right] \qquad \text{ET I 138(15)}$$

4. $\displaystyle\int_0^\infty \left(x^2 - ix\right)^{\nu-1} e^{-\mu x}\, dx = \frac{i\sqrt{\pi}\,e^{-\frac{i\mu}{2}}}{2\mu^{\nu-\frac12}}\,\Gamma(\nu)\,H^{(1)}_{\nu-\frac12}\left(\frac{\mu}{2}\right)$

$$\left[\mathrm{Re}\,\mu > 0, \quad \mathrm{Re}\,\nu > 0\right] \qquad \text{ET I 138(16)}$$

3.389

1. $\displaystyle\int_0^u x^{2\nu-1}\left(u^2 - x^2\right)^{\varrho-1} e^{\mu x}\, dx = \frac12\,\mathrm{B}(\nu,\varrho)u^{2\nu+2\varrho-2}\,{}_1F_2\left(\nu; \frac12, \nu+\varrho; \frac{\mu^2 u^2}{4}\right)$

$$+\frac{\mu}{2}\,\mathrm{B}\left(\nu+\frac12, \varrho\right)u^{2\nu+2\varrho-1}\,{}_1F_2\left(\nu+\frac12; \frac32, \nu+\varrho+\frac12; \frac{\mu^2 u^2}{4}\right)$$
$$\left[\mathrm{Re}\,\varrho > 0, \quad \mathrm{Re}\,\nu > 0\right] \qquad \text{ET II 188(21)}$$

2.[7] $\displaystyle\int_0^\infty x^{2\nu-1}\left(u^2 + x^2\right)^{\varrho-1} e^{-\mu x}\, dx = \frac{u^{2\nu+2\varrho-2}}{2\sqrt{\pi}\,\Gamma(1-\varrho)}\,G^{31}_{13}\left(\frac{\mu^2 u^2}{4}\,\bigg|\,\begin{matrix}1-\nu\\ 1-\varrho-\nu, 0, \frac12\end{matrix}\right)$

$$\left[|\arg u| < \frac{\pi}{2}, \quad \mathrm{Re}\,\mu > 0, \quad \mathrm{Re}\,\nu > 0\right]$$
$$\text{ET II 234(15)a}$$

3.[7] $\displaystyle\int_0^u x\left(u^2 - x^2\right)^{\nu-1} e^{\mu x}\, dx = \frac{u^{2\nu}}{2\nu} + \frac{\sqrt{\pi}}{2}\left(\frac{\mu}{2}\right)^{\frac12-\nu} u^{\nu+\frac12}\,\Gamma(\nu)\left[I_{\nu+\frac12}(\mu u) + \mathbf{L}_{\nu+\frac12}(\mu u)\right]$

$$\left[\mathrm{Re}\,\nu > 0\right] \qquad \text{ET II 188(19)a}$$

4. $\displaystyle\int_u^\infty x\left(x^2 - u^2\right)^{\nu-1} e^{-\mu x}\, dx = 2^{\nu-\frac12}\left(\sqrt{\pi}\right)^{-1}\mu^{\frac12-\nu} u^{\nu+\frac12}\,\Gamma(\nu)\,K_{\nu+\frac12}(u\mu)$

$$\left[\mathrm{Re}(u\mu) > 0\right] \qquad \text{ET II 203(16)a}$$

5. $\displaystyle\int_{-\infty}^\infty \frac{(ix)^{-\nu} e^{-ipx}\, dx}{\beta^2 + x^2} = \pi\beta^{-\nu-1} e^{-|p|\beta}$

$$\left[|\nu| < 1, \quad \mathrm{Re}\,\beta > 0, \quad \arg ix = \frac{\pi}{2}\,\mathrm{sign}\,x\right] \qquad \text{ET I 118(5)}$$

6. $\displaystyle\int_0^\infty \frac{x^\nu e^{-\mu x}}{\beta^2 + x^2}\, dx = \frac{1}{2}\, \Gamma(\nu)\beta^{\nu-1} \left[\exp\left(i\mu\beta + i\frac{(\nu-1)\pi}{2} \right) \right.$

$$\left. \times\, \Gamma(1-\nu, i\beta\mu) + \exp\left(-i\beta\mu - i\frac{(\nu-1)\pi}{2} \right) \Gamma(1-\nu, -i\beta\mu) \right]$$

$$[\operatorname{Re}\beta > 0, \quad \operatorname{Re}\mu > 0, \quad \operatorname{Re}\nu > -1] \quad \text{ET II 218(22)}$$

7. $\displaystyle\int_0^\infty \frac{x^{\nu-1} e^{-\mu x}\, dx}{1 + x^2} = \pi \operatorname{cosec}(\nu\pi) V_\nu(2\mu, 0)$ $\qquad [\operatorname{Re}\mu > 0, \quad \operatorname{Re}\nu > 0]$ $\qquad$ ET I 138(9)

8. $\displaystyle\int_{-\infty}^\infty \frac{(\beta + ix)^{-\nu} e^{-ipx}}{\gamma^2 + x^2}\, dx = \frac{\pi}{\gamma}(\beta + \gamma)^{-\nu} e^{-p\gamma}$

$$[\operatorname{Re}\nu > -1, \quad p > 0, \quad \operatorname{Re}\beta > 0, \quad \operatorname{Re}\gamma > 0] \quad \text{ET I 118(6)}$$

9.[6] $\displaystyle\int_{-\infty}^\infty \frac{(\beta - ix)^{-\nu} e^{-ipx}}{\gamma^2 + x^2}\, dx = \frac{\pi}{\gamma}(\beta + \gamma)^{-\nu} e^{\gamma p}$

$$[p < 0, \quad \operatorname{Re}\beta > 0, \quad \operatorname{Re}\gamma > 0, \quad \operatorname{Re}\nu > -1] \quad \text{ET I 118(7)}$$

3.391 $\displaystyle\int_0^\infty \left[\left(\sqrt{x+2\beta} + \sqrt{x}\right)^{2\nu} - \left(\sqrt{x+2\beta} - \sqrt{x}\right)^{2\nu} \right] e^{-\mu x}\, dx = 2^{\nu+1}\frac{\nu}{\mu}\beta^\nu e^{\beta\mu}\, K_\nu(\beta\mu)$

$$[|\arg\beta| < \pi, \quad \operatorname{Re}\mu > 0] \quad \text{ET I 140(30)}$$

3.392

1. $\displaystyle\int_0^\infty \left(x + \sqrt{1+x^2}\right)^\nu e^{-\mu x}\, dx = \frac{1}{\mu}\, S_{1,\nu}(\mu) + \frac{\nu}{\mu}\, S_{0,\nu}(\mu)$

$$[\operatorname{Re}\mu > 0] \quad \text{ET I 140(25)}$$

2. $\displaystyle\int_0^\infty \left(\sqrt{1+x^2} - x\right)^\nu e^{-\mu x}\, dx = \frac{1}{\mu}\, S_{1,\nu}(\mu) - \frac{\nu}{\mu}\, S_{0,\nu}(\mu)$

$$[\operatorname{Re}\mu > 0] \quad \text{ET I 140(26)}$$

3. $\displaystyle\int_0^\infty \frac{\left(x + \sqrt{1+x^2}\right)^\nu}{\sqrt{1+x^2}}\, e^{-\mu x}\, dx = \pi \operatorname{cosec}\nu\pi\, [\mathbf{J}_{-\nu}(\mu) - J_{-\nu}(\mu)]$

$$[\operatorname{Re}\mu > 0] \qquad \text{ET I 140(27), EH II 35(33)}$$

4. $\displaystyle\int_0^\infty \frac{\left(\sqrt{1+x^2} - x\right)^\nu}{\sqrt{1+x^2}}\, e^{-\mu x}\, dx = S_{0,\nu}(\mu) - \nu\, S_{-1,\nu}(\mu)$ $\qquad [\operatorname{Re}\mu > 0]$ $\qquad$ ET I 140(28)

3.393 $\displaystyle\int_0^\infty \frac{\left(x + \sqrt{x^2 + 4\beta^2}\right)^{2\nu}}{\sqrt{x^3 + 4\beta^2 x}}\, e^{-\mu x}\, dx$

$$= \frac{\sqrt{\mu\pi^3}}{2^{2\nu+3/2}\beta^{2\nu}}\left[J_{\nu+1/4}(\beta\mu)\, Y_{\nu-1/4}(\beta\mu) - J_{\nu-1/4}(\beta\mu)\, Y_{\nu+1/4}(\beta\mu) \right]$$

$$[\operatorname{Re}\beta > 0, \quad \operatorname{Re}\mu > 0] \quad \text{ET I 140(33)}$$

3.394 $\displaystyle\int_0^\infty \frac{\left(1 + \sqrt{1+x^2}\right)^{\nu+1/2}}{x^{\nu+1}\sqrt{1+x^2}}\, e^{-\mu x}\, dx = \sqrt{2}\,\Gamma(-\nu)\, D_\nu\left(\sqrt{2i\mu}\right) D_\nu\left(\sqrt{-2i\mu}\right)$

$$[\operatorname{Re}\mu \geq 0, \quad \operatorname{Re}\nu < 0] \quad \text{ET I 140(32)}$$

3.395

1.
$$\int_1^\infty \frac{\left(\sqrt{x^2-1}+x\right)^\nu + \left(\sqrt{x^2-1}+x\right)^{-\nu}}{\sqrt{x^2-1}} e^{-\mu x}\, dx = 2\, K_\nu(\mu)$$

$$[\operatorname{Re}\mu > 0] \qquad\qquad \text{ET I 140(29)}$$

2.
$$\int_1^\infty \frac{\left(x+\sqrt{x^2-1}\right)^{2\nu} + \left(x-\sqrt{x^2-1}\right)^{2\nu}}{\sqrt{x\left(x^2-1\right)}} e^{-\mu x}\, dx = \sqrt{\frac{2\mu}{\pi}}\, K_{\nu+1/4}\left(\frac{\mu}{2}\right) K_{\nu-1/4}\left(\frac{\mu}{2}\right)$$

$$[\operatorname{Re}\mu > 0] \qquad\qquad \text{ET I 140(34)}$$

3.
$$\int_0^\infty \frac{\left(x+\sqrt{x^2+1}\right)^\nu + \cos\nu\pi\left(x+\sqrt{x^2+1}\right)^{-\nu}}{\sqrt{x^2+1}} e^{-\mu x}\, dx = -\pi\left[\mathbf{E}_\nu(\mu) + Y_\nu(\mu)\right]$$

$$[\operatorname{Re}\mu > 0] \qquad\qquad \text{EH II 35(34)}$$

3.41–3.44 Combinations of rational functions of powers and exponentials

3.411

1.
$$\int_0^\infty \frac{x^{\nu-1}\, dx}{e^{\mu x}-1} = \frac{1}{\mu^\nu}\, \Gamma(\nu)\, \zeta(\nu) \qquad\qquad [\operatorname{Re}\mu > 0, \quad \operatorname{Re}\nu > 1] \qquad \text{FI II 792a}$$

2.
$$\int_0^\infty \frac{x^{2n-1}\, dx}{e^{px}-1} = (-1)^{n-1}\left(\frac{2\pi}{p}\right)^{2n} \frac{B_{2n}}{4n} \qquad\qquad [n=1,2,\ldots] \qquad \text{FI II 721a}$$

3.
$$\int_0^\infty \frac{x^{\nu-1}\, dx}{e^{\mu x}+1} = \frac{1}{\mu^\nu}\left(1-2^{1-\nu}\right)\Gamma(\nu)\,\zeta(\nu) \qquad\qquad [\operatorname{Re}\mu > 0, \quad \operatorname{Re}\nu > 0] \qquad \text{FI II 792a, WH}$$

4.
$$\int_0^\infty \frac{x^{2n-1}\, dx}{e^{px}+1} = \left(1-2^{1-2n}\right)\left(\frac{2\pi}{p}\right)^{2n} \frac{|B_{2n}|}{4n} \qquad\qquad [n=1,2,\ldots] \qquad \text{BI(83)(2), EH I 39(25)}$$

5.
$$\int_0^{\ln 2} \frac{x\, dx}{1-e^{-x}} = \frac{\pi^2}{12} \qquad\qquad\qquad \text{BI (104)(5)}$$

6.[8]
$$\int_0^\infty \frac{x^{\nu-1} e^{-\mu x}}{1-\beta e^{-x}}\, dx = \Gamma(\nu)\sum_{n=0}^\infty (\mu+n)^{-\nu}\beta^n = \Gamma(\nu)\,\Phi(\beta,\nu,\mu)$$

$$[\operatorname{Re}\mu > 0 \text{ and either } |\beta| \le 1, \quad \beta \ne 1, \quad \operatorname{Re}\nu > 0; \text{ or } \beta = 1, \quad \operatorname{Re}\nu > 1] \quad \text{EH I 27(3)}$$

7.
$$\int_0^\infty \frac{x^{\nu-1} e^{-\mu x}}{1-e^{-\beta x}}\, dx = \frac{1}{\beta^\nu}\,\Gamma(\nu)\,\zeta\left(\nu,\frac{\mu}{\beta}\right) \qquad\qquad [\operatorname{Re}\mu > 0, \quad \operatorname{Re}\nu > 1] \qquad \text{ET I 144(10)}$$

8.
$$\int_0^\infty \frac{x^{n-1} e^{-px}}{1+e^x}\, dx = (n-1)!\sum_{k=1}^\infty \frac{(-1)^{k-1}}{(p+k)^n} \qquad\qquad [p > -1; \quad n=1,2,\ldots] \qquad \text{BI (83)(9)}$$

9.
$$\int_0^\infty \frac{x e^{-x}\, dx}{e^x-1} = \frac{\pi^2}{6} - 1 \qquad\qquad (\text{cf. } \mathbf{4.231}\ 3) \qquad \text{BI (82)(1)}$$

10.
$$\int_0^\infty \frac{x e^{-2x}\, dx}{e^{-x}+1} = 1 - \frac{\pi^2}{12} \qquad\qquad (\text{cf. } \mathbf{4.251}\ 6) \qquad \text{BI (82)(2)}$$

11.
$$\int_0^\infty \frac{x e^{-3x}}{e^{-x}+1}\, dx = \frac{\pi^2}{12} - \frac{3}{4} \qquad\qquad (\text{cf. } \mathbf{4.251}\ 5) \qquad \text{BI (82)(3)}$$

12. $\int_0^\infty \frac{xe^{-2nx}}{1 + e^x}\, dx = -\frac{\pi^2}{12} + \sum_{k=1}^{2n-1} \frac{(-1)^{k-1}}{k^2}$ (cf. **4.251** 6)   BI (82)(5)

13. $\int_0^\infty \frac{xe^{-(2n-1)x}}{1 + e^x}\, dx = \frac{\pi^2}{12} + \sum_{k=1}^{2n} \frac{(-1)^k}{k^2}$ (cf. **4.251** 5)   BI (82)(4)

14.[7] $\int_0^\infty \frac{x^2 e^{-nx}}{1 - e^{-x}}\, dx = 2\sum_{k=n}^\infty \frac{1}{k^3} = 2\left(\zeta(3) - \sum_{k=1}^{n-1} \frac{1}{k^3}\right)$ $[n = 1, 2, \ldots]$ (cf. **4.261** 12)

     BI (82)(9)

15.[7] $\int_0^\infty \frac{x^2 e^{-nx}}{1 + e^{-x}}\, dx = 2\sum_{k=n}^\infty \frac{(-1)^{n+k}}{k^3} = (-1)^{n+1}\left(\frac{3}{2}\zeta(3) + 2\sum_{k=1}^{n-1} \frac{(-1)^k}{k^3}\right)$

     $[n = 1, 2, \ldots]$ (cf. **4.261** 11)

     LI (82)(10)

16. $\int_{-\infty}^\infty \frac{x^2 e^{-\mu x}}{1 + e^{-x}}\, dx = \pi^3 \csc^3 \mu\pi \left(2 - \sin^2 \mu\pi\right)$ $[0 < \operatorname{Re}\mu < 1]$ ET I 120(17)a

17. $\int_0^\infty \frac{x^3 e^{-nx}}{1 - e^{-x}}\, dx = \frac{\pi^4}{15} - 6\sum_{k=1}^{n-1} \frac{1}{k^4}$ (cf. **4.262** 5)   BI (82)(12)

18.[7] $\int_0^\infty \frac{x^3 e^{-nx}}{1 + e^{-x}}\, dx = 6\sum_{k=n}^\infty \frac{(-1)^{n+k}}{k^4} = (-1)^{n+1}\left(\frac{7}{120}\pi^4 - 6\sum_{k=1}^{n-1} \frac{(-1)^k}{k^4}\right)$

     (cf. **4.262** 4)   LI (82)(13)

19.[9] $\int_0^\infty e^{-px}\left(e^{-x} - 1\right)^n \frac{dx}{x} = -\sum_{k=0}^n (-1)^k \binom{n}{k} \ln(p + n - k)$ LI (89)(10)

20.[9] $\int_0^\infty e^{-px}\left(e^{-x} - 1\right)^n \frac{dx}{x^2} = \sum_{k=0}^n (-1)^k \binom{n}{k}(p + n - k)\ln(p + n - k)$ LI (89)(15)

21. $\int_0^\infty x^{n-1}\frac{1 - e^{-mx}}{1 - e^x}\, dx = (n-1)!\sum_{k=1}^m \frac{1}{k^n}$ (cf. **4.272** 11)   LI (83)(8)

22.[7] $\int_0^\infty \frac{x^{p-1}}{e^{rx} - q}\, dx = \frac{1}{qr^p}\Gamma(p)\sum_{k=1}^\infty \frac{q^k}{k^p} = \Gamma(p)r^{-p}\,\Phi(q, p, 1)$

     $[p > 0, \quad r > 0, \quad -1 < q < 1]$

     BI (83)(5)

23. $\int_{-\infty}^\infty \frac{xe^{\mu x}\, dx}{\beta + e^x} = \pi\beta^{\mu-1}\operatorname{cosec}(\mu\pi)\left[\ln\beta - \pi\cot(\mu\pi)\right]$ $[|\arg\beta| < \pi, \quad 0 < \operatorname{Re}\mu < 1]$

     BI (101)(5), ET I 120(16)a

24. $\int_{-\infty}^\infty \frac{xe^{\mu x}}{e^{\nu x} - 1}\, dx = \left(\frac{\pi}{\nu}\operatorname{cosec}\frac{\mu\pi}{\nu}\right)^2$ $[\operatorname{Re}\nu > \operatorname{Re}\mu > 0]$ (cf. **4.254** 2)

     LI (101)(3)

25. $\int_0^\infty x\frac{1 + e^{-x}}{e^x - 1}\, dx = \frac{\pi^2}{3} - 1$ (cf. **4.231** 4)   BI (82)(6)

26. $\displaystyle\int_0^\infty x\frac{1-e^{-x}}{1+e^{-3x}}e^{-x}\,dx = \frac{2\pi^2}{27}$ LI (82)(7)a

27. $\displaystyle\int_0^\infty \frac{1-e^{-\mu x}}{1+e^x}\frac{dx}{x} = \ln\left[\frac{\Gamma\left(\frac{\mu}{2}+1\right)}{\Gamma\left(\frac{\mu+1}{2}\right)}\sqrt{\pi}\right]$ $[\operatorname{Re}\mu > -1]$ BI (93)(4)

28. $\displaystyle\int_0^\infty \frac{e^{-\nu x}-e^{-\mu x}}{e^{-x}+1}\frac{dx}{x} = \ln\frac{\Gamma\left(\frac{\nu}{2}\right)\Gamma\left(\frac{\mu+1}{2}\right)}{\Gamma\left(\frac{\mu}{2}\right)\Gamma\left(\frac{\nu+1}{2}\right)}$ $[\operatorname{Re}\mu > 0, \quad \operatorname{Re}\nu > 0]$ BI (93)(6)

29. $\displaystyle\int_{-\infty}^\infty \frac{e^{px}-e^{qx}}{1+e^{rx}}\frac{dx}{x} = \ln\left[\tan\frac{p\pi}{2r}\cot\frac{q\pi}{2r}\right]$ $[|r|>|p|, \quad |r|>|q|, \quad rp>0, \quad rq>0]$

                                                                         BI (103)(3)

30. $\displaystyle\int_{-\infty}^\infty \frac{e^{px}-e^{qx}}{1-e^{rx}}\frac{dx}{x} = \ln\left[\sin\frac{p\pi}{r}\operatorname{cosec}\frac{q\pi}{r}\right]$ $[|r|>|p|, \quad |r|>|q|, \quad rp>0, \quad rq>0]$

                                                                          BI (103)(4)

31. $\displaystyle\int_0^\infty \frac{e^{-qx}+e^{(q-p)x}}{1-e^{-px}}x\,dx = \left(\frac{\pi}{p}\operatorname{cosec}\frac{q\pi}{p}\right)^2$ $[0<q<p]$ BI (82)(8)

32. $\displaystyle\int_0^\infty \frac{e^{-px}-e^{(p-q)x}}{e^{-qx}+1}\frac{dx}{x} = \ln\cot\frac{p\pi}{2q}$ $[0<p<q]$ BI (93)(7)

3.412 $\displaystyle\int_0^\infty \left\{\frac{a+be^{-px}}{ce^{px}+g+he^{-px}} - \frac{a+be^{-qx}}{ce^{qx}+g+he^{-qx}}\right\}\frac{dx}{x} = \frac{a+b}{c+g+h}\ln\frac{p}{q}$

                                                          $[p>0, \quad q>0]$ BI (96)(7)

3.413

1. $\displaystyle\int_0^\infty \frac{\left(1-e^{-\beta x}\right)\left(1-e^{-\gamma x}\right)e^{-\mu x}}{1-e^{-x}}\frac{dx}{x} = \ln\frac{\Gamma(\mu)\,\Gamma(\beta+\gamma+\mu)}{\Gamma(\mu+\beta)\,\Gamma(\mu+\gamma)}$

      $[\operatorname{Re}\mu > 0, \quad \operatorname{Re}\mu > -\operatorname{Re}\beta, \quad \operatorname{Re}\mu > -\operatorname{Re}\gamma, \quad \operatorname{Re}\mu > -\operatorname{Re}(\beta+\gamma)]$ (cf. **4.267** 25)

                                                                         BI (93)(13)

2. $\displaystyle\int_0^\infty \frac{\left\{1-e^{(q-p)x}\right\}^2}{e^{qx}-e^{(q-2p)x}}\frac{dx}{x} = \ln\operatorname{cosec}\frac{q\pi}{2p}$ $[0<q<p]$ BI (95)(6)

3. $\displaystyle\int_0^\infty \frac{e^{-px}-e^{-qx}}{1+e^{-x}}\frac{1+e^{-(2n+1)x}}{x}\,dx$

           $= \ln\left\{\dfrac{q(q+2)(q+4)\cdots(q+2n)(p+1)(p+3)\cdots(p+2n-1)}{p(p+2)(p+4)\cdots(p+2n)(q+1)(q+3)\cdots(q+2n-1)}\right\}$

                        $[\operatorname{Re}p > -2n, \quad \operatorname{Re}q > -2n]$ (cf. **4.267** 14)   BI (93)(11)

3.414 $\displaystyle\int_0^\infty \frac{\left(1-e^{-\beta x}\right)\left(1-e^{-\gamma x}\right)\left(1-e^{-\delta x}\right)e^{-\mu x}}{1-e^{-x}}\frac{dx}{x} = \ln\frac{\Gamma(\mu)\,\Gamma(\mu+\beta+\gamma)\,\Gamma(\mu+\beta+\delta)\,\Gamma(\mu+\gamma+\delta)}{\Gamma(\mu+\beta)\,\Gamma(\mu+\gamma)\,\Gamma(\mu+\delta)\,\Gamma(\mu+\beta+\gamma+\delta)}$

              $[2\operatorname{Re}\mu > |\operatorname{Re}\beta| + |\operatorname{Re}\gamma| + |\operatorname{Re}\delta|]$ (cf. **4.267** 31)   BI (93)(14), ET I 145(17)

3.415

1. $\displaystyle\int_0^\infty \frac{x\,dx}{(x^2+\beta^2)(e^{\mu x}-1)} = \frac{1}{2}\left[\ln\left(\frac{\beta\mu}{2\pi}\right) - \frac{\pi}{\beta\mu} - \psi\left(\frac{\beta\mu}{2\pi}\right)\right]$

                                         $[\operatorname{Re}\beta > 0, \quad \operatorname{Re}\mu > 0]$

                                                                  BI (97)(20), EH I 18(27)

2.7 $\displaystyle\int_0^\infty \frac{x\,dx}{(x^2+\beta^2)^2\,(e^{2\pi x}-1)}$ $\quad -\dfrac{1}{8\beta^3}-\dfrac{1}{4\beta^2}+\dfrac{1}{4\beta}\,\psi'(\beta)$

$$= \frac{1}{4\beta^4}\sum_{k=0}^\infty \frac{|B_{2k+2}|}{\beta^{2k}}$$

$$[\mathrm{Re}\,\beta > 0] \qquad \text{BI(97)(22), EH I 22(12)}$$

3.8 $\displaystyle\int_0^\infty \frac{x\,dx}{(x^2\beta^2)\,(e^{\mu x}+1)} = \frac{1}{2}\left[\psi\left(\frac{\beta\mu}{2\pi}+\frac{1}{2}\right)-\ln\left(\frac{\beta\mu}{2\pi}\right)\right]$

$$[\mathrm{Re}\,\beta > 0, \quad \mathrm{Re}\,\mu > 0]$$

4.8 $\displaystyle\int_0^\infty \frac{x\,dx}{(x^2+\beta^2)^2\,(e^{2\pi x}+1)} = \frac{1}{4\beta^2}-\frac{1}{4\beta}\,\psi'\left(\beta+\frac{1}{2}\right)$ $\quad [\mathrm{Re}\,\beta > 0, \quad \mathrm{Re}\,\mu > 0]$

3.416

1. $\displaystyle\int_0^\infty \frac{(1+ix)^{2n}-(1-ix)^{2n}}{i}\,\frac{dx}{e^{2\pi x}-1} = \frac{1}{2}\frac{2n-1}{2n+1}$ $\quad [n=1,2,\ldots]$ $\quad$ BI (88)(4)

2. $\displaystyle\int_0^\infty \frac{(1+ix)^{2n}-(1-ix)^{2n}}{i}\,\frac{dx}{e^{\pi x}+1} = \frac{1}{2n+1}$ $\quad [n=1,2,\ldots]$ $\quad$ BI (87)(1)

3.8 $\displaystyle\int_0^\infty \frac{(1+ix)^{2n-1}-(1-ix)^{2n-1}}{i}\,\frac{dx}{e^{\pi x}+1} = \frac{1}{2n}\left[1-2^{2^n}B_{2n}\right]$

$$[n=1,2,\ldots] \qquad \text{BI (87)(2)}$$

3.417

1. $\displaystyle\int_{-\infty}^\infty \frac{x\,dx}{a^2 e^x+b^2 e^{-x}} = \frac{\pi}{2ab}\ln\frac{b}{a}$ $\quad [ab>0] \quad$ (cf. **4.231** 8) $\quad$ BI (101)(1)

2. $\displaystyle\int_{-\infty}^\infty \frac{x\,dx}{a^2 e^x-b^2 e^{-x}} = \frac{\pi^2}{4ab}$ $\quad$ (cf. **4.231** 10) $\quad$ LI (101)(2)

3.418

1.6 $\displaystyle\int_0^\infty \frac{x\,dx}{e^x+e^{-x}-1} = \frac{1}{3}\left[\psi'\left(\frac{1}{3}\right)-\frac{2}{3}\pi^2\right] = 1.1719536193\ldots$ $\quad$ LI (88)(1)

2.6 $\displaystyle\int_0^\infty \frac{xe^{-x}\,dx}{e^x+e^{-x}-1} = \frac{1}{6}\left[\psi'\left(\frac{1}{3}\right)-\frac{5}{6}\pi^2\right] = 0.3118211319\ldots$ $\quad$ LI (88)(2)

3. $\displaystyle\int_0^{\ln 2} \frac{x\,dx}{e^x+2e^{-x}-2} = \frac{\pi}{8}\ln 2$ $\quad$ BI (104)(7)

3.419

1. $\displaystyle\int_{-\infty}^\infty \frac{x\,dx}{(\beta+e^x)\,(1+e^{-x})} = \frac{(\ln\beta)^2}{2(\beta-1)}$ $\quad [|\arg\beta|<\pi] \quad$ (cf. **4.232** 2)

$$\text{BI (101)(16)}$$

2. $\displaystyle\int_{-\infty}^\infty \frac{x\,dx}{(\beta+e^x)\,(1-e^{-x})} = \frac{\pi^2+(\ln\beta)^2}{2(\beta+1)}$ $\quad [|\arg\beta|<\pi] \quad$ (cf. **4.232** 3)

$$\text{BI (101)(17)}$$

3.

$$\int_{-\infty}^{\infty} \frac{x^2 \, dx}{(\beta + e^x)(1 - e^{-x})} = \frac{\left[\pi^2 + (\ln \beta)^2\right] \ln \beta}{3(\beta + 1)}$$

$$[|\arg \beta| < \pi] \qquad (\text{cf. } \mathbf{4.261} \ 4)$$

BI (102)(6)

4.

$$\int_{-\infty}^{\infty} \frac{x^3 \, dx}{(\beta + e^x)(1 - e^{-x})} = \frac{\left[\pi^2 + (\ln \beta)^2\right]^2}{4(\beta + 1)}$$

$$[|\arg \beta| < \pi] \qquad (\text{cf. } \mathbf{4.262} \ 3)$$

BI (102)(9)

5.

$$\int_{-\infty}^{\infty} \frac{x^4 \, dx}{(\beta + e^x)(1 - e^{-x})} = \frac{\left[\pi^2 + (\ln \beta)^2\right]^2}{15(\beta + 1)} \left[7\pi^2 + 3(\ln \beta)^2\right] \ln \beta$$

$$(\text{cf. } \mathbf{4.263} \ 1) \qquad \text{BI (102)(10)}$$

6.

$$\int_{-\infty}^{\infty} \frac{x^5 \, dx}{(\beta + e^x)(1 - e^{-x})} = \frac{\left[\pi^2 + (\ln \beta)^2\right]^2}{6(\beta + 1)} \left[3\pi^2 + (\ln \beta)^2\right]^2$$

$$(\text{cf. } \mathbf{4.264} \ 3) \qquad \text{BI (102)(11)}$$

7.

$$\int_{-\infty}^{\infty} \frac{(x - \ln \beta) x \, dx}{(\beta - e^x)(1 - e^{-x})} = \frac{-\left[4\pi^2 + (\ln \beta)^2\right] \ln \beta}{6(\beta - 1)}$$

$$[|\arg \beta| < \pi] \qquad (\text{cf. } \mathbf{4.257} \ 4)$$

BI (102)(7)

3.421

1.

$$\int_0^{\infty} \left(e^{-\nu x} - 1\right)^n \left(e^{-\rho x} - 1\right)^m e^{-\mu x} \frac{dx}{x^2} = \sum_{k=0}^{n} (-1)^k \binom{n}{k} \sum_{l=0}^{m} (-1)^l \binom{m}{l}$$

$$\times \left\{(m - l)\rho + (n - k)\nu + \mu\right\} \ln\left[(m - l)\rho + (n - k)\nu + \mu\right]$$

$$[\operatorname{Re} \nu > 0, \quad \operatorname{Re} \mu > 0, \quad \operatorname{Re} \rho > 0] \quad \text{BI (89)(17)}$$

2.

$$\int_0^{\infty} \left(1 - e^{-\nu x}\right)^n \left(1 - e^{-\rho x}\right) e^{-x} \frac{dx}{x^3} = \frac{1}{2} \sum_{k=0}^{n} (-1)^k \binom{n}{k} (\rho + k\nu + 1)^2$$

$$\times \ln(\rho + k\nu + 1) + \frac{1}{2} \sum_{k=1}^{n} (-1)^{k-1} \binom{n}{k} (k\nu + 1)^2 \ln(k\nu + 1)$$

$$[n \geq 2, \quad \operatorname{Re} \nu > 0, \quad \operatorname{Re} \rho > 0] \quad \text{BI (89)(31)}$$

3.

$$\int_{-\infty}^{\infty} \frac{x e^{-\mu x} \, dx}{(\beta + e^{-x})(\gamma + e^{-x})} = \frac{\pi \left(\beta^{\mu-1} \ln \beta - \gamma^{\mu-1} \ln \gamma\right)}{(\beta - \gamma) \sin \mu \pi} + \frac{\pi^2 \left(\beta^{\mu-1} - \gamma^{\mu-1}\right) \cos \mu \pi}{(\gamma - \beta) \sin^2 \mu \pi}$$

$$[|\arg \beta| < \pi, \quad |\arg \gamma| < \pi, \quad \beta \neq \gamma. \quad 0 < \operatorname{Re} \mu < 2] \quad \text{ET I 120(19)}$$

4.

$$\int_0^{\infty} \left(e^{-px} - e^{-qx}\right) \left(e^{-rx} - e^{-sx}\right) e^{-x} \frac{dx}{x} = \ln \frac{(p + s + 1)(q + r + 1)}{(p + r + 1)(q + s + 1)}$$

$$[p + s > -1, \quad p + r > -1, \quad q > p] \qquad (\text{cf. } \mathbf{4.267} \ 24) \quad \text{BI (89)(11)}$$

5. $\int_0^\infty \left(1 - e^{-px}\right) \left(1 - e^{-qx}\right) \left(1 - e^{-rx}\right) e^{-x} \dfrac{dx}{x}$

$$= (p+q+1)\ln(p+q+1)$$
$$+(p+r+1)\ln(p+r+1) + (q+r+1)\ln(q+r+1)$$
$$-(p+1)\ln(p+1) - (q+1)\ln(q+1) - (r+1)\ln(r+1)$$
$$-(p+q+r)\ln(p+q+r)$$

$[p > 0, \quad q > 0, \quad r > 0]$ (cf. **4.268** 3)   BI (89)(14)

3.422 $\int_{-\infty}^\infty \dfrac{x(x-a)e^{\mu x}\, dx}{(\beta - e^x)(1 - e^{-x})} = \dfrac{-\pi^2}{e^a - 1} \operatorname{cosec}^2 \mu\pi \left[(e^{\alpha\mu}+1)\ln\mu - 2\pi\cot\mu\pi\,(e^{\alpha\mu}-1)\right]$

$[a > 0, \quad |\arg\beta| < \pi, \quad |\operatorname{Re}\mu| < 1]$ (cf. **4.257** 5)   BI (102)(8)a

3.423

1. $\int_0^\infty \dfrac{x^{\nu-1}}{\left(e^x - 1\right)^2}\, dx = \Gamma(\nu)\left[\zeta(\nu-1) - \zeta(\nu)\right]$ $[\operatorname{Re}\nu > 2]$ ET I 313(10)

2.[6] $\int_0^\infty \dfrac{x^{\nu-1}e^{-\mu x}}{\left(e^x - 1\right)^2}\, dx = \Gamma(\nu)\left[\zeta(\nu-1, \mu+2) - (\mu+1)\,\zeta(\nu, \mu+2)\right]$

$[\operatorname{Re}\mu > -2, \quad \operatorname{Re}\nu > 2]$ ET I 313(11)

3.[8] $\int_0^\infty \dfrac{x^q e^{-px}\, dx}{\left(1 - ae^{-px}\right)^2} = \dfrac{\Gamma(q+1)}{ap^{q+1}} \sum_{k=1}^\infty \dfrac{a^k}{k^q}$ $[a < 1, \quad q > -1, \quad p > 0]$ BI (85)(13)

4.[7] $\int_0^\infty \dfrac{x^{\nu-1}e^{-\mu x}}{\left(1 - \beta e^{-x}\right)^2}\, dx = \Gamma(\nu)\left[\Phi(\beta; \nu-1; \mu) - (\mu-1)\,\Phi(\beta; \nu; \mu)\right]$

$[\operatorname{Re}\nu > 0, \quad \operatorname{Re}\mu > 0, \quad |\arg(1-\beta)| < \pi]$ (cf. **9.550**)   ET I 313(12)

5. $\int_{-\infty}^\infty \dfrac{xe^x\, dx}{\left(\beta + e^x\right)^2} = \dfrac{1}{\beta}\ln\beta$ $[|\arg\beta| < \pi]$ (cf. **4.231** 5)

BI (101)(10)

3.424

1.[7] $\int_0^\infty \dfrac{(1+a)e^x - a}{\left(1 - e^x\right)^2}\, e^{-ax}x^n\, dx = n!\,\zeta(n, a)$ $[a > -1, \quad n = 1, 2, \ldots]$ BI (85)(15)

2. $\int_0^\infty \dfrac{(1+a)e^x + a}{\left(1 + e^x\right)^2}\, e^{-ax}x^n\, dx = n! \sum_{k=1}^\infty \dfrac{(-1)^k}{(a+k)^n}$ $[a > -1, \quad n = 1, 2, \ldots]$ BI (85)(14)

3. $\int_{-\infty}^\infty \dfrac{a^2 e^x + b^2 e^{-x}}{\left(a^2 e^x - b^2 e^{-x}\right)^2}\, x^2\, dx = \dfrac{\pi^2}{2ab}$ $[ab > 0]$ BI (102)(3)a

4. $\int_{-\infty}^\infty \dfrac{a^2 e^x - b^2 e^{-x}}{\left(a^2 e^x + b^2 e^{-x}\right)^2}\, x^2\, dx = \dfrac{\pi}{ab}\ln\dfrac{b}{a}$ $[ab > 0]$ BI (102)(1)

5. $\displaystyle\int_0^\infty \frac{e^x - e^{-x} + 2}{(e^x - 1)^2} x^2 \, dx = \frac{2}{3}\pi^2 - 2$ BI (85)(7)

3.425

1.[7] $\displaystyle\int_{-\infty}^\infty \frac{xe^x \, dx}{(a^2 + b^2 e^{2x})^n} = \frac{\sqrt{\pi}\,\Gamma\left(n - \frac{1}{2}\right)}{4a^{2n-1} b\,\Gamma(n)} \left[2\ln\frac{a}{2b} - \boldsymbol{C} - \psi\left(n - \frac{1}{2}\right)\right]$

$$[ab > 0, \quad n > 0]$$

                                             BI(101)(13), LI(101)(13)

2.[7] $\displaystyle\int_{-\infty}^\infty \frac{\left(a^2 e^x - e^{-x}\right) x^2 \, dx}{(a^2 e^x + e^{-x})^{p+1}} = -\frac{1}{a^{p+1}} B\left(\frac{p}{2}, \frac{p}{2}\right)\ln a$ $[a > 0, \quad p > 0]$ BI (102)(5)

3.426

1. $\displaystyle\int_{-\infty}^\infty \frac{(e^x - ae^{-x}) x^2 \, dx}{(a + e^x)^2(1 + e^{-x})^2} = \frac{(\ln a)^2}{a - 1}$ BI (102)(12)

2. $\displaystyle\int_{-\infty}^\infty \frac{(e^x - ae^{-x}) x^2 \, dx}{(a + e^x)^2(1 - e^{-x})^2} = \frac{\pi^2 + (\ln a)^2}{a + 1}$ BI (102)(13)

3.427

1. $\displaystyle\int_0^\infty \left(\frac{e^{-x}}{x} + \frac{e^{-\mu x}}{e^{-x} - 1}\right) dx = \psi(\mu)$ $[\operatorname{Re}\mu > 0]$ (cf. **4.281** 4)   WH

2.[7] $\displaystyle\int_0^\infty \left(\frac{1}{1 - e^{-x}} - \frac{1}{x}\right) e^{-x} \, dx = \boldsymbol{C}$ (cf. **4.281** 1)   BI (94)(1)

3. $\displaystyle\int_0^\infty \left(\frac{1}{2} - \frac{1}{1 + e^{-x}}\right) \frac{e^{-2x}}{x} \, dx = \frac{1}{2}\ln\frac{\pi}{4}$ BI (94)(5)

4. $\displaystyle\int_0^\infty \left(\frac{1}{2} - \frac{1}{x} + \frac{1}{e^x - 1}\right) \frac{e^{-\mu x}}{x} \, dx = \ln\Gamma(\mu) - \left(\mu - \frac{1}{2}\right)\ln\mu + \mu - \frac{1}{2}\ln(2\pi)$

$$[\operatorname{Re}\mu > 0]$$

                                                          WH

5. $\displaystyle\int_0^\infty \left(\frac{1}{2}e^{-2x} - \frac{1}{e^x + 1}\right) \frac{dx}{x} = -\frac{1}{2}\ln\pi$ BI (94)(6)

6. $\displaystyle\int_0^\infty \left(\frac{e^{\mu x} - 1}{1 - e^{-x}} - \mu\right) \frac{e^{-x}}{x} \, dx = -\ln\Gamma(\mu) - \ln\sin(\pi\mu) + \ln\pi$

$$[\operatorname{Re}\mu < 1]$$

                                                      EH I 21(6)

7. $\displaystyle\int_0^\infty \left(\frac{e^{-\nu x}}{1 - e^{-x}} - \frac{e^{-\mu x}}{x}\right) dx = \ln\mu - \psi(\nu)$ (cf. **4.281** 5)   BI (94)(3)

8. $\displaystyle\int_0^\infty \left(\frac{n}{x} - \frac{e^{-\mu x}}{1 - e^{-x/n}}\right) e^{-x} \, dx = n\,\psi(n\mu + n) - n\ln n$

$$[\operatorname{Re}\mu > 0, \quad n = 1, 2, \ldots]$$

                                                     BI (94)(4)

9. $\displaystyle\int_0^\infty \left(\mu - \frac{1 - e^{-\mu x}}{1 - e^{-x}}\right) \frac{e^{-x}}{x} \, dx = \ln\Gamma(\mu + 1)$ $[\operatorname{Re}\mu > -1]$ WH

10. $\displaystyle\int_0^\infty \left(\nu e^{-x} - \frac{e^{-\mu x} - e^{-(\mu+\nu)x}}{e^x - 1}\right) \frac{dx}{x} = \ln \frac{\Gamma(\mu+\nu+1)}{\Gamma(\mu+1)}$

$\qquad\qquad\qquad\qquad\qquad\qquad$ [$\operatorname{Re}\mu > -1$, $\quad \operatorname{Re}\nu > 0$] $\qquad$ BI (94)(8)

11. $\displaystyle\int_0^\infty \left[(1 - e^x)^{-1} + x^{-1} - 1\right] e^{-xz}\, dx = \psi(z) - \ln z \qquad$ [$\operatorname{Re}z > 0$] $\qquad$ EH I 18(24)

3.428

1. $\displaystyle\int_0^\infty \left(\nu e^{-\mu x} - \frac{1}{\mu}e^{-x} - \frac{1}{\mu}\frac{e^{-1} - e^{-\mu\nu x}}{1 - e^{-x}}\right) \frac{dx}{x} = \frac{1}{\mu}\ln\Gamma(\mu\nu) - \nu\ln\mu$

$\qquad\qquad\qquad\qquad\qquad\qquad$ [$\operatorname{Re}\mu > 0$, $\quad \operatorname{Re}\nu > 0$] $\qquad$ BI (94)(18)

2. $\displaystyle\int_0^\infty \left(\frac{n-1}{2} + \frac{n-1}{1 - e^{-x}} + \frac{e^{(1-\mu)x}}{1 - e^{x/n}} + \frac{e^{-n\mu x}}{1 - e^{-x}}\right) e^{-x}\frac{dx}{x} = \frac{n-1}{2}\ln 2\pi - \left(n\mu + \frac{1}{2}\right)\ln n$

$\qquad\qquad\qquad\qquad\qquad\qquad$ [$\operatorname{Re}\mu > 0$, $\quad n = 1, 2, \ldots$] $\qquad$ BI (94)(14)

3. $\displaystyle\int_0^\infty \left(n\mu - \frac{n-1}{2} - \frac{n}{1 - e^{-x}} - \frac{e^{(1-\mu)x}}{1 - e^{x/n}}\right) \frac{e^{-x}}{x}\, dx = \sum_{k=0}^{n-1} \ln\Gamma\left(\mu - \frac{k}{n} + 1\right)$

$\qquad\qquad\qquad\qquad\qquad\qquad$ [$\operatorname{Re}\mu > 0$, $\quad n = 1, 2, \ldots$] $\qquad$ BI (94)(13)

4. $\displaystyle\int_0^\infty \left(\frac{e^{-\nu x}}{1 - e^x} - \frac{e^{-\mu\nu x}}{1 - e^{\mu x}} - \frac{e^x}{1 - e^x} + \frac{e^{\mu x}}{1 - e^{\mu x}}\right) \frac{dx}{x} = \nu\ln\mu$

$\qquad\qquad\qquad\qquad\qquad\qquad$ [$\operatorname{Re}\mu > 0$, $\quad \operatorname{Re}\nu > 0$] $\qquad$ LI (94)(15)

5. $\displaystyle\int_0^\infty \left[\frac{1}{e^x - 1} - \frac{\mu e^{-\mu x}}{1 - e^{-\mu x}} + \left(a\mu - \frac{\mu+1}{2}\right)e^{-\mu x} + (1 - a\mu)e^{-x}\right] \frac{dx}{x}$

$\qquad\qquad\qquad\qquad\qquad\qquad\qquad = \dfrac{\mu-1}{2}\ln(2\pi) + \left(\dfrac{1}{2} - a\mu\right)\ln\mu$

$\qquad\qquad\qquad\qquad\qquad\qquad$ [$\operatorname{Re}\mu > 0$] $\qquad$ BI (94)(16)

6. $\displaystyle\int_0^\infty \left[\frac{e^{-\nu x}}{1 - e^{-x}} - \frac{e^{-\mu\nu x}}{1 - e^{-\mu x}} - \frac{(\mu-1)e^{-\mu x}}{1 - e^{-\mu x}} - \frac{\mu-1}{2}e^{-\mu}x\right] \frac{dx}{x} = \frac{\mu-1}{2}\ln(2\pi) + \left(\frac{1}{2} - \mu\nu\right)\ln\mu$

$\qquad\qquad\qquad\qquad$ [$\operatorname{Re}\mu > 0$, $\quad \operatorname{Re}\nu > 0$] $\qquad$ (cf. **4.267** 37) $\quad$ BI (94)(17)

7. $\displaystyle\int_0^\infty \left[1 - e^{-x} - \frac{(1 - e^{-\nu x})(1 - e^{-\mu x})}{1 - e^{-x}}\right] \frac{dx}{x} = \ln\mathrm{B}(\mu, \nu)$

$\qquad\qquad\qquad\qquad\qquad\qquad$ [$\operatorname{Re}\mu > 0$, $\quad \operatorname{Re}\nu > 0$] $\qquad$ BI (94)(12)

3.429 $\displaystyle\int_0^\infty \left[e^{-x} - (1 + x)^{-\mu}\right] \frac{dx}{x} = \psi(\mu) \qquad$ [$\operatorname{Re}\mu > 0$] $\qquad$ NH 184(7)

3.431

1. $\displaystyle\int_0^\infty \left(e^{-\mu x} - 1 + \mu x - \frac{1}{2}\mu^2 x^2\right) x^{\nu-1}\, dx = \frac{-1}{\nu(\nu+1)(\nu+2)\mu^\nu}\Gamma(\nu+3)$

$\qquad\qquad\qquad\qquad\qquad\qquad$ [$\operatorname{Re}\mu > 0$, $\quad -2 > \operatorname{Re}\nu > -3$]

$\qquad\qquad\qquad\qquad\qquad\qquad\qquad\qquad\qquad\qquad\qquad$ LI (90)(5)

2. $\displaystyle\int_0^\infty \left[x^{-1} - \frac{1}{2} x^{-2}(x+2)\left(1 - e^{-x}\right) \right] e^{-px}\, dx = -1 + \left(p + \frac{1}{2}\right) \ln\left(1 + \frac{1}{p}\right)$

$$[\mathrm{Re}\, p > 0]$$
       ET I 144(6)

3.432

1. $\displaystyle\int_0^\infty x^{\nu-1} e^{-mx}\left(e^{-x} - 1\right)^n dx = \Gamma(\nu) \sum_{k=0}^n (-1)^k \binom{n}{k} \frac{1}{(n+m-k)^\nu}$

$$[n = 0, 1, \ldots, \mathrm{Re}\,\nu > 0]$$
       LI (90)(10)

2. $\displaystyle\int_0^\infty \left[x^{\nu-1} e^{-x} - e^{-\mu x}\left(1 - e^{-x}\right)^{\nu-1} \right] dx = \Gamma(\nu) - \frac{\Gamma(\mu)}{\Gamma(\mu+\nu)}$

$$[\mathrm{Re}\,\mu > 0, \quad \mathrm{Re}\,\nu > 0]$$
       LI (81)(14)

3.433 $\displaystyle\int_0^\infty x^{p-1} \left[e^{-x} + \sum_{k=1}^n (-1)^k \frac{x^{k-1}}{(k-1)!} \right] dx = \Gamma(p)$ $[-n < p < -n+1, \quad n = 0, 1, \ldots]$

       FI II 805

3.434

1. $\displaystyle\int_0^\infty \frac{e^{-\nu x} - e^{-\mu x}}{x^{\rho+1}}\, dx = \frac{\mu^\rho - \nu^\rho}{\rho}\, \Gamma(1-\rho)$ $[\mathrm{Re}\,\mu > 0, \quad \mathrm{Re}\,\nu > 0, \quad \mathrm{Re}\,\rho < 1]$

       BI (90)(6)

2. $\displaystyle\int_0^\infty \frac{e^{-\mu x} - e^{-\nu x}}{x}\, dx = \ln \frac{\nu}{\mu}$ $[\mathrm{Re}\,\mu > 0, \quad \mathrm{Re}\,\nu > 0]$

       FI II 634

3.435

1. $\displaystyle\int_0^\infty \left\{ (x+1)e^{-x} - e^{-\frac{x}{2}} \right\} \frac{dx}{x} = 1 - \ln 2$

       LI (89)(19)

2.[7] $\displaystyle\int_0^\infty \frac{1 - e^{-\mu x}}{x(x+\beta)}\, dx = \frac{1}{\beta}\left[\ln(\beta\mu \boldsymbol{C}) - e^{\beta\mu}\, \mathrm{Ei}(-\beta\mu) \right]$ $[|\arg\beta| < \pi, \quad \mathrm{Re}\,\mu > 0]$ ET II 217 (18)

3. $\displaystyle\int_0^\infty \left(\frac{1}{1+x} - e^{-x} \right) \frac{dx}{x} = \boldsymbol{C}$

       FI II 7 95, 802

4. $\displaystyle\int_0^\infty \left(e^{-\mu x} - \frac{1}{1+ax} \right) \frac{dx}{x} = \ln \frac{a}{\mu} - \boldsymbol{C}$ $[a > 0, \quad \mathrm{Re}\,\mu > 0]$ BI (92)(10)

3.436 $\displaystyle\int_0^\infty \left\{ \frac{e^{-npx} - e^{-nqx}}{n} - \frac{e^{-mpx} - e^{-mqx}}{m} \right\} \frac{dx}{x^2} = (q-p)\ln \frac{m}{n}$ $[p > 0, \quad q > 0]$ BI (89)(28)

3.437 $\displaystyle\int_0^\infty \left\{ pe^{-x} - \frac{1 - e^{-px}}{x} \right\} \frac{dx}{x} = p\ln p - p$ $[p > 0]$ BI (89)(24)

3.438

1. $\displaystyle\int_0^\infty \left\{ \left(\frac{1}{2} + \frac{1}{x} \right) e^{-x} - \frac{1}{x} e^{-\frac{x}{2}} \right\} \frac{dx}{x} = \frac{\ln 2 - 1}{2}$

       BI (89)(19)

2.[7] $\displaystyle\int_0^\infty \left\{ \frac{p^2}{6} e^{-x} - \frac{p^2}{2x} - \frac{p}{x^2} - \frac{1 - e^{-px}}{x^3} \right\} \frac{dx}{x} = \frac{p^2}{6} \ln p - \frac{11}{36} p^3$

$$[p > 0]$$
       BI (89)(33)

3. $\int_0^\infty \left(e^{-x} - e^{-2x} - \dfrac{1}{x} e^{-2x} \right) \dfrac{dx}{x} = 1 - \ln 2$　　　　　　　　　BI (89)(25)

4. $\int_0^\infty \left\{ \left(p - \dfrac{1}{2} \right) e^{-x} + \dfrac{x+2}{2x} \left(e^{-px} - e^{-\frac{x}{2}} \right) \right\} \dfrac{dx}{x} = \left(p - \dfrac{1}{2} \right) (\ln p - 1)$

$$[p > 0]$$　　　　　　　　　BI (89)(22)

3.439 $\int_0^\infty \left\{ (p-q)e^{-rx} + \dfrac{1}{mx} \left(e^{-mpx} - e^{-mqx} \right) \right\} \dfrac{dx}{x} = p \ln p - q \ln q - (p-q) \left(1 + \ln \dfrac{r}{m} \right)$

$$[p > 0, \quad q > 0, \quad r > 0] \quad \text{LI(89)(26), LI(89)(27)}$$

3.441 $\int_0^\infty \left\{ (p-r)e^{-qx} + (r-q)e^{-px} + (q-p)e^{-rx} \right\} \dfrac{dx}{x^2} = (r-q)p \ln p + (p-r)q \ln q + (q-p)r \ln r$

$$[p > 0, \quad q > 0, \quad r > 0] \qquad (\text{cf. } \mathbf{4.268}\ 6) \quad \text{BI (89)(18)}$$

3.442

1. $\int_0^\infty \left\{ 1 - \dfrac{x+2}{2x} \left(1 - e^{-x} \right) \right\} e^{-qx} \dfrac{dx}{x} = -1 + \left(q + \dfrac{1}{2} \right) \ln \dfrac{q+1}{q}$

$$[q > 0]$$　　　　　　　　　BI (89)(23)

2. $\int_0^\infty \left(\dfrac{e^{-x} - 1}{x} + \dfrac{1}{1+x} \right) \dfrac{dx}{x} = \boldsymbol{C} - 1$　　　　　　　　　BI (92)(16)

3. $\int_0^\infty \left(e^{-px} - \dfrac{1}{1+a^2 x^2} \right) \dfrac{dx}{x} = -\boldsymbol{C} + \ln \dfrac{a}{p}$　　　$[p > 0]$　　　BI (92)(11)

3.443

1. $\int_0^\infty \left\{ \dfrac{e^{-x} p^2}{2} - \dfrac{p}{x} + \dfrac{1 - e^{-px}}{x^2} \right\} \dfrac{dx}{x} = \dfrac{p^2}{2} \ln p - \dfrac{3}{4} p^2$　　$[p > 0]$　　BI (89)(32)

2. $\int_0^\infty \dfrac{(1 - e^{-px})^n e^{-qx}}{x^3} \, dx = \dfrac{1}{2} \sum_{k=2}^n (-1)^{k-1} \binom{n}{k} (q+kp)^2 \ln(q+kp)$

$$[n > 2, \quad q > 0, \quad pn + q > 0] \qquad (\text{cf. } \mathbf{4.268}\ 4) \quad \text{BI (89)(30)}$$

3. $\int_0^\infty \left(1 - e^{-px} \right)^2 e^{-qx} \dfrac{dx}{x^2} = (2p+q) \ln(2p+q) - 2(p+q) \ln(p+q) + q \ln q$

$$[q > 0, \quad 2p > -q] \qquad (\text{cf. } \mathbf{4.268}\ 2)$$
$$\text{BI (89)(13)}$$

3.45 Combinations of powers and algebraic functions of exponentials

3.451

1. $\int_0^\infty x e^{-x} \sqrt{1 - e^{-x}} \, dx = \dfrac{4}{3} \left(\dfrac{4}{3} - \ln 2 \right)$　　　　　　　　　BI (99)(1)

2. $\int_0^\infty x e^{-x} \sqrt{1 - e^{-2x}} \, dx = \dfrac{\pi}{4} \left(\dfrac{1}{2} + \ln 2 \right)$　　　$(\text{cf. } \mathbf{4.241}\ 9)$　　BI (99)(2)

3.452

1. $\displaystyle\int_0^\infty \frac{x\,dx}{\sqrt{e^x - 1}} = 2\pi \ln 2$ FI II 643a, BI(99)(4)

2. $\displaystyle\int_0^\infty \frac{x^2\,dx}{\sqrt{e^x - 1}} = 4\pi \left\{ (\ln 2)^2 + \frac{\pi^2}{12} \right\}$ BI (99)(5)

3. $\displaystyle\int_0^\infty \frac{xe^{-x}\,dx}{\sqrt{e^x - 1}} = \frac{\pi}{2}\left[2\ln 2 - 1\right]$ BI (99)(6)

4. $\displaystyle\int_0^\infty \frac{xe^{-x}\,dx}{\sqrt{e^{2x} - 1}} = 1 - \ln 2$ BI (99)(8)

5. $\displaystyle\int_0^\infty \frac{xe^{-2x}\,dx}{\sqrt{e^x - 1}} = \frac{3}{4}\pi \left(\ln 2 - \frac{7}{12} \right)$ BI (99)(7)

3.453

1. $\displaystyle\int_0^\infty \frac{xe^x}{a^2e^x - (a^2 - b^2)}\frac{dx}{\sqrt{e^x - 1}} = \frac{2\pi}{ab}\ln\left(1 + \frac{b}{a}\right)$ $[ab > 0]$ (cf. **4.298** 17)   BI (99)(16)

2. $\displaystyle\int_0^\infty \frac{xe^x\,dx}{[a^2e^x - (a^2 + b^2)]\sqrt{e^x - 1}} = \frac{2\pi}{ab}\arctan\frac{b}{a}$ $[ab > 0]$ (cf. **4.298** 18)   BI (99)(17)

3.454

1. $\displaystyle\int_0^\infty \frac{xe^{-2nx}\,dx}{\sqrt{e^{2x} + 1}} = \frac{(2n-1)!!}{(2n)!!}\frac{\pi}{2}\left\{\ln 2 + \sum_{k=1}^{2n}\frac{(-1)^k}{k}\right\}$ LI (99)(10)

2. $\displaystyle\int_0^\infty \frac{xe^{-(2n-1)x}\,dx}{\sqrt{e^{2x} - 1}} = -\frac{(2n-2)!!}{(2n-1)!!}\left\{\ln 2 + \sum_{k=1}^{2n-1}\frac{(-1)^k}{k}\right\}$ LI (99)(9)

3.455

1. $\displaystyle\int_0^\infty \frac{x^2e^x\,dx}{\sqrt{(e^x - 1)^3}} = 8\pi\ln 2$ BI (99)(11)

2. $\displaystyle\int_0^\infty \frac{x^3e^x\,dx}{\sqrt{(e^x - 1)^3}} = 24\pi\left[(\ln 2)^2 + \frac{\pi^2}{12}\right]$ BI (99)(12)

3.456

1. $\displaystyle\int_0^\infty \frac{x\,dx}{\sqrt[3]{e^{3x} - 1}} = \frac{\pi}{3\sqrt{3}}\left[\ln 3 + \frac{\pi}{3\sqrt{3}}\right]$ BI (99)(13)

2. $\displaystyle\int_0^\infty \frac{x\,dx}{\sqrt[3]{(e^{3x} - 1)^2}} = \frac{\pi}{3\sqrt{3}}\left[\ln 3 - \frac{\pi}{3\sqrt{3}}\right]$ (cf. **4.244** 3)   BI (99)(14)

3.457

1. $\displaystyle\int_0^\infty xe^{-x}\left(1 - e^{-2x}\right)^{n-1/2}\,dx = \frac{(2n-1)!!}{4\cdot(2n)!!}\pi\left[\boldsymbol{C} + \psi(n+1) + 2\ln 2\right]$

                                                 (cf. **4.241** 5)   BI (99)(3)

2. $\displaystyle\int_{-\infty}^{\infty} \frac{xe^x\,dx}{(a+e^x)^{n+3/2}} = \frac{2}{(2n+1)a^{n+1/2}}\left[\ln(4a) - 3\boldsymbol{C} - 2\,\psi(2n) - \psi(n)\right]$
                                                                              BI (101)(12)

3. $\displaystyle\int_{-\infty}^{\infty} \frac{x\,dx}{(a^2e^x + e^{-x})^{\mu}} = -\frac{1}{2a^{\mu}}\,\mathrm{B}\left(\frac{\mu}{2}, \frac{\mu}{2}\right)\ln a$ $[a>0, \quad \mathrm{Re}\,\mu > 0]$
                                                                              BI (101)(14)

3.458

1.[7] $\displaystyle\int_{0}^{\ln 2} xe^x(e^x - 1)^{p-1}\,dx = \frac{1}{p}\left[\ln 2 + \sum_{k=0}^{\infty} \frac{(-1)^{k-1}}{p+k+1}\right]$
                                                                              BI (104)(4)

2. $\displaystyle\int_{-\infty}^{\infty} \frac{xe^x\,dx}{(a+e^x)^{\nu+1}} = \frac{1}{\nu a^{\nu}}\left[\ln a - \boldsymbol{C} - \psi(\nu)\right]$ $[a>0]$

$\displaystyle\qquad\qquad = \frac{1}{\nu a^{\nu}}\left[\ln a - \sum_{k=1}^{\nu-1} \frac{1}{k}\right]$ $[a>0, \quad \nu = 1, 2, \ldots]$

                                                                              BI (101)(11)

3.46–3.48 Combinations of exponentials of more complicated arguments and powers

3.461

1. $\displaystyle\int_{u}^{\infty} \frac{e^{-p^2x^2}}{x^{2n}}\,dx = \frac{(-1)^n 2^{n-1} p^{2n-1}\sqrt{\pi}}{(2n-1)!!}\left[1 - \Phi(pu)\right]$

$\displaystyle\qquad\qquad + \frac{e^{-p^2u^2}}{2u^{2n-1}} \sum_{k=0}^{n-1} \frac{(-1)^k 2^{k+1}(pu)^{2k}}{(2n-1)(2n-3)\cdots(2n-2k-1)}$

                                                        $[p>0]$ NT 21(4)

2. $\displaystyle\int_{0}^{\infty} x^{2n}e^{-px^2}\,dx = \frac{(2n-1)!!}{2(2p)^n}\sqrt{\frac{\pi}{p}}$ $[p>0, \quad n=0,1,\ldots]$ FI II 743

3. $\displaystyle\int_{0}^{\infty} x^{2n+1}e^{-px^2}\,dx = \frac{n!}{2p^{n+1}}$ $[p>0]$ BI (81)(7)

4. $\displaystyle\int_{-\infty}^{\infty} (x+ai)^{2n}e^{-x^2}\,dx = \frac{(2n-1)!!}{2^n}\sqrt{\pi}\sum_{k=0}^{n} (-1)^k \frac{(2a)^{2k}n!}{(2k)!(n-k)!}$ BI (100)(12)

5.[3] $\displaystyle\int_{u}^{\infty} e^{-\mu x^2}\frac{dx}{x^2} = \frac{1}{u}e^{-\mu u^2} - \sqrt{\mu\pi}\left[1 - \Phi\left(\sqrt{\mu u}\right)\right]$ $\left[|\arg\mu| < \dfrac{\pi}{2}, \quad u>0\right]$ ET I 135(19)a

3.462

1. $\displaystyle\int_{0}^{\infty} x^{\nu-1}e^{-\beta x^2 - \gamma x}\,dx = (2\beta)^{-\nu/2}\,\Gamma(\nu)\exp\left(\frac{\gamma^2}{8\beta}\right)D_{-\nu}\left(\frac{\gamma}{\sqrt{2\beta}}\right)$

                                                        $[\mathrm{Re}\,\beta > 0, \quad \mathrm{Re}\,\nu > 0]$
                                                        EH II 119(3)a, ET I 313(13)

2.[8] $\displaystyle\int_{-\infty}^{\infty} x^n e^{-px^2 + 2qx}\,dx = \frac{1}{2^{n-1}p}\sqrt{\frac{\pi}{p}}\frac{d^{n-1}}{dq^{n-1}}\left(qe^{q^2/p}\right)$ $[p>0]$ BI (100)(8)

$\displaystyle\qquad\qquad = n!e^{q^2/p}\sqrt{\frac{\pi}{p}}\left(\frac{q}{p}\right)^n \sum_{k=0}^{\lfloor n/2\rfloor} \frac{1}{(n-2k)!(k)!}\left(\frac{p}{4q^2}\right)^k$ $[p>0]$ LI (100)(8)

3. $$\int_{-\infty}^{\infty} (ix)^{\nu} e^{-\beta^2 x^2 - iqx}\, dx = 2^{-\frac{\nu}{2}} \sqrt{\pi} \beta^{-\nu-1} \exp\left(-\frac{q^2}{8\beta^2}\right) D_{\nu}\left(\frac{q}{\beta\sqrt{2}}\right)$$

$$\left[\operatorname{Re}\beta > 0, \quad \operatorname{Re}\nu > -1, \quad \arg ix = \frac{\pi}{2}\operatorname{sign} x\right] \quad \text{ET I 121(23)}$$

4. $$\int_{-\infty}^{\infty} x^n \exp\left[-(x-\beta)^2\right] dx = (2i)^{-n}\sqrt{\pi}\, H_n(i\beta)$$ EH II 195(31)

5. $$\int_0^{\infty} xe^{-\mu x^2 - 2\nu x}\, dx = \frac{1}{2\mu} - \frac{\nu}{2\mu}\sqrt{\frac{\pi}{\mu}} e^{\frac{\nu^2}{\mu}}\left[1 - \Phi\left(\frac{\nu}{\sqrt{\mu}}\right)\right]$$

$$\left[|\arg \nu| < \frac{\pi}{2}, \quad \operatorname{Re}\mu > 0\right] \quad \text{ET I 146(31)a}$$

6. $$\int_{-\infty}^{\infty} xe^{-px^2 + 2qx}\, dx = \frac{q}{p}\sqrt{\frac{\pi}{p}} \exp\left(\frac{q^2}{p}\right)$$ $[\operatorname{Re}p > 0]$ BI (100)(7)

7. $$\int_0^{\infty} x^2 e^{-\mu x^2 - 2\nu x}\, dx = -\frac{\nu}{2\mu^2} + \sqrt{\frac{\pi}{\mu^5}}\frac{2\nu^2 + \mu}{4} e^{\frac{\nu^2}{\mu}}\left[1 - \Phi\left(\frac{\nu}{\sqrt{\mu}}\right)\right]$$

$$\left[|\arg \nu| < \frac{\pi}{2}, \quad \operatorname{Re}\mu > 0\right] \quad \text{ET I 146(32)}$$

8. $$\int_{-\infty}^{\infty} x^2 e^{-\mu x^2 + 2\nu x}\, dx = \frac{1}{2\mu}\sqrt{\frac{\pi}{\mu}}\left(1 + 2\frac{\nu^2}{\mu}\right) e^{\frac{\nu^2}{\mu}}$$ $[|\arg \nu| < \pi, \quad \operatorname{Re}\mu > 0]$ BI (100)(8)a

9.* $$\int_0^{\infty} x^{\alpha-1} e^{-px^{\mu}}\, dx = \frac{1}{\mu} p^{-\alpha/\mu}\, \Gamma\left(\frac{\alpha}{\mu}\right)$$ $[\mu > 0, \quad \operatorname{Re}\alpha > 0, \quad \operatorname{Re}p > 0]$

$$\text{PBM 346 (2.3.18(2))}$$

3.463 $$\int_0^{\infty} \left(e^{-x^2} - e^{-x}\right)\frac{dx}{x} = \frac{1}{2}\mathbf{C}$$ BI (89)(5)

3.464 $$\int_0^{\infty} \left(e^{-\mu x^2} - e^{-\nu x^2}\right)\frac{dx}{x^2} = \sqrt{\pi}\left(\sqrt{\nu} - \sqrt{\mu}\right)$$ $[\operatorname{Re}\mu > 0, \quad \operatorname{Re}\nu > 0]$ FI II 645

3.465 $$\int_0^{\infty} \left(1 + 2\beta x^2\right) e^{-\mu x^2}\, dx = \frac{\mu + \beta}{2}\sqrt{\frac{\pi}{\mu^3}}$$ $[\operatorname{Re}\mu > 0]$ ET I 136(24)a

3.466

1. $$\int_0^{\infty} \frac{e^{-\mu^2 x^2}}{x^2 + \beta^2}\, dx = [1 - \Phi(\beta\mu)]\frac{\pi}{2\beta} e^{\beta^2\mu^2}$$ $\left[\operatorname{Re}\beta > 0, \quad |\arg \mu| < \frac{\pi}{4}\right]$ NT 19(13)

2. $$\int_0^{\infty} \frac{x^2 e^{-\mu^2 x^2}}{x^2 + \beta^2}\, dx = \frac{\sqrt{\pi}}{2\mu} - \frac{\pi\beta}{2} e^{\mu^2\beta^2}[1 - \Phi(\beta\mu)]$$ $\left[\operatorname{Re}\beta > 0, \quad |\arg \mu| < \frac{\pi}{4}\right]$ ET II 217(16)

3. $$\int_0^1 \frac{e^{x^2} - 1}{x^2}\, dx = \sum_{k=1}^{\infty} \frac{1}{k!(2k-1)}$$ FI II 683

3.467 $$\int_0^{\infty} \left(e^{-x^2} - \frac{1}{1 + x^2}\right)\frac{dx}{x} = -\frac{1}{2}\mathbf{C}$$ BI (92)(12)

3.468

1. $$\int_{u\sqrt{2}}^{\infty} \frac{e^{-x^2}}{\sqrt{x^2 - u^2}}\frac{dx}{x} = \frac{\pi}{4u}[1 - \Phi(u)]^2$$ $[u > 0]$ NT 33(17)

2. $\displaystyle\int_0^\infty \frac{xe^{-\mu x^2}\,dx}{\sqrt{a^2+x^2}} = \frac{1}{2}\sqrt{\frac{\pi}{\mu}}e^{a^2\mu}\left[1-\Phi\left(a\sqrt{u}\right)\right]$ $\qquad$ $[\operatorname{Re}\mu > 0, \quad a > 0]$ $\qquad$ NT 19(11)

3.469

1. $\displaystyle\int_0^\infty e^{-\mu x^4 - 2\nu x^2}\,dx = \frac{1}{4}\sqrt{\frac{2\nu}{\mu}}\exp\left(\frac{\nu^2}{2\mu}\right)K_{\frac{1}{4}}\left(\frac{\nu^2}{2\mu}\right)$ $\qquad$ $[\operatorname{Re}\mu \geq 0]$ $\qquad$ ET I 146(23)

2. $\displaystyle\int_0^\infty \left(e^{-x^4} - e^{-x}\right)\frac{dx}{x} = \frac{3}{4}C$ $\qquad$ BI (89)(7)

3. $\displaystyle\int_0^\infty \left(e^{-x^4} - e^{-x^2}\right)\frac{dx}{x} = \frac{1}{4}C$ $\qquad$ BI (89)(6)

3.471

1. $\displaystyle\int_0^u \exp\left(-\frac{\beta}{x}\right)\frac{dx}{x^2} = \frac{1}{\beta}\exp\left(-\frac{\beta}{u}\right)$ $\qquad$ ET II 188(22)

2. $\displaystyle\int_0^u x^{\nu-1}(u-x)^{\mu-1}e^{-\frac{\beta}{x}}\,dx = \beta^{\frac{\nu-1}{2}}u^{\frac{2\mu+\nu-1}{2}}\exp\left(-\frac{\beta}{2u}\right)\Gamma(\mu)\,W_{\frac{1-2\mu-\nu}{2},\frac{\nu}{2}}\left(\frac{\beta}{u}\right)$
$\qquad\qquad$ $[\operatorname{Re}\mu > 0, \quad \operatorname{Re}\beta > 0, \quad u > 0]$
$\qquad\qquad$ ET II 187(18)

3. $\displaystyle\int_0^u x^{-\mu-1}(u-x)^{\mu-1}e^{-\frac{\beta}{x}}\,dx = \beta^{-\mu}u^{\mu-1}\,\Gamma(\mu)\exp\left(-\frac{\beta}{u}\right)$
$\qquad\qquad$ $[\operatorname{Re}\mu > 0, \quad u > 0]$ $\qquad$ ET II 187(16)

4. $\displaystyle\int_0^u x^{-2\mu}(u-x)^{\mu-1}e^{-\frac{\beta}{x}}\,dx = \frac{1}{\sqrt{\pi u}}\beta^{\frac{1}{2}-\mu}e^{-\frac{\beta}{2u}}\,\Gamma(\mu)\,K_{\mu-\frac{1}{2}}\left(\frac{\beta}{2u}\right)$
$\qquad\qquad$ $[u > 0, \quad \operatorname{Re}\beta > 0, \quad \operatorname{Re}\mu > 0]$
$\qquad\qquad$ ET II 187(17)

5. $\displaystyle\int_u^\infty x^{\nu-1}(x-u)^{\mu-1}e^{\frac{\beta}{x}}\,dx = \mathrm{B}(1-\mu-\nu,\mu)u^{\mu+\nu-1}\,{}_1F_1\left(1-\mu-\nu;1-\nu;\frac{\beta}{u}\right)$
$\qquad\qquad$ $[0 < \operatorname{Re}\mu < \operatorname{Re}(1-\nu), \quad u > 0]$
$\qquad\qquad$ ET II 203(15)

6. $\displaystyle\int_u^\infty x^{-2\mu}(x-u)^{\mu-1}e^{\frac{\beta}{x}}\,dx = \sqrt{\frac{\pi}{u}}\beta^{\frac{1}{2}-\mu}\,\Gamma(\mu)\exp\left(\frac{\beta}{2u}\right)I_{\mu-\frac{1}{2}}\left(\frac{\beta}{2u}\right)$
$\qquad\qquad$ $[\operatorname{Re}\mu > 0, \quad u > 0]$ $\qquad$ ET II 202(14)

7. $\displaystyle\int_0^\infty x^{\nu-1}(x+\gamma)^{\mu-1}e^{-\frac{\beta}{x}}\,dx = \beta^{\frac{\nu-1}{2}}\gamma^{\frac{\nu-1}{2}+\mu}\,\Gamma(1-\mu-\nu)e^{\frac{\beta}{2\gamma}}\,W_{\frac{\nu-1}{2}+\mu,-\frac{\nu}{2}}\left(\frac{\beta}{\gamma}\right)$
$\qquad\qquad$ $[|\arg\gamma| < \pi, \quad \operatorname{Re}(1-\mu) > \operatorname{Re}\nu > 0]$
$\qquad\qquad$ ET II 234(13)a

8. $\displaystyle\int_0^u x^{-2\mu}(u^2-x^2)^{\mu-1}e^{-\frac{\beta}{x}}\,dx = \frac{1}{\sqrt{\pi}}\left(\frac{2}{\beta}\right)^{\mu-\frac{1}{2}}u^{\mu-\frac{3}{2}}\,\Gamma(\mu)\,K_{\mu-\frac{1}{2}}\left(\frac{\beta}{u}\right)$
$\qquad\qquad$ $[\operatorname{Re}\beta > 0, \quad u > 0, \quad \operatorname{Re}\mu > 0]$
$\qquad\qquad$ ET II 188(23)a

9. $\int_0^\infty x^{\nu-1} e^{-\frac{\beta}{x}-\gamma x}\, dx = 2\left(\frac{\beta}{\gamma}\right)^{\frac{\nu}{2}} K_\nu\left(2\sqrt{\beta\gamma}\right)$ $[\operatorname{Re}\beta > 0, \quad \operatorname{Re}\gamma > 0]$

                           ET II 82(23)a, LET I 146(29)

10. $\int_0^\infty x^{\nu-1} \exp\left[\frac{i\mu}{2}\left(x - \frac{\beta^2}{x}\right)\right] dx = 2\beta^\nu e^{\frac{i\nu\pi}{2}} K_{-\nu}(\beta\mu)$

             $\left[\operatorname{Im}\mu > 0, \quad \operatorname{Im}\left(\beta^2\mu\right) < 0; \text{ note that } K_{-\nu} \equiv K_\nu\right]$ EH II 82(24)

11. $\int_0^\infty x^{\nu-1} \exp\left[\frac{i\mu}{2}\left(x + \frac{\beta^2}{x}\right)\right] dx = i\pi\beta^\nu e^{-\frac{i\nu\pi}{2}} H_{-\nu}^{(1)}(\beta\mu)$

                         $\left[\operatorname{Im}\mu > 0, \quad \operatorname{Im}\left(\beta^2\mu\right) > 0\right]$

                                            EH II 21(33)

12. $\int_0^\infty x^{\nu-1} \exp\left(-x - \frac{\mu^2}{4x}\right) dx = 2\left(\frac{\mu}{2}\right)^\nu K_{-\nu}(\mu)$

             $\left[|\arg\mu| < \frac{\pi}{2}, \operatorname{Re}\mu^2 > 0; \text{ note that } K_{-\nu} \equiv K_\nu\right]$ WA 203(15)

13. $\int_0^\infty \frac{x^{\nu-1} e^{-\frac{\beta}{x}}}{x+\gamma}\, dx = \gamma^{\nu-1} e^{\frac{\beta}{\gamma}} \Gamma(1-\nu)\, \Gamma\left(\nu, \frac{\beta}{\gamma}\right)$ $[|\arg\gamma| < \pi, \quad \operatorname{Re}\beta > 0, \quad \operatorname{Re}\nu < 1]$

                                            ET II 218(19)

14. $\int_0^1 \frac{\exp\left(1 - \frac{1}{x}\right) - x^\nu}{x(1-x)}\, dx = \psi(\nu)$ $[\operatorname{Re}\nu > 0]$ BI (80)(7)

15.* $\int_0^\infty x^{-\frac{1}{2}} e^{-\gamma x - \beta/x}\, dx = \sqrt{\frac{\pi}{\gamma}}\, e^{-2\sqrt{\beta\gamma}}$ $[\operatorname{Re}\beta \geq 0, \quad \operatorname{Re}\gamma > 0]$ ET 245 (5.6.1)

16.* $\int_0^\infty x^{n-\frac{1}{2}} e^{-px - q/x}\, dx = (-1)^n \sqrt{\pi}\, \frac{\partial^n}{\partial p^n}\left(p^{-1/2} e^{-2\sqrt{pq}}\right)$

                               $[\operatorname{Re}p > 0, \quad \operatorname{Re}q > 0]$

                                       PBM 344 (2.3.16(2))

3.472

1. $\int_0^\infty \left(\exp\left(-\frac{a}{x^2}\right) - 1\right) e^{-\mu x^2}\, dx = \frac{1}{2}\sqrt{\frac{\pi}{\mu}}\left[\exp\left(-2\sqrt{a\mu}\right) - 1\right]$

                          $[\operatorname{Re}\mu > 0, \quad \operatorname{Re}a > 0]$ ET I 146(30)

2. $\int_0^\infty x^2 \exp\left(-\frac{a}{x^2} - \mu x^2\right) dx = \frac{1}{4}\sqrt{\frac{\pi}{\mu^3}}\left(1 + 2\sqrt{a\mu}\right)\exp\left(-2\sqrt{a\mu}\right)$

                          $[\operatorname{Re}\mu > 0, \quad \operatorname{Re}a > 0]$ ET I 146(26)

3. $\int_0^\infty \exp\left(-\frac{a}{x^2} - \mu x^2\right) \frac{dx}{x^2} = \frac{1}{2}\sqrt{\frac{\pi}{a}}\exp\left(-2\sqrt{a\mu}\right)$ $[\operatorname{Re}\mu > 0, \quad a > 0]$ ET I 146(28)a

4. $\int_0^\infty \exp\left[-\frac{1}{2a}\left(x^2 + \frac{1}{x^2}\right)\right] \frac{dx}{x^4} = \sqrt{\frac{a\pi}{2}}(1+a)e^{-1/a}$ $[a > 0]$ BI (98)(14)

5.* $\int_0^\infty x^{-n-1/2} e^{-px - q/x}\, dx = (-1)^n \sqrt{\frac{\pi}{p}}\, \frac{\partial^n}{\partial q^n} e^{-2\sqrt{pq}}$ $[\operatorname{Re}p > 0, \quad \operatorname{Re}q > 0]$

                                       PBM 344 (2.3.16(3))

3.473 $\displaystyle\int_0^\infty \exp\left(-x^n\right) x^{(m+1/2)n-1}\, dx = \frac{(2m-1)!!}{2^m n}\sqrt{\pi}$ BI (98)(6)

3.474

1. $\displaystyle\int_0^1 \left\{\frac{n\exp\left(1-x^{-n}\right)}{1-x^n} - \frac{x^{np}}{1-x}\right\}\frac{dx}{x} = \frac{1}{n}\sum_{k=1}^{n}\psi\left(p+\frac{k-1}{n}\right)$

          $[p>0]$ BI (80)(8)

2. $\displaystyle\int_0^1 \left\{\frac{n\exp\left(1-x^{-n}\right)}{1-x^n} - \frac{\exp\left(1-\frac{1}{x}\right)}{1-x}\right\}\frac{dx}{x} = -\ln n$ BI (80)(9)

3.475

1.[7] $\displaystyle\int_0^\infty \left\{\exp\left(-x^2\right) - \frac{1}{1+x^{2n+1}}\right\}\frac{dx}{x} = -\frac{1}{2^n}\,\mathbf{C}$ $[n\in\mathbb{Z}]$ BI (92)(14)

2. $\displaystyle\int_0^\infty \left\{\exp\left(-x^{2^n}\right) - \frac{1}{1+x^2}\right\}\frac{dx}{x} = -2^{-n}\,\mathbf{C}$ BI (92)(13)

3. $\displaystyle\int_0^\infty \left\{\exp\left(-x^{2^n}\right) - e^{-x}\right\}\frac{dx}{x} = \left(1-2^{-n}\right)\mathbf{C}$ BI (89)(8)

3.476

1. $\displaystyle\int_0^\infty \left[\exp\left(-\nu x^p\right) - \exp\left(-\mu x^p\right)\right]\frac{dx}{x} = \frac{1}{p}\ln\frac{\mu}{\nu}$ $[\operatorname{Re}\mu>0, \quad \operatorname{Re}\nu>0]$ BI (89)(3)

2. $\displaystyle\int_0^\infty \left[\exp\left(-x^p\right) - \exp\left(-x^q\right)\right]\frac{dx}{x} = \frac{p-q}{pq}\,\mathbf{C}$ $[p>0, \quad q>0]$ BI (89)(9)

3.477

1.[10] $\displaystyle\int_{-\infty}^\infty \frac{e^{-a|x|}}{x-u}\, dx = e^{-au}\,\gamma(0,-au) - e^{au}\,\gamma(0,au)$ $[\operatorname{Re}a>0, \quad \operatorname{Im}u\neq 0, \quad \arg u\neq 0]$ MC

2.[8] $\displaystyle\int_{-\infty}^\infty \frac{\operatorname{sign}x\,\exp\left(-a|x|\right)}{x-u}\, dx = -\left[\exp\left(a|u|\right)\operatorname{Ei}\left(-a|u|\right) - \exp\left(-a|u|\right)\operatorname{Ei}\left(a|u|\right)\right]$

          $[a>0]$ ET II 251(36)

3.478

1. $\displaystyle\int_0^\infty x^{\nu-1}\exp\left(-\mu x^p\right)\, dx = \frac{1}{p}\mu^{-\frac{\nu}{p}}\,\Gamma\left(\frac{\nu}{p}\right)$ $[\operatorname{Re}\mu>0, \quad \operatorname{Re}\nu>0, \quad p>0]$

          BI(81)(8)a, ET I 313(15, 16)

2. $\displaystyle\int_0^\infty x^{\nu-1}\left[1 - \exp\left(-\mu x^p\right)\right]\, dx = -\frac{1}{|p|}\mu^{-\frac{\nu}{p}}\,\Gamma\left(\frac{\nu}{p}\right)$

          $[\operatorname{Re}\mu>0 \text{ and } -p<\operatorname{Re}\nu<0 \text{ for } p>0, \quad 0<\operatorname{Re}\nu<-p \text{ for } p<0]$ ET I 313(18, 19)

3. $\displaystyle\int_0^u x^{\nu-1}(u-x)^{\mu-1}\exp\left(\beta x^n\right)\, dx = \mathrm{B}(\mu,\nu)u^{\mu+\nu-1}\,{}_nF_n\left(\frac{\nu}{n},\frac{\nu+1}{n},\dots,\frac{\nu+n-1}{n}\right.$

          $\left.\frac{\mu+\nu}{n},\frac{\mu+\nu+1}{n},\dots,\frac{\mu+\nu+n-1}{n};\beta u^n\right)$

          $[\operatorname{Re}\mu>0, \quad \operatorname{Re}\nu>0, \quad n=2,3,\dots]$ ET II 187(15)

4. $\displaystyle\int_0^\infty x^{\nu-1} \exp\left(-\beta x^p - \gamma x^{-p}\right)\, dx = \frac{2}{p}\left(\frac{\gamma}{\beta}\right)^{\frac{\nu}{2p}} K_{\frac{\nu}{p}}\left(2\sqrt{\beta\gamma}\right)$

$$\qquad\qquad\qquad\qquad\qquad\qquad\qquad [\operatorname{Re}\beta > 0, \quad \operatorname{Re}\gamma > 0] \qquad \text{ET I 313(17)}$$

3.479

1. $\displaystyle\int_0^\infty \frac{x^{\nu-1} \exp\left(-\beta\sqrt{1+x}\right)}{\sqrt{1+x}}\, dx = \frac{2}{\sqrt{\pi}}\left(\frac{\beta}{2}\right)^{\frac{1}{2}-\nu} \Gamma(\nu)\, K_{\frac{1}{2}-\nu}(\beta)$

$$\qquad\qquad\qquad\qquad\qquad\qquad\qquad [\operatorname{Re}\beta > 0, \quad \operatorname{Re}\nu > 0] \qquad \text{ET I 313(14)}$$

2.[8] $\displaystyle\int_0^\infty \frac{x^{\nu-1} \exp\left(i\mu\sqrt{1+x^2}\right)}{\sqrt{1+x^2}}\, dx = i\frac{\sqrt{\pi}}{2}\left(\frac{\mu}{2}\right)^{\frac{1-\nu}{2}} \Gamma\left(\frac{\nu}{2}\right) H^{(1)}_{\frac{1-\nu}{2}}(\mu)$

$$\qquad\qquad\qquad\qquad\qquad\qquad\qquad [\operatorname{Im}\mu > 0, \quad \operatorname{Re}\nu > 2] \qquad \text{EH II 83(30)}$$

3.481

1. $\displaystyle\int_{-\infty}^\infty x e^x \exp\left(-\mu e^x\right)\, dx = -\frac{1}{\mu}\left(\boldsymbol{C} + \ln\mu\right) \qquad [\operatorname{Re}\mu > 0] \qquad \text{BI (100)(13)}$

2. $\displaystyle\int_{-\infty}^\infty x e^x \exp\left(-\mu e^{2x}\right)\, dx = -\frac{1}{4}\left[\boldsymbol{C} + \ln(4\mu)\right]\sqrt{\frac{\pi}{\mu}} \qquad [\operatorname{Re}\mu > 0] \qquad \text{BI (100)(14)}$

3.482

1.[3] $\displaystyle\int_0^\infty \exp\left(nx - \beta\sinh x\right)\, dx = \frac{1}{2}\left[S_n(\beta) - \pi\,\mathbf{E}_n(\beta) - \pi\,Y_n(\beta)\right]$

$$\qquad\qquad\qquad\qquad\qquad\qquad\qquad\qquad\qquad [\operatorname{Re}\beta > 0] \qquad \text{ET I 168(11)}$$

2. $\displaystyle\int_0^\infty \exp\left(-nx - \beta\sinh x\right)\, dx = (-1)^{n+1}\frac{1}{2}\left[S_n(\beta) + \pi\,\mathbf{E}_n(\beta) + \pi\,Y_n(\beta)\right]$

$$\qquad\qquad\qquad\qquad\qquad\qquad\qquad\qquad\qquad [\operatorname{Re}\beta > 0] \qquad \text{ET I 168(12)}$$

3. $\displaystyle\int_0^\infty \exp\left(-\nu x - \beta\sinh x\right)\, dx = \frac{\pi}{\sin\nu\pi}\left[\mathbf{J}_\nu(\beta) - J_\nu(\beta)\right]$

$$\qquad\qquad\qquad\qquad\qquad\qquad\qquad\qquad\qquad [\operatorname{Re}\beta > 0] \qquad \text{ET I 168(13)}$$

3.483 $\displaystyle\int_{-\infty}^\infty \frac{\exp\left(\nu\operatorname{arcsinh}x - iax\right)}{\sqrt{1+x^2}}\, dx = \begin{cases} 2\exp\left(-\dfrac{i\nu\pi}{2}\right) K_\nu(a) & \text{for } a > 0, \\[2mm] 2\exp\left(\dfrac{i\nu\pi}{2}\right) K_\nu(-a) & \text{for } a < 0 \end{cases} \qquad [|\operatorname{Re}\nu| < 1]$

$$\qquad\qquad\qquad\qquad\qquad\qquad\qquad\qquad\qquad\qquad\qquad\qquad \text{ET I 122(32)}$$

3.484 $\displaystyle\int_0^\infty \left[\left(1 + \frac{a}{qx}\right)^{qx} - \left(1 + \frac{a}{px}\right)^{px}\right]\frac{dx}{x} = (e^a - 1)\ln\frac{q}{p} \qquad [p > 0, \quad q > 0] \qquad \text{BI (89)(34)}$

3.485 $\displaystyle\int_0^{\pi/2} \exp\left(-\tan^2 x\right)\, dx = \frac{\pi e}{2}\left[1 - \Phi(1)\right]$

3.486[6] $\displaystyle\int_0^1 x^{-x}\, dx = \int_0^1 e^{-x\ln x}\, dx = \sum_{k=1}^\infty k^{-k} = 1.2912859970627\ldots \qquad\qquad \text{FI II 483}$

3.5 Hyperbolic Functions

3.51 Hyperbolic functions

3.511

1. $\displaystyle\int_0^\infty \frac{dx}{\cosh ax} = \frac{\pi}{2a}$ $[a > 0]$

2. $\displaystyle\int_0^\infty \frac{\sinh ax}{\sinh bx}\, dx = \frac{\pi}{2b} \tan \frac{a\pi}{2b}$ $[b > |a|]$ BI (27)(10)a

3. $\displaystyle\int_0^\infty \frac{\sinh ax}{\cosh bx}\, dx = \frac{\pi}{2b} \sec \frac{a\pi}{2b} - \frac{1}{b}\, \beta\left(\frac{a+b}{2b}\right)$ $[b > |a|]$ GW (351)(3b)

4. $\displaystyle\int_0^\infty \frac{\cosh ax}{\cosh bx}\, dx = \frac{\pi}{2b} \sec \frac{a\pi}{2b}$ $[b > |a|]$ BI (4)(14)a

5. $\displaystyle\int_0^\infty \frac{\sinh ax \cosh bx}{\sinh cx}\, dx = \frac{\pi}{2c}\, \frac{\sin \frac{a\pi}{c}}{\cos \frac{a\pi}{c} + \cos \frac{b\pi}{c}}$ $[c > |a| + |b|]$ BI (27)(11)

6. $\displaystyle\int_0^\infty \frac{\cosh ax \cosh bx}{\cosh cx}\, dx = \frac{\pi}{c}\, \frac{\cos \frac{a\pi}{2c} \cos \frac{b\pi}{2c}}{\cos \frac{a\pi}{c} + \cos \frac{b\pi}{c}}$ $[c > |a| + |b|]$ BI (27)(5)a

7. $\displaystyle\int_0^\infty \frac{\sinh ax \sinh bx}{\cosh cx}\, dx = \frac{\pi}{c}\, \frac{\sin \frac{a\pi}{2c} \sin \frac{b\pi}{2c}}{\cos \frac{a\pi}{c} + \cos \frac{b\pi}{c}}$ $[c > |a| + |b|]$ BI (27)(6)a

8. $\displaystyle\int_0^\infty \frac{dx}{\cosh^2 x} = \sqrt{\pi} \sum_{k=0}^\infty \frac{(-1)^k}{\sqrt{2k+1}}$ BI (98)(25)

9. $\displaystyle\int_{-\infty}^\infty \frac{\sinh^2 ax}{\sinh^2 x}\, dx = 1 - a\pi \cot a\pi$ $[a^2 < 1]$ BI (16)(3)a

10. $\displaystyle\int_0^\infty \frac{\sinh ax \sinh bx}{\cosh^2 bx}\, dx = \frac{a\pi}{2b^2} \sec \frac{a\pi}{2b}$ $[b > |a|]$ BI (27)(16)a

3.512

1. $\displaystyle\int_0^\infty \frac{\cosh 2\beta x}{\cosh^{2\nu} ax}\, dx = \frac{4^{\nu-1}}{a}\, B\left(\nu + \frac{\beta}{a}, \nu - \frac{\beta}{a}\right)$ $[\operatorname{Re}(\nu \pm \beta) > 0, \quad a > 0, \quad \beta > 0]$

      LI(27)(17)a, EH I 11(26)

2. $\displaystyle\int_0^\infty \frac{\sinh^\mu x}{\cosh^\nu x}\, dx = \frac{1}{2}\, B\left(\frac{\mu+1}{2}, \frac{\nu-\mu}{2}\right)$ $[\operatorname{Re}\mu > -1, \quad \operatorname{Re}(\mu - \nu) < 0]$

      EH I 11(23)

3.513

1. $\displaystyle\int_0^\infty \frac{dx}{a + b \sinh x} = \frac{1}{\sqrt{a^2 + b^2}} \ln \frac{a + b + \sqrt{a^2 + b^2}}{a + b - \sqrt{a^2 + b^2}}$ $[ab \neq 0]$ GW (351)(8)

2. $\displaystyle\int_0^\infty \frac{dx}{a + b \cosh x} = \frac{2}{\sqrt{b^2 - a^2}} \arctan \frac{\sqrt{b^2 - a^2}}{a + b}$ $[b^2 > a^2]$

$\displaystyle\qquad\qquad\qquad\quad = \frac{1}{\sqrt{a^2 - b^2}} \ln \frac{a + b + \sqrt{a^2 - b^2}}{a + b - \sqrt{a^2 - b^2}}$ $[b^2 < a^2]$

      GW (351)(7)

3. $$\int_0^\infty \frac{dx}{a\sinh x + b\cosh x} = \frac{2}{\sqrt{b^2-a^2}}\arctan\frac{\sqrt{b^2-a^2}}{a+b} \qquad [b^2 > a^2]$$

$$= \frac{1}{\sqrt{a^2-b^2}}\ln\frac{a+b+\sqrt{a^2-b^2}}{a+b-\sqrt{a^2-b^2}} \qquad [a^2 > b^2]$$

GW (351)(9)

4. $$\int_0^\infty \frac{dx}{a+b\cosh x + c\sinh x} = \frac{2}{\sqrt{b^2-a^2-c^2}}\left[\arctan\frac{\sqrt{b^2-a^2-c^2}}{a+b+c} + \epsilon\pi\right]$$

$$\left[\text{when } b^2 > a^2 + c^2; \text{ and } \begin{cases} \epsilon = 0 & \text{for } (b-a)(a+b+c) > 0 \\ |\epsilon| = 1 & \text{for } (b-a)(a+b+c) < 0 \\ \epsilon = 1 & \text{for } a < b+c \\ \epsilon = -1 & \text{for } a > b+c \end{cases}\right]$$

$$= \frac{1}{\sqrt{a^2-b^2+c^2}}\ln\frac{a+b+c+\sqrt{a^2-b^2+c^2}}{a+b+c-\sqrt{a^2-b^2+c^2}}$$

$$\left[b^2 < a^2 + c^2, \quad a^2 \neq b^2\right]$$

$$= \frac{1}{c}\ln\frac{a+c}{a}$$

$$[a = b \neq 0, \quad c \neq 0]$$

$$= \frac{2(a-b)}{c(a-b-c)}$$

$$\left[b^2 = a^2 + c^2, \quad c(a-b-c) < 0\right]$$

GW (351)(6)

3.514

1. $$\int_0^\infty \frac{dx}{\cosh ax + \cos t} = \frac{t}{a}\operatorname{cosec} t \qquad\qquad [0 < t < \pi, \quad a > 0] \qquad \text{BI (27)(22)a}$$

2. $$\int_0^\infty \frac{\cosh ax - \cos t_1}{\cosh bx - \cos t_2}\,dx = \frac{\pi}{b}\frac{\sin\frac{a(\pi t_2)}{b}}{\sin t_2 \sin\frac{a}{b}\pi} - \frac{\pi t_2}{b\sin t_2}\cos t_1$$

$$[0 < |a| < b, \quad 0 < t_2 < \pi] \qquad \text{BI (6)(20)a}$$

3. $$\int_0^\infty \frac{\cosh ax\, dx}{(\cosh x + \cos t)^2} = \frac{\pi\,(-\cos t\sin at + a\sin t\cos at)}{\sin^3 t\sin a\pi}$$

$$[0 < a^2 < 1, \quad 0 < t < \pi] \qquad \text{BI (6)(18)a}$$

4. $$\int_0^\infty \frac{\sinh ax\sinh bx}{(\cosh ax + \cos t)^2}\,dx = \frac{b\pi}{a^2}\operatorname{cosec} t\operatorname{cosec}\frac{b\pi}{a}\sin\frac{bt}{a} \qquad [0 < |b| < a, \quad 0 < t < \pi] \qquad \text{BI (27)(27)a}$$

3.515 $$\int_{-\infty}^\infty \left(1 - \frac{\sqrt{2}\cosh x}{\sqrt{\cosh 2x}}\right)dx = -\ln 2 \qquad\qquad \text{BI (21)(12)a}$$

3.516

1. $$\int_0^\infty \frac{dx}{\left(z+\sqrt{z^2-1}\cosh x\right)^\mu} = \frac{1}{2}\int_{-\infty}^\infty \frac{dx}{\left(z+\sqrt{z^2-1}\cosh x\right)^\mu} = Q_{\mu-1}(z)$$

$$[\operatorname{Re}\mu > -1]$$

For a suitable choice of a single-valued branch of the integrand, this formula is valid for arbitrary values of z in the z-plane cut from -1 to $+1$ provided $\mu < 0$. If $\mu > 0$, this formula ceases to be valid for points at which the denominator vanishes.

CO, WH

1. $$\int_0^\infty \frac{dx}{\left(\beta + \sqrt{\beta^2 - 1}\cosh x\right)^{n+1}} = Q_n(\beta)$$

EH II 181(32)

2. $$\int_0^\infty \frac{\cosh \gamma x\, dx}{\left(\beta + \sqrt{\beta^2 - 1}\cosh x\right)^{\nu+1}} = \frac{e^{-i\gamma\pi}\,\Gamma(\nu - \gamma + 1)\,Q_\nu^\gamma(\beta)}{\Gamma(\nu + 1)}$$

$[\operatorname{Re}(\nu \pm \gamma) > -1, \quad \nu \neq -1, -2, -3, \ldots]$

EH I 157(12)

3. $$\int_0^\infty \frac{\sinh^{2\mu} x\, dx}{\left(\beta + \sqrt{\beta^2 - 1}\cosh x\right)^{\nu+1}} = \frac{2^\mu e^{-i\mu\pi}\,\Gamma(\nu - 2\mu + 1)\,\Gamma\left(\mu + \frac{1}{2}\right)}{\sqrt{\pi}(\beta^2 - 1)^{\frac{\mu}{2}}\,\Gamma(\nu + 1)}\,Q_{\nu-\mu}^\mu(\beta)$$

$[\operatorname{Re}(\nu - 2\mu + 1) > 0, \quad \operatorname{Re}(\nu + 1) > 0]$

EH I 155(2)

3.517

1. $$\int_0^\infty \frac{\cosh\left(\gamma + \frac{1}{2}\right)x\, dx}{(\beta + \cosh x)^{\nu+\frac{1}{2}}} = \sqrt{\frac{\pi}{2}}(\beta^2 - 1)^{-\frac{\nu}{2}}\frac{\Gamma(\nu + \gamma + 1)\,\Gamma(\nu - \gamma)\,P_\gamma^{-\nu}(\beta)}{\Gamma\left(\nu + \frac{1}{2}\right)}$$

$[\operatorname{Re}(\nu - \gamma) > 0, \quad \operatorname{Re}(\nu + \gamma + 1) > 0]$

EH I 156(11)

2. $$\int_0^a \frac{\cosh\left(\gamma + \frac{1}{2}\right)x\, dx}{(\cosh a - \cosh x)^{\nu+\frac{1}{2}}} = \sqrt{\frac{\pi}{2}}\frac{\Gamma\left(\frac{1}{2} - \nu\right)}{\sinh^\nu a}\,P_\gamma^\nu(\cosh a)$$

$\left[\operatorname{Re}\nu < \frac{1}{2}, \quad a > 0\right]$ EH I 156(8)

3.518

1. $$\int_0^\infty \frac{\sinh^{2\mu} x\, dx}{(\cosh a + \sinh a \cosh x)^{\nu+1}} = \frac{2^\mu e^{-i\mu\pi}}{\sqrt{\pi}\sinh^\mu a}\frac{\Gamma(\nu - 2\mu + 1)\,\Gamma\left(\mu + \frac{1}{2}\right)}{\Gamma(\nu + 1)}\,Q_{\nu-\mu}^\mu(\cosh a)$$

$[\operatorname{Re}(\nu + 1) > 0, \quad \operatorname{Re}(\nu - 2\mu + 1) > 0, \quad a > 0]$ EH I 155(3)a

2.[10] $$\int_0^\infty \frac{\sinh^{2\mu+1} x\, dx}{(\beta + \cosh x)^{\nu+1}} = 2^\mu (\beta^2 - 1)^{\frac{\mu-\nu}{2}}\,\Gamma(\nu - 2\mu)\,\Gamma(\mu + 1)\,P_\mu^{\mu-\nu}(\beta)$$

    $[\operatorname{Re}(\nu - \mu) > \operatorname{Re}\mu > -1, \quad \beta \text{ does not lie on the ray } (-\infty, +1) \text{ of the real axis}]$ EH I 155(1)

3. $$\int_0^\infty \frac{\sinh^{2\mu-1} x \cosh x\, dx}{(1 + a\sinh^2 x)^\nu} = \frac{1}{2}a^{-\mu}\,\mathrm{B}(\mu, \nu - \mu)$$

$[\operatorname{Re}\nu > \operatorname{Re}\mu > 0, \quad a > 0]$ EH I 11(22)

4.[7] $$\int_0^\infty \frac{\sinh^{\mu-1} x(\cosh x + 1)^{\nu-1}\, dx}{(\beta + \cosh x)^\varrho} = 2^{\mu+\nu-\rho}\,\mathrm{B}\left(\frac{1}{2}\mu, \varrho + 2 - \mu - \nu\right)$$

$$\times\, {}_2F_1\left(\varrho, \varrho + 2 - \mu - \nu; 2 - \frac{1}{2}\mu - \nu; \frac{1}{2} - \frac{1}{2}\beta\right)$$

$[\operatorname{Re}\mu > 0, \quad \operatorname{Re}(\varrho - \mu - \nu) > -2, \quad |\arg(1 + \beta)| < \pi]$ EH I 115(11)

5.^{6}
$$\int_0^\infty \frac{\sinh^{\mu-1} x (\cosh x - 1)^{\nu-1}\, dx}{(\beta + \cosh x)^\varrho} = 2^{-(2-\mu-\nu+\varrho)}\; {}_2F_1\left(\varrho, 2 - \mu - \nu + \varrho; 1 + \varrho - \frac{\mu}{2}; \frac{1-\beta}{2}\right)$$
$$\times B\left(2 - \mu - \nu + \varrho, -1 + \nu + \frac{\mu}{2}\right)$$
$$[\beta \notin (-\infty, -1), \quad \mathrm{Re}(2+\varrho)\,\mathrm{Re}(\mu+\nu), \quad \mathrm{Re}(2\nu+\mu) > 2] \quad \text{EH I 115(10)}$$

6.^{7}
$$\int_0^\infty \frac{\sinh^{\mu-1} x \cosh^{\nu-1} x}{\left(\cosh^2 x - \beta\right)^\varrho}\, dx = {}_2F_1\left(\varrho, 1 + \varrho - \frac{\mu+\nu}{2}; 1 + \varrho - \frac{\nu}{2}; \beta\right) 2\,B\left(\frac{\mu}{2}, 1 + \varrho - \frac{\mu+\nu}{2}\right)$$
$$[\beta \notin (1, \infty), \quad \mathrm{Re}\,\mu > 0, \quad 2\,\mathrm{Re}(1+\varrho) > \mathrm{Re}(\mu+\nu)] \quad \text{EH I 115(9)}$$

3.519
$$\int_0^{\pi/2} \frac{\sinh\left[(r-p)\right]\tan x}{\sinh\left(r\tan x\right)}\, dx = \pi \sum_{k=1}^\infty \frac{1}{k\pi + r} \sin\frac{pk\pi}{r} \qquad \left[p^2 < r^2\right] \qquad \text{BI (274)(13)}$$

3.52–3.53 Combinations of hyperbolic functions and algebraic functions

3.521

1.
$$\int_0^\infty \frac{x\, dx}{\sinh ax} = \frac{\pi^2}{4a^2} \qquad\qquad [a > 0] \qquad\qquad \text{GW (352)(2b)}$$

2.
$$\int_0^\infty \frac{x\, dx}{\cosh x} = 2\,\boldsymbol{G} = \pi \ln 2 - 4\,L\left(\frac{\pi}{4}\right) = 1.831931188\ldots \qquad \text{LI III 225(103a), BI(84)(1)a}$$

3.
$$\int_1^\infty \frac{dx}{x \sinh ax} = -2 \sum_{k=0}^\infty \mathrm{Ei}[-(2k+1)a] \qquad [a > 0] \qquad \text{LI (104)(14)}$$

4.
$$\int_1^\infty \frac{dx}{x \cosh ax} = 2 \sum_{k=0}^\infty (-1)^{k+1}\, \mathrm{Ei}[-(2k+1)a] \qquad [a > 0] \qquad \text{LI (104)(13)}$$

3.522

1.
$$\int_0^\infty \frac{x\, dx}{(b^2 + x^2)\sinh ax} = \frac{\pi}{2ab} + \pi \sum_{k=1}^\infty \frac{(-1)^k}{ab + k\pi} \qquad [a > 0, \quad b > 0]$$

2.
$$\int_0^\infty \frac{x\, dx}{(b^2 + x^2)\sinh \pi x} = \frac{1}{2b} - \beta(b+1) \qquad [b > 0] \qquad \text{BI(97)(16), GW(352)(8)}$$

3.
$$\int_0^\infty \frac{dx}{(b^2 + x^2)\cosh ax} = \frac{2\pi}{b} \sum_{k=1}^\infty \frac{(-1)^{k-1}}{2ab + (2k-1)\pi} \qquad [a > 0, \quad b > 0] \qquad \text{BI (97)(5)}$$

4.
$$\int_0^\infty \frac{dx}{(b^2 + x^2)\cosh \pi x} = \frac{1}{b}\beta\left(b + \frac{1}{2}\right) \qquad [b > 0] \qquad \text{BI (97)(4)}$$

5.
$$\int_0^\infty \frac{x\, dx}{(1 + x^2)\sinh \pi x} = \ln 2 - \frac{1}{2} \qquad\qquad \text{BI (97)(7)}$$

6.
$$\int_0^\infty \frac{dx}{(1 + x^2)\cosh \pi x} = 2 - \frac{\pi}{2} \qquad\qquad \text{BI (97)(1)}$$

7.
$$\int_0^\infty \frac{x\, dx}{(1 + x^2)\sinh\frac{\pi x}{2}} = \frac{\pi}{2} - 1 \qquad\qquad \text{BI (97)(8)}$$

8. $$\int_0^\infty \frac{dx}{(1+x^2)\cosh\frac{\pi x}{2}} = \ln 2$$ BI (97)(2)

9. $$\int_0^\infty \frac{x\,dx}{(1+x^2)\sinh\frac{\pi x}{4}} = \frac{1}{\sqrt{2}}\left[\pi + 2\ln\left(\sqrt{2}+1\right)\right] - 2$$ BI (97)(9)

10. $$\int_0^\infty \frac{dx}{(1+x^2)\cosh\frac{\pi x}{4}} = \frac{1}{\sqrt{2}}\left[\pi - 2\ln\left(\sqrt{2}+1\right)\right]$$ BI (97)(3)

3.523

1. $$\int_0^\infty \frac{x^{\beta-1}}{\sinh ax}\,dx = \frac{2^\beta - 1}{2^{\beta-1}a^\beta}\,\Gamma(\beta)\,\zeta(\beta)$$ $[\operatorname{Re}\beta > 1, \quad a > 0]$ WH

2. $$\int_0^\infty \frac{x^{2n-1}}{\sinh ax}\,dx = \frac{2^{2n}-1}{2n}\left(\frac{\pi}{a}\right)^{2n}|B_{2n}|$$ $[a > 0, \quad n = 1, 2, \ldots]$

WH, GW(352)(2a)

3. $$\int_0^\infty \frac{x^{\beta-1}}{\cosh ax}\,dx = \frac{2}{(2a)^\beta}\,\Gamma(\beta)\,\Phi\left(-1, \beta, \frac{1}{2}\right)$$
$$= \frac{2}{(2a)^\beta}\,\Gamma(\beta)\sum_{k=0}^\infty (-1)^k\left(\frac{2}{2k+1}\right)^\beta$$

$[\operatorname{Re}\beta > 0, \quad a > 0]$ EH I 35, ET I 322(1)

4. $$\int_0^\infty \frac{x^{2n}}{\cosh ax}\,dx = \left(\frac{\pi}{2a}\right)^{2n+1}|E_{2n}|$$ $[a > 0]$ BI(84)(12)a, GW(352)(1a)

5. $$\int_0^\infty \frac{x^2\,dx}{\cosh x} = \frac{\pi^3}{8}$$ (cf. **4.261** 6) BI (84)(3)

6. $$\int_0^\infty \frac{x^3\,dx}{\sinh x} = \frac{\pi^4}{8}$$ (cf. **4.262** 1 and 2) BI (84)(5)

7. $$\int_0^\infty \frac{x^4\,dx}{\cosh x} = \frac{5}{32}\pi^5$$ BI (84)(7)

8. $$\int_0^\infty \frac{x^5}{\sinh x}\,dx = \frac{\pi^6}{4}$$ BI (84)(8)

9. $$\int_0^\infty \frac{x^6}{\cosh x}\,dx = \frac{61}{128}\pi^7$$ BI (84)(9)

10. $$\int_0^\infty \frac{x^7}{\sinh x}\,dx = \frac{17}{16}\pi^8$$ BI (84)(10)

11. $$\int_0^\infty \frac{\sqrt{x}\,dx}{\cosh x} = \sqrt{\pi}\sum_{k=0}^\infty (-1)^k \frac{1}{\sqrt{(2k+1)^3}}$$ BI (98)(7)a

12. $$\int_0^\infty \frac{dx}{\sqrt{x}\cosh x} = 2\sqrt{\pi}\sum_{k=0}^\infty \frac{(-1)^k}{\sqrt{2k+1}}$$ BI (98)(25)a

3.524

1. $$\int_0^\infty x^{\mu-1} \frac{\sinh \beta x}{\sinh \gamma x}\, dx = \frac{\Gamma(\mu)}{(2\gamma)^\mu} \left\{ \zeta\left[\mu, \frac{1}{2}\left(1 - \frac{\beta}{\gamma}\right)\right] - \zeta\left[\mu, \frac{1}{2}\left(1 + \frac{\beta}{\gamma}\right)\right] \right\}$$
$$[\operatorname{Re} \gamma > |\operatorname{Re} \beta|, \quad \operatorname{Re} \mu > -1]$$
ET I 323(10)

2. $$\int_0^\infty x^{2m} \frac{\sinh ax}{\sinh bx}\, dx = \frac{\pi}{2b} \frac{d^{2m}}{da^{2m}} \left(\tan \frac{a\pi}{2b}\right) \qquad [b > |a|] \qquad \text{BI (112)(20)a}$$

3. $$\int_0^\infty \frac{\sinh ax}{\sinh bx} \frac{dx}{x^p} = \Gamma(1-p) \sum_{k=0}^\infty \left\{ \frac{1}{[b(2k+1)-a]^{1-p}} - \frac{1}{[b(2k+1)+a]^{1-p}} \right\}$$
$$[b > |a|, \quad p < 1] \qquad \text{BI (131)(2)a}$$

4. $$\int_0^\infty x^{2m+1} \frac{\sinh ax}{\cosh bx}\, dx = \frac{\pi}{2b} \frac{d^{2m+1}}{da^{2m+1}} \left(\sec \frac{a\pi}{2b}\right) \qquad [b > |a|] \qquad \text{BI (112)(18)a}$$

5. $$\int_0^\infty x^{\mu-1} \frac{\cosh \beta x}{\sinh \gamma x}\, dx = \frac{\Gamma(\mu)}{(2\gamma)^\mu} \left\{ \zeta\left[\mu, \frac{1}{2}\left(1 - \frac{\beta}{\gamma}\right)\right] + \zeta\left[\mu, \frac{1}{2}\left(1 + \frac{\beta}{\gamma}\right)\right] \right\}$$
$$[\operatorname{Re} \gamma > |\operatorname{Re} \beta|, \quad \operatorname{Re} \mu > 1]$$
ET I 323(12)

6. $$\int_0^\infty x^{2m} \frac{\cosh ax}{\cosh bx}\, dx = \frac{\pi}{2b} \frac{d^{2m}}{da^{2m}} \left(\sec \frac{a\pi}{2b}\right) \qquad [b > |a|] \qquad \text{BI(112)(17)}$$

7. $$\int_0^\infty \frac{\cosh ax}{\cosh bx} \cdot \frac{dx}{x^p} = \Gamma(1-p) \sum_{k=0}^\infty (-1)^k \left\{ \frac{1}{[b(2k+1)-a]^{1-p}} + \frac{1}{[b(2k+1)+a]^{1-p}} \right\}$$
$$[b > |a|, \quad p < 1] \qquad \text{BI(131)(1)a}$$

8. $$\int_0^\infty x^{2m+1} \frac{\cosh ax}{\sinh bx}\, dx = \frac{\pi}{2b} \frac{d^{2m+1}}{da^{2m+1}} \left(\tan \frac{a\pi}{2b}\right) \qquad [b > |a|] \qquad \text{BI (112)(19)a}$$

9.[8] $$\int_0^\infty x^2 \frac{\sinh ax}{\sinh bx}\, dx = \frac{\pi^3}{4b^3} \sin \frac{a\pi}{2b} \sec^3 \frac{a\pi}{2b} \qquad [b > |a|] \qquad \text{BI (84)(18)}$$

10. $$\int_0^\infty x^4 \frac{\sinh ax}{\sinh bx}\, dx = 8 \left(\frac{\pi}{2b} \sec \frac{a\pi}{2b}\right)^5 \cdot \sin \frac{a\pi}{2b} \cdot \left(2 + \sin^2 \frac{a\pi}{2b}\right)$$
$$[b > |a|] \qquad \text{BI (82)(17)a}$$

11. $$\int_0^\infty x^6 \frac{\sinh ax}{\sinh bx}\, dx = 16 \left(\frac{\pi}{2b} \sec \frac{a\pi}{2b}\right)^7 \sin \frac{a\pi}{2b} \left(45 - 30 \cos^2 \frac{a\pi}{2b} + 2 \cos^4 \frac{a\pi}{2b}\right)$$
$$[b > |a|] \qquad \text{BI (82)(21)a}$$

12. $$\int_0^\infty x \frac{\sinh ax}{\cosh bx}\, dx = \frac{\pi^2}{4b^2} \sin \frac{a\pi}{2b} \sec^2 \frac{a\pi}{2b} \qquad [b > |a|] \qquad \text{BI (84)(15)a}$$

13. $$\int_0^\infty x^3 \frac{\sinh ax}{\cosh bx}\, dx = \left(\frac{\pi}{2b} \sec \frac{a\pi}{2b}\right)^4 \sin \frac{a\pi}{2b} \cdot \left(6 - \cos^2 \frac{a\pi}{2b}\right)$$
$$[b > |a|] \qquad \text{BI (82)(14)a}$$

14. $$\int_0^\infty x^5 \frac{\sinh ax}{\cosh bx}\, dx = \left(\frac{\pi}{2b} \sec \frac{a\pi}{2b}\right)^6 \sin \frac{a\pi}{2b} \left(120 - 60 \cos^2 \frac{a\pi}{2b} + \cos^4 \frac{a\pi}{2b}\right)$$
$$[b > |a|] \qquad \text{BI (82)(18)a}$$

15. $\int_0^\infty x^7 \frac{\sinh ax}{\cosh bx}\, dx = \left(\frac{\pi}{2b}\sec\frac{a\pi}{2b}\right)^8 \sin\frac{a\pi}{2b}\left(5040 - 4200\cos^2\frac{a\pi}{2b} + 546\cos^4\frac{a\pi}{2b} - \cos^6\frac{a\pi}{2b}\right)$

$\qquad\qquad\qquad [b > |a|]$ BI (82)(22)a

16. $\int_0^\infty x \frac{\cosh ax}{\sinh bx}\, dx = \left(\frac{\pi}{2b}\sec\frac{a\pi}{2b}\right)^2$ $\qquad [b > |a|]$ BI (84)(16)a

17. $\int_0^\infty x^3 \frac{\cosh ax}{\sinh bx}\, dx = 2\left(\frac{\pi}{2b}\sec\frac{a\pi}{2b}\right)^4\left(1 + 2\sin^2\frac{a\pi}{2b}\right)$ $\qquad [b > |a|]$ BI (82)(15)a

18. $\int_0^\infty x^5 \frac{\cosh ax}{\sinh bx}\, dx = 8\left(\frac{\pi}{2b}\sec\frac{a\pi}{2b}\right)^6\left(15 - 15\cos^2\frac{a\pi}{2b} + 2\cos^4\frac{a\pi}{2b}\right)$

$\qquad\qquad\qquad [b > |a|]$ BI (82)(19)a

19. $\int_0^\infty x^7 \frac{\cosh ax}{\sinh bx}\, dx = 16\left(\frac{\pi}{2b}\sec\frac{a\pi}{2b}\right)^8\left(315 - 420\cos^2\frac{a\pi}{2b} + 126\cos^4\frac{a\pi}{2b} - 4\cos^6\frac{a\pi}{2b}\right)$

$\qquad\qquad\qquad [b > |a|]$ BI(82)(23)a

20. $\int_0^\infty x^2 \frac{\cosh ax}{\cosh bx}\, dx = \frac{\pi^3}{8b^3}\left(2\sec^3\frac{a\pi}{2b} - \sec\frac{a\pi}{2b}\right)$ $\qquad [b > |a|]$ BI (84)(17)a

21. $\int_0^\infty x^4 \frac{\cosh ax}{\cosh bx}\, dx = \left(\frac{\pi}{2b}\sec\frac{a\pi}{2b}\right)^5\left(24 - 20\cos^2\frac{a\pi}{2b} + \cos^4\frac{a\pi}{2b}\right)$

$\qquad\qquad\qquad [b > |a|]$ BI (82)(16)a

22. $\int_0^\infty x^6 \frac{\cosh ax}{\cosh bx}\, dx = \left(\frac{\pi}{2b}\sec\frac{a\pi}{2b}\right)^7\left(720 - 840\cos^2\frac{a\pi}{2b} + 182\cos^4\frac{a\pi}{2b} - \cos^6\frac{a\pi}{2b}\right)$

$\qquad\qquad\qquad [b > |a|]$ BI (82)(20)a

23. $\int_0^\infty \frac{\sinh ax}{\cosh bx}\cdot\frac{dx}{x} = \ln\tan\left(\frac{a\pi}{4b} + \frac{\pi}{4}\right)$ $\qquad [b > |a|]$ BI (95)(3)a

3.525

1. $\int_0^\infty \frac{\sinh ax}{\sinh \pi x}\cdot\frac{dx}{1 + x^2} = -\frac{a}{2}\cos a + \frac{1}{2}\sin a\ln\left[2\left(1 + \cos a\right)\right]$

$\qquad\qquad\qquad [\pi \geq |a|]$ BI (97)(10)a

2. $\int_0^\infty \frac{\sinh ax}{\sinh \frac{\pi}{2}x}\cdot\frac{dx}{1 + x^2} = \frac{\pi}{2}\sin a + \frac{1}{2}\cos a\ln\frac{1 - \sin a}{1 + \sin a}$ $\qquad [\pi \geq 2|a|]$ BI (97)(11)a

3. $\int_0^\infty \frac{\cosh ax}{\sinh \pi x}\cdot\frac{x\, dx}{1 + x^2} = \frac{1}{2}\left(a\sin a - 1\right) + \frac{1}{2}\cos a\ln\left[2\left(1 + \cos a\right)\right]$

$\qquad\qquad\qquad [\pi > |a|]$ BI (97)(12)a

4. $\int_0^\infty \frac{\cosh ax}{\sinh \frac{\pi}{2}x}\cdot\frac{x\, dx}{1 + x^2} = \frac{\pi}{2}\cos a - 1 + \frac{1}{2}\sin a\ln\frac{1 + \sin a}{1 - \sin a}$

$\qquad\qquad\qquad \left[\frac{\pi}{2} > |a|\right]$ BI (97)(13)a

5. $\int_0^\infty \frac{\sinh ax}{\cosh \pi x}\cdot\frac{x\, dx}{1 + x^2} = -2\sin\frac{a}{2} + \frac{\pi}{2}\sin a - \cos a\ln\tan\frac{a + \pi}{4}$

$\qquad\qquad\qquad [\pi > |a|]$ GW (352)(12)

6. $\displaystyle\int_0^\infty \frac{\cosh ax}{\cosh \pi x}\cdot\frac{dx}{1+x^2} = 2\cos\frac{a}{2} - \frac{\pi}{2}\cos a - \sin a\ln\tan\frac{a+\pi}{4}$

                                                   $[\pi > |a|]$ GW (352)(11)

7. $\displaystyle\int_0^\infty \frac{\sinh ax}{\sinh bx}\cdot\frac{dx}{c^2+x^2} = \frac{\pi}{c}\sum_{k=1}^\infty \frac{\sin\frac{k(b-a)}{b}\pi}{bc+k\pi}$ $[b \ge |a|]$ BI (97)(18)

8. $\displaystyle\int_0^\infty \frac{\cosh ax}{\sinh bx}\cdot\frac{x\,dx}{c^2+x^2} = \frac{\pi}{2bc} + \pi\sum_{k=1}^\infty \frac{\cos\frac{k(b-a)}{b}\pi}{bc+k\pi}$ $[b > |a|]$ BI (97)(19)

3.526

1. $\displaystyle\int_0^\infty \frac{\sinh ax\cosh bx}{\cosh cx}\cdot\frac{dx}{x} = \frac{1}{2}\ln\left\{\tan\frac{(a+b+c)\pi}{4c}\cot\frac{(b+c-a)\pi}{4c}\right\}$

                                                   $[c > |a| + |b|]$ BI (93)(10)a

2. $\displaystyle\int_0^\infty \frac{\sinh^2 ax}{\sinh bx}\cdot\frac{dx}{x} = \frac{1}{2}\ln\sec\frac{a}{b}\pi$ $[b > |2a|]$ BI (95)(5)a

3. $\displaystyle\int_0^\infty \frac{x^{\mu-1}}{\sinh\beta x\cosh\gamma x}\,dx = \frac{\Gamma(\mu)}{(2\gamma)^\mu}\left\{\Phi\left[-1,\mu,\frac{1}{2}\left(1+\frac{\beta}{\gamma}\right)\right] + \Phi\left[-1,\mu,\frac{1}{2}\left(1-\frac{\beta}{\gamma}\right)\right]\right\}$

                                               $[\operatorname{Re}\gamma > |\operatorname{Re}\beta|, \quad \operatorname{Re}\mu > 0]$

                                                           ET I 323(11)

3.527

1. $\displaystyle\int_0^\infty \frac{x^{\mu-1}}{\sinh^2 ax}\,dx = \frac{4}{(2a)^\mu}\Gamma(\mu)\zeta(\mu-1)$ $[\operatorname{Re} a > 0, \quad \operatorname{Re}\mu > 2]$ BI (86)(7)a

2. $\displaystyle\int_0^\infty \frac{x^{2m}}{\sinh^2 ax}\,dx = \frac{\pi^{2m}}{a^{2m+1}}|B_{2m}|$ $[a > 0, \quad m = 1, 2, \ldots]$ BI(86)(5)a

3.[6] $\displaystyle\int_0^\infty \frac{x^{\mu-1}}{\cosh^2 ax}\,dx = \frac{4}{(2a)^\mu}\left(1 - 2^{2-\mu}\right)\Gamma(\mu)\zeta(\mu-1)$ $[\operatorname{Re} a > 0, \quad \operatorname{Re}\mu > 0, \quad \mu \ne 2]$

                                       $= \frac{1}{a^2}\ln 2$ $[\operatorname{Re} a > 0, \quad \mu = 2]$

                                                       BI (86)(6)a

4. $\displaystyle\int_0^\infty \frac{x\,dx}{\cosh^2 ax} = \frac{\ln 2}{a^2}$ $[a \ne 0]$ LO III 396

5. $\displaystyle\int_0^\infty \frac{x^{2m}}{\cosh^2 ax}\,dx = \frac{(2^{2m}-2)\pi^{2m}}{(2a)^{2m}a}|B_{2m}|$ $[a > 0, \quad m = 1, 2, \ldots]$ BI(86)(2)a

6. $\displaystyle\int_0^\infty x^{\mu-1}\frac{\sinh ax}{\cosh^2 ax}\,dx = \frac{2\,\Gamma(\mu)}{a^\mu}\sum_{k=0}^\infty \frac{(-1)^k}{(2k+1)^{\mu-1}}$ $[\operatorname{Re}\mu > 1, \quad a > 0]$ BI (86)(15)a

7. $\displaystyle\int_0^\infty \frac{x\sinh ax}{\cosh^2 ax}\,dx = \frac{\pi}{2a^2}$ $[a > 0]$ BI (86)(8)a

8. $\displaystyle\int_0^\infty x^{2m+1}\frac{\sinh ax}{\cosh^2 ax}\,dx = \frac{2m+1}{a}\left(\frac{\pi}{2a}\right)^{2m+1}|E_{2m}|$ $[a > 0, \quad m = 0, 1, \ldots]$ BI (86)(12)a

9. $\displaystyle\int_0^\infty x^{2m+1}\frac{\cosh ax}{\sinh^2 ax}\,dx = \frac{2^{2m+1}-1}{a^2(2a)^{2m}}(2m+1)!\,\zeta(2m+1)$

$\qquad\qquad\qquad\qquad\qquad\qquad\qquad\qquad\quad [a \neq 0, \quad m = 1, 2, \ldots]$ BI (86)(13)a

10. $\displaystyle\int_0^\infty x^{2m}\frac{\cosh ax}{\sinh^2 ax}\,dx = \frac{2^{2m}1}{a}\left(\frac{\pi}{a}\right)^{2m}|B_{2m}|$ $[a > 0, \quad m = 1, 2, \ldots]$ BI (86)(14)a

11.[8] $\displaystyle\int_0^\infty \frac{x \sinh ax}{\cosh^{2\mu+1} ax}\,dx = \frac{\sqrt{\pi}}{4\mu a^2}\frac{\Gamma(\mu)}{\Gamma\left(\mu+\frac{1}{2}\right)}$ $[\mu > 0, \quad a > 0]$ LI (86)(9)

12. $\displaystyle\int_{-\infty}^\infty \frac{x^2\,dx}{\sinh^2 x} = \frac{\pi^2}{3}$ BI (102)(2)a

13. $\displaystyle\int_0^\infty x^2\frac{\cosh ax}{\sinh^2 ax}\,dx = \frac{\pi^2}{2a^3}$ $[a > 0]$ BI (86)(11)a

14. $\displaystyle\int_0^\infty x^2\frac{\sinh ax}{\cosh^2 ax}\,dx = \frac{\ln 2}{2a^3}$ $[a \neq 0]$ BI (86)(10)a

15.[10] $\displaystyle\int_0^\infty \frac{\tanh\frac{x}{2}\,dx}{\cosh x} = \ln 2$ BI (93)(17)a

3.528

1. $\displaystyle\int_0^\infty \frac{(1+xi)^{2n-1}-(1-xi)^{2n-1}}{i\sinh\frac{\pi x}{2}}\,dx = 2$ BI (87)(8)

2. $\displaystyle\int_0^\infty \frac{(1+xi)^{2n}-(1-xi)^{2n}}{i\sinh\frac{\pi x}{2}}\,dx = (-1)^{n+1}2|E_{2n}|+2$ $[n = 0, 1, \ldots]$ BI (87)(7)

3.529

1. $\displaystyle\int_0^\infty \left(\frac{1}{\sinh x}-\frac{1}{x}\right)\frac{dx}{x} = -\ln 2$ BI (94)(10)a

2. $\displaystyle\int_0^\infty \frac{\cosh ax - 1}{\sinh bx}\cdot\frac{dx}{x} = -\ln\cos\frac{a\pi}{2b}$ $[b > |a|]$ GW (352)(66)

3. $\displaystyle\int_0^\infty \left(\frac{a}{\sinh ax}-\frac{b}{\sinh bx}\right)\frac{dx}{x} = (b-a)\ln 2$ BI (94)(11)a

3.531

1.[7] $\displaystyle\int_0^\infty \frac{x\,dx}{2\cosh x - 1} = \frac{4}{\sqrt{3}}\left[\frac{\pi}{3}\ln 2 - L\left(\frac{\pi}{3}\right)\right] = 1.1719536193\ldots$

$\qquad\qquad\qquad\qquad\qquad\qquad\qquad\qquad$ [see **8.26** for $L(x)$]   LI (88)(1)

2.[10] $\displaystyle\int_0^\infty \frac{x\,dx}{\cosh 2x + \cos 2t} = \frac{t\ln 2 - L(t)}{\sin 2t}$ LO III 402

3. $\displaystyle\int_0^\infty \frac{x^2\,dx}{\cosh x + \cos t} = \frac{t}{3}\cdot\frac{\pi^2 - t^2}{\sin t}$ $[0 < t < \pi]$ BI (88)(3)a

4. $\displaystyle\int_0^\infty \frac{x^4\,dx}{\cosh x + \cos t} = \frac{t}{15}\frac{\left(\pi^2-t^2\right)\left(7\pi^2-3t^2\right)}{\sin t}$ $[0 < t < \pi]$ BI (88)(4)a

$5.^3$ $\displaystyle\int_0^\infty \frac{x^{2m}\,dx}{\cosh x - \cos 2a\pi} = 2(2m)!\operatorname{cosec} 2a\pi \sum_{k=1}^\infty \frac{\sin 2ka\pi}{k^{2m+1}}$ $\left[0 < a < 1, \quad a \neq \tfrac{1}{2}\right]$

$\qquad\qquad\qquad = 2\left(2^{2m-1} - 1\right)\pi^{2m}|B_{2m}|$ $\left[a = \tfrac{1}{2}\right]$

BI (88)(5)a

$6.^3$ $\displaystyle\int_0^\infty \frac{x^{\mu-1}\,dx}{\cosh x - \cos t}$

$\qquad = \dfrac{i\,\Gamma(\mu)}{\sin t}\left[e^{-it}\,\Phi\left(e^{-it}, \mu, 1\right) - e^{it}\,\Phi\left(e^{it}, \mu, 1\right)\right]$ $[\operatorname{Re}\mu > 0, \quad 0 < t < 2\pi, \quad t \neq \pi]$ ET I 323(5)

$\qquad = \left(2 - 2^{3-\mu}\right)\Gamma(\mu)\,\zeta(\mu - 1)$ $[\mu \neq 2, \quad t = \pi]$

$\qquad = 2\ln 2$ $[\mu = 2, \quad t = \pi]$

$7.$ $\displaystyle\int_0^\infty \frac{x^\mu\,dx}{\cosh x + \cos t} = \frac{2\,\Gamma(\mu+1)}{\sin t}\sum_{k=1}^\infty (-1)^{k-1}\frac{\sin kt}{k^{\mu+1}}$ $[\mu > -1, \quad 0 < t < \pi]$ BII (96)(14)a

$8.$ $\displaystyle\int_0^u \frac{x\,dx}{\cosh 2x - \cos 2t} = \frac{1}{2}\operatorname{cosec} 2t\left[L(\theta + t) - L(\theta - t) - 2L(t)\right]$

$\qquad\qquad\qquad\qquad\qquad [\theta = \arctan(\tanh u \cot t), \quad t \neq n\pi]$

LO III 402

3.532

$1.$ $\displaystyle\int_0^\infty \frac{x^n\,dx}{a\cosh x + b\sinh x} = \frac{(2n)!}{a+b}\sum_{k=0}^\infty \frac{1}{(2k+1)^{n+1}}\left(\frac{b-a}{b+a}\right)^k$

$[a > 0, \quad b > 0, \quad n > -1]$ GW (352)(5)

$2.$ $\displaystyle\int_0^u \frac{x\cosh x\,dx}{\cosh 2x - \cos 2t} = \frac{1}{2}\operatorname{cosec} t\left\{L\left(\frac{\theta+t}{2}\right) - L\left(\frac{\theta-t}{2}\right) + L\left(\pi - \frac{\psi+t}{2}\right)\right.$

$\qquad\qquad\qquad\qquad\left. + L\left(\frac{\psi-t}{2}\right) - 2L\left(\frac{t}{2}\right) - 2L\left(\frac{\pi-t}{2}\right)\right\}$

$\qquad\qquad\left[\tan\frac{\theta}{2} = \tanh\frac{u}{2}\cot\frac{t}{2}, \quad \tan\frac{\psi}{2} = \coth\frac{u}{2}\cot\frac{t}{2}; \quad t \neq n\pi\right]$ LO III 288a

3.533

$1.$ $\displaystyle\int_0^\infty \frac{x\cosh x\,dx}{\cosh 2x - \cos 2t} = \operatorname{cosec} t\left[\frac{\pi}{2}\ln 2 - L\left(\frac{t}{2}\right) - L\left(\frac{(\pi-t)}{2}\right)\right]$

$\qquad\qquad\qquad\qquad\qquad [t \neq m\pi]$

LO III 403

$2.^6$ $\displaystyle\int_0^\infty x\frac{\sinh ax\,dx}{(\cosh ax - \cos t)^2} = \frac{\pi-t}{a^2}\operatorname{cosec} t$ $[a > 0, \quad 0 < t < \pi]$ (cf. **3.5141**)

BI (88)(11)a

$3.$ $\displaystyle\int_0^\infty x^3\frac{\sinh x\,dx}{(\cosh x + \cos t)^2} = \frac{t\left(\pi^2 - t^2\right)}{\sin t}$ $[0 < t < \pi]$ (cf. **3.531** 3)

BI (88)(13)

$4.^{10}$ $\displaystyle\int_0^\infty x^{2m+1} \frac{\sinh x \, dx}{(\cosh x - \cos 2a\pi)^2} = 2(2m+1)! \operatorname{cosec} 2a\pi \sum_{k=1}^{\infty} \frac{\sin 2ka\pi}{k^{2m+1}}$ $\quad \left[0 < a < 1, \quad a \ne \tfrac{1}{2}\right]$

$$= 2(2m+1)\left(2^{2m-1} - 1\right)\pi^{2m}|B_{2m}| \qquad \left[a = \tfrac{1}{2}\right]$$

BI (88)(14)

3.534

1. $\displaystyle\int_0^1 \sqrt{1 - x^2} \cosh ax \, dx = \frac{\pi}{2a} I_1(a)$ WA 94(9)

2. $\displaystyle\int_0^1 \frac{\cosh ax}{\sqrt{1 - x^2}} \, dx = \frac{\pi}{2} I_0(a)$ WA 94(9)

3.535 $\displaystyle\int_0^1 \frac{x}{\sqrt{\cosh 2a - \cosh 2ax}} \cdot \frac{dx}{\sinh ax} = \frac{\pi}{2\sqrt{2}a^2} \cdot \frac{\arcsin(\tanh a)}{\sinh a}$ $\quad [a > 0]$ BI (80)(11)

3.536

1. $\displaystyle\int_0^\infty \frac{x^2}{\cosh^2 x} \, dx = \frac{\sqrt{\pi}}{2} \sum_{k=0}^{\infty} \frac{(-1)^k}{\sqrt{(2k+1)^3}}$ BI (98)(7)

2. $\displaystyle\int_0^\infty \frac{x^2 \tanh x^2 \, dx}{\cosh^2 x} = \frac{\sqrt{\pi}}{2} \sum_{k=0}^{\infty} \frac{(-1)^k}{\sqrt{2k+1}}$ BI (98)(8)

3. $\displaystyle\int_0^\infty \sinh(\nu \operatorname{arcsinh} x) \frac{x^{\mu-1}}{\sqrt{1 + x^2}} \, dx = \frac{\sin\frac{\mu\pi}{2}\sin\frac{\nu\pi}{2}}{2^\mu \pi} \Gamma(\mu) \Gamma\left(\frac{1-\mu-\nu}{2}\right) \times \Gamma\left(\frac{1-\mu+\nu}{2}\right)$

$[-1 < \operatorname{Re}\mu < 1 - |\operatorname{Re}\nu|]$ ET I 324(14)

4. $\displaystyle\int_0^\infty \cosh(\nu \operatorname{arccosh} x) \frac{x^{\mu-1}}{\sqrt{1 + x^2}} \, dx = \frac{\cos\frac{\mu\pi}{2}\cos\frac{\nu\pi}{2}}{2^\mu \pi} \Gamma(\mu) \Gamma\left(\frac{1-\mu-\nu}{2}\right) \times \Gamma\left(\frac{1-\mu+\nu}{2}\right)$

$[0 < \operatorname{Re}\mu < 1 - |\operatorname{Re}\nu|]$ ET I 324(15)

3.54 Combinations of hyperbolic functions and exponentials

3.541

1. $\displaystyle\int_0^\infty e^{-\mu x} \sinh^\nu \beta x \, dx = \frac{1}{2^{\nu+1}\beta} \mathrm{B}\left(\frac{\mu}{2\beta} - \frac{\nu}{2}, \nu + 1\right)$ $\quad [\operatorname{Re}\beta > 0, \quad \operatorname{Re}\nu > -1, \operatorname{Re}\mu > \operatorname{Re}\beta\nu]$

EH I 11(25), ET I 163(5)

2. $\displaystyle\int_0^\infty e^{-\mu x} \frac{\sinh \beta x}{\sinh bx} \, dx = \frac{1}{2b}\left[\psi\left(\frac{1}{2} + \frac{\mu + \beta}{2b}\right) - \psi\left(\frac{1}{2} + \frac{\mu - \beta}{2b}\right)\right]$

$[\operatorname{Re}(\mu + b \pm \beta) > 0]$ EH I 16(14)a

3. $\displaystyle\int_{-\infty}^\infty e^{-\mu x} \frac{\sinh \mu x}{\sinh \beta x} \, dx = \frac{\pi}{2\beta} \tan \frac{\mu\pi}{\beta}$ $\quad [\operatorname{Re}\beta > 2|\operatorname{Re}\mu|]$ BI (18)(6)

4. $\displaystyle\int_0^\infty e^{-x} \frac{\sinh ax}{\sinh x} \, dx = \frac{1}{a} - \frac{\pi}{2} \cot \frac{a\pi}{2}$ $\quad [0 < a < 2]$ BI (4)(3)

5. $\displaystyle\int_0^\infty \frac{e^{-px}\,dx}{(\cosh px)^{2q+1}} = \frac{2^{2q-2}}{p}\,\mathrm{B}(q,q) - \frac{1}{2qp}$ $[p > 0, \quad q > 0]$ LI (27)(19)

6. $\displaystyle\int_0^\infty e^{-\mu x}\,\frac{dx}{\cosh x} = \beta\left(\frac{\mu+1}{2}\right)$ $[\mathrm{Re}\,\mu > -1]$ ET I 163(7)

7. $\displaystyle\int_0^\infty e^{-\mu x}\tanh x\,dx = \beta\left(\frac{\mu}{2}\right) - \frac{1}{\mu}$ $[\mathrm{Re}\,\mu > 0]$ ET I 163(9)

8. $\displaystyle\int_0^\infty \frac{e^{-\mu x}}{\cosh^2 x}\,dx = \mu\,\beta\left(\frac{\mu}{2}\right) - 1$ $[\mathrm{Re}\,\mu > 0]$ ET I 163(8)

9. $\displaystyle\int_0^\infty e^{-\mu x}\,\frac{\sinh \mu x}{\cosh^2 \mu x}\,dx = \frac{1}{\mu}\left(1 - \ln 2\right)$ $[\mathrm{Re}\,\mu > 0]$ LI (27)(15)

10. $\displaystyle\int_0^\infty e^{-qx}\,\frac{\sinh px}{\sinh qx}\,dx = \frac{1}{p} - \frac{\pi}{2q}\cot\frac{p\pi}{2q}$ $[0 < p < 2q]$ BI (27)(9)a

3.542

1. $\displaystyle\int_0^\infty e^{-\mu x}(\cosh \beta x - 1)^\nu\,dx = \frac{1}{2^\nu \beta}\,\mathrm{B}\left(\frac{\mu}{\beta} - \nu, 2\nu + 1\right)$

$$\left[\mathrm{Re}\,\beta > 0, \quad \mathrm{Re}\,\nu > -\frac{1}{2}, \quad \mathrm{Re}\,\mu > \mathrm{Re}\,\beta\nu\right] \quad \text{ET I 163(6)}$$

2. $\displaystyle\int_0^\infty e^{-\mu x}(\cosh x - \cosh u)^{\nu-1}\,dx = -i\sqrt{\frac{2}{\pi}}\,e^{i\pi\nu}\,\Gamma(\nu)\sinh^{\nu-\frac{1}{2}} u\, Q_{\mu-\frac{1}{2}}^{\frac{1}{2}-\nu}(\cosh u)$

$$[\mathrm{Re}\,\nu > 0, \quad \mathrm{Re}\,\mu > \mathrm{Re}\,\nu - 1]$$
$$\text{EH I 155(4), ET I 164(23)}$$

3.543

1. $\displaystyle\int_{-\infty}^\infty \frac{e^{-ibx}\,dx}{\sinh x + \sinh t} = -\frac{i\pi e^{itb}}{\sinh \pi b \cosh t}\left(\cosh \pi b - e^{-2itb}\right)$

$$[t > 0] \quad \text{ET I 121(30)}$$

2. $\displaystyle\int_0^\infty \frac{e^{-\mu x}}{\cosh x - \cos t}\,dx = 2\cosec t \sum_{k=1}^\infty \frac{\sin kt}{\mu + k}$ $[\mathrm{Re}\,\mu > -1, \quad t \neq 2n\pi]$ BI (6)(10)a

3. $\displaystyle\int_0^\infty \frac{1 - e^{-x}\cos t}{\cosh x - \cos t}\,e^{-(\mu-1)x}\,dx = 2\sum_{k=0}^\infty \frac{\cos kt}{\mu + k}$ $[\mathrm{Re}\,\mu > 0, \quad t \neq 2n\pi]$ BI (6)(9)a

4. $\displaystyle\int_0^\infty \frac{e^{px} - \cos t}{(\cosh px + \cos t)^2}\,dx = \frac{1}{p}\left(t\cosec t + \frac{1}{1 + \cos t}\right)$ $[p > 0]$ BI (27)(26)a

3.544 $\displaystyle\int_u^\infty \frac{\exp\left[-\left(n+\frac{1}{2}\right)x\right]}{\sqrt{2\left(\cosh x - \cosh u\right)}}\,dx = Q_n(\cosh u), \quad [u > 0]$ EH II 181(33)

3.545

1. $\displaystyle\int_0^\infty \frac{\sinh ax}{e^{px} + 1}\,dx = \frac{\pi}{2p}\cosec\frac{a\pi}{p} - \frac{1}{2a}$ $[p > a, \quad p > 0]$ BI (27)(3)

2. $\displaystyle\int_0^\infty \frac{\sinh ax}{e^{px} - 1}\, dx = \frac{1}{2a} - \frac{\pi}{2p} \cot \frac{a\pi}{p}$ $\qquad\qquad [p > a, \quad p > 0]$ $\qquad\qquad$ BI (27)(9)

3.546

1. $\displaystyle\int_0^\infty e^{-\beta x^2} \sinh ax\, dx = \frac{1}{2} \frac{\sqrt{\pi}}{\sqrt{\beta}} \exp \frac{a^2}{4\beta}\, \Phi\left(\frac{a}{2\sqrt{\beta}}\right)$ $\qquad [\operatorname{Re}\beta > 0]$ $\qquad\qquad$ ET I166(38)a

2. $\displaystyle\int_0^\infty e^{-\beta x^2} \cosh ax\, dx = \frac{1}{2} \sqrt{\frac{\pi}{\beta}} \exp \frac{a^2}{4\beta}$ $\qquad\qquad [\operatorname{Re}\beta > 0]$ $\qquad\qquad$ FI II 720a

3. $\displaystyle\int_0^\infty e^{-\beta x^2} \sinh^2 ax\, dx = \frac{1}{4} \sqrt{\frac{\pi}{\beta}} \left(\exp \frac{a^2}{\beta} - 1\right)$ $\qquad [\operatorname{Re}\beta > 0]$ $\qquad\qquad$ ET I 166(40)

4. $\displaystyle\int_0^\infty e^{-\beta x^2} \cosh^2 ax\, dx = \frac{1}{4} \sqrt{\frac{\pi}{\beta}} \left(\exp \frac{a^2}{\beta} + 1\right)$ $\qquad [\operatorname{Re}\beta > 0]$ $\qquad\qquad$ ET I 166(41)

3.547

1. $\displaystyle\int_0^\infty \exp\left(-\beta \sinh x\right) \sinh \gamma x\, dx = \frac{\pi}{2} \cot \frac{\gamma\pi}{2}\, [J_\gamma(\beta) - \mathbf{J}_\gamma(\beta)] - \frac{\pi}{2}\, [\mathbf{E}_\gamma(\beta) + Y_\gamma(\beta)] = \gamma\, S_{-1,\gamma}(\beta)$

$\qquad\qquad\qquad\qquad\qquad\qquad\qquad\qquad\qquad [\operatorname{Re}\beta > 0]$ $\qquad$ WA 341(5), ET I 168(14)a

2. $\displaystyle\int_0^\infty \exp\left(-\beta \cosh x\right) \sinh \gamma x \sinh x\, dx = \frac{\gamma}{\beta}\, K_\gamma(\beta)$

3. $\displaystyle\int_0^\infty \exp\left(-\beta \sinh x\right) \cosh \gamma x\, dx = \frac{\pi}{2} \tan \frac{\pi\gamma}{2}\, [\mathbf{J}_\gamma(\beta) - J_\gamma(\beta)] - \frac{\pi}{2}\, [\mathbf{E}_\gamma(\beta) + Y_\gamma(\beta)] = S_{0,\gamma}(\beta)$

$\qquad\qquad\qquad\qquad\qquad\qquad\qquad\qquad [\operatorname{Re}\beta > 0, \quad \gamma \text{ not an integer}]$
$\qquad\qquad\qquad\qquad\qquad\qquad\qquad\qquad$ ET I 168(16)a, WA 341(4), EH II 84(50)

4. $\displaystyle\int_0^\infty \exp\left(-\beta \cosh x\right) \cosh \gamma x\, dx = K_\gamma(\beta)$ $\qquad [\operatorname{Re}\beta > 0]$ $\qquad$ ET I 168(16)a, WA 201(5)

5. $\displaystyle\int_0^\infty \exp\left(-\beta \sinh x\right) \sinh \gamma x \cosh x\, dx = \frac{\gamma}{\beta}\, S_{0,\gamma}(\beta)$ $\qquad [\operatorname{Re}\beta > 0]$ $\qquad$ ET I 168(7), EH II 85(51)

6. $\displaystyle\int_0^\infty \exp\left(-\beta \sinh x\right) \sinh[(2n + 1)x] \cosh x\, dx = O_{2n+1}(\beta)$

$\qquad\qquad\qquad\qquad\qquad\qquad\qquad\qquad\qquad\qquad [\operatorname{Re}\beta > 0]$ $\qquad\qquad$ ET I 167(5)

7. $\displaystyle\int_0^\infty \exp\left(-\beta \sinh x\right) \cosh \gamma x \cosh x\, dx = \frac{1}{\beta}\, S_{1,\gamma}(\beta)$ $\qquad [\operatorname{Re}\beta > 0]$

8. $\displaystyle\int_0^\infty \exp\left(-\beta \sinh x\right) \cosh 2nx \cosh x\, dx = O_{2n}(\beta)$ $\qquad [\operatorname{Re}\beta > 0]$ $\qquad$ ET I 168(6)

9. $\displaystyle\int_0^\infty \exp\left(-\beta \cosh x\right) \sinh^{2\nu} x\, dx = \frac{1}{\sqrt{\pi}} \left(\frac{2}{\beta}\right)^\nu \Gamma\left(\nu + \frac{1}{2}\right) K_\nu(\beta)$

$\qquad\qquad\qquad\qquad\qquad\qquad\qquad [\operatorname{Re}\beta > 0, \quad \operatorname{Re}\nu > -\tfrac{1}{2}]$ $\qquad$ EH II 82(20)

10. $\displaystyle\int_0^\infty \exp\left[-2\left(\beta \coth x + \mu x\right)\right] \sinh^{2\nu} x\, dx = \frac{1}{4} \beta^{\frac{\nu-1}{2}} \Gamma(\mu - \nu)$

$\qquad\qquad\qquad\qquad\qquad\qquad \times \left[W_{-\mu+\frac{1}{2},\nu}(4\beta) - (\mu - \nu)\, W_{-\mu-\frac{1}{2},\nu}(4\beta)\right]$
$\qquad\qquad\qquad\qquad\qquad\qquad\qquad\qquad [\operatorname{Re}\beta > 0, \quad \operatorname{Re}\mu > \operatorname{Re}\nu]$ $\qquad$ ET I 165(31)

11. $\displaystyle\int_0^\infty \exp\left(-\frac{\beta^2}{2}\sinh x\right)\sinh^{\nu-1}x\,\cosh^\nu x\,dx = -\pi\,D_\nu\left(\beta e^{i\pi/4}\right)D_\nu\left(\beta e^{-i\pi/4}\right)$

$$\left[\operatorname{Re}\nu > 0, |\arg\beta| \le \frac{\pi}{4}\right] \qquad \text{EH II 120(10)}$$

12. $\displaystyle\int_0^\infty \frac{\exp\left(2\nu x - 2\beta\sinh x\right)}{\sqrt{\sinh x}}\,dx = \frac{1}{2}\sqrt{\pi^3\beta}\left[J_{\nu+\frac{1}{4}}(\beta)\,J_{\nu-\frac{1}{4}}(\beta) + Y_{\nu+\frac{1}{4}}(\beta)\,Y_{\nu-\frac{1}{4}}(\beta)\right]$

$$[\operatorname{Re}\beta > 0] \qquad \text{EH I 169(20)}$$

13. $\displaystyle\int_0^\infty \frac{\exp\left(-2\nu x - 2\beta\sinh x\right)}{\sqrt{\sinh x}}\,dx = \frac{1}{2}\sqrt{\pi^3\beta}\left[J_{\nu+\frac{1}{4}}(\beta)\,Y_{\nu-\frac{1}{4}}(\beta) - J_{\nu-\frac{1}{4}}(\beta)\,Y_{\nu+\frac{1}{4}}(\beta)\right]$

$$[\operatorname{Re}\beta > 0] \qquad \text{ET I 169(21)}$$

14. $\displaystyle\int_0^\infty \frac{\exp\left(-2\beta\sinh x\right)\sinh 2\nu x}{\sqrt{\sinh x}}\,dx = \frac{1}{4i}\sqrt{\frac{\pi^3\beta}{2}}\left\{e^{\nu\pi i}\,H_{\frac{1}{2}+\nu}^{(1)}(\beta)\,H_{\frac{1}{2}-\nu}^{(2)}(\beta)\right.$

$$\left. -e^{-\nu\pi i}\,H_{\frac{1}{2}-\nu}^{(1)}(\beta)\,H_{\frac{1}{2}+\nu}^{(2)}(\beta)\right\}$$

$$[\operatorname{Re}\beta > 0] \qquad \text{ET I 170(24)}$$

15. $\displaystyle\int_0^\infty \frac{\exp\left(-2\beta\sinh x\right)\cosh 2\nu x}{\sqrt{\sinh x}}\,dx = \frac{1}{4}\sqrt{\frac{\pi^3\beta}{2}}\left\{e^{\nu\pi i}\,H_{\frac{1}{2}+\nu}^{(1)}(\beta)\,H_{\frac{1}{2}-\nu}^{(2)}(\beta)\right.$

$$\left. +e^{-\nu\pi i}\,H_{\frac{1}{2}-\nu}^{(1)}(\beta)\,H_{\frac{1}{2}+\nu}^{(2)}(\beta)\right\}$$

$$[\operatorname{Re}\beta > 0] \qquad \text{ET I 170(25)}$$

16. $\displaystyle\int_0^\infty \frac{\exp\left(-2\beta\cosh x\right)\cosh 2\nu x}{\sqrt{\cosh x}}\,dx = \sqrt{\frac{\beta}{\pi}}\,K_{\nu+\frac{1}{4}}(\beta)\,K_{\nu-\frac{1}{4}}(\beta)$

$$[\operatorname{Re}\beta > 0] \qquad \text{ET I 170(26)}$$

17.[8] $\displaystyle\int_0^\infty \frac{\exp\left[-2\beta\left(\cosh x - 1\right)\right]\cosh 2\nu x}{\sqrt{\cosh x}}\,dx = \sqrt{\frac{\beta}{\pi}}\cdot e^{2\beta}\,K_{\nu+\frac{1}{4}}(\beta)\,K_{\nu-\frac{1}{4}}(\beta)$

$$[\operatorname{Re}\beta > 0] \qquad \text{ET I 170(27)}$$

18. $\displaystyle\int_0^\infty \frac{\cos\left[\left(\nu+\frac{1}{4}\right)\pi\right]\exp\left(-2\nu x - 2\beta\sinh x\right) + \sin\left[\left(\nu+\frac{1}{4}\right)\pi\right]\exp\left(2\nu x - 2\beta\sinh x\right)}{\sqrt{\sinh x}}\,dx$

$$= \frac{1}{2}\sqrt{\pi^3\beta}\left[J_{\frac{1}{4}+\nu}(\beta)\,J_{\frac{1}{4}-\nu}(\beta) + Y_{\frac{1}{4}+\nu}(\beta)\,Y_{\frac{1}{4}-\nu}(\beta)\right]$$

$$[\operatorname{Re}\beta > 0] \qquad \text{ET I 169(22)}$$

19. $\displaystyle\int_0^\infty \frac{\sin\left[\left(\nu+\frac{1}{4}\right)\pi\right]\exp\left(-2\nu x - 2\beta\sinh x\right) - \cos\left[\left(\nu+\frac{1}{4}\right)\pi\right]\exp\left(2\nu x - 2\beta\sinh x\right)}{\sqrt{\sinh x}}\,dx$

$$= \frac{1}{2}\sqrt{\pi^3\beta}\left[J_{\frac{1}{4}+\nu}(\beta)\,Y_{\frac{1}{4}-\nu}(\beta) - J_{\frac{1}{4}-\nu}(\beta)\,Y_{\frac{1}{4}+\nu}(\beta)\right]$$

$$[\operatorname{Re}\beta > 0] \qquad \text{ET I 169(23)}$$

20. $\displaystyle\int_0^\infty \frac{\exp\left[-\beta(\cosh x - 1)\right]\cosh\nu x\,\sinh x}{\sqrt{\cosh x\,(\cosh x - 1)}}\,dx = e^\beta\,K_\nu(\beta)$

$$[\operatorname{Re}\beta > 0] \qquad \text{ET I 169(19)}$$

3.548

1. $\int_0^\infty e^{-\mu x^4} \sinh ax^2 \, dx = \dfrac{\pi}{4} \sqrt{\dfrac{a}{2\mu}} \exp\left(\dfrac{a^2}{8\mu}\right) I_{\frac{1}{4}}\left(\dfrac{a^2}{8\mu}\right)$ $[\operatorname{Re}\mu > 0, \quad a \geq 0]$ ET I 166(42)

2. $\int_0^\infty e^{-\mu x^4} \cosh ax^2 \, dx = \dfrac{\pi}{4} \sqrt{\dfrac{a}{2\mu}} \exp\left(\dfrac{a^2}{8\mu}\right) I_{-\frac{1}{4}}\left(\dfrac{a^2}{8\mu}\right)$

$[\operatorname{Re}\mu > 0, \quad a > 0]$ ET I 166(43)

3.549

1. $\int_0^\infty e^{-\beta x} \sinh\left[(2n+1)\operatorname{arcsinh} x\right] \, dx = O_{2n+1}(\beta)$ $[\operatorname{Re}\beta > 0]$ (cf. **3.547** 6)

ET I 167(5)

2. $\int_0^\infty e^{-\beta x} \cosh\left(2n\operatorname{arcsinh} x\right) \, dx = O_{2n}(\beta)$ $[\operatorname{Re}\beta > 0]$ (cf. **3.547** 8)

ET I 168(6)

3. $\int_0^\infty e^{-\beta x} \sinh\left(\nu\operatorname{arcsinh} x\right) \, dx = \dfrac{\nu}{\beta} S_{0,\nu}(\beta)$ $[\operatorname{Re}\beta > 0]$ (cf. **3.547**5)   ET I 168(7)

4. $\int_0^\infty e^{-\beta x} \cosh\left(\nu\operatorname{arcsinh} x\right) \, dx = \dfrac{1}{\beta} S_{1,\nu}(\beta)$ $[\operatorname{Re}\beta > 0]$ (cf. **3.547** 7)

A number of other integrals containing hyperbolic functions and exponentials, depending on arcsinh x or arccosh x can be found by first making the substitution $x = \sinh t$ or $x = \cosh t$.

3.55–3.56 Combinations of hyperbolic functions, exponentials, and powers

3.551

1. $\int_0^\infty x^{\mu-1} e^{-\beta x} \sinh \gamma x \, dx = \dfrac{1}{2} \Gamma(\mu) \left[(\beta-\gamma)^{-\mu} - (\beta+\gamma)^{-\mu}\right]$

$[\operatorname{Re}\beta > -1, \quad \operatorname{Re}\beta > |\operatorname{Re}\gamma|]$

ET I 164(18)

2. $\int_0^\infty x^{\mu-1} e^{-\beta x} \cosh \gamma x \, dx = \dfrac{1}{2} \Gamma(\mu) \left[(\beta-\gamma)^{-\mu} + (\beta+\gamma)^{-\mu}\right]$

$[\operatorname{Re}\mu > 0, \quad \operatorname{Re}\beta > |\operatorname{Re}\gamma|]$

ET I 164(19)

3. $\int_0^\infty x^{\mu-1} e^{-\beta x} \coth x \, dx = \Gamma(\mu) \left[2^{1-\mu} \zeta\left(\mu, \dfrac{\beta}{2}\right) - \beta^{-\mu}\right]$

$[\operatorname{Re}\mu > 1, \quad \operatorname{Re}\beta > 0]$ ET I 164(21)

4. $\int_0^\infty x^n e^{-(p+mq)x} \sinh^m qx \, dx = 2^{-m} n! \sum_{k=0}^m \binom{m}{k} \dfrac{(-1)^k}{(p+2kq)^{n+1}}$

$[p > 0, \quad q > 0, \quad m < p + qm]$

LI (81)(4)

5. $\int_0^1 \dfrac{e^{-\beta x}}{x} \sinh \gamma x \, dx = \dfrac{1}{2} \left[\ln \dfrac{\beta+\gamma}{\beta-\gamma} \operatorname{Ei}(\gamma-\beta) - \operatorname{Ei}(-\gamma-\beta)\right]$

$[\beta > \gamma]$ BI (80)(4)

6. $\displaystyle\int_0^\infty \frac{e^{-\beta x}}{x} \sinh \gamma x \, dx = \frac{1}{2} \ln \frac{\beta + \gamma}{\beta - \gamma}$ $[\operatorname{Re}\beta > |\operatorname{Re}\gamma|]$ ET I 163(12)

7. $\displaystyle\int_1^\infty \frac{e^{-\beta x}}{x} \cosh \gamma x \, dx = \frac{1}{2}\left[-\operatorname{Ei}(\gamma - \beta) - \operatorname{Ei}(-\gamma - \beta)\right]$ $[\operatorname{Re}\beta > |\operatorname{Re}\gamma|]$ ET I 164(15)

8.[6] $\displaystyle\int_0^\infty x e^{-x} \coth x \, dx = \frac{\pi^2}{4} - 1$ BI (82)(6)

9. $\displaystyle\int_0^\infty e^{-\beta x} \tanh x \, \frac{dx}{x} = \ln \frac{\beta}{4} + 2 \ln \frac{\Gamma\left(\frac{\beta}{4}\right)}{\Gamma\left(\frac{\beta}{4} + \frac{1}{2}\right)}$ $[\operatorname{Re}\beta > 0]$ ET I 164(16)

10.[6] $\displaystyle\int_0^\infty x e^{-x} \coth(x/2) \, dx = \frac{\pi^2}{3} - 1$

3.552

1. $\displaystyle\int_0^\infty \frac{x^{\mu-1} e^{-\beta x}}{\sinh x} \, dx = 2^{1-\mu} \, \Gamma(\mu) \, \zeta\left[\mu, \frac{1}{2}(\beta + 1)\right]$ $[\operatorname{Re}\mu > 1, \quad \operatorname{Re}\beta > -1]$ ET I 164(20)

2. $\displaystyle\int_0^\infty \frac{x^{2m-1} e^{-ax}}{\sinh ax} \, dx = \frac{1}{2m} |B_{2m}| \left(\frac{\pi}{a}\right)^{2m}$ $[a > 0, \quad m = 1, 2, \ldots]$ EH I 38(24)a

3. $\displaystyle\int_0^\infty \frac{x^{\mu-1} e^{-x}}{\cosh x} \, dx = 2^{1-\mu}\left(1 - 2^{1-\mu}\right) \Gamma(\mu) \, \zeta(\mu)$ $[\operatorname{Re}\mu > 0, \quad \mu \neq 1]$

           $= \ln 2$ $[\text{if } \mu = 1]$

                                                                      EH I 32(5)

4. $\displaystyle\int_0^\infty \frac{x^{2m-1} e^{-ax}}{\cosh ax} \, dx = \frac{1 - 2^{1-2m}}{2m} |B_{2m}| \left(\frac{\pi}{a}\right)^{2m}$ $[a > 0, \quad m = 1, 2, \ldots]$ EH I 39(25)a

5. $\displaystyle\int_0^\infty \frac{x^2 e^{-2nx}}{\sinh x} \, dx = 4 \sum_{k=n}^\infty \frac{1}{(2k+1)^3}$ $[n = 0, 1, 2, \ldots]$ (cf. **4.261** 13)

                                                                    BI(84)(4)

6. $\displaystyle\int_0^\infty \frac{x^3 e^{-2nx}}{\sinh x} \, dx = \frac{\pi^4}{8} - 12 \sum_{k=1}^n \frac{1}{(2k+1)^4}$ $[n = 0, 1, \ldots]$ (cf. **4.262** 6)

                                                                    BI (84)(6)

3.553

1. $\displaystyle\int_0^\infty \frac{\sinh^2 ax \, e^{-x} \, dx}{\sinh x \quad x} = \frac{1}{2} \ln(a\pi \operatorname{cosec} a\pi)$ $[a < 1]$ BI (95)(7)

2. $\displaystyle\int_0^\infty \frac{\sinh^2 \frac{\pi}{2}}{\cosh x} \cdot \frac{e^{-x} \, dx}{x} = \frac{1}{2} \ln \frac{4}{\pi}$ (cf. **4.267** 2)   BI (95)(4)

3.554

1. $\displaystyle\int_0^\infty e^{-\beta x} (1 - \operatorname{sech} s) \, \frac{dx}{x} = 2 \ln \frac{\Gamma\left(\frac{\beta+3}{4}\right)}{\Gamma\left(\frac{\beta+1}{4}\right)} - \ln \frac{\beta}{4}$ $[\operatorname{Re}\beta > 0]$ ET I 164(17)

2. $\displaystyle\int_0^\infty e^{-\beta x}\left(\frac{1}{x}-\operatorname{cosech} x\right)\,dx = \psi\left(\frac{\beta+1}{2}\right)-\ln\frac{\beta}{2}$ $\qquad[\operatorname{Re}\beta>0]$ $\qquad$ ET I 163(10)

3. $\displaystyle\int_0^\infty\left[\frac{\sinh\left(\frac{1}{2}-\beta\right)x}{\sinh\frac{x}{2}}-(1-2\beta)e^{-x}\right]\frac{dx}{x} = 2\ln\Gamma(\beta)-\ln\pi+\ln\left(\sin\pi\beta\right)$

$\qquad\qquad\qquad\qquad\qquad\qquad\qquad\qquad\qquad\qquad\qquad\qquad [0<\operatorname{Re}\beta<1]$ $\qquad$ EH I 21(7)

4. $\displaystyle\int_0^\infty e^{-\beta x}\left(\frac{1}{x}-\coth x\right)\,dx = \psi\left(\frac{\beta}{2}\right)-\ln\frac{\beta}{2}+\frac{1}{\beta}$ $\qquad[\operatorname{Re}\beta>0]$ $\qquad$ ET I 163(11)

5. $\displaystyle\int_0^\infty\left\{-\frac{\sinh qx}{\sinh\frac{x}{2}}+2qe^{-x}\right\}\frac{dx}{x} = 2\ln\Gamma\left(q+\frac{1}{2}\right)+\ln\cos\pi q-\ln\pi$

$\qquad\qquad\qquad\qquad\qquad\qquad\qquad\qquad\qquad\qquad\qquad\qquad [q^2<\frac{1}{2}]$ $\qquad$ WH

6. $\displaystyle\int_0^\infty x^{\mu-1}e^{-\beta x}\left(\coth x-1\right)\,dx = 2^{1-\mu}\Gamma(\mu)\zeta\left(\mu,\frac{\beta}{2}+1\right)$

$\qquad\qquad\qquad\qquad\qquad\qquad\qquad\qquad\qquad [\operatorname{Re}\beta>0;\quad\operatorname{Re}\mu>1]$ $\qquad$ ET I 164(22)

3.555

1. $\displaystyle\int_0^\infty\frac{\sinh^2 ax}{1-e^{px}}\cdot\frac{dx}{x} = \frac{1}{4}\ln\left(\frac{p}{2a\pi}\sin\frac{2a\pi}{p}\right)$ $\qquad[0<2|a|<p]\qquad(\text{cf. }\mathbf{3.545}\ 2)$

$\qquad\qquad\qquad\qquad\qquad\qquad\qquad\qquad\qquad\qquad\qquad\qquad\qquad\qquad\qquad\qquad$ BI (93)(15)

2. $\displaystyle\int_0^\infty\frac{\sinh^2 ax}{e^x+1}\cdot\frac{dx}{x} = -\frac{1}{4}\ln\left(a\pi\cot a\pi\right)$ $\qquad[a<\frac{1}{2}]\qquad(\text{cf. }\mathbf{3.545}\ 1)$ $\qquad$ BI (93)(9)

3.556

1. $\displaystyle\int_{-\infty}^\infty x\frac{1-e^{px}}{\sinh x}\,dx = -\frac{\pi^2}{2}\tan^2\frac{p\pi}{2}$ $\qquad[p<1]\qquad(\text{cf. }\mathbf{4.255}\ 3)$ $\qquad$ BI (101)(4)

2. $\displaystyle\int_0^\infty\frac{1-e^{-px}}{\sinh x}\cdot\frac{1-e^{-(p+1)x}}{x}\,dx = 2p\ln 2$ $\qquad[p>-1]$ $\qquad$ BI (95)(8)

3.557

1. $\displaystyle\int_0^\infty\frac{e^{-px}-e^{-qx}}{\cosh x-\cos\frac{m}{n}\pi}\cdot\frac{dx}{x}$

$\displaystyle\qquad = 2\operatorname{cosec}\frac{m}{n}\pi\sum_{k=1}^{n-1}(-1)^{k-1}\sin\left(\frac{km}{n}\pi\right)\ln\frac{\Gamma\left(\frac{n+q+k}{2n}\right)\Gamma\left(\frac{p+k}{2n}\right)}{\Gamma\left(\frac{n+p+k}{2n}\right)\Gamma\left(\frac{q+k}{2n}\right)}$ $\qquad[m+n\text{ odd}]$

$\displaystyle\qquad = 2\operatorname{cosec}\frac{m}{n}\pi\sum_{k=1}^{\frac{n-1}{2}}(-1)^{k-1}\sin\left(\frac{km}{n}\pi\right)\ln\frac{\Gamma\left(\frac{n+q-k}{n}\right)\Gamma\left(\frac{p+k}{n}\right)}{\Gamma\left(\frac{n+p-k}{n}\right)\Gamma\left(\frac{q+k}{n}\right)}$ $\qquad[m+n\text{ even}]$

$\qquad\qquad\qquad\qquad\qquad\qquad\qquad\qquad\qquad\qquad\qquad [p>-1,\quad q>-1]$ $\qquad$ BI (96)(1)

2.
$$\int_0^\infty \frac{(1-e^{-x})^2}{\cosh x + \cos \frac{m}{n}\pi} \cdot \frac{dx}{x} = 2\,\mathrm{cosec}\,\frac{m}{n}\pi \sum_{k=1}^{n-1}(-1)^{k-1}\sin\left(\frac{km}{n}\pi\right)$$

$$\times \ln \frac{\left[\Gamma\left(\frac{n+k+1}{2n}\right)\right]^2 \Gamma\left(\frac{k+2}{2n}\right)\Gamma\left(\frac{k}{2n}\right)}{\left[\Gamma\left(\frac{k+1}{2n}\right)\right]^2 \Gamma\left(\frac{n+k}{2n}\right)\Gamma\left(\frac{n+k+2}{2n}\right)} \qquad [m+n \text{ odd}]$$

$$= 2\,\mathrm{cosec}\,\frac{m}{n}\pi \sum_{k=1}^{\frac{n-1}{2}}(-1)^{k-1}\sin\left(\frac{km}{n}\pi\right)$$

$$\times \ln \frac{\left[\Gamma\left(\frac{n-k+1}{n}\right)\right]^2 \Gamma\left(\frac{k+2}{n}\right)\Gamma\left(\frac{k}{n}\right)}{\left[\Gamma\left(\frac{k+1}{n}\right)\right]^2 \Gamma\left(\frac{n-k}{n}\right)\Gamma\left(\frac{n-k+2}{n}\right)} \qquad [m+n \text{ even}]$$

BI (96)(2)

3.
$$\int_0^\infty \left[e^{-x}\tan\frac{m}{2n}\pi - \frac{e^{-px}\sin\frac{m}{n}\pi}{\cosh x + \cos\frac{m}{n}\pi}\right]\cdot\frac{dx}{x}$$

$$= \tan\left(\frac{m}{2n}\pi\right)\ln(2n) + 2\sum_{k=1}^{n-1}(-1)^{k-1}\sin\left(\frac{km}{n}\pi\right)\ln\frac{\Gamma\left(\frac{p+n+k}{2n}\right)}{\Gamma\left(\frac{p+k}{2n}\right)} \qquad [m+n \text{ odd}]$$

$$= \tan\left(\frac{m}{2n}\pi\right)\ln n + 2\sum_{k=1}^{\frac{n-1}{2}}(-1)^{k-1}\sin\left(\frac{km}{n}\pi\right)\ln\frac{\Gamma\left(\frac{p+n-k}{n}\right)}{\Gamma\left(\frac{p+k}{n}\right)} \qquad [m+n \text{ even}]$$

BI (96)(3)

4.
$$\int_0^\infty \frac{1+e^{-x}}{\cosh x + \cos a}\cdot\frac{dx}{x^{1-p}} = 2\sec\frac{a}{2}\,\Gamma(p)\sum_{k=1}^\infty (-1)^{k-1}\frac{\cos\left(k-\frac{1}{2}\right)a}{k^p}$$

$$[p > 0] \qquad \text{LI (96)(5)}$$

5.
$$\int_0^\infty \frac{x^q e^{-\frac{x}{2}}\cosh\frac{x}{2}}{\cosh x + \cos\lambda}\,dx = \frac{\Gamma(q+1)}{\cos\frac{\lambda}{2}}\sum_{k=1}^\infty (-1)^{k-1}\frac{\cos\left(k-\frac{1}{2}\right)\lambda}{k^{q+1}}$$

$$[q > -1] \qquad \text{LI (96)(5)a}$$

6.
$$\int_0^\infty x\frac{e^{-x}-\cos a}{\cosh x - \cos a}\,dx = |a|\pi - \frac{a^2}{2} - \frac{\pi^2}{3} \qquad \text{BI (88)(8)}$$

7.
$$\int_0^\infty x^{2m+1}\frac{e^{-x}-\cos a\pi}{\cosh x - \cos a\pi}\,dx = 2\cdot(2m+1)!\sum_{k=1}^\infty \frac{\cos ka\pi}{k^{2m+2}} \qquad \text{BI (88)(6)}$$

3.558

1.
$$\int_0^\infty x\frac{1-e^{-nx}}{\sinh^2\frac{x}{2}}\,dx = \frac{2n\pi^2}{3} - 4\sum_{k=1}^{n-1}\frac{n-k}{k^2} \qquad \text{BI (85)(3)}$$

2. $$\int_0^\infty x \frac{1-(-1)^n e^{-nx}}{\cosh^2 \frac{x}{2}}\, dx = \frac{n\pi^2}{3} + 4\sum_{k=1}^{n-1}(-1)^k \frac{n-k}{k^2}$$ LI (85)(1)

3. $$\int_0^\infty x^2 \frac{1-e^{-nx}}{\sinh^2 \frac{x}{2}}\, dx = 8n\,\zeta(3) - 8\sum_{k=1}^{n-1} \frac{n-k}{k^3}$$ BI (85)(5)

4. $$\int_0^\infty x^2 e^x \frac{1-e^{-2nx}}{\sinh^2 x}\, dx = 8n\sum_{k=1}^{\infty} \frac{1}{(2k-1)^3} - 8\sum_{k=1}^{n-1} \frac{n-k}{(2k-1)^3}$$ LI (85)(6)

5. $$\int_0^\infty x^2 \frac{1+(-1)^n e^{-nx}}{\cosh^2 \frac{x}{2}}\, dx = 6n\,\zeta(3) - 8\sum_{k=1}^{n-1} \frac{n-k}{k^3}$$ LI (85)(4)

6. $$\int_0^\infty x^3 \frac{1-e^{-nx}}{\sinh^2 \frac{x}{2}}\, dx = \frac{4}{15}n\pi^4 - 24\sum_{k=1}^{n-1} \frac{n-k}{k^4}$$ BI (85)(9)

7. $$\int_0^\infty x^3 \frac{1+(-1)^n e^{-nx}}{\cosh^2 \frac{x}{2}}\, dx = \frac{7}{30}n\pi^4 + 24\sum_{k=1}^{n-1}(-1)^k \frac{n-k}{k^4}$$ BI (85)(8)

3.559 $$\int_0^\infty e^{-x}\left[a - \frac{1}{2} + \frac{(1-e^{-x})(1-ax) - xe^{-x}}{4\sinh^2 \frac{x}{2}} e^{(2-a)x}\right] \frac{dx}{x} = a - \frac{1}{2} + \ln\Gamma(a) - \frac{1}{2}\ln(2\pi)\qquad [a>0]$$

BI (96)(6)

3.561 $$\int_0^\infty \frac{e^{-2x}\tanh\frac{x}{2}}{x\cosh x}\, dx = 2\ln\frac{\pi}{2\sqrt{2}}$$ BI (93)(18)

3.562

1. $$\int_0^\infty x^{2\mu-1}e^{-\beta x^2}\sinh\gamma x\, dx = \frac{1}{2}\Gamma(2\mu)(2\beta)^{-\mu}\exp\left(\frac{\gamma^2}{8\beta}\right)\left[D_{-2\mu}\left(-\frac{\gamma}{\sqrt{2\beta}}\right) - D_{-2\mu}\left(\frac{\gamma}{\sqrt{2\beta}}\right)\right]$$

$$\left[\operatorname{Re}\mu > -\tfrac{1}{2},\quad \operatorname{Re}\beta > 0\right]\qquad \text{ET I 166(44)}$$

2. $$\int_0^\infty x^{2\mu-1}e^{-\beta x^2}\cosh\gamma x\, dx = \frac{1}{2}\Gamma(2\mu)(2\beta)^{-\mu}\exp\left(\frac{\gamma^2}{8\beta}\right)\left[D_{-2\mu}\left(-\frac{\gamma}{\sqrt{2\beta}}\right) + D_{-2\mu}\left(\frac{\gamma}{\sqrt{2\beta}}\right)\right]$$

$$\left[\operatorname{Re}\mu > 0,\quad \operatorname{Re}\beta > 0\right]\qquad \text{ET I 166(45)}$$

3. $$\int_0^\infty xe^{-\beta x^2}\sinh\gamma x\, dx = \frac{\gamma}{4\beta}\sqrt{\frac{\pi}{\beta}}\exp\left(\frac{\gamma^2}{4\beta}\right)\qquad [\operatorname{Re}\beta > 0]\qquad \text{BI(81)(12)a,ET I 165(34)}$$

4. $$\int_0^\infty xe^{-\beta x^2}\cosh\gamma x\, dx = \frac{\gamma}{4\beta}\sqrt{\frac{\pi}{\beta}}\exp\left(\frac{\gamma^2}{4\beta}\right)\Phi\left(\frac{\gamma}{2\sqrt{\beta}}\right) + \frac{1}{2\beta}$$

$$[\operatorname{Re}\beta > 0]\qquad \text{ET I 166(35)}$$

5. $$\int_0^\infty x^2 e^{-\beta x^2}\sinh\gamma x\, dx = \frac{\sqrt{\pi}\,(2\beta+\gamma^2)}{8\beta^2\sqrt{\beta}}\exp\left(\frac{\gamma^2}{4\beta}\right)\Phi\left(\frac{\gamma}{2\sqrt{\beta}}\right) + \frac{\gamma}{4\beta^2}$$

$$[\operatorname{Re}\beta > 0]\qquad \text{ET I 166(36)}$$

6. $$\int_0^\infty x^2 e^{-\beta x^2}\cosh\gamma x\, dx = \frac{\sqrt{\pi}\,(2\beta+\gamma^2)}{8\beta^2\sqrt{\beta}}\exp\left(\frac{\gamma^2}{4\beta}\right)\qquad [\operatorname{Re}\beta > 0]\qquad \text{ET I 166(37)}$$

3.6–4.1 Trigonometric Functions

3.61 Rational functions of sines and cosines and trigonometric functions of multiple angles

3.611

1. $\displaystyle\int_0^{2\pi} (1 - \cos x)^n \sin nx \, dx = 0$ BI (68)(10)

2. $\displaystyle\int_0^{2\pi} (1 - \cos x)^n \cos nx \, dx = (-1)^n \frac{\pi}{2^{n-1}}$ BI (68)(11)

3. $\displaystyle\int_0^{\pi} (\cos t + i \sin t \cos x)^n \, dx = \int_0^{\pi} (\cos t + i \sin t \cos x)^{-n-1} \, dx = \pi P_n (\cos t)$ EH I 158(23)a

3.612

1.[6] $\displaystyle\int_0^{\pi} \frac{\sin nx \cos mx}{\sin x} \, dx = 0$ for $n \le m$;

                     $= \pi$ for $n > m$,   if $m + n$ is odd and positive

                     $= 0$ for $n > m$,   if $m + n$ is even

                                                  LI (64)(3)

2. $\displaystyle\int_0^{\pi} \frac{\sin nx}{\sin x} \, dx = 0$ for n even

                  $= \pi$ for n odd

                                            BI (64)(1, 2)

3. $\displaystyle\int_0^{\pi/2} \frac{\sin(2n - 1)x}{\sin x} \, dx = \frac{\pi}{2}$ FI II 145

4. $\displaystyle\int_0^{\pi/2} \frac{\sin 2nx}{\sin x} \, dx = 2 \left(1 - \frac{1}{3} + \frac{1}{5} - \cdots + \frac{(-1)^{k-1}}{2n - 1} \right)$ GW (332)(21b)

5. $\displaystyle\int_0^{\pi} \frac{\sin 2nx}{\cos x} \, dx = 2 \int_0^{\pi/2} \frac{\sin 2nx}{\cos x} \, dx = (-1)^{n-1} 4 \left(1 - \frac{1}{3} + \frac{1}{5} - \cdots + \frac{(-1)^{n-1}}{2n - 1} \right)$ GW (332)(22a)

6. $\displaystyle\int_0^{\pi} \frac{\cos(2n + 1)x}{\cos x} \, dx = 2 \int_0^{\frac{\pi}{2}} \frac{\cos(2n + 1)x}{\cos x} \, dx = (-1)^n \pi$ GW (332)(22b)

7. $\displaystyle\int_0^{\pi/2} \frac{\sin 2nx \cos x}{\sin x} \, dx = \frac{\pi}{2}$ LI (45)(17)

3.613

1.[6] $\displaystyle\int_0^{\pi} \frac{\cos nx \, dx}{1 + a \cos x} = \frac{\pi}{\sqrt{1 - a^2}} \left(\frac{\sqrt{1 - a^2} - 1}{a} \right)^n$ $\left[a^2 < 1, \quad n \ge 0 \right]$ BI (64)(12)

2.[6] $\displaystyle\int_0^{\pi} \frac{\cos nx \, dx}{1 - 2a \cos x + a^2} = \frac{\pi a^n}{1 - a^2}$ $\left[a^2 < 1, \quad n \ge 0 \right]$

                       $= \dfrac{\pi}{(a^2 - 1) a^n}$ $\left[a^2 > 1, \quad n \ge 0 \right]$

                                                 BI (65)(3)

3. $\displaystyle\int_0^\pi \frac{\sin nx \sin x \, dx}{1 - 2a \cos x + a^2} = \frac{\pi}{2} a^{n-1}$ $\left[a^2 < 1, \quad n \geq 1\right]$

                                 $\displaystyle = \frac{\pi}{2a^{n+1}}$ $\left[a^2 > 1, \quad n \geq 1\right]$

                                                                   BI(65)(4), GW(332)(34a)

4.[10] $\displaystyle\int_0^\pi \frac{\cos nx \cos x \, dx}{1 - 2a \cos x + a^2} = \frac{\pi}{2} \cdot \frac{1 + a^2}{1 - a^2} a^{n-1}$ $\left[a^2 < 1, \quad n \geq 1\right]$

                                 $\displaystyle = \frac{\pi}{2a^{n+1}} \cdot \frac{a^2 + 1}{a^2 - 1}$ $\left[a^2 > 1, \quad n \geq 1\right]$

                                 $\displaystyle = \frac{\pi a}{1 - a^2}$ $\left[n = 0, \quad a^2 < 1\right]$

                                 $\displaystyle = \frac{\pi}{a\left(a^2 - 1\right)}$ $\left[n = 0, \quad a^2 > 1\right]$

                                                                   BI(65)(5), GW(332)(34b)

5. $\displaystyle\int_0^\pi \frac{\cos(2n - 1)x \, dx}{1 - 2a \cos 2x + a^2} = \int_0^\pi \frac{\cos 2nx \cos x \, dx}{1 - 2a \cos 2x + a^2} = 0$ $\left[a^2 \neq 1\right]$ BI (65)(9, 10)

6. $\displaystyle\int_0^\pi \frac{\cos(2n - 1)x \cos 2x \, dx}{1 - 2a \cos 2x + a^2} = 0$ $\left[a^2 \neq 1\right]$ BI (65)(12)

7. $\displaystyle\int_0^\pi \frac{\sin 2nx \sin x \, dx}{1 - 2a \cos 2x + a^2} = \int_0^\pi \frac{\sin(2n - 1)x \sin 2x \, dx}{1 - 2a \cos 2x + a^2} = 0$

                                                     $\left[a^2 \neq 1\right]$ BI (65)(6, 7)

8. $\displaystyle\int_0^\pi \frac{\sin(2n - 1)x \sin x \, dx}{1 - 2a \cos 2x + a^2} = \frac{\pi}{2} \cdot \frac{a^{n-1}}{1 + a}$ $\left[a^2 < 1\right]$

                                     $\displaystyle = \frac{\pi}{2} \cdot \frac{1}{(1 + a)a^n}$ $\left[a^2 > 1\right]$

                                                               BI (65)(8)

9. $\displaystyle\int_0^\pi \frac{\cos(2n - 1)x \cos x \, dx}{1 - 2a \cos 2x + a^2} = \frac{\pi}{2} \cdot \frac{a^{n-1}}{1 - a}$ $\left[a^2 < 1\right]$

                                     $\displaystyle = \frac{\pi}{2} \cdot \frac{1}{(a - 1)a^n}$ $\left[a^2 > 1\right]$

                                                              BI (65)(11)

10. $\displaystyle\int_0^\pi \frac{\sin nx - a \sin(n - 1)x}{1 - 2a \cos x + a^2} \sin mx \, dx = 0$ for $m < n$

                                     $\displaystyle = \frac{\pi}{2} a^{m-n}$ for $m \geq n$

                                                     $\left[a^2 < 1\right]$ LI (65)(13)

11.[6] $\displaystyle\int_0^\pi \frac{\cos nx - a \cos(n - 1)x}{1 - 2a \cos x + a^2} \cos mx \, dx = \frac{\pi}{2}\left(a^{|m|-n} - 1\right)$

                                                     $\left[a^2 < 1\right]$ BI (65)(14)

12. $\displaystyle\int_0^\pi \frac{\sin nx - a \sin[(n + 1)x]}{1 - 2a \cos x + a^2} \, dx = 0$ $\left[a^2 < 1\right]$ BI (68)(13)

13. $\displaystyle\int_0^\pi \frac{\cos nx - a \cos[(n + 1)x]}{1 - 2a \cos x + a^2} \, dx = \pi a^n$ $\left[a^2 < 1\right]$ BI (68)(14)

3.614[7]　$\displaystyle\int_0^\pi \frac{\sin x}{a^2 - 2ab\cos x + b^2} \cdot \frac{\sin px \cdot dx}{1 - 2a^p \cos px + a^{2p}}$

$$= \frac{\pi b^{p-1}}{2a^{p+1}(1 - b^p)} \qquad [0 < b \le a \le 1, \quad p = 1, 2, 3, \ldots]$$

$$= \frac{\pi a^{p-1}}{2b(b^p - a^{2p})} \qquad [0 < a \le 1, \quad a^2 < b, \quad p = 1, 2, 3, \ldots]$$

BI (66)(9)

3.615

1.　$\displaystyle\int_0^{\pi/2} \frac{\cos 2nx\, dx}{1 - a^2 \sin^2 x} = \frac{(-1)^n \pi}{2\sqrt{1 - a^2}} \left(\frac{1 - \sqrt{1 - a^2}}{a} \right)^{2n} \qquad [a^2 < 1]$ BI (47)(27)

2.　$\displaystyle\int_0^\pi \frac{\cos x \sin 2nx\, dx}{1 + (a + b\sin x)^2} = -\frac{\pi}{b} \sin \left\{ 2n \arctan \sqrt{\frac{s}{2}} \right\} \tan^{2n} \left(\frac{1}{2} \arccos \sqrt{\frac{s}{2a^2}} \right)$

3.　$\displaystyle\int_0^\pi \frac{\cos x \cos(2n+1)x\, dx}{1 + (a + b\sin x)^2} = \frac{\pi}{b} \cos \left\{ (2n+1) \arctan \sqrt{\frac{s}{2}} \right\} \tan^{2n+1} \left(\frac{1}{2} \arccos \sqrt{\frac{s}{2a^2}} \right)$

$$\text{where } s = -\left(1 + b^2 - a^2\right) + \sqrt{\left(1 + b^2 - a^2\right)^2 + 4a^2} \quad \text{BI (65)(21, 22)}$$

3.616

1.　$\displaystyle\int_0^\pi \left(1 - 2a\cos x + a^2\right)^n dx = \pi \sum_{k=0}^n \binom{n}{k}^2 a^{2k}$ BI (63)(1)

2.[10]　$\displaystyle\int_0^\pi \frac{dx}{\left(1 - 2a\cos x + a^2\right)^n} = \frac{1}{2} \int_0^{2\pi} \frac{dx}{\left(1 - 2a\cos x + a^2\right)^n}$

$$= \frac{\pi}{(1 - a^2)^n} \sum_{k=0}^{n-1} \frac{(n + k - 1)!}{(k!)^2 (n - k - 1)!} \left(\frac{a^2}{1 - a^2} \right)^k \qquad [a^2 < 1]$$

$$= \frac{\pi}{(a^2 - 1)^n} \sum_{k=0}^{n-1} \frac{(n + k - 1)!}{(k!)^2 (n - k - 1)!} \frac{1}{(a^2 - 1)^k} \qquad [a^2 > 1]$$

BI (331)(63)

3.　$\displaystyle\int_0^\pi \left(1 - 2a\cos x + a^2\right)^n \cos nx\, dx = (-1)^n \pi a^n$ BI (63)(2)

4.　$\displaystyle\int_0^\pi \left(1 - 2a\cos x + a^2\right)^n \cos mx\, dx$

$$= \frac{1}{2} \int_0^{2\pi} \left(1 - 2a\cos x + a^2\right)^n \cos mx\, dx$$

$$= 0 \qquad [n < m]$$

$$= \pi(-a)^m \left(1 + a^2\right)^{n-m} \sum_{k=0}^{[(n-m)/2]} \binom{n}{k} \binom{n - k}{m + k} \left(\frac{a}{1 + a^2} \right)^{2k} \qquad [n \ge m]$$

GW (332)(35a)

5.　$\displaystyle\int_0^{2\pi} \frac{\sin nx\, dx}{(1 - 2a\cos 2x + a^2)^m} = 0$ GW (332)(32a)

6. $$\int_0^\pi \frac{\sin x \, dx}{(1 - 2a \cos 2x + a^2)^m} = \frac{1}{2(m-1)a} \left[\frac{1}{(1-a)^{2m-2}} - \frac{1}{(1+a)^{2m-2}} \right] \qquad [a \neq 0, \quad \pm 1]$$

GW (332)(32c)

7. $$\int_0^\pi \frac{\cos nx \, dx}{(1 - 2a \cos x + a^2)^m} = \frac{1}{2} \int_0^{2\pi} \frac{\cos nx \, dx}{(1 - 2a \cos x + a^2)^m}$$

$$= \frac{a^{2m+n-2}\pi}{(1-a^2)^{2m-1}} \sum_{k=0}^{m-1} \binom{m+n-1}{k} \binom{2m-k-2}{m-1} \left(\frac{1-a^2}{a^2} \right)^k \qquad [a^2 < 1]$$

$$= \frac{\pi}{a^n(a^2-1)^{2m-1}} \sum_{k=0}^{m-1} \binom{m+n-1}{k} \binom{2m-k-2}{m-1} (a^2-1)^k \qquad [a^2 > 1]$$

GW (332)(31)

8. $$\int_0^{\pi/2} \frac{\cos 2nx \, dx}{(a^2 \cos^2 x + b^2 \sin^2 x)^{n+1}} = \binom{2n}{n} \frac{(b^2 - a^2)^n}{(2ab)^{2n+1}} \pi$$

$$[a > 0, \quad b > 0] \qquad \text{GW (332)(30b)}$$

3.617^{10} $$\int_0^\pi \frac{dx}{(1 - 2a \cos x + a^2)^{n+1/2}} = \frac{2}{|1+a|^{2n+1}} F_n \left(\frac{2\sqrt{|a|}}{|1+a|} \right), \qquad |a| \neq 1$$

with

$$F_n(k) = \int_0^{\pi/2} \frac{dx}{(1 - k^2 \sin^2 x)^{n+1/2}}$$

where the $F_n(k)$ satisfy the recurrence relation

$$F_{n+1}(k) = F_n(k) + \frac{k}{2n+1} \frac{dF_n(k)}{dk}, \qquad n = 0, 1, 2, \ldots$$

and

$$F_0(k) = K(k) \equiv \int_0^{\pi/2} \frac{dx}{(1 - k^2 \sin^2 x)^{1/2}}$$

is the complete elliptic integral of the first kind.
Introducing the complete elliptic integral of the second kind

$$E(k) = \int_0^{\pi/2} (1 - k^2 \sin^2 x)^{1/2} \, dx$$

the derivatives

$$\frac{dK}{dk} = \frac{E(k)}{k(1-k^2)} - \frac{K(k)}{k}, \qquad \frac{dE}{dk} = \frac{E(k) - K(k)}{k}$$

combined with the recurrence relation lead to

$$F_1((k)) = F_0(k) + k \frac{d F_0(k)}{dk}$$

$$= K(k) + \frac{E(k)}{1-k^2} - K(k) = \frac{E(k)}{1-k^2},$$

$$F_2(k) = \frac{E(k)}{1-k^2} + \frac{k}{3} \frac{d}{dk} \left[\frac{E(k)}{1-k^2} \right]$$

$$= \frac{1}{3(1-k^2)} \left[\left(\frac{4-2k^2}{1-k^2} \right) E(k) - K(k) \right]$$

3.62   Powers of trigonometric functions

3.621

1. $\displaystyle\int_0^{\pi/2} \sin^{\mu-1} x \, dx = \int_0^{\pi/2} \cos^{\mu-1} x \, dx = 2^{\mu-2} \, \mathrm{B}\left(\frac{\mu}{2}, \frac{\mu}{2}\right)$ FI II 789

2. $\displaystyle\int_0^{\pi/2} \sin^{3/2} x \, dx = \int_0^{\pi/2} \cos^{3/2} x \, dx = \frac{1}{6\sqrt{2\pi}}\left[\Gamma\left(\frac{1}{4}\right)\right]^2$

3. $\displaystyle\int_0^{\pi/2} \sin^{2m} x \, dx = \int_0^{\pi/2} \cos^{2m} x \, dx = \frac{(2m-1)!!}{(2m)!!}\frac{\pi}{2}$ FI II 151

4. $\displaystyle\int_0^{\pi/2} \sin^{2m+1} x \, dx = \int_0^{\pi/2} \cos^{2m+1} x \, dx = \frac{(2m)!!}{(2m+1)!!}$ FI II 151

5. $\displaystyle\int_0^{\pi/2} \sin^{\mu-1} x \cos^{\nu-1} x \, dx = \frac{1}{2}\, \mathrm{B}\left(\frac{\mu}{2}, \frac{\nu}{2}\right)$ $[\operatorname{Re}\mu > 0, \quad \operatorname{Re}\nu > 0]$

                                                          LO V 113(50), LO V 122, FI II 788

3.622

1. $\displaystyle\int_0^{\pi/2} \tan^{\pm\mu} x \, dx = \frac{\pi}{2}\sec\frac{\mu\pi}{2}$ $[|\operatorname{Re}\mu| < 1]$ BI (42)(1)

2. $\displaystyle\int_0^{\pi/4} \tan^{\mu} x \, dx = \frac{1}{2}\beta\left(\frac{\mu+1}{2}\right)$ $[\operatorname{Re}\mu > -1]$ BI (34)(1)

3. $\displaystyle\int_0^{\pi/4} \tan^{2n} x \, dx = (-1)^n \frac{\pi}{4} + \sum_{k=0}^{n-1} \frac{(-1)^k}{2n-2k-1}$ BI (34)(2)

4. $\displaystyle\int_0^{\pi/4} \tan^{2n+1} x \, dx = (-1)^{n+1}\frac{\ln 2}{2} + \sum_{k=0}^{n-1} \frac{(-1)^k}{2n-2k}$ BI (34)(3)

3.623

1. $\displaystyle\int_0^{\pi/2} \tan^{\mu-1} x \cos^{2\nu-2} x \, dx = \int_0^{\pi/2} \cot^{\mu-1} x \sin^{2\nu-2} x \, dx = \frac{1}{2}\, \mathrm{B}\left(\frac{\mu}{2}, \nu - \frac{\mu}{2}\right)$

                                         $[0 < \operatorname{Re}\mu < 2\operatorname{Re}\nu]$ BI(42)(6), BI(45)(22)

2.[6] $\displaystyle\int_0^{\pi/4} \tan^{\mu} x \sin^2 x \, dx = \frac{1+\mu}{4}\beta\left(\frac{\mu+1}{2}\right) - \frac{1}{4}$ $[\operatorname{Re}\mu > -1]$ BI (34)(4)

3.[6] $\displaystyle\int_0^{\pi/4} \tan^{\mu} x \cos^2 x \, dx = \frac{1-\mu}{4}\beta\left(\frac{\mu+1}{2}\right) + \frac{1}{4}$ $[\operatorname{Re}\mu > -1]$ BI (34)(5)

3.624

1. $\displaystyle\int_0^{\pi/4} \frac{\sin^p x}{\cos^{p+2} x} \, dx = \frac{1}{p+1}$ $[p > -1]$ GW (331)(34b)

2.[3] $\displaystyle\int_0^{\pi/2} \frac{\sin^{\mu-\frac{1}{2}} x}{\cos^{2\mu-1} x} \, dx = \int_0^{\pi/2} \frac{\cos^{\mu-\frac{1}{2}} x}{\sin^{2\mu-1} x} \, dx = \frac{1}{2}\left\{\frac{\Gamma\left(\frac{\mu}{2}+\frac{1}{4}\right)\Gamma(1-\mu)}{\Gamma\left(\frac{5}{4}-\frac{\mu}{2}\right)}\right\}$

                                             $\left[-\frac{1}{2} < \operatorname{Re}\mu < 1\right]$ LI (55)(12)

3. $\int_0^{\pi/4} \dfrac{\cos^{n-\frac{1}{2}} 2x}{\cos^{2n+1} x} \, dx = \dfrac{(2n-1)!!}{2 \cdot (2n)!!} \pi$ BI (38)(3)

4.[8] $\int_0^{\pi/4} \dfrac{\cos^\mu 2x}{\cos^{2(\mu+1)} x} \, dx = 2^{2\mu} \, \mathrm{B}(\mu+1, \mu+1)$ $[\operatorname{Re}\mu > -1]$ BI (35)(1)

5. $\int_0^{\pi/4} \dfrac{\sin^{2\mu-2} x}{\cos^\mu 2x} \, dx = 2^{1-2\mu} \, \mathrm{B}(2\mu-1, 1-\mu) = \dfrac{\Gamma\left(\mu-\frac{1}{2}\right)\Gamma(1-\mu)}{2\sqrt{\pi}}$

$$\left[\tfrac{1}{2} < \operatorname{Re}\mu < 1\right] \qquad \text{BI (35)(4)}$$

6.[6] $\int_0^{\pi/2} \left(\dfrac{\sin ax}{\sin x}\right)^2 dx = \dfrac{a\pi}{2} - \dfrac{1}{2} \sin \pi a \left[2a\, \beta(a) - 1\right],$ $[a > 0]$

3.625

1. $\int_0^{\pi/4} \dfrac{\sin^{2n-1} x \cos^p 2x}{\cos^{2p+2n+1} x} \, dx = \dfrac{(n-1)!}{2} \cdot \dfrac{\Gamma(p+1)}{\Gamma(p+n+1)}$

$$= \dfrac{(n-1)!}{2(p+n)(p+n-1)\cdots(p+1)} = \dfrac{1}{2}\, \mathrm{B}(n, p+1)$$

$$[p > -1] \qquad (\text{cf. } \mathbf{3.251}\ 1) \qquad \text{BI (35)(2)}$$

2. $\int_0^{\pi/4} \dfrac{\sin^{2n} x \cos^p 2x}{\cos^{2p+2n+2} x} \, dx = \tfrac{1}{2} \, \mathrm{B}\left(n+\tfrac{1}{2}, p+1\right)$ $[p > -1]$ $(\text{cf. } \mathbf{3.251}\ 1)$ BI (35)(3)

3. $\int_0^{\pi/4} \dfrac{\sin^{2n-1} x \cos^{m-\frac{1}{2}} 2x}{\cos^{2n+2m} x} \, dx = \dfrac{(2n-2)!!(2m-1)!!}{(2n+2m-1)!!}$ BI (38)(6)

4.[8] $\int_0^{\pi/4} \dfrac{\sin^{2n} x \cos^{m-\frac{1}{2}} 2x}{\cos^{2n+2m+1} x} \, dx = \dfrac{(2n-1)!!(2m-1)!!}{(2n+2m)!!} \cdot \dfrac{\pi}{2}$ BI (38)(7)

3.626

1. $\int_0^{\pi/4} \dfrac{\sin^{2n-1} x}{\cos^{2n+2} x} \sqrt{\cos 2x} \, dx = \dfrac{(2n-2)!!}{(2n+1)!!}$ $(\text{cf. } \mathbf{3.251}\ 1)$ BI (38)(4)

2. $\int_0^{\pi/4} \dfrac{\sin^{2n} x}{\cos^{2n+3} x} \sqrt{\cos 2x} \, dx = \dfrac{(2n-1)!!}{(2n+2)!!} \cdot \dfrac{\pi}{2}$ $(\text{cf. } \mathbf{3.251}\ 1)$ BI (38)(5)

3.627 $\int_0^{\pi/2} \dfrac{\tan^\mu x}{\cos^\mu x} \, dx = \int_0^{\pi/2} \dfrac{\cot^\mu x}{\sin^\mu x} \, dx = \dfrac{\Gamma(\mu)\,\Gamma\left(\frac{1}{2}-\mu\right)}{2^\mu \sqrt{\pi}} \sin \dfrac{\mu\pi}{2}$

$$\left[-1 < \operatorname{Re}\mu < \tfrac{1}{2}\right] \qquad \text{BI (55)(12)a}$$

3.628[8] $\int_0^{\frac{\pi}{2}} \sec^{2p} x \sin^{2p-1} x \, dx = \dfrac{1}{2p\pi} \Gamma(p+1)\Gamma\left(\tfrac{1}{2}-p\right)$ $\left[0 < p < \tfrac{1}{2}\right]$ WA 691

3.63 Powers of trigonometric functions and trigonometric functions of linear functions

3.631

1. $\int_0^\pi \sin^{\nu-1} x \sin ax \, dx = \dfrac{\pi \sin \frac{a\pi}{2}}{2^{\nu-1}\nu \, \mathrm{B}\left(\dfrac{\nu+a+1}{2}, \dfrac{\nu-a+1}{2}\right)}$

$$[\operatorname{Re}\nu > 0] \qquad \text{LO V 121(67a), WA 337a}$$

2.[7] $\displaystyle\int_0^{\pi/2} 2\sin^{\nu-2} x \sin \nu x \, dx = \frac{1}{1-\nu}\cos\frac{\nu\pi}{2}$ $\qquad [\operatorname{Re}\nu > 1]$ $\qquad$ GW(332)(16d), FI I 152

3.[6] $\displaystyle\int_0^{\pi} \sin^{\nu} x \sin \nu x \, dx = 2^{-\nu}\pi\sin\frac{\nu\pi}{2}$ $\qquad [\operatorname{Re}\nu > -1]$ $\qquad$ LO V 121(69)

4. $\displaystyle\int_0^{\pi} \sin^n x \sin 2mx \, dx = 0$ $\qquad$ GW (332)(11a)

5. $\displaystyle\int_0^{\pi} \sin^{2n} x \sin(2m+1)x \, dx = \int_0^{\pi/2} \sin^{2n} x \sin(2m+1)x \, dx$

$$= \frac{(-1)^m 2^{n+1} n!(2n-1)!!}{(2n-2m-1)!!(2m+2n+1)!!} \qquad [m \le n]^*$$

$$= \frac{(-1)^n 2^{n+1} n!(2m-2n-1)!!(2n-1)!!}{(2m+2n+1)!!} \qquad [m \ge n]^*$$

$\qquad$ GW (332)(11b)

6. $\displaystyle\int_0^{\pi} \sin^{2n+1} x \sin(2m+1)x \, dx = 2\int_0^{\pi/2} \sin^{2n+1} x \sin(2m+1)x \, dx$

$$= \frac{(-1)^m \pi}{2^{2n+1}}\binom{2n+1}{n-m} \qquad [n \ge m]$$

$$= 0 \qquad [n < m]$$

$\qquad$ BI(40)(12), GW(332)(11c)

7. $\displaystyle\int_0^{\pi} \sin^n x \cos(2m+1)x \, dx = 0$ $\qquad$ GW (332)(12a)

8. $\displaystyle\int_0^{\pi} \sin^{\nu-1} x \cos ax \, dx = \frac{\pi\cos\frac{a\pi}{2}}{2^{\nu-1}\nu\,\mathrm{B}\left(\dfrac{\nu+a+1}{2},\dfrac{\nu-a+1}{2}\right)}$

$\qquad [\operatorname{Re}\nu > 0]$ $\qquad$ LO V 121(68)a, WA 337a

9. $\displaystyle\int_0^{\pi/2} \cos^{\nu-1} x \cos ax \, dx = \frac{\pi}{2^{\nu}\nu\,\mathrm{B}\left(\dfrac{\nu+a+1}{2},\dfrac{\nu-a+1}{2}\right)}$

$\qquad [\operatorname{Re}\nu > 0]$ $\qquad$ GW (332)(9c)

10. $\displaystyle\int_0^{\pi/2} \sin^{\nu-2} x \cos \nu x \, dx = \frac{1}{\nu-1}\sin\frac{\nu\pi}{2}$ $\qquad [\operatorname{Re}\nu > 1]$ $\qquad$ GW(332)(16b), FI II 15 2

11. $\displaystyle\int_0^{\pi} \sin^{\nu} x \cos \nu x \, dx = \frac{\pi}{2^{\nu}}\cos\frac{\nu\pi}{2}$ $\qquad [\operatorname{Re}\nu > -1]$ $\qquad$ LO V 121(70)a

12. $\displaystyle\int_0^{\pi} \sin^{2n} x \cos 2mx \, dx = 2\int_0^{\pi/2} \sin^{2n} x \cos 2mx \, dx = \frac{(-1)^m}{2^{2n}}\binom{2n}{n-m}\pi \quad [n \ge m]$

$$= 0 \qquad [n < m]$$

$\qquad$ BI(40)(16), GW(332)(12b)

---

*In 3.631.5, for $m = n$ we should set $(2n-2m-1)!! = 1$

13.[7] $\displaystyle\int_0^\pi \sin^{2n+1} x \cos 2mx\, dx$

$$= 2\int_0^{\pi/2} \sin^{2n+1} x \cos 2mx\, dx = \frac{(-1)^m 2^{n+1} n!(2n+1)!!}{(2m-2n-3)!!(2m+2n+1)!!} \qquad [n \geq m-1]$$

$$= \frac{(-1)^{n+1} 2^{n+1} n!(2m-2n+3)!!(2n+1)!!}{(2m+2n+1)!!} \qquad [n < m-1]$$

$$\text{GW (332)(12c)}$$

14. $\displaystyle\int_0^{\pi/2} \cos^{\nu-2} x \sin \nu x\, dx = \frac{1}{\nu-1}$ $\qquad [\operatorname{Re}\nu > 1]$ $\qquad$ GW(332)(16c), FI II 152

15. $\displaystyle\int_0^\pi \cos^m x \sin nx\, dx = \left[1-(-1)^{m+n}\right]\int_0^{\pi/2} \cos^m x \sin nx\, dx$

$$= \left[1-(-1)^{m+n}\right]\left\{\sum_{k=0}^{r-1} \frac{m!}{(m-k)!}\frac{(m+n-2k-2)!!}{(m+n)!!} + s\frac{m!(n-m-2)!!}{(m+n)!!}\right\}$$

$$\left[r = \begin{cases} m & \text{if } m \leq n \\ n & \text{if } m \geq n \end{cases} \qquad s = \begin{cases} 2 & \text{if } n-m = 4l+2 > 0 \\ 1 & \text{if } n-m = 2l+1 > 0 \\ 0 & \text{if } n-m = 4l \text{ or } n-m < 0 \end{cases} \right] \qquad \text{GW (332)(13a)}$$

16. $\displaystyle\int_0^{\pi/2} \cos^n x \sin nx\, dx = \frac{1}{2^{n+1}}\sum_{k=1}^n \frac{2^k}{k}$ $\qquad$ FI II 153

17. $\displaystyle\int_0^\pi \cos^n x \cos mx\, dx = \left[1+(-1)^{m+n}\right]\int_0^{\pi/2} \cos^n x \cos mx\, dx$

$$= \left[1+(-1)^{m+n}\right]\begin{cases} s\dfrac{n!}{(m-n)(m-n+2)\cdots(m+n)} & \text{if } n < m \\[2mm] \dfrac{\pi}{2^{n+1}}\dbinom{n}{k} & \text{if } m \leq n \text{ and } n-m = 2k \\[2mm] \dfrac{n!}{(2k+1)!!(2m+2k+1)!!} & \text{if } m < n \text{ and } n-m = 2k+1 \end{cases}$$

$$\text{where } s = \begin{cases} 0 & \text{if } m-n = 2k \\ 1 & \text{if } m-n = 4k+1 \\ -1 & \text{if } m-n = 4k-1 \end{cases} \qquad \text{GW (332)(15a)}$$

18.[6] $\displaystyle\int_0^\pi \cos^m x \cos ax\, dx = \frac{(-1)^m \sin a\pi}{2^m(m+a)}\,{}_2F_1\left(-m, -\frac{a+m}{2}; 1-\frac{a+m}{2}; -1\right)$

$$[a \neq 0, \pm 1, \pm 2, \ldots] \qquad \text{WA 313}$$

19. $\displaystyle\int_0^{\pi/2} \cos^{\nu-2} x \cos \nu x\, dx = 0$ $\qquad [\operatorname{Re}\nu > 1]$ $\qquad$ GW(332)(16a), FI II 152

20.[10] $\displaystyle\int_0^{\pi/2} \cos^n x \cos nx\, dx = \frac{\pi}{2^{n+1}}$ $\qquad [\operatorname{Re}n > -1]$ $\qquad$ LO V 122(78), FI II 153

3.632

1. $\displaystyle\int_0^\pi \sin^{p-1} x \cos\left[a\left(\frac{\pi}{2} - x\right)\right] dx = 2^{p-1} \frac{\Gamma\left(\frac{p-a}{2}\right)\Gamma\left(\frac{p+a}{2}\right)}{\Gamma(p-a)\,\Gamma(p+a)}\,\Gamma(p)$

$$\left[p^2 < a^2\right] \qquad\qquad \text{BI (62)(11)}$$

2. $\displaystyle\int_{-\frac{\pi}{2}}^{\frac{\pi}{2}} \cos^{\nu-1} x \sin\left[a\left(x + \frac{\pi}{2}\right)\right] dx = \frac{\pi \sin\frac{a\pi}{2}}{2^{\nu-1}\nu\,\mathrm{B}\left(\dfrac{\nu+a+1}{2},\dfrac{\nu-a+1}{2}\right)}$

$$[\operatorname{Re}\nu > 0] \qquad\qquad \text{WA 337a}$$

3.[10] $\displaystyle\int_0^{\pi/2} \cos^p x \sin[(p+2n)x]\,dx = (-1)^{n-1} \sum_{k=0}^{n-1} \frac{(-1)^k 2^k}{p+k+1}\binom{n-1}{k}$

$$[n > 0] \qquad\qquad \text{LI (41)(12)}$$

4. $\displaystyle\int_{-\pi}^\pi \cos^{n-1} x \cos[m(x-a)]\,dx = \left[1 - (-1)^{n+m}\right] = \int_{-\frac{\pi}{2}}^{\frac{\pi}{2}} \cos^{n-1} x \cos[m(x-a)]\,dx$

$$= \frac{\left[1 - (-1)^{n+m}\right]\pi \cos ma}{2^{n-1} n\,\mathrm{B}\left(\dfrac{n+m+1}{2},\dfrac{n-m+1}{2}\right)}$$

$$[n \geq m] \qquad \text{LO V 123(80), LO V 139(94a)}$$

5. $\displaystyle\int_0^{\pi/2} \cos^{p+q-2} x \cos[(p-q)x]\,dx = \frac{\pi}{2^{p+q-1}(p+q-1)\,\mathrm{B}(p,q)}$

$$[p + q > 1] \qquad\qquad \text{WH}$$

3.633

1. $\displaystyle\int_0^{\pi/2} \cos^{p-1} x \sin ax \sin x\,dx = \frac{a\pi}{2^{p+1}p(p+1)\,\mathrm{B}\left(\dfrac{p+a}{2}+1,\dfrac{p-a}{2}+1\right)}$ LO V 150(110)

2. $\displaystyle\int_0^{\pi/2} \cos^n x \sin nx \sin 2mx\,dx = \int_0^{\pi/2} \cos^n x \cos nx \cos 2mx\,dx = \frac{\pi}{2^{n+2}}\binom{n}{m}$ BI (42)(19, 20)

3. $\displaystyle\int_0^{\pi/2} \cos^{n-1} x \cos[(n+1)x]\cos 2mx\,dx = \frac{\pi}{2^{n+1}}\binom{n-1}{m-1}$

$$[n > m - 1] \qquad\qquad \text{BI (42)(21)}$$

4. $\displaystyle\int_0^{\pi/2} \cos^{p+q} x \cos px \cos qx\,dx = \frac{\pi}{2^{p+q+2}}\left[1 + \frac{1}{(p+q+1)\,\mathrm{B}(p+1,q+1)}\right]$

$$[p + q > -1] \qquad\qquad \text{GW (332)(10c)}$$

5.[6] $\displaystyle\int_0^{\pi/2} \cos^{p+q} x \sin px \sin qx\,dx = \frac{\pi}{2^{p+q+2}} \sum_{k=1}^\infty \binom{p}{k}\binom{q}{k} = \frac{\pi}{2^{p+q+2}}\left[\frac{\Gamma(p+q+1)}{\Gamma(p+1)\,\Gamma(q+1)} - 1\right]$

$$[p + q > -1] \qquad\qquad \text{BI (42)(16)}$$

3.634

1. $$\int_0^{\pi/2} \sin^{\mu-1} x \cos^{\nu-1} x \sin(\mu+\nu)x \, dx = \sin \frac{\mu\pi}{2} \, B(\mu, \nu)$$

$$[\mathrm{Re}\,\mu > 0, \quad \mathrm{Re}\,\nu > 0]$$

BI(42)(23), FI II 814a

2. $$\int_0^{\pi/2} \sin^{\mu-1} x \cos^{\nu-1} x \cos(\mu+\nu)x \, dx = \cos \frac{\mu\pi}{2} \, B(\mu, \nu)$$

$$[\mathrm{Re}\,\mu > 0, \quad \mathrm{Re}\,\nu > 0]$$

BI(42)(24), FI II 814a

3. $$\int_0^{\pi/2} \cos^{p+n-1} x \sin px \cos[(n+1)x] \sin x \, dx = \frac{\pi}{2^{p+n+1}} \frac{\Gamma(p+n)}{n!\,\Gamma(p)}$$

$$[p > -n]$$

BI (42)(15)

3.635

1. $$\int_0^{\pi/4} \cos^{\mu-1} 2x \tan x \, dx = \frac{1}{4}\left[\psi\left(\frac{\mu+1}{2}\right) - \psi\left(\frac{\mu}{2}\right)\right] \qquad [\mathrm{Re}\,\mu > 0]$$

BI (34)(7)

2.[7] $$\int_0^{\pi/2} \cos^{p+2n} x \sin px \tan x \, dx = \frac{\pi}{2^{p+2n+1}\,\Gamma(p)} \sum_{k=0}^{\infty} \binom{n}{k} \frac{\Gamma(p+n-k)}{(n-k)!}$$

$$= \frac{p\pi}{2^{p+2+n+1}} \frac{\Gamma(p+2n)}{\Gamma(n+1)\,\Gamma(p+n+1)}$$

$$[p > -2n]$$

BI (42)(22)

3. $$\int_0^{\pi/2} \cos^{n-1} x \sin[(n+1)x] \cot x \, dx = \frac{\pi}{2}$$

BI (45)(18)

3.636

1. $$\int_0^{\pi/2} \tan^{\pm\mu} x \sin 2x \, dx = \frac{\mu\pi}{2} \operatorname{cosec} \frac{\mu\pi}{2} \qquad\qquad [0 < \mathrm{Re}\,\mu < 2]$$

BI (45)(20)a

2. $$\int_0^{\pi/2} \tan^{\pm\mu} x \cos 2x \, dx = \mp \frac{\mu\pi}{2} \sec \frac{\mu\pi}{2} \qquad\qquad [|\mathrm{Re}\,\mu| < 1]$$

BI (45)(21)

3. $$\int_0^{\pi/2} \frac{\tan^{2\mu} x}{\cos x} \, dx = \int_0^{\pi/2} \frac{\cot^{2\mu} x}{\sin x} \, dx = \frac{\Gamma\left(\mu+\frac{1}{2}\right)\Gamma(-\mu)}{2\sqrt{\mu}}$$

$$\left[-\tfrac{1}{2} < \mathrm{Re}\,\mu < 1\right] \qquad (\text{cf. } \mathbf{3.251}\ 1)$$

BI (45)(13, 14)

3.637

1. $$\int_0^{\pi/2} \tan^p x \sin^{q-2} x \sin qx \, dx = -\cos \frac{(p+q)\pi}{2} \, B(p+q-1, 1-p)$$

$$[p+q > 1 > p]$$

GW (332)(15d)

2. $$\int_0^{\pi/2} \tan^p x \sin^{q-2} x \cos qx \, dx = \sin \frac{(p+q)\pi}{2} \, B(p+q-1, 1-p)$$

$$[p+q > 1 > p]$$

GW (332)(15b)

3. $\displaystyle\int_0^{\pi/2} \cot^p x \cos^{q-2} x \sin qx \, dx = \cos\frac{p\pi}{2}\, \mathrm{B}(p+q-1, 1-p)$

$$[p+q > 1 > p] \qquad\qquad \text{GW (332)(15c)}$$

4. $\displaystyle\int_0^{\pi/2} \cot^p x \cos^{q-2} x \cos qx \, dx = \sin\frac{p\pi}{2}\, \mathrm{B}(p+q-1, 1-p)$

$$[p+q > 1 > p] \qquad\qquad \text{GW (332)(15a)}$$

3.638

1. $\displaystyle\int_0^{\pi/4} \frac{\sin^{2\mu} x \, dx}{\cos^{\mu+\frac{1}{2}} 2x \cos x} = \frac{\pi}{2}\sec\mu\pi$ $\left[|\mathrm{Re}\,\mu| < \tfrac{1}{2}\right]$ (cf. **3.192** 2)

$$\text{BI (38)(8)}$$

2. $\displaystyle\int_0^{\pi/4} \frac{\sin^{\mu-\frac{1}{2}} 2x \, dx}{\cos^\mu 2x \cos x} = \frac{2}{2\mu-1}\cdot\frac{\Gamma\left(\mu+\frac{1}{2}\right)\Gamma(1-\mu)}{\sqrt{\pi}}\sin\left(\frac{2\mu-1}{4}\pi\right)$

$$\left[-\tfrac{1}{2} < \mathrm{Re}\,\mu < 1\right] \qquad\qquad \text{BI (38)(17)}$$

3. $\displaystyle\int_0^{\pi/2} \frac{\cos^{p-1} x \sin px}{\sin x}\, dx = \frac{\pi}{2}$ $[p > 0]$ GW(332)(17), BI(45)(5)

3.64–3.65 Powers and rational functions of trigonometric functions

3.641

1. $\displaystyle\int_0^{\pi/2} \frac{\sin^{p-1} x \cos^{-p} x}{a\cos x + b\sin x}\, dx = \int_0^{\pi/2} \frac{\sin^{-p} x \cos^{p-1} x}{a\sin x + b\cos x}\, dx = \frac{\pi \cosec p\pi}{a^{1-p}b^p}$

$$[ab > 0, \quad 0 < p < 1] \qquad\qquad \text{GW (331)(62)}$$

2. $\displaystyle\int_0^{\pi/2} \frac{\sin^{1-p} x \cos^p x}{(\sin x + \cos x)^3}\, dx = \int_0^{\pi/2} \frac{\sin^p x \cos^{1-p} x}{(\sin x + \cos x)^3}\, dx = \frac{(1-p)p}{2}\pi \cosec p\pi$

$$[-1 < p < 2] \qquad\qquad \text{BI(48)(5)}$$

3.642

1. $\displaystyle\int_0^{\pi/2} \frac{\sin^{2\mu-1} x \cos^{2\nu-1} x \, dx}{\left(a^2\sin^2 x + b^2\cos^2 x\right)^{\mu+\nu}} = \frac{1}{2a^{2\mu}b^{2\nu}}\,\mathrm{B}(\mu,\nu)$ $[\mathrm{Re}\,\mu > 0, \quad \mathrm{Re}\,\nu > 0]$ BI (48)(28)

2. $\displaystyle\int_0^{\pi/2} \frac{\sin^{n-1} x \cos^{n-1} x \, dx}{\left(a^2\cos^2 x + b^2\sin^2 x\right)^n} = \frac{\mathrm{B}\left(\frac{n}{2}, \frac{n}{2}\right)}{2(ab)^n}$ $[ab > 0]$ GW (331)(59a)

3. $\displaystyle\int_0^{\pi/2} \frac{\sin^{2n} x \, dx}{\left(a^2\cos^2 x + b^2\sin^2 x\right)^{n+1}} = \frac{1}{2}\int_0^\pi \frac{\sin^{2n} x \, dx}{\left(a^2\cos^2 x + b^2\sin^2 x\right)^{n+1}}$

$$= \int_0^{\pi/2} \frac{\cos^{2n} x \, dx}{\left(a^2\sin^2 x + b^2\cos^2 x\right)^{n+1}} = \frac{1}{2}\int_0^\pi \frac{\cos^{2n} x \, dx}{\left(a^2\sin^2 x + b^2\cos^2 x\right)^{n+1}} = \frac{(2n-1)!!\pi}{2^{n+1}n!ab^{2n+1}}$$

$$[ab > 0] \qquad\qquad\qquad \text{GW (331)(58)}$$

4. $\displaystyle\int_0^{\pi/2} \frac{\cos^{p+2n} x \cos px \, dx}{\left(a^2 \cos^2 x + b^2 \sin^2 x\right)^{n+1}} = \pi \sum_{k=0}^{n} \binom{2n-k}{n}\binom{p+k-1}{k} \frac{b^{p-1}}{(2a)^{2n-k+1}(a+b)^{p+k}}$

$$\left[a > 0, \quad b > 0, \quad p > -2n - 1\right]$$

GW (332)(30)

3.643

1. $\displaystyle\int_0^{\pi/2} \frac{\cos^p x \cos px \, dx}{1 - 2a \cos 2x + a^2} = \frac{\pi}{2^{p+1}} \cdot \frac{(1+a)^{p-1}}{1-a}$ $\left[a^2 < 1, \quad p > -1\right]$ GW (332)(33c)

2. $\displaystyle\int_0^{\pi/2} \frac{\sin^{2n} x \cos^\mu x \cos \beta x}{(1 - 2a \cos 2x + a^2)^m}\, dx = \frac{(-1)^n \pi (1-a)^{2n-2m+1}}{2^{2m-\beta-1}(1+a)^{2m+\beta+1}} \sum_{k=0}^{m-1}\sum_{l=0}^{m-k-1} \binom{\beta}{k}\binom{2n}{l}$

$$\times \binom{2m-k-l-2}{m-1(-2)^l}(a-1)^k$$

$$\left[a^2 < 1, \quad \beta = 2m - 2n - \mu - 2, \quad \mu > -1\right] \quad \text{GW (332)(33)}$$

3.644

1. $\displaystyle\int_0^{\pi} \frac{\sin^m x}{p + q \cos x}\, dx = 2^{m-2}\frac{p}{q^2} \sum_{\nu=1}^{k}\left(\frac{p^2-q^2}{-4q^2}\right)^{\nu-1} B\left(\frac{m+1-2\nu}{2}, \frac{m+1-2\nu}{2}\right) + \left(\frac{p^2-q^2}{-q^2}\right)^k A$

$$\text{where } A = \begin{cases} \dfrac{\pi p}{q^2}\left(1 - \sqrt{1 - \dfrac{q^2}{p^2}}\right) & \text{if } m = 2k+2 \\[4mm] \dfrac{1}{q}\ln\dfrac{p+q}{p-q} & \text{if } m = 2k+1 \end{cases} \qquad \left[k \geq 1, \quad q \neq 0, \quad p^2 - q^2 \geq 0\right]$$

2. $\displaystyle\int_0^{\pi} \frac{\sin^m x}{1 + \cos x}\, dx = 2^{m-1} B\left(\frac{m-1}{2}, \frac{m+1}{2}\right)$ $[m \geq 2]$

3. $\displaystyle\int_0^{\pi} \frac{\sin^m x}{1 - \cos x}\, dx = 2^{m-1} B\left(\frac{m-1}{2}, \frac{m+1}{2}\right)$ $[m \geq 2]$

4. $\displaystyle\int_0^{\pi} \frac{\sin^2 x}{p + q \cos x}\, dx = \frac{p\pi}{q^2}\left(1 - \sqrt{1 - \frac{q^2}{p^2}}\right)$

5. $\displaystyle\int_0^{\pi} \frac{\sin^3 x}{p + q \cos x}\, dx = 2\frac{p}{q^2} + \frac{1}{q}\left(1 - \frac{p^2}{q^2}\right)\ln\frac{p+q}{p-q}$

3.645 $\displaystyle\int_0^{\pi} \frac{\cos^n x \, dx}{(a + b \cos x)^{n+1}} = \frac{\pi}{2^n (a+b)^n \sqrt{a^2 - b^2}} \sum_{k=0}^{n} (-1)^k \frac{(2n-2k-1)!!(2k-1)!!}{(n-k)!k!}\left(\frac{a+b}{a-b}\right)^k$

$$\left[a^2 > b^2\right]$$

LI (64)(16)

3.646

1. $\displaystyle\int_0^{\pi/2} \frac{\cos^n x \sin nx \sin 2x}{1 - 2a \cos 2x + a^2}\, dx = \frac{\pi}{4a}\left[\left(\frac{1+a}{2}\right)^n - \frac{1}{2^n}\right]$ $\left[a^2 < 1\right]$ BI (50)(6)

2. $\displaystyle\int_0^{\pi/2} \frac{1 - a \cos 2nx}{1 - 2a \cos 2nx + a^2}\cos^m x \cos mx \, dx = \frac{\pi}{2^{m+2}} \sum_{k=1}^{\infty} \binom{m}{kn}a^k + \frac{\pi}{2^{m+1}}$

$$\left[a^2 < 1\right]$$

LI (50)(7)

3.647 $\displaystyle\int_0^{\pi/2} \frac{\cos^p x \cos px\, dx}{a^2 \sin^2 x + b^2 \cos^2 x} = \frac{\pi}{2b} \cdot \frac{a^{p-1}}{(a+b)^p}$ $[p > -1, \quad a > 0, \quad b > 0]$ BI (47)(20)

3.648

1. $\displaystyle\int_0^{\pi/4} \frac{\tan^l x\, dx}{1 + \cos\frac{m}{n}\pi \sin 2x}$

$$= \frac{1}{2n}\operatorname{cosec}\frac{m}{n}\pi \sum_{k=0}^{n-1} (-1)^{k-1} \sin\frac{km}{n}\pi \left[\psi\left(\frac{n+l+k}{2n}\right) - \psi\left(\frac{l+k}{2n}\right)\right] \quad [m+n \text{ is odd}]$$

$$= \frac{1}{n}\operatorname{cosec}\frac{m}{n}\pi \sum_{k=0}^{\frac{n-1}{2}} (-1)^{k-1} \sin\frac{km}{n}\pi \left[\psi\left(\frac{n+l-k}{n}\right) - \psi\left(\frac{l+k}{n}\right)\right] \quad [m+n \text{ is even}]$$

$$[l \text{ is a natural number}] \qquad \text{BI (36)(5)}$$

2. $\displaystyle\int_0^{\pi/2} \frac{\tan^{\pm\mu} x\, dx}{1 + \cos t \sin 2x} = \pi \operatorname{cosec} t \sin\mu t \operatorname{cosec}(\mu\pi)$ $[|\operatorname{Re}\mu| < 1, \quad t^2 < \pi^2]$ BI (47)(4)

3.649

1. $\displaystyle\int_0^{\pi/2} \frac{\tan^{\pm\mu} x \sin 2x\, dx}{1 \mp 2a \cos 2x + a^2} = \frac{\pi}{4a}\operatorname{cosec}\frac{\mu\pi}{2}\left[1 - \left(\frac{1-a}{1+a}\right)^\mu\right]$ $[a^2 < 1]$

$$= \frac{\pi}{4a}\operatorname{cosec}\frac{\mu\pi}{2}\left[1 + \left(\frac{a-1}{a+1}\right)^\mu\right] \quad [a^2 > 1]$$

$$[-2 < \operatorname{Re}\mu < 1] \qquad \text{BI (50)(3)}$$

2. $\displaystyle\int_0^{\pi/2} \frac{\tan^{\pm\mu} x\,(1 \mp a \cos 2x)}{1 \mp 2a \cos 2x + a^2}\, dx = \frac{\pi}{4}\sec\frac{\mu\pi}{2}\left[1 + \left(\frac{1-a}{1+a}\right)^\mu\right]$ $[a^2 < 1]$

$$= \frac{\pi}{4}\sec\frac{\mu\pi}{2}\left[1 - \left(\frac{a-1}{a+1}\right)^\mu\right] \quad [a^2 > 1]$$

$$[|\operatorname{Re}\mu| < 1] \qquad \text{BI (50)(4)}$$

3.651

1. $\displaystyle\int_0^{\pi/4} \frac{\tan^\mu x\, dx}{1 + \sin x \cos x} = \frac{1}{3}\left[\psi\left(\frac{\mu+2}{3}\right) - \psi\left(\frac{\mu+1}{3}\right)\right]$ $[\operatorname{Re}\mu > -1]$ BI (36)(3)

2. $\displaystyle\int_0^{\pi/4} \frac{\tan^\mu x\, dx}{1 - \sin x \cos x} = \frac{1}{3}\left[\beta\left(\frac{\mu+2}{3}\right) + \beta\left(\frac{\mu+1}{3}\right)\right]$ $[\operatorname{Re}\mu > -1]$ BI (36)(4)a

3.652

1. $\displaystyle\int_0^{\pi/2} \frac{\tan^\mu x\, dx}{(\sin x + \cos x)\sin x} = \int_0^{\pi/2} \frac{\cot^\mu x\, dx}{(\sin x + \cos x)\cos x} = \pi\operatorname{cosec}\mu\pi$

$$[0 < \operatorname{Re}\mu < 1] \qquad \text{BI (49)(1)}$$

2. $\displaystyle\int_0^{\pi/2} \frac{\tan^\mu x\, dx}{(\sin x - \cos x)\sin x} = \int_0^{\pi/2} \frac{\cot^\mu x\, dx}{(\cos x - \sin x)\cos x} = -\pi\cot\mu\pi$

$$[0 < \operatorname{Re}\mu < 1] \qquad \text{BI (49)(2)}$$

3. $\displaystyle\int_0^{\pi/2} \frac{\cot^{\mu+\frac{1}{2}} x\, dx}{(\sin x + \cos x)\cos x} = \int_0^{\pi/2} \frac{\tan^{\mu-\frac{1}{2}} x\, dx}{(\sin x + \cos x)\cos x} = \pi\sec\mu\pi$

$$[|\operatorname{Re}\mu| < \tfrac{1}{2}] \qquad \text{BI (61)(1, 2)}$$

3.653

1. $$\int_0^{\pi/2} \frac{\tan^{1-2\mu} x \, dx}{a^2 \cos^2 x + b^2 \sin^2 x} = \int_0^{\pi/2} \frac{\cot^{1-2\mu} x \, dx}{a^2 \sin^2 x + b^2 \cos^2 x} = \frac{\pi}{2a^{2\mu} b^{2-2\mu} \sin \mu\pi}$$
$$[0 < \operatorname{Re}\mu < 1] \qquad \text{GW (331)(59b)}$$

2.[7] $$\int_0^{\pi/2} 2\frac{\tan^\mu x \, dx}{1 - a \sin^2 x} = \int_0^{\pi/2} \frac{\cot^\mu x \, dx}{1 - a \cos^2 x} = \frac{\pi \sec \frac{\mu\pi}{2}}{2\sqrt{(1-a)^{\mu+1}}}$$
$$[|\operatorname{Re}\mu| < 1, \quad a < 1] \qquad \text{BI (49)(6)}$$

3. $$\int_0^{\pi/2} \frac{\tan^{\pm\mu} x \, dx}{1 - \cos^2 t \sin^2 2x} = \frac{\pi}{2} \operatorname{cosec} t \sec \frac{\mu\pi}{2} \cos\left[\left(\frac{\pi}{2} - t\right)\mu\right]$$
$$[|\operatorname{Re}\mu| < 1, \quad t^2 < \pi^2]$$
$$\text{BI(49)(7), BI(47)(21)}$$

4. $$\int_0^{\pi/2} \frac{\tan^{\pm\mu} x \sin 2x}{1 - \cos^2 t \sin^2 2x} \, dx = \pi \operatorname{cosec} 2t \operatorname{cosec} \frac{\mu\pi}{2} \sin\left[\left(\frac{\pi}{2} - t\right)\mu\right]$$
$$[|\operatorname{Re}\mu| < 1, \quad t^2 < \pi^2] \qquad \text{BI (47)(22)a}$$

5. $$\int_0^{\pi/2} \frac{\tan^\mu x \sin^2 x \, dx}{1 - \cos^2 t \sin^2 2x} = \int_0^{\pi/2} \frac{\cot^\mu x \cos^2 x \, dx}{1 - \cos^2 t \sin^2 2x} = \frac{\pi}{2} \operatorname{cosec} 2t \sec \frac{\mu\pi}{2} \cos\left[\frac{\mu\pi}{2} - (\mu+1)t\right]$$
$$[|\operatorname{Re}\mu| < 1, \quad t^2 < \pi^2] \quad \text{BI(47)(23)a, BI(49)(10)}$$

6. $$\int_0^{\pi/2} \frac{\tan^\mu x \cos^2 x \, dx}{1 - \cos^2 t \sin^2 2x} = \int_0^{\pi/2} \frac{\cot^\mu x \sin^2 x \, dx}{1 - \cos^2 t \sin^2 2x} = \frac{\pi}{2} \operatorname{cosec} 2t \sec \frac{\mu\pi}{2} \cos\left[\frac{\mu\pi}{2} - (\mu-1)t\right]$$
$$[|\operatorname{Re}\mu| < 1, \quad t^2 < \pi^2] \quad \text{BI(47)(24)a, BI(49)(9)}$$

3.654

1. $$\int_0^{\pi/2} \frac{\tan^{\mu+1} x \cos^2 x \, dx}{(1 + \cos t \sin 2x)^2} = \int_0^{\pi/2} \frac{\cot^{\mu+1} x \sin^2 x \, dx}{(1 + \cos t \sin 2x)^2} = \frac{\pi (\mu \sin t \cos \mu t - \cos t \sin \mu t)}{2 \sin \mu\pi \sin^3 t}$$
$$[|\operatorname{Re}\mu| < 1, \quad t^2 < \pi^2]$$
$$\text{BI(48)(3), BI(49)(22)}$$

2. $$\int_0^{\pi/2} \frac{\tan^{\pm\mu} x \, dx}{(\sin x + \cos x)^2} = \frac{\mu\pi}{\sin \mu\pi} \qquad [0 < \operatorname{Re}\mu < 1] \qquad \text{BI (56)(9)a}$$

3. $$\int_0^{\pi/2} \frac{\tan^{\pm(\mu-1)x} \, dx}{\cos^2 x - \sin^2 x} = \pm\frac{\pi}{2} \cot \frac{\mu\pi}{2} \qquad [0 < \operatorname{Re}\mu < 2] \qquad \text{BI (45)(27, 29)}$$

3.655 $$\int_0^{\pi/2} \frac{\tan^{2\mu-1} x \, dx}{1 - 2a \left(\cos t_1 \sin^2 x + \cos t_2 \cos^2 x\right) + a^2} = \int_0^{\pi/2} \frac{\cot^{2\mu-1} x \, dx}{1 - 2a \left(\cos t_1 \cos^2 x + \cos t_2 \sin^2 x\right) + a^2}$$
$$= \frac{\pi \operatorname{cosec} \mu\pi}{(1 - 2a \cos t_2 + a^2)^\mu (1 - 2a \cos t_1 + a^2) 1 - \mu}$$
$$[0 < \operatorname{Re}\mu < 1, \quad t_1^2 < \pi^2, \quad t_2^2 < \pi^2] \quad \text{BI (50)(18)}$$

3.656

1.　　$\displaystyle\int_0^{\pi/4} \frac{\tan^\mu x\, dx}{1 - \sin^2 x \cos^2 x} = \frac{1}{12}\left\{-\psi\left(\frac{\mu+1}{6}\right) - \psi\left(\frac{\mu+2}{6}\right)\right.$

$$\left. + \psi\left(\frac{\mu+4}{6}\right) + \psi\left(\frac{\mu+5}{6}\right) + 2\psi\left(\frac{\mu+2}{3}\right) - 2\psi\left(\frac{\mu+1}{3}\right)\right\}$$

$$[\operatorname{Re}\mu > -1]\qquad(\text{cf. } \textbf{3.651} \text{ 1 and 2})\qquad\text{LI (36)(10)}$$

2.　　$\displaystyle\int_0^{\pi/2} \frac{\tan^{\mu-1} x \cos^2 x\, dx}{1 - \sin^2 x \cos^2 x} = \int_0^{\pi/2} \frac{\cot^{\mu-1} x \sin^2 x\, dx}{1 - \sin^2 x \cos^2 x} = \frac{\pi}{4\sqrt{3}}\operatorname{cosec}\frac{\mu\pi}{6}\operatorname{cosec}\left(\frac{2+\mu}{6}\pi\right)$

$$[0 < \operatorname{Re}\mu < 4]\qquad\qquad\text{LI (47)(26)}$$

3.66 Forms containing powers of linear functions of trigonometric functions

3.661

1.　　$\displaystyle\int_0^{2\pi} (a\sin x + b\cos x)^{2n+1}\, dx = 0$　　　　　　　　　　　　　　　　　　　BI (68)(9)

2.　　$\displaystyle\int_0^{2\pi} (a\sin x + b\cos x)^{2n}\, dx = \frac{(2n-1)!!}{(2n)!!}\cdot 2\pi(a^2+b^2)^n$　　　　　　BI (68)(8)

3.　　$\displaystyle\int_0^\pi (a + b\cos x)^n\, dx = \frac{1}{2}\int_0^{2\pi} (a + b\cos x)^n\, dx = \pi(a^2-b^2)^{\frac{n}{2}} P_n\left(\frac{a}{\sqrt{a^2-b^2}}\right)$

$$= \frac{\pi}{2^n}\sum_{k=0}^{\lfloor n/2\rfloor} \frac{(-1)^k(2n-2k)!}{k!(n-k)!(n-2k)!}a^{n-2k}(a^2-b^2)^k$$

$$[a^2 > b^2]\qquad\qquad\text{GW (332)(37a)}$$

4.　　$\displaystyle\int_0^\pi \frac{dx}{(a+b\cos x)^{n+1}} = \frac{1}{2}\int_0^{2\pi} \frac{dx}{(a+b\cos x)^{n+1}} = \frac{\pi}{(a^2-b^2)^{\frac{n+1}{2}}} P_n\left(\frac{a}{\sqrt{a^2-b^2}}\right)$

$$= \frac{\pi}{2^n(a+b)^n\sqrt{a^2-b^2}}\sum_{k=0}^{n} \frac{(2n-2k-1)!!(2k-1)!!}{(n-k)!k!}\cdot\left(\frac{a+b}{a-b}\right)^k$$

$$[a > |b|]\qquad\qquad\text{GW(332)(38), LI(64)(14)}$$

3.662

1.　　$\displaystyle\int_0^{\pi/2} (\sec x - 1)^\mu \sin x\, dx = \int_0^{\pi/2} (\operatorname{cosec} x - 1)^\mu \cos x\, dx = \mu\pi\operatorname{cosec}\mu\pi$

$$[|\operatorname{Re}\mu| < 1]\qquad\qquad\text{BI (55)(13)}$$

2.　　$\displaystyle\int_0^{\pi/2} (\operatorname{cosec} x - 1)^\mu \sin 2x\, dx = (1-\mu)\mu\pi\operatorname{cosec}\mu\pi\qquad[-1 < \operatorname{Re}\mu < 2]\qquad\text{BI (48)(7)}$

3.　　$\displaystyle\int_0^{\pi/2} (\sec x - 1)^\mu \tan x\, dx = \int_0^{\pi/2} (\operatorname{cosec} x - 1)^\mu \cot x\, dx = -\pi\operatorname{cosec}\mu\pi$

$$[-1 < \operatorname{Re}\mu < 0]\qquad\qquad\text{BI (46)(4,6)}$$

4.　　$\displaystyle\int_0^{\pi/4} (\cot x - 1)^\mu \frac{dx}{\sin 2x} = -\frac{\pi}{2}\operatorname{cosec}\mu\pi\qquad[-1 < \operatorname{Re}\mu < 0]\qquad\text{BI (38)(22)a}$

5. $\displaystyle\int_0^{\pi/4} (\cot x - 1)^\mu \frac{dx}{\cos^2 x} = \mu\pi \operatorname{cosec} \mu\pi$ $\qquad$ $[|\operatorname{Re}\mu| < 1]$ $\qquad$ BI (38)(11)a

3.663

1. $\displaystyle\int_0^u (\cos x - \cos u)^{\nu-\frac{1}{2}} \cos ax\, dx = \sqrt{\frac{\pi}{2}} \sin^\nu u\, \Gamma\left(\nu + \frac{1}{2}\right) P_{a-\frac{1}{2}}^{-\nu}(\cos u)$

$$\left[\operatorname{Re}\nu > -\tfrac{1}{2};\quad a > 0,\quad 0 < u < \pi\right]$$

EH I 159(27), ET I 22(28)

2. $\displaystyle\int_0^u (\cos x - \cos u)^{\nu-1} \cos[(\nu+\beta)x]\, dx = \frac{\sqrt{\pi}\,\Gamma(\beta+1)\,\Gamma(\nu)\,\Gamma(2\nu)\sin^{2\nu-1} u}{2^\nu\,\Gamma(\beta+2\nu)\,\Gamma\left(\nu+\frac{1}{2}\right)}\, C_\beta^\nu(\cos u)$

$$\left[\operatorname{Re}\nu > 0,\quad \operatorname{Re}\beta > -1,\quad 0 < u < \pi\right]$$

EH I 178(23)

3.664

1. $\displaystyle\int_0^\pi \left(z + \sqrt{z^2-1}\cos x\right)^q dx = \pi\, P_q(z)$

$$\left[\operatorname{Re} z > 0,\quad \arg\left(z + \sqrt{z^2-1}\cos x\right) = \arg z \text{ for } x = \frac{\pi}{2}\right]$$ SM 482

2. $\displaystyle\int_0^\pi \frac{dx}{\left(z + \sqrt{z^2-1}\cos x\right)^q} = \pi\, P_{q-1}(z)$

$$\left[\operatorname{Re} z > 0,\quad \arg\left(z + \sqrt{z^2-1}\cos x\right) = \arg z \text{ for } x = \frac{\pi}{2}\right]$$ WH

3. $\displaystyle\int_0^\pi \left(z + \sqrt{z^2-1}\cos x\right)^q \cos nx\, dx = \frac{\pi}{(q+1)(q+2)\cdots(q+n)}\, P_q^n(z)$

$$\left[\operatorname{Re} z > 0,\quad \arg\left(z + \sqrt{z^2-1}\cos x\right) = \arg z \text{ for } x = \frac{\pi}{2},\right.$$

$$\left. z \text{ lies outside the interval } (-1,1) \text{ of the real axis}\right]$$

WH, SM 483(15)

4. $\displaystyle\int_0^\pi \left(z + \sqrt{z^2-1}\cos x\right)^\mu \sin^{2\nu-1} x\, dx$

$$= \frac{2^{2\nu-1}\,\Gamma(\mu+1)[\Gamma(\nu)]^2}{\Gamma(2\nu+\mu)}\, C_\mu^\nu(z)$$

$$= \frac{\sqrt{\pi}\,\Gamma(\nu)\,\Gamma(2\nu)\,\Gamma(\mu+1)}{\Gamma(2\nu+\mu)\,\Gamma\left(\nu+\frac{1}{2}\right)}\, C_\mu^\nu(z) = 2^\nu\sqrt{\frac{\pi}{2}}(z^2-1)^{\frac{1}{4}-\frac{\nu}{2}}\,\Gamma(\nu)\, P_{\mu+\nu-\frac{1}{2}}^{\frac{1}{2}-\nu}(z)$$

$$[\operatorname{Re}\nu > 0]$$ EH I 155(6)a, EH I 178(22)

5. $\displaystyle\int_0^{2\pi} \left[\beta + \sqrt{\beta^2-1}\cos(a-x)\right]^\nu \left(\gamma + \sqrt{\gamma^2-1}\cos x\right)^{\nu-1} dx$

$$= 2\pi\, P_\nu\left(\beta\gamma - \sqrt{\beta^2-1}\sqrt{\gamma^2-1}\cos a\right)$$

$$[\operatorname{Re}\beta > 0,\quad \operatorname{Re}\gamma > 0]$$ EH I 157(18)

3.665

1. $\int_0^\pi \dfrac{\sin^{\mu-1} x \, dx}{(a + b\cos x)^\mu} = \dfrac{2^{\mu-1}}{\sqrt{(a^2 - b^2)^\mu}} \, \mathrm{B}\left(\dfrac{\mu}{2}, \dfrac{\mu}{2}\right)$ $[\mathrm{Re}\,\mu > 0, \quad 0 < b < a]$ FI II 790a

2. $\int_0^\pi \dfrac{\sin^{2\mu-1} x \, dx}{(1 + 2a\cos x + a^2)^\nu} = \mathrm{B}\left(\mu, \tfrac{1}{2}\right) F\left(\nu, \nu - \mu + \tfrac{1}{2}; \mu + \tfrac{1}{2}; a^2\right)$

                                                $[\mathrm{Re}\,\mu > 0, \quad |a| < 1]$ EH I 81(9)

3.666

1. $\int_0^\pi (\beta + \cos x)^{\mu-\nu-\frac{1}{2}} \sin^{2\nu} x \, dx = \dfrac{2^{\nu+\frac{1}{2}} e^{-i\mu\pi} (\beta^2 - 1)^{\frac{\mu}{2}} \Gamma\left(\nu + \tfrac{1}{2}\right) Q_{\nu-\frac{1}{2}}^\mu(\beta)}{\Gamma\left(\nu + \mu + \tfrac{1}{2}\right)}$

                                             $\left[\mathrm{Re}\left(\nu + \mu + \tfrac{1}{2}\right) > 0, \quad \mathrm{Re}\,\nu > -\tfrac{1}{2}\right]$
                                                                EH I 155(5)a

2.[6] $\int_0^\pi (\cosh \beta + \sinh \beta \cos x)^{\mu+\nu} \sin^{-2\nu} x \, dx = \dfrac{\sqrt{\pi}}{2^\nu} \sinh^\nu(\beta) \, \Gamma\left(\tfrac{1}{2} - \nu\right) P_\mu^\nu(\cosh \beta)$

                                                  $\left[\mathrm{Re}\,\nu < \tfrac{1}{2}\right]$ EH I 156(7)

3. $\int_0^\pi (\cos t + i\sin t \cos x)^\mu \sin^{2\nu-1} x \, dx = 2^{\nu-\frac{1}{2}} \sqrt{\pi} \sin^{\frac{1}{2}-\nu} t \, \Gamma(\nu) \, P_{\mu+\nu-\frac{1}{2}}^{\frac{1}{2}-\nu}(\cos t)$

                                                $\left[\mathrm{Re}\,\nu > 0, \quad t^2 < \pi^2\right]$ EH I 158(23)

4. $\int_0^{2\pi} [\cos t + i\sin t \cos(a - x)]^\nu \cos mx \, dx = \dfrac{i^{3m} 2\pi \, \Gamma(\nu + 1)}{\Gamma(\nu + m + 1)} \cos ma \, P_\nu^m(\cos t)$

                                                $\left[0 < t < \dfrac{\pi}{2}\right]$ EH I 159(25)

5.[10] $\int_0^{2\pi} [\cos t + i\sin t \cos(a - x)]^\nu \sin mx \, dx = \dfrac{i^{3m} 2\pi \, \Gamma(\nu + 1)}{\Gamma(\nu + m + 1)} \sin ma \, P_\nu^m(\cos t)$

                                                $\left[0 < t < \dfrac{\pi}{2}\right]$ EH I 159(26)

3.667

1. $\int_0^{\pi/4} \dfrac{\sin^{\mu-1} 2x \, dx}{(\cos x + \sin x)^{2\mu}} = \dfrac{\sqrt{\pi}}{2^{\mu+1}} \dfrac{\Gamma(\mu)}{\Gamma\left(\mu + \tfrac{1}{2}\right)}$ $[\mathrm{Re}\,\mu > 0]$ BI (37)(1)

2. $\int_0^{\pi/4} \dfrac{\sin^\mu x \, dx}{(\cos x - \sin x)^{\mu+1} \cos x} = -\pi \operatorname{cosec} \mu\pi$ $[-1 < \mathrm{Re}\,\mu < 0]$ (cf. **3.192** 2)

                                                                 BI (37)(16)

3. $\int_0^{\pi/4} \dfrac{(\cos x - \sin x)^\mu}{\sin^\mu x \sin 2x} \, dx = -\dfrac{\pi}{2} \operatorname{cosec} \mu\pi$ $[-1 < \mathrm{Re}\,\mu < 0]$ BI (35)(27)

4. $\int_0^{\pi/4} \dfrac{\sin^\mu x \, dx}{(\cos x - \sin x)^\mu \sin 2x} = \dfrac{\pi}{2} \operatorname{cosec} \mu\pi$ $[0 < \mathrm{Re}\,\mu < 1]$ LI (37)(20)a

5. $\int_0^{\pi/4} \dfrac{\sin^\mu x \, dx}{(\cos x - \sin x)^\mu \cos^2 x} = \mu\pi \operatorname{cosec} \mu\pi$ $[|\mathrm{Re}\,\mu| < 1]$ BI (37)(17)

6. $$\int_0^{\pi/4} \frac{\sin^\mu x \, dx}{(\cos x - \sin x)^{\mu-1} \cos^3 x} = \frac{1-\mu}{2} \mu\pi \operatorname{cosec} \mu\pi \qquad [|\operatorname{Re}\mu| < 1] \qquad \text{BI(35)(24), BI(37)(18)}$$

7. $$\int_0^{\pi/2} \frac{\sin^{\mu-1} x \cos^{\nu-1} x}{(\sin x + \cos x)^{\mu+\nu}} \, dx = \mathrm{B}(\mu, \nu) \qquad [\operatorname{Re}\mu > 0, \quad \operatorname{Re}\nu > 0] \qquad \text{BI (48)(8)}$$

3.668

1. $$\int_{-\frac{\pi}{4}}^{\frac{\pi}{4}} \left(\frac{\cos x + \sin x}{\cos x - \sin x}\right)^{\cos 2t} dx = \frac{\pi}{2\sin\left(\pi\cos^2 t\right)} \qquad \text{FI II 788}$$

2. $$\int_u^v \frac{(\cos u - \cos x)^{\mu-1}}{(\cos x - \cos v)^\mu} \cdot \frac{\sin x \, dx}{1 - 2a\cos x + a^2} = \frac{\left(1 - 2a\cos u + a^2\right)^{\mu-1}}{\left(1 - 2a\cos v + a^2\right)^\mu} \cdot \frac{\pi}{\sin\mu\pi}$$
$$\left[0 < \operatorname{Re}\mu < 1, \quad a^2 < 1\right] \qquad \text{BI (73)(2)}$$

3.669 $$\int_0^{\pi/2} \frac{\sin^{p-1} x \cos^{q-p-1} x \, dx}{(a\cos x + b\sin x)^q} = \int_0^{\pi/2} \frac{\sin^{q-p-1} x \cos^{p-1} x}{(a\sin x + b\cos x)^q} \, dx = \frac{\mathrm{B}(p, q-p)}{a^{q-p}b^p}$$
$$[q > p > 0, \quad ab > 0] \qquad \text{BI (331)(9)}$$

3.67 Square roots of expressions containing trigonometric functions

3.671

1. $$\int_0^{\pi/2} \sin^\alpha x \cos^\beta x \sqrt{1 - k^2\sin^2 x} \, dx = \frac{1}{2}\mathrm{B}\left(\frac{\alpha+1}{2}, \frac{\beta+1}{2}\right) F\left(\frac{\alpha+1}{2}, -\frac{1}{2}; \frac{\alpha+\beta+2}{2}; k^2\right)$$
$$[\alpha > -1, \quad \beta > -1, \quad |k| < 1]$$
$$\text{GW (331)(93)}$$

2. $$\int_0^{\pi/2} \frac{\sin^\alpha x \cos^\beta x}{\sqrt{1 - k^2\sin^2 x}} \, dx = \frac{1}{2}\mathrm{B}\left(\frac{\alpha+1}{2}, \frac{\beta+1}{2}\right) F\left(\frac{\alpha+1}{2}, \frac{1}{2}; \frac{\alpha+\beta+2}{2}; k^2\right)$$
$$[\alpha > -1, \quad \beta > -1, \quad |k| < 1]$$
$$\text{GW (331)(92)}$$

3. $$\int_0^\pi \frac{\sin^{2n} x \, dx}{\sqrt{1 - k^2\sin^2 x}} = \frac{\pi}{2^n} \sum_{j=0}^\infty \frac{(2j-1)!!\,(2n+2j-1)!!}{2^{2j}\,j!\,(n+j)!} k^{2j} \qquad [k^2 < 1]$$
$$= \frac{(2n-1)!!\pi}{2^n\sqrt{1-k^2}} \sum_{j=0}^\infty \frac{[(2j-1)!!]^2}{2^{2j}\,j!\,(n+j)!} \left(\frac{k^2}{k^2-1}\right)^j \qquad \left[k^2 < \frac{1}{2}\right]$$
$$\text{LI (67)(2)}$$

3.672

1. $$\int_0^{\pi/4} \frac{\sin^n x}{\cos^{n+1} x} \cdot \frac{dx}{\sqrt{\cos x\,(\cos x - \sin x)}} = 2 \cdot \frac{(2n)!!}{(2n+1)!!} \qquad \text{BI (39)(5)}$$

2. $$\int_0^{\pi/4} \frac{\sin^n x}{\cos^{n+1} x} \cdot \frac{dx}{\sqrt{\sin x\,(\cos x - \sin x)}} = \frac{(2n-1)!!}{(2n)!!}\pi \qquad \text{BI (39)(6)}$$

3.673 $$\int_u^{\frac{\pi}{2}} \frac{dx}{\sqrt{\sin x - \sin u}} = \sqrt{2}\,K\left(\sin\frac{\pi - 2u}{4}\right) \qquad \text{BI (74)(11)}$$

3.674

1.[8] $\displaystyle\int_0^{\frac{\pi}{2}} \frac{dx}{\sqrt{1 - (p^2/2)(1 - \cos 2x)}} = \boldsymbol{K}(p),$ $[1 > p > 0]$ BI (67)(5)

2. $\displaystyle\int_0^{\pi} \frac{\sin x \, dx}{\sqrt{1 - 2p\cos x + p^2}} = 2$ $[p^2 \le 1]$

                                        $= \dfrac{2}{p}$ $[p^2 \ge 1]$

                                                                   BI (67)(6)

3.[8] $\displaystyle\int_0^{\pi} \frac{\cos x \, dx}{\sqrt{1 - 2p\cos x + p^2}} = \frac{1}{p}\left[\frac{1 + p^2}{1 + p}\boldsymbol{K}\left(\frac{2\sqrt{p}}{1 + p}\right) - (1 + p)\boldsymbol{E}\left(\frac{2\sqrt{p}}{1 + p}\right)\right]$

                                                       $[p^2 < 1]$ BI (67)(7)

3.675

1. $\displaystyle\int_u^{\pi} \frac{\sin\left(n + \frac{1}{2}\right)x \, dx}{\sqrt{2(\cos u - \cos x)}} = \frac{\pi}{2} P_n(\cos u)$ WH

2. $\displaystyle\int_0^{u} \frac{\cos\left(n + \frac{1}{2}\right)x \, dx}{\sqrt{2(\cos x - \cos u)}} = \frac{\pi}{2} P_n(\cos u)$ FI II 684, WH

3.676

1. $\displaystyle\int_0^{\pi/2} \frac{\sin x \, dx}{\sqrt{1 + p^2\sin^2 x}} = \frac{1}{p}\arctan p$ BI (60)(5)

2. $\displaystyle\int_0^{\pi/2} \tan^2 x\sqrt{1 - p^2\sin^2 x} \, dx = \infty$ BI (53)(8)

3. $\displaystyle\int_0^{\pi/2} \frac{dx}{\sqrt{p^2\cos^2 x + q^2\sin^2 x}} = \frac{1}{p}\boldsymbol{K}\left(\frac{\sqrt{p^2 - q^2}}{p}\right)$ $[0 < q < p]$ FI II 165

3.677

1. $\displaystyle\int_0^{\pi/2} \frac{\sin^2 x \, dx}{\sqrt{1 + \sin^2 x}} = \sqrt{2}\boldsymbol{E}\left(\frac{\sqrt{2}}{2}\right) - \frac{1}{\sqrt{2}}\boldsymbol{K}\left(\frac{\sqrt{2}}{2}\right)$ BI (60)(2)

2. $\displaystyle\int_0^{\pi/2} \frac{\cos^2 x \, dx}{\sqrt{1 + \sin^2 x}} = \sqrt{2}\left[\boldsymbol{K}\left(\frac{\sqrt{2}}{2}\right) - \boldsymbol{E}\left(\frac{\sqrt{2}}{2}\right)\right]$ BI (60)(3)

3.678

1. $\displaystyle\int_0^{\pi/4} \left(\sec^{1/2} 2x - 1\right)\frac{dx}{\tan x} = \ln 2$ BI (38)(23)

2. $\displaystyle\int_0^{\pi/4} \frac{\tan^2 x \, dx}{\sqrt{1 - k^2\sin^2 2x}} = \sqrt{1 - k^2} - \boldsymbol{E}(k) + \frac{1}{2}\boldsymbol{K}(k)$ BI (39)(2)

3. $\displaystyle\int_0^{u} \sqrt{\frac{\cos 2x - \cos 2u}{\cos 2x + 1}} \, dx = \frac{\pi}{2}(1 - \cos u)$ $\left[u^2 < \dfrac{\pi^2}{4}\right]$ LI (74)(6)

4. $\int_0^{\pi/4} \frac{(\cos x - \sin x)^{n-\frac{1}{2}}}{\cos^{n+1} x} \sqrt{\operatorname{cosec} x}\, dx = \frac{(2n-1)!!}{(2n)!!}\pi$

BI (38)(24)

5. $\int_0^{\pi/4} \frac{(\cos x - \sin x)^{n-\frac{1}{2}}}{\cos^{n+1} x} \tan^m x \sqrt{\operatorname{cosec} x}\, dx = \frac{(2n-1)!!(2m-1)!!}{(2n+2m)!!}\pi$

BI (38)(25)

3.679

1. $\int_0^{\pi/2} \frac{\cos^2 x}{1-\cos^2\beta\cos^2 x} \cdot \frac{dx}{\sqrt{1-k^2\sin^2 x}}$
$$= \frac{1}{\sin\beta\cos\beta\sqrt{1-k'^2\sin^2\beta}}\left\{\frac{\pi}{2} - \boldsymbol{K}E(\beta,k') - \boldsymbol{E}F(\beta,k') + \boldsymbol{K}F(\beta,k')\right\}{}^{*}$$

MO 138

2. $\int_0^{\pi/2} \frac{\sin^2 x}{1-\left(1-k'^2\sin^2\beta\right)\sin^2 x} \cdot \frac{dx}{\sqrt{1-k^2\sin^2 x}}$
$$= \frac{1}{k'^2\sin\beta\cos\beta\sqrt{1-k'^2\sin^2\beta}}\left\{\frac{\pi}{2} - \boldsymbol{K}E(\beta,k') - \boldsymbol{E}F(\beta,k') + \boldsymbol{K}F(\beta,k')\right\}{}^{*}$$

MO 138

3. $\int_0^{\pi/2} \frac{\sin^2 x}{1-k^2\sin^2\beta\sin^2 x} \cdot \frac{dx}{\sqrt{1-k^2\sin^2 x}} = \frac{\boldsymbol{K}E(\beta,k) - \boldsymbol{E}F(\beta,k)}{k^2\sin\beta\cos\beta\sqrt{1-k^2\sin^2\beta}}$

MO 138

3.68 Various forms of powers of trigonometric functions

3.681

1. $\int_0^{\pi/2} \frac{\sin^{2\mu-1} x \cos^{2\nu-1} x\, dx}{\left(1-k^2\sin^2 x\right)^\varrho} = \frac{1}{2}\,\mathrm{B}(\mu,\nu)\,F\left(\varrho,\mu;\mu+\nu;k^2\right)$

[Re $\mu > 0$, Re $\nu > 0$]   EH I 115(7)

2. $\int_0^{\pi/2} \frac{\sin^{2\mu-1} x \cos^{2\nu-1} x\, dx}{\left(1-k^2\sin^2 x\right)^{\mu+\nu}} = \frac{\mathrm{B}(\mu,\nu)}{2(1-k^2)^\mu}$ [Re $\mu > 0$, Re $\nu > 0$]

EH I 10(20)

3. $\int_0^{\pi/2} \frac{\sin^\mu x\, dx}{\cos^{\mu-3} x \left(1-k^2\sin^2 x\right)^{\frac{\mu}{2}-1}}$
$$= \frac{\Gamma\left(\frac{\mu+1}{2}\right)\Gamma\left(2-\frac{\mu}{2}\right)}{k^3\sqrt{\pi(\mu-1)(\mu-3)(\mu-5)}}\left\{\frac{1+(\mu-3)k+k^2}{(1+k)^{\mu-3}} - \frac{1-(\mu-3)k+k^2}{(1-k)^{\mu-3}}\right\}$$

[$-1 < \operatorname{Re}\mu < 4$]   BI (54)(10)

4.[8] $\int_0^{\pi/2} \frac{\sin^{\mu+1} x\, dx}{\cos^\mu x \left(1-k^2\sin^2 x\right)^{\frac{\mu+1}{2}}} = \frac{(1-k)^{-\mu} - (1+k)^{-\mu}}{2k\mu\sqrt{\pi}}\Gamma\left(1+\frac{\mu}{2}\right)\Gamma\left(\frac{1-\mu}{2}\right)$

[$-2 < \operatorname{Re}\mu < 1$]   BI (61)(5)

---

*In 3.631.5, $k' = \sqrt{1-k^2}$

3.682 $\displaystyle\int_0^{\pi/2} \frac{\sin^\mu x \cos^\nu x}{(a - b\cos^2 x)^\varrho}\, dx = \frac{1}{2a^\varrho}\, \mathrm{B}\left(\frac{\mu+1}{2}, \frac{\nu+1}{2}\right) F\left(\frac{\nu+1}{2}, \varrho; \frac{\mu+\nu}{2}+1; \frac{b}{a}\right)$

$$[\mathrm{Re}\,\mu > -1, \quad \mathrm{Re}\,\nu > -1, \quad a > |b| \geq 0]$$

GW (331)(64)

3.683

1. $\displaystyle\int_0^{\pi/4} \left(\sin^n 2x - 1\right)\tan\left(\frac{\pi}{4}+x\right) dx = \int_0^{\pi/4} \left(\cos^n 2x - 1\right)\cot x\, dx = -\frac{1}{2}\sum_{k=1}^n \frac{1}{k}$

$$= -\frac{1}{2}\left[\boldsymbol{C} + \psi(n+1)\right]$$

$$[n \geq 0] \qquad \text{BI(34)(8), BI(35)(11)}$$

2. $\displaystyle\int_0^{\pi/4} \left(\sin^\mu 2x - 1\right)\operatorname{cosec}^\mu 2x \tan\left(\frac{\pi}{4}+x\right) dx = \int_0^{\pi/4} \left(\cos^\mu 2x - 1\right)\sec^\mu 2x \cot x\, dx$

$$= \frac{1}{2}\left[\boldsymbol{C} + \psi(1-\mu)\right]$$

$$[\mathrm{Re}\,\mu < 1] \qquad \text{BI (35)(20)}$$

3. $\displaystyle\int_0^{\frac{\pi}{4}} \left(\sin^{2\mu} 2x - 1\right)\operatorname{cosec}^\mu 2x \tan\left(\frac{\pi}{4}+x\right) dx = \int_0^{\pi/4} \left(\cos^{2\mu} 2x - 1\right)\sec^\mu 2x \cot x\, dx$

$$= -\frac{1}{2\mu} + \frac{\pi}{2}\cot\mu\pi$$

BI (35)(21)

4. $\displaystyle\int_0^{\pi/4} \left(1 - \sec^\mu 2x\right)\cot x\, dx = \int_0^{\pi/4} \left(1 - \operatorname{cosec}^\mu 2x\right)\tan\left(\frac{\pi}{4}+x\right) dx = \frac{1}{2}\left[\boldsymbol{C} + \psi(1-\mu)\right]$

$$[\mathrm{Re}\,\mu < 1] \qquad \text{BI (35)(13)}$$

3.684 $\displaystyle\int_0^{\pi/4} \frac{(\cot^\mu x - 1)\, dx}{(\cos x - \sin x)\sin x} = \int_0^{\pi/2} \frac{(\tan^\mu x - 1)\, dx}{(\sin x - \cos x)\cos x} = -\boldsymbol{C} - \psi(1-\mu) \qquad [\mathrm{Re}\,\mu < 1]$

BI (37)(9)

3.685

1. $\displaystyle\int_0^{\pi/4} \left(\sin^{\mu-1} 2x - \sin^{\nu-1} 2x\right)\tan\left(\frac{\pi}{4}+x\right) dx = \int_0^{\pi/4} \left(\cos^{\mu-1} 2x - \cos^{\nu-1} 2x\right)\cot x\, dx$

$$= \frac{1}{2}\left[\psi(\nu) - \psi(\mu)\right]$$

$$[\mathrm{Re}\,\mu > 0, \mathrm{Re}\,\nu > 0] \quad \text{BI(34)(9), BI(35)(12)}$$

2. $\displaystyle\int_0^{\pi/2} \left(\sin^{\mu-1} x - \sin^{\nu-1} x\right)\frac{dx}{\cos x} = \int_0^{\pi/2} \left(\cos^{\mu-1} x - \cos^{\nu-1} x\right)\frac{dx}{\sin x} = \frac{1}{2}\left[\psi\left(\frac{\nu}{2}\right) - \psi\left(\frac{\mu}{2}\right)\right]$

$$[\mathrm{Re}\,\mu > 0, \quad \mathrm{Re}\,\nu > 0] \qquad \text{BI (46)(2)}$$

3. $\displaystyle\int_0^{\pi/2} \left(\sin^\mu x - \operatorname{cosec}^\mu x\right)\frac{dx}{\cos x} = \int_0^{\pi/2} \left(\cos^\mu x - \sec^\mu x\right)\frac{dx}{\sin x} = -\frac{\pi}{2}\tan\frac{\mu\pi}{2}$

$$[|\mathrm{Re}\,\mu| < 1] \qquad \text{BI (46)(1, 3)}$$

4. $\int_0^{\pi/4} \left(\sin^\mu 2x - \operatorname{cosec}^\mu 2x\right) \cot\left(\frac{\pi}{4} + x\right) dx = \int_0^{\pi/4} \left(\cos^\mu 2x - \sec^\mu 2x\right) \tan x \, dx$
$$= \frac{1}{2\mu} - \frac{\pi}{2} \operatorname{cosec} \mu\pi$$
$$[|\operatorname{Re}\mu| < 1] \qquad\qquad \text{BI (35)(19, 22)}$$

5. $\int_0^{\pi/4} \left(\sin^\mu 2x - \operatorname{cosec}^\mu 2x\right) \tan\left(\frac{\pi}{4} + x\right) dx = \int_0^{\pi/4} \left(\cos^\mu 2x - \sec^\mu 2x\right) \cot x \, dx$
$$= -\frac{1}{2\mu} + \frac{\pi}{2} \cot \mu\pi$$
$$[|\operatorname{Re}\mu| < 1] \qquad\qquad \text{BI (35)(14)}$$

6. $\int_0^{\pi/4} \left(\sin^{\mu-1} 2x + \operatorname{cosec}^\mu 2x\right) \cot\left(\frac{\pi}{4} + x\right) dx$
$$= \int_0^{\pi/4} \left(\cos^{\mu-1} 2x + \sec^\mu 2x\right) \tan x \, dx = \frac{\pi}{4} \operatorname{cosec} \mu\pi$$
$$[0 < \operatorname{Re}\mu < 1] \qquad\qquad \text{BI (35)(18, 8)}$$

7. $\int_0^{\pi/4} \left(\sin^{\mu-1} 2x - \operatorname{cosec}^\mu 2x\right) \tan\left(\frac{\pi}{4} + x\right) dx = \int_0^{\pi/4} \left(\cos^{\mu-1} 2x - \sec^\mu 2x\right) \cot x \, dx = \frac{\pi}{2} \cot \mu\pi$
$$[0 < \operatorname{Re}\mu < 1] \qquad \text{BI(35)(7), LI(34)(10)}$$

3.686 $\int_0^{\pi/2} \frac{\tan x \, dx}{\cos^\mu x + \sec^\mu x} = \int_0^{\pi/2} \frac{\cot x \, dx}{\sin^\mu x + \operatorname{cosec}^\mu x} = \frac{\pi}{4\mu}$ \qquad BI(47)(28), BI(49)(14)

3.687

1. $\int_0^{\pi/2} \frac{\sin^{\mu-1} x + \sin^{\nu-1} x}{\cos^{\mu+\nu-1} x} dx = \int_0^{\pi/2} \frac{\cos^{\mu-1} x + \cos^{\nu-1} x}{\sin^{\mu+\nu-1} x} dx = \frac{\cos\left(\frac{\nu-\mu}{4}\pi\right)}{2\cos\left(\frac{\nu+\mu}{4}\pi\right)} B\left(\frac{\mu}{2}, \frac{\nu}{2}\right)$
$$[\operatorname{Re}\mu > 0, \quad \operatorname{Re}\nu > 0, \quad \operatorname{Re}(\mu+\nu) < 2]$$
$$\text{BI (46)(7)}$$

2. $\int_0^{\pi/2} \frac{\sin^{\mu-1} x - \sin^{\nu-1} x}{\cos^{\mu+\nu-1} x} dx = \int_0^{\pi/2} \frac{\cos^{\mu-1} x - \cos^{\nu-1} x}{\sin^{\mu+\nu-1} x} dx = \frac{\sin\left(\frac{\nu-\mu}{4}\pi\right)}{2\sin\left(\frac{\nu+\mu}{4}\pi\right)} B\left(\frac{\mu}{2}, \frac{\nu}{2}\right)$
$$[\operatorname{Re}\mu > 0, \quad \operatorname{Re}\nu > 0, \quad \operatorname{Re}(\mu+\nu) < 4]$$
$$\text{BI(46)(8)}$$

3. $\int_0^{\pi/2} \frac{\sin^\mu x + \sin^\nu x}{\sin^{\mu+\nu} x + 1} \cot x \, dx = \int_0^{\frac{\pi}{2}} \frac{\cos^\mu x + \cos^\nu x}{\cos^{\mu+\nu} x + 1} \tan x \, dx = \frac{\pi}{\mu+\nu} \sec\left(\frac{\mu-\nu}{\mu+\nu} \cdot \frac{\pi}{2}\right)$
$$[\operatorname{Re}\mu > 0, \quad \operatorname{Re}\nu > 0]$$
$$\text{BI (49)(15)a, BI (47)(29)}$$

4. $\int_0^{\pi/2} \frac{\sin^\mu x - \sin^\nu x}{\sin^{\mu+\nu} x - 1} \cot x \, dx = \int_0^{\frac{\pi}{2}} \frac{\cos^\mu x - \cos^\nu x}{\cos^{\mu+\nu} x - 1} \tan x \, dx = \frac{\pi}{\mu+\nu} \tan\left(\frac{\mu-\nu}{\mu+\nu} \cdot \frac{\pi}{2}\right)$
$$[\operatorname{Re}\mu > 0, \quad \operatorname{Re}\nu > 0]$$
$$\text{BI(149)(16)a, BI(47)(30)}$$

5. $\int_0^{\pi/2} \frac{\cos^\mu x + \sec^\mu x}{\cos^\nu x + \sec^\nu x} \tan x \, dx = \frac{\pi}{2\nu} \sec\left(\frac{\mu}{\nu} \cdot \frac{\pi}{2}\right)$ \qquad $[|\operatorname{Re}\nu| > |\operatorname{Re}\mu|]$ \qquad BI (49)(12)

6. $\int_0^{\pi/2} \dfrac{\cos^\mu x - \sec^\mu x}{\cos^\nu x - \sec^\nu x} \tan x \, dx = \dfrac{\pi}{2\nu} \tan\left(\dfrac{\mu}{\nu} \cdot \dfrac{\pi}{2}\right)$ $[|\operatorname{Re}\nu| > |\operatorname{Re}\mu|]$ BI (49)(13)

3.688

1. $\int_0^{\pi/4} \dfrac{\tan^\nu x - \tan^\mu x}{\cos x - \sin x} \cdot \dfrac{dx}{\sin x} = \psi(\mu) - \psi(\nu)$ $[\operatorname{Re}\mu > 0, \quad \operatorname{Re}\nu > 0]$ BI (37)(10)

2. $\int_0^{\pi/4} \dfrac{\tan^\mu x - \tan^{1-\mu} x}{\cos x - \sin x} \cdot \dfrac{dx}{\sin x} = \pi \cot \mu\pi$ $[0 < \operatorname{Re}\mu < 1]$ BI (37)(11)

3. $\int_0^{\pi/4} (\tan^\mu x + \cot^\mu x) \, dx = \dfrac{\pi}{2} \sec \dfrac{\mu\pi}{2}$ $[|\operatorname{Re}\mu| < 1]$ BI (35)(9)

4. $\int_0^{\pi/4} (\tan^\mu x - \cot^\mu x) \tan x \, dx = \dfrac{1}{\mu} - \dfrac{\pi}{2} \operatorname{cosec} \dfrac{\mu\pi}{2}$ $[0 < \operatorname{Re}\mu < 2]$ BI (35)(15)

5. $\int_0^{\pi/4} \dfrac{\tan^{\mu-1} x - \cot^{\mu-1} x}{\cos 2x} \, dx = \dfrac{\pi}{2} \cot \dfrac{\mu\pi}{2}$ $[|\operatorname{Re}\mu| < 2]$ BI (35)(10)

6. $\int_0^{\pi/4} \dfrac{\tan^\mu x - \cot^\mu x}{\cos 2x} \tan x \, dx = -\dfrac{1}{\mu} + \dfrac{\pi}{2} \cot \dfrac{\mu\pi}{2}$ $[-2 < \operatorname{Re}\mu < 0]$ BI (35)(23)

7. $\int_0^{\pi/4} \dfrac{\tan^\mu x + \cot^\mu x}{1 + \cos t \sin 2x} \, dx = \pi \operatorname{cosec} t \operatorname{cosec} \mu\pi \sin \mu t$ $[t \neq n\pi, \quad |\operatorname{Re}\mu| < 1]$ BI (36)(6)

8. $\int_0^{\pi/4} \dfrac{\tan^{\mu-1} x + \cot^\mu x}{(\sin x + \cos x) \cos x} \, dx = \pi \operatorname{cosec} \mu\pi$ $[0 < \operatorname{Re}\mu < 1]$ BI (37)(3)

9. $\int_0^{\pi/4} \dfrac{\tan^\mu x - \cot^\mu x}{(\sin x + \cos x) \cos x} \, dx = -\pi \operatorname{cosec} \mu\pi + \dfrac{1}{\mu}$ $[0 < \operatorname{Re}\mu < 1]$ BI (37)(4)

10. $\int_0^{\pi/4} \dfrac{\tan^\nu x - \cot^\mu x}{(\cos x - \sin x) \cos x} \, dx = \psi(1-\mu) - \psi(1+\nu)$ $[\operatorname{Re}\mu < 1, \quad \operatorname{Re}\nu > -1]$ BI (37)(5)

11. $\int_0^{\pi/4} \dfrac{\tan^{\mu-1} x - \cot^\mu x}{(\cos x - \sin x) \cos x} \, dx = \pi \cot \mu\pi$ $[0 < \operatorname{Re}\mu < 1]$ BI (37)(7)

12. $\int_0^{\pi/4} \dfrac{\tan^\mu x - \cot^\mu x}{(\cos x - \sin x) \cos x} \, dx = \pi \cot \mu\pi - \dfrac{1}{\mu}$ $[0 < \operatorname{Re}\mu < 1]$ BI (37)(8)

13. $\int_0^{\pi/4} \dfrac{1}{\tan^\mu x + \cot^\mu x} \cdot \dfrac{dx}{\sin 2x} = \dfrac{\pi}{8\mu}$ $[\operatorname{Re}\mu \neq 0]$ BI (37)(12)

14. $\int_0^{\pi/2} \dfrac{1}{(\tan^\mu x + \cot^\mu x)^\nu} \cdot \dfrac{dx}{\tan x} = \int_0^{\pi/2} \dfrac{1}{(\tan^\mu x + \cot^\mu x)^\nu} \cdot \dfrac{dx}{\sin 2x} = \dfrac{\sqrt{\pi}}{2^{2\nu+1}\mu} \dfrac{\Gamma(\nu)}{\Gamma\left(\nu + \frac{1}{2}\right)}$

                                            $[\nu > 0]$ BI(49)(25), BI(49)(26)

15. $\int_0^{\pi/4} (\tan^\mu x - \cot^\mu x)(\tan^\nu x - \cot^\nu x) \, dx = \dfrac{2\pi \sin \frac{\mu\pi}{2} \sin \frac{\nu\pi}{2}}{\cos \mu\pi + \cos \nu\pi}$

                                 $[|\operatorname{Re}\mu| < 1, \quad |\operatorname{Re}\nu| < 1]$ BI (35)(17)

16. $\displaystyle\int_0^{\pi/4} \left(\tan^\mu x + \cot^\mu x\right)\left(\tan^\nu x + \cot^\nu x\right)\,dx = \frac{2\pi \cos\frac{\mu\pi}{2}\cos\frac{\nu\pi}{2}}{\cos\mu\pi + \cos\nu\pi}$

$\qquad\qquad\qquad\qquad [|\mathrm{Re}\,\mu| < 1, \quad |\mathrm{Re}\,\nu| < 1] \qquad$ BI (35)(16)

17. $\displaystyle\int_0^{\pi/4} \frac{\left(\tan^\mu x - \cot^\mu x\right)\left(\tan^\nu x + \cot^\nu x\right)}{\cos 2x}\,dx = -\pi\frac{\sin\mu\pi}{\cos\mu\pi + \cos\nu\pi}$

$\qquad\qquad\qquad\qquad [|\mathrm{Re}\,\mu| < 1, \quad |\mathrm{Re}\,\nu| < 1] \qquad$ BI (35)(25)

18. $\displaystyle\int_0^{\pi/4} \frac{\tan^\nu x - \cot^\nu x}{\tan^\mu x - \cot^\mu x}\cdot\frac{dx}{\sin 2x} = \frac{\pi}{4\mu}\tan\frac{\nu\pi}{2\mu}$ $\qquad [0 < \mathrm{Re}\,\nu < 1] \qquad$ BI (37)(14)

19. $\displaystyle\int_0^{\pi/4} \frac{\tan^\nu x + \cot^\nu x}{\tan^\mu x + \cot^\mu x}\cdot\frac{dx}{\sin 2x} = \frac{\pi}{4\mu}\sec\frac{\nu\pi}{2\mu}$ $\qquad [0 < \mathrm{Re}\,\nu < 1] \qquad$ BI (37)(13)

20. $\displaystyle\int_0^{\pi/2} \frac{(1+\tan x)^\nu - 1}{(1+\tan x)^{\mu+\nu}}\frac{dx}{\sin x \cos x} = \psi(\mu+\nu) - \psi(\mu)$ $\qquad [\mu > 0, \quad \nu > 0] \qquad$ BI (49)(29)

3.689

1. $\displaystyle\int_0^{\pi/2} \frac{\left(\sin^\mu x + \mathrm{cosec}^\mu x\right)\cot x\,dx}{\sin^\nu x - 2\cos t + \mathrm{cosec}^\nu x} = \frac{\pi}{\nu}\,\mathrm{cosec}\,t\,\mathrm{cosec}\,\frac{\mu\pi}{\nu}\sin\frac{\mu t}{\nu}$

$\qquad\qquad\qquad\qquad\qquad [\mu < \nu] \qquad$ LI (50)(14)

2. $\displaystyle\int_0^{\pi/2} \frac{\sin^\mu x - 2\cos t_1 + \mathrm{cosec}^\mu x}{\sin^\nu x + 2\cos t_2 + \mathrm{cosec}^\nu x}\cdot\cot x\,dx = \frac{\pi}{\nu}\,\mathrm{cosec}\,t_2\,\mathrm{cosec}\,\frac{\mu\pi}{\nu}\sin\frac{\mu t_2}{\nu} - \frac{t_2}{\nu}\,\mathrm{cosec}\,t_2\cos t_1$

$[(\nu > \mu > 0) \text{ or } (\nu < \mu < 0) \text{ or } (\mu > 0, \nu < 0, \text{ and } \mu + \nu < 0) \text{ or } (\mu < 0, \nu > 0, \text{ and } \mu + \nu > 0)]$

$\qquad\qquad\qquad\qquad\qquad$ BI (50)(15)

3.69–3.71 Trigonometric functions of more complicated arguments

3.691

1. $\displaystyle\int_0^\infty \sin\left(ax^2\right)\,dx = \int_0^\infty \cos ax^2\,dx = \frac{1}{2}\sqrt{\frac{\pi}{2a}}$ $\qquad [a > 0] \qquad$ FI II 743a, ET I 64(7)a

2. $\displaystyle\int_0^1 \sin\left(ax^2\right)\,dx = \sqrt{\frac{\pi}{2a}}\,S\left(\sqrt{a}\right)$ $\qquad [a > 0]$

3. $\displaystyle\int_0^1 \cos\left(ax^2\right)\,dx = \sqrt{\frac{\pi}{2a}}\,C\left(\sqrt{a}\right)$ $\qquad [a > 0] \qquad$ ET I 8(5)a

4. $\displaystyle\int_0^\infty \sin\left(ax^2\right)\sin 2bx\,dx = \sqrt{\frac{\pi}{2a}}\left\{\cos\frac{b^2}{a}\,C\left(\frac{b}{\sqrt{a}}\right) + \sin\frac{b^2}{a}\,S\left(\frac{b}{\sqrt{a}}\right)\right\}$

$\qquad\qquad\qquad\qquad [a > 0, \quad b > 0] \qquad$ ET I 82(1)a

5. $\displaystyle\int_0^\infty \sin\left(ax^2\right)\cos 2bx\,dx = \frac{1}{2}\sqrt{\frac{\pi}{2a}}\left\{\cos\frac{b^2}{a} - \sin\frac{b^2}{a}\right\} = \frac{1}{2}\sqrt{\frac{\pi}{a}}\cos\left(\frac{b^2}{a} + \frac{\pi}{4}\right)$

$\qquad\qquad\qquad\qquad [a > 0, \quad b > 0]$

$\qquad\qquad\qquad\qquad$ ET I 82(18), BI(70)(13) GW(334)(5a)

6.
$$\int_0^\infty \cos ax^2 \sin 2bx \, dx = \sqrt{\frac{\pi}{2a}} \left\{ \sin \frac{b^2}{a} C\left(\frac{b}{\sqrt{a}}\right) - \cos \frac{b^2}{a} S\left(\frac{b}{\sqrt{a}}\right) \right\}$$
$$[a > 0, \quad b > 0] \qquad \text{ET I 83(3)a}$$

7.
$$\int_0^\infty \cos ax^2 \cos 2bx \, dx = \frac{1}{2}\sqrt{\frac{\pi}{2a}} \left\{ \cos \frac{b^2}{a} + \sin \frac{b^2}{a} \right\} \qquad [a > 0, \quad b > 0]$$
$$\text{GW(334)(5a), BI(70)(14), ET I 24(7)}$$

8.
$$\int_0^\infty (\cos ax + \sin ax) \sin \left(b^2 x^2\right) \, dx$$
$$= \frac{1}{2b}\sqrt{\frac{\pi}{2}} \left\{ \left(1 + 2C\left(\frac{a}{2b}\right)\right) \cos\left(\frac{a^2}{4b^2}\right) - \left(1 - 2S\left(\frac{a}{2b}\right)\right) \sin\left(\frac{a^2}{4b^2}\right) \right\}$$
$$[a > 0, \quad b > 0] \qquad \text{ET I 85(22)}$$

9.
$$\int_0^\infty (\cos ax + \sin ax) \cos \left(b^2 x^2\right) \, dx$$
$$= \frac{1}{2b}\sqrt{\frac{\pi}{2}} \left\{ \left(1 + 2C\left(\frac{a}{2b}\right)\right) \sin\left(\frac{a^2}{4b^2}\right) + \left(1 - 2S\left(\frac{a}{2b}\right)\right) \cos\left(\frac{a^2}{4b^2}\right) \right\}$$
$$[a > 0, \quad b > 0] \qquad \text{ET I 25(21)}$$

10.
$$\int_0^\infty \sin \left(a^2 x^2\right) \sin 2bx \sin 2cx \, dx = \frac{\sqrt{\pi}}{2a} \sin \frac{2bc}{a^2} \cos \left(\frac{b^2 + c^2}{a^2} - \frac{\pi}{4}\right)$$
$$[a > 0, \quad b > 0, \quad c > 0] \qquad \text{ET I 84(15)}$$

11.
$$\int_0^\infty \sin \left(a^2 x^2\right) \cos 2bx \cos 2cx \, dx = \frac{\sqrt{\pi}}{2a} \cos \frac{2bc}{a^2} \cos \left(\frac{b^2 + c^2}{a^2} + \frac{\pi}{4}\right)$$
$$[a > 0, \quad b > 0, \quad c > 0] \qquad \text{ET I 84(21)}$$

12.
$$\int_0^\infty \cos \left(a^2 x^2\right) \sin 2bx \sin 2cx \, dx = \frac{\sqrt{\pi}}{2a} \sin \frac{2bc}{a^2} \sin \left(\frac{b^2 + c^2}{a^2} - \frac{\pi}{4}\right)$$
$$[a > 0, \quad b > 0, \quad c > 0] \qquad \text{ET I 25(19)}$$

13.
$$\int_0^\infty \sin \left(ax^2\right) \cos \left(bx^2\right) \, dx = \frac{1}{4}\sqrt{\frac{\pi}{2}} \left(\frac{1}{\sqrt{a+b}} + \frac{1}{\sqrt{a-b}}\right) \qquad [a > b > 0]$$
$$= \frac{1}{4}\sqrt{\frac{\pi}{2}} \left(\frac{1}{\sqrt{b+a}} - \frac{1}{\sqrt{b-a}}\right) \qquad [b > a > 0]$$
$$\text{BI (177)(21)}$$

14.
$$\int_0^\infty \left(\sin^2 ax^2 - \sin^2 bx^2\right) \, dx = \frac{1}{8}\left(\sqrt{\frac{\pi}{b}} - \sqrt{\frac{\pi}{a}}\right) \qquad [a > 0, \quad b > 0] \qquad \text{BI (178)(1)}$$

15.
$$\int_0^\infty \left(\cos^2 ax^2 - \sin^2 bx^2\right) \, dx = \frac{1}{8}\left(\sqrt{\frac{\pi}{b}} + \sqrt{\frac{\pi}{a}}\right) \qquad [a > 0, \quad b > 0] \qquad \text{BI (178)(3)}$$

16.
$$\int_0^\infty \left(\cos^2 ax^2 - \cos^2 bx^2\right) \, dx = \frac{1}{8}\left(\sqrt{\frac{\pi}{a}} - \sqrt{\frac{\pi}{b}}\right) \qquad [a > 0, \quad b > 0] \qquad \text{BI (178)(5)}$$

17.
$$\int_0^\infty \left(\sin^4 ax^2 - \sin^4 bx^2\right) x = \frac{1}{64}\left(8 - \sqrt{2}\right)\left(\sqrt{\frac{\pi}{b}} - \sqrt{\frac{\pi}{a}}\right)$$
$$[a > 0, \quad b > 0] \qquad \text{BI (178)(2)}$$

18. $\int_0^\infty \left(\cos^4 ax^2 - \sin^4 bx^2\right) dx = \frac{1}{8}\left(\sqrt{\frac{\pi}{a}} + \sqrt{\frac{\pi}{b}}\right) + \frac{1}{32}\left(\sqrt{\frac{\pi}{2a}} - \sqrt{\frac{\pi}{2b}}\right)$

$[a > 0, \quad b > 0]$ BI (178)(4)

19. $\int_0^\infty \left(\cos^4 ax^2 - \cos^4 bx^2\right) dx = \frac{1}{64}\left(8 + \sqrt{2}\right)\left(\sqrt{\frac{\pi}{a}} - \sqrt{\frac{\pi}{b}}\right)$

$[a > 0, \quad b > 0]$ BI (178)(6)

20. $\int_0^\infty \sin^{2n} ax^2 \, dx = \int_0^\infty \cos^{2n} ax^2 \, dx = \infty$ BI (177)(5, 6)

21. $\int_0^\infty \sin^{2n+1}\left(ax^2\right) dx = \frac{1}{2^{2n+1}} \sum_{k=0}^n (-1)^{n+k} \binom{2n+1}{k} \sqrt{\frac{\pi}{2(2n-2k+1)a}}$

$[a > 0]$ BI (70)(9)

22. $\int_0^\infty \cos^{2n+1}\left(ax^2\right) dx = \frac{1}{2^{2n+1}} \sum_{k=0}^n \binom{2n+1}{k} \sqrt{\frac{\pi}{2(2n-2k+1)a}}$

$[a > 0]$ BI(177)(7)a, BI(70)(10)

3.692

1. $\int_0^\infty \left[\sin\left(a - x^2\right) + \cos\left(a - x^2\right)\right] dx = \sqrt{\frac{\pi}{a}} \sin a$ GW(333)(30c), BI(178)(7)a

2. $\int_0^\infty \cos\left(\frac{x^2}{2} - \frac{\pi}{8}\right) \cos ax \, dx = \sqrt{\frac{\pi}{2}} \cos\left(\frac{a^2}{2} - \frac{\pi}{8}\right)$ $[a > 0]$ ET I 24(8)

3. $\int_0^\infty \sin\left[a\left(1 - x^2\right)\right] \cos bx \, dx = -\frac{1}{2}\sqrt{\frac{\pi}{a}} \cos\left(a + \frac{b^2}{4a} + \frac{\pi}{4}\right)$

$[a > 0]$ ET I 23(2)

4. $\int_0^\infty \cos\left[a\left(1 - x^2\right)\right] \cos bx \, dx = \frac{1}{2}\sqrt{\frac{\pi}{a}} \sin\left(a + \frac{b^2}{4a} + \frac{\pi}{4}\right)$

$[a > 0]$ ET I 24(10)

5. $\int_0^\infty \sin\left(ax^2 + \frac{b^2}{a}\right) \cos 2bx \, dx = \int_0^\infty \cos\left(ax^2 + \frac{b^2}{a}\right) \cos 2bx \, dx = \frac{1}{2}\sqrt{\frac{\pi}{2a}}$

$[a > 0]$ BI (70)(19, 20)

6.[8] $\int_{-\infty}^\infty \left[\cos\sqrt{x^2 - 1} - \cos\sqrt{x^2 + 1}\right] dx = \sum_{n=0}^\infty \frac{\pi}{\left\{2^{4n+1}[(2n)!]^2\left(n + \frac{1}{2}\right)\right\}}$

3.693

1. $\int_0^\infty \sin\left(ax^2 + 2bx\right) dx = \sqrt{\frac{\pi}{2a}}\left\{\cos\frac{b^2}{a}\left(\frac{1}{2} - S_2\left(\frac{b^2}{a}\right)\right) - \sin\frac{b^2}{a}\left(\frac{1}{2} - C_2\left(\frac{b^2}{a}\right)\right)\right\}$

$[a > 0]$ BI (70)(3)

2. $\int_0^\infty \cos\left(ax^2 + 2bx\right)\, dx = \sqrt{\dfrac{\pi}{2a}}\left\{\cos\dfrac{b^2}{a}\left(\dfrac{1}{2} - C_2\left(\dfrac{b^2}{a}\right)\right) + \sin\dfrac{b^2}{a}\left(\dfrac{1}{2} - S_2\left(\dfrac{b^2}{a}\right)\right)\right\}$

$$[a > 0] \qquad\qquad \text{BI (70)(4)}$$

3.694

1. $\int_0^\infty \sin\left(ax^2 + 2bx + c\right)\, dx = \sqrt{\dfrac{\pi}{2a}}\cos\dfrac{b^2}{a}\left\{\left(\dfrac{1}{2} - C_2\left(\dfrac{b^2}{a}\right)\right)\sin c + \left(\dfrac{1}{2} - S_2\left(\dfrac{b^2}{a}\right)\right)\cos c\right\}$

$\qquad\qquad\qquad + \sqrt{\dfrac{\pi}{2a}}\sin\dfrac{b^2}{a}\left\{\left(\dfrac{1}{2} - S_2\left(\dfrac{b^2}{a}\right)\right)\sin c - \left(\dfrac{1}{2} - C_2\left(\dfrac{b^2}{a}\right)\right)\cos c\right\}$

$$[a > 0] \qquad\qquad \text{GW (334)(4a)}$$

2. $\int_0^\infty \cos\left(ax^2 + 2bx + c\right)\, dx = \sqrt{\dfrac{\pi}{2a}}\cos\dfrac{b^2}{a}\left\{\left(\dfrac{1}{2} - C_2\left(\dfrac{b^2}{a}\right)\right)\cos c - \left(\dfrac{1}{2} - S_2\left(\dfrac{b^2}{a}\right)\right)\sin c\right\}$

$\qquad\qquad\qquad + \sqrt{\dfrac{\pi}{2a}}\sin\dfrac{b^2}{a}\left\{\left(\dfrac{1}{2} - S_2\left(\dfrac{b^2}{a}\right)\right)\cos c + \left(\dfrac{1}{2} - C_2\left(\dfrac{b^2}{a}\right)\right)\sin c\right\}$

$$[a > 0] \qquad\qquad \text{GW (334)(4b)}$$

3.695

1. $\int_0^\infty \sin\left(a^3 x^3\right)\sin(bx)\, dx = \dfrac{\pi}{6a}\sqrt{\dfrac{b}{3a}}\left\{J_{\frac{1}{3}}\left(\dfrac{2b}{3a}\sqrt{\dfrac{b}{3a}}\right) + J_{-\frac{1}{3}}\left(\dfrac{2b}{3a}\sqrt{\dfrac{b}{3a}}\right) - \dfrac{\sqrt{3}}{\pi}K_{\frac{1}{3}}\left(\dfrac{2b}{3a}\sqrt{\dfrac{b}{3a}}\right)\right\}$

$$[a > 0, \quad b > 0] \qquad\qquad \text{ET I 83(5)}$$

2. $\int_0^\infty \cos\left(a^3 x^3\right)\cos(bx)\, dx = \dfrac{\pi}{6a}\sqrt{\dfrac{b}{3a}}\left\{J_{\frac{1}{3}}\left(\dfrac{2b}{3a}\sqrt{\dfrac{b}{3a}}\right) + J_{-\frac{1}{3}}\left(\dfrac{2b}{3a}\sqrt{\dfrac{b}{3a}}\right) + \dfrac{\sqrt{3}}{\pi}K_{\frac{1}{3}}\left(\dfrac{2b}{3a}\sqrt{\dfrac{b}{3a}}\right)\right\}$

$$[a > 0, \quad b > 0] \qquad\qquad \text{ET I 24(11)}$$

3.696

1. $\int_0^\infty \sin\left(ax^4\right)\sin\left(bx^2\right)\, dx = -\dfrac{\pi}{4}\sqrt{\dfrac{b}{2a}}\sin\left(\dfrac{b^2}{8a} - \dfrac{3}{8}\pi\right)J_{\frac{1}{4}}\left(\dfrac{b^2}{8a}\right)$

$$[a > 0, \quad b > 0] \qquad\qquad \text{ET I 83(2)}$$

2. $\int_0^\infty \sin\left(ax^4\right)\cos\left(bx^2\right)\, dx = -\dfrac{\pi}{4}\sqrt{\dfrac{b}{2a}}\sin\left(\dfrac{b^2}{8a} - \dfrac{\pi}{8}\right)J_{-\frac{1}{4}}\left(\dfrac{b^2}{8a}\right)$

$$[a > 0, \quad b > 0] \qquad\qquad \text{ET I 84(19)}$$

3. $\int_0^\infty \cos\left(ax^4\right)\sin\left(bx^2\right)\, dx = \dfrac{\pi}{4}\sqrt{\dfrac{b}{2a}}\cos\left(\dfrac{b^2}{8a} - \dfrac{3}{8}\pi\right)J_{\frac{1}{4}}\left(\dfrac{b^2}{8a}\right)$

$$[a > 0, \quad b > 0] \qquad \text{ET I 83(4), ET I 25(24)}$$

4. $\int_0^\infty \cos\left(ax^4\right)\cos\left(bx^2\right)\, dx = \dfrac{\pi}{4}\sqrt{\dfrac{b}{2a}}\cos\left(\dfrac{b^2}{8a} - \dfrac{\pi}{8}\right)J_{-\frac{1}{4}}\left(\dfrac{b^2}{8a}\right)$

$$[a > 0, \quad b > 0] \qquad\qquad \text{ET I 25(25)}$$

3.697 $\int_0^\infty \sin\left(\dfrac{a^2}{x}\right)\sin(bx)\, dx = \dfrac{a\pi}{2\sqrt{b}}J_1\left(2a\sqrt{b}\right) \qquad [a > 0, \quad b > 0]$ ET I 83(6)

3.698

1. $\int_0^\infty \sin\left(\dfrac{a^2}{x^2}\right) \sin\left(b^2 x^2\right) \, dx = \dfrac{1}{4b}\sqrt{\dfrac{\pi}{2}} \left[\sin 2ab - \cos 2ab + e^{-2ab}\right]$

$$[a > 0, \quad b > 0]$$ ET I 83(9)

2.[8] $\int_0^\infty \sin\left(\dfrac{a^2}{x^2}\right) \cos\left(b^2 x^2\right) \, dx = \dfrac{1}{4b}\sqrt{\dfrac{\pi}{2}} \left[\sin 2ab + \cos 2ab - e^{-2ab}\right]$ ET I 24(13)

3. $\int_0^\infty \cos\left(\dfrac{a^2}{x^2}\right) \sin\left(b^2 x^2\right) \, dx = \dfrac{1}{4b}\sqrt{\dfrac{\pi}{2}} \left[\sin 2ab + \cos 2ab + e^{-2ab}\right]$

$$[a > 0, \quad b > 0]$$ ET I 84(12)

4. $\int_0^\infty \cos\left(\dfrac{a^2}{x^2}\right) \cos\left(b^2 x^2\right) \, dx = \dfrac{1}{4b}\sqrt{\dfrac{\pi}{2}} \left[\cos 2ab - \sin 2ab + e^{-2ab}\right]$

$$[a > 0, \quad b > 0]$$ ET I 24(14)

3.699

1. $\int_0^\infty \sin\left(a^2 x^2 + \dfrac{b^2}{x^2}\right) \, dx = \dfrac{\sqrt{2\pi}}{4a} \left(\cos 2ab + \sin 2ab\right)$ $\quad [a > 0, \quad b > 0]$ BI (70)(27)

2. $\int_0^\infty \cos\left(a^2 x^2 + \dfrac{b^2}{x^2}\right) \, dx = \dfrac{\sqrt{2\pi}}{4a} \left(\cos 2ab - \sin 2ab\right)$ $\quad [a > 0, \quad b > 0]$ BI (70)(28)

3. $\int_0^\infty \sin\left(a^2 x^2 - 2ab + \dfrac{b^2}{x^2}\right) \, dx = \int_0^\infty \cos\left(a^2 x^2 - 2ab + \dfrac{b^2}{x^2}\right) \, dx = \dfrac{\sqrt{2\pi}}{4a}$

$$[a > 0, \quad b > 0]$$
BI(179)(11, 12)a, ET I 83(6)

4. $\int_0^\infty \sin\left(a^2 x^2 - \dfrac{b^2}{x^2}\right) \, dx = \dfrac{\sqrt{2\pi}}{4a} e^{-2ab}$ $\quad [a > 0, \quad b > 0]$ GW (334)(9b)a

5. $\int_0^\infty \cos\left(a^2 x^2 - \dfrac{b^2}{x^2}\right) \, dx = \dfrac{\sqrt{2\pi}}{4a} e^{-2ab}$ $\quad [a > 0, \quad b > 0]$ GW (334)(9b)a

3.711 $\int_0^u \sin\left(a\sqrt{u^2 - x^2}\right) \cos bx \, dx = \dfrac{\pi a u}{2\sqrt{a^2 + b^2}} J_1\left(u\sqrt{a^2 + b^2}\right)$ $\quad [a > 0, \quad b > 0, \quad u > 0]$

ET I 27(37)

3.712

1. $\int_0^\infty \sin\left(ax^p\right) \, dx = \dfrac{\Gamma\left(\frac{1}{p}\right) \sin\frac{\pi}{2p}}{p a^{\frac{1}{p}}}$ $\quad [a > 0, \quad p > 1]$ EH I 13(40)

2. $\int_0^\infty \cos\left(ax^p\right) \, dx = \dfrac{\Gamma\left(\frac{1}{p}\right) \cos\frac{\pi}{2p}}{p a^{\frac{1}{p}}}$ $\quad [a > 0, \quad p > 1]$ EH I 13(39)

3.713

1. $$\int_0^\infty \sin\left(ax^p + bx^q\right) dx = \frac{1}{p} \sum_{k=0}^\infty \frac{(-b)^k}{k!} a^{-\frac{kq+1}{p}} \Gamma\left(\frac{kq+1}{p}\right) \sin\left[\frac{k(q-p)+1}{2p}\pi\right]$$

$$[a > 0, \quad b > 0, \quad p > 0, \quad q > 0]$$

BI (70)(7)

2. $$\int_0^\infty \cos\left(ax^p + bx^q\right) dx = \frac{1}{p} \sum_{k=0}^\infty \frac{(-b)^k}{k!} a^{-(kq+1)/p} \Gamma\left(\frac{kq+1}{p}\right) \cos\left[\frac{k(q-p)+1}{2p}\pi\right]$$

$$[a > 0, \quad b > 0, \quad p > 0, \quad q > 0]$$

BI (70)(8)

3.714

1. $$\int_0^\infty \cos\left(z \sinh x\right) dx = K_0(z) \qquad\qquad [\operatorname{Re} z > 0] \qquad\qquad \text{WA 202(14)}$$

2. $$\int_0^\infty \sin\left(z \cosh x\right) dx = \frac{\pi}{2} J_0(z) \qquad\qquad [\operatorname{Re} z > 0] \qquad\qquad \text{MO 36}$$

3. $$\int_0^\infty \cos\left(z \cosh x\right) dx = -\frac{\pi}{2} Y_0(z) \qquad\qquad [\operatorname{Re} z > 0] \qquad\qquad \text{MO 37}$$

4. $$\int_0^\infty \cos\left(z \sinh x\right) \cosh \mu x\, dx = \cos\frac{\mu\pi}{2} K_\mu(z) \qquad [\operatorname{Re} z > 0, \quad |\operatorname{Re}\mu| < 1] \qquad \text{WA 202(13)}$$

5. $$\int_0^\pi \cos\left(z \cosh x\right) \sin^{2\mu} x\, dx = \sqrt{\pi}\left(\frac{2}{z}\right)^\mu \Gamma\left(\mu + \frac{1}{2}\right) I_\mu(z)$$

$$\left[\operatorname{Re} z > 0, \quad \operatorname{Re}\mu > -\tfrac{1}{2}\right] \qquad \text{WH}$$

3.715

1. $$\int_0^\pi \sin\left(z \sin x\right) \sin ax\, dx = \sin a\pi\, s_{0,a}(z) = \sin a\pi \sum_{k=1}^\infty \frac{(-1)^{k-1} z^{2k-1}}{(1^2 - a^2)(3^2 - a^2)\ldots\left[(2k-1)^2 - a^2\right]}$$

$$[a > 0] \qquad\qquad \text{WA 338(13)}$$

2. $$\int_0^\pi \sin\left(z \sin x\right) \sin nx\, dx = \frac{1}{2}\int_{-\pi}^\pi \sin\left(z \sin x\right) \sin nx\, dx$$

$$= [1 - (-1)^n]\int_0^{\pi/2} \sin\left(z \sin x\right) \sin nx\, dx = [1 - (-1)^n]\frac{\pi}{2} J_n(z)$$

$$[n = 0, \pm 1, \pm 2, \ldots] \quad \text{WA 30(6), GW(334)(153a)}$$

3. $$\int_0^{\pi/2} \sin\left(z \sin x\right) \sin 2x\, dx = \frac{2}{z^2}\left(\sin z - z \cos z\right) \qquad\qquad \text{LI (43)(14)}$$

4. $$\int_0^\pi \sin\left(z \sin x\right) \cos ax\, dx = (1 + \cos a\pi)\, s_{0,a}(z)$$

$$= (1 + \cos a\pi) \sum_{k=1}^\infty \frac{(-1)^{k-1} z^{2k-1}}{(1^2 - a^2)(3^2 - a^2)\ldots\left[(2k-1)^2 - a^2\right]}$$

$$[a > 0] \qquad\qquad \text{WA 338(14)}$$

5. $\displaystyle\int_0^\pi \sin\left(z\sin x\right)\cos[(2n+1)x]\,dx = 0$ GW (334)(53b)

6. $\displaystyle\int_0^\pi \cos\left(z\sin x\right)\sin ax\,dx = -a\left(1-\cos a\pi\right)s_{-1,a}(z)$

$$= -a\left(1-\cos a\pi\right)\left\{-\frac{1}{a^2}+\sum_{k=1}^\infty \frac{(-1)^{k-1}z^{2k}}{a^2\left(2^2-a^2\right)\left(4^2-a^2\right)\dots\left[(2k)^2-a^2\right]}\right\}$$

$[a>0]$ WA 338(12)

7. $\displaystyle\int_0^\pi \cos\left(z\sin x\right)\sin 2nx\,dx = 0$ GW (334)(54a)

8. $\displaystyle\int_0^\pi \cos\left(z\sin x\right)\cos ax\,dx = -a\sin a\pi\, s_{-1,a}(z)$

$$= -a\sin a\pi\left\{-\frac{1}{a^2}+\sum_{k=1}^\infty \frac{(-1)^{k-1}z^{2k}}{a^2\left(2^2-a^2\right)\left(4^2-a^2\right)\dots\left[(2k)^2-a^2\right]}\right\}$$

$[a>0]$ WA 338(11)

9. $\displaystyle\int_0^\pi \cos\left(z\sin x\right)\cos nx\,dx = \frac{1}{2}\int_{-\pi}^\pi \cos\left(z\sin x\right)\cos nx\,dx$

$$= [1+(-1)^n]\int_0^{\pi/2}\cos\left(z\sin x\right)\cos nx\,dx = [1+(-1)^n]\frac{\pi}{2}J_n(z)$$

GW (334)(54b)

10.[8] $\displaystyle\int_0^{\pi/2}\cos\left(z\sin x\right)\cos^{2n}x\,dx = \frac{\pi}{2}\frac{(2n-1)!!}{z^n}J_n(z)$ $[n=0,1,2,\dots]$ FI II 486, WA 35a

11. $\displaystyle\int_0^{\pi/2}\sin\left(z\cos x\right)\sin 2x\,dx = \frac{2}{z^2}\left(\sin z - z\cos z\right)$ LI (43)(15)

12.[8] $\displaystyle\int_0^{\pi/2}\sin\left(z\cos x\right)\cos ax\,dx = \cos\frac{a\pi}{2}s_{0,a}(z) = \frac{\pi}{4}\operatorname{cosec}\frac{a\pi}{2}\left[\mathbf{J}_a(z)-\mathbf{J}_{-a}(z)\right]$

$$= -\frac{\pi}{4}\sec\frac{a\pi}{4}\left[\mathbf{E}_a(z)+\mathbf{E}_{-a}(z)\right]$$

$$= \cos\frac{a\pi}{2}\sum_{k=1}^\infty \frac{(-1)^{k-1}z^{2k-1}}{\left(1^2-a^2\right)\left(3^2-a^2\right)\dots\left[(2k-1)^2-a^2\right]}$$

$[a>0]$ WA 339

13. $\displaystyle\int_0^\pi \sin\left(z\cos x\right)\cos nx\,dx = \frac{1}{2}\int_{-\pi}^\pi \sin\left(z\cos x\right)\cos nx\,dx = \pi\sin\frac{n\pi}{2}J_n(z)$ GW (334)(55b)

14. $\displaystyle\int_0^{\pi/2}\sin\left(z\cos x\right)\cos[(2n+1)x]\,dx = (-1)^n\frac{\pi}{2}J_{2n+1}(z)$ WA 30(8)

15. $\displaystyle\int_0^{\pi/2}\sin\left(a\cos x\right)\operatorname{tg}x\,dx = \operatorname{si}(a)+\frac{\pi}{2}$ $[a>0]$ BI (43)(17)

16. $\displaystyle\int_0^{\pi/2}\sin\left(z\cos x\right)\sin^{2\nu}x\,dx = \frac{\sqrt{\pi}}{2}\left(\frac{2}{z}\right)^\nu \Gamma\left(\nu+\frac{1}{2}\right)\mathbf{H}_\nu(z)$

$\left[\operatorname{Re}\nu>-\tfrac{1}{2}\right]$ WA 358(1)

$17.^7$ $\displaystyle\int_0^{\pi/2} \cos\left(z\cos x\right)\cos ax\, dx = -a\sin\frac{a\pi}{2}\, s_{-1,a}(z)$

$$= \frac{\pi}{4}\sec\frac{a\pi}{2}\left[\mathbf{J}_a(z) + \mathbf{J}_{-a}(z)\right] = \frac{\pi}{4}\operatorname{cosec}\frac{a\pi}{2}\left[\mathbf{E}_a(z) - \mathbf{E}_{-a}(z)\right]$$

$$= -a\sin\frac{a\pi}{2}\left\{-\frac{1}{a^2} + \sum_{k=1}^{\infty}\frac{(-1)^{k-1}z^{2k}}{a^2\left(2^2 - a^2\right)\left(4^2 - a^2\right)\ldots\left[(2k)^2 - a^2\right]}\right\}$$

$$[a > 0] \qquad\qquad\qquad \text{WA 339}$$

18. $\displaystyle\int_0^{\pi}\cos\left(z\cos x\right)\cos nx\, dx = \frac{1}{2}\int_{-\pi}^{\pi}\cos\left(z\cos x\right)\cos nx\, dx = \pi\cos\frac{n\pi}{2}\, J_n(z)$ GW (334)(56b)

19. $\displaystyle\int_0^{\pi/2}\cos\left(z\cos x\right)\cos 2nx\, dx = (-1)^n\cdot\frac{\pi}{2}\, J_{2n}(z)$ WA 30(9)

20. $\displaystyle\int_0^{\pi/2}\cos\left(z\cos x\right)\sin^{2\nu} x\, dx = \frac{\sqrt{\pi}}{2}\left(\frac{2}{z}\right)^{\nu}\Gamma\left(\nu + \frac{1}{2}\right)J_{\nu}(z)$

$$\left[\operatorname{Re}\nu > -\tfrac{1}{2}\right] \qquad\qquad \text{WA 35, WH}$$

21. $\displaystyle\int_0^{\pi}\cos\left(z\cos x\right)\sin^{2\mu} x\, dx = \sqrt{\pi}\left(\frac{2}{z}\right)^{\mu}\Gamma\left(\mu + \frac{1}{2}\right)J_{\mu}(z)$

$$\left[\operatorname{Re}\mu > -\tfrac{1}{2}\right] \qquad\qquad\qquad \text{WH}$$

3.716

1. $\displaystyle\int_0^{\pi/2}\sin\left(a\tan x\right)dx = \frac{1}{2}\left[e^{-a}\overline{\operatorname{Ei}(a)} - e^a\operatorname{Ei}(-a)\right]$ $[a > 0]$ (cf. **3.723** 1)   BI (43)(1)

2. $\displaystyle\int_0^{\pi/2}\cos\left(a\tan x\right)dx = \frac{\pi}{2}e^{-a}$ $[a \geq 0]$ BI (43)(2)

3. $\displaystyle\int_0^{\pi/2}\sin\left(a\tan x\right)\sin 2x\, dx = \frac{a\pi}{2}e^{-a}$ $[a \geq 0]$ BI (43)(7)

4. $\displaystyle\int_0^{\pi/2}\cos\left(a\tan x\right)\sin^2 x\, dx = \frac{1-a}{4}\pi e^{-a}$ $[a \geq 0]$ BI (43)(8)

5. $\displaystyle\int_0^{\pi/2}\cos\left(a\tan x\right)\cos^2 x\, dx = \frac{1+a}{4}\pi e^{-a}$ $[a \geq 0]$ BI (43)(9)

6. $\displaystyle\int_0^{\pi/2}\sin\left(a\tan x\right)\tan x\, dx = \frac{\pi}{2}e^{-a}$ $[a > 0]$ BI (43)(5)

7. $\displaystyle\int_0^{\pi/2}\cos\left(a\tan x\right)\tan x\, dx = -\frac{1}{2}\left[e^{-a}\overline{\operatorname{Ei}(a)} + e^a\operatorname{Ei}(-a)\right]$

$$[a > 0] \qquad\qquad (\text{cf. } \mathbf{3.723}\ 5) \qquad\qquad \text{BI (43)(6)}$$

8. $\displaystyle\int_0^{\pi/2}\sin\left(a\tan x\right)\sin^2 x\tan x\, dx = \frac{2-a}{4}\pi e^{-a}$ $[a > 0]$ BI (43)(11)

9. $\displaystyle\int_0^{\pi/2}\sin^2\left(a\tan x\right)dx = \frac{\pi}{4}\left(1 - e^{-2a}\right)$ $[a \geq 0]$ (cf. **3.742** 1)   BI (43)(3)

10. $\displaystyle\int_0^{\pi/2} \cos^2\left(a\tan x\right) dx = \frac{\pi}{4}\left(1 + e^{-2a}\right)$ $\qquad$ $[a \geq 0]$ $\qquad$ (cf. **3.742** 3) $\qquad$ BI (43)(4)

11. $\displaystyle\int_0^{\pi/2} \sin^2\left(a\tan x\right)\cot^2 x \, dx = \frac{\pi}{4}\left(e^{-2a} + 2a - 1\right)$ $\qquad$ $[a \geq 0]$ $\qquad$ BI (43)(19)

12. $\displaystyle\int_0^{\pi/2}\left[1 - \sec^2 x \cos\left(\tan x\right)\right]\frac{dx}{\tan x} = C$ $\qquad$ BI (51)(14)

13. $\displaystyle\int_0^{\pi/2} \sin\left(a\cot x\right)\sin 2x \, dx = \frac{a\pi}{2}e^{-a}$ $\qquad$ $[a \geq 0]$ $\qquad$ (cf. **3.716** 3.)

and in general, formulas **3.716** remain valid if we replace $\tan x$ in the argument of the sine or cosine with $\cot x$ if we also replace $\sin x$ with $\cos x$, $\cos x$ with $\sin x$, hence $\tan x$ with $\cot x$, $\cot x$ with $\tan x$, $\sec x$ with $\operatorname{cosec} x$, and $\operatorname{cosec} x$ with $\sec x$ in the factors. Analogously,

3.717 $\displaystyle\int_0^{\pi/2}\sin\left(a\operatorname{cosec} x\right)\sin\left(a\cot x\right)\frac{dx}{\cos x} = \int_0^{\pi/2}\sin\left(a\sec x\right)\sin\left(a\tan x\right)\frac{dx}{\sin x} = \frac{\pi}{2}\sin a$ $\qquad$ $[a \geq 0]$

$\qquad$ BI (52)(11, 12)

3.718

1. $\displaystyle\int_0^{\pi/2}\sin\left(\frac{\pi}{2}p - a\tan x\right)\tan^{p-1} x \, dx = \int_0^{\pi/2}\cos\left(\frac{\pi}{2}p - a\tan x\right)\tan^p x \, dx = \frac{\pi}{2}e^{-a}$

$\qquad$ $\left[p^2 < 1, \quad p \neq 0, \quad a \geq 0\right]$ $\qquad$ BI (44)(5, 6)

2. $\displaystyle\int_0^{\pi/2}\sin\left(a\tan x - \nu x\right)\sin^{\nu-2} x \, dx = 0$ $\qquad$ $[\operatorname{Re}\nu > 0, \quad a > 0]$ $\qquad$ NH 157(15)

3. $\displaystyle\int_0^{\pi/2}\sin\left(n\tan x + \nu x\right)\frac{\cos^{\nu-1} x}{\sin x}\, dx = \frac{\pi}{2}$ $\qquad$ $[\operatorname{Re}\nu > 0]$ $\qquad$ BI (51)(15)

4. $\displaystyle\int_0^{\pi/2}\cos\left(a\tan x - \nu x\right)\cos^{\nu-2} x \, dx = \frac{\pi e^{-a}a^{\nu-1}}{\Gamma(\nu)}$ $\qquad$ $[\operatorname{Re}\nu > 1, \quad a > 0]$

$\qquad$ LO V 153(112), NT 157(14)

5. $\displaystyle\int_0^{\pi/2}\cos\left(a\tan x + \nu x\right)\cos^{\nu} x \, dx = 2^{-\nu-1}\pi e^{-a}$ $\qquad$ $[\operatorname{Re}\nu > -1, \quad a \geq 0]$ $\qquad$ BI (44)(4)

6. $\displaystyle\int_0^{\pi/2}\cos\left(a\tan x - \gamma x\right)\cos^{\nu} x \, dx = \frac{\pi a^{\frac{\nu}{2}}}{2^{\frac{\nu}{2}+1}}\cdot\frac{W_{\frac{\gamma}{2},-\frac{\nu+1}{2}}(2a)}{\Gamma\left(1 + \frac{\gamma+\nu}{2}\right)}$

$\qquad$ $\left[a > 0, \quad \operatorname{Re}\nu > -1, \quad \frac{\nu+\gamma}{2} \neq -1, -2, \ldots\right]$ $\qquad$ EH I 274(13)a

7. $\displaystyle\int_0^{\pi/2}\frac{\sin nx - \sin\left(nx - a\tan x\right)}{\sin x}\cos^{n-1} x \, dx = \begin{cases}\pi/2 & [n = 0, \quad a > 0],\\ \pi\left(1 - e^{-a}\right) & [n = 1, \quad a \geq 0]\end{cases}$

$\qquad$ LO V 153(114)

3.719

1.[6] $\displaystyle\int_0^{\pi}\sin\left(\nu x - z\sin x\right) dx = \pi\,\mathbf{E}_{\nu}(z)$ $\qquad$ WA 336(2)

2. $\displaystyle\int_0^\pi \cos(nx - z\sin x)\, dx = \pi\, J_n(z)$ WH

3. $\displaystyle\int_0^\pi \cos(\nu x - z\sin x)\, dx = \pi\, \mathbf{J}_\nu(z)$ WA 336(1)

3.72–3.74 Combinations of trigonometric and rational functions

3.721

1. $\displaystyle\int_0^\infty \frac{\sin(ax)}{x}\, dx = \frac{\pi}{2}\,\mathrm{sign}\,a$ FI II 645

2. $\displaystyle\int_1^\infty \frac{\sin(ax)}{x}\, dx = -\,\mathrm{si}(a)$ BI 203(1)

3.[8] $\displaystyle\int_1^\infty \frac{\cos(ax)}{x}\, dx = -\,\mathrm{ci}(a)$ BI 203(5)

3.722

1. $\displaystyle\int_0^\infty \frac{\sin(ax)}{x+\beta}\, dx = \mathrm{ci}(a\beta)\sin(a\beta) - \cos(a\beta)\,\mathrm{si}(a\beta)$ $[|\arg\beta| < \pi, \quad a > 0]$

         BI(16)(1), FI II 646a

2.[10] $\displaystyle\int_{-\infty}^\infty \frac{\sin(ax)}{x+\beta}\, dx = \pi e^{iab}$ $[a > 0, \quad \mathrm{Im}\,\beta > 0]$

3. $\displaystyle\int_0^\infty \frac{\cos(ax)}{x+\beta}\, dx = -\sin(a\beta)\,\mathrm{si}(a\beta) - \cos(a\beta)\,\mathrm{ci}(a\beta)$ $[|\arg\beta| < \pi, \quad a > 0]$

         ET I 8(7), BI(160)(2)

4.[8] $\displaystyle\int_{-\infty}^\infty \frac{\cos(ax)}{x+\beta}\, dx = -i\pi e^{ia\beta}$ $[a > 0, \quad \mathrm{Im}\,\beta > 0]$

5.[10] $\displaystyle\int_0^\infty \frac{\sin(ax)}{\beta - x}\, dx = \sin(\beta a)\,\mathrm{ci}(\beta a) - \cos(\beta a)\,[\mathrm{si}(\beta a) + \pi]$

         $[a > 0, \quad \beta \text{ not real and positive}]$
         FI II 646, BI(161)(1)

6.[8] $\displaystyle\int_{-\infty}^\infty \frac{\sin(ax)}{\beta - x}\, dx = -\pi e^{ia\beta}$ $[a > 0, \quad \mathrm{Im}\,\beta > 0]$

7.[10] $\displaystyle\int_0^\infty \frac{\cos(ax)}{\beta - x}\, dx = -\cos(a\beta)\,\mathrm{ci}(a\beta) + \sin(a\beta)\,[\mathrm{si}(a\beta) + \pi]$

         $[a > 0, \quad \beta \text{ not real and positive}]$
         ET I 8(8), BI(161)(2)a

8.[7] $\displaystyle\int_{-\infty}^\infty \frac{\cos(ax)}{\beta - x}\, dx = -\pi e^{iab}$ $[a > 0, \quad \mathrm{Im}\,\beta > 0]$

3.723

1.[7] $\displaystyle\int_0^\infty \frac{\sin(ax)}{\beta^2 + x^2}\, dx = \frac{1}{2\beta}\left[e^{-a\beta}\overline{\mathrm{Ei}(a\beta)} - e^{a\beta}\,\mathrm{Ei}(-a\beta)\right]$ $[a > 0, \quad \beta > 0]$ ET I 65(14), BI(160)(3)

2. $$\int_0^\infty \frac{\cos(ax)}{\beta^2 + x^2} \, dx = \frac{\pi}{2\beta} e^{-a\beta}$$

    $[a \geq 0, \quad \operatorname{Re}\beta > 0]$

    FI II 741, 750, ET I 8(11), WH

3. $$\int_0^\infty \frac{x\sin(ax)}{\beta^2 + x^2} \, dx = \frac{\pi}{2} e^{-a\beta}$$

    $[a > 0, \quad \operatorname{Re}\beta > 0]$

    FI II 741, 750, ET I 65(15), WH

4. $$\int_{-\infty}^\infty \frac{x\sin(ax)}{\beta^2 + x^2} \, dx = \pi e^{-a\beta}$$

    $[a > 0, \quad \operatorname{Re}\beta > 0]$ BI (202)(10)

5.[7] $$\int_0^\infty \frac{x\cos(ax)}{\beta^2 + x^2} \, dx = -\frac{1}{2}\left[e^{-a\beta}\overline{\operatorname{Ei}(a\beta)} + e^{a\beta}\operatorname{Ei}(-a\beta) \right]$$

    $[a > 0, \quad \beta > 0]$ BI (160)(6)

6. $$\int_{-\infty}^\infty \frac{\sin[a(b - x)]}{c^2 + x^2} \, dx = \frac{\pi}{c} e^{-ac}\sin(ab)$$

    $[a > 0, \quad b > 0, \quad c > 0]$ LI (202)(9)

7. $$\int_{-\infty}^\infty \frac{\cos[a(b - x)]}{c^2 + x^2} \, dx = \frac{\pi}{c} e^{-ac}\cos(ab)$$

    $[a > 0, \quad b > 0, \quad c > 0]$ LI (202)(11)a

8. $$\int_0^\infty \frac{\sin(ax)}{\beta^2 - x^2} \, dx = \frac{1}{\beta}\left[\sin(a\beta)\operatorname{ci}(a\beta) - \cos(a\beta)\left(\operatorname{si}(a\beta) + \frac{\pi}{2}\right) \right]$$

    $[|\arg\beta| < \pi, \quad a > 0]$ BI (161)(3)

9. $$\int_0^\infty \frac{\cos(ax)}{b^2 - x^2} \, dx = \frac{\pi}{2b}\sin(ab)$$

    $[a > 0, \quad b > 0]$ BI(161)(5), ET I 9(15)

10. $$\int_0^\infty \frac{x\sin(ax)}{b^2 - x^2} \, dx = -\frac{\pi}{2}\cos(ab)$$

    $[a > 0]$ FI II 647, ET II 252(45)

11. $$\int_0^\infty \frac{x\cos(ax)}{\beta^2 + x^2} \, dx = \cos(a\beta)\operatorname{ci}(a\beta) + \sin(a\beta)\left[\operatorname{si}(a\beta) + \frac{\pi}{2}\right]$$

    $[|\arg\beta| < \pi, \quad a > 0]$ BI (161)(6)

12. $$\int_{-\infty}^\infty \frac{\sin(ax)}{x(x - b)} \, dx = \pi\frac{\cos(ab) - 1}{b}$$

    $[a > 0, \quad b > 0]$ ET II 252(44)

3.724

1. $$\int_{-\infty}^\infty \frac{b + cx}{p + 2qx + x^2} \sin(ax) \, dx = \left(\frac{cq - b}{\sqrt{p - q^2}}\sin(aq) + c\cos(aq) \right) \pi e^{-a\sqrt{p - q^2}}$$

    $[a > 0, \quad p > q^2]$ BI (202)(12)

2. $$\int_{-\infty}^\infty \frac{b + cx}{p + 2qx + x^2} \cos(ax) \, dx = \left(\frac{b - cq}{\sqrt{p - q^2}}\cos(aq) + c\sin(aq) \right) \pi e^{-a\sqrt{p - q^2}}$$

    $[a > 0, \quad p > q^2]$ BI (202)(13)

3. $$\int_{-\infty}^\infty \frac{\cos[(b - 1)t] - x\cos(bt)}{1 - 2x\cos t + x^2} \cos(ax) \, dx = \pi e^{-a\sin t}\sin(bt + a\cos t)$$

    $[a > 0, \quad t^2 < \pi^2]$ BI (202)(14)

3.725

1. $\displaystyle\int_0^\infty \frac{\sin(ax)\,dx}{x\left(\beta^2 + x^2\right)} = \frac{\pi}{2\beta^2}\left(1 - e^{-a\beta}\right)$ $[\operatorname{Re}\beta > 0, \quad a > 0]$ BI (172)(1)

2. $\displaystyle\int_0^\infty \frac{\sin(ax)\,dx}{x\left(b^2 - x^2\right)} = \frac{\pi}{2b^2}\left(1 - \cos(ab)\right)$ $[a > 0]$ BI (172)(4)

3. $\displaystyle\int_0^\infty \frac{\sin(ax)\cos(bx)}{x\left(x^2 + \beta^2\right)}\,dx = \frac{\pi}{2\beta^2}e^{-\beta b}\sinh(a\beta)$ $[0 < a < b]$

         $= -\dfrac{\pi}{2\beta^2}e^{-a\beta}\cosh(b\beta) + \dfrac{\pi}{2\beta^2}$ $[a > b > 0]$

                                                     ET I 19(4)

3.726

1.[7] $\displaystyle\int_0^\infty \frac{x\sin(ax)\,dx}{b^3 \pm b^2 x + b x^2 \pm x^3}$

         $= \pm\dfrac{1}{4b}\left[e^{-ab}\overline{\operatorname{Ei}(ab)} - e^{ab}\operatorname{Ei}(-ab) - 2\operatorname{ci}(ab)\sin(ab) + 2\cos(ab)\left(\operatorname{si}(ab) + \dfrac{\pi}{2}\right)\right]$

         $+ \dfrac{\pi e^{-ab} - \pi\cos(ab)}{4b}$

$[a > 0, \quad b > 0; \quad$ if the lower sign is taken, the above expression indicates the principal value$]$

                                       ET I 65(21)a, BI(176)(10, 13)

2.[7] $\displaystyle\int_0^\infty \frac{x^2\sin(ax)\,dx}{b^3 \pm b^2 x + b x^2 \pm x^3}$

         $= \dfrac{1}{4}\left[e^{ab}\operatorname{Ei}(-ab) - e^{-ab}\overline{\operatorname{Ei}(ab)} + 2\operatorname{ci}(ab)\sin(ab) - 2\cos(ab)\left(\operatorname{si}(ab) + \dfrac{\pi}{2}\right)\right]$

         $\pm\pi\left(e^{-ab} + \cos(ab)\right)$

$[a > 0, \quad b > 0; \quad$ if the lower sign is taken, the above expression indicates the principal value$]$

                                       ET I 66(22), BI(176)(11, 14)

3.727

1. $\displaystyle\int_0^\infty \frac{\cos(ax)}{b^4 + x^4}\,dx = \frac{\pi\sqrt{2}}{4b^3}\exp\left(-\frac{ab}{\sqrt{2}}\right)\left(\cos\frac{ab}{\sqrt{2}} + \sin\frac{ab}{\sqrt{2}}\right)$

                                 $[a > 0, \quad b > 0]$ BI(160)(25)a, ET I 9(19)

2.[8] $\displaystyle\int_0^\infty \frac{\sin(ax)}{b^4 - x^4}\,dx = \frac{1}{4b^3}\left[2\sin(ab)\operatorname{ci}(ab) - 2\cos(ab)\left(\operatorname{si}(ab) + \frac{\pi}{2}\right)\right.$

         $\left. + e^{-ab}\operatorname{Ei}(ab) - e^{ab}\operatorname{Ei}(-ab)\right]$

                                 $[a > 0, \quad b > 0]$ BI (161)(12)

3. $\displaystyle\int_0^\infty \frac{\cos(ax)}{b^4 - x^4}\,dx = \frac{\pi}{4b^3}\left[e^{-ab} + \sin(ab)\right]$

                 $[a > 0, \quad b > 0]$ (cf. **3.723** 2 and **3.723** 9)   BI (161)(16)

4. $\displaystyle\int_0^\infty \frac{x\sin(ax)}{b^4 + x^4}\,dx = \frac{\pi}{2b^2}\exp\left(-\frac{ab}{\sqrt{2}}\right)\sin\frac{ab}{\sqrt{2}}$ $[a > 0, \quad b > 0]$ BI (160)(23)a

5. $\displaystyle\int_0^\infty \frac{x\sin(ax)}{b^4 - x^4}\,dx = \frac{\pi}{4b^2}\left[e^{-ab} - \cos(ab)\right]$ $[a > 0, \quad b > 0]$ BI (161)(13)

6.7 $\displaystyle\int_0^\infty \frac{x\cos(ax)}{b^4-x^4}\,dx = \frac{1}{4b^2}\left[2\cos(ab)\,\mathrm{ci}(ab)+2\sin(ab)\left(\mathrm{si}(ab)+\frac{\pi}{2}\right)\right.$

$$\left.-\,e^{-ab}\overline{\mathrm{Ei}(ab)}-e^{ab}\,\mathrm{Ei}(-ab)\right]$$

$$[a>0,\quad b>0]\qquad (\text{cf. }\textbf{3.723}\ 5\text{ and }\textbf{3.723}\ 11)\quad \text{BI (161)(17)}$$

7. $\displaystyle\int_0^\infty \frac{x^2\cos(ax)}{b^4+x^4}\,dx = \frac{\pi\sqrt{2}}{4b}\exp\left(-\frac{ab}{\sqrt{2}}\right)\left(\cos\frac{ab}{\sqrt{2}}-\sin\frac{ab}{\sqrt{2}}\right)$

$$[a>0,\quad b>0]\qquad\qquad \text{BI (160)(26)a}$$

8.7 $\displaystyle\int_0^\infty \frac{x^2\sin(ax)\,dx}{b^4-x^4} = \frac{1}{4b}\left[2\sin(ab)\,\mathrm{ci}(ab)\right.$

$$\left.-2\cos(ab)\left(\mathrm{si}(ab)+\frac{\pi}{2}\right)-e^{-ab}\overline{\mathrm{Ei}(ab)}+e^{ab}\,\mathrm{Ei}(-ab)\right]$$

$$[a>0,\quad b>0]\qquad\qquad \text{BI (161)(14)}$$

9. $\displaystyle\int_0^\infty \frac{x^2\cos(ax)}{b^4-x^4}\,dx = \frac{\pi}{4b}\left(\sin(ab)-e^{-ab}\right)$ $[a>0,\quad b>0]$ BI (161)(18)

10. $\displaystyle\int_0^\infty \frac{x^3\sin(ax)}{b^4+x^4}\,dx = \frac{\pi}{2}\exp\left(-\frac{ab}{\sqrt{2}}\right)\cos\frac{ab}{\sqrt{2}}$ $[a>0,\quad b>0]$ BI (160)(24)

11. $\displaystyle\int_0^\infty \frac{x^3\sin(ax)}{b^4-x^4}\,dx = \frac{-\pi}{4}\left[e^{-ab}-\cos(ab)\right]$ $[a>0,\quad b>0]$ BI (161)(15)

12.7 $\displaystyle\int_0^\infty \frac{x^3\cos(ax)\,dx}{b^4-x^4} = \frac{1}{4}\left[2\cos(ab)\,\mathrm{ci}(ab)+2\sin(ab)\left(\mathrm{si}(ab)+\frac{\pi}{2}\right)\right.$

$$\left.+\,e^{-ab}\overline{\mathrm{Ei}(ab)}+e^{ab}\,\mathrm{Ei}(-ab)\right]$$

$$[a>0,\quad b>0]\qquad\qquad \text{BI(161)(19)}$$

13.* $\displaystyle\int_0^\infty \frac{x^3\sin ax}{(x^2+b^2)^3}\,dx = \frac{\pi e^{-ab}}{16b}\left(3a-ba^2\right)$

14.* $\displaystyle\int_0^\infty \frac{x^3\sin ax}{(x^2+b^2)^4}\,dx = \frac{\pi e^{-ab}a}{96b^3}\left(3+3ab-a^2b^2\right)$

3.728

1. $\displaystyle\int_0^\infty \frac{\cos(ax)\,dx}{(\beta^2+x^2)(\gamma^2+x^2)} = \frac{\pi\left(\beta e^{-a\gamma}-\gamma e^{-a\beta}\right)}{2\beta\gamma\left(\beta^2-\gamma^2\right)}$ $[a>0,\quad \mathrm{Re}\,\beta>0,\quad \mathrm{Re}\,\gamma>0]$

$$\text{BI (175)(1)}$$

2. $\displaystyle\int_0^\infty \frac{x\sin(ax)\,dx}{(\beta^2+x^2)(\gamma^2+x^2)} = \frac{\pi\left(e^{-a\beta}-e^{-a\gamma}\right)}{2\left(\gamma^2-\beta^2\right)}$ $[a>0,\quad \mathrm{Re}\,\beta>0,\quad \mathrm{Re}\,\gamma>0]$

$$\text{BI (174)(1)}$$

3. $\displaystyle\int_0^\infty \frac{x^2\cos(ax)\,dx}{(\beta^2+x^2)(\gamma^2+x^2)} = \frac{\pi\left(\beta e^{-a\beta}-\gamma e^{-a\gamma}\right)}{2\left(\beta^2-\gamma^2\right)}$ $[a>0,\quad \mathrm{Re}\,\beta>0,\quad \mathrm{Re}\,\gamma>0]$

$$\text{BI (175)(2)}$$

4. $\displaystyle\int_0^\infty \frac{x^3 \sin(ax)\,dx}{(\beta^2 + x^2)\,(\gamma^2 + x^2)} = \frac{\pi\,\left(\beta^2 e^{-a\beta} - \gamma^2 e^{-a\gamma}\right)}{2\,(\beta^2 - \gamma^2)}$ $[a > 0, \quad \mathrm{Re}\,\beta > 0, \quad \mathrm{Re}\,\gamma > 0]$

BI (174)(2)

5. $\displaystyle\int_0^\infty \frac{\cos(ax)\,dx}{(b^2 - x^2)\,(c^2 - x^2)} = \frac{\pi\,(b\sin(ac) - c\sin(ab))}{2bc\,(b^2 - c^2)}$ $[a > 0, \quad b > 0, \quad c > 0]$ BI (175)(3)

6. $\displaystyle\int_0^\infty \frac{x \sin(ax)\,dx}{(b^2 - x^2)\,(c^2 - x^2)} = \frac{\pi\,(\cos(ab) - \cos(ac))}{2\,(b^2 - c^2)}$ $[a > 0]$ BI (174)(3)

7. $\displaystyle\int_0^\infty \frac{x^2 \cos(ax)\,dx}{(b^2 - x^2)\,(c^2 - x^2)} = \frac{\pi\,(c\sin(ac) - b\sin(ab))}{2\,(b^2 - c^2)}$ $[a > 0, \quad b > 0, \quad c > 0]$ BI (175)(4)

8. $\displaystyle\int_0^\infty \frac{x^3 \sin(ax)\,dx}{(b^2 - x^2)\,(c^2 - x^2)} = \frac{\pi\,\left(b^2 \cos(ab) - c^2 \cos(ac)\right)}{2\,(b^2 - c^2)}$ $[a > 0, \quad b > 0, \quad c > 0]$ BI (174)(4)

9.* $\displaystyle\int_0^\infty \frac{x \sin ax}{(b^2 - x^2)\,(c^2 + x^2)}\,dx = \frac{\pi}{2}\frac{e^{-ac} - \cos ba}{a^2 + c^2}$ $[a > 0, \quad c > 0, \quad b \text{ real}]$

3.729

1. $\displaystyle\int_0^\infty \frac{\cos(ax)\,dx}{(b^2 + x^2)^2} = \frac{\pi}{4b^3}(1 + ab)e^{-ab}$ $[a > 0, \quad b > 0]$ BI (170)(7)

2. $\displaystyle\int_0^\infty \frac{x \sin(ax)\,dx}{(b^2 + x^2)^2} = \frac{\pi}{4b}ae^{-ab}$ $[a > 0, \quad b > 0]$ BI (170)(3)

3. $\displaystyle\int_0^\infty \cos(px)\frac{1 - x^2}{(1 + x^2)^2}\,dx = \frac{\pi p}{2}e^{-p}$ BI (43)(10)a

4. $\displaystyle\int_0^\infty \frac{x^3 \sin(ax)\,dx}{(b^2 + x^2)^2} = \frac{\pi}{4}(2 - ab)e^{-ab}$ $[a > 0, \quad b > 0]$ BI (170)(4)

3.731   **Notation:** $2A^2 = \sqrt{b^4 + c^2} + b^2, \quad 2B^2 = \sqrt{b^4 + c^2} - b^2,$

1. $\displaystyle\int_0^\infty \frac{\cos(ax)\,dx}{(x^2 + b^2)^2 + c^2} = \frac{\pi}{2c}\frac{e^{-aA}\,(B\cos(aB) + A\sin(aB))}{\sqrt{b^4 + c^2}}$

$[a > 0, \quad b > 0, \quad c > 0]$ BI (176)(3)

2. $\displaystyle\int_0^\infty \frac{x \sin(ax)\,dx}{(x^2 + b^2)^2 + c^2} = \frac{\pi}{2c}e^{-aA}\sin(aB)$ $[a > 0, \quad b > 0, \quad c > 0]$ BI (176)(1)

3. $\displaystyle\int_0^\infty \frac{\left(x^2 + b^2\right)\cos(ax)\,dx}{(x^2 + b^2)^2 + c^2} = \frac{\pi}{2}\frac{e^{-aA}\,(A\cos(aB) - B\sin(aB))}{\sqrt{b^4 + c^2}}$

$[a > 0, \quad b > 0, \quad c > 0]$ BI (176)(4)

4. $\displaystyle\int_0^\infty \frac{x\left(x^2 + b^2\right)\sin(ax)\,dx}{(x^2 + b^2)^2 + c^2} = \frac{\pi}{2}e^{-aA}\cos(aB)$ $[a > 0, \quad b > 0, \quad c > 0]$ BI (176)(2)

3.732

1. $\displaystyle\int_0^\infty \left[\frac{1}{\beta^2 + (\gamma - x)^2} - \frac{1}{\beta^2 + (\gamma + x)^2} \right] \sin(ax)\,dx = \frac{\pi}{\beta} e^{-a\beta} \sin(a\gamma)$

$\qquad\qquad\qquad\qquad\qquad\qquad [a > 0, \quad \operatorname{Re}\beta > 0, \quad \gamma + i\beta \text{ is not real}]$

$\qquad\qquad\qquad\qquad\qquad\qquad\qquad\qquad\qquad$ ET I 65(16)

2. $\displaystyle\int_0^\infty \left[\frac{1}{\beta^2 + (\gamma - x)^2} + \frac{1}{\beta^2 + (\gamma + x)^2} \right] \cos(ax)\,dx = \frac{\pi}{\beta} e^{-a\beta} \cos(a\gamma)$

$\qquad\qquad\qquad\qquad\qquad\qquad [a > 0, \quad |\operatorname{Im}\gamma| < \operatorname{Re}\beta] \qquad$ ET I 8(13)

3. $\displaystyle\int_0^\infty \left[\frac{\gamma + x}{\beta^2 + (\gamma + x)^2} - \frac{\gamma - x}{\beta^2 + (\gamma - x)^2} \right] \sin(ax)\,dx = \pi e^{-a\beta} \cos(a\gamma)$

$\qquad\qquad\qquad\qquad\qquad\qquad [a > 0, \quad \operatorname{Re}\beta > 0, \quad \gamma + i\beta \text{ is not real}]$

$\qquad\qquad\qquad\qquad\qquad\qquad\qquad\qquad\qquad$ LI (175)(17)

4. $\displaystyle\int_0^\infty \left[\frac{\gamma + x}{\beta^2 + (\gamma + x)^2} + \frac{\gamma - x}{\beta^2 + (\gamma - x)^2} \right] \cos(ax)\,dx = \pi e^{-a\beta} \sin(a\gamma)$

$\qquad\qquad\qquad\qquad\qquad\qquad [a > 0, \quad |\operatorname{Im} a| < \operatorname{Re}\beta] \qquad$ LI (176)(21)

3.733

1. $\displaystyle\int_0^\infty \frac{\cos(ax)\,dx}{x^4 + 2b^2 x^2 \cos 2t + b^4} = \frac{\pi}{2b^3} \exp(-ab\cos t) \frac{\sin(t + ab\sin t)}{\sin 2t}$

$\qquad\qquad\qquad\qquad\qquad\qquad \left[a > 0, \quad b > 0, \quad |t| < \frac{\pi}{2} \right] \qquad$ BI (176)(7)

2. $\displaystyle\int_0^\infty \frac{x \sin(ax)\,dx}{x^4 + 2b^2 x^2 \cos 2t + b^4} = \frac{\pi}{2b^2} \exp(-ab\cos t) \frac{\sin(ab\sin t)}{\sin 2t}$

$\qquad\qquad\qquad\qquad\qquad\qquad \left[a > 0, \quad b > 0, \quad |t| < \frac{\pi}{2} \right]$

$\qquad\qquad\qquad\qquad\qquad\qquad\qquad\qquad$ BI(176)(5), ET I 66(23)

3. $\displaystyle\int_0^\infty \frac{x^2 \cos(ax)\,dx}{x^4 + 2b^2 x^2 \cos 2t + b^4} = \frac{\pi}{2b} \exp(-ab\cos t) \frac{\sin(t - ab\sin t)}{\sin 2t}$

$\qquad\qquad\qquad\qquad\qquad\qquad \left[a > 0, \quad b > 0, \quad |t| < \frac{\pi}{2} \right] \qquad$ BI (176)(8)

4. $\displaystyle\int_0^\infty \frac{x^3 \sin(ax)\,dx}{x^4 + 2b^2 x^2 \cos 2t + b^4} = \frac{\pi}{2} \exp(-ab\cos t) \frac{\sin(2t - ab\sin t)}{\sin 2t}$

$\qquad\qquad\qquad\qquad\qquad\qquad \left[a > 0, \quad b > 0, \quad |t| < \frac{\pi}{2} \right] \qquad$ BI (176)(6)

5. $\displaystyle\int_0^\infty \frac{\sin(ax)\,dx}{x(x^4 + 2b^2 x^2 \cos 2t + b^4)} = \frac{\pi}{2b^4} \left[1 - \exp(-ab\cos t) \frac{\sin(2t + ab\sin t)}{\sin 2t} \right]$

$\qquad\qquad\qquad\qquad\qquad\qquad \left[a > 0, \quad b > 0, \quad |t| < \frac{\pi}{2} \right] \qquad$ BI (176)(22)

3.734

1. $\displaystyle\int_0^\infty \frac{\sin(ax)\,dx}{x(b^4 + x^4)} = \frac{\pi}{2b^4} \left[1 - \exp\left(-\frac{ab}{\sqrt{2}} \right) \cos \frac{ab}{\sqrt{2}} \right] \qquad [a > 0, \quad b > 0] \qquad$ BI (172)(7)

2. $\displaystyle\int_0^\infty \frac{\sin(ax)\,dx}{x\,(b^4-x^4)} = \frac{\pi}{4b^4}\left[2-e^{-ab}-\cos(ab)\right]$ $[a>0, \quad b>0]$ BI (172)(10)

3.735 $\displaystyle\int_0^\infty \frac{\sin(ax)\,dx}{x(b^2+x^2)^2} = \frac{\pi}{2b^4}\left[1-\frac{1}{2}e^{-ab}(2+ab)\right]$ $[a>0, \quad b>0]$ WH, BI (172)(22)

3.736

1. $\displaystyle\int_0^\infty \frac{\cos(ax)\,dx}{(b^2+x^2)(b^4-x^4)} = \frac{\pi}{8b^5}\left[\sin(ab)+(2+ab)e^{-ab}\right]$

                                                             $[a>0, \quad b>0]$ BI (176)(5)

2. $\displaystyle\int_0^\infty \frac{x\sin(ax)\,dx}{(b^2+x^2)(b^4-x^4)} = \frac{\pi}{8b^4}\left[(1+ab)e^{-ab}-\cos(ab)\right]$

                                                             $[a>0, \quad b>0]$ BI (174)(5)

3. $\displaystyle\int_0^\infty \frac{x^2\cos(ax)\,dx}{(b^2+x^2)(b^4-x^4)} = \frac{\pi}{8b^3}\left[\sin(ab)-abe^{-ab}\right]$ $[a>0, \quad b>0]$ BI (175)(6)

4. $\displaystyle\int_0^\infty \frac{x^3\sin(ax)\,dx}{(b^2+x^2)(b^4-x^4)} = \frac{\pi}{8b^2}\left[(1-ab)e^{-ab}-\cos(ab)\right]$

                                                             $[a>0, \quad b>0]$ BI (174)(6)

5. $\displaystyle\int_0^\infty \frac{x^4\cos(ax)\,dx}{(b^2+x^2)(b^4-x^4)} = \frac{\pi}{8b}\left[\sin(ab)+(ab-2)e^{-ab}\right]$

                                                             $[a>0, \quad b>0]$ BI (175)(7)

6. $\displaystyle\int_0^\infty \frac{x^5\sin(ax)\,dx}{(b^2+x^2)(b^4-x^4)} = \frac{\pi}{8}\left[(ab-3)e^{-ab}-\cos(ab)\right]$

                                                             $[a>0, \quad b>0]$ BI (174)(7)

3.737

1.[8] $\displaystyle\int_0^\infty \frac{\cos(ax)\,dx}{(b^2+x^2)^n} = \frac{\pi e^{-ab}}{(2b)^{2n-1}(n-1)!}\sum_{k=0}^{n-1}\frac{(2n-k-2)!(2ab)^k}{k!(n-k-1)!}$

$\displaystyle \qquad\qquad\qquad = \frac{(-1)^{n-1}\pi}{2b^{2n-1}(n-1)!}\left[\frac{d^{n-1}}{dp^{n-1}}\left(\frac{e^{-ab\sqrt{p}}}{\sqrt{p}}\right)\right]_{p=1}$

$\displaystyle \qquad\qquad\qquad = \frac{(-1)^{n-1}\pi}{2b^{2n-1}(n-1)!}\left[\frac{d^{n-1}}{dp^{n-1}}\left(\frac{e^{-abp}}{(1+p)^n}\right)\right]_{p=1}$

                                           $[a>0, \quad b>0]$ GW(333)(67b), WA 209, WA 192

2. $\displaystyle\int_0^\infty \frac{x\sin(ax)\,dx}{(x^2+\beta^2)^{n+1}} = \frac{\pi a e^{-a\beta}}{2^{2n}n!\beta^{2n-1}}\sum_{k=0}^{n-1}\frac{(2n-k-2)!(2a\beta)^k}{k!(n-k-1)!}$

$\displaystyle \qquad\qquad\qquad = \frac{\pi}{2}e^{-a\beta}$ $[n=0, \quad \beta\geq 0]$

                                           $[a>0, \quad \operatorname{Re}\beta>0]$ GW (333)(66c)

3. $\int_0^\infty \dfrac{\sin(ax)\,dx}{x(\beta^2 + x^2)^{n+1}} = \dfrac{\pi}{2\beta^{2n+2}} \left[1 - \dfrac{e^{-a\beta}}{2^n n!} F_n(a\beta) \right]$

$\quad \left[a > 0, \quad \operatorname{Re} \beta > 0, \quad F_0(z) = 1, \quad F_1(z) = z + 2, \dots, F_n(z) = (z + 2n)F_{n-1}(z) - zF'_{n-1}(z) \right]$

GW (333)(66e)

4. $\int_0^\infty \dfrac{x \sin(ax)\,dx}{(b^2 + x^2)^3} = \dfrac{\pi a}{16b^3}(1 + ab)e^{-ab}$ $\qquad [a > 0, \quad b > 0]$ $\quad$ BI(170)(5), ET I 67(35)a

5. $\int_0^\infty \dfrac{x \sin(ax)\,dx}{(b^2 + x^2)^4} = \dfrac{\pi a}{96b^5}\left(3 + 3ab + a^2b^2\right)e^{-ab}$ $\qquad [a > 0, \quad b > 0]$ $\quad$ BI(170)(6), ET I 67(35)a

6.* $\int_0^\infty \dfrac{x^3 \sin ax}{(x^2 + \beta^2)^{n+1}}\,dx = \dfrac{\pi e^{-a\beta}}{2^{2n}n!\beta^{2n-2}} \left[2^{n-1}(2n-3)!!(2 - \beta a) \right.$

$\qquad\qquad \left. - \sum_{k=1}^{n-1} \dfrac{(2n-k-2)!2^k(\beta a)^{k-1}}{k!(n-k-1)!}\left[k(k+1) - 2(k+1)\beta a + \beta^2 a^2 \right] \right]$

3.738

1. $\int_0^\infty \dfrac{x^{m-1}\sin(ax)}{x^{2n} + \beta^{2n}}\,dx = -\dfrac{\pi\beta^{m-2n}}{2n} \sum_{k=1}^n \exp\left[-a\beta \sin\dfrac{(2k-1)\pi}{2n} \right]$

$\qquad\qquad \times \cos\left\{ \dfrac{(2k-1)m\pi}{2n} + a\beta\cos\dfrac{(2k-1)\pi}{2n} \right\}$

$\qquad [m \text{ is even}], \quad \left[a > 0, \quad |\arg\beta| < \dfrac{\pi}{2n}, \quad 0 < m \le 2n \right]$ $\quad$ ET I 67(38)

2. $\int_0^\infty \dfrac{x^{m-1}\cos(ax)}{x^{2n} + \beta^{2n}}\,dx = \dfrac{\pi\beta^{m-2n}}{2n} \sum_{k=1}^n \exp\left[-a\beta\sin\dfrac{(2k-1)\pi}{2n} \right]$

$\qquad\qquad \times \sin\left\{ \dfrac{(2k-1)m\pi}{2n} + a\beta\cos\dfrac{(2k-1)\pi}{2n} \right\}$

$\qquad [m \text{ is odd}], \quad \left[a > 0, \quad |\arg\beta| < \dfrac{\pi}{2n}, \quad 0 < m < 2n+1 \right]$ $\quad$ BI(160)(29)a, ET I 10(29)

3.739

1. $\int_0^\infty \dfrac{\sin(ax)\,dx}{x\left(x^2 + 2^2\right)\left(x^2 + 4^2\right)\dots\left(x^2 + 4n^2\right)}$

$\qquad\qquad = \dfrac{\pi(-1)^n}{(2n)!2^{2n+1}} \left[2\sum_{k=0}^{n-1}(-1)^k \binom{2n}{k} e^{2(k-n)a} + (-1)^n \binom{2n}{n} \right]$

$\qquad\qquad [a > 0, \quad n \ge 0]$ $\quad$ LI(174)(8)

2. $\int_0^\infty \dfrac{\cos(ax)\,dx}{\left(x^2 + 1^2\right)\left(x^2 + 3^2\right)\dots\left[x^2 + (2n+1)^2\right]}$

$\qquad\qquad = \dfrac{(-1)^n}{(2n+1)!}\dfrac{\pi}{2^{2n+1}} \sum_{k=0}^n (-1)^k \binom{2n+1}{k} e^{(2k-2n-1)a}$ $\qquad [a \ge 0, \quad n \ge 0]$

$\qquad\qquad = \dfrac{\pi 2^{-2n-1}}{(2n+1)(n!)^2}$ $\qquad\qquad\qquad\qquad\qquad\qquad [a = 0, \quad n \ge 0]$

BI(175)(8)

3. $\int_0^\infty \dfrac{x \sin(ax)\, dx}{(x^2 + 1^2)(x^2 + 3^2)\ldots \left[x^2 + (2n+1)^2\right]}$

$$= \frac{\pi (-1)^n}{(2n+1)! \, 2^{2n+1}} \sum_{k=0}^n (-1)^k \binom{2n+1}{k} (2n - 2k + 1) e^{(2k - 2n - 1)a}$$

$$[a > 0, \quad n \geq 0] \qquad\qquad \text{LI (174)(9)}$$

4. $\int_0^\infty \dfrac{\cos ax \, dx}{(x^2 + 2^2)(x^2 + 4^2)\ldots (x^2 + 4n^2)} = \dfrac{\pi 2^{1-2n}}{(2n)!} \sum_{k=1}^n (-1)^k k \binom{2n}{n-k} e^{-2ak}$

$$[n \geq 1, \quad a \geq 0]$$

3.741

1. $\int_0^\infty \dfrac{\sin(ax)\sin(bx)}{x}\, dx = \dfrac{1}{4} \ln \left(\dfrac{a+b}{a-b}\right)^2$ $[a > 0, \quad b > 0, \quad a \neq b]$ FI II 647

2. $\int_0^\infty \dfrac{\sin(ax)\cos(bx)}{x}\, dx = \dfrac{\pi}{2}$ $[a > b \geq 0]$

$$= \frac{\pi}{4} \qquad\qquad\qquad\qquad [a = b > 0]$$

$$= 0 \qquad\qquad\qquad\qquad [b > a \geq 0]$$

$$\text{FI II 645}$$

3. $\int_0^\infty \dfrac{\sin(ax)\sin(bx)}{x^2}\, dx = \dfrac{a\pi}{2}$ $[0 < a \leq b]$

$$= \frac{b\pi}{2} \qquad\qquad\qquad [0 < b \leq a]$$

$$\text{BI (157)(1)}$$

3.742

1. $\int_0^\infty \dfrac{\sin(ax)\sin(bx)}{\beta^2 + x^2}\, dx = \dfrac{\pi}{4\beta}\left(e^{-|a-b|\beta} - e^{-(a+b)\beta}\right)$ $[a > 0, \quad b > 0, \quad \operatorname{Re}\beta > 0]$

$$= \frac{\pi}{2\beta} e^{-a\beta} \sinh b\beta \qquad\qquad [\beta > 0, \quad a \geq b \geq 0]$$

$$= \frac{\pi}{2\beta} e^{-b\beta} \sinh a\beta \qquad\qquad [\beta > 0, \quad b \geq a \geq 0]$$

$$\text{BI(162)(1)a, GW(333)(71a)}$$

2. $\int_0^\infty \dfrac{\sin(ax)\cos(bx)}{\beta^2 + x^2}\, dx = \dfrac{1}{4\beta} e^{-a\beta}\left\{e^{b\beta}\operatorname{Ei}\left[\beta(a-b)\right] + e^{-b\beta}\operatorname{Ei}\left[\beta(a+b)\right]\right\}$

$$- \frac{1}{4\beta} e^{a\beta}\left\{e^{b\beta}\operatorname{Ei}\left[-\beta(a+b)\right] + e^{-b\beta}\operatorname{Ei}\left[\beta(b-a)\right]\right\}$$

$$\text{BI (162)(3)}$$

3. $\int_0^\infty \dfrac{\cos(ax)\cos(bx)}{\beta^2 + x^2}\, dx = \dfrac{\pi}{4\beta}\left[e^{-|a-b|\beta} + e^{-(a+b)\beta}\right]$ $[a > 0, \quad b > 0, \quad \operatorname{Re}\beta > 0]$

$$= \frac{\pi}{2\beta} e^{-a\beta} \cosh b\beta \qquad\qquad [\beta > 0, \quad a \geq b \geq 0]$$

$$= \frac{\pi}{2\beta} e^{-b\beta} \cosh a\beta \qquad\qquad [\beta > 0, \quad b \geq a \geq 0]$$

$$\text{BI(163)(1)a, GW(333)(71c)}$$

4. $\int_0^\infty \dfrac{x\cos(ax)\cos(bx)}{\beta^2 + x^2}\, dx = -\dfrac{1}{4}e^{a\beta}\left\{e^{b\beta}\,\text{Ei}\left[-\beta(a+b)\right] + e^{-b\beta}\,\text{Ei}\left[\beta(b-a)\right]\right\}$
$$-\dfrac{1}{4}e^{-a\beta}\left\{e^{b\beta}\,\text{Ei}\left[\beta(a-b)\right] + e^{-b\beta}\,\text{Ei}\left[\beta(a+b)\right]\right\} \qquad [a \neq b]$$
$$= \infty \qquad\qquad\qquad\qquad [a = b]$$

BI (163)(2)

5. $\int_0^\infty \dfrac{x\sin(ax)\cos(bx)}{x^2 + \beta^2}\, dx = \dfrac{\pi}{2}e^{-a\beta}\cosh(b\beta) \qquad [0 < b < a]$
$$= \dfrac{\pi}{4}e^{-2a\beta} \qquad\qquad [0 < b = a]$$
$$= -\dfrac{\pi}{2}e^{-b\beta}\sinh(a\beta) \qquad [0 < a < b]$$

BI (162)(4)

6. $\int_0^\infty \dfrac{\sin(ax)\sin(bx)}{p^2 - x^2}\, dx = -\dfrac{\pi}{2p}\cos(ap)\sin(bp) \qquad [a > b > 0]$
$$= -\dfrac{\pi}{4p}\sin(2ap) \qquad\qquad [a = b > 0]$$
$$= -\dfrac{\pi}{2p}\sin(ap)\cos(bp) \qquad [b > a > 0]$$

BI (166)(1)

7. $\int_0^\infty \dfrac{\sin(ax)\cos(bx)}{p^2 - x^2}\, x\, dx = -\dfrac{\pi}{2}\cos(ap)\cos(bp) \qquad [a > b > 0]$
$$= -\dfrac{\pi}{4}\cos(2ap) \qquad\qquad [a = b > 0]$$
$$= \dfrac{\pi}{2}\sin(ap)\sin(bp) \qquad\quad [b > a > 0]$$

BI (166)(2)

8. $\int_0^\infty \dfrac{\cos(ax)\cos(bx)}{p^2 - x^2}\, dx = \dfrac{\pi}{2p}\sin(ap)\cos(bp) \qquad [a > b > 0]$
$$= \dfrac{\pi}{4p}\sin(2ap) \qquad\qquad [a = b > 0]$$
$$= \dfrac{\pi}{2p}\cos(ap)\sin(bp) \qquad [b > a > 0]$$

BI (166)(3)

3.743

1. $\int_0^\infty \dfrac{\sin(ax)}{\sin(bx)}\cdot\dfrac{dx}{x^2 + \beta^2} = \dfrac{\pi}{2\beta}\cdot\dfrac{\sinh(a\beta)}{\sinh(b\beta)} \qquad [0 < a < b, \quad \text{Re}\,\beta > 0]$ ET I 80(21)

2. $\int_0^\infty \dfrac{\sin(ax)}{\cos(bx)}\cdot\dfrac{x\,dx}{x^2 + \beta^2} = -\dfrac{\pi}{2}\cdot\dfrac{\sinh(a\beta)}{\cosh(b\beta)} \qquad [0 < a < b, \quad \text{Re}\,\beta > 0]$ ET I 81(30)

3. $\int_0^\infty \dfrac{\cos(ax)}{\sin(bx)}\cdot\dfrac{x\,dx}{x^2 + \beta^2} = \dfrac{\pi}{2}\cdot\dfrac{\cosh(a\beta)}{\sinh(b\beta)} \qquad [0 < a < b, \quad \text{Re}\,\beta > 0]$ ET I 23(37)

4. $\int_0^\infty \dfrac{\cos(ax)}{\cos(bx)}\cdot\dfrac{dx}{x^2 + \beta^2} = \dfrac{\pi}{2\beta}\cdot\dfrac{\cosh(a\beta)}{\cosh(b\beta)} \qquad [0 < a < b, \quad \text{Re}\,\beta > 0]$ ET I 23(36)

5.6 $\text{PV} \displaystyle\int_0^\infty \frac{\sin(ax)}{\sin x} \cdot \frac{dx}{b^2 - x^2} = 0$ if $0 \le a \le 1$

$\qquad\qquad\qquad\qquad\qquad = \dfrac{\pi}{b} \sin(a-1)b$ if $1 \le a \le 2$

$\qquad\qquad\qquad\qquad\qquad\qquad\qquad\qquad$ [b real, $b/\pi \notin \mathbb{Z}$]

3.744[3] $\displaystyle\int_0^\infty \frac{\sin(ax)}{\cos(bx)} \cdot \frac{dx}{x(x^2 + \beta^2)} = \frac{\pi}{2\beta^2} \cdot \frac{\sinh(a\beta)}{\cosh(b\beta)}$ $[0 < a < b, \quad \operatorname{Re}\beta > 0]$ ET I 82(32)

3.745[3] $\displaystyle\int_0^\infty \frac{\sin(ax)}{\cos(bx)} \cdot \frac{dx}{x(c^2 - x^2)} = 0$ $[0 < a < b, \quad c > 0]$ ET I 82(31)

3.746

1. $\displaystyle\int_0^\infty \frac{dx}{x^{n+1}} \prod_{k=0}^n \sin(a_k x) = \frac{\pi}{2} \prod_{k=1}^n a_k$ $\left[a_0 > \displaystyle\sum_{k=1}^n a_k, \quad a_k > 0 \right]$ FI II 646

2. $\displaystyle\int_0^\infty \frac{\sin(ax)}{x^{n+1}} dx \prod_{k=1}^n \sin(a_k x) \prod_{j=1}^m \cos(b_j x) = \frac{\pi}{2} \prod_{k=1}^n a_k$ $\left[a > \displaystyle\sum_{k=1}^n |a_k| + \sum_{j=1}^m |b_j| \right]$ WH

3.747

1.[7] $\displaystyle\int_0^{\pi/2} \frac{x^m}{\sin x} dx = \left(\frac{\pi}{2}\right)^m \left[\frac{1}{m} + \sum_{k=1}^\infty \frac{2^{2k-1} - 1}{4^{2k-1}(m + 2k)} \zeta(2k) \right] = 2\pi G - \frac{7}{2}\zeta(3)$

$\qquad\qquad\qquad\qquad\qquad\qquad\qquad\qquad\qquad\qquad\qquad$ [$m = 2$]   LI (206)(2)

2. $\displaystyle\int_0^{\pi/2} \frac{x\,dx}{\sin x} = \int_0^{\pi/2} \frac{\left(\frac{\pi}{2} - x\right)\,dx}{\cos x} = 2G$ BI(204)(18), BI(206)(1), GW(333)(32)

3. $\displaystyle\int_0^\infty \frac{x\,dx}{(x^2 + b^2)\sin(ax)} = \frac{\pi}{2\sinh(ab)}$ $[b > 0]$ GW (333)(79c)

4. $\displaystyle\int_0^\pi x \tan x\,dx = -\pi \ln 2$ BI (218)(4)

5. $\displaystyle\int_0^{\pi/2} x \tan x\,dx = \infty$ BI (205)(2)

6. $\displaystyle\int_0^{\pi/4} x \tan x\,dx = -\frac{\pi}{8} \ln 2 + \frac{1}{2} G = 0.1857845358\ldots$ BI (204)(1)

7. $\displaystyle\int_0^{\pi/2} x \cot x\,dx = \frac{\pi}{2} \ln 2$ FI II 623

8. $\displaystyle\int_0^{\pi/4} x \cot x\,dx = \frac{\pi}{8} \ln 2 + \frac{1}{2} G = 0.7301810584\ldots$ BI (204)(2)

9. $\displaystyle\int_0^{\pi/2} \left(\frac{\pi}{2} - x\right) \tan x\,dx = \frac{1}{2}\int_0^\pi \left(\frac{\pi}{2} - x\right) \tan x\,dx = \frac{\pi}{2} \ln 2$ GW(333)(33b), BI(218)(12)

10. $\displaystyle\int_0^\infty \tan ax \frac{dx}{x} = \frac{\pi}{2}$ $[a > 0]$ LO V 279(5)

11. $\displaystyle\int_0^{\pi/2} \frac{x \cot x}{\cos 2x} dx = \frac{\pi}{4} \ln 2$ BI (206)(12)

3.748

1. $$\int_0^{\pi/4} x^m \tan x \, dx = \frac{1}{2}\left(\frac{\pi}{4}\right)^m \sum_{k=1}^{\infty} \frac{(4^k-1)\,\zeta(2k)}{4^{2k-1}(m+2k)}$$ LI (204)(5)

2. $$\int_0^{\pi/2} x^p \cot x \, dx = \left(\frac{\pi}{2}\right)^p \left(\frac{1}{p} - 2\sum_{k=1}^{\infty} \frac{1}{4^k(p+2k)}\,\zeta(2k)\right)$$ LI (205)(7)

3. $$\int_0^{\pi/4} x^m \cot x \, dx = \frac{1}{2}\left(\frac{\pi}{4}\right)^m \left(\frac{2}{m} - \sum_{k=1}^{\infty} \frac{\zeta(2k)}{4^{2k-1}(m+2k)}\right)$$ LI (204)(6)

3.749

1. $$\int_0^{\infty} \frac{x\tan(ax)\,dx}{x^2+b^2} = \frac{\pi}{e^{2ab}+1}$$ $[a>0, \quad b>0]$ GW (333)(79a)

2. $$\int_0^{\infty} \frac{x\cot(ax)\,dx}{x^2+b^2} = \frac{\pi}{e^{2ab}-1}$$ $[a>0, \quad b>0]$ GW (333)(79b)

3. $$\int_0^{\infty} \frac{x\tan(ax)\,dx}{b^2-x^2} = \int_0^{\infty} \frac{x\cot(ax)\,dx}{b^2-x^2} = \int_0^{\infty} \frac{x\,\mathrm{cosec}(ax)\,dx}{b^2-x^2} = \infty$$ BI (161)(7, 8, 9)

3.75 Combinations of trigonometric and algebraic functions

3.751

1. $$\int_0^{\infty} \frac{\sin(ax)\,dx}{\sqrt{x+\beta}} = \sqrt{\frac{\pi}{2a}}\left[\cos(a\beta) - \sin(a\beta) + 2\,C\left(\sqrt{a\beta}\right)\sin(a\beta) - 2\,S\left(\sqrt{a\beta}\right)\cos(a\beta)\right]$$
$[a>0, \quad |\arg\beta| < \pi]$ ET I 65(12)a

2.[9] $$\int_0^{\infty} \frac{\cos(ax)\,dx}{\sqrt{x+\beta}} = \sqrt{\frac{\pi}{2a}}\left[\cos(a\beta) + \sin(a\beta) - 2\,C\left(\sqrt{a\beta}\right)\cos(a\beta) - 2\,S\left(\sqrt{a\beta}\right)\sin(a\beta)\right]$$
$[a>0, \quad |\arg\beta| < \pi]$ ET I 8(9)a

3. $$\int_u^{\infty} \frac{\sin(ax)}{\sqrt{x-u}}\,dx = \sqrt{\frac{\pi}{2a}}\left[\sin(au) + \cos(au)\right]$$ $[a>0, \quad u>0]$ ET I 65(13)

4. $$\int_u^{\infty} \frac{\cos(ax)}{\sqrt{x-u}}\,dx = \sqrt{\frac{\pi}{2a}}\left[\cos(au) - \sin(au)\right]$$ $[a>0, \quad u>0]$ ET I 8(10)

3.752

1.[8] $$\int_0^1 \sin(ax)\sqrt{1-x^2}\,dx = \sum_{k=0}^{\infty} \frac{(-1)^k a^{2k+1}}{(2k-1)!!(2k+3)!!} = \frac{\pi}{2a}\,\mathbf{H}_1(a)$$
$[a>0]$ BI (149)(6)

2. $$\int_0^1 \cos(ax)\sqrt{1-x^2}\,dx = \frac{\pi}{2a}\,J_1(a)$$ KU 65(6)a

3.753

1.[8] $$\int_0^1 \frac{\sin(ax)\,dx}{\sqrt{1-x^2}} = \sum_{k=0}^{\infty} \frac{(-1)^k a^{2k+1}}{[(2k+1)!!]^2} = \frac{\pi}{2}\,\mathbf{H}_0(a)$$ $[a>0]$ BI (149)(9)

2. $\displaystyle\int_0^1 \frac{\cos(ax)\,dx}{\sqrt{1-x^2}} = \frac{\pi}{2}\,J_0(a)$ WA 30(7)a

3. $\displaystyle\int_1^\infty \frac{\sin(ax)\,dx}{\sqrt{x^2-1}} = \frac{\pi}{2}\,J_0(a)$ $[a>0]$ WA 200(14)

4. $\displaystyle\int_1^\infty \frac{\cos(ax)}{\sqrt{x^2-1}}\,dx = -\frac{\pi}{2}\,Y_0(a)$ WA 200(15)

5. $\displaystyle\int_0^1 \frac{x\sin(ax)}{\sqrt{1-x^2}}\,dx = \frac{\pi}{2}\,J_1(a)$ $[a>0]$ WA 30(6)

3.754

1. $\displaystyle\int_0^\infty \frac{\sin(ax)\,dx}{\sqrt{\beta^2+x^2}} = \frac{\pi}{2}\left[I_0(a\beta) - \mathbf{L}_0(a\beta)\right]$ $[a>0, \quad \operatorname{Re}\beta>0]$ ET I 66(26)

2. $\displaystyle\int_0^\infty \frac{\cos(ax)\,dx}{\sqrt{\beta^2+x^2}} = K_0(a\beta)$ $[a>0, \quad \operatorname{Re}\beta>0]$

                                                       WA 191(1), GW(333)(78a)

3. $\displaystyle\int_0^\infty \frac{x\sin(ax)}{\sqrt{(\beta^2+x^2)^3}}\,dx = a\,K_0(a\beta)$ $[a>0, \quad \operatorname{Re}\beta>0]$ ET I 66(27)

3.755

1. $\displaystyle\int_0^\infty \frac{\sqrt{\sqrt{x^2+\beta^2}-\beta}\,\sin(ax)\,dx}{\sqrt{x^2+\beta^2}} = \sqrt{\frac{\pi}{2a}}\,e^{-a\beta}$ $[a>0]$ ET I 66(31)

2. $\displaystyle\int_0^\infty \frac{\sqrt{\sqrt{x^2+\beta^2}+\beta}\,\cos(ax)\,dx}{\sqrt{x^2+\beta^2}} = \sqrt{\frac{\pi}{2a}}\,e^{-a\beta}$ $[a>0, \quad \operatorname{Re}\beta>0]$ ET I 10(25)

3.756

1. $\displaystyle\int_0^\infty \frac{\sin(ax)}{x^{\frac{n}{2}-1}}\prod_{k=2}^n \sin(a_k x)\,dx = 0$ $\left[a_k>0, \quad a>\sum_{k=2}^n a_k\right]$ ET I 80(22)

2. $\displaystyle\int_0^\infty x^{\frac{n}{2}-1}\cos(ax)\prod_{k=1}^n \cos(a_k x)\,dx = 0$ $\left[a_k>0, \quad a>\sum_{k=1}^n a_k\right]$ ET I 22(26)

3.757

1. $\displaystyle\int_0^\infty \frac{\sin(ax)}{\sqrt{x}}\,dx = \sqrt{\frac{\pi}{2a}}$ BI (177)(1)

2. $\displaystyle\int_0^\infty \frac{\cos(ax)}{\sqrt{x}}\,dx = \sqrt{\frac{\pi}{2a}}$ BI (177)(2)

3.76–3.77 Combinations of trigonometric functions and powers

3.761

1. $\displaystyle\int_0^1 x^{\mu-1}\sin(ax)\,dx = \frac{-i}{2\mu}\left[\,_1F_1(\mu;\mu+1;ia) - \,_1F_1\left(\mu;\mu+1;-ia\right)\right]$

$$[a>0,\quad \operatorname{Re}\mu>-1,\quad \mu\neq 0]$$

ET I 68(2)a

2.[8] $\displaystyle\int_u^\infty x^{\mu-1}\sin x\,dx = \frac{i}{2}\left[e^{-\frac{\pi}{2}i\mu}\,\Gamma(\mu,iu) - e^{\frac{\pi}{2}i\mu}\,\Gamma(\mu,-iu)\right]$

$$[\operatorname{Re}\mu<1]$$

EH II 149(2)

3. $\displaystyle\int_1^\infty \frac{\sin(ax)}{x^{2n}}\,dx = \frac{a^{2n-1}}{(2n-1)!}\left[\sum_{k=1}^{2n-1}\frac{(2n-k-1)!}{a^{2n-k}}\sin\left(a+(k-1)\frac{\pi}{2}\right) + (-1)^n\operatorname{ci}(a)\right]$

$$[a>0]$$

LI (203)(15)

4. $\displaystyle\int_0^\infty x^{\mu-1}\sin(ax)\,dx = \frac{\Gamma(\mu)}{a^\mu}\sin\frac{\mu\pi}{2} = \frac{\pi\sec\frac{\mu\pi}{2}}{2a^\mu\,\Gamma(1-\mu)}$ $[a>0;\quad 0<|\operatorname{Re}\mu|<1]$

FI II 809a, BI(150)(1)

5.[10] $\displaystyle\int_0^\pi x^m\sin(nx)\,dx = \frac{(-1)^{n+1}}{n^{m+1}}\sum_{k=0}^{\lfloor m/2\rfloor}(-1)^k\frac{m!}{(m-2k)!}(n\pi)^{m-2k}$

$$-(-1)^{\lfloor m/2\rfloor}\frac{m!\left\lfloor m-2\lfloor\frac{m}{2}\rfloor-1\right\rfloor}{n^{m+1}}$$

GW(333)(6)

6.[8] $\displaystyle\int_0^1 x^{\mu-1}\cos(ax)\,dx = \frac{1}{2\mu}\left[\,_1F_1(\mu;u+1;ia) + \,_1F_1\left(\mu;u+1;-ia\right)\right]$

$$[a>0,\quad \operatorname{Re}\mu>0]$$

ET I 11(2)

7. $\displaystyle\int_u^\infty x^{\mu-1}\cos x\,dx = \frac{1}{2}\left[e^{-\frac{\pi}{2}i\mu}\,\Gamma(\mu,iu)s + e^{\frac{\pi}{2}i\mu}\,\Gamma(\mu,-iu)\right]$

$$[\operatorname{Re}\mu<1]$$

EH II 149(1)

8. $\displaystyle\int_1^\infty \frac{\cos(ax)}{x^{2n+1}}\,dx = \frac{a^{2n}}{(2n)!}\left[\sum_{k=1}^{2n}\frac{(2n-k)!}{a^{2n-k+1}}\cos\left(a+(k-1)\frac{\pi}{2}\right) + (-1)^{n+1}\operatorname{ci}(a)\right]$

$$[a>0]$$

LI (203)(16)

9.[8] $\displaystyle\int_0^\infty x^{\mu-1}\cos(ax)\,dx = \frac{\Gamma(\mu)}{a^\mu}\cos\frac{\mu\pi}{2} = \frac{\pi\operatorname{cosec}\frac{\mu\pi}{2}}{2a^\mu\,\Gamma(1-\mu)}$ $[a>0,\quad 0<\operatorname{Re}\mu<1]$

FI II 809a, BI(150)(2)

10. $\displaystyle\int_0^\pi x^m\cos(nx)\,dx = \frac{(-1)^n}{n^{m+1}}\sum_{k=0}^{\lfloor(m-1)/2\rfloor}(-1)^k\frac{m!}{(m-2k-1)!}(n\pi)^{m-2k-1}$

$$+(-1)^{\lfloor(m+1)/2\rfloor}\frac{2[(m+1)/2]-m}{n^{m+1}}\cdot m!$$

GW (333)(7)

11.
$$\int_0^{\pi/2} x^m \cos x \, dx = \sum_{k=0}^{\lfloor m/2 \rfloor} (-1)^k \frac{m!}{(m-2k)!} \left(\frac{\pi}{2}\right)^{m-2k} + (-1)^{\lfloor m/2 \rfloor} \left(2 \left\lfloor \frac{m}{2} \right\rfloor - m\right) m!$$

GW (333)(9c)

12.
$$\int_0^{2n\pi} x^m \cos kx \, dx = -\sum_{j=0}^{m-1} \frac{j!}{k^{j+1}} \binom{m}{j} (2n\pi)^{m-j} \cos \frac{j+1}{2}\pi$$

BI (226)(2)

3.762

1.
$$\int_0^\infty x^{\mu-1} \sin(ax) \sin(bx) \, dx = \frac{1}{2} \cos \frac{\mu\pi}{2} \Gamma(\mu) \left[|b-a|^{-\mu} - (b+a)^{-\mu} \right]$$

$$[a > 0, \quad b > 0, \quad a \neq b, \quad -2 < \operatorname{Re}\mu < 1]$$

(for $\mu = 0$, see **3.741** 1, for $\mu = -1$, see **3.741** 3)

BI(149)(7), ET I 321(40)

2.
$$\int_0^\infty x^{\mu-1} \sin(ax) \cos(bx) \, dx = \frac{1}{2} \sin \frac{\mu\pi}{2} \Gamma(\mu) \left[(a+b)^{-\mu} + |a-b|^{-\mu} \operatorname{sign}(a-b) \right]$$

$$[a > 0, \quad b > 0, \quad |\operatorname{Re}\mu| < 1] \qquad \text{(for } \mu = 0 \text{ see } \mathbf{3.741}\ 2) \quad \text{BI(159)(8)a, ET I 321(41)}$$

3.
$$\int_0^\infty x^{\mu-1} \cos(ax) \cos(bx) \, dx = \frac{1}{2} \cos \frac{\mu\pi}{2} \Gamma(\mu) \left[(a+b)^{-\mu} + |a-b|^{-\mu} \right]$$

$$[a > 0, \quad b > 0, \quad 0 < \operatorname{Re}\mu < 1]$$

ET I 20(17)

3.763

1.
$$\int_0^\infty \frac{\sin(ax) \sin(bx) \sin(cx)}{x^\nu} \, dx = \frac{1}{4} \cos \frac{\nu\pi}{2} \Gamma(1-\nu) \left\{ (c+a-b)^{\nu-1} - (c+a+b)^{\nu-1} \right.$$
$$\left. - |c-a+b|^{\nu-1} \operatorname{sign}(a-b-c) + |c-a-b|^{\nu-1} \operatorname{sign}(a+b-c) \right\}$$

$$[c > 0, \quad 0 < \operatorname{Re}\nu < 4, \quad \nu \neq 1, 2, 3, \quad a \geq b > 0] \quad \text{GW(333)(26a)a, ET I 79(13)}$$

2.
$$\int_0^\infty \frac{\sin(ax) \sin(bx) \sin(cx)}{x} \, dx = 0 \qquad\qquad\qquad [c < a-b \text{ and } c > a+b]$$

$$= \frac{\pi}{8} \qquad\qquad\qquad [c = a-b \text{ and } c = a+b]$$

$$= \frac{\pi}{4} \qquad\qquad\qquad [a-b < c < a+b]$$

$$[a \geq b > 0, \quad c > 0] \qquad\qquad \text{FI II 645}$$

3.
$$\int_0^\infty \frac{\sin(ax) \sin(bx) \sin(cx)}{x^2} \, dx = \frac{1}{4}(c+a+b) \ln(c+a+b)$$

$$- \frac{1}{4}(c+a-b) \ln(c+a-b) - \frac{1}{4}|c-a-b| \ln|c-a-b|$$

$$\times \operatorname{sign}(a+b-c) + \frac{1}{4}|c-a+b| \ln|c-a+b| \operatorname{sign}(a-b-c)$$

$$[a \geq b > 0, \quad c > 0] \quad \text{BI(157)(8)a, ET I 79(11)}$$

4. $\int_0^\infty \dfrac{\sin(ax)\sin(bx)\sin(cx)}{x^3}\,dx = \dfrac{\pi bc}{2}$ $[0 < c < a - b \text{ and } c > a + b]$

$\qquad\qquad\qquad\qquad\qquad = \dfrac{\pi bc}{2} - \dfrac{\pi(a - b - c)^2}{8}$ $[a - b < c < a + b]$

$\qquad\qquad\qquad\qquad\qquad\qquad\qquad\qquad\qquad [a \geq b > 0, \quad c > 0]$ BI(157)(20), ET I 79(12)

3.764

1. $\int_0^\infty x^p \sin(ax + b)\,dx = \dfrac{1}{a^{p+1}}\,\Gamma(1 + p)\cos\left(b + \dfrac{p\pi}{2}\right)$ $[a > 0, \quad -1 < p < 0]$ GW (333)(30a)

2. $\int_0^\infty x^p \cos(ax + b)\,dx = -\dfrac{1}{a^{p+1}}\,\Gamma(1 + p)\sin\left(b + \dfrac{\pi p}{2}\right)$

$\qquad\qquad\qquad\qquad\qquad\qquad\qquad\qquad\qquad [a > 0, \quad -1 < p < 0]$ GW (333)(30b)

3.765

1.[10] $\int_0^\infty \dfrac{\sin ax}{x^\nu(x + b)}\,dx$

$\qquad = a^{1+\nu} b \cos\dfrac{\pi\nu}{2}\,\Gamma(-1 - \nu)\ {}_1F_2\left(1;\ 1 + \dfrac{\nu}{2}, \dfrac{3}{2} + \dfrac{\nu}{2};\ -\dfrac{1}{4}a^2b^2\right)\mathrm{sign}(a)$

$\qquad -\dfrac{\pi\,\mathrm{cosec}(\pi\nu)\sin(ab)}{b^\nu} - a^\nu\,\Gamma(-\nu)\ {}_1F_2\left(1;\ 1 + \dfrac{\nu}{2}, 1 + \dfrac{\nu}{2};\ -\dfrac{1}{4}a^2b^2\right)\mathrm{sign}(a)\sin\dfrac{\pi\nu}{2}$

$\qquad\qquad\qquad\qquad\qquad [\mathrm{Im}\,a = 0, \quad -1 < \mathrm{Re}\,b < 2, \quad \arg b \neq \pi]$ MC

2. $\int_0^\infty \dfrac{\cos(ax)}{x^\nu(x + \beta)}\,dx = \dfrac{\Gamma(1 - \nu)}{2\beta^\nu}\left[e^{ia\beta}\,\Gamma(\nu, ia\beta) + e^{-ia\beta}\,\Gamma(\nu, -ia\beta)\right]$

$\qquad\qquad\qquad\qquad\qquad\qquad [a > 0, \quad |\mathrm{Re}\,\nu| < 1, \quad |\arg\beta| < \pi]$

$\qquad\qquad\qquad\qquad\qquad\qquad\qquad\qquad\qquad\qquad\qquad\qquad\qquad$ ET II 221(52)

3.766

1.[10] $\int_0^\infty \dfrac{x^{\mu-1}\sin ax}{1 + x^2}\,dx$

$\qquad = -a^{2-\mu}\,\Gamma(\mu - 2)\ {}_1F_2\left(1;\ \dfrac{3 - \mu}{2}, \dfrac{4 - \mu}{2};\ \dfrac{a^2}{4}\right)\mathrm{sign}(a)\sin\dfrac{\pi\mu}{2} + \dfrac{\pi}{2}\sec\dfrac{\pi\mu}{2}\sinh(a)$

$\qquad\qquad\qquad\qquad\qquad\qquad [\mathrm{Im}\,a = 0, \quad -1 < \mathrm{Re}\,\mu < 3]$ MC

2. $\int_0^\infty \dfrac{x^{\mu-1}\cos(ax)}{1 + x^2}\,dx = \dfrac{\pi}{2}\,\mathrm{cosec}\,\dfrac{\mu\pi}{2}\cosh a$

$\qquad\qquad + \dfrac{1}{2}\cos\dfrac{\mu\pi}{2}\,\Gamma(\mu)\left\{\exp\left[-a + i\pi(1 - \mu)\right]\gamma(1 - \mu, -a) - e^a\,\gamma(1 - \mu, a)\right\}$

$\qquad\qquad\qquad\qquad\qquad\qquad [a > 0, \quad 0 < \mathrm{Re}\,\mu < 3]$ ET I 319(24)

3.[9] $\int_0^\infty \dfrac{x^{2\mu+1}\sin(ax)\,dx}{x^2 + b^2} = -\dfrac{\pi}{2}b^{2\mu}\sec(\mu\pi)\sinh(ab)$

$\qquad\qquad + \dfrac{\sin(\mu\pi)}{2a^{2\mu}}\,\Gamma(2\mu)\left[\ {}_1F_1(1; 1 - 2\mu; ab) + {}_1F_1(1; 1 - 2\mu; -ab)\right]$

$\qquad\qquad\qquad\qquad [a > 0, \quad -\dfrac{3}{2} < \mathrm{Re}\,\mu < \dfrac{1}{2}]$ ET II 220(39)

$4.^9 \quad \int_0^\infty \frac{x^{2\mu+1}\cos(ax)\,dx}{x^2+b^2} = -\frac{\pi}{2}b^{2(\mu+\frac{1}{2})}\operatorname{cosec}\left[\left(\mu+\frac{1}{2}\right)\pi\right]\cosh(ab)$

$$+\frac{\cos\left[\left(\mu+\frac{1}{2}\right)\pi\right]}{2a^{2(\mu+\frac{1}{2})}}\Gamma\left[2\left(\mu+\frac{1}{2}\right)\right]\left\{{}_1F_1\left(1;1-2\left(\mu+\frac{1}{2}\right);ab\right)\right.$$

$$\left.+{}_1F_1\left(1;1-2\left(\mu+\frac{1}{2}\right);-ab\right)\right\}$$

$$\left[a>0,\quad -1<\operatorname{Re}\mu<\tfrac{1}{2}\right]\quad\text{ET II 221(56)}$$

3.767

1. $\quad \int_0^\infty \frac{x^{\beta-1}\sin\left(ax-\frac{\beta\pi}{2}\right)}{\gamma^2+x^2}\,dx = -\frac{\pi}{2}\gamma^{\beta-2}e^{-a\gamma} \qquad \left[a>0,\quad \operatorname{Re}\gamma>0,\quad 0<\operatorname{Re}\beta<2\right]$

$$\text{BI (160)(20)}$$

2. $\quad \int_0^\infty \frac{x^\beta\cos\left(ax-\frac{\beta\pi}{2}\right)}{\gamma^2+x^2}\,dx = \frac{\pi}{2}\gamma^{\beta-1}e^{-a\gamma} \qquad \left[a>0,\quad \operatorname{Re}\gamma>0,\quad |\operatorname{Re}\beta|<1\right]$

$$\text{BI (160)(21)}$$

3. $\quad \int_0^\infty \frac{x^{\beta-1}\sin\left(ax-\frac{\beta\pi}{2}\right)}{x^2-b^2}\,dx = \frac{\pi}{2}b^{\beta-2}\cos\left(ab-\frac{\pi\beta}{2}\right) \qquad \left[a>0,\quad b>0,\quad 0<\operatorname{Re}\beta<2\right]$

$$\text{BI (161)(11)}$$

4. $\quad \int_0^\infty \frac{x^\beta\cos\left(ax-\frac{\beta\pi}{2}\right)}{x^2-b^2}\,dx = -\frac{\pi}{2}b^{\beta-1}\sin\left(ab-\frac{\pi\beta}{2}\right) \qquad \left[a>0,\quad b>0,\quad |\beta|<1\right]$

$$\text{GW (333)(82)}$$

3.768

1. $\quad \int_u^\infty (x-u)^{\mu-1}\sin(ax)\,dx = \frac{\Gamma(\mu)}{a^\mu}\sin\left(au+\frac{\mu\pi}{2}\right) \qquad \left[a>0,\quad 0<\operatorname{Re}\mu<1\right]\quad\text{ET II 203(19)}$

2. $\quad \int_u^\infty (x-u)^{\mu-1}\cos(ax)\,dx = \frac{\Gamma(\mu)}{a^\mu}\cos\left(au+\frac{\mu\pi}{2}\right) \qquad \left[a>0,\quad 0<\operatorname{Re}\mu<1\right]\quad\text{ET II 204(24)}$

$3.^{10} \quad \int_0^1 (1-x)^\nu\sin(ax)\,dx = \frac{1}{a} - \frac{\Gamma(\nu+1)}{a^{\nu+1}}C_\nu(a) = a^{-\nu-1/2}s_{\nu+1/2,1/2}(a)$

$$=\sum_{n=0}^\infty \frac{(-1)^n a^{\nu+2n}}{\Gamma(\nu+2n+1)}$$

$$\left[a>0,\quad \operatorname{Re}\nu>-1\right]\qquad\text{ET I 11(3)a}$$

Here $C_\nu(a)$ is the Young's function given by:

$$C_\nu(a) = \frac{\frac{1}{2}a^\nu}{\Gamma(\nu+1)}\left[{}_1F_1(1;\nu+1;ia)+{}_1F_1\left(1;\nu+1;-ia\right)\right] = \sum_{n=0}^\infty \frac{(-1)^n a^{\nu+2n}}{\Gamma(\nu+2n+1)}$$

4.³ $\displaystyle\int_0^1 (1-x)^\nu \cos(ax)\,dx = \frac{i}{2} a^{-\nu-1} \left\{ \exp\left[\frac{i}{2}(\nu\pi - 2a)\right] \gamma(\nu+1, -ia) \right.$

$\displaystyle \left. - \exp\left[-\frac{i}{2}(\nu\pi - 2a)\right] \gamma(\nu+1, ia) \right\}$

$\displaystyle = \Gamma(\nu+1) \sum_{n=0}^{\infty} \frac{(-a^2)^n}{\Gamma(\nu+2+2n)}$

$[a > 0, \quad \operatorname{Re}\nu > -1]$ ET I 11(3)a

5. $\displaystyle\int_0^u x^{\nu-1}(u-x)^{\mu-1}\sin(ax)\,dx = \frac{u^{\mu+\nu-1}}{2i} B(\mu,\nu)\left[\,_1F_1(\nu; \mu+\nu; iau) - \,_1F_1(\nu; \mu+\nu; -iau)\right]$

$[a > 0, \quad \operatorname{Re}\mu > 0, \quad \operatorname{Re}\nu > -1, \quad \nu \neq 0]$ ET II 189(26)

6. $\displaystyle\int_0^u x^{\nu-1}(u-x)^{\mu-1}\cos(ax)\,dx = \frac{u^{\mu+\nu-1}}{2} B(\mu,\nu)\left[\,_1F_1(\nu; \mu+\nu; iau) + \,_1F_1(\nu; \mu+\nu; -iau)\right]$

$[a > 0, \quad \operatorname{Re}\mu > 0, \quad \operatorname{Re}\nu > 0]$

ET II 189(32)

7. $\displaystyle\int_0^u x^{\mu-1}(u-x)^{\mu-1}\sin(ax)\,dx = \sqrt{\pi}\left(\frac{u}{a}\right)^{\mu-1/2} \sin\frac{au}{2}\,\Gamma(\mu)\,J_{\mu-1/2}\left(\frac{au}{2}\right)$

$[\operatorname{Re}\mu > 0]$ ET II 189(25)

8. $\displaystyle\int_u^{\infty} x^{\mu-1}(x-u)^{\mu-1}\sin(ax)\,dx$

$\displaystyle = \frac{\sqrt{\pi}}{2}\left(\frac{u}{a}\right)^{\mu-1/2}\Gamma(\mu)\left[\cos\frac{au}{2}\,J_{1/2-\mu}\left(\frac{au}{2}\right) - \sin\frac{au}{2}\,Y_{1/2-\mu}\left(\frac{au}{2}\right)\right]$

$[a > 0, \quad 0 < \operatorname{Re}\mu < \tfrac{1}{2}]$ ET II 203(20)

9. $\displaystyle\int_0^u x^{\mu-1}(u-x)^{\mu-1}\cos(ax)\,dx = \sqrt{\pi}\left(\frac{u}{a}\right)^{\mu-\frac{1}{2}}\cos\frac{au}{2}\,\Gamma(\mu)\,J_{\mu-\frac{1}{2}}\left(\frac{au}{2}\right)$

$[\operatorname{Re}\mu > 0]$ ET II 189(31)

10. $\displaystyle\int_u^{\infty} x^{\mu-1}(x-u)^{\mu-1}\cos(ax)\,dx = -\frac{\sqrt{\pi}}{2}\left(\frac{u}{a}\right)^{\mu-\frac{1}{2}}\Gamma(\mu)\left[\sin\frac{au}{2}\,J_{\frac{1}{2}-\mu}\left(\frac{au}{2}\right) - \cos\frac{au}{2}\,Y_{\frac{1}{2}-\mu}\left(\frac{au}{2}\right)\right]$

$[a > 0, \quad 0 < \operatorname{Re}\mu < \tfrac{1}{2}]$ ET II 204(25)

11.³ $\displaystyle\int_0^1 x^{\nu-1}(1-x)^{\mu-1}\sin(ax)\,dx = -\frac{i}{2} B(\mu,\nu)\left[\,_1F_1(\nu; \nu+\mu; ia) - \,_1F_1(\nu; \nu+\mu; -ia)\right]$

$[\operatorname{Re}\mu > 0, \quad \operatorname{Re}\nu > -1, \quad \nu \neq 0]$

ET I 68 (5)a, ET I 317(5)

12.³ $\displaystyle\int_0^1 x^{\nu-1}(1-x)^{\mu-1}\cos(ax)\,dx = \frac{1}{2} B(\mu,\nu)\left[\,_1F_1(\nu; \nu+\mu; ia) + \,_1F_1(\nu; \nu+\mu; -ia)\right]$

$[\operatorname{Re}\mu > 0, \quad \operatorname{Re}\nu > 0]$ ET I 11(5)

13. $\displaystyle\int_0^1 x^{\mu}(1-x)^{\mu}\sin(2ax)\,dx = \frac{\sqrt{\pi}}{(2a)^{\mu+\frac{1}{2}}}\,\Gamma(\mu+1)\,J_{\mu+\frac{1}{2}}(a)\sin a$

$[a > 0, \quad \operatorname{Re}\mu > -1]$ ET I 68(4)

14. $\displaystyle\int_0^1 x^\mu (1-x)^\mu \cos(2ax)\, dx = \frac{\sqrt{\pi}}{(2a)^{\mu+\frac{1}{2}}} \Gamma(\mu+1)\, J_{\mu+\frac{1}{2}}(a) \cos a$

$$[a > 0, \quad \operatorname{Re}\mu > -1] \qquad \text{ET I 11(4)}$$

3.769

1. $\displaystyle\int_0^\infty \left[(\beta + ix)^{-\nu} - (\beta - ix)^{-\nu} \right] \sin(ax)\, dx = -\frac{\pi i a^{\nu-1} e^{-a\beta}}{\Gamma(\nu)}$

$$[a > 0, \quad \operatorname{Re}\beta > 0, \quad \operatorname{Re}\nu > 0]$$
$$\text{ET I 70(15)}$$

2. $\displaystyle\int_0^\infty \left[(\beta + ix)^{-\nu} + (\beta - ix)^{-\nu} \right] \cos(ax)\, dx = \frac{\pi a^{\nu-1} e^{-a\beta}}{\Gamma(\nu)}$

$$[a > 0, \quad \operatorname{Re}\beta > 0, \quad \operatorname{Re}\nu > 0]$$
$$\text{ET I 13(19)}$$

3. $\displaystyle\int_0^\infty x\left[(\beta + ix)^{-\nu} + (\beta - ix)^{-\nu} \right] \sin(ax)\, dx = -\frac{\pi a^{\nu-2}(\nu-1-a\beta)}{\Gamma(\nu)} e^{-a\beta}$

$$[a > 0, \quad \operatorname{Re}\beta > 0, \quad \operatorname{Re}\nu > 0]$$
$$\text{ET I 70(16)}$$

4. $\displaystyle\int_0^\infty x^{2n}\left[(\beta - ix)^{-\nu} - (\beta + ix)^{-\nu} \right] \sin(ax)\, dx = \frac{(-1)^n i}{\Gamma(\nu)} (2n)!\, \pi a^{\nu-2n-1} e^{-a\beta}\, L_{2n}^{\nu-2n-1}(a\beta)$

$$[a > 0, \quad \operatorname{Re}\beta > 0, \quad 0 \le 2n < \operatorname{Re}\nu]$$
$$\text{ET I 70(17)}$$

5. $\displaystyle\int_0^\infty x^{2n}\left[(\beta + ix)^{-\nu} + (\beta - ix)^{-\nu} \right] \cos(ax)\, dx = \frac{(-1)^n}{\Gamma(\nu)} (2n)!\, \pi a^{\nu-2n-1} e^{-a\beta}\, L_{2n}^{\nu-2n-1}(a\beta)$

$$[a > 0, \quad \operatorname{Re}\beta > 0, \quad 0 \le 2n < \operatorname{Re}\nu]$$
$$\text{ET I 13(20)}$$

6. $\displaystyle\int_0^\infty x^{2n+1}\left[(\beta + ix)^{-\nu} + (\beta - ix)^{-\nu} \right] \sin(ax)\, dx = \frac{(-1)^{n+1}}{\Gamma(\nu)} (2n+1)!\, \pi a^{\nu-2n-2} e^{-a\beta}\, L_{2n+1}^{\nu-2n-2}(a\beta)$

$$[a > 0, \quad \operatorname{Re}\beta > 0, \quad -1 \le 2n+1 < \operatorname{Re}\nu] \quad \text{ET I 70(18)}$$

7. $\displaystyle\int_0^\infty x^{2n+1}\left[(\beta + ix)^{-\nu} - (\beta - ix)^{-\nu} \right] \cos(ax)\, dx = \frac{(-1)^{n+1}}{\Gamma(\nu)} (2n+1)!\, \pi a^{\nu-2n-2} e^{-a\beta}\, L_{2n+1}^{\nu-2n-2}(a\beta)$

$$[a > 0, \quad \operatorname{Re}\beta > 0, \quad 0 \le 2n < \operatorname{Re}\nu - 1] \quad \text{ET I 13(21)}$$

3.771

1. $\displaystyle\int_0^\infty (\beta^2 + x^2)^{\nu-\frac{1}{2}} \sin(ax)\, dx = \frac{\sqrt{\pi}}{2}\left(\frac{2\beta}{a}\right)^\nu \Gamma\left(\nu + \frac{1}{2}\right)\left[I_{-\nu}(a\beta) - \mathbf{L}_\nu(a\beta)\right]$

$$\left[a > 0, \quad \operatorname{Re}\beta > 0, \quad \operatorname{Re}\nu < \tfrac{1}{2}, \quad \nu \ne -\tfrac{1}{2}, -\tfrac{3}{2}, -\tfrac{5}{2}, \dots\right] \quad \text{EH II 38a, ET I 68(6)}$$

2. $\displaystyle\int_0^\infty (\beta^2 + x^2)^{\nu-\frac{1}{2}} \cos(ax)\, dx = \frac{1}{\sqrt{\pi}}\left(\frac{2\beta}{a}\right)^\nu \cos(\pi\nu)\, \Gamma\left(\nu + \frac{1}{2}\right) K_{-\nu}(a\beta)$

$$\left[a > 0, \quad \operatorname{Re}\beta > 0, \quad \operatorname{Re}\nu < \frac{1}{2}\right]$$
$$\text{WA 191(1)a, GW(333)(78)a}$$

3. $\int_0^u x^{2\nu-1}\left(u^2 - x^2\right)^{\mu-1}\sin(ax)\,dx$

$$= \frac{a}{2}u^{2\mu+2\nu-1}\,\mathrm{B}\left(\mu, \nu + \frac{1}{2}\right)\ {}_1F_2\left(\nu + \frac{1}{2}; \frac{3}{2}, \mu + \nu + \frac{1}{2}; -\frac{a^2u^2}{4}\right)$$

$$\left[\operatorname{Re}\mu > 0, \quad \operatorname{Re}\nu > -\tfrac{1}{2}\right] \qquad \text{ET II 189(29)}$$

4. $\int_0^u x^{2\nu-1}\left(u^2 - x^2\right)^{\mu-1}\cos(ax)\,dx = \frac{1}{2}u^{2\mu+2\nu-2}\,\mathrm{B}(\mu, \nu)\ {}_1F_2\left(\nu; \frac{1}{2}, \mu + \nu; -\frac{a^2u^2}{4}\right)$

$$\left[\operatorname{Re}\mu > 0, \quad \operatorname{Re}\nu > 0\right] \qquad \text{ET II 190(35)}$$

5.[7] $\int_0^\infty x\left(x^2 + \beta^2\right)^{\nu-\frac{1}{2}}\sin(ax)\,dx = \frac{1}{\sqrt{\pi}}\beta\left(\frac{2\beta}{a}\right)^\nu \cos\nu\pi\,\Gamma\left(\nu + \frac{1}{2}\right)K_{\nu+1}(a\beta)$

$$= \sqrt{\pi}\beta\left(\frac{2\beta}{a}\right)^\nu \frac{1}{\Gamma\left(\frac{1}{2} - \nu\right)}K_{\nu+1}(a\beta)$$

$$\left[a > 0, \quad \operatorname{Re}\beta > 0, \quad \operatorname{Re}\nu < 0\right] \qquad \text{ET I 69(11)}$$

6. $\int_0^u \left(u^2 - x^2\right)^{\nu-\frac{1}{2}}\sin(ax)\,dx = \frac{\sqrt{\pi}}{2}\left(\frac{2u}{a}\right)^\nu \Gamma\left(\nu + \frac{1}{2}\right)\mathbf{H}_\nu(au)$

$$\left[a > 0, \quad u > 0, \quad \operatorname{Re}\nu > -\tfrac{1}{2}\right]$$
$$\text{ET I 69(7), WA 358(1)a}$$

7. $\int_u^\infty \left(x^2 - u^2\right)^{\nu-\frac{1}{2}}\sin(ax)\,dx = \frac{\sqrt{\pi}}{2}\left(\frac{2u}{a}\right)^\nu \Gamma\left(\nu + \frac{1}{2}\right)J_{-\nu}(au)$

$$\left[a > 0, \quad u > 0, \quad |\operatorname{Re}\nu| < \tfrac{1}{2}\right]$$
$$\text{EH II 81(12)a, ET I 69(8), WA 187(3)a}$$

8. $\int_0^u \left(u^2 - x^2\right)^{\nu-\frac{1}{2}}\cos(ax)\,dx = \frac{\sqrt{\pi}}{2}\left(\frac{2u}{a}\right)^\nu \Gamma\left(\nu + \frac{1}{2}\right)J_\nu(au)$

$$\left[a > 0, \quad u > 0, \quad \operatorname{Re}\nu > -\tfrac{1}{2}\right]$$
$$\text{ET I 11(8)}$$

9. $\int_u^\infty \left(x^2 - u^2\right)^{\nu-\frac{1}{2}}\cos(ax)\,dx = -\frac{\sqrt{\pi}}{2}\left(\frac{2u}{a}\right)^\nu \Gamma\left(\nu + \frac{1}{2}\right)Y_{-\nu}(au)$

$$\left[a > 0, \quad u > 0, \quad |\operatorname{Re}\nu| < \tfrac{1}{2}\right]$$
$$\text{WA 187(4)a, EH II 82(13)a, ET I 11(9)}$$

10. $\int_0^u x\left(u^2 - x^2\right)^{\nu-\frac{1}{2}}\sin(ax)\,dx = \frac{\sqrt{\pi}}{2}u\left(\frac{2u}{a}\right)^\nu \Gamma\left(\nu + \frac{1}{2}\right)J_{\nu+1}(au)$

$$\left[a > 0, \quad u > 0, \quad \operatorname{Re}\nu > -\tfrac{1}{2}\right]$$
$$\text{ET I 69(9)}$$

11. $\int_u^\infty x\left(x^2 - u^2\right)^{\nu-\frac{1}{2}}\sin(ax)\,dx = \frac{\sqrt{\pi}}{2}u\left(\frac{2u}{a}\right)^\nu \Gamma\left(\nu + \frac{1}{2}\right)Y_{-\nu-1}(au)$

$$\left[a > 0, \quad u > 0, \quad -\tfrac{1}{2} < \operatorname{Re}\nu < 0\right]$$
$$\text{ET I 69(10)}$$

12.^{7} $\displaystyle\int_0^u x\left(u^2-x^2\right)^{\nu-\frac{1}{2}}\cos(ax)\,dx = -\frac{u^{\nu+1}}{a^\nu}\,s_{(\nu-1)\nu+1}(au)$

$$= \frac{1}{2}\left(\nu+\frac{1}{2}\right)^{-1}u^{2\nu+1} - \frac{\sqrt{\pi}}{2}u\left(\frac{2u}{a}\right)^\nu\Gamma\left(\nu+\frac{1}{2}\right)\mathbf{H}_{\nu+1}(au)$$

$$\left[a>0,\quad u>0,\quad \mathrm{Re}\,\nu>-\tfrac{1}{2}\right]\quad \text{ET I 12(10)}$$

13. $\displaystyle\int_u^\infty x\left(x^2-u^2\right)^{\nu-1/2}\cos(ax)\,dx\;\frac{\sqrt{\pi}\,u}{2}\left(\frac{2u}{a}\right)^\nu\Gamma\left(\nu+\frac{1}{2}\right)J_{-\nu-1}(au)$

$$\left[a>0,\quad u>0,\quad 0<\mathrm{Re}\,\nu<\tfrac{1}{2}\right]\quad \text{ET I 12(11)}$$

3.772

1. $\displaystyle\int_0^\infty \left(x^2+2\beta x\right)^{\nu-1/2}\sin(ax)\,dx = \frac{\sqrt{\pi}}{2}\left(\frac{2\beta}{a}\right)^\nu\Gamma\left(\nu+\frac{1}{2}\right)\left[J_{-\nu}(a\beta)\cos(a\beta)+Y_{-\nu}(a\beta)\sin(a\beta)\right]$

$$\left[a>0,\quad |\arg\beta|<\pi,\quad \tfrac{1}{2}>\mathrm{Re}\,\nu>-\tfrac{3}{2}\right]\quad \text{ET I 69(12)}$$

2. $\displaystyle\int_0^\infty \left(x^2+2\beta x\right)^{\nu-1/2}\cos(ax)\,dx$

$$= -\frac{\sqrt{\pi}}{2}\left(\frac{2\beta}{a}\right)^\nu\Gamma\left(\nu+\frac{1}{2}\right)\left[Y_{-\nu}(a\beta)\cos(a\beta)-J_{-\nu}(a\beta)\sin(a\beta)\right]$$

$$\left[a>0,\quad |\mathrm{Re}\,\nu|<\tfrac{1}{2}\right]\quad \text{ET I 12(13)}$$

3. $\displaystyle\int_0^{2u} \left(2ux-x^2\right)^{\nu-1/2}\sin(ax)\,dx = \sqrt{\pi}\left(\frac{2u}{a}\right)^\nu\Gamma\left(\nu+\frac{1}{2}\right)\sin(au)\,J_\nu(au)$

$$\left[a>0,\quad u>0,\quad \mathrm{Re}\,\nu>-\tfrac{1}{2}\right]$$
$$\text{ET I 69(13)a}$$

4. $\displaystyle\int_{2u}^\infty \left(x^2-2ux\right)^{\nu-1/2}\sin(ax)\,dx = \frac{\sqrt{\pi}}{2}\left(\frac{2\beta}{a}\right)^\nu\Gamma\left(\nu+\frac{1}{2}\right)\left[J_{-\nu}(au)\cos(au)-Y_{-\nu}(au)\sin(au)\right]$

$$\left[a>0,\quad u>0,\quad |\mathrm{Re}\,\nu|<\tfrac{1}{2}\right]\quad \text{ET I 70(14)}$$

5. $\displaystyle\int_0^{2u} \left(2ux-x^2\right)^{\nu-1/2}\cos(ax)\,dx = \sqrt{\pi}\left(\frac{2u}{a}\right)^\nu\Gamma\left(\nu+\frac{1}{2}\right)J_\nu(au)\cos(au)$

$$\left[a>0,\quad u>0,\quad \mathrm{Re}\,\nu>-\tfrac{1}{2}\right]$$
$$\text{ET I 12(4)}$$

6. $\displaystyle\int_{2u}^\infty \left(x^2-2ux\right)^{\nu-1/2}\cos(ax)\,dx = -\frac{\sqrt{\pi}}{2}\left(\frac{2u}{a}\right)^\nu\Gamma\left(\nu+\frac{1}{2}\right)\left[J_{-\nu}(au)\sin(au)+Y_{-\nu}(au)\cos(au)\right]$

$$\left[a>0,\quad u>0,\quad |\mathrm{Re}\,\nu|<\tfrac{1}{2}\right]\quad \text{ET I 12(12)}$$

3.773

1.[8] $\displaystyle\int_0^\infty \frac{x^{2\nu}}{(x^2+\beta^2)^{\mu+1}} \sin(ax)\, dx$

$$= \frac{1}{2}\beta^{2\nu-2\mu}a\, B\left(1+\nu, \mu-\nu\right)\, {}_1F_2\left(\nu+1;\ \nu+1-\mu, \frac{3}{2};\ \frac{\beta^2 a^2}{4}\right)$$

$$+ \frac{\sqrt{\pi}a^{2\mu-2\nu+1}}{4^{\mu-\nu+1}}\frac{\Gamma(\nu-\mu)}{\Gamma\left(\mu-\nu+\frac{3}{2}\right)}\, {}_1F_2\left(\mu+1;\ \mu-\nu+\frac{3}{2}, \mu-\nu+1;\ \frac{\beta^2 a^2}{4}\right)$$

$$= \frac{\sqrt{\pi}}{2\,\Gamma(\mu+1)}\beta^{2\nu-2\mu-1}\, G^{21}_{13}\left(\frac{a^2\beta^2}{4}\ \middle|\ \begin{matrix}-\nu+\frac{1}{2}\\ \mu-\nu+\frac{1}{2}, \frac{1}{2}, 0\end{matrix}\right)$$

$$\left[a>0,\quad \operatorname{Re}\beta>0,\quad -1<\operatorname{Re}\nu<\operatorname{Re}\mu+1\right]\quad \text{ET I 71(28)a, ET II 234(17)}$$

2.[8] $\displaystyle\int_0^\infty \frac{x^{2m+1}\sin(ax)}{(z+x^2)^{n+1}}\, dx = \frac{(-1)^{n+m}}{n!}\cdot\frac{\pi}{2}\frac{d^n}{dz^n}\left(z^m e^{-a\sqrt{z}}\right)$

$$\left[a>0,\quad 0\le m\le n,\quad |\arg z|<\pi\right]$$
$$\text{ET I 68(39)}$$

3. $\displaystyle\int_0^\infty \frac{x^{2m+1}\sin(ax)\, dx}{(\beta^2+x^2)^{n+\frac{1}{2}}} = \frac{(-1)^{m+1}\sqrt{\pi}}{2^n\beta^n\,\Gamma\left(n+\frac{1}{2}\right)}\frac{d^{2m+1}}{da^{2m+1}}\left[a^n\, K_n(a\beta)\right]$

$$\left[a>0,\quad \operatorname{Re}\beta>0,\quad -1\le m\le n\right]$$
$$\text{ET I 67(37)}$$

4. $\displaystyle\int_0^\infty \frac{x^{2\nu}\cos(ax)\, dx}{(x^2+\beta^2)^{\mu+1}} = \frac{1}{2}\beta^{2\nu-2\mu-1}\, B\left(\nu+\frac{1}{2}, \mu-\nu+\frac{1}{2}\right)\, {}_1F_2\left(\nu+\frac{1}{2};\nu-\mu+\frac{1}{2}, \frac{1}{2};\frac{\beta^2 a^2}{4}\right)$

$$+ \frac{\sqrt{\pi}a^{2\mu-2\nu+1}}{4^{\mu-\nu+1}}\frac{\Gamma\left(\nu-\mu-\frac{1}{2}\right)}{\Gamma(\mu-\nu+1)}\, {}_1F_2\left(\mu+1;\mu-\nu+1, \mu-\nu+\frac{3}{2};\frac{\beta^2 a^2}{4}\right)$$

$$= \frac{\sqrt{\pi}}{2\,\Gamma(\mu+1)}\beta^{2\nu-2\mu-1}\, G^{21}_{13}\left(\frac{a^2\beta^2}{4}\ \middle|\ \begin{matrix}-\nu+\frac{1}{2}\\ \mu-\nu+\frac{1}{2}, 0, \frac{1}{2}\end{matrix}\right)$$

$$\left[a>0,\quad \operatorname{Re}\beta>0,\quad -\frac{1}{2}<\operatorname{Re}\nu<\operatorname{Re}\mu+1\right]\quad \text{ET I 14(29)a, ET II 235(19)}$$

5. $\displaystyle\int_0^\infty \frac{x^{2m}\cos(ax)\, dx}{(z+x^2)^{n+1}} = (-1)^{m+n}\frac{\pi}{2\cdot n!}\cdot\frac{d^n}{dz^n}\left(z^{m-\frac{1}{2}}e^{-a\sqrt{z}}\right)$

$$\left[a>0,\quad n+1>m\ge 0,\quad |\arg z|<\pi\right]$$
$$\text{ET I 10(28)}$$

6.[7] $\displaystyle\int_0^\infty \frac{x^{2m}\cos(ax)\, dx}{(\beta^2+x^2)^{n+\frac{1}{2}}} = \frac{(-1)^m\sqrt{\pi}}{2^n\beta^n\,\Gamma\left(n+\frac{1}{2}\right)}\cdot\frac{d^{2m}}{da^{2m}}\left\{a^n\, K_n(a\beta)\right\}$

$$\left[a>0,\quad \operatorname{Re}\beta>0,\quad 0\le m<n+\frac{1}{2}\right]$$
$$\text{ET I 14(28)}$$

3.774

1. $\displaystyle\int_0^\infty \frac{\sin(ax)\, dx}{\sqrt{x^2+b^2}\left(x+\sqrt{x^2+b^2}\right)^\nu} = \frac{\pi}{b^\nu \sin(\nu\pi)}\left[\sin\frac{\nu\pi}{2}\, I_\nu(ab) + \frac{i}{2}\, \mathbf{J}_\nu(iab) - \frac{i}{2}\, \mathbf{J}_\nu(-iab)\right]$

$$\left[a>0,\quad b>0,\quad \operatorname{Re}\nu>-1\right]$$
$$\text{ET I 70(19)}$$

2. $$\int_0^\infty \frac{\cos(ax)\,dx}{\sqrt{x^2+b^2}\left(x+\sqrt{x^2+b^2}\right)^\nu} = \frac{\pi}{b^\nu \sin(\nu\pi)}\left[\frac{1}{2}\mathbf{J}_\nu(iab) + \frac{1}{2}\mathbf{J}_\nu(-iab) - \cos\frac{\nu\pi}{2}I_\nu(ab)\right]$$
$$[a>0, \quad b>0, \quad \operatorname{Re}\nu>-1]$$
ET I 12(15)

3. $$\int_0^\infty \frac{\left(x+\sqrt{x^2+\beta^2}\right)^\nu}{\sqrt{x\left(x^2+\beta^2\right)}}\sin(ax)\,dx = \sqrt{\frac{a\pi}{2}}\beta^\nu I_{\frac{1}{4}-\frac{\nu}{2}}\left(\frac{a\beta}{2}\right)K_{\frac{1}{4}+\frac{\nu}{2}}\left(\frac{a\beta}{2}\right)$$
$$[a>0, \quad \operatorname{Re}\beta>0, \quad \operatorname{Re}\nu<\tfrac{3}{2}]$$
ET I 71(23)

4. $$\int_0^\infty \frac{\left(\sqrt{x^2+\beta^2}-x\right)^\nu}{\sqrt{x\left(x^2+\beta^2\right)}}\cos(ax)\,dx = \sqrt{\frac{a\pi}{2}}\beta^\nu I_{-\frac{1}{4}+\frac{\nu}{2}}\left(\frac{a\beta}{2}\right)K_{-\frac{1}{4}-\frac{\nu}{2}}\left(\frac{a\beta}{2}\right)$$
$$[a>0, \quad \operatorname{Re}\beta>0, \quad \operatorname{Re}\nu>-\tfrac{3}{2}]$$
ET I 12(17)

5. $$\int_0^\infty \frac{\left(\beta+\sqrt{x^2+\beta^2}\right)^\nu}{x^{\nu+\frac{1}{2}}\sqrt{x^2+\beta^2}}\sin(ax)\,dx = \frac{1}{\beta}\sqrt{\frac{2}{a}}\,\Gamma\left(\frac{3}{4}-\frac{\nu}{2}\right)W_{\frac{\nu}{2},\frac{1}{4}}(a\beta)M_{-\frac{\nu}{2},\frac{1}{4}}(a\beta)$$
$$[a>0, \quad \operatorname{Re}\beta>0, \quad \operatorname{Re}\nu<\tfrac{3}{2}]$$
ET I 71(27)

6. $$\int_0^\infty \frac{\left(\beta+\sqrt{x^2+\beta^2}\right)^\nu}{x^{\nu+\frac{1}{2}}\sqrt{\beta^2+x^2}}\cos(ax)\,dx = \frac{1}{\beta\sqrt{2a}}\,\Gamma\left(\frac{1}{4}-\frac{\nu}{2}\right)W_{\frac{\nu}{2},-\frac{1}{4}}(a\beta)M_{-\frac{\nu}{2},-\frac{1}{4}}(a\beta)$$
$$[a>0, \quad \operatorname{Re}\beta>0, \quad \operatorname{Re}\nu<\tfrac{1}{2}]$$
ET I 12(18)

3.775

1. $$\int_0^\infty \frac{\left(\sqrt{x^2+\beta^2}+x\right)^\nu - \left(\sqrt{x^2+\beta^2}-x\right)^\nu}{\sqrt{x^2+\beta^2}}\sin(ax)\,dx = 2\beta^\nu \sin\frac{\nu\pi}{2}K_\nu(a\beta)$$
$$[a>0, \quad \operatorname{Re}\beta>0, \quad |\operatorname{Re}\nu|<1]$$
ET I 70(20)

2. $$\int_0^\infty \frac{\left(\sqrt{x^2+\beta^2}+x\right)^\nu + \left(\sqrt{x^2+\beta^2}-x\right)^\nu}{\sqrt{x^2+\beta^2}}\cos(ax)\,dx = 2\beta^\nu \cos\frac{\nu\pi}{2}K_\nu(a\beta)$$
$$[a>0, \quad \operatorname{Re}\beta>0, \quad |\operatorname{Re}\nu|<1]$$
ET I 13(22)

3. $$\int_u^\infty \frac{\left(x+\sqrt{x^2-u^2}\right)^\nu + \left(x-\sqrt{x^2-u^2}\right)^\nu}{\sqrt{x^2-u^2}}\sin(ax)\,dx = \pi u^\nu\left[J_\nu(au)\cos\frac{\nu\pi}{2} - Y_\nu(au)\sin\frac{\nu\pi}{2}\right]$$
$$[a>0, \quad u>0, \quad |\operatorname{Re}\nu|<1]$$
ET I 70(22)

4. $$\int_u^\infty \frac{\left(x+\sqrt{x^2-u^2}\right)^\nu + \left(x-\sqrt{x^2-u^2}\right)^\nu}{\sqrt{x^2-u^2}}\cos(ax)\,dx = -\pi u^\nu\left[Y_\nu(au)\cos\frac{\nu\pi}{2} + J_\nu(au)\sin\frac{\nu\pi}{2}\right]$$
$$[a>0, \quad u>0, \quad |\operatorname{Re}\nu|<1]$$
ET I 13(25)

5. $\int_0^u \dfrac{\left(x + i\sqrt{u^2 - x^2}\right)^\nu + \left(x - i\sqrt{u^2 - x^2}\right)^\nu}{\sqrt{u^2 - x^2}}\, \sin(ax)\, dx = \dfrac{\pi}{2} u^\nu \operatorname{cosec} \dfrac{\nu\pi}{2}\left[\mathbf{J}_\nu(au) - \mathbf{J}_{-\nu}(au)\right]$

$$[a > 0, \quad u > 0]$$

ET I 70(21)

6. $\int_0^u \dfrac{\left(x + i\sqrt{u^2 - x^2}\right)^\nu + \left(x - i\sqrt{u^2 - x^2}\right)^\nu}{\sqrt{u^2 - x^2}}\, \cos(ax)\, dx = \dfrac{\pi}{2} u^\nu \sec \dfrac{\nu\pi}{2}\left[\mathbf{J}_\nu(au) + \mathbf{J}_{-\nu}(au)\right]$

$$[a > 0, \quad u > 0, \quad |\operatorname{Re}\nu| < 1]$$

ET I 13(24)

7.⁶ $\int_u^\infty \dfrac{\left(x + \sqrt{x^2 - u^2}\right)^\nu + \left(x - \sqrt{x^2 - u^2}\right)^\nu}{\sqrt{x(x^2 - u^2)}}\, \sin(ax)\, dx$

$$= -\sqrt{\left(\dfrac{\pi}{2}\right)^3} au^\nu \left[J_{1/4+\nu/2}\left(\dfrac{au}{2}\right) Y_{1/4-\nu/2}\left(\dfrac{au}{2}\right) + J_{1/4-\nu/2}\left(\dfrac{au}{2}\right) Y_{1/4+\nu/2}\left(\dfrac{au}{2}\right)\right]$$

$$[a > 0, \quad u > 0, \quad |\operatorname{Re}\nu| < \tfrac{3}{2}] \quad \text{ET I 71(25)}$$

8.⁶ $\int_u^\infty \dfrac{\left(x + \sqrt{x^2 - u^2}\right)^\nu + \left(x - \sqrt{x^2 - u^2}\right)^\nu}{\sqrt{x(x^2 - u^2)}}\, \cos(ax)\, dx$

$$= -\sqrt{\left(\dfrac{\pi}{2}\right)^3} au^\nu \left[J_{-1/4+\nu/2}\left(\dfrac{au}{2}\right) Y_{-1/4-\nu/2}\left(\dfrac{au}{2}\right) + J_{-1/4-\nu/2}\left(\dfrac{au}{2}\right) Y_{-1/4+\nu/2}\left(\dfrac{au}{2}\right)\right]$$

$$[a > 0, \quad u > 0, \quad |\operatorname{Re}\nu| < \tfrac{3}{2}] \quad \text{ET I 13(26)}$$

9. $\int_0^\infty \dfrac{\left(x + \beta + \sqrt{x^2 + 2\beta x}\right)^\nu + \left(x + \beta - \sqrt{x^2 + 2\beta x}\right)^\nu}{\sqrt{x^2 + 2\beta x}}\, \sin(ax)\, dx$

$$= \pi\beta^\nu \left[Y_\nu(\beta a)\sin\left(\beta a - \dfrac{\nu\pi}{2}\right) + J_\nu(\beta a)\cos\left(\beta a - \dfrac{\nu\pi}{2}\right)\right]$$

$$[a > 0, \quad |\arg\beta| < \pi, \quad |\operatorname{Re}\nu| < 1] \quad \text{ET I 71(26)}$$

10. $\int_0^\infty \dfrac{\left(x + \beta + \sqrt{x^2 + 2\beta x}\right)^\nu + \left(x + \beta - \sqrt{x^2 + 2\beta x}\right)^\nu}{\sqrt{x^2 + 2\beta x}}\, \cos(ax)\, dx$

$$= \pi\beta^\nu \left[J_\nu(\beta a)\sin\left(\beta a - \dfrac{\nu\pi}{2}\right) - Y_\nu(\beta a)\cos\left(\beta a - \dfrac{\nu\pi}{2}\right)\right]$$

$$[a > 0, \quad |\arg\beta| < \pi, \quad |\operatorname{Re}\nu| < 1] \quad \text{ET I 13(23)}$$

11. $\int_0^{2u} \dfrac{\left(\sqrt{2u + x} + i\sqrt{2u - x}\right)^{4\nu} + \left(\sqrt{2u + x} - i\sqrt{2u - x}\right)^{4\nu}}{\sqrt{4u^2 x - x^3}}\, \cos(ax)\, dx$

$$= (4u)^{2\nu} \pi^{3/2} \sqrt{\dfrac{a}{2}}\, J_{\nu-1/4}(au)\, J_{-\nu-1/4}(au)$$

$$[a > 0, \quad u > 0] \quad \text{ET I 14(27)}$$

3.776

1. $\int_0^\infty \dfrac{a^2(b + x)^2 + p(p + 1)}{(b + x)^{p+2}}\, \sin(ax)\, dx = \dfrac{a}{b^p}$ $[a > 0, \quad b > 0, \quad p > 0]$ BI (170)(1)

2. $\int_0^\infty \dfrac{a^2(b + x)^2 + p(p + 1)}{(b + x)^{p+2}}\, \cos(ax)\, dx = \dfrac{p}{b^{p+1}}$ $[a > 0, \quad b > 0, \quad p > 0]$ BI (170)(2)

3.78–3.81 Rational functions of x and of trigonometric functions

3.781

1. $\displaystyle\int_0^\infty \left(\frac{\sin x}{x} - \frac{1}{1+x}\right)\frac{dx}{x} = 1 - C$ (cf. **3.784** 4 and **3.781** 2)   BI (173)(7)

2. $\displaystyle\int_0^\infty \left(\cos x - \frac{1}{1+x}\right)\frac{dx}{x} = -C$ BI (173)(8)

3.782

1. $\displaystyle\int_0^u \frac{1-\cos x}{x}\,dx - \int_u^\infty \frac{\cos x}{x}\,dx = C + \ln u$ $[u > 0]$ GW (333)(31)

2. $\displaystyle\int_0^\infty \frac{1-\cos ax}{x^2}\,dx = \frac{a\pi}{2}$ $[a \geq 0]$ BI (158)(1)

3. $\displaystyle\int_{-\infty}^\infty \frac{1-\cos ax}{x(x-b)}\,dx = \pi\frac{\sin ab}{b}$ $[a > 0, \quad b \text{ real}, \quad b \neq 0]$ ET II 253(48)

3.783

1. $\displaystyle\int_0^\infty \left[\frac{\cos x - 1}{x^2} + \frac{1}{2(1+x)}\right]\frac{dx}{x} = \frac{1}{2}C - \frac{3}{4}$ BI (173)(19)

2. $\displaystyle\int_0^\infty \left(\cos x - \frac{1}{1+x^2}\right)\frac{dx}{x} = -C$ EH I 17, BI(273)(21)

3.784

1. $\displaystyle\int_0^\infty \frac{\cos ax - \cos bx}{x}\,dx = \ln\frac{b}{a}$ $[a > 0, \quad b > 0]$ FI II 635, GW(333)(20)

2. $\displaystyle\int_0^\infty \frac{a\sin bx - b\sin ax}{x^2}\,dx = ab\ln\frac{a}{b}$ $[a > 0, \quad b > 0]$ FI II 647

3. $\displaystyle\int_0^\infty \frac{\cos ax - \cos bx}{x^2}\,dx = \frac{(b-a)\pi}{2}$ $[a \geq 0, \quad b \geq 0]$ BI(158)(12), FI II 645

4. $\displaystyle\int_0^\infty \frac{\sin x - x\cos x}{x^2}\,dx = 1$ BI (158)(3)

5. $\displaystyle\int_0^\infty \frac{\cos ax - \cos bx}{x(x+\beta)}\,dx = \frac{1}{\beta}\left[\operatorname{ci}(a\beta)\cos a\beta + \operatorname{si}(a\beta)\sin a\beta - \operatorname{ci}(b\beta)\cos b\beta - \operatorname{si}(b\beta)\sin b\beta + \ln\frac{b}{a}\right]$

$[a > 0, \quad b > 0, \quad |\arg\beta| < \pi]$ ET II 221(49)

6. $\displaystyle\int_0^\infty \frac{\cos ax + x\sin ax}{1+x^2}\,dx = \pi e^{-a}$ $[a > 0]$ GW (333)(73)

7. $\displaystyle\int_0^\infty \frac{\sin ax - ax\cos ax}{x^3}\,dx = \frac{\pi}{4}a^2\operatorname{sign}a$ LI (158)(5)

8. $\displaystyle\int_0^\infty \frac{\cos ax - \cos bx}{x^2(x^2+\beta^2)}\,dx = \frac{\pi\left[(b-a)\beta + e^{-b\beta} - e^{-a\beta}\right]}{2\beta^3}$

$[a > 0, \quad b > 0, \quad |\arg\beta| < \pi]$
BI(173)(20)a, ET II 222(59)

$9.^{10}$ $\displaystyle\int_0^\infty \frac{\cos mx}{1 + a^2\, T_n(x)} = \frac{\pi}{2n\sqrt{1+a^2}} \sum_{k=1}^n e^{-m \sin u \sinh \phi} \left(\cos \beta \sin u \cosh \phi + \sin \beta \cos u \sinh \phi\right)$

$$[u = (2k-1)\,\pi/(2n), \quad \phi = \operatorname{arcsinh}(1/a), \quad \beta = m \cos u \cosh \phi, \quad 0 < |a| < 1]$$

3.785 $\displaystyle\int_0^\infty \frac{1}{x} \sum_{k=1}^n a_k \cos b_k x\, dx = -\sum_{k=1}^n a_k \ln b_k$ $\qquad \left[b_k > 0, \quad \displaystyle\sum_{k=1}^n a_k = 0\right]$ $\qquad$ FI II 649

3.786

1. $\displaystyle\int_0^\infty \frac{(1 - \cos ax)\sin bx}{x^2}\, dx = \frac{b}{2} \ln \frac{b^2 - a^2}{b^2} + \frac{a}{2} \ln \frac{a+b}{a-b}$

$$[a > 0, \quad b > 0] \qquad \text{ET I 81(29)}$$

2. $\displaystyle\int_0^\infty \frac{(1 - \cos ax)\cos bx}{x}\, dx = \ln \frac{\sqrt{|a^2 - b^2|}}{b}$ $\qquad [0 > 0, \quad b > 0, \quad a \neq b]$ $\qquad$ FI II 647

3. $\displaystyle\int_0^\infty \frac{(1 - \cos ax)\cos bx}{x^2}\, dx = \frac{\pi}{2}(a - b)$ $\qquad [a < b \leq a]$

$$= 0 \qquad\qquad\qquad [0 < a \leq b]$$

$$\text{ET I 20(16)}$$

3.787

1. $\displaystyle\int_0^\infty \frac{(\cos a - \cos nax)\sin mx}{x}\, dx = \frac{\pi}{2}(\cos a - 1)$ $\qquad [m > na > 0]$

$$= \frac{\pi}{2} \cos a \qquad\qquad [na > m]$$

$$\text{BI(155)(7)}$$

2. $\displaystyle\int_0^\infty \frac{\sin^2 ax - \sin^2 bx}{x}\, dx = \frac{1}{2} \ln \frac{a}{b}$ $\qquad [a > 0, \quad b > 0]$ $\qquad$ GW (333)(20b)

3. $\displaystyle\int_0^\infty \frac{x^3 - \sin^3 x}{x^5}\, dx = \frac{13}{32}\pi$ $\qquad\qquad$ BI (158)(6)

4. $\displaystyle\int_0^\infty \frac{(3 - 4\sin^2 ax)\sin^2 ax}{x}\, dx = \frac{1}{2} \ln 2$ $\qquad [a \text{ real}, a \neq 0]$ $\qquad$ HBI (155)(6)

3.788 $\displaystyle\int_0^{\pi/2} \left(\frac{1}{x} - \cot x\right) dx = \ln \frac{\pi}{2}$ $\qquad\qquad$ GW (333)(61)a

3.789 $\displaystyle\int_0^{\pi/2} \frac{4x^2 \cos x + (\pi - x)x}{\sin x}\, dx = \pi^2 \ln 2$ $\qquad\qquad$ LI (206)(10)

3.791

1. $\displaystyle\int_0^{\pi/2} \frac{x\, dx}{1 + \sin x} = \ln 2$ $\qquad\qquad$ GW (333)(55a)

2. $\displaystyle\int_0^\pi \frac{x \cos x}{1 + \sin x}\, dx = \pi \ln 2 - 4G$ $\qquad\qquad$ GW (333)(55c)

3. $\displaystyle\int_0^{\pi/2} \frac{x \cos x}{1 + \sin x}\, dx = \pi \ln 2 - 2G$ $\qquad\qquad$ GW (333)(55b)

4.
$$\int_0^\pi \frac{\left(\frac{\pi}{2} - x\right)\cos x}{1 - \sin x}\, dx = 2\int_0^{\pi/2} \frac{\left(\frac{\pi}{2} - x\right)\cos x}{1 - \sin x}\, dx = \pi\ln 2 + 4\boldsymbol{G} = 5.8414484669\ldots$$

BI(207)(3), GW(333)(56c)

5.
$$\int_0^{\pi/2} \frac{x^2\, dx}{1 - \cos x} = -\frac{\pi^2}{4} + \pi\ln 2 + 4\boldsymbol{G} = 3.3740473667\ldots$$

BI (207)(3)

6.
$$\int_0^\pi \frac{x^2\, dx}{1 - \cos x} = 4\pi\ln 2$$

BI (219)(1)

7.
$$\int_0^{\pi/2} \frac{x^{p+1}\, dx}{1 - \cos x} = -\left(\frac{\pi}{2}\right)^{p+1} + \left(\frac{\pi}{2}\right)^p (p + 1)\left\{\frac{2}{p} - \sum_{k=1}^\infty \frac{1}{4^{2k-1}(p + 2k)}\,\zeta(2k)\right\}$$

$[p > 0]$

LI (207)(4)

8.
$$\int_0^{\pi/2} \frac{x\, dx}{1 + \cos x} = \frac{\pi}{2} - \ln 2$$

GW (333)(55a)

9.
$$\int_0^{\pi/2} \frac{x\sin x\, dx}{1 - \cos x} = \frac{\pi}{2}\ln 2 + 2\boldsymbol{G}$$

GW (333)(56a)

10.
$$\int_0^\pi \frac{x\sin x\, dx}{1 - \cos x} = 2\pi\ln 2$$

GW (333)(56b)

11.
$$\int_0^\pi \frac{x - \sin x}{1 - \cos x}\, dx = \frac{\pi}{2} + \int_0^{\pi/2} \frac{x - \sin x}{1 - \cos x}\, dx = 2$$

GW (333)(57a)

12.
$$\int_0^{\pi/2} \frac{x\sin x}{1 + \cos x}\, dx = -\frac{\pi}{2}\ln 2 + 2\boldsymbol{G}$$

GW (333)(55b)

3.792

1.
$$\int_{-\pi}^\pi \frac{dx}{1 - 2a\cos x + a^2} = \frac{2\pi}{1 - a^2} \qquad [a^2 < 1]$$

FI II 485

2.
$$\int_0^{\pi/2} \frac{x\cos x\, dx}{1 + 2a\sin x + a^2} = \frac{\pi}{2a}\ln(1 + a) - \sum_{k=0}^\infty (-1)^k \frac{a^{2k}}{(2k + 1)^2}$$

$[a^2 < 1]$

LI (241)(2)

3.
$$\int_0^\pi \frac{x\sin x\, dx}{1 - 2a\cos x + a^2} = \frac{\pi}{a}\ln(1 + a) \qquad [a^2 < 1, \quad a \neq 0]$$
$$= \frac{\pi}{a}\ln\left(1 + \frac{1}{a}\right) \qquad [a^2 < 1]$$

BI (221)(2)

4.
$$\int_0^{2\pi} \frac{x\sin x\, dx}{1 - 2a\cos x + a^2} = \frac{2\pi}{a}\ln(1 - a) \qquad [a^2 < 1, \quad a \neq 0]$$
$$= \frac{2\pi}{a}\ln\left(1 - \frac{1}{a}\right) \qquad [a^2 > 1]$$

BI (223)(4)

5.
$$\int_0^{2\pi} \frac{x\sin nx\, dx}{1 - 2a\cos x + a^2} = \frac{2\pi}{1 - a^2}\left[(a^{-n} - a^n)\ln(1 - a) + \sum_{k=1}^{n-1} \frac{a^{-k} - a^k}{n - k}\right]$$

$[a^2 < 1, \quad a \neq 0]$

BI (223)(5)

6. $\int_0^\infty \dfrac{\sin x}{1 - 2a\cos x + a^2} \cdot \dfrac{dx}{x} = \dfrac{\pi}{4a}\left[\left|\dfrac{1+a}{1-a}\right| - 1\right]$ $[a \text{ real}, \quad a \neq 0, \quad a \neq 1]$

GW (333)(62b)

7.[8] $\int_0^\infty \dfrac{\sin bx}{1 - 2a\cos x + a^2} \cdot \dfrac{dx}{x} = \dfrac{\pi}{2}\dfrac{1 + a - 2a^{[b]+1}}{(1-a^2)(1-a)}$ $[b \neq 0, 1, 2, \ldots]$

$= \dfrac{\pi}{2}\dfrac{1 + a - a^b - a^{b+1}}{(1-a^2)(1-a)}$ $[b = 1, 2, \ldots]; \qquad [0 < a < 1]$

ET I 81(26)

8. $\int_0^\infty \dfrac{\sin x \cos bx}{1 - 2a\cos x + a^2} \cdot \dfrac{dx}{x} = \dfrac{\pi}{2(1-a)}a^{[b]}$ $[b \neq 0, 1, 2, \ldots]$

$= \dfrac{\pi}{2(1-a)}a^b + \dfrac{\pi}{4}a^{b-1}$ $[b = 1, 2, 3, \ldots];$

$[0 < a < 1, \quad b > 0]; (\text{for } b = 0, \text{ see } \mathbf{3.792}\ 6)$ ET I 19(5)

9. $\int_0^\infty \dfrac{(1 - a\cos x)\sin bx}{1 - 2a\cos x + a^2} \cdot \dfrac{dx}{x} = \dfrac{\pi}{2} \cdot \dfrac{1 - a^{[b]+1}}{1-a}$ $[b \neq 1, 2, 3, \ldots]$

$= \dfrac{\pi}{2} \cdot \dfrac{1 - a^b}{1-a} + \dfrac{\pi a^b}{4}$ $[b = 1, 2, 3, \ldots]$

$[0 < a < 1, \quad b > 0]$ ET I 82(33)

10.[3] $\int_0^\infty \dfrac{1}{1 - 2a\cos bx + a^2}\dfrac{dx}{\beta^2 + x^2} = \dfrac{\pi}{2\beta(1-a^2)}\dfrac{1 + ae^{-b\beta}}{1 - ae^{-b\beta}}$

$[a^2 < 1, \quad b \geq 0]$ BI (192)(1)

11. $\int_0^\infty \dfrac{1}{1 - 2a\cos bx + a^2}\dfrac{dx}{\beta^2 - x^2} = \dfrac{a\pi}{\beta(1-a^2)}\dfrac{\sin b\beta}{1 - 2a\cos b\beta + a^2}$

$[a^2 < 1, \quad b > 0]$ BI (193)(1)

12. $\int_0^\infty \dfrac{\sin bcx}{1 - 2a\cos bx + a^2}\dfrac{x\,dx}{\beta^2 + x^2} = \dfrac{\pi}{2}\dfrac{e^{-\beta bc} - a^c}{(1 - ae^{-b\beta})(1 - ae^{b\beta})}$

$[a^2 < 1, \quad b > 0, \quad c > 0]$ BI (192)(8)

13. $\int_0^\infty \dfrac{\sin bx}{1 - 2a\cos bx + a^2}\dfrac{x\,dx}{\beta^2 + x^2} = \dfrac{\pi}{2}\dfrac{1}{e^{b\beta} - a}$ $[a^2 < 1, \quad b > 0]$

$= \dfrac{\pi}{2a}\dfrac{1}{ae^{b\beta} - 1}$ $[a^2 > 1, \quad b > 0]$

BI (192)(2)

14. $\int_0^\infty \dfrac{\sin bcx}{1 - 2a\cos bx + a^2}\dfrac{x\,dx}{\beta^2 - x^2} = \dfrac{\pi}{2}\dfrac{a^c - \cos\beta bc}{1 - 2a\cos\beta b + a^2}$

$[a^2 < 1, \quad b > 0, \quad c > 0]$ BI (193)(5)

15. $\int_0^\infty \dfrac{\cos bcx}{1 - 2a\cos bx + a^2}\dfrac{dx}{\beta^2 - x^2} = \dfrac{\pi}{2\beta(1-a^2)}\dfrac{(1 - a^2)\sin\beta bc + 2a^{c+1}\sin\beta b}{1 - 2a\cos\beta b + a^2}$

$[a^2 < 1, \quad b > 0, \quad c > 0]$ BI (193)(9)

16. $\int_0^\infty \dfrac{1 - a\cos bx}{1 - 2a\cos bx + a^2}\dfrac{dx}{1 + x^2} = \dfrac{\pi}{2}\dfrac{e^b}{e^b - a}$ $[a^2 < 1, \quad b > 0]$ FI II 719

17. $\int_0^\infty \dfrac{\cos bx}{1 - 2a\cos x + a^2} \cdot \dfrac{dx}{x^2 + \beta^2} = \dfrac{\pi \left(e^{\beta - \beta b} + a e^{\beta b}\right)}{2\beta \left(1 - a^2\right)\left(e^\beta - a\right)}$

$$[0 \leq b < 1, \quad |a| < 1, \quad \operatorname{Re}\beta > 0]$$
$$\text{ET I 21(21)}$$

18. $\int_0^\infty \dfrac{\sin bx \sin x}{1 - 2a\cos x + a^2} \cdot \dfrac{dx}{x^2 + \beta^2}$

$$= \dfrac{\pi}{2\beta} \dfrac{\sinh b\beta}{e^\beta - a} \qquad\qquad [0 \leq b < 1]$$

$$= \dfrac{\pi}{4\beta\left(ae^\beta - 1\right)} \left[a^m e^{\beta(m+1-b)} - e^{(1-b)\beta}\right]$$

$$\qquad - \dfrac{\pi}{4\beta\left(ae^{-\beta} - 1\right)} \left[a^m e^{-(m+1-b)\beta} - e^{-(1-b)\beta}\right] \quad [m \leq b \leq m + 1]$$

$$[0 < a < 1, \quad \operatorname{Re}\beta > 0] \qquad \text{ET I 81(27)}$$

19. $\int_0^\infty \dfrac{(\cos x - a)\cos bx}{1 - 2a\cos x + a^2} \cdot \dfrac{dx}{x^2 + \beta^2} = \dfrac{\pi\cosh\beta b}{2\beta\left(e^\beta - a\right)} \qquad [0 \leq b < 1, \quad |a| < 1, \quad \operatorname{Re}\beta > 0]$

$$\text{ET I 21(23)}$$

20. $\int_0^\infty \dfrac{\sin x}{(1 - 2a\cos 2x + a^2)^{n+1}} \dfrac{dx}{x} = \int_0^\infty \dfrac{\tan x}{(1 - 2a\cos 2x + a^2)^{n+1}} \dfrac{dx}{x}$

$$= \int_0^\infty \dfrac{\tan x}{(1 - 2a\cos 4x + a^2)^{n+1}} \dfrac{dx}{x} = \dfrac{\pi}{2(1 - a^2)^{2n+1}} \sum_{k=0}^n \binom{n}{k}^2 a^{2k}$$
$$\text{BI (187)(14)}$$

3.793

1.3 $\int_0^{2\pi} \dfrac{\sin nx - a\sin[(n+1)x]}{1 - 2a\cos x + a^2} x\, dx = -2\pi a^n \left[\ln(1 - a) + \sum_{k=1}^n \dfrac{1}{ka^k}\right]$

$$[|a| < 1] \qquad\qquad \text{BI (223)(9)}$$

2. $\int_0^{2\pi} \dfrac{\cos nx - a\cos[(n+1)x]}{1 - 2a\cos x + a^2} x\, dx = 2\pi a^n \qquad [a^2 < 1] \qquad \text{BI (223)(13)}$

3.794

1.3 $\int_0^\pi \dfrac{x\, dx}{1 + a^2 + 2a\cos x} = \dfrac{\pi^2}{2\left(1 - a^2\right)} + \dfrac{4}{\left(1 - a^2\right)} \sum_{k=0}^\infty \dfrac{a^{2k+1}}{(2k+1)^2}$

$$[a^2 < 1]$$

2. $\int_0^{2\pi} \dfrac{x\sin nx}{1 \pm a\cos x}\, dx = \dfrac{2\pi}{\sqrt{1 - a^2}} \left[(\mp 1)^n \dfrac{\left(1 + \sqrt{1 - a^2}\right)^n - \left(1 - \sqrt{1 - a^2}\right)^n}{a^n}\right.$

$$\left. \times \ln\dfrac{2\sqrt{1 \pm a}}{\sqrt{1 + a} + \sqrt{1 - a}} + \sum_{k=0}^{n-1} \dfrac{(\mp 1)^k}{n - k} \dfrac{\left(1 + \sqrt{1 - a^2}\right)^k - \left(1 - \sqrt{1 - a^2}\right)^k}{a^k}\right]$$

$$[a^2 < 1] \qquad\qquad \text{BI (223)(2)}$$

3.3 $\int_0^{2\pi} \dfrac{x\cos nx}{1 \pm a\cos x}\, dx = \dfrac{2\pi^2}{\sqrt{1 - a^2}} \left(\dfrac{1 - \sqrt{1 - a^2}}{\mp a}\right)^n \qquad [a^2 < 1] \qquad \text{BI (223)(3)}$

4. $\displaystyle\int_0^\pi \frac{x \sin x\, dx}{a + b \cos x} = \frac{\pi}{b} \ln \frac{a + \sqrt{a^2 - b^2}}{2(a - b)}$ $\qquad [a > |b| > 0]$ $\qquad$ GW (333)(53a)

5. $\displaystyle\int_0^{2\pi} \frac{x \sin x\, dx}{a + b \cos x} = \frac{2\pi}{b} \ln \frac{a + \sqrt{a^2 - b^2}}{2(a + b)}$ $\qquad [a > |b| > 0]$ $\qquad$ GW (333)(53b)

6. $\displaystyle\int_0^\infty \frac{\sin x}{a \pm b \cos 2x} \cdot \frac{dx}{x} = \frac{\pi}{2\sqrt{a^2 - b^2}}$ $\qquad \left[a^2 > b^2\right]$

$\qquad\qquad\qquad\qquad\qquad\quad = 0$ $\qquad \left[a^2 < b^2\right]$

$\qquad\qquad\qquad\qquad\qquad\qquad\qquad\qquad\qquad\qquad$ BI (181)(1)

3.795 $\displaystyle\int_{-\infty}^{\infty} \frac{\left(b^2 + c^2 + x^2\right) x \sin ax - \left(b^2 - c^2 - x^2\right) c \sinh ac}{\left[x^2 + (b - c)^2\right]\left[x^2 + (b + c)^2\right](\cos ax + \cosh ac)}\, dx = \pi$ $\qquad [c > b > 0]$

$\qquad\qquad\qquad\qquad\qquad\qquad\qquad\qquad\qquad\qquad\qquad\qquad\qquad = \dfrac{2\pi}{e^{ab} + 1}$ $\qquad [b > c > 0]$

$\qquad\qquad\qquad\qquad\qquad\qquad\qquad\qquad\qquad\qquad\qquad\qquad\quad [a > 0]$ $\qquad$ BI (202)(18)

3.796

1. $\displaystyle\int_0^{\pi/2} \frac{\cos x \pm \sin x}{\cos x \mp \sin x} x\, dx = \mp \frac{\pi}{4} \ln 2 - \boldsymbol{G}$ $\qquad$ BI (207)(8, 9)

2. $\displaystyle\int_0^{\pi/4} \frac{\cos x - \sin x}{\cos x + \sin x} x\, dx = \frac{\pi}{4} \ln 2 - \frac{1}{2}\boldsymbol{G}$ $\qquad$ BI (204)(23)

3.797

1. $\displaystyle\int_0^{\pi/4} \left(\frac{\pi}{4} - x \tan x\right) \tan x\, dx = \frac{1}{2} \ln 2 + \frac{\pi^2}{32} - \frac{\pi}{4} + \frac{\pi}{8} \ln 2$ $\qquad$ BI (204)(8)

2. $\displaystyle\int_0^{\pi/4} \frac{\left(\frac{\pi}{4} - x\right) \tan x\, dx}{\cos 2x} = -\frac{\pi}{8} \ln 2 + \frac{1}{2}\boldsymbol{G}$ $\qquad$ BI (204)(19)

3. $\displaystyle\int_0^{\pi/4} \frac{\frac{\pi}{4} - x \tan x}{\cos 2x}\, dx = \frac{\pi}{8} \ln 2 + \frac{1}{2}\boldsymbol{G}$ $\qquad$ BI (204)(20)

3.798

1.[8] $\displaystyle\int_0^\infty \frac{\tan x}{a + b \cos 2x} \cdot \frac{dx}{x} = \frac{\pi}{2\sqrt{a^2 - b^2}}$ $\qquad [0 < b < a]$

$\qquad\qquad\qquad\qquad\qquad\qquad\quad = 0$ $\qquad [0 < a < b]$

$\qquad\qquad\qquad\qquad\qquad\qquad\qquad\qquad\qquad\qquad$ BI (181)(2)

2.[8] $\displaystyle\int_0^\infty \frac{\tan x}{a + b \cos 4x} \cdot \frac{dx}{x} = \frac{\pi}{2\sqrt{a^2 - b^2}}$ $\qquad [0 < b < a]$

$\qquad\qquad\qquad\qquad\qquad\qquad\quad = 0$ $\qquad [0 < a < b]$

$\qquad\qquad\qquad\qquad\qquad\qquad\qquad\qquad\qquad\qquad$ BI (181)(3)

3.799

1. $\displaystyle\int_0^{\pi/2} \frac{x\, dx}{(\sin x + a \cos x)^2} = \frac{a}{1 + a^2} \frac{\pi}{2} - \frac{\ln a}{1 + a^2}$ $\qquad [a > 0]$ $\qquad$ BI (208)(5)

2. $\displaystyle\int_0^{\pi/4} \frac{x\,dx}{(\cos x + a\sin x)^2} = \frac{1}{1+a^2}\ln\frac{1+a}{\sqrt{2}} + \frac{\pi}{4}\cdot\frac{1-a}{(1+a)(1+a^2)}$

$$[a > 0] \qquad\qquad\qquad \text{BI (204)(24)}$$

3. $\displaystyle\int_0^{\pi} \frac{a\cos x + b}{(a + b\cos x)^2}x^2\,dx = \frac{2\pi}{b}\ln\frac{2(a-b)}{a+\sqrt{a^2-b^2}} \qquad [a > |b| > 0] \qquad \text{GW (333)(58a)}$

3.811

1. $\displaystyle\int_0^{\pi} \frac{\sin x}{1 - \cos t_1\cos x}\cdot\frac{x\,dx}{1 - \cos t_2\cos x} = \pi\cosec\frac{t_1 + t_2}{2}\cosec\frac{t_1 - t_2}{2}\ln\frac{1 + \tan\dfrac{t_1}{2}}{1 + \tan\dfrac{t_2}{2}}$

$$(\text{cf. } \mathbf{3.794}\ 4) \qquad\qquad \text{BI (222)(5)}$$

2. $\displaystyle\int_0^{\pi/2} \frac{x\,dx}{(\cos x \pm \sin x)\sin x} = \frac{\pi}{4}\ln 2 + \mathbf{G} \qquad\qquad\qquad\qquad \text{BI (208))(16, 17)}$

3. $\displaystyle\int_0^{\pi/4} \frac{x\,dx}{(\cos x + \sin x)\sin x} = -\frac{\pi}{8}\ln 2 + \mathbf{G} \qquad\qquad\qquad\qquad \text{BI (204)(29)}$

4. $\displaystyle\int_0^{\pi/4} \frac{x\,dx}{(\cos x + \sin x)\cos x} = \frac{\pi}{8}\ln 2 \qquad\qquad\qquad\qquad\qquad \text{BI (204)(28)}$

5. $\displaystyle\int_0^{\pi/4} \frac{\sin x}{\sin x + \cos x}\frac{x\,dx}{\cos^2 x} = -\frac{\pi}{8}\ln 2 + \frac{\pi}{4} - \frac{1}{2}\ln 2 \qquad\qquad \text{BI (204)(30)}$

3.812

1. $\displaystyle\int_0^{\pi} \frac{x\sin x\,dx}{a + b\cos^2 x} = \frac{\pi}{\sqrt{ab}}\arctan\sqrt{\frac{b}{a}} \qquad\qquad [a > 0, \quad b > 0]$

$$= \frac{\pi}{2\sqrt{-ab}}\ln\frac{\sqrt{a}+\sqrt{-b}}{\sqrt{a}-\sqrt{-b}} \qquad\qquad [a > -b > 0]$$

$$\text{GW (333)(60a)}$$

2. $\displaystyle\int_0^{\pi/2} \frac{x\sin 2x\,dx}{1 + a\cos^2 x} = \frac{\pi}{a}\ln\frac{1+\sqrt{1+a}}{2} \qquad\qquad [a > -1, \quad a \neq 0] \qquad \text{BI (207)(10)}$

3. $\displaystyle\int_0^{\pi/2} \frac{x\sin 2x\,dx}{1 + a\sin^2 x} = \frac{\pi}{a}\ln\frac{2\left(1+a-\sqrt{1+a}\right)}{2} \qquad\qquad [a > -1, \quad a \neq 0] \qquad \text{BI (207)(2)}$

4.[7] $\displaystyle\int_0^{\pi} \frac{x\,dx}{a^2 - \cos^2 x} = \frac{\pi^2}{2a\sqrt{a^2-1}} \qquad\qquad [a^2 > 1]$

$$= 0 \qquad\qquad\qquad\qquad [0 < a^2 < 1]$$

$$[\text{divergent if } a = 0] \qquad \text{BI (219)(10)}$$

5.[7] $\displaystyle\int_0^{\pi} \frac{x\sin x\,dx}{a^2 - \cos^2 x} = \frac{\pi}{2a}\ln\left|\frac{1+a}{1-a}\right| \qquad\qquad [0 < a < 1] \qquad \text{divergent if } a = 0$

$$\text{BI (219)(13)}$$

6.[6] $\displaystyle\int_0^\pi \frac{x \sin 2x\, dx}{a^2 - \cos^2 x} = \pi \ln \left\{ 4 \left(1 - a^2 \right) \right\}$ $[0 \le a^2 < 1]$

$\displaystyle\qquad\qquad = 2\pi \ln \left[2 \left(1 - a^2 + a\sqrt{a^2 - 1} \right) \right]$ $[a^2 > 1]$

BI (219)(19)

7. $\displaystyle\int_0^{\pi/2} \frac{x \sin x\, dx}{\cos^2 t - \sin^2 x} = -2 \operatorname{cosec} t \sum_{k=0}^\infty \frac{\sin(2k+1)t}{(2k+1)^2}$ BI (207)(1)

8. $\displaystyle\int_0^\pi \frac{x \sin x\, dx}{1 - \cos^2 t \sin^2 x} = \pi(\pi - 2t) \operatorname{cosec} 2t$ BI (219)(12)

9. $\displaystyle\int_0^\pi \frac{x \cos x\, dx}{\cos^2 t - \cos^2 x} = 4 \operatorname{cosec} t \sum_{k=0}^\infty \frac{\sin(2k+1)t}{(2k+1)^2}$ BI (219)(17)

10. $\displaystyle\int_0^\pi \frac{x \sin x\, dx}{\tan^2 t + \cos^2 x} = \frac{\pi}{2}(\pi - 2t) \cot t$ BI (219)(14)

11. $\displaystyle\int_0^\infty \frac{x\,(a \cos x + b) \sin x\, dx}{\cot^2 t + \cos^2 x} = 2a\pi \ln \cos \frac{t}{2} + \pi bt \tan t$ BI (219)(18)

3.813

1. $\displaystyle\int_0^\pi \frac{x\, dx}{a^2 \cos^2 x + b^2 \sin^2 x} = \frac{1}{4} \int_0^{2\pi} \frac{x\, dx}{a^2 \cos^2 x + b^2 \sin^2 x} = \frac{\pi^2}{2ab}$

$[a > 0, \quad b > 0]$ GW (333)(36)

2. $\displaystyle\int_0^\infty \frac{1}{\beta^2 \sin^2 ax + \gamma^2 \cos^2 ax} \cdot \frac{dx}{x^2 + \delta^2} = \frac{\pi \sinh(2a\delta)}{4\delta \left(\beta^2 \sinh^2(a\delta) - \gamma^2 \cosh^2(a\delta) \right)} \left[\frac{\beta}{\gamma} - \frac{\gamma}{\beta} - \frac{2}{\sinh(2a\delta)} \right]$

$\displaystyle\left[\left| \arg \frac{\beta}{\gamma} \right| < \pi, \quad \operatorname{Re} \delta > 0, \quad a > 0 \right]$

GW(333)(81), ET II 222(63)

3. $\displaystyle\int_0^\infty \frac{\sin x\, dx}{x \left(a^2 \sin^2 x + b^2 \cos^2 x \right)} = \frac{\pi}{2ab}$ $[ab > 0]$ BI (181)(8)

4. $\displaystyle\int_0^\infty \frac{\sin^2 x\, dx}{x \left(a^2 \cos^2 x + b^2 \sin^2 x \right)} = \frac{\pi}{2b(a+b)}$ $[a > 0, \quad b > 0]$ BI (181)(11)

5. $\displaystyle\int_0^{\pi/2} \frac{x \sin 2x\, dx}{a^2 \cos^2 x + b^2 \sin^2 x} = \frac{\pi}{a^2 - b^2} \ln \frac{a+b}{2b}$ $[a > 0, \quad b > 0, \quad a \ne b]$ GW (333)(52a)

6. $\displaystyle\int_0^\pi \frac{x \sin 2x\, dx}{a^2 \cos^2 x + b^2 \sin^2 x} = \frac{2\pi}{a^2 - b^2} \ln \frac{a+b}{2a}$ $[a > 0, \quad b > 0, \quad a \ne b]$ GW (333)(52b)

7. $\displaystyle\int_0^\infty \frac{\sin 2x}{a^2 \cos^2 x + b^2 \sin^2 x} \cdot \frac{dx}{x} = \frac{\pi}{a(a+b)}$ $[a > 0, \quad b > 0]$ BI (182)(3)

8. $\displaystyle\int_0^\infty \frac{\sin 2ax}{\beta^2 \sin^2 ax + \gamma^2 \cos^2 ax} \cdot \frac{x\, dx}{x^2 + \delta^2} = \frac{\pi}{2 \left(\beta^2 \sinh^2(a\delta) - \gamma^2 \cosh^2(a\delta) \right)} \left[\frac{\beta - \gamma}{\beta + \gamma} - e^{-2a\delta} \right]$

$\displaystyle\left[a > 0, \quad \left| \arg \frac{\beta}{\gamma} \right| < \pi, \quad \operatorname{Re} \delta > 0 \right]$

ET II 222(64), GW(333)(80)

9. $\displaystyle\int_0^\infty \frac{(1-\cos x)\sin x}{a^2\cos^2 x + b^2\sin^2 x}\cdot\frac{dx}{x} = \frac{\pi}{2b(a+b)}$ $[a>0,\quad b>0]$ BI (182)(7)a

10. $\displaystyle\int_0^\infty \frac{\sin x\cos^2 x}{a^2\cos^2 x + b^2\sin^2 x}\cdot\frac{dx}{x} = \frac{\pi}{2a(a+b)}$ $[a>0,\quad b>0]$ BI (182)(4)

11. $\displaystyle\int_0^\infty \frac{\sin^3 x}{a^2\cos^2 x + b^2\sin^2 x}\cdot\frac{dx}{x} = \frac{\pi}{2b}\cdot\frac{2}{a+b}$ $[a>0,\quad b>0]$ BI (182)(1)

3.814

1. $\displaystyle\int_0^{\pi/2} \frac{(1-x\cot x)\,dx}{\sin^2 x} = \frac{\pi}{4}$ BI (206)(9)

2. $\displaystyle\int_0^{\pi/4} \frac{x\tan x\,dx}{(\sin x+\cos x)\cos x} = -\frac{\pi}{8}\ln 2 + \frac{\pi}{4} - \frac{1}{2}\ln 2$ BI (204)(30)

3. $\displaystyle\int_0^\infty \frac{\tan x}{a^2\cos^2 x + b^2\sin^2 x}\frac{dx}{x} = \frac{\pi}{2ab}$ $[a>0,\quad b>0]$ BI (181)(9)

4. $\displaystyle\int_0^{\pi/2} \frac{x\cot x\,dx}{a^2\cos^2 x + b^2\sin^2 x} = \frac{\pi}{2a^2}\ln\frac{a+b}{b}$ $[a>0,\quad b>0]$ LI (208)(20)

5. $\displaystyle\int_0^{\pi/2} \frac{\left(\frac{\pi}{2}-x\right)\tan x\,dx}{a^2\cos^2 x + b^2\sin^2 x} = \frac{1}{2}\int_0^\pi \frac{\left(\frac{\pi}{2}-x\right)\tan x\,dx}{a^2\cos^2 x + b^2\sin^2 x}$
$$= \frac{\pi}{2b^2}\ln\frac{a+b}{a}$$
     $[a>0,\quad b>0]$ GW (333)(59)

6. $\displaystyle\int_0^\infty \frac{\sin^2 x\tan x}{a^2\cos^2 x + b^2\sin^2 x}\cdot\frac{dx}{x} = \frac{\pi}{2b(a+b)}$ $[a>0,\quad b>0]$ BI (182)(6)

7. $\displaystyle\int_0^\infty \frac{\tan x}{a^2\cos^2 2x + b^2\sin^2 2x}\cdot\frac{dx}{x} = \frac{\pi}{2ab}$ $[a>0,\quad b>0]$ BI (181)(10)a

8. $\displaystyle\int_0^\infty \frac{\sin^2 2x\tan x}{a^2\cos^2 2x + b^2\sin^2 2x}\cdot\frac{dx}{x} = \frac{\pi}{2b}\cdot\frac{1}{a+b}$ $[a>0,\quad b>0]$ BI (182)(2)a

9. $\displaystyle\int_0^\infty \frac{\cos^2 2x\tan x}{a^2\cos^2 2x + b^2\sin^2 2x}\cdot\frac{dx}{x} = \frac{\pi}{2a}\cdot\frac{1}{a+b}$ $[a>0,\quad b>0]$ BI (182)(5)a

10. $\displaystyle\int_0^\infty \frac{\sin^2 x\cos x}{a^2\cos^2 2x + b^2\sin^2 2x}\cdot\frac{dx}{x\cos 4x} = -\frac{\pi}{8b}\frac{a}{a^2+b^2}$ $[a>0,\quad b>0]$ BI (186)(12)a

11. $\displaystyle\int_0^\infty \frac{\sin x}{a^2\cos^2 x + b^2\sin^2 x}\cdot\frac{dx}{x\cos 2x} = \frac{\pi}{2ab}\cdot\frac{b^2-a^2}{b^2+a^2}$ $[a>0,\quad b>0]$ BI (186)(4)a

12. $\displaystyle\int_0^\infty \frac{\sin x\cos x}{a^2\cos^2 x + b^2\sin^2 x}\cdot\frac{dx}{x\cos 2x} = \frac{\pi}{2a}\cdot\frac{b}{a^2+b^2}$ $[a>0,\quad b>0]$ BI (186)(7)a

13. $\displaystyle\int_0^\infty \frac{\sin x\cos^2 x}{a^2\cos^2 x + b^2\sin^2 x}\cdot\frac{dx}{x\cos 2x} = \frac{\pi}{2ab}\cdot\frac{b^2}{a^2+b^2}$ $[a>0,\quad b>0]$ BI (186)(8)a

14. $\displaystyle\int_0^\infty \frac{\sin^3 x}{a^2\cos^2 x + b^2\sin^2 x}\cdot\frac{dx}{x\cos 2x} = -\frac{\pi}{2b}\cdot\frac{a}{a^2+b^2}$ $[a>0,\quad b>0]$ BI (186)(10)

15. $\displaystyle\int_0^\infty \frac{1-\cos x}{a^2\cos^2 x + b^2\sin^2 x}\cdot\frac{dx}{x\sin x} = \frac{\pi}{2ab}$ $[a>0,\quad b>0]$ BI (186)(3)a

3.815

1. $\displaystyle\int_0^{\pi/2} \frac{x\sin 2x\,dx}{\left(1+a\sin^2 x\right)\left(1+b\sin^2 x\right)} = \frac{\pi}{a-b}\ln\left\{\frac{1+\sqrt{1+b}}{1+\sqrt{1+a}}\cdot\frac{\sqrt{1+a}}{\sqrt{1+b}}\right\}$

$[a>0,\quad b>0]$ (cf. **3.812** 3)

BI (208)(22)

2. $\displaystyle\int_0^{\pi/2} \frac{x\sin 2x\,dx}{\left(1+a\sin^2 x\right)\left(1+b\cos^2 x\right)} = \frac{\pi}{a+ab+b}\ln\frac{\left(1+\sqrt{1+n}\right)\sqrt{1+a}}{1+\sqrt{1+a}}$

$[a>0,\quad b>0]$ (cf. **3.812** 2 and 3)

BI (208)(24)

3. $\displaystyle\int_0^{\pi/2} \frac{x\sin 2x\,dx}{\left(1+a\cos^2 x\right)\left(1+b\cos^2 x\right)} = \frac{\pi}{a-b}\ln\frac{1+\sqrt{1+a}}{1+\sqrt{1+b}}$

$[a>0,\quad b>0]$ (cf. **3.812** 2)

BI (208)(23)

4. $\displaystyle\int_0^{\pi/2} \frac{x\sin 2x\,dx}{\left(1-\sin^2 t_1\cos^2 x\right)\left(1-\sin^2 t_2\cos^2 x\right)} = \frac{2\pi}{\cos^2 t_1 - \cos^2 t_2}\ln\frac{\cos\dfrac{t_1}{2}}{\cos\dfrac{t_2}{2}}$

$[-\pi < t_1 < \pi,\quad -\pi < t_2 < \pi]$

BI (208)(21)

3.816

1. $\displaystyle\int_0^{\pi} \frac{x^2\sin 2x}{\left(a^2-\cos^2 x\right)^2}\,dx = \pi^2\frac{\sqrt{a^2-1}-a}{a\left(a^2-1\right)}$ $[a>1]$ LI (220)(9)

2.[7] $\displaystyle\int_0^{\pi} \frac{\left(a^2-1-\sin^2 x\right)\cos x}{\left(a^2-\cos^2 x\right)^2}x^2\,dx = \frac{\pi}{2}\ln\left|\frac{1-a}{1+a}\right|$ $[a^2>1]$ (cf. **3.812** 5)   BI (220)(12)

3. $\displaystyle\int_0^{\pi} \frac{a\cos 2x - \sin^2 x}{\left(a+\sin^2 x\right)^2}x^2\,dx = -2\pi\ln\left[2\left(-a+\sqrt{a(a+1)}\right)\right]$

$[a>0]$ LI (220)(10)

4. $\displaystyle\int_0^{\pi} \frac{a\cos 2x + \sin^2 x}{\left(a-\sin^2 x\right)^2}x^2\,dx = \pi\ln(4a)$ $[a>1]$ (cf. **3.812** 6)   LI (220)(11)

3.817

1. $\displaystyle\int_0^\infty \frac{\sin x}{\left(a^2\cos^2 x + b^2\sin^2 x\right)^2}\cdot\frac{dx}{x} = \frac{\pi}{4}\cdot\frac{a^2+b^2}{a^3b^3}$ $[ab>0]$ BI (181)(12)

2. $\displaystyle\int_0^\infty \frac{\sin x\cos x}{\left(a^2\cos^2 x + b^2\sin^2 x\right)^2}\cdot\frac{dx}{x} = \frac{\pi}{4a^3b}$ $[ab>0]$ BI (182)(8)

3. $\displaystyle\int_0^\infty \frac{\sin^3 x}{\left(a^2\cos^2 x + b^2\sin^2 x\right)^2}\cdot\frac{dx}{x} = \frac{\pi}{4ab^3}$ $[ab>0]$ BI (181)(15)

4. $\displaystyle\int_0^\infty \frac{\sin x \cos^2 x}{\left(a^2\cos^2 x + b^2\sin^2 x\right)^2} \cdot \frac{dx}{x} = \frac{\pi}{4a^3 b}$ $[ab > 0]$ BI (182)(9)

5. $\displaystyle\int_0^\infty \frac{\tan x}{\left(a^2\cos^2 x + b^2\sin^2 x\right)^2} \cdot \frac{dx}{x} = \frac{\pi}{4}\cdot\frac{a^2 + b^2}{a^3 b^3}$ $[ab > 0]$ BI (181)(13)

6. $\displaystyle\int_0^\infty \frac{\tan x}{\left(a^2\cos^2 2x + b^2\sin^2 2x\right)^2} \cdot \frac{dx}{x} = \frac{\pi}{4}\frac{a^2 + b^2}{a^3 b^3}$ $[ab > 0]$ BI (181)(14)

7. $\displaystyle\int_0^\infty \frac{\sin^2 x \tan x}{\left(a^2\cos^2 x + b^2\sin^2 x\right)^2} \cdot \frac{dx}{x} = \frac{\pi}{4ab^3}$ $[ab > 0]$ BI (182)(11)

8. $\displaystyle\int_0^\infty \frac{\tan x \cos^2 2x}{\left(a^2\cos^2 2x + b^2\sin^2 2x\right)^2} \cdot \frac{dx}{x} = \frac{\pi}{4a^3 b}$ $[ab > 0]$ BI (182)(10)

3.818

1. $\displaystyle\int_0^\infty \frac{\sin x}{\left(a^2\cos^2 x + b^2\sin^2 x\right)^3} \cdot \frac{dx}{x} = \frac{\pi}{16}\cdot\frac{3a^4 + 2a^2 b^2 + 3b^4}{a^5 b^5}$

     $[ab > 0]$ BI (181)(16)

2. $\displaystyle\int_0^\infty \frac{\sin x \cos x}{\left(a^2\cos^2 x + b^2\sin^2 x\right)^3} \cdot \frac{dx}{x} = \frac{\pi}{16}\cdot\frac{a^2 + 3b^2}{a^5 b^3}$ $[ab > 0]$ BI (182)(13)

3. $\displaystyle\int_0^\infty \frac{\sin x \cos^2 x}{\left(a^2\cos^2 x + b^2\sin^2 x\right)^3} \cdot \frac{dx}{x} = \frac{\pi}{16}\cdot\frac{a^2 + 3b^2}{a^5 b^3}$ $[ab > 0]$ BI (182)(14)

4. $\displaystyle\int_0^\infty \frac{\sin^3 x}{\left(a^2\cos^2 x + b^2\sin^2 x\right)^3} \cdot \frac{dx}{x} = \frac{\pi}{16}\cdot\frac{3a^2 + b^2}{a^3 b^5}$ $[ab > 0]$ LI (181)(19)

5. $\displaystyle\int_0^\infty \frac{\sin^3 x \cos x}{\left(a^2\cos^2 2x + b^2\sin^2 2x\right)^3} \cdot \frac{dx}{x} = \frac{\pi}{64}\cdot\frac{3a^2 + b^2}{a^3 b^5}$ $[ab > 0]$ BI (182)(17)

6. $\displaystyle\int_0^\infty \frac{\tan x}{\left(a^2\cos^2 x + b^2\sin^2 x\right)^3} \cdot \frac{dx}{x} = \frac{\pi}{16}\cdot\frac{3a^4 + 2a^2 b^2 + 3b^4}{a^5 b^5}$

     $[ab > 0]$ BI (181)(17)

7. $\displaystyle\int_0^\infty \frac{\sin^2 x \tan x}{\left(a^2\cos^2 x + b^2\sin^2 x\right)^3} \cdot \frac{dx}{x} = \frac{\pi}{16}\cdot\frac{3a^2 + b^2}{a^3 b^5}$ $[ab > 0]$ BI (182)(16)

8. $\displaystyle\int_0^\infty \frac{\tan x}{\left(a^2\cos^2 2x + b^2\sin^2 2x\right)^3} \cdot \frac{dx}{x} = \frac{\pi}{16}\cdot\frac{3a^4 + 2a^2 b^2 + 3b^4}{a^5 b^5}$

     $[ab > 0]$ BI (181)(18)

9. $\displaystyle\int_0^\infty \frac{\tan x \cos^2 2x}{\left(a^2\cos^2 2x + b^2\sin^2 2x\right)^3} \cdot \frac{dx}{x} = \frac{\pi}{16}\cdot\frac{a^2 + 3b^2}{a^5 b^3}$ $[ab > 0]$ BI (182)(15)

3.819

1. $\int_0^\infty \dfrac{\sin x}{\left(a^2 \cos^2 x + b^2 \sin^2 x\right)^4} \cdot \dfrac{dx}{x} = \dfrac{\pi}{32} \cdot \dfrac{5a^6 + 3a^4 b^2 + 3a^2 b^4 + 5b^6}{a^7 b^7}$

$$[ab > 0] \qquad \text{BI (181)(20)}$$

2. $\int_0^\infty \dfrac{\sin x \cos x}{\left(a^2 \cos^2 x + b^2 \sin^2 x\right)^4} \cdot \dfrac{dx}{x} = \dfrac{\pi}{32} \cdot \dfrac{a^4 + 2a^2 b^2 + 5b^4}{a^7 b^5}$

$$[ab > 0] \qquad \text{BI (182)(18)}$$

3. $\int_0^\infty \dfrac{\sin x \cos^2 x}{\left(a^2 \cos^2 x + b^2 \sin^2 x\right)^4} \cdot \dfrac{dx}{x} = \dfrac{\pi}{32} \cdot \dfrac{a^4 + 2a^2 b^2 + 5b^4}{a^7 b^5}$

$$[ab > 0] \qquad \text{BI (182)(19)}$$

4. $\int_0^\infty \dfrac{\sin^3 x}{\left(a^2 \cos^2 x + b^2 \sin^2 x\right)^4} \cdot \dfrac{dx}{x} = \dfrac{\pi}{32} \cdot \dfrac{5a^4 + a^2 b^2 + b^4}{a^5 b^7}$

$$[ab > 0] \qquad \text{BI (181)(23)}$$

5. $\int_0^\infty \dfrac{\sin^3 x \cos x}{\left(a^2 \cos^2 x + b^2 \sin^2 x\right)^4} \cdot \dfrac{dx}{x} = \dfrac{\pi}{32} \cdot \dfrac{a^2 + b^2}{a^5 b^5} \qquad [ab > 0] \qquad \text{BI (182)(26)}$

6. $\int_0^\infty \dfrac{\sin x \cos^3 x}{\left(a^2 \cos^2 x + b^2 \sin^2 x\right)^4} \cdot \dfrac{dx}{x} = \dfrac{\pi}{32} \cdot \dfrac{a^2 + 5b^2}{a^7 b^3} \qquad [ab > 0] \qquad \text{BI (182)(23)}$

7. $\int_0^\infty \dfrac{\sin^3 x \cos^2 x}{\left(a^2 \cos^2 x + b^2 \sin^2 x\right)^4} \cdot \dfrac{dx}{x} = \dfrac{\pi}{32} \cdot \dfrac{a^2 + b^2}{a^5 b^5} \qquad [ab > 0] \qquad \text{BI (182)(27)}$

8. $\int_0^\infty \dfrac{\sin x \cos^4 x}{\left(a^2 \cos^2 x + b^2 \sin^2 x\right)^4} \cdot \dfrac{dx}{x} = \dfrac{\pi}{32} \cdot \dfrac{a^2 + 5b^2}{a^7 b^3} \qquad [ab > 0] \qquad \text{BI (182)(24)}$

9. $\int_0^\infty \dfrac{\sin^5 x}{\left(a^2 \cos^2 x + b^2 \sin^2 x\right)^4} \cdot \dfrac{dx}{x} = \dfrac{\pi}{32} \cdot \dfrac{5a^2 + b^2}{a^3 b^7} \qquad [ab > 0] \qquad \text{BI (181)(24)}$

10. $\int_0^\infty \dfrac{\sin^3 x \cos x}{\left(a^2 \cos^2 2x + b^2 \sin^2 2x\right)^4} \cdot \dfrac{dx}{x} = \dfrac{\pi}{128} \cdot \dfrac{5a^4 + 2a^2 b^2 + b^4}{a^5 b^7}$

$$[ab > 0] \qquad \text{BI (182)(22)}$$

11. $\int_0^\infty \dfrac{\sin^5 x \cos^3 x}{\left(a^2 \cos^2 2x + b^2 \sin^2 2x\right)^4} \cdot \dfrac{dx}{x} = \dfrac{\pi}{512} \cdot \dfrac{5a^2 + b^2}{a^3 b^7} \qquad [ab > 0] \qquad \text{BI (182)(30)}$

12. $\int_0^\infty \dfrac{\sin^2 x \tan x}{\left(a^2 \cos^2 x + b^2 \sin^2 x\right)^4} \cdot \dfrac{dx}{x} = \dfrac{\pi}{32} \cdot \dfrac{5a^4 + 2a^2 b^2 + b^4}{a^5 b^7}$

$$[ab > 0] \qquad \text{BI (182)(21)}$$

13. $\int_0^\infty \dfrac{\sin^4 x \tan x}{\left(a^2 \cos^2 x + b^2 \sin^2 x\right)^4} \cdot \dfrac{dx}{x} = \dfrac{\pi}{32} \cdot \dfrac{5a^2 + b^2}{a^3 b^7} \qquad [ab > 0] \qquad \text{BI (182)(29)}$

14. $\int_0^\infty \dfrac{\cos^2 2x \tan x}{\left(a^2 \cos^2 2x + b^2 \sin^2 2x\right)^4} \cdot \dfrac{dx}{x} = \dfrac{\pi}{32} \cdot \dfrac{a^4 + 2a^2 b^2 + 5b^4}{a^7 b^5}$

$\qquad\qquad\qquad\qquad\qquad\qquad\qquad\qquad\qquad\qquad [ab > 0]$ BI (182)(29)

15. $\int_0^\infty \dfrac{\sin^3 4x \tan x}{\left(a^2 \cos^2 2x + b^2 \sin^2 2x\right)^4} \cdot \dfrac{dx}{x} = \dfrac{\pi}{8} \cdot \dfrac{a^2 + b^2}{a^5 b^5}$ $[ab > 0]$ BI (182)(28)

16. $\int_0^\infty \dfrac{\cos^4 2x \tan x}{\left(a^2 \cos^2 2x + b^2 \sin^2 2x\right)^4} \cdot \dfrac{dx}{x} = \dfrac{\pi}{32} \cdot \dfrac{a^2 + 5b^2}{a^7 b^3}$ $[ab > 0]$ BI (182)(25)

3.82–3.83 Powers of trigonometric functions combined with other powers

3.821

1. $\int_0^\pi x \sin^p x \, dx = \dfrac{\pi^2}{2^{p+1}} \dfrac{\Gamma(p + 1)}{\left[\Gamma\left(\frac{p}{2} + 1\right)\right]^2}$ $[p > -1]$ BI(218)(7), LO V 121(71)

2. $\int_0^{r\pi} x \sin^n x \, dx = \dfrac{\pi^2}{2} \cdot \dfrac{(2m - 1)!!}{(2m)!!} r^2$ $[n = 2m]$

$\qquad\qquad\qquad = (-1)^{r+1} \pi \dfrac{(2m)!!}{(2m + 1)!!} r$ $[n = 2m + 1]$

$\qquad\qquad\qquad\qquad\qquad\qquad\qquad\qquad\qquad [r \text{ is a natural number}]$ GW (333)(8c)

3. $\int_0^{\pi/2} x \cos^n x \, dx = -\sum_{k=0}^{m-1} \dfrac{(n - 2k + 1)(n - 2k + 3) \cdots (n - 1)}{(n - 2k)(n - 2k + 2) \cdots n} \dfrac{1}{n - 2k}$

$\qquad\qquad\qquad\qquad + \begin{cases} \dfrac{\pi}{2} \cdot \dfrac{(2m - 2)!!}{(2m - 1)!!} & [n = 2m - 1] \\[2mm] \dfrac{\pi^2}{8} \cdot \dfrac{(2m - 1)!!}{(2m)!!} & [n = 2m] \end{cases}$

$\qquad\qquad\qquad\qquad\qquad\qquad\qquad\qquad\qquad\qquad\qquad$ GW (333)(9b)

4. $\int_0^\pi x \cos^{2m} x \, dx = \dfrac{\pi^2}{2} \dfrac{(2m - 1)!!}{(2m)!!}$ BI (218)(10)

5. $\int_{r\pi}^{s\pi} x \cos^{2m} x \, dx = \dfrac{\pi^2}{2} \left(s^2 - r^2\right) \dfrac{(2m - 1)!!}{(2m)!!}$ BI (226)(3)

6. $\int_0^\infty \dfrac{\sin^p x}{x} \, dx = \dfrac{\sqrt{\pi}}{2} \cdot \dfrac{\Gamma\left(\frac{p}{2}\right)}{\Gamma\left(\frac{p+1}{2}\right)} = 2^{p-2} \mathrm{B}\left(\dfrac{p}{2}, \dfrac{p}{2}\right)$

$\qquad\qquad [p \text{ is a fraction with odd numerator and denominator}]$ LO V 278, FI II 808

7. $\int_0^\infty \dfrac{\sin^{2n+1} x}{x} \, dx = \dfrac{(2n - 1)!!}{(2n)!!} \cdot \dfrac{\pi}{2}$ BI (151)(4)

8. $\int_0^\infty \dfrac{\sin^{2n} x}{x} \, dx = \infty$ BI (151)(3)

9. $\int_0^\infty \dfrac{\sin^2 ax}{x^2} \, dx = \dfrac{a\pi}{2}$ $[a > 0]$ LO V 307, 312, FI II 632

10. $$\int_0^\infty \frac{\sin^{2m} ax}{x^2}\, dx = \frac{(2m-3)!!}{(2m-2)!!} \cdot \frac{a\pi}{2} \qquad\qquad [a>0] \qquad\qquad \text{GW (333)(14b)}$$

11. $$\int_0^\infty \frac{\sin^{2m+1} ax}{x^3}\, dx = \frac{(2m-3)!!}{(2m)!!}(2m+1)\frac{a^2\pi}{4} \qquad\qquad [a>0] \qquad\qquad \text{GW (333)(14d)}$$

12. $$\int_0^\infty \frac{\sin^p x}{x^m}\, dx$$

$$= \frac{p}{m-1}\int_0^\infty \frac{\sin^{p-1} x}{x^{m-1}}\cos x\, dx \qquad\qquad\qquad\qquad [p>m-1>0]$$

$$= \frac{p(p-1)}{(m-1)(m-2)}\int_0^\infty \frac{\sin^{p-2} x}{x^{m-2}}\, dx - \frac{p^2}{(m-1)(m-2)}\int_0^\infty \frac{\sin^p x}{x^{m-2}}\, dx \quad [p>m-1>1]$$

$$\text{GW (333)(17)}$$

13. $$\int_0^\infty \frac{\sin^{2n} px}{\sqrt{x}}\, dx = \infty \qquad\qquad\qquad\qquad\qquad\qquad\qquad\qquad \text{BI (177)(5)}$$

14. $$\int_0^\infty \sin^{2n+1} px\, \frac{dx}{\sqrt{x}} = \frac{1}{2^{2n}}\sqrt{\frac{\pi}{2p}}\sum_{k=0}^n (-1)^k \binom{2n+1}{n+k+1}\frac{1}{\sqrt{2k+1}} \qquad \text{BI (177)(7)}$$

3.822

1. $$\int_0^{\pi/2} x^p \cos^m x\, dx = -\frac{p(p-1)}{m^2}\int_0^{\pi/2} x^{p-2}\cos^m x\, dx + \frac{m-1}{m}\int_0^{\pi/2} x^p \cos^{m-2} x\, dx$$

$$[m>1,\quad p>1] \qquad\qquad \text{GW (333)(9a)}$$

2. $$\int_0^\infty x^{-1/2}\cos^{2n+1}(px)\, dx = \frac{1}{2^{2n}}\sqrt{\frac{\pi}{2p}}\sum_{k=0}^n \binom{2n+1}{n+k+1}\frac{1}{\sqrt{2k+1}} \qquad \text{BI (177)(8)}$$

3.823 $$\int_0^\infty x^{\mu-1}\sin^2 ax\, dx = -\frac{\Gamma(\mu)\cos\frac{\mu\pi}{2}}{2^{\mu+1}a^\mu} \qquad\qquad [a>0,\quad -2<\operatorname{Re}\mu<0]$$

$$\text{ET I 319(15), GW(333)(19c)a}$$

3.824

1. $$\int_0^\infty \frac{\sin^2 ax}{x^2+\beta^2}\, dx = \frac{\pi}{4\beta}\left(1-e^{-2a\beta}\right) \qquad\qquad [a>0,\quad \operatorname{Re}\beta>0] \qquad \text{BI (160)(10)}$$

2. $$\int_0^\infty \frac{\cos^2 ax}{x^2+\beta^2}\, dx = \frac{\pi}{4\beta}\left(1+e^{-2a\beta}\right) \qquad\qquad [a>0,\quad \operatorname{Re}\beta>0] \qquad \text{BI (160)(11)}$$

3.[7] $$\int_0^\infty \sin^{2m} x\,\frac{dx}{a^2+x^2} = \frac{(-1)^m}{2^{2m+1}}\cdot\frac{\pi}{2}\left\{2^{2m}\sinh^{2m} a - 2\sum_{k=0}^m (-1)^k \binom{2m}{k}\sinh[2(m-k)a]\right\}$$

$$[a>0] \qquad\qquad \text{BI (160)(12)}$$

4.[7] $$\int_0^\infty \sin^{2m+1} x\,\frac{dx}{a^2+x^2} = \frac{(-1)^{m-1}}{2^{2m+2}a}\left\{e^{(2m+1)a}\sum_{k=0}^{2m+1}(-1)^k\binom{2m+1}{k}e^{-2ka}\operatorname{Ei}[(2k-2m-1)a]\right.$$

$$\left. + e^{-(2m+1)a}\sum_{k=0}^{2m+1}(-1)^{k-1}\binom{2m+1}{k}e^{2ka}\operatorname{Ei}[(2m+1-2k)a]\right\}$$

$$[a>0] \qquad\qquad \text{BI (160)(14)}$$

5.[7] $\quad \int_0^\infty \sin^{2m+1} x \frac{x\,dx}{a^2+x^2} = \frac{\pi}{2^{2m+1}} e^{-(2m+1)a} \sum_{k=0}^{m} (-1)^{m+k} \binom{2m+1}{k} e^{2ka}$

$$\left[|\arg a| < \frac{\pi}{2}\right], \quad m = 0, 1, 2, \ldots$$

6.[7] $\quad \int_0^\infty \cos^{2m} x \frac{dx}{a^2+x^2} = \frac{\pi}{2^{2m+1}a} \binom{2m}{m} + \frac{\pi}{2^{2m}} \sum_{k=1}^{m} \binom{2m}{m+k} e^{-2ka}$

$$[a > 0] \qquad\qquad \text{BI (160)(16)}$$

7. $\quad \int_0^\infty \cos^{2m+1} x \frac{dx}{a^2+x^2} = \frac{\pi}{2^{2m+1}a} \sum_{k=1}^{m} \binom{2m+1}{m+k+1} e^{-(2k+1)a}$

$$[a > 0] \qquad\qquad \text{BI (160)(17)}$$

8. $\quad \int_0^\infty \cos^{2m+1} x \frac{x\,dx}{a^2+x^2} = -\frac{e^{-(2m+1)a}}{2^{2m+2}} \sum_{k=0}^{2m+1} \binom{2m+1}{k} e^{2ka} \operatorname{Ei}[(2m-2k+1)a]$

$$\qquad\qquad\qquad - \frac{e^{(2m+1)a}}{2^{2m+2}} \sum_{k=0}^{2m+1} \binom{2m+1}{k} e^{-2ka} \operatorname{Ei}[(2k-2m-1)a]$$

$$\text{BI (160)(18)}$$

9. $\quad \int_0^\infty \frac{\cos^2 ax}{b^2-x^2} dx = \frac{\pi}{4b} \sin 2ab \qquad\qquad [a > 0, \quad b > 0] \qquad \text{BI (161)(10)}$

10. $\quad \int_0^\infty \frac{\sin^2 ax \cos^2 bx}{\beta^2+x^2} dx = \frac{\pi}{8\beta} \left[1 - \frac{1}{2}e^{-2(a+b)\beta} + e^{-2b\beta} - \frac{1}{2}e^{2(b-a)\beta} - e^{-2a\beta}\right] \quad [a > b]$

$$= \frac{\pi}{16\beta} \left[1 - e^{-4a\beta}\right] \qquad\qquad\qquad\qquad\qquad [a = b]$$

$$= \frac{\pi}{8\beta} \left[1 - \frac{1}{2}e^{-2(a+b)\beta} + e^{-2b\beta} - \frac{1}{2}e^{2(a-b)\beta} - e^{-2a\beta}\right] \quad [a < b]$$

$$[a > 0, \quad b > 0], \qquad (\text{cf. } \mathbf{3.824} \text{ 1 and 3}) \quad \text{BI (162)(6)}$$

11. $\quad \int_0^\infty \frac{x \sin 2ax \cos^2 bx}{\beta^2+x^2} dx = \frac{\pi}{8} \left[2e^{-2a\beta} + e^{-2(a+b)\beta} + e^{2(b-a)\beta}\right] \qquad [a > 0]$

$$= \frac{\pi}{8} \left[e^{-4a\beta} + 2e^{-2a\beta}\right] \qquad\qquad\qquad\qquad [a = b]$$

$$= \frac{\pi}{8} \left[2e^{-2a\beta} + e^{-2(a+b)\beta} - e^{2(a-b)\beta}\right] \qquad [a < b]$$

$$\text{LI (162)(5)}$$

3.825

1. $\quad \int_0^\infty \frac{\sin^2 ax\,dx}{(b^2+x^2)(c^2+x^2)} = \frac{\pi \left(b - c + ce^{-2ab} - be^{-2ac}\right)}{4bc(b^2-c^2)}$

$$[a > 0, \quad b > 0, \quad c > 0] \qquad \text{BI (174)(15)}$$

2. $\quad \int_0^\infty \frac{\cos^2 ax\,dx}{(b^2+x^2)(c^2+x^2)} = \frac{\pi \left(b - c + be^{-2ac} - ce^{-2ab}\right)}{4bc(b^2-c^2)}$

$$[a > 0, \quad b > 0, \quad c > 0] \qquad \text{BI (175)(14)}$$

$3.^3 \quad \displaystyle\int_0^\infty \frac{\sin^2 ax\, dx}{(b^2 - x^2)(c^2 - x^2)} = \frac{\pi\,(c \sin 2ab - b \sin 2ac)}{4bc\,(b^2 - c^2)} \qquad [a > 0, \quad b > 0, \quad c > 0, \quad b \neq c]$

LI (174)(16)

$4.^3 \quad \displaystyle\int_0^\infty \frac{\cos^2 ax\, dx}{(b^2 - x^2)(c^2 - x^2)} = \frac{\pi\,(b \sin 2ac - c \sin 2ab)}{4bc\,(b^2 - c^2)} \qquad [a > 0, \quad b > 0, \quad c > 0, \quad b \neq c]$

LI (175)(15)

3.826

1. $\quad \displaystyle\int_0^\infty \frac{\sin^2 ax\, dx}{x^2\,(b^2 + x^2)} = \frac{\pi}{4b^2}\left[2a - \frac{1}{b}\left(1 - e^{-2ab}\right)\right] \qquad [a > 0, \quad b > 0]$ BI (172)(13)

2. $\quad \displaystyle\int_0^\infty \frac{\sin^2 ax\, dx}{x^2\,(b^2 - x^2)} = \frac{\pi}{4b^2}\left(2a - \frac{1}{b}\sin 2ab\right) \qquad [a > 0, \quad b > 0]$ BII (172)(14)

3.827

$1.^8 \quad \displaystyle\int_0^\infty \frac{\sin^3 ax}{x^\nu}\, dx = \frac{3 - 3^{\nu-1}}{4}\, a^{\nu-1} \cos\frac{\nu\pi}{2}\,\Gamma(1 - \nu) \qquad [a < \operatorname{Re}\nu < 4, \nu \neq 1, 2, 3]$

GW (333)(19f)

$2.^8 \quad \displaystyle\int_0^\infty \frac{\sin^3 ax}{x}\, dx = \frac{\pi}{4}$ LO V 277

3. $\quad \displaystyle\int_0^\infty \frac{\sin^3 ax}{x^2}\, dx = \frac{3}{4}a \ln 3$ BI (156)(2)

$4.^8 \quad \displaystyle\int_0^\infty \frac{\sin^3 ax}{x^3}\, dx = \frac{3}{8}a^2 \pi$ BI(156)(7)a, LO V 312

5. $\quad \displaystyle\int_0^\infty \frac{\sin^4 ax}{x^2}\, dx = \frac{a\pi}{4} \qquad [a > 0]$ BI (156)(3)

6. $\quad \displaystyle\int_0^\infty \frac{\sin^4 ax}{x^3}\, dx = a^2 \ln 2$ BI (156)(8)

7. $\quad \displaystyle\int_0^\infty \frac{\sin^4 ax}{x^4}\, dx = \frac{a^3 \pi}{3} \qquad [a > 0]$ BI(156)(11), LO V 312

8. $\quad \displaystyle\int_0^\infty \frac{\sin^5 ax}{x^2}\, dx = \frac{5}{16}a\,(3 \ln 3 - \ln 5)$ BI (156)(4)

9. $\quad \displaystyle\int_0^\infty \frac{\sin^5 ax}{x^3}\, dx = \frac{5}{32}a^2 \pi \qquad [a > 0]$ BI (156)(9)

10. $\quad \displaystyle\int_0^\infty \frac{\sin^5 ax}{x^4}\, dx = \frac{5}{96}a^3\,(25 \ln 5 - 27 \ln 3)$ BI (156)(12)

11. $\quad \displaystyle\int_0^\infty \frac{\sin^5 ax}{x^5}\, dx = \frac{115}{384}a^4 \pi \qquad [a > 0]$ BI(156)(13), LO V 312

12. $\quad \displaystyle\int_0^\infty \frac{\sin^6 ax}{x^2}\, dx = \frac{3}{16}a\pi \qquad [a > 0]$ BI (156)(5)

13. $\quad \displaystyle\int_0^\infty \frac{\sin^6 ax}{x^3}\, dx = \frac{3}{16}a^2\,(8 \ln 2 - 3 \ln 3)$ BI (156)(10)

14. $\displaystyle\int_0^\infty \frac{\sin^6 ax}{x^5}\,dx = \frac{1}{16}a^4\left(27\ln 3 - 32\ln 2\right)$ BI (156)(14)

15. $\displaystyle\int_0^\infty \frac{\sin^6 ax}{x^6}\,dx = \frac{11}{40}a^5\pi$ $[a > 0]$ LO V 312

3.828   In **3.828** 1–21 the restrictions $a > 0$, $b > 0$, $c > 0$ apply.

1.[8] $\displaystyle\int_0^\infty \frac{\sin ax \sin bx}{x}\,dx = \frac{1}{2}\ln\left|\frac{a+b}{a-b}\right|$ $[a \neq b]$ FI II 647

2.[8] $\displaystyle\int_0^\infty \sin ax \sin bx\,\frac{dx}{x^2} = \frac{1}{2}\pi\min(a,b)$ BI (157)(1)

3.[8] $\displaystyle\int_0^\infty \frac{\sin^2 ax \sin bx}{x}\,dx = \frac{\pi}{4}$ $[b < 2a]$

                            $= \dfrac{\pi}{8}$ $[b = 2a]$

                            $= 0$ $[b > 2a]$

                                                BI (151)(10)

4.[8] $\displaystyle\int_0^\infty \frac{\sin^2 ax \cos bx}{x}\,dx = \frac{1}{4}\ln\frac{4a^2 - b^2}{b^2}$ $[2a \neq b]$ BI (151)(12)

5.[8] $\displaystyle\int_0^\infty \frac{\sin^2 ax \cos 2bx}{x^2}\,dx = \frac{1}{2}\pi\max(0, a - b)$

6. $\displaystyle\int_0^\infty \frac{\sin 2ax \cos^2 bx}{x}\,dx = \frac{\pi}{2}$ $[a > b]$

                            $= \dfrac{3}{8}\pi$ $[a = b]$

                            $= \dfrac{\pi}{4}$ $[a < b]$

                                              BI (151)(9)

7.[8] $\displaystyle\int_0^\infty \frac{\sin^2 ax \sin bx \sin cx}{x^2}\,dx = \frac{\pi}{16}\left(|b - 2a - c| - |2a - b - c| + 2c\right)$

                                      $[a > 0, \quad 0 < c \leq b]$

                                      BI(157)(9)a, ET I 79(15)

8.[8] $\displaystyle\int_0^\infty \frac{\sin^2 ax \sin bx \sin cx}{x}\,dx = \frac{1}{8}\ln\left|\frac{(b+c)^2(2a-b+c)(2a+b-c)}{(b-c)^2(2a+b+c)(2a-b-c)}\right|$

                     $[b \neq c, \quad 2a + c \neq b, \quad 2a + b \neq c, \quad 2a \neq b + c]$ LI (152)(2)

9. $\displaystyle\int_0^\infty \frac{\sin^2 ax \sin^2 bx}{x^2}\,dx = \frac{\pi}{4}a$ $[0 \leq a \leq b]$

                                $= \dfrac{\pi}{4}b$ $[0 \leq b \leq a]$

                                              BI (157)(3)

10.[8] $\displaystyle\int_0^\infty \frac{\sin^2 ax \sin^2 bx}{x^4}\,dx = \frac{1}{6}\pi\min\left(a^2, b^2\right)\left[3\max(a,b) - \min(a,b)\right]$ BI (157)(27)

$11.^8$ $\int_0^\infty \dfrac{\sin^2 ax \cos^2 bx}{x^2}\,dx = \dfrac{1}{4}\pi\left[a + \max(0,\,a-b)\right]$ BI (157)(6)

$12.$ $\int_0^\infty \dfrac{\sin^3 ax \sin 3bx}{x^4}\,dx = \dfrac{a^3\pi}{2}$ $[b > a]$

$\hphantom{12.\quad}= \dfrac{\pi}{16}\left[8a^3 - 9(a-b)^3\right]$ $[a \le 3b \le 3a]$ BI (157)(28)

$\hphantom{12.\quad}= \dfrac{9b\pi}{8}\left(a^2 - b^2\right)$ $[3b \le a]$ LI (157)(28)

$13.$ $\int_0^\infty \dfrac{\sin^3 ax \cos bx}{x}\,dx = 0$ $[b > 3a]$

$\hphantom{13.\quad}= -\dfrac{\pi}{16}$ $[b = 3a]$

$\hphantom{13.\quad}= -\dfrac{\pi}{8}$ $[3a > b > a]$

$\hphantom{13.\quad}= \dfrac{\pi}{16}$ $[b = a]$

$\hphantom{13.\quad}= \dfrac{\pi}{4}$ $[a > b]$

$[a > 0,\quad b > 0]$ BI (151)(15)

$14.^{10}$ $\int_0^\infty \dfrac{\sin^3 ax \cos 3bx}{x^2}\,dx = \dfrac{3}{16}\Bigg(a\log 81 - 2(a-3b)\log(a-3b) + 2(a-b)\log(a-b)$

$+\, 2(a+b)\log(a+b) - 2(a+3b)\log(a+3)\Bigg)$

$[\operatorname{Im} a = 0,\quad \operatorname{Im} b = 0]$ MC

$15.$ $\int_0^\infty \dfrac{\sin^3 ax \cos bx}{x^3}\,dx = \dfrac{\pi}{8}\left(3a^2 - b^2\right)$ $[b < a]$

$\hphantom{15.\quad}= \dfrac{\pi b^2}{4}$ $[a = b]$

$\hphantom{15.\quad}= \dfrac{\pi}{16}(3a - b)^2$ $[a < b < 3a]$

$\hphantom{15.\quad}= 0$ $[3a < b]$

$[a > 0,\quad b > 0]$ BI(157)(19), ET I 19(10)

$16.$ $\int_0^\infty \dfrac{\sin^3 ax \sin bx}{x^4}\,dx = \dfrac{b\pi}{24}\left(9a^2 - b^2\right)$ $[0 < b \le a]$

$\hphantom{16.\quad}= \dfrac{\pi}{48}\left[24a^3 - (3a - b)^3\right]$ $[0 < a \le b \le 3a]$

$\hphantom{16.\quad}= \dfrac{\pi a^3}{2}$ $[0 < 3a \le b]$

ET I 79(16)

17. $$\int_0^\infty \frac{\sin^3 ax \sin^2 bx}{x}\,dx = \frac{\pi}{8} \qquad [2b > 3a]$$

$$= \frac{5\pi}{32} \qquad [2b = 3a]$$

$$= \frac{3\pi}{16} \qquad [3a > 2b > a]$$

$$= \frac{3\pi}{32} \qquad [2b = a]$$

$$= 0 \qquad [a > 2b]$$

$$[a > 0, \quad b > 0] \qquad \text{BI (151)(14)}$$

18.8 $$\int_0^\infty \frac{\sin^2 ax \cos^3 bx}{x}\,dx = \frac{1}{16} \ln \left| \frac{(2a+b)^3 (b-2a)^3 (2a+3b)(3b-2a)}{9b^8} \right|$$

$$[2a \neq b, \quad 2a \neq 3b] \qquad \text{BI (151)(13)}$$

19.10 $$\int_0^\infty \frac{\sin^2 ax \sin^2 bx \sin^2 2cx}{x}\,dx$$

$$= \frac{\pi}{32} \Bigg(2\operatorname{sign}(a-c) - \operatorname{sign}(a-b-c) + 2\operatorname{sign}(b-c) - \operatorname{sign}(a+b-c)$$

$$+ 4\operatorname{sign}(c) - 2\operatorname{sign}(a+c) + \operatorname{sign}(a-b+c) - 2\operatorname{sign}(b+c) + \operatorname{sign}(a+b+c) \Bigg)$$

$$[\operatorname{Im} a = 0, \quad \operatorname{Im} b = 0, \quad \operatorname{Im} c = 0] \qquad \text{MC}$$

20. $$\int_0^\infty \frac{\sin^2 ax \sin^2 bx \sin 2cx\,dx}{x^2} = \frac{a-b-c}{16} \ln 4(a-b-c)^2$$

$$- \frac{a+b+c}{16} \ln 4(a+b+c)^2 + \frac{a+b-c}{16} \ln 4(a+b-c)^2$$

$$- \frac{a-b+c}{16} \ln 4(a-b+c)^2 + \frac{a+c}{8} \ln 4(a+c)^2$$

$$- \frac{a-c}{8} \ln 4(a-c)^2 + \frac{b+c}{8} \ln 4(b+c)^2 - \frac{b-c}{8} \ln 4(b-c)^2$$

$$- \frac{1}{2} c \ln 2c$$

$$[a > 0, \quad b > 0, \quad c > 0] \qquad \text{BI (157)(10)}$$

21.8 $$\int_0^\infty \frac{\sin^2 ax \sin^3 bx}{x^3}\,dx = \frac{3b^2 \pi}{16} \qquad [2a > 3b]$$

$$= \frac{a^2 \pi}{12} \qquad [2a = 3b]$$

$$= \frac{6b^2 - (3b - 2a)^2}{32}\pi \qquad [3b > 2a > b]$$

$$= \frac{a^2 \pi}{4} \qquad [b \geq 2a]$$

$$\text{BI (157)(18)}$$

3.829

1. $$\int_0^\infty \frac{x^n - \sin^n x}{x^{n+2}}\,dx = \frac{\pi}{2^n (n+1)!} \sum_{k=0}^{[(n-1)/2]} (-1)^k \binom{n}{k} (n-2k)^{n+1} \qquad \text{GW (333)(63)}$$

2. $\displaystyle\int_0^\infty \left(1 - \cos^{2m-1} x\right) \frac{dx}{x^2} = \int_0^\infty \left(1 - \cos^{2m} x\right) \frac{dx}{x^2} = \frac{m\pi}{2^{2m}} \binom{2m}{m}$ BI (158)(7, 8)

3.831

1. $\displaystyle\int_0^\infty \frac{\sin^{2n} ax - \sin^{2n} bx}{x} \, dx = \frac{(2n-1)!!}{(2n)!!} \ln \frac{b}{a}$ $[ab > 0, \quad n = 1, 2, \ldots]$ FI II 651

2. $\displaystyle\int_0^\infty \frac{\cos^{2n} ax - \cos^{2n} bx}{x} \, dx = \left[1 - \frac{(2n-1)!!}{(2n)!!}\right] \ln \frac{b}{a}$ $[ab > 0, \quad n = 0, 1, \ldots]$ FI II 651

3. $\displaystyle\int_0^\infty \frac{\cos^{2m+1} ax - \cos^{2m+1} bx}{x} \, dx = \ln \frac{b}{a}$ $[ab > 0, \quad m = 0, 1, \ldots]$ FI II

4. $\displaystyle\int_0^\infty \frac{\cos^m ax \cos max - \cos^m bx \cos mbx}{x} \, dx = \left(1 - \frac{1}{2^m}\right) \ln \frac{b}{a}$

                $[ab > 0, \quad m = 0, 1, \ldots]$ LI (155)(8)

3.832

1. $\displaystyle\int_0^{\pi/2} x \cos^{p-1} x \sin ax \, dx = \frac{\pi}{2^{p+1}} \Gamma(p) \frac{\psi\left(\dfrac{p+a+1}{2}\right) - \psi\left(\dfrac{p-a+1}{2}\right)}{\Gamma\left(\dfrac{p+a+1}{2}\right) \Gamma\left(\dfrac{p-a+1}{2}\right)}$

                $[p > 0, \quad -(p+1) < a < p+1]$

                BI (205)(6)

2.[3] $\displaystyle\int_0^\infty \sin^{2m+1} x \sin 2mx \frac{dx}{a^2 + x^2} = \frac{(-1)^m \pi}{2^{2m+1} a} \left[\left(1 - e^{-2a}\right)^{2m} - 1\right] \sinh a$

                $[a > 0, \quad m = 0, 1, \ldots]$ BI (162)(17)

3. $\displaystyle\int_0^\infty \sin^{2m-1} x \sin[(2m-1)x] \frac{dx}{a^2 + x^2} = \frac{(-1)^{m+1} \pi}{2^{2m} a} \left(1 - e^{-2a}\right)^{2m-1}$

                $[a > 0, \quad m = 1, 2, \ldots]$ BI (162)(11)

4. $\displaystyle\int_0^\infty \sin^{2m-1} x \sin[(2m+1)x] \frac{dx}{a^2 + x^2} = \frac{(-1)^{m-1} \pi}{2^{2m} a} e^{-2a} \left(1 - e^{-2a}\right)^{2m-1}$

                $[a > 0, \quad m = 1, 2, \ldots]$ BI (162)(12)

5. $\displaystyle\int_0^\infty \sin^{2m+1} x \sin[3(2m+1)x] \frac{dx}{a^2 + x^2} = \frac{(-1)^m \pi}{2a} e^{-3(2m+1)a} \sinh^{2m+1} a$

                $[a > 0]$ BI (162)(18)

6.[3] $\displaystyle\int_0^\infty \sin^{2m} x \sin[(2m-1)x] \frac{x \, dx}{a^2 + x^2} = \frac{(-1)^m \pi}{2^{2m+1}} e^a \left[\left(1 - e^{-2a}\right)^{2m} - \left(1 + e^{-2a}\right)\right]$

                $[a \geq 0, \quad m = 0, 1, \ldots]$ BI (162)(13)

7. $\displaystyle\int_0^\infty \sin^{2m} x \sin(2mx) \frac{x \, dx}{a^2 + x^2} = \frac{(-1)^m \pi}{2^{2m+1}} \left[\left(1 - e^{-2a}\right)^{2m} - 1\right]$

                $[a > 0, \quad m = 0, 1, \ldots]$ BI (162)(14)

8. $\displaystyle\int_0^\infty \sin^{2m} x \sin[(2m+2)x]\frac{x\,dx}{a^2+x^2} = \frac{(-1)^m \pi}{2^{2m+1}} e^{-2a}\left(1-e^{-2a}\right)^{2m}$

$$[a>0, \quad m=0,1,\ldots] \qquad \text{BI (162)(15)}$$

9. $\displaystyle\int_0^\infty \sin^{2m} x \sin 4mx \frac{x\,dx}{a^2+x^2} = \frac{(-1)^m \pi}{2} e^{-4ma}\sinh^{2m} a$

$$[a>0, \quad m=1,2,\ldots] \qquad \text{BI (162)(16)}$$

10. $\displaystyle\int_0^\infty \sin^{2m} x \cos x \frac{dx}{x^2} = \frac{(2m-3)!!}{(2m)!!}\cdot\frac{\pi}{2}$ $\qquad [m=1,2,\ldots]$ $\qquad \text{GW (333)(15a)}$

11. $\displaystyle\int_0^\infty \sin^{2m} x \cos[(2m-1)x]\frac{dx}{a^2+x^2} = \frac{(-1)^m \pi}{2^{2m}a}\left[\left(1-e^{-2a}\right)^{2m-1}-1\right]\sinh a$

$$[a>0, \quad m=1,2,\ldots] \qquad \text{BI (162)(25)}$$

12. $\displaystyle\int_0^\infty \sin^{2m} x \cos(2mx)\frac{dx}{a^2+x^2} = \frac{(-1)^m \pi}{2^{2m+1}a}\left(1-e^{-2a}\right)^{2m}$

$$[a>0, \quad m=0,1,\ldots] \qquad \text{BI (162)(26)}$$

13. $\displaystyle\int_0^\infty \sin^{2m} x \cos[(2m+2)x]\frac{dx}{a^2+x^2} = \frac{(-1)^m \pi}{2^{2m+1}a} e^{-2a}\left(1-e^{-2a}\right)^{2m}$

$$[a>0, \quad m=0,1,\ldots] \qquad \text{BI (162)(27)}$$

14. $\displaystyle\int_0^\infty \sin^{2m} x \cos 4mx \frac{dx}{a^2+x^2} = \frac{(-1)^m \pi}{2a} e^{-4ma}\sinh^{2m} a$

$$[a>0, \quad m=0,1,\ldots] \qquad \text{BI (162)(28)}$$

15. $\displaystyle\int_0^\infty \sin^{2m+1} x \cos x \frac{dx}{x} = \frac{(2m-1)!!}{(2m+2)!!}\cdot\frac{\pi}{2}$ $\qquad [m=0,1,\ldots]$ $\qquad \text{GW (333)(15)}$

16.³ $\displaystyle\int_0^\infty \sin^{2m+1} x \cos x \frac{dx}{x^3} = \frac{(2m-3)!!}{(2m)!!}\cdot\frac{\pi}{2}$ $\qquad [m=1,2,\ldots]$ $\qquad \text{GW (333)(15b)}$

17. $\displaystyle\int_0^\infty \sin^{2m-1} x \cos[(2m-1)x]\frac{x\,dx}{a^2+x^2} = \frac{(-1)^m \pi}{2^{2m}}\left[\left(1-e^{-2a}\right)^{2m-1}-1\right]$

$$[m=1,2,\ldots, \quad a>0] \qquad \text{BI (162)(23)}$$

18.³ $\displaystyle\int_0^\infty \sin^{2m+1} x \cos 2mx \frac{x\,dx}{a^2+x^2} = \frac{(-1)^{m-1}\pi}{2^{2m+2}}\left\{e^a\left[\left(1-e^{-2a}\right)^{2m+1}-1\right]-e^{-a}\right\}$

$$[m=0,1,\ldots, \quad a\geq 0] \qquad \text{BI (162)(29)}$$

19. $\displaystyle\int_0^\infty \sin^{2m-1} x \cos[(2m+1)x]\frac{x\,dx}{a^2+x^2} = \frac{(-1)^m \pi}{2^{2m}} e^{-2a}\left(1-e^{-2a}\right)^{2m-1}$

$$[m=1,2,\ldots, \quad a>0] \qquad \text{BI (162)(24)}$$

20. $\displaystyle\int_0^\infty \sin^{2m+1} x \cos[2(2m+1)x]\frac{x\,dx}{a^2+x^2} = \frac{(-1)^{m-1}\pi}{2} e^{-2(2m+1)a}\sinh^{2m+1} a$

$$[m=0,1,\ldots, \quad a>0] \qquad \text{BI (162)(30)}$$

21. $\displaystyle\int_0^\infty \cos^m x \sin mx \frac{x\,dx}{a^2+x^2} = \frac{1}{2^{m+1}a} \sum_{k=1}^m \binom{m}{k} \left[e^{-2ka}\,\mathrm{Ei}(2ka) - e^{2ka}\,\mathrm{Ei}(-2ka)\right]$

$$[a>0]$$
BI (162)(8)

22. $\displaystyle\int_0^\infty \cos^n sx \sin nsx \frac{x\,dx}{a^2+x^2} = \frac{\pi}{2^{n+1}} \left[\left(1+e^{-2as}\right)^n - 1\right]$

$$[s>0, \quad \mathrm{Re}\,a>0, \quad n\geq 0]$$
BI (163)(9)

23. $\displaystyle\int_0^\infty \cos^n sx \sin nsx \frac{x\,dx}{a^2-x^2} = \frac{\pi}{2} \left(2^{-n} - \cos^n as \cos nas\right)$

$$[n=0,1,\ldots]$$
BI (166)(10)

24. $\displaystyle\int_0^\infty \cos^{m-1} x \sin[(m+1)x] \frac{x\,dx}{a^2+x^2} = \frac{\pi}{2^m} e^{-2a} \left(1+e^{-2a}\right)^{m-1}$

$$[a>0, \quad m=1,2,\ldots]$$
BI (163)(6)

25. $\displaystyle\int_0^\infty \cos^m x \sin[(m+1)x] \frac{x\,dx}{a^2+x^2} = \frac{\pi}{2^{m+1}} e^{-a} \left(1+e^{-2a}\right)^m$

$$[m=0,1,\ldots, \quad a>0]$$
BI (163)(10)

26.[3] $\displaystyle\int_0^\infty \cos^m x \sin[(m-1)x] \frac{x\,dx}{a^2+x^2} = \frac{\pi}{2^m} \cosh a \left[\left(1+e^{-2a}\right)^{m-1} - 1\right]$

$$[m=0,1,\ldots, \quad a\geq 0]$$
BI (163)(7)

27. $\displaystyle\int_0^\infty \cos^m x \sin(3mx) \frac{x\,dx}{a^2+x^2} = \frac{\pi}{2} e^{-3a} \cosh^m a$ $\qquad [a>0, \quad m=1,2,\ldots]$ $\qquad$ BI (163)(11)

28. $\displaystyle\int_0^\infty \cos^n sx \cos nxs \frac{dx}{a^2+x^2} = \frac{\pi}{2^{n+1}a} \left(1+e^{-2as}\right)^n$ $\qquad [n=0,1,\ldots]$ $\qquad$ BI (163)(16)

29. $\displaystyle\int_0^\infty \cos^n sx \cos nsx \frac{dx}{a^2-x^2} = \frac{\pi}{2a} \cos^n as \sin nas$ $\qquad [n=0,1,\ldots]$

30. $\displaystyle\int_0^\infty \cos^{m-1} x \cos[(m+1)x] \frac{dx}{a^2+x^2} = \frac{\pi}{2^m a} e^{-2a} \left(1+e^{-2a}\right)^{m-1}$

$$[m=1,2,\ldots, \quad a>0]$$
BI (163)(14)

31. $\displaystyle\int_0^\infty \cos^m x \cos[(m-1)x] \frac{dx}{a^2+x^2} = \frac{\pi}{2^{m+1}a} e^a \left[\left(1+e^{-2a}\right)^m - \left(1-e^{-2a}\right)\right]$

$$[m=0,1,\ldots, \quad a>0]$$
BI (163)(15)

32. $\displaystyle\int_0^\infty \cos^m x \cos[(m+1)x] \frac{dx}{a^2+x^2} = \frac{\pi}{2^{m+1}a} e^{-a} \left(1+e^{-2a}\right)^m$

$$[m=0,1,\ldots, \quad a>0]$$
BI (163)(17)

33.
$$\int_0^\infty \sin^p x \cos x \frac{dx}{x^q} = \frac{p}{q-1}\int_0^\infty \frac{\sin^{p-1} x}{x^{q-1}}dx - \frac{p+1}{q-1}\int_0^\infty \frac{\sin^{p+1} x}{x^{q-1}}\,dx \qquad [p > q-1 > 0]$$

$$= \frac{p(p-1)}{(q-1)(q-2)}\int_0^\infty \sin^{p-2} x \cos x \frac{dx}{x^{q-2}}$$

$$- \frac{(p+1)^2}{(q-1)(q-2)}\int_0^\infty \sin^p x \cos x \frac{dx}{x^{q-2}} \qquad [p > q-1 > 1]$$

GW (333)(18)

34.
$$\int_0^\infty \cos^{2m} x \cos 2nx \sin x \frac{dx}{x}x = \int_0^\infty \cos^{2m-1} x \cos 2nx \sin x \frac{dx}{x}x = \frac{\pi}{2^{2m+1}}\binom{2m}{m+n}$$

BI (152)(5, 6)

35.
$$\int_0^\infty \cos^p ax \sin bx \cos x \frac{dx}{x} = \frac{\pi}{2} \qquad [b > ap, \quad p > -1]$$

BI (153)(12)

36.
$$\int_0^\infty \cos^p ax \sin pax \cos x \frac{dx}{x} = \frac{\pi}{2^{p+1}}(2^p - 1) \qquad [p > -1]$$

BI (153)(2)

37.
$$\int_0^\infty \frac{dx}{x^2}\left(\prod_{k=1}^n \cos^{p_k} a_k x\right)\sin bx \sin x = \frac{\pi}{2} \qquad \left[b > \sum_{k=1}^n a_k p_k, \quad a_k > 0, \quad p_k > 0\right]$$

BI (157)(15)

3.833

1.[10]
$$\int_0^\infty \sin^{2m+1} x \cos^{2n} x \frac{dx}{x} = \int_0^\infty \sin^{2m+1} x \cos^{2n-1} x \frac{dx}{x} = \frac{(2m-1)!!(2n-1)!!}{2^{m+n+1}(m+n)!}\pi$$

BI (151)(24, 25)

$$= \frac{1}{2}\,B\left(m+\frac{1}{2}, n+\frac{1}{2}\right)$$

GW (333)(24)

2.
$$\int_0^\infty \sin^{2m+1} 2x \cos^{2n-1} 2x \cos^2 x \frac{dx}{x} = \frac{\pi}{2}\cdot\frac{(2m-1)!!(2n-1)!!}{(2m+2n)!!}$$

LI (152)(4)

3.834

1.
$$\int_0^\infty \frac{\sin^{2m+1} x}{1-2a\cos x + a^2}\cdot\frac{dx}{x} = \frac{(-1)^m\pi(1+a)^{4m}}{2^{2m+2}a^{2m+1}}\left\{\left|\frac{1-a}{1+a}\right|^{2m-1}\right.$$

$$\left. - \sum_{k=0}^{2m}(-1)^k\binom{m-\frac{1}{2}}{k}\left(\frac{4a}{(1+a)^2}\right)^k\right\}$$

$$[|a| \neq 1]$$

GW (333)(62a)

2.
$$\int_0^\infty \frac{\sin^{2m+1} x \cos^n x}{(1-2a\cos x + a^2)^p}\cdot\frac{dx}{x}$$

$$= \frac{n!\pi}{2^{n+1}(2m+n+1)!(1+a)^{2p}}\sum_{k=0}^n \frac{(-1)^k(2m+2n-2k+1)!!(2m+2k-1)!!}{k!(n-k)!}$$

$$\times F\left(m+n-k+\frac{3}{2}, p; 2m+n+2; \frac{4a}{(1+a)^2}\right)$$

$$[a \neq \pm 1]$$

GW (333)(62)

3.835

1. $$\int_0^\infty \frac{\cos^{2m} x \cos 2mx \sin x}{a^2 \cos^2 x + b^2 \sin^2 x} \cdot \frac{dx}{x} = \frac{\pi}{2} \frac{b^{2m-1}}{a(a+b)^{2m}} \qquad [ab > 0] \qquad\qquad \text{BI (182)(31)a}$$

2. $$\int_0^\infty \frac{\cos^{2m-1} x \cos 2mx \sin x}{a^2 \cos^2 x + b^2 \sin^2 x} \cdot \frac{dx}{x} = \frac{\pi}{2a} \frac{b^{2m-1}}{(a+b)^{2m}} \qquad [ab > 0] \qquad\qquad \text{LI (182)(32)a}$$

3.836

1. $$\int_0^\infty \left(\frac{\sin x}{x}\right)^n \frac{\sin mx}{x} \, dx = \frac{\pi}{2} \qquad [m \geq n] \qquad\qquad \text{LI (159)(12)}$$

2.[8] $$\int_0^\infty \left(\frac{\sin x}{x}\right)^n \cos mx \, dx \frac{n\pi}{2^n} \sum_{k=0}^{\lfloor \frac{1}{2}(m+n) \rfloor} \frac{(-1)^k (n+m-2k)^{n-1}}{k!(n-k)!} \qquad [0 \leq m < n]$$

$$= 0 \qquad [m \geq n \geq 2]$$

$$= \frac{\pi}{4} \qquad [m = n = 1]$$

GI(159)(14), ET I 20(11)

3. $$\int_0^\infty \left(\frac{\sin x}{x}\right)^{n-1} \sin nx \cos x \frac{dx}{x} = \frac{\pi}{2} \qquad [n \geq 1] \qquad\qquad \text{BI (159)(20)}$$

4.[8] $$\int_0^\infty \left(\frac{\sin x}{x}\right)^n \frac{\sin(anx)}{x} \, dx = \frac{\pi}{2}\left[1 - \frac{1}{2^{n-1}n!} \sum_{k=0}^{\lfloor \frac{1}{2}n(1+a) \rfloor} (-1)^k \binom{n}{k}(n+an-2k)^n\right]$$

$$[\text{all real } a, \, n \geq 1] \qquad\qquad \text{ET I 20(11)}$$

5.[10] $$I_n(b) = \frac{2}{\pi} \int_0^\infty \left(\frac{\sin x}{x}\right)^n \cos bx \, dx = n(2^{n-1}n!)^{-1} \sum_{k=0}^{\lfloor r \rfloor} (-1)^k \binom{n}{k}(n-b-2k)^{n-1}$$

where $0 \leq b < n$, $n \geq 1$, $r = (n-b)/2$, and $\lfloor r \rfloor$ is the largest integer contained in r

LO V 340(14)

6. $$\int_0^\infty \left(\frac{\sin x}{x}\right)^n \cos anx \, dx = 0 \quad [a \leq -1 \text{ or } a \geq 1, n \geq 2; \text{ for } n = 1 \text{ see } \textbf{3.741 } 2]$$

3.837

1. $$\int_0^{\pi/2} \frac{x^2 \, dx}{\sin^2 x} = \pi \ln 2 \qquad\qquad \text{BI (206)(9)}$$

2. $$\int_0^{\pi/4} \frac{x^2 \, dx}{\sin^2 x} = -\frac{\pi^2}{16} + \frac{\pi}{4} \ln 2 + \boldsymbol{G} = 0.8435118417\ldots \qquad\qquad \text{BI (204)(10)}$$

3. $$\int_0^{\pi/4} \frac{x^2 \, dx}{\cos^2 x} = \frac{\pi^2}{16} + \frac{\pi}{4} \ln 2 - \boldsymbol{G} \qquad\qquad \text{GW (333)(35a)}$$

4. $$\int_0^{\pi/4} \frac{x^{p+1}}{\sin^2 x} \, dx = -\left(\frac{\pi}{4}\right)^{p+1} + (p+1)\left(\frac{\pi}{4}\right)^p \left\{\frac{1}{p} - \frac{1}{2} \sum_{k=1}^\infty \frac{1}{4^{2k-1}(p+2k)} \zeta(2k)\right\}$$

$$[p > 0] \qquad\qquad \text{LI (204)(14)}$$

5. $\displaystyle\int_0^{\pi/2} \frac{x^2 \cos x}{\sin^2 x}\, dx = -\frac{\pi^2}{4} + 4\boldsymbol{G} = 1.1964612764\ldots$ BI (206)(7)

6. $\displaystyle\int_0^{\pi/2} \frac{x^3 \cos x}{\sin^3 x}\, dx = -\frac{\pi^3}{16} + \frac{3}{2}\pi \ln 2$ BI (206)(8)

7. $\displaystyle\int_0^{\infty} \frac{\cos 2nx}{\cos x} \sin^{2n} x \frac{dx}{x^m} = 0$ $\left[n > \dfrac{m-1}{2}, \quad m > 0 \right]$ BI (180)(16)

8. $\displaystyle\int_0^{\infty} \frac{\cos 2nx}{\cos x} \sin^{2n+1} x \frac{dx}{x^m} = 0$ $\left[n > \dfrac{m-2}{2}, \quad m > 0 \right]$ BI (180)(17)

9. $\displaystyle\int_0^1 \frac{x\, dx}{\cos ax \cos[a(1-x)]} = \frac{1}{a} \operatorname{cosec} a \cdot \ln \sec a$ $\left[a < \dfrac{\pi}{2} \right]$ BI (149)(20)

10.^{3} $\displaystyle\int_0^{\pi} \frac{x \sin(2n+1)x}{\sin x}\, dx = \frac{1}{2}\pi^2$ $[n = 0, 1, 2, \ldots]$

11.^{3} $\displaystyle\int_0^{\pi} \frac{x \sin 2nx}{\sin x}\, dx = -4 \sum_{k=1}^{n} (2k-1)^{-2}$ $[n = 1, 2, 3, \ldots]$

3.838

1. $\displaystyle\int_0^{\pi/2} \frac{x \cos^{p-1} x}{\sin^{p+1} x}\, dx = \frac{\pi}{2p} \sec \frac{\pi p}{2}$ $[p < 1]$ BI (206)(13)a

2. $\displaystyle\int_0^{\pi/4} \frac{x \sin^{p-1} x}{\cos^{p+1} x}\, dx = \frac{\pi}{4p} - \frac{1}{2p} \beta\left(\frac{p+1}{2}\right)$ $[p > -1]$ LI (204)(15)

3. $\displaystyle\int_0^{\pi/4} \frac{x \sin^{2m-1} x}{\cos^{2m+1} x}\, dx = \frac{\pi}{8m}(1 - \cos m\pi) + \frac{1}{2m} \sum_{k=0}^{m-1} \frac{(-1)^{k-1}}{2m - 2k - 1}$ BI (204)(17)

4. $\displaystyle\int_0^{\pi/4} \frac{x \sin^{2m} x}{\cos^{2m+2} x}\, dx = \frac{1}{2(2m+1)}\left[\frac{\pi}{2} + (-1)^{m-1} \ln 2 + \sum_{k=0}^{m-1} \frac{(-1)^{k-1}}{m-k} \right]$ BI (204)(16)

3.839

1. $\displaystyle\int_0^{\pi/4} x \tan^2 x\, dx = \frac{\pi}{4} - \frac{\pi^3}{32} - \frac{1}{2}\ln 2$ BI (204)(3)

2. $\displaystyle\int_0^{\pi/4} x \tan^3 x\, dx = \frac{\pi}{4} - \frac{1}{2} + \frac{\pi}{8}\ln 2 - \frac{1}{2}\boldsymbol{G}$ BI (204)(7)

3. $\displaystyle\int_0^{\pi/4} \frac{x^2 \tan x}{\cos^2 x}\, dx = \frac{1}{2}\ln 2 - \frac{\pi}{4} + \frac{\pi^2}{16}$ (cf. **3.839** 1) BI (204)(13)

4. $\displaystyle\int_0^{\pi/4} \frac{x^2 \tan^2 x}{\cos^2 x}\, dx = \frac{1}{3}\left(1 - \frac{\pi}{4}\ln 2 - \frac{\pi}{2} + \frac{\pi^2}{16} + \boldsymbol{G} \right)$ (cf. **3.839** 2) BI (204)(12)

5. $\displaystyle\int_0^{\pi/2} x \cos^p x \tan x\, dx = \frac{\pi}{2^{p+1}p} \cdot \frac{\Gamma(p+1)}{\left[\Gamma\left(\frac{p}{2}+1\right)\right]^2}$ $[p > -1]$ BI (205)(3)

6. $\displaystyle\int_0^{\pi/2} x \sin^p x \cot x \, dx = \frac{\pi}{2p} - \frac{2^{p-1}}{p} \mathrm{B}\left(\frac{p+1}{2}, \frac{p+1}{2}\right)$

$\qquad\qquad\qquad\qquad\qquad\qquad\qquad [p > -1]$ \qquad\qquad BI (206)(11)

7. $\displaystyle\int_0^\infty \sin^{2n} x \tan x \frac{dx}{x} = \frac{\pi}{2} \cdot \frac{(2n-1)!!}{(2n)!!}$ \qquad\qquad GW (333)(16)

8. $\displaystyle\int_0^\infty \cos^s rx \tan qx \frac{dx}{x} = \frac{\pi}{2}$ \qquad\qquad $[s > -1]$ \qquad\qquad BI (151)(26)

9. $\displaystyle\int_0^\infty \frac{\cos[(2n-1)x]}{\cos x} \cdot \left(\frac{\sin x}{x}\right)^{2n} dx = (-1)^{n-1}\frac{2^{2n}-1}{(2n)!} \cdot 2^{2n-1}\pi|B_{2n}|$ \qquad BI (180)(15)

10. $\displaystyle\int_0^\infty \tan^r px \frac{dx}{q^2 + x^2} = \frac{\pi}{2q} \sec\frac{r\pi}{2} \tanh^r pq$ \qquad\qquad $[r^2 < 1]$ \qquad\qquad BI (160)(19)

3.84 Integrals containing $\sqrt{1 - k^2 \sin^2 x}$, $\sqrt{1 - k^2 \cos^2 x}$, and similar expressions

Notation: $k' = \sqrt{1 - k^2}$

3.841

1. $\displaystyle\int_0^\infty \sin x \sqrt{1 - k^2 \sin^2 x} \frac{dx}{x} = \boldsymbol{E}(k)$ \qquad\qquad BI (154)(8)

2. $\displaystyle\int_0^\infty \sin x \sqrt{1 - k^2 \cos^2 x} \frac{dx}{x} = \boldsymbol{E}(k)$ \qquad\qquad BI (154)(20)

3. $\displaystyle\int_0^\infty \tan x \sqrt{1 - k^2 \sin^2 x} \frac{dx}{x} = \boldsymbol{E}(k)$ \qquad\qquad BI (154)(9)

4. $\displaystyle\int_0^\infty \tan x \sqrt{1 - k^2 \cos^2 x} \frac{dx}{x} = \boldsymbol{E}(k)$ \qquad\qquad BI (154)(21)

3.842

1. $\displaystyle\int_0^\infty \frac{\sin x}{\sqrt{1 + \sin^2 x}} \frac{dx}{x} = \int_0^\infty \frac{\tan x}{\sqrt{1 + \sin^2 x}} \cdot \frac{dx}{x}$

$\displaystyle\qquad = \int_0^\infty \frac{\sin x}{\sqrt{1 + \cos^2 x}} \frac{dx}{x} = \int_0^\infty \frac{\tan x}{\sqrt{1 + \cos^2 x}} \frac{dx}{x} = \frac{1}{\sqrt{2}} \boldsymbol{K}\left(\frac{1}{\sqrt{2}}\right)$

$\qquad\qquad\qquad\qquad\qquad\qquad\qquad\qquad\qquad$ BI (183)(4, 5, 9, 10)

2. $\displaystyle\int_u^{\frac{\pi}{2}} \frac{x \cos x \, dx}{\sqrt{\sin^2 x - \sin^2 u}} = \frac{\pi}{2} \ln(1 + \cos u)$ \qquad\qquad BI (226)(4)

3. $\displaystyle\int_0^\infty \frac{\sin x}{\sqrt{1 - k^2 \sin^2 x}} \frac{dx}{x} = \int_0^\infty \frac{\tan x}{\sqrt{1 - k^2 \sin^2 x}} \frac{dx}{x}$

$\displaystyle\qquad = \int_0^\infty \frac{\sin x}{\sqrt{1 - k^2 \cos^2 x}} \frac{dx}{x} = \int_0^\infty \frac{\tan x}{\sqrt{1 - k^2 \cos^2 x}} \frac{dx}{x} = \boldsymbol{K}(k)$

$\qquad\qquad\qquad\qquad\qquad\qquad\qquad\qquad\qquad$ BI (183)(12, 13, 21, 22)

4. $\displaystyle\int_0^{\pi/2} \frac{x \sin x \cos x}{\sqrt{1 - k^2 \sin^2 x}} dx = \frac{1}{2k^2}\left[-\pi k' + 2\boldsymbol{E}(k)\right]$ \qquad\qquad BI (211)(1)

5. $\int_0^{\pi/2} \dfrac{x \sin x \cos x}{\sqrt{1 - k^2 \cos^2 x}}\, dx = \dfrac{1}{2k^2} \left[\pi - 2\, \boldsymbol{E}(k) \right]$ BI (214)(1)

6. $\int_0^{\alpha} \dfrac{x \sin x\, dx}{\cos^2 x \sqrt{\sin^2 \alpha - \sin^2 x}} = \dfrac{\pi \sin^2 \frac{\alpha}{2}}{\cos^2 \alpha}$ LO III 284

7. $\int_0^{\beta} \dfrac{x \sin x\, dx}{\left(1 - \sin^2 \alpha \sin^2 x\right)\sqrt{\sin^2 \beta - \sin^2 x}} = \dfrac{\pi \ln \dfrac{\cos \alpha + \sqrt{1 - \sin^2 \alpha \sin^2 \beta}}{2 \cos \beta \cos^2 \frac{\alpha}{2}}}{2 \cos \alpha \sqrt{1 - \sin^2 \alpha \sin^2 \beta}}$ LO III 284

3.843

1. $\int_0^{\infty} \tan x \sqrt{1 - k^2 \sin^2 2x}\, \dfrac{dx}{x} = \boldsymbol{E}(k)$ BI (154)(10)

2. $\int_0^{\infty} \tan x \sqrt{1 - k^2 \cos^2 2x}\, \dfrac{dx}{x} = \boldsymbol{E}(k)$ BI (154)(22)

3. $\int_0^{\infty} \dfrac{\tan x}{\sqrt{1 + \sin^2 2x}}\, \dfrac{dx}{x} = \int_0^{\infty} \dfrac{\tan x}{\sqrt{1 + \cos^2 2x}}\, \dfrac{dx}{x} = \dfrac{1}{\sqrt{2}} \boldsymbol{K}\left(\dfrac{1}{\sqrt{2}}\right)$ BI (183)(6, 11)

4. $\int_0^{\infty} \dfrac{\tan x}{\sqrt{1 - k^2 \sin^2 2x}}\, \dfrac{dx}{x} = \int_0^{\infty} \dfrac{\tan x}{\sqrt{1 - k^2 \cos^2 2x}}\, \dfrac{dx}{x} = \boldsymbol{K}(k)$ BI (183)(14, 23)

3.844

1. $\int_0^{\infty} \dfrac{\sin x \cos x}{\sqrt{1 - k^2 \cos^2 x}}\, \dfrac{dx}{x} = \dfrac{1}{k^2} \left[\boldsymbol{K}(k) - \boldsymbol{E}(k) \right]$ BI (185)(20)

2. $\int_0^{\infty} \dfrac{\sin x \cos^2 x}{\sqrt{1 - k^2 \cos^2 x}} \cdot \dfrac{dx}{x} = \dfrac{1}{k^2} \left[\boldsymbol{K}(k) - \boldsymbol{E}(k) \right]$ BI (185)(21)

3. $\int_0^{\infty} \dfrac{\sin x \cos^3 x}{\sqrt{1 - k^2 \cos^2 x}} \cdot \dfrac{dx}{x} = \dfrac{1}{3k^4} \left[\left(2 + k^2\right) \boldsymbol{K}(k) - 2\left(1 + k^2\right) \boldsymbol{E}(k) \right]$ BI (185)(22)

4. $\int_0^{\infty} \dfrac{\sin x \cos^4 x}{\sqrt{1 - k^2 \cos^2 x}} \cdot \dfrac{dx}{x} = \dfrac{1}{3k^4} \left[\left(2 + k^2\right) \boldsymbol{K}(k) - 2\left(1 + k^2\right) \boldsymbol{E}(k) \right]$ BI (185)(23)

5. $\int_0^{\infty} \dfrac{\sin^3 x \cos x}{\sqrt{1 - k^2 \cos^2 x}} \cdot \dfrac{dx}{x} = \dfrac{1}{3k^4} \left[\left(1 + k'^2\right) \boldsymbol{E}(k) - 2k'^2\, \boldsymbol{K}(k) \right]$ BI (185)(24)

6. $\int_0^{\infty} \dfrac{\sin^3 x \cos^2 x}{\sqrt{1 - k^2 \cos^2 x}} \cdot \dfrac{dx}{x} = \dfrac{1}{3k^4} \left[\left(1 + k'^2\right) \boldsymbol{E}(k) - 2k'^2\, \boldsymbol{K}(k) \right]$ BI (185)(25)

7. $\int_0^{\infty} \dfrac{\sin^2 x \tan x}{\sqrt{1 - k^2 \cos^2 x}} \cdot \dfrac{dx}{x} = \dfrac{1}{k^2} \left[\boldsymbol{E}(k) - k'^2\, \boldsymbol{K}(k) \right]$ BI (184)(16)

8. $\int_0^{\infty} \dfrac{\sin^4 x \tan x}{\sqrt{1 - k^2 \cos^2 x}} \cdot \dfrac{dx}{x} = \dfrac{1}{3k^4} \left[\left(2 + 3k^2\right) k'^2\, \boldsymbol{K}(k) - 2\left(k'^2 - k^2\right) \boldsymbol{E}(k) \right]$ BI (184)(18)

3.845

1. $\int_0^{\infty} \dfrac{\sin x \cos x}{\sqrt{1 + \cos^2 x}} \cdot \dfrac{dx}{x} = \sqrt{2} \left[\boldsymbol{E}\left(\dfrac{\sqrt{2}}{2}\right) - \dfrac{1}{2} \boldsymbol{K}\left(\dfrac{\sqrt{2}}{2}\right) \right]$ BI (185)(6)

2. $\int_0^\infty \frac{\sin x \cos^2 x}{\sqrt{1 + \cos^2 x}} \cdot \frac{dx}{x} = \sqrt{2} \left[E \left(\frac{\sqrt{2}}{2} \right) - \frac{1}{2} K \left(\frac{\sqrt{2}}{2} \right) \right]$ BI (185)(7)

3.[10] $\int_0^\infty \frac{\sin^2 x \tan x}{\sqrt{1 + \cos^2 x}} \cdot \frac{dx}{x} = \sqrt{2} \left[K \left(\frac{\sqrt{2}}{2} \right) - E \left(\frac{\sqrt{2}}{2} \right) \right]$ BU (184)(8)

3.846

1. $\int_0^\infty \frac{\sin x \cos x}{\sqrt{1 - k^2 \sin^2 x}} \cdot \frac{dx}{x} = \frac{1}{k^2} \left[E(k) - k'^2 K(k) \right]$ BI (185)(9)

2. $\int_0^\infty \frac{\sin x \cos^2 x}{\sqrt{1 - k^2 \sin^2 x}} \cdot \frac{dx}{x} = \frac{1}{k^2} \left[E(k) - k'^2 K(k) \right]$ BI (185)(10)

3. $\int_0^\infty \frac{\sin x \cos^3 x}{\sqrt{1 - k^2 \sin^2 x}} \cdot \frac{dx}{x} = \frac{1}{3k^4} \left[(2 - 3k^2) k'^2 K(k) - 2 \left(k'^2 - k^2 \right) E(k) \right]$ BI (185)(11)

4. $\int_0^\infty \frac{\sin x \cos^4 x}{\sqrt{1 - k^2 \sin^2 x}} \cdot \frac{dx}{x} = \frac{1}{3k^4} \left[(2 - 3k^2) k'^2 K(k) - 2 \left(k'^2 - k^2 \right) E(k) \right]$ BI (185)(12)

5. $\int_0^\infty \frac{\sin^3 x \cos x}{\sqrt{1 - k^2 \sin^2 x}} \cdot \frac{dx}{x} = \frac{1}{3k^4} \left[\left(1 + k'^2 \right) E(k) - 2k'^2 K(k) \right]$ BI (185)(13)

6. $\int_0^\infty \frac{\sin^3 x \cos^2 x}{\sqrt{1 - k^2 \sin^2 x}} \cdot \frac{dx}{x} = \frac{1}{3k^4} \left[\left(1 + k'^2 \right) E(k) - 2k'^2 K(k) \right]$ BI (185)(14)

7. $\int_0^\infty \frac{\sin^2 x \tan x}{\sqrt{1 - k^2 \sin^2 x}} \cdot \frac{dx}{x} = \frac{1}{k^2} \left[K(k) - E(k) \right]$ BI (184)(9)

8. $\int_0^\infty \frac{\sin^4 x \tan x}{\sqrt{1 - k^2 \sin^2 x}} \cdot \frac{dx}{x} = \frac{1}{3k^4} \left[(2 + k^2) K(k) - 2 \left(1 + k^2 \right) E(k) \right]$ BI (184)(11)

3.847 $\int_0^\infty \frac{\sin x \cos x}{\sqrt{1 + \sin^2 x}} \cdot \frac{dx}{x} = \int_0^\infty \frac{\sin x \cos^2 x}{\sqrt{1 + \sin^2 x}} \cdot \frac{dx}{x} = \sqrt{2} \left[K \left(\frac{\sqrt{2}}{2} \right) - E \left(\frac{\sqrt{2}}{2} \right) \right]$ BI (185)(3, 4)

3.848

1. $\int_0^\infty \frac{\sin^3 x \cos x}{\sqrt{1 - k^2 \sin^2 2x}} \cdot \frac{dx}{x} = \frac{1}{4k^2} \left[K(k) - E(k) \right]$ BI (185)(15)

2. $\int_0^\infty \frac{\cos^2 2x \tan x}{\sqrt{1 - k^2 \sin^2 2x}} \cdot \frac{dx}{x} = \frac{1}{k^2} \left[E(k) - k'^2 K(k) \right]$ BI (184)(12)

3. $\int_0^\infty \frac{\cos^4 2x \tan x}{\sqrt{1 - k^2 \sin^2 2x}} \cdot \frac{dx}{x} = \frac{1}{3k^4} \left[(2 - 3k^2) k'^2 K(k) - 2 \left(k'^2 - k^2 \right) E(k) \right]$ BI (184)(13)

4. $\int_0^\infty \frac{\sin^2 4x \tan x}{\sqrt{1 - k^2 \sin^2 2x}} \cdot \frac{dx}{x} = \frac{4}{3k^4} \left[\left(1 + k'^2 \right) E(k) - 2k'^2 K(k) \right]$ BI (184)(17)

5. $\int_0^\infty \frac{\sin^3 x \cos x}{\sqrt{1 - k^2 \cos^2 2x}} \cdot \frac{dx}{x} = \frac{1}{4k^2} \left[E(k) - k'^2 K(k) \right]$ BI (185)(26)

6. $\int_0^\infty \frac{\cos^2 2x \tan x}{\sqrt{1 - k^2 \cos^2 2x}} \cdot \frac{dx}{x} = \frac{1}{k^2} \left[K(k) - E(k) \right]$ BI (184)(19)

7. $\displaystyle\int_0^\infty \frac{\cos^4 2x \tan x}{\sqrt{1-k^2\cos^2 2x}} \cdot \frac{dx}{x} = \frac{1}{3k^4}\left[\left(2+k^2\right)\boldsymbol{K}(k) - 2\left(1+k^2\right)\boldsymbol{E}(k)\right]$ BI (184)(20)

3.849

1. $\displaystyle\int_0^\infty \frac{\sin^3 x \cos x}{\sqrt{1+\cos^2 2x}} \cdot \frac{dx}{x} = \frac{1}{2\sqrt{2}}\left[\boldsymbol{K}\left(\frac{\sqrt{2}}{2}\right) - \boldsymbol{E}\left(\frac{\sqrt{2}}{2}\right)\right]$ BI (185)(8)

2.^{10} $\displaystyle\int_0^\infty \frac{\sin^3 x \cos x}{\sqrt{1+\sin^2 2x}} \cdot \frac{dx}{x} = \frac{\sqrt{2}}{8}\left[2\boldsymbol{E}\left(\frac{\sqrt{2}}{2}\right) - \boldsymbol{K}\left(\frac{\sqrt{2}}{2}\right)\right]$ BI (185)(5)

3. $\displaystyle\int_0^\infty \frac{\cos^2 2x \tan x}{\sqrt{1+\sin^2 2x}} \cdot \frac{dx}{x} = \sqrt{2}\left[\boldsymbol{K}\left(\frac{\sqrt{2}}{2}\right) - \boldsymbol{E}\left(\frac{\sqrt{2}}{2}\right)\right]$ BI (184)(7)

3.85–3.88 Trigonometric functions of more complicated arguments combined with powers

3.851

1. $\displaystyle\int_0^\infty x\sin\left(ax^2\right)\sin(2bx)\,dx = \frac{b}{2a}\sqrt{\frac{\pi}{2a}}\left(\cos\frac{b^2}{a} + \sin\frac{b^2}{a}\right)$

                                                   $[a \geq 0, \quad b > 0]$ BI (150)(4)

2. $\displaystyle\int_0^\infty x\sin\left(ax^2\right)\cos(2bx)\,dx = \frac{1}{2a} - \frac{b}{a}\sqrt{\frac{\pi}{2a}}\left[\sin\frac{b^2}{a}\,C\left(\frac{b}{\sqrt{a}}\right) - \cos\frac{b^2}{a}\,S\left(\frac{b}{\sqrt{a}}\right)\right]$ BI (150)(5)a

3. $\displaystyle\int_0^\infty x\cos\left(ax^2\right)\sin(2bx)\,dx = \frac{b}{2a}\sqrt{\frac{\pi}{2a}}\left(\sin\frac{b^2}{a} - \cos\frac{b^2}{a}\right)$

                                          $[a > 0, \quad b > 0]$,   (cf. **3.691** 7)

                                                           BI (150)(7)

4. $\displaystyle\int_0^\infty x\cos\left(ax^2\right)\cos(2bx)\,dx = \frac{b}{a}\sqrt{\frac{\pi}{2a}}\left[\cos\frac{b^2}{a}\,C\left(\frac{b}{\sqrt{a}}\right) + \sin\frac{b^2}{a}\,S\left(\frac{b}{\sqrt{a}}\right)\right]$ BI (150)(6)a

5. $\displaystyle\int_0^\infty \sin\left(ax^2\right)\cos(bx)\frac{dx}{x^2} = \frac{b\pi}{2}\left\{S\left(\frac{b}{2\sqrt{a}}\right) - C\left(\frac{b}{2\sqrt{a}}\right) + \sqrt{a\pi}\sin\left(\frac{b^2}{4a} + \frac{\pi}{4}\right)\right\}$

                                        $[a > 0, \quad b > 0]$,   (cf. **3.691** 7)

                                                            ET I 23(3)a

3.852

1. $\displaystyle\int_0^\infty \frac{\sin\left(ax^2\right)}{x^2}\,dx = \sqrt{\frac{a\pi}{2}}$ $[a \geq 0]$ BI (177)(10)a

2. $\displaystyle\int_0^\infty \sin\left(ax^2\right)\cos\left(bx^2\right)\frac{dx}{x^2} = \frac{1}{2}\sqrt{\frac{\pi}{2}}\left(\sqrt{a+b} + \sqrt{a-b}\right)$ $[a > b > 0]$

                                 $= \dfrac{1}{2}\sqrt{\pi a}$ $[b = a \geq 0]$

                                 $= \dfrac{1}{2}\sqrt{\dfrac{\pi}{2}}\left(\sqrt{a+b} - \sqrt{b-a}\right)$

                                          $[b > a > 0]$,   (cf. **3.852** 1)   BI (177)(23)

3. $$\int_0^\infty \frac{\sin^2\left(a^2 x^2\right)}{x^4}\,dx = \frac{2\sqrt{\pi}}{3}a^3 \qquad\qquad [a \geq 0] \qquad\qquad \text{GW (333)(19e)}$$

4.[10] $$\int_0^\infty \frac{\sin^3\left(a^2 x^2\right)}{x^2}\,dx = \frac{a}{4}\sqrt{\frac{\pi}{2}}\left(3 - \sqrt{3}\right) \qquad\qquad \left[\operatorname{Im} a^2 = 0\right] \qquad\qquad \text{MC}$$

5. $$\int_0^\infty \left(\sin^2 x - x^2 \cos x^2\right)\frac{dx}{x^4} = \frac{1}{3}\sqrt{\frac{\pi}{2}} \qquad\qquad \text{BI (178)(8)}$$

6. $$\int_0^\infty \left\{\cos^2 x - \frac{1}{1+x^2}\right\}\frac{dx}{x} = -\frac{1}{2}C \qquad\qquad \text{BI (173)(22)}$$

3.853

1. $$\int_0^\infty \frac{\sin\left(ax^2\right)}{\beta^2 + x^2}\,dx = \frac{\pi}{2\beta}\left[\sqrt{2}\sin\left(a\beta^2 + \frac{\pi}{4}\right)C\left(\sqrt{a}\beta\right) - \sqrt{2}\cos\left(a\beta^2 + \frac{\pi}{4}\right)S\left(\sqrt{a}\beta\right) - \sin\left(a\beta^2\right)\right]$$
$$[a > 0, \quad \operatorname{Re}\beta > 0] \qquad\qquad \text{ET II 219(33)a}$$

2. $$\int_0^\infty \frac{\cos\left(ax^2\right)}{\beta^2 + x^2}\,dx = \frac{\pi}{2\beta}\left[\cos\left(a\beta^2\right) - \sqrt{2}\cos\left(a\beta^2 + \frac{\pi}{4}\right)C\left(\sqrt{a}\beta\right) - \sqrt{2}\sin\left(a\beta^2 + \frac{\pi}{4}\right)S\left(\sqrt{a}\beta\right)\right]$$
$$[a > 0, \quad \operatorname{Re}\beta > 0] \qquad\qquad \text{ET II 221(51)a}$$

3. $$\int_0^\infty \frac{x^2 \sin\left(ax^2\right)}{\beta^2 + x^2}\,dx$$
$$= \frac{\beta\pi}{2}\left[\sin\left(a\beta^2\right) - \sqrt{2}\sin\left(a\beta^2 + \frac{\pi}{4}\right)C\left(\sqrt{a}\beta\right) + \sqrt{2}\cos\left(a\beta^2 + \frac{\pi}{4}\right)S\left(\sqrt{a}\beta\right)\right]$$
$$- \frac{1}{2}\sqrt{\frac{\pi}{2a}}$$
$$[a > 0, \quad \operatorname{Re}\beta > 0] \qquad\qquad \text{ET II 219(32)a}$$

4. $$\int_0^\infty \frac{x^2 \cos\left(ax^2\right)}{\beta^2 + x^2}\,dx = \frac{1}{2}\sqrt{\frac{\pi}{2a}} - \frac{\beta\pi}{2}\left\{\cos\left(a\beta^2\right) - \sqrt{2}\cos\left(a\beta^2 + \frac{\pi}{4}\right)C\left(\sqrt{a}\beta\right)\right.$$
$$\left. - \sqrt{2}\sin\left(a\beta^2 + \frac{\pi}{4}\right)S\left(\sqrt{a}\beta\right)\right\}$$
$$[a > 0, \quad \operatorname{Re}\beta > 0] \qquad\qquad \text{ET II 221(50)a}$$

3.854

1. $$\int_0^\infty \left(\cos\left(ax^2\right) - \sin\left(ax^2\right)\right)\frac{dx}{x^4 + b^4} = \frac{\pi e^{-ab^2}}{2b^3\sqrt{2}} \qquad\qquad [a > 0, \quad b > 0]$$
$$\text{LI (178)(11)a, BI (168)(25)}$$

2. $$\int_0^\infty \left(\cos\left(ax^2\right) + \sin\left(ax^2\right)\right)\frac{x^2\,dx}{x^4 + b^4} = \frac{\pi e^{-ab^2}}{2b\sqrt{2}} \qquad\qquad [a > 0, b > 0] \qquad\qquad \text{LI (178)(12)}$$

3. $$\int_0^\infty \left(\cos\left(ax^2\right) + \sin\left(ax^2\right)\right)\frac{x^2\,dx}{\left(x^4 + b^4\right)^2} = \frac{\pi e^{-ab^2}}{4\sqrt{2}b^3}\left(a + \frac{1}{2b^2}\right)$$
$$[a > 0, \quad b > 0] \qquad\qquad \text{LI (178)(14)}$$

4. $$\int_0^\infty \left(\cos\left(ax^2\right) - \sin\left(ax^2\right)\right)\frac{x^4\,dx}{\left(x^4 + b^4\right)^2} = \frac{\pi e^{-ab^2}}{4\sqrt{2}b}\left(\frac{1}{2b^2} - a\right)$$
$$[a > 0, \quad b > 0] \qquad\qquad \text{BI (178)(15)}$$

3.855

1. $\int_0^\infty \dfrac{\sin\left(ax^2\right)}{\sqrt{\beta^2 + x^4}}\, dx = \dfrac{1}{2}\sqrt{\dfrac{a\pi}{2}}\, I_{\frac{1}{4}}\left(\dfrac{a\beta}{2}\right) K_{\frac{1}{4}}\left(\dfrac{a\beta}{2}\right)$ $[a > 0, \quad \operatorname{Re}\beta > 0]$ ET I 66(28)

2. $\int_0^\infty \dfrac{\cos\left(ax^2\right)}{\sqrt{\beta^2 + x^4}}\, dx = \dfrac{1}{2}\sqrt{\dfrac{a\pi}{2}}\, I_{-\frac{1}{4}}\left(\dfrac{a\beta}{2}\right) K_{\frac{1}{4}}\left(\dfrac{a\beta}{2}\right)$ $[a > 0, \quad \operatorname{Re}\beta > 0]$ ET I 9(22)

3. $\int_0^u \dfrac{\sin\left(a^2 x^2\right)}{\sqrt{u^4 - x^4}}\, dx = \dfrac{a}{4}\sqrt{\dfrac{\pi^3}{2}}\left[J_{\frac{1}{4}}\left(\dfrac{a^2}{u^2}2\right)\right]^2$ $[a > 0]$ ET I 66(29)

4. $\int_u^\infty \dfrac{\sin\left(a^2 x^2\right)}{\sqrt{x^4 - u^4}}\, dx = -\dfrac{a}{4}\sqrt{\dfrac{\pi^3}{2}}\, J_{\frac{1}{4}}\left(\dfrac{a^2 u^2}{2}\right) Y_{\frac{1}{4}}\left(\dfrac{a^2 u^2}{2}\right)$

                                                $[a > 0]$ ET I 66(30)

5. $\int_0^u \dfrac{\cos\left(a^2 x^2\right)}{\sqrt{u^4 - x^4}}\, dx = \dfrac{a}{4}\sqrt{\dfrac{\pi^3}{2}}\left[J_{-\frac{1}{4}}\left(\dfrac{a^2 u^2}{2}\right)\right]^2$ ET I 9(23)

6. $\int_u^\infty \dfrac{\cos\left(a^2 x^2\right)}{\sqrt{x^4 - u^4}}\, dx = -\dfrac{a}{4}\sqrt{\dfrac{\pi^3}{2}}\, J_{-\frac{1}{4}}\left(\dfrac{a^2 u^2}{2}\right) Y_{-\frac{1}{4}}\left(\dfrac{a^2 u^2}{2}\right)$ ET I 10(24)

3.856

1. $\int_0^\infty \dfrac{\left(\sqrt{\beta^4 + x^4} + x^2\right)^\nu}{\sqrt{\beta^4 + x^4}}\sin\left(a^2 x^2\right)\, dx = \dfrac{a}{2}\sqrt{\dfrac{\pi}{2}}\beta^{2\nu}\, I_{\frac{1}{4} - \frac{\nu}{2}}\left(\dfrac{a^2 \beta^2}{2}\right) K_{\frac{1}{4} + \frac{\nu}{2}}\left(\dfrac{a^2 \beta^2}{2}\right)$

                                       $\left[\operatorname{Re}\nu < \dfrac{3}{2}, \quad |\arg\beta| < \dfrac{\pi}{4}\right]$ ET I 71(23)

2. $\int_0^\infty \dfrac{\left(\sqrt{\beta^4 + x^4} + x^2\right)^\nu}{\sqrt{\beta^4 + x^4}}\cos\left(a^2 x^2\right)\, dx = \dfrac{a}{2}\sqrt{\dfrac{\pi}{2}}\beta^{2\nu}\, I_{-\frac{1}{4} - \frac{\nu}{2}}\left(\dfrac{a^2 \beta^2}{2}\right) K_{-\frac{1}{4} + \frac{\nu}{2}}\left(\dfrac{a^2 \beta^2}{2}\right)$

                                       $\left[\operatorname{Re}\nu < \dfrac{3}{2}, \quad |\arg\beta| < \dfrac{\pi}{4}\right]$ ET I 12(16)

3. $\int_0^\infty \dfrac{\left(\sqrt{\beta^4 + x^4} - x^2\right)^\nu}{\sqrt{\beta^4 + x^4}}\cos\left(a^2 x^2\right)\, dx = \dfrac{a}{2}\sqrt{\dfrac{\pi}{2}}\beta^{2\nu}\, I_{-\frac{1}{4} + \frac{\nu}{2}}\left(\dfrac{a^2 \beta^2}{2}\right) K_{-\frac{1}{4} - \frac{\nu}{2}}\left(\dfrac{a^2 \beta^2}{2}\right)$

                                       $\left[\operatorname{Re}\nu > -\dfrac{3}{2}, \quad |\arg\beta| < \dfrac{\pi}{4}\right]$

                                                                 ET I 12(17)

4. $\int_0^\infty \dfrac{\sin\left(a^2 x^2\right)\, dx}{\sqrt{\beta^4 + x^4}\sqrt{x^2 + \sqrt{\beta^4 + x^4}}} = \dfrac{\sinh\frac{a^2 \beta^2}{2}}{\sqrt{2}\beta^2}\, K_0\left(\dfrac{a^2 \beta^2}{2}\right)$

                                       $\left[|\arg\beta| < \dfrac{\pi}{4}\right]$ ET I 66(32)

5. $\int_0^\infty \dfrac{\cos\left(a^2 x^2\right)\, dx}{\sqrt{\beta^4 + x^4}\sqrt{\left(x^2 + \sqrt{\beta^4 + x^4}\right)^3}} = \dfrac{\sinh\frac{a^2 \beta^2}{2}}{2\sqrt{2}\beta^4}\, K_1\left(\dfrac{a^2 \beta^2}{2}\right)$

                                       $\left[|\arg\beta| < \dfrac{\pi}{4}\right]$ ET I 10(27)

6. $\int_0^\infty \dfrac{\sqrt{\sqrt{\beta^4 + x^4} + x^2}}{\sqrt{\beta^4 + x^4}} \sin\left(a^2 x^2\right)\, dx = \dfrac{\pi}{2\sqrt{2}} e^{-\frac{a^2\beta^2}{2}}\, I_0\left(\dfrac{a^2\beta^2}{2}\right)$

$$\left[|\arg\beta| < \dfrac{\pi}{4}\right]$$ ET I 67(33)

3.857

1. $\int_0^\infty \dfrac{x^2}{R_1 R_2}\sqrt{\dfrac{R_2 - R_1}{R_2 + R_1}}\, \sin\left(ax^2\right)\, dx = \dfrac{1}{2\sqrt{b}}\, K_0(ac)\sin ab$

$$\left[R_1 = \sqrt{c^2 + (b - x^2)^2}, \quad R_2 = \sqrt{c^2 + (b + x^2)^2}, \quad a > 0, \quad c > 0\right]$$ ET I 67(34)

2. $\int_0^\infty \dfrac{x^2}{R_1 R_2}\sqrt{\dfrac{R_2 + R_1}{R_2 - R_1}}\, \cos\left(ax^2\right)\, dx = \dfrac{1}{2\sqrt{b}}\, K_0(ac)\cos ab$

$$\left[R_1 = \sqrt{c^2 + (b - x^2)^2}, \quad R_2 = \sqrt{c^2 + (b + x^2)^2}, \quad a > 0, \quad c > 0\right]$$ ET I 10(26)

3.858

1. $\displaystyle\int_u^\infty \dfrac{\left(x^2 + \sqrt{x^4 - u^4}\right)^\nu + \left(x^2 - \sqrt{x^4 - u^4}\right)^\nu}{\sqrt{x^4 - u^4}}\, \sin\left(a^2 x^2\right)\, dx$

$$= -\dfrac{a}{4}\sqrt{\dfrac{\pi^3}{a}}\, u^{2\nu}\left[J_{\frac{1}{4} + \frac{\nu}{2}}\left(\dfrac{a^2 u^2}{2}\right) Y_{\frac{1}{4} - \frac{\nu}{2}}\left(\dfrac{a^2 u^2}{2}\right) + J_{\frac{1}{4} - \frac{\nu}{2}}\left(\dfrac{a^2 u^2}{2}\right) Y_{\frac{1}{4} + \frac{\nu}{2}}\left(\dfrac{a^2 u^2}{2}\right)\right]$$

$$\left[\operatorname{Re}\nu < \dfrac{3}{2}\right]$$ ET I 71(25)

2. $\displaystyle\int_u^\infty \dfrac{\left(x^2 + \sqrt{x^4 - u^4}\right)^\nu + \left(x^2 - \sqrt{x^4 - u^4}\right)^\nu}{\sqrt{x^4 - u^4}}\, \cos\left(a^2 x^2\right)\, dx$

$$= -\dfrac{a}{4}\sqrt{\dfrac{\pi^3}{a}}\, u^{2\nu}\left[J_{-\frac{1}{4} + \frac{\nu}{2}}\left(\dfrac{a^2 u^2}{2}\right) Y_{-\frac{1}{4} - \frac{\nu}{2}}\left(\dfrac{a^2 u^2}{2}\right) + J_{-\frac{1}{4} - \frac{\nu}{2}}\left(\dfrac{a^2 u^2}{2}\right) Y_{-\frac{1}{4} + \frac{\nu}{2}}\left(\dfrac{a^2 u^2}{2}\right)\right]$$

$$\left[\operatorname{Re}\nu < \dfrac{3}{2}\right]$$ ET I 13(26)

3.859 $\displaystyle\int_0^\infty \left[\cos\left(x^{2^n}\right) - \dfrac{1}{1 + x^{2^{n+1}}}\right] \dfrac{dx}{x} = -\dfrac{1}{2^n}\, C$ BI (173)(24)

3.861

1. $\displaystyle\int_0^\infty \sin^{2n+1}\left(ax^2\right) \dfrac{dx}{x^{2m}} = \pm\dfrac{\sqrt{\pi}\, a^{m - \frac{1}{2}}}{2^{2n - m + \frac{1}{2}}(2m - 1)!!} \sum_{k=1}^{n+1} (-1)^{k-1} \binom{2n+1}{n+k}(2k - 1)^{m - \frac{1}{2}}$

$$\left[\begin{array}{l}\text{the } + \text{ sign is taken when } m \equiv 0 \pmod 4 \text{ or } m \equiv 1 \pmod 4,\\ \text{the } - \text{ sign is taken when } m \equiv 2 \pmod 4 \text{ or } m \equiv 3 \pmod 4\end{array}\right]$$ BI (177)(19)a

2. $\displaystyle\int_0^\infty \sin^{2n}\left(ax^2\right) \dfrac{dx}{x^{2m}} = \pm\dfrac{\sqrt{\pi}\, a^{m - \frac{1}{2}}}{2^{2n - 2m + 1}(2m - 1)!!} \sum_{k=1}^{n} (-1)^k \binom{2n}{n+k} k^{m - \frac{1}{2}}$

$$\left[\begin{array}{l}\text{the } + \text{ sign is taken when } m \equiv 0 \pmod 4 \text{ or } m \equiv 3 \pmod 4,\\ \text{the } - \text{ sign is taken when } m \equiv 2 \pmod 4 \text{ or } m \equiv 1 \pmod 4\end{array}\right]$$ BI (177)(18)a, LI (177)(18)

3.862 $\displaystyle\int_0^\infty \left[\cos\left(ax^2\sqrt{n}\right) + \sin\left(ax^2\sqrt{n}\right)\right]\left(\dfrac{\sin^2 x}{x^2}\right)^n dx$

$$= \dfrac{\sqrt{\pi}}{(2n - 1)!!\sqrt{2}} \sum_{k=0}^{n} (-1)^k \binom{n}{k}\left(n - 2k + a\sqrt{n}\right)^{n - \frac{1}{2}}$$

$$\left[a > \sqrt{n} > 0\right]$$ BI (178)(9)

3.863

1. $$\int_0^\infty x^2 \cos\left(ax^4\right) \sin\left(2bx^2\right) dx = -\frac{\pi}{8}\sqrt{\frac{b^3}{a^3}}\left[\sin\left(\frac{b^2}{2a}-\frac{\pi}{8}\right) J_{-\frac{1}{4}}\left(\frac{b^2}{2a}\right) + \cos\left(\frac{b^2}{2a}-\frac{\pi}{8}\right) J_{\frac{3}{4}}\left(\frac{b^2}{2a}\right)\right]$$
    $$[a>0, \quad b>0]$$ ET I 25(22)

2. $$\int_0^\infty x^2 \cos\left(ax^4\right) \cos\left(2bx^2\right) dx = -\frac{\pi}{8}\sqrt{\frac{b^3}{a^3}}\left[\sin\left(\frac{b^2}{2a}+\frac{\pi}{8}\right) J_{-\frac{3}{4}}\left(\frac{b^2}{2a}\right) + \cos\left(\frac{b^2}{2a}+\frac{\pi}{8}\right) J_{-\frac{1}{4}}\left(\frac{b^2}{2a}\right)\right]$$
    $$[a>0, \quad b>0]$$ ET I 25(23)

3.864

1. $$\int_0^\infty \sin\frac{b}{x} \sin ax \frac{dx}{x} = \frac{\pi}{2}\, Y_0\left(2\sqrt{ab}\right) + K_0\left(2\sqrt{ab}\right) \qquad [a>0, \quad b>0]$$ WA 204(3)a

2. $$\int_0^\infty \cos\frac{b}{x} \cos ax \frac{dx}{x} = -\frac{\pi}{2}\, Y_0\left(2\sqrt{ab}\right) + K_0\left(2\sqrt{ab}\right)$$
    $$[a>0, \quad b>0]$$
    WA 204(4)a, ET I 24 (12)

3.865

1. $$\int_0^u \frac{\left(u^2-x^2\right)^{\mu-1}}{x^{2\mu}} \sin\frac{a}{x} dx = \frac{\sqrt{\pi}}{2}\left(\frac{2}{a}\right)^{\mu-\frac{1}{2}} u^{\mu-\frac{3}{2}}\, \Gamma(\mu)\, J_{\frac{1}{2}-\mu}\left(\frac{a}{u}\right)$$
    $$[a>0, \quad u>0, \quad 0<\operatorname{Re}\mu<1]$$
    ET II 189(30)

2. $$\int_u^\infty \frac{(x-u)^{\mu-1}}{x^{2\mu}} \sin\frac{a}{x} dx = \sqrt{\frac{\pi}{u}} a^{\frac{1}{2}-\mu}\, \Gamma(\mu) \sin\frac{a}{2u}\, J_{\mu-\frac{1}{2}}\left(\frac{a}{2u}\right)$$
    $$[a>0, \quad u>0, \quad \operatorname{Re}\mu>0]$$
    ET II 203(21)

3. $$\int_0^u \frac{\left(u^2-x^2\right)^{\mu-1}}{x^{2\mu}} \cos\frac{a}{x} dx = -\frac{\sqrt{\pi}}{2}\left(\frac{2}{a}\right)^{\mu-\frac{1}{2}} \Gamma(\mu) u^{\mu-\frac{3}{2}}\, Y_{\frac{1}{2}-\mu}\left(\frac{a}{u}\right)$$
    $$[a>0, \quad u>0, \quad 0<\operatorname{Re}\mu<1]$$
    ET II 190(36)

4. $$\int_u^\infty \frac{(x-u)^{\mu-1}}{x^{2\mu}} \cos\frac{a}{x} dx = \sqrt{\frac{\pi}{u}} a^{\frac{1}{2}-\mu}\, \Gamma(\mu) \cos\frac{a}{2u}\, J_{\mu-\frac{1}{2}}\left(\frac{a}{2u}\right)$$
    $$[a>0, \quad u>0, \quad \operatorname{Re}\mu>0]$$
    ET II 204(26)

3.866

1. $$\int_0^\infty x^{\mu-1} \sin\frac{b^2}{x} \sin\left(a^2 x\right) dx = \frac{\pi}{4}\left(\frac{b}{a}\right)^\mu \operatorname{cosec}\frac{\mu\pi}{2}\left[J_\mu(2ab) - J_{-\mu}(2ab) + I_{-\mu}(2ab) - I_\mu(2ab)\right]$$
    $$[a>0, \quad b>0, \quad |\operatorname{Re}\mu|<1]$$
    ET I 322(42)

2. $$\int_0^\infty x^{\mu-1} \sin\frac{b^2}{x} \cos\left(a^2 x\right) dx = \frac{\pi}{4}\left(\frac{b}{a}\right)^\mu \sec\frac{\mu\pi}{2}\left[J_\mu(2ab) + J_{-\mu}(2ab) + I_\mu(2ab) - I_{-\mu}(2ab)\right]$$
    $$[a>0, \quad b>0, \quad |\operatorname{Re}\mu|<1]$$
    ET I 322(43)

3. $\int_0^\infty x^{\mu-1} \cos \dfrac{b^2}{x} \cos\left(a^2 x\right)\, dx = \dfrac{\pi}{4} \left(\dfrac{b}{a}\right)^\mu \operatorname{cosec} \dfrac{\mu\pi}{2} \left[J_{-\mu}(2ab) - J_\mu(2ab) + I_{-\mu}(2ab) - I_\mu(2ab)\right]$

$$[a > 0, \quad b > 0, \quad |\operatorname{Re}\mu| < 1]$$

ET I 322(44)

3.867

1. $\int_0^1 \dfrac{\cos ax - \cos \frac{a}{x}}{1 - x^2}\, dx = \dfrac{1}{2}\int_0^\infty \dfrac{\cos ax - \cos \frac{a}{x}}{1 - x^2}\, dx = \dfrac{\pi}{2}\sin a$

$$[a > 0]$$

GW (334)(7a)

2. $\int_0^1 \dfrac{\cos ax + \cos \frac{a}{x}}{1 + x^2}\, dx = \dfrac{1}{2}\int_0^\infty \dfrac{\cos ax + \cos \frac{a}{x}}{1 + x^2}\, dx = \dfrac{\pi}{2}e^{-a}$

$$[a > 0]$$

GW (334)(7b)

3.868

1. $\int_0^\infty \sin\left(a^2 x + \dfrac{b^2}{x}\right) \dfrac{dx}{x} = \pi J_0(2ab)$ $\qquad [a > 0, \quad b > 0]$

GW (334)(11a), WA 200(16)

2. $\int_0^\infty \cos\left(a^2 x + \dfrac{b^2}{x}\right) \dfrac{dx}{x} = -\pi Y_0(2ab)$ $\qquad [a > 0, \quad b > 0]$ $\qquad$ GW (334)(11a)

3. $\int_0^\infty \sin\left(a^2 x - \dfrac{b^2}{x}\right) \dfrac{dx}{x} = 0$ $\qquad [a > 0, \quad b > 0]$ $\qquad$ GW (334)(11b)

4. $\int_0^\infty \cos\left(a^2 x - \dfrac{b^2}{x}\right) \dfrac{dx}{x} = 2 K_0(2ab)$ $\qquad [a > 0, \quad b > 0]$ $\qquad$ GW (334)(11b)

3.869

1. $\int_0^\infty \sin\left(ax - \dfrac{b}{x}\right) \dfrac{x\, dx}{\beta^2 + x^2} = \dfrac{\pi}{2}\exp\left(-\alpha\beta - \dfrac{b}{\beta}\right)$ $\qquad [a > 0, \quad b > 0, \quad \operatorname{Re}\beta > 0]$

ET II 220(42)

2. $\int_0^\infty \cos\left(ax - \dfrac{b}{x}\right) \dfrac{dx}{\beta^2 + x^2} = \dfrac{\pi}{2\beta}\exp\left(-a\beta - \dfrac{b}{\beta}\right)$ $\qquad [a > 0, \quad b > 0, \quad \operatorname{Re}\beta > 0]$

ET II 222(58)

3.871

1. $\int_0^\infty x^{\mu-1} \sin\left[a\left(x + \dfrac{b^2}{x}\right)\right] dx = \pi b^\mu \left[J_\mu(2ab)\cos\dfrac{\mu\pi}{2} - Y_\mu(2ab)\sin\dfrac{\mu\pi}{2}\right]$

$$[a > 0, \quad b > 0, \quad \operatorname{Re}\mu < 1]$$

ET I 319(17)

2. $\int_0^\infty x^{\mu-1} \cos\left[a\left(x + \dfrac{b^2}{x}\right)\right] dx = -\pi b^\mu \left[J_\mu(2ab)\sin\dfrac{\mu\pi}{2} + Y_\mu(2ab)\cos\dfrac{\mu\pi}{2}\right]$

$$[a > 0, \quad b > 0, \quad |\operatorname{Re}\mu| < 1]$$

ET I 321(35)

3. $\int_0^\infty x^{\mu-1} \sin\left[a\left(x - \dfrac{b^2}{x}\right)\right] dx = 2b^\mu K_\mu(2ab)\sin\dfrac{\mu\pi}{2}$ $\qquad [a > 0, \quad b > 0, \quad |\operatorname{Re}\mu| < 1]$

ET I 319(16)

4. $\int_0^\infty x^{\mu-1} \cos\left[a\left(x - \dfrac{b^2}{x}\right)\right] dx = 2b^\mu K_\mu(2ab) \cos\dfrac{\mu\pi}{2}$

$$[a > 0, \quad b > 0, \quad |\operatorname{Re}\mu| < 1]$$

ET I 321(36)

3.872

1. $\int_0^1 \sin\left[a\left(x + \dfrac{1}{x}\right)\right] \sin\left[a\left(x - \dfrac{1}{x}\right)\right] \dfrac{dx}{1 - x^2}$

$$= \dfrac{1}{2}\int_0^\infty \sin\left[a\left(x + \dfrac{1}{x}\right)\right] \sin\left[a\left(x - \dfrac{1}{x}\right)\right] \dfrac{dx}{1 - x^2} = -\dfrac{\pi}{4}\sin 2a$$

$$[a \geq 0]$$ BI (149)(15), GW (334)(8a)

2. $\int_0^1 \cos\left[a\left(x + \dfrac{1}{x}\right)\right] \cos\left[a\left(x - \dfrac{1}{x}\right)\right] \dfrac{dx}{1 + x^2}$

$$= \dfrac{1}{2}\int_0^\infty \cos\left[a\left(x + \dfrac{1}{x}\right)\right] \cos\left[a\left(x - \dfrac{1}{x}\right)\right] \dfrac{dx}{1 + x^2} = \dfrac{\pi}{4}e^{-2a}$$

$$[a \geq 0]$$ GW (334)(8b)

3.873

1. $\int_0^\infty \sin\dfrac{a^2}{x^2} \cos b^2 x^2 \dfrac{dx}{x^2} = \dfrac{\sqrt{\pi}}{4\sqrt{2}a}\left[\sin(2ab) + \cos(2ab) + e^{-2ab}\right]$

$$[a > 0, \quad b > 0]$$ ET I 24(15)

2. $\int_0^\infty \cos\dfrac{a^2}{x^2} \cos b^2 x^2 \dfrac{dx}{x^2} = \dfrac{\sqrt{\pi}}{4\sqrt{2}a}\left[\cos(2ab) - \sin(2ab) + e^{-2ab}\right]$

$$[a > 0, \quad b > 0]$$ ET I 24(16)

3.874

1. $\int_0^\infty \sin\left(a^2 x^2 + \dfrac{b^2}{x^2}\right) \dfrac{dx}{x^2} = \dfrac{\sqrt{\pi}}{2b}\sin\left(2ab + \dfrac{\pi}{4}\right)$ $[a > 0, \quad b > 0]$

BI (179)(6)a, GW(334)(10a)

2. $\int_0^\infty \cos\left(a^2 x^2 + \dfrac{b^2}{x^2}\right) \dfrac{dx}{x^2} = \dfrac{\sqrt{\pi}}{2b}\cos\left(2ab + \dfrac{\pi}{4}\right)$ $[a > 0, \quad b > 0]$

GI (179)(8)a, GW(334)(10a)

3. $\int_0^\infty \sin\left(a^2 x^2 - \dfrac{b^2}{x^2}\right) \dfrac{dx}{x^2} = -\dfrac{\sqrt{\pi}}{2\sqrt{2}b}e^{-2ab}$ $[a \geq 0, \quad b > 0]$ GW (335)(10b)

4. $\int_0^\infty \cos\left(a^2 x^2 - \dfrac{b^2}{x^2}\right) \dfrac{dx}{x^2} = \dfrac{\sqrt{\pi}}{2\sqrt{2}b}e^{-2ab}$ $[a \geq 0, \quad b > 0]$ GW (334)(10b)

5. $\int_0^\infty \sin\left(ax - \dfrac{b}{x}\right)^2 \dfrac{dx}{x^2} = \dfrac{\sqrt{2\pi}}{4b}$ $[a > 0, \quad b > 0]$ BI (179)(13)a

6. $\int_0^\infty \cos\left(ax - \dfrac{b}{x}\right)^2 \dfrac{dx}{x^2} = \dfrac{\sqrt{2\pi}}{4b}$ $[a > 0, \quad b > 0]$ BI (179)(14)a

3.875

1. $\int_u^\infty \dfrac{x \sin\left(p\sqrt{x^2 - u^2}\right)}{x^2 + a^2} \cos bx \, dx = \dfrac{\pi}{2} \exp\left(-p\sqrt{a^2 + u^2}\right) \cosh ab$

$$[0 < b < p] \qquad \text{ET I 27(39)}$$

2. $\int_u^\infty \dfrac{x \sin\left(p\sqrt{x^2 - u^2}\right)}{a^2 + x^2 - u^2} \cos bx \, dx = \dfrac{\pi}{2} e^{-ap} \cos\left(b\sqrt{u^2 - a^2}\right)$

$$[0 < b < p, \quad a > 0] \qquad \text{ET I 27(38)}$$

3.[6] $\int_0^\infty \dfrac{\sin\left(p\sqrt{a^2 + x^2}\right)}{\left(a^2 + x^2\right)^{3/2}} \cos bx \, dx = \dfrac{\pi p}{2a} e^{-ab} \qquad [0 < p < b, \quad a > 0] \qquad \text{ET I 26(29)}$

3.876

1. $\int_0^\infty \dfrac{\sin\left(p\sqrt{x^2 + a^2}\right)}{\sqrt{x^2 + a^2}} \cos bx \, dx = \dfrac{\pi}{2} J_0\left(a\sqrt{p^2 - b^2}\right) \qquad [0 < b < p]$

$$= 0 \qquad\qquad [b > p > 0]$$
$$[a > 0] \qquad \text{ET I 26(30)}$$

2. $\int_0^\infty \dfrac{\cos\left(p\sqrt{x^2 + a^2}\right)}{\sqrt{x^2 + a^2}} \cos bx \, dx = -\dfrac{\pi}{2} Y_0\left(a\sqrt{p^2 - b^2}\right) \qquad [0 < b < p]$

$$= K_0\left(a\sqrt{b^2 - p^2}\right) \qquad [b > p > 0]$$
$$[a > 0] \qquad \text{ET I 26(34)}$$

3. $\int_0^\infty \dfrac{\cos\left(p\sqrt{x^2 + a^2}\right)}{x^2 + c^2} \cos bx \, dx = \dfrac{\pi}{2c} e^{-bc} \cos\left(p\sqrt{a^2 - c^2}\right)$

$$[c > 0, \quad b > p] \qquad \text{ET I 26(33)}$$

4. $\int_0^\infty \dfrac{\sin\left(p\sqrt{x^2 + a^2}\right)}{\left(x^2 + c^2\right)\sqrt{x^2 + a^2}} \cos bx \, dx = \dfrac{\pi}{2c} \dfrac{e^{-bc} \sin\left(p\sqrt{a^2 - c^2}\right)}{\sqrt{a^2 - c^2}} \qquad [c \neq a]$

$$= \dfrac{\pi}{2} e^{-ba} \dfrac{p}{a} \qquad [c = a]$$
$$[b > p, \quad c > 0] \qquad \text{ET I 26(31)a}$$

5.[6] $\int_0^\infty \dfrac{\cos\left(p\sqrt{x^2 + a^2}\right)}{x^2 + a^2} \cos bx \, dx = \dfrac{\pi}{2a} e^{-ab} \qquad [b > p > 0; \quad a > 0] \qquad \text{ET I 27(35)a}$

6.[6] $\int_0^\infty \dfrac{x \cos\left(p\sqrt{x^2 + a^2}\right)}{x^2 + a^2} \sin bx \, dx = \dfrac{\pi}{2} e^{-ab} \qquad [a > 0, \quad b > p > 0] \qquad \text{ET I 85(29)a}$

7. $\int_0^u \dfrac{\cos\left(p\sqrt{u^2 - x^2}\right)}{\sqrt{u^2 - x^2}} \cos bx \, dx = \dfrac{\pi}{2} J_0\left(u\sqrt{b^2 + p^2}\right) \qquad \text{ET I 28(42)}$

8. $\int_u^\infty \dfrac{\cos\left(p\sqrt{x^2 - u^2}\right)}{\sqrt{x^2 - u^2}} \cos bx \, dx = K_0\left(u\sqrt{p^2 - b^2}\right) \qquad [0 < b < |p|]$

$$= -\dfrac{\pi}{2} Y_0\left(u\sqrt{b^2 - p^2}\right) \qquad [b > |p|]$$

$$\text{ET I 28(43)}$$

3.877

1. $\displaystyle\int_0^u \frac{\sin\left(p\sqrt{u^2-x^2}\right)}{\sqrt[4]{(u^2-x^2)^3}}\cos bx\,dx = \sqrt{\frac{\pi^3 p}{8}}\,J_{\frac{1}{4}}\left[\frac{u}{2}\left(\sqrt{b^2+p^2}-b\right)\right]J_{\frac{1}{4}}\left[\frac{u}{2}\left(\sqrt{b^2+p^2}+b\right)\right]$

$$[b>0, \quad p>0]\qquad\qquad\text{ET I 27(40)}$$

2. $\displaystyle\int_u^\infty \frac{\sin\left(p\sqrt{x^2-u^2}\right)}{\sqrt[4]{(x^2-u^2)^3}}\cos bx\,dx = -\sqrt{\frac{\pi^3 p}{8}}\,J_{\frac{1}{4}}\left[\frac{u}{2}\left(b-\sqrt{b^2-p^2}\right)\right]Z_{\frac{1}{4}}\left[\frac{u}{2}\left(b+\sqrt{b^2-p^2}\right)\right]$

$$[b>p>0]\qquad\qquad\text{ET I 27(41)}$$

3. $\displaystyle\int_0^u \frac{\cos\left(p\sqrt{u^2-x^2}\right)}{\sqrt[4]{(u^2-x^2)^3}}\cos bx\,dx = \sqrt{\frac{\pi^3 p}{8}}\,J_{-\frac{1}{4}}\left[\frac{u}{2}\left(\sqrt{p^2+b^2}-b\right)\right]J_{-\frac{1}{4}}\left[\frac{u}{2}\left(\sqrt{p^2+b^2}+b\right)\right]$

$$[u>0, \quad p>0]\qquad\qquad\text{ET I 28(44)}$$

4. $\displaystyle\int_u^\infty \frac{\cos\left(p\sqrt{x^2-u^2}\right)}{\sqrt[4]{(x^2-u^2)^3}}\cos bx\,dx = -\sqrt{\frac{\pi^3 p}{8}}\,J_{-\frac{1}{4}}\left[\frac{u}{2}\left(b-\sqrt{b^2-p^2}\right)\right]Y_{\frac{1}{4}}\left[\frac{u}{2}\left(b+\sqrt{b^2-p^2}\right)\right]$

$$[b>p>0]\qquad\qquad\text{ET I 28(45)}$$

3.878

1. $\displaystyle\int_0^\infty \frac{\sin\left(p\sqrt{x^4+a^4}\right)}{\sqrt{x^4+a^4}}\cos bx^2\,dx = \frac{1}{2}\sqrt{\left(\frac{\pi}{2}\right)^3}\,b\,J_{-\frac{1}{4}}\left[\frac{a^2}{2}\left(p-\sqrt{p^2-b^2}\right)\right]J_{\frac{1}{4}}\left[\frac{a^2}{2}\left(p+\sqrt{p^2-b^2}\right)\right]$

$$[p>b>0]\qquad\qquad\text{ET I 26(32)}$$

2. $\displaystyle\int_0^\infty \frac{\cos\left(p\sqrt{x^4+a^4}\right)}{\sqrt{x^4+a^4}}\cos bx^2\,dx$

$$= -\frac{1}{2}\sqrt{\left(\frac{\pi}{2}\right)^3}\,b\,J_{-\frac{1}{4}}\left[\frac{a^2}{2}\left(p-\sqrt{p^2-b^2}\right)\right]Y_{\frac{1}{4}}\left[\frac{a^2}{2}\left(p+\sqrt{p^2-b^2}\right)\right]$$
$$[a>0, \quad p>b>0]\qquad\qquad\text{ET I 27(36)}$$

3. $\displaystyle\int_0^u \frac{\cos\left(p\sqrt{u^4-x^4}\right)}{\sqrt{u^4-x^4}}\cos bx^2\,dx = \frac{1}{2}\sqrt{\left(\frac{\pi}{2}\right)^3}\,b\,J_{-\frac{1}{4}}\left[\frac{u^2}{2}\left(\sqrt{p^2+b^2}-p\right)\right]J_{-\frac{1}{4}}\left[\frac{u^2}{2}\left(\sqrt{p^2+b^2}+p\right)\right]$

$$[p>0, \quad b>0]\qquad\qquad\text{ET I 28(46)}$$

3.879 $\displaystyle\int_0^\infty \sin ax^p\,\frac{dx}{x} = \frac{\pi}{2p}$ $[a>0, \quad p>0]$ GW (334)(6)

3.881

1. $\displaystyle\int_0^{\pi/2} x\sin\left(a\tan x\right)\,dx = \frac{\pi}{4}e^{-a}\left[\mathbf{C}+\ln 2a - e^{2a}\,\text{Ei}(-2a)\right]$

$$[a>0]\qquad\qquad\text{BI (205)(9)}$$

2. $\displaystyle\int_0^\infty \sin\left(a\tan x\right)\frac{dx}{x} = \frac{\pi}{2}\left(1-e^{-a}\right)$ $[a>0]$ BI (151)(6)

3. $\displaystyle\int_0^\infty \sin\left(a\tan x\right)\cos x\frac{dx}{x} = \frac{\pi}{2}\left(1-e^{-a}\right)$ $[a>0]$ BI (151)(19)

4. $$\int_0^\infty \cos\left(a\tan x\right)\sin x\,\frac{dx}{x} = \frac{\pi}{2}e^{-a} \qquad\qquad [a>0] \qquad\qquad \text{BI (151)(20)}$$

5. $$\int_0^\infty \sin\left(a\tan x\right)\sin 2x\,\frac{dx}{x} = \frac{1+a}{2}\pi e^{-a} \qquad\qquad [a>0] \qquad\qquad \text{BI (152)(11)}$$

6. $$\int_0^\infty \cos\left(a\tan x\right)\sin^3 x\,\frac{dx}{x} = \frac{1-a}{4}\pi e^{-a} \qquad\qquad [a>0] \qquad\qquad \text{BI (151)(23)}$$

7. $$\int_0^\infty \sin\left(a\tan x\right)\tan\frac{x}{2}\cos^2 x\,\frac{dx}{x} = \frac{1+a}{4}\pi e^{-a} \qquad\qquad [a>0] \qquad\qquad \text{BI (152)(13)}$$

8. $$\int_0^{\pi/2} \cos\left(a\tan x\right)\frac{x\,dx}{\sin 2x} = -\frac{\pi}{4}\operatorname{Ei}(-a) \qquad\qquad [a>0] \qquad\qquad \text{BI (206)(15)}$$

9. $$\int_0^{\pi/2} \sin\left(a\cot x\right)\frac{x\,dx}{\sin^2 x} = \frac{1-e^{-a}}{2a}\pi \qquad\qquad [a>0] \qquad\qquad \text{LI (206)(14)}$$

10. $$\int_0^{\pi/2} x\cos\left(a\tan x\right)\tan x\,dx = -\frac{\pi}{4}e^{-a}\left[\boldsymbol{C}+\ln 2a + e^{2a}\operatorname{Ei}(-2a)\right]$$
    $$[a>0] \qquad\qquad \text{BI (205)(10)}$$

11. $$\int_0^\infty \cos\left(a\tan x\right)\tan x\,\frac{dx}{x} = \frac{\pi}{2}e^{-a} \qquad\qquad [a>0] \qquad\qquad \text{BI (151)(21)}$$

12. $$\int_0^\infty \cos\left(a\tan x\right)\sin^2 x\tan x\,\frac{dx}{x} = \frac{1-a}{16}\pi e^{-a} \qquad\qquad [a>0] \qquad\qquad \text{BI (152)(15)}$$

13. $$\int_0^\infty \sin\left(a\tan x\right)\tan^2 x\,\frac{dx}{x} = \frac{\pi}{2}e^{-a} \qquad\qquad [a>0] \qquad\qquad \text{BI (152)(9)}$$

14. $$\int_0^\infty \cos\left(a\tan 2x\right)\tan x\,\frac{dx}{x} = \frac{\pi}{2}e^{-a} \qquad\qquad [a>0] \qquad\qquad \text{BI (151)(22)}$$

15. $$\int_0^\infty \sin\left(a\tan 2x\right)\cos^2 2x\tan x\,\frac{dx}{x} = \frac{1+a}{4}\pi e^{-a} \qquad\qquad [a>0] \qquad\qquad \text{BI (152)(13)}$$

16. $$\int_0^\infty \sin\left(a\tan 2x\right)\tan x\tan 2x\,\frac{dx}{x} = \frac{\pi}{2}e^{-a} \qquad\qquad [a>0] \qquad\qquad \text{BI (152)(10)}$$

17. $$\int_0^\infty \sin\left(a\tan 2x\right)\tan x\cot 2x\,\frac{dx}{x} = \frac{\pi}{2}\left(1-e^{-a}\right) \qquad\qquad [a>0] \qquad\qquad \text{BI (180)(6)}$$

3.882

1. $$\int_0^\infty \sin\left(a\tan^2 x\right)\frac{x\,dx}{b^2+x^2} = \frac{\pi}{2}\left[\exp\left(-a\tanh b\right)-e^{-a}\right]$$
    $$[a>0,\quad b>0] \qquad\qquad \text{BI (160)(22)}$$

2. $$\int_0^\infty \cos\left(a\tan^2 x\right)\cos x\,\frac{dx}{b^2+x^2} = \frac{\pi}{2b}\left[\cosh b\exp\left(-a\tanh b\right)-e^{-a}\sinh b\right]$$
    $$[a>0,\quad b>0] \qquad\qquad \text{BI (163)(3)}$$

3. $$\int_0^\infty \cos\left(a\tan^2 x\right)\operatorname{cosec} 2x\,\frac{x\,dx}{b^2+x^2} = \frac{\pi}{2\sinh 2b}\exp\left(-a\tanh b\right)$$
    $$[a>0,\quad b>0] \qquad\qquad \text{BI (191)(10)}$$

4. $\displaystyle\int_0^\infty \cos\left(a\tan^2 x\right)\tan x\,\frac{x\,dx}{b^2+x^2} = \frac{\pi}{2\cosh b}\left[e^{-a}\cosh b - \exp\left(-a\tanh b\right)\sinh b\right]$

$$[a>0, \quad b>0] \qquad\qquad \text{BI (163)(4)}$$

5. $\displaystyle\int_0^\infty \cos\left(a\tan^2 x\right)\cot x\,\frac{x}{dx}b^2+x^2 = \frac{\pi}{2}\left[\coth b \exp\left(-a\tanh b\right) - e^{-a}\right]$

$$[a>0, \quad b>0] \qquad\qquad \text{BI (163)(5)}$$

6. $\displaystyle\int_0^\infty \cos\left(a\tan^2 x\right)\cot 2x\,\frac{x\,dx}{b^2+x^2} = \frac{\pi}{2}\left[\coth 2b \exp\left(-a\tanh b\right) - e^{-a}\right]$

$$[a>0, \quad b>0] \qquad\qquad \text{BI (191)(11)}$$

3.883

1. $\displaystyle\int_0^1 \cos\left(a\ln x\right)\frac{dx}{(1+x)^2} = \frac{a\pi}{2\sinh a\pi}$
$$\qquad\qquad\qquad\qquad\qquad\qquad \text{BI (404)(4)}$$

2. $\displaystyle\int_0^1 x^{\mu-1}\sin\left(\beta\ln x\right)\,dx = -\frac{\beta}{\beta^2+\mu^2}$
$$\qquad\qquad [\operatorname{Re}\mu > |\operatorname{Im}\beta|] \qquad\qquad \text{ET I 319(19)}$$

3. $\displaystyle\int_0^1 x^{\mu-1}\cos\left(\beta\ln x\right)\,dx = \frac{\mu}{\beta^2+\mu^2}$
$$\qquad\qquad [\operatorname{Re}\mu > |\operatorname{Im}\beta|] \qquad\qquad \text{ET I 321(38)}$$

3.884 $\displaystyle\int_{-\infty}^\infty \frac{\sin a\sqrt{|x|}}{x-b}\operatorname{sign} x\,dx = \cos a\sqrt{|b|} + \exp\left(-a\sqrt{|b|}\right)$

$$[a>0] \qquad\qquad \text{ET II 253(46)}$$

3.89–3.91 Trigonometric functions and exponentials

3.891

1. $\displaystyle\int_0^{2\pi} e^{imx}\sin nx\,dx = 0$
$$\qquad\qquad\qquad [m\neq n;\ \text{or}\ m=n=0]$$

$$= \pi i \qquad\qquad\qquad [m=n\neq 0]$$

2. $\displaystyle\int_0^{2\pi} e^{imx}\cos nx\,dx = 0$
$$\qquad\qquad\qquad [m\neq n]$$

$$= \pi \qquad\qquad\qquad [m=n\neq 0]$$

$$= 2\pi \qquad\qquad\qquad [m=n=0]$$

3.892

1. $\displaystyle\int_0^\pi e^{i\beta x}\sin^{\nu-1}x\,dx = \frac{\pi e^{i\beta\frac{\pi}{2}}}{2^{\nu-1}\nu\,\mathrm{B}\left(\dfrac{\nu+\beta+1}{2},\dfrac{\nu-\beta+1}{2}\right)}$

$$[\operatorname{Re}\nu > -1] \qquad\qquad \text{NH 158, EH I 12(29)}$$

2. $\int_{-\frac{\pi}{2}}^{\frac{\pi}{2}} e^{i\beta x} \cos^{\nu-1} x \, dx = \dfrac{\pi}{2^{\nu-1}\nu \, B\left(\dfrac{\nu+\beta+1}{2}, \dfrac{\nu-\beta+1}{2}\right)}$

$\qquad\qquad\qquad\qquad\qquad\qquad\qquad\qquad [\operatorname{Re}\nu > -1] \qquad\qquad \text{GW (335)(19)}$

$3.^6$ $\int_0^{\pi/2} e^{i2\beta x} \sin^{2\mu} x \cos^{2\nu} x \, dx = \dfrac{1}{2^{2\mu+2\nu+1}} \left\{ \exp\left[i\pi\left(\beta-\nu-\tfrac{1}{2}\right)\right] B\left(\beta-\mu-\nu, 2\nu+1\right) \right.$

$\qquad\qquad\qquad\qquad \times F\left(-2\mu, \beta-\mu-\nu; 1+\beta-\mu+\nu; -1\right) + \exp\left[i\pi\left(\mu+\tfrac{1}{2}\right)\right]$

$\qquad\qquad\qquad\qquad \left. \times \, B(\beta-\mu-\nu, 2\mu+1) \, F\left(-2\nu, \beta-\mu-\nu; 1+\beta+\mu-\nu; -1\right) \right\}$

$\qquad\qquad\qquad\qquad\qquad\qquad [\operatorname{Re}\mu > -\tfrac{1}{2}, \quad \operatorname{Re}\nu > -\tfrac{1}{2}] \qquad \text{EH I 80(6)}$

4. $\int_0^{\pi} e^{i2\beta x} \sin^{2\mu} x \cos^{2\nu} x \, dx = \dfrac{\pi \exp\left[i\pi(\beta-\nu)\right] F(-2\nu, \beta-\mu-\nu; 1+\beta+\mu-\nu; -1)}{4^{\mu+\nu}(2\mu+1)\, B(1-\beta+\mu+\nu, 1+\beta+\mu-\nu)}$

$\qquad\qquad\qquad\qquad\qquad\qquad\qquad\qquad\qquad\qquad\qquad\qquad \text{EH I 80(8)}$

5. $\int_0^{\pi/2} e^{i(\mu+\nu)x} \sin^{\mu-1} x \cos^{\nu-1} x \, dx = e^{i\mu\frac{\pi}{2}} B(\mu, \nu)$

$\qquad\qquad = \dfrac{1}{2^{\mu+\nu-1}} e^{i\mu\frac{\pi}{2}} \left\{ \dfrac{1}{\mu} \, F(1-\nu, 1; \mu+1; -1) + \dfrac{1}{\nu} \, F(1-\mu, 1; \nu+1; -1) \right\}$

$\qquad\qquad\qquad\qquad\qquad\qquad [\operatorname{Re}\mu > 0, \quad \operatorname{Re}\nu > 0] \qquad \text{EH I 80(7)}$

3.893

$1.^8$ $\int_0^{\infty} e^{-px} \sin(qx+\lambda)\, dx = \dfrac{1}{p^2+q^2}\left(q\cos\lambda + p\sin\lambda\right) \qquad [\operatorname{Re}p > 0] \qquad \text{BI (261)(3)}$

$2.^8$ $\int_0^{\infty} e^{-px} \cos(qx+\lambda)\, dx = \dfrac{1}{p^2+q^2}\left(p\cos\lambda - q\sin\lambda\right) \qquad [\operatorname{Re}p > 0] \qquad \text{BI (261)(4)}$

3. $\int_0^{\infty} e^{-x\cos t} \cos(t - x\sin t)\, dx = 1 \qquad\qquad\qquad\qquad\qquad\qquad\qquad \text{BI (261)(7)}$

$4.^8$ $\int_0^{\infty} \dfrac{e^{-\beta x} \sin ax}{\sin bx}\, dx = \operatorname{Re}\left\{ \dfrac{1}{2bi}\left[\psi\left(\dfrac{a+b}{2b} - i\dfrac{\beta}{2b}\right) - \psi\left(\dfrac{b-a}{2b} - i\dfrac{\beta}{2b}\right)\right]\right\}$

$\qquad\qquad\qquad\qquad\qquad\qquad [\operatorname{Re}\beta > 0, \quad b \neq 0] \qquad \text{GW (335)(15)}$

$5.^8$ $\int_0^{\infty} \dfrac{e^{-2px} \sin[(2n+1)x]}{\sin x}\, dx = \dfrac{1}{2p} + \sum_{k=1}^{n} \dfrac{p}{p^2+k^2} \qquad [\operatorname{Re}p > 0] \qquad \text{BI (267)(15)}$

$6.^8$ $\int_0^{\infty} \dfrac{e^{-px} \sin 2nx}{\sin x}\, dx = 2p \sum_{k=0}^{n-1} \dfrac{1}{p^2+(2k+1)^2} \qquad [\operatorname{Re}p > 0] \qquad \text{GW (335)(15c)}$

7. $\int_0^{\infty} e^{-px} \cos[(2n+1)x] \tan x \, dx = \dfrac{2n+1}{p^2+(2n+1)^2} + (-1)^n 2 \sum_{k=0}^{n-1} \dfrac{(-1)^k (2k+1)}{p^2+(2k+1)^2}$

$\qquad\qquad\qquad\qquad\qquad\qquad\qquad\qquad [p > 0] \qquad\qquad \text{LI (267)(16)}$

3.894 $\int_{-\pi}^{\pi} \left[\beta + \sqrt{\beta^2-1}\cos x\right]^{\nu} e^{inx} \, dx = \dfrac{2\pi\, \Gamma(\nu+1)\, P_\nu^m(\beta)}{\Gamma(\nu+m+1)}$

$\qquad\qquad\qquad\qquad\qquad\qquad [\operatorname{Re}\beta > 0] \qquad\qquad \text{ET I 157(15)}$

3.895

1.
$$\int_0^\infty e^{-\beta x} \sin^{2m} x \, dx = \frac{(2m)!}{\beta(\beta^2 + 2^2)(\beta^2 + 4^2) \cdots \left[\beta^2 + (2m)^2\right]}$$

$$[\operatorname{Re}\beta > 0] \qquad \text{FI II 615, WA 620a}$$

2.[10]
$$\int_0^\pi e^{-px} \sin^{2m} x \, dx = \frac{(2m)!\,(1 - e^{-p\pi})}{p\,(p^2 + 2^2)(p^2 + 4^2) \cdots \left[p^2 + (2m)^2\right]}$$

$$\text{GW (335)(4a)}$$

3.[10]
$$\int_0^{\pi/2} e^{-px} \sin^{2m} x \, dx$$

$$= \frac{(2m)!}{p\,(p^2 + 2^2)(p^2 + 4^2) \cdots \left[p^2 + (2m)^2\right]}$$

$$\times \left\{ 1 - e^{-\frac{p\pi}{2}} \left[1 + \frac{p^2}{2!} + \frac{p^2\,(p^2 + 2^2)}{4!} + \cdots + \frac{p^2\,(p^2 + 2^2) \cdots \left[p^2 + (2m-2)^2\right]}{(2m)!} \right] \right\}$$

$$\text{BI (270)(4)}$$

4.
$$\int_0^\infty e^{-\beta x} \sin^{2m+1} x \, dx = \frac{(2m+1)!}{(\beta^2 + 1^2)(\beta^2 + 3^2) \cdots \left[\beta^2 + (2m+1)^2\right]}$$

$$[\operatorname{Re}\beta > 0] \qquad \text{FI II 615, WA 620a}$$

5.[10]
$$\int_0^\pi e^{-px} \sin^{2m+1} x \, dx = \frac{(2m+1)!\,(1 + e^{-p\pi})}{(p^2 + 1^2)(p^2 + 3^2) \cdots \left[p^2 + (2m+1)^2\right]}$$

$$\text{GW (335)(4b)}$$

6.[8]
$$\int_0^{\pi/2} e^{-px} \sin^{2m+1} x \, dx$$

$$= \frac{(2m+1)!}{(p^2 + 1^2)(p^2 + 3^2) \cdots \left[p^2 + (2m+1)^2\right]}$$

$$\times \left\{ 1 - pe^{\frac{-p\pi}{2}} \left[1 + \frac{p^2 + 1^2}{3!} + \cdots + \frac{(p^2 + 1^2)(p^2 + 3^2) \cdots \left[p^2 + (2m-1)^2\right]}{(2m+1)!} \right] \right\}$$

$$\text{BI (270)(5)}$$

7.
$$\int_0^\infty e^{-px} \cos^{2m} x \, dx = \frac{(2m)!}{p\,(p^2 + 2^2) \cdots \left[p^2 + (2m)^2\right]}$$

$$\times \left\{ 1 + \frac{p^2}{2!} + \frac{p^2\,(p^2 + 2^2)}{4!} + \cdots + \frac{p^2\,(p^2 + 2^2) \cdots \left[p^2 + (2m-2)^2\right]}{(2m)!} \right\}$$

$$[p > 0] \qquad \text{BI (262)(3)}$$

8.[10] $\displaystyle\int_0^{\pi/2} e^{-px} \cos^{2m} x\, dx$

$$= \frac{(2m)!}{p\left(p^2 + 2^2\right)\cdots\left[p^2 + (2m)^2\right]}$$

$$\times\left\{-e^{-p\frac{\pi}{2}} + 1 + \frac{p^2}{2!} + \frac{p^2\left(p^2 + 2^2\right)}{4!} + \cdots + \frac{p^2\left(p^2 + 2^2\right)\cdots\left[p^2 + (2m-2)^2\right]}{(2m)!}\right\}$$

BI (270)(6)

9.[7] $\displaystyle\int_0^{\infty} e^{-px} \cos^{2m+1} x\, dx$

$$= \frac{(2m+1)!\,p}{\left(p^2 + 1^2\right)\left(p^2 + 3^2\right)\cdots\left[p^2 + (2m+1)^2\right]}$$

$$\times\left\{1 + \frac{p^2 + 1^2}{3!} + \frac{\left(p^2 + 1^2\right)\left(p^2 + 3^2\right)}{5!} + \cdots + \frac{\left(p^2 + 1^2\right)\left(p^2 + 3^2\right)\cdots\left[p^2 + (2m-1)^2\right]}{(2m+1)!}\right\}$$

$[p > 0]$

BI (262)(4)

10.[7] $\displaystyle\int_0^{\pi/2} e^{-px} \cos^{2m+1} x\, dx$

$$= \frac{(2m+1)!}{\left(p^2 + 1^2\right)\left(p^2 + 3^2\right)\cdots\left[p^2 + (2m+1)^2\right]}$$

$$\times\left\{e^{-p\frac{\pi}{2}} + p\left[1 + \frac{p^2 + 1^2}{3!} + \cdots + \frac{\left(p^2 + 1\right)\left(p^2 + 3^2\right)\cdots\left[p^2 + (2m-1)^2\right]}{(2m+1)!}\right]\right\}$$

$[p \neq 0]$

BI (270)(7)

11.[8] $\displaystyle\int_0^{\infty} e^{-\beta x} \sin^n ax \left\{\begin{matrix}\sin bx\\\cos bx\end{matrix}\right\} dx = \frac{2^{-n-2}}{a(n+1)} e^{\frac{1}{4}(1\mp 1 + 2n)\pi i}$

$$\times\left\{\left(\begin{matrix}\frac{b+na+i\beta}{2a}\\n+1\end{matrix}\right)^{-1} \pm (-1)^n \left(\begin{matrix}\frac{b+na-i\beta}{2a}\\n+1\end{matrix}\right)^{-1}\right\}$$

$[a > 0, \quad b > 0, \quad \operatorname{Re}\beta > 0]$

12.* $\displaystyle\int_0^{\infty} e^{-ax} \cos^2 mx\, dx = \frac{a^2 + 2m^2}{a\left(a^2 + 4m^2\right)}$

DW61 (861.06)

13.* $\displaystyle\int_0^{\infty} e^{-ax} \cos mx \cos nx\, dx = \frac{a\left(a^2 + m^2 + n^2\right)}{\left(a^2 + (m-n)^2\right)\left(a^2 + (m+n)^2\right)}$

DW61 (861.15)

14.* $\displaystyle\int_0^{\infty} e^{-ax} \sin mx \cos nx\, dx = \frac{m\left(a^2 + m^2 - n^2\right)}{\left(a^2 + (m-n)^2\right)\left(a^2 + (m+n)^2\right)}$

DW61 (861.14)

15.* $\displaystyle\int_0^{\infty} e^{-ax} \sin^2 mx\, dx = \frac{2m}{a\left(a^2 + 4m^2\right)}$ $[a > 0]$

DW61 (861.10)

16.* $\displaystyle\int_0^\infty e^{-ax}\sin mx\sin nx\,dx = \frac{2amn}{\left[a^2+(m-n)^2\right]\left[a^2+(m+n)^2\right]}$ DW61 (861.13)

3.896

1. $\displaystyle\int_{-\infty}^\infty e^{-q^2x^2}\sin[p(x+\lambda)]\,dx = \frac{\sqrt\pi}{q}e^{-\frac{p^2}{4q^2}}\sin p\lambda$ BI (269)(2)

2. $\displaystyle\int_{-\infty}^\infty e^{-q^2x^2}\cos[p(x+\lambda)]\,dx = \frac{\sqrt\pi}{q}e^{-\frac{p^2}{4q^2}}\cos p\lambda$ BI (269)(3)

3. $\displaystyle\int_0^\infty e^{-ax^2}\sin bx\,dx = \frac{b}{2a}\exp\left(-\frac{b^2}{4a}\right)\,{}_1F_1\left(\frac{1}{2};\frac{3}{2};\frac{b^2}{4a}\right)$

                   $= \dfrac{b}{2a}\,{}_1F_1\left(1;\dfrac{3}{2};-\dfrac{b^2}{4a}\right)$ ET I 73(18)

                   $= \dfrac{b}{2a}\displaystyle\sum_{k=1}^\infty\frac{1}{(2k-1)!!}\left(-\frac{b^2}{2a}\right)^{k-1}$ $[a>0]$ FI II 720

4. $\displaystyle\int_0^\infty e^{-\beta x^2}\cos bx\,dx = \frac{1}{2}\sqrt{\frac{\pi}{\beta}}\exp\left(-\frac{b^2}{4\beta}\right)$ $[\operatorname{Re}\beta>0]$ BI (263)(2)

3.897

1.[8] $\displaystyle\int_0^\infty e^{-\beta x^2-\gamma x}\sin bx\,dx = -\frac{i}{4}\sqrt{\frac{\pi}{\beta}}\left\{\exp\frac{(\gamma-ib)^2}{4\beta}\left[1-\Phi\left(\frac{\gamma-ib}{2\sqrt\beta}\right)\right]\right.$

                   $\left. -\exp\frac{(\gamma+ib)^2}{4\beta}\left[1-\Phi\left(\frac{\gamma+ib}{2\sqrt\beta}\right)\right]\right\}$

                            $[\operatorname{Re}\beta>0]$ ET I 74(27)

2. $\displaystyle\int_0^\infty e^{-\beta x^2-\gamma x}\cos bx\,dx = \frac{1}{4}\sqrt{\frac{\pi}{\beta}}\left\{\exp\frac{(\gamma-ib)^2}{4\beta}\left[1-\Phi\left(\frac{\gamma-ib}{2\sqrt\beta}\right)\right]\right.$

                   $\left. +\exp\frac{(\gamma+ib)^2}{4\beta}\left[1-\Phi\left(\frac{\gamma+ib}{2\sqrt\beta}\right)\right]\right\}$

                            $[\operatorname{Re}\beta>0]$ ET I 15(16)

3.898

1. $\displaystyle\int_0^\infty e^{-\beta x^2}\sin ax\sin bx\,dx = \frac{1}{4}\sqrt{\frac{\pi}{\beta}}\left\{e^{-\frac{(a-b)^2}{4\beta}}-e^{-\frac{(a+b)^2}{4\beta}}\right\}$

                            $[\operatorname{Re}\beta>0]$ BI (263)(4)

2. $\displaystyle\int_0^\infty e^{-\beta x^2}\cos ax\cos bx\,dx = \frac{1}{4}\sqrt{\frac{\pi}{\beta}}\left\{e^{-\frac{(a-b)^2}{4\beta}}+e^{-\frac{(a+b)^2}{4\beta}}\right\}$

                            $[\operatorname{Re}\beta>0]$ BI (263)(5)

3.[8] $\displaystyle\int_0^\infty e^{-px^2}\sin^2 ax\,dx = \frac{1}{4}\sqrt{\frac{\pi}{p}}\left(1-e^{-\frac{a^2}{p}}\right)$ $[\operatorname{Re}p>0]$ BI (263)(6)

3.899

1.[7] $\displaystyle\int_0^\infty \frac{e^{p^2 x^2} \sin[(2n+1)x]}{\sin x}\, dx = \frac{\sqrt{\pi}}{p}\left[\frac{1}{2} + \sum_{k=1}^n e^{-\left(\frac{k}{p}\right)^2}\right] \qquad [p > 0]$ BI (267)(17)

2. $\displaystyle\int_0^\infty \frac{e^{-p^2 x^2} \cos[(4n+1)x]}{\cos x}\, dx = \frac{\sqrt{\pi}}{p}\left[\frac{1}{2} + \sum_{k=0}^{2n} (-1)^k e^{-\left(\frac{k}{p}\right)^2}\right]$

$$[p > 0] \qquad \text{BI (267)(18)}$$

3. $\displaystyle\int_0^\infty \frac{e^{-px^2}\, dx}{1 - 2a\cos x + a^2} = \frac{\sqrt{\frac{\pi}{p}}}{1 - a^2}\left\{\frac{1}{2} + \sum_{k=1}^\infty a^k \exp\left(-\frac{k^2}{4p}\right)\right\} \qquad [a^2 < 1, \quad p > 0]$ EI (266)(1)

$$= \frac{\sqrt{\frac{\pi}{p}}}{a^2 - 1}\left\{\frac{1}{2} + \sum_{k=1}^\infty a^{-k} \exp\left(-\frac{k^2}{4p}\right)\right\} \qquad [a^2 > 1, \quad p > 0] \qquad \text{LI (266)(1)}$$

3.911

1. $\displaystyle\int_0^\infty \frac{\sin ax}{e^{\beta x} + 1}\, dx = \frac{1}{2a} - \frac{\pi}{2\beta \sinh \frac{a\pi}{\beta}} \qquad [a > 0, \quad \mathrm{Re}\,\beta > 0]$ BI (264)(1)

2. $\displaystyle\int_0^\infty \frac{\sin ax}{e^{\beta x} - 1}\, dx = \frac{\pi}{2\beta}\coth\left(\frac{\pi a}{\beta}\right) - \frac{1}{2a} \qquad [a > 0, \quad \mathrm{Re}\,\beta > 0]$ BI (264)(2), WH

3.[6] $\displaystyle\int_0^\infty \frac{\sin ax}{e^x - 1}\frac{x}{2}\, e^{\frac{x}{2}}\, dx = \frac{1}{2}\pi \tanh(a\pi) \qquad [a > 0]$ ET I 73(13)

4. $\displaystyle\int_0^\infty \frac{\sin ax}{1 - e^{-x}}e^{-nx}\, dx = \frac{\pi}{2} - \frac{1}{2a} + \frac{\pi}{e^{2\pi a} - 1} - \sum_{k=1}^{n-1} \frac{a}{a^2 + k^2}$

$$[a > 0] \qquad \text{BI (264)(8)}$$

5. $\displaystyle\int_0^\infty \frac{\sin ax}{e^{\beta x} - e^{\gamma x}}\, dx = \frac{1}{2i(\beta - \gamma)}\left[\psi\left(\frac{\beta + ia}{\beta - \gamma}\right) - \psi\left(\frac{\beta - ia}{\beta - \gamma}\right)\right]$

$$[\mathrm{Re}\,\beta > 0, \quad \mathrm{Re}\,\gamma > 0] \qquad \text{GW (335)(8)}$$

6. $\displaystyle\int_0^\infty \frac{\sin ax\, dx}{e^{\beta x}\left(e^{-x} - 1\right)} = \frac{i}{2}\left[\psi(\beta + ia) - \psi(\beta - ia)\right] \qquad [\mathrm{Re}\,\beta > -1]$ ET 73(15)

3.912

1. $\displaystyle\int_0^\infty e^{-\beta x}\left(1 - e^{-\gamma x}\right)^{\nu - 1}\sin ax\, dx = -\frac{i}{2\gamma}\left[\mathrm{B}\left(\nu, \frac{\beta - ia}{\gamma}\right) - \mathrm{B}\left(\nu, \frac{\beta + ia}{\gamma}\right)\right]$

$$[\mathrm{Re}\,\beta > 0, \quad \mathrm{Re}\,\gamma > 0, \quad \mathrm{Re}\,\nu > 0, \quad a > 0] \qquad \text{ET I 73(17)}$$

2. $\displaystyle\int_0^\infty e^{-\beta x}\left(1 - e^{-\gamma x}\right)^{\nu - 1}\cos ax\, dx = \frac{1}{2\gamma}\left[\mathrm{B}\left(\nu, \frac{\beta - ia}{\gamma}\right) + \mathrm{B}\left(\nu, \frac{\beta + ia}{\gamma}\right)\right]$

$$[\mathrm{Re}\,\beta > 0, \quad \mathrm{Re}\,\gamma > 0, \quad \mathrm{Re}\,\nu > 0, \quad a > 0] \qquad \text{ET I 15(10)}$$

3.913

1.
$$\int_{-\frac{\pi}{2}}^{\frac{\pi}{2}} e^{i\beta x} \cos^{\nu} x \left(\beta^2 e^{ix} + \nu^2 e^{-ix}\right)^{\mu} dx = \frac{\pi \, {}_2F_1\left(-\mu, \frac{\beta}{2} - \frac{\nu}{2} - \frac{\mu}{2}; 1 + \frac{\beta}{2} + \frac{\nu}{2} - \frac{\mu}{2}; \frac{\beta^2}{\nu^2}\right)}{2^{\nu}(\nu+1)\,\mathrm{B}\left(1 + \frac{\beta}{2} + \frac{\nu}{2} - \frac{\mu}{2}, 1 - \frac{\beta}{2} + \frac{\nu}{2} + \frac{\mu}{2}\right)}$$

$$[\operatorname{Re}\nu > -1, \quad |\nu| > |\beta|] \qquad \text{EH I 81(11)a}$$

2.
$$\int_{-\frac{\pi}{2}}^{\frac{\pi}{2}} e^{iux} \cos^{\mu} x \left(a^2 e^{ix} + b^2 e^{-ix}\right)^{\nu} dx$$

$$= \frac{\pi b^{2\nu} \, {}_2F_1\left(-\nu, \frac{u+\mu+\nu}{2}; 1 + \frac{\mu-\nu-u}{2}; \frac{a^2}{b^2}\right)}{2^{\mu}(\mu+1)\,\mathrm{B}\left(1 - \frac{u+\nu-\mu}{2}, 1 + \frac{u+\mu+\nu}{2}\right)} \qquad \left[\text{for } a^2 < b^2\right]$$

$$= \frac{\pi a^{2\nu} \, {}_2F_1\left(-\nu, \frac{u+\mu-\nu}{2}; 1 + \frac{\mu-\nu+u}{2}; \frac{b^2}{a^2}\right)}{2^{\mu}(\mu+1)\,\mathrm{B}\left(1 + \frac{u+\mu-\nu}{2}, 1 + \frac{\mu+\nu-u}{2}\right)} \qquad \left[\text{for } b^2 < a^2\right]$$

$$[\operatorname{Re}\mu > -1] \qquad \text{ET I 122(31)a}$$

3.914

1.
$$\int_0^{\infty} e^{-\beta\sqrt{\gamma^2+x^2}} \cos bx \, dx = \frac{\beta\gamma}{\sqrt{\beta^2+b^2}} K_1\left(\gamma\sqrt{\beta^2+b^2}\right)$$

$$[\operatorname{Re}\beta > 0, \quad \operatorname{Re}\gamma > 0] \qquad \text{ET I 16(26)}$$

2.*
$$\int_0^{\infty} \sqrt{\gamma^2+x^2}\, e^{-\beta\sqrt{\gamma^2+x^2}} \cos bx \, dx = \frac{\beta^2\gamma^2}{A^2} K_0(\gamma A) + \left(\frac{2\beta^2\gamma}{A^3} - \frac{\gamma}{A}\right) K_1(\gamma A)$$

$$\left[A = \sqrt{\beta^2+b^2}\right]$$

3.*
$$\int_0^{\infty} \left(\gamma^2+x^2\right) e^{-\beta\sqrt{\gamma^2+x^2}} \cos bx \, dx$$

$$= \left(-\frac{3\beta\gamma^2}{A^2} + \frac{4\beta^3\gamma^2}{A^4}\right) K_0(\gamma A) + \left(-\frac{6\beta\gamma}{A^3} + \frac{8\beta^3\gamma}{A^5} + \frac{\beta^3\gamma^3}{A^3}\right) K_1(\gamma A)$$

$$\left[A = \sqrt{\beta^2+b^2}\right]$$

4.*
$$\int_0^{\infty} \frac{e^{-\beta\sqrt{\gamma^2+x^2}}}{\sqrt{\gamma^2+x^2}} \cos bx \, dx = K_0\left(\gamma\sqrt{\beta^2+b^2}\right) \qquad [\operatorname{Re}\beta > 0, \quad \operatorname{Re}\gamma > 0, \quad b > 0]$$

$$\text{ET I 16(27)}$$

5.*
$$\int_0^{\infty} \left(\frac{1}{\beta(\gamma^2+x^2)^{3/2}} + \frac{1}{\gamma^2+x^2}\right) e^{-\beta\sqrt{\gamma^2+x^2}} \cos bx \, dx = \frac{1}{\beta\gamma}\sqrt{\beta^2+b^2}\, K_1\left(\gamma\sqrt{\beta^2+b^2}\right)$$

$$(6.726(4))$$

6.*
$$\int_0^{\infty} x e^{-\beta\sqrt{\gamma^2+x^2}} \sin bx \, dx = \frac{b\beta\gamma^2}{\beta^2+b^2} K_2\left(\gamma\sqrt{\beta^2+b^2}\right) \qquad \text{ET I 175(35)}$$

7.* $\displaystyle\int_0^\infty x\sqrt{\gamma^2+x^2}\,e^{-\beta\sqrt{\gamma^2+x^2}}\sin bx\,dx$

$$= \left(-\frac{b\gamma^2}{A^2}+\frac{4b\beta^2\gamma^2}{A^4}\right)K_0(\gamma A)+\left(-\frac{2b\gamma}{A^3}+\frac{8b\beta^2\gamma}{A^5}+\frac{b\beta^2\gamma^3}{A^3}\right)K_1(\gamma A)$$

$$\left[A=\sqrt{\beta^2+b^2}\right]$$

8.* $\displaystyle\int_0^\infty \left(\gamma^2+x^2\right)e^{-\beta\sqrt{\gamma^2+x^2}}x\sin bx\,dx = \left(-\frac{12b\beta\gamma^2}{A^4}+\frac{24b\beta^3\gamma^2}{A^6}+\frac{b\beta^3\gamma^4}{A^4}\right)K_0(\gamma A)$$

$$+\left(-\frac{24b\beta\gamma}{A^5}+\frac{48b\beta^3\gamma}{A^7}-\frac{3b\beta\gamma^3}{A^3}+\frac{8b\beta^3\gamma^3}{A^5}\right)K_1(\gamma A)$$

$$\left[A=\sqrt{\beta^2+b^2}\right]$$

9.* $\displaystyle\int_0^\infty \frac{xe^{-\beta\sqrt{\gamma^2+x^2}}}{\sqrt{\gamma^2+x^2}}\sin bx\,dx = \frac{\gamma b}{\sqrt{\beta^2+b^2}}K_1\left(\gamma\sqrt{\beta^2+b^2}\right)$ ET I 75(36)

10.* $\displaystyle\int_0^\infty \left(\frac{1}{\beta(\gamma^2+x^2)^{3/2}}+\frac{1}{\gamma^2+x^2}\right)e^{-\beta\sqrt{\gamma^2+x^2}}x\sin bx\,dx = \frac{b}{\beta}K_0\left(\gamma\sqrt{\beta^2+b^2}\right)$ (6.726(3))

3.915

1. $\displaystyle\int_0^\pi e^{a\cos x}\sin x\,dx = \frac{2}{a}\sinh a$ GW (337)(15c)

2. $\displaystyle\int_0^\pi e^{i\beta\cos x}\cos nx\,dx = i^n\pi J_n(\beta)$ EH II 81(2)

3.³ $\displaystyle\int_{-\frac{\pi}{2}}^{\frac{\pi}{2}} e^{i\beta\sin x}\cos^{2\nu}x\,dx = \sqrt{\pi}\left(\frac{2}{\beta}\right)^\nu\Gamma\left(\nu+\frac{1}{2}\right)J_\nu(\beta)$ $\left[\operatorname{Re}\nu>-\tfrac{1}{2}\right]$ EH II 81(6)

4. $\displaystyle\int_0^\pi e^{\pm\beta\cos x}\sin^{2\nu}x\,dx = \sqrt{\pi}\left(\frac{2}{\beta}\right)^\nu\Gamma\left(\nu+\frac{1}{2}\right)I_\nu(\beta)$ $\left[\operatorname{Re}\nu>-\tfrac{1}{2}\right]$ GW (337)(15b)

5. $\displaystyle\int_0^\pi e^{i\beta\cos x}\sin^{2\nu}x\,dx = \sqrt{\pi}\left(\frac{2}{\beta}\right)^\nu\Gamma\left(\nu+\frac{1}{2}\right)J_\nu(\beta)$ $\left[\operatorname{Re}\nu>-\tfrac{1}{2}\right]$ WA 34(2), WA 60(6)

3.916

1. $\displaystyle\int_0^{\pi/2} e^{-p^2\tan x}\frac{\sin\frac{x}{2}\sqrt{\cos x}}{\sin 2x}\,dx = \left[C(p)-\frac{1}{2}\right]^2+\left[S(p)-\frac{1}{2}\right]^2$ NT 33(18)a

2. $\displaystyle\int_0^{\pi/2} \frac{\exp(-p\tan x)\,dx}{\sin 2x+a\cos 2x+a} = -\frac{1}{2}e^{ap}\operatorname{Ei}(-ap)$ $[p>0]$,   (cf. **3552** 4 and 6)

BI (273)(11)

3. $\displaystyle\int_0^{\pi/2} \frac{\exp(-p\cot x)\,dx}{\sin 2x+a\cos 2x-a} = -\frac{1}{2}e^{-ap}\operatorname{Ei}(ap)$ $[p>0]$,   (cf. **3.552** 4 and 6)

BI (273)(12)

4. $\displaystyle\int_0^{\pi/2} \frac{\exp(-p\tan x)\sin 2x\,dx}{(1-a^2)-2a^2\cos 2x-(1+a^2)\cos^2 2x} = -\frac{1}{4}\left[e^{-ap}\operatorname{Ei}(ap)+e^{ap}\operatorname{Ei}(-ap)\right]$

$$[p>0]$$ BI (273)(13)

5. $\displaystyle\int_0^{\pi/2} \frac{\exp\left(-p\cot x\right)\sin 2x\,dx}{(1-a^2)+2a^2\cos 2x - (1+a^2)\cos^2 2x} = -\frac{1}{4}\left[e^{-ap}\operatorname{Ei}(ap) + e^{ap}\operatorname{Ei}(-ap)\right]$

$$[p > 0]$$ BI (273)(14)

3.917

1. $\displaystyle\int_0^{\pi/2} e^{-2\beta\cot x}\cos^{\nu-1/2}x\,\sin^{-(\nu+1)}x\,\sin\left[\beta - \left(\nu - \frac{1}{2}\right)x\right]dx = \frac{\sqrt{\pi}}{2(2\beta)^\nu}\,\Gamma\left(\nu + \frac{1}{2}\right)J_\nu(\beta)$

$$\left[\operatorname{Re}\nu > -\tfrac{1}{2}\right]$$ WA 186(7)

2. $\displaystyle\int_0^{\pi/2} e^{-2\beta\cot x}\cos^{\nu-1/2}x\,\sin^{-(\nu+1)x}\,\cos\left[\beta - \left(\nu - \frac{1}{2}\right)x\right]dx = \frac{\sqrt{\pi}}{2(2\beta)^\nu}\,\Gamma\left(\nu + \frac{1}{2}\right)Y_\nu(\beta)$

$$\left[\operatorname{Re}\nu > -\tfrac{1}{2}\right]$$ WA 186(8)

3.918

1. $\displaystyle\int_0^{\pi/2} \frac{\cos^\mu x}{\sin^{2\mu+2}x}\,e^{i\,\gamma(\beta-\mu x)-2\beta\cot x}\,dx = \frac{i\gamma}{2}\sqrt{\frac{\pi}{2\beta}}(2\beta)^{-\mu}\,\Gamma(\mu+1)\,H^{(\varepsilon)}_{\mu+\frac{1}{2}}(\beta)$

$$\left[\varepsilon = 1, 2, \quad \gamma = (-1)^{\varepsilon+1}, \quad \operatorname{Re}\beta > 0, \quad \operatorname{Re}\mu > -1\right]$$ GW (337)(16)

2. $\displaystyle\int_0^{\pi/2} \frac{\cos^\mu x\,\sin(\beta-\mu x)}{\sin^{2\mu+2}x}\,e^{-2\beta\cot x}\,dx = \frac{1}{2}\sqrt{\frac{\pi}{2\beta}}(2\beta)^{-\mu}\,\Gamma(\mu+1)\,J_{\mu+\frac{1}{2}}(\beta)$

$$[\operatorname{Re}\beta > 0, \quad \operatorname{Re}\mu > -1]$$ WH

3. $\displaystyle\int_0^{\pi/2} \frac{\cos^\mu x\,\cos(\beta-\mu x)}{\sin^{2\mu+2}x}\,e^{-2\beta\cot x}\,dx = -\frac{1}{2}\sqrt{\frac{\pi}{2\beta}}(2\beta)^{-\mu}\,\Gamma(\mu+1)\,Y_{\mu+\frac{1}{2}}(\beta)$

$$[\operatorname{Re}\beta > 0, \quad \operatorname{Re}\mu > -1]$$ GW (337)(17b)

3.919

1. $\displaystyle\int_0^{\pi/2} \frac{\sin 2nx}{\sin^{2n+2}x}\cdot\frac{dx}{\exp\left(2\pi\cot x\right)-1} = (-1)^{n-1}\frac{2n-1}{4(2n+1)}$ BI (275)(6), LI (275)(6)

2. $\displaystyle\int_0^{\pi/2} \frac{\sin 2nx}{\sin^{2n+2}x}\frac{dx}{\exp\left(\pi\cot x\right)-1} = (-1)^{n-1}\frac{n}{2n+1}$ BI (275)(7), LI (275)(7)

3.92 Trigonometric functions of more complicated arguments combined with exponentials

3.921[6]

1. $\displaystyle\int_0^\infty e^{-\gamma x}\cos ax^2\left(\cos\gamma x - \sin\gamma x\right)dx = \sqrt{\frac{\pi}{8a}}\exp\left(-\frac{\gamma^2}{2a}\right)$

$$[a > 0, \quad \operatorname{Re}\gamma \geq |\operatorname{Im}\gamma|]$$ ET I 26(28)

2.[10] $\displaystyle\int_0^{\pi/4}\prod_{n=1}^\infty \exp\left[-\frac{1}{n}\tan^{2n}x\right] = \frac{\pi}{2} - 1$

3.[10] $\displaystyle\int_0^{\pi/2} \exp\left[-\sum_{n=1}^{\infty} \frac{1}{n}\sin^{2n}x\right] = \int_0^{\pi/2}\exp\left[-\sum_{n=1}^{\infty}\frac{1}{n}\cos^{2n}x\right] = \frac{\pi}{4}$

3.922

1. $\displaystyle\int_0^{\infty} e^{-\beta x^2}\sin ax^2\,dx = \frac{1}{2}\int_{-\infty}^{\infty} e^{-\beta x^2}\sin ax^2\,dx = \sqrt{\frac{\pi}{8}}\sqrt{\frac{\sqrt{\beta^2+a^2}-\beta}{\beta^2+a^2}}$

$$= \frac{\sqrt{\pi}}{2\sqrt[4]{\beta^2+a^2}}\sin\left(\frac{1}{2}\arctan\frac{a}{\beta}\right)$$

$$[\mathrm{Re}\,\beta > 0,\quad a > 0]\quad\text{FI II 750, BI (263)(8)}$$

2. $\displaystyle\int_0^{\infty} e^{-\beta x^2}\cos ax^2\,dx = \frac{1}{2}\int_{-\infty}^{\infty} e^{-\beta x^2}\cos ax^2\,dx = \sqrt{\frac{\pi}{8}}\sqrt{\frac{\sqrt{\beta^2+a^2}+\beta}{\beta^2+a^2}}$

$$= \frac{\sqrt{\pi}}{2\sqrt[4]{\beta^2+a^2}}\cos\left(\frac{1}{2}\arctan\frac{a}{\beta}\right)$$

$$[\mathrm{Re}\,\beta > 0,\quad a > 0]\quad\text{FI II 750, BI (263)(9)}$$

[In formulas **3.922** 3 and 4, $a > 0$, $b > 0$, $\mathrm{Re}\,\beta > 0$, and

$$A = \frac{b^2}{4\left(a^2+\beta^2\right)},\qquad B = \sqrt{\frac{1}{2}\left(\sqrt{\beta^2+a^2}+\beta\right)},\qquad C = \sqrt{\frac{1}{2}\left(\sqrt{\beta^2+a^2}-\beta\right)}$$

If a is complex, then $\mathrm{Re}\,\beta > |\mathrm{Im}\,a|$.]

3. $\displaystyle\int_0^{\infty} e^{-\beta x^2}\sin ax^2\cos bx\,dx = -\frac{1}{2}\sqrt{\frac{\pi}{\beta^2+a^2}}\,e^{-A\beta}\left(B\sin Aa - C\cos Aa\right)$

$$= \frac{\sqrt{\pi}}{2\sqrt[4]{\beta^2+a^2}}\exp\left(-\frac{\beta b^2}{4\left(\beta^2+a^2\right)}\right)\sin\left\{\frac{1}{2}\arctan\frac{a}{\beta} - \frac{ab^2}{4\left(\beta^2+a^2\right)}\right\}$$

$$\text{LI (263)(10), GW (337)(5)}$$

4. $\displaystyle\int_0^{\infty} e^{-\beta x^2}\cos ax^2\cos bx\,dx = \frac{1}{2}\sqrt{\frac{\pi}{\beta^2+a^2}}\,e^{-A\beta}\left(B\cos Aa + C\sin Aa\right)$

$$= \frac{\sqrt{\pi}}{2\sqrt[4]{\beta^2+a^2}}\exp\left(-\frac{\beta b^2}{4\left(\beta^2+a^2\right)}\right)\cos\left\{\frac{1}{2}\arctan\frac{a}{\beta} - \frac{ab^2}{4\left(\beta^2+a^2\right)}\right\}$$

$$\text{LI (263)(11), GW (337)(5)}$$

3.923

1. $\displaystyle\int_{-\infty}^{\infty}\exp\left[-\left(ax^2+2bx+c\right)\right]\sin\left(px^2+2qx+r\right)\,dx$

$$= \frac{\sqrt{\pi}}{\sqrt[4]{a^2+p^2}}\exp\frac{a\left(b^2-ac\right)-\left(aq^2-2bpq+cp^2\right)}{a^2+p^2}$$

$$\times\sin\left\{\frac{1}{2}\arctan\frac{p}{a} - \frac{p\left(q^2-pr\right)-\left(b^2p-2abq+a^2r\right)}{a^2+p^2}\right\}$$

$$[a > 0]\qquad\text{GW (337)(3), BI (296)(6)}$$

2. $$\int_{-\infty}^{\infty} \exp\left[-\left(ax^2 + 2bx + c\right)\right] \cos\left(px^2 + 2qx + r\right) dx$$

$$= \frac{\sqrt{\pi}}{\sqrt[4]{a^2 + p^2}} \exp \frac{a\left(b^2 - ac\right) - \left(aq^2 - 2bpq + cp^2\right)}{a^2 + p^2}$$

$$\times \cos\left\{\frac{1}{2} \arctan \frac{p}{a} - \frac{p\left(q^2 - pr\right) - \left(b^2 p - 2abq + a^2 r\right)}{a^2 + p^2}\right\}$$

$$[a > 0] \qquad \text{GW (337)(3), BI (269)(7)}$$

3.924

1. $$\int_0^{\infty} e^{-\beta x^4} \sin bx^2 \, dx = \frac{\pi}{4}\sqrt{\frac{b}{2\beta}} \exp\left(-\frac{b^2}{8\beta}\right) I_{\frac{1}{4}}\left(\frac{b^2}{8\beta}\right)$$

$$[\operatorname{Re}\beta > 0, \quad b > 0] \qquad \text{ET 73(22)}$$

2. $$\int_0^{\infty} e^{-\beta x^4} \cos bx^2 \, dx = \frac{\pi}{4}\sqrt{\frac{b}{2\beta}} \exp\left(-\frac{b^2}{8\beta}\right) I_{-\frac{1}{4}}\left(\frac{b^2}{8\beta}\right)$$

$$[\operatorname{Re}\beta > 0, \quad b > 0] \qquad \text{ET I 15(12)}$$

3.925

1. $$\int_0^{\infty} e^{-\frac{p^2}{x^2}} \sin 2a^2 x^2 \, dx = \frac{1}{2}\int_{-\infty}^{\infty} e^{-\frac{p^2}{x^2}} \sin 2a^2 x^2 \, dx = \frac{\sqrt{\pi}}{4a} e^{-2ap}\left(\cos 2ap + \sin 2ap\right)$$

$$[a > 0, \quad b > 0] \qquad \text{BI (268)(12)}$$

2. $$\int_0^{\infty} e^{-\frac{p^2}{x^2}} \cos 2a^2 x^2 \, dx = \frac{1}{2}\int_{-\infty}^{\infty} e^{-\frac{p^2}{x^2}} \cos 2a^2 x^2 \, dx = \frac{\sqrt{\pi}}{4a} e^{-2ap}\left(\cos 2ap - \sin 2ap\right)$$

$$[a > 0, \quad b > 0] \qquad \text{BI (268)(13)}$$

3.926   Notation:

$$u = \sqrt{\frac{\sqrt{a^2 + \beta^2} + \beta}{2}}, \qquad v = \sqrt{\frac{\sqrt{a^2 + \beta^2} - \beta}{2}}$$

1. $$\int_0^{\infty} e^{-\left(\beta x^2 + \frac{\gamma}{x^2}\right)} \sin ax^2 \, dx = \frac{1}{2}\sqrt{\frac{\pi}{a^2 + \beta^2}} e^{-2u\sqrt{\gamma}}\left[v \cos\left(2v\sqrt{\gamma}\right) + u \sin\left(2v\sqrt{\gamma}\right)\right]$$

$$[\operatorname{Re}\beta > 0, \quad \operatorname{Re}\gamma > 0] \qquad \text{BI (268)(14)}$$

2. $$\int_0^{\infty} e^{-\left(\beta x^2 + \frac{\gamma}{x^2}\right)} \cos ax^2 \, dx = \frac{1}{2}\sqrt{\frac{\pi}{a^2 + \beta^2}} e^{-2u\sqrt{\gamma}}\left[u \cos\left(2v\sqrt{\gamma}\right) - v \sin\left(2v\sqrt{\gamma}\right)\right]$$

$$[\operatorname{Re}\beta > 0, \quad \operatorname{Re}\gamma > 0] \qquad \text{BI (268)(15)}$$

3.927 $$\int_0^{\infty} e^{-\frac{p}{x}} \sin^2 \frac{a}{x} \, dx = a \arctan \frac{2a}{p} + \frac{p}{4} \ln \frac{p^2}{p^2 + 4a^2} \qquad [a > 0, \quad p > 0] \qquad \text{LI (268)(4)}$$

3.928   Notation: In formulas **3.928** 1 and 2, $a^2 + p^2 > 0$, $r = \sqrt[4]{a^4 + p^4}$, $s = \sqrt[4]{b^4 + q^4}$, $A = \frac{1}{2}\arctan\frac{a^2}{p^2}$, and $B = \frac{1}{2}\arctan\frac{b^2}{q^2}$.

1. $$\int_0^\infty \exp\left[-\left(p^2x^2 + \frac{q^2}{x^2}\right)\right] \sin\left(a^2x^2 + \frac{b^2}{x^2}\right) dx = \frac{\sqrt{\pi}}{2r} e^{-2rs\cos(A+B)} \sin\left\{A + 2rs\sin(A+B)\right\}$$

BI (268)(22)

2. $$\int_0^\infty \exp\left[-\left(p^2x^2 + \frac{q^2}{x^2}\right)\right] \cos\left(a^2x^2 + \frac{b^2}{x^2}\right) dx = \frac{\sqrt{\pi}}{2r} e^{-2rs\cos(A+B)} \cos\left\{A + 2rs\sin(A+B)\right\}$$

BI (268)(23)

3.929 $$\int_0^\infty \left[e^{-x}\cos\left(p\sqrt{x}\right) + pe^{-x^2}\sin px\right] dx = 1$$ LI (268)(3)

3.93 Trigonometric and exponential functions of trigonometric functions

3.931

1. $$\int_0^{\pi/2} e^{-p\cos x} \sin\left(p\sin x\right) dx = \text{Ei}(-p) - \text{ci}(p)$$ NT 13(27)

2. $$\int_0^\pi e^{-p\cos x} \sin\left(p\sin x\right) dx = -\int_{-\pi}^0 e^{-p\cos x} \sin\left(p\sin x\right) dx = -2\,\text{shi}(p)$$ GW (337)(11b)

3. $$\int_0^{\pi/2} e^{-p\cos x} \cos\left(p\sin x\right) dx = -\,\text{si}(p)$$ NT 13(26)

4. $$\int_0^{\pi/2} e^{-p\cos x} \cos\left(p\sin x\right) dx = \frac{1}{2}\int_0^{2\pi} e^{-p\cos x} \cos\left(p\sin x\right) dx = \pi$$ GW (337)(11a)

3.932

1. $$\int_0^\pi e^{p\cos x} \sin\left(p\sin x\right) \sin mx\, dx = \frac{1}{2}\int_0^{2\pi} e^{p\cos x} \sin\left(p\sin x\right) \sin mx\, dx = \frac{\pi}{2}\cdot\frac{p^m}{m!}$$

BI (277)(7), GW (337)(13a)

2. $$\int_0^\pi e^{p\cos x} \cos\left(p\sin x\right) \cos mx\, dx = \frac{1}{2}\int_0^{2\pi} e^{p\cos x} \cos\left(p\sin x\right) \cos mx\, dx = \frac{\pi}{2}\cdot\frac{p^m}{m!}$$

BI (277)(8), GW (337)(13b)

3.933 $$\int_0^\pi e^{p\cos x} \sin\left(p\sin x\right) \text{cosec}\, x\, dx = \pi\sinh p$$ BI (278)(1)

3.934

1. $$\int_0^\pi e^{p\cos x} \sin\left(p\sin x\right) \tan\frac{x}{2}\, dx = \pi\left(1 - e^p\right)$$ BI (271)(8)

2. $$\int_0^\pi e^{p\cos x} \sin\left(p\sin x\right) \cot\frac{x}{2}\, dx = \pi\left(e^p - 1\right)$$ BI (272)(5)

3.935 $$\int_0^\pi e^{p\cos x} \cos\left(p\sin x\right) \frac{\sin 2nx}{\sin x}\, dx = \pi\sum_{k=0}^{n-1}\frac{p^{2k+1}}{(2k+1)!}\qquad [p > 0]$$ LI (278)(3)

3.936

1. $$\int_0^{2\pi} e^{p\cos x} \cos\left(p\sin x - mx\right) dx = 2\int_0^\pi e^{p\cos x} \cos\left(p\sin x - mx\right) dx = \frac{2\pi p^m}{m!}$$

BI (277)(9), GW (337)(14a)

2. $\displaystyle\int_0^{2\pi} e^{p\sin x}\sin\left(p\cos x + mx\right)\,dx = \frac{2\pi p^m}{m!}\sin\frac{m\pi}{2}$ $[p>0]$ GW (337)(14b)

3. $\displaystyle\int_0^{2\pi} e^{p\sin x}\cos\left(p\cos x + mx\right)\,dx = \frac{2\pi p^m}{m!}\cos\frac{m\pi}{2}$ $[p>0]$ GW (337)(14b)

4. $\displaystyle\int_0^{2\pi} e^{\cos x}\sin\left(mx - \sin x\right)\,dx = 0$ WH

5. $\displaystyle\int_0^{\pi} e^{\beta\cos x}\cos\left(ax + \beta\sin x\right)\,dx = \beta^{-a}\sin(a\pi)\,\gamma(a,\beta)$ EH II 137(2)

3.937   **Notation**: In formulas **3.937** 1 and 2, $(b-p)^2 + (a+q)^2 > 0$, $m = 0,1,2,\ldots$, $A = p^2 - q^2 + a^2 - b^2$, $B = 2(pq + ab)$, $C = p^2 + q^2 - a^2 - b^2$, and $D = 2(ap + bq)$.

1. $\displaystyle\int_0^{2\pi} \exp\left(p\cos x + q\sin x\right)\sin\left(a\cos x + b\sin x - mx\right)\,dx$

$$= i\pi\left[(b-p)^2 + (a+q)^2\right]^{-\frac{m}{2}}\left\{(A+iB)^{\frac{m}{2}}\,I_m\left(\sqrt{C-iD}\right) - (A-iB)^{\frac{m}{2}}\,I_m\left(\sqrt{C+iD}\right)\right\}$$

                                                                 GW (337)(9b)

2. $\displaystyle\int_0^{2\pi} \exp\left(p\cos x + q\sin x\right)\cos\left(a\cos x + b\sin x - mx\right)\,dx$

$$= \pi\left[(b-p)^2 + (a+q)^2\right]^{-\frac{m}{2}}\left\{(A+iB)^{\frac{m}{2}}\,I_m\left(\sqrt{C-iD}\right) + (A-iB)^{\frac{m}{2}}\,I_m\left(\sqrt{C+iD}\right)\right\}$$

                                                                 GW (337)(9a)

3. $\displaystyle\int_0^{2\pi} \exp\left(p\cos x + q\sin x\right)\sin\left(q\cos x - p\sin x + mx\right)\,dx = \frac{2\pi}{m!}\left(p^2 + q^2\right)^{\frac{m}{2}}\sin\left(m\arctan\frac{q}{p}\right)$

                                                                 GW (337)(12)

4. $\displaystyle\int_0^{2\pi} \exp\left(p\cos x + q\sin x\right)\cos\left(q\cos x - p\sin x + mx\right)\,dx = \frac{2\pi}{m!}\left(p^2 + q^2\right)^{\frac{m}{2}}\cos\left(m\arctan\frac{q}{p}\right)$

                                                                 GW (337)(12)

3.938

1. $\displaystyle\int_0^{\pi} e^{r(\cos px + \cos qx)}\sin\left(r\sin px\right)\sin\left(r\sin qx\right)\,dx = \frac{\pi}{2}\sum_{k=1}^{\infty}\frac{1}{\Gamma(pk+1)\,\Gamma(qk+1)}r^{(p+q)k}$

                                                                 BI (277)(14)

2. $\displaystyle\int_0^{\pi} e^{r(\cos px + \cos qx)}\cos\left(r\sin px\right)\cos\left(r\sin qx\right)\,dx = \frac{\pi}{2}\left(2 + \sum_{k=1}^{\infty}\frac{r^{(p+q)k}}{\Gamma(pk+1)\,\Gamma(qk+1)}\right)$

                                                                 BI (277)(15)

3.939

1. $\displaystyle\int_0^{\pi} e^{q\cos x}\frac{\sin rx}{1 - 2p^r\cos rx + p^{2r}}\sin\left(q\sin x\right)\,dx = \frac{\pi}{2pr}\sum_{k=1}^{\infty}\frac{(pq)^{kr}}{\Gamma(kr+1)}$

                                        $[r>0, \quad 0<p<1]$ BI (278)(15)

2.³ $\displaystyle\int_0^\pi e^{q\cos x}\frac{1-p^r\cos rx}{1-2p^r\cos rx+p^{2r}}\cos(q\sin x)\,dx = \frac{\pi}{2}\left[2+\sum_{k=1}^\infty\frac{(pq)^{kr}}{\Gamma(kr+1)}\right]$

$$[r>0,0<p<1]\qquad\text{BI (278)(16)}$$

3. $\displaystyle\int_0^{\pi/2}\frac{e^{p\cos 2x}\cos(p\sin 2x)\,dx}{\cos^2 x+q^2\sin^2 x} = \frac{\pi}{2q}\exp\left(p\frac{q-1}{q+1}\right)$ BI (273)(8)

3.94–3.97 Combinations involving trigonometric functions, exponentials, and powers

3.941

1. $\displaystyle\int_0^\infty e^{-px}\sin qx\,\frac{dx}{x} = \arctan\frac{q}{p}$ $[p>0]$ BI (365)(1)

2. $\displaystyle\int_0^\infty e^{-px}\cos qx\,\frac{dx}{x} = \infty$ BI (365)(2)

3.942

1. $\displaystyle\int_0^\infty e^{-px}\cos px\,\frac{x\,dx}{b^4+x^4} = \frac{\pi}{4b^2}\exp\left(-bp\sqrt{2}\right)$ $[p>0,\quad b>0]$ BI (386)(6)a

2. $\displaystyle\int_0^\infty e^{-px}\cos px\,\frac{x\,dx}{b^4-x^4} = \frac{\pi}{4b^2}e^{-bp}\sin bp$ $[p>0,\quad b>0]$ BI (386)(7)a

3.943 $\displaystyle\int_0^\infty e^{-\beta x}(1-\cos ax)\,\frac{dx}{x} = \frac{1}{2}\ln\frac{a^2+\beta^2}{\beta^2}$ $[\operatorname{Re}\beta>0]$ BI (367)(6)

3.944

1. $\displaystyle\int_0^u x^{\mu-1}e^{-\beta x}\sin\delta x\,dx = \frac{i}{2}(\beta+i\delta)^{-\mu}\gamma\left[\mu,(\beta+i\delta)u\right]-\frac{i}{2}(\beta-i\delta)^{-\mu}\gamma\left[\mu,(\beta-i\delta)u\right]$

$$[\operatorname{Re}\mu>-1]\qquad\text{ET I 318(8)}$$

2. $\displaystyle\int_u^\infty x^{\mu-1}e^{-\beta x}\sin\delta x\,dx = \frac{i}{2}(\beta+i\delta)^{-\mu}\Gamma\left[\mu,(\beta+i\delta)u\right]-\frac{i}{2}(\beta-i\delta)^{-\mu}\Gamma\left[\mu,(\beta-i\delta)u\right]$

$$[\operatorname{Re}\beta>|\operatorname{Im}\delta|]\qquad\text{ET I 318(9)}$$

3. $\displaystyle\int_0^u x^{\mu-1}e^{-\beta x}\cos\delta x\,dx = \frac{1}{2}(\beta+i\delta)^{-\mu}\gamma\left[\mu,(\beta+i\delta)u\right]+\frac{1}{2}(\beta-i\delta)^{-\mu}\gamma\left[\mu,(\beta-i\delta)u\right]$

$$[\operatorname{Re}\mu>0]\qquad\text{ET I 320(28)}$$

4. $\displaystyle\int_u^\infty x^{\mu-1}e^{-\beta x}\cos\delta x\,dx = \frac{1}{2}(\beta+i\delta)^{-\mu}\Gamma\left[\mu,(\beta+i\delta)u\right]+\frac{1}{2}(\beta-i\delta)^{-\mu}\Gamma\left[\mu,(\beta-i\delta)u\right]$

$$[\operatorname{Re}\beta>|\operatorname{Im}\delta|]\qquad\text{ET I 320(29)}$$

5. $\displaystyle\int_0^\infty x^{\mu-1}e^{-\beta x}\sin\delta x\,dx = \frac{\Gamma(\mu)}{(\beta^2+\delta^2)^{\frac{\mu}{2}}}\sin\left(\mu\arctan\frac{\delta}{\beta}\right)$

$$[\operatorname{Re}\mu>-1,\quad\operatorname{Re}\beta>|\operatorname{Im}\delta|]$$
$$\text{FI II 812, BI (361)(9)}$$

6. $$\int_0^\infty x^{\mu-1} e^{-\beta x} \cos \delta x \, dx = \frac{\Gamma(\mu)}{(\delta^2 + \beta^2)^{\frac{\mu}{2}}} \cos\left(\mu \arctan \frac{\delta}{\beta}\right)$$

$$[\operatorname{Re}\mu > 0, \quad \operatorname{Re}\beta > |\operatorname{Im}\delta|]$$
FI II 812, BI (361)(10)

7. $$\int_0^\infty x^{\mu-1} \exp\left(-ax \cos t\right) \sin\left(ax \sin t\right) dx = \Gamma(\mu) a^{-\mu} \sin(\mu t)$$

$$\left[\operatorname{Re}\mu > -1, \quad a > 0, \quad |t| < \frac{\pi}{2}\right]$$
EH I 13(36)

8. $$\int_0^\infty x^{\mu-1} \exp\left(-ax \cos t\right) \cos\left(ax \sin t\right) dx = \Gamma(\mu) a^{-\mu} \cos(\mu t)$$

$$\left[\operatorname{Re}\mu > -1, \quad a > 0, \quad |t| < \frac{\pi}{2}\right]$$
EH I 13(35)

9. $$\int_0^\infty x^{p-1} e^{-qx} \sin\left(qx \tan t\right) dx = \frac{1}{q^p} \Gamma(p) \cos^p t \sin pt \qquad \left[|t| < \frac{\pi}{2}, \quad q > 0\right]$$
LO V 288(16)

10. $$\int_0^\infty x^{p-1} e^{-qx} \cos\left(qx \tan t\right) dx = \frac{1}{q^p} \Gamma(p) \cos^p(t) \cos pt$$

$$\left[|t| < \frac{\pi}{2}, \quad q > 0\right]$$
LO V 288(15)

11. $$\int_0^\infty x^n e^{-\beta x} \sin bx \, dx = n!\left(\frac{\beta}{\beta^2 + b^2}\right)^{n+1} \sum_{0 \le 2k \le n} (-1)^k \binom{n+1}{2k+1}\left(\frac{b}{\beta}\right)^{2k+1}$$

$$= (-1)^n \frac{\partial^n}{\partial \beta^n}\left(\frac{b}{b^2 + \beta^2}\right)$$

$$[\operatorname{Re}\beta > 0, \quad b > 0] \quad \text{GW (336)(3), ET I 72(3)}$$

12. $$\int_0^\infty x^n e^{-\beta x} \cos bx \, dx = n!\left(\frac{\beta}{\beta^2 + b^2}\right)^{n+1} \sum_{0 \le 2k \le n+1} (-1)^k \binom{n+1}{2k}\left(\frac{b}{\beta}\right)^{2k}$$

$$= (-1)^n \frac{\partial^n}{\partial \beta^n}\left(\frac{\beta}{b^2 + \beta^2}\right)$$

$$[\operatorname{Re}\beta > 0, \quad b > 0] \quad \text{GW (336)(4), ET I 14(5)}$$

13. $$\int_0^\infty x^{n-1/2} e^{-\beta x} \sin bx \, dx = (-1)^n \sqrt{\frac{\pi}{2}} \frac{d^n}{d\beta^n}\left(\frac{\sqrt{\sqrt{\beta^2 + b^2} - \beta}}{\sqrt{\beta^2 + b^2}}\right)$$

$$[\operatorname{Re}\beta > 0, \quad b > 0] \qquad \text{ET I 72(6)}$$

14. $$\int_0^\infty x^{n-1/2} e^{-\beta x} \cos bx \, dx = (-1)^n \sqrt{\frac{\pi}{2}} \frac{d^n}{d\beta^n}\left(\frac{\sqrt{\sqrt{\beta^2 + b^2} + \beta}}{\sqrt{\beta^2 + b^2}}\right)$$

$$[\operatorname{Re}\beta > 0, \quad b > 0] \qquad \text{ET I 15(6)}$$

3.945

1. $$\int_0^\infty \left(e^{-\beta x}\sin ax - e^{-\gamma x}\sin bx\right)\frac{dx}{x^r}$$
$$= \Gamma(1-r)\left\{(b^2+\gamma^2)^{\frac{r-1}{2}}\sin\left[(r-1)\arctan\frac{b}{\gamma}\right] - (a^2+\beta^2)^{\frac{r-1}{2}}\sin\left[(r-1)\arctan\frac{a}{\beta}\right]\right\}$$
$$[\operatorname{Re}\beta > 0, \quad \operatorname{Re}\gamma > 0, \quad r < 2, \quad r \neq 1] \quad \text{BI}(371)(6)$$

2. $$\int_0^\infty \left(e^{-\beta x}\cos ax - e^{-\gamma x}\cos bx\right)\frac{dx}{x^r}$$
$$= \Gamma(1-r)\left\{(a^2+\beta^2)^{\frac{r-1}{2}}\cos\left[(r-1)\arctan\frac{a}{\beta}\right] - (b^2+\gamma^2)^{\frac{r-1}{2}}\cos\left[(r-1)\arctan\frac{b}{\gamma}\right]\right\}$$
$$[\operatorname{Re}\beta > 0, \quad \operatorname{Re}\gamma > 0, \quad r < 2, \quad r \neq 1] \quad \text{BI}(371)(7)$$

3. $$\int_0^\infty \left(ae^{-\beta x}\sin bx - be^{-\gamma x}\sin ax\right)\frac{dx}{x^2} = ab\left[\frac{1}{2}\ln\frac{a^2+\gamma^2}{b^2+\beta^2} + \frac{\gamma}{a}\operatorname{arccot}\frac{\gamma}{a} - \frac{\beta}{b}\operatorname{arccot}\frac{\beta}{b}\right]$$
$$[\operatorname{Re}\beta > 0, \quad \operatorname{Re}\gamma > 0] \quad \text{BI}(368)(22)$$

3.946

1. $$\int_0^\infty e^{-px}\sin^{2m+1}ax\frac{dx}{x} = \frac{(-1)^m}{2^{2m}}\sum_{k=0}^m (-1)^k\binom{2m+1}{k}\arctan\frac{(2m-2k+1)a}{p}$$
$$[m = 0, 1, \dots, \quad p > 0] \quad \text{GW}(336)(9a)$$

2. $$\int_0^\infty e^{-px}\sin^{2m}ax\frac{dx}{x} = \frac{(-1)^{m+1}}{2^{2m}}\sum_{k=0}^{m-1}(-1)^k\binom{2m}{k}\ln\left[p^2+(2m-2k)^2a^2\right] - \frac{1}{2^{2m}}\binom{2m}{m}\ln p$$
$$[m = 1, 2, \dots, \quad p > 0] \quad \text{GW}(336)(9b)$$

3.947

1. $$\int_0^\infty e^{-\beta x}\sin\gamma x\sin ax\frac{dx}{x} = \frac{1}{4}\ln\frac{\beta^2+(a+\gamma)^2}{\beta^2+(a-\gamma)^2} \qquad [\operatorname{Re}\beta > |\operatorname{Im}\gamma|, \quad a > 0] \qquad \text{BI}(365)(5)$$

2. $$\int_0^\infty e^{-px}\sin ax\sin bx\frac{dx}{x^2} = \frac{a}{2}\arctan\frac{2pb}{p^2+a^2-b^2} + \frac{b}{2}\arctan\frac{2pa}{p^2+b^2-a^2} + \frac{p}{4}\ln\frac{p^2+(a-b)^2}{p^2+(a+b)^2}$$
$$[p > 0] \qquad \text{BI}(368)(1), \text{FI II } 744$$

3. $$\int_0^\infty e^{-px}\sin ax\cos bx\frac{dx}{x} = \frac{1}{2}\arctan\frac{2pa}{p^2-a^2+b^2} + s\frac{\pi}{2}$$
$$[a \geq 0, \quad p > 0, \quad s = 0 \text{ for } p^2 - a^2 + b^2 \geq 0 \text{ and } s = 1 \text{ for } p^2 - a^2 + b^2 < 0] \quad \text{GW}(336)(10b)$$

3.948

1. $$\int_0^\infty e^{-\beta x}(\sin ax - \sin bx)\frac{dx}{x} = \arctan\frac{(a-b)\beta}{ab+\beta^2} \qquad [\operatorname{Re}\beta > 0], \qquad (\text{cf. } \mathbf{3.951}\ 2)$$
$$\text{BI}(367)(7)$$

2. $$\int_0^\infty e^{-\beta x}(\cos ax - \cos bx)\frac{dx}{x} = \frac{1}{2}\ln\frac{b^2+\beta^2}{a^2+\beta^2} \qquad [\operatorname{Re}\beta > 0], \qquad (\text{cf. } \mathbf{3.951}\ 3)$$
$$\text{BI}(367)(8), \text{FI II } 748a$$

3. $\int_0^\infty e^{-\beta x} \left(\cos ax - \cos bx\right) \dfrac{dx}{x^2} = \dfrac{\beta}{2} \ln \dfrac{a^2 + \beta^2}{b^2 + \beta^2} + b \arctan \dfrac{b}{\beta} - a \arctan \dfrac{a}{\beta}$

$$[\operatorname{Re} p > 0] \qquad\qquad \text{BI (368)(20)}$$

4. $\int_0^\infty e^{-\beta x} \left(\sin^2 ax - \sin^2 bx\right) \dfrac{dx}{x^2} = a \arctan \dfrac{2a}{p} - b \arctan \dfrac{2b}{p} - \dfrac{p}{4} \ln \dfrac{p^2 + 4a^2}{p^2 + 4b^2}$

$$[p > 0] \qquad\qquad \text{BI (368)(25)}$$

5. $\int_0^\infty e^{-\beta x} \left(\cos^2 ax - \cos^2 bx\right) \dfrac{dx}{x^2} = -a \arctan \dfrac{2a}{p} + b \arctan \dfrac{2b}{p} + \dfrac{p}{4} \ln \dfrac{p^2 + 4a^2}{p^2 + 4b^2}$

$$[p > 0] \qquad\qquad \text{BI (368)(26)}$$

3.949

1. $\int_0^\infty e^{-px} \sin ax \sin bx \sin cx \dfrac{dx}{x} = -\dfrac{1}{4} \arctan \dfrac{a+b+c}{p} + \dfrac{1}{4} \arctan \dfrac{a+b-c}{p} + \dfrac{1}{4} \arctan \dfrac{a-b+c}{p}$

$$+ \dfrac{1}{4} \arctan \dfrac{-a+b+c}{p}$$

$$[p > 0] \qquad\qquad \text{BI (365)(11)}$$

2.[8] $\int_0^\infty e^{-px} \sin^2 ax \sin bx \dfrac{dx}{x} = \dfrac{1}{2} \arctan \dfrac{b}{p} - \dfrac{1}{2} \left[\dfrac{1}{2} \arctan \dfrac{2pb}{p^2 + 4a^2 - b^2} + s \dfrac{\pi}{2} \right]$

$$\left[s = \begin{cases} 1 & \text{for } p^2 + 4a^2 - b^2 < 0 \\ 0 & \text{for } p^2 + 4a^2 - b^2 \geq 0 \end{cases} \right]$$

$$\text{BI (365)(8)}$$

3. $\int_0^\infty e^{-px} \sin^2 ax \cos bx \dfrac{dx}{x} = \dfrac{1}{8} \ln \dfrac{\left[p^2 + (2a+b)^2\right]\left[p^2 + (2a-b)^2\right]}{(p^2 + b^2)^2}$

$$[p > 0] \qquad\qquad \text{BI (365)(9)}$$

4.[8] $\int_0^\infty e^{-px} \sin ax \cos^2 bx \dfrac{dx}{x} = \dfrac{1}{2} \arctan \dfrac{a}{p} + \dfrac{1}{2} \left[\dfrac{1}{2} \arctan \dfrac{2pa}{p^2 + 4b^2 - a^2} + s \dfrac{\pi}{2} \right]$

$$\left[s = \begin{cases} 1 & \text{for } p^2 + 4b^2 - a^2 < 0 \\ 0 & \text{for } p^2 + 4b^2 - a^2 \geq 0 \end{cases} \right]$$

$$\text{BI (365)(10)}$$

5. $\int_0^\infty e^{-px} \sin^2 ax \sin bx \sin cx \dfrac{dx}{x} = \dfrac{1}{8} \ln \dfrac{p^2 + (b+c)^2}{p^2 + (b-c)^2}$

$$+ \dfrac{1}{16} \ln \dfrac{\left[p^2 + (2a-b+c)^2\right]\left[p^2 + (2a+b-c)^2\right]}{\left[p^2 + (2a+b+c)^2\right]\left[p^2 + (2a-b-c)^2\right]}$$

$$[p > 0] \qquad\qquad \text{BI (365)(15)}$$

3.951

1. $\int_0^\infty \left(1 - e^{-x}\right) \cos x \dfrac{dx}{x} = \ln \sqrt{2}$

$$\text{FI II 745}$$

2. $\int_0^\infty \dfrac{e^{-\gamma x} - e^{-\beta x}}{x} \sin bx \, dx = \arctan \dfrac{(\beta - \gamma)b}{b^2 + \beta \gamma}$

$$[\operatorname{Re} \beta > 0, \quad \operatorname{Re} \gamma \geq 0] \qquad\qquad \text{BI (367)(3)}$$

3. $\displaystyle\int_0^\infty \frac{e^{-\gamma x} - e^{-\beta x}}{x} \cos bx \, dx = \frac{1}{2} \ln \frac{b^2 + \beta^2}{b^2 + \gamma^2}$ $\qquad$ $[\operatorname{Re}\beta > 0, \quad \operatorname{Re}\gamma \geq 0]$ $\qquad$ BI (367)(4)

4. $\displaystyle\int_0^\infty \frac{e^{-\gamma x} - e^{-\beta x}}{x} \sin bx \, dx = \frac{b}{2} \ln \frac{b^2 + \beta^2}{b^2 + \gamma^2} + \beta \arctan \frac{b}{\beta} - \gamma \arctan \frac{b}{\gamma}$

$\qquad\qquad\qquad\qquad\qquad\qquad\qquad\qquad\qquad$ $[\operatorname{Re}\beta > 0, \quad \operatorname{Re}\gamma > 0]$ $\qquad$ BI (368)(21)a

5. $\displaystyle\int_0^\infty \frac{x}{e^{\beta x} - 1} \cos bx \, dx = \frac{1}{2b^2} - \frac{\pi^2}{2\beta^2} \operatorname{cosech}^2 \frac{b\pi}{\beta}$ $\qquad$ $[\operatorname{Re}\beta > 0]$ $\qquad$ ET I 15(18)

6. $\displaystyle\int_0^\infty \left(\frac{1}{e^x - 1} - \frac{1}{x}\right) \cos bx \, dx = \ln b - \frac{1}{2}\left[\psi(ib) + \psi(-ib)\right]$

$\qquad\qquad\qquad\qquad\qquad\qquad\qquad\qquad\qquad$ $[b > 0]$ $\qquad$ ET I 15(9)

7. $\displaystyle\int_0^\infty \frac{1 - \cos ax}{e^{2\pi x} - 1} \cdot \frac{dx}{x} = \frac{a}{4} + \frac{1}{2} \ln \frac{1 - e^{-a}}{a}$ $\qquad$ $[a > 0]$ $\qquad$ BI (387)(10)

8. $\displaystyle\int_0^\infty \left(e^{-\beta x} - e^{-\gamma x} \cos ax\right) \frac{dx}{x} = \frac{1}{2} \ln \frac{a^2 + \gamma^2}{\beta^2}$ $\qquad$ $[\operatorname{Re}\beta > 0, \quad \operatorname{Re}\gamma > 0]$ $\qquad$ BI (367)(10)

9. $\displaystyle\int_0^\infty \frac{\cos px - e^{-px}}{b^4 + x^4} \frac{dx}{x} = \frac{\pi}{2b^4} \exp\left(-\frac{1}{2}bp\sqrt{2}\right) \sin\left(\frac{1}{2}bp\sqrt{2}\right)$

$\qquad\qquad\qquad\qquad\qquad\qquad\qquad\qquad\qquad$ $[p > 0]$ $\qquad$ BI (390)(6)

10. $\displaystyle\int_0^\infty \left(\frac{1}{e^x - 1} - \frac{\cos x}{x}\right) dx = \boldsymbol{C}$ $\qquad$ NT 65(8)

11. $\displaystyle\int_0^\infty \left(ae^{-px} - \frac{e^{-qx}}{x} \sin ax\right) \frac{dx}{x} = \frac{a}{2} \ln \frac{a^2 + q^2}{p^2} + q \arctan \frac{a}{q} - a$

$\qquad\qquad\qquad\qquad\qquad\qquad\qquad\qquad\qquad$ $[p > 0, \quad q > 0]$ $\qquad$ BI (368)(24)

12. $\displaystyle\int_0^\infty \frac{x^{2m} \sin bx}{e^x - 1} \, dx = (-1)^m \frac{\partial^{2m}}{\partial b^{2m}} \left[\frac{\pi}{2} \coth b\pi - \frac{1}{2b}\right]$ $\qquad$ $[b > 0]$ $\qquad$ GW (336)(15a)

13. $\displaystyle\int_0^\infty \frac{x^{2m+1} \cos bx}{e^x - 1} \, dx = (-1)^m \frac{\partial^{2m+1}}{\partial b^{2m+1}} \left[\frac{\pi}{2} \coth b\pi - \frac{1}{2b}\right]$

$\qquad\qquad\qquad\qquad\qquad\qquad\qquad\qquad\qquad$ $[b > 0]$ $\qquad$ GW (336)(15b)

14. $\displaystyle\int_0^\infty \frac{x^{2m} \sin bx \, dx}{e^{(2n+1)cx} - e^{(2n-1)cx}} = (-1)^m \frac{\partial^{2m}}{\partial b^{2m}} \left[\frac{\pi}{4c} \tanh \frac{b\pi}{2c} - \sum_{k=1}^{n} \frac{b}{b^2 + (2k-1)^2 c^2}\right]$

$\qquad\qquad\qquad\qquad\qquad\qquad\qquad\qquad\qquad$ $[b > 0]$ $\qquad$ GW (336)(14a)

15. $\displaystyle\int_0^\infty \frac{x^{2m+1} \cos bx \, dx}{e^{(2n+1)cx} - e^{(2n-1)cx}} = (-1)^m \frac{\partial^{2m+1}}{\partial b^{2m+1}} \left[\frac{\pi}{4c} \tanh \frac{b\pi}{2c} - \sum_{k=1}^{n} \frac{b}{b^2 + (2k-1)^2 c^2}\right]$

$\qquad\qquad\qquad\qquad\qquad\qquad\qquad\qquad\qquad$ $[b > 0]$ $\qquad$ GW (336)(14b)

16. $\displaystyle\int_0^\infty \frac{x^{2m} \sin bx \, dx}{e^{(2n-2)cx}} = (-1)^m \frac{\partial^{2m}}{\partial b^{2m}} \left[\frac{\pi}{4c} \coth \frac{b\pi}{2c} - \frac{1}{2b} - \sum_{k=1}^{n-1} \frac{b}{b^2 + (2k)^2 c^2}\right]$

$\qquad\qquad\qquad\qquad\qquad\qquad\qquad\qquad\qquad$ $[b > 0, \quad c > 0]$ $\qquad$ GW (336)(14c)

17.
$$\int_0^\infty \frac{x^{2m+1}\cos bx\,dx}{e^{2ncx}-e^{(2n-2)cx}} = (-1)^m \frac{\partial^{2m+1}}{\partial b^{2m+1}}\left[\frac{\pi}{4c}\coth\frac{b\pi}{2c} - \frac{1}{2b} - \sum_{k=1}^{n-1}\frac{b}{b^2+(2k)^2 c^2}\right]$$

$$[b>0, \quad c>0] \qquad\qquad \text{GW (336)(14d)}$$

18.
$$\int_0^\infty \frac{\cos ax - \cos bx}{e^{(2m+1)px}-e^{(2m-1)px}}\frac{dx}{x} = \frac{1}{2}\ln\frac{\cosh\frac{b\pi}{2p}}{\cosh\frac{a\pi}{2p}} - \frac{1}{2}\sum_{k=1}^m \ln\frac{b^2+(2k-1)^2 p^2}{a^2+(2k-1)^2 p^2}$$

$$[p>0] \qquad\qquad \text{GW (336)(16a)}$$

19.
$$\int_0^\infty \frac{\cos ax - \cos bx}{e^{2mpx}-e^{(2m-2)px}}\frac{dx}{x} = \frac{1}{2}\ln\frac{a\sinh\frac{b\pi}{2p}}{b\sinh\frac{a\pi}{2p}} - \frac{1}{2}\sum_{k=1}^{m-1}\ln\frac{b^2+4k^2 p^2}{a^2+4k^2 p^2}$$

$$[p>0] \qquad\qquad \text{GW (336)(16b)}$$

20.
$$\int_0^\infty \frac{\sin x \sin bx}{1-e^x}\cdot\frac{dx}{x} = \frac{1}{4}\ln\frac{(b+1)\sinh[(b-1)\pi]}{(b-1)\sinh[(b+1)\pi]} \qquad [b^2\neq 1] \qquad\qquad \text{LO V 305}$$

21.
$$\int_0^\infty \frac{\sin^2 ax}{1-e^x}\cdot\frac{dx}{x} = \frac{1}{4}\ln\frac{2a\pi}{\sinh 2a\pi} \qquad\qquad \text{LO V 306, BI (387)(5)}$$

3.952

1.
$$\int_0^\infty xe^{-p^2 x^2}\sin ax\,dx = \frac{a\sqrt{\pi}}{4p^3}\exp\left(-\frac{a^2}{4p^2}\right) \qquad\qquad \text{BI (362)(1)}$$

2.
$$\int_0^\infty xe^{-p^2 x^2}\cos ax\,dx = \frac{1}{2p^2} - \frac{a}{4p^3}\sum_{k=0}^\infty \frac{(-1)^k k!}{(2k+1)!}\left(\frac{a}{p}\right)^{2k+1}$$

$$[a>0] \qquad\qquad \text{BI (362)(2)}$$

3.
$$\int_0^\infty x^2 e^{-p^2 x^2}\sin ax\,dx = \frac{a}{4p^4} + \frac{2p^2-a^2}{8p^5}\sum_{k=0}^\infty \frac{(-1)^k k!}{(2k+1)!}\left(\frac{a}{p}\right)^{2k+1}$$

$$[a>0] \qquad\qquad \text{BI (362)(4)}$$

4.
$$\int_0^\infty x^2 e^{-p^2 x^2}\cos ax\,dx = \sqrt{\pi}\frac{2p^2-a^2}{8p^5}\exp\left(-\frac{a^2}{4p^2}\right) \qquad\qquad \text{BI (362)(5)}$$

5.
$$\int_0^\infty x^3 e^{-p^2 x^2}\sin ax\,dx = \sqrt{\pi}\frac{6ap^2-a^3}{16p^7}\exp\left(-\frac{a^2}{4p^2}\right) \qquad\qquad \text{BI (362)(6)}$$

6.[3]
$$\int_0^\infty e^{-p^2 x^2}\sin ax\frac{dx}{x} = \frac{a\sqrt{\pi}}{2p}\sum_{k=0}^\infty \frac{(-1)^k}{k!(2k+1)}\left(\frac{a}{2p}\right)^{2k} = \frac{\pi}{2}\,\Phi\left(\frac{a}{2p}\right) \qquad\qquad \text{BI (365)(21)}$$

7.
$$\int_0^\infty x^{\mu-1}e^{-\beta x^2}\sin\gamma x\,dx = \frac{\gamma e^{-\frac{\gamma^2}{4\beta}}}{2\beta^{\frac{\mu+1}{2}}}\Gamma\left(\frac{1+\mu}{2}\right)\,{}_1F_1\left(1-\frac{\mu}{2};\frac{3}{2};\frac{\gamma^2}{4\beta}\right)$$

$$[\operatorname{Re}\beta>0, \quad \operatorname{Re}\mu>-1] \qquad \text{ET I 318(10)}$$

8.[10]
$$\int_0^\infty x^{\mu-1}e^{-\beta x^2}\cos ax\,dx = \frac{1}{2}\beta^{-\mu/2}\Gamma\left(\frac{\mu}{2}\right)e^{-a^2/4\beta}\,{}_1F_1\left(-\frac{\mu}{2}+\frac{1}{2};\frac{1}{2};\frac{a^2}{4\beta}\right)$$

$$[\operatorname{Re}\beta>0, \quad \operatorname{Re}\mu>0, \quad a>0]$$
$$\text{ET I 320(30)}$$

9. $$\int_0^\infty x^{2n} e^{-\beta^2 x^2} \cos ax \, dx = (-1)^n \frac{\sqrt{\pi}}{2^{n+1}\beta^{2n+1}} \exp\left(-\frac{a^2}{8\beta^2}\right) D_{2n}\left(\frac{a}{\beta\sqrt{2}}\right)$$

$$= (-1)^n \frac{\sqrt{\pi}}{(2\beta)^{2n+1}} \exp\left(-\frac{a^2}{4\beta^2}\right) H_{2n}\left(\frac{a}{2\beta}\right)$$

$$\left[|\arg\beta| < \frac{\pi}{4}, \quad a > 0\right] \quad \text{WH, ET I 15(13)}$$

10. $$\int_0^\infty x^{2n+1} e^{-\beta^2 x^2} \sin ax \, dx = (-1)^n \frac{\sqrt{\pi}}{2^{n+\frac{3}{2}}\beta^{2n+2}} \exp\left(-\frac{a^2}{8\beta^2}\right) D_{2n+1}\left(\frac{a}{\beta\sqrt{2}}\right)$$

$$= (-1)^n \frac{\sqrt{\pi}}{(2\beta)^{2n+2}} \exp\left(-\frac{a^2}{4\beta^2}\right) H_{2n+1}\left(\frac{a}{2\beta}\right)$$

$$\left[|\arg\beta| < \frac{\pi}{4}, \quad a > 0\right] \quad \text{WH, ET I 74(23)}$$

3.953

1. $$\int_0^\infty x^{\mu-1} e^{-\gamma x - \beta x^2} \sin ax \, dx$$

$$= -\frac{i}{2(2\beta)^{\frac{\mu}{2}}} \exp\left(\frac{\gamma^2 - a^2}{8\beta}\right) \Gamma(\mu) \left\{\exp\left(-\frac{ia\gamma}{4\beta}\right) D_{-\mu}\left(\frac{\gamma - ia}{\sqrt{2\beta}}\right) - \exp\left(\frac{ia\gamma}{4\beta}\right) D_{-\mu}\left(\frac{\gamma + ia}{\sqrt{2\beta}}\right)\right\}$$

$$[\operatorname{Re}\mu > -1, \quad \operatorname{Re}\beta > 0, \quad a > 0] \quad \text{ET I 318(11)}$$

2. $$\int_0^\infty x^{\mu-1} e^{-\gamma x - \beta x^2} \cos ax \, dx$$

$$= \frac{1}{2(2\beta)^{\frac{\mu}{2}}} \exp\left(\frac{\gamma^2 - a^2}{8\beta}\right) \Gamma(\mu) \left\{\exp\left(-\frac{ia\gamma}{4\beta}\right) D_{-\mu}\left(\frac{\gamma - ia}{\sqrt{2\beta}}\right) + \exp\left(\frac{ia\gamma}{4\beta}\right) D_{-\mu}\left(\frac{\gamma + ia}{\sqrt{2\beta}}\right)\right\}$$

$$[\operatorname{Re}\mu > 0, \quad \operatorname{Re}\beta > 0, \quad a > 0] \quad \text{ET I 16(18)}$$

3. $$\int_0^\infty x e^{-\gamma x - \beta x^2} \sin ax \, dx = \frac{i\sqrt{\pi}}{8\sqrt{\beta^3}} \left\{(\gamma - ia) \exp\left[-\frac{(\gamma - ia)^2}{4\beta}\right]\left[1 - \Phi\left(\frac{\gamma - ia}{2\sqrt{\beta}}\right)\right]\right.$$

$$\left. - (\gamma + ia) \exp\left[-\frac{(\gamma + ia)^2}{4\beta}\right]\left[1 - \Phi\left(\frac{\gamma + ia}{2\sqrt{\beta}}\right)\right]\right\}$$

$$[\operatorname{Re}\beta > 0, \quad a > 0] \quad \text{ET I 74(28)}$$

4. $$\int_0^\infty x e^{-\gamma x - \beta x^2} \cos ax \, dx = -\frac{\sqrt{\pi}}{8\sqrt{\beta^3}} \left\{(\gamma - ia) \exp\frac{(\gamma - ia)^2}{4\beta}\left[1 - \Phi\left(\frac{\gamma - ia}{2\sqrt{\beta}}\right)\right]\right.$$

$$\left. + (\gamma + ia) \exp\frac{(\gamma + ia)^2}{4\beta}\left[1 - \Phi\left(\frac{\gamma + ia}{2\sqrt{\beta}}\right)\right]\right\} + \frac{1}{2\beta}$$

$$[\operatorname{Re}\beta > 0, \quad a > 0] \quad \text{ET I 16(17)}$$

3.954

1. $$\int_0^\infty e^{-\beta x^2} \sin ax \frac{x \, dx}{\gamma^2 + x^2} = -\frac{\pi}{4} e^{\beta\gamma^2}\left[2\sinh a\gamma + e^{-\gamma a}\,\Phi\left(\gamma\sqrt{\beta} - \frac{a}{2\sqrt{\beta}}\right) - e^{\gamma a}\,\Phi\left(\Gamma\sqrt{\beta} + \frac{a}{2\sqrt{\beta}}\right)\right]$$

$$[\operatorname{Re}\beta > 0, \quad \operatorname{Re}\gamma > 0, \quad a > 0]$$

$$\text{ET I 74(26)a}$$

2. $\displaystyle\int_0^\infty e^{-\beta x^2}\cos ax\,\frac{dx}{\gamma^2+x^2} = \frac{\pi}{4\gamma}e^{\beta\gamma^2}\left[2\cosh a\gamma - e^{-\gamma a}\,\Phi\left(\gamma\sqrt{\beta}-\frac{a}{2\sqrt{\beta}}\right) - e^{\gamma a}\,\Phi\left(\gamma\sqrt{\beta}+\frac{a}{2\sqrt{\beta}}\right)\right]$

$[\operatorname{Re}\beta>0, \quad \operatorname{Re}\gamma>0, \quad a>0]$

ET I 15(15)

3.955 $\displaystyle\int_0^\infty x^\nu e^{-\frac{x^2}{2}}\cos\left(\beta x - \nu\frac{\pi}{2}\right)dx = \sqrt{\frac{\pi}{2}}\,e^{-\frac{\beta^2}{4}}\,D_\nu(\beta)$ $[\operatorname{Re}\nu>-1]$ EH II 120(4)

3.956 $\displaystyle\int_0^\infty e^{-x^2}\left(2x\cos x - \sin x\right)\sin x\,\frac{dx}{x^2} = \sqrt{\pi}\,\frac{e-1}{2e}$ BI (369)(19)

3.957

1. $\displaystyle\int_0^\infty x^{\mu-1}\exp\left(\frac{-\beta^2}{4x}\right)\sin ax\,dx$

$\displaystyle = \frac{i}{2^\mu}\beta^\mu a^{-\frac{\mu}{2}}\left[\exp\left(-\frac{i}{4}\mu\pi\right)K_\mu\left(\beta e^{\frac{\pi i}{4}}\sqrt{a}\right) - \exp\left(\frac{i}{4}\mu\pi\right)K_\mu\left(\beta e^{-\pi i/4}\sqrt{a}\right)\right]$

$[\operatorname{Re}\beta>0, \quad \operatorname{Re}\mu<1, \quad a>0]$ ET I 318(12)

2. $\displaystyle\int_0^\infty x^{\mu-1}\exp\left(\frac{-\beta^2}{4x}\right)\cos ax\,dx$

$\displaystyle = \frac{1}{2^\mu}\beta^\mu a^{-\frac{\mu}{2}}\left[\exp\left(-\frac{i}{4}\mu\pi\right)K_\mu\left(\beta e^{\pi i/4}\sqrt{a}\right) + \exp\left(\frac{i}{4}\mu\pi\right)K_\mu\left(\beta e^{-\pi i/4}\sqrt{a}\right)\right]$

$[\operatorname{Re}\beta>0, \quad \operatorname{Re}\mu<1, \quad a>0]$ ET I 320(32)a

3.958

1. $\displaystyle\int_{-\infty}^\infty x^n e^{-(ax^2+bx+c)}\sin(px+q)\,dx = -\left(\frac{-1}{2a}\right)^n\sqrt{\frac{\pi}{a}}\exp\left(\frac{b^2-p^2}{4a}-c\right)\sum_{k=0}^{\lfloor n/2\rfloor}\frac{n!}{(n-2k)!k!}a^k$

$\displaystyle \times \sum_{j=0}^{n-2k}\binom{n-2k}{j}b^{n-2k-j}p^j\sin\left(\frac{pb}{2a}-q+\frac{\pi}{2}j\right)$

$[a>0]$ GW (37)(1b)

2. $\displaystyle\int_{-\infty}^\infty x^n e^{-(ax^2+bx+c)}\cos(px+q)\,dx = \left(\frac{-1}{2a}\right)^n\sqrt{\frac{\pi}{a}}\exp\left(\frac{b^2-p^2}{4a}-c\right)\sum_{k=0}^{\lfloor n/2\rfloor}\frac{n!}{(n-2k)!k!}a^k$

$\displaystyle \times \sum_{j=0}^{n-2k}\binom{n-2k}{j}p^j\cos\left(\frac{pb}{2a}-q+\frac{\pi}{2}j\right)$

$[a>0]$ GW (337)(1a)

3.959 $\displaystyle\int_0^\infty x e^{-p^2 x^2}\tan ax\,dx = \frac{a\sqrt{\pi}}{p^3}\sum_{k=1}^\infty(-1)^k k\exp\left(-\frac{a^2 k^2}{p^2}\right)$

$[p>0]$ BI (362)(15)

3.961

1. $\displaystyle\int_0^\infty \exp\left(-\beta\sqrt{\gamma^2+x^2}\right)\sin ax\,\frac{x\,dx}{\sqrt{\gamma^2+x^2}} = \frac{a\gamma}{\sqrt{a^2+\beta^2}}K_1\left(\gamma\sqrt{a^2+\beta^2}\right)$

$[\operatorname{Re}\beta>0, \quad \operatorname{Re}\gamma>0, \quad a>0]$

ET I 75(36)

2. $\displaystyle\int_0^\infty \exp\left[-\beta\sqrt{\gamma^2+x^2}\right]\cos ax\,\frac{dx}{\sqrt{\gamma^2+x^2}} = K_0\left(\gamma\sqrt{a^2+\beta^2}\right)$

$[\operatorname{Re}\beta > 0, \quad \operatorname{Re}\gamma > 0, \quad a > 0]$

ET I 17(27)

3.962

1. $\displaystyle\int_0^\infty \frac{\sqrt{\sqrt{\gamma^2+x^2}-\gamma}\,\exp\left(-\beta\sqrt{\gamma^2+x^2}\right)}{\sqrt{\gamma^2+x^2}}\sin ax\,dx = \sqrt{\frac{\pi}{2}}\,\frac{a\exp\left(-\gamma\sqrt{a^2+\beta^2}\right)}{\sqrt{\beta^2+a^2}\sqrt{\beta+\sqrt{a^2+\beta^2}}}$

$[\operatorname{Re}\beta > 0, \quad \operatorname{Re}\gamma > 0, \quad a > 0]$

ET I 75(38)

2. $\displaystyle\int_0^\infty \frac{x\exp\left(-\beta\sqrt{\gamma^2+x^2}\right)}{\sqrt{\gamma^2+x^2}\sqrt{\sqrt{\gamma^2+x^2}-\gamma}}\cos ax\,dx = \sqrt{\frac{\pi}{2}}\,\frac{\sqrt{\beta+\sqrt{a^2+\beta^2}}}{\sqrt{a^2+\beta^2}}\exp\left[-\gamma\sqrt{a^2+\beta^2}\right]$

$[\operatorname{Re}\beta > 0, \quad \operatorname{Re}\gamma > 0, \quad a > 0]$

ET I 17(29)

3.963

1. $\displaystyle\int_0^\infty e^{-\tan^2 x}\frac{\sin x}{\cos^2 x}\frac{dx}{x} = \frac{\sqrt{\pi}}{2}$

BI (391)(1)

2. $\displaystyle\int_0^{\pi/2} e^{-p\tan x}\frac{x\,dx}{\cos^2 x} = \frac{1}{p}\left[\operatorname{ci}(p)\sin p - \cos p\operatorname{si}(p)\right]$ $\qquad [p>0]$ $\qquad$ (cf. **3.339**)

BI (396)(3)

3.[8] $\displaystyle\int_0^{\pi/2} xe^{-\tan^2 x}\sin 4x\frac{dx}{\cos^8 x} = -\frac{3}{2}\sqrt{\pi}$

BI (396)(5)

4.[8] $\displaystyle\int_0^{\pi/2} xe^{-\tan^2 x}\sin^3 2x\frac{dx}{\cos^8 x} = 2\sqrt{\pi}$

BI (396)(6)

3.964

1. $\displaystyle\int_0^{\pi/2} xe^{-p\tan x}\frac{p\sin x - \cos x}{\cos^3 x}\,dx = -\sin p\operatorname{si}(p) - \operatorname{ci}(p)\cos p$

$[p>0]$

LI (396)(4)

2. $\displaystyle\int_0^{\pi/2} xe^{-p\tan^2 x}\frac{p-\cos^2 x}{\cos^4 x\cot x}\,dx = \frac{1}{4}\sqrt{\frac{\pi}{p}}$ $\qquad [p>0]$

BI (396)(7)

3.[8] $\displaystyle\int_0^{\pi/2} xe^{-p\tan^2 x}\frac{p-2\cos^2 x}{\cos^6 x\cot x}\,dx = \frac{1+2p}{8p}\sqrt{\frac{\pi}{p}}$ $\qquad [p>0]$

BI (396)(8)

3.965

1. $\displaystyle\int_0^\infty xe^{-\beta x}\sin ax^2\sin\beta x\,dx = \frac{\beta}{4}\sqrt{\frac{\pi}{2a^3}}e^{-\frac{\beta^2}{2a}}$ $\qquad \left[|\arg\beta| < \frac{\pi}{4}, \quad a > 0\right]$

ET I 84(17)

2. $\displaystyle\int_0^\infty xe^{-\beta x}\cos ax^2\cos\beta x\,dx = \frac{\beta}{4}\sqrt{\frac{\pi}{2a^3}}e^{-\frac{\beta^2}{2a}}$ $\qquad [a>0, \quad \operatorname{Re}\beta > |\operatorname{Im}\beta|]$

ET 26(27)

3.966

1. $$\int_0^\infty x e^{-px} \cos\left(2x^2 + px\right)\, dx = 0 \qquad\qquad [p > 0] \qquad\qquad \text{BI (361)(16)}$$

2. $$\int_0^\infty x e^{-px} \cos\left(2x^2 - px\right)\, dx = \frac{p\sqrt{\pi}}{8} \exp\left(-\frac{1}{4}p^2\right) \qquad [p > 0] \qquad\qquad \text{BI (361)(17)}$$

3. $$\int_0^\infty x^2 e^{-px} \left[\sin\left(2x^2 + px\right) + \cos\left(2x^2 + px\right)\right]\, dx = 0 \quad [p > 0] \qquad\qquad \text{BI (361)(18)}$$

4. $$\int_0^\infty x^2 e^{-px} \left[\sin\left(2x^2 - px\right) - \cos\left(2x^2 - px\right)\right]\, dx = \frac{\sqrt{\pi}}{16}\left(2 - p^2\right) \exp\left(-\frac{1}{4}p^2\right) \qquad \text{BI (361)(19)}$$

5.[3] $$\int_0^\infty x^{\mu-1} e^{-x} \cos\left(x + ax^2\right)\, dx = \frac{e^{\frac{1}{4a}}\,\Gamma(\mu)}{(2a)^{\frac{\mu}{2}}} \cos\frac{\mu\pi}{4}\, D_{-\mu}\left(\frac{1}{\sqrt{a}}\right)$$
$$[\operatorname{Re}\mu > 0, \quad a > 0] \qquad\qquad \text{ET I 321(37)}$$

6.[6] $$\int_0^\infty x^{\mu-1} e^{-x} \sin\left(x + ax^2\right)\, dx = \frac{e^{\frac{1}{4a}}\,\Gamma(\mu)}{(2a)^{\frac{\mu}{2}}} \sin\frac{\mu\pi}{4}\, D_{-\mu}\left(\frac{1}{\sqrt{a}}\right)$$
$$[\operatorname{Re}\mu > -1, \quad a > 0] \qquad\qquad \text{ET I 319(18)}$$

3.967

1. $$\int_0^\infty e^{-\frac{\beta^2}{x^2}} \sin a^2 x^2\, \frac{dx}{x^2} = \frac{\sqrt{\pi}}{2\beta} e^{-\sqrt{2}a\beta} \sin\left(\sqrt{2}a\beta\right) \qquad [\operatorname{Re}\beta > 0, \quad a > 0]$$
$$\text{ET I 75(30)a, BI(369)(3)a}$$

2. $$\int_0^\infty e^{-\frac{\beta^2}{x^2}} \cos a^2 x^2\, \frac{dx}{x^2} = \frac{\sqrt{\pi}}{2\beta} e^{-\sqrt{2}a\beta} \cos\left(\sqrt{2}a\beta\right) \qquad [\operatorname{Re}\beta > 0, \quad a > 0]$$
$$\text{BI (369)(4), ET I 16(20)}$$

3. $$\int_0^\infty x^2 e^{-\beta x^2} \cos ax^2\, dx = \frac{\sqrt{\pi}}{4\sqrt[4]{(a^2 + \beta^2)^3}} \cos\left(\frac{3}{2}\arctan\frac{a}{\beta}\right)$$
$$[\operatorname{Re}\beta > 0] \qquad\qquad \text{ET I 14(3)a}$$

3.968

1. $$\int_0^\infty e^{-\beta x^2} \sin ax^4\, dx = -\frac{\pi}{8}\sqrt{\frac{\beta}{a}}\left[J_{\frac14}\left(\frac{\beta^2}{8a}\right)\cos\left(\frac{\beta^2}{8a}\right) + \frac{\pi}{8} + Y_{\frac14}\left(\frac{\beta^2}{8a}\right)\sin\left(\frac{\beta^2}{8a}\right) + \frac{\pi}{8}\right]$$
$$[\operatorname{Re}\beta > 0, \quad a > 0] \qquad\qquad \text{ET I 75(34)}$$

2. $$\int_0^\infty e^{-\beta x^2} \cos ax^4\, dx = \frac{\pi}{8}\sqrt{\frac{\beta}{a}}\left[J_{\frac14}\left(\frac{\beta^2}{8a}\right)\sin\left(\frac{\beta^2}{8a} + \frac{\pi}{8}\right) - Y_{\frac14}\left(\frac{\beta^2}{8a}\right)\cos\left(\frac{\beta^2}{8a}\right) + \frac{\pi}{8}\right]$$
$$[\operatorname{Re}\beta > 0, \quad a > 0] \qquad\qquad \text{ET I 16(24)}$$

3.969

1. $$\int_0^\infty e^{-p^2 x^4 + q^2 x^2}\left[2px \cos\left(2pqx^3\right) + q\sin\left(2pqx^3\right)\right]\, dx = \frac{\sqrt{\pi}}{2} \qquad\qquad \text{BI (363)(7)}$$

2. $\int_0^\infty e^{-p^2x^4+q^2x^2} \left[2px \sin\left(2pqx^3\right) - q\cos\left(2pqx^3\right)\right]\, dx = 0$ BI (363)(8)

3.971   **Notation**: In formulas **3.971** 1 and 2, $p \geq 0$, $q \geq 0$, $r = \sqrt[4]{a^2 + p^2}$, $s = \sqrt[4]{b^2 + q^2}$, $A = \arctan\frac{a}{p}$, and $B = \arctan\frac{b}{q}$.

1. $\int_0^\infty \exp\left(-px^2 - \frac{q}{x^2}\right) \sin\left(ax^2 + \frac{b}{x^2}\right) \frac{dx}{x^2} = \frac{1}{2} \int_{-\infty}^\infty \exp\left(-px^2 - \frac{q}{x^2}\right) \sin\left(ax^2 + \frac{b}{x^2}\right) \frac{dx}{x^2}$

$$= \frac{\sqrt{\pi}}{2s} \exp\left[-2rs\cos(A+B)\right] \sin\left[A + 2rs\sin(A+B)\right]$$

BI (369)(16, 17)

2. $\int_0^\infty \exp\left(-px^2 - \frac{q}{x^2}\right) \cos\left(ax^2 + \frac{b}{x^2}\right) \frac{dx}{x^2} = \frac{1}{2} \int_{-\infty}^\infty \exp\left(-px^2 - \frac{q}{x^2}\right) \cos\left(ax^2 + \frac{b}{x^2}\right) \frac{dx}{x^2}$

$$= \frac{\sqrt{\pi}}{2s} \exp\left[-2rs\cos(A+B)\right] \cos\left[A + 2rs\sin(A+B)\right]$$

BI (369)(15, 18)

3.972

1. $\int_0^\infty \exp\left[-\beta\sqrt{\gamma^4 + x^4}\right] \sin ax^2 \frac{dx}{\sqrt{\gamma^4 + x^4}}$

$$= \sqrt{\frac{a\pi}{8}}\, I_{1/4}\left[\frac{\gamma^2}{2}\left(\sqrt{\beta^2 + a^2} - \beta\right)\right] K_{1/4}\left[\frac{\gamma^2}{4}\left(\sqrt{\beta^2 + a^2} + \beta\right)\right]$$

$$\left[\operatorname{Re}\beta > 0, \quad |\arg\gamma| < \frac{\pi}{4}, \quad a > 0\right] \quad \text{ET I 75(37)}$$

2. $\int_0^\infty \exp\left[-\beta\sqrt{\gamma^4 + x^4}\right] \cos ax^2 \frac{dx}{\sqrt{\gamma^4 + x^4}}$

$$= \sqrt{\frac{a\pi}{8}}\, I_{-1/4}\left[\frac{\gamma^2}{2}\left(\sqrt{\beta^2 + a^2} - \beta\right)\right] K_{1/4}\left[\frac{\gamma^2}{4}\left(\sqrt{\beta^2 + a^2} + \beta\right)\right]$$

$$\left[\operatorname{Re}\beta > 0, \quad |\arg\gamma| < \frac{\pi}{4}, \quad a > 0\right] \quad \text{ET I 17(28)}$$

3.973

1. $\int_0^\infty \exp\left(p\cos ax\right) \sin\left(p\sin ax\right) \frac{dx}{x} = \frac{\pi}{2}\left(e^p - 1\right)$ $[p > 0, \quad a > 0]$ WH, FI II 725

2. $\int_0^\infty \exp\left(p\cos ax\right) \sin\left(p\sin ax + bx\right) \frac{x\, dx}{c^2 + x^2} = \frac{\pi}{2} \exp\left(-cb + pe^{-ac}\right)$

$$[a > 0, \quad b > 0, \quad c > 0, \quad p > 0]$$

BI (372)(3)

3. $\int_0^\infty \exp\left(p\cos ax\right) \cos\left(p\sin ax + bx\right) \frac{dx}{c^2 + x^2} = \frac{\pi}{2c} \exp\left(-cb + pe^{-ac}\right)$

$$[a > 0, \quad b > 0, \quad c > 0, \quad p > 0]$$

BI (372)(4)

4. $\int_0^\infty \exp\left(p\cos x\right) \sin\left(p\sin x + nx\right) \frac{dx}{x} = \frac{\pi}{2} e^p$ $[p > 0]$ BI (366)(2)

5. $$\int_0^\infty \exp\left(p\cos x\right)\sin\left(p\sin x\right)\cos nx\,\frac{dx}{x} = \frac{p^n}{n!}\cdot\frac{\pi}{4} + \frac{\pi}{2}\sum_{k=n+1}^\infty \frac{p^k}{k!}$$

$$[p>0] \qquad\qquad \text{LI (366)(3)}$$

6. $$\int_0^\infty \exp\left(p\cos x\right)\cos\left(p\sin x\right)\sin nx\,\frac{dx}{x} = \frac{\pi}{2}\sum_{k=0}^{n-1}\frac{p^k}{k!} + \frac{p^n}{n!}\frac{\pi}{4}$$

$$[p>0] \qquad\qquad \text{LI (366)(4)}$$

3.974

1. $$\int_0^\infty \exp\left(p\cos ax\right)\sin\left(p\sin ax\right)\operatorname{cosec} ax\,\frac{dx}{b^2+x^2} = \frac{\pi\left[e^p - \exp\left(pe^{-ab}\right)\right]}{2b\sinh ab}$$

$$[a>0, \quad b>0, \quad p>0] \qquad \text{BI (391)(4)}$$

2. $$\int_0^\infty \left[1 - \exp\left(p\cos ax\right)\cos\left(p\sin ax\right)\right]\operatorname{cosec} ax\,\frac{x\,dx}{b^2+x^2} = \frac{\pi\left[e^p - \exp\left(pe^{-ab}\right)\right]}{2\sinh ab}$$

$$[a>0, \quad b>0, \quad p>0] \qquad \text{BI (391)(5)}$$

3. $$\int_0^\infty \exp\left(p\cos ax\right)\sin\left(p\sin ax + ax\right)\operatorname{cosec} ax\,\frac{dx}{b^2+x^2} = \frac{\pi\left[e^p - \exp\left(pe^{-ab} - ab\right)\right]}{2b\sinh ab}$$

$$[a>0, \quad b>0, \quad p>0] \qquad \text{BI (391)(6)}$$

4. $$\int_0^\infty \exp\left(p\cos ax\right)\cos\left(p\sin ax + ax\right)\operatorname{cosec} ax\,\frac{x\,dx}{b^2+x^2} = \frac{\pi\left[e^p - \exp\left(pe^{-ab} - ab\right)\right]}{2\sinh ab}$$

$$[a>0, \quad b>0, \quad p>0] \qquad \text{BI (391)(7)}$$

5. $$\int_0^\infty \exp\left(p\cos ax\right)\sin\left(p\sin ax\right)\frac{x\,dx}{b^2-x^2} = \frac{\pi}{2}\left[1 - \exp\left(p\cos ab\right)\cos\left(p\sin ab\right)\right]$$

$$[p>0, \quad a>0] \qquad \text{BI (378)(1)}$$

6. $$\int_0^\infty \exp\left(p\cos ax\right)\cos\left(p\sin ax\right)\frac{dx}{b^2-x^2} = \frac{\pi}{2b}\exp\left(p\cos ab\right)\sin\left(p\sin ab\right)$$

$$[a>0, \quad b>0, \quad p>0] \qquad \text{BI (378)(2)}$$

7. $$\int_0^\infty \exp\left(p\cos ax\right)\sin\left(p\sin ax\right)\tan ax\,\frac{dx}{b^2+x^2} = \frac{\pi}{2b}\cdot\tanh ab\left[\exp\left(pe^{-ab}\right) - e^p\right]$$

$$[a>0, \quad b>0, \quad p>0] \qquad \text{BI (372)(14)}$$

8. $$\int_0^\infty \exp\left(p\cos ax\right)\sin\left(p\sin ax\right)\cot ax\,\frac{dx}{b^2+x^2} = \frac{\pi}{2b}\coth ab\left[e^p - \exp\left(pe^{-ab}\right)\right]$$

$$[a>0, \quad b>0, \quad p>0] \qquad \text{BI (372)(15)}$$

9. $$\int_0^\infty \exp\left(p\cos ax\right)\sin\left(p\sin ax\right)\operatorname{cosec} ax\,\frac{dx}{b^2-x^2} = \frac{\pi}{2b}\operatorname{cosec} ab\left[e^p - \exp\left(p\cos ab\right)\cos\left(p\sin ab\right)\right]$$

$$[a>0, \quad b>0, \quad p>0] \qquad \text{BI (391)(12)}$$

10. $$\int_0^\infty \left[1 - \exp\left(p\cos ax\right)\cos\left(p\sin ax\right)\right]\operatorname{cosec} ax\,\frac{x\,dx}{b^2-x^2} = -\frac{\pi}{2}\exp\left(p\cos ab\right)\sin\left(p\sin ab\right)\operatorname{cosec} ab$$

$$[a>0, \quad b>0, \quad p>0] \qquad \text{BI (391)(13)}$$

3.975

1. $\displaystyle\int_0^\infty \frac{\sin\left(\beta\arctan\frac{x}{\gamma}\right)}{\left(\gamma^2+x^2\right)^{\frac{\beta}{2}}}\cdot\frac{dx}{e^{2\pi x}-1}=\frac{1}{2}\,\zeta(\beta,\gamma)-\frac{1}{4\gamma^\beta}-\frac{\gamma^{1-\beta}}{2(\beta-1)}$

$\qquad\qquad\qquad\qquad\qquad\qquad\qquad$ [$\mathrm{Re}\,\beta>1,\quad \mathrm{Re}\,\gamma>0$]　　　WH, ET I 26(7)

2. $\displaystyle\int_0^\infty \frac{\sin\left(\beta\arctan x\right)}{\left(1+x^2\right)^{\frac{\beta}{2}}}\cdot\frac{dx}{e^{2\pi x}+1}=\frac{1}{2(\beta-1)}-\frac{\zeta(\beta)}{2^\beta}$　　　[$\mathrm{Re}\,\beta>1$]　　　　EH I 33(13)

3.976　$\displaystyle\int_0^\infty \left(1+x^2\right)^{\beta-\frac{1}{2}}e^{-px^2}\cos\left[2px+(2\beta-1)\arctan x\right]\,dx=\frac{e^{-p}}{2p^\beta}\sin\pi\beta\,\Gamma(\beta)$

$\qquad\qquad\qquad\qquad\qquad\qquad\qquad$ [$\mathrm{Re}\,\beta>0,\quad p>0$]　　　　　　　　WH

3.98–3.99 Combinations of trigonometric and hyperbolic functions

3.981

1. $\displaystyle\int_0^\infty \frac{\sin ax}{\sinh\beta x}\,dx=\frac{\pi}{2\beta}\tanh\frac{a\pi}{2\beta}$　　　　　　[$\mathrm{Re}\,\beta>0,\quad a>0$]　　　BI (264)(16)

2. $\displaystyle\int_0^\infty \frac{\sin ax}{\cosh\beta x}\,dx=-\frac{\pi}{2\beta}\tanh\frac{a\pi}{2\beta}-\frac{i}{2\beta}\left[\psi\left(\frac{\beta+ai}{4\beta}\right)-\psi\left(\frac{\beta-ai}{4\beta}\right)\right]$

$\qquad\qquad\qquad\qquad\qquad\qquad\qquad$ [$\mathrm{Re}\,\beta>0,\quad a>0$]

$\qquad\qquad\qquad\qquad\qquad\qquad\qquad\qquad$ GW (335)(12), ET I 88(1)

3. $\displaystyle\int_0^\infty \frac{\cos ax}{\cosh\beta x}\,dx=\frac{\pi}{2\beta}\,\mathrm{sech}\,\frac{a\pi}{2\beta}$　　　　　[$\mathrm{Re}\,\beta>0,\quad$ all real a]　　BI (264)(14)

4. $\displaystyle\int_0^\infty \sin ax\,\frac{\sinh\beta x}{\sinh\gamma x}\,dx=\frac{\pi}{2\gamma}\,\frac{\sinh\frac{a\pi}{\gamma}}{\cosh\frac{a\pi}{\gamma}+\cos\frac{\beta\pi}{\gamma}}+\frac{i}{2\gamma}\left[\psi\left(\frac{\beta+\gamma+ia}{2\gamma}\right)-\psi\left(\frac{\beta+\gamma-ia}{2\gamma}\right)\right]$

$\qquad\qquad\qquad\qquad\qquad\qquad\qquad$ [$|\mathrm{Re}\,\beta|<\mathrm{Re}\,\gamma,\quad a>0$]　　　ET I 88(5)

5. $\displaystyle\int_0^\infty \cos ax\,\frac{\sinh\beta x}{\sinh\gamma x}\,dx=\frac{\pi}{2\gamma}\,\frac{\sin\frac{\pi\beta}{\gamma}}{\cosh\frac{a\pi}{\gamma}+\cos\frac{\beta\pi}{\gamma}}$　　　[$|\mathrm{Re}\,\beta|<\mathrm{Re}\,\gamma$]　　BI (265)(7)

6. $\displaystyle\int_0^\infty \sin ax\,\frac{\sinh\beta x}{\cosh\gamma x}\,dx=\frac{\pi}{\gamma}\,\frac{\sin\frac{\beta\pi}{2\gamma}\sinh\frac{a\pi}{2\gamma}}{\cosh\frac{a\pi}{\gamma}+\cos\frac{\beta\pi}{\gamma}}$　　[$|\mathrm{Re}\,\beta|<\mathrm{Re}\,\gamma,\quad a>0$]　　BI (265)(2)

7. $\displaystyle\int_0^\infty \cos ax\,\frac{\sinh\beta x}{\cosh\gamma x}\,dx=\frac{1}{4\gamma}\left[\left\{\psi\left(\frac{3\gamma-\beta+ia}{4\gamma}\right)+\psi\left(\frac{3\gamma-\beta-ia}{4\gamma}\right)-\psi\left(\frac{3\gamma+\beta-ia}{4\gamma}\right)\right\}\right.$

$\qquad\qquad\qquad\qquad\left.-\psi\left(\frac{3\gamma+\beta+ia}{4\gamma}\right)+\frac{2\pi\sin\frac{\pi\beta}{\gamma}}{\cos\frac{\pi\beta}{\gamma}+\cosh\frac{\pi a}{\gamma}}\right]$

$\qquad\qquad\qquad\qquad\qquad$ [$|\mathrm{Re}\,\beta|<\mathrm{Re}\,\gamma,\quad a>0$]　　　　ET I 31(13)

8. $\displaystyle\int_0^\infty \sin ax \frac{\cosh \beta x}{\sinh \gamma x}\, dx = \frac{\pi}{2\gamma}\cdot\frac{\sinh\frac{\pi a}{\gamma}}{\cosh\frac{\pi a}{\gamma}+\cos\frac{\pi\beta}{\gamma}}$ $[|\operatorname{Re}\beta|<\operatorname{Re}\gamma,\quad a>0]$ BI (265)(4)

9. $\displaystyle\int_0^\infty \sin ax \frac{\cosh \beta x}{\cosh \gamma x}\, dx = \frac{i}{4\gamma}\left[\psi\left(\frac{3\gamma+\beta+ia}{4\gamma}\right)-\psi\left(\frac{3\gamma+\beta-ai}{4\gamma}\right)+\psi\left(\frac{3\gamma-\beta+ia}{4\gamma}\right)\right.$

$$\left.-\psi\left(\frac{3\gamma-\beta-ai}{4\gamma}\right)-\frac{2\pi i\sinh\frac{\pi a}{\gamma}}{\cosh\frac{a\pi}{\gamma}+\cos\frac{\beta\pi}{\gamma}}\right]$$

$[|\operatorname{Re}\beta|<\operatorname{Re}\gamma,\quad a>0]$ ET I 88(6)

10. $\displaystyle\int_0^\infty \cos ax \frac{\cosh \beta x}{\cosh \gamma x}\, dx = \frac{\pi}{\gamma}\frac{\cos\frac{\beta\pi}{2\gamma}\cosh\frac{a\pi}{2\gamma}}{\cosh\frac{a\pi}{\gamma}+\cos\frac{\beta\pi}{\gamma}}$ $[|\operatorname{Re}\beta|<\operatorname{Re}\gamma,\quad \text{all real } a]$ BI (265)(6)

11. $\displaystyle\int_0^{\pi/2} \cos^{2m} x \cosh \beta x\, dx = \frac{(2m)!\sinh\frac{\pi\beta}{2}}{\beta\left(\beta^2+2^2\right)\ldots\left[\beta^2+(2m)^2\right]}$

$[\operatorname{Re}\beta>0]$ WA 620a

12. $\displaystyle\int_0^{\pi/2} \cos^{2m-1} x \cosh \beta x\, dx = \frac{(2m-1)!\cosh\frac{\pi\beta}{2}}{\left(\beta^2+1^2\right)\left(\beta^2+3^2\right)\ldots\left[\beta^2+(2m+1)^2\right]}$

$[\operatorname{Re}\beta>0]$ WA 620a

3.982

1. $\displaystyle\int_0^\infty \frac{\cos ax}{\cosh^2\beta x}\, dx = \frac{a\pi}{2\beta^2\sinh\frac{a\pi}{2\beta}}$ $[\operatorname{Re}\beta>0,\quad a>0]$ BI (264)(16)

2. $\displaystyle\int_0^\infty \sin ax \frac{\sinh \beta x}{\cosh^2\gamma x}\, dx = \frac{\pi\left(a\sin\frac{\beta\pi}{2\gamma}\cosh\frac{a\pi}{2\gamma}-\beta\cos\frac{\beta\pi}{2\gamma}\sinh\frac{a\pi}{2\gamma}\right)}{\gamma^2\left(\cosh\frac{a\pi}{\gamma}-\cos\frac{\beta\pi}{\gamma}\right)}$

$[|\operatorname{Re}\beta|<2\operatorname{Re}\gamma,\quad a>0]$ ET I 88(9)

3.[8] $\displaystyle\int_0^\infty \frac{\sin^2 x \cos ax}{\sin^2 hx}\, dx = \frac{\pi}{4}\left\{\frac{a+2}{1-e^{-\pi(a+2)}}-\frac{2a}{1-e^{-\pi a}}+\frac{a-2}{1-e^{-\pi(a-2)}}\right\}=I(a)$

$$\left[I(0)=\frac{1}{2}\left(\pi\coth\pi-1\right),\quad I(\pm2)=\frac{1}{4}+\frac{\pi}{2}\left(\coth 2\pi-\coth\pi\right)\right]$$

3.983

1.[6] $\displaystyle\int_0^\infty \frac{\cos ax\,dx}{b\cosh\beta x + c} = \frac{\pi\sin\left(\frac{a}{\beta}\operatorname{arccosh}\frac{c}{b}\right)}{\beta\sqrt{c^2-b^2}\sinh\dfrac{a\pi}{\beta}}$ $[c>b>0]$

$\displaystyle\qquad\qquad = \frac{\pi\sinh\left(\frac{a}{\beta}\arccos\frac{c}{b}\right)}{\beta\sqrt{b^2-c^2}\sinh\dfrac{a\pi}{\beta}}$ $[b>|c|>0]$

$\qquad\qquad\qquad\qquad\qquad\qquad [\operatorname{Re}\beta>0,\quad a>0]$ GW (335)(13a)

2. $\displaystyle\int_0^\infty \frac{\cos ax\,dx}{\cosh\beta x+\cos\gamma} = \frac{\pi}{\beta}\,\frac{\sinh\dfrac{a\gamma}{\beta}}{\sin\gamma\sinh\dfrac{a\pi}{\beta}}$ $[\pi\operatorname{Re}\beta<\operatorname{Im}\overline{\beta}\gamma,\quad a>0]$ BI (267)(3)

3.[3] $\displaystyle\int_0^\infty \frac{\cos ax\,dx}{\cosh x-\cosh b} = -\pi\coth a\pi\,\frac{\sin ab}{\sinh b}$ $[a>0,\quad b>0]$ ET I 30(8)

4. $\displaystyle\int_0^\infty \frac{\cos ax\,dx}{1+2\cosh\left(\sqrt{\dfrac{2}{3}}\pi x\right)} = \frac{\sqrt{\dfrac{\pi}{2}}}{1+2\cosh\left(\sqrt{\dfrac{2}{3}}\pi a\right)}$ $[a>0]$ ET I 30(9)

5. $\displaystyle\int_0^\infty \frac{\sin ax\sinh\beta x}{\cosh\gamma x+\cos\delta}\,dx = \frac{\pi\left\{\sin\left[\frac{\beta}{\gamma}(\pi-\delta)\right]\sinh\left[\frac{a}{\gamma}(\pi+\delta)\right] - \sin\left[\frac{\beta}{\gamma}(\pi+\delta)\right]\sinh\left[\frac{a}{\gamma}(\pi-\delta)\right]\right\}}{\gamma\sin\delta\left(\cosh\dfrac{2\pi a}{\gamma}-\cos\dfrac{2\pi\beta}{\gamma}\right)}$

$\qquad\qquad [\pi\operatorname{Re}\gamma>|\operatorname{Re}\overline{\gamma}\delta|,\quad |\operatorname{Re}\beta|<\operatorname{Re}\gamma,\quad a>0]$ BI (267)(2)

6. $\displaystyle\int_0^\infty \frac{\cos ax\cosh\beta x}{\cosh\gamma x+\cos b}\,dx = \frac{\pi\left\{\cos\left[\frac{\beta}{\gamma}(\pi-b)\right]\cosh\left[\frac{a}{\gamma}(\pi+b)\right] - \cos\left[\frac{\beta}{\gamma}(\pi+b)\right]\cosh\left[\frac{a}{\gamma}(\pi-b)\right]\right\}}{\gamma\sin b\left(\cosh\dfrac{2\pi a}{\gamma}-\cos\dfrac{2\pi\beta}{\gamma}\right)}$

$\qquad\qquad [|\operatorname{Re}\beta|<\operatorname{Re}\gamma,\quad 0<b<\pi,\quad a<0]$
$\qquad\qquad\qquad\qquad\qquad\qquad\qquad\qquad$ BI (267)(6)

7. $\displaystyle\int_0^\infty \frac{\cos ax\,dx}{\left(\beta+\sqrt{\beta^2-1}\cosh x\right)^{\nu+1}} = \Gamma(\nu+1-ai)e^{a\pi}\frac{Q_\nu^{ai}(\beta)}{\Gamma(\nu+1)}$

$\qquad\qquad [\operatorname{Re}\nu>-1,\quad |\arg(\beta+1)|<\pi,\quad a>0]$
$\qquad\qquad\qquad\qquad\qquad\qquad\qquad\qquad$ ET I 30(10)

3.984

1.[6] $\displaystyle\lim_{c\uparrow 1}\int_0^\infty \frac{\sin ax\sinh cx}{\cosh x+\cos b}\,dx = \pi\,\frac{\cosh ab}{\sinh a\pi}$ $[|b|\le\pi,\quad a\text{ real}]$ BI (267)(1)

2.[6] $\displaystyle\lim_{c\uparrow 1}\int_0^\infty \frac{\cos ax\cosh cx}{\cosh x+\cos b}\,dx = -\pi\cot b\,\frac{\sinh ab}{\sinh a\pi}$ $[0<|b|<\pi,\quad a\text{ real}]$ BI (267)(5)

3.[8] $\displaystyle\int_0^\infty \frac{\sin ax\sinh\frac{x}{2}}{\cosh x+\cos\beta}\,dx = \frac{\pi\sinh a\beta}{2\sin\dfrac{\beta}{2}\cosh a\pi}$ $[\operatorname{Re}\beta<\pi,\quad a>0]$ ET I 80(10)

4. $\int_0^\infty \dfrac{\cos ax \cosh \frac{\beta}{2} x}{\cosh \beta x + \cosh \gamma} dx = \dfrac{\pi \cos \frac{a\gamma}{\beta}}{2\beta \cosh \frac{\gamma}{2} \cosh \frac{a\pi}{\beta}}$ $\left[\pi \operatorname{Re} \beta > \left| \operatorname{Im} \left(\overline{\beta} \gamma \right) \right| \right]$ ET I 31(16)

5. $\int_0^\infty \dfrac{\sin ax \sinh \beta x}{\cosh 2\beta x + \cos 2ax} dx = \dfrac{a\pi}{4 \left(a^2 + \beta^2 \right)}$ $[a > 0, \quad \operatorname{Re} \beta > 0]$ BI (267)(7)

6. $\int_0^\infty \dfrac{\cos ax \cosh \beta x}{\cosh 2\beta x + \cos 2ax} dx = \dfrac{\beta\pi}{4 \left(a^2 + \beta^2 \right)}$ $[\operatorname{Re} \beta > 0, \quad a > 0]$ BI (267)(8)

7.[8] $\int_0^\infty \dfrac{\sinh^{2\mu-1} x \cosh^{2\varrho-2\nu+1} x}{\left(\cosh^2 x - \beta \sinh^2 x \right)^\varrho} dx = \dfrac{1}{2} \operatorname{B}(\mu, \nu - \mu) \, {}_2F_1(\varrho, \mu; \nu; \beta)$

                                                  $[\operatorname{Re} \nu > \operatorname{Re} \mu > 0]$ EH I 115(12)

3.985

1. $\int_0^\infty \dfrac{\cos ax \, dx}{\cosh^\nu \beta x} = \dfrac{2^{\nu-2}}{\beta \, \Gamma(\nu)} \Gamma \left(\dfrac{\nu}{2} + \dfrac{ai}{2\beta} \right) \Gamma \left(\dfrac{\nu}{2} - \dfrac{ai}{2\beta} \right)$ $[\operatorname{Re} \beta > 0, \quad \operatorname{Re} \nu > 0, \quad a > 0]$

                                                               ET I 30(5)

2. $\int_0^\infty \dfrac{\cos ax \, dx}{\cosh^{2n} \beta x} = \dfrac{4^{n-1} \pi a}{2(2n-1)! \beta^2 \sinh \frac{a\pi}{2\beta}} \prod_{k=1}^{n-1} \left(\dfrac{a^2}{4\beta^2} + k^2 \right)$

$\qquad\qquad = \dfrac{\pi a \left(a^2 + 2^2 \beta^2 \right) \left(a^2 + 4^2 \beta^2 \right) \cdots \left[a^2 + (2n-2)^2 \beta^2 \right]}{2(2n-1)! \beta^{2n} \sinh \frac{a\pi}{2\beta}}$

                                                 $[n \geq 2, \quad a > 0]$ ET I 30(3)

3. $\int_0^\infty \dfrac{\cos ax \, dx}{\cosh^{2n+1} \beta x} = \dfrac{\pi 2^{2n-1}}{(2n)! \beta \cosh \frac{a\pi}{2\beta}} \prod_{k=1}^{n} \left[\dfrac{a^2}{4\beta^2} + \left(\dfrac{2k-1}{2} \right)^2 \right]$

$\qquad\qquad = \dfrac{\pi \left(a^2 + \beta^2 \right) \left(a^2 + 3^2 \beta^2 \right) \cdots \left[a^2 + (2n-1)^2 \beta^2 \right]}{2(2n)! \beta^{2n+1} \cosh \frac{a\pi}{2\beta}}$

                                     $[\operatorname{Re} \beta > 0, \quad n = 0, 1, \ldots, \text{ all real } a]$ ET I 30(4)

3.986

1. $\int_0^\infty \dfrac{\sin \beta x \sin \gamma x}{\cosh \delta x} dx = \dfrac{\pi}{\delta} \cdot \dfrac{\sinh \frac{\beta\pi}{2\delta} \sinh \frac{\gamma\pi}{2\delta}}{\cosh \frac{\beta}{\delta} \pi + \cosh \frac{\gamma}{\delta} \pi}$ $[|\operatorname{Im}(\beta + \gamma)| < \operatorname{Re} \delta]$ BI (264)(19)

2. $\int_0^\infty \dfrac{\sin \alpha x \cos \beta x}{\sinh \gamma x} dx = \dfrac{\pi \sinh \frac{\pi\alpha}{\gamma}}{2\gamma \left(\cosh \frac{\alpha\pi}{\gamma} + \cosh \frac{\beta\pi}{\gamma} \right)}$ $[|\operatorname{Im}(\alpha + \beta)| < \operatorname{Re} \gamma]$ LI (264)(20)

3. $\int_0^\infty \dfrac{\cos \beta x \cos \gamma x}{\cosh \delta x} dx = \dfrac{\pi}{\delta} \cdot \dfrac{\cosh \frac{\beta\pi}{2\delta} \cosh \frac{\gamma\pi}{2\delta}}{\cosh \frac{\beta\pi}{\delta} + \cosh \frac{\gamma\pi}{\delta}}$ $[|\operatorname{Im}(\beta + \gamma)| < \operatorname{Re} \delta]$ BI (264)(21)

4.³ $\quad \int_0^\infty \dfrac{\sin^2 \beta x}{\sinh^2 \pi x} \, dx = \dfrac{\beta}{\pi \left(e^{2\beta} - 1\right)} + \dfrac{\beta - 1}{2\pi} = \dfrac{\beta \coth \beta - 1}{2\pi}$

$$[|\mathrm{Im}\,\beta| < \pi] \qquad\qquad \text{EH I 44(3)}$$

3.987

1. $\quad \int_0^\infty \sin ax \left(1 - \tanh \beta x\right) dx = \dfrac{1}{a} - \dfrac{\pi}{2\beta \sinh \dfrac{a\pi}{2\beta}}$ $\qquad [\mathrm{Re}\,\beta > 0] \qquad$ ET I 88(4)a

2. $\quad \int_0^\infty \sin ax \left(\coth \beta x - 1\right) dx = \dfrac{\pi}{2\beta} \coth \dfrac{a\pi}{2\beta} - \dfrac{1}{a}$ $\qquad [\mathrm{Re}\,\beta > 0] \qquad$ ET I 88(3)

3.988

1. $\quad \int_0^{\pi/2} \dfrac{\cos ax \sinh \left(2b \cos x\right)}{\sqrt{\cos x}} \, dx = \dfrac{\pi}{2} \sqrt{\pi b} \, I_{\frac{\alpha}{2} + \frac{1}{4}}(b) \, I_{-\frac{\alpha}{2} + \frac{1}{4}}(b)$

$$[a > 0] \qquad\qquad \text{ET I 37(66)}$$

2. $\quad \int_0^{\pi/2} \dfrac{\cos ax \cosh \left(2b \cos x\right)}{\sqrt{\cos x}} \, dx = \dfrac{\pi}{2} \sqrt{\pi b} \, I_{\frac{a}{2} - \frac{1}{4}}(b) \, I_{-\frac{a}{2} - \frac{1}{4}}(b)$

$$[a > 0] \qquad\qquad \text{ET I 37(67)}$$

3. $\quad \int_0^\infty \dfrac{\cos ax \, dx}{\sqrt{\cosh x \cos b}} = \dfrac{\pi \, P_{-\frac{1}{2} + ia}\left(\cos b\right)}{\sqrt{2} \cosh a\pi}$ $\qquad [a > 0, \quad b > 0] \qquad$ ET I 30(7)

3.989

1. $\quad \int_0^\infty \dfrac{\sin \frac{a^2 x^2}{\pi} \sin bx}{\sinh ax} \, dx = \dfrac{\pi}{2a} \sin \dfrac{\pi b^2}{4a^2} \operatorname{cosech} \dfrac{\pi b}{2a}$ $\qquad [a > 0, \quad b > 0] \qquad$ ET I 93(44)

2. $\quad \int_0^\infty \dfrac{\cos \frac{a^2 x^2}{\pi} \sin bx}{\sinh ax} \, dx = \dfrac{\pi}{2a} \dfrac{\cosh \frac{\pi b}{a} - \cos \frac{\pi b^2}{4a^2}}{\sinh \dfrac{\pi b}{2a}}$ $\qquad [a > 0, \quad b > 0] \qquad$ ET I 93(45)

3. $\quad \int_0^\infty \dfrac{\sin \frac{x^2}{\pi} \cos ax}{\cosh x} \, dx = \dfrac{\pi}{2} \dfrac{\cos \frac{a^2 \pi}{4} - \frac{1}{\sqrt{2}}}{\cosh \dfrac{a\pi}{2}}$ $\qquad$ ET I 36(54)

4. $\quad \int_0^\infty \dfrac{\cos \frac{x^2}{\pi} \cos ax}{\cosh x} \, dx = \dfrac{\pi}{2} \cdot \dfrac{\sin \frac{a^2 \pi}{4} + \frac{1}{\sqrt{2}}}{\cosh \dfrac{a\pi}{2}}$ $\qquad$ ET I 36(55)

5. $\quad \int_0^\infty \dfrac{\sin \left(\pi a x^2\right) \cos bx}{\cosh \pi x} \, dx = -\sum_{k=0}^\infty \exp\left[-\left(k + \tfrac{1}{2}\right) b\right] \sin\left[\left(k + \tfrac{1}{2}\right)^2 \pi a\right]$

$$+ \dfrac{1}{\sqrt{a}} \sum_{k=0}^\infty \exp\left[-\dfrac{b\left(k + \frac{1}{2}\right)}{a}\right] \sin\left[\dfrac{\pi}{4} - \dfrac{b^2}{4\pi a} + \dfrac{\left(k + \frac{1}{2}\right)^2 \pi}{a}\right]$$

$$[a > 0, \quad b > 0] \qquad\qquad \text{ET I 36(56)}$$

6. $\displaystyle\int_0^\infty \frac{\cos\left(\pi a x^2\right)\cos bx}{\cosh \pi x}\,dx = \sum_{k=0}^\infty (-1)^k \exp\left[-\left(k+\frac{1}{2}\right)b\right]\cos\left[\left(k+\frac{1}{2}\right)^2 \pi a\right]$

$$+\frac{1}{\sqrt{a}}\sum_{k=0}^\infty \exp\left[-\frac{b\left(k+\frac{1}{2}\right)}{a}\right]\cos\left[\frac{\pi}{4}-\frac{b^2}{4\pi a}+\frac{\left(k+\frac{1}{2}\right)^2 \pi}{a}\right]$$

$$[a>0,\quad b>0] \qquad\qquad \text{ET I 36(57)}$$

3.991

1. $\displaystyle\int_0^\infty \sin \pi x^2 \sin ax \coth \pi x\,dx = \frac{1}{2}\tanh\frac{a}{2}\sin\left(\frac{\pi}{4}+\frac{a^2}{4\pi}\right)$ ET I 93(42)

2. $\displaystyle\int_0^\infty \cos \pi x^2 \sin ax \coth \pi x\,dx = \frac{1}{2}\tanh\frac{a}{2}\left[1\cos\left(\frac{\pi}{4}+\frac{a^2}{4\pi}\right)\right]$ ET I 93(43)

3.992

1. $\displaystyle\int_0^\infty \frac{\sin \pi x^2 \cos ax}{1+2\cosh\left(\frac{2}{\sqrt{3}}\pi x\right)}\,dx = -\sqrt{3}+\frac{\cos\left(\frac{\pi}{12}-\frac{a^2}{4\pi}\right)}{4\cosh\dfrac{a}{\sqrt{3}}-2}$ ET I 37(60)

2. $\displaystyle\int_0^\infty \frac{\cos \pi x^2 \cos ax}{1+2\cosh\left(\frac{2}{\sqrt{3}}\pi x\right)}\,dx = 1-\frac{\sin\left(\frac{\pi}{12}-\frac{a^2}{4\pi}\right)}{4\cosh\dfrac{a}{\sqrt{3}}-2}$ ET I 37(61)

3.993 $\displaystyle\int_0^\infty \frac{\sin^2 x+\cos x^2}{\cosh\left(\sqrt{\pi}x\right)}\cos ax\,dx = \frac{\sqrt{\pi}}{2}\cdot\frac{\sin^2 a+\cos a^2}{\cosh\left(\sqrt{\pi}a\right)}$ ET I 37(58)

3.994

1. $\displaystyle\int_0^\infty \frac{\sin\left(2a\cosh x\right)\cos bx}{\sqrt{\cosh x}}\,dx = -\frac{\pi}{4}\sqrt{a\pi}\left[J_{\frac{1}{4}+\frac{ib}{2}}(a)\,Y_{\frac{1}{4}-\frac{ib}{2}}(a)+J_{\frac{1}{4}-\frac{ib}{2}}(a)\,Y_{\frac{1}{4}+\frac{ib}{2}}(a)\right]$

$$[a>0,\quad b>0] \qquad\qquad \text{ET I 37(62)}$$

2. $\displaystyle\int_0^\infty \frac{\cos\left(2a\cosh x\right)\cos bx}{\sqrt{\cosh x}}\,dx = -\frac{\pi}{4}\sqrt{a\pi}\left[J_{-\frac{1}{4}+\frac{ib}{2}}(a)\,Y_{-\frac{1}{4}-\frac{ib}{2}}(a)+J_{-\frac{1}{4}-\frac{ib}{2}}(a)\,Y_{-\frac{1}{4}+\frac{ib}{2}}(a)\right]$

$$[a>0,\quad b>0] \qquad\qquad \text{ET I 37(63)}$$

3. $\displaystyle\int_0^\infty \frac{\sin\left(2a\sinh x\right)\sin bx}{\sqrt{\sinh x}}\,dx = -\frac{i}{2}\sqrt{\pi a}\left[I_{\frac{1}{4}-\frac{ib}{2}}(a)\,K_{-\frac{1}{4}+\frac{ib}{2}}(a)-I_{\frac{1}{4}+\frac{ib}{2}}(a)\,K_{\frac{1}{4}-\frac{ib}{2}}(a)\right]$

$$[a>0,\quad b>0] \qquad\qquad \text{ET I 93(47)}$$

4. $\displaystyle\int_0^\infty \frac{\cos\left(2a\sinh x\right)\sin bx}{\sqrt{\sinh x}}\,dx = -\frac{i}{2}\sqrt{\pi a}\left[I_{-\frac{1}{4}-\frac{ib}{2}}(a)\,K_{-\frac{1}{4}+\frac{ib}{2}}(a)-I_{-\frac{1}{4}+\frac{ib}{2}}(a)\,K_{-\frac{1}{4}-\frac{ib}{2}}(a)\right]$

$$[a>0,\quad b>0] \qquad\qquad \text{ET I 93(48)}$$

5. $\displaystyle\int_0^\infty \frac{\sin\left(2a\sinh x\right)\cos bx}{\sqrt{\sinh x}}\,dx = \frac{\sqrt{\pi a}}{2}\left[I_{\frac{1}{4}-\frac{ib}{2}}(a)\,K_{\frac{1}{4}+\frac{ib}{2}}(a)+I_{\frac{1}{4}+\frac{ib}{2}}(a)\,K_{\frac{1}{4}-\frac{ib}{2}}(a)\right]$

$$[a>0,\quad b>0] \qquad\qquad \text{ET I 37(64)}$$

6. $\int_0^\infty \dfrac{\cos(2a \sinh x) \cos bx}{\sqrt{\sinh x}} \, dx = \dfrac{\sqrt{\pi a}}{2} \left[I_{-\frac{1}{4}-\frac{ib}{2}}(a) \, K_{-\frac{1}{4}+\frac{ib}{2}}(a) + I_{-\frac{1}{4}+\frac{ib}{2}}(a) \, K_{-\frac{1}{4}-\frac{ib}{2}}(a) \right]$

$[a > 0, \quad b > 0]$ ET I 37(65)

7. $\int_0^\infty \sin(a \cosh x) \sin(a \sinh x) \dfrac{dx}{\sinh x} = \dfrac{\pi}{2} \sin a$ $[a > 0]$ BI (264)(22)

3.995

1. $\int_0^{\pi/2} \dfrac{\sin(2a \cos^2 x) \cosh(a \sin 2x)}{b^2 \cos^2 x + c^2 \sin^2 x} \, dx = \dfrac{\pi}{2bc} \sin \dfrac{2ac}{b+c}$

$[b > 0, \quad c > 0]$ BI (273)(9)

2. $\int_0^{\pi/2} \dfrac{\cos(2a \cos^2 x) \cosh(a \sin 2x)}{b^2 \cos^2 x + c^2 \sin^2 x} \, dx = \dfrac{\pi}{2bc} \cos \dfrac{2ac}{b+c}$

$[b > 0, \quad c > 0]$ BI (273)(10)

3.996

1. $\int_0^\infty \sin(a \sinh x) \sinh \beta x \, dx = \sin \dfrac{\beta \pi}{2} K_\beta(a)$ $[|\operatorname{Re} \beta| < 1, \quad a > 0]$ EH II 82(26)

2. $\int_0^\infty \cos(a \sinh x) \cosh \beta x \, dx = \cos \dfrac{\beta \pi}{2} K_\beta(a)$ $[|\operatorname{Re} \beta| < 1, \quad a > 0]$ WA 202(13)

3. $\int_0^{\pi/2} \cos(a \sin x) \cosh(\beta \cos x) \, dx = \dfrac{\pi}{2} J_0 \left(\sqrt{a^2 - \beta^2} \right)$ MO 40

4. $\int_0^\infty \sin\left(a \cosh x - \tfrac{1}{2}\beta\pi\right) \cosh \beta x \, dx = \dfrac{\pi}{2} J_\beta(a)$ $[|\operatorname{Re} \beta| < 1, \quad a > 0]$ WA 199(12)

5. $\int_0^\infty \cos\left(a \cosh x - \tfrac{1}{2}\beta\pi\right) \cosh \beta x \, dx = -\dfrac{\pi}{2} Y_\beta(a)$ $[|\operatorname{Re} \beta| < 1, \quad a > 0]$ WA 199(13)

3.997

1. $\int_0^{\pi/2} \sin^\nu x \sinh(\beta \cos x) \, dx = \dfrac{\sqrt{\pi}}{2} \left(\dfrac{2}{\beta} \right)^{\frac{\nu}{2}} \Gamma\left(\dfrac{\nu+1}{2} \right) \mathbf{L}_{\frac{\nu}{2}}(\beta)$

$[\operatorname{Re} \nu > -1]$ EH II 38(53)

2. $\int_0^\pi \sin^\nu x \cosh(\beta \cos x) \, dx = \sqrt{\pi} \left(\dfrac{2}{\beta} \right)^{\frac{\nu}{2}} \Gamma\left(\dfrac{\nu+1}{2} \right) I_{\frac{\nu}{2}}(\beta)$

$[\operatorname{Re} \nu > -1]$ WH

3. $\int_0^{\pi/2} \dfrac{dx}{\cosh(\tan x) \cos x \sqrt{\sin 2x}} = \sqrt{2\pi} \sum_{k=0}^\infty \dfrac{(-1)^k}{\sqrt{2k+1}}$ BI (276)(13)

4. $\int_0^{\pi/2} \dfrac{\tan^q x}{\cosh(\tan x) + \cos \lambda} \dfrac{dx}{\sin 2x} = \dfrac{\Gamma(q)}{\sin \lambda} \sum_{k=1}^\infty (-1)^{k-1} \dfrac{\sin k\lambda}{k^q}$

$[q > 0]$ BI (275)(20)

4.11–4.12 Combinations involving trigonometric and hyperbolic functions and powers

4.111

1.
$$\int_0^\infty \frac{\sin ax}{\sinh \beta x} \cdot x^{2m}\, dx = (-1)^m \frac{\pi}{2\beta} \cdot \frac{\partial^{2m}}{\partial a^{2m}}\left(\tanh \frac{a\pi}{2\beta}\right)$$
$$[\operatorname{Re}\beta > 0] \qquad (\text{cf. } \mathbf{3.981}\ 1)$$
$$\text{GW (336)(17a)}$$

2.
$$\int_0^\infty \frac{\cos ax}{\sinh \beta x} \cdot x^{2m+1}\, dx = (-1)^m \frac{\pi}{2\beta} \frac{\partial^{2m+1}}{\partial a^{2m+1}}\left(\tanh \frac{a\pi}{2\beta}\right)$$
$$[\operatorname{Re}\beta > 0] \qquad (\text{cf. } \mathbf{3.981}\ 1)$$
$$\text{GW (336)(17b)}$$

3.
$$\int_0^\infty \frac{\sin ax}{\cosh \beta x} \cdot x^{2m+1}\, dx = (-1)^{m+1} \frac{\pi}{2\beta} \cdot \frac{\partial^{2m+1}}{\partial a^{2m+1}}\left(\frac{1}{\cosh \dfrac{a\pi}{2\beta}}\right)$$
$$[\operatorname{Re}\beta > 0] \qquad (\text{cf. } \mathbf{3.981}\ 3)$$
$$\text{GW (336)(18b)}$$

4.
$$\int_0^\infty \frac{\cos ax}{\cosh \beta x} \cdot x^{2m}\, dx = (-1)^m \frac{\pi}{2\beta} \cdot \frac{\partial^{2m}}{\partial a^{2m}}\left(\frac{1}{\cosh \dfrac{a\pi}{2\beta}}\right)$$
$$[\operatorname{Re}\beta > 0] \qquad (\text{cf. } \mathbf{3.981}\ 3)$$
$$\text{GW (336)(18a)}$$

5.
$$\int_0^\infty x\frac{\sin 2ax}{\cosh \beta x}\, dx = \frac{\pi^2}{4\beta^2} \cdot \frac{\sinh \dfrac{a\pi}{\beta}}{\cosh^2 \dfrac{a\pi}{\beta}} \qquad [\operatorname{Re}\beta > 0, \quad a > 0] \qquad \text{BI (364)(6)a}$$

6.
$$\int_0^\infty x\frac{\cos 2ax}{\sinh \beta x}\, dx = \frac{\pi^2}{4\beta^2} \cdot \frac{1}{\cosh^2 \dfrac{a\pi}{\beta}} \qquad [\operatorname{Re}\beta > 0, \quad a > 0] \qquad \text{BI (364)(1)a}$$

7.
$$\int_0^\infty \frac{\sin ax}{\cosh \beta x}\frac{dx}{x} = 2\arctan\left(\exp \frac{\pi a}{2\beta}\right) - \frac{\pi}{2} \qquad [\operatorname{Re}\beta > 0, \quad a > 0]$$
$$\text{BI (387)(1), ET I 89(13), LI (298)(17)}$$

4.112

1.
$$\int_0^\infty \left(x^2 + \beta^2\right)\frac{\cos ax}{\cosh \dfrac{\pi x}{2\beta}}\, dx = \frac{2\beta^3}{\cosh^3 a\beta} \qquad [\operatorname{Re}\beta > 0, \quad a > 0] \qquad \text{ET I 32(19)}$$

2.
$$\int_0^\infty x\left(x^2 + 4\beta^2\right)\frac{\cos ax}{\sinh \dfrac{\pi x}{2\beta}}\, dx = \frac{6\beta^4}{\cosh^4 a\beta} \qquad [\operatorname{Re}\beta > 0, \quad a > 0] \qquad \text{ET I 32(20)}$$

4.113

1. $$\int_0^\infty \frac{\sin ax}{\sinh \pi x} \cdot \frac{dx}{x^2 + \beta^2} = -\frac{1}{2\beta^2} - \frac{\pi e^{-a\beta}}{\beta \sin \pi \beta}$$
$$+ \frac{1}{2\beta^2} \left[{}_2F_1\left(1, -\beta; 1-\beta; -e^{-a}\right) + {}_2F_1\left(1, \beta; 1+\beta : -e^{-a}\right) \right]$$
$$= \frac{1}{2\beta^2} - \frac{\pi e^{-a\beta}}{2\beta \sin \pi \beta} - \sum_{k=1}^\infty \frac{(-1)^k e^{-ak}}{k^2 - \beta^2}$$

$$[\operatorname{Re} \beta > 0, \quad \beta \neq 0, 1, 2, \dots, \quad a > 0] \quad \text{ET I 90(18)}$$

2. $$\int_0^\infty \frac{\sin ax}{\sinh \pi x} \cdot \frac{dx}{x^2 + m^2} = \frac{(-1)^m a e^{-ma}}{2m} + \frac{1}{2m} \sum_{k=1}^{m-1} \frac{(-1)^k e^{-ka}}{m-k} + \frac{(-1)^m e^{-ma}}{2m} \ln\left(1 + e^{-a}\right)$$
$$+ \frac{1}{2m!} \frac{d^{m-1}}{dz^{m-1}} \left[\frac{(1+z)^{m-1}}{z} \ln(1+z) \right]_{z=e^{-a}}$$

$$[a > 0] \qquad\qquad \text{ET I 89(17)}$$

3. $$\int_0^\infty \frac{\sin ax}{\sinh \pi x} \cdot \frac{dx}{1 + x^2} = \frac{1}{2} \int_{-\infty}^\infty \frac{\sin ax}{\sinh \pi x} \frac{dx}{1 + x^2} = -\frac{a}{2} \cosh a + \sinh a \ln\left(2 \cosh \frac{a}{2}\right) \qquad \text{GW (336)(21b)}$$

4. $$\int_0^\infty \frac{\sin ax}{\sinh \frac{\pi}{2}x} \cdot \frac{dx}{1 + x^2} = \frac{1}{2} \int_{-\infty}^\infty \frac{\sin ax}{\sinh \frac{\pi}{2}x} \cdot \frac{dx}{1 + x^2} = \frac{\pi}{2} \sinh a - \cosh a \arctan(\sinh a)$$

$$\text{GW (336)(21a)}$$

5. $$\int_0^\infty \frac{\sin ax}{\sinh \frac{\pi}{4}x} \cdot \frac{dx}{1 + x^2} = -\frac{\pi}{\sqrt{2}} e^{-a} + \frac{\sinh a}{\sqrt{2}} \ln \frac{2 \cosh a + \sqrt{2}}{2 \cosh a - \sqrt{2}} + \sqrt{2} \cosh a \arctan \frac{\sqrt{2}}{2 \sinh a}$$

$$[a > 0] \qquad\qquad \text{LI (389)(1)}$$

6. $$\int_0^\infty \frac{\sin ax}{\cosh \frac{\pi}{4}x} \cdot \frac{x\,dx}{1 + x^2} = \frac{\pi}{\sqrt{2}} e^{-a} + \frac{\sinh a}{\sqrt{2}} \ln \frac{2 \cosh a + \sqrt{2}}{2 \cosh a - \sqrt{2}} - \sqrt{2} \cosh a \arctan \left(\frac{1}{\sqrt{2} \sinh a} \right)$$

$$[a > 0] \qquad\qquad \text{BI (388)(1)}$$

7. $$\int_0^\infty \frac{\cos ax}{\sinh \pi x} \cdot \frac{x\,dx}{1 + x^2} = -\frac{1}{2} + \frac{a}{2} e^{-a} + \cosh a \ln\left(1 + e^{-a}\right)$$

$$[a > 0] \qquad\qquad \text{BI (389)(14), ET I 32(24)}$$

8. $$\int_0^\infty \frac{\cos ax}{\sinh \frac{\pi}{2}x} \cdot \frac{x\,dx}{1 + x^2} = 2 \sinh a \arctan\left(e^{-a}\right) + \frac{\pi}{2} e^{-a} - 1$$

$$[a > 0] \qquad\qquad \text{BI (389)(11)}$$

9. $$\int_0^\infty \frac{\cos ax}{\cosh \pi x} \cdot \frac{x\,dx}{x^2 + \beta^2} = \sum_{k=0}^\infty (-1)^k \frac{\left(k + \frac{1}{2}\right)^2 e^{-a\beta} - \beta e^{-\left(k+\frac{1}{2}\right)a}}{\beta \left[\left(k + \frac{1}{2}\right)^2 - \beta^2\right]}$$

$$[\operatorname{Re} \beta > 0, \quad a > 0] \qquad\qquad \text{ET I 32(26)}$$

10.
$$\int_0^\infty \frac{\cos ax}{\cosh \pi x} \cdot \frac{dx}{\left(m+\frac{1}{2}\right)^2 + x^2} = \frac{(-1)^m e^{-\left(m+\frac{1}{2}\right)a}}{2m+1}\left[a + \ln\left(1+e^{-a}\right)\right]$$
$$+ \frac{e^{-\frac{a}{2}}}{2m+1}\sum_{k=0}^{m-1}\frac{(-1)^k e^{-ak}}{k-m} + \frac{e^{-\frac{a}{2}}}{(2m+1)(m+1)}$$
$$\times\; _2F_1\left(1, m+1;\; m+2;\; -e^{-a}\right)$$
$$[a>0]$$
ET I 32(25)

11.
$$\int_0^\infty \frac{\cos ax}{\cosh \pi x} \cdot \frac{dx}{1+x^2} = 2\cosh\frac{a}{2} - \left[e^a \arctan\left(e^{-\frac{a}{2}}\right) + e^{-a}\arctan\left(e^{\frac{a}{2}}\right)\right]$$
$$[a>0]$$
ET I 32(21)

12.
$$\int_0^\infty \frac{\cos ax}{\cosh \frac{\pi}{2}x} \cdot \frac{dx}{1+x^2} = ae^{-a} + \cosh a \ln\left(1+e^{-2a}\right) \qquad [a>0]$$
BI (388)(6)

13.
$$\int_0^\infty \frac{\cos ax}{\cosh \frac{\pi}{4}x} \cdot \frac{dx}{1+x^2} = \frac{\pi}{\sqrt{2}}e^{-a} + \frac{2\sinh a}{\sqrt{2}}\arctan\left(\frac{1}{\sqrt{2}\sinh a}\right) - \frac{\cosh a}{\sqrt{2}}\ln\frac{2\cosh a + \sqrt{2}}{2\cosh a - \sqrt{2}}$$
$$[a>0]$$
BI (388)(5)

4.114

1.
$$\int_0^\infty \frac{\sin ax}{x}\frac{\sinh \beta x}{\sinh \gamma x}\,dx = \arctan\left(\tan\frac{\beta\pi}{2\gamma}\tanh\frac{a\pi}{2\gamma}\right) \qquad [|\operatorname{Re}\beta| < \operatorname{Re}\gamma, \quad a>0]$$
BI (387)(6)a

2.
$$\int_0^\infty \frac{\cos ax}{x}\frac{\sinh \beta x}{\cosh \gamma x}\,dx = \frac{1}{2}\ln\frac{\cosh\frac{a\pi}{2\gamma} + \sin\frac{\beta\pi}{2\gamma}}{\cosh\frac{a\pi}{2\gamma} - \sin\frac{\beta\pi}{2\gamma}} \qquad [|\operatorname{Re}\beta| < \operatorname{Re}\gamma]$$
ET I 33(34)

4.115

1.
$$\int_0^\infty \frac{x\sin ax}{x^2+b^2}\cdot\frac{\sinh \beta x}{\sinh \pi x}\,dx = \frac{\pi}{2}\frac{e^{-ab}\sin b\beta}{\sin b\pi} + \sum_{k=1}^\infty (-1)^k\frac{ke^{-ak}\sin k\beta}{k^2-b^2}$$
$$[0 < \operatorname{Re}\beta < \pi, \quad a>0, \quad b>0]$$
BI (389)(23)

2.
$$\int_0^\infty \frac{x\sin ax}{x^2+1}\cdot\frac{\sinh \beta x}{\sinh \pi x}\,dx = \frac{1}{2}e^{-a}\left(a\sin\beta - \beta\cos\beta\right) - \frac{1}{2}\sinh a\sin\beta\ln\left[1 + 2e^{-a}\cos\beta + e^{-2a}\right]$$
$$+ \cosh a\cos\beta\arctan\frac{\sin\beta}{e^a + \cos\beta}$$
$$[|\operatorname{Re}\beta| < \pi, \quad a>0]$$
LI (389)(10)

3.
$$\int_0^\infty \frac{x\sin ax}{x^2+1}\cdot\frac{\sinh \beta x}{\sinh \frac{\pi}{2}x}\,dx$$
$$= \frac{\pi}{2}e^{-a}\sin\beta + \frac{1}{2}\cos\beta\sinh a\ln\frac{\cosh a + \sin\beta}{\cosh a - \sin\beta} - \sin\beta\cosh a\arctan\left(\frac{\cos\beta}{\sinh a}\right)$$
$$\left[|\operatorname{Re}\beta| < \frac{\pi}{2}, \quad a>0\right]$$
BI (389)(8)

4.
$$\int_0^\infty \frac{\cos ax}{x^2+b^2}\cdot\frac{\sinh \beta x}{\sinh \pi x}\,dx = \frac{\pi}{2b}\cdot\frac{e^{-ab}\sin b\beta}{\sin b\pi} + \sum_{k=1}^\infty (-1)^k\frac{e^{-ak}\sin k\beta}{k^2-b^2}$$
$$[0 < \operatorname{Re}\beta < \pi, \quad a>0, \quad b>0]$$
BI (389)(22)

5. $\int_0^\infty \dfrac{\cos ax}{x^2+1} \cdot \dfrac{\sinh \beta x}{\sinh \pi x}\, dx = \dfrac{1}{2}e^{-a}\left(a \sin \beta - \beta \cos \beta\right) + \dfrac{1}{2}\cosh a \sin \beta \ln\left(1 + 2e^{-a}\cos \beta + e^{-2a}\right)$

$\qquad\qquad\qquad\qquad - \sinh a \cos \beta \arctan \dfrac{\sin \beta}{e^a + \cos \beta}$

$\qquad\qquad\qquad\qquad\qquad\qquad [|\operatorname{Re}\beta| < \pi, \quad a > 0, \quad b > 0] \quad$ BI (389)(20)a

6. $\int_0^\infty \dfrac{\cos ax}{x^2+1} \cdot \dfrac{\sinh \beta x}{\sinh \frac{\pi}{2}x}\, dx = \dfrac{\pi}{2}e^{-a}\sin \beta - \dfrac{1}{2}\cosh a \cos \beta \ln \dfrac{\cosh a + \sin \beta}{\cosh a - \sin \beta} + \sinh a \sin \beta \arctan \dfrac{\cos \beta}{\sinh a}$

$\qquad\qquad\qquad\qquad\qquad \left[|\operatorname{Re}\beta| < \dfrac{\pi}{2}, \quad a > 0, \quad b > 0\right]$

$\qquad\qquad\qquad\qquad\qquad\qquad\qquad\qquad\qquad$ BI (389)(18)

7. $\int_0^\infty \dfrac{\sin ax}{x^2+\frac{1}{4}} \cdot \dfrac{\sinh \beta x}{\cosh \pi x}\, dx = e^{-\frac{a}{2}}\left(a \sin \dfrac{\beta}{2} - \beta \cos \dfrac{\beta}{2}\right) - \sinh \dfrac{a}{2}\sin \dfrac{\beta}{2}\ln\left(1 + 2e^{-a}\cos \beta + e^{-2a}\right)$

$\qquad\qquad\qquad\qquad + \cosh \dfrac{a}{2}\cos \dfrac{\beta}{2}\arctan \dfrac{\sin \beta}{1 + e^{-a}\cos \beta}$

$\qquad\qquad\qquad\qquad\qquad\qquad [|\operatorname{Re}\beta| < \pi, \quad a > 0] \qquad\qquad$ ET I 91(26)

8. $\int_0^\infty \dfrac{\sin ax}{x^2+\beta^2} \cdot \dfrac{\cosh \gamma x}{\sinh \pi x}\, dx = \dfrac{1}{2\beta^2} - \dfrac{\pi}{2\beta}\cdot \dfrac{e^{-a\beta}\cos \beta\gamma}{\sin \beta\pi} + \sum_{k=1}^\infty (-1)^{k-1}\dfrac{e^{-ak}\cos k\gamma}{k^2 - \beta^2}$

$\qquad\qquad\qquad\qquad\qquad [0 \le \operatorname{Re}\beta, \quad |\operatorname{Re}\gamma| < \pi, \quad a > 0]$

$\qquad\qquad\qquad\qquad\qquad\qquad\qquad\qquad\qquad$ BI (389)(21)

9. $\int_0^\infty \dfrac{\sin ax}{x^2+1} \cdot \dfrac{\cosh \beta x}{\sinh \pi x}\, dx = -\dfrac{1}{2}e^{-a}\left(a \cos \beta + \beta \sin \beta\right) + \dfrac{1}{2}\sinh a \cos \beta \ln\left(1 + 2e^{-a}\cos \beta + e^{-2a}\right)$

$\qquad\qquad\qquad\qquad + \cosh a \sin \beta \arctan \dfrac{\sin \beta}{e^a + \cos \beta}$

$\qquad\qquad\qquad\qquad [|\operatorname{Re}\beta| < \pi, \quad a > 0] \quad$ ET I 91(25), LI (389)(9)

10. $\int_0^\infty \dfrac{\sin ax}{x^2+1} \cdot \dfrac{\cosh \beta x}{\sinh \frac{\pi}{2}x}\, dx = -\dfrac{\pi}{2}e^{-a}\cos \beta + \dfrac{1}{2}\sinh a \sin \beta \ln \dfrac{\cosh a + \sin \beta}{\cosh a - \sin \beta} + \cosh a \cos \beta \arctan \dfrac{\cos \beta}{\sinh a}$

$\qquad\qquad\qquad\qquad\qquad \left[|\operatorname{Re}\beta| < \dfrac{\pi}{2}, \quad a > 0\right] \qquad$ BI (389)(7)

11. $\int_0^\infty \dfrac{x \cos ax}{x^2+b^2} \cdot \dfrac{\cosh \beta x}{\sinh \pi x}\, dx = \dfrac{\pi}{2}\cdot \dfrac{e^{-ab}\cos b\beta}{\sin b\pi} + \sum_{k=1}^\infty (-1)^k \dfrac{ke^{-ak}\cos k\beta}{k^2 - b^2}$

$\qquad\qquad\qquad\qquad\qquad [|\operatorname{Re}\beta| < \pi, \quad a > 0] \qquad$ BI (389)(24)

12. $\int_0^\infty \dfrac{x \cos ax}{x^2+1} \cdot \dfrac{\cosh \beta x}{\sinh \pi x}\, dx = \dfrac{1}{2}e^{-a}\left(a \cos \beta + \beta \sin \beta\right)$

$\qquad\qquad\qquad\qquad - \dfrac{1}{2} + \dfrac{1}{2}\cosh a \cos \beta \ln\left[1 + 2e^{-a}\cos \beta + e^{-2a}\right]$

$\qquad\qquad\qquad\qquad + \sinh a \sin \beta \arctan \dfrac{\sin \beta}{e^a + \cos \beta}$

$\qquad\qquad\qquad\qquad\qquad\qquad [|\operatorname{Re}\beta| < \pi, \quad a > 0] \qquad$ BI (389)(19)

13. $\displaystyle\int_0^\infty \frac{x\cos ax}{x^2+1}\cdot\frac{\cosh\beta x}{\sinh\frac{\pi}{2}x}\,dx = -1 + \frac{\pi}{2}e^{-a}\cos\beta + \frac{1}{2}\cosh a\sin\beta\ln\frac{\cosh a+\sin\beta}{\cosh a-\sin\beta}$

$$+ \sinh a\cos\beta\arctan\frac{\cos\beta}{\sinh a}$$

$$\left[|\mathrm{Re}\,\beta| < \frac{\pi}{2},\quad a>0\right] \qquad \text{BI (389)(17)}$$

14. $\displaystyle\int_0^\infty \frac{\cos ax}{x^2+1}\cdot\frac{\cosh\beta x}{\cosh\frac{\pi}{2}x}\,dx = ae^{-a}\cos\beta + \beta e^{-a}\sin\beta + \sinh a\sin\beta\arctan\frac{e^{-2a}\sin 2\beta}{1+e^{-2a}\cos 2\beta}$

$$+\frac{1}{2}\cosh a\cos\beta\ln\left(1+2e^{-2a}\cos 2\beta+e^{-4a}\right)$$

$$\left[|\mathrm{Re}\,\beta| < \frac{\pi}{2}, a>0\right] \qquad \text{ET I 34(37)}$$

4.116

1.[6] $\displaystyle\int_0^\infty x\cos 2ax\tanh x\,dx$ the integral is divergent   BI (364)(2)

2. $\displaystyle\int_0^\infty \cos ax\tanh\beta x\,\frac{dx}{x} = \ln\coth\frac{a\pi}{4\beta}$ $[\mathrm{Re}\,\beta>0,\quad a>0]$ BI (387)(8)

4.117

1. $\displaystyle\int_0^\infty \frac{\sin ax}{1+x^2}\tanh\frac{\pi x}{2}\,dx = a\cosh a - \sinh a\ln\left(2\sinh a\right)$

$$[a>0] \qquad \text{BI (388)(3)}$$

2. $\displaystyle\int_0^\infty \frac{\sin ax}{1+x^2}\tanh\frac{\pi x}{4}\,dx = -\frac{\pi}{2}e^a + \sinh a\ln\coth\frac{a}{2} + 2\cosh a\arctan\left(e^a\right)$ BI (388)(4)

3. $\displaystyle\int_0^\infty \frac{\sin ax}{1+x^2}\coth\pi x\,dx = \frac{a}{2}e^{-a} - \sinh a\ln\left(1-e^{-a}\right)$ $[a>0]$ BI (389)(5)

4. $\displaystyle\int_0^\infty \frac{\sin ax}{1+x^2}\coth\frac{\pi}{2}x\,dx = \sinh a\ln\coth\frac{a}{2}$ $[a>0]$ BI (389)(6)

5. $\displaystyle\int_0^\infty \frac{x\cos ax}{1+x^2}\tanh\frac{\pi}{2}x\,dx = -ae^{-a} - \cosh a\ln\left(1-e^{-2a}\right)$

$$[a>0] \qquad \text{BI (388)(7)}$$

6. $\displaystyle\int_0^\infty \frac{x\cos ax}{1+x^2}\tanh\frac{\pi}{4}x\,dx - \frac{\pi}{2}e^a + \cosh a\ln\coth\frac{a}{2} + 2\sinh a\arctan\left(e^a\right)$

$$[a>0] \qquad \text{BI (388)(8)}$$

7. $\displaystyle\int_0^\infty \frac{x\cos ax}{1+x^2}\coth\pi x\,dx = -\frac{a}{2}e^{-a} - \frac{1}{2} - \cosh a\ln\left(1-e^{-a}\right)$ BI (389)(15)a, ET I 33(31)a

8. $\displaystyle\int_0^\infty \frac{x\cos ax}{1+x^2}\coth\frac{\pi}{2}x\,dx = -1 + \cosh a\ln\coth\frac{a}{2}$ $[a>0]$ BI (389)(12)

9. $\displaystyle\int_0^\infty \frac{x\cos ax}{1+x^2}\coth\frac{\pi}{4}x\,dx = -2 + \frac{\pi}{2}e^{-a} + \cosh a\ln\coth\frac{a}{2} + 2\sinh a\arctan\left(e^{-a}\right)$

$$[a>0] \qquad \text{BI (389)(13)}$$

4.118[8] $\int_0^\infty \dfrac{x \sin ax}{\cosh^2 x}\, dx = \dfrac{\pi}{2} \dfrac{1}{\sinh \frac{1}{2}\pi a} \left(\dfrac{1}{2}\pi a \coth \dfrac{1}{2}\pi a - 1 \right)$ ET I 89(14)

4.119 $\int_0^\infty \dfrac{1 - \cos px}{\sinh qx} \cdot \dfrac{dx}{x} = \ln \left(\cosh \dfrac{p\pi}{2q} \right)$ BI (387)(2)a

4.121

1. $\int_0^\infty \dfrac{\sin ax - \sin bx}{\cosh \beta x} \cdot \dfrac{dx}{x} = 2 \arctan \dfrac{\exp \dfrac{a\pi}{2\beta} - \exp \dfrac{b\pi}{2\beta}}{1 + \exp \dfrac{(a+b)\pi}{2\beta}}$

$[\operatorname{Re} \beta > 0]$ GW (336)(19b)

2. $\int_0^\infty \dfrac{\cos ax - \cos bx}{\sinh \beta x} \cdot \dfrac{dx}{x} = \ln \dfrac{\cosh \dfrac{b\pi}{2\beta}}{\cosh \dfrac{a\pi}{2\beta}}$ $[\operatorname{Re} \beta > 0]$ GW (336)(19a)

4.122

1.[6] $\int_0^\infty \dfrac{\cos \beta x \sin \gamma x}{\cosh \delta x} \cdot \dfrac{dx}{x} = \arctan \dfrac{\sinh \dfrac{\gamma \pi}{2\delta}}{\cosh \dfrac{\beta \pi}{2\delta}}$ $[\operatorname{Re} \delta > |\operatorname{Im} \beta| + |\operatorname{Im} \gamma|]$ ET I 93(46)a

2. $\int_0^\infty \sin^2 ax \dfrac{\cosh \beta x}{\sinh x} \cdot \dfrac{dx}{x} = \dfrac{1}{4} \ln \dfrac{\cosh 2a\pi + \cos \beta \pi}{1 + \cos \beta \pi}$ $[|\operatorname{Re} \beta| < 1]$ BI (387)(7)

4.123

1. $\int_0^\infty \dfrac{\sin x}{\cosh ax + \cos x} \cdot \dfrac{x\, dx}{x^2 - \pi^2} = \arctan \dfrac{1}{a} - \dfrac{1}{a}$ BI (390)(1)

2. $\int_0^\infty \dfrac{\sin x}{\cosh ax - \cos x} \cdot \dfrac{x\, dx}{x^2 - \pi^2} = \dfrac{a}{1 + a^2} - \arctan \dfrac{1}{a}$ BI (390)(2)

3. $\int_0^\infty \dfrac{\sin 2x}{\cosh 2ax - \cos 2x} \cdot \dfrac{x\, dx}{x^2 - \pi^2} = \dfrac{1}{2a} \cdot \dfrac{1 + 2a^2}{1 + a^2} - \arctan \dfrac{1}{a}$ BI (390)(4)

4. $\int_0^\infty \dfrac{\cosh ax \sin x}{\cosh 2ax - \cos 2x} \cdot \dfrac{x\, dx}{x^2 - \pi^2} = \dfrac{-1}{2a(1 + a^2)}$ LI (390)(3)

5. $\int_0^\infty \dfrac{\cos ax}{\cosh \pi x + \cos \pi \beta} \cdot \dfrac{dx}{x^2 + \gamma^2} = \dfrac{\pi e^{-a\gamma}}{2\gamma(\cos \gamma \pi + \cos \beta \pi)}$

$+ \dfrac{1}{\sinh \beta \pi} \displaystyle\sum_{k=0}^\infty \left\{ \dfrac{\exp[-(2k+1-\beta)a]}{\gamma^2 - (2k+1-\beta)^2} - \dfrac{\exp[-(2k+1+\beta)a]}{\gamma^2 - (2k+1+\beta)^2} \right\}$

$[0 < \operatorname{Re} \beta < 1, \quad \operatorname{Re} \gamma > 0, \quad a > 0]$ ET I 33(27)

6. $\int_0^\infty \dfrac{\sin ax \sinh bx}{\cos 2ax + \cosh 2bx} x^{p-1}\, dx = \dfrac{\Gamma(p)}{(a^2 + b^2)^{\frac{p}{2}}} \sin\left(p \arctan \dfrac{a}{b} \right) \displaystyle\sum_{k=0}^\infty \dfrac{(-1)^k}{(2k+1)^p}$

$[p > 0]$ BI (364)(8)

7.　$\int_0^\infty \sin ax^2 \dfrac{\sin \frac{\pi x}{2} \sinh \frac{\pi x}{2}}{\cos \pi x + \cosh \pi x} \cdot x\, dx = \dfrac{1}{4}\left[\dfrac{\partial\, \vartheta_1(z\mid q)}{\partial z}\right]_{z=0,\, q=e^{-2a}}$

$[a > 0]$　　　　　　　ET I 93(49)

4.124

1.　$\int_0^1 \dfrac{\cos px \cosh\left(q\sqrt{1-x^2}\right)}{\sqrt{1-x^2}}\, dx = \dfrac{\pi}{2}\, J_0\left(\sqrt{p^2-q^2}\right)$　　　　　　　MO (40)

2.　$\int_u^\infty \cos ax \cosh\sqrt{\beta\left(u^2-x^2\right)} \cdot \dfrac{dx}{\sqrt{u^2-x^2}} = \dfrac{\pi}{2}\, J_0\left(\dfrac{u}{\sqrt{a^2-\beta^2}}\right)$　　　　　　　ET I 34(38)

4.125

1.　$\int_0^\infty \sinh\left(a\sin x\right)\cos\left(a\cos x\right)\sin x \sin 2nx\, \dfrac{dx}{x} = \dfrac{(-1)^{n-1}a^{2n-1}}{(2n-1)!}\dfrac{\pi}{8}\left[1 + \dfrac{a^2}{2n(2n+1)}\right]$

LI (367)(14)

2.　$\int_0^\infty \cosh\left(a\sin x\right)\cos\left(a\cos x\right)\sin x \cos(2n-1)x\, \dfrac{dx}{x} = \dfrac{(-1)^{n-1}a^{2(n-1)}}{[2(n-1)]!}\dfrac{\pi}{8}\left[1 - \dfrac{a^2}{2n(2n-1)}\right]$

LI (367)(15)

3.　$\int_0^\infty \sinh\left(a\sin x\right)\cos\left(a\cos x\right)\cos x \cos 2nx\, \dfrac{dx}{x} = \dfrac{\pi}{2}\displaystyle\sum_{k=n+1}^\infty \dfrac{(-1)^k a^{2k+1}}{(2k+1)!} + \dfrac{(-1)^n a^{2n+1}}{(2n+1)!}\dfrac{3\pi}{8}$

$\qquad\qquad + \dfrac{(-1)^{n-1}a^{2n-1}}{(2n-1)!}\dfrac{\pi}{8}$

LI (367)(21)

4.126

1.　$\int_0^\infty \sin\left(a\cos bx\right)\sinh\left(a\sin bx\right)\dfrac{x\, dx}{c^2-x^2} = \dfrac{\pi}{2}\left[\cos\left(a\cos bc\right)\cosh\left(a\sin bc\right) - 1\right]$

$[b > 0]$　　　　　　　BI (381)(2)

2.　$\int_0^\infty \sin\left(a\cos bx\right)\cosh\left(a\sin bx\right)\dfrac{dx}{c^2-x^2} = \dfrac{\pi}{2c}\cos\left(a\cos bc\right)\sinh\left(a\sin bc\right)$

$[b > 0, \quad c > 0]$　　　　　　　BI (381)(1)

3.　$\int_0^\infty \cos\left(a\cos bx\right)\sinh\left(a\sin bx\right)\dfrac{x\, dx}{c^2-x^2} = \dfrac{\pi}{2}\left[a\cos bc - \sin\left(a\cos bc\right)\cosh\left(a\sin bc\right)\right]$

$[b > 0]$　　　　　　　BI (381)(4)

4.　$\int_0^\infty \cos\left(a\cos bx\right)\cosh\left(a\sin bx\right)\dfrac{dx}{c^2-x^2} = -\dfrac{\pi}{2c}\sin\left(a\cos bc\right)\sinh\left(a\sin bc\right)$

$[b > 0]$　　　　　　　BI (381)(3)

4.13 Combinations of trigonometric and hyperbolic functions and exponentials

4.131

1.
$$\int_0^\infty \sin ax \sinh^\nu \gamma x e^{-\beta x}\, dx = -\frac{i\,\Gamma(\nu+1)}{2^{\nu+2}\gamma}\left\{\frac{\Gamma\left(\dfrac{\beta-\nu\gamma-ai}{2\gamma}\right)}{\Gamma\left(\dfrac{\beta+\nu\gamma-ai}{2\gamma}+1\right)} - \frac{\Gamma\left(\dfrac{\beta-\nu\gamma+ai}{2\gamma}\right)}{\Gamma\left(\dfrac{\beta+\gamma\nu+ai}{2\gamma}+1\right)}\right\}$$

$$[\operatorname{Re}\nu > -2, \quad \operatorname{Re}\gamma > 0, \quad |\operatorname{Re}(\gamma\nu)| < \operatorname{Re}\beta] \qquad \text{ET I 91(30)a}$$

2.
$$\int_0^\infty \cos ax \sinh^\nu \gamma x e^{-\beta x}\, dx = \frac{\Gamma(\nu+1)}{2^{\nu+2}\gamma}\left\{\frac{\Gamma\left(\dfrac{\beta-\nu\gamma-ai}{2\gamma}\right)}{\Gamma\left(\dfrac{\beta+\gamma\nu-ai}{2\gamma}+1\right)} - \frac{\Gamma\left(\dfrac{\beta-\nu\gamma+ai}{2\gamma}\right)}{\Gamma\left(\dfrac{\beta+\nu\gamma+ai}{2\gamma}+1\right)}\right\}$$

$$[\operatorname{Re}\nu > -1, \quad \operatorname{Re}\gamma > 0, \quad |\operatorname{Re}(\gamma\nu)| < \operatorname{Re}\beta] \qquad \text{ET I 34(40)a}$$

3.
$$\int_0^\infty e^{-\beta x}\frac{\sin ax}{\sinh \gamma x}\, dx = \sum_{k=1}^\infty \frac{2a}{a^2 + [\beta + (2k-1)\gamma]^2} \qquad \text{BI (264)(9)a}$$

$$= \frac{1}{2\gamma i}\left[\psi\left(\frac{\beta+\gamma+ia}{2\gamma}\right) - \psi\left(\frac{\beta+\gamma-ia}{2\gamma}\right)\right] \qquad [\operatorname{Re}\beta > |\operatorname{Re}\gamma|] \qquad \text{ET I 91(28)}$$

4.
$$\int_0^\infty e^{-x}\frac{\sin ax}{\sinh x}\, dx = \frac{\pi}{2}\coth\frac{a\pi}{2} - \frac{1}{a} \qquad \text{ET I 91(29)}$$

4.132

1.
$$\int_0^\infty \frac{\sin ax \sinh \beta x}{e^{\gamma x}-1}\, dx = -\frac{a}{2(a^2+\beta^2)} + \frac{\pi}{2\gamma}\cdot\frac{\sinh\frac{2\pi a}{\gamma}}{\cosh\frac{2\pi a}{\gamma} - \cos\frac{2\pi\beta}{\gamma}}$$

$$+ \frac{i}{2\gamma}\left[\psi\left(\frac{\beta}{\gamma}+i\frac{a}{\gamma}+1\right) - \psi\left(\frac{\beta}{\gamma}-i\frac{a}{\gamma}+1\right)\right]$$

$$[\operatorname{Re}\gamma > |\operatorname{Re}\beta|, a > 0] \qquad \text{ET I 92(33)}$$

2.
$$\int_0^\infty \frac{\sin ax \cosh \beta x}{e^{\gamma x}-1}\, dx = -\frac{a}{2(a^2+\beta^2)} + \frac{\pi}{2\gamma}\cdot\frac{\sinh\frac{2\pi a}{\gamma}}{\cosh\frac{2\pi a}{\gamma} - \cos\frac{2\pi\beta}{\gamma}}$$

$$[\operatorname{Re}\gamma > |\operatorname{Re}\beta|] \qquad \text{BI (265)(5)a, ET I 92(34)}$$

3.
$$\int_0^\infty \frac{\sin ax \cosh \beta x}{e^{\gamma x}+1}\, dx = \frac{a}{2(a^2+\beta^2)} - \frac{\pi}{\gamma}\cdot\frac{\sinh\frac{a\pi}{\gamma}\cos\frac{\beta\pi}{\gamma}}{\cosh\frac{2a\pi}{\gamma} - \cos\frac{2\beta\pi}{\gamma}}$$

$$[\operatorname{Re}\gamma > |\operatorname{Re}\beta|] \qquad \text{ET I 92(35)}$$

4.
$$\int_0^\infty \frac{\cos ax \sinh \beta x}{e^{\gamma x}-1}\, dx = \frac{\beta}{2(a^2+\beta^2)} - \frac{\pi}{2\gamma}\cdot\frac{\sin\frac{2\pi\beta}{\gamma}}{\cosh\frac{2a\pi}{\gamma} - \cos\frac{2\beta\pi}{\gamma}}$$

$$[\operatorname{Re}\gamma > |\operatorname{Re}\beta|] \qquad \text{LI (265)(8)}$$

5. $$\int_0^\infty \frac{\cos ax \sinh \beta x}{e^{\gamma x} + 1} \, dx = -\frac{\beta}{2(a^2 + \beta^2)} + \frac{\pi}{\gamma} \frac{\sin \frac{\pi \beta}{\gamma} \cosh \frac{\pi a}{\gamma}}{\cosh \frac{2a\pi}{\gamma} - \cos \frac{2\beta \pi}{\gamma}}$$

$$[\operatorname{Re} \gamma > |\operatorname{Re} \beta|] \qquad \text{ET I 34(39)}$$

4.133

1. $$\int_0^\infty \sin ax \sinh \beta x \exp\left(-\frac{x^2}{4\gamma}\right) dx = \sqrt{\pi\gamma} \exp \gamma \left(\beta^2 - a^2\right) \sin(2a\beta\gamma)$$

$$[\operatorname{Re} \gamma > 0] \qquad \text{ET I 92(37)}$$

2. $$\int_0^\infty \cos ax \cosh \beta x \exp\left(-\frac{x^2}{4\gamma}\right) dx = \sqrt{\pi\gamma} \exp \gamma \left(\beta^2 - a^2\right) \cos(2a\beta\gamma)$$

$$[\operatorname{Re} \gamma > 0] \qquad \text{ET I 35(41)}$$

4.134

1. $$\int_0^\infty e^{-\beta x^2} \left(\cosh x - \cos x\right) dx = \sqrt{\frac{\pi}{\beta}} \cosh \frac{1}{4\beta} \qquad [\operatorname{Re} \beta > 0] \qquad \text{ME 24}$$

2. $$\int_0^\infty e^{-\beta x^2} \left(\cosh x - \cos x\right) dx = \sqrt{\frac{\pi}{\beta}} \sinh \frac{1}{4\beta} \qquad [\operatorname{Re} \beta > 0] \qquad \text{ME 24}$$

4.135

1. $$\int_0^\infty \sin ax^2 \cosh 2\gamma x e^{-\beta x^2} \, dx = \frac{1}{2} \sqrt[4]{\frac{\pi^2}{a^2 + \beta^2}} \exp\left(-\frac{\beta\gamma^2}{a^2 + \beta^2}\right) \sin\left(\frac{a\gamma^2}{a^2 + \beta^2} + \frac{1}{2} \arctan \frac{a}{\beta}\right)$$

$$[\operatorname{Re} \beta > 0] \qquad \text{LI (268)(7)}$$

2. $$\int_0^\infty \cos ax^2 \cosh 2\gamma x e^{-\beta x^2} \, dx = \frac{1}{2} \sqrt[4]{\frac{\pi^2}{a^2 + \beta^2}} \exp\left(-\frac{\beta\gamma^2}{a^2 + \beta^2}\right) \cos\left(\frac{a\gamma^2}{a^2 + \beta^2} + \frac{1}{2} \arctan \frac{a}{\beta}\right)$$

$$[\operatorname{Re} \beta > 0] \qquad \text{LI (268)(8)}$$

4.136

1. $$\int_0^\infty \left(\sinh^2 x + \sin x^2\right) e^{-\beta x^4} \, dx = \frac{\sqrt{2}\pi}{4\sqrt{\beta}} I_{\frac{1}{4}}\left(\frac{1}{8\beta}\right) \cosh \frac{1}{8\beta}$$

$$[\operatorname{Re} \beta > 0] \qquad \text{ME 24}$$

2. $$\int_0^\infty \left(\sinh^2 x - \sin x^2\right) e^{-\beta x^4} \, dx = \frac{\sqrt{2}\pi}{4\sqrt{\beta}} I_{\frac{1}{4}}\left(\frac{1}{8\beta}\right) \sinh \frac{1}{8\beta}$$

$$[\operatorname{Re} \beta > 0] \qquad \text{ME 24}$$

3. $$\int_0^\infty \left(\cosh^2 x + \cos x^2\right) e^{-\beta x^4} \, dx = \frac{\sqrt{2}\pi}{4\sqrt{\beta}} I_{-\frac{1}{4}}\left(\frac{1}{8\beta}\right) \cosh \frac{1}{8\beta}$$

$$[\operatorname{Re} \beta > 0] \qquad \text{ME 24}$$

4. $\displaystyle\int_0^\infty \left(\cosh^2 x - \cos x^2\right) e^{-\beta x^4}\, dx = \frac{\sqrt{2}\pi}{4\sqrt{\beta}}\, I_{-\frac{1}{4}}\left(\frac{1}{8\beta}\right) \sinh\frac{1}{8\beta}$

$$[\operatorname{Re}\beta > 0] \qquad\qquad \text{ME 24}$$

4.137

1. $\displaystyle\int_0^\infty \sin 2x^2 \sinh 2x^2 e^{-\beta x^4}\, dx = \frac{\pi}{\sqrt[4]{128\beta^2}}\, J_{-\frac{1}{4}}\left(\frac{1}{\beta}\right)\cos\left(\frac{1}{\beta}+\frac{\pi}{4}\right)$

$$[\operatorname{Re}\beta > 0] \qquad\qquad \text{MI 32}$$

2. $\displaystyle\int_0^\infty \sin 2x^2 \cosh 2x^2 e^{-\beta x^4}\, dx = \frac{\pi}{\sqrt[4]{128\beta^2}}\, J_{\frac{1}{4}}\left(\frac{1}{\beta}\right)\cos\left(\frac{1}{\beta}-\frac{\pi}{4}\right)$

$$[\operatorname{Re}\beta > 0] \qquad\qquad \text{MI 32}$$

3. $\displaystyle\int_0^\infty \cos 2x^2 \sinh 2x^2 e^{-\beta x^4}\, dx = \frac{-\pi}{\sqrt[4]{128\beta^2}}\, J_{\frac{1}{4}}\left(\frac{1}{\beta}\right)\sin\left(\frac{1}{\beta}-\frac{\pi}{4}\right)$

$$[\operatorname{Re}\beta > 0] \qquad\qquad \text{MI 32}$$

4. $\displaystyle\int_0^\infty \cos 2x^2 \cosh 2x^2 e^{-\beta x^4}\, dx = \frac{\pi}{\sqrt[4]{128\beta^2}}\, J_{-\frac{1}{4}}\left(\frac{1}{\beta}\right)\sin\left(\frac{1}{\beta}+\frac{\pi}{4}\right)$

$$[\operatorname{Re}\beta > 0] \qquad\qquad \text{MI 32}$$

4.138

1. $\displaystyle\int_0^\infty \left(\sin^2 2x \cosh 2x^2 + \cos 2x^2 \sinh 2x^2\right) e^{-\beta x^4}\, dx = \frac{\pi}{\sqrt[4]{32\beta^2}}\, J_{\frac{1}{4}}\left(\frac{1}{\beta}\right)\cos\left(\frac{1}{\beta}\right)$

$$[\operatorname{Re}\beta > 0] \qquad\qquad \text{MI 32}$$

2. $\displaystyle\int_0^\infty \left(\sin^2 2x \cosh 2x^2 - \cos 2x^2 \sinh 2x^2\right) e^{-\beta x^4}\, dx = \frac{\pi}{\sqrt[4]{32\beta^2}}\, J_{\frac{1}{4}}\left(\frac{1}{\beta}\right)\sin\left(\frac{1}{\beta}\right)$

$$[\operatorname{Re}\beta > 0] \qquad\qquad \text{MI 32}$$

3. $\displaystyle\int_0^\infty \left(\cos^2 2x \cosh 2x^2 + \sin 2x^2 \sinh 2x^2\right) e^{-\beta x^4}\, dx = \frac{\pi}{\sqrt[4]{32\beta^2}}\, J_{-\frac{1}{4}}\left(\frac{1}{\beta}\right)\cos\left(\frac{1}{\beta}\right)$

$$[\operatorname{Re}\beta > 0] \qquad\qquad \text{MI 32}$$

4. $\displaystyle\int_0^\infty \left(\cos^2 2x \cosh 2x^2 - \sin 2x^2 \sinh 2x^2\right) e^{-\beta x^4}\, dx = \frac{\pi}{\sqrt[4]{32\beta^2}}\, J_{-\frac{1}{4}}\left(\frac{1}{\beta}\right)\sin\left(\frac{1}{\beta}\right)$

$$[\operatorname{Re}\beta > 0] \qquad\qquad \text{MI 32}$$

4.14 Combinations of trigonometric and hyperbolic functions, exponentials, and powers

4.141

1. $\displaystyle\int_0^\infty x e^{-\beta x^2} \cosh x \sin x\, dx = \frac{1}{4}\sqrt{\frac{\pi}{\beta^3}}\left(\cos\frac{1}{2\beta}+\sin\frac{1}{2\beta}\right)$

$$[\operatorname{Re}\beta > 0] \qquad\qquad \text{MI 32}$$

2. $\displaystyle\int_0^\infty x e^{-\beta x^2}\sinh x \cos x\, dx = \frac{1}{4}\sqrt{\frac{\pi}{\beta^3}}\left(\cos\frac{1}{2\beta}-\sin\frac{1}{2\beta}\right)$

$$[\operatorname{Re}\beta > 0]\qquad\qquad \text{MI 32}$$

3. $\displaystyle\int_0^\infty x^2 e^{-\beta x^2}\cosh x \cos x\, dx = \frac{1}{4}\sqrt{\frac{\pi}{\beta^3}}\left(\cos\frac{1}{2\beta}-\frac{1}{\beta}\sin\frac{1}{2\beta}\right)$

$$[\operatorname{Re}\beta > 0]\qquad\qquad \text{MI 32}$$

4. $\displaystyle\int_0^\infty x^2 e^{-\beta x^2}\sinh x \sin x\, dx = \frac{1}{4}\sqrt{\frac{\pi}{\beta^3}}\left(\sin\frac{1}{2\beta}+\frac{1}{\beta}\cos\frac{1}{2\beta}\right)$

$$[\operatorname{Re}\beta > 0]\qquad\qquad \text{MI 32}$$

4.142

1. $\displaystyle\int_0^\infty x e^{-\beta x^2}\left(\sinh x + \sin x\right) dx = \frac{1}{2}\sqrt{\frac{\pi}{\beta^3}}\cosh\frac{1}{4\beta}$ $[\operatorname{Re}\beta > 0]$ ME 24

2. $\displaystyle\int_0^\infty x e^{-\beta x^2}\left(\sinh x - \sin x\right) dx = \frac{1}{2}\sqrt{\frac{\pi}{\beta^3}}\sinh\frac{1}{4\beta}$ $[\operatorname{Re}\beta > 0]$ ME 24

3. $\displaystyle\int_0^\infty x^2 e^{-\beta x^2}\left(\cosh x + \cos x\right) dx = \frac{1}{2}\sqrt{\frac{\pi}{\beta^3}}\left(\cosh\frac{1}{4\beta}+\frac{1}{2\beta}\sinh\frac{1}{4\beta}\right)$

$$[\operatorname{Re}\beta > 0]\qquad\qquad \text{ME 24}$$

4. $\displaystyle\int_0^\infty x^2 e^{-\beta x^2}\left(\cosh x - \cos x\right) dx = \frac{1}{2}\sqrt{\frac{\pi}{\beta^3}}\left(\sinh\frac{1}{4\beta}+\frac{1}{2\beta}\cosh\frac{1}{4\beta}\right)$

$$[\operatorname{Re}\beta > 0]\qquad\qquad \text{ME 24}$$

4.143

1. $\displaystyle\int_0^\infty x e^{-\beta x^2}\left(\cosh x \sin x + \sinh x \cos x\right) dx = \frac{1}{2\beta}\sqrt{\frac{\pi}{\beta}}\cos\frac{1}{2\beta}$

$$[\operatorname{Re}\beta > 0]\qquad\qquad \text{MI 32}$$

2. $\displaystyle\int_0^\infty x e^{-\beta x^2}\left(\cosh x \sin x - \sinh x \cos x\right) dx = \frac{1}{2\beta}\sqrt{\frac{\pi}{\beta}}\sin\frac{1}{2\beta}$

$$[\operatorname{Re}\beta > 0]\qquad\qquad \text{MI 32}$$

4.144 $\displaystyle\int_0^\infty e^{-x^2}\sinh x^2 \cos ax\,\frac{dx}{x^2} = \sqrt{\frac{\pi}{2}}e^{-\frac{a^2}{8}}-\frac{\pi a}{4}\left[1-\Phi\left(\frac{a}{\sqrt{8}}\right)\right]$

$$[a > 0]\qquad\qquad \text{ET I 35(44)}$$

4.145

1. $\displaystyle\int_0^\infty x e^{-\beta x^2}\cosh\left(2ax\sin t\right)\sin\left(2ax\cos t\right) dx = \frac{a}{2}\sqrt{\frac{\pi}{\beta^3}}\exp\left(-\frac{a^2}{\beta}\cos 2t\right)\cos\left(t-\frac{a^2}{\beta}\sin 2t\right)$

$$[\operatorname{Re}\beta > 0]\qquad\qquad \text{BI (363)(5)}$$

2. $\displaystyle\int_0^\infty x e^{-\beta x^2}\sinh\left(2ax\sin t\right)\cos\left(2ax\cos t\right) dx = \frac{a}{2}\sqrt{\frac{\pi}{\beta^3}}\exp\left(-\frac{a^2}{\beta}\cos 2t\right)\sin\left(t-\frac{a^2}{\beta}\sin 2t\right)$

$$[\operatorname{Re}\beta > 0]\qquad\qquad \text{BI (363)(6)}$$

4.146^{10}

1.^{8} $\displaystyle\int_0^\infty e^{-\beta x^2}\sinh ax \sin bx\,dx = \frac{1}{2}\sqrt{\frac{\pi}{\beta}}\exp\left(\frac{a^2-b^2}{4\beta}\right)\sin\frac{ab}{2\beta}$

$$[\operatorname{Re}\beta > 0]$$

2.^{8} $\displaystyle\int_0^\infty e^{-\beta x^2}\cosh ax \cos bx\,dx = \frac{1}{2}\sqrt{\frac{\pi}{\beta}}\exp\left(\frac{a^2-b^2}{4\beta}\right)\cos\frac{ab}{2\beta}$

$$[\operatorname{Re}\beta > 0]$$

3. $\displaystyle\int_0^\infty xe^{-\beta x^2}\cosh ax \sin ax\,dx = \frac{a}{4\beta}\sqrt{\frac{\pi}{\beta}}\left(\cos\frac{a^2}{2\beta}+\sin\frac{a^2}{2\beta}\right)$

$$[\operatorname{Re}\beta > 0]$$

4. $\displaystyle\int_0^\infty xe^{-\beta x^2}\sinh ax \cos ax\,dx = \frac{a}{4\beta}\sqrt{\frac{\pi}{\beta}}\left(\cos\frac{a^2}{2\beta}-\sin\frac{a^2}{2\beta}\right)$

$$[\operatorname{Re}\beta > 0]$$

5.^{8} $\displaystyle\int_0^\infty x^2e^{-\beta x^2}\cosh ax \sin ax\,dx = \frac{1}{4}\sqrt{\frac{\pi}{\beta^3}}\left(\sin\frac{a^2}{2\beta}+\frac{a^2}{\beta}\cos\frac{a^2}{2\beta}\right)$

$$[\operatorname{Re}\beta > 0]$$

6.^{8} $\displaystyle\int_0^\infty x^2e^{-\beta x^2}\cosh ax \cos ax\,dx = \frac{1}{4}\sqrt{\frac{\pi}{\beta^3}}\left(\cos\frac{a^2}{2\beta}-\frac{a^2}{\beta}\sin\frac{a^2}{2\beta}\right)$

$$[\operatorname{Re}\beta > 0]$$

4.2–4.4 Logarithmic Functions

4.21 Logarithmic functions

4.211

1. $\displaystyle\int_e^\infty \frac{dx}{\ln\frac{1}{x}} = -\infty$ BI (33)(9)

2. $\displaystyle\int_0^u \frac{dx}{\ln x} = \operatorname{li} u$ FI III 653, FI II 606

4.212

1.^{7} $\displaystyle\int_0^1 \frac{dx}{a+\ln x} = e^{-a}\operatorname{Ei}(a)$ $[a>0]$ BI (31)(4)

2. $\displaystyle\int_0^1 \frac{dx}{a-\ln x} = -e^{a}\operatorname{Ei}(-a)$ $[a>0]$ BI (31)(5)

3.^{7} $\displaystyle\int_0^1 \frac{dx}{(a+\ln x)^2} = -\frac{1}{a}+e^{-a}\operatorname{Ei}(a)$ $[a\geq 0]$ BI (31)(14)

4. $\displaystyle\int_0^1 \frac{dx}{(a - \ln x)^2} = \frac{1}{a} + e^a \operatorname{Ei}(-a)$ $\qquad [a > 0]$ $\qquad$ BI (31)(16)

5.[8] $\displaystyle\int_0^1 \frac{\ln x \, dx}{(a + \ln x)^2} = 1 + (1 - a)e^{-a} \operatorname{Ei}(a)$ $\qquad [a \geq 0]$ $\qquad$ BI (31)(15)

6. $\displaystyle\int_0^1 \frac{\ln x \, dx}{(a - \ln x)^2} = 1 + (1 + a)e^a \operatorname{Ei}(-a)$ $\qquad [a > 0]$ $\qquad$ BI (31)(17)

7. $\displaystyle\int_1^e \frac{\ln x \, dx}{(1 + \ln x)^2} = \frac{e}{2} - 1$ $\qquad$ BI (33)(10)

8.[7] $\displaystyle\int_0^1 \frac{dx}{(a + \ln x)^n} = \frac{1}{(n-1)!} e^{-a} \operatorname{Ei}(a) - \frac{1}{(n-1)!} \sum_{k=1}^{n-1} (n - k - 1)! a^{k-n}$

$\qquad\qquad\qquad\qquad\qquad\qquad\qquad\qquad\qquad [a \geq 0]$ $\qquad$ BI (31))(22)

9. $\displaystyle\int_0^1 \frac{dx}{(a - \ln x)^n} = \frac{(-1)^n}{(n-1)!} e^a \operatorname{Ei}(-a) + \frac{(-1)^{n-1}}{(n-1)!} \sum_{k=1}^{n-1} (n - k - 1)!(-a)^{k-n}$

$\qquad\qquad\qquad\qquad\qquad\qquad\qquad\qquad\qquad [a > 0, \quad n \text{ odd}]$ $\qquad$ BI (31)(23)

In integrals of the form $\displaystyle\int \frac{(\ln x)^m}{[a^n + (\ln x)^n]^l} \, dx$ it is convenient to make the substitution $x = e^{-t}$.

Results **4.212** 3, **4.212** 5, and **4.212** 8 [for $n > 1$] and **4.213** 6, **4.213** 8 below are divergent but may be considered to be valid if defined as follows:

$$\int_0^a \frac{f(z) \, dz}{(z - z_0)^n} = \frac{1}{(n-1)!} \left(\frac{d}{dz_0}\right)^{n-1} \left[\operatorname{PV} \int_0^a \frac{f(z)}{z - z_0} \, dz\right]$$

where $a > z_0 > 0, n = 1, 2, 3, \ldots$ and PV indicates the Cauchy principal value.

4.213

1. $\displaystyle\int_0^1 \frac{dx}{a^2 + (\ln x)^2} = \frac{1}{a} \left[\operatorname{ci}(a) \sin a - \operatorname{si}(a) \cos a\right]$ $\qquad [a > 0]$ $\qquad$ BI (31)(6)

2.[7] $\displaystyle\int_0^1 \frac{dx}{a^2 - (\ln x)^2} = \frac{1}{2a} \left[e^{-a}\overline{\operatorname{Ei}}(a) - e^a \operatorname{Ei}(-a)\right]$ $\qquad [a > 0], \quad$ (cf. **4.212** 1 and 2)

$\qquad\qquad\qquad\qquad\qquad\qquad\qquad\qquad\qquad\qquad\qquad\qquad\qquad\qquad\quad$ BI (31)(8)

3. $\displaystyle\int_0^1 \frac{\ln x \, dx}{a^2 + (\ln x)^2} = \operatorname{ci}(a) \cos a + \operatorname{si}(a) \sin a$ $\qquad [a > 0]$ $\qquad$ BI (31)(7)

4.[7] $\displaystyle\int_0^1 \frac{\ln x \, dx}{a^2 - (\ln x)^2} = -\frac{1}{2} \left[e^{-a}\overline{\operatorname{Ei}}(a) + e^a \operatorname{Ei}(-a)\right]$ $\qquad [a > 0], \quad$ (cf. **4.212** 1 and 2)

$\qquad\qquad\qquad\qquad\qquad\qquad\qquad\qquad\qquad\qquad\qquad\qquad\qquad\qquad\quad$ BI (31)(9)

5. $\displaystyle\int_0^1 \frac{dx}{\left[a^2 + (\ln x)^2\right]^2} = \frac{1}{2a^3} \left[\operatorname{ci}(a) \sin a - \operatorname{si}(a) \cos a\right] - \frac{1}{2a^2} \left[\operatorname{ci}(a) \cos a + \operatorname{si}(a) \sin a\right]$

$\qquad\qquad\qquad\qquad\qquad\qquad\qquad\qquad\qquad [a > 0]$ $\qquad$ LI (31)(18)

6.[8] $\displaystyle\int_0^1 \frac{dx}{\left[a^2 - (\ln x)^2\right]^2}$ $\qquad\qquad\qquad\qquad$ is divergent

7. $\displaystyle\int_0^1 \frac{\ln x\, dx}{\left[a^2+(\ln x)^2\right]^2} = \frac{1}{2a}\left[\mathrm{ci}(a)\sin a - \mathrm{si}(a)\cos a\right] - \frac{1}{2a^2}$

$$[a>0] \qquad\qquad \text{BI (31)(19)}$$

8.[8] $\displaystyle\int_0^1 \frac{\ln x\, dx}{\left[a^2-(\ln x)^2\right]^2}$ \hfill is divergent

4.214

1. $\displaystyle\int_0^1 \frac{dx}{a^4-(\ln x)^4} = -\frac{1}{4a^3}\left[e^a\,\mathrm{Ei}(-a) - e^{-a}\overline{\mathrm{Ei}}(a) - 2\,\mathrm{ci}(a)\sin a + 2\,\mathrm{si}(a)\cos a\right]$

$$[a>0] \qquad\qquad \text{BI (31)(10)}$$

2. $\displaystyle\int_0^1 \frac{\ln x\, dx}{a^4-(\ln x)^4} = -\frac{1}{4a^2}\left[e^a\,\mathrm{Ei}(-a) + e^{-a}\overline{\mathrm{Ei}}(a) - 2\,\mathrm{ci}(a)\cos a - 2\,\mathrm{si}(a)\sin a\right]$

$$[a>0] \qquad\qquad \text{BI (31)(11)}$$

3. $\displaystyle\int_0^1 \frac{(\ln x)^2\, dx}{a^4-(\ln x)^4} = -\frac{1}{4a}\left[e^a\,\mathrm{Ei}(-a) - e^{-a}\overline{\mathrm{Ei}}(a) + 2\,\mathrm{ci}(a)\sin a - 2\,\mathrm{si}(a)\cos a\right]$

$$[a>0] \qquad\qquad \text{BI (31)(12)}$$

4.[7] $\displaystyle\int_0^1 \frac{(\ln x)^3\, dx}{a^4-(\ln x)^4} = -\frac{1}{4}\left[e^a\,\mathrm{Ei}(-a) + e^{-a}\overline{\mathrm{Ei}}(a) + 2\,\mathrm{ci}(a)\cos a + 2\,\mathrm{si}(a)\sin a\right]$

$$[a>0] \qquad\qquad \text{BI (31)(13)}$$

4.215

1. $\displaystyle\int_0^1 \left(\ln\frac{1}{x}\right)^{\mu-1} dx = \Gamma(\mu)$ \hfill $[\mathrm{Re}\,\mu>0]$ \hfill FI II 778

2. $\displaystyle\int_0^1 \frac{dx}{\left(\ln\dfrac{1}{x}\right)^{\mu}} = \frac{\pi}{\Gamma(\mu)}\,\mathrm{cosec}\,\mu\pi$ \hfill $[\mathrm{Re}\,\mu<1]$ \hfill BI (31)(1)

3. $\displaystyle\int_0^1 \sqrt{\ln\frac{1}{x}}\, dx = \frac{\sqrt{\pi}}{2}$ \hfill BI (32)(1)

4. $\displaystyle\int_0^1 \frac{dx}{\sqrt{\ln\dfrac{1}{x}}} = \sqrt{\pi}$ \hfill BI (32)(3)

4.216 $\displaystyle\int_0^{1/e} \frac{dx}{\sqrt{(\ln x)^2-1}} = K_0(1)$ \hfill GW (32)(2)

4.22   Logarithms of more complicated arguments

4.221

1. $$\int_0^1 \ln x \ln(1-x)\, dx = 2 - \frac{\pi^2}{6}$$ BI (30)(7)

2. $$\int_0^1 \ln x \ln(1+x)\, dx = 2 - \frac{\pi^2}{12} - 2\ln 2$$ BI (30)(8)

3. $$\int_0^1 \ln \frac{1-ax}{1-a} \frac{dx}{\ln x} = -\sum_{k=1}^{\infty} a^k \frac{\ln(1+k)}{k}$$ $[a<1]$ BI (31)(3)

4.222

1. $$\int_0^{\infty} \ln \frac{a^2+x^2}{b^2+x^2}\, dx = (a-b)\pi$$ $[a>0, \quad b>0]$ GW (322)(20)

2. $$\int_0^{\infty} \ln x \ln \frac{a^2+x^2}{b^2+x^2}\, dx = \pi(b-a) + \pi \ln \frac{a^a}{b^b}$$ $[a>0, \quad b>0]$ BI (33)(1)

3. $$\int_0^{\infty} \ln x \ln \left(1 + \frac{b^2}{x^2}\right) dx = \pi b\, (\ln b - 1)$$ $[b>0]$ BI (33)(2)

4. $$\int_0^{\infty} \ln \left(1 + a^2 x^2\right) \ln \left(1 + \frac{b^2}{x^2}\right) dx = 2\pi \left[\frac{1+ab}{a} \ln(1+ab) - b\right]$$

   $[a>0, \quad b>0]$ BI (33)(3)

5. $$\int_0^{\infty} \ln \left(a^2 + x^2\right) \ln \left(1 + \frac{b^2}{x^2}\right) dx = 2\pi \left[(a+b)\ln(a+b) - a\ln a - b\right]$$

   $[a>0, \quad b>0]$ BI (33)(4)

6. $$\int_0^{\infty} \ln \left(1 + \frac{a^2}{x^2}\right) \ln \left(1 + \frac{b^2}{x^2}\right) dx = 2\pi \left[(a+b)\ln(a+b) - a\ln a - b\ln b\right]$$

   $[a>0, \quad b>0]$ BI (33)(5)

7. $$\int_0^{\infty} \ln \left(a^2 + \frac{1}{x^2}\right) \ln \left(1 + \frac{b^2}{x^2}\right) dx = 2\pi \left[\frac{1+ab}{a} \ln(1+ab) - b\ln b\right]$$

   $[a>0, \quad b>0]$ BI (33)(7)

4.223

1. $$\int_0^{\infty} \ln \left(1 + e^{-x}\right) dx = \frac{\pi^2}{12}$$ BI (256)(10)

2. $$\int_0^{\infty} \ln \left(1 - e^{-x}\right) dx = -\frac{\pi^2}{6}$$ BI (256)(11)

3. $$\int_0^{\infty} \ln \left(1 + 2e^{-x} \cos t + e^{-2x}\right) dx = \frac{\pi^2}{6} - \frac{t^2}{2}$$ $[|t|<\pi]$ BI (256)(18)

4.224

1. $$\int_0^u \ln \sin x\, dx = L\left(\frac{\pi}{2} - u\right) - L\left(\frac{\pi}{2}\right)$$ LO III 186(15)

2. $\displaystyle\int_0^{\pi/4} \ln \sin x \, dx = -\frac{\pi}{4} \ln 2 - \frac{1}{2} \boldsymbol{G}$

BI (285)(1)

3. $\displaystyle\int_0^{\pi/2} \ln \sin x \, dx = \frac{1}{2} \int_0^{\pi} \ln \sin x \, dx = -\frac{\pi}{2} \ln 2$

FI II 629,643

4. $\displaystyle\int_0^{u} \ln \cos x \, dx = -L(u)$

LO III 184(10)

5. $\displaystyle\int_0^{\pi/4} \ln \cos x \, dx = -\frac{\pi}{4} \ln 2 + \frac{1}{2} \boldsymbol{G}$

BI (286)(1)

6. $\displaystyle\int_0^{\pi/2} \ln \cos x \, dx = -\frac{\pi}{2} \ln 2$

BI 306(1)

7. $\displaystyle\int_0^{\pi/2} (\ln \sin x)^2 \, dx = \frac{\pi}{2} \left[(\ln 2)^2 + \frac{\pi^2}{12} \right]$

BI (305)(19)

8. $\displaystyle\int_0^{\pi/2} (\ln \cos x)^2 \, dx = \frac{\pi}{2} \left[(\ln 2)^2 + \frac{\pi^2}{12} \right]$

BI (306)(14)

9.[8] $\displaystyle\int_0^{\pi} \ln (a + b \cos x) \, dx = \pi \ln \frac{a + \sqrt{a^2 - b^2}}{2}$ $\qquad [a \geq |b| > 0]$

GW (322)(15)

10. $\displaystyle\int_0^{\pi} \ln (1 \pm \sin x) \, dx = -\pi \ln 2 \pm 4 \boldsymbol{G}$

GW (322)(16a)

11.[7] $\displaystyle\int_0^{\pi/2} \ln (1 + a \sin x) \, dx = \frac{\pi}{2} \ln \frac{a}{2} + 2\boldsymbol{G} + 2 \sum_{k=1}^{\infty} \frac{b^k}{k} \sum_{n=1}^{k} \frac{(-1)^{n+1}}{2n - 1}$ $\quad [a > 0] \qquad b = \frac{1 - a}{1 + a}$

$\qquad\qquad\qquad\qquad = -\frac{\pi}{2} \ln 2 + 2 \boldsymbol{G}$ $\qquad\qquad\qquad [a = 1]$

12. $\displaystyle\int_0^{\pi} \ln (1 + a \cos x) \, dx = \pi \ln \frac{1 + \sqrt{1 - a^2}}{2}$ $\qquad [a^2 \leq 1]$

BI (330)(1)

13. $\displaystyle\int_0^{\pi/2} \ln (1 + 2a \sin x + a^2) \, dx = \sum_{k=0}^{\infty} \frac{2^{2k} (k!)^2}{(2k + 1) \cdot (2k + 1)!!} \left(\frac{2a}{1 + a^2} \right)^{2k+1}$

$\qquad\qquad\qquad\qquad\qquad\qquad\qquad\qquad [a^2 \leq 1]$

BI (308)(24)

14.[8] $\displaystyle\int_0^{n\pi} \ln (a^2 - 2ab \cos x + b^2) \, dx = 2\pi \ln [\max (|a|, |b|)]$

$\qquad\qquad\qquad\qquad\qquad\qquad\qquad\qquad [ab > 0]$

FI II 142, 163, 688

15.[8] $\displaystyle\int_0^{n\pi} \ln (1 - 2a \cos x + a^2) \, dx = 0$ $\qquad [a^2 \leq 1]$

$\qquad\qquad\qquad\qquad = n\pi \ln a^2$ $\qquad\qquad [a^2 \geq 1]$

4.225

1. $\displaystyle\int_0^{\pi/4} \ln (\cos x - \sin x) \, dx = -\frac{\pi}{8} \ln 2 - \frac{1}{2} \boldsymbol{G}$

GW (322)(9b)

2. $\int_0^{\pi/4} \ln\left(\cos x + \sin x\right) dx = \dfrac{1}{2}\int_0^{\pi/2} \ln\left(\cos x + \sin x\right) dx = -\dfrac{\pi}{8}\ln 2 + \dfrac{1}{2}\,\mathbf{G}$ GW (322)(9a)

3. $\int_0^{2\pi} \ln\left(1 + a\sin x + b\cos x\right) dx = 2\pi\ln\dfrac{1+\sqrt{1-a^2-b^2}}{2}$

$$\left[a^2 + b^2 < 1\right] \qquad\qquad \text{BI (332)(2)}$$

4. $\int_0^{2\pi} \ln\left(1 + a^2 + b^2 + 2a\sin x + 2b\cos x\right) dx = 0 \qquad \left[a^2 + b^2 \le 1\right]$

$$= 2\pi\ln\left(a^2 + b^2\right) \quad \left[a^2 + b^2 \ge 1\right]$$

$$\text{BI (322)(3)}$$

4.226

1. $\int_0^{\pi/2} \ln\left(a^2 - \sin^2 x\right)^2 dx = -2\pi\ln 2 \qquad\qquad\qquad \left[a^2 \le 1\right]$

$$= 2\pi\ln\dfrac{a + \sqrt{a^2 - 1}}{2} = 2\pi\left(\operatorname{arccosh} a - \ln 2\right) \quad \left[a > 1\right]$$

$$\text{FI II 644, 687}$$

2. $\int_0^{\pi/2} \ln\left(1 + a\sin^2 x\right) dx = \dfrac{1}{2}\int_0^{\pi} \ln\left(1 + a\sin^2 x\right) dx = \int_0^{\pi/2} \ln\left(1 + a\cos^2 x\right) dx$

$$= \dfrac{1}{2}\int_0^{\pi} \ln\left(1 + a\cos^2 x\right) dx = \pi\ln\dfrac{1 + \sqrt{1+a}}{2}$$

$$\left[a \ge -1\right] \qquad \text{BI (308)(15), GW(322)(12)}$$

3. $\int_0^{u} \ln\left(1 - \sin^2\alpha\sin^2 x\right) dx = (\pi - 2\theta)\ln\cot\dfrac{\alpha}{2} + 2u\ln\left(\dfrac{1}{2}\sin\alpha\right) - \dfrac{\pi}{2}\ln 2$

$$+ L(\theta + u) - L(\theta - u) + L\left(\dfrac{\pi}{2} - 2u\right)$$

$$\left[\cot\theta = \cos\alpha\tan u; \quad -\pi \le \alpha \le \pi, \quad -\dfrac{\pi}{2} \le u \le \dfrac{\pi}{2}\right] \quad \text{LO III 287}$$

4. $\int_0^{\pi/2} \ln\left[1 - \cos^2 x\left(\sin^2\alpha - \sin^2\beta\sin^2 x\right)\right] dx = \pi\ln\left[\dfrac{1}{2}\left(\cos^2\dfrac{\alpha}{2} + \sqrt{\cos^4\dfrac{\alpha}{2} + \sin^2\dfrac{\beta}{2}\cos^2\dfrac{\beta}{2}}\right)\right]$

$$\left[\alpha > \beta > 0\right] \qquad\qquad \text{LO III 283}$$

5. $\int_0^{u} \ln\left(1 - \dfrac{\sin^2 x}{\sin^2 a}\right) dx = -u\ln\sin^2\alpha - L\left(\dfrac{\pi}{2} - \alpha + u\right) + L\left(\dfrac{\pi}{2} - \alpha - u\right)$

$$\left[-\dfrac{\pi}{2} \le u \le \dfrac{\pi}{2}, \quad |\sin u| \le |\sin\alpha|\right]$$

$$\text{LO III 287}$$

6. $\int_0^{\pi/2} \ln\left(a^2\cos^2 x + b^2\sin^2 x\right) dx = \dfrac{1}{2}\int_0^{\pi} \ln\left(a^2\cos^2 x + b^2\sin^2 x\right) dx = \pi\ln\dfrac{a+b}{2}$

$$\left[a > 0, \quad b > 0\right] \qquad\qquad \text{GW (322)(13)}$$

7. $\int_0^{\pi/2} \ln \frac{1 + \sin t \cos^2 x}{1 - \sin t \cos^2 x}\, dx = \pi \ln \dfrac{1 + \sin \dfrac{t}{2}}{\cos \dfrac{t}{2}} = \pi \ln \cot \dfrac{\pi - t}{4}$

$$\left[|t| < \frac{\pi}{2}\right] \qquad \text{LO III 283}$$

4.227

1. $\int_0^u \ln \tan x\, dx = L(u) + L\left(\frac{\pi}{2} - u\right) - L\left(\frac{\pi}{2}\right)$ — LO III 186(16)

2. $\int_0^{\pi/4} \ln \tan x\, dx = -\int_{\frac{\pi}{4}}^{\frac{\pi}{2}} \ln \tan x\, dx = -\boldsymbol{G}$ — BI (286)(11)

3. $\int_0^{\pi/2} \ln\left(a \tan x\right)\, dx = \frac{\pi}{2} \ln a$ $\qquad\qquad [a > 0]$ — BI (307)(2)

4.[7] $\int_0^{\pi/4} (\ln \tan x)^n\, dx = n!(-1)^n \sum_{k=0}^{\infty} \frac{(-1)^k}{(2k+1)^{n+1}}$
$$= \frac{1}{2}\left(\frac{\pi}{2}\right)^{n+1} |E_n| \qquad\qquad [n \text{ even}]$$
— BI (286)(21)

5.[7] $\int_0^{\pi/2} (\ln \tan x)^{2n}\, dx = 2(2n)! \sum_{k=0}^{\infty} \frac{(-1)^k}{(2k+1)^{2n+1}} = \left(\frac{\pi}{2}\right)^{2n+1} |E_{2n}|$ — BI (307)(15)

6. $\int_0^{\pi/2} (\ln \tan x)^{2n+1}\, dx = 0$ — BI (307)(14)

7. $\int_0^{\pi/4} (\ln \tan x)^2\, dx = \frac{\pi^3}{16}$ — BI (286)(16)

8. $\int_0^{\pi/4} (\ln \tan x)^4\, dx = \frac{5}{64}\pi^5$ — BI (286)(19)

9. $\int_0^{\pi/4} \ln\left(1 + \tan x\right)\, dx = \frac{\pi}{8} \ln 2$ — BI (287)(1)

10. $\int_0^{\pi/2} \ln\left(1 + \tan x\right)\, dx = \frac{\pi}{4} \ln 2 + \boldsymbol{G}$ — BI (308)(9)

11. $\int_0^{\pi/4} \ln\left(1 - \tan x\right)\, dx = \frac{\pi}{8} \ln 2 - \boldsymbol{G}$ — BI (287)(2)

12. $\int_0^{\pi/2} \ln^{(1-\tan x)}\, dx = \frac{\pi}{2} e \ln 2 - 2\boldsymbol{G}$ — BI (308)(10)

13. $\int_0^{\pi/4} \ln\left(1 + \cot x\right)\, dx = \frac{\pi}{8} \ln 2 + \boldsymbol{G}$ — BI (287)(3)

14. $\int_0^{\pi/4} \ln\left(\cot x - 1\right)\, dx = \frac{\pi}{8} \ln 2$ — BI (287)(4)

15. $\displaystyle\int_0^{\pi/4} \ln\left(\tan x + \cot x\right)\, dx = \frac{1}{2}\int_0^{\pi/2} \ln\left(\tan x + \cot x\right)\, dx = \frac{\pi}{2}\ln 2$ BI (287)(5), BI (308)(11)

16. $\displaystyle\int_0^{\pi/4} \ln^2\left(\cot x - \tan x\right)\, dx = \frac{1}{2}\int_0^{\pi/2} \ln^2\left(\cot x - \tan x\right)\, dx = \frac{\pi}{2}\ln 2$ BI (287)(6), BI (308)(12)

17. $\displaystyle\int_0^{\pi/2} \ln\left(a^2 + b^2 \tan^2 x\right)\, dx = \frac{1}{2}\int_0^{\pi} \ln\left(a^2 + b^2 \tan^2 x\right)\, dx = \pi\ln(a + b)$

$$[a > 0, \quad b > 0]$$ GW (322)(17)

4.228

1. $\displaystyle\int_0^{\pi/2} \ln\left(\sin t \sin x + \sqrt{1 - \cos^2 t \sin^2 x}\right)\, dx = \frac{\pi}{2}\ln 2 - 2L\left(\frac{t}{2}\right) - 2L\left(\frac{\pi - t}{2}\right)$ LO III 290

2. $\displaystyle\int_0^{u} \ln\left(\cos x + \sqrt{\cos^2 x - \cos^2 t}\right)\, dx = -\left(\frac{\pi}{2} - t - \varphi\right)\ln\cos t + \frac{1}{2}L(u + \varphi) - \frac{1}{2}L(u - \varphi) - L(\varphi)$

$$\left[\cos\varphi = \frac{\sin u}{\sin t} \quad 0 \le u \le t \le \frac{\pi}{2}\right]$$ LO III 290

3. $\displaystyle\int_0^{t} \ln\left(\cos x + \sqrt{\cos^2 x - \cos^2 t}\right)\, dx = -\left(\frac{\pi}{2} - t\right)\ln\cos t$ LO III 285

4. $\displaystyle\int_0^{u} \ln\frac{\sin u + \sin t \cos x \sqrt{\sin^2 u - \sin^2 x}}{\sin u - \sin t \cos x \sqrt{\sin^2 u - \sin^2 x}}\, dx = \pi\ln\left[\tan\frac{t}{2}\sin u + \sqrt{\tan^2\frac{t}{2}\sin^2 u + 1}\right]$

$$[t > 0, \quad u > 0]$$ LO III 283

5. $\displaystyle\int_0^{\pi/4} \sqrt{\ln\cot x}\, dx = \frac{\sqrt{\pi}}{2}\sum_{k=0}^{\infty}\frac{(-1)^k}{\sqrt{(2k + 1)^3}}$ BI (297)(9)

6. $\displaystyle\int_0^{\pi/4} \frac{dx}{\sqrt{\ln\cot x}} = \sqrt{\pi}\sum_{k=0}^{\infty}\frac{(-1)^k}{\sqrt{2k + 1}}$ BI (304)(24)

7. $\displaystyle\int_0^{\pi/4} \ln\left(\sqrt{\tan x} + \sqrt{\cot x}\right)\, dx = \frac{1}{2}\int_0^{\pi/2} \ln\left(\sqrt{\tan x} + \sqrt{\cot x}\right)\, dx = \frac{\pi}{8}\ln 2 + \frac{1}{2}\boldsymbol{G}$

     BI (287)(7), BI (308)(22)

8. $\displaystyle\int_0^{\pi/4} \ln^2\left(\sqrt{\cot x} - \sqrt{\tan x}\right)\, dx = \frac{1}{2}\int_0^{\pi/2} \ln^2\left(\sqrt{\cot x} - \sqrt{\tan x}\right)\, dx = \frac{\pi}{4}\ln 2 - \boldsymbol{G}$

     BI (287)(8), BI (308)(23)

4.229

1. $\displaystyle\int_0^{1} \ln\left(\ln\frac{1}{x}\right)\, dx = -\boldsymbol{C}$ FI II 807

2. $\displaystyle\int_0^{1} \frac{dx}{\ln\left(\ln\frac{1}{x}\right)} = 0$ BI (31)(2)

3. $\displaystyle\int_0^1 \ln\left(\ln\frac{1}{x}\right) \frac{dx}{\sqrt{\ln\frac{1}{x}}} = -\left(C + 2\ln 2\right)\sqrt{\pi}$ BI (32)(4)

4. $\displaystyle\int_0^1 \ln\left(\ln\frac{1}{x}\right)\left(\ln\frac{1}{x}\right)^{u-1} dx = \psi(\mu)\,\Gamma(\mu)$ $[\mathrm{Re}\,\mu > 0]$ BI (30)(10)

   If the integrand contains $(\ln\ln\frac{1}{x})$, it is convenient to make the substitution $\ln\frac{1}{x} = u$, i.e., $x = e^{-u}$.

5.[7] $\displaystyle\int_0^1 \ln\left(a + \ln x\right) dx = \ln a - e^{-a}\,\mathrm{Ei}(a)$ $[a > 0]$ BI (30)(5)

6. $\displaystyle\int_0^1 \ln\left(a - \ln x\right) dx = \ln a - e^{a}\,\mathrm{Ei}(-a)$ $[a > 0]$ BI (30)(6)

7. $\displaystyle\int_{\pi/4}^{\pi/2} \ln\ln\tan x\, dx = \frac{\pi}{2}\ln\left\{\frac{\Gamma\left(\frac{3}{4}\right)}{\Gamma\left(\frac{1}{4}\right)}\sqrt{2\pi}\right\}$ BI (308)(28)

4.23 Combinations of logarithms and rational functions

4.231

1. $\displaystyle\int_0^1 \frac{\ln x}{1 + x}\, dx = -\frac{\pi^2}{12}$ FI II 483a

2. $\displaystyle\int_0^1 \frac{\ln x}{1 - x}\, dx = -\frac{\pi^2}{6}$ FI II 714

3. $\displaystyle\int_0^1 \frac{x\ln x}{1 - x}\, dx = 1 - \frac{\pi^2}{6}$ BI (108)(7)

4. $\displaystyle\int_0^1 \frac{1 + x}{1 - x}\ln x\, dx = 1 - \frac{\pi^2}{3}$ BI (108)(9)

5.[7] $\displaystyle\int_0^\infty \frac{\ln x\, dx}{(x + a)^2} = \frac{\ln a}{a}$ $[0 < a < 1]$ BI (139)(1)

6. $\displaystyle\int_0^1 \frac{\ln x}{(1 + x)^2}\, dx = -\ln 2$ BI (111)(1)

7.[7] $\displaystyle\int_0^\infty \ln x \frac{dx}{(a^2 + b^2 x^2)^n} = \frac{\Gamma\left(n - \frac{1}{2}\right)\sqrt{\pi}}{4(n-1)!\,a^{2n-1}b}\left[2\ln\frac{a}{2b} - C - \psi\left(n - \frac{1}{2}\right)\right]$

                                                                                $[a > 0, \quad b > 0]$ LI (139)(3)

8. $\displaystyle\int_0^\infty \frac{\ln x\, dx}{a^2 + b^2 x^2} = \frac{\pi}{2ab}\ln\frac{a}{b}$ $[ab > 0]$ BI (135)(6)

9. $\displaystyle\int_0^\infty \frac{\ln px}{q^2 + x^2}\, dx = \frac{\pi}{2q}\ln pq$ $[p > 0, \quad q > 0]$ BI (135)(4)

10. $\displaystyle\int_0^\infty \frac{\ln x\, dx}{a^2 - b^2 x^2} = -\frac{\pi^2}{4ab}$ $[ab > 0]$ LI (324)(7b)

11. $\int_0^a \dfrac{\ln x \, dx}{x^2 + a^2} = \dfrac{\pi \ln a}{4a} - \dfrac{G}{a}$ $\qquad [a > 0]$ $\qquad$ GW (324)(7b)

12. $\int_0^1 \dfrac{\ln x}{1 + x^2} \, dx = -\int_1^\infty \dfrac{\ln x}{1 + x^2} \, dx = -G$ $\qquad$ FI II 482, 614

13. $\int_0^1 \dfrac{\ln x \, dx}{1 - x^2} = -\dfrac{\pi^2}{8}$ $\qquad$ BI (108)(11)

14. $\int_0^1 \dfrac{x \ln x}{1 + x^2} \, dx = -\dfrac{\pi^2}{48}$ $\qquad$ GW (324)(7b)

15. $\int_0^1 \dfrac{x \ln x}{1 - x^2} \, dx = -\dfrac{\pi^2}{24}$

16. $\int_0^1 \ln x \dfrac{1 - x^{2n+2}}{(1 - x^2)^2} \, dx = -\dfrac{(n+1)\pi^2}{8} + \sum_{k=1}^n \dfrac{n - k + 1}{(2k - 1)^2}$ $\qquad$ BI (111)(5)

17. $\int_0^1 \ln x \dfrac{1 + (-1)^n x^{n+1}}{(1 + x)^2} \, dx = -\dfrac{(n+1)\pi^2}{12} - \sum_{k=1}^n (-1)^k \dfrac{n - k + 1}{k^2}$ $\qquad$ BI (111)(2)

18. $\int_0^1 \ln x \dfrac{1 - x^{n+1}}{(1 - x)^2} \, dx = -\dfrac{(n+1)\pi^2}{6} + \sum_{k=1}^n \dfrac{n - k + 1}{k^2}$ $\qquad$ BI (111)(3)

4.232

1. $\int_u^v \dfrac{\ln x \, dx}{(x + u)(x + v)} = \dfrac{\ln uv}{2(v - u)} \ln \dfrac{(u + v)^2}{4uv}$ $\qquad$ BI (145)(32)

2. $\int_0^\infty \dfrac{\ln x \, dx}{(x + \beta)(x + \gamma)} = \dfrac{(\ln \beta)^2 - (\ln \gamma)^2}{2(\beta - \gamma)}$ $\qquad [|\arg \beta| < \pi, \quad |\arg \gamma| < \pi]$

$\qquad$ ET II 218(24)

3. $\int_0^\infty \dfrac{\ln x}{x + a} \dfrac{dx}{x - 1} = \dfrac{\pi^2 + (\ln a)^2}{2(a + 1)}$ $\qquad [a > 0]$ $\qquad$ BI (140)(10)

4.233

1.[3] $\int_0^1 \dfrac{\ln x \, dx}{1 + x + x^2} = \dfrac{2}{9}\left[\dfrac{2\pi^2}{3} - \psi'\left(\dfrac{1}{3}\right)\right] = -0.7813024129\ldots$ $\qquad$ LI (113)(1)

2.[3] $\int_0^1 \dfrac{\ln x \, dx}{1 - x + x^2} = \dfrac{1}{3}\left[\dfrac{2\pi^2}{3} - \psi'\left(\dfrac{1}{3}\right)\right] = -1.17195361934\ldots$ $\qquad$ LI (113)(2)

3.[7] $\int_0^1 \dfrac{x \ln x \, dx}{1 + x + x^2} = -\dfrac{1}{9}\left[\dfrac{2\pi^2}{6} - \psi'\left(\dfrac{1}{3}\right)\right] = -0.15766014917\ldots$ $\qquad$ LI (113)(2)

4.[3] $\int_0^1 \dfrac{x \ln x \, dx}{1 - x + x^2} = \dfrac{1}{6}\left[\dfrac{5\pi^2}{6} - \psi'\left(\dfrac{1}{3}\right)\right] = -0.3118211319\ldots$ $\qquad$ LI (113)(4)

5. $\int_0^\infty \dfrac{\ln x \, dx}{x^2 + 2xa \cos t + a^2} = \dfrac{t \ln a}{a \sin t}$ $\qquad [a > 0, \quad 0 < t < \pi]$ $\qquad$ GW (324)(13c)

4.234

1. $$\int_1^\infty \frac{\ln x\, dx}{(1+x^2)^2} = \ln 2 \qquad\qquad\qquad\qquad \text{BI (144)(18)a}$$

2. $$\int_0^1 \frac{x\ln x\, dx}{(1+x^2)^2} = -\frac{1}{4}\ln 2 \qquad\qquad\qquad\qquad \text{BI (111)(4)}$$

3. $$\int_0^\infty \frac{1+x^2}{(1-x^2)^2}\ln x\, dx = 0 \qquad\qquad\qquad\qquad \text{BI (142)(2)a}$$

4. $$\int_0^\infty \frac{1-x^2}{(1+x^2)^2}\ln x\, dx = -\frac{\pi}{2} \qquad\qquad\qquad\qquad \text{BI (142)(1)a}$$

5. $$\int_0^1 \frac{x^2\ln x\, dx}{(1-x^2)(1+x^4)} = -\frac{\pi^2}{16\left(2+\sqrt{2}\right)} \qquad\qquad\qquad\qquad \text{BI (112)(21)}$$

6. $$\int_0^\infty \frac{\ln x\, dx}{(a^2+b^2x^2)(1+x^2)} = \frac{b\pi}{2a(b^2-a^2)}\ln\frac{a}{b} \qquad [ab>0] \qquad \text{BI (317)(16)a}$$

7. $$\int_0^\infty \frac{\ln x}{x^2+a^2}\cdot\frac{dx}{1+b^2x^2} = \frac{\pi}{2(1-a^2b^2)}\left(\frac{1}{a}\ln a + b\ln b\right)$$
$$\qquad\qquad\qquad\qquad\qquad [a>0,\quad b>0] \qquad \text{LI (140)(12)}$$

8. $$\int_0^\infty \frac{x^2\ln x\, dx}{(a^2+b^2x^2)(1+x^2)} = \frac{a\pi}{2b(b^2-a^2)}\ln\frac{b}{a} \qquad [ab>0] \qquad \text{LI (140)(12), BI (317)(15)a}$$

4.235

1. $$\int_0^\infty \ln x\frac{(1-x)x^{n-2}}{1-x^{2n}}\, dx = -\frac{\pi^2}{4n^2}\tan^2\frac{\pi}{2n} \qquad [n>1] \qquad \text{BI (135)(10)}$$

2. $$\int_0^\infty \ln x\frac{(1-x^2)\,x^{m-1}}{1-x^{2n}}\, dx = -\frac{\pi^2\sin\left(\dfrac{m+1}{n}\right)\pi\sin\dfrac{\pi}{n}}{4n^2\sin^2\dfrac{m\pi}{2n}\sin^2\left(\dfrac{m+2}{2n}\pi\right)} \qquad \text{LI (135)(12)}$$

3. $$\int_0^\infty \ln x\frac{(1-x^2)\,x^{n-2}}{1-x^{2n}}\, dx = -\frac{\pi^2}{4n^2}\tan^2\frac{\pi}{n} \qquad [n>2] \qquad \text{BI (135)(11)}$$

4. $$\int_0^1 \ln x\frac{x^{m-1}+x^{n-m-1}}{1-x^n}\, dx = -\frac{\pi^2}{n^2\sin^2\left(\dfrac{m}{n}\pi\right)} \qquad [n>m] \qquad \text{BI (108)(15)}$$

4.236

1. $$\int_0^1 \left\{\frac{1+(p-1)\ln x}{1-x} + \frac{x\ln x}{(1-x)^2}\right\} x^{p-1}\, dx = -1 + \psi'(p)$$
$$\qquad\qquad\qquad\qquad\qquad [p>0] \qquad \text{BI (111)(6)a, GW (326)(13)}$$

2. $$\int_0^1 \left[\frac{1}{1-x} + \frac{x\ln x}{(1-x)^2}\right] dx = \frac{\pi^2}{6} - 1 \qquad\qquad\qquad \text{GW (326)(13a)}$$

4.24 Combinations of logarithms and algebraic functions

4.241

1. $$\int_0^1 \frac{x^{2n}\ln x}{\sqrt{1-x^2}}\,dx = \frac{(2n-1)!!}{(2n)!!}\cdot\frac{\pi}{2}\left(\sum_{k=1}^{2n}\frac{(-1)^{k-1}}{k}-\ln 2\right)$$ BI (118)(5)a

2. $$\int_0^1 \frac{x^{2n+1}\ln x}{\sqrt{1-x^2}}\,dx = \frac{(2n)!!}{(2n+1)!!}\left(\ln 2+\sum_{k=1}^{2n+1}\frac{(-1)^k}{k}\right)$$ BI (118)(5)a

3. $$\int_0^1 x^{2n}\sqrt{1-x^2}\ln x\,dx = \frac{(2n-1)!!}{(2n+2)!!}\cdot\frac{\pi}{2}\left(\sum_{k=1}^{2n}\frac{(-1)^{k-1}}{k}-\frac{1}{2n+2}-\ln 2\right)$$

                                         LI (117)(4), GW (324)(53a)

4. $$\int_0^1 x^{2n+1}\sqrt{1-x^2}\ln x\,dx = \frac{(2n)!!}{(2n+3)!!}\left(\ln 2+\sum_{k=1}^{2n+1}\frac{(-1)^k}{k}-\frac{1}{2n+3}\right)$$

                                         BI (117)(5), GW (324)(53b)

5. $$\int_0^1 \ln x\cdot\sqrt{(1-x^2)^{2n-1}}\,dx = -\frac{(2n-1)!!}{4\cdot(2n)!!}\pi\left[\psi(n+1)+C+\ln 4\right]$$ BI (117)(3)

6. $$\int_0^{\sqrt{\frac{1}{2}}} \frac{\ln x\,dx}{\sqrt{1-x^2}} = -\frac{\pi}{4}\ln 2-\frac{1}{2}G$$ BI (145)(1)

7. $$\int_0^1 \frac{\ln x\,dx}{\sqrt{1-x^2}} = -\frac{\pi}{2}\ln 2$$ FI II 614, 643

8. $$\int_1^\infty \frac{\ln x\,dx}{x^2\sqrt{x^2-1}} = 1-\ln 2$$ BI (144)(17)

9. $$\int_0^1 \sqrt{1-x^2}\ln x\,dx = -\frac{\pi}{8}-\frac{\pi}{4}\ln 2$$ BI (117)(1), GW (324)(53c)

10. $$\int_0^1 x\sqrt{1-x^2}\ln x\,dx = \tfrac{1}{3}\ln 2-\tfrac{4}{9}$$ BI (117)(2)

11. $$\int_0^1 \frac{\ln x\,dx}{\sqrt{x(1-x^2)}} = -\frac{\sqrt{2\pi}}{8}\left[\Gamma\left(\frac{1}{4}\right)\right]^2$$ GW (324)(54a)

4.242

1. $$\int_0^\infty \frac{\ln x\,dx}{\sqrt{(a^2+x^2)(x^2+b^2)}} = \frac{1}{2a}K\left(\frac{\sqrt{a^2-b^2}}{a}\right)\ln ab$$

                                 $[a>b>0]$ BY (800.04)

2. $$\int_0^b \frac{\ln x\,dx}{\sqrt{(a^2+x^2)(b^2-x^2)}} = \frac{1}{2\sqrt{a^2+b^2}}\left[K\left(\frac{b}{\sqrt{a^2+b^2}}\right)\ln ab-\frac{\pi}{2}K\left(\frac{a}{\sqrt{a^2+b^2}}\right)\right]$$

                                 $[a>0,\quad b>0]$ BY (800.02)

3. $\int_b^\infty \dfrac{\ln x\, dx}{\sqrt{(x^2 + a^2)(x^2 - b^2)}} = \dfrac{1}{2\sqrt{a^2 + b^2}}\left[\mathbf{K}\left(\dfrac{a}{\sqrt{a^2 + b^2}}\right)\ln ab + \dfrac{\pi}{2}\,\mathbf{K}\left(\dfrac{b}{\sqrt{a^2 + b^2}}\right)\right]$

$[a > 0, \quad b > 0]$ BY (800.06)

4. $\int_0^b \dfrac{\ln x\, dx}{\sqrt{(a^2 - x^2)(b^2 - x^2)}} = \dfrac{1}{2a}\left[\mathbf{K}\left(\dfrac{b}{a}\right)\ln ab - \dfrac{\pi}{2}\,\mathbf{K}\left(\dfrac{\sqrt{a^2 - b^2}}{a}\right)\right]$

$[a > b > 0]$ BY (800.01)

5. $\int_b^a \dfrac{\ln x\, dx}{\sqrt{(a^2 - x^2)(x^2 - b^2)}} = \dfrac{1}{2a}\,\mathbf{K}\left(\dfrac{\sqrt{a^2 - b^2}}{a}\right)\ln ab$ BY (800.03)

6. $\int_a^\infty \dfrac{\ln x\, dx}{\sqrt{(x^2 - a^2)(x^2 - b^2)}} = \dfrac{1}{2a}\left[\mathbf{K}\left(\dfrac{b}{a}\right)\ln ab + \dfrac{\pi}{2}\,\mathbf{K}\left(\dfrac{\sqrt{a^2 - b^2}}{a}\right)\right]$

$[a > b > 0]$ BY (800.05)

4.243 $\int_0^1 \dfrac{x\ln x}{\sqrt{1 - x^4}}\, dx = -\dfrac{\pi}{8}\ln 2$ GW (324)(56b)

4.244

1. $\int_0^1 \dfrac{\ln x\, dx}{\sqrt[3]{x(1 - x^2)^2}} = -\dfrac{1}{8}\left[\Gamma\left(\dfrac{1}{3}\right)\right]^3$ GW (324)(54b)

2. $\int_0^1 \dfrac{\ln x\, dx}{\sqrt[3]{1 - x^3}} = -\dfrac{\pi}{3\sqrt{3}}\left(\ln 3 + \dfrac{\pi}{3\sqrt{3}}\right)$ BI (118)(7)

3. $\int_0^1 \dfrac{x\ln x\, dx}{\sqrt[3]{(1 - x^3)^2}} = \dfrac{\pi}{3\sqrt{3}}\left(\dfrac{\pi}{3\sqrt{3}} - \ln 3\right)$ BI (118)(8)

4.245

1. $\int_0^1 \dfrac{x^{4n+1}\ln x}{\sqrt{1 - x^4}}\, dx = \dfrac{(2n - 1)!!}{(2n)!!}\cdot\dfrac{\pi}{8}\left(\sum_{k=1}^{2n}\dfrac{(-1)^{k-1}}{k} - \ln 2\right)$ GW (324)(56a)

2. $\int_0^1 \dfrac{x^{4n+3}\ln x}{\sqrt{1 - x^4}}\, dx = \dfrac{(2n)!!}{4\cdot(2n + 1)!!}\left(\ln 2 + \sum_{k=1}^{2n+1}\dfrac{(-1)^k}{k}\right)$ GW (324)(56c)

4.246 $\int_0^1 (1 - x^2)^{n - \frac{1}{2}}\ln x\, dx = -\dfrac{(2n - 1)!!}{(2n)!!}\cdot\dfrac{\pi}{4}\left[2\ln 2 + \sum_{k=1}^n \dfrac{1}{k}\right]$ GW (324)(55)

4.247

1.[6] $\int_0^1 \dfrac{\ln x}{\sqrt[n]{1 - x^{2n}}}\, dx = -\dfrac{\pi\,\mathrm{B}\left(\dfrac{1}{2n},\dfrac{1}{2n}\right)}{8n^2\sin\dfrac{\pi}{2n}}$ $[n > 1]$ GW (324)(54c)a

2.[6] $\int_0^1 \dfrac{\ln x\, dx}{\sqrt[n]{x^{n-1}(1 - x^2)}} = -\dfrac{\pi\,\mathrm{B}\left(\dfrac{1}{2n},\dfrac{1}{2n}\right)}{8\sin\dfrac{\pi}{2n}}$ GW (324)(54)

4.25 Combinations of logarithms and powers

4.251

1. $\displaystyle\int_0^\infty \frac{x^{\mu-1}\ln x}{\beta+x}\,dx = \frac{\pi\beta^{\mu-1}}{\sin\mu\pi}\left(\ln\beta - \pi\cot\mu\pi\right)$ $\qquad[|\arg\beta| < \pi, \quad 0 < \operatorname{Re}\mu < 1]$

BI (135)(1)

2. $\displaystyle\int_0^\infty \frac{x^{\mu-1}\ln x}{a-x}\,dx = \pi a^{\mu-1}\left(\cot\mu\pi\ln a - \frac{\pi}{\sin^2\mu\pi}\right)$ $\qquad[a>0, \quad 0<\operatorname{Re}\mu<1]$ ET I 314(5)

3.[10] $\displaystyle\int_0^1 \frac{x^{\mu-1}\ln x}{x+1}\,dx = \beta'(\mu)$ $\qquad[\operatorname{Re}\mu>0]$ GW (324)(6), ET I 314(3)

4. $\displaystyle\int_0^1 \frac{x^{\mu-1}\ln x}{1-x}\,dx = -\psi'(\mu) = -\zeta(2,\mu)$ $\qquad[\operatorname{Re}\mu>0]$ BI (108)(8)

5. $\displaystyle\int_0^1 \ln x \frac{x^{2n}}{1+x}\,dx = -\frac{\pi^2}{12} + \sum_{k=1}^{2n}\frac{(-1)^{k-1}}{k^2}$

BI (108)(4)

6. $\displaystyle\int_0^1 \ln x \frac{x^{2n-1}}{1+x}\,dx = \frac{\pi^2}{12} + \sum_{k=1}^{2n-1}\frac{(-1)^k}{k^2}$

BI (108)(5)

4.252

1. $\displaystyle\int_0^\infty \frac{x^{\mu-1}\ln x}{(x+\beta)(x+\gamma)}\,dx = \frac{\pi}{(\gamma-\beta)\sin\mu\pi}\left[\beta^{\mu-1}\ln\beta - \gamma^{\mu-1}\ln\gamma - \pi\cot\mu\pi\left(\beta^{\mu-1}-\gamma^{\mu-1}\right)\right]$
$\qquad[|\arg\beta|<\pi, \quad |\arg\gamma|<\pi, \quad 0<\operatorname{Re}\mu<2, \quad \mu\neq1]$ BI (140)(9)a, ET 314(6)

2. $\displaystyle\int_0^\infty \frac{x^{\mu-1}\ln x\,dx}{(x+\beta)(x-1)} = \frac{\pi}{(\beta+1)\sin^2\mu\pi}\left[\pi - \beta^{\mu-1}\left(\sin\mu\pi\ln\beta - \pi\cos\mu\pi\right)\right]$
$\qquad[|\arg\beta|<\pi, \quad 0<\operatorname{Re}\mu<2, \quad \mu\neq1]$
BI (140)(11)

3. $\displaystyle\int_0^\infty \frac{x^{p-1}\ln x}{1-x^2}\,dx = -\frac{\pi^2}{4}\operatorname{cosec}^2\frac{p\pi}{2}$ $\qquad[0<p<2]$ (see also **4.254** 2)

4.[6] $\displaystyle\int_0^\infty \frac{x^{\mu-1}\ln x}{(x+a)^2}\,dx = \frac{(1-\mu)a^{\mu-2}\pi}{\sin\mu\pi}\left(\ln a - \pi\cot\mu\pi + \frac{1}{\mu-1}\right)$
$\qquad[|\arg a|<\pi \quad 0<\operatorname{Re}\mu<2 \quad (\mu\neq1)]$
GW (324)(13b)

4.253

1.[8] $\displaystyle\int_0^1 x^{\mu-1}(1-x^r)^{\nu-1}\ln x\,dx = \frac{1}{r^2}\operatorname{B}\left(\frac{\mu}{r},\nu\right)\left[\psi\left(\frac{\mu}{r}\right) - \psi\left(\frac{\mu}{r}+\nu\right)\right]$
$\qquad[\operatorname{Re}\mu>0, \quad \operatorname{Re}\nu>0, \quad r>0]$
GW (324)(3b)a, BI (107)(5)a

2. $\displaystyle\int_0^1 \frac{x^{p-1}}{(1-x)^{p+1}}\ln x\,dx = -\frac{\pi}{p}\operatorname{cosec}p\pi$ $\qquad[0<p<1]$ bi (319)(10)a

3. $\int_u^\infty \dfrac{(x-u)^{\mu-1}\ln x\, dx}{x^\lambda} = u^{\mu-\lambda}\, \mathrm{B}(\lambda-\mu,\mu)\,[\ln u + \psi(\lambda) - \psi(\lambda-\mu)]$

                                                       $[0 < \mathrm{Re}\,\mu < \mathrm{Re}\,\lambda]$ ET II 203(18)

4.[8] $\int_0^1 \ln x \left(\dfrac{x}{a^2+x^2}\right)^p \dfrac{dx}{x} = \dfrac{\ln a}{2a^p}\, \mathrm{B}\left(\dfrac{p}{2},\dfrac{p}{2}\right)$ $[a > 0, \quad p > 0]$ BI (140)(6)

5. $\int_1^\infty (x-1)^{p-1}\ln x\, dx = \dfrac{\pi}{p}\, \mathrm{cosec}\,\pi p$ $[-1 < p < 0]$ BI (289)(12)a

6.[7] $\int_0^\infty \ln x \dfrac{dx}{(a+x)^{\mu+1}} = \dfrac{1}{\mu a^\mu}\,(\ln a - \boldsymbol{C} - \psi(\mu))$ $[\mathrm{Re}\,\mu > 0, \quad a \neq 0, \quad |\arg a| < \pi]$

                                                                    NT 68(7)

7.[7] $\int_0^\infty \ln x \dfrac{dx}{(a+x)^{n+\frac{1}{2}}} = \dfrac{2}{(2n-1)a^{n-\frac{1}{2}}}\left(\ln a + 2\ln 2 - 2\sum_{k=1}^{n-1}\dfrac{1}{2k-1}\right)$

                                                  $[|\arg a| < \pi, \quad n = 1, 2, \ldots]$ BI (142)(5)

4.254

1. $\int_0^1 \dfrac{x^{p-1}\ln x}{1-x^q}\, dx = -\dfrac{1}{q^2}\,\psi'\left(\dfrac{p}{q}\right)$ $[p > 0, \quad q > 0]$ GW (324)(5)

2. $\int_0^\infty \dfrac{x^{p-1}\ln x}{1-x^q}\, dx = -\dfrac{\pi^2}{q^2 \sin^2 \frac{p\pi}{q}}$ $[0 < p < q]$ BI (135)(8)

3. $\int_0^\infty \dfrac{\ln x}{x^q-1}\dfrac{dx}{x^p} = \dfrac{\pi^2}{q^2 \sin^2 \frac{p-1}{q}\pi}$ $[p < 1, \quad p+q > 1]$ BI (140)(2)

4.[3] $\int_0^1 \dfrac{x^{p-1}\ln x}{1+x^q}\, dx = \dfrac{1}{q^2}\,\beta'\left(\dfrac{p}{q}\right)$ $[p > 0, \quad q > 0]$ GW (324)(7)

5. $\int_0^\infty \dfrac{x^{p-1}\ln x}{1+x^q}\, dx = -\dfrac{\pi^2}{q^2}\dfrac{\cos\frac{p\pi}{q}}{\sin^2\frac{p\pi}{q}}$ $[0 < p < q]$ BI (135)(7)

6. $\int_0^1 \dfrac{x^{q-1}\ln x}{1-x^{2q}}\, dx = -\dfrac{\pi^2}{8q^2}$ $[q > 0]$ BI (108)(12)

4.255

1. $\int_0^1 \ln x \dfrac{(1-x^2)\,x^{p-2}}{1+x^{2p}}\, dx = -\left(\dfrac{\pi}{2p}\right)^2 \dfrac{\sin\frac{\pi}{2p}}{\cos^2\frac{\pi}{2p}}$ $[p > 1]$ BI (108)(13)

2. $\int_0^1 \ln x \dfrac{(1+x^2)\,x^{p-2}}{1-x^{2p}}\, dx = -\left(\dfrac{\pi}{2p}\right)^2 \sec^2 \dfrac{\pi}{2p}$ $[p > 1]$ BI (108)(14)

3. $\int_0^\infty \ln x \frac{1 - x^p}{1 - x^2}\, dx = \frac{\pi^2}{4} \tan^2 \frac{p\pi}{2}$ $[p < 1]$ BI (140)(3)

4.256 $\int_0^1 \ln \frac{1}{x} \frac{x^{\mu-1}\, dx}{\sqrt[n]{(1 - x^n)^{n-m}}} = \frac{1}{n^2}\, B\left(\frac{\mu}{n}, \frac{m}{n}\right)\left[\psi\left(\frac{\mu + m}{n}\right) - \psi\left(\frac{\mu}{n}\right)\right]$

                                                  $[\operatorname{Re}\mu > 0]$ LI (118)(12)

4.257

1. $\int_0^\infty \frac{x^\nu \ln \frac{x}{\beta}\, dx}{(x + \beta)(x + \gamma)} = \frac{\pi\left[\gamma^\nu \ln \frac{\gamma}{\beta} + \pi\left(\beta^\nu - \gamma^\nu\right)\cot \nu\pi\right]}{\sin \nu\pi(\gamma - \beta)}$

                                  $[|\arg \beta| < \pi, \quad |\arg \gamma| < \pi, \quad |\operatorname{Re}\nu| < 1]$

                                                        ET II 219(30)

2. $\int_0^\infty \ln \frac{x}{q} \left(\frac{x^p}{q^{2p} + x^{2p}}\right)\frac{dx}{x} = 0$ $[q > 0]$ BI (140)(4)a

3. $\int_0^\infty \ln \frac{x}{q} \left(\frac{x^p}{q^{2p} + x^{2p}}\right)^r \frac{dx}{q^2 + x^2} = 0$ $[q > 0]$ BI (140)(4)a

4. $\int_0^\infty \ln x \ln \frac{x}{a} \frac{dx}{(x - 1)(x - a)} = \frac{\left[4\pi^2 + (\ln a)^2\right]\ln a}{6(a - 1)}$ $[a > 0]$ (for $a = 1$ see **4.261** 5)

                                                                    BI (141)(5)

5. $\int_0^\infty \ln x \ln \frac{x}{a} \frac{x^p\, dx}{(x - 1)(x - a)} = \frac{\pi^2\left[(a^p + 1)\ln a - 2\pi\left(a^p - 1\right)\cot p\pi\right]}{(a - 1)\sin^2 p\pi}$

                                               $[p^2 < 1, \quad a > 0]$ BI (141)(6)

4.26-4.27 Combinations involving powers of the logarithm and other powers

4.261

1.[7] $\int_0^1 (\ln x)^2 \frac{dx}{1 + 2x \cos t + x^2} = \frac{t\left(\pi^2 - t^2\right)}{6 \sin t}$ $[0 \le t \le \pi]$ BI (113)(7)

2. $\int_0^1 \frac{(\ln x)^2\, dx}{x^2 - x + 1} = \frac{1}{2}\int_0^\infty \frac{(\ln x)^2\, dx}{x^2 - x + 1} = \frac{10\pi^3}{81\sqrt{3}}$ GW (324)(16c)

3. $\int_0^1 \frac{(\ln x)^2\, dx}{x^2 + x + 1} = \frac{1}{2}\int_0^\infty \frac{(\ln x)^2\, dx}{x^2 + x + 1} = \frac{8\pi^3}{81\sqrt{3}}$ GW (324)(16b)

4. $\int_0^\infty (\ln x)^2 \frac{dx}{(x - 1)(x + a)} = \frac{\left[\pi^2 + (\ln a)^2\right]\ln a}{3(1 + a)}$ $[a > 0]$ BI (141)(1)

5. $\int_0^\infty (\ln x)^2 \frac{dx}{(1 - x)^2} = \frac{2}{3}\pi^2$ BI (139)(4)

6. $\int_0^1 (\ln x)^2 \frac{dx}{1 + x^2} = \frac{\pi^3}{16}$ BI (109)(3)

7. $\int_0^1 (\ln x)^2 \dfrac{1+x^2}{1+x^4}\,dx = \dfrac{1}{2}\int_0^\infty (\ln x)^2 \dfrac{1+x^2}{1+x^4}\,dx = \dfrac{3\sqrt{2}}{64}\pi^3$ BI (109)(5), BI (135)(13)

8.³ $\int_0^1 (\ln x)^2 \dfrac{1-x}{1-x^6}\,dx = \dfrac{1}{36}\left(\dfrac{4\sqrt{3}\pi^3}{27} - \psi''\left(\dfrac{1}{3}\right)\right)$

9. $\int_0^1 (\ln x)^2 \dfrac{dx}{\sqrt{1-x^2}} = \dfrac{\pi}{2}\left[(\ln 2)^2 + \dfrac{\pi^2}{12}\right]$ BI (118)(13)

10. $\int_0^\infty (\ln x)^2 \dfrac{x^{\mu-1}}{1+x}\,dx = \dfrac{\pi^3\left(2-\sin^2\mu\pi\right)}{\sin^3\mu\pi}$ $[0 < \operatorname{Re}\mu < 1]$ ET I 315(10)

11.⁷ $\int_0^1 (\ln x)^2 \dfrac{x^n\,dx}{1+x} = 2\sum_{k=n}^\infty \dfrac{(-1)^{n+k}}{(k+1)^3} = (-1)^n\left(\dfrac{3}{2}\zeta(3) + 2\sum_{k=1}^n \dfrac{(-1)^k}{k^3}\right)$

     $[n = 0, 1, \ldots]$ BI (109)(1)

12.⁷ $\int_0^1 (\ln x)^2 \dfrac{x^n\,dx}{1-x} = 2\sum_{k=n}^\infty \dfrac{1}{(k+1)^3} = 2\left(\zeta(3) - \sum_{k=1}^n \dfrac{1}{k^3}\right)$

     $[n = 0, 1, \ldots]$ BI (109)(2)

13.⁷ $\int_0^1 (\ln x)^2 \dfrac{x^{2n}\,dx}{1-x^2} = 2\sum_{k=n}^\infty \dfrac{1}{(2k+1)^3} = \dfrac{7}{2}\zeta(3) - 2\sum_{k=1}^n \dfrac{1}{(2k-1)^3}$

     $[n = 0, 1, \ldots]$ BI (109)(4)

14. $\int_0^\infty (\ln x)^2 \dfrac{x^{p-1}\,dx}{x^2+2x\cos t+1} = \dfrac{\pi\sin(1-p)t}{\sin t \sin p\pi}\left\{\pi^2 - t^2 + 2\pi\cot p\pi\left[\pi\cot p\pi + t\cot(1-p)t\right]\right\}$

     $[0 < t < \pi, \quad 0 < p < 2 \quad (p \neq 1)]$

     GW (324)(17)

15. $\int_0^1 (\ln x)^2 \dfrac{x^{2n}\,dx}{\sqrt{1-x^2}} = \dfrac{(2n-1)!!}{2\cdot(2n)!!}\pi\left\{\dfrac{\pi^2}{12} + \sum_{k=1}^{2n}\dfrac{(-1)^k}{k^2} + \left[\sum_{k=1}^{2n}\dfrac{(-1)^k}{k} + \ln 2\right]^2\right\}$ GW (324)(60a)

16. $\int_0^1 (\ln x)^2 \dfrac{x^{2n+1}\,dx}{\sqrt{1-x^2}} = \dfrac{(2n)!!}{(2n+1)!!}\left\{-\dfrac{\pi^2}{12} - \sum_{k=1}^{2n+1}\dfrac{(-1)^k}{k^2} + \left[\sum_{k=1}^{2n+1}\dfrac{(-1)^k}{k} + \ln 2\right]^2\right\}$

     GW (324)(60b)

17.⁷ $\int_0^1 (\ln x)^2 x^{\mu-1}(1-x)^{\nu-1}\,dx = \mathrm{B}(\mu,\nu)\left\{\left[\psi(\mu) - \psi(\nu+\mu)\right]^2 + \psi'(\mu) - \psi'(\mu+\nu)\right\}$

     $[\operatorname{Re}\mu > 0, \quad \operatorname{Re}\nu > 0]$ ET I 315(11)

18. $\int_0^1 (\ln x)^2 \dfrac{1-x^{n+1}}{(1-x)^2}\,dx = 2(n+1)\zeta(3) - 2\sum_{k=1}^n \dfrac{n-k+1}{k^3}$ LI (111)(8)

19. $\int_0^1 (\ln x)^2 \dfrac{1+(-1)^n x^{n+1}}{(1+x)^2}\,dx = \dfrac{3}{2}(n+1)\zeta(3) - 2\sum_{k=1}^n (-1)^{k-1}\dfrac{n-k+1}{k^3}$ LI (111)(7)

$20.^7$ $\displaystyle\int_0^1 (\ln x)^2 \frac{1 - x^{2n+2}}{(1 - x^2)^2}\, dx = \frac{7}{4}(n+1)\,\zeta(3) - 2\sum_{k=1}^{n} \frac{n-k+1}{(2k-1)^3}$

$$[n = 0, 1, \ldots] \qquad \text{LI (111)(9)}$$

21. $\displaystyle\int_0^1 (\ln x)^2 x^{p-1}(1 - x^r)^{q-1}\, dx = \frac{1}{r^3}\, \mathrm{B}\left(\frac{p}{r}, q\right)\left\{\psi'\left(\frac{p}{r}\right) - \psi'\left(\frac{p}{r}+q\right) + \left[\psi\left(\frac{p}{r}\right) - \psi\left(\frac{p}{r}+q\right)\right]^2\right\}$

$$[p > 0, \quad q > 0, \quad r > 0] \qquad \text{GW (324)(8a)}$$

4.262

1. $\displaystyle\int_0^1 (\ln x)^3 \frac{dx}{1 + x} = -\frac{7}{120}\pi^4$ BI (109)(9)

2. $\displaystyle\int_0^1 (\ln x)^3 \frac{dx}{1 - x} = -\frac{\pi^4}{15}$ BI (109)(11)

3. $\displaystyle\int_0^\infty (\ln x)^3 \frac{dx}{(x + a)(x - 1)} = \frac{\left[\pi^2 + (\ln a)^2\right]^2}{4(a + 1)}$ $[a > 0]$ BI (141)(2)

4. $\displaystyle\int_0^1 (\ln x)^3 \frac{x^n\, dx}{1 + x} = (-1)^{n+1}\left[\frac{7\pi^4}{120} - 6\sum_{k=0}^{n-1} \frac{(-1)^k}{(k+1)^4}\right]$ $[n = 1, 2, \ldots]$ BI (109)(10)

5. $\displaystyle\int_0^1 (\ln x)^3 \frac{x^n\, dx}{1 - x} = -\frac{\pi^4}{15} + 6\sum_{k=0}^{n-1} \frac{1}{(k+1)^4}$ $[n = 1, 2, \ldots]$ BI (109)(12)

6. $\displaystyle\int_0^1 (\ln x)^3 \frac{x^{2n}\, dx}{1 - x^2} = -\frac{\pi^4}{16} + 6\sum_{k=0}^{n-1} \frac{1}{(2k+1)^4}$ $[n = 1, 2, \ldots]$ BI (109)(14)

7. $\displaystyle\int_0^1 (\ln x)^3 \frac{1 - x^{n+1}}{(1 - x)^2}\, dx = -\frac{(n+1)\pi^4}{15} + 6\sum_{k=1}^{n} \frac{n-k+1}{k^4}$ BI (111)(11)

8. $\displaystyle\int_0^1 (\ln x)^3 \frac{1 + (-1)^n x^{n+1}}{(1 + x)^2}\, dx = -\frac{7(n+1)\pi^4}{120} + 6\sum_{k=1}^{n} (-1)^{k-1}\frac{n-k+1}{k^4}$ BI (111)(10)

9. $\displaystyle\int_0^1 (\ln x)^3 \frac{1 - x^{2n+2}}{(1 - x^2)^2}\, dx = -\frac{(n+1)\pi^4}{16} + 6\sum_{k=1}^{n} \frac{n-k+1}{(2k-1)^4}$ BI (111)(12)

4.263

$1.^8$ $\displaystyle\int_0^\infty (\ln x)^4 \frac{dx}{(x - 1)(x + a)} = \frac{\ln a \left[\pi^2 + (\ln a)^2\right]\left[7\pi^2 + 3(\ln a)^2\right]}{15(1 + a)}$

$$[a > 0] \qquad \text{BI (141)(3)}$$

2. $\displaystyle\int_0^1 (\ln x)^4 \frac{dx}{1 + x^2} = \frac{5\pi^5}{64}$ BI (109)(17)

3. $\displaystyle\int_0^1 (\ln x)^4 \frac{dx}{1 + 2x\cos t + x^2} = \frac{t\left(\pi^2 - t^2\right)\left(7\pi^2 - 3t^2\right)}{30\sin t}$ $[|t| < \pi]$ BI (113)(8)

4.264

1. $\displaystyle\int_0^1 (\ln x)^5 \frac{dx}{1+x} = -\frac{31\pi^6}{252}$ BI (109)(20)

2. $\displaystyle\int_0^1 (\ln x)^5 \frac{dx}{1-x} = -\frac{8\pi^6}{63}$ BI (109)(21)

3. $\displaystyle\int_0^\infty (\ln x)^5 \frac{dx}{(x-1)(x+a)} = \frac{\left[\pi^2 + (\ln a)^2\right]^2 \left[3\pi^2 + (\ln a)^2\right]}{6(1+a)}$

$$[a > 0]$$ BI (141)(4)

4.265 $\displaystyle\int_0^1 (\ln x)^6 \frac{dx}{1+x^2} = \frac{61\pi^7}{256}$ BI (109)(25)

4.266

1. $\displaystyle\int_0^1 (\ln x)^7 \frac{dx}{1+x} = -\frac{127\pi^8}{240}$ BI (109)(28)

2. $\displaystyle\int_0^1 (\ln x)^7 \frac{dx}{1-x} = -\frac{8\pi^8}{15}$ BI (109)(29)

4.267

1. $\displaystyle\int_0^1 \frac{1-x}{1+x} \frac{dx}{\ln x} = \ln\frac{2}{\pi}$ BI (127)(3)

2. $\displaystyle\int_0^1 \frac{(1-x)^2}{1+x^2} \frac{dx}{\ln x} = \ln\frac{\pi}{4}$ BI (128)(2)

3.[8] $\displaystyle\int_0^1 \frac{(1-x)^2}{1+2x\cos\dfrac{mx}{n}+x^2} \cdot \frac{dx}{\ln x}$

$$= \frac{1}{\sin\dfrac{m\pi}{n}} \sum_{k=1}^{n-1} (-1)^k \sin\frac{km\pi}{n} \ln \frac{\left\{\Gamma\left(\dfrac{n+k+1}{2n}\right)\right\}^2 \Gamma\left(\dfrac{k+2}{2n}\right)\Gamma\left(\dfrac{k}{2n}\right)}{\left\{\Gamma\left(\dfrac{k+1}{2n}\right)\right\}^2 \Gamma\left(\dfrac{n+k}{2n}\right)\Gamma\left(\dfrac{n+k+2}{2n}\right)}$$

$$[m+n \text{ is odd}]$$

$$= \frac{1}{\sin\dfrac{m\pi}{n}} \sum_{k=1}^{\left\lfloor\frac{1}{2}(n-1)\right\rfloor} (-1)^k \sin\frac{km\pi}{n} \ln \frac{\left\{\Gamma\left(\dfrac{n-k+1}{n}\right)\right\}^2 \Gamma\left(\dfrac{k+2}{n}\right)\Gamma\left(\dfrac{k}{n}\right)}{\left\{\Gamma\left(\dfrac{k+1}{n}\right)\right\}^2 \Gamma\left(\dfrac{n-k}{n}\right)\Gamma\left(\dfrac{n-k+2}{n}\right)}$$

$$[m+n \text{ is even}]$$

$$[m < n]$$ BI (130)(3)

4. $\displaystyle\int_0^1 \frac{1-x}{1+x} \cdot \frac{1}{1+x^2} \cdot \frac{dx}{\ln x} = -\frac{\ln 2}{2}$ BI (130)(16)

5. $\displaystyle\int_0^1 \frac{1-x}{1+x} \cdot \frac{x^2}{1+x^2} \cdot \frac{dx}{\ln x} = \ln\frac{2\sqrt{2}}{\pi}$ BI (130)(17)

6. $\displaystyle\int_0^1 (4-x)^p \frac{dx}{\ln x} = \sum_{k=1}^{\infty} (-1)^k \binom{p}{k} \ln(1+k)$ $[p \geq 1]$ BI (123)(2)

7. $\displaystyle\int_0^1 \left(\frac{1-x^p}{1-x} - p \right) \frac{dx}{\ln x} = \ln \Gamma(p+1)$ GW (326)(10)

8. $\displaystyle\int_0^1 \frac{x^{p-1} - x^{q-1}}{\ln x} dx = \ln \frac{p}{q}$ $[p > 0, \quad q > 0]$ FI II 647

9. $\displaystyle\int_0^1 \frac{x^{p-1} - x^{q-1}}{\ln x} \cdot \frac{dx}{1+x} = \ln \frac{\Gamma\left(\frac{q}{2}\right) \Gamma\left(\frac{p+1}{2}\right)}{\Gamma\left(\frac{p}{2}\right) \Gamma\left(\frac{q+1}{2}\right)}$ $[p > 0, \quad q > 0]$ FI II 186

10. $\displaystyle\int_0^1 \frac{x^{p-1} - x^{-p}}{(1+x)\ln x} dx = \frac{1}{2}\int_0^{\infty} \frac{x^{p-1} - x^{-p}}{(1+x)\ln x} dx = \ln\left(\tan \frac{p\pi}{2} \right)$

$[0 < p < 1]$ FI II 816

11. $\displaystyle\int_0^1 (x^p - x^q) x^{r-1} \frac{dx}{\ln x} = \ln \frac{p+r}{r+q}$ $[r > 0, \quad p > 0, \quad q > 0]$ LI (123)(5)

12. $\displaystyle\int_0^1 \frac{x^p - x^q}{(1-ax)^n} \frac{dx}{x \ln x} = \ln \frac{p}{q} + \sum_{k=1}^{\infty} \binom{n+k-1}{k} a^k \ln \frac{p+k}{q+k}$

$[p > 0, \quad q > 0, \quad a^2 < 1]$ BI (130)(15)

13. $\displaystyle\int_0^1 (x^p - 1)(x^q - 1) \frac{dx}{\ln x} = \ln \frac{p+q+1}{(p+1)(q+1)}$ $[p > -1, \quad q > -1, \quad p+q > -1]$

GW (324)(19b)

14. $\displaystyle\int_0^1 \frac{x^p - x^q}{1+x} \cdot \frac{1+x^{2n+1}}{x \ln x} dx = \ln \frac{\Gamma\left(\frac{p}{2}+n+1\right) \Gamma\left(\frac{q+1}{2}+n\right) \Gamma\left(\frac{p+1}{2}\right) \Gamma\left(\frac{q}{2}\right)}{\Gamma\left(\frac{q}{2}+n+1\right) \Gamma\left(\frac{p+1}{2}+n\right) \Gamma\left(\frac{q+1}{2}\right) \Gamma\left(\frac{p}{2}\right)}$

$[p > 0, \quad q > 0]$ BI (127)(7)

15. $\displaystyle\int_0^1 \frac{x^p - x^q}{1-x} \cdot \frac{1-x^r}{\ln x} dx = \ln \frac{\Gamma(q+1) \Gamma(p+r+1)}{\Gamma(p+1) \Gamma(q+r+1)}$

$[p > -1, \quad q > -1, \quad p+r > -1, \quad q+r > -1]$ GW (324)(23)

16. $\displaystyle\int_0^1 \frac{x^{p-1} - x^{q-1}}{(1+x^r)\ln x} dx = \ln \frac{\Gamma\left(\frac{p+r}{2r}\right) \Gamma\left(\frac{q}{2r}\right)}{\Gamma\left(\frac{q+r}{2r}\right) \Gamma\left(\frac{p}{2r}\right)}$ $[p > 0, \quad q > 0, \quad r > 0]$ GW (324)(21)

17. $\displaystyle\int_0^1 \frac{1 - x^{2p-2q}}{1+x^{2p}} \frac{x^{q-1} dx}{\ln x} = \ln \tan \frac{q\pi}{4p}$ $[0 < q < p]$ (see also **3.524** 27)

BI (128)(6)

18. $\displaystyle\int_0^\infty \frac{x^{p-1} - x^{q-1}}{(1+x^r)\ln x}\,dx = \ln\left(\tan\frac{p\pi}{2r}\cot\frac{q\pi}{2r}\right)$ $\qquad [0 < p < r, \quad 0 < q < r]$

$\qquad\qquad\qquad\qquad\qquad\qquad\qquad\qquad\qquad$ GW (324)(22), BI (143)(2)

19. $\displaystyle\int_0^\infty \frac{x^{p-1} - x^{q-1}}{(1-x^r)\ln x}\,dx = \ln\left(\frac{\sin\frac{p\pi}{r}}{\sin\frac{q\pi}{r}}\right)$ $\qquad [0 < p < r, \quad 0 < q < r]$ $\qquad$ BI (143)(4)

20. $\displaystyle\int_0^1 \frac{x^{p-1} - x^{q-1}}{1 - x^{2n}}\cdot\frac{1-x^2}{\ln x}\,dx = \ln\frac{\Gamma\left(\dfrac{p+2}{2n}\right)\Gamma\left(\dfrac{q}{2n}\right)}{\Gamma\left(\dfrac{q+2}{2n}\right)\Gamma\left(\dfrac{p}{2n}\right)}$ $\qquad [p > 0, \quad q > 0]$ $\qquad$ BI (128)(11)

21. $\displaystyle\int_0^1 \frac{x^{p-1} - x^{q-1}}{1 + x^{2(2n+1)}}\frac{1+x^2}{\ln x}\,dx = \ln\frac{\Gamma\left(\dfrac{p+4n+4}{4(2n+1)}\right)\Gamma\left(\dfrac{q+2}{4(2n+1)}\right)\Gamma\left(\dfrac{p+4n+2}{4(2n+1)}\right)\Gamma\left(\dfrac{q}{4(2n+1)}\right)}{\Gamma\left(\dfrac{q+4n+4}{4(2n+1)}\right)\Gamma\left(\dfrac{p+2}{4(2n+1)}\right)\Gamma\left(\dfrac{q+4n+2}{4(2n+1)}\right)\Gamma\left(\dfrac{p}{4(2n+1)}\right)}$

$\qquad\qquad\qquad\qquad\qquad\qquad\qquad\qquad\qquad [p > 0, \quad q > 0]$ $\qquad$ BI (128)(7)

22. $\displaystyle\int_0^\infty \frac{x^{p-1} - x^{q-1}}{1 + x^{2(2n+1)}}\cdot\frac{1+x^2}{\ln x}\,dx = \ln\left\{\tan\frac{p\pi}{4(2n+1)}\cdot\tan\frac{(p+2)\pi}{4(2n+1)}\cdot\cot\frac{q\pi}{4(2n+1)}\cdot\cot\frac{(q+2)\pi}{4(2n+1)}\right\}$

$\qquad\qquad\qquad\qquad\qquad\qquad [0 < p < 4n, \quad 0 < q < 4n]$ $\qquad$ BI (143)(5)

23. $\displaystyle\int_0^\infty \frac{x^{p-1} - x^{q-1}}{1 - x^{2n}}\frac{1-x^2}{\ln x}\,dx = \ln\frac{\sin\dfrac{p\pi}{2n}\cdot\sin\dfrac{(q+2)\pi}{2n}}{\sin\dfrac{q\pi}{2n}\cdot\sin\dfrac{(p+2)\pi}{2n}}$

$\qquad\qquad\qquad\qquad\qquad\qquad [0 < p < 2n, \quad 0 < q < 2n]$ $\qquad$ BI (143)(6)

24. $\displaystyle\int_0^1 (1 - x^p)(1 - x^q)\frac{x^{r-1}\,dx}{\ln x} = \ln\frac{(p+q+r)r}{(p+r)(q+r)}$ $\qquad [p > 0, \quad q > 0, \quad r > 0]$ $\qquad$ BI (123)(8)

25. $\displaystyle\int_0^1 (1 - x^p)(1 - x^q)\frac{x^{r-1}\,dx}{(1-x)\ln x} = \ln\frac{\Gamma(p+r)\,\Gamma(q+r)}{\Gamma(p+q+r)\,\Gamma(r)}$

$\qquad\qquad [r > 0, \quad r + p > 0, \quad r + q > 0, \quad r + p + q > 0]$ $\qquad$ FI II 815a

26. $\displaystyle\int_0^1 (1 - x^p)(1 - x^q)(1 - x^r)\frac{dx}{\ln x} = \ln\frac{(p+q+1)(q+r+1)(r+p+1)}{(p+q+r+1)(p+1)(q+1)(r+1)}$

$\qquad [p > -1, \quad q > -1, \quad r > -1, \quad p+q > -1, \quad p+r > -1, \quad q+r > -1, \quad p+q+r > -1]$

$\qquad\qquad\qquad\qquad\qquad\qquad\qquad\qquad\qquad GW (324)(19c)$

27. $\displaystyle\int_0^1 (1 - x^p)(1 - x^q)(1 - x^r)\frac{dx}{(1-x)\ln x} = \ln\frac{\Gamma(p+1)\,\Gamma(q+1)\,\Gamma(r+1)\,\Gamma(p+q+r+1)}{\Gamma(p+q+1)\,\Gamma(p+r+1)\,\Gamma(q+r+1)}$

$\qquad [p > -1, \quad q > -1, \quad r > -1, \quad p+q > -1, \quad p+r > -1, \quad q+r > -1, \quad p+q+r > -1]$

$\qquad\qquad\qquad\qquad\qquad\qquad\qquad\qquad\qquad FI II 815$

28. $\displaystyle\int_0^1 (1 - x^p)(1 - x^q)(1 - x^r)\frac{x^{s-1}\,dx}{\ln x} = \ln\frac{(p+q+s)(p+r+s)(q+r+s)s}{(p+s)(q+s)(r+s)(p+q+r+s)}$

$\qquad\qquad\qquad\qquad [p > 0, \quad q > 0, \quad r > 0, \quad s > 0]$

$\qquad\qquad\qquad\qquad\qquad\qquad\qquad\qquad\qquad BI (123)(10)$

29. $$\int_0^1 (1 - x^p)(1 - x^q) \frac{x^{s-1}\, dx}{(1 - x^r) \ln x} = \ln \frac{\Gamma\left(\dfrac{p+s}{r}\right) \Gamma\left(\dfrac{q+s}{r}\right)}{\Gamma\left(\dfrac{s}{r}\right) \Gamma\left(\dfrac{p+q+s}{r}\right)}$$

$$[p > 0, \quad q > 0, \quad r > 0, \quad s > 0]$$
GW (324)(23a)

30. $$\int_0^\infty (1 - x^p)(1 - x^q) \frac{x^{s-1}\, dx}{(1 - x^{p+q+2s}) \ln x} = 2 \int_0^1 (1 - x^p)(1 - x^q) \frac{x^{s-1}\, dx}{(1 - x^{p+q+2s}) \ln x}$$

$$= 2 \ln \left\{ \sin \frac{s\pi}{p+q+2s} \operatorname{cosec} \frac{(p+s)\pi}{p+q+2s} \right\}$$

$$[s > 0, \quad s + p > 0, \quad s + p + q > 0] \quad \text{GW (324)(23b)a}$$

31. $$\int_0^1 (1 - x^p)(1 - x^q)(1 - x^r) \frac{x^{s-1}\, dx}{(1 - x) \ln x} = \ln \frac{\Gamma(p+s)\,\Gamma(q+s)\,\Gamma(r+s)\,\Gamma(p+q+r+s)}{\Gamma(p+q+s)\,4\,\Gamma(p+r+s)\,\Gamma(q+r+s)\,\Gamma(s)}$$

$$[p > 0, \quad q > 0, \quad r > 0, \quad s > 0]^* \quad \text{BI (127)(11)}$$

32. $$\int_0^1 (1 - x^p)(1 - x^q)(1 - x^r) \frac{x^{s-1}\, dx}{(1 - x^t) \ln x}$$

$$= \ln \frac{\Gamma\left(\dfrac{p+s}{t}\right) \Gamma\left(\dfrac{q+s}{t}\right) \Gamma\left(\dfrac{r+s}{t}\right) \Gamma\left(\dfrac{p+q+r+s}{t}\right)}{\Gamma\left(\dfrac{p+q+s}{t}\right) \Gamma\left(\dfrac{q+r+s}{t}\right) \Gamma\left(\dfrac{p+r+s}{t}\right) \Gamma\left(\dfrac{s}{t}\right)}$$

$$[p > 0, \quad q > 0, \quad r > 0, \quad s > 0, \quad t > 0]^* \quad \text{GW (324)(23b)}$$

33. $$\int_0^1 \left\{ \frac{x^p - x^{p+q}}{1 - x} - q \right\} \frac{dx}{\ln x} = \ln \frac{\Gamma(p+q+1)}{\Gamma(p+1)} \qquad [p > -1, \quad p+q > -1] \qquad \text{BI (127)(19)}$$

34. $$\int_0^1 \left\{ \frac{x^\mu - x}{x - 1} - x(\mu - 1) \right\} \frac{dx}{x \ln x} = \ln \Gamma(\mu) \qquad [\operatorname{Re} \mu > 0] \qquad \text{WH, BI (127)(18)}$$

35. $$\int_0^1 \left\{ 1 - x - \frac{(1 - x^p)(1 - x^q)}{1 - x} \right\} \frac{dx}{x \ln x} = -\ln \{ \mathrm{B}(p, q) \}$$

$$[p > 0, \quad q > 0] \qquad \text{BI (130)(18)}$$

36. $$\int_0^1 \left\{ \frac{x^{p-1}}{1 - x} - \frac{x^{pq-1}}{1 - x^q} - \frac{1}{x(1 - x)} + \frac{1}{x(1 - x^q)} \right\} \frac{dx}{\ln x} = q \ln p$$

$$[p > 0] \qquad \text{BI (130)(20)}$$

37. $$\int_0^1 \left\{ \frac{x^{q-1}}{1 - x} - \frac{x^{pq-1}}{1 - x^p} - \frac{p-1}{1 - x^p} x^{p-1} - \frac{p-1}{2} x^{p-1} \right\} \frac{dx}{\ln x} = \frac{1 - p}{2} \ln(2\pi) + \left(pq - \frac{1}{2} \right) \ln p$$

$$[p > 0, \quad q > 0] \qquad \text{BI (130)(22)}$$

38. $$\int_0^1 \frac{(1 - x^p)(1 - x^q) - (1 - x)^2}{x(1 - x) \ln x}\, dx = \ln \mathrm{B}(p, q) \qquad [p > 0, \quad q > 0] \qquad \text{GW (324)(24)}$$

---

*In 4.267.31 the restrictions can be somewhat weakened by writing, for example, $s > 0$, $p + s > 0$, $q + s > 0$, $r + s > 0$, $p + q + s > 0$, $p + r + s > 0$, $q + r + s > 0$, $p + q + r + s > 0$, in **4.267** 31 and 32.

39.[6] $\displaystyle\int_0^1 (x^p - 1)^n \frac{dx}{\ln x} = \sum_{k=0}^n \binom{n}{n-k} (-1)^{n-k} \ln(pk+1)$ $[n > 0, \quad pn > -1]$

GW (324)(19d), BI (123)(12)a

40.[6] $\displaystyle\int_0^1 \frac{(1-x^p)^n}{1-x} \frac{dx}{\ln x} = \sum_{k=0}^n (-1)^{k-1} \ln \Gamma[(n-k)p+1]$ $[n > 1, \quad pn > -1]$ BI (127)(12)

41. $\displaystyle\int_0^1 (x^p - 1)^n x^{q-1} \frac{dx}{\ln x} = \sum_{k=0}^n (-1)^k \binom{n}{k} \ln[q + (n-k)p]$

$[n > 0, \quad q > 0, \quad pn > -q]$

BI (123)(12)

42.[6] $\displaystyle\int_0^1 (1-x^p)^n x^{q-1} \frac{dx}{(1-x)\ln x} = \sum_{k=0}^n (-1)^{k-1} \ln \Gamma[(n-k)p+q]$

$[n > 1, \quad q > 0, \quad pn > -q]$

BI (127)(13)

43.[10] $\displaystyle\int_0^1 (x^p - 1)^n (x^q - 1)^m \frac{x^{r-1} dx}{\ln x} = \sum_{j=0}^n (-1)^j \binom{n}{j} \sum_{k=0}^m (-1)^k \binom{m}{k} \ln[r + (m-k)q + (n-j)p]$

$[n \geq 0, \quad m \geq 0, \quad n+m > 0, \quad r > 0, \quad pn + qm + r > 0]$ BI (123)(16)

4.268

1. $\displaystyle\int_0^1 \frac{(x^p - x^q)(1-x^r)}{(\ln x)^2} dx = (p+1)\ln(p+1) - (q+1)\ln(q+1)$

$-(p+r+1)\ln(p+r+1) + (q+r+1)\ln(q+r+1)$

$[p > -1, \quad q > -1, \quad p+r > -1, \quad q+r > -1]$ GW (324)(26)

2. $\displaystyle\int_0^1 (x^p - x^q)^2 \frac{dx}{(\ln x)^2} = (2p+1)\ln(2p+1) + (2q+1)\ln(2q+1) - 2(p+q+1)\ln(p+q+1)$

$\left[p > -\frac{1}{2}, \quad q > -\frac{1}{2}\right]$ GW (324)(26a)

3. $\displaystyle\int_0^1 (1-x^p)(1-x^q)(1-x^r) \frac{dx}{(\ln x)^2}$

$= (p+q+1)\ln(p+q+1) + (q+r+1)\ln(q+r+1) + (p+r+1)\ln(p+r+1)$

$-(p+1)\ln(p+1) - (q+1)\ln(q+1) - (r+1)\ln(r+1) - (p+q+r)\ln(p+q+r)$

$[p > -1, \quad q > -1, \quad r > -1, \quad p+q > -1, \quad p+r > -1, \quad q+r > -1, \quad p+q+r > 0]$

BI (124)(4)

4. $\displaystyle\int_0^1 (1-x^p)^n x^{q-1} \frac{dx}{(\ln x)^2} = \frac{1}{2} \sum_{k=0}^n (-1)^k \binom{n}{k} (pk+q)^2 \ln(pk+q)$

$\left[q > 0, \quad p > -\frac{q}{n}\right]$ BI (124)(14)

5. $\displaystyle\int_0^1 (1-x^p)^n (1-x^q)^m x^{r-1} \frac{dx}{(\ln x)^2} = \left(\sum_{j=0}^{n} (-1)^j \binom{n}{j}\right) \left(\sum_{k=0}^{m} (-1)^k \binom{m}{k}\right)$

$$\times [(m-k)q + (n-j)p + r] \ln[(m-k)q + (n-j)p + r]$$

$$[r > 0, \quad mq + r > 0, \quad np + r > 0, \quad mq + np + r > 0] \quad \text{BI (124)(8)}$$

6. $\displaystyle\int_0^1 \left[(q-r)x^{p-1} + (r-p)x^{q-1} + (p-q)x^{r-1}\right] \frac{dx}{(\ln x)^2}$

$$= (q-r)p \ln p + (r-p)q \ln q + (p-q)r \ln r$$

$$[p > 0, \quad q > 0, \quad r > 0] \quad \text{BI (124)(9)}$$

7. $\displaystyle\int_0^1 \left[\frac{x^{p-1}}{(p-q)(p-r)(p-s)} + \frac{x^{q-1}}{(q-p)(q-r)(q-s)} + \frac{x^{r-1}}{(r-p)(r-q)(r-s)} + \right.$

$$\left. + \frac{x^{s-1}}{(s-p)(s-q)(s-r)}\right] \frac{dx}{(\ln x)^2} = \frac{1}{2}\left[\frac{p^2 \ln p}{(p-q)(p-r)(p-s)} + \frac{q^2 \ln q}{(q-p)(q-r)(q-s)} \right.$$

$$\left. + \frac{r^2 \ln r}{(r-p)(r-q)(r-s)} + \frac{s^2 \ln s}{(s-p)(s-q)(s-r)}\right]$$

$$[p > 0, \quad q > 0, \quad r > 0, \quad s > 0] \quad \text{BI (124)(16)}$$

4.269

1. $\displaystyle\int_0^1 \sqrt{\ln \frac{1}{x}} \cdot \frac{dx}{1+x^2} = \frac{\sqrt{\pi}}{2} \sum_{k=0}^{\infty} \frac{(-1)^k}{\sqrt{(2k+1)^3}}$ BI (115)(33)

2. $\displaystyle\int_0^1 \frac{dx}{\sqrt{\ln \frac{1}{x}} \cdot (1+x)^2} = \sqrt{\pi} \sum_{k=0}^{\infty} \frac{(-1)^k}{\sqrt{2k+1}}$ BI (133)(2)

3. $\displaystyle\int_0^1 \sqrt{\ln \frac{1}{x}} \cdot x^{p-1} \, dx = \frac{1}{2}\sqrt{\frac{\pi}{p^3}}$ $[p > 0]$ GW (324)(1c)

4. $\displaystyle\int_0^1 \frac{x^{p-1}}{\sqrt{\ln \frac{1}{x}}} \, dx = \sqrt{\frac{\pi}{p}}$ $[p > 0]$ BI (133)(1)

5. $\displaystyle\int_0^1 \frac{\sin t - x^n \sin[(n+1)t] + x^{n+1} \sin nt}{1 - 2x \cos t + x^2} \cdot \frac{dx}{\sqrt{\ln \frac{1}{x}}} = \sqrt{\pi} \sum_{k=1}^{n} \frac{\sin kt}{\sqrt{k}}$

$$[|t| < \pi] \quad \text{BI (133)(5)}$$

6. $\displaystyle\int_0^1 \frac{\cos t - x - x^{n-1} \cos nt + x^n \cos[(n-1)t]}{1 - 2x \cos t + x^2} \cdot \frac{dx}{\sqrt{\ln \frac{1}{x}}} = \sqrt{\pi} \sum_{k=1}^{n-1} \frac{\cos kt}{\sqrt{k}}$

$$[|t| < \pi] \quad \text{BI (133)(6)}$$

7. $\int_u^v \dfrac{dx}{x \cdot \sqrt{\ln \dfrac{x}{u} \ln \dfrac{v}{x}}} = \pi$ $[uv > 0]$ BI (145)(37)

4.271

1. $\int_0^1 (\ln x)^{2n} \dfrac{dx}{1+x} = \dfrac{2^{2n}-1}{2^{2n}} \cdot (2n)!\, \zeta(2n+1)$ BI (110)(1)

2. $\int_0^1 (\ln x)^{2n-1} \dfrac{dx}{1+x} = \dfrac{1-2^{2n-1}}{2n}\pi^{2n}|B_{2n}|$ $[n = 1, 2, \ldots]$ BI (110)(2)

3. $\int_0^1 (\ln x)^{2n-1} \dfrac{dx}{1-x} = -\dfrac{1}{n}2^{2n-2}\pi^{2n}|B_{2n}|$ $[n = 1, 2, \ldots]$ BI (110)(5), GW(324)(9a)

4. $\int_0^1 (\ln x)^{p-1} \dfrac{dx}{1-x} = e^{i(p-1)\pi}\,\Gamma(p)\,\zeta(p)$ $[p > 1]$ GW (324)(9b)

5. $\int_0^1 (\ln x)^n \dfrac{dx}{1+x^2} = (-1)^n n! \sum_{k=0}^{\infty} \dfrac{(-1)^k}{(2k+1)^{n+1}}$ BI (110)(11)

6. $\int_0^1 (\ln x)^{2n} \dfrac{dx}{1+x^2} = \dfrac{1}{2}\int_0^{\infty} (\ln x)^{2n} \dfrac{dx}{1+x^2} = \dfrac{\pi^{2n+1}}{2^{2n+2}}|E_{2n}|$ GW (324)(10)a

7. $\int_0^{\infty} \dfrac{(\ln x)^{2n+1}}{1+bx+x^2}\, dx = 0$ $[|b| < 2]$ BI (135)(2)

8. $\int_0^1 (\ln x)^{2n} \dfrac{dx}{1-x^2} = \dfrac{2^{2n+1}-1}{2^{2n+1}} \cdot (2n)!\, \zeta(2n+1)$ $[n = 1, 2, \ldots]$ BI (110)(12)

9. $\int_0^{\infty} (\ln x)^{2n} \dfrac{dx}{1-x^2} = 0$ BI (312)(7)a

10. $\int_0^1 (\ln x)^{2n-1} \dfrac{dx}{1-x^2} = \dfrac{1}{2}\int_0^{\infty} (\ln x)^{2n-1} \dfrac{dx}{1-x^2} = \dfrac{1-2^{2n}}{4n}\pi^{2n}|B_{2n}|$

$[n = 1, 2, \ldots]$ BI (290)(17)a, BI(312)(6)a

11. $\int_0^1 (\ln x)^{2n-1} \dfrac{x\, dx}{1-x^2} = -\dfrac{1}{4n}\pi^{2n}|B_{2n}|$ $[n = 1, 2, \ldots]$ BI (290)(19)a

12. $\int_0^1 (\ln x)^{2n} \dfrac{1+x^2}{(1-x^2)^2}\, dx = \dfrac{2^{2n}-1}{2}\pi^{2n}|B_{2n}|$ $[n = 1, 2, \ldots]$ BI (296)(17)a

13. $\int_0^1 (\ln x)^{2n+1} \dfrac{(\cos 2a\pi - x)\, dx}{1 - 2x\cos 2a\pi + x^2} = -(2n+1)! \sum_{k=1}^{\infty} \dfrac{\cos 2ak\pi}{k^{2n+2}}$

[a is not an integer] LI (113)(10)

14.[6] $\int_0^{\infty} (\ln x)^n \dfrac{x^{\nu-1}\, dx}{a^2 + 2ax\cos t + x^2} = -\pi \csc t \dfrac{d^n}{d\nu^n}\left[a^{\nu-2}\dfrac{\sin(\nu-1)t}{\sin \nu\pi}\right]$

$[a > 0, \quad 0 < \operatorname{Re}\nu < 2, \quad 0 < |t| < \pi]$
ET I 315(12)

15. $\int_0^1 (\ln x)^n \dfrac{x^{p-1}}{1-x^q}\, dx = -\dfrac{1}{q^{n+1}}\psi^{(n)}\left(\dfrac{p}{q}\right)$ $[p > 0, \quad q > 0]$ GW (324)(9)

16.³ $\displaystyle\int_0^1 (\ln x)^n \frac{x^{p-1}}{1+x^q}\,dx = \frac{1}{q^{n+1}} \beta^{(n)}\left(\frac{p}{q}\right)$ $[p > 0, \quad q > 0]$ GW (324)(10)

4.272

1. $\displaystyle\int_0^1 \frac{\left[\ln\left(\frac{1}{x}\right)\right]^{q-1} dx}{1 + 2x\cos t + x^2} = \operatorname{cosec} t\, \Gamma(q) \sum_{k=1}^{\infty} (-1)^{k-1}\frac{\sin kt}{k^q}$ $[|t| < \pi, q < 1]$ LI (130)(1)

2. $\displaystyle\int_0^1 \left(\ln\frac{1}{x}\right)^{q-1} \frac{(1+x)\,dx}{1 + 2x\cos t + x^2} = \sec\frac{t}{2}\cdot\Gamma(q)\sum_{k=1}^{\infty}(-1)^{k-1}\frac{\cos\left[\left(k-\frac{1}{2}\right)t\right]}{k^q}$

$\left[|t| < \pi, \quad q < \frac{1}{2}\right]$ LI (130)(5)

3.⁹ $\displaystyle\int_0^1 \left[\ln\left(\frac{1}{x}\right)\right]^{\mu} \frac{x^{\nu-1}\,dx}{1 - 2ax\cos t + x^2 a^2} = \frac{\Gamma(\mu+1)}{a\sin t}\sum_{k=1}^{\infty}\frac{a^k \sin kt}{(\nu+k-1)^{\mu+1}}$

$[a > 0, \quad \operatorname{Re}\mu > 0, \quad \operatorname{Re}\nu > 0, \quad -\pi < t < \pi]$ BI (140)(14)a

4. $\displaystyle\int_0^1 \left(\ln\frac{1}{x}\right)^{r-1} \frac{\cos\lambda - px}{1 + p^2 x^2 - 2px\cos\lambda}x^{q-1}\,dx = \Gamma(r)\sum_{k=1}^{\infty}\frac{p^{k-1}\cos k\lambda}{(q+k-1)^r}$

$[r > 0, \quad q > 0]$ BI (113)(11)

5. $\displaystyle\int_1^{\infty} (\ln x)^p \frac{dx}{x^2} = \Gamma(1+p)$ $[p > -1]$ BI (149)(1)

6. $\displaystyle\int_0^1 \left(\ln\frac{1}{x}\right)^{\mu-1} x^{\nu-1}\,dx = \frac{1}{\nu^{\mu}}\Gamma(\mu)$ $[\operatorname{Re}\mu > 0, \quad \operatorname{Re}\nu > 0]$ BI (107)(3)

7. $\displaystyle\int_0^1 \left(\ln\frac{1}{x}\right)^{n-\frac{1}{2}} x^{\nu-1}\,dx = \frac{(2n-1)!!}{(2\nu)^n}\sqrt{\frac{\pi}{\nu}}$ $[\operatorname{Re}\nu > 0]$ BI (107)(2)

8. $\displaystyle\int_0^1 \left(\ln\frac{1}{x}\right)^{n-1}\frac{x^{\nu-1}}{1+x}\,dx = (n-1)!\sum_{k=0}^{\infty}\frac{(-1)^k}{(\nu+k)^n}$ $[\operatorname{Re}\nu > 0]$ BI (110)(4)

9. $\displaystyle\int_0^1 \left(\ln\frac{1}{x}\right)^{n-1}\frac{x^{\nu-1}}{1-x}\,dx = (n-1)!\,\zeta(n,\nu)$ $[\operatorname{Re}\nu > 0]$ BI (110)(7)

10. $\displaystyle\int_0^1 \left(\ln\frac{1}{x}\right)^{\mu-1}(x-1)^n\left(a+\frac{nx}{x-1}\right)x^{a-1}\,dx = \Gamma(\mu)\sum_{k=0}^{n}\frac{(-1)^k n(n-1)\dots(n-k+1)}{(a+n-k)^{\mu-1}k!}$

$[\operatorname{Re}\mu > 0]$ LI (110)(10)

11. $\displaystyle\int_0^1 \left(\ln\frac{1}{x}\right)^{n-1}\frac{1-x^m}{1-x}\,dx = (n-1)!\sum_{k=1}^{m}\frac{1}{k^n}$ LI (110)(9)

12. $\displaystyle\int_0^1 \left(\ln\frac{1}{x}\right)^{\mu-1}\frac{x^{\nu-1}\,dx}{1-x^2} = \Gamma(\mu)\sum_{k=0}^{\infty}\frac{1}{(\nu+2k)^{\mu}} = \frac{1}{2^{\mu}}\Gamma(\mu)\,\zeta\left(\mu, \frac{\nu}{2}\right)$

$[\operatorname{Re}\mu > 0, \quad \operatorname{Re}\nu > 0]$ BI (110)(13)

13. $\int_0^1 \dfrac{x^q - x^{-q}}{1 - x^2}\left(\ln\dfrac{1}{x}\right)^p dx = \Gamma(p+1)\sum_{k=1}^{\infty}\left\{\dfrac{1}{(2k+q-1)^{p+1}} - \dfrac{1}{(2k-q-1)^{p+1}}\right\}$

$\qquad\qquad\qquad\qquad\qquad\qquad\qquad\qquad\qquad [p > -1, \quad q^2 < 1]$ LI (326)(12)a

14. $\int_0^1 \left(\ln\dfrac{1}{x}\right)^{r-1}\dfrac{x^{p-1}\,dx}{(1+x^q)^s} = \Gamma(r)\sum_{k=0}^{\infty}\binom{-s}{k}\dfrac{1}{(p+kq)^r}$ $\quad [p>0, \quad q>0, \quad r>0, \quad 0<s<r+2]$

$\qquad\qquad\qquad\qquad\qquad\qquad\qquad\qquad\qquad\qquad\qquad\qquad$ GW (324)(11)

15. $\int_0^1 \left(\ln\dfrac{1}{x}\right)^n (1+x^q)^m x^{p-1}\,dx = n!\sum_{k=0}^{m}\binom{m}{k}\dfrac{1}{(p+kq)^{n+1}}$

$\qquad\qquad\qquad\qquad\qquad\qquad\qquad\qquad\qquad [p>0, \quad q>0]$ BI (107)(6)

16. $\int_0^1 \left(\ln\dfrac{1}{x}\right)^n (1-x^q)^m x^{p-1}\,dx = n!\sum_{k=0}^{m}\binom{m}{k}\dfrac{(-1)^k}{(p+kq)^{n+1}}$

$\qquad\qquad\qquad\qquad\qquad\qquad\qquad\qquad\qquad [p>0, \quad q>0]$ BI (107)(7)

17. $\int_0^1 \left(\ln\dfrac{1}{x}\right)^{p-1}\dfrac{x^{q-1}\,dx}{1-ax^q} = \dfrac{1}{aq^p}\Gamma(p)\sum_{k=1}^{\infty}\dfrac{a^k}{k^p}$ $\quad [p>0, \quad q>0, \quad a<1]$ LI (110)(8)

18. $\int_0^1 \left(\ln\dfrac{1}{x}\right)^{2-\frac{1}{n}}\left(x^{p-1}-x^{q-1}\right)dx = \dfrac{n}{n-1}\Gamma\left(\dfrac{1}{n}\right)\left(q^{1-\frac{1}{n}}-p^{1-\frac{1}{n}}\right)$

$\qquad\qquad\qquad\qquad\qquad\qquad\qquad\qquad\qquad [q>p>0]$ BI (133)(4)

19. $\int_0^1 \left(\ln\dfrac{1}{x}\right)^{2n-1}\dfrac{x^p - x^{-p}}{1-x^q}x^{q-1}\,dx = \dfrac{1}{p^{2n}}\sum_{k=n}^{\infty}\left(\dfrac{2p\pi}{q}\right)^k\dfrac{|B_{2k}|}{2k\cdot(2k-2n)!}$

$\qquad\qquad\qquad\qquad\qquad\qquad\qquad\qquad\qquad \left[p < \dfrac{q}{2}\right]$ LI (110)(16)

4.273 $\int_u^v \left(\ln\dfrac{x}{u}\right)^{p-1}\left(\ln\dfrac{v}{x}\right)^{q-1}\dfrac{dx}{x} = B(p,q)\left(\ln\dfrac{v}{u}\right)^{p+q-1}$ $\quad [p>0, \quad q>0, \quad uv>0]$ BI (145)(36)

4.274 $\int_0^{\frac{1}{e}} \dfrac{\sqrt[q]{x}\,dx}{x\sqrt{-(1+\ln x)}} = \dfrac{\sqrt{q\pi}}{\sqrt[q]{e}}$ $\qquad\qquad\qquad [q>0]$ BI (145)(4)

4.275

1. $\int_0^1 \left[\left(\ln\dfrac{1}{x}\right)^{q-1} - x^{p-1}(1-x)^{q-1}\right]dx = \dfrac{\Gamma(q)}{\Gamma(p+q)}\left[\Gamma(p+q) - \Gamma(p)\right]$

$\qquad\qquad\qquad\qquad\qquad\qquad\qquad\qquad\qquad [p>0, \quad q>0]$ BI (107)(8)

2. $\int_0^1 \left[x - \left(\dfrac{1}{1-\ln x}\right)^q\right]\dfrac{dx}{x\ln x} = -\psi(q)$ $\qquad\qquad [q>0]$ BI (126)(5)

4.28 Combinations of rational functions of $\ln x$ and powers

4.281

1. $\displaystyle\int_0^1 \left[\frac{1}{\ln x} + \frac{1}{1-x}\right] dx = C$ BI (127)(15)

2. $\displaystyle\int_1^\infty \frac{dx}{x^2(\ln p - \ln x)} = \frac{1}{p}\,\mathrm{li}(p)$ LA 281(30)

3. $\displaystyle\int_0^1 \frac{x^{p-1}\,dx}{q \pm \ln x} = \pm e^{\mp pq}\,\mathrm{Ei}\,(\pm pq)$ $[p > 0, \quad q > 0]$ LI (144)(11,12)

4. $\displaystyle\int_0^1 \left[\frac{1}{\ln x} + \frac{x^{\mu-1}}{1-x}\right] dx = -\psi(\mu)$ $[\mathrm{Re}\,\mu > 0]$ WH

5. $\displaystyle\int_0^1 \left[\frac{x^{p-1}}{\ln x} + \frac{x^{q-1}}{1-x}\right] dx = \ln p - \psi(q)$ $[p > 0, \quad q > 0]$ BI (127)(17)

6. $\displaystyle\int_0^1 \left[\frac{1}{1-x^2} + \frac{1}{2x\ln x}\right]\frac{dx}{\ln x} = \frac{\ln 2}{2}$ LI (130)(19)

7. $\displaystyle\int_0^1 \left[q - \frac{1}{2} + \frac{(1-x)(1+q\ln x) + x\ln x}{(1-x)^2} x^{q-1}\right]\frac{dx}{\ln x} = \frac{1}{2} - q - \ln\Gamma(q) + \frac{\ln 2\pi}{2}$

                 $[q > 0]$ BI (128)(15)

4.282

1. $\displaystyle\int_0^1 \frac{\ln x}{4\pi^2 + (\ln x)^2}\cdot\frac{dx}{1-x} = \frac{1}{4} - \frac{1}{2}C$ BI (129)(1)

2. $\displaystyle\int_0^1 \frac{1}{a^2 + (\ln x)^2}\cdot\frac{dx}{1+x^2} = \frac{1}{2a}\,\beta\left(\frac{2a+\pi}{4\pi}\right)$ $\left[a > -\frac{\pi}{2}\right]$ BI (129)(9)

3. $\displaystyle\int_0^1 \frac{1}{\pi^2 + (\ln x)^2}\frac{dx}{1+x^2} = \frac{4-\pi}{4\pi}$ BI (129)(6)

4. $\displaystyle\int_0^1 \frac{\ln x}{\pi^2 + (\ln x)^2}\cdot\frac{dx}{1-x^2} = \frac{1}{2}\left(\frac{1}{2} - \ln 2\right)$ BI (129)(10)

5. $\displaystyle\int_0^1 \frac{\ln x}{a^2 + (\ln x)^2}\cdot\frac{x\,dx}{1-x^2} = \frac{1}{2}\left[\frac{\pi}{2a} + \ln\frac{\pi}{a} + \psi\left(\frac{a}{\pi}\right)\right]$ $[a > 0]$ BI (129)(14)

6. $\displaystyle\int_0^1 \frac{\ln x}{\pi^2 + (\ln x)^2}\cdot\frac{x\,dx}{1-x^2} = \frac{1}{2}\left(\frac{1}{2} - C\right)$ BI (129)(13)

7. $\displaystyle\int_0^1 \frac{1}{\pi^2 + 4(\ln x)^2}\cdot\frac{dx}{1+x^2} = \frac{\ln 2}{4\pi}$ BI (129)(7)

8. $\displaystyle\int_0^1 \frac{\ln x}{\pi^2 + 4(\ln x)^2}\cdot\frac{dx}{1-x^2} = \frac{2-\pi}{16}$ BI (129)(11)

9.[10] $\displaystyle\int_0^1 \frac{1}{\pi^2 + 16(\ln x)^2}\cdot\frac{dx}{1+x^2} = \frac{1}{8\pi\sqrt{2}}\left[\pi + 2\ln\left(\sqrt{2}-1\right)\right]$ BI (129)(8)

10. $\int_0^1 \frac{\ln x}{\pi^2 + 16(\ln x)^2} \cdot \frac{dx}{1 - x^2} = -\frac{\pi}{32\sqrt{2}} + \frac{1}{16} + \frac{1}{16\sqrt{2}} \ln\left(\sqrt{2} - 1\right)$ BI (129)(12)

11. $\int_0^1 \frac{\ln x}{\left[a^2 + (\ln x)^2\right]^2} \frac{dx}{1 - x} = -\frac{\pi^2}{a^4} \sum_{k=1}^{\infty} |B_{2k}| \left(\frac{2\pi}{a}\right)^{2k-2}$ BI (129)(4)

12. $\int_0^1 \frac{\ln x}{\left[a^2 + (\ln x)^2\right]^2} \frac{x\,dx}{1 - x^2} = -\frac{\pi^2}{4a^4} \sum_{k=1}^{\infty} |B_{2k}| \left(\frac{\pi}{a}\right)^{2k-2}$ BI (129)(16)

13. $\int_0^1 \frac{x^p - x^{-p}}{x^2 - 1} \frac{dx}{q^2 + (\ln x)^2} = \frac{2\pi}{q} \sum_{k=1}^{\infty} (-1)^{k-1} \frac{\sin kp\pi}{2q + k\pi} \qquad \left[p^2 < 1\right]$ BI (132)(13)a

4.283

1. $\int_0^1 \left(\frac{x - 1}{\ln x} - x\right) \frac{dx}{\ln x} = \ln 2 - 1$ BI (132)(17)a

2. $\int_0^1 \left(\frac{1}{\ln x} + \frac{1}{1 - x} - \frac{1}{2}\right) \frac{dx}{\ln x} = \frac{\ln 2\pi}{2} - 1$ BI (127)(20)

3. $\int_0^1 \left(\frac{1}{\ln x} + \frac{x}{1 - x} + \frac{x}{2}\right) \frac{dx}{x \ln x} = \frac{\ln 2\pi}{2}$ BI (127)(23)

4. $\int_0^1 \left[\frac{1}{(\ln x)^2} - \frac{x}{(1 - x)^2}\right] dx = \boldsymbol{C} - \frac{1}{2}$ GW (326)(8a)

5. $\int_0^1 \left(\frac{1}{1 - x^2} + \frac{1}{2 \ln x} - \frac{1}{2}\right) \frac{dx}{\ln x} = \frac{\ln 2 - 1}{2}$ BI (128)(14)

6. $\int_0^1 \left(\frac{1}{\ln x} + \frac{1}{2} \cdot \frac{1 + x}{1 - x} - \ln x\right) \frac{dx}{\ln x} = \frac{\ln 2\pi}{2}$ BI (127)(22)

7. $\int_0^1 \left[\frac{1}{1 - \ln x} - x\right] \frac{dx}{x \ln x} = -\boldsymbol{C}$ GW (326)(11a)

8. $\int_0^1 \left[\frac{x^q - 1}{x(\ln x)^2} - \frac{q}{\ln x}\right] dx = q \ln q - q \qquad [q > 0]$ BI (126)(2)

9. $\int_0^1 \left[x + \frac{1}{a \ln x - 1}\right] \frac{dx}{x \ln x} = \ln \frac{a}{q} + \boldsymbol{C} \qquad [a > 0, \quad q > 0]$ BI (126)(8)

10. $\int_0^1 \left[\frac{1}{\ln x} + \frac{1 + x}{2(1 - x)}\right] \frac{x^{p-1}}{\ln x}\,dx = -\ln\Gamma(p) + \left(p - \frac{1}{2}\right)\ln p - p + \frac{\ln 2\pi}{2}$

$[p > 0]$ GW (326)(9)

11. $\int_0^1 \left[p - 1 - \frac{1}{1 - x} + \left(\frac{1}{2} - \frac{1}{\ln x}\right) x^{p-1}\right] \frac{dx}{\ln x} = \left(\frac{1}{2} - p\right)\ln p + p - \frac{\ln 2\pi}{2}$

$[p > 0]$ BI (127)(25)

12.
$$\int_0^1 \left[-\frac{1}{(\ln x)^2} + \frac{(p-2)x^p - (p-1)x^{p-1}}{(1-x)^2} \right] dx = -\psi(p) + p - \frac{3}{2}$$
$$[p > 0] \qquad\qquad \text{GW (326)(8)}$$

13.
$$\int_0^1 \left[\left(p - \frac{1}{2}\right) x^3 + \frac{1}{2}\left(1 - \frac{1}{\ln x}\right)\left(x^{2p-1} - 1\right) \right] \frac{dx}{\ln x} = \left(\frac{1}{2} - p\right)(\ln p - 1)$$
$$[p > 0] \qquad\qquad \text{BI (132)(23)a}$$

14.
$$\int_0^1 \left[\left(q - \frac{1}{2}\right) \frac{x^{p-1} - x^{r-1}}{\ln x} + \frac{p x^{pq-1}}{1 - x^p} - \frac{r x^{rq-1}}{1 - x^r} \right] \frac{dx}{\ln x} = (p-r)\left[\frac{1}{2} - q - \ln\Gamma(q) + \frac{\ln 2\pi}{2}\right]$$
$$[q > 0] \qquad\qquad \text{BI (132)(13)}$$

4.284

1.
$$\int_0^1 \left[\frac{x^q - 1}{x(\ln x)^3} - \frac{q}{x(\ln x)^2} - \frac{q^2}{2\ln x} \right] dx = \frac{q^2}{2}\ln q - \frac{3}{4}q^2$$
$$[q > 0] \qquad\qquad \text{BI (126)(3)}$$

2.
$$\int_0^1 \left[\frac{x^q - 1}{x(\ln x)^4} - \frac{q}{x(\ln x)^3} - \frac{q^2}{2x(\ln x)^2} - \frac{q^3}{6\ln x} \right] dx = \frac{q^3}{6}\ln q - \frac{11}{36}q^3$$
$$[q > 0] \qquad\qquad \text{BI (126)(4)}$$

4.285
$$\int_0^1 \frac{x^{p-1}\,dx}{(q + \ln x)^n} = \frac{p^{n-1}}{(n-1)!}e^{-pq}\,\text{Ei}(pq) - \frac{1}{(n-1)!q^{n-1}}\sum_{k=1}^{n-1}(n-k-1)!(pq)^{k-1}$$
$$[p > 0, \quad q < 0] \qquad\qquad \text{BI (125)(21)}$$

In integrals of the form $\int \dfrac{x^a (\ln x)^n \, dx}{[b \pm (\ln x)^m]^l}$, we should make the substitution $x = e^t$ or $x = e^{-t}$ and then seek the resulting integrals in **3.351–3.356**.

4.29–4.32 Combinations of logarithmic functions of more complicated arguments and powers

4.291

1.
$$\int_0^1 \frac{\ln(1+x)}{x}\,dx = \frac{\pi^2}{12} \qquad\qquad \text{FI II 483}$$

2.
$$\int_0^1 \frac{\ln(1-x)}{x}\,dx = -\frac{\pi^2}{6} \qquad\qquad \text{FI II 714}$$

3.
$$\int_0^{1/2} \frac{\ln(1-x)}{x}\,dx = \frac{1}{2}(\ln 2)^2 - \frac{\pi^2}{12} \qquad\qquad \text{BI (145)(2)}$$

4.
$$\int_0^1 \ln\left(1 - \frac{x}{2}\right)\frac{dx}{x} = \frac{1}{2}(\ln 2)^2 - \frac{\pi^2}{12} \qquad\qquad \text{BI (114)(18)}$$

5.
$$\int_0^1 \frac{\ln\dfrac{1+x}{2}}{1-x}\,dx = \frac{1}{2}(\ln 2)^2 - \frac{\pi^2}{12} \qquad\qquad \text{BI (115)(1)}$$

6. $\displaystyle\int_0^1 \frac{\ln(1+x)}{1+x}\, dx = \frac{1}{2}(\ln 2)^2$

BI (114)(14)a

7.7 $\displaystyle\int_0^\infty \frac{\ln(1+ax)}{1+x^2}\, dx = \frac{\pi}{4}\ln\left(1+a^2\right) - \int_0^a \frac{\ln u\, du}{1+u^2}$ $[a>0]$

GI II (2209)

8. $\displaystyle\int_0^1 \frac{\ln(1+x)}{1+x^2}\, dx = \frac{\pi}{8}\ln 2$

FI II 157

9. $\displaystyle\int_0^\infty \frac{\ln(1+x)}{1+x^2}\, dx = \frac{\pi}{4}\ln 2 + \boldsymbol{G}$

BI (136)(1)

10. $\displaystyle\int_0^1 \frac{\ln(1-x)}{1+x^2}\, dx = \frac{\pi}{8}\ln 2 - \boldsymbol{G}$

BI (114)(17)

11. $\displaystyle\int_1^\infty \frac{\ln(x-1)}{1+x^2}\, dx = \frac{\pi}{8}\ln 2$

BI (144)(4)

12. $\displaystyle\int_0^1 \frac{\ln(1+x)}{x(1+x)}\, dx = \frac{\pi^2}{12} - \frac{1}{2}(\ln 2)^2$

BI (144)(4)

13. $\displaystyle\int_0^\infty \frac{\ln(1+x)}{x(1+x)}\, dx = \frac{\pi^2}{6}.$

BI (141)(9)a

14. $\displaystyle\int_0^1 \frac{\ln(1+x)}{(ax+b)^2}\, dx = \frac{1}{a(a-b)}\ln\frac{a+b}{b} + \frac{2\ln 2}{b^2-a^2}$ $[a\neq b,\quad ab>0]$

$\displaystyle\qquad\qquad = \frac{1}{2a^2}\left(1-\ln 2\right)$ $[a=b]$

LI (114)(5)a

15. $\displaystyle\int_0^\infty \frac{\ln(1+x)}{(ax+b)^2}\, dx = \frac{\ln\dfrac{a}{b}}{a(a-b)}$ $[ab>0]$

BI (139)(5)

16. $\displaystyle\int_0^1 \ln(a+x)\frac{dx}{a+x^2} = \frac{1}{2\sqrt{a}}\operatorname{arccot}\sqrt{a}\ln[(1+a)a]$ $[a>0]$

BI (114)(20)

17. $\displaystyle\int_0^\infty \ln(a+x)\frac{dx}{(b+x)^2} = \frac{a\ln a - b\ln b}{b(a-b)}$ $[a>0,\quad b>0,\quad a\neq b]$

LI (139)(6)

18. $\displaystyle\int_0^a \frac{\ln(1+ax)}{1+x^2}\, dx = \frac{1}{2}\arctan a\ln\left(1+a^2\right)$

GI II (2195)

19. $\displaystyle\int_0^1 \frac{\ln(1+ax)}{1+ax^2}\, dx = \frac{1}{2\sqrt{a}}\arctan\sqrt{a}\ln(1+a)$ $[a>0]$

BI (114)(21)

20. $\displaystyle\int_0^1 \frac{\ln(ax+b)}{(1+x)^2}\, dx = \frac{1}{a-b}\left[\frac{1}{2}(a+b)\ln(a+b) - b\ln b - a\ln 2\right]$

$[a>0,\quad b>0,\quad a\neq b]$ BI (114)(22)

21. $\displaystyle\int_0^\infty \frac{\ln(ax+b)}{(1+x)^2}\, dx = \frac{1}{a-b}\left[a\ln a - b\ln b\right]$ $[a>0,\quad b>0]$

BI (139)(8)

22. $\displaystyle \int_0^\infty \ln(a+x) \frac{x\,dx}{(b^2+x^2)^2} = \frac{1}{2(a^2+b^2)}\left(\ln b + \frac{a\pi}{2b} + \frac{a^2}{b^2}\ln a\right)$

$[a>0, \quad b>0]$ BI (139)(9)

23. $\displaystyle \int_0^1 \ln(1+x)\frac{1+x^2}{(1+x)^4}\,dx = -\frac{1}{3}\ln 2 + \frac{23}{72}$ LI (114)(12)

24. $\displaystyle \int_0^1 \ln(1+x)\frac{1+x^2}{a^2+x^2}\cdot\frac{dx}{1+a^2x^2} = \frac{1}{2a(1+a^2)}\left[\frac{\pi}{2}\ln\left(1+a^2\right) - 2\arctan a\cdot\ln a\right]$

$[a>0]$ LI (114)(11)

25. $\displaystyle \int_0^1 \ln(1+x)\frac{1-x^2}{(ax+b)^2\,(bx+a)^2}\,dx = \frac{1}{a^2-b^2}\left\{\frac{1}{a-b}\left[\frac{a+b}{ab}\ln(a+b) - \frac{1}{a}\ln b - \frac{1}{b}\ln a\right]\right.$

$\left.+ \frac{4\ln 2}{b^2-a^2}\right\}$

$[a>0, \quad b>0, \quad a^2\neq b^2]$ LI (114)(13)

26. $\displaystyle \int_0^\infty \ln(1+x)\frac{1-x^2}{(ax+b)^2}\cdot\frac{dx}{(bx+a)^2} = \frac{1}{ab(a^2-b^2)}\ln\frac{b}{a}$

$[a>0, \quad b>0]$ LI (139)(14)

27. $\displaystyle \int_0^1 \ln(1+ax)\frac{1-x^2}{(1+x^2)^2}\,dx = \frac{1}{2}\frac{(1+a)^2}{1+a^2}\ln(1+a) - \frac{1}{2}\cdot\frac{a}{1+a^2}\ln 2 - \frac{\pi}{4}\cdot\frac{a^2}{1+a^2}$

$[a>-1]$ BI (114)(23)

28. $\displaystyle \int_0^\infty \ln(a+x)\frac{b^2-x^2}{(b^2+x^2)^2}\,dx = \frac{1}{a^2+b^2}\left(a\ln\frac{b}{a} - \frac{b\pi}{2}\right)$

$[a>0, \quad b>0]$ BI (139)(11)

29. $\displaystyle \int_0^\infty \ln^2(a-x)\frac{b^2-x^2}{(b^2+x^2)^2}\,dx = \frac{2}{a^2+b^2}\left(a\ln\frac{a}{b} - \frac{b\pi}{2}\right)$

$[a>0, \quad b>0]$ BI (139)(12)

30. $\displaystyle \int_0^\infty \ln^2(a-x)\frac{x\,dx}{(b^2+x^2)^2} = \frac{1}{a^2+b^2}\left(\ln b - \frac{a\pi}{2b} + \frac{a^2}{b^2}\ln a\right)$

$[a>0, \quad b>0]$ BI (139)(10)

4.292

1. $\displaystyle \int_0^1 \frac{\ln(1\pm x)}{\sqrt{1-x^2}}\,dx = -\frac{\pi}{2}\ln 2 \pm 2\mathbf{G}$ GW (325)(20)

2. $\displaystyle \int_0^1 \frac{x\ln(1\pm x)}{\sqrt{1-x^2}}\,dx = -1 \pm \frac{\pi}{2}$ GW (325)(22c)

3. $\displaystyle \int_{-a}^a \frac{\ln(1+bx)}{\sqrt{a^2-x^2}}\,dx = \pi\ln\frac{1+\sqrt{1-a^2b^2}}{2}$ $\left[0\leq|b|\leq\frac{1}{a}\right]$

BI (145)(16, 17)a, GW (325)(21e)

4.
$$\int_0^1 \frac{x \ln(1 + ax)}{\sqrt{1 - x^2}}\, dx = -1 + \frac{\pi}{2} \cdot \frac{1 - \sqrt{1 - a^2}}{a} + \frac{\sqrt{1 - a^2}}{a} \arcsin a \quad [|a| \leq 1]$$
$$= -1 + \frac{\pi}{2a} + \frac{\sqrt{a^2 - 1}}{a} \ln\left(a + \sqrt{a^2 - 1}\right) \qquad [a \geq 1]$$

GW (325)(22)

5.
$$\int_0^1 \frac{\ln(1 + ax)}{x\sqrt{1 - x^2}}\, dx = \frac{1}{2} \arcsin a\,(\pi - \arcsin a) = \frac{\pi^2}{8} - \frac{1}{2}(\arccos a)^2$$

$$[|a| \leq 1] \qquad \text{BI (120)(4), GW (325)(21a)}$$

4.293

1.
$$\int_0^1 x^{\mu-1} \ln(1 + x)\, dx = \frac{1}{\mu}\left[\ln 2 - \beta(\mu + 1)\right] \qquad [\operatorname{Re}\mu > -1] \qquad \text{BI (106)(4)a}$$

2.[6]
$$\int_1^\infty x^{\mu-1} \ln(1 + x)\, dx = \frac{-1}{\mu}\left[\beta(-\mu) + \ln 2\right] \qquad [\operatorname{Re}\mu < 0] \qquad \text{ET I 315(17)}$$

3.
$$\int_0^\infty x^{\mu-1} \ln(1 + x)\, dx = \frac{\pi}{\mu \sin \mu\pi} \qquad [-1 < \operatorname{Re}\mu < 0] \qquad \text{GW (325)(3)a}$$

4.
$$\int_0^1 x^{2n-1} \ln(1 + x)\, dx = \frac{1}{2n} \sum_{k=1}^{2n} \frac{(-1)^{k-1}}{k} \qquad \text{GW (325)(2b)}$$

5.
$$\int_0^1 x^{2n} \ln(1 + x)\, dx = \frac{1}{2n+1}\left[\ln 4 + \sum_{k=1}^{2n+1} \frac{(-1)^k}{k}\right] \qquad \text{GW (325)(2c)}$$

6.
$$\int_0^1 x^{n-\frac{1}{2}} \ln(1 + x)\, dx = \frac{2 \ln 2}{2n+1} + \frac{(-1)^n \cdot 4}{2n+1}\left[\pi - \sum_{k=0}^n \frac{(-1)^k}{2k+1}\right] \qquad \text{GW (325)(2f)}$$

7.
$$\int_0^\infty x^{\mu-1} \ln|1 - x|\, dx = \frac{\pi}{\mu} \cot(\mu\pi) \qquad [-1 < \operatorname{Re}\mu < 0]$$

$$\text{BI (134)(4), ET I 315(18)}$$

8.
$$\int_0^1 x^{\mu-1} \ln(1 - x)\, dx = -\frac{1}{\mu}\left[\psi(\mu + 1) - \psi(1)\right] \qquad [\operatorname{Re}\mu > -1] \qquad \text{ET I 316(19)}$$

9.[7]
$$\int_1^\infty x^{\mu-1} \ln(x - 1)\, dx = \frac{1}{\mu}\left[\pi \cot(\mu\pi) + \psi(\mu + 1) + C\right]$$

$$[\operatorname{Re}\mu < 0] \qquad \text{ET I 316(20)}$$

10.
$$\int_0^\infty x^{\mu-1} \ln(1 + \gamma x)\, dx = \frac{\pi}{\mu\gamma^\mu \sin \mu\pi} \qquad [-1 < \operatorname{Re}\mu < 0, \quad |\arg \gamma| < \pi]$$

$$\text{BI (134)(3)}$$

11.
$$\int_0^\infty \frac{x^{\mu-1} \ln(1 + x)}{1 + x}\, dx = \frac{\pi}{\sin \mu\pi}\left[C + \psi(1 - \mu)\right] \qquad [-1 < \operatorname{Re}\mu < 1] \qquad \text{ET I 316(21)}$$

12.
$$\int_0^1 \frac{\ln(1 + x)}{(1 + x)^{\mu+1}}\, dx = -\frac{\ln 2}{2^\mu \mu} + \frac{2^\mu - 1}{2^\mu \mu^2} \qquad \text{BI (114)(6)}$$

13.
$$\int_0^1 \frac{x^{\mu-1} \ln(1 - x)}{(1 - x)^{1-\nu}}\, dx = B(\mu, \nu)\left[\psi(\nu) - \psi(\mu + \nu)\right] \qquad [\operatorname{Re}\mu > 0, \quad \operatorname{Re}\nu > 0] \qquad \text{ET I 316(122)}$$

14. $\displaystyle\int_0^\infty \frac{x^{\mu-1}\ln(\gamma+x)}{(\gamma+x)^\nu}\,dx = \gamma^{\mu-\nu}\,\mathrm{B}(\mu,\nu-\mu)\left[\psi(\nu)-\psi(\nu-\mu)+\ln\gamma\right]$

$$[0 < \operatorname{Re}\mu < \operatorname{Re}\nu] \qquad\qquad \text{ET I 316(23)}$$

4.294

1. $\displaystyle\int_0^1 \ln(1+x)\frac{(p-1)x^{p-1}-px^{-p}}{x}\,dx = 2\ln 2 - \frac{\pi}{\sin p\pi}$

$$[0 < p < 1] \qquad\qquad \text{BI (114)(2)}$$

2. $\displaystyle\int_0^1 \ln(1+x)\frac{1+x^{2n+1}}{1+x}\,dx = 2\ln 2\sum_{k=0}^{n}\frac{1}{2k+1} - \sum_{j=1}^{2n+1}\frac{1}{j}\sum_{k=1}^{j}\frac{(-1)^{k-1}}{k}$ BI (114)(7)

3. $\displaystyle\int_0^1 \ln(1+x)\frac{1-x^{2n}}{1+x}\,dx = 2\ln 2\cdot\sum_{k=0}^{n-1}\frac{1}{2k+1} - \sum_{j=1}^{2n}\frac{1}{j}\sum_{k=1}^{j}\frac{(-1)^{k-1}}{k}$ BI (114)(8)

4. $\displaystyle\int_0^1 \ln(1+x)\frac{1-x^{2n}}{1-x}\,dx = 2\ln 2\cdot\sum_{k=0}^{n-1}\frac{1}{2k+1} + \sum_{i=1}^{2n}\frac{(-1)^j}{j}\sum_{k=1}^{j}\frac{(-1)^{k-1}}{k}$ BI (114)(9)

5. $\displaystyle\int_0^1 \ln(1+x)\frac{1-x^{2n+1}}{1-x}\,dx = 2\ln 2\sum_{k=0}^{n}\frac{1}{2k+1} + \sum_{j=1}^{2n+1}\frac{(-1)^j}{j}\sum_{k=1}^{j}\frac{(-1)^{k-1}}{k}$ BI (114)(10)

6. $\displaystyle\int_0^1 \ln(1-x)\frac{1-(-1)^n x^n}{1-x}\,dx = \sum_{j=1}^{n}\frac{(-1)^j}{j}\sum_{k=1}^{j}\frac{1}{k}$ BI (114)(15)

7. $\displaystyle\int_0^1 \ln(1-x)\frac{1-x^n}{1-x}\,dx = -\sum_{j=1}^{n}\frac{1}{j}\sum_{k=1}^{j}\frac{1}{k}$ BI (114)(16)

8. $\displaystyle\int_0^\infty \ln^2(1-x)x^p\,dx = \frac{2\pi}{p+1}\cot p\pi$ $[-2 < p < -1]$ BI (134)(13)a

9. $\displaystyle\int_0^1 [\ln(1+x)]^n (1+x)^r\,dx = (-1)^{n-1}\frac{n!}{(r+1)^{n+1}} + 2^{r+1}\sum_{k=0}^{n}\frac{(-1)^k n!(\ln 2)^{n-k}}{(n-k)!(r+1)^{k+1}}$ LI (106)(34)a

10. $\displaystyle\int_0^1 [\ln(1-x)]^n (1-x)^r\,dx = (-1)^n\frac{n!}{(r+1)^{n+1}}$ $[r > -1]$ BI (106)(35)a

11. $\displaystyle\int_0^1 \left(\ln\frac{1}{1-x^2}\right)^n x^{2q-1}\,dx = \frac{n!}{2}\zeta(n+1,q+1)$ $[-1 < q < 0]$ BI (311)(15)a

12. $\displaystyle\int_0^1 (\ln x)^{2n}\ln\left(1-x^2\right)\frac{dx}{x} = -\frac{\pi^{2n+2}}{2(n+1)(2n+1)}|B_{2n+2}|$ BI (309)(5)a

13.[6] $\displaystyle\int_0^1 \left[\ln\frac{1}{x}\right]^m \ln\left(1-x^2\right)\,dx = -\sum_{n=1}^{\infty}\frac{\Gamma(m+1)}{n(2n+1)^{m+1}}$ $[m+1 > 0, \quad n+1 > 0]$

4.295

1. $\displaystyle\int_0^\infty \ln\left(\mu x^2+\beta\right)\frac{dx}{\gamma+x^2} = \frac{\pi}{\sqrt{\gamma}}\ln\left(\sqrt{\mu\gamma}+\sqrt{\beta}\right)$ $[\operatorname{Re}\beta > 0, \quad \operatorname{Re}\mu > 0, \quad |\arg\gamma| < \pi]$

$$\text{ET II 218(27)}$$

2. $\displaystyle\int_0^1 \ln\left(1+x^2\right) \frac{dx}{x^2} = \frac{\pi}{2} - \ln 2$ GW (325)(2g)

3. $\displaystyle\int_0^\infty \ln\left(1+x^2\right) \frac{dx}{x^2} = \pi$ GW (325)(4c)

4. $\displaystyle\int_0^\infty \ln\left(1+x^2\right) \frac{dx}{(a+x)^2} = \frac{2a}{1+a^2}\left(\frac{\pi}{2a}+\ln a\right)$ $[a>0]$ BI (319)(6)a

5. $\displaystyle\int_0^1 \ln\left(1+x^2\right) \frac{dx}{1+x^2} = \frac{\pi}{2}\ln 2 - \boldsymbol{G}$ BI (114)(24)

6. $\displaystyle\int_1^\infty \ln\left(1+x^2\right) \frac{dx}{1+x^2} = \frac{\pi}{2}\ln 2 + \boldsymbol{G}$ BI (114)(5)

7. $\displaystyle\int_0^\infty \ln\left(a^2+b^2x^2\right) \frac{dx}{c^2+g^2x^2} = \frac{\pi}{cg}\ln\frac{ag+bc}{g}$ $[a>0,\quad b>0,\quad c>0,\quad g>0]$

BI (136)(11-14)a

8. $\displaystyle\int_0^\infty \ln\left(a^2+b^2x^2\right) \frac{dx}{c^2-g^2x^2} = -\frac{\pi}{cg}\arctan\frac{bc}{ag}$ $[a>0,\quad b>0,\quad c>0,\quad g>0]$

BI (136)(15)a

9. $\displaystyle\int_0^\infty \frac{\ln\left(1+p^2x^2\right) - \ln\left(1+q^2x^2\right)}{x^2}\,dx = \pi(p-q)$ $[p>0,\quad q>0]$ FI II 645

10. $\displaystyle\int_0^1 \ln\frac{1+a^2x^2}{1+a^2}\frac{dx}{1-x^2} = -(\arctan a)^2$ BI (115)(2)

11. $\displaystyle\int_0^1 \ln\left(1-x^2\right) \frac{dx}{x} = -\frac{\pi^2}{12}$

12. $\displaystyle\int_0^\infty \ln^2\left(1-x^2\right) \frac{dx}{x^2} = 0$ BI (142)(9)a

13. $\displaystyle\int_0^1 \ln\left(1-x^2\right) \frac{dx}{1+x^2} = \frac{\pi}{4}\ln 2 - \boldsymbol{G}$ GW (325)(17)

14. $\displaystyle\int_1^\infty \ln\left(x^2-1\right) \frac{dx}{1+x^2} = \frac{\pi}{4}\ln 2 + \boldsymbol{G}$ BI (144)(6)

15. $\displaystyle\int_0^\infty \ln^2\left(a^2-x^2\right) \frac{dx}{b^2+x^2} = \frac{\pi}{b}\ln\left(a^2+b^2\right)$ $[b>0]$ BI (136)(16)

16. $\displaystyle\int_0^\infty \ln^2\left(a^2-x^2\right) \frac{b^2-x^2}{(b^2+x^2)^2}\,dx = -\frac{2b\pi}{a^2+b^2}$ $[b>0]$ BI (136)(20)

17. $\displaystyle\int_0^1 \ln\left(1+x^2\right) \frac{dx}{x\left(1+x^2\right)} = \frac{1}{2}\left[\frac{\pi^2}{12}-\frac{1}{2}(\ln 2)^2\right]$ BI (114)(25)

18. $\displaystyle\int_0^\infty \ln\left(1+x^2\right) \frac{dx}{x\left(1+x^2\right)} = \frac{\pi^2}{12}$ BI (141)(9)

19. $\displaystyle\int_0^1 \ln\left(\cos^2 t + x^2 \sin^2 t\right) \frac{dx}{1-x^2} = -t^2$ BI (114)(27)a

20. $\displaystyle\int_0^\infty \ln\left(a^2 + b^2 x^2\right) \frac{dx}{(c+gx)^2} = \frac{2\ln b}{cg} + \frac{b^2}{a^2 g^2 + b^2 c^2}\left(\frac{a}{b}\pi + 2\frac{c}{g}\ln\frac{c}{g} + 2\frac{a^2 g}{b^2 c}\ln\frac{a}{b}\right)$

$\qquad\qquad\qquad\qquad\qquad\qquad [a > 0, \quad b > 0, \quad c > 0, \quad g > 0]$

$\qquad\qquad\qquad\qquad\qquad\qquad\qquad\qquad\qquad$ BI (139)(16)a

21. $\displaystyle\int_0^1 \ln\left(a^2 + b^2 x^2\right) \frac{dx}{(c+gx)^2}$

$\qquad = \dfrac{2}{c(c+g)}\ln a + \dfrac{b^2}{a^2 g^2 + b^2 c^2}\left[\dfrac{2a}{b}\operatorname{arccot}\dfrac{a}{b} + \dfrac{cb^2 - ga^2}{b^2(c+g)}\ln\dfrac{a^2 + b^2}{a^2} - 2\dfrac{c}{g}\ln\dfrac{c+g}{c}\right]$

$\qquad\qquad\qquad [a > 0, \quad b > 0, \quad c > 0, \quad g > 0]$ BI (114)(28)a

22. $\displaystyle\int_0^\infty \frac{\ln\left(1 + p^2 x^2\right)}{r^2 + q^2 x^2}\,dx = \int_0^\infty \frac{\ln\left(p^2 + x^2\right)}{q^2 + r^2 x^2}\,dx = \frac{\pi}{qr}\ln\frac{q+pr}{q}$

$\qquad\qquad\qquad\qquad\qquad\qquad [qr > 0, \quad p > 0]$

$\qquad\qquad\qquad\qquad\qquad$ FI II 745a, BI (318)(1)a, BI (318)(4)a

23. $\displaystyle\int_0^\infty \frac{\ln\left(1 + a^2 x^2\right)}{b^2 + c^2 x^2}\,\frac{dx}{d^2 + g^2 x^2} = \frac{\pi}{b^2 g^2 - c^2 d^2}\left[\frac{g}{d}\ln\left(1 + \frac{ad}{g}\right) - \frac{c}{b}\ln\left(1 + \frac{ab}{c}\right)\right]$

$\qquad [a > 0, \quad b > 0, \quad c > 0, \quad d > 0, \quad g > 0, \quad b^2 g^2 \neq c^2 d^2]$ BI (141)(10)

24. $\displaystyle\int_0^\infty \frac{\ln\left(1 + a^2 x^2\right)}{b^2 + c^2 x^2}\,\frac{x^2\,dx}{d^2 + g^2 x^2} = \frac{\pi}{b^2 g^2 - c^2 d^2}\left[\frac{b}{c}\ln\left(1 + \frac{ab}{c}\right) - \frac{d}{g}\ln\left(1 + \frac{ad}{g}\right)\right]$

$\qquad [a > 0, \quad b > 0, \quad c > 0, \quad d > 0, \quad g > 0, \quad b^2 g^2 \neq c^2 d^2]$ BI (141)(11)

25. $\displaystyle\int_0^\infty \ln\left(a^2 + b^2 x^2\right) \frac{dx}{(c^2 + g^2 x^2)^2} = \frac{\pi}{2c^3 g}\left(\ln\frac{ag+bc}{g} - \frac{bc}{ag+bc}\right)$

$\qquad\qquad\qquad\qquad\qquad\qquad [a > 0, \quad b > 0, \quad c > 0, \quad g > 0]$

$\qquad\qquad\qquad\qquad\qquad\qquad\qquad\qquad\qquad$ GW (325)(18a)

26. $\displaystyle\int_0^\infty \ln\left(a^2 + b^2 x^2\right) \frac{x^2\,dx}{(c^2 + g^2 x^2)^2} = \frac{\pi}{2cg^3}\left(\ln\frac{ag+bc}{g} + \frac{bc}{ag+bc}\right)$

$\qquad\qquad\qquad\qquad\qquad\qquad [a > 0, \quad b > 0, \quad c > 0, \quad g > 0]$

$\qquad\qquad\qquad\qquad\qquad\qquad\qquad\qquad\qquad$ GW (325)(18b)

27. $\displaystyle\int_0^1 \ln\left(1 + ax^2\right)\sqrt{1 - x^2}\,dx = \frac{\pi}{2}\left\{\ln\frac{1 + \sqrt{1+a}}{2} + \frac{1}{2}\frac{1 - \sqrt{1+a}}{1 + \sqrt{1+a}}\right\}$

$\qquad\qquad\qquad\qquad\qquad\qquad\qquad [a > 0]$ $\qquad\qquad\qquad\qquad$ BI (117)(6)

28. $\displaystyle\int_0^1 \ln\left(1 + a - ax^2\right)\sqrt{1 - x^2}\,dx = \frac{\pi}{2}\left\{\ln\frac{1 + \sqrt{1+a}}{2} - \frac{1}{2}\frac{1 - \sqrt{1+a}}{1 + \sqrt{1+a}}\right\}$

$\qquad\qquad\qquad\qquad\qquad\qquad\qquad [a > 0]$ $\qquad\qquad\qquad\qquad$ BI (117)(7)

29. $\displaystyle\int_0^1 \ln\left(1 - a^2 x^2\right)\frac{dx}{\sqrt{1 - x^2}} = \pi\ln\frac{1 + \sqrt{1 - a^2}}{2}$ $\qquad [a^2 < 1]$ $\qquad\qquad$ BI (119)(1)

30.[6] $\displaystyle\int_0^1 \ln\left(1 - a^2 x^2\right)\frac{dx}{x\sqrt{1 - x^2}} = -\left(\arccos|a| - \frac{\pi}{2}\right)^2$ $\qquad\qquad\qquad$ LI (120)(11)

31. $\int_0^1 \ln\left(1-x^2\right) \dfrac{dx}{\sqrt{\left(1-x^2\right)\left(1-k^2x^2\right)}} = \ln\dfrac{k'}{k}\,\boldsymbol{K}(k) - \dfrac{\pi}{2}\,\boldsymbol{K}(k')$　　　　　BI (120)(12)

32. $\int_0^1 \ln\left(1\pm kx^2\right) \dfrac{dx}{\sqrt{\left(1-x^2\right)\left(1-k^2x^2\right)}} = \dfrac{1}{2}\ln\dfrac{2\pm 2k}{\sqrt{k}}\,\boldsymbol{K}(k) - \dfrac{\pi}{8}\,\boldsymbol{K}(k')$　　　BI (120)(8), BI (120)(14)

33. $\int_0^1 \dfrac{\ln\left(1-k^2x^2\right)}{\sqrt{\left(1-x^2\right)\left(1-k^2x^2\right)}}\,dx = \ln k'\,\boldsymbol{K}(k)$　　　　　　　　　　BI (119)(27)

34. $\int_0^1 \ln\left(1-k^2x^2\right)\sqrt{\dfrac{1-k^2x^2}{1-x^2}}\,dx = \left(2-k^2\right)\boldsymbol{K}(k) - \left(2-\ln k'\right)\boldsymbol{E}(k)$　　　　BI (119)(3)

35. $\int_0^1 \sqrt{\dfrac{1-x^2}{1-k^2x^2}}\,\ln\left(1-k^2x^2\right)\,dx = \dfrac{1}{k^2}\left(1+k'^2 - k'^2\ln k'\right)\boldsymbol{K}(k) - \left(2-\ln k'\right)\boldsymbol{E}(k)$　　BI (119)(7)

36. $\int_{-1}^1 \ln\left(1-x^2\right)\dfrac{dx}{(a+bx)\sqrt{1-x^2}} = \dfrac{2\pi}{\sqrt{a^2-b^2}}\ln\dfrac{\sqrt{a^2-b^2}}{a+\sqrt{a^2-b^2}}$

　　　　　　　　　　　　　　　　　　　　　　$[a>0,\quad b>0,\quad a\neq b]$　　BI (145)(15)

37.[8] $\int_0^1 \ln\left(1-x^2\right)\left(px^{p-1}-qx^{q-1}\right)dx = \psi\left(\dfrac{q}{2}+1\right) + \psi\left(\dfrac{p}{2}+1\right)$

　　　　　　　　　　　　　　　　　　　　　　$[p>-2,\quad q>-2]$　　BI (106)(15)

38. $\int_0^1 \ln\left(1+ax^2\right)\dfrac{dx}{\sqrt{1-x^2}} = \pi\ln\dfrac{1+\sqrt{1+a}}{2}$　　　　$[a\geq -1]$　　GW (325)(21b)

39. $\int_0^1 \ln\left(1+x^2\right)x^{\mu-1}\,dx = \dfrac{1}{\mu}\left[\ln 2 - \beta\left(\dfrac{\mu}{2}+1\right)\right]$　　　$[\operatorname{Re}\mu > -2]$　　BI (106)(12)

40. $\int_0^\infty \ln\left(1+x^2\right)x^{\mu-1}\,dx = \dfrac{\pi}{\mu\sin\dfrac{\mu\pi}{2}}$　　　　　$[-2<\operatorname{Re}\mu<0]$

　　　　　　　　　　　　　　　　　　　　　　　BI (311)(4)a, ET I 315(15)

41. $\int_0^\infty \ln\left(1+x^2\right)\dfrac{x^{\mu-1}\,dx}{1+x}$

　　　$= \dfrac{\pi}{\sin\mu\pi}\left\{\ln 2 - (1-\mu)\sin\dfrac{\mu\pi}{2}\,\beta\left(\dfrac{1-\mu}{2}\right) - (2-\mu)\cos\dfrac{\mu\pi}{2}\,\beta\left(\dfrac{2-\mu}{2}\right)\right\}$

　　　　　　　　　　　　　　　$[-2<\operatorname{Re}\mu<1]$　　　　ET I 316(25)

4.296

1. $\int_0^1 \ln\left(1+2x\cos t+x^2\right)\dfrac{dx}{x} = \dfrac{\pi^2}{6} - \dfrac{t^2}{2}$　　　　　BI (114)(34)

2. $\int_{-\infty}^\infty \ln\left(a^2-2ax\cos t+x^2\right)\dfrac{dx}{1+x^2} = \pi\ln\left(1+2a|\sin t|+a^2\right)$　　BI (145)(28)

3. $\int_0^\infty \ln\left(1+2x\cos t+x^2\right)x^{\mu-1}\,dx = \dfrac{2\pi}{\mu}\dfrac{\cos\mu t}{\sin\mu\pi}$　　$[|t|<\pi,\quad -1<\operatorname{Re}\mu<0]$　　ET I 316(27)

4. $\displaystyle\int_0^\infty \ln\left(\frac{x^2 + 2ax\cos t + a^2}{x^2 - 2ax\cos t + a^2}\right) \frac{x\,dx}{x^2 + b^2} = \frac{1}{2}\pi^2 - \pi t + \pi\arctan\frac{(a^2 - b^2)\cos t}{(a^2 + b^2)\sin t + 2ab}$

$$[a > 0, \quad b > 0, \quad 0 < t < \pi]$$

4.297

1. $\displaystyle\int_0^1 \ln\frac{ax + b}{bx + a}\frac{dx}{(1 + x)^2} = \frac{1}{a - b}\left[(a + b)\ln\frac{a + b}{2} - a\ln a - b\ln b\right]$

$$[a > 0, \quad b > 0] \qquad\qquad \text{BI (115)(16)}$$

2. $\displaystyle\int_0^\infty \ln\frac{ax + b}{bx + a}\frac{dx}{(1 + x)^2} = 0 ,$ $[ab > 0]$ BI (139)(23)

3. $\displaystyle\int_0^1 \ln\frac{1 - x}{x}\frac{dx}{1 + x^2} = \frac{\pi}{8}\ln 2$ BI (115)(5)

4. $\displaystyle\int_0^1 \ln\frac{1 + x}{1 - x}\frac{dx}{1 + x^2} = G$ BI (115)(17)

5. $\displaystyle\int_0^\infty \ln^2\left(\frac{1 + x}{1 - x}\right)\frac{dx}{x(1 + x^2)} = \frac{\pi^2}{2}$ BI (141)(13)

6. $\displaystyle\int_u^v \ln\frac{v + x}{u + x}\frac{dx}{x} = \frac{1}{2}\left(\ln\frac{v}{u}\right)^2$ $[uv > 0]$ BI (145)(33)

7. $\displaystyle\int_0^\infty \frac{b\ln(1 + ax) - a\ln(1 + bx)}{x^2}\,dx = ab\ln\frac{b}{a}$ $[a > 0, \quad b > 0]$ FI II 647

8. $\displaystyle\int_0^1 \ln\frac{1 + ax}{1 - ax}\frac{dx}{x\sqrt{1 - x^2}} = \pi\arcsin a$ $[|a| \leq 1]$ GW (325)(21c), BI (122)(2)

9. $\displaystyle\int_u^v \ln\left(\frac{1 + ax}{1 - ax}\right)\frac{dx}{\sqrt{(x^2 - u^2)(v^2 - x^2)}} = \frac{\pi}{v}F\left(\arcsin av, \frac{u}{v}\right)$

$$[|av| < 1] \qquad\qquad \text{BI (145)(35)}$$

10.[8] $\displaystyle\text{PV}\int_0^1 \ln\left|\frac{a + y}{a - y}\right|\frac{dy}{y\sqrt{1 - y^2}} = \frac{\pi^2}{2}$ $[0 < a \leq 1]$

4.298

1. $\displaystyle\int_0^\infty \ln\frac{1 + x^2}{x}\frac{x^{2n-1}}{1 + x}\,dx = \frac{\ln 2}{2n} + \frac{1}{4n^2} - \frac{1}{2n}\beta(2n + 1)$ BI (137)(1)

2. $\displaystyle\int_0^\infty \ln\frac{1 + x^2}{x}\frac{x^{2n}}{1 + x}\,dx = \frac{\ln 2}{2n} + \frac{1}{4n^2} - \frac{1}{2n}\beta(2n + 1)$ BI (137)(3)

3. $\displaystyle\int_0^\infty \ln\frac{1 + x^2}{x}\frac{x^{2n-1}}{1 - x}\,dx = \frac{\ln 2}{2n} + \frac{1}{4n^2} - \frac{1}{2n}\beta(2n + 1)$ BI (137)(2)

4. $\displaystyle\int_0^\infty \ln\frac{1 + x^2}{x}\frac{x^{2n}}{1 - x}\,dx = -\frac{\ln 2}{2n} - \frac{1}{4n^2} + \frac{1}{2n}\beta(2n + 1)$ BI (137)(4)

5. $\displaystyle\int_0^\infty \ln\frac{1 + x^2}{x}\frac{x^{2n-1}}{1 + x^2}\,dx = \frac{\ln 2}{2n} + \frac{1}{4n^2} - \frac{1}{2n}\beta(2n + 1)$ BI (137)(10)

6. $\int_0^1 \ln \frac{1+x^2}{x} x^{2n}\, dx = \frac{1}{2n+1}\left\{(-1)^n \frac{\pi}{2} + \ln 2 - \frac{1}{2n+1} + 2\sum_{k=0}^{n-1} \frac{(-1)^k}{2n-2k-1}\right\}$ BI (294)(8)

7. $\int_0^1 \ln \frac{1+x^2}{x} x^{2n-1}\, dx = \frac{1}{2n}\left\{(-1)^{n+1}\ln 2 + \ln 2 - \frac{1}{2n} + (-1)^{n+1}\sum_{k=1}^{n-1}\frac{(-1)^k}{k}\right\}$ BI (294)(9)a

8. $\int_0^1 \ln \frac{1+x^2}{x}\frac{dx}{1+x^2} = \frac{\pi}{2}\ln 2$ BI (115)(7)

9. $\int_0^\infty \ln \frac{1+x^2}{x}\frac{dx}{1+x^2} = \pi \ln 2$ BI (137)(8)

10. $\int_0^\infty \ln \frac{1+x^2}{x}\frac{dx}{1-x^2} = 0$ BI (137)(9)

11. $\int_0^1 \ln \frac{1-x^2}{x}\frac{dx}{1+x^2} = \frac{\pi}{4}\ln 2$ BI (115)(9)

12. $\int_1^\infty \ln \frac{1+x^2}{x+1}\frac{dx}{1+x^2} = \frac{3\pi}{8}\ln 2$ BI (144)(8)

13. $\int_0^1 \ln \frac{1+x^2}{x+1}\frac{dx}{1+x^2} = \frac{3\pi}{8}\ln 2 - \boldsymbol{G}$ BI (115)(18)

14. $\int_1^\infty \ln \frac{1+x^2}{x-1}\frac{dx}{1+x^2} = \frac{3\pi}{8}\ln 2 + \boldsymbol{G}$ BI (144)(9)

15. $\int_0^1 \ln \frac{1+x^2}{1-x}\frac{dx}{1+x^2} = \frac{3\pi}{8}\ln 2$ BI (115)(19)

16. $\int_0^\infty \ln \frac{1+x^2}{x^2}\frac{x\, dx}{1+x^2} = \frac{\pi^2}{12}$ BI (138)(3)

17. $\int_0^\infty \ln \frac{a^2+b^2x^2}{x^2}\frac{dx}{c^2+g^2x^2} = \frac{\pi}{cg}\ln \frac{ag+bc}{c}$ $[a>0,\quad b>0,\quad c>0,\quad g>0]$
       BI (138)(6, 7, 9, 10)a

18. $\int_0^\infty \ln \frac{a^2+b^2x^2}{x^2}\frac{dx}{c^2-g^2x^2} = \frac{1}{cg}\arctan \frac{ag}{bc}$ $[a>0, b>0, c>0, g>0]$
       BI (138)(8, 11)a

19. $\int_0^\infty \ln \frac{1+x^2}{x^2}\frac{x^2\, dx}{(1+x^2)^2} = \frac{\pi}{4}(\ln 4 - 1)$ BI (139)(21)

20. $\int_0^1 \ln^2\left(\frac{1-x^2}{x^2}\right)\sqrt{1-x^2}\, dx = \pi$ FI II 643a

21. $\int_0^1 \ln \frac{1+2x\cos t + x^2}{(1+x)^2}\frac{dx}{x} = \frac{1}{2}\int_0^\infty \ln \frac{1+2x\cos t + x^2}{(1+x)^2}\frac{dx}{x} = -\frac{t^2}{2}$
       $[|t|<\pi]$ BI (115)(23), BI (134)(15)

22. $\int_0^\infty \ln \frac{1+2x\cos t + x^2}{(1+x)^2} x^{p-1}\, dx = -\frac{2\pi(1-\cos pt)}{p\sin p\pi}$ $[0<|p|<1,\quad |t|<\pi]$ BI (134)(17)

23. $\displaystyle\int_0^1 \ln\frac{1+x^2\sin t}{1-x^2\sin t}\frac{dx}{\sqrt{1-x^2}} = \pi\ln\cot\left(\frac{\pi-t}{4}\right)$ $[|t|<\pi]$ GW (325)(21d)

4.299

1. $\displaystyle\int_0^\infty \ln\frac{(x+1)\left(x+a^2\right)}{(x+a)^2}\frac{dx}{x} = (\ln a)^2$ $[a>0]$ BI (134)(14)

2. $\displaystyle\int_0^1 \ln\frac{(1-ax)\left(1+ax^2\right)}{\left(1-ax^2\right)^2}\frac{dx}{1+ax^2} = \frac{1}{2\sqrt{a}}\arctan\sqrt{a}\ln(1+a)$

                                                 $[a>0]$ BI (115)(25)

3. $\displaystyle\int_0^1 \ln\frac{\left(1-a^2x^2\right)\left(1+ax^2\right)}{\left(1-ax^2\right)^2}\frac{dx}{1+ax^2} = \frac{1}{\sqrt{a}}\arctan\sqrt{a}\ln(1+a)$

                                                   $[a>0]$ BI (115)(26)

4. $\displaystyle\int_0^1 \ln\frac{(x+1)\left(x+a^2\right)}{(x+a)^2}x^{\mu-1}\,dx = \frac{\pi(a^\mu-1)^2}{\mu\sin\mu\pi}$ $[a>0,\quad\operatorname{Re}\mu>0]$ BI (134)(16)

4.311

1.[3] $\displaystyle\int_0^\infty \ln\left(a^3-x^3\right)\frac{dx}{x^3}$ integral divergent   BI (134)(7)

2. $\displaystyle\int_0^\infty \ln\left(1+x^3\right)\frac{dx}{1-x+x^2} = \frac{2\pi}{\sqrt{3}}\ln 3$ LI (136)(8)

3. $\displaystyle\int_0^\infty \ln\left(1+x^3\right)\frac{dx}{1+x^3} = \frac{\pi}{\sqrt{3}}\ln 3 - \frac{\pi^2}{9}$ LI (136)(6)

4. $\displaystyle\int_0^\infty \ln\left(1+x^3\right)\frac{x\,dx}{1+x^3} = \frac{\pi}{\sqrt{3}}\ln 3 + \frac{\pi^2}{9}$ LI (136)(7)

5. $\displaystyle\int_0^\infty \ln\left(1+x^3\right)\frac{1-x}{1+x^3}\,dx = -\frac{2}{9}\pi^2$ BI (136)(9)

6.[8] $\displaystyle\int_0^\infty \left|1-\frac{x^3}{a^3}\right|\frac{dx}{x^3} = -\frac{\pi\sqrt{3}}{6a^2}$

4.312

1. $\displaystyle\int_0^\infty \ln\frac{1+x^3}{x^3}\frac{dx}{1+x^3} = \frac{\pi}{\sqrt{3}}\ln 3 + \frac{\pi^2}{9}$ BI (138)(12)

2. $\displaystyle\int_0^\infty \ln\frac{1+x^3}{x^3}\frac{x\,dx}{1+x^3} = \frac{\pi}{\sqrt{3}}\ln 3 - \frac{\pi^2}{9}$ BI (138)(13)

4.313

1. $\displaystyle\int_0^\infty \ln x\ln\left(1+a^2x^2\right)\frac{dx}{x^2} = \pi a\left(1-\ln a\right)$ $[a>0]$ BI (134)(18)

2. $\displaystyle\int_0^\infty \ln\left(1+c^2x^2\right)\ln\left(a^2+b^2x^2\right)\frac{dx}{x^2} = 2\pi\left[\left(c+\frac{b}{a}\right)\ln(b+ac) - \frac{b}{a}\ln b - c\ln c\right]$

                                               $[a>0,\quad b>0,\quad c>0]$

                                                                 BI (134)(20, 21)a

3. $\int_0^\infty \ln\left(1 + c^2 x^2\right) \ln\left(a^2 + \dfrac{b^2}{x^2}\right) \dfrac{dx}{x^2} = 2\pi\left[\dfrac{a + bc}{b}\ln(a + bc) - \dfrac{a}{b}\ln a - c\right]$

$\qquad\qquad\qquad\qquad\qquad\qquad\qquad\qquad [a > 0, \quad a + bc > 0] \qquad$ BI (134)(22, 23)a

4. $\int_0^\infty \ln x \ln \dfrac{1 + a^2 x^2}{1 + b^2 x^2} \dfrac{dx}{x^2} = \pi(a - b) + \pi \ln \dfrac{b^b}{a^a} \qquad [a > 0, \quad b > 0] \qquad$ BI (134)(24)

5. $\int_0^\infty \ln x \ln \dfrac{a^2 + 2bx + x^2}{a^2 - 2bx + x^2} \dfrac{dx}{x} = 2\pi \ln a \arcsin \dfrac{b}{a} \qquad [a \geq |b|] \qquad$ BI (134)(25)

6. $\int_0^\infty \ln(1 + x) \dfrac{x \ln x - x - a}{(x + a)^2} \dfrac{dx}{x} = \dfrac{(\ln a)^2}{2(a - 1)} \qquad [a > 0] \qquad$ BI (141)(7)

7. $\int_0^\infty \ln^2(1 - x) \dfrac{x \ln x - x - a}{(x + a)^2} \dfrac{dx}{x} = \dfrac{\pi^2 + (\ln a)^2}{1 + a} \qquad [a > 0] \qquad$ LI (141)(8)

4.314

1. $\int_0^1 \ln(1 + ax) \dfrac{x^{p-1} - x^{q-1}}{\ln x} dx = \sum_{k=1}^\infty \dfrac{a^k}{k} \ln \dfrac{p + k}{q + k} + \ln \dfrac{p}{q}$

$\qquad\qquad\qquad\qquad\qquad\qquad\qquad [a > 0, \quad p > 0, \quad q > 0] \qquad$ BI (123)(18)

2. $\int_0^\infty \left[\dfrac{(q - 1)x}{(1 + x)^2} - \dfrac{1}{x + 1} + \dfrac{1}{(1 + x)^q}\right] \dfrac{dx}{x \ln(1 + x)} = \ln \Gamma(q)$

$\qquad\qquad\qquad\qquad\qquad\qquad\qquad\qquad\qquad [q > 0] \qquad$ BI (143)(7)

3. $\int_0^1 \dfrac{x \ln x + 1 - x}{x(\ln x)^2} \ln(1 + x) dx = \ln \dfrac{4}{\pi} \qquad$ BI (126)(12)

4. $\int_0^1 \dfrac{\ln\left(1 - x^2\right) dx}{x\left(q^2 + (\ln x)^2\right)} = -\dfrac{\pi}{q} \ln \Gamma\left(\dfrac{q + \pi}{\pi}\right) + \dfrac{\pi}{2q} \ln 2q + \ln \dfrac{q}{\pi} - 1$

$\qquad\qquad\qquad\qquad\qquad\qquad\qquad\qquad\qquad [q > 0] \qquad$ LI (327)(12)a

4.315

1. $\int_0^1 \ln(1 + x)(\ln x)^{n-1} \dfrac{dx}{x} = (-1)^{n-1}(n - 1)!\left(1 - \dfrac{1}{2^n}\right)\zeta(n + 1) \qquad$ BI (116)(3)

2. $\int_0^1 \ln(1 + x)(\ln x)^{2n} \dfrac{dx}{x} = \dfrac{2^{2n+1} - 1}{(2n + 1)(2n + 2)}\pi^{2n+2}|B_{2n+2}| \qquad$ BI (116)(1)

3. $\int_0^1 \ln(1 - x)(\ln x)^{n-1} \dfrac{dx}{x} = (-1)^n(n - 1)! \zeta(n + 1) \qquad$ BI (116)(4)

4. $\int_0^1 \ln(1 - x)(\ln x)^{2n} \dfrac{dx}{x} = -\dfrac{2^{2n}}{(n + 1)(2n + 1)}\pi^{2n+2}|B_{2n+2}| \qquad$ BI (116)(2)

4.316

1. $\int_0^1 \ln\left(1 - ax^r\right)\left(\ln \dfrac{1}{x}\right)^p \dfrac{dx}{x} = -\dfrac{1}{r^{p+1}}\Gamma(p + 1)\sum_{k=1}^\infty \dfrac{a^k}{k^{p+2}}$

$\qquad\qquad\qquad\qquad\qquad\qquad\qquad [p > -1, \quad a < 1, \quad r > 0] \qquad$ BI (116)(7)

2. $\displaystyle\int_0^1 \ln\left(1 - 2ax\cos t + a^2 x^2\right)\left(\ln\frac{1}{x}\right)^p \frac{dx}{x} = -2\,\Gamma(p+1)\sum_{k=1}^{\infty}\frac{a^k\cos kt}{k^{p+2}}$ LI (116)(8)

4.317

1. $\displaystyle\int_0^{\infty} \ln\frac{\sqrt{1+x^2}+a}{\sqrt{1+x^2}-a}\frac{dx}{\sqrt{1+x^2}} = \pi\arcsin a$ $[|a| < 1]$ BI (142)(11)

2. $\displaystyle\int_0^1 \ln\frac{\sqrt{1-a^2x^2}-x\sqrt{1-a^2}}{1-x}\frac{dx}{x} = \frac{1}{2}(\arcsin a)^2$ BI (115)(32)

3. $\displaystyle\int_0^1 \ln\frac{1+\cos t\sqrt{1-x^2}}{1-\cos t\sqrt{1-x^2}}\frac{dx}{x^2+\tan^2 v} = \pi\cot t\frac{\cos\dfrac{v-t}{2}}{\sin\dfrac{v+t}{2}}$ BI (115)(30)

4. $\displaystyle\int_0^1 \ln^2\left(\frac{x+\sqrt{1-x^2}}{x-\sqrt{1-x^2}}\right)\frac{x\,dx}{1-x^2} = \frac{\pi^2}{2}$ BI (115)(31)

5. $\displaystyle\int_0^1 \ln\left\{\sqrt{1+kx}+\sqrt{1-kx}\right\}\frac{dx}{\sqrt{(1-x^2)(1-k^2x^2)}} = \frac{1}{4}\ln(4k)\,\boldsymbol{K}(k) + \frac{\pi}{8}\,\boldsymbol{K}(k')$ BI (121)(8)

6. $\displaystyle\int_0^1 \ln\left\{\sqrt{1+kx}-\sqrt{1-kx}\right\}\frac{dx}{\sqrt{(1-x^2)(1-k^2x^2)}} = \frac{1}{4}\ln(4k)\,\boldsymbol{K}(k) + \frac{3}{8}\pi\,\boldsymbol{K}(k')$ BI (121)(9)

7. $\displaystyle\int_0^1 \ln\left\{1+\sqrt{1-k^2x^2}\right\}\frac{dx}{\sqrt{(1-x^2)(1-k^2x^2)}} = \frac{1}{2}\ln k\,\boldsymbol{K}(k) + \frac{\pi}{4}\,\boldsymbol{K}(k')$ BI (121)(6)

8. $\displaystyle\int_0^1 \ln\left\{1-\sqrt{1-k^2x^2}\right\}\frac{dx}{\sqrt{(1-x^2)(1-k^2x^2)}} = \frac{1}{2}\ln k\,\boldsymbol{K}(k) - \frac{3}{4}\pi\,\boldsymbol{K}(k')$ BI (121)(7)

9. $\displaystyle\int_0^1 \ln\frac{1+p\sqrt{1-x^2}}{1-p\sqrt{1-x^2}}\frac{dx}{1-x} = \pi\arcsin p$ $[p^2 < 1]$ BI (115)(29)

10. $\displaystyle\int_0^1 \ln\frac{1+q\sqrt{1-k^2x^2}}{1-q\sqrt{1-k^2x^2}}\frac{dx}{\sqrt{(1-x^2)(1-k^2x^2)}} = \pi\,F\left(\arcsin q, k'\right)$

                                                             $[q^2 < 1]$ BI (122)(15)

11.[10] $\displaystyle\int_{-\infty}^{\infty} \ln\left|\frac{1+2\sqrt{1+x^2}}{1-2\sqrt{1+x^2}}\right|\frac{dx}{\sqrt{1+x^2}} = \frac{\pi^2}{3}$

4.318

1. $\displaystyle\int_0^1 \frac{\ln(1-x^q)}{1+(\ln x)^2}\frac{dx}{x} = \pi\left[\ln\Gamma\left(\frac{q}{2\pi}+1\right) - \frac{\ln q}{2} + \frac{q}{2\pi}\left(\ln\frac{q}{2\pi}-1\right)\right]$

                                                        $[q > 0]$ BI (126)(11)

2. $\displaystyle\int_0^{\infty} \ln(1+x^r)\left[\frac{(p-r)x^p-(q-r)x^q}{\ln x} + \frac{x^q-x^p}{(\ln x)^2}\right]\frac{dx}{x^{r+1}} = r\ln\left(\tan\frac{q\pi}{2r}\cot\frac{p\pi}{2r}\right)$

                                                   $[p < r, \quad q < r]$ BI (143)(9)

In integrals containing $\ln(a+bx^r)$, it is useful to make the substitution $x^r = t$ and then to seek the resulting integral in the tables. For example,

$$\int_0^\infty x^{p-1} \ln\left(1 + x^r\right) dx = \frac{1}{r}\int_0^\infty t^{\frac{p}{r}-1} \ln(1+t)\, dt = \frac{\pi}{p\sin\dfrac{p\pi}{r}} \qquad \text{(see 4.293 3)}$$

4.319

1. $$\int_0^\infty \ln\left(1 - e^{-2a\pi x}\right) \frac{dx}{1+x^2} = -\pi\left[\frac{1}{2}\ln 2a\pi + a\left(\ln a - 1\right) - \ln\Gamma(a+1)\right]$$

$$[a > 0] \qquad \text{BI (354)(6)}$$

2. $$\int_0^\infty \ln\left(1 + e^{-2a\pi x}\right) \frac{dx}{1+x^2} = \pi\left[\ln\Gamma(2a) - \ln\Gamma(a) + a\left(1 - \ln a\right) - \left(2a - \frac{1}{2}\right)\ln 2\right]$$

$$[a > 0] \qquad \text{BI (354)(7)}$$

3. $$\int_0^\infty \ln\frac{a + be^{-px}}{a + be^{-qx}} \frac{dx}{x} = \ln\frac{a}{a+b}\ln\frac{p}{q} \qquad \left[\frac{b}{a} > -1, \quad pq > 0\right]$$

$$\text{FI II 635, BI (354)(1)}$$

4.321

1. $$\int_{-\infty}^\infty x\ln\cosh x\, dx = 0 \qquad\qquad \text{BI (358)(2)a}$$

2. $$\int_{-\infty}^\infty \ln\cosh x\frac{dx}{1-x^2} = 0 \qquad\qquad \text{BI (138)(20)a}$$

4.322

1.[7] $$\int_0^\pi \ln\sin xx\, dx = \frac{1}{2}\int_0^\pi \ln\cos^2 x\, dx = -\frac{\pi^2}{2}\ln 2 \qquad \text{BI (432)(1, 2) FI II 643}$$

2. $$\int_0^\infty \frac{\ln\sin^2 ax}{b^2 + x^2}\, dx = \frac{\pi}{b}\ln\frac{1 - e^{-2ab}}{2} \qquad [a > 0, \quad b > 0] \qquad \text{GW (338)(28b)}$$

3. $$\int_0^\infty \frac{\ln\cos^2 ax}{b^2 + x^2}\, dx = \frac{\pi}{b}\ln\frac{1 + e^{-2ab}}{2} \qquad [a > 0, \quad b > 0] \qquad \text{GW (338)(28a)}$$

4. $$\int_0^\infty \frac{\ln\sin^2 ax}{b^2 - x^2}\, dx = -\frac{\pi^2}{2b} + a\pi \qquad [a > 0, \quad b > 0] \qquad \text{BI (418)(1)}$$

5. $$\int_0^\infty \frac{\ln\cos^2 ax}{b^2 - x^2}\, dx = a\pi \qquad [a > 0] \qquad \text{BI (418)(2)}$$

6. $$\int_0^\infty \frac{\ln\cos^2 x}{x^2}\, dx = -\pi \qquad\qquad \text{FI II 686}$$

7.[7] $$\int_0^{\pi/4} \ln\sin xx^{\mu-1}\, dx = -\frac{1}{2\mu}\left(\frac{\pi}{4}\right)^\mu \left[\ln 2 + \frac{2}{\mu} - \sum_{k=1}^\infty \frac{\zeta(2k)}{4^{2k-1}(\mu + 2k)}\right]$$

$$[\operatorname{Re}\mu > 0] \qquad \text{LI (425)(1)}$$

8.[7] $$\int_0^{\pi/2} \ln\sin xx^{\mu-1}\, dx = -\frac{1}{\mu}\left(\frac{\pi}{2}\right)^\mu \left[\frac{1}{\mu} - 2\sum_{k=1}^\infty \frac{\zeta(2k)}{4^k(\mu + 2k)}\right]$$

$$[\operatorname{Re}\mu > 0] \qquad \text{LI (430)(1)}$$

9. $\displaystyle\int_0^{\pi/2} \ln\left(1 - \cos x\right) x^{\mu-1}\, dx = -\frac{1}{\mu}\left(\frac{\pi}{2}\right)^{\mu}\left[\frac{2}{\mu} - \sum_{k=1}^{\infty} \frac{\zeta(2k)}{4^{2k-1}(\mu+2k)}\right]$

$$[\operatorname{Re}\mu > 0] \qquad\qquad \text{LI (430)(2)}$$

10. $\displaystyle\int_0^{\infty} \ln\left(1 \pm 2p\cos\beta x + p^2\right) \frac{dx}{q^2 + x^2} = \frac{\pi}{q}\ln\left(1 \pm pe^{-\beta q}\right) \qquad [p^2 < 1]$

$$= \frac{\pi}{q}\ln\left(p \pm e^{-\beta q}\right) \qquad [p^2 > 1]$$

$$\text{FI II 718a}$$

4.323

1.[7] $\displaystyle\int_0^{\pi} x\ln\tan^2 dx = 0$ BI (432)(3)

2. $\displaystyle\int_0^{\infty} \frac{\ln\tan^2 ax}{b^2 + x^2}\, dx = \frac{\pi}{b}\ln\tanh ab$ $[a > 0, \quad b > 0]$ GW (338)(28c)

3. $\displaystyle\int_0^{\infty} \ln\left(\frac{1 + \tan x}{1 - \tan x}\right)^2 \frac{dx}{x} = \frac{\pi^2}{2}$ GW (338)(26)

4.324

1. $\displaystyle\int_0^{\infty} \ln\left(\frac{1 + \sin x}{1 - \sin x}\right)^2 \frac{dx}{x} = \pi^2$ GW (338)(25)

2. $\displaystyle\int_0^{\infty} \ln\frac{1 + 2a\cos px + a^2}{1 + 2a\cos qx + a^2} \frac{dx}{x} = \ln(1 + a)\ln\frac{q^2}{p^2}$ $[-1 < a \leq 1]$

$$= \ln\left(1 + \frac{1}{a}\right)\ln\frac{q^2}{p^2} \qquad [a < -1 \text{ or } a \geq 1]$$

$$\text{GW (338)(27)}$$

3. $\displaystyle\int_0^{\infty} \ln\left(a^2\sin^2 px + b^2\cos^2 px\right) \frac{dx}{c^2 + x^2} = \frac{\pi}{c}\left[\ln\left(a\sinh cp + b\cosh cp\right) - cp\right]$

$$[a > 0, \quad b > 0, \quad c > 0, \quad p > 0]$$
$$\text{GW (338)(29)}$$

4.325

1.[3] $\displaystyle\int_0^1 \ln\ln\left(\frac{1}{x}\right) \frac{dx}{1 + x} = -\boldsymbol{C}\ln 2 + \sum_{k=2}^{\infty} (-1)^k \frac{\ln k}{k} = -\boldsymbol{C}\ln 2 + 0.159868905\cdots = -\frac{1}{2}(\ln 2)^2$

$$\text{GW (325)(25a)}$$

2. $\displaystyle\int_0^1 \ln\ln\left(\frac{1}{x}\right) \frac{dx}{x + e^{i\lambda}} = \sum_{k=1}^{\infty} \frac{(-1)^k}{k} e^{-ik\lambda}\left(\boldsymbol{C} + \ln k\right)$ GW (325)(26)

3. $\displaystyle\int_0^1 \ln\ln\left(\frac{1}{x}\right) \frac{dx}{(1 + x)^2} = \int_1^{\infty} \ln\ln x \frac{dx}{(1 + x)^2} = \frac{1}{2}\left[\psi\left(\frac{1}{2}\right) + \ln 2\pi\right] = \frac{1}{2}\left(\ln\frac{\pi}{2} - \boldsymbol{C}\right)$

$$\text{BI (147)(7)}$$

4. $\displaystyle\int_0^1 \ln\ln\left(\frac{1}{x}\right) \frac{dx}{1 + x^2} = \int_1^{\infty} \ln\ln x \frac{dx}{1 + x^2} = \frac{\pi}{2}\ln\frac{\sqrt{2\pi}\,\Gamma\left(\frac{3}{4}\right)}{\Gamma\left(\frac{1}{4}\right)}$ BI (148)(1)

5. $\displaystyle\int_0^1 \ln\ln\left(\frac{1}{x}\right)\frac{dx}{1+x+x^2} = \int_1^\infty \ln\ln x \frac{dx}{1+x+x^2} = \frac{\pi}{\sqrt{3}}\ln\frac{\sqrt[3]{2\pi}\,\Gamma\left(\frac{2}{3}\right)}{\Gamma\left(\frac{1}{3}\right)}$ BI (148)(2)

6. $\displaystyle\int_0^1 \ln\ln\left(\frac{1}{x}\right)\frac{dx}{1-x+x^2} = \int_1^\infty \ln\ln x \frac{dx}{1-x+x^2} = \frac{2\pi}{\sqrt{3}}\left[\frac{5}{6}\ln 2\pi - \ln\Gamma\left(\frac{1}{6}\right)\right]$ BI (148)(5)

7. $\displaystyle\int_0^1 \ln\ln\left(\frac{1}{x}\right)\frac{dx}{1+2x\cos t + x^2} = \int_1^\infty \ln\ln x \frac{dx}{1+2x\cos t + x^2} = \frac{\pi}{2\sin t}\ln\frac{(2\pi)^{t/\pi}\,\Gamma\left(\frac{1}{2}+\frac{t}{2\pi}\right)}{\Gamma\left(\frac{1}{2}-\frac{t}{2\pi}\right)}$

                                                                BI (147)(9)

8. $\displaystyle\int_0^1 \ln\ln\frac{1}{x}x^{\mu-1}\,dx = -\frac{1}{\mu}\left(C + \ln\mu\right)$ $[\mathrm{Re}\,\mu > 0]$ BI (147)(1)

9. $\displaystyle\int_1^\infty \ln\ln x \frac{x^{n-2}\,dx}{1+x^2+x^4+\cdots+x^{2n-2}}$

$$= \frac{\pi}{2n}\tan\frac{\pi}{2n}\ln 2\pi + \frac{\pi}{n}\sum_{k=1}^{n-1}(-1)^{k-1}\sin\frac{k\pi}{n}\ln\frac{\Gamma\left(\dfrac{n+k}{2n}\right)}{\Gamma\left(\dfrac{k}{2n}\right)} \qquad [n \text{ is even}]$$

$$= \frac{\pi}{2n}\tan\frac{\pi}{2n}\ln\pi + \frac{\pi}{n}\sum_{k=1}^{\frac{n-1}{2}}(-1)^{k-1}\sin\frac{k\pi}{n}\ln\frac{\Gamma\left(\dfrac{n-k}{n}\right)}{\Gamma\left(\dfrac{k}{n}\right)} \qquad [n \text{ is odd}]$$

                                                                BI (148)(4)

10. $\displaystyle\int_0^1 \ln\ln\left(\frac{1}{x}\right)\frac{dx}{(1+x^2)\sqrt{\ln\dfrac{1}{x}}} = \int_1^\infty \ln\ln x \frac{dx}{(1+x^2)\sqrt{\ln x}}$

$$= \sqrt{\pi}\sum_{k=0}^\infty \frac{(-1)^{k+1}}{\sqrt{2k+1}}\left[\ln(2k+1) + 2\ln 2 + C\right]$$

                                                              BI (147)(4)

11. $\displaystyle\int_0^1 \ln\ln\left(\frac{1}{x}\right)\frac{x^{\mu-1}\,dx}{\sqrt{\ln\dfrac{1}{x}}} = -\left(C + \ln 4\mu\right)\sqrt{\frac{\pi}{\mu}}$ $[\mathrm{Re}\,\mu > 0]$ BI (147)(3)

12. $\displaystyle\int_0^1 \ln\ln\left(\frac{1}{x}\right)\left(\ln\frac{1}{x}\right)^{\mu-1}x^{\nu-1}\,dx = \frac{1}{\nu^\mu}\Gamma(\mu)\left[\psi(\mu) - \ln(\nu)\right]$

                                                     $[\mathrm{Re}\,\mu > 0, \quad \mathrm{Re}\,\nu > 0]$ BI (147)(2)

4.326

1. $\displaystyle\int_0^1 \ln\left(a - \ln x\right)x^{\mu-1}\,dx = \frac{1}{\mu}\left[\ln a - e^{a\mu}\,\mathrm{Ei}(-a\mu)\right]$ $[\mathrm{Re}\,\mu > 0, \quad a > 0]$ BI (107)(23)

2. $\displaystyle\int_0^{\frac{1}{e}} \ln\left(2\ln\frac{1}{x} - 1\right)\frac{x^{2\mu-1}}{\ln x}\,dx = -\frac{1}{2}\left[\mathrm{Ei}(-\mu)\right]^2$ $[\mathrm{Re}\,\mu > 0]$ BI (145)(5)

4.327

1. $$\int_0^1 \ln\left[a^2 + (\ln x)^2\right] \frac{dx}{1 + x^2} = \pi \ln \frac{2\,\Gamma\left(\dfrac{2a + 3\pi}{4\pi}\right)}{\Gamma\left(\dfrac{2a + \pi}{4\pi}\right)} + \frac{\pi}{2} \ln \frac{\pi}{2}$$

$$\left[a > -\frac{\pi}{2}\right]$$ BI (147)(10)

2. $$\int_0^1 \ln\left[a^2 + 4(\ln x)^2\right] \frac{dx}{1 + x^2} = \pi \ln \frac{2\,\Gamma\left(\dfrac{a + 3\pi}{4\pi}\right)}{\Gamma\left(\dfrac{a + \pi}{4\pi}\right)} + \frac{\pi}{2} \ln \pi$$

$$[a > -\pi]$$ BI (147)(16)a

3. $$\int_0^\infty \ln\left[a^2 + (\ln x)^2\right] x^{\mu-1}\, dx = \frac{2}{\mu}\left[-\cos a\mu\, \mathrm{ci}(a\mu) - \sin a\mu\, \mathrm{si}(a\mu) + \ln a\right]$$

$$[a > 0, \quad \mathrm{Re}\,\mu > 0]$$ GW (325)(28)

If the integrand contains a logarithm whose argument also contains a logarithm, for example, if the integrand contains $\ln\ln\dfrac{1}{x}$, it is useful to make the substitution $\ln x = t$ and then seek the transformed integral in the tables.

4.33–4.34 Combinations of logarithms and exponentials

4.331

1. $$\int_0^\infty e^{-\mu x} \ln x\, dx = -\frac{1}{\mu}\left(\mathbf{C} + \ln\mu\right)$$ $[\mathrm{Re}\,\mu > 0]$ BI (256)(2)

2. $$\int_1^\infty e^{-\mu x} \ln x\, dx = -\frac{1}{\mu}\,\mathrm{Ei}(-\mu)$$ $[\mathrm{Re}\,\mu > 0]$ BI (260)(5)

3. $$\int_0^1 e^{\mu x} \ln x\, dx = -\frac{1}{\mu}\int_0^1 \frac{e^{\mu x} - 1}{x}\, dx$$ $[\mu \neq 0]$ GW (324)(81a)

4.332

1. $$\int_0^\infty \frac{\ln x\, dx}{e^x + e^{-x} - 1} = \frac{2\pi}{\sqrt{3}}\left[\frac{5}{6}\ln 2\pi - \ln\Gamma\left(\frac{1}{6}\right)\right]$$ (cf. **4.325** 6)   BI (257)(6)

2. $$\int_0^\infty \frac{\ln x\, dx}{e^x + e^{-x} + 1} = \frac{\pi}{\sqrt{3}}\ln\left[\frac{\Gamma\left(\frac{2}{3}\right)}{\Gamma\left(\frac{1}{3}\right)}\sqrt{2\pi}\right]$$ (cf. **4.325** 5)   BI (257)(7)a, LI (260)(3)

4.333 $$\int_0^\infty e^{-\mu x^2} \ln x\, dx = -\frac{1}{4}\left(\mathbf{C} + \ln 4\mu\right)\sqrt{\frac{\pi}{\mu}}$$ $[\mathrm{Re}\,\mu > 0]$ BI (256)(8), FI II 807a

4.334 $$\int_0^\infty \frac{\ln x\, dx}{e^{x^2} + 1 + e^{-x^2}} = \frac{1}{2}\sqrt{\frac{\pi}{3}}\sum_{k=1}^\infty (-1)^k\,\frac{\mathbf{C} + \ln 4k}{\sqrt{k}}\sin\frac{k\pi}{3}$$ BI (357)(13)

4.335

1. $$\int_0^\infty e^{-\mu x}(\ln x)^2\, dx = \frac{1}{\mu}\left[\frac{\pi^2}{6} + (\mathbf{C} + \ln\mu)^2\right]$$ $[\mathrm{Re}\,\mu > 0]$ ET I 149(13)

2. $\int_0^\infty e^{-x^2} (\ln x)^2 \, dx = \frac{\sqrt{\pi}}{8} \left[(\boldsymbol{C} + 2\ln 2)^2 + \frac{\pi^2}{2} \right]$

FI II 808

3.[7] $\int_0^\infty e^{-\mu x} (\ln x)^3 \, dx = -\frac{1}{\mu} \left[(\boldsymbol{C} + \ln \mu)^3 + \frac{\pi^2}{2} (\boldsymbol{C} + \ln \mu) - \psi''(1) \right]$

MI 26

4.336

1.[7] $\text{PV} \int_0^\infty \frac{e^{-x}}{\ln x} \, dx = -0.154479567$

BI (260)(9)

2. $\int_0^\infty \frac{e^{-\mu x} \, dx}{\pi^2 + (\ln x)^2} = \nu'(\mu) - e^\mu$ $\qquad$ [$\text{Re}\,\mu > 0$]

MI 26

4.337

1. $\int_0^\infty e^{-\mu x} \ln(\beta + x) dx = \frac{1}{\mu} \left[\ln \beta - e^{\mu\beta} \, \text{Ei}(-\beta\mu) \right]$ $\qquad$ [$|\arg \beta| < \pi, \quad \text{Re}\,\mu > 0$]

BI (256)(3)

2. $\int_0^\infty e^{-\mu x} \ln(1 + \beta x) dx = -\frac{1}{\mu} e^{\frac{\mu}{\beta}} \, \text{Ei} \left(-\frac{\mu}{\beta} \right)$ $\qquad$ [$|\arg \beta| < \pi, \quad \text{Re}\,\mu > 0$]

ET I 148(4)

3. $\int_0^\infty e^{-\mu x} \ln|a - x| \, dx = \frac{1}{\mu} \left[\ln a - e^{-a\mu} \, \text{Ei}(a\mu) \right]$ $\qquad$ [$a > 0, \quad \text{Re}\,\mu > 0$]

BI (256)(4)

4.[7] $\int_0^\infty e^{-\mu x} \ln \left| \frac{\beta}{\beta - x} \right| \, dx = \frac{1}{\mu} \left[e^{-\beta\mu} \, \text{Ei}(\beta\mu) \right]$ $\qquad$ [$\text{Re}\,\mu > 0$]

MI 26

4.338

1. $\int_0^\infty e^{-\mu x} \ln \left(\beta^2 + x^2 \right) \, dx = \frac{2}{\mu} \left[\ln \beta - \text{ci}(\beta\mu) \cos(\beta\mu) - \text{si}(\beta\mu) \sin(\beta\mu) \right]$

$\qquad\qquad$ [$\text{Re}\,\beta > 0, \quad \text{Re}\,\mu > 0$]

BI (256)(6)

2. $\int_0^\infty e^{-\mu x} \ln^2 \left(x^2 - \beta^2 \right) \, dx = \frac{2}{\mu} \left[\ln^2 \beta - e^{\beta\mu} \, \text{Ei}(-\beta\mu) - e^{\beta\mu} \, \text{Ei}(\beta\mu) \right]$

$\qquad\qquad$ [$\text{Im}\,\beta > 0, \quad \text{Re}\,\mu > 0$]

BI (256)(5)

4.339 $\int_0^\infty e^{-\mu x} \ln \left| \frac{x + 1}{x - 1} \right| \, dx = \frac{1}{\mu} \left[e^{-\mu} (\ln 2\mu + \gamma) - e^\mu \, \text{Ei}(-2\mu) \right]$

$\qquad\qquad$ [$\text{Re}\,\mu > 0$]

MI 27

4.341 $\int_0^\infty e^{-\mu x} \ln \frac{\sqrt{x + ai} + \sqrt{x - ai}}{\sqrt{2a}} \, dx = \frac{\pi}{4\mu} \left[\mathbf{H}_0(a\mu) - Y_0(a\mu) \right]$

$\qquad\qquad$ [$a > 0, \quad \text{Re}\,\mu > 0$]

ET I 149(20)

4.342

1. $\int_0^\infty e^{-2nx} \ln(\sinh x) \, dx = \frac{1}{2n} \left[\frac{1}{n} + \ln 2 - 2\,\beta(2n + 1) \right]$

BI (256)(17)

2. $\int_0^\infty e^{-\mu x} \ln(\cosh x) \, dx = \frac{1}{\mu} \left[\beta \left(\frac{\mu}{2} \right) - \frac{1}{\mu} \right]$ $\qquad$ [$\text{Re}\,\mu > 0$]

ET I 165(32)

3. $\int_0^\infty e^{-\mu x} [\ln(shx) - \ln x] \, dx = \frac{1}{\mu} \left[\ln \frac{\mu}{2} - \frac{1}{2\mu} - \psi \left(\frac{\mu}{2} \right) \right]$

$\qquad\qquad$ [$\text{Re}\,\mu > 0$]

ET I 165(33)

4.343 $\int_0^\pi e^{\mu \cos x} \left[\ln \left(2\mu \sin^2 x \right) + \boldsymbol{C} \right] \, dx = -\pi K_0(\mu)$

WA 95(16)

4.35–4.36 Combinations of logarithms, exponentials, and powers

4.351

1. $\displaystyle\int_0^1 (1-x)e^{-x}\ln x\,dx = \frac{1-e}{e}$ BI (352)(1)

2. $\displaystyle\int_0^1 e^{\mu x}\left(\mu x^2 + 2x\right)\ln x\,dx = \frac{1}{\mu^2}\left[(1-\mu)e^{\mu}-1\right]$ BI (352)(2)

3. $\displaystyle\int_1^{\infty} \frac{e^{-\mu x}\ln x}{1+x}\,dx = \frac{1}{2}e^{\mu}[\mathrm{Ei}(-\mu)]^2$ $[\mathrm{Re}\,\mu > 0]$ NT 32(10)

4.352

1. $\displaystyle\int_0^{\infty} x^{\nu-1}e^{-\mu x}\ln x\,dx = \frac{1}{\mu^{\nu}}\,\Gamma(\nu)\left[\psi(\nu) - \ln\mu\right]$ $[\mathrm{Re}\,\mu > 0, \quad \mathrm{Re}\,\nu > 0]$

                                                                    BI (353)(3), ET I 315(10)a

2. $\displaystyle\int_0^{\infty} x^n e^{-\mu x}\ln x\,dx = \frac{n!}{\mu^{n+1}}\left[1 + \frac{1}{2} + \frac{1}{3} + \cdots + \frac{1}{n} - C - \ln\mu\right]$

                                                         $[\mathrm{Re}\,\mu > 0]$ ET I 148(7)

3. $\displaystyle\int_0^{\infty} x^{n-\frac{1}{2}} e^{-\mu x}\ln x\,dx = \sqrt{\pi}\,\frac{(2n-1)!!}{2^n \mu^{n+\frac{1}{2}}}\left[2\left(1 + \frac{1}{3} + \frac{1}{5} + \cdots + \frac{1}{2n-1}\right) - C - \ln 4\mu\right]$

                                                         $[\mathrm{Re}\,\mu > 0]$ ET I 148(10)

4. $\displaystyle\int_0^{\infty} x^{\mu-1} e^{-x}\ln x\,dx = \Gamma'(\mu)$ $[\mathrm{Re}\,\mu > 0]$ GW (324)(83a)

4.353

1. $\displaystyle\int_0^{\infty} (x-\nu)x^{\nu-1}e^{-x}\ln x\,dx = \Gamma(\nu)$ $[\mathrm{Re}\,\nu > 0]$ GW (324)(84)

2. $\displaystyle\int_0^{\infty} \left(\mu x - n - \frac{1}{2}\right)x^{n-\frac{1}{2}}e^{-\mu x}\ln x\,dx = \frac{(2n-1)!!}{(2\mu)^n}\sqrt{\frac{\pi}{\mu}}$

                                                         $[\mathrm{Re}\,\mu > 0]$ BI (357)(2)

3. $\displaystyle\int_0^1 (\mu x + n + 1)x^n e^{\mu x}\ln x\,dx = e^{\mu}\sum_{k=0}^{n}(-1)^{k-1}\frac{n!}{(n-k)!\mu^{k+1}} + (-1)^n\frac{n!}{\mu^{n+1}}$

                                                         $[\mu \neq 0]$ GW (324)(82)

4.354

1.[6] $\displaystyle\int_0^{\infty} \frac{x^{\nu-1}\ln x}{e^x + 1}\,dx = \Gamma(\nu)\sum_{k=1}^{\infty}\frac{(-1)^{k-1}}{k^{\nu}}\left[\psi(\nu) - \ln k\right]$ $[\mathrm{Re}\,\nu > 0]$

                             $= -\frac{1}{2}(\ln 2)^2$ $[\text{for } \nu = 1]$

                                                                     GW (324)(86a)

2.[7] $\displaystyle\int_0^{\infty} \frac{x^{\nu-1}\ln x}{(e^x + 1)^2}\,dx = \Gamma(\nu)\sum_{k=2}^{\infty}\frac{(-1)^k(k-1)}{k^{\nu}}\left[\psi(\nu) - \ln k\right]$

                                                         $[\mathrm{Re}\,\nu > 1]$ GW (324)(86b)

3. $\int_0^\infty \dfrac{(x-\nu)e^x - \nu}{(e^x+1)^2} x^{\nu-1} \ln x \, dx = \Gamma(\nu) \sum_{k=1}^\infty \dfrac{(-1)^{k-1}}{k^\nu}$ $\qquad$ [Re $\nu > 0$] $\qquad$ GW (324)(87a)

4. $\int_0^\infty \dfrac{(x-2n)e^x - 2n}{(e^x+1)^2} x^{2n-1} \ln x \, dx = \dfrac{2^{2n-1}-1}{2n} \pi^{2n} |B_{2n}|$

$\qquad\qquad\qquad\qquad\qquad\qquad\qquad\qquad\qquad$ [$n = 1, 2, \ldots$] $\qquad$ GW (324)(87b)

5. $\int_0^\infty \dfrac{x^{\nu-1} \ln x}{(e^x+1)^n} \, dx = (-1)^n \dfrac{\Gamma(\nu)}{(n-1)!} \sum_{k=n}^\infty \dfrac{(-1)^k (k-1)!}{(k-n)! k^\nu} [\psi(\nu) - \ln k]$

$\qquad\qquad\qquad\qquad\qquad\qquad\qquad\qquad\qquad$ [Re $\nu > 0$] $\qquad$ GW (324)(86c)

4.355

1. $\int_0^\infty x^2 e^{-\mu x^2} \ln x \, dx = \dfrac{1}{8\mu} (2 - \ln 4\mu - \boldsymbol{C}) \sqrt{\dfrac{\pi}{\mu}}$ $\qquad$ [Re $\mu > 0$] $\qquad$ BI (357)(1)a

2. $\int_0^\infty x \left(\mu x^2 - \nu x - 1\right) e^{-\mu x^2 + 2\nu x} \ln x \, dx = \dfrac{1}{4\mu} + \dfrac{\nu}{4\mu} \sqrt{\dfrac{\pi}{\mu}} \exp\left(\dfrac{\nu^2}{\mu}\right) \left[1 + \Phi\left(\dfrac{\nu}{\sqrt{\mu}}\right)\right]$

$\qquad\qquad\qquad\qquad\qquad\qquad\qquad\qquad\qquad$ [Re $\mu > 0$] $\qquad$ BI (358)(1)

3. $\int_0^\infty \left(\mu x^2 - n\right) x^{2n-1} e^{-\mu x^2} \ln x \, dx = \dfrac{(n-1)!}{4\mu^n}$ $\qquad$ [Re $\mu > 0$] $\qquad$ BI (353)(4)

4. $\int_0^\infty \left(2\mu x^2 - 2n - 1\right) x^{2n} e^{-\mu x^2} \ln x \, dx = \dfrac{(2n-1)!!}{2(2\mu)^n} \sqrt{\dfrac{\pi}{\mu}}$

$\qquad\qquad\qquad\qquad\qquad\qquad\qquad\qquad\qquad$ [Re $\mu > 0$] $\qquad$ BI (353)(5)

4.356

1. $\int_0^\infty \exp\left[-\mu\left(\dfrac{x}{a} + \dfrac{a}{x}\right)\right] \ln x \, \dfrac{dx}{x} = 2 \ln a \, K_0(2\mu)$ $\qquad$ [$a > 0$, $\quad$ Re $\mu > 0$] $\qquad$ GW (324)(91)

2. $\int_0^\infty \exp\left(-ax - \dfrac{b}{x}\right) \ln x \left[2ax^2 - (2n+1)x - 2b\right] x^{n-\frac{1}{2}} \, dx$

$$= 2\left(\dfrac{b}{a}\right)^{\frac{n}{2}} \sqrt{\dfrac{\pi}{a}} e^{-2\sqrt{ab}} \sum_{k=0}^\infty \dfrac{(n+k)!}{(n-k)!(2k)!! \left(2\sqrt{ab}\right)^k}$$

$\qquad\qquad\qquad\qquad\qquad\qquad\qquad\qquad\qquad$ [$a > 0$, $\quad b > 0$] $\qquad$ BI (357)(4)

3. $\int_0^\infty \exp\left(-ax - \dfrac{b}{x}\right) \ln x \left[2ax^2 + (2n-1)x - 2b\right] \dfrac{dx}{x^{n+\frac{3}{2}}}$

$$= 2\left(\dfrac{a}{b}\right)^{\frac{n}{2}} \sqrt{\dfrac{\pi}{a}} e^{-2\sqrt{ab}} \sum_{k=0}^\infty \dfrac{(n+k-1)!}{(n-k-1)!(2k)!! \left(2\sqrt{ab}\right)^k}$$

$\qquad\qquad\qquad\qquad\qquad\qquad\qquad\qquad\qquad$ [$a > 0$, $\quad b > 0$] $\qquad$ BI (357)(11)

For $n = \frac{1}{2}$:

4. $\int_0^\infty \exp\left(-ax - \dfrac{a}{x}\right) \ln x \dfrac{ax^2 - b}{x^2} \, dx = 2 K_0\left(2\sqrt{ab}\right)$ $\qquad$ [$a > 0$, $\quad b > 0$] $\qquad$ GW (324)(92c)

For $n = 0$:

5. $\int_0^\infty \exp\left(-ax - \dfrac{b}{x}\right) \ln x \dfrac{2ax^2 - x - 2b}{x\sqrt{x}}\, dx = 2\sqrt{\dfrac{\pi}{a}} e^{-2\sqrt{ab}}$

$$[a > 0, \quad b > 0]$$

BI (357)(7), GW(324)(92a)

For $n = -1$:

6. $\int_0^\infty \exp\left(-ax - \dfrac{b}{x}\right) \ln x \dfrac{2ax^2 - 3x - 2b}{\sqrt{x}}\, dx = \dfrac{1 + 2\sqrt{ab}}{a}\sqrt{\dfrac{\pi}{a}} e^{-2\sqrt{ab}}$

$$[a > 0, \quad b > 0]$$

LI (357)(6), GW (324)(92b)

7.[9] $\int_0^\infty \exp\left(-ax - \dfrac{b}{x}\right) \ln x \left(a - \dfrac{b}{x^2}\right) dx = K_0\left(2\sqrt{ab}\right)$

$$[a > 0, \quad b > 0]$$

8.[9] $\int_0^\infty \exp\left(-ax - \dfrac{b}{x}\right) \ln x \left[2ax^2 - (2n+1)x - 2b\right] x^{n-\frac{3}{2}}\, dx$

$$= 4\left(\dfrac{b}{a}\right)^{(2n+1)/4} K_{n+\frac{1}{2}}\left(2\sqrt{ab}\right)$$

$$= 2\left(\dfrac{b}{a}\right)^{\frac{n}{2}}\sqrt{\dfrac{\pi}{a}} e^{-2\sqrt{ab}} \sum_{k=0}^{n} \dfrac{(n+k)!}{(n-k)!(2k)!!\left(2\sqrt{ab}\right)^k}$$

$$[n = 0, 1, \ldots, a > 0, \quad b > 0]$$

9.[9] $\int_0^\infty \exp\left(-ax - \dfrac{b}{x}\right) \ln\left[(ax^2 - b)\cos(\alpha \ln x) + \alpha x \sin(\alpha \ln x)\right] \dfrac{dx}{x^2}$

$$= 2\cos\left(\alpha \ln \sqrt{b/a}\right) K_{i\alpha}\left(2\sqrt{ab}\right)$$

$$[a > 0, \quad b > 0, \quad -\infty < \alpha < \infty]$$

10.[9] $\int_0^\infty \exp\left(-ax - \dfrac{b}{x}\right) \ln x \left[(ax^2 - b)\sin(\alpha \ln x) - \alpha x \cos(\alpha \ln x)\right] \dfrac{dx}{x^2}$

$$= 2\sin\left(\alpha \ln \sqrt{b/a}\right) K_{i\alpha}\left(2\sqrt{ab}\right)$$

$$[a > 0, \quad b > 0, \quad -\infty < \alpha < \infty]$$

11.[9] $q\int_0^\infty x^\alpha \ln x \left[a - \dfrac{\alpha}{x} - \dfrac{b}{x^2}\right] \exp\left(-ax - \dfrac{b}{x}\right) dx = 2\left(\dfrac{b}{a}\right)^{\alpha/2} K_\alpha\left(2\sqrt{ab}\right)$

$$[a > 0, \quad b > 0, \quad -\infty < \alpha < \infty]$$

4.357

1. $\int_0^\infty \exp\left(-\dfrac{1 + x^4}{2ax^2}\right) \ln x \dfrac{1 + ax^2 - x^4}{x^2}\, dx = -\dfrac{\sqrt{2a^3\pi}}{2\sqrt[a]{e}}$

$$[a > 0]$$

BI (357)(8)

2. $\int_0^\infty \exp\left(-\dfrac{1 + x^4}{2ax^2}\right) \ln x \dfrac{x^4 + ax^2 - 1}{x^4}\, dx = \dfrac{\sqrt{2a^3\pi}}{2\sqrt[a]{e}} \quad [a > 0]$

BI (357)(9)

3. $\quad \int_0^\infty \exp\left(-\dfrac{1+x^4}{2ax^2}\right) \ln x \dfrac{x^4 + 3ax - 1}{x^6} \, dx = \dfrac{(1+a)\sqrt{2a^3\pi}}{2\sqrt[a]{e}}$

$$[a > 0] \qquad\qquad \text{BI (357)(10)}$$

4.358

1.[6] $\quad \int_1^\infty x^{\nu-1} e^{-\mu x} (\ln x)^m \, dx = \dfrac{\partial^m}{\partial \nu^m} \left\{\mu^{-\nu} \, \Gamma(\nu, \mu)\right\}$ $\qquad [m = 0, 1, \ldots, \quad \operatorname{Re}\mu > 0, \quad \operatorname{Re}\nu > 0]$

$$\text{MI 26}$$

2. $\quad \int_0^\infty x^{\nu-1} e^{-\mu x} (\ln x)^2 \, dx = \dfrac{\Gamma(\nu)}{\mu^\nu} \left\{[\psi(\nu) - \ln\mu]^2 + \zeta(2, \nu)\right\}$

$$[\operatorname{Re}\mu > 0, \quad \operatorname{Re}\nu > 0] \qquad \text{MI 26}$$

3.[9] $\quad \int_0^\infty x^{\nu-1} e^{-\mu x} (\ln x)^3 \, dx = \dfrac{\Gamma(\nu)}{\mu^\nu} \left\{[\psi(\nu) - \ln\mu]^3 + 3\,\zeta(2, \nu)\,[\psi(\nu) - \ln\mu] - 2\,\zeta(3, \nu)\right\}$

$$[\operatorname{Re}\mu > 0, \quad \operatorname{Re}\nu > 0] \qquad \text{MI 26}$$

4.[7] $\quad \int_0^\infty x^{\nu-1} e^{-\mu x} (\ln x)^4 \, dx = \dfrac{\Gamma(\nu)}{\nu} \left\{ [\psi(\nu) - \ln\mu]^4 + 6\,\zeta(2, \nu)[\psi(\nu) - \ln\mu]^2 \right.$

$$\left. -8\,\zeta(3, \nu)\,[\psi(\nu) - \ln\mu] + 3[\zeta(2, \nu)]^2 + 6\,\zeta(4, \nu) \right\}$$

$$[\operatorname{Re}\mu > 0, \quad \operatorname{Re}\nu > 0]$$

5.[3] $\quad \int_0^\infty x^{\nu-1} e^{-\mu x} (\ln x)^n \, dx = \dfrac{\partial^n}{\partial \nu^n} \left\{\mu^{-\nu} \, \Gamma(\nu)\right\}$ $\qquad [n = 0, 1, 2, \ldots]$

4.359

1. $\quad \int_0^\infty e^{-\mu x} \dfrac{x^{p-1} - x^{q-1}}{\ln x} \, dx = \dfrac{1}{\mu}[\lambda(\mu, p - 1) - \lambda(\mu, q - 1)]$

$$[\operatorname{Re}\mu > 0, \quad p > 0, \quad q > 0] \qquad \text{MI 27}$$

2. $\quad \int_0^1 e^{\mu x} \dfrac{x^{p-1} - x^{q-1}}{\ln x} \, dx = \sum_{k=0}^\infty \dfrac{\mu^k}{k!} \ln \dfrac{p + k}{q + k}$ $\qquad [\operatorname{Re}\mu > 0, \quad p > 0, \quad q > 0] \qquad \text{BI (352)(9)}$

4.361

1. $\quad \int_0^\infty \dfrac{(x+1)e^{-\mu x}}{\pi^2 + (\ln x)^2} \, dx = \nu'(\mu) - \nu''(\mu)$ $\qquad [\operatorname{Re}\mu > 0] \qquad\qquad \text{MI 27}$

2. $\quad \int_0^\infty \dfrac{e^{-\mu x} \, dx}{x\left[\pi^2 + (\ln x)^2\right]} = e^\mu - \nu(\mu)$ $\qquad [\operatorname{Re}\mu > 0] \qquad\qquad \text{MI 27}$

4.362

1. $\quad \int_0^1 x e^x \ln(1 - x) \, dx = 1 - e$

$$\text{BI (352)(5)a}$$

2. $\displaystyle\int_1^\infty e^{-\mu x}\ln(2x-1)\frac{dx}{x}=\frac{1}{2}\left[\mathrm{Ei}\left(-\frac{\mu}{2}\right)\right]^2$ $[\mathrm{Re}\,\mu>0]$ ET I 148(8)

4.363

1. $\displaystyle\int_0^\infty e^{-\mu x}\ln(a+x)\frac{\mu(x+a)\ln(x+a)-2}{x+a}\,dx$

$$=\frac{1}{4}\int_0^\infty e^{-\mu x}\ln^2(a-x)\frac{\mu(x-a)\ln^2(x-a)-4}{x-a}\,dx=(\ln a)^2$$

$$[\mathrm{Re}\,\mu>0,\quad a>0]\qquad \text{BI (354)(4, 5)}$$

2. $\displaystyle\int_0^1 x(1-x)(2-x)e^{-(1-x)^2}\ln(1-x)\,dx=\frac{1-e}{4e}$ BI (352)(4)

4.364

1. $\displaystyle\int_0^\infty e^{-\mu x}\ln[(x+a)(x+b)]\frac{dx}{x+a+b}=e^{(a+b)\mu}\{\mathrm{Ei}(-a\mu)\,\mathrm{Ei}(-b\mu)-\ln(ab)\,\mathrm{Ei}[-(a+b)\mu]\}$

$$[a>0,\quad b>0,\quad \mathrm{Re}\,\mu>0]\quad \text{BI (354)(11)}$$

2. $\displaystyle\int_0^\infty e^{-\mu x}\ln(x+a+b)\left(\frac{1}{x+a}+\frac{1}{x+b}\right)dx$

$$=(1+\ln a\ln b)\ln(a+b)+e^{-(a+b)\mu}\{\mathrm{Ei}(-\alpha\mu)\,\mathrm{Ei}(-b\mu)\}$$

$$+(1-\ln(ab))\,\mathrm{Ei}[-(a+b)\mu]$$

$$[a>0,\quad b>0,\quad \mathrm{Re}\,\mu>0]\quad \text{BI (354)(12)}$$

4.365 $\displaystyle\int_0^\infty\left[e^{-x}-\frac{x}{(1+x)^{p+1}\ln(1+x)}\right]\frac{dx}{x}=\ln p$ $[p>0]$ BI (354)(15)

4.366

1. $\displaystyle\int_0^\infty e^{-\mu x}\ln\left(1+\frac{x^2}{a^2}\right)\frac{dx}{x}=[\mathrm{ci}(a\mu)]^2+[\mathrm{si}(a\mu)]^2$ $[\mathrm{Re}\,\mu>0]$ NT 32(11)a

2. $\displaystyle\int_0^\infty e^{-\mu x}\ln\left|1-\frac{x^2}{a^2}\right|\frac{dx}{x}=\mathrm{Ei}(a\mu)\,\mathrm{Ei}(-a\mu)$ $[\mathrm{Re}\,\mu>0]$ ME 18

3. $\displaystyle\int_0^\infty xe^{-\mu x^2}\ln\left|\frac{1+x^2}{1-x^2}\right|dx=\frac{1}{\mu}\left[\cosh\mu\sinh(i\mu)-\sinh\mu\cosh(i\mu)\right]$

$$[\mathrm{Re}\,\mu>0]\,;\qquad (\text{cf. } \textbf{4.339})\qquad \text{MI 27}$$

4.367 $\displaystyle\int_0^\infty xe^{-\mu x^2}\ln\frac{x+\sqrt{x^2+2\beta}}{\sqrt{2\beta}}\,dx=\frac{e^{\beta\mu}}{4\mu}K_0(\beta\mu)$ $[|\arg\beta|<\pi,\quad \mathrm{Re}\,\mu>0]$ ET I 149(19)

4.368 $\displaystyle\int_0^{2u} e^{-\mu x^2}\ln\frac{x^2(4u^2-x^2)}{u^4}\frac{dx}{\sqrt{4u^2-x^2}}=\frac{\pi}{2}e^{-2u^2\mu}\left[\frac{\pi}{2}\,Y_0\left(2iu^2\mu\right)-(\boldsymbol{C}-\ln 2)\,J_0\left(2iu^2\mu\right)\right]$

$$[\mathrm{Re}\,\mu>0]\qquad \text{ET I 149(21)a}$$

4.369

1. $\displaystyle\int_0^\infty x^{\nu-1}e^{-\mu x}\left[\psi(\nu)-\ln x\right]dx=\frac{\Gamma(\nu)\ln\mu}{\mu^\nu}$ $[\mathrm{Re}\,\nu>0]$ ET I 149(12)

2.　$\int_0^\infty x^n e^{-\mu x} \left\{ \left[\ln x - \frac{1}{2} \psi(n+1) \right]^2 - \frac{1}{2} \psi'(n+1) \right\} dx$

$$= \frac{n!}{\mu^{n+1}} \left\{ \left[\ln \mu - \frac{1}{2} \psi(n+1) \right]^2 + \frac{1}{2} \psi'(n+1) \right\}$$

$[\operatorname{Re} \mu > 0]$　　　　　　　　MI 26

4.37 Combinations of logarithms and hyperbolic functions

4.371

1.　$\int_0^\infty \frac{\ln x}{\cosh x} dx = \pi \ln \left[\frac{\sqrt{2\pi}\,\Gamma\left(\frac{3}{4}\right)}{\Gamma\left(\frac{1}{4}\right)} \right]$　　　　　　　　LI (260)(1)a

2.　$\int_0^\infty \frac{\ln x\, dx}{\cosh x + \cos t} = \frac{\pi}{\sin t} \ln \frac{(2\pi)^{t/\pi}\,\Gamma\left(\dfrac{\pi + t}{2\pi}\right)}{\Gamma\left(\dfrac{\pi - t}{2\pi}\right)}$　　$[t^2 < \pi^2]$　　BI (257)(7)a

3.　$\int_0^\infty \frac{\ln x\, dx}{\cosh^2 x} = \psi\left(\frac{1}{2}\right) + \ln \pi = \ln \pi - 2\ln 2 - C$　　　　BI (257)(4)a

4.372

1.　$\int_1^\infty \ln x \frac{\sinh mx}{\sinh nx} dx$

$$= \frac{\pi}{2n} \tan \frac{m\pi}{2n} \ln 2\pi + \frac{\pi}{n} \sum_{k=1}^{n-1} (-1)^{k-1} \sin \frac{km\pi}{n} \ln \frac{\Gamma\left(\dfrac{n+k}{2n}\right)}{\Gamma\left(\dfrac{k}{2n}\right)}$$　　$]m+n$ is odd$]$

$$= \frac{\pi}{2n} \tan \frac{m\pi}{2n} \ln \pi + \frac{\pi}{n} \sum_{k=1}^{\frac{n-1}{2}} (-1)^{k-1} \sin \frac{km\pi}{n} \ln \frac{\Gamma\left(\dfrac{n-k}{n}\right)}{\Gamma\left(\dfrac{k}{n}\right)}$$　　$[m+n$ is even$]$

BI (148)(3)a

2.　$\int_1^\infty \ln x \frac{\cosh mx}{\cosh nx} dx$

$$= \frac{\pi}{2n} \frac{\ln 2\pi}{\cos \dfrac{m\pi}{2n}} + \frac{\pi}{n} \sum_{k=1}^{n} (-1)^{k-1} \cos \frac{(2k-1)m\pi}{2n} \ln \frac{\Gamma\left(\dfrac{2n+2k-1}{4n}\right)}{\Gamma\left(\dfrac{2k-1}{4n}\right)}$$　　$[m+n$ is odd$]$

$$= \frac{\pi}{2n} \frac{\ln \pi}{\cos \dfrac{m\pi}{2n}} + \frac{\pi}{n} \sum_{k=1}^{\frac{n-1}{2}} (-1)^{k-1} \cos \frac{(2k-1)m\pi}{2n} \ln \frac{\Gamma\left(\dfrac{2n-2k+1}{2n}\right)}{\Gamma\left(\dfrac{2k-1}{2n}\right)}$$　　$[m+n$ is even$]$

BI (148)(6)a

4.373

1. $$\int_0^\infty \frac{\ln\left(a^2 + x^2\right)}{\cosh bx}\, dx = \frac{\pi}{b}\left[2\ln\frac{2\,\Gamma\left(\dfrac{2ab + 3\pi}{4\pi}\right)}{\Gamma\left(\dfrac{2ab + \pi}{4\pi}\right)} - \ln\frac{2b}{\pi}\right]$$

$$\left[b > 0, \quad a > -\frac{\pi}{2b}\right].$$ BI (258)(11)a

2. $$\int_0^\infty \ln\left(1 + x^2\right)\frac{dx}{\cosh\dfrac{\pi x}{2}} = 2\ln\frac{4}{\pi}$$ BI (258)(1)a

3. $$\int_0^\infty \ln\left(a^2 + x^2\right)\frac{\sinh\left(\dfrac{2}{3}\pi x\right)}{\sinh \pi x}\, dx = 2\sin\frac{\pi}{3}\ln\frac{6\,\Gamma\left(\dfrac{a+4}{6}\right)\Gamma\left(\dfrac{a+5}{6}\right)}{\Gamma\left(\dfrac{a+1}{6}\right)\Gamma\left(\dfrac{a+2}{6}\right)}$$

$$[a > -1].$$ BI (258)(12)

4. $$\int_0^\infty \ln\left(1 + x^2\right)\frac{dx}{\sinh^2 ax} = \frac{2}{a}\left[\ln\frac{a}{\pi} + \frac{\pi}{2a} - \psi\left(\frac{\pi + a}{\pi}\right)\right]$$

$$[a > 0]$$ BI (258)(5)

5. $$\int_0^\infty \ln\left(1 + x^2\right)\frac{\cosh\left(\dfrac{\pi}{2}x\right)}{\sinh^2\left(\dfrac{\pi}{2}x\right)}\, dx = \frac{2\pi - 4}{\pi}$$ BI (258)(3)

6. $$\int_0^\infty \ln\left(1 + x^2\right)\frac{\cosh\left(\dfrac{\pi}{4}x\right)}{\sinh^2\left(\dfrac{\pi}{4}x\right)}\, dx = 4\sqrt{2} - \frac{16}{\pi} + \frac{8\sqrt{2}}{\pi}\ln\left(\sqrt{2} + 1\right)$$ BI (258)(2)

4.374

1. $$\int_0^\infty \ln\left(\cos^2 t + e^{-2x}\sin^2 t\right)\frac{dx}{\sinh x} = -2t^2$$ BI (259)(10)a

2. $$\int_0^\infty \ln\left(a + be^{-2x}\right)\frac{dx}{\cosh^2 x} = \frac{2}{(b-a)}\left[\frac{a+b}{2}\ln(a+b) - a\ln a - b\ln 2\right]$$

$$[a > 0, \quad a + b > 0]$$ LI (259)(14)

4.375

1. $$\int_0^\infty \ln\cosh\frac{x}{2}\frac{dx}{\cosh x} = \boldsymbol{G} + \frac{\pi}{4}\ln 2$$ BI (259)(11)

2. $$\int_0^\infty \ln\coth x\frac{dx}{\cosh x} = \frac{\pi}{2}\ln 2$$ BI (259)(16)

4.376

1. $$\int_0^\infty \frac{\ln x}{\sqrt{x}\cosh x}\, dx = 2\sqrt{\pi}\sum_{k=0}^\infty \frac{(-1)^{k+1}}{\sqrt{2k+1}}\left\{\ln(2k+1) + 2\ln 2 + \boldsymbol{C}\right\}$$ BI (147)(4)

2. $\displaystyle\int_0^\infty \ln x \frac{(\mu + 1)\cosh x - x \sinh x}{\cosh^2 x} x^\mu \, dx = 2\,\Gamma(\mu + 1) \sum_{k=0}^\infty \frac{(-1)^{k+1}}{(2k+1)^{\mu+1}}$

$$[\mathrm{Re}\,\mu > -1] \qquad\qquad \text{BI (356)(10)}$$

3. $\displaystyle\int_0^\infty \ln x \frac{(n + 1)\cosh x - x \sinh x}{\cosh^2 x} x^n \, dx = \frac{(-1)^n}{2^n} \beta^{(n)}\left(\frac{1}{2}\right)$

4. $\displaystyle\int_0^\infty \ln 2x \frac{n \sinh 2ax - ax}{\sinh^2 ax} x^{2n-1} \, dx = -\frac{1}{n}\left(\frac{\pi}{a}\right)^{2n} |B_{2n}|$

$$[n = 1, 2, \ldots] \qquad\qquad \text{BI (356)(9)a}$$

5. $\displaystyle\int_0^\infty \ln x \frac{ax \cosh ax - (2n+1)\sinh ax}{\sinh^2 ax} x^{2n} \, dx = 2\frac{2^{2n+1}-1}{(2a)^{2n+1}} (2n)!\,\zeta(2n+1)$ BI (356)(14)

6. $\displaystyle\int_0^\infty \ln x \frac{ax \cosh ax - 2n \sinh ax}{\sinh^2 ax} x^{2n-1} \, dx = \frac{2^{2n-1}-1}{2n} |B_{2n}|\left(\frac{\pi}{a}\right)^{2n}$

$$[n = 1, 2, \ldots, a > 0] \qquad\qquad \text{BI (356)(15)}$$

7. $\displaystyle\int_0^\infty \ln \frac{(2n+1)\cosh ax - ax \sinh ax}{\cosh^2 ax} x^{2n} \, dx = -\left(\frac{\pi}{2a}\right)^{2n+1} |E_{2n}|$

$$[a > 0] \qquad\qquad \text{BI (356)(11)}$$

8.[6] $\displaystyle\int_0^\infty \ln x \frac{2ax \sinh ax - (2n+1)\cosh ax}{\cosh^3 ax} x^{2n} \, dx = \begin{cases} \dfrac{2}{a}\left(2^{2n-1}-1\right)\left(\dfrac{\pi}{2a}\right)^{2n} |B_{2n}| & n = 1, 2, \ldots \\[2ex] \dfrac{1}{a} & n = 0 \end{cases}$

$$[a > 0] \qquad\qquad \text{BI (356)(2)}$$

9.[6] $\displaystyle\int_0^\infty \ln x \frac{2ax \cosh ax - (2n+1)\sinh ax}{\sinh^3 ax} x^{2n} \, dx = \frac{1}{a}\left(\frac{\pi}{a}\right)^{2n} |B_{2n}|$

$$[a > 0, \quad n = 1, 2, \ldots] \qquad\qquad \text{BI (356)(6)a}$$

10. $\displaystyle\int_0^\infty \ln x \frac{x \sinh x - 6 \sinh^2\left(\frac{x}{2}\right) - 6 \cos^2 \frac{t}{2}}{(\cosh x + \cos t)^2} x^2 \, dx = \frac{(\pi - t^2)\,t}{3 \sin t}$

$$[0 < t < \pi] \qquad\qquad \text{BI (356)(16)a}$$

11. $\displaystyle\int_0^\infty \ln\left(1 + x^2\right) \frac{\cosh \pi x + \pi x \sinh \pi x}{\cosh^2 \pi x} \frac{dx}{x^2} = 4 - \pi$ BI (356)(12)

12. $\displaystyle\int_0^\infty \ln\left(1 + 4x^2\right) \frac{\cosh \pi x + \pi x \sinh \pi x}{\cosh^2 \pi x} \frac{dx}{x^2} = 4 \ln 2$ BI (356)(13)

4.377 $\displaystyle\int_0^\infty \ln 2x \frac{ax - n\left(1 - e^{-2ax}\right)}{\sinh^2 ax} x^{2n-1} \, dx = \frac{1}{2n}\left(\frac{\pi}{a}\right)^{2n} |B_{2n}|$

$$[n = 1, 2, \ldots] \qquad\qquad \text{LI (356)(8)a}$$

4.38–4.41 Logarithms and trigonometric functions

4.381

1. $\displaystyle\int_0^1 \ln x \sin ax \, dx = -\frac{1}{a}\left[C + \ln a - \operatorname{ci}(a)\right]$ $[a > 0]$ GW (338)(2a)

2. $\displaystyle\int_0^1 \ln x \cos ax \, dx = -\frac{1}{a}\left[\operatorname{si}(a) + \frac{\pi}{2}\right]$ $[a > 0]$ BI (284)(2)

3. $\displaystyle\int_0^{2\pi} \ln x \sin nx \, dx = -\frac{1}{n}\left[C + \ln(2n\pi) - \operatorname{ci}(2n\pi)\right]$ GW (338)(1a)

4. $\displaystyle\int_0^{2\pi} \ln x \cos nx \, dx = -\frac{1}{n}\left[\operatorname{si}(2n\pi) + \frac{\pi}{2}\right]$ GW (338)(1b)

4.382

1. $\displaystyle\int_0^\infty \ln\left|\frac{x+a}{x-a}\right| \sin bx \, dx = \frac{\pi}{b}\sin ab$ $[a < 0, \quad b > 0]$ ET I 77(11)

2.[10] $\displaystyle\int_0^\infty \ln\left|\frac{x+a}{x-a}\right| \cos bx \, dx = \frac{2}{b}\left[\cos(ab)\left\{\operatorname{si}(ab) + \frac{\pi}{2}\right\} - \sin(ab)\operatorname{ci}(ab)\right]$

$[a > 0, \quad b > 0]$ ET I 18(9)

3. $\displaystyle\int_0^\infty \ln\frac{a^2 + x^2}{b^2 + x^2} \cos cx \, dx = \frac{\pi}{c}\left(e^{-bc} - e^{-ac}\right)$ $[a > 0, \quad b > 0, \quad c > 0]$

FI III 648a, BI (337)(5)

4. $\displaystyle\int_0^\infty \ln\frac{x^2 + x + a^2}{x^2 - x + a^2} \sin bx \, dx = \frac{2\pi}{b}\exp\left(-b\sqrt{a^2 - \frac{1}{4}}\right)\sin\frac{b}{2}$

$[b > 0]$ ET I 77(12)

5. $\displaystyle\int_0^\infty \ln\frac{(x+\beta)^2 + \gamma^2}{(x-\beta)^2 + \gamma^2} \sin bx \, dx = \frac{2\pi}{b}e^{-\gamma b}\sin \beta b$ $[\operatorname{Re}\gamma > 0, \quad |\operatorname{Im}\beta| \le \operatorname{Re}\gamma, \quad b > 0]$

ET I 77(13)

4.383

1. $\displaystyle\int_0^\infty \ln\left(1 + e^{-\beta x}\right)\cos bx \, dx = \frac{\beta}{2b^2} - \frac{\pi}{2b\sinh\left(\dfrac{\pi b}{\beta}\right)}$ $[\operatorname{Re}\beta > 0, \quad b > 0]$ ET I 18(13)

2. $\displaystyle\int_0^\infty \ln\left(1 - e^{-\beta x}\right)\cos bx \, dx = \frac{\beta}{2b^2} - \frac{\pi}{2b}\coth\left(\frac{\pi b}{\beta}\right)$ $[\operatorname{Re}\beta > 0, \quad b > 0]$ ET I 18(14)

4.384

1. $\displaystyle\int_0^1 \ln(\sin\pi x)\sin 2n\pi x \, dx = 0$ GW (338)(3a)

2.[7] $\displaystyle\int_0^1 \ln(\sin\pi x)\sin(2n+1)\pi x \, dx = 2\int_0^{1/2} \ln(\sin\pi x)\sin(2n+1)\pi x \, dx$

$$= \frac{2}{(2n+1)\pi}\left[\ln 2 - \frac{1}{2n+1} - 2\sum_{k=1}^n \frac{1}{2k-1}\right]$$

GW (338)(3b)

3.^{6} $\int_0^1 \ln(\sin \pi x) \cos 2n\pi x \, dx = 2 \int_0^{1/2} \ln(\sin \pi x) \cos 2n\pi x \, dx$

$$= -\ln 2 \qquad\qquad [n = 0]$$

$$= -\frac{1}{2n} \qquad\qquad [n > 0]$$

GW (338)(3c)

4. $\int_0^1 \ln(\sin \pi x) \cos(2n+1)\pi x \, dx = 0$ GW (338)(3d)

5. $\int_0^{\pi/2} \ln \sin x \sin x \, dx = \ln 2 - 1$ BI (305)(4)

6. $\int_0^{\pi/2} \ln \sin x \cos x \, dx = -1$ BI (305)(5)

7. $\int_0^{\pi/2} \ln \sin x \cos 2nx \, dx = \begin{cases} -\dfrac{\pi}{4n}, & \text{for } n > 0 \\ -\dfrac{\pi}{2} \ln 2, & \text{for } n = 0 \end{cases}$ LI (305)(6)

8. $\int_0^{\pi} \ln \sin x \cos[2m(x-n)] \, dx = -\dfrac{\pi \cos 2mn}{2m}$ LI (330)(8)

9. $\int_0^{\pi/2} \ln \sin x \sin^2 x \, dx = \dfrac{\pi}{8}(1 - \ln 4)$ BI (305)(7)

10. $\int_0^{\pi/2} \ln \sin x \cos^2 x \, dx = -\dfrac{\pi}{8}(1 + \ln 4)$ BI (305)(8)

11. $\int_0^{\pi/2} \ln \sin x \sin x \cos^2 x \, dx = \dfrac{1}{9}(\ln 8 - 4)$ BI (305)(9)

12. $\int_0^{\pi/2} \ln \sin x \tan x \, dx = -\dfrac{\pi^2}{24}$ BI (305)(11)

13. $\int_0^{\pi/2} \ln 2x \sin x \, dx = \int_0^{\pi/2} \ln \sin 2x \cos x \, dx = 2(\ln 2 - 1)$ BI (305)(16, 17)

14. $\int_0^{\pi} \dfrac{\ln(1 + p \cos x)}{\cos x} \, dx = \pi \arcsin p \qquad [p^2 < 1]$ FI II 484

15. $\int_0^{\pi} \ln \sin x \dfrac{dx}{1 - 2a \cos x + a^2} = \dfrac{\pi}{1 - a^2} \ln \dfrac{1 - a^2}{2} \qquad [a^2 < 1]$

$$= \dfrac{\pi}{a^2 - 1} \ln \dfrac{a^2 - 1}{2a^2} \qquad [a^2 > 1]$$

BI (331)(8)

16. $\int_0^{\pi} \ln \sin bx \dfrac{dx}{1 - 2a \cos x + a^2} = \dfrac{\pi}{1 - a^2} \ln \dfrac{1 - a^{2b}}{2} \qquad [a^2 < 1]$ BI (331)(10)

17. $\int_0^{\pi} \ln \cos bx \dfrac{dx}{1 - 2a \cos x + a^2} = \dfrac{\pi}{1 - a^2} \ln \dfrac{1 + a^{2b}}{2} \qquad [a^2 < 1]$ BI (331)(11)

18. $\int_0^{\pi/2} \ln \sin x \dfrac{dx}{1 - 2a \cos 2x + a^2} = \dfrac{1}{2} \int_0^{\pi} \ln \sin x \dfrac{dx}{1 - 2a \cos 2x + a^2}$

$\qquad\qquad\qquad\qquad\qquad\quad = \dfrac{\pi}{2(1 - a^2)} \ln \dfrac{1 - a}{2} \qquad\qquad [a^2 < 1]$

$\qquad\qquad\qquad\qquad\qquad\quad = \dfrac{\pi}{2(a^2 - 1)} \ln \dfrac{a - 1}{2a} \qquad\qquad [a^2 > 1]$

$\qquad\qquad\qquad\qquad\qquad\qquad\qquad\qquad\qquad\qquad$ BI (321)(1), BI (331)(13)

19. $\int_0^{\pi} \ln \sin bx \dfrac{dx}{1 - 2a \cos 2x + a^2} = \dfrac{\pi}{1 - a^2} \ln \dfrac{1 - a^b}{2} \qquad [a^2 < 1]$ $\qquad$ BI (331)(18)

20. $\int_0^{\pi} \ln \cos bx \dfrac{dx}{1 - 2a \cos 2x + a^2} = \dfrac{\pi}{1 - a^2} \ln \dfrac{1 + a^b}{2} \qquad [a^2 < 1]$ $\qquad$ BI (331)(21)

21. $\int_0^{\pi/2} \dfrac{\ln \cos x \, dx}{1 - 2p \cos 2x + p^2} = \dfrac{\pi}{2(1 - p^2)} \ln \dfrac{1 + p}{2} \qquad [p^2 < 1]$

$\qquad\qquad\qquad\qquad\qquad\quad = \dfrac{\pi}{2(p^2 - 1)} \ln \dfrac{p + 1}{2p} \qquad [p^2 > 1]$

$\qquad\qquad\qquad\qquad\qquad\qquad\qquad\qquad\qquad\qquad$ BI (321)(8)

22. $\int_0^{\pi} \ln \sin x \dfrac{\cos x \, dx}{1 - 2a \cos x + a^2} = \dfrac{\pi}{2a} \dfrac{1 + a^2}{1 - a^2} \ln(1 - a^2) - \dfrac{a\pi \ln 2}{1 - a^2} \qquad [a^2 < 1]$

$\qquad\qquad\qquad\qquad\qquad\qquad = \dfrac{\pi}{2a} \dfrac{a^2 + 1}{a^2 - 1} \ln \dfrac{a^2 - 1}{a^2} - \dfrac{\pi \ln 2}{a(a^2 - 1)} \qquad [a^2 > 1]$

$\qquad\qquad\qquad\qquad\qquad\qquad\qquad\qquad\qquad\qquad$ LI (331)(9)

23. $\int_0^{\pi} \ln \sin bx \dfrac{\cos x \, dx}{1 - 2a \cos 2x + a^2} = \int_0^{\pi} \ln \cos bx \dfrac{\cos x \, dx}{1 - 2a \cos 2x + a^2} = 0$

$\qquad\qquad\qquad\qquad\qquad\qquad\quad [0 < a < 1] \qquad\quad$ BI (331)(19, 22)

24. $\int_0^{\pi} \ln \sin x \dfrac{\cos^2 x \, dx}{1 - 2a \cos 2x + a^2} = \dfrac{\pi}{4a} \dfrac{1 + a}{1 - a} \ln(1 - a) - \dfrac{\pi \ln 2}{2(1 - a)} \qquad [0 < a < 1]$

$\qquad\qquad\qquad\qquad\qquad\qquad = \dfrac{\pi}{4a} \dfrac{a + 1}{a - 1} \ln \dfrac{a - 1}{a} - \dfrac{\pi \ln 2}{2a(a - 1)} \qquad [a > 1]$

$\qquad\qquad\qquad\qquad\qquad\qquad\qquad\qquad\qquad\qquad$ BI (331)(16)

25. $\int_0^{\pi/2} \ln \sin x \dfrac{\cos 2x \, dx}{1 - 2a \cos 2x + a^2} = \dfrac{1}{2} \int_0^{\pi} \ln \sin x \dfrac{\cos 2x \, dx}{1 - 2a \cos 2x + a^2}$

$\qquad\qquad\qquad\qquad\qquad\quad = \dfrac{\pi}{2a(1 - a^2)} \left\{ \dfrac{1 + a^2}{2} \ln(1 - a) - a^2 \ln 2 \right\} \qquad [a^2 < 1]$

$\qquad\qquad\qquad\qquad\qquad\quad = \dfrac{\pi}{2a(a^2 - 1)} \left\{ \dfrac{1 + a^2}{2} \ln \dfrac{a - 1}{a} - \ln 2 \right\} \qquad [a^2 > 1]$

$\qquad\qquad\qquad\qquad\qquad\qquad$ BI (321)(2), BI (331)(15), LI (321))(2)

26. $\int_0^{\pi/2} \ln \cos x \dfrac{\cos 2x \, dx}{1 - 2a \cos 2x + a^2} = \dfrac{\pi}{2a(1 - a^2)} \left\{ \dfrac{1 + a^2}{2} \ln(1 + a) - a^2 \ln 2 \right\} \qquad [a^2 < 1]$

$\qquad\qquad\qquad\qquad\qquad\quad = \dfrac{\pi}{2a(a^2 - 1)} \left\{ \dfrac{1 + a^2}{2} \ln \dfrac{1 + a}{a} - \ln 2 \right\} \qquad [a^2 > 1]$

$\qquad\qquad\qquad\qquad\qquad\qquad\qquad\qquad\qquad\qquad$ BI (321)(9)

4.385

1. $\displaystyle\int_0^\pi \ln \sin x \frac{dx}{a + b \cos x} = \frac{\pi}{\sqrt{a^2 - b^2}} \ln \frac{\sqrt{a^2 - b^2}}{a + \sqrt{a^2 - b^2}}$ $[a > 0, \quad a > b]$ BI (331)(6)

2. $\displaystyle\int_0^{\pi/2} \ln \sin x \frac{dx}{(a \sin x \pm b \cos x)^2} = \int_0^{\pi/2} \ln \cos x \frac{dx}{(a \cos x \pm b \sin x)^2}$

$$= \frac{1}{b(a^2 + b^2)} \left(\mp a \ln \frac{a}{b} - \frac{b\pi}{2} \right)$$

$[a > 0, \quad b > 0]$ BI (319)(1,6)a

3. $\displaystyle\int_0^{\pi/2} \frac{\ln \sin x \, dx}{a^2 \sin^2 x + b^2 \cos^2 x} = \int_0^{\pi/2} \frac{\ln \cos x \, dx}{b^2 \sin^2 x + a^2 \cos^2 x} = \frac{\pi}{2ab} \ln \frac{b}{a + b}$

$[a > 0, \quad b > 0]$ BI (317)(4, 10)

4. $\displaystyle\int_0^{\pi/2} \ln \sin x \frac{\sin 2x \, dx}{\left(a \sin^2 x + b \cos^2 x\right)^2} = \int_0^{\pi/2} \ln \cos x \frac{\sin 2x \, dx}{\left(b \sin^2 x + a \cos^2 x\right)^2}$

$$= \frac{1}{2b(b - a)} \ln \frac{a}{b}$$

$[a > 0, \quad b > 0]$ BI (319)(3, 7), LI (319)(3)

5. $\displaystyle\int_0^{\pi/2} \ln \sin x \frac{a^2 \sin^2 x - b^2 \cos^2 x}{\left(a^2 \sin^2 x + b^2 \cos^2 x\right)^2} \, dx = \int_0^{\pi/2} \ln \cos x \frac{a^2 \cos^2 x - b^2 \sin^2 x}{\left(a^2 \cos^2 x + b^2 \sin^2 x\right)^2} \, dx$

$$= \frac{\pi}{2b(a + b)}$$

$[a > 0, \quad b > 0]$ LI (319)(2, 8)

4.386

1. $\displaystyle\int_0^{\pi/2} \ln \sin x \frac{\sin x}{\sqrt{1 + \sin^2 x}} \, dx = \int_0^{\pi/2} \frac{\cos x \ln \cos x}{\sqrt{1 + \cos^2 x}} \, dx = -\frac{\pi}{8} \ln 2$ BI (322)Zsurround1, 6

2. $\displaystyle\int_0^{\pi/2} \frac{\sin^3 x \ln \sin x}{\sqrt{1 + \sin^2 x}} \, dx = \int_0^{\pi/2} \frac{\cos^3 x \ln \cos x}{\sqrt{1 + \cos^2 x}} \, dx = \frac{\ln 2 - 1}{4}$ BI (322)(2, 7)

3. $\displaystyle\int_0^{\pi/2} \ln \sin x \frac{dx}{\sqrt{1 - k^2 \sin^2 x}} = -\frac{1}{2} \boldsymbol{K}(k) \ln k - \frac{\pi}{4} \boldsymbol{K}(k')$ BI (322)(3)

4. $\displaystyle\int_0^{\pi/2} \frac{\ln \cos x \, dx}{\sqrt{1 - k^2 \sin^2 x}} = \frac{1}{2} \boldsymbol{K}(k) \ln \frac{k'}{k} - \frac{\pi}{4} \boldsymbol{K}(k')$ BI (322)(9)

4.387

1. $\displaystyle\int_0^{\pi/2} \ln \sin x \sin^\mu x \cos^\nu x \, dx = \int_0^{\pi/2} \ln \cos x \cos^\mu x \sin^\nu x \, dx$

$$= \frac{1}{4} \mathrm{B}\left(\frac{\mu + 1}{2}, \frac{\nu + 1}{2} \right) \left[\psi\left(\frac{\mu + 1}{2} \right) - \psi\left(\frac{\mu + \nu + 2}{2} \right) \right]$$

$[\operatorname{Re} \mu > -1, \quad \operatorname{Re} \nu > -1]$ GW (338)(6c)

2. $$\int_0^{\pi/2} \ln \sin x \, \sin^{\mu-1} x \, dx = \frac{\sqrt{\pi} \, \Gamma\left(\dfrac{\mu}{2}\right)}{4 \, \Gamma\left(\dfrac{\mu+1}{2}\right)} \left[\psi\left(\frac{\mu}{2}\right) - \psi\left(\frac{\mu+1}{2}\right) \right]$$

$$[\operatorname{Re}\mu > 0] \qquad\qquad \text{GW (338)(6a)}$$

3. $$\int_0^{\pi/2} \ln \sin x \, \cos^{\nu-1} x \, dx = \frac{\sqrt{\pi} \, \Gamma\left(\dfrac{\nu}{2}\right)}{4 \, \Gamma\left(\dfrac{\nu+1}{2}\right)} \left[\psi\left(\frac{\nu}{2}\right) - \psi\left(\frac{\nu+1}{2}\right) \right]$$

$$[\operatorname{Re}\nu > 0] \qquad\qquad \text{GW (338)(6b)}$$

4. $$\int_0^{\pi/2} \ln \sin x \, \sin^{2n} x \, dx = \frac{(2n-1)!!}{(2n)!!} \frac{\pi}{2} \left\{ \sum_{k=1}^{2n} \frac{(-1)^{k+1}}{k} - \ln 2 \right\} \qquad \text{FI II 811}$$

5. $$\int_0^{\pi/2} \ln \sin x \, \sin^{2n+1} x \, dx = \frac{(2n)!!}{(2n+1)!!} \left\{ \sum_{k=1}^{2n+1} \frac{(-1)^k}{k} + \ln 2 \right\} \qquad \text{BI (305)(13)}$$

6. $$\int_0^{\pi/2} \ln \sin x \, \cos^{2n} x \, dx = -\frac{(2n-1)!!}{(2n)!!} \frac{\pi}{4} \left[\sum_{k=1}^{n} \frac{1}{k} + \ln 4 \right]$$

$$= -\frac{(2n-1)!!}{(2n)!!} \frac{\pi}{4} \left[\boldsymbol{C} + \psi(n+1) + \ln 4 \right]$$

$$\text{BI (305)(14)}$$

7. $$\int_0^{\pi/2} \ln \sin x \, \cos^{2n+1} x \, dx = -\frac{(2n)!!}{(2n+1)!!} \sum_{k=0}^{n} \frac{1}{2k+1}$$

$$= -\frac{(2n)!!}{2(2n+1)!!} \left[\psi\left(n+\frac{3}{2}\right) - \psi\left(\frac{1}{2}\right) \right]$$

$$\text{GW (338)(7b)}$$

8. $$\int_0^{\pi/2} \ln \cos x \, \sin^{2n} x \, dx = -\frac{(2n-1)!!}{2^{n+1} \cdot n!} \frac{\pi}{2} \left\{ \boldsymbol{C} + 2 \ln 2 + \psi(n+1) \right\}$$

$$\text{BI (306)(8)}$$

9. $$\int_0^{\pi/2} \ln \cos x \, \cos^{2n} x \, dx = -\frac{(2n-1)!!}{2^n n!} \frac{\pi}{2} \left(\ln 2 + \sum_{k=1}^{2n} \frac{(-1)^k}{k} \right)$$

$$\text{BI (306)(10)}$$

10. $$\int_0^{\pi/2} \ln \cos x \, \cos^{2n} x \, dx = \frac{2^{n-1}(n-1)!}{(2n-1)!!} \left[\ln 2 + \sum_{k=1}^{2n-1} \frac{(-1)^k}{k} \right]$$

$$\text{BI (306)(9)}$$

4.388

1. $$\int_0^{\pi/4} \ln \sin x \, \frac{\sin^{2n} x}{\cos^{2n+2} x} \, dx = \frac{1}{2n+1} \left[\frac{1}{2} \ln 2 + (-1)^n \frac{\pi}{4} + \sum_{k=0}^{n-1} \frac{(-1)^k}{2n-2k-1} \right] \qquad \text{BI (288)(1)}$$

2. $$\int_0^{\pi/4} \ln \sin x \, \frac{\sin^{2n-1} x}{\cos^{2n+1} x} \, dx = \frac{1}{4n} \left[-\ln 2 + (-1)^n \ln 2 + \sum_{k=1}^{n-1} \frac{(-1)^k}{n-k} \right] \qquad \text{LI (288)(2)}$$

3. $\displaystyle\int_0^{\pi/4} \ln\cos x\, \frac{\sin^{2n} x}{\cos^{2n+2} x}\, dx = \frac{1}{2n+1}\left[-\frac{1}{2}\ln 2 + (-1)^{n+1}\frac{\pi}{4} + \sum_{k=0}^n \frac{(-1)^{k-1}}{2n-2k+1}\right]$ BI (288)(10)

4. $\displaystyle\int_0^{\pi/4} \ln\cos x\, \frac{\sin^{2n-1} x}{\cos^{2n+1} x}\, dx = \frac{1}{4n}\left[-\ln 2 + (-1)^n \ln 2 + \sum_{k=0}^{n-1}\frac{(-1)^k}{n-k}\right]$ BI (288)(11)

5. $\displaystyle\int_0^{\pi/2} \ln\sin x\, \frac{\sin^{p-1} x}{\cos^{p+1} x}\, dx = -\frac{\pi}{2p}\operatorname{cosec}\frac{p\pi}{2}$ $[0 < p < 2]$ BI (310)(4)

6. $\displaystyle\int_0^{\pi/2} \ln\sin x\, \frac{dx}{\tan^{p-1} x \sin 2x} = \frac{1}{4}\frac{\pi}{p-1}\sec\frac{p\pi}{2}$ $[p^2 < 1]$ BI (310)(3)

4.389

1. $\displaystyle\int_0^{\pi} \ln\sin x \sin^{2n} 2x \cos 2x\, dx = -\frac{(2n-1)!!}{(2n)!!}\frac{\pi}{4n+2}$ BI (330)(9)

2. $\displaystyle\int_0^{\pi/4} \ln\sin x \cos^n 2x \sin 2x\, dx = -\frac{1}{4(n+1)}\left\{\boldsymbol{C} + \psi(n+2) + \ln 2\right\}$ BI (285)(2)

3. $\displaystyle\int_0^{\pi/4} \ln\cos x \cos^{\mu-1} 2x \tan 2x\, dx = \frac{1}{4(1-\mu)}\beta(\mu)$
$[\operatorname{Re}\mu > 0]$ BI (286)(2)

4. $\displaystyle\int_0^{\pi/2} \ln\sin x \sin^{\mu-1} x \cos x\, dx = \int_0^{\pi/2} \ln\cos x \cos^{\mu-1} x \sin x\, dx = -\frac{1}{\mu^2}$
$[\operatorname{Re}\mu > 0]$ BI (306)(11)

5.³ $\displaystyle\int_{-\frac{\pi}{2}}^{\frac{\pi}{2}} \ln\cos x \cos^p x \cos px\, dx = \frac{\pi}{2^{p+1}}\left[\boldsymbol{C} + \psi(p+1) - 2\ln 2\right]$

$[p > -1]$

6. $\displaystyle\int_0^{\pi/2} \ln\cos x \cos^{p-1} x \sin px \sin x\, dx = \frac{\pi}{2^{p+2}}\left[\boldsymbol{C} + \psi(p) - \frac{1}{p} - 2\ln 2\right]$
$[p > 0]$ BI (306)(12)

4.391

1. $\displaystyle\int_0^{\pi/4} (\ln\cos 2x)^n \cos^{p-1} 2x \tan x\, dx = \int_0^{\pi/4} (\ln\sin 2x)^n \sin^{p-1} 2x \tan\left(\frac{\pi}{4} - x\right) dx = \frac{1}{2}\beta^{(n)}(p)$
$[p > 0]$ BI (286)(10), BI (285)(18)

2. $\displaystyle\int_0^{\pi/4} (\ln\sin 2x)^n \sin^{p-1} 2x \tan\left(\frac{\pi}{4} + x\right) dx = \frac{(-1)^n n!}{2}\zeta(n+1, p)$ BI (285)(17)

3. $\displaystyle\int_0^{\pi/4} (\ln\cos 2x)^{2n-1} \tan x\, dx = \frac{1 - 2^{2n-1}}{4n}\pi^{2n}|B_{2n}|$ $[n = 1, 2, \dots]$ BI (286)Zsurround7

4. $\displaystyle\int_0^{\pi/4} (\ln\cos 2x)^{2n} \tan x\, dx = \frac{2^{2n}1}{2^{2n+1}}(2n)!\,\zeta(2n+1)$ BI (286)(8)

4.392

1. $\int_0^{\pi/4} \ln(\sin x \cos x) \dfrac{\sin^{2n} x}{\cos^{2n+2} x}\, dx = \dfrac{1}{2n+1}\left[(-1)^{n+1}\dfrac{\pi}{2} - \ln 2 + \dfrac{1}{2n+1} + 2\sum_{k=0}^{n-1}\dfrac{(-1)^{k-1}}{2n-2k-1}\right]$

BI (294)(8)

2. $\int_0^{\pi/4} \ln(\sin x \cos x) \dfrac{\sin^{2n-1} x}{\cos^{2n+1} x}\, dx = \dfrac{1}{2n}\left[(-1)^n \ln 2 - \ln 2 + \dfrac{1}{2n} + (-1)^n\sum_{k=1}^{n-1}\dfrac{(-1)^k}{k}\right]$

BI (294)(9)

4.393

1. $\int_0^{\pi/2} \ln\tan x \sin x\, dx = \ln 2$

BI (307)(3)

2. $\int_0^{\pi/2} \ln\tan x \cos x\, dx = -\ln 2$

BI (307)(4)

3. $\int_0^{\pi/2} \ln\tan x \sin^2 x\, dx = -\int_0^{\pi/2} \ln\tan x \cos^2 x\, dx = \dfrac{\pi}{4}$

BI (307)(5, 6)

4. $\int_0^{\pi/4} \dfrac{\ln\tan x}{\cos 2x}\, dx = -\dfrac{\pi^2}{8}$

GW (338)(10b)a

5. $\int_0^{\pi/2} \sin x \ln\cot\dfrac{x}{2}\, dx = \ln 2$

LO III 290

4.394

1. $\int_0^{\pi/2} \dfrac{\ln\tan x\, dx}{1 - 2a\cos 2x + a^2} = \dfrac{\pi}{2(1-a^2)}\ln\dfrac{1-a}{1+a}$ $[a^2 < 1]$

$\qquad\qquad\qquad\qquad\qquad = \dfrac{\pi}{2(a^2-1)}\ln\dfrac{a-1}{a+1}$ $[a^2 > 1]$

BI (321)(15)

2. $\int_0^{\pi/2} \dfrac{\ln\tan x \cos 2x\, dx}{1 - 2a\cos 2x + a^2} = \dfrac{\pi}{4a}\dfrac{1+a^2}{1-a^2}\ln\dfrac{1-a}{1+a}$ $[a^2 < 1]$

$\qquad\qquad\qquad\qquad\qquad = \dfrac{\pi}{4a}\dfrac{a^2+1}{a^2-1}\ln\dfrac{a-1}{a+1}$ $[a^2 > 1]$

BI (321)(16)

3. $\int_0^{\pi} \dfrac{\ln\tan bx\, dx}{1 - 2a\cos 2x + a^2} = \dfrac{\pi}{1-a^2}\ln\dfrac{1-a^b}{1+a^b}$ $[0 < a < 1, \quad b > 0]$ BI (331)(24)

4. $\int_0^{\pi} \dfrac{\ln\tan bx \cos x\, dx}{1 - 2a\cos 2x + a^2} = 0$ $[0 < a < 1]$ BI (331)(25)

5. $\int_0^{\pi/4} \ln\tan x \dfrac{\cos 2x\, dx}{1 - a\sin 2x} = -\dfrac{\arcsin a}{4a}(\pi + \arcsin a)$ $[a^2 \le 1]$ BI (291)(2,3)

6. $\int_0^{\pi/4} \ln\tan x \dfrac{\cos 2x\, dx}{1 - a^2\sin^2 2x} = -\dfrac{\pi}{4a}\arcsin a$ $[a^2 < 1]$ BI (291)(9)

7. $\int_0^{\pi/4} \ln \tan x \dfrac{\cos 2x \, dx}{1 + a^2 \sin^2 2x} = -\dfrac{\pi}{4a} \operatorname{arcsinh} a = -\dfrac{\pi}{4a} \ln\left(a + \sqrt{1 + a^2}\right)$

$$\left[a^2 < 1\right] \qquad \text{BI (291)(10)}$$

8. $\int_0^u \dfrac{\sin x \ln \cot \dfrac{x}{2}}{1 - \cos^2 \alpha \sin^2 x} \, dx = \operatorname{cosec} 2\alpha \left\{ \dfrac{\pi}{2} \ln 2 + L(\varphi - \alpha) - L(\varphi + \alpha) - L\left(\dfrac{\pi}{2} - 2\alpha\right) \right\}$

$$\left[\tan \varphi = \cot \alpha \cos u; \quad 0 < u < \pi\right]$$
$$\text{LO III 290}$$

9. $\int_0^{\pi/4} \dfrac{\ln \tan x \sin 2x \, dx}{1 - \cos^2 t \sin^2 2x} = \operatorname{cosec} 2t \left[L\left(\dfrac{\pi}{2} - t\right) - \left(\dfrac{\pi}{2} - t\right) \ln 2 \right]$ $\qquad$ LO III 290a

4.395

1. $\int_0^{\pi/2} \dfrac{\ln \tan x \, dx}{\sqrt{1 - k^2 \sin^2 x}} = -\ln k' \, \boldsymbol{K}(k)$ $\qquad$ BI (322)(11)

2. $\int_u^{\pi/4} \dfrac{\ln \tan x \sin 4x \, dx}{\left(\sin^2 u + \tan^2 v \sin^2 2x\right) \sqrt{\sin^2 2x - \sin^2 u}} = -\dfrac{\pi}{2} \dfrac{\cos^2 v}{\sin u \sin v} \ln \dfrac{\sin v + \sqrt{1 - \cos^2 u \cos^2 v}}{\sin u \, (1 + \sin v)}$

$$\left[0 < u < \dfrac{\pi}{2}, \quad 0 < v < \dfrac{\pi}{2}\right] \qquad \text{LO III 285a}$$

4.396

1. $\int_0^{\pi/2} \ln(a \tan x) \sin^{\mu-1} 2x \, dx = 2^{\mu-2} \ln a \dfrac{\left\{\Gamma\left(\dfrac{a}{2}\right)\right\}^2}{\Gamma(a)}$ $\qquad [a > 0, \quad \operatorname{Re} \mu > 0]$ $\qquad$ LI (307)(8)

2. $\int_0^{\pi/2} \ln \tan x \cos^{2(\mu-1)x} \, dx = -\dfrac{\sqrt{\pi}}{4} \dfrac{\Gamma\left(u - \dfrac{1}{2}\right)}{\Gamma(\mu)} \left[\boldsymbol{C} + \psi\left(\dfrac{2\mu - 1}{2}\right) + \ln 4\right]$

$$\left[\operatorname{Re} \mu > \tfrac{1}{2}\right] \qquad \text{BI (307)(9)}$$

3. $\int_0^{\pi/2} \ln \tan x \cos^{q-1} x \cot x \sin[(q+1)x] \, dx = -\dfrac{\pi}{2} \left[\boldsymbol{C} + \psi(q+1)\right]$

$$[q > -1] \qquad \text{BI (307)(11)}$$

4. $\int_0^{\pi/2} \ln \tan x \cos^{q-1} x \cos[(q+1)x] \, dx = -\dfrac{\pi}{2q}$ $\qquad [q > 0]$ $\qquad$ BI (307)(10)

5. $\int_0^{\pi/4} (\ln \tan x)^n \tan^p x \, dx = \dfrac{1}{2^{n+1}} B^{(n)}\left(\dfrac{p+1}{2}\right)$ $\qquad [p > -1]$ $\qquad$ LI (286)(22)

6. $\int_0^{\pi/2} (\ln \tan x)^{2n-1} \dfrac{dx}{\cos 2x} = \dfrac{1 - 2^{2n}}{2n} \pi^{2n} |B_{2n}|$ $\qquad [n = 1, 2, \ldots]$ $\qquad$ BI (312)(6)

7. $\int_0^{\pi/4} \ln \tan x \tan^{2n+1} x \, dx = \dfrac{(-1)^{n+1}}{4} \left[\dfrac{\pi^2}{12} + \sum_{k=1}^{n} \dfrac{(-1)^k}{k^2}\right]$ $\qquad$ GW (338)(8a)

4.397

1. $\int_0^{\pi/2} \ln\left(1 + p\sin x\right)\dfrac{dx}{\sin x} = \dfrac{\pi^2}{8} - \dfrac{1}{2}\left(\arccos p^2\right)$ $\left[p^2 < 1\right]$ BI (313)(1)

2. $\int_0^{\pi/2} \ln\left(1 + p\cos x\right)\dfrac{dx}{\cos x} = \dfrac{\pi^2}{8} - \dfrac{1}{2}\left(\arccos p\right)^2$ $\left[p^2 < 1\right]$ BI (313)(8)

3. $\int_0^{\pi} \ln\left(1 + p\cos x\right)\dfrac{dx}{\cos x} = \pi\arcsin p$ $\left[p^2 < 1\right]$ BI (331)(1)

4. $\int_0^{\pi/2} \dfrac{\cos x \ln\left(1 + \cos\alpha\cos x\right)}{1 - \cos^2\alpha\cos^2 x}\, dx = \dfrac{L\left(\dfrac{\pi}{2} - \alpha\right) - \alpha\ln\sin\alpha}{\sin\alpha\cos\alpha}$

                                                       $\left[0 < \alpha < \dfrac{\pi}{2}\right]$ LO III 291

5. $\int_0^{\pi/2} \dfrac{\cos x \ln\left(1 - \cos\alpha\cos x\right)}{1 - \cos^2\alpha\cos^2 x}\, dx = \dfrac{L\left(\dfrac{\pi}{2} - \alpha\right) + (\pi - \alpha)\ln\sin\alpha}{\sin\alpha\cos\alpha}$

                                                       $\left[0 < \alpha < \dfrac{\pi}{2}\right]$ LO III 291

6. $\int_0^{\pi} \ln\left(1 - 2a\cos x + a^2\right)\cos nx\, dx$

$$= \frac{1}{2}\int_0^{2\pi} \ln\left(1 - 2a\cos x + a^2\right)\cos nx\, dx$$

$$= -\frac{\pi}{n}a^n \qquad\qquad \left[a^2 < 1\right] \quad \text{BI (330)(11), BI (332)(5)}$$

$$= -\frac{\pi}{na^n} \qquad\qquad \left[a^2 > 1\right] \qquad \text{GW (338)(13a)}$$

7. $\int_0^{\pi} \ln\left(1 - 2a\cos x + a^2\right)\sin nx\sin x\, dx = \dfrac{1}{2}\int_0^{2\pi} \ln\left(1 - 2a\cos x + a^2\right)\sin nx\sin x\, dx$

$$= \frac{\pi}{2}\left(\frac{a^{n+1}}{n+1} - \frac{a^{n-1}}{n-1}\right)$$

$$\left[a^2 > 1\right] \qquad \text{BI (330)(10), BI (332)(4)}$$

8. $\int_0^{\pi} \ln\left(1 - 2a\cos x + a^2\right)\sin nx\sin x\, dx = \dfrac{1}{2}\int_0^{2\pi} \ln\left(1 - 2a\cos x + a^2\right)\cos nx\cos x\, dx$

$$= -\frac{\pi}{2}\left(\frac{a^{n+1}}{n+1} + \frac{a^{n-1}}{n-1}\right)$$

                                                   BI (330)(12), BI (332)(6)

9. $\int_0^{\pi} \ln\left(1 - 2a\cos 2x + a^2\right)\cos(2n-1)x\, dx = 0$ $\left[a^2 < 1\right]$ BI (330)(15)

10. $\int_0^{\pi} \ln\left(1 - 2a\cos 2x + a^2\right)\sin 2nx\sin x\, dx = 0$ $\left[a^2 < 1\right]$ BI (330)(13)

11. $\displaystyle\int_0^\pi \ln\left(1 - 2a\cos 2x + a^2\right)\sin(2n-1)x\sin x\,dx = \frac{\pi}{2}\left(\frac{a^n}{n} - \frac{a^{n-1}}{n-1}\right)$

$\left[a^2 < 1\right]$ BI (330)(14)

12. $\displaystyle\int_0^\pi \ln\left(1 - 2a\cos 2x + a^2\right)\cos 2nx\cos x\,dx = 0$ $\left[a^2 < 1\right]$ BI (330)(16)

13. $\displaystyle\int_0^\pi \ln\left(1 - 2a\cos 2x + a^2\right)\cos(2n-1)x\cos x\,dx = -\frac{\pi}{2}\left(\frac{a^n}{n} + \frac{a^{n-1}}{n-1}\right)$

$\left[a^2 < 1\right]$ BI (330)(17)

14. $\displaystyle\int_0^{\pi/2} \ln\left(1 + 2a\cos 2x + a^2\right)\sin^2 x\,dx = -\frac{a\pi}{4}$ $\left[a^2 < 1\right]$

$\displaystyle = \frac{\pi\ln a^2}{4} - \frac{\pi}{4a}$ $\left[a^2 > 1\right]$

BI (309)(22), LI (309)(22)

15. $\displaystyle\int_0^{\pi/2} \ln\left(1 + 2a\cos 2x + a^2\right)\cos^2 x\,dx = \frac{a\pi}{4}$ $\left[a^2 < 1\right]$

$\displaystyle = \frac{\pi\ln a^2}{4} + \frac{\pi}{4a}$ $\left[a^2 > 1\right]$

BI (309)(23), LI (309)(23)

16. $\displaystyle\int_0^\pi \frac{\ln\left(1 - 2a\cos x + a^2\right)}{1 - 2b\cos x + b^2}\,dx = \frac{2\pi\ln(1 - ab)}{1 - b^2}$ $\left[a^2 \leq 1, \quad b^2 < 1\right]$ BI (331)(26)

4.398

1. $\displaystyle\int_0^\pi \ln\frac{1 + 2a\cos x + a^2}{1 - 2a\cos x + a^2}\sin(2n+1)x\,dx = (-1)^n\frac{2\pi a^{2n+1}}{2n+1}$

$\left[a^2 < 1\right]$ BI (330)(18)

2. $\displaystyle\int_0^{2\pi} \ln\frac{1 - 2a\cos x + a^2}{1 - 2a\cos nx + a^2}\cos mx\,dx = 2\pi\left(\frac{n}{m}a^{m/n} - \frac{a^m}{m}\right)$ $\left[a^2 \leq 1\right]$

$\displaystyle = 2\pi\left(\frac{n}{m}a^{-m/n} - \frac{a^{-m}}{m}\right)$ $\left[a^2 \geq 1\right]$

BI (332)(9)

3. $\displaystyle\int_0^\pi \ln\frac{1 + 2a\cos 2x + a^2}{1 + 2a\cos 2nx + a^2}\cot x\,dx = 0$ BI (331)(5), LI(331)(5)

4.399

1. $\displaystyle\int_0^{\pi/2} \ln\left(1 + a\sin^2 x\right)\sin^2 x\,dx = \frac{\pi}{2}\left(\ln\frac{1 + \sqrt{1+a}}{2} - \frac{1}{2}\frac{1 - \sqrt{1+a}}{1 + \sqrt{1+a}}\right)$

$\left[a > -1\right]$ BI (309)(14)

2. $\displaystyle\int_0^{\pi/2} \ln\left(1 + a\sin^2 x\right)\cos^2 x\,dx = \frac{\pi}{2}\left(\ln\frac{1 + \sqrt{1+a}}{2} + \frac{1}{2}\frac{1 - \sqrt{1+a}}{1 + \sqrt{1+a}}\right)$

$\left[a > -1\right]$ BI (309)(15)

3.
$$\int_0^{\pi/2} \frac{\ln\left(1 - \cos^2 \beta \cos^2 x\right)}{1 - \cos^2 \alpha \cos^2 x}\, dx = -\frac{\pi}{\sin \alpha} \ln \frac{1 + \sin \alpha}{\sin \alpha + \sin \beta}$$

$$\left[0 < \beta < \frac{\pi}{2}, \quad 0 < \alpha < \frac{\pi}{2}\right] \qquad \text{LO III 285}$$

4.411

1.
$$\int_0^{\pi} \ln \frac{1 + \sin x}{1 + \cos \lambda \sin x} \frac{dx}{\sin x} = \lambda^2 \qquad\qquad \left[\lambda^2 < \pi^2\right] \qquad\qquad \text{BI (331)(2)}$$

2.
$$\int_0^{\pi/2} \ln \frac{p + q \sin ax}{p - q \sin ax} \frac{dx}{\sin ax} = \int_0^{\pi/2} \ln \frac{p + q \cos ax}{p - q \cos ax} \frac{dx}{\cos ax} = \int_0^{\pi/2} \ln \frac{p + q \tan ax}{p - q \tan ax} \frac{dx}{\tan ax} = \pi \arcsin \frac{q}{p}$$
$$[p > q > 0]$$

FI II 695a, BI (315)(5, 13,17)a

3.
$$\int_0^{\pi/2} \frac{\cos x}{1 - \cos^2 \alpha \cos^2 x} \ln \frac{1 + \cos \beta \cos x}{1 - \cos \beta \cos x}\, dx = \frac{2\pi}{\sin 2\alpha} \ln \frac{\cos \dfrac{\alpha - \beta}{2}}{\sin \dfrac{\alpha + \beta}{2}}$$

$$\left[0 < \alpha \le \beta < \frac{\pi}{2}\right] \qquad \text{LO III 284}$$

4.412

1.
$$\int_0^{\pi/4} \ln \tan\left(\frac{\pi}{4} \pm x\right) \frac{dx}{\sin 2x} = \pm \frac{\pi^2}{8} \qquad\qquad \text{BI (293)(1)}$$

2.
$$\int_0^{\pi/4} \ln \tan\left(\frac{\pi}{4} \pm x\right) \frac{dx}{\tan 2x} = \pm \frac{\pi^2}{16} \qquad\qquad \text{BI (293)(2)}$$

3.
$$\int_0^{\pi/4} \ln \tan\left(\frac{\pi}{4} \pm x\right) (\ln \tan x)^{2n} \frac{dx}{\sin 2x} = \pm \frac{2^{2n+2} - 1}{4(n+1)(2n+1)} \pi^{2n+2} |B_{2n+2}| \qquad \text{BI (294)(24)}$$

4.
$$\int_0^{\pi/4} \ln \tan\left(\frac{\pi}{4} \pm x\right) (\ln \tan x)^{2n-1} \frac{dx}{\sin 2x} = \pm \frac{1 - 2^{2n+1}}{2^{2n+2} n} (2n)!\, \zeta(2n+1) \qquad \text{BI (294)(25)}$$

5.
$$\int_0^{\pi/4} \ln \tan\left(\frac{\pi}{4} \pm x\right) (\ln \sin 2x)^{n-1} \frac{dx}{\tan 2x} = \frac{(-1)^{n-1}}{2} (n-1)!\, \zeta(n+1) \qquad \text{LI (294)(20)}$$

4.413

1.
$$\int_0^{\pi/2} \ln\left(p^2 + q^2 \tan^2 x\right) \frac{dx}{a^2 \sin^2 x + b^2 \cos^2 x} = \frac{\pi}{ab} \ln \frac{ap + bq}{a}$$
$$[a > 0, \quad b > 0, \quad p > 0, \quad q > 0]$$

BI (318)(1–4)a

2.
$$\int_0^{\pi/2} \ln\left(1 + q^2 \tan^2 x\right) \frac{1}{p^2 \sin^2 x + r^2 \cos^2 x} \frac{dx}{s^2 \sin^2 x + t^2 \cos^2 x}$$
$$= \frac{\pi}{p^2 t^2 - s^2 r^2} \left\{ \frac{p^2 - r^2}{pr} \ln\left(1 + \frac{qr}{p}\right) + \frac{t^2 - s^2}{st} \ln\left(1 + \frac{qt}{s}\right) \right\}$$
$$[q > 0, \quad p > 0, \quad r > 0, \quad s > 0, \quad t > 0] \qquad \text{BI (320)(18)}$$

3. $\displaystyle\int_0^{\pi/2} \ln\left(1 + q^2 \tan^2 x\right) \frac{\sin^2 x}{p^2 \sin^2 x + r^2 \cos^2 x} \frac{dx}{s^2 \sin^2 x + t^2 \cos^2 x}$

$$= \frac{\pi}{p^2 t^2 - s^2 r^2} \left\{ \frac{t}{s} \ln\left(1 + \frac{qr}{p}\right) - \frac{r}{p} \ln\left(1 + \frac{qt}{s}\right) \right\}$$

$$[q > 0, \quad p > 0, \quad r > 0, \quad s > 0, \quad t > 0] \quad \text{BI (320)(20)}$$

4. $\displaystyle\int_0^{\pi/2} \ln\left(1 + q^2 \tan^2 x\right) \frac{\cos^2 x}{p^2 \sin^2 x + r^2 \cos^2 x} \frac{dx}{s^2 \sin^2 x + t^2 \cos^2 x}$

$$= \frac{\pi}{p^2 t^2 - s^2 r^2} \left\{ \frac{p}{r} \ln\left(1 + \frac{qr}{p}\right) - \frac{s}{t} \ln\left(1 + \frac{qt}{s}\right) \right\}$$

$$[q > 0, \quad p > 0, \quad r > 0, \quad s > 0, \quad t > 0] \quad \text{BI (320)(21)}$$

5. $\displaystyle\int_0^{\pi} \frac{\ln \tan rx \, dx}{1 - 2p \cos x + p^2} = \frac{\pi}{1 - p^2} \ln \frac{1 - p^{2r}}{1 + p^{2r}} \qquad \left[p^2 < 1\right] \qquad \text{BI (331)(12)}$

4.414

1. $\displaystyle\int_0^{\pi/2} \ln\left(1 - k^2 \sin^2 x\right) \frac{dx}{\sqrt{1 - k^2 \sin^2 x}} = \ln k' \, \boldsymbol{K}(k) \qquad \text{BI (323)(1)}$

2. $\displaystyle\int_0^{\pi/2} \ln\left(1 - k^2 \sin^2 x\right) \frac{\sin^2 x \, dx}{\sqrt{1 - k^2 \sin^2 x}} = \frac{1}{k^2} \left\{ \left(k^2 - 2 + \ln k'\right) \boldsymbol{K}(k) + (2 - \ln k') \boldsymbol{E}(k) \right\}$

$$\text{BI (323)(3)}$$

3. $\displaystyle\int_0^{\pi/2} \ln\left(1 - k^2 \sin^2 x\right) \frac{\cos^2 x}{dx} \sqrt{1 - k^2 \sin^2 x} = \frac{1}{k^2} \left[\left(1 + k'^2 - k'^2 \ln k'\right) \boldsymbol{K}(k) - (2 - \ln k') \boldsymbol{E}(k) \right]$

$$\text{BI (323)(6)}$$

4. $\displaystyle\int_0^{\pi/2} \ln\left(1 - k^2 \sin^2 x\right) \frac{dx}{\sqrt{\left(1 - k^2 \sin^2 x\right)^3}} = \frac{1}{k'^2} \left[\left(k^2 - 2\right) \boldsymbol{K}(k) + (2 + \ln k') \boldsymbol{E}(k) \right]$

$$\text{BI (323)(9)}$$

5. $\displaystyle\int_0^{\pi/2} \ln\left(1 - k^2 \sin^2 x\right) \frac{\sin^2 x}{dx} \sqrt{\left(1 - k^2 \sin^2 x\right)^3}$

$$= \frac{1}{k^2 k'^2} \left[(2 + \ln k') \boldsymbol{E}(k) - \left(1 + k'^2 + k'^2 \ln k'\right) \boldsymbol{K}(k) \right]$$

$$\text{BI (323)(10)}$$

6. $\displaystyle\int_0^{\pi/2} \ln\left(1 - k^2 \sin^2 x\right) \frac{\cos^2 x \, dx}{\sqrt{\left(1 - k^2 \sin^2 x\right)^3}} = \frac{1}{k^2} \left[\left(1 + k'^2 + \ln k'\right) \boldsymbol{K}(k) - (2 + \ln k') \boldsymbol{E}(k) \right]$

$$\text{BI (323)(16)}$$

7. $\displaystyle\int_0^{\pi/2} \ln\left(1 - k^2 \sin^2 x\right) \sqrt{1 - k^2 \sin^2 x} \, dx = \left(1 + k'^2\right) \boldsymbol{K}(k) - (2 - \ln k') \boldsymbol{E}(k) \qquad \text{BI (324)(18)}$

8. $$\int_0^{\pi/2} \ln\left(1 - k^2 \sin^2 x\right) \sin^2 x \sqrt{1 - k^2 \sin^2 x}\, dx = \frac{1}{9k^2}\left\{\left(-2 + 11k^2 - 6k^4 + 3k'^2 \ln k'\right) \boldsymbol{K}(k)\right.$$

$$\left. + \left[2 - 10k^2 - 3\left(1 - 2k^2\right)\ln k'\right] \boldsymbol{E}(k)\right\}$$

BI (324)(20)

9. $$\int_0^{\pi/2} \ln\left(1 - k^2 \sin^2 x\right) \cos^2 x \sqrt{1 - k^2 \sin^2 x}\, dx = \frac{1}{9k^2}\left\{\left(2 + 7k^2 - 3k^4 - 3k'^2 \ln k'\right) \boldsymbol{K}(k)\right.$$

$$\left. - \left[2 + 8k^2 - 3\left(1 + k^2\right)\ln k'\right] \boldsymbol{E}(k)\right\}$$

BI (324)(21), LI (324)(21)

10. $$\int_0^{\pi/2} \ln\left(1 - k^2 \sin^2 x\right) \frac{\sin x \cos x\, dx}{\sqrt{\left(1 - k^2 \sin^2 x\right)^{2n+1}}} = \frac{2}{(2n-1)^2 k^2}\left\{\left[1 + (2n-1)\ln k'\right] k'^{1-2n} - 1\right\}$$

BI (324)(17)

4.415

1. $$\int_0^\infty \ln x \sin ax^2\, dx = -\frac{1}{4}\sqrt{\frac{\pi}{2a}}\left(\ln 4a + \boldsymbol{C} - \frac{\pi}{2}\right) \qquad [a > 0]$$ GW (338)(19)

2. $$\int_0^\infty \ln x \cos ax^2\, dx = -\frac{1}{4}\sqrt{\frac{\pi}{2a}}\left(\ln 4a + \boldsymbol{C} - \frac{\pi}{2}\right) \qquad [a > 0]$$ GW (338)(19)

4.416

1. $$\int_0^{\pi/2} \frac{\cos x \ln\left(1 + \sqrt{\sin^2 \beta - \cos^2 \beta \tan^2 \alpha \sin^2 x}\right)}{1 - \sin^2 \alpha \cos^2 x}\, dx$$

$$= \operatorname{cosec} 2\alpha \left\{(2\alpha + 2\gamma - \pi)\ln \cos \beta + 2\,L(\alpha) - 2\,L(\gamma) + L(\alpha + \gamma) - L(\alpha - \gamma)\right\}$$

$$\left[\cos \gamma = \frac{\sin \alpha}{\sin \beta};\quad 0 < \alpha < \beta < \frac{\pi}{2}\right]$$ LO III 291

2. $$\int_0^{\pi/2} \frac{\cos x \ln\left(1 - \sqrt{\sin^2 \beta - \cos^2 \beta \tan^2 \alpha \sin^2 x}\right)}{1 - \sin^2 \alpha \cos^2 x}\, dx$$

$$= \operatorname{cosec} 2\alpha \left\{(\pi + 2\alpha - 2\gamma)\ln \cos \beta + 2\,L(\alpha) + 2\,L(\gamma) - L(\alpha + \gamma) + L(\alpha - \gamma)\right\}$$

$$\left[\cos \gamma = \frac{\sin \alpha}{\sin \beta};\quad 0 < \alpha < \beta < \frac{\pi}{2}\right]$$ LO III 291

3. $$\int_\beta^{\pi/2} \frac{\ln\left(\sin x + \sqrt{\sin^2 x - \sin^2 \beta}\right)}{1 - \cos^2 \alpha \cos^2 x}\, dx$$

$$= -\operatorname{cosec} \alpha \left\{\arctan\left(\frac{\tan \beta}{\sin \alpha}\right)\ln \sin \beta + \frac{\pi}{2}\ln \frac{1 + \sin \alpha}{\sin \alpha + \sqrt{1 - \cos^2 \alpha \cos^2 \beta}}\right\}$$

$$\left[0 < \alpha < \pi,\quad 0 < \beta < \frac{\pi}{2}\right]$$ LO III 285

4.7 $\int_0^{\pi/4} \ln \tan x (\ln \cos 2x)^{n-1} \tan 2x \, dx = \frac{1}{2}(-1)^n (n-1)! \left(1 - 2^{-(n+1)}\right) \zeta(n+1)$

BI (287)(20)

4.42–4.43 Combinations of logarithms, trigonometric functions, and powers

4.421

1. $\int_0^\infty \ln x \sin ax \frac{dx}{x} = -\frac{\pi}{2} \left(\boldsymbol{C} + \ln a\right)$ $\qquad\qquad [a > 0]$ $\qquad\qquad$ FI II 810a

2. $\int_0^\infty \ln ax \sin bx \frac{x \, dx}{\beta^2 + x^2} = \frac{\pi}{2} e^{-b\beta'} \ln(a\beta') - \frac{\pi}{4} \left[e^{b\beta'} \operatorname{Ei}(-b\beta') + e^{-b\beta'} \operatorname{Ei}(b\beta')\right]$

$\qquad\qquad [\beta' = \beta \operatorname{sign} \beta; \quad a > 0, \quad b > 0]$
ET I 76(5), NT 27(10)a

3. $\int_0^\infty \ln ax \cos bx \frac{\beta' \, dx}{\beta^2 + x^2} = \frac{\pi}{2} e^{-b\beta'} \ln(a\beta') + \frac{\pi}{4} \left[e^{b\beta'} \operatorname{Ei}(-b\beta') - e^{-b\beta'} \operatorname{Ei}(b\beta')\right]$

$\qquad\qquad [\beta' = \beta \operatorname{sign} \beta; \quad a > 0, \quad b > 0]$
ET I 17(3), NT 27(11)a

4. $\int_0^\infty \ln ax \sin bx \frac{x \, dx}{x^2 - c^2} = \frac{\pi}{2} \left\{-\operatorname{si}(bc) \sin bc + \cos bc \left[\ln ac - ci(bc)\right]\right\}$

$\qquad\qquad [a > 0, \quad b > 0, \quad c > 0]$ $\qquad$ BI (422)(5)

5. $\int_0^\infty \ln ax \cos bx \frac{dx}{x^2 - c^2} = \frac{\pi}{2c} \left\{\sin bc \left[ci(bc) - \ln ac\right] - \cos bc \operatorname{si}(bc)\right\}$

$\qquad\qquad [a > 0, \quad b > 0, \quad c > 0]$ $\qquad$ BI (422)(6)

4.422

1. $\int_0^\infty \ln x \sin ax \, x^{\mu-1} \, dx = \frac{\Gamma(\mu)}{a^\mu} \sin \frac{\mu\pi}{2} \left[\psi(\mu) - \ln a + \frac{\pi}{2} \cot \frac{\mu\pi}{2}\right]$

$\qquad\qquad [a > 0, \quad |\operatorname{Re}\mu| < 1]$ $\qquad$ BI (411)(5)

2. $\int_0^\infty \ln x \cos ax \, x^{\mu-1} \, dx = \frac{\Gamma(\mu)}{a^\mu} \cos \frac{\mu\pi}{2} \left[\psi(\mu) - \ln a - \frac{\pi}{2} \tan \frac{\mu\pi}{2}\right]$

$\qquad\qquad [a > 0, \quad 0 < \operatorname{Re}\mu < 1]$ $\qquad$ BI (411)(6)

4.423

1. $\int_0^\infty \ln x \frac{\cos ax - \cos bx}{x} \, dx = \ln \frac{a}{b} \left(\boldsymbol{C} + \frac{1}{2} \ln ab\right)$ $\qquad [a > 0, \quad b > 0]$ $\qquad$ GW (338)(21a)

2. $\int_0^\infty \ln x \frac{\cos ax - \cos bx}{x^2} \, dx = \frac{\pi}{2} \left[(a - b)(\boldsymbol{C} - 1) + a \ln a - b \ln b\right]$

$\qquad\qquad [a > 0, \quad b > 0]$ $\qquad$ GW (338)(21b)

3. $\int_0^\infty \ln x \frac{\sin^2 ax}{x^2} \, dx = -\frac{a\pi}{2} \left(\boldsymbol{C} + \ln 2a - 1\right)$ $\qquad\qquad [a > 0]$ $\qquad$ GW (338)(20b)

4.424

1. $\displaystyle \int_0^\infty (\ln x)^2 \sin ax \frac{dx}{x} = \frac{\pi}{2} \boldsymbol{C}^2 + \frac{\pi^3}{24} + \pi \boldsymbol{C} \ln a + \frac{\pi}{2}(\ln a)^2$

$$[a > 0] \qquad\qquad \text{ET I 77(9), FI II 810a}$$

2.[6] $\displaystyle \int_0^\infty (\ln x)^2 \sin ax\, x^{\mu-1}\, dx = \frac{\Gamma(\mu)}{a^\mu} \sin \frac{\mu\pi}{2} \left[\psi'(\mu) + \psi^2(\mu) + \pi\,\psi(\mu) \cot \frac{\mu\pi}{2} - 2\,\psi(\mu) \ln a \right.$

$$\left. - \pi \ln a \cot \frac{\mu\pi}{2} + (\ln a)^2 - \frac{1}{4}\pi^2 \right]$$

$$[a > 0, \quad 0 < \operatorname{Re}\mu < 1] \qquad \text{ET I 77(10)}$$

4.425

1. $\displaystyle \int_0^\infty \ln(1 + x) \cos ax \frac{dx}{x} = \frac{1}{2} \left\{ [\operatorname{si}(a)]^2 + [\operatorname{ci}(a)]^2 \right\}$ $[a > 0]$ ET I 18(8)

2. $\displaystyle \int_0^\infty \ln^2 \left(\frac{b + x}{b - x} \right) \cos ax \frac{dx}{x} = -2\pi \operatorname{si}(ab)$ $[a \geq 0, \quad b > 0]$ ET I 18(11)

3. $\displaystyle \int_0^\infty \ln\left(1 + b^2 x^2\right) \sin ax \frac{dx}{x} = -\pi \operatorname{Ei}\left(-\frac{a}{b}\right)$ $[a > 0, \quad b > 0]$

$$\text{GW (338)(24), ET I 77(14)}$$

4. $\displaystyle \int_0^1 \ln\left(1 - x^2\right) \cos\left(p \ln x\right) \frac{dx}{x} = \frac{1}{2p^2} + \frac{\pi}{2p} \coth \frac{p\pi}{2}$ LI (309)(1)a

4.426

1. $\displaystyle \int_0^\infty \ln \frac{b^2 + x^2}{c^2 + x^2} \sin ax\, x\, dx = \frac{\pi}{a^2} \left[(1 + ac)e^{-ac} - (1 + ab)e^{-ab} \right]$

$$[b \geq 0, \quad c \geq 0, \quad a > 0] \qquad \text{GW (338)(23)}$$

2. $\displaystyle \int_0^\infty \ln \frac{b^2 x^2 + p^2}{c^2 x^2 + p^2} \sin ax \frac{dx}{x} = \pi \left[\operatorname{Ei}\left(-\frac{ap}{c}\right) - \operatorname{Ei}\left(-\frac{ap}{b}\right) \right]$

$$[b > 0, \quad c > 0, \quad p > 0, \quad a > 0]$$
$$\text{ET I 77(15)}$$

4.427 $\displaystyle \int_0^\infty \ln\left(x + \sqrt{\beta^2 + x^2}\right) \frac{\sin ax}{\sqrt{\beta^2 + x^2}}\, dx = \frac{\pi}{2}\, K_0(a\beta) + \frac{\pi}{2} \ln(\beta) \left[I_0(a\beta) - \mathbf{L}(a\beta) \right]$

$$[\operatorname{Re}\beta > 0, \quad a > 0] \qquad \text{ET I 77(16)}$$

4.428

1. $\displaystyle \int_0^\infty \ln \cos^2 ax \frac{\cos bx}{x^2}\, dx = \pi b \ln 2 - a\pi$ $[a > 0, \quad b > 0]$ ET I 22(29)

2. $\displaystyle \int_0^\infty \ln\left(4 \cos^2 ax\right) \frac{\cos bx}{x^2 + c^2}\, dx = \frac{\pi}{c} \cosh(bc) \ln\left(1 + e^{-2ac}\right)$

$$\left[a < b < 2a < \frac{\pi}{c} \right] \qquad \text{ET I 22(30)}$$

3. $\displaystyle \int_0^\infty \ln \cos^2 ax \frac{\sin bx}{x\left(1 + x^2\right)}\, dx = \pi \ln\left(1 + e^{-2a}\right) \sinh b - \pi \ln 2 \left(1 - e^{-b}\right)$

$$[a > 0, \quad b > 0] \qquad \text{ET I 82(36)}$$

4. $\displaystyle\int_0^\infty \ln \cos^2 ax \, \frac{\cos bx}{x^2 (1 + x^2)} \, dx = -\pi \ln \left(1 + e^{-2a}\right) \cosh b + \left(b + e^{-b}\right) \pi \ln 2 - a\pi$

$$[a > 0, \quad b > 0]$$ ET I 22(31)

4.429 $\displaystyle\int_0^1 \frac{(1 + x)x}{\ln x} \sin (\ln x) \, dx = \frac{\pi}{4}$ BI (326)(2)a

4.431

1. $\displaystyle\int_0^\infty \ln (2 \pm 2 \cos x) \frac{\sin bx}{x^2 + c^2} x \, dx = - \pi \sinh(bc) \ln \left(1 \pm e^{-c}\right)$

$$[b > 0, \quad c > 0]$$ ET I 22(32)

2. $\displaystyle\int_0^\infty \ln (2 \pm 2 \cos x) \frac{\cos bx}{x^2 + c^2} \, dx = \frac{\pi}{c} \cosh(bc) \ln \left(1 \pm e^{-c}\right)$

$$[b > 0, \quad c > 0]$$ ET I 22(32)

3. $\displaystyle\int_0^\infty \ln \left(1 + 2a \cos x + a^2\right) \frac{\sin bx}{x} \, dx = -\frac{\pi}{2} \sum_{k=1}^{[b]} \frac{(-a)^k}{k} \left[1 + \operatorname{sign}(b - k)\right]$

$$[0 < a < 1, \quad b > 0]$$ ET I 82(25)

4. $\displaystyle\int_0^\infty \ln \left(1 - 2a \cos x + a^2\right) \frac{\cos bx}{x^2 + c^2} \, dx = \frac{\pi}{c} \ln \left(1 - ae^{-c}\right) \cosh(bc) + \frac{\pi}{c} \sum_{k=1}^{\lfloor b \rfloor} \frac{a^k}{k} \sinh[c(b - k)]$

$$[|a| < 1, \quad b > 0, \quad c > 0]$$ ET I 22(33)

4.432

1. $\displaystyle\int_0^\infty \ln \left(1 - k^2 \sin^2 x\right) \frac{\sin x}{\sqrt{1 - k^2 \sin^2 x}} \frac{dx}{x} = \int_0^\infty \ln \left(1 - k^2 \cos^2 x\right) \frac{\sin x}{\sqrt{1 - k^2 \cos^2 x}} \frac{dx}{x} = \ln k' \, \boldsymbol{K}(k)$

BI ((412, 414))(4)

2. $\displaystyle\int_0^{\pi/2} \ln \left(1 - k^2 \sin^2 x\right) \frac{\sin x \cos x}{\sqrt{1 - k^2 \sin^2 x}} x \, dx$

$$= \frac{1}{k^2} \left\{\pi k' \left(1 - \ln k'\right) + \left(2 - k^2\right) \boldsymbol{K}(k) - \left(4 - \ln k'\right) \boldsymbol{E}(k)\right\}$$

BI (426)(3)

3. $\displaystyle\int_0^{\pi/2} \ln \left(1 - k^2 \cos^2 x\right) \frac{\sin x \cos x}{\sqrt{1 - k^2 \cos^2 x}} x \, dx = \frac{1}{k^2} \left\{-\pi - \left(2 - k^2\right) \boldsymbol{K}(k) + \left(4 - \ln k'\right) \boldsymbol{E}(k)\right\}$

BI (426)(6)

4. $\displaystyle\int_0^\infty \ln \left(1 - k^2 \sin^2 x\right) \frac{\sin x \cos x}{\sqrt{1 - k^2 \sin^2 x}} \frac{dx}{x} = \frac{1}{k^2} \left\{\left(2 - k^2 - k'^2 \ln k'\right) \boldsymbol{K}(k) - \left(2 - \ln k'\right) \boldsymbol{E}(k)\right\}$

BI (412)(5)

5. $\displaystyle\int_0^\infty \ln \left(1 - k^2 \cos^2 x\right) \frac{\sin x \cos x}{\sqrt{1 - k^2 \cos^2 x}} \frac{dx}{x} = \frac{1}{k^2} \left\{\left(k^2 - 2 + \ln k'\right) \boldsymbol{K}(k) + \left(2 - \ln k'\right) \boldsymbol{E}(k)\right\}$

BI (414)(5)

6. $\displaystyle\int_0^\infty \ln\left(1 \pm k\sin^2 x\right) \frac{\sin x}{\sqrt{1 - k^2\sin^2 x}} \frac{dx}{x} = \int_0^\infty \ln\left(1 \pm k\cos^2 x\right) \frac{\sin x}{\sqrt{1 - k^2\cos^2 x}} \frac{dx}{x}$

$$= \int_0^\infty \ln\left(1 \pm k\sin^2 x\right) \frac{\tan x}{\sqrt{1 - k^2\sin^2 x}} \frac{dx}{x}$$

$$= \int_0^\infty \ln\left(1 \pm k\cos^2 x\right) \frac{\tan x}{\sqrt{1 - k^2\cos^2 x}} \frac{dx}{x}$$

$$= \int_0^\infty \ln\left(1 \pm k\sin^2 2x\right) \frac{\tan x}{\sqrt{1 - k^2\sin^2 2x}} \frac{dx}{x}$$

$$= \int_0^\infty \ln\left(1 \pm k^2\cos^2 2x\right) \frac{\tan x}{\sqrt{1 - k^2\cos^2 2x}} \frac{dx}{x}$$

$$= \frac{1}{2}\ln\frac{2\left(1 \pm k\right)}{\sqrt{k}}\, \boldsymbol{K}(k) - \frac{\pi}{8}\, \boldsymbol{K}\left(k'\right)$$

$$\text{BI (413)(1–6), BI (415)(1–6)}$$

7. $\displaystyle\int_0^\infty \ln\left(1 - k^2\sin^2 x\right) \frac{\sin^3 x}{\sqrt{1 - k^2\sin^2 x}} \frac{dx}{x} = \frac{1}{k^2}\left\{\left(k^2 - 2 + \ln k'\right)\boldsymbol{K}(k) + \left(2 - \ln k'\right)\boldsymbol{E}(k)\right\}$

$$\text{BI (412)(6)}$$

8. $\displaystyle\int_0^\infty \ln\left(1 - k^2\cos^2 x\right) \frac{\sin^3 x}{\sqrt{1 - k^2\cos^2 x}} \frac{dx}{x} = \frac{1}{k^2}\left\{\left(2 - k^2 - k'^2\ln k'\right)\boldsymbol{K}(k) - \left(2 - \ln k'\right)\boldsymbol{E}(k)\right\}$

$$\text{BI (414)(6)a}$$

9. $\displaystyle\int_0^\infty \ln\left(1 - k^2\sin^2 x\right) \frac{\sin x\cos^2 x}{\sqrt{1 - k^2\sin^2 x}} \frac{dx}{x} = \frac{1}{k^2}\left\{\left(2 - k^2 - k'^2\ln k'\right)\boldsymbol{K}(k) - \left(2 - \ln k'\right)\boldsymbol{E}(k)\right\}$

$$\text{BI (412)(7)}$$

10. $\displaystyle\int_0^\infty \ln\left(1 - k^2\cos^2 x\right) \frac{\sin x\cos^2 x}{\sqrt{1 - k^2\cos^2 x}} \frac{dx}{x} = \frac{1}{k^2}\left\{\left(k^2 - 2 + \ln k'\right)\boldsymbol{K}(k) + \left(2 - \ln k'\right)\boldsymbol{E}(k)\right\}$

$$\text{BI (414)(7)}$$

11. $\displaystyle\int_0^\infty \ln\left(1 - k^2\sin^2 x\right) \frac{\tan x}{\sqrt{1 - k^2\sin^2 x}} \frac{dx}{x} = \int_0^\infty \ln\left(1 - k^2\cos^2 x\right) \frac{\tan x}{\sqrt{1 - k^2\cos^2 x}} \frac{dx}{x} = \ln k'\, \boldsymbol{K}(k)$

$$\text{BI ((412, 414))(9)}$$

12. $\displaystyle\int_0^\infty \ln\left(1 - k^2\sin^2 x\right) \frac{\sin^2 x\tan x}{\sqrt{1 - k^2\sin^2 x}} \frac{dx}{x} = \frac{1}{k^2}\left\{\left(k^2 - 2 + \ln k'\right)\boldsymbol{K}(k) + \left(2 - \ln k'\right)\boldsymbol{E}(k)\right\}$

$$\text{BI (412)(8)}$$

13. $\displaystyle\int_0^\infty \ln\left(1 - k^2\cos^2 x\right) \frac{\sin^2 x\tan x}{\sqrt{1 - k^2\cos^2 x}} \frac{dx}{x} = \frac{1}{k^2}\left\{\left(2 - k^2 - k'^2\ln k'\right)\boldsymbol{K}(k) - \left(2 - \ln k'\right)\boldsymbol{E}(k)\right\}$

$$\text{BI (414)(8)}$$

14. $\displaystyle\int_0^\infty \ln\left(1 - k^2\sin^2 x\right) \frac{\sin^2 x}{\sqrt{\left(1 - k^2\sin^2 x\right)^3}} \frac{dx}{x} = \int_0^\infty \ln\left(1 - k^2\cos^2 x\right) \frac{\sin x}{\sqrt{\left(1 - k^2\cos^2 x\right)^3}} \frac{dx}{x}$

$$= \frac{1}{k'^2}\left\{\left(k^2 - 2\right)\boldsymbol{K}(k) + \left(2 + \ln k'\right)\boldsymbol{E}(k)\right\}$$

$$\text{BI ((412, 414))(13)}$$

15. $\int_0^{\pi/2} \ln\left(1 - k^2 \sin^2 x\right) \dfrac{\sin x \cos x}{\sqrt{\left(1 - k^2 \sin^2 x\right)^3}} x\, dx = \dfrac{1}{k^2} \left\{ (1 + \ln k') \dfrac{\pi}{k'} - (2 + \ln k')\, \boldsymbol{K}(k) \right\}$

BI (426)(9)

16. $\int_0^{\pi/2} \ln\left(1 - k^2 \cos^2 x\right) \dfrac{\sin x \cos x}{\sqrt{\left(1 - k^2 \cos^2 x\right)^3}} x\, dx = \dfrac{1}{k^2} \left\{ -\pi + (2 + \ln k')\, \boldsymbol{K}(k) \right\}$ BI (426)(15)

17. $\int_0^\infty \ln\left(1 - k^2 \sin^2 x\right) \dfrac{\sin x \cos x}{\sqrt{\left(1 - k^2 \sin^2 x\right)^3}} \dfrac{dx}{x} = \int_0^\infty \ln\left(1 - k^2 \cos^2 x\right) \dfrac{\sin^3 x}{\sqrt{\left(1 - k^2 \cos^2 x\right)^3}} \dfrac{dx}{x}$

$$= \dfrac{1}{k^2} \left\{ \left(2 - k^2 + \ln k'\right) \boldsymbol{K}(k) - (2 + \ln k')\, \boldsymbol{E}(k) \right\}$$

BI (412)(14), BI(414)(15)

18. $\int_0^\infty \ln\left(1 - k^2 \sin^2 x\right) \dfrac{\sin^3 x}{\sqrt{\left(1 - k^2 \sin^2 x\right)^3}} \dfrac{dx}{x}$

$$= \int_0^\infty \ln\left(1 - k^2 \cos^2 x\right) \dfrac{\sin x \cos x}{\sqrt{\left(1 - k^2 \cos^2 x\right)^3}} \dfrac{dx}{x}$$

$$= \dfrac{1}{k^2 k'^2} \left\{ (2 + \ln k')\, \boldsymbol{E}(k) - \left(2 - k^2 + k'^2 \ln k'\right) \boldsymbol{K}(k) \right\}$$

BI (412)(15), BI(414)(14)

19. $\int_0^\infty \ln\left(1 - k^2 \sin^2 x\right) \dfrac{\sin x \cos^2 x}{\sqrt{\left(1 - k^2 \sin^2 x\right)^3}} \dfrac{dx}{x} = \int_0^\infty \ln\left(1 - k^2 \cos^2 x\right) \dfrac{\sin^2 x \tan x}{\sqrt{\left(1 - k^2 \cos^2 x\right)^3}} \dfrac{dx}{x}$

$$= \dfrac{1}{k^2} \left\{ \left(2 - k^2 + \ln k'\right) \boldsymbol{K}(k) - (2 + \ln k')\, \boldsymbol{E}(k) \right\}$$

BI (412)(16), BI(414)(17)

20. $\int_0^\infty \ln\left(1 - k^2 \sin^2 x\right) \dfrac{\sin^2 x \tan x}{\sqrt{\left(1 - k^2 \sin^2 x\right)^3}} \dfrac{dx}{x}$

$$= \int_0^\infty \ln\left(1 - k^2 \cos^2 x\right) \dfrac{\sin x \cos^2 x}{\sqrt{\left(1 - k^2 \cos^2 x\right)^3}} \dfrac{dx}{x}$$

$$= \dfrac{1}{k^2 k'^2} \left\{ (2 + \ln k')\, \boldsymbol{E}(k) - \left(2 - k^2 + k'^2 \ln k'\right) \boldsymbol{K}(k) \right\}$$

BI (412)(17), BI(414)(16)

21. $\int_0^\infty \ln\left(1 - k^2 \sin^2 x\right) \dfrac{\tan x}{\sqrt{\left(1 - k^2 \sin^2 x\right)^3}} \dfrac{dx}{x} = \int_0^\infty \ln\left(1 - k^2 \cos^2 x\right) \dfrac{\tan x}{\sqrt{\left(1 - k^2 \cos^2 x\right)^3}} \dfrac{dx}{x}$

$$= \dfrac{1}{k'^2} \left\{ \left(k^2 - 2\right) \boldsymbol{K}(k) + (2 + \ln k')\, \boldsymbol{E}(k) \right\}$$

BI ((412, 414))(18)

22. $\int_0^\infty \ln\left(1 - k^2 \sin^2 x\right) \sqrt{1 - k^2 \sin^2 x}\, \sin x \dfrac{dx}{x} = \int_0^\infty \ln\left(1 - k^2 \cos^2 x\right) \sqrt{1 - k^2 \cos^2 x}\, \sin x \dfrac{dx}{x}$

$$= \left(2 - k^2\right) \boldsymbol{K}(k) - (2 - \ln k')\, \boldsymbol{E}(k)$$

BI ((412, 414))(1)

23. $\int_0^{\pi/2} \ln\left(1 - k^2 \sin^2 x\right) \sqrt{1 - k^2 \sin^2 x} \sin x \cos x \cdot x \, dx$

$$= \frac{1}{27k^2} \left\{ 3\pi k'^3 \left(1 - 3\ln k'\right) + \left(22k'^2 + 6k^4 - 3k'^2 \ln k'\right) K(k) \right\}$$
$$- \left(2 - k^2\right) \left(14 - 6\ln k'\right) E(k)$$

BI (426)(1)

24. $\int_0^{\pi/2} \ln\left(1 - k^2 \cos^2 x\right) \sqrt{1 - k^2 \cos^2 x} \sin x \cos x \cdot x \, dx$

$$= \frac{1}{27k^2} \left\{ -3\pi - \left(22k'^2 + 6k^4 - 3k'^2 \ln k'\right) K(k) + \left(2 - k^2\right) \left(14 - 6\ln k'\right) E(k) \right\}$$

BI (426)(2)

25. $\int_0^\infty \ln\left(1 - k^2 \sin^2 x\right) \sqrt{1 - k^2 \sin^2 x} \tan x \frac{dx}{x} = \int_0^\infty \ln\left(1 - k^2 \cos^2 x\right) \sqrt{1 - k^2 \cos^2 x} \tan x \frac{dx}{x}$

$$= \left(2 - k^2\right) K(k) - \left(2 - \ln k'\right) E(k)$$

((412,414))(2)

26. $\int_0^\infty \ln\left(\sin^2 x + k' \cos^2 x\right) \frac{\sin x}{\sqrt{1 - k^2 \cos^2 x}} \frac{dx}{x} = \int_0^\infty \ln\left(\sin^2 x + k' \cos^2 x\right) \frac{\tan x}{\sqrt{1 - k^2 \cos^2 x}} \frac{dx}{x}$

$$= \int_0^\infty \ln\left(\sin^2 2x + k' \cos^2 2x\right) \frac{\tan x}{\sqrt{1 - k^2 \cos^2 2x}} \frac{dx}{x}$$

$$= \frac{1}{2} \ln \left[\frac{2\left(\sqrt{k'}\right)^3}{1 + k'} \right] K(k)$$

BI (415)(19–21)

4.44 Combinations of logarithms, trigonometric functions, and exponentials

4.441

1.[7] $\int_0^\infty e^{-qx} \sin px \ln x \, dx = \frac{1}{p^2 + q^2} \left[q \arctan \frac{p}{q} - pC - \frac{p}{c} \ln\left(p^2 - q^2\right) \right]$

$[q > 0, \quad p > 0]$ BI (467)(1)

2. $\int_0^\infty e^{-qx} \cos px \ln x \, dx = -\frac{1}{p^2 + q^2} \left[\frac{q}{2} \ln\left(p^2 + q^2\right) + p \arctan \frac{p}{q} + qC \right]$

$[q > 0]$ BI (467)(2)

4.442 $\int_0^{\pi/2} \frac{e^{-p \tan x} \ln \cos x \, dx}{\sin x \cos x} = -\frac{1}{2}[\operatorname{ci}(p)]^2 + \frac{1}{2}[\operatorname{si}(p)]^2$ $[\operatorname{Re} p > 0]$ NT 32(11)

4.5 Inverse Trigonometric Functions

4.51 Inverse trigonometric functions

4.511 $\displaystyle\int_0^\infty \operatorname{arccot} px \operatorname{arccot} qx \, dx = \frac{\pi}{2}\left\{\frac{1}{p}\ln\left(1+\frac{p}{q}\right) + \frac{1}{q}\ln\left(1+\frac{q}{p}\right)\right\}$

$$[p > 0, \quad q > 0] \qquad\qquad \text{BI (77)(8)}$$

4.512 $\displaystyle\int_0^\pi \arctan\left(\cos x\right) \, dx = 0$ BI (345)(1)

4.52 Combinations of arcsines, arccosines, and powers

4.521

1. $\displaystyle\int_0^1 \frac{\arcsin x}{x} \, dx = \frac{\pi}{2}\ln 2$ FI II 614, 623

2. $\displaystyle\int_0^1 \frac{\arccos x}{1 \pm x} \, dx = \mp\frac{\pi}{2}\ln 2 + 2\boldsymbol{G}$ BI (231)(7, 8)

3. $\displaystyle\int_0^1 \arcsin x \frac{x}{1+qx^2} \, dx = \frac{\pi}{2q}\ln\frac{2\sqrt{1+q}}{1+\sqrt{1+q}}$ $[q > -1]$ BI (231)(1)

4. $\displaystyle\int_0^1 \arcsin x \frac{x}{1-p^2x^2} \, dx = \frac{\pi}{2p^2}\ln\frac{1+\sqrt{1-p^2}}{2\sqrt{1-p^2}}$ $[p^2 < 1]$ LI (231)(3)

5. $\displaystyle\int_0^1 \arccos x \frac{dx}{\sin^2\lambda - x^2} = 2\operatorname{cosec}\lambda \sum_{k=0}^\infty \frac{\sin[(2k+1)\lambda]}{(2k+1)^2}$ BI (231)(10)

6. $\displaystyle\int_0^1 \arcsin x \frac{dx}{x\left(1+qx^2\right)} = \frac{\pi}{2}\ln\frac{1+\sqrt{1+q}}{\sqrt{1+q}}$ $[q > -1]$ BI (235)(10)

7. $\displaystyle\int_0^1 \arcsin x \frac{x}{\left(1+qx^2\right)^2} \, dx = \frac{\pi}{4q}\frac{\sqrt{1+q}-1}{1+q}$ $[q > -1]$ BI (234)(2)

8. $\displaystyle\int_0^1 \arccos x \frac{x}{\left(1+qx^2\right)^2} \, dx = \frac{\pi}{4q}\frac{\sqrt{1+q}-1}{1+q}$ $[q > -1]$ BI (234)(4)

4.522

1. $\displaystyle\int_0^1 x\sqrt{1-k^2x^2}\,\arccos x \, dx = \frac{1}{9k^2}\left[\frac{3}{2}\pi + k'^2\,\boldsymbol{K}(k) - 2\left(1+k'^2\right)\boldsymbol{E}(k)\right]$ BI (236)(9)

2. $\displaystyle\int_0^1 x\sqrt{1-k^2x^2}\,\arcsin x \, dx = \frac{1}{9k^2}\left[-\frac{3}{2}\pi k'^3 - k'^2\,\boldsymbol{K}(k) + 2\left(1+k'^2\right)\boldsymbol{E}(k)\right]$ BI (236)(1)

3. $\displaystyle\int_0^1 x\sqrt{k'^2+k^2x^2}\,\arcsin x \, dx = \frac{1}{9k^2}\left[\frac{3}{2}\pi + k'^2\,\boldsymbol{K}(k) - 2\left(1+k'^2\right)\boldsymbol{E}(k)\right]$ BI(236)(5)

4. $\displaystyle\int_0^1 \frac{x\arcsin x}{\sqrt{1-k^2x^2}} \, dx = \frac{1}{k^2}\left[-\frac{\pi}{2}k' + \boldsymbol{E}(k)\right]$ BI (237)(1)

5. $\displaystyle\int_0^1 \frac{x\arccos x}{\sqrt{1-k^2x^2}} \, dx = \frac{1}{k^2}\left[\frac{\pi}{2} - \boldsymbol{E}(k)\right]$ BI (240)(1)

6. $\int_0^1 \dfrac{x \arcsin x}{\sqrt{k'^2 + k^2 x^2}} \, dx = \dfrac{1}{k^2} \left[\dfrac{\pi}{2} - \boldsymbol{E}(k) \right]$ BI (238)(1)

7. $\int_0^1 \dfrac{x \arccos x}{\sqrt{k'^2 + k^2 x^2}} \, dx = \dfrac{1}{k^2} \left[-\dfrac{\pi}{2} k' + \boldsymbol{E}(k) \right]$ BI (241)(1)

8. $\int_0^1 \dfrac{x \arcsin x \, dx}{(x^2 - \cos^2 \lambda)\sqrt{1 - x^2}} = \dfrac{2}{\sin \lambda} \sum_{k=0}^{\infty} \dfrac{\sin[(2k+1)\lambda]}{(2k+1)^2}$ BI (243)(11)

9. $\int_0^1 \dfrac{x \arcsin kx}{\sqrt{(1-x^2)(1-k^2 x^2)}} \, dx = -\dfrac{\pi}{2k} \ln k'$ BI (239)(1)

10. $\int_0^1 \dfrac{x \arccos kx}{\sqrt{(1-x^2)(1-k^2 x^2)}} \, dx = \dfrac{\pi}{2k} \ln(1+k)$ BI (242)(1)

4.523

1. $\int_0^1 x^{2n} \arcsin x \, dx = \dfrac{1}{2n+1} \left[\dfrac{\pi}{2} - \dfrac{2^n n!}{(2n+1)!!} \right]$ BI (229)(1)

2. $\int_0^1 x^{2n-1} \arcsin x \, dx = \dfrac{\pi}{4n} \left[1 - \dfrac{(2n-1)!!}{2^n n!} \right]$ BI (229)(2)

3. $\int_0^1 x^{2n} \arccos x \, dx = \dfrac{2^n n!}{(2n+1)(2n+1)!!}$ BI (229)(4)

4. $\int_0^1 x^{2n-1} \arccos x \, dx = \dfrac{\pi}{4n} \dfrac{(2n-1)!!}{2^n n!}$ BI (229)(5)

5. $\int_{-1}^1 (1-x^2)^n \arccos x \, dx = \pi \dfrac{2^n n!}{(2n+1)!!}$ BI (254)(2)

6. $\int_{-1}^1 (1-x^2)^{n-\frac{1}{2}} \arccos x \, dx = \dfrac{\pi^2}{2} \dfrac{(2n-1)!!}{2^n n!}$ BI (254)(3)

4.524

1. $\int_0^1 (\arcsin x)^2 \dfrac{dx}{x^2 \sqrt{1-x^2}} = \pi \ln 2$ BI (243)(13)

2. $\int_0^1 (\arccos x)^2 \dfrac{dx}{\left(\sqrt{1-x^2}\right)^3} = \pi \ln 2$ BI (244)(9)

4.53–4.54 Combinations of arctangents, arccotangents, and powers

4.531

1. $\int_0^1 \dfrac{\arctan x}{x} \, dx = \int_1^{\infty} \dfrac{\operatorname{arccot} x}{x} \, dx = \boldsymbol{G}$ FI II 482, BI (253)(8)

2. $\int_0^{\infty} \dfrac{\operatorname{arccot} x}{1 \pm x} \, dx = \pm \dfrac{\pi}{4} \ln 2 + \boldsymbol{G}$ BI (248)(6, 7)

3. $\int_0^1 \dfrac{\operatorname{arccot} x}{x(1+x)}\, dx = -\dfrac{\pi}{8}\ln 2 + \boldsymbol{G}$ BI (235)(11)

4. $\int_0^\infty \dfrac{\arctan x}{1-x^2}\, dx = -\boldsymbol{G}.$ BI (248)(2)

5. $\int_0^1 \arctan qx \dfrac{dx}{(1+px)^2} = \dfrac{1}{2}\dfrac{q}{p^2+q^2}\ln\dfrac{(1+p)^2}{1+q^2} + \dfrac{q^2-p}{(1+p)(p^2+q^2)}\arctan q$

$[p > -1]$ BI (243)(7)

6. $\int_0^1 \operatorname{arccot} qx \dfrac{dx}{(1+px)^2} = \dfrac{1}{2}\dfrac{q}{p^2+q^2}\ln\dfrac{1+q^2}{(1+p)^2} + \dfrac{p}{p^2+q^2}\arctan q + \dfrac{1}{1+p}\operatorname{arccot} q$

$[p > -1]$ BI (234)(10)

7. $\int_0^1 \dfrac{\arctan x}{x(1+x^2)}\, dx = \dfrac{\pi}{8}\ln 2 + \dfrac{1}{2}\boldsymbol{G}$ BI (235)(12)

8. $\int_0^\infty \dfrac{x \arctan x}{1+x^4}\, dx = \dfrac{\pi^2}{16}$ BI (248)(3)

9. $\int_0^\infty \dfrac{x \arctan x}{1-x^4}\, dx = -\dfrac{\pi}{8}\ln 2$ BI (248)(4)

10. $\int_0^\infty \dfrac{x \arctan x}{1-x^4}\, dx = \dfrac{\pi}{8}\ln 2$ BI (248)(12)

11. $\int_0^\infty \dfrac{\operatorname{arccot} x}{x\sqrt{1+x^2}}\, dx = \int_0^\infty \dfrac{\operatorname{arccot} x}{\sqrt{1+x^2}}\, dx = 2\boldsymbol{G}$ BI (251)(3, 10)

12. $\int_0^1 \dfrac{\arctan x}{x\sqrt{1-x^2}}\, dx = \dfrac{\pi}{2}\ln\left(1+\sqrt{2}\right)$ FI II 694

13. $\int_0^1 \dfrac{x \arctan x\, dx}{\sqrt{(1+x^2)(1+k'^2 x^2)}} = \dfrac{1}{k^2}\left[F\left(\dfrac{\pi}{4}, k\right) - \dfrac{\pi}{2\sqrt{2\left(1+k'^2\right)}}\right]$ BI (294)(14)

4.532

1. $\int_0^1 x^p \arctan x\, dx = \dfrac{1}{2(p+1)}\left[\dfrac{\pi}{2} - \beta\left(\dfrac{p}{2}+1\right)\right]$ $[p > -2]$ BI (229)(7)

2. $\int_0^\infty x^p \arctan x\, dx = \dfrac{\pi}{2(p+1)}\operatorname{cosec}\dfrac{p\pi}{2}$ $[-1 > p > -2]$ BI (246)(1)

3. $\int_0^1 x^p \operatorname{arccot} x\, dx = \dfrac{1}{2(p+1)}\left[\dfrac{\pi}{2} + \beta\left(\dfrac{p}{2}+1\right)\right]$ $[p > -1]$ BI (229)(8)

4. $\int_0^\infty x^p \operatorname{arccot} x\, dx = -\dfrac{\pi}{2(p+1)}\operatorname{cosec}\dfrac{p\pi}{2}$ $[-1 < p < 0]$ BI (246)(2)

5. $\int_0^\infty \left(\dfrac{x^p}{1+x^{2p}}\right)^{2q} \arctan x \dfrac{dx}{x} = \dfrac{\sqrt{\pi^3}}{2^{2q+2}p}\dfrac{\Gamma(q)}{\Gamma\left(q+\frac{1}{2}\right)}$ $[q > 0]$ BI (250)(10)

4.533

1. $\displaystyle\int_0^\infty (1 - x\operatorname{arccot} x)\, dx = \frac{\pi}{4}$ BI (246)(3)

2. $\displaystyle\int_0^1 \left(\frac{\pi}{4} - \arctan x\right)\frac{dx}{1 - x} = -\frac{\pi}{8}\ln 2 + \boldsymbol{G}$ BI (232)(2)

3. $\displaystyle\int_0^1 \left(\frac{\pi}{4} - \arctan x\right)\frac{1 + x}{1 - x}\frac{dx}{1 + x^2} = \frac{\pi}{8}\ln 2 + \frac{1}{2}\boldsymbol{G}$ BI (235)(25)

4. $\displaystyle\int_0^1 \left(x\operatorname{arccot} x - \frac{1}{x}\arctan x\right)\frac{dx}{1 - x^2} = -\frac{\pi}{4}\ln 2$ BI (232)(1)

4.534 $\displaystyle\int_0^\infty (\arctan x)^2\,\frac{dx}{x^2\sqrt{1 + x^2}} = \int_0^\infty (\operatorname{arccot} x)^2\,\frac{x\,dx}{\sqrt{1 + x^2}} = -\frac{\pi^2}{4} + 4\boldsymbol{G}$ BI (251)(9, 17)

4.535

1. $\displaystyle\int_0^1 \frac{\arctan px}{1 + p^2 x}\, dx = \frac{1}{2p^2}\arctan p\ln\left(1 + p^2\right)$ BI (231)(19)

2. $\displaystyle\int_0^1 \frac{\operatorname{arccot} px}{1 + p^2 x}\, dx = \frac{1}{p^2}\left\{\frac{\pi}{4} + \frac{1}{2}\operatorname{arccot} p\right\}\ln\left(1 + p^2\right)$ $[p > 0]$ BI (231)(24)

3. $\displaystyle\int_0^\infty \frac{\arctan qx}{(p + x)^2}\, dx = -\frac{q}{1 + p^2 q^2}\left(\ln pq - \frac{\pi}{2}pq\right)$ $[p > 0, \quad q > 0]$ BI (249)(1)

4. $\displaystyle\int_0^\infty \frac{\operatorname{arccot} qx}{(p + x)^2}\, dx = \frac{q}{1 + p^2 q^2}\left(\ln pq + \frac{\pi}{2pq}\right)$ $[p > 0, \quad q > 0]$ BI (249)(8)

5. $\displaystyle\int_0^\infty \frac{x\operatorname{arccot} px}{q^2 + x^2}\, dx = \frac{\pi}{2}\ln\frac{1 + pq}{pq}$ $[p > 0, \quad q > 0]$ BI (248)(9)

6. $\displaystyle\int_0^\infty \frac{x\operatorname{arccot} px\,dx}{x^2 - q^2} = \frac{\pi}{4}\ln\frac{1 + p^2 q^2}{p^2 q^2}$ $[p > 0, \quad q > 0]$ BI (248)(10)

7. $\displaystyle\int_0^\infty \frac{\arctan px}{x\left(1 + x^2\right)}\, dx = \frac{\pi}{2}\ln(1 + p)$ $[p \geq 0]$ FI II 745

8. $\displaystyle\int_0^\infty \frac{\arctan px}{x\left(1 - x^2\right)}\, dx = \frac{\pi}{4}\ln\left(1 + p^2\right)$ $[p \geq 0]$ BI (250)(6)

9. $\displaystyle\int_0^\infty \arctan qx\frac{dx}{x\left(p^2 + x^2\right)} = \frac{\pi}{2p^2}\ln(1 + pq)$ $[p > 0, \quad q \geq 0]$ BI (250)(3)

10. $\displaystyle\int_0^\infty \arctan qx\frac{dx}{x\left(1 - p^2 x^2\right)} = \frac{\pi}{4}\ln\frac{p^2 + q^2}{p^2}$ $[p \geq 0]$ BI (250)(6)

11. $\displaystyle\int_0^\infty \frac{x\arctan qx}{\left(p^2 + x^2\right)^2}\, dx = \frac{\pi q}{4p(1 + pq)}$ $[p > 0, \quad q \geq 0]$ BI (252)(12)a

12. $\displaystyle\int_0^\infty \frac{x\operatorname{arccot} qx}{\left(p^2 + x^2\right)^2}\, dx = \frac{\pi}{4p^2(1 + pq)}$ $[p > 0, \quad q \geq 0]$ BI (252)(20)a

13. $\displaystyle\int_0^1 \frac{\arctan qx}{x\sqrt{1 - x^2}}\, dx = \frac{\pi}{2}\ln\left(q + \sqrt{1 + q^2}\right)$ BI (244)(11)

14.[9] $\displaystyle\int_{-\infty}^{\infty} \frac{x\arctan(\alpha x)\, dx}{(x^2+\beta^2)(x^2+\gamma^2)} = \begin{cases} \dfrac{\pi}{\beta^2-\gamma^2}\log\left(\dfrac{1+|\alpha\beta|}{1+|\alpha\gamma|}\right)\operatorname{sign}(\alpha) & (\alpha,\beta,\gamma\ \text{real};\quad \beta\neq\gamma) \\ \dfrac{\pi\alpha}{2|\beta|(1+|\alpha\beta|)} & \beta=\gamma \end{cases}$

15.[9] $\displaystyle\int_{-\infty}^{\infty} \frac{x\arctan(\alpha/x)\, dx}{(x^2+\beta^2)(x^2+\gamma^2)} = \begin{cases} \dfrac{\pi}{\beta^2-\gamma^2}\log\left(\dfrac{1+|\alpha/\gamma|}{1+|\alpha/\beta|}\right)\operatorname{sign}(\alpha) & (\alpha,\beta,\gamma\ \text{real};\quad \beta\neq\gamma) \\ \dfrac{\pi\alpha}{2\beta^2(|\beta|+|\alpha|)} & (\beta=\gamma) \end{cases}$

4.536

1. $\displaystyle\int_0^{\infty} \arctan qx \arcsin x\, \frac{dx}{x^2} = \frac{1}{2}q\pi\ln\frac{1+\sqrt{1+q^2}}{\sqrt{1+q^2}} + \frac{\pi}{2}\ln\left(q+\sqrt{1+q^2}\right) - \frac{\pi}{2} - \arctan q$

BI (230)(7)

2. $\displaystyle\int_0^{\infty} \frac{\arctan px - \arctan qx}{x}\, dx = \frac{\pi}{2}\ln\frac{p}{q}$ $\qquad [p>0, \quad q>0]$ $\qquad$ FI II 635

3. $\displaystyle\int_0^{\infty} \frac{\arctan px \arctan qx}{x^2}\, dx = \frac{\pi}{2}\ln\frac{(p+q)^{p+q}}{p^p q^q}$ $\qquad [p>0, \quad q>0]$ $\qquad$ FI II 745

4.537

1.[8] $\displaystyle\int_0^1 \arctan\left(\sqrt{1-x^2}\right)\frac{dx}{1-x^2\cos^2\lambda} = \frac{\pi}{2\cos\lambda}\ln\left[\cos\left(\frac{\pi-4\lambda}{8}\right)\operatorname{cosec}\left(\frac{\pi+4\lambda}{8}\right)\right]$ $\qquad$ BI (245)(9)

2. $\displaystyle\int_0^1 \arctan\left(p\sqrt{1-x^2}\right)\frac{dx}{1-x^2} = \frac{1}{2}\pi\ln\left(p+\sqrt{1+p^2}\right)$

$\qquad\qquad\qquad [p>0]$ $\qquad$ BI (245)(10)

3. $\displaystyle\int_0^1 \arctan\left(\tan\lambda\sqrt{1-k^2}x^2\right)\sqrt{\frac{1-x^2}{1-k^2x^2}}\, dx = \frac{\pi}{2k^2}\left[E(\lambda,k)-k'^2\,F(\gamma,k)\right]$
$\qquad\qquad\qquad\qquad\qquad\qquad\qquad\qquad -\dfrac{\pi}{2k^2}\cot\gamma\left(1-\sqrt{1-k^2\sin^2\gamma}\right)$

BI (245)(12)

4. $\displaystyle\int_0^1 \arctan\left(\tan\lambda\sqrt{1-k^2x^2}\right)\sqrt{\frac{1-k^2x^2}{1-x^2}}\, dx = \frac{\pi}{2}\,E(\lambda,k) - \frac{\pi}{2}\cot\lambda\left(1-\sqrt{1-k^2\sin^2\lambda}\right)$

BI (245)(11)

5. $\displaystyle\int_0^1 \frac{\arctan\left(\tan\lambda\sqrt{1-k^2x^2}\right)}{\sqrt{(1-x^2)(1-k^2x^2)}}\, dx = \frac{\pi}{2}\,F(\lambda,k)$ $\qquad$ BI (245)(13)

4.538

1. $\displaystyle\int_0^{\infty} \arctan x^2\, \frac{dx}{1+x^2} = \int_0^{\infty} \arctan x^3\, \frac{dx}{1+x^2}$ $\qquad$ BI (252)(10, 11)

$\qquad\qquad = \displaystyle\int_0^{\infty} \operatorname{arccot} x^2\, \frac{dx}{1+x^2} = \int_0^{\infty} \operatorname{arccot} x^3\, \frac{dx}{1+x^2} = \frac{\pi^2}{8}$ $\qquad$ BI (252)(18, 19)

2. $\displaystyle\int_0^\infty \frac{1-x^2}{x^2} \arctan x^2 \, dx = \frac{\pi}{2}\left(\sqrt{2}-1\right)$ BI (244)(10)a

4.539 $\displaystyle\int_0^\infty x^{s-1} \arctan\left(ae^{-x}\right)dx = 2^{-s-1}\,\Gamma(s)a\,\Phi\left(-a^2,s+1,\tfrac{1}{2}\right)$ ET I 222(47)

4.541 $\displaystyle\int_0^\infty \arctan\left(\frac{p\sin qx}{1+p\cos qx}\right)\frac{x\,dx}{1+x^2} = \frac{\pi}{2}\ln\left(1+pe^{-q}\right) \quad [p>-e^q]$ BI (341)(14)a

4.55 Combinations of inverse trigonometric functions and exponentials

4.551

1.[9] $\displaystyle\int_0^1 (\arcsin x)\, e^{-bx}\, dx = \frac{\pi}{2b}\left[I_0(b) - \mathbf{L}_0(b)\right] - \frac{\pi e^{-b}}{2b}$ ET I 160(1)

2. $\displaystyle\int_0^1 x\,(\arcsin x)\, e^{-bx}\, dx = \frac{\pi}{2b^2}\left[\mathbf{L}_0(b) - I_0(b) + b\,\mathbf{L}_1(b) - b\,I_1(b)\right] + \frac{1}{b}$ ET I 161(2)

3.[9] $\displaystyle\int_0^\infty \left(\arctan\frac{x}{a}\right)e^{-bx}\, dx = \frac{1}{b}\left[\mathrm{ci}(ab)\sin(ab) - \mathrm{si}(ab)\cos(ab)\right]$

                                               $[\mathrm{Re}\,b>0]$ ET I 161(3)

4.[9] $\displaystyle\int_0^\infty \left(\mathrm{arccot}\frac{x}{a}\right)e^{-bx}\, dx = \frac{1}{b}\left[\frac{\pi}{2} - \mathrm{ci}(ab)\sin(ab) + \mathrm{si}(ab)\cos(ab)\right]$

                                               $[\mathrm{Re}\,b>0]$ ET I 161(4)

4.552 $\displaystyle\int_0^\infty \frac{\arctan\dfrac{x}{q}}{e^{2\pi x}-1}\, dx = \frac{1}{2}\left[\ln\Gamma(q) - \left(q-\frac{1}{2}\right)\ln q + q - \frac{1}{2}\ln 2\pi\right]$

                                                   $[q>0]$ WH

4.553 $\displaystyle\int_0^\infty \left(\frac{2}{\pi}\mathrm{arccot}\, x - e^{-px}\right)\frac{dx}{x} = \mathbf{C} + \ln p \qquad [p>0]$ NT 66(12)

4.56   A combination of the arctangent and a hyperbolic function

4.561 $\displaystyle\int_{-\infty}^\infty \frac{\arctan e^{-x}}{\cosh^{2q} px}\, dx = \frac{1}{2}\int_{-\infty}^\infty \frac{\Pi(x)}{\cosh^{2q} px}\, dx = \frac{\sqrt{\pi^3}}{4p}\frac{\Gamma(q)}{\Gamma\left(q+\frac{1}{2}\right)}$

                                                   $[q>0]$ LI (282)(10)

4.57 Combinations of inverse and direct trigonometric functions

4.571 $\displaystyle\int_0^{\pi/2} \arcsin\left(k\sin x\right)\frac{\sin x\,dx}{\sqrt{1-k^2\sin^2 x}} = -\frac{\pi}{2k}\ln k'$ BI (344)(2)

4.572 $\displaystyle\int_0^\infty \left(\frac{2}{\pi}\mathrm{arccot}\, x - \cos px\right)dx = \mathbf{C} + \ln p \qquad [p>0]$ NT 66(12)

4.573

1. $\displaystyle\int_0^\infty \mathrm{arccot}\, qx\,\sin px\, dx = \frac{\pi}{2p}\left(1 - e^{-\frac{p}{q}}\right) \qquad [p>0,\quad q>0]$ BI (347)(1)a

2. $$\int_0^\infty \operatorname{arccot} qx \cos px \, dx = \frac{1}{2p} \left[e^{-\frac{p}{q}} \operatorname{Ei} \left(\frac{p}{q} \right) - e^{\frac{p}{q}} \operatorname{Ei} \left(-\frac{p}{q} \right) \right]$$

$$[p > 0, \quad q > 0] \qquad\qquad \text{BI (347)(2)a}$$

3. $$\int_0^\infty \operatorname{arccot} rx \frac{\sin px \, dx}{1 \pm 2q \cos px + q^2} = \pm \frac{\pi}{2pq} \ln \frac{1 \pm q}{1 \pm qe^{-\frac{p}{r}}} \qquad [p^2 < 1, \quad r > 0, \quad p > 0]$$

$$= \pm \frac{\pi}{2pq} \ln \frac{q \pm 1}{q \pm e^{-\frac{p}{r}}} \qquad [q^2 > 1, \quad r > 0, \quad p > 0]$$

$$\text{BI (347)(10)}$$

4. $$\int_0^\infty \operatorname{arccot} px \frac{\tan x \, dx}{q^2 \cos^2 x + r^2 \sin^2 x} = \frac{\pi}{2r^2} \ln \left(1 + \frac{r}{q} \tanh \frac{1}{p} \right)$$

$$[p > 0, \quad q > 0, \quad r > 0] \qquad \text{BI (347)(9)}$$

4.574

1. $$\int_0^\infty \arctan \left(\frac{2a}{x} \right) \sin(bx) \, dx = \frac{\pi}{b} e^{-ab} \sinh(ab) \qquad [\operatorname{Re} a > 0, \quad b > 0] \qquad \text{ET I 87(8)}$$

2.[7] $$\int_0^\infty \arctan \frac{a}{x} \cos(bx) \, dx = \frac{1}{2b} \left[e^{-ab} \operatorname{Ei}(ab) - e^{ab} \operatorname{Ei}(-ab) \right]$$

$$[a > 0, \quad b > 0] \qquad\qquad \text{ET I 29(7)}$$

3. $$\int_0^\infty \arctan \left[\frac{2ax}{x^2 + c^2} \right] \sin(bx) \, dx = \frac{\pi}{b} e^{-b\sqrt{a^2+c^2}} \sinh(ab)$$

$$[b > 0] \qquad\qquad \text{ET I 87(9)}$$

4. $$\int_0^\infty \arctan \left(\frac{2}{x^2} \right) \cos(bx) \, dx = \frac{\pi}{b} e^{-b} \sin b \qquad\qquad [b > 0] \qquad\qquad \text{ET I 29(8)}$$

4.575

1. $$\int_0^\pi \arctan \frac{p \sin x}{1 - p \cos x} \sin nx \, dx = \frac{\pi}{2n} p^n \qquad\qquad [p^2 < 1] \qquad\qquad \text{BI (345)(4)}$$

2. $$\int_0^\pi \arctan \frac{p \sin x}{1 - p \cos x} \sin nx \cos x \, dx = \frac{\pi}{4} \left(\frac{p^{n+1}}{n+1} + \frac{p^{n-1}}{n-1} \right)$$

$$[p^2 < 1] \qquad\qquad \text{BI (345)(5)}$$

3. $$\int_0^\pi \arctan \frac{p \sin x}{1 - p \cos x} \cos nx \sin x \, dx = \frac{\pi}{4} \left(\frac{p^{n+1}}{n+1} - \frac{p^{n-1}}{n-1} \right)$$

$$[p^2 < 1] \qquad\qquad \text{BI (345)(6)}$$

4.576

1. $$\int_0^\pi \arctan \frac{p \sin x}{1 - p \cos x} \frac{dx}{\sin x} = \frac{\pi}{2} \ln \frac{1 + p}{1 - p} \qquad\qquad [p^2 < 1] \qquad\qquad \text{BI(346)(1)}$$

2. $$\int_0^\pi \arctan \frac{p \sin x}{1 - p \cos x} \frac{dx}{\tan x} = -\frac{\pi}{2} \ln \left(1 - p^2 \right) \qquad [p^2 < 1] \qquad\qquad \text{BI(346)(3)}$$

4.577

1. $\displaystyle\int_0^{\pi/2} \arctan\left(\tan\lambda\sqrt{1-k^2\sin^2 x}\right) \frac{\sin^2 x \, dx}{\sqrt{1-k^2\sin^2 x}}$

$$= \frac{\pi}{2k^2}\left[F(\lambda,k) - E(\lambda,k) + \cot\lambda\left(1 - \sqrt{1-k^2\sin^2\lambda}\right)\right]$$

BI (344)(4)

2. $\displaystyle\int_0^{\pi/2} \arctan\left(\tan\lambda\sqrt{1-k^2\sin^2 x}\right) \frac{\cos^2 x \, dx}{\sqrt{1-k^2\sin^2 x}}$

$$= \frac{\pi}{2k^2}\left[E(\lambda,k) - k'^2\, F(\lambda,k) + \cot\lambda\left(\sqrt{1-k^2\sin^2\lambda} - 1\right)\right]$$

BI (344)(5)

4.58 A combination involving an inverse and a direct trigonometric function and a power

4.581[10] $\displaystyle\int_0^\infty \arctan x \cos px\, \frac{dx}{x} = \int_0^\infty \arctan\frac{x}{p}\cos x\, \frac{dx}{x} = -\frac{\pi}{2}\,\mathrm{Ei}(-p)$

$$[\mathrm{Re}(p) > 0] \qquad \text{ET I 29(3), NT 25(13)}$$

4.59 Combinations of inverse trigonometric functions and logarithms

4.591

1. $\displaystyle\int_0^1 \arcsin x \ln x \, dx = 2 - \ln 2 - \frac{1}{2}\pi$ BI (339)(1)

2. $\displaystyle\int_0^1 \arccos x \ln x \, dx = \ln 2 - 2$ BI (339)(2)

4.592 $\displaystyle\int_0^1 \arccos x \, \frac{dx}{\ln x} = -\sum_{k=0}^\infty \frac{(2k-1)!!}{2^k k!}\frac{\ln(2k+2)}{2k+1}$ BI (339)(8)

4.593

1. $\displaystyle\int_0^1 \arctan x \ln x \, dx = \frac{1}{2}\ln 2 - \frac{\pi}{4} + \frac{1}{48}\pi^2$ BI (339)(3)

2. $\displaystyle\int_0^1 \mathrm{arccot}\, x \ln x \, dx = -\frac{1}{48}\pi^2 - \frac{\pi}{4} - \frac{1}{2}\ln 2$ BI (339)(4)

4.594 $\displaystyle\int_0^1 \arctan x (\ln x)^{n-1}(\ln x + n)\, dx = \frac{n!}{(-2)^{n+1}}\left(2^{-n} - 1\right)\zeta(n+1)$ BI (339)(7)

4.6 Multiple Integrals

4.60 Change of variables in multiple integrals

4.601

1.
$$\iint\limits_{(\sigma)} f(x,y)\,dx\,dy = \iint\limits_{(\sigma')} f\left[\varphi(u,v),\psi(u,v)\right]|\Delta|\,du\,dv$$

where $x = \varphi(u,v), y = \psi(u,v)$, and $\Delta = \dfrac{\partial\varphi}{\partial u}\dfrac{\partial\psi}{\partial v} - \dfrac{\partial\psi}{\partial u}\dfrac{\partial\varphi}{\partial v} \equiv \dfrac{D(\varphi,\psi)}{D(u,v)}$ is the Jacobian determinant of the functions φ and ψ.

2.
$$\iiint\limits_{(V)} f(x,y,z)\,dx\,dy\,dz = \iiint\limits_{(V')} f\left[\varphi(u,v,w),\psi(u,v,w),\mathrm{chi}(u,v,w)\right]|\Delta|\,du\,dv\,dw$$

where $x = \varphi(u,v,w), y = \psi(u,v,w)$, and $z = \mathrm{chi}(u,v,w)$ and where

$$\Delta = \begin{vmatrix} \dfrac{\partial\varphi}{\partial u} & \dfrac{\partial\varphi}{\partial v} & \dfrac{\partial\varphi}{\partial w} \\ \dfrac{\partial\psi}{\partial u} & \dfrac{\partial\psi}{\partial v} & \dfrac{\partial\psi}{\partial w} \\ \dfrac{\partial\chi}{\partial u} & \dfrac{\partial\chi}{\partial v} & \dfrac{\partial\chi}{\partial w} \end{vmatrix} \equiv \dfrac{D(\varphi,\psi,\chi)}{D(u,v,w)}$$

is the Jacobian determinant of the functions φ, ψ, and χ.

Here, we assume, both in (**4.601** 1) and in (**4.601** 2) that

(a)   the functions φ, ψ, and χ and also their first partial derivatives are continuous in the region of integration;

(b)   the Jacobian does not change sign in this region;

(c)   there exists a one-to-one correspondence between the old variables x, y, z and the new ones u, v, w in the region of integration;

(d)   when we change from the variables x, y, z to the variables u, v, w, the region V (resp. σ) is mapped into the region V' (resp. σ').

4.602   Transformation to polar coordinates:

$$x = r\cos\varphi, \quad y = r\sin\varphi; \quad \frac{D(x,y)}{D(r,\varphi)} = r$$

4.603   Transformation to spherical coordinates:

$$x = r\sin\theta\cos\varphi, \quad y = r\sin\theta\sin\varphi, \quad z = r\cos\theta, \quad \frac{D(x,y,z)}{D(r,\theta,\varphi)} = r^2\sin\theta$$

4.61 Change of the order of integration and change of variables

4.611

1.
$$\int_0^\alpha dx \int_0^x f(x,y)\,dy = \int_0^\alpha dy \int_y^\alpha f(x,y)\,dx$$

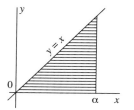

2. $\displaystyle \int_0^\alpha dx \int_0^{\frac{\beta}{\alpha}x} f(x,y)\, dy = \int_0^\beta dy \int_{\frac{\alpha}{\beta}y}^\alpha f(x,y)\, dx$

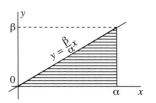

4.612

1. $\displaystyle \int_0^R dx \int_0^{\sqrt{R^2-x^2}} f(x,y)\, dy = \int_0^R dy \int_0^{\sqrt{R^2-y^2}} f(x,y)\, dx$

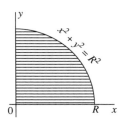

2. $\displaystyle \int_0^{2p} dx \int_0^{q/p\sqrt{2px-x^2}} f(x,y)\, dy = \int_0^q dy \int_{p\left[1-\sqrt{1-(y/q)^2}\right]}^{p\left[1+\sqrt{1-(y/q)^2}\right]} f(x,y)\, dx$

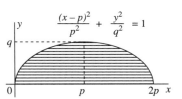

4.613

1. $\displaystyle \int_0^\alpha dx \int_0^{\beta/(\beta+x)} f(x,y)\, dy = \int_0^{\beta/(\beta+\alpha)} dy \int_0^\alpha f(x,y)\, dx$
$$+ \int_{\beta/(\beta+\alpha)}^1 dy \int_0^{\beta(1-y)/y} f(x,y)\, dx$$

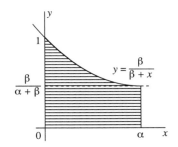

2. $\displaystyle \int_0^\alpha dx \int_{\beta x}^{\delta - \nu x} f(x,y)\, dy = \int_0^{\alpha\beta} dy \int_0^{y/\beta} f(x,y)\, dx$
$$+ \int_{\alpha\beta}^\delta dy \int_0^{(\delta-y)/\gamma} f(x,y)\, dx$$

$$\left[\alpha = \frac{\delta}{\beta+\gamma}, \quad a > 0, \quad \beta > 0, \quad \gamma > 0 \right]$$

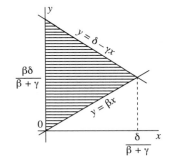

3. $\displaystyle\int_0^{2\alpha} dx \int_{x^2/4\alpha}^{3\alpha-x} f(x,y)\, dy = \int_0^{\alpha} dy \int_0^{2\sqrt{\alpha y}} f(x,y)\, dx +$

$$+ \int_{\alpha}^{3\alpha} dy \int_0^{3\alpha-y} f(x,y)\, dx$$

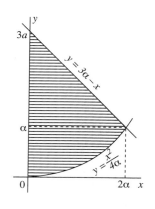

4. $\displaystyle\int_0^{R} dx \int_{\sqrt{R^2-x^2}}^{x+2R} f(x,y)\, dy = \int_0^{R} dy \int_{\sqrt{R^2-y^2}}^{R} f(x,y)\, dx$

$$+ \int_{R}^{2R} dy \int_0^{R} f(x,y)\, dx$$

$$+ \int_{2R}^{3R} dy \int_{y-2R}^{R} f(x,y)\, dx$$

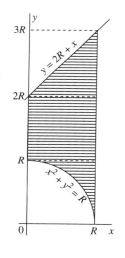

4.614 $\displaystyle\int_0^{\pi/2} d\varphi \int_0^{2R\cos\varphi} f(r,\varphi)\, dr = \int_0^{2R} dr \int_0^{\arccos \frac{r}{2R}} f(r,\varphi)\, d\varphi$

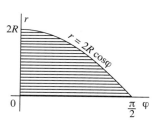

4.615 $\displaystyle\int_0^{R} dx \int_0^{\sqrt{R^2-x^2}} f(x,y)\, dy = \int_0^{\pi/2} d\varphi \int_0^{R} f(r\cos\varphi r, \sin\varphi)\, r\, dr$

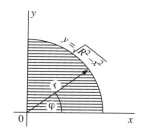

4.616
$$\int_0^{2R} dx \int_0^{\sqrt{2R-x^2}} f(x,y)\,dy = \int_0^{\pi/2} d\varphi \int_0^{2R\cos\varphi} f\left(r\cos\varphi r, \sin\varphi\right) r\,dr$$

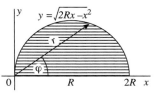

4.617
$$\int_\alpha^\beta dx \int_{\varphi_1(x)}^{\varphi_2(x)} f(x,y)dy = \int_0^\beta dx \int_0^{\varphi_2(x)} f(x,y)dy - \int_0^\beta dx \int_0^{\varphi_1(x)} f(x,y)dy - \int_0^\alpha dx \int_0^{\varphi_2(x)} f(x,y)dy$$
$$+ \int_0^\alpha dx \int_0^{\varphi_1(x)} f(x,y)\,dy$$
$$[\varphi_1(x) \le \varphi_2(x) \text{ for } \alpha \le x \le \beta]$$

4.618
$$\int_0^\gamma dx \int_0^{\varphi(x)} f(x,y)\,dy = \int_0^\gamma dx \int_0^1 f\left[x, z\varphi(x)\right]\varphi(x)dz \quad [y = z\varphi(x)]$$
$$= \gamma \int_0^1 dz \int_0^{\varphi(\gamma z)} f(\gamma z, y)\,dy \quad [x = \gamma z]$$

4.619
$$\int_{x_0}^{x_1} dx \int_{y_0}^{y_1} f(x,y)\,dy = \int_{x_0}^{x_1} dx \int_0^1 (y_1 - y_0)\, f\left[x, y_0 + (y_1 - y_0)\,t\right]\,dt$$
$$[y = y_0 + (y_1 - y_0)\,t]$$

4.62   Double and triple integrals with constant limits

4.620   General formulas

1.
$$\int_0^\pi d\omega \int_0^\infty f'\left(p\cosh x + q\cos\omega \sinh x\right)\sinh x\,dx = -\frac{\pi\,\text{sign}\,p}{\sqrt{p^2 - q^2}} f\left(\text{sign}\,p\sqrt{p^2 - q^2}\right)$$
$$\left[p^2 > q^2, \quad \lim_{x\to+\infty} f(x) = 0\right] \qquad \text{LO III 389}$$

2.
$$\int_0^{2\pi} d\omega \int_0^\infty f'\left[p\cosh x + (q\cos\omega + r\sin\omega)\sinh x\right]\sinh x\,dx$$
$$= -\frac{2\pi\,\text{sign}\,p}{\sqrt{p^2 - q^2 - r^2}} f\left(\text{sign}\,p\sqrt{p^2 - q^2 - r^2}\right)$$
$$\left[p^2 > q^2 + r^2, \quad \lim_{x\to+\infty} f(x) = 0\right] \qquad \text{LO III 390}$$

3.
$$\int_0^\pi \int_0^\pi \frac{dx\,dy}{\sin x \sin^2 y} f'\left[\frac{p - q\cos x}{\sin x \sin y} + r\cot y\right] = -\frac{2\pi\,\text{sign}\,p}{\sqrt{p^2 - q^2 - r^2}} f\left(\text{sign}\,p\sqrt{p^2 - q^2 - r^2}\right)$$
$$\left[p^2 > q^2 + r^2, \quad \lim_{x\to+\infty} f(x) = 0\right]$$
$$\text{LO III 280}$$

4. $\int_{-\infty}^{\infty} dx \int_{-\infty}^{\infty} f'\left(p\cosh x\cosh y + q\sinh x\cosh y + r\sinh y\right)\cosh y\,dy$

$$= -\frac{2\pi\,\mathrm{sign}\,p}{\sqrt{p^2 - q^2 - r^2}} f\left(\mathrm{sign}\,p\sqrt{p^2 - q^2 - r^2}\right)$$

$$\left[p^2 > q^2 + r^2, \quad \lim_{x\to+\infty} f(x) = 0\right] \quad \text{LO III 390}$$

5. $\int_0^{\infty} dx \int_0^{\pi} f\left(p\cosh x + q\cos\omega\sinh x\right)\sinh^2 x\sin\omega\,d\omega = 2\int_0^{\infty} f\left(\mathrm{sign}\,p\sqrt{p^2 - q^2}\cosh x\right)\sinh^2 x\,dx$

$$\left[\lim_{x\to+\infty} f(x) = 0\right] \quad \text{LO III 391}$$

6. $\int_0^{\infty} dx \int_0^{2\pi} d\omega \int_0^{\pi} f\left[p\cosh x + (q\cos\omega + r\sin\omega)\sin\theta\sinh x\right]\sinh^2 x\sin\theta\,d\theta$

$$= 4\int_0^{\infty} f\left(\mathrm{sign}\,p\sqrt{p^2 - q^2 - r^2}\cosh x\right)\sinh^2 x\,dx$$

$$\left[p^2 > q^2 + r^2, \quad \lim_{x\to+\infty} f(x) = 0\right] \quad \text{LO III 390}$$

7. $\int_0^{\infty} dx \int_0^{2\pi} d\omega \int_0^{\pi} f\left\{p\cosh x + \left[(q\cos\omega + r\sin\omega)\sin\theta + s\cosh\theta\right]\sinh x\right\}\sinh^2 x\sin\theta\,d\theta$

$$= 4\pi\int_0^{\infty} f\left(\mathrm{sign}\,p\sqrt{p^2 - q^2 - r^2 - s^2}\cosh x\right)\sinh^2 x\,dx$$

$$\left[p^2 > q^2 + r^2 + s^2, \quad \lim_{x\to+\infty} f(x) = 0\right] \quad \text{LO III 391}$$

4.621

1. $\int_0^{\pi/2}\int_0^{\pi/2} \frac{\sin y\sqrt{1 - k^2\sin^2 x\sin^2 y}}{1 - k^2\sin^2 y}\,dx\,dy = \frac{\pi}{2\sqrt{1 - k^2}}$ LO I 252(90)

2. $\int_0^{\pi/2}\int_0^{\pi/2} \frac{\cos y\sqrt{1 - k^2\sin^2 x\sin^2 y}}{1 - k^2\sin^2 y}\,dx\,dy = \boldsymbol{K}(k)$ LO I 252(91)

3. $\int_0^{\pi/2}\int_0^{\pi/2} \frac{\sin\alpha\sin y\,dx\,dy}{\sqrt{1 - \sin^2\alpha\sin^2 x\sin^2 y}} = \frac{\pi\alpha}{2}$ LO I 253

4.622

1. $\int_0^{\pi}\int_0^{\pi}\int_0^{\pi} \frac{dx\,dy\,dz}{1 - \cos x\cos y\cos z} = 4\pi\,\boldsymbol{K}^2\left(\frac{\sqrt{2}}{2}\right)$ MO 137

2. $\int_0^{\pi}\int_0^{\pi}\int_0^{\pi} \frac{dx\,dy\,dz}{3 - \cos y\cos z - \cos x\cos z - \cos x\cos y} = \sqrt{3}\pi\,\boldsymbol{K}^2\left(\sin\frac{\pi}{12}\right)$ MO 137

3. $\int_0^{\pi}\int_0^{\pi}\int_0^{\pi} \frac{dx\,dy\,dz}{3 - \cos x - \cos y - \cos z} = 4\pi\left[18 + 12\sqrt{2} - 10\sqrt{3} - 7\sqrt{6}\right]\boldsymbol{K}^2\left[\left(2 - \sqrt{3}\right)\left(\sqrt{3} - \sqrt{2}\right)\right]$

MO 137

4.623[3] $\int_0^{\infty}\int_0^{\infty} \varphi\left(a^2 x^2 + b^2 y^2\right)\,dx\,dy = \frac{\pi}{2ab}\int_0^{\infty} \varphi\left(x^2\right)x\,dx$

4.624 $\displaystyle\int_0^\pi \int_0^{2\pi} f\left(\alpha\cos\theta + \beta\sin\theta\cos\psi + \gamma\sin\theta\sin\psi\right)\sin\theta\,d\theta\,d\psi$

$$= 2\pi\int_0^\pi f\left(R\cos p\right)\sin p\,dp = 2\pi\int_{-1}^1 f(Rt)\,dt$$

$$\left[R = \sqrt{\alpha^2 + \beta^2 + \gamma^2}\right]$$

4.625[8] $\displaystyle p_l(a,b) = \int_0^a dx\int_0^b dy\left(x^2 + y^2 + 1\right)^{-3/2} P_l\left(1/\sqrt{x^2 + y^2 + 1}\right)$

Then, for even and odd subscripts:

- $\displaystyle p_{2l}(a,b) = \frac{1}{l(2l+1)2^{2l}}\,\frac{ab}{\sqrt{a^2 + b^2 + 1}}\sum_{k=0}^{l-1}\frac{(-1)^{l-k-1}2^{2k}\binom{2l+2k}{l+k}\binom{l+k}{l-k-1}}{\binom{2k}{k}(2k+1)}$

$$\times(2l + 2k + 1)\sum_{j=0}^k \frac{\binom{2j}{j}}{2^{2j}}\,\frac{1}{(a^2 + b^2 + 1)^j}\left(\frac{1}{(a^2+1)^{k-j+1}} + \frac{1}{(b^2+1)^{k-j+1}}\right)$$

- $\displaystyle p_{2l+1}(a,b) = \frac{1}{2^{2l+1}(2l+1)}\sum_{k=0}^l \frac{(-1)^{l+k}}{2^{2k}}\binom{l}{k}\binom{l+k+1}{k}\binom{2l+2k+1}{l+k}$

$$\times\left\{\frac{1}{(b^2+1)^k}\,\frac{b}{\sqrt{b^2+1}}\arctan^{-1}\frac{a}{\sqrt{b^2+1}} + \frac{1}{(a^2+1)^k}\,\frac{a}{\sqrt{a^2+1}}\arctan^{-1}\frac{b}{\sqrt{a^2+1}}\right.$$

$$\left. +ab\sum_{j=1}^k \frac{2^{2j-1}}{j\binom{2j}{j}}\cdot\frac{1}{(a^2+b^2+1)^j}\left(\frac{1}{(a^2+1)^{k-j+1}} + \frac{1}{(b^2+1)^{k-j+1}}\right)\right\}$$

4.63–4.64 Multiple integrals

4.631 $\displaystyle\int_p^x dt_{n-1}\int_p^{t_{n-1}} dt_{n-2}\cdots\int_p^{t_1} f(t)\,dt = \frac{1}{(n-1)!}\int_p^x (x-t)^{n-1}f(t)\,dt,$

where $f(t)$ is continuous on the interval $[p,q]$ and $p \le x \le q$.   FI II 692

4.632

1. $\displaystyle\idotsint_{\substack{x_1\ge 0, x_2\ge 0,\ldots,x_n\ge 0 \\ x_1+x_2+\cdots+x_n\le h}} dx_1\,dx_2\cdots dx_n = \frac{h^n}{n!}$

[the volume of an n-dimensional simplex]   FI III 472

2. $\displaystyle\idotsint_{x_1^2+x_2^2+\cdots+x_n^2\le R^2} dx_1\,dx_2\cdots dx_n = \frac{\sqrt{\pi^n}}{\Gamma\left(\dfrac{n}{2}+1\right)}R^n$ [the volume of an n-dimensional sphere]

FI III 473

4.633 $\displaystyle\idotsint_{x_1^2+x_2^2+\cdots+x_n^2\le 1}\frac{dx_1\,dx_2\cdots dx_n}{\sqrt{1 - x_1^2 - x_2^2 - \cdots - x_n^2}} = \frac{\pi^{(n+1)/2}}{\Gamma\left(\dfrac{n+1}{2}\right)}$ $[n > 1]$

[Half-area of the surface of an $(n+1)$-dimensional sphere $x_1^2 + x_2^2 + \cdots + x_{n+1}^2 = 1$]   FI III 474

4.634[8]
$$\underset{\substack{x_1 \geq 0, x_2 \geq 0, \ldots, x_n \geq 0 \\ \left(\frac{x_1}{q_1}\right)^{\alpha_1} + \left(\frac{x_2}{q_2}\right)^{\alpha_2} + \cdots + \left(\frac{x_n}{q_n}\right)^{\alpha_n} \leq 1}}{\iint \cdots \int} x_1^{p_1-1} x_2^{p_2-1} \cdots x_n^{p_n-1} \, dx_1 \, dx_2 \ldots dx_n$$

$$= \frac{q_1^{p_1} q_2^{p_2} \cdots q_n^{p_n}}{\alpha_1 \alpha_2 \ldots \alpha_n} \frac{\Gamma\left(\dfrac{p_1}{\alpha_1}\right) \Gamma\left(\dfrac{p_2}{\alpha_2}\right) \ldots \Gamma\left(\dfrac{p_n}{\alpha_n}\right)}{\Gamma\left(\dfrac{p_1}{\alpha_1} + \dfrac{p_2}{\alpha_2} + \cdots + \dfrac{p_n}{\alpha_n} + 1\right)}$$

$$[\alpha_i > 0, \quad p_i > 0, \quad q_i > 0, \quad i = 1, 2, \ldots, n] \quad \text{FI III 477}$$

4.635

1.[8]
$$\underset{\substack{x_1 \geq 0, x_2 \geq 0, \ldots, x_n \geq 0 \\ \left(\frac{x_1}{q_1}\right)^{\alpha_1} + \left(\frac{x_2}{q_2}\right)^{\alpha_2} + \cdots + \left(\frac{x_n}{q_n}\right)^{\alpha_n} \geq 1}}{\iint \cdots \int} f\left[\left(\frac{x_1}{q_1}\right)^{\alpha_1} + \left(\frac{x_2}{q_2}\right)^{\alpha_2} + \cdots + \left(\frac{x_n}{q_n}\right)^{\alpha_n}\right]$$

$$\times x_1^{p_1-1} x_2^{p_2-1} \cdots x_n^{p_n-1} \, dx_1 \, dx_2 \cdots dx_n$$

$$= \frac{q_1^{p_1} q_2^{p_2} \cdots q_n^{p_n}}{\alpha_1 \alpha_2 \cdots \alpha_n} \frac{\Gamma\left(\dfrac{p_1}{\alpha_1}\right) \Gamma\left(\dfrac{p_2}{\alpha_2}\right) \ldots \Gamma\left(\dfrac{p_n}{\alpha_n}\right)}{\Gamma\left(\dfrac{p_1}{\alpha_1} + \dfrac{p_2}{\alpha_2} + \cdots + \dfrac{p_n}{\alpha_n}\right)} \int_1^\infty f(x) x^{\frac{p_1}{\alpha_1} + \frac{p_2}{\alpha_2} + \cdots + \frac{p_n}{\alpha_n} - 1} \, dx$$

under the assumption that the integral on the right converges absolutely.   FI III 487

2.[8]
$$\underset{\substack{x_1 \geq 0, x_2 \geq 0, \cdots, x_n \geq 0 \\ \left(\frac{x_1}{q_1}\right)^{\alpha_1} + \left(\frac{x_2}{q_2}\right)^{\alpha_2} + \cdots + \left(\frac{x_n}{q_n}\right)^{\alpha_n} \leq 1}}{\iint \cdots \int} f\left[\left(\frac{x_1}{q_1}\right)^{\alpha_1} + \left(\frac{x_2}{q_2}\right)^{\alpha_2} + \cdots + \left(\frac{x_n}{q_n}\right)^{\alpha_n}\right]$$

$$\times x_1^{p_1-1} x_2^{p_2-1} \cdots x_n^{p_n-1} \, dx_1 \, dx_2 \cdots dx_n$$

$$= \frac{q_1^{p_1} q_2^{p_2} \cdots q_n^{p_n}}{\alpha_1 \alpha_2 \cdots \alpha_n} \frac{\Gamma\left(\dfrac{p_1}{\alpha_1}\right) \Gamma\left(\dfrac{p_2}{\alpha_2}\right) \ldots \Gamma\left(\dfrac{p_n}{\alpha_n}\right)}{\Gamma\left(\dfrac{p_1}{\alpha_1} + \dfrac{p_2}{\alpha_2} + \cdots + \dfrac{p_n}{\alpha_n}\right)} \int_0^1 f(x) x^{\frac{p_1}{\alpha_1} + \frac{p_2}{\alpha_2} + \cdots + \frac{p_n}{\alpha_n} - 1} \, dx$$

under the assumptions that the one-dimensional integral on the right converges absolutely and that the numbers q_i, α_i, and p_i are positive.   FI III 479

In particular,

3.
$$\underset{\substack{x_1 \geq 0, x_2 \geq 0, \ldots, x_n \geq 0 \\ x_1 + x_2 + \cdots + x_n \leq 1}}{\iint \cdots \int} x_1^{p_1-1} x_2^{p_2-1} \ldots x_n^{p_n-1} e^{-q(x_1+x_2+\cdots+x_n)} \, dx_1 \, dx_2 \ldots dx_n$$

$$= \frac{\Gamma(p_1) \Gamma(p_2) \ldots \Gamma(p_n)}{\Gamma(p_1 + p_2 + \cdots + p_n)} \int_0^1 x^{p_1+p_2+\cdots+p_n-1} e^{-qx} \, dx$$

$$[n > 0, \quad p_1 > 0, \quad p_2 > 0, \ldots, p_n > 0]$$

4.8

$$\iint \cdots \int_{\substack{x_1 \geq 0, x_2 \geq 0, \cdots, x_n \geq 0 \\ x_1^{\alpha_1} + x_2^{\alpha_2} + \cdots + x_n^{\alpha_n} \leq 1}} \frac{x_1^{p_1-1} x_2^{p_2-1} \cdots x_n^{p_n-1}}{(1 - x_1^{\alpha_1} - x_2^{\alpha_2} - \cdots - x_n^{\alpha_n})^{\mu}} \, dx_1 \, dx_2 \ldots dx_n$$

$$= \frac{\Gamma(1-\mu)}{\alpha_1 \alpha_2 \ldots \alpha_n} \frac{\Gamma\left(\frac{p_1}{\alpha_1}\right) \Gamma\left(\frac{p_2}{\alpha_2}\right) \ldots \Gamma\left(\frac{p_n}{\alpha_n}\right)}{\Gamma\left(1 - \mu + \frac{p_1}{\alpha_1} + \frac{p_2}{\alpha_2} + \cdots + \frac{p_n}{\alpha_n}\right)}$$

$$[p_1 > 0, \quad p_2 > 0, \ldots, p_n > 0, \quad \mu < 1] \quad \textsf{FI III 480}$$

4.636

1.8

$$\iint \cdots \int_{\substack{x_1 \geq 0, x_2 \geq 0, \ldots, x_n \geq 0 \\ x_1^{\alpha_1} + x_2^{\alpha_2} + \cdots + x_n^{\alpha_n} \geq 1}} \frac{x_1^{p_1-1} x_2^{p_2-1} \cdots x_n^{p_n-1}}{(x_1^{\alpha_1} + x_2^{\alpha_2} + \cdots + x_n^{\alpha_n})^{\mu}} \, dx_1 \, dx_2 \ldots dx_n$$

$$= \frac{1}{\alpha_1 \alpha_2 \ldots \alpha_n \left(\mu - \frac{p_1}{\alpha_1} - \frac{p_2}{\alpha_2} - \cdots - \frac{p_n}{\alpha_n}\right)} \frac{\Gamma\left(\frac{p_1}{\alpha_1}\right) \Gamma\left(\frac{p_2}{\alpha_2}\right) \ldots \Gamma\left(\frac{p_n}{\alpha_n}\right)}{\Gamma\left(\frac{p_1}{\alpha_1} + \frac{p_2}{\alpha_2} + \cdots + \frac{p_n}{\alpha_n}\right)}$$

$$\left[p_1 > 0, \quad p_2 > 0, \ldots, p_n > 0; \quad \mu > \frac{p_1}{\alpha_1} + \frac{p_2}{\alpha_2} + \cdots + \frac{p_n}{\alpha_n}\right] \quad \textsf{FI III 488}$$

2.8

$$\iint \cdots \int_{\substack{x_1 \geq 0, x_2 \geq 0, \cdots, x_n \geq 0 \\ x_1^{\alpha_1} + x_2^{\alpha_2} + \cdots + x_n^{\alpha_n} \leq 1}} \frac{x_1^{p_1-1} x_2^{p_2-1} \cdots x_n^{p_n-1}}{(x_1^{\alpha_1} + x_2^{\alpha_2} + \cdots + x_n^{\alpha_n})^{\mu}} \, dx_1 \, dx_2 \ldots dx_n$$

$$= \frac{1}{\alpha_1 \alpha_2 \ldots \alpha_n \left(\frac{p_1}{\alpha_1} + \frac{p_2}{\alpha_2} + \cdots + \frac{p_n}{\alpha_n} - \mu\right)} \frac{\Gamma\left(\frac{p_1}{\alpha_1}\right) \Gamma\left(\frac{p_2}{\alpha_2}\right) \ldots \Gamma\left(\frac{p_n}{\alpha_n}\right)}{\Gamma\left(\frac{p_1}{\alpha_1} + \frac{p_2}{\alpha_2} + \cdots + \frac{p_n}{\alpha_n}\right)}$$

$$\left[\mu < \frac{p_1}{\alpha_1} + \frac{p_2}{\alpha_2} + \cdots + \frac{p_n}{\alpha_n}\right] \quad \textsf{FI III 480}$$

3.8

$$\iint \cdots \int_{\substack{x_1 \geq 0, x_2 \geq 0, \ldots, x_n \geq 0 \\ x_1^{\alpha_1} + x_2^{\alpha_2} + \cdots + x_n^{\alpha_n} \leq 1}} x_1^{p_1-1} x_2^{p_2-1} \cdots x_n^{p_n-1} \sqrt{\frac{1 - x_1^{\alpha_1} - x_2^{\alpha_2} - \cdots - x_n^{\alpha_n}}{1 + x_1^{\alpha_1} + x_2^{\alpha_2} + \cdots + x_n^{\alpha_n}}} \, dx_1 \, dx_2 \ldots dx_n$$

$$= \frac{\sqrt{\pi}}{2} \frac{\Gamma\left(\frac{p_1}{\alpha_1}\right) \Gamma\left(\frac{p_1}{\alpha_2}\right) \ldots \Gamma\left(\frac{p_n}{\alpha_n}\right)}{\alpha_1 \alpha_2 \ldots \alpha_n} \frac{1}{\Gamma(m)} \left\{ \frac{\Gamma\left(\frac{m}{2}\right)}{\Gamma\left(\frac{m+1}{2}\right)} - \frac{\Gamma\left(\frac{m+1}{2}\right)}{\Gamma\left(\frac{m+2}{2}\right)} \right\},$$

where $m = \dfrac{p_1}{\alpha_1} + \dfrac{p_2}{\alpha_2} + \cdots + \dfrac{p_n}{\alpha_n}$. $\textsf{FI III 480}$

4.637[8]
$$\iint \cdots \int_{\substack{x_1 \geq 0, \ x_2 \geq 0, \ldots, x_n \geq 0 \\ x_1 + x_2 + \cdots + x_n \leq 1}} f(x_1 + x_2 + \cdots + x_n) \frac{x_1^{p_1-1} x_2^{p_2-1} \ldots x_n^{p_n-1} \, dx_1 \, dx_2 \ldots dx_n}{(q_1 x_1 + q_2 x_2 + \cdots + q_n x_n + r)^{p_1 + p_2 + \cdots + p_n}}$$

$$= \frac{\Gamma(p_1) \Gamma(p_2) \ldots \Gamma(p_n)}{\Gamma(p_1 p_2 + \ldots p_n)} \int_0^1 f(x) \frac{x^{p_1 p_2 + \cdots p_n - 1}}{(q_1 x + r)^{p_1} (q_2 x + r)^{p_2} \ldots (q_n x + r)^{p_n}} \, dx,$$
$$[q_1 \geq 0, \quad q_2 \geq 0, \ldots, q_n \geq 0; \quad r > 0]$$

where $f(x)$ is continuous on the interval $(0, 1)$.

4.638

1.
$$\int_0^\infty \int_0^\infty \cdots \int_0^\infty \frac{x_1^{p_1-1} x_2^{p_2-1} \ldots x_n^{p_n-1} e^{-(q_1 x_1 + q_2 x_2 + \cdots + q_n x_n)}}{(r_0 + r_1 x_1 + r_2 x_2 + \cdots + r_n x_n)^s} \, dx_1 \, dx_2 \ldots dx_n$$

$$= \frac{\Gamma(p_1) \Gamma(p_2) \ldots \Gamma(p_n)}{\Gamma(s)} \int_0^\infty \frac{e^{r_0 x} x^{s-1} \, dx}{(q_1 r_1 x)^{p_1} (q_1 r_2 x)^{p_2} \ldots (q_n r_n x)^{p_n}}$$

where $p_i, q_i, r_i,$ and s are positive. This result is also valid for $r_0 = 0$ provided $p_1 + p_2 + \cdots + p_n > s$.

2.
$$\int_0^\infty \int_0^\infty \cdots \int_0^\infty \frac{x_1^{p_1-1} x_2^{p_2-1} \ldots x_n^{p_n-1}}{(r_0 + r_1 x_1 + r_2 x_2 + \cdots + r_n x_n)^s} \, dx_1 \, dx_2 \ldots dx_n$$

$$= \frac{\Gamma(p_1) \Gamma(p_2) \ldots \Gamma(p_n) \Gamma(s p_1 p_2 - \cdots - p_n)}{r_1^{p_1} r_2^{p_2} \cdots r_n^{p_n} r_0^{s - p_1 - p_2 - \cdots - p_n} \Gamma(s)}$$
$$[p_i > 0, \quad r_i > 0, \quad s > 0]$$

3.[8]
$$\int_0^\infty \int_0^\infty \cdots \int_0^\infty \frac{x_1^{p_1-1} x_2^{p_2-1} \ldots x_n^{p_n-1}}{[1 + (r_1 x_1)^{q_1} + (r_2 x_2)^{q_2} + \cdots + (r_n x_n)^{q_n}]^s} \, dx_1 \, dx_2 \ldots dx_n$$

$$= \frac{\Gamma\left(\frac{p_1}{q_1}\right) \Gamma\left(\frac{p_2}{q_2}\right) \ldots \Gamma\left(\frac{p_n}{q_n}\right)}{q_1 q_2 \ldots q_n r_1^{p_1 q_1} r_2^{p_2 q_2} \ldots r_n^{p_n q_n}} \frac{\Gamma\left(s - \frac{p_1}{q_1} - \frac{p_2}{q_2} - \cdots - \frac{p_n}{q_n}\right)}{\Gamma(s)}$$
$$[p_i > 0, \quad q_i > 0, \quad r_i > 0, \quad s > 0]$$

4.639

1.
$$\iint \cdots \int_{x_1^2 + x_2^2 + \cdots + x_n^2 \leq 1} (p_1 x_1 + p_2 x_2 + \cdots + p_n x_n)^{2m} \, dx_1 \, dx_2 \ldots dx_n$$

$$= \frac{(2m-1)!!}{2^m} \frac{\sqrt{\pi^n}}{\Gamma\left(\frac{n}{2} + m + 1\right)} \left(p_1^2 + p_2^2 + \cdots + p_n^2\right)^m$$

FI III 482

2.
$$\iint \cdots \int_{x_1^2 + x_2^2 + \cdots + x_n^2 \leq 1} (p_1 x_1 + p_2 x_2 + \cdots + p_n x_n)^{2m+1} \, dx_1 \, dx_2 \ldots dx_n = 0$$

FI III 483

4.641

1.
$$\iint \cdots \int_{x_1^2 + x_2^2 + \cdots + x_n^2 \leq 1} e^{p_1 x_1 p_2 x_2 + \cdots p_n x_n} \, dx_1 \, dx_2 \ldots dx_n$$

$$= \sqrt{\pi^n} \sum_{k=0}^\infty \frac{1}{k! \, \Gamma\left(\frac{n}{2} + k + 1\right)} \left(\frac{p_1^2 + p_2^2 + \cdots + p_n^2}{4}\right)^k$$

FI III 483

2.
$$\iint \cdots \int_{x_1^2+x_2^2+\cdots+x_{2n}^2 \leq 1} e^{p_1 x_1 p_2 x_2 + \cdots p_{2n} x_{2n}}\, dx_1\, dx_2 \ldots dx_{2n} = \frac{(2\pi)^n\, I_n\left(\sqrt{p_1^2+p_2^2+\cdots+p_{2n}^2}\right)}{\left(p_1^2+p_2^2+\cdots+p_{2n}^2\right)^{n/2}}$$

FI III 483a

4.642
$$\iint \cdots \int_{x_1^2+x_2^2+\cdots+x_n^2 \leq R^2} f\left(\sqrt{x_1^2+x_2^2+\cdots+x_n^2}\right) dx_1\, dx_2 \ldots dx_n = \frac{2\sqrt{\pi^n}}{\Gamma\left(\dfrac{n}{2}\right)} \int_0^R x^{n-1} f(x)\, dx,$$

where $f(x)$ is a function that is continuous on the interval $(0, R)$. FI III 485

4.643 $\displaystyle \int_0^1 \int_0^1 \cdots \int_0^1 f(x_1 x_2 \cdots x_n)(1-x_1)^{p_1-1}(1-x_2)^{p_2-1}\ldots(1-x_n)^{p_n-1}$

$$\times x_2^{p_1} x_3^{p_1+p_2} \cdots x_n^{p_1+p_2+\cdots+p_{n-1}}\, dx_1\, dx_2 \ldots dx_n$$

$$= \frac{\Gamma(p_1)\Gamma(p_2)\ldots\Gamma(p_n)}{\Gamma(p_1+p_2+\cdots+p_n)} \int_0^1 f(x)(1-x)^{p_1+p_2+\cdots+p_n-1}\, dx$$

under the assumption that the integral on the right converges absolutely. FI III 488

4.644 $\displaystyle \overbrace{\iint \cdots \int}^{n-1}_{x_1^2+x_2^2+\cdots+x_n^2=1} f(p_1 x_1 + p_2 x_2 + \cdots + p_n x_n) \frac{dx_1\, dx_2 \ldots dx_{n-1}}{|x_n|}$

$$= 2 \iint \cdots \int_{x_1^2+x_2^2+\cdots+x_{n-1}^2 \leq 1} f(p_1 x_1 + p_2 x_2 + \cdots + p_n x_n) \frac{dx_1\, dx_2 \cdots dx_{n-1}}{\sqrt{1-x_1^2-x_2^2-\cdots-x_{n-1}^2}}$$

$$= \frac{2\sqrt{\pi^{n-1}}}{\Gamma\left(\dfrac{n-1}{2}\right)} \int_0^\pi f\left(\sqrt{p_1^2+p_2^2+\cdots+p_n^2}\cos x\right)\sin^{n-2} x\, dx \qquad [n \geq 3]$$

where $f(x)$ is continuous on the interval $\left\{-\sqrt{p_1^2+p_2^2+\cdots+p_n^2}, \sqrt{p_1^2+p_2^2+\cdots+p_n^2}\right\}$ FI III 489

4.645 Suppose that two functions $f(x_1, x_2, \ldots, x_n)$ and $g(x_1, x_2, \ldots, x_n)$ are continuous in a closed bounded region D and that the smallest and greatest values of the function g in D are m and M respectively. Let $\varphi(u)$ denote a function that is continuous for $m \leq u \leq M$. We denote by $\psi(u)$ the integral

1. $\displaystyle \psi(u) = \iint \cdots \int_{m \leq g(x_1, x_2, \ldots, x_n) \leq u} f(x_1, x_2, \ldots, x_n)\, dx_1\, dx_2 \ldots dx_n,$

over that portion of the region D on which the inequality $m \leq g(x_1, x_2, \ldots, x_n) \leq u$ is satisfied. Then

2. $\displaystyle \iint \cdots \int_{m \leq g(x_1, x_2, \ldots, x_n) \leq M} f(x_1, x_2, \ldots, x_n)\, \varphi[g(x_1, x_2, \ldots, x_n)]\, dx_1\, dx_2 \ldots dx_n$

$$= (S) \int_m^M \varphi(u)\, d\psi(u) = (R) \int_m^M \varphi(u) \frac{d\psi(u)}{du}\, du$$

where the middle integral must be understood in the sense of Stieltjes. If the derivative $\frac{d\psi}{du}$ exists and is continuous, the Riemann integral on the right exists.

M may be $+\infty$ in formulas **4.645** 2, in which case $\int_m^{+\infty}$ should be understood to mean $\displaystyle \lim_{M \to +\infty} \int_m^M$.

4.646[8] $\displaystyle\iint\cdots\int_{\substack{x_1\geq0,x_2\geq0,\ldots,x_n\geq0 \\ x_1+x_2+\cdots+x_n\leq1}} \frac{x_1^{p_1-1}x_2^{p_2-1}\ldots x_n^{p_n-1}}{(q_1x_1+q_2x_2+\cdots+q_nx_n)^r}\,dx_1\,dx_2\ldots dx_n$

$$=\frac{\Gamma(p_1)\,\Gamma(p_2)\ldots\Gamma(p_n)}{\Gamma(p_1+p_2+\cdots+p_n-r+1)\,\Gamma(r)}\int_0^\infty\frac{x^{r-1}\,dx}{(1+q_1x)^{p_1}(1+q_2x)^{p_2}\cdots(1+q_nx)^{p_n}}$$

$$=[p_1>0,\quad p_2>0,\ldots,p_n>0,\quad q_1>0,\quad q_2>0,\ldots,q_n>0,\quad p_1+p_2+\cdots+p_n>r>0]$$

FI III 493

4.647 $\displaystyle\iint\cdots\int_{0\leq x_1^2+x_2^2+\cdots+x_n^2\leq1}\exp\left\{\frac{p_1x_1+p_2x_2+\cdots+p_nx_n}{\sqrt{x_1^2+x_2^2+\cdots+x_n^2}}\right\}dx_1\,dx_2\ldots dx_n$

$$=\frac{2\sqrt{\pi^n}}{n(p_1^2+p_2^2+\cdots+p_n^2)^{\frac{n}{4}-\frac{1}{2}}}\,I_{\frac{n}{2}-1}\left(\sqrt{p_1^2+p_2^2+\cdots+p_n^2}\right)$$

FI III 495

4.648[8] $\displaystyle\int_0^\infty\int_0^\infty\cdots\int_0^\infty\exp\left[-\left(x_1+x_2+\cdots+x_n+\frac{\lambda^{n+1}}{x_1x_2\ldots x_n}\right)\right]$

$$\times xc_1^{\frac{1}{n+1}-1}x_2^{\frac{2}{n+1}-1}\ldots x_n^{\frac{n}{n+1}-1}\,dx_1\,dx_2\cdots dx_n$$

$$=\frac{1}{\sqrt{n+1}}(2\pi)^{\frac{n}{2}}e^{-(n+1)\lambda}$$

FI III 496

5 Indefinite Integrals of Special Functions

5.1 Elliptic Integrals and Functions

Notation: $k' = \sqrt{1 - k^2}$ (cf. 8.1).

5.11 Complete elliptic integrals

5.111

1. $$\int K(k)k^{2p+3}\,dk = \frac{1}{(2p+3)^2}\left\{4(p+1)^2\int K(k)k^{2p+1}\,dk + k^{2p+2}\left[E(k) - (2p+3)\,K(k)k'^2\right]\right\}$$

BY (610.04)

2. $$\int E(k)k^{2p+3}\,dk = \frac{1}{4p^2 + 16p + 15}\left\{4(p+1)^2\int E(k)k^{2p+1}\,dk\right.$$
$$\left. - E(k)k^{2p+2}\left[(2p+3)k'^2 - 2\right] - k^{2p+2}k'^2\,K(k)\right\}$$

BY (611.04)

5.112

1. $$\int K(k)\,dk = \frac{\pi k}{2}\left[1 + \sum_{j=1}^{\infty}\frac{[(2j)!]^2 k^{2j}}{(2j+1)2^{4j}(j!)^4}\right]$$

BY (610.00)

2.[6] $$\int E(k)\,dk = \frac{\pi k}{2}\left[1 - \sum_{j=1}^{\infty}\frac{[(2j)!]^2 k^{2j}}{(4j^2 - 1)\,2^{4j}(j!)^4}\right]$$

BY (611.00)

3. $$\int K(k)k\,dk = E(k) - k'^2\,K(k)$$

BY (610.01)

4. $$\int E(k)k\,dk = \frac{1}{3}\left[(1 + k^2)\,E(k) - k'^2\,K(k)\right]$$

BY (611.01)

5. $$\int K(k)k^3\,dk = \frac{1}{9}\left[(4 + k^2)\,E(k) - k'^2\,(4 + 3k^2)\,K(k)\right]$$

BY (610.02)

6. $\int \boldsymbol{E}(k)k^3 \, dk = \dfrac{1}{45} \left[\left(4 + k^2 + 9k^4 \right) \boldsymbol{E}(k) - k'^2 \left(4 + 3k^2 \right) \boldsymbol{K}(k) \right]$ BY 611.02)

7. $\int \boldsymbol{K}(k)k^5 \, dk = \dfrac{1}{225} \left[\left(64 + 16k^2 + 9k^4 \right) \boldsymbol{E}(k) - k'^2 \left(64 + 48k^2 + 45k^4 \right) \boldsymbol{K}(k) \right]$ BY (610.03)

8. $\int \boldsymbol{E}(k)k^5 \, dk = \dfrac{1}{1575} \left[\left(64 + 16k^2 + 9k^4 + 225k^6 \right) \boldsymbol{E}(k) - k'^2 \left(64 + 48k^2 + 45k^4 \right) \boldsymbol{K}(k) \right]$

                                                                     BY (611.03)

9. $\int \dfrac{\boldsymbol{K}(k)}{k^2} \, dk = -\dfrac{\boldsymbol{E}(k)}{k}$ BY (612.05)

10. $\int \dfrac{\boldsymbol{E}(k)}{k^2} \, dk = \dfrac{1}{k} \left[k'^2 \boldsymbol{K}(k) - 2 \boldsymbol{E}(k) \right]$ BY (612.02)

11. $\int \dfrac{\boldsymbol{E}(k)}{k'^2} \, dk = k \, \boldsymbol{K}(k)$ BY (612.01)

12. $\int \dfrac{\boldsymbol{E}(k)}{k^4} \, dk = \dfrac{1}{9k^3} \left[2 \left(k^2 - 2 \right) \boldsymbol{E}(k) + k'^2 \boldsymbol{K}(k) \right]$ BY (612.03)

13. $\int \dfrac{k \, \boldsymbol{E}(k)}{k'^2} \, dk = \boldsymbol{K}(k) - \boldsymbol{E}(k)$ BY (612.04)

5.113

1. $\int \left[\boldsymbol{K}(k) - \boldsymbol{E}(k) \right] \dfrac{dk}{k} = -\boldsymbol{E}(k)$ BY (612.06)

2. $\int \left[\boldsymbol{E}(k) - k'^2 \boldsymbol{K}(k) \right] \dfrac{dk}{k} = 2 \boldsymbol{E}(k) - k'^2 \boldsymbol{K}(k)$ BY (612.09)

3. $\int \left[\left(1 + k^2 \right) \boldsymbol{K}(k) - \boldsymbol{E}(k) \right] \dfrac{dk}{k} = -k'^2 \boldsymbol{K}(k)$ BY (612.12)

4. $\int \left[\boldsymbol{K}(k) - \boldsymbol{E}(k) \right] \dfrac{dk}{k^2} = \dfrac{1}{k} \left[\boldsymbol{E}(k) - k'^2 \boldsymbol{K}(k) \right]$ BY (612.07)

5. $\int \left[\boldsymbol{E}(k) - k'^2 \boldsymbol{K}(k) \right] \dfrac{dk}{k^2 k'^2} = \dfrac{1}{k} \left[\boldsymbol{K}(k) - \boldsymbol{E}(k) \right]$

6. $\int \left[\left(1 + k^2 \right) \boldsymbol{E}(k) - k'^2 \boldsymbol{K}(k) \right] \dfrac{dk}{k k'^4} = \dfrac{\boldsymbol{E}(k)}{k'^2}$ BY (612.13)

5.114 $\int \dfrac{k \, \boldsymbol{K}(k) \, dk}{\left[\boldsymbol{E}(k) - k'^2 \boldsymbol{K}(k) \right]^2} = \dfrac{1}{k'^2 \boldsymbol{K}(k) - \boldsymbol{E}(k)}$ BY (612.11)

5.115

1. $\int \Pi \left(\dfrac{\pi}{2}, r^2, k \right) k \, dk = \left(k^2 - r^2 \right) \Pi \left(\dfrac{\pi}{2}, r^2, k \right) - \boldsymbol{K}(k) + \boldsymbol{E}(k)$ BY (612.14)

2. $\int \left[\boldsymbol{K}(k) - \Pi \left(\dfrac{\pi}{2}, r^2, k \right) \right] k \, dk = k^2 \boldsymbol{K}(k) - \left(k^2 - r^2 \right) \Pi \left(\dfrac{\pi}{2}, r^2, k \right)$ BY (612.15)

3. $\int \left[\dfrac{\boldsymbol{E}(k)}{k'^2} + \Pi \left(\dfrac{\pi}{2}, r^2, k \right) \right] k \, dk = \left(k^2 - r^2 \right) \Pi \left(\dfrac{\pi}{2}, r^2, k \right)$ BY (612.16)

5.12   Elliptic integrals

5.121 $\displaystyle\int_0^x \frac{F(x,k)\,dx}{\sqrt{1-k^2\sin^2 x}} = \frac{[F(x,k)]^2}{2} \qquad \left[0 < x \le \frac{\pi}{2}\right]$ BY (630.01)

5.122 $\displaystyle\int_0^x E(x,k)\sqrt{1-k^2\sin^2 x}\,dx = \frac{[\boldsymbol{E}(x,k)]^2}{2}$ BY (630.32)

5.123

1. $\displaystyle\int_0^x F(x,k)\sin x\,dx = -\cos x\, F(x,k) + \frac{1}{k}\arcsin(k\sin x)$ BY (630.11)

2. $\displaystyle\int_0^x F(x,k)\cos x\,dx = \sin x\, F(x,k) + \frac{1}{k}\operatorname{arccosh}\sqrt{\frac{1-k^2\sin^2 x}{k'^2}} - \frac{1}{k}\operatorname{arccosh}\left(\frac{1}{k'}\right)$ BY (630.21)

5.124

1. $\displaystyle\int_0^x E(x,k)\sin x\,dx = -\cos x\, E(x,k) + \frac{1}{2k}\left[k\sin x\sqrt{1-k^2\sin^2 x} + \arcsin(k\sin x)\right].$

                  BY (630.12)

2. $\displaystyle\int_0^x E(x,k)\cos x\,dx = \sin x\, E(x,k) + \frac{1}{2k}\left[k\cos x\sqrt{1-k^2\sin^2 x}\right.$

         $\displaystyle\left. - k'^2\operatorname{arccosh}\sqrt{\frac{1-k^2\sin^2 x}{k'^2}} - k + k'^2\operatorname{arccosh}\left(\frac{1}{k'}\right)\right]$

                  BY (630.22)

5.125

1. $\displaystyle\int_0^x \Pi\left(x,\alpha^2,k\right)\sin x\,dx$

   $\displaystyle= -\cos x\,\Pi\left(x,\alpha^2,k\right) + \frac{1}{\sqrt{k^2-\alpha^2}}\arctan\left[\sqrt{\frac{k^2-\alpha^2}{1-k^2\sin^2 x}}\sin x\right] \qquad \left[\alpha^2 < k^2\right]$

   $\displaystyle= -\cos x\,\Pi\left(x,\alpha^2,k\right) + \frac{1}{\sqrt{\alpha^2-k^2}}\operatorname{arctanh}\left[\sqrt{\frac{\alpha^2-k^2}{1-k^2\sin^2 x}}\sin x\right] \qquad \left[\alpha^2 > k^2\right]$

                  BY (630.13)

2. $\displaystyle\int_0^x \Pi\left(x,\alpha^2,k\right)\cos x\,dx = \sin x\,\Pi\left(x,\alpha^2,k\right) - f - f_0$

  where

$$f = \frac{1}{2\sqrt{(1-\alpha^2)(\alpha^2-k^2)}} \arctan\left[\frac{2(1-\alpha^2)(\alpha^2-k^2) + (1-\alpha^2\sin^2 x)(2k^2-\alpha^2-\alpha^2 k^2)}{2\alpha^2\sqrt{(1-\alpha^2)(\alpha^2-k^2)}\cos x\sqrt{1-k^2\sin^2 x}}\right]$$

$$\text{for } (1-\alpha^2)(\alpha^2-k^2) > 0;$$

$$= \frac{1}{2\sqrt{(\alpha^2-1)(\alpha^2-k^2)}} \ln\left[\frac{2(\alpha^2-1)(\alpha^2-k^2) + (1-\alpha^2\sin^2 x)(\alpha^2+\alpha^2 k^2 - 2k^2)}{1-\alpha^2\sin^2 x}\right.$$

$$\left. + \frac{2\alpha^2\sqrt{(\alpha^2-1)(\alpha^2-k^2)}\cos x\sqrt{1-k^2\sin^2 x}}{1-\alpha^2\sin^2 x}\right]$$

$$\text{for } (1-\alpha^2)(\alpha^2-k^2) < 0,$$

f_0 is the value of f at $x=0$ BY (630.23)

Integration with respect to the modulus

5.126 $\displaystyle\int F(x,k)k\,dk = E(x,k) - k'^2 F(x,k) + \left(\sqrt{1-k^2\sin^2 x}-1\right)\cot x$ BY (613.01)

5.127 $\displaystyle\int E(x,k)k\,dk = \frac{1}{3}\left[(1+k^2)E(x,k) - k'^2 F(x,k) + \left(\sqrt{1-k^2\sin^2 x}-1\right)\cot x\right].$ BY (613.02)

5.128 $\displaystyle\int \Pi\left(x,r^2,k\right)k\,dk = \left(k^2-r^2\right)\Pi\left(x,r^2,k\right) - F(x,k) + E(x,k) + \left(\sqrt{1-k^2\sin^2 x}-1\right)\cot x$

BY (613.03)

5.13 Jacobian elliptic functions

5.131

1. $\displaystyle\int \operatorname{sn}^m u\,du = \frac{1}{m+1}\left[\operatorname{sn}^{m+1}u\operatorname{cn}u\operatorname{dn}u + (m+2)\left(1+k^2\right)\int \operatorname{sn}^{m+2}u\,du\right.$

$$\left. - (m+3)k^2\int \operatorname{sn}^{m+4}u\,du\right]$$

SI 259, PE(567)

2. $\displaystyle\int \operatorname{cn}^m u\,du = \frac{1}{(m+1)k'^2}\left[-\operatorname{cn}^{m+1}u\operatorname{sn}u\operatorname{dn}u\right.$

$$\left. + (m+2)\left(1-2k^2\right)\int \operatorname{cn}^{m+2}u\,du + (m+3)k^2\int \operatorname{cn}^{m+4}u\,du\right]$$

PE (568)

3. $\displaystyle \int \mathrm{dn}^{\,m} u\, du = \frac{1}{(m+1)k'^2} \left[k^2 \,\mathrm{dn}^{\,m+1} u \,\mathrm{sn}\, u \,\mathrm{cn}\, u \right.$

$$\left. + (m+2)\left(2 - k^2\right) \int \mathrm{dn}^{\,m+2} u\, du - (m+3) \int \mathrm{dn}^{\,m+4} u\, du \right]$$

PE (569)

By using formulas **5.131**, we can reduce the integrals $\int \mathrm{sn}^{\,m} u\, du$, $\int \mathrm{cn}^{\,m} u\, du$, and $\int \mathrm{dn}^{\,m} u\, du$ to the integrals **5.132**, **5.133** and **5.134**.

5.132

1. $\displaystyle \int \frac{du}{\mathrm{sn}\, u} = \ln \frac{\mathrm{sn}\, u}{\mathrm{cn}\, u + \mathrm{dn}\, u}$ H 87(164)

$\displaystyle \qquad\qquad = \ln \frac{\mathrm{dn}\, u - \mathrm{cn}\, u}{\mathrm{sn}\, u}$ SI 266(4)

2. $\displaystyle \int \frac{du}{\mathrm{cn}\, u} = \frac{1}{k'} \ln \frac{k'\,\mathrm{sn}\, u + \mathrm{dn}\, u}{\mathrm{cn}\, u}$ SI 266(5)

3. $\displaystyle \int \frac{du}{\mathrm{dn}\, u} = \frac{1}{k'} \arctan \frac{k'\,\mathrm{sn}\, u - \mathrm{cn}\, u}{k'\,\mathrm{sn}\, u + \mathrm{cn}\, u}$ H 88(166)

$\displaystyle \qquad\qquad = \frac{1}{k'} \arccos \frac{\mathrm{cn}\, u}{\mathrm{dn}\, u}$ JA

$\displaystyle \qquad\qquad = \frac{1}{ik'} \ln \frac{\mathrm{cn}\, u + ik'\,\mathrm{sn}\, u}{\mathrm{dn}\, u}$ SI 266(6)

$\displaystyle \qquad\qquad = \frac{1}{k'} \arcsin \frac{k'\,\mathrm{sn}\, u}{\mathrm{dn}\, u}$ JA

5.133

1. $\displaystyle \int \mathrm{sn}\, u\, du = \frac{1}{k} \ln \left(\mathrm{dn}\, u - k\,\mathrm{cn}\, u \right)$ H 87(161)

$\displaystyle \qquad\qquad = \frac{1}{k} \operatorname{arccosh} \frac{\mathrm{dn}\, u - k^2 \,\mathrm{cn}\, u}{1 - k^2}$ JA

$\displaystyle \qquad\qquad = \frac{1}{k} \operatorname{arcsinh} \left(k \frac{\mathrm{dn}\, u - \mathrm{cn}\, u}{1 - k^2} \right);$ JA

$\displaystyle \qquad\qquad = -\frac{1}{k} \ln \left(\mathrm{dn}\, u + k\,\mathrm{cn}\, u \right)$ SI 365(1)

2. $\displaystyle \int \mathrm{cn}\, u\, du = \frac{1}{k} \arccos \left(\mathrm{dn}\, u \right);$ H 87(162)

$\displaystyle \qquad\qquad = \frac{i}{k} \ln \left(\mathrm{dn}\, u - ik\,\mathrm{sn}\, u \right);$ SI 265(2)a, ZH 87(162)

$\displaystyle \qquad\qquad = \frac{1}{k} \arcsin \left(k\,\mathrm{sn}\, u \right)$ JA

3. $\displaystyle\int \mathrm{dn}\, u\, du = \arcsin\left(\mathrm{sn}\, u\right);$ H 87(163)

$\qquad = \mathrm{am}\, u = i \ln\left(\mathrm{cn}\, u - i\, \mathrm{sn}\, u\right)$ SI 266(3), ZH 87(163)

5.134

1. $\displaystyle\int \mathrm{sn}^2 u\, du = \frac{1}{k^2}\left[u - E\left(\mathrm{am}\, u, k\right)\right]$ PE (564)

2. $\displaystyle\int \mathrm{cn}^2 u\, du = \frac{1}{k^2}\left[E\left(\mathrm{am}\, u, k\right) - k'^2 u\right]$ PE (565)

3. $\displaystyle\int \mathrm{dn}^2 u\, du = E\left(\mathrm{am}\, u, k\right)$ PE (566)

5.135

1. $\displaystyle\int \frac{\mathrm{sn}\, u}{\mathrm{cn}\, u}\, du = \frac{1}{k'} \ln \frac{\mathrm{dn}\, u + k'}{\mathrm{cn}\, u}$ SI 266(7)

$\qquad = \dfrac{1}{2k'} \ln \dfrac{\mathrm{dn}\, u + k'}{\mathrm{dn}\, u - k'}$ H 88(167)

2. $\displaystyle\int \frac{\mathrm{sn}\, u}{\mathrm{dn}\, u}\, du = \frac{i}{kk'} \ln \frac{ik' - k\, \mathrm{cn}\, u}{\mathrm{dn}\, u}$ SI 266(8)

$\qquad = \dfrac{1}{kk'} \operatorname{arccot} \dfrac{k\, \mathrm{cn}\, u}{k'}$

3. $\displaystyle\int \frac{\mathrm{cn}\, u}{\mathrm{sn}\, u}\, du = \ln \frac{1 - \mathrm{dn}\, u}{\mathrm{sn}\, u}$ SI 266(10)

$\qquad = \dfrac{1}{2} \ln \dfrac{1 - \mathrm{dn}\, u}{1 + \mathrm{dn}\, u}$ H 88(168)

4. $\displaystyle\int \frac{\mathrm{cn}\, u}{\mathrm{dn}\, u}\, du = -\frac{1}{k} \ln \frac{1 - k\, \mathrm{sn}\, u}{\mathrm{dn}\, u}$ SI 266(9)

$\qquad = \dfrac{1}{2k} \ln \dfrac{1 + k\, \mathrm{sn}\, u}{1 - k\, \mathrm{sn}\, u}$

5. $\displaystyle\int \frac{\mathrm{dn}\, u}{\mathrm{cn}\, u}\, du = \frac{1}{2} \ln \frac{1 + \mathrm{sn}\, u}{1 - \mathrm{sn}\, u}$ H 88(172)

$\qquad = \ln \dfrac{1 + \mathrm{sn}\, u}{\mathrm{cn}\, u}$ JA

6. $\displaystyle\int \frac{\mathrm{dn}\, u}{\mathrm{sn}\, u}\, du = \frac{1}{2} \ln \frac{1 - \mathrm{cn}\, u}{1 + \mathrm{cn}\, u}$ H 87(170)

5.136

1. $\displaystyle\int \operatorname{sn} u \operatorname{cn} u \, du = -\frac{1}{k^2} \operatorname{dn} u$

2. $\displaystyle\int \operatorname{sn} u \operatorname{dn} u \, du = -\operatorname{cn} u$

3. $\displaystyle\int \operatorname{cn} u \operatorname{dn} u \, du = \operatorname{sn} u$

5.137

1. $\displaystyle\int \frac{\operatorname{sn} u}{\operatorname{cn}^2 u} \, du = \frac{1}{k'^2} \frac{\operatorname{dn} u}{\operatorname{cn} u}$ H 88(173)

2. $\displaystyle\int \frac{\operatorname{sn} u}{\operatorname{dn}^2 u} \, du = -\frac{1}{k'^2} \frac{\operatorname{cn} u}{\operatorname{dn} u}$ H 88(175)

3. $\displaystyle\int \frac{\operatorname{cn} u}{\operatorname{sn}^2 u} \, du = -\frac{\operatorname{dn} u}{\operatorname{sn} u}$ H 88(174)

4. $\displaystyle\int \frac{\operatorname{cn} u}{\operatorname{dn}^2 u} \, du = \frac{\operatorname{sn} u}{\operatorname{dn} u}$ H 88(177)

5. $\displaystyle\int \frac{\operatorname{dn} u}{\operatorname{sn}^2 u} \, du = -\frac{\operatorname{cn} u}{\operatorname{sn} u}$ H 88(176)

6. $\displaystyle\int \frac{\operatorname{dn} u}{\operatorname{cn}^2 u} \, du = \frac{\operatorname{sn} u}{\operatorname{cn} u}$ H 88(178)

5.138

1. $\displaystyle\int \frac{\operatorname{cn} u}{\operatorname{sn} u \operatorname{dn} u} \, du = \ln \frac{\operatorname{sn} u}{\operatorname{dn} u}$ H 88(183)

2. $\displaystyle\int \frac{\operatorname{sn} u}{\operatorname{cn} u \operatorname{dn} u} \, du = \frac{1}{k'^2} \ln \frac{\operatorname{dn} u}{\operatorname{cn} u}$ H 88(182)

3. $\displaystyle\int \frac{\operatorname{dn} u}{\operatorname{sn} u \operatorname{cn} u} \, du = \ln \frac{\operatorname{sn} u}{\operatorname{cn} u}$ H 88(184)

5.139

1. $\displaystyle\int \frac{\operatorname{cn} u \operatorname{dn} u}{\operatorname{sn} u} \, du = \ln snu.$ H 88(179)

2. $\displaystyle\int \frac{\operatorname{sn} u \operatorname{dn} u}{\operatorname{cn} u} \, du = \ln \frac{1}{\operatorname{cn} u}$ H 88(180)

3. $\displaystyle\int \frac{\operatorname{sn} u \operatorname{cn} u}{\operatorname{dn} u} \, du = -\frac{1}{k^2} \ln \operatorname{dn} u$ H 88(181)

5.14 Weierstrass elliptic functions

5.141

1. $$\int \wp(u)\, du = -\zeta(u)$$

2. $$\int \wp^2(u)\, du = \frac{1}{6}\,\wp'(u) + \frac{1}{12}g_2 u. \qquad\qquad\qquad\qquad\qquad\qquad\qquad \text{H } 120(192)$$

3. $$\int \wp^3(u)\, du = \frac{1}{120}\,\wp'''(u) - \frac{3}{20}g_2\,\zeta(u) + \frac{1}{10}g_3 u. \qquad\qquad\qquad \text{H } 120(193)$$

4.[8] $$\int \frac{du}{\wp(u) - \wp(v)} = \frac{1}{\wp'(v)}\left[2u\,\zeta(v) + \ln\frac{\sigma(u-v)}{\sigma(u+v)}\right] \qquad [\wp(v) \neq e_1, e_2, e_3] \qquad (\text{see } \mathbf{8.162})$$

$$\text{H } 120(194)$$

5. $$\int \frac{\alpha\wp(u) + \beta}{\gamma\wp(u) + \delta}\, du = \frac{au}{\gamma} + \frac{\alpha\delta - \beta\gamma}{\gamma^2\wp'(v)}\left[\ln\frac{\sigma(u+v)}{\sigma(u-v)} - 2u\,\zeta(v)\right]$$

$$\text{where } v = \wp^{-1}\left(\frac{-\delta}{\gamma}\right) \qquad \text{H } 120(195)$$

5.2 The Exponential Integral Function

5.21 The exponential integral function

5.211 $$\int_x^\infty \mathrm{Ei}(-\beta x)\,\mathrm{Ei}(-\gamma x)\, dx = \left(\frac{1}{\beta} + \frac{1}{\gamma}\right)\mathrm{Ei}[-(\beta + \gamma)x]$$

$$-x\,\mathrm{Ei}(-\beta x)\,\mathrm{Ei}(-\gamma x) - \frac{e^{-\beta x}}{\beta}\mathrm{Ei}(-\gamma x) - \frac{e^{-\gamma x}}{\gamma}\mathrm{Ei}(-\beta x)$$

$$[\mathrm{Re}(\beta + \gamma) > 0] \qquad \text{NT } 53(2)$$

5.22 Combinations of the exponential integral function and powers

5.221

1. $$\int_x^\infty \frac{\mathrm{Ei}[-a(x+b)]}{x^{n+1}}\, dx = \left[\frac{1}{x^n} - \frac{(-1)^n}{b^n}\right]\frac{\mathrm{Ei}[-a(x+b)]}{n} + \frac{e^{-ab}}{n}\sum_{k=0}^{n-1}\frac{(-1)^{n-k-1}}{b^{n-k}}\int_x^\infty \frac{e^{-ax}}{x^{k+1}}\, dx$$

$$[a > 0, \quad b > 0] \qquad \text{NT } 52(3)$$

2. $$\int_x^\infty \frac{\mathrm{Ei}[-a(x+b)]}{x^2}\, dx = \left(\frac{1}{x} + \frac{1}{b}\right)\mathrm{Ei}[-a(x+b)] - \frac{e^{-ab}\,\mathrm{Ei}(-ax)}{b}$$

$$[a > 0, b > 0] \qquad \text{NT } 52(4)$$

5.23 Combinations of the exponential integral and the exponential

5.231

1. $$\int_0^x e^x\,\mathrm{Ei}(-x)\, dx = -\ln x - \boldsymbol{C} + e^x\,\mathrm{Ei}(-x) \qquad\qquad\qquad\qquad \text{ET II } 308(11)$$

1. $\int_0^x e^{-\beta x}\,\mathrm{Ei}(-\alpha x)\,dx = -\dfrac{1}{\beta}\left\{e^{-\beta x}\,\mathrm{Ei}(-\alpha x) + \ln\left(1 + \dfrac{\beta}{\alpha}\right) - \mathrm{Ei}[-(\alpha+\beta)x]\right\}$ ET II 308(12)

5.3 The Sine Integral and the Cosine Integral

5.31

1. $\int \cos\alpha x\,\mathrm{ci}(\beta x)\,dx = \dfrac{\sin\alpha x\,\mathrm{ci}(\beta x)}{\alpha} - \dfrac{\mathrm{si}(\alpha x + \beta x) + \mathrm{si}(\alpha x - \beta x)}{2\alpha}$ NT 49(1)

2. $\int \sin\alpha x\,\mathrm{ci}(\beta x)\,dx = -\dfrac{\cos\alpha x\,\mathrm{ci}(\beta x)}{\alpha} + \dfrac{\mathrm{ci}(\alpha x + \beta x) + \mathrm{ci}(\alpha x - \beta x)}{2\alpha}$ NT 49(2)

5.32

1. $\int \cos\alpha x\,\mathrm{si}(\beta x)\,dx = \dfrac{\sin\alpha x\,\mathrm{si}(\beta x)}{\alpha} + \dfrac{\mathrm{ci}(\alpha x + \beta x) - \mathrm{ci}(\alpha x - \beta x)}{2\alpha}$ NT 49(3)

2. $\int \sin\alpha x\,\mathrm{si}(\beta x)\,dx = -\dfrac{\cos\alpha x\,\mathrm{si}(\beta x)}{\alpha} + \dfrac{\mathrm{si}(\alpha x + \beta x) - \mathrm{si}(\alpha x - \beta x)}{2\alpha}$ NT 49(4)

5.33

1. $\int \mathrm{ci}(\alpha x)\,\mathrm{ci}(\beta x)\,dx = x\,\mathrm{ci}(\alpha x)\,\mathrm{ci}(\beta x) + \dfrac{1}{2\alpha}\left(\mathrm{si}(\alpha x + \beta x) + \mathrm{si}(\alpha x - \beta x)\right)$
$$+\dfrac{1}{2\beta}\left(\mathrm{si}(\alpha x + \beta x) + \mathrm{si}(\beta x - \alpha x)\right) - \dfrac{1}{\alpha}\sin\alpha x\,\mathrm{ci}(\beta x) - \dfrac{1}{\beta}\sin\beta x\,\mathrm{ci}(\alpha x)$$
NT 53(5)

2. $\int \mathrm{si}(\alpha x)\,\mathrm{si}(\beta x)\,dx = x\,\mathrm{si}(\alpha x)\,\mathrm{si}(\beta x) - \dfrac{1}{2\beta}\left(\mathrm{si}(\alpha x + \beta x) + \mathrm{si}(\alpha x - \beta x)\right)$
$$-\dfrac{1}{2\alpha}\left(\mathrm{si}(\alpha x + \beta x) + \mathrm{si}(\beta x + \alpha x)\right) + \dfrac{1}{\alpha}\cos\alpha x\,\mathrm{si}(\beta x) + \dfrac{1}{\beta}\cos\beta x\,\mathrm{si}(\alpha x)$$
NT 54(6)

3. $\int \mathrm{si}(\alpha x)\,\mathrm{ci}(\beta x)\,dx = x\,\mathrm{si}(\alpha x)\,\mathrm{ci}(\beta x) + \dfrac{1}{\alpha}\cos\alpha x\,\mathrm{ci}(\beta x)$
$$-\dfrac{1}{\beta}\sin\beta x\,\mathrm{si}(\alpha x) - \left(\dfrac{1}{2\alpha} + \dfrac{1}{2\beta}\right)\mathrm{ci}(\alpha x + \beta x) - \left(\dfrac{1}{2\alpha} - \dfrac{1}{2\beta}\right)\mathrm{ci}(\alpha x - \beta x)$$
NT 54(10)

5.34

1. $\int_x^\infty \mathrm{si}[a(x+b)]\dfrac{dx}{x^2} = \left(\dfrac{1}{x} + \dfrac{1}{b}\right)\mathrm{si}[a(x+b)] - \dfrac{\cos ab\,\mathrm{si}(ax) + \sin ab\,\mathrm{ci}(ax)}{b}$
$$[a > 0, \quad b > 0]$$
NT 52(6)

2. $\int_x^\infty \mathrm{ci}[a(x+b)]\dfrac{dx}{x^2} = \left(\dfrac{1}{x} + \dfrac{1}{b}\right)\mathrm{ci}[a(x+b)] + \dfrac{\sin ab\,\mathrm{si}(ax) - \cos ab\,\mathrm{ci}(ax)}{b}$
$$[a > 0, \quad b > 0]$$
NT 52(5)

5.4 The Probability Integral and Fresnel Integrals

5.41 $\displaystyle\int \Phi(\alpha x)\, dx = x\,\Phi(\alpha x) + \frac{e^{-\alpha^2 x^2}}{\alpha\sqrt{\pi}}$ NT 12(20)a

5.42 $\displaystyle\int S(\alpha x)\, dx = x\,S(\alpha x) + \frac{\cos^2 \alpha x^2}{\alpha\sqrt{2\pi}}$ NT 12(22)a

5.43 $\displaystyle\int C(\alpha x)\, dx = x\,C(\alpha x) - \frac{\sin^2 \alpha x^2}{\alpha\sqrt{2\pi}}$ NT 12(21)a

5.5 Bessel Functions

Notation: Z and $\mathfrak{Z}$ denote any of J, N, $H^{(1)}$, $H^{(2)}$. In formulae 5.52–5.56, $Z_p(x)$ and $\mathfrak{Z}_p(x)$ are arbitrary Bessel functions of the first, second, or third kinds.

5.51 $\displaystyle\int J_p(x)\, dx = 2\sum_{k=0}^{\infty} J_{p+2k+1}(x)$ JA, MO 30

5.52

1. $\displaystyle\int x^{p+1}\, Z_p(x)\, dx = x^{p+1}\, Z_{p+1}(x)$ WA 132(1)

2. $\displaystyle\int x^{-p+1}\, Z_p(x)\, dx = -x^{-p+1}\, Z_{p+1}(x)$ WA 132(2)

5.53[10] $\displaystyle\int \left[(\alpha^2 - \beta^2)\, x - \frac{p^2 - q^2}{x} \right] Z_p(\alpha x)\, \mathfrak{Z}_q(\beta x)\, dx$

$$= \alpha x\, Z_{p+1}(\alpha x)\, \mathfrak{Z}_q(\beta x) - \beta x\, Z_p(\alpha x)\, \mathfrak{Z}_{q+1}(\beta x) - (p - q)\, Z_p(\alpha x)\, \mathfrak{Z}_q(\beta x)$$

$$= \beta x\, Z_p(\alpha x)\, \mathfrak{Z}_{q-1}(\beta x) - \alpha x\, Z_{p-1}(\alpha x)\, \mathfrak{Z}_q(\beta x) + (p - q)\, Z_p(\alpha x)\, \mathfrak{Z}_q(\beta x)$$

JA, MO 30, WA 134(7)

5.54

1.[10] $\displaystyle\int x\, Z_p(\alpha x)\, \mathfrak{Z}_p(\beta x)\, dx = \frac{\alpha x\, Z_{p+1}(\alpha x)\, \mathfrak{Z}_p(\beta x) - \beta x\, Z_p(\alpha x)\, \mathfrak{Z}_{p+1}(\beta x)}{\alpha^2 - \beta^2}$

$$= \frac{\beta x\, Z_p(\alpha x)\, \mathfrak{Z}_{p-1}(\beta x) - \alpha x\, Z_{p-1}(\alpha x)\, \mathfrak{Z}_p(\beta x)}{\alpha^2 - \beta^2}$$

WA 134(8)

2. $\displaystyle\int x[Z_p(\alpha x)]^2\, dx = \frac{x^2}{2} \left\{ [Z_p(\alpha x)]^2 - Z_{p-1}(\alpha x)\, Z_{p+1}(\alpha x) \right\}$ WA 135(11)

5.55[10] $\displaystyle\int \frac{1}{x}\, Z_p(\alpha x)\, \mathfrak{Z}_q(\alpha x)\, dx = \alpha x \frac{Z_p(\alpha x)\, \mathfrak{Z}_{q+1}(\alpha x) - Z_{p+1}(\alpha x)\, \mathfrak{Z}_q(\alpha x)}{p^2 - q^2} + \frac{Z_p(\alpha x)\, \mathfrak{Z}_q(\alpha x)}{p + q}$

$$= \alpha x \frac{Z_{p-1}(\alpha x)\, \mathfrak{Z}_q(\alpha x) - Z_p(\alpha x)\, \mathfrak{Z}_{q-1}(\alpha x)}{p^2 - q^2} - \frac{Z_p(\alpha x)\, \mathfrak{Z}_q(\alpha x)}{p + q}$$

WA 135(13)

5.56

1. $\displaystyle\int Z_1(x)\, dx = -\, Z_0(x)$ JA

2. $\displaystyle\int x\, Z_0(x)\, dx = x\, Z_1(x)$ JA

6–7 Definite Integrals of Special Functions

6.1 Elliptic Integrals and Functions

Notation: $k' = \sqrt{1 - k^2}$ (cf. 8.1).

6.11 Forms containing $F(x, k)$

6.111 $\displaystyle\int_0^{\pi/2} F(x, k) \cot x \, dx = \frac{\pi}{4} \boldsymbol{K}(k') + \frac{1}{2} \ln k \, \boldsymbol{K}(k)$ BI (350)(1)

6.112

1. $\displaystyle\int_0^{\pi/2} F(x, k) \frac{\sin x \cos x}{1 + k \sin^2 x} \, dx = \frac{1}{4k} \boldsymbol{K}(k) \ln \frac{(1 + k)\sqrt{k}}{2} + \frac{\pi}{16k} \boldsymbol{K}(k')$ BI (350)(6)

2. $\displaystyle\int_0^{\pi/2} F(x, k) \frac{\sin x \cos x}{1 - k \sin^2 x} \, dx = \frac{1}{4k} \boldsymbol{K}(k) \ln \frac{2}{(1 - k)\sqrt{k}} - \frac{\pi}{16k} \boldsymbol{K}(k')$ BI (350)(7)

3. $\displaystyle\int_0^{\pi/2} F(x, k) \frac{\sin x \cos x}{1 - k^2 \sin^2 x} \, dx = -\frac{1}{2k^2} \ln k' \, \boldsymbol{K}(k)$ BI (350)(2)a, BY(802.12)a

6.113

1. $\displaystyle\int_0^{\pi/2} F(x, k') \frac{\sin x \cos x \, dx}{\cos^2 x + k \sin^2 x} = \frac{1}{4(1 - k)} \ln \frac{2}{(1 + k)\sqrt{k}} \boldsymbol{K}(k')$ BI (350)(5)

2. $\displaystyle\int_0^{\pi/2} F(x, k) \frac{\sin x \cos x}{1 - k^2 \sin^2 t \sin^2 x} \cdot \frac{dx}{\sqrt{1 - k^2 \sin^2 x}}$

$$= -\frac{1}{k^2 \sin t \cos t} \left[\boldsymbol{K}(k) \arctan (k' \tan t) - \frac{\pi}{2} F(t, k) \right]$$
BI (350)(12)

6.114 $\displaystyle\int_u^v F(x, k) \frac{dx}{\sqrt{(\sin^2 x - \sin^2 u)(\sin^2 v - \sin^2 x)}} = \frac{1}{2 \cos u \sin v} \boldsymbol{K}(k) \, \boldsymbol{K}\left(\sqrt{1 - \tan^2 u \cot^2 v}\right)$

$\qquad\qquad\qquad\qquad\qquad\qquad\qquad\qquad [k^2 = 1 - \cot^2 u \cdot \cot^2 v]$ BI (351)(9)

6.115 $\displaystyle\int_0^1 F(\arcsin x, k) \frac{x \, dx}{1 + kx^2} = \frac{1}{4k} \boldsymbol{K}(k) \ln \frac{(1 + k)\sqrt{k}}{2} + \frac{\pi}{16k} \boldsymbol{K}(k')$

$\qquad\qquad\qquad\qquad\qquad\qquad$ (cf. **6.112** 2)   BI (466)(1)

This and similar formulas can be obtained from formulas **6.111–6.113** by means of the substitution $x = \arcsin t$.

6.12 Forms containing $E(x, k)$

6.121 $\displaystyle\int_0^{\pi/2} E(x, k) \frac{\sin x \cos x}{1 - k^2 \sin^2 x} \, dx = \frac{1}{2k^2} \left\{ \left(1 + k'^2\right) K(k) - (2 + \ln k') E(k) \right\}$ BI (350)(4)

6.122 $\displaystyle\int_0^{\pi/2} E(x, k) \frac{dx}{\sqrt{1 - k^2 \sin^2 x}} = \frac{1}{2} \left\{ E(k) K(k) - \ln k' \right\}$ BI (350)(10), BY (630.02)

6.123 $\displaystyle\int_0^{\pi/2} E(x, k) \frac{\sin x \cos x}{1 - k^2 \sin^2 t \sin^2 x} \cdot \frac{dx}{\sqrt{1 - k^2 \sin^2 x}}$

$\displaystyle = -\frac{1}{k^2 \sin t \cos t} \left[E(k) \arctan (k' \tan t) - \frac{\pi}{2} E(t, k) + \frac{\pi}{2} \cot t \left(1 - \sqrt{1 - k^2 \sin^2 t}\right) \right]$

                    BI (350)(13)

6.124 $\displaystyle\int_u^v E(x, k) \frac{dx}{\sqrt{\left(\sin^2 x - \sin^2 u\right)\left(\sin^2 v - \sin^2 x\right)}} = \frac{1}{2 \cos u \sin v} E(k) K\left(\sqrt{1 - \frac{tg^2 u}{tg^2 v}}\right)$

$\displaystyle + \frac{k^2 \sin v}{2 \cos u} K\left(\sqrt{1 - \frac{\sin^2 2u}{\sin^2 2v}}\right)$

$\left[k^2 = 1 - \cot^2 u \cot^2 v\right]$ BI (351)(10)

6.13 Integration of elliptic integrals with respect to the modulus

6.131 $\displaystyle\int_0^1 F(x, k) k \, dk = \frac{1 - \cos x}{\sin x} = \tan \frac{x}{2}$ BY (616.03)

6.132 $\displaystyle\int_0^1 E(x, k) k \, dk = \frac{\sin^2 x + 1 - \cos x}{3 \sin x}$ BY (616.04)

6.133 $\displaystyle\int_0^1 \Pi\left(x, r^2, k\right) k \, dk = \tan \frac{x}{2} - r \ln \sqrt{\frac{1 + r \sin x}{1 - r \sin x}} - r^2 \Pi\left(x, r^2, 0\right)$ BY (616.05)

6.14–6.15 Complete elliptic integrals

6.141

1. $\displaystyle\int_0^1 K(k) \, dk = 2G$ FI II 755

2. $\displaystyle\int_0^1 K(k') \, dk = \frac{\pi^2}{4}$ BY (615.03)

6.142 $\displaystyle\int_0^1 \left(K(k) - \frac{\pi}{2}\right) \frac{dk}{k} = \pi \ln 2 - 2G$ BY (615.05)

6.143[7] $\displaystyle\int_0^1 K(k) \frac{dk}{k'} = K^2\left(\frac{\sqrt{2}}{2}\right) = \frac{1}{16\pi} \Gamma^4\left(\frac{1}{4}\right)$ BY (615.08)

6.144 $\displaystyle\int_0^1 K(k) \frac{dk}{1 + k} = \frac{\pi^2}{8}$ BY (615.09)

6.145 $\displaystyle\int_0^1 \left(\boldsymbol{K}(k') - \ln \frac{4}{k} \right) \frac{dk}{k} = \frac{1}{12} \left[24(\ln 2)^2 - \pi^2 \right]$ BY (615.13)

6.146 $\displaystyle n^2 \int_0^1 k^n \, \boldsymbol{K}(k) \, dk = (n-1)^2 \int_0^1 k^{n-2} \, \boldsymbol{K}(k) \, dk + 1$ BY (615.12)

6.147 $\displaystyle n \int_0^1 k^n \, \boldsymbol{K}(k') \, dk = (n-1) \int_0^1 k^{n-2} \, \boldsymbol{E}(k) \, dk$ $[n > 1]$ (see **6.152**) BY (615.11)

6.148

1. $\displaystyle\int_0^1 \boldsymbol{E}(k) \, dk = \frac{1}{2} + \boldsymbol{G}$ BY (615.02)

2. $\displaystyle\int_0^1 \boldsymbol{E}(k') \, dk = \frac{\pi^2}{8}$ BY (615.04)

6.149

1. $\displaystyle\int_0^1 \left(\boldsymbol{E}(k) - \frac{\pi}{2} \right) \frac{dk}{k} = \pi \ln 2 - 2\boldsymbol{G} + 1 - \frac{\pi}{2}$ BY (615.06)

2. $\displaystyle\int_0^1 (\boldsymbol{E}(k') - 1) \frac{dk}{k} = 2 \ln 2 - 1$ BY (615.07)

6.151 $\displaystyle\int_0^1 \boldsymbol{E}(k) \frac{dk}{k'} = \frac{1}{8} \left[4 \boldsymbol{K}^2 \left(\frac{\sqrt{2}}{2} \right) + \frac{\pi^2}{\boldsymbol{K}^2 \left(\frac{\sqrt{2}}{2} \right)} \right]$ BY (615.10)

6.152 $\displaystyle (n+2) \int_0^1 k^n \, \boldsymbol{E}(k') \, dk = (n+1) \int_0^1 k^n \, \boldsymbol{K}(k') \, dk$ $[n > 1]$ (see **6.147**) BY (615.14)

6.153[6] $\displaystyle\int_0^a \frac{\boldsymbol{K}(k) k \, dk}{k'^2 \sqrt{a^2 - k^2}} = \frac{\pi}{4} \frac{1}{\sqrt{1 - a^2}} \ln \left(\frac{1+a}{1-a} \right)$ $[0 < a < 1]$ LO I 252

6.154 $\displaystyle\int_0^{\pi/2} \frac{\boldsymbol{E}(p \sin x)}{1 - p^2 \sin^2 x} \sin x \, dx = \frac{\pi}{2\sqrt{1 - p^2}}$ $[p^2 > 1]$ FI II 489

6.16 The theta function

6.161

1. $\displaystyle\int_0^\infty x^{s-1} \vartheta_2 \left(0 \mid ix^2 \right) dx = 2^s \left(1 - 2^{-s} \right) \pi^{-\frac{s}{2}} \Gamma \left(\tfrac{1}{2} s \right) \zeta(s)$

   $[\operatorname{Re} s > 2]$ ET I 339(20)

2. $\displaystyle\int_0^\infty x^{s-1} \left[\vartheta_3 \left(0 \mid ix^2 \right) - 1 \right] dx = \pi^{-\frac{s}{2}} \Gamma \left(\tfrac{1}{2} s \right) \zeta(s)$ $[\operatorname{Re} s > 2]$ ET I 339(21)

3. $\displaystyle\int_0^\infty x^{s-1} \left[1 - \vartheta_4 \left(0 \mid ix^2 \right) \right] dx = \left(1 - 2^{1-s} \right) \pi^{-\frac{1}{2}s} \Gamma \left(\tfrac{1}{2} s \right) \zeta(s)$

   $[\operatorname{Re} s > 2]$ ET I 339(22)

4. $\displaystyle\int_0^\infty x^{s-1}\left[\vartheta_4\left(0\mid ix^2\right)+\vartheta_2\left(0\mid ix^2\right)-\vartheta_3\left(0\mid ix^2\right)\right]\,dx=-\left(2^s-1\right)\left(2^{1-s}-1\right)\pi^{-\frac12 s}\,\Gamma\left(\tfrac12 s\right)\zeta(s)$

ET I 339(24)

6.162

1. $\displaystyle\int_0^\infty e^{-ax}\,\vartheta_4\left(\frac{b\pi}{2l}\,\middle|\,\frac{i\pi x}{l^2}\right)\,dx=\frac{l}{\sqrt a}\cosh\left(b\sqrt a\right)\operatorname{cosec}\left(l\sqrt a\right)$

[Re $a>0$, $|b|\le l$]   ET I 224(1)a

2. $\displaystyle\int_0^\infty e^{-ax}\,\vartheta_1\left(\frac{b\pi}{2l}\,\middle|\,\frac{i\pi x}{l^2}\right)\,dx=-\frac{l}{\sqrt a}\sinh\left(b\sqrt a\right)\operatorname{sech}\left(l\sqrt a\right)$

[Re $a>0$, $|b|\le l$]   ET I 224(2)a

3. $\displaystyle\int_0^\infty e^{-ax}\,\vartheta_2\left(\frac{(1+b)\pi}{2l}\,\middle|\,\frac{i\pi x}{l^2}\right)\,dx=-\frac{l}{\sqrt a}\sinh\left(b\sqrt a\right)\operatorname{sech}\left(l\sqrt a\right)$

[Re $a>0$, $|b|\le l$]   ET I 224(3)a

4. $\displaystyle\int_0^\infty e^{-ax}\,\vartheta_3\left(\frac{(1+b)\pi}{2l}\,\middle|\,\frac{i\pi x}{l^2}\right)\,dx=\frac{l}{\sqrt a}\cosh\left(b\sqrt a\right)\operatorname{cosech}\left(l\sqrt a\right)$

[Re $a>0$, $|b|\le l$]   ET I 224(4)a

6.163[10]

1. $\displaystyle\int_0^\infty e^{-(a-\mu)x}\,\vartheta_3\left(\pi\sqrt\mu x\mid i\pi x\right)\,dx=\frac{1}{2\sqrt a}\left[\coth\left(\sqrt a+\sqrt\mu\right)+\coth\left(\sqrt a-\sqrt\mu\right)\right]$

[Re $a>0$]   ET I 224(7)a

2.[10] $\displaystyle\int_0^\infty \vartheta_3\left(i\pi kx\mid i\pi x\right)e^{-\left(k^2+l^2\right)x}\,dx=\frac{\sinh 2l}{l\left(\cosh 2l-\cos 2k\right)}$

6.164 $\displaystyle\int_0^\infty\left[\vartheta_4\left(0\middle|ie^{2x}\right)+\vartheta_2(0)ie^{2x}-\vartheta_3\left(0\mid ie^{2x}\right)\right]e^{\frac12 x}\cos(ax)\,dx$

$$=\frac12\left(2^{\frac12+ia}-1\right)\left(1-2^{\frac12-ia}\right)\pi^{-\frac14-\frac12 ia}\,\Gamma\left(\tfrac14+\tfrac12 ia\right)\zeta\left(\tfrac12+ia\right)$$

[$a>0$]   ET I 61(11)

6.165 $\displaystyle\int_0^\infty e^{\frac12 x}\left[\vartheta_3\left(0\mid ie^{2x}\right)-1\right]\cos(ax)\,dx$

$$=\frac{2}{1+4a^2}\left\{1+\left[\left(a^2+\tfrac14\right)\pi^{-\frac12 ia-\frac14}\,\Gamma\left(\tfrac12 ia+\tfrac14\right)\zeta\left(ia+\tfrac12\right)\right]\right\}$$

[$a>0$]   ET I 61(12)

6.17[10] Generalized elliptic integrals

1.   Set

$$\Omega_j(k)\equiv\int_0^\pi\left[1-k^2\cos\phi\right]^{-\left(j+\frac12\right)}\,d\phi,$$

$$\alpha_m(j)=\frac{\pi}{(64)^m}\,\frac{j!}{(2j)!}\,\frac{(4m+2j)!}{(2m+j)!}\left(\frac{1}{m!}\right)^2,\qquad\lambda=\frac{\pi}{2}\sqrt{\frac{(2j+1)k^2}{1-k^2}},$$

then

$$\Omega_j(k) = \sum_{m=0}^{\infty} \alpha_m(j) k^{4m} = \sqrt{\frac{\pi}{(2j+1)k^2}} \left(1 - k^2\right)^{-j} \left[\operatorname{erf} \lambda + \frac{1}{2}(2j+1)^{-1} \left(1 + \frac{1}{2k^2}\right) \right.$$
$$\times \left\{ \operatorname{erf} \lambda - \left(\frac{2}{\sqrt{\pi}}\right) \left(\lambda e^{-\lambda^2}\right) \left(1 + \frac{2}{3}\lambda^2\right) \right\} - \frac{1}{12}(2j+1)^{-2} \left(16 + \frac{13}{k^2} + \frac{1}{k^4}\right)$$
$$\times \left. \left\{ \operatorname{erf} \lambda - \left(\frac{2}{\sqrt{\pi}}\right) \left(\lambda e^{-\lambda^2}\right) \left(1 + \frac{2}{3}\lambda^2 + \frac{4}{15}\lambda^4\right) \right\} + \cdots \right]$$

while for large λ

$$\lim_{j\to\infty} \Omega_j(k) = \sqrt{\frac{\pi}{(2j+1)}} k^2 \left(1 - k^2\right)^{-j}$$
$$\times \left[1 + \frac{1}{2}(2j+1)^{-1} \left\{ 1 + \frac{1}{2k^2} \right\} - \frac{4}{3}(2j+1)^{-2} \left\{ 1 + \frac{13}{16k^2} + \frac{1}{16k^4} \right\} + \cdots \right]$$

2. Set

$$R_\mu(k, \alpha, \delta) = \int_0^\pi \frac{\cos^{2\alpha-1}(\theta/2) \sin^{2\delta-2\alpha-1}(\theta/2)\, d\theta}{[1 - k^2 \cos\theta]^{\mu+\frac{1}{2}}},$$
$$0 < k < 1, \quad \operatorname{Re}\delta > \operatorname{Re}\alpha > 0, \quad \operatorname{Re}\mu > -1/2,$$
$$M_\nu(\mu, \alpha, \delta) = \frac{(-1)^\nu 2^\nu \left(\mu + \frac{1}{2}\right)_\nu}{\nu!} \frac{\Gamma(\alpha)\,\Gamma(\delta - \alpha + \nu)}{\Gamma(\delta + \nu)},$$
$$\text{with } (\lambda)_\nu = \Gamma(\lambda + \nu)/\Gamma(\lambda), \text{ and}$$
$$W_\nu(\mu, \alpha, \delta) = \frac{2^\nu \left(\mu + \frac{1}{2}\right)_\nu}{\nu!} \frac{\Gamma(\alpha + \nu)\,\Gamma(\delta - \alpha)}{\Gamma(\delta + \nu)},$$

then:

• for small k;

$$R_\mu(k, \alpha, \delta) = \left(1 - k^2\right)^{-\left(\mu+\frac{1}{2}\right)} \sum_{\nu=0}^{\infty} \left[k^2/\left(1 - k^2\right)\right]^\nu M_\nu(\mu, \alpha, \delta)$$
$$= \left(1 + k^2\right)^{-\left(\mu+\frac{1}{2}\right)} \sum_{\nu=0}^{\infty} \left[k^2/\left(1 + k^2\right)\right]^\nu W_\nu(\mu, \alpha, \delta),$$

• for k^2 close to 1;

$R_\mu(k, \alpha, \delta)$

$$= \left[\Gamma(\delta - \alpha) \Gamma\left(\mu + \alpha - \delta + \tfrac{1}{2}\right) \Gamma\left(\mu + \tfrac{1}{2}\right) \right] \left(2k^2\right)^{\alpha - \delta} \left(1 - k^2\right)^{\delta - \alpha - \mu - \frac{1}{2}}$$

$$\times \left\{ \Gamma\left(\delta - \alpha - \mu - \tfrac{1}{2}\right) \Gamma(\alpha) \left[\Gamma\left(\delta - \mu - \tfrac{1}{2}\right) \left(2k^2\right)^{\mu + \frac{1}{2}} \right] \right\}$$

$$\left[\operatorname{Re}\left(\mu + \alpha - \delta + \tfrac{1}{2}\right) \text{ not an integer} \right]$$

$$= \left[2^{\mu + \frac{1}{2}} k^{2\mu + 1} \Gamma\left(\mu + \tfrac{1}{2}\right) \Gamma(1 - \alpha) \right]$$

$$\times \sum_{n=0}^{\infty} \left[\Gamma(\delta - \alpha + n) \Gamma(1 - \alpha + n) \Gamma\left(\alpha - \delta + \mu - n + \tfrac{1}{2}\right) n! \right] \left[2k^2 / \left(1 - k^2\right) \right]^{\alpha - \delta + \mu - n + \frac{1}{2}}$$

$$\left[\alpha - \delta + \mu + \tfrac{1}{2} = m, \text{ with } m \text{ a non-negative integer} \right]$$

6.2–6.3 The Exponential Integral Function and Functions Generated by It

6.21 The logarithm integral

6.211 $\displaystyle\int_0^1 \operatorname{li}(x)\, dx = -\ln 2$ BI (79)(5)

6.212

1. $\displaystyle\int_0^1 \operatorname{li}\left(\frac{1}{x}\right) x\, dx = 0$ BI (255)(1)

2. $\displaystyle\int_0^1 \operatorname{li}(x) x^{p-1}\, dx = -\frac{1}{p} \ln(p+1)$ $[p > -1]$ BI (255)(2)

3. $\displaystyle\int_0^1 \operatorname{li}(x) \frac{dx}{x^{q+1}} = \frac{1}{q} \ln(1 - q)$ $[q < 1]$ BI (255)(3)

4. $\displaystyle\int_1^\infty \operatorname{li}(x) \frac{dx}{x^{q+1}} = -\frac{1}{q} \ln(q - 1)$ $[q > 1]$ BI (255)(4)

6.213

1. $\displaystyle\int_0^1 \operatorname{li}\left(\frac{1}{x}\right) \sin(a \ln x)\, dx = \frac{1}{1 + a^2}\left(a \ln a - \frac{\pi}{2}\right)$ $[a > 0]$ BI (475)(1)

2. $\displaystyle\int_1^\infty \operatorname{li}\left(\frac{1}{x}\right) \sin(a \ln x)\, dx = -\frac{1}{1 + a^2}\left(\frac{\pi}{2} + a \ln a\right)$ $[a > 0]$ BI (475)(9)

3. $\displaystyle\int_0^1 \operatorname{li}\left(\frac{1}{x}\right) \cos(a \ln x)\, dx = -\frac{1}{1 + a^2}\left(\ln a + \frac{\pi}{2} a\right)$ $[a > 0]$ BI (475)(2)

4. $\displaystyle\int_1^\infty \operatorname{li}\left(\frac{1}{x}\right) \cos(a \ln x)\, dx = \frac{1}{1 + a^2}\left(\ln a - \frac{\pi}{2} a\right)$ $[a > 0]$ BI (475)(10)

5. $\displaystyle\int_0^1 \operatorname{li}(x) \sin(a \ln x) \frac{dx}{x} = \frac{\ln\left(1 + a^2\right)}{2a}$ $[a > 0]$ BI(479)(1), ET I 98(20)a

6. $\displaystyle\int_0^1 \operatorname{li}(x)\cos(a\ln x)\,\frac{dx}{x} = -\frac{\arctan a}{a}$ BI (479)(2)

7. $\displaystyle\int_0^1 \operatorname{li}(x)\sin(a\ln x)\,\frac{dx}{x^2} = \frac{1}{1+a^2}\left(a\ln a + \frac{\pi}{2}\right)$ $[a>0]$ BI (479)(3)

8. $\displaystyle\int_1^\infty \operatorname{li}(x)\sin(a\ln x)\,\frac{dx}{x^2} = \frac{1}{1+a^2}\left(\frac{\pi}{2} - a\ln a\right)$ $[a>0]$ BI (479)(13)

9. $\displaystyle\int_0^1 \operatorname{li}(x)\cos(a\ln x)\,\frac{dx}{x^2} = \frac{1}{1+a^2}\left(\ln a - \frac{\pi}{2}a\right)$ $[a>0]$ BI (479)(4)

10. $\displaystyle\int_1^\infty \operatorname{li}(x)\cos(a\ln x)\,\frac{dx}{x^2} = -\frac{1}{1+a^2}\left(\ln a + \frac{\pi}{2}a\right)$ $[a>0]$ BI (479)(14)

11. $\displaystyle\int_0^1 \operatorname{li}(x)\sin(a\ln x)\,x^{p-1}\,dx = \frac{1}{a^2+p^2}\left\{\frac{a}{2}\ln\left[(1+p)^2 + a^2\right] - p\arctan\frac{a}{1+p}\right\}$

$$[p>0] \qquad\qquad\qquad \text{BI (477)(1)}$$

12. $\displaystyle\int_0^1 \operatorname{li}(x)\cos(a\ln x)\,x^{p-1}\,dx = -\frac{1}{a^2+p^2}\left\{a\arctan\frac{a}{1+p} + \frac{p}{2}\ln\left[(1+p)^2 + a^2\right]\right\}$

$$[p>0] \qquad\qquad\qquad \text{BI (477)(2)}$$

6.214

1. $\displaystyle\int_0^1 \operatorname{li}\left(\frac{1}{x}\right)\left(\ln\frac{1}{x}\right)^{p-1}dx = -\pi\cot p\pi \cdot \Gamma(p)$ $[0<p<1]$ BI (340)(1)

2. $\displaystyle\int_1^\infty \operatorname{li}\left(\frac{1}{x}\right)(\ln x)^{p-1}\,dx = -\frac{\pi}{\sin p\pi}\,\Gamma(p)$ $[0<p<1]$ BI (340)(9)

6.215

1. $\displaystyle\int_0^1 \operatorname{li}(x)\frac{x^{p-1}}{\sqrt{\ln\left(\dfrac{1}{x}\right)}}\,dx = -2\sqrt{\frac{\pi}{p}}\operatorname{arcsinh}\sqrt{p} = -2\sqrt{\frac{\pi}{p}}\ln\left(\sqrt{p} + \sqrt{p+1}\right)$

$$[p>0] \qquad\qquad\qquad \text{BI (444)(3)}$$

2. $\displaystyle\int_0^1 \operatorname{li}(x)\frac{dx}{x^{p+1}\sqrt{\ln\left(\dfrac{1}{x}\right)}} = -2\sqrt{\frac{\pi}{p}}\arcsin\sqrt{p}$ $[1>p>0]$ BI (444)(4)

6.216

1. $\displaystyle\int_0^1 \operatorname{li}(x)\left[\ln\left(\frac{1}{x}\right)\right]^{p-1}\frac{ax}{x} = -\frac{1}{p}\Gamma(p)$ $[0<p\le1]$ BI (444)(1)

2. $\displaystyle\int_0^1 \operatorname{li}(x)\left[\ln\left(\frac{1}{x}\right)\right]^{p-1}\frac{dx}{x^2} = -\frac{\pi\,\Gamma(p)}{\sin p\pi}$ $[0<p<1]$ BI (444)(2)

6.22–6.23 The exponential integral function

6.221 $\int_0^p \mathrm{Ei}(\alpha x)\,dx = p\,\mathrm{Ei}(\alpha p) + \dfrac{1 - e^{\alpha p}}{\alpha}$ NT 11(7)

6.222 $\int_0^\infty \mathrm{Ei}(-px)\,\mathrm{Ei}(-qx)\,dx = \left(\dfrac{1}{p} + \dfrac{1}{q}\right)\ln(p + q) - \dfrac{\ln q}{p} - \dfrac{\ln p}{q}$

$[p > 0, \quad q > 0]$ FI II 653, NT 53(3)

6.223 $\int_0^\infty \mathrm{Ei}(-\beta x)x^{\mu-1}\,dx = -\dfrac{\Gamma(\mu)}{\mu\beta^\mu}$ $[\mathrm{Re}\,\beta \geq 0, \quad \mathrm{Re}\,\mu > 0]$

NT 55(7), ET I 325(10)

6.224

1. $\int_0^\infty \mathrm{Ei}(-\beta x)e^{-\mu x}\,dx = -\dfrac{1}{\mu}\ln\left(1 + \dfrac{\mu}{\beta}\right)$ $[\mathrm{Re}(\beta + \mu) \geq 0, \quad \mu > 0]$

$\qquad\qquad\qquad\qquad = -1/\beta$ $[\mu = 0]$

FI II 652, NT 48(8)

2. $\int_0^\infty \mathrm{Ei}(ax)e^{-\mu x}\,dx = -\dfrac{1}{\mu}\ln\left(\dfrac{\mu}{a} - 1\right)$ $[a > 0, \quad \mathrm{Re}\,\mu > 0, \quad \mu > a]$

ET I 178(23)a, BI (283)(3)

6.225

1. $\int_0^\infty \mathrm{Ei}\left(-x^2\right)e^{-\mu x^2}\,dx = -\sqrt{\dfrac{\pi}{\mu}}\,\mathrm{arcsinh}\,\sqrt{\mu} = -\sqrt{\dfrac{\pi}{\mu}}\ln\left(\sqrt{\mu} + \sqrt{1 + \mu}\right)$

$[\mathrm{Re}\,\mu > 0]$ BI (283)(5), ET I 178(25)a

2. $\int_0^\infty \mathrm{Ei}\left(-x^2\right)e^{px^2}\,dx = -\sqrt{\dfrac{\pi}{p}}\,\arcsin\sqrt{p}$ $[1 > p > 0]$ NT 59(9)a

6.226

1. $\int_0^\infty \mathrm{Ei}\left(-\dfrac{1}{4x}\right)e^{-\mu x}\,dx = -\dfrac{2}{\mu}K_0\left(\sqrt{\mu}\right)$ $[\mathrm{Re}\,\mu > 0]$ MI 34

2. $\int_0^\infty \mathrm{Ei}\left(\dfrac{a^2}{4x}\right)e^{-\mu x}\,dx = -\dfrac{2}{\mu}K_0\left(a\sqrt{\mu}\right)$ $[a > 0, \quad \mathrm{Re}\,\mu > 0]$ MI 34

3. $\int_0^\infty \mathrm{Ei}\left(-\dfrac{1}{4x^2}\right)e^{-\mu x^2}\,dx = \sqrt{\dfrac{\pi}{\mu}}\,\mathrm{Ei}\left(-\sqrt{\mu}\right)$ $[\mathrm{Re}\,\mu > 0]$ MI 34

4. $\int_0^\infty \mathrm{Ei}\left(-\dfrac{1}{4x^2}\right)e^{-\mu x^2 + \frac{1}{4x^2}}\,dx = \sqrt{\dfrac{\pi}{\mu}}\left[\cos\sqrt{\mu}\,\mathrm{ci}\,\sqrt{\mu} - \sin\sqrt{\mu}\,\mathrm{si}\,\sqrt{\mu}\right]$

$[\mathrm{Re}\,\mu > 0]$ MI 34

6.227

1. $\int_0^\infty \mathrm{Ei}(-x)e^{-\mu x}x\,dx = \dfrac{1}{\mu(\mu + 1)} - \dfrac{1}{\mu^2}\ln(1 + \mu)$ $[\mathrm{Re}\,\mu > 0]$ MI 34

2. $\displaystyle\int_0^\infty \left[\frac{e^{-ax}\,\mathrm{Ei}(ax)}{x-b} - \frac{e^{ax}\,\mathrm{Ei}(-ax)}{x+b} \right] dx = 0$ $\qquad [a > 0, \quad b < 0]$

$\qquad\qquad\qquad\qquad\qquad\qquad = \pi^2 e^{-ab}$ $\qquad [a > 0, \quad b > 0]$

$\qquad\qquad\qquad\qquad\qquad\qquad\qquad\qquad\qquad\qquad\qquad$ ET II 253(1)a

6.228

1. $\displaystyle\int_0^\infty \mathrm{Ei}(-x)e^x x^{\nu-1}\, dx = -\frac{\pi\,\Gamma(\nu)}{\sin\nu\pi}$ $\qquad [0 < \mathrm{Re}\,\nu < 1]$ $\qquad$ ET II 308(13)

2. $\displaystyle\int_0^\infty \mathrm{Ei}(-\beta x)e^{-\mu x} x^{\nu-1}\, dx = -\frac{\Gamma(\nu)}{\nu(\beta+\mu)^\nu}\, {}_2F_1\left(1,\nu;\nu+1;\frac{\mu}{\beta+\mu}\right)$

$\qquad\qquad\qquad\qquad [|\arg\beta| < \pi, \quad \mathrm{Re}(\beta+\mu) > 0, \quad \mathrm{Re}\,\nu > 0]$ $\quad$ ET II 308(14)

6.229 $\displaystyle\int_0^\infty \mathrm{Ei}\left(-\frac{1}{4x^2}\right)\exp\left(-\mu x^2 + \frac{1}{4x^2}\right)\frac{dx}{x^2} = 2\sqrt{\pi}\left(\cos\sqrt{\mu}\,\mathrm{si}\,\sqrt{\mu} - \sin\sqrt{\mu}\,\mathrm{ci}\,\sqrt{\mu}\right)$

$\qquad\qquad\qquad\qquad\qquad\qquad\qquad\qquad\qquad\qquad [\mathrm{Re}\,\mu > 0]$ $\qquad$ MI 34

6.231 $\displaystyle\int_{-\ln a}^\infty \left[\mathrm{Ei}(-a) - \mathrm{Ei}\left(-e^{-x}\right)\right]e^{-\mu x}\, dx = \frac{1}{\mu}\gamma(\mu, a)$ $\qquad [a < 1, \quad \mathrm{Re}\,\mu > 0]$ $\qquad$ MI 34

6.232

1. $\displaystyle\int_0^\infty \mathrm{Ei}(-ax)\sin bx\, dx = -\frac{\ln\left(1+\dfrac{b^2}{a^2}\right)}{2b}$ $\qquad [a > 0, \quad b > 0]$ $\qquad$ BI (473)(1)a

2. $\displaystyle\int_0^\infty \mathrm{Ei}(-ax)\cos bx\, dx = -\frac{1}{b}\arctan\frac{b}{a}$ $\qquad [a > 0, \quad b > 0]$ $\qquad$ BI (473)(2)a

6.233

1. $\displaystyle\int_0^\infty \mathrm{Ei}(-x)e^{-\mu x}\sin\beta x\, dx = -\frac{1}{\beta^2+\mu^2}\left\{\frac{\beta}{2}\ln\left[(1+\mu)^2+\beta^2\right] - \mu\arctan\frac{\beta}{1+\mu}\right\}$

$\qquad\qquad\qquad\qquad\qquad\qquad\qquad\qquad\qquad [\mathrm{Re}\,\mu > |\mathrm{Im}\,\beta|]$ $\qquad$ BI (473)(7)a

2. $\displaystyle\int_0^\infty \mathrm{Ei}(-x)e^{-\mu x}\cos\beta x\, dx = -\frac{1}{\beta^2+\mu^2}\left\{\frac{\mu}{2}\ln\left[(1+\mu)^2+\beta^2\right] + \beta\arctan\frac{\beta}{1+\mu}\right\}$

$\qquad\qquad\qquad\qquad\qquad\qquad\qquad\qquad\qquad [\mathrm{Re}\,\mu > |\mathrm{Im}\,\beta|]$ $\qquad$ BI (473)(8)a

6.234 $\displaystyle\int_0^\infty \mathrm{Ei}(-x)\ln x\, dx = C + 1$ $\qquad\qquad\qquad\qquad\qquad\qquad$ NT 56(10)

6.24–6.26 The sine integral and cosine integral functions

6.241

1. $\displaystyle\int_0^\infty \mathrm{si}(px)\,\mathrm{si}(qx)\, dx = \frac{\pi}{2p}$ $\qquad [p \geq q]$ $\qquad$ BI II 653, NT 54(8)

2. $\displaystyle\int_0^\infty \mathrm{ci}(px)\,\mathrm{ci}(qx)\, dx = \frac{\pi}{2p}$ $\qquad [p \geq q]$ $\qquad$ FI II 653, NT 54(7)

3. $\displaystyle\int_0^\infty \operatorname{si}(px)\operatorname{ci}(qx)\,dx = \frac{1}{4q}\ln\left(\frac{p+q}{p-q}\right)^2 + \frac{1}{4p}\ln\frac{(p^2-q^2)^2}{q^4}$ $[p \neq q]$

$\displaystyle\qquad\qquad\qquad\qquad\qquad = \frac{1}{q}\ln 2$ $[p = q]$

FI II 653, NT 54(10, 12)

6.242 $\displaystyle\int_0^\infty \frac{\operatorname{ci}(ax)}{\beta + x}\,dx = -\frac{1}{2}\left\{[\operatorname{si}(a\beta)]^2 + [\operatorname{ci}(a\beta)]^2\right\}$ $[a>0, \quad |\arg\beta| < \pi]$ ET II 224(1)

6.243

1. $\displaystyle\int_{-\infty}^\infty \frac{\operatorname{si}(a|x|)}{x-b}\operatorname{sign}x\,dx = \pi\operatorname{ci}(a|b|)$ $[a>0, \quad b>0]$ ET II 253(3)

2. $\displaystyle\int_{-\infty}^\infty \frac{\operatorname{ci}(a|x|)}{x-b}\,dx = -\pi\operatorname{sign}b\cdot\operatorname{si}(a|b|)$ $[a>0]$ ET II 253(2)

6.244

1.[8] $\displaystyle\int_0^\infty \operatorname{si}(px)\frac{x\,dx}{q^2+x^2} = \frac{\pi}{2}\operatorname{Ei}(-pq)$ $[p>0, \quad q>0]$ BI (255)(6)

2.[8] $\displaystyle\int_0^\infty \operatorname{si}(px)\frac{x\,dx}{q^2-x^2} = -\frac{\pi}{2}\operatorname{ci}(pq)$ $[p>0, \quad q>0]$ BI (255)(6)

6.245

1. $\displaystyle\int_0^\infty \operatorname{ci}(px)\frac{dx}{q^2+x^2} = \frac{\pi}{2q}\operatorname{Ei}(-pq)$ $[p>0, \quad q>0]$ BI (255)(7)

2. $\displaystyle\int_0^\infty \operatorname{ci}(px)\frac{dx}{q^2-x^2} = \frac{\pi}{2q}\operatorname{si}(pq)$ $[p>0, \quad q>0]$ BI (255)(8)

6.246

1. $\displaystyle\int_0^\infty \operatorname{si}(ax)x^{\mu-1}\,dx = -\frac{\Gamma(\mu)}{\mu a^\mu}\sin\frac{\mu\pi}{2}$ $[a>0, \quad 0<\operatorname{Re}\mu<1]$

NT 56(9), ET I 325(12)a

2. $\displaystyle\int_0^\infty \operatorname{ci}(ax)x^{\mu-1}\,dx = -\frac{\Gamma(\mu)}{\mu a^\mu}\cos\frac{\mu\pi}{2}$ $[a>0, \quad 0<\operatorname{Re}\mu<1]$

NT 56(8), ET I 325(13)a

6.247

1. $\displaystyle\int_0^\infty \operatorname{si}(\beta x)e^{-\mu x}\,dx = -\frac{1}{\mu}\arctan\frac{\mu}{\beta}$ $[\operatorname{Re}\mu>0]$ NT 49(12), ET I 177(18)

2. $\displaystyle\int_0^\infty \operatorname{ci}(\beta x)e^{-\mu x}\,dx = -\frac{1}{\mu}\ln\sqrt{1+\frac{\mu^2}{\beta^2}}$ $[\operatorname{Re}\mu>0]$ NT 49(11), ET I 178(19)a

6.248

1.[8] $\displaystyle\int_0^\infty \operatorname{si}(x)e^{-\mu x^2}x\,dx = \frac{\pi}{4\mu}\left[\Phi\left(\frac{1}{2\sqrt{\mu}}\right) - 1\right]$ $[\operatorname{Re}\mu>0]$ MI 34

2. $\displaystyle\int_0^\infty \operatorname{ci}(x) e^{-\mu x^2}\,dx = \frac{1}{4}\sqrt{\frac{\pi}{\mu}}\,\operatorname{Ei}\left(-\frac{1}{4\mu}\right)$ $[\operatorname{Re}\mu > 0]$ MI 34

6.249 $\displaystyle\int_0^\infty \left[\operatorname{si}\left(x^2\right) + \frac{\pi}{2}\right] e^{-\mu x}\,dx = \frac{\pi}{\mu}\left\{\left[S\left(\frac{\mu^2}{4}\right) - \frac{1}{2}\right]^2 + \left[C\left(\frac{\mu^2}{4}\right) - \frac{1}{2}\right]^2\right\}$

$\qquad\qquad\qquad\qquad\qquad\qquad\qquad\qquad\qquad\qquad$ $[\operatorname{Re}\mu > 0]$ ME 26

6.251

1. $\displaystyle\int_0^\infty \operatorname{si}\left(\frac{1}{x}\right) e^{-\mu x}\,dx = \frac{2}{\mu}\operatorname{kei}\left(2\sqrt{\mu}\right)$ $[\operatorname{Re}\mu > 0]$ MI 34

2. $\displaystyle\int_0^\infty \operatorname{ci}\left(\frac{1}{x}\right) e^{-\mu x}\,dx = -\frac{2}{\mu}\operatorname{ker}\left(2\sqrt{\mu}\right)$ $[\operatorname{Re}\mu > 0]$ MI 34

6.252

1. $\displaystyle\int_0^\infty \sin px\,\operatorname{si}(qx)\,dx = -\frac{\pi}{2p}$ $\left[p^2 > q^2\right]$

$\qquad\qquad\qquad\qquad\qquad = -\frac{\pi}{4p}$ $\left[p^2 = q^2\right]$

$\qquad\qquad\qquad\qquad\qquad = 0$ $\left[p^2 < q^2\right]$

$\qquad\qquad\qquad\qquad\qquad\qquad\qquad\qquad\qquad\qquad$ FI II 652, NT 50(8)

2.[6] $\displaystyle\int_0^\infty \cos px\,\operatorname{si}(qx)\,dx = -\frac{1}{4p}\ln\left(\frac{p+q}{p-q}\right)^2$ $\left[p \neq 0, \quad p^2 \neq q^2\right]$

$\qquad\qquad\qquad\qquad\qquad = \frac{1}{q}$ $\left[p = 0\right]$

$\qquad\qquad\qquad\qquad\qquad\qquad\qquad\qquad\qquad\qquad$ FI II 652, NT 50(10)

3. $\displaystyle\int_0^\infty \sin px\,\operatorname{ci}(qx)\,dx = -\frac{1}{4p}\ln\left(\frac{p^2}{q^2} - 1\right)^2$ $\left[p \neq 0, \quad p^2 \neq q^2\right]$

$\qquad\qquad\qquad\qquad\qquad = 0$ $\left[p = 0\right]$

$\qquad\qquad\qquad\qquad\qquad\qquad\qquad\qquad\qquad\qquad$ FI II 652, NT 50(9)

4. $\displaystyle\int_0^\infty \cos px\,\operatorname{ci}(qx)\,dx = -\frac{\pi}{2p}$ $\left[p^2 > q^2\right]$

$\qquad\qquad\qquad\qquad\qquad = -\frac{\pi}{4p}$ $\left[p^2 = q^2\right]$

$\qquad\qquad\qquad\qquad\qquad = 0$ $\left[p^2 < q^2\right]$

$\qquad\qquad\qquad\qquad\qquad\qquad\qquad\qquad\qquad\qquad$ FI II 654, NT 50(7)

6.253 $\displaystyle\int_0^\infty \frac{\operatorname{si}(ax)\sin bx}{1 - 2r\cos x + r^2}\,dx = -\frac{\pi\left(r^m + r^{m+1}\right)}{4b(1-r)\left(1-r^2\right)}$ $[b = a - m]$

$\qquad\qquad\qquad\qquad\qquad\qquad = -\frac{\pi\left(2 + 2r - r^m - r^{m+1}\right)}{4b(1-r)\left(1-r^2\right)}$ $[b = a + m]$

$\qquad\qquad\qquad\qquad\qquad\qquad = -\frac{\pi r^{m+1}}{2b(1-r)\left(1-r^2\right)}$ $[a - m - 1 < b < a - m]$

$\qquad\qquad\qquad\qquad\qquad\qquad = -\frac{\pi\left(1 + r - r^{m+1}\right)}{2b(1-r)\left(1-r^2\right)}$ $[a + m < b < a + m + 1]$

$\qquad\qquad\qquad\qquad\qquad\qquad\qquad\qquad\qquad\qquad$ ET I 97(10)

6.254

1. $\int_0^\infty \left[\mathrm{si}(ax) + \dfrac{\pi}{2} \right] \sin bx \, \dfrac{dx}{x} = \dfrac{1}{2} \left[L_2 \left(\dfrac{a}{b} \right) - L_2 \left(-\dfrac{a}{b} \right) \right]$

$\qquad\qquad\qquad\qquad\qquad\qquad\qquad [a > 0, \quad b > 0] \qquad\qquad$ ET I 97(12)

2. $\int_0^\infty \left[\mathrm{si}(ax) + \dfrac{\pi}{2} \right] \cos bx \cdot \dfrac{dx}{x} = \dfrac{\pi}{2} \ln \dfrac{a}{b} \qquad\qquad [a > 0, \quad b > 0] \qquad\qquad$ ET I 41(11)

6.255

1. $\int_{-\infty}^\infty [\cos ax \, \mathrm{ci}\,(a|x|) + \sin\,(a|x|)\,\mathrm{si}\,(a|x|)] \, \dfrac{dx}{x - b} = -\pi \,[\mathrm{sign}\, b \cos ab \, \mathrm{si}\,(a|b|) - \sin ab \, \mathrm{ci}\,(a|b|)]$

$\qquad\qquad\qquad\qquad\qquad\qquad\qquad [a > 0] \qquad\qquad$ ET II 253(4)

2. $\int_{-\infty}^\infty [\sin ax \, \mathrm{ci}\,(a|x|) - \mathrm{sign}\, x \cos ax \, \mathrm{si}\,(a|x|)] \, \dfrac{dx}{x - b} = -\pi \,[\sin\,(a|b|)\,\mathrm{si}\,(a|b|) + \cos ab \, \mathrm{ci}\,(a|b|)]$

$\qquad\qquad\qquad\qquad\qquad\qquad\qquad [a > 0] \qquad\qquad$ ET II 253(5)

6.256 $\int_0^\infty \left[\mathrm{si}^2(x) + \mathrm{ci}^2(x) \right] \cos ax \, dx = \dfrac{\pi}{a} \ln(1 + a) \qquad [a > 0] \qquad\qquad$ ET I 42(18)

6.257 $\int_0^\infty \mathrm{si} \left(\dfrac{a}{x} \right) \sin bx \, dx = -\dfrac{\pi}{2b} J_0 \left(2\sqrt{ab} \right) \qquad\qquad [b > 0] \qquad\qquad$ ET I 96(9)

6.258

1. $\int_0^\infty \left[\mathrm{si}(ax) + \dfrac{\pi}{2} \right] \sin bx \, \dfrac{dx}{x^2 + c^2}$

$\qquad = \dfrac{\pi}{4c} \left\{ e^{-bc} \left[\mathrm{Ei}(bc) - \mathrm{Ei}(-ac) \right] + e^{bc} \left[\mathrm{Ei}(-ac) - \mathrm{Ei}(-bc) \right] \right\} \quad [0 < b \le a, \quad c > 0]$

$\qquad = \dfrac{\pi}{4c} e^{-bc} \left[\mathrm{Ei}(ac) - \mathrm{Ei}(-ac) \right] \qquad\qquad\qquad\qquad\qquad\qquad [0 < a \le b, \quad c > 0]$

$\qquad\qquad\qquad\qquad\qquad\qquad\qquad\qquad\qquad\qquad\qquad$ BI (460)(1)

2. $\int_0^\infty \left[\mathrm{si}(ax) + \dfrac{\pi}{2} \right] \cos bx \, \dfrac{x \, dx}{x^2 + c^2}$

$\qquad = -\dfrac{\pi}{4} \left\{ e^{-bc} \left[\mathrm{Ei}(bc) - \mathrm{Ei}(-ac) \right] + e^{bc} \left[\mathrm{Ei}(-bc) - \mathrm{Ei}(-ac) \right] \right\} \quad [0 < b \le a, \quad c > 0]$

$\qquad = \dfrac{\pi}{4} e^{-bc} \left[\mathrm{Ei}(-ac) - \mathrm{Ei}(ac) \right] \qquad\qquad\qquad\qquad\qquad\qquad [0 < a \le b, \quad c > 0]$

$\qquad\qquad\qquad\qquad\qquad\qquad\qquad\qquad\qquad\qquad\qquad$ BI (460)(2, 5)

6.259

1. $\int_0^\infty \mathrm{si}(ax) \sin bx \, \dfrac{dx}{x^2 + c^2} = \dfrac{\pi}{2c} \mathrm{Ei}(-ac) \sinh(bc) \qquad\qquad [0 < b \le a, \quad c > 0]$

$\qquad = \dfrac{\pi}{4c} e^{-cb} \left[\mathrm{Ei}(-bc) + \mathrm{Ei}(bc) - \mathrm{Ei}(-ac) - \mathrm{Ei}(ac) \right]$

$\qquad\qquad + \dfrac{\pi}{2c} \mathrm{Ei}(-bc) \sinh(bc) \qquad\qquad\qquad\qquad\qquad [0 < a \le b, \quad c > 0]$

$\qquad\qquad\qquad\qquad\qquad\qquad\qquad\qquad\qquad\qquad\qquad$ ET I 96(8)

2. $\int_0^\infty \mathrm{ci}(ax) \sin bx \, \dfrac{x \, dx}{x^2 + c^2} = -\dfrac{\pi}{2} \sinh(bc) \, \mathrm{Ei}(-ac) \qquad\qquad [0 < b \le a, \quad c > 0]$

$\qquad = -\dfrac{\pi}{2} \sinh(bc) \, \mathrm{Ei}(-bc) + \dfrac{\pi}{4} e^{-bc} [\mathrm{Ei}(-bc) + \mathrm{Ei}(bc)$

$\qquad\qquad - \mathrm{Ei}(-ac) - \mathrm{Ei}(ac)] \qquad\qquad\qquad\qquad\qquad [0 < a \le b, \quad c > 0]$

$\qquad\qquad\qquad\qquad\qquad\qquad\qquad\qquad$ BI (460)(3)a, ET I 97(15)a

3. $\displaystyle\int_0^\infty \operatorname{ci}(ax)\cos bx\,\frac{dx}{x^2+c^2}$

$$= \frac{\pi}{2c}\cosh bc\,\operatorname{Ei}(-ac) \qquad\qquad [0 < b \le a, \quad c > 0]$$

$$= \frac{\pi}{4c}\left\{e^{-bc}\left[\operatorname{Ei}(ac)+\operatorname{Ei}(-ac)-\operatorname{Ei}(bc)\right]+e^{bc}\operatorname{Ei}(-bc)\right\} \qquad [0 < a \le b, \quad c > 0]$$

BI (460)(4), ET I 41(15)

6.261

1. $\displaystyle\int_0^\infty \operatorname{si}(bx)\cos ax\,e^{-px}\,dx = -\frac{1}{2\left(a^2+p^2\right)}\left[\frac{a}{2}\ln\frac{p^2+(a+b)^2}{p^2+(a-b)^2}+p\arctan\frac{2bp}{b^2-a^2-p^2}\right]$

$[a > 0, \quad b > 0, \quad p > 0]$ ET I 40(8)

2. $\displaystyle\int_0^\infty \operatorname{si}(\beta x)\cos ax\,e^{-\mu x}\,dx = -\frac{\arctan\dfrac{\mu+ai}{\beta}}{2(\mu+ai)} - \frac{\arctan\dfrac{\mu-ai}{\beta}}{2(\mu-ai)}$

$[a > 0, \quad \operatorname{Re}\mu > |\operatorname{Im}\beta|]$ ET I 40(9)

6.262

1. $\displaystyle\int_0^\infty \operatorname{ci}(bx)\sin ax\,e^{-\mu x}\,dx = \frac{1}{2\left(a^2+\mu^2\right)}\left\{\mu\arctan\frac{2a\mu}{\mu^2+b^2-a^2}-\frac{a}{2}\ln\frac{\left(\mu^2+b^2-a^2\right)^2+4a^2\mu^2}{b^4}\right\}$

$[a > 0, \quad b > 0, \quad \operatorname{Re}\mu > 0]$
ET I 98(16)a

2. $\displaystyle\int_0^\infty \operatorname{ci}(bx)\cos ax\,e^{-px}\,dx = \frac{-1}{2\left(a^2+p^2\right)}\left\{\frac{p}{2}\ln\frac{\left[\left(b^2+p^2-a^2\right)^2+4a^2p^2\right]}{b^4}+a\arctan\frac{2ap}{b^2+p^2-a^2}\right\}$

$[a > 0, \quad b > 0, \quad \operatorname{Re}p > 0]$ ET I 41(16)

3. $\displaystyle\int_0^\infty \operatorname{ci}(\beta x)\cos ax\,e^{-\mu x}\,dx = \frac{-\ln\left[1+\dfrac{(\mu+ai)^2}{\beta^2}\right]}{4(\mu+ai)} - \frac{\ln\left[1+\dfrac{(\mu-ai)^2}{\beta^2}\right]}{4(\mu-ai)}$

$[a > 0, \quad \operatorname{Re}\mu > |\operatorname{Im}\beta|]$ ET I 41(17)

6.263

1. $\displaystyle\int_0^\infty \left[\operatorname{ci}(x)\cos x+\operatorname{si}(x)\sin x\right]e^{-\mu x}\,dx = \frac{-\dfrac{\pi}{2}-\mu\ln\mu}{1+\mu^2} \qquad [\operatorname{Re}\mu > 0]$ ME 26a, ET I 178(21)a

2. $\displaystyle\int_0^\infty \left[\operatorname{si}(x)\cos x-\operatorname{ci}(x)\sin x\right]e^{-\mu x}\,dx = \frac{-\dfrac{\pi}{2}\mu+\ln\mu}{1+\mu^2} \qquad [\operatorname{Re}\mu > 0]$ ME 26a, ET I 178(20)a

3. $\displaystyle\int_0^\infty \left[\sin x-x\operatorname{ci}(x)\right]e^{-\mu x}\,dx = \frac{\ln\left(1+\mu^2\right)}{2\mu^2} \qquad [\operatorname{Re}\mu > 0]$ ME 26

6.264

1. $\displaystyle\int_0^\infty \operatorname{si}(x)\ln x\,dx = \boldsymbol{C}+1$ NT 46(10)

2. $\displaystyle\int_0^\infty \operatorname{ci}(x)\ln x\,dx = \frac{\pi}{2}$ NT 56(11)

6.27 The hyperbolic sine integral and hyperbolic cosine integral functions

6.271

1. $\displaystyle\int_0^\infty \operatorname{shi}(x)e^{-\mu x}\,dx = \frac{1}{2\mu}\ln\frac{\mu+1}{\mu-1} = \frac{1}{\mu}\operatorname{arccoth}\mu$ $[\operatorname{Re}\mu > 1]$ MI 34

2. $\displaystyle\int_0^\infty \chi(x)e^{-\mu x}\,dx = -\frac{1}{2\mu}\ln\left(\mu^2-1\right)$ $[\operatorname{Re}\mu > 1]$ MI 34

6.272 $\displaystyle\int_0^\infty \chi(x)e^{-px^2}\,dx = \frac{1}{4}\sqrt{\frac{\pi}{p}}\operatorname{Ei}\left(\frac{1}{4p}\right)$ $[p > 0]$ MI 35

6.273

1. $\displaystyle\int_0^\infty \left[\cosh x\,\operatorname{shi}(x) - \sinh x\,\chi(x)\right]e^{-\mu x}\,dx = \frac{\ln\mu}{\mu^2-1}$ $[\operatorname{Re}\mu > 0]$ MI 35

2. $\displaystyle\int_0^\infty \left[\cosh x\,\chi(x) + \sinh x\,\operatorname{shi}(x)\right]e^{-\mu x}\,dx = \frac{\mu\ln\mu}{1-\mu^2}$ $[\operatorname{Re}\mu > 2]$ MI 35

6.274 $\displaystyle\int_0^\infty \left[\cosh x\,\operatorname{shi}(x) - \sinh x\,\chi(x)\right]e^{-\mu x^2}\,dx = \frac{1}{4}\sqrt{\frac{\pi}{\mu}}e^{\frac{1}{4\mu}}\operatorname{Ei}\left(-\frac{1}{4\mu}\right)$

$[\operatorname{Re}\mu > 0]$ MI 35

6.275 $\displaystyle\int_0^\infty \left[x\operatorname{chi}(x) - \sinh x\right]e^{-\mu x}\,dx = -\frac{\ln\left(\mu^2-1\right)}{2\mu^2}$ $[\operatorname{Re}\mu > 1]$ MI 35

6.276 $\displaystyle\int_0^\infty \left[\cosh x\,\operatorname{chi}(x) + \sinh x\,\operatorname{shi}(x)\right]e^{-\mu x^2}x\,dx = \frac{1}{8}\sqrt{\frac{\pi}{\mu^3}}\exp\left(\frac{1}{4\mu}\right)\operatorname{Ei}\left(-\frac{1}{4\mu}\right)$

$[\operatorname{Re}\mu > 0]$ MI 35

6.277

1. $\displaystyle\int_0^\infty \left[\operatorname{chi}(x) + \operatorname{ci}(x)\right]e^{-\mu x}\,dx = -\frac{\ln\left(\mu^4-1\right)}{2\mu}$ $[\operatorname{Re}\mu > 1]$ MI 34

2. $\displaystyle\int_0^\infty \left[\operatorname{chi}(x) - \operatorname{ci}(x)\right]e^{-\mu x}\,dx = \frac{1}{2\mu}\ln\frac{\mu^2+1}{\mu^2-1}$ $[\operatorname{Re}\mu > 1]$ MI 35

6.28–6.31 The probability integral

6.281

1.[6] $\displaystyle\int_0^\infty \left[1 - \Phi(px)\right]x^{2q-1}\,dx = \frac{\Gamma\left(q+\frac{1}{2}\right)}{2\sqrt{\pi}qp^{2q}}$ $[\operatorname{Re}q > 0, \quad \operatorname{Re}p > 0]$

NT 56(12), ET II 306(1)a

2.[6] $\displaystyle\int_0^\infty \left[1 - \Phi\left(at^\alpha \pm \frac{b}{t^\alpha}\right)\right]dt = \frac{2b}{\sqrt{\pi}}\left(\frac{b}{a}\right)^{\frac{1-\alpha}{2\alpha}}\left[K_{\frac{1+\alpha}{2\alpha}}(2ab) \pm K_{\frac{1-\alpha}{2\alpha}}(2ab)\right]e^{\pm 2ab}$

$[a > 0, \quad b > 0, \quad \alpha \neq 0]$

6.282

1. $\int_0^\infty \Phi(qt)e^{-pt}\,dt = \frac{1}{p}\left[1 - \Phi\left(\frac{p}{2q}\right)\right]\exp\left(\frac{p^2}{4q^2}\right)$ $\left[\operatorname{Re}p > 0, \quad |\arg q| < \frac{\pi}{4}\right]$

                                                                        MO 175, EH II 148(11)

2. $\int_0^\infty \left[\Phi\left(x + \frac{1}{2}\right) - \Phi\left(\frac{1}{2}\right)\right]e^{-\mu x + \frac{1}{4}}\,dx = \frac{1}{(\mu+1)(\mu+2)}\exp\frac{(\mu+1)^2}{4}\left[1 - \Phi\left(\frac{\mu+1}{2}\right)\right]$

                                                                                                   ME 27

6.283

1. $\int_0^\infty e^{\beta x}\left[1 - \Phi\left(\sqrt{\alpha}x\right)\right]\,dx = \frac{1}{\beta}\left[\frac{\sqrt{\alpha}}{\sqrt{\alpha - \beta}} - 1\right]$ $[\operatorname{Re}\alpha > 0, \quad \operatorname{Re}\beta < \operatorname{Re}\alpha]$ ET II 307(5)

2. $\int_0^\infty \Phi\left(\sqrt{qt}\right)e^{-pt}\,dt = \frac{\sqrt{q}}{p}\frac{1}{\sqrt{p+q}}$ $[\operatorname{Re}p > 0, \quad \operatorname{Re}(q + p) > 0]$

                                                                                                 EH II 148(12)

6.284 $\int_0^\infty \left[1 - \Phi\left(\frac{q}{2\sqrt{x}}\right)\right]e^{-px}\,dx = \frac{1}{p}e^{-q\sqrt{p}}$ $\left[\operatorname{Re}p > 0, \quad |\arg q| < \frac{\pi}{4}\right]$

                                                                            EF 147(235), EH II 148(13)

6.285

1. $\int_0^\infty [1 - \Phi(x)]e^{-\mu^2 x^2}\,dx = \frac{\arctan\mu}{\sqrt{\pi}\mu}$ $[\operatorname{Re}\mu > 0]$ MI 37

2. $\int_0^\infty \Phi(iat)e^{-a^2t^2 - st}\,dt = \frac{-1}{2ai\sqrt{\pi}}\exp\left(\frac{s^2}{4a^2}\right)\operatorname{Ei}\left(-\frac{s^2}{4a^2}\right)$

                                                               $\left[\operatorname{Re}s > 0, \quad |\arg a| < \frac{\pi}{4}\right]$

                                                                                         EH II 148(14)a

6.286

1. $\int_0^\infty [1 - \Phi(\beta x)]e^{\mu^2 x^2}x^{\nu - 1}\,dx = \frac{\Gamma\left(\dfrac{\nu+1}{2}\right)}{\sqrt{\pi}\nu\beta^\nu}\ {}_2F_1\left(\frac{\nu}{2}, \frac{\nu+1}{2}; \frac{\nu}{2} + 1; \frac{\mu^2}{\beta^2}\right)$

                                                            $[\operatorname{Re}^2\beta > \operatorname{Re}\mu^2, \quad \operatorname{Re}\nu > 0]$

                                                                                 ET II 306(2)

2. $\int_0^\infty \left[1 - \Phi\left(\frac{\sqrt{2}x}{2}\right)\right]e^{\frac{x^2}{2}}x^{\nu - 1}\,dx = 2^{\frac{\nu}{2} - 1}\sec\frac{\nu\pi}{2}\Gamma\left(\frac{\nu}{2}\right)$

                                                                  $[0 < \operatorname{Re}\nu < 1]$ ET I 325(9)

6.287

1. $\int_0^\infty \Phi(\beta x)e^{-\mu x^2}x\,dx = \frac{\beta}{2\mu\sqrt{\mu + \beta^2}}$ $[\operatorname{Re}\mu > -\operatorname{Re}\beta^2, \quad \operatorname{Re}\mu > 0]$

                                                                                  ME 27a, ET I 176(4)

2. $\displaystyle\int_0^\infty [1 - \Phi(\beta x)] e^{-\mu x^2} x\, dx = \frac{1}{2\mu}\left(1 - \frac{\beta}{\sqrt{\mu + \beta^2}}\right)$ $\quad\left[\operatorname{Re}\mu > -\operatorname{Re}\beta^2, \quad \operatorname{Re}\mu > 0\right]$

NT 49(14), ET I 177(9)

6.288 $\displaystyle\int_0^\infty \Phi(iax)e^{-\mu x^2} x\, dx = \frac{ai}{2\mu\sqrt{\mu - a^2}}$ $\quad\left[a > 0, \quad \operatorname{Re}\mu > \operatorname{Re}a^2\right]$ MI 37a

6.289

1. $\displaystyle\int_0^\infty \Phi(\beta x)e^{(\beta^2 - \mu^2)x^2} x\, dx = \frac{\beta}{2\mu\,(\mu^2 - \beta^2)}$ $\quad\left[\operatorname{Re}^2\mu > \operatorname{Re}\beta^2, \quad |\arg\mu| < \frac{\pi}{4}\right]$

ET I 176(5)

2. $\displaystyle\int_0^\infty [1 - \Phi(\beta x)] e^{(\beta^2 - \mu^2)x^2} x\, dx = \frac{1}{2\mu(\mu + \beta)}$ $\quad\left[\operatorname{Re}^2\mu > \operatorname{Re}\beta^2, \quad \arg\mu < \frac{\pi}{4}\right]$

ET I 177(10)

3. $\displaystyle\int_0^\infty \Phi\left(\sqrt{b - a}\,x\right) e^{-(a+\mu)x^2} x\, dx = \frac{\sqrt{b - a}}{2(\mu + a)\sqrt{\mu + b}}$ $\quad[\operatorname{Re}\mu > -a > 0, \quad b > a]$ ME 27

6.291 $\displaystyle\int_0^\infty \Phi(ix)e^{-(\mu x + x^2)} x\, dx = \frac{i}{\sqrt{\pi}}\left[\frac{1}{\mu} + \frac{\mu}{4}\operatorname{Ei}\left(-\frac{\mu^2}{4}\right)\right]$ $\quad[\operatorname{Re}\mu > 0]$ MI 37

6.292 $\displaystyle\int_0^\infty [1 - \Phi(x)] e^{-\mu^2 x^2} x^2\, dx = \frac{1}{2\sqrt{\pi}}\left\{\frac{\arctan\mu}{\mu^3} - \frac{1}{\mu^2\,(\mu^2 + 1)}\right\}$

$\left[|\arg\mu| < \frac{\pi}{4}\right]$ MI 37

6.293 $\displaystyle\int_0^\infty \Phi(x)e^{-\mu x^2}\frac{dx}{x} = \frac{1}{2}\ln\frac{\sqrt{\mu + 1} + 1}{\sqrt{\mu + 1} - 1} = \operatorname{arccoth}\sqrt{\mu + 1}$

$[\operatorname{Re}\mu > 0]$
MI 37a

6.294

1. $\displaystyle\int_0^\infty \left[1 - \Phi\left(\frac{\beta}{x}\right)\right] e^{-\mu^2 x^2} x\, dx = \frac{1}{2\mu^2}\exp(-2\beta\mu)$ $\quad\left[|\arg\beta| < \frac{\pi}{4}, \quad |\arg\mu| < \frac{\pi}{4}\right]$

ET I 177(11)

2. $\displaystyle\int_0^\infty \left[1 - \Phi\left(\frac{1}{x}\right)\right] e^{-\mu^2 x^2}\frac{dx}{x} = -\operatorname{Ei}(-2\mu)$ $\quad\left[|\arg\mu| < \frac{\pi}{4}\right]$ MI 37

6.295

1. $\displaystyle\int_0^\infty \left[1 - \Phi\left(\frac{1}{x}\right)\right]\exp\left(-\mu^2 x^2 + \frac{1}{x^2}\right) dx = \frac{1}{\sqrt{\pi}\mu}[\sin 2\mu\operatorname{ci}(2\mu) - \cos 2\mu\operatorname{si}(2\mu)]$

$\left[|\arg\mu| < \frac{\pi}{4}\right]$ MI 37

2. $\displaystyle\int_0^\infty \left[1 - \Phi\left(\frac{1}{x}\right)\right]\exp\left(-\mu^2 x^2 + \frac{1}{x^2}\right) x\, dx = \frac{\pi}{2\mu}[\mathbf{H}_1(2\mu) - Y_1(2\mu)] - \frac{1}{\mu^2}$

$\left[|\arg\mu| < \frac{\pi}{4}\right]$ MI 37

3. $\displaystyle\int_0^\infty \left[1 - \Phi\left(\frac{1}{x}\right)\right]\exp\left(-\mu^2 x^2 + \frac{1}{x^2}\right)\frac{dx}{x} = \frac{\pi}{2}[\mathbf{H}_0(2\mu) - Y_0(2\mu)]$

$\left[|\arg\mu| < \frac{\pi}{4}\right]$ MI 37

6.296 $\displaystyle\int_0^\infty \left\{ (x^2 + a^2)\left[1 - \Phi\left(\frac{a}{\sqrt{2}x}\right)\right] - \sqrt{\frac{2}{\pi}}\, ax \cdot e^{-\frac{a^2}{2x^2}} \right\} e^{-\mu^2 x^2} x\, dx = \frac{1}{2\mu^4} e^{-a\mu\sqrt{2}}$

$$\left[|\arg\mu| < \frac{\pi}{4}, \quad a > 0\right] \qquad \text{MI 38a}$$

6.297

1. $\displaystyle\int_0^\infty \left[1 - \Phi\left(\gamma x + \frac{\beta}{x}\right)\right] e^{(\gamma^2 - \mu)x^2} x\, dx = \frac{1}{2\sqrt{\mu}\,(\sqrt{\mu} + \gamma)} \exp\left[-2\left(\beta\gamma + \beta\sqrt{\mu}\right)\right]$

$$[\operatorname{Re}\beta > 0, \quad \operatorname{Re}\mu > 0] \qquad \text{ET I 177(12)a}$$

2. $\displaystyle\int_0^\infty \left[1 - \Phi\left(\frac{b + 2ax^2}{2x}\right)\right] \exp\left[-\left(\mu^2 - a^2\right)x^2 + ab\right] x\, dx = \frac{e^{-b\mu}}{2\mu(\mu + a)}$

$$[a > 0, \quad b > 0, \quad \operatorname{Re}\mu > 0] \qquad \text{MI 38}$$

3. $\displaystyle\int_0^\infty \left\{ \left[1 - \Phi\left(\frac{b - 2ax^2}{2x}\right)\right] e^{-ab} + \left[1 - \Phi\left(\frac{b + 2ax^2}{2x}\right)\right] e^{ab} \right\} e^{-\mu x^2} x\, dx = \frac{1}{\mu} \exp\left(-b\sqrt{a^2 + \mu}\right)$

$$[a > 0, \quad b > 0, \quad \operatorname{Re}\mu > 0] \qquad \text{MI 38}$$

6.298 $\displaystyle\int_0^\infty \left\{ 2\cosh ab - e^{-ab}\,\Phi\left(\frac{b - 2ax^2}{2x}\right) - e^{ab}\,\Phi\left(\frac{b + 2ax^2}{2x}\right) \right\} e^{-(\mu - a^2)x^2} x\, dx = \frac{1}{\mu - a^2}\exp\left(-b\sqrt{\mu}\right)$

$$[a > 0, \quad b > 0, \quad \operatorname{Re}\mu > 0] \qquad \text{MI 38}$$

6.299 $\displaystyle\int_0^\infty \cosh(2\nu t)\exp\left[(a\cosh t)^2\right]\left[1 - \Phi\left(a\cosh t\right)\right] dt = \frac{1}{2\cos(\nu\pi)}\exp\left(\tfrac{1}{2}a^2\right) K_\nu\left(a^2\right)$

$$\left[\operatorname{Re}a > 0, \quad -\tfrac{1}{2} < \operatorname{Re}\nu < \tfrac{1}{2}\right]$$
$$\text{ET II 308(10)}$$

6.311 $\displaystyle\int_0^\infty \left[1 - \Phi(ax)\right]\sin bx\, dx = \frac{1}{b}\left(1 - e^{-\frac{b^2}{4a^2}}\right) \qquad [a > 0, \quad b > 0] \qquad \text{ET I 96(4)}$

6.312 $\displaystyle\int_0^\infty \Phi(ax)\sin bx^2\, dx = \frac{1}{4\sqrt{2\pi b}}\left(\ln\frac{b + a^2 + a\sqrt{2b}}{b + a^2 - a\sqrt{2b}} + 2\arctan\frac{a\sqrt{2b}}{b - a^2}\right)$

$$[a > 0, \quad b > 0] \qquad \text{ET I 96(3)}$$

6.313

1. $\displaystyle\int_0^\infty \sin(\beta x)\left[1 - \Phi\left(\sqrt{\alpha x}\right)\right] dx = \frac{1}{\beta} - \left(\frac{\frac{\alpha}{2}}{\alpha^2 + \beta^2}\right)^{\frac{1}{2}}\left[(\alpha^2 + \beta^2)^{\frac{1}{2}} - \alpha\right]^{-\frac{1}{2}}$

$$[\operatorname{Re}\alpha > |\operatorname{Im}\beta|] \qquad \text{ET II 307(6)}$$

2. $\displaystyle\int_0^\infty \cos(\beta x)\left[1 - \Phi\left(\sqrt{\alpha x}\right)\right] dx = \left(\frac{\frac{\alpha}{2}}{\alpha^2 + \beta^2}\right)^{\frac{1}{2}}\left[(\alpha^2 + \beta^2)^{\frac{1}{2}} + \alpha\right]^{-\frac{1}{2}}$

$$[\operatorname{Re}\alpha > |\operatorname{Im}\beta|] \qquad \text{ET II 307(7)}$$

6.314

1. $\displaystyle\int_0^\infty \sin(bx)\left[1 - \Phi\left(\sqrt{\frac{a}{x}}\right)\right] dx = b^{-1}\exp\left[-(2ab)^{\frac{1}{2}}\right]\cos\left[(2ab)^{\frac{1}{2}}\right]$

$$[\operatorname{Re}a > 0, \quad b > 0] \qquad \text{ET II 307(8)}$$

2.
$$\int_0^\infty \cos(bx)\left[1 - \Phi\left(\sqrt{\frac{a}{x}}\right)\right] dx = -b^{-1}\exp\left[-(2ab)^{\frac{1}{2}}\right]\sin\left[(2ab)^{\frac{1}{2}}\right]$$

$$[\operatorname{Re} a > 0, \quad b > 0] \qquad \text{ET II 307(9)}$$

6.315

1.
$$\int_0^\infty x^{\nu-1}\sin(\beta x)\left[1 - \Phi(\alpha x)\right] dx = \frac{\Gamma\left(1 + \frac{1}{2}\nu\right)\beta}{\sqrt{\pi}(\nu+1)\alpha^{\nu+1}} \, {}_2F_2\left(\frac{\nu+1}{2}, \frac{\nu}{2}+1; \frac{3}{2}, \frac{\nu+3}{2}; -\frac{\beta^2}{4\alpha^2}\right)$$

$$[\operatorname{Re}\alpha > 0, \quad \operatorname{Re}\nu > -1] \qquad \text{ET II 307(3)}$$

2.
$$\int_0^\infty x^{\nu-1}\cos(\beta x)\left[1 - \Phi(\alpha x)\right] dx = \frac{\Gamma\left(\frac{1}{2} + \frac{1}{2}\nu\right)}{\sqrt{\pi}\nu\alpha^\nu} \, {}_2F_2\left(\frac{\nu}{2}, \frac{\nu+1}{2}; \frac{1}{2}, \frac{\nu}{2}+1; -\frac{\beta^2}{4\alpha^2}\right)$$

$$[\operatorname{Re}\alpha > 0, \quad \operatorname{Re}\nu > 0] \qquad \text{ET II 307(4)}$$

3.
$$\int_0^\infty [1 - \Phi(ax)]\cos bx \cdot x \, dx = \frac{1}{2a^2}\exp\left(-\frac{b^2}{4a^2}\right) - \frac{1}{b^2}\left[1 - \exp\left(-\frac{b^2}{4a^2}\right)\right]$$

$$[a > 0, \quad b > 0] \qquad \text{ET I 40(5)}$$

4.
$$\int_0^\infty [\Phi(ax) - \Phi(bx)]\cos px \frac{dx}{x} = \frac{1}{2}\left[\operatorname{Ei}\left(-\frac{p^2}{4b^2}\right) - \operatorname{Ei}\left(\frac{p^2}{4a^2}\right)\right]$$

$$[a > 0, \quad b > 0, \quad p > 0] \qquad \text{ET I 40(6)}$$

5.
$$\int_0^\infty x^{-\frac{1}{2}}\Phi\left(a\sqrt{x}\right)\sin bx \, dx = \frac{1}{2\sqrt{2\pi b}}\left\{\ln\left[\frac{b + a\sqrt{2b} + a^2}{b - a\sqrt{2b} + a^2}\right] + 2\arctan\left[\frac{a\sqrt{2b}}{b - a^2}\right]\right\}$$

$$[a > 0, \quad b > 0] \qquad \text{ET I 96(3)}$$

6.316
$$\int_0^\infty e^{\frac{1}{2}x^2}\left[1 - \Phi\left(\frac{x}{\sqrt{2}}\right)\right]\sin bx \, dx = \sqrt{\frac{\pi}{2}}e^{\frac{b^2}{2}}\left[1 - \Phi\left(\frac{b}{\sqrt{2}}\right)\right]$$

$$[b > 0] \qquad \text{ET I 96(5)}$$

6.317[6]
$$\int_0^\infty e^{-a^2x^2}\Phi(iax)\sin bx \, dx = \frac{i}{a}\frac{\sqrt{\pi}}{2}e^{-\frac{b^2}{4a^2}} \qquad [b > 0] \qquad \text{ET I 96(2)}$$

6.318
$$\int_0^\infty [1 - \Phi(x)]\operatorname{si}(2px) \, dx = \frac{2}{\pi p}\left(1 - e^{-p^2}\right) - \frac{2}{\sqrt{\pi}}(1 - \Phi(p))$$

$$[p > 0] \qquad \text{NT 61(13)a}$$

6.32 Fresnel integrals

6.321

1.
$$\int_0^\infty \left[\frac{1}{2} - S(px)\right]x^{2q-1} \, dx = \frac{\sqrt{2}\,\Gamma\left(q + \frac{1}{2}\right)\sin\dfrac{2q+1}{4}\pi}{4\sqrt{\pi}q p^{2q}}$$

$$\left[0 < \operatorname{Re} q < \frac{3}{2}, \quad p > 0\right] \qquad \text{NT 56(14)a}$$

2.
$$\int_0^\infty \left[\frac{1}{2} - C(px)\right]x^{2q-1} \, dx = \frac{\sqrt{2}\,\Gamma\left(q + \frac{1}{2}\right)\cos\dfrac{2q+1}{4}\pi}{4\sqrt{\pi}q p^{2q}}$$

$$\left[0 < \operatorname{Re} q < \frac{3}{2}, \quad p > 0\right] \qquad \text{NT 56(13)a}$$

6.322

1. $\displaystyle\int_0^\infty S(t)e^{-pt}\,dt = \frac{1}{p}\left\{\cos\frac{p^2}{4}\left[\frac{1}{2} - C\left(\frac{p}{2}\right)\right] + \sin\frac{p^2}{4}\left[\frac{1}{2} - S\left(\frac{p}{2}\right)\right]\right\}$ MO 173a

2. $\displaystyle\int_0^\infty C(t)e^{-pt}\,dt = \frac{1}{p}\left\{\cos\frac{p^2}{4}\left[\frac{1}{2} - S\left(\frac{p}{2}\right)\right] - \sin\frac{p^2}{4}\left[\frac{1}{2} - C\left(\frac{p}{2}\right)\right]\right\}$ MO 172a

6.323

1. $\displaystyle\int_0^\infty S\left(\sqrt{t}\right)e^{-pt}\,dx = \frac{\left(\sqrt{p^2+1} - p\right)^{\frac{1}{2}}}{2p\sqrt{p^2+1}}$ EF 122(58)a

2. $\displaystyle\int_0^\infty C\left(\sqrt{t}\right)e^{-pt}\,dt = \frac{\left(\sqrt{p^2+1} + p\right)^{\frac{1}{2}}}{2p\sqrt{p^2+1}}$ EF 122(58)a

6.324

1. $\displaystyle\int_0^\infty \left[\frac{1}{2} - S(x)\right]\sin 2px\,dx = \frac{1 + \sin p^2 - \cos p^2}{4p}$ $[p > 0]$ NT 61(12)a

2. $\displaystyle\int_0^\infty \left[\frac{1}{2} - C(x)\right]\sin 2px\,dx = \frac{1 - \sin p^2 - \cos p^2}{4p}$ $[p > 0]$ NT 61(11)a

6.325

1. $\displaystyle\int_0^\infty S(x)\sin b^2 x^2\,dx = \frac{\sqrt{\pi}}{b}2^{-\frac{5}{2}}$ $\left[0 < b^2 < 1\right]$

                   $= 0$ $\left[b^2 > 1\right]$

                                                ET I 98(21)a

2. $\displaystyle\int_0^\infty C(x)\cos b^2 x^2\,dx = \frac{\sqrt{\pi}}{b}2^{-\frac{5}{2}}$ $\left[0 < b^2 < 1\right]$

                   $= 0$ $\left[b^2 > 1\right]$

                                                ET I 42(22)

6.326

1. $\displaystyle\int_0^\infty \left[\frac{1}{2} - S(x)\right]\operatorname{si}(2px)\,dx = \left(\frac{\pi}{8}\right)^{1/2}(S(p) + C(p) - 1) - \frac{1 + \sin p^2 - \cos p^2}{4p}$

                                     $[p > 0]$ NT 61(15)a

2. $\displaystyle\int_0^\infty \left[\frac{1}{2} - C(x)\right]\operatorname{si}(2px)\,dx = \left(\frac{\pi}{8}\right)^{1/2}(S(p) - C(p)) - \frac{1 - \sin p^2 - \cos p^2}{4p}$

                                     $[p > 0]$ NT 61(14)a

6.4 The Gamma Function and Functions Generated by It

6.41 The gamma function

6.411 $\displaystyle\int_{-\infty}^{\infty} \Gamma(\alpha + x)\,\Gamma(\beta - x)\,dx$

$$= -i\pi 2^{1-\alpha-\beta}\,\Gamma(\alpha+\beta) \qquad [\text{Re}(\alpha+\beta)<1, \quad \text{Im}\,\alpha, \quad \text{Im}\,\beta > 0]\quad \text{ET II 297(3)}$$

$$= i\pi 2^{1-\alpha-\beta}\,\Gamma(\alpha+\beta) \qquad [\text{Re}(\alpha+\beta)<1, \quad \text{Im}\,\alpha, \quad \text{Im}\,\beta < 0]\quad \text{ET II 297(2)}$$

$$= 0 \qquad [\text{Re}(\alpha+\beta)<1, \quad \text{Im}\,\alpha, \quad \text{Im}\,\beta < 0]$$

ET II 297(1)

6.412 $\displaystyle\int_{-i\infty}^{i\infty} \Gamma(\alpha+s)\,\Gamma(\beta+s)\,\Gamma(\gamma-s)\,\Gamma(\delta-s)\,ds = 2\pi i \frac{\Gamma(\alpha+\gamma)\,\Gamma(\alpha+\delta)\,\Gamma(\beta+\gamma)\,\Gamma(\beta+\delta)}{\Gamma(\alpha+\beta+\gamma+\delta)}$

$$[\text{Re}\,\alpha, \quad \text{Re}\,\beta, \quad \text{Re}\,\gamma, \quad \text{Re}\,\delta > 0]$$

ET II 302(32)

6.413

1. $\displaystyle\int_0^{\infty} |\Gamma(a+ix)\,\Gamma(b+ix)|^2\,dx = \frac{\sqrt{\pi}\,\Gamma(a)\,\Gamma\left(a+\frac{1}{2}\right)\Gamma(b)\,\Gamma\left(b+\frac{1}{2}\right)\Gamma(a+b)}{2\,\Gamma\left(a+b+\frac{1}{2}\right)}$

$$[a>0, \quad b>0] \qquad \text{ET II 302(27)}$$

2. $\displaystyle\int_0^{\infty} \left|\frac{\Gamma(a+ix)}{\Gamma(b+ix)}\right|^2\,dx = \frac{\sqrt{\pi}\,\Gamma(a)\,\Gamma\left(a+\frac{1}{2}\right)\Gamma\left(b-a-\frac{1}{2}\right)}{2\,\Gamma(b)\,\Gamma\left(b-\frac{1}{2}\right)\Gamma(b-a)}$

$$\left[0<a<b-\tfrac{1}{2}\right] \qquad \text{ET II 302(28)}$$

6.414

1. $\displaystyle\int_{-\infty}^{\infty} \frac{\Gamma(\alpha+x)}{\Gamma(\beta+x)}\,dx = 0 \qquad [\text{Im}\,\alpha \neq 0, \quad \text{Re}(\alpha-\beta) < -1]$

ET II 297(4)

2. $\displaystyle\int_{-\infty}^{\infty} \frac{dx}{\Gamma(\alpha+x)\,\Gamma(\beta-x)} = \frac{2^{\alpha+\beta-2}}{\Gamma(\alpha+\beta-1)} \qquad [\text{Re}(\alpha+\beta) > 1] \qquad \text{ET II 297(5)}$

3. $\displaystyle\int_{-\infty}^{\infty} \frac{\Gamma(\gamma+x)\,\Gamma(\delta+x)}{\Gamma(\alpha+x)\,\Gamma(\beta+x)}\,dx = 0$

$$[\text{Re}(\alpha+\beta-\gamma-\delta) > 1, \quad \text{Im}\,\gamma, \quad \text{Im}\,\delta > 0] \qquad \text{ET II 299(18)}$$

4. $\displaystyle\int_{-\infty}^{\infty} \frac{\Gamma(\gamma+x)\,\Gamma(\delta+x)}{\Gamma(\alpha+x)\,\Gamma(\beta+x)}\,dx = \frac{\pm 2\pi^2 i\,\Gamma(\alpha+\beta-\gamma-\delta-1)}{\sin[\pi(\gamma-\delta)]\,\Gamma(\alpha-\gamma)\,\Gamma(\alpha-\delta)\,\Gamma(\beta-\gamma)\,\Gamma(\beta-\delta)}$

$[\text{Re}(\alpha+\beta-\gamma-\delta) > 1, \quad \text{Im}\,\gamma < 0, \quad \text{Im}\,\delta < 0.$ In the numerator, we take the plus sign if $\text{Im}\,\gamma > \text{Im}\,\delta$ and the minus sign if $\text{Im}\,\gamma < \text{Im}\,\delta.]$ ET II 300(19)

5. $\displaystyle\int_{-\infty}^{\infty} \frac{\Gamma(\alpha-\beta-\gamma+x+1)\,dx}{\Gamma(\alpha+x)\,\Gamma(\beta-x)\,\Gamma(\gamma+x)} = \frac{\pi \exp\left(\pm\frac{1}{2}\pi(\delta-\gamma)i\right)}{\Gamma(\beta+\gamma-1)\,\Gamma\left(\frac{1}{2}(\alpha+\beta)\right)\Gamma\left(\frac{1}{2}(\gamma-\delta+1)\right)}$

$[\text{Re}(\beta+\gamma) > 1, \quad \delta = \alpha-\beta-\gamma+1, \quad \text{Im}\,\delta \neq 0.$ The sign is plus in the argument if the exponential for $\text{Im}\,\delta > 0$ and minus for $\text{Im}\,\delta < 0].$ ET II 300(20)

6.
$$\int_{-\infty}^{\infty} \frac{dx}{\Gamma(\alpha + x)\,\Gamma(\beta - x)\,\Gamma(\gamma + x)\,\Gamma(\delta - x)} = \frac{\Gamma(\alpha + \beta + \gamma + \delta - 3)}{\Gamma(\alpha + \beta - 1)\,\Gamma(\beta + \gamma - 1)\,\Gamma(\gamma + \delta - 1)\,\Gamma(\delta + \alpha - 1)}$$

$$[\operatorname{Re}(\alpha + \beta + \gamma + \delta) > 3] \qquad \text{ET II 300(21)}$$

6.415

1.
$$\int_{-\infty}^{-\infty} \frac{R(x)\,dx}{\Gamma(\alpha + x)\,\Gamma(\beta - x)\,\Gamma(\gamma + x)\,\Gamma(\delta - x)}$$

$$= \frac{\Gamma(\alpha + \beta + \gamma + \delta - 3)}{\Gamma(\alpha + \beta - 1)\,\Gamma(\beta + \gamma - 1)\,\Gamma(\gamma + \delta - 1)\,\Gamma(\delta + \alpha - 1)} \int_0^1 R(t)\,dt$$

$$[\operatorname{Re}(\alpha + \beta + \gamma + \delta) > 3, \quad R(x + 1) = R(x)] \quad \text{ET II 301(24)}$$

2.
$$\int_{-\infty}^{\infty} \frac{R(x)\,dx}{\Gamma(\alpha + x)\,\Gamma(\beta - x)\,\Gamma(\gamma + x)\,\Gamma(\delta - x)} = \frac{\int_0^1 R(t)\cos\left[\frac{1}{2}\pi(2t + \alpha - \beta)\right]\,dt}{\Gamma\left(\dfrac{\alpha + \beta}{2}\right)\Gamma\left(\dfrac{\gamma + \delta}{2}\right)\Gamma(\alpha + \delta - 1)}$$

$$[\alpha + \delta = \beta + \gamma, \quad \operatorname{Re}(\alpha + \beta + \gamma + \delta) > 2, \quad R(x + 1) = -R(x)] \quad \text{ET II 301(25)}$$

6.42 Combinations of the gamma function, the exponential, and powers

6.421

1.
$$\int_{-\infty}^{\infty} \Gamma(\alpha + x)\,\Gamma(\beta - x)\,\exp\left[2(\pi n + \theta)xi\right]\,dx = 2\pi i\,\Gamma(\alpha + \beta)(2\cos\theta)^{-\alpha - \beta}\exp[(\beta - \alpha)i\theta]$$

$$\times \left[\eta_n(\beta)\exp(2n\pi\beta i) - \eta_n(-\alpha)\exp(-2n\pi\alpha i)\right]$$

$$\left[\operatorname{Re}(\alpha + \beta) < 1, \quad -\frac{\pi}{2} < \theta < \frac{\pi}{2}, \quad n \text{ an integer}, \quad \eta_n(\xi) = \begin{cases} 0 & \text{if } \left(\frac{1}{2} - n\right)\operatorname{Im}\xi > 0 \\ \operatorname{sign}\left(\frac{1}{2} - n\right) & \text{if } \left(\frac{1}{2} - n\right)\operatorname{Im}\xi < 0 \end{cases}\right]$$

$$\text{ET II 298(7)}$$

2.
$$\int_{-\infty}^{\infty} \frac{e^{\pi i c x}\,dx}{\Gamma(\alpha + x)\,\Gamma(\beta - x)\,\Gamma(\gamma + kx)\,\Gamma(\delta - kx)} = 0$$

$$[\operatorname{Re}(\alpha + \beta + \gamma + \delta) > 2, \quad c \text{ and } k \text{ are real}, \quad |c| > |k| + 1] \quad \text{ET II 301(26)}$$

3.
$$\int_{-\infty}^{\infty} \frac{\Gamma(\alpha + x)}{\Gamma(\beta + x)}\exp[(2\pi n + \pi - 2\theta)xi]\,dx$$

$$= 2\pi i\,\operatorname{sign}\left(n + \frac{1}{2}\right)\frac{(2\cos\theta)^{\beta - \alpha - 1}}{\Gamma(\beta - \alpha)}\exp[-(2\pi n + \pi - \theta)\alpha i + \theta i(\beta - 1)]$$

$$\left[\operatorname{Re}(\beta - \alpha) > 0, \quad -\frac{\pi}{2} < \theta < \frac{\pi}{2}, \quad n \text{ is an integer}, \quad \left(n + \frac{1}{2}\right)\operatorname{Im}\alpha < 0\right] \quad \text{ET II 298(8)}$$

4.
$$\int_{-\infty}^{\infty} \frac{\Gamma(\alpha + x)}{\Gamma(\beta + x)}\exp[(2\pi n + \pi - 2\theta)xi]\,dx = 0$$

$$\left[\operatorname{Re}(\beta - \alpha) > 0, \quad -\frac{\pi}{2} < \theta < \frac{\pi}{2}, \quad n \text{ is an integer}, \quad \left(n + \frac{1}{2}\right)\operatorname{Im}\alpha > 0\right] \quad \text{ET II 297(6)}$$

6.422

1. $$\int_{-i\infty}^{i\infty} \Gamma(s - k - \lambda)\,\Gamma\left(\lambda + \mu - s + \tfrac{1}{2}\right)\Gamma\left(\lambda - \mu - s + \tfrac{1}{2}\right)z^s\,ds$$

    $$= 2\pi i\,\Gamma\left(\tfrac{1}{2} - k - \mu\right)\Gamma\left(\tfrac{1}{2} - k + \mu\right)z^\lambda e^{\frac{z}{2}}\,W_{k,\mu}(z)$$

    $$\left[\operatorname{Re}(k + \lambda) < 0, \quad \operatorname{Re}\lambda > |\operatorname{Re}\mu| - \tfrac{1}{2}, \quad |\arg z| < \tfrac{3}{2}\pi\right]\quad \text{ET II 302(29)}$$

2. $$\int_{\gamma-i\infty}^{\gamma+i\infty} \Gamma(\alpha + s)\,\Gamma(-s)\,\Gamma(1 - c - s)x^s\,ds = 2\pi i\,\Gamma(\alpha)\,\Gamma(\alpha - c + 1)\Psi(\alpha, c; x)$$

    $$\left[-\operatorname{Re}\alpha < \gamma < \min\left(0, 1 - \operatorname{Re}c\right), \quad -\tfrac{3}{2}\pi < \arg x < \tfrac{3}{2}\pi\right]\quad \text{EH I 256(5)}$$

3. $$\int_{\gamma-i\infty}^{\gamma+i\infty} \Gamma(-s)\,\Gamma(\beta + s)t^s\,ds = 2\pi i\,\Gamma(\beta)(1 + t)^{-\beta} \qquad \left[0 > \gamma > \operatorname{Re}(1 - \beta), \quad |\arg t| < \pi\right]$$

    $$\text{EH I 256, BU 75}$$

4. $$\int_{-\infty i}^{\infty i} \Gamma\left(\frac{t - p}{2}\right)\Gamma(-t)\left(\sqrt{2}\right)^{t-p-2}z^t\,dt = 2\pi i e^{\frac{1}{4}z^2}\,\Gamma(-p)\,D_p(z)$$

    $$\left[|\arg z| < \tfrac{3}{4}\pi, \quad p \text{ is not a positive integer}\right]\quad \text{WH}$$

5. $$\int_{-i\infty}^{i\infty} \Gamma(s)\,\Gamma\left(\tfrac{1}{2}\nu + \tfrac{1}{4} - s\right)\Gamma\left(\tfrac{1}{2}\nu - \tfrac{1}{4} - s\right)\left(\frac{z^2}{2}\right)^s\,ds$$

    $$= 2\pi i \cdot 2^{\frac{1}{4} - \frac{1}{2}\nu}z^{-\frac{1}{2}}e^{\frac{3}{4}z^2}\,\Gamma\left(\tfrac{1}{2}\nu + \tfrac{1}{4}\right)\Gamma\left(\tfrac{1}{2}\nu - \tfrac{1}{4}\right)D_\nu(z)$$

    $$\left[|\arg z| < \tfrac{3}{4}\pi, \quad \nu \neq \tfrac{1}{2}, \ -\tfrac{1}{2}, \ -\tfrac{3}{2}, \ldots\right]\quad \text{EH II 120}$$

6.[3] $$\int_{c-i\infty}^{c+i\infty} \left(\tfrac{1}{2}x\right)^{-s}\Gamma\left(\tfrac{1}{2}\nu + \tfrac{1}{2}s\right)\left[\Gamma\left(1 + \tfrac{1}{2}\nu - \tfrac{1}{2}s\right)\right]^{-1}\,ds = 4\pi i\,J_\nu(x)$$

    $$\left[x > 0, \ -\operatorname{Re}\nu < c < 1\right]\quad \text{EH II 21(34)}$$

7. $$\int_{-c-i\infty}^{-c+i\infty} \Gamma(-\nu - s)\,\Gamma(-s)\left(-\tfrac{1}{2}iz\right)^{\nu+2s}\,ds = -2\pi^2 e^{\frac{1}{2}i\nu\pi}\,H_\nu^{(1)}(z)$$

    $$\left[|\arg(-iz)| < \frac{\pi}{2}, \quad 0 < \operatorname{Re}\nu < c\right]$$

    $$\text{EH II 83(34)}$$

8. $$\int_{-c-i\infty}^{-c+i\infty} \Gamma(-\nu - s)\,\Gamma(-s)\left(\tfrac{1}{2}iz\right)^{\nu+2s}\,ds = 2\pi^2 e^{-\frac{1}{2}i\nu\pi}\,H_\nu^{(2)}(z)$$

    $$\left[|\arg(iz)| < \frac{\pi}{2}, \quad 0 < \operatorname{Re}\nu < c\right]$$

    $$\text{EH II 83(35)}$$

9. $$\int_{-i\infty}^{i\infty} \Gamma(-s)\frac{\left(\tfrac{1}{2}x\right)^{\nu+2s}}{\Gamma(\nu + s + 1)}\,ds = 2\pi i\,J_\nu(x) \qquad \left[x > 0, \quad \operatorname{Re}\nu > 0\right]\quad \text{EH II 83(36)}$$

10. $$\int_{-i\infty}^{i\infty} \Gamma(-s)\,\Gamma(-2\nu - s)\,\Gamma\left(\nu + s + \tfrac{1}{2}\right)(-2iz)^s\,ds = -\pi^{\frac{5}{2}}e^{-i(z-\nu\pi)}\sec(\nu\pi)(2z)^{-\nu}\,H_\nu^{(1)}(z)$$

    $$\left[|\arg(-iz)| < \tfrac{3}{2}\pi, \quad 2\nu \neq \pm 1, \ \pm 3\ldots\right]$$

    $$\text{EH II 83(37)}$$

11. $\int_{-i\infty}^{i\infty} \Gamma(-s)\,\Gamma(-2\nu - s)\,\Gamma\left(\nu + s + \tfrac{1}{2}\right)(2iz)^s\,ds = \pi^{\frac{5}{2}}e^{i(z-\nu\pi)}\sec(\nu\pi)(2z)^{-\nu}\,H_\nu^{(2)}(z)$

$$\left[|\arg(iz)| < \tfrac{3}{2}\pi, \quad 2\nu \neq \pm 1, \quad \pm 3\dots\right]$$
EH II 84(38)

12. $\int_{-i\infty}^{i\infty} \Gamma(s)\,\Gamma\left(\tfrac{1}{2} - s - \nu\right)\Gamma\left(\tfrac{1}{2} - s + \nu\right)(2z)^s\,ds = 2^{\frac{3}{2}}\pi^{\frac{3}{2}}iz^{\frac{1}{2}}e^z\sec(\nu\pi)\,K_\nu(z)$

$$\left[|\arg z| < \tfrac{3}{2}\pi, \quad 2\nu \neq \pm 1, \quad \pm 3,\dots\right]$$
EH II 84(39)

13. $\int_{-\frac{1}{2}-i\infty}^{-\frac{1}{2}+i\infty} \dfrac{\Gamma(-s)}{s\,\Gamma(1+s)}\,x^{2s}\,ds = 4\pi\int_{2x}^{\infty}\dfrac{J_0(t)}{t}\,dt \qquad [x > 0]$ MO 41

14. $\int_{-i\infty}^{i\infty} \dfrac{\Gamma(\alpha + s)\,\Gamma(\beta + s)\,\Gamma(-s)}{\Gamma(\gamma + s)}(-z)^s\,ds = 2\pi i\dfrac{\Gamma(\alpha)\,\Gamma(\beta)}{\Gamma(\gamma)}\,F(\alpha, \beta; \gamma; z)$

[For $\arg(-z) < \pi$, the path of integration must separate the poles of the integrand at the points $s = 0, 1, 2, 3, \dots$ from the poles $s = -\alpha - n$ and $s = -\beta - n$ (for $n = 0, 1, 2, \dots$)].

15. $\int_{\delta - i\infty}^{\delta + i\infty} \dfrac{\Gamma(\alpha + s)\,\Gamma(-s)}{\Gamma(\gamma + s)}(-z)^s\,ds = \dfrac{2\pi i\,\Gamma(\alpha)}{\Gamma(\gamma)}\,{}_1F_1(\alpha; \gamma; z)$

$$\left[-\tfrac{\pi}{2} < \arg(-z) < \tfrac{\pi}{2}, \quad 0 > \delta > -\operatorname{Re}\alpha, \quad \gamma \neq 0, 1, 2, \dots\right]$$ EH I 62(15), EH I 256(4)

16. $\int_{-i\infty}^{i\infty} \left[\dfrac{\Gamma\left(\tfrac{1}{2} - s\right)}{\Gamma(s)}\right]^2 z^s\,ds = 2\pi iz^{\frac{1}{2}}\left[2\pi^{-1}K_0\left(4z^{\frac{1}{4}}\right) - Y_0\left(4z^{\frac{1}{4}}\right)\right]$

$$[z > 0]$$
ET II 303(33)

17. $\int_{-i\infty}^{i\infty} \dfrac{\Gamma\left(\lambda + \mu - s + \tfrac{1}{2}\right)\Gamma\left(\lambda - \mu - s + \tfrac{1}{2}\right)}{\Gamma(\lambda - k - s + 1)}z^s\,ds = 2\pi iz^\lambda e^{-\frac{z}{2}}\,W_{k,\mu}(z)$

$$\left[\operatorname{Re}\lambda > |\operatorname{Re}\mu| - \tfrac{1}{2}, \quad |\arg z| < \tfrac{\pi}{2}\right]$$
ET II 302(30)

18. $\int_{-i\infty}^{i\infty} \dfrac{\Gamma(k - \lambda + s)\,\Gamma\left(\lambda + \mu - s + \tfrac{1}{2}\right)}{\Gamma\left(\mu - \lambda + s + \tfrac{1}{2}\right)}z^s\,ds = 2\pi i\dfrac{\Gamma\left(k + \mu + \tfrac{1}{2}\right)}{\Gamma(2\mu + 1)}z^\lambda e^{-\frac{z}{2}}\,M_{k,\mu}(z)$

$$\left[\operatorname{Re}(k - \lambda) > 0, \quad \operatorname{Re}(\lambda + \mu) > -\tfrac{1}{2}, \quad |\arg z| < \tfrac{\pi}{2}\right]$$ ET II 302(31)

19.
$$\int_{-i\infty}^{i\infty} \frac{\prod_{j=1}^{m} \Gamma(b_j - s) \prod_{j=1}^{n} \Gamma(1 - a_j + s)}{\prod_{j=m+1}^{q} \Gamma(1 - b_j + s) \prod_{j=n+1}^{p} \Gamma(a_j - s)} z^s \, ds = 2\pi i \, G_{mn}^{pq} \left(z \left| \begin{array}{c} a_1, \ldots, a_p \\ b_1, \ldots, b_q \end{array} \right. \right)$$

$$\left[p + q < 2(m + n); \quad |\arg z| < \left(m + n - \tfrac{1}{2}p - \tfrac{1}{2}q \right) \pi; \right.$$

$$\left. \operatorname{Re} a_k < 1, \quad k = 1, \ldots, n; \quad \operatorname{Re} b_j > 0, \quad j = 1, \ldots, m \right]$$

ET II 303(34)

6.423

1. $\displaystyle\int_0^\infty e^{-\alpha x} \frac{dx}{\Gamma(1 + x)} = \nu\left(e^{-\alpha}\right)$ MI 39, EH III 222(16)

2. $\displaystyle\int_0^\infty e^{-\alpha x} \frac{dx}{\Gamma(x + \beta + 1)} = e^{\beta \alpha} \nu\left(e^{-\alpha}, \beta\right)$ MI 39, EH III 222(16)

3. $\displaystyle\int_0^\infty e^{-\alpha x} \frac{x^m}{\Gamma(x + 1)} \, dx = \mu\left(e^{-\alpha}, m\right) \Gamma(m + 1)$ $[\operatorname{Re} m > -1]$ MI 39, EH III 222(17)

4. $\displaystyle\int_0^\infty e^{-\alpha x} \frac{x^m}{\Gamma(x + n + 1)} \, dx = e^{n\alpha} \mu\left(e^{-\alpha}, m, n\right) \Gamma(m + 1)$ MI 39, EH III 222(17)

6.424 $\displaystyle\int_{-\infty}^\infty \frac{R(x) \exp[(2\pi n + \theta)xi] \, dx}{\Gamma(\alpha + x) \Gamma(\beta - x)} = \frac{\left[2 \cos\left(\frac{\theta}{2}\right) \right]^{\alpha + \beta - 2}}{\Gamma(\alpha + \beta - 1)} \exp\left[\frac{1}{2}\theta(\beta - \alpha)i \right] \int_0^1 R(t) \exp(2\pi n t i) \, dt$

$[\operatorname{Re}(\alpha + \beta) > 1, \quad -\pi < \theta < \pi, \quad n \text{ is an integer}, \quad R(x + 1) = R(x)]$ ET II 299(16)

6.43 Combinations of the gamma function and trigonometric functions

6.431

1. $\displaystyle\int_{-\infty}^{-\infty} \frac{\sin rx \, dx}{\Gamma(p + x) \Gamma(q - x)} = \frac{\left(2 \cos \frac{r}{2} \right)^{p+q-2} \sin \frac{r(q-p)}{2}}{\Gamma(p + q - 1)}$ $[|r| < \pi]$

$$= 0 \qquad [|r| > \pi]$$

$[r \text{ is real}; \quad \operatorname{Re}(p + q) > 1]$ MO 10a, ET II 298(9, 10)

2. $\displaystyle\int_{-\infty}^\infty \frac{\cos rx \, dx}{\Gamma(p + x) \Gamma(q - x)} = \frac{\left(2 \cos \frac{r}{2} \right)^{p+q-2} \cos \frac{r(q-p)}{2}}{\Gamma(p + q - 1)}$ $[|r| < \pi]$

$$= 0 \qquad [|r| > \pi]$$

$[r \text{ is real}; \quad \operatorname{Re}(p + q) > 1]$ MO 10a, ET II 299(13, 14)

6.432 $\displaystyle\int_{-\infty}^{\infty} \frac{\sin(m\pi x)}{\sin(\pi x)} \frac{dx}{\Gamma(\alpha+x)\,\Gamma(\beta-x)} = 0$ $\qquad$ [m is an even integer]

$$= \frac{2^{\alpha+\beta-2}}{\Gamma(\alpha+\beta-1)} \qquad \text{[m is an odd integer]}$$

$$[\operatorname{Re}(\alpha+\beta) > 1] \qquad \text{ET II 298(11, 12)}$$

6.433

1. $\displaystyle\int_{-\infty}^{\infty} \frac{\sin \pi x \, dx}{\Gamma(\alpha+x)\,\Gamma(\beta-x)\,\Gamma(\gamma+x)\,\Gamma(\delta-x)} = \frac{\sin\left[\dfrac{\pi}{2}(\beta-\alpha)\right]}{2\,\Gamma\left(\dfrac{\alpha+\beta}{2}\right)\Gamma\left(\dfrac{\gamma+\delta}{2}\right)\Gamma(\alpha+\delta-1)}$

$$[\alpha+\delta=\beta+\gamma, \quad \operatorname{Re}(\alpha+\beta+\gamma+\delta) > 2] \qquad \text{ET II 300(22)}$$

2. $\displaystyle\int_{-\infty}^{\infty} \frac{\cos \pi x \, dx}{\Gamma(\alpha+x)\,\Gamma(\beta-x)\,\Gamma(\gamma+x)\,\Gamma(\delta-x)} = \frac{\cos\left[\dfrac{\pi}{2}(\beta-\alpha)\right]}{2\,\Gamma\left(\dfrac{\alpha+\beta}{2}\right)\Gamma\left(\dfrac{\gamma+\delta}{2}\right)\Gamma(\alpha+\delta-1)}$

$$[\alpha+\delta=\beta+\gamma, \quad \operatorname{Re}(\alpha+\beta+\gamma+\delta) > 2] \qquad \text{ET II 301(23)}$$

6.44 The logarithm of the gamma function*

6.441

1. $\displaystyle\int_p^{p+1} \ln\Gamma(x)\,dx = \frac{1}{2}\ln 2\pi + p\ln p - p$ $\qquad\qquad$ FI II 784

2. $\displaystyle\int_0^1 \ln\Gamma(x)\,dx = \int_0^1 \ln\Gamma(1-x)\,dx = \frac{1}{2}\ln 2\pi$ $\qquad\qquad$ FI II 783

3. $\displaystyle\int_0^1 \ln\Gamma(x+q)\,dx = \frac{1}{2}\ln 2\pi + q\ln q - q$ $\qquad$ [$q \geq 0$] $\qquad$ NH 89(17), ET II 304(40)

4. $\displaystyle\int_0^z \ln\Gamma(x+1)\,dx = \frac{z}{2}\ln 2\pi - \frac{z(z+1)}{2} + z\ln\Gamma(z+1) - \ln G(z+1)$,

$\qquad$ where $G(z+1) = (2\pi)^{\frac{z}{2}} \exp\left(-\dfrac{z(z+1)}{2} - \dfrac{Cz^2}{2}\right) \displaystyle\prod_{k=1}^{\infty}\left\{\left(1+\dfrac{z}{k}\right)^k \exp\left(-z+\dfrac{z^2}{2k}\right)\right\}$ $\quad$ WH

5. $\displaystyle\int_0^n \ln\Gamma(\alpha+x)\,dx = \sum_{k=0}^{n-1}(a+k)\ln(a+k) - na + \frac{1}{2}n\ln(2\pi) - \frac{1}{2}n(n-1)$

$$[a \geq 0; \quad n = 1, 2, \ldots] \qquad \text{ET II 304(41)}$$

6.442 $\displaystyle\int_0^1 \exp(2\pi nxi)\ln\Gamma(a+x)\,dx = (2\pi ni)^{-1}\left[\ln a - \exp(-2\pi nai)\,\text{Ei}(2\pi nai)\right]$

$$[a > 0; \quad n = \pm 1, \pm 2, \ldots] \qquad \text{ET II 304(38)}$$

---

*Here, we are violating our usual order of presentation of the formulas in order to make it easier to examine the integrals involving the gamma function

6.443

1. $\displaystyle\int_0^1 \ln\Gamma(x)\sin 2\pi nx\, dx = \frac{1}{2\pi n}\left[\ln(2\pi n) + \boldsymbol{C}\right]$ NH 203(5), ET II 304(42)

2. $\displaystyle\int_0^1 \ln\Gamma(x)\sin(2n+1)\pi x\, dx = \frac{1}{(2n+1)\pi}\left[\ln\left(\frac{\pi}{2}\right) + 2\left(1 + \frac{1}{3} + \cdots + \frac{1}{2n-1}\right) + \frac{1}{2n+1}\right]$

ET II 305(43)

3. $\displaystyle\int_0^1 \ln\Gamma(x)\cos 2\pi nx\, dx = \frac{1}{4n}$ NH 203(6), ET II 305(44)

4.[8] $\displaystyle\int_0^1 \ln\Gamma(x)\cos(2n+1)\pi x\, dx = \frac{2}{\pi^2}\left[\frac{1}{(2n+1)^2}\left(\boldsymbol{C} + \ln 2\pi\right) + 2\sum_{k=2}^{\infty}\frac{\ln k}{4k^2 - (2n+1)^2}\right]$ NH 203(6)

5. $\displaystyle\int_0^1 \sin(2\pi nx)\ln\Gamma(a+x)\, dx = -(2\pi n)^{-1}\left[\ln a + \cos(2\pi na)\,\mathrm{ci}(2\pi na) - \sin(2\pi na)\,\mathrm{si}(2\pi na)\right]$

$[a > 0;\quad n = 1, 2, \ldots]$ ET II 304(36)

6. $\displaystyle\int_0^1 \cos(2\pi nx)\ln\Gamma(a+x)\, dx = -(2\pi n)^{-1}\left[\sin(2\pi na)\,\mathrm{ci}(2\pi na) + \cos(2\pi na)\,\mathrm{si}(2\pi na)\right]$

$[a > 0;\quad n = 1, 2, \ldots]$ ET II 304(37)

6.45 The incomplete gamma function

6.451

1. $\displaystyle\int_0^\infty e^{-\alpha x}\,\gamma(\beta, x)\, dx = \frac{1}{\alpha}\,\Gamma(\beta)(1+\alpha)^{-\beta}$ $[\beta > 0]$ MI 39

2. $\displaystyle\int_0^\infty e^{-\alpha x}\,\Gamma(\beta, x)\, dx = \frac{1}{\alpha}\,\Gamma(\beta)\left[1 - \frac{1}{(\alpha+1)^\beta}\right]$ $[\beta > 0]$ MI 39

6.452

1. $\displaystyle\int_0^\infty e^{-\mu x}\gamma\left(\nu, \frac{x^2}{8a^2}\right)dx = \frac{1}{\mu}2^{-\nu-1}\,\Gamma(2\nu)e^{(a\mu)^2}\,D_{-2\nu}(2a\mu)$

$\left[|\arg a| < \dfrac{\pi}{4}, \quad \mathrm{Re}\,\nu > -\dfrac{1}{2}, \quad \mathrm{Re}\,\mu > 0\right]$
ET I 179(36)

2. $\displaystyle\int_0^\infty e^{-\mu x}\gamma\left(\frac{1}{4}, \frac{x^2}{8a^2}\right)dx = \frac{2^{\frac{3}{4}}\sqrt{a}}{\sqrt{\mu}}e^{(a\mu)^2}\,K_{\frac{1}{4}}\left(a^2\mu^2\right)$ $\left[|\arg a| < \dfrac{\pi}{4}, \quad \mathrm{Re}\,\mu > 0\right]$ ET I 179(35)

6.453 $\displaystyle\int_0^\infty e^{-\mu x}\,\Gamma\left(\nu, \frac{a}{x}\right)dx = 2a^{\frac{1}{2}\nu}\mu^{\frac{1}{2}\nu-1}\,K_\nu\left(2\sqrt{\mu a}\right)$ $\left[|\arg a| < \dfrac{\pi}{2}, \quad \mathrm{Re}\,\mu > 0\right]$ ET I 179(32)

6.454 $\displaystyle\int_0^\infty e^{-\beta x}\gamma\left(\nu, \alpha\sqrt{x}\right)dx = 2^{-\frac{1}{2}\nu}\alpha^\nu\beta^{-\frac{1}{2}\nu-1}\,\Gamma(\nu)\exp\left(\frac{\alpha^2}{8\beta}\right)D_{-\nu}\left(\frac{\alpha}{\sqrt{2\beta}}\right)$

$[\mathrm{Re}\,\beta > 0, \quad \mathrm{Re}\,\nu > 0]$
ET II 309(19), MI 39a

6.455

1. $\displaystyle\int_0^\infty x^{\mu-1}e^{-\beta x}\,\Gamma(\nu,\alpha x)\,dx = \frac{\alpha^\nu\,\Gamma(\mu+\nu)}{\mu(\alpha+\beta)^{\mu+\nu}}\;{}_2F_1\left(1,\mu+\nu;\mu+1;\frac{\beta}{\alpha+\beta}\right)$

$$[\operatorname{Re}(\alpha+\beta) > 0, \quad \operatorname{Re}\mu > 0, \quad \operatorname{Re}(\mu+\nu) > 0] \quad\text{ET II 309(16)}$$

2. $\displaystyle\int_0^\infty x^{\mu-1}e^{-\beta x}\,\gamma(\nu,\alpha x)\,dx = \frac{\alpha^\nu\,\Gamma(\mu+\nu)}{\nu(\alpha+\beta)^{\mu+\nu}}\;{}_2F_1\left(1,\mu+\nu;\nu+1;\frac{\alpha}{\alpha+\beta}\right)$

$$[\operatorname{Re}(\alpha+\beta) > 0, \quad \operatorname{Re}\beta > 0, \quad \operatorname{Re}(\mu+\nu) > 0] \quad\text{ET II 308(15)}$$

6.456

1. $\displaystyle\int_0^\infty e^{-\alpha x}(4x)^{\nu-\frac12}\gamma\left(\nu,\frac{1}{4x}\right)dx = \sqrt{\pi}\,\frac{\gamma(2\nu,\sqrt{\alpha})}{\alpha^{\nu+\frac12}}$ MI 39a

2. $\displaystyle\int_0^\infty e^{-\alpha x}(4x)^{\nu-\frac12}\,\Gamma\left(\nu,\frac{1}{4x}\right)dx = \frac{\sqrt{\pi}\,\Gamma(2\nu,\sqrt{\alpha})}{\alpha^{\nu+\frac12}}$ MI 39a

6.457

1. $\displaystyle\int_0^\infty e^{-\alpha x}\frac{(4x)^\nu}{\sqrt{x}}\gamma\left(\nu+1,\frac{1}{4x}\right)dx = \sqrt{\pi}\,\frac{\gamma(2\nu+1,\sqrt{\alpha})}{\alpha^{\nu+\frac12}}$ MI 39

2. $\displaystyle\int_0^\infty e^{-\alpha x}\frac{(4x)^\nu}{\sqrt{x}}\,\Gamma\left(\nu+1,\frac{1}{4x}\right)dx = \sqrt{\pi}\,\frac{\Gamma(2\nu+1,\sqrt{\alpha})}{\alpha^{\nu+\frac12}}$ MI 39

6.458 $\displaystyle\int_0^\infty x^{1-2\nu}\exp\left(\alpha x^2\right)\sin(bx)\,\Gamma\left(\nu,\alpha x^2\right)dx = \pi^{\frac12}2^{-\nu}\alpha^{\nu-1}\,\Gamma\left(\tfrac32-\nu\right)\exp\left(\frac{b^2}{8\alpha}\right)D_{2\nu-2}\left[\frac{b}{(2\alpha)^{\frac12}}\right]$

$$\left[|\arg\alpha| < \frac{3\pi}{2}, \quad 0 < \operatorname{Re}\nu < 1\right]$$

$$\text{ET II 309(18)}$$

6.46–6.47 The function $\psi(x)$

6.461 $\displaystyle\int_1^x \psi(t)\,dt = \ln\Gamma(x)$

6.462 $\displaystyle\int_0^1 \psi(\alpha+x)\,dx = \ln\alpha$ $[\alpha > 0]$ ET II 305(1)

6.463 $\displaystyle\int_0^\infty x^{-\alpha}\left[C+\psi(1+x)\right] = -\pi\operatorname{cosec}(\pi\alpha)\,\zeta(\alpha)$ $[1 < \operatorname{Re}\alpha < 2]$ ET II 305(6)

6.464 $\displaystyle\int_0^1 e^{2\pi nxi}\,\psi(\alpha+x)\,dx = e^{-2\pi n\alpha i}\operatorname{Ei}(2\pi n\alpha i)$ $[\alpha > 0; \quad n = \pm i, \pm 2, \ldots]$ ET II 305(2)

6.465

1.[8] $\displaystyle\int_0^1 \psi(x)\sin\pi x\,dx = -\frac{2}{\pi}\left[C+\ln 2\pi + 2\sum_{k=2}^\infty \frac{\ln k}{4k^2-1}\right]$

$$\text{(see } \mathbf{6.443}\ 4) \qquad\qquad\qquad \text{NH 204}$$

2. $\displaystyle\int_0^1 \psi(x)\sin(2\pi nx)\,dx = -\frac{1}{2}\pi$ $[n = 1, 2, \ldots]$ ET II 305(3)

6.466 $\displaystyle\int_0^\infty [\psi(\alpha + ix) - \psi(\alpha - ix)]\sin xy\,dx = i\pi e^{-\alpha y}\left(1 - e^{-y}\right)^{-1}$

$[\alpha > 0, \quad y > 0]$ ET I 96(1)

6.467

1. $\displaystyle\int_0^1 \sin(2\pi nx)\,\psi(\alpha + x)\,dx = \sin(2\pi n\alpha)\operatorname{ci}(2\pi n\alpha) + \cos(2\pi n\alpha)\operatorname{si}(2\pi n\alpha)$

$[\alpha \geq 0; \quad n = 1, 2, \ldots]$ ET II 305(4)

2. $\displaystyle\int_0^1 \cos(2\pi nx)\,\psi(\alpha + x)\,dx = \sin(2\pi n\alpha)\operatorname{si}(2\pi n\alpha) - \cos(2\pi n\alpha)\operatorname{ci}(2\pi n\alpha)$

$[\alpha > 0; \quad n = 1, 2, \ldots]$ ET II 305(5)

6.468 $\displaystyle\int_0^1 \psi(x)\sin^2\pi x\,dx = -\frac{1}{2}\left[\boldsymbol{C} + \ln(2\pi)\right]$ NH 204

6.469

1. $\displaystyle\int_0^1 \psi(x)\sin\pi x\cos\pi x\,dx = -\frac{\pi}{4}$ NH 204

2.[8] $\displaystyle\int_0^1 \psi(x)\sin\pi x\sin(n\pi x)\,dx = \frac{n}{1 - n^2}$ $[n \text{ is even}]$

$\displaystyle\qquad\qquad\qquad\qquad\qquad\qquad = \frac{1}{2}\ln\frac{n-1}{n+1}$ $[n > 1 \text{ is odd}]$

NH 204(8)a

6.471

1. $\displaystyle\int_0^\infty x^{-\alpha}\left[\ln x - \psi(1 + x)\right]\,dx = \pi\operatorname{cosec}(\pi\alpha)\,\zeta(\alpha)$ $[0 < \operatorname{Re}\alpha < 1]$ ET II 306(7)

2. $\displaystyle\int_0^\infty x^{-\alpha}\left[\ln(1 + x) - \psi(1 + x)\right]\,dx = \pi\operatorname{cosec}(\pi\alpha)\left[\zeta(\alpha) - (\alpha - 1)^{-1}\right]$

$[0 < \operatorname{Re}\alpha < 1]$ ET II 306(8)

3. $\displaystyle\int_0^\infty [\psi(x + 1) - \ln x]\cos(2\pi xy)\,dx = \frac{1}{2}\left[\psi(y + 1) - \ln y\right]$ ET II 306(12)

6.472

1. $\displaystyle\int_0^\infty x^{-\alpha}\left[(1 + x)^{-1} - \psi'(1 + x)\right]\,dx = -\pi\alpha\operatorname{cosec}(\pi\alpha)\left[\zeta(1 + \alpha) - \alpha^{-1}\right]$

$[|\operatorname{Re}\alpha| < 1]$ ET II 306(9)

2. $\displaystyle\int_0^\infty x^{-\alpha}\left[x^{-1} - \psi'(1 + x)\right]\,dx = -\pi\alpha\operatorname{cosec}(\pi\alpha)\,\zeta(1 + \alpha)$

$[-2 < \operatorname{Re}\alpha < 0]$ ET II 306(10)

6.473 $\displaystyle\int_0^\infty x^{-\alpha}\psi^{(n)}(1 + x)\,dx = (-1)^{n-1}\frac{\pi\,\Gamma(\alpha + n)}{\Gamma(\alpha)\sin\pi\alpha}\,\zeta(\alpha + n)$

$[n = 1, 2, \ldots; \quad 0 < \operatorname{Re}\alpha < 1]$

ET II 306(11)

6.5–6.7 Bessel Functions

6.51 Bessel functions

6.511

1. $\displaystyle\int_0^\infty J_\nu(bx)\,dx = \frac{1}{b}$ $[\mathrm{Re}\,\nu > -1, \quad b > 0]$ ET II 22(3)

2. $\displaystyle\int_0^\infty Y_\nu(bx)\,dx = -\frac{1}{b}\tan\left(\frac{\nu\pi}{2}\right)$ $[|\mathrm{Re}\,\nu| < 1, \quad b > 0]$

                                                    WA 432(7), ET II 96(1)

3. $\displaystyle\int_0^a J_\nu(x)\,dx = 2\sum_{k=0}^\infty J_{\nu+2k+1}(a)$ $[\mathrm{Re}\,\nu > -1]$ ET II 333(1)

4. $\displaystyle\int_0^a J_{\frac{1}{2}}(t)\,dt = 2\,S\left(\sqrt{a}\right)$ WA 599(4)

5. $\displaystyle\int_0^a J_{-\frac{1}{2}}(t)\,dt = 2\,C\left(\sqrt{a}\right)$ WA 599(3)

6. $\displaystyle\int_0^a J_0(x)\,dx = a\,J_0(a) + \frac{\pi a}{2}\left[J_1(a)\,\mathbf{H}_0(a) - J_0(a)\,\mathbf{H}_1(a)\right]$

                                          $[a > 0]$ ET II 7(2)

7. $\displaystyle\int_0^a J_1(x)\,dx = 1 - J_0(a)$ $[a > 0]$ ET II 18(1)

8. $\displaystyle\int_a^\infty J_0(x)\,dx = 1 - a\,J_0(a) + \frac{\pi a}{2}\left[J_0(a)\,\mathbf{H}_1(a) - J_1(a)\,\mathbf{H}_0(a)\right]$

                                          $[a > 0]$ ET II 7(3)

9. $\displaystyle\int_a^\infty J_1(x)\,dx = J_0(a)$ $[a > 0]$ ET II 18(2)

10. $\displaystyle\int_a^b Y_\nu(x)\,dx = 2\sum_{n=0}^\infty\left[Y_{\nu+2n+1}(b) - Y_{\nu+2n+1}(a)\right]$ ET II 339(46)

11. $\displaystyle\int_0^a I_\nu(x)\,dx = 2\sum_{n=0}^\infty(-1)^n\,I_{\nu+2n+1}(a)$ $[\mathrm{Re}\,\nu > -1]$ ET II 364(1)

6.512

1. $\displaystyle\int_0^\infty J_\mu(ax)\,J_\nu(bx)\,dx = b^\nu a^{-\nu-1}\frac{\Gamma\left(\dfrac{\mu+\nu+1}{2}\right)}{\Gamma(\nu+1)\,\Gamma\left(\dfrac{\mu-\nu+1}{2}\right)}\,F\left(\frac{\mu+\nu+1}{2},\frac{\nu-\mu+1}{2};\nu+1;\frac{b^2}{a^2}\right)$

                               $[a > 0, \quad b > 0, \quad \mathrm{Re}(\mu+\nu) > -1, \quad b > a$

                   For $a < b$, the positions of μ and ν should be reversed.]

                                                      ET II 48(6)

$2.^7$ $\displaystyle\int_0^\infty J_{\nu+n}(\alpha t)\, J_{\nu-n-1}(\beta t)\, dt = \frac{\beta^{\nu-n-1}\,\Gamma(\nu)}{\alpha^{\nu-n}\, n!\,\Gamma(\nu-n)}\, F\left(\nu, -n; \nu-n; \frac{\beta^2}{\alpha^2}\right)$ $\qquad [0 < \beta < \alpha]$

$\displaystyle\hspace{6cm} = (-1)^n\, \frac{1}{2\alpha}$ $\qquad [0 < \beta = \alpha]$

$\displaystyle\hspace{6cm} = 0$ $\qquad [0 < \alpha < \beta]$

$\hspace{8cm} [\mathrm{Re}(\nu) > 0]$ $\qquad$ MO 50

$3.^8$ $\displaystyle\int_0^\infty J_\nu(\alpha x)\, J_{\nu-1}(\beta x)\, dx = \frac{\beta^{\nu-1}}{\alpha^\nu}$ $\qquad [\beta < \alpha]$

$\displaystyle\hspace{5.3cm} = \frac{1}{2\beta}$ $\qquad [\beta = \alpha]$

$\displaystyle\hspace{5.3cm} = 0$ $\qquad [\beta > \alpha]$

$\hspace{6cm} [\mathrm{Re}\,\nu > 0]$ $\qquad$ WA 444(8), KU (40)a

$4.$ $\displaystyle\int_0^\infty J_{\nu+2n+1}(ax)\, J_\nu(bx)\, dx = b^\nu a^{-\nu-1}\, P_n^{(\nu,0)}\left(1 - \frac{2b^2}{a^2}\right)$ $\qquad [\mathrm{Re}\,\nu > -1-n, \quad 0 < b < a]$

$\displaystyle\hspace{6cm} = 0$ $\qquad [\mathrm{Re}\,\nu > -1-n, \quad 0 < a < b]$

$\hspace{10cm}$ ET II 47(5)

$5.$ $\displaystyle\int_0^\infty J_{\nu+n}(ax)\, Y_{\nu-n}(ax)\, dx = (-1)^{n+1}\, \frac{1}{2a}$ $\qquad \left[\mathrm{Re}\,\nu > -\tfrac{1}{2}, \quad a > 0, \quad n = 0, 1, 2, \ldots\right]$

$\hspace{10cm}$ ET II 347(57)

$6.$ $\displaystyle\int_0^\infty J_1(bx)\, Y_0(ax)\, dx = -\frac{b^{-1}}{\pi}\, \ln\left(1 - \frac{b^2}{a^2}\right)$ $\qquad [0 < b < a]$ $\qquad$ ET II 21(31)

$7.$ $\displaystyle\int_0^a J_\nu(x)\, J_{\nu+1}(x)\, dx = \sum_{n=0}^\infty\, [J_{\nu+n+1}(a)]^2$ $\qquad [\mathrm{Re}\,\nu > -1]$ $\qquad$ ET II 338(37)

$8.^9$ $\displaystyle\int_0^\infty k\, J_n(ka)\, J_n(kb)\, dk = \frac{1}{a}\, \delta(b-a)$ $\qquad [n = 0, 1, \ldots]$ $\qquad$ JAC 110

6.513

$1.$ $\displaystyle\int_0^\infty [J_\mu(ax)]^2\, J_\nu(bx)\, dx = a^{2\mu} b^{-2\mu-1}\, \frac{\Gamma\left(\dfrac{1+\nu+2\mu}{2}\right)}{[\Gamma(\mu+1)]^2\, \Gamma\left(\dfrac{1+\nu-2\mu}{2}\right)}$

$\displaystyle\hspace{3cm} \times \left[F\left(\frac{1-\nu+2\mu}{2}, \frac{1+\nu+2\mu}{2}; \mu+1; \frac{1 - \sqrt{1 - \dfrac{4a^2}{b^2}}}{2}\right)\right]^2$

$\hspace{6cm} [\mathrm{Re}\,\nu + \mathrm{Re}\,2\mu > -1, \quad 0 < 2a < b]$ $\qquad$ ET II 52(33)

$2.$ $\displaystyle\int_0^\infty [J_\mu(ax)]^2\, K_\nu(bx)\, dx = \frac{b^{-1}}{2}\, \Gamma\left(\frac{2\mu+\nu+1}{2}\right)\, \Gamma\left(\frac{2\mu-\nu+1}{2}\right)\left[P_{\frac{1}{2}\nu-\frac{1}{2}}^{-\mu}\left(\sqrt{1 + \frac{4a^2}{b^2}}\right)\right]^2$

$\hspace{6cm} [2\,\mathrm{Re}\,\mu > |\mathrm{Re}\,\nu| - 1, \quad \mathrm{Re}\,b > 2|\mathrm{Im}\,a|]$

$\hspace{10cm}$ ET II 138(18)

3. $\displaystyle\int_0^\infty I_\mu(ax)\,K_\mu(ax)\,J_\nu(bx)\,dx = \frac{e^{\mu\pi i}\,\Gamma\left(\dfrac{\nu+2\mu+1}{2}\right)}{b\,\Gamma\left(\dfrac{\nu-2\mu+1}{2}\right)}\,P_{\frac12\nu-\frac12}^{-\mu}\left(\sqrt{1+\frac{4a^2}{b^2}}\right)Q_{\frac12\nu-\frac12}^{-\mu}\left(\sqrt{1+\frac{4a^2}{b^2}}\right)$

$$[\operatorname{Re} a>0, \quad b>0, \quad \operatorname{Re}\nu>-1, \quad \operatorname{Re}(\nu+2\mu)>-1] \quad \text{ET II 65(20)}$$

4. $\displaystyle\int_0^\infty J_\mu(ax)\,J_{-\mu}(ax)\,K_\nu(bx)\,dx = \frac{\pi}{2b}\sec\left(\frac{\nu\pi}{2}\right)P_{\frac12\nu-\frac12}^{\mu}\left(\sqrt{1+\frac{4a^2}{b^2}}\right)P_{\frac12\nu-\frac12}^{-\mu}\left(\sqrt{1+\frac{4a^2}{b^2}}\right)$

$$[|\operatorname{Re}\nu|<1, \quad \operatorname{Re} b>2|\operatorname{Im} a|]$$
$$\text{ET II 138(21)}$$

5. $\displaystyle\int_0^\infty [K_\mu(ax)]^2\,J_\nu(bx)\,dx = \frac{e^{2\mu\pi i}\,\Gamma\left(\dfrac{1+\nu+2\mu}{2}\right)}{b\,\Gamma\left(\dfrac{1+\nu-2\mu}{2}\right)}\left[Q_{\frac12\nu-\frac12}^{-\mu}\left(\sqrt{1+\frac{4a^2}{b^2}}\right)\right]^2$

$$\left[\operatorname{Re} a>0, \quad b>0, \quad \operatorname{Re}\left(\tfrac12\nu\pm\mu\right)>-\tfrac12\right] \quad \text{ET II 66(28)}$$

6. $\displaystyle\int_0^z J_\mu(x)\,J_\nu(z-x)\,dx = 2\sum_{k=0}^\infty (-1)^k\,J_{\mu+\nu+2k+1}(z)$ $[\operatorname{Re}\mu>-1, \quad \operatorname{Re}\nu>-1]$ WA 414(2)

7. $\displaystyle\int_0^z J_\mu(x)\,J_{-\mu}(z-x)\,dx = \sin z$ $[-1<\operatorname{Re}\mu<1]$ WA 415(4)

8. $\displaystyle\int_0^z J_\mu(x)\,J_{1-\mu}(z-x)\,dx = J_0(z)-\cos(z)$ $[-1<\operatorname{Re}\mu<2]$ WA 415(4)

6.514

1. $\displaystyle\int_0^\infty J_\nu\left(\frac{a}{x}\right)J_\nu(bx)\,dx = b^{-1}J_{2\nu}\left(2\sqrt{ab}\right)$ $\left[a>0, \quad b>0, \quad \operatorname{Re}\nu>-\tfrac12\right]$

$$\text{ET II 57(9)}$$

2. $\displaystyle\int_0^\infty J_\nu\left(\frac{a}{x}\right)Y_\nu(bx)\,dx = b^{-1}\left[Y_{2\nu}\left(2\sqrt{ab}\right)+\frac{2}{\pi}K_{2\nu}\left(\sqrt{2ab}\right)\right]$

$$\left[a>0, \quad b>0, \quad -\tfrac12<\operatorname{Re}\nu<\tfrac32\right]$$
$$\text{ET II 110(12)}$$

3. $\displaystyle\int_0^\infty J_\nu\left(\frac{a}{x}\right)K_\nu(bx)\,dx = b^{-1}e^{\frac12 i(\nu+1)\pi}K_{2\nu}\left[2e^{\frac14 i\pi}\sqrt{ab}\right]+b^{-1}e^{-\frac12 i(\nu+1)\pi}K_{2\nu}\left[2e^{-\frac14\pi i}\sqrt{ab}\right]$

$$\left[a>0, \quad \operatorname{Re} b>0, \quad |\operatorname{Re}\nu|<\tfrac52\right]$$
$$\text{ET II 141(31)}$$

4. $\displaystyle\int_0^\infty Y_\nu\left(\frac{a}{x}\right)J_\nu(bx)\,dx = -\frac{2b^{-1}}{\pi}\left[K_{2\nu}\left(2\sqrt{ab}\right)-\frac{\pi}{2}Y_{2\nu}\left(2\sqrt{ab}\right)\right]$

$$\left[a>0, \quad b>0, \quad |\operatorname{Re}\nu|<\tfrac12\right]$$
$$\text{ET II 62(37)a}$$

5. $\displaystyle\int_0^\infty Y_\nu\left(\frac{a}{x}\right)Y_\nu(bx)\,dx = -b^{-1}J_{2\nu}\left(2\sqrt{ab}\right)$ $\left[a>0, \quad b>0, \quad |\operatorname{Re}\nu|<\tfrac12\right]$

$$\text{ET II 110(14)}$$

6. $$\int_0^\infty Y_\nu\left(\frac{a}{x}\right) K_\nu(bx)\, dx = -b^{-1} e^{\frac{1}{2}\nu\pi i} K_{2\nu}\left(2e^{\frac{1}{4}\pi i}\sqrt{ab}\right) - b^{-1} e^{-\frac{1}{2}\nu\pi i} K_{2\nu}\left(2e^{-\frac{1}{4}\pi i}\sqrt{ab}\right)$$
$$\left[a>0, \quad \operatorname{Re} b>0, \quad |\operatorname{Re}\nu|<\tfrac{5}{2}\right]$$
ET II 143(37)

7. $$\int_0^\infty K_\nu\left(\frac{a}{x}\right) Y_\nu(bx)\, dx = -2b^{-1}\left[\sin\left(\frac{3\nu\pi}{2}\right)\operatorname{ker}_{2\nu}\left(2\sqrt{ab}\right) + \cos\left(\frac{3\nu\pi}{2}\right)\operatorname{kei}_{2\nu}\left(2\sqrt{ab}\right)\right]$$
$$\left[\operatorname{Re} a>0, \quad b>0, \quad |\operatorname{Re}\nu|<\tfrac{1}{2}\right]$$
ET II 113(28)

8. $$\int_0^\infty K_\nu\left(\frac{a}{x}\right) K_\nu(bx)\, dx = \pi b^{-1} K_{2\nu}\left(2\sqrt{ab}\right) \qquad \left[\operatorname{Re} a>0, \quad \operatorname{Re} b>0\right] \qquad \text{ET II 146(54)}$$

6.515

1. $$\int_0^\infty J_\mu\left(\frac{a}{x}\right) Y_\mu\left(\frac{a}{x}\right) K_0(bx)\, dx = -2b^{-1} J_{2\mu}\left(2\sqrt{ab}\right) K_{2\mu}\left(2\sqrt{ab}\right)$$
$$\left[a>0, \quad \operatorname{Re} b>0\right]$$
ET II 143(42)

2. $$\int_0^\infty \left[K_\mu\left(\frac{a}{x}\right)\right]^2 K_0(bx)\, dx = 2\pi b^{-1} K_{2\mu}\left(2e^{\frac{1}{4}\pi i}\sqrt{ab}\right) K_{2\mu}\left(2e^{-\frac{1}{4}\pi i}\sqrt{ab}\right)$$
$$\left[\operatorname{Re} a>0, \quad \operatorname{Re} b>0\right]$$
ET II 147(59)

3. $$\int_0^\infty H_\mu^{(1)}\left(\frac{a^2}{x}\right) H_\mu^{(2)}\left(\frac{a^2}{x}\right) J_0(bx)\, dx = 16\pi^{-2} b^{-1} \cos\mu\pi\, K_{2\mu}\left(2e^{\pi i/4} a\sqrt{b}\right) K_{2\mu}\left(2e^{-\pi i/4} a\sqrt{b}\right)$$
$$\left[|\arg a|<\frac{\pi}{4}, \quad b>0, \quad |\operatorname{Re}\mu|<\tfrac{1}{4}\right]$$
ET II 17(36)

6.516

1. $$\int_0^\infty J_{2\nu}\left(a\sqrt{x}\right) J_\nu(bx)\, dx = b^{-1} J_\nu\left(\frac{a^2}{4b}\right) \qquad \left[a>0, \quad b>0, \quad \operatorname{Re}\nu>-\tfrac{1}{2}\right]$$
ET II 58(16)

2. $$\int_0^\infty J_{2\nu}\left(a\sqrt{x}\right) Y_\nu(bx)\, dx = -b^{-1} \mathbf{H}_\nu\left(\frac{a^2}{4b}\right) \qquad \left[a>0, \quad b>0, \quad \operatorname{Re}\nu>-\tfrac{1}{2}\right]$$
ET II 111(18)

3. $$\int_0^\infty J_{2\nu}\left(a\sqrt{x}\right) K_\nu(bx)\, dx = \frac{\pi}{2} b^{-1}\left[I_\nu\left(\frac{a^2}{4b}\right) - \mathbf{L}_\nu\left(\frac{a^2}{4b}\right)\right]$$
$$\left[\operatorname{Re} b>0, \quad \operatorname{Re}\nu>-\tfrac{1}{2}\right] \qquad \text{ET II 144(45)}$$

4.[10] $$\int_0^\infty Y_{2\nu}\left(a\sqrt{x}\right) J_\nu(bx)\, dx = \frac{1}{b} J_\nu\left(\frac{a^2}{4b}\right) \cot(2\pi\nu) - \frac{1}{2b} J_{-\nu}\left(\frac{a^2}{4b}\right) \operatorname{cosec}(2\pi\nu)$$
$$- \frac{2^{3\nu-3} a^{2-2\nu} b^{\nu-2}}{\pi^{3/2}} \Gamma\left(\nu-\tfrac{1}{2}\right) {}_1F_2\left(1; \frac{3}{2}, \frac{3}{2}-\nu; \frac{a^4}{64b^2}\right)$$
$$\left[a>0, \quad b>0\right] \qquad \text{MC}$$

5. $\displaystyle\int_0^\infty Y_{2\nu}\left(a\sqrt{x}\right) Y_\nu(bx)\, dx$

$$= \frac{b^{-1}}{2}\left[\sec(\nu\pi)\, J_{-\nu}\left(\frac{a^2}{4b}\right) + \operatorname{cosec}(\nu\pi)\, \mathbf{H}_{-\nu}\left(\frac{a^2}{4b}\right) - 2\cot(2\nu\pi)\, \mathbf{H}_\nu\left(\frac{a^2}{4b}\right)\right]$$

$$\left[a > 0, \quad b > 0, \quad |\mathrm{Re}\,\nu| < \tfrac{1}{2}\right] \quad \text{ET II 111(19)}$$

6. $\displaystyle\int_0^\infty Y_{2\nu}\left(a\sqrt{x}\right) K_\nu(bx)\, dx = \frac{\pi b^{-1}}{2}\left[\operatorname{cosec}(2\nu\pi)\, \mathbf{L}_{-\nu}\left(\frac{a^2}{4b}\right) - \cot(2\nu\pi)\, \mathbf{L}_\nu\left(\frac{a^2}{4b}\right)\right.$

$$\left. - \tan(\nu\pi)\, I_\nu\left(\frac{a^2}{4b}\right) - \frac{\sec(\nu\pi)}{\pi}\, K_\nu\left(\frac{a^2}{4b}\right)\right]$$

$$\left[\mathrm{Re}\,b > 0, \quad |\mathrm{Re}\,\nu| < \tfrac{1}{2}\right] \quad \text{ET II 144(46)}$$

7. $\displaystyle\int_0^\infty K_{2\nu}\left(a\sqrt{x}\right) J_\nu(bx)\, dx = \frac{1}{4}\pi b^{-1}\sec(\nu\pi)\left[\mathbf{H}_{-\nu}\left(\frac{a^2}{4b}\right) - Y_{-\nu}\left(\frac{a^2}{4b}\right)\right]$

$$\left[\mathrm{Re}\,a > 0, \quad b > 0, \quad \mathrm{Re}\,\nu > -\tfrac{1}{2}\right]$$
$$\text{ET II 70(22)}$$

8. $\displaystyle\int_0^\infty K_{2\nu}\left(a\sqrt{x}\right) Y_\nu(bx)\, dx$

$$= -\frac{1}{4}\pi b^{-1}\left[\sec(\nu\pi)\, J_{-\nu}\left(\frac{a^2}{4b}\right) - \operatorname{cosec}(\nu\pi)\, \mathbf{H}_{-\nu}\left(\frac{a^2}{4b}\right) + 2\operatorname{cosec}(2\nu\pi)\, \mathbf{H}_\nu\left(\frac{a^2}{4b}\right)\right]$$

$$\left[\mathrm{Re}\,a > 0, \quad b > 0, \quad |\mathrm{Re}\,\nu| < \tfrac{1}{2}\right] \quad \text{ET II 114(34)}$$

9. $\displaystyle\int_0^\infty K_{2\nu}\left(a\sqrt{x}\right) K_\nu(bx)\, dx = \frac{\pi b^{-1}}{4\cos(\nu\pi)}\left\{K_\nu\left(\frac{a^2}{4b}\right) + \frac{\pi}{2\sin(\nu\pi)}\left[\mathbf{L}_{-\nu}\left(\frac{a^2}{4b}\right) - \mathbf{L}_\nu\left(\frac{a^2}{4b}\right)\right]\right\}$

$$\left[\mathrm{Re}\,b > 0, \quad |\mathrm{Re}\,\nu| < \tfrac{1}{2}\right] \quad \text{ET II 147(63)}$$

10. $\displaystyle\int_0^\infty I_{2\nu}\left(a\sqrt{x}\right) K_\nu(bx)\, dx = \frac{\pi b^{-1}}{2}\left[I_\nu\left(\frac{a^2}{4b}\right) + \mathbf{L}_\nu\left(\frac{a^2}{4b}\right)\right]$

$$\left[\mathrm{Re}\,b > 0, \quad \mathrm{Re}\,\nu > -\tfrac{1}{2}\right] \quad \text{ET II 147(60)}$$

6.517 $\displaystyle\int_0^z J_0\left(\sqrt{z^2 - x^2}\right) dx = \sin z$ MO 48

6.518 $\displaystyle\int_0^\infty K_{2\nu}(2z\sinh x)\, dx = \frac{\pi^2}{8\cos\nu\pi}\left(J_\nu^{\,2}(z) + N_\nu^{\,2}(z)\right)$

$$\left[\mathrm{Re}\,z > 0, \quad -\tfrac{1}{2} < \mathrm{Re}\,\nu < \tfrac{1}{2}\right] \quad \text{MO 45}$$

6.519

1. $\displaystyle\int_0^{\pi/2} J_{2\nu}(2z\cos x)\, dx = \frac{\pi}{2} J_\nu^{\,2}(z)$ $\left[\mathrm{Re}\,\nu > -\tfrac{1}{2}\right]$ WH

2. $\displaystyle\int_0^{\pi/2} J_{2\nu}(2z\sin x)\, dx = \frac{\pi}{2} J_\nu^{\,2}(z)$ $\left[\mathrm{Re}\,\nu > -\tfrac{1}{2}\right]$ WA 42(1)a

6.52 Bessel functions combined with x and x^2

6.521

1. $\displaystyle\int_0^1 x\, J_\nu(\alpha x)\, J_\nu(\beta x)\, dx = 0$ $[\alpha \neq \beta]$

$\displaystyle = \frac{1}{2}\{J_{\nu+1}(\alpha)\}^2$ $[\alpha = \beta]$

$[J_\nu(\alpha) = J_\nu(\beta) = 0, \quad \nu > -1]$ **WH**

2.[10] $\displaystyle\int_0^\infty x\, K_\nu(ax)\, J_\nu(bx)\, dx = \frac{b^\nu}{a^\nu\,(b^2 + a^2)}$ $[\operatorname{Re} a > 0, \quad b > 0, \quad \operatorname{Re}\nu > -1]$

ET II 63(2)

3. $\displaystyle\int_0^\infty x\, K_\nu(ax)\, K_\nu(bx)\, dx = \frac{\pi(ab)^{-\nu}\left(a^{2\nu} - b^{2\nu}\right)}{2\sin(\nu\pi)\,(a^2 - b^2)}$ $[|\operatorname{Re}\nu| < 1, \quad \operatorname{Re}(a + b) > 0]$

ET II 145(48)

4. $\displaystyle\int_0^a x\, J_\nu(\lambda x)\, K_\nu(\mu x)\, dx = \left(\mu^2 + \lambda^2\right)^{-1}\left[\left(\frac{\lambda}{\mu}\right)^\nu + \lambda a\, J_{\nu+1}(\lambda a)\, K_\nu(\mu a) - \mu a\, J_\nu(\lambda a)\, K_{\nu+1}(\mu a)\right]$

$[\operatorname{Re}\nu > -1]$ **ET II 367(26)**

6.522   **Notation:** $\ell_1 = \dfrac{1}{2}\left[\sqrt{(b+c)^2 + a^2} - \sqrt{(b-c)^2 + a^2}\right],\ \ell_2 = \dfrac{1}{2}\left[\sqrt{(b+c)^2 + a^2} + \sqrt{(b-c)^2 + a^2}\right]$

1.[8] $\displaystyle\int_0^\infty x[J_\mu(ax)]^2\, K_\nu(bx)\, dx = \Gamma\left(\mu + \tfrac{1}{2}\nu + 1\right)\Gamma\left(\mu - \tfrac{1}{2}\nu + 1\right) b^{-2}$

$$\times\left(1 + 4a^2b^{-2}\right)^{-\frac{1}{2}} P_{\frac{1}{2}\nu}^{-\mu}\left[\left(1 + 4a^2b^{-2}\right)^{\frac{1}{2}}\right] P_{-\frac{1}{2}\nu}^{-\mu}\left[\left(1 + 4a^2b^{-2}\right)^{\frac{1}{2}}\right]$$

$[\operatorname{Re} b > 2|\operatorname{Im} a|, \quad 2\operatorname{Re}\mu > |\operatorname{Re}\nu| - 2]$ **ET II 138(19)**

2. $\displaystyle\int_0^\infty x[K_\mu(ax)]^2\, J_\nu(bx)\, dx = \frac{2e^{2\mu\pi i}\,\Gamma\left(1 + \tfrac{1}{2}\nu + \mu\right)}{b\left(4a^2 + b^2\right)^{\frac{1}{2}}\Gamma\left(\tfrac{1}{2}\nu - \mu\right)}$

$$\times Q_{\frac{1}{2}\nu}^{-\mu}\left(\sqrt{1 + 4a^2b^{-2}}\right) Q_{\frac{1}{2}\nu-1}^{-\mu}\left(\sqrt{1 + 4a^2b^{-2}}\right)$$

$\left[b > 0, \quad \operatorname{Re} a > 0, \quad \operatorname{Re}\left(\tfrac{1}{2}\nu \pm \mu\right) > -1\right]$ **ET II 66(27)a**

3.[10] $\displaystyle\int_0^\infty x\, K_0(ax)\, J_\nu(bx)\, J_\nu(cx)\, dx = r_1^{-1} r_2^{-1} (r_2 r_1)^\nu (r_2 r_1)^{-\nu} = \frac{\ell_1^\nu}{\ell_2^\nu\,(\ell_2^2 - \ell_1^2)},$

$$\left[r_1 = \sqrt{a^2 + (b-c)^2}, \quad r_2 = \sqrt{a^2 + (b+c)^2}, \quad c > 0, \quad \operatorname{Re}\nu > -1, \quad \operatorname{Re} a > |\operatorname{Im} b|\right]$$

ET II 63(6)

4.[10] $\displaystyle\int_0^\infty x\, I_0(ax)\, K_0(bx)\, J_0(cx)\, dx = \left(a^4 + b^4 + c^4 - 2a^2b^2 + 2a^2c^2 + 2b^2c^2\right)^{-\frac{1}{2}}$

$[\operatorname{Re} b > \operatorname{Re} a, \quad c > 0]$ **ET II 16(27)**

alternatively, with a and c interchanged

$\displaystyle\int_0^\infty x\, I_0(cx)\, K_0(bx)\, J_0(ax)\, dx = \frac{1}{\ell_2^2 - \ell_1^2}$ $[\operatorname{Re} b > \operatorname{Re} c, \quad a > 0]$

5.^{10} $\displaystyle\int_0^\infty x\, J_0(ax)\, K_0(bx)\, J_0(cx)\, dx = \left(a^4 + b^4 + c^4 - 2a^2 c^2 + 2a^2 b^2 + 2b^2 c^2\right)^{-\frac{1}{2}}$

$$[\operatorname{Re} b > |\operatorname{Im} a|, \quad c > 0] \qquad \text{ET II 15(25)}$$

alternatively, with a and b interchanged

$\displaystyle\int_0^\infty x\, J_0(bx)\, K_0(ax)\, J_0(cx)\, dx = \frac{1}{\ell_2^2 - \ell_1^2}$ $[\operatorname{Re} a > |\operatorname{Im} b|, \quad c > 0]$

6. $\displaystyle\int_0^\infty x\, J_0(ax)\, Y_0(ax)\, J_0(bx)\, dx = 0$ $[0 < b < 2a]$

$$= -2\pi^{-1} b^{-1} \left[b^2 - 4a^2\right]^{-\frac{1}{2}} \quad [0 < 2a < b < \infty]$$

$$\text{ET II 15(21)}$$

7. $\displaystyle\int_0^\infty x\, J_\mu(ax)\, J_{\mu+1}(ax)\, K_\nu(bx)\, dx = \Gamma\left(\mu + \frac{3+\nu}{2}\right)\Gamma\left(\mu + \frac{3-\nu}{2}\right) b^{-2}\left(1 + 4a^2 b^{-2}\right)^{-\frac{1}{2}}$

$$\times P_{-\mu}^{\frac{1}{2}\nu - \frac{1}{2}}\left[\sqrt{1 + 4a^2 b^{-2}}\right] P_{-\mu-1}^{\frac{1}{2}\nu - \frac{1}{2}}\left[\sqrt{1 + 4a^2 b^{-2}}\right]$$

$$[\operatorname{Re} b > 2|\operatorname{Im} a|, \quad 2\operatorname{Re}\mu > |\operatorname{Re}\nu| - 3] \quad \text{ET II 138(20)}$$

8. $\displaystyle\int_0^\infty x\, K_{\mu-\frac{1}{2}}(ax)\, K_{\mu+\frac{1}{2}}(ax)\, J_\nu(bx)\, dx$

$$= -\frac{2e^{2\mu\pi i}\,\Gamma\left(\frac{1}{2}\nu + \mu + 1\right)}{b\,\Gamma\left(\frac{1}{2}\nu - \mu\right)\left(b^2 + 4a^2\right)^{\frac{1}{2}}}\, Q_{\frac{1}{2}\nu - \frac{1}{2}}^{-\mu+\frac{1}{2}}\left[\left(1 + 4a^2 b^{-2}\right)^{\frac{1}{2}}\right] Q_{\frac{1}{2}\nu - \frac{1}{2}}^{-\mu-\frac{1}{2}}\left[\left(1 + 4a^2 b^{-2}\right)^{\frac{1}{2}}\right]$$

$$\left[b > 0, \quad \operatorname{Re} a > 0, \quad \operatorname{Re}\nu > -1, \quad |\operatorname{Re}\mu| < 1 + \tfrac{1}{2}\operatorname{Re}\nu\right] \quad \text{ET II 67(29)a}$$

9.^{8} $\displaystyle\int_0^\infty x\, I_{\frac{1}{2}\nu}(ax)\, K_{\frac{1}{2}\nu}(ax)\, J_\nu(bx)\, dx = b^{-1}\left(b^2 + 4a^2\right)^{-\frac{1}{2}}$

$$[b > 0, \quad \operatorname{Re} a > 0, \quad \operatorname{Re}\nu > -1]$$

$$\text{ET II 65(16)}$$

10. $\displaystyle\int_0^\infty x\, J_{\frac{1}{2}\nu}(ax)\, Y_{\frac{1}{2}\nu}(ax)\, J_\nu(bx)\, dx$

$$= 0 \qquad\qquad\qquad [a > 0, \quad \operatorname{Re}\nu > -1, \quad 0 < b < 2a]$$

$$= -2\pi^{-1} b^{-1}\left(b^2 - 4a^2\right)^{-\frac{1}{2}} \quad [a > 0, \quad \operatorname{Re}\nu > -1, \quad 2a < b < \infty]$$

$$\text{ET II 55(48)}$$

11.^{8} $\displaystyle\int_0^\infty x\, J_{\frac{1}{2}(\nu+n)}(ax)\, J_{\frac{1}{2}(\nu-n)}(ax)\, J_\nu(bx)\, dx$

$$= 2\pi^{-1} b^{-1}\left(4a^2 - b^2\right)^{-\frac{1}{2}} T_n\left(\frac{b}{2a}\right) \quad [a > 0, \quad \operatorname{Re}\nu > -1, \quad 0 < b < 2a]$$

$$= 0 \qquad\qquad\qquad\qquad\qquad\qquad\quad [a > 0, \quad \operatorname{Re}\nu > -1, \quad 2a < b]$$

$$\text{ET II 52(32)}$$

12. $\displaystyle\int_0^\infty x\, I_{\frac{1}{2}(\nu-\mu)}(ax)\, K_{\frac{1}{2}(\nu+\mu)}(ax)\, J_\nu(bx)\, dx = 2^{-\mu} a^{-\mu} b^{-1}\left(b^2 + 4a^2\right)^{-\frac{1}{2}}\left[b + \left(b^2 + 4a^2\right)^{\frac{1}{2}}\right]^\mu$

$$[b > 0, \quad \operatorname{Re} a > 0, \quad \operatorname{Re}\nu > -1, \quad \operatorname{Re}(\nu - \mu) > -2] \quad \text{ET II 66(23)}$$

13.^{8} $\displaystyle\int_0^\infty x\,J_\mu\,(xa\sin\varphi)\,K_{\nu-\mu}\,(ax\cos\varphi\cos\psi)\,J_\nu\,(xa\sin\psi)\,dx = \frac{(\sin\varphi)^\mu(\sin\psi)^\nu(\cos\varphi)^{\nu-\mu}(\cos\psi)^{\mu-\nu}}{a^2\left(1-\sin^2\varphi\sin^2\psi\right)}$

$$\left[a>0,\quad 0<\varphi<\frac{\pi}{2},\quad 0<\psi<\frac{\pi}{2},\quad \operatorname{Re}\mu>-1,\quad \operatorname{Re}\nu>-1\right]\quad \text{ET II 64(10)}$$

14.^{8} $\displaystyle\int_0^\infty x\,J_\mu\,(xa\sin\varphi\cos\psi)\,J_{\nu-\mu}(ax)\,J_\nu\,(xa\cos\varphi\sin\psi)\,dx$

$$= -2\pi^{-1}a^{-2}\sin(\mu\pi)(\sin\varphi)^\mu(\sin\psi)^\nu(\cos\varphi)^{-\nu}(\cos\psi)^{-\mu}[\cos(\varphi+\psi)\cos(\varphi-\psi)]^{-1}$$

$$\left[a>0,\quad 0<\varphi<\frac{\pi}{2},\quad 0<\psi<\tfrac{1}{2}\pi,\quad \operatorname{Re}\nu>-1\right]\quad \text{ET II 54(39)}$$

15.^{10} $\displaystyle\int_0^\infty x^{\nu+1}\,J_\nu\,(bx)\,K_\nu(ax)\,J_\nu(cx)\,dx = \frac{2^{3\nu}(abc)^\nu\,\Gamma\left(\nu+\frac{1}{2}\right)}{\sqrt{\pi}(\ell_2^2-\ell_1^2)^{2\nu+1}}$

$$[\operatorname{Re}a>|\operatorname{Im}b|,\quad c>0]$$

16.^{10} $\displaystyle\int_0^\infty x^{\nu+1}\,I_\nu(cx)\,K_\nu(bx)\,J_\nu(ax)\,dx = \frac{2^{3\nu}(abc)^\nu\,\Gamma\left(\nu+\frac{1}{2}\right)}{\sqrt{\pi}(\ell_2^2-\ell_1^2)^{2\nu+1}}$

$$[\operatorname{Re}b>|\operatorname{Im}a|+|\operatorname{Im}c|]$$

17.^{10} $\displaystyle\int_0^\infty t^{\nu-\mu-\rho+1}\,J_\mu(ct)\,J_\nu(bt)\,K_\rho(at)\,dx$

$$= \frac{2^{1+\nu-\mu-\rho}}{c^\mu b^\nu a^\rho\,\Gamma\left(\mu-\nu+\rho\right)}\int_0^{\ell_1}\frac{x^{1+2\nu-2\rho}\left[(\ell_1^2-x^2)\,(\ell_2^2-x^2)\right]^{\mu-\nu+\rho-1}}{(b^2-x^2)^{\mu-\nu}}\,dt$$

$$\ell_1 = \frac{1}{2}\left[\sqrt{(b+c)^2+a^2}-\sqrt{(b-c)^2+a^2}\right],\quad \ell_2 = \frac{1}{2}\left[\sqrt{(b+c)^2+a^2}+\sqrt{(b-c)^2+a^2}\right]$$

$$[\operatorname{Re}a>|\operatorname{Im}b|,\quad c>0]$$

18.^{10} $\displaystyle\int_0^\infty t^{\mu-\nu+\rho+1}\,J_\mu(ct)\,J_\nu(bt)\,K_\rho(at)\,dx$

$$= \frac{2^{1+\mu-\nu+\rho}a^\rho}{c^\mu b^\nu\,\Gamma\left(\nu-\mu-\rho\right)}\int_0^{\ell_1}\frac{x^{1+2\mu+2\rho}\left[(\ell_1^2-x^2)\,(\ell_2^2-x^2)\right]^{\nu-\mu-\rho-1}}{(c^2-x^2)^{\nu-\mu}}\,dt$$

$$\ell_1 = \frac{1}{2}\left[\sqrt{(b+c)^2+a^2}-\sqrt{(b-c)^2+a^2}\right],\quad \ell_2 = \frac{1}{2}\left[\sqrt{(b+c)^2+a^2}+\sqrt{(b-c)^2+a^2}\right]$$

$$[\operatorname{Re}a>|\operatorname{Im}b|,\quad c>0]$$

6.523 $\displaystyle\int_0^\infty x\left[2\pi^{-1}K_0(ax)-Y_0(ax)\right]K_0(bx)\,dx = 2\pi^{-1}\left[\left(a^2+b^2\right)^{-1}+\left(b^2-a^2\right)^{-1}\right]\ln\frac{b}{a}$

$$[\operatorname{Re}b>|\operatorname{Im}a|,\quad \operatorname{Re}(a+b)>0]$$
$$\text{ET II 145(50)}$$

6.524

1. $\displaystyle\int_0^\infty x\,J_\nu^2(ax)\,J_\nu(bx)\,Y_\nu(bx)\,dx = 0$ $\qquad\qquad\left[0<a<b,\quad \operatorname{Re}\nu>-\tfrac{1}{2}\right]$

$$= -(2\pi ab)^{-1}\qquad\qquad\left[0<b<a,\quad \operatorname{Re}\nu>-\tfrac{1}{2}\right]$$

$$\text{ET II 352(14)}$$

2. $\displaystyle\int_0^\infty x[J_0(ax)\,K_0(bx)]^2\,dx = \frac{\pi}{8ab}-\frac{1}{4ab}\arcsin\left(\frac{b^2-a^2}{b^2+a^2}\right)$

$$[a>0,\quad b>0]\qquad\qquad \text{ET II 373(9)}$$

6.525   **Notation:** $\ell_1 = \dfrac{1}{2}\left[\sqrt{(b+c)^2 + a^2} - \sqrt{(b-c)^2 + a^2}\right]$, $\ell_2 = \dfrac{1}{2}\left[\sqrt{(b+c)^2 + a^2} + \sqrt{(b-c)^2 + a^2}\right]$

1.[10] $\displaystyle\int_0^\infty x^2 J_1(ax) K_0(bx) J_0(cx)\,dx = 2a\left(a^2 + b^2 - c^2\right)\left[\left(a^2 + b^2 + c^2\right)^2 - 4a^2 c^2\right]^{-\frac{3}{2}}$

$$[c > 0, \quad \operatorname{Re} b \geq |\operatorname{Im} a|, \quad \operatorname{Re} a > 0]$$
$$\text{ET II 15(26)}$$

alternatively, with a and b interchanged

$$\int_0^\infty x^2 J_1(bx) K_0(ax) J_0(cx)\,dx = \frac{2b\left(a^2 + b^2 - c^2\right)}{\left(\ell_2^2 - \ell_1^2\right)^3} \qquad [\operatorname{Re} a > |\operatorname{Im} b|, \quad \operatorname{Re} b > 0, \quad c > 0]$$

2.[10] $\displaystyle\int_0^\infty x^2 I_0(ax) K_1(bx) J_0(cx)\,dx = 2b\left(b^2 + c^2 - a^2\right)\left[\left(a^2 + b^2 + c^2\right)^2 - 4a^2 b^2\right]^{-\frac{3}{2}}$

$$[\operatorname{Re} b > |\operatorname{Re} a|, \quad c > 0] \qquad \text{ET II 16(28)}$$

3.[10] $\displaystyle\int_0^\infty x^2 I_0(cx) K_0(bx) J_0(ax)\,dx = \frac{2b\left(a^2 + b^2 - c^2\right)}{\left(\ell_2^2 - \ell_1^2\right)^3} \qquad [\operatorname{Re} a > |\operatorname{Im} b|, \quad c > 0]$

6.526

1. $\displaystyle\int_0^\infty x J_{\frac{1}{2}\nu}\left(ax^2\right) J_\nu(bx)\,dx = (2a)^{-1} J_{\frac{1}{2}\nu}\left(\frac{b^2}{4a}\right)$

$$[a > 0, \quad b > 0, \quad \operatorname{Re}\nu > -1] \quad \text{ET II 56(1)}$$

2. $\displaystyle\int_0^\infty x J_{\frac{1}{2}\nu}\left(ax^2\right) Y_\nu(bx)\,dx$

$$= (4a)^{-1}\left[Y_{\frac{1}{2}\nu}\left(\frac{b^2}{4a}\right) - \tan\left(\frac{\nu\pi}{2}\right) J_{\frac{1}{2}\nu}\left(\frac{b^2}{4a}\right) + \sec\left(\frac{\nu\pi}{2}\right)\mathbf{H}_{-\frac{1}{2}\nu}\left(\frac{b^2}{4a}\right)\right]$$
$$[a > 0, \quad b > 0, \quad \operatorname{Re}\nu > -1] \quad \text{ET II 109(9)}$$

3. $\displaystyle\int_0^\infty x J_{\frac{1}{2}\nu}\left(ax^2\right) K_\nu(bx)\,dx = \frac{\pi}{8a\cos\left(\frac{\nu\pi}{2}\right)}\left[\mathbf{H}_{-\frac{1}{2}\nu}\left(\frac{b^2}{4a}\right) - Y_{-\frac{1}{2}\nu}\left(\frac{b^2}{4a}\right)\right]$

$$[a > 0, \quad \operatorname{Re} b > 0, \quad \operatorname{Re}\nu > -1]$$
$$\text{ET II 140(27)}$$

4. $\displaystyle\int_0^\infty x Y_{\frac{1}{2}\nu}\left(ax^2\right) J_\nu(bx)\,dx = -(2a)^{-1}\mathbf{H}_{\frac{1}{2}\nu}\left(\frac{b^2}{4a}\right) \qquad [a > 0, \quad \operatorname{Re} b > 0, \quad \operatorname{Re}\nu > -1]$

$$\text{ET II 61(35)}$$

5. $\displaystyle\int_0^\infty x Y_{\frac{1}{2}\nu}\left(ax^2\right) K_\nu(bx)\,dx$

$$= \frac{\pi}{4a\sin(\nu\pi)}\left[\cos\left(\frac{\nu\pi}{2}\right)\mathbf{H}_{-\frac{1}{2}\nu}\left(\frac{b^2}{4a}\right) - \sin\left(\frac{\nu\pi}{2}\right) J_{-\frac{1}{2}\nu}\left(\frac{b^2}{4a}\right) - \mathbf{H}_{\frac{1}{2}\nu}\left(\frac{b^2}{4a}\right)\right]$$
$$[a > 0, \quad \operatorname{Re} b > 0, \quad |\operatorname{Re}\nu| < 1] \quad \text{ET II 141(28)}$$

6. $\displaystyle\int_0^\infty x K_{\frac{1}{2}\nu}\left(ax^2\right) J_\nu(bx)\,dx = \frac{\pi}{4a}\left[I_{\frac{1}{2}\nu}\left(\frac{b^2}{4a}\right) - \mathbf{L}_{\frac{1}{2}\nu}\left(\frac{b^2}{4a}\right)\right]$

$$[\operatorname{Re} a > 0, \quad b > 0, \quad \operatorname{Re}\nu > -1]$$
$$\text{ET II 68(9)}$$

7. $$\int_0^\infty x\, K_{\frac{1}{2}\nu}\left(ax^2\right) Y_\nu(bx)\, dx = \frac{\pi}{4a}\left[\operatorname{cosec}(\nu\pi)\, \mathbf{L}_{-\frac{1}{2}\nu}\left(\frac{b^2}{4a}\right) - \cot(\nu\pi)\, \mathbf{L}_{\frac{1}{2}\nu}\left(\frac{b^2}{4a}\right)\right.$$

$$\left. - \tan\left(\frac{\nu\pi}{2}\right) I_{\frac{1}{2}\nu}\left(\frac{b^2}{4a}\right) - \frac{1}{\pi}\sec\left(\frac{\nu\pi}{2}\right) K_{\frac{1}{2}\nu}\left(\frac{b^2}{4a}\right)\right]$$

$$[\operatorname{Re} a > 0, \quad b > 0, \quad |\operatorname{Re}\nu| < 1] \quad \text{ET II 112(25)}$$

8. $$\int_0^\infty x\, K_{\frac{1}{2}\nu}\left(ax^2\right) K_\nu(bx)\, dx$$

$$= \frac{\pi}{8a}\left\{\sec\left(\frac{\nu\pi}{2}\right) K_{\frac{1}{2}\nu}\left(\frac{b^2}{4a}\right) + \pi\operatorname{cosec}(\nu\pi)\left[\mathbf{L}_{-\frac{1}{2}\nu}\left(\frac{b^2}{4a}\right) - \mathbf{L}_{\frac{1}{2}\nu}\left(\frac{b^2}{4a}\right)\right]\right\}$$

$$[\operatorname{Re} a > 0, \quad |\operatorname{Re}\nu| < 1] \quad \text{ET II 146(52)}$$

6.527

1. $$\int_0^\infty x^2\, J_{2\nu}(2ax)\, J_{\nu-\frac{1}{2}}\left(x^2\right) dx = \frac{1}{2}a\, J_{\nu+\frac{1}{2}}\left(a^2\right) \qquad [a > 0, \quad \operatorname{Re}\nu > -\tfrac{1}{2}] \qquad \text{ET II 355(33)}$$

2. $$\int_0^\infty x^2\, J_{2\nu}(2ax)\, J_{\nu+\frac{1}{2}}\left(x^2\right) dx = \frac{1}{2}a\, J_{\nu-\frac{1}{2}}\left(a^2\right) \qquad [a > 0, \quad \operatorname{Re}\nu > -2] \qquad \text{ET II 355(35)}$$

3. $$\int_0^\infty x^2\, J_{2\nu}(2ax)\, Y_{\nu+\frac{1}{2}}\left(x^2\right) dx = -\frac{1}{2}a\, \mathbf{H}_{\nu-\frac{1}{2}}\left(a^2\right) \qquad [a > 0, \quad \operatorname{Re}\nu > -2] \qquad \text{ET II 355(36)}$$

6.528 $$\int_0^\infty x\, K_{\frac{1}{4}\nu}\left(\frac{x^2}{4}\right) I_{\frac{1}{4}\nu}\left(\frac{x^2}{4}\right) J_\nu(bx)\, dx = K_{\frac{1}{4}\nu}\left(\frac{x^2}{4}\right) I_{\frac{1}{4}\nu}\left(\frac{b^2}{4}\right)$$

$$[b > 0, \quad \nu > -1] \qquad \text{MO 183a}$$

6.529

1. $$\int_0^\infty x\, J_\nu\left(2\sqrt{ax}\right) K_\nu\left(2\sqrt{ax}\right) J_\nu(bx)\, dx = \frac{1}{2}b^{-2}e^{-\frac{2a}{b}} \qquad [\operatorname{Re} a > 0, \quad b > 0, \quad \operatorname{Re}\nu > -1]$$

$$\text{ET II 70(23)}$$

2. $$\int_0^a x\, J_\lambda(2a)\, I_\lambda(2x)\, J_\mu\left(2\sqrt{a^2 - x^2}\right) I_\mu\left(2\sqrt{a^2 - x^2}\right) dx$$

$$= \frac{a^{2\lambda+2\mu+2}}{2\,\Gamma(\lambda+1)\,\Gamma(\mu+1)\,\Gamma(\lambda+\mu+2)}$$

$$\times\, {}_1F_4\left(\frac{\lambda+\mu+1}{2}; \lambda+1, \mu+1, \lambda+\mu+1, \frac{\lambda+\mu+3}{2}; -a^4\right)$$

$$[\operatorname{Re}\lambda > -1, \quad \operatorname{Re}\mu > -1] \quad \text{ET II 376(31)}$$

6.53–6.54 Combinations of Bessel functions and rational functions

6.531

1.[10] $$\int_0^\infty \frac{Y_\nu(bx)}{x+a}\, dx$$

$$= -\pi J_\nu(ab)\cot(\pi\nu)\operatorname{cosec}(\pi\nu) - \pi J_{-\nu}(ab)\operatorname{cosec}^2(\pi\nu) + \frac{1}{\nu}\cot\frac{\pi\nu}{2}\, {}_1F_2\left(1; \frac{2-\nu}{2}, \frac{2+\nu}{2}; -\frac{a^2b^2}{4}\right)$$

$$+ \frac{ab}{\nu^2-1}\, {}_1F_2\left(1; \frac{3-\nu}{2}, \frac{3+\nu}{2}; -\frac{a^2b^2}{4}\right)\tan\frac{\pi\nu}{2}$$

$$[\operatorname{Re}\nu < 1, \quad \arg a \neq \pi, \quad b > 0] \qquad \text{MC}$$

2. $$\int_0^\infty \frac{Y_\nu(bx)}{x-a}\,dx = \pi\left\{\cot(\nu\pi)\left[Y_\nu(ab) + \mathbf{E}_\nu(ab)\right] + \mathbf{J}_\nu(ab) + 2[\cot(\nu\pi)]^2\left[\mathbf{J}_\nu(ab) - J_\nu(ab)\right]\right\}$$

$$[b>0, \quad a>0, \quad |\operatorname{Re}\nu| < 1]$$

ET II 98(9)

3. $$\int_0^\infty \frac{K_\nu(bx)}{x+a}\,dx = \frac{\pi^2}{2}[\operatorname{cosec}(\nu\pi)]^2\left[I_\nu(ab) + I_{-\nu}(ab) - e^{-\frac{1}{2}i\nu\pi}\,\mathbf{J}_\nu(iab) - e^{\frac{1}{2}i\nu\pi}\,\mathbf{J}_{-\nu}(iab)\right]$$

$$[\operatorname{Re}b>0, \quad |\arg a| < \pi, \quad |\operatorname{Re}\nu| < 1]$$

ET II 128(5)

6.532

1. $$\int_0^\infty \frac{J_\nu(x)}{x^2+a^2}\,dx = \frac{\pi\left[\mathbf{J}_\nu(a) - J_\nu(a)\right]}{a\sin(\nu\pi)}$$

$$[\operatorname{Re}a>0, \quad \operatorname{Re}\nu > -1]$$

ET II 340(2)

2. $$\int_0^\infty \frac{Y_\nu(x)}{x^2+a^2}\,dx = \frac{1}{\cos\dfrac{\nu\pi}{2}}\left[-\frac{\pi}{2a}\tan\left(\frac{\nu\pi}{2}\right)I_\nu(ab) - \frac{1}{a}K_\nu(ab)\right.$$

$$\left. +\frac{b\sin\left(\dfrac{\nu\pi}{2}\right)}{1-\nu^2}\,{}_1F_2\left(1; \frac{3-\nu}{2}, \frac{3+\nu}{2}; \frac{a^2b^2}{4}\right)\right]$$

$$[b>0, \quad \operatorname{Re}a>0, \quad |\operatorname{Re}\nu| < 1]\quad \text{ET II 99(13)}$$

3. $$\int_0^\infty \frac{Y_\nu(bx)}{x^2-a^2}\,dx = \frac{\pi}{2a}\left\{J_\nu(ab) + \tan\left(\frac{\nu\pi}{2}\right)\left\{\tan\left(\frac{\nu\pi}{2}\right)\left[\mathbf{J}_\nu(ab) - J_\nu(ab)\right] - \mathbf{E}_\nu(ab) - Y_\nu(ab)\right\}\right\}$$

$$[b>0, \quad a>0, \quad |\operatorname{Re}\nu| < 1]$$

ET II 101(21)

4. $$\int_0^\infty \frac{x\,J_0(ax)}{x^2+k^2}\,dx = K_0(ak)$$ $$[a>0, \quad \operatorname{Re}k>0]$$ WA 466(5)

5. $$\int_0^\infty \frac{Y_0(ax)}{x^2+k^2}\,dx = -\frac{K_0(ak)}{k}$$ $$[a>0, \quad \operatorname{Re}k>0]$$ WA 466(6)

6. $$\int_0^\infty \frac{J_0(ax)}{x^2+k^2}\,dx = \frac{\pi}{2k}\left[I_0(ak) - \mathbf{L}_0(ak)\right]$$ $$[a>0, \quad \operatorname{Re}k>0]$$ WA 467(7)

6.533

1. $$\int_0^z J_p(x)\,J_q(z-x)\,\frac{dx}{x} = \frac{J_{p+q}(z)}{p}$$ $$[\operatorname{Re}p>0, \quad \operatorname{Re}q>-1]$$ WA 415(3)

2. $$\int_0^z \frac{J_p(x)}{x}\,\frac{J_q(z-x)}{z-x}\,dx = \left(\frac{1}{p} + \frac{1}{q}\right)\frac{J_{p+q}(z)}{z}$$ $$[\operatorname{Re}p>0, \quad \operatorname{Re}q>0]$$ WA 415(5)

3. $$\int_0^\infty [J_0(ax) - 1]\,J_1(bx)\,\frac{dx}{x} = -\frac{b}{4}\left[1 + 2\ln\frac{a}{b}\right]$$ $$[0<b<a]$$

$$= -\frac{a^2}{4b}$$ $$[0<a<b]$$

ET II 21(28)a

4. $$\int_0^\infty [1 - J_0(ax)]\,J_0(bx)\,\frac{dx}{x} = 0$$ $$[0<a<b]$$

$$= \ln\frac{a}{b}$$ $$[0<b<a]$$

ET II 14(16)

6.534 $\displaystyle\int_0^\infty \frac{x^3 J_0(x)}{x^4 - a^4}\, dx = \frac{1}{2} K_0(a) - \frac{1}{4}\pi\, Y_0(a)$ $\qquad [a > 0]$ $\qquad$ ET II 340(5)

6.535 $\displaystyle\int_0^\infty \frac{x}{x^2 + a^2}\left[J_\nu(x)\right]^2 dx = I_\nu(a)\, K_\nu(a)$ $\qquad [\operatorname{Re} a > 0, \quad \operatorname{Re}\nu > -1]$ $\qquad$ ET II 342(26)

6.536 $\displaystyle\int_0^\infty \frac{x^3 J_0(bx)}{x^4 + a^4}\, dx = \ker(ab)$ $\qquad \left[b > 0, \quad |\arg a| < \tfrac{1}{4}\pi\right]$

ET II 8(9), MO 46a

6.537 $\displaystyle\int_0^\infty \frac{x^2 J_0(bx)}{x^4 + a^4}\, dx = -\frac{1}{a^2}\operatorname{kei}(ab)$ $\qquad \left[b > 0, \quad |\arg a| < \dfrac{\pi}{4}\right]$ $\qquad$ MO 46a

6.538

1. $\displaystyle\int_0^\infty J_1(ax)\, J_1(bx)\, \frac{dx}{x^2} = \frac{a+b}{\pi}\left[E\left(\frac{2i\sqrt{ab}}{|b-a|}\right) - K\left(\frac{2i\sqrt{ab}}{|b-a|}\right)\right]$

$\qquad\qquad [a > 0, \quad b > 0]$ $\qquad$ ET II 21(30)

2.[8] $\displaystyle\int_0^\infty x^{-1} J_{\nu+2n+1}(x)\, J_{\nu+2m+1}(x)\, dx = 0$ $\qquad [m \neq n \text{ with } m, n \text{ integers}, \nu > -1]$

$\qquad\qquad\qquad = (4n + 2\nu + 2)^{-1}$ $\qquad [m = n, \quad \nu > -1]$

EH II 64

6.539

1. $\displaystyle\int_a^b \frac{dx}{x[J_\nu(x)]^2} = \frac{\pi}{2}\left[\frac{Y_\nu(b)}{J_\nu(b)} - \frac{Y_\nu(a)}{J_\nu(a)}\right]$ $\qquad [J_\nu(x) \neq 0 \text{ for } x \in [a, b]]$ $\quad$ ET II 338(41)

2. $\displaystyle\int_a^b \frac{dx}{x[Y_\nu(x)]^2} = \frac{\pi}{2}\left[\frac{J_\nu(a)}{Y_\nu(a)} - \frac{J_\nu(b)}{Y_\nu(b)}\right]$ $\qquad [Y_\nu(x) \neq 0 \text{ for } x \in [a, b]]$

ET II 339(49)

3. $\displaystyle\int_a^b \frac{dx}{x\, J_\nu(x)\, Y_\nu(x)} = \frac{\pi}{2}\ln\left[\frac{J_\nu(a)\, Y_\nu(b)}{J_\nu(b)\, Y_\nu(a)}\right]$ $\qquad$ ET II 339(50)

6.541

1. $\displaystyle\int_0^\infty x\, J_\nu(ax)\, J_\nu(bx)\, \frac{dx}{x^2 + c^2} = I_\nu(bc)\, K_\nu(ac)$ $\qquad [0 < b < a, \quad \operatorname{Re} c > 0, \quad \operatorname{Re}\nu > -1]$

$\qquad\qquad\qquad\qquad\qquad\qquad\quad = I_\nu(ac)\, K_\nu(bc)$ $\qquad [0 < a < b, \quad \operatorname{Re} c > 0, \quad \operatorname{Re}\nu > -1]$

ET II 49(10)

2.[8] $\displaystyle\int_0^\infty x^{1-2n} J_\nu(ax)\, J_\nu(bx)\, \frac{dx}{x^2 + c^2}$

$\displaystyle = \left(-\frac{1}{c^2}\right)^n\left[I_\nu(bc)\, K_\nu(ac) - \frac{1}{2}\left(\frac{b}{a}\right)^\nu \frac{\pi}{\sin(\pi\nu)}\sum_{p=0}^{n-1}\frac{(a^2c^2/4)^p}{p!\,\Gamma(1-\nu+p)}\sum_{k=0}^{n-1-p}\frac{(b^2c^2/4)^k}{k!\,\Gamma(1-\nu+k)}\right]$

$\qquad\qquad\qquad\qquad\qquad\qquad\qquad\qquad\qquad\qquad [0 < b < a]$

$\displaystyle = \left(-\frac{1}{c^2}\right)^n\left[I_\nu(bc)\, K_\nu(ac) - \frac{1}{2\nu}\left(\frac{b}{a}\right)^\nu \sum_{p=0}^{n-1}\frac{(a^2c^2/4)^p}{p!\,(1-\nu)_p}\sum_{k=0}^{n-1-p}\frac{(b^2c^2/4)^k}{k!\,(1+\nu)_k}\right]$

$\qquad [n = 1, 2, \ldots, \quad \operatorname{Re}\nu > n - 1, \quad \operatorname{Re} c > 0, \quad 0 < b < a]$

$3.^8$ $\displaystyle\int_0^\infty \frac{x^{\alpha-1}}{(x^2+z^2)^\rho}\, J_\mu(cx)\, J_\nu(cx)\, dx = \frac{1}{2}\left(\frac{c}{2}\right)^{2\rho-\alpha}$

$$\times\, \Gamma\left[\begin{matrix}(\mu+\nu+\alpha)/2-\rho,\, 1+2\rho-\alpha\\(\mu-\nu-\alpha)/2+\rho+1,\,(\mu+\nu-\alpha)/2+\rho+1,\,(\nu-\mu-\alpha)/2+\rho+1\end{matrix}\right]$$

$$\times\, {}_3F_4\left(\frac{1-\alpha}{2}+\rho,\, 1-\frac{\alpha}{2}+\rho,\,\rho;\,\rho+1-\frac{\mu+\nu+\alpha}{2},\,\rho+1+\frac{\mu-\nu-\alpha}{2},\right.$$

$$\left.\rho+1+\frac{\mu+\nu-\alpha}{2},\,\rho+1+\frac{\nu-\mu-\alpha}{2};\,c^2z^2\right)+\frac{z^{\alpha-2\rho}}{2}\left(\frac{cz}{2}\right)^{\mu+\nu},$$

$$\Gamma\left[\begin{matrix}\rho-(\alpha+\mu+\nu)/2,\,(\alpha+\mu+\nu)/2\\\rho,\,\mu+1,\,\nu+1\end{matrix}\right]\,{}_3F_4\left(\frac{1+\mu+\nu}{2},\, 1+\frac{\mu+\nu}{2}\right.$$

$$\left.\frac{\alpha+\mu+\nu}{2};\, 1-\rho+\frac{\alpha+\mu+\nu}{2},\,\mu+1,\,\nu+1,\,\mu+\nu+1;\,c^2z^2\right)$$

$$\left[\Gamma\left[\begin{matrix}a_1,\ldots,a_p\\b_1,\ldots,b_q\end{matrix}\right]=\frac{\Gamma(a_1)\ldots\Gamma(a_p)}{\Gamma(b_1)\ldots\Gamma(b_q)},\quad c>0,\quad \operatorname{Re}z>0,\quad \operatorname{Re}(\alpha+\mu+\nu)>0;\quad \operatorname{Re}(\alpha-2\rho)>1\right]$$

6.542 $\displaystyle\int_0^\infty \frac{J_\nu(ax)\,Y_\nu(bx)-J_\nu(bx)\,Y_\nu(ax)}{x\left\{[J_\nu(bx)]^2+[Y_\nu(bx)]^2\right\}}\, dx = -\frac{\pi}{2}\left(\frac{b}{a}\right)^\nu$ $[0<b<a]$ ET II 352(16)

6.543 $\displaystyle\int_0^\infty J_\mu(bx)\left\{\cos\left[\frac{1}{2}(\nu-\mu)\pi\right]J_\nu(ax)-\sin\left[\frac{1}{2}(\nu-\mu)\pi\right]Y_\nu(ax)\right\}\frac{x\,dx}{x^2+r^2}=I_\mu(br)\,K_\nu(ar)$

$$[\operatorname{Re}r>0,\quad a\geq b>0,\quad \operatorname{Re}\mu>|\operatorname{Re}\nu|-2]$$

6.544

1. $\displaystyle\int_0^\infty J_\nu\left(\frac{a}{x}\right)Y_\nu\left(\frac{x}{b}\right)\frac{dx}{x^2}=-\frac{1}{a}\left[\frac{2}{\pi}K_{2\nu}\left(\frac{2\sqrt{a}}{\sqrt{b}}\right)-Y_{2\nu}\left(\frac{2\sqrt{a}}{\sqrt{b}}\right)\right]$

$$\left[a>0,\quad b>0,\quad |\operatorname{Re}\nu|<\tfrac{1}{2}\right]$$
EI II 357(47)

2. $\displaystyle\int_0^\infty J_\nu\left(\frac{a}{x}\right)J_\nu\left(\frac{x}{b}\right)\frac{dx}{x^2}=\frac{1}{a}J_{2\nu}\left(\frac{2\sqrt{a}}{\sqrt{b}}\right)$ $\left[a>0,\quad b>0,\quad \operatorname{Re}\nu>-\tfrac{1}{2}\right]$

ET II 57(10)

3. $\displaystyle\int_0^\infty J_\nu\left(\frac{a}{x}\right)K_\nu\left(\frac{x}{b}\right)\frac{dx}{x^2}=\frac{1}{a}e^{\frac{1}{2}i\nu\pi}K_{2\nu}\left(\frac{2\sqrt{a}}{\sqrt{b}}e^{\frac{1}{4}i\pi}\right)+\frac{1}{a}e^{-\frac{1}{2}i\nu\pi}K_{2\nu}\left(\frac{2\sqrt{a}}{\sqrt{b}}e^{-\frac{1}{4}i\pi}\right)$

$$\left[\operatorname{Re}b>0,\quad a>0,\quad |\operatorname{Re}\nu|<\tfrac{1}{2}\right]$$
ET II 142(32)

4. $\displaystyle\int_0^\infty Y_\nu\left(\frac{a}{x}\right)J_\nu\left(\frac{x}{b}\right)\frac{dx}{x^2}=\frac{2}{a\pi}\left[K_{2\nu}\left(\frac{2\sqrt{a}}{\sqrt{b}}\right)+\frac{\pi}{2}Y_{2\nu}\left(\frac{2\sqrt{a}}{\sqrt{b}}\right)\right]$

$$\left[a>0,\quad b>0,\quad |\operatorname{Re}\nu|<\tfrac{1}{2}\right]$$
ET II 62(38)

5. $\displaystyle\int_0^\infty Y_\nu\left(\frac{a}{x}\right)K_\nu\left(\frac{x}{b}\right)\frac{dx}{x^2}=\frac{4}{a}\left[e^{\frac{1}{2}i(\nu+1)\pi}K_{2\nu}\left(\frac{2\sqrt{a}}{\sqrt{b}}e^{\frac{1}{4}i\pi}\right)+e^{-\frac{1}{2}i(\nu+1)\pi}K_{2\nu}\left(\frac{2\sqrt{a}}{\sqrt{b}}e^{-\frac{1}{4}i\pi}\right)\right]$

$$\left[\operatorname{Re}b>0,\quad a>0,\quad |\operatorname{Re}\nu|<\tfrac{1}{2}\right]$$
ET II 143(38)

6. $\int_0^\infty K_\nu \left(\dfrac{a}{x}\right) J_\nu \left(\dfrac{x}{b}\right) \dfrac{dx}{x^2} = \dfrac{i}{a} \left[e^{\frac{1}{2}\nu\pi i} K_{2\nu} \left(e^{\frac{1}{4}\pi i} \dfrac{2\sqrt{a}}{\sqrt{b}} \right) - e^{-\frac{1}{2}\nu\pi i} K_{2\nu} \left(e^{-\frac{1}{4}\pi i} \dfrac{2\sqrt{a}}{\sqrt{b}} \right) \right]$

$$\left[\operatorname{Re} a > 0, \quad b > 0, \quad |\operatorname{Re}\nu| < \tfrac{5}{2} \right]$$
ET II 70(19)

7. $\int_0^\infty K_\nu \left(\dfrac{a}{x}\right) Y_\nu \left(\dfrac{x}{b}\right) \dfrac{dx}{x^2} = \dfrac{2}{a} \left[\sin \left(\dfrac{3}{2}\pi\nu\right) \operatorname{kei}_{2\nu} \left(\dfrac{2\sqrt{a}}{\sqrt{b}}\right) - \cos \left(\dfrac{3}{2}\pi\nu\right) \operatorname{ker}_{2\nu} \left(\dfrac{2\sqrt{a}}{\sqrt{b}}\right) \right]$

$$\left[\operatorname{Re} a > 0, \quad b > 0, \quad |\operatorname{Re}\nu| < \tfrac{5}{2} \right]$$
ET II 113(29)

8. $\int_0^\infty K_\nu \left(\dfrac{a}{x}\right) K_\nu \left(\dfrac{x}{b}\right) \dfrac{dx}{x^2} = \dfrac{\pi}{a} K_{2\nu} \left(\dfrac{2\sqrt{a}}{\sqrt{b}}\right)$ $\left[\operatorname{Re} a > 0, \quad \operatorname{Re} b > 0 \right]$ ET II 146(55)

6.55 Combinations of Bessel functions and algebraic functions

6.551[10]

1. $\int_0^1 x^{1/2} J_\nu(xy)\, dx = \sqrt{2} y^{-3/2} \dfrac{\Gamma\left(\frac{3}{4} + \frac{1}{2}\nu\right)}{\Gamma\left(\frac{1}{4} + \frac{1}{2}\nu\right)}$

$$+ y^{-1/2} \left[\left(\nu - \tfrac{1}{2}\right) J_\nu(y)\, S_{-1/2,\nu-1}(y) - J_{\nu-1}(y)\, S_{1/2,\nu}(y) \right]$$
$$\left[y > 0, \quad \operatorname{Re}\nu > -\tfrac{3}{2} \right]$$
ET II 21(1)

2. $\int_1^\infty x^{1/2} J_\nu(xy)\, dx = y^{-1/2} \left[J_{\nu-1}(y)\, S_{1/2,\nu}(y) + \left(\tfrac{1}{2} - \nu\right) J_\nu(y)\, S_{-1/2,\nu-1}(y) \right]$

$$\left[y > 0 \right]$$
ET II 22(2)

6.552

1. $\int_0^\infty J_\nu(xy) \dfrac{dx}{(x^2 + a^2)^{1/2}} = I_{\nu/2}\left(\tfrac{1}{2}ay\right) K_{\nu/2}\left(\tfrac{1}{2}ay\right)$ $\left[\operatorname{Re} a > 0, \quad y > 0, \quad \operatorname{Re}\nu > -1 \right]$

ET II 23(11), WA 477(3), MO 44

2. $\int_0^\infty Y_\nu(xy) \dfrac{dx}{(x^2 + a^2)^{1/2}} = -\dfrac{1}{\pi} \sec\left(\tfrac{1}{2}\nu\pi\right) K_{\nu/2}\left(\tfrac{1}{2}ay\right) \left[K_{\nu/2}\left(\tfrac{1}{2}ay\right) + \pi \sin\left(\tfrac{1}{2}\nu\pi\right) I_{\nu/2}\left(\tfrac{1}{2}ay\right) \right]$

$$\left[y > 0, \quad \operatorname{Re} a > 0, \quad |\operatorname{Re}\nu| < 1 \right]$$
ET II 100(18)

3. $\int_0^\infty K_\nu(xy) \dfrac{dx}{(x^2 + a^2)^{1/2}} = \dfrac{\pi^2}{8} \sec\left(\tfrac{1}{2}\nu\pi\right) \left\{ \left[J_{\nu/2}\left(\tfrac{1}{2}ay\right) \right]^2 + \left[Y_{\nu/2}\left(\tfrac{1}{2}ay\right) \right]^2 \right\}$

$$\left[\operatorname{Re} a > 0, \quad \operatorname{Re} y > 0, \quad |\operatorname{Re}\nu| < 1 \right]$$
ET II 128(6)

4. $\int_0^1 J_\nu(xy) \dfrac{dx}{(1 - x^2)^{1/2}} = \dfrac{\pi}{2} \left[J_{\nu/2}\left(\tfrac{1}{2}y\right) \right]^2$ $\left[y > 0, \quad \operatorname{Re}\nu > -1 \right]$ ET II 24(22)a

5. $\int_0^1 Y_0(xy) \dfrac{dx}{(1 - x^2)^{1/2}} = \dfrac{\pi}{2} J_0\left(\tfrac{1}{2}y\right) Y_0\left(\tfrac{1}{2}y\right)$ $\left[y > 0 \right]$ ET II 102(26)a

6. $\int_1^\infty J_\nu(xy) \dfrac{dx}{(x^2 - 1)^{1/2}} = -\dfrac{\pi}{2} J_{\nu/2}\left(\tfrac{1}{2}y\right) Y_{\nu/2}\left(\tfrac{1}{2}y\right)$ $\left[y > 0 \right]$ ET II 24(23)a

7. $\displaystyle\int_1^\infty Y_\nu(xy)\frac{dx}{(x^2-1)^{1/2}} = \frac{\pi}{4}\left\{\left[J_{\nu/2}\left(\tfrac{1}{2}y\right)\right]^2 - \left[Y_{\nu/2}\left(\tfrac{1}{2}y\right)\right]^2\right\}$

$$[y > 0]$$ ET II 102(27)

6.553 $\displaystyle\int_0^\infty x^{-1/2}I_\nu(x)K_\nu(x)K_\mu(2x)\,dx = \frac{\Gamma\left(\tfrac{1}{4}+\tfrac{1}{2}\mu\right)\Gamma\left(\tfrac{1}{4}-\tfrac{1}{2}\mu\right)\Gamma\left(\tfrac{1}{4}+\nu+\tfrac{1}{2}\mu\right)\Gamma\left(\tfrac{1}{4}+\nu-\tfrac{1}{2}\mu\right)}{4\Gamma\left(\tfrac{3}{4}+\nu+\tfrac{1}{2}\mu\right)\Gamma\left(\tfrac{3}{4}+\nu-\tfrac{1}{2}\mu\right)}$

$$\left[|\operatorname{Re}\mu| < \tfrac{1}{2}, \quad 2\operatorname{Re}\nu > |\operatorname{Re}\mu| - \tfrac{1}{2}\right]$$
ET II 372(2)

6.554

1. $\displaystyle\int_0^\infty x\,J_0(xy)\frac{dx}{(a^2+x^2)^{1/2}} = y^{-1}e^{-ay}$ $[y > 0, \quad \operatorname{Re}a > 0]$ ET II 7(4)

2. $\displaystyle\int_0^1 x\,J_0(xy)\frac{dx}{(1-x^2)^{1/2}} = y^{-1}\sin y$ $[y > 0]$ ET II 7(5)a

3. $\displaystyle\int_1^\infty x\,J_0(xy)\frac{dx}{(x^2-1)^{1/2}} = y^{-1}\cos y$ $[y > 0]$ ET II 7(6)a

4. $\displaystyle\int_0^\infty x\,J_0(xy)\frac{dx}{(x^2+a^2)^{3/2}} = a^{-1}e^{-ay}$ $[y > 0, \quad \operatorname{Re}a > 0]$ ET II 7(7)a

5.[8] $\displaystyle\int_0^\infty \frac{x^{\nu+1}J_\nu(ax)}{(x^4+4k^4)^{\nu+1/2}}\,dx = \frac{\left(\tfrac{1}{2}a\right)^\nu\sqrt{2\pi}}{(2k)^{2\nu}\Gamma\left(\nu+\tfrac{1}{2}\right)}J_\nu(ak)K_\nu(ak)$

$$\left[a > 0, \quad k > 0, \quad \operatorname{Re}\nu - \tfrac{1}{2}\right] \quad \text{WA 473(1)}$$

6.555 $\displaystyle\int_0^\infty x^{1/2}J_{2\nu-1}\left(ax^{1/2}\right)Y_\nu(xy)\,dx = -\frac{a}{2y^2}\mathbf{H}_{\nu-1}\left(\frac{a^2}{4y}\right)$

$$\left[a > 0, \quad y > 0, \quad \operatorname{Re}\nu > -\tfrac{1}{2}\right]$$
ET II 111(17)

6.556 $\displaystyle\int_0^\infty J_\nu\left[a\left(x^2+1\right)^{1/2}\right]\frac{dx}{\sqrt{x^2+1}} = -\frac{\pi}{2}J_{\nu/2}\left(\frac{a}{2}\right)Y_{\nu/2}\left(\frac{a}{2}\right)$ $[\operatorname{Re}\nu > -1, \quad a > 0]$ MO 46

6.56–6.58 Combinations of Bessel functions and powers

6.561

1. $\displaystyle\int_0^1 x^\nu J_\nu(ax)\,dx = 2^{\nu-1}a^{-\nu}\pi^{\frac{1}{2}}\Gamma\left(\nu+\tfrac{1}{2}\right)\left[J_\nu(a)\mathbf{H}_{\nu-1}(a) - \mathbf{H}_\nu(a)J_{\nu-1}(a)\right]$

$$\left[\operatorname{Re}\nu > -\tfrac{1}{2}\right]$$
ET II 333(2)a

2. $\displaystyle\int_0^1 x^\nu Y_\nu(ax)\,dx = 2^{\nu-1}a^{-\nu}\pi^{\frac{1}{2}}\Gamma\left(\nu+\tfrac{1}{2}\right)\left[Y_\nu(a)\mathbf{H}_{\nu-1}(a) - \mathbf{H}_\nu(a)Y_{\nu-1}(a)\right]$

$$\left[\operatorname{Re}\nu > -\tfrac{1}{2}\right]$$
ET II 338(43)a

3. $\displaystyle\int_0^1 x^\nu I_\nu(ax)\,dx = 2^{\nu-1}a^{-\nu}\pi^{\frac{1}{2}}\Gamma\left(\nu+\tfrac{1}{2}\right)\left[I_\nu(a)\mathbf{L}_{\nu-1}(a) - \mathbf{L}_\nu(a)I_{\nu-1}(a)\right]$

$$\left[\operatorname{Re}\nu > -\tfrac{1}{2}\right]$$
ET II 364(2)a

4. $$\int_0^1 x^\nu K_\nu(ax)\,dx = 2^{\nu-1}a^{-\nu}\pi^{\frac{1}{2}}\,\Gamma\left(\nu+\tfrac{1}{2}\right)\left[K_\nu(a)\,\mathbf{L}_{\nu-1}(a) + \mathbf{L}_\nu(a)\,K_{\nu-1}(a)\right]$$

$$\left[\operatorname{Re}\nu > -\tfrac{1}{2}\right] \qquad \text{ET II 367(21)a}$$

5. $$\int_0^1 x^{\nu+1} J_\nu(ax)\,dx = a^{-1} J_{\nu+1}(a) \qquad [\operatorname{Re}\nu > -1] \qquad \text{ET II 333(3)a}$$

6. $$\int_0^1 x^{\nu+1} Y_\nu(ax)\,dx = a^{-1} Y_{\nu+1}(a) + 2^{\nu+1}a^{-\nu-2}\pi^{-1}\Gamma(\nu+1)$$

$$[\operatorname{Re}\nu > -1] \qquad \text{ET II 339(44)a}$$

7. $$\int_0^1 x^{\nu+1} I_\nu(ax)\,dx = a^{-1} I_{\nu+1}(a) \qquad [\operatorname{Re}\nu > -1] \qquad \text{ET II 365(3)a}$$

8. $$\int_0^1 x^{\nu+1} K_\nu(ax)\,dx = 2^\nu a^{-\nu-2}\Gamma(\nu+1) - a^{-1} K_{\nu+1}(a)$$

$$[\operatorname{Re}\nu > -1] \qquad \text{ET II 367(22)a}$$

9. $$\int_0^1 x^{1-\nu} J_\nu(ax)\,dx = \frac{a^{\nu-2}}{2^{\nu-1}\Gamma(\nu)} - a^{-1} J_{\nu-1}(a) \qquad \text{ET II 333(4)a}$$

10. $$\int_0^1 x^{1-\nu} Y_\nu(ax)\,dx = \frac{a^{\nu-2}\cot(\nu\pi)}{2^{\nu-1}\Gamma(\nu)} - a^{-1} Y_{\nu-1}(a) \qquad [\operatorname{Re}\nu < 1] \qquad \text{ET II 339(45)a}$$

11. $$\int_0^1 x^{1-\nu} I_\nu(ax)\,dx = a^{-1} I_{\nu-1}(a) - \frac{a^{\nu-2}}{2^{\nu-1}\Gamma(\nu)} \qquad \text{ET II 365(4)a}$$

12. $$\int_0^1 x^{1-\nu} K_\nu(ax)\,dx = 2^{-\nu}a^{\nu-2}\Gamma(1-\nu) - a^{-1} K_{\nu-1}(a)$$

$$[\operatorname{Re}\nu < 1] \qquad \text{ET II 367(23)a}$$

13.[7] $$\int_0^1 x^\mu J_\nu(ax)\,dx = \frac{2^\mu\,\Gamma\left(\frac{\nu+\mu+1}{2}\right)}{a^{\mu+1}\,\Gamma\left(\frac{\nu-\mu+1}{2}\right)} + a^{-\mu}\left\{(\mu+\nu-1)\,J_\nu(a)\,S_{\mu-1,\nu-1}(a) - J_{\nu-1}(a)\,S_{\mu,\nu}(a)\right\}$$

$$[a > 0,\quad \operatorname{Re}(\mu+\nu) > -1] \qquad \text{ET II 22(8)a}$$

14. $$\int_0^\infty x^\mu J_\nu(ax)\,dx = 2^\mu a^{-\mu-1}\frac{\Gamma\left(\frac{1}{2}+\frac{1}{2}\nu+\frac{1}{2}\mu\right)}{\Gamma\left(\frac{1}{2}+\frac{1}{2}\nu-\frac{1}{2}\mu\right)} \qquad \left[-\operatorname{Re}\nu-1 < \operatorname{Re}\mu < \tfrac{1}{2},\quad a > 0\right]$$

$$\text{EH II 49(19)}$$

15. $$\int_0^\infty x^\mu Y_\nu(ax)\,dx = 2^\mu \cot\left[\tfrac{1}{2}(\nu+1-\mu)\pi\right] a^{-\mu-1}\frac{\Gamma\left(\frac{1}{2}+\frac{1}{2}\nu+\frac{1}{2}\mu\right)}{\Gamma\left(\frac{1}{2}+\frac{1}{2}\nu-\frac{1}{2}\mu\right)}$$

$$\left[|\operatorname{Re}\nu|-1 < \mu < \tfrac{1}{2},\quad a > 0\right]$$

$$\text{ET II 97(3)a}$$

16. $$\int_0^\infty x^\mu K_\nu(ax)\,dx = 2^{\mu-1}a^{-\mu-1}\,\Gamma\left(\frac{1+\mu+\nu}{2}\right)\Gamma\left(\frac{1+\mu-\nu}{2}\right)$$

$$[\operatorname{Re}(\mu+1\pm\nu) > 0,\quad \operatorname{Re}a > 0]$$

$$\text{EH II 51(27)}$$

17. $$\int_0^\infty \frac{J_\nu(ax)}{x^{\nu-q}}\,dx = \frac{\Gamma\left(\frac{1}{2}q+\frac{1}{2}\right)}{2^{\nu-q}a^{q-\nu+1}\,\Gamma\left(\nu-\frac{1}{2}q+\frac{1}{2}\right)} \qquad \left[-1 < \mathrm{Re}\,q < \mathrm{Re}\,\nu - \frac{1}{2}\right]$$

WA 428(1), KU 144(5)

18. $$\int_0^\infty \frac{Y_\nu(x)}{x^{\nu-\mu}}\,dx = \frac{\Gamma\left(\frac{1}{2}+\frac{1}{2}\mu\right)\Gamma\left(\frac{1}{2}+\frac{1}{2}\mu-\nu\right)\sin\left(\frac{1}{2}\mu-\nu\right)\pi}{2^{\nu-\mu}\pi}$$

$$\left[|\mathrm{Re}\,\nu| < \mathrm{Re}(1+\mu-\nu) < \frac{3}{2}\right]$$

WA 430(5)

19.* $$\int_0^1 x^{2m+n+1/2}\,K_{n+1/2}(\alpha x)\,dx = \sqrt{\frac{\pi}{2}}\sum_{k=0}^n \frac{(n+k)!}{k!(n-k)!}\,\frac{\gamma(2m+n-k+1,\alpha)}{\alpha^{2m+n+3/2}2^k}$$

STR

6.562

1. $$\int_0^\infty x^\mu\,Y_\nu(bx)\frac{dx}{x+a} = (2a)^\mu\pi^{-1}\left\{\sin\left[\frac{1}{2}\pi(\mu-\nu)\right]\Gamma\left[\frac{1}{2}(\mu+\nu+1)\right]\Gamma\left[\frac{1}{2}(1+\mu-\nu)\right]S_{-\mu,\nu}(ab)\right.$$

$$\left.-2\cos\left[\frac{1}{2}\pi(\mu-\nu)\right]\Gamma\left(1+\frac{1}{2}\mu+\frac{1}{2}\nu\right)\Gamma\left(1+\frac{1}{2}\mu-\frac{1}{2}\nu\right)S_{-\mu-1,\nu}(ab)\right\}$$

$$\left[b>0,\quad |\arg a| < \pi,\quad \mathrm{Re}\,(\mu\pm\nu) > -1,\quad \mathrm{Re}\,\mu < \frac{3}{2}\right] \quad \text{ET II 98(8)}$$

2. $$\int_0^\infty \frac{x^\nu\,J_\nu(ax)}{x+k}\,dx = \frac{\pi k^\nu}{2\cos\nu\pi}\left[\mathbf{H}_{-\nu}(ak)-Y_{-\nu}(ak)\right] \qquad \left[-\frac{1}{2} < \mathrm{Re}\,\nu < \frac{3}{2},\quad a>0,\quad |\arg k| < \pi\right]$$

WA 479(7)

3. $$\int_0^\infty x^\mu\,K_\nu(bx)\frac{dx}{x+a}$$

$$= 2^{\mu-2}\,\Gamma\left[\frac{1}{2}(\mu+\nu)\right]\Gamma\left[\frac{1}{2}(\mu-\nu)\right]b^{-\mu}\,{}_1F_2\left(1;1-\frac{\mu+\nu}{2},1-\frac{\mu-\nu}{2};\frac{a^2b^2}{4}\right)$$

$$-2^{\mu-3}\,\Gamma\left[\frac{1}{2}(\mu-\nu-1)\right]\Gamma\left[\frac{1}{2}(\mu+\nu-1)\right]ab^{1-\mu}\,{}_1F_2\left(1;\frac{3-\mu-\nu}{2},\frac{3-\mu+\nu}{2};\frac{a^2b^2}{4}\right)$$

$$-\pi a^\mu\,\mathrm{cosec}[\pi(\mu-\nu)]\left\{K_\nu(ab)+\pi\cos(\mu\pi)\,\mathrm{cosec}[\pi(\nu+\mu)]\,I_\nu(ab)\right\}$$

$$\left[\mathrm{Re}\,b>0,\quad |\arg a|<\pi,\quad \mathrm{Re}\,\mu > |\mathrm{Re}\,\nu|-1\right] \quad \text{ET II 127(4)}$$

6.563 $$\int_0^\infty x^{\varrho-1}\,J_\nu(bx)\frac{dx}{(x+a)^{1+\mu}} = \frac{\pi a^{\varrho-\mu-1}}{\sin[(\varrho+\nu-\mu)\pi]\,\Gamma(\mu+1)}$$

$$\times\left\{\sum_{m=0}^\infty \frac{(-1)^m\left(\frac{1}{2}ab\right)^{\nu+2m}\Gamma(\varrho+\nu+2m)}{m!\,\Gamma(\nu+m+1)\,\Gamma(\varrho+\nu-\mu+2m)}\right.$$

$$\left.-\sum_{m=0}^\infty \frac{\left(\frac{1}{2}ab\right)^{\mu+1-\varrho+m}\Gamma(\mu+m+1)}{m!\,\Gamma\left[\frac{1}{2}(\mu+\nu-\varrho+m+3)\right]}\,\frac{\sin\left[\frac{1}{2}(\varrho+\nu-\mu-m)\pi\right]}{\Gamma\left[\frac{1}{2}(\mu-\nu-\varrho+m+3)\right]}\right\}$$

$$\left[b>0,\quad |\arg a|<\pi,\quad \mathrm{Re}(\varrho+\nu)>0,\quad \mathrm{Re}(\varrho-\mu)<\frac{5}{2}\right] \quad \text{ET II 23(10), WA 479}$$

6.564

1. $$\int_0^\infty x^{\nu+1}\,J_\nu(bx)\frac{dx}{\sqrt{x^2+a^2}} = \sqrt{\frac{2}{\pi b}}\,a^{\nu+\frac{1}{2}}\,K_{\nu+\frac{1}{2}}(ab) \qquad \left[\mathrm{Re}\,a>0,\quad b>0,\quad -1<\mathrm{Re}\,\nu<\frac{1}{2}\right]$$

ET II 23(15)

2. $$\int_0^\infty x^{1-\nu} J_\nu(bx) \frac{dx}{\sqrt{x^2 + a^2}} = \sqrt{\frac{\pi}{2b}} a^{\frac{1}{2}-\nu} \left[I_{\nu-\frac{1}{2}}(ab) - \mathbf{L}_{\nu-\frac{1}{2}}(ab) \right]$$

$$\left[\operatorname{Re} a > 0, \quad b > 0, \quad \operatorname{Re} \nu > -\tfrac{1}{2} \right]$$
ET II 23(16)

6.565

1. $$\int_0^\infty x^{-\nu} \left(x^2 + a^2\right)^{-\nu-\frac{1}{2}} J_\nu(bx)\, dx = 2^\nu a^{-2\nu} b^\nu \frac{\Gamma(\nu+1)}{\Gamma(2\nu+1)} I_\nu\left(\frac{ab}{2}\right) K_\nu\left(\frac{ab}{2}\right)$$

$$\left[\operatorname{Re} a > 0, \quad b > 0, \quad \operatorname{Re} \nu > -\tfrac{1}{2} \right]$$
WA 477(4), ET II 23(17)

2. $$\int_0^\infty x^{\nu+1} \left(x^2 + a^2\right)^{-\nu-\frac{1}{2}} J_\nu(bx)\, dx = \frac{\sqrt{\pi}\, b^{\nu-1}}{2^\nu e^{ab}\, \Gamma\left(\nu+\frac{1}{2}\right)}$$

$$\left[\operatorname{Re} a > 0, \quad b > 0, \quad \operatorname{Re} \nu > -\tfrac{1}{2} \right]$$
ET II 24(18)

3. $$\int_0^\infty x^{\nu+1} \left(x^2 + a^2\right)^{-\nu-\frac{3}{2}} J_\nu(bx)\, dx = \frac{b^\nu \sqrt{\pi}}{2^{\nu+1} a e^{ab}\, \Gamma\left(\nu+\frac{3}{2}\right)}$$

$$\left[\operatorname{Re} a > 0, \quad b > 0, \quad \operatorname{Re} \nu > -1 \right]$$
ET II 24(19)

4. $$\int_0^\infty \frac{J_\nu(bx) x^{\nu+1}}{(x^2 + a^2)^{\mu+1}}\, dx = \frac{a^{\nu-\mu} b^\mu}{2^\mu\, \Gamma(\mu+1)} K_{\nu-\mu}(ab)$$

$$\left[-1 < \operatorname{Re} \nu < \operatorname{Re}\left(2\mu + \tfrac{3}{2}\right), \quad a > 0, \quad b > 0 \right] \quad \text{MO 43}$$

5. $$\int_0^\infty x^{\nu+1} \left(x^2 + a^2\right)^\mu Y_\nu(bx)\, dx = 2^{\nu-1} \pi^{-1} a^{2\mu+2} (1+\mu)^{-1}\, \Gamma(\nu) b^{-\nu}$$

$$\times {}_1F_2\left(1; 1-\nu, 2+\mu; \frac{a^2 b^2}{4}\right) - 2^\mu a^{\mu+\nu+1} [\sin(\nu\pi)]^{-1}$$

$$\times \Gamma(\mu+1) b^{-1-\mu} \left[I_{\mu+\nu+1}(ab) - 2\cos(\mu\pi)\, K_{\mu+\nu+1}(ab) \right]$$

$$\left[b > 0, \quad \operatorname{Re} a > 0, \quad -1 < \operatorname{Re} \nu < -2\operatorname{Re}\mu \right] \quad \text{ET II 100(19)}$$

6.[10] $$\int_0^\infty x^{1-\nu} \left(x^2 + a^2\right)^\mu Y_\nu(bx)\, dx = \frac{2^\mu a^{1+\mu-\nu} b^{-1-\mu} \pi}{\Gamma(-\mu)} I_{-1-\mu+\nu}(ab) \cot[\pi(\mu-\nu)] \operatorname{cosec}(\pi\mu)$$

$$- \frac{2^\mu a^{1+\mu-\nu} b^{-1-\mu} \pi}{\Gamma(-\mu)} I_{1+\mu-\nu}(ab) \operatorname{cosec}[\pi(\mu-\nu)] \operatorname{cosec}(\pi\nu)$$

$$+ \frac{2^{-1-\nu} a^{2+2\mu} b^\nu}{(1+\mu)\pi} \cos(\pi\nu)\, \Gamma(-\mu)\, {}_1F_2\left(1;\, 2+\mu, 1+\nu;\, \frac{a^2 b^2}{4}\right)$$

$$\left[\operatorname{Re} \nu < 1, \quad \operatorname{Re}(\nu - 2\mu) > -3, \quad \arg a^2 \neq \pi, \quad b > 0 \right] \quad \text{MC}$$

7. $$\int_0^\infty x^{1+\nu} \left(x^2 + a^2\right)^\mu K_\nu(bx)\, dx = 2^\nu\, \Gamma(\nu+1) a^{\nu+\mu+1} b^{-1-\mu} S_{\mu-\nu,\mu+\nu+1}(ab)$$

$$\left[\operatorname{Re} a > 0, \quad \operatorname{Re} b > 0, \quad \operatorname{Re} \nu > -1 \right]$$
ET II 128(8)

8.
$$
\int_0^\infty \frac{x^{\varrho-1} J_\nu(ax)}{(x^2 + k^2)^{\mu+1}} \, dx = \frac{a^\nu k^{\varrho+\nu-2\mu-2} \, \Gamma\left(\frac{1}{2}\varrho + \frac{1}{2}\nu\right) \Gamma\left(\mu + 1 - \frac{1}{2}\varrho - \frac{1}{2}\nu\right)}{2^{\nu+1} \, \Gamma(\mu+1) \, \Gamma(\nu+1)}
$$
$$
\times \, {}_1F_2\left(\frac{\varrho+\nu}{2}; \frac{\varrho+\nu}{2} - \mu, \nu+1; \frac{a^2 k^2}{4}\right)
$$
$$
+ \frac{a^{2\mu+2-\varrho} \, \Gamma\left(\frac{1}{2}\nu + \frac{1}{2}\varrho - \mu - 1\right)}{2^{2\mu+3-\varrho} \, \Gamma\left(\mu + 2 + \frac{1}{2}\nu - \frac{1}{2}\varrho\right)}
$$
$$
\times \, {}_1F_2\left(\mu+1; \mu+2 + \frac{\nu-\varrho}{2}, \mu+2 - \frac{\nu+\varrho}{2}; \frac{a^2 k^2}{4}\right)
$$
$$
\left[a > 0, \quad -\operatorname{Re}\nu < \operatorname{Re}\varrho < 2\operatorname{Re}\mu + \tfrac{7}{2}\right] \quad \text{WA 477(1)}
$$

6.566

1.
$$
\int_0^\infty x^\mu \, Y_\nu(bx) \frac{dx}{x^2 + a^2} = 2^{\mu-2} \pi^{-1} b^{1-\mu}
$$
$$
\times \cos\left[\frac{\pi}{2}(\mu - \nu + 1)\right] \Gamma\left(\frac{1}{2}\mu + \frac{1}{2}\nu - \frac{1}{2}\right) \Gamma\left(\frac{1}{2}\mu - \frac{1}{2}\nu - \frac{1}{2}\right)
$$
$$
\times \, {}_1F_2\left(1; 2 - \frac{\mu+1+\nu}{2}, 2 - \frac{\mu+1-\nu}{2}; \frac{a^2 b^2}{4}\right)
$$
$$
- \frac{1}{2} \pi a^{\mu-1} \operatorname{cosec}\left[\frac{\pi}{2}(\mu + \nu + 1)\right] \cot\left[\frac{\pi}{2}(\mu - \nu + 1)\right] I_\nu(ab)
$$
$$
- a^{\mu-1} \operatorname{cosec}\left[\frac{\pi}{2}(\mu - \nu + 1)\right] K_\nu(ab)
$$
$$
\left[b > 0, \quad \operatorname{Re} a > 0, \quad |\operatorname{Re}\nu| - 1 < \operatorname{Re}\mu < \tfrac{5}{2}\right] \quad \text{ET II 100(17)}
$$

2.
$$
\int_0^\infty x^{\nu+1} J_\nu(ax) \frac{dx}{x^2 + b^2} = b^\nu K_\nu(ab) \qquad \left[a > 0, \quad \operatorname{Re} b > 0, \quad -1 < \operatorname{Re}\nu < \tfrac{3}{2}\right]
$$
$$
\text{EH II 96(58)}
$$

3.
$$
\int_0^\infty x^\nu K_\nu(ax) \frac{dx}{x^2 + b^2} = \frac{\pi^2 b^{\nu-1}}{4 \cos\nu\pi} \left[\mathbf{H}_{-\nu}(ab) - Y_{-\nu}(ab)\right]
$$
$$
\left[a > 0, \quad \operatorname{Re} b > 0, \quad \operatorname{Re}\nu > -\tfrac{1}{2}\right]
$$
$$
\text{WA 468(9)}
$$

4.
$$
\int_0^\infty x^{-\nu} K_\nu(ax) \frac{dx}{x^2 + b^2} = \frac{\pi^2}{4 b^{\nu+1} \cos\nu\pi} \left[\mathbf{H}_\nu(ab) - Y_\nu(ab)\right]
$$
$$
\left[a > 0, \quad \operatorname{Re} b > 0, \quad \operatorname{Re}\nu < \tfrac{1}{2}\right]
$$
$$
\text{WA 468(10)}
$$

5.
$$
\int_0^\infty x^{-\nu} J_\nu(ax) \frac{dx}{x^2 + b^2} = \frac{\pi}{2 b^{\nu+1}} \left[I_\nu(ab) - \mathbf{L}_\nu(ab)\right] \qquad \left[a > 0, \quad \operatorname{Re} b > 0, \quad \operatorname{Re}\nu > -\tfrac{5}{2}\right]
$$
$$
\text{WA 468(11)}
$$

6.567

1.
$$
\int_0^1 x^{\nu+1} \left(1 - x^2\right)^\mu J_\nu(bx) \, dx = 2^\mu \, \Gamma(\mu+1) b^{-(\mu+1)} J_{\nu+\mu+1}(b)
$$
$$
\left[b > 0, \quad \operatorname{Re}\nu > -1, \quad \operatorname{Re}\mu > -1\right]
$$
$$
\text{ET II 26(33)a}
$$

2. $\int_0^1 x^{\nu+1} (1-x^2)^\mu Y_\nu(bx) \, dx$

$$= b^{-(\mu+1)} \left[2^\mu \Gamma(\mu+1) Y_{\mu+\nu+1}(b) + 2^{\nu+1} \pi^{-1} \Gamma(\nu+1) S_{\mu-\nu,\mu+\nu+1}(b) \right]$$

$$[b > 0, \quad \operatorname{Re}\mu > -1, \quad \operatorname{Re}\nu > -1] \quad \text{ET II 103(35)a}$$

3. $\int_0^1 x^{1-\nu} (1-x^2)^\mu J_\nu(bx) \, dx = \dfrac{2^{1-\nu} S_{\nu+\mu,\mu-\nu+1}(b)}{b^{\mu+1} \Gamma(\nu)}$ $\qquad [b > 0, \quad \operatorname{Re}\mu > -1]$ $\qquad\qquad$ ET II 25(31)a

4. $\int_0^1 x^{1-\nu} (1-x^2)^\mu Y_\nu(bx) \, dx = b^{-(\mu+1)} \left[2^{1-\nu} \pi^{-1} \cos(\nu\pi) \Gamma(1-\nu) \right.$

$$\times \left. S_{\mu+\nu,\mu-\nu+1}(b) - 2^\mu \operatorname{cosec}(\nu\pi) \Gamma(\mu+1) J_{\mu-\nu+1}(b) \right]$$

$$[b > 0, \quad \operatorname{Re}\mu > -1, \quad \operatorname{Re}\nu < 1] \quad \text{ET II 104(37)a}$$

5. $\int_0^1 x^{1-\nu} (1-x^2)^\mu K_\nu(bx) \, dx = 2^{-\nu-2} b^\nu (\mu+1)^{-1} \Gamma(-\nu) \, {}_1F_2 \left(1; \nu+1, \mu+2; \dfrac{b^2}{4} \right)$

$$+ \pi 2^{\mu-1} b^{-(\mu+1)} \operatorname{cosec}(\nu\pi) \Gamma(\mu+1) I_{\mu-\nu+1}(b)$$

$$[\operatorname{Re}\mu > -1, \quad \operatorname{Re}\nu < 1] \quad \text{ET II 129(12)a}$$

6. $\int_0^1 x^{1-\nu} J_\nu(bx) \dfrac{dx}{\sqrt{1-x^2}} = \sqrt{\dfrac{\pi}{2b}} \, \mathbf{H}_{\nu-\frac{1}{2}}(b)$ $\qquad\qquad [b > 0]$ $\qquad\qquad$ ET II 24(24)a

7. $\int_0^1 x^{1+\nu} Y_\nu(bx) \dfrac{dx}{\sqrt{1-x^2}} = \sqrt{\dfrac{\pi}{2b}} \operatorname{cosec}(\nu\pi) \left[\cos(\nu\pi) J_{\nu+\frac{1}{2}}(b) - \mathbf{H}_{-\nu-\frac{1}{2}}(b) \right]$

$$[b > 0, \quad \operatorname{Re}\nu > -1] \quad \text{ET II 102(28)a}$$

8. $\int_0^1 x^{1-\nu} Y_\nu(bx) \dfrac{dx}{\sqrt{1-x^2}} = \sqrt{\dfrac{\pi}{2b}} \left\{ \cot(\nu\pi) \left[\mathbf{H}_{\nu-\frac{1}{2}}(b) - Y_{\nu-\frac{1}{2}}(b) \right] - J_{\nu-\frac{1}{2}}(b) \right\}$

$$[b > 0, \quad \operatorname{Re}\nu < 1] \quad \text{ET II 102(30)a}$$

9. $\int_0^1 x^\nu (1-x^2)^{\nu-\frac{1}{2}} J_\nu(bx) \, dx = 2^{\nu-1} \sqrt{\pi} b^{-\nu} \Gamma\left(\nu+\tfrac{1}{2}\right) \left[J_\nu\left(\dfrac{b}{2}\right) \right]^2$

$$[b > 0, \quad \operatorname{Re}\nu > -\tfrac{1}{2}] \quad \text{ET II 24(25)a}$$

10. $\int_0^1 x^\nu (1-x^2)^{\nu-\frac{1}{2}} Y_\nu(bx) \, dx = 2^{\nu-1} \sqrt{\pi} b^{-\nu} \Gamma\left(\nu+\dfrac{1}{2}\right) J_\nu\left(\dfrac{b}{2}\right) Y_\nu\left(\dfrac{b}{2}\right)$

$$[b > 0, \quad \operatorname{Re}\nu > -\tfrac{1}{2}] \quad \text{ET II 102(31)a}$$

11. $\int_0^1 x^\nu (1-x^2)^{\nu-\frac{1}{2}} K_\nu(bx) \, dx = 2^{\nu-1} \sqrt{\pi} b^{-\nu} \Gamma\left(\nu+\dfrac{1}{2}\right) I_\nu\left(\dfrac{b}{2}\right) K_\nu\left(\dfrac{b}{2}\right)$

$$[\operatorname{Re}\nu > -\tfrac{1}{2}] \quad \text{ET II 129(10)a}$$

12. $\int_0^1 x^\nu (1-x^2)^{\nu-\frac{1}{2}} I_\nu(bx) \, dx = 2^{-\nu-1} \sqrt{\pi} b^{-\nu} \Gamma\left(\nu+\dfrac{1}{2}\right) \left[I_\nu\left(\dfrac{b}{2}\right) \right]^2$ $\qquad\qquad$ ET II 365(5)a

13. $\int_0^1 x^{\nu+1} (1-x^2)^{-\nu-\frac{1}{2}} J_\nu(bx) \, dx = 2^{-\nu} \dfrac{b^{\nu-1}}{\sqrt{\pi}} \Gamma\left(\dfrac{1}{2}-\nu\right) \sin b$

$$[b > 0, \quad |\operatorname{Re}\nu| < \tfrac{1}{2}] \quad \text{ET II 25(27)a}$$

14. $$\int_1^\infty x^\nu \left(x^2 - 1\right)^{\nu - \frac{1}{2}} Y_\nu(bx)\, dx = 2^{\nu-2}\sqrt{\pi}\, b^{-\nu}\, \Gamma\left(\nu + \frac{1}{2}\right)\left[J_\nu\left(\frac{b}{2}\right) J_{-\nu}\left(\frac{b}{2}\right) - Y_\nu\left(\frac{b}{2}\right) Y_{-\nu}\left(\frac{b}{2}\right)\right]$$

$$\left[|\operatorname{Re}\nu| < \tfrac{1}{2}, \quad b > 0\right] \qquad \text{ET II 103(32)a}$$

15. $$\int_1^\infty x^\nu \left(x^2 - 1\right)^{\nu - \frac{1}{2}} K_\nu(bx)\, dx = \frac{2^{\nu-1}}{\sqrt{\pi}}\, b^{-\nu}\, \Gamma\left(\nu + \frac{1}{2}\right)\left[K_\nu\left(\frac{b}{2}\right)\right]^2$$

$$\left[\operatorname{Re} b > 0, \quad \operatorname{Re}\nu > -\tfrac{1}{2}\right] \qquad \text{ET II 129(11)a}$$

16. $$\int_1^\infty x^{-\nu}\left(x^2 - 1\right)^{-\nu - \frac{1}{2}} J_\nu(bx)\, dx = -2^{-\nu-1}\sqrt{\pi}\, b^\nu\, \Gamma\left(\frac{1}{2} - \nu\right) J_\nu\left(\frac{b}{2}\right) Y_\nu\left(\frac{b}{2}\right)$$

$$\left[b > 0, \quad |\operatorname{Re}\nu| < \tfrac{1}{2}\right] \qquad \text{ET II 25(26)a}$$

17.[8] $$\int_1^\infty x^{-\nu+1}\left(x^2 - 1\right)^{\nu - \frac{1}{2}} J_\nu(bx)\, dx = \frac{2^\nu}{\sqrt{\pi}}\, b^{-\nu-1}\, \Gamma\left(\frac{1}{2} + \nu\right)\cos b$$

$$\left[b > 0, \quad |\operatorname{Re}\nu| < \tfrac{1}{2}\right] \qquad \text{ET II 25(28)}$$

6.568

1. $$\int_0^\infty x^\nu\, Y_\nu(bx)\, \frac{dx}{x^2 - a^2} = \frac{\pi}{2} a^{\nu-1}\, J_\nu(ab) \qquad\qquad \left[a > 0, \quad b > 0, \quad -\tfrac{1}{2} < \operatorname{Re}\nu < \tfrac{5}{2}\right]$$

$$\text{ET II 101(22)}$$

2. $$\int_0^\infty x^\mu\, Y_\nu(bx)\, \frac{dx}{x^2 - a^2} = \frac{\pi}{2} a^{\mu-1}\, J_\nu(ab) + 2^\mu \pi^{-1} a^{\mu-1} \cos\left[\frac{\pi}{2}(\mu - \nu + 1)\right]$$

$$\times\, \Gamma\left(\frac{\mu - \nu + 1}{2}\right) \Gamma\left(\frac{\mu + \nu + 1}{2}\right) S_{-\mu,\nu}(ab)$$

$$\left[a > 0, \quad b > 0, \quad |\operatorname{Re}\nu| - 1 < \operatorname{Re}\mu < \tfrac{5}{2}\right] \qquad \text{ET II (101)(25)}$$

6.569 $\displaystyle\int_0^1 x^\lambda (1 - x)^{\mu-1}\, J_\nu(ax)\, dx$

$$= \frac{\Gamma(\mu)\, \Gamma(1 + \lambda + \nu)\, 2^{-\nu} a^\nu}{\Gamma(\nu + 1)\, \Gamma(1 + \lambda + \mu + \nu)}$$

$$\times\, {}_2F_3\left(\frac{\lambda + 1 + \nu}{2}, \frac{\lambda + 2 + \nu}{2}; \nu + 1, \frac{\lambda + 1 + \mu + \nu}{2}, \frac{\lambda + 2 + \mu + \nu}{2}; -\frac{a^2}{4}\right)$$

$$\left[\operatorname{Re}\mu > 0, \quad \operatorname{Re}(\lambda + \nu) > -1\right] \qquad \text{ET II 193(56)a}$$

6.571

1. $$\int_0^\infty \left[\left(x^2 + a^2\right)^{\frac{1}{2}} \pm x\right]^\mu J_\nu(bx)\, \frac{dx}{\sqrt{x^2 + a^2}} = a^\mu\, I_{\frac{1}{2}(\nu\mp\mu)}\left(\frac{ab}{2}\right) K_{\frac{1}{2}(\nu\pm\mu)}\left(\frac{ab}{2}\right)$$

$$\left[\operatorname{Re} a > 0, \quad b > 0, \quad \operatorname{Re}\nu > -1, \quad \operatorname{Re}\mu < \tfrac{3}{2}\right] \qquad \text{ET II 26(38)}$$

2. $$\int_0^\infty \left[\left(x^2 + a^2\right)^{\frac{1}{2}} - x\right]^\mu Y_\nu(bx)\, \frac{dx}{\sqrt{x^2 + a^2}}$$

$$= a^\mu\left[\cot(\nu\pi)\, I_{\frac{1}{2}(\mu+\nu)}\left(\frac{ab}{2}\right) K_{\frac{1}{2}(\mu-\nu)}\left(\frac{ab}{2}\right) - \operatorname{cosec}(\nu\pi)\, I_{\frac{1}{2}(\mu-\nu)}\left(\frac{ab}{2}\right) K_{\frac{1}{2}(\mu+\nu)}\left(\frac{ab}{2}\right)\right]$$

$$\left[\operatorname{Re} a > 0, \quad b > 0, \quad \operatorname{Re}\mu > -\tfrac{3}{2}, \quad |\operatorname{Re}\nu| < 1\right] \qquad \text{ET II 104(40)}$$

3. $\int_0^\infty \left[(x^2 + a^2)^{\frac{1}{2}} + x \right]^\mu K_\nu(bx) \dfrac{dx}{\sqrt{x^2 + a^2}}$

$$= \frac{\pi^2}{4} a^\mu \operatorname{cosec}(\nu\pi) \left[J_{\frac{1}{2}(\nu-\mu)}\left(\frac{ab}{2}\right) Y_{-\frac{1}{2}(\nu+\mu)}\left(\frac{ab}{2}\right) - Y_{\frac{1}{2}(\nu-\mu)}\left(\frac{ab}{2}\right) J_{-\frac{1}{2}(\nu+\mu)}\left(\frac{ab}{2}\right) \right]$$

$$[\operatorname{Re} a > 0, \quad \operatorname{Re} b > 0] \qquad \text{ET II 130(15)}$$

6.572

1. $\int_0^\infty x^{-\mu} \left[(x^2 + a^2)^{\frac{1}{2}} + a \right]^\mu J_\nu(bx) \dfrac{dx}{\sqrt{x^2 + a^2}} = \dfrac{\Gamma\left(\frac{1+\nu-\mu}{2}\right)}{ab\,\Gamma(\nu+1)} W_{\frac{1}{2}\mu, \frac{1}{2}\nu}(ab)\, M_{-\frac{1}{2}\mu, \frac{1}{2}\nu}(ab)$

$$[\operatorname{Re} a > 0, \quad b > 0, \quad \operatorname{Re}(\nu - \mu) > -1]$$
$$\text{ET II 26(40)}$$

2. $\int_0^\infty x^{-\mu} \left[(x^2 + a^2)^{\frac{1}{2}} + a \right]^\mu K_\nu(bx) \dfrac{dx}{\sqrt{x^2 + a^2}}$

$$= \dfrac{\Gamma\left(\frac{1+\nu-\mu}{2}\right) \Gamma\left(\frac{1-\nu-\mu}{2}\right)}{2ab} W_{\frac{1}{2}\mu, \frac{1}{2}\nu}(iab)\, W_{\frac{1}{2}\mu, \frac{1}{2}\nu}(-iab)$$

$$[\operatorname{Re} a > 0, \quad \operatorname{Re} b > 0, \quad \operatorname{Re}\mu + |\operatorname{Re}\nu| < 1] \quad \text{ET II 130(18), BU 87(6a)}$$

3. $\int_0^\infty x^{-\mu} \left[(x^2 + a^2)^{\frac{1}{2}} - a \right]^\mu Y_\nu(bx) \dfrac{dx}{\sqrt{x^2 + a^2}}$

$$= -\frac{1}{ab} W_{-\frac{1}{2}\mu, \frac{1}{2}\nu}(ab) \left\{ \frac{\Gamma\left(\frac{1+\nu+\mu}{2}\right)}{\Gamma(\nu+1)} \tan\left(\frac{\nu-\mu}{2}\pi\right) M_{\frac{1}{2}\mu, \frac{1}{2}\nu}(ab) \right.$$

$$\left. + \sec\left(\frac{\nu-\mu}{2}\pi\right) W_{\frac{1}{2}\mu, \frac{1}{2}\nu}(ab) \right\}$$

$$\left[\operatorname{Re} a > 0, \quad b > 0, \quad |\operatorname{Re}\nu| < \tfrac{1}{2} + \tfrac{1}{2}\operatorname{Re}\mu\right] \quad \text{ET II 105(42)}$$

6.573

1. $\int_0^\infty x^{\nu-M+1} J_\nu(bx) \prod_{i=1}^k J_{\mu_i}(a_i x)\, dx = 0 \qquad\qquad M = \sum_{i=1}^k \mu_i$

$$\left[a_i > 0, \quad \sum_{i=1}^k a_i < b < \infty, \quad -1 < \operatorname{Re}\nu < \operatorname{Re} M + \tfrac{1}{2}k - \tfrac{1}{2} \right] \quad \text{ET II 54(42)}$$

2. $\int_0^\infty x^{\nu-M-1} J_\nu(bx) \prod_{i=1}^k J_{\mu_i}(a_i x)\, dx = 2^{\nu-M-1} b^{-\nu} \Gamma(\nu) \prod_{i=1}^k \dfrac{a_i^{\mu_i}}{\Gamma(1+\mu_i)}, \qquad M = \sum_{i=1}^k \mu_i$

$$\left[a_i > 0, \quad \sum_{i=1}^k a_i < b < \infty, \quad 0 < \operatorname{Re}\nu < \operatorname{Re} M + \tfrac{1}{2}k + \tfrac{3}{2} \right] \quad \text{WA 460(16)a, ET II 54(43)}$$

6.574

1.^{8} $$\int_0^\infty J_\nu(\alpha t)\, J_\mu(\beta t) t^{-\lambda}\, dt = \frac{\alpha^\nu\, \Gamma\left(\dfrac{\nu+\mu-\lambda+1}{2}\right)}{2^\lambda \beta^{\nu-\lambda+1}\, \Gamma\left(\dfrac{-\nu+\mu+\lambda+1}{2}\right)\Gamma(\nu+1)}$$

$$\times F\left(\frac{\nu+\mu-\lambda+1}{2}, \frac{\nu-\mu-\lambda+1}{2}; \nu+1; \frac{\alpha^2}{\beta^2}\right)$$

$$[\operatorname{Re}(\nu+\mu-\lambda+1) > 0, \quad \operatorname{Re}\lambda > -1, \quad 0 < \alpha < \beta] \quad \text{WA 439(2)a, MO 49}$$

If we reverse the positions of ν and μ and at the same time reverse the positions of α and β, the function on the right hand side of this equation will change. Thus, the right hand side represents a function of $\dfrac{\alpha}{\beta}$ that is not analytic at $\dfrac{\alpha}{\beta} = 1$.

For $\alpha = \beta$, we have the following equation

2. $$\int_0^\infty J_\nu(\alpha t)\, J_\mu(\alpha t) t^{-\lambda}\, dt = \frac{\alpha^{\lambda-1}\, \Gamma(\lambda)\, \Gamma\left(\dfrac{\nu+\mu-\lambda+1}{2}\right)}{2^\lambda\, \Gamma\left(\dfrac{-\nu+\mu+\lambda+1}{2}\right)\Gamma\left(\dfrac{\nu+\mu+\lambda+1}{2}\right)\Gamma\left(\dfrac{\nu-\mu+\lambda+1}{2}\right)}$$

$$[\operatorname{Re}(\nu+\mu+1) > \operatorname{Re}\lambda > 0, \quad \alpha > 0]$$

$$\text{MO 49, WA 441(2)a}$$

If $\mu - \nu + \lambda + 1$ (or $\nu - \mu + \lambda + 1$) is a negative integer, the right hand side of equation **6.574** 1. (or **6.574** 3.) vanishes. The cases in which the hypergeometric function F in **6.574** 3 (or **6.574** 1.) can be reduced to an elementary function are then especially important.

6.575

1. $$\int_0^\infty J_{\nu+1}(\alpha t)\, J_\mu(\beta t) t^{\mu-\nu}\, dt = 0 \qquad\qquad\qquad [\alpha < \beta]$$

$$= \frac{\left(\alpha^2-\beta^2\right)^{\nu-\mu}\beta^\mu}{2^{\nu-\mu}\alpha^{\nu+1}\, \Gamma(\nu-\mu+1)} \quad [\alpha \geq \beta]$$

$$[\operatorname{Re}\mu > \operatorname{Re}(\nu+1) > 0] \qquad\qquad \text{MO 51}$$

2. $$\int_0^\infty \frac{J_\nu(x)\, J_\mu(x)}{x^{\nu+\mu}}\, dx = \frac{\sqrt{\pi}\, \Gamma(\nu+\mu)}{2^{\nu+\mu}\, \Gamma\left(\nu+\mu+\tfrac{1}{2}\right)\Gamma\left(\nu+\tfrac{1}{2}\right)\Gamma\left(\mu+\tfrac{1}{2}\right)}$$

$$[\operatorname{Re}(\nu+\mu) > 0] \qquad \text{KU 147(17), WA 434(1)}$$

6.576

1. $$\int_0^\infty x^{\mu-\nu+1}\, J_\mu(x)\, K_\nu(x)\, dx = \tfrac{1}{2}\, \Gamma(\mu-\nu+1) \qquad [\operatorname{Re}\mu > -1, \quad \operatorname{Re}(\mu-\nu) > -1]$$

$$\text{ET II 370(47)}$$

2. $\displaystyle\int_0^\infty x^{-\lambda} J_\nu(ax) J_\nu(bx)\, dx = \frac{2^\nu b^\nu \Gamma\left(\nu + \dfrac{1-\lambda}{2}\right)}{2^\lambda (a+b)^{2\nu-\lambda+1} \Gamma(\nu+1) \Gamma\left(\dfrac{1+\lambda}{2}\right)}$

$$\times F\left(\nu + \frac{1-\lambda}{2}, \nu + \frac{1}{2}; 2\nu+1; \frac{4ab}{(a+b)^2}\right)$$

$\qquad\qquad\qquad [a>0, \quad b>0, \quad 2\operatorname{Re}\nu + 1 > \operatorname{Re}\lambda > -1]$ **ET II 47(4)**

3. $\displaystyle\int_0^\infty x^{-\lambda} K_\mu(ax) J_\nu(bx)\, dx = \frac{b^\nu \Gamma\left(\dfrac{\nu-\lambda+\mu+1}{2}\right) \Gamma\left(\dfrac{\nu-\lambda-\mu+1}{2}\right)}{2^{\lambda+1} a^{\nu-\lambda+1} \Gamma(1+\nu)}$

$$\times F\left(\frac{\nu-\lambda+\mu+1}{2}, \frac{\nu-\lambda-\mu+1}{2}; \nu+1; -\frac{b^2}{a^2}\right)$$

$\qquad [\operatorname{Re}(a \pm ib) > 0, \quad \operatorname{Re}(\nu-\lambda+1) > |\operatorname{Re}\mu|]$ **EH II 52(31), ET II 63(4), WA 449(1)**

4. $\displaystyle\int_0^\infty x^{-\lambda} K_\mu(ax) K_\nu(bx)\, dx = \frac{2^{-2-\lambda} a^{-\nu+\lambda-1} b^\nu}{\Gamma(1-\lambda)} \Gamma\left(\frac{1-\lambda+\mu+\nu}{2}\right) \Gamma\left(\frac{1-\lambda-\mu+\nu}{2}\right)$

$$\times \Gamma\left(\frac{1-\lambda+\mu-\nu}{2}\right) \Gamma\left(\frac{1-\lambda-\mu-\nu}{2}\right)$$

$$\times F\left(\frac{1-\lambda+\mu+\nu}{2}, \frac{1-\lambda-\mu+\nu}{2}; 1-\lambda; 1-\frac{b^2}{a^2}\right)$$

$\qquad\qquad [\operatorname{Re}a + b > 0, \quad \operatorname{Re}\lambda < 1 - |\operatorname{Re}\mu| - |\operatorname{Re}\nu|]$ **ET II 145(49), EH II 93(36)**

5. $\displaystyle\int_0^\infty x^{-\lambda} K_\mu(ax) I_\nu(bx)\, dx = \frac{b^\nu \Gamma\left(\frac{1}{2} - \frac{1}{2}\lambda + \frac{1}{2}\mu + \frac{1}{2}\nu\right) \Gamma\left(\frac{1}{2} - \frac{1}{2}\lambda - \frac{1}{2}\mu + \frac{1}{2}\nu\right)}{2^{\lambda+1} \Gamma(\nu+1) a^{-\lambda+\nu+1}}$

$$\times F\left(\frac{1}{2} - \frac{1}{2}\lambda + \frac{1}{2}\mu + \frac{1}{2}\nu, \frac{1}{2} - \frac{1}{2}\lambda - \frac{1}{2}\mu + \frac{1}{2}\nu; \nu+1; \frac{b^2}{a^2}\right)$$

$\qquad\qquad\qquad [\operatorname{Re}(\nu+1-\lambda \pm \mu) > 0, \quad a > b]$ **EH II 93(35)**

6. $\displaystyle\int_0^\infty x^{-\lambda} Y_\mu(ax) J_\nu(bx)\, dx = \frac{2}{\pi} \sin\frac{\pi(\nu-\mu-\lambda)}{2} \int_0^\infty x^{-\lambda} K_\mu(ax) I_\nu(bx)\, dx$

$\qquad [a>b, \quad \operatorname{Re}\lambda > -1, \quad \operatorname{Re}(\nu-\lambda+1 \pm \mu) > 0]$ $\qquad$ (see **6.576** 5) **EH II 93(37)**

7.[8] $\displaystyle\int_0^\infty x^{\mu+\nu+1} J_\mu(ax) K_\nu(bx)\, dx = 2^{\mu+\nu} a^\mu b^\nu \frac{\Gamma(\mu+\nu+1)}{(a^2+b^2)^{\mu+\nu+1}}$

$\qquad\qquad\qquad\qquad [\operatorname{Re}\mu > |\operatorname{Re}\nu| - 1, \quad \operatorname{Re}b > |\operatorname{Im}a|]$

$\qquad\qquad\qquad\qquad\qquad\qquad\qquad\qquad\qquad$ **ET 137(16), EH II 93(36)**

6.577

1.[8] $\displaystyle\int_0^\infty x^{\nu-\mu+1+2n} J_\mu(ax) J_\nu(bx) \frac{dx}{x^2+c^2} = (-1)^n c^{\nu-\mu+2n} I_\mu(ac) K_\nu(bc)$

$\qquad [a>0, \quad b>a, \quad \operatorname{Re}c>0, \quad 2+\operatorname{Re}\mu-2n > \operatorname{Re}\nu > -1-n, \quad n \geq 0 \text{ an integer}]$ **ET II 49(13)**

$2.^8 \qquad \int_0^\infty x^{\mu-\nu+1+2n} J_\mu(ax) J_\nu(bx) \frac{dx}{x^2+c^2} = (-1)^n c^{\mu-\nu+2n} I_\nu(bc) K_\mu(ac)$

$$[b>0, \quad a>b, \quad \operatorname{Re}\nu - 2n+2 > \operatorname{Re}\mu > -n-1, \quad n \geq 0 \text{ an integer}] \quad \text{ET II 49(15)}$$

6.578

1. $\int_0^\infty x^{\varrho-1} J_\lambda(ax) J_\mu(bx) J_\nu(cx)\, dx = \dfrac{2^{\varrho-1} a^\lambda b^\mu c^{-\lambda-\mu-\varrho}\, \Gamma\left(\frac{\lambda+\mu+\nu+\varrho}{2}\right)}{\Gamma(\lambda+1)\,\Gamma(\mu+1)\,\Gamma\left(1 - \frac{\lambda+\mu-\nu+\varrho}{2}\right)}$

$$\times F_4\left(\frac{\lambda+\mu-\nu+\varrho}{2}, \frac{\lambda+\mu+\nu+\varrho}{2}; \lambda+1, \mu+1; \frac{a^2}{c^2}, \frac{b^2}{c^2}\right)$$

$$\left[\operatorname{Re}(\lambda+\mu+\nu+\varrho) > 0, \quad \operatorname{Re}\varrho < \frac{5}{2}, \quad a>0, \quad b>0, \quad c>0, \quad c>a+b\right] \quad \text{ET II 351(9)}$$

2. $\int_0^\infty x^{\varrho-1} J_\lambda(ax) J_\mu(bx) K_\nu(cx)\, dx$

$$= \frac{2^{\varrho-2} a^\lambda b^\mu c^{-\varrho-\lambda-\mu}}{\Gamma(\lambda+1)\,\Gamma(\mu+1)}\, \Gamma\left(\frac{\varrho+\lambda+\mu-\nu}{2}\right) \Gamma\left(\frac{\varrho+\lambda+\mu+\nu}{2}\right)$$

$$\times F_4\left(\frac{\varrho+\lambda+\mu-\nu}{2}, \frac{\varrho+\lambda+\mu+\nu}{2}; \lambda+1, \mu+1; -\frac{a^2}{c^2}, -\frac{b^2}{c^2}\right)$$

$$[\operatorname{Re}(\varrho+\lambda+\mu) > |\operatorname{Re}\nu|, \quad \operatorname{Re}c > |\operatorname{Im}a| + |\operatorname{Im}b|] \quad \text{ET II 373(8)}$$

3. $\int_0^\infty x^{\lambda-\mu-\nu+1} J_\nu(ax) J_\mu(bx) J_\lambda(cx)\, dx = 0$

$$\left[\operatorname{Re}\lambda > -1, \quad \operatorname{Re}(\lambda-\mu-\nu) < \tfrac{1}{2}, \quad c>b>0, \quad 0<a<c-b\right] \quad \text{ET II 53(36)}$$

4. $\int_0^\infty x^{\lambda-\mu-\nu-1} J_\nu(ax) J_\mu(bx) J_\lambda(cx)\, dx = \dfrac{2^{\lambda-\mu-\nu-1} a^\nu b^\mu\, \Gamma(\lambda)}{c^\lambda\, \Gamma(\mu+1)\,\Gamma(\nu+1)}$

$$\left[\operatorname{Re}\lambda > 0, \quad \operatorname{Re}(\lambda-\mu-\nu) < \tfrac{5}{2}, \quad c>b>0, \quad 0<a<c-b\right] \quad \text{ET II 53(37)}$$

5. $\int_0^\infty x^{1+\mu} Y_\mu(ax) J_\nu(bx) J_\nu(cx)\, dx = 0 \qquad\qquad [0<b<c, \quad 0<a<c-b]$

$$\text{ET II 352(13)}$$

$6.^8 \qquad \int_0^\infty x^{\mu+1} K_\mu(ax) J_\nu(bx) J_\nu(cx)\, dx = \dfrac{1}{\sqrt{2\pi}} a^\mu b^{-\mu-1} c^{-\mu-1} e^{-\left(\mu+\frac{1}{2}\right)\pi i} \left(\mu^2-1\right)^{-\frac{1}{2}\mu-\frac{1}{4}} Q_{\nu-\frac{1}{2}}^{\mu+\frac{1}{2}}(\mu)$

$$\left[2bc\mu = a^2+b^2+c^2, \quad \operatorname{Re}a > |\operatorname{Im}b| + |\operatorname{Im}c|, \quad \operatorname{Re}\nu > -1, \quad \operatorname{Re}(\mu+\nu) > -1\right]$$

$$\text{WA 452(2), ET II 64(12)}$$

$7.^8 \qquad \int_0^\infty x^{\mu+1} I_\nu(ax) K_\mu(bx) J_\nu(cx)\, dx = \dfrac{1}{\sqrt{2\pi}} a^{-\mu-1} b^\mu c^{-\mu-1} e^{-\left(\mu-\frac{1}{2}\nu+\frac{1}{4}\right)\pi i} \left(\nu^2+1\right)^{-\frac{1}{2}\mu-\frac{1}{4}} Q_{\nu-\frac{1}{2}}^{\mu+\frac{1}{2}}(i\nu),$

$$2ac\nu = b^2 - a^2 + c^2 \qquad [\operatorname{Re}b > |\operatorname{Re}a| + |\operatorname{Im}c|; \quad \operatorname{Re}\nu > -1, \quad \operatorname{Re}(\mu+\nu) > -1] \quad \text{ET II 66(22)}$$

8.8 $\quad \displaystyle\int_0^\infty x^{1-\mu} J_\mu(ax) J_\nu(bx) J_\nu(cx)\,dx$

$$= \sqrt{\frac{2}{\pi^3}}\, a^{-\mu}(bc)^{\mu-1}(\sinh\mu)^{\mu-\frac12}\sin[(\mu-\nu)\pi]e^{\left(\mu-\frac12\right)\pi i}\, Q_{\nu-\frac12}^{\frac12-\mu}(\cosh\mu) \qquad [a>b+c]$$

$$= \frac{1}{\sqrt{2\pi}}\, a^{-\mu}(bc)^{\mu-1}(\sin\nu)^{\mu-\frac12}\, P_{\nu-\frac12}^{\frac12-\mu}(\cos\nu) \qquad\qquad [|b-c|<a<b+c]$$

$$= 0 \qquad\qquad\qquad\qquad [0<a<|b-c|]$$

$$\left[2bc\cosh\mu = a^2-b^2-c^2,\quad 2bc\cos\nu = b^2+c^2-a^2,\quad b>0,\quad c>0;\quad \mathrm{Re}\,\nu>-1,\mathrm{Re}\,\mu>-\tfrac12\right]$$

9. $\quad \displaystyle\int_0^\infty J_\nu(ax) J_\nu(bx) J_\nu(cx) x^{1-\nu}\,dx = 0 \qquad\qquad [0<c\le|a-b|\ \text{or}\ c\ge a+b]$

$$= \frac{2^{\nu-1}\Delta^{2\nu-1}}{(abc)^\nu\,\Gamma\left(\nu+\frac12\right)\Gamma\left(\frac12\right)} \qquad [|a-b|<c<a+b]$$

$$\Delta = \tfrac14\sqrt{\left[c^2-(a-b)^2\right]\left[(a+b)^2-c^2\right]},\quad [a>0,\quad b>0,\quad c>0;\quad \mathrm{Re}\,\nu>-\tfrac12]$$

($\Delta>0$ is equal to the area of a triangle whose sides are a, b, and c.)

10.8 $\quad \displaystyle\int_0^\infty x^{\nu+1} K_\mu(ax) K_\mu(bx) J_\nu(cx)\,dx = \frac{\sqrt{\pi}c^\nu\,\Gamma(\nu+\mu+1)\,\Gamma(\nu-\mu+1)}{2^{\frac23}(ab)^{\nu+1}(\mu^2-1)^{\frac12\nu+\frac14}}\, P_{\mu-\frac12}^{-\nu-\frac12}(\mu)$

$$\left[2ab\mu = a^2+b^2+c^2,\quad \mathrm{Re}(a+b)>|\mathrm{Im}\,c|,\quad \mathrm{Re}(\nu\pm\mu)>-1,\quad \mathrm{Re}\,\nu>-1\right] \quad \text{ET II 67(30)}$$

11.8 $\quad \displaystyle\int_0^\infty x^{\nu+1} K_\mu(ax) I_\mu(bx) J_\nu(cx)\,dx = \frac{(ab)^{-\nu-1}c^\nu e^{-\left(\nu+\frac12\right)\pi i}\, Q_{\mu-\frac12}^{\nu+\frac12}(\mu)}{\sqrt{2\pi}(\mu^2-1)^{\frac12\nu+\frac14}}\qquad 2ab\mu = a^2+b^2+c^2$

$$[\mathrm{Re}\,a>|\mathrm{Re}\,b|+|\mathrm{Im}\,c|;\quad \mathrm{Re}\,\nu>-1,\quad \mathrm{Re}(\mu+\nu)>-1] \quad \text{ET II 66(24)}$$

12.8 $\quad \displaystyle\int_0^\infty x^{\nu+1}[J_\nu(ax)]^2\, Y_\nu(bx)\,dx = 0 \qquad\qquad \left[0<b<2a,\quad |\mathrm{Re}\,\nu|<\tfrac12\right]$

$$= \frac{2^{3\nu+1}a^{2\nu}b^{-\nu-1}}{\sqrt{\pi}\,\Gamma\left(\frac12-\nu\right)}\left(b^2-4a^2\right)^{-\nu-\frac12} \qquad \left[0<2a<b,\quad |\mathrm{Re}\,\nu|<\tfrac12\right]$$

$$\text{ET II 109(3)}$$

13. $\quad \displaystyle\int_0^\infty x^{\nu+1} J_\nu(ax) Y_\nu(ax) J_\nu(bx)\,dx$

$$= 0 \qquad\qquad \left[a>0,\quad |\mathrm{Re}\,\nu|<\tfrac12,\quad 0<b<2a\right]$$

$$= -\frac{2^{3\nu+1}a^{2\nu}b^{-\nu-1}}{\sqrt{\pi}\,\Gamma\left(\frac12-\nu\right)}\left(b^2-4a^2\right)^{-\nu-\frac12} \qquad \left[a>0,\quad 2a<b<\infty,\quad |\mathrm{Re}\,\nu|<\tfrac12\right]$$

$$\text{ET II 55(49)}$$

14. $\quad \displaystyle\int_0^\infty x^{\nu+1} J_\mu(xa\sin\psi) J_\nu(xa\sin\varphi) K_\mu(xa\cos\varphi\cos\psi)\,dx$

$$= \frac{2^\nu\,\Gamma(\mu+\nu+1)(\sin\varphi)^\nu\left(\cos\frac\alpha2\right)^{2\nu+1}}{a^{\nu+2}(\cos\psi)^{2\nu+2}}\, P_\nu^{-\mu}(\cos\alpha)$$

$$\left[\tan\tfrac12\alpha = \tan\psi\cos\varphi,\quad a>0,\quad \frac\pi2>\varphi>0,\quad 0<\psi<\frac\pi2,\quad \mathrm{Re}\,\nu>-1,\quad \mathrm{Re}(\mu+\nu)>-1\right]$$

$$\text{ET II 64(11)}$$

15. $$\int_0^\infty x^{\nu+1} J_\nu(ax) K_\nu(bx) J_\nu(cx)\,dx = \frac{2^{3\nu}(abc)^\nu \Gamma\left(\nu+\tfrac12\right)}{\sqrt{\pi}\left[(a^2+b^2+c^2)^2 - 4a^2c^2\right]^{\nu+\frac12}}$$

$$\left[\operatorname{Re} b > |\operatorname{Im} a|, \quad c > 0, \quad \operatorname{Re}\nu > -\tfrac12\right]$$

ET II 63(8)

16.8 $$\int_0^\infty x^{\nu+1} I_\nu(ax) K_\nu(bx) J_\nu(cx)\,dx = \frac{2^{3\nu}(abc)^\nu \Gamma\left(\nu+\tfrac12\right)}{\sqrt{\pi}\left[(b^2-a^2+c^2)^2 + 4a^2c^2\right]^{\nu+\frac12}}$$

$$\left[\operatorname{Re} b > |\operatorname{Re} a| + |\operatorname{Im} c|; \quad \operatorname{Re}\nu > -\tfrac12\right]$$

ET II 65(18)

6.579

1. $$\int_0^\infty x^{2\nu+1} J_\nu(ax)\,Y_\nu(ax)\,J_\nu(bx)\,Y_\nu(bx)\,dx$$

$$= \frac{a^{2\nu}\,\Gamma(3\nu+1)}{2\pi b^{4\nu+2}\,\Gamma\left(\tfrac12-\nu\right)\Gamma\left(2\nu+\tfrac32\right)}\, F\left(\nu+\tfrac12, 3\nu+1; 2\nu+\tfrac32; \frac{a^2}{b^2}\right)$$

$$\left[0 < a < b, \quad -\tfrac13 < \operatorname{Re}\nu < \tfrac12\right]$$ EH II 94(45), ET II 352(15)

2. $$\int_0^\infty x^{2\nu+1} J_\nu(ax)\,K_\nu(ax)\,J_\nu(bx)\,K_\nu(bx)\,dx$$

$$= \frac{2^{\nu-3} a^{2\nu}\,\Gamma\left(\tfrac{\nu+1}{2}\right)\Gamma\left(\nu+\tfrac12\right)\Gamma\left(\tfrac{3\nu+1}{2}\right)}{\sqrt{\pi}b^{4\nu+2}\,\Gamma(\nu+1)}\, F\left(\nu+\tfrac12, \frac{3\nu+1}{2}; 2\nu+1; 1-\frac{a^4}{b^4}\right)$$

$$\left[0 < a < b, \quad \operatorname{Re}\nu > -\tfrac13\right]$$ ET II 373(10)

3. $$\int_0^\infty x^{1-2\nu}[J_\nu(ax)]^4\,dx = \frac{\Gamma(\nu)\,\Gamma(2\nu)}{2\pi\left[\Gamma\left(\nu+\tfrac12\right)\right]^2 \Gamma(3\nu)}$$ $[\operatorname{Re}\nu > 0]$ ET II 342(25)

4. $$\int_0^\infty x^{1-2\nu}[J_\nu(ax)]^2[J_\nu(bx)]^2\,dx = \frac{a^{2\nu-1}\,\Gamma(\nu)}{2\pi b\,\Gamma\left(\nu+\tfrac12\right)\Gamma\left(2\nu+\tfrac12\right)}\, F\left(\nu, \frac12-\nu; 2\nu+\frac12; \frac{a^2}{b^2}\right)$$

ET II 351(10)

6.581

1. $$\int_0^a x^{\lambda-1} J_\mu(x)\,J_\nu(a-x)\,dx = 2^\lambda \sum_{m=0}^\infty \frac{(-1)^m \Gamma(\lambda+\mu+m)\,\Gamma(\lambda+m)}{m!\,\Gamma(\lambda)\,\Gamma(\mu+m+1)}\, J_{\lambda+\mu+\nu+2m}(a)$$

$$\left[\operatorname{Re}(\lambda+\mu) > 0, \quad \operatorname{Re}\nu > -1\right]$$

ET II 354(25)

2.8 $$\int_0^a x^{\lambda-1}(a-x)^{-1} J_\mu(x)\,J_\nu(a-x)\,dx$$

$$= \frac{2^\lambda}{a\nu} \sum_{m=0}^\infty \frac{(-1)^m \Gamma(\lambda+\mu+m)\,\Gamma(\lambda+m)}{m!\,\Gamma(\lambda)\,\Gamma(\mu+m+1)}(\lambda+\mu+\nu+2m)\, J_{\lambda+\mu+\nu+2m}(a)$$

$$\left[\operatorname{Re}(\lambda+\mu) > 0, \quad \operatorname{Re}\nu > 0\right]$$ ET II 354(27)

3. $$\int_0^a x^\mu(a-x)^\nu J_\mu(x)\,J_\nu(a-x)\,dx = \frac{\Gamma\left(\mu+\tfrac12\right)\Gamma\left(\nu+\tfrac12\right)}{\sqrt{2\pi}\,\Gamma(\mu+\nu+1)}a^{\mu+\nu+\frac12}\, J_{\mu+\nu+\frac12}(a)$$

$$\left[\operatorname{Re}\mu > -\tfrac12, \quad \operatorname{Re}\nu > -\tfrac12\right]$$

ET II 354(28), EH II 46(6)

4. $\displaystyle\int_0^a x^\mu (a-x)^{\nu+1} J_\mu(x) J_\nu(a-x)\,dx = \frac{\Gamma\left(\mu+\frac{1}{2}\right)\Gamma\left(\nu+\frac{3}{2}\right)}{\sqrt{2\pi}\,\Gamma(\mu+\nu+2)} a^{\mu+\nu+\frac{3}{2}} J_{\mu+\nu+\frac{1}{2}}(a)$

$$\left[\operatorname{Re}\nu > -1, \quad \operatorname{Re}\mu > -\tfrac{1}{2}\right]$$

ET II 354(29)

5. $\displaystyle\int_0^a x^\mu (a-x)^{-\mu-1} J_\mu(x) J_\nu(a-x)\,dx = \frac{2^\mu\,\Gamma\left(\mu+\frac{1}{2}\right)\Gamma(\nu-\mu)}{\sqrt{\pi}\,\Gamma(\mu+\nu+1)} a^\mu J_\nu(a)$

$$\left[\operatorname{Re}\nu > \operatorname{Re}\mu > -\tfrac{1}{2}\right]$$ ET II 355(30)

6.582 $\displaystyle\int_0^\infty x^{\mu-1}|x-b|^{-\mu} K_\mu\left(|x-b|\right) K_\nu(x)\,dx = \frac{1}{\sqrt{\pi}}(2b)^{-\mu}\,\Gamma\left(\tfrac{1}{2}-\mu\right)\Gamma(\mu+\nu)\,\Gamma(\mu-\nu)\,K_\nu(b)$

$$\left[b>0, \quad \operatorname{Re}\mu < \tfrac{1}{2}, \quad \operatorname{Re}\mu > |\operatorname{Re}\nu|\right]$$
ET II 374(14)

6.583 $\displaystyle\int_0^\infty x^{\mu-1}(x+b)^{-\mu} K_\mu(x+b) K_\nu(x)\,dx = \frac{\sqrt{\pi}\,\Gamma(\mu+\nu)\,\Gamma(\mu-\nu)}{2^\mu b^\mu\,\Gamma\left(\mu+\frac{1}{2}\right)} K_\nu(b)$

$$\left[|\arg b| < \pi, \quad \operatorname{Re}\mu > |\operatorname{Re}\nu|\right]$$
ET II 374(15)

6.584

1.[8] $\displaystyle\int_0^\infty \frac{x^{\varrho-1}\left[H_\nu^{(1)}(ax) - e^{\varrho\pi i} H_\nu^{(1)}\left(axe^{\pi i}\right)\right]}{(x^2-r^2)^{m+1}}\,dx = \frac{\pi i}{2^m m!}\left(\frac{d}{r\,dr}\right)^m \left[r^{\varrho-2} H_\nu^{(1)}(ar)\right]$

$$\left[m = 0,1,2,\ldots, \quad \operatorname{Im}r > 0, \quad a>0, \quad |\operatorname{Re}\nu| < \operatorname{Re}\varrho < 2m+\tfrac{7}{2}\right]$$ WA 465

2.[8] $\displaystyle\int_0^\infty \left[\cos\tfrac{1}{2}(\varrho-\nu)\pi\, J_\nu(ax) + \sin\tfrac{1}{2}(\varrho-\nu)\pi\, Y_\nu(ax)\right] \frac{x^{\varrho-1}}{(x^2+k^2)^{m+1}}\,dx$

$$= \frac{(-1)^{m+1}}{2^m\cdot m!}\left(\frac{d}{k\,dk}\right)^m \left[k^{\varrho-2} K_\nu(ak)\right]$$

$$\left[m = 0,1,2,\ldots, \quad \operatorname{Re}k > 0, \quad a>0, \quad |\operatorname{Re}\nu| < \operatorname{Re}\varrho < 2m+\tfrac{7}{2}\right]$$ WA 466(2)

3. $\displaystyle\int_0^\infty \left\{\cos\nu\pi\, J_\nu(ax) - \sin\nu\pi\, Y_\nu(ax)\right\} \frac{x^{1-\nu}\,dx}{(x^2+k^2)^{m+1}} = \frac{a^m K_{\nu+m}(ak)}{2^m\cdot m!k^{\nu+m}}$

$$\left[m = 0,1,2,\ldots, \quad \operatorname{Re}k > 0, \quad a>0, \quad -2m-\tfrac{3}{2} < \operatorname{Re}\nu < 1\right]$$ WA 466(3)

4. $\displaystyle\int_0^\infty \left\{\cos\left[\left(\tfrac{1}{2}\varrho-\tfrac{1}{2}\nu-\mu\right)\pi\right] J_\nu(ax) + \sin\left[\left(\tfrac{1}{2}\varrho-\tfrac{1}{2}\nu-\mu\right)\pi\right] Y_\nu(ax)\right\} \frac{x^{\varrho-1}}{(x^2+k^2)^{\mu+1}}\,dx$

$$= \frac{\pi k^{\varrho-2\mu-2}}{2\sin\nu\pi\cdot\Gamma(\mu+1)}\left[\frac{\left(\tfrac{1}{2}ak\right)^\nu\Gamma\left(\tfrac{1}{2}\varrho+\tfrac{1}{2}\nu\right)}{\Gamma(\nu+1)\,\Gamma\left(\tfrac{1}{2}\varrho+\tfrac{1}{2}\nu-\mu\right)}\,{}_1F_2\left(\frac{\varrho+\nu}{2};\frac{\varrho+\nu}{2}-\mu,\nu+1;\frac{a^2k^2}{4}\right)\right.$$

$$\left.- \frac{\left(\tfrac{1}{2}ak\right)^{-\nu}\Gamma\left(\tfrac{1}{2}\varrho-\tfrac{1}{2}\nu\right)}{\Gamma(1-\nu)\,\Gamma\left(\tfrac{1}{2}\varrho-\tfrac{1}{2}\nu-\mu\right)}\,{}_1F_2\left(\frac{\varrho-\nu}{2};\frac{\varrho-\nu}{2}-\mu,1-\nu;\frac{a^2k^2}{4}\right)\right]$$

$$\left[a>0, \quad \operatorname{Re}k > 0, \quad |\operatorname{Re}\nu| < \operatorname{Re}\varrho < 2\operatorname{Re}\mu+\tfrac{7}{2}\right]$$ WA 407(1)

5.8 $\displaystyle\int_0^\infty \left[\prod_{j=1}^n J_{\mu_j}(b_n x)\right] \left\{\cos\left[\frac{1}{2}\left(\varrho + \sum_j \mu_j - \nu\right)\pi\right] J_\nu(ax)\right.$

$\displaystyle\left. + \sin\left[\frac{1}{2}\left(\varrho + \sum_j \mu_j - \nu\right)\pi\right] Y_\nu(ax)\right\} \frac{x^{\varrho-1}}{x^2+k^2}\,dx$

$\displaystyle = -\left[\prod_{j=1}^n I_{\mu_j}(b_n k)\right] K_\nu(ak)k^{\varrho-2}$

$\displaystyle\left[\operatorname{Re} k > 0, \quad a > \sum_j |\operatorname{Re} b_j|, \quad \operatorname{Re}\left(\varrho + \sum_j \mu_j\right) > |\operatorname{Re}\nu|\right]$ WA 472(9)

6.59 Combinations of powers and Bessel functions of more complicated arguments

6.591

1. $\displaystyle\int_0^\infty x^{2\nu+\frac{1}{2}} J_{\nu+\frac{1}{2}}\left(\frac{a}{x}\right) K_\nu(bx)\,dx = \sqrt{2\pi}\,b^{-\nu-1}a^{\nu+\frac{1}{2}} J_{1+2\nu}\left(\sqrt{2ab}\right) K_{1+2\nu}\left(\sqrt{2ab}\right)$

$[a>0, \quad \operatorname{Re} b > 0, \quad \operatorname{Re}\nu > -1]$

ET II 142(35)

2. $\displaystyle\int_0^\infty x^{2\nu+\frac{1}{2}} Y_{\nu+\frac{1}{2}}\left(\frac{a}{x}\right) K_\nu(bx)\,dx = \sqrt{2\pi}\,b^{-\nu-1}a^{\nu+\frac{1}{2}} Y_{2\nu+1}\left(\sqrt{2ab}\right) K_{2\nu+1}\left(\sqrt{2ab}\right)$

$[a>0, \quad \operatorname{Re} b > 0, \quad \operatorname{Re}\nu > -1]$

ET II 143(41)

3. $\displaystyle\int_0^\infty x^{2\nu+\frac{1}{2}} K_{\nu+\frac{1}{2}}\left(\frac{a}{x}\right) K_\nu(bx)\,dx = \sqrt{2\pi}\,b^{-\nu-1}a^{\nu+\frac{1}{2}} K_{2\nu+1}\left(e^{\frac{1}{4}i\pi}\sqrt{2ab}\right) K_{2\nu+1}\left(e^{-\frac{1}{4}i\pi}\sqrt{2ab}\right)$

$[\operatorname{Re} a > 0, \quad \operatorname{Re} b > 0]$ ET II 146(56)

4. $\displaystyle\int_0^\infty x^{-2\nu+\frac{1}{2}} J_{\nu-\frac{1}{2}}\left(\frac{a}{x}\right) K_\nu(bx)\,dx = \sqrt{2\pi}\,b^{\nu-1}a^{\frac{1}{2}-\nu} K_{2\nu-1}\left(\sqrt{2ab}\right)$

$\displaystyle\times\left[\sin(\nu\pi) J_{2\nu-1}\left(\sqrt{2ab}\right) + \cos(\nu\pi) Y_{2\nu-1}\left(\sqrt{2ab}\right)\right]$

$[a>0, \quad \operatorname{Re} b > 0, \quad \operatorname{Re}\nu < 1]$ ET II 142(34)

5. $\displaystyle\int_0^\infty x^{-2\nu+\frac{1}{2}} Y_{\nu-\frac{1}{2}}\left(\frac{a}{x}\right) K_\nu(bx)\,dx = -\sqrt{\frac{\pi}{2}}\,b^{\nu-1}a^{\frac{1}{2}-\nu}\sec(\nu\pi) K_{2\nu-1}\left(\sqrt{2ab}\right)$

$\displaystyle\times\left[J_{2\nu-1}\left(\sqrt{2ab}\right) - J_{1-2\nu}\left(\sqrt{2ab}\right)\right]$

$[a>0, \quad \operatorname{Re}\nu < 1]$ ET II 143(40)

6. $\displaystyle\int_0^\infty x^{-2\nu+\frac{1}{2}} J_{\frac{1}{2}-\nu}\left(\frac{a}{x}\right) J_\nu(bx)\,dx$

$\displaystyle = -\frac{1}{2}i\operatorname{cosec}(2\nu\pi)b^{\nu-1}a^{\frac{1}{2}-\nu}\left[e^{2\nu\pi i} J_{1-2\nu}(u) J_{2\nu-1}(v) - e^{-2\nu\pi i} J_{2\nu-1}(u) J_{1-2\nu}(v)\right]$

$\displaystyle\left[u = \left(\tfrac{1}{2}ab\right)^{\frac{1}{2}}e^{\frac{1}{4}\pi i}, \quad v = \left(\tfrac{1}{2}ab\right)^{\frac{1}{2}}e^{-\frac{1}{4}\pi i}, \quad a>0, \quad b>0, \quad -\tfrac{1}{2} < \operatorname{Re}\nu < 3\right]$ ET II 58(12)

7. $\int_0^\infty x^{-2\nu+\frac{1}{2}} K_{\nu-\frac{1}{2}}\left(\frac{a}{x}\right) Y_\nu(bx)\,dx = \sqrt{2\pi}b^{\nu-1}a^{\frac{1}{2}-\nu}\, Y_{2\nu-1}\left(\sqrt{2ab}\right) K_{2\nu-1}\left(\sqrt{2ab}\right)$

$$\left[b>0, \quad \operatorname{Re} a>0, \quad \operatorname{Re}\nu>\tfrac{1}{6}\right]$$

ET II 113(30)

8. $\int_0^\infty x^{\varrho-1} J_\mu(ax) J_\nu\left(\frac{b}{x}\right)\,dx = \dfrac{a^{\nu-\varrho}b^\nu\,\Gamma\left(\frac{1}{2}\mu+\frac{1}{2}\varrho-\frac{1}{2}\nu\right)}{2^{2\nu-\varrho+1}\,\Gamma(\nu+1)\,\Gamma\left(\frac{1}{2}\mu+\frac{1}{2}\nu-\frac{1}{2}\varrho+1\right)}$

$$\times\; {}_0F_3\left(\nu+1,\frac{\nu-\mu-\varrho}{2}+1,\frac{\nu+\mu-\varrho}{2}+1;\frac{a^2b^2}{16}\right)$$

$$+\dfrac{a^\mu b^{\mu+\varrho}\,\Gamma\left(\frac{1}{2}\nu-\frac{1}{2}\mu-\frac{1}{2}\varrho\right)}{2^{2\mu+\varrho+1}\,\Gamma(\mu+1)\,\Gamma\left(\frac{1}{2}\mu+\frac{1}{2}\nu+\frac{1}{2}\varrho+1\right)}$$

$$\times\; {}_0F_3\left(\mu+1,\frac{\mu-\nu+\varrho}{2}+1,\frac{\nu+\mu+\varrho}{2}+1;\frac{a^2b^2}{16}\right)$$

$$\left[a>0, \quad b>0, \quad -\operatorname{Re}\left(\mu+\tfrac{3}{2}\right)<\operatorname{Re}\varrho<\operatorname{Re}\left(\nu+\tfrac{3}{2}\right)\right]\quad \text{WA 480(1)}$$

6.592

1. $\int_0^\infty x^\lambda(1-x)^{\mu-1} Y_\nu\left(a\sqrt{x}\right)\,dx = 2^{-\nu}a^\nu\cot(\nu\pi)\dfrac{\Gamma(\mu)\,\Gamma\left(\lambda+1+\frac{1}{2}\nu\right)}{\Gamma(1+\nu)\,\Gamma\left(\lambda+1+\mu+\frac{1}{2}\nu\right)}$

$$\times\; {}_1F_2\left(\lambda+1+\tfrac{1}{2}\nu;1+\nu,\lambda+1+\mu+\tfrac{1}{2}\nu;-\frac{a^2}{4}\right)$$

$$-2^\nu a^{-\nu}\operatorname{cosec}(\nu\pi)\dfrac{\Gamma(\mu)\,\Gamma\left(\lambda+1-\frac{1}{2}\nu\right)}{\Gamma(1-\nu)\,\Gamma\left(\lambda+1+\mu-\frac{1}{2}\nu\right)}$$

$$\times\; {}_1F_2\left(\lambda-\tfrac{1}{2}\nu+1;1-\nu,\lambda+1+\mu-\tfrac{1}{2}\nu;-\frac{a^2}{4}\right)$$

$$\left[\operatorname{Re}\lambda>-1+\tfrac{1}{2}|\operatorname{Re}\nu|, \quad \operatorname{Re}\mu>0\right]\quad \text{ET II 197(76)a}$$

2.^{10} $\int_0^1 x^\lambda(1-x)^{\mu-1} K_\nu\left(a\sqrt{x}\right)\,dx$

$$= 2^{-\nu-1}a^{-\nu}\dfrac{\Gamma(\nu)\,\Gamma(\mu)\,\Gamma\left(\lambda+1-\frac{1}{2}\nu\right)}{\Gamma\left(\lambda+1+\mu-\frac{1}{2}\nu\right)}\; {}_1F_2\left(\lambda+1-\frac{1}{2}\nu;1-\nu,\lambda+1+\mu-\frac{1}{2}\nu;\frac{a^2}{4}\right)$$

$$+2^{-1-\nu}a^\nu\dfrac{\Gamma(-\nu)\,\Gamma\left(\lambda+1+\frac{1}{2}\nu\right)\Gamma(\mu)}{\Gamma\left(\lambda+1+\mu+\frac{1}{2}\nu\right)}\; {}_1F_2\left(\lambda+1+\frac{1}{2}\nu;1+\nu,\lambda+1+\mu+\frac{1}{2}\nu;\frac{a^2}{4}\right)$$

$$= \dfrac{2^{\nu-1}}{a^\nu}\,\Gamma(\mu)\,G_{13}^{21}\left(\frac{a^2}{4}\left|\begin{matrix}\frac{\nu}{2}-\lambda\\ \nu,0,\frac{\nu}{2}-\lambda-\mu\end{matrix}\right.\right)$$

OB 159 (3.16)

$$\left[\operatorname{Re}\lambda>-1+\tfrac{1}{2}|\operatorname{Re}\nu|, \quad \operatorname{Re}\mu>0\right]\quad \text{ET II 198(87)a}$$

3. $\int_1^\infty x^\lambda(x-1)^{\mu-1} J_\nu\left(a\sqrt{x}\right)\,dx = 2^{2\lambda}a^{-2\lambda}\,G_{13}^{20}\left(\frac{a^2}{4}\left|\begin{matrix}0\\ -\mu,\lambda+\frac{1}{2}\nu,\lambda-\frac{1}{2}\nu\end{matrix}\right.\right)\Gamma(\mu)$

$$\left[a>0, \quad 0<\operatorname{Re}\mu<\tfrac{1}{4}-\operatorname{Re}\lambda\right]$$

ET II 205(36)a

4. $\displaystyle\int_1^\infty x^\lambda (x-1)^{\mu-1} K_\nu\left(a\sqrt{x}\right) dx = \Gamma(\mu) 2^{2\lambda-1} a^{-2\lambda}\, G_{13}^{30}\left(\frac{a^2}{4}\,\bigg|\,{0 \atop -\mu,\,\frac{1}{2}\nu+\lambda,\,-\frac{1}{2}\nu+\lambda}\right)$

                                     $[\operatorname{Re} a > 0, \quad \operatorname{Re}\mu > 0]$ ET II 209(60)a

5. $\displaystyle\int_0^1 x^{-\frac{1}{2}}(1-x)^{-\frac{1}{2}} J_\nu\left(a\sqrt{x}\right) dx = \pi\left[J_{\frac{1}{2}\nu}\left(\frac{1}{2}a\right)\right]^2$ $[\operatorname{Re}\nu > -1]$ ET II 194(59)a

6. $\displaystyle\int_0^1 x^{-\frac{1}{2}}(1-x)^{-\frac{1}{2}} I_\nu\left(a\sqrt{x}\right) dx = \pi\left[I_{\frac{1}{2}\nu}\left(\frac{1}{2}a\right)\right]^2$ $[\operatorname{Re}\nu > -1]$ ET II 197(79)

7. $\displaystyle\int_0^1 x^{-\frac{1}{2}}(1-x)^{-\frac{1}{2}} K_\nu\left(a\sqrt{x}\right) dx = \frac{1}{2}\pi\sec\left(\frac{1}{2}\nu\pi\right)\left[I_{\frac{\nu}{2}}\left(\frac{a}{2}\right) + I_{-\frac{\nu}{2}}\left(\frac{a}{2}\right)\right] K_{\frac{\nu}{2}}\left(\frac{a}{2}\right)$

                                     $[|\operatorname{Re}\nu| < 1]$ ET II 198(85)a

8. $\displaystyle\int_1^\infty x^{-\frac{1}{2}}(x-1)^{-\frac{1}{2}} K_\nu\left(a\sqrt{x}\right) dx = \left[K_{\frac{\nu}{2}}\left(\frac{a}{2}\right)\right]^2$ $[\operatorname{Re} a > 0]$ ET II 208(56)a

9. $\displaystyle\int_0^1 x^{-\frac{1}{2}}(1-x)^{-\frac{1}{2}} Y_\nu\left(a\sqrt{x}\right) dx = \pi\left\{\cot(\nu\pi)\left[J_{\frac{\nu}{2}}\left(\frac{a}{2}\right)\right]^2 - \operatorname{cosec}(\nu\pi)\left[J_{-\frac{\nu}{2}}\left(\frac{a}{2}\right)\right]^2\right\}$

                                     $[|\operatorname{Re}\nu| < 1]$ ET II 195(68)a

10. $\displaystyle\int_1^\infty x^{-\frac{1}{2}\nu}(x-1)^{\mu-1} J_\nu\left(a\sqrt{x}\right) dx = \Gamma(\mu) 2^\mu a^{-\mu} J_{\nu-\mu}(a)$

                                     $\left[a > 0, \quad 0 < \operatorname{Re}\mu < \frac{1}{2}\operatorname{Re}\nu + \frac{3}{4}\right]$
                                                     ET II 205(34)a

11. $\displaystyle\int_1^\infty x^{-\frac{1}{2}\nu}(x-1)^{\mu-1} J_{-\nu}\left(a\sqrt{x}\right) dx = \Gamma(\mu) 2^\mu a^{-\mu}\left[\cos(\nu\pi) J_{\nu-\mu}(a) - \sin(\nu\pi) Y_{\nu-\mu}(a)\right]$

                                     $\left[a > 0, \quad 0 < \operatorname{Re}\mu < \frac{1}{2}\operatorname{Re}\nu + \frac{3}{4}\right]$
                                                     ET II 205(35)a

12. $\displaystyle\int_1^\infty x^{-\frac{1}{2}\nu}(x-1)^{\mu-1} K_\nu\left(a\sqrt{x}\right) dx = \Gamma(\mu) 2^\mu a^{-\mu} K_{\nu-\mu}(a)$

                                     $[\operatorname{Re} a > 0, \quad \operatorname{Re}\mu > 0]$ ET II 209(59)a

13. $\displaystyle\int_1^\infty x^{-\frac{1}{2}\nu}(x-1)^{\mu-1} Y_\nu\left(a\sqrt{x}\right) dx = 2^\mu a^{-\mu} Y_{\nu-\mu}(a)\,\Gamma(\mu)$

                                     $\left[a > 0, \quad 0 < \operatorname{Re}\mu < \frac{1}{2}\operatorname{Re}\nu + \frac{3}{4}\right]$
                                                     ET II 206(40)a

14. $\displaystyle\int_1^\infty x^{-\frac{1}{2}\nu}(x-1)^{\mu-1} H_\nu^{(1)}\left(a\sqrt{x}\right) dx = 2^\mu a^{-\mu} H_{\nu-\mu}^{(1)}(a)\,\Gamma(\mu)$

                                     $[\operatorname{Re}\mu > 0, \quad \operatorname{Im} a > 0]$ ET II 206(45)a

15. $\displaystyle\int_1^\infty x^{-\frac{1}{2}\nu}(x-1)^{\mu-1} H_\nu^{(2)}\left(a\sqrt{x}\right) dx = 2^\mu a^{-\mu} H_{\nu-\mu}^{(2)}(a)\,\Gamma(\mu)$

                                     $[\operatorname{Re}\mu > 0, \quad \operatorname{Im} a < 0]$ ET II 207(48)a

16. $\displaystyle\int_0^1 x^{-\frac{1}{2}\nu}(1-x)^{\mu-1} J_\nu\left(a\sqrt{x}\right) dx = \frac{2^{2-\nu} a^{-\mu}}{\Gamma(\nu)}\, s_{\mu+\nu-1,\mu-\nu}(a)$

                                     $[\operatorname{Re}\mu > 0]$ ET II 194(64)a

17. $\int_0^1 x^{-\frac{1}{2}\nu}(1-x)^{\mu-1}\,Y_\nu\left(a\sqrt{x}\right)\,dx = \dfrac{2^{2-\nu}a^{-\mu}\cot(\nu\pi)}{\Gamma(\nu)}\,s_{\mu+\nu-1,\mu-\nu}(a)$

$$-2^\mu a^{-\mu}\,\operatorname{cosec}(\nu\pi)\,J_{\mu-\nu}(a)\,\Gamma(\mu)$$

$$[\operatorname{Re}\mu>0,\quad \operatorname{Re}\nu<1]\qquad \text{ET II 196(75)a}$$

6.593

1. $\int_0^\infty \sqrt{x}\,J_{2\nu-1}\left(a\sqrt{x}\right)\,J_\nu(bx)\,dx = \dfrac{1}{2}ab^{-2}\,J_{\nu-1}\left(\dfrac{a^2}{4b}\right)\qquad \left[b>0,\quad \operatorname{Re}\nu>-\tfrac{1}{2}\right]\qquad \text{ET II 58(15)}$

2. $\int_0^\infty \sqrt{x}\,J_{2\nu-1}\left(a\sqrt{x}\right)\,K_\nu(bx)\,dx = \dfrac{\pi a}{4b^2}\left[I_{\nu-1}\left(\dfrac{a^2}{4b}\right) - \mathbf{L}_{\nu-1}\left(\dfrac{a^2}{4b}\right)\right]$

$$\left[\operatorname{Re}b>0,\quad \operatorname{Re}\nu>-\tfrac{1}{2}\right]\qquad \text{ET II 144(44)}$$

6.594

1. $\int_0^\infty x^\nu\,I_{2\nu-1}\left(a\sqrt{x}\right)\,J_{2\nu-1}\left(a\sqrt{x}\right)\,K_\nu(bx)\,dx = \sqrt{\pi}2^{-\nu}a^{2\nu-1}b^{-2\nu-\frac{1}{2}}\,J_{\nu-\frac{1}{2}}\left(\dfrac{a^2}{2b}\right)$

$$[\operatorname{Re}b>0,\quad \operatorname{Re}\nu>0]\qquad \text{ET II 148(65)}$$

2. $\int_0^\infty x^\nu\,I_{2\nu-1}\left(a\sqrt{x}\right)\,Y_{2\nu-1}\left(a\sqrt{x}\right)\,K_\nu(bx)\,dx$

$$= \sqrt{\pi}2^{-\nu-1}a^{2\nu-1}b^{-2\nu-\frac{1}{2}}\operatorname{cosec}(\nu\pi)$$

$$\times \left[\mathbf{H}_{\frac{1}{2}-\nu}\left(\dfrac{a^2}{2b}\right) + \cos(\nu\pi)\,J_{\nu-\frac{1}{2}}\left(\dfrac{a^2}{2b}\right) + \sin(\nu\pi)\,Y_{\nu-\frac{1}{2}}\left(\dfrac{a^2}{2b}\right)\right]$$

$$[\operatorname{Re}b>0,\quad \operatorname{Re}\nu>0]\qquad \text{ET II 148(66)}$$

3. $\int_0^\infty x^\nu\,J_{2\nu-1}\left(a\sqrt{x}\right)\,K_{2\nu-1}\left(a\sqrt{x}\right)\,K_\nu(bx)\,dx$

$$= \pi^2 2^{-\nu-2}a^{2\nu-1}b^{-2\nu-\frac{1}{2}}\operatorname{cosec}(\nu\pi)\left[\mathbf{H}_{\frac{1}{2}-\nu}\left(\dfrac{a^2}{2b}\right) - Y_{\frac{1}{2}-\nu}\left(\dfrac{a^2}{2b}\right)\right]$$

$$[\operatorname{Re}b>0,\quad \operatorname{Re}\nu>0]\qquad \text{ET II 148(67)}$$

6.595

1. $\int_0^\infty x^{\nu+1}\,J_\nu(cx)\prod_{i=1}^n z_i^{-\mu_i}\,J_{\mu_i}(a_i z_i)\,dx = 0 \qquad\qquad z_i=\sqrt{x^2+b_i^2}$

$$\left[a_i>0,\quad \operatorname{Re}b_i>0,\quad \sum_{i=1}^n a_i<c;\quad \operatorname{Re}\left(\dfrac{1}{2}n+\sum_{i=1}^n \mu_i-\dfrac{1}{2}\right)>\operatorname{Re}\nu>-1\right]$$

$$\text{EH II 52(33), ET II 60(26)}$$

2. $\int_0^\infty x^{\nu-1}\,J_\nu(cx)\prod_{i=1}^n z_i^{-\mu_i}\,J_{\mu_i}(a_i z_i)\,dx = 2^{\nu-1}\,\Gamma(\nu)c^{-\nu}\prod_{i=1}^n \left[b_i^{-\mu_i}\,J_{\mu_i}(a_i b_i)\right]\qquad z_i=\sqrt{x^2+b_i^2}$

$$\left[a_i>0,\quad \operatorname{Re}b_i>0,\quad \sum_{i=1}^n a_i<c,\quad \operatorname{Re}\left(\dfrac{1}{2}n+\sum_{i=1}^n \mu_i+\dfrac{3}{2}\right)>\operatorname{Re}\nu>0\right]$$

$$\text{EH II 52(34), ET II 60(27)}$$

6.596

1. $$\int_0^\infty J_\nu\left(\alpha\sqrt{x^2+z^2}\right)\frac{x^{2\mu+1}}{\sqrt{(x^2+z^2)^\nu}}\,dx = \frac{2^\mu\,\Gamma(\mu+1)}{\alpha^{\mu+1}z^{\nu-\mu-1}}\,J_{\nu-\mu-1}(\alpha z)$$

$$\left[\alpha>0,\quad \operatorname{Re}\left(\frac{1}{2}\nu-\frac{1}{4}\right)>\operatorname{Re}\mu>-1\right]$$

WA 457(5)

2. $$\int_0^\infty \frac{J_\nu\left(\alpha\sqrt{t^2+1}\right)}{\sqrt{t^2+1}}\,dt = -\frac{\pi}{2}\,J_{\frac{\nu}{2}}\left(\frac{\alpha}{2}\right)Y_{\frac{\nu}{2}}\left(\frac{\alpha}{2}\right)\qquad [\operatorname{Re}\nu>-1,\quad \alpha>0]$$

MO 46

3. $$\int_0^\infty K_\nu\left(\alpha\sqrt{x^2+z^2}\right)\frac{x^{2\mu+1}}{\sqrt{(x^2+z^2)^\nu}}\,dx = \frac{2^\mu\,\Gamma(\mu+1)}{\alpha^{\mu+1}z^{\nu-\mu-1}}\,K_{\nu-\mu-1}(\alpha z)$$

$$[\alpha>0,\quad \operatorname{Re}\mu>-1]$$

WA 457(6)

4.[8] $$\int_0^\infty J_\nu(\beta x)\frac{J_{\mu-1}\left\{\alpha\sqrt{x^2+z^2}\right\}}{(x^2+z^2)^{\frac{1}{2}\mu+\frac{1}{2}}}x^{\nu+1}\,dx = \frac{\alpha^{\mu-1}z^\nu}{2^{\mu-1}\,\Gamma(\mu)}\,K_\nu(\beta z)$$

$$[\alpha<\beta,\quad \operatorname{Re}(\mu+2)>\operatorname{Re}\nu>-1]$$

ET II 59(19)

5.[8] $$\int_0^\infty J_\nu(\beta x)\frac{J_\mu\left\{\alpha\sqrt{x^2+z^2}\right\}}{\sqrt{(x^2+z^2)^\mu}}x^{\nu-1}\,dx = \frac{2^{\nu-1}\,\Gamma(\nu)}{\beta^\nu}\frac{J_\mu(\alpha z)}{z^\mu}$$

$$[\operatorname{Re}(\mu+2)>\operatorname{Re}\nu>0,\quad \beta>\alpha>0]$$

WA 459(12)

6.[6] $$\int_0^\infty J_\nu(\beta x)\frac{J_\mu\left(\alpha\sqrt{x^2+z^2}\right)}{\sqrt{(x^2+z^2)^\mu}}x^{\nu+1}\,dx$$

$$= 0 \qquad\qquad\qquad\qquad\qquad [0<\alpha<\beta]$$

$$= \frac{\beta^\nu}{\alpha^\mu}\left(\frac{\sqrt{\alpha^2-\beta^2}}{z}\right)^{\mu-\nu-1}J_{\mu-\nu-1}\left\{z\sqrt{\alpha^2-\beta^2}\right\}\quad [\alpha>\beta>0]$$

$$[\operatorname{Re}\mu>\operatorname{Re}\nu>-1]\qquad\text{WA 415(1)}$$

7.[8] $$\int_0^\infty J_\nu(\beta x)\frac{K_\mu\left(\alpha\sqrt{x^2+z^2}\right)}{\sqrt{(x^2+z^2)^\mu}}x^{\nu+1}\,dx = \frac{\beta^\nu}{\alpha^\mu}\left(\frac{\sqrt{\alpha^2+\beta^2}}{z}\right)^{\mu-\nu-1}K_{\mu-\nu-1}\left(z\sqrt{\alpha^2+\beta^2}\right)$$

$$\left[\alpha>0,\quad \beta>0,\quad \operatorname{Re}\nu>-1,\quad |\arg z|<\frac{\pi}{2}\right]\qquad\text{KU 151(31), WA 416(2)}$$

8.[8] $$\int_0^\infty J_\nu(ux)\,K_\mu\left(v\sqrt{x^2-y^2}\right)(x^2-y^2)^{-\frac{\mu}{2}}x^{\nu+1}\,dx = \frac{\pi}{2}\exp\left[-i\pi\left(\mu-\nu-\frac{1}{2}\right)\right]\cdot\frac{u^\nu}{v^\mu}$$

$$\cdot\left[\frac{\sqrt{u^2+v^2}}{y}\right]^{\mu-\nu-1}H_{\mu-\nu-1}^{(2)}\left(y\sqrt{u^2+v^2}\right)$$

$$\left[\operatorname{Re}\mu<1,\quad \operatorname{Re}\nu>-1, u>0,\quad v>0, y>0;\quad (x^2-y^2)^{\frac{1}{2}\alpha}=e^{\frac{1}{2}\alpha\pi i}(y^2-n^2)^{\frac{1}{2}\alpha}\text{if }x<y\right]$$

9.^{8} $\displaystyle\int_0^\infty J_\nu(ux)\, H_\mu^{(2)}\left(v\sqrt{x^2+y^2}\right)\left(x^2+y^2\right)^{-\frac{\mu}{2}} x^{\nu+1}\, dx$

$$= \frac{u^\nu}{v^\mu}\left[\frac{\sqrt{v^2-u^2}}{y}\right]^{\mu-\nu-1} H_{\mu-\nu-1}^{(2)}\left(y\sqrt{v^2-u^2}\right)$$

$$[u < v]$$

$$\left[\mathrm{Re}\,\mu > \mathrm{Re}\,\nu > -1, \quad u > 0, \quad v > 0, \quad y > 0;, \quad \arg\sqrt{v^2-u^2} = 0,\ \text{for}\ v > u\right.$$

$$\left.\arg\left(v^2-u^2\right)^\sigma = -\pi\sigma\ \text{for}\ v < u,\ \text{where}\ \sigma = \frac{1}{2}\ \text{or}\ \sigma = \frac{\mu-\nu-1}{2}\right]$$

MO 43

10.^{8} $\displaystyle\int_0^\infty J_\nu(\beta x)\, J_\mu\left(\alpha\sqrt{x^2+z^2}\right) J_\mu\left(\gamma\sqrt{x^2+z^2}\right)\frac{x^{\nu-1}}{(x^2+z^2)^\mu}\, dx = \frac{2^{\nu-1}\,\Gamma(\nu)}{\beta^\nu}\frac{J_\mu(\alpha z)}{z^\mu}\frac{J_\mu(\gamma z)}{z^\mu}$

$$\left[\alpha > 0;\quad \beta > \alpha+\gamma;\quad \gamma > 0, \quad \mathrm{Re}\left(2\mu+\tfrac{5}{2}\right) > \mathrm{Re}\,\nu > 0\right]\quad \text{WA 459(14)}$$

11.^{8} $\displaystyle\int_0^\infty J_\nu(\beta t)\, t^{\nu-1}\prod_{k=1}^n J_\mu\left(\alpha_k\sqrt{t^2+x^2}\right)\sqrt{(t^2+x^2)^{-n\mu}}\, dt = 2^{\nu-1}\beta^{-\nu}\,\Gamma(\nu)\prod_{k=1}^n\left[x^{-\mu}\,J_\mu\left(\alpha_k x\right)\right]$

$$\left[x > 0, \quad \alpha_1 > 0, \quad \alpha_2 > 0,\ldots,\alpha_n > 0, \quad \beta > \prod_{k=1}^n \alpha_k;\quad \mathrm{Re}\left(n\mu+\frac{1}{2}n+\frac{1}{2}\right) > \mathrm{Re}\,\nu > 0\right]$$

MO 43

12.^{8} $\displaystyle\int_0^\infty \frac{J_\mu^2\left(\sqrt{a^2+x^2}\right)}{(a^2+x^2)^\nu}\, x^{2\nu-2}\, dx = \frac{\Gamma\left(\nu-\frac{1}{2}\right)}{2a^{\nu+1}\sqrt{\pi}}\,\mathbf{H}_\nu(2a) \qquad \left[\mathrm{Re}\,\nu > \tfrac{1}{2}\right]$ 　　　　　　WA 457(8)

6.597 $\displaystyle\int_0^\infty t^{\nu+1}\, J_\mu\left[b\left(t^2+y^2\right)^{\frac{1}{2}}\right]\left(t^2+y^2\right)^{-\frac{1}{2}\mu}\left(t^2+\beta^2\right)^{-1} J_\nu(at)\, dt$

$$= \beta^\nu\, J_\mu\left[b\left(y^2-\beta^2\right)^{\frac{1}{2}}\right]\left(y^2-\beta^2\right)^{-\frac{1}{2}\mu} K_\nu(a\beta)$$

$$[a \geq b, \quad \mathrm{Re}\,\beta > 0, \quad -1 < \mathrm{Re}\,\nu < 2 + \mathrm{Re}\,\mu]\quad \text{EH II 95(56)}$$

6.598 $\displaystyle\int_0^1 x^{\frac{\mu}{2}}(1-x)^{\frac{\nu}{2}}\, J_\mu\left(a\sqrt{x}\right) J_\nu\left(b\sqrt{1-x}\right)\, dx = 2a^\mu b^\nu\left(a^2+b^2\right)^{-\frac{1}{2}(\nu+\mu+1)} J_{\nu+\mu+1}\left(\sqrt{a^2+b^2}\right)$

$$[\mathrm{Re}\,\nu > -1, \quad \mathrm{Re}\,\mu > -1]\qquad \text{EH II 46a}$$

6.61 Combinations of Bessel functions and exponentials

6.611

1. $\displaystyle\int_0^\infty e^{-\alpha x}\, J_\nu(\beta x)\, dx = \frac{\beta^{-\nu}\left[\sqrt{\alpha^2+\beta^2}-\alpha\right]^\nu}{\sqrt{\alpha^2+\beta^2}}$ 　　　　　　$[\mathrm{Re}\,\nu > -1, \quad \mathrm{Re}\left(\alpha\pm i\beta\right) > 0]$

EH II 49(18), WA 422(8)

2. $\displaystyle\int_0^\infty e^{-\alpha x}\, Y_\nu(\beta x)\, dx = \left(\alpha^2 + \beta^2\right)^{-\frac{1}{2}} \operatorname{cosec}(\nu\pi)$
$$\times \left\{ \beta^\nu \left[\left(\alpha^2 + \beta^2\right)^{\frac{1}{2}} + \alpha \right]^{-\nu} \cos(\nu\pi) - \beta^{-\nu} \left[\left(\alpha^2 + \beta^2\right)^{\frac{1}{2}} + \alpha \right]^\nu \right\}$$
$$[\operatorname{Re}\alpha > 0, \quad \beta > 0, \quad |\operatorname{Re}\nu| < 1] \quad \text{MO 179, ET II 105(1)}$$

3. $\displaystyle\int_0^\infty e^{-\alpha x}\, K_\nu(\beta x)\, dx = \frac{\pi}{\beta \sin(\nu\pi)} \frac{\sin(\nu\theta)}{\sin\theta}$
$$\left[\cos\theta = \frac{\alpha}{\beta}; \quad \theta \to \frac{\pi}{2} \quad \text{for } \beta \to \infty \right]$$
$$\text{ET II 131(22)}$$
$$= \frac{\pi \operatorname{cosec}(\nu\pi)}{2\sqrt{\alpha^2 - \beta^2}} \left[\beta^{-\nu}\left(\alpha + \sqrt{\alpha^2 - \beta^2}\right)^\nu - \beta^\nu \left(\sqrt{\alpha^2 - \beta^2} + \alpha\right)^{-\nu} \right]$$
$$[|\operatorname{Re}\nu| < 1, \quad \operatorname{Re}(\alpha + \beta) > 0]$$
$$\text{ET I 197(24), MO 180}$$

4.[8] $\displaystyle\int_0^\infty e^{-\alpha x}\, I_\nu(\beta x)\, dx = \frac{\beta^{-\nu}\left[\alpha - \sqrt{\alpha^2 - \beta^2}\right]^\nu}{\sqrt{\alpha^2 - \beta^2}} \qquad [\operatorname{Re}\nu > -1, \quad \operatorname{Re}\alpha > |\operatorname{Re}\beta|]$
$$\text{MO 180, ET I 195(1)}$$

5. $\displaystyle\int_0^\infty e^{-\alpha x}\, H_\nu^{(1,2)}(\beta x)\, dx = \frac{\left(\sqrt{\alpha^2 + \beta^2} - \alpha\right)^\nu}{\beta^\nu \sqrt{\alpha^2 + \beta^2}} \left\{ 1 \pm \frac{i}{\sin(\nu\pi)} \left[\cos(\nu\pi) - \frac{\left(\alpha + \sqrt{\alpha^2 + \beta^2}\right)^{2\nu}}{b^{2\nu}} \right] \right\}$

$[-1 < \operatorname{Re}\nu < 1$; a plus sign corresponds to the function $H_\nu^{(1)}$, a minus sign to the function $H_\nu^{(2)}]$.
$$\text{MO 180, ET I188(54, 55)}$$

6. $\displaystyle\int_0^\infty e^{-\alpha x}\, H_0^{(1)}(\beta x)\, dx = \frac{1}{\sqrt{\alpha^2 + \beta^2}} \left\{ 1 - \frac{2i}{\pi} \ln\left[\frac{\alpha}{\beta} + \sqrt{1 + \left(\frac{\alpha}{\beta}\right)^2} \right] \right\}$
$$[\operatorname{Re}\alpha > |\operatorname{Im}\beta|] \quad \text{MO 180, ET I 188(53)}$$

7. $\displaystyle\int_0^\infty e^{-\alpha x}\, H_0^{(2)}(\beta x)\, dx = \frac{1}{\sqrt{\alpha^2 + \beta^2}} \left\{ 1 + \frac{2i}{\pi} \ln\left[\frac{\alpha}{\beta} + \sqrt{1 + \left(\frac{\alpha}{\beta}\right)^2} \right] \right\}$
$$[\operatorname{Re}\alpha > |\operatorname{Im}\beta|] \quad \text{MO 180, ET I 188(53)}$$

8. $\displaystyle\int_0^\infty e^{-\alpha x}\, Y_0(\beta x)\, dx = \frac{-2}{\pi\sqrt{\alpha^2 + \beta^2}} \ln\frac{\alpha + \sqrt{\alpha^2 + \beta^2}}{\beta}$
$$[\operatorname{Re}\alpha > |\operatorname{Im}\beta|] \quad \text{MO 47, ET I 187(44)}$$

9. $\displaystyle\int_0^\infty e^{-\alpha x}\, K_0(\beta x)\, dx$

$$= \frac{\arccos \frac{\alpha}{\beta}}{\sqrt{\beta^2 - \alpha^2}} \qquad\qquad [0 < \alpha < \beta, \quad \text{Re}(\alpha + \beta) > 0] \quad \text{WA 424, ET II 131(22)}$$

$$= \frac{1}{\sqrt{\alpha^2 - \beta^2}} \ln\left(\frac{\alpha}{\beta} + \sqrt{\frac{\alpha^2}{\beta^2} - 1}\right) \qquad [0 \le \beta < \alpha, \quad \text{Re}(\alpha + \beta) > 0]$$

MO 48

$10.^{10}$ $\displaystyle\int_a^b \alpha\,d\alpha \int_0^\infty dk\, J_1(k\alpha) e^{-k|\beta|} = \int_a^b \left(1 - \frac{|\beta|}{\sqrt{\alpha^2 + \beta^2}}\right) d\alpha$

(see 3.241 6)

6.612

1. $\displaystyle\int_0^\infty e^{-2\alpha x} J_0(x)\, Y_0(x)\, dx = \frac{\boldsymbol{K}\left[\alpha\left(\alpha^2 + 1\right)^{-\frac{1}{2}}\right]}{\pi(\alpha^2 + 1)^{\frac{1}{2}}}$ $\qquad [\text{Re}\,\alpha > 0]$ $\qquad$ ET II 347(58)

2. $\displaystyle\int_0^\infty e^{-2\alpha x} I_0(x)\, K_0(x)\, dx = \frac{1}{2}\boldsymbol{K}\left[\left(1 - \alpha^2\right)^{\frac{1}{2}}\right]$ $\qquad [0 < \alpha < 1]$

$$\qquad\qquad\qquad\qquad = \frac{1}{2\alpha}\boldsymbol{K}\left[\left(1 - \frac{1}{\alpha^2}\right)^{\frac{1}{2}}\right] \qquad [1 < \alpha < \infty]$$

ET II 370(48)

3. $\displaystyle\int_0^\infty e^{-\alpha x} J_\nu(\beta x)\, J_\nu(\gamma x)\, dx = \frac{1}{\pi\sqrt{\gamma\beta}}\, Q_{\nu - \frac{1}{2}}\left(\frac{\alpha^2 + \beta^2 + \gamma^2}{2\beta\gamma}\right)$

$$\left[\text{Re}\left(\alpha \pm i\beta \pm i\gamma\right) > 0, \quad \gamma > 0, \quad \text{Re}\,\nu > -\tfrac{1}{2}\right] \quad \text{WA 426(2), ET II 50(17)}$$

4. $\displaystyle\int_0^\infty e^{-\alpha x}[J_0(\beta x)]^2\, dx = \frac{2}{\pi\sqrt{\alpha^2 + 4\beta^2}}\boldsymbol{K}\left(\frac{2\beta}{\sqrt{\alpha^2 + 4\beta^2}}\right)$ $\qquad$ MO 178

5. $\displaystyle\int_0^\infty e^{-2\alpha x} J_1^2(\beta x)\, dx = \frac{\left(2\alpha^2 + \beta^2\right)\boldsymbol{K}\left(\frac{\beta}{\sqrt{\alpha^2 + \beta^2}}\right) - 2\left(\alpha^2 + \beta^2\right)\boldsymbol{E}\left(\frac{\beta}{\sqrt{\alpha^2 + \beta^2}}\right)}{\pi\beta^2\sqrt{\alpha^2 + \beta^2}}$ $\qquad$ WA 428(3)

6. $\displaystyle\int_0^\infty e^{-3x} I_l(x)\, I_m(x)\, I_n(x)\, dx = r_1 g + \frac{r_2}{\pi^2 g} + r_3$

where

$$g = \frac{\sqrt{3} - 1}{96\pi^3}\, \Gamma^2\left(\frac{1}{24}\right)\Gamma^2\left(\frac{11}{24}\right)$$

and

(lmn)	r_1	r_2	r_3
000	1	0	0
100	1	0	$-1/3$
110	$5/12$	$-1/2$	0
111	$-1/8$	$3/4$	0
200	$10/3$	2	-2
210	$3/8$	$-9/4$	$1/3$
211	$-2/3$	2	0
220	$73/36$	$-29/6$	0
221	$-15/16$	$21/8$	0
222	$5/8$	$-27/20$	0
300	$35/2$	21	-13
310	$-79/36$	$-85/6$	4
311	$-11/4$	$21/2$	$-2/3$
320	$319/48$	$-119/8$	$-1/3$
321	$-125/36$	$269/30$	0
322	$35/16$	$-213/40$	0
330	$50/3$	$-1046/25$	0
331	$-35/3$	$148/5$	0
332	$35/9$	$-1012/105$	0
333	$-35/16$	$1587/280$	0
400	$994/9$	$542/3$	-92
410	$-515/16$	$-879/8$	$115/3$
411	$-9/2$	$357/5$	-12
420	$12907/120$	$-13903/10$	-6
421	$-229/16$	$1251/40$	1
422	$35/3$	$-1024/35$	0
430	$2641/48$	$-28049/200$	$1/3$
431	$-1505/36$	$118051/1050$	0

(lmn)	r_1	r_2	r_3
432	$525/32$	$-4617/112$	0
433	$-595/72$	$8809/420$	0
440	$6025/36$	$-620161/1470$	0
441	$-29175/224$	$131379/400$	0
442	$2975/48$	$-31231/200$	0
443	$-539/32$	$119271/2800$	0
444	$77/8$	$-186003/7700$	0
500	$9287/12$	$3005/2$	$-2077/3$
510	$-189029/180$	$-138331/50$	348
511	$275/4$	$5751/10$	-150
520	$2897/16$	$-15123/20$	$-229/3$
521	$-937/12$	$27059/30$	24
522	$509/8$	$-4209/28$	0
530	$3589/18$	$-1993883/3075$	0
531	$-1329/8$	$297981/700$	$-4/3$
532	$2555/36$	$-187777/1050$	0
533	$-2233/48$	$164399/1400$	0
540	$18471/32$	$-28493109/19600$	$-1/3$
541	$-1390/3$	$286274/245$	0
542	$7777/32$	$-1715589/2800$	0
543	$-5621/72$	$4550057/23100$	0
544	$1155/32$	$-560001/6160$	0
550	$197045/108$	$-101441689/22050$	0
551	$-12023/8$	$18569853/4900$	0
552	$1683/2$	$-5718309/2695$	0
553	$-5159/16$	$2504541/3080$	0
554	$24563/312$	$-1527851/77000$	0
555	$-9251/208$	$12099711/107800$	0

6.613[7] $\displaystyle\int_0^\infty e^{-x^2} J_{\nu+\frac{1}{2}}\left(\frac{x^2}{2}\right) dx = \frac{\Gamma(\nu+1)}{\sqrt{\pi}} D_{-\nu-1}\left(ze^{\frac{\pi}{4}i}\right) D_{-\nu-1}\left(ze^{-\frac{\pi i}{4}}\right)$ $[\operatorname{Re}\nu > -1]$ MO 122

6.614

1. $\displaystyle\int_0^\infty e^{-\alpha x} J_\nu\left(\beta\sqrt{x}\right) dx = \frac{\beta}{4}\sqrt{\frac{\pi}{\alpha^3}} \exp\left(-\frac{\beta^2}{8\alpha}\right)\left[I_{\frac{1}{2}(\nu-1)}\left(\frac{\beta^2}{8\alpha}\right) - I_{\frac{1}{2}(\nu+1)}\left(\frac{\beta^2}{8\alpha}\right)\right]$

$\displaystyle = \frac{1}{\alpha} e^{-\beta^2/4\alpha}$ $[\nu = 0]$

    MO 178

2. $\displaystyle\int_0^\infty e^{-\alpha x} Y_{2\nu}\left(2\sqrt{\beta x}\right) dx = \frac{e^{-\frac{1}{2}\frac{\beta}{\alpha}}}{\sqrt{\alpha\beta}}\left\{\cot(\nu\pi)\frac{\Gamma(\nu+1)}{\Gamma(2\nu+1)} M_{\frac{1}{2},\nu}\left(\frac{\beta}{\alpha}\right) - \operatorname{cosec}(\nu\pi)\, W_{\frac{1}{2},nu}\left(\frac{\beta}{\alpha}\right)\right\}$

$[\operatorname{Re}\alpha > 0, \quad |\operatorname{Re}\nu| < 1]$ ET I 188(50)a

3. $\displaystyle\int_0^\infty e^{-\alpha x} I_{2\nu}\left(2\sqrt{\beta x}\right) dx = \frac{e^{\frac{1}{2}\frac{\beta}{\alpha}}}{\sqrt{\alpha\beta}}\frac{\Gamma(\nu+1)}{\Gamma(2\nu+1)} M_{-\frac{1}{2},\nu}\left(\frac{\beta}{\alpha}\right)$

$[\operatorname{Re}\alpha > 0, \quad \operatorname{Re}\nu > -1]$ ET I 197(20)a

4. $$\int_0^\infty e^{-\alpha x} K_{2\nu}\left(2\sqrt{\beta x}\right) dx = \frac{e^{\frac{1}{2}\frac{\beta}{\alpha}}}{2\sqrt{\alpha\beta}} \Gamma(\nu+1)\,\Gamma(1-\nu)\,W_{-\frac{1}{2},\nu}\left(\frac{\beta}{\alpha}\right)$$

$$[\operatorname{Re}\alpha > 0, \quad |\operatorname{Re}\nu| < 1] \qquad \text{ET I 199(37)a}$$

5. $$\int_0^\infty e^{-\alpha x} K_1\left(\beta\sqrt{x}\right) dx = \frac{\beta}{8}\sqrt{\frac{\pi}{\alpha^3}} \exp\left(\frac{\beta^2}{8\alpha}\right)\left[K_1\left(\frac{\beta^2}{8\alpha}\right) - K_0\left(\frac{\beta^2}{8\alpha}\right)\right]$$

MO 181

6.615 $$\int_0^\infty e^{-\alpha x} J_\nu\left(2\beta\sqrt{x}\right) J_\nu\left(2\gamma\sqrt{x}\right) dx = \frac{1}{\alpha} I_\nu\left(\frac{2\beta\gamma}{\alpha}\right) \exp\left(-\frac{\beta^2+\gamma^2}{\alpha}\right) \qquad [\operatorname{Re}\nu > -1]$$

MO 178

6.616

1. $$\int_0^\infty e^{-\alpha x} J_0\left(\beta\sqrt{x^2+2\gamma x}\right) dx = \frac{1}{\sqrt{\alpha^2+\beta^2}} \exp\left[\gamma\left(\alpha - \sqrt{\alpha^2+\beta^2}\right)\right]$$

MO 179

2. $$\int_1^\infty e^{-\alpha x} J_0\left(\beta\sqrt{x^2-1}\right) dx = \frac{1}{\sqrt{\alpha^2+\beta^2}} \exp\left(-\sqrt{\alpha^2+\beta^2}\right)$$

MO 179

3. $$\int_{-\infty}^\infty e^{itx} H_0^{(1)}\left(r\sqrt{\alpha^2-t^2}\right) dt = -2i\frac{e^{i\alpha\sqrt{r^2+x^2}}}{\sqrt{r^2+x^2}}$$

$$\left[0 \le \arg\sqrt{\alpha^2-t^2} < \pi, \quad 0 \le \arg\alpha < \pi; \quad r \text{ and } x \text{ are real}\right] \quad \text{MO 49}$$

4. $$\int_{-\infty}^\infty e^{-itx} H_0^{(2)}\left(r\sqrt{\alpha^2-t^2}\right) dt = 2i\frac{e^{-i\alpha\sqrt{r^2+x^2}}}{\sqrt{r^2+x^2}}$$

$$\left[-\pi < \arg\sqrt{\alpha^2-t^2} \le 0, \quad -\pi < \arg\alpha \le 0, \quad r \text{ and } x \text{ are real}\right] \quad \text{MO 49}$$

5.[3] $$\int_{-1}^1 e^{-ax} I_0\left(b\sqrt{1-x^2}\right) dx = 2(a^2+b^2)^{-1/2} \sinh\sqrt{a^2+b^2}$$

$$[a > 0, \quad b > 0]$$

6.[8] $$\int_0^\infty e^{-xy} J_0\left[y\sqrt{1-x^2}\right]/(\alpha+y)\,dy = \sum_{n=0}^\infty n!\frac{P_n(x)}{\alpha^{n+1}}$$

6.617

1. $$\int_0^\infty K_{q-p}(2z\sinh x)\,e^{(p+q)x}\,dx = \frac{\pi^2}{4\sin[(p-q)\pi]}\left[J_p(z)\,Y_q(z) - J_q(z)\,Y_p(z)\right]$$

$$[\operatorname{Re}z > 0, \quad -1 < \operatorname{Re}(p-q) < 1]$$

MO 44

2. $$\int_0^\infty K_0(2z\sinh x)\,e^{-2px}\,dx = -\frac{\pi}{4}\left\{J_p(z)\frac{\partial Y_p(z)}{\partial p} - Y_p(z)\frac{\partial J_p(z)}{\partial p}\right\}$$

$$[\operatorname{Re}z > 0]$$

MO 44

6.618

1. $$\int_0^\infty e^{-\alpha x^2} J_\nu(\beta x)\,dx = \frac{\sqrt{\pi}}{2\sqrt{\alpha}} \exp\left(-\frac{\beta^2}{8\alpha}\right) I_{\frac{1}{2}\nu}\left(\frac{\beta^2}{8\alpha}\right) \qquad [\operatorname{Re}\alpha > 0, \quad \beta > 0, \quad \operatorname{Re}\nu > -1]$$

WA 432(5), ET II 29(8)

2. $\int_0^\infty e^{-\alpha x^2} Y_\nu(\beta x)\,dx = -\dfrac{\sqrt{\pi}}{2\sqrt{\alpha}}\exp\left(-\dfrac{\beta^2}{8\alpha}\right)\left[\tan\dfrac{\nu\pi}{2}\,I_{\frac{1}{2}\nu}\left(\dfrac{\beta^2}{8\alpha}\right) + \dfrac{1}{\pi}\sec\left(\dfrac{\nu\pi}{2}\right)K_{\frac{1}{2}\nu}\left(\dfrac{\beta^2}{8\alpha}\right)\right]$

$$[\operatorname{Re}\alpha > 0, \quad \beta > 0, \quad |\operatorname{Re}\nu| < 1]$$

WA 432(6), ET II 106(3)

3. $\int_0^\infty e^{-\alpha x^2} K_\nu(\beta x)\,dx = \dfrac{1}{4}\sec\left(\dfrac{\nu\pi}{2}\right)\dfrac{\sqrt{\pi}}{\sqrt{\alpha}}\exp\left(\dfrac{\beta^2}{8\alpha}\right)K_{\frac{1}{2}\nu}\left(\dfrac{\beta^2}{8\alpha}\right)$

$$[\operatorname{Re}\alpha > 0, \quad |\operatorname{Re}\nu| < 1]$$

EH II 51(28), ET II 132(24)

4. $\int_0^\infty e^{-\alpha x^2} I_\nu(\beta x)\,dx = \dfrac{\sqrt{\pi}}{2\sqrt{\alpha}}\exp\left(\dfrac{\beta^2}{8\alpha}\right)I_{\frac{1}{2}\nu}\left(\dfrac{\beta^2}{8\alpha}\right)$ $[\operatorname{Re}\nu > -1, \quad \operatorname{Re}\alpha > 0]$ EH II 92(27)

5. $\int_0^\infty e^{-\alpha x^2} J_\mu(\beta x) J_\nu(\beta x)\,dx$

$$= 2^{-\nu-\mu-1}\alpha^{-\frac{\nu+\mu+1}{2}}\beta^{\nu+\mu}\frac{\Gamma\left(\frac{\mu+\nu+1}{2}\right)}{\Gamma(\mu+1)\,\Gamma(\nu+1)}$$

$$\times\; {}_3F_3\left(\frac{\nu+\mu+1}{2},\frac{\nu+\mu+2}{2},\frac{\nu+\mu+1}{2};\mu+1,\nu+1,\nu+\mu+1;-\frac{\beta^2}{\alpha}\right)$$

$$[\operatorname{Re}(\nu+\mu) > -1, \quad \operatorname{Re}\alpha > 0] \quad \text{EH II 50(21)a}$$

6.62–6.63 Combinations of Bessel functions, exponentials, and powers

6.621   **Notation**:

$$\ell_1 = \frac{1}{2}\left[\sqrt{(a+\rho)^2+z^2} - \sqrt{(a-\rho)^2+z^2}\right], \quad \ell_2 = \frac{1}{2}\left[\sqrt{(a+\rho)^2+z^2} + \sqrt{(a-\rho)^2+z^2}\right]$$

1. $\int_0^\infty e^{-\alpha x} J_\nu(\beta x) x^{\mu-1}\,dx$

$$= \frac{\left(\frac{\beta}{2\alpha}\right)^\nu \Gamma(\nu+\mu)}{\alpha^\mu\,\Gamma(\nu+1)}\,F\left(\frac{\nu+\mu}{2},\frac{\nu+\mu+1}{2};\nu+1;-\frac{\beta^2}{\alpha^2}\right)$$

WA 421(2)

$$= \frac{\left(\frac{\beta}{2\alpha}\right)^\nu \Gamma(\nu+\mu)}{\alpha^\mu\,\Gamma(\nu+1)}\left(1+\frac{\beta^2}{\alpha^2}\right)^{\frac{1}{2}-\mu}F\left(\frac{\nu-\mu+1}{2},\frac{\nu-\mu}{2}+1;\nu+1;-\frac{\beta^2}{\alpha^2}\right)$$

WA 421(3)

$$= \frac{\left(\frac{\beta}{2}\right)^\nu \Gamma(\nu+\mu)}{\sqrt{(\alpha^2+\beta^2)^{\nu+\mu}}\,\Gamma(\nu+1)}\,F\left(\frac{\nu+\mu}{2},\frac{1-\mu+\nu}{2};\nu+1;\frac{\beta^2}{\alpha^2+\beta^2}\right)$$

$$[\operatorname{Re}(\nu+\mu) > 0, \quad \operatorname{Re}(\alpha+i\beta) > 0, \quad \operatorname{Re}(\alpha-i\beta) > 0]$$

WA 421(3)

$$= \left(\alpha^2+\beta^2\right)^{-\frac{1}{2}\mu}\Gamma(\nu+\mu)\,P_{\mu-1}^{-\nu}\left[\alpha\left(\alpha^2+\beta^2\right)^{-\frac{1}{2}}\right]$$

$$[\alpha > 0, \quad \beta > 0, \quad \operatorname{Re}(\nu+\mu) > 0]$$

ET II 29(6)

2. $\displaystyle\int_0^\infty e^{-\alpha x}\, Y_\nu(\beta x) x^{\mu-1}\, dx$

$$= \cot\nu\pi \frac{\left(\frac{\beta}{2}\right)^\nu \Gamma(\nu+\mu)}{\sqrt{(\alpha^2+\beta^2)^{\nu+\mu}}\,\Gamma(\nu+1)}\, F\left(\frac{\nu+\mu}{2},\frac{\nu-\mu+1}{2};\nu+1;\frac{\beta^2}{\alpha^2+\beta^2}\right)$$

$$-\csc\nu\pi \frac{\left(\frac{\beta}{2}\right)^{-\nu}\Gamma(\mu-\nu)}{\sqrt{(\alpha^2+\beta^2)^{\mu-\nu}}\,\Gamma(1-\nu)}\, F\left(\frac{\mu-\nu}{2},\frac{1-\nu-\mu}{2};1-\nu;\frac{\beta^2}{\alpha^2+\beta^2}\right)$$

$$[\operatorname{Re}\mu \ge |\operatorname{Re}\nu|, \quad \operatorname{Re}(\alpha\pm i\beta)>0]$$
WA 421(4)

$$= -\frac{2}{\pi}\,\Gamma(\nu+\mu)\left(\beta^2+\alpha^2\right)^{-\frac12\mu}\, Q_{\mu-1}^{-\nu}\left[\alpha\left(\alpha^2+\beta^2\right)^{-\frac12}\right]$$

$$[\alpha>0,\quad \beta>0,\quad \operatorname{Re}\mu>|\operatorname{Re}\nu|]$$
ET II 105(2)

3. $\displaystyle\int_0^\infty x^{\mu-1} e^{-\alpha x}\, K_\nu(\beta x)\, dx = \frac{\sqrt{\pi}(2\beta)^\nu}{(\alpha+\beta)^{\mu+\nu}}\, \frac{\Gamma(\mu+\nu)\,\Gamma(\mu-\nu)}{\Gamma\left(\mu+\frac12\right)}\, F\left(\mu+\nu,\nu+\frac12;\mu+\frac12;\frac{\alpha-\beta}{\alpha+\beta}\right)$

$$[\operatorname{Re}\mu>|\operatorname{Re}\nu|,\quad \operatorname{Re}(\alpha+\beta)>0]$$
ET II 131(23)a, EH II 50(26)

4. $\displaystyle\int_0^\infty x^{m+1} e^{-\alpha x}\, J_\nu(\beta x)\, dx = (-1)^{m+1}\beta^{-\nu}\frac{d^{m+1}}{d\alpha^{m+1}}\left[\frac{\left(\sqrt{\alpha^2+\beta^2}-\alpha\right)^\nu}{\sqrt{\alpha^2+\beta^2}}\right]$

$$[\beta>0,\quad \operatorname{Re}\nu>-m-2]\qquad \text{ET II 28(3)}$$

5.[10] $\displaystyle\int_0^\infty e^{-zx}\, J_1(ax)\, J_{1/2}(\rho x)\, x^{-3/2}\, dx$

$$= \frac{1}{a}\sqrt{\frac{2}{\pi\rho}}\left\{\frac{\ell_1}{2}\sqrt{a^2-\ell_1^2}+\frac{a^2}{2}\arcsin\left(\frac{\ell_1}{2}\right)+z\left[\sqrt{\rho^2-\ell_1^2}-\rho\right]\right\}$$

$$[\arg a>0,\quad \arg\rho>0,\quad \arg z>0]$$

6.[10] $\displaystyle\int_0^\infty e^{-zx}\, J_1(ax)\, J_{1/2}(\rho x)\, x^{-1/2}\, dx = \frac{1}{a}\sqrt{\frac{2}{\pi\rho}}\left[\rho-\sqrt{\rho^2-\ell_1^2}\right]$

$$[\arg a>0,\quad \arg\rho>0,\quad \arg z>0]$$

7.[10] $\displaystyle\int_0^\infty e^{-zx}\, J_1(ax)\, J_{1/2}(\rho x)\, x^{1/2}\, dx = \frac{1}{a}\sqrt{\frac{2}{\pi\rho}}\frac{\ell_1\sqrt{a^2-\ell_1^2}}{\ell_2^2-\ell_1^2}$

$$[\arg a>0,\quad \arg\rho>0,\quad \arg z>0]$$

8.[10] $\displaystyle\int_0^\infty e^{-zx}\, J_1(ax)\, J_{3/2}(\rho x)\, x^{1/2}\, dx = \sqrt{\frac{2}{\pi}}\frac{\ell_1^2\sqrt{\rho^2-\ell_1^2}}{\rho^{3/2}a\left(\ell_2^2-\ell_1^2\right)}$

$$[\arg a>0,\quad \arg\rho>0,\quad \arg z>0]$$

9.[10] $\displaystyle\int_0^\infty e^{-zx}\, J_1(ax)\, J_{3/2}(\rho x)\, x^{-3/2}\, dx = \frac{1}{\sqrt{2\pi}}\frac{1}{\rho^{3/2}a}\left[a^2\arcsin\left(\frac{\ell_1}{a}\right)-\ell_1\sqrt{a^2-\ell_1^2}\right]$

$$[\arg a>0,\quad \arg\rho>0,\quad \arg z>0]$$

$10.^{10}$ $\displaystyle\int_0^\infty e^{-zx} J_1(ax) J_{5/2}(\rho x) x^{-1/2}\,dx = \frac{1}{\sqrt{2\pi}}\frac{z}{\rho^{5/2}a}\left[\ell_1\sqrt{a^2-\ell_1^2}+\frac{2a^2\ell_1}{\sqrt{a^2-\ell_1^2}}-3a^2\arcsin\left(\frac{\ell_1}{a}\right)\right]$

$$[\arg a > 0, \quad \arg \rho > 0, \quad \arg z > 0]$$

$11.^{10}$ $\displaystyle\int_0^\infty e^{-zx} J_1(ax) J_{5/2}(\rho x) x^{-3/2}\,dx$

$$= \frac{1}{\sqrt{2\pi}}\frac{1}{\rho^{5/2}a}\left[\frac{\ell_1}{\sqrt{a^2-\ell_1^2}}\left(\frac{7a^2}{8}-a^2z^2-\frac{\ell_1^4}{4}-\frac{5a^2\ell_1^2}{8}\right)\right.$$

$$\left.-\frac{1}{2}\left(\ell_1^2+\ell_2^2\right)\ell_1\sqrt{a^2-\ell_1^2}+\arcsin\left(\frac{\ell_1}{a}\right)\left(\frac{3}{2}a^2z^2+\frac{1}{2}a^2\rho^2-\frac{3a^4}{8}\right)\right]$$

$$[\arg a > 0, \quad \arg \rho > 0, \quad \arg z > 0]$$

$12.^{10}$ $\displaystyle\int_0^\infty e^{-zx} J_1(ax) J_{5/2}(\rho x) x^{-5/2}\,dx$

$$= \frac{1}{\sqrt{2\pi}}\frac{1}{\rho^{5/2}a}\left\{\frac{2\left[\rho^{5/2}-\left(\rho^2-\ell_1^2\right)^{5/2}\right]}{15}+za^2\arcsin\left(\frac{\ell_1}{a}\right)\left[\frac{3a^2}{8}-\frac{\rho^2}{2}-\frac{z^2}{2}\right]\right.$$

$$\left.+z\ell_1\sqrt{a^2-\ell_1^2}\left[\frac{\rho^2}{2}-\frac{3a^2}{8}+\frac{z^2}{6}-\frac{\ell_1^2}{4}\right]+\frac{z^3a^2\ell_1}{3\sqrt{a^2-\ell_1^2}}\right\}$$

$$[\arg a > 0, \quad \arg \rho > 0, \quad \arg z > 0]$$

$13.^{10}$ $\displaystyle\int_0^\infty e^{-zx} J_2(ax) J_{3/2}(\rho x) x^{1/2}\,dx = \sqrt{\frac{2}{\pi}}a^2\rho^{3/2}\frac{\sqrt{\ell_2^2-\rho^2}}{(\ell_2^2-\ell_1^2)\ell_2^4}$

$$[\arg a > 0, \quad \arg \rho > 0, \quad \arg z > 0]$$

$14.^{10}$ $\displaystyle\int_0^\infty e^{-zx} J_2(ax) J_{3/2}(\rho x) x^{-1/2}\,dx = \sqrt{\frac{2}{\pi}}\frac{\rho^{3/2}}{a^2}\left[\frac{2}{3}-\frac{\sqrt{\rho^2-\ell_1^2}}{\rho}+\frac{(\rho^2-\ell_1^2)^{3/2}}{3\rho^3}\right]$

$$[\arg a > 0, \quad \arg \rho > 0, \quad \arg z > 0]$$

$15.^{10}$ $\displaystyle\int_0^\infty e^{-zx} J_3(ax) J_{1/2}(\rho x) x^{-1/2}\,dx$

$$= \sqrt{\frac{2}{\pi\rho}}\frac{1}{3a^3}\left\{\rho\left[3a^2-4\rho^2+12z^2\right]-\sqrt{\rho^2-\ell_1^2}\left\{12\ell_2^2-16\rho^2+4\ell_1^2-3a^2\right\}\right\}$$

$$[\arg a > 0, \quad \arg \rho > 0, \quad \arg z > 0]$$

$16.^{10}$ $\displaystyle\int_0^\infty e^{-zx} J_3(ax) J_{3/2}(\rho x) x^{1/2}\,dx$

$$= \sqrt{\frac{2}{\pi}}\rho^{3/2}\left\{\frac{4}{a^3}\left[\frac{2}{3}-\frac{\sqrt{\rho^2-\ell_1^2}}{\rho}+\frac{(\rho^2-\ell_1^2)^{3/2}}{3\rho^2}\right]-\frac{a\sqrt{\ell_2^2-a^2}}{(\ell_2^2-\ell_1^2)\ell_2^3}\right\}$$

$$[\arg a > 0, \quad \arg \rho > 0, \quad \arg z > 0]$$

$17.^{10}$ $\displaystyle\int_0^\infty e^{-zx} J_3(ax) J_{3/2}(\rho x) x^{-1/2}\,dx = \sqrt{\frac{2}{\pi}}\frac{\rho^{3/2}}{3a^3}\left[\sqrt{\ell_2^2-\rho^2}\left(\frac{4\rho^2\left(2\rho^2-\ell_1^2\right)-\ell_1^4}{\rho^4}\right)-8z\right]$

$$[\arg a > 0, \quad \arg \rho > 0, \quad \arg z > 0]$$

$18.^{10}$ $\int_0^\infty e^{-zx} J_3(ax) J_{3/2}(\rho x) x^{-3/2}\, dx$

$$= \sqrt{\frac{2}{\pi}} \frac{\rho^{3/2}}{3a^3} \left\{ a^2 - \frac{4}{5}\rho^2 + 4z^2 - \sqrt{\rho^2 - \ell_1^2} \left[\frac{4\ell_2^2}{\rho} - \frac{24\rho}{5} + \frac{8\ell_1^2}{5\rho} - \frac{a^2}{\rho} + \frac{\ell_1^4}{5\rho^3} \right] \right\}$$

$$[\arg a > 0, \quad \arg \rho > 0, \quad \arg z > 0]$$

$19.^{10}$ $\int_0^\infty e^{-zx} J_3(ax) J_{3/2}(\rho x) x^{-5/2}\, dx$

$$= -\sqrt{\frac{2}{\pi}} \frac{\rho^{3/2}}{3a^3} \left\{ \left(a^2 - \frac{4}{5}\rho^2 \right) z + \frac{4z^3}{3} \right.$$

$$+ \sqrt{\ell_2^2 - \rho^2} \left[a^2 + \frac{32}{15}\rho^2 - \frac{12}{5}\ell_1^2 - \frac{4}{3}\ell_2^2 + \frac{2\ell_1^4}{5\rho^2} + \frac{a^4\ell_1^2}{16\rho^4} + \frac{a^2\ell_1^2}{24\rho^4} + \frac{\ell_1^6}{30\rho^4} \right]$$

$$\left. - \frac{a^6}{16\rho^3} \arcsin\left(\frac{\rho}{\ell_2} \right) \right\}$$

$$[\arg a > 0, \quad \arg \rho > 0, \quad \arg z > 0]$$

6.622

1. $\int_0^\infty \left(J_0(x) - e^{-\alpha x} \right) \frac{dx}{x} = \ln 2\alpha$ $\qquad [\alpha > 0]$ $\qquad$ NT 66(13)

2. $\int_0^\infty \frac{e^{i(u+x)}}{u+x} J_0(x)\, dx = \frac{\pi}{2} i\, H_0^{(1)}(u)$ $\qquad$ MO 44

$3.^8$ $\int_0^\infty e^{-x \cosh \alpha} I_\nu(x) x^{\mu-1}\, dx = \sqrt{\frac{2}{\pi}} e^{-\left(\mu-\frac{1}{2}\right)\pi i} \dfrac{Q_{\nu-\frac{1}{2}}^{\mu-\frac{1}{2}}(\cosh \alpha)}{\sinh^{\mu-\frac{1}{2}} \alpha}$

$$[\operatorname{Re}(\mu + \nu) > 0, \quad \operatorname{Re}(\cosh \alpha) > 1]$$

$$\text{WA 388(6)a}$$

6.623

1. $\int_0^\infty e^{-\alpha x} J_\nu(\beta x) x^\nu\, dx = \dfrac{(2\beta)^\nu \Gamma\left(\nu + \frac{1}{2}\right)}{\sqrt{\pi}(\alpha^2 + \beta^2)^{\nu+\frac{1}{2}}}$ $\qquad \left[\operatorname{Re}\nu > -\frac{1}{2}, \quad \operatorname{Re}\alpha > |\operatorname{Im}\beta|\right]$

$$\text{WA 422(5)}$$

2. $\int_0^\infty e^{-\alpha x} J_\nu(\beta x) x^{\nu+1}\, dx = \dfrac{2\alpha(2\beta)^\nu \Gamma\left(\nu + \frac{3}{2}\right)}{\sqrt{\pi}(\alpha^2 + \beta^2)^{\nu+\frac{3}{2}}}$ $\qquad \left[\operatorname{Re}\nu > -1, \quad \operatorname{Re}\alpha > |\operatorname{Im}\beta|\right]$

$$\text{WA 422(6)}$$

3. $\int_0^\infty e^{-\alpha x} J_\nu(\beta x) \dfrac{dx}{x} = \dfrac{\left(\sqrt{\alpha^2 + \beta^2} - \alpha\right)^\nu}{\nu \beta^\nu}$

$$[\operatorname{Re}\nu > 0; \quad \operatorname{Re}\alpha > |\operatorname{Im}\beta|] \qquad (\text{cf. } \mathbf{6.611}\ 1) \quad \text{WA 422(7)}$$

6.624

1. $\int_0^\infty x e^{-\alpha x} K_0(\beta x)\, dx = \dfrac{1}{\alpha^2 - \beta^2} \left\{ \dfrac{\alpha}{\sqrt{\alpha^2 - \beta^2}} \ln\left[\dfrac{\alpha}{\beta} + \sqrt{\left(\dfrac{\alpha}{\beta}\right)^2 - 1} \right] - 1 \right\}$ $\qquad$ MO 181

2. $\displaystyle\int_0^\infty \sqrt{x}\,e^{-\alpha x}\,K_{\pm\frac12}(\beta x)\,dx = \sqrt{\frac{\pi}{2\beta}}\,\frac{1}{\alpha+\beta}$ MO 181

3. $\displaystyle\int_0^\infty e^{-tz}(z^2-1)^{-1/2}\,K_\mu(t)t^\nu\,dt = \frac{\Gamma(\nu-\mu+1)}{(z^2-1)^{-\frac12(\nu+1)}}\,e^{i\mu\pi}\,Q_\nu^\mu(z)$

$$[\mathrm{Re}\,(\nu\pm\mu)>-1]$$ EH II 57(7)

4. $\displaystyle\int_0^\infty e^{-tz}(z^2-1)^{-1/2}\,I_{-\mu}(t)t^\nu\,dt = \frac{\Gamma(-\nu-\mu)}{(z^2-1)^{\frac12\nu}}\,P_\nu^\mu(z)$ $[\mathrm{Re}(\nu+\mu)<0]$ EH II 57(8)

5. $\displaystyle\int_0^\infty e^{-tz}(z^2-1)^{-\frac12}\,I_\mu(t)t^\nu\,dt = \frac{\Gamma(\nu+\mu+1)}{(z^2-1)^{-\frac12(\nu+1)}}\,P_\nu^{-\mu}(z)$

$$[\mathrm{Re}(\nu+\mu)>-1]$$ EH II 57(9)

6. $\displaystyle\int_0^\infty e^{-t\cos\theta}\,J_\mu\,(t\sin\theta)\,t^\nu\,dt = \Gamma(\nu+\mu+1)\,P_\nu^{-\mu}\,(\cos\theta)$

$$\left[\mathrm{Re}(\nu+\mu)>-1,\quad 0\le\theta<\tfrac12\pi\right]$$
EH II 57(10)

7. $\displaystyle\int_0^\infty \frac{J_\nu(bx)x^\nu}{e^{\pi x}-1}\,dx = \frac{(2b)^\nu\,\Gamma\left(\nu+\frac12\right)}{\sqrt{\pi}}\sum_{n=1}^\infty \frac{1}{(n^2\pi^2+b^2)^{\nu+\frac12}}$

$$[\mathrm{Re}\,\nu>0,\quad |\mathrm{Im}\,b|<\pi]$$ WA 423(9)

6.625

1. $\displaystyle\int_0^1 x^{\lambda-\nu-1}(1-x)^{\mu-1}e^{\pm i\alpha x}\,J_\nu(\alpha x)\,dx = \frac{2^{-\nu}\alpha^\nu\,\Gamma(\lambda)\,\Gamma(\mu)}{\Gamma(\lambda+\mu)\,\Gamma(\nu+1)}\,{}_2F_2\left(\lambda,\nu+\frac12;\lambda+\mu,2\nu+1;\pm2i\alpha\right)$

$$[\mathrm{Re}\,\lambda>0,\quad \mathrm{Re}\,\mu>0]$$ ET II 194(58)a

2. $\displaystyle\int_0^1 x^\nu(1-x)^{\mu-1}e^{\pm i\alpha x}\,J_\nu(\alpha x)\,dx \overset{\cdot}{=} \frac{(2\alpha)^\nu\,\Gamma(\mu)\,\Gamma\left(\nu+\frac12\right)}{\sqrt{\pi}\,\Gamma(\mu+2\nu+1)}\,{}_1F_1\left(\nu+\frac12;\mu+2\nu+1;\pm2i\alpha\right)$

$$\left[\mathrm{Re}\,\mu>0,\quad \mathrm{Re}\,\nu>-\tfrac12\right]$$ ET II 194(57)a

3. $\displaystyle\int_0^1 x^\nu(1-x)^{\mu-1}e^{\pm\alpha x}\,J_\nu(\alpha x)\,dx = \frac{(2\alpha)^\nu\,\Gamma\left(\nu+\frac12\right)\Gamma(\mu)}{\sqrt{\pi}\,\Gamma(\mu+2\nu+1)}\,{}_1F_1\left(\nu+\frac12;\mu+2\nu+1;\pm2\alpha\right)$

$$\left[\mathrm{Re}\,\mu>0,\quad \mathrm{Re}\,\nu>-\tfrac12\right]$$
BU 9(16a), ET II 197(77)a

4. $\displaystyle\int_0^1 x^{\lambda-1}(1-x)^{\mu-1}e^{\pm\alpha x}\,I_\nu(\alpha x)\,dx = \frac{\left(\frac12\alpha\right)^\nu\,\Gamma(\lambda+\nu)\,\Gamma(\mu)}{\Gamma(\nu+1)\,\Gamma(\lambda+\mu+\nu)}$

$$\times\,{}_2F_2\left(\nu+\frac12,\lambda+\nu;2\nu+1,\mu+\lambda+\nu;\pm2\alpha\right)$$

$$[\mathrm{Re}\,\mu>0,\quad \mathrm{Re}(\lambda+\nu)>0]$$ ET II 197(78)a

5. $\displaystyle\int_0^1 x^{\mu-\kappa}(1-x)^{2\kappa-1}\,I_{\mu-\kappa}\left(\frac12 xz\right)e^{-\frac12 xz}\,dx = \frac{\Gamma(2\kappa)}{\sqrt{\pi}\,\Gamma(1+2\mu)}\,e^{\frac{x}{2}}z^{-\kappa-\frac12}\,M_{\kappa,u}(z)$

$$\left[\mathrm{Re}\left(\kappa-\tfrac12-\mu\right)<0,\quad \mathrm{Re}\,\kappa>0\right]$$
BU 129(14a)

6. $\int_1^\infty x^{-\lambda}(x-1)^{\mu-1}e^{-\alpha x}\,I_\nu(\alpha x)\,dx = \dfrac{(2\alpha)^\lambda\,\Gamma(\mu)}{\sqrt{\pi}}\,G_{23}^{21}\left(2\alpha\left|\begin{matrix}\frac{1}{2}-\lambda,0\\-\mu,\nu-\lambda,-\nu-\lambda\end{matrix}\right.\right)$

$\left[0<\operatorname{Re}\mu<\tfrac{1}{2}+\operatorname{Re}\lambda,\quad\operatorname{Re}\alpha>0\right]$
ET II 207(50)a

7. $\int_1^\infty x^{-\lambda}(x-1)^{\mu-1}e^{-\alpha x}\,K_\nu(\alpha x)\,dx = \Gamma(\mu)\sqrt{\pi}(2\alpha)^\lambda\,G_{23}^{30}\left(2\alpha\left|\begin{matrix}0,\frac{1}{2}-\lambda\\-\mu,\nu-\lambda,-\nu-\lambda\end{matrix}\right.\right)$

$\left[\operatorname{Re}\mu>0,\quad\operatorname{Re}\alpha>0\right]$ ET II 208(55)a

8. $\int_1^\infty x^{-\nu}(x-1)^{\mu-1}e^{-\alpha x}\,I_\nu(\alpha x)\,dx = \dfrac{(2\alpha)^{\nu-\mu}\,\Gamma\left(\frac{1}{2}-\mu+\nu\right)\Gamma(\mu)}{\sqrt{\pi}\,\Gamma(1-\mu+2\nu)}$

$\times\,{}_1F_1\left(\dfrac{1}{2}-\mu+\nu;1-\mu+2\nu;-2\alpha\right)$

$\left[0<\operatorname{Re}\mu<\tfrac{1}{2}+\operatorname{Re}\nu,\quad\operatorname{Re}\alpha>0\right]$ ET II 207(49)a

9. $\int_1^\infty x^{-\nu}(x-1)^{\mu-1}e^{-\alpha x}\,K_\nu(\alpha x)\,dx = \sqrt{\pi}\,\Gamma(\mu)(2\alpha)^{-\frac{1}{2}\mu-\frac{1}{2}}e^{-\alpha}\,W_{-\frac{1}{2}\mu,\nu-\frac{1}{2}\mu}(2\alpha)$

$\left[\operatorname{Re}\mu>0,\quad\operatorname{Re}\alpha>0\right]$ ET II 208(53)a

10. $\int_1^\infty x^{-\mu-\frac{1}{2}}(x-1)^{\mu-1}e^{-\alpha x}\,K_\nu(\alpha x)\,dx = \sqrt{\pi}\,\Gamma(\mu)(2\alpha)^{-\frac{1}{2}}e^{-\alpha}\,W_{-\mu,\nu}(2\alpha)$

$\left[\operatorname{Re}\mu>0,\quad\operatorname{Re}\alpha>0\right]$ ET II 207(51)a

11.[3] $\int_{-1}^1 (1-x^2)^{-1/2}xe^{-ax}\,I_1\left(b\sqrt{1-x^2}\right)dx = \dfrac{2}{b}\left\{\sinh a - a(a^2+b^2)^{-1/2}\sinh\sqrt{a^2+b^2}\right\}$

$\left[a>0,\quad b>0\right]$

6.626

1. $\int_0^\infty x^{\lambda-1}e^{-\alpha x}\,J_\mu(\beta x)\,J_\nu(\gamma x)\,dx = \dfrac{\beta^\mu\gamma^\nu}{\Gamma(\nu+1)}2^{-\nu-\mu}\alpha^{-\lambda-\mu-\nu}\sum_{m=0}^\infty\dfrac{\Gamma(\lambda+\mu+\nu+2m)}{m!\,\Gamma(\mu+m+1)}$

$\times\,F\left(-m,-\mu-m;\nu+1;\dfrac{\gamma^2}{\beta^2}\right)\left(-\dfrac{\beta^2}{4\alpha^2}\right)^m$

$\left[\operatorname{Re}(\lambda+\mu+\nu)>0,\quad\operatorname{Re}(\alpha\pm i\beta\pm i\gamma)>0\right]$ EH II 48(15)

2. $\int_0^\infty e^{-2\alpha x}\,J_\nu(\beta x)\,J_\mu(\beta x)x^{\nu+\mu}\,dx = \dfrac{\Gamma\left(\nu+\mu+\frac{1}{2}\right)\beta^{\nu+\mu}}{\sqrt{\pi^3}}$

$\times\int_0^{\frac{\pi}{2}}\dfrac{\cos^{\nu+\mu}\varphi\cos(\nu-\mu)\varphi}{(\alpha^2+\beta^2\cos^2\varphi)^{\nu+\mu}\sqrt{\alpha^2+\beta^2\cos^2\varphi}}\,d\varphi$

$\left[\operatorname{Re}\alpha>|\operatorname{Im}\beta|,\quad\operatorname{Re}(\nu+\mu)>-\tfrac{1}{2}\right]$ WA 427(1)

3. $\int_0^\infty e^{-2\alpha x}\,J_0(\beta x)\,J_1(\beta x)x\,dx = \dfrac{\boldsymbol{K}\left(\frac{\beta}{\sqrt{\alpha^2+\beta^2}}\right)-\boldsymbol{E}\left(\frac{\beta}{\sqrt{\alpha^2+\beta^2}}\right)}{2\pi\beta\sqrt{\alpha^2+\beta^2}}$ WA 427(2)

4. $$\int_0^\infty e^{-2\alpha x} I_0(\beta x)\, I_1(\beta x)\, x\, dx = \frac{1}{2\pi\beta}\left\{\frac{\alpha}{\alpha^2-\beta^2}\, \boldsymbol{E}\left(\frac{\beta}{\alpha}\right) - \frac{1}{\alpha}\, \boldsymbol{K}\left(\frac{\beta}{\alpha}\right)\right\}$$

$$[\mathrm{Re}\,\alpha > \mathrm{Re}\,\beta] \qquad\qquad \text{WA 428(5)}$$

5.[10] $$\int_0^\infty x^{\nu-\mu+2n} e^{-zx}\, J_\mu(\alpha x)\, J_\nu(\rho x)\, dx = \frac{1}{\sqrt{\pi}}\left(\frac{a}{2}\right)^{\mu-\nu-2n-1}\left(\frac{\rho}{a}\right)^\nu$$

$$\times \frac{1}{\Gamma\left(\mu-\nu-n+\frac{1}{2}\right)}\sum_{q=0}^\infty \frac{\Gamma\left(\nu+n+q+\frac{1}{2}\right)\left(\nu-\mu+n+\frac{1}{2}\right)_q}{q!\,\Gamma\left(\nu+q+\frac{1}{2}\right)}$$

$$\times a^{-2q}\int_0^{\ell_1/\rho}\frac{dx}{\sqrt{1-x^2}}\, x^{2\nu+2q}\left(\rho^2+\frac{z^2}{1-x^2}\right)^q$$

where $\ell_1 = \frac{1}{2}\left[\sqrt{(a+\rho)^2+z^2} - \sqrt{(a-\rho)^2+z^2}\right]$ $\qquad \left[\mu>\nu+2n,\quad n=0,1,\dots,\quad \nu>-\tfrac{1}{2}\right]$

6.627 $$\int_0^\infty \frac{x^{-1/2}}{x+a}\, e^{-x}\, K_\nu(x)\, dx = \frac{\pi e^a\, K_\nu(a)}{\sqrt{a}\cos(\nu\pi)} \qquad \left[|\arg a|<\pi,\quad |\mathrm{Re}\,\nu|<\tfrac{1}{2}\right] \qquad \text{ET II 368(29)}$$

6.628

1. $$\int_0^\infty e^{-x\cos\beta}\, J_{-\nu}(x\sin\beta)\, x^\mu\, dx = \Gamma(\mu-\nu+1)\, P_\mu^\nu(\cos\beta)$$

$$\left[0<\beta<\frac{\pi}{2},\quad \mathrm{Re}(\mu-\nu)>-1\right]$$

$$\text{WA 424(3), WH}$$

2. $$\int_0^\infty e^{-x\cos\beta}\, Y_\nu(x\sin\beta)\, x^\mu\, dx = -\frac{\sin\mu\pi}{\sin(\mu+\nu)\pi}\frac{\Gamma(\mu-\nu+1)}{\pi}$$

$$\times\left[Q_\mu^\nu(\cos\beta+0\cdot i)\, e^{\frac{1}{2}\nu\pi i} + Q_\mu^\nu(\cos\beta-0\cdot i)\, e^{-\frac{1}{2}\nu\pi i}\right]$$

$$\left[\mathrm{Re}(\mu+\nu)>-1,\quad 0<\beta<\frac{\pi}{2}\right] \quad \text{WA 424(4)}$$

3. $$\int_0^1 e^{\frac{xu}{2}}(1-x)^{2\nu-1}\, x^{\mu-\nu}\, J_{\mu-\nu}\left(\frac{ixu}{2}\right) dx = 2^{2(\nu-\mu)} e^{\frac{\pi}{2}(\mu-\nu)i}\frac{\mathrm{B}(2\nu,2\mu-2\nu+1)}{\Gamma(\mu-\nu+1)}\frac{e^{\frac{u}{2}}}{u^{\nu+\frac{1}{2}}}\, M_{\nu,\mu}(u)$$

$$\text{MO 118a}$$

4.[8] $$\int_0^\infty e^{-x\cosh\alpha}\, I_\nu(x\sinh\alpha)\, x^\mu\, dx = \Gamma(\nu+\mu+1)\, P_\mu^{-\nu}(\cosh\alpha)$$

$$\left[\mathrm{Re}(\mu+\nu)>-1,\quad |\mathrm{Im}\,\alpha|<\tfrac{1}{2}\pi\right]$$

$$\text{WA 423(1)}$$

5. $$\int_0^\infty e^{-x\cosh\alpha}\, K_\nu(x\sinh\alpha)\, x^\mu\, dx = \frac{\sin\mu\pi}{\sin(\nu+\mu)\pi}\, \Gamma(\mu-\nu+1)\, Q_\mu^\nu(\cosh\alpha)$$

$$[\mathrm{Re}(\mu+1)>|\mathrm{Re}\,\nu|] \qquad\qquad \text{WA 423(2)}$$

6. $$\int_0^\infty e^{-x\cosh\alpha}\, I_\nu(x)\, x^{\mu-1}\, dx = \frac{\cos\nu\pi}{\sin(\mu+\nu)\pi}\frac{Q_{\mu-\frac{1}{2}}^{\nu-\frac{1}{2}}(\cosh\alpha)}{\sqrt{\frac{\pi}{2}}(\sinh\alpha)^{\mu-\frac{1}{2}}}$$

$$[\mathrm{Re}(\mu+\nu)>0,\quad \mathrm{Re}(\cosh\alpha)>1]$$

$$\text{WA 424(6)}$$

7.

$$\int_0^\infty e^{-x\cosh\alpha} K_\nu(x) x^{\mu-1}\, dx = \sqrt{\frac{\pi}{2}}\, \Gamma(\mu-\nu)\, \Gamma(\mu+\nu) \frac{P_{\nu-\frac{1}{2}}^{\frac{1}{2}-\mu}(\cosh\alpha)}{(\sinh\alpha)^{\mu-\frac{1}{2}}}$$

$$[\operatorname{Re}\mu > |\operatorname{Re}\nu|, \quad \operatorname{Re}(\cosh\alpha) > -1]$$

WA 424(7)

6.629⁸ $\displaystyle\int_0^\infty x^{-1/2} e^{-x\alpha\cos\varphi\cos\psi}\, J_\mu(\alpha x \sin\varphi)\, J_\nu(\alpha x \sin\psi)\, dx$

$$= \Gamma\left(\mu+\nu+\tfrac{1}{2}\right) \alpha^{-\frac{1}{2}}\, P_{\nu-\frac{1}{2}}^{-\mu}(\cos\varphi)\, P_{\mu-\frac{1}{2}}^{-\nu}(\cos\psi)$$

$$\left[\alpha > 0, \quad 0 < \varphi < \frac{\pi}{2}, \quad 0 < \psi < \frac{\pi}{2}, \quad \operatorname{Re}(\mu+\nu) > -\frac{1}{2}\right]$$ ET II 50(19)

6.631

1.

$$\int_0^\infty x^\mu e^{-\alpha x^2} J_\nu(\beta x)\, dx = \frac{\beta^\nu\, \Gamma\left(\frac{1}{2}\nu + \frac{1}{2}\mu + \frac{1}{2}\right)}{2^{\nu+1}\alpha^{\frac{1}{2}(\mu+\nu+1)}\, \Gamma(\nu+1)}\ {}_1F_1\left(\frac{\nu+\mu+1}{2}; \nu+1; -\frac{\beta^2}{4\alpha}\right)$$

BU 8(15)

$$= \frac{\Gamma\left(\frac{1}{2}\nu + \frac{1}{2}\mu + \frac{1}{2}\right)}{\beta\alpha^{\frac{1}{2}\mu}\, \Gamma(\nu+1)} \exp\left(-\frac{\beta^2}{8\alpha}\right) M_{\frac{1}{2}\mu,\frac{1}{2}\nu}\left(\frac{\beta^2}{4\alpha}\right)$$

$$[\operatorname{Re}\alpha > 0, \quad \operatorname{Re}(\mu+\nu) > -1]$$

EH II 50(22), ET II 30(14), BU 14(13b)

2. $\displaystyle\int_0^\infty x^\mu e^{-\alpha x^2}\, Y_\nu(\beta x)\, dx$

$$= -\alpha^{-\frac{1}{2}\mu}\beta^{-1} \sec\left(\frac{\nu-\mu}{2}\pi\right) \exp\left(-\frac{\beta^2}{8\alpha}\right)$$

$$\times \left\{\frac{\Gamma\left(\frac{1}{2} + \frac{1}{2}\mu + \frac{1}{2}\nu\right)}{\Gamma(1+\nu)} \sin\left(\frac{\nu-\mu}{2}\pi\right) M_{\frac{1}{2}\mu,\frac{1}{2}\nu}\left(\frac{\beta^2}{4\alpha}\right) + W_{\frac{1}{2}\mu,\frac{1}{2}\nu}\left(\frac{\beta^2}{4\alpha}\right)\right\}$$

$$[\operatorname{Re}\alpha > 0, \quad \operatorname{Re}\mu > |\operatorname{Re}\nu| - 1, \quad \beta > 0]$$ ET II 106(4)

3.

$$\int_0^\infty x^\mu e^{-\alpha x^2} K_\nu(\beta x)\, dx = \frac{1}{2}\alpha^{-\frac{1}{2}\mu}\beta^{-1} \Gamma\left(\frac{1+\nu+\mu}{2}\right) \Gamma\left(\frac{1-\nu+\mu}{2}\right) \exp\left(\frac{\beta^2}{8\alpha}\right) W_{-\frac{1}{2}\mu,\frac{1}{2}\nu}\left(\frac{\beta^2}{4\alpha}\right)$$

$$[\operatorname{Re}\mu > |\operatorname{Re}\nu| - 1]$$ ET II 132(25)

4.

$$\int_0^\infty x^{\nu+1} e^{-\alpha x^2} J_\nu(\beta x)\, dx = \frac{\beta^\nu}{(2\alpha)^{\nu+1}} \exp\left(-\frac{\beta^2}{4\alpha}\right) \qquad [\operatorname{Re}\alpha > 0, \quad \operatorname{Re}\nu > -1]$$

WA 43(4), ET II 29(10)

5.

$$\int_0^\infty x^{\nu-1} e^{-\alpha x^2} J_\nu(\beta x)\, dx = 2^{\nu-1}\beta^{-\nu} \gamma\left(\nu, \frac{\beta^2}{4\alpha}\right) \qquad [\operatorname{Re}\alpha > 0, \quad \operatorname{Re}\nu > 0]$$ ET II 30(11)

6.

$$\int_0^\infty x^{\nu+1} e^{\pm i\alpha x^2} J_\nu(\beta x)\, dx = \frac{\beta^\nu}{(2\alpha)^{\nu+1}} \exp\left[\pm i\left(\frac{\nu+1}{2}\pi - \frac{\beta^2}{4\alpha}\right)\right]$$

$$[\alpha > 0, \quad -1 < \operatorname{Re}\nu < \tfrac{1}{2}, \quad \beta > 0]$$

ET II 30(12)

7. $\displaystyle\int_0^\infty x e^{-\alpha x^2} J_\nu(\beta x)\, dx = \frac{\sqrt{\pi}\beta}{8\alpha^{\frac{3}{2}}} \exp\left(-\frac{\beta^2}{8\alpha}\right) \left[I_{\frac{1}{2}\nu-\frac{1}{2}}\left(\frac{\beta^2}{8\alpha}\right) - I_{\frac{1}{2}\nu+\frac{1}{2}}\left(\frac{\beta^2}{8\alpha}\right) \right]$

$$[\operatorname{Re}\alpha > 0, \quad \operatorname{Re}\nu > -2] \qquad \text{ET II 29(9)}$$

8. $\displaystyle\int_0^1 x^{n+1} e^{-\alpha x^2} I_n(2\alpha x)\, dx = \frac{1}{4\alpha} \left[e^\alpha - e^{-\alpha} \sum_{r=-n}^n I_r(2\alpha) \right]$

$$[n = 0, 1, \ldots] \qquad \text{ET II 365(8)a}$$

9. $\displaystyle\int_1^\infty x^{1-n} e^{-\alpha x^2} I_n(2\alpha x)\, dx = \frac{1}{4\alpha} \left[e^\alpha - e^{-\alpha} \sum_{r=1-n}^{n-1} I_r(2\alpha) \right]$

$$[n = 1, 2, \ldots] \qquad \text{ET II 367(20)a}$$

10. $\displaystyle\int_0^\infty e^{-x^2} x^{2n+\mu+1} J_\mu\left(2x\sqrt{z}\right)\, dx = \frac{n!}{2} e^{-z} z^{\frac{1}{2}\mu} L_n^\mu(z) \qquad [n = 0, 1, \ldots; \quad n + \operatorname{Re}\mu > -1]$

$$\text{BU 135(5)}$$

6.632 $\displaystyle\int_0^\infty x^{-\frac{1}{2}} \exp\left[-\left(x^2 + a^2 - 2ax\cos\varphi\right)^{\frac{1}{2}}\right] \left[x^2 + a^2 - 2ax\cos\varphi\right]^{-\frac{1}{2}} K_\nu(x)\, dx$

$$= \pi a^{-\frac{1}{2}} \sec(\nu\pi)\, P_{\nu-\frac{1}{2}}\left(-\cos\varphi\right) K_\nu(a)$$

$$\left[|\arg a| + |\operatorname{Re}\varphi| < \pi, \quad |\operatorname{Re}\nu| < \tfrac{1}{2}\right] \quad \text{ET II 368(32)}$$

6.633

1. $\displaystyle\int_0^\infty x^{\lambda+1} e^{-\alpha x^2} J_\mu(\beta x) J_\nu(\gamma x)\, dx = \frac{\beta^\mu \gamma^\nu \alpha^{-\frac{\mu+\nu+\lambda+2}{2}}}{2^{\nu+\mu+1} \Gamma(\nu+1)} \sum_{m=0}^\infty \frac{\Gamma\left(m + \frac{1}{2}\nu + \frac{1}{2}\mu + \frac{1}{2}\lambda + 1\right)}{m!\, \Gamma(m+\mu+1)} \left(-\frac{\beta^2}{4\alpha}\right)^m$

$$\times F\left(-m, -\mu - m; \nu + 1; \frac{\gamma^2}{\beta^2}\right)$$

$$[\operatorname{Re}\alpha > 0, \operatorname{Re}(\mu+\nu+\lambda) > -2, \beta > 0, \quad \gamma > 0] \quad \text{EH II 49(20)a, ET II 51(24)a}$$

2. $\displaystyle\int_0^\infty e^{-\varrho^2 x^2} J_p(\alpha x) J_p(\beta x) x\, dx = \frac{1}{2\varrho^2} \exp\left(-\frac{\alpha^2+\beta^2}{4\varrho^2}\right) I_p\left(\frac{\alpha\beta}{2\varrho^2}\right)$

$$\left[\operatorname{Re} p > -1, \quad |\arg \varrho| < \frac{\pi}{4}, \quad \alpha > 0, \quad \beta > 0\right] \quad \text{KU 146(16)a, WA 433(1)}$$

3. $\displaystyle\int_0^\infty x^{2\nu+1} e^{-\alpha x^2} J_\nu(x) Y_\nu(x)\, dx = -\frac{1}{2\sqrt{\pi}} \alpha^{-\frac{3}{2}\nu-\frac{1}{2}} \exp\left(-\frac{1}{2\alpha}\right) W_{\frac{1}{2}\nu, \frac{1}{2}\nu}\left(\frac{1}{\alpha}\right)$

$$[\operatorname{Re}\alpha > 0, \quad \operatorname{Re}\nu > -\tfrac{1}{2}] \qquad \text{ET II 347(59)}$$

4. $\displaystyle\int_0^\infty x e^{-\alpha x^2} I_\nu(\beta x) J_\nu(\gamma x)\, dx = \frac{1}{2\alpha} \exp\left(\frac{\beta^2-\gamma^2}{4\alpha}\right) J_\nu\left(\frac{\beta\gamma}{2\alpha}\right)$

$$[\operatorname{Re}\alpha > 0, \quad \operatorname{Re}\nu > -1] \qquad \text{ET II 63(1)}$$

5. $\int_0^\infty x^{\lambda-1} e^{-\alpha x^2} J_\mu(\beta x) J_\nu(\beta x)\, dx$

$$= 2^{-\nu-\mu-1} \alpha^{-\frac{1}{2}(\nu+\lambda+\mu)} \beta^{\nu+\mu} \frac{\Gamma\left(\frac{1}{2}\lambda + \frac{1}{2}\mu + \frac{1}{2}\nu\right)}{\Gamma(\mu+1)\Gamma(\nu+1)}$$

$$\times \, _3F_3\left[\frac{\nu}{2} + \frac{\mu}{2} + \frac{1}{2}, \frac{\nu}{2} + \frac{\mu}{2} + 1, \frac{\nu+\mu+\lambda}{2}; \mu+1, \nu+1, \mu+\nu+1; -\frac{\beta^2}{\alpha}\right]$$

$$[\operatorname{Re}(\nu+\lambda+\mu) > 0, \quad \operatorname{Re}\alpha > 0] \quad \text{WA 434, EH II 50(21)}$$

6.634 $\int_0^\infty x e^{-\frac{x^2}{2a}} [I_\nu(x) + I_{-\nu}(x)] K_\nu(x)\, dx = a e^a K_\nu(a) \qquad [\operatorname{Re}a > 0, \quad -1 < \operatorname{Re}\nu < 1]$

$$\text{ET II 371(49)}$$

6.635

1. $\int_0^\infty x^{-1} e^{-\frac{\alpha}{x}} J_\nu(\beta x)\, dx = 2 J_\nu\left(\sqrt{2\alpha\beta}\right) K_\nu\left(\sqrt{2\alpha\beta}\right)$

$$[\operatorname{Re}\alpha > 0, \quad \beta > 0] \qquad \text{ET II 30(15)}$$

2. $\int_0^\infty x^{-1} e^{-\frac{\alpha}{x}} Y_\nu(\beta x)\, dx = 2 Y_\nu\left(\sqrt{2\alpha\beta}\right) K_\nu\left(\sqrt{2\alpha\beta}\right)$

$$[\operatorname{Re}\alpha > 0, \quad \beta > 0] \qquad \text{ET II 106(5)}$$

3. $\int_0^\infty x^{-1} e^{-\frac{\alpha}{x} - \beta x} J_\nu(\gamma x)\, dx = 2 J_\nu\left\{\sqrt{2\alpha}\left[\sqrt{\beta^2 + \gamma^2} - \beta\right]^{\frac{1}{2}}\right\} K_\nu\left\{\sqrt{2\alpha}\left[\sqrt{\beta^2 + \gamma^2} + \beta\right]^{\frac{1}{2}}\right\}$

$$[\operatorname{Re}\alpha > 0, \quad \operatorname{Re}\beta > 0, \quad \gamma > 0]$$
$$\text{ET II 30(16)}$$

6.636 $\int_0^\infty x^{-\frac{1}{2}} e^{-\alpha\sqrt{x}} J_\nu(\beta x)\, dx = \frac{\sqrt{2}}{\sqrt{\pi\beta}} \Gamma\left(\nu + \frac{1}{2}\right) D_{-\nu-\frac{1}{2}}\left(2^{-\frac{1}{2}} \alpha e^{\frac{1}{4}\pi i} \beta^{-\frac{1}{2}}\right) D_{-\nu-\frac{1}{2}}\left(2^{-\frac{1}{2}} \alpha e^{-\frac{1}{4}\pi i} \beta^{-\frac{1}{2}}\right)$

$$\left[\operatorname{Re}\alpha > 0, \quad \beta > 0, \quad \operatorname{Re}\nu > -\frac{1}{2}\right]$$
$$\text{ET II 30(17)}$$

6.637

1. $\int_0^\infty (\beta^2 + x^2)^{-\frac{1}{2}} \exp\left[-\alpha(\beta^2 + x^2)^{\frac{1}{2}}\right] J_\nu(\gamma x)\, dx$

$$= I_{\frac{1}{2}\nu}\left\{\frac{1}{2}\beta\left[(\alpha^2 + \gamma^2)^{\frac{1}{2}} - \alpha\right]\right\} K_{\frac{1}{2}\nu}\left\{\frac{1}{2}\beta\left[(\alpha^2 + \gamma^2)^{\frac{1}{2}} + \alpha\right]\right\}$$

$$[\operatorname{Re}\alpha > 0, \quad \operatorname{Re}\beta > 0, \quad \gamma > 0, \quad \operatorname{Re}\nu > -1] \quad \text{ET II 31(20)}$$

2. $\int_0^\infty (\beta^2 + x^2)^{-\frac{1}{2}} \exp\left[-\alpha(\beta^2 + x^2)^{\frac{1}{2}}\right] Y_\nu(\gamma x)\, dx$

$$= -\sec\left(\frac{\nu\pi}{2}\right) K_{\frac{1}{2}\nu}\left\{\frac{1}{2}\beta\left[(\alpha^2 + \gamma^2)^{\frac{1}{2}} + \alpha\right]\right\}$$

$$\times \left(\frac{1}{\pi} K_{\frac{1}{2}\nu}\left\{\frac{1}{2}\beta\left[(\alpha^2 + \gamma^2)^{\frac{1}{2}} + \alpha\right]\right\} + \sin\left(\frac{\nu\pi}{2}\right) I_{\frac{1}{2}\nu}\left\{\frac{1}{2}\beta\left[(\alpha^2 + \gamma^2)^{\frac{1}{2}} - \alpha\right]\right\}\right)$$

$$[\operatorname{Re}\alpha > 0, \quad \operatorname{Re}\beta > 0, \quad \gamma > 0, \quad |\operatorname{Re}\nu| < 1] \quad \text{ET II 106(6)}$$

3. $$\int_0^\infty \left(x^2 + \beta^2\right)^{-\frac{1}{2}} \exp\left[-\alpha\left(x^2 + \beta^2\right)^{\frac{1}{2}}\right] K_\nu(\gamma x)\, dx$$

$$= \frac{1}{2} \sec\left(\frac{\nu\pi}{2}\right) K_{\frac{1}{2}\nu}\left(\frac{1}{2}\beta\left[\alpha + \left(\alpha^2 - \gamma^2\right)^{\frac{1}{2}}\right]\right) K_{\frac{1}{2}\nu}\left(\frac{1}{2}\beta\left[\alpha - \left(\alpha^2 - \gamma^2\right)^{\frac{1}{2}}\right]\right)$$

$$[\operatorname{Re}\alpha > 0, \quad \operatorname{Re}\beta > 0, \quad \operatorname{Re}(\gamma + \beta) > 0, \quad |\operatorname{Re}\nu| < 1] \quad \text{ET II 132(26)}$$

6.64 Combinations of Bessel functions of more complicated arguments, exponentials, and powers

6.641 $$\int_0^\infty \sqrt{x}\, e^{-\alpha x} J_{\pm\frac{1}{4}}\left(x^2\right) dx = \frac{\sqrt{\pi\alpha}}{4}\left[\mathbf{H}_{\mp\frac{1}{4}}\left(\frac{\alpha^2}{4}\right) - Y_{\mp\frac{1}{4}}\left(\frac{\alpha^2}{4}\right)\right]$$ MI 42

6.642

1.[10] $$\int_0^\infty x^{-1} e^{-\alpha x} Y_\nu\left(\frac{2}{x}\right) dx = 2 K_\nu\left(2\sqrt{a}\right) Y_\nu\left(2\sqrt{a}\right)$$

$$[\operatorname{Re} a > 0] \qquad\qquad \text{MC}$$

2. $$\int_0^\infty x^{-1} e^{-\alpha x} H_\nu^{(1,2)}\left(\frac{2}{x}\right) dx = H_\nu^{(1,2)}\left(\sqrt{\alpha}\right) K_\nu\left(\sqrt{\alpha}\right)$$ MI 44, EH II 91(26)

6.643

1. $$\int_0^\infty x^{\mu-\frac{1}{2}} e^{-\alpha x} J_{2\nu}\left(2\beta\sqrt{x}\right) dx = \frac{\Gamma\left(\mu + \nu + \frac{1}{2}\right)}{\beta\,\Gamma(2\nu+1)} e^{-\frac{\beta^2}{2\alpha}} \alpha^{-\mu} M_{\mu,\nu}\left(\frac{\beta^2}{\alpha}\right)$$

$$\left[\operatorname{Re}\left(\mu + \nu + \tfrac{1}{2}\right) > 0\right]$$

$$\text{BU 14(13a), MI 42a}$$

2. $$\int_0^\infty x^{\mu-\frac{1}{2}} e^{-\alpha x} I_{2\nu}\left(2\beta\sqrt{x}\right) dx = \frac{\Gamma\left(\mu + \nu + \frac{1}{2}\right)}{\Gamma(2\nu+1)} \beta^{-1} e^{\frac{\beta^2}{2\alpha}} \alpha^{-\mu} M_{-\mu,\nu}\left(\frac{\beta^2}{\alpha}\right)$$

$$\left[\operatorname{Re}\left(\mu + \nu + \tfrac{1}{2}\right) > 0\right] \qquad \text{MI 45}$$

3. $$\int_0^\infty x^{\mu-\frac{1}{2}} e^{-\alpha x} K_{2\nu}\left(2\beta\sqrt{x}\right) dx = \frac{\Gamma\left(\mu + \nu + \frac{1}{2}\right)\Gamma\left(\mu - \nu + \frac{1}{2}\right)}{2\beta} e^{\frac{\beta^2}{2\alpha}} \alpha^{-\mu} W_{-\mu,\nu}\left(\frac{\beta^2}{\alpha}\right)$$

$$\left[\operatorname{Re}\left(\mu + \nu + \tfrac{1}{2}\right) > 0\right], \qquad (\text{cf. } \mathbf{6.631}\ 3)$$

$$\text{MI 47a}$$

4. $$\int_0^\infty x^{n+\frac{1}{2}\nu} e^{-\alpha x} J_\nu\left(2\beta\sqrt{x}\right) dx = n!\,\beta^\nu e^{-\frac{\beta^2}{\alpha}} \alpha^{-n-\nu-1} L_n^\nu\left(\frac{\beta^2}{\alpha}\right)$$

$$[n + \nu > -1] \qquad\qquad \text{MO 178a}$$

5. $$\int_0^\infty x^{-\frac{1}{2}} e^{-\alpha x} Y_{2\nu}\left(\beta\sqrt{x}\right) dx = -\sqrt{\frac{\pi}{\alpha}} \frac{\exp\left(-\frac{\beta^2}{8\alpha}\right)}{\cos(\nu\pi)}\left[\sin(\nu\pi) I_\nu\left(\frac{\beta^2}{8\alpha}\right) + \frac{1}{\pi} K_\nu\left(\frac{\beta^2}{8\alpha}\right)\right]$$

$$\left[|\operatorname{Re}\nu| < \tfrac{1}{2}\right] \qquad\qquad \text{MI 44}$$

6. $$\int_0^\infty x^{\frac{1}{2}m} e^{-\alpha x} K_m\left(2\sqrt{x}\right) dx = \frac{\Gamma(m+1)}{2\alpha}\left(\frac{1}{\alpha}\right)^{\frac{1}{2}m-\frac{1}{2}} e^{\frac{1}{2\alpha}} W_{-\frac{1}{2}(m+1),-\frac{1}{2}m}\left(\frac{1}{\alpha}\right)$$ MI 48a

6.644 $\displaystyle\int_0^\infty e^{-\beta x} J_{2\nu}\left(2a\sqrt{x}\right) J_\nu(bx)\,dx = \exp\left(-\frac{a^2\beta}{\beta^2+b^2}\right) J_\nu\left(\frac{a^2 b}{\beta^2+b^2}\right)\frac{1}{\sqrt{\beta^2+b^2}}$

$$\left[\operatorname{Re}\beta>0,\quad b>0,\quad \operatorname{Re}\nu>-\tfrac{1}{2}\right]$$

ET II 58(17)

6.645

1. $\displaystyle\int_1^\infty \left(x^2-1\right)^{-\frac{1}{2}} e^{-\alpha x} J_\nu\left(\beta\sqrt{x^2-1}\right) dx = I_{\frac{1}{2}\nu}\left[\frac{1}{2}\left(\sqrt{\alpha^2+\beta^2}-\alpha\right)\right] K_{\frac{1}{2}\nu}\left[\frac{1}{2}\left(\sqrt{\alpha^2+\beta^2}+\alpha\right)\right]$

MO 179a

2. $\displaystyle\int_1^\infty \left(x^2-1\right)^{\frac{1}{2}\nu} e^{-\alpha x} J_\nu\left(\beta\sqrt{x^2-1}\right) dx = \sqrt{\frac{2}{\pi}}\beta^\nu\left(\alpha^2+\beta^2\right)^{-\frac{1}{2}\nu-\frac{1}{4}} K_{\nu+\frac{1}{2}}\left(\sqrt{\alpha^2+\beta^2}\right)$

MO 179a

3.[3] $\displaystyle\int_{-1}^1 \left(1-x^2\right)^{-1/2} e^{-ax} I_1\left(b\sqrt{1-x^2}\right) dx = \frac{2}{b}\left(\cosh\sqrt{a^2+b^2}-\cosh a\right)$

$$\left[a>0,\quad b>0\right]$$

6.646

1. $\displaystyle\int_1^\infty \left(\frac{x-1}{x+1}\right)^{\frac{1}{2}\nu} e^{-\alpha x} J_\nu\left(\beta\sqrt{x^2-1}\right) dx = \frac{\exp\left(-\sqrt{\alpha^2+\beta^2}\right)}{\sqrt{\alpha^2+\beta^2}}\left(\frac{\beta}{\alpha+\sqrt{\alpha^2+\beta^2}}\right)^\nu$

$$\left[\operatorname{Re}\nu>-1\right]$$ EF 89(52), MO 179

2. $\displaystyle\int_1^\infty \left(\frac{x-1}{x+1}\right)^{\frac{1}{2}\nu} e^{-\alpha x} I_\nu\left(\beta\sqrt{x^2-1}\right) dx = \frac{\exp\left(-\sqrt{\alpha^2-\beta^2}\right)}{\sqrt{\alpha^2-\beta^2}}\left(\frac{\beta}{\alpha+\sqrt{\alpha^2-\beta^2}}\right)^\nu$

$$\left[\operatorname{Re}\nu>-1,\quad \alpha>\beta\right]$$ MO 180

3.[7] $\displaystyle\int_b^\infty e^{-pt}\left(\frac{t-b}{t+b}\right)^{\nu/2} K_\nu\left[a\left(t^2-b^2\right)^{1/2}\right] dt = \frac{\Gamma(\nu+1)}{2s a^\nu}\left[x^\nu e^{-bx}\Gamma(-\nu,bx)-y^\nu e^{bs}\Gamma(-\nu,by)\right]$

where $x=p-s,\quad y=p+s,\quad s=\left(p^2-a^2\right)^{1/2}$ $\left[\operatorname{Re}(p+a)>0,\quad |\operatorname{Re}(\nu)|<1\right].$

ME 39a

6.647

1. $\displaystyle\int_0^\infty x^{-\lambda-\frac{1}{2}}(\beta+x)^{\lambda-\frac{1}{2}} e^{-\alpha x} K_{2\mu}\left[\sqrt{x(\beta+x)}\right] dx$

$$= \frac{1}{\beta} e^{\frac{1}{2}\alpha\beta}\,\Gamma\left(\tfrac{1}{2}-\lambda+\mu\right)\Gamma\left(\tfrac{1}{2}-\lambda-\mu\right) W_{\lambda,\mu}(z_1)\,W_{\lambda,\mu}(z_2)$$

$$z_1=\tfrac{1}{2}\beta\left(\alpha+\sqrt{\alpha^2-1}\right),\quad z_2=\tfrac{1}{2}\beta\left(\alpha-\sqrt{\alpha^2-1}\right)$$

$$\left[|\arg\beta|<\pi,\quad \operatorname{Re}\alpha>-1,\quad \operatorname{Re}\lambda+|\operatorname{Re}\mu|<\tfrac{1}{2}\right]$$ ET II 377(37)

2. $\int_0^\infty (\alpha + x)^{-\frac{1}{2}} x^{-\frac{1}{2}} e^{-x \cosh t} K_\nu \left[\sqrt{x(\alpha + x)} \right] dx$

$$= \frac{1}{2} \sec \left(\frac{\nu \pi}{2} \right) e^{\frac{1}{2} \alpha \cosh t} K_{\frac{1}{2}\nu} \left(\frac{1}{4} \alpha e^t \right) K_{\frac{1}{2}\nu} \left(\frac{1}{4} \alpha e^{-t} \right)$$

$$[-1 < \operatorname{Re} \nu < 1] \qquad \text{ET II 377(36)}$$

3.[7] $\int_0^\alpha x^{\lambda - \frac{1}{2}} (\alpha - x)^{-\lambda - \frac{1}{2}} e^{-x \sinh t} I_{2\mu} \left[\sqrt{x(\alpha - x)} \right] dx$

$$= e^{-(a/2)\sinh t} \frac{2\,\Gamma\left(\frac{1}{2} + \lambda + \mu\right) \Gamma\left(\frac{1}{2} - \lambda + \mu\right)}{\alpha[\Gamma(2\mu + 1)]^2} M_{\lambda, \mu} \left(\frac{1}{2} \alpha e^t \right) M_{-\lambda, \mu} \left(\frac{1}{2} \alpha e^{-t} \right)$$

$$\left[\operatorname{Re} \mu > |\operatorname{Re} \lambda| - \tfrac{1}{2} \right] \qquad \text{ET II 377(32)}$$

6.648 $\int_{-\infty}^\infty e^{\varrho x} \left(\frac{\alpha + \beta e^x}{\alpha e^x + \beta} \right)^\nu K_{2\nu} \left[\left(\alpha^2 + \beta^2 + 2\alpha\beta \cosh x \right)^{\frac{1}{2}} \right] dx = 2\,K_{\nu+\varrho}(\alpha)\,K_{\nu-\varrho}(\beta)$

$$[\operatorname{Re} \alpha > 0, \quad \operatorname{Re} \beta > 0] \qquad \text{ET II 379(45)}$$

6.649

1. $\int_0^\infty K_{\mu-\nu}(2z \sinh x) e^{(\nu+\mu)x} dx = \frac{\pi^2}{4 \sin[(\nu - \mu)\pi]} [J_\nu(z)\,Y_\mu(z) - J_\mu(z)\,Y_\nu(z)]$

$$[\operatorname{Re} z > 0, \quad -1 < \operatorname{Re}(\nu - \mu) < 1]$$

$$\text{MO 44}$$

2. $\int_0^\infty J_{\nu+\mu}(2x \sinh t) e^{(\nu-\mu)t} dt = K_\nu(x)\,I_\mu(x)$

$$\left[\operatorname{Re}(\nu - \mu) < \tfrac{3}{2}, \quad \operatorname{Re}(\nu + \mu) > -1, \quad x > 0 \right] \quad \text{EH II 97(68)}$$

3. $\int_0^\infty Y_{\nu-\mu}(2x \sinh t) e^{-(\nu+\mu)t} dt = \frac{1}{\sin[\pi(\mu - \nu)]} \{ I_\mu(x)\,K_\nu(x) - \cos[(\nu - \mu)\pi]\,I_\nu(x)\,K_\mu(x) \}$

$$\left[|\operatorname{Re}(\nu - \mu)| < 1, \quad \operatorname{Re}(\nu + \mu) > -\tfrac{1}{2}, \quad x > 0 \right] \quad \text{EH II 97(73)}$$

4. $\int_0^\infty K_0(2z \sinh x) e^{-2\nu x} dx = -\frac{\pi}{4} \left\{ J_\nu(z) \frac{\partial\,Y_\nu(z)}{\partial \nu} - Y_\nu(z) \frac{\partial\,J_\nu(z)}{\partial \nu} \right\}$

6.65 Combinations of Bessel and exponential functions of more complicated arguments and powers

6.651

1. $\int_0^\infty x^{\lambda + \frac{1}{2}} e^{-\frac{1}{4}\alpha^2 x^2} I_\mu \left(\frac{1}{4} \alpha^2 x^2 \right) J_\nu(\beta x) dx$

$$= \frac{1}{\sqrt{2\pi}} 2^{\lambda+1} \beta^{-\lambda - \frac{3}{2}} G_{23}^{21} \left(\frac{\beta^2}{2\alpha^2} \left| \begin{matrix} 1 - \mu, 1 + \mu \\ h, \frac{1}{2}, k \end{matrix} \right. \right)$$

$$h = \frac{3}{4} + \frac{1}{2}\lambda + \frac{1}{2}\nu, \quad k = \frac{3}{4} + \frac{1}{2}\lambda - \frac{1}{2}\nu$$

$$\left[|\arg \alpha| < \frac{\pi}{4}, \quad \beta > 0, \quad -\frac{3}{2} - \operatorname{Re}(2\mu + \nu) < \operatorname{Re} \lambda < 0 \right] \quad \text{ET II 68(8)}$$

2.
$$\int_0^\infty x^{\lambda+\frac{1}{2}} e^{-\frac{1}{4}\alpha^2 x^2} K_\mu\left(\tfrac{1}{4}\alpha^2 x^2\right) J_\nu(\beta x)\, dx$$
$$= \sqrt{\frac{\pi}{2}}\, 2^{\lambda+1} \beta^{-\lambda-\frac{3}{2}}\, G_{23}^{12}\left(\frac{\beta^2}{2\alpha^2} \left| \begin{array}{l} 1-\mu, 1+\mu \\ h, \frac{1}{2}, k \end{array}\right.\right)$$

$$h = \tfrac{3}{4} + \tfrac{1}{2}\lambda + \tfrac{1}{2}\nu, \quad k = \tfrac{3}{4} + \tfrac{1}{2}\lambda - \tfrac{1}{2}\nu$$

$$\left[|\arg\alpha| < \frac{\pi}{4}, \quad \operatorname{Re}(\lambda+\nu\pm 2\mu) > -\tfrac{3}{2}\right] \quad \textbf{ET II 69(15)}$$

3.
$$\int_0^\infty x^{2\mu-\nu+1} e^{-\frac{1}{4}\alpha x^2} I_\mu\left(\tfrac{1}{4}\alpha x^2\right) J_\nu(\beta x)\, dx$$
$$= 2^{\mu-\nu+\frac{1}{2}} (\pi\alpha)^{-\frac{1}{2}} \Gamma\left(\frac{1}{2}+\mu\right) \frac{\beta^{\nu-2\mu-1}}{\Gamma\left(\frac{1}{2}-\mu+\nu\right)}\, {}_1F_1\left(\frac{1}{2}+\mu; \frac{1}{2}-\mu+\nu; -\frac{\beta^2}{2\alpha}\right)$$
$$\left[\operatorname{Re}\alpha > 0, \quad \beta > 0, \quad \operatorname{Re}\nu > 2\operatorname{Re}\mu + \tfrac{1}{2} > -\tfrac{1}{2}\right] \quad \textbf{ET II 68(6)}$$

4.
$$\int_0^\infty x^{2\mu+\nu+1} e^{-\frac{1}{4}\alpha^2 x^2} K_\mu\left(\tfrac{1}{4}\alpha^2 x^2\right) J_\nu(\beta x)\, dx$$
$$= \sqrt{\pi}\, 2^\mu \alpha^{-2\mu-2\nu-2} \beta^\nu \frac{\Gamma(1+2\mu+\nu)}{\Gamma\left(\mu+\nu+\frac{3}{2}\right)}\, {}_1F_1\left(1+2\mu+\nu; \mu+\nu+\frac{3}{2}; -\frac{\beta^2}{2\alpha^2}\right)$$
$$\left[|\arg\alpha| < \tfrac{1}{4}\pi, \quad \operatorname{Re}\nu > -1, \quad \operatorname{Re}(2\mu+\nu) > -1, \quad \beta > 0\right] \quad \textbf{ET II 69(13)}$$

5.
$$\int_0^\infty x^{2\mu+\nu+1} e^{-\frac{1}{2}\alpha x^2} I_\mu\left(\tfrac{1}{2}\alpha x^2\right) K_\nu(\beta x)\, dx$$
$$= \frac{2^{\mu-\frac{1}{2}}}{\sqrt{\pi}} \beta^{-\mu-\frac{3}{2}} \alpha^{-\frac{1}{2}\mu-\frac{1}{2}\nu-\frac{1}{4}} \Gamma(2\mu+\nu+1)\, \Gamma\left(\mu+\tfrac{1}{2}\right) \exp\left(\frac{\beta^2}{8\alpha}\right) W_{k,m}\left(\frac{\beta^2}{4\alpha}\right)$$
$$2k = -3\mu - \nu - \tfrac{1}{2}, \quad 2m = \mu+\nu+\tfrac{1}{2}$$

$$\left[\operatorname{Re}\alpha > 0, \quad \operatorname{Re}\mu > -\tfrac{1}{2}, \quad \operatorname{Re}(2\mu+\nu) > -1\right] \quad \textbf{ET II 146(53)}$$

6.
$$\int_0^\infty x e^{-\frac{1}{4}\alpha x^2} J_{\frac{1}{2}\nu}\left(\tfrac{1}{4}\beta x^2\right) J_\nu(\gamma x)\, dx = 2\left(\alpha^2+\beta^2\right)^{-\frac{1}{2}} \exp\left(-\frac{\alpha\gamma^2}{\alpha^2+\beta^2}\right) J_{\frac{1}{2}\nu}\left(\frac{\beta\gamma^2}{\alpha^2+\beta^2}\right)$$
$$\left[\gamma > 0, \quad \operatorname{Re}\alpha > |\operatorname{Im}\beta|, \quad \operatorname{Re}\nu > -1\right]$$
$$\textbf{ET II 56(2)}$$

7.
$$\int_0^\infty x e^{-\frac{1}{4}\alpha x^2} I_{\frac{1}{2}\nu}\left(\tfrac{1}{4}\alpha x^2\right) J_\nu(\beta x)\, dx = \left(\frac{1}{2}\pi\alpha\right)^{-\frac{1}{2}} \beta^{-1} \exp\left(-\frac{\beta^2}{2\alpha}\right)$$
$$\left[\operatorname{Re}\alpha > 0, \quad \beta > 0, \quad \operatorname{Re}\nu > -1\right]$$
$$\textbf{ET II 67(3)}$$

8.
$$\int_0^\infty x^{1-\nu} e^{-\frac{1}{4}\alpha^2 x^2} I_\nu\left(\tfrac{1}{4}\alpha^2 x^2\right) J_\nu(\beta x)\, dx = \sqrt{\frac{2}{\pi}} \frac{\beta^{\nu-1}}{\alpha} \exp\left(-\frac{\beta^2}{4\alpha^2}\right) D_{-2\nu}\left(\frac{\beta}{\alpha}\right)$$
$$\left[|\arg\alpha| < \tfrac{1}{4}\pi, \quad \beta > 0, \quad \operatorname{Re}\nu > -\tfrac{1}{2}\right]$$
$$\textbf{ET II 67(1)}$$

9.
$$\int_0^\infty x^{-\nu-1} e^{-\frac{1}{4}\alpha^2 x^2} I_{\nu+1}\left(\tfrac{1}{4}\alpha^2 x^2\right) J_\nu(\beta x)\, dx = \sqrt{\frac{2}{\pi}} \beta^\nu \exp\left(-\frac{\beta^2}{4\alpha^2}\right) D_{-2\nu-3}\left(\frac{\beta}{\alpha}\right)$$
$$\left[|\arg\alpha| < \tfrac{1}{4}\pi, \quad \operatorname{Re}\nu > -1, \quad \beta > 0\right]$$
$$\textbf{ET II 67(2)}$$

6.652 $\displaystyle\int_0^\infty x^{2\nu} e^{-\left(\frac{x^2}{8}+\alpha x\right)} I_\nu\left(\frac{x^2}{8}\right) dx = \frac{\Gamma(4\nu+1)}{2^{4\nu}\,\Gamma(\nu+1)}\frac{e^{\frac{\alpha^2}{2}}}{\alpha^{\nu+1}}\, W_{-\frac{3}{2}\nu,\frac{1}{2}\nu}\left(\alpha^2\right)$

$\left[\operatorname{Re}\left(\nu+\frac{1}{4}\right)>0\right]$ MI 45

6.653

1. $\displaystyle\int_0^\infty \exp\left[-\frac{1}{2}x - \frac{1}{2x}\left(a^2+b^2\right)\right] I_\nu\left(\frac{ab}{x}\right)\frac{dx}{x} = 2\,I_\nu(a)\,K_\nu(b)$ $[0<a<b]$

$= 2\,K_\nu(a)\,I_\nu(b)$ $[0<b<a]$

$[\operatorname{Re}\nu>-1]$ WA 482(2)a, EH II 53(37), WA 482(3)a

2. $\displaystyle\int_0^\infty \exp\left[-\frac{1}{2}x - \frac{1}{2x}\left(z^2+w^2\right)\right] K_\nu\left(\frac{zw}{x}\right)\frac{dx}{x} = 2\,K_\nu(z)\,K_\nu(w)$

$\left[|\arg z|<\pi,\quad |\arg w|<\pi,\quad \arg(z+w)|<\frac{1}{4}\pi\right]$ WA 483(1), EH II 53(36)

6.654 $\displaystyle\int_0^\infty x^{-\frac{1}{2}} e^{-\frac{\beta^2}{8x}-\alpha x} K_\nu\left(\frac{\beta^2}{8x}\right) dx = \sqrt{4\pi}\,\alpha^{-\frac{1}{2}} K_{2\nu}\left(\beta\sqrt{\alpha}\right)$ ME 39

6.655 $\displaystyle\int_0^\infty x\left(\beta^2+x^2\right)^{-\frac{1}{2}} \exp\left(-\frac{\alpha^2\beta}{\beta^2+x^2}\right) J_\nu\left(\frac{\alpha^2 x}{\beta^2+x^2}\right) J_\nu(\gamma x)\,dx = \gamma^{-1} e^{-\beta\gamma} J_{2\nu}\left(2\alpha\sqrt{\gamma}\right)$

$\left[\operatorname{Re}\beta>0,\quad \gamma>0,\quad \operatorname{Re}\nu>-\frac{1}{2}\right]$

ET II 58(14)

6.656

1. $\displaystyle\int_0^\infty e^{-(\xi-z)\cosh t}\, J_{2\nu}\left[2(z\xi)^{\frac{1}{2}}\sinh t\right] dt = I_\nu(z)\,K_\nu(\xi)$

$\left[\operatorname{Re}\nu>-\frac{1}{2},\quad \operatorname{Re}(\xi-z)>0\right]$

EH II 98(78)

2. $\displaystyle\int_0^\infty e^{-(\xi+z)\cosh t}\, K_{2\nu}\left[2(z\xi)^{\frac{1}{2}}\sinh t\right] dt = \frac{1}{2}K_\nu(z)\,K_\nu(\xi)\sec(\nu\pi)$

$\left[|\operatorname{Re}\nu|<\frac{1}{2},\quad \operatorname{Re}\left(z^{\frac{1}{2}}+\xi^{\frac{1}{2}}\right)^2\ge 0\right]$

EH II 98(79)

6.66 Combinations of Bessel, hyperbolic, and exponential functions

Bessel and hyperbolic functions

6.661

1. $\displaystyle\int_0^\infty \sinh(ax)\,K_\nu(bx)\,dx = \frac{\pi}{2}\frac{\operatorname{cosec}\left(\frac{\nu\pi}{2}\right)\sin\left[\nu\arcsin\left(\frac{a}{b}\right)\right]}{\sqrt{b^2-a^2}}$

$[\operatorname{Re}b>|\operatorname{Re}a|,\quad |\operatorname{Re}\nu|<2]$

ET II 133(32)

2. $\displaystyle\int_0^\infty \cosh(ax)\,K_\nu(bx)\,dx = \frac{\pi\cos\left[\nu\arcsin\left(\frac{a}{b}\right)\right]}{2\sqrt{b^2-a^2}\cos\left(\frac{\nu\pi}{2}\right)}$ $[\operatorname{Re}b>|\operatorname{Re}a|,\quad |\operatorname{Re}\nu|<1]$

ET II 134(33)

6.662   Notation:

$$\ell_1 = \frac{1}{2}\left[\sqrt{(b+c)^2+a^2}-\sqrt{(b-c)^2+a^2}\right], \qquad \ell_2 = \frac{1}{2}\left[\sqrt{(b+c)^2+a^2}+\sqrt{(b-c)^2+a^2}\right]$$

1.[10] $$\int_0^\infty \cosh(\beta x)\, K_0(\alpha x)\, J_0(\gamma x)\, dx = \frac{\boldsymbol{K}(k)}{\sqrt{u+v}}$$

$$u = \frac{1}{2}\left\{\sqrt{(\alpha^2+\beta^2+\gamma^2)^2-4\alpha^2\beta^2}\right\} + \alpha^2 - \beta^2 - \gamma^2$$

$$v = \frac{1}{2}\left\{\sqrt{(\alpha^2+\beta^2+\gamma^2)^2-4\alpha^2\beta^2}\right\} - \alpha^2 + \beta^2 + \gamma^2$$

$$k^2 = v(u+v)^{-1} \qquad [\operatorname{Re}\alpha > |\operatorname{Re}\beta|, \quad \gamma > 0]$$

ET II 15(23)

alternatively, with $a = \gamma$, $b = \beta$, $c = \alpha$,

$$\int_0^\infty \cosh(bx)\, K_0(cx)\, J_0(ax)\, dx = \frac{\boldsymbol{K}(k)}{\sqrt{\ell_2^2-\ell_1^2}}$$

$$k^2 = \frac{\ell_2^2-c^2}{\ell_2^2-\ell_1^2}, \qquad [\operatorname{Re}c > |\operatorname{Re}b|, \quad a > 0]$$

2.[10] $$\int_0^\infty \sinh(\beta x)\, K_1(\alpha x)\, J_0(\gamma x)\, dx = a^{-1}\left[u\,\boldsymbol{E}(k) - \boldsymbol{K}(k)\,\boldsymbol{E}(u) + \frac{\boldsymbol{K}(k)\operatorname{sn}u\operatorname{dn}u}{\operatorname{cn}u}\right]$$

$$\operatorname{cn}^2 u = 2\gamma^2\left\{\left[(\alpha^2+\beta^2+\gamma^2)^2-4\alpha^2\beta^2\right]^{\frac{1}{2}} - \alpha^2 + \beta^2 + \gamma^2\right\}^{-1}$$

$$k^2 = \frac{1}{2}\left\{1 - (\alpha^2-\beta^2-\gamma^2)\left[(\alpha^2+\beta^2+\gamma^2)^2-4\alpha^2\beta^2\right]^{-\frac{1}{2}}\right\}$$

$$[\operatorname{Re}\alpha > |\operatorname{Re}\beta|, \quad \gamma > 0]$$

ET II 15(24)

alternatively, with $a = \gamma$, $b = \beta$, $c = \alpha$,

$$\int_0^\infty \sinh(bx)\, K_1(cx)\, J_0(ax)\, dx = c^{-1}\left[u\,\boldsymbol{E}(k) - \boldsymbol{K}(k)\,\boldsymbol{E}(u) + \frac{\boldsymbol{K}(k)\operatorname{sn}u\operatorname{dn}u}{\operatorname{cn}u}\right]$$

$$\operatorname{cn}^2 u = \frac{a^2}{\ell_2^2-c^2}, \quad k^2 = \frac{\ell_2^2-c^2}{\ell_2^2-\ell_1^2} \qquad [\operatorname{Re}c > |\operatorname{Re}b|, \quad a > 0]$$

6.663

1. $$\int_0^\infty K_{\nu\pm\mu}(2z\cosh t)\cosh[(\mu\mp\nu)t]\, dt = \frac{1}{2}K_\mu(z)\,K_\nu(z)$$

$$[\operatorname{Re}z > 0] \qquad\qquad \text{WA 484(1), EH II 54(39)}$$

2. $$\int_0^\infty Y_{\mu+\nu}(2z\cosh t)\cosh[(\mu-\nu)t]\, dt = \frac{\pi}{4}\left[J_\mu(z)\,J_\nu(z) - Y_\mu(z)\,Y_\nu(z)\right]$$

$$[z > 0] \qquad\qquad \text{EH II 96(64)}$$

3. $\displaystyle\int_0^\infty J_{\mu+\nu}\left(2z\cosh t\right)\cosh[(\mu-\nu)t]\,dt = -\frac{\pi}{4}\left[J_\mu(z)\,Y_\nu(z) + J_\nu(z)\,Y_\mu(z)\right]$

$$[z > 0]$$ EH II 97(65)

4. $\displaystyle\int_0^\infty J_{\mu+\nu}\left(2z\sinh t\right)\cosh[(\mu-\nu)t]\,dt = \frac{1}{2}\left[I_\nu(z)\,K_\mu(z) + I_\mu(z)\,K_\nu(z)\right]$

$$\left[\operatorname{Re}(\nu+\mu) > -1, \quad |\operatorname{Re}(\mu-\nu)| < \tfrac{3}{2}, \quad z > 0\right]$$ EH II 97(71)

5. $\displaystyle\int_0^\infty J_{\mu+\nu}\left(2z\sinh t\right)\sinh[(\mu-\nu)t]\,dt = \frac{1}{2}\left[I_\nu(z)\,K_\mu(z) - I_\mu(z)\,K_\nu(z)\right]$

$$\left[\operatorname{Re}(\nu+\mu) > -1, \quad |\operatorname{Re}(\mu-\nu)| < \tfrac{3}{2}, \quad z > 0\right]$$ EH II 97(72)

6.664

1. $\displaystyle\int_0^\infty J_0\left(2z\sinh t\right)\sinh(2\nu t)\,dt = \frac{\sin(\nu\pi)}{\pi}[K_\nu(z)]^2$ $\left[|\operatorname{Re}\nu| < \tfrac{3}{4}, \quad z > 0\right]$ EH II 97(69)

2. $\displaystyle\int_0^\infty Y_0\left(2z\sinh t\right)\cosh(2\nu t)\,dt = -\frac{\cos(\nu\pi)}{\pi}[K_\nu(z)]^2$ $\left[|\operatorname{Re}\nu| < \tfrac{3}{4}, \quad z > 0\right]$ EH II 97(70)

3. $\displaystyle\int_0^\infty Y_0\left(2z\sinh t\right)\sinh(2\nu t)\,dt = \frac{1}{\pi}\left[I_\nu(z)\frac{\partial K_\nu(z)}{\partial \nu} - K_\nu(z)\frac{\partial I_\nu(z)}{\partial \nu}\right] - \frac{1}{\pi}\cos(\nu\pi)[K_\nu(z)]^2$

$$\left[|\operatorname{Re}\nu| < \tfrac{3}{4}, \quad z > 0\right]$$ EH II 97(75)

4. $\displaystyle\int_0^\infty K_0\left(2z\sinh t\right)\cosh 2\nu t\,dt = \frac{\pi^2}{8}\left\{J_\nu{}^2(z) + N_\nu{}^2(z)\right\}$

$$[\operatorname{Re} z > 0]$$ MO 44

5. $\displaystyle\int_0^\infty K_{2\mu}\left(z\sinh 2t\right)\coth^{2\nu} t\,dt = \frac{1}{4z}\,\Gamma\left(\frac{1}{2}+\mu-\nu\right)\Gamma\left(\frac{1}{2}-\mu-\nu\right)W_{\nu,\mu}(iz)\,W_{\nu,\mu}(-iz)$

$$\left[|\arg z| \le \frac{\pi}{2}, \quad |\operatorname{Re}\mu| + \operatorname{Re}\nu < \tfrac{1}{2}\right]$$
MO 119

6. $\displaystyle\int_0^\infty \cosh(2\mu x)\,K_{2\nu}\left(2a\cosh x\right)\,dx = \frac{1}{2}\,K_{\mu+\nu}(a)\,K_{\mu-\nu}(a)$

$$[\operatorname{Re} a > 0]$$ ET II 378(42)

6.665 $\displaystyle\int_0^\infty \operatorname{sech} x\,\cosh(2\lambda x)\,I_{2\mu}\left(a\operatorname{sech} x\right)\,dx = \frac{\Gamma\left(\frac{1}{2}+\lambda+\mu\right)\Gamma\left(\frac{1}{2}-\lambda+\mu\right)}{2a[\Gamma(2\mu+1)]^2}\,M_{\lambda,\mu}(a)\,M_{-\lambda,\mu}(a)$

$$\left[|\operatorname{Re}\lambda| - \operatorname{Re}\mu < \tfrac{1}{2}\right]$$ ET II 378(43)

Bessel, hyperbolic, and algebraic functions

6.666 $\displaystyle\int_0^\infty x^{\nu+1}\sinh(\alpha x)\operatorname{cosech}(\pi x)\,J_\nu(\beta x)\,dx = \frac{2}{\pi}\sum_{n=1}^\infty (-1)^{n-1}n^{\nu+1}\sin(n\alpha)\,K_\nu(n\beta)$

$$[|\operatorname{Re}\alpha| < \pi, \quad \operatorname{Re}\nu > -1]$$
ET II 41(3), WA 469(12)

6.667

1.^{3} $\quad \displaystyle\int_0^a \frac{\cosh\left(\sqrt{a^2-x^2}\right)\sinh t\, I_{2\nu}(x)}{\sqrt{a^2-x^2}}\, dx = \frac{\pi}{2}\, I_\nu\left(\frac{1}{2}ae^t\right) I_\nu\left(\frac{1}{2}ae^{-t}\right)$

$$\left[\operatorname{Re}\nu > -\tfrac{1}{2}\right] \qquad \text{ET II 365(10)}$$

2. $\quad \displaystyle\int_0^a \frac{\cosh\left(\sqrt{a^2-x^2}\,\sinh t\right) K_{2\nu}(x)}{\sqrt{a^2-x^2}}\, dx = \frac{\pi^2}{4}\operatorname{cosec}(\nu\pi)\left[I_{-\nu}\left(ae^t\right)I_{-\nu}\left(ae^{-t}\right) - I_\nu\left(ae^t\right)I_\nu\left(ae^{-t}\right)\right]$

$$\left[|\operatorname{Re}\nu| < \tfrac{1}{2}\right] \qquad \text{ET II 367(25)}$$

Exponential, hyperbolic, and Bessel functions

6.668 **Notation**:

$$\ell_1 = \frac{1}{2}\left[\sqrt{(b+c)^2+a^2} - \sqrt{(b-c)^2+a^2}\right], \qquad \ell_2 = \frac{1}{2}\left[\sqrt{(b+c)^2+a^2} + \sqrt{(b-c)^2+a^2}\right]$$

1.^{10} $\quad \displaystyle\int_0^\infty e^{-\alpha x}\sinh(\beta x)\, J_0(\gamma x)\, dx = (\alpha\beta)^{\frac{1}{2}} r_1^{-1} r_2^{-1} (r_2 - r_1)^{\frac{1}{2}} (r_2 + r_1)^{-\frac{1}{2}}$

$\qquad r_1 = \sqrt{\gamma^2 + (\beta-\alpha)^2}, \qquad r_2 = \sqrt{\gamma^2 + (\beta+\alpha)^2}, \qquad [\operatorname{Re}\alpha > |\operatorname{Re}\beta|, \quad \gamma > 0] \quad \text{ET II 12(52)}$

alternatively, with $a = \gamma$, $b = \beta$, $c = \alpha$,

$$\int_0^\infty e^{-cx}\sinh(bx)\, J_0(ax)\, dx = \frac{\ell_1}{\ell_2^2 - \ell_1^2}$$

$$[\operatorname{Re}c > |\operatorname{Re}b|, \quad a > 0]$$

2.^{10} $\quad \displaystyle\int_0^\infty e^{-\alpha x}\cosh(\beta x)\, J_0(\gamma x)\, dx = (\alpha\beta)^{\frac{1}{2}} r_1^{-1} r_2^{-1} (r_2 - r_1)^{\frac{1}{2}} (r_2 + r_1)^{-\frac{1}{2}}$

$\qquad r_1 = \sqrt{\gamma^2 + (\beta-\alpha)^2}, \qquad r_2 = \sqrt{\gamma^2 + (\beta+\alpha)^2}, \qquad [\operatorname{Re}\alpha > |\operatorname{Re}\beta|, \quad \gamma > 0] \quad \text{ET II 12(54)}$

alternatively, with $a = \gamma$, $b = \beta$, $c = \alpha$,

$$\int_0^\infty e^{-cx}\cosh(bx)\, J_0(ax)\, dx = \frac{\ell_2}{\ell_2^2 - \ell_1^2}$$

$$[\operatorname{Re}c > |\operatorname{Re}b|, \quad a > 0]$$

6.669

1. $\quad \displaystyle\int_0^\infty \left[\coth\left(\frac{1}{2}x\right)\right]^{2\lambda} e^{-\beta\cosh x}\, J_{2\mu}(\alpha\sinh x)\, dx = \frac{\Gamma\left(\frac{1}{2}-\lambda+\mu\right)}{\alpha\,\Gamma(2\mu+1)}\, M_{-\lambda,\mu}\left[(\alpha^2+\beta^2)^{\frac{1}{2}} - \beta\right]$

$$\times\, W_{\lambda,\mu}\left[(\alpha^2+\beta^2)^{\frac{1}{2}} + \beta\right]$$

$$\left[\operatorname{Re}\beta > |\operatorname{Re}\alpha|, \quad \operatorname{Re}(\mu-\lambda) > -\tfrac{1}{2}\right] \quad \text{BU 86(5b)a, ET II 363(34)}$$

2. $\quad \displaystyle\int_0^\infty \left[\coth\left(\frac{1}{2}x\right)\right]^{2\lambda} e^{-\beta\cosh x}\, Y_{2\mu}(\alpha\sinh x)\, dx$

$$= -\frac{\sec[(\mu+\lambda)\pi]}{\alpha}\, W_{\lambda,\mu}\left(\sqrt{\alpha^2+\beta^2} + \beta\right) W_{-\lambda,\mu}\left(\sqrt{\alpha^2+\beta^2} - \beta\right)$$

$$-\frac{\tan[(\mu+\lambda)\pi]\,\Gamma\left(\frac{1}{2}-\lambda+\mu\right)}{\alpha\,\Gamma(2\mu+1)}\, W_{\lambda,\mu}\left(\sqrt{\alpha^2+\beta^2} + \beta\right) M_{-\lambda,\mu}\left(\sqrt{\alpha^2+\beta^2} - \beta\right)$$

$$\left[\operatorname{Re}\beta > |\operatorname{Re}\alpha|, \quad \operatorname{Re}\lambda < \tfrac{1}{2} - |\operatorname{Re}\mu|\right] \quad \text{ET II 363(35)}$$

3. $\int_0^\infty e^{-\frac{1}{2}(a_1a_2)t\cosh x} \left[\coth\left(\frac{1}{2}x\right)\right]^{2\nu} K_{2\mu}\left(t\sqrt{a_1a_2}\sinh x\right) dx$

$$= \frac{\Gamma\left(\frac{1}{2}+\mu-\nu\right)\Gamma\left(\frac{1}{2}-\mu-\nu\right)}{2t\sqrt{a_1a_2}} W_{\nu,\mu}(a_1t)\, W_{\nu,\mu}(a_2t)$$

$$\left[\operatorname{Re}\nu < \operatorname{Re}\frac{1\pm 2\mu}{2}, \quad \operatorname{Re}\left[t(\sqrt{a_1}+\sqrt{a_2})^2\right] > 0\right] \quad \text{BU 85(4a)}$$

4. $\int_0^\infty e^{-\frac{1}{2}(a_1a_2)t\cosh x} \left[\coth\left(\frac{x}{2}\right)\right]^{2\nu} I_{2\mu}\left(t\sqrt{a_1a_2}\sinh x\right) dx = \frac{\Gamma\left(\frac{1}{2}+\mu-\nu\right)}{t\sqrt{a_1a_2}\,\Gamma(1+2\mu)} W_{\nu,\mu}(a_1t)\, M_{\nu,\mu}(a_2t)$

$$\left[\operatorname{Re}\left(\tfrac{1}{2}+\mu-\nu\right) > 0, \quad \operatorname{Re}\mu > 0, \quad a_1 > a_2\right] \quad \text{BU 86(5c)}$$

5. $\int_{-\infty}^\infty e^{2\nu s - \frac{x-y}{2}\tanh s} I_{2\mu}\left(\frac{\sqrt{xy}}{\cosh s}\right) \frac{ds}{\cosh s} = \frac{\Gamma\left(\frac{1}{2}+\mu+\nu\right)\Gamma\left(\frac{1}{2}+\mu-\nu\right)}{\sqrt{xy}[\Gamma(1+2\mu)]^2} M_{\nu,\mu}(x)\, M_{-\nu,\mu}(y)$

$$\left[\operatorname{Re}\left(\pm\nu+\tfrac{1}{2}+\mu\right) > 0\right] \quad \text{BU 83(3a)a}$$

6. $\int_{-\infty}^\infty e^{2\nu s - \frac{x+y}{2}\tanh s} J_{2\mu}\left(\frac{\sqrt{xy}}{\cosh s}\right) \frac{ds}{\cosh s} = \frac{\Gamma\left(\frac{1}{2}+\mu+\nu\right)\Gamma\left(\frac{1}{2}+\mu-\nu\right)}{\sqrt{xy}[\Gamma(1+2\mu)]^2} M_{\nu,\mu}(x)\, M_{\nu,\mu}(y)$

$$\left[\operatorname{Re}\left(\mp\nu+\tfrac{1}{2}+\mu\right) > 0\right] \quad \text{BU 84(3b)a}$$

6.67–6.68 Combinations of Bessel and trigonometric functions

6.671

1. $\int_0^\infty J_\nu(\alpha x)\sin\beta x\, dx = \dfrac{\sin\left(\nu\arcsin\frac{\beta}{\alpha}\right)}{\sqrt{\alpha^2-\beta^2}}$ $[\beta < \alpha]$

$$= \infty \text{ or } 0 \quad\quad [\beta = \alpha]$$

$$= \frac{\alpha^\nu\cos\frac{\nu\pi}{2}}{\sqrt{\beta^2-\alpha^2}\left(\beta+\sqrt{\beta^2-\alpha^2}\right)^\nu} \quad\quad [\beta > \alpha]$$

$$[\operatorname{Re}\nu > -2] \quad\quad\quad \text{WA 444(4)}$$

2. $\int_0^\infty J_\nu(\alpha x)\cos\beta x\, dx = \dfrac{\cos\left(\nu\arcsin\frac{\beta}{\alpha}\right)}{\sqrt{\alpha^2-\beta^2}}$ $[\beta < \alpha]$

$$= \infty \text{ or } 0 \quad\quad [\beta = \alpha]$$

$$= \frac{-\alpha^\nu\sin\frac{\nu\pi}{2}}{\sqrt{\beta^2-\alpha^2}\left(\beta+\sqrt{\beta^2-\alpha^2}\right)^\nu} \quad\quad [\beta > \alpha]$$

$$[\operatorname{Re}\nu > -1] \quad\quad\quad \text{WA 444(5)}$$

3. $\displaystyle\int_0^\infty Y_\nu(ax)\sin(bx)\,dx$

$$= \cot\left(\frac{\nu\pi}{2}\right)\left(a^2 - b^2\right)^{-\frac{1}{2}}\sin\left[\nu\arcsin\left(\frac{b}{a}\right)\right] \qquad [0 < b < a, |\mathrm{Re}\,\nu| < 2]$$

$$= \frac{1}{2}\operatorname{cosec}\left(\frac{\nu\pi}{2}\right)\left(b^2 - a^2\right)^{-\frac{1}{2}}$$

$$\times\left\{a^{-\nu}\cos(\nu\pi)\left[b - \left(b^2 - a^2\right)^{\frac{1}{2}}\right]^\nu - a^\nu\left[b - \left(b^2 - a^2\right)^{\frac{1}{2}}\right]^{-\nu}\right\} \qquad [0 < a < b, \quad |\mathrm{Re}\,\nu| < 2]$$

ET I 103(33)

4. $\displaystyle\int_0^\infty Y_\nu(ax)\cos(bx)\,dx$

$$= \frac{\tan\left(\frac{\nu\pi}{2}\right)}{\left(a^2 - b^2\right)^{\frac{1}{2}}}\cos\left[\nu\arcsin\left(\frac{b}{a}\right)\right] \qquad [0 < b < a, \quad |\mathrm{Re}\,\nu| < 1]$$

$$= -\sin\left(\frac{\nu\pi}{2}\right)\left(b^2 - a^2\right)^{-\frac{1}{2}}\left\{a^{-\nu}\left[b - \left(b^2 - a^2\right)^{\frac{1}{2}}\right]^\nu + \cot(\nu\pi)\right.$$

$$\left. + a^\nu\left[b - \left(b^2 - a^2\right)^{\frac{1}{2}}\right]^{-\nu}\operatorname{cosec}(\nu\pi)\right\} \qquad [0 < a < b, \quad |\mathrm{Re}\,\nu| < 1]$$

ET I 47(29)

5. $\displaystyle\int_0^\infty K_\nu(ax)\sin(bx)\,dx$

$$= \frac{1}{4}\pi a^{-\nu}\operatorname{cosec}\left(\frac{\nu\pi}{2}\right)\left(a^2 + b^2\right)^{-\frac{1}{2}}\left\{\left[\left(b^2 + a^2\right)^{\frac{1}{2}} + b\right]^\nu - \left[\left(b^2 + a^2\right)^{\frac{1}{2}} - b\right]^\nu\right\}$$

$$[\mathrm{Re}\,a > 0, \quad b > 0, \quad |\mathrm{Re}\,\nu| < 2, \quad \nu \neq 0] \quad \text{ET I 105(48)}$$

6. $\displaystyle\int_0^\infty K_\nu(ax)\cos(bx)\,dx$

$$= \frac{\pi}{4}\left(b^2 + a^2\right)^{-\frac{1}{2}}\sec\left(\frac{\nu\pi}{2}\right)\left\{a^{-\nu}\left[b + \left(b^2 + a^2\right)^{\frac{1}{2}}\right]^\nu + a^\nu\left[b + \left(b^2 + a^2\right)^{\frac{1}{2}}\right]^{-\nu}\right\}$$

$$[\mathrm{Re}\,a > 0, b > 0, |\mathrm{Re}\,\nu| < 1] \quad \text{ET I 49(40)}$$

7. $\displaystyle\int_0^\infty J_0(ax)\sin(bx)\,dx = 0 \qquad\qquad\qquad [0 < b < a]$

$$= \frac{1}{\sqrt{b^2 - a^2}} \qquad\qquad\qquad [0 < a < b]$$

ET I 99(1)

8. $\displaystyle\int_0^\infty J_0(ax)\cos(bx)\,dx = \frac{1}{\sqrt{a^2 - b^2}} \qquad\qquad [0 < b < a]$

$$= \infty \qquad\qquad\qquad [a = b]$$

$$= 0 \qquad\qquad\qquad [0 < a < b]$$

ET I 43(1)

9. $\displaystyle\int_0^\infty J_{2n+1}(ax)\sin(bx)\,dx = (-1)^n \frac{1}{\sqrt{a^2-b^2}}\,T_{2n+1}\left(\frac{b}{a}\right)$ $[0<b<a]$

$$= 0 \qquad\qquad [0<a<b]$$

ET I 99(2)

10. $\displaystyle\int_0^\infty J_{2n}(ax)\cos(bx)\,dx = (-1)^n \frac{1}{\sqrt{a^2-b^2}}\,T_{2n}\left(\frac{b}{a}\right)$ $[0<b<a]$

$$= 0 \qquad\qquad [0<a<b]$$

ET I 43(2)

11. $\displaystyle\int_0^\infty Y_0(ax)\sin(bx)\,dx = \frac{2\arcsin\left(\frac{b}{a}\right)}{\pi\sqrt{a^2-b^2}}$ $[0<b<a]$

$$= \frac{2}{\pi}\frac{1}{\sqrt{b^2-a^2}}\ln\left[\frac{b}{a}-\sqrt{\frac{b^2}{a^2}-1}\right] \qquad [0<a<b]$$

ET I 103(31)

12. $\displaystyle\int_0^\infty Y_0(ax)\cos(bx)\,dx = 0$ $[0<b<a]$

$$= -\frac{1}{\sqrt{b^2-a^2}} \qquad [0<a<b]$$

ET I 47(28)

13. $\displaystyle\int_0^\infty K_0(\beta x)\sin\alpha x\,dx = \frac{1}{\sqrt{\alpha^2+\beta^2}}\ln\left(\frac{\alpha}{\beta}+\sqrt{\frac{\alpha^2}{\beta^2}+1}\right)$

$[\alpha>0,\quad\beta>0]$ WA 425(11)a, MO 48

14.[8] $\displaystyle\int_0^\infty K_0(\beta x)\cos\alpha x\,dx = \frac{\pi}{2\sqrt{\alpha^2+\beta^2}}$ $[\alpha>0]$ WA 425(10)a, MO 48

6.672

1. $\displaystyle\int_0^\infty J_\nu(ax)\,J_\nu(bx)\sin(cx)\,dx$

$$= 0 \qquad\qquad [\operatorname{Re}\nu>-1,\quad 0<c<b-a,\quad 0<a<b]$$

$$= \frac{1}{2\sqrt{ab}}\,P_{\nu-\frac{1}{2}}\left(\frac{b^2+a^2-c^2}{2ab}\right) \qquad [\operatorname{Re}\nu>-1,\quad b-a<c<b+a,\quad 0<a<b]$$

$$= -\frac{\cos(\nu\pi)}{\pi\sqrt{ab}}\,Q_{\nu-\frac{1}{2}}\left(-\frac{b^2+a^2-c^2}{2ab}\right) \qquad [\operatorname{Re}\nu>-1,\quad b+a<c,\quad 0<a<b]$$

ET I 102(27)

2. $\displaystyle\int_0^\infty J_\nu(x)\,J_{-\nu}(x)\cos(bx)\,dx = \frac{1}{2}\,P_{\nu-\frac{1}{2}}\left(\frac{1}{2}b^2-1\right)$ $[0<b<2]$

$$= 0 \qquad\qquad [2<b]$$

ET I 46(21)

3. $\int_0^\infty K_\nu(ax)\, K_\nu(bx)\cos(cx)\, dx = \dfrac{\pi^2}{4\sqrt{ab}}\sec(\nu\pi)\, P_{\nu-\frac{1}{2}}\left[\left(a^2+b^2+c^2\right)(2ab)^{-1}\right]$

$\left[\operatorname{Re}(a+b)>0,\quad c>0,\quad |\operatorname{Re}\nu|<\tfrac{1}{2}\right]$

ET I 50(51)

4. $\int_0^\infty K_\nu(ax)\, I_\nu(bx)\cos(cx)\, dx = \dfrac{1}{2\sqrt{ab}}\, Q_{\nu-\frac{1}{2}}\left(\dfrac{a^2+b^2+c^2}{2ab}\right)$

$\left[\operatorname{Re}a>|\operatorname{Re}b|,\quad c>0,\quad \operatorname{Re}\nu>-\tfrac{1}{2}\right]$

ET I 49(47)

5. $\int_0^\infty \sin(2ax)[J_\nu(x)]^2\, dx = \dfrac{1}{2}\, P_{\nu-\frac{1}{2}}\left(1-2a^2\right)$ $\qquad[0<a<1,\quad \operatorname{Re}\nu>-1]$

$\qquad\qquad\qquad\qquad\quad = \dfrac{1}{\pi}\cos(\nu\pi)\, Q_{\nu-\frac{1}{2}}\left(2a^2-1\right)\quad[a>1,\quad \operatorname{Re}\nu>-1]$

ET II 343(30)

6. $\int_0^\infty \cos(2ax)[J_\nu(x)]^2\, dx = \dfrac{1}{\pi}\, Q_{\nu-\frac{1}{2}}\left(1-2a^2\right)$ $\qquad\left[0<a<1,\quad \operatorname{Re}\nu>-\tfrac{1}{2}\right]$

$\qquad\qquad\qquad\qquad\quad = -\dfrac{1}{\pi}\sin(\nu\pi)\, Q_{\nu-\frac{1}{2}}\left(2a^2-1\right)\quad\left[a>1,\quad \operatorname{Re}\nu>-\tfrac{1}{2}\right]$

ET II 344(32)

7. $\int_0^\infty \sin(2ax)\, J_0(x)\, Y_0(x)\, dx = 0$ $\qquad\qquad\qquad[0<a<1]$

$\qquad\qquad\qquad\qquad\quad = -\dfrac{K\left[\left(1-a^{-2}\right)^{\frac{1}{2}}\right]}{\pi a}\qquad[a>1]$

ET II 348(60)

8. $\int_0^\infty K_0(ax)\, I_0(bx)\cos(cx)\, dx = \dfrac{1}{\sqrt{c^2+(a+b)^2}}\, K\left\{\dfrac{2\sqrt{ab}}{\sqrt{c^2+(a+b)^2}}\right\}$

$\left[\operatorname{Re}a>|\operatorname{Re}b|,\quad c>0\right]\qquad$ ET I 49(46)

9. $\int_0^\infty \cos(2ax)\, J_0(x)\, Y_0(x)\, dx = -\dfrac{1}{\pi}\, K(a)$ $\qquad[0<a<1]$

$\qquad\qquad\qquad\qquad\quad = -\dfrac{1}{\pi a}\, K\left(\dfrac{1}{a}\right)\qquad[a>1]$

ET II 348(61)

10. $\int_0^\infty \cos(2ax)[Y_0(x)]^2\, dx = \dfrac{1}{\pi}\, K\left(\sqrt{1-a^2}\right)$ $\qquad[0<a<1]$

$\qquad\qquad\qquad\qquad\quad = \dfrac{2}{\pi a}\, K\left(\sqrt{1-\dfrac{1}{a^2}}\right)\qquad[a>1]$

ET II 348(62)

6.673

1. $\displaystyle\int_0^\infty \left[J_\nu(ax) \cos\left(\frac{\nu\pi}{2}\right) - Y_\nu(ax) \sin\left(\frac{\nu\pi}{2}\right) \right] \sin(bx)\, dx$

$$= 0 \qquad\qquad\qquad\qquad\qquad\qquad [0 < b < a, \quad |\operatorname{Re}\nu| < 2]$$

$$= \frac{1}{2a^\nu \sqrt{b^2 - a^2}} \left\{ \left[b + (b^2 - a^2)^{\frac{1}{2}} \right]^\nu + \left[b - (b^2 - a^2)^{\frac{1}{2}} \right]^\nu \right\} \quad [0 < a < b, \quad |\operatorname{Re}\nu| < 2]$$

$$\text{ET I } 104(39)$$

2. $\displaystyle\int_0^\infty \left[Y_\nu(ax) \cos\left(\frac{\nu\pi}{2}\right) + J_\nu(ax) \sin\left(\frac{\nu\pi}{2}\right) \right] \cos(bx)\, dx$

$$= 0 \qquad\qquad\qquad\qquad\qquad\qquad [0 < b < a, \quad |\operatorname{Re}\nu| < 1]$$

$$= -\frac{1}{2a^\nu \sqrt{b^2 - a^2}} \left\{ \left[b + (b^2 - a^2)^{\frac{1}{2}} \right]^\nu + \left[b - (b^2 - a^2)^{\frac{1}{2}} \right]^\nu \right\} \quad [0 < a < b, \quad |\operatorname{Re}\nu| < 1]$$

$$\text{ET I } 48(32)$$

6.674

1. $\displaystyle\int_0^a \sin(a - x)\, J_\nu(x)\, dx = a\, J_{\nu+1}(a) - 2\nu \sum_{n=0}^\infty (-1)^n\, J_{\nu+2n+2}(a)$

$$[\operatorname{Re}\nu > -1] \qquad\qquad \text{ET II } 334(12)$$

2. $\displaystyle\int_0^a \cos(a - x)\, J_\nu(x)\, dx = a\, J_\nu(a) - 2\nu \sum_{n=0}^\infty (-1)^n\, J_{\nu+2n+1}(a)$

$$[\operatorname{Re}\nu > -1] \qquad\qquad \text{ET II } 336(23)$$

3. $\displaystyle\int_0^a \sin(a - x)\, J_{2n}(x)\, dx = a\, J_{2n+1}(a) + (-1)^n 2n \left[\cos a - J_0(a) - 2 \sum_{m=1}^n (-1)^m\, J_{2m}(a) \right]$

$$[n = 0, 1, 2, \ldots] \qquad\qquad \text{ET II } 334(10)$$

4. $\displaystyle\int_0^a \cos(a - x)\, J_{2n}(x)\, dx = a\, J_{2n}(a) - (-1)^n 2n \left[\sin a - 2 \sum_{m=0}^{n-1} (-1)^m\, J_{2m+1}(a) \right]$

$$[n = 0, 1, 2, \ldots] \qquad\qquad \text{ET II } 335(21)$$

5. $\displaystyle\int_0^a \sin(a - x)\, J_{2n+1}(x)\, dx = a\, J_{2n+2}(a) + (-1)^n (2n+1) \left[\sin a - 2 \sum_{m=0}^n (-1)^m\, J_{2m+1}(a) \right]$

$$[n = 0, 1, 2, \ldots] \qquad\qquad \text{ET II } 334(11)$$

6. $\displaystyle\int_0^a \cos(a - x)\, J_{2n+1}(x)\, dx = a\, J_{2n+1}(a) + (-1)^n (2n+1) \left[\cos a - J_0(a) - 2 \sum_{m=1}^n (-1)^m\, J_{2m}(a) \right]$

$$[n = 0, 1, 2, \ldots] \qquad\qquad \text{ET II } 336(22)$$

7. $\displaystyle\int_0^z \sin(z - x)\, J_0(x)\, dx = z\, J_1(z)$

$$\text{WA } 415(2)$$

8. $\displaystyle\int_0^z \cos(z - x)\, J_0(x)\, dx = z\, J_0(z)$

$$\text{WA } 415(1)$$

6.675

1. $$\int_0^\infty J_\nu\left(a\sqrt{x}\right)\sin(bx)\,dx = \frac{a\sqrt{\pi}}{4b^{\frac{3}{2}}}\left[\cos\left(\frac{a^2}{8b}-\frac{\nu\pi}{4}\right)J_{\frac{1}{2}\nu-\frac{1}{2}}\left(\frac{a^2}{8b}\right)-\sin\left(\frac{a^2}{8b}-\frac{\nu\pi}{4}\right)J_{\frac{1}{2}\nu+\frac{1}{2}}\left(\frac{a^2}{8b}\right)\right]$$

$$[a>0, \quad b>0, \quad \operatorname{Re}\nu>-4]$$

ET I 110(23)

2. $$\int_0^\infty J_\nu\left(a\sqrt{x}\right)\cos(bx)\,dx$$

$$= -\frac{a\sqrt{\pi}}{4b^{\frac{3}{2}}}\left[\sin\left(\frac{a^2}{8b}-\frac{\nu\pi}{4}\right)J_{\frac{1}{2}\nu-\frac{1}{2}}\left(\frac{a^2}{8b}\right)+\cos\left(\frac{a^2}{8b}-\frac{\nu\pi}{4}\right)J_{\frac{1}{2}\nu+\frac{1}{2}}\left(\frac{a^2}{8b}\right)\right]$$

$$[a>0, \quad b>0, \quad \operatorname{Re}\nu>-2] \quad \text{ET I 53(22)a}$$

3. $$\int_0^\infty J_0\left(a\sqrt{x}\right)\sin(bx)\,dx = \frac{1}{b}\cos\left(\frac{a^2}{4b}\right) \qquad [a>0, \quad b>0] \qquad\qquad \text{ET I 110(22)}$$

4. $$\int_0^\infty J_0\left(a\sqrt{x}\right)\cos(bx)\,dx = \frac{1}{b}\sin\left(\frac{a^2}{4b}\right) \qquad [a>0, \quad b>0] \qquad\qquad \text{ET I 53(21)}$$

6.676

1. $$\int_0^\infty J_\nu\left(a\sqrt{x}\right)J_\nu\left(b\sqrt{x}\right)\sin(cx)\,dx = \frac{1}{c}J_\nu\left(\frac{ab}{2c}\right)\cos\left(\frac{a^2+b^2}{4c}-\frac{\nu\pi}{2}\right)$$

$$[a>0, \quad b>0, \quad c>0, \quad \operatorname{Re}\nu>-2]$$

ET I 111(29)a

2. $$\int_0^\infty J_\nu\left(a\sqrt{x}\right)J_\nu\left(b\sqrt{x}\right)\cos(cx)\,dx = \frac{1}{c}J_\nu\left(\frac{ab}{2c}\right)\sin\left(\frac{a^2+b^2}{4c}-\frac{\nu\pi}{2}\right)$$

$$[a>0, \quad b>0, \quad c>0, \quad \operatorname{Re}\nu>-1]$$

ET I 54(27)

3. $$\int_0^\infty J_0\left(a\sqrt{x}\right)K_0\left(a\sqrt{x}\right)\sin(bx)\,dx = \frac{1}{2b}K_0\left(\frac{a^2}{2b}\right) \qquad [\operatorname{Re}a>0, \quad b>0] \qquad \text{ET I 111(31)}$$

4. $$\int_0^\infty J_0\left(\sqrt{ax}\right)K_0\left(\sqrt{ax}\right)\cos(bx)\,dx = \frac{\pi}{4b}\left[I_0\left(\frac{a}{2b}\right)-\mathbf{L}_0\left(\frac{a}{2b}\right)\right]$$

$$[\operatorname{Re}a>0, \quad b>0] \qquad\qquad \text{ET I 54(29)}$$

5. $$\int_0^\infty K_0\left(\sqrt{ax}\right)Y_0\left(\sqrt{ax}\right)\cos(bx)\,dx = -\frac{1}{2b}K_0\left(\frac{a}{2b}\right) \qquad [\operatorname{Re}\sqrt{a}>0, \quad b>0] \qquad \text{ET I 54(30)}$$

6. $$\int_0^\infty K_0\left(\sqrt{ax}e^{\frac{1}{4}\pi i}\right)K_0\left(\sqrt{ax}e^{-\frac{1}{4}\pi i}\right)\cos(bx)\,dx = \frac{\pi^2}{8b}\left[\mathbf{H}_0\left(\frac{a}{2b}\right)-Y_0\left(\frac{a}{2b}\right)\right]$$

$$[\operatorname{Re}a>0, b>0] \qquad\qquad \text{ET I 54(31)}$$

6.677

1. $$\int_a^\infty J_0\left(b\sqrt{x^2-a^2}\right)\sin(cx)\,dx = 0 \qquad [0<c<b]$$

$$= \frac{\cos\left(a\sqrt{c^2-b^2}\right)}{\sqrt{c^2-b^2}} \qquad [0<b<c]$$

ET I 113(47)

2. $\displaystyle\int_a^\infty J_0\left(b\sqrt{x^2-a^2}\right)\cos(cx)\,dx = \frac{\exp\left(-a\sqrt{b^2-c^2}\right)}{\sqrt{b^2-c^2}}$ $[0<c<b]$

$$= \frac{-\sin\left(a\sqrt{c^2-b^2}\right)}{\sqrt{c^2-b^2}} \quad [0<b<c]$$

ET I 57(48)a

3.[6] $\displaystyle\int_0^\infty J_0\left(\alpha\sqrt{x^2+z^2}\right)\cos\beta x\,dx = \frac{\cos z\sqrt{\alpha^2-\beta^2}}{\sqrt{\alpha^2-\beta^2}}$ $[0<\beta<\alpha,\quad z>0]$

$$= 0 \quad [0<\alpha<\beta,\quad z>0]$$

MO 47a

4. $\displaystyle\int_0^\infty Y_0\left(\alpha\sqrt{x^2+z^2}\right)\cos\beta x\,dx = \frac{1}{\sqrt{\alpha^2-\beta^2}}\sin\left(z\sqrt{\alpha^2-\beta^2}\right)$ $[0<\beta<\alpha,\quad z>0]$

$$= -\frac{1}{\sqrt{\beta^2-\alpha^2}}\exp\left(-z\sqrt{\beta^2-\alpha^2}\right) \quad [0<\alpha<\beta,\quad z>0]$$

MO 47a

5. $\displaystyle\int_0^\infty K_0\left[\alpha\sqrt{x^2+\beta^2}\right]\cos(\gamma x)\,dx = \frac{\pi}{2\sqrt{\alpha^2+\gamma^2}}\exp\left(-\beta\sqrt{\alpha^2+\gamma^2}\right)$

$$[\operatorname{Re}\alpha>0,\quad \operatorname{Re}\beta>0,\quad \gamma>0]$$

ET I 56(43)

6. $\displaystyle\int_0^a J_0\left(b\sqrt{a^2-x^2}\right)\cos(cx)\,dx = \frac{\sin\left(a\sqrt{b^2+c^2}\right)}{\sqrt{b^2+c^2}}$ $[b>0]$ MO 48a, ET I 57(47)

7. $\displaystyle\int_0^\infty J_0\left(b\sqrt{x^2-a^2}\right)\cos(cx)\,dx = \frac{\cosh\left(a\sqrt{b^2-c^2}\right)}{\sqrt{b^2-c^2}}$ $[0<c<b,\quad a>0]$

$$= 0 \quad [0<b<c,\quad a>0]$$

ET I 57(49)

8. $\displaystyle\int_0^\infty H_0^{(1)}\left(\alpha\sqrt{\beta^2-x^2}\right)\cos(\gamma x)\,dx = -i\frac{\exp\left(i\beta\sqrt{\alpha^2+\gamma^2}\right)}{\sqrt{\alpha^2+\gamma^2}}$

$$\left[\pi>\arg\sqrt{\beta^2-x^2}\geq 0,\quad \alpha>0,\quad \gamma>0\right] \quad \text{ET I 59(59)}$$

9. $\displaystyle\int_0^\infty H_0^{(2)}\left(\alpha\sqrt{\beta^2-x^2}\right)\cos(\gamma x)\,dx = \frac{i\exp\left(-i\beta\sqrt{\alpha^2+\gamma^2}\right)}{\sqrt{\alpha^2+\gamma^2}}$

$$\left[-\pi<\arg\sqrt{\beta^2-x^2}\leq 0,\quad \alpha>0,\quad \gamma>0\right] \quad \text{ET I 58(58)}$$

6.678 $\displaystyle\int_0^\infty \left[K_0\left(2\sqrt{x}\right)+\frac{\pi}{2}Y_0\left(2\sqrt{x}\right)\right]\sin(bx)\,dx = \frac{\pi}{2b}\sin\left(\frac{1}{b}\right)$ $[b>0]$ ET I 111(34)

6.679

1. $\displaystyle\int_0^\infty J_{2\nu}\left[2b\sinh\left(\frac{x}{2}\right)\right]\sin(bx)\,dx = -i\left[I_{\nu-ib}(a)\,K_{\nu+ib}(a) - I_{\nu+ib}(a)\,K_{\nu-ib}(a)\right]$

$$[a>0,\quad b>0,\quad \operatorname{Re}\nu>-1]$$

ET I 115(59)

2. $\int_0^\infty J_{2\nu}\left[2a\sinh\left(\dfrac{x}{2}\right)\right]\cos(bx)\,dx = I_{\nu-ib}(a)\,K_{\nu+ib}(a) + I_{\nu+ib}(a)\,K_{\nu-ib}(a)$

$$\left[a>0,\quad b>0,\quad \operatorname{Re}\nu>-\tfrac{1}{2}\right]$$

ET I 59(64)

3. $\int_0^\infty J_{2\nu}\left[2a\cosh\left(\dfrac{x}{2}\right)\right]\cos(bx)\,dx = -\dfrac{\pi}{2}\left[J_{\nu+ib}(a)\,Y_{\nu-ib}(a) + J_{\nu-ib}(a)\,Y_{\nu+ib}(a)\right]$ ET I 59(63)

4. $\int_0^\infty J_0\left[2a\sinh\left(\dfrac{x}{2}\right)\right]\sin(bx)\,dx = \dfrac{2}{\pi}\sinh(\pi b)[K_{ib}(a)]^2$

$$\left[a>0,\quad b>0\right]$$ ET I 115(58)

5. $\int_0^\infty J_0\left[2a\sinh\left(\dfrac{x}{2}\right)\right]\cos(bx)\,dx = \left[I_{ib}(a) + I_{-ib}(a)\right]K_{ib}(a)$

$$\left[a>0,\quad b>0\right]$$ ET I 59(62)

6. $\int_0^\infty Y_0\left[2a\sinh\left(\dfrac{x}{2}\right)\right]\cos(bx)\,dx = -\dfrac{2}{\pi}\cosh(\pi b)[K_{ib}(a)]^2$

$$\left[a>0,\quad b>0\right]$$ ET I 59(65)

7. $\int_0^\infty K_0\left[2a\sinh\left(\dfrac{x}{2}\right)\right]\cos(bx)\,dx = \dfrac{\pi^2}{4}\left\{[J_{ib}(a)]^2 + [Y_{ib}(a)]^2\right\}$

$$\left[\operatorname{Re}a>0,\quad b>0\right]$$ ET I 59(66)

6.681

1. $\int_0^{\frac{\pi}{2}}\cos(2\mu x)\,J_{2\nu}\,(2a\cos x)\,dx = \dfrac{\pi}{2}\,J_{\nu+\mu}(a)\,J_{\nu-\mu}(a)\qquad\left[\operatorname{Re}\nu>-\tfrac{1}{2}\right]$ ET II 361(23)

2. $\int_0^{\frac{\pi}{2}}\cos(2\mu x)\,Y_{2\nu}\,(2a\cos x)\,dx = \dfrac{\pi}{2}\left[\cot(2\nu\pi)\,J_{\nu+\mu}(a)\,J_{\nu-\mu}(a) - \operatorname{cosec}(2\nu\pi)\,J_{\mu-\nu}(a)\,J_{-\mu-\nu}(a)\right]$

$$\left[|\operatorname{Re}\nu|<\tfrac{1}{2}\right]$$ ET II 361(24)

3. $\int_0^{\frac{\pi}{2}}\cos(2\mu x)\,I_{2\nu}\,(2a\cos x)\,dx = \dfrac{\pi}{2}\,I_{\nu-\mu}(a)\,I_{\nu+\mu}(a)\qquad\left[\operatorname{Re}\nu>-\tfrac{1}{2}\right]$ ET I 59(61)

4. $\int_0^{\frac{\pi}{2}}\cos(\nu x)\,K_\nu\,(2a\cos x)\,dx = \dfrac{\pi}{2}\,I_0(a)\,K_\nu(a)\qquad\left[\operatorname{Re}\nu<1\right]$ WA 484(3)

5. $\int_0^\pi J_0\,(2z\cos x)\cos 2nx\,dx = (-1)^n\pi J_n^2(z).$ MO 45

6. $\int_0^\pi J_0\,(2z\sin x)\cos 2nx\,dx = \pi J_n^2(z).$ WA 43(3), MO 45

7. $\int_0^{\frac{\pi}{2}}\cos(2n\pi)\,Y_0\,(2a\sin x)\,dx = \dfrac{\pi}{2}\,J_n(a)\,Y_n(a)\qquad[n=0,1,2,\ldots]$ ET II 360(16)

8. $\int_0^\pi \sin(2\mu x)\,J_{2\nu}\,(2a\sin x)\,dx = \pi\sin(\mu\pi)\,J_{\nu-\mu}(a)\,J_{\nu+\mu}(a)$

$$\left[\operatorname{Re}\nu>-1\right]$$ ET II 360(13)

9. $\int_0^\pi \cos(2\mu x) \, J_{2\nu} \, (2a \sin x) \, dx = \pi \cos(\mu\pi) \, J_{\nu-\mu}(a) \, J_{\nu+\mu}(a)$

$$\left[\operatorname{Re}\nu > -\tfrac{1}{2}\right] \qquad\qquad \text{ET II 360(14)}$$

10. $\int_0^{\frac{\pi}{2}} J_{\nu+\mu} \, (2z \cos x) \cos[(\nu - \mu)x] \, dx = \dfrac{\pi}{2} \, J_\nu(z) \, J_\mu(z) \qquad [\operatorname{Re}(\nu + \mu) > -1]$ MO 42

11. $\int_0^{\frac{\pi}{2}} \cos[(\mu - \nu)x] \, I_{\mu+\nu} \, (2a \cos x) \, dx = \dfrac{\pi}{2} \, I_\mu(a) \, I_\nu(a) \qquad [\operatorname{Re}(\mu + \nu) > -1]$

$$\text{WA 484(2), ET II 378(39)}$$

12. $\int_0^{\frac{\pi}{2}} \cos[(\mu - \nu)x] \, K_{\mu+\nu} \, (2a \cos x) \, dx = \dfrac{\pi^2}{4} \operatorname{cosec}[(\mu + \nu)\pi] \, [I_{-\mu}(a) \, I_{-\nu}(a) - I_\mu(a) \, I_\nu(a)]$

$$[|\operatorname{Re}(\mu + \nu)| < 1] \qquad\qquad \text{ET II 378(40)}$$

13.[8] $\int_0^{\frac{\pi}{2}} K_{\nu-m} \, (2a \cos x) \cos[(m + \nu)x] \, dx = (-1)^m \dfrac{\pi}{2} \, I_m(a) \, K_\nu(a)$

$$[|\operatorname{Re}(\nu - m)| < 1] \qquad\qquad \text{WA 485(4)}$$

6.682

1.[7] $\int_0^{\frac{\pi}{2}} J_{\nu-\frac{1}{2}} \, (x \sin t) \sin^{\nu+\frac{1}{2}} t \, dt = \sqrt{\dfrac{\pi}{2x}} \, J_\nu(x)$

[ν may be zero, a natural number, one half, or a natural number plus one half; $x > 0$]   MO 42a

2. $\int_0^{\frac{\pi}{2}} J_\nu \, (z \sin x) \sin^\nu x \cos^{2\nu} x \, dx = 2^{\nu-1} \sqrt{\pi} \, \Gamma\left(\nu + \dfrac{1}{2}\right) z^{-\nu} J_\nu^{\,2} \left(\dfrac{z}{2}\right)$

$$\left[\operatorname{Re}\nu > -\tfrac{1}{2}\right] \qquad\qquad \text{MO 42a}$$

6.683

1. $\int_0^{\frac{\pi}{2}} J_\nu \, (z \sin x) \, I_\mu \, (z \cos x) \tan^{\nu+1} x \, dx = \dfrac{\left(\dfrac{z}{2}\right)^\nu \Gamma\left(\dfrac{\mu - \nu}{2}\right)}{\Gamma\left(\dfrac{\mu + \nu}{2} + 1\right)} \, J_\mu(z)$

$$[\operatorname{Re}\nu > \operatorname{Re}\mu > -1] \qquad\qquad \text{WA 407(4)}$$

2. $\int_0^{\frac{\pi}{2}} J_\nu \, (z_1 \sin x) \, J_\mu \, (z_2 \cos x) \sin^{\nu+1} x \cos^{\mu+1} x \, dx = \dfrac{z_1^{\,\nu} z_2^{\,\mu} \, J_{\nu+\mu+1} \left(\sqrt{z_1^2 + z_2^2}\right)}{\sqrt{(z_1^2 + z_2^2)^{\nu+\mu+1}}}$

$$[\operatorname{Re}\nu > -1, \quad \operatorname{Re}\mu > -1] \qquad\qquad \text{WA 410(1)}$$

3. $\int_0^{\frac{\pi}{2}} J_\nu \, (z \cos^2 x) \, J_\mu \, (z \sin^2 x) \sin x \cos x \, dx = \dfrac{1}{z} \sum_{k=0}^{\infty} (-1)^k \, J_{\nu+\mu+2k+1}(z)$

$$[\operatorname{Re}\nu > -1, \quad \operatorname{Re}\mu > -1] \qquad \text{(see also \textbf{6.513} 6)} \qquad \text{WA 414(1)}$$

4. $\int_0^{\frac{\pi}{2}} J_\mu \, (z \sin \theta) \, (\sin \theta)^{1-\mu} (\cos \theta)^{2\nu+1} \, d\theta = \dfrac{s_{\mu+\nu,\nu-\mu+1}(z)}{2^{\mu-1} z^{\nu+1} \Gamma(\mu)}$

$$[\operatorname{Re}\nu > -1] \qquad\qquad \text{WA 407(2)}$$

5. $\int_0^{\frac{\pi}{2}} J_\mu \left(z \sin \theta\right) (\sin \theta)^{1-\mu} \, d\theta = \dfrac{\mathbf{H}_{\mu - \frac{1}{2}}(z)}{\sqrt{\dfrac{2z}{\pi}}}$ WA 407(3)

6. $\int_0^{\frac{\pi}{2}} J_\mu \left(a \sin \theta\right) (\sin \theta)^{\mu+1} (\cos \theta)^{2\varrho+1} \, d\theta = 2^\varrho \, \Gamma(\varrho+1) a^{-\varrho-1} \, J_{\varrho+\mu+1}(a)$

$$[\operatorname{Re} \varrho > -1, \quad \operatorname{Re} \mu > -1]$$
$$\text{WA 406(1)}, \quad \text{EH II 46(5)}$$

7. $\int_0^{\frac{\pi}{2}} J_\nu \left(2z \sin \theta\right) (\sin \theta)^\nu (\cos \theta)^{2\nu} \, d\theta$

$$= \frac{1}{2} \sum_{m=0}^\infty \frac{(-1)^m z^{\nu+2m} \, \Gamma\left(\nu+m+\frac{1}{2}\right) \Gamma\left(\nu+\frac{1}{2}\right)}{m! \, \Gamma(\nu+m+1) \, \Gamma(2\nu+m+1)}$$
$$= \frac{1}{2} z^{-\nu} \sqrt{\pi} \, \Gamma\left(\nu+\frac{1}{2}\right) \left[J_\nu(z)\right]^2 \qquad \left[\operatorname{Re} \nu > -\tfrac{1}{2}\right]$$
$$\text{EH II 47(10)}$$

8. $\int_0^{\frac{\pi}{2}} J_\nu \left(z \sin \theta\right) (\sin \theta)^{\nu+1} (\cos \theta)^{-2\nu} \, d\theta = 2^{-\nu} \dfrac{z^{\nu-1}}{\sqrt{\pi}} \, \Gamma\left(\dfrac{1}{2} - \nu\right) \sin z$

$$\left[-1 < \operatorname{Re} \nu < \tfrac{1}{2}\right] \qquad \text{EH II 68(39)}$$

9. $\int_0^{\frac{\pi}{2}} J_\nu \left(z \sin^2 \theta\right) J_\nu \left(z \cos^2 \theta\right) (\sin \theta)^{2\nu+1} (\cos \theta)^{2\nu+1} \, d\theta = \dfrac{\Gamma\left(\frac{1}{2}+\nu\right) J_{2\nu+\frac{1}{2}}(z)}{2^{2\nu+\frac{3}{2}} \Gamma(\nu+1) \sqrt{z}}$

$$\left[\operatorname{Re} \nu > -\tfrac{1}{2}\right] \qquad \text{WA 409(1)}$$

10. $\int_0^{\frac{\pi}{2}} J_\mu \left(z \sin^2 \theta\right) J_\nu \left(z \cos^2 \theta\right) \sin^{2\mu+1} \theta \cos^{2\nu+1} \theta \, d\theta = \dfrac{\Gamma\left(\mu+\frac{1}{2}\right) \Gamma\left(\nu+\frac{1}{2}\right) J_{\mu+\nu+\frac{1}{2}}(z)}{2\sqrt{\pi} \, \Gamma(\mu+\nu+1) \sqrt{2z}}$

$$\left[\operatorname{Re} \mu > -\tfrac{1}{2}, \quad \operatorname{Re} \nu > -\tfrac{1}{2}\right] \qquad \text{WA 417(1)}$$

6.684

1.[8] $\int_0^\pi (\sin x)^{2\nu} \dfrac{J_\nu \left(\sqrt{\alpha^2 + \beta^2 - 2\alpha\beta \cos x}\right)}{\left(\sqrt{\alpha^2 + \beta^2 - 2\alpha\beta \cos x}\right)^\nu} \, dx = 2^\nu \sqrt{\pi} \, \Gamma\left(\nu+\dfrac{1}{2}\right) \dfrac{J_\nu(\alpha)}{\alpha^\nu} \dfrac{J_\nu(\beta)}{\beta^\nu}$

$$\left[\operatorname{Re} \nu > -\tfrac{1}{2}\right] \qquad \text{ET II 362(27)}$$

2. $\int_0^\pi (\sin x)^{2\nu} \dfrac{Y_\nu \left(\sqrt{\alpha^2 + \beta^2 - 2\alpha\beta \cos x}\right)}{\left(\sqrt{\alpha^2 + \beta^2 - 2\alpha\beta \cos x}\right)^\nu} \, dx = 2^\nu \sqrt{\pi} \, \Gamma\left(\nu+\dfrac{1}{2}\right) \dfrac{J_\nu(\alpha)}{\alpha^\nu} \dfrac{Y_\nu(\beta)}{\beta^\nu}$

$$\left[|\alpha| < |\beta|, \quad \operatorname{Re} \nu > -\tfrac{1}{2}\right] \qquad \text{ET II 362(28)}$$

6.685 $\int_0^{\frac{\pi}{2}} \sec x \cos(2\lambda x) \, K_{2\mu}(a \sec x) \, dx = \dfrac{\pi}{2a} \, W_{\lambda,\mu}(a) \, W_{-\lambda,\mu}(a)$ $[\operatorname{Re} a > 0]$ ET II 378(41)

6.686

1. $\int_0^\infty \sin\left(ax^2\right) J_\nu(bx) \, dx = -\dfrac{\sqrt{\pi}}{2\sqrt{a}} \sin\left(\dfrac{b^2}{8a} - \dfrac{\nu+1}{4}\pi\right) J_{\frac{1}{2}\nu}\left(\dfrac{b^2}{8a}\right)$

$$[a > 0, b > 0, \operatorname{Re} \nu > -3] \qquad \text{ET II 34(13)}$$

2. $\displaystyle\int_0^\infty \cos\left(ax^2\right) J_\nu(bx)\, dx = \frac{\sqrt{\pi}}{2\sqrt{a}}\cos\left(\frac{b^2}{8a}-\frac{\nu+1}{4}\pi\right) J_{\frac{1}{2}\nu}\left(\frac{b^2}{8a}\right)$

$$[a>0,\quad b>0,\quad \operatorname{Re}\nu>-1]$$

ET II 38(38)

3. $\displaystyle\int_0^\infty \sin\left(ax^2\right) Y_\nu(bx)\, dx$

$$= -\frac{\sqrt{\pi}}{4\sqrt{a}}\sec\left(\frac{\nu\pi}{2}\right)$$
$$\times\left[\cos\left(\frac{b^2}{8a}-\frac{3\nu+1}{4}\pi\right) J_{\frac{1}{2}\nu}\left(\frac{b^2}{8a}\right) - \sin\left(\frac{b^2}{8a}+\frac{\nu-1}{4}\pi\right) Y_{\frac{1}{2}\nu}\left(\frac{b^2}{8a}\right)\right]$$
$$[a>0,\quad b>0,\quad -3<\operatorname{Re}\nu<3]\quad\text{ET II 107(7)}$$

4. $\displaystyle\int_0^\infty \cos\left(ax^2\right) Y_\nu(bx)\, dx$

$$= \frac{\sqrt{\pi}}{4\sqrt{a}}\sec\left(\frac{\nu\pi}{2}\right)$$
$$\times\left[\sin\left(\frac{b^2}{8a}-\frac{3\nu+1}{4}\pi\right) J_{\frac{1}{2}\nu}\left(\frac{b^2}{8a}\right) + \cos\left(\frac{b^2}{8a}+\frac{\nu-1}{4}\pi\right) Y_{\frac{1}{2}\nu}\left(\frac{b^2}{8a}\right)\right]$$
$$[a>0,\quad b>0,\quad -1<\operatorname{Re}\nu<1]\quad\text{ET II 107(8)}$$

5. $\displaystyle\int_0^\infty \sin\left(ax^2\right) J_1(bx)\, dx = \frac{1}{b}\sin\frac{b^2}{4a}$ $[a>0,\quad b>0]$ ET II 19(16)

6. $\displaystyle\int_0^\infty \cos\left(ax^2\right) J_1(bx)\, dx = \frac{2}{b}\sin^2\left(\frac{b^2}{8a}\right)$ $[a>0,\quad b>0]$ ET II 20(20)

7. $\displaystyle\int_0^\infty \sin^2\left(ax^2\right) J_1(bx)\, dx = \frac{1}{2b}\cos\left(\frac{b^2}{8a}\right)$ $[a>0,\quad b>0]$ ET II 19(17)

6.687 $\displaystyle\int_0^\infty \cos\left(\frac{x^2}{2a}\right) K_{2\nu}\left(xe^{i\frac{\pi}{4}}\right) K_{2\nu}\left(xe^{-i\frac{\pi}{4}}\right) dx$

$$= \frac{\Gamma\left(\frac{1}{4}+\nu\right)\Gamma\left(\frac{1}{4}-\nu\right)\sqrt{\pi}}{8\sqrt{a}} W_{\frac{1}{4},\nu}\left(ae^{i\frac{\pi}{2}}\right) W_{\frac{1}{4},\nu}\left(ae^{-i\frac{\pi}{2}}\right)$$
$$\left[a>0,\quad |\operatorname{Re}\nu|<\tfrac{1}{4}\right]\quad\text{ET II 372(1)}$$

6.688

1. $\displaystyle\int_0^{\frac{\pi}{2}} J_\nu\left(\mu z\sin t\right)\cos\left(\mu x\cos t\right) dt = \frac{\pi}{2} J_{\frac{\nu}{2}}\left(\mu\frac{\sqrt{x^2+z^2}+x}{2}\right) J_{\frac{\nu}{2}}\left(\mu\frac{\sqrt{x^2+z^2}-x}{2}\right)$

$$[\operatorname{Re}\nu>-1,\quad \operatorname{Re} z>0]\quad\text{MO 46}$$

2. $\displaystyle\int_0^{\frac{\pi}{2}} (\sin x)^{\nu+1}\cos\left(\beta\cos x\right) J_\nu\left(\alpha\sin x\right) dx = 2^{-\frac{1}{2}}\sqrt{\pi}\alpha^\nu\left(\alpha^2+\beta^2\right)^{-\frac{1}{2}\nu-\frac{1}{4}} J_{\nu+\frac{1}{2}}\left[\left(\alpha^2+\beta^2\right)^{\frac{1}{2}}\right]$

$$[\operatorname{Re}\nu>-1]\quad\text{ET II 361(19)}$$

3. $\displaystyle\int_0^{\frac{\pi}{2}} \cos\left[(z-\zeta)\cos\theta\right] J_{2\nu}\left[2\sqrt{z\zeta}\sin\theta\right] d\theta = \frac{\pi}{2} J_\nu(z) J_\nu(\zeta)$

$$\left[\operatorname{Re}\nu>-\tfrac{1}{2}\right]\quad\text{EH II 47(8)}$$

6.69–6.74 Combinations of Bessel and trigonometric functions and powers

6.691 $\displaystyle\int_0^\infty x \sin(bx)\, K_0(ax)\, dx = \frac{\pi b}{2}\left(a^2 + b^2\right)^{-\frac{3}{2}}$ \qquad $[\operatorname{Re} a > 0, \quad b > 0]$ \hfill ET I 105(47)

6.692

1. $\displaystyle\int_0^\infty x\, K_\nu(ax)\, I_\nu(bx) \sin(cx)\, dx = -\frac{1}{2}(ab)^{-\frac{3}{2}} c\left(u^2 - 1\right)^{-\frac{1}{2}} Q^1_{\nu - \frac{1}{2}}(u), \qquad u = (2ab)^{-1}\left(a^2 + b^2 + c^2\right)$

$\left[\operatorname{Re} a > |\operatorname{Re} b|, \quad c > 0, \quad \operatorname{Re} \nu > -\frac{3}{2}\right]$

\hfill ET I 106(54)

2. $\displaystyle\int_0^\infty x\, K_\nu(ax)\, K_\nu(bx) \sin(cx)\, dx = \frac{\pi}{4}(ab)^{-\frac{3}{2}} c\left(u^2 - 1\right)^{-\frac{1}{2}} \Gamma\left(\tfrac{3}{2} + \nu\right) \Gamma\left(\tfrac{3}{2} - \nu\right) P^{-1}_{\nu - \frac{1}{2}}(u)$

$u = (2ab)^{-1}\left(a^2 + b^2 + c^2\right)$ \qquad $\left[\operatorname{Re}(a + b) > 0, \quad c > 0, \quad |\operatorname{Re}\nu| < \tfrac{3}{2}\right]$ \quad ET I 107(61)

6.693

1. $\displaystyle\int_0^\infty J_\nu(\alpha x) \sin \beta x\, \frac{dx}{x} = \frac{1}{\nu}\sin\left(\nu \arcsin \frac{\beta}{\alpha}\right)$ \hspace{2cm} $[\beta \le \alpha]$

$\displaystyle = \frac{\alpha^\nu \sin \frac{\nu\pi}{2}}{\nu\left(\beta + \sqrt{\beta^2 - \alpha^2}\right)^\nu}$ \hspace{2cm} $[\beta \ge \alpha]$

$[\operatorname{Re}\nu > -1]$ \hfill WA 443(2)

2.[8] $\displaystyle\int_0^\infty J_\nu(\alpha x) \cos \beta x\, \frac{dx}{x} = \frac{1}{\nu}\cos\left(\nu \arcsin \frac{\beta}{\alpha}\right)$ \hspace{2cm} $[\beta \le \alpha]$

$\displaystyle = \frac{\alpha^\nu \cos \frac{\nu\pi}{2}}{\nu\left(\beta + \sqrt{\beta^2 - \alpha^2}\right)^\nu}$ \hspace{1cm} $[\beta \ge \alpha]$ \qquad $[\operatorname{Re}\nu > 0]$

\hfill WA 443(3)

3. $\displaystyle\int_0^\infty Y_\nu(ax) \sin(bx)\, \frac{dx}{x} = -\frac{1}{\nu}\tan\left(\frac{\nu\pi}{2}\right)\sin\left[\nu \arcsin\left(\frac{b}{a}\right)\right]$

$[0 < b < a, \quad |\operatorname{Re}\nu| < 1]$

$\displaystyle = \frac{1}{2\nu}\sec\left(\frac{\nu\pi}{2}\right)\left\{a^{-\nu}\cos(\nu\pi)\left[b - \left(b^2 - a^2\right)^{\frac{1}{2}}\right]^\nu - a^\nu\left[b - \left(b^2 - a^2\right)^{\frac{1}{2}}\right]^{-\nu}\right\}$

$[0 < a < b, \quad |\operatorname{Re}\nu| < 1]$

\hfill ET I 103(35)

4. $\displaystyle\int_0^\infty J_\nu(ax) \sin(bx)\, \frac{dx}{x^2}$

$\displaystyle = \frac{\sqrt{a^2 - b^2}\,\sin\left[\nu \arcsin\left(\frac{b}{a}\right)\right]}{\nu^2 - 1} - \frac{b\cos\left[\nu \arcsin\left(\frac{b}{a}\right)\right]}{\nu\left(\nu^2 - 1\right)}$ \qquad $[0 < b < a, \quad \operatorname{Re}\nu > 0]$

$\displaystyle = \frac{-a^\nu \cos\left(\frac{\nu\pi}{2}\right)\left[b + \nu\sqrt{b^2 - a^2}\right]}{\nu\left(\nu^2 - 1\right)\left[b + \sqrt{b^2 - a^2}\right]^\nu}$ \qquad $[0 < a < b, \quad \operatorname{Re}\nu > 0]$

\hfill ET I 99(6)

5. $\displaystyle\int_0^\infty J_\nu(ax)\cos(bx)\frac{dx}{x^2}$

$$= \frac{a\cos\left[(\nu-1)\arcsin\left(\frac{b}{a}\right)\right]}{2\nu(\nu-1)} + \frac{a\cos\left[(\nu+1)\arcsin\left(\frac{b}{a}\right)\right]}{2\nu(\nu+1)} \qquad [0<b<a, \quad \operatorname{Re}\nu>1]$$

$$= \frac{a^\nu\sin\left(\frac{\nu\pi}{2}\right)}{2\nu(\nu-1)\left[b+\sqrt{b^2-a^2}\right]^{\nu-1}} - \frac{a^{\nu+2}\sin\left(\frac{\nu\pi}{2}\right)}{2\nu(\nu+1)\left[b+\sqrt{b^2-a^2}\right]^{\nu+1}} \qquad [0<a<b, \quad \operatorname{Re}\nu>1]$$

ET I 44(6)

6. $\displaystyle\int_0^\infty J_0(\alpha x)\sin x\,\frac{dx}{x} = \frac{\pi}{2}$ $[0<\alpha<1]$

$$= \operatorname{arccosec}\alpha \qquad\qquad [\alpha>1]$$

WH

7. $\displaystyle\int_0^\infty J_0(x)\sin\beta x\,\frac{dx}{x} = \frac{\pi}{2}$ $[\beta>1]$

$$= \arcsin\beta \qquad\qquad [\beta^2<1]$$

$$= -\frac{\pi}{2} \qquad\qquad [\beta<-1]$$

8. $\displaystyle\int_0^\infty [J_0(x)-\cos\alpha x]\frac{dx}{x} = \ln 2\alpha$ NT 66(13)

9. $\displaystyle\int_0^z J_\nu(x)\sin(z-x)\frac{dx}{x} = \frac{2}{\nu}\sum_{k=0}^\infty (-1)^k J_{\nu+2k+1}(z)$ $[\operatorname{Re}\nu>0]$ WA 416(4)

10. $\displaystyle\int_0^z J_\nu(x)\cos(z-x)\frac{dx}{x} = \frac{1}{\nu}J_\nu(z) + \frac{2}{\nu}\sum_{k=1}^\infty (-1)^k J_{\nu+2k}(z)$

$$[\operatorname{Re}\nu>0] \qquad\qquad \text{WA 416(5)}$$

6.694[10] $\displaystyle\int_0^\infty \left[\frac{J_1(ax)}{x}\right]^2 \sin(bx)\,dx$

$$= \frac{1}{2}b - \left(\frac{4a}{3\pi}\right)\left[\left(1+\frac{b^2}{4a^2}\right)\boldsymbol{E}\left(\frac{b}{2a}\right) + \left(1-\frac{b^2}{4a^2}\right)\boldsymbol{K}\left(\frac{b}{2a}\right)\right] \qquad [0\le b\le 2a] \quad \text{ET I 102(22)}$$

$$= \frac{1}{2}b - \frac{2b}{3\pi}\left[\left(1+\frac{b^2}{4a^2}\right)\boldsymbol{E}\left(\frac{2a}{b}\right) - \left(1-\left(\frac{4a^2}{b^2}\right)^{-1}\right)\boldsymbol{K}\left(\frac{2a}{b}\right)\right] \qquad [0\le 2a\le b]$$

6.695

1. $\displaystyle\int_0^\infty \frac{\sin\alpha x}{\beta^2+x^2}J_0(ux)\,dx = \frac{\sinh\alpha\beta}{\beta}K_0(\beta u)$ $[\alpha>0, \quad \operatorname{Re}\beta>0, \quad u>\alpha]$ MO 46

2. $\displaystyle\int_0^\infty \frac{\cos\alpha x}{\beta^2+x^2}J_0(ux)\,dx = \frac{\pi}{2}\frac{e^{-\alpha\beta}}{\beta}I_0(\beta u)$ $[\alpha>0, \quad \operatorname{Re}\beta>0, \quad -\alpha<u<\alpha]$

MO 46

3. $\displaystyle\int_0^\infty \frac{x}{x^2+\beta^2}\sin(\alpha x)J_0(\gamma x)\,dx = \frac{\pi}{2}e^{-\alpha\beta}I_0(\gamma\beta)$ $[\alpha>0, \quad \operatorname{Re}\beta>0, \quad 0<\gamma<\alpha]$

ET II 10(36)

4. $\displaystyle\int_0^\infty \frac{x}{x^2 + \beta^2} \cos(\alpha x)\, J_0(\gamma x)\, dx = \cosh(\alpha\beta)\, K_0(\beta\gamma)$ $[\alpha > 0, \quad \operatorname{Re}\beta > 0, \quad \alpha < \gamma]$

ET II 11(45)

6.696 $\displaystyle\int_0^\infty [1 - \cos(\alpha x)]\, J_0(\beta x)\, \frac{dx}{x} = \operatorname{arccosh}\left(\frac{\alpha}{\beta}\right)$ $[0 < \beta < \alpha]$

$= 0$ $[0 < \alpha < \beta]$

ET II 11(43)

6.697

1. $\displaystyle\int_{-\infty}^\infty \frac{\sin[\alpha(x+\beta)]}{x+\beta}\, J_0(x)\, dx = 2\int_0^\alpha \frac{\cos\beta u}{\sqrt{1-u^2}}\, du$ $[0 \le \alpha \le 1]$ WA 463(2)

$= \pi\, J_0(\beta)$ $[1 \le \alpha < \infty]$ WA 463(1), ET II 345(42)

2. $\displaystyle\int_0^\infty \frac{\sin(x+t)}{x+t}\, J_0(t)\, dt = \frac{\pi}{2}\, J_0(x)$ $[x > 0]$ WA 475(4)

3. $\displaystyle\int_0^\infty \frac{\cos(x+t)}{x+t}\, J_0(t)\, dt = -\frac{\pi}{2}\, Y_0(x)$ $[x > 0]$ WA 475(5)

4. $\displaystyle\int_{-\infty}^\infty \frac{|x|}{x+\beta}\, \sin[\alpha(x+\beta)]\, J_0(bx)\, dx = 0$ $[0 \le \alpha < b]$ WA 464(5), ET II 345(43)a

5. $\displaystyle\int_{-\infty}^\infty \frac{\sin[\alpha(x+\beta)]}{x+\beta}\left[J_{n+\frac{1}{2}}(x)\right]^2 dx = \pi\left[J_{n+\frac{1}{2}}(\beta)\right]^2$ $[2 \le \alpha < \infty, \quad n = 0, 1, \ldots]$

ET II 346(45)

6. $\displaystyle\int_{-\infty}^\infty \frac{\sin[\alpha(x+\beta)]}{x+\beta}\, J_{n+\frac{1}{2}}(x)\, J_{-n-\frac{1}{2}}(x)\, dx = \pi\, J_{n+\frac{1}{2}}(\beta)\, J_{-n-\frac{1}{2}}(\beta)$

$[2 \le \alpha < \infty, \quad n = 0, 1, \ldots]$

ET II 346(46)

7. $\displaystyle\int_{-\infty}^\infty \frac{J_\mu[a(z+x)]}{(z+x)^\mu}\, \frac{J_\nu[a(\zeta+x)]}{(\zeta+x)^\nu}\, dx = \frac{\Gamma(\mu+\nu)\sqrt{\pi}\sqrt{\frac{2}{a}}}{\Gamma\left(\mu+\frac{1}{2}\right)\Gamma\left(\nu+\frac{1}{2}\right)} \cdot \frac{J_{\mu+\nu-\frac{1}{2}}[a(z-\zeta)]}{(z-\zeta)^{\mu+\nu-\frac{1}{2}}}$

$[\operatorname{Re}(\mu+\nu) > 0]$ WA 463(3)

6.698

1. $\displaystyle\int_0^\infty \sqrt{x}\, J_{\nu+\frac{1}{4}}(ax)\, J_{-\nu+\frac{1}{4}}(ax)\, \sin(bx)\, dx = \sqrt{\frac{2}{\pi b}}\, \frac{\cos\left[2\nu \arccos\left(\frac{b}{2a}\right)\right]}{\sqrt{4a^2 - b^2}}$ $[0 < b < 2a]$

$= 0$ $[0 < 2a < b]$

ET I 102(26)

2. $\displaystyle\int_0^\infty \sqrt{x}\, J_{\nu-\frac{1}{4}}(ax)\, J_{-\nu-\frac{1}{4}}(ax)\, \cos(bx)\, dx = \sqrt{\frac{2}{\pi b}}\, \frac{\cos\left[2\nu \arccos\left(\frac{b}{2a}\right)\right]}{\sqrt{4a^2 - b^2}}$ $[0 < b < 2a]$

$= 0$ $[0 < 2a < b]$

ET I 46(24)

3. $\displaystyle\int_0^\infty \sqrt{x}\, I_{\frac{1}{4}-\nu}\left(\frac{1}{2}ax\right) K_{\frac{1}{4}+\nu}\left(\frac{1}{2}ax\right)\sin(bx)\,dx = \sqrt{\frac{\pi}{2b}}\, a^{-2\nu}\frac{\left(b+\sqrt{a^2+b^2}\right)^{2\nu}}{\sqrt{a^2+b^2}}$

$$\left[\operatorname{Re} a > 0,\quad b > 0,\quad \operatorname{Re}\nu < \tfrac{5}{4}\right]$$

ET I 106(56)

4. $\displaystyle\int_0^\infty \sqrt{x}\, I_{-\frac{1}{4}-\nu}\left(\frac{1}{2}ax\right) K_{-\frac{1}{4}+\nu}\left(\frac{1}{2}ax\right)\cos(bx)\,dx = \sqrt{\frac{\pi}{2b}}\, a^{-2\nu}\frac{\left(b+\sqrt{a^2+b^2}\right)^{2\nu}}{\sqrt{a^2+b^2}}$

$$\left[\operatorname{Re} a > 0,\quad b > 0,\quad \operatorname{Re}\nu < \tfrac{3}{4}\right]$$

ET I 50(49)

6.699

1. $\displaystyle\int_0^\infty x^\lambda J_\nu(ax)\sin(bx)\,dx = 2^{1+\lambda}a^{-(2+\lambda)}b\frac{\Gamma\left(\frac{2+\lambda+\nu}{2}\right)}{\Gamma\left(\frac{\nu-\lambda}{2}\right)}F\left(\frac{2+\lambda+\nu}{2},\frac{2+\lambda-\nu}{2};\frac{3}{2};\frac{b^2}{a^2}\right)$

$$\left[0 < b < a,\quad -\operatorname{Re}\nu - 1 < 1 + \operatorname{Re}\lambda < \tfrac{3}{2}\right]$$

$$= \left(\frac{1}{2}a\right)^\nu b^{-(\nu+\lambda+1)}\frac{\Gamma(\nu+\lambda+1)}{\Gamma(\nu+1)}\sin\left[\pi\left(\frac{1+\lambda+\nu}{2}\right)\right]$$

$$\times F\left(\frac{2+\lambda+\nu}{2},\frac{1+\lambda+\nu}{2};\nu+1;\frac{a^2}{b^2}\right)$$

$$\left[0 < a < b,\quad -\operatorname{Re}\nu - 1 < 1 + \operatorname{Re}\lambda < \tfrac{3}{2}\right]$$

ET I 100(11)

2. $\displaystyle\int_0^\infty x^\lambda J_\nu(ax)\cos(bx)\,dx$

$$= \frac{2^\lambda a^{-(1+\lambda)}\Gamma\left(\frac{1+\lambda+\nu}{2}\right)}{\Gamma\left(\frac{\nu-\lambda+1}{2}\right)}F\left(\frac{1+\lambda+\nu}{2},\frac{1+\lambda-\nu}{2};\frac{1}{2};\frac{b^2}{a^2}\right)$$

$$\left[0 < b < a,\quad -\operatorname{Re}\nu < 1 + \operatorname{Re}\lambda < \tfrac{3}{2}\right]$$

$$= \frac{\left(\frac{a}{2}\right)^\nu b^{-(\nu+1+\lambda)}\Gamma(1+\lambda+\nu)\cos\left[\frac{\pi}{2}(1+\lambda+\nu)\right]}{\Gamma(\nu+1)}F\left(\frac{1+\lambda+\nu}{2},\frac{2+\lambda+\nu}{2};\nu+1;\frac{a^2}{b^2}\right)$$

$$\left[0 < a < b,\quad -\operatorname{Re}\nu < 1 + \operatorname{Re}\lambda < \tfrac{3}{2}\right]$$

ET I 45(13)

3. $\displaystyle\int_0^\infty x^\lambda K_\mu(ax)\sin(bx)\,dx = \frac{2^\lambda b\,\Gamma\left(\frac{2+\mu+\lambda}{2}\right)\Gamma\left(\frac{2+\lambda-\mu}{2}\right)}{a^{2+\lambda}}F\left(\frac{2+\mu+\lambda}{2},\frac{2+\lambda-\mu}{2};\frac{3}{2};-\frac{b^2}{a^2}\right)$

$$\left[\operatorname{Re}\left(-\lambda\pm\mu\right) < 2,\quad \operatorname{Re} a > 0,\quad b > 0\right]$$

ET I 106(50)

4. $\displaystyle\int_0^\infty x^\lambda K_\mu(ax)\cos(bx)\,dx = 2^{\lambda-1}a^{-\lambda-1}\Gamma\left(\frac{\mu+\lambda+1}{2}\right)\Gamma\left(\frac{1+\lambda-\mu}{2}\right)$

$$\times F\left(\frac{\mu+\lambda+1}{2},\frac{1+\lambda-\mu}{2};\frac{1}{2};-\frac{b^2}{a^2}\right)$$

$$\left[\operatorname{Re}\left(-\lambda\pm\mu\right) < 1,\quad \operatorname{Re} a > 0,\quad b > 0\right]\quad \text{ET I 49(42)}$$

5. $$\int_0^\infty x^\nu \sin(ax) J_\nu(bx)\,dx = \frac{\sqrt{\pi}\,2^\nu b^\nu \left(a^2 - b^2\right)^{-\nu-\frac{1}{2}}}{\Gamma\left(\frac{1}{2}-\nu\right)} \qquad \left[0 < b < a, \quad -1 < \operatorname{Re}\nu < \tfrac{1}{2}\right]$$

$$= 0 \qquad \left[0 < a < b, \quad -1 < \operatorname{Re}\nu < \tfrac{1}{2}\right]$$

ET II 32(4)

6. $$\int_0^\infty x^\nu \cos(ax) J_\nu(bx)\,dx = -2^\nu \frac{\sin(\nu\pi)}{\sqrt{\pi}} \Gamma\left(\frac{1}{2}+\nu\right) b^\nu \left(a^2-b^2\right)^{-\nu-\frac{1}{2}} \qquad \left[0 < b < a, \quad |\operatorname{Re}\nu| < \tfrac{1}{2}\right]$$

$$= 2^\nu \frac{b^\nu}{\sqrt{\pi}} \Gamma\left(\frac{1}{2}+\nu\right) \left(b^2-a^2\right)^{-\nu-\frac{1}{2}} \qquad \left[0 < a < b, \quad |\operatorname{Re}\nu| < \tfrac{1}{2}\right]$$

ET II 36(29)

7. $$\int_0^\infty x^{\nu+1} \sin(ax) J_\nu(bx)\,dx$$

$$= -2^{1+\nu} a \frac{\sin(\nu\pi)}{\sqrt{\pi}} b^\nu \Gamma\left(\nu+\frac{3}{2}\right) \left(a^2-b^2\right)^{-\nu-\frac{3}{2}} \qquad \left[0 < b < a, \quad -\tfrac{3}{2} < \operatorname{Re}\nu < -\tfrac{1}{2}\right]$$

$$= -\frac{2^{1+\nu}}{\sqrt{\pi}} a b^\nu \Gamma\left(\nu+\frac{3}{2}\right) \left(b^2-a^2\right)^{-\nu-\frac{3}{2}} \qquad \left[0 < a < b, \quad -\tfrac{3}{2} < \operatorname{Re}\nu < -\tfrac{1}{2}\right]$$

ET II 32(3)

8. $$\int_0^\infty x^{\nu+1} \cos(ax) J_\nu(bx)\,dx = 2^{1+\nu}\sqrt{\pi}\,ab^\nu \frac{\left(a^2-b^2\right)^{-\nu-\frac{3}{2}}}{\Gamma\left(-\frac{1}{2}-\nu\right)} \qquad \left[0 < b < a, \quad -1 < \operatorname{Re}\nu < -\tfrac{1}{2}\right]$$

$$= 0 \qquad \left[0 < a < b, \quad -1 < \operatorname{Re}\nu < -\tfrac{1}{2}\right]$$

ET II 36(28)

9. $$\int_0^1 x^\nu \sin(ax) J_\nu(ax)\,dx = \frac{1}{2\nu+1}\left[\sin a J_\nu(a) - \cos a\, J_{\nu+1}(a)\right]$$

$$\left[\operatorname{Re}\nu > -1\right] \qquad \text{ET II 334(9)a}$$

10. $$\int_0^1 x^\nu \cos(ax) J_\nu(ax)\,dx = \frac{1}{2\nu+1}\left[\cos a J_\nu(a) + \sin a\, J_{\nu+1}(a)\right]$$

$$\left[\operatorname{Re}\gamma > -\tfrac{1}{2}\right] \qquad \text{ET II 335(20)}$$

11. $$\int_0^\infty x^{1+\nu} K_\nu(ax) \sin(bx)\,dx = \sqrt{\pi}\,(2a)^\nu \Gamma\left(\frac{3}{2}+\nu\right) b\left(b^2+a^2\right)^{-\frac{3}{2}-\nu}$$

$$\left[\operatorname{Re} a > 0, \quad b > 0, \quad \operatorname{Re}\nu > -\tfrac{3}{2}\right]$$

ET I 105(49)

12. $$\int_0^\infty x^\mu K_\mu(ax) \cos(bx)\,dx = \frac{1}{2}\sqrt{\pi}\,(2a)^\mu \Gamma\left(\mu+\frac{1}{2}\right)\left(b^2+a^2\right)^{-\mu-\frac{1}{2}}$$

$$\left[\operatorname{Re} a > 0, \quad b > 0, \quad \operatorname{Re}\mu > -\tfrac{1}{2}\right]$$

ET I 49(41)

13. $$\int_0^\infty x^\nu Y_{\nu-1}(ax) \sin(bx)\,dx = 0 \qquad \left[0 < b < a, \quad |\operatorname{Re}\nu| < \tfrac{1}{2}\right]$$

$$= \frac{2^\nu \sqrt{\pi}\,a^{\nu-1}b}{\Gamma\left(\frac{1}{2}-\nu\right)}\left(b^2-a^2\right)^{-\nu-\frac{1}{2}} \qquad \left[0 < a < b, \quad |\operatorname{Re}\nu| < \tfrac{1}{2}\right]$$

ET I 104(36)

14. $\displaystyle\int_0^\infty x^\nu\, Y_\nu(ax)\cos(bx)\,dx = 0$ $\left[0 < b < a, \quad |\operatorname{Re}\nu| < \tfrac{1}{2}\right]$

$$= -2^\nu\sqrt{\pi}a^\nu\,\frac{\left(b^2 - a^2\right)^{-\nu-\frac{1}{2}}}{\Gamma\left(\tfrac{1}{2} - \nu\right)} \qquad \left[0 < a < b, \quad |\operatorname{Re}\nu| < \tfrac{1}{2}\right]$$

ET I 47(30)

6.711

1. $\displaystyle\int_0^\infty x^{\nu-\mu}\, J_\mu(ax)\, J_\nu(bx)\sin(cx)\,dx = 0$ $\left[0 < c < b - a, \quad -1 < \operatorname{Re}\nu < 1 + \operatorname{Re}\mu\right]$

ET I 103(28)

2. $\displaystyle\int_0^\infty x^{\nu-\mu+1}\, J_\mu(ax)\, J_\nu(bx)\cos(cx)\,dx = 0$

$$\left[0 < c < b - a, \quad a > 0, \quad b > 0, \quad -1 < \operatorname{Re}\nu < \operatorname{Re}\mu\right] \quad \text{ET I 47(25)}$$

3. $\displaystyle\int_0^\infty x^{\nu-\mu-2}\, J_\mu(ax)\, J_\nu(bx)\sin(cx)\,dx = 2^{\nu-\mu-1}a^\mu b^{-\nu}\,\frac{c\,\Gamma(\nu)}{\Gamma(\mu+1)}$

$$\left[0 < a, \quad 0 < b, \quad 0 < c < b - a, \quad 0 < \operatorname{Re}\nu < \operatorname{Re}\mu + 3\right] \quad \text{ET I 103(29)}$$

4. $\displaystyle\int_0^\infty x^{\varrho-\mu-1}\, J_\mu(ax)\, J_\varrho(bx)\cos(cx)\,dx = 2^{\varrho-\mu-1}b^{-\varrho}a^\mu\,\frac{\Gamma(\varrho)}{\Gamma(\mu+1)}$

$$\left[b > 0, \quad a > 0, \quad 0 < c < b - a, \quad 0 < \operatorname{Re}\varrho < \operatorname{Re}\mu + 2\right] \quad \text{ET I 47(26)}$$

5. $\displaystyle\int_0^\infty x^{1-2\nu}\sin(2ax)\, J_\nu(x)\, Y_\nu(x)\,dx = -\frac{\Gamma\left(\tfrac{3}{2} - \nu\right)a}{2\,\Gamma\left(2\nu - \tfrac{1}{2}\right)\Gamma(2 - \nu)}\,F\left(\frac{3}{2} - \nu, \frac{3}{2} - 2\nu; 2 - \nu; a^2\right)$

$$\left[0 < \operatorname{Re}\nu < \tfrac{3}{2}, \quad 0 < a < 1\right]$$

ET II 348(63)

6.[10] $\displaystyle\int_0^\infty \arg\sin(zx)x^{\nu-\mu-4}\, J_\mu(ax)\, J_\nu(\rho x)\,dx = z\frac{\Gamma(\nu)\,a^\mu\rho^{-\nu}}{2^{\mu-\nu+3}\,\Gamma(\mu+1)}\left[\frac{\rho^2}{\nu-1} - \frac{a^2}{\mu+1} - \frac{2z^2}{3}\right]$

7.[10] $\displaystyle\int_0^\infty \cos(zx)x^{\nu-\mu-3}\, J_\mu(ax)\, J_\nu(\rho x)\,dx = \frac{\Gamma(\nu)\,a^\mu\rho^{-\nu}}{2^{\mu-\nu+3}\,\Gamma(\mu+1)}\left[\frac{\rho^2}{\nu-1} - \frac{a^2}{\mu+1} - 2z^2\right]$

6.712

1. $\displaystyle\int_0^\infty x^\nu\left[J_\nu(ax)\cos(ax) + Y_\nu(ax)\sin(ax)\right]\sin(bx)\,dx = \frac{\sqrt{\pi}(2a)^\nu}{\Gamma\left(\tfrac{1}{2} - \nu\right)}\left(b^2 + 2ab\right)^{-\nu-\frac{1}{2}}$

$$\left[b > 0, \quad -1 < \operatorname{Re}\nu < \tfrac{1}{2}\right] \quad \text{ET I 104(40)}$$

2. $\displaystyle\int_0^\infty x^\nu\left[Y_\nu(ax)\cos(ax) - J_\nu(ax)\sin(ax)\right]\cos(bx)\,dx = -\frac{\sqrt{\pi}(2a)^\nu}{\Gamma\left(\tfrac{1}{2} - \nu\right)}\left(b^2 + 2ab\right)^{-\nu-\frac{1}{2}}$

ET I 48(35)

3. $\displaystyle\int_0^\infty x^\nu\left[J_\nu(ax)\cos(ax) - Y_\nu(ax)\sin(ax)\right]\sin(bx)\,dx$

$$= 0 \qquad\qquad \left[0 < b < 2a, \quad -1 < \operatorname{Re}\nu < \tfrac{1}{2}\right]$$

$$= \frac{2^\nu\sqrt{\pi}b^\nu}{\Gamma\left(\tfrac{1}{2} - \nu\right)}\left(b^2 - 2ab\right)^{-\nu-\frac{1}{2}} \qquad \left[2a < b, \quad -1 < \operatorname{Re}\nu < \tfrac{1}{2}\right]$$

ET I 104(41)

4. $\displaystyle\int_0^\infty x^\nu \left[J_\nu(ax) \sin(ax) + Y_\nu(ax) \cos(ax) \right] \cos(bx)\, dx$

$$= 0 \qquad\qquad\qquad\qquad \left[0 < b < 2a, \quad |\operatorname{Re}\nu| < \tfrac{1}{2} \right]$$

$$= -\frac{\sqrt{\pi}(2a)^\nu}{\Gamma\left(\tfrac{1}{2} - \nu\right)} \left(b^2 - 2ab \right)^{-\nu - \frac{1}{2}} \qquad \left[0 < 2a < b, \quad |\operatorname{Re}\nu| < \tfrac{1}{2} \right]$$

ET I 48(33)

6.713

1. $\displaystyle\int_0^\infty x^{1-2\nu} \sin(2ax) \left\{ [J_\nu(x)]^2 - [Y_\nu(x)]^2 \right\} dx$

$$= \frac{\sin(2\nu\pi)\, \Gamma\left(\tfrac{3}{2} - \nu\right) \Gamma\left(\tfrac{3}{2} - 2\nu\right) a}{\pi\, \Gamma(2 - \nu)} F\left(\frac{3}{2} - \nu, \frac{3}{2} - 2\nu; 2 - \nu; a^2 \right)$$

$$\left[0 < \operatorname{Re}\nu < \tfrac{3}{4}, \quad 0 < a < 1 \right] \quad \text{ET II 348(64)}$$

2. $\displaystyle\int_0^\infty x^{2-2\nu} \sin(2ax) \left[J_\nu(x) J_{\nu-1}(x) - Y_\nu(x) Y_{\nu-1}(x) \right] dx$

$$= -\frac{\sin(2\nu\pi)\, \Gamma\left(\tfrac{3}{2} - \nu\right) \Gamma\left(\tfrac{5}{2} - 2\nu\right) a}{\pi\, \Gamma(2 - \nu)} F\left(\frac{3}{2} - \nu, \frac{5}{2} - 2\nu; 2 - \nu; a^2 \right)$$

$$\left[\tfrac{1}{2} < \operatorname{Re}\nu < \tfrac{5}{4}, \quad 0 < a < 1 \right] \quad \text{ET II 348(65)}$$

3. $\displaystyle\int_0^\infty x^{2-2\nu} \sin(2ax) \left[J_\nu(x) Y_{\nu-1}(x) + Y_\nu(x) J_{\nu-1}(x) \right] dx$

$$= -\frac{\Gamma\left(\tfrac{3}{2} - \nu\right) a}{\Gamma\left(2\nu - \tfrac{3}{2} \right) \Gamma(2 - \nu)} F\left(\frac{3}{2} - \nu, \frac{5}{2} - 2\nu; 2 - \nu; a^2 \right)$$

$$\left[\tfrac{1}{2} < \operatorname{Re}\nu < \tfrac{5}{2}, \quad 0 < a < 1 \right] \quad \text{ET II 349(66)}$$

6.714

1. $\displaystyle\int_0^\infty \sin(2ax)[x^\nu J_\nu(x)]^2\, dx$

$$= \frac{a^{-2\nu}\, \Gamma\left(\tfrac{1}{2} + \nu\right)}{2\sqrt{\pi}\, \Gamma(1 - \nu)} F\left(\frac{1}{2} + \nu, \frac{1}{2}; 1 - \nu; a^2 \right) \qquad\qquad \left[0 < a < 1, \quad |\operatorname{Re}\nu| < \tfrac{1}{2} \right]$$

$$= \frac{a^{-4\nu-1}\, \Gamma\left(\tfrac{1}{2} + \nu\right)}{2\, \Gamma(1 + \nu)\, \Gamma\left(\tfrac{1}{2} - 2\nu\right)} F\left(\frac{1}{2} + \nu, \frac{1}{2} + 2\nu; 1 + \nu; \frac{1}{a^2} \right) \qquad \left[a > 1, \quad |\operatorname{Re}\nu| < \tfrac{1}{2} \right]$$

ET II 343(31)

2. $\displaystyle\int_0^\infty \cos(2ax)[x^\nu J_\nu(x)]^2\, dx$

$$= \frac{a^{-2\nu}\, \Gamma(\nu)}{2\sqrt{\pi}\, \Gamma\left(\tfrac{1}{2} - \nu\right)} F\left(\nu + \frac{1}{2}, \frac{1}{2}; 1 - \nu; a^2 \right)$$

$$+ \frac{\Gamma(-\nu)\, \Gamma\left(\tfrac{1}{2} + 2\nu\right)}{2\pi\, \Gamma\left(\tfrac{1}{2} - \nu\right)} F\left(\frac{1}{2} + \nu, \frac{1}{2} + 2\nu; 1 + \nu; a^2 \right) \qquad \left[0 < a < 1, \quad -\tfrac{1}{4} < \operatorname{Re}\nu < \tfrac{1}{2} \right]$$

$$= -\frac{\sin(\nu\pi) a^{-4\nu-1}\, \Gamma\left(\tfrac{1}{2} + 2\nu\right)}{\Gamma(1 + \nu)\, \Gamma\left(\tfrac{1}{2} - \nu\right)} F\left(\frac{1}{2} + \nu, \frac{1}{2} + 2\nu; 1 + \nu; \frac{1}{a^2} \right) \qquad \left[a > 1, \quad -\tfrac{1}{4} < \operatorname{Re}\nu < \tfrac{1}{2} \right]$$

ET II 344(33)

6.715

1. $$\int_0^\infty \frac{x^\nu}{x+\beta} \sin(x+\beta) J_\nu(x)\, dx = \frac{\pi}{2} \sec(\nu\pi)\beta^\nu J_{-\nu}(\beta)$$

$$\left[|\arg\beta| < \pi, \quad |\operatorname{Re}\nu| < \tfrac{1}{2}\right] \qquad \text{ET II 340(8)}$$

2. $$\int_0^\infty \frac{x^\nu}{x+\beta} \cos(x+\beta) J_\nu(x)\, dx = -\frac{\pi}{2} \sec(\nu\pi)\beta^\nu Y_{-\nu}(\beta)$$

$$\left[|\arg\beta| < \pi, \quad |\operatorname{Re}\nu| < \tfrac{1}{2}\right] \qquad \text{ET II 340(9)}$$

6.716

1. $$\int_0^a x^\lambda \sin(a-x) J_\nu(x)\, dx = 2a^{\lambda+1} \sum_{n=0}^\infty \frac{(-1)^n \,\Gamma(\nu-\lambda+2n)\,\Gamma(\nu+\lambda+1)}{\Gamma(\nu-\lambda)\,\Gamma(\nu+\lambda+3+2n)}(\nu+2n+1)\, J_{\nu+2n+1}(a)$$

$$[\operatorname{Re}(\lambda+\nu) > -1] \qquad \text{ET II 335(16)}$$

2. $$\int_0^a x^\lambda \cos(a-x) J_\nu(x)\, dx = \frac{a^{\lambda+1} J_\nu(a)}{\lambda+\nu+1} + 2a^{\lambda+1}$$
$$\times \sum_{n=1}^\infty \frac{(-1)^n \,\Gamma(\nu-\lambda+2n-1)\,\Gamma(\nu+\lambda+1)}{\Gamma(\nu-\lambda)\,\Gamma(\nu+\lambda+2n+2)}(\nu+2n)\, J_{\nu+2n}(a)$$

$$[\operatorname{Re}(\lambda+\nu) > -1] \qquad \text{ET II 335(26)}$$

6.717 $$\int_{-\infty}^\infty \frac{\sin[a(x+\beta)]}{x^\nu(x+\beta)} J_{\nu+2n}(x)\, dx = \pi\beta^{-\nu} J_{\nu+2n}(\beta)$$

$$\left[1 \le a < \infty, n = 0, 1, 2, \ldots; \quad \operatorname{Re}\nu > -\tfrac{3}{2}\right] \qquad \text{ET II 345(44)}$$

6.718

1. $$\int_0^\infty \frac{x^\nu}{x^2+\beta^2} \sin(\alpha x) J_\nu(\gamma x)\, dx = \beta^{\nu-1} \sinh(\alpha\beta)\, K_\nu(\beta\gamma)$$

$$\left[0 < \alpha \le \gamma, \quad \operatorname{Re}\beta > 0, \quad -1 < \operatorname{Re}\nu < \tfrac{3}{2}\right] \qquad \text{ET II 33(8)}$$

2. $$\int_0^\infty \frac{x^{\nu+1}}{x^2+\beta^2} \cos(\alpha x) J_\nu(\gamma x)\, dx = \beta^\nu \cosh(\alpha\beta)\, K_\nu(\beta\gamma)$$

$$\left[0 < \alpha \le \gamma, \quad \operatorname{Re}\beta > 0, \quad -1 < \operatorname{Re}\nu < \tfrac{1}{2}\right] \qquad \text{ET II 37(33)}$$

3. $$\int_0^\infty \frac{x^{1-\nu}}{x^2+\beta^2} \sin(\alpha x) J_\nu(\gamma x)\, dx = \frac{\pi}{2}\beta^{-\nu} e^{-\alpha\beta} I_\nu(\beta\gamma) \qquad \left[0 < \gamma \le \alpha, \quad \operatorname{Re}\beta > 0, \quad \operatorname{Re}\nu > -\tfrac{1}{2}\right]$$

$$\text{ET II 33(9)}$$

4. $$\int_0^\infty \frac{x^{-\nu}}{x^2+\beta^2} \cos(\alpha x) J_\nu(\gamma x)\, dx = \frac{\pi}{2}\beta^{-\nu-1} e^{-\alpha\beta} I_\nu(\beta\gamma)$$

$$\left[0 < \gamma \le \alpha, \operatorname{Re}\beta > 0, \operatorname{Re}\nu > -\tfrac{3}{2}\right]$$
$$\text{ET II 37(34)}$$

6.719

1.[6] $$\int_0^\alpha \frac{\sin(\beta x)}{\sqrt{\alpha^2-x^2}} J_\nu(x)\, dx = \pi \sum_{n=0}^\infty (-1)^n J_{2n+1}(\alpha\beta)\, J_{\frac{1}{2}\nu+n+\frac{1}{2}}\left(\tfrac{1}{2}\alpha\right) J_{\frac{1}{2}\nu-n-\frac{1}{2}}\left(\tfrac{1}{2}\alpha\right)$$

$$[\operatorname{Re}\nu > -2] \qquad \text{ET II 335(17)}$$

2. $\displaystyle\int_0^\alpha \frac{\cos(\beta x)}{\sqrt{\alpha^2 - x^2}} J_\nu(x)\,dx = \frac{\pi}{2}\, J_0(\alpha\beta)\left[J_{\frac{1}{2}\nu}\left(\tfrac{1}{2}\alpha\right)\right]^2 + \pi\sum_{n=1}^\infty (-1)^n J_{2n}(\alpha\beta)\, J_{\frac{1}{2}\nu+n}\left(\tfrac{1}{2}\alpha\right) J_{\frac{1}{2}\nu-n}\left(\tfrac{1}{2}\alpha\right)$

$$[\operatorname{Re}\nu > -1] \qquad \text{ET II 336(27)}$$

6.721

1. $\displaystyle\int_0^\infty \sqrt{x}\, J_{\frac{1}{4}}\left(a^2 x^2\right)\sin(bx)\,dx = 2^{-3/2} a^{-2}\sqrt{\pi b}\, J_{\frac{1}{4}}\left(\frac{b^2}{4a^2}\right)$

$$[b > 0] \qquad \text{ET I 108(1)}$$

2. $\displaystyle\int_0^\infty \sqrt{x}\, J_{-\frac{1}{4}}\left(a^2 x^2\right)\cos(bx)\,dx = 2^{-3/2} a^{-2}\sqrt{\pi b}\, J_{-\frac{1}{4}}\left(\frac{b^2}{4a^2}\right)$

$$[b > 0] \qquad \text{ET I 51(1)}$$

3. $\displaystyle\int_0^\infty \sqrt{x}\, Y_{\frac{1}{4}}\left(a^2 x^2\right)\sin(bx)\,dx = -2^{-3/2}\sqrt{\pi b}\, a^{-2}\, \mathbf{H}_{\frac{1}{4}}\left(\frac{b^2}{4a^2}\right)$

$$\text{ET I 108(7)}$$

4. $\displaystyle\int_0^\infty \sqrt{x}\, Y_{-\frac{1}{4}}\left(a^2 x^2\right)\cos(bx)\,dx = -2^{-3/2}\sqrt{\pi b}\, a^{-2}\, \mathbf{H}_{-\frac{1}{4}}\left(\frac{b^2}{4a^2}\right)$

$$\text{ET I 52(7)}$$

5. $\displaystyle\int_0^\infty \sqrt{x}\, K_{\frac{1}{4}}\left(a^2 x^2\right)\sin(bx)\,dx = 2^{-5/2}\sqrt{\pi^3 b}\, a^{-2}\left[I_{\frac{1}{4}}\left(\frac{b^2}{4a^2}\right) - \mathbf{L}_{\frac{1}{4}}\left(\frac{b^2}{4a^2}\right)\right]$

$$\left[|\arg a| < \frac{\pi}{4}, \quad b > 0\right] \qquad \text{ET I 109(11)}$$

6. $\displaystyle\int_0^\infty \sqrt{x}\, K_{-\frac{1}{4}}\left(a^2 x^2\right)\cos(bx)\,dx = 2^{-5/2}\sqrt{\pi^3 b}\, a^{-2}\left[I_{-\frac{1}{4}}\left(\frac{b^2}{4a^2}\right) - \mathbf{L}_{-\frac{1}{4}}\left(\frac{b^2}{4a^2}\right)\right]$

$$[b > 0] \qquad \text{ET I 52(10)}$$

6.722

1. $\displaystyle\int_0^\infty \sqrt{x}\, K_{\frac{1}{8}+\nu}\left(a^2 x^2\right) I_{\frac{1}{8}-\nu}\left(a^2 x^2\right)\sin(bx)\,dx = \sqrt{2\pi}\, b^{-3/2}\frac{\Gamma\left(\frac{5}{8} - \nu\right)}{\Gamma\left(\frac{5}{4}\right)}\, W_{\nu,\frac{1}{8}}\left(\frac{b^2}{8a^2}\right) M_{-\nu,\frac{1}{8}}\left(\frac{b^2}{8a^2}\right)$

$$\left[\operatorname{Re}\nu < \frac{5}{8}, \quad |\arg a| < \frac{\pi}{4}, \quad b > 0\right]$$
$$\text{ET I 109(13)}$$

2.[10] $\displaystyle\int_0^\infty \sqrt{x}\, J_{-\frac{1}{8}-\nu}\left(a^2 x^2\right) J_{-\frac{1}{8}+\nu}\left(a^2 x^2\right)\cos(bx)\,dx$

$$= \frac{\sqrt{\pi}}{2^{3/4} a^{3/2}}\frac{\Gamma\left(\frac{1}{4}\right)}{\Gamma\left(\frac{3}{4}\right)\Gamma\left(\frac{5}{8}-\nu\right)\Gamma\left(\frac{5}{8}+\nu\right)}\, {}_2F_3\left(\frac{3}{8}-\nu, \frac{3}{8}+\nu; \frac{3}{8}, \frac{3}{4}, \frac{7}{8}; -\left(\frac{b}{4a}\right)^4\right)$$

$$- \frac{1}{a^2}\sqrt{\frac{2b}{\pi}}\cos(\pi\nu)\, {}_2F_3\left(\frac{1}{2}-\nu, \frac{1}{2}+\nu; \frac{1}{2}, \frac{7}{8}, \frac{9}{8}; -\left(\frac{b}{4a}\right)^4\right)$$

$$- \frac{b^{5/2}\nu}{15 a^4}\sqrt{\frac{2}{\pi}}\sin(\pi\nu)\, {}_2F_3\left(1-\nu, 1+\nu; \frac{11}{8}, \frac{3}{2}, \frac{13}{8}; -\left(\frac{b}{4a}\right)^4\right)$$

$$[a^2 > 0, \quad \operatorname{Im} b = 0] \qquad \text{MC}$$

3. $\displaystyle\int_0^\infty \sqrt{x}\, J_{\frac{1}{8}-\nu}\left(a^2 x^2\right) J_{\frac{1}{8}+\nu}\left(a^2 x^2\right) \sin(bx)\, dx$

$$= \sqrt{\frac{2}{\pi}}\, b^{-3/2}\left[e^{\pi i/8}\, W_{\nu,\frac{1}{8}}\left(\frac{b^2 e^{\pi i/2}}{8a^2}\right) W_{-\nu,\frac{1}{8}}\left(\frac{b^2 e^{\pi i/2}}{8a^2}\right)\right.$$

$$\left. + e^{-i\pi/8}\, W_{\nu,\frac{1}{8}}\left(\frac{b^2 e^{-\pi i/2}}{8a^2}\right) W_{-\nu,\frac{1}{8}}\left(\frac{b^2 e^{-\frac{\pi i}{2}}}{8a^2}\right)\right]$$

$$[b>0] \hspace{3cm} \text{ET I 108(6)}$$

4. $\displaystyle\int_0^\infty \sqrt{x}\, K_{\frac{1}{8}-\nu}\left(a^2 x^2\right) I_{-\frac{1}{8}-\nu}\left(a^2 x^2\right) \cos(bx)\, dx$

$$= \sqrt{2\pi}\, b^{-3/2}\frac{\Gamma\left(\frac{3}{8}-\nu\right)}{\Gamma\left(\frac{3}{4}\right)} W_{\nu,-\frac{1}{8}}\left(\frac{b^2}{8a^2}\right) M_{-\nu,-\frac{1}{8}}\left(\frac{b^2}{8a^2}\right)$$

$$\left[\operatorname{Re}\nu<\tfrac{3}{8}, \quad b>0\right] \hspace{2cm} \text{ET I 52(12)}$$

6.723 $\displaystyle\int_0^\infty x\, J_\nu\left(x^2\right)\left[\sin(\nu\pi)\, J_\nu\left(x^2\right)-\cos(\nu\pi)\, Y_\nu\left(x^2\right)\right] J_{4\nu}(4ax)\, dx = \frac{1}{4}\, J_\nu\left(a^2\right) J_{-\nu}\left(a^2\right)$

$$[a>0, \quad \operatorname{Re}\nu>-1] \hspace{2cm} \text{ET II 375(20)}$$

6.724

1. $\displaystyle\int_0^\infty x^{2\lambda}\, J_{2\nu}\left(\frac{a}{x}\right) \sin(bx)\, dx$

$$= \frac{\sqrt{\pi}\, a^{2\nu}\, \Gamma(\lambda-\nu+1) b^{2\nu-2\lambda-1}}{4^{2\nu-\lambda}\, \Gamma(2\nu+1)\, \Gamma\left(\nu-\lambda+\frac{1}{2}\right)}\, {}_0F_3\left(2\nu+1,\nu-\lambda,\nu-\lambda+\frac{1}{2};\frac{a^2 b^2}{16}\right)$$

$$+ \frac{a^{2\lambda+2}\, \Gamma(\nu-\lambda-1) b}{2^{2\lambda+3}\, \Gamma(\nu+\lambda+2)}\, {}_0F_3\left(\frac{3}{2},\lambda-\nu+2,\lambda+\nu+2;\frac{a^2 b^2}{16}\right)$$

$$\left[-\tfrac{5}{4}<\operatorname{Re}\lambda<\operatorname{Re}\nu, \quad a>0, \quad b>0\right] \hspace{1cm} \text{ET I 109(15)}$$

2. $\displaystyle\int_0^\infty x^{2\lambda}\, J_{2\nu}\left(\frac{a}{x}\right) \cos(bx)\, dx$

$$= 4^{\lambda-2\nu}\sqrt{\pi}\, a^{2\nu} b^{2\nu-2\lambda-1}\frac{\Gamma\left(\lambda-\nu+\frac{1}{2}\right)}{\Gamma(2\nu+1)\, \Gamma(\nu-\lambda)}\, {}_0F_3\left(2\nu+1,\nu-\lambda+\frac{1}{2},\nu-\lambda;\frac{a^2 b^2}{16}\right)$$

$$+ 4^{-\lambda-1} a^{2\lambda+1}\frac{\Gamma\left(\nu-\lambda-\frac{1}{2}\right)}{\Gamma\left(\nu+\lambda+\frac{3}{2}\right)}\, {}_0F_3\left(\frac{1}{2},\lambda-\nu+\frac{3}{2},\nu+\lambda+\frac{3}{2};\frac{a^2 b^2}{16}\right)$$

$$\left[-\tfrac{3}{4}<\operatorname{Re}\lambda<\operatorname{Re}\nu-\tfrac{1}{2}, \quad a>0, \quad b>0\right] \hspace{0.7cm} \text{ET I 53(14)}$$

6.725

1. $\displaystyle\int_0^\infty \frac{\sin(bx)}{\sqrt{x}}\, J_\nu\left(a\sqrt{x}\right) dx = -\sqrt{\frac{\pi}{b}}\, \sin\left(\frac{a^2}{8b}-\frac{\nu\pi}{4}-\frac{\pi}{4}\right) J_{\frac{\nu}{2}}\left(\frac{a^2}{8b}\right)$

$$[\operatorname{Re}\nu>-3, \quad a>0, \quad b>0]$$
$$\text{ET I 110(27)}$$

2. $\displaystyle\int_0^\infty \frac{\cos(bx)}{\sqrt{x}}\, J_\nu\left(a\sqrt{x}\right) dx = \sqrt{\frac{\pi}{b}}\, \cos\left(\frac{a^2}{8b}-\frac{\nu\pi}{4}-\frac{\pi}{4}\right) J_{\frac{1}{2}\nu}\left(\frac{a^2}{8b}\right)$

$$[\operatorname{Re}\nu>-1, \quad a>0, \quad b>0]$$
$$\text{ET I 54(25)}$$

3. $\displaystyle\int_0^\infty x^{\frac{1}{2}\nu} J_\nu\left(a\sqrt{x}\right) \sin(bx)\, dx = 2^{-\nu} a^\nu b^{-\nu-1} \cos\left(\frac{a^2}{4b} - \frac{\nu\pi}{2}\right)$

$$\left[-2 < \operatorname{Re}\nu < \tfrac{1}{2}, \quad a > 0, \quad b > 0\right]$$

ET I 110(28)

4. $\displaystyle\int_0^\infty x^{\frac{1}{2}\nu} J_\nu\left(a\sqrt{x}\right) \cos(bx)\, dx = 2^{-\nu} b^{-\nu-1} a^\nu \sin\left(\frac{a^2}{4b} - \frac{\nu\pi}{2}\right)$

$$\left[-1 < \operatorname{Re}\nu < \tfrac{1}{2}, \quad a > 0, \quad b > 0\right]$$

ET I 54(26)

6.726

1. $\displaystyle\int_0^\infty x\left(x^2 + b^2\right)^{-\frac{1}{2}\nu} J_\nu\left(a\sqrt{x^2 + b^2}\right) \sin(cx)\, dx$

$$= \sqrt{\frac{\pi}{2}}\, a^{-\nu} b^{-\nu+\frac{3}{2}} c\left(a^2 - c^2\right)^{\frac{1}{2}\nu-\frac{3}{4}} J_{\nu-\frac{3}{2}}\left(b\sqrt{a^2 - c^2}\right) \quad \left[0 < c < a, \quad \operatorname{Re}\nu > \tfrac{1}{2}\right]$$

$$= 0 \qquad\qquad\qquad \left[0 < a < c, \quad \operatorname{Re}\nu > \tfrac{1}{2}\right]$$

ET I 111(37)

2. $\displaystyle\int_0^\infty \left(x^2 + b^2\right)^{-\frac{1}{2}\nu} J_\nu\left(a\sqrt{x^2 + b^2}\right) \cos(cx)\, dx$

$$= \sqrt{\frac{\pi}{2}}\, a^{-\nu} b^{-\nu+\frac{1}{2}} \left(a^2 - c^2\right)^{\frac{1}{2}\nu-\frac{1}{4}} J_{\nu-\frac{1}{2}}\left(b\sqrt{a^2 - c^2}\right) \quad \left[0 < c < a, \quad b > 0, \quad \operatorname{Re}\nu > -\tfrac{1}{2}\right]$$

$$= 0 \qquad\qquad\qquad \left[0 < a < c, \quad b > 0, \quad \operatorname{Re}\nu > -\tfrac{1}{2}\right]$$

ET I 55(37)

3. $\displaystyle\int_0^\infty x\left(x^2 + b^2\right)^{\frac{1}{2}\nu} K_{\pm\nu}\left(a\sqrt{x^2 + b^2}\right) \sin(cx)\, dx$

$$= \sqrt{\frac{\pi}{2}}\, a^\nu b^{\nu+\frac{3}{2}} c\left(a^2 + c^2\right)^{-\frac{1}{2}\nu-\frac{3}{4}} K_{-\nu-\frac{3}{2}}\left(b\sqrt{a^2 + c^2}\right)$$

$$\left[\operatorname{Re}a > 0, \quad \operatorname{Re}b > 0, \quad c > 0\right] \quad \text{ET I 113(45)}$$

4. $\displaystyle\int_0^\infty \left(x^2 + b^2\right)^{\mp\frac{1}{2}\nu} K_\nu\left(a\sqrt{x^2 + b^2}\right) \cos(cx)\, dx$

$$= \sqrt{\frac{\pi}{2}}\, a^{\mp\nu} b^{\frac{1}{2}\mp\nu} \left(a^2 + c^2\right)^{\pm\frac{1}{2}\nu-\frac{1}{4}} K_{\pm\nu-\frac{1}{2}}\left(b\sqrt{a^2 + c^2}\right)$$

$$\left[\operatorname{Re}a > 0, \quad \operatorname{Re}b > 0, \quad c > 0\right] \quad \text{ET I 56(45)}$$

5. $\displaystyle\int_0^\infty \left(x^2 + a^2\right)^{-\frac{1}{2}\nu} Y_\nu\left(b\sqrt{x^2 + a^2}\right) \cos(cx)\, dx$

$$= \sqrt{\frac{a\pi}{2}}\, (ab)^{-\nu} \left(b^2 - c^2\right)^{\frac{1}{2}\nu-\frac{1}{4}} Y_{\nu-\frac{1}{2}}\left(a\sqrt{b^2 - c^2}\right) \quad \left[0 < c < b, \quad a > 0, \quad \operatorname{Re}\nu > -\tfrac{1}{2}\right]$$

$$= -\sqrt{\frac{2a}{\pi}}\, (ab)^{-\nu} \left(c^2 - b^2\right)^{\frac{1}{2}\nu-\frac{1}{4}} K_{\nu-\frac{1}{2}}\left(a\sqrt{c^2 - b^2}\right) \quad \left[0 < b < c, \quad a > 0, \quad \operatorname{Re}\nu > -\tfrac{1}{2}\right]$$

ET I 56(41)

6.727

1.9 $\displaystyle\int_0^a \frac{\cos(cx)}{\sqrt{a^2-x^2}} J_\nu\left(b\sqrt{a^2-x^2}\right) dx = \frac{\pi}{2} J_{\frac{1}{2}\nu}\left[\frac{a}{2}\left(\sqrt{b^2+c^2}-c\right)\right] J_{\frac{1}{2}\nu}\left[\frac{a}{2}\left(\sqrt{b^2+c^2}+c\right)\right]$

$$[\operatorname{Re}\nu > -1, \quad c > 0, \quad a > 0]$$

ET I 113(48)

2. $\displaystyle\int_a^\infty \frac{\sin(cx)}{\sqrt{x^2-a^2}} J_\nu\left(b\sqrt{x^2-a^2}\right) dx = \frac{\pi}{2} J_{\frac{1}{2}\nu}\left[\frac{a}{2}\left(c-\sqrt{c^2+b^2}\right)\right] J_{-\frac{1}{2}\nu}\left[\frac{a}{2}\left(c+\sqrt{c^2+b^2}\right)\right]$

$$[0 < b < c, \quad a > 0, \quad \operatorname{Re}\nu > -1]$$

ET I 113(49)

3. $\displaystyle\int_a^\infty \frac{\cos(cx)}{\sqrt{x^2-a^2}} J_\nu\left(b\sqrt{x^2-a^2}\right) dx = -\frac{\pi}{2} J_{\frac{1}{2}\nu}\left[\frac{a}{2}\left(c-\sqrt{c^2-b^2}\right)\right] Y_{-\frac{1}{2}\nu}\left[\frac{a}{2}\left(c+\sqrt{c^2-b^2}\right)\right]$

$$[0 < b < c, \quad a > 0, \quad \operatorname{Re}\nu > -1]$$

ET I 58(54)

4.8 $\displaystyle\int_0^a \left(a^2-x^2\right)^{\frac{1}{2}\nu}\cos x\, I_\nu\left(\sqrt{a^2-x^2}\right) dx = \frac{\sqrt{\pi}\, a^{2\nu+1}}{2^{\nu+1}\Gamma\left(\nu+\frac{3}{2}\right)}$

$$\left[\operatorname{Re}\nu > -\tfrac{1}{2}\right]$$

WA 409(2)

6.728

1. $\displaystyle\int_0^\infty x\sin\left(ax^2\right) J_\nu(bx)\, dx$

$$= \frac{\sqrt{\pi}\,b}{8a^{3/2}}\left[\cos\left(\frac{b^2}{8a}-\frac{\nu\pi}{4}\right) J_{\frac{1}{2}\nu-\frac{1}{2}}\left(\frac{b^2}{8a}\right) - \sin\left(\frac{b^2}{8a}-\frac{\nu\pi}{4}\right) J_{\frac{1}{2}\nu+\frac{1}{2}}\left(\frac{b^2}{8a}\right)\right]$$

$$[a > 0, \quad b > 0, \quad \operatorname{Re}\nu > -4]\quad\text{ET II 34(14)}$$

2. $\displaystyle\int_0^\infty x\cos\left(ax^2\right) J_\nu(bx)\, dx$

$$= \frac{\sqrt{\pi}\,b}{8a^{3/2}}\left[\cos\left(\frac{b^2}{8a}-\frac{\nu\pi}{4}\right) J_{\frac{1}{2}\nu+\frac{1}{2}}\left(\frac{b^2}{8a}\right) + \sin\left(\frac{b^2}{8a}-\frac{\nu\pi}{4}\right) J_{\frac{1}{2}\nu-\frac{1}{2}}\left(\frac{b^2}{8a}\right)\right]$$

$$[a > 0, \quad b > 0, \quad \operatorname{Re}\nu > -2]\quad\text{ET II 38(39)}$$

3. $\displaystyle\int_0^\infty J_0(\beta x)\sin\left(\alpha x^2\right) x\, dx = \frac{1}{2\alpha}\cos\frac{\beta^2}{4\alpha}$ $\qquad [\alpha > 0, \quad \beta > 0]$ $\qquad$ MO 47

4. $\displaystyle\int_0^\infty J_0(\beta x)\cos\left(\alpha x^2\right) x\, dx = \frac{1}{2\alpha}\sin\frac{\beta^2}{4\alpha}$ $\qquad [\alpha > 0, \quad \beta > 0]$ $\qquad$ MO 47

5. $\displaystyle\int_0^\infty x^{\nu+1}\sin\left(ax^2\right) J_\nu(bx)\, dx = \frac{b^\nu}{2^{\nu+1}a^{\nu+1}}\cos\left(\frac{b^2}{4a}-\frac{\nu\pi}{2}\right)$

$$\left[a > 0, \quad b > 0, \quad -2 < \operatorname{Re}\nu < \tfrac{1}{2}\right]$$

ET II 34(15)

6. $\displaystyle\int_0^\infty x^{\nu+1}\cos\left(ax^2\right) J_\nu(bx)\, dx = \frac{b^\nu}{2^{\nu+1}a^{\nu+1}}\sin\left(\frac{b^2}{4a}-\frac{\nu\pi}{2}\right)$

$$\left[a > 0, \quad b > 0, \quad -1 < \operatorname{Re}\nu < \tfrac{1}{2}\right]$$

ET II 38(40)

6.729

1. $$\int_0^\infty x \sin\left(ax^2\right) J_\nu(bx) J_\nu(cx)\, dx = \frac{1}{2a} \cos\left(\frac{b^2+c^2}{4a} - \frac{\nu\pi}{2}\right) J_\nu\left(\frac{bc}{2a}\right)$$

$$[a > 0, \quad b > 0, \quad c > 0, \quad \operatorname{Re}\nu > -2]$$
ET II 51(26)

2. $$\int_0^\infty x \cos\left(ax^2\right) J_\nu(bx) J_\nu(cx)\, dx = \frac{1}{2a} \sin\left(\frac{b^2+c^2}{4a} - \frac{\nu\pi}{2}\right) J_\nu\left(\frac{bc}{2a}\right)$$

$$[a > 0, \quad b > 0, \quad c > 0, \quad \operatorname{Re}\nu > -1]$$
ET II 51(27)

6.731

1. $$\int_0^\infty x \sin\left(ax^2\right) J_\nu\left(bx^2\right) J_{2\nu}(2cx)\, dx$$

$$= \frac{1}{2\sqrt{b^2-a^2}} \sin\left(\frac{ac^2}{b^2-a^2}\right) J_\nu\left(\frac{bc^2}{b^2-a^2}\right) \quad [0 < a < b, \quad \operatorname{Re}\nu > -1]$$

$$= \frac{1}{2\sqrt{a^2-b^2}} \cos\left(\frac{ac^2}{a^2-b^2}\right) J_\nu\left(\frac{bc^2}{a^2-b^2}\right) \quad [0 < b < a, \quad \operatorname{Re}\nu > -1]$$
$ETII356(41)a$

2.[10] $$\int_0^\infty x \cos\left(ax^2\right) J_\nu\left(bx^2\right) J_{2\nu}(2cx)\, dx$$

$$= \frac{1}{2\sqrt{b^2-a^2}} \cos\left(\frac{ac^2}{b^2-a^2}\right) J_\nu\left(\frac{bc^2}{b^2-a^2}\right) \quad \left[0 < a < b, \quad \operatorname{Re}\nu > -\tfrac{1}{2}\right]$$

$$= \frac{1}{2\sqrt{a^2-b^2}} \sin\left(\frac{ac^2}{a^2-b^2}\right) J_\nu\left(\frac{bc^2}{a^2-b^2}\right) \quad \left[0 < b < a, \quad \operatorname{Re}\nu > -\tfrac{1}{2}\right]$$
ET II 356(42)a

6.732 $$\int_0^\infty x^2 \cos\left(\frac{x^2}{2a}\right) Y_1(x) K_1(x)\, dx = -a^3 K_0(a) \qquad [a > 0]$$ ET II 371(52)

6.733

1. $$\int_0^\infty \sin\left(\frac{a}{2x}\right) [\sin x\, J_0(x) + \cos x\, Y_0(x)] \frac{dx}{x} = \pi J_0\left(\sqrt{a}\right) Y_0\left(\sqrt{a}\right)$$

$$[a > 0]$$
ET II 346(51)

2. $$\int_0^\infty \cos\left(\frac{a}{2x}\right) [\sin x\, Y_0(x) - \cos x\, J_0(x)] \frac{dx}{x} = \pi J_0\left(\sqrt{a}\right) Y_0\left(\sqrt{a}\right)$$

$$[a > 0]$$
ET II 347(52)

3. $$\int_0^\infty x \sin\left(\frac{a}{2x}\right) K_0(x)\, dx = \frac{\pi a}{2} J_1\left(\sqrt{a}\right) K_1\left(\sqrt{a}\right) \qquad [a > 0]$$ ET II 368(34)

4. $$\int_0^\infty x \cos\left(\frac{a}{2x}\right) K_0(x)\, dx = -\frac{\pi a}{2} Y_1\left(\sqrt{a}\right) K_1\left(\sqrt{a}\right) \qquad [a > 0]$$ ET II 369(35)

6.734 $$\int_0^\infty \cos\left(a\sqrt{x}\right) K_\nu(bx) \frac{dx}{\sqrt{x}}$$

$$= \frac{\pi}{2\sqrt{b}} \sec(\nu\pi) \left[D_{\nu-\frac{1}{2}}\left(\frac{a}{\sqrt{2b}}\right) D_{-\nu-\frac{1}{2}}\left(-\frac{a}{\sqrt{2b}}\right) + D_{\nu-\frac{1}{2}}\left(-\frac{a}{\sqrt{2b}}\right) D_{-\nu-\frac{1}{2}}\left(\frac{a}{\sqrt{2b}}\right)\right]$$

$$\left[\operatorname{Re} b > 0, \quad |\operatorname{Re}\nu| < \tfrac{1}{2}\right]$$
ET II 132(27)

6.735

1. $$\int_0^\infty x^{1/4} \sin\left(2a\sqrt{x}\right) J_{-\frac{1}{4}}(x)\, dx = \sqrt{\pi} a^{3/2} J_{\frac{3}{4}}\left(a^2\right) \qquad [a>0] \qquad \text{ET II 341(10)}$$

2. $$\int_0^\infty x^{1/4} \cos\left(2a\sqrt{x}\right) J_{\frac{1}{4}}(x)\, dx = \sqrt{\pi} a^{3/2} J_{-\frac{3}{4}}\left(a^2\right) \qquad [a>0] \qquad \text{ET II 341(12)}$$

3. $$\int_0^\infty x^{1/4} \sin\left(2a\sqrt{x}\right) J_{\frac{3}{4}}(x)\, dx = \sqrt{\pi} a^{3/2} J_{-\frac{1}{4}}\left(a^2\right) \qquad [a>0] \qquad \text{ET II 341(11)}$$

4. $$\int_0^\infty x^{1/4} \cos\left(2a\sqrt{x}\right) J_{-\frac{3}{4}}(x)\, dx = \sqrt{\pi} a^{3/2} J_{\frac{1}{4}}\left(a^2\right) \qquad [a>0] \qquad \text{ET II 341(13)}$$

6.736

1. $$\int_0^\infty x^{-1/2} \sin\cos\left(4a\sqrt{x}\right) J_0(x)\, dx = -2^{-3/2}\sqrt{\pi}\left[\cos\left(a^2-\frac{\pi}{4}\right) J_0\left(a^2\right) - \sin\left(a^2-\frac{\pi}{4}\right) Y_0\left(a^2\right)\right]$$
$$[a>0] \qquad \text{ET II 341(18)}$$

2. $$\int_0^\infty x^{-1/2} \cos x \cos\left(4a\sqrt{x}\right) J_0(x)\, dx = -2^{-3/2}\sqrt{\pi}\left[\sin\left(a^2-\frac{\pi}{4}\right) J_0\left(a^2\right) + \cos\left(a^2-\frac{\pi}{4}\right) Y_0\left(a^2\right)\right]$$
$$[a>0] \qquad \text{ET II 342(22)}$$

3. $$\int_0^\infty x^{-1/2} \sin x \sin\left(4a\sqrt{x}\right) J_0(x)\, dx = \sqrt{\frac{\pi}{2}}\cos\left(a^2+\frac{\pi}{4}\right) J_0\left(a^2\right)$$
$$[a>0] \qquad \text{ET II 341(16)}$$

4. $$\int_0^\infty x^{-1/2} \cos x \sin\left(4a\sqrt{x}\right) J_0(x)\, dx = \sqrt{\frac{\pi}{2}}\cos\left(a^2-\frac{\pi}{4}\right) J_0\left(a^2\right)$$
$$[a>0] \qquad \text{ET II 342(20)}$$

5. $$\int_0^\infty x^{-1/2} \sin x \cos\left(4a\sqrt{x}\right) Y_0(x)\, dx = 2^{-3/2}\sqrt{\pi}\left[3\sin\left(a^2-\frac{\pi}{4}\right) J_0\left(a^2\right) - \cos\left(a^2-\frac{\pi}{4}\right) Y_0\left(a^2\right)\right]$$
$$[a>0] \qquad \text{ET II 347(55)}$$

6. $$\int_0^\infty x^{-1/2} \cos x \cos\left(4a\sqrt{x}\right) Y_0(x)\, dx$$
$$= -2^{-3/2}\sqrt{\pi}\left[3\cos\left(a^2-\frac{\pi}{4}\right) J_0\left(a^2\right) + \sin\left(a^2-\frac{\pi}{4}\right) Y_0\left(a^2\right)\right]$$
$$[a>0] \qquad \text{ET II 347(56)}$$

6.737

1. $$\int_0^\infty \frac{\sin\left(a\sqrt{x^2+b^2}\right)}{\sqrt{x^2+b^2}} J_\nu(cx)\, dx = \frac{\pi}{2} J_{\frac{1}{2}\nu}\left[\frac{b}{2}\left(a-\sqrt{a^2-c^2}\right)\right] J_{-\frac{1}{2}\nu}\left[\frac{b}{2}\left(a+\sqrt{a^2-c^2}\right)\right]$$
$$[a>0, \quad \operatorname{Re} b>0, \quad c>0, \quad a>c, \quad \operatorname{Re}\nu>-1] \quad \text{ET II 35(19)}$$

2. $$\int_0^\infty \frac{\cos\left(a\sqrt{x^2+b^2}\right)}{\sqrt{x^2+b^2}} J_\nu(cx)\, dx = -\frac{\pi}{2} J_{\frac{1}{2}\nu}\left[\frac{b}{2}\left(a-\sqrt{a^2-c^2}\right)\right] Y_{-\frac{1}{2}\nu}\left[\frac{b}{2}\left(a+\sqrt{a^2-c^2}\right)\right]$$
$$[a>0, \quad \operatorname{Re} b>0, \quad c>0, \quad a>c, \quad \operatorname{Re}\nu>-1] \quad \text{ET II 39(44)}$$

3. $$\int_0^a \frac{\cos\left(b\sqrt{a^2-x^2}\right)}{\sqrt{a^2-x^2}} J_\nu(cx)\,dx = \frac{\pi}{2} J_{\frac{1}{2}\nu}\left[\frac{a}{2}\left(\sqrt{b^2+c^2}-b\right)\right] J_{\frac{1}{2}\nu}\left[\frac{a}{2}\left(\sqrt{b^2+c^2}+b\right)\right]$$

$$[c>0,\quad \mathrm{Re}\,\nu > -1] \qquad \text{ET II 39(47)}$$

4. $$\int_0^a x^{\nu+1}\frac{\cos\left(\sqrt{a^2-x^2}\right)}{\sqrt{a^2-x^2}} I_\nu(x)\,dx = \frac{\sqrt{\pi}\,a^{2\nu+1}}{2^{\nu+1}\,\Gamma\left(\nu+\frac{3}{2}\right)} \qquad [\mathrm{Re}\,\nu > -1] \qquad \text{ET II 365(9)}$$

5. $$\int_0^\infty x^{\nu+1}\frac{\sin\left(a\sqrt{b^2+x^2}\right)}{\sqrt{b^2+x^2}} J_\nu(cx)\,dx$$

$$= \sqrt{\frac{\pi}{2}}\,b^{\frac{1}{2}+\nu}c^\nu\left(a^2-c^2\right)^{-\frac{1}{4}-\frac{1}{2}\nu} J_{-\nu-\frac{1}{2}}\left(b\sqrt{a^2-c^2}\right) \qquad \left[0<c<a,\quad \mathrm{Re}\,b>0,\quad -1<\mathrm{Re}\,\nu<\tfrac{1}{2}\right]$$

$$= 0 \qquad \left[0<a<c,\quad \mathrm{Re}\,b>0,\quad -1<\mathrm{Re}\,\nu<\tfrac{1}{2}\right]$$

$$\text{ET II 35(20)}$$

6. $$\int_0^\infty x^{\nu+1}\frac{\cos\left(a\sqrt{x^2+b^2}\right)}{\sqrt{x^2+b^2}} J_\nu(cx)\,dx = -\sqrt{\frac{\pi}{2}}\,b^{\frac{1}{2}+\nu}c^\nu\left(a^2-c^2\right)^{-\frac{1}{4}-\frac{1}{2}\nu} Y_{-\nu-\frac{1}{2}}\left(b\sqrt{a^2-c^2}\right)$$

$$\left[0<c<a,\quad \mathrm{Re}\,b>0,\quad -1<\mathrm{Re}\,\nu<\frac{1}{2}\right]$$

$$= \sqrt{\frac{2}{\pi}}\,b^{\frac{1}{2}+\nu}c^\nu\left(c^2-a^2\right)^{-\frac{1}{4}-\frac{1}{2}\nu} K_{\nu+\frac{1}{2}}\left(b\sqrt{c^2-a^2}\right)$$

$$\left[0<a<c,\quad \mathrm{Re}\,b>0,\quad -1<\mathrm{Re}\,\nu<\frac{1}{2}\right]$$

$$\text{ET II 39(45)}$$

6.738

1. $$\int_0^a x^{\nu+1}\sin\left(b\sqrt{a^2-x^2}\right) J_\nu(x)\,dx = \sqrt{\frac{\pi}{2}}\,a^{\nu+\frac{3}{2}}b\left(1+b^2\right)^{-\frac{1}{2}\nu-\frac{3}{4}} J_{\nu+\frac{3}{2}}\left(a\sqrt{1+b^2}\right)$$

$$[\mathrm{Re}\,\nu > -1] \qquad \text{ET II 335(19)}$$

2. $$\int_0^\infty x^{\nu+1}\cos\left(a\sqrt{x^2+b^2}\right) J_\nu(cx)\,dx$$

$$= \sqrt{\frac{\pi}{2}}\,ab^{\nu+\frac{3}{2}}c^\nu\left(a^2-c^2\right)^{-\frac{1}{2}\nu-\frac{3}{4}}\left[\cos(\pi\nu)\,J_{\nu+\frac{3}{2}}\left(b\sqrt{a^2-c^2}\right)-\sin(\pi\nu)\,Y_{\nu+\frac{3}{2}}\left(b\sqrt{a^2-c^2}\right)\right]$$

$$\left[0<c<a,\quad \mathrm{Re}\,b>0,\quad -1<\mathrm{Re}\,\nu<-\tfrac{1}{2}\right]$$

$$= 0$$

$$\left[0<a<c,\quad \mathrm{Re}\,b>0,\quad -1<\mathrm{Re}\,\nu<-\tfrac{1}{2}\right]$$

$$\text{ET II 39(43)}$$

6.739 $$\int_0^t x^{-1/2}\frac{\cos\left(b\sqrt{t-x}\right)}{\sqrt{t-x}} J_{2\nu}\left(a\sqrt{x}\right)\,dx = \pi J_\nu\left[\frac{\sqrt{t}}{2}\left(\sqrt{a^2+b^2}+b\right)\right] J_\nu\left[\frac{\sqrt{t}}{2}\left(\sqrt{a^2+b^2}-b\right)\right]$$

$$[\mathrm{Re}\,\nu > -\tfrac{1}{2}] \qquad \text{EH II 47(7)}$$

6.741

1. $$\int_0^1 \frac{\cos\left(\mu\arccos x\right)}{\sqrt{1-x^2}} J_\nu(ax)\,dx = \frac{\pi}{2} J_{\frac{1}{2}(\mu+\nu)}\left(\frac{a}{2}\right) J_{\frac{1}{2}(\nu-\mu)}\left(\frac{a}{2}\right)$$

$$[\mathrm{Re}(\mu+\nu) > -1,\quad a>0] \qquad \text{ET II 41(54)}$$

2. $\displaystyle\int_0^1 \frac{\cos\left[(\nu+1)\arccos x\right]}{\sqrt{1-x^2}} J_\nu(ax)\,dx = \sqrt{\frac{\pi}{a}}\cos\left(\frac{a}{2}\right) J_{\nu+\frac{1}{2}}\left(\frac{a}{2}\right)$

$$[\operatorname{Re}\nu > -1, \quad a > 0] \qquad \text{ET II 40(53)}$$

3. $\displaystyle\int_0^1 \frac{\cos\left[(\nu-1)\arccos x\right]}{\sqrt{1-x^2}} J_\nu(ax)\,dx = \sqrt{\frac{\pi}{a}}\sin\left(\frac{a}{2}\right) J_{\nu-\frac{1}{2}}\left(\frac{a}{2}\right)$

$$[\operatorname{Re}\nu > 0, \quad a > 0] \qquad \text{ET II 40(52)a}$$

6.75 Combinations of Bessel, trigonometric, and exponential functions and powers

6.751 **Notation:** $\ell_1 = \dfrac{1}{2}\left[\sqrt{(b+c)^2 + a^2} - \sqrt{(b-c)^2 + a^2}\right]$, $\ell_2 = \dfrac{1}{2}\left[\sqrt{(b+c)^2 + a^2} + \sqrt{(b-c)^2 + a^2}\right]$

1. $\displaystyle\int_0^\infty e^{-\frac{1}{2}ax}\sin(bx) I_0\left(\frac{1}{2}ax\right)dx = \frac{1}{\sqrt{2b}}\frac{1}{\sqrt{b^2+a^2}}\sqrt{b+\sqrt{b^2+a^2}}$

$$[\operatorname{Re}a > 0, \quad b > 0] \qquad \text{ET I 105(44)}$$

2. $\displaystyle\int_0^\infty e^{-\frac{1}{2}ax}\cos(bx) I_0\left(\frac{1}{2}ax\right)dx = \frac{a}{\sqrt{2b}}\frac{1}{\sqrt{a^2+b^2}\sqrt{b+\sqrt{a^2+b^2}}}$

$$[\operatorname{Re}a > 0, \quad b > 0] \qquad \text{ET I 48(38)}$$

3.10 $\displaystyle\int_0^\infty e^{-bx}\cos(ax) J_0(cx)\,dx = \frac{\left[\sqrt{(b^2+c^2-a^2)^2 + 4a^2b^2} + b^2 + c^2 - a^2\right]^{1/2}}{\sqrt{2}\sqrt{(b^2+c^2-a^2)^2 + 4a^2b^2}}$

$$[c > 0] \qquad \text{ET II 11(46)}$$

alternatively, with a and b interchanged,

$$\int_0^\infty e^{-ax}\cos(bx) J_0(cx)\,dx = \frac{\sqrt{\ell_2^2 - b^2}}{\ell_2^2 - \ell_1^2} \qquad [c > 0]$$

6.752

1.10 $\displaystyle\int_0^\infty e^{-ax} J_0(bx)\sin(cx)\frac{dx}{x} = \arcsin\left(\frac{2c}{\sqrt{a^2+(c+b)^2} + \sqrt{a^2+(c-b)^2}}\right) = \arcsin\left(\frac{c}{\ell_2}\right)$

$$[\operatorname{Re}a > |\operatorname{Im}b|, \quad c > 0] \qquad \text{ET I 101(17)}$$

2.10 $\displaystyle\int_0^\infty e^{-ax} J_1(cx)\sin(bx)\frac{dx}{x} = \frac{b}{c}(1-r) = \frac{b - \sqrt{b^2 - \ell_1^2}}{c},$

$$\left[b^2 = \frac{c^2}{1-r^2} - \frac{a^2}{r^2}, \quad c > 0\right]$$

$$\text{ET II 19(15)}$$

Notation: For integrals 6.752 3–6.752 5 we define the auxiliary functions

$$\ell_1(a) \equiv \ell_1(a, \rho, z) = \frac{1}{2}\left[\sqrt{(a+\rho)^2 + z^2} - \sqrt{(a-\rho)^2 + z^2}\right]$$

$$\ell_2(a) \equiv \ell_1(a, \rho, z) = \frac{1}{2}\left[\sqrt{(a+\rho)^2 + z^2} + \sqrt{(a-\rho)^2 + z^2}\right]$$

when $a \geq 0$, $\rho \geq 0$, and $z \geq 0$.

$3.^{10}$ $\sqrt{\dfrac{\pi}{2}} \displaystyle\int_0^\infty e^{-zx} J_{\nu+1/2}(ax) J_{\nu+1}(\rho x) \sqrt{x} \, dx$

$$= a^{-\nu-3/2} \rho^{-\nu-1} \frac{\ell_1^{2\nu+2}}{\sqrt{\rho^2 - \ell_1^2}} \frac{a\left(\rho^2 - \ell_1^2\right)}{\ell_1\left(\ell_2^2 - \ell_1^2\right)}$$

$$= a^{\nu+1/2} \frac{\rho^{\nu+1}}{\ell_2^{2\nu+2}} \frac{\sqrt{\ell_2^2 - a^2}}{\ell_2^2 - \ell_1^2} \qquad [\operatorname{Re} z > |\operatorname{Im} a| + |\operatorname{Im} \rho|]$$

$4.^{10}$ $\sqrt{\dfrac{\pi}{2}} \displaystyle\int_0^\infty e^{-zx} J_{\nu+1/2}(ax) J_\nu(\rho x) \dfrac{dx}{\sqrt{x}}$

$$= a^{\nu+1/2} \rho^\nu \int_0^{1/\ell_2} \frac{1}{\ell_2^{2\nu}} \frac{1}{\sqrt{1 - a^2/\ell_2^2}} \, d\left(\frac{1}{\ell_2}\right)$$

$$= a^{-\nu-1/2} \rho^\nu \int_0^{a/\ell_2} x^{2\nu} \frac{dx}{\sqrt{1 - x^2}} \qquad \left[\nu > -\tfrac{1}{2}, \quad \operatorname{Re} z > |\operatorname{Im} a| + |\operatorname{Im} \rho|\right]$$

$5.^{10}$ $\displaystyle\int_0^\infty e^{-zx} \sin(ax) J_1(\rho x) \dfrac{dx}{x^2} = \dfrac{\sqrt{\ell_2^2 - a^2}\left(a - \sqrt{a^2 - \ell_1^2}\right)^2}{2a\rho} + \dfrac{\rho}{2} \arcsin\left(\dfrac{a}{\ell_2}\right)$

$$[\operatorname{Re} z > |\operatorname{Im} a| + |\operatorname{Im} \rho|]$$

6.753

$1.^8$ $\displaystyle\int_0^\infty \dfrac{\sin(xa \sin\psi)}{x} e^{-xa \cos\varphi \cos\psi} J_\nu(xa \sin\varphi) \, dx = \nu^{-1}\left(\tan\dfrac{\varphi}{2}\right)^\nu \sin(\nu\psi)$

$$\left[\operatorname{Re}\nu > -1, \quad a > 0, \quad 0 < \varphi < \frac{\pi}{2}, \quad 0 < \psi < \frac{\pi}{2}\right] \qquad \text{ET II 33(10)}$$

$2.$ $\displaystyle\int_0^\infty \dfrac{\cos(xa \sin\psi)}{x} e^{-xa \cos\varphi \cos\psi} J_\nu(xa \sin\varphi) \, dx = \nu^{-1}\left(\tan\dfrac{\varphi}{2}\right)^\nu \cos(\nu\psi)$

$$\left[\operatorname{Re}\nu > 0, \quad a > 0, \quad 0 < \varphi, \quad \psi < \frac{\pi}{2}\right]$$
$$\text{ET II 38(35)}$$

$3.^8$ $\displaystyle\int_0^\infty x^{\nu+1} e^{-sx} \sin(bx) J_\nu(ax) \, dx = -\dfrac{2(2a)^\nu}{\sqrt{\pi}} \Gamma(\nu + \tfrac{3}{2}) R^{-2\nu-3} \left[b \cos(\nu + \tfrac{3}{2})\varphi + s \sin(\nu + \tfrac{3}{2})\varphi\right]$

$$\left[\operatorname{Re}\nu > -\tfrac{3}{2}, \quad \operatorname{Re} s > |\operatorname{Im} a| + |\operatorname{Im} b|, \right.$$

$$\left. R^4 = \left(s^2 + a^2 - b^2\right)^2 + 4b^2 s^2, \quad \varphi = \arg\left(s^2 + a^2 - b^2 - 2ibs\right)\right]$$

$4.^8$ $\displaystyle\int_0^\infty x^{\nu+1} e^{-sx} \cos(bx) J_\nu(ax) \, dx = \dfrac{2(2a)^\nu}{\sqrt{\pi}} \Gamma(\nu + \tfrac{3}{2}) R^{-2\nu-3} \left[s \cos(\nu + \tfrac{3}{2})\varphi - b \sin(\nu + \tfrac{3}{2})\varphi\right],$

$$\left[\operatorname{Re}\nu > -1, \quad \operatorname{Re} s > |\operatorname{Im} a| + |\operatorname{Im} b|, \right.$$

$$\left. R^4 = \left(s^2 + a^2 - b^2\right)^2 + 4b^2 s^2, \quad \varphi = \arg\left(s^2 + a^2 - b^2 - 2ibs\right)\right]$$

$5.^{10}$ $\displaystyle\int_0^\infty x^\nu e^{-ax\cos\varphi\cos\psi}\sin{(ax\sin\psi)}\,J_\nu\,(ax\sin\varphi)\,dx$

$$= 2^\nu \frac{\Gamma\left(\nu+\frac{1}{2}\right)}{\sqrt{\pi}}a^{-\nu-1}(\sin\varphi)^\nu\left(\cos^2\psi+\sin^2\psi\cos^2\varphi\right)^{-\nu-\frac{1}{2}}\sin\left[\left(\nu+\tfrac{1}{2}\right)\beta\right]$$

$$\tan\frac{\beta}{2}=\tan\psi\cos\varphi \qquad \left[a>0,\quad 0<\varphi<\frac{\pi}{2},\quad 0<\psi<\frac{\pi}{2},\quad \operatorname{Re}\nu>-1\right]\quad\text{ET II 34(12)}$$

6. $\displaystyle\int_0^\infty x^\nu e^{-ax\cos\varphi\cos\psi}\cos{(ax\sin\psi)}\,J_\nu\,(ax\sin\varphi)\,dx$

$$= 2^\nu \frac{\Gamma\left(\nu+\frac{1}{2}\right)}{\sqrt{\pi}}a^{-\nu-1}(\sin\varphi)^\nu\left(\cos^2\psi+\sin^2\psi\cos^2\varphi\right)^{-\nu-\frac{1}{2}}\cos\left[\left(\nu+\tfrac{1}{2}\right)\beta\right]$$

$$\tan\frac{\beta}{2}=\tan\psi\cos\varphi \qquad \left[a>0,\quad 0<\varphi,\quad \psi<\frac{\pi}{2},\quad \operatorname{Re}\nu>-\frac{1}{2}\right]\quad\text{ET II 38(37)}$$

6.754

1. $\displaystyle\int_0^\infty e^{-x^2}\sin(bx)\,I_0\,(x^2)\,dx = \frac{\sqrt{\pi}}{2^{3/2}}e^{-\frac{b^2}{8}}I_0\left(\frac{b^2}{8}\right)\qquad [b>0]\qquad\qquad\text{ET I 108(9)}$

2. $\displaystyle\int_0^\infty e^{-ax}\cos{(x^2)}\,J_0\,(x^2)\,dx = \frac{1}{4}\sqrt{\frac{\pi}{2}}\left[J_0\left(\frac{a^2}{16}\right)\cos\left(\frac{a^2}{16}-\frac{\pi}{4}\right)-Y_0\left(\frac{a^2}{16}\right)\cos\left(\frac{a^2}{16}+\frac{\pi}{4}\right)\right]$

$$[a>0]\qquad\qquad\text{MI 42}$$

3. $\displaystyle\int_0^\infty e^{-ax}\sin{(x^2)}\,J_0\,(x^2)\,dx = \frac{1}{4}\sqrt{\frac{\pi}{2}}\left[J_0\left(\frac{a^2}{16}\right)\sin\left(\frac{a^2}{16}-\frac{\pi}{4}\right)-Y_0\left(\frac{a^2}{16}\right)\sin\left(\frac{a^2}{16}+\frac{\pi}{4}\right)\right]$

$$[a>0]\qquad\qquad\text{MI 42}$$

6.755

1. $\displaystyle\int_0^\infty x^{-\nu}e^{-x}\sin{\left(4a\sqrt{x}\right)}\,I_\nu(x)\,dx = \left(2^{3/2}a\right)^{\nu-1}e^{-a^2}\,W_{\frac{1}{2}-\frac{3}{2}\nu,\,\frac{1}{2}-\frac{1}{2}\nu}\left(2a^2\right)$

$$[a>0,\quad \operatorname{Re}\nu>0]\qquad\qquad\text{ET II 366(14)}$$

2. $\displaystyle\int_0^\infty x^{-\nu-\frac{1}{2}}e^{-x}\cos{\left(4a\sqrt{x}\right)}\,I_\nu(x)\,dx = 2^{\frac{3}{2}\nu-1}a^{\nu-1}e^{-a^2}\,W_{-\frac{3}{2}\nu,\,\frac{1}{2}\nu}\left(2a^2\right)$

$$[a>0,\quad \operatorname{Re}\nu>-\tfrac{1}{2}]\qquad\qquad\text{ET II 366(16)}$$

3. $\displaystyle\int_0^\infty x^{-\nu}e^{x}\sin{\left(4a\sqrt{x}\right)}\,K_\nu(x)\,dx = \left(2^{3/2}a\right)^{\nu-1}\frac{\Gamma\left(\frac{3}{2}-2\nu\right)}{\Gamma\left(\frac{1}{2}+\nu\right)}\pi\, e^{a^2}\,W_{\frac{3}{2}\nu-\frac{1}{2},\,\frac{1}{2}-\frac{1}{2}\nu}\left(2a^2\right)$

$$[a>0,\quad 0<\operatorname{Re}\nu<\tfrac{3}{4}]\qquad\qquad\text{ET II 369(38)}$$

4. $\displaystyle\int_0^\infty x^{-\nu-\frac{1}{2}}e^{x}\cos{\left(4a\sqrt{x}\right)}\,K_\nu(x)\,dx = 2^{\frac{3}{2}\nu-1}\pi a^{\nu-1}\frac{\Gamma\left(\frac{1}{2}-2\nu\right)}{\Gamma\left(\frac{1}{2}+\nu\right)}e^{a^2}\,W_{\frac{3}{2}\nu,\,-\frac{1}{2}\nu}\left(2a^2\right)$

$$[a>0,\quad -\tfrac{1}{2}<\operatorname{Re}\nu<\tfrac{1}{4}]$$

$$\text{ET II 369(42)}$$

5. $\displaystyle\int_0^\infty x^{\varrho-\frac{3}{2}}e^{-x}\sin{\left(4a\sqrt{x}\right)}\,K_\nu(x)\,dx = \frac{\sqrt{\pi}a\,\Gamma(\varrho+\nu)\,\Gamma(\varrho-\nu)}{2^{\varrho-2}\,\Gamma\left(\varrho+\frac{1}{2}\right)}\,{}_2F_2\left(\varrho+\nu,\varrho-\nu;\frac{3}{2},\varrho+\frac{1}{2};-2a^2\right)$

$$[\operatorname{Re}\varrho>|\operatorname{Re}\nu|]\qquad\qquad\text{ET II 369(39)}$$

6. $\displaystyle\int_0^\infty x^{\varrho-1}e^{-x}\cos\left(4a\sqrt{x}\right)K_\nu(x)\,dx = \frac{\sqrt{\pi}\,\Gamma(\varrho+\nu)\,\Gamma(\varrho-\nu)}{2^\varrho\,\Gamma\left(\varrho+\frac{1}{2}\right)}\;{}_2F_2\left(\varrho+\nu,\varrho-\nu;\frac{1}{2},\varrho+\frac{1}{2};-2a^2\right)$

$$[\operatorname{Re}\varrho > |\operatorname{Re}\nu|]\qquad\text{ET II 370(43)}$$

7. $\displaystyle\int_0^\infty x^{-1/2}e^{-x}\cos\left(4a\sqrt{x}\right)I_0(x)\,dx = \frac{1}{\sqrt{2\pi}}e^{-a^2}K_0\left(a^2\right)$

$$[a>0]\qquad\text{ET II 366(15)}$$

8. $\displaystyle\int_0^\infty x^{-1/2}e^{x}\cos\left(4a\sqrt{x}\right)K_0(x)\,dx = \sqrt{\frac{\pi}{2}}e^{a^2}K_0\left(a^2\right)\qquad[a>0]\qquad\text{ET II 369(40)}$

9. $\displaystyle\int_0^\infty x^{-1/2}e^{-x}\cos\left(4a\sqrt{x}\right)K_0(x)\,dx = \frac{1}{\sqrt{2}}\pi^{3/2}e^{-a^2}I_0\left(a^2\right)$

$$\text{ET II 369(41)}$$

6.756

1. $\displaystyle\int_0^\infty x^{-\frac{1}{2}}e^{-a\sqrt{x}}\sin\left(a\sqrt{x}\right)J_\nu(bx)\,dx$

$$= \frac{i}{\sqrt{2\pi b}}\Gamma\left(\nu+\frac{1}{2}\right)D_{-\nu-\frac{1}{2}}\left(\frac{a}{\sqrt{b}}\right)\left[D_{-\nu-\frac{1}{2}}\left(\frac{ia}{\sqrt{b}}\right)-D_{-\nu-\frac{1}{2}}\left(-\frac{ia}{\sqrt{b}}\right)\right]$$

$$[a>0,\quad b>0,\quad \operatorname{Re}\nu>-1]\quad\text{ET II 34(17)}$$

2. $\displaystyle\int_0^\infty x^{-\frac{1}{2}}e^{-a\sqrt{x}}\cos\left(a\sqrt{x}\right)J_\nu(bx)\,dx$

$$= \frac{1}{\sqrt{2\pi b}}\Gamma\left(\nu+\frac{1}{2}\right)D_{-\nu-\frac{1}{2}}\left(\frac{a}{\sqrt{b}}\right)\left[D_{-\nu-\frac{1}{2}}\left(\frac{ia}{\sqrt{b}}\right)+D_{-\nu-\frac{1}{2}}\left(-\frac{ia}{\sqrt{b}}\right)\right]$$

$$[a>0,\quad b>0,\quad \operatorname{Re}\nu>-\tfrac{1}{2}]\quad\text{ET II 39(42)}$$

3. $\displaystyle\int_0^\infty x^{-1/2}e^{-a\sqrt{x}}\sin\left(a\sqrt{x}\right)J_0(bx)\,dx = \frac{1}{2b}a\,I_{\frac{1}{4}}\left(\frac{a^2}{4b}\right)K_{\frac{1}{4}}\left(\frac{a^2}{4b}\right)$

$$\left[|\arg a|<\frac{\pi}{4},\quad b>0\right]\qquad\text{ET II 11(40)}$$

4. $\displaystyle\int_0^\infty x^{-1/2}e^{-a\sqrt{x}}\cos\left(a\sqrt{x}\right)J_0(bx)\,dx = \frac{a}{2b}\,I_{-\frac{1}{4}}\left(\frac{a^2}{4b}\right)K_{\frac{1}{4}}\left(\frac{a^2}{4b}\right)$

$$\left[|\arg a|<\frac{\pi}{4},\quad b>0\right]\qquad\text{ET II 12(49)}$$

6.757

1. $\displaystyle\int_0^\infty e^{-bx}\sin\left[a\left(1-e^{-x}\right)\right]J_\nu\left(ae^{-x}\right)\,dx$

$$= 2\sum_{n=0}^\infty \frac{(-1)^n\,\Gamma(\nu-b+2n+1)\,\Gamma(\nu+b)}{\Gamma(\nu-b+1)\,\Gamma(\nu+b+2n+2)}(\nu+2n-1)\,J_{\nu+2n+1}(a)$$

$$[\operatorname{Re}b>-\operatorname{Re}\nu]\qquad\text{ET I 193(26)}$$

2. $\displaystyle\int_0^\infty e^{-bx}\cos\left[a\left(1-e^{-x}\right)\right]J_\nu\left(ae^{-x}\right)\,dx$

$$= \frac{J_\nu(a)}{\nu+b}+\sum_{n=0}^\infty 2(-1)^n\frac{\Gamma(\nu-b+2n)\,\Gamma(\nu+b)}{\Gamma(\nu-b+1)\,\Gamma(\nu+b+2n+1)}(\nu+2n)\,J_{\nu+2n}(a)$$

$$[\operatorname{Re}b>-\operatorname{Re}\nu]\qquad\text{ET I 193(27)}$$

6.758 $\displaystyle\int_{-\frac{\pi}{2}}^{\frac{\pi}{2}} e^{i(\mu-\nu)\theta}(\cos\theta)^{\nu+\mu}(\lambda z)^{-\nu-\mu}\, J_{\nu+\mu}(\lambda z)\, d\theta$

$$= \pi(2az)^{-\mu}(2bz)^{-\nu} J_\mu(az)\, J_\nu(bz);\ \lambda = \sqrt{2\cos\theta\,(a^2 e^{i\theta}+b^2 e^{-i\theta})}$$

$$\lambda = \sqrt{2\cos\theta\,(a^2 e^{i\theta}+b^2 e^{-i\theta})} \qquad [\mathrm{Re}(\nu+\mu)>-1] \quad \text{EH II 48(12)}$$

6.76 Combinations of Bessel, trigonometric, and hyperbolic functions

6.761 $\displaystyle\int_0^\infty \cosh x \cos(2a\sinh x)\, J_\nu(be^x)\, J_\nu(be^{-x})\, dx = \dfrac{J_{2\nu}\left(2\sqrt{b^2-a^2}\right)}{2\sqrt{b^2-a^2}}$ $[0<a<b,\quad \mathrm{Re}\,\nu>-1]$

$$= 0 \qquad\qquad [0<b<a,\quad \mathrm{Re}\,\nu>-1]$$

ET II 359(10)

6.762 $\displaystyle\int_0^\infty \cosh x \sin(2a\sinh x)\left[J_\nu(be^x)\, Y_\nu(be^{-x}) - Y_\nu(be^x)\, J_\nu(be^{-x})\right]\, dx$

$$= 0 \qquad\qquad\qquad\qquad\qquad\qquad\qquad\qquad\quad [0<a<b,\quad |\mathrm{Re}\,\nu|<\tfrac{1}{2}]$$

$$= -\frac{2}{\pi}\cos(\nu\pi)\left(a^2-b^2\right)^{-1/2} K_{2\nu}\left[2\left(a^2-b^2\right)^{1/2}\right] \qquad [0<b<a,\quad |\mathrm{Re}\,\nu|<\tfrac{1}{2}]$$

ET II 360(12)

6.763 $\displaystyle\int_0^\infty \cosh x \cos(2a\sinh x)\, Y_\nu(be^x)\, Y_\nu(be^{-x})\, dx$

$$= -\frac{1}{2}\left(b^2-a^2\right)^{-1/2} J_{2\nu}\left[2\left(b^2-a^2\right)^{1/2}\right] \qquad [0<a<b,\quad |\mathrm{Re}\,\nu|<1]$$

$$= \frac{2}{\pi}\cos(\nu\pi)\left(a^2-b^2\right)^{-1/2} K_{2\nu}\left[2\left(a^2-b^2\right)^{1/2}\right] \qquad [0<b<a,\quad |\mathrm{Re}\,\nu|<1]$$

ET II 360(11)

6.77 Combinations of Bessel functions and the logarithm, or arctangent

6.771 $\displaystyle\int_0^\infty x^{\mu+\frac{1}{2}}\ln x\, J_\nu(ax)\, dx = \dfrac{2^{\mu-\frac{1}{2}}\,\Gamma\left(\frac{\mu+\nu}{2}+\frac{3}{4}\right)}{\Gamma\left(\frac{\nu-\mu}{2}+\frac{1}{4}\right) a^{\mu+\frac{3}{2}}}\left[\psi\left(\frac{\mu+\nu}{2}+\frac{3}{4}\right) + \psi\left(\frac{\nu-\mu}{2}+\frac{1}{4}\right) - \ln\frac{a^2}{4}\right]$

$$\left[a>0,\quad -\mathrm{Re}\,\nu-\tfrac{3}{2}<\mathrm{Re}\,\mu<0\right]$$

ET II 32(25)

6.772

1. $\displaystyle\int_0^\infty \ln x\, J_0(ax)\, dx = -\frac{1}{a}\left[\ln(2a)+C\right]$ WA 430(4)a, ET II 10(27)

2. $\displaystyle\int_0^\infty \ln x\, J_1(ax)\, dx = -\frac{1}{a}\left[\ln\left(\frac{a}{2}\right)+C\right]$ ET II 19(11)

3. $\displaystyle\int_0^\infty \ln\left(a^2+x^2\right) J_1(bx)\, dx = \frac{2}{b}\left[K_0(ab)+\ln a\right]$ ET II 19(12)

4. $\displaystyle\int_0^\infty J_1(tx)\ln\sqrt{1+t^4}\, dt = \frac{2}{x}\ker x$ MO 46

6.773 $\displaystyle\int_0^\infty \frac{\ln\left(x+\sqrt{x^2+a^2}\right)}{\sqrt{x^2+a^2}}\, J_0(bx)\, dx = \left[\frac{1}{2}K_0^2\left(\frac{ab}{2}\right) + \ln a\, I_0\left(\frac{ab}{2}\right)K_0\left(\frac{ab}{2}\right)\right]$

$[a>0, \quad b>0]$ ET II 10(28)

6.774 $\displaystyle\int_0^\infty \ln\frac{\sqrt{x^2+a^2}+x}{\sqrt{x^2+a^2}-x}\, J_0(bx)\frac{dx}{\sqrt{x^2+a^2}} = K_0^2\left(\frac{ab}{2}\right)$ $[\operatorname{Re} a>0, \quad b>0]$ ET II 10(29)

6.775 $\displaystyle\int_0^\infty x\left[\ln\left(1+\sqrt{a^2+x^2}\right)-\ln x\right]J_0(bx)\, dx = \frac{1}{b^2}\left(1-e^{-ab}\right)$

$[\operatorname{Re} a>0, \quad b>0]$ ET II 12(55)

6.776 $\displaystyle\int_0^\infty x\ln\left(1+\frac{a^2}{x^2}\right)J_0(bx)\, dx = \frac{2}{b}\left[\frac{1}{b}-a\, K_1(ab)\right]$ $[\operatorname{Re} a>0, \quad b>0]$ ET II 10(30)

6.777 $\displaystyle\int_0^\infty J_1(tx)\arctan t^2\, dt = -\frac{2}{x}\operatorname{kei} x$ MO 46

6.78 Combinations of Bessel and other special functions

6.781 $\displaystyle\int_0^\infty \operatorname{si}(ax)\, J_0(bx)\, dx = -\frac{1}{b}\arcsin\left(\frac{b}{a}\right)$ $[0<b<a]$

$\displaystyle = 0$ $[0<a<b]$

ET II 13(6)

6.782

1. $\displaystyle\int_0^\infty \operatorname{Ei}(-x)\, J_0\left(2\sqrt{zx}\right)dx = \frac{e^{-z}-1}{z}$ NT 60(4)

2. $\displaystyle\int_0^\infty \operatorname{si}(x)\, J_0\left(2\sqrt{zx}\right)dx = -\frac{\sin z}{z}$ NT 60(6)

3. $\displaystyle\int_0^\infty \operatorname{ci}(x)\, J_0\left(2\sqrt{zx}\right)dx = \frac{\cos z-1}{z}$ NT 60(5)

4. $\displaystyle\int_0^\infty \operatorname{Ei}(-x)\, J_1\left(2\sqrt{zx}\right)\frac{dx}{\sqrt{x}} = \frac{\operatorname{Ei}(-z)-\boldsymbol{C}-\ln z}{\sqrt{z}}$ NT 60(7)

5. $\displaystyle\int_0^\infty \operatorname{si}(x)\, J_1\left(2\sqrt{zx}\right)\frac{dx}{\sqrt{x}} = -\frac{\frac{\pi}{2}-\operatorname{si}(z)}{\sqrt{z}}$ NT 60(9)

6. $\displaystyle\int_0^\infty \operatorname{ci}(z)\, J_1\left(2\sqrt{zx}\right)\frac{dx}{\sqrt{x}} = \frac{\operatorname{ci}(z)-\boldsymbol{C}-\ln z}{\sqrt{z}}$ NT 60(8)

7. $\displaystyle\int_0^\infty \operatorname{Ei}(-x)\, Y_0\left(2\sqrt{zx}\right)dx = \frac{\boldsymbol{C}+\ln z-e^2\operatorname{Ei}(-z)}{\pi z}$ NT 63(5)

6.783

1. $\displaystyle\int_0^\infty x\operatorname{si}\left(a^2x^2\right)J_0(bx)\, dx = -\frac{2}{b^2}\sin\left(\frac{b^2}{4a^2}\right)$ $[a>0]$ ET II 13(7)a

2. $\displaystyle\int_0^\infty x\operatorname{ci}\left(a^2x^2\right)J_0(bx)\, dx = \frac{2}{b^2}\left[1-\cos\left(\frac{b^2}{4a^2}\right)\right]$ $[a>0]$ ET II 13(8)a

3. $\displaystyle\int_0^\infty \operatorname{ci}\left(a^2x^2\right)J_0(bx)\, dx = \frac{1}{b}\left[\operatorname{ci}\left(\frac{b^2}{4a^2}\right)+\ln\left(\frac{b^2}{4a^2}\right)+2\boldsymbol{C}\right]$

$[a>0]$ ET II 13(8)a

4. $\displaystyle\int_0^\infty \operatorname{si}\left(a^2 x^2\right) J_1(bx)\, dx = \frac{1}{b}\left[-\operatorname{si}\left(\frac{b^2}{4a^2}\right) - \frac{\pi}{2}\right]$ $[a > 0]$ ET II 20(25)a

6.784

1. $\displaystyle\int_0^\infty x^{\nu+1}\left[1 - \Phi(ax)\right] J_\nu(bx)\, dx = a^{-\nu}\frac{\Gamma\left(\nu + \frac{3}{2}\right)}{b^2\,\Gamma(\nu+2)}\exp\left(-\frac{b^2}{8a^2}\right) M_{\frac{1}{2}\nu + \frac{1}{2},\,\frac{1}{2}\nu + \frac{1}{2}}\left(\frac{b^2}{4a^2}\right)$

$\left[|\arg a| < \dfrac{\pi}{4}, \quad b > 0, \quad \operatorname{Re}\nu > -1\right]$

ET II 92(22)

2. $\displaystyle\int_0^\infty x^\nu\left[1 - \Phi(ax)\right] J_\nu(bx)\, dx = \sqrt{\frac{2}{\pi}}\,\frac{a^{\frac{1}{2}-\nu}\,\Gamma\left(\nu + \frac{1}{2}\right)}{b^{3/2}\,\Gamma\left(\nu + \frac{3}{2}\right)}\exp\left(-\frac{b^2}{8a^2}\right) M_{\frac{1}{2}\nu - \frac{1}{4},\,\frac{1}{2}\nu + \frac{1}{4}}\left(\frac{b^2}{4a^2}\right)$

$\left[|\arg a| < \dfrac{\pi}{4}, \quad \operatorname{Re}\nu > -\frac{1}{2}, \quad b > 0\right]$

ET II 92(23)

6.785 $\displaystyle\int_0^\infty \frac{\exp\left(\frac{a^2}{2x} - x\right)}{x}\left[1 - \Phi\left(\frac{a}{\sqrt{2x}}\right)\right] K_\nu(x)\, dx = \frac{\pi^{5/2}}{4}\sec(\nu\pi)\left\{\left[J_\nu(a)\right]^2 + \left[Y_\nu(a)\right]^2\right\}$

$\left[\operatorname{Re} a > 0, \quad |\operatorname{Re}\nu| < \frac{1}{2}\right]$ ET II 370(46)

6.786 $\displaystyle\int_0^\infty x^{\nu - 2\mu + 2n + 2}e^{x^2}\,\Gamma\left(\mu, x^2\right) Y_\nu(bx)\, dx$

$\displaystyle = (-1)^n \frac{\Gamma\left(\frac{3}{2} - \mu + \nu + n\right)\Gamma\left(\frac{3}{2} - \mu + n\right)}{b\,\Gamma(1-\mu)}\exp\left(\frac{b^2}{8}\right) W_{\mu - \frac{1}{2}\nu - n - 1,\,\frac{1}{2}\nu}\left(\frac{b^2}{4}\right)$

$\left[n \text{ is an integer}, \quad b > 0, \quad \operatorname{Re}(\nu - \mu + n) > -\frac{3}{2}, \quad \operatorname{Re}(-\mu + n) > -\frac{3}{2}, \quad \operatorname{Re}\nu < \frac{1}{2} - 2n\right]$

ET II 108(2)

6.787 $\displaystyle\int_0^\infty \frac{x^{\nu + 2n - \frac{1}{2}}}{\mathrm{B}(a + x, a - x)} J_\nu(bx)\, dx = 0$

$\left[\pi \le b < \infty, \quad -1 < \operatorname{Re}\nu < 2a - 2n - \frac{7}{2}\right]$ ET II 92(21)

6.79 Integration of Bessel functions with respect to the order

6.791

1. $\displaystyle\int_{-\infty}^\infty K_{ix+iy}(a)\, K_{ix+iz}(b)\, dx = \pi\, K_{iy-iz}(a + b)$ $[|\arg a| + |\arg b| < \pi]$ ET II 382(21)

2. $\displaystyle\int_{-\infty}^\infty J_{\nu - x}(a)\, J_{\mu + x}(a)\, dx = J_{\mu + \nu}(2a)$ $[\operatorname{Re}(\mu + \nu) > 1]$ ET II 379(1)

3. $\int_{-\infty}^{\infty} J_{\kappa+x}(a) \, J_{\lambda-x}(a) \, J_{\mu+x}(a) \, J_{\nu-x}(a) \, dx$

$$= \frac{\Gamma(\kappa+\lambda+\mu+\nu+1)}{\Gamma(\kappa+\lambda+1)\,\Gamma(\lambda+\mu+1)\,\Gamma(\mu+\nu+1)\,\Gamma(\nu+\kappa+1)}$$

$$\times \; {}_4F_5\left(\frac{\kappa+\lambda+\mu+\nu+1}{2}, \frac{\kappa+\lambda+\mu+\nu+1}{2}, \frac{\kappa+\lambda+\mu+\nu}{2}+1, \frac{\kappa+\lambda+\mu+\nu}{2}+1; \right.$$

$$\left. \kappa+\lambda+\mu+\nu+1, \kappa+\lambda+1, \lambda+\mu+1, \mu+\nu+1, \nu+\kappa+1; -4a^2 \right)$$

$[\operatorname{Re}(\kappa+\lambda+\mu+\nu) > -1]$ ET II 379(3)

6.792

1. $\int_{-\infty}^{\infty} e^{\pi x} \, K_{ix+iy}(a) \, K_{ix+iz}(b) \, dx = \pi e^{-\pi z} \, K_{i(y-z)}(a-b)$

$[a > b > 0]$ ET II 382(22)

2. $\int_{-\infty}^{\infty} e^{i\varrho x} \, K_{\nu+ix}(\alpha) \, K_{\nu-ix}(\beta) \, dx = \pi \left(\frac{\alpha e^{\varrho} + \beta}{\alpha + \beta e^{\varrho}} \right)^{\nu} K_{2\nu}\left(\sqrt{\alpha^2 + \beta^2 + 2\alpha\beta \cosh \varrho} \right)$

$[|\arg \alpha| + |\arg \beta| + |\operatorname{Im} \varrho| < \pi]$
ET II 382(23)

3. $\int_{-\infty}^{\infty} e^{(\pi-\gamma)x} \, K_{ix+iy}(a) \, K_{ix+iz}(b) \, dx = \pi e^{-\beta y - \alpha z} \, K_{iy-iz}(c)$

     $[0 < \gamma < \pi, \quad a > 0, \quad b > 0, \quad c > 0, \quad \alpha, \beta, \gamma$—the angles of the triangle with sides a, b, $c]$

ET II 382(24), EH II 55(44)a

4. $\int_{-\infty}^{\infty} e^{-cxi} \, H_{\nu-ix}^{(2)}(a) \, H_{\nu+ix}^{(2)}(b) \, dx = 2i \left(\frac{h}{k} \right)^{2\nu} H_{2\nu}^{(2)}(hk)$

     $h = \sqrt{ae^{\frac{1}{2}c} + be^{-\frac{1}{2}c}}, \quad k = \sqrt{ae^{-\frac{1}{2}c} + be^{\frac{1}{2}c}}$ $[a, b > 0, \quad \operatorname{Im} ce = 0]$ ET II 380(11)

5. $\int_{-\infty}^{\infty} a^{-\mu-x} b^{-\nu+x} e^{cxi} \, J_{\mu+x}(a) \, J_{\nu-x}(b) \, dx$

$$= \left[\frac{2\cos\left(\frac{c}{2}\right)}{a^2 e^{-\frac{1}{2}ci} + b^2 e^{\frac{1}{2}ci}} \right]^{\frac{1}{2}\mu+\frac{1}{2}\nu} \exp\left[\frac{c}{2}(\nu-\mu)i \right] J_{\mu+\nu}\left\{ \left[2\cos\left(\frac{c}{2}\right) \left(a^2 e^{-\frac{1}{2}ci} + b^2 e^{\frac{1}{2}ci} \right) \right]^{1/2} \right\}$$

$[a > 0, \quad b > 0, \quad |c| < \pi, \quad \operatorname{Re}(\mu+\nu) > 1]$

$$= 0$$

$[a > 0, \quad b > 0, \quad |c| \geq \pi, \quad \operatorname{Re}(\mu+\nu) > 1]$

EH II 54(41), ET II 379(2)

6.793

1. $\int_{-\infty}^{\infty} e^{-cxi} \left[J_{\nu-ix}(a) \, Y_{\nu+ix}(b) + Y_{\nu-ix}(a) \, J_{\nu+ix}(b) \right] dx = -2 \left(\frac{h}{k} \right)^{2\nu} J_{2\nu}(hk)$

     $h = \sqrt{ae^{\frac{1}{2}c} + be^{-\frac{1}{2}c}}, \quad k = \sqrt{ae^{-\frac{1}{2}c} + be^{\frac{1}{2}c}}$ $[a, b > 0, \quad \operatorname{Im} c = 0]$ ET II 380(9)

2. $\displaystyle\int_{-\infty}^{\infty} e^{-cxi}\left[J_{\nu-ix}(a)\,J_{\nu+ix}(b) - Y_{\nu-ix}(a)\,Y_{\nu+ix}(b)\right]dx = 2\left(\frac{h}{k}\right)^{2\nu} Y_{2\nu}(hk)$

$$h = \sqrt{ae^{\frac{1}{2}c} + be^{-\frac{1}{2}c}}, \qquad k = \sqrt{ae^{-\frac{1}{2}c} + be^{\frac{1}{2}c}} \qquad [a, b > 0, \quad \operatorname{Im} c = 0] \quad \text{ET II 380(10)}$$

3.^{10} $\displaystyle\int_{-\infty}^{\infty} e^{i\gamma x}\operatorname{sech}(\pi x)\left[J_{-ix}(\alpha)\,J_{ix}(\beta) - J_{ix}(\alpha)\,J_{-ix}(\beta)\right]dx = 2i\,\mathrm{H}(\sigma)\,\operatorname{sgn}(\beta - \alpha)\,J_0\left(\sigma^{1/2}\right)$

$$\left[\alpha, \beta, \gamma \in \mathbb{R}, \quad \alpha, \beta > 0, \quad \sigma = \alpha^2 + \beta^2 - 2\alpha\beta\cosh\gamma, \quad \mathrm{H}(\sigma)\ \text{the Heaviside step function}\right]$$

6.794

1. $\displaystyle\int_0^{\infty} K_{ix}(a)\,K_{ix}(b)\,\cosh[(\pi - \varphi)x]\,dx = \frac{\pi}{2}\,K_0\left(\sqrt{a^2 + b^2 - 2ab\cos\varphi}\right)$ EH II 55(42)

2. $\displaystyle\int_0^{\infty} \cosh\left(\frac{\pi}{2}x\right) K_{ix}(a)\,dx = \frac{\pi}{2}$ $[a > 0]$ ET II 382(19)

3. $\displaystyle\int_0^{\infty} \cosh(\varrho x)\,K_{ix+\nu}(a)\,K_{-ix+\nu}(a)\,dx = \frac{\pi}{2}\,K_{2\nu}\left[2a\cos\left(\frac{\varrho}{2}\right)\right]$

$$[2|\arg a| + |\operatorname{Re}\varrho| < \pi] \quad \text{ET II 383(28)}$$

4. $\displaystyle\int_{-\infty}^{\infty} \operatorname{sech}\left(\frac{\pi}{2}x\right) J_{ix}(a)\,dx = 2\sin a$ $[a > 0]$ ET II 380(6)

5. $\displaystyle\int_{-\infty}^{\infty} \operatorname{cosech}\left(\frac{\pi}{2}x\right) J_{ix}(a)\,dx = -2i\cos a$ $[a > 0]$ ET II 380(7)

6. $\displaystyle\int_0^{\infty} \operatorname{sech}(\pi x)\left\{[J_{ix}(a)]^2 + [Y_{ix}(a)]^2\right\}dx = -Y_0(2a) - \mathbf{E}_0(2a)$

$$[a > 0] \quad \text{ET II 380(12)}$$

7. $\displaystyle\int_0^{\infty} x\sinh\left(\frac{\pi}{2}x\right) K_{ix}(a)\,dx = \frac{\pi a}{2}$ $[a > 0]$ ET II 382(20)

8. $\displaystyle\int_0^{\infty} x\tanh(\pi x)\,K_{ix}(\beta)\,K_{ix}(\alpha)\,dx = \frac{\pi}{2}\sqrt{\alpha\beta}\,\frac{\exp(-\beta - \alpha)}{\alpha + \beta}$

$$[|\arg\beta| < \pi, \quad |\arg\alpha| < \pi] \quad \text{ET II 175(4)}$$

9. $\displaystyle\int_0^{\infty} x\sinh(\pi x)\,K_{2ix}(\alpha)\,K_{ix}(\beta)\,dx = \frac{\pi^{3/2}\alpha}{2^{5/2}\sqrt{\beta}}\exp\left(-\beta - \frac{\alpha^2}{8\beta}\right)$

$$\left[\beta > 0, \quad |\arg\alpha| < \frac{\pi}{4}\right] \quad \text{ET II 175(5)}$$

10. $\displaystyle\int_0^{\infty} \frac{x\sinh(\pi x)}{x^2 + n^2}\,K_{ix}(\alpha)\,K_{ix}(\beta)\,dx = \frac{\pi^2}{2}\,I_n(\beta)\,K_n(\alpha)$ $[0 < \beta < \alpha;\quad n = 0, 1, 2, \ldots]$

$$= \frac{\pi^2}{2}\,I_n(\alpha)\,K_n(\beta) \qquad [0 < \alpha < \beta;\quad n = 0, 1, 2, \ldots]$$

$$\text{ET II 176(8)}$$

11. $\displaystyle\int_0^{\infty} x\sinh(\pi x)\,K_{ix}(\alpha)\,K_{ix}(\beta)\,K_{ix}(\gamma)\,dx = \frac{\pi^2}{4}\exp\left[-\frac{\gamma}{2}\left(\frac{\alpha}{\beta} + \frac{\beta}{\alpha} + \frac{\alpha\beta}{\gamma^2}\right)\right]$

$$\left[|\arg\alpha| + |\arg\beta| < \frac{\pi}{2}, \quad \gamma > 0\right]$$

$$\text{ET II 176(9)}$$

12. $\int_0^\infty x \sinh\left(\frac{\pi}{2}x\right) K_{\frac{1}{2}ix}(\alpha) K_{\frac{1}{2}ix}(\beta) K_{ix}(\gamma) dx = \frac{\pi^2 \gamma}{2\sqrt{\gamma^2 + 4\alpha\beta}} \exp\left[-\frac{(\alpha+\beta)\sqrt{\gamma^2+4\alpha\beta}}{2\sqrt{\alpha\beta}}\right]$

$$[|\arg \alpha| + |\arg \beta| < \pi, \quad \gamma > 0]$$

ET II 176(10)

13. $\int_0^\infty x \sinh(\pi x) K_{\frac{1}{2}ix+\lambda}(\alpha) K_{\frac{1}{2}ix-\lambda}(\alpha) K_{ix}(\gamma) dx = 0 \qquad [0 < \gamma < 2\alpha]$

$$= \frac{\pi^2 \gamma}{2^{2\lambda+1}\alpha^{2\lambda}z}\left[(\gamma+z)^{2\lambda} + (\gamma-z)^{2\lambda}\right]$$

$$z = \sqrt{\gamma^2 - 4\alpha^2} \qquad [0 < 2\alpha < \gamma] \quad \text{ET II 176(11)}$$

6.795

1. $\int_0^\infty \cos(bx) K_{ix}(a) dx = \frac{\pi}{2}e^{-a\cosh b} \qquad \left[|\operatorname{Im} b| < \frac{\pi}{2}, \quad a > 0\right]$

EH II 55(46), ET II 175(2)

2. $\int_0^\infty J_x(ax) J_{-x}(ax) \cos(\pi x) dx = \frac{1}{4}(1-a^2)^{-1/2} \qquad [|a| < 1]$ ET II 380(4)

3. $\int_0^\infty x \sin(ax) K_{ix}(b) dx = \frac{\pi b}{2}\sinh a \exp(-b\cosh a) \qquad \left[|\operatorname{Im} a| < \frac{\pi}{2}, \quad b > 0\right]$ ET II 175(1)

4. $\int_{-\infty}^{-\infty} \frac{\sin[(\nu+ix)\pi]}{n+\nu+ix} K_{\nu+ix}(a) K_{\nu-ix}(b) dx = \pi^2 I_n(a) K_{n+2\nu}(b) \qquad [0 < a < b; \quad n = 0, 1, \ldots]$

$$= \pi^2 K_{n+2\nu}(a) I_n(b) \qquad [0 < b < a; \quad n = 0, 1, \ldots]$$

ET II 382(25)

5. $\int_0^\infty x \sin\left(\frac{1}{2}\pi x\right) K_{\frac{1}{2}ix}(a) K_{ix}(b) dx = \frac{\pi^{3/2}b}{\sqrt{2a}}\exp\left(-a - \frac{b^2}{8a}\right)$

$$\left[|\arg a| < \frac{\pi}{2}, \quad b > 0\right] \qquad \text{ET II 175(6)}$$

6.796

1. $\int_{-\infty}^\infty \frac{e^{\frac{1}{2}\pi x}\cos(bx)}{\sinh(\pi x)} J_{ix}(a) dx = -i\exp(ia\cosh b) \qquad [a > 0, \quad b > 0]$ ET II 380(8)

2. $\int_0^\infty \cos(bx) \cosh\left(\frac{1}{2}\pi x\right) K_{ix}(a) dx = \frac{\pi}{2}\cos(a\sinh b)$ EH II 55(47)

3. $\int_0^\infty \sin(bx) \sinh\left(\frac{1}{2}\pi x\right) K_{ix}(a) dx = \frac{\pi}{2}\sin(a\sinh b)$ EH II 55(48)

4. $\int_0^\infty \cos(bx) \cosh(\pi x)[K_{ix}(a)]^2 dx = -\frac{\pi^2}{4}Y_0\left[2a\sinh\left(\frac{b}{2}\right)\right]$

$$[a > 0, \quad b > 0] \qquad \text{ET II 383(27)}$$

5. $\int_0^\infty \sin(bx) \sinh(\pi x)[K_{ix}(a)]^2 dx = \frac{\pi^2}{4}J_0\left[2a\sinh\left(\frac{b}{2}\right)\right]$

$$[a > 0, \quad b > 0] \qquad \text{ET II 382(26)}$$

6.797

1. $$\int_0^\infty x e^{\pi x} \sinh(\pi x)\, \Gamma(\nu + ix)\, \Gamma(\nu - ix)\, H_{ix}^{(2)}(a)\, H_{ix}^{(2)}(b)\, dx$$

$$= i 2^\nu \sqrt{\pi}\, \Gamma\left(\tfrac{1}{2} + \nu\right) (ab)^\nu (a+b)^{-\nu}\, K_\nu(a+b)$$

$$[a > 0, \quad b > 0, \quad \operatorname{Re}\nu > 0] \quad \text{ET II 381(14)}$$

2. $$\int_0^\infty x e^{\pi x} \sinh(\pi x) \cosh(\pi x)\, \Gamma(\nu+ix)\, \Gamma(\nu-ix)\, H_{ix}^{(2)}(a)\, H_{ix}^{(2)}(b)\, dx = \frac{i\pi^{3/2} 2^\nu}{\Gamma\left(\tfrac{1}{2} - \nu\right)} (b-a)^{-\nu}\, H_\nu^{(2)}(b-a)$$

$$\left[0 < a < b, \quad 0 < \operatorname{Re}\nu < \tfrac{1}{2}\right]$$

$$\text{ET II 381(15)}$$

3. $$\int_0^\infty x e^{\pi x} \sinh(\pi x)\, \Gamma\left(\frac{\nu + ix}{2}\right) \Gamma\left(\frac{\nu - ix}{2}\right) H_{ix}^{(2)}(a)\, H_{ix}^{(2)}(b)\, dx$$

$$= i\pi 2^{2-\nu}(ab)^\nu \left(a^2 + b^2\right)^{-\frac{1}{2}\nu}\, H_\nu^{(2)}\left(\sqrt{a^2 + b^2}\right)$$

$$[a > 0, \quad b > 0, \quad \operatorname{Re}\nu > 0] \quad \text{ET II 381(16)}$$

4. $$\int_0^\infty x \sinh(\pi x)\, \Gamma(\lambda+ix)\, \Gamma(\lambda-ix)\, K_{ix}(a)\, K_{ix}(b)\, dx = 2^{\nu-1}\pi^{3/2}(ab)^\lambda (a+b)^{-\lambda}\, \Gamma\left(\lambda + \tfrac{1}{2}\right) K_\lambda(a+b)$$

$$[|\arg a| < \pi, \quad \operatorname{Re}\lambda > 0, \quad b > 0]$$

$$\text{ET II 176(12)}$$

5. $$\int_0^\infty x \sinh(2\pi x)\, \Gamma(\lambda + ix)\, \Gamma(\lambda - ix)\, K_{ix}(a)\, K_{ix}(b)\, dx = \frac{2^\lambda \pi^{\frac{5}{2}}}{\Gamma\left(\tfrac{1}{2} - \lambda\right)} \left(\frac{ab}{|b-a|}\right)^\lambda K_\lambda\left(|b-a|\right)$$

$$\left[a > 0, \quad 0 < \operatorname{Re}\lambda < \tfrac{1}{2}, \quad b > 0\right]$$

$$\text{ET II 176(13)}$$

6. $$\int_0^\infty x \sinh(\pi x)\, \Gamma\left(\lambda + \tfrac{1}{2}ix\right) \Gamma\left(\lambda - \tfrac{1}{2}ix\right) K_{ix}(a)\, K_{ix}(b)\, dx = 2\pi^2 \left(\frac{ab}{2\sqrt{a^2 + b^2}}\right) K_{2\lambda}\left(\sqrt{a^2 + b^2}\right)$$

$$\left[|\arg a| < \frac{\pi}{2}, \quad \operatorname{Re}\lambda > 0, \quad b > 0\right]$$

$$\text{ET II 177(14)}$$

7. $$\int_0^\infty \frac{x \tanh(\pi x)\, K_{ix}(a)\, K_{ix}(b)}{\Gamma\left(\tfrac{3}{4} + \tfrac{1}{2}ix\right) \Gamma\left(\tfrac{3}{4} - \tfrac{1}{2}ix\right)}\, dx = \frac{1}{2}\sqrt{\frac{\pi ab}{a^2 + b^2}}\, \exp\left(-\sqrt{a^2 + b^2}\right)$$

$$\left[|\arg a| < \frac{\pi}{2}, \quad b > 0\right], \quad\quad \text{ET II 177(15)}$$

6.8 Functions Generated by Bessel Functions

6.81 Struve functions

6.811

1. $$\int_0^\infty \mathbf{H}_\nu(bx)\, dx = -\frac{\cot\left(\frac{\nu\pi}{2}\right)}{b} \qquad\qquad [-2 < \operatorname{Re}\nu < 0, \quad b > 0] \qquad \text{ET II 158(1)}$$

2. $\displaystyle\int_0^\infty \mathbf{H}_\nu\left(\frac{a^2}{x}\right)\mathbf{H}_\nu(bx)\,dx = -\frac{J_{2\nu}\left(2a\sqrt{b}\right)}{b}$ $\qquad\left[a>0,\quad b>0,\quad \mathrm{Re}\,\nu>-\frac{3}{2}\right]$

ET II 170(37)

3. $\displaystyle\int_0^\infty \mathbf{H}_{\nu-1}\left(\frac{a^2}{x}\right)\mathbf{H}_\nu(bx)\,\frac{dx}{x} = -\frac{1}{a\sqrt{b}}J_{2\nu-1}\left(2a\sqrt{b}\right)$ $\qquad\left[a>0,\quad b>0,\quad \mathrm{Re}\,\nu>-\frac{1}{2}\right]$

ET II 170(38)

6.812

1. $\displaystyle\int_0^\infty \frac{\mathbf{H}_1(bx)\,dx}{x^2+a^2} = \frac{\pi}{2a}\left[I_1(ab)-\mathbf{L}_1(ab)\right]$ $\qquad[\mathrm{Re}\,a>0,\quad b>0]$ $\qquad$ ET II 158(6)

2. $\displaystyle\int_0^\infty \frac{\mathbf{H}_\nu(bx)}{x^2+a^2}\,dx = -\frac{\pi}{2a\sin\left(\frac{\nu\pi}{2}\right)}\mathbf{L}_\nu(ab)+\frac{b\cot\left(\frac{\nu\pi}{2}\right)}{1-\nu^2}\,{}_1F_2\left(1;\frac{3-\nu}{2};\frac{3+\nu}{2};\frac{a^2b^2}{2}\right)$

$\qquad\left[\mathrm{Re}\,a>0,\quad b>0,\quad |\mathrm{Re}\,\nu|<2\right]$

ET II 159(7)

6.813

1. $\displaystyle\int_0^\infty x^{s-1}\,\mathbf{H}_\nu(ax)\,dx = \frac{2^{s-1}\,\Gamma\left(\frac{s+\nu}{2}\right)}{a^s\,\Gamma\left(\frac{1}{2}\nu-\frac{1}{2}s+1\right)}\tan\left(\frac{s+\nu}{2}\pi\right)$

$\qquad\left[a>0,\quad -1-\mathrm{Re}\,\nu<\mathrm{Re}\,s<\min\left(\frac{3}{2},1-\mathrm{Re}\,\nu\right)\right]$ $\quad$ WA 429(2), ET I 335(52)

2. $\displaystyle\int_0^\infty x^{-\nu-1}\,\mathbf{H}_\nu(x)\,dx = \frac{2^{-\nu-1}\pi}{\Gamma(\nu+1)}$ $\qquad\left[\mathrm{Re}\,\nu>-\frac{3}{2}\right]$ $\qquad$ ET II 383(2)

3. $\displaystyle\int_0^\infty x^{-\mu-\nu}\,\mathbf{H}_\mu(x)\,\mathbf{H}_\nu(x)\,dx = \frac{2^{-\mu-\nu}\sqrt{\pi}\,\Gamma(\mu+\nu)}{\Gamma\left(\mu+\frac{1}{2}\right)\Gamma\left(\nu+\frac{1}{2}\right)\Gamma\left(\mu+\nu+\frac{1}{2}\right)}$

$\qquad[\mathrm{Re}(\mu+\nu)>0]$ $\quad$ WA 435(2), ET II 384(8)

4. $\displaystyle\int_0^1 x^{\nu+1}\,\mathbf{H}_\nu(ax)\,dx = \frac{1}{a}\,\mathbf{H}_{\nu+1}(a)$ $\qquad\left[a>0,\quad \mathrm{Re}\,\nu>-\frac{3}{2}\right]$ $\qquad$ ET II 158(2)a

5. $\displaystyle\int_0^1 x^{1-\nu}\,\mathbf{H}_\nu(ax)\,dx = \frac{a^{\nu-1}}{2^{\nu-1}\sqrt{\pi}\,\Gamma\left(\nu+\frac{1}{2}\right)} - \frac{1}{a}\,\mathbf{H}_{\nu-1}(a)$

$\qquad[a>0]$ $\qquad$ ET II 158(3)a

6.814

1. $\displaystyle\int_0^\infty \frac{x^{\nu+1}\,\mathbf{H}_\nu(bx)}{(x^2+a^2)^{1-\mu}}\,dx = \frac{2^{\mu-1}\pi a^{\mu+\nu}b^{-\mu}}{\Gamma(1-\mu)\cos[(\mu+\nu)\pi]}\left[I_{-\mu-\nu}(ab)-\mathbf{L}_{\mu+\nu}(ab)\right]$

$\qquad\left[\mathrm{Re}\,a>0,\quad b>0,\quad \mathrm{Re}\,\nu>-\frac{3}{2},\quad \mathrm{Re}(\mu+\nu)<\frac{1}{2},\quad \mathrm{Re}(2\mu+\nu)<\frac{3}{2}\right]$ $\quad$ ET II 159(8)

6.815

1. $\displaystyle\int_0^1 x^{\frac{1}{2}\nu}(1-x)^{\mu-1}\,\mathbf{H}_\nu\left(a\sqrt{x}\right)\,dx = 2^\mu a^{-\mu}\,\Gamma(\mu)\,\mathbf{H}_{\mu+\nu}(a)$

$\qquad\left[\mathrm{Re}\,\nu>-\frac{3}{2},\quad \mathrm{Re}\,\mu>0\right]$ $\quad$ ET II 199(88)a

2. $\int_0^1 x^{\lambda - \frac{1}{2}\nu - \frac{3}{2}}(1-x)^{\mu-1}\,\mathbf{H}_\nu\left(a\sqrt{x}\right)\,dx = \dfrac{\mathrm{B}(\lambda,\mu)a^{\nu+1}}{2^\nu\sqrt{\pi}\,\Gamma\left(\nu+\frac{3}{2}\right)}\,{}_2F_3\left(1,\lambda;\frac{3}{2},\nu+\frac{3}{2},\lambda+\mu;-\frac{a^2}{4}\right)$

$$[\operatorname{Re}\lambda > 0,\quad \operatorname{Re}\mu > 0] \qquad \text{ET II 199(89)a}$$

6.82 Combinations of Struve functions, exponentials, and powers

6.821

$1.^6$ $\int_0^\infty e^{-\alpha x}\,\mathbf{H}_{-n-\frac{1}{2}}(\beta x)\,dx = (-1)^n\beta^{n+\frac{1}{2}}\left(\alpha+\sqrt{\alpha^2+\beta^2}\right)^{-n-\frac{1}{2}}\dfrac{1}{\sqrt{\alpha^2+\beta^2}}$

$$[\operatorname{Re}\alpha > |\operatorname{Im}\beta|] \qquad \text{ET I 206(6)}$$

$2.^6$ $\int_0^\infty e^{-\alpha x}\,\mathbf{L}_{-n-\frac{1}{2}}(\beta x)\,dx = \beta^{n+\frac{1}{2}}\left(\alpha+\sqrt{\alpha^2-\beta^2}\right)^{-n-\frac{1}{2}}\dfrac{1}{\sqrt{\alpha^2-\beta^2}}$

$$[\operatorname{Re}\alpha > |\operatorname{Re}\beta|] \qquad \text{ET I 208(26)}$$

3. $\int_0^\infty e^{-\alpha x}\,\mathbf{H}_0(\beta x)\,dx = \dfrac{2}{\pi}\dfrac{\ln\left(\frac{\sqrt{\alpha^2+\beta^2}+\beta}{\alpha}\right)}{\sqrt{\alpha^2+\beta^2}}$ $[\operatorname{Re}\alpha > |\operatorname{Im}\beta|]$ ET II 205(1)

4. $\int_0^\infty e^{-\alpha x}\,\mathbf{L}_0(\beta x)\,dx = \dfrac{2}{\pi}\dfrac{\arcsin\left(\frac{\beta}{\alpha}\right)}{\sqrt{\alpha^2+\beta^2}}$ $[\operatorname{Re}\alpha > |\operatorname{Re}\beta|]$ ET II 207(18)

6.822 $\int_0^\infty e^{(\nu+1)x}\,\mathbf{H}_\nu(a\sinh x)\,dx = \sqrt{\dfrac{\pi}{a}}\,\operatorname{cosec}(\nu\pi)\left[\sinh\left(\frac{a}{2}\right)I_{\nu+\frac{1}{2}}\left(\frac{a}{2}\right) - \cosh\left(\frac{a}{2}\right)I_{-\nu-\frac{1}{2}}\left(\frac{a}{2}\right)\right]$

$$[\operatorname{Re}a > 0,\quad -2 < \operatorname{Re}\nu < 0]$$
$$\text{ET II 385(11)}$$

6.823

1. $\int_0^\infty x^\lambda e^{-\alpha x}\,\mathbf{H}_\nu(bx)\,dx = \dfrac{b^{\nu+1}\,\Gamma(\lambda+\nu+2)}{2^\nu a^{\lambda+\nu+2}\sqrt{\pi}\,\Gamma\left(\nu+\frac{3}{2}\right)}\,{}_3F_2\left(1,\frac{\lambda+\nu}{2}+1,\frac{\lambda+\nu+3}{2};\frac{3}{2},\nu+\frac{3}{2};-\frac{b^2}{a^2}\right)$

$$[\operatorname{Re}a > 0,\quad b > 0,\quad \operatorname{Re}(\lambda+\nu) > -2]$$
$$\text{ET II 161(19)}$$

2. $\int_0^\infty x^\nu e^{-\alpha x}\,\mathbf{L}_\nu(\beta x)\,dx = \dfrac{(2\beta)^\nu\,\Gamma\left(\nu+\frac{1}{2}\right)}{\sqrt{\pi}\left(\sqrt{\alpha^2-\beta^2}\right)^{2\nu+1}} - \dfrac{\Gamma(2\nu+1)\left(\frac{\beta}{\alpha}\right)^\nu}{\sqrt{\frac{\pi}{2}}\,\alpha\left(\beta^2-\alpha^2\right)^{\frac{1}{2}\nu+\frac{1}{4}}}\,P_{-\nu-\frac{1}{2}}^{-\nu-\frac{1}{2}}\left(\frac{\beta}{\alpha}\right)$

$$[\operatorname{Re}\alpha > |\operatorname{Re}\beta|,\quad \operatorname{Re}\nu > -\tfrac{1}{2}]$$
$$\text{ET I 209(35)a}$$

6.824

1. $\int_0^\infty t^\nu e^{-at}\,\mathbf{L}_{2\nu}\left(2\sqrt{t}\right)\,dt = \dfrac{1}{a^{2\nu+1}}e^{\frac{1}{a}}\,\Phi\left(\frac{1}{\sqrt{a}}\right)$ MI 51

2. $\displaystyle\int_0^\infty t^\nu e^{-at}\, \mathbf{L}_{-2\nu}\left(\sqrt{t}\right) dt = \frac{1}{\Gamma\left(\frac{1}{2} - 2\nu\right) a^{2\nu+1}} e^{\frac{1}{a}} \gamma\left(\frac{1}{2} - 2\nu, \frac{1}{a}\right)$ MI 51

6.825 $\displaystyle\int_0^\infty x^{s-1} e^{-\alpha^2 x^2}\, \mathbf{H}_\nu(\beta x)\, dx = \frac{\beta^{\nu+1}\, \Gamma\left(\frac{1}{2} + \frac{s}{2} + \frac{\nu}{2}\right)}{2^{\nu+1}\sqrt{\pi}\,\alpha^{\nu+s+1}\, \Gamma\left(\nu + \frac{3}{2}\right)}\; {}_2F_2\left(1, \frac{\nu+s+1}{2}; \frac{3}{2}, \nu + \frac{3}{2}; -\frac{\beta^2}{4\alpha^2}\right)$

$$\left[\operatorname{Re} s > -\operatorname{Re}\nu - 1, \quad |\arg \alpha| < \frac{\pi}{4}\right]$$

ET I 335(51)a, ET II 162(20)

6.83 Combinations of Struve and trigonometric functions

6.831 $\displaystyle\int_0^\infty x^{-\nu} \sin(ax)\, \mathbf{H}_\nu(bx)\, dx = 0$ $\left[0 < b < a, \quad \operatorname{Re}\nu > -\frac{1}{2}\right]$

$$= \sqrt{\pi}\, 2^{-\nu} b^{-\nu} \frac{\left(b^2 - a^2\right)^{\nu - \frac{1}{2}}}{\Gamma\left(\nu + \frac{1}{2}\right)} \qquad \left[0 < a < b, \quad \operatorname{Re}\nu > -\frac{1}{2}\right]$$

ET II 162(21)

6.832 $\displaystyle\int_0^\infty \sqrt{x}\, \sin(ax)\, \mathbf{H}_{\frac{1}{4}}\left(b^2 x^2\right) dx = -2^{-3/2}\sqrt{\pi}\, \frac{\sqrt{a}}{b^2}\, Y_{\frac{1}{4}}\left(\frac{a^2}{4b^2}\right)$

$$[a > 0]$$ ET I 109(14)

6.84–6.85 Combinations of Struve and Bessel functions

6.841 $\displaystyle\int_0^\infty \mathbf{H}_{\nu-1}(ax)\, Y_\nu(bx)\, dx = -a^{\nu-1} b^{-\nu}$ $\left[0 < b < a, \quad |\operatorname{Re}\nu| < \frac{1}{2}\right]$

$$= 0 \qquad \left[0 < a < b, \quad |\operatorname{Re}\nu| < \frac{1}{2}\right]$$

ET II 114(36)

6.842 $\displaystyle\int_0^\infty \left[\mathbf{H}_0(ax) - Y_0(ax)\right] J_0(bx)\, dx = \frac{4}{\pi(a+b)}\, K\left(\frac{|a-b|}{a+b}\right)$

$$[a > 0, \quad b > 0]$$ ET II 15(22)

6.843

1. $\displaystyle\int_0^\infty J_{2\nu}\left(a\sqrt{x}\right) \mathbf{H}_\nu(bx)\, dx = -\frac{1}{b}\, Y_\nu\left(\frac{a^2}{4b}\right)$ $\left[a > 0, \quad b > 0, \quad -1 < \operatorname{Re}\nu < \frac{5}{4}\right]$

ET II 164(10)

2. $\displaystyle\int_0^\infty K_{2\nu}\left(2a\sqrt{x}\right) \mathbf{H}_\nu(bx)\, dx = \frac{2^\nu}{\pi b}\, \Gamma(\nu+1)\, S_{-\nu-1,\nu}\left(\frac{a^2}{b}\right)$

$$\left[\operatorname{Re} a > 0, \quad b > 0, \quad \operatorname{Re}\nu > -1\right]$$ ET II 168(27)

6.844 $\displaystyle\int_0^\infty \left[\cos\left(\frac{\mu-\nu}{2}\pi\right) J_\mu\left(a\sqrt{x}\right) - \sin\left(\frac{\mu-\nu}{2}\pi\right) Y_\mu\left(a\sqrt{x}\right)\right] K_\mu\left(a\sqrt{x}\right) \mathbf{H}_\nu(bx)\, dx$

$$= \frac{1}{a^2}\, W_{\frac{1}{2}\nu, \frac{1}{2}\mu}\left(\frac{a^2}{2b}\right) W_{-\frac{1}{2}\nu, \frac{1}{2}\mu}\left(\frac{a^2}{2b}\right)$$

$$\left[|\arg a| < \frac{\pi}{4}, \quad b > 0, \quad \operatorname{Re}\nu > |\operatorname{Re}\mu| - 2\right] \quad \text{ET II 169(35)}$$

6.845

1. $\displaystyle\int_0^\infty \left[\mathbf{H}_{-\nu}\left(\frac{a}{x}\right) - Y_{-\nu}\left(\frac{a}{x}\right)\right] J_\nu(bx)\, dx = \frac{4}{\pi b}\cos(\nu\pi)\, K_{2\nu}\left(2\sqrt{ab}\right)$

$$\left[|\arg a| < \pi,\quad b > 0,\quad |\operatorname{Re}\nu| < \tfrac{1}{2}\right]$$
$$\text{ET II 73(7)}$$

2. $\displaystyle\int_0^\infty \left[J_{-\nu}\left(\frac{a^2}{x}\right) + \sin(\nu\pi)\,\mathbf{H}_\nu\left(\frac{a^2}{x}\right)\right] \mathbf{H}_\nu(bx)\, dx = \frac{1}{b}\left[\frac{2}{\pi}K_{2\nu}\left(2a\sqrt{b}\right) - Y_{2\nu}\left(2a\sqrt{b}\right)\right]$

$$\left[a > 0,\quad b > 0,\quad -\tfrac{3}{2} < \operatorname{Re}\nu < 0\right]$$
$$\text{ET II 170(39)}$$

6.846 $\displaystyle\int_0^\infty \left[\frac{2}{\pi}K_{2\nu}\left(2a\sqrt{x}\right) + Y_{2\nu}\left(2a\sqrt{x}\right)\right] \mathbf{H}_\nu(bx)\, dx = \frac{1}{b} J_\nu\left(\frac{a^2}{b}\right)$

$$\left[a > 0,\quad b > 0,\quad |\operatorname{Re}\nu| < \tfrac{1}{2}\right]$$
$$\text{ET II 169(30)}$$

6.847 $\displaystyle\int_0^\infty \left[\cos\frac{\nu\pi}{2} J_\nu(ax) + \sin\frac{\nu\pi}{2}\,\mathbf{H}_\nu(ax)\right] \frac{dx}{x^2 + k^2} = \frac{\pi}{2k}\left[I_\nu(ak) - \mathbf{L}_\nu(ak)\right]$

$$\left[a > 0,\quad \operatorname{Re}k > 0,\quad -\tfrac{1}{2} < \operatorname{Re}\nu < 2\right]$$
$$\text{ET II 384(5)a, WA 467(8)}$$

6.848

1. $\displaystyle\int_0^\infty x\left[I_\nu(ax) - \mathbf{L}_{-\nu}(ax)\right] J_\nu(bx)\, dx = \frac{2}{\pi}\left(\frac{a}{b}\right)^{\nu-1}\cos(\nu\pi)\frac{1}{a^2 + b^2}$

$$\left[\operatorname{Re}a > 0,\quad b > 0,\quad -1 < \operatorname{Re}\nu < -\tfrac{1}{2}\right]$$
$$\text{ET II 74(12)}$$

2. $\displaystyle\int_0^\infty x\left[\mathbf{H}_{-\nu}(ax) - Y_{-\nu}(ax)\right] J_\nu(bx)\, dx = 2\frac{\cos(\nu\pi)}{a^\nu \pi}b^{\nu-1}\frac{1}{a + b}$

$$\left[|\arg a| < \pi,\quad -\tfrac{1}{2} < \operatorname{Re}\nu,\quad b > 0\right]$$
$$\text{ET II 73(5)}$$

6.849

1. $\displaystyle\int_0^\infty x K_\nu(ax)\,\mathbf{H}_\nu(bx)\, dx = a^{-\nu-1}b^{\nu+1}\frac{1}{a^2 + b^2}$ $\left[\operatorname{Re}a > 0,\quad b > 0,\quad \operatorname{Re}\nu > -\tfrac{3}{2}\right]$

$$\text{ET II 164(12)}$$

2. $\displaystyle\int_0^\infty x[K_\mu(ax)]^2\,\mathbf{H}_0(bx)\, dx = -2^{-\mu-1}\pi a^{-2\mu}\frac{\left[(z+b)^{2\mu} + (z-b)^{2\mu}\right]}{bz}\sec(\mu\pi),$

$$z = \sqrt{4a^2 + b^2}\qquad \left[\operatorname{Re}a > 0,\quad b > 0,\quad |\operatorname{Re}\mu| < \tfrac{3}{2}\right]\quad \text{ET II 166(18)}$$

6.851

1. $\displaystyle\int_0^\infty x\left\{\left[J_{\frac{1}{2}\nu}(ax)\right]^2 - \left[Y_{\frac{1}{2}\nu}(ax)\right]^2\right\} \mathbf{H}_\nu(bx)\, dx$

$$= 0 \qquad\qquad \left[0 < b < 2a,\quad -\tfrac{3}{2} < \operatorname{Re}\nu < 0\right]$$

$$= \frac{4}{\pi b}\frac{1}{\sqrt{b^2 - 4a^2}} \qquad \left[0 < 2a < b,\quad -\tfrac{3}{2} < \operatorname{Re}\nu < 0\right]$$
$$\text{ET II 164(7)}$$

2. $\int_0^\infty x^{\nu+1} \left\{ [J_\nu(ax)]^2 - [Y_\nu(ax)]^2 \right\} \mathbf{H}_\nu(bx)\, dx$

$$= 0 \qquad\qquad\qquad \left[0 < b < 2a, \quad -\tfrac{3}{4} < \operatorname{Re}\nu < 0 \right]$$

$$= \frac{2^{3\nu+2} a^{2\nu} b^{-\nu-1}}{\sqrt{\pi}\, \Gamma\left(\tfrac{1}{2} - \nu\right)} \left(b^2 - 4a^2\right)^{-\nu-\frac{1}{2}} \qquad \left[0 < 2a < b, \quad -\tfrac{3}{4} < \operatorname{Re}\nu < 0 \right]$$

ET II 163(6)

6.852

1. $\int_0^\infty x^{1-\mu-\nu}\, J_\nu(x)\, \mathbf{H}_\mu(x)\, dx = \dfrac{(2\nu-1)2^{-\mu-\nu}}{(\mu+\nu-1)\,\Gamma\left(\mu+\tfrac{1}{2}\right)\Gamma\left(\nu+\tfrac{1}{2}\right)}$

$$\left[\operatorname{Re}\nu > \tfrac{1}{2}, \quad \operatorname{Re}(\mu+\nu) > 1 \right]$$

ET II 383(4)

2. $\int_0^\infty x^{\mu-\nu+1}\, Y_\mu(ax)\, \mathbf{H}_\nu(bx)\, dx$

$$= 0 \qquad\qquad \left[0 < b < a, \quad \operatorname{Re}(\nu-\mu) > 0, \quad -\tfrac{3}{2} < \operatorname{Re}\mu < \tfrac{1}{2} \right]$$

$$= \frac{2^{1+\mu-\nu} a^\mu b^{-\nu}}{\Gamma(\nu-\mu)} \left(b^2 - a^2\right)^{\nu-\mu-1} \qquad \left[0 < a < b, \quad \operatorname{Re}(\nu-\mu) > 0, \quad -\tfrac{3}{2} < \operatorname{Re}\mu < \tfrac{1}{2} \right]$$

ET II 163(3)

3. $\int_0^\infty x^{\mu+\nu+1}\, K_\mu(ax)\, \mathbf{H}_\nu(bx)\, dx = \dfrac{2^{\mu+\nu+1} b^{\nu+1}}{\sqrt{\pi}\, a^{\mu+2\nu+3}}\, \Gamma\left(\mu+\nu+\tfrac{3}{2}\right) F\left(1, \mu+\nu+\tfrac{3}{2}; \tfrac{3}{2}; -\tfrac{b^2}{a^2}\right)$

$$\left[\operatorname{Re}a > 0, \quad b > 0, \quad \operatorname{Re}\nu > -\tfrac{3}{2}, \quad \operatorname{Re}(\mu+\nu) > -\tfrac{3}{2} \right] \quad \text{ET II 165(13)}$$

6.853

1. $\int_0^\infty x^{1-\mu} \left[\sin(\mu\pi)\, J_{\mu+\nu}(ax) + \cos(\mu\pi)\, Y_{\mu+\nu}(ax) \right] \mathbf{H}_\nu(bx)\, dx$

$$= 0 \qquad \left[0 < b < a, \quad 1 < \operatorname{Re}\mu < \tfrac{3}{2}, \quad \operatorname{Re}\nu > -\tfrac{3}{2}, \quad \operatorname{Re}(\nu-\mu) < \tfrac{1}{2} \right]$$

$$= \frac{b^\nu \left(b^2 - a^2\right)^{\mu-1}}{2^{\mu-1} a^{\mu+\nu}\, \Gamma(\mu)} \qquad \left[0 < a < b, \quad 1 < \operatorname{Re}\mu < \tfrac{3}{2}, \quad \operatorname{Re}\nu > -\tfrac{3}{2}, \quad \operatorname{Re}(\nu-\mu) < \tfrac{1}{2} \right]$$

ET II 163(4)

2. $\int_0^\infty x^{\lambda+\frac{1}{2}} \left[I_\mu(ax) - \mathbf{L}_{-\mu}(ax) \right] J_\nu(bx)\, dx$

$$= 2^{\lambda+\frac{1}{2}} \frac{\cos(\mu\pi)}{\pi}\, b^{-\lambda-\frac{3}{2}}\, G_{33}^{22}\left(\frac{b^2}{a^2} \left|\begin{array}{c} \tfrac{1+\mu}{2}, 1-\tfrac{\mu}{2}, 1+\tfrac{\mu}{2} \\ \tfrac{3}{4}+\tfrac{\lambda+\nu}{2}, \tfrac{1+\mu}{2}, \tfrac{3}{4}+\tfrac{\lambda-\nu}{2} \end{array}\right.\right)$$

$$\left[\operatorname{Re}a > 0, \quad b > 0, \quad \operatorname{Re}(\mu+\nu+\lambda) > -\tfrac{3}{2}, \quad -\operatorname{Re}\nu-\tfrac{5}{2} < \operatorname{Re}(\lambda-\mu) < 1 \right] \quad \text{ET II 76(21)}$$

3. $\displaystyle\int_0^\infty x^{\lambda+\frac{1}{2}} \left[\mathbf{H}_\mu(ax) - Y_\mu(ax)\right] J_\nu(bx)\, dx$

$$= 2^{\lambda+\frac{1}{2}} \frac{\cos(\mu\pi)}{\pi^2} b^{-\lambda-\frac{3}{2}}\, G_{33}^{23} \left(\frac{b^2}{a^2} \left| \begin{array}{l} \frac{1-\mu}{2}, 1-\frac{\mu}{2}, 1+\frac{\mu}{2} \\ \frac{3}{4}+\frac{\lambda+\nu}{2}, \frac{1-\mu}{2}, \frac{3}{4}+\frac{\lambda-\nu}{2} \end{array} \right. \right)$$

$$\left[b > 0, \quad |\arg a| < \pi, \quad \operatorname{Re}(\lambda+\mu) < 1, \quad \operatorname{Re}(\lambda+\nu) + \tfrac{3}{2} > |\operatorname{Re}\mu| \right] \quad \text{ET II 73(6)}$$

4. $\displaystyle\int_0^\infty \sqrt{x} \left[I_{\nu-\frac{1}{2}}(ax) - \mathbf{L}_{\nu-\frac{1}{2}}(ax) \right] J_\nu(bx)\, dx = \sqrt{\frac{2}{\pi}}\, a^{\nu-\frac{1}{2}} b^{-\nu} \frac{1}{\sqrt{a^2+b^2}}$

$$\left[\operatorname{Re} a > 0, \quad b > 0, \quad |\operatorname{Re}\nu| < \tfrac{1}{2} \right]$$

$$\text{ET II 74(11)}$$

5. $\displaystyle\int_0^\infty x^{\mu-\nu+1} \left[I_\mu(ax) - \mathbf{L}_\mu(ax) \right] J_\nu(bx)\, dx = \frac{2^{\mu-\nu+1} a^{\mu-1} b^{\nu-2\mu-1}}{\sqrt{\pi}\, \Gamma\left(\nu-\mu+\frac{1}{2}\right)}\, F\left(1, \frac{1}{2}; \nu-\mu+\frac{1}{2}; -\frac{b^2}{a^2}\right)$

$$\left[-1 < 2\operatorname{Re}\mu + 1 < \operatorname{Re}\nu + \tfrac{1}{2}, \quad \operatorname{Re} a > 0, \quad b > 0 \right] \quad \text{ET II 74(13)}$$

6. $\displaystyle\int_0^\infty x^{\mu-\nu+1} \left[I_\mu(ax) - \mathbf{L}_{-\mu}(ax) \right] J_\nu(bx)\, dx = \frac{2^{\mu-\nu+1} a^{-\mu-1} b^{\nu-1}}{\Gamma\left(\frac{1}{2}-\mu\right)\Gamma\left(\frac{1}{2}+\nu\right)}\, F\left(1, \frac{1}{2}+\mu; \frac{1}{2}+\nu; -\frac{b^2}{a^2}\right)$

$$\left[\operatorname{Re} a > 0, \quad \operatorname{Re}\nu > -\tfrac{1}{2}, \quad \operatorname{Re}\mu > -1, \quad b > 0 \right] \quad \text{ET II 75(18)}$$

6.854

1. $\displaystyle\int_0^\infty x\, \mathbf{H}_{\frac{1}{2}\nu}\left(ax^2\right) K_\nu(bx)\, dx = \frac{\Gamma\left(\frac{1}{2}\nu+1\right)}{2^{1-\frac{1}{2}\nu} a\pi}\, S_{-\frac{1}{2}\nu-1, \frac{1}{2}\nu}\left(\frac{b^2}{4a}\right)$

$$\left[a > 0, \quad \operatorname{Re} b > 0, \quad \operatorname{Re}\nu > -2 \right]$$

$$\text{ET II 150(75)}$$

2. $\displaystyle\int_0^\infty x\, \mathbf{H}_{\frac{1}{2}\nu}\left(ax^2\right) J_\nu(bx)\, dx = -\frac{1}{2a}\, Y_{\frac{1}{2}\nu}\left(\frac{b^2}{4a}\right) \qquad \left[a > 0, \quad b > 0, \quad -2 < \operatorname{Re}\nu < \tfrac{3}{2} \right]$

$$\text{ET II 73(3)}$$

6.855

1. $\displaystyle\int_0^\infty x^{2\nu+\frac{1}{2}} \left[I_{\nu+\frac{1}{2}}\left(\frac{a}{x}\right) - \mathbf{L}_{\nu+\frac{1}{2}}\left(\frac{a}{x}\right) \right] J_\nu(bx)\, dx = 2^{\frac{3}{2}} \frac{a^{\nu+\frac{1}{2}}}{\sqrt{\pi}\, b^{\nu+1}}\, J_{2\nu+1}\left(\sqrt{2ab}\right) K_{2\nu+1}\left(\sqrt{2ab}\right)$

$$\left[\operatorname{Re} a > 0, \quad b > 0, \quad -1 < \operatorname{Re}\nu < \tfrac{1}{2} \right]$$

$$\text{ET II 76(22)}$$

2. $\displaystyle\int_0^\infty \left[\mathbf{H}_{-\nu-1}\left(\frac{a}{x}\right) - Y_{-\nu-1}\left(\frac{a}{x}\right) \right] J_\nu(bx)\, \frac{dx}{x} = -\frac{4}{\pi\sqrt{ab}} \cos(\nu\pi)\, K_{-2\nu-1}\left(2\sqrt{ab}\right)$

$$\left[|\arg a| < \pi, \quad b > 0, \quad |\operatorname{Re}\nu| < \tfrac{1}{2} \right]$$

$$\text{ET II 74(8)}$$

3. $\displaystyle\int_0^\infty x^{2\nu+\frac{1}{2}} \left[\mathbf{H}_{\nu+\frac{1}{2}}\left(\frac{a}{x}\right) - Y_{\nu+\frac{1}{2}}\left(\frac{a}{x}\right) \right] J_\nu(bx)\, dx$

$$= -2^{5/2} \pi^{-3/2} a^{\nu+\frac{1}{2}} b^{-\nu-1} \sin(\nu\pi)\, K_{2\nu+1}\left(\sqrt{2ab}\, e^{\frac{1}{4}\pi i}\right) K_{2\nu+1}\left(\sqrt{2ab}\, e^{-\frac{1}{4}\pi i}\right)$$

$$\left[|\arg a| < \pi, \quad b > 0, \quad -1 < \operatorname{Re}\nu < -\tfrac{1}{6} \right] \quad \text{ET II 74(9)}$$

6.856 $\displaystyle\int_0^\infty x\, Y_\nu\left(a\sqrt{x}\right) K_\nu\left(a\sqrt{x}\right) \mathbf{H}_\nu(bx)\, dx = \frac{1}{2b^2}\exp\left(-\frac{a^2}{2b}\right)$

$$\left[b>0,\quad |\arg a|<\frac{\pi}{4},\quad \operatorname{Re}\nu>-\frac{3}{2}\right]$$

ET II 169(32)

6.857

1. $\displaystyle\int_0^\infty x\exp\left(\frac{a^2x^2}{8}\right) K_{\frac{1}{2}\nu}\left(\frac{a^2x^2}{8}\right)\mathbf{H}_\nu(bx)\, dx$

$$= \frac{2}{\sqrt{\pi}} a^{-\frac{\nu}{2}-1} b^{\frac{\nu}{2}-1}\cos\left(\frac{\nu\pi}{2}\right)\Gamma\left(-\frac{1}{2}\nu\right)\exp\left(\frac{b^2}{2a^2}\right) W_{k,m}\left(\frac{b^2}{a^2}\right)$$

$$k=\tfrac{1}{4}\nu,\qquad m=\tfrac{1}{2}+\tfrac{1}{4}\nu\qquad \left[|\arg a|<\tfrac{3}{4}\pi,\quad b>0,\quad -\tfrac{3}{2}<\operatorname{Re}\nu<0\right]\quad \text{ET II 167(24)}$$

2. $\displaystyle\int_0^\infty x^{\sigma-2}\exp\left(-\frac{1}{2}a^2x^2\right) K_\mu\left(\frac{1}{2}a^2x^2\right)\mathbf{H}_\nu(bx)\, dx$

$$= \frac{\sqrt{\pi}}{2^{\nu+2}} a^{-\nu-\sigma} b^{\nu+1}\frac{\Gamma\left(\frac{\nu+\sigma}{2}+\mu\right)\Gamma\left(\frac{\nu+\sigma}{2}-\mu\right)}{\Gamma\left(\frac{3}{2}\right)\Gamma\left(\nu+\frac{3}{2}\right)\Gamma\left(\frac{\nu+\sigma}{2}\right)}$$

$$\times {}_3F_3\left(1,\frac{\nu+\sigma}{2}+\mu,\frac{\nu+\sigma}{2}-\mu;\frac{3}{2},\nu+\frac{3}{2},\frac{\nu+\sigma}{2};-\frac{b^2}{4a^2}\right)$$

$$\left[b>0,\quad |\arg a|<\frac{\pi}{4},\quad \operatorname{Re}(\sigma+\nu)>2|\operatorname{Re}\mu|\right]\quad \text{ET II 167(23)}$$

6.86 Lommel functions

6.861

1. $\displaystyle\int_0^\infty x^{\lambda-1} S_{\mu,\nu}(x)\, dx = \frac{\Gamma\left[\frac{1}{2}(1+\lambda+\mu)\right]\Gamma\left[\frac{1}{2}(1-\lambda-\mu)\right]\Gamma\left[\frac{1}{2}(1+\mu+\nu)\right]\Gamma\left[\frac{1}{2}(1+\mu-\nu)\right]}{2^{2-\lambda-\mu}\Gamma\left[\frac{1}{2}(\nu-\lambda)+1\right]\Gamma\left[1-\frac{1}{2}(\lambda+\nu)\right]}$

$$\left[-\operatorname{Re}\mu<\operatorname{Re}\lambda+1<\tfrac{5}{2}\right]\quad \text{ET II 385(17)}$$

6.862

1. $\displaystyle\int_0^u x^{\lambda-\frac{1}{2}\mu-\frac{1}{2}}(u-x)^{\sigma-1} s_{\mu,\nu}\left(a\sqrt{x}\right) dx$

$$= \Gamma(\sigma)\frac{a^{\mu+1} u^{\lambda+\sigma}\Gamma(\lambda+1)}{(\mu-\nu+1)(\mu+\nu+1)\Gamma(\lambda+\sigma+1)}$$

$$\times {}_2F_3\left(1,1+\lambda;\frac{\mu-\nu+3}{2},\frac{\mu+\nu+3}{2},\lambda+\sigma+1;-\frac{a^2u}{4}\right)$$

$$[\operatorname{Re}\lambda>-1,\quad \operatorname{Re}\sigma>0]\quad \text{ET II 199(92)}$$

2. $\displaystyle\int_u^\infty x^{\frac{1}{2}\nu}(x-u)^{\mu-1} s_{\lambda,\nu}\left(a\sqrt{x}\right) dx = \frac{\mathrm{B}\left[\mu,\frac{1}{2}(1-\lambda-\nu)-\mu\right] u^{\frac{1}{2}\mu+\frac{1}{2}\nu}}{a^\mu} S_{\lambda+\mu,\mu+\nu}\left(a\sqrt{u}\right)$

$$\left[|\arg\left(a\sqrt{u}\right)|<\pi,\quad 0<2\operatorname{Re}\mu<1-\operatorname{Re}(\lambda+\nu)\right]\quad \text{ET II 211(71)}$$

6.863 $\displaystyle\int_0^\infty \sqrt{x}\, e^{-\alpha x} s_{\mu,\frac{1}{4}}\left(\frac{x^2}{2}\right) dx = 2^{-2\mu-1}\sqrt{\alpha}\,\Gamma\left(2\mu+\frac{3}{2}\right) S_{-\mu-1,\frac{1}{4}}\left(\frac{\alpha^2}{2}\right)$

$$\left[\operatorname{Re}\alpha>0,\quad \operatorname{Re}\mu>-\tfrac{3}{4}\right]\quad \text{ET I 209(38)}$$

6.864 $\displaystyle\int_0^\infty \exp[(\mu+1)x]\, s_{\mu,\nu}\,(a\sinh x)\,dx = 2^{\mu-2}\pi\,\mathrm{cosec}(\mu\pi)\,\Gamma(\varrho)\,\Gamma(\sigma)$

$$\times \left[I_\varrho\left(\frac{a}{2}\right) I_\sigma\left(\frac{a}{2}\right) - I_{-\varrho}\left(\frac{a}{2}\right) I_{-\sigma}\left(\frac{a}{2}\right) \right]$$

$$2\varrho = \mu+\nu+1, \quad 2\sigma = \mu-\nu+1 \qquad [a > 0, \quad -2 < \mathrm{Re}\,\mu < 0] \quad \text{ET II 386(22)}$$

6.865 $\displaystyle\int_0^\infty \sqrt{\sinh x}\,\cosh(\nu x)\, S_{\mu,\frac{1}{2}}\,(a\cosh x)\,dx = \frac{\mathrm{B}\left(\frac{1}{4}-\frac{\mu+\nu}{2},\,\frac{1}{4}-\frac{\mu-\nu}{2}\right)}{\sqrt{a}\,2^{\mu+\frac{3}{2}}}\, S_{\mu+\frac{1}{2},\nu}(a)$

$$\left[|\arg a| < \pi, \quad \mathrm{Re}\,\mu + |\mathrm{Re}\,\nu| < \tfrac{1}{2} \right]$$
$$\text{ET II 388(31)}$$

6.866

1. $\displaystyle\int_0^\infty x^{-\mu-1}\cos(ax)\,s_{\mu,\nu}(x)\,dx$

$$= 0 \qquad\qquad\qquad\qquad\qquad\qquad\qquad\qquad [a > 1]$$

$$= 2^{\mu-\frac{1}{2}}\sqrt{\pi}\,\Gamma\left(\frac{\mu+\nu+1}{2}\right)\Gamma\left(\frac{\mu-\nu+1}{2}\right)(1-a^2)^{\frac{1}{2}\mu+\frac{1}{4}}\,P_{\nu-\frac{1}{2}}^{\mu-\frac{1}{2}}(a) \quad [0 < a < 1]$$
$$\text{ET II 386(18)}$$

2. $\displaystyle\int_0^\infty x^{-\mu}\sin(ax)\,S_{\mu,\nu}(x)\,dx = 2^{-\mu-\frac{1}{2}}\sqrt{\pi}\,\Gamma\left(1-\frac{\mu+\nu}{2}\right)\Gamma\left(1-\frac{\mu-\nu}{2}\right)(a^2-1)^{\frac{1}{2}\mu-\frac{1}{4}}\,P_{\nu-\frac{1}{2}}^{\mu-\frac{1}{2}}(a)$

$$[a > 1, \quad \mathrm{Re}\,\mu < 1 - |\mathrm{Re}\,\nu|]$$
$$\text{ET II 387(23)}$$

6.867

1. $\displaystyle\int_0^{\pi/2}\cos(2\mu x)\, S_{2\mu-1,2\nu}\,(a\cos x)\,dx$

$$= \frac{\pi 2^{2\mu-3}a^{2\mu}\,\mathrm{cosec}(2\nu\pi)}{\Gamma(1-\mu-\nu)\,\Gamma(1-\mu+\nu)}\left[J_{\mu+\nu}\left(\frac{a}{2}\right) Y_{\mu-\nu}\left(\frac{a}{2}\right) - J_{\mu-\nu}\left(\frac{a}{2}\right) Y_{\mu+\nu}\left(\frac{a}{2}\right) \right]$$
$$[\mathrm{Re}\,\mu > -2, \quad |\mathrm{Re}\,\nu| < 1] \quad \text{ET II 388(29)}$$

2. $\displaystyle\int_0^{\pi/2}\cos\left[(\mu+1)x\right]\,s_{\mu,\nu}\,(a\cos x)\,dx = 2^{\mu-2}\pi\,\Gamma(\varrho)\,\Gamma(\sigma)\, J_\varrho\left(\frac{a}{2}\right) J_\sigma\left(\frac{a}{2}\right)$

$$2\varrho = \mu+\nu+1, \quad 2\sigma = \mu-\nu+1 \qquad [\mathrm{Re}\,\mu > -2] \quad \text{ET II 386(21)}$$

6.868 $\displaystyle\int_0^{\pi/2}\frac{\cos(2\mu x)}{\cos x}\, S_{2\mu,2\nu}\,(a\sec x)\,dx = \frac{\pi 2^{2\mu-1}}{a}\, W_{\mu,\nu}\left(ae^{i\frac{\pi}{2}}\right) W_{\mu,\nu}\left(ae^{-i\frac{\pi}{2}}\right)$

$$[|\arg a| < \pi, \quad \mathrm{Re}\,\mu < 1] \quad \text{ET II 388(30)}$$

6.869

1. $\displaystyle\int_0^\infty x^{1-\mu-\nu}\, J_\nu(ax)\, S_{\mu,-\mu-2\nu}(x)\,dx = \frac{\sqrt{\pi}\,a^{\nu-1}\,\Gamma(1-\mu-\nu)}{2^{\mu+2\nu}\,\Gamma\left(\nu+\frac{1}{2}\right)}(a^2-1)^{\frac{1}{2}(\mu+\nu-1)}\,P_{\mu+\nu}^{\mu+\nu-1}(a)$

$$[a > 1, \quad \mathrm{Re}\,\nu > -\tfrac{1}{2}, \quad \mathrm{Re}(\mu+\nu) < 1]$$
$$\text{ET II 388(28)}$$

2. $\displaystyle\int_0^\infty x^{-\mu} J_\nu(ax)\, s_{\nu+\mu,-\nu+\mu+1}(x)\, dx$

$$= 2^{\nu-1}\,\Gamma(\nu)a^{-\nu}\left(1-a^2\right)^\mu \qquad \left[0<a<1,\quad \operatorname{Re}\mu>-1,\quad -1e<\operatorname{Re}\nu<\tfrac{3}{2}\right]$$

$$= 0 \qquad\qquad\qquad\qquad \left[1<a,\quad \operatorname{Re}\mu>-1,\quad -1<\operatorname{Re}\nu<\tfrac{3}{2}\right]$$

ET II 388(28)

3. $\displaystyle\int_0^\infty x\, K_\nu(bx)\, s_{\mu,\frac{1}{2}\nu}\left(ax^2\right) dx = \frac{1}{4a}\,\Gamma\left(\mu+\frac{1}{2}\nu+1\right)\Gamma\left(\mu-\frac{1}{2}\nu+1\right)S_{-\mu-1,\frac{1}{2}\nu}\left(\frac{b^2}{4a}\right)$

$$\left[\operatorname{Re}\mu>\tfrac{1}{2}|\operatorname{Re}\nu|-2,\quad a>0,\quad \operatorname{Re}b>0\right] \qquad \text{ET II 151(78)}$$

6.87 Thomson functions

6.871

1. $\displaystyle\int_0^\infty e^{-\beta x}\,\operatorname{ber} x\, dx = \frac{\left(\sqrt{\beta^4+1}+\beta^2\right)^{1/2}}{\sqrt{2\left(\beta^4+1\right)}}$

ME 40

2. $\displaystyle\int_0^\infty e^{-\beta x}\,\operatorname{bei} x\, dx = \frac{\left(\sqrt{\beta^4+1}-\beta^2\right)^{1/2}}{\sqrt{2\left(\beta^4+1\right)}}$

ME 40

6.872

1. $\displaystyle\int_0^\infty e^{-\beta x}\,\operatorname{ber}_\nu\left(2\sqrt{x}\right) dx = \frac{1}{2\beta}\sqrt{\frac{\pi}{\beta}}\left[J_{\frac{1}{2}(\nu-1)}\left(\frac{1}{2\beta}\right)\cos\left(\frac{1}{2\beta}+\frac{3\nu\pi}{4}\right)\right.$

$$\left.- J_{\frac{1}{2}(\nu+1)}\left(\frac{1}{2\beta}\right)\cos\left(\frac{1}{2\beta}+\frac{3\nu+6}{4}\pi\right)\right]$$

MI 49

2. $\displaystyle\int_0^\infty e^{-\beta x}\,\operatorname{bei}_\nu\left(2\sqrt{x}\right) dx = \frac{1}{2\beta}\sqrt{\frac{\pi}{\beta}}\left[J_{\frac{1}{2}(\nu-1)}\left(\frac{1}{2\beta}\right)\sin\left(\frac{1}{2\beta}+\frac{3\nu}{4}\pi\right)\right.$

$$\left.- J_{\frac{1}{2}(\nu+1)}\left(\frac{1}{2\beta}\right)\sin\left(\frac{1}{2\beta}+\frac{3\nu+6}{4}\pi\right)\right]$$

MI 49

3. $\displaystyle\int_0^\infty e^{-\beta x}\,\operatorname{ber}\left(2\sqrt{x}\right) dx = \frac{1}{\beta}\cos\frac{1}{\beta}$

ME 40

4. $\displaystyle\int_0^\infty e^{-\beta x}\,\operatorname{bei}\left(2\sqrt{x}\right) dx = \frac{1}{\beta}\sin\frac{1}{\beta}$

ME 40

5. $\displaystyle\int_0^\infty e^{-\beta x}\,\operatorname{ker}\left(2\sqrt{x}\right) dx = -\frac{1}{2\beta}\left[\cos\frac{1}{\beta}\operatorname{ci}\frac{1}{\beta}+\sin\frac{1}{\beta}\operatorname{si}\frac{1}{\beta}\right]$

MI 50

6. $\displaystyle\int_0^\infty e^{-\beta x}\,\operatorname{kei}\left(2\sqrt{x}\right) dx = -\frac{1}{2\beta}\left[\sin\frac{1}{\beta}\operatorname{ci}\frac{1}{\beta}-\cos\frac{1}{\beta}\operatorname{si}\frac{1}{\beta}\right]$

MI 50

7. $\int_0^\infty e^{-\beta x} \, \mathrm{ber}_\nu \left(2\sqrt{x} \right) \mathrm{bei}_\nu \left(2\sqrt{x} \right) \, dx = \frac{1}{2\beta} J_\nu \left(\frac{2}{\beta} \right) \sin \left(\frac{2}{\beta} + \frac{3\nu\pi}{2} \right)$

$$[\mathrm{Re}\,\nu > -1]$$ MI 49

6.873 $\int_0^\infty \left[\mathrm{ber}_\nu^2 \left(2\sqrt{x} \right) + \mathrm{bei}_\nu^2 \left(2\sqrt{x} \right) \right] e^{-\beta x} \, dx = \frac{1}{\beta} I_\nu \left(\frac{2}{\beta} \right)$

$$[\mathrm{Re}\,\nu > -1]$$ ME 40

6.874

1. $\int_0^\infty \frac{e^{-\beta x}}{\sqrt{x}} \, \mathrm{ber}_{2\nu} \left(2\sqrt{2x} \right) \, dx = \sqrt{\frac{\pi}{\beta}} \, J_\nu \left(\frac{1}{\beta} \right) \cos \left(\frac{1}{\beta} - \frac{3\pi}{4} + \frac{3\nu\pi}{2} \right)$

$$\left[\mathrm{Re}\,\nu > -\tfrac{1}{2} \right]$$ MI 49

2. $\int_0^\infty \frac{e^{-\beta x}}{\sqrt{x}} \, \mathrm{bei}_{2\nu} \left(2\sqrt{2x} \right) \, dx = \sqrt{\frac{\pi}{\beta}} \, J_\nu \left(\frac{1}{\beta} \right) \sin \left(\frac{1}{\beta} - \frac{3\pi}{4} + \frac{3\nu\pi}{2} \right)$

$$\left[\mathrm{Re}\,\nu > -\tfrac{1}{2} \right]$$ MI 49

3. $\int_0^\infty x^{\frac{\nu}{2}} \, \mathrm{ber}_\nu \left(\sqrt{x} \right) e^{-\beta x} \, dx = \frac{2^{-\nu}}{\beta^{1+\nu}} \cos \left(\frac{1}{4\beta} + \frac{3\nu\pi}{4} \right)$ $[\mathrm{Re}\,\nu > -1]$ ME 40

4. $\int_0^\infty x^{\frac{\nu}{2}} \, \mathrm{bei}_\nu \left(\sqrt{x} \right) e^{-\beta x} \, dx = \frac{2^{-\nu}}{\beta^{1+\nu}} \sin \left(\frac{1}{4\beta} + \frac{3\nu\pi}{4} \right)$ $[\mathrm{Re}\,\nu > -1]$ ME 40

6.875

1. $\int_0^\infty e^{-\beta x} \left[\mathrm{ker} \left(2\sqrt{x} \right) - \frac{1}{2} \ln x \, \mathrm{ber} \left(2\sqrt{x} \right) \right] \, dx = \frac{1}{\beta} \left[\ln \beta \cos \frac{1}{\beta} + \frac{\pi}{4} \sin \frac{1}{\beta} \right]$ MI 50

2. $\int_0^\infty e^{-\beta x} \left[\mathrm{kei} \left(2\sqrt{x} \right) - \frac{1}{2} \ln x \, \mathrm{bei} \left(2\sqrt{x} \right) \right] \, dx = \frac{1}{\beta} \left[\ln \beta \sin \frac{1}{\beta} - \frac{\pi}{4} \cos \frac{1}{\beta} \right]$ MI 50

6.876

1. $\int_0^\infty x \, \mathrm{kei}\, x \, J_1(ax) \, dx = -\frac{1}{2a} \arctan a^2$ $[a > 0]$ ET II 21(32)

2. $\int_0^\infty x \, \mathrm{ker}\, x \, J_1(ax) \, dx = \frac{1}{2a} \ln \sqrt{(1 + a^4)}$ $[a > 0]$ ET II 21(33)

6.9 Mathieu Functions

Notation: $k^2 = q$. For definition of the coefficients $A_p^{(m)}$ and $B_p^{(m)}$ see section 8.6.

6.91 Mathieu functions

6.911

1. $\int_0^{2\pi} \mathrm{ce}_m(z, q) \, \mathrm{ce}_p(z, q) \, dz = 0$ $[m \neq p]$ MA

2. $\int_0^{2\pi} [\mathrm{ce}_{2n}(z, q)]^2 \, dz = 2\pi \left[A_0^{(2n)} \right]^2 + \pi \sum_{r=1}^\infty \left[A_{2r}^{(2n)} \right]^2 = \pi$ MA

3. $$\int_0^{2\pi} \left[ce_{2n+1}(z,q)\right]^2 dz = \pi \sum_{r=0}^{\infty} \left[A_{2r+1}^{(2n+1)}\right]^2 = \pi \qquad \text{MA}$$

4. $$\int_0^{2\pi} se_m(z,q)\, se_p(z,q)\, dz = 0 \qquad [m \neq p] \qquad \text{MA}$$

5. $$\int_0^{2\pi} \left[se_{2n+1}(z,q)\right]^2 dz = \pi \sum_{r=0}^{\infty} \left[B_{2r+1}^{(2n+1)}\right]^2 = \pi \qquad \text{MA}$$

6. $$\int_0^{2\pi} \left[se_{2n+2}(z,q)\right]^2 dz = \pi \sum_{r=0}^{\infty} \left[B_{2r+2}^{(2n+2)}\right]^2 = \pi \qquad \text{MA}$$

7. $$\int_0^{2\pi} se_m(z,q)\, ce_p(z,q)\, dz = 0 \qquad [m = 1,2,\ldots; \quad p = 1,2,\ldots] \qquad \text{MA}$$

6.92 Combinations of Mathieu, hyperbolic, and trigonometric functions

6.921

1. $$\int_0^{\pi} \cosh\left(2k\cos u \sinh z\right) ce_{2n}(u,q)\, du = \frac{\pi A_0^{(2n)}}{ce_{2n}\left(\frac{\pi}{2},q\right)} (-1)^n\, Ce_{2n}(z,-q)$$
$$[q > 0] \qquad \text{MA}$$

2. $$\int_0^{\pi} \cosh\left(2k\sin u \cosh z\right) ce_{2n}(u,q)\, du = \frac{\pi A_0^{(2n)}}{ce_{2n}(0,q)} (-1)^n\, Ce_{2n}(z,-q)$$
$$[q > 0] \qquad \text{MA}$$

3. $$\int_0^{\pi} \sinh\left(2k\sin u \cosh z\right) se_{2n+1}(u,q)\, du = \frac{\pi k B_1^{(2n+1)}}{se'_{2n+1}(0,q)} (-1)^n\, Ce_{2n+1}(z,-q)$$
$$[q > 0] \qquad \text{MA}$$

4. $$\int_0^{\pi} \sinh\left(2k\cos u \sinh z\right) ce_{2n+1}(u,q)\, du = \frac{\pi k A_1^{(2n+1)}}{ce'_{2n+1}\left(\frac{\pi}{2},q\right)} (-1)^{n+1}\, Se_{2n+1}(z,-q)$$
$$[q > 0] \qquad \text{MA}$$

5. $$\int_0^{\pi} \sinh\left(2k\sin u \sin z\right) se_{2n+1}(u,q)\, du = \frac{\pi k B_1^{(2n+1)}}{se'_{2n+1}(0,q)} se_{2n+1}(z,q)$$
$$[q > 0] \qquad \text{MA}$$

6.922

1. $$\int_0^{\pi} \cos u \cosh z \cos\left(2k\sin u \sinh z\right) ce_{2n+1}(u,q)\, du = \frac{\pi A_1^{(2n+1)}}{2\, ce_{2n+1}(0,q)} Ce_{2n+1}(z,q)$$
$$[q > 0] \qquad \text{MA}$$

2. $$\int_0^{\pi} \sin u \sinh z \cos\left(2k\cos u \cosh z\right) se_{2n+1}(u,q)\, du = \frac{\pi B_1^{(2n+1)}}{2\, se_{2n+1}\left(\frac{\pi}{2},q\right)} Se_{2n+1}(z,q)$$
$$[q > 0] \qquad \text{MA}$$

3. $\quad \displaystyle\int_0^\pi \sin u \sinh z \sin\left(2k\cos u \cosh z\right) \mathrm{se}_{2n+2}(u,q)\,du = -\frac{\pi k B_2^{(2n+2)}}{2\,\mathrm{se}'_{2n+2}\left(\frac{\pi}{2},q\right)} \mathrm{Se}_{2n+2}(z,q)$

$\qquad\qquad\qquad\qquad\qquad\qquad\qquad\qquad\qquad\qquad [q>0]$ 　　　　MA

4. $\quad \displaystyle\int_0^\pi \cos u \cosh z \sin\left(2k\sin u \sinh z\right) \mathrm{se}_{2n+2}(u,q)\,du = \frac{\pi k B_2^{(2n+2)}}{2\,\mathrm{se}'_{2n+2}(0,q)} \mathrm{Se}_{2n+2}(z,q)$

$\qquad\qquad\qquad\qquad\qquad\qquad\qquad\qquad\qquad\qquad [q>0]$ 　　　　MA

5. $\quad \displaystyle\int_0^\pi \sin u \cosh z \cosh\left(2k\cos u \sinh z\right) \mathrm{se}_{2n+1}(u,q)\,du = \frac{\pi B_1^{(2n+1)}}{2\,\mathrm{se}_{2n+1}\left(\frac{\pi}{2},q\right)} (-1)^n \mathrm{Ce}_{2n+1}(z,-q)$

$\qquad\qquad\qquad\qquad\qquad\qquad\qquad\qquad\qquad\qquad [q>0]$ 　　　　MA

6. $\quad \displaystyle\int_0^\pi \cos u \sinh z \cosh\left(2k\sin u \cosh z\right) \mathrm{ce}_{2n+1}(u,q)\,du = \frac{\pi A_1^{(2n+1)}}{2\,\mathrm{ce}_{2n+1}(0,q)} (-1)^n \mathrm{Se}_{2n+1}(z,-q)$

$\qquad\qquad\qquad\qquad\qquad\qquad\qquad\qquad\qquad\qquad [q>0]$ 　　　　MA

7. $\quad \displaystyle\int_0^\pi \sin u \cosh z \sinh\left(2k\cos u \sinh z\right) \mathrm{se}_{2n+2}(u,q)\,du = \frac{\pi k B_2^{(2n+2)}}{2\,\mathrm{se}'_{2n+2}\left(\dfrac{\pi}{2},q\right)} (-1)^{n+1} \mathrm{Se}_{2n+2}(z,-q)$

$\qquad\qquad\qquad\qquad\qquad\qquad\qquad\qquad\qquad\qquad [q>0]$ 　　　　MA

8. $\quad \displaystyle\int_0^\pi \cos u \sinh z \sinh\left(2k\sin u \cosh z\right) \mathrm{se}_{2n+2}(u,q)\,du = \frac{\pi k B_2^{(2n+2)}}{2\,\mathrm{se}'_{2n+2}(0,q)} (-1)^n \mathrm{Se}_{2n+2}(z,-q)$

$\qquad\qquad\qquad\qquad\qquad\qquad\qquad\qquad\qquad\qquad [q>0]$ 　　　　MA

6.923

1. $\quad \displaystyle\int_0^\infty \sin\left(2k\cosh z \cosh u\right) \sinh z \sinh u \,\mathrm{Se}_{2n+1}(u,q)\,du = -\frac{\pi B_1^{(2n+1)}}{4\,\mathrm{se}_{2n+1}\left(\frac{1}{2}\pi,q\right)} \mathrm{Se}_{2n+1}(z,q)$

$\qquad\qquad\qquad\qquad\qquad\qquad\qquad\qquad\qquad\qquad [q>0]$ 　　　　MA

2. $\quad \displaystyle\int_0^\infty \cos\left(2k\cosh z \cosh u\right) \sinh z \sinh u \,\mathrm{Se}_{2n+1}(u,q)\,du = -\frac{\pi B_1^{(2n+1)}}{4\,\mathrm{se}_{2n+1}\left(\frac{1}{2}\pi,q\right)} \mathrm{Gey}_{2n+1}(z,q)$

$\qquad\qquad\qquad\qquad\qquad\qquad\qquad\qquad\qquad\qquad [q>0]$ 　　　　MA

3. $\quad \displaystyle\int_0^\infty \sin\left(2k\cosh z \cosh u\right) \sinh z \sinh u \,\mathrm{Se}_{2n+2}(u,q)\,du = -\frac{k\pi B_2^{(2n+2)}}{4\,\mathrm{se}'_{2n+2}\left(\frac{1}{2}\pi,q\right)} \mathrm{Gey}_{2n+2}(z,q)$

$\qquad\qquad\qquad\qquad\qquad\qquad\qquad\qquad\qquad\qquad [q>0]$ 　　　　MA

4. $\quad \displaystyle\int_0^\infty \cos\left(2k\cosh z \cosh u\right) \sinh z \sinh u \,\mathrm{Se}_{2n+2}(u,q)\,du = -\frac{k\pi B_2^{(2n+2)}}{4\,\mathrm{se}_{2n+2}\left(\frac{1}{2}\pi,q\right)} \mathrm{Se}_{2n+2}(z,q)$

$\qquad\qquad\qquad\qquad\qquad\qquad\qquad\qquad\qquad\qquad [q>0]$ 　　　　MA

5. $\quad \displaystyle\int_0^\infty \sin\left(2k\cosh z \cosh u\right) \mathrm{Ce}_{2n}(u,q)\,du = \frac{\pi A_0^{(2n)}}{2\,\mathrm{ce}_{2n}\left(\frac{1}{2}\pi,q\right)} \mathrm{Ce}_{2n}(z,q)$

$\qquad\qquad\qquad\qquad\qquad\qquad\qquad\qquad\qquad\qquad [q>0]$ 　　　　MA

6. $$\int_0^\infty \cos\left(2k \cosh z \cosh u\right) \mathrm{Ce}_{2n}(u,q)\, du = -\frac{\pi A_0^{(2n)}}{2\, \mathrm{ce}_{2n}\left(\frac{1}{2}\pi, q\right)}\, \mathrm{Fey}_{2n}(z,q)$$

$$[q > 0] \qquad \text{MA}$$

7. $$\int_0^\infty \sin\left(2k \cosh z \cosh u\right) \mathrm{Ce}_{2n+1}(u,q)\, du = \frac{k\pi A_1^{(2n+1)}}{2\, \mathrm{ce}'_{2n+1}\left(\frac{1}{2}\pi, q\right)}\, \mathrm{Fey}_{2n+1}(z,q)$$

$$[q > 0] \qquad \text{MA}$$

8. $$\int_0^\infty \cos\left(2k \cosh z \cosh u\right) \mathrm{Ce}_{2n+1}(u,q)\, du = \frac{k\pi A_1^{(2n+1)}}{2\, \mathrm{ce}'_{2n+1}\left(\frac{1}{2}\pi, q\right)}\, \mathrm{Ce}_{2n+1}(z,q)$$

$$[q > 0] \qquad \text{MA}$$

6.924

1. $$\int_0^\pi \cos\left(2k \cos u \cos z\right) \mathrm{ce}_{2n}(u,q)\, du = \frac{\pi A_0^{(2n)}}{\mathrm{ce}_{2n}\left(\frac{1}{2}\pi, q\right)}\, \mathrm{ce}_{2n}(z,q)$$

$$[q > 0] \qquad \text{MA}$$

2. $$\int_0^\pi \sin\left(2k \cos u \cos z\right) \mathrm{ce}_{2n+1}(u,q)\, du = -\frac{\pi k A_1^{(2n+1)}}{\mathrm{ce}'_{2n+1}\left(\frac{1}{2}\pi, q\right)}\, \mathrm{ce}_{2n+1}(z,q)$$

$$[q > 0] \qquad \text{MA}$$

3. $$\int_0^\pi \cos\left(2k \cos u \cosh z\right) \mathrm{ce}_{2n}(u,q)\, du = \frac{\pi A_0^{(2n)}}{\mathrm{ce}_{2n}\left(\frac{1}{2}\pi, q\right)}\, \mathrm{Ce}_{2n}(z,q)$$

$$[q > 0] \qquad \text{MA}$$

4. $$\int_0^\pi \cos\left(2k \sin u \sinh z\right) \mathrm{ce}_{2n}(u,q)\, du = \frac{\pi A_0^{(2n)}}{\mathrm{ce}_{2n}(0, q)}\, \mathrm{Ce}_{2n}(z,q)$$

$$[q > 0] \qquad \text{MA}$$

5. $$\int_0^\pi \sin\left(2k \cos u \cosh z\right) \mathrm{ce}_{2n+1}(u,q)\, du = -\frac{\pi k A_1^{(2n+1)}}{\mathrm{ce}'_{2n+1}\left(\frac{1}{2}\pi, q\right)}\, \mathrm{Ce}_{2n+1}(z,q)$$

$$[q > 0] \qquad \text{MA}$$

6. $$\int_0^\pi \sin\left(2k \sin u \sinh z\right) \mathrm{se}_{2n+1}(u,q)\, du = \frac{\pi k B_1^{(2n+1)}}{\mathrm{se}''_{2n+1}(0, q)}\, \mathrm{Se}_{2n+1}(z,q)$$

$$[q > 0] \qquad \text{MA}$$

6.925   **Notation:** $z_1 = 2k\sqrt{\cosh^2 \xi - \sin^2 \eta}$, and $\tan\alpha = \tanh\xi \tan\eta$

1. $$\int_0^{2\pi} \sin\left[z_1 \cos(\theta - \alpha)\right] \mathrm{ce}_{2n}(\theta, q)\, d\theta = 0.$$

$$\text{MA}$$

2. $$\int_0^{2\pi} \cos\left[z_1 \cos(\theta - \alpha)\right] \mathrm{ce}_{2n}(\theta, q)\, d\theta = \frac{2\pi A_0^{(2n)}}{\mathrm{ce}_{2n}(0, q)\, \mathrm{ce}_{2n}\left(\frac{1}{2}\pi, q\right)}\, \mathrm{Ce}_{2n}(\xi, q)\, \mathrm{ce}_{2n}(\eta, q)$$

$$\text{MA}$$

3. $\displaystyle \int_0^{2\pi} \sin\left[z_1 \cos(\theta - \alpha)\right] \mathrm{ce}_{2n+1}(\theta, q)\, d\theta = -\frac{2\pi k A_1^{(2n+1)}}{\mathrm{ce}_{2n+1}(0, q)\, \mathrm{ce}'_{2n+1}\left(\frac{1}{2}\pi, q\right)} \mathrm{Ce}_{2n+1}(\xi, q)\, \mathrm{ce}_{2n+1}(\eta, q)$

MA

4. $\displaystyle \int_0^{2\pi} \cos\left[z_1 \cos(\theta - \alpha)\right] \mathrm{ce}_{2n+1}(\theta, q)\, d\theta = 0$

MA

5. $\displaystyle \int_0^{2\pi} \sin\left[z_1 \cos(\theta - \alpha)\right] \mathrm{se}_{2n+1}(\theta, q)\, d\theta = \frac{2\pi k B_1^{(2n+1)}}{\mathrm{se}_{2n+1}(0, q)\, \mathrm{se}_{2n+1}\left(\frac{1}{2}\pi, q\right)} \mathrm{Se}_{2n+1}(\xi, q)\, \mathrm{se}_{2n+1}(\eta, q)$

MA

6. $\displaystyle \int_0^{2\pi} \cos\left[z_1 \cos(\theta - \alpha)\right] \mathrm{se}_{2n+1}(\theta, q)\, d\theta = 0$

MA

7. $\displaystyle \int_0^{2\pi} \sin\left[z_1 \cos(\theta - \alpha)\right] \mathrm{se}_{2n+2}(\theta, q)\, d\theta = 0$

MA

8. $\displaystyle \int_0^{2\pi} \cos\left[z_1 \cos(\theta - \alpha)\right] \mathrm{se}_{2n+2}(\theta, q)\, d\theta = \frac{2\pi k^2 B_2^{(2n+2)}}{\mathrm{se}'_{2n+2}(0, q)\, \mathrm{se}'_{2n+2}\left(\frac{1}{2}\pi, q\right)} \mathrm{Se}_{2n+2}(\xi, q)\, \mathrm{se}_{2n+2}(\eta, q)$

MA

6.926 $\displaystyle \int_0^{\pi} \sin u \sin z \sin\left(2k \cos u \cos z\right) \mathrm{se}_{2n+2}(u, q)\, du = -\frac{\pi k B_2^{(2n+2)}}{2\, \mathrm{se}'_{2n+2}\left(\frac{\pi}{2}, q\right)} \mathrm{se}_{2n+2}(z, q)$

$$[q > 0]$$

MA

6.93 Combinations of Mathieu and Bessel functions

6.931

1. $\displaystyle \int_0^{\pi} J_0\left\{k\left[2\left(\cos 2u + \cos 2z\right)\right]^{1/2}\right\} \mathrm{ce}_{2n}(u, q)\, du = \frac{\pi\left[A_0^{(2n)}\right]^2}{\mathrm{ce}_{2n}(0, q)\, \mathrm{ce}_{2n}\left(\frac{\pi}{2}, q\right)} \mathrm{ce}_{2n}(z, q)$

MA

2. $\displaystyle \int_0^{2\pi} Y_0\left\{k\left[2\left(\cos 2u + \cosh 2z\right)\right]^{1/2}\right\} \mathrm{ce}_{2n}(u, q)\, du = \frac{2\pi\left[A_0^{(2n)}\right]^2}{\mathrm{ce}_{2n}(0, q)\, \mathrm{ce}_{2n}\left(\frac{\pi}{2}, q\right)} \mathrm{Fey}_{2n}(z, q)$

MA

6.94* Relationships between eigenfunctions of the Helmholtz equation in different coordinate systems

Notation: Particular solutions of the Helmholtz equation in three-dimensional infinite space

$$\nabla^2 \Psi + k^2 \Psi = 0$$

in Cartesian (x, y, z), spherical (r, θ, ϕ), and cylindrical (ρ, z, ϕ) coordinates are

$$\Psi_{k_x k_y k_z}(x, y, z) \propto \exp\left[i\left(k_x x + k_y y + k_z z\right)\right] \quad \text{with} \quad k^2 = k_x^2 + k_y^2 + k_z^2$$

$$\Psi_{lm}(r, \theta, \phi) \propto \exp(im\phi)\sqrt{\frac{k}{r}}\, Z_{l+1/2}(kr)\, P_l^m(\cos\theta)$$

$$\Psi_{mk_z}(\rho, z, \phi) \propto \exp\left[i\left(m\phi + k_z z\right)\right] Z_{l+1/2}\left(\rho\sqrt{k^2 - k_z^2}\right)$$

with $P_l^m(\cos\theta)$ the associated Legendre function, Z is any Bessel function, $m = 0, 1, \ldots, l$; $l \in \mathbb{N}$, $r^2 = \rho^2 + z^2$, $\rho = r\sin\theta$, $z = r\cos\theta$, $\phi = \operatorname{arccot}(x/y)$, and $k_t^2 = k^2 - k_z^2$.

6.941

1. $$\int_{-k}^{k} e^{i\rho z} J_m\left(\rho\sqrt{k^2 - \rho^2}\right) P_l^m\left(\frac{p}{k}\right) dp = i^{l-m}\sqrt{\frac{2\pi k}{r}} \, J_{l+1/2}(kr) P_l^m\left(\frac{z}{r}\right)$$

$$[\rho > 0, \quad l \geq m \geq 0]$$

2. $$\int_{-\infty}^{\infty} e^{-i\rho z} J_{l+1/2}(kr) P_l^m\left(\frac{z}{r}\right) dz = i^{m-l}\sqrt{\frac{2\pi r}{k}} \, J_m\left(\rho\sqrt{k^2 - \rho^2}\right) P_l^m\left(\frac{\rho}{k}\right)$$

$$[\rho > 0, \quad l \geq m \geq 0]$$

3. $$\int_0^{\infty} J_m\left(\rho k_t\right) \cos\left[k_x x + m\arcsin\left(\frac{x}{\rho}\right)\right] dx$$
$$= \frac{(-1)^m}{\sqrt{k_t^2 - k_x^2}} \cos\left[y\sqrt{k_t^2 - k_x^2} + m\arccos\left(\frac{k_x}{k_t}\right)\right] \qquad [k_x^2 < k_t^2]$$
$$= 0 \qquad\qquad [k_x^2 > k_t^2]$$

4. $$\int_0^{\infty} Y_m\left(\rho k_t\right) \cos\left[k_x x + m\arcsin\left(\frac{x}{\rho}\right)\right] dx$$
$$= \frac{(-1)^m}{\sqrt{k_t^2 - k_x^2}} \sin\left[y\sqrt{k_t^2 - k_x^2} + m\arccos\left(\frac{k_x}{k_t}\right)\right] \qquad [k_x^2 < k_t^2]$$
$$= \frac{(-1)^m}{\sqrt{k_x^2 - k_t^2}} \exp\left[-y\sqrt{k_x^2 - k_t^2} - m\operatorname{sign}(k_x)\operatorname{arccosh}\left(\frac{|k_x|}{k_t}\right)\right] \qquad [k_x^2 > k_t^2]$$

5. $$\int_{-\infty}^{\infty} H_{l+1/2}^{(j)}(kr) P_l^m\left(\frac{z}{r}\right) e^{-ik_z z} dx = i^{m-l}\sqrt{\frac{2\pi r}{k}} \, H_m^{(j)}\left(\rho\sqrt{k^2 - k_z^2}\right) P_l^m\left(\frac{k_z}{k}\right)$$

$$[\rho > 0]$$

The result is true for $j = 1$ if $\pi > \arg\sqrt{k^2 - k_z^2} \geq 0$, for $j = 2$ if $-\pi < \arg\sqrt{k^2 - k_z^2} \leq 0$.

6. $$\int_{-\infty}^{\infty} H_m^{(j)}\left(\rho\sqrt{k^2 - k_z^2}\right) P_l^m\left(\frac{k_z}{k}\right) e^{ik_z z} dk_z = i^{l-m}\sqrt{\frac{2\pi k}{r}} \, H_{l+1/2}^{(j)}(kr) P_l^m\left(\frac{z}{r}\right)$$

The result is true for $j = 1$ if $\pi > \arg\sqrt{k^2 - k_z^2} \geq 0$, for $j = 2$ if $-\pi < \arg\sqrt{k^2 - k_z^2} \leq 0$.

7. $$\int_{-\infty}^{\infty} J_{l+1/2}(kr) P_l^m\left(\frac{z}{r}\right) e^{-ik_z z} dz = i^{m-l}\sqrt{\frac{2\pi r}{k}} \, J_m\left(\rho\sqrt{k^2 - k_z^2}\right) P_l^m\left(\frac{k_z}{k}\right) \qquad [k_z^2 < k^2]$$
$$= 0 \qquad\qquad [k_z^2 > k^2]$$

8. $$\int_{-k}^{k} J_m\left(\rho\sqrt{k^2 - k_z^2}\right) P_l^m\left(\frac{k_z}{k}\right) e^{ik_z z} dk_z = i^{l-m}\sqrt{\frac{2\pi k}{r}} \, J_{l+1/2}(kr) P_l^m\left(\frac{z}{r}\right)$$

9. $$\int_{-\infty}^{\infty} Y_{l+1/2}(kr) P_l^m\left(\frac{z}{r}\right) e^{-ik_z z} dz = i^{m-l}\sqrt{\frac{2\pi r}{k}} \, Y_m\left(\rho\sqrt{k^2 - k_z^2}\right) P_l^m\left(\frac{k_z}{k}\right) \qquad [k_z^2 < k^2]$$
$$= -2i^{m-l}\sqrt{\frac{2r}{k\pi}} \, K_m\left(\rho\sqrt{k_z^2 - k^2}\right) P_l^m\left(\frac{k_z}{k}\right) \qquad [k_z^2 > k^2]$$

10.　$i^{l-m} \displaystyle\int_{-k}^{k} Y_m \left(\rho \sqrt{k^2 - k_z^2} \right) P_l^m \left(\frac{k_z}{k} \right) e^{ik_z z} \, dk_z$

$$- \frac{4}{\pi} \int_k^{\infty} \cos \left[k_z z + \tfrac{1}{2}\pi (m - l) \right] P_l^m \left(\frac{k_z}{k} \right) K_m \left(\rho \sqrt{k_z^2 - k^2} \right) e^{ik_z z} \, dk_z$$

$$= \sqrt{\frac{2\pi k}{r}} \, Y_{l+1/2}(kr) P_l^m \left(\frac{z}{r} \right)$$

7.1–7.2 Associated Legendre Functions

7.11 Associated Legendre functions

7.111 $\displaystyle\int_{\cos\varphi}^{1} P_\nu(x)\,dx = \sin\varphi\,P_\nu^{-1}(\cos\varphi)$ MO 90

7.112

1. $\displaystyle\int_{-1}^{1} P_n^m(x)\,P_k^m(x)\,dx = 0$ $[n \neq k]$

   $\displaystyle = \frac{2}{2n+1}\frac{(n+m)!}{(n-m)!}$ $[n = k]$

SM III 185, WH

2. $\displaystyle\int_{-1}^{1} Q_n^m(x)\,P_k^m(x)\,dx = (-1)^m \frac{1-(-1)^{n+k}(n+m)!}{(k-n)(k+n+1)(n-m)!}$ EH I 171(18)

3. $\displaystyle\int_{-1}^{1} P_\nu(x)\,P_\sigma(x)\,dx$

   $\displaystyle = \frac{2\pi\sin\pi(\sigma-\nu)+4\sin(\pi\nu)\sin(\pi\sigma)\left[\psi(\nu+1)-\psi(\sigma+1)\right]}{\pi^2(\sigma-\nu)(\sigma+\nu+1)}$ $[\sigma+\nu+1\neq 0]$ EH I 170(7)

   $\displaystyle = \frac{\pi^2-2(\sin\pi\nu)^2\,\psi'(\nu+1)}{\pi^2\left(\nu+\dfrac{1}{2}\right)}$ $[\sigma=\nu]$ EH I 170(9)a

4. $\displaystyle\int_{-1}^{1} Q_\nu(x)\,Q_\sigma(x)\,dx = \frac{\left[\psi(\nu+1)-\psi(\sigma+1)\right]\left[1+\cos(\pi\sigma)\cos(\nu\pi)\right]-\frac{\pi}{2}\sin\pi(\nu-\sigma)}{(\sigma-\nu)(\sigma+\nu+1)}$

   $[\sigma+\nu+1\neq 0;\quad \nu,\quad \sigma\neq -1,-2,-3,\ldots]$

EH I 170(11)

   $\displaystyle = \frac{\frac{1}{2}\pi^2-\psi'(\nu+1)\left[1+(\cos\nu\pi)^2\right]}{2\nu+1}$

   $[\nu=\sigma,\quad \nu\neq -1,-2,-3,\ldots]$

EH I 170(12)

5. $\displaystyle\int_{-1}^{1} P_\nu(x)\,Q_\sigma(x)\,dx = \frac{1-\cos\pi(\sigma-\nu)-2\pi^{-1}\sin(\pi\nu)\cos(\pi\sigma)\left[\psi(\nu+1)-\psi(\sigma+1)\right]}{(\nu-\sigma)(\nu+\sigma+1)}$

   $[\operatorname{Re}\nu>0,\quad \operatorname{Re}\sigma>0,\quad \sigma\neq\nu]$

EH I 170(13)

   $\displaystyle = -\frac{\sin(2\nu\pi)\,\psi'(\nu+1)}{\pi(2\nu+1)}$

   $[\operatorname{Re}\nu>0,\quad \sigma=\nu]$

EH I 171(14)

7.113   **Notation:** $\displaystyle A = \frac{\Gamma\left(\frac{1}{2}+\frac{\nu}{2}\right)\Gamma\left(1+\frac{\sigma}{2}\right)}{\Gamma\left(\frac{1}{2}+\frac{\sigma}{2}\right)\Gamma\left(1+\frac{\nu}{2}\right)}$

1. $\displaystyle\int_{0}^{1} P_\nu(x)\,P_\sigma(x)\,dx = \frac{A\sin\frac{\pi\sigma}{2}\cos\frac{\pi\nu}{2}-A^{-1}\sin\frac{\pi\nu}{2}\cos\frac{\pi\sigma}{2}}{\frac{1}{2}\pi(\sigma-\nu)(\sigma+\nu+1)}$ EH I 171(15)

2. $\displaystyle\int_0^1 Q_\nu(x)\,Q_\sigma(x)\,dx = \frac{\psi(\nu+1) - \psi(\sigma+1) - \frac{\pi}{2}\left[\left(A - A^{-1}\right)\sin\frac{\pi(\sigma+\nu)}{2}\,\left(A + A^{-1}\right)\sin\frac{\pi(\sigma-\nu)}{2}\right]}{(\sigma-\nu)(\sigma+\nu+1)}$

$\qquad\qquad\qquad\qquad\qquad\qquad\qquad$ [$\operatorname{Re}\nu > 0, \quad \operatorname{Re}\sigma > 0$] $\qquad$ EH I 171(16)

3. $\displaystyle\int_0^1 P_\nu(x)\,Q_\sigma(x)\,dx = \frac{A^{-1}\cos\frac{\pi(\nu-\sigma)}{2} - 1}{(\sigma-\nu)(\sigma+\nu+1)}$ $\qquad$ [$\operatorname{Re}\nu > 0, \quad \operatorname{Re}\sigma > 0$] $\qquad$ EH I 171(17)

7.114

1. $\displaystyle\int_1^\infty P_\nu(x)\,Q_\sigma(x)\,dx = \frac{1}{(\sigma-\nu)(\sigma+\nu+1)}$ $\qquad$ [$\operatorname{Re}(\sigma-\nu) > 0, \quad \operatorname{Re}(\sigma+\nu) > -1$]

$\qquad\qquad\qquad\qquad\qquad\qquad\qquad\qquad\qquad\qquad\qquad\qquad\qquad$ ET II 324(19)

2. $\displaystyle\int_1^\infty Q_\nu(x)\,Q_\sigma(x)\,dx = \frac{\psi(\sigma+1) - \psi(\nu+1)}{(\sigma-\nu)(\sigma+\nu+1)}$

$\qquad\qquad\qquad\qquad\qquad\qquad$ [$\operatorname{Re}(\nu+\sigma) > -1; \quad \sigma,\nu \neq -1,-2,-3,\ldots$] $\quad$ EH I 170(5)

3. $\displaystyle\int_1^\infty [Q_\nu(x)]^2\,dx = \frac{\psi'(\nu+1)}{2\nu+1}$ $\qquad$ [$\operatorname{Re}\nu > -\frac{1}{2}$] $\qquad$ EH I 170(6)

7.115 $\displaystyle\int_1^\infty Q_\nu(x)\,dx = \frac{1}{\nu(\nu+1)}$ $\qquad$ [$\operatorname{Re}\nu > 0$] $\qquad$ ET II 324(18)

7.12–7.13 Combinations of associated Legendre functions and powers

7.121 $\displaystyle\int_{\cos\varphi}^1 x\,P_\nu(x)\,dx = \frac{-\sin\varphi}{(\nu-1)(\nu+2)}\left[\sin\varphi P_\nu(\cos\varphi) + \cos\varphi\,P_\nu^1(\cos\varphi)\right]$ $\qquad$ MO 90

7.122

1. $\displaystyle\int_0^1 \frac{[P_n^m(x)]^2}{1-x^2}\,dx = \frac{1}{2m}\frac{(n+m)!}{(n-m)!}$ $\qquad$ [$0 < m \leq n$] $\qquad$ MO 74

2. $\displaystyle\int_0^1 [P_\nu^\mu(x)]^2\,\frac{dx}{1-x^2} = -\frac{\Gamma(1+\mu+\nu)}{2\mu\,\Gamma(1-\mu+\nu)}$ $\qquad$ [$\operatorname{Re}\mu < 0, \quad \nu+\mu$ is a positive integer]

$\qquad\qquad\qquad\qquad\qquad\qquad\qquad\qquad\qquad\qquad\qquad\qquad\qquad$ EH I 172(26)

3. $\displaystyle\int_0^1 \left[P_\nu^{n-\nu}(x)\right]^2\frac{dx}{1-x^2} = -\frac{n!}{2(n-\nu)\,\Gamma(1-n+2\nu)}$ $\qquad$ [$n = 0,1,2,\ldots; \quad \operatorname{Re}\nu > n$]

$\qquad\qquad\qquad\qquad\qquad\qquad\qquad\qquad\qquad\qquad\qquad\qquad\qquad$ ET II 315(9)

7.123 $\displaystyle\int_{-1}^1 P_n^m(x)\,P_n^k(x)\frac{dx}{1-x^2} = 0$ $\qquad$ [$0 \leq m \leq n, \quad 0 \leq k \leq n; \quad m \neq k$]

$\qquad\qquad\qquad\qquad\qquad\qquad\qquad\qquad\qquad\qquad\qquad\qquad\qquad$ MO 74

7.124 $\displaystyle\int_{-1}^1 x^k(z-x)^{-1}\left(1-x^2\right)^{\frac{1}{2}m}P_n^m(x)\,dx = (-2)^m\left(z^2-1\right)^{\frac{1}{2}m}Q_n^m(z)\cdot z^k$

$\qquad\qquad\qquad\qquad\qquad\qquad\qquad$ [$m \leq n; k = 0,1,\ldots,n-m;$

z is in the complex plane with a cut along the interval $(-1,1)$ on the real axis]

$\qquad\qquad\qquad\qquad\qquad\qquad\qquad\qquad\qquad\qquad\qquad\qquad\qquad$ ET II 279(26)

7.125 $\int_{-1}^{1} \left(1-x^2\right)^{\frac{1}{2}m} P_k^m(x)\, P_l^m(x)\, P_n^m(x)\, dx = (-1)^m \pi^{-3/2} \dfrac{(k+m)!(l+m)!\,(n+m)!(s-m)!}{(k-m)!(l-m)!(n-m)!(s-k)!}$

$$\times \frac{\Gamma\left(m+\tfrac{1}{2}\right)\Gamma\left(t-k+\tfrac{1}{2}\right)\Gamma\left(t-l+\tfrac{1}{2}\right)\Gamma\left(t-n+\tfrac{1}{2}\right)}{(s-l)!(s-n)!\,\Gamma\left(s+\tfrac{3}{2}\right)}$$

$[2s = k+l+n+m$ and $2t = k+l-n-m$ are both even

$l \geq m, \quad m \leq k-l-m \leq n \leq k+l+m]$

ET II 280(32)

7.126

1. $\int_0^1 P_\nu(x) x^\sigma\, dx = \dfrac{\sqrt{\pi}\, 2^{-\sigma-1}\, \Gamma(1+\sigma)}{\Gamma\left(1+\tfrac{1}{2}\sigma-\tfrac{1}{2}\nu\right)\Gamma\left(\tfrac{1}{2}\sigma+\tfrac{1}{2}\nu+\tfrac{3}{2}\right)}$ $[\operatorname{Re}\sigma > -1]$ EH I 171(23)

2. $\int_0^1 x^\sigma P_\nu^m(x)\, dx = \dfrac{(-1)^m \pi^{1/2} 2^{-2m-1} \Gamma\left(\tfrac{1+\sigma}{2}\right) \Gamma(1+m+\nu)}{\Gamma\left(\tfrac{1}{2}+\tfrac{1}{2}m\right)\Gamma\left(\tfrac{3}{2}+\tfrac{\sigma}{2}+\tfrac{m}{2}\right)\Gamma(1-m+\nu)}$

$$\times {}_3F_2\left(\frac{m+\nu+1}{2}, \frac{m-\nu}{2}, \frac{m}{2}+1; m+1, \frac{3+\sigma+m}{2}; 1\right)$$

$[\operatorname{Re}\sigma > -1; \quad m = 0,1,2,\ldots]$ ET II 313(2)

3. $\int_0^1 x^\sigma P_\nu^\mu(x)\, dx = \dfrac{\pi^{1/2} 2^{2\mu-1} \Gamma\left(\tfrac{1+\sigma}{2}\right)}{\Gamma\left(\tfrac{1-\mu}{2}\right)\Gamma\left(\tfrac{3+\sigma-\mu}{2}\right)} {}_3F_2\left(\frac{\nu-\mu+1}{2}, -\frac{\mu+\nu}{2}, 1-\frac{\mu}{2}; 1-\mu, \frac{3+\sigma-\mu}{2}; 1\right)$

$[\operatorname{Re}\sigma > -1, \quad \operatorname{Re}\mu < 2]$ ET II 313(3)

4. $\int_1^\infty x^{\mu-1} Q_\nu(ax)\, dx = e^{\mu\pi i}\, \Gamma(\mu) a^{-\mu}\left(a^2-1\right)^{\frac{1}{2}\mu} Q_\nu^{-\mu}(a)$

$[|\arg(a-1)| < \pi, \quad \operatorname{Re}\mu > 0, \quad \operatorname{Re}(\nu-\mu) > -1]$ ET II 325(26)

7.127 $\int_{-1}^{1}(1+x)^\sigma P_\nu(x)\, dx = \dfrac{2^{\sigma+1}[\Gamma(\sigma+1)]^2}{\Gamma(\sigma+\nu+2)\,\Gamma(1+\sigma-\nu)}$ $[\operatorname{Re}\sigma > -1]$ ET II 316(15)

7.128

1. $\int_{-1}^{1}(1-x)^{-\frac{1}{2}\mu}(1+x)^{\frac{1}{2}\mu-\frac{1}{2}}(z+x)^{\mu-\frac{3}{2}} P_\nu^\mu(x)\, dx$

$$= -\frac{\Gamma\left(\mu-\tfrac{1}{2}\right)(z-1)^{\mu-\frac{1}{2}}(z+1)^{-1/2}}{\pi^{1/2} e^{2\mu\pi i}\,\Gamma(\mu+\nu)\,\Gamma(\mu-\nu-1)}$$

$$\times \left\{ Q_\nu^\mu\left[\left(\frac{1+z}{2}\right)^{1/2}\right] Q_{-\nu-1}^{\mu-1}\left[\left(\frac{1+z}{2}\right)^{1/2}\right] + Q_\nu^{\mu-1}\left[\left(\frac{1+z}{2}\right)^{1/2}\right] Q_{-\nu-1}^\mu\left[\left(\frac{1+z}{2}\right)^{1/2}\right] \right\}$$

$[-\tfrac{1}{2} < \operatorname{Re}\mu < 1,$

z is in the complex plane with a cut along the interval $(-1,1)$ of the real axis]

ET II 317(20)

2. $\int_{-1}^{1}(1-x)^{-\frac{1}{2}\mu}(1+x)^{\frac{1}{2}\mu-\frac{1}{2}}(z+x)^{\mu-\frac{1}{2}} P_\nu^\mu(x)\, dx$

$$= \frac{2e^{-2\mu\pi i}\,\Gamma\left(\tfrac{1}{2}+\mu\right)}{\pi^{1/2}\,\Gamma(\mu-\nu)\,\Gamma(\mu+\nu+1)}(z-1)^\mu Q_\nu^\mu\left[\left(\frac{1+z}{2}\right)^{1/2}\right] Q_{-\nu-1}^\mu\left[\left(\frac{1+z}{2}\right)^{1/2}\right]$$

$[-\tfrac{1}{2} < \operatorname{Re}\mu < 1,$

z is in the complex plane with a cut along the interval $(-1,1)$ of the real axis]

ET II 316(18)

7.129 $\quad \displaystyle\int_{-1}^{1} P_{\nu}(x)\, P_{\lambda}(x)(1+x)^{\lambda+\nu}\, dx = \frac{2^{\lambda+\nu+1}[\Gamma(\lambda+\nu+1)]^4}{[\Gamma(\lambda+1)\,\Gamma(\nu+1)]^2\,\Gamma(2\lambda+2\nu+2)}$

$$[\operatorname{Re}(\nu+\lambda+1) > 0] \qquad \text{EH I 172(30)}$$

7.131

1. $\quad \displaystyle\int_{1}^{\infty} (x-1)^{-\frac{1}{2}\mu}(x+1)^{\frac{1}{2}\mu-\frac{1}{2}}(z+x)^{\mu-\frac{1}{2}}\, P_{\nu}^{\mu}(x)\, dx$

$$= \pi^{1/2}\frac{\Gamma(-\mu-\nu)\,\Gamma(1-\mu+\nu)}{\Gamma\left(\frac{1}{2}-\mu\right)}(z-1)^{\mu}\left\{P_{\nu}^{\mu}\left[\left(\frac{1+z}{2}\right)^{1/2}\right]\right\}^2$$

$$[\operatorname{Re}(\mu+\nu) < 0, \quad \operatorname{Re}(\mu-\nu) < 1, \quad |\arg(z+1)| < \pi] \quad \text{ET II 321(6)}$$

2. $\quad \displaystyle\int_{1}^{\infty} (x-1)^{-\frac{1}{2}\mu}(x+1)^{\frac{1}{2}\mu-\frac{1}{2}}(z+x)^{\mu-\frac{3}{2}}\, P_{\nu}^{\mu}(x)\, dx$

$$= \frac{\pi^{1/2}\,\Gamma(1-\mu-\nu)\,\Gamma(2-\mu+\nu)\,(z-1)^{\mu-\frac{1}{2}}(z+1)^{-1/2}}{\Gamma\left(\frac{3}{2}-\mu\right)}P_{\nu}^{\mu}\left[\left(\frac{1+z}{2}\right)^{1/2}\right]P_{\nu}^{\mu-1}\left[\left(\frac{1+z}{2}\right)^{1/2}\right]$$

$$[\operatorname{Re}\mu < 1, \quad \operatorname{Re}(\mu+\nu) < 1, \quad \operatorname{Re}(\mu-\nu) < 2, \quad |\arg(1+z)| < \pi] \quad \text{ET II 321(7)}$$

7.132

1. $\quad \displaystyle\int_{-1}^{1}\left(1-x^2\right)^{\lambda-1} P_{\nu}^{\mu}(x)\, dx = \frac{\pi 2^{\mu}\,\Gamma\left(\lambda+\frac{1}{2}\mu\right)\Gamma\left(\lambda-\frac{1}{2}\mu\right)}{\Gamma\left(\lambda+\frac{1}{2}\nu+\frac{1}{2}\right)\Gamma\left(\lambda-\frac{1}{2}\nu\right)\Gamma\left(-\frac{1}{2}\mu+\frac{1}{2}\nu+1\right)\Gamma\left(-\frac{1}{2}\mu-\frac{1}{2}\nu+\frac{1}{2}\right)}$

$$[2\operatorname{Re}\lambda > |\operatorname{Re}\mu|] \qquad \text{ET II 316(6)}$$

2. $\quad \displaystyle\int_{1}^{\infty}\left(x^2-1\right)^{\lambda-1} P_{n}^{\mu}(x)\, dx = \frac{2^{\mu-1}\,\Gamma\left(\lambda-\frac{1}{2}\mu\right)\Gamma\left(1-\lambda+\frac{1}{2}\nu\right)\Gamma\left(\frac{1}{2}-\lambda-\frac{1}{2}\nu\right)}{\Gamma\left(1-\frac{1}{2}\mu+\frac{1}{2}\nu\right)\Gamma\left(\frac{1}{2}-\frac{1}{2}\mu-\frac{1}{2}\nu\right)\Gamma\left(1-\lambda-\frac{1}{2}\mu\right)}$

$$[\operatorname{Re}\lambda > \operatorname{Re}\mu, \quad \operatorname{Re}(1-2\lambda-\nu) > 0, \quad \operatorname{Re}(2-2\lambda+\nu) > 0] \quad \text{ET II 320(2)}$$

3.[9] $\quad \displaystyle\int_{1}^{\infty}\left(x^2-1\right)^{\lambda-1} Q_{\nu}^{\mu}(x)\, dx = e^{\mu\pi i}\frac{\Gamma\left(\frac{1}{2}+\frac{1}{2}\nu+\frac{1}{2}\mu\right)\Gamma\left(1-\lambda+\frac{1}{2}\nu\right)\Gamma\left(\lambda+\frac{1}{2}\mu\right)\Gamma\left(\lambda-\frac{1}{2}\mu\right)}{2^{2-\mu}\,\Gamma\left(1+\frac{1}{2}\nu-\frac{1}{2}\mu\right)\Gamma\left(\frac{1}{2}+\lambda+\frac{1}{2}\nu\right)}$

$$[|\operatorname{Re}\mu| < 2\operatorname{Re}\lambda < \operatorname{Re}\nu+2]$$

$$\text{ET II 324(23)}$$

4. $\quad \displaystyle\int_{0}^{1} x^{\sigma}\left(1-x^2\right)^{-\frac{1}{2}\mu} P_{\nu}^{\mu}(x)\, dx = \frac{2^{\mu-1}\,\Gamma\left(\frac{1}{2}+\frac{1}{2}\sigma\right)\Gamma\left(1+\frac{1}{2}\sigma\right)}{\Gamma\left(1+\frac{1}{2}\sigma-\frac{1}{2}\nu-\frac{1}{2}\mu\right)\Gamma\left(\frac{1}{2}\sigma+\frac{1}{2}\nu-\frac{1}{2}\mu+\frac{3}{2}\right)}$

$$[\operatorname{Re}\mu < 1, \quad \operatorname{Re}\sigma > -1] \qquad \text{EH I 172(24)}$$

5. $\quad \displaystyle\int_{0}^{1} x^{\sigma}\left(1-x^2\right)^{\frac{1}{2}m} P_{\nu}^{m}(x)\, dx = \frac{(-1)^m 2^{-m-1}\,\Gamma\left(\frac{1}{2}+\frac{1}{2}\sigma\right)\Gamma\left(1+\frac{1}{2}\sigma\right)\Gamma(1+m+\nu)}{\Gamma(1-m+\nu)\,\Gamma\left(1+\frac{1}{2}\sigma+\frac{1}{2}m-\frac{1}{2}\nu\right)\Gamma\left(\frac{3}{2}+\frac{1}{2}\sigma+\frac{1}{2}m+\frac{1}{2}\nu\right)}$

$$[\operatorname{Re}\sigma > -1, \quad m \text{ is a positive integer}] \quad \text{EH I 172(25), ET II 313(4)}$$

6. $\quad \displaystyle\int_{0}^{1}\left(1-x^2\right)^{\eta} P_{\nu}^{\mu}(x)\, dx = \frac{2^{\mu-1}\,\Gamma\left(1+\eta-\frac{1}{2}\mu\right)\Gamma\left(\frac{1}{2}+\frac{1}{2}\sigma\right)}{\Gamma(1-\mu)\,\Gamma\left(\frac{3}{2}+\eta+\frac{1}{2}\sigma-\frac{1}{2}\mu\right)}$

$$\times\; {}_3F_2\left(\frac{\nu-\mu+1}{2}, -\frac{\mu+\nu}{2}, 1+\eta-\frac{\mu}{2}; 1-\mu, \frac{3+\sigma-\mu}{2}+\eta; 1\right)$$

$$\left[\operatorname{Re}\left(\eta-\tfrac{1}{2}\mu\right) > -1, \operatorname{Re}\sigma > -1\right] \quad \text{ET II 314(6)}$$

7. $$\int_1^\infty x^{-\varrho}\left(x^2-1\right)^{-\frac{1}{2}\mu} P_\nu^\mu(x)\,dx = \frac{2^{\varrho+\mu-2}\,\Gamma\left(\frac{\varrho+\mu+\nu}{2}\right)\Gamma\left(\frac{\varrho+\mu-\nu-1}{2}\right)}{\sqrt{\pi}\,\Gamma(\varrho)}$$

$$[\operatorname{Re}\mu < 1, \quad \operatorname{Re}(\varrho+\mu+\nu) > 0, \quad \operatorname{Re}(\varrho+\mu-\nu) > 1] \quad \text{ET II 320(3)}$$

7.133

1. $$\int_u^\infty Q_\nu(x)(x-u)^{\mu-1}\,dx = \Gamma(\mu)e^{\mu\pi i}\left(u^2-1\right)^{\frac{1}{2}\mu} Q_\nu^{-\mu}(u)$$

$$[|\arg(u-1)| < \pi, \quad 0 < \operatorname{Re}\mu < 1+\operatorname{Re}\nu] \quad \text{MO 90a}$$

2. $$\int_u^\infty \left(x^2-1\right)^{\frac{1}{2}\lambda} Q_\nu^{-\lambda}(x)(x-u)^{\mu-1}\,dx = \Gamma(\mu)e^{\mu\pi i}\left(u^2-1\right)^{\frac{1}{2}\lambda+\frac{1}{2}\mu} Q_\nu^{-\lambda-\mu}(u)$$

$$[|\arg(u-1)| < \pi, \quad 0 < \operatorname{Re}\mu < 1+\operatorname{Re}(\nu-\lambda)] \quad \text{ET II 204(30)}$$

7.134

1. $$\int_1^\infty (x-1)^{\lambda-1}\left(x^2-1\right)^{\frac{1}{2}\mu} P_\nu^\mu(x)\,dx = \frac{2^{\lambda+\mu}\,\Gamma(\lambda)\,\Gamma(-\lambda-\mu-\nu)\,\Gamma(1-\lambda-\mu+\nu)}{\Gamma(1-\mu+\nu)\,\Gamma(-\mu-\nu)\,\Gamma(1-\lambda-\mu)}$$

$$[\operatorname{Re}\lambda > 0, \quad \operatorname{Re}(\lambda+\mu+\nu) < 0, \quad \operatorname{Re}(\lambda+\mu-\nu) < 1] \quad \text{ET II 321(4)}$$

2. $$\int_1^\infty (x-1)^{\lambda-1}\left(x^2-1\right)^{-\frac{1}{2}\mu} P_\nu^\mu(x)\,dx = -\frac{2^{\lambda-\mu}\sin\pi\nu\,\Gamma(\lambda-\mu)\,\Gamma(-\lambda+\mu-\nu)\,\Gamma(1-\lambda+\mu+\nu)}{\pi\,\Gamma(1-\lambda)}$$

$$[\operatorname{Re}(\lambda-\mu) > 0, \quad \operatorname{Re}(\mu-\lambda-\nu) > 0, \quad \operatorname{Re}(\mu-\lambda+\nu) > -1] \quad \text{ET II 321(5)}$$

7.135

1. $$\int_{-1}^1 \left(1-x^2\right)^{-\frac{1}{2}\mu}(z-x)^{-1} P_{\mu+n}^\mu(x)\,dx = 2e^{-i\mu\pi}\left(z^2-1\right)^{-\frac{1}{2}\mu} Q_{\mu+n}^\mu(z)$$

$[n = 0,1,2,\ldots, \quad \operatorname{Re}\mu+n > -1, z$ is in the complex plane with a cut along the interval $(-1,1)$ of the real axis.]   ET II 316(17)

2. $$\int_1^\infty (x-1)^{\lambda-1}\left(x^2-1\right)^{\mu/2}(x+z)^{-\rho} P_\nu^\mu(x)\,dx$$

$$= \frac{2^{\lambda+\mu-\rho}\,\Gamma(\lambda-\rho)\,\Gamma(\rho-\lambda-\mu-\nu)\,\Gamma(\rho-\lambda-\mu+\nu+1)}{\Gamma(1-\mu+\nu)\,\Gamma(-\mu-\nu)\,\Gamma(1+\rho-\lambda-\mu)}$$

$$\times {}_3F_2\left(\rho, \rho-\lambda-\mu-\nu, \rho-\lambda-\mu+\nu+1; \rho-\lambda+1, \rho-\lambda-\mu+1; \frac{1+z}{2}\right)$$

$$+ \frac{\Gamma(\rho-\lambda)\,\Gamma(\lambda)}{\Gamma(\rho)\,\Gamma(1-\mu)}2^\mu(z+1)^{\lambda-\rho}\,{}_3F_2\left(\lambda, -\mu-\nu, 1-\mu+\nu; 1-\mu, 1-\rho+\lambda; \frac{1+z}{2}\right)$$

$$[\operatorname{Re}\lambda > 0, \quad \operatorname{Re}(\rho-\lambda-\mu-\nu) > 0, \quad \operatorname{Re}(\rho-\lambda-\mu+\nu+1) > 0, \quad |\arg(z+1)| < \pi]$$

ET II 322(9)

3. $\displaystyle\int_1^\infty (x-1)^{\lambda-1}\left(x^2-1\right)^{-\mu/2}(x+z)^{-\rho}\,P_\nu^\mu(x)\,dx$

$$= -\frac{\sin(\nu\pi)\,\Gamma(\lambda-\mu-\rho)\,\Gamma\left(\rho-\lambda+\mu-\nu\right)\Gamma(\rho-\lambda+\mu+\nu+1)}{2^{\rho-\lambda+\mu}\pi\,\Gamma\left(1+\rho-\lambda\right)}$$

$$\times\ {}_3F_2\left(\rho,\rho-\lambda+\mu-\nu,\rho-\lambda+\mu+\nu+1;1+\rho-\lambda,1+\rho-\lambda+\mu;\frac{1+z}{2}\right)$$

$$+\frac{\Gamma(\lambda-\mu)\,\Gamma\left(\rho-\lambda+\mu\right)}{\Gamma(\rho)\,\Gamma(1-\mu)}(z+1)^{\lambda-\rho-\mu}$$

$$\times\ {}_3F_2\left(\lambda-\mu,-\nu,\nu+1;1+\lambda-\mu-\rho,1-\mu;\frac{1+z}{2}\right)$$

$$[\mathrm{Re}(\lambda-\mu)>0,\quad \mathrm{Re}\left(\rho-\lambda+\mu-\nu\right)>0,\quad \mathrm{Re}(\rho-\lambda+\mu+\nu+1)>0,\quad |\arg(z+1)|<\pi]$$

ET II 322(10)

7.136

1. $\displaystyle\int_{-1}^1 \left(1-x^2\right)^{\lambda-1}\left(1-a^2x^2\right)^{\mu/2}P_\nu(ax)\,dx$

$$=\frac{\pi 2^\mu\,\Gamma(\lambda)}{\Gamma\left(\frac12+\lambda\right)\Gamma\left(\frac12-\frac12\mu-\frac12\nu\right)\Gamma\left(1-\frac12\mu+\frac12\nu\right)}\ {}_2F_1\left(-\frac{\mu+\nu}{2},\frac{1-\mu+\nu}{2};\frac12+\lambda;a^2\right)$$

$$[\mathrm{Re}\,\lambda>0,\quad -1<a<1]\quad \text{ET II 318(31)}$$

2. $\displaystyle\int_1^\infty \left(x^2-1\right)^{\lambda-1}\left(a^2x^2-1\right)^{\mu/2}P_\nu^\mu(ax)\,dx$

$$=\frac{\Gamma(\lambda)\,\Gamma\left(1-\lambda-\frac12\mu+\frac12\nu\right)\Gamma\left(\frac12-\lambda-\frac12\mu-\frac12\nu\right)}{\Gamma\left(1-\frac12\mu+\frac12\nu\right)\Gamma\left(\frac12-\frac12\nu-\frac12\mu\right)\Gamma(1-\lambda-\mu)}$$

$$\times 2^{\mu-1}a^{\mu-\nu-1}\ {}_2F_1\left(\frac{1-\mu+\nu}{2},1-\lambda-\frac{\mu-\nu}{2};1-\lambda-\mu;1-\frac{1}{a^2}\right)$$

$$[\mathrm{Re}\,a>0,\quad \mathrm{Re}\,\lambda>0,\quad \mathrm{Re}(\nu-\mu-2\lambda)>-2,\quad \mathrm{Re}(2\lambda+\mu+\nu)<1]\quad \text{ET II 325(25)}$$

3. $\displaystyle\int_1^\infty \left(x^2-1\right)^{\lambda-1}\left(a^2x^2-1\right)^{-\frac12\mu}Q_\nu^\mu(ax)\,dx=\frac{\Gamma\left(\frac{\mu+\nu+1}{2}\right)\Gamma(\lambda)\,\Gamma\left(1-\lambda+\frac{\mu+\nu}{2}\right)2^{\mu-2}e^{\mu\pi i}a^{-\mu-\nu-1}}{\Gamma\left(\nu+\frac32\right)}$

$$\times\ {}_2F_1\left(\frac{\mu+\nu+1}{2},1-\lambda+\frac{\mu+\nu}{2};\nu+\frac32;a^{-2}\right)$$

$$[|\arg(a-1)|<\pi,\quad \mathrm{Re}\,\lambda>0,\quad \mathrm{Re}(2\lambda-\mu-\nu)<2]\quad \text{ET II 325(27)}$$

7.137

1. $\displaystyle\int_1^\infty x^{-\frac12\mu-\frac12}(x-1)^{-\mu-\frac12}(1+ax)^{\frac12\mu}\,Q_\nu^\mu(1+2ax)\,dx$

$$=\pi^{-1/2}e^{-\mu\pi i}\,\Gamma\left(\tfrac12-\mu\right)a^{\frac12\mu}\left\{Q_\nu^\mu\left[(1+a)^{1/2}\right]\right\}^2$$

$$\left[|\arg a|<\pi,\quad \mathrm{Re}\,\mu<\tfrac12,\quad \mathrm{Re}(\mu+\nu)>-1\right]\quad \text{ET II 325(28)}$$

2. $\displaystyle\int_1^\infty x^{-\frac12\mu-\frac12}(x-1)^{-\mu-\frac32}(1+ax)^{\frac12\mu}\,Q_\nu^\mu(1+2ax)\,dx$

$$=-\pi^{-1/2}e^{-\mu\pi i}\,\Gamma\left(-\mu-\tfrac12\right)a^{\frac12\mu+\frac12}\left(1+a^2\right)^{-1/2}Q_\nu^{\mu+1}\left[(1+a)^{1/2}\right]Q_\nu^\mu\left[(1+a)^{1/2}\right]$$

$$\left[|\arg a|<\pi,\quad \mathrm{Re}\,\mu<-\tfrac12,\quad \mathrm{Re}(\mu+\nu+2)>0\right]\quad \text{ET II 326(29)}$$

3. $\int_0^1 x^{-\frac{1}{2}\mu-\frac{1}{2}}(1-x)^{-\mu-\frac{1}{2}}(1+ax)^{\frac{1}{2}\mu}P_\nu^\mu(1+2ax)\,dx = \pi^{1/2}\Gamma\left(\frac{1}{2}-\mu\right)a^{\frac{1}{2}\mu}\left\{P_\nu^\mu\left[(1+a)^{1/2}\right]\right\}^2$

$$\left[\operatorname{Re}\mu<\tfrac{1}{2},\quad |\arg a|<\pi\right]\quad \text{ET II 319(32)}$$

4. $\int_0^1 x^{-\frac{1}{2}\mu-\frac{1}{2}}(1-x)^{-\mu-\frac{3}{2}}(1+ax)^{\frac{1}{2}\mu}P_\nu^\mu(1+2ax)\,dx$

$$= \pi^{1/2}\Gamma\left(-\tfrac{1}{2}-\mu\right)a^{\frac{1}{2}\mu+\frac{1}{2}}P_\nu^{\mu+1}\left[(1+a)^{1/2}\right]P_\nu^\mu\left[(1+a)^2\right]$$

$$\left[\operatorname{Re}\mu<-\tfrac{1}{2},\quad |\arg a|<\pi\right]\quad \text{ET II 319(33)}$$

5. $\int_0^1 x^{\frac{1}{2}\mu-\frac{1}{2}}(1-x)^{\mu-\frac{1}{2}}(1+ax)^{-\frac{1}{2}\mu}P_\nu^\mu(1+2ax)\,dx$

$$= \pi^{1/2}\Gamma\left(\tfrac{1}{2}+\mu\right)a^{-\frac{1}{2}\mu}P_\nu^\mu\left[(1+a)^{1/2}\right]P_\nu^{-\mu}\left[(1+a)^{1/2}\right]$$

$$\left[\operatorname{Re}\mu>-\tfrac{1}{2},\quad |\arg a|<\pi\right]\quad \text{ET II 319(34)}$$

6. $\int_0^1 x^{\frac{1}{2}\mu-\frac{1}{2}}(1-x)^{\mu-\frac{3}{2}}(1+ax)^{-\frac{1}{2}\mu}P_\nu^\mu(1+2ax)\,dx$

$$= \frac{1}{2}\pi^{1/2}\Gamma\left(\mu-\tfrac{1}{2}\right)a^{\frac{1}{2}-\frac{1}{2}\mu}(1+a)^{-1/2}\left\{P_\nu^{1-\mu}\left[(1+a)^{1/2}\right]P_\nu^\mu\left[(1+a)^{1/2}\right]\right\}$$

$$+(\mu+\nu)(1-\mu+\nu)P_\nu^{-\mu}\left[(1+a)^{1/2}\right]P_\nu^\mu\left[(1+a)^{1/2}\right]$$

$$\left[\operatorname{Re}\mu>\tfrac{1}{2},\quad |\arg a|<\pi\right]\quad \text{ET II 319(35)}$$

7. $\int_0^1 x^{-\frac{\mu}{2}-\frac{1}{2}}(1-x)^{-\mu-\frac{1}{2}}(1+ax)^{\frac{1}{2}\mu}Q_\nu^\mu(1+2ax)\,dx$

$$= \pi^{1/2}\Gamma\left(\tfrac{1}{2}-\mu\right)a^{\frac{1}{2}\mu}P_\nu^\mu\left[(1+a)^{1/2}\right]Q_\nu^\mu\left[(1+a)^{1/2}\right]$$

$$\left[\operatorname{Re}\mu<\tfrac{1}{2},\quad |\arg a|<\pi\right]\quad \text{ET II 320(38)}$$

8. $\int_0^1 x^{-\frac{\mu}{2}-\frac{1}{2}}(1-x)^{-\mu-\frac{3}{2}}(1+ax)^{\frac{1}{2}\mu}Q_\nu^\mu(1+2ax)\,dx$

$$= \frac{1}{2}\pi^{1/2}\Gamma\left(-\mu-\tfrac{1}{2}\right)(1+a)^{-1/2}a^{\frac{1}{2}\mu+\frac{1}{2}}$$

$$\times\left\{P_\nu^{\mu+1}\left[(1+a)^{1/2}\right]Q_\nu^\mu\left[(1+a)^{1/2}\right]+P_\nu^\mu\left[(1+a)^{1/2}\right]Q_\nu^{\mu+1}\left[(1+a)^{1/2}\right]\right\}$$

$$\left[\operatorname{Re}\mu<-\tfrac{1}{2},\quad |\arg a|<\pi\right]\quad \text{ET II 320(39)}$$

9. $\int_0^y (y-x)^{\mu-1}\left[x\left(1+\tfrac{1}{2}\gamma x\right)\right]^{-\frac{1}{2}\lambda}P_\nu^\lambda(1+\gamma x)\,dx$

$$= \Gamma(\mu)\left(\frac{2}{\gamma}\right)^{\frac{1}{2}\mu}\left[y\left(1+\frac{1}{2}\gamma y\right)\right]^{\frac{1}{2}\mu-\frac{1}{2}\lambda}P_\nu^{\lambda-\mu}(1+\gamma y)$$

$$\left[\operatorname{Re}\lambda<1,\quad \operatorname{Re}\mu>0,\quad |\arg\gamma y|<\pi\right]\quad \text{ET II 193(52)}$$

10. $\int_0^y (y-x)^{\mu-1}x^{\sigma+\frac{1}{2}\lambda-1}\left(1+\tfrac{1}{2}\gamma x\right)^{-\frac{1}{2}\lambda}P_\nu^\lambda(1+\gamma x)\,dx$

$$= \frac{\left(\frac{\gamma}{2}\right)^{-\frac{1}{2}\lambda}\Gamma(\sigma)\Gamma(\mu)y^{\sigma+\mu-1}}{\Gamma(1-\lambda)\Gamma(\sigma+\mu)}\,{}_3F_2\left(-\nu,1+\nu,\sigma;1-\lambda,\sigma+\mu;-\frac{1}{2}\gamma y\right)$$

$$\left[\operatorname{Re}\sigma>0,\quad \operatorname{Re}\mu>0,\quad |\gamma y|<1\right]\quad \text{ET II 193(53)}$$

11. $$\int_0^y (y-x)^{\mu-1}[x(1-x)]^{-\frac{1}{2}\lambda}\,P_\nu^\lambda(1-2x)\,dx = \Gamma(\mu)[y(1-y)]^{\frac{1}{2}\mu-\frac{1}{2}\lambda}\,P_\nu^{\lambda-\mu}(1-2y)$$

$$[\operatorname{Re}\lambda < 1, \quad \operatorname{Re}\mu > 0, \quad 0 < y < 1]$$
ET II 193(54)

12. $$\int_0^y (y-x)^{\mu-1}x^{\sigma+\frac{1}{2}\lambda-1}(1-x)^{-\frac{1}{2}\lambda}\,P_\nu^\lambda(1-2x)\,dx$$

$$= \frac{\Gamma(\mu)\,\Gamma(\sigma)y^{\sigma+\mu-1}}{\Gamma(\sigma+\mu)\,\Gamma(1-\lambda)}\,{}_3F_2\left(-\nu, 1+\nu, \sigma; 1-\lambda, \sigma+\mu; y\right)$$

$$[\operatorname{Re}\sigma > 0, \quad \operatorname{Re}\mu > 0, \quad 0 < y < 1] \quad \text{ET II 193(155)}$$

7.138 $$\int_0^\infty (a+x)^{-\mu-\nu-2}\,P_\mu\left(\frac{a-x}{a+x}\right)P_\nu\left(\frac{a-x}{a+x}\right)dx = \frac{a^{-\mu-\nu-1}[\Gamma(\mu+\nu+1)]^4}{[\Gamma(\mu+1)\,\Gamma(\nu+1)]^2\,\Gamma(2\mu+2\nu+2)}$$

$$[|\arg a| < \pi, \quad \operatorname{Re}(\mu+\nu) > -1]$$
ET II 326(3)

7.14 Combinations of associated Legendre functions, exponentials, and powers

7.141

1. $$\int_1^\infty e^{-ax}(x-1)^{\lambda-1}\left(x^2-1\right)^{\frac{1}{2}\mu}P_\nu^\mu(x)\,dx = \frac{a^{-\lambda-\mu}e^{-a}}{\Gamma(1-\mu+\nu)\,\Gamma(-\mu-\nu)}\,G_{23}^{31}\left(2a\,\middle|\,\begin{matrix}1+\mu, 1\\ \lambda+\mu, -\nu, 1+\nu\end{matrix}\right)$$

$$[\operatorname{Re}a > 0, \quad \operatorname{Re}\lambda > 0] \quad \text{ET II 323(13)}$$

2. $$\int_1^\infty e^{-ax}(x-1)^{\lambda-1}\left(x^2-1\right)^{\frac{1}{2}\mu}Q_\nu^\mu(x)\,dx$$

$$= \frac{\Gamma(\nu+\mu+1)e^{\mu\pi i}}{2\,\Gamma(\nu-\mu+1)}a^{-\lambda-\mu}e^{-a}\,G_{23}^{22}\left(2a\,\middle|\,\begin{matrix}1+\mu, 1\\ \lambda+\mu, \nu+1, -\nu\end{matrix}\right)$$

$$[\operatorname{Re}a > 0, \quad \operatorname{Re}\lambda > 0, \quad \operatorname{Re}(\lambda+\mu) > 0] \quad \text{ET II 325(24)}$$

3. $$\int_1^\infty e^{-ax}(x-1)^{\lambda-1}\left(x^2-1\right)^{-\frac{1}{2}\mu}P_\nu^\mu(x)\,dx = -\pi^{-1}\sin(\nu\pi)a^{\mu-\lambda}e^{-a}\,G_{23}^{31}\left(2a\,\middle|\,\begin{matrix}1, 1-\mu\\ \lambda-\mu, 1+\nu, -\nu\end{matrix}\right)$$

$$[\operatorname{Re}a > 0, \quad \operatorname{Re}(\lambda-\mu) > 0]$$
ET II 323(15)

4. $$\int_1^\infty e^{-ax}(x-1)^{\lambda-1}\left(x^2-1\right)^{-\frac{1}{2}\mu}Q_\nu^\mu(x)\,dx = \frac{1}{2}e^{\mu\pi i}a^{\mu-\lambda}e^{-a}\,G_{23}^{22}\left(2a\,\middle|\,\begin{matrix}1-\mu, 1\\ \lambda-\mu, \nu+1, -\nu\end{matrix}\right)$$

$$[\operatorname{Re}a > 0, \quad \operatorname{Re}\lambda > 0, \quad \operatorname{Re}(\lambda-\mu) > 0]$$
ET II 323(14)

5. $$\int_1^\infty e^{-ax}\left(x^2-1\right)^{-\frac{1}{2}\mu}P_\nu^\mu(x)\,dx = 2^{1/2}\pi^{-1/2}a^{\mu-\frac{1}{2}}K_{\nu+\frac{1}{2}}(a)$$

$$[\operatorname{Re}a > 0, \quad \operatorname{Re}\mu < 1]$$
ET II 323(11), MO 90

7.142 $$\int_1^\infty e^{-\frac{1}{2}ax}\left(\frac{x+1}{x-1}\right)^{\frac{1}{2}\mu}P_{\nu-\frac{1}{2}}^\mu(x)\,dx = \frac{2}{a}\,W_{\mu,\nu}(a) \qquad \left[\operatorname{Re}\mu < 1, \quad \nu-\frac{1}{2} \neq 0, \pm 1, \pm 2, \dots\right]$$
BU 79(34), MO 118

7.143

1. $$\int_0^\infty [x(1+x)]^{-\frac{1}{2}\mu} e^{-\beta x}\, P_\nu^\mu(1+2x)\, dx = \frac{\beta^{\mu-\frac{1}{2}}}{\sqrt{\pi}} e^{\frac{1}{2}\beta}\, K_{\nu+\frac{1}{2}}\left(\frac{\beta}{2}\right)$$

$$[\operatorname{Re}\mu < 1, \quad \operatorname{Re}\beta > 0] \qquad \text{ET I 179(1)}$$

2. $$\int_0^\infty \left(1+\frac{1}{x}\right)^{\frac{1}{2}\mu} e^{-\beta x}\, P_\nu^\mu(1+2x)\, dx = \frac{e^{\frac{1}{2}\beta}}{\beta}\, W_{\mu,\nu+\frac{1}{2}}(\beta)$$

$$[\operatorname{Re}\mu < 1, \quad \operatorname{Re}\beta > 0] \qquad \text{ET I 179(2)}$$

7.144

1. $$\int_0^\infty e^{-\beta x} x^{\lambda+\frac{1}{2}\mu-1}(x+2)^{\frac{1}{2}\mu}\, Q_\nu^\mu(1+x)\, dx$$

$$= \frac{\Gamma(\nu+\mu+1)}{\Gamma(\nu-\mu+1)}\left\{ \frac{\sin(\nu\pi)}{2\beta^{\lambda+\mu}\sin(\mu\pi)}\, E\left(-\nu,\nu+1,\lambda+\mu;\mu+1:2\beta\right) \right.$$
$$\left. - \frac{\sin[(\mu+\nu)\pi]}{2^{1-\mu}\beta^\lambda\sin(\mu\pi)}\, E\left(\nu-\mu+1,-\nu-\mu,\lambda:1-\mu:2\beta\right) \right\}$$

$$[\operatorname{Re}\beta > 0, \quad \operatorname{Re}\lambda > 0, \quad \operatorname{Re}(\lambda+\mu) > 0] \quad \text{ET I 181(16)}$$

2. $$\int_0^\infty e^{-\beta x} x^{\lambda-\frac{1}{2}\mu-1}(x+2)^{\frac{1}{2}\mu}\, Q_\nu^\mu(1+x)\, dx = -\frac{\sin(\nu\pi)}{2\beta^{\lambda-\mu}\sin(\mu\pi)}\, E(-\nu,\nu+1,\lambda-\mu:1-\mu:2\beta)$$
$$- \frac{\sin[(\mu-\nu)\pi]}{2^{1+\mu}\beta^\lambda\sin(\mu\pi)}\, E(\mu+\nu+1,\mu-\nu,\lambda:1+\mu:2\beta)$$

$$[\operatorname{Re}\beta > 0, \quad \operatorname{Re}\lambda > 0, \quad \operatorname{Re}(\lambda-\mu)] > 0 \quad \text{ET I 181(17)}$$

7.145

1. $$\int_0^\infty \frac{e^{-\beta x}}{1+x}\, P_\nu\left[\frac{1}{(1+x)^2}-1\right] dx = \frac{e^\beta}{\beta}\, W_{\nu+\frac{1}{2},0}(\beta)\, W_{-\nu-\frac{1}{2},0}(\beta)$$

$$[\operatorname{Re}\beta > 0] \qquad \text{ET I 180(6)}$$

2. $$\int_0^\infty x^{-1} e^{-\beta x}\, Q_{-\frac{1}{2}}\left(1+2x^{-2}\right) dx = \frac{\pi^2}{8}\left\{ \left[J_0\left(\frac{1}{2}\beta\right)\right]^2 + \left[Y_0\left(\frac{1}{2}\beta\right)\right]^2 \right\}$$

$$[\operatorname{Re}\beta > 0] \qquad \text{ET II 327(5)}$$

3. $$\int_0^\infty x^{-1} e^{-ax}\, Q_\nu\left(1+2x^{-2}\right) dx = \frac{1}{2}[\Gamma(\nu+1)]^2 a^{-1}\, W_{-\nu-\frac{1}{2},0}(ai)\, W_{-\nu-\frac{1}{2},0}(-ai)$$

$$[\operatorname{Re}a > 0, \quad \operatorname{Re}\nu > -1] \qquad \text{ET II 327(6)}$$

7.146

1. $$\int_0^\infty x^{-\frac{1}{2}\mu} e^{-\beta x}\, P_\nu^\mu\left(\sqrt{1+x}\right) dx = 2^\mu \beta^{\frac{1}{2}\mu-\frac{5}{4}} e^{\frac{\beta}{2}}\, W_{\frac{1}{2}\mu+\frac{1}{4},\frac{1}{2}\nu+\frac{1}{4}}(\beta)$$

$$[\operatorname{Re}\mu < 1, \quad \operatorname{Re}\beta > 0] \qquad \text{ET I 180(7)}$$

2. $$\int_0^\infty x^{-\frac{1}{2}\mu} \frac{e^{-\beta x}}{\sqrt{1+x}}\, P_\nu^\mu\left(\sqrt{1+x}\right) dx = 2^\mu \beta^{\frac{1}{2}\mu-\frac{3}{4}} e^{\frac{\beta}{2}}\, W_{\frac{1}{2}\mu+\frac{1}{4},\frac{1}{2}\nu+\frac{1}{4}}(\beta)$$

$$[\operatorname{Re}\mu < 1, \operatorname{Re}\beta > 0] \qquad \text{ET I 180(8)a}$$

3. $\int_0^\infty \sqrt{x}\, e^{-\beta x}\, P_\nu^{1/4}\left(\sqrt{1+x^2}\right) P_\nu^{-1/4}\left(\sqrt{1+x^2}\right) dx = \frac{1}{2}\sqrt{\frac{\pi}{2\beta}}\, H_{\nu+\frac{1}{2}}^{(1)}\left(\frac{1}{2}\beta\right) H_{\nu+\frac{1}{2}}^{(2)}\left(\frac{1}{2}\beta\right)$

$$[\operatorname{Re}\beta > 0] \qquad\qquad \text{ET I 180(9)}$$

7.147 $\int_0^\infty x^{\lambda-1}\left(x^2+a^2\right)^{\frac{1}{2}\nu} e^{-\beta x}\, P_\nu^\mu\left[\frac{x}{\left(x^2+a^2\right)^{1/2}}\right] dx$

$$= \frac{2^{-\nu-2} a^{\lambda+\nu}}{\pi\,\Gamma(-\mu-\nu)}\, G_{24}^{32}\left(\frac{a^2\beta^2}{4} \left|\begin{array}{c} 1-\frac{\lambda}{2}, \frac{1-\lambda}{2} \\ 0, \frac{1}{2}, -\frac{\lambda+\mu+\nu}{2}, -\frac{\lambda-\mu+\nu}{2} \end{array}\right.\right)$$

$$[a>0, \quad \operatorname{Re}\beta > 0, \quad \operatorname{Re}\lambda > 0] \quad \text{ET II 327(7)}$$

7.148 $\int_{-1}^1 (1-x)^{-\frac{1}{2}\mu}(1+x)^{\frac{1}{2}\mu+\nu-1}\exp\left(-\frac{1-x}{1+x}y\right) P_\nu^\mu(x)\, dx = 2^\nu y^{\frac{1}{2}\mu+\nu-\frac{1}{2}} e^{\frac{1}{2}y}\, W_{\frac{1}{2}\mu-\nu-\frac{1}{2},\frac{1}{2}\mu}(y)$

$$[\operatorname{Re}y > 0] \qquad\qquad \text{ET II 317(21)}$$

7.149 $\int_1^\infty \left(\alpha^2+\beta^2+2\alpha\beta x\right)^{-1/2}\exp\left[-\left(\alpha^2+\beta^2+2\alpha\beta x\right)^{1/2}\right] P_\nu(x)\, dx$

$$= 2\pi^{-1}(\alpha\beta)^{-1/2}\, K_{\nu+\frac{1}{2}}(\alpha)\, K_{\nu+\frac{1}{2}}(\beta)$$

$$[\operatorname{Re}\alpha > 0, \quad \operatorname{Re}\beta > 0] \qquad \text{ET II 323(16)}$$

7.15 Combinations of associated Legendre and hyperbolic functions

7.151

1. $\int_0^\infty (\sinh x)^{\alpha-1}\, P_\nu^{-\mu}(\cosh x)\, dx = \frac{2^{-1-\mu}\,\Gamma\left(\frac{1}{2}\alpha+\frac{1}{2}\mu\right)\Gamma\left(\frac{1}{2}\nu-\frac{1}{2}\alpha+1\right)\Gamma\left(\frac{1}{2}-\frac{1}{2}\alpha-\frac{1}{2}\nu\right)}{\Gamma\left(\frac{1}{2}\mu+\frac{1}{2}\nu+1\right)\Gamma\left(\frac{1}{2}+\frac{1}{2}\mu-\frac{1}{2}\nu\right)\Gamma\left(1+\frac{1}{2}\mu-\frac{1}{2}\alpha\right)}$

$$[\operatorname{Re}(\alpha+\mu) > 0, \quad \operatorname{Re}(\nu-\alpha+2) > 0, \quad \operatorname{Re}(1-\alpha-\nu) > 0] \quad \text{EH I 172(28)}$$

2. $\int_0^\infty (\sinh x)^{\alpha-1}\, Q_\nu^\mu(\cosh x)\, dx \; \frac{e^{i\mu\pi} 2^{\mu-\alpha}\,\Gamma\left(\frac{1}{2}+\frac{1}{2}\nu+\frac{1}{2}\mu\right)\Gamma\left(1+\frac{1}{2}\nu-\frac{1}{2}\alpha\right)}{\Gamma\left(1+\frac{1}{2}\nu-\frac{1}{2}\mu\right)\Gamma\left(\frac{1}{2}+\frac{1}{2}\nu+\frac{1}{2}\alpha\right)}$

$$\times \Gamma\left(\frac{1}{2}\alpha+\frac{1}{2}\mu\right)\Gamma\left(\frac{1}{2}\alpha-\frac{1}{2}\mu\right)$$

$$[\operatorname{Re}(\alpha\pm\mu) > 0, \quad \operatorname{Re}(\nu-\alpha+2) > 0] \quad \text{EH I 172(29)}$$

7.152 $\int_0^\infty e^{-\alpha x}\sinh^{2\mu}\left(\frac{1}{2}x\right) P_{2n}^{-2\mu}\left[\cosh\left(\frac{1}{2}x\right)\right] dx = \frac{\Gamma\left(2\mu+\frac{1}{2}\right)\Gamma(\alpha-n-\mu)\Gamma\left(\alpha+n-\mu+\frac{1}{2}\right)}{4^\mu\sqrt{\pi}\,\Gamma(\alpha+n+\mu+1)\Gamma\left(\alpha-n+\mu+\frac{1}{2}\right)}$

$$\left[\operatorname{Re}\alpha > n+\operatorname{Re}\mu, \quad \operatorname{Re}\mu > -\frac{1}{4}\right]$$

$$\text{ET I 181(15)}$$

7.16 Combinations of associated Legendre functions, powers, and trigonometric functions

7.161

1. $\displaystyle\int_0^1 x^{\lambda-1}\left(1-x^2\right)^{-\frac{1}{2}\mu}\sin(ax)\,P_\nu^\mu(x)\,dx$

$$= \frac{\pi^{1/2}2^{\mu-\lambda-1}\,\Gamma\left(\lambda+1\right)a}{\Gamma\left(1+\frac{\lambda-\mu-\nu}{2}\right)\Gamma\left(\frac{3+\lambda-\mu+\nu}{2}\right)}$$

$$\times\, {}_2F_3\left(\frac{1+\lambda}{2},1+\frac{\lambda}{2};\frac{3}{2},1+\frac{\lambda-\mu-\nu}{2},\frac{3+\lambda-\mu+\nu}{2};-\frac{a^2}{4}\right)$$

$$[\operatorname{Re}\lambda>-1,\quad \operatorname{Re}\mu<1]\qquad \text{ET II 314(7)}$$

2. $\displaystyle\int_0^1 x^{\lambda-1}\left(1-x^2\right)^{-\frac{1}{2}\mu}\cos(ax)\,P_\nu^\mu(x)\,dx$

$$= \frac{\pi^{1/2}2^{\mu-\lambda}\,\Gamma(\lambda)}{\Gamma\left(1+\dfrac{\lambda-\mu+\nu}{2}\right)\Gamma\left(\dfrac{1+\lambda-\mu-\nu}{2}\right)}$$

$$\times\, {}_2F_3\left(\frac{\lambda}{2},\frac{\lambda+1}{2};\frac{1}{2},\frac{1+\lambda-\mu-\nu}{2},1+\frac{\lambda-\mu+\nu}{2};-\frac{a^2}{4}\right)$$

$$[\operatorname{Re}\lambda>0,\quad \operatorname{Re}\mu<1]\qquad \text{ET II 314(8)}$$

3. $\displaystyle\int_0^\infty \left(x^2-1\right)^{\frac{1}{2}\mu}\sin(ax)\,P_\nu^\mu(x)\,dx = \frac{2^\mu\pi^{1/2}a^{-\mu-\frac{1}{2}}}{\Gamma\left(\frac{1}{2}-\frac{1}{2}\mu-\frac{1}{2}\nu\right)\Gamma\left(1-\frac{1}{2}\mu+\frac{1}{2}\nu\right)}S_{\mu+\frac{1}{2},\nu+\frac{1}{2}}(a)$

$$\left[a>0,\quad \operatorname{Re}\mu<\tfrac{3}{2},\quad \operatorname{Re}(\mu+\nu)<1\right]$$
$$\text{ET II 320(1)}$$

7.162

1. $\displaystyle\int_a^\infty P_\nu\left(2x^2a^{-2}-1\right)\sin(bx)\,dx = -\frac{\pi a}{4\cos(\nu\pi)}\left\{\left[J_{\nu+\frac{1}{2}}\left(\frac{ab}{2}\right)\right]^2-\left[J_{-\nu-\frac{1}{2}}\left(\frac{ab}{2}\right)\right]^2\right\}$

$$[a>0,\quad b>0,\quad -1<\operatorname{Re}\nu<0]$$
$$\text{ET II 326(1)}$$

2. $\displaystyle\int_a^\infty P_\nu\left(2x^2a^{-2}-1\right)\cos(bx)\,dx$

$$= -\frac{\pi}{4}a\left[J_{\nu+\frac{1}{2}}\left(\frac{ab}{2}\right)J_{-\nu-\frac{1}{2}}\left(\frac{ab}{2}\right)-Y_{\nu+\frac{1}{2}}\left(\frac{ab}{2}\right)Y_{-\nu-\frac{1}{2}}\left(\frac{ab}{2}\right)\right]$$

$$[a>0,\quad b>0,\quad -1<\operatorname{Re}\nu<0]\qquad \text{ET II 326(2)}$$

3. $\displaystyle\int_0^\infty \left(x^2+2\right)^{-1/2}\sin(ax)\,P_\nu^{-1}\left(x^2+1\right)\,dx = 2^{-1/2}\pi^{-1}a\sin(\nu\pi)\left[K_{\nu+\frac{1}{2}}\left(2^{-1/2}a\right)\right]^2$

$$[a>0,\quad -2<\operatorname{Re}\nu<1]\qquad \text{ET I 98(22)}$$

4. $\displaystyle\int_0^\infty \left(x^2+2\right)^{-1/2}\sin(ax)\,Q_\nu^1\left(x^2+1\right)\,dx = -2^{-3/2}\pi a\,K_{\nu+\frac{1}{2}}\left(2^{-1/2}a\right)I_{\nu+\frac{1}{2}}\left(2^{-1/2}a\right)$

$$\left[a>0,\quad \operatorname{Re}\nu>-\tfrac{3}{2}\right]\qquad \text{ET 98(23)}$$

5. $\displaystyle\int_0^\infty \cos(ax)\, P_\nu\left(1+x^2\right)\, dx = -\frac{\sqrt{2}}{\pi}\sin(\nu\pi)\left[K_{\nu+\frac{1}{2}}\left(\frac{a}{\sqrt{2}}\right)\right]^2$

$$[a>0,\quad -1<\operatorname{Re}\nu<0]\qquad \text{ET I 42(23)}$$

6. $\displaystyle\int_0^\infty \cos(ax)\, Q_\nu\left(1+x^2\right)\, dx = \frac{\pi}{\sqrt{2}}\, K_{\nu+\frac{1}{2}}\left(\frac{a}{\sqrt{2}}\right) I_{\nu+\frac{1}{2}}\left(\frac{a}{\sqrt{2}}\right)$

$$[a>0,\quad \operatorname{Re}\nu>-1]\qquad \text{ET I 42(24)}$$

7. $\displaystyle\int_0^1 \cos(ax)\, P_\nu\left(2x^2-1\right)\, dx = \frac{\pi}{2}\, J_{\nu+\frac{1}{2}}\left(\frac{a}{2}\right) J_{-\nu-\frac{1}{2}}\left(\frac{a}{2}\right)$

$$[a>0]\qquad \text{ET I 42(25)}$$

7.163

1. $\displaystyle\int_a^\infty \left(x^2-a^2\right)^{\frac{1}{2}\nu-\frac{1}{4}}\sin(bx)\, P_0^{\frac{1}{2}-\nu}\left(ax^{-1}\right)\, dx = b^{-\nu-\frac{1}{2}}\cos\left(ab-\frac{\nu\pi}{2}+\frac{\pi}{4}\right)$

$$[a>0,\quad |\operatorname{Re}\nu|<\tfrac{1}{2}]\qquad \text{ET I 98(24)}$$

2. $\displaystyle\int_0^1 x^{-1}\cos(ax)\, P_\nu\left(2x^{-2}-1\right)\, dx = -\frac{1}{2}\pi\,\operatorname{cosec}(\nu\pi)\ {}_1F_1\left((\nu+1;1;ai)\right)\ {}_1F_1\left(\nu+1;1;-ai\right)$

$$[a>0,\quad -1<\operatorname{Re}\nu<0]\qquad \text{ET II 327(4)}$$

7.164

1. $\displaystyle\int_0^\infty x^{1/2}\sin(bx)\left[P_\nu^{-1/4}\left(\sqrt{1+a^2x^2}\right)\right]^2 dx = \frac{\sqrt{\frac{2}{\pi}}\,a^{-1}b^{-1/2}}{\Gamma\left(\frac{5}{4}+\nu\right)\Gamma\left(\frac{1}{4}-\nu\right)}\left[K_{\nu+\frac{1}{2}}\left(\frac{b}{2a}\right)\right]^2$

$$\left[\operatorname{Re}a>0,\quad b>0,\quad -\tfrac{5}{4}<\operatorname{Re}\nu<\tfrac{1}{4}\right]$$
$$\text{ET II 327(8)}$$

2. $\displaystyle\int_0^\infty x^{1/2}\sin(bx)\, P_\nu^{-1/4}\left(\sqrt{1+a^2x^2}\right) Q_{\nu-1}^{-1/4}\left(\sqrt{1+a^2x^2}\right) dx$

$$= \frac{\sqrt{\frac{\pi}{2}}\,e^{-\frac{1}{4}\pi i}\,\Gamma\left(\nu+\frac{5}{4}\right)}{ab^{\frac{1}{2}}\,\Gamma\left(\nu+\frac{3}{4}\right)}\, I_{\nu+\frac{1}{2}}\left(\frac{b}{2a}\right) K_{\nu+\frac{1}{2}}\left(\frac{b}{2a}\right)$$
$$\left[\operatorname{Re}a>0,\quad b>0,\quad \operatorname{Re}\nu>-\tfrac{5}{4}\right]\qquad \text{ET II 328(9)}$$

3. $\displaystyle\int_0^\infty x^{1/2}\sin(bx)\, P_\nu^{-1/4}\left(\sqrt{1+a^2x^2}\right) P_{\nu-1}^{-1/4}\left(\sqrt{1+a^2x^2}\right) \frac{dx}{\sqrt{1+a^2x^2}}$

$$= \frac{a^{-2}b^{1/2}}{\sqrt{2\pi}\,\Gamma\left(\frac{5}{4}+\nu\right)\Gamma\left(\frac{5}{4}-\nu\right)}\, K_{\nu-\frac{1}{2}}\left(\frac{b}{2a}\right) K_{\nu+\frac{1}{2}}\left(\frac{b}{2a}\right)$$
$$\left[\operatorname{Re}a>0,\quad b>0,\quad -\tfrac{5}{4}<\operatorname{Re}\nu<\tfrac{5}{4}\right]\qquad \text{ET II 328(10)}$$

4. $\displaystyle\int_0^\infty x^{1/2}\sin(bx)\, P_\nu^{1/4}\left(\sqrt{1+a^2x^2}\right) P_\nu^{-3/4}\left(\sqrt{1+a^2x^2}\right) \frac{dx}{\sqrt{1+a^2x^2}}$

$$= \frac{a^{-2}b^{1/2}}{\sqrt{2\pi}\,\Gamma\left(\frac{7}{4}+\nu\right)\Gamma\left(\frac{3}{4}-\nu\right)}\left[K_{\nu+\frac{1}{2}}\left(\frac{b}{2a}\right)\right]^2$$
$$\left[\operatorname{Re}a>0,\quad b>0,\quad -\tfrac{7}{4}<\operatorname{Re}\nu<\tfrac{3}{4}\right]\qquad \text{ET II 328(11)}$$

5. $\int_0^\infty x^{1/2} \cos(bx) \left[P_\nu^{1/4} \left(\sqrt{1 + a^2 x^2} \right) \right]^2 dx = \dfrac{a^{-1} \left(\frac{\pi b}{2} \right)^{-1/2}}{\Gamma \left(\frac{3}{4} + \nu \right) \Gamma \left(-\frac{1}{4} - \nu \right)} \left[K_{\nu + \frac{1}{2}} \left(\dfrac{b}{2a} \right) \right]^2$

$$\left[\operatorname{Re} a > 0, \quad b > 0, \quad -\tfrac{3}{4} < \operatorname{Re} \nu < -\tfrac{1}{4} \right]$$

ET II 328(12)

6. $\int_0^\infty x^{1/2} \cos(bx) \, P_\nu^{1/4} \left(\sqrt{1 + a^2 x^2} \right) Q_\nu^{1/4} \left(\sqrt{1 + a^2 x^2} \right) dx$

$$= \dfrac{\sqrt{\frac{\pi}{2}} e^{\frac{1}{4} \pi i} \Gamma \left(\nu + \frac{3}{4} \right)}{a b^{1/2} \Gamma \left(\nu + \frac{5}{4} \right)} I_{\nu + \frac{1}{2}} \left(\dfrac{b}{2a} \right) K_{\nu + \frac{1}{2}} \left(\dfrac{b}{2a} \right)$$

$$\left[\operatorname{Re} a > 0, \quad b > 0, \quad \operatorname{Re} \nu > -\tfrac{3}{4} \right] \quad \text{ET II 328(13)}$$

7. $\int_0^\infty x^{1/2} \cos(bx) \, P_\nu^{-1/4} \left(\sqrt{1 + a^2 x^2} \right) P_\nu^{3/4} \left(\sqrt{1 + a^2 x^2} \right) \dfrac{dx}{\sqrt{1 + a^2 x^2}}$

$$= \dfrac{a^{-2} b^{1/2}}{\sqrt{2\pi} \, \Gamma \left(\frac{5}{4} + \nu \right) \Gamma \left(\frac{1}{4} - \nu \right)} \left[K_{\nu + \frac{1}{2}} \left(\dfrac{b}{2a} \right) \right]^2$$

$$\left[\operatorname{Re} a > 0, \quad b > 0, \quad -\tfrac{5}{4} < \operatorname{Re} \nu < \tfrac{1}{4} \right] \quad \text{ET II 328(14)}$$

8. $\int_0^\infty x^{1/2} \cos(bx) \, P_\nu^{1/4} \left(\sqrt{1 + a^2 x^2} \right) P_{\nu - 1}^{1/4} \left(\sqrt{1 + a^2 x^2} \right) \dfrac{dx}{\sqrt{1 + a^2 x^2}}$

$$= \dfrac{a^{-2} b^{1/2}}{\sqrt{2\pi} \, \Gamma \left(\frac{3}{4} + \nu \right) \Gamma \left(\frac{3}{4} - \nu \right)} K_{\nu - \frac{1}{2}} \left(\dfrac{b}{2a} \right) K_{\nu + \frac{1}{2}} \left(\dfrac{b}{2a} \right)$$

$$\left[\operatorname{Re} a > 0, \quad b > 0, \quad |\operatorname{Re} \nu| < \tfrac{3}{4} \right] \quad \text{ET II 329(15)}$$

7.165 $\int_0^\infty \cos(ax) \, P_\nu (\cosh x) \, dx$

$$= -\dfrac{\sin(\nu\pi)}{4\pi^2} \, \Gamma \left(\dfrac{1 + \nu + i\alpha}{2} \right) \Gamma \left(\dfrac{1 + \nu - i\alpha}{2} \right) \Gamma \left(-\dfrac{\nu + i\alpha}{2} \right) \Gamma \left(-\dfrac{\nu - i\alpha}{2} \right)$$

$$[a > 0, \quad -1 < \operatorname{Re} \nu < 0] \quad \text{ET II 329(18)}$$

7.166 $\int_0^\pi P_\nu^{-\mu} (\cos \varphi) \sin^{\alpha - 1} \varphi \, d\varphi = \dfrac{2^{-\mu} \pi \, \Gamma \left(\frac{1}{2}\alpha + \frac{1}{2}\mu \right) \Gamma \left(\frac{1}{2}\alpha - \frac{1}{2}\mu \right)}{\Gamma \left(\frac{1}{2} + \frac{1}{2}\alpha + \frac{1}{2}\nu \right) \Gamma \left(\frac{1}{2}\alpha - \frac{1}{2}\nu \right) \Gamma \left(\frac{1}{2}\mu + \frac{1}{2}\nu + 1 \right) \Gamma \left(\frac{1}{2}\mu - \frac{1}{2}\nu + \frac{1}{2} \right)}$

$$[\operatorname{Re} (\alpha \pm \mu) > 0] \quad \text{MO 90, EH I 172(27)}$$

7.167 $\int_0^a P_\nu^{-\mu} (\cos x) \, P_\nu^{-\eta} [\cos(a - x)] \left[\dfrac{\sin(a - x)}{\sin x} \right]^\eta \dfrac{dx}{\sin x} = \dfrac{2^\eta \, \Gamma(\mu - \eta) \, \Gamma \left(\eta + \frac{1}{2} \right) (\sin a)^\eta}{\sqrt{\pi} \, \Gamma(\eta + \mu + 1)} P_\nu^{-\mu} (\cos a)$

$$\left[\operatorname{Re} \mu > \operatorname{Re} \eta > -\tfrac{1}{2} \right] \quad \text{ET II 329(16)}$$

7.17 A combination of an associated Legendre function and the probability integral

7.171 $\int_1^\infty (x^2 - 1)^{-\frac{1}{2}\mu} \exp \left(a^2 x^2 \right) [1 - \Phi(ax)] \, P_\nu^\mu(x) \, dx$

$$= \pi^{-1} 2^{\mu - 1} \Gamma \left(\dfrac{1 + \mu + \nu}{2} \right) \Gamma \left(\dfrac{\mu - \nu}{2} \right) a^{\mu - \frac{3}{2}} e^{\frac{a^2}{2}} \, W_{\frac{1}{4} - \frac{1}{2}\mu, \frac{1}{4} + \frac{1}{2}\nu} \left(a^2 \right)$$

$$[\operatorname{Re} a > 0, \quad \operatorname{Re} \mu < 1, \quad \operatorname{Re} (\mu + \nu) > -1, \quad \operatorname{Re}(\mu - \nu) > 0]$$

ET II 324(17)

7.18 Combinations of associated Legendre and Bessel functions

7.181

1. $$\int_1^\infty P_{\nu-\frac{1}{2}}(x)x^{1/2}\,Y_\nu(ax)\,dx = 2^{-1/2}a^{-1}\left[\cos\left(\tfrac{1}{2}a\right)J_\nu\left(\tfrac{1}{2}a\right) - \sin\left(\tfrac{1}{2}a\right)Y_\nu\left(\tfrac{1}{2}a\right)\right]$$
$$\left[a>0,\quad \operatorname{Re}\nu<\tfrac{1}{2}\right] \qquad \text{ET II 108(3)a}$$

2. $$\int_1^\infty P_{\nu-\frac{1}{2}}(x)x^{1/2}\,J_\nu(ax)\,dx = -\frac{1}{\sqrt{2a}}\left[\cos\left(\tfrac{1}{2}a\right)Y_\nu\left(\tfrac{1}{2}a\right) + \sin\left(\tfrac{1}{2}a\right)J_\nu\left(\tfrac{1}{2}a\right)\right]$$
$$\left[|\operatorname{Re}\nu|<\tfrac{1}{2}\right] \qquad \text{ET II 344(36)a}$$

7.182

1. $$\int_1^\infty x^\nu\left(x^2-1\right)^{\frac{1}{2}\lambda-\frac{1}{2}}P_\lambda^{\lambda-1}(x)\,J_\nu(ax)\,dx = \frac{2^{\lambda+\nu}a^{-\lambda}\,\Gamma\left(\tfrac{1}{2}+\nu\right)}{\pi^{1/2}\,\Gamma(1-\lambda)}S_{\lambda-\nu,\lambda+\nu}(a)$$
$$\left[a>0,\quad \operatorname{Re}\nu<\tfrac{5}{2},\quad \operatorname{Re}(2\lambda+\nu)<\tfrac{3}{2}\right]$$
$$\text{ET II 345(38)a}$$

2. $$\int_1^\infty x^{\frac{1}{2}-\mu}\left(x^2-1\right)^{-\frac{1}{2}\mu}P_{\nu-\frac{1}{2}}^\mu(x)\,J_\nu(ax)\,dx$$
$$= -2^{-3/2}\pi^{1/2}a^{\mu-\frac{1}{2}}\left[J_{\mu-\frac{1}{2}}\left(\frac{a}{2}\right)Y_\nu\left(\frac{a}{2}\right) + Y_{\mu-\frac{1}{2}}\left(\frac{a}{2}\right)J_\nu\left(\frac{a}{2}\right)\right]$$
$$\left[-\tfrac{1}{4}<\operatorname{Re}\mu<1,\quad a>0,\quad |\operatorname{Re}\nu|<\tfrac{1}{2}+2\operatorname{Re}\mu\right] \quad \text{ET II 344(37)a}$$

3. $$\int_1^\infty x^{\frac{1}{2}-\mu}\left(x^2-1\right)^{-\frac{1}{2}\mu}P_{\nu-\frac{1}{2}}^\mu(x)\,Y_\nu(ax)\,dx$$
$$= 2^{-3/2}\pi^{1/2}a^{\mu-\frac{1}{2}}\left[J_\nu\left(\frac{a}{2}\right)J_{\mu-\frac{1}{2}}\left(\frac{a}{2}\right) - Y_\nu\left(\frac{a}{2}\right)Y_{\mu-\frac{1}{2}}\left(\frac{a}{2}\right)\right]$$
$$\left[-\tfrac{1}{4}<\operatorname{Re}\mu<1,\quad a>0,\quad \operatorname{Re}(2\mu-\nu)>-\tfrac{1}{2}\right] \quad \text{ET II 349(67)a}$$

4. $$\int_0^1 x^{\frac{1}{2}-\mu}\left(1-x^2\right)^{-\frac{1}{2}\mu}P_\nu^\mu(x)\,J_{\nu+\frac{1}{2}}(ax)\,dx = \sqrt{\frac{\pi}{2}}a^{\mu-\frac{1}{2}}\,J_{\frac{1}{2}-\mu}\left(\tfrac{1}{2}a\right)J_{\nu+\frac{1}{2}}\left(\tfrac{1}{2}a\right)$$
$$\left[\operatorname{Re}\mu<1,\quad \operatorname{Re}(\mu-\nu)<2\right]$$
$$\text{ET II 337(33)a}$$

5. $$\int_1^\infty x^{\frac{1}{2}-\mu}\left(x^2-1\right)^{-\frac{1}{2}\mu}P_{\nu-\frac{1}{2}}^\mu(x)\,K_\nu(ax)\,dx = (2\pi)^{-1/2}a^{\mu-\frac{1}{2}}\,K_\nu\left(\tfrac{1}{2}a\right)K_{\mu-\frac{1}{2}}\left(\tfrac{1}{2}a\right)$$
$$\left[\operatorname{Re}\mu<1,\quad \operatorname{Re}a>0\right] \qquad \text{ET II 135(5)a}$$

6. $$\int_1^\infty x^{\mu+\frac{1}{2}}\left(x^2-1\right)^{-\frac{1}{2}\mu}P_{\nu-\frac{1}{2}}^\mu(x)\,K_\nu(ax)\,dx = \sqrt{\frac{\pi}{2}}a^{-3/2}e^{-\frac{1}{2}a}\,W_{\mu,\nu}(a)$$
$$\left[\operatorname{Re}\mu<1,\quad \operatorname{Re}a>0\right] \qquad \text{ET II 135(3)a}$$

7. $$\int_1^\infty x^{\mu-\frac{3}{2}}\left(x^2-1\right)^{-\frac{1}{2}\mu}P_{\nu-\frac{1}{2}}^\mu(x)\,K_\nu(ax)\,dx = \sqrt{\frac{\pi}{2}}a^{-1/2}e^{-\frac{1}{2}a}\,W_{\mu-1,\nu}(a)$$
$$\left[\operatorname{Re}\mu<1,\quad \operatorname{Re}a>0\right] \qquad \text{ET II 135(4)a}$$

8. $$\int_1^\infty x^{\mu-\frac{1}{2}}\left(x^2-1\right)^{-\frac{1}{2}\mu}P_{\nu-\frac{3}{2}}^\mu(x)\,K_\nu(ax)\,dx = \sqrt{\frac{\pi}{2}}a^{-1}e^{-\frac{1}{2}a}\,W_{\mu-\frac{1}{2},\nu-\frac{1}{2}}(a)$$
$$\left[\operatorname{Re}\mu<1\right] \qquad \text{ET II 135(6)a}$$

9. $\displaystyle\int_1^\infty x^{1/2}\left(x^2-1\right)^{\frac{1}{2}\nu-\frac{1}{4}} P_\mu^{\frac{1}{2}-\nu}\left(2x^2-1\right) K_\nu(ax)\,dx = \pi^{-1/2}a^{-\nu}2^{\nu-1}\left[K_{\mu+\frac{1}{2}}\left(\dfrac{a}{2}\right)\right]^2$

$\left[\operatorname{Re}\nu > -\tfrac{1}{2}, \quad \operatorname{Re}a > 0\right]$ ET II 136(11)a

10. $\displaystyle\int_1^\infty x^{1/2}\left(x^2-1\right)^{\frac{1}{2}\nu-\frac{1}{4}} P_\mu^{\frac{1}{2}-\nu}\left(2x^2-1\right) Y_\nu(ax)\,dx$

$= \pi^{1/2}2^{\nu-2}a^{-\nu}\left[J_{\mu+\frac{1}{2}}\left(\dfrac{a}{2}\right)J_{-\mu-\frac{1}{2}}\left(\dfrac{a}{2}\right) - Y_{\mu+\frac{1}{2}}\left(\dfrac{a}{2}\right)Y_{-\mu-\frac{1}{2}}\left(\dfrac{a}{2}\right)\right]$

$\left[\operatorname{Re}\nu > -\tfrac{1}{2}, \quad a > 0, \quad \operatorname{Re}\nu + |2\operatorname{Re}\mu + 1| < \tfrac{3}{2}\right]$ ET II 108(5)a

11. $\displaystyle\int_1^\infty x^{1/2}\left(x^2-1\right)^{\frac{1}{2}\nu-\frac{1}{4}} P_\mu^{\frac{1}{2}-\nu}\left(2x^2-1\right) J_\nu(ax)\,dx$

$= -2^{\nu-2}a^{-\nu}\pi^{1/2}\sec(\mu\pi)\left\{\left[J_{\mu+\frac{1}{2}}\left(\dfrac{a}{2}\right)\right]^2 - \left[J_{-\mu-\frac{1}{2}}\left(\dfrac{a}{2}\right)\right]^2\right\}$

$\left[\operatorname{Re}\nu > -\tfrac{1}{2}, \quad a > 0, \quad \operatorname{Re}\nu - \tfrac{3}{2} < 2\operatorname{Re}\mu < \tfrac{1}{2} - \operatorname{Re}\nu\right]$ ET II 345(39)a

12. $\displaystyle\int_1^\infty x\left(x^2-1\right)^{-\frac{1}{2}\nu} P_\mu^\nu\left(2x^2-1\right) K_\nu(ax)\,dx = 2^{-\nu}a^{\nu-1} K_{\mu+1}(a)$

$\left[\operatorname{Re}a > 0, \quad \operatorname{Re}\nu < 1\right]$ ET II 136(10)a

13. $\displaystyle\int_0^\infty x\left(x^2+a^2\right)^{\frac{1}{2}\nu} P_\mu^\nu\left(1+2x^2a^{-2}\right) K_\nu(xy)\,dx = 2^{-\nu}ay^{-\nu-1} S_{2\nu,2\mu+1}(ay)$

$\left[\operatorname{Re}a > 0, \quad \operatorname{Re}y > 0, \quad \operatorname{Re}\nu < 1\right]$
ET II 135(7)

14. $\displaystyle\int_0^\infty x\left(x^2+a^2\right)^{\frac{1}{2}\nu}\left[(\mu-\nu) P_\mu^\nu\left(1+2x^2a^{-2}\right) + (\mu+\nu) P_{-\mu}^\nu\left(1+2x^2a^{-2}\right)\right] K_\nu(xy)\,dx$

$= 2^{1-\nu}\mu y^{-\nu-2} S_{2\nu+1,2\mu}(ay)$

$\left[\operatorname{Re}a > 0, \quad \operatorname{Re}y > 0, \quad \operatorname{Re}\nu < 1\right]$ ET II 136(8)

15. $\displaystyle\int_0^\infty x\left(x^2+a^2\right)^{\frac{1}{2}\nu-1}\left[P_\mu^\nu\left(1+2x^2a^{-2}\right) + P_{-\mu}^\nu\left(1+2x^2a^{-2}\right)\right] K_\nu(xy)\,dx = 2^{1-\nu}y^{-\nu} S_{2\nu-1,2\mu}(ay)$

$\left[\operatorname{Re}a > 0, \quad \operatorname{Re}y > 0, \quad \operatorname{Re}\nu < 1\right]$
ET II 136(9)

16. $\displaystyle\int_0^\infty x^{1/2}\left(x^2+2\right)^{-\frac{1}{2}\nu-\frac{1}{4}} P_\mu^{-\nu-\frac{1}{2}}\left(x^2+1\right) J_\nu(xy)\,dx = \dfrac{y^{-1/2}2^{\frac{1}{2}-\nu}\pi^{-1/2}\left[K_{\mu+\frac{1}{2}}\left(2^{-1/2}y\right)\right]^2}{\Gamma\left(\nu+\mu+\frac{3}{2}\right)\Gamma\left(\nu-\mu+\frac{1}{2}\right)}$

$\left[-\tfrac{3}{2} - \operatorname{Re}\nu < \operatorname{Re}\mu < \operatorname{Re}\nu + \tfrac{1}{2}, \quad y > 0\right]$
ET II 44(1)

17. $\displaystyle\int_0^\infty x^{1/2}\left(x^2+2\right)^{-\frac{1}{2}\nu-\frac{1}{4}} Q_\mu^{\nu+\frac{1}{2}}\left(x^2+1\right) J_\nu(xy)\,dx$

$= 2^{-\nu-\frac{1}{2}}\pi^{1/2}e^{\left(\nu+\frac{1}{2}\right)\pi i}y^\nu K_{\mu+\frac{1}{2}}\left(2^{-1/2}y\right) I_{\mu+\frac{1}{2}}\left(2^{-1/2}y\right)$

$\left[\operatorname{Re}\nu > -1, \quad \operatorname{Re}(2\mu+\nu) > -\tfrac{5}{2}, \quad y > 0\right]$ ET II 46(12)

7.183 $\displaystyle\int_0^\infty x^{1-\mu}\left(1+a^2x^2\right)^{-\frac{1}{2}\mu-\frac{1}{4}} Q_{\nu-\frac{1}{2}}^{\mu+\frac{1}{2}}\left(\pm iax\right) J_\nu(xy)\,dx$

$$= i(2\pi)^{1/2} e^{i\pi\left(\mu\mp\frac{1}{2}\nu\mp\frac{1}{4}\right)} a^{-1} y^{\mu-1} I_\nu\left(\tfrac{1}{2}a^{-1}y\right) K_\mu\left(\tfrac{1}{2}a^{-1}y\right)$$

$$\left[-\tfrac{3}{4}-\tfrac{1}{2}\operatorname{Re}\nu<\operatorname{Re}\mu<1+\operatorname{Re}\nu,\quad y>0,\quad \operatorname{Re} a>0\right]\quad\text{ET II 46(11)}$$

7.184

1. $\displaystyle\int_1^\infty x^{1/2}\left(x^2-1\right)^{\frac{1}{2}\mu-\frac{1}{4}} P_{-\frac{1}{2}+\nu}^{-\frac{1}{2}-\mu}\left(x^{-1}\right) J_\nu(xa)\,dx = 2^{1/2}a^{-1-\mu}\pi^{-1/2}\cos\left[a+\tfrac{1}{2}(\nu-\mu)\pi\right]$

$$\left[|\operatorname{Re}\mu|<\tfrac{1}{2},\quad \operatorname{Re}\nu>-1,\quad a>0\right]$$
$$\text{ET II 44(2)a}$$

2. $\displaystyle\int_1^\infty x^{-\nu}\left(x^2-1\right)^{\frac{1}{4}-\frac{1}{2}\nu} P_\mu^{\nu-\frac{1}{2}}\left(2x^{-2}-1\right) K_\nu(ax)\,dx$

$$= \pi^{1/2}2^{-\nu}a^{-2+\nu} W_{\mu+\frac{1}{2},\nu-\frac{1}{2}}(a)\, W_{-\mu-\frac{1}{2},\nu-\frac{1}{2}}(a)$$
$$\left[\operatorname{Re}\nu<\tfrac{3}{2},\quad a>0\right]\qquad\text{ET II 370(45)a}$$

3. $\displaystyle\int_0^\infty x^\nu\left(1+x^2\right)^{\frac{1}{4}+\frac{\nu}{2}} Q_\mu^{\nu+\frac{1}{2}}\left(1+\frac{2}{x^2}\right) J_\nu(ax)\,dx$

$$= -ie^{i\pi\nu}\pi^{-\frac{1}{2}}2^\nu a^{-\nu-2}\left[\Gamma\left(\tfrac{3}{2}+\mu+\nu\right)\right]^2\Gamma\left(\tfrac{1}{2}+\nu-\mu\right)$$

$$\times W_{-\mu-\frac{1}{2},\nu+\frac{1}{2}}(a)\left[\frac{\cos(\mu\pi)}{\Gamma(2+2\nu)} M_{\mu+\frac{1}{2},\nu+\frac{1}{2}}(a)+\frac{\sin(\mu\pi)}{\Gamma\left(\nu+\mu+\frac{3}{2}\right)} W_{\mu+\frac{1}{2},\nu+\frac{1}{2}}(a)\right]$$

$$\left[a>0,\quad \operatorname{Re}(\mu+\nu)>-\tfrac{3}{2},\quad \operatorname{Re}(\mu-\nu)<\tfrac{1}{2}\right]\quad\text{ET II 46(14)}$$

4. $\displaystyle\int_0^1 x^\nu\left(1-x^2\right)^{\frac{1}{2}\nu+\frac{1}{4}} P_\mu^{-\nu-\frac{1}{2}}\left(2x^{-2}-1\right) J_\nu(xy)\,dx$

$$= 2^{\nu+\frac{1}{2}}y^\nu\frac{\Gamma\left(\tfrac{3}{2}+\mu+\nu\right)\Gamma\left(\tfrac{1}{2}+\nu-\mu\right)}{(2\pi)^{1/2}\left[\Gamma\left(\tfrac{3}{2}+\nu\right)\right]^2}$$

$$\times {}_1F_1\left(\nu+\mu+\frac{3}{2};2\nu+2;iy\right){}_1F_1\left(\nu+\mu+\frac{3}{2};2\nu+2;-iy\right)$$

$$\left[y>0,\quad -\tfrac{3}{2}-\operatorname{Re}\nu<\operatorname{Re}\mu<\operatorname{Re}\nu+\tfrac{1}{2}\right]\quad\text{ET II 45(3)}$$

5. $\displaystyle\int_0^\infty x^{-\nu}\left(x^2+a^2\right)^{\frac{1}{4}-\frac{1}{2}\nu} Q_\mu^{\frac{1}{2}-\nu}\left(1+2a^2x^{-2}\right) K_\nu(xy)\,dx$

$$= ie^{-i\pi\nu}\pi^{1/2}2^{-\nu-1}a^{-\nu-\frac{1}{2}}y^{\nu-2}\left[\Gamma\left(\tfrac{3}{2}+\mu-\nu\right)\right]^2 W_{-\mu-\frac{1}{2},\nu-\frac{1}{2}}(iay)\, W_{-\mu-\frac{1}{2},\nu-\frac{1}{2}}(-iay)$$

$$\left[\operatorname{Re} a>0,\quad \operatorname{Re} y>0,\quad \operatorname{Re}\mu>-\tfrac{3}{2},\quad \operatorname{Re}(\mu-\nu)>-\tfrac{3}{2}\right]\quad\text{ET II 137(13)}$$

6. $\displaystyle\int_0^\infty x^{-\nu}\left(x^2+1\right)^{\frac{1}{4}-\frac{1}{2}\nu} Q_\mu^{\frac{1}{2}-\nu}\left(1+2x^{-2}\right) J_\nu(ax)\,dx$

$$= 2^{-\nu}a^{-\nu-2}\frac{ie^{-i\nu\pi}\pi^{1/2}\Gamma\left(\tfrac{3}{2}+\mu-\nu\right)}{\Gamma(2\nu)} M_{\mu+\frac{1}{2},\nu-\frac{1}{2}}(a)\, W_{-\mu-\frac{1}{2},\nu-\frac{1}{2}}(a)$$

$$\left[a>0,\quad 0<\operatorname{Re}\nu<\operatorname{Re}\mu+\tfrac{3}{2}\right]\quad\text{ET II 47(15)a}$$

7. $\displaystyle\int_0^\infty x^{-\nu}\left(x^2+a^2\right)^{\frac14-\frac12\nu}Q_{-\frac12}^{\frac12-\nu}\left(1+2a^2x^{-2}\right)K_\nu(xy)\,dx$

$$= ie^{-i\pi\nu}\pi^{3/2}2^{-\nu-3}a^{\frac12-\nu}y^{\nu-1}[\Gamma(1-\nu)]^2\times\left\{\left[J_{\nu-\frac12}\left(\frac{ay}{2}\right)\right]^2+\left[Y_{\nu-\frac12}\left(\frac{ay}{2}\right)\right]^2\right\}$$

$$\left[\operatorname{Re}a>0,\quad\operatorname{Re}y>0,\quad\operatorname{Re}\nu<1\right]\quad\text{ET II 136(12)}$$

7.185 $\displaystyle\int_0^\infty x^{1/2}Q_{\nu-\frac12}\left[\left(a^2+x^2\right)x^{-1}\right]J_\nu(xy)\,dx=2^{-1/2}\pi y^{-1}\exp\left[-\left(a^2-\tfrac14\right)^{1/2}y\right]J_\nu\left(\tfrac12y\right)$

$$\left[\operatorname{Re}\nu>-\tfrac12,\quad y>0\right]\qquad\text{ET II 46(10)}$$

7.186 $\displaystyle\int_0^\infty x\left(1+x^2\right)^{-\nu-1}P_\nu\left(\frac{1-x^2}{1+x^2}\right)J_0(xy)\,dx=y^{2\nu}[2^\nu\,\Gamma(\nu+1)]^{-2}K_0(y)$

$$\left[\operatorname{Re}\nu>0\right]\qquad\text{ET II 13(10)}$$

7.187

1. $\displaystyle\int_0^\infty xP_\mu^\nu\left(\sqrt{1+x^2}\right)K_\nu(xy)\,dx=y^{-3/2}S_{\nu+\frac12,\mu+\frac12}(y)$

$$\left[\operatorname{Re}\nu<1,\quad\operatorname{Re}y>0\right]\qquad\text{ET II 137(14)}$$

2. $\displaystyle\int_0^\infty x\left[P_{\lambda-\frac12}\left(\sqrt{1+a^2x^2}\right)\right]^2J_0(xy)\,dx=2\pi^{-2}y^{-1}a^{-1}\cos(\lambda\pi)\left[K_\lambda\left(\frac{y}{2a}\right)\right]^2$

$$\left[\operatorname{Re}a>0,\quad|\operatorname{Re}\lambda|<\tfrac14,\quad y>0\right]$$
$$\text{ET II 13(11)}$$

3. $\displaystyle\int_0^\infty x\left(1+x^2\right)^{-1/2}P_\mu^\nu\left(\sqrt{1+x^2}\right)K_\nu(xy)\,dx=y^{-1/2}S_{\nu-\frac12,\mu+\frac12}(y)$

$$\left[\operatorname{Re}\nu<1,\quad\operatorname{Re}y>0\right]\qquad\text{ET II 137(15)}$$

4. $\displaystyle\int_0^\infty xP_\mu^{-\frac12\nu}\left(\sqrt{1+a^2x^2}\right)Q_\mu^{-\frac12\nu}\left(\sqrt{1+a^2x^2}\right)J_\nu(xy)\,dx$

$$=\frac{y^{-1}e^{-\frac12\nu\pi i}\,\Gamma\left(1+\mu+\frac12\nu\right)}{a\,\Gamma\left(1+\mu-\frac12\nu\right)}I_{\mu+\frac12}\left(\frac{y}{2a}\right)K_{\mu+\frac12}\left(\frac{y}{2a}\right)$$

$$\left[\operatorname{Re}a>0,\quad y>0,\quad\operatorname{Re}\mu>-\tfrac34,\quad\operatorname{Re}\nu>-1\right]\quad\text{ET II 47(16)}$$

5. $\displaystyle\int_0^\infty xP_{\sigma-\frac12}^\mu\left(\sqrt{1+a^2x^2}\right)Q_{\sigma-\frac12}^\mu\left(\sqrt{1+a^2x^2}\right)J_0(xy)\,dx$

$$=y^{-2}e^{\mu\pi i}\frac{\Gamma\left(\frac12+\sigma-\mu\right)}{\Gamma(1+2\sigma)}W_{\mu,\sigma}\left(\frac{y}{a}\right)M_{-\mu,\sigma}\left(\frac{y}{a}\right)$$

$$\left[\operatorname{Re}a>0,\quad y>0,\quad\operatorname{Re}\sigma>-\tfrac14,\quad\operatorname{Re}\mu<1\right]\quad\text{ET II 14(15)}$$

6. $\displaystyle\int_0^\infty xP_{\sigma-\frac12}^\mu\left(\sqrt{1+a^2x^2}\right)P_{\sigma-\frac12}^{-\mu}\left(\sqrt{1+a^2x^2}\right)J_0(xy)\,dx$

$$=2\pi^{-1}y^{-2}\cos(\sigma\pi)W_{\mu,\sigma}\left(\frac{y}{a}\right)W_{-\mu,\sigma}\left(\frac{y}{a}\right)$$

$$\left[\operatorname{Re}a>0,\quad y>0,\quad|\operatorname{Re}\sigma|<\tfrac14\right]\quad\text{ET II 14(14)}$$

7. $\displaystyle\int_0^\infty x\left\{P_{\sigma-\frac12}^\mu\left(\sqrt{1+a^2x^2}\right)\right\}^2J_0(xy)\,dx=-i\pi^{-1}y^{-2}W_{\mu,\sigma}\left(\frac{y}{a}\right)\left[W_{\mu,\sigma}\left(e^{\pi i}\frac{y}{a}\right)-W_{\mu,\sigma}\left(e^{-\pi i}\frac{y}{a}\right)\right]$

$$\left[\operatorname{Re}a>0,\quad y>0,\quad|\operatorname{Re}\sigma|<\tfrac14,\quad\operatorname{Re}\mu<1\right]\quad\text{ET II 14(13)}$$

8.
$$\int_0^\infty x\left(1+a^2x^2\right)^{-1/2} P_\mu^{-\frac{1}{2}-\frac{1}{2}\nu}\left(\sqrt{1+a^2x^2}\right) P_\mu^{\frac{1}{2}-\frac{1}{2}\nu}\left(\sqrt{1+a^2x^2}\right) J_\nu(xy)\,dx$$

$$= \frac{\left[K_{\mu+\frac{1}{2}}\left(\frac{y}{2a}\right)\right]^2}{\pi a^2\,\Gamma\left(\frac{\nu}{2}+\mu+\frac{3}{2}\right)\Gamma\left(\frac{\nu}{2}-\mu+\frac{1}{2}\right)}$$

$$\left[\operatorname{Re}a>0,\quad y>0,\quad -\tfrac{5}{4}<\operatorname{Re}\mu<\tfrac{1}{4}\right] \quad \textbf{ET II 46(9)}$$

9.
$$\int_0^\infty x\left\{P_\mu^{-\frac{1}{2}\nu}\left(\sqrt{1+a^2x^2}\right)\right\}^2 J_\nu(xy)\,dx = \frac{2\left[K_{\mu+\frac{1}{2}}\left(\frac{y}{2a}\right)\right]^2 y^{-1}}{\pi a\,\Gamma\left(1+\mu+\frac{1}{2}\nu\right)\Gamma\left(\frac{1}{2}\nu-\mu\right)}$$

$$\left[\operatorname{Re}a>0,\quad y>0,\quad -\tfrac{3}{4}<\operatorname{Re}\mu<-\tfrac{1}{4},\quad \operatorname{Re}\nu>-1\right] \quad \textbf{ET II 45(7)}$$

10.
$$\int_0^\infty x\left(1+a^2x^2\right)^{-1/2} P_\mu^{-\frac{1}{2}\nu}\left(\sqrt{1+a^2x^2}\right) P_{\mu+1}^{-\frac{1}{2}\nu}\left(\sqrt{1+a^2x^2}\right) J_\nu(xy)\,dx$$

$$= \frac{K_{\mu+\frac{1}{2}}\left(\frac{y}{2a}\right) K_{\mu+\frac{3}{2}}\left(\frac{y}{2a}\right)}{\pi a^2\,\Gamma\left(2+\frac{1}{2}\nu+\mu\right)\Gamma\left(\frac{1}{2}\nu-\mu\right)}$$

$$\left[\operatorname{Re}a>0,\quad y>0,\quad -\tfrac{7}{4}<\operatorname{Re}\mu<-\tfrac{1}{4}\right] \quad \textbf{ET II 45(8)}$$

7.188

1.
$$\int_0^\infty x\left(a^2+x^2\right)^{-\frac{1}{2}\mu} P_{\mu-1}^{-\nu}\left[\frac{a}{\sqrt{a^2+x^2}}\right] J_\nu(xy)\,dx = \frac{y^{\mu-2}e^{-ay}}{\Gamma(\mu+\nu)}$$

$$\left[\operatorname{Re}a>0,\quad y>0,\quad \operatorname{Re}\nu>-1,\quad \operatorname{Re}\mu>\tfrac{1}{2}\right] \quad \textbf{ET II 45(4)}$$

2.
$$\int_0^\infty x^{\nu+1}\left(x^2+a^2\right)^{\frac{1}{2}\nu} P_\nu\left(\frac{x^2+2a^2}{2a\sqrt{x^2+a^2}}\right) J_\nu(xy)\,dx = \frac{(2a)^{\nu+1}y^{-\nu-1}}{\pi\,\Gamma(-\nu)}\left[K_{\nu+\frac{1}{2}}\left(\frac{ya}{2}\right)\right]^2$$

$$\left[\operatorname{Re}a>0,\quad -1<\operatorname{Re}\nu<0,\quad y>0\right]$$
$$\textbf{ET II 45(5)}$$

3.
$$\int_0^\infty x^{1-\nu}\left(x^2+a^2\right)^{-\frac{1}{2}\nu} P_{\nu-1}\left(\frac{x^2+2a^2}{2a\sqrt{x^2+a^2}}\right) J_\nu(xy)\,dx = \frac{(2a)^{1-\nu}y^{\nu-1}}{\Gamma(\nu)}I_{\nu-\frac{1}{2}}\left(\frac{ay}{2}\right)K_{\nu-\frac{1}{2}}\left(\frac{ay}{2}\right)$$

$$\left[\operatorname{Re}a>0,\quad y>0,\quad 0<\operatorname{Re}\nu<1\right]$$
$$\textbf{ET II 45(6)}$$

7.189

1.
$$\int_0^\infty (a+x)^\mu e^{-x} P_\nu^{-2\mu}\left(1+\frac{2x}{a}\right) I_\mu(x)\,dx = 0$$

$$\left[-\tfrac{1}{2}<\operatorname{Re}\mu<0,\quad -\tfrac{1}{2}+\operatorname{Re}\mu<\operatorname{Re}\nu<-\tfrac{1}{2}-\operatorname{Re}\mu\right] \quad \textbf{ET II 366(18)}$$

2.
$$\int_0^\infty (x+a)^{-\mu}e^{-x} P_\nu^{-2\mu}\left(1+\frac{2x}{a}\right) I_\mu(x)\,dx$$

$$= \frac{2^{\mu-1}\,\Gamma\left(\mu+\nu+\frac{1}{2}\right)\Gamma\left(\mu-\nu-\frac{1}{2}\right)e^a}{\pi^{1/2}\,\Gamma(2\mu+\nu+1)\Gamma(2\mu-\nu)}W_{\frac{1}{2}-\mu,\frac{1}{2}+\nu}(2a)$$

$$\left[|\arg a|<\pi,\quad \operatorname{Re}\mu>\left|\operatorname{Re}\nu+\tfrac{1}{2}\right|\right] \quad \textbf{ET II 367(19)}$$

3. $$\int_0^\infty x^{-\mu} e^x \, P_\nu^{2\mu} \left(1 + \frac{2x}{a}\right) K_\mu(x + a) \, dx$$

$$= \pi^{-1/2} 2^{\mu-1} \cos(\mu\pi) \, \Gamma\left(\mu + \nu + \tfrac{1}{2}\right) \Gamma\left(\mu - \nu + \tfrac{1}{2}\right) W_{\frac{1}{2}-\mu, \frac{1}{2}+\nu}(2a)$$

$$\left[|\arg a| < \pi, \quad \operatorname{Re}\mu > \left|\operatorname{Re}\nu + \tfrac{1}{2}\right|\right] \quad \text{ET II 373(11)}$$

4. $$\int_0^\infty x^{-\frac{1}{2}\mu}(x + a)^{-1/2} e^{-x} \, P_{\nu-\frac{1}{2}}^\mu \left(\frac{a - x}{a + x}\right) K_\nu(a + x) \, dx = \sqrt{\frac{\pi}{2}} \, a^{-\frac{1}{2}\mu} \, \Gamma(\mu, 2a)$$

$$[a > 0, \quad \operatorname{Re}\mu < 1] \quad \text{ET II 374(12)}$$

5. $$\int_0^\infty (\sinh x)^{\mu+1} (\cosh x)^{-2\mu-\frac{3}{2}} \, P_\nu^{-\mu}[\cosh(2x)] \, I_{\mu-\frac{1}{2}}(a \operatorname{sech} x) \, dx$$

$$= \frac{2^{\mu-\frac{1}{2}} \Gamma(\mu - \nu) \Gamma(\mu + \nu + 1)}{\pi^{1/2} a^{\mu+\frac{3}{2}} [\Gamma(\mu + 1)]^2} \, M_{\nu+\frac{1}{2},\mu}(a) \, M_{-\nu-\frac{1}{2},\mu}(a)$$

$$[\operatorname{Re}\mu > \operatorname{Re}\nu, \quad \operatorname{Re}\mu > -\operatorname{Re}\nu - 1] \quad \text{ET II 378(44)}$$

7.19 Combinations of associated Legendre functions and functions generated by Bessel functions

7.191

1. $$\int_a^\infty x^{1/2} \left(x^2 - a^2\right)^{-\frac{1}{4}-\frac{1}{2}\nu} \, P_\mu^{\nu+\frac{1}{2}} \left(2x^2 a^{-2} - 1\right) [\mathbf{H}_\nu(x) - Y_\nu(x)] \, dx$$

$$= 2^{-\nu-2} \pi^{1/2} a \operatorname{cosec}(\mu\pi) \cos(\nu\pi) \left\{ \left[Y_\nu\left(\tfrac{1}{2}a\right)\right]^2 - \left[J_\nu\left(\tfrac{1}{2}a\right)\right]^2 \right\}$$

$$\left[-1 < \operatorname{Re}\mu < 0, \quad \operatorname{Re}\nu < \tfrac{1}{2}\right] \quad \text{ET II 384(6)}$$

2. $$\int_0^\infty x^{1/2} \left(x^2 - a^2\right)^{-1/4-\nu/2} \, P_\mu^{\nu+1/2} \left(2x^2 a^{-2} - 1\right) [I_{-\nu}(x) - \mathbf{L}_\nu(x)] \, dx$$

$$= 2^{-\nu-1} \pi^{1/2} a \operatorname{cosec}(2\mu\pi) \cos(\nu\pi) \left\{ \left[I_\nu\left(\tfrac{1}{2}a\right)\right]^2 - \left[I_{-\nu}\left(\tfrac{1}{2}a\right)\right]^2 \right\}$$

$$\left[-1 < \operatorname{Re}\mu < 0, \quad \operatorname{Re}\nu < \tfrac{1}{2}\right] \quad \text{ET II 385(15)}$$

7.192

1. $$\int_0^1 x^{(\nu-\mu-1)/2} \left(1 - x^2\right)^{(\nu-\mu-2)/4} \, P_{\nu-1/2}^{(\mu-\nu+2)/2}(x) \, S_{\mu,\nu}(ax) \, dx$$

$$= 2^{\mu-3/2} \pi^{1/2} a^{-(\nu-\mu-1)/2} \, \Gamma\left(\frac{\mu + \nu + 3}{4}\right) \Gamma\left(\frac{\mu - 3\nu + 3}{4}\right) \cos\left(\frac{\mu - \nu}{2}\pi\right)$$

$$\times \left[J_\nu\left(\tfrac{1}{2}a\right) Y_{-(\mu-\nu+1)/2}\left(\tfrac{1}{2}a\right) - Y_\nu\left(\tfrac{1}{2}a\right) J_{-(\mu-\nu+1)/2}\left(\tfrac{1}{2}a\right)\right]$$

$$[\operatorname{Re}(\mu - \nu) < 0, \quad a > 0, \quad |\operatorname{Re}(\mu + \nu)| < 1, \quad \operatorname{Re}(\mu - 3\nu) < 1] \quad \text{ET II 387(24)a}$$

2. $\int_1^\infty x^{1/2}(x^2-1)^{-\beta/2} P_\nu^\beta(x) S_{\mu,1/2}(ax)\,dx$

$$= \frac{2^{-3/2+\beta-\mu}a^{\beta-1}\,\Gamma\left(\frac{\beta-\mu+\nu}{2}+\frac14\right)\Gamma\left(\frac{\beta-\mu-\nu}{2}-\frac14\right)}{\pi^{1/2}\,\Gamma\left(\frac12-\mu\right)} S_{\mu-\beta+1,\nu+1/2}(a)$$

$$\left[\operatorname{Re}\beta<1,\quad a>0,\quad \operatorname{Re}(\mu+\nu-\beta)<-\tfrac12,\quad \operatorname{Re}(\mu-\nu-\beta)<\tfrac12\right] \quad \text{ET II 387(25)a}$$

7.193

1. $\int_1^\infty x^{-\nu}(x^2-1)^{1/4-\nu/2} P_{\mu/2-\nu/2}^{\nu-1/2}\left(2x^{-2}-1\right) S_{\mu,\nu}(ax)\,dx$

$$= \frac{2^{\mu-\nu}a^{\nu-2}\pi^{1/2}\,\Gamma\left(\frac{3\nu-\mu-1}{2}\right)}{\Gamma\left(\frac{1+\nu-\mu}{2}\right)} W_{\rho,\sigma}\left(ae^{i\pi/2}\right) W_{\rho,\sigma}\left(ae^{-i\pi/2}\right)$$

$$\rho=\tfrac12(\mu+1-\nu),\quad \sigma=\nu-\tfrac12,\quad \left[\operatorname{Re}(\mu-\nu)<0,\quad a>0,\quad \operatorname{Re}\nu<\tfrac32,\quad \operatorname{Re}(3\nu-\mu)>1\right]$$

$$\text{ET II 387(27)a}$$

2. $\int_1^\infty x(x^2-1)^{-\nu/2} P_\lambda^\nu\left(2x^2-1\right) S_{\mu,\nu}(ax)\,dx$

$$= \frac{a^{\nu-1}\,\Gamma\left(\frac{\nu-\mu+1}{2}+\lambda\right)\Gamma\left(\frac{\nu-\mu-1}{2}-\lambda\right)}{2\,\Gamma\left(\frac{1-\mu-\nu}{2}\right)\Gamma\left(\frac{1-\mu+\nu}{2}\right)} S_{\mu-\nu+1,2\lambda+1}(a)$$

$$\left[\operatorname{Re}\nu<1,\quad a>0,\quad \operatorname{Re}(\mu-\nu+\lambda)<-1,\quad \operatorname{Re}(\mu-\nu+\lambda)<0\right] \quad \text{ET II 387(26)a}$$

7.21 Integration of associated Legendre functions with respect to the order

7.211

1. $\int_0^\infty P_{-x-\frac12}(\cos\theta)\,dx = \dfrac{1}{2}\operatorname{cosec}\left(\dfrac12\theta\right)$ $[0<\theta<\pi]$ ET II 329(19)

2. $\int_{-\infty}^\infty P_x(\cos\theta)\,dx = \operatorname{cosec}\left(\dfrac12\theta\right)$ $[0<\theta<\pi]$ ET II 329(20)

7.212 $\int_0^\infty x^{-1}\tanh(\pi x)\,P_{-\frac12+ix}(\cosh a)\,dx = 2e^{-\frac12 a}\,\boldsymbol{K}\left(e^{-a}\right)$

$$[a>0] \quad \text{ET II 330(22)}$$

7.213 $\int_0^\infty \dfrac{x\tanh(\pi x)}{a^2+x^2}\,P_{-\frac12+ix}(\cosh b)\,dx = Q_{a-\frac12}(\cosh b) \quad [\operatorname{Re}a>0]$ ET II 387(23)

7.214 $\int_0^\infty \sinh(\pi x)\cos(ax)\,P_{-\frac12+ix}(b)\,dx = \dfrac{1}{\sqrt{2(b+\cosh a)}}$

$$[a>0,\quad |b|<1] \quad \text{ET I 42(27)}$$

7.215 $\int_0^\infty \cos(bx)\,P_{-\frac12+ix}^\mu(\cosh a)\,dx = 0$ $[0<a<b]$

$$= \frac{\sqrt{\frac{\pi}{2}}(\sinh a)^\mu}{\Gamma\left(\frac12-\mu\right)(\cosh a-\cosh b)^{\mu+\frac12}} \quad [0<b<a]$$

$$\text{ET II 330(21)}$$

7.216 $\int_0^\infty \cos(bx)\,\Gamma(\mu+ix)\,\Gamma(\mu-ix)\,P_{-\frac{1}{2}+ix}^{\frac{1}{2}-\mu}(\cosh a)\,dx = \dfrac{\sqrt{\frac{\pi}{2}}\,\Gamma(\mu)(\sinh a)^{\mu-\frac{1}{2}}}{(\cosh a + \cosh b)^\mu}$

$$[a>0,\quad b>0,\quad \mathrm{Re}\,\mu>0]$$

ET II 330(24)

7.217

1. $\int_{-\infty}^\infty \left(\nu-\dfrac{1}{2}+ix\right)\Gamma\left(\dfrac{1}{2}-ix\right)\Gamma\left(2\nu-\dfrac{1}{2}+ix\right)P_{\nu+ix-1}^{\frac{1}{2}-\nu}(\cos\theta)\,I_{\nu-\frac{1}{2}+ix}(a)\,K_{\nu-\frac{1}{2}+ix}(b)\,dx$

$$= \sqrt{2\pi}(\sin\theta)^{\nu-\frac{1}{2}}\left(\frac{ab}{\omega}\right)^\nu K_\nu(\omega)$$

$$\left[\omega = \left(a^2+b^2+2ab\cos\theta\right)^{1/2}\right]\quad \text{ET II 383(29)}$$

2. $\int_0^\infty xe^{\pi x}\tanh(\pi x)\,P_{-\frac{1}{2}+ix}(-\cos\theta)\,H_{ix}^{(2)}(ka)\,H_{ix}^{(2)}(kb)\,dx = -\dfrac{2(ab)^{1/2}}{\pi R}e^{-ikR};$

$$R = \left(a^2+b^2-2ab\cos\theta\right)^{1/2}\qquad [a>0,\quad b>0,\quad 0<\theta<\pi,\quad \mathrm{Im}\,k\le 0]\quad \text{ET II 381(17)}$$

3. $\int_0^\infty xe^{\pi x}\sinh(\pi x)\,\Gamma(\nu+ix)\,\Gamma(\nu-ix)\,P_{-\frac{1}{2}+ix}^{\frac{1}{2}-\nu}(-\cos\theta)\,H_{ix}^{(2)}(a)\,H_{ix}^{(2)}(b)\,dx$

$$= i(2\pi)^{1/2}(\sin\theta)^{\nu-\frac{1}{2}}\left(\frac{ab}{R}\right)^\nu H_\nu^{(2)}(R)$$

$$R = \left(a^2+b^2-2ab\cos\theta\right)^{1/2}\qquad [a>0,\quad b>0,\quad 0<\theta<\pi,\quad \mathrm{Re}\,\nu>0]\quad \text{ET II 381 (18)}$$

4. $\int_0^\infty x\sinh(\pi x)\,\Gamma(\lambda+ix)\,\Gamma(\lambda-ix)\,K_{ix}(a)\,K_{ix}(b)\,P_{-\frac{1}{2}+ix}^{\frac{1}{2}-\lambda}(\beta)\,dx = \dfrac{\pi^{1/2}}{\sqrt{2}}\left(\frac{ab}{z}\right)^\lambda(\beta^2-1)^{\frac{1}{2}\lambda-\frac{1}{4}}K_\lambda(z)$

$$z = \sqrt{a^2+b^2+2ab\beta}\qquad \left[|\arg a|<\dfrac{\pi}{2},\quad |\arg(\beta-1)|<\pi,\quad \mathrm{Re}\,\lambda>0\right]\quad \text{ET II 177(16)}$$

7.22 Combinations of Legendre polynomials, rational functions, and algebraic functions

7.221

1. $\int_{-1}^1 P_n(x)\,P_m(x)\,dx = 0 \qquad [m\ne n]$

$$= \dfrac{2}{2n+1}\qquad [m=n]$$

WH, EH I 170(8, 10)

2.[6] $\int_0^1 P_n(x)\,P_m(x)\,dx = \dfrac{1}{2n+1} \qquad [m=n]$

$$= 0 \qquad [n-m\text{ is even},\quad m\ne n]$$

$$= \dfrac{(-1)^{\frac{1}{2}(m+n-1)}m!\,n!}{2^{m+n-1}(m-n)(n+m+1)\left[\left(\frac{n}{2}\right)!\left(\frac{m-1}{2}\right)!\right]^2}\qquad [n\text{ is even},\,m\text{ is odd}]$$

WH

3. $\displaystyle\int_0^{2\pi} P_{2n}(\cos\varphi)\,d\varphi = 2\pi\left[\dbinom{2n}{n}2^{-2n}\right]^2.$ MO 70, EH II 183(50)

7.222

1. $\displaystyle\int_{-1}^1 x^m\,P_n(x)\,dx = 0$ $[m<n]$

2. $\displaystyle\int_{-1}^1 (1+x)^{m+n}\,P_m(x)\,P_n(x)\,dx = \frac{2^{m+n+1}[(m+n)!]^4}{(m!n!)^2(2m+2n+1)!}$ ET II 277(15)

3. $\displaystyle\int_{-1}^1 (1+x)^{m-n-1}\,P_m(x)\,P_n(x)\,dx = 0$ $[m>n]$ ET II 278(16)

4. $\displaystyle\int_{-1}^1 (1-x^2)^n\,P_{2m}(x)\,dx = \frac{2n^2}{(n-m)(2m+2n+1)}\int_{-1}^1 (1-x^2)^{n-1}\,P_{2m}(x)\,dx$

                                              $[m<n]$ WH

5. $\displaystyle\int_0^1 x^2\,P_{n+1}(x)\,P_{n-1}(x)\,dx = \frac{n(n+1)}{(2n-1)(2n+1)(2n+3)}$ WH

7.223 $\displaystyle\int_{-1}^1 \frac{1}{z-x}\{P_n(x)\,P_{n-1}(x) - P_{n-1}(x)\,P_n(z)\}\,dx = -\frac{2}{n}$ WH

7.224   [z belongs to the complex plane with a discontinuity along the interval from -1 to $+1$.]

1. $\displaystyle\int_{-1}^1 (z-x)^{-1}\,P_n(x)\,dx = 2\,Q_n(z)$ ET II 277(7)

2. $\displaystyle\int_{-1}^1 x(z-x)^{-1}\,P_0(x)\,dx = 2\,Q_1(z)$ ET II 277(8)

3. $\displaystyle\int_{-1}^1 x^{n+1}(z-x)^{-1}\,P_n(x)\,dx = 2z^{n+1}\,Q_n(z) - \frac{2^{n+1}(n!)^2}{(2n+1)!}$ ET II 277(9)

4. $\displaystyle\int_{-1}^1 x^m(z-x)^{-1}\,P_n(x)\,dx = 2z^m\,Q_n(z)$ $[m\le n]$ ET II 277(10)a

5. $\displaystyle\int_{-1}^1 (z-x)^{-1}\,P_m(x)\,P_n(x)\,dx = 2\,P_m(z)\,Q_n(z)$ $[m\le n]$ ET II 278(18)a

6. $\displaystyle\int_{-1}^1 (z-x)^{-1}\,P_n(x)\,P_{n+1}(x)\,dx = 2\,P_{n+1}(z)\,Q_n(z) - \frac{2}{n+1}$ ET II 278(19)

7. $\displaystyle\int_{-1}^1 x(z-x)^{-1}\,P_m(x)\,P_n(x)\,dx = 2z\,P_m(z)\,Q_n(z)$ $[m<n]$ ET II 278(21)

8. $\displaystyle\int_{-1}^1 x(z-x)^{-1}[P_n(x)]^2\,dx = 2z\,P_n(z)\,Q_n(z) - \frac{2}{2n+1}$ ET II 278(20)

7.225

1. $\displaystyle\int_{-1}^x (x-t)^{-1/2}\,P_n(t)\,dt = \left(n+\frac{1}{2}\right)^{-1}(1+x)^{-1/2}\,[T_n(x) + T_{n+1}(x)]$ EH II 187(43)

2. $\displaystyle\int_x^1 (t-x)^{-1/2} P^{-1/2} P_n(t)\, dt = \left(n+\frac{1}{2}\right)^{-1} (1-x)^{-1/2} \left[T_n(x) - T_{n+1}(x)\right]$ EH II 187(44)

3. $\displaystyle\int_{-1}^1 (1-x)^{-1/2} P_n(x)\, dx = \frac{2^{3/2}}{2n+1}$ EH II 183(49)

4. $\displaystyle\int_{-1}^1 (\cosh 2p - x)^{-1/2} P_n(x)\, dx = \frac{2\sqrt{2}}{2n+1} \exp[-(2n+1)p]$

                                                         $[p > 0]$ WH

5.[10] $\displaystyle\frac{1}{2}\int_{-1}^1 \frac{P_\ell(z)\, dz}{\sqrt{(xy-z)^2 - (x^2-1)(y^2-1)}} = P_\ell(x)\, Q_\ell(y) \quad (1 < x \le y)$

                                                                    $= P_\ell(y)\, Q_\ell(x) \quad (1 < y \le x)$

7.226

1. $\displaystyle\int_{-1}^1 (1-x^2)^{-1/2} P_{2m}(x)\, dx = \left[\frac{\Gamma\left(\frac{1}{2}+m\right)}{m!}\right]^2$ ET II 276(4)

2. $\displaystyle\int_{-1}^1 x(1-x^2)^{-1/2} P_{2m+1}(x)\, dx = \frac{\Gamma\left(\frac{1}{2}+m\right)\Gamma\left(\frac{3}{2}+m\right)}{m!(m+1)!}$ ET II 276(5)

3. $\displaystyle\int_{-1}^1 (1+px^2)^{-m-3/2} P_{2m}(x)\, dx = \frac{2}{2m+1}(-p)^m(1+p)^{-m-1/2}$

                                                            $[|p| < 1]$ MO 71

7.227 $\displaystyle\int_0^1 x(a^2+x^2)^{-1/2} P_n(1-2x^2)\, dx = \frac{\left[a+(a^2+1)^{1/2}\right]^{-2n-1}}{2n+1}$

                                                     $[\operatorname{Re} a > 0]$ ET II 278(23)

7.228[6] $\displaystyle\frac{1}{2}\Gamma(1+\mu)\int_{-1}^1 P_l(x)(z-x)^{-\mu-1}\, dx = (z^2-1)^{-\mu/2} e^{-i\pi\mu}\, Q_l^\mu(z)$

                                               $[l = 0, 1, 2, \ldots, \quad |\arg(z-1)| < \pi]$

7.23 Combinations of Legendre polynomials and powers

7.231

1. $\displaystyle\int_0^1 x^\lambda P_{2m}(x)\, dx = \frac{(-1)^m \Gamma\left(m-\frac{1}{2}\lambda\right)\Gamma\left(\frac{1}{2}+\frac{1}{2}\lambda\right)}{2\,\Gamma\left(-\frac{1}{2}\lambda\right)\Gamma\left(m+\frac{3}{2}+\frac{1}{2}\lambda\right)}$ $[\operatorname{Re}\lambda > -1]$ EH II 183(51)

2.[6] $\displaystyle\int_0^1 x^\lambda P_{2m+1}(x)\, dx = \frac{(-1)^m \Gamma\left(m+\frac{1}{2}-\frac{1}{2}\lambda\right)\Gamma\left(1+\frac{1}{2}\lambda\right)}{2\,\Gamma\left(\frac{1}{2}-\frac{1}{2}\lambda\right)\Gamma\left(m+2+\frac{1}{2}\lambda\right)}$

                                                     $[\operatorname{Re}\lambda > -2]$ EH II 183(52)

7.232

1. $$\int_{-1}^{1} (1-x)^{a-1} P_m(x) P_n(x) \, dx$$

$$= \frac{2^a \, \Gamma(a) \, \Gamma(n-a+1)}{\Gamma(1-a) \, \Gamma(n+a+1)} \, {}_4F_3\left(-m, m+1, a, a; 1, a+n+1, a-n; 1\right)$$

$$[\operatorname{Re} a > 0] \qquad\qquad \text{ET II 278(17)}$$

2. $$\int_{-1}^{1} (1-x)^{a-1}(1+x)^{b-1} P_n(x) \, dx = \frac{2^{a+b-1} \, \Gamma(a) \, \Gamma(b)}{\Gamma(a+b)} \, {}_3F_2(-n, 1+n, a; 1, a+b; 1)$$

$$[\operatorname{Re} a > 0, \quad \operatorname{Re} b > 0] \qquad \text{ET II 276(6)}$$

3. $$\int_{0}^{1} (1-x)^{\mu-1} P_n(1-\gamma x) \, dx = \frac{\Gamma(\mu) n!}{\Gamma(\mu+n+1)} \, P_n^{(\mu,-\mu)}(1-\gamma)$$

$$[\operatorname{Re} \mu > 0] \qquad\qquad \text{ET II 190(37)a}$$

4. $$\int_{0}^{1} (1-x)^{\mu-1} x^{\nu-1} P_n(1-\gamma x) \, dx = \frac{\Gamma(\mu) \, \Gamma(\nu)}{\Gamma(\mu+\nu)} \, {}_3F_2\left(-n, n+1, \nu; 1, \mu+\nu; \frac{1}{2}\gamma\right)$$

$$[\operatorname{Re} \mu > 0, \quad \operatorname{Re} \nu > 0] \qquad \text{ET II 190(38)}$$

7.233 $$\int_{0}^{1} x^{2\mu-1} P_n\left(1-2x^2\right) \, dx = \frac{(-1)^n [\Gamma(\mu)]^2}{2 \, \Gamma(\mu+n+1) \, \Gamma(\mu-n)}$$

$$[\operatorname{Re} \mu > 0] \qquad\qquad \text{ET II 278(22)}$$

7.24 Combinations of Legendre polynomials and other elementary functions

7.241 $$\int_{0}^{\infty} P_n(1-x) e^{-ax} \, dx = e^{-a} a^n \left(\frac{1}{a}\frac{d}{da}\right)^n \left(\frac{e^a}{a}\right)$$

$$= a^n \left(1 + \frac{1}{2}\frac{d}{da}\right)^n \left(\frac{1}{a^{n+1}}\right)$$

$$[\operatorname{Re} a > 0] \qquad\qquad \text{ET I 171(2)}$$

7.242 $$\int_{0}^{\infty} P_n\left(e^{-x}\right) e^{-ax} \, dx = \frac{(a-1)(a-2)\cdots(a-n+1)}{(a+n)(a+n-2)\cdots(a-n+2)}$$

$$[n \geq 2, \quad \operatorname{Re} a > 0] \qquad \text{ET I 171(3)}$$

7.243

1. $$\int_{0}^{\infty} P_{2n}(\cosh x) e^{-ax} \, dx = \frac{\left(a^2-1^2\right)\left(a^2-3^2\right)\cdots\left[a^2-(2n-1)^2\right]}{a\left(a^2-2^2\right)\left(a^2-4^2\right)\cdots\left[a^2-(2n)^2\right]}$$

$$[\operatorname{Re} a > 2n] \qquad\qquad \text{ET I 171(6)}$$

2. $$\int_{0}^{\infty} P_{2n+1}(\cosh x) e^{-ax} \, dx = \frac{a\left(a^2-2^2\right)\left(a^2-4^2\right)\cdots\left[a^2-(2n)^2\right]}{\left(a^2-1\right)\left(a^2-3^2\right)\cdots\left[a^2-(2n+1)^2\right]}$$

$$[\operatorname{Re} a > 2n+1] \qquad\qquad \text{ET I 171(7)}$$

3. $$\int_0^\infty P_{2n}\left(\cos x\right) e^{-ax}\,dx = \frac{\left(a^2+1^2\right)\left(a^2+3^2\right)\cdots\left[a^2+(2n-1)^2\right]}{a\left(a^2+2^2\right)\left(a^2+4^2\right)\cdots\left[a^2+(2n)^2\right]}$$

$$[\operatorname{Re} a > 0] \qquad\qquad \text{ET I 171(4)}$$

4. $$\int_0^\infty P_{2n+1}\left(\cos x\right) e^{-ax}\,dx = \frac{a\left(a^2+2^2\right)\left(a^2+4^2\right)\cdots\left[a^2+(2n)^2\right]}{\left(a^2+1^2\right)\left(a^2+3^2\right)\cdots\left[a^2+(2n+1)^2\right]}$$

$$[\operatorname{Re} a > 0] \qquad\qquad \text{ET I 171(5)}$$

5.* $$\int_{-1}^1 e^{ix\alpha}\,P_n(x)\,dx = i^n \sqrt{\frac{2\pi}{\alpha}}\,J_{n+\frac{1}{2}}(\alpha) \qquad\qquad [n=0,1,2,\ldots] \qquad \text{GH2 24 (171.10)}$$

7.244

1. $$\int_0^1 P_n\left(1-2x^2\right)\sin ax\,dx = \frac{\pi}{2}\left[J_{n+\frac{1}{2}}\left(\frac{a}{2}\right)\right]^2 \qquad\qquad [a>0] \qquad\qquad \text{ET I 94(2)}$$

2. $$\int_0^1 P_n\left(1-2x^2\right)\cos ax\,dx = \frac{\pi}{2}(-1)^n\,J_{n+\frac{1}{2}}\left(\frac{a}{2}\right)J_{-n-\frac{1}{2}}\left(\frac{a}{2}\right)$$

$$[a>0] \qquad\qquad \text{ET I 38(1)}$$

7.245

1. $$\int_0^{2\pi} P_{2m+1}\left(\cos\theta\right)\cos\theta\,d\theta = \frac{\pi}{2^{4m+1}}\binom{2m}{m}\binom{2m+2}{m+1}. \qquad\qquad \text{MO 70, EH II 183(5)}$$

2. $$\int_0^\pi P_m\left(\cos\theta\right)\sin n\theta\,d\theta = \frac{2(n-m+1)(n-m+3)\cdots(n+m-1)}{(n-m)(n-m+2)\cdots(n+m)} \qquad [n>m \text{ and } n+m \text{ is odd}]$$

$$= 0 \qquad\qquad [n\le m \text{ or } n+m \text{ is even}]$$

$$\text{MO 71}$$

3.[10] $$\int_0^{2\pi} P_{2n+1}\left(\sin\alpha\sin\phi\right)\sin\phi\,d\phi = (-1)^{n+1}\frac{2\sqrt{\pi}\,\Gamma\left(n+\frac{3}{2}\right)}{(2n+1)\,\Gamma\left(n+2\right)}\,P_{2n+1}^1\left(\cos\alpha\right)$$

$$\left[\alpha \ne \tfrac{1}{2}(2n+1)\pi, \quad n \text{ an integer}\right]$$

4.* $$\int_{-1}^1 \cos(\alpha x)\,P_n(x)\,dx = 0 \qquad\qquad [n \text{ is odd}]$$

$$= (-1)^v\sqrt{\frac{2\pi}{\alpha}}\,J_{2v+\frac{1}{2}}(\alpha) \qquad\qquad [n=2v \text{ is even}]$$

$$\text{GH2 24 (171.10a)}$$

7.246 $$\int_0^\pi P_n\left(1-2\sin^2 x\sin^2\theta\right)\sin x\,dx = \frac{2\sin(2n+1)\theta}{(2n+1)\sin\theta} \qquad\qquad \text{MO 71}$$

7.247 $$\int_0^1 P_{2n+1}(x)\sin ax\,\frac{dx}{\sqrt{x}} = (-1)^{n+1}\sqrt{\frac{\pi}{2a}}\,J_{2n+\frac{3}{2}}(a) \qquad [a>0] \qquad\qquad \text{ET I 94(1)}$$

7.248

1. $$\int_{-1}^1 \left(a^2+b^2-2abx\right)^{-1/2}\sin\left[\lambda\left(a^2+b^2-2abx\right)^{1/2}\right]P_n(x)\,dx = \pi(ab)^{-1/2}\,J_{n+\frac{1}{2}}(a\lambda)\,J_{n+\frac{1}{2}}(b\lambda)$$

$$[a>0, \quad b>0] \qquad\qquad \text{ET II 277(11)}$$

2. $\int_{-1}^{1} \left(a^2 + b^2 - 2abx\right)^{-1/2} \cos\left[\lambda\left(a^2 + b^2 - 2abx\right)^{1/2}\right] P_n(x)\, dx = -\pi(ab)^{-1/2}\, J_{n+\frac{1}{2}}(a\lambda)\, Y_{n+\frac{1}{2}}(b\lambda)$

$$[0 \le a \le b] \qquad\qquad \text{ET II 277(12)}$$

7.249

1. $\int_{-1}^{1} P_n(x) \arcsin x\, dx = 0 \qquad\qquad [n \text{ is even}]$

$$= \pi \left\{ \frac{(n-2)!!}{2^{\frac{1}{2}(n+1)}\left(\dfrac{n+1}{2}\right)!} \right\}^2 \qquad [n \text{ is odd}]$$

$$\text{WH}$$

2. $P_n(x) = \dfrac{1}{t} \displaystyle\sum_{t=0}^{t-1} \left(x + \sqrt{x^2 - 1}\cos\dfrac{2\pi r}{t}\right)^n \qquad\qquad [t > n]$

7.25 Combinations of Legendre polynomials and Bessel functions

7.251

1. $\int_{0}^{1} x\, P_n\left(1 - 2x^2\right) Y_\nu(xy)\, dx = \pi^{-1}y^{-1}\left[S_{2n+1}(y) + \pi\, Y_{2n+1}(y)\right]$

$$[n = 0, 1, \ldots; \quad y > 0, \quad \nu > 0]$$
$$\text{ET II 108(1)}$$

2. $\int_{0}^{1} x\, P_n\left(1 - 2x^2\right) K_0(xy)\, dx = y^{-1}\left[(-1)^{n+1} K_{2n+1}(y) + \dfrac{i}{2} S_{2n+1}(iy)\right]$

$$[y > 0] \qquad\qquad \text{ET II 134(1)}$$

3. $\int_{0}^{1} x\, P_n\left(1 - 2x^2\right) J_0(xy)\, dx = y^{-1} J_{2n+1}(y) \qquad [y > 0] \qquad \text{ET II 13(1)}$

4. $\int_{0}^{1} x\, P_n\left(1 - 2x^2\right) [J_0(ax)]^2\, dx = \dfrac{1}{2(2n+1)}\left\{[J_n(a)]^2 + [J_{n+1}(a)]^2\right\} \qquad \text{ET II 338(39)a}$

5. $\int_{0}^{1} x\, P_n\left(1 - 2x^2\right) J_0(ax)\, Y_0(ax)\, dx = \dfrac{1}{2(2n+1)}\left[J_n(a)\, Y_n(a) + J_{n+1}(a)\, Y_{n+1}(a)\right]$

$$\text{ET II 339(48)a}$$

6. $\int_{0}^{1} x^2\, P_n\left(1 - 2x^2\right) J_1(xy)\, dx = y^{-1}(2n+1)^{-1}\left[(n+1) J_{2n+2}(y) - n J_{2n}(y)\right]$

$$[y > 0] \qquad\qquad \text{ET II 20(23)}$$

7. $\int_{0}^{1} x^{\mu-1}\, P_n\left(2x^2 - 1\right) J_\nu(ax)\, dx = \dfrac{2^{-\nu-1}a^\nu\left[\Gamma\left(\frac{1}{2}\mu + \frac{1}{2}\nu\right)\right]^2}{\Gamma(\nu + 1)\,\Gamma\left(\frac{1}{2}\mu + \frac{1}{2}\nu + n + 1\right)\Gamma\left(\frac{1}{2} + \frac{1}{2}\nu - n\right)}$

$$\times\ {}_2F_3\left(\dfrac{\mu + \nu}{2}, \dfrac{\mu + \nu}{2}; \nu + 1, \dfrac{\mu + \nu}{2} + n + 1, \dfrac{\mu + \nu}{2} - n; -\dfrac{a^2}{4}\right)$$

$$[a > 0, \quad \operatorname{Re}(\mu + \nu) > 0] \qquad \text{ET II 337(32)a}$$

7.252 $\displaystyle\int_0^1 e^{-ax}\,P_n(1-2x)\,I_0(ax)\,dx = \frac{e^{-a}}{2n+1}\left[I_n(a)+I_{n+1}(a)\right]$

$$[a>0]$$ ET II 366(11)a

7.253 $\displaystyle\int_0^{\pi/2} \sin(2x)\,P_n(\cos 2x)\,J_0(a\sin x)\,dx = a^{-1}J_{2n+1}(a)$ ET II 361(20)

7.254 $\displaystyle\int_0^1 x\,P_n\left(1-2x^2\right)\left[I_0(ax)-\mathbf{L}_0(ax)\right]dx = (-1)^n\left[I_{2n+1}(a)-\mathbf{L}_{2n+1}(a)\right]$

$$[a>0]$$ ET II 385(14)a

7.3–7.4 Orthogonal Polynomials

7.31 Combinations of Gegenbauer polynomials $C_n^\nu(x)$ and powers

7.311

1. $\displaystyle\int_{-1}^1 \left(1-x^2\right)^{\nu-\frac12}C_n^\nu(x)\,dx = 0$ $\left[n>0,\quad \operatorname{Re}\nu>-\tfrac12\right]$ ET II 280(1)

2. $\displaystyle\int_0^1 x^{n+2\varrho}\left(1-x^2\right)^{\nu-\frac12}C_n^\nu(x)\,dx = \frac{\Gamma(2\nu+n)\,\Gamma(2\varrho+n+1)\,\Gamma\left(\nu+\frac12\right)\Gamma\left(\varrho+\frac12\right)}{2^{n+1}\,\Gamma(2\nu)\,\Gamma(2\varrho+1)\,n!\,\Gamma(n+\nu+\varrho+1)}$

$$\left[\operatorname{Re}\varrho>-\tfrac12,\quad \operatorname{Re}\nu>-\tfrac12\right]$$ ET II 280(2)

3. $\displaystyle\int_{-1}^1 (1-x)^{\nu-\frac12}(1+x)^{\beta}C_n^\nu(x)\,dx = \frac{2^{\beta+\nu+\frac12}\,\Gamma(\beta+1)\,\Gamma\left(\nu+\frac12\right)\Gamma(2\nu+n)\,\Gamma\left(\beta-\nu+\frac32\right)}{n!\,\Gamma(2\nu)\,\Gamma\left(\beta-\nu-n+\frac32\right)\Gamma\left(\beta+\nu+n+\frac32\right)}$

$$\left[\operatorname{Re}\beta>-1,\quad \operatorname{Re}\nu>-\tfrac12\right]$$ ET II 280(3)

4. $\displaystyle\int_{-1}^1 (1-x)^{\alpha}(1+x)^{\beta}C_n^\nu(x)\,dx = \frac{2^{\alpha+\beta+1}\,\Gamma(\alpha+1)\,\Gamma(\beta+1)\,\Gamma(n+2\nu)}{n!\,\Gamma(2\nu)\,\Gamma(\alpha+\beta+2)}$

$$\times\,{}_3F_2\left(-n,n+2\nu,\alpha+1;\nu+\frac12,\alpha+\beta+2;1\right)$$

$$\left[\operatorname{Re}\alpha>-1,\quad \operatorname{Re}\beta>-1\right]$$ ET II 281(4)

7.312   In the following integrals, z belongs to the complex plane with a cut along the interval of the real axis from -1 to 1.

1. $\displaystyle\int_{-1}^1 x^m(z-x)^{-1}\left(1-x^2\right)^{\nu-\frac12}C_n^\nu(x)\,dx = \frac{\pi^{1/2}2^{\frac32-\nu}}{\Gamma(\nu)}e^{-\left(\nu-\frac12\right)\pi i}z^m\left(z^2-1\right)^{\frac12\nu-\frac14}Q_{n+\nu-\frac12}^{\nu-\frac12}(z)$

$$\left[m\le n,\quad \operatorname{Re}\nu>-\tfrac12\right]$$ ET II 281(5)

2. $\displaystyle\int_{-1}^1 x^{n+1}(z-x)^{-1}\left(1-x^2\right)^{\nu-\frac12}C_n^\nu(x)\,dx = \frac{\pi^{1/2}2^{\frac32-\nu}}{\Gamma(\nu)}e^{-\left(\nu-\frac12\right)\pi i}z^{n+1}\left(z^2-1\right)^{\frac12\nu-\frac14}Q_{n+\nu-\frac12}^{\nu-\frac12}(z)$

$$-\frac{\pi 2^{1-2\nu-n}n!}{\Gamma(\nu)\,\Gamma(\nu+n+1)}$$

$$\left[\operatorname{Re}\nu>-\tfrac12\right]$$ ET II 281(6)

3.[6] $\displaystyle\int_{-1}^1 (z-x)^{-1}\left(1-x^2\right)^{\nu-\frac12}C_m^\nu(x)\,C_n^\nu(x)\,dx = \frac{\pi^{1/2}2^{\frac32-\nu}}{\Gamma(\nu)}e^{-\left(\nu-\frac12\right)\pi i}\left(z^2-1\right)^{\frac12\nu-\frac14}C_m^\nu(z)\,Q_{n+\nu-\frac12}^{\nu-\frac12}(z)$

$$\left[m\le n,\quad \operatorname{Re}\nu>-\tfrac12\right]$$ ET II 283(17)

7.313

1. $\displaystyle \int_{-1}^{1} \left(1-x^2\right)^{\nu-\frac{1}{2}} C_m^\nu(x)\, C_n^\nu(x)\, dx = 0$ $\left[m \neq n, \quad \operatorname{Re}\nu > -\tfrac{1}{2}\right]$

ET II 282(12), MO 98a, EH I 177(16)

2. $\displaystyle \int_{-1}^{1} \left(1-x^2\right)^{\nu-\frac{1}{2}} [C_n^\nu(x)]^2\, dx = \frac{\pi 2^{1-2\nu}\, \Gamma(2\nu+n)}{n!(n+\nu)[\Gamma(\nu)]^2}$ $\left[\operatorname{Re}\nu > -\tfrac{1}{2}\right]$

ET II 281(8), MO 98a, EH I 177(17)

7.314

1. $\displaystyle \int_{-1}^{1} (1-x)^{\nu-\frac{3}{2}}(1+x)^{\nu-\frac{1}{2}} [C_n^\nu(x)]^2\, dx = \frac{\pi^{1/2}\, \Gamma\left(\nu-\frac{1}{2}\right)\Gamma(2\nu+n)}{n!\, \Gamma(\nu)\, \Gamma(2\nu)}$

$$\left[\operatorname{Re}\nu > \tfrac{1}{2}\right] \qquad \text{ET II 281(9)}$$

2. $\displaystyle \int_{-1}^{1} (1-x)^{\nu-\frac{1}{2}}(1+x)^{2\nu-1} [C_n^\nu(x)]^2\, dx = \frac{2^{3\nu-\frac{1}{2}}[\Gamma(2\nu+n)]^2\, \Gamma\left(2n+\nu+\frac{1}{2}\right)}{(n!)^2\, \Gamma(2\nu)\, \Gamma\left(3\nu+2n+\frac{1}{2}\right)}$

$$\left[\operatorname{Re}\nu > 0\right] \qquad \text{ET II 282(10)}$$

3. $\displaystyle \int_{-1}^{1} (1-x)^{3\nu+2n-\frac{3}{2}}(1+x)^{\nu-\frac{1}{2}} [C_n^\nu(x)]^2\, dx$

$$= \frac{\pi^{1/2}\left[\Gamma\left(\nu+\frac{1}{2}\right)\right]^2 \Gamma\left(\nu+2n+\frac{1}{2}\right)\Gamma(2\nu+2n)\,\Gamma\left(3\nu+2n-\frac{1}{2}\right)}{2^{2\nu+2n}\left[n!\,\Gamma\left(\nu+n+\frac{1}{2}\right)\Gamma(2\nu)\right]^2 \Gamma\left(2\nu+2n+\frac{1}{2}\right)}$$

$$\left[\operatorname{Re}\nu > \tfrac{1}{6}\right] \qquad \text{ET II 282(11)}$$

4. $\displaystyle \int_{-1}^{1} (1-x)^{\nu-\frac{1}{2}}(1+x)^{\nu+m-n-\frac{3}{2}} C_m^\nu(x)\, C_n^\nu(x)\, dx$

$$= (-1)^m \frac{2^{2-2\nu-m+n}\pi^{3/2}\,\Gamma(2\nu+n)}{m!(n-m)![\Gamma(\nu)]^2\,\Gamma\left(\frac{1}{2}+\nu+m\right)} \frac{\Gamma\left(\nu-\frac{1}{2}+m-n\right)\Gamma\left(\frac{1}{2}-\nu+m-n\right)}{\Gamma\left(\frac{1}{2}-\nu-n\right)\Gamma\left(\frac{1}{2}+m-n\right)}$$

$$\left[\operatorname{Re}\nu > -\tfrac{1}{2}; \quad n \geq m\right] \qquad \text{ET II 282(13)a}$$

5. $\displaystyle \int_{-1}^{1} (1-x)^{2\nu-1}(1+x)^{\nu-\frac{1}{2}} C_m^\nu(x)\, C_n^\nu(x)\, dx$

$$= \frac{2^{3\nu-\frac{1}{2}}\,\Gamma\left(\nu+\frac{1}{2}\right)\Gamma(2\nu+m)\,\Gamma(2\nu+n)}{m!\, n!\, \Gamma(2\nu)\, \Gamma\left(\frac{1}{2}-\nu\right)} \frac{\Gamma\left(\nu+\frac{1}{2}+m+n\right)\Gamma\left(\frac{1}{2}-\nu+n-m\right)}{\Gamma\left(\nu+\frac{1}{2}+n-m\right)\Gamma\left(3\nu+\frac{1}{2}+m+n\right)}$$

$$\left[\operatorname{Re}\nu > 0\right] \qquad \text{ET II 282(14)}$$

6. $\displaystyle \int_{-1}^{1} (1-x)^{\nu-\frac{1}{2}}(1+x)^{3\nu+m+n-\frac{3}{2}} C_m^\nu(x)\, C_n^\nu(x)\, dx$

$$= \frac{2^{4\nu+m+n-1}\left[\Gamma\left(\nu+\frac{1}{2}\right)\Gamma(2\nu+m+n)\right]^2}{\Gamma\left(\nu+m+\frac{1}{2}\right)\Gamma\left(\nu+n+\frac{1}{2}\right)\Gamma(2\nu+m)} \frac{\Gamma\left(\nu+m+n+\frac{1}{2}\right)\Gamma\left(3\nu+m+n-\frac{1}{2}\right)}{\Gamma(2\nu+n)\,\Gamma(4\nu+2m+2n)}$$

$$\left[\operatorname{Re}\nu > \tfrac{1}{6}\right] \qquad \text{ET II 282(15)}$$

7. $$\int_{-1}^{1} (1-x)^{\alpha}(1+x)^{\nu-\frac{1}{2}} C_m^{\mu}(x)\, C_n^{\nu}(x)\, dx$$

$$= \frac{2^{\alpha+\nu+\frac{1}{2}}\,\Gamma(\alpha+1)\,\Gamma\left(\nu+\frac{1}{2}\right)\Gamma\left(\nu-\alpha+n-\frac{1}{2}\right)}{m!\,n!\,\Gamma\left(\nu-\alpha-\frac{1}{2}\right)\Gamma\left(\nu-\alpha+n+\frac{3}{2}\right)}\,\frac{\Gamma(2\mu+m)\,\Gamma(2\nu+n)}{\Gamma(2\mu)\,\Gamma(2\nu)}$$

$$\times\ {}_4F_3\left(-m,m+2\mu,\alpha+1,\alpha-\nu+\frac{3}{2};\mu+\frac{1}{2},\nu+\alpha+n+\frac{3}{2},\alpha-\nu-n+\frac{3}{2};1\right)$$

$$\left[\operatorname{Re}\alpha > -1,\quad \operatorname{Re}\nu > -\tfrac{1}{2}\right]\quad \text{ET II 283(16)}$$

7.315 $$\int_{-1}^{1}\left(1-x^2\right)^{\frac{1}{2}\nu-1} C_{2n}^{\nu}(ax)\, dx = \frac{\pi^{1/2}\,\Gamma\left(\frac{1}{2}\nu\right)}{\Gamma\left(\frac{1}{2}\nu+\frac{1}{2}\right)}\, C_n^{\frac{1}{2}\nu}\left(2a^2-1\right)$$

$$[\operatorname{Re}\nu > 0]\qquad \text{ET II 283(19)}$$

7.316 $$\int_{-1}^{1}\left(1-x^2\right)^{\nu-1} C_n^{\nu}(\cos\alpha\cos\beta + x\sin\alpha\sin\beta)\, dx = \frac{2^{2\nu-1}n!\,[\Gamma(\nu)]^2}{\Gamma(2\nu+n)}\, C_n^{\nu}(\cos\alpha)\, C_n^{\nu}(\cos\beta)$$

$$[\operatorname{Re}\nu > 0]\qquad \text{ET II 283(20)}$$

7.317

1. $$\int_{0}^{1}(1-x)^{\mu-1}x^{\lambda-\frac{1}{2}}\, C_n^{\lambda}(1-\gamma x)\, dx = \frac{\Gamma(2\lambda+n)\,\Gamma\left(\lambda+\frac{1}{2}\right)\Gamma(\mu)}{\Gamma(2\lambda)\,\Gamma\left(\lambda+\mu+n+\frac{1}{2}\right)}\, P_n^{(\alpha,\beta)}(1-\gamma)$$

$$\alpha = \lambda+\mu-\tfrac{1}{2},\qquad \beta = \lambda-\mu-\tfrac{1}{2}\qquad \left[\operatorname{Re}\lambda > -1,\ \lambda \neq 0,\ -\tfrac{1}{2},\ \operatorname{Re}\mu > 0\right]\quad \text{ET II 190(39)a}$$

2. $$\int_{0}^{1}(1-x)^{\mu-1}x^{\nu-1}\, C_n^{\lambda}(1-\gamma x)\, dx = \frac{\Gamma(2\lambda+n)\,\Gamma(\mu)\,\Gamma(\nu)}{n!\,\Gamma(2\lambda)\,\Gamma(\mu+\nu)}$$

$$\times\ {}_3F_2\left(-n,n+2\lambda,\nu;\lambda+\frac{1}{2},\mu+\nu;\frac{\gamma}{2}\right)$$

$$[2\lambda \neq 0,-1,-2,\dots,\quad \operatorname{Re}\mu > 0,\quad \operatorname{Re}\nu > 0]\quad \text{ET II 191(40)a}$$

7.318 $$\int_{0}^{1} x^{2\nu}\left(1-x^2\right)^{\sigma-1} C_n^{\nu}\left(1-x^2 y\right)\, dx = \frac{\Gamma(2\nu+n)\,\Gamma\left(\nu+\frac{1}{2}\right)\Gamma(\sigma)}{2\,\Gamma(2\nu)\,\Gamma\left(n+\nu+\sigma+\frac{1}{2}\right)}\, P_n^{(\alpha,\beta)}(1-y),$$

$$\alpha = \nu+\sigma-\tfrac{1}{2},\qquad \beta = \nu-\sigma-\tfrac{1}{2}\qquad \left[\operatorname{Re}\nu > -\tfrac{1}{2},\quad \operatorname{Re}\sigma > 0\right]\quad \text{ET II 283(21)}$$

7.319

1. $$\int_{0}^{1}(1-x)^{\mu-1}x^{\nu-1}\, C_{2n}^{\lambda}\left(\gamma x^{1/2}\right)\, dx = (-1)^n\frac{\Gamma(\lambda+n)\,\Gamma(\mu)\,\Gamma(\nu)}{n!\,\Gamma(\lambda)\,\Gamma(\mu+\nu)}\ {}_3F_2\left(-n,n+\lambda,\nu;\frac{1}{2},\mu+\nu;\gamma^2\right)$$

$$[\operatorname{Re}\mu > 0,\quad \operatorname{Re}\nu > 0]\qquad \text{ET II 191(41)a}$$

2. $$\int_{0}^{1}(1-x)^{\mu-1}x^{\nu-1}\, C_{2n+1}^{\lambda}\left(\gamma x^{1/2}\right)\, dx = \frac{(-1)^n 2\gamma\,\Gamma(\mu)\,\Gamma(\lambda+n+1)\,\Gamma\left(\nu+\frac{1}{2}\right)}{n!\,\Gamma(\lambda)\,\Gamma\left(\mu+\nu+\frac{1}{2}\right)}$$

$$\times\ {}_3F_2\left(-n,n+\lambda+1,\nu+\frac{1}{2};\frac{3}{2},\mu+\nu+\frac{1}{2};\gamma^2\right)$$

$$\left[\operatorname{Re}\mu > 0,\quad \operatorname{Re}\nu > -\tfrac{1}{2}\right]\qquad \text{ET II 191(42)}$$

7.32 Combinations of Gegenbauer polynomials $C_n^\nu(x)$ and some elementary functions

7.321 $\qquad \int_{-1}^{1} \left(1 - x^2\right)^{\nu - \frac{1}{2}} e^{iax} \, C_n^\nu(x) \, dx = \dfrac{\pi 2^{1-\nu} i^n \, \Gamma(2\nu + n)}{n! \, \Gamma(\nu)} a^{-\nu} \, J_{\nu+n}(a)$

$$\left[\operatorname{Re}\nu > -\tfrac{1}{2}\right] \qquad \text{ET II 281(7), MO 99a}$$

7.322 $\qquad \int_{0}^{2a} \left[x(2a - x)\right]^{\nu - \frac{1}{2}} C_n^\nu\left(\dfrac{x}{a} - 1\right) e^{-bx} \, dx = (-1)^n \dfrac{\pi \, \Gamma(2\nu + n)}{n! \, \Gamma(\nu)} \left(\dfrac{a}{2b}\right)^\nu e^{-ab} \, I_{\nu+n}(ab)$

$$\left[\operatorname{Re}\nu > -\tfrac{1}{2}\right] \qquad \text{ET I 171(9)}$$

7.323

1. $\qquad \int_{0}^{\pi} C_n^\nu\left(\cos\varphi\right) \left(\sin\varphi\right)^{2\nu} \, d\varphi = 0 \qquad\qquad\qquad [n = 1, 2, 3, \ldots]$

$$= 2^{-2\nu} \pi \, \Gamma(2\nu + 1)\left[\Gamma(1 + \nu)\right]^{-2} \quad [n = 0]$$

$$\text{EH I 177(18)}$$

2. $\qquad \int_{0}^{\pi} C_n^\nu\left(\cos\psi\cos\psi' + \sin\psi\sin\psi'\cos\varphi(\sin\varphi)^{2\nu-1} \, d\varphi\right)$

$$= 2^{2\nu - 1} n! \left[\Gamma(\nu)\right]^2 \, C_n^\nu(\cos\psi) \, C_n^\nu(\cos\psi')\left[\Gamma(2\nu + n)\right]^{-1}$$

$$\left[\operatorname{Re}\nu > 0\right] \qquad \text{EH I 177(20)}$$

7.324

1. $\qquad \int_{0}^{1} \left(1 - x^2\right)^{\nu - \frac{1}{2}} C_{2n+1}^\nu(x) \sin ax \, dx = (-1)^n \pi \dfrac{\Gamma(2n + 2\nu + 1) \, J_{2n+\nu+1}(a)}{(2n + 1)! \, \Gamma(\nu)(2a)^\nu}$

$$\left[\operatorname{Re}\nu > -\tfrac{1}{2}, \quad a > 0\right] \qquad \text{ET I 94(4)}$$

2. $\qquad \int_{0}^{1} \left(1 - x^2\right)^{\nu - \frac{1}{2}} C_{2n}^\nu(x) \cos ax \, dx = \dfrac{(-1)^n \pi \, \Gamma(2n + 2\nu) \, J_{\nu+2n}(a)}{(2n)! \, \Gamma(\nu)(2a)^\nu}$

$$\left[\operatorname{Re}\nu > -\tfrac{1}{2}, \quad a > 0\right] \qquad \text{ET I 38(3)a}$$

7.33 Combinations of the polynomials $C_n^\nu(x)$ and Bessel functions. Integration of Gegenbauer functions with respect to the index.

7.331

1. $\qquad \int_{1}^{\infty} x^{2n+1-\nu} \left(x^2 - 1\right)^{\nu - 2n - \frac{1}{2}} C_{2n}^{\nu-2n}\left(\dfrac{1}{x}\right) J_\nu(xy) \, dx$

$$= (-1)^n 2^{2n-\nu+1} y^{-\nu+2n-1}\left[(2n)!\right]^{-1} \Gamma(2\nu - 2n)\left[\Gamma(\nu - 2n)\right]^{-1} \cos y$$

$$\left[y > 0, \quad 2n - \tfrac{1}{2} < \operatorname{Re}\nu < 2n + \tfrac{1}{2}\right] \qquad \text{ET II 44(10)a}$$

7.332

1.
$$\int_0^\infty x^{\nu+1}\left(x^2+\beta^2\right)^{-\frac{1}{2}\nu-\frac{3}{4}} C_{2n+1}^{\nu+\frac{1}{2}}\left[\left(x^2+\beta^2\right)^{-1/2}\beta\right] J_{\nu+\frac{3}{2}+2n}\left[\left(x^2+\beta^2\right)^{1/2}a\right] J_\nu(xy)\, dx$$
$$= (-1)^n 2^{1/2}\pi^{-1/2}a^{\frac{1}{2}-\nu}y^\nu\left(a^2-y^2\right)^{-1/2}\sin\left[\beta\left(a^2-y^2\right)^{1/2}\right] C_{2n+1}^{\nu+\frac{1}{2}}\left[\left(1-\frac{y^2}{a^2}\right)^{1/2}\right]$$
$$[0 < y < a]$$
$$= 0$$
$$[a < y < \infty] \qquad [a > 0, \quad \mathrm{Re}\,\beta > 0, \quad \mathrm{Re}\,\nu > -1]$$
ET II 59(23)

2.
$$\int_0^\infty x^{\nu+1}\left(x^2+\beta^2\right)^{-\frac{1}{2}\nu-\frac{3}{4}} C_{2n}^{\nu+\frac{1}{2}}\left[\beta\left(x^2+\beta^2\right)^{-1/2}\right] J_{\nu+\frac{1}{2}+2n}\left[\left(x^2+\beta^2\right)^{1/2}a\right] J_\nu(xy)\, dx$$
$$= (-1)^n 2^{1/2}\pi^{-1/2}a^{\frac{1}{2}-\nu}y^\nu\left(a^2-y^2\right)^{-1/2}\cos\left[\beta\left(a^2-y^2\right)^{1/2}\right] C_{2n}^{\nu+\frac{1}{2}}\left[\left(1-\frac{y^2}{a^2}\right)^{1/2}\right]$$
$$[0 < y < a]$$
$$= 0$$
$$[a < y < \infty] \qquad [a > 0, \quad \mathrm{Re}\,\beta > 0, \quad \mathrm{Re}\,\nu > -1]$$
ET II 59(24)

7.333

1.
$$\int_0^\pi (\sin x)^{\nu+1}\cos\left(a\cos\theta\cos x\right) C_n^{\nu+\frac{1}{2}}(\cos x) J_\nu\left(a\sin\theta\sin x\right)\, dx$$
$$= (-1)^{\frac{n}{2}}\left(\frac{2\pi}{a}\right)^{1/2}(\sin\theta)^\nu C_n^{\nu+\frac{1}{2}}(\cos\theta) J_{\nu+\frac{1}{2}+n}(a) \qquad [n = 0, 2, 4, \ldots]$$
$$= 0 \qquad [n = 1, 3, 5, \ldots]$$
$$[[\mathrm{Re}\,\nu > -1]] \qquad \text{WA 414(2)a}$$

2.
$$\int_0^\pi (\sin x)^{\nu+1}\sin\left(a\cos\theta\cos x\right) C_n^{\nu+\frac{1}{2}}(\cos x) J_\nu\left(a\sin\theta\sin x\right)\, dx$$
$$= 0 \qquad [n = 0, 2, 4, \ldots]$$
$$= (-1)^{\frac{n-1}{2}}\left(\frac{2\pi}{a}\right)^{1/2}(\sin\theta)^\nu C_n^{\nu+\frac{1}{2}}(\cos\theta) J_{\nu+\frac{1}{2}+n}(a) \qquad [n = 1, 3, 5, \ldots]$$
$$[\mathrm{Re}\,\nu > -1] \qquad \text{WA 414(3)a}$$

7.334

1.
$$\int_0^\pi (\sin x)^{2\nu} C_n^\nu(\cos x)\frac{J_\nu(\omega)}{\omega^\nu}\, dx = \frac{\pi\,\Gamma(2\nu+n)}{2^{\nu-1}n!\,\Gamma(\nu)}\frac{J_{\nu+n}(\alpha)}{\alpha^\nu}\frac{J_{\nu+n}(\beta)}{\beta^\nu},$$
$$\omega = \left(\alpha^2+\beta^2-2\alpha\beta\cos x\right)^{1/2} \qquad [n = 0, 1, 2, \ldots; \quad \mathrm{Re}\,\nu > -\tfrac{1}{2}] \quad \text{ET II 362(29)}$$

2.
$$\int_0^\pi (\sin x)^{2\nu} C_n^\nu(\cos x)\frac{Y_\nu(\omega)}{\omega^\nu}\, dx = \frac{\pi\,\Gamma(2\nu+n)}{2^{\nu-1}n!\,\Gamma(\nu)}\frac{J_{\nu+n}(\alpha)}{\alpha^\nu}\frac{Y_{\nu+n}(\beta)}{\beta^\nu},$$
$$\omega = \left(\alpha^2+\beta^2-2\alpha\beta\cos x\right)^{1/2} \qquad [|\alpha| < |\beta|, \quad \mathrm{Re}\,\nu - \tfrac{1}{2}] \quad \text{ET II 362(30)}$$

Integration of Gegenbauer functions with respect to the index

7.335 $\displaystyle\int_{c-i\infty}^{c+i\infty} [\sin(\alpha\pi)]^{-1} t^\alpha\, C_\alpha^\nu(z)\, d\alpha = -2i\left(1 + 2tz + t^2\right)^{-\nu}$

$$[-2 < \operatorname{Re}\nu < c < 0, \quad |\arg(z \pm 1)| < \pi]$$

EH I 178(25)

7.336 $\displaystyle\int_{-\infty}^{\infty} \operatorname{sech}(\pi x)\left(\nu - \frac{1}{2} + ix\right) K_{\nu - \frac{1}{2} + ix}(a)\, I_{\nu - \frac{1}{2} + ix}(b)\, C_{-\frac{1}{2} + ix}^\nu(-\cos\varphi)\, dx$

$$= \frac{2^{-\nu+1}(ab)^\nu}{\Gamma(\nu)}\omega^{-\nu}\, K_\nu(\omega)$$

$$\omega = \sqrt{a^2 + b^2 - 2ab\cos\varphi} \qquad \text{EH II 55(45)}$$

7.34 Combinations of Chebyshev polynomials and powers

7.341 $\displaystyle\int_{-1}^{1} [T_n(x)]^2\, dx = 1 - \left(4n^2 - 1\right)^{-1}$ ET II 271(6)

7.342 $\displaystyle\int_{-1}^{1} U_n\left[x\left(1 - y^2\right)^{1/2}\left(1 - z^2\right)^{1/2} + yz\right] dx = \frac{2}{n+1}\, U_n(y)\, U_n(z)$

$$[|y| < 1, \quad |z| < 1] \qquad \text{ET II 275(34)}$$

7.343

1. $\displaystyle\int_{-1}^{1} T_n(x)\, T_m(x)\frac{dx}{\sqrt{1 - x^2}} = 0$ $[m \neq n]$

$$= \frac{\pi}{2} \qquad [m = n \neq 0]$$

$$= \pi \qquad [m = n = 0]$$

MO 104

2. $\displaystyle\int_{-1}^{1} \sqrt{1 - x^2}\, U_n(x)\, U_m(x)\, dx = 0$ $[m \neq n]$ ET II 274(28)

$$= \frac{\pi}{2} \quad [m = n] \qquad \text{ET II 274(27), MO 105a}$$

7.344

1. $\displaystyle\int_{-1}^{1} (y - x)^{-1}\left(1 - y^2\right)^{-1/2} T_n(y)\, dy = \pi\, U_{n-1}(x)$ $[n = 1, 2, \ldots]$ EH II 187(47)

2. $\displaystyle\int_{-1}^{1} (y - x)^{-1}\left(1 - y^2\right)^{1/2} U_{n-1}(y)\, dy = -\pi\, T_n(x)$ $[n = 1, 2, \ldots]$ EH II 187(48)

7.345

1. $\displaystyle\int_{-1}^{1} (1 - x)^{-1/2}(1 + x)^{m - n - \frac{3}{2}}\, T_m(x)\, T_n(x)\, dx = 0$ $[m > n]$ ET II 272(10)

2. $\displaystyle\int_{-1}^{1} (1 - x)^{-1/2}(1 + x)^{m + n - \frac{3}{2}}\, T_m(x)\, T_n(x)\, dx = \frac{\pi(2m + 2n - 2)!}{2^{m+n}(2m - 1)!(2n - 1)!}$

$$[m + n \neq 0] \qquad \text{ET II 272(11)}$$

3. $\displaystyle\int_{-1}^{1} (1-x)^{1/2}(1+x)^{m+n+\frac{3}{2}}\, U_m(x)\, U_n(x)\, dx = \frac{\pi(2m+2n+2)!}{2^{m+n+2}(2m+1)!(2n+1)!}$ ET II 274(31)

4. $\displaystyle\int_{-1}^{1} (1-x)^{1/2}(1+x)^{m-n-\frac{1}{2}}\, U_m(x)\, U_n(x)\, dx = 0 \qquad [m>n]$ ET II 274(30)

5. $\displaystyle\int_{-1}^{1} (1-x)(1+x)^{1/2}\, U_m(x)\, U_n(x)\, dx = \frac{2^{5/2}(m+1)(n+1)}{\left(m+n+\frac{3}{2}\right)\left(m+n+\frac{5}{2}\right)\left[1-4(m-n)^2\right]}$

ET II 274(29)

6. $\displaystyle\int_{-1}^{1} (1+x)^{-1/2}(1-x)^{\alpha-1}\, T_m(x)\, T_n(x)\, dx$

$$= \frac{\pi^{1/2}2^{\alpha-\frac{1}{2}}\,\Gamma(\alpha)\,\Gamma\left(n-\alpha+\frac{1}{2}\right)}{\Gamma\left(\frac{1}{2}-\alpha\right)\Gamma\left(\alpha+n+\frac{1}{2}\right)}\ {}_4F_3\left(-m,m,\alpha,\alpha+\frac{1}{2};\frac{1}{2},\alpha+n+\frac{1}{2},\alpha-n+\frac{1}{2};1\right)$$

$$[\operatorname{Re}\alpha>0] \qquad \text{ET II 272(12)}$$

7. $\displaystyle\int_{-1}^{1} (1+x)^{1/2}(1-x)^{\alpha-1}\, U_m(x)\, U_n(x)\, dx$

$$= \frac{\pi^{1/2}2^{\alpha-\frac{1}{2}}(m+1)(n+1)\,\Gamma(\alpha)\,\Gamma\left(n-\alpha+\frac{3}{2}\right)}{\Gamma\left(\frac{3}{2}-\alpha\right)\Gamma\left(\frac{3}{2}+\alpha+n\right)}$$

$$\times\ {}_4F_3\left(-m,m+2,\alpha,\alpha-\frac{1}{2};\frac{3}{2},\alpha+n+\frac{3}{2},\alpha-n-\frac{1}{2};1\right)$$

$$[\operatorname{Re}\alpha>0] \qquad \text{ET II 275(32)}$$

7.346 $\displaystyle\int_{0}^{1} x^{s-1}\, T_n(x)\frac{dx}{\sqrt{1-x^2}} = \frac{\pi}{s2^s\,\mathrm{B}\left(\frac{1}{2}+\frac{1}{2}s+\frac{1}{2}n,\,\frac{1}{2}+\frac{1}{2}s-\frac{1}{2}n\right)}$

$$[\operatorname{Re}s>0] \qquad \text{ET II 324(2)}$$

7.347

1. $\displaystyle\int_{-1}^{1} (1-x)^{\alpha}(1+x)^{\beta}\, T_n(x)\, dx = \frac{2^{\alpha+\beta+2n+1}(n!)^2\,\Gamma(\alpha+1)\,\Gamma(\beta+1)}{(2n)!\,\Gamma(\alpha+\beta+2)}$

$$\times\ {}_3F_2\left(-n,n,\alpha+1;\frac{1}{2},\alpha+\beta+2;1\right)$$

$$[\operatorname{Re}\alpha>-1,\quad \operatorname{Re}\beta>-1] \quad \text{ET II 271(2)}$$

2. $\displaystyle\int_{-1}^{1} (1-x)^{\alpha}(1+x)^{\beta}\, U_n(x)\, dx = \frac{2^{\alpha+\beta+2n+2}[(n+1)!]^2\,\Gamma(\alpha+1)\,\Gamma(\beta+1)}{(2n+2)!\,\Gamma(\alpha+\beta+2)}$

$$\times\ {}_3F_2\left(-n,n+1,\alpha+1;\frac{3}{2},\alpha+\beta+2;1\right)$$

ET II 273(22)

7.348 $\displaystyle\int_{-1}^{1} \left(1-x^2\right)^{-1/2}\, U_{2n}(xz)\, dx = \pi\, P_n\left(2z^2-1\right) \qquad [|z|<1]$ ET II 275(33)

7.349 $\displaystyle\int_{-1}^{1} \left(1-x^2\right)^{-1/2}\, T_n\left(1-x^2 y\right)\, dx = \frac{1}{2}\pi\left[P_n(1-y)+P_{n-1}(1-y)\right]$ ET II 222(14)

7.35 Combinations of Chebyshev polynomials and some elementary functions

7.351 $\qquad \int_0^1 x^{-1/2}\left(1-x^2\right)^{-\frac{1}{2}} e^{-\frac{2a}{x}}\, T_n(x)\, dx = \pi^{1/2}\, D_{n-\frac{1}{2}}\left(2a^{1/2}\right) D_{-n-\frac{1}{2}}\left(2a^{1/2}\right)$

$$[\operatorname{Re} a > 0] \qquad\qquad\qquad \text{ET II 272(13)}$$

7.352

1. $\displaystyle\int_0^\infty \frac{x\, U_n\left[a\left(a^2+x^2\right)^{-1/2}\right]}{\left(a^2+x^2\right)^{\frac{1}{2}n+1}\left(e^{\pi x}+1\right)}\, dx = \frac{a^{-n}}{2n} - 2^{-n-1}\,\zeta\left(n+1, \frac{a+1}{2}\right)$

$$[\operatorname{Re} a > 0] \qquad\qquad\qquad \text{ET II 275(39)}$$

2. $\displaystyle\int_0^\infty \frac{x\, U_n\left[a\left(a^2+x^2\right)^{-1/2}\right]}{\left(a^2+x^2\right)^{\frac{1}{2}n+1}\left(e^{2\pi x}-1\right)}\, dx = \frac{1}{2}\,\zeta(n+1,a) - \frac{a^{-n-1}}{4} - \frac{a^{-n}}{2n}$

$$[\operatorname{Re} a > 0] \qquad\qquad\qquad \text{ET II 276(40)}$$

7.353

1. $\displaystyle\int_0^\infty \left(a^2+x^2\right)^{-\frac{1}{2}n}\operatorname{sech}\left(\frac{1}{2}\pi x\right) T_n\left[a\left(a^2+x^2\right)^{-1/2}\right]\, dx = 2^{1-2n}\left[\zeta\left(n, \frac{a+1}{4}\right) - \zeta\left(n, \frac{a+3}{4}\right)\right]$

$$= 2^{1-n}\,\Phi\left(-1, n, \frac{a+1}{2}\right)$$

$$[\operatorname{Re} a > 0] \qquad\qquad\qquad \text{ET II 273(19)}$$

2. $\displaystyle\int_0^\infty \left(a^2+x^2\right)^{-\frac{1}{2}n}\left[\cosh\left(\frac{1}{2}\pi x\right)\right]^{-2} T_n\left[a\left(a^2+x^2\right)^{-1/2}\right]\, dx = \pi^{-1}n\, 2^{1-n}\,\zeta\left(n+1, \frac{a+1}{2}\right)$

$$[\operatorname{Re} a > 0] \qquad\qquad\qquad \text{ET II 273(20)}$$

7.354

1. $\displaystyle\int_{-1}^1 \sin(xyz)\cos\left[\left(1-x^2\right)^{1/2}\left(1-y^2\right)^{1/2}z\right] T_{2n+1}(x)\, dx = (-1)^n\pi\, T_{2n+1}(y)\, J_{2n+1}(x)$

$$\text{ET II 271(4)}$$

2. $\displaystyle\int_{-1}^1 \sin(xyz)\sin\left[\left(1-x^2\right)^{1/2}\left(1-y^2\right)^{1/2}z\right] U_{2n+1}(x)\, dx = (-1)^n\pi\left(1-y^2\right)^{1/2} U_{2n+1}(y)\, J_{2n+2}(z)$

$$\text{ET II 274(25)}$$

3. $\displaystyle\int_{-1}^1 \cos(xyz)\cos\left[\left(1-x^2\right)^{1/2}\left(1-y^2\right)^{1/2}z\right] T_{2n}(x)\, dx = (-1)^n\pi\, T_{2n}(y)\, J_{2n}(z) \qquad\qquad \text{ET II 271(5)}$

4. $\displaystyle\int_{-1}^1 \cos(xyz)\sin\left[\left(1-x^2\right)^{1/2}\left(1-y^2\right)^{1/2}z\right] U_{2n}(x)\, dx = (-1)^n\pi\left(1-y^2\right)^{1/2} U_{2n}(y)\, J_{2n+1}(z)$

$$\text{ET II 274(24)}$$

7.355

1. $\displaystyle\int_0^1 T_{2n+1}(x)\sin ax\,\frac{dx}{\sqrt{1-x^2}} = (-1)^n\,\frac{\pi}{2}\, J_{2n+1}(a) \qquad\qquad [a > 0] \qquad\qquad \text{ET I 94(3)a}$

2. $\displaystyle\int_0^1 T_{2n}(x)\cos ax\,\frac{dx}{\sqrt{1-x^2}} = (-1)^n\,\frac{\pi}{2}\, J_{2n}(a) \qquad\qquad [a > 0] \qquad\qquad \text{ET I 38(2)a}$

7.36 Combinations of Chebyshev polynomials and Bessel functions

7.361 $\int_0^1 \left(1 - x^2\right)^{-1/2} T_n(x)\, J_\nu(xy)\, dx = \frac{1}{2}\pi\, J_{\frac{1}{2}(\nu+n)}\left(\frac{1}{2}y\right) J_{\frac{1}{2}(\nu-n)}\left(\frac{1}{2}y\right)$

$$[y > 0, \quad \operatorname{Re}\nu > -n - 1] \qquad \text{ET II 42(1)}$$

7.362 $\int_1^\infty \left(x^2 - 1\right)^{-\frac{1}{2}} T_n\left(\frac{1}{x}\right) K_{2\mu}(ax)\, dx = \frac{\pi}{2a}\, W_{\frac{1}{2}n,\mu}(a)\, W_{-\frac{1}{2}n,\mu}(a)$

$$[\operatorname{Re} a > 0] \qquad \text{ET II 366(17)a}$$

7.37–7.38 Hermite polynomials

7.371 $\int_0^x H_n(y)\, dy = [2(n+1)]^{-1} [H_{n+1}(x) - H_{n+1}(0)]$ EH II 194(27)

7.372 $\int_{-1}^1 \left(1 - t^2\right)^{\alpha-\frac{1}{2}} H_{2n}\left(\sqrt{x}\,t\right) dx = \dfrac{(-1)^n \pi^{1/2}(2n)!\, \Gamma\left(\alpha + \frac{1}{2}\right) L_n^\alpha(x)}{\Gamma(n+\alpha+1)}$

$$\left[\operatorname{Re} a > -\tfrac{1}{2}\right] \qquad \text{EH II 195(34)}$$

7.373

1. $\int_0^x e^{-y^2} H_n(y)\, dy = H_{n-1}(0) - e^{-x^2} H_{n-1}(x)$ [see **8.956**]   EH II 194(26)

2. $\int_{-\infty}^\infty e^{-x^2} H_{2m}(xy)\, dx = \sqrt{\pi}\,\dfrac{(2m)!}{m!}\left(y^2 - 1\right)^m$ EH II 195(28)

7.374

1. $\int_{-\infty}^\infty e^{-x^2} H_n(x) H_m(x)\, dx = 0 \qquad [m \neq n]$ SM II 567

$$= 2^n \cdot n!\sqrt{\pi} \qquad [m = n]$$

SM II 568

2. $\int_{-\infty}^\infty e^{-2x^2} H_m(x) H_n(x)\, dx = (-1)^{\frac{1}{2}(m+n)} 2^{\frac{m+n-1}{2}} \Gamma\left(\dfrac{m+n+1}{2}\right)$

$$[m + n \text{ is even}] \qquad \text{ET II 289(10)a}$$

3. $\int_{-\infty}^\infty e^{-x^2} H_m(ax) H_n(x)\, dx = 0 \qquad [m < n]$ ET II 290(20)a

4. $\int_{-\infty}^\infty e^{-x^2} H_{2m+n}(ax) H_n(x)\, dx = \sqrt{\pi}\, 2^n \dfrac{(2m+n)!}{m!}\left(a^2 - 1\right)^m a^n$ ET II 291(21)a

5. $\int_{-\infty}^\infty e^{-2\alpha^2 x^2} H_m(x) H_n(x)\, dx = 2^{\frac{m+n-1}{2}}\alpha^{-m-n-1}\left(1 - 2\alpha^2\right)^{\frac{m+n}{2}} \Gamma\left(\dfrac{m+n+1}{2}\right)$

$$\times\ {}_2F_1\left(-m, n; \dfrac{1-m-n}{2}; \dfrac{\alpha^2}{2\alpha^2 - 1}\right)$$

$$\left[\operatorname{Re}\alpha^2 > 0, \quad \alpha^2 \neq \tfrac{1}{2}, \quad m+n \text{ is even}\right] \qquad \text{ET II 289(12)a}$$

6. $\int_{-\infty}^\infty e^{-(x-y)^2} H_n(x)\, dx = \pi^{1/2} y^n 2^n$ ET II 288(2)a, EH II 195(31)

7. $\displaystyle\int_{-\infty}^{\infty} e^{-(x-y)^2} H_m(x)\,H_n(x)\,dx = 2^n\pi^{1/2}m!\,y^{n-m}\,L_m^{n-m}\left(-2y^2\right)$

$$[m \le n] \qquad \text{BU 148(15), ET II 289(13)a}$$

8. $\displaystyle\int_{-\infty}^{\infty} e^{-(x-y)^2} H_n(\alpha x)\,dx = \pi^{1/2}\left(1-\alpha^2\right)^{\frac{n}{2}} H_n\left[\frac{\alpha y}{(1-\alpha^2)^{1/2}}\right]$ ET II 290(17)a

9. $\displaystyle\int_{-\infty}^{\infty} e^{-(x-y)^2} H_m\left(\alpha x\right) H_n(\alpha x)\,dx$

$$= \pi^{1/2}\sum_{k=0}^{\min(m,n)} 2^k k!\binom{m}{k}\binom{n}{k}\left(1-\alpha^2\right)^{\frac{m+n}{2}-k} H_{m+n-2k}\left[\frac{\alpha y}{(1-\alpha^2)^{1/2}}\right]$$

$$\text{ET II 291(26)a}$$

10. $\displaystyle\int_{-\infty}^{\infty} e^{-\frac{(x-y)^2}{2u}} H_n(x)\,dx = (2\pi u)^{1/2}(1-2u)^{\frac{n}{2}} H_n\left[y(1-2u)^{-1/2}\right]$

$$\left[0 \le u < \tfrac{1}{2}\right] \qquad \text{EH II 195(30)}$$

7.375

1. $\displaystyle\int_{-\infty}^{\infty} e^{-2x^2} H_k(x)\,H_m(x)\,H_n(x)\,dx = \pi^{-1}2^{\frac{1}{2}(m+n+k-1)}\,\Gamma(s-k)\,\Gamma(s-m)\,\Gamma(s-n)$

$$2s = k+m+n+1 \qquad [k+m+n \text{ is even}] \quad \text{ET II 290(14)a}$$

2. $\displaystyle\int_{-\infty}^{\infty} e^{-x^2} H_k(x)\,H_m(x)\,H_n(x)\,dx = \frac{2^{\frac{m+n+k}{2}}\pi^{1/2}k!\,m!\,n!}{(s-k)!\,(s-m)!\,(s-n)!},$

$$2s = m+n+k \qquad [k+m+n \text{ is even}]$$
$$\text{ET II 290(15)a}$$

7.376

1. $\displaystyle\int_{-\infty}^{\infty} e^{ixy}e^{-\frac{x^2}{2}} H_n(x)\,dx = (2\pi)^{1/2}e^{-\frac{y^2}{2}} H_n(y)i^n$ MO 165a

2. $\displaystyle\int_{0}^{\infty} e^{-2\alpha x^2} x^\nu H_{2n}(x)\,dx = (-1)^n 2^{2n-\frac{3}{2}-\frac{1}{2}\nu}\frac{\Gamma\left(\frac{\nu+1}{2}\right)\Gamma\left(n+\frac{1}{2}\right)}{\sqrt{\pi}\,\alpha^{\frac{1}{2}(\nu+1)}} F\left(-n,\frac{\nu+1}{2};\frac{1}{2};\frac{1}{2\alpha}\right)$

$$[\operatorname{Re}\alpha > 0, \quad \operatorname{Re}\nu > -1] \qquad \text{BU 150(18a)}$$

3. $\displaystyle\int_{0}^{\infty} e^{-2\alpha x^2} x^\nu H_{2n+1}(x)\,dx = (-1)^n 2^{2n-\frac{1}{2}\nu}\frac{\Gamma\left(\frac{\nu}{2}+1\right)\Gamma\left(n+\frac{3}{2}\right)}{\sqrt{\pi}\,\alpha^{\frac{1}{2}\nu+1}} F\left(-n,\frac{\nu}{2}+1;\frac{3}{2};\frac{1}{2\alpha}\right)$

$$[\operatorname{Re}\alpha > 0, \quad \operatorname{Re}\nu > -2] \qquad \text{BU 150(18b)}$$

7.377[8] $\displaystyle\int_{-\infty}^{\infty} e^{-x^2} H_m(x+y)\,H_n(x+z)\,dx = 2^n\pi^{1/2}m!\,z^{n-m}\,L_m^{n-m}(-2yz)$

$$[m \le n] \qquad \text{ET II 292(30)a}$$

7.378 $\displaystyle\int_{0}^{\infty} x^{\alpha-1}e^{-\beta x} H_n(x)\,dx = 2^n\sum_{m=0}^{\lfloor\frac{n}{2}\rfloor}\frac{n!\,\Gamma(\alpha+n-2m)}{m!\,(n-2m)!}(-1)^m 2^{-2m}\beta^{2m-\alpha-n}$

$$[\operatorname{Re}\alpha > 0, \text{ if } n \text{ is even}; \quad \operatorname{Re}\alpha > -1, \text{ if } n \text{ is odd}; \quad \operatorname{Re}\beta > 0] \quad \text{ET I 172(11)a}$$

7.379

1. $$\int_{-\infty}^{\infty} x e^{-x^2} H_{2m+1}(xy)\, dx = \pi^{1/2} \frac{(2m+1)!}{m!} y \left(y^2 - 1\right)^m$$ EH II 195(28)

2. $$\int_{-\infty}^{\infty} x^n e^{-x^2} H_n(xy)\, dx = \pi^{1/2} n! P_n(y).$$ EH II 195(29)

7.381 $$\int_{-\infty}^{\infty} (x \pm ic)^{\nu} e^{-x^2} H_n(x)\, dx = 2^{n-1-\nu} \pi^{1/2} \frac{\Gamma\left(\frac{n-\nu}{2}\right)}{\Gamma(-\nu)} \exp\left[\pm \tfrac{1}{2}\pi(\nu+n)i\right]$$

$$[c > 0]$$ ET II 288(3)a

7.382 $$\int_{0}^{\infty} x^{-1}\left(x^2 + a^2\right)^{-1} e^{-x^2} H_{2n+1}(x)\, dx = (-2)^n \pi^{1/2} a^{-2}\left[2^{\nu} n! - (2n+1)! e^{\frac{1}{2}a^2} D_{-2n-2}\left(a\sqrt{2}\right)\right]$$

ET II 288(4)a

7.383

1. $$\int_{0}^{\infty} e^{-xp} H_{2n+1}\left(\sqrt{x}\right) dx = (-1)^n 2^n (2n+1)!! \pi^{1/2} (p-1)^n p^{-n-\frac{3}{2}}$$

$$[\operatorname{Re} p > 0]$$ EF 151(261)a, ET I 172(12)a

2. $$\int_{0}^{\infty} e^{-(b-\beta x)} H_{2n+1}\left(\sqrt{(\alpha-\beta)x}\right) dx = (-1)^n \sqrt{\pi}\sqrt{\alpha-\beta}\frac{(2n+1)!}{n!}\frac{(b-\alpha)^n}{(b-\beta)^{n+\frac{3}{2}}}$$

$$[\operatorname{Re}(b-\beta) > 0]$$ ET I 172(15)a

3. $$\int_{0}^{\infty} \frac{1}{\sqrt{x}} e^{-(b-\beta)x} H_{2n}\left(\sqrt{(\alpha-\beta)x}\right) dx = (-1)^n \sqrt{\pi}\frac{(2n)!}{n!}\frac{(b-\alpha)^n}{(b-\beta)^{n+\frac{1}{2}}}$$

$$[\operatorname{Re}(b-\beta) > 0]$$ ET I 172(16)a

4. $$\int_{0}^{\infty} x^{a-\frac{1}{2}n-1} e^{-bx} H_n\left(\sqrt{x}\right) dx = 2^n\, \Gamma(a) b^{-a}\, {}_2F_1\left(-\tfrac{1}{2}n, \tfrac{1}{2} - \tfrac{1}{2}n; 1-a; b\right)$$

$$\left[\operatorname{Re} a > \tfrac{1}{2}n, \text{ if } n \text{ is even,}\quad \operatorname{Re} a > \tfrac{1}{2}n - \tfrac{1}{2}, \text{ if } n \text{ is odd,}\quad \operatorname{Re} b > 0,\right.$$

$$\left.\text{If } a \text{ is even, only the first } 1 + \left\lfloor \frac{n}{2} \right\rfloor \text{ terms are kept in the series for } {}_2F_1\right]$$

ET I 172(14)a

5. $$\int_{0}^{\infty} x^{-1/2} e^{-px} H_{2n}\left(\sqrt{x}\right) dx = (-1)^n 2^n (2n-1)!! \pi^{1/2} (p-1)^n p^{-n-\frac{1}{2}}$$ MO 177a

7.384 $$\int_{0}^{\infty} \frac{1}{\sqrt{x}} e^{-bx}\left[H_n\left(\frac{\alpha+\sqrt{x}}{\lambda}\right) + H_n\left(\frac{a-\sqrt{x}}{\lambda}\right)\right] dx = \sqrt{\frac{2\pi}{b}}\left(1 - \lambda^{-2}b^{-1}\right)^{\frac{n}{2}} H_n\left(\frac{\alpha}{\sqrt{\lambda^2 - \frac{1}{b}}}\right)$$

$$[\operatorname{Re} b > 0]$$ ET I 173(17)a

7.385

1. $$\int_{0}^{\infty} \frac{e^{-bx}}{\sqrt{e^x - 1}} H_{2n}\left[\sqrt{s\left(1 - e^{-x}\right)}\right] dx = (-1)^n 2^{2n} \sqrt{\pi}\frac{(2n)!\,\Gamma\left(b+\frac{1}{2}\right)}{\Gamma(n+b+1)} L_n^n(s)$$

$$\left[\operatorname{Re} b > -\tfrac{1}{2}\right]$$ ET I 174(23)a

2. $\displaystyle\int_0^\infty e^{-bx} H_{2n+1}\left[\sqrt{s}\sqrt{1-e^{-x}}\right] dx = (-1)^n 2^{2n}\sqrt{\pi s}\,\frac{(2n+1)!\,\Gamma(b)}{\Gamma\left(n+b+\frac{3}{2}\right)}\,L_n^b(s)$

$$[\operatorname{Re} b > 0] \hspace{4cm} \text{ET I 174(24)a}$$

7.386 $\displaystyle\int_0^\infty x^{-\frac{n+1}{2}} e^{-\frac{q^2}{4x}} H_n\left(\frac{q}{2\sqrt{x}}\right) e^{-px}\, dx = 2^n \pi^{1/2} p^{\frac{n-1}{2}} e^{-q\sqrt{p}}$ EF 129(117)

7.387

1. $\displaystyle\int_0^\infty e^{-x^2}\sinh\left(\sqrt{2}\beta x\right) H_{2n+1}(x)\, dx = 2^{n-\frac{1}{2}}\pi^{1/2}\beta^{2n+1} e^{\frac{1}{2}\beta^2}$ ET II 289(7)a

2. $\displaystyle\int_0^\infty e^{-x^2}\cosh\left(\sqrt{2}\beta x\right) H_{2n}(x)\, dx = 2^{n-1}\pi^{1/2}\beta^{2n} e^{\frac{1}{2}\beta^2}$ ET II 289(8)a

7.388

1. $\displaystyle\int_0^\infty e^{-x^2}\sin\left(\sqrt{2}\beta x\right) H_{2n+1}(x)\, dx = (-1)^n 2^{n-\frac{1}{2}}\pi^{1/2}\beta^{2n+1} e^{-\frac{1}{2}\beta^2}$ ET II 288(5)a

2. $\displaystyle\int_0^\infty e^{-x^2}\sin\left(\sqrt{2}\beta x\right) H_{2n+1}(ax)\, dx = (-1)^n 2^{-1}\pi^{1/2}\left(a^2-1\right)^{n+\frac{1}{2}} e^{-\frac{1}{2}\beta^2} H_{2n+1}\left(\frac{a\beta}{\sqrt{2}(a^2-1)^{1/2}}\right)$

$$\text{ET II 290(18)a}$$

3. $\displaystyle\int_0^\infty e^{-x^2}\cos\left(\sqrt{2}\beta x\right) H_{2n}(x)\, dx = (-1)^n 2^{n-1}\pi^{1/2}\beta^{2n} e^{-\frac{1}{2}\beta^2}$ ET II 289(6)a

4. $\displaystyle\int_0^\infty e^{-x^2}\cos\left(\sqrt{2}\beta x\right) H_{2n}(ax)\, dx = 2^{-1}\pi^{1/2}\left(1-a^2\right)^n e^{-\frac{1}{2}\beta^2} H_{2n}\left[\frac{a\beta}{\sqrt{2}(a^2-1)^{1/2}}\right]$

$$\text{ET II 290(19)a}$$

5. $\displaystyle\int_0^\infty e^{-y^2}[H_n(y)]^2 \cos\left(\sqrt{2}\beta y\right) dy = \pi^{1/2} 2^{n-1} n!\, e^{-\frac{\beta^2}{2}} L_n\left(\beta^2\right)$ EH II 195(33)

6. $\displaystyle\int_0^\infty e^{-x^2}\sin(bx) H_n(x) H_{n+2m+1}(x)\, dx = 2^n (-1)^m \sqrt{\frac{\pi}{2}} n!\, b^{2m+1} e^{-\frac{b^2}{4}} L_n^{2m+1}\left(\frac{b^2}{2}\right)$

$$[b > 0] \hspace{4cm} \text{ET I 39(11)a}$$

7. $\displaystyle\int_0^\infty e^{-x^2}\cos(bx) H_n(x) H_{n+2m}(x)\, dx = 2^{n-\frac{1}{2}}\sqrt{\frac{\pi}{2}} n!\, (-1)^m b^{2m} e^{-\frac{b^2}{4}} L_n^{2m}\left(\frac{b^2}{2}\right)$

$$[b > 0] \hspace{4cm} \text{ET I 39(11)a}$$

7.389 $\displaystyle\int_0^\pi (\cos x)^n H_{2n}\left[a(1-\sec x)^{1/2}\right] dx = 2^{-n}(-1)^n \pi\,\frac{(2n)!}{(n!)^2}[H_n(a)]^2$ ET II 292(31)

7.39 Jacobi polynomials

7.391

1. $$\int_{-1}^{1} (1-x)^\alpha (1+x)^\beta P_n^{(\alpha,\beta)}(x) P_m^{(\alpha,\beta)}(x)\, dx$$

$$= 0 \qquad\qquad [m \neq n, \quad \operatorname{Re}\alpha > -1, \quad \operatorname{Re}\beta > -1]$$

$$= \frac{2^{\alpha+\beta+1}\,\Gamma(\alpha+n+1)\,\Gamma(\beta+n+1)}{n!\,(\alpha+\beta+1+2n)\,\Gamma(\alpha+\beta+n+1)} \qquad [m = n, \quad \operatorname{Re}\alpha > -1, \quad \operatorname{Re}\beta > -1]$$

ET II 285(5, 9)

2. $$\int_{-1}^{1} (1-x)^\varrho (1+x)^\sigma P_n^{(\alpha,\beta)}(x)\, dx = \frac{2^{\varrho+\sigma+1}\,\Gamma(\varrho+1)\,\Gamma(\sigma+1)\,\Gamma(n+1+\alpha)}{n!\,\Gamma(\varrho+\sigma+2)\,\Gamma(1+\alpha)}$$

$$\times\; {}_3F_2\left(-n, \alpha+\beta+n+1, \varrho+1; \alpha+1, \varrho+\sigma+2; 1\right)$$

$$[\operatorname{Re}\varrho > -1, \quad \operatorname{Re}\sigma > -1] \qquad \text{ET II 284(3)}$$

3.[6] $$\int_{-1}^{1} (1-x)^\alpha (1+x)^\sigma P_n^{(\alpha,\beta)}(x)\, dx = \frac{2^{\alpha+\sigma+1}\,\Gamma(\sigma+1)\,\Gamma(\alpha+1)\,\Gamma(\sigma-\beta+1)}{n!\,\Gamma(\sigma-\beta-n+1)\,\Gamma(\alpha+\sigma+n+2)}$$

$$[\operatorname{Re}\alpha > -1, \quad \operatorname{Re}\sigma > -1] \qquad \text{ET II 284(1)}$$

4. $$\int_{-1}^{1} (1-x)^\varrho (1+x)^\beta P_n^{(\alpha,\beta)}(x)\, dx = \frac{2^{\beta+\varrho+1}\,\Gamma(\varrho+1)\,\Gamma(\beta+n+1)\,\Gamma(\alpha-\varrho+n)}{n!\,\Gamma(\alpha-\varrho)\,\Gamma(\beta+\varrho+n+2)}$$

$$[\operatorname{Re}\varrho > -1, \quad \operatorname{Re}\beta > -1] \qquad \text{ET II 284(2)}$$

5. $$\int_{-1}^{1} (1-x)^{\alpha-1} (1+x)^\beta \left[P_n^{(\alpha,\beta)}(x) \right]^2 dx = \frac{2^{\alpha+\beta}\,\Gamma(\alpha+n+1)\,\Gamma(\beta+n+1)}{n!\,\alpha\,\Gamma(\alpha+\beta+n+1)}$$

$$[\operatorname{Re}\alpha > 0, \quad \operatorname{Re}\beta > -1] \qquad \text{ET II 285(6)}$$

6. $$\int_{-1}^{1} (1-x)^{2\alpha} (1+x)^\beta \left[P_n^{(\alpha,\beta)}(x) \right]^2 dx = \frac{2^{4\alpha+\beta+1}\,\Gamma\left(\alpha+\frac{1}{2}\right) \left[\Gamma(\alpha+n+1)\right]^2 \Gamma(\beta+2n+1)}{\sqrt{\pi}\,(n!)^2\,\Gamma(\alpha+1)\,\Gamma(2\alpha+\beta+2n+2)}$$

$$\left[\operatorname{Re}\alpha > -\tfrac{1}{2}, \quad \operatorname{Re}\beta > -1\right] \qquad \text{ET II 285(7)}$$

7. $$\int_{-1}^{1} (1-x)^\varrho (1+x)^\beta P_n^{(\alpha,\beta)}(x) P_n^{(\varrho,\beta)}(x)\, dx$$

$$= \frac{2^{\varrho+\beta+1}\,\Gamma(\varrho+n+1)\,\Gamma(\beta+n+1)\,\Gamma(\alpha+\beta+2n+1)}{n!\,\Gamma(\beta+\varrho+2n+2)\,\Gamma(\alpha+\beta+n+1)}$$

$$[\operatorname{Re}\varrho > -1, \quad \operatorname{Re}\beta > -1] \qquad \text{ET II285(10)}$$

8. $$\int_{-1}^{1} (1-x)^{\varrho-1} (1+x)^\beta P_n^{(\alpha,\beta)}(x) P_n^{(\varrho,\beta)}(x)\, dx = \frac{2^{\varrho+\beta}\,\Gamma(\alpha+n+1)\,\Gamma(\beta+n+1)\,\Gamma(\varrho)}{n!\,\Gamma(\alpha+1)\,\Gamma(\varrho+\beta+n+1)}$$

$$[\operatorname{Re}\beta > -1, \quad \operatorname{Re}\varrho > 0] \qquad \text{ET II 286(11)}$$

9.[7] $$\int_{-1}^{1} (1-x)^\alpha (1+x)^\sigma P_n^{(\alpha,\beta)}(x) P_m^{(\alpha,\sigma)}(x)\, dx$$

$$= \frac{2^{\alpha+\sigma+1}\,\Gamma(\alpha+n+1)\,\Gamma(\alpha+\beta+m+n+1)\,\Gamma(\sigma+m+1)\,\Gamma(\sigma-\beta+1)}{m!\,(n-m)!\,\Gamma(\alpha+\beta+n+1)\,\Gamma(\alpha+\sigma+m+n+2)\,\Gamma(\alpha-\beta+m-n+1)}$$

$$[\operatorname{Re}\alpha > -1, \quad \operatorname{Re}\sigma > -1] \qquad \text{ET II 286(12)}$$

$10.^6$ $\displaystyle\int_{-1}^{1} (1-x)^\varrho (1+x)^\beta \, P_n^{(\alpha,\beta)}(x) \, P_m^{(\varrho,\beta)}(x) \, dx$

$$= \frac{2^{\beta+\varrho+1}\, \Gamma(\alpha+\beta+m+n+1)\, \Gamma(\beta+n+1)\, \Gamma(\varrho+m+1)}{m!(n-m)!\, \Gamma(\alpha+\beta+n+1)\, \Gamma(\beta+\varrho+m+n+2)} \frac{\Gamma(\alpha-\varrho-m+n)}{\Gamma(\alpha-\varrho)}$$

$$[\operatorname{Re}\beta > -1, \quad \operatorname{Re}\varrho > -1] \quad \text{ET II 287(16)}$$

11. $\displaystyle\int_0^x (1-y)^\alpha (1+y)^\beta \, P_n^{(\alpha,\beta)}(y) \, dy = \frac{1}{2n}\left[P_{n-1}^{(\alpha+1,\beta+1)}(0) - (1-x)^{\alpha+1}(1+x)^{\beta+1}\, P_{n-1}^{(\alpha+1,\beta+1)}(x) \right]$

$$\text{EH II 173(38)}$$

7.392

1. $\displaystyle\int_0^1 x^{\lambda-1}(1-x)^{\mu-1}\, P_n^{(\alpha,\beta)}(1-\gamma x)\, dx$

$$= \frac{\Gamma(\alpha+n+1)\,\Gamma(\lambda)\,\Gamma(\mu)}{n!\,\Gamma(\alpha+1)\,\Gamma(\lambda+\mu)}\; {}_3F_2\left(-n, n+\alpha+\beta+1, \lambda; \alpha+1, \lambda+\mu; \frac{1}{2}\gamma\right)$$

$$[\operatorname{Re}\lambda > 0, \quad \operatorname{Re}\mu > 0] \quad \text{ET II 192(46)a}$$

2. $\displaystyle\int_0^1 x^{\lambda-1}(1-x)^{\mu-1}\, P_n^{(\alpha,\beta)}(\gamma x - 1)\, dx$

$$= (-1)^n \frac{\Gamma(\beta+n+1)\,\Gamma(\lambda)\,\Gamma(\mu)}{n!\,\Gamma(\beta+1)\,\Gamma(\lambda+\mu)}\; {}_3F_2\left(-n, n+\alpha+\beta+1, \lambda; \beta+1, \lambda+\mu; \frac{1}{2}\gamma\right)a$$

$$[\operatorname{Re}\lambda > 0, \quad \operatorname{Re}\mu > 0] \quad \text{ET II 192(47)a}$$

3. $\displaystyle\int_0^1 x^{\alpha}(1-x)^{\mu-1}\, P_n^{(\alpha,\beta)}(1-\gamma x)\, dx = \frac{\Gamma(\alpha+n+1)\,\Gamma(\mu)}{\Gamma(\alpha+\mu+n+1)}\, P_n^{(\alpha+\mu,\beta-\mu)}(1-\gamma)$

$$[\operatorname{Re}a > -1, \quad \operatorname{Re}\mu > 0] \quad \text{ET II 191(43)a}$$

4. $\displaystyle\int_0^1 x^{\beta}(1-x)^{\mu-1}\, P_n^{(\alpha,\beta)}(\gamma x - 1)\, dx = \frac{\Gamma(\beta+n+1)\,\Gamma(\mu)}{\Gamma(\beta+\mu+n+1)}\, P_n^{(\alpha-\mu,\beta+\mu)}(\gamma-1)$

$$[\operatorname{Re}\beta > -1, \quad \operatorname{Re}\mu > 0] \quad \text{ET II 191(44)a}$$

7.393

1. $\displaystyle\int_0^1 (1-x^2)^\nu \sin bx \, P_{2n+1}^{(\nu,\nu)}(x)\, dx = \frac{(-1)^n \sqrt{\pi}\, \Gamma(2n+\nu+2)\, J_{2n+\nu+\frac{3}{2}}(b)}{2^{\frac{1}{2}-\nu}(2n+1)!\, b^{\nu+\frac{1}{2}}}$

$$[b > 0, \quad \operatorname{Re}\nu > -1] \quad \text{ET I 94(5)}$$

2. $\displaystyle\int_0^1 (1-x^2)^\nu \cos bx \, P_{2n}^{(\nu,\nu)}(x)\, dx = \frac{(-1)^n 2^{\nu-\frac{1}{2}}\sqrt{\pi}\, \Gamma(2n+\nu+1)\, J_{2n+\nu+\frac{1}{2}}(b)}{(2n)!\, b^{\nu+\frac{1}{2}}}$

$$[b > 0, \quad \operatorname{Re}\nu > -1] \quad \text{ET I 38(4)}$$

7.41–7.42   Laguerre polynomials

7.411

1. $\displaystyle\int_0^t L_n(x)\, dx = L_n(t) - L_{n+1}(t)/(n+1).$ MO 110

2. $\quad \int_0^t L_n^\alpha(x)\, dx = L_n^\alpha(t) - L_{n+1}^\alpha(t) - \binom{n+\alpha}{n} + \binom{n+1+\alpha}{n+1}$

               EH II 189(16)a

3. $\quad \int_0^t L_{n-1}^{\alpha+1}(x)\, dx = -L_n^\alpha(t) + \binom{n+\alpha}{n}$

               EH II 189(15)a

4. $\quad \int_0^t L_m(x)\, L_n(t-x)\, dx = L_{m+n}(t) - L_{m+n+1}(t)$

               EH II 191(31)

5. $\quad \sum_{k=0}^\infty \left[\int_0^t \frac{L_k(x)}{k!}\, dx \right]^2 = e^t - 1 \qquad\qquad [t \geq 0]$

               MO 110

7.412

1. $\quad \int_0^1 (1-x)^{\mu-1} x^\alpha\, L_n^\alpha(ax)\, dx = \frac{\Gamma(\alpha+n+1)\,\Gamma(\mu)}{\Gamma(\alpha+\mu+n+1)}\, L_n^{\alpha+\mu}(a)$

               $[\operatorname{Re}\alpha > -1, \quad \operatorname{Re}\mu > 0]$

               EH II 191(30)a, BU 129(14c)

2. $\quad \int_0^1 (1-x)^{\mu-1} x^{\lambda-1}\, L_n^\alpha(\beta x)\, dx = \frac{\Gamma(\alpha+n+1)\,\Gamma(\lambda)\,\Gamma(\mu)}{n!\,\Gamma(\alpha+1)\,\Gamma(\lambda+\mu)}\, {}_2F_2(-n, \lambda; \alpha+1, \lambda+\mu : \beta)$

               $[\operatorname{Re}\lambda > 0, \quad \operatorname{Re}\mu > 0] \qquad$ ET II 192(50)a

7.413 $\quad \int_0^1 x^\alpha (1-x)^\beta\, L_m^\alpha(xy)\, L_n^\beta[(1-x)y]\, dx = \frac{(m+n)!\,\Gamma(\alpha+m+1)\,\Gamma(\beta+n+1)}{m!\,n!\,\Gamma(\alpha+\beta+m+n+2)}\, L_{m+n}^{\alpha+\beta+1}(y)$

               $[\operatorname{Re}\alpha > -1, \quad \operatorname{Re}\beta > -1] \qquad$ ET II 293(7)

7.414

1. $\quad \int_0^\infty e^{-x}\, L_n^\alpha(x)\, dx = e^{-y}\left[L_n^\alpha(y) - L_{n-1}^\alpha(y) \right]$

               EH II 191(29)

2. $\quad \int_0^\infty e^{-bx}\, L_n(\lambda x)\, L_n(\mu x)\, dx = \frac{(b-\lambda-\mu)^n}{b^{n+1}}\, P_n\left[\frac{b^2 - (\lambda+\mu)b + 2\lambda\mu}{b(b-\lambda-\mu)} \right]$

               $[\operatorname{Re}b > 0] \qquad$ ET I 175(34)

3.[8] $\quad \int_0^\infty e^{-x} x^\alpha\, L_n^\alpha(x)\, L_m^\alpha(x)\, dx = 0 \qquad\qquad [m \neq n, \quad \operatorname{Re}\alpha > -1] \qquad$ BU 115(8), ET II 293(3)

               $= \frac{\Gamma(\alpha+n+1)}{n!} \qquad\quad [m = n, \quad \operatorname{Re}\alpha > 0] \qquad$ BU 115(8), ET II 292(2)

4. $\quad \int_0^\infty e^{-bx} x^\alpha\, L_n^\alpha(\lambda x)\, L_m^\alpha(\mu x)\, dx = \frac{\Gamma(m+n+\alpha+1)}{m!\,n!}\, \frac{(b-\lambda)^n (b-\mu)^m}{b^{m+n+\alpha+1}}$

               $\times F\left[-m, -n; -m-n-\alpha, \frac{b(b-\lambda-\mu)}{(b-\lambda)(b-\mu)} \right]$

               $[\operatorname{Re}\alpha > -1, \quad \operatorname{Re}b > 0] \qquad$ ET I 175(35)

4(1)[9]. $\quad \int_0^\infty e^{-x} x^{\alpha+1/2}\, L_n^\alpha(x)\, L_m^\alpha(x)\, dx = \frac{\Gamma(\alpha+n+1)^2\,\Gamma(\alpha+m+1)\,\Gamma\left(\alpha+\frac{3}{2}\right)\,\Gamma\left(m-\frac{1}{2}\right)}{n!\,m!\,\Gamma(\alpha+1)\,\Gamma\left(-\frac{1}{2}\right)}$

               $\times {}_3F_2\left(-n, \alpha+\frac{3}{2}, \frac{3}{2}; \alpha+1, \frac{3}{2}-m; 1 \right)$

5. $\quad \int_0^\infty e^{-bx}\, L_n^a(x)\, dx = \sum_{m=0}^n \binom{a+m-1}{m} \frac{(b-1)^{n-m}}{b^{n-m+1}} \qquad [\operatorname{Re}b > 0] \qquad$ ET I 174(27)

6. $\displaystyle\int_0^\infty e^{-bx} L_n(x)\, dx = (b-1)^n b^{-n-1}$ $[\operatorname{Re} b > 0]$ ET I 174(25)

7. $\displaystyle\int_0^\infty e^{-st} t^\beta\, L_n^\alpha(t)\, dt = \frac{\Gamma(\beta+1)\,\Gamma(\alpha+n+1)}{n!\,\Gamma(\alpha+1)} s^{-\beta-1}\, F\left(-n, \beta+1; \alpha+1; \frac{1}{s}\right)$

$$[\operatorname{Re}\beta > -1, \quad \operatorname{Re} s > 0]$$

                                      BU 119(4b), EH II 191(133)

8. $\displaystyle\int_0^\infty e^{-st} t^\alpha\, L_n^\alpha(t)\, dt = \frac{\Gamma(\alpha+n+1)(s-1)^n}{n!\,s^{\alpha+n+1}}$ $[\operatorname{Re}\alpha > -1, \quad \operatorname{Re} s > 0]$

                                      EH II 191(32), MO 176a

9. $\displaystyle\int_0^\infty e^{-x} x^{\alpha+\beta}\, L_m^\alpha(x)\, L_n^\beta(x)\, dx = (-1)^{m+n}(\alpha+\beta)!\binom{\alpha+m}{n}\binom{\beta+n}{m}$

$$[\operatorname{Re}(\alpha+\beta) > -1]$$ ET II 293(4)

10.[6] $\displaystyle\int_0^\infty e^{-bx} x^{2a} [L_n^a(x)]^2\, dx = \frac{2^{2a}\,\Gamma\left(a+\tfrac{1}{2}\right)\Gamma\left(n+\tfrac{1}{2}\right)}{\pi(n!)^2 b^{2a+1}}$

$$\times\, F\left(-n, a+\frac{1}{2}; \frac{1}{2}-n; \left(1-\frac{2}{b}\right)^2\right)\Gamma(a+n+1)$$

$$\left[\operatorname{Re} a > -\frac{1}{2}, \quad \operatorname{Re} b > 0\right]$$ ET I 174(30)

11. $\displaystyle\int_0^\infty e^{-x} x^{\gamma-1}\, L_n^\mu(x)\, dx = \frac{\Gamma(\gamma)\,\Gamma(1+\mu+n-\gamma)}{n!\,\Gamma(1+\mu-\gamma)}$ $[\operatorname{Re}\gamma > 0]$ BU 120(4b)

12. $\displaystyle\int_0^\infty e^{-x\left(s+\frac{a_1+a_2}{2}\right)} x^{\mu+\beta}\, L_k^\mu(a_1 x)\, L_k^\mu(a_2 x)\, dx$

$$= \frac{\Gamma(1+\mu+\beta)\,\Gamma(1+\mu+k)}{k!\,k!\,\Gamma(1+\mu)}\left\{\frac{d^k}{dh^k}\left[\frac{F\left(\frac{1+\mu+\beta}{2}, 1+\frac{\mu+\beta}{2}; 1+\mu; \frac{A^2}{B^2}\right)}{(1-h)^{1+\mu} B^{1+\mu+\beta}}\right]\right\}_{h=0}$$

$$A^2 = \frac{4a_1 a_2 h}{(1-h)^2}; \qquad B = s + \frac{a_1+a_2}{2}\frac{1+h}{1-h}$$

$$\left[\operatorname{Re}\left(s+\frac{a_1+a_2}{2}\right) > 0, \quad a_1 > 0, \quad a_2 > 0, \quad \operatorname{Re}(\mu+\beta) > -1\right]$$ BU 142(19)

13. $\displaystyle\int_0^\infty \exp\left[-x\left(s+\frac{a_1+a_2}{2}\right)\right] x^\mu\, L_k^\mu(a_1 x)\, L_k^\mu(a_2 x)\, dx = \frac{\Gamma(1+\mu+k)}{b_0^{1+\mu+k}}\cdot\frac{b_0^k}{k!}\cdot P_k^{(\mu,0)}\left(\frac{b_1^2}{b_0 b_2}\right)$

$$b_0 = s + \frac{a_1+a_2}{2}, \quad b_1^2 = b_0 b_2 + 2a_1 a_2, \quad b_2 = s - \frac{a_1+a_2}{2}$$

$$\left[\operatorname{Re}\mu > -1, \quad \operatorname{Re}\left(s+\frac{a_1+a_2}{2}\right) > 0\right]$$

                                      BU 144(22)

7.415 $\displaystyle\int_0^1 (1-x)^{\mu-1} x^{\lambda-1} e^{-\beta x}\, L_n^\alpha(\beta x)\, dx = \frac{\Gamma(\alpha+n+1)}{n!\,\Gamma(\alpha+1)}\, \mathrm{B}(\lambda,\mu)\, {}_2F_2\left(\alpha+n+1, \lambda; \alpha+1, \lambda+\mu; -\beta\right)$

$$[\operatorname{Re}\lambda > 0, \quad \operatorname{Re}\mu > 0]$$ ET II 193(51)a

7.416 $\int_{-\infty}^{\infty} x^{m-n} \exp\left[-\frac{1}{2}(x-y)^2\right] L_n^{m-n}\left(x^2\right) dx = \frac{(2\pi)^{1/2}}{n!} i^{n-m} 2^{-\frac{n+m}{2}} H_n\left(\frac{iy}{\sqrt{2}}\right) H_m\left(\frac{iy}{\sqrt{2}}\right)$

BU 149(15b), ET II 293(8)a

7.417

1. $\int_0^{\infty} x^{\nu-2n-1} e^{-ax} \sin(bx) L_{2n}^{\nu-2n-1}(ax) dx = (-1)^n i\, \Gamma(\nu) \dfrac{b^{2n}\left[(a-ib)^{-\nu} - (a+ib)^{-\nu}\right]}{2(2n)!}$

$[b > 0, \quad \operatorname{Re} a > 0, \quad \operatorname{Re}\nu > 2n]$
ET I 95(12)

2. $\int_0^{\infty} x^{\nu-2n-2} e^{-ax} \sin(bx) L_{2n+1}^{\nu-2n-2}(ax) dx = (-1)^{n+1}\, \Gamma(\nu) \dfrac{b^{2n+1}\left[(a+ib)^{-\nu} + (a-ib)^{-\nu}\right]}{2(2n+1)!}$

$[b > 0, \quad \operatorname{Re} a > 0, \quad \operatorname{Re}\nu > 2n+1]$
ET I 95(13)

3. $\int_0^{\infty} x^{\nu-2n} e^{-ax} \cos(bx) L_{\nu-2n}^{2n-1}(ax) dx = i(-1)^{n+1}\, \Gamma(\nu) \dfrac{b^{2n-1}\left[(a-ib)^{-\nu} - (a+ib)^{-\nu}\right]}{2(2n-1)!}$

$[b > 0, \quad \operatorname{Re} a > 0, \quad \operatorname{Re}\nu > 2n-1]$
ET I 39(12)

4. $\int_0^{\infty} x^{\nu-2n-1} e^{-ax} \cos(bx) L_{2n}^{\nu-2n-1}(ax) dx = (-1)^n\, \Gamma(\nu) \dfrac{b^{2n}\left[(a+ib)^{-\nu} + (a-ib)^{-\nu}\right]}{2(2n)!}$

$[b > 0, \quad \operatorname{Re}\nu > 2n, \quad \operatorname{Re} a > 0]$
ET I 39(13)

7.418

1. $\int_0^{\infty} e^{-\frac{1}{2}x^2} \sin(bx) L_n\left(x^2\right) dx = (-1)^n \dfrac{i}{2} n! \dfrac{1}{\sqrt{2\pi}} \left\{[D_{-n-1}(ib)]^2 - [D_{-n-1}(-ib)]^2\right\}$

$[b > 0]$ ET I 95(14)

2. $\int_0^{\infty} e^{-\frac{1}{2}x^2} \cos(bx) L_n\left(x^2\right) dx = \sqrt{\dfrac{\pi}{2}} (n!)^{-1} e^{-\frac{1}{2}b^2} 2^{-n} \left[H_n\left(\dfrac{b}{\sqrt{2}}\right)\right]^2$

$[b > 0]$ ET I 39(14)

3. $\int_0^{\infty} x^{2n+1} e^{-\frac{1}{2}x^2} \sin(bx) L_n^{n+\frac{1}{2}}\left(\dfrac{1}{2}x^2\right) dx = \sqrt{\dfrac{\pi}{2}} b^{2n+1} e^{-\frac{1}{2}b^2} L_n^{n+\frac{1}{2}}\left(\dfrac{b^2}{2}\right)$

$[b > 0]$ ET I 95(15)

4. $\int_0^{\infty} x^{2n} e^{-\frac{1}{2}x^2} \cos(bx) L_n^{n-\frac{1}{2}}\left(\dfrac{1}{2}x^2\right) dx = \sqrt{\dfrac{\pi}{2}} b^{2n} e^{-\frac{1}{2}b^2} L_n^{n+\frac{1}{2}}\left(\dfrac{1}{2}b^2\right)$

$[b > 0]$ ET I 39(16)

5. $\int_0^{\infty} x e^{-\frac{1}{2}x^2} L_n^{\alpha}\left(\dfrac{1}{2}x^2\right) L_{\tilde{n}}^{\frac{1}{2}-\alpha}\left(\dfrac{1}{2}x^2\right) \sin(xy) dx = \left(\dfrac{\pi}{2}\right)^{1/2} y e^{-\frac{1}{2}y^2} L_n^{\alpha}\left(\dfrac{1}{2}y^2\right) L_{\tilde{n}}^{\frac{1}{2}-\alpha}\left(\dfrac{1}{2}y^2\right)$

ET II 294(11)

6. $\int_0^\infty e^{-\frac{1}{2}x^2} L_n^\alpha \left(\frac{1}{2}x^2\right) L_n^{-\frac{1}{2}-\alpha} \left(\frac{1}{2}x^2\right) \cos(xy)\, dx = \left(\frac{\pi}{2}\right)^{1/2} e^{-\frac{1}{2}y^2} L_n^\alpha \left(\frac{1}{2}y^2\right) L_n^{-\alpha-\frac{1}{2}} \left(\frac{1}{2}y^2\right)$

$$\text{ET II 294(12)}$$

7.419 $\int_0^\infty x^{n+2\nu-\frac{1}{2}} \exp[-(1+a)x]\, L_n^{2\nu}(ax)\, K_\nu(x)\, dx$

$$= \frac{\pi^{1/2}\, \Gamma\left(n+\nu+\frac{1}{2}\right) \Gamma\left(n+3\nu+\frac{1}{2}\right)}{2^{n+2\nu+\frac{1}{2}} n!\, \Gamma(2\nu+1)}\, F\left(n+\nu+\frac{1}{2}, n+3\nu+\frac{1}{2}; 2\nu+1; -\frac{1}{2}a\right)$$

$$\left[\operatorname{Re} a > -2, \quad \operatorname{Re}(n+\nu) > -\frac{1}{2}, \quad \operatorname{Re}(n+3\nu) > -\frac{1}{2}\right] \quad \text{ET II 370(44)}$$

7.421

1. $\int_0^\infty x e^{-\frac{1}{2}\alpha x^2} L_n\left(\frac{1}{2}\beta x^2\right) J_0(xy)\, dx = \frac{(\alpha-\beta)^n}{\alpha^{n+1}} e^{-\frac{1}{2\alpha}y^2} L_n\left[\frac{\beta y^2}{2\alpha(\beta-\alpha)}\right]$

$$[y > 0, \quad \operatorname{Re}\alpha > 0] \qquad \text{ET II 13(4)a}$$

2. $\int_0^\infty x e^{-x^2} L_n\left(x^2\right) J_0(xy)\, dx = \frac{2^{-2n-1}}{n!} y^{2n} e^{-\frac{1}{4}y^2}$

$$\text{ET II 13(5)}$$

3. $\int_0^\infty x^{2n+\nu+1} e^{-\frac{1}{2}x^2} L_n^{\nu+n}\left(\frac{1}{2}x^2\right) J_\nu(xy)\, dx = y^{2n+\nu} e^{-\frac{1}{2}y^2} L_n^{\nu+n}\left(\frac{1}{2}y^2\right)$

$$[y > 0, \quad \operatorname{Re}\nu > -1] \qquad \text{MO 183}$$

4. $\int_0^\infty x^{\nu+1} e^{-\beta x^2} L_n^\nu\left(\alpha x^2\right) J_\nu(xy)\, dx = 2^{-\nu-1}\beta^{-\nu-n-1}(\beta-\alpha)^n y^\nu e^{-\frac{y^2}{4\beta}} L_n^\nu\left[\frac{\alpha y^2}{4\beta(\alpha-\beta)}\right]$

$$\text{ET II 43(5)}$$

5. $\int_0^\infty e^{-\frac{1}{2q}x^2} x^{\nu+1} L_n^\nu\left[\frac{x^2}{2q(1-q)}\right] J_\nu(xy)\, dx = \frac{q^{n+\nu+1}}{(q-1)^n} e^{-\frac{qy^2}{2}} y^\nu L_n^\nu\left(\frac{y^2}{2}\right)$

$$[\nu > 0] \qquad \text{MO 183}$$

7.422

1. $\int_0^\infty x^{\nu+1} e^{-\beta x^2} \left[L_n^{\frac{1}{2}\nu}\left(\alpha x^2\right)\right]^2 J_\nu(xy)\, dx$

$$= \frac{y^\nu}{\pi n!}\, \Gamma\left(n+1+\frac{1}{2}\nu\right) (2\beta)^{-\nu-1} e^{-\frac{y^2}{4\beta}}$$

$$\times \sum_{l=0}^n \frac{(-1)^l \Gamma\left(n-l+\frac{1}{2}\right) \Gamma\left(l+\frac{1}{2}\right)}{\Gamma\left(l+1+\frac{1}{2}\nu\right)(n-l)!} \left(\frac{2\alpha-\beta}{\beta}\right)^{2l} L_{2l}^\nu\left[\frac{\alpha y^2}{2\beta(2\alpha-\beta)}\right]$$

$$[y > 0, \quad \operatorname{Re}\beta > 0, \quad \operatorname{Re}\nu > -1] \quad \text{ET II 43(7)}$$

2.[9] $\int_0^\infty x^{\nu+1} e^{-\alpha x^2} L_m^{\nu-\sigma}\left(\alpha x^2\right) L_n^\sigma\left(\alpha x^2\right) J_\nu(xy)\, dx$

$$= (-1)^{m+n}(2\alpha)^{-\nu-1} y^\nu e^{-\frac{y^2}{4\alpha}} L_n^{m-n-\sigma}\left(\frac{y^2}{4\alpha}\right) L_m^{n-m+\sigma-\nu}\left(\frac{y^2}{4\alpha}\right)$$

$$[y > 0, \quad \operatorname{Re}\alpha > 0, \quad \operatorname{Re}\nu > -1, \quad n \neq 0, \quad \sigma \neq 0, \quad \alpha \neq 1] \quad \text{ET II 43(8)}$$

7.423

1.　$\int_0^\infty e^{-\frac{1}{2}x^2} L_n\left(\frac{1}{2}x^2\right) H_{2n+1}\left(\frac{x}{2\sqrt{2}}\right) \sin(xy)\, dx = \left(\frac{\pi}{2}\right)^{1/2} e^{-\frac{1}{2}y^2} L_n\left(\frac{1}{2}y^2\right) H_{2n+1}\left(\frac{y}{2\sqrt{2}}\right)$

ET II 294(13)a

2.　$\int_0^\infty e^{-\frac{1}{2}x^2} L_n\left(\frac{1}{2}x^2\right) H_{2n}\left(\frac{x}{2\sqrt{2}}\right) \cos(xy)\, dx = \left(\frac{\pi}{2}\right)^{1/2} e^{-\frac{1}{2}y^2} L_n\left(\frac{1}{2}y^2\right) H_{2n}\left(\frac{y}{2\sqrt{2}}\right)$

ET II 294(14)a

7.5 Hypergeometric Functions

7.51 Combinations of hypergeometric functions and powers

7.511　$\int_0^\infty F(a,b;c;-z)z^{-s-1}\, dx = \dfrac{\Gamma(a+s)\,\Gamma(b+s)\,\Gamma(c)\,\Gamma(-s)}{\Gamma(a)\,\Gamma(b)\,\Gamma(c+s)}$

$[c \neq 0, -1, -2, \ldots, \quad \operatorname{Re} s < 0, \quad \operatorname{Re}(a+s) > 0, \quad \operatorname{Re}(b+s) > 0]$　EH I 79(4)

7.512

1.　$\int_0^1 x^{\alpha-\gamma}(1-x)^{\gamma-\beta-1} F(\alpha,\beta;\gamma;x)\, dx = \dfrac{\Gamma\left(1+\frac{\alpha}{2}\right)\Gamma(\gamma)\,\Gamma(\alpha-\gamma+1)\,\Gamma\left(\gamma-\frac{\alpha}{2}-\beta\right)}{\Gamma(1+\alpha)\,\Gamma\left(1+\frac{\alpha}{2}-\beta\right)\Gamma\left(\gamma-\frac{\alpha}{2}\right)}$

$\left[\operatorname{Re}\alpha+1 > \operatorname{Re}\gamma > \operatorname{Re}\beta, \quad \operatorname{Re}\left(\gamma-\frac{\alpha}{2}-\beta\right) > 0\right]$　ET II 398(1)

2.　$\int_0^1 x^{\varrho-1}(1-x)^{\beta-\gamma-n} F(-n,\beta;\gamma;x)\, dx = \dfrac{\Gamma(\gamma)\,\Gamma(\varrho)\,\Gamma(\beta-\gamma+1)\,\Gamma(\gamma-\varrho+n)}{\Gamma(\gamma+n)\,\Gamma(\gamma-\varrho)\,\Gamma(\beta-\gamma+\varrho+1)}$

$[n = 0, 1, 2 \ldots; \quad \operatorname{Re}\varrho > 0, \quad \operatorname{Re}(\beta-\gamma) > n-1]$　ET II 398(2)

3.　$\int_0^1 x^{\varrho-1}(1-x)^{\beta-\varrho-1} F(\alpha,\beta;\gamma;x)\, dx = \dfrac{\Gamma(\gamma)\,\Gamma(\varrho)\,\Gamma(\beta-\varrho)\,\Gamma(\gamma-\alpha-\varrho)}{\Gamma(\beta)\,\Gamma(\gamma-\alpha)\,\Gamma(\gamma-\varrho)}$

$[\operatorname{Re}\varrho > 0, \quad \operatorname{Re}(\beta-\varrho) > 0, \quad \operatorname{Re}(\gamma-\alpha-\varrho) > 0]$　ET II 399(3)

4.　$\int_0^1 x^{\gamma-1}(1-x)^{\varrho-1} F(\alpha,\beta;\gamma;x)\, dx = \dfrac{\Gamma(\gamma)\,\Gamma(\varrho)\,\Gamma(\gamma+\varrho-\alpha-\beta)}{\Gamma(\gamma+\varrho-\alpha)\,\Gamma(\gamma+\varrho-\beta)}$

$[\operatorname{Re}\gamma > 0, \quad \operatorname{Re}\varrho > 0, \quad \operatorname{Re}(\gamma+\varrho-\alpha-\beta) > 0]$　ET II 399(4)

5.　$\int_0^1 x^{\varrho-1}(1-x)^{\sigma-1} F(\alpha,\beta;\gamma;x)\, dx = \dfrac{\Gamma(\varrho)\,\Gamma(\sigma)}{\Gamma(\varrho+\sigma)}\, {}_3F_2(\alpha,\beta,\varrho;\gamma,\varrho+\sigma;1)$

$[\operatorname{Re}\varrho > 0, \quad \operatorname{Re}\sigma > 0, \quad \operatorname{Re}(\gamma+\sigma-\alpha-\beta) > 0]$　ET II 399(5)

6.[10]　$\int_0^1 x^{\lambda-1}(1-x)^{\beta-\lambda-1} F\left(\alpha,\beta;\lambda;\frac{zx}{b}\right) dx = \mathrm{B}(\lambda,\beta-\lambda)(1-z/b)^{-\alpha}$

BU 9

7. $\int_0^1 x^{\gamma-1}(1-x)^{\delta-\gamma-1} F(\alpha,\beta;\gamma;xz) F(\delta-\alpha,\delta-\beta;\delta-\gamma;(1-x)\zeta)\, dx$

$$= \frac{\Gamma(\gamma)\,\Gamma(\delta-\gamma)}{\Gamma(\delta)}(1-\zeta)^{2\alpha-\delta} F(\alpha,\beta;\delta;z+\zeta-z\zeta)$$

$$[0<\operatorname{Re}\gamma<\operatorname{Re}\delta, \quad |\arg(1-z)|<\pi, \quad |\arg(1-\zeta)|<\pi] \quad \textbf{ET II 400(11)}$$

8. $\int_0^1 x^{\gamma-1}(1-x)^{\epsilon-1}(1-xz)^{-\delta} F(\alpha,\beta;\gamma;xz) F\left[\delta,\beta-\gamma;\epsilon;\dfrac{(1-x)z}{(1-xz)}\right] dx$

$$= \frac{\Gamma(\gamma)\,\Gamma(\epsilon)}{\Gamma(\gamma+\epsilon)} F(\alpha+\delta,\beta;\gamma+\epsilon;z)$$

$$[\operatorname{Re}\gamma>0, \quad \operatorname{Re}\epsilon>0, \quad |\arg(z-1)|<\pi] \quad \textbf{ET II 400(12), Eh I 78(3)}$$

9. $\int_0^1 x^{\gamma-1}(1-x)^{\varrho-1}(1-zx)^{-\sigma} F(\alpha,\beta;\gamma;x)\, dx$

$$= \frac{\Gamma(\gamma)\,\Gamma(\varrho)\,\Gamma(\gamma+\varrho-\alpha-\beta)}{\Gamma(\gamma+\varrho-\alpha)\,\Gamma(\gamma+\varrho-\beta)}(1-z)^{-\sigma}$$

$$\times {}_3F_2\left(\varrho,\sigma,\gamma+\varrho-\alpha-\beta;\gamma+\varrho-\alpha,\gamma+\varrho-\beta;\frac{z}{z-1}\right)$$

$$[\operatorname{Re}\gamma>0, \quad \operatorname{Re}\varrho>0, \quad \operatorname{Re}(\gamma+\varrho-\alpha-\beta)>0, \quad |\arg(1-z)|<\pi] \quad \textbf{ET II 399(6)}$$

10. $\int_0^\infty x^{\gamma-1}(x+z)^{-\sigma} F(\alpha,\beta;\gamma;-x)\, dx = \dfrac{\Gamma(\gamma)\,\Gamma(\alpha-\gamma+\sigma)\,\Gamma(\beta-\gamma+\sigma)}{\Gamma(\sigma)\,\Gamma(\alpha+\beta-\gamma+\sigma)}$

$$\times F(\alpha-\gamma+\sigma,\beta-\gamma+\sigma;\alpha+\beta-\gamma+\sigma;1-z)$$

$$[\operatorname{Re}\gamma>0, \quad \operatorname{Re}(\alpha-\gamma+\sigma)>0, \quad \operatorname{Re}(\beta-\gamma+\sigma)>0, \quad |\arg z|<\pi] \quad \textbf{ET II 400(10)}$$

11. $\int_0^1 (1-x)^{\mu-1}x^{\nu-1}\, {}_pF_q(a_1,\ldots,a_p;\nu,b_2,\ldots,b_q;ax)\, dx$

$$= \frac{\Gamma(\mu)\,\Gamma(\nu)}{\Gamma(\mu+\nu)}\, {}_pF_q(a_1,\ldots,a_p;\mu+\nu,b_2,\ldots,b_q;a)$$

$$[\operatorname{Re}\mu>0, \quad \operatorname{Re}\nu>0, \quad p\le q+1; \text{ if } p=q+1, \text{ then } |a|<1] \quad \textbf{ET II 200(94)}$$

12. $\int_0^1 (1-x)^{\mu-1}x^{\nu-1}\, {}_pF_q(a_1,\ldots,a_p;b_1,\ldots,b_q;ax)\, dx$

$$= \frac{\Gamma(\mu)\,\Gamma(\nu)}{\Gamma(\mu+\nu)}\, {}_{p+1}F_{q+1}(\nu,a_1,\ldots,a_p;\mu+\nu,b_1,\ldots,b_q;a)$$

$$[\operatorname{Re}\mu>0, \quad \operatorname{Re}\nu>0, \quad p\le q+1, \text{ if } p=q+1, \text{ then } |a|<1] \quad \textbf{ET II 200(95)}$$

7.513 $\int_0^1 x^{s-1}\left(1-x^2\right)^\nu F(-n,a;b;x^2)\, dx = \dfrac{1}{2}\operatorname{B}\left(\nu+1,\dfrac{s}{2}\right) {}_3F_2\left(-n,a,\dfrac{s}{2};b,\nu+1+\dfrac{s}{2};1\right)$

$$[\operatorname{Re}s>0, \quad \operatorname{Re}\nu>-1] \quad \textbf{ET I 336(4)}$$

7.52 Combinations of hypergeometric functions and exponentials

7.521
$$\int_0^\infty e^{-st} \, _pF_q(a_1, \ldots, a_p; b_1, \ldots, b_q, t) \, dt = \frac{1}{s} \, _{p+1}F_q\left(1, a_1, \ldots, a_p; b_1, \ldots, b_q, s^{-1}\right)$$

$$[p \leq q] \qquad \text{EH I 192}$$

7.522

1. $$\int_0^\infty e^{-\lambda x} x^{\gamma-1} \, _2F_1(\alpha, \beta; \delta; -x) \, dx = \frac{\Gamma(\delta)\lambda^{-\gamma}}{\Gamma(\alpha)\,\Gamma(\beta)} \, E(\alpha, \beta, \gamma; \delta : \lambda)$$

$$[\operatorname{Re}\lambda > 0, \quad \operatorname{Re}\gamma > 0] \qquad \text{EH I 205(10)}$$

2.⁶ $$\int_0^\infty e^{-bx} x^{a-1} F\left(\frac{1}{2}+\nu, \frac{1}{2}-\nu; a; -\frac{x}{2}\right) dx = 2^a e^b \frac{1}{\sqrt{\pi}} \Gamma(a)(2b)^{\frac{1}{2}-a} K_\nu(b)$$

$$[\operatorname{Re}a > 0, \quad \operatorname{Re}b > 0] \qquad \text{ET I 212(1)}$$

3. $$\int_0^\infty e^{-bx} x^{\gamma-1} F(2\alpha, 2\beta; \gamma; -\lambda x) \, dx = \Gamma(\gamma)b^{-\gamma} \left(\frac{b}{\lambda}\right)^{\alpha+\beta-\frac{1}{2}} e^{\frac{b}{2\lambda}} W_{\frac{1}{2}-\alpha-\beta, \alpha-\beta}\left(\frac{b}{\lambda}\right)$$

$$[\operatorname{Re}b > 0, \quad \operatorname{Re}\gamma > 0, \quad |\arg\lambda| < \pi]$$
$$\text{BU 78(30), ET I 212(4)}$$

4.⁶ $$\int_0^\infty e^{-xt} t^{b-1} F(a, a-c+1; b; -t) \, dt = x^{a-b} \Gamma(b) \Psi(a, c; x)$$

$$[\operatorname{Re}b > 0, \quad \operatorname{Re}x > 0] \qquad \text{EH I 273(11)}$$

5. $$\int_0^\infty e^{-x} x^{s-1} \, _pF_q(a_1, \ldots, a_p, b_1, \ldots, b_q; ax) \, dx = \Gamma(s) \, _{p+1}F_q(s, a_1, \ldots, a_p; b_1, \ldots, b_q; a)$$

$$[p < q, \quad \operatorname{Re}s > 0] \qquad \text{ET I 337(11)}$$

6. $$\int_0^\infty x^{\beta-1} e^{-\mu x} \, _2F_2(-n, n+1; 1, \beta; x) \, dx = \Gamma(\beta)\mu^{-\beta} P_n\left(1 - \frac{2}{\mu}\right)$$

$$[\operatorname{Re}\mu > 0, \quad \operatorname{Re}\beta > 0] \qquad \text{ET I 218(6)}$$

7. $$\int_0^\infty x^{\beta-1} e^{-\mu x} \, _2F_2\left(-n, n; \beta, \frac{1}{2}; x\right) dx = \Gamma(\beta)\mu^{-\beta} \cos\left[2n \arcsin\left(\frac{1}{\sqrt{\mu}}\right)\right]$$

$$[\operatorname{Re}\mu > 0, \quad \operatorname{Re}\beta > 0] \qquad \text{ET I 218(7)}$$

8. $$\int_0^\infty x^{\varrho_n-1} e^{-\mu x} \, _mF_n(a_1, \ldots, a_m; \varrho_1, \ldots, \varrho_n; \lambda x) \, dx$$

$$= \Gamma(\varrho_n) \mu^{-\varrho_n} \, _mF_{n-1}\left(a_1, \ldots, a_m; \varrho_1, \ldots, \varrho_{n-1}; \frac{\lambda}{\mu}\right)$$

$$[m \leq n; \quad \operatorname{Re}\varrho_n > 0, \quad \operatorname{Re}\mu > 0, \text{ if } m < n; \operatorname{Re}\mu > \operatorname{Re}\lambda, \text{ if } m = n] \quad \text{ET I 219(16)a}$$

9. $$\int_0^\infty x^{\sigma-1} e^{-\mu x} \, _mF_n(a_1, \ldots, a_m; \varrho_1, \ldots, \varrho_n; \lambda x) \, dx$$

$$= \Gamma(\sigma)\mu^{-\sigma} \, _{m+1}F_n\left(a_1, \ldots, a_m, \sigma; \varrho_1, \ldots, \varrho_n; \frac{\lambda}{\mu}\right)$$

$$[m \leq n, \quad \operatorname{Re}\sigma > 0, \quad \operatorname{Re}\mu > 0, \text{ if } m < n; \operatorname{Re}\mu > \operatorname{Re}\lambda, \text{ if } m = n] \quad \text{ET I 219(17)}$$

7.523 $\displaystyle\int_0^1 x^{\gamma-1}(1-x)^{\varrho-1}e^{-xz}\,F(\alpha,\beta;\gamma;x)\,dx$

$$= \frac{\Gamma(\gamma)\,\Gamma(\varrho)\,\Gamma(\gamma+\varrho-\alpha-\beta)}{\Gamma(\gamma+\varrho-\alpha)\,\Gamma(\gamma+\varrho-\beta)}e^{-z}\,{}_2F_2\left(\varrho,\gamma+\varrho-\alpha-\beta;\gamma+\varrho-\alpha,\gamma+\varrho-\beta;z\right)$$

$$[\operatorname{Re}\gamma>0,\quad \operatorname{Re}\varrho>0,\quad \operatorname{Re}(\gamma+\varrho-\alpha-\beta)>0]\quad \text{ET II 400(8)}$$

7.524

1. $\displaystyle\int_0^\infty e^{-\lambda x}\,F\left(\alpha,\beta;\tfrac{1}{2};-x^2\right)dx = \lambda^{\alpha+\beta-1}\,S_{1-\alpha-\beta,\alpha-\beta}(\lambda)$

$$[\operatorname{Re}\lambda>0]\qquad\qquad \text{ET II 401(13)}$$

2. $\displaystyle\int_0^\infty e^{-st}\,{}_pF_q\left(a_1,\ldots,a_p;b_1,\ldots,b_q;t^2\right)dx = s^{-1}\,{}_{p+2}F_q\left(a_1,\ldots,a_p,1,\tfrac{1}{2};b_1,\ldots,b_q;\tfrac{4}{s^2}\right)$

$$[p<q]\qquad\qquad \text{MO 176}$$

3. $\displaystyle\int_0^\infty e^{-st}\,{}_0F_q\left(\tfrac{1}{q},\tfrac{2}{q},\ldots,\tfrac{q-1}{q},1;\tfrac{t^q}{q^q}\right)dt = s^{-1}\exp\left(s^{-q}\right)$

$$\text{MO 176}$$

7.525

1. $\displaystyle\int_0^\infty x^{\sigma-1}e^{-\mu x}\,{}_mF_n\left(a_1,\ldots,a_m;\varrho_1,\ldots,\varrho_n;(\lambda x)^k\right)dx$

$$= \Gamma(\sigma)\mu^{-\sigma}\,{}_{m+k}F_n\left(a_1,\ldots,a_m,\frac{\sigma}{k},\frac{\sigma+1}{k},\ldots,\frac{\sigma+k-1}{k};\varrho_1,\ldots,\varrho_n;\left(\frac{k\lambda}{\mu}\right)^k\right)$$

$$\left[m+k\le n+1,\quad \operatorname{Re}\sigma>0;\quad \operatorname{Re}\mu>0,\text{ if }m+k\le n;\right.$$

$$\left.\operatorname{Re}\left(\mu+k\lambda e^{\frac{2\pi i}{k}}\right)>0;\quad r=0,1,\ldots,k-1\text{ for }m+k=n+1\right]$$

$$\text{ET I 220(19)}$$

2. $\displaystyle\int_0^\infty xe^{-\lambda x}\,F\left(\alpha,\beta;\tfrac{3}{2};-x^2\right)dx = \lambda^{\alpha+\beta-2}\,S_{1-\alpha-\beta,\alpha-\beta}(\lambda)$

$$[\operatorname{Re}\lambda>0]\qquad\qquad \text{ET II 401(14)}$$

7.526

1. $\displaystyle\int_{\gamma-i\infty}^{\gamma+i\infty} e^{st}s^{-b}\,F\left(a,b;a+b-c+1;1-\tfrac{1}{s}\right)dx = 2\pi i\frac{\Gamma(a+b-c+1)}{\Gamma(b)\,\Gamma(b-c+1)}t^{b-1}\,\Psi(a;c;t)$

$$\left[\operatorname{Re}b>0,\quad \operatorname{Re}(b-c)>-1,\quad \gamma>\tfrac{1}{2}\right]$$

$$\text{EH I 273(12)}$$

2. $\displaystyle\int_0^\infty e^{-t}t^{\gamma-1}(x+t)^{-\alpha}(y+t)^{-\alpha'}\,F\left[a,a';\gamma;\frac{t(x+y+t)}{(x+t)(y+t)}\right]dt = \Gamma(\gamma)\Psi(a,c;x)\Psi(a',c;y),$

$$\gamma=a+a'-c+1\qquad [\operatorname{Re}\gamma>0,\quad xy\ne0]\quad \text{EH I 287(21)}$$

3. $$\int_0^\infty x^{\gamma-1}(x+y)^{-\alpha}(x+z)^{-\beta}e^{-x}\,F\left[\alpha,\beta;\gamma;\frac{x(x+y+z)}{(x+y)(x+z)}\right]\,dx$$

$$= \Gamma(\gamma)(zy)^{-\frac{1}{2}-\mu}e^{\frac{y+z}{2}}\,W_{\nu,\mu}(y)\,W_{\lambda,\mu}(z)$$

$$2\nu = 1 - \alpha + \beta - \gamma;\quad 2\lambda = 1 + \alpha - \beta - \gamma;\quad 2\mu = \alpha + \beta - \gamma$$

$$[\operatorname{Re}\gamma > 0,\quad |\arg y| < \pi,\quad |\arg z| < \pi]$$

ET II 401(15)

7.527

1. $$\int_0^\infty \left(1 - e^{-x}\right)^{\lambda-1}e^{-\mu x}\,F\left(\alpha,\beta;\gamma;\delta e^{-x}\right)\,dx = \mathrm{B}(\mu,\lambda)\,{}_3F_2(\alpha,\beta,\mu;\gamma,\mu+\lambda;\delta)$$

$$[\operatorname{Re}\lambda > 0,\quad \operatorname{Re}\mu > 0,\quad |\arg(1-\delta)| < \pi]\quad \text{ET I 213(9)}$$

2. $$\int_0^\infty \left(1 - e^{-x}\right)^{\mu}e^{-\alpha x}\,F\left(-n,\mu+\beta+n;\beta;e^{-x}\right)\,dx = \frac{\mathrm{B}(\alpha,\mu+n+1)\,\mathrm{B}(\alpha,\beta+n-\alpha)}{\mathrm{B}(\alpha,\beta-\alpha)}$$

$$[\operatorname{Re}\alpha > 0,\quad \operatorname{Re}\mu > -1]\quad \text{ET I 213(10)}$$

3. $$\int_0^\infty \left(1 - e^{-x}\right)^{\gamma-1}e^{-\mu x}\,F\left(\alpha,\beta;\gamma;1-e^{-x}\right)\,dx = \frac{\Gamma(\mu)\,\Gamma(\gamma-\alpha-\beta+\mu)\,\Gamma(\gamma)}{\Gamma(\gamma-\alpha+\mu)\,\Gamma(\gamma-\beta+\mu)}$$

$$[\operatorname{Re}\mu > 0,\quad \operatorname{Re}\mu > \operatorname{Re}(\alpha+\beta-\gamma),\quad \operatorname{Re}\gamma > 0]\quad \text{ET I 213(11)}$$

4. $$\int_0^\infty \left(1 - e^{-x}\right)^{\gamma-1}e^{-\mu x}\,F\left[\alpha,\beta;\gamma;\delta\left(1-e^{-x}\right)\right]\,dx = \mathrm{B}(\mu,\gamma)\,F(\alpha,\beta;\mu+\gamma;\delta)$$

$$[\operatorname{Re}\mu > 0,\quad \operatorname{Re}\gamma > 0,\quad |\arg(1-\delta)| < \pi]\quad \text{ET I 213(12)}$$

7.53 Hypergeometric and trigonometric functions

7.531

1. $$\int_0^\infty x\sin\mu x\,F\left(\alpha,\beta;\frac{3}{2};-c^2x^2\right)\,dx = 2^{-\alpha-\beta+1}\pi c^{-\alpha-\beta}\mu^{\alpha+\beta-2}\frac{K_{\alpha-\beta}\left(\frac{\mu}{c}\right)}{\Gamma(\alpha)\,\Gamma(\beta)}$$

$$\left[\mu > 0,\quad \operatorname{Re}\alpha > \tfrac{1}{2},\quad \operatorname{Re}\beta > \tfrac{1}{2}\right]$$

ET I 115(6)

2. $$\int_0^\infty \cos\mu x\,F\left(\alpha,\beta;\frac{1}{2};-c^2x^2\right)\,dx = 2^{-\alpha-\beta+1}\pi c^{-\alpha-\beta}\mu^{\alpha+\beta-1}\frac{K_{\alpha-\beta}\left(\frac{\mu}{c}\right)}{\Gamma(\alpha)\,\Gamma(\beta)}$$

$$[\mu > 0,\quad \operatorname{Re}\alpha > 0,\quad \operatorname{Re}\beta > 0,\quad c > 0]$$

ET I 61(9)

7.54 Combinations of hypergeometric and Bessel functions

7.541 $$\int_0^\infty x^{\alpha+\beta-2\nu-1}(x+1)^{-\nu}e^{xz}\,K_\nu[(x+1)z]\,F\left(\alpha,\beta;\alpha+\beta-2\nu;-x\right)\,dx$$

$$= \pi^{-\frac{1}{2}}\cos(\nu\pi)\,\Gamma\left(\tfrac{1}{2}-\alpha+\nu\right)\,\Gamma\left(\tfrac{1}{2}-\beta+\nu\right)\,\Gamma(\gamma)(2z)^{-\frac{1}{2}-\frac{1}{2}\gamma}\,W_{\frac{1}{2}\gamma,\frac{1}{2}(\beta-\alpha)}(2z)$$

$$\gamma = \alpha+\beta-2\nu\qquad \left[\operatorname{Re}(\alpha+\beta-2\nu) > 0,\quad \operatorname{Re}\left(\tfrac{1}{2}-\alpha+\nu\right) > 0,\quad \operatorname{Re}\left(\tfrac{1}{2}-\beta+\nu\right) > 0,\quad |\arg z| < \tfrac{3}{2}\pi\right]$$

ET II 401(16)

7.542

1.
$$\int_0^\infty x^{\sigma-1} \, {}_pF_{p-1}\left(a_1,\ldots,a_p; b_1,\ldots,b_{p-1}; -\lambda x^2\right) Y_\nu(xy)\, dx$$

$$= \frac{\Gamma(b_1)\ldots\Gamma(b_{p-1})}{2\lambda^{\frac{1}{2}\sigma}\,\Gamma(a_1)\ldots\Gamma(a_p)} G^{p+2,1}_{p+2,p+3}\left(\frac{y^2}{4\lambda}\,\middle|\,\begin{matrix}b_0^*,\ldots,b_{p+1}^*,\\ h,k,a_1^*,\ldots,a_p^*,l\end{matrix}\right)$$

$$a_j^* = a_j - \frac{\sigma}{2},\quad j=1,\ldots,p;\quad b_0^* = 1 - \frac{\sigma}{2};\quad b_j^* = b_j - \frac{\sigma}{2},$$

$$j=1,\ldots,p-1; h = \frac{\nu}{2},\quad k = -\frac{\nu}{2},\quad l = -\frac{1+\nu}{2}$$

$$\left[|\arg\lambda| < \pi,\quad \operatorname{Re}\sigma > |\operatorname{Re}\nu|,\quad \operatorname{Re} a_j > \tfrac{1}{2}\operatorname{Re}\sigma - \tfrac{3}{4},\quad y > 0\right]$$

ET II 118(53)

2.
$$\int_0^\infty x^{\sigma-1} \, {}_pF_p\left(a_1,\ldots,a_p; b_1,\ldots,b_p; -\lambda x^2\right) Y_\nu(xy)\, dx$$

$$= \frac{\Gamma(b_1)\ldots\Gamma(b_p)}{2\lambda^{\frac{1}{2}\sigma}\,\Gamma(a_1)\ldots\Gamma(a_p)} G^{p+2,1}_{p+2,p+3}\left(\frac{y^2}{4\lambda}\,\middle|\,\begin{matrix}b_0^*,\ldots,b_p^*,l\\ h,k,a_1^*,\ldots,a_p^*,l\end{matrix}\right)$$

$$b_0^* = 1 - \frac{\sigma}{2};\quad a_j^* = a_j - \frac{\sigma}{2},\quad b_{j*} = b_j - \frac{\sigma}{2};\quad j=1,\ldots,p;\quad h = \frac{\nu}{2},\quad k = -\frac{\nu}{2},\quad l = -\frac{1+\nu}{2}$$

$$\left[\operatorname{Re}\lambda > 0,\quad \operatorname{Re}\sigma > |\operatorname{Re}\nu|,\quad \operatorname{Re} a_j > \tfrac{1}{2}\operatorname{Re}\sigma - \tfrac{3}{4},\quad y > 0\right]$$

ET II 119(54)

3.
$$\int_0^\infty x^{\sigma-1} \, {}_pF_q\left(a_1,\ldots,a_p; b_1,\ldots,b_q; -\lambda x^2\right) Y_\nu(xy)\, dx$$

$$= -\pi^{-1}2^{\sigma-1}y^{-\sigma}\cos\left[\frac{\pi}{2}(\sigma-\nu)\right]\Gamma\left(\frac{\sigma+\nu}{2}\right)\Gamma\left(\frac{\sigma-\nu}{2}\right)$$

$$\times\; {}_{p+2}F_q\left(a_1,\ldots,a_p,\frac{\sigma+\nu}{2},\frac{\sigma-\nu}{2}; b_1,\ldots,b_q; -\frac{4\lambda}{y^2}\right)$$

$$\left[y > 0,\quad p \le q-1,\quad \operatorname{Re}\sigma > |\operatorname{Re}\nu|\right]\quad \text{ET II 119(55)}$$

4.
$$\int_0^\infty x^{\sigma-1} \, {}_pF_q\left(a_1,\ldots,a_p; b_1,\ldots,b_q; -\lambda x^2\right) K_\nu(xy)\, dx$$

$$= 2^{\sigma-2}y^{-\sigma}\,\Gamma\left(\frac{\sigma+\nu}{2}\right)\Gamma\left(\frac{\sigma-\nu}{2}\right)\; {}_{p+2}F_q\left(a_1,\ldots,a_p,\frac{\sigma+\nu}{2},\frac{\sigma-\nu}{2}; b_1,\ldots,b_q; \frac{4\lambda}{y^2}\right)$$

$$\left[\operatorname{Re} y > 0,\quad p \le q-1,\quad \operatorname{Re}\sigma > |\operatorname{Re}\nu|\right]\quad \text{ET II 153(88)}$$

5.
$$\int_0^\infty x^{2\varrho} \, {}_pF_p\left(a_1,\ldots,a_p; b_1,\ldots,b_p; -\lambda x^2\right) J_\nu(xy)\, dx$$

$$= \frac{2^{2\varrho}\,\Gamma(b_1)\ldots\Gamma(b_p)}{y^{2\varrho+1}\,\Gamma(a_1)\ldots\Gamma(a_p)} G^{p+1,1}_{p+1,p+2}\left(\frac{y^2}{4\lambda}\,\middle|\,\begin{matrix}1,&b_1,\ldots,b_p\\ h,&a_1,\ldots,a_p,&k\end{matrix}\right)$$

$$h = \frac{1}{2} + \varrho + \frac{1}{2}\nu,\qquad k = \frac{1}{2} + \varrho - \frac{1}{2}\nu$$

$$\left[y > 0,\quad \operatorname{Re}\lambda > 0,\quad -1 - \operatorname{Re}\nu < 2\operatorname{Re}\varrho < \tfrac{1}{2} + 2\operatorname{Re} a_r,\quad r=1,\ldots,p\right]\quad \text{ET II 91(18)}$$

6. $\int_0^\infty x^{2\varrho}\,_{m+1}F_m\left(a_1,\ldots,a_{m+1};b_1,\ldots,b_m;-\lambda^2 x^2\right)J_\nu(xy)\,dx$

$$= \frac{2^{2\varrho}\,\Gamma\left(b_1\right)\ldots\Gamma\left(b_m\right)y^{-2\varrho-1}}{\Gamma\left(a_1\right)\ldots\Gamma\left(a_{m+1}\right)}\,G_{m+1,m+3}^{m+2,1}\left(\frac{y^2}{4\lambda^2}\,\middle|\,\begin{array}{llll}1, & b_1,\ldots,b_m, & \\ h, & a_1,\ldots,a_{m+1}, & k\end{array}\right)$$
$$h=\tfrac{1}{2}+\varrho+\tfrac{1}{2}\nu, \qquad k=\tfrac{1}{2}+\varrho-\tfrac{1}{2}\nu,$$

$$\left[y>0, \quad \operatorname{Re}\lambda>0, \quad \operatorname{Re}(2\varrho+\nu)>-1, \quad \operatorname{Re}\left(\varrho-a_r\right)<\tfrac{1}{4}; \quad r=1,\ldots,m+1\right] \quad \textbf{ET II 91(19)}$$

7. $\int_0^\infty x^\delta\,F\left(\alpha,\beta;\gamma;-\lambda^2 x^2\right)J_\nu(xy)\,dx$

$$= \frac{2^\delta\,\Gamma(\gamma)}{\Gamma(\alpha)\,\Gamma(\beta)}y^{-\delta-1}\,G_{24}^{22}\left(\frac{y^2}{4\lambda^2}\,\middle|\,\begin{array}{cccc}1-\alpha, & 1-\beta & & \\ \frac{1+\delta+\nu}{2}, & 0, & 1-\gamma, & \frac{1+\delta-\nu}{2}\end{array}\right)$$
$$\left[y>0, \quad \operatorname{Re}\lambda>0, \quad -1-\operatorname{Re}\nu-2\min\left(\operatorname{Re}\alpha, \quad \operatorname{Re}\beta\right)<\operatorname{Re}\delta<-\tfrac{1}{2}\right] \quad \textbf{ET II 82(9)}$$

8. $\int_0^\infty x^\delta\,F\left(\alpha,\beta;\gamma;-\lambda^2 x^2\right)J_\nu(xy)\,dx = \frac{2^\delta y^{-\delta-1}\,\Gamma(\gamma)}{\Gamma(\alpha)\,\Gamma(\beta)}\,G_{24}^{31}\left(\frac{y^2}{4\lambda^2}\,\middle|\,\begin{array}{cccc}1, & \gamma & & \\ \frac{1+\delta+\nu}{2}, & \alpha, & \beta, & \frac{1+\delta-\nu}{2}\end{array}\right)$

$$\left[y>0, \quad \operatorname{Re}\lambda>0, \quad -\operatorname{Re}\nu-1<\operatorname{Re}\delta<2\max\left(\operatorname{Re}\alpha, \quad \operatorname{Re}\beta\right)-\tfrac{1}{2}\right] \quad \textbf{ET II 81(6)}$$

9. $\int_0^\infty x^{\nu+1}\,F\left(\alpha,\beta;\gamma;-\lambda^2 x^2\right)J_\nu(xy)\,dx = \frac{2^{\nu+1}\,\Gamma(\gamma)}{\Gamma(\alpha)\,\Gamma(\beta)}y^{-\nu-2}\,G_{13}^{30}\left(\frac{y^2}{4\lambda^2}\,\middle|\,\begin{array}{ccc}\gamma & & \\ \nu+1, & \alpha, & \beta\end{array}\right)$

$$\left[y>0, \quad \operatorname{Re}\lambda>0, \quad -1<\operatorname{Re}\nu<2\max\left(\operatorname{Re}\alpha, \quad \operatorname{Re}\beta\right)-\tfrac{3}{2}\right] \quad \textbf{ET II 81(5)}$$

10. $\int_0^\infty x^{\nu+1}\,F\left(\alpha,\beta;\nu+1;-\lambda^2 x^2\right)J_\nu(xy)\,dx = \frac{2^{\nu-\alpha-\beta+2}\,\Gamma(\nu+1)}{\lambda^{\alpha+\beta}\,\Gamma(\alpha)\,\Gamma(\beta)}y^{\alpha+\beta-\nu-2}\,K_{\alpha-\beta}\left(\frac{y}{\lambda}\right)$

$$\left[y>0, \quad \operatorname{Re}\lambda>0, \quad -1<\operatorname{Re}\nu<2\max\left(\operatorname{Re}\alpha, \quad \operatorname{Re}\beta\right)-\tfrac{3}{2}\right] \quad \textbf{ET II 81(3)}$$

11. $\int_0^\infty x^{\nu+1}\,F\left(\alpha,\beta;\nu+1;-\lambda^2 x^2\right)K_\nu(xy)\,dx = 2^{\nu+1}\lambda^{-\alpha-\beta}y^{\alpha+\beta-\nu-2}\,\Gamma(\nu+1)\,S_{1-\alpha-\beta,\alpha-\beta}\left(\frac{y}{\lambda}\right)$

$$\left[\operatorname{Re}y>0, \quad \operatorname{Re}\lambda>0, \quad \operatorname{Re}\nu>-1\right]$$
$$\textbf{ET II 152(86)}$$

12. $\int_0^\infty x^{\nu+1}F\left(\alpha,\beta;\frac{\beta+\nu}{2}+1;-\lambda^2 x^2\right)J_\nu(xy)\,dx = \frac{\Gamma\left(\frac{\beta+\nu+2}{2}\right)y^{\beta-1}\lambda^{-\nu-\beta-1}}{\pi^{\frac{1}{2}}\,\Gamma(\alpha)\,\Gamma(\beta)2^{\beta-1}}K_{\frac{1}{2}(\nu-\beta+1)}\left(\frac{y}{2\lambda}\right)^2$

$$\left[y>0, \quad -1<\operatorname{Re}\nu<\left(2\max\left(\operatorname{Re}\alpha,\operatorname{Re}\beta\right)-\tfrac{3}{2}\right)\right] \quad \textbf{ET II 81(4)}$$

13. $\int_0^\infty x^{\sigma+\frac{1}{2}}\,F\left(\alpha,\beta;\gamma;-\lambda^2 x^2\right)Y_\nu(xy)\,dx = \frac{\lambda^{-\sigma-1}y^{-\frac{1}{2}}\,\Gamma(\gamma)}{\sqrt{2}\,\Gamma(\alpha)\,\Gamma(\beta)}\,G_{35}^{41}\left(\frac{y^2}{4\lambda^2}\,\middle|\,\begin{array}{ll}1-p,\gamma-p,l \\ h, \quad k,\alpha-p,\beta-p,l\end{array}\right)$

$$h=\tfrac{1}{4}+\tfrac{1}{2}\nu, \quad k=\tfrac{1}{4}-\tfrac{1}{2}\nu, \quad l=-\tfrac{1}{4}-\tfrac{1}{2}\nu, \quad p=\tfrac{1}{2}+\tfrac{1}{2}\sigma$$

$$\left[y>0, \quad \operatorname{Re}\lambda>0, \quad \operatorname{Re}\sigma>|\operatorname{Re}\nu|-\tfrac{3}{2}, \quad \operatorname{Re}\sigma<2\operatorname{Re}\alpha, \quad \operatorname{Re}\sigma<2\operatorname{Re}\beta\right] \quad \textbf{ET II 118(52)}$$

14. $\displaystyle\int_0^\infty x^{\nu+2} F\left(\frac{1}{2}, \frac{1}{2}-\nu; \frac{3}{2}; -\lambda^2 x^2\right) Y_\nu(xy)\, dx = \frac{2^\nu y^{-\nu-1}}{\pi^{\frac{1}{2}} \lambda^2 \Gamma\left(\frac{1}{2}-\nu\right)} K_\nu\left(\frac{y}{2\lambda}\right) K_{\nu+1}\left(\frac{y}{2\lambda}\right)$

$$\left[y > 0, \quad \operatorname{Re}\lambda > 0, \quad -\tfrac{3}{2} < \operatorname{Re}\nu < -\tfrac{1}{2}\right]$$
ET II 117(49)

15. $\displaystyle\int_0^\infty x^{\nu+2} F\left(1, 2\nu + \frac{3}{2}; \nu+2; -\lambda^2 x^2\right) Y_\nu(xy)\, dx = \pi^{-\frac{1}{2}} 2^{-\nu} \lambda^{-2\nu-3} \frac{\Gamma(\nu+2)}{\Gamma\left(2\nu+\frac{3}{2}\right)} \left[K_\nu\left(\frac{y}{2\lambda}\right)\right]^2$

$$\left[y > 0, \quad \operatorname{Re}\lambda > 0, \quad -\tfrac{1}{2} < \operatorname{Re}\nu < \tfrac{1}{2}\right]$$
ET II 117(50)

16. $\displaystyle\int_0^\infty x^{\nu+2} F\left(1, \mu + \nu + \frac{3}{2}; \frac{3}{2}; -\lambda^2 x^2\right) Y_\nu(xy)\, dx = \frac{\pi^{\frac{1}{2}} 2^{-\mu-\nu-1} \lambda^{-\mu-2\nu-3} y^{\mu+\nu}}{\Gamma\left(\mu+\nu+\frac{3}{2}\right)} K_\mu\left(\frac{y}{\lambda}\right)$

$$\left[y > 0, \quad \operatorname{Re}\lambda > 0, \quad -\tfrac{3}{2} < \operatorname{Re}\nu < \tfrac{1}{2}, \quad \operatorname{Re}(2\mu+\nu) > -\tfrac{3}{2}\right] \quad \text{ET II 118(51)}$$

17. $\displaystyle\int_0^\infty x^{2\alpha+\nu} F\left(\alpha - \nu - \frac{1}{2}, \alpha; 2\alpha; -\lambda^2 x^2\right) J_\nu(xy)\, dx$

$$= \frac{i\,\Gamma\left(\frac{1}{2}+\alpha\right) \Gamma\left(\frac{1}{2}+\alpha+\nu\right)}{\pi 2^{1-\nu-2\alpha} \lambda^{2\alpha-1} y^{\nu+2}} W_{\frac{1}{2}-\alpha, -\frac{1}{2}-\nu}\left(\frac{y}{\lambda}\right) \left[W_{\frac{1}{2}-\alpha, -\frac{1}{2}-\nu}\left(e^{-i\pi}\frac{y}{\lambda}\right) - W_{\frac{1}{2}-\alpha, -\frac{1}{2}-\nu}\left(e^{i\pi}\frac{y}{\lambda}\right)\right]$$

$$\left[y > 0, \quad \operatorname{Re}\lambda > 0, \quad \operatorname{Re}\nu < -\tfrac{1}{2}, \quad \operatorname{Re}(\alpha+\nu) > -\tfrac{1}{2}\right] \quad \text{ET II 80(1)}$$

18. $\displaystyle\int_0^\infty x^{2\alpha-\nu} F\left(\nu + \alpha - \frac{1}{2}, \alpha; 2\alpha; -\lambda^2 x^2\right) J_\nu(xy)\, dx$

$$= \frac{2^{2\alpha-\nu} \Gamma\left(\frac{1}{2}+\alpha\right) y^{\nu-2}}{\lambda^{2\alpha-1} \Gamma(2\nu)} M_{\alpha-\frac{1}{2}, \nu-\frac{1}{2}}\left(\frac{y}{\lambda}\right) W_{\frac{1}{2}-\alpha, \nu-\frac{1}{2}}\left(\frac{y}{\lambda}\right)$$
ET II 80(2)

7.543

1. $\displaystyle\int_0^\infty x^{-2\alpha-1} F\left(\frac{1}{2}+\alpha, 1+\alpha; 1+2\alpha; -\frac{4\lambda^2}{x^2}\right) J_\nu(xy)\, dx = \lambda^{-2\alpha} I_{\frac{1}{2}\nu+\alpha}(\lambda y)\, K_{\frac{1}{2}\nu-\alpha}(\lambda y)$

$$\left[y > 0, \quad \operatorname{Re}\lambda > 0, \quad \operatorname{Re}\nu > -1, \quad \operatorname{Re}\alpha > -\tfrac{1}{2}\right] \quad \text{ET II 81(7)}$$

2. $\displaystyle\int_0^\infty x^{\nu+1-4\alpha} F\left(\alpha, \alpha + \frac{1}{2}; \nu+1; -\frac{\lambda^2}{x^2}\right) J_\nu(xy)\, dx$

$$= \frac{\Gamma(\nu)}{\Gamma(2\alpha)} 2^\nu \lambda^{1-2\alpha} y^{2\alpha-\nu-1} I_\nu\left(\frac{1}{2}\lambda y\right) K_{2\alpha-\nu-1}\left(\frac{1}{2}\lambda y\right)$$

$$\left[y > 0, \quad \operatorname{Re}\lambda > 0, \quad \operatorname{Re}\alpha - 1 < \operatorname{Re}\nu < 4\operatorname{Re}\alpha - \tfrac{3}{2}\right] \quad \text{ET II 81(8)}$$

7.544 $\displaystyle\int_0^\infty x^{\nu+1}(1+x)^{-2\alpha} F\left[\alpha, \nu + \frac{1}{2}; 2\nu+1; \frac{4x}{(1+x)^2}\right] J_\nu(xy)\, dx$

$$= \frac{\Gamma(\nu+1)\,\Gamma(\nu-\alpha+1)}{\Gamma(\alpha)} 2^{2\nu-2\alpha+1} y^{2(\alpha-\nu-1)} J_\nu(y)$$

$$\left[y > 0, \quad -1 < \operatorname{Re}\nu < 2\operatorname{Re}\alpha - \tfrac{3}{2}\right] \quad \text{ET II 82(10)}$$

7.6 Confluent Hypergeometric Functions

7.61 Combinations of confluent hypergeometric functions and powers

7.611

1. $\displaystyle\int_0^\infty x^{-1}\, W_{k,\mu}(x)\, dx = \frac{\pi^{\frac{3}{2}} 2^k \sec(\mu\pi)}{\Gamma\left(\frac{3}{4}-\frac{1}{2}k+\frac{1}{2}\mu\right)\Gamma\left(\frac{3}{4}-\frac{1}{2}k-\frac{1}{2}\mu\right)}$

$$\left[|\mathrm{Re}\,\mu|<\tfrac{1}{2}\right] \qquad\qquad \text{ET II 406(22)}$$

2. $\displaystyle\int_0^\infty x^{-1}\, M_{k,\mu}(x)\, W_{\lambda,\mu}(x)\, dx = \frac{\Gamma(2\mu+1)}{(k-\lambda)\,\Gamma\left(\frac{1}{2}+\mu-\lambda\right)}$

$$\left[\mathrm{Re}\,\mu>-\tfrac{1}{2},\quad \mathrm{Re}(k-\lambda)>0\right]$$
$$\text{BU 116(11), ET II 409(39)}$$

3. $\displaystyle\int_0^\infty x^{-1}\, W_{k,\mu}(x)\, W_{\lambda,\mu}(x)\, dx$

$$= \frac{1}{(k-\lambda)\sin(2\mu\pi)}\left[\frac{1}{\Gamma\left(\frac{1}{2}-k+\mu\right)\Gamma\left(\frac{1}{2}-\lambda-\mu\right)} - \frac{1}{\Gamma\left(\frac{1}{2}-k-\mu\right)\Gamma\left(\frac{1}{2}-\lambda+\mu\right)}\right]$$
$$\left[|\mathrm{Re}\,\mu|<\tfrac{1}{2}\right] \qquad \text{BU 116(12), ET II 409(40)}$$

4. $\displaystyle\int_0^\infty \{W_{\kappa,\mu}(z)\}^2\,\frac{dz}{z} = \frac{\pi}{\sin 2\pi\mu}\,\frac{\psi\left(\frac{1}{2}+\mu-\kappa\right)-\psi\left(\frac{1}{2}-\mu-\kappa\right)}{\Gamma\left(\frac{1}{2}+\mu-\kappa\right)\Gamma\left(\frac{1}{2}-\mu-\kappa\right)}$

$$\left[|\mathrm{Re}\,\mu|<\tfrac{1}{2}\right] \qquad\qquad \text{BU 117(12a)}$$

5. $\displaystyle\int_0^\infty \frac{1}{z}\,[W_{\kappa,0}(z)]^2\, dx = \frac{\psi'\left(\frac{1}{2}-\kappa\right)}{\left[\Gamma\left(\frac{1}{2}-\kappa\right)\right]^2}$

$$\qquad\qquad\qquad\qquad \text{BU 117(12b)}$$

6. $\displaystyle\int_0^\infty x^{\varrho-1}\, W_{k,\mu}(x)\, W_{-k,\mu}(x)\, dx = \frac{\Gamma(\varrho+1)\,\Gamma\left(\frac{1}{2}\varrho+\frac{1}{2}+\mu\right)\Gamma\left(\frac{1}{2}\varrho+\frac{1}{2}-\mu\right)}{2\,\Gamma\left(1+\frac{1}{2}\varrho+k\right)\Gamma\left(1+\frac{1}{2}\varrho-k\right)}$

$$\left[\mathrm{Re}\,\varrho>2|\mathrm{Re}\,\mu|-1\right] \qquad \text{ET II 409(41)}$$

7. $\displaystyle\int_0^\infty x^{\varrho-1}\, W_{k,\mu}(x)\, W_{\lambda,\nu}(x)\, dx$

$$= \frac{\Gamma(1+\mu+\nu+\varrho)\,\Gamma(1-\mu+\nu+\varrho)\,\Gamma(-2\nu)}{\Gamma\left(\frac{1}{2}-\lambda-\nu\right)\Gamma\left(\frac{3}{2}-k+\nu+\varrho\right)} \times$$

$$\times\; _3F_2\left(1+\mu+\nu+\varrho, 1-\mu+\nu+\varrho, \frac{1}{2}-\lambda-\nu; 1+2\nu, \frac{3}{2}-k+\nu+\varrho; 1\right)$$

$$+ \frac{\Gamma(1+\mu-\nu+\varrho)\,\Gamma(1-\mu-\nu+\varrho)\,\Gamma(2\nu)}{\Gamma\left(\frac{1}{2}-\lambda+\nu\right)\Gamma\left(\frac{3}{2}-k-\nu+\varrho\right)}$$

$$\times\; _3F_2\left(1+\mu-\nu+\varrho, 1-\mu-\nu+\varrho, \frac{1}{2}-\lambda-\nu; 1-2\nu, \frac{3}{2}-k-\nu+\varrho; 1\right)$$

$$\left[|\mathrm{Re}\,\mu|+|\mathrm{Re}\,\nu|<\mathrm{Re}\,\varrho+1\right] \quad \text{ET II 410(42)}$$

7.612

1. $\displaystyle\int_0^\infty t^{b-1}\, _1F_1(a;c;-t)\, dt = \frac{\Gamma(b)\,\Gamma(c)\,\Gamma(a-b)}{\Gamma(a)\,\Gamma(c-b)} \qquad \left[0<\mathrm{Re}\,b<\mathrm{Re}\,a\right] \qquad \text{EH I 285(10)}$

2. $\displaystyle\int_0^\infty t^{b-1}\Psi(a,c;t)\,dt = \frac{\Gamma(b)\,\Gamma(a-b)\,\Gamma(b-c+1)}{\Gamma(a)\,\Gamma(a-c+1)}$ $[0 < \operatorname{Re}b < \operatorname{Re}a \quad \operatorname{Re}c < \operatorname{Re}b + 1]$

EH I 285(11)

7.613

1. $\displaystyle\int_0^t x^{\gamma-1}(t-x)^{c-\gamma-1}\,{}_1F_1(a;\gamma;x)\,dx = t^{c-1}\frac{\Gamma(\gamma)\,\Gamma(c-\gamma)}{\Gamma(c)}\,{}_1F_1(a;c;t)$

$[\operatorname{Re}c > \operatorname{Re}\gamma > 0]$

BU 9(16)a, EH I 271(16)

2. $\displaystyle\int_0^t x^{\beta-1}(t-x)^{\gamma-1}\,{}_1F_1(t;\beta;x)\,dx = \frac{\Gamma(\beta)\,\Gamma(\gamma)}{\Gamma(\beta+\gamma)}t^{\beta+\gamma-1}\,{}_1F_1(t;\beta+\gamma;t)$

$[\operatorname{Re}\beta > 0, \quad \operatorname{Re}\gamma > 0]$ ET II 401(1)

3. $\displaystyle\int_0^1 x^{\lambda-1}(1-x)^{2\mu-\lambda}\,{}_1F_1\left(\frac{1}{2}+\mu-\nu;\lambda;xz\right)dx = \mathrm{B}(\lambda,1+2\mu-\lambda)e^{\frac{1}{2}z}z^{-\frac{1}{2}-\mu}\,M_{\nu,\mu}(z)$

$[\operatorname{Re}\lambda > 0, \quad \operatorname{Re}(2\mu-\lambda) > -1]$

BU 14(14)

4. $\displaystyle\int_0^t x^{\beta-1}(t-x)^{\delta-1}\,{}_1F_1(t;\beta;x)\,{}_1F_1(\gamma;\delta;t-x)\,dx = \frac{\Gamma(\beta)\,\Gamma(\delta)}{\Gamma(\beta+\delta)}t^{\beta+\delta-1}\,{}_1F_1(t+\gamma;\beta+\delta;t)$

$[\operatorname{Re}\beta > 0, \quad \operatorname{Re}\delta > 0]$

ET II 402(2), EH I 271(15)

5. $\displaystyle\int_0^t x^{\mu-\frac{1}{2}}(t-x)^{\nu-\frac{1}{2}}\,M_{k,\mu}(x)\,M_{\lambda,\nu}(t-x)\,dx = \frac{\Gamma(2\mu+1)\,\Gamma(2\nu+1)}{\Gamma(2\mu+2\nu+2)}t^{\mu+\nu}\,M_{k+\lambda,\mu+\nu+\frac{1}{2}}(t)$

$\left[\operatorname{Re}\mu > -\dfrac{1}{2}, \quad \operatorname{Re}\nu > -\dfrac{1}{2}\right]$

BU 128(14), ET II 402(7)

6. $\displaystyle\int_0^1 x^{\beta-1}(1-x)^{\sigma-\beta-1}\,{}_1F_1(\alpha;\beta;\lambda x)\,{}_1F_1[\sigma-\alpha;\sigma-\beta;\mu(1-x)]\,dx$

$$= \frac{\Gamma(\beta)\,\Gamma(\sigma-\beta)}{\Gamma(\sigma)}e^{\lambda}\,{}_1F_1(\alpha;\sigma;\mu-\lambda)$$

$[0 < \operatorname{Re}\beta < \operatorname{Re}\sigma]$ ET II 402(3)

7.62–7.63 Combinations of confluent hypergeometric functions and exponentials

7.621

1. $\displaystyle\int_0^\infty e^{-st}t^\alpha\,M_{\mu,\nu}(t)\,dt = \frac{\Gamma\left(\alpha+\nu+\frac{3}{2}\right)}{\left(\frac{1}{2}+s\right)^{\alpha+\nu+\frac{3}{2}}}\,F\left(\alpha+\nu+\frac{3}{2},-\mu+\nu+\frac{1}{2};2\nu+1;\frac{2}{2s+1}\right)$

$\left[\operatorname{Re}\left(\alpha+\mu+\frac{3}{2}\right) > 0, \quad \operatorname{Re}s > \frac{1}{2}\right]$

BU 118(1), MO 176a, EH I 270(12)a

2. $\displaystyle\int_0^\infty e^{-st}t^{\mu-\frac{1}{2}}\,M_{\lambda,\mu}(qt)\,dt = q^{\mu+\frac{1}{2}}\,\Gamma(2\mu+1)\left(s-\tfrac{1}{2}q\right)^{\lambda-\mu-\frac{1}{2}}\left(s+\tfrac{1}{2}q\right)^{-\lambda-\mu-\frac{1}{2}}$

$\left[\operatorname{Re}\mu > -\dfrac{1}{2}, \quad \operatorname{Re}s > \dfrac{|\operatorname{Re}q|}{2}\right]$

BU 119(4c), MO 176a, EH I 271(13)a

3. $\displaystyle\int_0^\infty e^{-st} t^\alpha \, W_{\lambda,\mu}(qt) \, dt = \frac{\Gamma\left(\alpha+\mu+\frac{3}{2}\right)\Gamma\left(\alpha-\mu+\frac{3}{2}\right) q^{\mu+\frac{1}{2}}}{\Gamma(\alpha-\lambda+2)} \left(s+\frac{1}{2}q\right)^{-\alpha-\mu-\frac{3}{2}}$

$$\times F\left(\alpha+\mu+\frac{3}{2}, \mu-\lambda+\frac{1}{2}; \alpha-\lambda+2; \frac{2s-q}{2s+q}\right)$$

$$\left[\operatorname{Re}\left(\alpha\pm\mu+\frac{3}{2}\right) > 0, \quad \operatorname{Re} s > -\frac{q}{2}, \quad q > 0\right] \qquad \text{EH I 271(14)a, BU 121(6), MO 176}$$

4. $\displaystyle\int_0^\infty e^{-st} t^{b-1} \, {}_1F_1(a; c; kt) \, dt = \Gamma(b) s^{-b} \, F\left(a, b; c; ks^{-1}\right) \qquad\qquad\qquad [|s| > |k|]$

$$= \Gamma(b)(s-k)^{-b} \, F\left(c-a, b; c; \frac{k}{k-s}\right) \qquad [|s-k > |k||]$$

$$[\operatorname{Re} b > 0, \quad \operatorname{Re} s > \max(0, \operatorname{Re} k)] \qquad \text{EH I 269(5)}$$

5. $\displaystyle\int_0^\infty t^{c-1} \, {}_1F_1(a; c; t) e^{-st} \, dt = \Gamma(c) s^{-c} \left(1 - s^{-1}\right)^{-a} \qquad [\operatorname{Re} c > 0, \quad \operatorname{Re} s > 1] \qquad \text{EH I 270(6)}$

6. $\displaystyle\int_0^\infty t^{b-1} \Psi(a, c; t) \, e^{-st} \, dt = \frac{\Gamma(b)\,\Gamma(b-c+1)}{\Gamma(a+b-c+1)} \, F(b, b-c+1; a+b-c+1; 1-s)$

$$[\operatorname{Re} b > 0, \quad \operatorname{Re} c < \operatorname{Re} b + 1, \quad |1-s| < 1]$$

$$= \frac{\Gamma(b)\,\Gamma(b-c+1)}{\Gamma(a+b-c+1)} s^{-b} \, F\left(a, b; a+b-c+1; 1-s^{-1}\right)$$

$$\left[\operatorname{Re} s > \frac{1}{2}\right]$$

$$\text{EH I 270(7)}$$

7. $\displaystyle\int_0^\infty e^{-\frac{b}{2}x} x^{\nu-1} \, M_{\kappa,\mu}(bx) \, dx = \frac{\Gamma(1+2\mu)\,\Gamma(\kappa-\nu)\,\Gamma\left(\frac{1}{2}+\mu+\nu\right)}{\Gamma\left(\frac{1}{2}+\mu+\kappa\right)\Gamma\left(\frac{1}{2}+\mu-\nu\right)} b^\nu$

$$\left[\operatorname{Re}\left(\nu+\tfrac{1}{2}+\mu\right) > 0, \quad \operatorname{Re}(\kappa-\nu) > 0\right]$$
$$\text{BU 119(3)a, ET I 215(11)a}$$

8. $\displaystyle\int_0^\infty e^{-sx} \, M_{\kappa,\mu}(x) \frac{dx}{x} = \frac{2\,\Gamma(1+2\mu) e^{-i\pi\kappa}}{\Gamma\left(\frac{1}{2}+\mu+\kappa\right)} \left(\frac{s-\frac{1}{2}}{s+\frac{1}{2}}\right)^{\frac{\kappa}{2}} Q^\kappa_{\mu-\frac{1}{2}}(2s)$

$$\left[\operatorname{Re}\left(\tfrac{1}{2}+\mu\right) > 0, \quad \operatorname{Re} s > \tfrac{1}{2}\right]$$
$$\text{BU 119(4a)}$$

9. $\displaystyle\int_0^\infty e^{-sx} \, W_{\kappa,\mu}(x) \frac{dx}{x} = \frac{\pi}{\cos\left(\frac{\pi\mu}{2}\right)} \left(\frac{s-\frac{1}{2}}{s+\frac{1}{2}}\right)^{\frac{\kappa}{2}} P^\kappa_{\mu-\frac{1}{2}}(2s)$

$$\left[\operatorname{Re}\left(\tfrac{1}{2}\pm\mu\right) > 0, \quad \operatorname{Re} s > -\tfrac{1}{2}\right]$$
$$\text{BU 121(7)}$$

10. $\displaystyle\int_0^\infty x^{k+2\mu-1} e^{-\frac{3}{2}x} \, W_{k,\mu}(x) \, dx = \frac{\Gamma\left(k+\mu+\frac{1}{2}\right)\Gamma\left[\frac{1}{4}(2k+6\mu+5)\right]}{\left(k+3\mu+\frac{1}{2}\right)\Gamma\left[\frac{1}{4}(2\mu-2k+3)\right]}$

$$\left[\operatorname{Re}(k+\mu) > -\tfrac{1}{2}, \quad \operatorname{Re}(k+3\mu) > -\tfrac{1}{2}\right]$$
$$\text{BU 122(8a), ET II 406(23)}$$

11. $\displaystyle\int_0^\infty e^{-\frac{1}{2}x} x^{\nu-1} \, W_{\kappa,\mu}(x) \, dx = \frac{\Gamma\left(\nu+\frac{1}{2}-\mu\right)\Gamma\left(\nu+\frac{1}{2}+\mu\right)}{\Gamma(\nu-\kappa+1)}$

$$\left[\operatorname{Re}\left(\nu+\tfrac{1}{2}\pm\mu\right) > 0\right] \qquad \text{BU 122(8b)}$$

12. $\displaystyle\int_0^\infty e^{\frac{1}{2}x} x^{\nu-1}\, W_{\kappa,\mu}(x)\, dx = \Gamma\left(-\kappa-\mu\right)\frac{\Gamma\left(\frac{1}{2}+\mu+\nu\right)\Gamma\left(\frac{1}{2}-\mu+\nu\right)}{\Gamma\left(\frac{1}{2}-\mu-\kappa\right)\Gamma\left(\frac{1}{2}+\mu-\kappa\right)}$

$$\left[\operatorname{Re}\left(\nu+\tfrac{1}{2}\pm\mu\right)>0,\quad \operatorname{Re}\left(\kappa+\nu\right)<0\right]$$

BU 122(8c)a

7.622

1. $\displaystyle\int_0^\infty e^{-st} t^{c-1}\, {}_1F_1(a;c;t)\, {}_1F_1(\alpha;c;\lambda t)\, dt$

$$= \Gamma(c)(s-1)^{-a}(s-\lambda)^{-\alpha} s^{a+\alpha-c}\, F\left[a,\alpha;c;\lambda(s-1)^{-1}(s-\lambda)^{-1}\right]$$

$$\left[\operatorname{Re} c > 0,\quad \operatorname{Re} s > \operatorname{Re}\lambda + 1\right]\quad \text{EH I 287(22)}$$

2. $\displaystyle\int_0^\infty e^{-t} t^\varrho\, {}_1F_1(a;c;t)\Psi\left(a';c';\lambda t\right)\, dt$

$$= C\frac{\Gamma(c)\,\Gamma(\beta)}{\Gamma(\gamma)}\lambda^\sigma\, F\left(c-a,\beta;\gamma;1-\lambda^{-1}\right),$$

$$\varrho = c-1,\quad \sigma=-c,\quad \beta=c-c'+1,\quad \gamma=c-a+a'-c'+1,\quad C=\frac{\Gamma\left(a'-a\right)}{\Gamma\left(a'\right)},\ \text{or}$$

$$\varrho = c+c'-2,\quad \sigma=1-c-c',\quad \beta=c+c'-1,\quad \gamma=a'-a+c,\quad C=\frac{\Gamma\left(a'-a-c'+1\right)}{\Gamma\left(a'-c'+1\right)}$$

EH I 287(24)

3. $\displaystyle\int_0^\infty x^{\nu-1} e^{-bx}\, M_{\lambda_1,\mu_1-\frac{1}{2}}(a_1 x)\dots M_{\lambda_n,\mu_n-\frac{1}{2}}(a_n x)\, dx$

$$= a_1^{\mu_1}\dots a_n^{\mu_n} (b+A)^{-\nu-M}\,\Gamma(\nu+M)$$

$$\times F_A\left(\nu+M;\mu_1-\lambda_1,\dots,\mu_n-\lambda_n;2\mu_1,\dots,2\mu_n;\frac{a_1}{b+A},\dots,\frac{a_n}{b+A}\right),$$

$$M=\mu_1+\dots+\mu_n,\qquad A=\tfrac{1}{2}\left(a_1+\dots+a_n\right)$$

$$\left[\operatorname{Re}(\nu+M)>0,\quad \operatorname{Re}\left(b\pm\tfrac{1}{2}a_1\pm\dots+\tfrac{1}{2}a_n\right)>0\right]\quad \text{ET I 216(14)}$$

7.623

1. $\displaystyle\int_0^\infty e^{-x} x^{c+n-1}(x+y)^{-1}\, {}_1F_1(a;c;x)\, dx = (-1)^n\,\Gamma(c)\,\Gamma(1-a)\,y^{c+n-1}\Psi(c-a,c;y)$

$$\left[-\operatorname{Re} c < n < 1-\operatorname{Re} a,\quad n=0,1,2,\dots,\quad |\arg y|<\pi\right]\quad \text{EH I 285(16)}$$

2. $\displaystyle\int_0^t x^{-1}(t-x)^{k-1} e^{\frac{1}{2}(t-x)}\, M_{k,\mu}(x)\, dx = \frac{\Gamma(k)\,\Gamma(2\mu+1)}{\Gamma\left(k+\mu+\frac{1}{2}\right)}\pi^{\frac{1}{2}} t^{k-\frac{1}{2}} l_\mu\left(\frac{1}{2}t\right)$

$$\left[\operatorname{Re} k > 0,\quad \operatorname{Re}\mu > -\tfrac{1}{2}\right]\quad \text{ET II 402(5)}$$

3. $\displaystyle\int_0^t x^{k-1}(t-x)^{\lambda-1} e^{\frac{1}{2}(t-x)}\, M_{k+\lambda,\mu}(x)\, dx = \frac{\Gamma(\lambda)\,\Gamma\left(k+\mu+\frac{1}{2}\right) t^{k+\lambda-1}}{\Gamma\left(k+\lambda+\mu+\frac{1}{2}\right)} M_{k,\mu}(t)$

$$\left[\operatorname{Re}(k+\mu)>-\tfrac{1}{2},\quad \operatorname{Re}\lambda>0\right]$$

ET II 402(6)

4. $\displaystyle\int_0^t x^{-k-\lambda-1}(t-x)^{\lambda-1}e^{\frac{1}{2}x}\,W_{k,\mu}(x)\,dx = \frac{\Gamma(\lambda)\,\Gamma\left(\frac{1}{2}-k-\lambda+\mu\right)\Gamma\left(\frac{1}{2}-k-\lambda-\mu\right)}{t^{k+1}\,\Gamma\left(\frac{1}{2}-k+\mu\right)\Gamma\left(\frac{1}{2}-k-\mu\right)}\,W_{k+\lambda,\mu}(t)$

$$\left[\operatorname{Re}\lambda>0,\quad \operatorname{Re}(k+\lambda)<\tfrac{1}{2}-|\operatorname{Re}\mu|\right]$$

ET II 405(21)

5. $\displaystyle\int_1^\infty (x-1)^{\mu-1}x^{\lambda-\frac{1}{2}}e^{\frac{1}{2}ax}\,W_{k,\lambda}(ax)\,dx = \frac{\Gamma(\mu)\,\Gamma\left(\frac{1}{2}-k-\lambda-\mu\right)}{\Gamma\left(\frac{1}{2}-k-\lambda\right)}a^{-\frac{1}{2}\mu}e^{\frac{1}{2}a}\,W_{k+\frac{1}{2}\mu,\lambda+\frac{1}{2}\mu}(a)$

$$\left[|\arg(a)|<\tfrac{3}{2}\pi,\quad 0<\operatorname{Re}\mu<\tfrac{1}{2}-\operatorname{Re}(k+\lambda)\right]\quad \text{ET II 211(72)a}$$

6. $\displaystyle\int_1^\infty (x-1)^{\mu-1}x^{\lambda-\frac{1}{2}}e^{-\frac{1}{2}ax}\,W_{k,\lambda}(ax)\,dx = a^{-\frac{1}{2}\mu}\Gamma(\mu)e^{-\frac{1}{2}a}\,W_{k-\frac{1}{2}\mu,\lambda-\frac{1}{2}\mu}(a)$

$$\left[\operatorname{Re}\mu>0,\quad \operatorname{Re}a>0\right]\quad \text{ET II 211(74)a}$$

7. $\displaystyle\int_1^\infty (x-1)^{\mu-1}x^{k-\mu-1}e^{-\frac{1}{2}ax}\,W_{k,\lambda}(ax)\,dx = \Gamma(\mu)e^{-\frac{1}{2}a}\,W_{k-\mu,\lambda}(a)$

$$\left[\operatorname{Re}\mu>0,\quad \operatorname{Re}a>0\right]\quad \text{ET II 211(73)a}$$

8. $\displaystyle\int_0^1 (1-x)^{\mu-1}x^{k-\mu-1}e^{-\frac{1}{2}ax}\,W_{k,\lambda}(ax)\,dx$

$$= \Gamma(\mu)e^{-\frac{1}{2}a}\sec[(k-\mu-\lambda)\pi]$$

$$\times\left\{\sin(\mu\pi)\frac{\Gamma\left(k-\mu+\lambda+\frac{1}{2}\right)}{\Gamma(2\lambda+1)}\,M_{k-\mu,\lambda}(a)+\cos[(k-\lambda)\pi]\,W_{k-\mu,\lambda}(a)\right\}$$

$$\left[0<\operatorname{Re}\mu<\operatorname{Re}k-|\operatorname{Re}\lambda|+\tfrac{1}{2}\right]\quad \text{ET II 200(93)a}$$

7.624

1. $\displaystyle\int_0^\infty x^{\varrho-1}\left[x^{\frac{1}{2}}+(a+x)^{\frac{1}{2}}\right]^{2\sigma}e^{-\frac{1}{2}x}\,M_{k,\mu}(x)\,dx$

$$= \frac{-\sigma\,\Gamma(2\mu+1)a^\sigma}{\pi^{\frac{1}{2}}\Gamma\left(\frac{1}{2}+k+\mu\right)}\,G_{34}^{23}\left(a\left|\begin{array}{l}\frac{1}{2},1,1-k+\varrho\\ \frac{1}{2}+\mu+\varrho,-\sigma,\sigma,\frac{1}{2}-\mu+\varrho\end{array}\right.\right)$$

$$\left[|\arg a|<\pi,\quad \operatorname{Re}(\mu+\varrho)>-\tfrac{1}{2},\quad \operatorname{Re}(k-\varrho-\sigma)>0\right]\quad \text{ET II 403(8)}$$

2. $\displaystyle\int_0^\infty x^{\varrho-1}\left[x^{\frac{1}{2}}+(a+x)^{\frac{1}{2}}\right]^{2\sigma}e^{-\frac{1}{2}x}\,W_{k,\mu}(x)\,dx = -\pi^{-\frac{1}{2}}\sigma a^\sigma\,G_{34}^{32}\left(a\left|\begin{array}{l}\frac{1}{2},1,1-k+\varrho\\ \frac{1}{2}+\mu+\varrho,\frac{1}{2}-\mu+\varrho,-\sigma,\sigma\end{array}\right.\right)$

$$\left[|\arg a|<\pi,\quad \operatorname{Re}\varrho>|\operatorname{Re}\mu|-\tfrac{1}{2}\right]\quad \text{ET II 406(24)}$$

3. $\displaystyle\int_0^\infty x^{\varrho-1}\left[x^{\frac{1}{2}}+(a+x)^{\frac{1}{2}}\right]^{2\sigma}e^{-\frac{1}{2}x}\,W_{k,\mu}(x)\,dx$

$$= -\frac{\sigma\pi^{-\frac{1}{2}}a^\sigma}{\Gamma\left(\frac{1}{2}-k+\mu\right)\Gamma\left(\frac{1}{2}-k-\mu\right)}\,G_{34}^{33}\left(a\left|\begin{array}{l}\frac{1}{2},1,1+k+\varrho\\ \frac{1}{2}+\mu+\varrho,\frac{1}{2}-\mu+\varrho,-\sigma,\sigma\end{array}\right.\right)$$

$$\left[|\arg a|<\pi,\quad \operatorname{Re}\varrho>|\operatorname{Re}\mu|-\tfrac{1}{2},\quad \operatorname{Re}(k+\varrho+\sigma)<0\right]\quad \text{ET II 406(25)}$$

4. $\int_0^\infty x^{\varrho-1}(a+x)^{-\frac{1}{2}}\left[x^{\frac{1}{2}}+(a+x)^{\frac{1}{2}}\right]^{2\sigma}e^{-\frac{1}{2}x}\,M_{k,\mu}(x)\,dx$

$$= \frac{\Gamma(2\mu+1)a^\sigma}{\pi^{\frac{1}{2}}\,\Gamma\left(\frac{1}{2}+k+\mu\right)}\,G_{34}^{23}\left(a\left|\begin{array}{c}0,\frac{1}{2},\frac{1}{2}-k-\varrho\\-\sigma,\varrho+\mu,\varrho-\mu,\sigma\end{array}\right.\right)$$

$$\left[|\arg a|<\pi,\quad \mathrm{Re}(\varrho+\mu)>-\tfrac{1}{2},\quad \mathrm{Re}(k-\varrho-\sigma)>-\tfrac{1}{2}\right]\quad \textbf{ET II 403(9)}$$

5. $\int_0^\infty x^{\varrho-1}(a+x)^{-\frac{1}{2}}\left[x^{\frac{1}{2}}+(a+x)^{\frac{1}{2}}\right]^{2\sigma}e^{-\frac{1}{2}x}\,W_{k,\mu}(x)\,dx$

$$= \frac{\pi^{-\frac{1}{2}}a^\sigma}{\Gamma\left(\frac{1}{2}-k+\mu\right)\Gamma\left(\frac{1}{2}-k-\mu\right)}\,G_{34}^{33}\left(a\left|\begin{array}{c}0,\frac{1}{2},\frac{1}{2}+k+\varrho\\-\sigma,\varrho+\mu,\varrho-\mu,\sigma\end{array}\right.\right)$$

$$\left[|\arg a|<\pi,\quad \mathrm{Re}\,\varrho>|\mathrm{Re}\,\mu|-\tfrac{1}{2},\quad \mathrm{Re}(k+\varrho+\sigma)<\tfrac{1}{2}\right]\quad \textbf{ET II 406(26)}$$

6. $\int_0^\infty x^{\varrho-1}(a+x)^{-\frac{1}{2}}\left[x^{\frac{1}{2}}+(a+x)^{\frac{1}{2}}\right]^{2\sigma}e^{-\frac{1}{2}x}\,W_{k,\mu}(x)\,dx = \pi^{-\frac{1}{2}}a^\sigma\,G_{34}^{32}\left(a\left|\begin{array}{c}0,\frac{1}{2},\frac{1}{2}-k+\varrho\\-\sigma,\varrho+\mu,\varrho-\mu,\sigma\end{array}\right.\right)$

$$\left[|\arg a|<\pi,\quad \mathrm{Re}\,\varrho>|\mathrm{Re}\,\mu|-\tfrac{1}{2}\right]\quad \textbf{ET II 406(27)}$$

7.625

1. $\int_0^\infty x^{\varrho-1}\exp\left[-\tfrac{1}{2}(\alpha+\beta)x\right]M_{k,\mu}(\alpha x)\,W_{\lambda,\nu}(\beta x)\,dx$

$$= \frac{\Gamma(1+\mu+\nu+\varrho)\,\Gamma(1+\mu-\nu+\varrho)}{\Gamma\left(\frac{3}{2}-\lambda+\mu+\varrho\right)}\alpha^{\mu+\frac{1}{2}}\beta^{-\mu-\varrho-\frac{1}{2}}$$

$$\times\,{}_3F_2\left(\frac{1}{2}+k+\mu,1+\mu+\nu+\varrho,1+\mu-\nu+\varrho;2\mu+1,\frac{3}{2}-\lambda+\mu+\varrho;-\frac{\alpha}{\beta}\right)$$

$$\left[\mathrm{Re}\,\alpha>0,\quad \mathrm{Re}\,\beta>0,\quad \mathrm{Re}(\varrho+\mu)>|\mathrm{Re}\,\nu|-1\right]\quad \textbf{ET II 410(43)}$$

2. $\int_0^\infty x^{\varrho-1}\exp\left[-\frac{1}{2}(\alpha+\beta)x\right]W_{k,\mu}(\alpha x)\,W_{\lambda,\nu}(\beta x)\,dx$

$$= \beta^{-\varrho}\left[\Gamma\left(\tfrac{1}{2}-k+\mu\right)\Gamma\left(\tfrac{1}{2}-k-\mu\right)\Gamma\left(\tfrac{1}{2}-\lambda+\nu\right)\Gamma\left(\tfrac{1}{2}-\lambda-\nu\right)\right]^{-1}$$

$$\times\,G_{33}^{33}\left(\frac{\beta}{\alpha}\left|\begin{array}{c}\frac{1}{2}+\mu,\frac{1}{2}-\mu,1+\lambda+\varrho\\\frac{1}{2}+\nu+\varrho,\frac{1}{2}-\nu+\varrho,-k\end{array}\right.\right)$$

$$\left[|\mathrm{Re}\,\mu|+|\mathrm{Re}\,\nu|<\mathrm{Re}\,\varrho+1,\quad \mathrm{Re}(k+\lambda+\varrho)<0\right]\quad \textbf{ET II 410(44)a}$$

3. $\int_0^\infty x^{\varrho-1}\exp\left[-\frac{1}{2}(\alpha+\beta)x\right]W_{k,\mu}(\alpha x)\,W_{\lambda,\nu}(\beta x)\,dx = \beta^{-\varrho}\,G_{33}^{22}\left(\frac{\beta}{\alpha}\left|\begin{array}{c}\frac{1}{2}+\mu,\frac{1}{2}-\nu,1-\lambda+\varrho\\\frac{1}{2}+\nu+\varrho,\frac{1}{2}-\nu+\varrho,k\end{array}\right.\right)$

$$\textbf{ET II 411(46)}$$

4. $\int_0^\infty x^{\varrho-1}\exp\left[-\frac{1}{2}(\alpha-\beta)x\right]W_{k,\mu}(\alpha x)\,W_{\lambda,\nu}(\beta x)\,dx$

$$= \beta^{-\varrho}\left[\Gamma\left(\tfrac{1}{2}-\lambda+\nu\right)\Gamma\left(\tfrac{1}{2}-\lambda-\nu\right)\right]^{-1}G_{33}^{23}\left(\frac{\beta}{\alpha}\left|\begin{array}{c}\frac{1}{2}+\mu,\frac{1}{2}-\mu,1+\lambda+\varrho\\\frac{1}{2}+\nu+\varrho,\frac{1}{2}-\nu+\varrho,k\end{array}\right.\right)$$

$$\left[\mathrm{Re}\,\alpha>0,\quad |\mathrm{Re}\,\mu|+|\mathrm{Re}\,\nu|<\mathrm{Re}\,\varrho+1\right]\quad \textbf{ET II 411(45)}$$

7.626

1. $\displaystyle\int_0^1 \left[\frac{k}{x} - \frac{1}{4}(\xi + \eta)\exp\left[-\frac{1}{2}(\xi + \eta)x\right]x\right]x^c \; {}_1F_1(a; c; \xi x)\; {}_1F_1(a; c; \eta x)\, dx$

$$= 0 \qquad\qquad [\xi \neq \eta, \quad \operatorname{Re} c > 0]$$

$$= \frac{a}{\xi}e^{-\xi}[\,{}_1F_1(a + 1; c; \xi)]^2 \qquad [\xi = \eta, \quad \operatorname{Re} c > 0]$$

[where ξ and η are any two zeros of the function $\,{}_1F_1(a; c; x)$]   **EH I 285**

2. $\displaystyle\int_1^\infty \left[\frac{k}{x} - \frac{1}{4}(\xi + \eta)\right]e^{-\frac{1}{2}(\xi+\eta)x}x^c\Psi(a, c; \xi x)\,\Psi(a, c; \eta x)\, dx = 0 \qquad\qquad [\xi \neq \eta]\,;$

$$= -\xi^{-1}e^{-\xi}[\Psi(a - 1, c; \xi)]^2 \qquad [\xi = \eta]$$

[where ξ and η are any two zeros of the function $\Psi(a, c; x)$]   **EH I 286**

7.627

1. $\displaystyle\int_0^\infty x^{2\lambda-1}(a + x)^{-\mu-\frac{1}{2}}e^{\frac{1}{2}x}\, W_{k,\mu}(a + x)\, dx = \frac{\Gamma(2\lambda)\,\Gamma\left(\frac{1}{2} - k + \mu - 2\lambda\right)}{\Gamma\left(\frac{1}{2} - k + \mu\right)}a^{\lambda-\mu-\frac{1}{2}}\, W_{k+\lambda,\mu-\lambda}(a)$

$$\left[|\arg a| < \pi, \quad 0 < 2\operatorname{Re}\lambda < \frac{1}{2} - \operatorname{Re}(k + \mu)\right] \quad \textbf{ET II 411(50)}$$

2. $\displaystyle\int_0^\infty x^{2\lambda-1}(a + x)^{-\mu-\frac{1}{2}}e^{-\frac{1}{2}x}\, M_{k,\mu}^{-\frac{1}{2}x}(a + x)\, dx$

$$= \frac{\Gamma(2\lambda)\,\Gamma(2\mu + 1)\,\Gamma\left(k + \mu - 2\lambda + \frac{1}{2}\right)}{\Gamma\left(k + \mu + \frac{1}{2}\right)\Gamma(1 - 2\lambda + 2\mu)}a^{\lambda-\mu-\frac{1}{2}}\, M_{k-\lambda,\mu-\lambda}(a)$$
$$\left[\operatorname{Re}\lambda > 0, \quad \operatorname{Re}(k + \mu - 2\lambda) > -\tfrac{1}{2}\right] \quad \textbf{ET II 405(20)}$$

3. $\displaystyle\int_0^\infty x^{2\lambda-1}(a + x)^{-\mu-\frac{1}{2}}e^{-\frac{1}{2}x}\, W_{k,\mu}(a + x)\, dx = \Gamma(2\lambda)a^{\lambda-\mu-\frac{1}{2}}\, W_{k-\lambda,\mu-\lambda}(a)$

$$[|\arg a| < \pi, \quad \operatorname{Re}\lambda > 0] \quad \textbf{ET II 411(47)}$$

4. $\displaystyle\int_0^\infty x^{\lambda-1}(a + x)^{k-\lambda-1}e^{-\frac{1}{2}x}\, W_{k,\mu}(a + x)\, dx = \Gamma(\lambda)a^{k-1}\, W_{k-\lambda,\mu}(a)$

$$[|\arg a| < \pi, \quad \operatorname{Re}\lambda > 0] \quad \textbf{ET II 411(48)}$$

5. $\displaystyle\int_0^\infty x^{\varrho-1}(a + x)^{-\sigma}e^{-\frac{1}{2}x}\, W_{k,\mu}(a + x)\, dx = \Gamma(\varrho)a^\varrho e^{\frac{1}{2}a}\, G_{23}^{30}\left(a \left| \begin{matrix} 0, 1 - k - \sigma \\ -\varrho, \frac{1}{2} + \mu - \sigma, \frac{1}{2} - \mu - \sigma \end{matrix}\right.\right)$

$$[|\arg a| < \pi, \quad \operatorname{Re}\varrho > 0] \quad \textbf{ET II 411(49)}$$

6. $\displaystyle\int_0^\infty x^{\varrho-1}(a + x)^{-\sigma}e^{\frac{1}{2}x}\, W_{k,\mu}(a + x)\, dx$

$$= \frac{\Gamma(\varrho)a^\varrho e^{-\frac{1}{2}a}}{\Gamma\left(\frac{1}{2} - k + \mu\right)\Gamma\left(\frac{1}{2} - k - \mu\right)}\, G_{23}^{31}\left(a \left| \begin{matrix} k - \sigma + 1, 0 \\ -\varrho, \frac{1}{2} + \mu - \sigma, \frac{1}{2} - \mu - \sigma \end{matrix}\right.\right)$$
$$[|\arg a| < \pi, \quad 0 < \operatorname{Re}\varrho < \operatorname{Re}(\sigma - k)] \quad \textbf{ET II 412(51)}$$

7. $\displaystyle\int_0^\infty e^{-\frac{1}{2}(a+x)}\frac{(a+x)^{2\kappa-1}}{(ax)^\kappa}\,W_{\kappa,\mu}(x)\frac{dx}{x} = \frac{\Gamma\left(\frac{1}{2}-\mu-\kappa\right)\Gamma\left(\frac{1}{2}+\mu-\kappa\right)}{a\,\Gamma(1-2\kappa)}\,W_{\kappa,\mu}(a)$

$$\left[\operatorname{Re}\left(\tfrac{1}{2}\pm\mu-\kappa\right)>0\right] \qquad \text{BU 126(7a)}$$

8. $\displaystyle\int_0^\infty e^{-\frac{1}{2}x}x^{\gamma+\alpha-1}\,M_{\kappa,\mu}(x)\frac{dx}{(x+a)^\alpha}$

$$= \frac{\Gamma(1+2\mu)\,\Gamma\left(\frac{1}{2}+\mu+\gamma\right)\Gamma(\kappa-\gamma)}{\Gamma\left(\frac{1}{2}+\mu-\gamma\right)\Gamma\left(\frac{1}{2}+\mu+\kappa\right)}\;{}_2F_2\left(\alpha,\kappa-\gamma;\frac{1}{2}+\mu-\gamma,\frac{1}{2}-\mu-\gamma;a\right)$$

$$+\frac{\Gamma\left(\alpha+\gamma+\frac{1}{2}+\mu\right)\Gamma\left(-\gamma-\frac{1}{2}-\mu\right)}{\Gamma(\alpha)}a^{\gamma+\frac{1}{2}+\mu}$$

$$\times\;{}_2F_2\left(\alpha+\gamma+\mu+\frac{1}{2},\kappa+\mu+\frac{1}{2};1+2\mu,\frac{3}{2}+\mu+\gamma;a\right)$$

$$\left[\operatorname{Re}\left(\gamma+\alpha+\tfrac{1}{2}+\mu\right)>0,\quad \operatorname{Re}(\gamma-\kappa)<0\right] \qquad \text{BU 126(8)a}$$

9. $\displaystyle\int_0^\infty e^{-\frac{1}{2}x}x^{n+\mu+\frac{1}{2}}\,M_{\kappa,\mu}(x)\frac{dx}{x+a} = (-1)^{n+1}a^{n+\mu+\frac{1}{2}}e^{\frac{1}{2}a}\Gamma(1+2\mu)\Gamma\left(\frac{1}{2}-\mu+\kappa\right)W_{-\kappa,\mu}(a)$

$$\left[n=0,1,2,\dots,\quad \operatorname{Re}\left(\mu+1+\frac{n}{2}\right)>0,\quad \operatorname{Re}\left(\kappa-\mu-\frac{1}{2}\right)<n,\quad |\arg a|<\pi\right] \qquad \text{BU 127(10a)a}$$

7.628

1. $\displaystyle\int_0^\infty e^{-st}e^{-t^2}t^{2c-2}\,{}_1F_1\left(a;c;t^2\right)dt = 2^{1-2c}\Gamma(2c-1)\Psi\left(c-\frac{1}{2},a+\frac{1}{2};\frac{1}{4}s^2\right)$

$$\left[\operatorname{Re}c>\tfrac{1}{2},\quad \operatorname{Re}s>0\right] \qquad \text{EH I 270(11)}$$

2. $\displaystyle\int_0^\infty t^{2\nu-1}e^{-\frac{1}{2a}t^2}e^{-st}\,M_{-3\nu,\nu}\left(\frac{t^2}{a}\right)dt = \frac{1}{2\sqrt{\pi}}\Gamma(4\nu+1)a^{-\nu}s^{-4\nu}e^{as^2/8}K_{2\nu}\left(\frac{as^2}{8}\right)$

$$\left[\operatorname{Re}a>0,\quad \operatorname{Re}\nu>-\tfrac{1}{4},\quad \operatorname{Re}s>0\right]$$
$$\text{ET I 215(12)}$$

3. $\displaystyle\int_0^\infty t^{2\mu-1}e^{-\frac{1}{2a}t^2}e^{-st}\,M_{\lambda,\mu}\left(\frac{t^2}{a}\right)dt$

$$= 2^{-3\mu-\lambda}\Gamma(4\mu+1)a^{\frac{1}{2}(\lambda+\mu-1)}s^{\lambda-\mu-1}e^{\frac{as^2}{8}}W_{-\frac{1}{2}(\lambda+3\mu),\frac{1}{2}(\lambda-\mu)}\left(\frac{as^2}{4}\right)$$

$$\left[\operatorname{Re}a>0,\quad \operatorname{Re}\mu>-\tfrac{1}{4},\quad \operatorname{Re}s>0\right] \qquad \text{ET I 215(13)}$$

7.629

1.[8] $\displaystyle\int_0^\infty t^k\exp\left(\frac{a}{2t}\right)e^{-st}\,W_{k,\mu}\left(\frac{a}{t}\right)dt = 2^{1-2k}\sqrt{a}\,s^{-k-\frac{1}{2}}S_{2k,2\mu}\left(2\sqrt{as}\right)$

$$\left[|\arg a|<\pi,\quad \operatorname{Re}(k\pm\mu)>-\tfrac{1}{2},\quad \operatorname{Re}s>0\right] \qquad \text{ET I 217(21)}$$

2. $\displaystyle\int_0^\infty t^{-k}\exp\left(-\frac{a}{2t}\right)e^{-st}\,W_{k,\mu}\left(\frac{a}{t}\right)dt = 2\sqrt{a}\,s^{k-\frac{1}{2}}K_{2\mu}\left(2\sqrt{as}\right)$

$$\left[\operatorname{Re}a>0,\quad \operatorname{Re}s>0\right] \qquad \text{ET I 217(22)}$$

7.631

1.
$$\int_0^\infty x^{\varrho-1} \exp\left[\frac{1}{2}\left(\alpha^{-1}x - \beta x^{-1}\right)\right] W_{k,\mu}\left(\alpha^{-1}x\right) W_{\lambda,\nu}\left(\beta x^{-1}\right) dx$$
$$= \beta^\varrho \left[\Gamma\left(\tfrac{1}{2} - k + \mu\right) \Gamma\left(\tfrac{1}{2} - k - \mu\right)\right]^{-1}$$
$$\times G_{24}^{41}\left(\frac{\beta}{\alpha} \middle| \begin{matrix} 1+k, & 1-\lambda-\varrho \\ \tfrac{1}{2}+\mu, & \tfrac{1}{2}-\mu, & \tfrac{1}{2}+\nu-\varrho, & \tfrac{1}{2}-\nu-\varrho \end{matrix}\right)$$
$$\left[|\arg\alpha| < \tfrac{3}{2}\pi, \quad \operatorname{Re}\beta > 0, \quad \operatorname{Re}(k+\varrho) < -|\operatorname{Re}\nu| - \tfrac{1}{2}\right] \quad \textsf{ET II 412(55)}$$

2.
$$\int_0^\infty x^{\varrho-1} \exp\left[\frac{1}{2}\left(\alpha^{-1}x - \beta x^{-1}\right)\right] W_{k,\mu}\left(\alpha^{-1}x\right) W_{\lambda,\nu}\left(\beta x^{-1}\right) dx$$
$$= \beta^\varrho \left[\Gamma\left(\tfrac{1}{2} - k + \mu\right) \Gamma\left(\tfrac{1}{2} - k - \mu\right) \Gamma\left(\tfrac{1}{2} - \lambda + \nu\right) \Gamma\left(\tfrac{1}{2} - \lambda - \nu\right)\right]^{-1}$$
$$\times G_{24}^{42}\left(\frac{\beta}{\alpha} \middle| \begin{matrix} 1+k, & 1+\lambda-\varrho \\ \tfrac{1}{2}+\mu, & \tfrac{1}{2}-\mu, & \tfrac{1}{2}+\nu-\varrho, & \tfrac{1}{2}-\nu-\varrho \end{matrix}\right)$$
$$\left[|\arg\alpha| < \tfrac{3}{2}\pi, \quad |\arg\beta| < \tfrac{3}{2}\pi, \quad \operatorname{Re}(\lambda-\varrho) < \tfrac{1}{2} - |\operatorname{Re}\mu|, \quad \operatorname{Re}(k+\varrho) < \tfrac{1}{2} - |\operatorname{Re}\nu|\right]$$
$$\textsf{ET II 412(57)}$$

3.
$$\int_0^\infty x^{\varrho-1} \exp\left[\frac{1}{2}\left(\alpha^{-1}x + \beta x^{-1}\right)\right] W_{k,\mu}\left(\alpha^{-1}x\right) W_{\lambda,\nu}\left(\beta x^{-1}\right) dx$$
$$= \beta^\varrho G_{24}^{40}\left(\frac{\beta}{\alpha} \middle| \begin{matrix} 1-k, & 1-\lambda-\varrho \\ \tfrac{1}{2}+\mu, & \tfrac{1}{2}-\mu, & \tfrac{1}{2}+\nu-\varrho, & \tfrac{1}{2}-\nu-\varrho \end{matrix}\right)$$
$$\left[\operatorname{Re}\alpha > 0, \quad \operatorname{Re}\beta > 0\right] \quad \textsf{ET II 412(54)}$$

7.632 $\displaystyle\int_0^\infty e^{-st}\left(e^t - 1\right)^{\mu-\frac{1}{2}} \exp\left(-\frac{1}{2}\lambda e^t\right) M_{k,\mu}\left(\lambda e^t - \lambda\right) dt$
$$= \frac{\Gamma(2\mu+1)\,\Gamma\left(\tfrac{1}{2} + k - \mu + s\right)}{\Gamma(s+1)} W_{-k-\frac{1}{2}s,\mu-\frac{1}{2}s}(\lambda)$$
$$\left[\operatorname{Re}\mu > -\tfrac{1}{2}, \quad \operatorname{Re}s > \operatorname{Re}(\mu-k) - \tfrac{1}{2}\right] \quad \textsf{ET I 216(15)}$$

7.64 Combinations of confluent hypergeometric and trigonometric functions

7.641 $\displaystyle\int_0^\infty \cos(ax)\, {}_1F_1(\nu+1;1;ix)\, {}_1F_1(\nu+1;1;-ix)\, dx$
$$= -a^{-1}\sin(\nu\pi)\, P_\nu\left(2a^{-2} - 1\right) \quad [0 < a < 1];$$
$$= 0 \quad\quad\quad\quad [1 < a < \infty]$$
$$[-1 < \operatorname{Re}\nu < 0] \quad\quad \textsf{ET II 402(4)}$$

7.642 $\displaystyle\int_0^\infty \cos(2xy)\, {}_1F_1\left(a;c;-x^2\right) dx = \frac{1}{2}\pi^{\frac{1}{2}}\frac{\Gamma(c)}{\Gamma(a)}y^{2a-1}e^{-y^2}\Psi\left(c - \tfrac{1}{2}, a + \tfrac{1}{2}; y^2\right)$ **EH I 285(12)**

7.643

1.
$$\int_0^\infty x^{4\nu} e^{-\frac{1}{2}x^2} \sin(bx)\, {}_1F_1\left(\frac{1}{2} - 2\nu; 2\nu+1; \frac{1}{2}x^2\right) dx = \sqrt{\frac{\pi}{2}}b^{4\nu}c^{-\frac{1}{2}b^2}\, {}_1F_1\left(\frac{1}{2} - 2\nu; 1+2\nu; \frac{1}{2}b^2\right)$$
$$\left[b > 0, \quad \operatorname{Re}\nu > -\tfrac{1}{4}\right] \quad \textsf{ET I 115(5)}$$

2. $\int_0^\infty x^{2\nu-1} e^{-\frac{1}{4}x^2} \sin(bx)\, M_{3\nu,\nu}\left(\frac{1}{2}x^2\right) dx = \sqrt{\frac{\pi}{2}}\, b^{2\nu-1} e^{-\frac{1}{4}b^2}\, M_{3\nu,\nu}\left(\frac{1}{2}b^2\right)$

$$\left[b > 0, \quad \operatorname{Re}\nu > -\tfrac{1}{4}\right] \qquad \text{ET I 116(10)}$$

3. $\int_0^\infty x^{-2\nu-1} e^{\frac{1}{4}x^2} \cos(bx)\, W_{3\nu,\nu}\left(\frac{1}{2}x^2\right) dx = \sqrt{\frac{\pi}{2}}\, b^{-2\nu-1} e^{\frac{1}{4}b^2}\, W_{3\nu,\nu}\left(\frac{1}{2}b^2\right)$

$$\left[\operatorname{Re}\nu < \tfrac{1}{4}, \quad b > 0\right] \qquad \text{ET I 61(7)}$$

4. $\int_0^\infty x^{-2\nu} e^{\frac{1}{4}x^2} \sin(bx)\, W_{3\nu-1,\nu}\left(\frac{1}{2}x^2\right) dx = \sqrt{\frac{\pi}{2}}\, b^{-2\nu} e^{\frac{1}{4}b^2}\, W_{3\nu-1,\nu}\left(\frac{1}{2}b^2\right)$

$$\left[\operatorname{Re}\nu < \tfrac{1}{2}, \quad b > 0\right] \qquad \text{ET I 116(9)}$$

7.644

1. $\int_0^\infty x^{-\mu-\frac{1}{2}} e^{-\frac{1}{2}x} \sin\left(2ax^{\frac{1}{2}}\right) M_{k,\mu}(x)\, dx = \pi^{\frac{1}{2}} a^{k+\mu-1} \dfrac{\Gamma(3-2\mu)}{\Gamma\left(\frac{1}{2}+k+\mu\right)} \exp\left(-\dfrac{a^2}{2}\right) W_{\varrho,\sigma}\left(a^2\right),$

$$2\varrho = k - 3\mu + 1, \qquad 2\sigma = k + \mu - 1 \quad [a > 0, \quad \operatorname{Re}(k+\mu) > 0] \quad \text{ET II 403(10)}$$

2. $\int_0^\infty x^{\varrho-1} \sin\left(cx^{\frac{1}{2}}\right) e^{-\frac{1}{2}x}\, W_{k,\mu}(x)\, dx = \dfrac{c\,\Gamma(1+\mu+\varrho)\,\Gamma(1-\mu+\varrho)}{\Gamma\left(\frac{3}{2}-k+\varrho\right)}$

$$\times\ {}_2F_2\left(1+\mu+\varrho, 1-\mu+\varrho; \frac{3}{2}, \frac{3}{2}-k+\varrho; -\frac{c^2}{4}\right)$$

$$\left[\operatorname{Re}\varrho > |\operatorname{Re}\mu| - 1\right] \qquad \text{ET II 407(28)}$$

3. $\int_0^\infty x^{\varrho-1} \sin\left(cx^{\frac{1}{2}}\right) e^{\frac{1}{2}x}\, W_{k,\mu}(x)\, dx$

$$= \dfrac{\pi^{\frac{1}{2}}}{\Gamma\left(\frac{1}{2}-k+\mu\right)\Gamma\left(\frac{1}{2}-k-\mu\right)}\, G_{23}^{22}\left(\dfrac{c^2}{4}\,\middle|\, \begin{matrix} \frac{1}{2}+\mu-\varrho, \frac{1}{2}-\mu-\varrho \\ \frac{1}{2}, -k-\varrho, 0 \end{matrix}\right)$$

$$\left[c > 0, \quad \operatorname{Re}\varrho > |\operatorname{Re}\mu| - 1, \quad \operatorname{Re}(k+\varrho) < \tfrac{1}{2}\right] \qquad \text{ET II 407(29)}$$

4. $\int_0^\infty x^{\varrho-1} \cos\left(cx^{\frac{1}{2}}\right) e^{-\frac{1}{2}x}\, W_{k,\mu}(x)\, dx = \dfrac{\Gamma\left(\frac{1}{2}+\mu+\varrho\right)\Gamma\left(\frac{1}{2}-\mu+\varrho\right)}{\Gamma(1-k+\varrho)}$

$$\times\ {}_2F_2\left(\frac{1}{2}+\mu+\varrho, \frac{1}{2}-\mu+\varrho; \frac{1}{2}, 1-k+\varrho; -\frac{c^2}{4}\right)$$

$$\left[\operatorname{Re}\varrho > |\operatorname{Re}\mu| - \tfrac{1}{2}\right] \qquad \text{ET II 407(30)}$$

5. $\int_0^\infty x^{\varrho-1} \cos\left(cx^{\frac{1}{2}}\right) e^{\frac{1}{2}x}\, W_{k,\mu}(x)\, dx$

$$= \dfrac{\pi^{\frac{1}{2}}}{\Gamma\left(\frac{1}{2}-k+\mu\right)\Gamma\left(\frac{1}{2}-k-\mu\right)}\, G_{23}^{22}\left(\dfrac{c^2}{4}\,\middle|\, \begin{matrix} \frac{1}{2}+\mu-\varrho, \frac{1}{2}-\mu-\varrho \\ 0, -k-\varrho, \frac{1}{2} \end{matrix}\right)$$

$$\left[c > 0, \quad \operatorname{Re}\varrho > |\operatorname{Re}\mu| - \tfrac{1}{2}, \quad \operatorname{Re}(k+\varrho) < \tfrac{1}{2}\right] \qquad \text{ET II 407(31)}$$

7.65 Combinations of confluent hypergeometric functions and Bessel functions

7.651

1. $\displaystyle\int_0^\infty J_\nu(xy)\, M_{-\frac{1}{2}\mu,\frac{1}{2}\nu}(ax)\, W_{\frac{1}{2}\mu,\frac{1}{2}\nu}(ax)\, dx$

$$= ay^{-\mu-1}\frac{\Gamma(\nu+1)}{\Gamma\left(\frac{1}{2}-\frac{1}{2}\mu+\frac{1}{2}\nu\right)}\left[a+\left(a^2+y^2\right)^{\frac{1}{2}}\right]^\mu\left(a^2+y^2\right)^{-\frac{1}{2}}$$

$$\left[y>0,\quad \mathrm{Re}\,\nu>-1,\quad \mathrm{Re}\,\mu<\tfrac{1}{2},\quad \mathrm{Re}\,a>0\right]\quad \text{ET II 85(19)}$$

2. $\displaystyle\int_0^\infty M_{k,\frac{1}{2}\nu}(-iax)\, M_{-k,\frac{1}{2}\nu}(-iax)\, J_\nu(xy)\, dx$

$$= \frac{ae^{-\frac{1}{2}(\nu+1)\pi i}\left[\Gamma(1+\nu)\right]^2}{\Gamma\left(\frac{1}{2}+k+\frac{1}{2}\nu\right)\Gamma\left(\frac{1}{2}-k+\frac{1}{2}\nu\right)}y^{-1-2k}$$

$$\times\left(a^2-y^2\right)^{-\frac{1}{2}}\left\{\left[a+\left(a^2-y^2\right)^{\frac{1}{2}}\right]^{2k}+\left[a-\left(a^2-y^2\right)^{\frac{1}{2}}\right]^{2k}\right\}\quad [0<y<a];$$

$$= 0 \qquad\qquad\qquad [a<y<\infty]$$

$$\left[a>0,\quad \mathrm{Re}\,\nu>-1,\quad |\mathrm{Re}\,k|<\tfrac{1}{4}\right]\quad \text{ET II 85(18)}$$

7.652 $\displaystyle\int_0^\infty M_{-\mu,\frac{1}{2}\nu}\left\{a\left[\left(b^2+x^2\right)^{\frac{1}{2}}-b\right]\right\}W_{\mu,\frac{1}{2}\nu}\left\{a\left[\left(b^2+x^2\right)^{\frac{1}{2}}+b\right]\right\}J_\nu(xy)\, dx$

$$= \frac{ay^{-2\mu-1}\,\Gamma(1+\nu)\left[\left(a^2+y^2\right)^{\frac{1}{2}}+a\right]^{2\mu}}{\Gamma\left(\frac{1}{2}+\frac{1}{2}\nu-\mu\right)\left(A^2+Y^2\right)^{\frac{1}{2}}}\exp\left[-b\left(a^2+y^2\right)^{\frac{1}{2}}\right]$$

$$\left[y>0,\quad \mathrm{Re}\,\nu>-1,\quad \mathrm{Re}\,\mu<\tfrac{1}{4},\quad \mathrm{Re}\,a>0,\quad \mathrm{Re}\,b>0\right]\quad \text{ET II 87(29)}$$

7.66 Combinations of confluent hypergeometric functions, Bessel functions, and powers

7.661

1. $\displaystyle\int_0^\infty x^{-1}\, W_{k,\mu}(ax)\, M_{-k,\mu}(ax)\, J_0(xy)\, dx$

$$= e^{-ik\pi}\frac{\Gamma(1+2\mu)}{\Gamma\left(\frac{1}{2}+\mu+k\right)}P_{\mu-\frac{1}{2}}^k\left[\left(1+\frac{y^2}{a^2}\right)^{\frac{1}{2}}\right]Q_{\mu-\frac{1}{2}}^k\left[\left(1+\frac{y^2}{a^2}\right)^{\frac{1}{2}}\right]$$

$$\left[y>0,\quad \mathrm{Re}\,a>0,\quad \mathrm{Re}\,\mu>-\tfrac{1}{2},\quad \mathrm{Re}\,k<\tfrac{3}{4}\right]\quad \text{ET II 18(44)}$$

2. $\displaystyle\int_0^\infty x^{-1}\, W_{k,\mu}(ax)\, W_{-k,\mu}(ax)\, J_0(xy)\, dx = \frac{1}{2}\pi\cos(\mu\pi)\, P_{\mu-\frac{1}{2}}^k\left[\left(1+\frac{y^2}{a^2}\right)^{\frac{1}{2}}\right]P_{\mu-\frac{1}{2}}^{-k}\left[\left(1+\frac{y^2}{a^2}\right)^{\frac{1}{2}}\right]$

$$\left[y>0,\quad \mathrm{Re}\,a>0,\quad |\mathrm{Re}\,\mu|<\tfrac{1}{2}\right]$$
$$\text{ET II 18(45)}$$

3. $\displaystyle\int_0^\infty x^{2\mu-\nu}\, W_{k,\mu}(ax)\, M_{-k,\mu}(ax)\, J_\nu(xy)\, dx$

$$= 2^{2\mu-\nu+2k} a^{2k} y^{\nu-2\mu-2k-1} \frac{\Gamma(2\mu+1)}{\Gamma\left(\nu-k-\mu+\frac{1}{2}\right)}$$

$$\times\ _3F_2\left(\frac{1}{2}-k, 1-k, \frac{1}{2}-k+\mu; 1-2k, \frac{1}{2}-k-\mu+\nu; -\frac{y^2}{a^2}\right)$$

$$\left[y>0,\quad \operatorname{Re}\mu>-\tfrac{1}{2},\quad \operatorname{Re} a>0,\quad \operatorname{Re}(2\mu+2k-\nu)<\tfrac{1}{2}\right]\quad \text{ET II 85(20)}$$

4. $\displaystyle\int_0^\infty x^{2\varrho-\nu}\, W_{k,\mu}(iax)\, W_{k,\mu}(-iax)\, J_\nu(xy)\, dx$

$$= 2^{2\varrho-\nu} y^{\nu-2\varrho-1} \pi^{-\frac{1}{2}} \left[\Gamma\left(\frac{1}{2}-k+\mu\right)\Gamma\left(\frac{1}{2}-k-\mu\right)\right]^{-1} G^{24}_{44}\left(\frac{y^2}{a^2}\,\middle|\,\begin{matrix}\frac{1}{2}, 0, \frac{1}{2}-\mu, \frac{1}{2}+\mu\\ \varrho+\frac{1}{2}, -k, k, \varrho-\nu+\frac{1}{2}\end{matrix}\right)$$

$$\left[y>0,\quad \operatorname{Re} a>0,\quad \operatorname{Re}\varrho>|\operatorname{Re}\mu|-1,\quad \operatorname{Re}(2\varrho+2k-\nu)<\tfrac{1}{2}\right]\quad \text{ET II 86(23)a}$$

5. $\displaystyle\int_0^\infty x^{2\varrho-\nu}\, W_{k,\mu}(ax)\, M_{-k,\mu}(ax)\, J_\nu(xy)\, dx$

$$= \frac{2^{2\varrho-\nu}\Gamma(2\mu+1)}{\pi^{\frac{1}{2}}\Gamma\left(\frac{1}{2}-k+\mu\right)} y^{\nu-2\varrho-1} G^{23}_{44}\left(\frac{y^2}{a^2}\,\middle|\,\begin{matrix}\frac{1}{2}, 0, \frac{1}{2}-\mu, \frac{1}{2}+\mu\\ \varrho+\frac{1}{2}, -k, k, \varrho-\nu+\frac{1}{2}\end{matrix}\right)$$

$$\left[y>0,\quad \operatorname{Re} a>0,\quad \operatorname{Re}\varrho>-1,\quad \operatorname{Re}(\varrho+\mu)>-1,\quad \operatorname{Re}(2e+2k+\nu)<\tfrac{1}{2}\right]\quad \text{ET II 86(21)a}$$

6. $\displaystyle\int_0^\infty x^{2\varrho-\nu}\, W_{k,\mu}(ax)\, W_{-k,\mu}(ax)\, J_\nu(xy)\, dx$

$$= \frac{\Gamma(\varrho+1+\mu)\,\Gamma(\varrho+1-\mu)\,\Gamma(2\varrho+2)}{\Gamma\left(\frac{3}{2}+k+\varrho\right)\Gamma\left(\frac{3}{2}-k+\varrho\right)\Gamma(1+\nu)} y^\nu 2^{-\nu-1} a^{-2\varrho-1}$$

$$\times\ _4F_3\left(\varrho+1, \varrho+\frac{3}{2}, \varrho+1+\mu, \varrho+1-\mu; \frac{3}{2}+k+\varrho, \frac{3}{2}-k+\varrho, 1+\nu; -\frac{y^2}{a^2}\right)$$

$$\left[y>0,\quad \operatorname{Re}\varrho>|\operatorname{Re}\mu|-1,\quad \operatorname{Re} a>0\right]\quad \text{ET II 86(22)a}$$

7.662

1. $\displaystyle\int_0^\infty x^{-1} M_{-\mu,\frac{1}{4}\nu}\left(\frac{1}{2}x^2\right) W_{\mu,\frac{1}{4}\nu}\left(\frac{1}{2}x^2\right) J_\nu(xy)\, dx = \frac{\Gamma\left(1+\frac{1}{2}\nu\right)}{\Gamma\left(\frac{1}{2}+\frac{1}{4}\nu-\mu\right)} I_{\frac{1}{4}\nu-\mu}\left(\frac{1}{4}y^2\right) K_{\frac{1}{4}\nu+\mu}\left(\frac{1}{4}y^2\right)$

$$\left[y>0,\quad \operatorname{Re}\nu>-1\right]\quad \text{ET II 86(24)}$$

2. $\displaystyle\int_0^\infty x^{-1} M_{\alpha-\beta,\frac{1}{4}\nu-\gamma}\left(\frac{1}{2}x^2\right) W_{\alpha+\beta,\frac{1}{4}\nu+\gamma}\left(\frac{1}{2}x^2\right) J_\nu(xy)\, dx$

$$= \frac{\Gamma\left(1+\frac{1}{2}\nu-2\gamma\right)}{\Gamma\left(1+\frac{1}{2}\nu-2\beta\right)} y^{-2} M_{\alpha-\gamma,\frac{1}{4}\nu-\beta}\left(\frac{1}{2}y^2\right) W_{\alpha+\gamma,\frac{1}{4}\nu+\beta}\left(\frac{1}{2}y^2\right)$$

$$\left[y>0,\quad \operatorname{Re}\beta<\tfrac{1}{8},\quad \operatorname{Re}\nu>-1,\quad \operatorname{Re}(\nu-4\gamma)>-2\right]\quad \text{ET II 86(25)}$$

3. $\displaystyle\int_0^\infty x^{-1} M_{k,0}\left(iax^2\right) M_{k,0}\left(-iax^2\right) K_0(xy)\, dx = \frac{\pi}{16}\left\{\left[J_k\left(\frac{y^2}{8a}\right)\right]^2 + \left[Y_k\left(\frac{y^2}{8a}\right)\right]^2\right\}$

$$\left[a>0\right]\quad \text{ET II 152(83)}$$

4.　$\displaystyle\int_0^\infty x^{-1} M_{k,\mu}\left(iax^2\right) M_{k,\mu}\left(-iax^2\right) K_0(xy)\, dx = ay^{-2}[\Gamma(2\mu+1)]^2\, W_{-\mu,k}\left(\frac{iy^2}{4a}\right) W_{-\mu,k}\left(-\frac{iy^2}{4a}\right)$

$$\left[a>0, \quad \operatorname{Re} y>0, \quad \operatorname{Re}\mu>-\tfrac{1}{2}\right]$$

ET II 152(84)

7.663

1.　$\displaystyle\int_0^\infty x^{2\varrho}\, {}_1F_1\left(a;b;-\lambda x^2\right) J_\nu(xy)\, dx = \frac{2^{2\varrho}\,\Gamma(b)}{\Gamma(a)y^{2\varrho+1}}\, G_{23}^{21}\left(\frac{y^2}{4\lambda}\;\middle|\;\begin{matrix}1,b\\ \tfrac{1}{2}+\varrho+\tfrac{1}{2}\nu, a, \tfrac{1}{2}+\varrho-\tfrac{1}{2}\nu\end{matrix}\right)$

$$\left[y>0, \quad -1-\operatorname{Re}\nu<2\operatorname{Re}\varrho<\tfrac{1}{2}+2\operatorname{Re}a, \quad \operatorname{Re}\lambda>0\right]\quad \text{ET II 88(6)}$$

2.　$\displaystyle\int_0^\infty x^{\nu+1}\, {}_1F_1\left(2a-\nu;a+1;-\frac{1}{2}x^2\right) J_\nu(xy)\, dx = \frac{2^{\nu-a+\frac{1}{2}}\,\Gamma(a+1)}{\pi^{\frac{1}{2}}\,\Gamma(2a-\nu)}\, y^{2a-\nu-1} e^{-\frac{1}{4}y^2}\, K_{a-\nu-\frac{1}{2}}\left(\frac{1}{4}y^2\right)$

$$\left[y>0, \quad \operatorname{Re}\nu>-1, \quad \operatorname{Re}(4a-3\nu)>\tfrac{1}{2}\right]\quad \text{ET II 87(1)}$$

3.　$\displaystyle\int_0^\infty x^a\, {}_1F_1\left(a;\frac{1+a+\nu}{2};-\frac{1}{2}x^2\right) J_\nu(xy)\, dx = y^{a-1}\, {}_1F_1\left(a;\frac{1+a+\nu}{2};-\frac{y^2}{2}\right)$

$$\left[y>0, \quad \operatorname{Re}a>-\tfrac{1}{2}, \quad \operatorname{Re}(a+\nu)>-1\right]$$

ET II 87(2)

4.　$\displaystyle\int_0^\infty x^{\nu+1-2a}\, {}_1F_1\left(a;1+\nu-a;-\frac{1}{2}x^2\right) J_\nu(xy)\, dx$

$$= \frac{\pi^{\frac{1}{2}}\,\Gamma(1+\nu-a)}{\Gamma(a)}\, 2^{-2a+\nu+\frac{1}{2}}\, y^{2a-\nu-1} e^{-\frac{1}{4}y^2}\, I_{a-\frac{1}{2}}\left(\frac{1}{4}y^2\right)$$

$$\left[y>0, \quad \operatorname{Re}a-1<\operatorname{Re}\nu<4\operatorname{Re}a-\tfrac{1}{2}\right]\quad \text{ET II 87(3)}$$

5.　$\displaystyle\int_0^\infty x\, {}_1F_1\left(\lambda;1;-x^2\right) J_0(xy)\, dx = \left[2^{2\lambda-1}\,\Gamma(\lambda)\right]^{-1} y^{2\lambda-2} e^{-\frac{1}{4}y^2}$

$$\left[y>0, \quad \operatorname{Re}\lambda>0\right]\quad \text{ET II 18(46)}$$

6.　$\displaystyle\int_0^\infty x^{\nu+1}\, {}_1F_1\left(a;b;-\lambda x^2\right) J_\nu(xy)\, dx$

$$= \frac{2^{1-a}\,\Gamma(b)}{\Gamma(a)\lambda^{\frac{1}{2}a+\frac{1}{2}\nu}}\, y^{a-2} e^{-\frac{y^2}{8\lambda}}\, W_{k,\mu}\left(\frac{y^2}{4\lambda}\right), \qquad 2k=a-2b+\nu+2, \qquad 2\mu=a-\nu-1$$

$$\left[y>0, \quad -1<\operatorname{Re}\nu<2\operatorname{Re}a-\tfrac{1}{2}, \quad \operatorname{Re}\lambda>0\right]\quad \text{ET II 88(4)}$$

7.　$\displaystyle\int_0^\infty x^{2b-\nu-1}\, {}_1F_1\left(a;b;-\lambda x^2\right) J_\nu(xy)\, dx = \frac{2^{2b-2a-\nu-1}\,\Gamma(b)}{\Gamma(a-b+\nu+1)}\, \lambda^{-a} y^{2a-2b+\nu}$

$$\times\, {}_1F_1\left(a;1+a-b+\nu;-\frac{y^2}{4\lambda}\right)$$

$$\left[y>0, \quad 0<\operatorname{Re}b<\tfrac{3}{4}+\operatorname{Re}\left(a+\tfrac{1}{2}\nu\right), \quad \operatorname{Re}\lambda>0\right]\quad \text{ET II 88(5)}$$

7.664

1.　$\displaystyle\int_0^\infty x\, W_{\frac{1}{2}\nu,\mu}\left(\frac{a}{x}\right) W_{-\frac{1}{2}\nu,\mu}\left(\frac{a}{x}\right) K_\nu(xy)\, dx = 2ay^{-1}\, K_{2\mu}\left[(2ay)^{\frac{1}{2}} e^{\frac{1}{4}i\pi}\right] K_{2\mu}\left[(2ay)^{\frac{1}{2}} e^{-\frac{1}{4}i\pi}\right]$

$$\left[\operatorname{Re}y>0, \quad \operatorname{Re}a>0\right]\quad \text{ET II 152(85)}$$

2. $$\int_0^\infty x \, W_{\frac{1}{2}\nu,\mu}\left(\frac{2}{x}\right) W_{-\frac{1}{2}\nu,\mu}\left(\frac{2}{x}\right) J_\nu(xy) \, dx$$
$$= -4y^{-1}\left\{\sin\left[(\mu - \tfrac{1}{2}\nu)\,\pi\right] J_{2\mu}\left(2y^{\frac{1}{2}}\right) + \cos\left[(\mu - \tfrac{1}{2}\nu)\,\pi\right] Y_{2\mu}\left(2y^{\frac{1}{2}}\right)\right\} K_{2\mu}\left(2y^{\frac{1}{2}}\right)$$
$$[y > 0, \quad \operatorname{Re}(\nu \pm 2\mu) > -1] \quad \text{ET II 87(27)}$$

3. $$\int_0^\infty x \, W_{\frac{1}{2}\nu,\mu}\left(\frac{2}{x}\right) W_{-\frac{1}{2}\nu,\mu}\left(\frac{2}{x}\right) Y_\nu(xy) \, dx$$
$$= 4y^{-1}\left\{\left\{\cos\left[(\mu - \tfrac{1}{2}\nu)\,\pi\right] J_{2\mu}\left(2y^{\frac{1}{2}}\right) - \sin\left[(\mu - \tfrac{1}{2}\nu)\,\pi\right] Y_{2\mu}\left(2y^{\frac{1}{2}}\right)\right\} K_{2\mu}\left(2y^{\frac{1}{2}}\right)\right\}$$
$$[y > 0, \quad |\operatorname{Re}\mu| < \tfrac{1}{4}] \quad \text{ET II 117(48)}$$

4. $$\int_0^\infty x \, W_{-\frac{1}{2}\nu,\mu}\left(\frac{2}{x}\right) M_{\frac{1}{2}\nu,\mu}\left(\frac{2}{x}\right) J_\nu(xy) \, dx = \frac{4\,\Gamma(1+2\mu)y^{-1}}{\Gamma\left(\frac{1}{2}+\frac{1}{2}\nu+\mu\right)} J_{2\mu}\left(2y^{\frac{1}{2}}\right) K_{2\mu}\left(2y^{\frac{1}{2}}\right)$$
$$[y > 0, \quad \operatorname{Re}\nu > -1, \quad \operatorname{Re}\mu > -\tfrac{1}{4}]$$
$$\text{ET II 86(26)}$$

5. $$\int_0^\infty x \, W_{-\frac{1}{2}\nu,\mu}\left(\frac{ia}{x}\right) W_{-\frac{1}{2}\nu,\mu}\left(-\frac{ia}{x}\right) J_\nu(xy) \, dx$$
$$= 4ay^{-1}\left[\Gamma\left(\tfrac{1}{2}+\mu+\tfrac{1}{2}\nu\right)\Gamma\left(\tfrac{1}{2}-\mu+\tfrac{1}{2}\nu\right)\right]^{-1} K_\mu\left[(2iay)^{\frac{1}{2}}\right] K_\mu\left[(-2iay)^{\frac{1}{2}}\right]$$
$$[y > 0, \quad \operatorname{Re}a > 0, \quad |\operatorname{Re}\mu| < \tfrac{1}{2}, \quad \operatorname{Re}\nu > -1] \quad \text{ET II 87(28)}$$

7.665

1. $$\int_0^\infty x^{-\frac{1}{2}} J_\nu\left(ax^{\frac{1}{2}}\right) K_{\frac{1}{2}\nu-\mu}\left(\frac{1}{2}x\right) M_{k,\mu}(x) \, dx$$
$$= \frac{\Gamma(2\mu+1)}{a\,\Gamma\left(k+\frac{1}{2}\nu+1\right)} W_{\frac{1}{2}(k-\mu),\frac{1}{2}k-\frac{1}{4}\nu}\left(\frac{a^2}{2}\right) M_{\frac{1}{2}(k+\mu),\frac{1}{2}k+\frac{1}{4}\nu}\left(\frac{a^2}{2}\right)$$
$$\left[a > 0, \quad \operatorname{Re}k > -\tfrac{1}{4}, \quad \operatorname{Re}\mu > -\tfrac{1}{2}, \quad \operatorname{Re}\nu > -1\right] \quad \text{ET II 405(18)}$$

2. $$\int_0^\infty x^{\frac{1}{2}c+\frac{1}{2}c'-1}\Psi(a,c;x) \, {}_1F_1\left(a';c';-x\right) J_{c+c'-2}\left[2(xy)^{\frac{1}{2}}\right] dx$$
$$= \frac{\Gamma(c')}{\Gamma(a+a')} y^{\frac{1}{2}c+\frac{1}{2}c'-1}\Psi\left(c'-a',c+c'-a-a';y\right) {}_1F_1\left(a';a+a';-y\right)$$
$$\left[\operatorname{Re}c' > 0, \quad 1 < \operatorname{Re}(c+c') < 2\operatorname{Re}(a+a') + \tfrac{1}{2}\right] \quad \text{EH I 287(23)}$$

7.666 $\displaystyle\int_0^\infty x^{\frac{1}{2}c-\frac{1}{2}} \, {}_1F_1\left(a;c;-2x^{\frac{1}{2}}\right)\Psi\left(a,c;2x^{\frac{1}{2}}\right) J_{c-1}\left[2(xy)^{\frac{1}{2}}\right] dx$
$$= 2^{-c}\frac{\Gamma(c)}{\Gamma(a)} y^{a-\frac{1}{2}c-\frac{1}{2}}\left[1+(1+y)^{\frac{1}{2}}\right]^{c-2a}(1+y)^{-\frac{1}{2}}$$
$$\left[\operatorname{Re}c > 2, \quad \operatorname{Re}(c-2a) < \tfrac{1}{2}\right] \quad \text{EH I 285(13)}$$

7.67 Combinations of confluent hypergeometric functions, Bessel functions, exponentials, and powers

7.671

1.
$$\int_0^\infty x^{k-\frac{3}{2}} \exp\left[-\frac{1}{2}(a+1)x\right] K_\nu\left(\frac{1}{2}ax\right) M_{k,\nu}(x)\, dx$$
$$= \frac{\pi^{\frac{1}{2}}\,\Gamma(k)\,\Gamma(k+2\nu)}{a^{k+\nu}\,\Gamma\left(k+\nu+\frac{1}{2}\right)}\ {}_2F_1\left(k, k+2\nu; 2\nu+1; -a^{-1}\right)$$
$$\left[\operatorname{Re} a > 0, \quad \operatorname{Re} k > 0, \quad \operatorname{Re}(k+2\nu) > 0\right] \quad \textbf{ET II 405(17)}$$

2.
$$\int_0^\infty x^{-k-\frac{3}{2}} \exp\left[-\frac{1}{2}(a-1)x\right] K_\mu\left(\frac{1}{2}ax\right) W_{k,\mu}(x)\, dx$$
$$= \frac{\pi\,\Gamma(-k)\,\Gamma(2\mu-k)\,\Gamma(-2\mu-k)}{\Gamma\left(\frac{1}{2}-k\right)\Gamma\left(\frac{1}{2}+\mu-k\right)\Gamma\left(\frac{1}{2}-\mu-k\right)}2^{2k+1}a^{k-\nu}\ {}_2F_1\left(-k, 2\mu-k; -2k; 1-a^{-1}\right)$$
$$\left[\operatorname{Re} a > 0, \quad \operatorname{Re} k < 2\operatorname{Re}\mu < -\operatorname{Re} k\right] \quad \textbf{ET II 408(36)}$$

7.672

1.
$$\int_0^\infty x^{2\varrho} e^{-\frac{1}{2}ax^2} M_{k,\mu}\left(ax^2\right) J_\nu(xy)\, dx$$
$$= \frac{\Gamma(2\mu+1)}{\Gamma\left(\mu+k+\frac{1}{2}\right)} 2^{2\varrho} y^{-2\varrho-1}\, G_{23}^{21}\left(\frac{y^2}{4a}\ \middle|\ \begin{matrix} \frac{1}{2}-\mu, \frac{1}{2}+\mu \\ \frac{1}{2}+\varrho+\frac{1}{2}\nu, k, \frac{1}{2}+\varrho-\frac{1}{2}\nu \end{matrix}\right)$$
$$\left[y > 0, \quad -1-\operatorname{Re}\left(\tfrac{1}{2}\nu+\mu\right) < \operatorname{Re}\varrho < \operatorname{Re} k - \tfrac{1}{4}, \quad \operatorname{Re} a > 0\right] \quad \textbf{ET II 83(10)}$$

2.
$$\int_0^\infty x^{2\varrho} e^{-\frac{1}{2}ax^2} W_{k,\mu}\left(ax^2\right) J_\nu(xy)\, dx$$
$$= \frac{\Gamma\left(1+\mu+\frac{1}{2}\nu+\varrho\right)\Gamma\left(1-\mu+\frac{1}{2}\nu+\varrho\right)2^{-\nu-1}}{\Gamma(\nu+1)\Gamma\left(\frac{3}{2}-k+\frac{1}{2}\nu+\varrho\right)}a^{-\frac{1}{2}\nu-\varrho-\frac{1}{2}}y^\nu$$
$$\times\ {}_2F_2\left(\lambda+\mu, \lambda-\mu; \nu+1, \frac{1}{2}-k+\lambda; -\frac{y^2}{4a}\right),$$
$$\lambda = 1+\tfrac{1}{2}\nu+\varrho \qquad \left[y > 0, \quad \operatorname{Re} a > 0, \quad \operatorname{Re}\left(\varrho\pm\mu+\tfrac{1}{2}\nu\right) > -1\right] \quad \textbf{ET II 85(16)}$$

3.
$$\int_0^\infty x^{2\varrho} e^{\frac{1}{2}ax^2} W_{k,\mu}\left(ax^2\right) J_\nu(xy)\, dx = \frac{2^{2\varrho} y^{-2\varrho-1}}{\Gamma\left(\frac{1}{2}+\mu-k\right)\Gamma\left(\frac{1}{2}-\mu-k\right)}$$
$$\times G_{23}^{22}\left(\frac{y^2}{4a}\ \middle|\ \begin{matrix} \frac{1}{2}-\mu, & \frac{1}{2}+\mu \\ \frac{1}{2}+\varrho+\frac{1}{2}\nu, & -k, & \frac{1}{2}+\varrho-\frac{1}{2}\nu \end{matrix}\right)$$
$$\left[y > 0, \quad |\arg a| < \pi, \quad -1-\operatorname{Re}\left(\tfrac{1}{2}\nu\pm\mu\right) < \operatorname{Re}\varrho < -\tfrac{1}{4}-\operatorname{Re} k\right] \quad \textbf{ET II 85(17)}$$

4.
$$\int_0^\infty x^{2\lambda+\frac{1}{2}} e^{-\frac{1}{4}x^2} M_{k,\mu}\left(\frac{1}{2}x^2\right) Y_\nu(xy)\, dx = \frac{2^\lambda y^{-1/2}\,\Gamma(2\mu+1)}{\Gamma\left(\frac{1}{2}+k+\mu\right)}\, G_{34}^{31}\left(\frac{y^2}{2}\ \middle|\ \begin{matrix} -\mu-\lambda, & \mu-\lambda, & \\ h, & \kappa, & -\lambda-\frac{1}{2}, & l \end{matrix}\right)$$
$$h = \tfrac{1}{4}+\tfrac{1}{2}\nu, \quad \kappa = \tfrac{1}{4}-\tfrac{1}{2}\nu, \quad l = -\tfrac{1}{4}-\tfrac{1}{2}\nu$$
$$\left[y > 0, \quad \operatorname{Re}(k-\lambda) > 0, \quad \operatorname{Re}\left(2\lambda+2\mu\pm\nu\right) > -\tfrac{5}{2}\right] \quad \textbf{ET II 116(45)}$$

5.
$$\int_0^\infty x^{2\lambda+\frac{1}{2}} e^{\frac{1}{4}x^2} W_{k,\mu}\left(\frac{1}{2}x^2\right) Y_\nu(xy)\, dx$$

$$= 2^\lambda \left[\Gamma\left(\frac{1}{2}-k+\mu\right)\Gamma\left(\frac{1}{2}-k-\mu\right)\right]^{-1} G_{34}^{32}\left(\frac{y^2}{2}\left|\begin{matrix}-\mu-\lambda, & \mu-\lambda, & l \\ h, & \kappa, & -\frac{1}{2}-k-\lambda, & l\end{matrix}\right.\right) y^{-1/2},$$

$$h = \tfrac{1}{4}+\tfrac{1}{2}\nu, \quad \kappa = \tfrac{1}{4}-\tfrac{1}{2}\nu, \quad l = -\tfrac{1}{4}-\tfrac{1}{2}\nu$$

$$\left[y>0, \quad \operatorname{Re}(k+\lambda)<0, \quad \operatorname{Re}(2\lambda\pm2\mu\pm\nu)>-\tfrac{5}{2}\right] \quad \text{ET II 117(47)}$$

6.
$$\int_0^\infty x^{-1/2} e^{-\frac{1}{2}x^2} M_{\frac{1}{2}\nu-\frac{1}{4},\frac{1}{2}\nu+\frac{1}{4}}\left(x^2\right) J_\nu(xy)\, dx = (2\nu+1)2^{-\nu}y^{\nu-1}\left[1-\Phi\left(\frac{1}{2}y\right)\right]$$

$$\left[y>0, \quad \operatorname{Re}\nu>-\tfrac{1}{2}\right] \quad \text{ET II 82(1)}$$

7.
$$\int_0^\infty x^{-1} e^{-\frac{1}{2}x^2} M_{\frac{1}{2}\nu+\frac{1}{2},\frac{1}{2}\nu+\frac{1}{2}}\left(x^2\right) J_\nu(xy)\, dx = \frac{\Gamma(\nu+2)y^\nu}{\Gamma\left(\nu+\frac{3}{2}\right)2^\nu}\left[1-\Phi\left(\frac{1}{2}y\right)\right]$$

$$\left[y>0, \operatorname{Re}\nu>-1\right] \quad \text{ET II 82(2)}$$

8.
$$\int_0^\infty e^{-\frac{1}{4}x^2} M_{k,\frac{1}{2}\nu}\left(\frac{1}{2}\right) x^2 J_\nu(xy)\, dx = \frac{2^{-k}\,\Gamma(\nu+1)}{\Gamma\left(k+\frac{1}{2}\nu+\frac{1}{2}\right)} y^{2k-1} e^{-\frac{1}{2}y^2}$$

$$\left[y>0, \quad \operatorname{Re}\nu>-1, \quad \operatorname{Re}k<\tfrac{1}{2}\right]$$
$$\text{ET II 83(7)}$$

9.
$$\int_0^\infty x^{\nu-2\mu} e^{-\frac{1}{4}x^2} M_{k,\mu}\left(\frac{1}{2}\right) x^2 J_\nu(xy)\, dx$$

$$= 2^{\frac{1}{2}\left(\frac{1}{2}-k-3\mu+\nu\right)} \frac{\Gamma(2\mu+1)}{\Gamma\left(\mu+k+\frac{1}{2}\right)} y^{k+\mu-\frac{3}{2}} e^{-\frac{1}{4}y^2} W_{\alpha,\beta}\left(\frac{1}{2}y^2\right),$$

$$2\alpha = k-3\mu+\nu+\tfrac{1}{2}, \qquad 2\beta = k+\mu-\nu-\tfrac{1}{2}$$

$$\left[y>0, \quad -1<\operatorname{Re}\nu<2\operatorname{Re}(k+\mu)-\tfrac{1}{2}\right] \quad \text{ET II 83(9)}$$

10.
$$\int_0^\infty x^{\nu-2\mu} e^{\frac{1}{4}x^2} W_{k,\pm\mu}\left(\frac{1}{2}x^2\right) J_\nu(xy)\, dx = \frac{\Gamma(1+\nu-2\mu)}{\Gamma(1+2\beta)} 2^{\beta-\mu} y^{k+\mu-\frac{3}{2}} e^{-\frac{1}{4}y^2} M_{\alpha,\beta}\left(\frac{1}{2}y^2\right)$$

$$2\alpha = \tfrac{1}{2}+k+\nu-3\mu, \quad 2\beta = \tfrac{1}{2}-k+\nu-\mu$$

$$\left[y>0, \quad \operatorname{Re}\nu>-1, \quad \operatorname{Re}(\nu-2\mu)>-1\right]$$
$$\text{ET II 84(14)}$$

11.
$$\int_0^\infty x^{\nu-2\mu} e^{-\frac{1}{4}x^2} W_{k,\pm\mu}\left(\frac{1}{2}x^2\right) J_\nu(xy)\, dx$$

$$= \frac{\Gamma(1+\nu-2\mu)}{\Gamma\left(\frac{1}{2}+\mu-k\right)} 2^{\frac{1}{2}\left(\frac{1}{2}+k-3\mu+\nu\right)} y^{\mu-k-\frac{3}{2}} e^{\frac{1}{4}y^2} W_{\alpha,\beta}\left(\frac{1}{2}y^2\right),$$

$$2\alpha = k+3\mu-\nu-\tfrac{1}{2}, \qquad 2\beta = k-\mu+\nu+\tfrac{1}{2}$$

$$\left[y>0, \quad \operatorname{Re}\nu>-1, \quad \operatorname{Re}(\nu-2\mu)>-1, \quad \operatorname{Re}\left(k-\mu+\tfrac{1}{2}\nu\right)<-\tfrac{1}{4}\right] \quad \text{ET II 84(15)}$$

12. $\int_0^\infty x^{2\mu-\nu} e^{-\frac{1}{4}x^2} M_{k,\mu}\left(\frac{1}{2}x^2\right) J_\nu(xy)\,dx$

$$= \frac{\Gamma(2\mu+1)}{\Gamma\left(\frac{1}{2}+k-\mu+\nu\right)} 2^{\frac{1}{2}\left(\frac{1}{2}-k+3\mu-\nu\right)} y^{k-\mu-\frac{3}{2}} e^{-\frac{1}{4}y^2} M_{\alpha,\beta}\left(\frac{1}{2}y^2\right)$$

$$2\alpha = \frac{1}{2}+k+3\mu-\nu, \qquad 2\beta = -\frac{1}{2}+k-\mu+\nu$$

$$\left[y>0, \quad -\frac{1}{2}<\operatorname{Re}\mu<\operatorname{Re}\left(k+\frac{1}{2}\nu\right)-\frac{1}{4}\right] \quad \text{ET II 83(8)}$$

13. $\int_0^\infty x^{2\mu-\nu} e^{-\frac{1}{4}x^2} M_{k,\mu}\left(\frac{1}{2}x^2\right) Y_\nu(xy)\,dx$

$$= \pi^{-1} 2^{\mu+\beta} y^{k-\mu-\frac{3}{2}} e^{-\frac{1}{4}y^2} \Gamma(2\mu+1)$$

$$\times \Gamma\left(\frac{1}{2}-k-\mu\right)\left\{\cos[(\nu-2\mu)\pi]\frac{\Gamma(2\mu-\nu-1)}{\Gamma(2\beta+1)} M_{\alpha,\beta}\left(\frac{1}{2}y^2\right)\right.$$

$$\left.- \sin[(\nu+k-\mu)\pi] W_{\alpha,\beta}\left(\frac{1}{2}y^2\right)\right\}$$

$$2\alpha = 3\mu-\nu+k+\frac{1}{2}, \qquad 2\beta = \mu-\nu-k+\frac{1}{2}$$

$$\left[y>0, \quad -1<2\operatorname{Re}\mu<\operatorname{Re}(2k+\nu)+\frac{1}{2}, \quad \operatorname{Re}(2\mu-\nu)>-1\right] \quad \text{ET II 116(44)}$$

14. $\int_0^\infty x^{2\mu+\nu} e^{-\frac{1}{4}x^2} M_{k,\mu}\left(\frac{1}{2}x^2\right) Y_\nu(xy)\,dx$

$$= \pi^{-1} 2^{\mu+\beta} y^{k-\mu-\frac{3}{2}} \Gamma(2\mu+1)$$

$$\times \Gamma\left(\frac{1}{2}-\mu-k\right) e^{-\frac{1}{4}y^2}\left\{\cos(2\mu\pi)\frac{\Gamma(2\mu+\nu+1)}{\Gamma\left(\mu+\nu-k+\frac{3}{2}\right)} M_{\alpha,\beta}\left(\frac{1}{2}y^2\right)\right.$$

$$\left.+ \sin[(\mu-k)\pi] W_{\alpha,\beta}\left(\frac{1}{2}y^2\right)\right\}$$

$$2\alpha = 3\mu+\nu+k+\frac{1}{2}, \qquad 2\beta = \mu+\nu-k+\frac{1}{2}$$

$$\left[y>0, \quad -1<2\operatorname{Re}\mu<\operatorname{Re}(2k-\nu)+\frac{1}{2}, \quad \operatorname{Re}(2\mu+\nu)>-1\right] \quad \text{ET II 116(43)}$$

15. $\int_0^\infty x^{2\mu+\nu} e^{-\frac{1}{2}ax^2} M_{k,\mu}\left(ax^2\right) K_\nu(xy)\,dx = 2^{\mu-k-\frac{1}{2}} a^{\frac{1}{4}-\frac{1}{2}(\mu+\nu+k)} y^{k-\mu-\frac{3}{2}}$

$$\times \Gamma(2\mu+1)\Gamma(2\mu+\nu+1)\exp\left(\frac{y^2}{8a}\right) W_{\kappa,m}\left(\frac{y^2}{4a}\right),$$

$$2\kappa = -3\mu-\nu-k-\frac{1}{2}, \qquad 2m = \mu+\nu-k+\frac{1}{2}$$

$$\left[\operatorname{Re}y>0, \quad \operatorname{Re}a>0, \quad \operatorname{Re}\mu>-\frac{1}{2}, \quad \operatorname{Re}(2\mu+\nu)>-1\right] \quad \text{ET II 152(82)}$$

7.673

1.[10] $\displaystyle\int_0^\infty e^{-\frac{1}{2}ax} x^{\frac{1}{2}(\mu-\nu-1)} M_{\kappa,\frac{1}{2}\mu}(ax) J_\nu\left(2\sqrt{bx}\right) dx$

$$= \left(\frac{b}{a}\right)^{\frac{\kappa-1}{2}-\frac{1+\mu}{4}} a^{-\frac{1}{2}(\mu+1-\nu)} \Gamma(1+\mu) e^{-\frac{b}{2a}} \frac{1}{\Gamma\left(1+\dfrac{\kappa+\nu}{2}-\dfrac{1+\mu}{4}\right)}$$

$$\times M_{\frac{1}{2}(\kappa-\nu-1)+\frac{3}{4}(1+\mu),\frac{\kappa+\nu}{2}-\frac{1+\mu}{4}}\left(\frac{b}{a}\right)$$

$$\left[\operatorname{Re}(1+\mu)>0, \quad \operatorname{Re}\left(\kappa+\frac{\nu-\mu}{2}\right)>-\frac{3}{4}, \quad \operatorname{Im} b = 0\right] \quad \text{BU 128(12)a}$$

2. $\displaystyle\int_0^\infty e^{\frac{1}{2}ax} x^{\frac{1}{2}(\nu-1\mp\mu)} W_{\kappa,\frac{1}{2}\mu}(ax) J_\nu\left(2\sqrt{bx}\right) dx = a^{-\frac{1}{2}(\nu+1\mp\mu)} \frac{\Gamma\left(\nu+1\mp\mu\right) e^{\frac{b}{2a}}}{\Gamma\left(\frac{1\pm\mu}{2}-\kappa\right)} \left(\frac{a}{b}\right)^{\frac{1}{2}(\kappa+1)+\frac{1}{4}(1\mp\nu)}$

$$\times W_{\frac{1}{2}(\kappa+1-\nu)-\frac{3}{4}(1\mp\mu),\frac{1}{2}(\kappa+\nu)+\frac{1}{4}(1\mp\nu)}\left(\frac{b}{a}\right)$$

$$\left[\operatorname{Re}\left(\frac{\nu\mp\mu}{2}+\kappa\right)<\frac{3}{4}, \quad \operatorname{Re}\nu>-1\right] \quad \text{BU 128(13)}$$

7.674

1. $\displaystyle\int_0^\infty x^{\varrho-1} e^{-\frac{1}{2}\kappa} J_{\lambda+\nu}\left(ax^{1/2}\right) J_{\lambda-\nu}\left(ax^{1/2}\right) W_{k,\mu}(x) dx$

$$= \frac{\left(\frac{1}{2}a\right)^{2\lambda} \Gamma\left(\frac{1}{2}+\lambda+\mu+\varrho\right)\Gamma\left(\frac{1}{2}+\lambda-\mu+\varrho\right)}{\Gamma(1+\lambda+\nu)\Gamma(1+\lambda-\nu)\Gamma(1+\lambda-k+\varrho)}$$

$$\times {}_4F_4\left(1+\lambda,\frac{1}{2}+\lambda,\frac{1}{2}+\lambda+\mu+\varrho,\frac{1}{2}+\lambda-\mu+\varrho;1+\lambda+\nu,\right.$$

$$\left.1+\lambda-\nu,1+2\lambda,1+\lambda-k+\varrho;-a^2\right)$$

$$\left[|\operatorname{Re}\mu|<\operatorname{Re}(\lambda+\varrho)+\frac{1}{2}\right] \quad \text{ET II 409(37)}$$

2. $\displaystyle\int_0^\infty x^{\varrho-1} e^{-\frac{1}{2}\kappa} I_{\lambda+\nu}\left(ax^{1/2}\right) K_{\lambda-\nu}\left(ax^{1/2}\right) W_{k,\mu}(x) dx$

$$= \frac{\pi^{-1/2}}{2} G_{45}^{24}\left(a^2 \left|\begin{array}{c} 0,\frac{1}{2},\frac{1}{2}+\mu-\varrho,\frac{1}{2}-\mu-\varrho \\ \lambda,\nu,-\lambda,-\nu,k-\varrho \end{array}\right.\right)$$

$$\left[|\operatorname{Re}\mu|<\operatorname{Re}(\lambda+\varrho)+\frac{1}{2}, \quad |\operatorname{Re}\mu|<\operatorname{Re}(\nu+\varrho)+\frac{1}{2}\right] \quad \text{ET II 409(38)}$$

Combinations of Struve functions and confluent hypergeometric functions

7.675

1. $\displaystyle\int_0^\infty x^{2\lambda+\frac{1}{2}} e^{-\frac{1}{4}x^2} M_{k,\mu}\left(\frac{1}{2}x^2\right) \mathbf{H}_\nu(xy) dx = \frac{2^{-\lambda}\Gamma(2\mu+1)}{y^{1/2}\Gamma\left(\frac{1}{2}+k+\mu\right)} G_{34}^{22}\left(\frac{y^2}{2} \left|\begin{array}{c} l,-\mu-\lambda,mu-\lambda \\ l,k-\lambda-\frac{1}{2},h,\kappa \end{array}\right.\right)$

$$h = \frac{1}{4}+\frac{1}{2}\nu, \quad \kappa = \frac{1}{4}-\frac{1}{2}\nu, \quad l = \frac{3}{4}+\frac{1}{2}\nu$$

$$\left[\operatorname{Re}(2\lambda+2\mu+\nu)>-\frac{7}{2}, \quad \operatorname{Re}(k-\lambda)>0, \quad y>0, \quad \operatorname{Re}(2\lambda-2k+\nu)<-\frac{1}{2}\right]$$

$$\text{ET II 171(42)}$$

2. $$\int_0^\infty x^{2\lambda+\frac{1}{2}} e^{-\frac{1}{4}x^2} W_{k,\mu}\left(\frac{1}{2}x^2\right) \mathbf{H}_\nu(xy)\,dx$$

$$= 2^{\frac{1}{4}-\lambda-\frac{1}{2}\nu}\pi^{-1/2}y^{\nu+1}\frac{\Gamma\left(\frac{7}{4}+\frac{1}{2}\nu+\lambda+\mu\right)\Gamma\left(\frac{7}{4}+\frac{1}{2}\nu+\lambda-\mu\right)}{\Gamma\left(\nu+\frac{3}{2}\right)\Gamma\left(\frac{9}{4}+\lambda-k-\frac{1}{2}\nu\right)}$$

$$\times\ {}_3F_3\left(1,\frac{7}{4}+\frac{\nu}{2}+\lambda+\mu,\frac{7}{4}+\frac{\nu}{2}+\lambda-\mu;\frac{3}{2},\nu+\frac{3}{2},\frac{9}{4}+\lambda-k+\frac{\nu}{2};-\frac{y^2}{2}\right)$$

$$\left[\operatorname{Re}(2\lambda+\nu)>2|\operatorname{Re}\mu|-\tfrac{7}{4},\quad y>0\right]\quad \text{ET II 171(43)}$$

3. $$\int_0^\infty x^{2\lambda+\frac{1}{2}} e^{\frac{1}{4}x^2} W_{k,\mu}\left(\frac{1}{2}x^2\right) \mathbf{H}_\nu(xy)\,dx$$

$$= \left[2^\lambda\,\Gamma\left(\frac{1}{2}-k+\mu\right)\Gamma\left(\frac{1}{2}-k-\mu\right)\right]^{-1} y^{-1/2}\,G^{23}_{34}\left(\frac{y^2}{2}\left|\begin{array}{c}l,-\mu-\lambda,\mu-\lambda\\ l,-k-\lambda-\frac{1}{2},h,\kappa\end{array}\right.\right)$$

$$h=\tfrac{1}{4}+\tfrac{1}{2}\nu,\quad \kappa=\tfrac{1}{4}-\tfrac{1}{2}\nu,\quad l=\tfrac{3}{4}+\tfrac{1}{2}\nu$$

$$\left[y>0,\quad \operatorname{Re}(2\lambda+\nu)>2|\operatorname{Re}\mu|-\tfrac{7}{2},\quad \operatorname{Re}(2k+2\lambda+\nu)<-\tfrac{1}{2},\quad \operatorname{Re}(k+\lambda)<0\right]\quad \text{ET II 172(46)a}$$

4. $$\int_0^\infty e^{\frac{1}{2}x^2} W_{-\frac{1}{2}\nu-\frac{1}{2},\frac{1}{2}\nu}\left(x^2\right)\mathbf{H}_\nu(xy)\,dx = 2^{-\nu-1}y^\nu\pi e^{\frac{1}{4}y^2}\left[1-\Phi\left(\frac{y}{2}\right)\right]$$

$$\left[y>0,\quad \operatorname{Re}\nu>-1\right]\qquad \text{ET II 171(44)}$$

7.68 Combinations of confluent hypergeometric functions and other special functions

Combinations of confluent hypergeometric functions and associated Legendre functions

7.681

1. $$\int_0^\infty x^{-1/2}(a+x)^\mu e^{-\frac{1}{2}x} P_\nu^{-2\mu}\left(1+2\frac{x}{a}\right) M_{k,\mu}(x)\,dx$$

$$= -\frac{\sin(\nu\pi)}{\pi\,\Gamma(k)}\,\Gamma(2\mu+1)\,\Gamma\left(k-\mu+\nu+\tfrac{1}{2}\right)\Gamma\left(k-\mu-\nu-\tfrac{1}{2}\right)e^{\frac{1}{2}a}\,W_{\varrho,\sigma}(a),$$

$$\varrho=\tfrac{1}{2}-k+\mu,\qquad \sigma=\tfrac{1}{2}+\nu$$

$$\left[|\arg a|<\pi,\quad \operatorname{Re}\mu>-\tfrac{1}{2},\quad \operatorname{Re}(k-\mu)>|\operatorname{Re}\nu+\tfrac{1}{2}|\right]\quad \text{ET II 403(11)}$$

2. $$\int_0^\infty x^{-1/2}(a+x)^{-\mu} e^{-\frac{1}{2}x} P_\nu^{-2\mu}\left(1+2\frac{x}{a}\right) M_{k,\mu}(x)\,dx$$

$$= \frac{\Gamma(2\mu+1)\,\Gamma\left(k+\mu+\nu+\tfrac{1}{2}\right)\Gamma\left(k+\mu-\nu-\tfrac{1}{2}\right)e^{\frac{1}{2}a}}{\Gamma\left(k+\mu+\tfrac{1}{2}\right)\Gamma(2\mu+\nu+1)\,\Gamma(2\mu-\nu)}\,W_{\frac{1}{2}-k-\mu,\frac{1}{2}+\nu}(a)$$

$$\left[|\arg a|<\pi,\quad \operatorname{Re}\mu>-\tfrac{1}{2},\quad \operatorname{Re}(k+\mu)>|\operatorname{Re}\nu+\tfrac{1}{2}|\right]\quad \text{ET II 403(12)}$$

3. $$\int_0^\infty x^{-\frac{1}{2}-\frac{1}{2}\mu-\nu}(a+x)^{\frac{1}{2}\mu} e^{-\frac{1}{2}x} P^\mu_{k+\nu-\frac{3}{2}}\left(1+2\frac{x}{a}\right) W_{k,\nu}(x)\,dx$$

$$= \frac{\Gamma(1-\mu-2\nu)}{\Gamma\left(\frac{3}{2}-k-\mu-\nu\right)}a^{-\frac{1}{4}+\frac{1}{2}k-\frac{1}{2}\nu}e^{\frac{1}{2}a}\,W_{\varrho,\sigma}(a)$$

$$2\varrho=\tfrac{1}{2}+2\mu+\nu-k,\quad 2\sigma=k+3\nu-\tfrac{3}{2}$$

$$\left[|\arg a|<\pi,\quad \operatorname{Re}\mu<1,\quad \operatorname{Re}(\mu+2\nu)<1\right]$$

$$\text{ET II 407(32)}$$

4. $\int_0^\infty x^{-\frac{1}{2}-\frac{1}{2}\mu-\nu}(a+x)^{-\frac{1}{2}\mu}e^{-\frac{1}{2}x}P_{k+\mu+\nu-\frac{3}{2}}^{\mu}\left(1+2\dfrac{x}{a}\right)W_{k,\nu}(x)\,dx$

$$= \frac{\Gamma(1-\mu-2\nu)}{\Gamma\left(\frac{3}{2}-k-\mu-\nu\right)}a^{-\frac{1}{2}+\frac{1}{2}k-\frac{1}{2}\nu}e^{\frac{1}{2}a}W_{\varrho,\sigma}(a)$$

$$2\varrho = \tfrac{1}{2}-k+\nu, \qquad 2\sigma = k+2\mu+3\nu-\tfrac{3}{2}$$

$$\left[|\arg a|<\pi, \quad \operatorname{Re}\mu<1, \quad \operatorname{Re}(\mu+2\nu)<1\right]$$

ET II 408(33)

5. $\int_0^\infty x^{\mu-\frac{1}{4}k-\frac{1}{2}\nu-\frac{1}{2}}(a+x)^{\frac{1}{2}\nu}e^{-\frac{1}{2}x}Q_{\mu-k+\frac{3}{2}}^{\nu}\left(1+2\dfrac{x}{a}\right)M_{k,\nu}(x)\,dx$

$$= \frac{e^{\nu\pi i}\,\Gamma(1+2\mu-\nu)\,\Gamma(1+2\mu)\,\Gamma\left(\frac{5}{2}-k+\mu+\nu\right)}{2\,\Gamma\left(\frac{1}{2}+k+\mu\right)}a^{\frac{1}{4}(\kappa+2\mu-2\nu+5)}e^{\frac{1}{2}a}W_{\varrho,\sigma}(a)$$

$$2\varrho = \tfrac{1}{2}-k-\mu+2\nu, \qquad 2\sigma = k-3\mu-\tfrac{3}{2}$$

$$\left[|\arg a|<\pi, \quad \operatorname{Re}\mu>-\tfrac{1}{2}, \quad \operatorname{Re}(2\mu-\nu)>-1\right]$$

ET II 404(14)

7.682

1. $\int_0^\infty x^{-1/2}e^{-\frac{1}{2}x}P_\nu^{-2\mu}\left[\left(1+\dfrac{x}{a}\right)^{1/2}\right]M_{k,\mu}(x)\,dx$

$$= \frac{\Gamma(2\mu+1)\,\Gamma\left(k+\frac{1}{2}\nu\right)\,\Gamma\left(k-\frac{1}{2}\nu-\frac{1}{2}\right)e^{\frac{1}{2}a}}{2^{2\mu}a^{1/4}\,\Gamma\left(k+\mu+\frac{1}{2}\right)\,\Gamma\left(\mu+\frac{1}{2}\nu+\frac{1}{2}\right)\,\Gamma\left(\mu-\frac{1}{2}\nu\right)}W_{\frac{3}{4}-k,\frac{1}{4}+\frac{1}{2}\nu}(a)$$

$$\left[|\arg a|<\pi, \quad \operatorname{Re}k>\tfrac{1}{2}\operatorname{Re}\nu-\tfrac{1}{2}, \quad \operatorname{Re}k>-\tfrac{1}{2}\operatorname{Re}\nu\right] \quad \text{ET II 404(13)}$$

2. $\int_0^\infty x^{\frac{1}{2}(k+\mu+\nu)-1}(a+x)^{-1/2}e^{-\frac{1}{2}x}Q_{k-\mu-\nu-1}^{1-k+\mu-\nu}\left[\left(1+\dfrac{x}{a}\right)^{1/2}\right]M_{k,\mu}(x)\,dx$

$$= e^{(1-k+\mu-\nu)\pi i}2^{\mu-k-\nu}a^{\frac{1}{2}(k+\mu-1)}\frac{\Gamma\left(\frac{1}{2}-\nu\right)\Gamma(1+2\mu)\Gamma(k+\mu+\nu)}{\Gamma\left(k+\mu+\frac{1}{2}\right)}e^{\frac{1}{2}a}W_{\varrho,\sigma}(a),$$

$$\varrho = \tfrac{1}{2}-k-\tfrac{1}{2}\nu, \qquad \sigma = \mu+\tfrac{1}{2}\nu$$

$$\left[|\arg a|<\pi, \quad \operatorname{Re}\mu>-\tfrac{1}{2}, \quad \operatorname{Re}(k+\mu+\nu)>0\right] \quad \text{ET II 404(15)}$$

3. $\int_0^\infty x^{\nu-\frac{1}{2}}e^{-\frac{1}{2}x}Q_{2k-2\nu-3}^{2\mu-2\nu}\left[\left(1+\dfrac{x}{a}\right)^{1/2}\right]M_{k,\mu}(x)\,dx$

$$= e^{2(\mu-\nu)\pi i}2^{2\mu-2\nu-1}a^{\frac{1}{2}(k+\mu-1)}e^{\frac{1}{2}a}\frac{\Gamma(2\mu+1)\,\Gamma(\nu+1)\,\Gamma\left(k+\mu-2\nu-\frac{1}{2}\right)}{\Gamma\left(k+\mu+\frac{1}{2}\right)}W_{\varrho,\sigma}(a),$$

$$2\varrho = 1-k+\mu-2\nu, \qquad 2\sigma = k-\mu-2\nu-2$$

$$\left[|\arg a|<\pi, \quad \operatorname{Re}\mu>-\tfrac{1}{2}, \quad \operatorname{Re}\nu>-1, \quad \operatorname{Re}(k+\mu-2\nu)>\tfrac{1}{2}\right] \quad \text{ET II 404(16)}$$

4. $\int_0^\infty x^{-\frac{1}{2}-\frac{1}{2}\mu-\nu}e^{-\frac{1}{2}x}P_{2k+\mu+2\nu-3}^{\mu}\left[\left(1+\dfrac{x}{a}\right)^{\frac{1}{2}}\right]W_{k,\nu}(x)\,dx$

$$= \frac{2^\mu\,\Gamma(1-\mu-2\nu)}{\Gamma\left(\frac{3}{2}-k-\mu-\nu\right)}a^{-\frac{1}{2}+\frac{1}{2}k-\frac{1}{2}\nu}e^{\frac{1}{2}a}W_{\varrho,\sigma}(a),$$

$$2\varrho = 1-k+\mu+\nu, \qquad 2\sigma = k+\mu+3\nu-2$$

$$\left[|\arg a|<\pi, \quad \operatorname{Re}\mu<1, \quad \operatorname{Re}(\mu+2\nu)<1\right]$$

ET II 408(34)

5.^{8} $\quad \int_0^\infty x^{-\frac{1}{2}-\frac{1}{2}\mu-\nu}(a+x)^{-1/2}e^{-\frac{1}{2}x}\,P_{2k+\mu+2\nu-2}^\mu\left[\left(1+\frac{x}{a}\right)^{1/2}\right]W_{k,\nu}(x)\,dx$

$$= \frac{2^\mu\,\Gamma(1-\mu-2\nu)}{\Gamma\left(\frac{3}{2}-k-\mu-\nu\right)}a^{-\frac{1}{2}+\frac{1}{2}k-\frac{1}{2}\nu}e^{\frac{1}{2}a}\,W_{\varrho,\sigma}(a), \quad 2\varrho=\mu+\nu-k, \quad 2\sigma=k+\mu+3\nu-1$$

$$[|\arg a|<\pi, \quad \operatorname{Re}\mu>0, \quad \operatorname{Re}\nu>0] \quad \text{ET II 408(35)}$$

A combination of confluent hypergeometric functions and orthogonal polynomials

7.683^8 $\quad \int_0^1 e^{-\frac{1}{2}ax}x^\alpha(1-x)^{\frac{\mu-\alpha}{2}-1}\,L_n^\alpha(ax)\,M_{\alpha-\frac{1+\alpha}{2},\frac{\mu-\alpha-1}{1}}\,[a(1-x)]\,dx$

$$= \frac{\Gamma(\mu-\alpha)}{\Gamma(1+\mu)}\frac{\Gamma(1+n+\alpha)}{n!}a^{-\frac{1+\alpha}{2}}\,M_{\alpha+n,\frac{\mu}{2}}(a)$$

$$[\operatorname{Re}a>-1, \quad \operatorname{Re}(\mu-\alpha)>0, \quad n=0,1,2,\ldots] \quad \text{BU 129(14b)}$$

A combination of hypergeometric and confluent hypergeometric functions

7.684 $\quad \int_0^\infty x^{\varrho-1}e^{-\frac{1}{2}x}\,M_{\gamma+\varrho,\beta+\varrho+\frac{1}{2}}(x)\ _2F_1\left(\alpha,\beta;\gamma;-\frac{\lambda}{x}\right)dx$

$$= \frac{\Gamma(\alpha+\beta+2\varrho)\,\Gamma(2\beta+2\varrho)\,\Gamma(\gamma)}{\Gamma(\beta)\,\Gamma(\beta+\gamma+2\varrho)}\lambda^{\frac{1}{2}\beta+\varrho-\frac{1}{2}}e^{\frac{1}{2}\lambda}\,W_{k,\mu}(\lambda);$$

$$k=\tfrac{1}{2}-\alpha-\tfrac{1}{2}\beta-\varrho, \qquad \mu=\tfrac{1}{2}\beta+\varrho$$

$$[|\arg\lambda|<\pi, \quad \operatorname{Re}(\beta+\varrho)>0, \quad \operatorname{Re}(\alpha+\beta+2\varrho)>0, \quad \operatorname{Re}\gamma>0]$$

$$\text{ET II 405(19)}$$

7.69 Integration of confluent hypergeometric functions with respect to the index

7.691 $\quad \int_{-\infty}^\infty \operatorname{sech}(\pi x)\,W_{ix,0}(\alpha)\,W_{-ix,0}(\beta)\,dx = 2\frac{(\alpha\beta)^{1/2}}{\alpha+\beta}\exp\left[-\frac{1}{2}(\alpha+\beta)\right]$ $\qquad$ ET II 414(61)

7.692 $\quad \int_{-i\infty}^{i\infty}\Gamma(-a)\,\Gamma(c-a)\Psi(a,c;x)\Psi(c-a,c;y)\,da = 2\pi i\,\Gamma(c)\Psi(c,2c;x+y).$ $\qquad$ EH I 285(15)

7.693

1. $\quad \int_{-\infty}^\infty \Gamma(ix)\,\Gamma(2k+ix)\,W_{k+ix,k-\frac{1}{2}}(\alpha)\,W_{-k-ix,k-\frac{1}{2}}(\beta)\,dx$

$$= 2\pi^{1/2}\,\Gamma(2k)(\alpha\beta)^k(\alpha+\beta)^{\frac{1}{2}-2k}\,K_{2k-\frac{1}{2}}\left(\frac{\alpha+\beta}{2}\right)$$

$$\text{ET II 414(62)}$$

2. $\quad \int_{-i\infty}^{i\infty}\Gamma\left(\tfrac{1}{2}+\nu+\mu+x\right)\Gamma\left(\tfrac{1}{2}+\nu+\mu-x\right)\Gamma\left(\tfrac{1}{2}+\nu-\mu+x\right)\Gamma\left(\tfrac{1}{2}+\nu-\mu-x\right)$

$$\times M_{\mu+ix,\nu}(\alpha)\,M_{\mu-ix,\nu}(\beta)\,dx$$

$$= \frac{2\pi(\alpha\beta)^{\nu+\frac{1}{2}}[\Gamma(2\nu+1)]^2\,\Gamma(2\nu+2\mu+1)\,\Gamma(2\nu-2\mu+1)}{(\alpha+\beta)^{2\nu+1}\,\Gamma(4\nu+2)}\,M_{2\mu,2\nu+\frac{1}{2}}(\alpha+\beta)$$

$$[\operatorname{Re}\nu>|\operatorname{Re}\mu|-\tfrac{1}{2}] \qquad \text{ET II 413(59)}$$

7.694
$$\int_{-\infty}^{\infty} e^{-2\varrho x i}\, \Gamma\left(\tfrac{1}{2}+\nu+ix\right) \Gamma\left(\tfrac{1}{2}+\nu-ix\right) M_{ix,\nu}(\alpha)\, M_{ix,\nu}(\beta)\, dx$$

$$= \frac{2\pi(\alpha\beta)^{1/2}}{\cosh\varrho} \exp\left[-(\alpha+\beta)\tanh\varrho\right] J_{2\nu}\left(\frac{2\alpha^{1/2}\beta^{1/2}}{\cosh\varrho}\right)$$
$$\left[|\mathrm{Im}\,\varrho| < \tfrac{1}{2}\pi, \quad \mathrm{Re}\,\nu > -\tfrac{1}{2}\right]$$

7.7 Parabolic Cylinder Functions

7.71 Parabolic cylinder functions

7.711

1.
$$\int_{-\infty}^{\infty} D_n(x)\, D_m(x)\, dx = 0 \qquad\qquad [m \neq n]$$
$$= n!(2\pi)^{1/2} \qquad\qquad [m = n]$$

WH

2.
$$\int_{0}^{\infty} D_\mu\left(\pm t\right) D_\nu(t)\, dt = \frac{\pi 2^{\frac{1}{2}(\mu+\nu+1)}}{\mu-\nu}\left[\frac{1}{\Gamma\left(\tfrac{1}{2}-\tfrac{1}{2}\mu\right)\Gamma\left(-\tfrac{1}{2}\nu\right)} \mp \frac{1}{\Gamma\left(\tfrac{1}{2}-\tfrac{1}{2}\nu\right)\Gamma\left(-\tfrac{1}{2}\mu\right)}\right]$$
$$[\text{when the lower sign is taken, } \mathrm{Re}\,\mu > \mathrm{Re}\,\nu] \quad \text{BU 11 117(13a), EH II 122(21)}$$

3.
$$\int_{0}^{\infty} [D_\nu(t)]^2\, dt = \pi^{1/2} 2^{-3/2} \frac{\psi\left(\tfrac{1}{2}-\tfrac{1}{2}\nu\right) - \psi\left(-\tfrac{1}{2}\nu\right)}{\Gamma(-\nu)} \qquad \text{BU 117(13b)a, EH II 122(22)a}$$

7.72 Combinations of parabolic cylinder functions, powers, and exponentials

7.721

1.
$$\int_{-\infty}^{\infty} e^{-\frac{1}{4}x^2}(x-z)^{-1} D_n(x)\, dx = \pm i e^{\mp n\pi i}(2\pi)^{1/2} n!\, e^{-\frac{1}{4}z^2} D_{-n-1}\left(\mp iz\right)$$

[The upper or lower sign is taken according as the imaginary part of z is positive or negative.]
WH

2.
$$\int_{1}^{\infty} x^\nu (x-1)^{\frac{1}{2}\mu-\frac{1}{2}\nu-1} \exp\left[-\frac{(x-1)^2 a^2}{4}\right] D_\mu(ax)\, dx = 2^{\mu-\nu-2} a^{\frac{\mu}{2}-\frac{\nu}{2}-1} \Gamma\left(\frac{\mu-\nu}{2}\right) D_\nu(a)$$
$$[\mathrm{Re}(\mu-\nu) > 0] \qquad \text{ET II 395(4)a}$$

7.722

1.
$$\int_{0}^{\infty} e^{-\frac{3}{4}x^2} x^\nu D_{\nu+1}(x)\, dx = 2^{-\frac{1}{2}-\frac{1}{2}\nu} \Gamma(\nu+1) \sin\frac{1}{4}(1-\nu)\pi$$
$$[\mathrm{Re}\,\nu > -1] \qquad \text{WH}$$

2.
$$\int_{0}^{\infty} e^{-\frac{1}{4}x^2} x^{\mu-1} D_{-\nu}(x)\, dx = \frac{\pi^{1/2} 2^{-\frac{1}{2}\mu-\frac{1}{2}\nu} \Gamma(\mu)}{\Gamma\left(\tfrac{1}{2}\mu+\tfrac{1}{2}\nu+\tfrac{1}{2}\right)} \qquad [\mathrm{Re}\,\mu > 0] \qquad \text{EH II 122(20)}$$

3.
$$\int_{0}^{\infty} e^{-\frac{3}{4}x^2} x^\nu D_{\nu-1}(x)\, dx = 2^{-\frac{1}{2}\nu-1} \Gamma(\nu) \sin\frac{1}{4}\pi\nu \qquad [\mathrm{Re}\,\nu > -1] \qquad \text{ET II 395(2)}$$

7.723

1. $$\int_0^\infty e^{-\frac{1}{4}x^2} x^\nu \left(x^2 + y^2\right)^{-1} D_\nu(x)\, dx = \left(\frac{\pi}{2}\right)^{1/2} \Gamma(\nu+1) y^{\nu-1} e^{\frac{1}{4}y^2} D_{-\nu-1}(y)$$

$$[\operatorname{Re} y > 0, \quad \operatorname{Re} \nu > -1]$$

EH II 121(18)a, ET II 396(6)a

2. $$\int_0^\infty e^{-\frac{1}{4}x^2} x^{\nu-1} \left(x^2 + y^2\right)^{-1/2} D_\nu(x)\, dx = y^{\nu-1} \Gamma(\nu) e^{\frac{1}{4}y^2} D_{-\nu}(y)$$

$$[\operatorname{Re} y > 0, \quad \operatorname{Re} \nu > 0] \qquad \text{ET II 396(7)}$$

3. $$\int_0^1 x^{2\nu-1} \left(1 - x^2\right)^{\lambda-1} e^{\frac{a^2 x^2}{4}} D_{-2\lambda-2\nu}(ax)\, dx = \frac{\Gamma(\lambda)\,\Gamma(2\nu)}{\Gamma(2\lambda+2\nu)} 2^{\lambda-1} e^{\frac{a^2}{4}} D_{-2\nu}(a)$$

$$[\operatorname{Re} \lambda > 0, \quad \operatorname{Re} \nu > 0] \qquad \text{ET II 395(3)a}$$

7.724 $$\int_{-\infty}^\infty e^{-\frac{(x-y)^2}{2\mu}} e^{\frac{1}{4}x^2} D_\nu(x)\, dx = (2\pi\mu)^{1/2} (1-\mu)^{\frac{1}{2}\nu} e^{\frac{y^2}{4-4\mu}} D_\nu\left[y(1-\mu)^{-1/2}\right] \qquad [0 < \operatorname{Re} \mu < 1]$$

EH II 121(15)

7.725

1. $$\int_0^\infty e^{-pt} (2t)^{\frac{\nu-1}{2}} e^{-\frac{t}{2}} D_{-\nu-2}\left(\sqrt{2t}\right) dt = \left(\frac{\pi}{2}\right)^{1/2} \frac{\left(\sqrt{p+1}-1\right)^{\nu+1}}{(\nu+1) p^{\nu+1}}$$

$$[\operatorname{Re} \nu > -1] \qquad \text{MO 175}$$

2. $$\int_0^\infty e^{-pt} (2t)^{\frac{\nu-1}{2}} e^{-\frac{t}{2}} D_{-\nu}\left(\sqrt{2t}\right) dt = \left(\frac{\pi}{2}\right)^{1/2} \frac{\left(\sqrt{p+1}-1\right)^\nu}{p^\nu \sqrt{p+1}}$$

$$[\operatorname{Re} \nu > -1] \qquad \text{MO 175}$$

3. $$\int_0^\infty e^{-bx} D_{2n+1}\left(\sqrt{2x}\right) dx = (-2)^n \Gamma\left(n + \frac{3}{2}\right) \left(b - \frac{1}{2}\right)^n \left(b + \frac{1}{2}\right)^{-n-\frac{3}{2}}$$

$$\left[\operatorname{Re} b > -\frac{1}{2}\right] \qquad \text{ET I 210(3)}$$

4. $$\int_0^\infty \left(\sqrt{x}\right)^{-1} e^{-bx} D_{2n}\left(\sqrt{2x}\right) dx = (-2)^n \Gamma\left(n + \frac{1}{2}\right) \left(b - \frac{1}{2}\right)^n \left(b + \frac{1}{2}\right)^{-n-\frac{1}{2}}$$

$$\left[\operatorname{Re} b > -\frac{1}{2}\right] \qquad \text{ET I 210(5)}$$

5. $$\int_0^\infty x^{-\frac{1}{2}(\nu+1)} e^{-sx} D_\nu\left(\sqrt{x}\right) dx = \sqrt{\pi}\left(1 + \sqrt{\frac{1}{2} + 2s}\right)^\nu \frac{1}{\sqrt{\frac{1}{4} + s}}$$

$$\left[\operatorname{Re} s > -\frac{1}{4}, \quad \operatorname{Re} \nu < 1\right] \qquad \text{ET I 210(7)}$$

6. $$\int_0^\infty e^{-zt} t^{-1+\frac{\beta}{2}} D_{-\nu}\left[2(kt)^{1/2}\right] dt = \frac{2^{1-\beta-\frac{\nu}{2}} \pi^{1/2} \Gamma(\beta)}{\Gamma\left(\frac{1}{2}\nu + \frac{1}{2}\beta + \frac{1}{2}\right)} (z+k)^{-\frac{\beta}{2}} F\left(\frac{\nu}{2}, \frac{\beta}{2}; \frac{\nu+\beta+1}{2}; \frac{z-k}{z+k}\right)$$

$$\left[\operatorname{Re}(z+k) > 0, \quad \operatorname{Re}\frac{z}{k} > 0\right]$$

EH II 121(11)

7.726 $$\int_{-\infty}^\infty e^{ixy - \frac{(1+\lambda)x^2}{4}} D_\nu\left[x(1-\lambda)^{1/2}\right] dx = (2\pi)^{1/2} \lambda^{\frac{1}{2}\nu} e^{-\frac{(1+\lambda)y^2}{4\lambda}} D_\nu\left[i\left(\lambda^{-1}-1\right)^{1/2} y\right] \qquad [\operatorname{Re} \lambda > 0]$$

EH II 121(16)

7.727
$$\int_0^\infty \frac{e^{\frac{1}{2}x}e^{-bx}}{(e^x-1)^{\mu+\frac{1}{2}}} \exp\left(-\frac{a}{1-e^{-x}}\right) D_{2\mu}\left(\frac{2\sqrt{a}}{\sqrt{1-e^{-x}}}\right) dx = e^{-a}2^{b+\mu}\,\Gamma(b+\mu)\,D_{-2b}\left(2\sqrt{a}\right)$$
$$[\operatorname{Re} a > 0, \quad \operatorname{Re} b > -\operatorname{Re}\mu]$$
ET I 211(13)

7.728 $\displaystyle\int_0^\infty (2t)^{-\frac{\nu}{2}} e^{-pt} e^{-\frac{q^2}{8t}}\, D_{\nu-1}\left(\frac{q}{\sqrt{2t}}\right) dt = \left(\frac{\pi}{2}\right)^{\frac{1}{2}} p^{\frac{1}{2}\nu-1}e^{-q\sqrt{p}}$ MO 175

7.73 Combinations of parabolic cylinder and hyperbolic functions

7.731

1. $\displaystyle\int_0^\infty \cosh(2\mu x) \exp\left[-(a\sinh x)^2\right] D_{2k}\left(2a\cosh x\right) dx = 2^{k-\frac{3}{2}}\pi^{1/2}a^{-1}\,W_{k,\mu}\left(2a^2\right)$
$$[\operatorname{Re}^2 a > 0] \qquad\qquad \text{ET II 398(20)}$$

2. $\displaystyle\int_0^\infty \cosh(2\mu x) \exp\left[(a\sinh x)^2\right] D_{2k}\left(2a\cosh x\right) dx = \frac{\Gamma(\mu-k)\,\Gamma(-\mu-k)}{2^{k+\frac{5}{2}}a\,\Gamma(-2k)}\,W_{k+\frac{1}{2},\mu}\left(2a^2\right)$
$$\left[|\arg a| < \frac{3\pi}{4}, \quad \operatorname{Re} k + |\operatorname{Re}\mu| < 0\right]$$
ET II 398(21)

7.74 Combinations of parabolic cylinder and trigonometric functions

7.741

1. $\displaystyle\int_0^\infty \sin(bx)\left\{[D_{-n-1}(ix)]^2 - [D_{-n-1}(-ix)]^2\right\} dx = (-1)^{n+1}\frac{i}{n!}\pi\sqrt{2\pi}e^{-\frac{1}{2}b^2}\,L_n\left(b^2\right)$
$$[b > 0] \qquad\qquad \text{ET I 115(3)}$$

2. $\displaystyle\int_0^\infty e^{-\frac{1}{4}x^2}\sin(bx)\,D_{2n+1}(x)\,dx = (-1)^n\sqrt{\frac{\pi}{2}}b^{2n+1}e^{-\frac{1}{2}b^2}$
$$[b > 0] \qquad\qquad \text{ET I 115(1)}$$

3. $\displaystyle\int_0^\infty e^{-\frac{1}{4}x^2}\cos(bx)\,D_{2n}(x)\,dx = (-1)^n\sqrt{\frac{\pi}{2}}b^{2n}e^{-\frac{1}{2}b^2}$ $[b > 0]$ ET I 60(2)

4. $\displaystyle\int_0^\infty e^{-\frac{1}{4}x^2}\sin(bx)\left[D_{2\nu-\frac{1}{2}}(x) - D_{2\nu-\frac{1}{2}}(-x)\right] dx = \sqrt{2\pi}\sin\left[\left(\nu-\frac{1}{4}\right)\pi\right]b^{2\nu-\frac{1}{2}}e^{-\frac{1}{2}b^2}$
$$\left[\operatorname{Re}\nu > \frac{1}{4}, \quad b > 0\right] \qquad \text{ET I 115(2)}$$

5. $\displaystyle\int_0^\infty e^{-\frac{1}{2}x^2}\cos(bx)\left[D_{2\nu-\frac{1}{2}}(x) + D_{2\nu-\frac{1}{2}}(-x)\right] dx = \frac{2^{\frac{1}{4}-2\nu}\sqrt{\pi}b^{2\nu-\frac{1}{2}}e^{-\frac{1}{4}b^2}}{\operatorname{cosec}\left[\left(\nu+\frac{1}{4}\right)\pi\right]}$
$$\left[\operatorname{Re}\nu > \frac{1}{4}, \quad b > 0\right] \qquad \text{ET I 61(4)}$$

7.742

1. $\displaystyle\int_0^\infty x^{2\varrho-1}\sin(ax)e^{-\frac{x^2}{4}}\,D_{2\nu}(x)\,dx = 2^{\nu-\varrho-\frac{1}{2}}\pi^{1/2}a\frac{\Gamma(2\varrho+1)}{\Gamma(\varrho-\nu+1)}$
$$\times\ {}_2F_2\left(\varrho+\frac{1}{2},\varrho+1;\frac{3}{2},\varrho-\nu+1;-\frac{a^2}{2}\right)$$
$$\left[\operatorname{Re}\varrho > -\frac{1}{2}\right] \qquad\qquad \text{ET II 396(8)}$$

2. $$\int_0^\infty x^{2\varrho-1} \sin(ax) e^{\frac{x^2}{4}} D_{2\nu}(x)\, dx = \frac{2^{\varrho-\nu-2}}{\Gamma(-2\nu)} G_{23}^{22}\left(\frac{a^2}{2}\left|\begin{array}{c} \frac{1}{2}-\varrho, 1-\varrho \\ -\varrho-\nu, \frac{1}{2}, 0 \end{array}\right.\right)$$

$$\left[a > 0, \quad \mathrm{Re}\,\varrho > -\tfrac{1}{2}, \quad \mathrm{Re}(\varrho+\nu) < \tfrac{1}{2}\right]$$
ET II 396(9)

3. $$\int_0^\infty x^{2\varrho-1} \cos(ax) e^{-\frac{x^2}{4}} D_{2\nu}(x)\, dx = \frac{2^{\nu-\varrho}\Gamma(2\varrho)\pi^{1/2}}{\Gamma\left(\varrho-\nu+\frac{1}{2}\right)}\, {}_2F_2\left(\varrho, \varrho+\frac{1}{2}; \frac{1}{2}, \varrho-\nu+\frac{1}{2}; -\frac{a^2}{2}\right)$$

$$[\mathrm{Re}\,\varrho > 0]$$
ET II 396(10)a

4. $$\int_0^\infty x^{2\varrho-1} \cos(ax) e^{\frac{x^2}{4}} D_{2\nu}(x)\, dx = \frac{2^{\varrho-\nu-2}}{\Gamma(-2\nu)} G_{23}^{22}\left(\frac{a^2}{2}\left|\begin{array}{c} \frac{1}{2}-\varrho, 1-\varrho \\ -\varrho-\nu, 0, \frac{1}{2} \end{array}\right.\right)$$

$$\left[a > 0, \quad \mathrm{Re}\,\varrho > 0, \quad \mathrm{Re}(\varrho+\nu) < \tfrac{1}{2}\right]$$
ET II 396(11)

7.743 $$\int_0^{\pi/2} (\cos x)^{-\mu-2}(\sin x)^{-\nu} D_\nu(a\sin x) D_\mu(a\cos x)\, dx = -\left(\tfrac{1}{2}\pi\right)^{1/2}(1+\mu)^{-1} D_{\mu+\nu+1}(a)$$

$$[\mathrm{Re}\,\nu < 1, \quad \mathrm{Re}\,\mu < -1]\qquad \text{ET II 397(19)}$$

7.744

1. $$\int_0^\infty \sin(bx)\left[D_{-\nu-\frac{1}{2}}\left(\sqrt{2x}\right) - D_{-\nu-\frac{1}{2}}\left(-\sqrt{2x}\right)\right] D_{\nu-\frac{1}{2}}\left(\sqrt{2x}\right) dx$$

$$= -\sqrt{2\pi}\sin\left[\left(\tfrac{1}{4}+\tfrac{1}{2}\nu\right)\pi\right] b^{-\nu-\frac{1}{2}} \frac{\left(1+\sqrt{1+b^2}\right)^\nu}{\sqrt{1+b^2}}$$

$$[b > 0]\qquad \text{ET I 115(4)}$$

2. $$\int_0^\infty \cos(bx)\left[D_{-2\nu-\frac{1}{2}}\left(\sqrt{2x}\right) + D_{-2\nu-\frac{1}{2}}\left(-\sqrt{2x}\right)\right] D_{2\nu-\frac{1}{2}}\left(\sqrt{2x}\right) dx$$

$$= -\frac{\sqrt{\pi}\sin\left[\left(\nu-\tfrac{1}{4}\right)\pi\right]\left(1+\sqrt{1+b^2}\right)^{2\nu}}{\sqrt{1+b^2}\, b^{2\nu+\frac{1}{2}}}$$

$$[b > 0]\qquad \text{ET I 60(3)}$$

7.75 Combinations of parabolic cylinder and Bessel functions

7.751

1. $$\int_0^\infty [D_n(ax)]^2 J_1(xy)\, dx = (-1)^{n-1} y^{-1}\left[D_n\left(\frac{y}{a}\right)\right]^2 \qquad [y > 0]\qquad \text{ET II 20(24)}$$

2. $$\int_0^\infty J_0(xy) D_n(ax) D_{n+1}(ax)\, dx = (-1)^n y^{-1} D_n\left(\frac{y}{a}\right) D_{n+1}\left(\frac{y}{a}\right)$$

$$[y > 0, \quad |\arg a| < \tfrac{1}{4}\pi]\qquad \text{ET II 17(42)}$$

3. $$\int_0^\infty J_0(xy) D_\nu(x) D_{\nu+1}(x)\, dx = 2^{-1} y^{-1}\left[D_\nu(-y) D_{\nu+1}(y) - D_{\nu+1}(-y) D_\nu(y)\right] \qquad \text{ET II 397(17)a}$$

7.752

1. $$\int_0^\infty x^\nu e^{-\frac{1}{4}x^2} D_{2\nu-1}(x) J_\nu(xy)\, dx = -\frac{1}{2}\sec(\nu\pi) y^{\nu-1} e^{-\frac{1}{4}y^2}\left[D_{2\nu-1}(y) - D_{2\nu-1}(-y)\right]$$

$$\left[y > 0, \quad \mathrm{Re}\,\nu > -\tfrac{1}{2}\right]$$
ET II 76(1), MO 183

2. $\displaystyle\int_0^\infty x^\nu e^{\frac{1}{4}x^2} D_{2\nu-1}(x) J_\nu(xy)\, dx = 2^{\frac{1}{2}-\nu}\pi \sin(\nu\pi) y^{-\nu}\Gamma(2\nu) e^{\frac{1}{4}y^2} K_\nu\left(\tfrac{1}{4}y^2\right)$

$$\left[y > 0, \quad -\tfrac{1}{2} < \operatorname{Re}\nu < \tfrac{1}{2}\right] \qquad \text{ET II 77(4)}$$

3. $\displaystyle\int_0^\infty x^{\nu+1} e^{-\frac{1}{4}x^2} D_{2\nu}(x) J_\nu(xy)\, dx = \frac{1}{2}\sec(\nu\pi) y^{\nu-1} e^{-\frac{1}{4}y^2}\left[D_{2\nu+1}(y) - D_{2\nu+1}(-y)\right]$

$$\left[y > 0, \quad \operatorname{Re}\nu > -1\right] \qquad \text{ET II 78(13)}$$

4. $\displaystyle\int_0^\infty x^\nu e^{-\frac{1}{4}x^2} D_{2\nu+1}(x) J_\nu(xy)\, dx = \frac{1}{2}\sec(\nu\pi) e^{-\frac{1}{4}y^2} y^\nu\left[D_{2\nu}(y) + D_{2\nu}(-y)\right]$

$$\left[y > 0, \quad \operatorname{Re}\nu > -\tfrac{1}{2}\right] \qquad \text{ET II 77(5)}$$

5. $\displaystyle\int_0^\infty x^{\nu+1} e^{-\frac{1}{4}x^2} D_{2\nu+2}(x) J_\nu(xy)\, dx = -\tfrac{1}{2}\sec(\nu\pi) y^\nu e^{-\frac{1}{4}y^2}\left[D_{2\nu+2}(y) + D_{2\nu+2}(-y)\right]$

$$\left[\operatorname{Re}\nu > -1, \quad y > 0\right] \qquad \text{ET II 78(16)}$$

6. $\displaystyle\int_0^\infty x^{\nu+1} e^{\frac{1}{4}x^2} D_{2\nu+2}(x) J_\nu(xy)\, dx = \pi^{-1}\sin(\nu\pi)\Gamma(2\nu+3) y^{-\nu-2} e^{\frac{1}{4}y^2} K_{\nu+1}\left(\tfrac{1}{4}y^2\right)$

$$\left[y > 0, \quad -1 < \operatorname{Re}\nu < -\tfrac{5}{6}\right]$$
$$\text{ET II 78(19)}$$

7. $\displaystyle\int_0^\infty x^\nu e^{-\frac{1}{4}x^2} D_{-2\nu}(x) J_\nu(xy)\, dx = 2^{-1/2}\pi^{1/2} y^{-\nu} e^{-\frac{1}{4}y^2} I_\nu\left(\tfrac{1}{4}y^2\right)$

$$\left[y > 0, \quad \operatorname{Re}\nu > -\tfrac{1}{2}\right] \qquad \text{ET II 77(8)}$$

8. $\displaystyle\int_0^\infty x^\nu e^{\frac{1}{4}x^2} D_{-2\nu}(x) J_\nu(xy)\, dx = y^{\nu-1} e^{\frac{1}{4}y^2} D_{-2\nu}(y) \qquad \left[\operatorname{Re}\nu > -\tfrac{1}{2}, \quad y > 0\right]$

$$\text{ET II 77(9), EH II 121(17)}$$

9. $\displaystyle\int_0^\infty x^\nu e^{\frac{1}{4}x^2} D_{-2\nu-2}(x) J_\nu(xy)\, dx = (2\nu+1)^{-1} y^\nu e^{\frac{1}{4}y^2} D_{-2\nu-1}(y)$

$$\left[y > 0, \quad \operatorname{Re}\nu > -\tfrac{1}{2}\right] \qquad \text{ET II 77(10)}$$

10. $\displaystyle\int_0^\infty x^\nu e^{-\frac{1}{4}a^2 x^2} D_{2\mu}(ax) J_\nu(xy)\, dx = \frac{2^{\mu-\frac{1}{2}}\Gamma\left(\nu+\frac{1}{2}\right) y^\nu}{\Gamma(\nu-\mu+1) a^{1+2\nu}}\,{}_1F_1\left(\nu+\frac{1}{2}; \nu-\mu+1; -\frac{y^2}{2a^2}\right)$

$$\left[y > 0, \quad |\arg a| < \tfrac{1}{4}\pi, \quad \operatorname{Re}\nu > -\tfrac{1}{2}\right]$$
$$\text{ET II 77(11)}$$

11. $\displaystyle\int_0^\infty x^\nu e^{\frac{1}{4}a^2 x^2} D_{2\mu}(ax) J_\nu(xy)\, dx = \frac{\Gamma\left(\frac{1}{2}+\nu\right) a^{2k} 2^{m+\mu}}{\Gamma\left(\frac{1}{2}-\mu\right) y^{\mu+\frac{3}{2}}} e^{\frac{y^2}{4a^2}} W_{k,m}\left(\frac{y^2}{4a^2}\right)$

$$2k = \tfrac{1}{2}+\mu-\nu, \quad 2m = \tfrac{1}{2}+\mu+\nu$$
$$\left[y > 0, \quad |\arg a| < \tfrac{1}{4}\pi, \quad -\tfrac{1}{2} < \operatorname{Re}\nu < \operatorname{Re}\left(\tfrac{1}{2}-2\mu\right)\right]$$
$$\text{ET II 78(12)}$$

12. $\displaystyle\int_0^\infty x^{\nu+1} e^{-\frac{1}{4}a^2 x^2} D_{2\mu}(ax) J_\nu(xy)\, dx = \frac{2^\mu \Gamma\left(\nu+\frac{3}{2}\right) y^\nu}{\Gamma\left(\nu-\mu+\frac{3}{2}\right) a^{2\nu+2}}\,{}_1F_1\left(\nu+\frac{3}{2}; \nu-\mu+\frac{3}{2}; -\frac{y^2}{2a^2}\right)$

$$\left[y > 0, \quad |\arg a| < \tfrac{1}{4}\pi, \quad \operatorname{Re}\nu > -1\right]$$
$$\text{ET II 79(23)}$$

13. $\displaystyle\int_0^\infty x^{\nu+1} e^{\frac{1}{4} a^2 x^2} D_{2\mu}(ax) J_\nu(xy)\, dx = \frac{\Gamma\left(\frac{3}{2}+\nu\right) 2^{\frac{1}{2}+m+\mu} a^{2k+1}}{\Gamma(-\mu) y^{\mu+2}} e^{\frac{y^2}{4a^2}} W_{k,m}\left(\frac{y^2}{2a^2}\right)$

$$2k = \mu - \nu - 1, \quad 2m = \mu + \nu + 1$$

$$\left[y > 0, \quad |\arg a| < \tfrac{3}{4}\pi, \quad -1 < \operatorname{Re}\nu < -\tfrac{1}{2} - 2\operatorname{Re}\mu \right]$$

ET II 79(24)

14. $\displaystyle\int_0^\infty x^{\lambda+\frac{1}{2}} e^{\frac{1}{4} a^2 x^2} D_\mu(ax) J_\nu(xy)\, dx = \frac{2^{\lambda-\frac{1}{2}\mu} \pi^{-\frac{1}{2}}}{\Gamma(-\mu) y^{\lambda+\frac{3}{2}}} G_{23}^{22}\left(\frac{y^2}{2a^2} \left|\begin{array}{l} \frac{1}{2}, 1 \\ \frac{3}{4}+\frac{\lambda+\nu}{2}, -\frac{\mu}{2}, \frac{3}{4}+\frac{\lambda-\nu}{2} \end{array}\right.\right)$

$$\left[y > 0, \quad |\arg a| < \tfrac{3}{4}\pi, \quad \operatorname{Re}\mu < -\operatorname{Re}\lambda < \operatorname{Re}\nu + \tfrac{3}{2} \right] \quad \text{ET II 80(26)}$$

15. $\displaystyle\int_0^\infty x^{\nu+1} e^{\frac{1}{4} x^2} D_{-2\nu-1}(x) J_\nu(xy)\, dx = (2\nu+1) y^{\nu-1} e^{\frac{1}{4} y^2} D_{-2\nu-2}(y)$

$$\left[y > 0, \quad \operatorname{Re}\nu > -\tfrac{1}{2} \right] \qquad \text{ET II 79(20)}$$

16. $\displaystyle\int_0^\infty x^{\nu+1} e^{-\frac{1}{4} x^2} D_{-2\nu-3}(x) J_\nu(xy)\, dx = 2^{-1/2} \pi^{1/2} y^{-\nu-2} e^{-\frac{1}{4} y^2} I_{\nu+1}\left(\tfrac{1}{4} y^2\right)$

$$\left[y > 0, \quad \operatorname{Re}\nu > -1 \right] \qquad \text{ET II 79(21)}$$

17. $\displaystyle\int_0^\infty x^{\nu+1} e^{\frac{1}{4} x^2} D_{-2\nu-3}(x) J_\nu(xy)\, dx = y^\nu e^{\frac{1}{4} y^2} D_{-2\nu-3}(y)$

$$\left[y > 0, \quad \operatorname{Re}\nu > -1 \right] \qquad \text{ET II 79(22)}$$

18. $\displaystyle\int_0^\infty x^\nu e^{\frac{1}{4} a^2 x^2} D_{\frac{1}{2}\nu-\frac{1}{2}}(ax) Y_\nu(xy)\, dx = -\pi^{-1} 2^{\frac{3}{4}\nu+\frac{3}{4}} a^{-\nu} y^{-1} \Gamma(\nu+1) e^{\frac{y^2}{4a^2}} W_{-\frac{1}{2}\nu-\frac{1}{2}, \frac{1}{2}\nu}\left(\frac{y^2}{2a^2}\right)$

$$\left[y > 0, \quad |\arg a| < \tfrac{3}{4}\pi, \quad -\tfrac{1}{2} < \operatorname{Re}\nu < \tfrac{2}{3} \right] \quad \text{ET II 115(39)}$$

7.753

1. $\displaystyle\int_0^\infty x^{\nu-\frac{1}{2}} e^{-(x+a)^2} I_{\nu-\frac{1}{2}}(2ax) D_\nu(2x)\, dx = \frac{1}{2} \pi^{-1/2} \Gamma(\nu) a^{\nu-\frac{1}{2}} D_{-\nu}(2a)$

$$\left[\operatorname{Re} a > 0, \quad \operatorname{Re}\nu > 0 \right] \qquad \text{ET II 397(12)}$$

2. $\displaystyle\int_0^\infty x^{\nu-\frac{3}{2}} e^{-(x+a)^2} I_{\nu-\frac{3}{2}}(2ax) D_\nu(2x)\, dx = \frac{1}{2} \pi^{-1/2} \Gamma(\nu) a^{\nu-\frac{3}{2}} D_{-\nu}(2a)$

$$\left[\operatorname{Re} a > 0, \quad \operatorname{Re}\nu > 1 \right] \qquad \text{ET II 397(13)}$$

7.754

1. $\displaystyle\int_0^\infty x^\nu e^{-\frac{1}{4} x^2} \left\{ [1 \mp 2\cos(\nu\pi)] D_{2\nu-1}(x) - D_{2\nu-1}(-x) \right\} J_\nu(xy)\, dx$

$$= \pm y^{\nu-1} e^{-\frac{1}{4} y^2} \left\{ [1 \mp 2\cos(\nu\pi)] D_{2\nu-1}(y) - D_{2\nu-1}(-y) \right\}$$

$$\left[y > 0, \quad \operatorname{Re}\nu > -\tfrac{1}{2} \right] \qquad \text{ET II 76(2, 3)}$$

2. $\displaystyle\int_0^\infty x^\nu e^{-\frac{1}{4} x^2} \left\{ [1 \mp 2\cos(\nu\pi)] D_{2\nu+1}(x) - D_{2\nu+1}(-x) \right\} J_\nu(xy)\, dx$

$$= \mp y^\nu e^{-\frac{1}{4} y^2} \left\{ [1 \mp 2\cos(\nu\pi)] D_{2\nu}(y) + D_{2\nu}(-y) \right\}$$

$$\left[y > 0, \quad \operatorname{Re}\nu > -\tfrac{1}{2} \right] \qquad \text{ET II 77(6, 7)}$$

3. $\displaystyle\int_0^\infty x^{\nu+1} e^{-\frac{1}{4}x^2} \left\{ [1 \pm 2\cos(\nu\pi)] D_{2\nu}(x) + D_{2\nu}(-x) \right\} J_\nu(xy)\, dx$

$$= \pm y^{\nu-1} e^{-\frac{1}{4}y^2} \left\{ [1 \pm 2\cos(\nu\pi)] D_{2\nu+1}(y) - D_{2\nu+1}(-y) \right\}$$

$$[y > 0, \quad \mathrm{Re}\,\nu > -1] \qquad \text{ET II 78(14, 15)}$$

4. $\displaystyle\int_0^\infty x^{\nu+1} e^{-\frac{1}{4}x^2} \left\{ [1 \mp 2\cos(\nu\pi)] D_{2\nu+2}(x) + D_{2\nu+2}(-x) \right\} J_\nu(xy)\, dx$

$$= \pm y^{\nu} e^{-\frac{1}{4}y^2} \left\{ [1 \mp 2\cos(\nu\pi)] D_{2\nu+2}(y) + D_{2\nu+2}(-y) \right\}$$

$$[y > 0, \quad \mathrm{Re}\,\nu > -1] \qquad \text{ET II 78(17, 18)}$$

7.755

1. $\displaystyle\int_0^\infty x^{-1/2} D_\nu\left(\sqrt{ax}\right) D_{-\nu-1}\left(\sqrt{ax}\right) J_0(xy)\, dx$

$$= 2^{-3/2} \pi a^{-1/2}\, P_{-\frac{1}{4}}^{\frac{1}{2}\nu+\frac{1}{4}} \left[\left(1 + \frac{4y^2}{a^2}\right)^{1/2} \right] P_{\frac{1}{4}}^{\frac{1}{2}\nu-\frac{1}{4}} \left[\left(1 + \frac{4y^2}{a^2}\right)^{1/2} \right]$$

$$[y > 0, \mathrm{Re}\,a > 0] \qquad \text{ET II 17(43)}$$

2. $\displaystyle\int_0^\infty x^{1/2} D_{-\frac{1}{2}-\nu}\left(ae^{\frac{1}{4}\pi i} x^{1/2}\right) D_{-\frac{1}{2}-\nu}\left(ae^{-\frac{1}{4}\pi i} x^{1/2}\right) J_\nu(xy)\, dx$

$$= 2^{-\nu} \pi^{1/2} y^{-\nu-1} \left(a^2 + 2y\right)^{-1/2} \left[\Gamma\left(\nu + \tfrac{1}{2}\right)\right]^{-1} \left[\left(a^2 + 2y\right)^{1/2} - a\right]^{2\nu}$$

$$[y > 0, \quad \mathrm{Re}\,a > 0, \quad \mathrm{Re}\,\nu > -\tfrac{1}{2}] \qquad \text{ET II 80(27)}$$

3. $\displaystyle a\int_0^\infty D_{-\frac{1}{2}-\nu}\left(ae^{\frac{1}{4}\pi i} x^{-1/2}\right) D_{-\frac{1}{2}-\nu}\left(ae^{-\frac{1}{4}\pi i} x^{-1/2}\right) J_\nu(xy)\, dx$

$$= 2^{1/2} \pi^{1/2} y^{-1} \left[\Gamma\left(\nu + \tfrac{1}{2}\right)\right]^{-1} \exp\left[-a(2y)^{1/2}\right]$$

$$[y > 0, \quad \mathrm{Re}\,a > 0, \quad e\,\mathrm{Re}\,\nu > -\tfrac{1}{2}] \qquad \text{ET II 80(28)a}$$

4. $\displaystyle\int_0^\infty x^{1/2} D_{\nu-\frac{1}{2}}\left(ax^{-1/2}\right) D_{-\nu-\frac{1}{2}}\left(ax^{-1/2}\right) Y_\nu(xy)\, dx$

$$= y^{-3/2} \exp\left(-ay^{1/2}\right) \sin\left[ay^{1/2} - \tfrac{1}{2}\left(\nu - \tfrac{1}{2}\right)\pi\right]$$

$$[y > 0, \quad |\arg a| < \tfrac{1}{4}\pi] \qquad \text{ET II 115(40)}$$

5. $\displaystyle\int_0^\infty x^{1/2} D_{\nu-\frac{1}{2}}\left(ax^{-1/2}\right) D_{-\nu-\frac{1}{2}}\left(ax^{-1/2}\right) K_\nu(xy)\, dx = 2^{-1} y^{-3/2} \pi \exp\left[-a(2y)^{1/2}\right]$

$$\left[\mathrm{Re}\,y > 0, \quad |\arg a| < \tfrac{1}{4}\pi\right] \qquad \text{ET II 151(81)}$$

Combinations of parabolic cylinder and Struve functions

7.756 $\displaystyle\int_0^\infty x^{-\nu} e^{-\frac{1}{4}x^2} \left[D_\mu(x) - D_\mu(-x)\right] \mathbf{H}_\nu(xy)\, dx$

$$= \frac{2^{3/2}\,\Gamma\left(\tfrac{1}{2}\mu + \tfrac{1}{2}\right)}{\Gamma\left(\tfrac{1}{2}\mu + \nu + 1\right)} y^{\mu+\nu} \sin\left(\tfrac{1}{2}\mu\pi\right) \, {}_1F_1\left(\tfrac{1}{2}\mu + \tfrac{1}{2}; \tfrac{1}{2}\mu + \nu + 1; -\tfrac{1}{2}y^2\right)$$

$$[y > 0, \quad \mathrm{Re}(\mu + \nu) > -\tfrac{3}{2}, \quad \mathrm{Re}\,\mu > -1] \qquad \text{ET II 171(41)}$$

7.76 Combinations of parabolic cylinder functions and confluent hypergeometric functions

7.761

1. $\displaystyle \int_0^\infty e^{\frac{1}{4}t^2} t^{2c-1} D_{-\nu}(t) \, {}_1F_1\left(a; c; -\frac{1}{2}pt^2\right) dt$

$$= \frac{\pi^{1/2}}{2^{c+\frac{1}{2}\nu}} \frac{\Gamma(2c)\,\Gamma\left(\frac{1}{2}\nu - c + a\right)}{\Gamma\left(\frac{1}{2}\nu\right)\Gamma\left(a + \frac{1}{2} + \frac{1}{2}\nu\right)} F\left(a, c + \frac{1}{2}; a + \frac{1}{2} + \frac{1}{2}\nu; 1 - p\right)$$

$$\left[|1-p| < 1, \quad \operatorname{Re} c > 0, \quad \operatorname{Re}\nu > 2\operatorname{Re}(c-a)\right] \quad \text{EH II 121(12)}$$

2. $\displaystyle \int_0^\infty e^{\frac{1}{4}t^2} t^{2c-2} D_{-\nu}(t) \, {}_1F_1\left(a; c; -\frac{1}{2}pt^2\right) dt$

$$= \frac{\pi^{1/2}}{2^{c+\frac{1}{2}\nu-\frac{1}{2}}} \frac{\Gamma(2c-1)\,\Gamma\left(\frac{1}{2}\nu + \frac{1}{2} - c + a\right)}{\Gamma\left(\frac{1}{2} + \frac{1}{2}\nu\right)\Gamma\left(a + \frac{1}{2}\nu\right)} F\left(a, c - \frac{1}{2}; a + \frac{1}{2}\nu; 1 - p\right)$$

$$\left[|1-p| < 1, \quad \operatorname{Re} c > \frac{1}{2}, \quad \operatorname{Re}\nu > 2\operatorname{Re}(c-a) - 1\right] \quad \text{EH II 121(13)}$$

7.77 Integration of a parabolic cylinder function with respect to the index

7.771 $\displaystyle \int_0^\infty \cos(ax) \, D_{x-\frac{1}{2}}(\beta) \, D_{-x-\frac{1}{2}}(\beta) \, dx = \frac{1}{2}\left(\frac{\pi}{\cos a}\right)^{1/2} \exp\left(-\frac{\beta^2 \cos a}{2}\right) \quad \left[|a| < \frac{1}{2}\pi\right]$

$$= 0 \qquad\qquad \left[|a| > \frac{1}{2}\pi\right]$$

$$\text{ET II 298(22)}$$

7.772

1. $\displaystyle \int_{-\frac{1}{2}-i\infty}^{-\frac{1}{2}+i\infty} \left[\frac{\left(\tan\frac{1}{2}\varphi\right)^\nu}{\cos\frac{1}{2}\varphi} D_\nu\left(-e^{\frac{1}{4}i\pi}\xi\right) D_{-\nu-1}\left(e^{\frac{1}{4}i\pi}\eta\right) \right.$

$$\left. + \frac{\left(\cot\frac{1}{2}\varphi\right)^\nu}{\sin\frac{1}{2}\varphi} D_{-\nu-1}\left(e^{\frac{1}{4}i\pi}\xi\right) D_\nu\left(-e^{\frac{1}{4}i\pi}\eta\right) \right] \frac{d\nu}{\sin\nu\pi}$$

$$= -2i(2\pi)^{1/2} \exp\left[-\frac{1}{4}i\left(\xi^2 - \eta^2\right)\cos\varphi - \frac{1}{2}i\xi\eta\sin\varphi\right]$$

$$\text{EH II 125(7)}$$

2. $\displaystyle \int_{-\frac{1}{2}-i\infty}^{-\frac{1}{2}+i\infty} \frac{\left(\tan\frac{1}{2}\varphi\right)^\nu}{\cos\frac{1}{2}\varphi} D_\nu\left(-e^{\frac{1}{4}i\pi}\zeta\right) D_{-\nu-1}\left(e^{\frac{1}{4}i\pi}\eta\right) \frac{d\nu}{\sin\nu\pi}$

$$= -2i\, D_0\left[e^{\frac{1}{4}i\pi}\left(\zeta\cos\frac{1}{2}\varphi + \eta\sin\frac{1}{2}\varphi\right)\right] D_{-1}\left[e^{\frac{1}{4}i\pi}\left(\eta\cos\frac{1}{2}\varphi - \zeta\sin\frac{1}{2}\varphi\right)\right]$$

$$\text{EH II 125(8)}$$

7.773

1. $\displaystyle \int_{c-i\infty}^{c+i\infty} D_\nu(z) t^\nu \, \Gamma(-\nu) \, d\nu = 2\pi i e^{-\frac{1}{4}z^2 - zt - \frac{1}{2}t^2}$ $\qquad \left[c < 0, \quad |\arg t| < \frac{\pi}{4}\right] \qquad \text{EH II 126(10)}$

2.

$$\int_{c-i\infty}^{c+i\infty} \left[D_\nu(x)\, D_{-\nu-1}(iy) + D_\nu(-x)\, D_{-\nu-1}(iy) \right] \frac{t^{-\nu-1}\, d\nu}{\sin(-\nu\pi)}$$

$$= \frac{2\pi i}{\left(\frac{\pi}{2}\right)^{1/2}} \left(1 + t^2\right)^{-\frac{1}{2}} \exp\left[\frac{1}{4}\frac{1 - t^2}{1 + t^2}\left(x^2 + y^2\right) + i\frac{txy}{1 + t^2} \right]$$

$$\left[-1 < c < 0, \quad |\arg t| < \frac{1}{2}\pi \right] \quad \text{EH II 126(11)}$$

7.774 $\int_{c-i\infty}^{c+i\infty} D_\nu\left[k^{\frac{1}{2}}(1+i)\xi \right] D_{-\nu-1}\left[k^{1/2}(1+i)\eta \right] \Gamma\left(-\tfrac{1}{2}\nu\right) \Gamma\left(\tfrac{1}{2} + \tfrac{1}{2}\nu\right)\, d\nu$

$$= 2^{1/2}\pi^2\, H_0^{(2)}\left[\tfrac{1}{2}k\left(\xi^2 + \eta^2\right)\right]$$

$$\left[-1 < c < 0, \quad \operatorname{Re} ik \geq 0 \right] \quad \text{EH II 125(9)}$$

7.8 Meijer's and MacRobert's Functions (G and E)

7.81 Combinations of the functions G and E and the elementary functions

7.811

1.

$$\int_0^\infty G_{p,q}^{m,n}\left(\eta x \left| \begin{matrix} a_1, \ldots, a_p \\ b_1, \ldots, b_q \end{matrix} \right. \right) G_{\sigma,\tau}^{\mu,\nu}\left(\omega x \left| \begin{matrix} c_1, \ldots, c_\sigma \\ d, \ldots, d_\tau \end{matrix} \right. \right) dx$$

$$= \frac{1}{\eta} G_{q+\sigma,p+\tau}^{n+\mu,m+\nu}\left(\frac{\omega}{\eta} \left| \begin{matrix} -b_1, \ldots, -b_m, c_1, \ldots, c_\sigma, -b_{m+1}, \ldots, -b_q \\ -a_1, \ldots, -a_n, d_1, \ldots, d_\tau, -a_{n+1}, \ldots, -a_p \end{matrix} \right. \right)$$

subject to the following constraints

- $m, n, p, q, \mu, \nu, \sigma, \tau$ are integers;
- $1 \leq n \leq p < q < p + \tau - \sigma$
- $\frac{1}{2}p + \frac{1}{2}q - n < m \leq q, \quad 0 \leq \nu \leq \sigma, \quad \frac{1}{2}\sigma + \frac{1}{2}\tau - \nu < \mu \leq \tau$;
- $\operatorname{Re}(b_j + d_k) > -1 \quad (j = 1, \ldots, m; k = 1, \ldots, \mu)$,
- $\operatorname{Re}(a_j + c_k) < 1 \quad (j = 1, \ldots, n; k = 1, \ldots, \tau)$;
- $\omega \neq 0, \quad \eta \neq 0, \quad |\arg \eta| < \left(m + n - \frac{1}{2}p - \frac{1}{2}q\right)\pi, \quad |\arg \omega| < \left(\mu + \nu - \frac{1}{2}\sigma - \frac{1}{2}\tau\right)\pi$
- The following must not be integers:

$$\begin{aligned}
b_j - b_k \quad &(j = 1, \ldots, m; k = 1, \ldots, m; j \neq k), \\
a_j - a_k \quad &(j = 1, \ldots, n; k = 1, \ldots, n; j \neq k), \\
d_j - d_k \quad &(j = 1, \ldots, \mu; k = 1, \ldots, \mu; j \neq k), \\
a_j + d_k \quad &(j = 1, \ldots, n; k = 1, \ldots, n);
\end{aligned}$$

- The following must not be positive integers:

$$\begin{aligned}
a_j - b_k \quad &(j = 1, \ldots, n; k = 1, \ldots, m), \\
c_j - d_k \quad &(j = 1, \ldots, \nu; k = 1, \ldots, \mu);
\end{aligned}$$

Formula **7.811** 1 also holds for four sets of restrictions. See C. S. Meijer, Neue Integraldarstellungen für Whittakersche Funktionen, Nederl. Akad. Wetensch. Proc. **44** (1941), 82–92.

ET II 422(14)

Hereafter, $G_{p,q}^{m,n}$ will be written as G_{pq}^{mn}, and commas will only be inserted in entries like $G_{p+1,q+1}^{m,n+1}$ where their omission could cause ambiguity.

2. $\displaystyle\int_0^1 x^{\varrho-1}(1-x)^{\sigma-1}\,G_{pq}^{mn}\left(\alpha x\,\bigg|\begin{array}{c}a_1,\ldots,a_p\\b_1,\ldots,b_q\end{array}\right)\,dx = \Gamma(\sigma)\,G_{p+1,q+1}^{m,n+1}\left(\alpha\,\bigg|\begin{array}{c}1-\varrho,a_1,\ldots,a_p\\b_1,\ldots,b_q,1-\varrho-\sigma\end{array}\right)$

where

- $(p+q) < 2(m+n)$;
- $|\arg a| < \left(m+n-\frac{1}{2}p-\frac{1}{2}q\right)\pi$;
- $\operatorname{Re}(\varrho+b_j) > 0,\ j=1,\ldots,m$;
- $\operatorname{Re}\sigma > 0$,
- either

$$p+q \le 2(m+n),\quad |\arg\alpha| \le \left(m+n-\tfrac{1}{2}\varrho-\tfrac{1}{2}q\right)\pi,$$
$$\operatorname{Re}(\varrho+b_j) > 0;\quad j=1,\ldots,m;\quad \operatorname{Re}\sigma > 0,$$
$$\operatorname{Re}\left[\sum_{j=1}^p a_j - \sum_{j=1}^q b_j + (p-q)\left(\varrho-\frac{1}{2}\right)\right] > -\frac{1}{2},$$

   or

$$p < q \quad (\text{or } p \le q \text{ for } |\alpha| < 1),\quad \operatorname{Re}(p+b_j) > 0;\quad j=1,\ldots,m;\quad \operatorname{Re}\sigma > 0$$

ET II 417(1)

3. $\displaystyle\int_1^\infty x^{-\varrho}(x-1)^{\sigma-1}\,G_{pq}^{mn}\left(\alpha x\,\bigg|\begin{array}{c}a_1,\ldots,a_p\\b_1,\ldots,b_q\end{array}\right)\,dx = \Gamma(\sigma)\,G_{p+1,q+1}^{m+1,n}\left(\alpha\,\bigg|\begin{array}{c}a_1,\ldots,a_p,\varrho\\\varrho-\sigma,b_1,\ldots,b_q\end{array}\right)$

where

- $p+q < 2(m+n)$
- $|\arg\alpha| < \left(m+n-\frac{1}{2}p-\frac{1}{2}q\right)\pi$
- $\operatorname{Re}(\varrho-\sigma-a_j) > -1;\quad j=1,\ldots,n$
- $\operatorname{Re}\sigma > 0$
- either

$$p+q \le 2(m+n),\quad |\arg\alpha| \le \left(m+n-\tfrac{1}{2}p-\tfrac{1}{2}q\right)\pi,$$
$$\operatorname{Re}(\varrho-\sigma-a_j) > -1;\quad j=1,\ldots,n;\quad \operatorname{Re}\sigma > 0,$$
$$\operatorname{Re}\left[\sum_{j=1}^p a_j - \sum_{j=1}^q b_j + (q-p)\left(\varrho-\sigma+\frac{1}{2}\right)\right] > -\frac{1}{2},$$

   or

$$q < p \quad (\text{or } q \le p \text{ for } |\alpha| > 1),\quad \operatorname{Re}(\varrho-\sigma-a_j) > -1;\quad j=1,\ldots,n;\quad \operatorname{Re}\sigma > 0$$

ET II 417(2)

4. $\displaystyle\int_0^\infty x^{\varrho-1}\,G_{pq}^{mn}\left(\alpha x\,\bigg|\begin{array}{c}a_1,\ldots,a_p\\b_1,\ldots,b_q\end{array}\right)\,dx = \frac{\prod_{j=1}^m \Gamma(b_j+\varrho)\prod_{j=1}^n \Gamma(1-a_j-\varrho)}{\prod_{j=m+1}^q \Gamma(1-b_j-\varrho)\prod_{j=n+1}^p \Gamma(a_j+\varrho)}\,\alpha^{-\varrho}$

$p+q < 2(m+n),\quad |\arg\alpha| < \left(m+n-\frac{1}{2}p-\frac{1}{2}q\right)\pi,\quad -\min_{1\le j\le m}\operatorname{Re}b_j < \operatorname{Re}\varrho < 1-\max_{1\le j\le n}\operatorname{Re}a_j$

ET II 418(3)a, ET I 337(14)

5. $$\int_0^\infty x^{\varrho-1}(x+\beta)^{-\sigma} G_{pq}^{mn}\left(\alpha x \left|\begin{matrix} a_1,\dots,a_p \\ b_1,\dots,b_q \end{matrix}\right.\right) dx = \frac{\beta^{\varrho-\sigma}}{\Gamma(\sigma)} G_{p+1,q+1}^{m+1,n+1}\left(\alpha\beta \left|\begin{matrix} 1-\varrho,a_1,\dots,a_p \\ \sigma-\varrho,b_1,\dots,b_q \end{matrix}\right.\right)$$

where

- $p + q < 2(m+n)$
- $|\arg\alpha| < \left(m+n-\frac{1}{2}p-\frac{1}{2}q\right)\pi$
- $|\arg\beta| < \pi$
- $\operatorname{Re}(\varrho + b_j) > 0, \quad j = 1,\dots,m$
- $\operatorname{Re}(\varrho - \sigma + a_j) < 1, \quad j = 1,\dots,n$
- either

$$p \le q, \quad p+q \le 2(m+n), \quad |\arg\alpha| \le \left(m+n-\tfrac{1}{2}p-\tfrac{1}{2}q\right)\pi, \quad |\arg\beta| < \pi$$
$$\operatorname{Re}(\varrho = b_j) > 0, \quad j=1,\dots,m, \quad \operatorname{Re}(\varrho-\sigma+a_j) < 1, \quad j=1,\dots,n,$$
$$\operatorname{Re}\left[\sum_{j=1}^p a_j - \sum_{j=1}^q b_j - (q-p)\left(\varrho-\sigma-\frac{1}{2}\right)\right] > 1,$$

or

$$p \ge q, \quad p+q \le 2(m+n), \quad |\arg\alpha| \le \left(m+n-\frac{1}{2}p-\frac{1}{2}q\right)\pi, \quad |\arg\beta| < \pi,$$
$$\operatorname{Re}(\varrho+b_j) > 0, \quad j=1,\dots,m, \quad \operatorname{Re}(\varrho-\sigma+a_j) < 1, \quad j=1,\dots,n,$$
$$\operatorname{Re}\left[\sum_{j=1}^p a_j - \sum_{j=1}^q b_j + (p-q)\left(\varrho-\frac{1}{2}\right)\right] > 1$$

ET II 418(4)

7.812

1. $$\int_0^1 x^{\beta-1}(1-x)^{\gamma-\beta-1} E\left(a_1,\dots,a_p : \varrho_1,\dots,\varrho_q; \frac{z}{x^m}\right) dx$$
$$= \Gamma(\gamma-\beta)m^{\beta-\gamma} E\left(a_1,\dots,a_{p+m} : \varrho_1,\dots,\varrho_{q+m} : z\right)$$
$$a_{p+k} = \frac{\beta+k-1}{m}, \quad \varrho_{q+k} = \frac{\gamma+k-1}{m}, \quad k = 1,\dots,m$$
$$[\operatorname{Re}\gamma > \operatorname{Re}\beta > 0, \quad m = 1,2,\dots] \quad \text{ET II 414(2)}$$

2. $$\int_0^\infty x^{\varrho-1}(1+x)^{-\sigma} E\left[a_1,\dots,a_p : \varrho_1,\dots,\varrho_q : (1+x)z\right] dx$$
$$= \Gamma(\varrho) E\left(a_1,\dots,a_p,\sigma-\varrho;\varrho_1,\dots,\varrho_q,\sigma;z\right)$$
$$[\operatorname{Re}\sigma > \operatorname{Re}\varrho > 0] \quad \text{ET II 415(3)}$$

3. $\displaystyle\int_0^\infty (1+x)^{-\beta} x^{s-1}\, G^{mn}_{pq}\left(\frac{ax}{1+x}\ \middle|\ \begin{matrix} a_1,\ldots,a_p \\ b_1,\ldots,b_q \end{matrix}\right)\, dx = \Gamma(\beta - s)\, G^{m,n+1}_{p+1,q+1}\left(a\ \middle|\ \begin{matrix} 1-s,a_1,\ldots,a_p \\ b_1,\ldots,b_q,1-\beta \end{matrix}\right)$

$$\left[-\min\operatorname{Re} b_k < \operatorname{Re} s < \operatorname{Re}\beta, \quad 1 \le k \le m; \quad (p+q) < 2(m+n),\right.$$

$$\left.|\arg a| < \left(m+n-\tfrac{1}{2}p-\tfrac{1}{2}q\right)\pi\right]$$

ET I 338(19)

7.813

1. $\displaystyle\int_0^\infty x^{-\varrho} e^{-\beta x}\, G^{mn}_{pq}\left(\alpha x\ \middle|\ \begin{matrix} a_1,\ldots,a_p \\ b_1,\ldots,b_q \end{matrix}\right)\, dx = \beta^{\varrho-1}\, G^{m,n+1}_{p+1,q}\left(\frac{\alpha}{\beta}\ \middle|\ \begin{matrix} \varrho,a_1,\ldots,a_p \\ b_1,\ldots,b_q \end{matrix}\right)$

$$\left[p+q < 2(m+n), \quad |\arg\alpha| < \left(m+n-\tfrac{1}{2}p-\tfrac{1}{2}q\right)\pi,\right.$$

$$\left.|\arg\beta| < \tfrac{1}{2}\pi, \quad \operatorname{Re}(b_j - \varrho) > -1, \quad j = 1,\ldots,m\right]$$

ET II 419(5)

2. $\displaystyle\int_0^\infty e^{-\beta x}\, G^{mn}_{pq}\left(\alpha x^2\ \middle|\ \begin{matrix} a_1,\ldots,a_p \\ b_1,\ldots,b_q \end{matrix}\right)\, dx = \pi^{-1/2}\beta^{-1}\, G^{m,n+2}_{p+2,q}\left(\frac{4\alpha}{\beta^2}\ \middle|\ \begin{matrix} 0,\tfrac{1}{2},a_1,\ldots,a_p \\ b_1,\ldots,b_q \end{matrix}\right)$

$$\left[p+q < 2(m+n), \quad |\arg\alpha| < \left(m+n-\tfrac{1}{2}p-\tfrac{1}{2}q\right)\pi,\right.$$

$$\left.|\arg\beta| < \tfrac{1}{2}\pi, \quad \operatorname{Re} b_j > -\tfrac{1}{2}; \quad j = 1,\ldots,m\right]$$

ET II 419(6)

7.814

1. $\displaystyle\int_0^\infty x^{\beta-1} e^{-x}\, E\left(a_1,\ldots,a_p : \varrho_1,\ldots,\varrho_q : xz\right)\, dx$

$$= \pi\cosec(\beta\pi)\left[E\left(a_1,\ldots,a_p : 1-\beta,\varrho_1,\ldots,\varrho_q : e^{\pm i\pi}z\right)\right.$$

$$\left. - z^{-\beta} E\left(a_1+\beta,\ldots,a_p+\beta : 1+\beta,\varrho_1+\beta,\ldots,\varrho_l+\beta : e^{\pm i\pi}z\right)\right]$$

$[p \ge q+1,\ \operatorname{Re}(a_r+\beta) > 0,\ r = 1,\ldots,p,\ |\arg z| < \pi.$ The formula holds also for $p < q+1$, provided the integral converges].   ET II 415(4)

2. $\displaystyle\int_0^\infty x^{\beta-1} e^{-x}\, E\left(a_1,\ldots,a_p : \varrho_1,\ldots,\varrho_q : x^{-m}z\right)\, dx$

$$= (2\pi)^{\frac{1}{2}-\frac{1}{2}m} m^{\beta-\frac{1}{2}}\, E\left(a_1,\ldots,a_{p+m} : \varrho_1,\ldots,\varrho_q : m^{-m}z\right)$$

$$\left[\operatorname{Re}\beta > 0, \quad a_{p+k} = \frac{\beta+k-1}{m}, \quad k = 1,\ldots,m; \quad m = 1,2,\ldots\right] \quad \text{ET II 415(5)}$$

7.815

1. $$\int_0^\infty \sin(cx)\, G_{pq}^{mn}\left(\alpha x^2 \left|\begin{matrix} a_1,\ldots,a_p \\ b_1,\ldots,b_q \end{matrix}\right.\right) dx = \sqrt{\pi}\,c^{-1}\, G_{p+2,q}^{m,n+1}\left(\frac{4\alpha}{c^2}\left|\begin{matrix} 0,a_1,\ldots,a_p,\frac{1}{2} \\ b_1,\ldots,b_q \end{matrix}\right.\right)$$

$$\left[p+q < 2(m+n), \quad |\arg\alpha| < \left(m+n-\tfrac{1}{2}p-\tfrac{1}{2}q\right)\pi,\right.$$
$$\left. c > 0, \quad \operatorname{Re} b_j > -1, \quad j=1,2,\ldots,m, \quad \operatorname{Re} a_j < \tfrac{1}{2}, \quad j=1,\ldots,n\right]$$

ET II 420(7)

2. $$\int_0^\infty \cos(cx)\, G_{pq}^{mn}\left(\alpha x^2 \left|\begin{matrix} a_1,\ldots,a_p \\ b_1,\ldots,b_q \end{matrix}\right.\right) dx = \pi^{1/2}c^{-1}\, G_{p+2,q}^{m,n+1}\left(\frac{4\alpha}{c^2}\left|\begin{matrix} \frac{1}{2},a_1,\ldots,a_p,0 \\ b_1,\ldots,b_q \end{matrix}\right.\right)$$

$$\left[p+q < 2(m+n), \quad |\arg\alpha| < \left(m+n-\tfrac{1}{2}p-\tfrac{1}{2}q\right)\pi,\right.$$
$$\left. c > 0, \quad \operatorname{Re} b_j > -\tfrac{1}{2}, \quad j=1,\ldots,m, \quad \operatorname{Re} a_j < \tfrac{1}{2}, \quad j=1,\ldots,n\right]$$

ET II 420(8)

7.82 Combinations of the functions G and E and Bessel functions

7.821

1. $$\int_0^\infty x^{-\varrho}\, J_\nu\left(2\sqrt{x}\right) G_{pq}^{mn}\left(\alpha x \left|\begin{matrix} a_1,\ldots,a_p \\ b_1,\ldots,b_q \end{matrix}\right.\right) dx = G_{p+2,q}^{m,n+1}\left(\alpha \left|\begin{matrix} \varrho-\frac{1}{2}\nu,a_1,\ldots,a_p,\varrho+\frac{1}{2}\nu \\ b_1,\ldots,b_q \end{matrix}\right.\right)$$

$$\left[p+q < 2(m+n), \quad |\arg\alpha| < \left(m+n-\tfrac{1}{2}p-\tfrac{1}{2}q\right)\pi\right.$$

$$\left. -\tfrac{3}{4} + \max_{1\le j\le n}\operatorname{Re} a_j < \operatorname{Re}\varrho < 1+\tfrac{1}{2}\operatorname{Re}\nu + \min_{1\le j\le m}\operatorname{Re} b_j\right]$$

ET II 420(9)

2. $$\int_0^\infty x^{-\varrho}\, Y_\nu\left(2\sqrt{x}\right) G_{pq}^{mn}\left(\alpha x \left|\begin{matrix} a_1,\ldots,a_p \\ b_1,\ldots,b_q \end{matrix}\right.\right) dx$$

$$= G_{p+3,q+1}^{m,n+2}\left(\alpha \left|\begin{matrix} \varrho-\frac{1}{2}\nu,\varrho+\frac{1}{2}\nu,a_1,\ldots,a_p,\varrho+\frac{1}{2}+\frac{1}{2}\nu \\ b_1,\ldots,b_q,\varrho+\frac{1}{2}+\frac{1}{2}\nu \end{matrix}\right.\right)$$

$$\left[p+q < 2(m+n), \quad |\arg\alpha| < \left(m+n-\tfrac{1}{2}p-\tfrac{1}{2}q\right)\pi,\right.$$

$$\left. -\tfrac{3}{4} + \max_{1\le j\le n}\operatorname{Re} a_j < \operatorname{Re}\varrho < \min_{1\le j\le m}\operatorname{Re} b_j + \tfrac{1}{2}|\operatorname{Re}\nu| + 1\right]$$

ET II 420(10)

3. $$\int_0^\infty x^{-\varrho}\, K_\nu\left(2\sqrt{x}\right) G_{pq}^{mn}\left(\alpha x \left|\begin{matrix} a_1,\ldots,a_p \\ b_1,\ldots,b_q \end{matrix}\right.\right) dx = \frac{1}{2}\, G_{p+2,q}^{m,n+2}\left(\alpha \left|\begin{matrix} \varrho-\frac{1}{2}\nu,\varrho+\frac{1}{2}\nu,a_1,\ldots,a_p \\ b_1,\ldots,b_q \end{matrix}\right.\right)$$

$$\left[p+q < 2(m+n), \quad |\arg\alpha| < \left(m+n-\tfrac{1}{2}p-\tfrac{1}{2}q\right)\pi,\right.$$

$$\left. \operatorname{Re}\varrho < 1 - \tfrac{1}{2}|\operatorname{Re}\nu| + \min_{1\le j\le m}\operatorname{Re} b_j\right]$$

ET II 421(11)

7.822

1. $$\int_0^\infty x^{2\varrho}\, J_\nu(xy)\, G_{pq}^{mn}\left(\lambda x^2\, \middle|\, \begin{matrix} a_1,\ldots,a_p \\ b_1,\ldots,b_q \end{matrix}\right)\, dx = \frac{2^{2\varrho}}{y^{2\varrho+1}}\, G_{p+2,q}^{m,n+1}\left(\frac{4\lambda}{y^2}\, \middle|\, \begin{matrix} h,a_1,\ldots,a_p,k \\ b_1,\ldots,b_q \end{matrix}\right)$$

    $$h = \tfrac{1}{2} - \varrho - \tfrac{1}{2}\nu, \quad k = \tfrac{1}{2} - \varrho + \tfrac{1}{2}\nu$$

    $$\left[p + q < 2(m+n), \quad |\arg \lambda| < \left(m+n-\tfrac{1}{2}p-\tfrac{1}{2}q\right)\pi, \quad \mathrm{Re}\left(b_j + \varrho + \tfrac{1}{2}\nu\right) > -\tfrac{1}{2}, \right.$$

    $$\left. j = 1,2,\ldots,m, \quad \mathrm{Re}\left(a_j + \varrho\right) < \tfrac{3}{4}, \quad j = 1,\ldots,n, \quad y > 0 \right]$$

    ET II 91(20)

2. $$\int_0^\infty x^{1/2}\, Y_\nu(xy)\, G_{pq}^{mn}\left(\lambda x^2\, \middle|\, \begin{matrix} a_1,\ldots,a_p \\ b_1,\ldots,b_q \end{matrix}\right)\, dx$$

    $$= (2\lambda)^{-1/2} y^{-1/2}\, G_{q+1,p+3}^{n+2,m}\left(\frac{y^2}{4\lambda}\, \middle|\, \begin{matrix} \tfrac{1}{2}-b_1,\ldots,\tfrac{1}{2}-b_q,l \\ h,k,\tfrac{1}{2}-a_1,\ldots,\tfrac{1}{2}-a_p,l \end{matrix}\right)$$

    $$h = \tfrac{1}{4} + \tfrac{1}{2}\nu, \quad k = \tfrac{1}{4} - \tfrac{1}{2}\nu, \quad l = -\tfrac{1}{4} - \tfrac{1}{2}\nu$$

    $$\left[p + q < 2(m+n), \quad |\arg \lambda| < \left(m+n-\tfrac{1}{2}p-\tfrac{1}{2}q\right)\pi, \quad y > 0, \right.$$

    $$\left. \mathrm{Re}\, a_j < 1, \quad j = 1,\ldots,n, \quad \mathrm{Re}\left(b_j \pm \tfrac{1}{2}\nu\right) > -\frac{3}{4}, \quad j = 1,\ldots,m \right]$$

    ET II 119(56)

3. $$\int_0^\infty x^{1/2}\, K_\nu(xy)\, G_{pq}^{mn}\left(\lambda x^2\, \middle|\, \begin{matrix} a_1,\ldots,a_p \\ b_1,\ldots,b_q \end{matrix}\right)\, dx$$

    $$= 2^{-3/2}\lambda^{-1/2} y^{-1/2}\, G_{q,p+2}^{n+2,m}\left(\frac{y^2}{4\lambda}\, \middle|\, \begin{matrix} \tfrac{1}{2}-b_1,\ldots,\tfrac{1}{2}-b_q \\ h,k,\tfrac{1}{2}-a_1,\ldots,\tfrac{1}{2}-a_p \end{matrix}\right)$$

    $$h = \tfrac{1}{4} + \tfrac{1}{2}\nu, \quad k = \tfrac{1}{4} - \tfrac{1}{2}\nu$$

    $$\left[\mathrm{Re}\, y > 0, \quad p + q < 2(m+n), \quad |\arg \lambda| < \left(m+n-\tfrac{1}{2}p-\tfrac{1}{2}q\right)\pi, \right.$$

    $$\left. \mathrm{Re}\, b_j > \tfrac{1}{2}|\mathrm{Re}\, \nu| - \tfrac{3}{4}, \quad j = 1,\ldots,m \right]$$

    ET II 153(90)

7.823

1. $$\int_0^\infty x^{\beta-1}\, J_\nu(x)\, E\left(a_1,\ldots,a_p : \varrho_1,\ldots,\varrho_q : x^{-2m}z\right)\, dx$$

    $$= (2\pi)^{-m}(2m)^{\beta-1}\left\{\exp\left[\tfrac{1}{2}\pi(\beta-\nu-1)i\right] E\left[a_1,\ldots,a_{p+2m} : \varrho_1,\ldots,\varrho_q : (2m)^{-2m}ze^{-m\pi i}\right]\right.$$

    $$\left. + \exp\left[-\tfrac{1}{2}\pi(\beta-\nu-1)i\right] E\left[a_1,\ldots,a_{p+2m} : \varrho_1,\ldots,\varrho_q : (2m)^{-2m}ze^{m\pi i}\right]\right\},$$

    $$a_{p+k} = \frac{\beta+\nu+2k-2}{2m}, \quad a_{p+m+k} = \frac{\beta-\nu+2k-2}{2m}, \quad m = 1,2,\ldots,; \quad k = 1,\ldots,m$$

    $$\left[\mathrm{Re}(\beta+\nu) > 0, \quad \mathrm{Re}\left(2a_r m - \beta\right) > -\tfrac{3}{2}, \quad r = 1,\ldots,p\right] \quad \text{ET II 415(7)}$$

2. $\displaystyle\int_0^\infty x^{\beta-1} K_\nu(x)\, E\left(a_1,\ldots,a_p : \varrho_1,\ldots,\varrho_q : x^{-2m}z\right)\,dx$

$$= (2\pi)^{1-m} 2^{\beta-2} m^{\beta-1}$$

$$\times E\left[a_1,\ldots,a_{p+2m} : \varrho_1,\ldots,\varrho_q : (2m)^{-2m}z\right],$$

$$a_{p+k} = \frac{\beta+\nu+2k-2}{2m}, \quad a_{p+m+k} = \frac{\beta-\nu+2k-2}{2m}, \quad k = 1,2,\ldots,m$$

$$\left[\operatorname{Re}\beta > |\operatorname{Re}\nu|, \quad m = 1,2,\ldots\right]$$

ET II 416(8)

7.824

1. $\displaystyle\int_0^\infty x^{1/2}\, \mathbf{H}_\nu(xy)\, G_{pq}^{mn}\left(\lambda x^2 \left|\begin{array}{c} a_1,\ldots,a_p \\ b_1,\ldots,b_q \end{array}\right.\right)\,dx$

$$= (2\lambda y)^{-1/2}\, G_{q+1,p+3}^{n+1,m+1}\left(\frac{y^2}{4\lambda}\left|\begin{array}{c} l, \frac{1}{2}-b_1,\ldots,\frac{1}{2}-b_q \\ l, \frac{1}{2}-a_1,\ldots,\frac{1}{2}-a_p, h, k \end{array}\right.\right)$$

$$h = \frac{1}{4}+\frac{\nu}{2}, \quad k = \frac{1}{4}-\frac{\nu}{2}, \quad l = \frac{3}{4}+\frac{\nu}{2}$$

$$\left[p+q < 2(m+n), \quad |\arg\lambda| < \left(m+n-\tfrac{1}{2}p-\tfrac{1}{2}q\right)\pi, \quad y > 0,\right.$$

$$\left.\operatorname{Re} a_j < \min\left(1, \tfrac{3}{4}-\tfrac{1}{2}\nu\right), \quad j = 1,\ldots,n, \quad \operatorname{Re}(2b_j+\nu) > -\tfrac{5}{2}, \quad j = 1,\ldots,m\right]$$

ET II 172(47)

2. $\displaystyle\int_0^\infty x^{-\varrho}\, \mathbf{H}_\nu\left(2\sqrt{x}\right) G_{pq}^{mn}\left(\alpha x \left|\begin{array}{c} a_1,\ldots,a_p \\ b_1,\ldots,b_q \end{array}\right.\right)\,dx$

$$= G_{p+3,q+1}^{m+1,n+1}\left(\alpha\left|\begin{array}{c} \varrho-\frac{1}{2}-\frac{1}{2}\nu, a_1,\ldots,a_p, \varrho+\frac{1}{2}\nu, \varrho-\frac{1}{2}\nu \\ \varrho-\frac{1}{2}-\frac{1}{2}\nu, b_1,\ldots,b_q \end{array}\right.\right)$$

$$\left[p+q < 2(m+n), \quad |\arg\alpha| < \left(m+n-\tfrac{1}{2}p-\tfrac{1}{2}q\right)\pi,\right.$$

$$\left.\max\left(-\frac{3}{4}, \operatorname{Re}\frac{\nu-1}{2}\right) + \max_{1\le j\le n}\operatorname{Re} a_j < \operatorname{Re}\varrho < \min_{1\le j\le m}\operatorname{Re} b_j + \tfrac{1}{2}\operatorname{Re}\nu + \tfrac{3}{2}\right]$$

ET II 421(12)

7.83 Combinations of the functions G and E and other special functions

7.831 $\displaystyle\int_1^\infty x^{-\varrho}(x-1)^{\sigma-1} F(k+\sigma-\varrho, \lambda+\sigma-\varrho; \sigma; 1-x)\, G_{pq}^{mn}\left(\alpha x \left|\begin{array}{c} a_1,\ldots,a_p \\ b_1,\ldots,b_q \end{array}\right.\right)\,dx$

$$= \Gamma(\sigma)\, G_{p+2,q+2}^{m+2,n}\left(\alpha\left|\begin{array}{c} a_1,\ldots,a_p, k+\lambda+\sigma-\varrho, \varrho \\ k, \lambda, b_1,\ldots,b_q \end{array}\right.\right)$$

where

ET II 421(13)

- $\displaystyle\operatorname{Re}\left[\sum_{j=1}^p a_j - \sum_{j=1}^q b_j + (q-p)\left(k+\frac{1}{2}\right)\right] > -\frac{1}{2},$

- Re $\left[\displaystyle\sum_{j=1}^{p} a_j - \sum_{j=1}^{q} b_j + (q - p)\left(\lambda + \dfrac{1}{2}\right) \right] > -\dfrac{1}{2}$
- either

$$p + q < 2(m + n), \quad |\arg \alpha| < \left(m + n - \tfrac{1}{2}p - \tfrac{1}{2}q\right)\pi,$$
$$\mathrm{Re}\,\sigma > 0, \quad \mathrm{Re}\,k \geq \mathrm{Re}\,\lambda > \mathrm{Re}\,a_j - 1, \quad j = 1, \ldots, n,$$

or

$$p + q \leq 2(m + n), \quad |\arg \alpha| \leq \left(m + n - \tfrac{1}{2}p - \tfrac{1}{2}q\right)\pi,$$
$$\mathrm{Re}\,\sigma > 0, \quad \mathrm{Re}\,k \geq \mathrm{Re}\,\lambda > \mathrm{Re}\,a_j - 1, \quad j = 1, \ldots, n,$$

7.832 $\displaystyle\int_0^\infty x^{\beta-1} e^{-\frac{1}{2}x}\, W_{\kappa,\mu}(x)\, E\left(a_1, \ldots, a_p : \varrho_1, \ldots, \varrho_q : x^{-m} z\right)\, dx$

$$= (2\pi)^{\frac{1}{2} - \frac{1}{2}m}\, m^{\beta + \kappa - \frac{1}{2}}\, E\left(a_1, \ldots, a_{p+2m} : \varrho_1, \ldots, \varrho_{q+m} : m^{-m} z\right),$$

$$a_{p+k} = \frac{\beta + k + \mu - \frac{1}{2}}{m}, \quad a_{p+m+k} = \frac{\beta - \mu + k - \frac{1}{2}}{m}, \quad \varrho_{q+k} = \frac{\beta - \kappa + k}{m}, \qquad k = 1, \ldots, m$$

$$\left[\mathrm{Re}\,\beta > |\mathrm{Re}\,\mu| - \tfrac{1}{2}, \quad m = 1, 2, \ldots\right] \quad \text{ET II 416(10)}$$

8–9 Special Functions

8.1 Elliptic integrals and functions

8.11 Elliptic integrals

8.110

1. Every integral of the form $\int R\left(x, \sqrt{P(x)}\right) dx$, where $P(x)$ is a third- or fourth-degree polynomial, can be reduced to a linear combination of integrals leading to elementary functions and the following three integrals:

$$\int \frac{dx}{\sqrt{(1 - x^2)(1 - k^2 x^2)}}, \qquad \int \frac{\sqrt{1 - k^2 x^2}}{\sqrt{1 - x^2}} dx, \qquad \int \frac{dx}{(1 - nx^2)\sqrt{(1 - x^2)(1 - k^2 x^2)}},$$

which are called respectively *elliptic integrals of the first, second, and third kind in the Legendre normal form*. The results of this reduction for the more frequently encountered integrals are given in formulas **3.13–3.17**. The number k is called the *modulus** of these integrals, the number $k' = \sqrt{1 - k^2}$ is called the complementary modulus, and the number n is called the parameter of the integral of the third kind. BY (110.04)

2. By means of the substitution $x = \sin\varphi$, elliptic integrals can be reduced to the normal trigonometric forms

$$\int \frac{d\varphi}{\sqrt{1 - k^2 \sin^2 \varphi}}, \qquad \int \sqrt{1 - k^2 \sin^2 \varphi}\, d\varphi, \qquad \int \frac{d\varphi}{(1 - n\sin^2 \varphi)\sqrt{1 - k^2 \sin^2 \varphi}} \qquad \text{BY (110.04)}$$

The results of reducing integrals of trigonometric functions to normal form are given in **2.58–2.62**.

3. Elliptic integrals from 0 to $\frac{\pi}{2}$ are called *complete elliptic integrals*.

8.111

Notations:

1. $\Delta\varphi = \sqrt{1 - k^2 \sin^2 \varphi}; \quad k' = \sqrt{1 - k^2}; \quad k^2 < 1.$
2. The elliptic integral of the first kind:

---

*The quantity k is sometimes called the *module* of the functions.

$$F(\varphi, k) = \int_0^\varphi \frac{d\alpha}{\sqrt{1 - k^2 \sin^2 \alpha}} = \int_0^{\sin \varphi} \frac{dx}{\sqrt{(1 - x^2)(1 - k^2 x^2)}}.$$

3.   The elliptic integral of the second kind:

$$E(\varphi, k) = \int_0^\varphi \sqrt{1 - k^2 \sin^2 \alpha}\, d\alpha = \int_0^{\sin \varphi} \frac{\sqrt{1 - k^2 x^2}}{\sqrt{1 - x^2}}\, dx \qquad \text{FI II 135}$$

4.   The elliptic integral of the third kind:

$$\Pi(\varphi, n, k) = \int_0^\varphi \frac{d\alpha}{(1 - n \sin^2 \alpha)\sqrt{1 - k^2 \sin^2 \alpha}} = \frac{\int_0^{\sin \varphi} dx}{(1 - nx^2)\sqrt{(1 - x^2)(1 - k^2 x^2)}} \qquad \text{BY (110.04)}$$

5. $$D(\varphi, k) = \frac{F(\varphi, k) - E(\varphi, k)}{k^2} = \int_0^\varphi \frac{\sin^2 \alpha\, d\alpha}{\sqrt{1 - k^2 \sin^2 \alpha}} = \int_0^{\sin \varphi} \frac{x^2\, dx}{\sqrt{(1 - x^2)(1 - k^2 x^2)}}$$

8.112 Complete elliptic integrals

1. $$\boldsymbol{K}(k) = F\left(\frac{\pi}{2}, k\right) = \boldsymbol{K}'(k')$$

2. $$\boldsymbol{E}(k) = E\left(\frac{\pi}{2}, k\right) = \boldsymbol{E}'(k')$$

3. $$\boldsymbol{K}'(k) = F\left(\frac{\pi}{2}, k'\right) = \boldsymbol{K}(k')$$

4. $$\boldsymbol{E}'(k) = E\left(\frac{\pi}{2}, k'\right) = \boldsymbol{E}(k')$$

5. $$\boldsymbol{D} = D\left(\frac{\pi}{2}, k\right) = \frac{\boldsymbol{K} - \boldsymbol{E}}{k^2}$$

In writing complete elliptic integrals, the modulus k, which acts as an independent variable, is often omitted and we write

$$\boldsymbol{K}\,(\equiv \boldsymbol{K}(k)), \quad \boldsymbol{K}'\,(\equiv \boldsymbol{K}'(k)), \quad \boldsymbol{E}'\,(\equiv \boldsymbol{E}(k)), \quad \boldsymbol{K}'\,(\equiv \boldsymbol{E}'(k))$$

Series representations

8.113

1. $$\boldsymbol{K} = \frac{\pi}{2}\left\{1 + \left(\frac{1}{2}\right)^2 k^2 + \left(\frac{1 \cdot 3}{2 \cdot 4}\right)^2 k^4 + \cdots + \left[\frac{(2n-1)!!}{2^n n!}\right]^2 k^{2n} + \ldots\right\} = \frac{\pi}{2} F\left(\frac{1}{2}, \frac{1}{2}; 1; k^2\right)$$

FI II 487, WH

2. $$\boldsymbol{K} = \frac{\pi}{1 + k'}\left\{1 + \left(\frac{1}{2}\right)^2\left(\frac{1 - k'}{1 + k'}\right)^2 + \left(\frac{1 \cdot 3}{2 \cdot 4}\right)^2\left(\frac{1 - k'}{1 + k'}\right)^4 + \cdots + \left[\frac{(2n-1)!!}{2^n n!}\right]^2\left(\frac{1 - k'}{1 + k'}\right)^{2n} + \ldots\right\}$$

DW

3. $$\boldsymbol{K} = \ln\frac{4}{k'} + \left(\frac{1}{2}\right)^2\left(\ln\frac{4}{k'} - \frac{2}{1 \cdot 2}\right)k'^2 + \left(\frac{1 \cdot 3}{2 \cdot 4}\right)^2\left(\ln\frac{4}{k'} - \frac{2}{1 \cdot 2} - \frac{2}{3 \cdot 4}\right)k'^4$$
$$+ \left(\frac{1 \cdot 3 \cdot 5}{2 \cdot 4 \cdot 6}\right)^2\left(\ln\frac{4}{k'} - \frac{2}{1 \cdot 2} - \frac{2}{3 \cdot 4} - \frac{2}{5 \cdot 6}\right)k'^6 + \ldots$$

DW

See also **8.197** 1 and **8.197** 2.

8.114

$1.^6$ $\qquad E = \dfrac{\pi}{2}\left\{1 - \dfrac{1}{2^2}k^2 - \dfrac{1^2 \cdot 3}{2^2 \cdot 4^2}k^4 - \cdots - \left[\dfrac{(2n-1)!!}{2^n n!}\right]^2 \dfrac{k^{2n}}{2n-1} - \cdots\right\} = \dfrac{\pi}{2} F\left(-\dfrac{1}{2}, \dfrac{1}{2}; 1; k^2\right)$

FI II 487

$2.\qquad E = \dfrac{(1+k')\pi}{4}\left\{1 + \dfrac{1}{2^2}\left(\dfrac{1-k'}{1+k'}\right)^2 + \dfrac{1^2}{2^2 \cdot 4^2}\left(\dfrac{1-k'}{1+k'}\right)^4 + \cdots + \left[\dfrac{(2n-3)!!}{2^n n!}\right]^2 \left(\dfrac{1-k'}{1+k'}\right)^{2n} + \cdots\right\}$

DW

$3.\qquad E = 1 + \dfrac{1}{2}\left(\ln\dfrac{4}{k'} - \dfrac{1}{1 \cdot 2}\right)k'^2 + \dfrac{1^2 \cdot 3}{2^2 \cdot 4}\left(\ln\dfrac{4}{k'} - \dfrac{2}{1 \cdot 2} - \dfrac{1}{3 \cdot 4}\right)k'^4$

$\qquad\qquad + \dfrac{1^2 \cdot 3^2 \cdot 5}{2^2 \cdot 4^2 \cdot 6}\left(\ln\dfrac{4}{k'} - \dfrac{2}{1 \cdot 2} - \dfrac{2}{3 \cdot 4} - \dfrac{1}{5 \cdot 6}\right)k'^6 + \cdots$

DW

8.115 $\quad D = \pi\left\{\dfrac{1}{1}\left(\dfrac{1}{2}\right)^2 + \dfrac{2}{3}\left(\dfrac{1 \cdot 3}{2 \cdot 4}\right)^2 k^2 + \cdots + \dfrac{n}{2n-1}\left[\dfrac{(2n-1)!!}{2^n n!}\right]^2 k^{2(n-1)} + \cdots\right\}$ ZH 43(158)

8.116 $\quad \displaystyle\int_0^{\frac{\pi}{2}} \dfrac{\sqrt{1 - k^2 \sin^2\varphi}}{1 - n^2 \sin^2\varphi}\, d\varphi = \sqrt{n'^2 - k'^2}\left(\dfrac{\arccos\frac{1}{n'}}{n'\sqrt{n'^2 - 1}} + \mathbf{R},\right)$ ZH 44(163)

where

$\mathbf{R} = \dfrac{k'^2}{2}\left(p + \dfrac{1}{2}\right)\dfrac{1}{n'^3} + \dfrac{k'^4}{16}\left[-1 + \left(p + \dfrac{1}{4}\right)\dfrac{1}{n'^3}\left(1 + \dfrac{6}{n'^2}\right)\right]$

$\quad + \dfrac{k'^6}{16}\left[-\dfrac{7}{16} - \dfrac{1}{n'^2} + \left(p + \dfrac{1}{6}\right)\dfrac{1}{n'^3}\left(\dfrac{3}{8} + \dfrac{1}{n'^2} + \dfrac{5}{n'^4}\right)\right]$

$\quad + \dfrac{15k'^8}{256}\left[-\dfrac{37}{144} - \dfrac{21}{40n'^2} - \dfrac{1}{n'^4} + \left(p + \dfrac{1}{8}\right)\dfrac{1}{n'^3}\left(\dfrac{5}{24} + \dfrac{9}{20n'^2} + \dfrac{1}{n'^4} + \dfrac{14}{3n'^6}\right)\right] + \cdots,$

$\qquad\qquad p = \ln\dfrac{4}{k'}, \quad k' = 4e^{-p}, \quad k'^2 = 1 - k^2, \quad n'^2 = 1 - n^2$ ZH 44(163)

Trigonometric series

8.117 For *small* values of k and φ, we may use the series

$1.\qquad F(\varphi, k) = \dfrac{2}{\pi}K\varphi - \sin\varphi\cos\varphi\left(a_0 + \dfrac{2}{3}a_1 \sin^2\varphi + \dfrac{2 \cdot 4}{3 \cdot 5}a_2 \sin^4\varphi + \cdots\right),$

where

$\qquad\qquad a_0 = \dfrac{2}{\pi}K - 1; \quad a_n = a_{n-1} - \left[\dfrac{(2n-1)!!}{2^n n!}\right]^2 k^{2n}$ ZH 10(19)

$2.\qquad E(\varphi, k) = \dfrac{2}{\pi}E\varphi + \sin\varphi\cos\varphi\left(b_0 + \dfrac{2}{3}b_1 \sin^2\varphi + \dfrac{2 \cdot 4}{3 \cdot 5}b_2 \sin^4\varphi + \cdots\right),$

where

$$b_0 = 1 - \frac{2}{\pi} \boldsymbol{E}, \quad b_n = b_{n-1} - \left[\frac{(2n-1)!!}{2^n n!}\right]^2 \frac{k^{2n}}{2n-1}$$

ZH 27(86)

8.118 For k close to 1, we may use the series

1. $$F(\varphi, k) = \frac{2}{\pi} \boldsymbol{K}' \ln \tan \left(\frac{\varphi}{2} + \frac{\pi}{4}\right) - \frac{\tan \varphi}{\cos \varphi} \left(a_0' - \frac{2}{3} a_1' \tan^2 \varphi + \frac{2 \cdot 4}{3 \cdot 5} a_2' \tan^4 \varphi - \ldots\right),$$

where

$$a_0' = \frac{2}{\pi} \boldsymbol{K}' - 1; \quad a_n' = a_{n-1} - \left[\frac{(2n-1)!!}{2^n n!}\right]^2 k'^{2n}$$

ZH 10(23)

2. $$E(\varphi, k) = \frac{2}{\pi} \left(\boldsymbol{K}' - \boldsymbol{E}'\right) \ln \tan \left(\frac{\varphi}{2} + \frac{\pi}{2}\right)$$
$$+ \frac{\tan \varphi}{\cos \varphi} \left(b_1' - \frac{2}{3} b_2' \tan^2 \varphi + \frac{2 \cdot 4}{3 \cdot 5} b_3' \tan^4 \varphi - \ldots\right) + \frac{1}{\sin \varphi} \left[1 - \cos \varphi \sqrt{1 - k^2 \sin^2 \varphi}\right],$$

where

$$b_0' = \frac{2}{\pi} \left(\boldsymbol{K}' - \boldsymbol{E}'\right), b_n' = b_{n-1}' - \left[\frac{(2n-3)!!}{2^{n-1}(n-1)!}\right]^2 \left(\frac{2n-1}{2n}\right) k'^{2n}$$

ZH 27(90)

For the expansion of complete elliptic integrals in Legendre polynomials, see **8.928**.

8.119 Representation in the form of an infinite product:

1. $$\boldsymbol{K}(k) = \frac{\pi}{2} \prod_{n=1}^{\infty} (1 + k_n), \qquad \text{where}$$

$$k_n = \frac{1 - \sqrt{1 - k_{n-1}^2}}{1 + \sqrt{1 - k_{n-1}^2}}; \qquad k_0 = k$$

FI II 166

See also **8.197**.

8.12 Functional relations between elliptic integrals

8.121

1. $F(-\varphi, k) = -F(\varphi, k)$ JA

2. $E(-\varphi, k) = -E(\varphi, k)$ JA

3. $F(n\pi \pm \varphi, k) = 2n \boldsymbol{K}(k) \pm F(\varphi, k)$ JA

4. $E(n\pi \pm \varphi, k) = 2n \boldsymbol{E}(k) \pm E(\varphi, k)$ JA

8.122 $\boldsymbol{E}(k) \boldsymbol{K}'(k) + \boldsymbol{E}'(k) \boldsymbol{K}(k) - \boldsymbol{K}(k) \boldsymbol{K}'(k) = \dfrac{\pi}{2}$ FI II 691, 791

8.123

1. $$\frac{\partial F}{\partial k} = \frac{1}{k'^2} \left(\frac{E - k'^2 F}{k} - \frac{k \sin \varphi \cos \varphi}{\sqrt{1 - k^2 \sin^2 \varphi}}\right)$$ MO 138, BY (710.07)

2. $\dfrac{d\,\boldsymbol{K}(k)}{dk} = \dfrac{\boldsymbol{E}(k)}{kk'^2} - \dfrac{\boldsymbol{K}(k)}{k}$ FI II 691

3. $\dfrac{\partial E}{\partial k} = \dfrac{E - F}{k}$ MO 138

4. $\dfrac{d\,\boldsymbol{E}(k)}{dk} = \dfrac{\boldsymbol{E}(k) - \boldsymbol{K}(k)}{k}$ FI II 690

8.124

1.   The functions $\boldsymbol{K}$ and $\boldsymbol{K}'$ satisfy the equation

$$\frac{d}{dk}\left\{kk'^2\frac{du}{dk}\right\} - ku = 0. \qquad\qquad \text{WH}$$

2.   The functions $\boldsymbol{E}$ and $\boldsymbol{E}' - \boldsymbol{K}'$ satisfy the equation

$$k'^2\frac{d}{dk}\left(k\frac{du}{dk}\right) + ku = 0. \qquad\qquad \text{WH}$$

8.125

1. $F\left(\psi, \dfrac{1-k'}{1+k'}\right) = (1+k')\,F(\varphi, k)$ $[\tan(\psi - \varphi) = k'\tan\varphi]$ MO 130

2. $E\left(\psi, \dfrac{1-k'}{1+k'}\right) = \dfrac{2}{1+k'}\left[E(\varphi, k) + k'\,F(\varphi, k)\right] - \dfrac{1-k'}{1+k'}\sin\psi$

                                                        $[\tan(\psi - \varphi) = k'\tan\varphi]$ MO 131

3. $F\left(\psi, \dfrac{2\sqrt{k}}{1+k}\right) = (1+k)\,F(\varphi, k)$ $\left[\sin\psi = \dfrac{(1+k)\sin\varphi}{1+k\sin^2\varphi}\right]$

4. $E\left(\psi, \dfrac{2\sqrt{k}}{1+k}\right) = \dfrac{1}{1+k}\left[2E(\varphi, k) - k'^2\,F(\varphi, k) + 2k\dfrac{\sin\varphi\cos\varphi}{1+k\sin^2\varphi}\sqrt{1 - k^2\sin^2\varphi}\right]$

                                                        $\left[\sin\psi = \dfrac{(1+k)\sin\varphi}{1+k\sin^2\varphi}\right]$ MO 131

8.126   In particular,

1. $\boldsymbol{K}\left(\dfrac{1-k'}{1+k'}\right) = \dfrac{1+k'}{2}\,\boldsymbol{K}(k).$ MO 130

2. $\boldsymbol{E}\left(\dfrac{1-k'}{1+k'}\right) = \dfrac{1}{1+k'}\left[\boldsymbol{E}(k) + k'\,\boldsymbol{K}(k)\right]$ MO 130

3. $\boldsymbol{K}\left(\dfrac{2\sqrt{k}}{1+k}\right) = (1+k)\,\boldsymbol{K}(k).$ MO 130

4. $\boldsymbol{E}\left(\dfrac{2\sqrt{k}}{1+k}\right) = \dfrac{1}{1+k}\left[2\,\boldsymbol{E}(k) - k'^2\,\boldsymbol{K}(k)\right]$ MO 130

8.127

k_1	$\sin\varphi_1$	$\cos\varphi_1$	$F(\varphi_1, k_1)$	$E(\varphi_1, k_1)$
$i\dfrac{k}{k'}$	$k'\dfrac{\sin\varphi}{\Delta\varphi}$	$\dfrac{\cos\varphi}{\Delta\varphi}$	$k'\,F(\varphi, k)$	$\dfrac{1}{k'}\left[E(\varphi, k) - \dfrac{k^2\sin\varphi\cos\varphi}{\Delta\varphi}\right]$
k'	$-i\tan\varphi$	$\sec\varphi$	$-i\,F(\varphi, k)$	$ia\left[E(\varphi, k) - F(\varphi, k) - \Delta\varphi\tan\varphi\right]$
$\dfrac{1}{k}$	$k\sin\varphi$	$\Delta\varphi$	$k\,F(\varphi, k)$	$\dfrac{1}{k}\left[E(\varphi, k) - k'^2\,F(\varphi, k)\right]$
$\dfrac{1}{k'}$	$-k'\tan\varphi$	$\dfrac{\Delta\varphi}{\cos\varphi}$	$-ik'\,F(\varphi, k)$	$\dfrac{i}{k'}\left[E(\varphi, k) - k'^2\,F(\varphi, k) - \Delta\varphi\tan\varphi\right]$
$\dfrac{k'}{ik}$	$\dfrac{-ik\sin\varphi}{\Delta\varphi}$	$\dfrac{1}{\Delta\varphi}$	$-ik\,F(\varphi, k)$	$\dfrac{i}{k}\left[E(\varphi, k) - F(\varphi, k) - \dfrac{k^2\sin\varphi\cos\varphi}{\Delta\varphi}\right]$

(see **8.111** 1)   MO 131

8.128   In particular,

1. $K\left(i\dfrac{k}{k'}\right) = k'\,K(k)$ $[\operatorname{Im}(k) < 0]$ MO 130

2. $K\left(i\dfrac{k}{k'}\right) = k'\left[K'(k') - i\,K(k)\right]$ $[\operatorname{Im}(k) < 0]$ MO 130

3. $K\left(\dfrac{1}{k}\right) = k\left[K(k) + i\,K'(k)\right]$ $[\operatorname{Im}(k) < 0]$ MO 130

For integrals of elliptic integrals, see **6.11–6.15**. For indefinite integrals of complete elliptic integrals, see **5.11**.

8.129   Special values:

1. $K\left(\sin\dfrac{\pi}{4}\right) = K\left(\dfrac{\sqrt{2}}{2}\right) = K'\left(\dfrac{\sqrt{2}}{2}\right) = \sqrt{2}\displaystyle\int_0^1 \dfrac{dt}{\sqrt{1-t^4}} = \dfrac{1}{4\sqrt{\pi}}\left[\Gamma\left(\dfrac{1}{4}\right)\right]^2$ MO 130

2. $K'\left(\sqrt{2}-1\right) = \sqrt{2}\,K\left(\sqrt{2}-1\right)$ MO 130

3. $K'\left(\sin\dfrac{\pi}{12}\right) = \sqrt{3}\,K\left(\sin\dfrac{\pi}{12}\right)$ MO 130

4. $K'\left(\tan^2\dfrac{\pi}{8}\right) = K'\left(\dfrac{2-\sqrt{2}}{2+\sqrt{2}}\right) = 2\,K\left(\tan^2\dfrac{\pi}{8}\right)$ MO 130

8.13   Elliptic functions

8.130   Definition and general properties.

1.   A single-valued function $f(z)$ of a complex variable, which is not a constant, is said to be elliptic if it has two periods $2\omega_1$ and $2\omega_2$, that is

$$f\left(z + 2m\omega_1 + 2n\omega_2\right) = f(z) \qquad [m, n \text{ integers}]$$

The ratio of the periods of an analytic function cannot be a real number. For an elliptic function $f(z)$, the z-plane can be partitioned into parallelograms—the period parallelograms—the vertices of which are the points $z_0 + 2m\omega_1 + 2n\omega_2$. At corresponding points of these parallelograms, the function $f(z)$ has the same value.   ZH 117, SI 299

2.   Suppose that α is the angle between the sides a and b of one of the period parallelograms. Then,

$$\tau = \frac{\omega_1}{\omega_2} = \frac{a}{b}e^{i\alpha}, \quad q = e^{i\pi\tau} = e^{-\frac{a}{b}\pi\sin\alpha}\left[\cos\left(\frac{a}{b}\pi\cos\alpha\right) + i\sin\left(\frac{a}{b}\pi\cos\alpha\right)\right]$$

3.   The *derivative* of an elliptic function is also an elliptic function with the same periods.

                                                          SM III 598

4.   A non-constant elliptic function has a finite number of poles in a period parallelogram: it can have no more than two simple and one second-order pole in such a parallelogram. Suppose that these poles lie at the points $a_1, a_2, \ldots, a_n$ and that their orders are $\alpha_1, \alpha_2, \ldots, \alpha_n$. Suppose that the zeros of an analytic function that occur in a single parallelogram are $b_1, b_2, \ldots, b_m$ and that the orders of the zeros are $\beta_1, \beta_2, \ldots, \beta_m$, respectively. Then,

$$\gamma = \alpha_1 + \alpha_2 + \cdots + \alpha_n = \beta_1 + \beta_2 + \cdots + \beta_m \qquad \text{ZH 118}$$

The number γ representing this sum is called the *order* of the elliptic function.

5.   The sum of the residues of an elliptic function with respect to all the poles belonging to a period parallelogram is equal to zero.

6.   The difference between the sum of all the zeros and the sum of all the poles of an elliptic function that are located in a period parallelogram is equal to one of its periods.

7.   Every two elliptic functions with the same periods are related by an algebraic relationship.

                                                              GO II 151

8.[7]   A non-constant single-valued function which is not constant cannot have more than two periods.

                                                              GO II 147

9.   An elliptic function of order γ assumes *an arbitrary value* γ times in a period parallelogram.

                                                            SM 601, SI 301

8.14   Jacobian elliptic functions

8.141   Consider the upper limit φ of the integral

$$u = \int_0^\varphi \frac{d\alpha}{\sqrt{1 - k^2\sin^2\alpha}}$$

as a function of u. Using the notation

$$\varphi = \operatorname{am} u$$

we call this upper limit the *amplitude*. The quantity u is called the *argument*, and its dependence on φ is written

$$u = \operatorname{arg}\varphi$$

8.142   The amplitude is an *infinitely-many-valued* function of u and has a period of $4\boldsymbol{K}i$. The *branch points* of the amplitude correspond to the values of the argument

$$u = 2m\boldsymbol{K} + (2n+1)\boldsymbol{K'}i, \qquad\qquad \text{ZH 67–69}$$

where m and n are arbitrary integers (see also **8.151**).

8.143 The first two of the following functions

$$\operatorname{sn} u = \sin \varphi = \sin \operatorname{am} u, \qquad \operatorname{cn} u = \cos \varphi = \cos \operatorname{am} u,$$

$$\operatorname{dn} u = \Delta\varphi = \sqrt{1 - k^2 \sin^2 \varphi} = \frac{d\varphi}{du}$$

are called, respectively, the *sine-amplitude* and the *cosine-amplitude* while the third may be called the *delta amplitude*. All these elliptic functions were exhibited by Jacobi and they bear his name.   SI 16

The Jacobian elliptic functions are *doubly-periodic* functions and have *two simple poles* in a period parallelogram.   ZH 69

8.144

1. $u = \displaystyle\int_0^{\operatorname{sn} u} \frac{dt}{\sqrt{(1-t^2)(1-k^2t^2)}}$ SI 21(23)

2. $u = \displaystyle\int_1^{\operatorname{cn} u} \frac{dt}{\sqrt{(1-t^2)(k'^2 + k^2t^2)}}$ SI 21(23)

3. $u = \displaystyle\int_1^{\operatorname{dn} u} \frac{dt}{\sqrt{(1-t^2)(t^2 - k'^2)}}$ SI 21(23)

8.145 Power series representations:

1. $\operatorname{sn} u = u - \dfrac{1+k^2}{3!}u^3 + \dfrac{1+14k^2+k^4}{5!}u^5 - \dfrac{1+135k^2+135k^2+k^6}{7!}u^7$
$$+ \frac{1+1228k^2+5478k^4+1228k^6+k^8}{9!}u^9 - \dots$$
$$[|u| < |\boldsymbol{K'}|] \qquad\qquad \text{ZH 81(97)}$$

2. $\operatorname{cn} u = 1 - \dfrac{1}{2!}u^2 + \dfrac{1+4k^2}{4!}u^4 - \dfrac{1+44k^2+16k^4}{6!}u^6 + \dfrac{1+408k^2+912k^4+64k^6}{8!}u^8 - \dots$
$$[|u| < |\boldsymbol{K'}|] \qquad\qquad \text{ZH 81(98)}$$

3. $\operatorname{dn} u =$
$$1 - \frac{k^2}{2!}u^2 + \frac{k^2(4+k^2)}{4!}u^4 - \frac{k^2(16+44k^2+k^4)}{6!}u^6 + \frac{k^2(64+912k^2+408k^4+k^6)}{8!}u^8 - \dots$$
$$[|u| < |\boldsymbol{K'}|] \qquad\qquad \text{ZH 81(99)}$$

4. $\operatorname{am} u$
$$= u - \frac{k^2}{3!}u^3 + \frac{k^2(4+k^2)}{5!}u^5 - \frac{k^2(16+44k^2+k^4)}{7!}u^7 + \frac{k^2(64+912k^2+408k^4+k^6)}{9!}u^9 - \dots$$
$$[|u| < |\boldsymbol{K'}|] \qquad\qquad \text{LA 380(4)}$$

8.146 Representation as a trigonometric series or a product $\left(q = e^{-\frac{\pi k'}{K}}\right)$*

---

*The expansions 1–22 are valid in every strip of the form $\left|\operatorname{Im} \dfrac{\pi u}{2\boldsymbol{K}}\right| < \dfrac{1}{2}\pi \operatorname{Im}\tau$. The expansions 23–25 are valid in an arbitrary bounded portion of u.

1. $\operatorname{sn} u = \dfrac{2\pi}{k\boldsymbol{K}} \displaystyle\sum_{n=1}^{\infty} \dfrac{q^{n-\frac{1}{2}}}{1 - q^{2n-1}} \sin(2n-1)\dfrac{\pi u}{2\boldsymbol{K}}$ WH, ZH 84(108)

2. $\operatorname{cn} u = \dfrac{2\pi}{k\boldsymbol{K}} \displaystyle\sum_{n=1}^{\infty} \dfrac{q^{n-\frac{1}{2}}}{1 + q^{2n-1}} \cos(2n-1)\dfrac{\pi u}{2\boldsymbol{K}}$ WH, ZH 84(109)

3. $\operatorname{dn} u = \dfrac{\pi}{2\boldsymbol{K}} + \dfrac{2\pi}{\boldsymbol{K}} \displaystyle\sum_{n=1}^{\infty} \dfrac{q^n}{1 + q^{2n}} \cos\dfrac{n\pi u}{\boldsymbol{K}}$ WH, ZH 84(110)

4. $am\,u = \dfrac{\pi u}{2\boldsymbol{K}} + 2 \displaystyle\sum_{n=1}^{\infty} \dfrac{1}{n} \dfrac{q^n}{1 + q^{2n}} \sin\dfrac{n\pi u}{\boldsymbol{K}}$ WH

5. $\dfrac{1}{\operatorname{sn} u} = \dfrac{\pi}{2\boldsymbol{K}} \left[\dfrac{1}{\sin\frac{\pi u}{2\boldsymbol{K}}} + 4 \displaystyle\sum_{n=1}^{\infty} \dfrac{q^{2n-1}}{1 - q^{2n-1}} \sin(2n-1)\dfrac{\pi u}{2\boldsymbol{K}} \right]$ LA 369(3)

6. $\dfrac{1}{\operatorname{cn} u} = \dfrac{\pi}{2k'\boldsymbol{K}} \left[\dfrac{1}{\cos\frac{\pi u}{2\boldsymbol{K}}} + 4 \displaystyle\sum_{n=1}^{\infty} (-1)^n \dfrac{q^{2n-1}}{1 + q^{2n-1}} \cos(2n-1)\dfrac{\pi u}{2\boldsymbol{K}} \right]$ LA 369(3)

7. $\dfrac{1}{\operatorname{dn} u} = \dfrac{\pi}{2k'\boldsymbol{K}} \left[1 + 4 \displaystyle\sum_{n=1}^{\infty} (-1)^n \dfrac{q^n}{1 + q^{2n}} \cos\dfrac{n\pi u}{\boldsymbol{K}} \right]$ LA 369(3)

8. $\dfrac{\operatorname{sn} u}{\operatorname{cn} u} = \dfrac{\pi}{2k'\boldsymbol{K}} \left[\tan\dfrac{\pi u}{2\boldsymbol{K}} + 4 \displaystyle\sum_{n=1}^{\infty} (-1)^n \dfrac{q^{2n}}{1 + q^{2n}} \sin\dfrac{n\pi u}{\boldsymbol{K}} \right]$ LA 369(4)

9. $\dfrac{\operatorname{sn} u}{\operatorname{dn} u} = -\dfrac{2\pi}{kk'\boldsymbol{K}} \displaystyle\sum_{n=1}^{\infty} (-1)^n \dfrac{q^{n-\frac{1}{2}}}{1 + q^{2n-1}} \sin(2n-1)\dfrac{\pi u}{2\boldsymbol{K}}$ LA 369(4)

10. $\dfrac{\operatorname{cn} u}{\operatorname{sn} u} = \dfrac{\pi}{2\boldsymbol{K}} \left[\cot\dfrac{\pi u}{2\boldsymbol{K}} - 4 \displaystyle\sum_{n=1}^{\infty} \dfrac{q^{2n}}{1 + q^{2n}} \sin\dfrac{\pi n u}{\boldsymbol{K}} \right]$ LA 369(5)

11. $\dfrac{\operatorname{cn} u}{\operatorname{dn} u} = -\dfrac{2\pi}{k\boldsymbol{K}} \displaystyle\sum_{n=1}^{\infty} (-1)^n \dfrac{q^{n-\frac{1}{2}}}{1 - q^{2n-1}} \cos(2n-1)\dfrac{\pi u}{2\boldsymbol{K}}$ LA 369(5)

12. $\dfrac{\operatorname{dn} u}{\operatorname{sn} u} = \dfrac{\pi}{2\boldsymbol{K}} \left[\dfrac{1}{\sin\frac{\pi u}{2\boldsymbol{K}}} - 4 \displaystyle\sum_{n=1}^{\infty} \dfrac{q^{2n-1}}{1 + q^{2n-1}} \sin(2n-1)\dfrac{\pi u}{2\boldsymbol{K}} \right]$ LA 369(6)

13. $\dfrac{\operatorname{dn} u}{\operatorname{cn} u} = \dfrac{\pi}{2\boldsymbol{K}} \left[\dfrac{1}{\cos\frac{\pi u}{2\boldsymbol{K}}} - 4 \displaystyle\sum_{n=1}^{\infty} (-1)^n \dfrac{q^{2n-1}}{1 - q^{2n-1}} \cos(2n-1)\dfrac{\pi u}{2\boldsymbol{K}} \right]$ LA 369(6)

14. $\dfrac{\operatorname{cn} u \operatorname{dn} u}{\operatorname{sn} u} = \dfrac{\pi}{2\boldsymbol{K}} \left[\cot\dfrac{\pi u}{2\boldsymbol{K}} - 4 \displaystyle\sum_{n=1}^{\infty} \dfrac{q^n}{1 + q^n} \sin\dfrac{n\pi u}{\boldsymbol{K}} \right]$ LA 369(7)

15. $\dfrac{\operatorname{sn} u \operatorname{dn} u}{\operatorname{cn} u} = \dfrac{\pi}{2\boldsymbol{K}} \left\{ \tan\dfrac{\pi u}{2\boldsymbol{K}} + 4 \displaystyle\sum_{n=1}^{\infty} \dfrac{q^n}{1 + (-1)^n q^n} \sin\dfrac{n\pi u}{\boldsymbol{K}} \right\}$ LA 369(7)

16. $\dfrac{\operatorname{sn} u \operatorname{cn} u}{\operatorname{dn} u} = \dfrac{4\pi^2}{k^2\boldsymbol{K}} \displaystyle\sum_{n=1}^{\infty} \dfrac{q^{2n-1}}{1 - q^{2(2n-1)}} \sin(2n-1)\dfrac{\pi u}{\boldsymbol{K}}$ LA 369(7)

17. $\dfrac{\operatorname{sn} u}{\operatorname{cn} u \operatorname{dn} u} = \dfrac{\pi}{2\left(1-k^2\right)K}\left[\tan\dfrac{\pi u}{2K} + 4\sum_{n=1}^{\infty}(-1)^n\dfrac{q^n}{1-q^n}\sin\dfrac{n\pi u}{K}\right]$ LA 369(8)

18. $\dfrac{\operatorname{cn} u}{\operatorname{sn} u \operatorname{dn} u} = \dfrac{\pi}{2K}\left[\cot\dfrac{\pi u}{2K} - 4\sum_{n=1}^{\infty}\dfrac{(-1)^n q^n}{1+(-1)^n q^n}\sin\dfrac{n\pi u}{K}\right]$ LA 369(8)

19. $\dfrac{\operatorname{dn} u}{\operatorname{sn} u \operatorname{cn} u} = \dfrac{\pi}{K}\left[\dfrac{1}{\sin\frac{\pi u}{K}} + 4\sum_{n=1}^{\infty}\dfrac{q^{2(2n-1)}}{1-q^{2(2n-1)}}\sin(2n-1)\dfrac{\pi u}{K}\right]$ LA 369(8)

20. $\ln snu = \ln\dfrac{2K}{\pi} + \ln\sin\dfrac{\pi u}{2K} - 4\sum_{n=1}^{\infty}\dfrac{1}{n}\dfrac{q^n}{1+q^n}\sin^2\dfrac{n\pi u}{2K}$ LA 369(2)

21. $\ln\operatorname{cn} u = \ln\cos\dfrac{\pi u}{2K} - 4\sum_{n=1}^{\infty}\dfrac{1}{n}\dfrac{q^n}{1+(-1)^n q^n}\sin^2\dfrac{n\pi u}{2K}$ LA 369(2)

22. $\ln\operatorname{dn} u = -8\sum_{n=1}^{\infty}\dfrac{1}{2n-1}\dfrac{q^{2n-1}}{1-q^{2(2n-1)}}\sin^2(2n-1)\dfrac{\pi u}{2K}$ LA 369(2)

23. $\operatorname{sn} u = \dfrac{2\sqrt[4]{q}}{\sqrt{k}}\sin\dfrac{\pi u}{2K}\prod_{n=1}^{\infty}\dfrac{1-2q^{2n}\cos\frac{\pi u}{K}+q^{4n}}{1-2q^{2n-1}\cos\frac{\pi u}{K}+q^{4n-2}}$ ZH 86(145)

24. $\operatorname{cn} u = \dfrac{2\sqrt{k'}\sqrt[4]{q}}{\sqrt{k}}\cos\dfrac{\pi u}{2K}\prod_{n=1}^{\infty}\dfrac{1+2q^{2n}\cos\frac{\pi u}{K}+q^{4n}}{1-2q^{2n-1}\cos\frac{\pi u}{K}+q^{4n-2}}$ ZH 86(146)

25. $\operatorname{dn} u = \sqrt{k'}\prod_{n=1}^{\infty}\dfrac{1+2q^{2n-1}\cos\frac{\pi u}{K}+q^{4n-2}}{1-2q^{2n-1}\cos\frac{\pi u}{K}+q^{4n-2}}$ ZH 86(147)

26. $\operatorname{sn}^3 u = \sum_{n=0}^{\infty}\left[\dfrac{1+k^2}{2k^3} - \dfrac{(2n+1)^2}{2k^3}\dfrac{\pi^2}{4K^2}\right]\dfrac{2\pi q^{n+\frac{1}{2}}\sin(2n+1)\frac{\pi u}{2K}}{K\left(1-q^{2n+1}\right)}$

$$\left[\left|\operatorname{Im}\dfrac{u}{2K}\right| < \operatorname{Im}\tau\right]$$ MO 147

27. $\dfrac{1}{\operatorname{sn}^2 u} = \dfrac{\pi^2}{4K^2}\operatorname{cosec}^2\dfrac{\pi u}{2K} + \dfrac{K-E}{K} - \dfrac{2\pi^2}{K^2}\sum_{n=1}^{\infty}\dfrac{nq^{2n}\cos\frac{n\pi u}{K}}{1-q^{2n}}$

$$\left[\left|\operatorname{Im}\dfrac{u}{2K}\right| < \dfrac{1}{2}\operatorname{Im}\tau\right]$$ MO 148

8.147

1. $\operatorname{sn} u = \dfrac{\pi}{2kK}\sum_{n=-\infty}^{\infty}\dfrac{1}{\sin\frac{\pi}{2K}\left[u-(2n-1)iK'\right]}$ MO 149

2. $\operatorname{cn} u = \dfrac{\pi i}{2kK}\sum_{n=-\infty}^{\infty}\dfrac{(-1)^n}{\sin\frac{\pi}{2K}\left[u-(2n-1)iK'\right]}$ MO 150

3. $\operatorname{dn} u = \dfrac{\pi i}{2K}\sum_{n=-\infty}^{\infty}\dfrac{(-1)^n}{\tan\frac{\pi}{2K}\left[u-(2n-1)iK'\right]}$ MO150

8.148   The Weierstrass expansions of the functions $\operatorname{sn} u$, $\operatorname{cn} u$, $\operatorname{dn} u$:

$$\operatorname{sn} u = \frac{B}{A}, \qquad \operatorname{cn} u = \frac{C}{A}, \qquad \operatorname{dn} u = \frac{D}{A}, \qquad\qquad \text{ZH 82–83(105,106,107)}$$

where

$$A = 1 - \sum_{n=1}^{\infty} (-1)^{n+1} a_{n+1} \frac{u^{2n+2}}{(2n+2)!} \qquad\qquad B = \sum_{n=0}^{\infty} (-1)^n b_n \frac{u^{2n+1}}{(2n+1)!}$$

$$C = \sum_{n=0}^{\infty} (-1)^n c_n \frac{u^{2n}}{(2n)!} \qquad\qquad D = \sum_{n=0}^{\infty} (-1)^n d_n \frac{u^{2n}}{(2n)!}$$

and

$a_2 = 2k^2, \quad a_3 = 8\left(k^2 + k^4\right), \quad a_4 = 32\left(k^2 + k^6\right) + 68k^4, \quad a_5 = 128\left(k^2 + k^8\right) + 480\left(k^4 + k^6\right),$

$a_6 = 512\left(k^2 + k^{10}\right) + 3008\left(k^4 + k^8\right) + 5400k^6, \quad \ldots$

$b_0 = 1, \quad b_1 = 1 + k^2, \quad b_2 = 1 + k^4 + 4k^2, \quad b_3 = 1 + k^6 + 9\left(k^2 + k^4\right),$

$b_4 = 1 + k^8 + 16\left(k^2 + k^6\right) - 6k^4, \quad b_5 = 1 + k^{10} + 25\left(k^2 + k^8\right) - 494\left(k^4 + k^6\right),$

$b_6 = 1 + k^{12} + 36\left(k^2 + k^{10}\right) - 5781\left(k^4 + k^8\right) - 12184k^6, \quad \ldots$

$c_0 = 1, \quad c_1 = 1, \quad c_2 = 1 + 2k^2, \quad c_3 = 1 + 6k^2 + 8k^4, \quad c_4 = 1 + 12k^2 + 60k^4 + 32k^6,$

$c_5 = 1 + 20k^2 + 348k^4 + 448k^6 + 128k^8, \quad c_6 = 1 + 30k^2 + 2372k^4 + 4600k^6 + 2880k^8 + 512k^{10}, \quad \ldots$

$d_0 = 1, \quad d_1 = k^2, \quad d_2 = 2k^2 + k^4, \quad d_3 = 8k^2 + 6k^4 + k^6, \quad d_4 = 32k^2 + 60k^4 + 12k^4 + k^8,$

$d_5 = 128k^2 + 448k^4 + 348k^6 + 20k^8 + k^{10},$

$d_6 = 512k^2 + 2880k^4 + 4600k^6 + 2372k^8 + 30k^{10} + k^{12}, \quad \ldots$

8.15 Properties of Jacobian elliptic functions and functional relationships between them

8.151   The periods, zeros, poles, and residues of Jacobian elliptic functions:

1.

	Periods	Zeros	Poles	Residues
$\operatorname{sn} u$	$4m\mathbf{K} + 2n\mathbf{K}'i$	$2m\mathbf{K} + 2n\mathbf{K}'i$	$2m\mathbf{K} + (2n+1)\mathbf{K}'i$	$(-1)^m \dfrac{1}{k}$
$\operatorname{cn} u$	$4m\mathbf{K} + 2n\left(\mathbf{K} + \mathbf{K}'i\right)$	$(2m+1)\mathbf{K} + 2n\mathbf{K}'i$	$2m\mathbf{K} + (2n+1)\mathbf{K}'i$	$(-1)^{m-1}\dfrac{i}{k}$
$\operatorname{dn} u$	$2m\mathbf{K} + 4n\mathbf{K}'i$	$(2m+1)\mathbf{K} + (2n+1)\mathbf{K}'i$	$2m\mathbf{K} + (2n+1)\mathbf{K}'i$	$(-1)^{n-1}i$

SM 630, ZH 69–72

2.

$u^* = u + K$	$u + iK$	$u + K + iK'$	$u + 2K$	$u + 2iK'$	$u + 2K + 2iK'$
$\operatorname{sn} u^* = \dfrac{\operatorname{cn} u}{\operatorname{dn} u}$	$\dfrac{1}{k \operatorname{sn} u}$	$\dfrac{1}{k} \dfrac{\operatorname{dn} u}{\operatorname{cn} u}$	$-\operatorname{sn} u$	$\operatorname{sn} u$	$-\operatorname{sn} u$
$\operatorname{cn} u^* = -k' \dfrac{\operatorname{sn} u}{\operatorname{dn} u}$	$-\dfrac{i}{k} \dfrac{\operatorname{dn} u}{\operatorname{sn} u}$	$-\dfrac{ik'}{k \operatorname{cn} u}$	$-\operatorname{cn} u$	$-\operatorname{cn} u$	$\operatorname{cn} u$
$\operatorname{dn} u^* = k' \dfrac{1}{\operatorname{dn} u}$	$-i\dfrac{\operatorname{cn} u}{\operatorname{sn} u}$	$ik' \dfrac{\operatorname{sn} u}{\operatorname{cn} u}$	$\operatorname{dn} u$	$-\operatorname{dn} u$	$-\operatorname{dn} u$

SM 630

3.

$u^* = 0$	$-u$	$\frac{1}{2} K$	$\frac{1}{2}(K + iK')$	$\frac{1}{2} iK'$	$u + 2mK + 2nK'i$
$\operatorname{sn} u^* = 0$	$-\operatorname{sn} u$	$\dfrac{1}{\sqrt{1 + k'}}$	$\dfrac{\sqrt{1 + k} + i\sqrt{1 - k}}{\sqrt{2k}}$	$\dfrac{i}{\sqrt{k}}$	$(-1)^m \operatorname{sn} u$
$\operatorname{cn} u^* = 1$	$\operatorname{cn} u$	$\dfrac{\sqrt{k'}}{\sqrt{1 + k'}}$	$\dfrac{(1 - i)\sqrt{k'}}{\sqrt{2k}}$	$\dfrac{\sqrt{1 + k}}{\sqrt{k}}$	$(-1)^{m+n} \operatorname{cn} u$
$\operatorname{dn} u^* = 1$	$\operatorname{dn} u$	$\sqrt{k'}$	$\dfrac{\sqrt{k'}\left(\sqrt{1 + k'} - i\sqrt{1 - k'}\right)}{\sqrt{2}}$	$\sqrt{1 + k}$	$(-1)^n \operatorname{dn} u$

SI 19, SI 18(13), WH,   WH   WH   WH

8.152   Transformation formulas

u_1	l_1	$sn(u_1,k_1)$	$cn(u_1,k_1)$	$dn(u_1,k_1)$
ku	$\dfrac{1}{k}$	$k\,\operatorname{sn}(u,k)$	$\operatorname{dn}(u,k)$	$\operatorname{cn}(u,k)$
iu	k'	$i\dfrac{\operatorname{sn}(u,k)}{\operatorname{cn}(u,k)}$	$\dfrac{1}{\operatorname{cn}(u,k)}$	$\dfrac{\operatorname{dn}(u,k)}{\operatorname{cn}(u,k)}$
$k'u$	$i\dfrac{k}{k'}$	$k'\dfrac{\operatorname{sn}(u,k)}{\operatorname{dn}(u,k)}$	$\dfrac{\operatorname{cn}(u,k)}{\operatorname{dn}(u,k)}$	$\dfrac{1}{\operatorname{dn}(u,k)}$
iku	$i\dfrac{k'}{k}$	$ik\dfrac{\operatorname{sn}(u,k)}{\operatorname{dn}(u,k)}$	$\dfrac{1}{\operatorname{dn}(u,k)}$	$\dfrac{\operatorname{cn}(u,k)}{\operatorname{dn}(u,k)}$
$ik'u$	$\dfrac{1}{k'}$	$ik'\dfrac{\operatorname{sn}(u,k)}{\operatorname{cn}(u,k)}$	$\dfrac{\operatorname{dn}(u,k)}{\operatorname{cn}(u,k)}$	$\dfrac{1}{\operatorname{cn}(u,k)}$
$(1+k)u$	$\dfrac{2\sqrt{k}}{1+k}$	$\dfrac{(1+k)\,\operatorname{sn}(u,k)}{1+k\,\operatorname{sn}^2(u,k)}$	$\dfrac{\operatorname{cn}(u,k)\,\operatorname{dn}(u,k)}{1+k\,\operatorname{sn}^2(u,k)}$	$\dfrac{1-k\,\operatorname{sn}^2(u,k)}{1+k\,\operatorname{sn}^2(u,k)}$
$(1+k')\,u$	$\dfrac{1-k'}{1+k'}$	$(1+k')\dfrac{\operatorname{sn}(u,k)\,\operatorname{cn}(u,k)}{\operatorname{dn}(u,k)}$	$\dfrac{1-(1+k')\,\operatorname{sn}^2(u,k)}{\operatorname{dn}(u,k)}$	$\dfrac{1-(1-k')\,\operatorname{sn}^2(u,k)}{\operatorname{dn}(u,k)}$
$\dfrac{\left(1+\sqrt{k'}\right)^2}{2}u$	$\left(\dfrac{1-\sqrt{k'}}{1+\sqrt{k'}}\right)^2$	$\dfrac{k^2\,\operatorname{sn}(u,k)\,d\,\operatorname{cn}(u,k)[k'+\operatorname{dn}(u,k)]}{\sqrt{k_1}\,[1+\operatorname{dn}(u,k)]}$	$\dfrac{\operatorname{dn}(u,k)-\sqrt{k'}}{1-\sqrt{k'}} \times \sqrt{\dfrac{2(1+k')}{[1+\operatorname{dn}(u,k)][k'+\operatorname{dn}(u,k)]}}$	$\dfrac{\sqrt{1+k_1}\left(\operatorname{dn}(u,k)+\sqrt{k'}\right)}{\sqrt{[1+\operatorname{dn}(u,k)][k'+\operatorname{dn}(u,k)]}}$

8.153

1. $\operatorname{sn}(iu, k) = i\dfrac{\operatorname{sn}(u, k')}{\operatorname{cn}(u, k')}$ SI 50(64)

2. $\operatorname{cn}(iu, k) = \dfrac{1}{\operatorname{cn}(u, k')}$ SI 50(65)

3. $\operatorname{dn}(iu, k) = \dfrac{\operatorname{dn}(u, k')}{\operatorname{cn}(u, k')}$ SI 50(65)

4. $\operatorname{sn}(u, k) = k^{-1}\operatorname{sn}\left(ku, k^{-1}\right)$

5. $\operatorname{cn}(u, k) = \operatorname{dn}\left(ku, k^{-1}\right)$

6. $\operatorname{dn}(u, k) = \operatorname{cn}\left(ku, k^{-1}\right)$

7. $\operatorname{sn}(u, ik) = \dfrac{1}{\sqrt{1 + k^2}}\dfrac{\operatorname{sn}\left(u\sqrt{1 + k^2}\right), k\left(1 + k^2\right)^{-1/2}}{\operatorname{dn}\left(u\sqrt{1 + k^2}\right), k(1 + k^2)^{-1/2}}$

8. $\operatorname{cn}(u, ik) = \dfrac{\operatorname{sn}\left(u\left(1 + k^2\right)\right)^{F}FFRAC12, k\left(1 + k^2\right)^{-1/2}}{\operatorname{dn}\left(u\left(1 + k^2\right)\right)^{1/2}, k(1 + k^2)^{-1/2}}$

9. $\operatorname{dn}(u, ik) = \dfrac{1}{\operatorname{dn}\left(u\left(1 + k^2\right)\right)^{F}FFRAC12, k\left(1 + k^2\right)^{-1/2}}$

Functional relations

8.154

1. $\operatorname{sn}^2 u = \dfrac{1 - \operatorname{cn} 2u}{1 + \operatorname{dn} 2u}$ MO 146

2. $\operatorname{cn}^2 u = \dfrac{\operatorname{cn} 2u + \operatorname{dn} 2u}{1 + \operatorname{dn} 2u}$ MO 146

3. $\operatorname{dn}^2 u = \dfrac{\operatorname{dn} 2u + k^2 \operatorname{cn} 2u + k'^2}{1 + \operatorname{dn} 2u}$ MO 146

4. $\operatorname{sn}^2 u + \operatorname{cn}^2 u = 1.$ SI 16(9)

5. $\operatorname{dn}^2 u + k^2 \operatorname{sn}^2 u = 1.$ SI 16(9)

8.155

1. $\dfrac{1 - \operatorname{dn} 2u}{1 + \operatorname{dn} 2u} = k^2 \dfrac{\operatorname{sn}^2 u \operatorname{cn}^2 u}{\operatorname{dn}^2 u}$ MO 146

2. $\dfrac{1 - \operatorname{cn} 2u}{1 + \operatorname{cn} 2u} = \dfrac{\operatorname{sn}^2 u \operatorname{dn}^2 u}{\operatorname{cn}^2 u}$ MO 146

8.156

1. $\operatorname{sn}(u \pm v) = \dfrac{\operatorname{sn} u \operatorname{cn} v \operatorname{dn} v \pm \operatorname{sn} v \operatorname{cn} u \operatorname{dn} u}{1 - k^2 \operatorname{sn}^2 u \operatorname{sn}^2 v}$ SI 46(56)

2. $\operatorname{cn}(u \pm v) = \dfrac{\operatorname{cn} u \operatorname{cn} v \mp \operatorname{sn} u \operatorname{sn} v \operatorname{dn} u \operatorname{dn} v}{1 - k^2 \operatorname{sn}^2 u \operatorname{sn}^2 v}$ SI 46(57)

3. $\mathrm{dn}\,(u \pm v) = \dfrac{\mathrm{dn}\,u\,\mathrm{dn}\,v \mp k^2\,\mathrm{sn}\,u\,\mathrm{sn}\,v\,\mathrm{cn}\,u\,\mathrm{cn}\,v}{1 - k^2\,\mathrm{sn}^2 u\,\mathrm{sn}^2 v}$ SI 46(58)

8.157

1. $\mathrm{sn}\,\dfrac{u}{2} = \pm\dfrac{1}{k}\sqrt{\dfrac{1 - \mathrm{dn}\,u}{1 + \mathrm{cn}\,u}} = \pm\sqrt{\dfrac{1 - \mathrm{cn}\,u}{1 + \mathrm{dn}\,u}}$ SI 47(61), SU 67(15)

2. $\mathrm{cn}\,\dfrac{u}{2} = \pm\sqrt{\dfrac{\mathrm{cn}\,u + \mathrm{dn}\,u}{1 + \mathrm{dn}\,u}} = \pm\dfrac{k'}{k}\sqrt{\dfrac{1 - \mathrm{dn}\,u}{\mathrm{dn}\,u - \mathrm{cn}\,u}}$ SI 48(62), SI 67(16)

3. $\mathrm{dn}\,\dfrac{u}{2} = \pm\sqrt{\dfrac{\mathrm{cn}\,u + \mathrm{dn}\,u}{1 + \mathrm{cn}\,u}} = \pm k'\sqrt{\dfrac{1 - \mathrm{cn}\,u}{\mathrm{dn}\,u + \mathrm{cn}\,u}}$ SI 48(63), SI 67(17)

8.158

1. $\dfrac{d}{du}\,\mathrm{sn}\,u = \mathrm{cn}\,u\,\mathrm{dn}\,u$ SI 21(21)

2. $\dfrac{d}{du}\,\mathrm{cn}\,u = -\,\mathrm{sn}\,u\,\mathrm{dn}\,u$ SI 21(21)

3.[8] $\dfrac{d}{du}\,\mathrm{dn}\,u = -k^2\,\mathrm{dn}\,u\,\mathrm{cn}\,u$ SI 21(21)

8.159   Jacobian elliptic functions are solutions of the following differential equations:

1. $\dfrac{d}{du}\,\mathrm{sn}\,u = \sqrt{(1 - \mathrm{sn}^2 u)\,(1 - k^2\,\mathrm{sn}^2 u)}$ SI 21(22)

2. $\dfrac{d}{du}\,\mathrm{cn}\,u = -\sqrt{(1 - \mathrm{cn}^2 u)\,(k'^2 + k^2\,\mathrm{cn}^2 u)},$ SI 21(22)

3. $\dfrac{d}{du}\,\mathrm{dn}\,u = -\sqrt{(1 - \mathrm{dn}^2 u)\,(\mathrm{dn}^2 u - k'^2)}$ SI 21(22)

For the indefinite integrals of Jacobi's elliptic functions, see **5.13**.

8.16   The Weierstrass function $\wp(u)$

8.160   The Weierstrass elliptic function $\wp(u)$ is defined by

1. $\wp(u) = \dfrac{1}{u^2} + \sum_{m,n}{}' \left\{ \dfrac{1}{(u - 2m\omega_1 - 2n\omega_2)^2} - \dfrac{1}{(2m\omega_1 + 2n\omega_2)^2} \right\},$ SI 307(6)

where the symbol $\sum'$ means that the summation is made over all combinations of integers m and n except for the combination $m = n = 0$; $2\omega_1$ and $2\omega_2$ are the periods of the function $\wp(u)$. Obviously,

2. $\wp\,(u + 2m\omega_1 + 2n\omega_2) = \wp(u)$ and $\mathrm{Im}\left(\dfrac{\omega_1}{\omega_2}\right) \neq 0,$

3. $\dfrac{d}{du}\,\wp(u) = -2\sum_{m,n} \dfrac{1}{(u - 2m\omega_1 - 2n\omega_2)^3},$

where the summation is made over all integral values of m and n.

The series **8.160** 1 and **8.160** 3 converge everywhere except at the poles, that is, at the points $2m\omega_1 + 2n\omega_2$ (where m and n are integers).

4.   The function $\wp(u)$ is a *doubly periodic function* and has *one second-order pole* in a period parallelogram.

SI 306

8.161   The function $\wp(u)$ satisfies the differential equation

1. $$\left[\frac{d\,\wp(u)}{du}\right]^2 = 4\,\wp^3(u) - g_2\,\wp(u) - g_3,$$

SI 142, 310, WH

    where

2. $$g_2 = 60 \sum_{m,n}{}' (m\omega_1 + n\omega_2)^{-4}; \qquad g_3 = 140 \sum_{m,n}{}' (m\omega_1 + n\omega_2)^{-6}$$

WH, SI 310

    The functions g_2 and g_3 are called the *invariants* of the function $\wp(u)$.

8.162 $$u = \int_{\wp(u)}^{\infty} \frac{dz}{\sqrt{4z^3 - g_2 z - g_3}} = \int_{\wp(u)}^{\infty} \frac{dz}{\sqrt{4\,(z-e_1)\,(z-e_2)\,(z-e_3)}},$$

    where e_1, e_2, and e_3 are the roots of the equation $4z^3 - g_2 z - g_3 = 0$; that is,

$$e_1 + e_2 + e_3 = 0, \quad e_1 e_2 + e_2 e_3 + e_3 e_1 = -\frac{g_2}{4}, \quad e_1 e_2 e_3 = \frac{g_3}{4}$$

SI 142, 143, 144

8.163 $\wp(\omega_1) = e_1$, $\wp(\omega_1) + \omega_2 = e_2$, $\wp(\omega_2) = e_3$. Here, it is assumed that if e_1, e_2, and e_3 lie on a straight line in the complex plane, e_2 lies between e_1 and e_3.

8.164 The number $\Delta = g_2^3 - 27 g_3^2$ is called the *discriminant* of the function $\wp(u)$. If $\Delta > 0$, all roots e_1, e_2, and e_3 of the equation $4z^3 - g_2 z - g_3 = 0$ (where g_2 and g_3 are real numbers) are *real*. In this case, the roots e_1, e_2, and e_3 are numbered in such a way that $e_1 > e_2 > e_3$.

1.   If $\Delta > 0$, then

$$\omega_1 = \int_{e_1}^{\infty} \frac{dz}{\sqrt{4z^3 - g_2 z - g_3}}, \quad qquad\omega_2 = i \int_{-\infty}^{e_3} \frac{dz}{\sqrt{g_3 + g_2 z - 4z^3}},$$

    where ω_1 is real and ω_2 is a purely imaginary number. Here, the values of the radical in the integrand are chosen in such a way that ω_1 and $\dfrac{\omega_2}{i}$ will be positive.

2.   If $\Delta < 0$, the root e_2 of the equation $4z^3 - g_2 z - g_3 = 0$ is *real* and the remaining two roots (e_1 and e_3) are *complex conjugates*. Suppose that $e_1 = \alpha = i\beta$, and $e_3 = \alpha - i\beta$. In this case, it is convenient to take

$$\omega' = \int_{e_1}^{\infty} \frac{dz}{\sqrt{4z^3 - g_2 z - g_3}} \quad \text{and} \quad \omega'' = \int_{e_3}^{\infty} \frac{dz}{\sqrt{4z^3 - g_2 z - g_3}}$$

    as basic semiperiods.

    In the first integral, the integration is taken over a path lying entirely in the upper half-plane and in the second over a path lying entirely in the lower-half plane.

SI 151(21, 22)

8.165 Series representation:

1. $$\wp(u) = \frac{1}{u^2} + \frac{g_2 u^2}{4 \cdot 5} + \frac{g_3 u^4}{4 \cdot 7} + \frac{g_2^2 u^6}{2^4 \cdot 3 \cdot 5^2} + \frac{3 g_2 g_3 u^8}{2^4 \cdot 5 \cdot 7 \cdot 11} + \dots$$

WH

8.166 Functional relations

1. $\wp(u) = \wp(-u), \quad \wp'(u) = -\wp'(-u)$

2. $\qquad \wp(u + v) = -\wp(u) - \wp(v) + \dfrac{1}{4}\left[\dfrac{\wp'(u) - \wp'(v)}{\wp(u) - \wp(v)}\right]^2$ SI 163(32)

8.167 $\qquad \wp\left(u; g_2, g_3\right) = \mu^2\, \wp\left(\mu u; \dfrac{g_2}{\mu^4}, \quad \dfrac{g_3}{\mu^6}\right)$ (the formula for homogeneity)

SI 149(13)

The special case: $\mu = i$.

1. $\qquad \wp\left(u; g_2, g_3\right) = -\wp\left(iu; g_2, -g_3\right)$

8.168 An arbitrary elliptic function can be expressed in terms of the elliptic function $\wp(u)$ having the same periods as the original function and its derivative $\wp'(u)$. This expression is rational with respect to $\wp(u)$ and linear with respect to $\wp'(u)$.

8.169 A connection with the Jacobian elliptic functions. For $\Delta > 0$ (see **8.164** 1).

1. $\qquad \wp\left(\dfrac{u}{\sqrt{e_1 - e_2}}\right) = e_1 + (e_1 - e_3)\dfrac{\operatorname{cn}^2(u; k)}{\operatorname{sn}^2(u; k)}$

$\qquad\qquad\qquad\quad = e_2 + (e_1 - e_3)\dfrac{\operatorname{dn}^2(u; k)}{\operatorname{sn}^2(u; k)}$

$\qquad\qquad\qquad\quad = e_3 + (e_1 - e_3)\dfrac{1}{\operatorname{sn}^2(u; k)}$

SI 145(5), ZH 120(197–199)a

2. $\qquad \omega_1 = \dfrac{K}{\sqrt{e_1 - e_3}}, \qquad \omega_2 = \dfrac{iK'}{\sqrt{e_1 - e_3}},$ SI 154(29)

where

3. $\qquad k = \sqrt{\dfrac{e_2 - e_3}{e_1 - e_3}}, \qquad k' = \sqrt{\dfrac{e_1 - e_2}{e_1 - e_3}}$ SI 145(7)

For $\Delta < 0$ (see **8.164** 2)

4. $\qquad \wp\left(\dfrac{u}{\sqrt[4]{9\alpha^2 + \beta^2}}\right) = e_2 + \sqrt{9\alpha^2 + \beta^2}\,\dfrac{1 + \operatorname{cn}(2u; k)}{1 - \operatorname{cn}(2u; k)};$ SI 147(12)

5. $\qquad \omega' = \dfrac{K - iK'}{2\sqrt{9\alpha^2 + \beta^2}}, \qquad \omega'' = \dfrac{K + iK'}{\sqrt[4]{9\alpha^2 + \beta^2}},$ SI 153(28)

where

6. $\qquad k = \sqrt{\dfrac{1}{2} - \dfrac{3e_2}{\sqrt{9\alpha^2 + \beta^2}}}; \qquad k' = \sqrt{\dfrac{1}{2} + \dfrac{3e_2}{\sqrt{9\alpha^2 + \beta^2}}}$ SI 147

For $\Delta = 0$, all the roots e_1, e_2, and e_3 are real and if $g_2 g_3 \neq 0$, two of them are equal to each other. If $e_1 = e_2 \neq e_3$, then

7. $\qquad \wp(u) = \dfrac{3g_3}{g_2} - \dfrac{9g_3}{2g_2}\coth^2\left(u\sqrt{-\dfrac{9g_3}{2g_2}}\right)$ SI 148

If $e_1 \neq e_2 = e_3$, then

8. $\qquad \wp(u) = -\dfrac{3g_3}{2g_2} + \dfrac{9g_3}{2g_2}\dfrac{1}{\sin^2\left(u\sqrt{\dfrac{9g_3}{2g_2}}\right)}$ SI 149

If $g_2 = g_3 = 0$, then $e_1 = e_2 = e_3 = 0$, and

9. $\wp(u) = \dfrac{1}{u^2}$

SI 149

8.17 The functions $\zeta(u)$ and $\sigma(u)$

8.171 Definitions:

1. $\zeta(u) = \dfrac{1}{u} - \displaystyle\int_0^u \left(\wp(z) - \dfrac{1}{z^2} \right) dz$

SI 181(45)

2. $\sigma(u) = u \exp\left\{ \displaystyle\int_0^u \left(\wp(z) - \dfrac{1}{z^2} \right) dz \right\}$

SI 181(46)

8.172 Series and infinite-product representation

1. $\zeta(u) = \dfrac{1}{u} + \displaystyle\sum_{m,n}{}' \left(\dfrac{1}{u - 2m\omega_1 - 2n\omega_2} + \dfrac{1}{2m\omega_1 + 2n\omega_2} + \dfrac{u}{(2m\omega_1 - 2n\omega_2)^2} \right)$

SI 307(8)

2. $\sigma(u) = u \displaystyle\prod_{mn,}{}' \left(1 - \dfrac{u}{2m\omega_1 + 2n\omega_2} \right) \exp\left\{ \dfrac{u}{2m\omega_1 + 2n\omega_2} + \dfrac{u^2}{2(2m\omega_1 + 2n\omega_2)^2} \right\}$

SI 308(9)

8.173

1. $\zeta(u) = u - \dfrac{g_2 u^3}{2^2 \cdot 3 \cdot 5} - \dfrac{g_3 u^5}{2^2 \cdot 5 \cdot 7} - \dfrac{g_2^2 u^7}{2^4 \cdot 3 \cdot 5^2 \cdot 7} - \dfrac{3 g_2 g_3 u^9}{2^4 \cdot 5 \cdot 7 \cdot 9 \cdot 11} - \cdots$

SI 181(49)

2. $\sigma(u) = u - \dfrac{g_2 u^5}{2^4 \cdot 3 \cdot 5} - \dfrac{g_3 u^7}{2^3 \cdot 3 \cdot 5 \cdot 7} - \dfrac{g_2^2 u^9}{2^9 \cdot 3^2 \cdot 5 \cdot 7} - \dfrac{3 g_2 g_3 u^{11}}{2^7 \cdot 3^2 \cdot 5^2 \cdot 7 \cdot 11} - \cdots$

SI 181(49)

8.174 $\zeta(u) = \dfrac{\zeta(\omega_1)}{\omega_1} u + \dfrac{\pi}{2\omega_1} \cot \dfrac{\pi u}{2\omega_1} + \dfrac{\pi}{2\omega_1} \displaystyle\sum_{n=1}^{\infty} \left\{ \cot\left(\dfrac{\pi u}{2\omega_1} + n\pi \dfrac{\omega_2}{\omega_1} \right) \right.$

$\left. + \cot\left(\dfrac{\pi u}{2\omega_1} - n\pi \dfrac{\omega_2}{\omega_1} \right) \right\}$

MO 154

$= \dfrac{\zeta(\omega_1)}{\omega_1} u + \dfrac{\pi}{2\omega_1} \cot \dfrac{\pi u}{2\omega_1} + \dfrac{2\pi}{\omega_1} \displaystyle\sum_{n=1}^{\infty} \dfrac{q^{2n}}{1 - q^{2n}} \sin \dfrac{\pi n u}{\omega_1}$

MO 155

Functional relations and properties

8.175 $\zeta(u) = -\zeta(-u), \quad \sigma(u) = -\sigma(-u)$

SI 181

8.176

1. $\zeta(u + 2\omega_1) = \zeta(u) + 2\zeta(\omega_1)$

SI 184(57)

2. $\zeta(u + 2\omega_2) = \zeta(u) + 2\zeta(\omega_2)$

SI 184(57)

3. $\sigma(u + 2\omega_1) = -\sigma(u) \exp\left\{ 2(u + \omega_1)\zeta(\omega_1) \right\}.$

SI 185(60)

4. $\sigma(u + 2\omega_2) = -\sigma(u) \exp\left\{ 2(u + \omega_2)\zeta(\omega_2) \right\}.$

SI 185(60)

5. $\omega_2 \zeta(\omega_1) - \omega_1 \zeta(\omega_2) = \dfrac{\pi}{2} i$ SI 186(62)

8.177

1. $\zeta(u+v) - \zeta(u) - \zeta(v) = \dfrac{1}{2} \dfrac{\wp'(u) - \wp'(v)}{\wp(u) - \wp(v)}$ SI 182(53)

2. $\wp(u) - \wp(v) = -\dfrac{\sigma(u-v)\,\sigma(u+v)}{\sigma^2(u)\,\sigma^2(v)}$ SI 183(54)

3. $\zeta(u-v) + \zeta(u+v) - 2\zeta(u) = \dfrac{\wp'(u)}{\wp(u) - \wp(v)}$ SI 182(51)

8.178

1. $\zeta(u; \omega_1, \omega_2) = t\,\zeta(tu; t\omega_1, t\omega_2)$ MO 154

2.[8] $\sigma(u; \omega_1, \omega_2) = t^{-1}\,\sigma(tu; t\omega_1, t\omega_2)$ MO 156

For the indefinite integrals of Weierstrass elliptic functions, see **5.14**.

8.18–8.19   Theta functions

8.180   *Theta functions* are defined as the sums (for $|q| < 1$) of the following series:

1. $\vartheta_4(u) = \displaystyle\sum_{n=-\infty}^{\infty} (-1)^n q^{n^2} e^{2nui} = 1 + 2\sum_{n=1}^{\infty} (-1)^n q^{n^2} \cos 2nu$ WH

2. $\vartheta_1(u) = \dfrac{1}{i} \displaystyle\sum_{n=-\infty}^{\infty} (-1)^n q^{\left(n+\frac{1}{2}\right)^2} e^{(2n+1)ui} = 2\sum_{n=1}^{\infty} (-1)^{n+1} q^{\left(n-\frac{1}{2}\right)^2} \sin(2n-1)u.$ WH

3. $\vartheta_2(u) = \displaystyle\sum_{n=-\infty}^{\infty} q^{\left(n+\frac{1}{2}\right)^2} e^{(2n+1)ui} = 2\sum_{n=1}^{\infty} q^{\left(n+\frac{1}{2}\right)^2} \cos(2n-1)u.$ WH

4. $\vartheta_3(u) = \displaystyle\sum_{n=-\infty}^{\infty} q^{n^2} e^{2nui} = 1 + 2\sum_{n=1}^{\infty} q^{n^2} \cos 2nu$ WH

The notations $\vartheta(u, q)$ and $\vartheta(u \mid \tau)$, where τ and q are related by $q = e^{i\pi\tau}$, are also used. Here, q is called the *nome* of the theta function and τ its *parameter*.

8.181   Representation of theta functions in terms of infinite products

1. $\vartheta_4(u) = \displaystyle\prod_{n=1}^{\infty} \left(1 - 2q^{2n-1} \cos 2u + q^{2(2n-1)}\right)\left(1 - q^{2n}\right)$ SI 200(9), ZH 90(9)

2. $\vartheta_3(u) = \displaystyle\prod_{n=1}^{\infty} \left(1 + 2q^{2n-1} \cos 2u + q^{2(2n-1)}\right)\left(1 - q^{2n}\right)$ SI 200(9), ZH 90(9)

3. $\vartheta_1(u) = 2\sqrt[4]{q} \sin u \displaystyle\prod_{n=1}^{\infty} \left(1 - 2q^{2n} \cos 2u + q^{4n}\right)\left(1 - q^{2n}\right)$ SI 200(9), ZH 90(9)

4.[8] $\vartheta_2(u) = 2\sqrt[4]{q} \cos u \displaystyle\prod_{n=1}^{\infty} \left(1 + 2q^{2n} \cos 2u + q^{4n}\right)\left(1 - q^{2n}\right)$ SI 200(0), ZH 90(9)

Functional relations and properties

8.182   Quasiperiodicity.   Suppose that $q = e^{\pi\tau i} (\operatorname{Im}\tau > 0)$. Then, theta functions that are periodic functions of u are called *quasiperiodic functions* of τ and u. This property follows from the equations

1. $\vartheta_4(u + \pi) = \vartheta_4(u)$ SI 200(10)

2. $\vartheta_4(u + \tau\pi) = -\dfrac{1}{q} e^{-2iu}\, \vartheta_4(u)$ SI 200(10)

3. $\vartheta_1(u + \pi) = -\vartheta_1(u)$ SI 200(10)

4. $\vartheta_1(u + \tau\pi) = -\dfrac{1}{q} e^{-2iu}\, \vartheta_1(u)$ SI 200(10)

5. $\vartheta_2(u + \pi) = -\vartheta_2(u)$ SI 200(10)

6. $\vartheta_2(u + \tau\pi) = \dfrac{1}{q} e^{-2iu}\, \vartheta_2(u)$ SI 200(10)

7. $\vartheta_3(u + \pi) = \vartheta_3(u)$ SI 200(10)

8. $\vartheta_3(u + \tau\pi) = \dfrac{1}{q} e^{-2iu}\, \vartheta_3(u)$ SI 200(10)

9.[10] $\vartheta_4(u + \pi) = \vartheta_4(u)$ LW 6(1.3.5)

8.183

1. $\vartheta_4\left(u + \tfrac{1}{2}\pi\right) = \vartheta_3(u)$ WH

2. $\vartheta_1\left(u + \tfrac{1}{2}\pi\right) = \vartheta_2(u)$ WH

3. $\vartheta_2\left(u + \tfrac{1}{2}\pi\right) = -\vartheta_1(u)$ WH

4. $\vartheta_3\left(u + \tfrac{1}{2}\pi\right) = \vartheta_4(u)$ WH

5. $\vartheta_4\left(u + \tfrac{1}{2}\pi\tau\right) = iq^{-1/4} e^{-iu}\, \vartheta_1(u)$ WH

6. $\vartheta_1\left(u + \tfrac{1}{2}\pi\tau\right) = iq^{-1/4} e^{-iu}\, \vartheta_4(u)$ WH

7. $\vartheta_2\left(u + \tfrac{1}{2}\pi\tau\right) = q^{-1/4} e^{-iu}\, \vartheta_3(u)$ WH

8. $\vartheta_3\left(u + \tfrac{1}{2}\pi\tau\right) = q^{-1/4} e^{-iu}\, \vartheta_2(u)$ WH

8.184   Even and odd theta functions

1. $\vartheta_1(-u) = -\vartheta_1(u)$ WH

2. $\vartheta_2(-u) = \vartheta_2(u)$ WH

3. $\vartheta_3(-u) = \vartheta_3(u)$ WH

4. $\vartheta_4(-u) = \vartheta_4(u)$ WH

8.185 $\vartheta_4^4(u) + \vartheta_2^4(u) = \vartheta_1^4(u) + \vartheta_3^4(u)$ WH

8.186[7]   Considering the theta functions as functions of two independent variables u and τ, we have

$$\pi i \frac{\partial^2 \vartheta_k(u \mid \tau)}{\partial u^2} + 4 \frac{\partial \vartheta_k(u \mid \tau)}{\partial \tau} = 0 \qquad [k = 1, 2, 3, 4] \qquad \text{WH}$$

8.187   We denote the partial derivatives of the theta functions with respect to u by a prime and consider them as functions of the single argument u. Then,

1. $\vartheta_1'(0) = \vartheta_2(0)\,\vartheta_3(0)\,\vartheta_4(0)$ WH

2. $\dfrac{\vartheta_1'''(0)}{\vartheta_1'(0)} = \dfrac{\vartheta_2''(0)}{\vartheta_2(0)} + \dfrac{\vartheta_3''(0)}{\vartheta_3(0)} + \dfrac{\vartheta_4''(0)}{\vartheta_4(0)}$ WH

8.188 $\vartheta_1(u)\,\vartheta_2(u)\,\vartheta_3(u)\,\vartheta_4(0) = \frac{1}{2}\,\vartheta_1(2u)\,\vartheta_2(0)\,\vartheta_3(0)\,\vartheta_4(0)$ WH

8.189   The zeros of the theta functions:

1.[8] $\vartheta_4(u) = 0$ for $u = 2m\dfrac{\pi}{2} + (2n - 1)\dfrac{\pi\tau}{2}$ SI 201

2.[10] $\vartheta_1(u) = 0$ for $u = 2m\dfrac{\pi}{2} + 2n\dfrac{\pi\tau}{2}$ SI 201

3. $\vartheta_2(u) = 0$ for $u = (2m - 1)\dfrac{\pi}{2} + 2n\dfrac{\pi\tau}{2}$ SI 201

4. $\vartheta_3(u) = 0$ for $u = (2m - 1)\dfrac{\pi}{2} + (2n - 1)\dfrac{\pi\tau}{2}$ $[m$ and n are integers or zero$]$ SI 201

For integrals of theta functions, see **6.16**.

8.191   Connections with the Jacobian elliptic functions:

    For $\tau = i\dfrac{K'}{K}$, i.e. for $q = \exp\left(-\pi\dfrac{K'}{K}\right)$,

1. $\operatorname{sn} u = \dfrac{1}{\sqrt{k}} \dfrac{\vartheta_1\left(\dfrac{\pi u}{2K}\right)}{\vartheta_4\left(\dfrac{\pi u}{2K}\right)} = \dfrac{1}{\sqrt{k}} \dfrac{H(u)}{\Theta(u)}$ SI 206(22), SI 209(35)

2. $\operatorname{cn} u = \sqrt{\dfrac{k'}{k}} \dfrac{\vartheta_2\left(\dfrac{\pi u}{2K}\right)}{\vartheta_4\left(\dfrac{\pi u}{2K}\right)} = \sqrt{\dfrac{k'}{k}} \dfrac{H_1(u)}{\Theta(u)}$ SI 207(23), SI 209(35)

3. $\operatorname{dn} u = \sqrt{k'} \dfrac{\vartheta_3\left(\dfrac{\pi u}{2K}\right)}{\vartheta_4\left(\dfrac{\pi u}{2K}\right)} = \sqrt{k'} \dfrac{\Theta_1(u)}{\Theta(u)}$ SI 207(24), SI 209(35)

8.192   Series representation of the functions H, H_1, Θ, Θ_1.

    In these formulas, $q = \exp\left(-\pi\dfrac{K'}{K}\right)$.

1. $\Theta(u) = \vartheta_4\left(\dfrac{\pi u}{2K}\right) = 1 + 2\displaystyle\sum_{n=1}^{\infty}(-1)^n q^{n^2} \cos\dfrac{n\pi u}{K}$ SI 207(25), SI 212(42)

2. $H(u) = \vartheta_1\left(\dfrac{\pi u}{2K}\right) = 2\displaystyle\sum_{n=1}^{\infty}(-1)^{n+1} \sqrt[4]{q^{(2n+1)^2}} \sin(2n - 1)\dfrac{\pi u}{2K}$ SI 207(25), SI 212(43)

3. $\Theta_1(u) = \vartheta_3\left(\dfrac{\pi u}{2K}\right) = 1 + 2\displaystyle\sum_{n=1}^{\infty}q^{n^2} \cos\dfrac{n\pi u}{K}$ SI 207(25), SI 212(45)

4. $H_1(u) = \vartheta_2\left(\dfrac{\pi u}{2K}\right) = 2\displaystyle\sum_{n=1}^{\infty}\sqrt[4]{q^{(2n-1)^2}}\cos(2n-1)\dfrac{\pi u}{2K}$ SI 207(25), SI 212(44)

8.193   Connections with the Weierstrass elliptic functions

1. $\wp(u) = e_1 + \left[\dfrac{H_1\left(u\sqrt{\lambda}\right) H'(0)}{H_1(0)\, H\left(u\sqrt{\lambda}\right)}\right]^2 \lambda = e_2 + \left[\dfrac{\Theta_1\left(u\sqrt{\lambda}\right) H'(0)}{\Theta_1(0)\, H'\left(u\sqrt{\lambda}\right)}\right]^2 \lambda = e_3 + \left[\dfrac{\Theta\left(u\sqrt{\lambda}\right) H'(0)}{\Theta(0)\, H'\left(u\sqrt{\lambda}\right)}\right]^2 \lambda$

                                                                      SI 235(77,78)

2. $\zeta(u) = \dfrac{\eta_1 u}{\omega_1} + \sqrt{\lambda}\,\dfrac{H'\left(u\sqrt{\lambda}\right)}{H\left(u\sqrt{\lambda}\right)}$ SI 234(73)

3. $\sigma(u) = \dfrac{1}{\sqrt{\lambda}}\exp\left(\dfrac{\eta_1 u^2}{2\omega_1}\right)\dfrac{H\left(u\sqrt{\lambda}\right)}{H'(0)}$ SI 234(72)

    where

$$\lambda = e_1 - e_3; \qquad \eta_1 = \zeta(\omega_1) = -\dfrac{\omega_1 \lambda}{3}\dfrac{H'''(0)}{H'(0)} \qquad\qquad \text{SI 236}$$

8.194   The connection with elliptic integrals:

1. $E(u,k) = u - u\dfrac{\Theta''(0)}{\Theta(0)} + \dfrac{\Theta'(u)}{\Theta(u)}$ SI 228(65)

2. $\Pi\left(u, -k^2\sin^2 a, k\right) = \displaystyle\int_0^u \dfrac{d\varphi}{1 - k^2\sin^2 a\,\operatorname{sn}^2\varphi} = u + \dfrac{\operatorname{sn} a}{\operatorname{cn} a\,\operatorname{dn} a}\left[\dfrac{\Theta'_{(a)}}{\Theta(a)}u + \dfrac{1}{2}\ln\dfrac{\Theta(u-a)}{\Theta(u+a)}\right]$

                                                                          SI 228(65)

q-series and products, $\left[q = \exp\left(-\pi\dfrac{K'}{K}\right)\right]$

8.195 $\dfrac{\pi}{2}\left[1 + 2\displaystyle\sum_{n=1}^{\infty}q^{n^2}\right]^2 = K = \dfrac{\pi}{2}\,\Theta^2(K)$ (cf. **8.197** 1)   SI 219

8.196 $E = K - K\dfrac{\Theta''(0)}{\Theta(0)} = K - \dfrac{2\pi^2}{K}\dfrac{\displaystyle\sum_{n=1}^{\infty}(-1)^{n+1}n^2 q^{n^2}}{1 + 2\displaystyle\sum_{n=1}^{\infty}(-1)^n q^{n^2}}$ SI 230(67)

8.197

1. $1 + 2\displaystyle\sum_{n=1}^{\infty}q^{n^2} = \sqrt{\dfrac{2K}{\pi}} = \vartheta_3(0)$ (cf. **8.195**)   WH

2. $\displaystyle\sum_{n=1}^{\infty}q^{\left(\frac{2n-1}{2}\right)^2} = \sqrt{\dfrac{kK}{2\pi}} = \dfrac{1}{2}\vartheta_2(0)$ WH

3. $\displaystyle 4\sqrt{q}\,\prod_{n=1}^{\infty}\left(\frac{1+q^{2n}}{1+q^{2n-1}}\right)^4 = k$ SI 206(17, 18)

4. $\displaystyle \prod_{n=1}^{\infty}\left(\frac{1-q^{2n-1}}{1+q^{2n-1}}\right)^4 = k'$ SI 206(19, 20)

5. $\displaystyle 2\sqrt[4]{q}\,\prod_{n=1}^{\infty}\left(\frac{1-q^{2n}}{1-q^{2n-1}}\right)^2 = 2\sqrt{k}\,\frac{\boldsymbol{K}}{\pi}$ WH

6. $\displaystyle \prod_{n=1}^{\infty}\left(\frac{1-q^{2n}}{1+q^{2n}}\right)^2 = 2\sqrt{k'}\,\frac{\boldsymbol{K}}{\pi}$ WH

8.198

1. $\displaystyle \lambda = \frac{1}{2}\frac{1-\sqrt{k'}}{1+\sqrt{k'}} = \frac{\displaystyle\sum_{n=0}^{\infty} q^{(2n+1)^2}}{\displaystyle 1 + 2\sum_{n=1}^{\infty} q^{4n^2}}$ [for $0 < k < 1$, we have $0 < \lambda < \frac{1}{2}$]   WH

     The series

2. $q = \lambda + 2\lambda^5 + 15\lambda^9 + 150\lambda^{13} + 1707\lambda^{17} + \ldots$ WH

     is used to determine q from the given modulus k.

8.199[10]   Identities involving products of theta functions

1. $\vartheta_1(x,q)\,\vartheta_1(y,q) = \vartheta_3\left(x+y,q^2\right)\vartheta_2\left(x-y,q^2\right) - \vartheta_2\left(x+y,q^2\right)\vartheta_3\left(x-y,q^2\right)$ LW 7(1.4.7)

2. $\vartheta_1(x,q)\,\vartheta_2(y,q) = \vartheta_1\left(x+y,q^2\right)\vartheta_4\left(x-y,q^2\right) + \vartheta_4\left(x+y,q^2\right)\vartheta_1\left(x-y,q^2\right)$ LW 8(1.4.8)

3. $\vartheta_2(x,q)\,\vartheta_2(y,q) = \vartheta_2\left(x+y,q^2\right)\vartheta_3\left(x-y,q^2\right) + \vartheta_3\left(x+y,q^2\right)\vartheta_2\left(x-y,q^2\right)$ LW 8(1.4.9)

4. $\vartheta_3(x,q)\,\vartheta_3(y,q) = \vartheta_3\left(x+y,q^2\right)\vartheta_3\left(x-y,q^2\right) + \vartheta_2\left(x+y,q^2\right)\vartheta_2\left(x-y,q^2\right)$ LW 8(1.4.10)

5. $\vartheta_3(x,q)\,\vartheta_4(y,q) = \vartheta_4\left(x+y,q^2\right)\vartheta_4\left(x-y,q^2\right) - \vartheta_1\left(x+y,q^2\right)\vartheta_1\left(x-y,q^2\right)$ LW 8(1.4.11)

6. $\vartheta_4(x,q)\,\vartheta_4(y,q) = \vartheta_3\left(x+y,q^2\right)\vartheta_3\left(x-y,q^2\right) - \vartheta_2\left(x+y,q^2\right)\vartheta_2\left(x-y,q^2\right)$ LW 8(1.4.12)

7. $\vartheta_1(x+y)\,\vartheta_1(x-y)\,\vartheta_4^2(0) = \vartheta_3^2(x)\,\vartheta_2^2(y) - \vartheta_2^2(x)\,\vartheta_3^2(y) = \vartheta_1^2(x)\,\vartheta_4^2(y) - \vartheta_4^2(x)\,\vartheta_1^2(y)$ LW 8(1.4.16)

8. $\vartheta_2(x+y)\,\vartheta_2(x-y)\,\vartheta_4^2(0) = \vartheta_4^2(x)\,\vartheta_2^2(y) - \vartheta_1^2(x)\,\vartheta_3^2(y) = \vartheta_2^2(x)\,\vartheta_4^2(y) - \vartheta_3^2(x)\,\vartheta_1^2(y)$ LW 8(1.4.17)

9. $\vartheta_3(x+y)\,\vartheta_3(x-y)\,\vartheta_4^2(0) = \vartheta_4^2(x)\,\vartheta_3^2(y) - \vartheta_1^2(x)\,\vartheta_2^2(y) = \vartheta_3^2(x)\,\vartheta_4^2(y) - \vartheta_2^2(x)\,\vartheta_1^2(y)$ LW 8(1.4.18)

10. $\vartheta_4(x+y)\,\vartheta_4(x-y)\,\vartheta_4^2(0) = \vartheta_4^2(x)\,\vartheta_4^2(y) - \vartheta_1^2(x)\,\vartheta_1^2(y)$ LW 8(1.4.15)

11. $\vartheta_4(x+y)\,\vartheta_4(x-y)\,\vartheta_4^2(0) = \vartheta_3^2(x)\,\vartheta_3^2(y) - \vartheta_2^2(x)\,\vartheta_2^2(y) = \vartheta_4^2(x)\,\vartheta_4^2(y) - \vartheta_1^2(x)\,\vartheta_1^2(y)$ LW 9(1.4.19)

12. $\vartheta_1(x+y)\,\vartheta_1(x-y)\,\vartheta_3^2(0) = \vartheta_1^2(x)\,\vartheta_3^2(y) - \vartheta_3^2(x)\,\vartheta_1^2(y) = \vartheta_4^2(x)\,\vartheta_2^2(y) - \vartheta_2^2(x)\,\vartheta_4^2(y)$ LW 9(1.4.23)

13. $\vartheta_2(x+y)\,\vartheta_2(x-y)\,\vartheta_3^2(0) = \vartheta_2^2(x)\,\vartheta_3^2(y) - \vartheta_4^2(x)\,\vartheta_1^2(y) = \vartheta_3^2(x)\,\vartheta_2^2(y) - \vartheta_1^2(x)\,\vartheta_4^2(y)$ LW 9(1.4.24)

14. $\vartheta_3(x+y)\,\vartheta_3(x-y)\,\vartheta_3^2(0) = \vartheta_1^2(x)\,\vartheta_1^2(y) + \vartheta_3^2(x)\,\vartheta_3^2(y) = \vartheta_2^2(x)\,\vartheta_2^2(y) + \vartheta_4^2(x)\,\vartheta_4^2(y)$ LW 9(1.4.25)

15. $\vartheta_4(x+y)\,\vartheta_4(x-y)\,\vartheta_3^2(0) = \vartheta_1^2(x)\,\vartheta_2^2(y) + \vartheta_3^2(x)\,\vartheta_4^2(y) = \vartheta_2^2(x)\,\vartheta_1^2(y) + \vartheta_4^2(x)\,\vartheta_3^2(y)$ LW 9(1.4.26)

16. $\vartheta_1(x+y)\,\vartheta_1(x-y)\,\vartheta_2^2(0) = \vartheta_1^2(x)\,\vartheta_2^2(y) - \vartheta_2^2(x)\,\vartheta_1^2(y) = \vartheta_4^2(x)\,\vartheta_3^2(y) - \vartheta_3^2(x)\,\vartheta_4^2(y)$ LW 9(1.4.30)

17. $\vartheta_2(x+y)\,\vartheta_2(x-y)\,\vartheta_2^2(0) = \vartheta_2^2(x)\,\vartheta_2^2(y) - \vartheta_1^2(x)\,\vartheta_1^2(y) = \vartheta_3^2(x)\,\vartheta_3^2(y) - \vartheta_4^2(x)\,\vartheta_4^2(y)$ LW 10(1.4.31)

18. $\vartheta_3(x+y)\,\vartheta_3(x-y)\,\vartheta_2^2(0) = \vartheta_3^2(x)\,\vartheta_2^2(y) + \vartheta_4^2(x)\,\vartheta_1^2(y) = \vartheta_2^2(x)\,\vartheta_3^2(y) + \vartheta_1^2(x)\,\vartheta_4^2(y)$ LW 10(1.4.32)

19. $\vartheta_4(x+y)\,\vartheta_4(x-y)\,\vartheta_2^2(0) = \vartheta_4^2(x)\,\vartheta_2^2(y) + \vartheta_3^2(x)\,\vartheta_1^2(y) = \vartheta_1^2(x)\,\vartheta_3^2(y) + \vartheta_2^2(x)\,\vartheta_4^2(y)$ LW 10(1.4.33)

20. $\vartheta_3^2(x)\,\vartheta_3^2(0) = \vartheta_4^2(x)\,\vartheta_4^2(0) + \vartheta_2^2(x)\,\vartheta_2^2(0)$ LW 11(1.4.49)

21. $\vartheta_4^2(x)\,\vartheta_3^2(0) = \vartheta_1^2(x)\,\vartheta_2^2(0) + \vartheta_3^2(x)\,\vartheta_4^2(0)$ LW 11(1.4.50)

22. $\vartheta_4^2(x)\,\vartheta_2^2(0) = \vartheta_1^2(x)\,\vartheta_3^2(0) + \vartheta_2^2(x)\,\vartheta_4^2(0)$ LW 11(1.4.51)

23. $\vartheta_3^2(x)\,\vartheta_2^2(0) = \vartheta_1^2(x)\,\vartheta_4^2(0) + \vartheta_2^2(x)\,\vartheta_3^2(0)$ LW 11(1.4.52)

24.[8] $\vartheta_3^4(x) = \vartheta_2^4(0) + \vartheta_4^4(0)$ LW 11(1.4.53)

8.199(1)[10]   Theta functions as infinite products

1. $\vartheta_1(z) = 2q^{\frac{1}{4}}\sin z\left(\prod_{n=1}^{\infty}\left(1-q^{2n}\right)\left(1-2q^{2n}\cos 2z + q^{4n}\right)\right)$ LW 15(1.6.23)

2. $\vartheta_2(z) = 2q^{\frac{1}{4}}\cos z\left(\prod_{n=1}^{\infty}\left(1-q^{2n}\right)\left(1+2q^{2n}\cos 2z + q^{4n}\right)\right)$ LW 15(1.5.24)

3. $\vartheta_3(z) = \prod_{n=1}^{\infty}\left(1-q^{2n}\right)\left(1+2q^{2n-1}\cos 2z + q^{4n-2}\right)$ LW 15(1.6.25)

4. $\vartheta_4(z) = \prod_{n=1}^{\infty}\left(1-q^{2n}\right)\left(1-2q^{2n-1}\cos 2z + q^{4n-2}\right)$ LW 15(1.6.26)

8.199(2)[10]   Derivatives of ratios of theta functions

1. $\dfrac{d}{dx}\left(\vartheta_1\,/\,\vartheta_4\right) = \vartheta_4^2(0)\,\vartheta_2(x)\,\vartheta_3(x)/\,\vartheta_4^2(x)$ LW 19(1.9.3)

2. $\dfrac{d}{dx}\left(\vartheta_2\,/\,\vartheta_4\right) = -\,\vartheta_3^2(0)\,\vartheta_1(x)\,\vartheta_3(x)/\,\vartheta_4^2(x)$ LW 19(1.9.6)

3. $\dfrac{d}{dx}\left(\vartheta_3\,/\,\vartheta_4\right) = -\,\vartheta_2^2(0)\,\vartheta_1(x)\,\vartheta_2(x)/\,\vartheta_4^2(x)$ LW 19(1.9.7)

4. $\dfrac{d}{dx}\left(\vartheta_1\,/\,\vartheta_3\right) = \vartheta_3^2(0)\,\vartheta_2(x)\,\vartheta_4(x)/\,\vartheta_3^2(x)$ LW 19(1.9.8)

5. $\dfrac{d}{dx}\left(\vartheta_2\,/\,\vartheta_3\right) = -\,\vartheta_4^2(0)\,\vartheta_1(x)\,\vartheta_4(x)/\,\vartheta_3^2(x)$ LW 19(1.9.9)

6. $\dfrac{d}{dx}\left(\vartheta_1\,/\,\vartheta_2\right) = \vartheta_2^2(0)\,\vartheta_3(x)\,\vartheta_4(x)/\,\vartheta_2^2(x)$ LW 19(1.9.10)

7. $\dfrac{d}{dx}\left(\vartheta_4\,/\,\vartheta_1\right) = -\,\vartheta_4^2(0)\,\vartheta_2(x)\,\vartheta_3(x)/\,\vartheta_1^2(x)$ LW 19(1.9.11)

8. $\dfrac{d}{dx}\left(\vartheta_4\,/\,\vartheta_2\right) = \vartheta_3^2(0)\,\vartheta_1(x)\,\vartheta_3(x)/\,\vartheta_2^2(x)$ LW 20(1.9.12)

9. $\dfrac{d}{dx}\left(\vartheta_4 / \vartheta_3\right) = \vartheta_2^2(0)\,\vartheta_1(x)\,\vartheta_2(x)/\vartheta_3^2(x)$ LW 20(1.9.13)

10. $\dfrac{d}{dx}\left(\vartheta_3 / \vartheta_1\right) = -\,\vartheta_3^2(0)\,\vartheta_2(x)\,\vartheta_4(x)/\vartheta_1^2(x)$ LW 20(1.9.14)

11. $\dfrac{d}{dx}\left(\vartheta_3 / \vartheta_2\right) = \vartheta_4^2(0)\,\vartheta_1(x)\,\vartheta_4(x)/\vartheta_2^2(x)$ LW 20(1.9.15)

12. $\dfrac{d}{dx}\left(\vartheta_2 / \vartheta_1\right) = -\,\vartheta_2^2(0)\,\vartheta_3(x)\,\vartheta_4(x)/\vartheta_1^2(x)$ LW 20(1.9.16)

8.199(3)[10]   Derivatives of theta functions

1. $\dfrac{d}{du}\ln\vartheta_1(u) = \cot u + 4\sin 2u\displaystyle\sum_{n=1}^{\infty}\frac{q^{2n}}{1 - 2q^{2n}\cos 2u + q^{4n}}$

2. $\dfrac{d}{du}\ln\vartheta_2(u) = -\tan u - 4\sin 2u\displaystyle\sum_{n=1}^{\infty}\frac{q^{2n}}{1 + 2q^{2n}\cos 2u + q^{4n}}$

3. $\dfrac{d}{du}\ln\vartheta_3(u) = -4\sin 2u\displaystyle\sum_{n=1}^{\infty}\frac{q^{2n-1}}{1 + 2q^{2n}\cos 2u + q^{4n-2}}$

4. $\dfrac{d}{du}\ln\vartheta_4(u) = 4\sin 2u\displaystyle\sum_{n=1}^{\infty}\frac{q^{2n-1}}{1 - 2q^{2n}\cos 2u + q^{4n-2}}$

5. $\dfrac{d^2}{du^2}\ln\vartheta_2(u) = -\displaystyle\sum_{n=-\infty}^{\infty}\operatorname{sech}^2\left\{i(u + n\pi\tau)\right\}$

8.2 The Exponential Integral Function and Functions Generated by It

8.21   The exponential integral function Ei(x)

8.211

1. $\operatorname{Ei}(x) = -\displaystyle\int_{-x}^{\infty}\frac{e^{-t}}{t}\,dt = \int_{-\infty}^{x}\frac{e^{t}}{t}\,dt = \operatorname{li}\left(e^{x}\right)$ $[x < 0]$

2.[8] $\operatorname{Ei}(x) = -\displaystyle\lim_{\varepsilon\to+0}\left[\int_{-x}^{-\varepsilon}\frac{e^{-t}}{t}\,dt + \int_{\varepsilon}^{\infty}\frac{e^{-t}}{t}\,dt\right] = \operatorname{PV}\int_{-\infty}^{x}\frac{e^{t}}{t}\,dt$

                                                    $[x > 0]$

3.[7] $\operatorname{Ei}(x) = \tfrac{1}{2}\left\{\operatorname{Ei}(x + i0) + \operatorname{Ei}(x - i0)\right\}$ $[x > 0]$ ET I 386

8.212

1.[8] $\operatorname{Ei}(-x) = \boldsymbol{C} + \ln x + \displaystyle\int_{0}^{x}\frac{e^{-t} - 1}{t}\,dt$ $[x > 0]$ NT 11(1)

          $= \boldsymbol{C} + e^{-x}\ln x + \displaystyle\int_{0}^{x}e^{-t}\ln t\,dt$ $[x > 0]$ NT 11(10)

$2.^7 \quad \mathrm{Ei}(x) = e^x \left[\dfrac{1}{x} + \displaystyle\int_0^\infty \dfrac{e^{-t}\,dt}{(x-t)^2} \right]$ $\qquad [x > 0] \qquad$ (cf. **8.211** 1)

$3. \quad \mathrm{Ei}(-x) = e^{-x} \left[-\dfrac{1}{x} + \displaystyle\int_0^\infty \dfrac{e^{-t}\,dt}{(x+t)^2} \right]$ $\qquad [x > 0] \qquad$ (cf. **8.211** 1) $\qquad$ LA 281(28)

$4. \quad \mathrm{Ei}\,(\pm x) = \pm e^{\pm x} \displaystyle\int_0^1 \dfrac{dt}{x \pm \ln t}$ $\qquad [x > 0] \qquad$ (cf. **8.211** 1)

$5. \quad \mathrm{Ei}\,(\pm xy) = \pm e^{\pm xy} \displaystyle\int_0^\infty \dfrac{e^{-xt}}{y \mp t}\,dt$ $\qquad [\mathrm{Re}\,y > 0, \quad x > 0] \qquad$ NT 19(11)

$6. \quad \mathrm{Ei}\,(\pm x) = -e^{\pm x} \displaystyle\int_0^\infty \dfrac{e^{-it}}{t \pm ix}\,dt$ $\qquad [x > 0] \qquad$ NT 23(2, 3)

$7.^8 \quad \mathrm{Ei}(xy) = e^{xy} \displaystyle\int_0^1 \dfrac{t^{y-1}}{x + \ln t}\,dt$ $\qquad$ LA 282(44)a

$8. \quad \mathrm{Ei}(-xy) = -e^{-xy} \displaystyle\int_0^1 \dfrac{t^{y-1}}{x - \ln t}\,dt$ $\qquad$ LA 282(45)a

$\qquad\qquad = x^{-1} e^{-xy} \left[\displaystyle\int_0^1 \dfrac{t^{x-1}}{(y - \ln t)^2}\,dt - y^{-1} \right]$ $\qquad [x > 0, \quad y > 0] \qquad$ LA 283(47)a

$9. \quad \mathrm{Ei}(x) = e^x \displaystyle\int_1^\infty \dfrac{1}{x - \ln t} \dfrac{dt}{t^2}$ $\qquad [x > 0] \qquad$ LA 283(48)

$10. \quad \mathrm{Ei}(-x) = -e^{-x} \displaystyle\int_1^\infty \dfrac{1}{x + \ln t} \dfrac{dt}{t^2}$ $\qquad [x > 0] \qquad$ LA 283(48)

$11. \quad \mathrm{Ei}(-x) = -e^{-x} \displaystyle\int_0^\infty \dfrac{t \cos t + x \sin t}{t^2 + x^2}\,dt$ $\qquad [x > 0] \qquad$ NT 23(6)

$12. \quad \mathrm{Ei}(-x) = -e^{-x} \displaystyle\int_0^\infty \dfrac{t \cos t - x \sin t}{t^2 + x^2}\,dt$ $\qquad [x < 0] \qquad$ NT 23(6)

$13. \quad \mathrm{Ei}(-x) = \dfrac{2}{\pi} \displaystyle\int_0^\infty \dfrac{\cos t}{t} \arctan \dfrac{t}{x}\,dt$ $\qquad [\mathrm{Re}\,x > 0] \qquad$ NT 25(13)

$14. \quad \mathrm{Ei}(-x) = \dfrac{2e^{-x}}{\pi} \displaystyle\int_0^\infty \dfrac{x \cos t - t \sin t}{t^2 + x^2} \ln t\,dt$ $\qquad [x > 0] \qquad$ NT 26(7)

$15. \quad \mathrm{Ei}(x) = 2 \ln x - \dfrac{2e^x}{\pi} \displaystyle\int_0^\infty \dfrac{x \cos t + t \sin t}{t^2 + x^2} \ln t\,dt$ $\qquad [x > 0] \qquad$ NT 27(8)

$16. \quad \mathrm{Ei}(-x) = -x \displaystyle\int_1^\infty e^{-tx} \ln t\,dt$ $\qquad [x > 0] \qquad$ NT 32(12)

See also **3.327**, **3.881** 8, **3.916** 2 and 3, **4.326** 1, **4.326** 2, **4.331** 2, **4.351** 3, **4.425** 3, **4.581**. For integrals of the exponential integral function, see **6.22**–**6.23**, **6.78**.

Series and asymptotic representations

8.213

1. $\operatorname{li}(x) = C + \ln\left(-\ln x\right) + \sum_{k=1}^{\infty} \frac{(\ln x)^k}{k \cdot k!}$ $[0 < x < 1]$ NT 3(9)

2. $\operatorname{li}(x) = C + \ln \ln x + \sum_{k=1}^{\infty} \frac{(\ln x)^k}{k \cdot k!}$ $[x > 1]$ NT 3(10)

8.214

1. $\operatorname{Ei}(x) = C + \ln(-x) + \sum_{k=1}^{\infty} \frac{x^k}{k \cdot k!}$ $[x < 0]$

2. $\operatorname{Ei}(x) = C + \ln x + \sum_{k=1}^{\infty} \frac{x^k}{k \cdot k!}$ $[x > 0]$

3. $\operatorname{Ei}(x) - \operatorname{Ei}(-x) = 2x \sum_{k=0}^{\infty} \frac{x^{2k}}{(2k+1)(2k+1)!}$ $[x > 0]$ NT 39(13)

8.215⁷ $\operatorname{Ei}(z) = \frac{e^z}{z}\left[\sum_{k=0}^{n} \frac{k!}{z^k} + R_n(z)\right]$ $|R_n(z)| = O\left(|z|^{-n-1}\right)$

         $[z \to \infty, \quad |\arg(-z)| \le \pi - \delta; \quad \delta > 0 \text{ small}]$, $|R_n(z)| \le (n+1)!|z|^{-n-1}$ $[\operatorname{Re} z \le 0]$

8.216⁷ $\operatorname{Ei}(nx) - \operatorname{Ei}(-nx) = e^{nx'}\left(\frac{1}{nx} + \frac{1}{n^2 x^2} + \frac{k_n}{n^3 x^3}\right)$,

         where $x' = x \operatorname{sign} \operatorname{Re}(x)$, $k_n = O(1)$, and $n \to \infty$ NT 39(15)

8.217   Functional relations:

1. $e^{x'} \operatorname{Ei}(-x') - e^{-x'} \operatorname{Ei}(x') = -2 \int_0^{\infty} \frac{x' \sin t}{t^2 + x^2} \, dt$ NT 24(11)

         $= \frac{4}{\pi} \int_0^{\infty} \frac{x' \cos t}{t^2 + x^2} \ln t \, dt - 2e^{-x'} \ln x'$ $[x' = x \operatorname{sign} \operatorname{Re} x]$ NT 27(9)

2. $e^{x'} \operatorname{Ei}(-x') + e^{-x'} \operatorname{Ei}(x') = -2 \int_0^{\infty} \frac{t \cos t}{t^2 + x^2} \, dt = 2e^{-x'} \ln x' - \frac{4}{\pi} \int_0^{\infty} \frac{t \sin t}{t^2 + x^2} \ln t \, dt$

                                          $[x' = x \operatorname{sign} \operatorname{Re} x]$ NT 24(10), NT 27(10)

3. $\operatorname{Ei}(-x) - \operatorname{Ei}\left(-\frac{1}{x}\right) = \frac{2}{\pi} \int_0^{\infty} \frac{\cos t}{t} \arctan \frac{t\left(x - \frac{1}{x}\right)}{1 + t^2} \, dt$

                                       $[\operatorname{Re} x > 0]$ NT 25(14)

4. $\operatorname{Ei}(-\alpha x) \operatorname{Ei}(-\beta x) - \ln(\alpha\beta) \operatorname{Ei}[-(\alpha + \beta)x] = e^{-(\alpha+\beta)x} \int_0^{\infty} \frac{e^{-tx} \ln[(\alpha + t)(\beta + t)]}{t + \alpha + \beta} \, dt$ NT 32(9)

See also **3.723** 1 and 5, **3.742** 2 and 4, **3.824** 4, **4.573** 2.

- For a connection with a confluent hypergeometric function, see **9.237**.

- For integrals of the exponential integral function, see **5.21**, **5.22**, **5.23**, **6.22**, and **6.23**.

8.218   Two numerical values:

1. $\mathrm{Ei}(-1) = -0.219\ 383\ 934\ 395\ 520\ 273\ 665\ldots$ NT 89

2. $\mathrm{Ei}(1) = 1.895\ 117\ 816\ 355\ 936\ 755\ 478\ldots$ NT 89

8.22   The hyperbolic sine integral $\mathrm{shi}\,x$ and the hyperbolic cosine integral $\mathrm{chi}\,x$

8.221

1. $\mathrm{shi}\,x = \displaystyle\int_0^x \frac{\sinh t}{t}\,dt = -i\left[\frac{\pi}{2} + \mathrm{si}(ix)\right]$ (see **8.230** 1)   EH II 146(17)

2. $\mathrm{chi}\,x = C + \ln x + \displaystyle\int_0^x \frac{\cosh t - 1}{t}\,dt$ EH II 146(18)

8.23 The sine integral and the cosine integral: $\mathrm{si}\,x$ and $\mathrm{ci}\,x$

8.230

1.^{10} $\mathrm{si}(x) = -\displaystyle\int_x^\infty \frac{\sin t}{t}\,dt = -\frac{\pi}{2} + \mathrm{Si}(x),\ \text{where}\ \mathrm{Si}(x) = \int_0^x \frac{\sin t}{t}\,dt$ NT 11(3)

2.^{10} $\mathrm{ci}(x) = -\displaystyle\int_x^\infty \frac{\cos t}{t}\,dt = C + \ln x + \int_0^x \frac{\cos t - 1}{t}\,dt$ [$\mathrm{ci}(x)$ is also written $\mathrm{Ci}(x)$]   NT 11(2)

8.231

1. $\mathrm{si}(xy) = -\displaystyle\int_x^\infty \frac{\sin ty}{t}\,dt$ NT 18(7)

2. $\mathrm{ci}(xy) = -\displaystyle\int_x^\infty \frac{\cos ty}{t}\,dt$ NT 18(6)

3. $\mathrm{si}(x) = -\displaystyle\int_0^{\pi/2} e^{-x\cos t} \cos\left(x\sin t\right)\,dt$ NT 13(26)

8.232

1. $\mathrm{si}(x) = -\dfrac{\pi}{2} + \displaystyle\sum_{k=1}^\infty \frac{(-1)^{k+1} x^{2k-1}}{(2k-1)(2k-1)!}$ NT 7(4)

2.^{7} $\mathrm{ci}(x) = C + \ln(x) + \displaystyle\sum_{k=1}^\infty (-1)^k \frac{x^{2k}}{2k(2k)!}$ NT 7(3)

8.233

1. $\mathrm{ci}(x) \pm i\,\mathrm{si}(x) = \mathrm{Ei}\left(\pm ix\right)$ NT 6a

2. $\mathrm{ci}(x) - \mathrm{ci}\left(xe^{\pm\pi i}\right) = \mp\pi i$ NT 7(5)

3. $\mathrm{si}(x) + \mathrm{si}(-x) = -\pi$ NT 7(7)

8.234

1.[7] $\text{Ei}(-x) - \text{ci}(x) = \int_0^{\pi/2} e^{-x\cos\varphi} \sin\left(s\sin\varphi\right) d\varphi$ NT 13(27)

2. $[\text{ci}(x)]^2 + [\text{si}(x)]^2 = -2 \int_0^{\pi/2} \dfrac{\exp\left(-x\tan\varphi\right)\ln\cos\varphi}{\sin\varphi\cos\varphi} d\varphi$

$$[\text{Re}\,x > 0] \qquad (\text{see also } \mathbf{4.366})$$

NT 32(11)

See also **3.341**, **3.351** 1 and 2, **3.354** 1 and 2, **3.721** 2 and 3, **3.722** 1, 3, 5 and 7, **3.723** 8 and 11, **4.338** 1, **4.366** 1.

8.235

1. $\lim\limits_{x\to+\infty} \left(x^\varrho \, \text{si}(x)\right) = 0, \quad \lim\limits_{x\to+\infty} \left(x^\varrho \, \text{ci}(x)\right) = 0$ $[\varrho < 1]$ NT 38(5)

2. $\lim\limits_{x\to-\infty} \text{si}(x) = -\pi, \quad \lim\limits_{x\to-\infty} \text{ci}(x) = \pm\pi i$ NT 38(6)

- For integrals of the sine integral and cosine integral, see **6.24–6.26**, **6.781**, **6.782**, and **6.783**.
- For indefinite integrals of the sine integral and cosine integral, see **5.3**.

8.24 The logarithm integral $\text{li}(x)$

8.240

1. $\text{li}(x) = \int_0^x \dfrac{dt}{\ln t} = \text{Ei}\left(\ln x\right)$ $[x < 1]$ JA

2. $\text{li}(x) = \lim\limits_{\varepsilon\to 0} \left[\int_0^{1-\varepsilon} \dfrac{dt}{\ln t} + \int_{1+\varepsilon}^x \dfrac{dt}{\ln t} \right] = \text{Ei}\left(\ln x\right)$ $[x > 1]$ JA

3. $\text{li}\left\{\exp\left(-xe^{\pm\pi i}\right)\right\} = \text{Ei}\left(-xe^{\pm i\pi}\right) = \text{Ei}\left(x \mp i0\right) = \text{Ei}(x) \pm i\pi = \text{li}\left(e^x\right) \pm i\pi$

$$[x > 0] \qquad \text{JA, NT 2(6)}$$

Integral representations

8.241

1. $\text{li}(x) = \int_{-\infty}^{\ln x} \dfrac{e^t}{t} dt = x\ln\ln\dfrac{1}{x} - \int_{-\ln x}^\infty e^{-t}\ln t\, dt$ $[x < 1]$ LA 281(33)

2. $\text{li}(x) = x\int_0^1 \dfrac{dt}{\ln x + \ln t}$ LA 280(22)

      $= \dfrac{x}{\ln x} + x\int_0^1 \dfrac{dt}{(\ln x + \ln t)^2}$ LA 280(29)

      $= x\int_1^\infty \dfrac{1}{\ln x - \ln t}\dfrac{dt}{t^2}$ $[x < 1]$ LA 280(30)

3. $\text{li}\left(a^x\right) = \dfrac{1}{\ln a}\int_{-\infty}^x \dfrac{a^t}{t} dt$ $[x > 0]$

For integrals of the logarithm integral, see **6.21**

8.25 The probability integral, the Fresnel integrals $\Phi(x)$, $S(x)$, $C(x)$, the error function $\operatorname{erf}(x)$, and the complementary error function $\operatorname{erfc}(x)$

8.250 Definition:

1. $\Phi(x) = \dfrac{2}{\sqrt{\pi}} \displaystyle\int_0^x e^{-t^2}\, dt$ (also called the error function and denoted by $\operatorname{erf}(x)$)

2. $S(x) = \dfrac{2}{\sqrt{2\pi}} \displaystyle\int_0^x \sin t^2\, dt$

3. $C(x) = \dfrac{2}{\sqrt{2\pi}} \displaystyle\int_0^x \cos t^2\, dt$

4.[10] $\operatorname{erfc}(x) = 1 - \operatorname{erf}(x) = 1 - \Phi(x)$ (called the complementary error function)

Integral representations

8.251

1. $\Phi(x) = \dfrac{1}{\sqrt{\pi}} \displaystyle\int_0^{x^2} \dfrac{e^{-t}}{\sqrt{t}}\, dt$ (see also **3.361** 1)

2. $S(x) = \dfrac{1}{\sqrt{2\pi}} \displaystyle\int_0^{x^2} \dfrac{\sin t}{\sqrt{t}}\, dt$

3. $C(x) = \dfrac{1}{\sqrt{2\pi}} \displaystyle\int_0^{x^2} \dfrac{\cos t}{\sqrt{t}}\, dt$

8.252

1. $\Phi(xy) = \dfrac{2y}{\sqrt{\pi}} \displaystyle\int_0^x e^{-t^2 y^2}\, dt$

2. $S(xy) = \dfrac{2y}{\sqrt{2\pi}} \displaystyle\int_0^x \sin\left(t^2 y^2\right)\, dt$

3. $C(xy) = \dfrac{2y}{\sqrt{2\pi}} \displaystyle\int_0^x \cos\left(t^2 y^2\right)\, dt$

4. $\Phi(xy) = 1 - \dfrac{2}{\sqrt{\pi}} e^{-x^2 y^2} \displaystyle\int_0^\infty \dfrac{e^{-t^2 y^2}\, ty\, dt}{\sqrt{t^2 + x^2}}$ $\left[\operatorname{Re}^2 y > 0\right]$ NT 19(11)a

            $= 1 - \dfrac{2x}{\pi} e^{-x^2 y^2} \displaystyle\int_0^\infty \dfrac{e^{-t^2 y^2}\, dt}{t^2 + x^2}$ $\left[\operatorname{Re}^2 y > 0\right]$ NT 19(13)a

5.[7] $\Phi\left(\dfrac{-y}{2xi}\right) - \Phi\left(\dfrac{y}{2xi}\right) = \dfrac{4xi e^{\frac{y^2}{4x^2}}}{\sqrt{\pi}} \displaystyle\int_0^\infty e^{-t^2 y^2} \sin(ty)\, dt$ $\left[\operatorname{Re} x^2 > 0\right]$ NT 28(3)a

6.[8] $\Phi\left(\dfrac{y}{2x}\right) = 1 - \dfrac{2}{\sqrt{\pi}} x e^{-\frac{y^2}{4}} \displaystyle\int_0^\infty e^{-t^2 x^2 - ty}\, dt$ $\left[\operatorname{Re} x^2 > 0\right]$ NT 27(1)a

See also **3.322**, **3.362** 2, **3.363**, **3.468**, **3.897**, **6.511** 4 and 5.

8.253[8]   Series representations:

1. $\Phi(x) = \dfrac{2}{\sqrt{\pi}} e^{-x^2} x_1 F_1\left(1; \dfrac{3}{2}; x^2\right) = \dfrac{2}{\sqrt{\pi}} \displaystyle\sum_{k=1}^{\infty} (-1)^{k+1} \dfrac{x^{2k-1}}{(2k-1)(k-1)!}$ NT 7(9)a

        $= \dfrac{2}{\sqrt{\pi}} e^{-x^2} \displaystyle\sum_{k=0}^{\infty} \dfrac{2^k x^{2k+1}}{(2k+1)!!}$ NT 10(11)a

2. $S(x) = \dfrac{2}{\sqrt{2\pi}}\left(x \sin x^2 \, F\left(1; \dfrac{5}{4}, \dfrac{3}{4}; -\dfrac{1}{4}x^2\right) - \dfrac{2}{3}x^3 \cos x^2 \, F\left(1; \dfrac{7}{4}, \dfrac{5}{4}; -\dfrac{1}{4}x^2\right)\right)$

        $\dfrac{2}{\sqrt{2\pi}} \displaystyle\sum_{k=0}^{\infty} \dfrac{(-1)^k x^{4k+3}}{(2k+1)!(4k+3)}$ NT 8(14)a

        $= \dfrac{2}{\sqrt{2\pi}} \left\{ \sin^2 x \displaystyle\sum_{k=0}^{\infty} \dfrac{(-1)^k 2^{2k} x^{4k+1}}{(4k+1)!!} - \cos x^2 \displaystyle\sum_{k=0}^{\infty} \dfrac{(-1)^k 2^{2k+1} x^{4k+3}}{(4k+3)!!} \right\}$ NT 10(13)a

3. $C(x) = \dfrac{2}{\sqrt{2\pi}}\left(\dfrac{2}{3}x^3 \sin x^2 \, F\left(1; \dfrac{7}{4}, \dfrac{5}{4}; -\dfrac{1}{4}x^2\right) - x \cos x^2 \, F\left(1; \dfrac{5}{4}, \dfrac{3}{4}; -\dfrac{1}{4}x^2\right)\right)$

        $= \dfrac{2}{\sqrt{2\pi}} \displaystyle\sum_{k=0}^{\infty} \dfrac{(-1)^k x^{4k+1}}{(2k)!(4k+1)}$ NT 8(13)a

        $= \dfrac{2}{\sqrt{2\pi}} \left\{ \sin^2 x \displaystyle\sum_{k=0}^{\infty} \dfrac{(-1)^k 2^{2k+1} x^{4k+3}}{(4k+3)!!} + \cos x^2 \displaystyle\sum_{k=0}^{\infty} \dfrac{(-1)^k 2^{2k} x^{4k+1}}{(4k+1)!!} \right\}$ NT 10(12)a

For the expansions in Bessel functions, see **8.515** 2, **8.515** 3.

Asymptotic representations

8.254[8] $\Phi(z) = 1 - \dfrac{e^{-z^2}}{\sqrt{\pi} z}\left[\displaystyle\sum_{k=0}^{n} (-1)^k \dfrac{(2k-1)!!}{(2z^2)^k} + O\left(|z|^{-2n-z}\right) \right]$,

                                     $[z \to \infty, \quad |\arg(-z)| \leq \pi - \delta; \quad \delta > 0 \text{ small}]$

      where

      $|R_n| < \dfrac{\Gamma\left(n + \frac{1}{2}\right)}{|x|^{n+\frac{1}{2}}} \cos\dfrac{\varphi}{2}, \quad x = |x|e^{i\varphi} \text{ and } \varphi^2 < \pi^2$ NT 37(10)a

8.255

1. $S(x) = \dfrac{1}{2} - \dfrac{1}{\sqrt{2\pi}x} \cos x^2 + O\left(\dfrac{1}{x^2}\right)$ $[x \to \infty]$ MO 127a

2. $C(x) = \dfrac{1}{2} + \dfrac{1}{\sqrt{2\pi}x} \sin x^2 + O\left(\dfrac{1}{x^2}\right)$ $[x \to \infty]$ MO 127a

8.256   Functional relations:

1. $C(z) + i\,S(z) = \sqrt{\dfrac{i}{2}}\, \Phi\left(\dfrac{z}{\sqrt{i}}\right) = \dfrac{2}{\sqrt{2\pi}} \displaystyle\int_0^z e^{it^2}\,dt$

2. $\quad C(z) - i\, S(z) = \dfrac{1}{\sqrt{2i}}\, \Phi\left(z\sqrt{i}\right) = \dfrac{2}{\sqrt{2\pi}} \displaystyle\int_0^z e^{-it^2}\, dt$

3. $\quad \left[\cos^2 u\, C(u) + \sin u^2\, S(u)\right] = \dfrac{1}{2}\left[\cos^2 u + \sin u^2\right] + \sqrt{\dfrac{2}{\pi}} \displaystyle\int_0^\infty e^{-2ut} \sin t^2\, dt$

$$[\operatorname{Re} u \geq 0] \qquad\qquad \text{NT 28(6)a}$$

4. $\quad \left[\cos^2 u\, S(u) - \sin u^2\, C(u)\right] = \dfrac{1}{2}\left[\cos^2 u - \sin u^2\right] - \sqrt{\dfrac{2}{\pi}} \displaystyle\int_0^\infty e^{-2ut} \cos t^2\, dt$

$$[\operatorname{Re} u \geq 0] \qquad\qquad \text{NT 28(5)a}$$

5. $\quad \left[C(x) - \dfrac{1}{2}\right]^2 + \left[S(x) - \dfrac{1}{2}\right]^2 + \dfrac{2}{\pi} \displaystyle\int_0^{\pi/2} \dfrac{\exp\left(-x^2 \tan\varphi\right) \sin\frac{\varphi}{2} \sqrt{\cos\varphi}}{\sin 2\varphi}\, d\varphi$

$$(\text{see also } \mathbf{6.322}) \qquad\qquad \text{NT 33(18)a}$$

- For a connection with a confluent hypergeometric function, see **9.236**.
- For a connection with a parabolic cylinder function, see **9.254**.

8.257

1. $\quad \displaystyle\lim_{x \to +\infty} \left(x^\varrho \left[S(x) - \tfrac{1}{2}\right]\right) = 0$ $\qquad\qquad\qquad [\varrho < 1] \qquad\qquad$ NT 38(11)

2. $\quad \displaystyle\lim_{x \to +\infty} \left(x^\varrho \left[C(x) - \tfrac{1}{2}\right]\right) = 0$ $\qquad\qquad\qquad [\varrho < 1] \qquad\qquad$ NT 38(11)

3. $\quad \displaystyle\lim_{x \to +\infty} S(x) = \dfrac{1}{2}$ $\qquad\qquad\qquad\qquad\qquad\qquad\qquad$ NT 38(12)a

4. $\quad \displaystyle\lim_{x \to +\infty} C(x) = \dfrac{1}{2}$ $\qquad\qquad\qquad\qquad\qquad\qquad\qquad$ NT 38(12)a

- For integrals of the probability integral, see **6.28–6.31**.
- For integrals of Fresnel's sine integral and cosine integral, see **6.32**.

8.258[10]   Integrals involving the complementary error function

1. $\quad \displaystyle\int_0^\infty \operatorname{erfc}^2(x) e^{-\beta x^2}\, dx = \dfrac{1}{\sqrt{\beta\pi}}\left(-\arccos\left(\dfrac{1}{1+\beta}\right) + 2\arctan\left(\sqrt{\beta}\right)\right)$

$$[\beta > 0]$$

2. $\quad \displaystyle\int_0^\infty x\, \operatorname{erfc}^2(x) e^{-\beta x^2}\, dx = \dfrac{1}{2\beta}\left(1 - \dfrac{4}{\pi} \dfrac{\arctan\left(\sqrt{1+\beta}\right)}{\sqrt{1+\beta}}\right)$

$$[\beta > 0]$$

3. $\quad \displaystyle\int_0^\infty x^3\, \operatorname{erfc}^2(x) e^{-\beta x^2}\, dx = \dfrac{1}{2\beta^2}\left(1 - \dfrac{4}{\pi} \dfrac{\arctan\left(\sqrt{1+\beta}\right)}{\sqrt{1+\beta}}\right)$

$$+ \dfrac{1}{\beta\pi}\left(\dfrac{1}{(1+\beta)(\beta^2 + 2\beta + 2)} - \dfrac{\arctan\left(\sqrt{1+\beta}\right)}{(1+\beta)^{\frac{3}{2}}}\right)$$

$$[\beta > 0]$$

4. $\displaystyle\int_0^\infty x\,\mathrm{erfc}\left(\sqrt{x}\right)e^{-\beta x}\,dx = \frac{1}{\beta^2}\left[1 - \frac{1 + \frac{3}{2}\beta}{(1+\beta)^{\frac{3}{2}}}\right]$ $[\beta > 0]$

5. $\displaystyle\int \sqrt{x}\,\mathrm{erfc}\left(\sqrt{x}\right)e^{-\beta x}\,dx = \frac{1}{\sqrt{\pi}}\left(\frac{1}{2}\frac{\arctan\left(\sqrt{\beta}\right)}{\beta^{\frac{3}{2}}} - \frac{1}{2\beta(1+\beta)}\right)$

$$[\beta > 0]$$

8.26 Lobachevskiy's function $L(x)$

8.260   Definition:

$$L(x) = -\int_0^x \ln\cos t\,dt \qquad\qquad \text{LO III 184(10)}$$

For integral representations of the function $L(x)$, see also **3.531** 8, **3.532** 2, **3.533**, and **4.224**.

8.261   Representation in the form of a series:

$$L(x) = x\ln 2 - \frac{1}{2}\sum_{k=1}^\infty (-1)^{k-1}\frac{\sin 2kx}{k^2} \qquad\qquad \text{LO III 185(11)}$$

8.262   Functional relationships:

1. $L(-x) = -L(x)$ $\left[-\dfrac{\pi}{2} \le x \le \dfrac{\pi}{2}\right]$ LO III 185(13)

2. $L(\pi - x) = \pi\ln 2 - L(x)$ LO III 286

3. $L(\pi + x) = \pi\ln 2 + L(x)$ LO III 286

4. $L(x) - L\left(\dfrac{\pi}{2} - x\right) = \left(x - \dfrac{\pi}{4}\right)\ln 2 - \dfrac{1}{2}L\left(\dfrac{\pi}{2} - 2x\right)$ $\left[0 \le x < \dfrac{\pi}{4}\right]$ LO III 186(14)

8.3 Euler's Integrals of the First and Second Kinds and Functions Generated by Them

8.31 The gamma function (Euler's integral of the second kind): $\Gamma(z)$

8.310   Definition:

1. $\displaystyle\Gamma(z) = \int_0^\infty e^{-t}t^{z-1}\,dt$ $[\mathrm{Re}\,z > 0]$ (Euler)   FI II 777(6)

     Generalization:

2. $\displaystyle\Gamma(z) = -\frac{1}{2i\sin\pi z}\int_C (-t)^{z-1}e^{-t}\,dt$

     for z not an integer. The contour C is shown in the drawing.   WH

$\Gamma(z)$ is an analytic function z with simple poles at the points $z = -l$ (for $l = 0, 1, 2,\dots$) to which correspond to residues $\dfrac{(-1)^l}{l!}$. $\Gamma(z)$ satisfies the relation $\Gamma(1) = 1$.   WH, MO 1

Integral representations

8.311 $\quad \Gamma(z) = \dfrac{1}{e^{2\pi i z} - 1} \displaystyle\int_\infty^{(0+)} e^{-t} t^{z-1}\, dt$ $\hfill$ MO 2

8.312

1. $\quad \Gamma(z) = \displaystyle\int_0^1 \left(\ln \frac{1}{t} \right)^{z-1} dt$ $\hfill$ [Re $z > 0$] $\hfill$ FI II 778

2. $\quad \Gamma(z) = x^z \displaystyle\int_0^\infty e^{-xt} t^{z-1}\, dt$ $\hfill$ [Re $z > 0$, $\quad$ Re $x > 0$] $\hfill$ FI II 779(8)

3. $\quad \Gamma(z) = \dfrac{2a^z e^a}{\sin \pi z} \displaystyle\int_0^\infty e^{-at^2} \left(1 + t^2\right)^{z - \frac{1}{2}} \cos \left[2at + (2z - 1) \arctan t\right] dt$

$\hfill$ [$a > 0$] $\hfill$ WH

4. $\quad \Gamma(z) = \dfrac{1}{2 \sin \pi z} \displaystyle\int_0^\infty e^{-t^2} t^{z-1} \left(1 + t^2\right)^{\frac{z}{2}} \left\{ 3 \sin \left[t + z \operatorname{arccot}(-t)\right] + \sin \left[t + (z - 2) \operatorname{arccot}(-t)\right] \right\} dt$

$\hfill$ [arccot denotes an obtuse angle] $\hfill$ WH

5. $\quad \Gamma(y) = x^y e^{-i\beta y} \displaystyle\int_0^\infty t^{y-1} \exp\left(-x t e^{-i\beta}\right) dt$

$\hfill$ $\left[x, y, \beta \text{ real}, \quad x > 0, \quad y > 0, \quad |\beta| < \dfrac{\pi}{2} \right]$ $\hfill$ MO 8

6. $\quad \Gamma(z) = \dfrac{b^z}{2 \sin \pi z} \displaystyle\int_{-\infty}^\infty e^{bti} (it)^{z-1}\, dt$ $\hfill$ [$b > 0$, $\quad 0 < $ Re $z < 1$] $\hfill$ NH 154(3)

7. $\quad \Gamma(z) = \dfrac{\left(\sqrt{a^2 + b^2}\right)^z}{\cos\left(z \arctan \frac{b}{a}\right)} \displaystyle\int_0^\infty e^{-at} \cos(bt) t^{z-1}\, dt$ $\hfill$ NH 152(1)a

$\quad\quad\quad = \dfrac{\left(\sqrt{a^2 + b^2}\right)^z}{\sin\left(z \arctan \frac{b}{a}\right)} \displaystyle\int_0^\infty e^{-at} \sin(bt) t^{z-1}\, dt$ $\hfill$ NH 152(2)

$\hfill$ [$a > 0$, $\quad b \geq 0$, $\quad$ Re $z > 0$]

8. $\quad \Gamma(z) = \dfrac{b^z}{\cos \frac{\pi z}{2}} \displaystyle\int_0^\infty \cos(bt) t^{z-1}\, dt$

$\quad\quad\quad = \dfrac{b^z}{\sin \frac{\pi z}{2}} \displaystyle\int_0^\infty \sin(bt) t^{z-1}\, dt$

$\hfill$ [$b > 0$, $\quad 0 < $ Re $z < 1$] $\hfill$ NH 152(5)

9. $\quad \Gamma(z) = \displaystyle\int_0^\infty e^{-t} (t - z) t^{z-1} \ln t\, dt$

$\hfill$ [Re $z > 0$] $\hfill$ NH 173(7)

10. $\quad \Gamma(z) = \displaystyle\int_{-\infty}^\infty \exp\left(zt - e^t\right) dt$

$\hfill$ [Re $z > 0$] $\hfill$ NH 145(14)

11. $\quad \Gamma(z) \cos \alpha x = \lambda^x \displaystyle\int_0^\infty t^{x-1} e^{-\lambda t \cos \alpha} \cos\left(\lambda t \sin \alpha\right) dt$

$\hfill$ $\left[\lambda > 0, \quad x > 0, \quad -\dfrac{\pi}{2} < \alpha < \dfrac{\pi}{2} \right]$ $\hfill$ WH

12. $\Gamma(x)\sin\alpha x = \lambda^x \int_0^\infty t^{x-1}e^{-\lambda t\cos\alpha}\sin(\lambda t\sin\alpha)\,dt$

$$\left[\lambda > 0, \quad x > 0, \quad -\frac{\pi}{2} < \alpha < \frac{\pi}{2}\right] \quad \text{WH}$$

13. $\Gamma(-z) = \int_0^\infty \left[\dfrac{e^{-t} - \displaystyle\sum_{k=0}^n (-1)^k \dfrac{t^k}{k!}}{t^{z+1}}\right] dt$ $[n = \lfloor\operatorname{Re} z\rfloor]$ MO 2

8.313 $\Gamma\left(\dfrac{z+1}{v}\right) = vu^{\frac{z+1}{v}} \int_0^\infty \exp\left(-ut^v\right) t^z\,dt$ $[\operatorname{Re} u > 0, \quad \operatorname{Re} v > 0, \quad \operatorname{Re} z > -1]$

JA, MO 7a

8.315

1.[7] $\dfrac{1}{\Gamma(z)} = \dfrac{i}{2\pi} \displaystyle\int_C (-t)^{-z} e^{-t}\,dt$ [For C see **8.310** 2]

2.[8] $\displaystyle\int_{-\infty}^\infty \dfrac{e^{bti}}{(a+it)^z}\,dt = \dfrac{2\pi e^{-ab} b^{z-1}}{\Gamma(z)}$

$\displaystyle\int_{-\infty}^\infty \dfrac{e^{-bti}}{(a+it)^z}\,dt = 0$ $\left[\operatorname{Re} a > 0, \quad b > 0, \quad \operatorname{Re} z > 0, \quad |\arg(a+it)| < \tfrac{1}{2}\pi\right]$

3. $\dfrac{1}{\Gamma(z)} = a^{1-z}\dfrac{e^a}{\pi}\displaystyle\int_0^{\pi/2} \cos(a\tan\theta - z\theta)\cos^{z-2}\theta\,d\theta$ $[\operatorname{Re} z > 1]$ NH 157(14)

See also **3.324** 2, **3.326**, **3.328**, **3.381** 4, **3.382** 2, **3.389** 2, **3.433**, **3.434**, **3.478** 1, **3.551** 1, 2, **3.827** 1, **4.267** 7, **4.272**, **4.353** 1, **4.369** 1, **6.214**, **6.223**, **6.246**, **6.281**.

8.32 Representation of the gamma function as series and products

8.321   Representation in the form of a series:

1.[6] $\Gamma(z+1) = \displaystyle\sum_{k=0}^\infty c_k z^k$

$$\left[c_0 = 1, \quad c_{n+1} = \frac{\sum_{k=0}^n (-1)^{k+1} s_{k+1} c_{n-k}}{n+1}; \quad s_1 = C, \quad s_n = \zeta(n) \text{ for } n \geq 2, \quad |z| < 1\right]$$

NH 40(1, 3)

2. $\dfrac{1}{\Gamma(z+1)} = \displaystyle\sum_{k=0}^\infty d_k z^k$

$$\left[d_0 = 1, \quad d_{n+1} = \frac{\sum_{k=0}^n (-1)^k c_{k+1} d_{n-k}}{n+1}; \quad s_1 = C, \quad s_n = \zeta(n) \text{ for } n \geq 2\right] \quad \text{NH 41(4, 6)}$$

Infinite-product representation

8.322 $\quad \Gamma(z) = e^{-Cz} \dfrac{1}{z} \displaystyle\prod_{k=1}^{\infty} \dfrac{e^{\frac{z}{k}}}{1 + \frac{z}{k}} \qquad [\mathrm{Re}\, z > 0]$ SM 269

$\qquad\qquad = \dfrac{1}{z} \displaystyle\prod_{k=1}^{\infty} \dfrac{\left(1 + \frac{1}{k}\right)^z}{1 + \frac{z}{k}} \qquad [\mathrm{Re}\, z > 0]$ WH

$\qquad\qquad = \displaystyle\lim_{n\to\infty} \dfrac{n^z}{z} \displaystyle\prod_{k=1}^{n} \dfrac{k}{z + k} \qquad [\mathrm{Re}\, z > 0]$ SM 267(130)

8.323[7] $\quad \Gamma(z) = 2z^z e^{-z} \displaystyle\prod_{k=1}^{\infty} \sqrt[2^k]{\mathrm{B}\left(2^{k-1} z, \tfrac{1}{2}\right)}$ NH 98(12)

8.324[7] $\quad \Gamma(1 + z) = 4^z \displaystyle\prod_{k=1}^{\infty} \dfrac{\Gamma\left(\dfrac{1}{2} + \dfrac{z}{2^k}\right)}{\sqrt{\pi}}$ MO 3

8.325

1. $\quad \dfrac{\Gamma(\alpha)\,\Gamma(\beta)}{\Gamma(\alpha + \gamma)\,\Gamma(\beta - \gamma)} = \displaystyle\prod_{k=0}^{\infty} \left[\left(1 + \dfrac{\gamma}{\alpha + k}\right)\left(1 - \dfrac{\gamma}{\beta + k}\right)\right]$ NH 62(2)

2. $\quad \dfrac{e^{Cx}\,\Gamma(z+1)}{\Gamma(z - x + 1)} = \displaystyle\prod_{k=1}^{\infty} \left[\left(1 - \dfrac{x}{z + k}\right) e^{\frac{x}{k}}\right] \qquad [z \neq 0, -1, -2, \ldots; \quad \mathrm{Re}\, z > 0, \quad \mathrm{Re}(z - x) > 0]$

3.[7] $\quad \dfrac{\sqrt{\pi}}{\Gamma\left(1 + \frac{z}{2}\right)\Gamma\left(\frac{1}{2} - \frac{z}{2}\right)} = \displaystyle\prod_{k=1}^{\infty} \left(1 - \dfrac{z}{2k - 1}\right)\left(1 + \dfrac{z}{2k}\right)$ MO 2

8.326

1. $\quad \dfrac{\dfrac{[\Gamma(x)]^2}{\Gamma(2x)}}{\mathrm{B}(x + iy, x - iy)} = \left|\dfrac{\Gamma(x)}{\Gamma(x - iy)}\right|^2 = \displaystyle\prod_{k=0}^{\infty} \left(1 + \dfrac{y^2}{(x + k)^2}\right)$

$\qquad\qquad\qquad\qquad\qquad\qquad\qquad\qquad [x, y \text{ real}, \quad x \neq 0, \quad -1, -2, \ldots]$

     LO V, NH 63(4)

2. $\quad \dfrac{\Gamma(x + iy)}{\Gamma(x)} = \dfrac{x e^{-iCy}}{x + iy} \displaystyle\prod_{n=1}^{\infty} \dfrac{\exp\left(\frac{iy}{n}\right)}{1 + \frac{iy}{x + n}} \qquad [x, r \text{ real}, \quad x \neq 0, \quad -1, -2, \ldots]$ MO 2

8.327   Asymptotic representation for large values of $|z|$:

$$\Gamma(z) = z^{z - \frac{1}{2}} e^{-z} \sqrt{2\pi} \left\{1 + \dfrac{1}{12z} + \dfrac{1}{288z^2} - \dfrac{139}{51840z^3} - \dfrac{571}{2488320z^4} + O\left(z^{-5}\right)\right\}$$

$\qquad\qquad\qquad\qquad\qquad\qquad\qquad [|\arg z| < \pi]$ WH

For z real and positive, the remainder of the series is less than the last term that is retained.

8.328

1. $\quad \displaystyle\lim_{|y|\to\infty} |\Gamma(x + iy)| e^{\frac{\pi}{2}|y|} |y|^{\frac{1}{2} - x} = \sqrt{2\pi} \qquad [x \text{ and } y \text{ are real}]$ MO 6

2. $\quad \displaystyle\lim_{|z|\to\infty} \dfrac{\Gamma(z + a)}{\Gamma(z)} e^{-a \ln z} = 1$ MO 6

8.33   Functional relations involving the gamma function

8.331 $\Gamma(x+1) = x\,\Gamma(x)$

8.332

1. $|\Gamma(iy)|^2 = \dfrac{\pi}{y \sinh \pi y}$ $[y \text{ is real}]$ MO 3

2. $\left|\Gamma\left(\frac{1}{2}+iy\right)\right|^2 = \dfrac{\pi}{\cosh \pi y}$ $[y \text{ is real}]$

3. $\Gamma(1+ix)\,\Gamma(1-ix) = \dfrac{\pi x}{\sinh x\pi}$ $[x \text{ is real}]$ LO V

4. $\Gamma(1+x+iy)\,\Gamma(1-x+iy)\,\Gamma(1+x-iy)\,\Gamma(1-x-iy) = \dfrac{2\pi^2\left(x^2+y^2\right)}{\cosh 2y\pi - \cos 2x\pi}$

                                                            $[x \text{ and } y \text{ are real}]$ LO V

8.333 $[\Gamma(n+1)]^n = G(n+1)\displaystyle\prod_{k=1}^{n} k^k,$

    where n is a natural number and

$$G(z+1) = (2\pi)^{\frac{z}{2}} \exp\left[-\frac{z(z+1)}{2} - \frac{C}{2}z^2\right] \prod_{n=1}^{\infty}\left\{\left(1+\frac{z}{n}\right)^n \exp\left(-z+\frac{z^2}{2n}\right)\right\}$$ WH

8.334

1. $\displaystyle\prod_{k=1}^{n} \frac{1}{\Gamma\left(-z\exp\frac{2\pi ki}{n}\right)} = -z^n \prod_{k=1}^{\infty}\left[1-\left(\frac{z}{k}\right)^n\right]$ $[n=2,3,3\ldots]$ MO 2

2. $\Gamma\left(\frac{1}{2}+x\right)\Gamma\left(\frac{1}{2}-x\right) = \dfrac{\pi}{\cos \pi x}$

3. $\Gamma(1-x)\,\Gamma(x) = \dfrac{\pi}{\sin \pi x}$ FI II 430

Special cases

8.335[7] $\Gamma(nx) = (2\pi)^{\frac{1-n}{2}}\, n^{nx-\frac{1}{2}} \displaystyle\prod_{k=0}^{n-1} \Gamma\left(x+\frac{k}{n}\right)$ $[\text{product theorem}]$ FI II 782a, WH

1. $\Gamma(2x) = \dfrac{2^{2x-1}}{\sqrt{\pi}}\,\Gamma(x)\,\Gamma\left(x+\frac{1}{2}\right)$ $[\text{doubling formula}]$

2. $\Gamma(3x) = \dfrac{3^{3x-\frac{1}{2}}}{2\pi}\,\Gamma(x)\,\Gamma\left(x+\frac{1}{3}\right)\Gamma\left(x+\frac{2}{3}\right)$

3. $\displaystyle\prod_{k=1}^{n-1} \Gamma\left(\frac{k}{n}\right)\Gamma\left(1-\frac{k}{n}\right) = \dfrac{(2\pi)^{n-1}}{n}$ WH

4.[10] $\displaystyle\sum_{n=0}^{\infty} \frac{\Gamma^2\left(n-\frac{1}{2}\right)}{4(n!)^2\,\Gamma^2\left(-\frac{1}{2}\right)} = \frac{1}{4} + \frac{1}{16} + \frac{1}{256} + \frac{1}{1024} + \frac{25}{65536} + \cdots = \frac{1}{\pi}$

8.336 $\Gamma\left(-\dfrac{yz+xi}{2y}\right)\Gamma(1-z) = (2i)^{z+1}y\,\Gamma\left(1+\dfrac{yz-xi}{2y}\right)\displaystyle\int_{0}^{\infty} e^{-tx}\sin^z(ty)\,dt$

                                               $[\text{Re}(yi) > 0, \quad \text{Re}(x-yzi) > 0]$

                                                                      NH 133(10)

- For a connection with the psi function, **8.361** 1.
- For a connection with the beta function, see **8.384** 1.
- For integrals of the gamma function, see **8.412** 4, **8.414**, **9.223**, **9.242** 3, **9.242** 4.

8.337

1. $\left[\Gamma'(x)\right]^2 < \Gamma(x)\,\Gamma''(x)$ $[x > 0]$ MO 1

2.   For $x > 0$, $\min \Gamma(1 + x) = 0.88560\ldots$ is attained when $x = 0.46163\ldots$ JA

Particular values

8.338

1. $\Gamma(1) = \Gamma(2) = 1$

2. $\Gamma\left(\frac{1}{2}\right) = \sqrt{\pi}$

3. $\Gamma\left(-\frac{1}{2}\right) = -2\sqrt{\pi}$

4. $\left[\Gamma\left(\frac{1}{4}\right)\right]^4 = 16\pi^2 \prod_{k=1}^{\infty} \frac{(4k-1)^2 \left[(4k+1)^2 - 1\right]}{\left[(4k-1)^2 - 1\right](4k+1)^2}$ MO 1a

5. $\prod_{k=1}^{8} \Gamma\left(\frac{k}{3}\right) = \frac{640}{3^6}\left(\frac{\pi}{\sqrt{3}}\right)^3$ WH

8.339   For n a natural number

1. $\Gamma(n) = (n-1)!$

2. $\Gamma\left(n + \frac{1}{2}\right) = \frac{\sqrt{\pi}}{2^n}(2n-1)!!$

3. $\Gamma\left(\frac{1}{2} - n\right) = (-1)^n \frac{2^n \sqrt{\pi}}{(2n-1)!!}$

4. $\dfrac{\Gamma\left(p + n + \frac{1}{2}\right)}{\Gamma\left(p - n + \frac{1}{2}\right)} = \dfrac{\left(4p^2 - 1^2\right)\left(4p^2 - 3^2\right)\ldots\left[4p^2 - (2n-1)^2\right]}{2^{2n}}$ WA 221

8.34   The logarithm of the gamma function

8.341   Integral representation:

1. $\ln \Gamma(z) = \left(z - \frac{1}{2}\right)\ln z - z + \frac{1}{2}\ln 2\pi + \int_0^\infty \left(\frac{1}{2} - \frac{1}{t} + \frac{1}{e^t - 1}\right)\frac{e^{-tz}}{t}\,dt$

                                            $[\operatorname{Re} z > 0]$ WH

2.[7] $\ln \Gamma(z) = z \ln z - z - \frac{1}{2}\ln z + \ln\sqrt{2\pi} + 2\int_0^\infty \frac{\arctan\frac{t}{z}}{e^{2\pi t} - 1}\,dt$

       $\left[\operatorname{Re} z > 0 \text{ and } \arctan w = \int_0^\omega \frac{du}{1 + u^2} \text{ is taken over a rectangular path in the w-plane}\right]$ WH

3. $\ln \Gamma(z) = \int_0^\infty \left\{ \dfrac{e^{-zt} - e^{-t}}{1 - e^{-t}} + (z-1)e^{-t} \right\} \dfrac{dt}{t}$ [Re $z > 0$]   WH

4. $\ln \Gamma(z) = \int_0^\infty \left\{ (z-1)e^{-t} + \dfrac{(1+t)^{-z} - (1+t)^{-1}}{\ln(1+t)} \right\} \dfrac{dt}{t}$

    [Re $z > 0$]   WH

5. $\ln \Gamma(x) = \dfrac{\ln \pi - \ln \sin \pi x}{2} + \dfrac{1}{2}\int_0^\infty \left\{ \dfrac{\sinh\left(\frac{1}{2} - x\right)t}{\sinh \frac{t}{2}} - (1-2x)e^{-t} \right\} \dfrac{dt}{t}$

    [$0 < x < 1$]   WH

6. $\ln \Gamma(z) = \int_0^1 \left\{ \dfrac{t^z - t}{t - 1} - t(z-1) \right\} \dfrac{dt}{t \ln t}$ [Re $z > 0$]   WH

7. $\ln \Gamma(z) = \int_0^\infty \left[(z-1)e^{-t} + \dfrac{e^{-tz} - e^{-t}}{1 - e^{-t}} \right] \dfrac{dt}{t}$ [Re $z > 0$]   NH 187(7)

See also **3.427** 9, **3.554** 5.

8.342   Series representations:

1. $\ln \Gamma(z+1)$

    $= \dfrac{1}{2}\left[\ln\left(\dfrac{\pi 2}{\sin \pi z} \right) - \ln \dfrac{1+z}{1-z} \right] + (1 - \boldsymbol{C})z + \displaystyle\sum_{k=1}^\infty \dfrac{1 - \zeta(2k+1)}{2k+1} z^{2k+1}$

    $= -\boldsymbol{C}z + \displaystyle\sum_{k=2}^\infty (-1)^k \dfrac{z^k}{k} \zeta(k)$ [$|z| < 1$] NH 38(16, 12)

2. $\ln \Gamma(1+x) = \dfrac{1}{2} \ln \dfrac{\pi x}{\sin \pi x} - \boldsymbol{C}x - \displaystyle\sum_{n=1}^\infty \dfrac{x^{2n+1}}{2n+1} \zeta(2n+1)$

    [$|x| < 1$]   NH 38(14)

8.343

1. $\ln \Gamma(x) = \ln \sqrt{2\pi} + \displaystyle\sum_{n=1}^\infty \left\{ \dfrac{1}{2n} \cos 2n\pi x + \dfrac{1}{n\pi} \left(\boldsymbol{C} + \ln 2n\pi \right) \sin 2n\pi x \right\}$

    [$0 < x < 1$]   FI III 558

2. $\ln \Gamma(z) = z \ln z - z - \dfrac{1}{2} \ln z + \ln \sqrt{2\pi} + \dfrac{1}{2} \displaystyle\sum_{m=1}^\infty \dfrac{m}{(m+1)(m+2)} \displaystyle\sum_{n=1}^\infty \dfrac{1}{(z+n)^{m+1}}$

    [$|\arg z| < \pi$]   MO 9

8.344[7]   Asymptotic expansion for large values of $|z|$:

$$\ln \Gamma(z) = z \ln z - z - \dfrac{1}{2} \ln z + \ln \sqrt{2\pi} + \sum_{k=1}^{n-1} \dfrac{B_{2k}}{2k(2k-1)z^{2k-1}} + R_n(z),$$

where

$$|R_n(z)| < \frac{|B_{2n}|}{2n(2n-1)|z|^{2n-1}\cos^{2n-1}\left(\frac{1}{2}\arg z\right)} \qquad \text{MO5}$$

For integrals of $\ln\Gamma(x)$, see **6.44**.

8.35 The incomplete gamma function

8.350 Definition:

1. $\gamma(\alpha, x) = \int_0^x e^{-t} t^{\alpha-1}\, dt$ $[\operatorname{Re}\alpha > 0]$ EH II 133(1), NH 1(1)

2.[7] $\Gamma(\alpha, x) = \int_0^\infty e^{-t} t^{\alpha-1}\, dt$ EH II 133(2), NH 2(2), LE 339

8.351

1. $\gamma^*(\alpha, x) = \dfrac{x^{-\alpha}}{\Gamma(\alpha)}\,\gamma(\alpha, x)$ is an analytic function with respect to α and x EH II 133(5)

2.   Another definition of $\Gamma(\alpha, x)$, that is also suitable for the case $\operatorname{Re}\alpha \le 0$:

$$\gamma(\alpha, x) = \frac{x^\alpha}{\alpha} e^{-x}\, \Phi(1, 1+\alpha; x) = \frac{x^\alpha}{\alpha}\, \Phi(a, 1+a; -x) \qquad \text{EH II 133(3)}$$

3.   For fixed x, $\Gamma(\alpha, x)$ is an entire function of α. For non-integral α, $\Gamma(\alpha, x)$ is a multiple-valued function of x with a branch point at $x = 0$.

4.   A second definition of $\Gamma(\alpha, x)$:

$$\Gamma(\alpha, x) = x^\alpha e^{-x} \Psi(1, 1+\alpha; x) = e^{-x}\Psi(1-\alpha, 1-\alpha; x) \qquad \text{EH II 133(4)}$$

8.352 Special cases:

1. $\gamma(1+n, x) = n!\left[1 - e^{-x}\left(\displaystyle\sum_{m=0}^n \frac{x^m}{m!}\right)\right]$ $[n = 0, 1, \ldots]$

                                                           EH II 136(17, 16), NH 6(11)

2. $\Gamma(1+n, x) = n!e^{-x}\displaystyle\sum_{m=0}^n \frac{x^m}{m!}$ $[n = 0, 1, \ldots]$ EH II 136(16, 18)

3.[7] $\Gamma(-n, x) = \dfrac{(-1)^n}{n!}\left[\operatorname{Ei}(-x) - e^{-x}\displaystyle\sum_{m=0}^{n-1} (-1)^m \frac{m!}{x^{m+1}}\right]$ $[n = 1, 2, \ldots]$ EH II 137(20), NH 4(4)

8.353 Integral representations:

1. $\gamma(\alpha, x) = x^\alpha \operatorname{cosec}\pi\alpha \displaystyle\int_0^\pi e^x \cos\theta \cos(\alpha\theta + x\sin\theta)\, d\theta$ $[x \ne 0, \quad \operatorname{Re}\alpha > 0, \quad \alpha \ne 1, 2, \ldots]$

                                                            EH II 137(2)

2. $\gamma(\alpha, x) = x^{\frac{1}{2}\alpha}\displaystyle\int_0^\infty e^{-t} t^{\frac{1}{2}\alpha-1} J_\alpha\left(2\sqrt{xt}\right) dt$ $[\operatorname{Re}\alpha > 0]$ EH II 138(4)

3. $\Gamma(\alpha, x) = \dfrac{\rho^{-x} x^{\alpha}}{\Gamma(1-\alpha)} \displaystyle\int_0^\infty \dfrac{e^{-t} t^{-\alpha}}{x+t}\, dt$ $[\operatorname{Re}\alpha < 1, \quad x > 0]$

EH II 137(3), NH 19(12)

4. $\Gamma(\alpha, x) = \dfrac{2 x^{\frac{1}{2}\alpha} e^{-x}}{\Gamma(1-\alpha)} \displaystyle\int_0^\infty e^{-t} t^{-\frac{1}{2}\alpha} K_\alpha \left[2\sqrt{xt}\,\right] dt$ $[\operatorname{Re}\alpha < 1]$ EH II 138(5)

5. $\Gamma(\alpha, xy) = y^\alpha e^{-xy} \displaystyle\int_0^\infty e^{-ty}(t+x)^{\alpha-1}\, dt$

$[\operatorname{Re} y > 0, \quad x > 0, \quad \operatorname{Re}\alpha > 1]$ (See also **3.936** 5, **3.944** 1–4)   NH 19(10)

For integrals of the gamma function, see **6.45**.

8.354 Series representations:

1. $\gamma(\alpha, x) = \displaystyle\sum_{n=0}^\infty \dfrac{(-1)^n x^{\alpha+n}}{n!(\alpha+n)}$ EH II 135(4)

2. $\Gamma(\alpha, x) = \Gamma(\alpha) - \displaystyle\sum_{n=0}^\infty \dfrac{(-1)^n x^{\alpha+n}}{n!(\alpha+n)}$ $[\alpha \neq 0, -1, -2, \ldots]$

EH II 135(5), LE 340(2)

3. $\Gamma(\alpha, x) - \Gamma(\alpha, x+y) = \gamma(\alpha, x+y) - \gamma(\alpha, x)$

$$= e^{-x} x^{\alpha-1} \sum_{k=0}^\infty \dfrac{(-1)^k \left[1 - e^{-y} e_k(y)\right] \Gamma(1-\alpha+k)}{x^k\, \Gamma(1-\alpha)}$$

$$e_k(x) = \sum_{m=0}^k \dfrac{x^m}{m!} \qquad [|y| < |x|] \quad \text{EH II 139(2)}$$

4. $\gamma(\alpha, x) = \Gamma(\alpha) e^{-x} x^{\frac{1}{2}\alpha} \displaystyle\sum_{n=0}^\infty x^{\frac{1}{2}n} I_{n+\alpha}\left(2\sqrt{x}\right) \sum_{m=0}^n \dfrac{(-1)^m}{m!}$ $[x \neq 0, \quad \alpha \neq 0, \quad -1, -2, \ldots]$

EH II 139(3)

5. $\Gamma(\alpha, x) = e^{-x} x^\alpha \displaystyle\sum_{n=0}^\infty \dfrac{L_n^\alpha(x)}{n+1}$ $[x > 0]$ EH II 140(5)

8.355 $\Gamma(\alpha, x)\, \gamma(\alpha, y) = e^{-x-y}(xy)^\alpha \displaystyle\sum_{n=0}^\infty \dfrac{n!\, \Gamma(\alpha)}{(n+1)\, \Gamma(\alpha+n+1)} L_n^\alpha(x) L_n^\alpha(y)$

$[y > 0, \quad x \geq y, \quad \alpha \neq 0, -1, \ldots]$

EH II 139(4)

8.356 Functional relations:

1. $\nu(\alpha+1, x) = \alpha\, \gamma(\alpha, x) - x^\alpha e^{-x}$ EH II 134(2)

2. $\Gamma(\alpha+1, x) = \alpha\, \Gamma(\alpha, x) + x^\alpha e^{-x}$ EH II 134(3)

3. $\Gamma(\alpha, x) + \gamma(\alpha, x) = \Gamma(\alpha)$ EH II 134(1)

4. $\dfrac{d\, \gamma(\alpha, x)}{dx} = -\dfrac{d\, \Gamma(\alpha, x)}{dx} = x^{\alpha-1} e^{-x}$ EH II 135(8)

5. $\dfrac{\Gamma(\alpha + n, x)}{\Gamma(\alpha + n)} = \dfrac{\Gamma(\alpha, x)}{\Gamma(\alpha)} + e^{-x} \sum\limits_{s=0}^{n-1} \dfrac{x^{\alpha+s}}{\Gamma(\alpha + s + 1)}$ NH 4(3)

6. $\Gamma(\alpha)\,\Gamma(\alpha + n, x) - \Gamma(\alpha + n)\,\Gamma(\alpha, x) = \Gamma(\alpha + n)\,\gamma(\alpha, x) - \Gamma(\alpha)\,\Gamma(\alpha + n, x)$ NH 5

8.357   Asymptotic representation for large values of $|x|$:

1. $\Gamma(\alpha, x) = x^{\alpha-1} e^{-x} \left[\sum\limits_{m=0}^{M-1} \dfrac{(-1)^m\,\Gamma(1 - \alpha + m)}{x^m\,\Gamma(1 - \alpha)} + O\left(|x|^{-M}\right) \right]$

$$\left[|x| \to \infty, -\frac{3\pi}{2} < \arg x < \frac{3\pi}{2}, \quad M = 1, 2, \ldots \right]$$ EH II 135(6), NH 37(7), LE 340(3)

8.358   Representation as a continued fraction:

$$\Gamma(\alpha, x) = \cfrac{e^{-x} x^\alpha}{x + \cfrac{1 - \alpha}{1 + \cfrac{1}{x + \cfrac{2 - \alpha}{1 + \cfrac{2}{x + \cfrac{3 - \alpha}{1 + \ldots}}}}}}$$ EH II 136(13), NH 42(9)

8.359   Relationships with other functions:

1. $\Gamma(0, x) = -\operatorname{Ei}(-x)$ EH II 143(1)

2. $\Gamma\left(0, \ln \dfrac{1}{x}\right) = -\operatorname{li}(x)$ EH II 143(2)

3. $\Gamma\left(\frac{1}{2}, x^2\right) = \sqrt{\pi} - \sqrt{\pi}\,\Phi(x)$ EH II 147(2)

4. $\Gamma\left(\frac{1}{2}, x^2\right) = \sqrt{\pi}\,\Phi(x)$ EH II 147(1)

8.36   The psi function $\psi(x)$

8.360   Definition:

1. $\psi(x) = \dfrac{d}{dx} \ln \Gamma(x)$

8.361   Integral representations:

1.[8] $\psi(z) = \dfrac{d \ln \Gamma(z)}{dz} = \displaystyle\int_0^\infty \left(\dfrac{e^{-t}}{t} - \dfrac{e^{-zt}}{1 - e^{-t}} \right) dt$ $[\operatorname{Re} z > 0]$ NH 183(1), WH

2. $\psi(z) = \displaystyle\int_0^\infty \left\{ e^{-t} - \dfrac{1}{(1 + t)^z} \right\} \dfrac{dt}{t}$ $[\operatorname{Re} z > 0]$ NH 184(7), WH

3. $\psi(z) = \ln z - \dfrac{1}{2z} - 2 \displaystyle\int_0^\infty \dfrac{t\,dt}{(t^2 + z^2)(e^{2\pi t} - 1)}$ $[\operatorname{Re} z > 0]$ WH

4. $\psi(z) = \int_0^1 \left(\dfrac{1}{-\ln t} - \dfrac{t^{z-1}}{1-t} \right) dt$ [Re $z > 0$]   WH

5. $\psi(z) = \int_0^\infty \dfrac{e^{-t} - e^{-zt}}{1 - e^{-t}} dt - \boldsymbol{C},$ WH

6. $\psi(z) = \int_0^\infty \left\{ (1+t)^{-1} - (1+t)^{-z} \right\} \dfrac{dt}{t} - \boldsymbol{C},$ [Re $z > 0$]   WH

7. $\psi(z) = \int_0^1 \dfrac{t^{z-1} - 1}{t - 1} dt - \boldsymbol{C}$ FI II 796, WH

8. $\psi(z) = \ln z + \int_0^\infty e^{-tz} \left[\dfrac{1}{t} - \dfrac{1}{1 - e^{-t}} \right] dt$ [Re $z > 0$]   MO 4

See also **3.244** 3, **3.311** 6, **3.317** 1, **3.457**, **3.458** 2, **3.471** 14, **4.253** 1 and 6, **4.275** 2, **4.281** 4, **4.482** 5.
For integrals of the psi function, see **6.46**, **6.47**.

Series representation

8.362

1. $\psi(x) = -\boldsymbol{C} - \displaystyle\sum_{k=0}^\infty \left(\dfrac{1}{x+k} - \dfrac{1}{k+1} \right)$ FI II 799(26), KU 26(1)

     $= -\boldsymbol{C} - \dfrac{1}{x} + x \displaystyle\sum_{k=1}^\infty \dfrac{1}{k(x+k)}$ FI II 495

2. $\psi(x) = \ln x - \displaystyle\sum_{k=0}^\infty \left[\dfrac{1}{x+k} - \ln\left(1 + \dfrac{1}{x+k} \right) \right]$ MO 4

3. $\psi(x) = -\boldsymbol{C} + \dfrac{\pi^2}{6}(x-1) - (x-1) \displaystyle\sum_{k=1}^\infty \left(\dfrac{1}{k+1} - \dfrac{1}{x+k} \right) \displaystyle\sum_{n=0}^{k-1} \dfrac{1}{x+n}$ NH 54(12)

8.363

1. $\psi(x+1) = -\boldsymbol{C} + \displaystyle\sum_{k=2}^\infty (-1)^k \zeta(k) x^{k-1}$ NH 37(5)

2. $\psi(x+1) = \dfrac{1}{2x} - \dfrac{\pi}{2} \cot \pi x - \dfrac{x^2}{1-x^2} - \boldsymbol{C} + \displaystyle\sum_{k=1}^\infty [1 - \zeta(2k+1)] x^{2k}$ NH 38(10)

3. $\psi(x) - \psi(y) = \displaystyle\sum_{k=0}^\infty \left(\dfrac{1}{y+k} - \dfrac{1}{x+k} \right)$

           (see also **3.219**, **3.231** 5, **3.311** 7, **3.688** 20, **4.253** 1, **4.295** 37)   NH 99(3)

4. $\psi(x+iy) - \psi(x-iy) = \displaystyle\sum_{k=0}^\infty \dfrac{2yi}{y^2 + (x+k)^2}$

5. $\psi\left(\dfrac{p}{q}\right) = -C + \displaystyle\sum_{k=0}^{\infty}\left(\dfrac{1}{k+1} - \dfrac{q}{p+kq}\right)$ (see also **3.244** 3)   NH 29(1)

6.[8] $\psi\left(\dfrac{p}{q}\right) = -C - \ln(2q) - \dfrac{\pi}{2}\cot\dfrac{p\pi}{q} + 2\displaystyle\sum_{k=1}^{\left[\frac{q+1}{2}\right]-1}\left[\cos\dfrac{2kp\pi}{q}\ln\sin\dfrac{k\pi}{q}\right]$

$$[q = 2,3,\ldots,p = 1,2,\ldots,q-1]$$

MO 4, EH I 19(29)

7. $\psi\left(\dfrac{p}{q}\right) - \psi\left(\dfrac{p-1}{q}\right) = q\displaystyle\sum_{n=2}^{\infty}\sum_{k=0}^{\infty}\dfrac{1}{(p+kq)^n - 1}$ NH 59(3)

8. $\psi^{(n)}(x) = (-1)^{n+1}n!\displaystyle\sum_{k=0}^{\infty}\dfrac{1}{(x+k)^{n+1}} = (-1)^{n+1}n!\,\zeta(n+1,x)$ NH 37(1)

Infinite-product representation

8.364

1. $e^{\psi(x)} = x\displaystyle\prod_{k=0}^{\infty}\left(1 + \dfrac{1}{x+k}\right)e^{-\frac{1}{x+k}}$ NH 65(12)

2. $e^{y\,\psi(x)} = \dfrac{\Gamma(x+y)}{\Gamma(x)}\displaystyle\prod_{k=0}^{\infty}\left(1 + \dfrac{y}{x+k}\right)e^{-\frac{y}{x+k}}$ NH 65(11)

See also **8.37**.

- For a connection with Riemann's zeta function, see **9.533** 2.
- For a connection with the gamma function, see **4.325** 12 and **4.352** 1.
- For a connection with the beta function, see **4.253** 1.
- For series of psi functions, see **8.403** 2, **8.446**, and **8.447** 3 (Bessel functions), **8.761** (derivatives of associated Legendre functions with respect to the degree), **9.153**, **9.154** (hypergeometric function), **9.237** (confluent hypergeometric function).
- For integrals containing psi functions, see **6.46–6.47**.

8.365   Functional relations:

1. $\psi(x+1) = \psi(x) + \dfrac{1}{x}$ JA

2. $\psi\left(\dfrac{x+1}{2}\right) - \psi\left(\dfrac{x}{2}\right) = 2\,\beta(x)$ (cf. **8.37** 0)

3. $\psi(x+n) = \psi(x) + \displaystyle\sum_{k=0}^{n-1}\dfrac{1}{x+k}$ GA 154(64)a

4. $\psi(n+1) = -C + \displaystyle\sum_{k=1}^{n}\dfrac{1}{k}$ MO 4

5. $\displaystyle\lim_{n\to\infty}\left[\psi(z+n) - \ln n\right] = 0$ MO 3

6. $\psi(nz) = \dfrac{1}{n}\sum\limits_{k=0}^{n-1}\psi\left(z + \dfrac{k}{n}\right) + \ln n$ $[n = 2, 3, 4, \ldots]$ MO 3

7. $\psi(x - n) = \psi(x) - \sum\limits_{k=1}^{n}\dfrac{1}{x - k}$

8. $\psi(1 - z) = \psi(z) + \pi\cot\pi z$ GA 155(68)a

9. $\psi\left(\frac{1}{2} + z\right) = \psi\left(\frac{1}{2} - z\right) + \pi\tan\pi z$ JA

10. $\psi\left(\frac{3}{4} - n\right) = \psi\left(\frac{1}{4} + n\right) + \pi$ $[n = 0, \quad \pm 1, \quad \pm 2, \ldots]$

8.366   Particular values

1. $\psi(1) = -\boldsymbol{C}$ (cf. **8.367** 1)

2. $\psi\left(\frac{1}{2}\right) = -\boldsymbol{C} - 2\ln 2 = -1.963\,510\,026\ldots$ GA 155a

3. $\psi\left(\frac{1}{2} \pm n\right) = -\boldsymbol{C} + 2\left[\sum\limits_{k=1}^{n}\dfrac{1}{2k - 1} - \ln 2\right]$ JA

4. $\psi\left(\frac{1}{4}\right) = -\boldsymbol{C} - \dfrac{\pi}{2} - 3\ln 2$ GA 157a

5. $\psi\left(\frac{3}{4}\right) = -\boldsymbol{C} + \dfrac{\pi}{2} - 3\ln 2$ GA 157a

6. $\psi\left(\frac{1}{3}\right) = -\boldsymbol{C} - \dfrac{\pi}{2}\sqrt{\frac{1}{3}} - \frac{3}{2}\ln 3$ GA 157a

7. $\psi\left(\frac{2}{3}\right) = -\boldsymbol{C} + \dfrac{\pi}{2}\sqrt{\frac{1}{3}} - \frac{3}{2}\ln 3$ GA 157a

8. $\psi'(1) = \dfrac{\pi^2}{6} = 1.644\,934\,066\,848\ldots$ JA

9. $\psi'\left(\frac{1}{2}\right) = \dfrac{\pi^2}{2} = 4.934\,802\,200\,5\ldots$ JA

10. $\psi'(-n) = \infty$ $[n$ is a natural number$]$ JA

11. $\psi'(n) = \dfrac{\pi^2}{6} - \sum\limits_{k=1}^{n-1}\dfrac{1}{k^2}$ $[n$ is a natural number$]$ JA

12. $\psi'\left(\frac{1}{2} + n\right) = \dfrac{\pi^2}{2} - 4\sum\limits_{k=1}^{n}\dfrac{1}{(2k - 1)^2}$ $[n$ is a natural number$]$ JA

13. $\psi'\left(\frac{1}{2} - n\right) = \dfrac{\pi^2}{2} + 4\sum\limits_{k=1}^{n}\dfrac{1}{(2k - 1)^2}$ $[n$ is a natural number$]$ JA

8.367   Euler's constant (also denoted by γ):

1. $\boldsymbol{C} = -\psi(1) = 0.577\,215\,664\,90\ldots$ FI II 319, 795

2. $\boldsymbol{C} = \lim\limits_{n\to\infty}\left[\sum\limits_{k=1}^{n-1}\dfrac{1}{k} - \ln n\right]$ FI II 801a

3. $\qquad C = \lim_{x \to 1+0} \left[\zeta(x) - \frac{1}{x-1} \right]$ FI II 804

Integral representations:

4. $\qquad C = -\int_0^\infty e^{-t} \ln t \, dt$ FI II 807

5. $\qquad C = -\int_0^1 \ln \left(\ln \frac{1}{t} \right) dt$ FI II 807

6. $\qquad C = \int_0^1 \left[\frac{1}{\ln t} + \frac{1}{1-t} \right] dt$ DW

7. $\qquad C = -\int_0^\infty \left[\cos t - \frac{1}{1+t} \right] \frac{dt}{t}$ MO 10

8. $\qquad C = 1 - \int_0^\infty \left[\frac{\sin t}{t} - \frac{1}{1+t} \right] \frac{dt}{t}$ MO 10

9. $\qquad C = -\int_0^\infty \left[e^{-t} - \frac{1}{1+t} \right] \frac{dt}{t}$ FI II 795, 802

10. $\qquad C = -\int_0^\infty \left[e^{-t} - \frac{1}{1+t^2} \right] \frac{dt}{t}$ DW, MO 10

11. $\qquad C = \int_0^\infty \left[\frac{1}{e^t - 1} - \frac{1}{te^t} \right] dt$ DW

12. $\qquad C = \int_0^1 \left(1 - e^{-t} \right) \frac{dt}{t} - \int_1^\infty \frac{e^{-t}}{t} dt$ FI II 802

See also **8.361** 5–**8.361** 7, **3.311** 6, **3.435** 3 and 4, **3.476** 2, **3.481** 1 and 2, **3.951** 10, **4.283** 9, **4.331** 1, **4.421** 1, **4.424** 1, **4.553**, **4.572**, **6.234**, **6.264** 1, **6.468**.

13. $\qquad$ Asymptotic expansions

$$C = \sum_{k=1}^{n-1} \frac{1}{k} - \ln n + \frac{1}{2n} + \frac{1}{12n^2} - \frac{1}{120n^4} + \frac{1}{252n^6} - \frac{1}{240n^8} + \cdots$$

$$\cdots + \frac{B_{2r}}{2r} \frac{1}{n^{2r}} + \frac{B_{2r+2}}{2(r+1)} \frac{\theta}{n^{2r+2}}$$

$[0 < \theta < 1]$ FI II 827

8.37 The function $\beta(x)$

Definition:

8.370 $\qquad \beta(x) = \frac{1}{2} \left[\psi \left(\frac{x+1}{2} \right) - \psi \left(\frac{x}{2} \right) \right]$ NH 16(13)

8.371 Integral representations:

1.[3] $\qquad \beta(x) = \int_0^1 \frac{t^{x-1}}{1+t} dt$ $\qquad\qquad [\operatorname{Re} x > 0]$ WH

2. $\qquad \beta(x) = \int_0^\infty \frac{e^{-xt}}{1+e^{-t}} dt$ $\qquad\qquad [\operatorname{Re} x > 0]$ MO 4

3. $$\beta\left(\frac{x+1}{2}\right) = \int_0^\infty \frac{e^{-xt}}{\cosh t}\, dt \qquad\qquad [\operatorname{Re} x > -1]$$

See also **3.241** 1, **3.251** 7, **3.522** 2 and 4, **3.623** 2 and 3, **4.282** 2, **4.389** 3, **4.532** 1 and 3.

Series representation

8.372

1.[7] $$\beta(x) = \sum_{k=0}^\infty \frac{(-1)^k}{x+k} \qquad\qquad [-x \notin \mathbb{N}] \qquad\qquad \text{NH 37, 101(1)}$$

2.[7] $$\beta(x) = \sum_{k=0}^\infty \frac{1}{(x+2k)(x+2k+1)} \qquad\qquad [-x \notin \mathbb{N}] \qquad\qquad \text{NH 101(2)}$$

3.[8] $$\beta(x) = \frac{1}{2}\sum_{k=0}^\infty \frac{k!}{x(x+1)\dots(x+k)}\frac{1}{2^k} \qquad\qquad [-x \notin \mathbb{N}]$$
$$[\beta \text{ has simple poles at } x = -n \text{ with residue } (-1)^n] \quad \text{NH 246(7)}$$

8.373

1.[6] $$\beta(x+1) = \ln 2 + \sum_{k=1}^\infty (-1)^k\left(1 - 2^{-k}\right)\zeta(k+1)x^k \qquad [|x| < 1] \qquad\qquad \text{NH 37(5)}$$

2.[6] $$\beta(x+1) = \ln 2 - 1 + \frac{1}{2x} - \frac{\pi}{2\sin \pi x} + \frac{1}{1-x^2} - \sum_{k=1}^\infty \left[1 - \left(1 - 2^{-2k}\right)\zeta(2k+1)\right]x^{2k}$$
$$[0 < |x| < 2; \quad x \neq \pm 1] \qquad\qquad \text{NH 38(11)}$$

8.374 $$\frac{d^n}{dx^n}\beta(x) = (-1)^n n! \sum_{k=0}^\infty \frac{(-1)^k}{(x+k)^{n+1}} \qquad\qquad [-x \in \mathbb{N}] \qquad\qquad \text{NH 37(2)}$$

8.375   Representation in the form of a finite sum:

1.[6] $$\beta\left(\frac{p}{q}\right) = \frac{\pi}{2\sin\frac{p\pi}{q}} - \sum_{k=0}^{\left\lfloor\frac{q-1}{2}\right\rfloor} \cos\frac{p(2k+1)\pi}{q}\ln\sin\frac{(2k+1)\pi}{2q}$$
$$[q = 2, 3, \dots, p = 1, 2, 3, \dots, q-1] \qquad (\text{see also } \mathbf{8.362}\ 5\text{--}7) \quad \text{NH 23(9)}$$

2. $$\beta(n) = (-1)^{n+1}\ln 2 + \sum_{k=1}^{n-1}\frac{(-1)^{k+n+1}}{k}$$

Functional relations

8.376 $$\sum_{k=0}^{2n}(-1)^k\,\beta\left(\frac{x+k}{2n+1}\right) = (2n+1)\,\beta(x) \qquad\qquad\qquad\qquad \text{NH 19}$$

8.377 $$\sum_{k=1}^{n}\beta\left(2^k x\right) = \psi\left(2^n x\right) - \psi(x) - n\ln 2 \qquad\qquad\qquad\qquad \text{NH 20(10)}$$

8.38 The beta function (Euler's integral of the first kind): $\mathrm{B}(x, y)$

Integral representation

8.380

1. $\mathrm{B}(x, y) = \displaystyle\int_0^1 t^{x-1}(1-t)^{y-1} \, dt^*$

            $= 2 \displaystyle\int_0^1 t^{2x-1}\left(1-t^2\right)^{y-1} \, dt$ $[\mathrm{Re}\, x > 0, \quad \mathrm{Re}\, y > 0]$ FI II 774(1)

2. $\mathrm{B}(x, y) = 2 \displaystyle\int_0^{\pi/2} \sin^{2x-1}\varphi \cos^{2y-1}\varphi \, d\varphi$ $[\mathrm{Re}\, x > 0, \quad \mathrm{Re}\, y > 0]$ KU 10

3. $\mathrm{B}(x, y) = \displaystyle\int_0^\infty \frac{t^{x-1}}{(1+t)^{x+y}} \, dt = 2 \int_0^\infty \frac{t^{2x-1}}{(1+t^2)^{x+y}} \, dt$ $[\mathrm{Re}\, x > 0, \quad \mathrm{Re}\, y > 0]$ FI II 775

4. $\mathrm{B}(x, y) = 2^{2-y-x} \displaystyle\int_{-1}^1 \frac{(1+t)^{2x-1}(1-t)^{2y-1}}{(1+t^2)^{x+y}} \, dt$ $[\mathrm{Re}\, x > 0, \quad \mathrm{Re}\, y > 0]$ MO 7

5. $\mathrm{B}(x, y) = \displaystyle\int_0^1 \frac{t^{x-1} + t^{y-1}}{(1+t)^{x+y}} \, dt = \int_1^\infty \frac{t^{x-1} + t^{y-1}}{(1+t)^{x+y}} \, dt$ $[\mathrm{Re}\, x > 0, \quad \mathrm{Re}\, y > 0]$ BI (1)(15)

6. $\mathrm{B}(x, y) = \dfrac{1}{2^{x+y-1}} \displaystyle\int_0^1 \left[(1+t)^{x-1}(1-t)^{y-1} + (1+t)^{y-1}(1-t)^{x-1}\right] \, dt$

                                                $[\mathrm{Re}\, x > 0, \quad \mathrm{Re}\, y > 0]$ BI (1)(15)

7. $\mathrm{B}(x, y) = z^y(1+z)^x \displaystyle\int_0^1 \frac{t^{x-1}(1-t)^{y-1}}{(t+z)^{x+y}}$

                $[\mathrm{Re}\, x > 0, \quad \mathrm{Re}\, y > 0, \quad 0 > z > -1, \quad \mathrm{Re}(x+y) < 1]$ NH 163(8)

8. $\mathrm{B}(x, y) = z^y(1+z)^x \displaystyle\int_0^{\pi/2} \frac{\cos^{2x-1}\varphi \sin^{2y-1}\varphi}{(z + \cos^2\varphi)^{x+y}} \, d\varphi$

                $[\mathrm{Re}\, x > 0, \quad \mathrm{Re}\, y > 0, \quad 0 > z > -1, \quad \mathrm{Re}(x+y) < 1]$ NH 163(8)

      See also **3.196** 3, **3.198**, **3.199**, **3.215**, **3.238** 3, **3.251** 1–3, 11, **3.253**, **3.312** 1, **3.512** 1 and 2, **3.541** 1, **3.542** 1, **3.621** 5, **3.623** 1, **3.631** 1, 8, 9, **3.632** 2, **3.633** 1, 4, **3.634** 1, 2, **3.637**, **3.642** 1, **3.667** 8, **3.681** 2.

9. $\mathrm{B}(x, x) = \dfrac{1}{2^{2x-2}} \displaystyle\int_0^1 \left(1-t^2\right)^{x-1} \, dt = \frac{1}{2^{2x-1}} \int_0^1 \frac{(1-t)^{x-1}}{\sqrt{t}} \, dt$

      See **8.384** 4, **8.382** 3, and also **3.621** 1, **3.642** 2, **3.665** 1, **3.821** 6, **3.839** 6.

10. $\mathrm{B}(x+y, x-y) = 4^{1-x} \displaystyle\int_0^\infty \frac{\cosh 2yt}{\cosh^{2x} t} \, dt$ $[\mathrm{Re}\, x > |\mathrm{Re}\, y|, \quad \mathrm{Re}\, x > 0]$ MO 9

11. $\mathrm{B}\left(x, \dfrac{y}{z}\right) = z \displaystyle\int_0^1 \left(1-t^z\right)^{x-1} t^{y-1} \, dt$ $\left[\mathrm{Re}\, z > 0, \quad \mathrm{Re}\, \dfrac{y}{z} > 0, \quad \mathrm{Re}\, x > 0\right]$

                                                                     FI II 787a

---

*This equation is used as the definition of the function $\mathrm{B}(x, y)$.

8.381

1. $\displaystyle\int_{-\infty}^{\infty} \frac{dt}{(a+it)^x (b-it)^y} = \frac{2\pi (a+b)^{1-x-y}}{(x+y-1)\,\mathrm{B}(x,y)}$

$$[a>0, \quad b>0; \quad x \text{ and } y \text{ are real}, \quad x+y>1] \quad \text{MO 7}$$

2. $\displaystyle\int_{-\infty}^{\infty} \frac{dt}{(a-it)^x (b-it)^y} = 0$

$$[a>0, \quad b>0; \quad x \text{ and } y \text{ are real}, \quad x+y>1] \quad \text{MO 7}$$

3. $\displaystyle\mathrm{B}(x+iy, x-iy) = 2^{1-2x}\alpha e^{-2i\gamma y}\int_{-\infty}^{\infty} \frac{e^{2i\alpha y t}\,dt}{\cosh^{2x}(\alpha t - \gamma)}$

$$[y, \alpha, \gamma \text{ are real}, \quad \alpha > 0; \quad \operatorname{Re}x > 0]$$
$$\text{MI 8a}$$

For an integral representation of $\ln \mathrm{B}(x,y)$, see **3.428** 7.

4. $\displaystyle\frac{1}{\mathrm{B}(x,y)} = \frac{2^{x+y-1}(x+y-1)}{\pi}\int_0^{\pi/2} \cos[(x-y)t]\cos^{x+y-2}t\,dt$ NH 158(5)a

$\displaystyle = \frac{2^{x+y-2}(x+y-1)}{\pi\cos\left[(x-y)\frac{\pi}{2}\right]}\int_0^\pi \cos[(x-y)t]\sin^{x+y-2}t\,dt$ NH 159(8)a

$\displaystyle = \frac{2^{x+y-2}(x+y-1)}{\pi\sin\left[(x-y)\frac{\pi}{2}\right]}\int_0^\pi \sin[(x-y)t]\sin^{x+y-2}t\,dt$ NH 159(9)a

Series representation

8.382

1. $\displaystyle\mathrm{B}(x,y) = \frac{1}{y}\sum_{n=0}^{\infty} (-1)^n y \frac{(y-1)\ldots(y-n)}{n!(x+n)}$ $[y>0]$ WH

2. $\displaystyle\ln\mathrm{B}\left(\frac{1+x}{2}, \frac{1}{2}\right)\ln\sqrt{2\pi} + \frac{1}{2}\left[\ln\left(\frac{\tan\frac{\pi x}{2}}{x}\right) - \ln\left(\frac{1+x}{1-x}\right)\right] + \sum_{k=0}^{\infty}\frac{1-(1-2^{-2k})\,\zeta(2k+1)}{2k+1}x^{2k+1}$

$$[|x|<2] \quad \text{NH 39(17)}$$

3. $\displaystyle\mathrm{B}\left(z, \frac{1}{2}\right) = \sum_{k=1}^{\infty}\frac{(2k-1)!!}{2^k k!}\frac{1}{z+k} + \frac{1}{z}$ (see also **8.384** and **8.380** 9)   WH

8.383   Infinite-product representation:

$$(x+y+1)\,\mathrm{B}(x+1, y+1) = \prod_{k=1}^{\infty}\frac{k(x+y+k)}{(x+k)(y+k)} \quad [x, \quad y \neq -1, \quad -2, \ldots] \quad \text{MO 2}$$

8.384   Functional relations involving the beta function:

1. $\displaystyle\mathrm{B}(x,y) = \frac{\Gamma(x)\,\Gamma(y)}{\Gamma(x+y)} = \mathrm{B}(y,x)$ FI II 779

2. $\mathrm{B}(x,y)\,\mathrm{B}(x+y, z) = \mathrm{B}(y,z)\,\mathrm{B}(y+z, x)$ MO 6

3. $\displaystyle\sum_{k=0}^{\infty} \mathrm{B}(x, y+k) = \mathrm{B}(x-1, y)$ WH

4. $\mathrm{B}(x, x) = 2^{1-2x}\, \mathrm{B}\left(\frac{1}{2}, x\right)$ (see also **8.380** 9 and **8.382** 3)

                                                                      FI II 784

5. $\mathrm{B}(x, x)\, \mathrm{B}\left(x + \frac{1}{2}, x + \frac{1}{2}\right) = \dfrac{\pi}{2^{4x-1} x}$ WH

6. $\dfrac{1}{\mathrm{B}(n, m)} = m\dbinom{n + m - 1}{n - 1} = n\dbinom{n + m - 1}{m - 1}$ [m and n are natural numbers]

For a connection with the psi function, see **4.253** 1.

8.39 The incomplete beta function $\mathrm{B}_x(p, q)$

8.391[7] $\displaystyle \mathrm{B}_x(p, q) = \int_0^x t^{p-1}(1-t)^{q-1}\, dt = \frac{x^p}{p}\, {}_2F_1\left((p, 1-q; p+1; x)\right)$ ET I 373

8.392 $\displaystyle I_x(p, q) = \frac{\mathrm{B}_x(p, q)}{\mathrm{B}(p, q)}$ ET II 429

8.4–8.5 Bessel Functions and Functions Associated with Them

8.40 Definitions

8.401   Bessel functions $Z_\nu(z)$ are solutions of the differential equation

$$\frac{d^2 Z_\nu}{dz^2} + \frac{1}{z}\frac{d Z_\nu}{dz} + \left(1 - \frac{\nu^2}{z^2}\right) Z_\nu = 0 \qquad\text{KU 37(1)}$$

     Special types of Bessel functions are what are called Bessel functions of the first kind $J_\nu(z)$, Bessel functions of the second kind $Y_\nu(z)$ (also called Neumann functions and often written $Y\nu(z)$), and Bessel functions of the third kind $H_\nu^{(1)}(z)$ and $H_\nu^{(2)}(z)$ (also called Hankel's functions).

8.402 $\displaystyle J_\nu(z) = \frac{z^\nu}{2^\nu}\sum_{k=0}^{\infty}(-1)^k\frac{z^{2k}}{2^{2k} k!\,\Gamma(\nu + k + 1)}$ [$|\arg z| < \pi$]   KU 55(1)

8.403

1. $Y_\nu(z) = \dfrac{1}{\sin \nu\pi}\left[\cos\nu\pi\, J_\nu(z) - J_{-\nu}(z)\right]$ [for non-integral ν, $|\arg z| < \pi$]

                                                                   KU 41(3)

2. $$\pi\, Y_n(z) = J_n(z) \ln \frac{z}{2} - \sum_{k=0}^{n-1} \frac{(n-k-1)!}{k!} \left(\frac{z}{2}\right)^{2k-n}$$

$$- \sum_{k=0}^{\infty} (-1)^k \frac{1}{k!\,(k+n)!} \left(\frac{z}{2}\right)^{n+2k} [\psi(k+1) + \psi(k+n+1)]$$

KU 43(10)

$$= J_n(z) \left(\ln \frac{z}{2} + C\right) - \sum_{k=0}^{n-1} \frac{(n-k-1)!}{k!} \left(\frac{z}{2}\right)^{2k-n}$$

$$- \left(\frac{z}{2}\right)^n \frac{1}{n!} \sum_{k=1}^{n} \frac{1}{k} - \sum_{k=1}^{\infty} \frac{(-1)^k \left(\frac{z}{2}\right)^{n+2k}}{k!\,(k+n)!} \left[\sum_{m=1}^{n+k} \frac{1}{m} + \sum_{m=1}^{k} \frac{1}{m}\right]$$

$[n+1$ a natural number, $|\arg z| < \pi]$
KU 44, WA 75(3)a

8.404

1. $Y_{-n}(z) = (-1)^n\, Y_n(z)$ $[n$ is a natural number$]$ KU 41(2)

2. $J_{-n}(z) = (-1)^n\, J_n(z)$ $[n$ is a natural number$]$ KU 41(2)

8.405[7]

1. $H_\nu^{(1)}(z) = J_\nu(z) + i\, Y_\nu(z)$ KU 44(1)

2. $H_\nu^{(2)}(z) = J_\nu(z) - i\, Y_\nu(z)$ KU 44(1)

   In all relationships that hold for an arbitrary Bessel function $Z_\nu(z)$, that is, for the functions $J_\nu(z)$, $Y_\nu(z)$, and linear combinations of them, for example, $H_\nu^{(1)}(z)$ and $H_\nu^{(2)}(z)$, we shall write simply the letter Z instead of the letters J, N, $H^{(1)}$, and $H^{(2)}$.

Modified Bessel functions of imaginary argument $I_\nu(z)$ and $K_\nu(z)$

8.406

1. $I_\nu(z) = e^{-\frac{\pi}{2}\nu i}\, J_\nu\left(e^{\frac{\pi}{2}i}z\right)$ $\left[-\pi < \arg z \le \dfrac{\pi}{2}\right]$ WA 92

2. $I_\nu(z) = e^{\frac{3}{2}\pi\nu i}\, J_\nu\left(e^{-\frac{3}{2}\pi i}z\right)$ $\left[\dfrac{\pi}{2} < \arg z \le \pi\right]$ WA 92

        For integral ν,

3. $I_n(z) = i^{-n}\, J_n(iz)$ KU 46(1)

8.407

1.[8] $K_\nu(z) = \dfrac{\pi i}{2} e^{\frac{\pi}{2}\nu i}\, H_\nu^{(1)}\left(ze^{\frac{1}{2}\pi i}\right)$ $\left[-\pi < \arg z \le \tfrac{1}{2}\pi\right]$

2.[8] $K_\nu(z) = \dfrac{-\pi i}{2} e^{-\frac{\pi}{2}\nu i}\, H_{-\nu}^{(2)}\left(ze^{-\frac{1}{2}\pi i}\right)$ $\left[-\tfrac{1}{2}\pi < \arg z \le \pi\right]$ WA 92(8)

   For the differential equation defining these functions, see **8.494**.

8.41 Integral representations of the functions $J_\nu(z)$ and $N_\nu(z)$

8.411

1.[7] $J_n(z) \quad \dfrac{1}{2\pi} \displaystyle\int_{-\pi}^{\pi} e^{-ni\theta + iz\sin\theta}\, d\theta$

$$= \frac{1}{\pi} \int_0^{\pi} \cos\left(n\theta - z\sin\theta\right) d\theta \qquad [n = 0, 1, 2, \ldots] \qquad\qquad \text{WH}$$

2. $J_{2n}(z) = \dfrac{1}{\pi} \displaystyle\int_0^{\pi} \cos 2n\theta \cos\left(z\sin\theta\right) d\theta = \dfrac{2}{\pi} \displaystyle\int_0^{\pi/2} \cos 2n\theta \cos\left(z\sin\theta\right) d\theta$

$$[n \text{ an integer}] \qquad\qquad \text{WA 30(7)}$$

3. $J_{2n+1}(z) \quad \dfrac{1}{\pi} \displaystyle\int_0^{\pi} \sin(2n+1)\theta \sin\left(z\sin\theta\right) d\theta$

$$= \frac{2}{\pi} \int_0^{\pi/2} \sin(2n+1)\theta \sin\left(z\sin\theta\right) d\theta \qquad [n \text{ an integer}] \qquad \text{WA 30(6)}$$

4. $J_\nu(z) = 2\dfrac{\left(\frac{z}{2}\right)^\nu}{\Gamma\left(\nu + \frac{1}{2}\right)\Gamma\left(\frac{1}{2}\right)} \displaystyle\int_0^{\pi/2} \sin^{2\nu}\theta \cos\left(z\cos\theta\right) d\theta$

$$\left[\operatorname{Re}\nu > -\tfrac{1}{2}\right] \qquad\qquad \text{WH}$$

5. $J_\nu(z) = \dfrac{\left(\frac{z}{2}\right)^\nu}{\Gamma\left(\nu + \frac{1}{2}\right)\Gamma\left(\frac{1}{2}\right)} \displaystyle\int_0^{\pi} \sin^{2\nu}\theta \cos\left(z\cos\theta\right) d\theta \qquad \left[\operatorname{Re}\nu > -\tfrac{1}{2}\right]$

6. $J_\nu(z) = \dfrac{\left(\frac{z}{2}\right)^\nu}{\Gamma\left(\nu + \frac{1}{2}\right)\Gamma\left(\frac{1}{2}\right)} \displaystyle\int_{-\pi/2}^{\pi/2} \cos\left(z\sin\theta\right) \cos^{2\nu}\theta\, d\theta$

$$\left[\operatorname{Re}\nu > -\tfrac{1}{2}\right] \qquad \text{KU 65(5), WA 35(4)a}$$

7. $J_\nu(z) = \dfrac{\left(\frac{z}{2}\right)^\nu}{\Gamma\left(\nu + \frac{1}{2}\right)\Gamma\left(\frac{1}{2}\right)} \displaystyle\int_0^{\pi} e^{\pm iz\cos\varphi} \sin^{2\nu}\varphi\, d\varphi \qquad \left[\operatorname{Re}\left(\nu + \tfrac{1}{2}\right) > 0\right] \qquad \text{WH}$

8. $J_\nu(z) = \dfrac{\left(\frac{z}{2}\right)^\nu}{\Gamma\left(\nu + \frac{1}{2}\right)\Gamma\left(\frac{1}{2}\right)} \displaystyle\int_{-1}^{1} \left(1 - t^2\right)^{\nu - \frac{1}{2}} \cos zt\, dt \qquad \left[\operatorname{Re}\nu > -\tfrac{1}{2}\right] \qquad \text{KU 65(6), WH}$

9. $J_\nu(x) = 2\dfrac{\left(\frac{x}{2}\right)^{-\nu}}{\Gamma\left(\frac{1}{2} - \nu\right)\Gamma\left(\frac{1}{2}\right)} \displaystyle\int_1^{\infty} \dfrac{\sin xt}{(t^2 - 1)^{\nu + \frac{1}{2}}}\, dt \qquad \left[-\tfrac{1}{2} < \operatorname{Re}\nu < \tfrac{1}{2}, \quad x > 0\right] \qquad \text{MO 37}$

10. $J_\nu(z) = \dfrac{\left(\frac{z}{2}\right)^\nu}{\Gamma\left(\nu + \frac{1}{2}\right)\Gamma\left(\frac{1}{2}\right)} \displaystyle\int_{-1}^{1} e^{izt} \left(1 - t^2\right)^{\nu - \frac{1}{2}}\, dt \qquad \left[\operatorname{Re}\nu > -\tfrac{1}{2}\right] \qquad \text{WA 34(3)}$

11. $J_\nu(x) = \dfrac{2}{\pi} \displaystyle\int_0^{\infty} \sin\left(x\cosh t - \dfrac{\nu\pi}{2}\right) \cosh \nu t\, dt \qquad\qquad \text{WA 199(12)}$

12. $J_\nu(z) = \dfrac{2^{\nu+1}z^\nu}{\Gamma\left(\nu + \frac{1}{2}\right)\Gamma\left(\frac{1}{2}\right)} \displaystyle\int_0^{\pi/2} \dfrac{\left(\cos^{\nu - \frac{1}{2}}\theta\right) \sin\left(z - \nu\theta + \frac{1}{2}\theta\right)}{\sin^{2\nu+1}\theta}\, e^{-2z\cot\theta}\, d\theta$

$$\left[|\arg z| < \frac{\pi}{2}, \quad \operatorname{Re}\left(\nu + \tfrac{1}{2}\right) > 0\right] \qquad \text{WH}$$

13.[10] $J_\nu(z) = \dfrac{1}{\pi} \displaystyle\int_0^\pi \cos(\nu\theta - z\sin\theta)\, d\theta - \dfrac{\sin\nu\pi}{\pi} \int_0^\infty e^{-\nu\theta - z\sinh\theta}\, d\theta$

$$[\operatorname{Re} z > 0] \qquad\qquad \text{WA 195(4)}$$

14. $J_\nu(z) = \dfrac{e^{\pm\nu\pi i}}{\pi} \left[\displaystyle\int_0^\pi \cos(\nu\theta + z\sin\theta)\, d\theta - \sin\nu\pi \int_0^\infty e^{-\nu\theta + z\sinh\theta}\, d\theta \right]$

$$\left[\text{for } \frac{\pi}{2} < |\arg z| < \pi, \text{ with the upper sign taken for } |\arg z| > \frac{\pi}{2} \right.$$

$$\left. \text{and the lower sign taken for } |\arg z| < -\frac{\pi}{2} \right]$$

$$\text{WH}$$

8.412

1. $J_\nu(z) = \dfrac{1}{2\pi i} \displaystyle\int_{-\infty}^{(0+)} t^{-\nu-1} \exp\left[\frac{z}{2}\left(t - \frac{1}{t} \right) \right] dt$ $\left[|\arg z| < \dfrac{\pi}{2} \right]$ WH, WA 195(2)

2. $J_\nu(z) = \dfrac{z^\nu}{2^{\nu+1}\pi i} \displaystyle\int_{-\infty}^{(0+)} t^{-\nu-1} \exp\left(t - \frac{z^2}{4t} \right) dt$ WA 195(1)

3.[8] $J_\nu(z) = \dfrac{z^\nu}{2^{\nu+1}\pi i} \displaystyle\sum_{k=1}^\infty \frac{(-1)^k z^{2k}}{2^{2k} k!} \int_{-\infty}^{(0+)} e^t t^{-\nu-k-1}\, dt$ WA 195(1)

4. $J_\nu(x) = \dfrac{1}{2\pi i} \displaystyle\int_{-i\infty}^{i\infty} \frac{\Gamma(-t)}{\Gamma(\nu+t+1)} \left(\frac{x}{2} \right)^{\nu+2t} dt$ $[\operatorname{Re}\nu > 0, \quad x > 0]$ WA 214(7)

5.[7] $J_\nu(z) = \dfrac{\Gamma\left(\frac{1}{2} - \nu\right) \left(\frac{z}{2}\right)^\nu}{2\pi i\, \Gamma\left(\frac{1}{2}\right)} \displaystyle\int_A^{(1+,-1-)} (t^2 - 1)^{\nu - \frac{1}{2}} \cos(zt)\, dt$

$$\left[\nu \neq \frac{1}{2}, \frac{3}{2}, \ldots; \text{ The point } A \text{ falls to the right of the point } t = 1, \right.$$

$$\left. \text{and } \arg(t-1) = \arg(t+1) = 0 \text{ at the point } A \right]$$

$$\text{WH}$$

6.[8] $J_\nu(z) = \dfrac{1}{2\pi} \displaystyle\int_{-\pi+\infty i}^{\pi+\infty i} e^{-iz\sin\theta + i\nu\theta}\, d\theta$ $[\operatorname{Re} z > 0]$

The path of integration being taken around the
semi-infinite strip $y \geq 0, -\pi \leq x \leq \pi$.

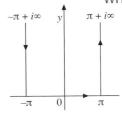

8.413[8] $\dfrac{J_\nu\left(\sqrt{z^2 + \zeta^2}\right)}{(z^2 - \zeta^2)^{\frac{\nu}{2}}} = \dfrac{1}{\pi(z+\zeta)^\nu} \left\{ \displaystyle\int_0^\infty e^{\zeta\cos t} \cos(z\sin t - \nu t)\, dt \right.$

$$\left. - \sin\nu\pi \int_0^\infty \exp(-z\sinh t - \zeta\cosh t - \nu t)\, dt \right\}$$

$$[\operatorname{Re}(z+\zeta) > 0] \qquad\qquad \text{MO 40}$$

8.414 $\displaystyle\int_{2x}^\infty \frac{J_0(t)}{t}\, dt = \frac{1}{4\pi} \int_{-\frac{1}{2}-i\infty}^{-\frac{1}{2}+i\infty} \frac{\Gamma(-t)}{t\,\Gamma(1+t)} x^{2t}\, dt$ $[x > 0]$ MO 41

See **3.715** 2, 9, 10, 13, 14, 19–21, **3.865** 1, 2, 4, **3.996** 4.

- For an integral representation of $J_0(z)$ see **3.714** 2, **3.753** 2, 3, and **4.124**.
- For an integral representation of $J_1(z)$ see **3.697**, **3.711**, **3.752** 2, and **3.753** 5.

8.415

1. $Y_0(x) = \dfrac{4}{\pi^2} \displaystyle\int_0^1 \dfrac{\arcsin t}{\sqrt{1-t^2}} \sin(xt)\, dt - \dfrac{4}{\pi^2} \int_1^\infty \dfrac{\ln\left(t + \sqrt{t^2-1}\right)}{\sqrt{t^2-1}} \sin(xt)\, dt$

$$[x > 0] \qquad\qquad \text{MO 37}$$

2. $Y_\nu(x) = -2 \dfrac{\left(\frac{x}{2}\right)^{-\nu}}{\Gamma\left(\frac{1}{2} - \nu\right)\Gamma\left(\frac{1}{2}\right)} \displaystyle\int_1^\infty \dfrac{\cos xt}{(t^2 - 1)^{\nu + \frac{1}{2}}}\, dt \qquad \left[-\tfrac{1}{2} < \operatorname{Re}\nu < \tfrac{1}{2}, \quad x > 0\right]$

$$\text{KU 89(28)a, MO 38}$$

3. $Y_\nu(x) = -\dfrac{2}{\pi} \displaystyle\int_0^\infty \cos\left(x\cosh t - \dfrac{\nu\pi}{2}\right)\cosh\nu t\, dt \qquad [-1 < \operatorname{Re}\nu < 1, \quad x > 0] \qquad \text{WA 199(13)}$

4.[8] $Y_\nu(z) = \dfrac{1}{\pi} \displaystyle\int_0^\pi \sin(z\sin\theta - \nu\theta)\, d\theta - \dfrac{1}{\pi} \int_0^\infty \left(e^{\nu t} + e^{-\nu t}\cos\nu\pi\right) e^{-z\sinh t}\, dt$

$$[\operatorname{Re} z > 0] \qquad\qquad \text{WA 197(1)}$$

5. $Y_\nu(z) = \dfrac{2\left(\frac{z}{2}\right)^\nu}{\Gamma\left(\nu + \frac{1}{2}\right)\Gamma\left(\frac{1}{2}\right)} \left[\displaystyle\int_0^{\pi/2} \sin(z\sin\theta)\cos^{2\nu}\theta\, d\theta - \int_0^\infty e^{-z\sinh\theta}\cosh^{2\nu}\theta\, d\theta \right]$

$$\left[\operatorname{Re}\nu > -\tfrac{1}{2}, \quad \operatorname{Re} z > 0\right] \qquad \text{WA 181(5)a}$$

6. $Y_\nu(z) = -\dfrac{2^{\nu+1}z^\nu}{\Gamma\left(\nu + \frac{1}{2}\right)\Gamma\left(\frac{1}{2}\right)} \displaystyle\int_0^{\frac{\pi}{2}} \dfrac{\cos^{\nu - \frac{1}{2}}\theta \cos\left(z - \nu\theta + \frac{1}{2}\theta\right)}{\sin^{2\nu+1}\theta} e^{-2z\cot\theta}\, d\theta$

$$\left[|\arg z| < \dfrac{\pi}{2}, \quad \operatorname{Re}\left(\nu + \tfrac{1}{2}\right) > 0\right]$$
$$\text{WA 186(8)}$$

For an integral representation of $Y_0(z)$, see **3.714** 3, **3.753** 4, **3.864**. See also **3.865** 3.

8.42 Integral representations of the functions $H_\nu^{(1)}(z)$ and $H_\nu^{(2)}(z)$

8.421

1. $\begin{aligned} H_\nu^{(1)}(x) &= \dfrac{e^{-\frac{\nu\pi i}{2}}}{\pi i} \int_{-\infty}^\infty e^{ix\cosh t - \nu t}\, dt \\ &= \dfrac{2e^{-\frac{\nu\pi i}{2}}}{\pi i} \int_0^\infty e^{ix\cosh t}\cosh\nu t\, dt \end{aligned}$

$$[-1 < \operatorname{Re}\nu < 1, \quad x > 0] \qquad \text{WA 199(10)}$$

2. $\begin{aligned} H_\nu^{(2)}(x) &= -\dfrac{e^{\frac{\nu\pi i}{2}}}{\pi i} \int_{-\infty}^\infty e^{-ix\cosh t - \nu t}\, dt \\ &= -\dfrac{2e^{\frac{\nu\pi i}{2}}}{\pi i} \int_0^\infty e^{-ix\cosh t}\cosh\nu t\, dt \end{aligned}$

$$[-1 < \operatorname{Re}\nu < 1, \quad x > 0] \qquad \text{WA 199(11)}$$

3. $H_\nu^{(1)}(z) = -\dfrac{2^{\nu+1} i z^\nu}{\Gamma\left(\nu + \frac{1}{2}\right)\Gamma\left(\frac{1}{2}\right)} \displaystyle\int_0^{\pi/2} \dfrac{\cos^{\nu-\frac{1}{2}} t \; e^{i\left(z - \nu t + \frac{t}{2}\right)}}{\sin^{2\nu+1} t} \exp\left(-2z \cot t\right)\, dt$

$$\left[\operatorname{Re}\nu > -\tfrac{1}{2}, \quad \operatorname{Re} z > 0\right] \qquad \text{WA 186(5)}$$

4. $H_\nu^{(2)}(z) = \dfrac{2^{\nu+1} i z^\nu}{\Gamma\left(\nu + \frac{1}{2}\right)\Gamma\left(\frac{1}{2}\right)} \displaystyle\int_0^{\pi/2} \dfrac{\cos^{\nu-\frac{1}{2}} t \; e^{-i\left(z - \nu t + \frac{t}{2}\right)}}{\sin^{2\nu+1} t} \exp\left(-2z \cot t\right)\, dt$

$$\left[\operatorname{Re}\nu > -\tfrac{1}{2}, \quad \operatorname{Re} z > 0\right] \qquad \text{WA 186(6)}$$

5. $H_\nu^{(1)}(x) = -\dfrac{2i\left(\frac{x}{2}\right)^{-\nu}}{\sqrt{\pi}\,\Gamma\left(\frac{1}{2} - \nu\right)} \displaystyle\int_1^\infty \dfrac{e^{ixt}}{(t^2 - 1)^{\nu+\frac{1}{2}}}\, dt$ $\left[-\tfrac{1}{2} < \operatorname{Re}\nu < \tfrac{1}{2}, \quad x > 0\right]$ WA 87(1)

6. $H_\nu^{(2)}(x) = \dfrac{2i\left(\frac{x}{2}\right)^{-\nu}}{\sqrt{\pi}\,\Gamma\left(\frac{1}{2} - \nu\right)} \displaystyle\int_1^\infty \dfrac{e^{-ixt}}{(t^2 - 1)^{\nu+\frac{1}{2}}}\, dt$ $\left[-\tfrac{1}{2} < \operatorname{Re}\nu < \tfrac{1}{2}, \quad x > 0\right]$ WA 187(2)

7. $H_\nu^{(1)}(z) = -\dfrac{i}{\pi} e^{-\frac{1}{2} i\nu\pi} \displaystyle\int_0^\infty \exp\left[\frac{1}{2} iz\left(t + \frac{1}{t}\right)\right] t^{-\nu-1}\, dt$

$$\left[0 < \arg z < \pi; \text{ or } \arg z = 0 \text{ and } -1 < \operatorname{Re}\nu < 1\right] \quad \text{MO 38}$$

8. $H_\nu^{(1)}(xz) = -\dfrac{i}{\pi} e^{-\frac{1}{2} i\nu\pi} z^\nu \displaystyle\int_0^\infty \exp\left[\frac{1}{2} ix\left(t + \frac{z^2}{t}\right)\right] t^{-\nu-1}\, dt$

$$\left[0 < \arg z < \frac{\pi}{2}, \quad x > 0, \quad \operatorname{Re}\nu > -1; \text{ or } \arg z = \frac{\pi}{2}, \quad x > 0 \text{ and } -1 < \operatorname{Re}\nu < 1\right] \quad \text{MO 38}$$

9. $H_\nu^{(1)}(xz) = \sqrt{\dfrac{2}{\pi z}}\, \dfrac{x^\nu \exp\left[i\left(xz - \frac{\pi}{2}\nu - \frac{\pi}{4}\right)\right]}{\Gamma\left(\nu + \frac{1}{2}\right)} \displaystyle\int_0^\infty \left(1 + \frac{it}{2z}\right)^{\nu-\frac{1}{2}} t^{\nu-\frac{1}{2}} e^{-xt}\, dt$

$$\left[\operatorname{Re}\nu > -\tfrac{1}{2}, \quad -\tfrac{1}{2}\pi < \arg z < \tfrac{3}{2}\pi, \quad x > 0\right] \quad \text{MO 39}$$

10. $H_\nu^{(1)}(z) = \dfrac{-2i e^{-i\nu\pi}\left(\frac{z}{2}\right)^\nu}{\sqrt{\pi}\,\Gamma\left(\nu + \frac{1}{2}\right)} \displaystyle\int_0^\infty e^{iz\cosh t}\sinh^{2\nu} t\, dt$

$$\left[0 < \arg z < \pi, \quad \operatorname{Re}\nu > -\tfrac{1}{2} \text{ or } \arg z = 0 \text{ and } -\tfrac{1}{2} < \operatorname{Re}\nu < \tfrac{1}{2}\right] \quad \text{MO 38}$$

11. $H_0^{(1)}(x) = -\dfrac{i}{\pi} \displaystyle\int_{-\infty}^\infty \dfrac{\exp\left(i\sqrt{x^2 + t^2}\right)}{\sqrt{x^2 + t^2}}\, dt$ $\left[x > 0\right]$ MO 38

8.422

1. $H_\nu^{(1)}(z) = \dfrac{\Gamma\left(\frac{1}{2} - \nu\right)\left(\frac{z}{2}\right)^\nu}{\pi i\,\Gamma\left(\frac{1}{2}\right)} \displaystyle\int_{1+\infty i}^{(1+)} e^{izt}(t^2 - 1)^{\nu-\frac{1}{2}}\, dt$ $\left[-\pi < \arg z < 2\pi\right]$ WA 183(4)

2. $H_\nu^{(2)}(z) = \dfrac{\Gamma\left(\frac{1}{2} - \nu\right)\left(\frac{z}{2}\right)^\nu}{\pi i\,\Gamma\left(\frac{1}{2}\right)} \displaystyle\int_{-1+\infty i}^{(-1-)} e^{izt}(t^2 - 1)^{\nu-\frac{1}{2}}\, dt$

$$\left[-2\pi < \arg z < \pi\right]$$

The paths of integration are shown in the drawing.

8.423

1. $H_\nu^{(1)}(z) = -\dfrac{1}{\pi} \displaystyle\int_{-\infty i}^{-\pi+\infty i} e^{-iz\sin\theta + i\nu\theta}\, d\theta$ $[\operatorname{Re} z > 0]$ WA 197(2)a

2. $H_\nu^{(2)}(z) = -\dfrac{1}{\pi} \displaystyle\int_{\pi+\infty i}^{-\infty i} e^{-iz\sin\theta + i\nu\theta}\, d\theta$ $[\operatorname{Re} z > 0]$ WA 197(3)a

The path of integration for **8.423** 1 is shown in the left hand drawing and for **8.423** 2 in the right hand drawing.

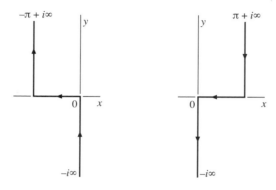

8.424

1. $H_\nu^{(1)}(z)\, J_\nu(\zeta) = \dfrac{1}{\pi i} \displaystyle\int_0^{\gamma+i\infty} \exp\left[\dfrac{1}{2}\left(t - \dfrac{z^2+\zeta^2}{t}\right)\right] I_\nu\left(\dfrac{z\zeta}{t}\right) \dfrac{dt}{t}$

                             $[\gamma > 0, \quad \operatorname{Re}\nu > -1, \quad |\zeta| < |z|]$ MO 45

2. $H_\nu^{(2)}(z)\, J_\nu(\zeta) = \dfrac{i}{\pi} \displaystyle\int_0^{\gamma-i\infty} \exp\left[\dfrac{1}{2}\left(t - \dfrac{z^2+\zeta^2}{t}\right)\right] I_\nu\left(\dfrac{z\zeta}{t}\right) \dfrac{dt}{t}$

                             $[\gamma > 0, \quad \operatorname{Re}\nu > -1, \quad |\zeta| < |z|]$ MO 45

8.43 Integral representations of the functions $I_\nu(z)$ and $K_\nu(z)$

The function $I_\nu(z)$

8.431

1. $I_\nu(z) = \dfrac{\left(\frac{z}{2}\right)^\nu}{\Gamma\left(\nu+\frac{1}{2}\right)\Gamma\left(\frac{1}{2}\right)} \displaystyle\int_{-1}^1 (1-t^2)^{\nu-\frac{1}{2}} e^{\pm zt}\, dt$ $\left[\operatorname{Re}\left(\nu+\frac{1}{2}\right) > 0\right]$ WA 94(9)

2. $I_\nu(z) = \dfrac{\left(\frac{z}{2}\right)^\nu}{\Gamma\left(\nu+\frac{1}{2}\right)\Gamma\left(\frac{1}{2}\right)} \displaystyle\int_{-1}^1 (1-t^2)^{\nu-\frac{1}{2}} \cosh zt\, dt$ $\left[\operatorname{Re}\left(\nu+\frac{1}{2}\right) > 0\right]$ WA 94(9)

3. $I_\nu(z) = \dfrac{\left(\frac{z}{2}\right)^\nu}{\Gamma\left(\nu+\frac{1}{2}\right)\Gamma\left(\frac{1}{2}\right)} \displaystyle\int_0^\pi e^{\pm z\cos\theta} \sin^{2\nu}\theta\, d\theta$ $\left[\operatorname{Re}\left(\nu+\frac{1}{2}\right) > 0\right]$ WA 94(9)

4. $I_\nu(z) = \dfrac{\left(\frac{z}{2}\right)^\nu}{\Gamma\left(\nu+\frac{1}{2}\right)\Gamma\left(\frac{1}{2}\right)} \displaystyle\int_0^\pi \cosh(z\cos\theta) \sin^{2\nu}\theta\, d\theta$ $\left[\operatorname{Re}\left(\nu+\frac{1}{2}\right) > 0\right]$ WA 94(9)

5. $$I_\nu(z) = \frac{1}{\pi} \int_0^\pi e^{z\cos\theta} \cos\nu\theta \, d\theta - \frac{\sin\nu\pi}{\pi} \int_0^\infty e^{-z\cosh t - \nu t} \, dt$$

$$\left[|\arg z| \le \frac{\pi}{2}, \quad \operatorname{Re}\nu > 0 \right] \qquad \text{WA 201(4)}$$

See also **3.383** 2, **3.387** 1, **3.471** 6, **3.714** 5.

For an integral representation of $I_0(z)$ and $I_1(z)$, see **3.366** 1, **3.534 3.856** 6.

The function $K_\nu(z)$

8.432

1. $$K_\nu(z) = \int_0^\infty e^{-z\cosh t} \cosh\nu t \, dt$$

$$\left[|\arg z| < \frac{\pi}{2} \text{ or } \operatorname{Re} z = 0 \text{ and } \nu = 0 \right]$$

MO 39

2. $$K_\nu(z) = \frac{\left(\frac{z}{2}\right)^\nu \Gamma\left(\frac{1}{2}\right)}{\Gamma\left(\nu + \frac{1}{2}\right)} \int_0^\infty e^{-z\cosh t} \sinh^{2\nu} t \, dt$$

$$\left[\operatorname{Re}\nu > -\frac{1}{2}, \quad \operatorname{Re} z > 0; \text{ or } \operatorname{Re} z = 0 \text{ and } -\frac{1}{2} < \operatorname{Re}\nu < \frac{1}{2} \right] \quad \text{WA 190(5), WH}$$

3. $$K_\nu(z) = \frac{\left(\frac{z}{2}\right)^\nu \Gamma\left(\frac{1}{2}\right)}{\Gamma\left(\nu + \frac{1}{2}\right)} \int_1^\infty e^{-zt} \left(t^2 - 1\right)^{\nu - \frac{1}{2}} \, dt$$

$$\left[\operatorname{Re}\left(\nu + \frac{1}{2}\right) > 0, \quad |\arg z| < \frac{\pi}{2}; \text{ or } \operatorname{Re} z = 0 \text{ and } \nu = 0 \right] \quad \text{WA 190(4)}$$

4. $$K_\nu(x) = \frac{1}{\cos\frac{\nu\pi}{2}} \int_0^\infty \cos\left(x\sinh t\right) \cosh\nu t \, dt$$

$$[x > 0, \quad -1 < \operatorname{Re}\nu < 1] \qquad \text{WA 202(13)}$$

5. $$K_\nu(xz) = \frac{\Gamma\left(\nu + \frac{1}{2}\right)(2z)^\nu}{x^\nu \Gamma\left(\frac{1}{2}\right)} \int_0^\infty \frac{\cos xt \, dt}{(t^2 + z^2)^{\nu + \frac{1}{2}}}$$

$$\left[\operatorname{Re}\left(\nu + \frac{1}{2}\right) \ge 0, \quad x > 0, \quad |\arg z| < \frac{\pi}{2} \right]$$

WA 191(1)

6.[7] $$K_\nu(z) = \frac{1}{2}\left(\frac{z}{2}\right)^\nu \int_0^\infty \frac{e^{-t - \frac{z^2}{4t}} \, dt}{t^{\nu + 1}}$$

$$\left[|\arg z| < \frac{\pi}{2}, \quad \operatorname{Re} z^2 > 0 \right] \qquad \text{WA 203(15)}$$

7.[7] $$K_\nu(xz) = \frac{z^\nu}{2} \int_0^\infty \exp\left[-\frac{x}{2}\left(t + \frac{z^2}{t}\right) \right] t^{-\nu - 1} \, dt$$

$$\left[|\arg z| < \frac{\pi}{4} \text{ or } |\arg z| = \frac{\pi}{4} \text{ and } \operatorname{Re}\nu < 1 \right] \quad \text{MO 39}$$

8. $$K_\nu(xz) = \sqrt{\frac{\pi}{2z}} \frac{x^\nu e^{-xz}}{\Gamma\left(\nu + \frac{1}{2}\right)} \int_0^\infty e^{-xt} t^{\nu - \frac{1}{2}} \left(1 + \frac{t}{2z}\right)^{\nu - \frac{1}{2}} \, dt$$

$$\left[|\arg z| < \pi, \quad \operatorname{Re}\nu > -\frac{1}{2}, x > 0 \right]$$

MO 39

9. $$K_\nu(xz) = \frac{\sqrt{\pi}}{\Gamma\left(\nu + \frac{1}{2}\right)} \left(\frac{x}{2z}\right)^\nu \int_0^\infty \frac{\exp\left(-x\sqrt{t^2 + z^2}\right)}{\sqrt{t^2 + z^2}} t^{2\nu} \, dt$$

$$\left[\operatorname{Re}\nu > -\frac{1}{2}, \quad \operatorname{Re} z > 0, \quad \operatorname{Re}\sqrt{t^2 + z^2} > 0, \quad x > 0 \right] \quad \text{MO 39}$$

See also **3.383** 3, **3.387** 3, 6, **3.388** 2, **3.389** 4, **3.391**, **3.395** 1, **3.471** 9, **3.483**, **3.547** 2, **3.856**, **3.871** 3, 4, **7.141** 5.

8.433 $K_{\frac{1}{3}}\left(\dfrac{2x\sqrt{x}}{3\sqrt{3}}\right) = \dfrac{3}{\sqrt{x}}\displaystyle\int_0^\infty \cos\left(t^3 + xt\right)\,dt$ KU 98(31), WA 211(2)

For an integral representation of $K_0(z)$, see **3.754** 2, **3.864**, **4.343**, **4.356**, **4.367**.

8.44 Series representation

The function $J_\nu(z)$

8.440 $J_\nu(z) = \left(\dfrac{z}{2}\right)^\nu \displaystyle\sum_{k=0}^\infty \dfrac{(-1)^k}{k!\,\Gamma(\nu+k+1)} \left(\dfrac{z}{2}\right)^{2k}$ $[|\arg z| < \pi]$

8.441 Special cases:

1. $J_0(z) = \displaystyle\sum_{k=0}^\infty (-1)^k \dfrac{z^{2k}}{2^{2k}(k!)^2}$

2. $J_1(z) = -J_0'(z) = \dfrac{z}{2} \displaystyle\sum_{k=0}^\infty \dfrac{(-1)^k z^{2k}}{2^{2k} k!(k+1)!}$

3. $J_{\frac{1}{3}}(z) = \dfrac{1}{\Gamma\left(\frac{4}{3}\right)} \sqrt[3]{\dfrac{z}{2}} \displaystyle\sum_{k=0}^\infty (-1)^k \dfrac{\left(z\sqrt{3}\right)^{2k}}{2^{2k} k!\cdot 1\cdot 4\cdot 7 \cdot\,\cdots\,(3k+1)}$

4. $J_{-\frac{1}{3}}(z) = \dfrac{1}{\Gamma\left(\frac{2}{3}\right)} \sqrt[3]{\dfrac{2}{z}} \left\{ 1 + \displaystyle\sum_{k=1}^\infty (-1)^k \dfrac{\left(z\sqrt{3}\right)^{2k}}{2^{2k} k!\cdot 2\cdot 5\cdot 8 \cdot\,\cdots\,(3k-1)} \right\}$

For the expansion of $J_\nu(z)$ in Laguerre polynomials, see **8.975** 3.

8.442

1.[7] $J_\nu(z) J_\mu(z) = \displaystyle\sum_{m=0}^\infty \dfrac{(-1)^m \left(\frac{1}{2}z\right)^{\mu+\nu+2m} (\mu+\nu+m+1)_m}{m!\,\Gamma(\mu+m+1)\,\Gamma(\nu+m+1)}$

2.[8] $J_\nu(az) J_\mu(bz) = \dfrac{\left(\frac{az}{2}\right)^\nu \left(\frac{bz}{2}\right)^\mu}{\Gamma(\mu+1)} \displaystyle\sum_{k=0}^\infty \dfrac{(-1)^k \left(\frac{az}{2}\right)^{2k} F\left(-k, -\nu-k; \mu-1; \frac{b^2}{a^2}\right)}{k!\,\Gamma(\nu+k+1)}$ MO 28

The function $Y_\nu(z)$

8.443 $Y_\nu(z)\quad \dfrac{1}{\sin\nu\pi} \left\{ \cos\nu\pi \left(\dfrac{z}{2}\right)^\nu \displaystyle\sum_{k=0}^\infty (-1)^k \dfrac{z^{2k}}{2^{2k} k!\,\Gamma(\nu+k+1)} \right.$

$\left. - \left(\dfrac{z}{2}\right)^{-\nu} \displaystyle\sum_{k=0}^\infty (-1)^k \dfrac{z^{2k}}{2^{2k} k!\,\Gamma(k-\nu+1)} \right\}$

          $[\nu \neq \text{an integer}]$ (cf. **8.403** 1)

For $\nu+1$ a natural number, see **8.403** 2.; for ν a negative integer see **8.404** 1

8.444 Special cases,

1. $\pi\, Y_0(z) = 2\, J_0(z) \left(\ln \frac{z}{2} + C \right) - 2 \sum_{k=1}^{\infty} \frac{(-1)^k}{(k!)^2} \left(\frac{z}{2} \right)^{2k} \sum_{m=1}^{k} \frac{1}{m}$ KU 44

2. $\pi\, Y_1(z) = 2\, J_1(z) \left(\ln \frac{z}{2} + C \right) - \frac{2}{z} - \sum_{k=1}^{\infty} \frac{(-1)^{k+1} \left(\frac{z}{2} \right)^{2k-1}}{k!(k-1)!} \left\{ \frac{1}{k} + 2 \sum_{m=1}^{k-1} \frac{1}{m} \right\}$

The functions $I_\nu(z)$ and $K_\nu(z)$

8.445 $I_\nu(z) = \sum_{k=0}^{\infty} \frac{1}{k!\, \Gamma(\nu + k + 1)} \left(\frac{z}{2} \right)^{\nu + 2k}$ WH

8.446[8] $K_n(z) = \frac{1}{2} \sum_{k=0}^{n-1} (-1)^k \frac{(n-k-1)!}{k! \left(\frac{z}{2} \right)^{n-2k}}$

$$+(-1)^{n+1} \sum_{k=0}^{\infty} \frac{\left(\frac{z}{2} \right)^{n+2k}}{k!(n+k)!} \left[\ln \frac{z}{2} - \frac{1}{2}\, \psi(k+1) - \frac{1}{2}\, \psi(n+k+1) \right]$$

        WA 95(15)

$$= (-1)^{n+1}\, I_n(z) \left(\ln \frac{1}{2} z + C \right) + \frac{1}{2}(-1)^n \sum_{l=0}^{\infty} \frac{\left(\frac{z}{2} \right)^{n+2l}}{l!(n+l)!} \left(\sum_{k=1}^{l} \frac{1}{k} + \sum_{k=1}^{n+l} \frac{1}{k} \right)$$

$$+\frac{1}{2} \sum_{l=0}^{n-1} \frac{(-1)^l (n-l-1)!}{l!} \left(\frac{z}{2} \right)^{2l-n}$$

        $[n+1$ is a natural number$]$
        MO 29

8.447   Special cases:

1. $I_0(z) = \sum_{k=0}^{\infty} \frac{\left(\frac{z}{2} \right)^{2k}}{(k!)^2}$

2. $I_1(z) = I_0'(z) = \sum_{k=0}^{\infty} \frac{\left(\frac{z}{2} \right)^{2k+1}}{k!(k+1)!}$

3. $K_0(z) = -\ln \frac{z}{2}\, I_0(z) + \sum_{k=0}^{\infty} \frac{z^{2k}}{2^{2k}(k!)^2}\, \psi(k+1)$ WA 95(14)

8.45   Asymptotic expansions of Bessel functions

8.451   For large values of $|z|$ *

---

*An estimate of the remainders in formulas **8.451** is given in **8.451** 7 and **8.451** 8.

1. $$J_{\pm\nu}(z) = \sqrt{\frac{2}{\pi z}} \left\{ \cos\left(z \mp \frac{\pi}{2}\nu - \frac{\pi}{4}\right) \left[\sum_{k=0}^{n-1} \frac{(-1)^k}{(2z)^{2k}} \frac{\Gamma\left(\nu + 2k + \frac{1}{2}\right)}{(2k)!\,\Gamma\left(\nu - 2k + \frac{1}{2}\right)} + R_1 \right] \right.$$
$$\left. - \sin\left(z \mp \frac{\pi}{2}\nu - \frac{\pi}{4}\right) \left[\sum_{k=0}^{n-1} \frac{(-1)^k}{(2z)^{2k+1}} \frac{\Gamma\left(\nu + 2k + \frac{3}{2}\right)}{(2k+1)!\,\Gamma\left(\nu - 2k - \frac{1}{2}\right)} + R_2 \right] \right\}$$
$$[|\arg z| < \pi] \qquad \text{(see } \mathbf{8.339}\ 4) \qquad \text{WA } 222(1, 3)$$

2. $$Y_{\pm\nu}(z) \sqrt{\frac{2}{\pi z}} \left\{ \sin\left(z \mp \frac{\pi}{2}\nu - \frac{\pi}{4}\right) \left[\sum_{k=0}^{n-1} \frac{(-1)^k}{(2z)^{2k}} \frac{\Gamma\left(\nu + 2k + \frac{1}{2}\right)}{(2k)!\,\Gamma\left(\nu - 2k + \frac{1}{2}\right)} + R_1 \right] \right.$$
$$\left. + \cos\left(z \mp \frac{\pi}{2}\nu - \frac{\pi}{4}\right) \left[\sum_{k=0}^{n-1} \frac{(-1)^k}{(2z)^{2k+1}} \frac{\Gamma\left(\nu + 2k + \frac{3}{2}\right)}{(2k+1)!\,\Gamma\left(\nu - 2k - \frac{1}{2}\right)} + R_2 \right] \right\}$$
$$[|\arg z| < \pi] \qquad \text{(see } \mathbf{8.339}\ 4) \qquad \text{WA } 222(2, 4, 5)$$

3. $$H_\nu^{(1)}(z) = \sqrt{\frac{2}{\pi z}} e^{i\left(z - \frac{\pi}{2}\nu - \frac{\pi}{4}\right)} \left[\sum_{k=0}^{n-1} \frac{(-1)^k}{(2iz)^k} \frac{\Gamma\left(\nu + k + \frac{1}{2}\right)}{k!\,\Gamma\left(\nu - k + \frac{1}{2}\right)} + \vartheta_1 \frac{(-1)^n}{(2iz)^n} \frac{\Gamma\left(\nu + n + \frac{1}{2}\right)}{k!\,\Gamma\left(\nu - n + \frac{1}{2}\right)} \right]$$
$$\left[\operatorname{Re}\nu > -\tfrac{1}{2}, \quad |\arg z| < \pi \right] \qquad \text{(see } \mathbf{8.339}\ 4) \qquad \text{WA } 221(5)$$

4. $$H_\nu^{(2)}(z) = \sqrt{\frac{2}{\pi z}} e^{-i\left(z - \frac{\pi}{2}\nu - \frac{\pi}{4}\right)} \left[\sum_{k=0}^{n-1} \frac{1}{(2iz)^k} \frac{\Gamma\left(\nu + k + \frac{1}{2}\right)}{k!\,\Gamma\left(\nu - k + \frac{1}{2}\right)} + \vartheta_2 \frac{1}{(2iz)^n} \frac{\Gamma\left(\nu + n + \frac{1}{2}\right)}{k!\,\Gamma\left(\nu - n + \frac{1}{2}\right)} \right]$$
$$\left[\operatorname{Re}\nu > -\tfrac{1}{2}, \quad |\arg z| < \pi \right] \qquad \text{(see } \mathbf{8.339}\ 4) \qquad \text{WA } 221(6)$$

For indices of the form $\nu = \frac{2n-1}{2}$ (where n is a natural number), the series **8.451** terminate. In this case, the closed formulas **8.46** are valid for all values.

5. $$I_\nu(z) \sim \frac{e^z}{\sqrt{2\pi z}} \sum_{k=0}^{\infty} \frac{(-1)^k}{(2z)^k} \frac{\Gamma\left(\nu + k + \frac{1}{2}\right)}{k!\,\Gamma\left(\nu - k + \frac{1}{2}\right)}$$
$$+ \frac{\exp\left[-z \pm \left(\nu + \frac{1}{2}\right)\pi i\right]}{\sqrt{2\pi z}} \sum_{k=0}^{\infty} \frac{1}{(2z)^k} \frac{\Gamma\left(\nu + k + \frac{1}{2}\right)}{k!\,\Gamma\left(\nu - k + \frac{1}{2}\right)}$$

$\left[\text{The } + \text{ sign is taken for } -\frac{1}{2}\pi < \arg z < \frac{3}{2}\pi, \text{ the } - \text{ sign for } -\frac{3}{2}\pi < \arg z < \frac{1}{2}\pi\right]$ (see **8.339** 4)]
$$\text{WA } 226(2,3)$$

6. $$K_\nu(z) = \sqrt{\frac{\pi}{2z}} e^{-z} \left[\sum_{k=0}^{n-1} \frac{1}{(2z)^k} \frac{\Gamma\left(\nu + k + \frac{1}{2}\right)}{k!\,\Gamma\left(\nu - k + \frac{1}{2}\right)} + \vartheta_3 \frac{\Gamma\left(\nu + n + \frac{1}{2}\right)}{(2z)^n n!\,\Gamma\left(\nu - n + \frac{1}{2}\right)} \right]$$
$$\text{(see } \mathbf{8.339}\ 4) \qquad \text{WA } 231, 245(9)$$

An estimate of the remainders of the asymptotic series in formulas **8.451**:

7. $$|R_1| < \left| \frac{\Gamma\left(\nu + 2n + \frac{1}{2}\right)}{(2z)^{2n}(2n)!\,\Gamma\left(\nu - 2n + \frac{1}{2}\right)} \right| \qquad\qquad \left[n > \frac{\nu}{2} - \frac{1}{4} \right] \qquad\qquad \text{WA } 231$$

8. $$|R_2| < \left| \frac{\Gamma\left(\nu + 2n + \frac{3}{2}\right)}{(2z)^{2n+1}(2n+1)!\,\Gamma\left(\nu - 2n - \frac{1}{2}\right)} \right| \qquad\qquad \left[n \geq \frac{\nu}{2} - \frac{3}{4} \right] \qquad\qquad \text{WA } 231$$

---

*The contradiction that this condition contains at first glance is explained by the so-called Stokes phenomenon (see Watson, G.N., *A Treatise on the Theory of Bessel Functions*, 2nd Edition, Cambridge Univ. Press, 1944, page 201).

For $-\dfrac{\pi}{2} < \arg z < \dfrac{3}{2}\pi$, ν real, and $n + \frac{1}{2} > |\nu|$ WA 245

$$|\theta_1| < \begin{cases} 1, & \text{if } \operatorname{Im} z \geq 0 \\ |\sec (\arg z)|, & \text{if } \operatorname{Im} z \leq 0 \end{cases}$$

For $-\dfrac{3}{2}\pi < \arg z < \dfrac{\pi}{2}$, ν real, and $n + \frac{1}{2} > |\nu|$ WA 246

$$|\theta_2| < \begin{cases} 1, & \text{if } \operatorname{Im} z \leq 0 \\ |\sec (\arg z)|, & \text{if } \operatorname{Im} z \geq 0 \end{cases}$$

For ν real,   WA 245

$$|\theta_3| < \begin{cases} 1 & \text{if } \operatorname{Re} z \geq 0 \\ |\operatorname{cosec} (\arg z)|, & \text{if } \operatorname{Re} z < 0 \end{cases}$$
$$\operatorname{Re} \theta_3 \geq 0, \qquad \text{if } \operatorname{Re} z \geq 0$$

For ν and z real and $n \geq \nu - \frac{1}{2}$,   WA 231

$$0 \leq |\theta_3| \leq 1$$

In particular, it follows from **8.451** 7 and **8.451** 8 that for real positive values of z and ν, the errors $|R_1|$ and $|R_2|$ are less than the absolute value of the first discarded term. For values of $|\arg z|$ close to π, the series **8.451** 1 and **8.451** 2 may not be suitable for calculations. In particular, the error for $|\arg z| > \pi$ can be greater in absolute value than the first discarded term.

"Approximation by tangents"

8.452   For large values of the index (where the argument is less than the index).

Suppose that $x > 0$ and $\nu > 0$. Let us set $\nu/x = \cos \beta$. Then, for large values of ν, the following expansions are valid:

1. $J_\nu \left(\dfrac{\nu}{\cosh \alpha} \right) \sim \dfrac{\exp (\nu \tanh \alpha - \nu \alpha)}{\sqrt{2\nu \pi \tanh \alpha}} \left\{ 1 + \dfrac{1}{\nu} \left(\dfrac{1}{8} \coth \alpha - \dfrac{5}{24} \coth^3 \alpha \right) \right.$

$$\left. + \dfrac{1}{\nu^2} \left(\dfrac{9}{128} \coth^2 \alpha - \dfrac{231}{576} \coth^4 \alpha + \dfrac{1155}{3456} \coth^6 \alpha \right) + \dots \right\}$$

                                                          WA 269(3)

2. $Y_\nu \left(\dfrac{\nu}{\cosh \alpha} \right) \sim -\dfrac{\exp (\nu \alpha - \nu \tanh \alpha)}{\sqrt{\frac{\pi}{2}\nu \tanh \alpha}} \left\{ 1 - \dfrac{1}{\nu} \left(\dfrac{1}{8} \coth \alpha - \dfrac{5}{24} \coth^3 \alpha \right) \right.$

$$\left. + \dfrac{1}{\nu^2} \left(\dfrac{9}{128} \coth^2 \alpha - \dfrac{231}{576} \coth^4 \alpha + \dfrac{1155}{3456} \coth^6 \alpha \right) + \dots \right\}$$

                                                          WA 270(5)

8.453   For large values of the index (where the argument is greater than the index).

Suppose that $x > 0$ and $\nu > 0$. Let us set $\nu/x = \cos\beta$. Then, for large values of ν, the following expansions are valid:

1. $$J_\nu\,(\nu\sec\beta) \sim \sqrt{\frac{2}{\nu\pi\tan\beta}}\left\{\left[1 - \frac{1}{\nu^2}\left(\frac{9}{128}\cot^2\beta + \frac{231}{576}\cot^4\beta\right.\right.\right.$$
$$\left.\left.+\ \frac{1155}{3456}\cot^6\beta\right) + \ldots\right]\cos\left(\nu\tan\beta - \nu\beta - \frac{\pi}{4}\right)$$
$$\left.+\ \left[\frac{1}{\nu}\left(\frac{1}{8}\cot\beta + \frac{5}{24}\cot^3\beta\right) - \ldots\right]\sin\left(\nu\tan\beta - \nu\beta - \frac{\pi}{4}\right)\right]$$

WA 271(4)

2. $$Y_\nu\,(\nu\sec\beta) \sim \sqrt{\frac{2}{\nu\pi\tan\beta}}\left\{\left[1 - \frac{1}{\nu^2}\left(\frac{9}{128}\cot^2\beta + \frac{231}{576}\cot^4\beta\right.\right.\right.$$
$$\left.\left.+\ \frac{1155}{3456}\cot^6\beta\right) + \ldots\right]\sin\left(\nu\tan\beta - \nu\beta - \frac{\pi}{4}\right)$$
$$\left.-\ \frac{1}{\nu}\left(\frac{1}{8}\cot\beta + \frac{5}{24}\cot^3\beta\right) - \ldots\right]\cos\left(\nu\tan\beta - \nu\beta - \frac{\pi}{4}\right)\right]$$

WA 271(5)

3. $$H_\nu^{(1)}\,(\nu\sec\beta) \sim \frac{\exp\left[\nu i\,(\tan\beta - \beta) - \frac{\pi}{4}i\right]}{\sqrt{\frac{\pi}{2}\nu\tan\beta}}\left\{1 - \frac{i}{\nu}\left(\frac{1}{8}\cot\beta + \frac{5}{24}\cot^3\beta\right)\right.$$
$$\left.-\ \frac{1}{\nu^2}\left(\frac{9}{128}\cot^2\beta + \frac{231}{576}\cot^4\beta + \frac{1155}{3456}\cot^6\beta\right) + \ldots\right\}$$

WA 271(1)

4. $$H_\nu^{(2)}\,(\nu\sec\beta) \sim \frac{\exp\left[-\nu i\,(\tan\beta - \beta) + \frac{\pi}{4}i\right]}{\sqrt{\frac{\pi}{2}\nu\tan\beta}}\left\{1 + \frac{i}{\nu}\left(\frac{1}{8}\cot\beta + \frac{5}{24}\cot^3\beta\right)\right.$$
$$\left.-\ \frac{1}{\nu^2}\left(\frac{9}{128}\cot^2\beta + \frac{231}{576}\cot^4\beta + \frac{1155}{3456}\cot^6\beta\right) + \ldots\right\}$$

WA 271(2)

Formulas **8.453** are not valid when $|x - \nu|$ is of a size comparable to $x^{\frac{1}{3}}$. For arbitrary small (and also large) values of $|x - \nu|$, we may use the following formulas:

8.454   Suppose that $x > 0$ and $\nu > 0$, we set

$$w = \sqrt{\frac{x^2}{\nu^2} - 1};$$

Then,

1. $$H_\nu^{(1)}(x) = \frac{w}{\sqrt{3}}\exp\left\{\left[\frac{\pi}{6} + \nu\left(w - \frac{w^3}{3} - \arctan w\right)\right]i\right\}H_{\frac{1}{3}}^{(1)}\left(\frac{\nu}{3}w^3\right) + O\left(\frac{1}{|\nu|}\right)$$

2. $H_\nu^{(2)}(x) = \dfrac{w}{\sqrt{3}} \exp\left\{\left[-\dfrac{\pi}{6} - \nu\left(w - \dfrac{w^3}{3} - \arctan w\right)\right]i\right\} H_{\frac{1}{3}}^{(2)}\left(\dfrac{\nu}{3} w^3\right) + O\left(\dfrac{1}{|\nu|}\right)$ MO 34

The absolute value of the error $O\left(\dfrac{1}{|\nu|}\right)$ is then less than $24\sqrt{2}\left|\dfrac{1}{\nu}\right|$.

8.455   For x real and ν a natural number ($\nu = n$), if $n \gg 1$, the following approximations are valid:

1.[7] $J_n(x) \approx \dfrac{1}{\pi}\sqrt{\dfrac{2(n-x)}{3x}}\, K_{\frac{1}{3}}\left\{\dfrac{[2(n-x)]^{\frac{3}{2}}}{3\sqrt{x}}\right\}$

$\qquad\qquad\qquad\qquad\qquad\qquad\qquad\qquad\qquad$ $[n > x]$ (see also **8.433**)

$\qquad\qquad\qquad\qquad\qquad\qquad\qquad\qquad\qquad\qquad\qquad\qquad$ WA 276(1)

$\qquad\qquad \approx \dfrac{1}{2}e^{\frac{2}{3}\pi i}\sqrt{\dfrac{2(n-x)}{3x}}\, H_{\frac{1}{3}}^{(1)}\left\{\dfrac{i}{3}\dfrac{[2(n-x)]^{\frac{3}{2}}}{\sqrt{x}}\right\}$

$\qquad\qquad\qquad\qquad\qquad\qquad\qquad\qquad\qquad\qquad$ $[n > x]$

$\qquad\qquad\qquad\qquad\qquad\qquad\qquad\qquad\qquad\qquad\qquad$ MO 34

$\qquad\qquad \approx \dfrac{1}{\sqrt{3}}\sqrt{\dfrac{2(x-n)}{3x}}\left\{J_{\frac{1}{3}}\left[\dfrac{\{2(x-n)\}^{\frac{3}{2}}}{3\sqrt{x}}\right] + J_{-\frac{1}{3}}\left[\dfrac{\{2(x-n)\}^{\frac{3}{2}}}{3\sqrt{x}}\right]\right\}$

$\qquad\qquad\qquad\qquad\qquad\qquad\qquad\qquad$ (see also **8.441** 3, **8.441** 4)

$\qquad\qquad\qquad\qquad\qquad\qquad\qquad\qquad\qquad\qquad\qquad$ WA 276(2)

2. $Y_n(x) \approx \sqrt{\dfrac{2(x-n)}{3x}}\left\{J_{-\frac{1}{3}}\left[\dfrac{\{2(x-n)\}^{\frac{3}{2}}}{3\sqrt{x}}\right] - J_{\frac{1}{3}}\left[\dfrac{\{2(x-n)\}^{\frac{3}{2}}}{3\sqrt{x}}\right]\right\}$

$\qquad\qquad\qquad\qquad\qquad\qquad\qquad\qquad$ $[x > n]$ WA 276(3)

An estimate of the error in formulas **8.455** has not yet been achieved.

8.456 $J_\nu^2(z) + N_\nu^2(z) \approx \dfrac{2}{\pi z}\sum_{k=0}^{\infty}\dfrac{(2k-1)!!}{2^k z^{2k}}\dfrac{\Gamma\left(\nu+k+\frac{1}{2}\right)}{k!\,\Gamma\left(\nu-k+\frac{1}{2}\right)}$ $[|\arg z| < \pi]$ (see also **8.479** 1)

$\qquad\qquad\qquad\qquad\qquad\qquad\qquad\qquad\qquad\qquad\qquad\qquad$ WA 250(5)

8.457 $J_\nu^2(x) + J_{\nu+1}^2(x) \approx \dfrac{2}{\pi x}$ $[x \gg |\nu|]$ WA 223

8.46   Bessel functions of order equal to an integer plus one-half

The function $J_\nu(z)$

8.461

1. $J_{n+\frac{1}{2}}(z) = \sqrt{\dfrac{2}{\pi z}}\left\{\sin\left(z - \dfrac{\pi}{2}n\right)\sum_{k=0}^{\lfloor\frac{n}{2}\rfloor}\dfrac{(-1)^k(n+2k)!}{(2k)!(n-2k)!}(2z)^{2k}\right.$

$\qquad\qquad\qquad\qquad + \cos\left(z - \dfrac{\pi}{2}n\right)\left.\sum_{k=0}^{\lfloor\frac{n-1}{2}\rfloor}\dfrac{(-1)^k(n+2k+1)!}{(2k+1)!(n-2k-1)!}(2z)^{2k+1}\right\}$

$\qquad\qquad\qquad\qquad$ $[n+1$ is a natural number$]$ (cf. **8.451** 1)   KU 59(6), WA 66(2)

2.
$$J_{-n-\frac{1}{2}}(z) = \sqrt{\frac{2}{\pi z}} \left\{ \cos\left(z + \frac{\pi}{2}n\right) \sum_{k=0}^{\lfloor \frac{n}{2} \rfloor} \frac{(-1)^k (n+2k)!}{(2k)!(n-2k)!(2z)^{2k}} \right.$$

$$\left. - \sin\left(z + \frac{\pi}{2}n\right) \sum_{k=0}^{\lfloor \frac{n-1}{2} \rfloor} \frac{(-1)^k (n+2k+1)!}{(2k+1)!(n-2k-1)!(2z)^{2k+1}} \right\}$$

$$[n+1 \text{ is a natural number}] \qquad (\text{cf. } \mathbf{8.451}\ 1) \quad \text{KU 58(7), WA 67(5)}$$

8.462

1.
$$J_{n+\frac{1}{2}}(z) = \frac{1}{\sqrt{2\pi z}} \left\{ e^{iz} \sum_{k=0}^{n} \frac{i^{-n+k-1}(n+k)!}{k!(n-k)!(2z)^k} + e^{-iz} \sum_{k=0}^{n} \frac{(-i)^{-n+k-1}(n+k)!}{k!(n-k)!(2z)^k} \right\}$$

$$[n+1 \text{ is a natural number}]$$
$$\text{KU 59(6), WA 66(1)}$$

2.
$$J_{-n-\frac{1}{2}}(z) = \frac{1}{\sqrt{2\pi z}} \left\{ e^{iz} \sum_{k=0}^{n} \frac{i^{n+k}(n+k)!}{k!(n-k)!(2z)^k} + e^{-iz} \sum_{k=0}^{n} \frac{(-i)^{n+k}(n+k)!}{k!(n-k)!(2z)^k} \right\}$$

$$[n+1 \text{ is a natural number}]$$
$$\text{KU 59(7), WA 67(4)}$$

8.463

1.
$$J_{n+\frac{1}{2}}(z) = (-1)^n z^{n+\frac{1}{2}} \sqrt{\frac{2}{\pi}} \frac{d^n}{(z\,dz)^n} \left(\frac{\sin z}{z}\right)$$
$$\text{KU 58(4)}$$

2.
$$J_{-n-\frac{1}{2}}(z) = z^{n+\frac{1}{2}} \sqrt{\frac{2}{\pi}} \frac{d^n}{(z\,dz)^n} \left(\frac{\cos z}{z}\right)$$
$$\text{KU 58(5)}$$

8.464 Special cases:

1.
$$J_{\frac{1}{2}}(z) = \sqrt{\frac{2}{\pi z}} \sin z$$
$$\text{DW}$$

2.
$$J_{-\frac{1}{2}}(z) = \sqrt{\frac{2}{\pi z}} \cos z$$
$$\text{DW}$$

3.
$$J_{\frac{3}{2}}(z) = \sqrt{\frac{2}{\pi z}} \left(\frac{\sin z}{z} - \cos z\right)$$
$$\text{DW}$$

4.
$$J_{-\frac{3}{2}}(z) = \sqrt{\frac{2}{\pi z}} \left(-\sin z - \frac{\cos z}{z}\right)$$
$$\text{DW}$$

5.[8]
$$J_{\frac{5}{2}}(z) = \sqrt{\frac{2}{\pi z}} \left\{ \left(\frac{3}{z^2} - 1\right) \sin z - \frac{3}{z} \cos z \right\}$$
$$\text{DW}$$

6.
$$J_{-\frac{5}{2}}(z) = \sqrt{\frac{2}{\pi z}} \left\{ \frac{3}{z} \sin z + \left(\frac{3}{z^2} - 1\right) \cos z \right\}$$
$$\text{DW}$$

The function $Y_{n+\frac{1}{2}}(z)$

8.465

1. $Y_{n+\frac{1}{2}}(z) = (-1)^{n-1} J_{-n-\frac{1}{2}}(z)$ JA

2. $Y_{-n-\frac{1}{2}}(z) = (-1)^{n} J_{n+\frac{1}{2}}(z)$ JA

The functions $H_{n+\frac{1}{2}}^{(1,2)}(z)$, $I_{n+\frac{1}{2}}(z)$, $K_{n+\frac{1}{2}}(z)$

8.466

1. $H_{n-\frac{1}{2}}^{(1)}(z) = \sqrt{\dfrac{2}{\pi z}} i^{-n} e^{iz} \sum_{k=0}^{n-1} (-1)^{k} \dfrac{(n+k-1)!}{k!(n-k-1)!} \dfrac{1}{(2iz)^{k}}$

$$\text{(cf. } \mathbf{8.451}\ 3)$$

2. $H_{n-\frac{1}{2}}^{(2)}(z) = \sqrt{\dfrac{2}{\pi z}} i^{n} e^{-iz} \sum_{k=0}^{n-1} \dfrac{(n+k-1)!}{k!(n-k-1)!} \dfrac{1}{(2iz)^{k}}$ $\text{(cf. } \mathbf{8.451}\ 4)$

8.467 $I_{\pm\left(n+\frac{1}{2}\right)}(z) = \dfrac{1}{\sqrt{2\pi z}} \left[e^{z} \sum_{k=0}^{n} \dfrac{(-1)^{k}(n+k)!}{k!(n-k)!(2z)^{k}} \pm (-1)^{n+1} e^{-z} \sum_{k=0}^{n} \dfrac{(n+k)!}{k!(n-k)!(2z)^{k}} \right]$

$$\text{(cf. } \mathbf{8.451}\ 5) \qquad\qquad \text{KU 60a}$$

8.468 $K_{n+\frac{1}{2}}(z) = \sqrt{\dfrac{\pi}{2z}} e^{-z} \sum_{k=0}^{n} \dfrac{(n+k)!}{k!(n-k)!(2z)^{k}}$ $\text{(cf. } \mathbf{8.451}\ 6)$ KU 60

8.469   Special cases:

1. $Y_{\frac{1}{2}}(z) = -\sqrt{\dfrac{2}{\pi z}} \cos z$

2. $Y_{-\frac{1}{2}}(z) = \sqrt{\dfrac{2}{\pi z}} \sin z$

3. $K_{\pm\frac{1}{2}}(z) = \sqrt{\dfrac{\pi}{2z}} e^{-z}$ WA 95(13)

4. $H_{\frac{1}{2}}^{(1)}(z) = \sqrt{\dfrac{2}{\pi z}} \dfrac{e^{iz}}{i}$ MO 27

5. $H_{\frac{1}{2}}^{(2)}(z) = \sqrt{\dfrac{2}{\pi z}} \dfrac{e^{-iz}}{-i}$ MO 27

6. $H_{-\frac{1}{2}}^{(1)}(z) = \sqrt{\dfrac{2}{\pi z}} e^{iz}$ MO 27

7. $H_{-\frac{1}{2}}^{(2)}(z) = \sqrt{\dfrac{2}{\pi z}} e^{-iz}$ MO 27

8.47–8.48 Functional relations

8.471[8]   Recursion formulas:

1. $z\,Z_{\nu-1}(z) + z\,Z_{\nu+1}(z) = 2\nu\,Z_\nu(z)$ KU 56(13), WA 56(1), WA 79(1), WA 88(3)

2. $Z_{\nu-1}(z) - Z_{\nu+1}(z) = 2\dfrac{d}{dz}\,Z_\nu(z)$ KU 56(12), WA 56(2), WA 79(2), We 88(4)

Sonin and Nielsen, in their construction of the theory of Bessel functions, defined Bessel functions as analytic functions of z that satisfy the recursion relations **8.471**. Z denotes J, N, $H^{(1)}$, $H^{(2)}$ or any linear combination of these functions, the coefficients in which are independent of z and ν.

8.472   Consequences of the recursion formulas for Z defined as above:

1. $z\dfrac{d}{dz}\,Z_\nu(z) + \nu\,Z_\nu(z) = z\,Z_{\nu-1}(z)$ KU 56(11), WA 56(3), WA 79(3), WA 88(5)

2. $z\dfrac{d}{dz}\,Z_\nu(z) - \nu\,Z_\nu(z) = -z\,Z_{\nu+1}(z)$ KU 56(10), WA 56(4), WA 79(4), WA 88(6)

3. $\left(\dfrac{d}{z\,dz}\right)^m \left(z^\nu\,Z_\nu(z)\right) = z^{\nu-m}\,Z_{\nu-m}(z)$ KU 56(8), WA 57(5), WA 89(9)

4. $\left(\dfrac{d}{z\,dz}\right)^m \left(z^{-\nu}\,Z_\nu(z)\right) = (-1)^m z^{-\nu-m}\,Z_{\nu+m}(z)$ WA 89(10), Ku 55(5), WA 57(6)

5. $Z_{-n}(z) = (-1)^n\,Z_n(z)$ [n is a natural number]   (cf. **8.404**)

8.473   Special cases:

1. $J_2(z) = \dfrac{2}{z}\,J_1(z) - J_0(z)$

2. $Y_2(z) = \dfrac{2}{z}\,Y_1(z) - Y_0(z)$

3. $H_2^{(1,2)}(z) = \dfrac{2}{z}\,H_1^{(1,2)}(z) - H_0^{(1,2)}(z)$

4. $\dfrac{d}{dz}\,J_0(z) = -J_1(z)$

5. $\dfrac{d}{dz}\,Y_0(z) = -Y_1(z)$

6. $\dfrac{d}{dz}\,H_0^{(1,2)}(z) = -H_1^{(1,2)}(z)$

8.474[8]   Each of the pairs of functions $J_\nu(z)$ and $J_{-\nu}(z)$ (for $\nu \neq 0, \pm1, \pm2,\ldots$), $J_\nu(z)$ and $Y_\nu(z)$, and $H_\nu^{(1)}(z)$ and $H_\nu^{(2)}(z)$, which are solutions of equation **8.401**, and also the pair $I_\nu(z)$ and $K_\nu(z)$ is a pair of linearly independent functions. The Wronskians of these pairs are, respectively,

$$-\frac{2}{\pi z}\sin\nu\pi, \quad \frac{2}{\pi z}, \quad -\frac{4i}{\pi z}, \quad -\frac{1}{z} \qquad \text{KU 52(10, 11, 12), WA 90(1, 4)}$$

8.475[6] The functions $J_\nu(z)$, and $Y_\nu(z)$, $H_\nu^{(1,2)}(z)$, $I_\nu(z)$, $K_\nu(z)$ with the exception of $J_n(z)$ and $I_n(z)$ for n an integer are *non-single-valued*: $z = 0$ is a branch point for these functions. The branches of these functions that lie on opposite sides of the cut $(-\infty, 0)$ are connected by the relations

8.476

1. $\quad J_\nu\left(e^{m\pi i}z\right) = e^{m\nu\pi i}\, J_\nu(z)$ $\hfill$ WA 90(1)

2. $\quad Y_\nu\left(e^{m\pi i}z\right) = e^{-m\nu\pi i}\, Y_\nu(z) + 2i\sin m\nu\pi \cot\nu\pi\, J_\nu(z)$ $\hfill$ WA 90(3)

3. $\quad Y_{-\nu}\left(e^{m\pi i}z\right) = e^{-m\nu\pi i}\, Y_{-\nu}(z) + 2i\sin m\nu\pi \operatorname{cosec}\nu\pi\, J_\nu(z)$ $\hfill$ WA 90(4)

4. $\quad I_\nu\left(e^{m\pi i}z\right) = e^{m\nu\pi i}\, I_\nu(z)$ $\hfill$ WA 95(17)

5. $\quad K_\nu\left(e^{m\pi i}z\right) = e^{-m\nu\pi i}\, K_\nu(z) - i\pi\dfrac{\sin m\nu\pi}{\sin\nu\pi}\, I_\nu(z)$ $\quad$ [ν not an integer] $\hfill$ WA 95(18)

6. $\quad H_\nu^{(1)}\left(e^{m\pi i}z\right) = e^{-m\nu\pi i}\, H_\nu^{(1)}(z) - 2e^{-\nu\pi i}\dfrac{\sin m\nu\pi}{\sin\nu\pi}\, J_\nu(z)$

$\qquad\qquad = \dfrac{\sin(1-m)\nu\pi}{\sin\nu\pi}\, H_\nu^{(1)}(z) - e^{-\nu\pi i}\dfrac{\sin m\nu\pi}{\sin\nu\pi}\, H_\nu^{(2)}(z)$

$\hfill$ WA 95(5)

7. $\quad H_\nu^{(2)}\left(e^{m\pi i}z\right) = e^{-m\nu\pi i}\, H_\nu^{(2)}(z) + 2e^{\nu\pi i}\dfrac{\sin m\nu\pi}{\sin\nu\pi}\, J_\nu(z)$

$\qquad\qquad = \dfrac{\sin(1+m)\nu\pi}{\sin\nu\pi}\, H_\nu^{(2)}(z) + e^{\nu\pi i}\dfrac{\sin m\nu\pi}{\sin\nu\pi}\, H_\nu^{(1)}(z)$

$\hfill$ [m an integer] $\hfill$ WA 90(6)

8. $\quad H_\nu^{(1)}\left(e^{i\pi}z\right) = -H_{-\nu}^{(2)}(z) = -e^{-i\pi\nu}\, H_\nu^{(2)}(z)$ $\hfill$ MO 26

9. $\quad H_\nu^{(2)}\left(e^{-i\pi}z\right) = -H_{-\nu}^{(1)}(z) = -e^{i\pi\nu}\, H_\nu^{(1)}(z)$ $\hfill$ MO 26

10.[8] $\quad \overline{H_\nu^{(2)}}(z) = H_{\bar\nu}^{(1)}\left(\bar z\right)$ $\hfill$ MO 26

8.477

1. $\quad J_\nu(z)\, Y_{\nu+1}(z) - J_{\nu+1}(z)\, Y_\nu(z) = -\dfrac{2}{\pi z}$ $\hfill$ WA 91(12)

2. $\quad I_\nu(z)\, K_{\nu+1}(z) + I_{\nu+1}(z)\, K_\nu(z) = \dfrac{1}{z}$ $\hfill$ WA 95(20)

See also **3.863**.

- For a connection with Legendre functions, see **8.722**.
- For a connection with the polynomials $C_n^\lambda(t)$, see **8.936** 4.
- For a connection with a confluent hypergeometric function, see **9.235**.

8.478 For $\nu > 0$ and $x > 0$, the product

$$x\left[J_\nu^{\,2}(x) + N_\nu^{\,2}(x)\right],$$

considered as a function of x, decreases monotonically, if $\nu > \frac{1}{2}$ and increases monotonically if $0 < \nu < \frac{1}{2}$.

$\hfill$ MO 35

8.479

1. $\dfrac{1}{\sqrt{x^2 - \nu^2}} > \dfrac{\pi}{2} \left[J_\nu{}^2(x) + N_\nu{}^2(x) \right] \geq \dfrac{1}{x}$ $\left[x \geq \nu \geq \frac{1}{2} \right]$ MO 35

2. $|J_n(nz)| \leq 1$

$\left[\left| \dfrac{z \exp \sqrt{1 - z^2}}{1 + \sqrt{1 - z^2}} \right| < 1, n \text{ a natural number} \right]$ MO 35

Relations between Bessel functions of the first, second, and third kinds

8.481 $J_\nu(z) = \dfrac{Y_{-\nu}(z) - Y_\nu(z) \cos \nu\pi}{\sin \nu\pi} = H_\nu^{(1)}(z) - i\, Y_\nu(z)$

$= H_\nu^{(2)}(z) + i\, Y_\nu(z) = \dfrac{1}{2} \left(H_\nu^{(1)}(z) + H_\nu^{(2)}(z) \right)$

(cf. **8.403** 1, **8.405**)   WA 89(1), JA

8.482 $Y_\nu(z) = \dfrac{J_\nu(z) \cos \nu\pi - J_{-\nu}(z)}{\sin \nu\pi} = i\, J_\nu(z) - i\, H_\nu^{(1)}(z)$

$= i\, H_\nu^{(2)}(z) - i\, J_\nu(z) = \dfrac{i}{2} \left(H_\nu^{(2)}(z) - H_\nu^{(1)}(z) \right)$

(cf. **8.403** 1, **8.405**)   WA 89(3), JA

8.483

1. $H_\nu^{(1)}(z) = \dfrac{J_{-\nu}(z) - e^{-\nu\pi i} J_\nu(z)}{i \sin \nu\pi} = \dfrac{Y_{-\nu}(z) - e^{-\nu\pi i} Y_\nu(z)}{\sin \nu\pi} = J_\nu(z) + i\, Y_\nu(z)$ WA 89(5)

2. $H_\nu^{(2)}(z) = \dfrac{e^{\nu\pi i} J_\nu(z) - J_{-\nu}(z)}{i \sin \nu\pi} = \dfrac{Y_{-\nu}(z) - e^{\nu\pi i} Y_\nu(z)}{\sin \nu\pi} = J_\nu(z) - i\, Y_\nu(z)$ WA 89(6)

(cf. **8.405**)   WA 89(6)

8.484

1. $H_{-\nu}^{(1)}(z) = e^{\nu\pi i} H_\nu^{(1)}(z)$ WA 89(7)

2. $H_{-\nu}^{(2)}(z) = e^{-\nu\pi i} H_\nu^{(2)}(z)$ WA 89(7)

8.485[7] $K_\nu(z) = \dfrac{\pi}{2} \dfrac{I_{-\nu}(z) - I_\nu(z)}{\sin \nu\pi}$ [ν not an integer]   (see also **8.407**)

WA 92(6)

8.486   Recursion formulas for the functions $I_\nu(z)$ and $K_\nu(z)$ and their consequences:

1. $z\, I_{\nu-1}(z) - z\, I_{\nu+1}(z) = 2\nu\, I_\nu(z)$ WA 93(1)

2. $I_{\nu-1}(z) + I_{\nu+1}(z) = 2\dfrac{d}{dz} I_\nu(z)$ WA 93(2)

3. $z\dfrac{d}{dz} I_\nu(z) + \nu\, I_\nu(z) = z\, I_{\nu-1}(z)$ WA 93(3)

4. $z\dfrac{d}{dz} I_\nu(z) - \nu\, I_\nu(z) = z\, I_{\nu+1}(z)$ WA 93(4)

5. $\left(\dfrac{d}{z\, dz} \right)^m \{ z^\nu I_\nu(z) \} = z^{\nu - m} I_{\nu - m}(z)$ WA 93(5)

6. $\left(\dfrac{d}{z\,dz}\right)^m \left\{z^{-\nu}\,I_\nu(z)\right\} = z^{-\nu-m}\,I_{\nu+m}(z)$ WA 93(6)

7. $I_{-n}(z) = l_n(z)$ [n a natural number]   WA 93(8)

8. $I_2(z) = -\dfrac{2}{z}l_1(z) + I_0(z)$

9. $\dfrac{d}{dz}\,I_0(z) = I_1(z)$ WA 93(7)

10. $z\,K_{\nu-1}(z) - z\,K_{\nu+1}(z) = -2\nu\,K_\nu(z)$ WA 93(1)

11. $K_{\nu-1}(z) + K_{\nu+1}(z) = -2\dfrac{d}{dz}\,K_\nu(z)$ WA 93(2)

12. $z\dfrac{d}{dz}\,K_\nu(z) + \nu\,K_\nu(z) = -z\,K_{\nu-1}(z)$ WA 93(3)

13. $z\dfrac{d}{dz}\,K_\nu(z) - \nu\,K_\nu(z) = -z\,K_{\nu+1}(z)$ WA 93(4)

14. $\left(\dfrac{d}{z\,dz}\right)^m \left\{z^\nu\,K_\nu(z)\right\} = (-1)^m z^{\nu-m}\,K_{\nu-m}(z)$ WA 93(5)

15. $\left(\dfrac{d}{z\,dz}\right)^m \left\{z^{-\nu}\,K_\nu(z)\right\} = (-1)^m z^{-\nu-m}\,K_{\nu+m}(z)$ WA 93(6)

16. $K_{-\nu}(z) = K_\nu(z)$ WA 93(8)

17. $K_2(z) = \dfrac{2}{z}\,K_1(z) + K_0(z)$

18. $\dfrac{d}{dz}\,K_0(z) = -K_1(z)$ WA 93(7)

19.* $\dfrac{\partial J_\nu(z)}{\partial \nu} = \left[\ln\dfrac{z}{2} - \psi(\nu+1)\right] J_\nu(z) + \dfrac{(z/2)^{\nu+1}}{\Gamma(\nu+1)}\sum_{n=0}^{\infty}\dfrac{(z/2)^n\,J_{n+1}(z)}{n!\,(\nu+n+1)^2}$ LUKE 360

8.486(1)[7]   Differentiation with respect to order

1. $\dfrac{\partial J_\nu(z)}{\partial \nu} = J_\nu(z)\ln\left(\dfrac{1}{2}z\right) - \sum_{k=0}^{\infty}(-1)^k\left(\dfrac{1}{2}z\right)^{\nu+2k}\dfrac{\psi(\nu+k+1)}{k!\,\Gamma(\nu+k+1)}$

                                            [$\nu \neq n$ or $n+\frac{1}{2}$, n integer]   MS 3.1.3

2. $\dfrac{\partial J_{-\nu}(z)}{\partial \nu} = -J_{-\nu}(z)\ln\left(\dfrac{1}{2}z\right) + \sum_{k=0}^{\infty}(-1)^k\left(\dfrac{1}{2}z\right)^{-\nu+2k}\dfrac{\psi(-\nu+k+1)}{k!\,\Gamma(-\nu+k+1)}$

                                            [$\nu \neq n$ or $n+\frac{1}{2}$, n integer]   MS 3.1.3

3. $\dfrac{\partial Y_\nu(z)}{\partial \nu} = \cot\pi\nu\dfrac{\partial J_\nu(z)}{\partial \nu} - \operatorname{cosec}\pi\nu\dfrac{\partial J_{-\nu}(z)}{\partial \nu} - \pi\operatorname{cosec}\pi\nu\,Y_\nu(z)$

                                            [$\nu \neq n$ or $n+\frac{1}{2}$, n integer]   MS 3.1.3

4. $\dfrac{\partial I_\nu(z)}{\partial \nu} = I_\nu(z)\ln\left(\dfrac{1}{2}z\right) - \sum_{k=0}^{\infty}\left(\dfrac{1}{2}z\right)^{\nu+2k}\dfrac{\psi(\nu+k+1)}{k!\,\Gamma(\nu+k+1)}$ [$\nu \neq n$ or $n+\frac{1}{2}$, n integer]

                                                                   MS 3.1.3

5. $\dfrac{\partial\, K_\nu(z)}{\partial \nu} = -\pi \cot \pi\nu\, K_\nu(z) + \dfrac{1}{2}\pi \operatorname{cosec} \pi\nu \left[\dfrac{\partial\, I_{-\nu}(z)}{\partial \nu} - \dfrac{\partial\, I_\nu(z)}{\partial \nu}\right]$

$$\left[\nu \neq n \text{ or } n+\tfrac{1}{2}, \quad n \text{ integer}\right] \quad \text{MS 3.1.3}$$

6. $\left[\dfrac{\partial\, J_\nu(z)}{\partial \nu}\right]_{\nu=\pm n} = \dfrac{1}{2}\pi(\pm1)^n\, Y_n(z) \pm (\pm1)^n \dfrac{1}{2}n! \displaystyle\sum_{k=0}^{n-1} \dfrac{\left(\frac{1}{2}z\right)^{k-n} J_k(z)}{k!(n-k)}$ $[n=0,1,\ldots]$ MS 3.2.3

7. $\left[\dfrac{\partial\, Y_\nu(z)}{\partial \nu}\right]_{\nu=\pm n} = -\dfrac{1}{2}\pi(\pm1)^n\, J_n(z) \pm (\pm1)^n \dfrac{1}{2}n! \displaystyle\sum_{k=0}^{n-1} \dfrac{\left(\frac{1}{2}z\right)^{k-n} Y_k(z)}{k!(n-k)}$ $[n=0,1,\ldots]$

$$\text{MS 3.2.3}$$

8. $\left[\dfrac{\partial\, I_\nu(z)}{\partial \nu}\right]_{\nu=\pm n} = (-1)^{n+1}\, K_n(z) \pm (-1)^n \dfrac{1}{2}n! \displaystyle\sum_{k=0}^{n-1} \dfrac{(-1)^k \left(\frac{1}{2}z\right)^{k-n} I_k(z)}{k!(n-k)}$ $[n=0,1,\ldots]$

$$\text{MS 3.2.3}$$

9. $\left[\dfrac{\partial\, K_\nu(z)}{\partial \nu}\right]_{\nu=\pm n} = \pm\dfrac{1}{2}n! \displaystyle\sum_{k=0}^{n-1} \dfrac{\left(\frac{1}{2}z\right)^{k-n} K_k(z)}{k!(n-k)}$ $[n=0,1,\ldots]$ MS 3.2.3

10. $(-1)^n \left[\dfrac{\partial}{\partial \nu} I_\nu(z)\right]_{\nu=n} = -K_n(z) + \dfrac{1}{2}n! \displaystyle\sum_{k=0}^{n-1} \dfrac{(-1)^k \left(\frac{1}{2}z\right)^{k-n} I_k(z)}{k!(n-k)}$

$$\left[n=0,1,\ldots\right] \quad \text{AS 9.6.44}$$

11. $\left[\dfrac{\partial\, K_\nu(z)}{\partial \nu}\right]_{\nu=n} = \dfrac{1}{2}n! \displaystyle\sum_{k=0}^{n-1} \dfrac{\left(\frac{1}{2}z\right)^{k-n} I_k(z)}{k!(n-k)}$ $[n=0,1,\ldots]$ AS 9.6.45

Special cases

12. $\left[\dfrac{\partial\, J_\nu(z)}{\partial \nu}\right]_{\nu=0} = \tfrac{1}{2}\pi\, Y_0(z)$ MS 3.2.3

13. $\left[\dfrac{\partial\, Y_\nu(z)}{\partial \nu}\right]_{\nu=0} = -\tfrac{1}{2}\pi\, J_0(z)$ MS 3.2.3

14. $\left[\dfrac{\partial\, I_\nu(z)}{\partial \nu}\right]_{\nu=0} = -K_0(z)$ MS 3.2.3

15. $\left[\dfrac{\partial\, K_\nu(z)}{\partial \nu}\right]_{\nu=0} = 0$ MS 3.2.3

16. $\left[\dfrac{\partial\, J_\nu(x)}{\partial \nu}\right]_{\nu=\frac{1}{2}} = \left(\tfrac{1}{2}\pi x\right)^{-1/2} \left[\sin x\, \operatorname{Ci}(3x) - \cos x\, \operatorname{Si}(2x)\right]$ MS 3.3.3

17. $\left[\dfrac{\partial\, J_\nu(x)}{\partial \nu}\right]_{\nu=-\frac{1}{2}} = \left(\tfrac{1}{2}\pi x\right)^{-1/2} \left[\cos x\, \operatorname{Ci}(2x) + \sin x\, \operatorname{Si}(2x)\right]$ MS 3.3.3

18. $\left[\dfrac{\partial\, Y_\nu(x)}{\partial \nu}\right]_{\nu=\frac{1}{2}} = \left(\tfrac{1}{2}\pi x\right)^{-1/2} \left\{\cos x\, \operatorname{Ci}(2x) + \sin x\, \left[\operatorname{Si}(2x) - \pi\right]\right\}$ MS 3.3.3

19. $\left[\dfrac{\partial\, Y_\nu(x)}{\partial \nu}\right]_{\nu=-\frac{1}{2}} = -\left(\tfrac{1}{2}\pi x\right)^{-1/2} \left\{\sin x\, \operatorname{Ci}(2x) - \cos x\, \left[\operatorname{Si}(2x) - \pi\right]\right\}$ MS 3.3.3

20. $\left[\dfrac{\partial I_\nu(x)}{\partial \nu}\right]_{\nu=\pm\frac{1}{2}} = (2\pi x)^{-1/2}\left[e^x\,\mathrm{Ei}(-2x) \mp e^{-x}\overline{\mathrm{Ei}}(2x)\right]$ MS 3.3.3

21. $\left[\dfrac{\partial K_\nu(x)}{\partial \nu}\right]_{\nu=\pm\frac{1}{2}} = \mp\left(\dfrac{\pi}{2x}\right)^{\frac{1}{2}} e^x\,\mathrm{Ei}(-2x)$ MS 3.3.3

8.487   Continuity with respect to the order*:

1. $\displaystyle\lim_{\nu\to n} Y_\nu(z) = Y_n(z)$ $[n \text{ an integer}]$ WA 76

2. $\displaystyle\lim_{\nu\to n} H_\nu^{(1,2)}(z) = H_n^{(1,2)}(z)$ $[n \text{ an integer}]$ WA 183

3. $\displaystyle\lim_{\nu\to n} K_\nu(z) = K_n(z)$ $[n \text{ an integer}]$ WA 92

8.49   Differential equations leading to Bessel functions

See also **8.401**

8.491

1. $\dfrac{1}{z}\dfrac{d}{dz}(zu') + \left(\beta^2 - \dfrac{\nu^2}{z^2}\right)u = 0$ $u = Z_\nu(\beta z)$ JA

2. $\dfrac{1}{z}\dfrac{d}{dz}(zu') + \left[(\beta\gamma z^{\gamma-1})^2 - \left(\dfrac{\nu\gamma}{z}\right)^2\right]u = 0$ $u = Z_\nu(\beta z^\gamma)$ JA

3. $u'' + \dfrac{1-2\alpha}{z}u' + \left[(\beta\gamma z^{\gamma-1})^2 + \dfrac{\alpha^2 - \nu^2\gamma^2}{z^2}\right]u = 0$ $u = z^\alpha Z_\nu(\beta z^\gamma)$ JA

4. $u'' + \left[(\beta\gamma z^{\gamma-1})^2 - \dfrac{4\nu^2\gamma^2 - 1}{4z^2}\right]u = 0$ $u = \sqrt{z}\,Z_\nu(\beta z^\gamma)$ JA

5. $u'' + \left(\beta^2 - \dfrac{4\nu^2 - 1}{4z^2}\right)u = 0$ $u = \sqrt{z}\,Z_\nu(\beta z)$ JA

6. $u'' + \dfrac{1-2\alpha}{z}u' + \left(\beta^2 + \dfrac{\alpha^2 - \nu^2}{z^2}\right)u = 0$ $u = z^\alpha Z_\nu(\beta z)$ JA

7. $u'' + bz^m u = 0$ $u = \sqrt{z}\,Z_{\frac{1}{m+2}}\left(\dfrac{2\sqrt{b}}{m+2}z^{\frac{m+2}{2}}\right)$

                                                                       JA 111(5)

8. $u'' + \dfrac{1}{z}u' + 4\left(z^2 - \dfrac{\nu^2}{z^2}\right)u = 0$ $u = Z_\nu(z^2)$ WA 111(6)

9. $u'' + \dfrac{1}{z}u' + \dfrac{1}{4z}\left(1 - \dfrac{\nu^2}{z}\right)u = 0$ $u = Z_\nu(\sqrt{z})$ WA 111(7)

10. $u'' + \dfrac{1-\nu}{z}u' + \dfrac{1}{4}\dfrac{u}{z} = 0$ $u = z^{\frac{\nu}{2}} Z_\nu(\sqrt{z})$ WA 111(9)a

11. $u'' + \beta^2\gamma^2 z^{2\beta-2}u = 0$ $u = z^{1/2} Z_{\frac{1}{2\beta}}(\gamma z^\beta)$ WA 110(3)

---

*The continuity of the functions $J_\nu(z)$ and $I_\nu(z)$ follows directly from the series representations of these functions.

12. $z^2 u'' + (2\alpha - 2\beta\nu + 1)zu' + \left[\beta^2\gamma^2 z^{2\beta} + \alpha(\alpha - 2\beta\nu)\right] u = 0$

$$u = z^{\beta\nu-\alpha} \, Z_\nu\left(\gamma z^\beta\right) \qquad \text{WA 112(21)}$$

8.492

1. $u'' + \left(e^{2z} - \nu^2\right) u = 0$ $\qquad\qquad\qquad\qquad u = Z_\nu\left(e^z\right)$ $\qquad\qquad$ WA 112(22)

2. $u'' + \dfrac{e^{2/z} - \nu^2}{z^4} u = 0$ $\qquad\qquad\qquad u = z \, Z_\nu\left(e^{1/z}\right)$ $\qquad\quad$ WA 112(22)

8.493

1. $u'' + \left(\dfrac{1}{z} - 2\tan z\right) u' - \left(\dfrac{\nu^2}{z^2} + \dfrac{\tan z}{z}\right) u = 0$ $\qquad u = \sec z \, Z_\nu(z)$ $\qquad$ JA

2. $u'' + \left(\dfrac{1}{z} + 2\cot z\right) u' - \left(\dfrac{\nu^2}{z^2} - \dfrac{\cot z}{z}\right) u = 0$ $\qquad u = \operatorname{cosec} z \, Z_\nu(z)$ $\quad$ JA

8.494

1. $u'' + \dfrac{1}{z} u' - \left(1 + \dfrac{\nu^2}{z^2}\right) u = 0$ $\qquad u = Z_\nu(iz) = C_1 \, I_\nu(z) + C_2 \, K_\nu(z)$ $\quad$ JA

2. $u'' + \dfrac{1}{z} u' - \left[\dfrac{1}{z} + \left(\dfrac{\nu}{2z}\right)^2\right] u = 0$ $\qquad\qquad u = Z_\nu\left(2i\sqrt{z}\right)$ $\qquad\qquad$ JA

3. $u'' + u' + \dfrac{1}{z^2}\left(\dfrac{1}{4} - \nu^2\right) u = 0$ $\qquad\qquad u = \sqrt{z} e^{-\frac{z}{2}} \, Z_\nu\left(\dfrac{iz}{2}\right)$ $\qquad$ JA

4.[10] $u'' + \left(\dfrac{2\nu + 1}{z} - k\right) u' - \dfrac{2\nu + 1}{2z} k u = 0$ $\qquad u = z^{-\nu} e^{\frac{1}{2}kx} \, Z_\nu\left(\dfrac{ikz}{2}\right)$ $\quad$ JA

5. $u'' + \dfrac{1 - \nu}{z} u' - \dfrac{1}{4} \dfrac{u}{z} = 0$ $\qquad\qquad u = z^{\frac{\nu}{2}} \, Z_\nu\left(i\sqrt{z}\right)$ $\qquad\quad$ WA 111(8)

6. $u'' \pm \dfrac{u}{\sqrt{z}} = 0$

$$u = \sqrt{z} \, Z_{\frac{2}{3}}\left(\tfrac{4}{3} z^{\frac{3}{4}}\right), \qquad u = \sqrt{z} \, Z_{\frac{2}{3}}\left(\tfrac{4}{3} i z^{\frac{3}{4}}\right) \quad \text{WA 111(10)}$$

7. $u'' \pm zu = 0$

$$u = \sqrt{z} \, Z_{\frac{1}{3}}\left(\tfrac{2}{3} z^{\frac{3}{2}}\right), \qquad u = \sqrt{z} \, Z_{\frac{1}{3}}\left(\tfrac{2}{3} i z^{\frac{3}{2}}\right) \quad \text{WA 111(10)}$$

8. $u'' - \left(c^2 + \dfrac{\nu(\nu + 1)}{z^2}\right) u = 0$ $\qquad\qquad u = \sqrt{z} \, Z_{\nu+\frac{1}{2}}(icz)$ $\qquad$ WA 108(1)

9. $u'' - \dfrac{2\nu}{z} u' - c^2 u = 0$ $\qquad\qquad\qquad u = z^{\nu+\frac{1}{2}} \, Z_{\nu+\frac{1}{2}}(icz)$ $\qquad$ WA 109(3, 4)

10. $u'' - c^2 z^{2\nu-2} u = 0$ $\qquad\qquad\qquad u = \sqrt{z} \, Z_{\frac{1}{2\nu}}\left(i\dfrac{c}{\nu} z^\nu\right)$ $\qquad$ WA 109(5, 6)

8.495

1. $u'' + \dfrac{1}{z} u' + \left(i - \dfrac{\nu^2}{z^2}\right) u = 0$ $\qquad\qquad u = Z_\nu\left(z\sqrt{i}\right)$ $\qquad\qquad$ JA

2. $\quad u'' + \left(\dfrac{1}{z} \mp 2i \right) u' - \left(\dfrac{\nu^2}{z^2} \pm \dfrac{i}{z} \right) u = 0$ $\qquad\qquad u = e^{\pm iz} Z_\nu(z)$ $\qquad$ JA

3. $\quad u'' + \dfrac{1}{z} u' + se^{i\alpha} u = 0$ $\qquad\qquad\qquad\qquad u = Z_0 \left(\sqrt{s} z e^{\frac{i}{2}\alpha} \right)$ $\qquad$ JA

4. $\quad u'' + \left(se^{i\alpha} + \dfrac{1}{4z^2} \right) u = 0$ $\qquad\qquad\qquad u = \sqrt{z}\, Z_0 \left(\sqrt{s} z e^{\frac{i}{2}\alpha} \right)$ $\qquad$ JA

8.496

1. $\quad \dfrac{d^2}{dz^2} \left(z^4 \dfrac{d^2 u}{dz^2} \right) - z^2 u = 0$ $\qquad\qquad u = \dfrac{1}{z} \left\{ Z_2 \left(2\sqrt{z} \right) + \overline{Z_2 \left(2i\sqrt{z} \right)} \right\}$

$\hfill$ WA 122(7)

2. $\quad \dfrac{d^2}{dz^2} \left(z^{\frac{16}{5}} \dfrac{d^2 u}{dz^2} \right) - z^{\frac{8}{5}} u = 0$ $\qquad u = z^{-7/10} \left\{ Z_{\frac{5}{6}} \left(\dfrac{5}{3} z^{\frac{3}{5}} \right) + \overline{Z_{\frac{5}{6}} \left(\dfrac{5}{3} i z^{\frac{3}{5}} \right)} \right\}$

$\hfill$ WA 122(8)

3. $\quad \dfrac{d^2}{dz^2} \left(z^{12} \dfrac{d^2 u}{dz^2} \right) - z^6 u = 0$

$\qquad\qquad\qquad u = z^{-4} \left\{ Z_{10} \left(2z^{-1/2} \right) + \overline{Z_{10} \left(2i z^{-1/2} \right)} \right\}$ $\quad$ WA 122(9)

4. $\quad \dfrac{d^4 u}{dz^4} + \dfrac{2}{z} \dfrac{d^3 u}{dz^3} - \dfrac{2\nu^2 + 1}{z^2} \dfrac{d^2 u}{dz^2} + \dfrac{2\nu^2 + 1}{z^3} \dfrac{du}{dz} + \left(\dfrac{\nu^4 - 4\nu^2}{z^4} - 1 \right) u = 0,$

$\quad u = A_1 J_\nu(z) + A_2 Y_\nu(z) + A_3 I_\nu(z) + A_4 K_\nu(z),$ where A_1, A_2, A_3, A_4 are constants $\quad$ MO 29

8.51–8.52 Series of Bessel functions

8.511 Generating functions for Bessel functions:

1. $\quad \exp \dfrac{1}{2} \left(t - \dfrac{1}{t} \right) z = J_0(z) + \sum\limits_{k=1}^{\infty} \left[t^k + (-t)^{-k} \right] J_k(z) = \sum\limits_{k=-\infty}^{\infty} J_k(z) t^k$

$\hfill [|z| < |t|] \qquad$ KU 119(12)

2. $\quad \exp \left(t - \dfrac{1}{t} \right) z = \left\{ \sum\limits_{k=-\infty}^{\infty} t^k J_k(z) \right\} \left\{ \sum\limits_{m=-\infty}^{\infty} t^m J_m(z) \right\}$ $\hfill$ WA 40

3. $\quad \exp \left(\pm iz \sin \varphi \right) = J_0(z) + 2 \sum\limits_{k=1}^{\infty} J_{2k}(z) \cos 2k\varphi \pm 2i \sum\limits_{k=0}^{\infty} J_{2k+1}(z) \sin(2k+1)\varphi$ $\hfill$ KU 120(13)

4. $\quad \exp \left(iz \cos \varphi \right) = \sqrt{\dfrac{\pi}{2z}} \sum\limits_{k=0}^{\infty} (2k+1) i^k J_{k+\frac{1}{2}}(z) P_k \left(\cos \varphi \right)$ $\hfill$ WA 401(1)

$\qquad\qquad\qquad = \sum\limits_{k=-\infty}^{\infty} i^k J_k(z) e^{ik\varphi}$ $\hfill$ MO 27

$\qquad\qquad\qquad = J_0(z) + 2 \sum\limits_{k=1}^{\infty} i^k J_k(z) \cos k\varphi$ $\hfill$ MO 27

5. $\sqrt{\dfrac{i}{\pi}} e^{iz\cos 2\varphi} \displaystyle\int_{-\infty}^{\sqrt{2z}\cos\varphi} e^{-it^2}\,dt = \dfrac{1}{2}J_0(z) + \sum_{k=1}^{\infty} e^{\frac{1}{4}k\pi i}\,J_{\frac{k}{2}}(z)\cos k\varphi$ MO 28

The series $\sum J_k(z)$

8.512

1. $J_0(z) + 2\displaystyle\sum_{k=1}^{\infty} J_{2k}(z) = 1$ WA 44

2. $\displaystyle\sum_{k=0}^{\infty} \dfrac{(n+2k)(n+k-1)!}{k!}\,J_{n+2k}(z) = \left(\dfrac{z}{2}\right)^n$ $[n=1,2,\ldots]$ WA 45

3. $\displaystyle\sum_{k=0}^{\infty} \dfrac{(4k+1)(2k-1)!!}{2^k k!}\,J_{2k+\frac{1}{2}}(z) = \sqrt{\dfrac{2z}{\pi}}$

8.513

Notation: In formulas **8.513** $Q_k^{(p)} = \displaystyle\sum_{m=0}^{\left\lfloor \frac{k-1}{2}\right\rfloor} \dfrac{(-1)^m \binom{k}{m}(k-2m)^p}{2^k k!}$

1. $\displaystyle\sum_{k=1}^{\infty} (2k)^{2p}\,J_{2k}(z) = \sum_{k=0}^{p} Q_{2k}^{(2p)}\,z^{2k}$ $[p=1,2,3,\ldots]$ WA 46(1)

2. $\displaystyle\sum_{k=0}^{\infty} (2k+1)^{2p+1}\,J_{2k+1}(z) = \sum_{k=0}^{p} Q_{2k+1}^{(2p+1)}\,z^{2k+1}$ $[p=0,1,2,3,\ldots]$ WA 46(2)

In particular:

3. $\displaystyle\sum_{k=0}^{\infty} (2k+1)^3\,J_{2k+1}(z) = \dfrac{1}{2}\left(z+z^3\right)$ WA 47(4)

4. $\displaystyle\sum_{k=1}^{\infty} (2k)^2\,J_{2k}(z) = \dfrac{1}{2}z^2$ WA 47(4)

5. $\displaystyle\sum_{k=1}^{\infty} 2k(2k+1)(2k+2)\,J_{2k+1}(z) = \dfrac{1}{2}z^3$ WA 47(4)

8.514

1. $\displaystyle\sum_{k=0}^{\infty} (-1)^k\,J_{2k+1}(z) = \dfrac{\sin z}{2}$ WH

2. $J_0(z) + 2\displaystyle\sum_{k=1}^{\infty} (-1)^k\,J_{2k}(z) = \cos z$ WH

3. $\displaystyle\sum_{k=1}^{\infty} (-1)^{k+1}(2k)^2\,J_{2k}(z) = \dfrac{z\sin z}{2}$ WA 32(9)

4. $$\sum_{k=0}^{\infty} (-1)^k (2k+1)^2 \, J_{2k+1}(z) = \frac{z \cos z}{2}$$
WA 32(10)

5. $$J_0(z) + 2 \sum_{k=1}^{\infty} J_{2k}(z) \cos 2k\theta = \cos(z \sin \theta)$$
KU 120(14), WA 32

6. $$\sum_{k=0}^{\infty} J_{2k+1}(z) \sin(2k+1)\theta = \frac{\sin(z \sin \theta)}{2}$$
KU 120(15), WA 32

7. $$\sum_{k=0}^{\infty} J_{2k+1}(x) = \frac{1}{2} \int_0^x J_0(t) \, dt \qquad\qquad [x \text{ is real}]$$
WA 638

8.515

1. $$\sum_{k=0}^{\infty} \frac{(-1)^k t^k}{k!} \left(\frac{2z+t}{2z} \right)^k J_{\nu+k}(z) = \left(\frac{z}{z+t} \right)^{\nu} J_{\nu}(z+t)$$
AD (9140)

2. $$\sum_{k=1}^{\infty} J_{2k-\frac{1}{2}} \left(x^2 \right) = S(x)$$
MO 127a

3. $$\sum_{k=0}^{\infty} J_{2k+\frac{1}{2}} \left(x^2 \right) = C(x)$$
MO 127a

8.516 $$\sum_{k=0}^{\infty} \frac{(2n+2k)(2n+k-1)!}{k!} J_{2n+2k} \left(2z \sin \theta \right) = (z \sin \theta)^{2n}$$
WA 47

The series $\sum A_k \, J_k(kx)$ and $\sum A_k \, J'_k(kx)$

8.517

1. $$\sum_{k=1}^{\infty} J_k(kz) = \frac{z}{2(1-z)} \qquad\qquad \left[\left| \frac{z \exp \sqrt{1-z^2}}{1 + \sqrt{1-z^2}} \right| < 1 \right]$$
WA 615(1)

2. $$\sum_{k=1}^{\infty} (-1)^k J_k(kz) = -\frac{z}{2(1+z)} \qquad\qquad \left[\left| \frac{z \exp \sqrt{1-z^2}}{1 + \sqrt{1-z^2}} \right| < 1 \right]$$
WA 622(1)

3. $$\sum_{k=1}^{\infty} J_{2k}(2kz) = \frac{z^2}{2 \left(1-z^2 \right)} \qquad\qquad \left[\left| \frac{z \exp \sqrt{1-z^2}}{1 + \sqrt{1-z^2}} \right| < 1 \right]$$
MO 58

8.518

1. $$\sum_{k=1}^{\infty} \frac{J'_{(kx)}}{k} = \frac{1}{2} + \frac{x}{4} \qquad\qquad [0 \le x < 1]$$
MO 58

2. $$\sum_{k=1}^{\infty} (-1)^{k-1} \frac{J'_{(kx)}}{k} = \frac{1}{2} - \frac{x}{4} \qquad\qquad [0 \le x < 1]$$
MO 58

3. $$\sum_{k=1}^{\infty} k \, J'_k(kx) = \frac{1}{2(1-x)^2} \qquad\qquad [0 \le x < 1]$$
MO 58

4. $\quad \displaystyle\sum_{k=1}^{\infty} (-1)^{k-1} J_k'(kx) k = \frac{1}{2(1+x)^2}$ $\qquad\qquad [0 \le x < 1]$ $\qquad\qquad$ MO 58

The series $\sum A_k J_0(kx)$

8.519 If, on the interval $[0 \le x \le \pi]$, a function $f(x)$ possesses a continuous derivative with respect to x that is of bounded variation, then

1. $\quad f(x) = \dfrac{a_0}{2} + \displaystyle\sum_{k=1}^{\infty} a_k J_0(kx)$ $\qquad\qquad [0 < x < \pi]$

   where

2. $\quad a_0 = 2f(0) + \dfrac{2}{\pi} \displaystyle\int_0^\pi du \int_0^{\pi/2} u f'(u \sin \varphi) \, d\varphi$

3. $\quad a_n = \dfrac{2}{\pi} \displaystyle\int_0^\pi du \int_0^{\pi/2} u f'(u \sin \varphi) \cos n u \, d\varphi$ $\qquad\qquad$ WH

8.521 Examples:

1. $\quad \displaystyle\sum_{k=1}^{\infty} J_0(kx) = -\frac{1}{2} + \frac{1}{x} + 2 \sum_{m=1}^{n} \frac{1}{\sqrt{x^2 - 4m^2\pi^2}}$ $\qquad [2n\pi < x < 2(n+1)\pi]$ $\qquad$ MO 59

2. $\quad \displaystyle\sum_{k=1}^{\infty} (-1)^{k+1} J_0(kx) = \frac{1}{2}$ $\qquad\qquad [0 < x < \pi]$ $\qquad\qquad$ KU 124(12)

3. $\quad \displaystyle\sum_{k=1}^{\infty} \frac{1}{(2k-1)^2} J_0\{(2k-1)x\} \quad \dfrac{\pi^2}{8} - \dfrac{|x|}{2}$ $\qquad\qquad [-\pi < x < \pi]$ $\qquad$ KU 124

   $\qquad\qquad\qquad\qquad = \dfrac{\pi^2}{8} + \sqrt{x^2 - \pi^2} - \dfrac{x}{2} - \pi \arccos \dfrac{\pi}{x}$ $\qquad [\pi < x < 2\pi]$ $\qquad$ MO 59

4. $\quad \displaystyle\sum_{k=1}^{\infty} e^{-kz} J_0\left(k\sqrt{x^2 + y^2}\right)$

   $\qquad = \dfrac{1}{r} - \dfrac{1}{2} + \displaystyle\sum_{k=1}^{\infty} \left\{ \dfrac{1}{\sqrt{(2ki\pi + z)^2 + x^2 + y^2}} - \dfrac{1}{\sqrt{(2ki\pi - z)^2 + x^2 + y^2}} \right\}$

   $\qquad = \dfrac{1}{r} - \dfrac{1}{2} + \displaystyle\sum_{k=1}^{\infty} \dfrac{1}{(2k)!} B_{2k} r^{2k-1} P_{2k-1}\left(\dfrac{z}{r}\right)$ $\qquad\qquad [0 < r < 2\pi]$ MO 59

where $r = \sqrt{x^2 + y^2 + z^2}$ and where the radical indicates the square root with a positive real part. In formula **8.521** 4, the first equation holds when x and y are real and $\operatorname{Re} z > 0$; the second equation holds when x, y, and z are all real.

The series $\sum A_k \, Z_0(kx) \sin kx$ and $\sum A_k \, Z_0(kx) \cos kx$

8.522

1. $$\sum_{k=1}^{\infty} J_0(kx) \cos kxt = -\frac{1}{2} + \sum_{l=1}^{m} \frac{1}{\sqrt{x^2 - (2\pi l + tx)^2}} + \frac{1}{x\sqrt{1 - t^2}} + \sum_{l=1}^{n} \frac{1}{\sqrt{x^2 - (2\pi l - tx)^2}}$$

MO 59

2. $$\sum_{k=1}^{\infty} J_0(kx) \sin kxt = \frac{1}{2\pi} \left\{ \sum_{l=1}^{n} \frac{1}{l} - \sum_{l=1}^{m} \frac{1}{l} \right\} + \sum_{l=m+1}^{\infty} \left\{ \frac{1}{\sqrt{(2\pi l + tx)^2 - x^2}} - \frac{1}{2\pi l} \right\}$$
$$- \sum_{l=n+1}^{\infty} \left\{ \frac{1}{\sqrt{(2\pi l - tx)^2 - x^2}} - \frac{1}{2\pi l} \right\}$$

MO 59

3. $$\sum_{k=1}^{\infty} Y_0(kx) \cos kxt = -\frac{1}{\pi} \left(\boldsymbol{C} + \ln \frac{x}{4\pi} \right) + \frac{1}{2\pi} \left\{ \sum_{l=1}^{m} \frac{1}{l} + \sum_{l=1}^{n} \frac{1}{l} \right\}$$
$$- \sum_{l=m+1}^{\infty} \left\{ \frac{1}{\sqrt{(2\pi l + tx)^2 - x^2}} - \frac{1}{2\pi l} \right\}$$
$$- \sum_{l=n+1}^{\infty} \left\{ \frac{1}{\sqrt{(2\pi l - tx)^2 - x^2}} - \frac{1}{2\pi l} \right\}$$

MO 60

In formulas **8.522**, $x > 0, 0 \le t < 1, 2\pi m < x(1 - t) < 2(m + 1)\pi, 2n\pi < x(1 + t) < 2(n + 1)\pi, m + 1$ and $n + 1$ are natural numbers.

8.523

1. $$\sum_{k=1}^{\infty} (-1)^k J_0(kx) \cos kxt = -\frac{1}{2} + \sum_{l=1}^{m} \frac{1}{\sqrt{x^2 - [(2l - 1)\pi + tx]^2}} + \sum_{l=1}^{n} \frac{1}{\sqrt{x^2 - [(2l - 1)\pi - tx]^2}}$$

MO 60

2. $$\sum_{k=1}^{\infty} (-1)^k J_0(kx) \sin kxt \quad \frac{1}{2\pi} \left\{ \sum_{l=1}^{n} \frac{1}{l} - \sum_{l=1}^{m} \frac{1}{l} \right\} + \sum_{l=m+1}^{\infty} \left\{ \frac{1}{\sqrt{[(2l - 1)\pi + tx]^2 - x^2}} - \frac{1}{2l\pi} \right\}$$
$$- \sum_{l=n+1}^{\infty} \left\{ \frac{1}{\sqrt{[(2l - 1)\pi - tx]^2 - x^2}} - \frac{1}{2l\pi} \right\}$$

MO 60

3. $\sum_{k=1}^{\infty} (-1)^k \, Y_0(kx) \cos kxt$

$-\frac{1}{\pi} \left(C + \ln \frac{x}{4\pi} \right) + \frac{1}{2\pi} \left\{ \sum_{l=1}^{m} \frac{1}{l} + \sum_{l=1}^{n} \frac{1}{l} \right\}$

$- \sum_{l=m+1}^{\infty} \left\{ \frac{1}{\sqrt{[(2l-1)\pi + tx]^2 - x^2}} - \frac{1}{2l\pi} \right\}$

$- \sum_{l=n+1}^{\infty} \left\{ \frac{1}{\sqrt{[(2l-1)\pi - tx]^2 - x^2}} - \frac{1}{2l\pi} \right\}$

MO 60

In formulas **8.523**, $x > 0, 0 \le t < 1$, $(2m-1)\pi < x(1-t) < (2m+1)\pi$, $(2n-1)\pi < x(1+t) < (2n+1)\pi$, m and n are natural numbers.

8.524[6]

1. $\sum_{k=1}^{\infty} J_0(kx) \cos kxt = -\frac{1}{2} + \sum_{l=m+1}^{n} \frac{1}{\sqrt{x^2 - (2l\pi - tx)^2}}$ MO 60

2. $\sum_{k=1}^{\infty} J_0(kx) \sin kxt \sum_{l=0}^{m} \frac{1}{\sqrt{(2l\pi - tx)^2 - x^2}} + \sum_{l=1}^{\infty} \left\{ \frac{1}{\sqrt{(2l\pi + tx)^2 - x^2}} - \frac{1}{2l\pi} \right\}$

$- \sum_{l=n+1}^{\infty} \left\{ \frac{1}{\sqrt{(2l\pi - tx)^2 - x^2}} - \frac{1}{2l\pi} \right\} + \frac{1}{2\pi} \sum_{l=1}^{n} \frac{1}{l}$

MO 60

3.[6] $\sum_{k=1}^{\infty} Y_0(kx) \cos kxt$ $-\frac{1}{\pi} \left(C + \ln \frac{x}{4\pi} \right) - \sum_{l=0}^{m} \frac{1}{\sqrt{(2\pi l - tx)^2 - x^2}} + \frac{1}{2\pi} \sum_{l=1}^{n} \frac{1}{l}$

$- \sum_{l=1}^{\infty} \left\{ \frac{1}{\sqrt{(2l\pi + tx)^2 - x^2}} - \frac{1}{2l\pi} \right\}$

$- \sum_{l=n+1}^{\infty} \left\{ \frac{1}{\sqrt{(2l\pi - tx)^2 - x^2}} - \frac{1}{2l\pi} \right\}$

MO 61

In formulas **8.524**, $x > 0, t > 1, 2m\pi < x(t-1) < 2(m+1)\pi$, $2n\pi < x(t+1) < 2(n+1)\pi, m+1$ and $n+1$ are natural numbers.

8.525

1. $\sum_{k=1}^{\infty} (-1)^k J_0(kx) \cos kxt = -\frac{1}{2} + \sum_{l=m+1}^{n} \frac{1}{\sqrt{x^2 - [(2l-1)\pi - tx]^2}}$ MO 61

2. $$\sum_{k=1}^{\infty} (-1)^k J_0(kx) \sin kxt = \sum_{l=1}^{m} \frac{1}{\sqrt{[(2l-1)\pi - tx]^2 - x^2}} + \frac{1}{2\pi} \sum_{l=1}^{n} \frac{1}{l}$$

$$+ \sum_{l=1}^{\infty} \left\{ \frac{1}{\sqrt{[(2l-1)\pi + tx]^2 - x^2}} - \frac{1}{2l\pi} \right\}$$

$$- \sum_{l=n+1}^{\infty} \left\{ \frac{1}{\sqrt{[(2l-1)\pi - tx]^2 - x^2}} - \frac{1}{2l\pi} \right\}$$

MO 61

3. $$\sum_{k=1}^{\infty} (-1)^k Y_0(kx) \cos kxt = -\frac{1}{\pi} \left(C + \ln \frac{x}{4\pi} \right) + \frac{1}{2\pi} \sum_{l=1}^{n} \frac{1}{l}$$

$$- \sum_{l=1}^{m} \frac{1}{\sqrt{[(2l-1)\pi - tx]^2 - x^2}}$$

$$- \sum_{l=1}^{\infty} \left\{ \frac{1}{\sqrt{[(2l-1)\pi + tx]^2 - x^2}} - \frac{1}{2l\pi} \right\}$$

$$- \sum_{l=n+1}^{\infty} \left\{ \frac{1}{\sqrt{[(2l-1)\pi - tx]^2 - x^2}} - \frac{1}{2l\pi} \right\}$$

MO 61

In formulas **8.525**, $x > 0, t > 1$, $(2m-1)\pi < x(t-1) < (2m+1)\pi$, $(2n-1)\pi < x(t+1) < (2n+1)\pi$, m and n are natural numbers.

8.526

1. $$\sum_{k=1}^{\infty} K_0(kx) \cos kxt = \frac{1}{2} \left(C + \ln \frac{x}{4\pi} \right) + \frac{\pi}{2x\sqrt{1+t^2}} + \frac{\pi}{2} \sum_{l=1}^{\infty} \left\{ \frac{1}{\sqrt{x^2 + (2l\pi - tx)^2}} - \frac{1}{2l\pi} \right\}$$

$$+ \frac{\pi}{2} \sum_{l=1}^{\infty} \left\{ \frac{1}{\sqrt{x^2 + (2l\pi + tx)^2}} - \frac{1}{2l\pi} \right\}$$

MO 61

2. $$\sum_{k=1}^{\infty} (-1)^k K_0(kx) \cos kxt = \frac{1}{2} \left(C + \ln \frac{x}{4\pi} \right) + \frac{\pi}{2} \sum_{l=1}^{\infty} \left\{ \frac{1}{\sqrt{x^2 + [(2l-1)\pi - xt]^2}} - \frac{1}{2l\pi} \right\}$$

$$+ \frac{\pi}{2} \sum_{l=1}^{\infty} \left\{ \frac{1}{\sqrt{x^2 + [(2l-1)\pi + xt]^2}} - \frac{1}{2l\pi} \right\}$$

$$[x > 0, \quad t \text{ real}] \qquad (\text{see also } \mathbf{8.66}) \quad \text{MO } 62$$

8.53 Expansion in products of Bessel functions

"Summation theorems"

8.530   Suppose that $r > 0, \varrho > 0, \varphi > 0$, and $R = \sqrt{r^2 + \varrho^2 - 2r\varrho\cos\varphi}$; that is, suppose that r, ϱ, and R are the sides of a triangle such that the angle between the sides r and ϱ is equal to φ. Suppose also that $\varrho <r$ and that ψ is the angle opposite the side ϱ, so that

1. $$0 < \psi < \frac{\pi}{2}, \quad e^{2i\psi} = \frac{r - \varrho e^{-i\varphi}}{r - \varrho e^{i\varphi}}$$

When these conditions are satisfied, we have the "summation theorem" for Bessel functions:

1. $$e^{i\nu\psi} Z_\nu(mR) = \sum_{k=-\infty}^{\infty} J_k(m\varrho) Z_{\nu+k}(mr) e^{ik\varphi} \qquad \text{[m is an arbitrary complex number]}$$

WA 394(6)

For $Z_\nu = J_\nu$ and ν an integer, the restriction $\varrho <r$ is superfluous.

MO 31

8.531   Special cases:

1. $$J_0(mR) = J_0(m\varrho) J_0(mr) + 2\sum_{k=1}^{\infty} J_k(m\varrho) J_k(mr) \cos k\varphi$$

WA 391(1)

2. $$H_0^{(1,2)}(mR) = J_0(m\varrho) H_0^{(1,2)}(mr) + 2\sum_{k=1}^{\infty} J_k(m\varrho) H_k^{(1,2)}(mr) \cos k\varphi$$

MO 31

3. $$J_0(z\sin\alpha) = J_0^2\left(\frac{z}{2}\right) + 2\sum_{k=1}^{\infty} J_k^2\left(\frac{z}{2}\right) \cos 2k\alpha$$

$$= \sqrt{\frac{2\pi}{z}} \sum_{k=0}^{\infty} \left(2k + \frac{1}{2}\right) \frac{(2k-1)!!}{2^k k!} J_{2k+\frac{1}{2}}(z) P_{2k}(\cos\alpha)$$

MO 31

8.532   The term "summation theorem" is also applied to the formula

1. $$\frac{Z_\nu(mR)}{R^\nu} = 2^\nu m^{-\nu} \Gamma(\nu) \sum_{k=0}^{\infty} (\nu + k) \frac{J_{\nu+k}(m\varrho)}{\varrho^\nu} \frac{Z_{\nu+k}(mr)}{r^\nu} C_k^\nu(\cos\varphi)$$

$[\nu \neq -1, -2, -3, \ldots;$ the conditions on r, ϱ, R, φ, and m are the same as in formula **8.530**; for $Z_\nu = J_\nu$ and ν an integer, formula **8.532** 1 is valid for arbitrary r, ϱ, and $\varphi]$.

WA 398(4)

8.533   Special cases:

1. $$\frac{e^{imR}}{R} = \frac{\pi i}{2\sqrt{r\varrho}} \sum_{k=0}^{\infty} (2k + 1) J_{k+\frac{1}{2}}(m\varrho) H_{k+\frac{1}{2}}^{(1)}(mr) P_k(\cos\varphi)$$

MO 31

2. $$\frac{e^{-imR}}{R} = -\frac{\pi i}{2\sqrt{r\varrho}} \sum_{k=0}^{\infty} (2k + 1) J_{k+\frac{1}{2}}(m\varrho) H_{k+\frac{1}{2}}^{(2)}(mr) P_k(\cos\varphi)$$

MO 31

8.534   A degenerate addition theorem $(r \to \infty)$:

$$e^{im\varrho \cos \varphi} = \sqrt{\frac{\pi}{2m\varrho}} \sum_{k=0}^{\infty} i^k (2k+1) J_{k+\frac{1}{2}}(m\varrho) P_k(\cos \varphi) \qquad \text{WA 401(1)}$$

$$= 2^\nu \Gamma(\nu) \sum_{k=0}^{\infty} (\nu + k) i^k (m\varrho)^{-\nu} J_{\nu+k}(m\varrho) C_k^\nu (\cos \varphi) \qquad [\nu \neq 0, -1, -2, \ldots] \qquad \text{WA 401(2)}$$

8.535   The term "product theorem" is also applied to the formula

$$Z_\nu(\lambda z) = \lambda^\nu \sum_{k=0}^{\infty} \frac{1}{k!} Z_{\nu+k}(z) \left(\frac{1-\lambda^2}{2} z \right)^k \qquad \left[|1-\lambda|^2 < 1 \right]$$

For $Z_\nu = J_\nu$, it is valid for all values of λ and z. $\qquad$ MO 32

8.536

1. $\quad \displaystyle\sum_{k=0}^{\infty} \frac{(2n+2k)(2n+k-1)!}{k!} J_{n+k}^2(z) = \frac{(2n)!}{(n!)^2} \left(\frac{z}{2} \right)^{2n} \qquad [n > 0] \qquad$ WA 47(1)

2. $\quad \displaystyle 2\sum_{k=n}^{\infty} \frac{k \, \Gamma(n+k)}{\Gamma(k-n+1)} J_k^2(z) = \frac{(2n)!}{(n!)^2} \left(\frac{z}{2} \right)^{2n} \qquad [n > 0] \qquad$ WA 47(2)

3. $\quad \displaystyle J_0^2(z) + 2\sum_{k=1}^{\infty} J_k^2(z) = 1 \qquad$ WA 41(3)

8.537

1. $\quad \displaystyle\sum_{k=-\infty}^{\infty} Z_{\nu-k}(t) J_k(z) = Z_\nu(z+t) \qquad [|z| < |t|] \qquad$ WA 158(2)

2. $\quad \displaystyle\sum_{k=-\infty}^{\infty} J_k(z) J_{n-k}(z) = J_n(2z) \qquad$ WA 41

8.538

1. $\quad \displaystyle\sum_{k=-\infty}^{\infty} (-1)^k J_{-\nu+k}(t) J_k(z) = J_{-\nu}(z+t) \qquad [|z| < |t|] \qquad$ WA 159

2. $\quad \displaystyle\sum_{k=-\infty}^{\infty} Z_{\nu+k}(t) J_k(z) = Z_\nu(t-z) \qquad [|z| < |t|] \qquad$ WA 159(5)

8.54 The zeros of Bessel functions

8.541   For arbitrary real ν, the function $J_\nu(z)$ has infinitely many real zeros. For $\nu > -1$, all its zeros are real. $\qquad$ WA 526, 530

A Bessel function $Z_\nu(z)$ has no multiple zeros except possibly the coordinate origin. $\qquad$ WA 528

8.542   All zeros of the function $Y_0(z)$ with positive real parts are real. $\qquad$ WA 531

8.543   If $-(2s+2) < \nu < -(2s+1)$, where s is a natural number or 0, then $J_\nu(z)$ has exactly $4s+2$ complex roots, two of which are purely imaginary. If $-(2s+1) < \nu < -2s$, where s is a natural number, then the function $J_\nu(z)$ has exactly $4s$ complex zeros none of which are purely imaginary. $\qquad$ WA 532

8.544 If x_ν and x'_ν are, respectively, the smallest positive zeros of the functions $J_\nu(z)$ and $J'_\nu(z)$ for $\nu > 0$, then $x_\nu > \nu$ and $x'_\nu > \nu$. Suppose also that y_ν is the smallest positive zero of the function $Y_\nu(z)$. Then, $x_\nu < y_\nu < x'_\nu$.   WA 534, 536

Suppose that $z_{\nu,m}$ (for $m = 1, 2, 3, \dots$) are the zeros of the function $z^{-\nu} J_\nu(z)$, numbered in order of the absolute value of their real parts. Here, we assume that $\nu \neq -1, -2, -3, \dots$. Then, for arbitrary z

$$J_\nu(z) = \frac{\left(\frac{z}{2}\right)^\nu}{\Gamma(\nu+1)} \prod_{m=1}^{\infty} \left(1 - \frac{z^2}{z_{\nu,m}^2}\right)$$ WA 550

8.545[8] The number of zeros of the function $z^{-\nu} J_\nu(z)$ that occur between the imaginary axis and the line on which

$$\operatorname{Re} z = \left(m + \tfrac{1}{2} \operatorname{Re} \nu + \tfrac{1}{4}\right) \pi,$$ WA 497

is exactly m.

8.546 For $\nu \geq 0$, the number of zeros of the function $K_\nu(z)$ that occur in the region $\operatorname{Re} z < 0$, $|\arg z| < \pi$ is equal to the even number closest to $\nu - \tfrac{1}{2}$.   WA 562

8.547 Large zeros of the functions $J_\nu(z) \cos \alpha - Y_\nu(z) \sin \alpha$, where ν and α are real numbers, are given by the asymptotic expansion

$$x_{\nu,m} \sim \left(m + \frac{1}{2}\nu - \frac{1}{4}\right) \pi - \alpha - \frac{4\nu^2 - 1}{8\left[\left(m + \frac{1}{2}\nu - \frac{1}{4}\right)\pi - \alpha\right]}$$
$$- \frac{\left(4\nu^2 - 1\right)\left(28\nu^2 - 31\right)}{384\left[\left(m + \frac{1}{2}\nu - \frac{1}{4}\right)\pi - \alpha\right]^3} - \cdots$$ KU 109(24), WA 558

8.548 In particular, large zeros of the function $J_0(z)$ are given by the expansion

$$x_{0,m} \sim \frac{\pi}{4}(4m-1) + \frac{1}{2\pi(4m-1)} - \frac{31}{6\pi^3(4m-1)^3} + \frac{3779}{15\pi^5(4m-1)^5} - \cdots$$ KU 109(25), WA 556

This series is suitable for calculating all (except the smallest x_{01}) zeros of the function $J_0(z)$ correctly to at least five digits.

8.549 To calculate the roots $x_{\nu,m}$ of the function $J_\nu(z)$ of smallest absolute value, we may use the identity

$$\sum_{m=1}^{\infty} \frac{1}{x_{\nu,m}^{16}} = \frac{429\nu^5 + 7640\nu^4 + 53752\nu^3 + 185430\nu^2 + 311387\nu + 202738}{2^{16}(\nu+1)^8(\nu+2)^4(\nu+3)^2(\nu+4)^2(\nu+5)(\nu+6)(\nu+7)(\nu+8)}$$ KU 112(27)a, WA 554

8.55 Struve functions

8.550 Definitions:

1. $\mathbf{H}_\nu(z) = \sum_{m=0}^{\infty} (-1)^m \frac{\left(\frac{z}{2}\right)^{2m+\nu+1}}{\Gamma\left(m + \frac{3}{2}\right)\Gamma\left(\nu + m + \frac{3}{2}\right)}$ WA 358(2)

2. $\mathbf{L}_\nu(z) = -ie^{-i\nu\frac{\pi}{2}} \mathbf{H}_\nu\left(ze^{i\frac{\pi}{2}}\right) = \sum_{m=0}^{\infty} \frac{\left(\frac{z}{2}\right)^{2m+\nu+1}}{\Gamma\left(m + \frac{3}{2}\right)\Gamma\left(\nu + m + \frac{3}{2}\right)}$ WA 360(11)

8.551 Integral representations:

1. $\mathbf{H}_\nu(z) = \frac{2\left(\frac{z}{2}\right)^\nu}{\sqrt{\pi}\,\Gamma\left(\nu + \frac{1}{2}\right)} \int_0^1 \left(1 - t^2\right)^{\nu - \frac{1}{2}} \sin zt \, dt = \frac{2\left(\frac{z}{2}\right)^\nu}{\sqrt{\pi}\,\Gamma\left(\nu + \frac{1}{2}\right)} \int_0^{\pi/2} \sin\left(z \cos \varphi\right)(\sin \varphi)^{2\nu} \, d\varphi$

$$\left[\operatorname{Re} \nu > -\tfrac{1}{2}\right]$$ WA 358(1)

2.　$\mathbf{L}_\nu(z) = \dfrac{2\left(\frac{z}{2}\right)^\nu}{\sqrt{\pi}\,\Gamma\left(\nu + \frac{1}{2}\right)} \displaystyle\int_0^{\pi/2} \sinh\left(z\cos\varphi\right)\left(\sin\varphi\right)^{2\nu} d\varphi$

$$\left[\operatorname{Re}\nu > -\tfrac{1}{2}\right] \qquad\qquad \text{WA 360(11)}$$

8.552　Special cases:

1.[6]　$\mathbf{H}_n(z) = \dfrac{1}{\pi}\displaystyle\sum_{m=0}^{\left\lfloor\frac{n-1}{2}\right\rfloor} \dfrac{\Gamma\left(m+\frac{1}{2}\right)\left(\frac{z}{2}\right)^{n-2m-1}}{\Gamma\left(n+\frac{1}{2}-m\right)} - \mathbf{E}_n(z)$　$[n = 1, 2, \ldots]$　EH II 40(66), WA 337(1)

2.[6]　$\mathbf{H}_{-n}(z) = (-1)^{n+1}\dfrac{1}{\pi}\displaystyle\sum_{m=0}^{\left\lfloor\frac{n-1}{2}\right\rfloor} \dfrac{\Gamma\left(n-m-\frac{1}{2}\right)\left(\frac{z}{2}\right)^{-n+2m+1}}{\Gamma\left(m+\frac{3}{2}\right)} - \mathbf{E}_{-n}(z)$

$$[n = 1, 2, \ldots] \qquad\qquad \text{EH II 40(67), WA 337(2)}$$

3.　$\mathbf{H}_{n+\frac{1}{2}}(z) = Y_{n+\frac{1}{2}}(z) + \dfrac{1}{\pi}\displaystyle\sum_{m=0}^{n} \dfrac{\Gamma\left(m+\frac{1}{2}\right)\left(\frac{z}{2}\right)^{-2m+n-\frac{1}{2}}}{\Gamma(n+1-m)}$

$$[n = 0, 1, \ldots] \qquad\qquad \text{EH II 39(64)}$$

4.　$\mathbf{H}_{-\left(n+\frac{1}{2}\right)}(z) = (-1)^n J_{n+\frac{1}{2}}(z)$　$[n = 0, 1, \ldots]$　EH II 39(65)

5.　$\mathbf{L}_{-\left(n+\frac{1}{2}\right)}(z) = I_{n+\frac{1}{2}}(z)$　$[n = 0, 1, \ldots]$　EH II 39(65)

6.　$\mathbf{H}_{\frac{1}{2}}(z) = \dfrac{\sqrt{2}}{\sqrt{\pi z}}\left(1 - \cos z\right)$　EH II 39, WA 364(3)

7.　$\mathbf{H}_{\frac{3}{2}}(z) = \left(\dfrac{z}{2\pi}\right)^{1/2}\left(1 + \dfrac{2}{z^2}\right) - \left(\dfrac{2}{\pi z}\right)^{1/2}\left(\sin z + \dfrac{\cos z}{z}\right)$　WA 364(3)

8.553　Functional relations:

1.　$\mathbf{H}_\nu\left(ze^{im\pi}\right) = e^{i\pi(\nu+1)m}\,\mathbf{H}_\nu(z)$　$[m = 1, 2, 3, \ldots]$　WA 362(5)

2.　$\dfrac{d}{dz}\left[z^\nu\,\mathbf{H}_\nu(z)\right] = z^\nu\,\mathbf{H}_{\nu-1}(z)$　WA 358

3.　$\dfrac{d}{dz}\left[z^{-\nu}\,\mathbf{H}_\nu(z)\right] = 2^{-\nu}\pi^{-1/2}\left[\Gamma\left(\nu+\frac{3}{2}\right)\right]^{-1} - z^{-\nu}\,\mathbf{H}_{\nu+1}(z)$　WA 359

4.　$\mathbf{H}_{\nu-1}(z) + \mathbf{H}_{\nu+1}(z) = 2\nu z^{-1}\,\mathbf{H}_\nu(z) + \pi^{-1/2}\left(\dfrac{z}{2}\right)^\nu\left[\Gamma\left(\nu+\frac{3}{2}\right)\right]^{-1}$　WA 359(5)

5.　$\mathbf{H}_{\nu-1}(z) - \mathbf{H}_{\nu+1}(z) = 2\,\mathbf{H}_\nu'(z) - \pi^{-1/2}\left(\dfrac{z}{2}\right)^\nu\left[\Gamma\left(\nu+\frac{3}{2}\right)\right]^{-1}$　WA 359(6)

8.554　Asymptotic representations:

$$\mathbf{H}_\nu(\xi) = Y_\nu(\xi) + \dfrac{1}{\pi}\sum_{m=0}^{p-1}\dfrac{\Gamma\left(m+\frac{1}{2}\right)\left(\frac{\xi}{2}\right)^{-2m+\nu-1}}{\Gamma\left(\nu+\frac{1}{2}-m\right)} + O\left(|\xi|^{\nu-2p-1}\right)$$

$$[|\arg\xi| < \pi] \qquad\qquad \text{EH II 39(63), WA 363(2)}$$

For the asymptotic representation of $Y_\nu(\xi)$, see **8.451** 2.

8.555　The differential equation for Struve functions:

$$z^2 y'' + z y' + \left(z^2 - \nu^2\right) y = \frac{1}{\sqrt{\pi}} \frac{4\left(\frac{z}{2}\right)^{\nu+1}}{\Gamma\left(\nu + \frac{1}{2}\right)}$$

WA 359(10)

8.56 Thomson functions and their generalizations

$\mathrm{ber}_\nu(z)$, $\mathrm{bei}_\nu(z)$, $\mathrm{her}_\nu(z)$, $\mathrm{hei}_\nu(z)$, $\mathrm{ker}_\nu(z)$, $\mathrm{kei}_\nu(z)$

8.561

1. $\mathrm{ber}_\nu(z) + i\,\mathrm{bei}_\nu(z) = J_\nu\left(z e^{\frac{3}{4}\pi i}\right)$ WA 96(6)

2. $\mathrm{ber}_\nu(z) - i\,\mathrm{bei}_\nu(z) = J_\nu\left(z e^{-\frac{3}{4}\pi i}\right).$ WA 96(6)

8.562

1. $\mathrm{her}_\nu(z) + i\,\mathrm{hei}_\nu(z) = H_{(1)}^\nu\left(z e^{\frac{3}{4}\pi i}\right)$ (see also **8.567**)   WA 96(7)

2. $\mathrm{her}_\nu(z) - i\,\mathrm{hei}_\nu(z) = H_{(1)}^\nu\left(z e^{-\frac{3}{4}\pi i}\right)$ (see also **8.567**)   WA 96(7)

8.563

1. $\mathrm{ber}_0(z) \equiv \mathrm{ber}(z); \quad \mathrm{bei}_0(z) \equiv \mathrm{bei}(z)$ WA 96(8)

2. $\mathrm{ker}(z) \equiv -\dfrac{\pi}{2}\,\mathrm{hei}_0(z); \quad \mathrm{kei}(z) \equiv \dfrac{\pi}{2}\,\mathrm{hei}_0(z)$ WA 96(8)

For integral representations, see **6.251**, **6.536**, **6.537**, **6.772** 4, **6.777**.

Series representation

8.564

1. $\mathrm{ber}(z) = \displaystyle\sum_{k=0}^\infty \frac{(-1)^k z^{4k}}{2^{4k}[(2k)!]^2}$ WA 96(3)

2. $\mathrm{bei}(z) = \displaystyle\sum_{k=0}^\infty \frac{(-1)^k z^{4k+2}}{2^{4k+2}[(2k+1)!]^2}$ WA 96(4)

3. $\mathrm{ker}(z) = \left(\ln\dfrac{2}{z} - C\right)\mathrm{ber}(z) + \dfrac{\pi}{4}\,\mathrm{bei}(z) + \displaystyle\sum_{k=1}^\infty (-1)^k \frac{z^{4k}}{2^{4k}[(2k)!]^2}\sum_{m=1}^{2k}\frac{1}{m}$ WA 96(9)a, DW

4. $\mathrm{kei}(z) = \left(\ln\dfrac{2}{z} - C\right)\mathrm{bei}(z) - \dfrac{\pi}{4}\,\mathrm{ber}(z) + \displaystyle\sum_{k=0}^\infty (-1)^k \frac{z^{4k+2}}{2^{4k+2}[(2k+1)!]^2}\sum_{m=1}^{2k+1}\frac{1}{m}$ WA 96(10)a, DW

8.565 $\mathrm{ber}_\nu^2(z) + \mathrm{bei}_\nu^2(z) = \displaystyle\sum_{k=0}^\infty \frac{(z/2)^{2\nu+4k}}{k!\,\Gamma(\nu+k+1)\,\Gamma(\nu+2k+1)}$ WA 163(6)

Asymptotic representation

8.566

1. $\mathrm{ber}(z) = \dfrac{e^{\alpha(z)}}{\sqrt{2\pi z}}\cos\beta(z)$ $\left[|\arg z| < \dfrac{\pi}{4}\right]$ WA 227(1)

2. $\mathrm{bei}(z) = \dfrac{e^{\alpha(z)}}{\sqrt{2\pi z}} \sin \beta(z)$ $\left[|\arg z| < \dfrac{\pi}{4} \right]$ WA 227(1)

3. $\mathrm{ker}(z) = \sqrt{\dfrac{\pi}{2z}}\, e^{\alpha(-z)} \cos \beta(-z)$ $\left[|\arg z| < \dfrac{5}{4}\pi \right]$ WA 227(2)

4. $\mathrm{kei}(z) = \sqrt{\dfrac{\pi}{2z}}\, e^{\alpha(-z)} \sin \beta(-z)$ $\left[|\arg z| < \dfrac{5}{4}\pi \right],$ WA 227(2)

where

$$\alpha(z) \sim \frac{z}{\sqrt{2}} + \frac{1}{8z\sqrt{2}} - \frac{25}{384 z^3 \sqrt{2}} - \frac{13}{128 z^4} - \cdots,$$

$$\beta(z) \sim \frac{z}{\sqrt{2}} - \frac{\pi}{8} - \frac{1}{8z\sqrt{2}} - \frac{1}{16 z^2} - \frac{25}{384 z^3 \sqrt{2}} + \cdots$$

8.567   Functional relations

1. $\mathrm{ker}(z) + i\,\mathrm{kei}(z) = K_0\left(z\sqrt{i}\right)$ (see **8.562**)   WA 96(5), DW

2. $\mathrm{ker}(z) - i\,\mathrm{kei}(z) = K_0\left(z\sqrt{-i}\right)$ (see **8.562**)   WA 96(5), DW

For integrals of Thomson's functions, see **6.87**.

8.57   Lommel functions

8.570   Definitions of the Lommel functions $s_{\mu,\nu}(z)$ and $S_{\mu,\nu}(z)$:

1. $s_{\mu,\nu}(z) = \dfrac{(-1)^m z^{\mu+1+2m}}{\left[(\mu+1)^2 - \nu^2\right]\left[(\mu+3)^2 - \nu^2\right]\dots\left[(\mu+2m+1)^2 - \nu^2\right]}$

$$= z^{\mu-1} \sum_{m=0}^{\infty} \frac{(-1)^m \left(\frac{z}{2}\right)^{2m+2} \Gamma\left(\frac{1}{2}\mu - \frac{1}{2}\nu + \frac{1}{2}\right) \Gamma\left(\frac{1}{2}\mu + \frac{1}{2}\nu + \frac{1}{2}\right)}{\Gamma\left(\frac{1}{2}\mu - \frac{1}{2}\nu + m + \frac{3}{2}\right) \Gamma\left(\frac{1}{2}\mu + \frac{1}{2}\nu + m + \frac{3}{2}\right)}$$

$\qquad\qquad\qquad$ $[\mu \pm \nu$ is not a negative odd integer$]$ EH II 40(69), WA 377(2)

2.[8] $S_{\mu,\nu}(z) = s_{\mu,\nu}(z) + \left[2^{\mu-1}\,\Gamma\left(\frac{1}{2}\mu - \frac{1}{2}\nu + \frac{1}{2}\right)\Gamma\left(\frac{1}{2}\mu + \frac{1}{2}\nu + \frac{1}{2}\right)\right]$

$$\times \frac{\cos\left[\frac{1}{2}(\mu-\nu)\pi\right] J_{-\nu}(z) - \cos\left[\frac{1}{2}(\mu+\nu)\pi\right] J_{\nu}(z)}{\sin \nu\pi}$$

$\qquad\qquad\qquad$ $[\mu \pm \nu$ is a positive odd integer, ν is not an integer$]$

$\qquad\qquad\qquad\qquad\qquad\qquad\qquad$ EH II 40(71), WA 379(2)

$$= s_{\mu,\nu}(z) + 2^{\mu-1}\,\Gamma\left(\frac{1}{2}\mu - \frac{1}{2}\nu + \frac{1}{2}\right)\Gamma\left(\frac{1}{2}\mu + \frac{1}{2}\nu + \frac{1}{2}\right)$$

$$\times \left\{ \sin\left[\frac{1}{2}(\mu-\nu)\pi\right] J_{\nu}(z) - \cos\left[\frac{1}{2}(\mu-\nu)\pi\right] Y_{\nu}(z) \right\}$$

$\qquad\qquad\qquad$ $[\mu \pm \nu$ is a positive odd integer, ν is an integer$]$

$\qquad\qquad\qquad\qquad\qquad\qquad\qquad$ EH II 41(71), WA 379(3)

Integral representations

8.571 $s_{\mu,\nu}(z) = \dfrac{\pi}{2} \left[Y_\nu(z) \displaystyle\int_0^z z^\mu \, J_\nu(z)\, dz - J_\nu(z) \int_0^z z^\mu \, Y_\nu(z)\, dz \right]$ WA 378(9)

8.572 $s_{\mu,\nu}(z)$

$$= 2^\mu \left(\frac{z}{2}\right)^{\frac{1}{2}(1+\nu+\mu)} \Gamma\left(\frac{1}{2} + \frac{1}{2}\mu - \frac{1}{2}\nu\right) \int_0^{\pi/2} J_{\frac{1}{2}(1+\mu-\nu)}(z\sin\theta)\,(\sin\theta)^{\frac{1}{2}(1+\nu-\mu)}(\cos\theta)^{\nu+\mu}\, d\theta$$

$$[\operatorname{Re}(\nu+\mu+1) > 0] \qquad \text{EH II 42(86)}$$

8.573 Special cases:

1. $S_{1,2n}(z) = z O_{2n}(z)$ WA 382(1)

2. $S_{0,2n+1}(z) = \dfrac{z}{2n+1} O_{2n+1}(z)$ WA 382(1)

3. $S_{-1,2n}(z) = \dfrac{1}{4n} S_{2n}(z)$ WA 382(2)

4. $S_{0,2n+1}(z) = \dfrac{1}{2} S_{2n+1}(z)$ WA 382(2)

5. $S_{\nu,\nu}(z) = \Gamma\left(\nu + \dfrac{1}{2}\right)\sqrt{\pi}\, 2^{\nu-1}\, \mathbf{H}_\nu(z)$ EH II 42(84)

6. $S_{\nu,\nu}(z) = [\mathbf{H}_\nu(z) - Y_\nu(z)]\, 2^{\nu-1}\sqrt{\pi}\,\Gamma\left(\nu + \tfrac{1}{2}\right)$ EH II 42(84)

8.574 Connections with other special functions:

1. $\mathbf{J}_\nu(z) = \dfrac{1}{\pi}\sin(\nu\pi)\left[s_{0,\nu}(z) - \nu\, s_{-1,\nu}(z)\right]$ EH II 41(82)

2. $\mathbf{E}_\nu(z) = -\dfrac{1}{\pi}\left[(1 + \cos\nu\pi)\, s_{0,\nu}(z) + \nu\,(1 - \cos\nu\pi)\, s_{-1,\nu}(z)\right]$ EH II 42(83)

A connection with a hypergeometric function

3. $s_{\mu,\nu}(z) = \dfrac{z^{\mu+1}}{(\mu - \nu + 1)(\mu + \nu + 1)}\; {}_1F_2\left(1; \dfrac{\mu - \nu + 3}{2}, \dfrac{\mu + \nu + 3}{2}; -\dfrac{z^2}{4}\right)$

EH II 40(69), WA 378(10)

8.575 Functional relations:

1. $s_{\mu+2,\nu}(z) = z^{\mu+1} - \left[(\mu+1)^2 - \nu^2\right] s_{\mu,\nu}(z)$ EH II 41(73), WA 380(1)

2.[8] $s'_{\mu,\nu}(z) + \left(\dfrac{\nu}{z}\right) s_{\mu,\nu}(z) = (\mu + \nu - 1)\, s_{\mu-1,\nu-1}(z)$ EH II 41(74), WA 380(2)

3. $s'_{\mu,\nu}(z) - \left(\dfrac{\nu}{z}\right) s_{\mu,\nu}(z) = (\mu - \nu - 1)\, s_{\mu-1,\nu+1}(z)$ EH II 41(75), WA 380(3)

4. $\left(2\dfrac{\nu}{z}\right) s_{\mu,\nu}(z) = (\mu + \nu - 1)\, s_{\mu-1,\nu-1}(z) - (\mu - \nu - 1)\, s_{\mu-1,\nu+1}(z)$ EH II 41(76), WA 380(4)

5.[8] $2\, s'_{\mu,\nu}(z) = (\mu + \nu - 1)\, s_{\mu-1,\nu-1}(z) + (\mu - \nu - 1)\, s_{\mu-1,\nu+1}(z)$ EH II 41(77), WA 380(5)

In formulas **8.575** 1–5, $s_{\mu,\nu}(z)$ can be replaced with $S_{\mu,\nu}(z)$.

8.576   Asymptotic expansion of $S_{\mu,\nu}(z)$. In the case in which $\mu \pm \nu$ is not a positive odd integer, the following asymptotic expansion is valid for $S_{\mu,\nu}(z)$:

$$S_{\mu,\nu}(z) = z^{\mu-1} \sum_{m=0}^{p-1} \frac{(-1)^m \, \Gamma\left(\frac{1}{2} - \frac{1}{2}\mu + \frac{1}{2}\nu + m\right) \Gamma\left(\frac{1}{2} - \frac{1}{2}\mu - \frac{1}{2}\nu + m\right)}{\left(\frac{z}{2}\right)^m \Gamma\left(\frac{1}{2} - \frac{1}{2}\mu + \frac{1}{2}\nu\right)} \frac{}{\Gamma\left(\frac{1}{2} - \frac{1}{2}\mu - \frac{1}{2}\nu\right)} + O\left(z^{\mu-2p}\right) \qquad \text{WA 385}$$

8.577   Lommel functions satisfy the following differential equation:

$$z^2 w'' + z w' + \left(z^2 - \nu^2\right) w = z^{\mu+1} \qquad \text{WA 377(1), EH II 40(68)}$$

8.578   Lommel functions of two variables $U_\nu(w,z)$ and $V_\nu(w,z)$:

Definition

1. $U_\nu(w,z) = \displaystyle\sum_{m=0}^{\infty} (-1)^m \left(\frac{w}{z}\right)^{\nu+2m} J_{\nu+2m}(z)$ EH II 42(87), WA 591(5)

2. $V_\nu(w,z) = \cos\left[\dfrac{1}{2}\left(w + \dfrac{z^2}{w} + \nu\pi\right)\right] + U_{-\nu+2}(w,z)$ EH II 42(88), WA 591(6)

      Particular values:

3. $U_0(z,z) = V_0(z,z) = \frac{1}{2}\left\{J_0(z) + \cos z\right\}$ WA 591(9)

4. $U_1(z,z) = -V_1(z,z) = \frac{1}{2}\sin z$ WA 591(10)

5. $U_{2n}(z,z) = \dfrac{(-1)^n}{2}\left\{\cos z - \displaystyle\sum_{m=0}^{n-1}(-1)^m \varepsilon_{2m} J_{2m}(z)\right\}$

$$[n \geq 1], \quad \varepsilon_m = \begin{cases} 2, & m > 0, \\ 1, & m = 0 \end{cases} \qquad \text{WA 591(11)}$$

6. $U_{2n+1}(z,z) = \dfrac{(-1)^n}{2}\left\{\sin z - \displaystyle\sum_{m=0}^{n-1}(-1)^m \varepsilon_{2m+1} J_{2m+1}(z)\right\}$

$$[n \geq 0], \quad \varepsilon_m = \begin{cases} 2, & m > 0, \\ 1, & m = 0 \end{cases} \qquad \text{WA 591(12)}$$

7. $V_n(w,z) = (-1)^n \, U_n\left(\dfrac{z^2}{w}, z\right)$

8. $U_\nu(w,0) = \dfrac{\left(\frac{w}{2}\right)^{1/2}}{\Gamma(\nu-1)} \, S_{\nu-\frac{3}{2},\frac{1}{2}}\left(\dfrac{w}{2}\right)$ WA 593(9)

9. $V_{-\nu+2}(w,0) = \dfrac{\left(\frac{w}{2}\right)^{1/2}}{\Gamma(\nu-1)} \, S_{\nu-\frac{3}{2},\frac{1}{2}}\left(\dfrac{w}{2}\right)$ WA 593(10)

8.579   Functional relations:

1. $2\dfrac{\partial}{\partial w} U_\nu(w,z) = U_{\nu-1}(w,z) + \left(\dfrac{z}{w}\right)^2 U_{\nu+1}(w,z)$ WA 593(2)

2. $2\dfrac{\partial}{\partial w} V_\nu(w,z) = V_{\nu+1}(w,z) + \left(\dfrac{z}{w}\right)^2 V_{\nu-1}(w,z)$ WA 593(4)

3.   The function $U_\nu(w,z)$ is a particular solution of the differential equation

$$\frac{\partial^2 U}{\partial z^2} - \frac{1}{z}\frac{\partial U}{\partial z} + \frac{z^2 U}{w^2} = \left(\frac{w}{z}\right)^{\nu-2} J_\nu(z) \qquad \text{WA 592(2)}$$

4.   The function $V_\nu(w, z)$ is a particular solution of the differential equation

$$\frac{\partial^2 V}{\partial z^2} - \frac{1}{z}\frac{\partial V}{\partial z} + \frac{z^2 V}{w^2} = \left(\frac{w}{z}\right)^{-\nu} J_{-\nu+2}(z) \qquad \text{WA 592(3)}$$

8.58 Anger and Weber functions $\mathbf{J}_\nu(z)$ and $\mathbf{E}_\nu(z)$

8.580   Definitions:

1.   The Anger function $\mathbf{J}_\nu(z)$:

$$\mathbf{J}_\nu(z) = \frac{1}{\pi}\int_0^\pi \cos(\nu\theta - z\sin\theta)\, d\theta \qquad \text{WA 336(1), EH II 35(32)}$$

2.   The Weber function $\mathbf{E}_\nu(z)$:

$$\mathbf{E}_\nu(z) = \frac{1}{\pi}\int_0^\pi \sin(\nu\theta - z\sin\theta)\, d\theta \qquad \text{WA 336(2), EH II 35(32)}$$

8.581   Series representations:

1. $\mathbf{J}_\nu(z) = \cos\dfrac{\nu\pi}{2}\displaystyle\sum_{n=0}^\infty \frac{(-1)^n\left(\frac{z}{2}\right)^{2n}}{\Gamma\left(n+1+\frac{1}{2}\nu\right)\Gamma\left(n+1-\frac{1}{2}\nu\right)}$

$$+\sin\frac{\nu\pi}{2}\sum_{n=0}^\infty \frac{(-1)^n\left(\frac{z}{2}\right)^{2n+1}}{\Gamma\left(n+\frac{3}{2}+\frac{1}{2}\nu\right)\Gamma\left(n+\frac{3}{2}-\frac{1}{2}\nu\right)}$$

EH II 36(36), WA 337(3)

2. $\mathbf{E}_\nu(z) = \sin\dfrac{\nu\pi}{2}\displaystyle\sum_{n=0}^\infty \frac{(-1)^n\left(\frac{z}{2}\right)^{2n}}{\Gamma\left(n+1+\frac{1}{2}\nu\right)\Gamma\left(n+1-\frac{1}{2}\nu\right)}$

$$-\cos\frac{\nu\pi}{2}\sum_{n=0}^\infty \frac{(-1)^n\left(\frac{z}{2}\right)^{2n+1}}{\Gamma\left(n+\frac{3}{2}+\frac{1}{2}\nu\right)\Gamma\left(n+\frac{3}{2}-\frac{1}{2}\nu\right)}$$

EH II 36(37), WA 338(4)

8.582   Functional relations:

1.[6] $2\,\mathbf{J}'_\nu(z) = \mathbf{J}_{\nu-1}(z) - \mathbf{J}_{\nu+1}(z)$ \qquad EH II 36(40), WA 340(2)

2.[6] $2\,\mathbf{E}'_\nu(z) = \mathbf{E}_{\nu-1}(z) - \mathbf{E}_{\nu+1}(z)$ \qquad EH II 36(41), WA 340(6)

3.[6] $\mathbf{J}_{\nu-1}(z) + \mathbf{J}_{\nu+1}(z) = 2\nu z^{-1}\,\mathbf{J}_\nu(z) - 2(\pi z)^{-1}\sin(\nu\pi)$ \qquad EH II 36(42), WA 340(1)

4.[6] $\mathbf{E}_{\nu-1}(z) + \mathbf{E}_{\nu+1}(z) = 2\nu z^{-1}\,\mathbf{E}_\nu(z) - 2(\pi z)^{-1}(1 - \cos\nu\pi)$ \qquad EH II 36(43), WA 340(5)

8.583 Asymptotic expansions:

1.[6] $\mathbf{J}_\nu(z)$

$$= J_\nu(z) + \frac{\sin \nu\pi}{\pi z} \left[\sum_{n=0}^{p-1} (-1)^n 2^{2n} \frac{\Gamma\left(n + \frac{1+\nu}{2}\right) \Gamma\left(n + \frac{1-\nu}{2}\right)}{\Gamma\left(\frac{1+\nu}{2}\right) \Gamma\left(\frac{1-\nu}{2}\right)} z^{-2n} \right.$$

$$\left. + O\left(|z|^{-2p}\right) - \nu \sum_{n=0}^{p-1} (-1)^n 2^{2n} \frac{\Gamma\left(n + 1 + \frac{1}{2}\nu\right) \Gamma\left(n + 1 - \frac{1}{2}\nu\right)}{\Gamma\left(1 + \frac{1}{2}\nu\right) \Gamma\left(1 - \frac{1}{2}\nu\right)} z^{-2n-1} + \nu\, O\left(|z|^{-2p-1}\right) \right]$$

$$[|\arg z| < \pi] \qquad \text{EH II 37(47), WA 344(1)}$$

2. $\mathbf{E}_\nu(z) = -\,Y_\nu(z)$

$$- \frac{1 + \cos(\nu\pi)}{\pi z} \left[\sum_{n=0}^{p-1} (-1)^n 2^{2n} \frac{\Gamma\left(n + \frac{1+\nu}{2}\right) \Gamma\left(n + \frac{1-\nu}{2}\right)}{\Gamma\left(\frac{1+\nu}{2}\right) \Gamma\left(\frac{1-\nu}{2}\right)} z^{-2n} + O\left(|z|^{-2p}\right) \right]$$

$$- \frac{\nu\,(1 - \cos \nu\pi)}{z\pi} \left[\sum_{n=0}^{p-1} (-1)^n 2^{2n} \frac{\Gamma\left(n + 1 + \frac{1}{2}\nu\right) \Gamma\left(n + 1 - \frac{1}{2}\nu\right)}{\Gamma\left(1 + \frac{1}{2}\nu\right) \Gamma\left(1 - \frac{1}{2}\nu\right)} z^{-2n-1} + O\left(|z|^{-2p-1}\right) \right]$$

$$\text{WA344(2), EH II 37(48)}$$

For the asymptotic expansion of $J_\nu(z)$ and $Y_\nu(z)$, see **8.451**.

8.584   The Anger and Weber functions satisfy the differential equation

$$y'' + z^{-1}y' + \left(1 - \frac{\nu^2}{z^2}\right) y = f(\nu, z),$$

where $f(\nu, z) = \dfrac{z - \nu}{\pi z^2} \sin \nu\pi$ for $\mathbf{J}_\nu(z)$ WA 341(9), EH II 37(44)

and $f(\nu, z) = -\dfrac{1}{\pi z^2} \left[z + \nu + (z - \nu) \cos \nu\pi\right]$ for $\mathbf{E}_\nu(z)$ EH II 37(45), WA 341(10)

8.59 Neumann's and Schläfli's polynomials: $O_n(z)$ and $S_n(z)$

8.590   Definition of Neumann's polynomials

1. $O_n(z) = \dfrac{1}{4} \sum_{m=0}^{\lfloor \frac{n}{2} \rfloor} \dfrac{n(n - m - 1)!}{m!} \left(\dfrac{z}{2}\right)^{2m-n-1}$ $[n \geq 1]$ WA 299(2), EH II 33(6)

2. $O_{-n}(z) = (-1)^n O_n(z)$ $[n \geq 1]$ WA 303(8)

3. $O_0(z) = \dfrac{1}{z}$ WA 299(3), EH II 33(7)

4. $O_1(z) = \dfrac{1}{z^2}$ EH II 33(7)

5. $O_2(z) = \dfrac{1}{z} + \dfrac{4}{z^3}$ EH II 33(7)

In general, $O_n(z)$ is a polynomial in z^{-1} of degree $n + 1$.

8.591   Functional relations:

1. $O_0'(z) = -\,O_1(z)$ EH II 33(9), WA 301(3)

2. $2\,O'_n(z) = O_{n-1}(z) - O_{n+1}(z)$ $[n \geq 1]$ EH II 33(10), WA 301(2)

3. $(n-1)\,O_{n+1}(z) + (n+1)\,O_{n-1}(z) - 2z^{-1}\left(n^2 - 1\right)O_n(z) = 2nz^{-1}\left(\sin n\frac{\pi}{2}\right)^2$

                                               $[n \geq 1]$ EH II 33(11), WA 301(1)

4. $nz\,O_{n-2}(z) - \left(n^2 - 1\right)O_n(z) = (n-1)z\,O'_n(z) + n\left(\sin n\frac{\pi}{2}\right)^2$ EH II 33(12), WA 303(4)

5. $nz\,O_{n+1}(z) - \left(n^2 - 1\right)O_n(z) = -(n+1)z\,O'_n(z) + n\left(\sin n\frac{\pi}{2}\right)^2$ EH II 33(13), WA 303(5)a

8.592   The generating function:

$$\frac{1}{z - \xi} = J_0(\xi)z^{-1} + 2\sum_{n=1}^{\infty} J_n(\xi)\,O_n(z) \qquad [|\xi| < |z|] \qquad\qquad \text{EH II 32(1), WA 298(1)}$$

8.593   The integral representation:

$$O_n(z) = \int_0^{\infty} \frac{\left[u + \sqrt{u^2 + z^2}\right]^n + \left[u - \sqrt{u^2 + z^2}\right]^n}{2z^{n+1}}\,e^{-u}\,du$$

See also **3.547** 6, 8, **3.549** 1, 2.   EH II 32(3), WA 305(1)

8.594   The inequality

$$|O_n(z)| \leq 2^{n-1}n!|z|^{-n-1}e^{\frac{1}{4}|z|^2} \qquad [n > 1] \qquad\qquad \text{EH II 33(8), WA 300(8)}$$

8.595   Neumann's polynomial $O_n(z)$ satisfies the differential equation

$$z^2\frac{d^2y}{dz^2} + 3z\frac{dy}{dz} + \left(z^2 + 1 - n^2\right)y = z\left(\cos n\frac{\pi}{2}\right)^2 + n\left(\sin n\frac{\pi}{2}\right)^2 \qquad \text{EH II 33(14), WA 303(1)}$$

8.596   Schläfli's polynomials $S_n(z)$. These are the functions that satisfy the formulas

1. $S_0(z) = 0$ EH II 34(18), WA 312(2)

2. $S_n(z) = \frac{1}{n}\left[2zO_n(z) - 2\left(\cos n\frac{\pi}{2}\right)^2\right]$ $[n \geq 1]$ EH II 34(19), WA 312(3)

        $= \sum_{m=0}^{\left\lfloor\frac{n}{2}\right\rfloor} \frac{(n - m - 1)!}{m!}\left(\frac{z}{2}\right)^{2m-n}$ $[n \geq 1]$ EH II 34(18)

3. $S_{-n}(z) = (-1)^{n+1}\,S_n(z)$ WA 313(6)

8.597   Functional relations:

1. $S_{n-1}(z) + S_{n+1}(z) = 4\,O_n(z)$ WA 313(7)

Other functional relations may be obtained from **8.591** by replacing $O_n(z)$ with the expression for $S_n(z)$ given by **8.596** 2.

8.6 Mathieu Functions

8.60 Mathieu's equation

$$\frac{d^2y}{dz^2} + \left(a - 2k^2\cos 2z\right)y = 0, \quad k^2 = q \qquad\qquad \text{MA}$$

8.61 Periodic Mathieu functions

8.610   In general, Mathieu's equation **8.60** does not have periodic solutions. If k is a real number, there exist infinitely many *eigenvalues a*, not identically equal to zero, corresponding to the periodic solutions

$$y(z) = y(2\pi + z),$$

If k is nonzero, there are no other linearly independent periodic solutions. Periodic solutions of Mathieu's equations are called *Mathieu's periodic functions* or *Mathieu functions of the first kind*, or, more simply, *Mathieu functions*.

8.611   Mathieu's equation has four series of distinct periodic solutions:

1. $\displaystyle \mathrm{ce}_{2n}(z, q) = \sum_{r=0}^{\infty} A_{2r}^{(2n)} \cos 2rz$ MA

2. $\displaystyle \mathrm{ce}_{2n+1}(z, q) = \sum_{r=0}^{\infty} A_{2r+1}^{(2n+1)} \cos(2r + 1)z$ MA

3. $\displaystyle \mathrm{se}_{2n+1}(z, q) = \sum_{r=0}^{\infty} B_{2r+1}^{(2n+1)} \sin(2r + 1)z$ MA

4. $\displaystyle \mathrm{se}_{2n+2}(z, q) = \sum_{r=0}^{\infty} B_{2r+2}^{(2n+2)} \sin(2r + 2)z$ MA

5.   The coefficients A and B depend on q. The eigenvalues a of the functions ce_{2n}, ce_{2n+1}, se_{2n}, se_{2n+1} are denoted by a_{2n}, a_{2n+1}, b_{2n}, b_{2n+1}.

8.612   The solutions of Mathieu's equation are normalized so that

$$\int_0^{2\pi} y^2 \, dx = \pi$$ MO 65

8.613

1. $\displaystyle \lim_{q \to 0} \mathrm{ce}_0(x) = \frac{1}{\sqrt{2}}$

2. $\displaystyle \lim_{q \to 0} \mathrm{ce}_n(x) = \cos nx$ $[n \neq 0]$

3. $\displaystyle \lim_{q \to 0} \mathrm{se}_n(x) = \sin nx$ MO 65

8.62 Recursion relations for the coefficients $A_{2r}^{(2n)}$, $A_{2r+1}^{(2n+1)}$, $B_{2r+1}^{(2n+1)}$, $B_{2r+2}^{(2n+2)}$

8.621

1. $a A_0^{(2n)} - q A_2^{(2n)} = 0$ MA

2. $(a - 4) A_2^{(2n)} - q \left(A_4^{(2n)} + 2A_0^{(2n)} \right) = 0$ MA

3. $\left(a - 4r^2 \right) A_{2r}^{(2n)} - q \left(A_{2r+2}^{(2n)} + A_{2r-2}^{(2n)} \right) = 0$ $[r \geq 2]$ MA

8.622

1. $(a - 1 - q)A_1^{(2n+1)} - qA_3^{(2n+1)} = 0$ MA

2. $\left[a - (2r+1)^2\right] A_{2r+1}^{(2n+1)} - q\left(A_{2r+3}^{(2n+1)} + A_{2r-1}^{(2n+1)}\right) = 0$

                                         $[r \geq 1]$ MA

8.623

1. $(a - 1 + q)B_1^{(2n+1)} - qB_3^{(2n+1)} = 0$ MA

2. $\left[a - (2r+1)^2\right] B_{2r+1}^{(2n+1)} - q\left(B_{2r+3}^{(2n+1)} + B_{2r-1}^{(2n+1)}\right) = 0$

                                         $[r \geq 1]$ MA

8.624

1. $(a - 4)B_2^{(2n+2)} - qB_4^{(2n+2)} = 0$ MA

2. $\left(a - 4r^2\right) B_{2r}^{(2n+2)} - q\left(B_{2r+2}^{(2n+2)} - B_{2r-2}^{(2n+2)}\right) = 0$ $[r \geq 2]$ MA

8.625 We can determine the coefficients A and B from equations **8.612**, **8.613** and **8.621-8.624** provided a is known. Suppose, for example, that we need to determine the coefficients $A_{2r}^{(2n)}$ for the function $\mathrm{ce}_{2n}(z, q)$. From the recursion formulas, we have

1.
$$
\begin{vmatrix}
a & -q & 0 & 0 & 0 & \cdots \\
-2q & a-4 & -q & 0 & 0 & \cdots \\
0 & -q & a-16 & -q & 0 & \cdots \\
0 & 0 & -q & a-36 & -q & \\
0 & 0 & 0 & -q & a-64 & \\
\vdots & \vdots & \vdots & & & \ddots
\end{vmatrix} = 0
$$
                                                             ST

For given q in equation **8.625** 1, we may determine the eigenvalues

2. $a = A_0, A_2, A_4, \ldots$ $[|a_0| \leq |a_2| \leq |a_4| \leq \ldots]$

If we now set $a = A_{2n}$, we can determine the coefficients $A_{2r}^{(2n)}$ from the recursion formulas **8.621** up to a proportionality coefficient. This coefficient is determined from the formula

3. $2\left[A_0^{(2n)}\right]^2 + \displaystyle\sum_{r=1}^{\infty} \left[A_{2r}^{(2n)}\right]^2 = 1,$ MA

which follows from the conditions of normalization.

8.63 Mathieu functions with a purely imaginary argument

8.630 If, in equation **8.60**, we replace z with iz, we arrive at the differential equation

1. $\dfrac{d^2 y}{dz^2} + (-a + 2q \cosh 2x)\, y = 0$

We can find the solutions of this equation if we replace the argument z with iz in the functions $\mathrm{ce}_n(z, q)$ and $\mathrm{se}_n(z, q)$. The functions obtained in this way are called *associated Mathieu functions of the first kind* and are denoted as follows:

1. $\mathrm{Ce}_{2n}(z, q), \quad \mathrm{Ce}_{2n+1}(z, q), \quad \mathrm{Se}_{2n+1}(z, q), \quad \mathrm{Se}_{2n+2}(z, q)$

8.631

1. $$\mathrm{Ce}_{2n}(z, q) = \sum_{r=0}^{\infty} A_{2r}^{(2n)} \cosh 2rz$$ MA

2. $$\mathrm{Ce}_{2n+1}(z, q) = \sum_{r=0}^{\infty} A_{2r+1}^{(2n+1)} \cosh(2r+1)z$$ MA

3. $$\mathrm{Se}_{2n+1}(z, q) = \sum_{r=0}^{\infty} B_{2r+1}^{(2n+1)} \sinh(2r+1)z$$ MA

4. $$\mathrm{Se}_{2n+2}(z, q) = \sum_{r=0}^{\infty} B_{2r+2}^{(2n+2)} \sinh(2r+2)z$$ MA

8.64 Non-periodic solutions of Mathieu's equation

Along with each periodic solution of equation **8.60**, there exists a second non-periodic solution that is linearly independent. The non-periodic solutions are denoted as follows:

$$\mathrm{fe}_{2n}(z, q), \quad \mathrm{fe}_{2n+1}(z, q), \quad \mathrm{ge}_{2n+1}(z, q), \quad \mathrm{ge}_{2n+2}(z, q)$$

Analogously, the second solutions of equation **8.630** 1 are denoted by

$$\mathrm{Fe}_{2n}(z, q), \quad \mathrm{Fe}_{2n+1}(z, q), \quad \mathrm{ge}_{2n+1}(z, q), \quad \mathrm{ge}_{2n+2}(z, q)$$

8.65 Mathieu functions for negative q

8.651   If we replace the argument z in equation **8.60** with $\pm\left(\dfrac{\pi}{2} \pm z\right)$, we get the equation

$$\frac{d^2 y}{dz^2} + (a + 2q \cos 2z)\, y = 0$$ MA

This equation has the following solutions:

8.652

1. $\mathrm{ce}_{2n}(z, -q) = (-1)^n \, \mathrm{ce}_{2n}\left(\frac{1}{2}\pi - z, q\right)$ MA

2. $\mathrm{ce}_{2n+1}(z, -q) = (-1)^n \, \mathrm{se}_{2n+1}\left(\frac{1}{2}\pi - z, q\right)$ MA

3. $\mathrm{se}_{2n+1}(z, -q) = (-1)^n \, \mathrm{ce}_{2n+1}\left(\frac{1}{2}\pi - z, q\right)$ MA

4. $\mathrm{se}_{2n+2}(z, -q) = (-1)^n \, \mathrm{se}_{2n+2}\left(\frac{1}{2}\pi - z, q\right)$ MA

5. $\mathrm{fe}_{2n}(z, -q) = (-1)^{n+1} \, \mathrm{fe}_{2n}\left(\frac{1}{2}\pi - z, q\right)$ MA

6. $\mathrm{fe}_{2n+1}(z, -q) = (-1)^n \, \mathrm{ge}_{2n+1}\left(\frac{1}{2}\pi - z, q\right)$ MA

7. $\mathrm{ge}_{2n+1}(z, -q) = (-1)^n \, \mathrm{fe}_{2n+1}\left(\frac{1}{2}\pi - z, q\right)$ MA

8. $\mathrm{ge}_{2n+2}(z, -q) = (-1)^n \, \mathrm{ge}_{2n+2}\left(\frac{1}{2}\pi - z, q\right)$ MA

8.653   Analogously, if we replace z with $\dfrac{\pi}{2}i + z$ in equation **8.630** 1 we get the equation

$$\frac{d^2y}{dz^2} - (a + 2q\cosh z)\,y = 0$$

It has the following solutions:

8.654

1. $\text{Ce}_{2n}(z, -q) = (-1)^n\,\text{Ce}_{2n}\left(\frac{\pi}{2}i + z, q\right)$ MA

2. $\text{Ce}_{2n+1}(z, -q) = (-1)^{n+1}i\,\text{Se}_{2n+1}\left(\frac{1}{2}\pi i + z, q\right)$ MA

3. $\text{Se}_{2n+1}(z, -q) = (-1)^{n+1}i\,\text{Ce}_{2n+1}\left(\frac{1}{2}\pi i + z, q\right)$ MA

4. $\text{Se}_{2n+2}(z, -q) = (-1)^{n+1}\,\text{Se}_{2n+2}\left(\frac{1}{2}\pi i + z, q\right)$ MA

5. $\text{Fe}_{2n}(z, -q) = (-1)^n\,\text{Fe}_{2n}\left(\frac{1}{2}\pi i + z, q\right)$ MA

6. $\text{Fe}_{2n+1}(z, -q) = (-1)^{n+1}i\,\text{ge}_{2n+1}\left(\frac{1}{2}\pi i + z, q\right)$ MA

7. $\text{ge}_{2n+1}(z, -q) = (-1)^{n+1}i\,\text{Fe}_{2n+1}\left(\frac{1}{2}\pi i + z, q\right)$ MA

8. $\text{ge}_{2n+2}(z, -q) = (-1)^{n+1}\,\text{ge}_{2n+2}\left(\frac{1}{2}\pi i + z, q\right)$ MA

8.66 Representation of Mathieu functions as series of Bessel functions

8.661

1. $\text{ce}_{2n}(z, q) = \dfrac{\text{ce}_{2n}\left(\frac{\pi}{2}, q\right)}{A_0^{(2n)}} \displaystyle\sum_{r=0}^{\infty} (-1)^r A_{2r}^{(2n)}\, J_{2r}(2k\cos z)$ MA

     $= \dfrac{\text{ce}_{2n}(0, q)}{A_0^{(2n)}} \displaystyle\sum_{r=0}^{\infty} (-1)^r A_{2r}^{(2n)}\, I_{2r}(2k\sin z)$ MA

2. $\text{ce}_{2n+1}(z, q) = -\dfrac{\text{ce}'_{2n+1}\left(\frac{\pi}{2}, q\right)}{kA_1^{(2n+1)}} \displaystyle\sum_{r=0}^{\infty} (-1)^r A_{2r+1}^{(2n+1)}\, J_{2r+1}(2k\cos z)$ MA

     $= \dfrac{\text{ce}_{2n+1}(0, q)}{kA_1(2n+1)} \cot z \displaystyle\sum_{r=0}^{\infty} (-1)^r (2r+1) A_{2r+1}^{(2n+1)}\, I_{2r+1}(2k\sin z)$ MA

3. $\text{se}_{2n+1}(z, q) = \dfrac{\text{se}_{2n+1}\left(\frac{\pi}{2}, q\right)}{kB_1^{(2n+1)}} \tan z \displaystyle\sum_{r=0}^{\infty} (-1)^r (2r+1) B_{2r+1}^{(2n+1)}\, J_{2r+1}(2k\cos z)$ MA

     $= \dfrac{\text{se}'_{2n+1}(0, q)}{kB_1^{(2n+1)}} \displaystyle\sum_{r=0}^{\infty} (-1)^r B_{2r+1}^{(2n+1)}\, I_{2r+1}(2k\sin z)$ MA

4. $\text{se}_{2n+2}(z, q) = \dfrac{-\text{se}'_{2n+2}\left(\frac{\pi}{2}, q\right)}{k^2 B_2^{(2n+2)}} \tan z \displaystyle\sum_{r=0}^{\infty} (-1)^r (2r+2) B_{2r+2}^{(2n+2)}\, J_{2r+2}(2k\cos z)$ MA

     $= \dfrac{\text{se}'_{2n+2}(0, q)}{k^2 B_2^{(2n+2)}} \cot z \displaystyle\sum_{r=0}^{\infty} (-1)^r (2r+2) B_{2r+2}^{(2n+2)}\, I_{2r+2}(2k\sin z)$ MA

8.662

1. $\quad \mathrm{fe}_{2n}(z,q) = -\dfrac{\pi\,\mathrm{fe}'_{2n}(0,q)}{2\,\mathrm{ce}_{2n}\left(\frac{\pi}{2},q\right)} \displaystyle\sum_{r=0}^{\infty} (-1)^r A_{2r}^{(2n)}\,\mathrm{Im}\left[J_r\left(ke^{iz}\right) Y_r\left(ke^{-iz}\right)\right]$ MA

2. $\quad \mathrm{fe}_{2n+1}(z,q) = \dfrac{\pi k\,\mathrm{fe}'_{2n+1}(0,q)}{2\,\mathrm{ce}'_{2n+1}\left(\frac{\pi}{2},q\right)}$

$\qquad\qquad\qquad \times \displaystyle\sum_{r=0}^{\infty} (-1)^r A_{2r+1}^{(2n+1)}\,\mathrm{Im}\left[J_r\left(ke^{iz}\right) Y_{r+1}\left(ke^{-iz}\right) + J_{r+1}\left(ke^{iz}\right) Y_r\left(ke^{-iz}\right)\right]$

    MA

3. $\quad \mathrm{ge}_{2n+1}(z,q) = -\dfrac{\pi k\,\mathrm{ge}_{2n+1}(0,q)}{2\,\mathrm{se}_{2n+1}\left(\frac{\pi}{2},q\right)}$

$\qquad\qquad\qquad \times \displaystyle\sum_{r=0}^{\infty} (-1)^r B_{2r+1}^{(2n+1)}\,\mathrm{Re}\left[J_r\left(ke^{iz}\right) Y_{r+1}\left(ke^{-iz}\right) - J_{r+1}\left(ke^{iz}\right) Y_r\left(ke^{-iz}\right)\right]$

    MA

4. $\quad \mathrm{ge}_{2n+2}(z,q) = -\dfrac{\pi k^2\,\mathrm{ge}_{2n+2}(0,q)}{2\,\mathrm{se}'_{2n+2}\left(\frac{1}{2}\pi,q\right)}$

$\qquad\qquad\qquad \times \displaystyle\sum_{r=0}^{\infty} (-1)^r\,\mathrm{Re}\left[J_k\left(ke^{iz}\right) Y_{r+2}\left(ke^{-iz}\right) - J_{r+2}\left(ke^{iz}\right) Y_r\left(ke^{-iz}\right)\right]$

    MA

The expansions of the functions Fe_n and Ge_n as series of the functions Y_ν are denoted, respectively, by Fey_n and Gey_n and the expansions of these functions as series of the functions K_ν are denoted, respectively, by Fek_n and Gek_n.

8.663

1. $\quad \mathrm{Fey}_{2n}(z,q) = \dfrac{\mathrm{ce}_{2n}(0,q)}{A_0^{(2n)}} \displaystyle\sum_{r=0}^{\infty} A_{2r}^{(2n)}\,Y_{2r}(2k\sinh z)$

$\qquad\qquad\qquad\qquad\qquad\qquad\qquad\qquad\qquad\qquad k^2 = q\,[|\sinh z| > 1, \quad \mathrm{Re}\,z > 0]$

    MA

$\qquad\qquad = \dfrac{\mathrm{ce}_{2n}\left(\frac{\pi}{2},q\right)}{A_0^{(2n)}} \displaystyle\sum_{r=0}^{\infty} (-1)^r A_{2r}^{(2n)}\,Y_{2r}(2k\cosh z)$

$\qquad\qquad\qquad\qquad\qquad\qquad\qquad\qquad\qquad\qquad\qquad\qquad [|\cosh z| > 1]$

    MA

$\qquad\qquad = \dfrac{\mathrm{ce}_{2n}(0,q)\,\mathrm{ce}_{2n}\left(\frac{\pi}{2},q\right)}{\left[A_0^{(2n)}\right]^2} \displaystyle\sum_{r=0}^{\infty} (-1)^r A_{2r}^{(2n)}\,J_r\left(ke^{-z}\right) Y_r\left(ke^z\right)$

    MA

2. $\mathrm{Fey}_{2n+1}(z, q) = \dfrac{\mathrm{ce}_{2n+1}(0, q)\coth z}{kA_1(2n+1)} \displaystyle\sum_{r=0}^{\infty} (2r+1)A_{2r+1}^{(2n+1)}\, Y_{2r+1}(2k\sinh z),$

$$k^2 = q, \quad [|\sinh z| > 1, \quad \mathrm{Re}\, z > 0]$$
MA

$$= -\dfrac{\mathrm{ce}_{2n+1}'\left(\frac{\pi}{2}, q\right)}{kA_1^{(2n+1)}} \sum_{r=0}^{\infty} (-1)^r A_{2r+1}^{(2n+1)}\, Y_{2r+1}(2k\cosh z)$$

$$[|\cosh z| > 1]$$
MA

$$= -\dfrac{\mathrm{ce}_{2n+1}(0, q)\,\mathrm{ce}_{2n+1}'\left(\frac{\pi}{2}, q\right)}{k\left[A_1^{(2n+1)}\right]^2}$$

$$\times \sum_{r=0}^{\infty} (-1)^r A_{2r+1}^{(2n+1)} \left[J_r\left(ke^{-z}\right) Y_{r+1}\left(ke^z\right) + J_{r+1}\left(ke^{-z}\right) Y_r\left(ke^z\right)\right]$$
MA

3. $\mathrm{Gey}_{2n+1}(z, q) = \dfrac{\mathrm{se}_{2n+1}'(0, q)}{kB_1^{(2n+1)}} \displaystyle\sum_{r=0}^{\infty} B_{2r+1}^{(2n+1)}\, Y_{2r+1}(2k\sinh z)$

$$[|\sinh z| > 1, \quad \mathrm{Re}\, z > 0]$$
MA

$$= \dfrac{\mathrm{se}_{2n+1}\left(\frac{\pi}{2}, q\right)}{kB_1^{(2n+1)}}\tanh z \sum_{r=0}^{\infty} (-1)^r (2r+1)B_{2r+1}^{(2n+1)}\, Y_{2r+1}(2k\cosh z)$$

$$[|\cosh z| > 1]$$
MA

$$= \dfrac{\mathrm{se}_{2n+1}(0, q)\,\mathrm{se}_{2n+1}\left(\frac{\pi}{2}, q\right)}{k\left[B_1^{(2n+1)}\right]^2} \sum_{r=0}^{\infty} (-1)^r B_{2r+1}^{(2n+1)}$$

$$\times \left[J_r\left(ke^{-z}\right) Y_{r+1}\left(ke^z\right)\right] J_{r+1}\left(ke^{-z}\right) Y_r\left(ke^z\right)$$
MA

4. $$\text{Gey}_{2n+2}(z, q) = \frac{\text{se}'_{2n+2}(0, q)}{k^2 B_2^{(2n+2)}} \coth z \sum_{r=0}^{\infty} (2r+2) B_{2r+2}^{(2n+2)} Y_{2r+2}(2k \sinh z)$$

$$[|\sinh z| > 1, \quad \text{Re } z > 0]$$
MA

$$= -\frac{\text{se}'_{2n+2}\left(\frac{\pi}{2}, q\right)}{k^2 B_2^{(2n+2)}} \tanh z \sum_{r=0}^{\infty} (-1)^r (2r+2) B_{2r+2}^{(2n+2)} Y_{2r+2}(2k \cosh z)$$

$$[|\cosh z| > 1]$$
MA

$$= \frac{\text{se}'_{2n+2}(0, q) \, \text{se}'_{2n+2}\left(\frac{\pi}{2}, q\right)}{k^2 \left[B_2^{(2n+2)}\right]^2} \sum_{r=0}^{\infty} (-1)^r B_{2r+2}^{(2n+2)}$$

$$\times \left[J_r\left(ke^{-z}\right) Y_{r+2}\left(ke^z\right)\right] - J_{r+2}\left(ke^{-z}\right) Y_r\left(ke^z\right)$$

MA

8.664

1. $$\text{Fek}_{2n}(z, q) = \frac{\text{ce}_{2n}(0, q)}{\pi A_0^{(2n)}} \sum_{r=0}^{\infty} (-1)^r A_{2r}^{(2n)} K_{2r}(-2ik \sinh z)$$

$$k^2 = q, \qquad [|\sinh z| > 1, \quad \text{Re } z > 0]$$
MA

2. $$\text{Fek}_{2n+1}(z, q) = \frac{\text{ce}_{2n+1}(0, q)}{\pi k A_1^{(2n+1)}} \coth z \sum_{r=0}^{\infty} (-1)^r (2r+1) A_{2r+1}^{(2n+1)} K_{2r+1}(-2ik \sinh z)$$

$$k^2 = q \qquad [|\sinh z| > 1, \quad \text{Re } z > 0]$$
MA

3. $$\text{Gek}_{2n+1}(z, q) = \frac{\text{se}_{2n+1}\left(\frac{\pi}{2}, q\right)}{\pi k B_1^{(2n+1)}} \tanh z \sum_{r=0}^{\infty} (2r+1) B_{2r+1}^{(2n+1)} K_{2r+1}(-2ik \cosh z)$$
MA

4. $$\text{Gek}_{2n+2}(z, q) = \frac{\text{se}'_{2n+2}\left(\frac{\pi}{2}, q\right)}{\pi k^2 B_2^{(2n+2)}} \tanh z \sum_{r=0}^{\infty} (2r+2) B_{2r+2}^{(2n+2)} K_{2r+2}(-2ik \cosh z)$$
MA

8.67 The general theory

If $i\mu$ is not an integer, the general solution of equation **8.60** can be found in the form
8.671

1. $$y = Ae^{\mu z} \sum_{r=-\infty}^{\infty} c_{2r} e^{2rzi} + Be^{-\mu z} \sum_{r=-\infty}^{\infty} c_{2r} e^{-2rzi}$$
MA

The coefficients c_{2r} can be determined from the homogeneous system of linear algebraic equations

2. $$c_{2r} + \xi_{2r}(c_{2r+2} + c_{2r-2}) = 0, r = \ldots, -2, -1, 0, 1, 2, 2 \ldots,$$
MA

where

$$\xi_{2r} = \frac{q}{(2r - i\mu)^2 - a}$$

The condition that this system be compatible yields an equation that μ must satisfy:

$$
3.^7 \quad \Delta(i\mu) = \begin{vmatrix} \cdot & \cdot & \cdot & \cdot & \cdot & \cdot & \cdot & \cdot & \cdot \\ \cdot & \xi_{-4} & 1 & \xi_{-4} & 0 & 0 & 0 & 0 & \cdot \\ \cdot & 0 & \xi_{-2} & 1 & \xi_{-2} & 0 & 0 & 0 & \cdot \\ \cdot & 0 & 0 & \xi_0 & 1 & \xi_0 & 0 & 0 & \cdot \\ \cdot & 0 & 0 & 0 & \xi_2 & 1 & \xi_2 & 0 & \cdot \\ \cdot & \cdot & \cdot & \cdot & \cdot & \cdot & \cdot & \cdot & \cdot \end{vmatrix} = 0 \qquad\qquad \text{MA}
$$

This equation can also be written in the form

4. $\cosh \mu\pi = 1 - 2\Delta(0)\sin^2\left(\dfrac{\pi\sqrt{a}}{2}\right)$, where $\Delta(0)$ is the value that is assumed by the determinant of the preceding article if we set $\mu = 0$ in the expressions for ξ_{2r}.

5.   If the pair (a, q) is such that $|\cosh \mu\pi| < 1$, then $\mu = i\beta, \operatorname{Im}\beta = 0$, and the solution **8.671** 1 is bounded on the real axis.

6.   If $|\cosh \mu\pi| > 1$, μ may be real or complex and the solution **8.671** 1 will not be bounded on the real axis.

7.   If $\cosh \mu\pi = \pm 1$, then $i\mu$ will be an integer. In this case, one of the solutions will be of period π or 2π (depending on whether n is even or odd). The second solution is non-periodic (see **8.61** and **8.64**).

8.7–8.8 Associated Legendre Functions

8.70 Introduction

8.700   An *associated Legendre function* is a solution of the differential equation

1. $$\left(1 - z^2\right)\frac{d^2 u}{dz^2} - 2z\frac{du}{dz} + \left[\nu(\nu + 1) - \frac{\mu^2}{1 - z^2}\right]u = 0,$$

in which ν and μ are arbitrary complex constants.

This equation is a special case of (Riemann's) hypergeometric equation (see **9.151**). The points

$$+1, -1, \infty$$

are, in general, its *singular points*, specifically, its ordinary branch points.

We are interested, on the one hand, in solutions of the equation that correspond to real values of the independent variable z that lie in the interval $[-1, 1]$ and, on the other hand, in solutions corresponding to an arbitrary complex number z such that $\operatorname{Re} z > 1$. These are multiple-valued in the z-plane. To separate these functions into single-valued branches, we make a cut along the real axis from $-\infty$ to $+1$. We are also interested in those solutions of equation **8.700** 1 for which ν or μ or both are integers. Of special significance is the case in which $\mu = 0$.

8.701   In connection with this, we shall use the following notations:

The letter z will denote *an arbitrary complex variable*; the letter x will denote a *real* variable that varies over the interval $[-1, +1]$. We shall sometimes set $x = \cos \varphi$, where φ is a real number.

We shall use the symbols $P_\nu^\mu(z)$, $Q_\nu^\mu(z)$ to denote those solutions of equation **8.700** 1, that are single-valued and regular for $|z| < 1$ and, in particular, uniquely determined for $z = x$.

We shall use the symbols $P_\nu^\mu(z)$, $Q_\nu^\mu(z)$ to denote those solutions of equation **8.700** 1 that are single-valued and regular for Re $z > 1$. When these functions cannot be unrestrictedly extended without violating their single-valuedness we make a cut along the real axis to the left of the point $z = 1$. The values of the functions $P_\nu^\mu(z)$ and $Q_\nu^\mu(z)$ on the upper and lower boundaries of that portion of the cuts lying between the points -1 and $+1$ are denoted respectively by

$$P_\nu^\mu (x \pm i0), \quad Q_\nu^\mu (x \pm i0)$$

The letters n and m denote natural numbers or zero. The letters ν and μ denote arbitrary complex numbers unless the contrary is stated.

The upper index will be omitted when it is equal to zero. That is, we set

$$P_\nu^0(z) = P_\nu(z), \quad Q_\nu^0(z) = Q_\nu(z)$$

The *linearly independent* functions

8.702 $P_\nu^\mu(z) = \dfrac{1}{\Gamma(1-\mu)} \left(\dfrac{z+1}{z-1} \right)^{\frac{\mu}{2}} F \left(-\nu, \nu+1; \quad 1-\mu; \quad \dfrac{1-z}{2} \right)$

$$\left[\arg \frac{z+1}{z-1} = 0, \text{ if } z \text{ is real and greater than 1 and} \right] \qquad \text{MO 80, WH}$$

8.703 $Q_\nu^\mu(z) = \dfrac{e^{\mu \pi i} \Gamma(\nu+\mu+1) \Gamma\left(\frac{1}{2}\right)}{2^{\nu+1} \Gamma\left(\nu+\frac{3}{2}\right)} \left(z^2 - 1 \right)^{\frac{\mu}{2}} z^{-\nu-\mu-1} F \left(\dfrac{\nu+\mu+2}{2}, \dfrac{\nu+\mu+1}{2}; \nu + \dfrac{3}{2}; \dfrac{1}{z^2} \right)$

$[\arg\left(z^2 - 1\right) = 0$ when z is real and greater than 1; $\arg z = 0$ when z is real and greater than zero] which are solutions of the differential equation **8.700** 1, are called *associated Legendre functions* (or *spherical functions*) of *the first* and *second kinds* respectively. They are uniquely defined, respectively, in the intervals $|1 - z| < 2$ and $|z| > 1$ with the portion of the real axis that lies between $-\infty$ and $+1$ excluded. They can be extended by means of hypergeometric series to the entire z-plane where the above-mentioned cut was made. These expressions for $P_\nu^\mu(z)$ and $Q_\nu^\mu(z)$ lose their meaning when $1 - \mu$ and $\nu + \frac{3}{2}$ are non-positive integers respectively.   MO 80

When z is a real number lying on the interval $[-1, +1]$, so that $(z = x = \cos \varphi)$, we take the following functions as linearly independent solutions of the equation

8.704 $P_\nu^\mu(x) = \frac{1}{2} \left[e^{\frac{1}{2}\mu\pi i} P_\nu^\mu (\cos \varphi + i0) + e^{-\frac{1}{2}\mu\pi i} P_\nu^\mu (\cos \varphi - i0) \right]$ EH I 143(1)

$$= \frac{1}{\Gamma(1-\mu)} \left(\frac{1+x}{1-x} \right)^{\frac{\mu}{2}} F \left(-\nu, \nu+1; 1-\mu; \frac{1-x}{2} \right) \qquad \text{EH I 143(6)}$$

8.705 $Q_\nu^\mu(x) = \frac{1}{2} e^{-\mu\pi i} \left[e^{-\frac{1}{2}\mu\pi i} Q_\nu^\mu (x+i0) + e^{\frac{1}{2}\mu\pi i} Q_\nu^\mu (x-i0) \right]$ EH I 143(2)

$$= \frac{\pi}{2 \sin \mu\pi} \left[P_\nu^\mu(x) \cos \mu\pi - \frac{\Gamma(\nu+\mu+1)}{\Gamma(\nu-\mu+1)} P_\nu^{-\mu}(x) \right] \qquad (\text{cf. } \mathbf{8.732}\ 5)$$

If $\mu = \pm m$ is an integer, the last equation loses its meaning. In this case, we get the following formulas by passing to the limit:

8.706

1. $Q_\nu^m(x) = (-1)^m \left(1 - x^2 \right)^{\frac{m}{2}} \dfrac{d^m}{dx^m} Q_\nu(x)$ $(\text{cf. } \mathbf{8.752}\ 1)$ EH I 149(7)

2. $Q_\nu^{-m}(x) = (-1)^m \dfrac{\Gamma(\nu - m + 1)}{\Gamma(\nu + m + 1)} Q_\nu^m(x)$ EH I 144(18)

The functions $Q_\nu^\mu(z)$ are not defined when $\nu + \mu$ is equal to a negative integer. Therefore, we must exclude the cases when $\nu + \mu = -1, -2, -3, \ldots$ for these formulas.

The functions

$$P_\nu^{\pm\mu}(\pm z), \quad Q_\nu^{\pm\mu}(\pm z), \quad P_{-\nu-1}^{\pm\mu}(\pm z), \quad Q_{-\nu-1}^{\pm\mu}(\pm z)$$

are *linearly independent solutions* of the differential equation for $\nu + \mu \neq 0, \pm 1, \pm 2, \ldots$.

8.707   Nonetheless, two linearly independent solutions can always be found. Specifically, for $\nu \pm \mu$ not an integer, the differential equation **8.700** 1 has the following solutions:

1. $P_\nu^{\pm\mu}(\pm z), \quad Q_\nu^{\pm\mu}(\pm z), \quad P_{-\nu-1}^{\pm\mu}(\pm z), \quad Q_{-\nu-1}^{\pm\mu}(\pm z)$

respectively, for $z = x = \cos\varphi$,

2. $P_\nu^{\pm\mu}(\pm x), \quad Q_\nu^{\pm\mu}(\pm x), \quad P_{-\nu-1}^{\pm\mu}(\pm x), \quad Q_{-\nu-1}^{\pm\mu}(\pm x)$

If $\nu \pm \mu$ is not an integer, the solutions

3. $P_\nu^\mu(z), \quad Q_\nu^\mu(z)$, respectively, and $P_\nu^\mu(x), \quad Q_\nu^\mu(x)$

are linearly independent. If $\nu \pm \mu$ is an integer but μ itself is not an integer, the following functions are linearly independent solutions of equation **8.700** 1:

4. $P_\nu^\mu(z), \quad P_\nu^{-\mu}(z)$, respectively, and $P_\nu^\mu(x), \quad P_\nu^{-\mu}(x)$

If $\mu = \pm m, \nu = n$, or $\nu = -n - 1$, the following functions are linearly independent solutions of equation **8.700** 1 for $n \geq$ m:

5. $P_n^m(z), \quad Q_n^m(z)$, respectively, and $P_n^m(x), \quad Q_n^m(x)$,

and for $n <$ m, the following functions will be linearly independent solutions

6. $P_n^{-m}(z), \quad Q_n^m(z)$, respectively, and $P_n^{-m}(x), \quad Q_n^m(x)$

8.71 Integral representations

8.711

1. $P_\nu^{-\mu}(z) = \dfrac{(z^2 - 1)^{\frac{\mu}{2}}}{2^\mu \sqrt{\pi}\, \Gamma\left(\mu + \frac{1}{2}\right)} \displaystyle\int_{-1}^1 \dfrac{\left(1 - t^2\right)^{\mu - \frac{1}{2}}}{\left(z + t\sqrt{z^2 - 1}\right)^{\mu - \nu}}\, dt \qquad \left[\operatorname{Re}\mu > -\frac{1}{2}, \quad |\arg(z \pm 1)| < \pi\right]$

MO 88

2. $P_\nu^m(z) = \dfrac{(\nu + 1)(\nu + 2)\ldots(\nu + m)}{\pi} \displaystyle\int_0^\pi \left[z + \sqrt{z^2 - 1}\cos\varphi\right]^\nu \cos m\varphi\, d\varphi$

   $= (-1)^m \dfrac{\nu(\nu - 1)\ldots(\nu - m + 1)}{\pi} \displaystyle\int_0^\pi \dfrac{\cos m\varphi\, d\varphi}{\left[z + \sqrt{z^2 - 1}\cos\varphi\right]^{\nu+1}}$

   $\left[|\arg z| < \dfrac{\pi}{2}, \quad \arg\left(z + \sqrt{z^2 - 1}\cos\varphi\right) = \arg z \text{ for } \varphi = \dfrac{\pi}{2}\right]$ (cf. **8.822** 1)   SM 483(15), WH

3. $Q_\nu^\mu(z) = \sqrt{\pi}\, \dfrac{e^{\mu\pi i}\,\Gamma(\nu + \mu + 1)}{2^\mu\,\Gamma\left(\mu + \frac{1}{2}\right)\Gamma(\nu - \mu + 1)} (z^2 - 1)^{\frac{\mu}{2}} \displaystyle\int_0^\infty \dfrac{\sinh^{2\mu} t\, dt}{\left(z + \sqrt{z^2 - 1}\cosh t\right)^{\nu+\mu+1}}$

   $[\operatorname{Re}(\nu \pm \mu) > -1, \quad |\arg(z \pm 1)| < \pi]$ (cf. **8.822** 2)   MO 88

4. $\quad Q_\nu^\mu(z) = \dfrac{e^{\mu\pi i}\,\Gamma(\nu+1)}{\Gamma(\nu-\mu+1)} \displaystyle\int_0^\infty \dfrac{\cosh\mu t\, dt}{\left(z+\sqrt{z^2-1}\cosh t\right)^{\nu+1}}$

$$[\operatorname{Re}(\nu+\mu) > -1, \nu \neq -1,-2,-3,\ldots, \quad |\arg(z\pm 1)| < \pi] \quad \text{WH, MO 88}$$

5. $\quad \displaystyle\int_{-1}^1 P_l^2(x)\,P_l^0(x)\,dx = -\dfrac{l!}{(l-2)!}\dfrac{1}{2l+1} = -\dfrac{l(l-1)}{2l+1}$

8.712 $\quad Q_\nu^\mu(z) = \dfrac{e^{\mu\pi i}\,\Gamma(\nu+\mu+1)}{2^{\nu+1}\,\Gamma(\nu+1)}\left(z^2-1\right)^{-\frac{\mu}{2}}\displaystyle\int_{-1}^1 \left(1-t^2\right)^\nu (z-t)^{-\nu-\mu-1}\,dt$

$$[\operatorname{Re}(\nu+\mu)>-1, \quad \operatorname{Re}\mu>-1, \quad |\arg(z\pm 1)|<\pi] \qquad \text{(cf. \textbf{8.821} 2)} \quad \text{MO 88a, EH I 155(5)a}$$

8.713

1. $\quad Q_\nu^\mu(z) = \dfrac{e^{\mu\pi i}\,\Gamma\left(\mu+\dfrac{1}{2}\right)}{\sqrt{2\pi}}\left(z^2-1\right)^{\frac{\mu}{2}}\left\{\displaystyle\int_0^\pi \dfrac{\cos\left(\nu+\frac{1}{2}\right)t\,dt}{(z-\cos t)^{\mu+\frac{1}{2}}} - \cos\nu\pi\displaystyle\int_0^\infty \dfrac{e^{-\left(\nu+\frac{1}{2}\right)t}\,dt}{(z+\cosh t)^{\mu+\frac{1}{2}}}\right\}$

$$\left[\operatorname{Re}\mu>-\tfrac{1}{2}, \quad \operatorname{Re}(\nu+\mu)>-1, \quad |\arg(z\pm 1)|<\pi\right] \quad \text{MO 89}$$

2. $\quad P_\nu^{-\mu}(z) = \dfrac{\left(z^2-1\right)^{\frac{\mu}{2}}}{2^\nu\,\Gamma(\mu-\nu)\,\Gamma(\nu+1)}\displaystyle\int_0^\infty \dfrac{\sinh^{2\nu+1}t}{(z+\cosh t)^{\nu+\mu+1}}\,dt$

$$[\operatorname{Re}z>-1, \quad |\arg(z\pm 1)|<\pi, \quad \operatorname{Re}(\nu+1)>0, \quad \operatorname{Re}(\mu-\nu)>0] \quad \text{MO 89}$$

3. $\quad P_\nu^{-\mu}(z) = \sqrt{\dfrac{2}{\pi}}\dfrac{\Gamma\left(\mu+\frac{1}{2}\right)\left(z^2-1\right)^{\frac{\mu}{2}}}{\Gamma(\nu+\mu+1)\,\Gamma(\mu-\nu)}\displaystyle\int_0^\infty \dfrac{\cosh\left(\nu+\frac{1}{2}\right)t\,dt}{(z+\cosh t)^{\mu+\frac{1}{2}}}$

$$[\operatorname{Re}z>-1, \quad |\arg(z\pm 1)|<\pi, \quad \operatorname{Re}(\nu+\mu)>-1, \quad \operatorname{Re}(\mu-\nu)>0] \quad \text{MO 89}$$

8.714

1. $\quad P_\nu^\mu(\cos\varphi) = \sqrt{\dfrac{2}{\pi}}\dfrac{\sin^\mu\varphi}{\Gamma\left(\frac{1}{2}-\mu\right)}\displaystyle\int_0^\varphi \dfrac{\cos\left(\nu+\frac{1}{2}\right)t\,dt}{(\cos t-\cos\varphi)^{\mu+\frac{1}{2}}} \qquad \left[0<\varphi<\pi, \quad \operatorname{Re}\mu<\tfrac{1}{2}\right]; \quad \text{(cf. \textbf{8.823})}$

$$\text{MO 87}$$

2. $\quad P_\nu^{-\mu}(\cos\varphi) = \dfrac{\Gamma(2\mu+1)\sin^\mu\varphi}{2^\mu\,\Gamma(\mu+1)\,\Gamma(\nu+\mu+1)\,\Gamma(\mu-\nu)}\displaystyle\int_0^\infty \dfrac{t^{\nu+\mu}\,dt}{(1+2t\cos\varphi+t^2)^{\mu+\frac{1}{2}}}$

$$[\operatorname{Re}(\nu+\mu)>-1, \quad \operatorname{Re}(\mu-\nu)>0] \quad \text{MO 89}$$

3. $\quad Q_\nu^\mu(\cos\varphi) = \dfrac{1}{2^{\mu+1}}\dfrac{\Gamma(\nu+\mu+1)}{\Gamma(\nu-\mu+1)}\dfrac{\sin^\mu\varphi}{\Gamma\left(\mu+\frac{1}{2}\right)}$

$$\times\displaystyle\int_0^\infty \left[\dfrac{\sinh^{2\mu}t}{(\cos\varphi+i\sin\varphi\cosh t)^{\nu+\mu+1}} + \dfrac{\sinh^{2\mu}t}{(\cos\varphi-i\sin\varphi\cosh t)^{\nu+\mu+1}}\right]dt$$

$$\left[\operatorname{Re}(\nu+\mu+1)>0, \quad \operatorname{Re}(\nu-\mu+1)>0, \quad \operatorname{Re}\mu>-\tfrac{1}{2}\right] \quad \text{MO 89}$$

4. $$P_\nu^\mu (\cos \varphi) = \frac{i}{2^\mu} \frac{\Gamma(\nu + \mu + 1)}{\Gamma(\nu - \mu + 1)} \frac{\sin^\mu \varphi}{\Gamma\left(\mu + \frac{1}{2}\right)}$$

$$\times \int_0^\infty \left[\frac{\sinh^{2\mu} t}{(\cos \varphi + i \sin \varphi \cosh t)^{\nu + \mu + 1}} - \frac{\sinh^{2\mu} t}{(\cos \varphi - i \sin \varphi \cosh t)^{\nu + \mu + 1}} \right] dt$$

$$\left[\operatorname{Re}(\nu \pm \mu + 1) > 0, \quad \operatorname{Re}\mu > -\tfrac{1}{2} \right] \quad \text{MO 89}$$

8.715

1. $$P_\nu^\mu (\cosh \alpha) = \frac{\sqrt{2} \sinh^\mu \alpha}{\sqrt{\pi}\, \Gamma\left(\frac{1}{2} - \mu\right)} \int_0^\alpha \frac{\cosh\left(\nu + \frac{1}{2}\right) t\, dt}{(\cosh \alpha - \cosh t)^{\mu + \frac{1}{2}}}$$

$$\left[\alpha > 0, \quad \operatorname{Re}\mu < \tfrac{1}{2} \right] \qquad \text{MO 87}$$

2. $$Q_\nu^\mu (\cosh \alpha) = \sqrt{\frac{\pi}{2}} \frac{e^{\mu \pi i} \sinh^\mu \alpha}{\Gamma\left(\frac{1}{2} - \mu\right)} \int_\alpha^\infty \frac{e^{-\left(\nu + \frac{1}{2}\right)t}\, dt}{(\cosh t - \cosh \alpha)^{\mu + \frac{1}{2}}}$$

$$\left[\alpha > 0, \quad \operatorname{Re}\mu < \tfrac{1}{2}, \quad \operatorname{Re}(\nu + \mu) > -1 \right]$$
$$\text{MO 87}$$

See also **3.277** 1, 4, 5, 7, **3.318**, **3.516** 3, **3.518** 1, 2, **3.542** 2, **3.663** 1, **3.894**, **3.988** 3, **6.622** 3, **6.628** 1, 4–7, and also **8.742**.

8.72 Asymptotic series for large values of $|\nu|$

8.721[6]   For real values of μ, $|\nu| \gg 1$, $|\nu| \gg |\mu|$, $|\arg \nu| < \pi$, we have:

1. $$P_\nu^\mu (\cos \varphi) = \frac{2}{\sqrt{\pi}} \Gamma(\nu + \mu + 1) \sum_{k=0}^\infty \frac{\Gamma\left(\mu + k + \frac{1}{2}\right)}{\Gamma\left(\mu - k + \frac{1}{2}\right)} \frac{\cos\left[\left(\nu + k + \frac{1}{2}\right)\varphi + \frac{\pi}{4}(2k - 1) + \frac{\mu \pi}{2}\right]}{k!\, \Gamma\left(\nu + k + \frac{3}{2}\right) (2 \sin \varphi)^{k + \frac{1}{2}}}$$

$$\left[\nu + \mu \neq -1, -2, -3, \ldots; \quad \nu \neq -\tfrac{3}{2}, -\tfrac{5}{2}, \tfrac{7}{2} \ldots; \quad \text{for } \frac{\pi}{6} < \varphi < \frac{5\pi}{6} \right.$$

This series also converges for complex values of ν and μ.
In the remaining cases, it is an asymptotic expansion for

$$\left. |\nu| \gg |\mu|, |\nu| \gg 1, \text{if } \nu > 0, \mu > 0 \text{ and } 0 < \varepsilon \leq \varphi \leq \pi - \varepsilon \right]$$

$$\text{MO 92}$$

2.[6] $$Q_\nu^\mu (\cos \varphi) = \sqrt{\pi}\, \Gamma(\nu + \mu + 1)$$

$$\times \sum_{k=0}^\infty (-1)^k \frac{\Gamma\left(\mu + k + \frac{1}{2}\right)}{\Gamma\left(\mu - k + \frac{1}{2}\right)} \frac{\cos\left[\left(\nu + k + \frac{1}{2}\right)\varphi - \frac{\pi}{4}(2k - 1) + \frac{\mu \pi}{2}\right]}{k!\, \Gamma\left(\nu + k + \frac{3}{2}\right) (2 \sin \varphi)^{k + \frac{1}{2}}}$$

$$\left[\nu + \mu \neq -1, -2, -3, \ldots; \quad \nu \neq -\tfrac{3}{2}, -\tfrac{5}{2}, -\tfrac{7}{2}, \ldots; \quad \text{for } \frac{\pi}{6} < \varphi < \frac{5}{6}\pi \right.$$

This series also converges for complex values of ν and μ.
In the remaining cases, it is an asymptotic expansion for

$$\left. |\nu| \gg |\mu|, \quad |\nu| \gg 1, \text{if } \nu > 0, \quad \mu > 0, \quad 0 < \varepsilon \leq \varphi \leq \pi - \varphi \right]$$

EH I 147(6), MO 92

3. $$P_\nu^\mu (\cos \varphi) = \frac{2}{\sqrt\pi} \frac{\Gamma(\nu + \mu + 1)}{\Gamma\left(\nu + \frac{3}{2}\right)} \frac{\cos \left[\left(\nu + \frac{1}{2}\right) \varphi - \frac{\pi}{4} + \frac{\mu\pi}{2}\right]}{\sqrt{2 \sin \varphi}} \left[1 + O\left(\frac{1}{\nu}\right)\right]$$

$$\left[0 < \varepsilon \le \varphi \le \pi - \varepsilon, \quad |\nu| \gg \frac{1}{\varepsilon}\right] \quad \text{MO 92}$$

For $\nu > 0, \mu > 0$ and $\nu > \mu$, it follows from formulas **8.721** 1 and **8.721** 2 that

4. $$\nu^{-\mu} P_\nu^\mu (\cos \varphi) = \sqrt{\frac{2}{\nu\pi \sin \varphi}} \cos \left[\left(\nu + \frac{1}{2}\right) \varphi - \frac{\pi}{4} + \frac{\mu\pi}{2}\right] + O\left(\frac{1}{\sqrt{\nu^3}}\right)$$

5. $$\nu^{-\mu} Q_\nu^\mu (\cos \varphi) = \sqrt{\frac{\pi}{2\nu \sin \varphi}} \cos \left[\left(\nu + \frac{1}{2}\right) \varphi + \frac{\pi}{4} + \frac{\mu\pi}{2}\right] O\left(\frac{1}{\sqrt{\nu^3}}\right)$$

$$\left[0 < \varepsilon \le \varphi \le \pi - \varepsilon; \quad \nu \gg \frac{1}{\varepsilon}\right] \quad \text{MO 92}$$

8.722   If φ is sufficiently close to 0 or π that $\nu\varphi$ or $\nu(\pi - \varphi)$ is small in comparison with 1, the asymptotic formulas **8.721** become unsuitable. In this case, the following asymptotic representation is applicable for $\mu \le 0, \nu \gg 1$, and *small* values of φ:

1. $$\left[\left(\nu + \frac{1}{2}\right) \cos \frac{\varphi}{2}\right]^\mu P_\nu^{-\mu} (\cos \varphi) = J_\mu(\eta) + \sin^2 \frac{\varphi}{2} \left[\frac{J_{\mu+1}(\eta)}{2\eta} - J_{\mu+2}(\eta) + \frac{\eta}{6} J_{\mu+3}(\eta)\right] + O\left(\sin^4 \frac{\varphi}{2}\right)$$

where $\eta = (2\nu + 1) \sin \frac{\varphi}{2}$. In particular, it follows that

1. $$\lim_{\nu \to \infty} \nu^\mu P_\nu^{-\mu} \left(\cos \frac{x}{\nu}\right) = J_\mu(x) \qquad\qquad [x \ge 0, \mu \ge 0] \qquad \text{MO 93}$$

8.723   We can see how the functions $P_\nu^\mu(z)$ and $Q_\nu^\mu(z)$ behave for large $|\nu|$ and real values of $z > \frac{3}{2\sqrt2}$:

1. $$P_\nu^\mu (\cosh \alpha) = \frac{2^\mu}{\sqrt\pi} \left\{ \frac{\Gamma\left(-\nu - \frac{1}{2}\right)}{\Gamma(-\nu - \mu)} \frac{e^{(\mu-\nu)\alpha} \sinh^\mu \alpha}{(e^{2\alpha} - 1)^{\mu + \frac{1}{2}}} F\left(\mu + \frac{1}{2}, -\mu + \frac{1}{2}; \nu + \frac{3}{2}; \frac{1}{1 - e^{2\alpha}}\right) \right.$$
$$\left. + \frac{\Gamma\left(\nu + \frac{1}{2}\right)}{\Gamma(\nu - \mu + 1)} \frac{e^{(\nu+\mu+1)\alpha} \sinh^\mu \alpha}{(e^{2\alpha} - 1)^{\mu + \frac{1}{2}}} F\left(\mu + \frac{1}{2}, -\mu + \frac{1}{2}; -\nu + \frac{1}{2}; \frac{1}{1 - e^{2\alpha}}\right) \right\}$$

$$\left[\nu \ne \pm\frac{1}{2}, \pm\frac{3}{2}, \pm\frac{5}{2}, \dots; \quad a > \frac{1}{2} \ln 2\right] \quad \text{MO 94}$$

2. $$Q_\nu^\mu (\cosh \alpha) = e^{\mu\pi i} 2^\mu \sqrt\pi \frac{\Gamma(\nu + \mu + 1)}{\Gamma\left(\nu + \frac{3}{2}\right)} \frac{e^{-(\nu+\mu+1)\alpha}}{(1 - e^{-2\alpha})^{\mu + \frac{1}{2}}} \sinh^\mu \alpha$$
$$\times F\left(\mu + \frac{1}{2}, -\mu + \frac{1}{2}; \nu + \frac{3}{2}; \frac{1}{1 - e^{2\alpha}}\right)$$

$$\left[\mu + \nu + 1 \ne 0, -1, -2, \dots; \quad \alpha > \frac{1}{2} \ln 2\right] \quad \text{MO 94}$$

See also **8.776**.

8.724   For the inequalities in **8.776** 1–4, ν and μ are arbitrary real numbers satisfying the inequalities $\nu \ge 1, \nu - \mu + 1 > 0$, and $\mu \ge 0$:

1. $$\left|P_\nu^{\pm\mu} (\cos \varphi)\right| < \sqrt{\frac{8}{\nu\pi}} \frac{\Gamma(\nu \pm \mu + 1)}{\Gamma(\nu + 1)} \frac{1}{\sin^{\mu + \frac{1}{2}} \varphi} \qquad\qquad\qquad \text{MO 91-92}$$

2. $\left|Q_\nu^{\pm\mu}\left(\cos\varphi\right)\right| < \sqrt{\dfrac{2\pi}{\nu}}\dfrac{\Gamma\left(\nu\pm\mu+1\right)}{\Gamma(\nu+1)}\dfrac{1}{\sin^{\mu+\frac{1}{2}}\varphi}$ MO 91-92

3. $\left|P_\nu^{\pm\mu}\left(\cos\varphi\right)\right| < \dfrac{2}{\sqrt{\nu\pi}}\dfrac{\Gamma\left(\nu\pm\mu+1\right)}{\Gamma(\nu+1)}\dfrac{1}{\sin^{\mu+\frac{1}{2}}\varphi}$ MO 91-92

4. $\left|Q_\nu^{\pm\mu}\left(\cos\varphi\right)\right| < \sqrt{\dfrac{\pi}{\nu}}\dfrac{\Gamma\left(\nu\pm\mu+1\right)}{\Gamma(\nu+1)}\dfrac{1}{\sin^{\mu+\frac{1}{2}}\varphi}$ MO 91-92

5.[8] $\left|\sqrt{\sin\varphi}\,P_n^m\left(\cos\varphi\right)\right| < \dfrac{\Gamma\left(n+\frac{1}{2}\right)}{\Gamma(n-m+1)}2^{(m+n)^2/n}\sup\limits_{0<t<\infty}\left|\sqrt{t}\,J_m(t)\right|$

$$[\text{uniformly } 0 \le m \le n]$$

8.725[10]   For fixed z and ν and $\operatorname{Re}\mu \to \infty$, with z not on the real axis between $-\infty$ and -1 and $+\infty$ and $+1$, the following are asymptotic expansions in which the upper and lower signs are taken according to whether $\operatorname{Im}z$ is greater than or less than 0.

1. $$P_\nu^\mu(z) = \frac{\Gamma(\nu+\mu+1)\,\Gamma(\mu-\nu)}{\pi\,\Gamma(\mu+1)}\left(\frac{z+1}{z-1}\right)^{\frac{1}{2}\mu}\sin\mu\pi\left[F\left(-\nu,\nu+1;1+\mu;\frac{1}{2}+\frac{1}{2}z\right)\right.$$

$$\left.-\frac{\sin\nu\pi}{\sin\mu\pi}e^{\mp i\mu\pi}\left(\frac{z-1}{z+1}\right)^\mu F\left(-\nu,\nu+1;1+\mu;\frac{1}{2}-\frac{1}{2}z\right)\right]$$

AS 8.10.1

2. $$Q_\nu^\mu(z) = \frac{1}{2}e^{i\mu\pi}\frac{\Gamma(\nu+\mu+1)}{\Gamma(\mu+1)}\left(\frac{z+1}{z-1}\right)^{\frac{1}{2}\mu}\Gamma(\mu-\nu)\left[F\left(-\nu,\nu+1;1+\mu;\frac{1}{2}+\frac{1}{2}z\right)\right.$$

$$\left.-e^{\mp i\nu\pi}\left(\frac{z-1}{z+1}\right)^\mu F\left(-\nu,\nu+1;1+\mu;\frac{1}{2}-\frac{1}{2}z\right)\right]$$

AS 8.10.2

3. $$Q_\nu^{-\mu}(z) = \frac{e^{-i\mu\pi}\operatorname{cosec}\left[\pi(\nu-\mu)\right]}{2\pi\,\Gamma(1+\mu)}\left[e^{\mp i\nu\pi}\left(\frac{z+1}{z-1}\right)^{-\frac{1}{2}\mu}F\left(-\nu,\nu+1;1+\mu;\frac{1}{2}-\frac{1}{2}z\right)\right.$$

$$\left.-\left(\frac{z-1}{z+1}\right)^{-\frac{1}{2}\mu}F\left(-\nu,\nu+1;1+\mu;\frac{1}{2}+\frac{1}{2}z\right)\right]$$

AS 8.10.3

8.73–8.74 Functional relations

8.731

1. $\left(z^2-1\right)\dfrac{d\,P_\nu^\mu(z)}{dz} = (\nu-\mu+1)\,P_{\nu+1}^\mu(z) - (\nu+1)z\,P_\nu^\mu(z)$

(cf. **8.832** 1, **8.914** 2)

EH I 161(10), MO 81

$1(1)^9.$ $\left(z^2 - 1\right) \dfrac{d\, P_\nu^\mu(z)}{dz} = \nu z\, P_\nu^\mu(z) - (\nu + \mu)\, P_{\nu-1}^\mu(z)$ AS 8.5.4

$1(2)^9.$ $\left(z^2 - 1\right) \dfrac{d\, P_\nu^\mu(z)}{dz} = (\nu + \mu)(\nu - \mu + 1)\left(z^2 - 1\right) P_\nu^{\mu-1}(z) - \mu z\, P_\nu^\mu(z)$ AS 8.5.2

2. $(2\nu + 1)z\, P_\nu^\mu(z) = (\nu - \mu + 1)\, P_{\nu+1}^\mu(z) + (\nu + \mu)\, P_{\nu-1}^\mu(z)$
                                         (cf. **8.832** 2, **8.914** 1)

                                                 EH I 160(2), MO 81

3. $P_\nu^{\mu+2}(z) + 2(\mu + 1)\dfrac{z}{\sqrt{z^2 - 1}}\, P_\nu^{\mu+1}(z) = (\nu - \mu)(\nu + \mu + 1)\, P_\nu^\mu(z)$ MO 82, EH I 160(1)

$3(1)^9.$ $P_\nu^{\mu+1}(z) = \left(z^2 - 1\right)^{-1/2}\left[(\nu - \mu)z\, P_\nu^\mu(z) - (\nu + \mu)\, P_{\nu-1}^\mu(z)\right]$ AS 8.5.1

4. $P_{\nu+1}^\mu(z) - P_{\nu-1}^\mu(z) = (2\nu + 1)\sqrt{z^2 - 1}\, P_\nu^{\mu-1}(z)$ EH I 160(3), MO 82

$4(1)^9.$ $(\nu - \mu + 1)\, P_{\nu+1}^\mu(z) = (2\nu + 1)z\, P_\nu^\mu(z) - (\nu + \mu)\, P_{\nu-1}^\mu(z)$ AS 334(8.5.3)

$4(2)^9.$ $P_{\nu+1}^\mu(z) = P_{\nu-1}^\mu(z) + (2\nu + 1)\left(z^2 - 1\right)^{1/2} P_\nu^{\mu-1}(z)$ AS 334(8.5.5)

5. $P_{-\nu-1}^\mu(z) = P_\nu^\mu(z)$ (cf. **8.820**, **8.832** 4)

                                                 EH I 140(1), MO 82

8.732

1. $\left(z^2 - 1\right) \dfrac{d\, Q_\nu^\mu(z)}{dz} = (\nu - \mu + 1)\, Q_{\nu+1}^\mu(z) - (\nu + 1)z\, Q_\nu^\mu(z)$

                                       (cf. **8.832** 3)   MO 82

$2.^{10}$ $(2\nu + 1)z\, Q_\nu^\mu(z) = (\nu - \mu + 1)\, Q_{\nu+1}^\mu(z) + (\nu + \mu)\, Q_{\nu-1}^\mu(z)$

                                       (cf. **8.832** 4)   MO 82

3. $Q_\nu^{\mu+2}(z) + 2(\mu + 1)\dfrac{z}{\sqrt{z^2 - 1}}\, Q_\nu^{\mu+1}(z) = (\nu - \mu)(\nu + \mu + 1)\, Q_\nu^\mu(z)$ MO 82

4. $Q_{\nu-1}^\mu(z) - Q_{\nu+1}^\mu(z) = -(2\nu + 1)\sqrt{z^2 - 1}\, Q_\nu^{\mu-1}(z)$ MO 82a

5. $e^{-\mu\pi i}\, Q_\nu^\mu(x \pm i0) = e^{\pm \frac{1}{2}\mu\pi i}\left[Q_\nu^\mu(x) \mp i\dfrac{\pi}{2}\, P_\nu^\mu(x)\right]$ MO 83

8.733

1. $\left(1 - x^2\right) \dfrac{d\, P_\nu^\mu(x)}{dx} = P_\nu^\mu(x) - (\nu - \mu + 1)\, P_{\nu+1}^\mu(x)$ (cf. **8.731** 1)

                     $= -\nu x\, P_\nu^\mu(x) + (\nu + \mu)\, P_{\nu-1}^\mu(x)$

                     $= -\sqrt{1 - x^2}\, P_\nu^{\mu+1}(x) - \mu x\, P_\nu^\mu(x);$

                     $= (\nu - \mu + 1)(\nu + \mu)\sqrt{1 - x^2}\, P_\nu^{\mu-1}(x) + \mu x\, P_\nu^\mu(x)$

                                                       MO 82

2. $(2\nu + 1)x\, P_\nu^\mu(x) = (\nu - \mu + 1)\, P_{\nu+1}^\mu(x) + (\nu + \mu)\, P_{\nu-1}^\mu(x)$

                                       (cf. **8.731** 2)   MO 82

3. $P_\nu^{\mu+2}(x) + 2(\mu + 1)\dfrac{x}{\sqrt{1 - x^2}}\, P_\nu^{\mu+1}(x) + (\nu - \mu)(\nu + \mu + 1)\, P_{(x)}^{\mu n} = 0$

                                       (cf. **8.731** 3)   MO 82

4. $P^\mu_{\nu-1}(x) - P^\mu_{\nu+1}(x) = (2\nu + 1)\sqrt{1-x^2}\, P^{\mu-1}_\nu(x)$ (cf. **8.731** 4)   MO 82

5. $P^\mu_{-\nu-1}(x) = P^\mu_\nu(x)$ (cf. **8.731** 5)

8.734

1. $(\nu + \mu + 1)z\, Q^\nu_\mu(z) + \sqrt{z^2-1}\, Q^{\mu+1}_\nu(z) = (\nu - \mu + 1)\, Q^\mu_{\nu+1}(z)$ MO 82

2. $(\nu + \mu)\, Q^\mu_{\nu-1}(z) + \sqrt{z^2-1}\, Q^{\mu+1}_\nu(z) = (\nu - \mu)z\, Q^\mu_\nu(z)$ MO 82

3. $Q^\mu_{\nu-1}(z) - z\, Q^\mu_\nu(z) = -(\nu - \mu + 1)\sqrt{z^2-1}\, Q^{\mu-1}_\nu(z)$ MO 82

4. $z\, Q^\mu_\nu(z) - Q^\mu_{\nu+1}(z) = -(\nu + \mu)\sqrt{z^2-1}\, Q^{\mu-1}_\nu(z)$ MO 82

5. $(\nu + \mu)(\nu + \mu + 1)\, Q^\mu_{\nu-1}(z) + (2\nu + 1)\sqrt{z^2-1}\, Q^{\mu+1}_\nu(z) = (\nu - \mu)(\nu - \mu + 1)\, Q^\mu_{\nu+1}(z)$ MO 82

8.735

1. $(\nu + \mu + 1)x\, P^\mu_\nu(x) + \sqrt{1-x^2}\, P^{\mu+1}_\nu(x) = (\nu - \mu + 1)\, P^\mu_{\nu+1}(x)$ MO 83

2. $(\nu - \mu)x\, P^\mu_\nu(x) - (\nu + \mu)\, P^\mu_{\nu-1}(x) = \sqrt{1-x^2}\, P^{\mu+1}_\nu(x)$ MO 83

3. $P^\mu_{\nu-1}(x) - x\, P^\mu_\nu(x) = (\nu - \mu + 1)\sqrt{1-x^2}\, P^{\mu-1}_\nu(x)$ MO 83

4. $x\, P^\mu_\nu(x) - P^\mu_{\nu+1}(x) = (\nu + \mu)\sqrt{1-x^2}\, P^{\mu-1}_\nu(x)$ MO 83

5. $(\nu - \mu)(\nu - \mu + 1)\, P^\mu_{\nu+1}(x) = (\nu + \mu)(\nu + \mu + 1)\, P^\mu_{\nu-1}(x) + (2\nu + 1)\sqrt{1-x^2}\, P^{\mu+1}_\nu(x)$ MO 83

8.736

1. $P^{-\mu}_\nu(z) = \dfrac{\Gamma(\nu - \mu + 1)}{\Gamma(\nu + \mu + 1)} \left[P^\mu_\nu(z) - \dfrac{2}{\pi} e^{-\mu\pi i} \sin \mu\pi\, Q^\mu_\nu(z) \right]$ MO 83

2. $P^\mu_\nu(-z) = e^{\nu\pi i} P^\mu_\nu(z) - \dfrac{2}{\pi} \sin[(\nu + \mu)\pi] e^{-\mu\pi i}\, Q^\mu_\nu(z)$ [Im $z < 0$]   (cf. **8.833** 1)   MO 83

3. $P^\mu_\nu(-z) = e^{-\nu\pi i} P^\mu_\nu(z) - \dfrac{2}{\pi} \sin[(\nu + \mu)\pi] e^{-\mu\pi i}\, Q^\mu_\nu(z)$

[Im $z > 0$]   (cf. **8.833** 2)   MO 83

4. $Q^{-\mu}_\nu(z) = e^{-2\mu\pi i} \dfrac{\Gamma(\nu - \mu + 1)}{\Gamma(\nu + \mu + 1)}\, Q^\mu_\nu(z)$ MO 82

5. $Q^\mu_\nu(-z) = -e^{-\nu\pi i}\, Q^\mu_\nu(z)$ [Im $z < 0$]   MO 82

6. $Q^\mu_\nu(-z) = -e^{\nu\pi i}\, Q^\mu_\nu(z)$ [Im $z > 0$]   MO 82

7.[6] $Q^\mu_\nu(z) \sin[(\nu + \mu)\pi] - Q^\mu_{-\nu-1}(z) \sin[(\nu - \mu)\pi] = \pi e^{\mu\pi i} \cos \nu\pi\, P^\mu_\nu(z)$ MO 83

8.737

1. $P^{-\mu}_\nu(x) = \dfrac{\Gamma(\nu - \mu + 1)}{\Gamma(\nu + \mu + 1)} \left[\cos \mu\pi\, P^\mu_\nu(x) - \dfrac{2}{\pi} \sin(\mu\pi)\, Q^\mu_\nu(x) \right]$ MO 84

2. $P^\mu_\nu(-x) = \cos[(\nu + \mu)\pi]\, P^\mu_\nu(x) - \dfrac{2}{\pi} \sin[(\nu + \mu)\pi]\, Q^\mu_\nu(x)$ MO 84

3. $Q^\mu_\nu(-x) = -\cos[(\nu + \mu)\pi]\, Q^\mu_\nu(x) - \dfrac{\pi}{2} \sin[(\nu + \mu)\pi]\, P^\mu_\nu(x)$ MO 83, EH I 144(15)

4. $Q^\mu_{-\nu-1}(x) = \dfrac{\sin[(\nu+\mu)\pi]}{\sin[(\nu-\mu)\pi]} \, Q^\mu_\nu(x) - \dfrac{\pi \cos\nu\pi \cos\mu\pi}{\sin[(\nu-\mu)\pi]} \, P^\mu_\nu(x)$ MO 84

8.738

1.[7] $Q^\mu_\nu(i \cot\varphi) = \exp\left[i\pi\left(\mu - \dfrac{1}{2}\right)\right] \sqrt{\pi}\,\Gamma(\nu+\mu+1) \sqrt{\dfrac{1}{2}\sin\varphi}\; P^{-\nu-\frac{1}{2}}_{-\mu-\frac{1}{2}}(\cos\varphi)$

$$\left[0 < \varphi < \frac{\pi}{2}\right] \qquad \text{MO 83}$$

2.[6] $P^\mu_\nu(i\cot\varphi) = \sqrt{\dfrac{2}{\pi}}\,\exp\left[i\pi\left(\nu + \dfrac{1}{4}\right)\right] \dfrac{\sqrt{\sin\varphi}}{\Gamma(-\nu-\mu)}\, Q^{-\nu-\frac{1}{2}}_{-\mu-\frac{1}{2}}(\cos\varphi - i0)$

$$\left[0 < \varphi < \frac{\pi}{2}\right] \qquad \text{MO 83}$$

8.739 $e^{-\mu\pi i}\, Q^\mu_\nu(\cosh\alpha) = \dfrac{\sqrt{\pi}\,\Gamma(\nu+\mu+1)}{\sqrt{2\sinh\alpha}}\, P^{-\nu-\frac{1}{2}}_{-\mu-\frac{1}{2}}(\coth\alpha)$ $[\operatorname{Re}(\cosh\alpha) > 0]$ MO 83

8.741

1. $P^{-\mu}_\nu(x)\dfrac{d\,P^\mu_\nu(x)}{dx} - P^\mu_\nu(x)\dfrac{d\,P^{-\mu}_\nu(x)}{dx} = \dfrac{2\sin\mu\pi}{\pi(1-x^2)}$ MO 83

2. $P^\mu_\nu(x)\dfrac{d\,Q^\mu_\nu(x)}{dx} - Q^\mu_\nu(x)\dfrac{d\,P^\mu_\nu(x)}{dx} = \dfrac{2^{2\mu}}{1-x^2}\dfrac{\Gamma\left(\frac{\nu+\mu+1}{2}\right)\Gamma\left(\frac{\nu+\mu}{2}+1\right)}{\Gamma\left(\frac{\nu-\mu+1}{2}\right)\Gamma\left(\frac{\nu-\mu}{2}+1\right)}$ MO 83

8.742

1. $\dfrac{\Gamma(\nu-\mu-1)}{\Gamma(\nu+\mu+1)}\left\{\cos\mu\pi\, P^\mu_\nu(\cos\varphi) - \dfrac{2}{\pi}\sin\mu\pi\, Q^\mu_\nu(\cos\varphi)\right\} = \sqrt{\dfrac{2}{\pi}}\,\dfrac{\operatorname{cosec}^\mu\varphi}{\Gamma\left(\mu+\frac{1}{2}\right)}\displaystyle\int_0^\varphi \dfrac{\cos\left(\nu+\frac{1}{2}\right)t\,dt}{(\cos t - \cos\varphi)^{\frac{1}{2}-\mu}}$

$$\left[\operatorname{Re}\mu > -\frac{1}{2}\right] \qquad \text{MO 88}$$

2. $\dfrac{\Gamma(\nu-\mu+1)}{\Gamma(\nu+\mu+1)}\left\{\cos\nu\pi\, P^\mu_\nu(\cos\varphi) - \dfrac{2}{\pi}\sin\nu\pi\, Q^\mu_\nu(\cos\varphi)\right\}$

$$= \sqrt{\frac{2}{\pi}}\,\frac{\operatorname{cosec}^\mu\varphi}{\Gamma\left(\mu+\frac{1}{2}\right)}\int_\varphi^\pi \frac{\cos\left[\left(\nu+\frac{1}{2}\right)(t-\pi)\right]dt}{(\cos\varphi - \cos t)^{\frac{1}{2}-\mu}}$$

$$\left[\operatorname{Re}\mu > -\frac{1}{2}\right] \qquad \text{MO 88}$$

3. $P^\mu_\nu(\cos\varphi)\cos(\nu+\mu)\pi - \dfrac{2}{\pi}\, Q^\mu_\nu(\cos\varphi)\sin(\nu+\mu)\pi = \sqrt{\dfrac{2}{\pi}}\,\dfrac{\sin^\mu\varphi}{\Gamma\left(\frac{1}{2}-\mu\right)}\displaystyle\int_\varphi^\pi \dfrac{\cos\left[\left(\nu+\frac{1}{2}\right)(t-\pi)\right]dt}{(\cos\varphi - \cos t)^{\mu+\frac{1}{2}}}$

$$\left[\operatorname{Re}\mu < \frac{1}{2}\right] \qquad \text{MO 88}$$

4. $\cos\mu\pi\, P^\mu_\nu(\cos\varphi) - \dfrac{2}{\pi}\sin\mu\pi\, Q^\mu_\nu(\cos\varphi)$

$$= \frac{1}{2^\mu\sqrt{\pi}}\,\frac{\Gamma(\nu+\mu+1)}{\Gamma(\nu-\mu+1)}\,\frac{\sin^\mu\varphi}{\Gamma\left(\mu+\frac{1}{2}\right)}\int_0^\pi \frac{\sin^{2\mu}t\,dt}{(\cos\varphi \pm i\sin\varphi\cos t)^{\nu-\mu}}$$

$$\left[\operatorname{Re}\mu > -\frac{1}{2},\quad 0 < \varphi < \pi\right] \qquad \text{MO 38}$$

For integrals of Legendre functions, see **7.11–7.21**.

8.75 Special cases and particular values

8.751

1. $P_\nu^m(x) = (-1)^m \dfrac{\Gamma(\nu + m + 1)(1 - x^2)^{\frac{m}{2}}}{2^m \, \Gamma(\nu - m + 1)m!} \, F\left(m - \nu, m + \nu + 1; m + 1; \dfrac{1 - x}{2}\right)$ MO 84

2. $P_\nu^m(z) = \dfrac{\Gamma(\nu + m + 1)(z^2 - 1)^{\frac{m}{2}}}{2^m m! \, \Gamma(\nu - m + 1)} \, F\left(m - \nu, m + \nu + 1; m + 1; \dfrac{1 - z}{2}\right)$ MO 84

3.[8] $Q_{n + \frac{1}{2}}^\mu(z) = \dfrac{e^{\mu \pi i} \, \Gamma\left(\mu + n + \frac{3}{2}\right)}{2^{n + \frac{3}{2}}(n + 1)!}(z^2 - 1)^{\frac{\mu}{2}} \pi^{1/2} z^{-n - \mu - 3/2} \, F\left(\dfrac{\mu + n + \frac{5}{2}}{2}, \dfrac{\mu + n + \frac{3}{2}}{2}; n + 2; \dfrac{1}{z^2}\right)$

MO 84

8.752

1. $P_\nu^m(x) = (-1)^m (1 - x^2)^{\frac{m}{2}} \dfrac{d^m}{dx^m} \, P_\nu(x)$ WH, MO 84, EH I 148(6)

2. $P_\nu^{-m}(x) = (-1)^m \dfrac{\Gamma(\nu - m + 1)}{\Gamma(\nu + m + 1)} \, P_\nu^m(x) = (1 - x^2)^{-\frac{m}{2}} \displaystyle\int_x^1 \cdots \int_x^1 P_\nu(x)(dx)^m$

$[m \geq 1]$ HO 99a, MO 85, EH I 149(10)a

3. $P_\nu^{-m}(z) = (z^2 - 1)^{-\frac{m}{2}} \displaystyle\int_1^z \cdots \int_1^z P_\nu(z)(dz)^m$ $[m \geq 1]$ MO 85, EH I 149(8)

4. $Q_\nu^m(z) = (z^2 - 1)^{\frac{m}{2}} \dfrac{d^m}{dz^m} \, Q_\nu(z)$ WH, MO 85, EH I 148(5)

5. $Q_\nu^{-m}(z) = (-1)^m (z^2 - 1)^{-\frac{m}{2}} \displaystyle\int_z^\infty \cdots \int_z^\infty Q_\nu(z)(dz)^m$

$[m \geq 1]$ MO 85, EH I 149(9)

Special values of the indices

8.753

1. $P_0^\mu(\cos \varphi) = \dfrac{1}{\Gamma(1 - \mu)} \cot^\mu \dfrac{\varphi}{2}$ MO 84

2. $P_\nu^{-1}(\cos \varphi) = -\dfrac{1}{\nu(\nu + 1)} \dfrac{d \, P_\nu(\cos \varphi)}{d\varphi}$ MO 84

3. $P_n^m(z) \equiv 0,$ $\qquad\qquad\qquad P_n^m(x) \equiv 0$ for $m > n$ MO 85

8.754

1. $P_{\nu - \frac{1}{2}}^{1/2}(\cosh \alpha) = \sqrt{\dfrac{2}{\pi \sinh \alpha}} \cosh \nu\alpha$ MO 85

2. $P_{\nu - \frac{1}{2}}^{1/2}(\cos \varphi) = \sqrt{\dfrac{2}{\pi \sin \varphi}} \cos \nu\varphi$ MO 85

3. $P_{\nu-\frac{1}{2}}^{-1/2}(\cos\varphi) = \sqrt{\dfrac{2}{\pi\sin\varphi}}\dfrac{\sin\nu\varphi}{\nu}$ MO 85

4. $Q_{\nu-\frac{1}{2}}^{1/2}(\cosh\alpha) = i\sqrt{\dfrac{\pi}{2\sinh\alpha}}e^{-\nu\alpha}$ MO 85

8.755

1. $P_{\nu}^{-\nu}(\cos\varphi) = \dfrac{1}{\Gamma(1+\nu)}\left(\dfrac{\sin\varphi}{2}\right)^{\nu}$ MO 85

2. $P_{\nu}^{-\nu}(\cosh\alpha) = \dfrac{1}{\Gamma(1+\nu)}\left(\dfrac{\sinh\alpha}{2}\right)^{\nu}$ MO 85

Special values of Legendre functions

8.756

1. $P_{\nu}^{\mu}(0) = \dfrac{2^{\mu}\sqrt{\pi}}{\Gamma\left(\frac{\nu-\mu}{2}+1\right)\Gamma\left(\frac{-\nu-\mu+1}{2}\right)}$ MO 84

2. $\dfrac{d\,P_{\nu}^{\mu}(0)}{dx} = \dfrac{2^{\mu+1}\sin\frac{1}{2}(\nu+\mu)\pi\,\Gamma\left(\frac{\nu+\mu}{2}+1\right)}{\sqrt{\pi}\,\Gamma\left(\frac{\nu-\mu+1}{2}\right)}$ MO 84

3. $Q_{\nu}^{\mu}(0) = -2^{\mu-1}\sqrt{\pi}\sin\dfrac{1}{2}(\nu+\mu)\pi\dfrac{\Gamma\left(\frac{\nu+\mu+1}{2}\right)}{\Gamma\left(\frac{\nu-\mu}{2}+1\right)}$ MO84

4. $\dfrac{d\,Q_{\nu}^{\mu}(0)}{dx} = 2^{\mu}\sqrt{\pi}\cos\dfrac{1}{2}(\nu+\mu)\pi\dfrac{\Gamma\left(\frac{\nu+\mu}{2}+1\right)}{\Gamma\left(\frac{\nu-\mu+1}{2}\right)}$ MO 84

8.76   Derivatives with respect to the order

8.761 $\dfrac{\partial P_{\nu}^{-\mu}(x)}{\partial\nu} = \dfrac{1}{\Gamma(\mu+1)}\left(\dfrac{1-x}{1+x}\right)^{\frac{\mu}{2}}\sum_{n=1}^{\infty}\dfrac{(-\nu)(1-\nu)\ldots(n-1-\nu)(\nu+1)(\nu+2)\ldots(\nu+n)}{(\mu+1)(\mu+2)\ldots(\mu+n)1\cdot2\ldots n}$

$$\times\left[\psi(\nu+n+1)-\psi(\nu-n+1)\right]\left(\dfrac{1-x}{2}\right)^{n}$$

$$[\nu\neq0,\pm1,\pm2,\ldots;\quad\operatorname{Re}\mu>-1]\quad\text{MO 94}$$

8.762

1. $\left[\dfrac{\partial P_{\nu}(\cos\varphi)}{\partial\nu}\right]_{\nu=0} = 2\ln\cos\dfrac{\varphi}{2}$ MO 94

2. $\left[\dfrac{\partial P_{\nu}^{-1}(\cos\varphi)}{\partial\nu}\right]_{\nu=0} = -\tan\dfrac{\varphi}{2} - 2\cot\dfrac{\varphi}{2}\ln\cos\dfrac{\varphi}{2}$ MO 94

3. $\left[\dfrac{\partial P_{\nu}^{-1}(\cos\varphi)}{\partial\nu}\right]_{\nu=1} = -\dfrac{1}{2}\tan\dfrac{\varphi}{2}\sin^{2}\dfrac{\varphi}{2} + \sin\varphi\ln\cos\dfrac{\varphi}{2}$ MO 94

- For a connection with the polynomials $C_{n}^{\lambda}(x)$, see **8.936**.
- For a connection with a hypergeometric function, see **8.77**.

8.77 Series representation

For a representation in the form of a series, see **8.721**. It is also possible to represent associated Legendre functions in the form of a series by expressing them in terms of a hypergeometric function.

8.771

1. $\quad P_\nu^\mu(z) = \left(\dfrac{z+1}{z-1}\right)^{\frac{\mu}{2}} \dfrac{1}{\Gamma(1-\mu)} F\left(-\nu, \nu+1; 1-\mu; \dfrac{1-z}{2}\right)$ MO 15

2.[8] $\quad Q_\nu^\mu(z) = \dfrac{e^{\mu\pi i}}{2^{\nu+1}} \dfrac{\Gamma(\nu+\mu+1)}{\Gamma\left(\nu+\frac{3}{2}\right)} \dfrac{\Gamma\left(\frac{1}{2}\right)\left(z^2-1\right)^{\frac{\mu}{2}}}{z^{\nu+\mu+1}} F\left(\dfrac{\nu+\mu}{2}+1, \dfrac{\nu+\mu+1}{2}; \nu+\dfrac{3}{2}; \dfrac{1}{z^2}\right)$ MO 15

See also **8.702**, **8.703**, **8.704**, **8.723**, **8.751**, **8.772**.

The analytic continuation for $|z| > 1$

The formulas are consequences of theorems on the analytic continuation of hypergeometric series (see **9.154** and **9.155**):

8.772

1. $\quad P_\nu^\mu(z) = \dfrac{\sin(\nu+\mu)\pi \, \Gamma(\nu+\mu+1)}{2^{\nu+1}\sqrt{\pi}\cos\nu\pi \, \Gamma\left(\nu+\frac{3}{2}\right)}\left(z^2-1\right)^{\frac{\mu}{2}} z^{-\nu-\mu-1} F\left(\dfrac{\nu+\mu}{2}+1, \dfrac{\nu+\mu+1}{2}; \nu+\dfrac{3}{2}; \dfrac{1}{z^2}\right)$

   $\qquad + \dfrac{2^\nu \, \Gamma\left(\nu+\frac{1}{2}\right)}{\sqrt{\pi}\,\Gamma(\nu-\mu+1)}\left(z^2-1\right)^{\frac{\mu}{2}} z^{\nu-\mu} F\left(\dfrac{\mu-\nu+1}{2}, \dfrac{\mu-\nu}{2}; \dfrac{1}{2}-\nu; \dfrac{1}{z^2}\right)$

   $\qquad\qquad\qquad \left[2\nu \neq \pm 1, \pm 3, \pm 5, \ldots; \quad |z| > 1; \quad |\arg(z\pm 1)| < \pi\right]$ MO 85

2. $\quad P_\nu^\mu(z) = \dfrac{\Gamma\left(-\nu-\frac{1}{2}\right)\left(z^2-1\right)^{-\frac{\nu+1}{2}}}{2^{\nu+1}\sqrt{\pi}\,\Gamma(-\nu-\mu)} F\left(\dfrac{\nu-\mu+1}{2}, \dfrac{\nu+\mu+1}{2}; \nu+\dfrac{3}{2}; \dfrac{1}{1-z^2}\right)$

   $\qquad + \dfrac{2^\nu \, \Gamma\left(\nu+\frac{1}{2}\right)}{\sqrt{\pi}\,\Gamma(\nu-\mu+1)}\left(z^2-1\right)^{\frac{\nu}{2}} F\left(\dfrac{\mu-\nu}{2}, -\dfrac{\mu+\nu}{2}; \dfrac{1}{2}-\nu; \dfrac{1}{1-z^2}\right)$

   $\qquad\qquad\qquad \left[2\nu \neq \pm 1, \pm 3, \pm 5; \ldots; \quad |1-z^2| > 1; \quad |\arg(z\pm 1)| < \pi\right]$ MO 85

3. $\quad P_\nu^\mu(z) = \dfrac{1}{\Gamma(1-\mu)}\left(\dfrac{z-1}{z+1}\right)^{-\frac{\mu}{2}}\left(\dfrac{z+1}{2}\right)^\nu F\left(-\nu, -\nu-\mu; 1-\mu; \dfrac{z-1}{z+1}\right)$

   $\qquad\qquad\qquad\qquad\qquad \left[\left|\dfrac{z-1}{z+1}\right| < 1\right]$ MO 86

8.773

1. $\quad Q_\nu^\mu(z) = e^{\mu\pi i} \dfrac{\sqrt{\pi}\,\Gamma(\nu+\mu+1)}{2^{\nu+1}\Gamma\left(\nu+\frac{3}{2}\right)}\left(z^2-1\right)^{-\frac{\nu+1}{2}} F\left(\dfrac{\nu+\mu+1}{2}, \dfrac{\nu-\mu+1}{2}; \nu+\dfrac{3}{2}; \dfrac{1}{1-z^2}\right)$

   $\qquad\qquad \left[\nu+\mu \neq -1, -2, -3, \ldots; \quad |\arg(z\pm 1)| < \pi; \quad |1-z^2| > 1\right]$ MO 86

2. $\quad Q_\nu^\mu(z) = \dfrac{1}{2}e^{\mu\pi i}\left\{\Gamma(\mu)\left(\dfrac{z+1}{z-1}\right)^{\frac{\mu}{2}} F\left(-\nu, \nu+1; 1-\mu; \dfrac{1-z}{2}\right)\right.$

   $\qquad + \left.\dfrac{\Gamma(-\mu)\,\Gamma(\nu+\mu+1)}{\Gamma(\nu-\mu+1)}\left(\dfrac{z-1}{z+1}\right)^{\frac{\mu}{2}} F\left(-\nu, \nu+1; \quad 1+\mu; \dfrac{1-z}{2}\right)\right\}$

   $\qquad\qquad\qquad\qquad \left[|\arg(z\pm 1)| < \pi, \quad |1-z| < 2\right]$ MO 86

8.774 $\quad P_\nu^\mu\left(i\cot\varphi\right) = \sqrt{\dfrac{\sin\varphi}{2\pi}}\dfrac{\Gamma\left(-\nu-\frac{1}{2}\right)}{\Gamma(-\nu-\mu)}e^{-i(\nu+1)\frac{\pi}{2}}\left(\tan\dfrac{\varphi}{2}\right)^{\nu+\frac{1}{2}}F\left(\dfrac{1}{2}+\mu,\dfrac{1}{2}-\mu;\nu+\dfrac{3}{2};\sin^2\dfrac{\varphi}{2}\right)$

$\qquad\qquad\qquad + \sqrt{\dfrac{\sin\varphi}{2\pi}}\dfrac{\Gamma\left(\nu+\frac{1}{2}\right)}{\Gamma(\nu-\mu+1)}e^{i\nu\frac{\pi}{2}}\left(\cot\dfrac{\varphi}{2}\right)^{\nu+\frac{1}{2}}F\left(\dfrac{1}{2}+\mu,\dfrac{1}{2}-\mu;\dfrac{1}{2}-\nu;\sin^2\dfrac{\varphi}{2}\right)$

$$\left[2\nu\neq\pm1,\pm3,\pm5,\dots,\quad 0<\varphi<\frac{\pi}{2}\right]\quad\text{MO 86}$$

8.775

1.[6] $\quad P_\nu^\mu(x) = \dfrac{2^\mu\cos\left(\frac{1}{2}(\nu+\mu)\pi\right)\Gamma\left(\frac{\nu+\mu+1}{2}\right)}{\sqrt{\pi}\,\Gamma\left(\frac{\nu-\mu}{2}+1\right)}\left(1-x^2\right)^{\frac{\mu}{2}}F\left(\dfrac{\nu+\mu+1}{2},\dfrac{\mu-\nu}{2};\dfrac{1}{2};x^2\right)$

$\qquad\qquad + \dfrac{2^{\mu+1}\sin\left(\frac{1}{2}(\nu+\mu)\pi\right)\Gamma\left(\frac{\nu+\mu}{2}+1\right)}{\sqrt{\pi}}\dfrac{1}{\Gamma\left(\frac{\nu-\mu+1}{2}\right)}x\left(1-x^2\right)^{\frac{\mu}{2}}F\left(\dfrac{\nu+\mu}{2}+1,\dfrac{-\nu+\mu+1}{2};\dfrac{3}{2};x^2\right)$

$$\text{MO 87}$$

2.[6] $\quad Q_\nu^\mu(x) = -\dfrac{\sqrt{\pi}}{2^{1-\mu}}\dfrac{\sin\left(\frac{1}{2}(\nu+\mu)\pi\right)\Gamma\left(\frac{\nu+\mu+1}{2}\right)}{\Gamma\left(\frac{\nu-\mu}{2}+1\right)}\left(1-x^2\right)^{\frac{\mu}{2}}F\left(\dfrac{\nu+\mu+1}{2},\dfrac{\mu-\nu}{2};\dfrac{1}{2};x^2\right)$

$\qquad\qquad + 2^\mu\sqrt{\pi}\dfrac{\cos\left(\frac{1}{2}(\nu+\mu)\pi\right)\Gamma\left(\frac{\nu+\mu}{2}+1\right)}{\Gamma\left(\frac{\nu-\mu+1}{2}\right)}x\left(1-x^2\right)^{\frac{\mu}{2}}F\left(\dfrac{\nu+\mu}{2}+1,\dfrac{\mu-\nu+1}{2};\dfrac{3}{2};x^2\right)$

$$\text{MO 87}$$

8.776 For $|z|\gg1$

1. $\quad P_\nu^\mu(z) = \left\{\dfrac{2^\nu\Gamma\left(\nu+\frac{1}{2}\right)}{\sqrt{\pi}\,\Gamma(\nu-\mu+1)}z^\nu + \dfrac{\Gamma\left(-\nu-\frac{1}{2}\right)}{2^{\nu+1}\sqrt{\pi}\,\Gamma(-\nu-\mu)}z^{-\nu-1}\right\}\left(1+O\left(\dfrac{1}{z^2}\right)\right)$

$$\left[2\nu\neq\pm1,\pm3,\pm5,\dots,\quad |\arg z|<\pi\right]$$
$$\text{MO 87}$$

2. $\quad Q_\nu^\mu(z) = \sqrt{\pi}\dfrac{e^{\mu\pi i}}{2^{\nu+1}}\dfrac{\Gamma(\mu+\nu+1)}{\Gamma\left(\nu+\frac{3}{2}\right)}z^{-\nu-1}\left(1+O\left(\dfrac{1}{z^2}\right)\right)$

$$\left[2\nu\neq-3,-5,-7,\dots;\quad |\arg z|<\pi\right]$$
$$\text{MO 87}$$

8.777 Set $\zeta = z+\sqrt{z^2-1}$. The variable ζ is uniquely defined by this equation on the entire z-plane in which a cut is made from $-\infty$ to $+1$. Here, we are considering that branch of the variable ζ for which values of ζ exceeding 1 correspond to real values of z exceeding 1. In this case,

1. $\quad P_\nu^\mu(z) = \dfrac{2^\mu\Gamma\left(-\nu-\frac{1}{2}\right)}{\sqrt{\pi}\,\Gamma(-\nu-\mu)}\dfrac{(z^2-1)^{\frac{\mu}{2}}}{\zeta^{\nu+\mu+1}}F\left(\dfrac{1}{2}+\mu,\nu+\mu+1;\nu+\dfrac{3}{2};\dfrac{1}{\zeta^2}\right)$

$\qquad\qquad + \dfrac{2^\mu}{\sqrt{\pi}}\dfrac{\Gamma\left(\nu+\frac{1}{2}\right)}{\Gamma(\nu-\mu+1)}\dfrac{(z^2-1)^{\frac{\mu}{2}}}{\zeta^{\mu-\nu}}F\left(\dfrac{1}{2}+\mu,\mu-\nu;\dfrac{1}{2}-\nu;\dfrac{1}{\zeta^2}\right)$

$$\left[2\nu\neq\pm1,\pm3,\pm5,\dots;\quad |\arg(z-1)|<\pi\right]\quad\text{MO 86}$$

2. $\quad Q_\nu^\mu(z) = 2^\mu e^{\mu\pi i}\sqrt{\pi}\dfrac{\Gamma(\nu+\mu+1)}{\Gamma\left(\nu+\frac{3}{2}\right)}\dfrac{(z^2-1)^{\frac{\mu}{2}}}{\zeta^{\nu+\mu+1}}F\left(\dfrac{1}{2}+\mu,\nu+\mu+1;\nu+\dfrac{3}{2};\dfrac{1}{\zeta^2}\right)$

$$\left[|\arg(z-1)|<\pi\right]\quad\text{MO 86}$$

8.78 The zeros of associated Legendre functions

8.781 The function $P_\nu^{-\mu}(\cos\varphi)$, considered as a function of ν has infinitely many zeros for $\mu \geq 0$. These are all simple and real. If a number ν_0 is a zero of the function $P_\nu^{-\mu}(\cos\varphi)$, the number $-\nu_0 - 1$ is also a zero of this function.

MO 91

8.782 If ν and μ are both real and $\mu \leq 0$, or if ν and μ are integers, the function $P_\nu^\mu(t)$ has no *real* zeros exceeding 1. If ν and μ are both real with $\nu < \mu < 0$, the function $P_\nu^\mu(t)$ has no real zeros exceeding 1 when $\sin\mu\pi\sin(\mu - \nu)\pi > 0$, but does have one such zero when $\sin\mu\pi\sin(\mu - \nu)\pi < 0$. Finally, if $\mu \leq \nu$, the function $P_\nu^\mu(t)$ has no zeros exceeding 1 for $\lfloor\mu\rfloor$ even but does have one zero for $\lfloor\mu\rfloor$ odd.

8.783 If $\nu > -\frac{3}{2}$ and $\nu + \mu + 1 > 0$, the function $Q_\nu^\mu(t)$ has no real zeros exceeding 1.   MO 91

8.784 The function $P_{-\frac{1}{2}+i\lambda}(z)$ has infinitely many zeros for real λ. All these zeros are *real* and *greater than unity*.

8.785 For n a natural number, the function $P_n(x)$ has exactly n real zeros which lie in the closed interval $-1, +1$.

8.786 The function $Q_n(z)$ has no zeros for which $|\arg(z - 1)| < \pi$ if n is a natural number. The function $Q_n(\cos\varphi)$ has exactly $n + 1$ zeros in the interval $0 \leq \varphi \leq \pi$.

MO 91

8.787 The following approximate formula can be used to calculate the values of ν for which the equation $P_\nu^{-\mu}(\cos\varphi) = 0$ holds for given small values of φ:

$$\nu + \frac{1}{2} = -\frac{j_\mu}{2\sin\frac{\varphi}{2}}\left\{1 - \frac{\sin^2\frac{\varphi}{2}}{6}\left(1 - \frac{4\mu^2 - 1}{j_\mu{}^2}\right) + O\left(\sin^4\frac{\varphi}{2}\right)\right\}$$

MO 93

Here, j_μ denotes an arbitrary nonzero root of the equation $J_\mu(z) = 0$ (for $\mu \geq 0$). If φ is close to π then, instead of this formula, we can use the following formulas:

1. $\nu \approx \mu + k + \dfrac{\Gamma(2\mu + k + 1)}{\Gamma(\mu)\,\Gamma(\mu + 1)\,\Gamma(k + 1)}\left(\dfrac{\pi - \varphi}{3}\right)^{2\mu}$ $[\mu > 0, \quad k = 0, 1, 2, \ldots]$ MO 93

2. $\nu \approx k + \dfrac{1}{2\ln\left(\frac{2}{\pi - \varphi}\right)}$ $[\mu = 0, \quad k = 0, 1, 2, \ldots]$ MO 93

8.79 Series of associated Legendre functions

8.791

1. $\dfrac{1}{z - t} = \sum_{k=0}^{\infty}(2k + 1)\,P_k(t)\,Q_k(z)$ $\left[\left|t + \sqrt{t^2 - 1}\right| < \left|z + \sqrt{z^2 - 1}\right|\right]$

Here, t must lie inside an ellipse passing through the point z with foci at the points ± 1.

2. $\dfrac{1}{\sqrt{1 - 2tz + t^2}}\ln\dfrac{z - t + \sqrt{1 - 2tz + t^2}}{\sqrt{z^2 - 1}} = \sum_{k=0}^{\infty} t^k\,Q_k(z)$

           $[\operatorname{Re} z > 1, \quad |t| < 1]$ MO 78

8.792 $P_\nu^{-\alpha}(\cos\varphi)\,P_\nu^{-\beta}(\cos\psi) = \dfrac{\sin\nu\pi}{\pi}\sum_{k=0}^{\infty}(-1)^k\left[\dfrac{1}{\nu - k} - \dfrac{1}{\nu + k + 1}\right]P_k^{-\alpha}(\cos\varphi)\,P_k^{-\beta}(\cos\psi)$

           $[a \geq 0, \quad \beta \geq 0, \quad \nu\text{ real}, \quad -\pi < \varphi \pm \psi < \pi]$ MO 94

8.793 $P_\nu^{-\mu}(\cos\varphi) = \dfrac{\sin\nu\pi}{\pi}\displaystyle\sum_{k=0}^{\infty}(-1)^k\left(\dfrac{1}{\nu-k}-\dfrac{1}{\nu+k+1}\right)P_k^{-\mu}(\cos\varphi)$ $[\mu\geq 0,\quad 0<\varphi<\pi]$

MO 94

Addition theorems

8.794

1. $P_\nu(\cos\psi_1\cos\psi_2+\sin\psi_1\sin\psi_2\cos\varphi)$

$$= P_\nu(\cos\psi_1)\,P_\nu(\cos\psi_2)+2\sum_{k=1}^{\infty}(-1)^k\,P_\nu^{-k}(\cos\psi_1)\,P_\nu^k(\cos\psi_2)\cos k\varphi$$

$$= P_\nu(\cos\psi_1)\,P_\nu(\cos\psi_2)+2\sum_{k=1}^{\infty}\frac{\Gamma(\nu-k+1)}{\Gamma(\nu+k+1)}\,P_\nu^k\left(\cos\psi_1\,P_\nu^k(\cos\psi_2)\cos k\varphi\right)$$

$[0\leq\psi_1<\pi,\quad 0\leq\psi_2<\pi,\quad \psi_1+\psi_2<\pi,\quad \varphi\text{ real}]$ (cf. **8.814**, **8.844** 1)   MO 90

2. $Q_\nu(\cos\psi_1)\cos\psi_2+\sin\psi_1\sin\psi_2\cos\varphi$

$$= P_\nu(\cos\psi_1)\,Q_\nu(\cos\psi_2)+2\sum_{k=1}^{\infty}(-1)^k\,P_\nu^{-k}(\cos\psi_1)\,Q_\nu\,k(\cos\psi_2)\cos k\varphi$$

$\left[0<\psi_1<\dfrac{\pi}{2},\quad 0<\psi_2<\pi,\quad 0<\psi_1+\psi_2<\pi;\quad \varphi\text{ real}\right]$ (cf. **8.844** 3)   MO 90

8.795

1. $P_\nu\left(z_1z_2-\sqrt{z_1^2-1}\sqrt{z_2^2-1}\cos\varphi\right)=P_\nu(z_1)\,P_\nu(z_2)+2\displaystyle\sum_{k=1}^{\infty}(-1)^k\,P_\nu^k(z_1)\,P_\nu^{-k}(z_2)\cos k\varphi$

$[\text{Re }z_1>0,\quad \text{Re }z_2>0,\quad |\arg(z_1-1)|<\pi,\quad |\arg(z_2-1)|<\pi]$ MO 91

2. $Q_\nu\left(x_1x_2-\sqrt{x_1^2-1}\sqrt{x_2^2-1}\cos\varphi\right)=P_\nu(x_1)\,Q_\nu(x_2)+2\displaystyle\sum_{k=1}^{\infty}(-1)^k\,P_\nu^{-k}(x_1)\,Q_\nu^k(x_2)\cos k\varphi$

$[1<x_1<x_2,\quad \nu\neq -1,-2,-3,\dots,\quad \varphi\text{ real}]$ MO 91

3. $Q_n\left(x_1x_2+\sqrt{x_1^2+1}\sqrt{x_2^2+1}\cosh\alpha\right)=\displaystyle\sum_{k=n+1}^{\infty}\frac{1}{(k-n-1)!(k+n)!}\,Q_n^k(ix_1)\,Q_n^k(ix_2)\,e^{-k\alpha}$

$[x_1>0,\quad x_2>0,\quad \alpha>0]$ MO 91

8.796 $P_\nu(-\cos\psi_1\cos\psi_2-\sin\psi_1\sin\psi_2\cos\varphi)=P_\nu(-\cos\psi_1)\,P_\nu(\cos\psi_2)+2\displaystyle\sum_{k=1}^{\infty}(-1)^k\frac{\Gamma(\nu+k+1)}{\Gamma(\nu-k+1)}$

$$\times P_\nu^{-k}(-\cos\psi_1)\,P_\nu^{-k}(\cos\psi_2)\cos k\varphi$$

$[0<\psi_2<\psi_1<\pi,\quad \varphi\text{ real}]$ (cf. **8.844** 2)   MO 91

See also **8.934** 3.

8.81 Associated Legendre functions with integral indices

8.810   For *integral* values of ν and μ, the differential equation **8.700** 1. (with $|\nu| > |\mu|$) has a simple solution in the real domain, namely:

$$u = P_n^m(x) = (-1)^m \left(1 - x^2\right)^{\frac{m}{2}} \frac{d^m}{dx^m} P_n(x)$$

The functions $P_n^m(x)$ are called *associated Legendre functions* (or *spherical functions*) *of the first kind*. The number n is called the *degree* and the number m is called the *order* of the function $P_n^m(x)$. The functions

$$\cos m\vartheta \, P_n^m(\cos\varphi), \quad \sin m\vartheta \, P_n^m(\cos\varphi),$$

which depend on the angles φ and ϑ, are also called Legendre functions of the first kind, or, more specifically, *tesseral harmonics* for $m <$n and *sectoral harmonics* for $m = n$. These last functions are periodic with respect to the angles φ and ϑ. Their periods are respectively π and 2π. They are single-valued and continuous everywhere on the surface of the unit sphere $x_1^2 + x_2^2 + x_3^2 = 1$ (where $x_1 = \sin\varphi\cos\vartheta$, $x_2 = \sin\varphi\sin\vartheta$, $x_3 = \cos\varphi$) and they are solutions of the differential equation

$$\frac{1}{\sin\varphi} \frac{\partial}{\partial\varphi} \left(\sin\varphi \frac{\partial Y}{\partial\varphi}\right) + \frac{1}{\sin^2\varphi} \frac{\partial^2 Y}{\partial\vartheta^2} + n(n+1)Y = 0$$

8.811[7]   The integral representation

$$P_n^m(\cos\varphi) = \frac{(-1)^m (n+m)!}{\Gamma\left(m + \frac{1}{2}\right)(n-m)!} \sqrt{\frac{2}{\pi}} \sin^{-m}\varphi \int_0^\varphi (\cos t - \cos\varphi)^{m-\frac{1}{2}} \cos\left(n + \tfrac{1}{2}\right) t \, dt \qquad \text{MO 75}$$

8.812   The series representation:

$$P_n^m(x) = \frac{(-1)^m (n+m)!}{2^m m!(n-m)!} \left(1 - x^2\right)^{\frac{m}{2}} \left\{1 - \frac{(n-m)(m+n+1)}{1!(m+1)} \frac{1-x}{2}\right.$$
$$\left. + \frac{(n-m)(n-m+1)(m+n+1)(m+n+2)}{2!(m+1)(m+2)} \left(\frac{1-x}{2}\right)^2 - \cdots\right\} \qquad \text{MO 73}$$

$$= \frac{(-1)^m (2n-1)!!}{(n-m)!} \left(1 - x^2\right)^{\frac{m}{2}} \left\{x^{n-m} - \frac{(n-m)(n-m-1)}{2(2n-1)} x^{n-m-2}\right.$$
$$\left. + \frac{(n-m)(n-m-1)(n-m-2)(n-m-3)}{2 \cdot 4(2n-1)(2n-3)} x^{n-m-4} - \cdots\right\} \qquad \text{MO 73}$$

$$= \frac{(-1)^m (2n-1)!!}{(n-m)!} \left(1 - x^2\right)^{\frac{m}{2}} x^{n-m} F\left(\frac{m-n}{2}, \frac{m-n+1}{2}; \frac{1}{2} - n; \frac{1}{x^2}\right) \qquad \text{MO 73}$$

8.813   Special cases:

1. $P_1^1(x) = -\left(1 - x^2\right)^{1/2} = -\sin\varphi$ MO 73

2. $P_2^1(x) = -3\left(1 - x^2\right)^{1/2} x = -\frac{3}{2}\sin 2\varphi$ MO 73

3. $P_2^2(x) = 3\left(1 - x^2\right) = \frac{3}{2}\left(1 - \cos 2\varphi\right)$ MO 73

4. $P_3^1(x) = -\frac{3}{2}\left(1 - x^2\right)^{1/2}\left(5x^2 - 1\right) = -\frac{3}{8}\left(\sin\varphi + 5\sin 3\varphi\right)$ MO 73

5. $P_3^2(x) = 15\left(1 - x^2\right) x = \frac{15}{4}\left(\cos\varphi - \cos 3\varphi\right)$ MO 73

6. $P_3^3(x) = -15\left(1 - x^2\right)^{3/2} = -\frac{15}{4}\left(3\sin\varphi - \sin 3\varphi\right)$ MO 73

Functional relations

For recursion formulas, see **8.731**.

8.814 $P_n \left(\cos \varphi_1 \cos \varphi_2 + \sin \varphi_1 \sin \varphi_2 \cos \Theta \right)$

$$= P_n \left(\cos \varphi_1 \right) P_n \left(\cos \varphi_2 \right) + 2 \sum_{m=1}^{n} \frac{(n-m)!}{(n+m)!} \, P_n^m \left(\cos \varphi_1 \right) P_n^m \left(\cos \varphi_2 \right) \cos m\Theta$$

$$[0 \le \varphi_1 \le \pi, \quad 0 \le \varphi_2 \le \pi] \qquad (\text{``addition theorem''}) \quad \text{MO 74}$$

8.815   If

$$Y_{n_1}(\varphi, \vartheta) = A_0 \, P_{n_1}(\cos \varphi) + \sum_{m=1}^{n_1} \left(a_m \cos m\vartheta + b_m \sin m\vartheta \right) P_{n_1}^m (\cos \varphi),$$

$$Z_{n_2}(\varphi, \vartheta) = \alpha_0 \, P_{n_2}(\cos \varphi) + \sum_{m=1}^{n_2} \left(\alpha_m \cos m\vartheta + \beta_m \sin m\vartheta \right) P_{n_2}^m (\cos \varphi),$$

then

$$\int_0^{2\pi} d\vartheta \int_0^{\pi} \sin \varphi \, d\varphi \, Y_{n_1}(\varphi, \vartheta) \, Y_{n_2}(\varphi, \vartheta) = 0,$$

$$\int_0^{2\pi} d\vartheta \int_0^{\pi} \sin \varphi \, d\varphi \, Y_n(\varphi, \vartheta) \, P_n \left[\cos \varphi \cos \psi + \sin \varphi \sin \psi \cos(\vartheta - \theta) \right] = \frac{4\pi}{2n+1} \, Y_n(\psi, \theta) \qquad \text{MO 75}$$

8.816 $\left(\cos \varphi + i \sin \varphi \cos \vartheta \right)^n = P_n(\cos \varphi) + 2 \sum_{m=1}^{n} (-1)^m \frac{n!}{(n+m)!} \cos m\vartheta \, P_n^m (\cos \varphi)$ MO 75

For integrals of the functions, $P_n^m(x)$, see **7.112 1**, **7.122 1**.

8.82–8.83   Legendre functions

8.820   The differential equation

$$\frac{d}{dz} \left[(1 - z^2) \frac{du}{dz} \right] + \nu(\nu + 1) u = 0 \qquad (\text{cf. } \textbf{8.700 1}),$$

where the parameter ν can be an arbitrary number, has the following two linearly independent solutions:

1. $P_\nu(z) = F \left(-\nu, \nu + 1; 1; \dfrac{1 - z}{2} \right)$

2. $Q_\nu(z) = \dfrac{\Gamma(\nu + 1) \Gamma \left(\frac{1}{2} \right)}{2^{\nu+1} \Gamma \left(\nu + \frac{3}{2} \right)} z^{-\nu - 1} F \left(\dfrac{\nu + 2}{2}, \dfrac{\nu + 1}{2}; \dfrac{2\nu + 3}{2}; \dfrac{1}{z^2} \right)$ SM 518(137)

     The functions $P_\nu(z)$ and $Q_\nu(z)$ are called *Legendre functions of the first* and *second kind* respectively. If ν is not an integer, the function $P_\nu(z)$ has *singularities* at $z = -1$ and $z = \infty$. However, if $\nu = n = 0, 1, 2, \ldots$, the function $P_\nu(z)$ becomes the *Legendre polynomial* $P_n(z)$ (see **8.91**) For $\nu = -n = -1, -2, \ldots$, we have

$$P_{-n-1}(z) = P_n(z)$$

3.   If $\nu \ne 0, 1, 2, \ldots$, the function $Q_\nu(z)$ has singularities at the points $z = \pm 1$ and $z = \infty$. These points are branch points of the function. On the other hand, if $\nu = n = 0, 1, 2, \ldots$, the function $Q_n(z)$ is single-valued for $|z| > 1$ and regular for $z = \infty$.

4.   In the right half-plane,

$$P_\nu(z) = \left(\frac{1+z}{2}\right)^\nu F\left(-\nu, -\nu; 1; \frac{z-1}{z+1}\right) \qquad [\operatorname{Re} z > 0]$$

5. The function $P_\nu(z)$ is uniquely determined by equations **8.820** 1 and **8.820** 4 within a circle of radius 2 with its center at the point $z = 1$ in the right half-plane.

    For $z = x = \cos\varphi$, a solution of equation **8.820** is the function

6. $$P_\nu(x) = P_\nu(\cos\varphi) = F\left(-\nu, \nu+1; 1; \sin^2\frac{\varphi}{2}\right);$$

    In general,

7. $$P_\nu(z) = P_{-\nu-1}(z) = P_\nu(x) = P_{-\nu-1}(x), \text{ for } z = x$$

8. The function $Q_\nu(z)$ for $|z| > 1$ is uniquely determined by equation **8.820** 2 everywhere in the z-plane in which a cut is made from the point $z = -\infty$ to the point $z = 1$. By means of a hypergeometric series, the function can be continued analytically inside the unit circle. On the cut $(-1 \leq x \leq +1)$ of the real axis, the function $Q_\nu(x)$ is determined by the equation

9. $$Q_\nu(x) = \tfrac{1}{2}\left[Q_\nu(x+i0) + Q_\nu(x-i0)\right] \hspace{4cm} \text{HO 52(53), WH}$$

Integral representations

8.821

1. $$P_\nu(z) = \frac{1}{2\pi i}\int_A^{(1+,z+)} \frac{(t^2-1)^\nu}{2^\nu(t-z)^{\nu+1}}\, dt$$

    Here, A is a point on the real axis to the right of the point $t = 1$ and to the right of z if z is real. At the point A, we set

    $$\arg(t-1) = \arg(t+1) = 0 \text{ and } [|\arg(t-z)| < \pi] \hspace{3cm} \text{WH}$$

2. $$Q_\nu(z) = \frac{1}{4i\sin\nu\pi}\int_A^{(1-,1+)} \frac{(t^2-1)^\nu}{2^\nu(z-t)^{\nu+1}}\, dt$$

    [ν is not an integer; the point A is at the end of the major axis of an ellipse to the right of $t = 1$ drawn in the t-plane with foci at the points ± 1 and with a minor axis sufficiently small that the point z lies outside it. The contour begins at the point A, follows the path $(1-, -1+)$ and returns to A; $|\arg z| \leq \pi$ and $|\arg(z-t)| \to \arg z$ as $t \to 0$ on the contour; $\arg(t+1) = \arg(t-1) = 0$ at the point A; z does not lie on the real axis between -1 and 1.]

    For $\nu = n$ an integer,

3. $$Q_n(z) = \frac{1}{2^{n+1}}\int_{-1}^1 (1-t^2)^n (z-t)^{-n-1}\, dt \hspace{3cm} \text{SM 517(134), WH}$$

8.822

1. $$P_\nu(z) = \frac{1}{\pi}\int_0^\pi \frac{d\varphi}{\left(z+\sqrt{z^2-1}\cos\varphi\right)^{\nu+1}} = \frac{1}{\pi}\int_0^\pi \left(z+\sqrt{z^2-1}\cos\varphi\right)^\nu d\varphi$$

    $$\left[\operatorname{Re} z > 0 \text{ and } \arg\left\{z+\sqrt{z^2-1}\cos\varphi\right\} = \arg z \text{ for } \varphi = \frac{\pi}{2}\right] \hspace{1cm} \text{WH}$$

2. $\quad Q_\nu(z) = \displaystyle\int_0^\infty \frac{d\varphi}{\left(z + \sqrt{z^2 - 1}\,\cosh\varphi\right)^{\nu+1}},$

$\left[\mathrm{Re}\,\nu > -1; \quad \text{if } \nu \text{ is not an integer}, \left\{\left(z + \sqrt{z^2 - 1}\right)\cosh\varphi\right\} \text{ for } \varphi = 0 \text{ has its principal value}\right]$

WH

8.823 $\quad P_\nu(\cos\theta) = \dfrac{2}{\pi}\displaystyle\int_0^\theta \frac{\cos\left(\nu + \frac{1}{2}\right)\varphi}{\sqrt{2(\cos\varphi - \cos\theta)}}\,d\varphi$

WH

8.824 $\quad Q_n(z) = 2^n n! \displaystyle\int_z^\infty \cdots \int_z^\infty \frac{(dz)^{n+1}}{(z^2 - 1)^{n+1}} = 2^n \int_z^\infty \frac{(t - z)^n}{(t^2 - 1)^{n+1}}\,dt$

$\qquad = \dfrac{(-1)^n}{(2n-1)!!}\dfrac{d^n}{dz^n}\left[(z^2 - 1)^n \displaystyle\int_z^\infty \frac{dt}{(t^2 - 1)^{n+1}}\right]$ $\qquad\qquad$ [$\mathrm{Re}\,z > 1$]

WH, MO 78

8.825 $\quad Q_n(z) = \dfrac{1}{2}\displaystyle\int_{-1}^1 \frac{P_n(t)}{z - t}\,dt$ $\qquad$ [$|\arg(z - 1)| < \pi$]

WH, MO 78

See also **6.622** 3, **8.842**.

8.826 Fourier series:

1. $\quad P_n(\cos\varphi) = \dfrac{2^{n+2}}{\pi}\dfrac{n!}{(2n+1)!!}\left[\sin(n+1)\varphi + \dfrac{1}{1}\dfrac{n+1}{2n+3}\sin(n+3)\varphi\right.$

$\qquad\qquad + \left.\dfrac{1 \cdot 3(n+1)(n+2)}{1 \cdot 2(2n+3)(2n+5)}\sin(n+5)\varphi + \ldots\right]$

$\qquad\qquad\qquad\qquad\qquad\qquad\qquad\qquad$ [$0 < \varphi < \pi$] $\qquad$ MO 79

2. $\quad Q_n(\cos\varphi) = 2^{n+1}\dfrac{n!}{(2n+1)!!}\left[\cos(n+1)\varphi + \dfrac{1}{1}\dfrac{n+1}{2n+3}\cos(n+3)\varphi\right.$

$\qquad\qquad + \left.\dfrac{1 \cdot 3}{1 \cdot 2}\dfrac{(n+1)(n+2)}{(2n+3)(2n+5)}\cos(n+5)\varphi + \ldots\right]$

$\qquad\qquad\qquad\qquad\qquad\qquad\qquad\qquad$ [$0 < \varphi < \pi$] $\qquad$ MO 79

The expressions for Legendre functions in terms of a hypergeometric function (see **8.820**) provide other series representations of these functions.

Special cases and particular values

8.827

1. $\quad Q_0(x) = \dfrac{1}{2}\ln\dfrac{1 + x}{1 - x} = \mathrm{arctanh}\,x$

JA

2. $\quad Q_1(x) = \dfrac{x}{2}\ln\dfrac{1 + x}{1 - x} - 1$

JA

3. $\quad Q_2(x) = \dfrac{1}{4}(3x^2 - 1)\ln\dfrac{1 + x}{1 - x} - \dfrac{3}{2}x$

JA

4. $\quad Q_3(x) = \dfrac{1}{4}(5x^3 - 3x)\ln\dfrac{1 + x}{1 - x} - \dfrac{5}{2}x^2 + \dfrac{2}{3}$

JA

5. $Q_4(x) = \dfrac{1}{16}\left(35x^4 - 30x^2 + 3\right)\ln\dfrac{1+x}{1-x} - \dfrac{35}{8}x^3 + \dfrac{55}{24}x$ JA

6. $Q_5(x) = \dfrac{1}{16}\left(63x^5 - 70x^3 + 15x\right)\ln\dfrac{1+x}{1-x} - \dfrac{63}{8}x^4 + \dfrac{49}{8}x^2 - \dfrac{8}{15}$ JA

8.828

1. $P_\nu(1) = 1$ MO 79

2. $P_\nu(0) = -\dfrac{1}{2}\dfrac{\sin\nu\pi}{\sqrt{\pi^3}}\,\Gamma\left(\dfrac{\nu+1}{2}\right)\Gamma\left(-\dfrac{\nu}{2}\right)$ MO 79

8.829 $Q_\nu(0) = \dfrac{1}{4\sqrt{\pi}}\,(1 - \cos\nu\pi)\,\Gamma\left(\dfrac{\nu+1}{2}\right)\Gamma\left(-\dfrac{\nu}{2}\right)$ MO 79

Functional relationships

8.831

1. $Q_\nu(x) = \dfrac{\pi}{2\sin\nu\pi}\left[\cos\nu\pi\, P_\nu(x) - P_\nu(-x)\right]$ $[\nu \neq 0,\quad \pm 1, \pm 2, \dots]$ MO 76

2. $Q_n(x) = \dfrac{1}{2}\, P_n(x)\ln\dfrac{1+x}{1-x} - W_{n-1}(x)$ $[n = 0, 1, 2, \dots]$,

    where

3. $W_{n-1}(x) = \displaystyle\sum_{k=0}^{\lfloor\frac{n-1}{2}\rfloor} \dfrac{2(n-2k)-1}{(2k+1)(n-k)}\, P_{n-2k-1}(x) = \sum_{k=1}^{n} \dfrac{1}{k}\, P_{k-1}(x)\, P_{n-k}(x)$

    and

4. $W_{-1}(x) \equiv 0$ (see also **8.839**)   SM 516(131), MO 76

5. $\displaystyle\sum_{k=0}^{\infty}(-1)^k\left(\dfrac{1}{\nu-k} - \dfrac{1}{\nu+k+1}\right)P_k(\cos\varphi) = \dfrac{\pi}{\sin\nu\pi}\, P_\nu(\cos\varphi)$

                                              $[\nu \text{ not an integer};\quad 0 \leq \varphi < \pi]$ MO 77

6. $\displaystyle\sum_{k=0}^{\infty}(-1)^k\left(\dfrac{1}{\nu-k} - \dfrac{1}{\nu+k+1}\right)P_k(\cos\varphi)\, P_k(\cos\psi) = \dfrac{\pi}{\sin\nu\pi}\, P_\nu(\cos\varphi)\, P_\nu(\cos\psi)$

                  $[\nu \text{ not an integer},\quad -\pi < \varphi + \psi < \pi,\quad -\pi < \varphi - \psi < \pi]$ MO 77

See also **8.521** 4.

8.832

1. $\left(z^2 - 1\right)\dfrac{d}{dz}\, P_\nu(z) = (\nu+1)\left[P_{\nu+1}(z) - z\, P_\nu(z)\right]$ WH

2. $(2\nu+1)z\, P_\nu(z) = (\nu+1)\, P_{\nu+1}(z) + \nu\, P_{\nu-1}(z)$ WH

3. $\left(z^2 - 1\right)\dfrac{d}{dz}\, Q_\nu(z) = (\nu+1)\left[Q_{\nu+1}(z) - z\, Q_\nu(z)\right]$ WH

4. $(2\nu+1)z\, Q_\nu(z)\, Q_{\nu+1}(z) + \nu\, Q_{\nu-1}(z)$ WH

8.833

1. $\quad P_\nu(-z) = e^{\nu\pi i}\, P_\nu(z) - \dfrac{2}{\pi}\sin\nu\pi\, Q_\nu(z)$ $\qquad$ [Im $z < 0$] $\qquad$ MO 77

2. $\quad P_\nu(-z) = e^{-\nu\pi i}\, P_\nu(z) - \dfrac{2}{\pi}\sin\nu\pi\, Q_\nu(z)$ $\qquad$ [Im $z > 0$] $\qquad$ MO 77

3. $\quad Q_\nu(-z) = -e^{-\nu\pi i}\, Q_\nu(z)$ $\qquad$ [Im $z < 0$] $\qquad$ MO 77

4. $\quad Q_\nu(-z) = -e^{\nu\pi i}\, Q_\nu(z)$ $\qquad$ [Im $z > 0$] $\qquad$ MO 77

8.834

1. $\quad Q_\nu\left(x \pm i0\right) = Q_\nu(x) \mp \dfrac{\pi i}{2}\, P_\nu(x)$ $\qquad$ MO 77

2. $\quad Q_n(z) = \dfrac{1}{2}\, P_n(z)\ln\dfrac{z+1}{z-1} - W_{n-1}(z)$ $\qquad$ (see **8.831** 3) $\qquad$ MO 77

8.835

1. $\quad Q_\nu(z) - Q_{-\nu-1}(z) = \pi\cot\nu\pi\, P_\nu(z)$ $\qquad$ [$\sin\nu\pi \neq 0$] $\qquad$ MO 77

2. $\quad Q_{-\nu-1}\left(\cos\varphi\right) = Q_\nu\left(\cos\varphi\right) - \pi\cot\nu\pi\, P_\nu\left(\cos\varphi\right)$ $\qquad$ [$\sin\nu\pi \neq 0$] $\qquad$ MO 77

3. $\quad Q_\nu\left(-\cos\varphi\right) = -\cos\nu\pi\, Q_\nu\left(\cos\varphi\right) - \dfrac{\pi}{2}\sin\nu\pi\, P_\nu\left(\cos\varphi\right)$ $\qquad$ MO 77

8.836

1. $\quad Q_n(z) = \dfrac{1}{2^n n!}\dfrac{d^n}{dz^n}\left[\left(z^2-1\right)^n \ln\dfrac{z+1}{z-1}\right] - \dfrac{1}{2}\, P_n(z)\ln\dfrac{z+1}{z-1}$ $\qquad$ MO 79

2. $\quad Q_n(x) = \dfrac{1}{2^n n!}\dfrac{d^n}{dx^n}\left[\left(x^2-1\right)^n \ln\dfrac{1+x}{1-x}\right] - \dfrac{1}{2}\, P_n(x)\ln\dfrac{1+x}{1-x}$ $\qquad$ MO 79

8.837

1. $\quad P_\nu(x) = P_\nu\left(\cos\varphi\right) = F\left(-\nu, \nu+1; 1; \sin^2\dfrac{\varphi}{2}\right)$ $\qquad$ (cf. **8.820** 6) $\qquad$ MO 76

2. $\quad P_\nu(z) = \dfrac{\tan\nu\pi}{2^{\nu+1}\sqrt{\pi}}\dfrac{\Gamma(\nu+1)}{\Gamma\left(\nu+\frac{3}{2}\right)}z^{-\nu-1}\, F\left(\dfrac{\nu}{2}+1, \dfrac{\nu+1}{2}; \nu+\dfrac{3}{2}; \dfrac{1}{z^2}\right)$

$\qquad\qquad + \dfrac{2^\nu}{\sqrt{\pi}}\dfrac{\Gamma\left(\nu+\frac{1}{2}\right)}{\Gamma(\nu+1)}z^\nu\, F\left(\dfrac{1-\nu}{2}, -\dfrac{\nu}{2}; \dfrac{1}{2}-\nu; \dfrac{1}{z^2}\right)$

$\qquad$ MO 78

See also **8.820**.

For integrals of Legendre functions, see **7.1**–**7.2**.

8.838 Inequalities ($0 \leq \varphi \leq \pi$, $\nu > 1$, and C_0 is a number that does not depend on the values of ν or φ):

1. $\quad \left|P_\nu\left(\cos\varphi\right) - P_{\nu+2}\left(\cos\varphi\right)\right| \leq 2C_0\sqrt{\dfrac{1}{\nu\pi}}$ $\qquad$ MO 78

2. $\quad \left|Q_\nu\left(\cos\varphi\right) - Q_{\nu+2}\left(\cos\varphi\right)\right| < C_0\sqrt{\dfrac{\pi}{\nu}}$ $\qquad$ MO 78

With regard to the zeros of Legendre functions of the second kind, see **8.784**, **8.785**, and **8.786**. For the expansion of Legendre functions in series of associated Legendre functions, see **8.794**, **8.795**, and **8.796**.

8.839 A differential equation leading to the functions W_{n-1} (see **8.831** 3):

$$\left(1 - x^2\right) \frac{d^2 W_{n-1}}{dx^2} - 2x \frac{dW_{n-1}}{dx} + (n+1)nW_{n-1} = 2\frac{d\,P_\nu}{dx} \qquad \text{MO 76}$$

8.84 Conical functions

8.840 Let us set

$$\nu = -\tfrac{1}{2} + i\lambda,$$

where λ is a real parameter, in the defining differential equation **8.700** 1 for associated Legendre functions. We then obtain the differential equation of the so-called conical functions. A conical function is a special case of the associated Legendre function. However, the Legendre functions

$$P_{-\frac{1}{2}+i\lambda}(x), \quad Q_{-\frac{1}{2}+i\lambda}(x)$$

have certain peculiarities that make us distinguish them as a special class—the class of conical functions. The most important of these peculiarities is the following

8.841 The functions

$$P_{-\frac{1}{2}+i\lambda}\left(\cos\varphi\right) = 1 + \frac{4\lambda^2 + 1^2}{2^2} \sin^2\frac{\varphi}{2} + \frac{\left(4\lambda^2 + 1^2\right)\left(4\lambda^2 + 3^2\right)}{2^2 4^2} \sin^4\frac{\varphi}{2} + \dots$$

are real for real values of φ. Also,

$$P_{-\frac{1}{2}+i\lambda}(x) \equiv P_{-\frac{1}{2}-i\lambda}(x) \qquad \text{MO 95}$$

8.842 Integral representations:

1. $$P_{-\frac{1}{2}+i\lambda}\left(\cos\varphi\right) = \frac{2}{\pi}\int_0^\varphi \frac{\cosh\lambda u\, du}{\sqrt{2\left(\cos u - \cos\varphi\right)}} = \frac{2}{\pi}\cosh\lambda\pi \int_0^\infty \frac{\cos\lambda u\, du}{\sqrt{2\left(\cos\varphi + \cosh u\right)}} \qquad \text{MO 95}$$

2.[6] $$Q_{-\frac{1}{2}\mp\lambda i}\left(\cos\varphi\right) = \pm i \sinh\lambda\pi \int_0^\infty \frac{\cos\lambda u\, du}{\sqrt{2\left(\cosh u + \cos\varphi\right)}} + \int_0^\infty \frac{\cos\lambda u\, du}{\sqrt{2\left(\cosh u - \cos\varphi\right)}} \qquad \text{MO 95}$$

Functional relations

(See also **8.73**)

8.843 $$P_{-\frac{1}{2}+i\lambda}\left(-\cos\varphi\right) = \frac{\cosh\lambda\pi}{\pi} \left[Q_{-\frac{1}{2}+i\lambda}\left(\cos\varphi\right) + Q_{-\frac{1}{2}-i\lambda}\left(\cos\varphi\right)\right] \qquad \text{MO 95}$$

8.844

1. $$P_{-\frac{1}{2}+i\lambda}\left(\cos\psi\cos\vartheta + \sin\psi\sin\vartheta\cos\varphi\right)$$

$$= P_{-\frac{1}{2}+i\lambda}\left(\cos\psi\right) P_{-\frac{1}{2}+i\lambda}\left(\cos\vartheta\right) + 2\sum_{k=1}^\infty \frac{(-1)^k 2^{2k}\, P^k_{-\frac{1}{2}+i\lambda}\left(\cos\psi\right) P^k_{-\frac{1}{2}+i\lambda}\left(\cos\vartheta\right)\cos k\varphi}{\left(4\lambda^2 + 1^2\right)\left(4\lambda^2 + 3^2\right)\cdots\left[4\lambda^2 + (2k-1)^2\right]}$$

$$\left[0 < \vartheta < \frac{\pi}{2}, \quad 0 < \psi < \pi, \quad 0 < \psi + \vartheta < \pi\right] \qquad \text{(cf. } \textbf{8.794} \text{ 1)} \quad \text{MO 95}$$

2. $$P_{-\frac{1}{2}+i\lambda}\left(-\cos\psi\cos\vartheta - \sin\psi\sin\vartheta\cos\varphi\right)$$

$$= P_{-\frac{1}{2}+i\lambda}\left(\cos\psi\right) P_{-\frac{1}{2}+i\lambda}\left(-\cos\vartheta\right) + 2\sum_{k=1}^\infty \frac{(-1)^k 2^{2k}\, P^k_{-\frac{1}{2}+i\lambda}\left(\cos\psi\right) P^k_{-\frac{1}{2}+i\lambda}\left(-\cos\vartheta\right)\cos k\varphi}{\left(4\lambda^2 + 1\right)\left(4\lambda^2 + 3^2\right)\cdots\left[4\lambda^2 + (2k-1)^2\right]}$$

$$\left[0 < \psi < \frac{\pi}{2} < \vartheta, \quad \psi + \vartheta < \pi\right] \qquad \text{(cf. } \textbf{8.796}\text{)} \quad \text{MO 95}$$

3. $Q_{-\frac{1}{2}+i\lambda}\left(\cos\psi\cos\vartheta+\sin\psi\sin\vartheta\cos\varphi\right)$

$$= P_{-\frac{1}{2}+i\lambda}\left(\cos\psi\right)Q_{-\frac{1}{2}+i\lambda}\left(\cos\vartheta\right)+2\sum_{k=1}^{\infty}\frac{(-1)^k 2^{2k}\,P^k_{-\frac{1}{2}+i\lambda}\left(\cos\psi\right)Q^k_{-\frac{1}{2}+i\lambda}\left(\cos\vartheta\right)\cos k\varphi}{\left(4\lambda^2+1\right)\left(4\lambda^2+3^2\right)\cdots\left[4\lambda^2+(2k-1)^2\right]}$$

$$\left[0<\psi<\frac{\pi}{2}<\vartheta,\quad\psi+\vartheta<\pi\right]\qquad\text{(cf. }\textbf{8.794}\,2)\qquad\text{MO 96}$$

Regarding the zeros of conical functions, see **8.784**.

8.85   Toroidal functions*

8.850   Solutions of the differential equation

1. $\dfrac{d^2u}{d\eta^2}+\dfrac{\cosh\eta}{\sinh\eta}\dfrac{du}{d\eta}-\left(n^2-\dfrac{1}{4}+\dfrac{m^2}{\sinh^2\eta}\right)u=0,$

are called toroidal functions. They are equivalent (under a coordinate transformation) to associated Legendre functions. In particular, the functions

$$P^m_{n-\frac{1}{2}}\left(\cosh\eta\right),\quad Q^m_{n-\frac{1}{2}}\left(\sinh\eta\right)\qquad\qquad\text{MO 96}$$

are solutions of equation **8.850** 1.

The following formulas, obtained from the formulas obtained earlier for associated Legendre functions, are valid for toroidal functions:

8.851   Integral representations:

1. $P^m_{n-\frac{1}{2}}\left(\cosh\eta\right)=\dfrac{\Gamma\left(n+m+\frac{1}{2}\right)}{\Gamma\left(n-m+\frac{1}{2}\right)}\dfrac{(\sinh\eta)^m}{2^m\sqrt{\pi}\,\Gamma\left(m+\frac{1}{2}\right)}\displaystyle\int_0^\pi\dfrac{\sin^{2m}\varphi\,d\varphi}{\left(\cosh\eta+\sinh\eta\cos\varphi\right)^{n+m+\frac{1}{2}}}$

$$=\dfrac{(-1)^m}{2\pi}\dfrac{\Gamma\left(n+\frac{1}{2}\right)}{\Gamma\left(n-m+\frac{1}{2}\right)}\int_0^{2\pi}\dfrac{\cos m\varphi\,d\varphi}{\left(\cosh\eta+\sinh\eta\cos\varphi\right)^{n+\frac{1}{2}}}$$

$$\text{MO 96}$$

2. $Q^m_{n-\frac{1}{2}}\left(\cosh\eta\right)=(-1)^m\dfrac{\Gamma\left(n+\frac{1}{2}\right)}{\Gamma\left(n-m+\frac{1}{2}\right)}\displaystyle\int_0^\infty\dfrac{\cosh mt\,dt}{\left(\cosh\eta+\sinh\eta\cosh t\right)^{n+\frac{1}{2}}}\qquad [n\geq m]$

$$=(-1)^m\dfrac{\Gamma\left(n+m+\frac{1}{2}\right)}{\Gamma\left(n+\frac{1}{2}\right)}\int_0^{\ln\coth\frac{\eta}{2}}\left(\cosh\eta-\sinh\eta\cosh t\right)^{n-\frac{1}{2}}\cosh mt\,dt$$

$$\text{MO 96}$$

8.852   Functional relations:

1. $Q^m_{n-\frac{1}{2}}\left(\cosh\eta\right)=(-1)^m\dfrac{2^m\,\Gamma\left(n+m+\frac{1}{2}\right)\sqrt{\pi}}{\Gamma(n+1)}\sinh^m\left(\eta e^{-\left(n+m+\frac{1}{2}\right)\eta}\right)$

$$\times F\left(m+\tfrac{1}{2},n+m+\tfrac{1}{2};n+1;e^{-2\eta}\right)$$

$$\text{MO 96}$$

---

*Sometimes called *torus functions*

2. $\qquad P_{n-\frac{1}{2}}^{-m}(\cosh\eta) = \dfrac{2^{-2m}}{\Gamma(m+1)}\left(1-e^{-2\eta}\right)^m e^{-\left(n+\frac{1}{2}\right)\eta} F\left(m+\tfrac{1}{2}, n+m+\tfrac{1}{2}; 2m+1; 1-e^{-2\eta}\right)$

MO 96

8.853   An asymptotic representation $P_{n-\frac{1}{2}}(\cosh\eta)$ for large values of n:

$$P_{n-\frac{1}{2}}(\cosh\eta) = \frac{\Gamma(n)e^{\left(n-\frac{1}{2}\right)\eta}}{\sqrt{\pi}\,\Gamma\left(n+\frac{1}{2}\right)}$$
$$\times\left[\frac{2\,\Gamma^2\left(n+\frac{1}{2}\right)}{\pi n!\,\Gamma(n)}\ln(4e^\eta)\,e^{-2n\eta}\,F\left(\frac{1}{2}, n+\frac{1}{2}; n+1; e^{-2\eta}\right)+A+B\right],$$

where

$$A = 1+\frac{1}{2^2}\frac{1\cdot(2n-1)}{1\cdot(n-1)}e^{-2\eta}+\frac{1}{2^4}\frac{1\cdot3\cdot(2n-1)(2n-3)}{1\cdot2\cdot(n-1)(n-2)}e^{-4\eta}+\cdots+\frac{1}{2^{2n-2}}\left(\frac{(2n-1)!!}{(n-1)!}\right)^2 e^{-2(n-1)\eta}$$

$$B = \frac{\Gamma\left(n+\frac{1}{2}\right)}{\sqrt{\pi^3}\,\Gamma(n)}\sum_{k=1}^{\infty}\frac{\Gamma\left(k+\frac{1}{2}\right)\Gamma\left(n+k+\frac{1}{2}\right)}{\Gamma(n+k+1)\,\Gamma(k+1)}\left(u_{n+k}+u_k-v_{n+k-\frac{1}{2}}-v_{k-\frac{1}{2}}\right)e^{-2(n+k)\eta}$$

Here,

$$u_r = \sum_{s=1}^{r}\frac{1}{s}, \quad v_{r-\frac{1}{2}} = \sum_{s=1}^{r}\frac{2}{2s-1} \qquad [r\text{ is a natural number}]$$

MO 97

8.9 Orthogonal Polynomials

8.90 Introduction

8.901   Suppose that $w(x)$ is a nonnegative real function of a real variable x. Let (a,b) be a fixed interval on the x-axis. Let us suppose further that, for $n = 0, 1, 2, \ldots$, the integral

$$\int_a^b x^n w(x)\,dx$$

exists and that the integral

$$\int_a^b w(x)\,dx$$

is positive. In this case, there exists a sequence of polynomials $p_0(x), p_1(x), \ldots, p_n(x), \ldots$, that is uniquely determined by the following conditions:

1. $p_n(x)$ is a polynomial of degree n and the coefficient of x^n in this polynomial is positive.

2.   The polynomials $p_0(x), p_1(x), \ldots$ are orthonormal; that is,

$$\int_a^b p_n(x)p_m(x)w(x)\,dx = \begin{cases} 0 & \text{for } n \neq m, \\ 1 & \text{for } n = m. \end{cases}$$

We say that the polynomials $p_n(x)$ constitute *a system of orthogonal polynomials on the interval* (a,b) *with the weight function* $w(x)$.

8.902   If q_n is the coefficient of x^n in the polynomial $p_n(x)$, then

1. $\displaystyle\sum_{k=0}^{n} p_k(x)p_k(y) = \frac{q_n}{q_{n+1}} \frac{p_{n+1}(x)p_n(y) - p_n(x)p_{n+1}(y)}{x - y}$ (Darboux–Christoffel formula)

EH II 159(10)

2. $\displaystyle\sum_{k=0}^{n} [p_k(x)]^2 = \frac{q_n}{q_{n+1}} \left[p_n(x)p'_{n+1}x) - p'_n x)p_{n+1}(x) \right]$ EH II 159(11)

8.903   Between any three consecutive orthogonal polynomials, there is a dependence

$$p_n(x) = (A_n x + B_n)\, p_{n-1}(x) - C_n p_{n-2}(x) \qquad [n = 2, 3, 4, \ldots]$$

In this formula, A_n, B_n, and C_n are constants and

$$A_n = \frac{q_n}{q_{n-1}}, \quad C_n = \frac{q_n q_{n-2}}{q_{n-1}^2}$$

MO 102

8.904   Examples of normalized systems of orthogonal polynomials:

Notation and name		Interval	Weight
$\left(n + \tfrac{1}{2}\right)^{1/2} P_n(x)$	see **8.91**	$(-1, +1)$	1
$2^\lambda \Gamma(\lambda) \left[\dfrac{(n + \lambda)\, n!}{2\pi\, \Gamma\,(2\lambda + n)}\right]^{1/2} C_n^\lambda(x)$	see **8.93**	$(-1, +1)$	$\left(1 - x^2\right)^{\lambda - \frac{1}{2}}$
$\sqrt{\dfrac{\varepsilon_n}{\pi}}\, T_n(x), \quad \varepsilon_0 = 1, \varepsilon_n = 2 \text{ for } n = 1, 2, 3, \ldots$	see **8.94**	$(-1, +1)$	$\left(1 - x^2\right)^{-1/2}$
$2^{-\frac{n}{2}} \pi^{-1/4} (n!)^{-1/2} H_n(x)$	see **8.95**	$(-\infty, \infty)$	e^{-x^2}
$\left[\dfrac{\Gamma(n+1)\,\Gamma(\alpha + \beta + 1 + n)(\alpha + \beta + 1 + 2n)}{\Gamma(\alpha + 1 + n)\,\Gamma(\beta + 1 + n)2^{\alpha + \beta + 1}}\right]^{1/2} P_n^{(\alpha, \beta)}(x)$	see **8.96**	$(-1, +1)$	$(1 - x)^\alpha (1 + x)^\beta$
$\left[\dfrac{\Gamma(n+1)}{\Gamma\,(\alpha + n + 1)}\right]^{1/2} (-1)^n L_n^\alpha(x)$	see **8.97**	$(0, \infty)$	$x^\alpha e^{-x}$

Cf. **7.221** 1, **7.313**, **7.343**, **7.374** 1, **7.391** 1, **7.414** 3.

8.91   Legendre polynomials

8.910   Definition.   The Legendre polynomials $P_n(z)$ are polynomials satisfying equation **8.700** 1 with $\mu = 0$ and $\nu = n$: that is, they satisfy the differential equation

1. $\left(1 - z^2\right) \dfrac{d^2 u}{dz^2} - 2z \dfrac{du}{dz} + n(n+1)u = 0$

This equation has a polynomial solution if, and only if, n is an integer. Thus, Legendre polynomials constitute a special type of associated Legendre function.

Legendre polynomials of degree n are of the form

2. $P_n(z) = \dfrac{1}{2^n n!} \dfrac{d^n}{dz^n} \left(z^2 - 1\right)^n$

8.911   Legendre polynomials written in expanded form:

1.
$$P_n(z) = \frac{1}{2^n} \sum_{k=0}^{\lfloor \frac{n}{2} \rfloor} \frac{(-1)^k (2n-2k)!}{k!(n-k)!(n-2k)!} z^{n-2k}$$
$$= \frac{(2n)!}{2^n (n!)^2} \left(z^n - \frac{n(n-1)}{2(2n-1)} z^{n-2} + \frac{n(n-1)(n-2)(n-3)}{2 \cdot 4 (2n-1)(2n-3)} z^{n-4} - \cdots \right)$$
$$= \frac{(2n-1)!!}{n!} z^n F\left(-\frac{n}{2}, \frac{1-n}{2}; \frac{1}{2} - n; \frac{1}{z^2} \right)$$

HO 13, AD (9001), MO 69

2.
$$P_{2n}(z) = (-1)^n \frac{(2n-1)!!}{2^n n!} \left(1 - \frac{2n(2n+1)}{2!} z^2 + \frac{2n(2n-2)(2n+1)(2n+3)}{4!} z^4 - \cdots \right)$$
$$= (-1)^n \frac{(2n-1)!!}{2^n n!} F\left(-n, n + \frac{1}{2}; \frac{1}{2}; z^2 \right)$$

AD (9002), MO 69

3.
$$P_{2n+1}(z) = (-1)^n \frac{(2n+1)!!}{2^n n!} \left(z - \frac{2n(2n+3)}{3!} z^3 + \frac{2n(2n-2)(2n+3)(2n+5)}{5!} z^5 - \cdots \right)$$
$$= (-1)^n \frac{(2n+1)!!}{2^n n!} z F\left(-n, n + \frac{3}{2}; \frac{3}{2}; z^2 \right)$$

AD (9002), MO 69

4.
$$P_n(\cos\varphi) = \frac{(2n-1)!!}{2^n n!} \left(\cos n\varphi + \frac{1}{1} \frac{n}{2n-1} \cos(n-2)\varphi \right.$$
$$+ \frac{1 \cdot 3}{1 \cdot 2} \frac{n(n-1)}{(2n-1)(2n-3)} \cos(n-4)\varphi$$
$$\left. + \frac{1 \cdot 3 \cdot 5}{1 \cdot 2 \cdot 3} \frac{n(n-1)(n-2)}{(2n-1)(2n-3)(2n-5)} \cos(n-6)\varphi - \cdots \right)$$

WH

5.
$$P_{2n}(\cos\varphi) = (-1)^n \frac{(2n-1)!!}{2^n n!}$$
$$\times \left\{ \sin^{2n}\varphi - \frac{(2n)^2}{2!} \sin^{2n-2}\varphi \cos^2\varphi + \cdots + (-1)^n \frac{2^n n!}{(2n-1)!!} \cos^{2n}\varphi \right\}$$

AD (9011)

6.
$$P_{2n+1}(\cos\varphi) = (-1)^n \frac{(2n+1)!!}{2^n n!} \cos\varphi$$
$$\times \left\{ \sin^{2n}\varphi - \frac{(2n)^2}{3!} \sin^{2n-2}\varphi \cos^2\varphi + \cdots + (-1)^n \frac{2^n n!}{(2n+1)!!} \cos^{2n}\varphi \right\}$$

AD (9012)

7.
$$P_n(z) = \sum_{k=0}^{n} \frac{(-1)^k (n+k)!}{(n-k)!(k!)^2 2^{k+1}} \left[(1-z)^k + (-1)^n (1+z)^k \right]$$

WH

8.912   Special cases:

1. $P_0(x) = 1$

JA

2. $P_1(x) = x = \cos\varphi$ JA

3. $P_2(x) = \dfrac{1}{2}\left(3x^2 - 1\right) = \dfrac{1}{4}\left(3\cos 2\varphi + 1\right)$ JA

4. $P_3(x) = \dfrac{1}{2}\left(5x^3 - 3x\right) = \dfrac{1}{8}\left(5\cos 3\varphi + 3\cos\varphi\right)$ JA

5. $P_4(x) = \dfrac{1}{8}\left(35x^4 - 30x^2 + 3\right) = \dfrac{1}{64}\left(35\cos 4\varphi + 20\cos 2\varphi + 9\right)$ JA

6. $P_5(x) = \dfrac{1}{8}\left(63x^5 - 70x^3 + 15x\right) = \dfrac{1}{128}\left(63\cos 5\varphi + 35\cos 3\varphi + 30\cos\varphi\right)$ JA

7.[10] $P_6(x) = \dfrac{1}{16}\left(231x^6 - 315x^4 + 105x^2 - 5\right) = \dfrac{1}{512}\left(231\cos 6\varphi + 126\cos 4\varphi + 105\cos 2\varphi + 50\right)$

8. $P_7(x) = \dfrac{1}{16}\left(429x^7 - 693x^5 + 315x^3 - 35x\right)$

          $= \dfrac{1}{1024}\left(429\cos 7\varphi + 231\cos 5\varphi + 189\cos 3\varphi + 175\cos\varphi\right)$

9. $P_8(x) = \dfrac{1}{128}\left(6435x^8 - 12012x^6 + 6930x^4 - 1260x^2 + 35\right)$

          $= \dfrac{1}{16384}\left(6435\cos 8\varphi - 3432\cos 6\varphi + 2772\cos 4\varphi - 2520\cos 2\varphi + 1225\right)$

8.913   Integral representations:

1. $P_n(\cos\varphi) = \dfrac{2}{\pi}\displaystyle\int_\varphi^\pi \dfrac{\sin\left(n + \frac{1}{2}\right)t}{\sqrt{2\left(\cos\varphi - \cos t\right)}}\,dt$ WH

     See also **3.611** 3, **3.661** 3, 4.

2.[7]   Schläfli's integral formula:

$$P_n(z) = \frac{1}{2\pi i}\int_C \frac{\left(t^2 - 1\right)^n}{2^n(t - z)^{n+1}}\,dt,$$

     with C a simple contour containing z.   SA 175(9)

3.[10]   Laplace integral formula:

$$P_n(z) = \frac{1}{\pi}\int_0^\pi \left[x + \left(x^2 - 1\right)^{1/2}\cos\varphi\right]^n d\varphi \qquad [|x| \le 1]$$
       SA 180(19)

Functional relations

8.914   Recurrence formulas:

1. $(n+1)\,P_{n+1}(z) - (2n+1)z\,P_n(z) + n\,P_{n-1}(z) = 0$ WH

2. $\left(z^2 - 1\right)\dfrac{d\,P_n}{dz} = n\left[z\,P_n(z) - P_{n-1}(z)\right] = \dfrac{n(n+1)}{2n+1}\left[P_{n+1}(z) - P_{n-1}(z)\right]$ WH

8.915

1.[10] $$\sum_{k=0}^{n}(2k+1)\,P_k(x)\,P_k(y) = (n+1)\frac{P_n(x)\,P_{n+1}(y) - P_n(y)\,P_{n+1}(x)}{y-x}$$

(Christoffel summation formula)

MO 70

1(1).[10] $$(y-x)\sum_{k=0}^{n}(2k+1)\,P_k(x)\,Q_k(y) = 1 - (n+1)\left[P_{n+1}(x)\,Q_n(y) - P_n(x)\,Q_{n+1}(y)\right]$$

AS 335(8.9.2)

2.[7] $$\sum_{k=0}^{\lfloor\frac{n-1}{2}\rfloor}(2n-4k-1)\,P_{n-2k-1}(z) = P'_n(z)$$ (summation theorem)

MO 70

3.[7] $$\sum_{k=0}^{\lfloor\frac{n-2}{2}\rfloor}(2n-4k-3)\,P_{n-2k-2}(z) = z\,P'_n(z) - n\,P_n(z)$$

SM 491(42), WH

4.[10] $$\sum_{k=1}^{\lfloor\frac{n}{2}\rfloor}(2n-4k+1)[k(2n-2k+1)-2]\,P_{n-2k}(z) = z^2\,P''_n(z) - n(n-1)\,P_n(z)$$

WH

5.[10] $$\sum_{k=0}^{m}\frac{a_{m-k}a_k a_{n-k}}{A_{n+m-k}}\left(\frac{2n+2m-4k+1}{2n+2m-2k+1}\right)P_{n+m-2k}(z) = P_n(z)\,P_m(z)$$

$$\left[a_k = \frac{(2k-1)!!}{k!}, \quad m \le n\right]$$ AD (9036)

8.916

1. $$P_n(\cos\varphi) = \frac{(2n-1)!!}{2^n n!}e^{\mp in\varphi}\,F\left(\frac{1}{2}, -n; \frac{1}{2}-n; e^{\pm 2i\varphi}\right)$$

MO 69

2. $$P_n(\cos\varphi) = F\left(n+1, -n; 1; \sin^2\frac{\varphi}{2}\right)$$

MO 69

3. $$P_n(\cos\varphi) = (-1)^n\,F\left(n+1, -n; 1; \cos^2\frac{\varphi}{2}\right)$$

WH

4. $$P_n(\cos\varphi) = \cos^n\varphi\,F\left(-\frac{1}{2}n, \frac{1}{2}-\frac{1}{2}n; 1; -\tan^2\varphi\right)$$

HO 23

5. $$P_n(\cos\varphi) = \cos^{2n}\frac{\varphi}{2}\,F\left(-n, -n; 1; -\tan^2\frac{\varphi}{2}\right)$$

HO 23, 29, WH

See also **8.911** 1, **8.911** 2, **8.911** 3. For a connection with other functions, see **8.936** 3, **8.836**, **8.962** 2.

- For integrals of Legendre polynomials, see **7.22–7.25**.
- For the zeros of Legendre polynomials, see **8.785**.

8.917 Inequalities:

1. $$P_0(x) < P_1(x) < P_2(x) < \cdots < P_n(x) < \ldots$$ $[x > 1]$

MO 71

2. For $x > -1$, $P_0(x) + P_1(x) + \cdots + P_n(x) > 0$.

MO 71

3. $[P_n(\cos\varphi)]^2 > \dfrac{\sin(2n+1)\varphi}{(2n+1)\sin\varphi}$ MO 71

4. $\sqrt{n\sin\varphi}\,|P_n(\cos\varphi)| \le 1.$ MO 71

5. $|P_n(\cos\varphi)| \le 1.$ WH

6.[10]   Let $n \ge 2$.   The successive relative maxima of $|P_n(x)|$, when x decreases from 1 to 0, form a decreasing sequence. More precisely, if $\mu_1, \mu_2, \ldots, \mu_{\lfloor n/2 \rfloor}$ denote these maxima corresponding to decreasing values of x, we have

$$1 > \mu_1 > \mu_2 > \cdots > \mu_{\lfloor n/2 \rfloor} \qquad\qquad\qquad \text{SZ 162(7.3.1)}$$

7.[10]   Let $n \ge 2$.   The successive relative maxima of $(\sin\theta)^{1/2}|P_n(\cos\theta)|$ when θ increases from 0 to $\pi/2$, form an increasing sequence.   SZ 163(7.3.2)

8.[10]   We have

$$(\sin\theta)^{1/2}|P_n(\cos\theta)| < (2/\pi)^{1/2} n^{-1/2} \qquad [0 \le q\theta \le q\pi] \qquad \text{SZ 163(7.3.8)}$$

Here the constant $(2/\pi)^{1/2}$ cannot be replaced by a smaller one.

9.[10] $\displaystyle\max_{0 \le q\theta \le q\pi} (\sin\theta)^{1/2}|P_n(\cos\theta)| \cong (2/\pi)^{1/2} n^{-\frac{1}{2}}$ $[n \to \infty]$ SZ 164(7.3.12)

10.[10]   Stieltjes' first theorem:

$$|P_n(\cos\theta)| \le \left(\frac{2}{\pi}\right)^{1/2} \frac{4}{\sqrt{n\sin\theta}} \qquad [n = 1, 2, \ldots, 0 < \theta < \pi] \qquad \text{SA 197(8)}$$

11.[10]   Stieltjes' second theorem:

$$|P_n(x) - P_{n+2}(x)| < \frac{4}{\sqrt{\pi}\sqrt{n+2}} \qquad [|x| \le 1] \qquad \text{SA 199(15)}$$

12.[10] $\left|\dfrac{d\,P_n(x)}{dx}\right| < \dfrac{2}{\sqrt{\pi}} \dfrac{\sqrt{n}}{1-x^2}$ $[|x| < 1, \quad n = 1, 2, \ldots]$ SA 201(18)

13.[10] $|P_{n+1}(x) + P_n(x)| < 6\left(\dfrac{2}{\pi n}\right)^{\frac{1}{2}}(1-x)^{-1/2}$ $[|x| < 1, \quad n = 0, 1, \ldots]$ SA 201(19)

8.918[10]   Asymptotic approximations:

1. $P_n(\cos\theta) = \left(\dfrac{2}{\pi n\sin\varphi}\right)^{1/2} \cos\left[\left(n+\dfrac{1}{2}\right)\theta - \dfrac{\pi}{4}\right] + O\left(n^{-3/2}\right)$

$$[\varepsilon \le \theta \le \pi - \varepsilon, \quad 0 < \varepsilon < \pi/2m] \qquad \text{(Laplace's formula)} \qquad \text{SA 208(1)}$$

2. $P_n(\cos\theta) = \left(\dfrac{2}{\pi n\sin\theta}\right)^{1/2} \left\{\left(1 - \dfrac{1}{4n}\right)\cos\left[\left(n+\dfrac{1}{2}\right)\theta - \dfrac{\pi}{4}\right] + \dfrac{1}{8n}\cos\theta\sin\left[\left(n+\dfrac{1}{2}\right)\theta - \dfrac{\pi}{4}\right]\right\}$

$$+ O\left(n^{-5/2}\right)$$

$$[\varepsilon \le \theta \le \pi - \varepsilon, \quad 0 < \varepsilon < \pi/2] \qquad \text{(Bonnet–Heine formula)} \qquad \text{SA 208(2)}$$

8.919[10] Series of products of Legendre and Chebyshev polynomials

1.
$$2\int_{-1}^{1} T_n(x)\, P_n(x)\, dx = \sum_{i,j=0}^{i+j=n} \int_{-1}^{1} P_i(x)\, P_j(x)\, P_n(x)\, dx$$

8.92 Series of Legendre polynomials

8.921   The generating function:

$$\frac{1}{\sqrt{1-2tz+t^2}} = \sum_{k=0}^{\infty} t^k\, P_k(z) \qquad \left[|t| < \min\left|z \pm \sqrt{z^2-1}\right|\right] \qquad \text{SM 489(31), WH}$$

$$= \sum_{k=0}^{\infty} \frac{1}{t^{k+1}}\, P_k(z) \qquad \left[|t| > \max\left|z \pm \sqrt{z^2-1}\right|\right] \qquad \text{MO 70}$$

8.922

1.
$$z^{2n} = \frac{1}{2n+1}\, P_0(z) + \sum_{k=1}^{\infty}(4k+1)\frac{2n(2n-2)\dots(2n-2k+2)}{(2n+1)(2n+3)\dots(2n+2k+1)}\, P_{2k}(z) \qquad \text{MO 72}$$

2.
$$z^{2n+1} = \frac{3}{2n+3}\, P_1(z) + \sum_{k=1}^{\infty}(4k+3)\frac{2n(2n-2)\dots(2n-2k+2)}{(2n+3)(2n+5)\dots(2n+2k+3)}\, P_{2k+1}(z) \qquad \text{MO 72}$$

3.
$$\frac{1}{\sqrt{1-x^2}} = \frac{\pi}{2}\sum_{k=0}^{\infty}(4k+1)\left\{\frac{(2k-1)!!}{2^k k!}\right\}^2 P_{2k}(x) \qquad [|x| < 1, \quad (-1)!! \equiv 1]$$

$$\text{MO 72, LA 385(15)}$$

4.
$$\frac{x}{\sqrt{1-x^2}} = \frac{\pi}{2}\sum_{k=0}^{\infty}(4k+3)\frac{(2k-1)!!(2k+1)!!}{2^{2k+1}k!(k+1)!}\, P_{2k+1}(x)$$

$$[|x| < 1, \quad (-1)!! \equiv 1] \qquad \text{LA 385(17)}$$

5.
$$\sqrt{1-x^2} = \frac{\pi}{2}\left\{\frac{1}{2} - \sum_{k=1}^{\infty}(4k+1)\frac{(2k-3)!!(2k-1)!!}{2^{2k+1}k!(k+1)!}\, P_{2k}(x)\right\}$$

$$[|x| < 1, \quad (-1)!! \equiv 1] \qquad \text{LA 385(18)}$$

6.[10]
$$\sqrt{\frac{1-x}{2}} = \frac{2}{3}\, P_0(x) - 2\sum_{n=1}^{\infty}\frac{1}{(2n-1)(2n+3)}\, P_n(x) \qquad [-1 \le x \le 1]$$

7.[10]
$$\frac{1-\rho^2}{(1-2\rho x+\rho^2)^{1/2}} = 1 + \sum_{n=0}^{\infty}(2n+1)\rho^n\, P_n(x), \qquad [|\rho| < 1, \quad |x| \le 1] \qquad \text{SA 170(4)}$$

8.923 $\arcsin x = \dfrac{\pi}{2}\displaystyle\sum_{k=1}^{\infty}\left\{\frac{(2k-1)!!}{2^k k!}\right\}^2 [P_{2k+1}(x) - P_{2k-1}(x)] + \pi x/2$

$$[|x| < 1, \quad (-1)!! \equiv 1] \qquad \text{WH}$$

8.924

1. $$-\frac{1+\cos n\pi}{2\left(n^2-1\right)}\,P_0\left(\cos\theta\right)-\frac{1+\cos n\pi}{2}\sum_{k=0}^{\infty}\frac{(4k+5)n^2\left(n^2-2^2\right)\ldots\left[n^2-(2k)^2\right]}{\left(n^2-1^2\right)\left(n^2-3^2\right)\ldots\left[n^2-(2k+3)^2\right]}\,P_{2k+2}\left(\cos\theta\right)$$

$$-\frac{3\left(1-\cos n\pi\right)}{2\left(n^2-2^2\right)}\,P_1\left(\cos\theta\right)$$

$$-\frac{1-\cos n\pi}{2}\sum_{k=1}^{\infty}\frac{(4k+3)\left(n^2-1^2\right)\ldots\left[n^2-(2k-1)^2\right]}{\left(n^2-2^2\right)\left(n^2-4^2\right)\ldots\left[n^2-(2k+2)^2\right]}\,P_{2k+1}\left(\cos\theta\right)=\cos n\theta$$

AD (9060.1)

2. $$\frac{-\sin n\pi}{2\left(n^2-1\right)}\,P_0\left(\cos\theta\right)-\frac{\sin n\pi}{2}\sum_{k=0}^{\infty}\frac{(4k+5)n^2\left(n^2-2^2\right)\ldots\left[n^2-(2k)^2\right]}{\left(n^2-1^2\right)\left(n^2-3^2\right)\ldots\left[n^2-(2k+3)^2\right]}\,P_{2k+2}\left(\cos\theta\right)$$

$$+\frac{3\sin n\pi}{2\left(n^2-2^2\right)}\,P_1\left(\cos\theta\right)$$

$$+\frac{\sin n\pi}{2}\sum_{k=1}^{\infty}\frac{(4k+3)\left(n^2-1^2\right)\left(n^2-3^2\right)\ldots\left[n^2-(2k-1)^2\right]}{\left(n^2-2^2\right)\left(n^2-4^2\right)\ldots\left[n^2-(2k+2)^2\right]}\,P_{2k+1}\left(\cos\theta\right)=\sin n\theta$$

AD (9060.2)

3.[3] $$\frac{2^{n-1}n!}{(2n-1)!!}\,P_n\left(\cos\theta\right)-n\sum_{k=1}^{\lfloor n/2\rfloor}(2n-4k+1)\frac{2^{n-2k-1}(n-k-1)!(2k-3)!!}{(2n-2k+1)!!k!}\,P_{n-2k}\left(\cos\theta\right)$$

$$=\cos n\theta$$

AD (9061.1)

4. $$\frac{(2n-1)!!P_{n-1}\left(\cos\theta\right)}{2^{n-1}(n-1)!}-\frac{n}{2^{n+1}}\sum_{k=0}^{\infty}\frac{(2n+2k-1)!!(2k-1)!!\left(2n+4k+3\right)}{2^{2k}(n+k+1)!(k+1)!}\,P_{n+2k+1}\left(\cos\theta\right)$$

$$=\frac{4\sin n\theta}{\pi}$$

AD (9061.2)

8.925

1. $$\sum_{k=1}^{\infty}\frac{4k-1}{2^{2k}(2k-1)^2}\left[\frac{(2k-1)!!}{k!}\right]^2 P_{2k-1}\left(\cos\theta\right)=1-\frac{2\theta}{\pi}$$

2. $$\sum_{k=1}^{\infty}\frac{4k+1}{2^{2k+1}(2k-1)(k+1)}\left[\frac{(2k-1)!!}{k!}\right]^2 P_{2k}\left(\cos\theta\right)=\frac{1}{2}-\frac{2\sin\theta}{\pi} \qquad \text{AD (9062.2)}$$

3. $$\sum_{k=1}^{\infty}\frac{k(4k-1)}{2^{2k-1}(2k-1)}\left[\frac{(2k-1)!!}{k!}\right]^2 P_{2k-1}\left(\cos\theta\right)=\frac{2\cot\theta}{\pi} \qquad \text{AD (9062.3)}$$

4. $$\sum_{k=1}^{\infty}\frac{4k+1}{2^{2k}}\left[\frac{(2k-1)!!}{k!}\right]^2 P_{2k}\left(\cos\theta\right)=\frac{2}{\pi\sin\theta}-1 \qquad \text{AD (9062.4)}$$

8.926

1. $$\sum_{n=1}^{\infty} \frac{1}{n} P_n(\cos\theta) = \ln \frac{2\tan\frac{\pi-\theta}{4}}{\sin\theta} = -\ln\sin\frac{\theta}{2} - \ln\left(1 + \sin\frac{\theta}{2}\right) \qquad \text{AD (9063.2)}$$

2. $$\sum_{n=1}^{\infty} \frac{1}{n+1} P_n(\cos\theta) = \ln \frac{1 + \sin\frac{\theta}{2}}{\sin\frac{\theta}{2}} - 1 \qquad \text{AD (9063.1)}$$

8.927 $$\sum_{k=0}^{\infty} \cos\left(k + \tfrac{1}{2}\right)\beta\, P_k(\cos\varphi) = \frac{1}{\sqrt{2(\cos\beta - \cos\varphi)}} \qquad [0 \le \beta < \varphi < \pi]$$

$$= 0 \qquad [0 < \varphi < \beta < \pi]$$

$$\text{MO 72}$$

8.928

1. $$\sum_{n=1}^{\infty} \frac{(-1)^n (4k+1)[(2n-1)!!]^3}{2^{3n}(n!)^3} P_{2n}(\cos\theta) = \frac{4\,K(\sin\theta)}{\pi^2} - 1 \qquad \text{AD (9064.1)}$$

2. $$\sum_{n=1}^{\infty} (-1)^{n+1} \frac{(4n+1)[(2n-1)!!]^3}{(2n-1)(2n+2)2^{3n}(n!)^3} P_{2n}(\cos\theta) = \frac{4\,E(\sin\theta)}{\pi^2} - \frac{1}{2} \qquad \text{AD (9064.2)}$$

- For series of products of Bessel functions and Legendre polynomials, see **8.511** 4, **8.531** 3, **8.533** 1, **8.533** 2, and **8.534**.
- For series of products of Legendre and Chebyshev polynomials, see **8.919**.

8.93 Gegenbauer polynomials $C_n^\lambda(t)$

8.930 Definition.   The polynomials $C_n^\lambda(t)$ of degree n are the coefficients of α^n in the power-series expansion of the function

$$\left(1 - 2t\alpha + \alpha^2\right)^{-\lambda} = \sum_{n=0}^{\infty} C_n^\lambda(t)\alpha^n \qquad \text{WH}$$

Thus, the polynomials $C_n^\lambda(t)$ are a *generalization of the Legendre polynomials*.

1.[10] $C_0^\lambda(t) = 1$

2.[10] $C_1^\lambda(t) = 2\lambda t$

3.[10] $C_2^\lambda(t) = 2\lambda(\lambda+1)t^2 - \lambda$

4.[10] $C_3^\lambda(t) = \frac{1}{3}\lambda\left(4\lambda^2 + 12\lambda + 8\right)t^3 - 2\lambda(\lambda+1)t$

5.[10] $C_4^\lambda(t) = \frac{2}{3}\lambda\left(\lambda^3 + 6\lambda^2 + 11\lambda + 6\right)t^4 - 2\lambda\left(\lambda^3 + 3\lambda + 2\right)t^2 + \frac{1}{2}\lambda(\lambda+1)$

6.[10] $C_5^\lambda(t) = \frac{1}{15}\lambda\left(4\lambda^4 + 40\lambda^3 + 140\lambda^2 + 200\lambda + 96\right)t^5$

$$- \frac{1}{3}\lambda\left(4\lambda^3 + 24\lambda^2 + 44\lambda + 24\right)t^3 + \lambda\left(\lambda^2 + 3\lambda + 2\right)t$$

$$7.^{10} \quad C_6^\lambda(t) = \frac{1}{45}\lambda\left(\lambda^5 + 60\lambda^4 + 340\lambda^3 + 900\lambda^2 + 1096\lambda + 480\right)t^6$$

$$-\frac{1}{3}\lambda\left(2\lambda^4 + 20\lambda^3 + 70\lambda^2 + 100\lambda + 48\right)t^4$$

$$+\lambda\left(\lambda^3 + 6\lambda^2 + 11\lambda + 6\right)t^2 + \frac{1}{6}\lambda\left(\lambda^2 + 3\lambda + 2\right)$$

8.931   Integral representation:

$$C_n^\lambda(t) = \frac{1}{\sqrt{\pi}}\frac{\Gamma(2\lambda + n)}{n!\,\Gamma(2\lambda)}\frac{\Gamma\left(\frac{2\lambda+1}{2}\right)}{\Gamma(\lambda)}\int_0^\pi \left(t + \sqrt{t^2 - 1}\cos\varphi\right)^n \sin^{2\lambda-1}\varphi\,d\varphi \qquad \text{MO 99}$$

See also **3.252** 11, **3.663** 2, **3.664** 4.

Functional relations

8.932   Expressions in terms of hypergeometric functions:

1. $C_n^\lambda(t) = \dfrac{\Gamma(2\lambda + n)}{\Gamma(n + 1)\,\Gamma(2\lambda)}\,F\left(2\lambda + n, -n; \lambda + \dfrac{1}{2}; \dfrac{1 - t}{2}\right)^*$ MO 97

$$= \frac{2^n\,\Gamma(\lambda + n)}{n!\,\Gamma(\lambda)}t^n\,F\left(-\frac{n}{2}, \frac{1 - n}{2}; 1 - \lambda - n; \frac{1}{t^2}\right) \qquad \text{MO 99}$$

2. $C_{2n}^\lambda(t) = \dfrac{(-1)^n}{(\lambda + n)\,\mathrm{B}(\lambda, n + 1)}\,F\left(-n, n + \lambda; \dfrac{1}{2}; t^2\right)$ MO 99

3. $C_{2n+1}^\lambda(t) = \dfrac{(-1)^n 2t}{\mathrm{B}(\lambda, n + 1)}\,F\left(-n, n + \lambda + 1; \dfrac{3}{2}; t^2\right)$ MO 99

8.933   Recursion formulas:

1. $(n + 2)\,C_{n+2}^\lambda(t) = 2(\lambda + n + 1)t\,C_{n+1}^\lambda(t) - (2\lambda + n)\,C_n^\lambda(t)$ Mo 98

2. $n\,C_n^\lambda(t) = 2\lambda\left[t\,C_{n-1}^{\lambda+1}(t) - C_{n-2}^{\lambda+1}(t)\right]$ WH

3. $(2\lambda + n)\,C_n^\lambda(t) = 2\lambda\left[C_n^{\lambda+1}(t) - t\,C_{n-1}^{\lambda+1}(t)\right]$ WH

4. $n\,C_n^\lambda(t) = (2\lambda + n - 1)t\,C_{n-1}^\lambda(t) - 2\lambda\left(1 - t^2\right)C_{n-2}^{\lambda+1}(t)$ WH

8.934

1. $C_n^\lambda(t) = \dfrac{(-1)^n}{2^n}\dfrac{\Gamma(2\lambda + n)\,\Gamma\left(\frac{2\lambda+1}{2}\right)}{\Gamma(2\lambda)\,\Gamma\left(\frac{2\lambda+1}{2} + n\right)}\dfrac{\left(1 - t^2\right)^{\frac{1}{2}-\lambda}}{n!}\dfrac{d^n}{dt^n}\left[\left(1 - t^2\right)^{\lambda+n-\frac{1}{2}}\right]$ WH

2. $C_n^\lambda\left(\cos\varphi\right) = \displaystyle\sum_{\substack{k,l=0 \\ k+l=n}}^n \dfrac{\Gamma(\lambda + k)\,\Gamma(\lambda + l)}{k!\,l!\,[\Gamma(\lambda)]^2}\cos(k - l)\varphi$ MO 99

---

*Equation 8.932.1 defines the generalized functions $C_n^\lambda(t)$, where the subscript n can be an arbitrary number

3. $C_n^\lambda (\cos\psi\cos\vartheta + \sin\psi\sin\vartheta\cos\varphi)$

$$= \frac{\Gamma(2\lambda-1)}{[\Gamma(\lambda)]^2}\sum_{k=0}^{n}\frac{2^{2k}(n-k)![\Gamma(\lambda+k)]^2}{\Gamma(2\lambda+n+k)}(2\lambda+2k-1)\sin^k\psi\,\sin^k\vartheta$$

$$\times C_{n-k}^{\lambda+k}(\cos\psi)\,C_{n-k}^{\lambda+k}(\cos\vartheta)\,C_k^{\lambda-\frac{1}{2}}(\cos\varphi)$$

$\left[\psi,\vartheta,\varphi\ \text{real};\quad \lambda\neq\tfrac{1}{2}\right]$ ["summation theorem"]   (see also **8.794–8.796**)   WH

4. $\lim\limits_{\lambda\to 0}\Gamma(\lambda)\,C_n^\lambda(\cos\varphi)=\dfrac{2\cos n\varphi}{n}$ MO 98

For orthogonality, see **8.904, 7.313**.

8.935 Derivatives:

1. $\dfrac{d^k}{dt^k}\,C_n^\lambda(t)=2^k\dfrac{\Gamma(\lambda+k)}{\Gamma(\lambda)}\,C_{n-k}^{\lambda+k}(t)$ MO 99

In particular,

2. $\dfrac{d\,C_n^\lambda(t)}{d=}2\lambda\,C_{n-1}^{\lambda+1}(t)$ WH

For integrals of the polynomials $C_n^\lambda(x)$ see **7.31–7.33**.

8.936 Connections with other functions:

1. $C_n^\lambda(t)=\dfrac{\Gamma(2\lambda+n)\,\Gamma\left(\lambda+\frac{1}{2}\right)}{\Gamma(2\lambda)\,\Gamma(n+1)}\left\{\dfrac{1}{4}\left(t^2-1\right)\right\}^{\frac{1}{4}-\frac{\lambda}{2}}P_{\lambda+n-\frac{1}{2}}^{\frac{1}{2}-\lambda}(t)$ MO 98

2. $C_{n-m}^{m+\frac{1}{2}}(t)=\dfrac{1}{(2m-1)!!}\dfrac{d^m\,P_n(t)}{dt^m}=(-1)^m\dfrac{(1-t^2)^{-\frac{m}{2}}\,m!\,2^m}{(2m)!}\,P_n^m(t)$

$[m+1\text{ a natural number}]$ MO 98, WH

3. $C_n^{1/2}(t)=P_n(t)$

4. $J_{\lambda-\frac{1}{2}}\left(r\sin\vartheta\sin\alpha\right)\left(r\sin\vartheta\sin\alpha\right)^{-\lambda+\frac{1}{2}}e^{-ir\cos\vartheta\cos\alpha}$

$$=\sqrt{2}\,\frac{\Gamma(\lambda)}{\Gamma\left(\lambda+\frac{1}{2}\right)}\sum_{k=0}^{\infty}(\lambda+k)i^{-k}\frac{\mathbf{J}_{\lambda+k}(r)\,C_k^\lambda(\cos\vartheta)\,C_k^\lambda(\cos\alpha)}{r^\lambda\,C_k^\lambda(1)}$$

MO 99

5. $\lim\limits_{\lambda\to\infty}\lambda^{-\frac{n}{2}}\,C_n^{\frac{\lambda}{2}}\left(t\sqrt{\dfrac{2}{\lambda}}\right)=\dfrac{2^{-\frac{n}{2}}}{n!}\,H_n(t)$ MO 99a

See also **8.932**.

8.937 Special cases and particular values:

1. $C_n^1(\cos\varphi)=\dfrac{\sin(n+1)\varphi}{\sin\varphi}$ MO 99

2. $C_0^0(\cos\varphi)=1$ MO 98

3. $C_0^\lambda(t)\equiv 1$ MO 98

4. $\quad C_n^\lambda(1) \equiv \begin{pmatrix} 2\lambda + n - 1 \\ n \end{pmatrix}$ MO 98

8.938 A differential equation leading to the polynomials $C_n^\lambda(t)$:

$$y'' + \frac{(2\lambda + 1)t}{t^2 - 1} y' - \frac{n(2\lambda + n)}{t^2 - 1} y = 0 \qquad \text{(cf. 9.174)}$$ WH

For series of products of Bessel functions and the polynomials $C_n^\lambda(x)$, see **8.532, 8.534**.

8.939[10] Differentiation and Rodrigues' formulas and orthogonality relation

1. $\quad \dfrac{d}{dt} C_n^\lambda(t) = 2\lambda\, C_{n-1}^{\lambda+1}(t)$ MS 5.3.2

2. $\quad \dfrac{d^m}{dt^m} C_n^\lambda(t) = 2^m \lambda(\lambda+1)(\lambda+2)\ldots(\lambda+m-1)\, C_{n-m}^{\lambda+m}(t)$ MS 5.3.2

3. $\quad \dfrac{d}{dt} C_{n-1}^\lambda(t) = t\dfrac{d}{dt} C_n^\lambda(t) - n\, C_n^\lambda(t)$ MS 5.3.2

4. $\quad \dfrac{d}{dt} C_{n+1}^\lambda(t) = t\dfrac{d}{dt} C_n^\lambda(t) + (2\lambda+n)\, C_n^\lambda(t)$ MS 5.3.2

5. $\quad \left(1 - t^2\right)\dfrac{d}{dt} C_n^\lambda(t) = (n+2\lambda-1)\, C_{n-1}^\lambda(t) - nt\, C_n^\lambda(t) = (n+2\lambda)t\, C_n^\lambda(t) - (n+1)\, C_{n+1}^\lambda(t)$

$$= 2\lambda\left(1 - t^2\right) C_{n-1}^{\lambda+1}(t)$$ MS 5.3.2

6. $\quad \dfrac{d}{dt}\left[C_{n+1}^\lambda(t) - C_{n-1}^\lambda(t)\right] = 2(n+\lambda)\, C_n^\lambda(t)$ MS 5.3.2

7. $\quad C_n^\lambda(t) = \dfrac{(-1)^n 2\lambda(2\lambda+1)(2\lambda+2)\ldots(2\lambda+n-1)\left(1-t^2\right)^{\frac{1}{2}-\lambda}}{2^n n!\left(\lambda+\frac{1}{2}\right)\left(\lambda+\frac{3}{2}\right)\ldots\left(\lambda+n-\frac{1}{2}\right)} \dfrac{d^n}{dt^n}\left[\left(1-t^2\right)^{n+\lambda-\frac{1}{2}}\right]$

$$= \dfrac{(-1)^n \, \Gamma\left(\lambda+\frac{1}{2}\right)\Gamma(n+2\lambda)\left(1-t^2\right)^{\frac{1}{2}-\lambda}}{2^n n!\,\Gamma(2\lambda)\,\Gamma\left(n+\lambda+\frac{1}{2}\right)} \dfrac{d^n}{dt^n}\left[\left(1-t^2\right)^{n+\lambda-\frac{1}{2}}\right]$$

[Rodrigues' formula]   MS 5.3.2

8. $\quad \displaystyle\int_{-1}^1 C_n^\lambda(t)\, C_m^\lambda(t)\left(1-t^2\right)^{\lambda-\frac{1}{2}} dt = 0 \qquad n \neq m$

$$= \frac{\pi 2^{1-2\lambda}\,\Gamma(n+2\lambda)}{n!(\lambda+n)[\Gamma(\lambda)]^2} \qquad n = m$$

$(\lambda \neq 0)$,   [Orthogonality relation]   MS 5.3.2

8.94 The Chebyshev polynomials $T_n(x)$ and $U_n(x)$

8.940 Definition

1. Chebyshev's polynomials of the first kind

$$T_n(x) = \cos\left(n\arccos x\right) = \frac{1}{2}\left[\left(x + i\sqrt{1-x^2}\right)^n + \left(x - i\sqrt{1-x^2}\right)^n\right]$$

$$= x^n - \binom{n}{2}x^{n-2}\left(1-x^2\right) + \binom{n}{4}x^{n-4}\left(1-x^2\right)^2 - \binom{n}{6}x^{n-6}\left(1-x^2\right)^3 + \ldots$$

NA 66, 71

2.   Chebyshev's polynomials of the second kind:

$$U_n(x) = \frac{\sin\left[(n+1)\arccos x\right]}{\sin\left[\arccos x\right]} = \frac{1}{2i\sqrt{1-x^2}}\left[\left(x+i\sqrt{1-x^2}\right)^{n+1} - \left(x-i\sqrt{1-x^2}\right)^{n+1}\right]$$

$$= \binom{n+1}{1}x^n - \binom{n+1}{3}x^{n-2}\left(1-x^2\right) + \binom{n+1}{5}x^{n-4}\left(1-x^2\right)^2 - \dots$$

Functional relations

8.941   Recursion formulas:

1. $T_{n+1}(x) - 2x\,T_n(x) + T_{n-1}(x) = 0$ NA 358

2. $U_{n+1}(x) - 2x\,U_n(x) + U_{n-1}(x) = 0$

3. $T_n(x) = U_n(x) - x\,U_{n-1}(x)$ EH II 184(3)

4. $\left(1-x^2\right)U_{n-1}(x) = x\,T_n(x) - T_{n+1}(x)$ EH II 184(4)

For the orthogonality, see **7.343** and **8.904**.

8.942   Relations with other functions:

1. $T_n(x) = F\left(n, -n; \dfrac{1}{2}; \dfrac{1-x}{2}\right)$ MO 104

2. $T_n(x) = (-1)^n \dfrac{\sqrt{1-x^2}}{(2n-1)!!}\dfrac{d^n}{dx^n}\left(1-x^2\right)^{n-\frac{1}{2}}$ MO 104

3. $U_n(x) = \dfrac{(-1)^n(n+1)}{\sqrt{1-x^2}(2n+1)!!}\dfrac{d^n}{dx^n}\left(1-x^2\right)^{n+\frac{1}{2}}$ EH II 185(15)

See also **8.962** 3.

8.943[10]   Special cases

1. $T_0(x) = 1$

2. $T_1(x) = x$

3. $T_2(x) = 2x^2 - 1$

4. $T_3(x) = 4x^3 - 3x$

5. $T_4(x) = 8x^4 - 8x^2 + 1$

6. $T_5(x) = 16x^5 - 20x^3 + 5x$

7. $T_6(x) = 32x^6 - 48x^4 + 18x^2 - 1$

8. $T_7(x) = 64x^7 - 112x^5 + 56x^3 - 7x$

9. $T_8(x) = 128x^8 - 256x^6 + 160x^4 - 32x^2 + 1$

10. $U_0(x) = 1$

11. $U_1(x) = 2x$

12. $U_2(x) = 4x^2 - 1$

13. $U_3(x) = 8x^3 - 4x$

14. $U_4(x) = 16x^4 - 12x^2 + 1$

15. $U_5(x) = 32x^5 - 32x^3 + 6x$

16. $U_6(x) = 64x^6 - 80x^4 + 24x^2 - 1$

17. $U_7(x) = 128x^7 - 192x^5 + 80x^3 - 8x$

18. $U_8(x) = 256x^8 - 448x^6 + 240x^4 - 40x^2 + 1$

8.944   Particular values:

1. $T_n(1) = 1$ 5. $U_{2n+1}(0) = 0$

2. $T_n(-1) = (-1)^n$ 6. $U_{2n}(0) = (-1)^n$

3. $T_{2n}(0) = (-1)^n$

4. $T_{2n+1}(0) = 0$

8.945   The generating function:

1. $\dfrac{1 - t^2}{1 - 2tx + t^2} = T_0(x) + 2 \sum\limits_{k=1}^{\infty} T_k(x) t^k$ MO 104

2. $\dfrac{1}{1 - 2tx + t^2} = \sum\limits_{k=0}^{\infty} U_k(x) t^k$ MO 104a, EH II 186(31)

8.946   Zeros.   The polynomials $T_n(x)$ and $U_n(x)$ only have real simple zeros. All these zeros lie in the interval $(-1, +1)$.

8.947   The functions $T_n(x)$ and $\sqrt{1 - x^2}\, U_{n-1}(x)$ are two linearly independent solutions of the differential equation

$$\left(1 - x^2\right) \frac{d^2 y}{dx^2} - x \frac{dy}{dx} + n^2 y = 0 \qquad\qquad \text{NA 69(58)}$$

8.948   Of all polynomials of degree n with leading coefficient equal to 1, the one that deviates the least from zero on the interval $[-1, +1]$ is the polynomial $2^{-n+1}\, T_n(x)$.

8.949[10]   Differentiation and Rodrigues' formulas and orthogonality relations

1. $\dfrac{d}{dx}\, T_n(x) = n\, U_{n-1}(x)$ MS 5.7.2

2. $\dfrac{d^m}{dx^m}\, T_n(x) = 2^{m-1}\, \Gamma(m) n\, C_{n-m}^m(x)$ MS 5.7.2

3. $\left(1 - x^2\right) \dfrac{d}{dx}\, T_n(x) = n\left[T_{n-1}(x) - x\, T_n(x)\right] = n\left[x\, T_n(x) - T_{n+1}(x)\right]$ MS 5.7.2

4. $\dfrac{d}{dx}\, U_n(x) = 2\, C_{n-1}^2(x)$ MS 5.7.2

5. $\dfrac{d^m}{dx^m}\, U_n(x) = 2^m\, m!\, C_{n-m}^{m+1}(x)$ MS 5.7.2

6. $\left(1 - x^2\right) \dfrac{d}{dx} U_n(x) = (n + 1) U_{n-1}(x) - nx\, U_n(x) = (n + 2)x\, U_n(x) - (n + 1) U_{n+1}(x)$

MS 5.7.2

7. $T_n(x) = \dfrac{(-1)^n \pi^{1/2} \left(1 - x^2\right)^{c\frac{1}{2}}}{2^{n+1}\, \Gamma\left(n + \frac{1}{2}\right)} \dfrac{d^n}{dx^n}\left[\left(1 - x^2\right)^{n - \frac{1}{2}}\right]$ [Rodrigues' formula]   MS 5.7.2

8. $U_n(x) = \dfrac{(-1)^n \pi^{1/2}(n + 1)\left(1 - x^2\right)^{-1/2}}{2^{n+1}\, \Gamma\left(n + \frac{3}{2}\right)} \dfrac{d^n}{dx^n}\left[\left(1 - x^2\right)^{n + \frac{1}{2}}\right]$

[Rodrigues' formula]   MS 5.7.2

9. $\displaystyle\int_{-1}^{1} T_m(x)\, T_n(x)\left(1 - x^2\right)^{-1/2} dx = \begin{cases} 0, & m \neq n \\ \pi/2, & m = n \neq 0 \\ \pi, & m = n = 0 \end{cases}$

[Orthogonality relation]   MS 5.7.2

10. $\displaystyle\int_{-1}^{1} U_m(x)\, U_n(x)\left(1 - x^2\right)^{-1/2} dx = \begin{cases} 0, & m \neq n \\ \pi/8, & m = n \end{cases}$ [Orthogonality relation]   MS 5.7.2

8.95 The Hermite polynomials $H_n(x)$

8.950   Definition

1. $H_n(x) = (-1)^n e^{x^2} \dfrac{d^n}{dx^n}\left(e^{-x^2}\right)$ SM 567(14)

or

2. $H_n(x) = 2^n x^n - 2^{n-1}\dbinom{n}{2} x^{n-2} + 2^{n-2} \cdot 1 \cdot 3 \cdot \dbinom{n}{4} x^{n-4} - 2^{n-3} \cdot 1 \cdot 3 \cdot 5 \cdot \dbinom{n}{6} x^{n-6} + \ldots$ MO 105a

3.[10] $H_0(x) = 1$

4.[10] $H_1(x) = 2x$

5.[10] $H_2(x) = 4x^2 - 2$

6.[10] $H_3(x) = 8x^3 - 12x$

7.[10] $H_4(x) = 16x^4 - 48x^2 + 12$

8.[10] $H_5(x) = 32x^5 - 160x^3 + 120x$

9.[10] $H_6(x) = 64x^6 - 480x^4 + 720x^2 - 120$

10.[10] $H_7(x) = 128x^7 - 1344x^5 + 3360x^3 - 1680x$

11.[10] $H_8(x) = 256x^8 - 3584x^6 + 13440x^4 - 13440x^2 + 1680$

8.951   The integral representation:

$$H_n(x) = \frac{2^n}{\sqrt{\pi}} \int_{-\infty}^{\infty} (x + it)^n e^{-t^2}\, dt$$ MO 106a

Functional relations

8.952　Recursion formulas:

1.　$\dfrac{d\,H_n(x)}{dx} = 2n\,H_{n-1}(x)$　　　　　　　　　　　　　　SM 569(22)

2.　$H_{n+1}(x) = 2x\,H_n(x) - 2n\,H_{n-1}(x)$　　　　　　　　　　SM 570(23)

　　For the orthogonality, see **7.374** 1 and **8.904**.

3.[10]　$n\,H_n(x) = -n\,H'_{n-1}(x) + x\,H'_n(x)$　　　　　　　　　　MS 5.6.2

4.[10]　$H_n(x) = 2x\,H_{n-1}(x) - H'_{n-1}(x)$　　　　　　　　　　　MS 5.6.2

8.953　The connection with other functions:

1.　$H_{2n}(x) = (-1)^n \dfrac{(2n)!}{n!}\,\Phi\left(-n,\tfrac{1}{2};x^2\right)$　　　　　　　　　MO 106a

2.　$H_{2n+1}(x) = (-1)^n 2\dfrac{(2n+1)!}{n!}\,x\,\Phi\left(-n,\tfrac{3}{2};x^2\right)$　　　　MO 106a

- For a connection with the polynomials $C_n^\lambda(x)$, see **8.936** 5.
- For a connection with the Laguerre polynomials, see **8.972** 2 and **8.972** 3.
- For a connection with functions of a parabolic cylinder, see **9.253**.

8.954　Inequalities:

1.[10]　$|H_n(x)| \le 2^{\frac{n}{2}-\lfloor\frac{n}{2}\rfloor}\dfrac{n!}{\lfloor n/2\rfloor!}e^{2x\sqrt{\lfloor n/2\rfloor}}$　　　　　　　　　MO 106a

2.[10]　$|H_n(x)| < k\sqrt{n!}\,2^{n/2}e^{x^2/2}, \quad k \approx 1.086435$　　　　SA 324

8.955　Asymptotic representation:

1.　$H_{2n}(x) = (-1)^n 2^n (2n-1)!!\,e^{x^2/2}\left[\cos\left(\sqrt{4n+1}x\right) + O\left(\dfrac{1}{\sqrt[4]{n}}\right)\right]$　　　SM 579

2.　$H_{2n+1}(x) = (-1)^n 2^{n+\frac{1}{2}}(2n-1)!!\sqrt{2n+1}\,e^{x^2/2}\left[\sin\left(\sqrt{4n+3}x\right) + O\left(\dfrac{1}{\sqrt[4]{n}}\right)\right]$　　SM 579

8.956　Special cases and particular values:

1.　$H_0(x) = 1$

2.　$H_1(x) = 2x$

3.　$H_2(x) = 4x^2 - 2$

4.　$H_3(x) = 8x^3 - 12x$

5.　$H_4(x) = 16x^4 - 48x^2 + 12$

6.　$H_{2n}(0) = (-1)^n 2^n (2n-1)!!$　　　　　　　　　　　　　SM 570(24)

7.　$H_{2n+1}(0) = 0$

Series of Hermite polynomials

8.957   The generating function:

1. $\exp\left(-t^2 + 2tx\right) = \sum_{k=0}^{\infty} \frac{t^k}{k!} H_k(x)$ SM 569(21)

2. $\frac{1}{e} \sinh 2x = \sum_{k=0}^{\infty} \frac{1}{(2k+1)!} H_{2k+1}(x)$ MO 106a

3. $\frac{1}{e} \cosh 2x = \sum_{k=0}^{\infty} \frac{1}{(2k)!} H_{2k}(x)$ MO 106a

4. $e \sin 2x = \sum_{k=0}^{\infty} (-1)^k \frac{1}{(2k+1)!} H_{2k+1}(x)$ MO 106a

5. $e \cos 2x = \sum_{k=0}^{\infty} (-1)^k \frac{1}{(2k)!} H_{2k}(x)$ MO 106a

8.958   "The summation theorem":

1. $\dfrac{\left(\sum\limits_{k=1}^{r} a_k^2\right)^{\frac{n}{2}}}{n!} H_n\left(\dfrac{\sum\limits_{k=1}^{r} a_k x_k}{\sqrt{\sum a_k^2}}\right) = \sum_{m_1+m_2+\cdots+m_r=n} \prod_{k=1}^{r} \left\{\dfrac{a_k^{m_k}}{m_k!} H_{m_k}(x_k)\right\}$ MO 106a

2.   A special case:

$$2^{\frac{n}{2}} H_n(x+y) = \sum_{k=0}^{n} \binom{n}{k} H_{n-k}\left(x\sqrt{2}\right) H_k\left(y\sqrt{2}\right)$$ MO 107a

8.959   Hermite polynomials satisfy the differential equation

1. $\dfrac{d^2 u_n}{dx^2} - 2x \dfrac{du_n}{dx} + 2n u_n = 0;$ SM 566(9)

A second solution of this differential equation is provided by the functions (A and B are arbitrary constants):

2. $u_{2n} = Ax\, \Phi\left(\frac{1}{2} - n; \frac{3}{2}; x^2\right),$

3. $u_{2n+1} = B\, \Phi\left(-\frac{1}{2} - n; \frac{1}{2}; x^2\right)$ MO 107

8.959(1)^{10}   Rodrigues' formula and orthogonality relation

1. $H_n(x) = (-1)^n e^{x^2} \dfrac{d^n}{dx^n}\left[e^{-x^2}\right]$ [Rodrigues' formula]   MS 5.6.2

2. $\displaystyle\int_{-\infty}^{\infty} e^{-x^2} H_m(x) H_n(x)\, dx = \begin{cases} 0 & \text{for } m \neq n \\ \pi^{1/2} 2^n n! & \text{for } m = n \end{cases}$ MS 5.6.2

8.96   Jacobi's polynomials

8.960   Definition

1. $P_n^{(\alpha,\beta)}(x) = \dfrac{(-1)^n}{2^n n!}(1-x)^{-\alpha}(1+x)^{-\beta}\dfrac{d^n}{dx^n}\left[(1-x)^{\alpha+n}(1+x)^{\beta+n}\right]$ EH II 169(10), CO

$$= \frac{1}{2^n}\sum_{m=0}^{n}\binom{n+\alpha}{m}\binom{n+\beta}{n-m}(x-1)^{n-m}(x+1)^m \qquad \text{EH II 169(2)}$$

8.961   Functional relations:

1. $P_n^{(\alpha,\beta)}(-x) = (-1)^n\, P_n^{(\alpha,\beta)}(x)$ EH II 169(13)

2. $2(n+1)(n+\alpha+\beta+1)(2n+\alpha+\beta)\, P_{n+1}^{(\alpha,\beta)}(x)$

$$= (2n+\alpha+\beta+1)\left[(2n+\alpha+\beta)(2n+\alpha+\beta+2)x+\alpha^2-\beta^2\right]P_n^{(\alpha,\beta)}(x)$$

$$-2(n+\alpha)(n+\beta)(2n+\alpha+\beta+2)\,P_{n-1}^{(\alpha,\beta)}(x)$$

         EH II 169(11)

3. $(2n+\alpha+\beta)\left(1-x^2\right)\dfrac{d}{dx}\,P_n^{(\alpha,\beta)}(x) = n[(\alpha-\beta)-(2n+\alpha+\beta)x]\,P_n^{(\alpha,\beta)}(x)$

$$+2(n+\alpha)(n+\beta)\,P_{n-1}^{(\alpha,\beta)}(x)$$

         EH II 170(15)

4. $\dfrac{d^m}{dx^m}\left[P_{n-1}^{(\alpha,\beta)}(x)\right] = \dfrac{1}{2^m}\dfrac{\Gamma(n+m+\alpha+\beta+1)}{\Gamma(n+\alpha+\beta+1)}\,P_{n-m}^{(\alpha,\beta)}(x)$

$$[m=1,2,\dots,n] \qquad \text{EH II 170(17)}$$

5. $\left(n+\tfrac{1}{2}\alpha+\tfrac{1}{2}\beta+1\right)(1-x)\,P_n^{(\alpha+1,\beta)}(x) = (n+\alpha+1)\,P_n^{(\alpha,\beta)}(x)-(n+1)\,P_{n+1}^{(\alpha,\beta)}(x)$ EH II 173(32)

6. $\left(n+\tfrac{1}{2}\alpha+\tfrac{1}{2}\beta+1\right)(1+x)\,P_n^{(\alpha,\beta+1)}(x) = (n+\beta+1)\,P_n^{(\alpha,\beta)}(x)+(n+1)\,P_{n+1}^{(\alpha,\beta)}(x)$ EH II 173(33)

7. $(1-x)\,P_n^{(\alpha+1,\beta)}(x)+(1+x)\,P_n^{(\alpha,\beta+1)}(x) = 2\,P_n^{(\alpha,\beta)}(x)$ EH II 173(34)

8. $(2n+\alpha+\beta)\,P_n^{(\alpha-1,\beta)}(x) = (n+\alpha+\beta)\,P_n^{(\alpha,\beta)}(x)-(n+\beta)\,P_{n-1}^{(\alpha,\beta)}(x)$ EH II 173(35)

9. $(2n+\alpha+\beta)\,P_n^{(\alpha,\beta-1)}(x) = (n+\alpha+\beta)\,P_n^{(\alpha,\beta)}(x)+(n+\alpha)\,P_{n-1}^{(\alpha,\beta)}(x)$ EH II 173(36)

10. $P_n^{(\alpha,\beta-1)}(x)-P_n^{(\alpha-1,\beta)}(x) = P_{n-1}^{(\alpha,\beta)}(x)$ EH II 173(37)

8.962   Connections with other functions:

1. $$P_n^{(\alpha,\beta)}(x) = \frac{(-1)^n \, \Gamma(n+1+\beta)}{n! \, \Gamma(1+\beta)} \, F\left(n+\alpha+\beta+1, -n; 1+\beta; \frac{1+x}{2}\right)$$ CO, EH II 170(16)

$$= \frac{\Gamma(n+1+\alpha)}{n! \, \Gamma(1+\alpha)} \, F\left(n+\alpha+\beta+1, -n; 1+\alpha; \frac{1-x}{2}\right)$$ EH II 170(16)

$$= \frac{\Gamma(n+1+\alpha)}{n! \, \Gamma(1+\alpha)} \left(\frac{1+x}{2}\right)^n F\left(-n, -n-\beta; \alpha+1; \frac{x-1}{x+1}\right)$$ EH II 170(16)

$$= \frac{\Gamma(n+1+\beta)}{n! \, \Gamma(1+\beta)} \left(\frac{x-1}{2}\right)^n F\left(-n, -n-\alpha; \beta+1; \frac{x+1}{x-1}\right)$$ EH II 170(16)

2. $$P_n(x) = P_n^{(0,0)}(x)$$ CO, EH II 179(3)

3. $$T_n(x) = \frac{2^{2n}(n!)^2}{(2n)!} \, P_n^{(-\frac{1}{2},-\frac{1}{2})}(x)$$ CO, EH II 184(5)a

4. $$C_n^\nu(x) = \frac{\Gamma(n+2\nu)\,\Gamma\left(\nu+\frac{1}{2}\right)}{\Gamma(2\nu)\,\Gamma\left(n+\nu+\frac{1}{2}\right)} \, P_n^{(\nu-1/2,\nu-1/2)}(x)$$ MO 108a, EH II 174(4)

8.963 The generating function:

$$\sum_{n=0}^{\infty} P_n^{(\alpha,\beta)}(x) z^n = 2^{\alpha+\beta} R^{-1}(1-z+R)^{-\alpha}(1+z+R)^{-\beta},$$

$$R = \sqrt{1-2xz+z^2} \qquad [|z| < 1]$$
EH II 172(29)

8.964 The Jacobi polynomials constitute the *unique* rational solution of the differential (hypergeometric) equation

$$\left(1-x^2\right)y'' + [\beta - \alpha - (\alpha+\beta+2)x]y' + n(n+\alpha+\beta+1)y = 0$$ EH II 169(14)

8.965 Asymptotic representation

$$P_n^{(\alpha,\beta)}(\cos\theta) =$$
$$\frac{\cos\left\{\left[n+\frac{1}{2}(\alpha+\beta+1)\right]\theta - \left(\frac{1}{2}\alpha+\frac{1}{4}\right)\pi\right\}}{\sqrt{\pi n}\left(\sin\frac{1}{2}\theta\right)^{\alpha+\frac{1}{2}}\left(\cos\frac{1}{2}\theta\right)^{\beta+\frac{1}{2}}} + O\left(n^{-3/2}\right) \qquad [\operatorname{Im}\alpha = \operatorname{Im}\beta = 0, \quad 0 < \theta < \pi]$$ EH II 198(10)

8.966 A limit relationship:

$$\lim_{n\to\infty}\left[n^{-\alpha} P_n^{(\alpha,\beta)}\left(\cos\frac{z}{n}\right)\right] = \left(\frac{z}{2}\right)^{-\alpha} J_\alpha(z)$$ EH II 173(41)

8.967 If $\alpha > -1$ and $\beta > -1$, all the zeros of the polynomial $P_n^{(\alpha,\beta)}(x)$ are simple and they lie in the interval $(-1,1)$.

8.97 The Laguerre polynomials

8.970 Definition.

1. $$L_n^\alpha(x) = \frac{1}{n!} e^x x^{-\alpha} \frac{d^n}{dx^n}\left(e^{-x} x^{n+\alpha}\right) \qquad \text{[Rodrigues' formula]}$$ EH II 188(5), MO 108

$$= \sum_{m=0}^{n} (-1)^m \binom{n+\alpha}{n-m} \frac{x^m}{m!}$$ MO 109, EH II 188(7)

2. $$L_n^0(x) = L_n(x)$$ ET I 369

$3.^{10}$ $L_0^\alpha(x) = 1$

$4.^{10}$ $L_1^\alpha(x) = -x + \alpha + 1$

$5.^{10}$ $L_2^\alpha(x) = \dfrac{1}{2} \left[x^2 - 2(\alpha + 2)x + (\alpha + 1)(\alpha + 2) \right]$

$6.^{10}$ $L_3^\alpha(x) = -\dfrac{1}{6} \left[x^3 - 3(\alpha + 3)x^2 + 3(\alpha + 2)(\alpha + 3)x - (\alpha + 1)(\alpha + 2)(\alpha + 3) \right]$

$7.^{10}$ $L_4^\alpha(x) = \dfrac{1}{24} \Big[x^4 - 4(\alpha + 4)x^3 + 6(\alpha + 3)(\alpha + 4)x^2 - 4(\alpha + 2)(\alpha + 3)(\alpha + 4)x$

$$+ (\alpha + 1)(\alpha + 2)(\alpha + 3)(\alpha + 4) \Big]$$

$8.^{10}$ $L_5^\alpha(x) = -\dfrac{1}{120} \Big[x^5 - 5(\alpha + 5)x^4 + 10(\alpha + 4)(\alpha + 5)x^3 - 10(\alpha + 3)(\alpha + 4)(\alpha + 5)x^2$

$$+ 5(\alpha + 2)(\alpha + 3)(\alpha + 4)(\alpha + 5)x - (\alpha + 1)(\alpha + 2)(\alpha + 3)(\alpha + 4)(\alpha + 5) \Big]$$

8.971   Functional relations:

1. $\dfrac{d}{dx} \left[L_n^\alpha(x) - L_{n+1}^\alpha(x) \right] = L_n^\alpha(x)$ EH II 189(16)

2. $\dfrac{d}{dx} L_n^\alpha(x) = -L_{n-1}^\alpha(x) = \dfrac{n\,L_n^\alpha(x) - (n + \alpha)\,L_{n-1}^\alpha(x)}{x}$ EH II 189(15), SM 575(42)a

3. $x\dfrac{d}{dx} L_n^\alpha(x) = n\,L_n^\alpha(x) - (n + \alpha)\,L_{n-1}^\alpha(x)$

$$= (n + 1)\,L_{n+1}^\alpha(x) - (n + \alpha + 1 - x)\,L_n^\alpha(x)$$

                                                     EH II 189(12), MO 109

4. $x\,L_n^{\alpha+1}(x) = (n + \alpha + 1)\,L_n^\alpha(x) - (n + 1)\,L_{n+1}^\alpha(x)$

$$= (n + \alpha)\,L_{n-1}^\alpha(x) - (n - x)\,L_n^\alpha(x)$$

                                                   SM 575(43)a, EH II 190(23)

5. $L_n^{\alpha-1}(x) = L_n^\alpha(x) - L_{n-1}^\alpha(x)$ SM 575(44)a, EH II 190(24)

6. $(n + 1)\,L_{n+1}^\alpha(x) - (2n + \alpha + 1 - x)\,L_n^\alpha(x) + (n + \alpha)\,L_{n-1}^\alpha(x) = 0$

                                            $[n = 1, 2, \ldots]$ MO 109, EH II 190(25, 24)

$7.^{10}$ $(n + \alpha)\,L_n^{\alpha-1}(x) = (n + 1)\,L_{n+1}^\alpha(x) - (n + 1 - x)\,L_n^\alpha(x)$ MS 5.5.2

$8.^{10}$ $n\,L_n^\alpha(x) = (2n + \alpha - 1 - x)\,L_{n-1}^\alpha(x) - (n + \alpha - 1)\,L_{n-2}^\alpha(x)$

                                                   $[n = 2, 3, \ldots]$ MS 5.5.2

8.972   Connections with other functions:

1. $L_n^\alpha(x) = \dbinom{n + \alpha}{n} \Phi(-n, \alpha + 1; x)$ MO 109, FI II 189(14)

2. $H_{2n}(x) = (-1)^n 2^{2n} n!\, L_n^{-1/2}\left(x^2\right)$ EH II 193(2), SM 576(47)

3. $H_{2n+1}(x) = (-1)^n 2^{2n+1} n! x\, L_n^{1/2}\left(x^2\right)$ EH II 193(3), SM 577(48)

8.973   Special cases:

1. $L_0^\alpha(x) = 1$ EH II 188(6)

2. $L_1^\alpha(x) = \alpha + 1 - x$ EH II 188(6)

3. $L_n^\alpha(0) = \begin{pmatrix} n + \alpha \\ n \end{pmatrix}$ EH II 189(13)

4. $L_n^{-n}(x) = (-1)^n \dfrac{x^n}{n!}$ MO 109

5. $L_1(x) = 1 - x$

6. $L_2(x) = 1 - 2x + \dfrac{x^2}{2}$ MO 109

8.974   Finite sums:

1. $\displaystyle\sum_{m=0}^{n} \frac{m!}{\Gamma(m + \alpha + 1)} L_m^\alpha(x)\, L_m^\alpha(y) = \frac{(n+1)!}{\Gamma(n + \alpha + 1)(x - y)} \left[L_n^\alpha(x)\, L_{n+1}^\alpha(y) - L_{n+1}^\alpha(x)\, L_n^\alpha(y) \right]$

       EH II 188(9)

2. $\displaystyle\sum_{m=0}^{n} \frac{\Gamma(\alpha - \beta + m)}{\Gamma(\alpha - \beta) m!} L_{n-m}^\beta(x) = L_n^\beta(x + y)$ MO 110, EH II 192(39)

3. $\displaystyle\sum_{m=0}^{n} L_m^\alpha(x) = L_n^{\alpha+1}(x)$ EH II 192(38)

4. $\displaystyle\sum_{m=0}^{n} L_m^\alpha(x)\, L_{n-m}^\beta(x) = L_n^{\alpha+\beta+1}(x + y)$ EH II 192(41)

8.975   Arbitrary functions:

1. $(1 - z)^{-\alpha - 1} \exp \dfrac{xz}{z - 1} = \displaystyle\sum_{n=0}^{\infty} L_n^\alpha(x) z^n$ $[|z| < 1]$ EH II 189(17), MO 109

2. $e^{-xz}(1 + z)^\alpha = \displaystyle\sum_{n=0}^{\infty} L_n^{\alpha - n}(x) z^n$ $[|z| < 1]$ MO 110, EH II 189(19)

3. $J_\alpha\left(2\sqrt{xz}\right) e^z (xz)^{-\frac{1}{2}\alpha} = \displaystyle\sum_{n=0}^{\infty} \frac{z^n}{\Gamma(n + \alpha + 1)} L_n^\alpha(x)$ $[\alpha > -1]$ EH II 189(18), MO 109

8.976   Other series of Laguerre polynomials:

1. $\displaystyle\sum_{n=0}^{\infty} n! \frac{L_n^\alpha(x)\, L_n^\alpha(y) z^n}{\Gamma(n + \alpha + 1)} = \frac{(xyz)^{-\frac{1}{2}\alpha}}{1 - z} \exp\left(-z \frac{x + y}{1 - z}\right) I_\alpha\left(2 \frac{\sqrt{xyz}}{1 - z}\right)$

                                  $[|z| < 1]$ EH II 189(20)

2. $$\sum_{n=0}^{\infty} \frac{L_n^\alpha(x)}{n+1} = e^x x^{-\alpha}\, \Gamma(\alpha, x) \qquad\qquad [\alpha > -1, \quad x > 0] \qquad\qquad \text{EH II 215(19)}$$

3.[6] $$L_n^\alpha(x)^2 = \frac{\Gamma(n+\alpha+1)}{2^{2n}n!} \sum_{k=0}^{n} \binom{2n-2k}{n-k} \frac{(2k)!}{k!} \frac{1}{\Gamma(\alpha+k+1)} L_{2k}^{2\alpha}(2x) \qquad\qquad \text{MO 110}$$

4.[6] $$L_n^\alpha(x)\, L_n^\alpha(y) = \frac{\Gamma(1+\alpha+n)}{n!} \sum_{k=0}^{n} \frac{L_{n-k}^{\alpha+2k}(x+y)}{\Gamma(1+\alpha+k)} \frac{(xy)^k}{k!} \qquad\qquad \text{MO 110, EH II 192(42)}$$

8.977 Summation theorems:

1. $$L_n^{\alpha_1+\alpha_2+\cdots+\alpha_k+k-1}(x_1+x_2+\cdots+x_k) = \sum_{i_1+i_2+\cdots+i_2=n} L_{i_1}^{\alpha_1}(x_1)\, L_{i_2}^{\alpha_2}(x_2) \cdots L_{i_k}^{\alpha_k}(x_k) \qquad \text{MO 110}$$

2. $$L_n^\alpha(x+y) = e^y \sum_{k=0}^{\infty} \frac{(-1)^k}{k!} y^k\, L_n^{\alpha+k}(x) \qquad\qquad \text{MO 110}$$

8.978 Limit relations and asymptotic behavior:

1. $$L_n^\alpha(x) = \lim_{\beta \to \infty} P_n^{(\alpha,\beta)}\left(1 - \frac{2x}{\beta}\right) \qquad\qquad \text{EH II 191(35)}$$

2. $$\lim_{n \to \infty}\left[n^{-\alpha} L_n^\alpha\left(\frac{x}{n}\right)\right] = x^{-\frac{1}{2}\alpha}\, J_\alpha\left(2\sqrt{x}\right) \qquad\qquad \text{EH II 191(36)}$$

3. $$L_n^\alpha(x) = \frac{1}{\sqrt{\pi}} e^{\frac{1}{2}x} x^{-\frac{1}{2}\alpha-\frac{1}{4}} n^{\frac{1}{2}\alpha-\frac{1}{4}} \cos\left[2\sqrt{nx} - \frac{\alpha\pi}{2} - \frac{\pi}{4}\right] + O\left(n^{\frac{1}{2}\alpha-\frac{3}{4}}\right)$$

$$[\operatorname{Im}\alpha = 0, \quad x > 0] \qquad\qquad \text{EH II 199(1)}$$

8.979 Laguerre polynomials satisfy the following differential equation:

$$x\frac{d^2u}{dx^2} + (\alpha - x + 1)\frac{du}{dx} + nu = 0 \qquad\qquad \text{EH II 188(10), SM 574(34)}$$

8.980[10] Orthogonality relation

$$\int_0^\infty e^{-x} x^\alpha\, L_n^\alpha(x)\, L_m^{alpha}(x)\, dx = \begin{cases} 0, & m \neq n \\ \Gamma(1+\alpha)\binom{n+\alpha}{n}, & m = n \end{cases} \qquad\qquad \text{MS 5.5.2}$$

8.981[10] Behavior of relative maxima of $|L_n^\alpha(x)|$

1. Let α be arbitrary and real. The sequence formed by the relative maxima of $|L_n^\alpha(x)|$ and by the value of this function at $x = 0$, is decreasing for $x < \alpha + \frac{1}{2}$, and increasing for $x > \alpha + \frac{1}{2}$. The successive relative maxima of $|L_n^\alpha(x)|$ form a decreasing sequence for $x \leq 0$, and an increasing sequence for $x \geq 0$.
    SZ 174(7.6.1)

2. Let α be an arbitrary real number. The successive relative maxima of

$$e^{-x/2} x^{(\alpha+1)/2} |L_n^\alpha(x)| \quad \text{and} \quad e^{-x/2} x^{\alpha/2 + \frac{1}{4}} |L_n^\alpha(x)|$$

form an increasing sequence provided $x > x_0$. In the first case

$$x_0 = \begin{cases} 0 & \text{if } \alpha^2 \leq 1, \\ \dfrac{\alpha^2 - 1}{2n + \alpha + 1} & \text{if } \alpha^2 > 1 \end{cases}$$

In the second case

$$x_0 = \begin{cases} 0 & \text{if } \alpha^2 \leq q\frac{1}{4}, \\ \left(\alpha^2 - \frac{1}{4}\right)^{\frac{1}{2}} & \text{if } \alpha^2 > \frac{1}{4} \end{cases} \qquad \text{SZ 174(7.6.2)}$$

In the first case we take n so large that $2n + \alpha + 1 > 0$.

8.982^{10}   Asymptotic and limiting behavior of $L_n^\alpha(x)$

1.   Let α be arbitrary and real, c and w fixed positive constants, and let $n \to \infty$. Then

$$L_n^\alpha(x) = \begin{cases} x^{-\alpha/2 - \frac{1}{4}} O\left(n^{\alpha/2 - \frac{1}{4}}\right) & \text{if } cn^{-1} \leq qx \leq qw \\ O(n^\alpha) & \text{if } 0 \leq qx \leq qcn^{-1} \end{cases}$$

These bounds are precise as regards their orders in n. For $\alpha \geq q - \frac{1}{2}$, both bounds hold in both intervals, that is,

$$L_n^\alpha(x) = \begin{cases} x^{-\alpha/2 - \frac{1}{4}} O\left(n^{\alpha/2 - \frac{1}{4}}\right), \\ O(n^\alpha), \end{cases} \quad 0 < x \leq qw, \quad \alpha \geq q - \frac{1}{2} \qquad \text{SZ 175(7.6.4)}$$

2.   Let α be arbitrary and real. Then for an arbitrary complex z

$$\lim_{n \to \infty} n^{-\alpha} L_n^\alpha(x) = z^{-\alpha/2} J_\alpha\left(2z^{1/2}\right), \qquad \text{SZ 191(8.1.3)}$$

uniformly if z is bounded.

9.1 Hypergeometric Functions

9.10   Definition

9.100   A *hypergeometric series* is a series of the form

$$F(\alpha, \beta; \gamma; z) = 1 + \frac{\alpha \cdot \beta}{\gamma \cdot 1} z + \frac{\alpha(\alpha+1)\beta(\beta+1)}{\gamma(\gamma+1) \cdot 1 \cdot 2} z^2 + \frac{\alpha(\alpha+1)(\alpha+2)\beta(\beta+1)(\beta+2)}{\gamma(\gamma+1)(\gamma+2) \cdot 1 \cdot 2 \cdot 3} z^3 + \dots$$

9.101   A hypergeometric series terminates if α or β is equal to a negative integer or to zero. For $\gamma = -n \, (n = 0, 1, 2, \dots)$, the hypergeometric series is indeterminate if neither α nor β is equal to $-m$ (where $m < n$ and m is a natural number). However,

1. $\displaystyle \lim_{\gamma \to -n} \frac{F(\alpha, \beta; \gamma; z)}{\Gamma(\gamma)} = \frac{\alpha(\alpha+1) \dots (\alpha+n)\beta(\beta+1) \dots (\beta+n)}{(n+1)!}$

$$\times z^{n+1} F(\alpha+n+1, \beta+n+1; n+2; z)$$

EH I 62(16)

9.102   If we exclude these values of the parameters α, β, γ, a hypergeometric series converges in the unit circle $|z| < 1$. F then has a branch point at $z = 1$. Then we have the following conditions for convergence on the unit circle:

1. $1 > \operatorname{Re}(\alpha + \beta - \gamma) \geq 0$.   The series converges throughout the entire unit circle except at the point $z = 1$.

2. $\operatorname{Re}(\alpha + \beta - \gamma) < 0$.   The series converges (absolutely) throughout the entire unit circle.

3. $\operatorname{Re}(\alpha + \beta - \gamma) \geq 1$.   The series diverges on the entire unit circle.   FI II 410, WH

9.11   Integral representations

9.111 $\displaystyle F(\alpha, \beta; \gamma; z) = \frac{1}{\mathrm{B}(\beta, \gamma - \beta)} \int_0^1 t^{\beta-1}(1-t)^{\gamma-\beta-1}(1-tz)^{-\alpha} \, dt$ $[\operatorname{Re}\gamma > \operatorname{Re}\beta > 0]$ WH

9.112[8] $\displaystyle F\left(p, n+p; n+1; z^2\right) = \frac{z^{-n}}{2\pi} \frac{\Gamma(p) n!}{\Gamma(p+n)} \int_0^{2\pi} \frac{\cos nt \, dt}{(1 - 2z\cos t + z^2)^p}$

$[n = 0, 1, 2, \dots; \quad p \neq 0, -1, -2, \dots; \quad |z| < 1]$ WH, MO 16

9.113 $\displaystyle F(\alpha, \beta; \gamma; z) = \frac{\Gamma(\gamma)}{\Gamma(\alpha)\,\Gamma(\beta)} \frac{1}{2\pi i} \int_{-\infty i}^{\infty i} \frac{\Gamma(\alpha+t)\,\Gamma(\beta+t)\,\Gamma(-t)}{\Gamma(\gamma+t)} (-z)^t \, dt$

Here, $|\arg(-z)| < \pi$ and the path of integration is chosen in such a way that the poles of the functions $\Gamma(\alpha+t)$ and $\Gamma(\beta+t)$ lie to the left of the path of integration and the poles of the function $\Gamma(-t)$ lie to the right of it.

9.114 $\displaystyle F\left(-m, -\frac{p+m}{2}; 1 - \frac{p+m}{2}; -1\right) = \frac{(-2)^m (p+m)}{\sin p\pi} \int_0^{\pi} \cos^m \varphi \cos p\varphi \, d\varphi$

$[m + 1 \text{ is a natural number}; \quad p \neq 0, \quad \pm 1, \dots]$ EH I 80(8), MO 16

See also **3.194** 1, 2, 5, **3.196** 1, **3.197** 6, 9, **3.259** 3, **3.312** 3, **3.518** 4–6, **3.665** 2, **3.671** 1, 2, **3.681** 1, **3.984** 7.

9.12 Representation of elementary functions in terms of a hypergeometric functions

9.121

1.[8] $\quad F(-n, \beta; \beta; -z) = (1 + z)^n$ EH I 101(4), GA 127 Ia

2. $\quad F\left(-\dfrac{n}{2}, -\dfrac{n-1}{2}; \dfrac{1}{2}; \dfrac{z^2}{t^2}\right) = \dfrac{(t+z)^n + (t-z)^n}{2t^n}$ GA 127 II

3. $\quad \lim\limits_{\omega \to \infty} F\left(-n, \omega; 2\omega; -\dfrac{z}{t}\right) = \left(1 + \dfrac{z}{2t}\right)^n$ GA 127 IIIa

4. $\quad F\left(-\dfrac{n-1}{2}, -\dfrac{n-2}{2}; \dfrac{3}{2}; \dfrac{z^2}{t^2}\right) = \dfrac{(t+z)^n - (t-z)^n}{2nzt^{n-1}}$ GA 127 IV

5. $\quad F\left(1-n, 1; 2; -\dfrac{z}{t}\right) = \dfrac{(t+z)^n - t^n}{nzt^{n-1}}$ GA 127 V

6. $\quad F(1, 1; 2; -z) = \dfrac{\ln(1+z)}{z}$ GA 127 VI

7. $\quad F\left(\dfrac{1}{2}, 1; \dfrac{3}{2}; z^2\right) = \dfrac{\ln \frac{1+z}{1-z}}{2z}$ GA 127 VII

8. $\quad \lim\limits_{k \to \infty} F\left(1, k; 1; \dfrac{z}{k}\right) = 1 + z \lim\limits_{k \to \infty} F\left(1, k; 2; \dfrac{z}{k}\right)$

$\qquad\qquad = 1 + z + \dfrac{z^2}{2} \lim\limits_{k \to \infty} F\left(1, k; 3; \dfrac{z}{k}\right) = \cdots = e^z$

GA 127 VIII

9. $\quad \lim\limits_{\substack{k \to \infty \\ k' \to \infty}} F\left(k, k'; \dfrac{1}{2}; \dfrac{z^2}{4kk'}\right) = \dfrac{e^z + e^{-z}}{2} = \cosh z$ GA 127 IX

10. $\quad \lim\limits_{\substack{k \to \infty \\ k' \to \infty}} F\left(k, k'; \dfrac{3}{2}; \dfrac{z^2}{4kk'}\right) = \dfrac{e^z - e^{-z}}{2z} = \dfrac{\sinh z}{z}$ GA 127 X

11. $\quad \lim\limits_{\substack{k \to \infty \\ k' \to \infty}} F\left(k, k'; \dfrac{3}{2}; -\dfrac{z^2}{4kk'}\right) = \dfrac{\sin z}{z}$ GA 127 XI

12. $\quad \lim\limits_{\substack{k \to \infty \\ k' \to \infty}} F\left(k, k'; \dfrac{1}{2}; -\dfrac{z^2}{4kk'}\right) = \cos z$ GA 127 XII

13. $\quad F\left(\dfrac{1}{2}, \dfrac{1}{2}; \dfrac{3}{2}; \sin^2 z\right) = \dfrac{z}{\sin z}$ GA 127 XIII

14. $\quad F\left(1, 1; \dfrac{3}{2}; \sin^2 z\right) = \dfrac{z}{\sin z \cos z}$ GA 127 XIV

15. $\quad F\left(\dfrac{1}{2}, 1; \dfrac{3}{2}; -\tan^2 z\right) = \dfrac{z}{\tan z}$ GA 127 XV

16. $\quad F\left(\dfrac{n+1}{2}, -\dfrac{n-1}{2}; \dfrac{3}{2}; \sin^2 z\right) = \dfrac{\sin nz}{n \sin z}$ GA 127 XVI

17. $\quad F\left(\dfrac{n+2}{2}, -\dfrac{n-2}{2}; \dfrac{3}{2}; \sin^2 z\right) = \dfrac{\sin nz}{n \sin z \cos z}$ GA 127 XVII

18. $F\left(-\dfrac{n-2}{2}, -\dfrac{n-1}{2}; \dfrac{3}{2}; -\tan^2 z\right) = \dfrac{\sin nz}{n \sin z \cos^{n-1} z}$ GA 127 XVIII

19. $F\left(\dfrac{n+2}{2}, \dfrac{n+1}{2}; \dfrac{3}{2}; -\tan^2 z\right) = \dfrac{\sin nz \cos^{n+1} z}{n \sin z}$ GA 127 XIX

20. $F\left(\dfrac{n}{2}, -\dfrac{n}{2}; \dfrac{1}{2}; \sin^2 z\right) = \cos nz$ EH I 101(11), GA 127 XX

21. $F\left(\dfrac{n+1}{2}, -\dfrac{n-1}{2}; \dfrac{1}{2}; \sin^2 z\right) = \dfrac{\cos nz}{\cos z}$ EH I 101(11), GA 127 XXI

22. $F\left(-\dfrac{n}{2}, -\dfrac{n-1}{2}; \dfrac{1}{2}; -\tan^2 z\right) = \dfrac{\cos nz}{\cos^n z}$ EH I 101(11), GA 127 XXII

23. $F\left(\dfrac{n+1}{2}, \dfrac{n}{2}; \dfrac{1}{2}; -\tan^2 z\right) = \cos nz \cos^n z$ GA 127 XXIII

24. $F\left(\dfrac{1}{2}, 1; 2; 4z(1-z)\right) = \dfrac{1}{1-z}$ $\left[|z| \le \tfrac{1}{2}; \quad |z(1-z)| \le \tfrac{1}{4}\right]$

25. $F\left(\dfrac{1}{2}, 1; 1; \sin^2 z\right) = \sec z$

26. $F\left(\dfrac{1}{2}, \dfrac{1}{2}; \dfrac{3}{2}; z^2\right) = \dfrac{\arcsin z}{z}$ (cf. **9.121** 13)

27. $F\left(\dfrac{1}{2}, 1; \dfrac{3}{2}; -z^2\right) = \dfrac{\arctan z}{z}$ (cf. **9.121** 15)

28. $F\left(\dfrac{1}{2}, \dfrac{1}{2}; \dfrac{3}{2}; -z^2\right) = \dfrac{\operatorname{arcsinh} z}{z}$ (cf. **9.121** 26)

29. $F\left(\dfrac{1+n}{2}, \dfrac{1-n}{2}; \dfrac{3}{2}; z^2\right) = \dfrac{\sin(n \arcsin z)}{nz}$ (cf. **9.121** 16)

30. $F\left(1+\dfrac{n}{2}, 1-\dfrac{n}{2}; \dfrac{3}{2}; z^2\right) = \dfrac{\sin(n \arcsin z)}{nz\sqrt{1-z^2}}$ (cf. **9.121** 17)

31. $F\left(\dfrac{n}{2}, -\dfrac{n}{2}; \dfrac{1}{2}; z^2\right) = \cos(n \arcsin z)$ (cf. **9.121** 20)

32. $F\left(\dfrac{1+n}{2}, \dfrac{1-n}{2}; \dfrac{1}{2}; z^2\right) = \dfrac{\cos(n \arcsin z)}{\sqrt{1-z^2}}$ (cf. **9.121** 21)

The representation of special functions in terms of a hypergeometric function:

- for complete elliptic integrals, see **8.113** 1 and **8.114** 1;
- for integrals of Bessel functions, see **6.574** 1, 3, **6.576** 2–5, **6.621** 1–3;
- for Legendre polynomials, see **8.911** and **8.916**. (All these hypergeometric series terminate; that is, these series are finite sums);
- for Legendre functions, see **8.820** and **8.837**;
- for associated Legendre functions, see **8.702**, **8.703**, **8.751**, **8.77**, **8.852**, and **8.853**;
- for Chebyshev polynomials, see **8.942** 1;
- for Jacobi's polynomials, see **8.962**;

- for Gegenbauer polynomials, see **8.932**;
- for integrals of parabolic cylinder functions, see **7.725** 6.

9.122   Particular values:

1. $F(\alpha, \beta; \gamma; 1) = \dfrac{\Gamma(\gamma)\,\Gamma(\gamma - \alpha - \beta)}{\Gamma(\gamma - \alpha)\,\Gamma(\gamma - \beta)}$ $[\operatorname{Re}\gamma > \operatorname{Re}(\alpha + \beta)]$

GA 147(48), FI II 793

2. $F(\alpha, \beta; \gamma; 1) = F(-\alpha, -\beta; \gamma - \alpha - \beta; 1)$ $[\operatorname{Re}\gamma > \operatorname{Re}(\alpha + \beta)]$ GA 148(49)

$\qquad\qquad = \dfrac{1}{F(-\alpha, \beta; \gamma - \alpha; 1)}$ $[\operatorname{Re}\gamma > \operatorname{Re}(\alpha + \beta)]$ GA 148(50)

$\qquad\qquad = \dfrac{1}{F(\alpha, -\beta; \gamma - \beta; 1)}$ $[\operatorname{Re}\gamma > \operatorname{Re}(\alpha + \beta)]$ GA 148(51)

3. $F\left(1, 1; \dfrac{3}{2}; \dfrac{1}{2}\right) = \dfrac{\pi}{2}$ (cf. **9.121** 14)

9.13 Transformation formulas and the analytic continuation of functions defined by hypergeometric series

9.130   The series $F(\alpha, \beta; \gamma; z)$ defines an analytic function that, speaking generally, has singularities at the points $z = 0$, 1, and ∞ (In the general case, there are branch points). We make a cut in the z-plane along the real axis from $z = 1$ to $z = \infty$; that is, we require that $|\arg(-z)| < \pi$ for $|z| \geq 1$. Then, the series $f(\alpha, \beta; \gamma; z)$ will, in the cut plane, yield a single-valued analytic continuation which we can obtain by means of the formulas below (provided $\gamma + 1$ is not a natural number and $\alpha - \beta$ and $\gamma - \alpha - \beta$ are not integers). These formulas make it possible to calculate the values of F in the given region even in the case in which $|z| > 1$. There are other closely related transformation formulas that can also be used to get the analytic continuation when the corresponding relationships hold between α, β, γ.

Transformation formulas

9.131

1. $F(\alpha, \beta; \gamma; z) = (1 - z)^{-\alpha}\, F\left(\alpha, \gamma - \beta; \gamma; \dfrac{z}{z - 1}\right)$ GA 218(91)

$\qquad\qquad = (1 - z)^{-\beta}\, F\left(\beta, \gamma - \alpha; \gamma; \dfrac{z}{z - 1}\right)$ GA 218(92)

$\qquad\qquad = (1 - z)^{\gamma - \alpha - \beta}\, F(\gamma - \alpha; \gamma - \beta; \gamma; z)$

2. $F(\alpha, \beta; \gamma; z) = \dfrac{\Gamma(\gamma)\,\Gamma(\gamma - \alpha - \beta)}{\Gamma(\gamma - \alpha)\,\Gamma(\gamma - \beta)}\, F(\alpha, \beta; \alpha + \beta - \gamma + 1; 1 - z)$

$\qquad\qquad + (1 - z)^{\gamma - \alpha - \beta}\dfrac{\Gamma(\gamma)\,\Gamma(\alpha + \beta - \gamma)}{\Gamma(\alpha)\,\Gamma(\beta)}\, F(\gamma - \alpha, \gamma - \beta; \gamma - \alpha - \beta + 1; 1 - z)$

EH I 94, MO 13

9.132

1. $F(\alpha, \beta; \gamma; z) = \dfrac{(1-z)^{-\alpha}\,\Gamma(\gamma)\,\Gamma(\beta-\alpha)}{\Gamma(\beta)\,\Gamma(\gamma-\alpha)}\,F\left(\alpha, \gamma-\beta; \alpha-\beta+1; \dfrac{1}{1-z}\right)$

$\qquad\qquad + (1-z)^{-\beta}\dfrac{\Gamma(\gamma)\,\Gamma(\alpha-\beta)}{\Gamma(\alpha)\,\Gamma(\gamma-\beta)}\,F\left(\beta, \gamma-\alpha; \beta-\alpha+1; \dfrac{1}{1-z}\right)$

$\qquad\qquad\qquad\qquad\qquad\qquad\qquad\qquad\qquad\qquad\qquad\qquad\qquad\qquad\qquad$ MO 13

2.[8] $F(\alpha, \beta; \gamma; z) = \dfrac{\Gamma(\gamma)\,\Gamma(\beta-\alpha)}{\Gamma(\beta)\,\Gamma(\gamma-\alpha)}(-z)^{-\alpha}\left(\alpha, \alpha+1-\gamma; \alpha+1-\beta; \dfrac{1}{z}\right)$

$\qquad\qquad + \dfrac{\Gamma(\gamma)\,\Gamma(\alpha-\beta)}{\Gamma(\alpha)\,\Gamma(\gamma-\beta)}(-z)^{-\beta}\,F\left(\beta, \beta+1-\gamma; \beta+1-\alpha; \dfrac{1}{z}\right)$

$\qquad\qquad\qquad\qquad [|\arg z| < \pi, \quad \alpha-\beta \neq \pm m, \quad m = 0, 1, 2, \dots]$ GA 220(93)

9.133 $F\left(2\alpha, 2\beta; \alpha+\beta+\tfrac{1}{2}; z\right) = F\left(\alpha, \beta; \alpha+\beta+\tfrac{1}{2}; 4z(1-z)\right)$

$\qquad\qquad\qquad\qquad\qquad\left[|z| \leq \tfrac{1}{2}, \quad |z(1-z)| \leq \tfrac{1}{4}\right]$ WH

9.134

1. $F(\alpha, \beta; 2\beta; z) = \left(1-\dfrac{z}{2}\right)^{-\alpha}F\left(\dfrac{\alpha}{2}, \dfrac{\alpha+1}{2}; \beta+\dfrac{1}{2}; \left(\dfrac{z}{2-z}\right)^2\right)$ MO 13, EH I 111(4)

2. $F(2\alpha, 2\alpha+1-\gamma; \gamma; z) = (1+z)^{-2\alpha}F\left(\alpha, \alpha+\dfrac{1}{2}; \gamma; \dfrac{4z}{(1+z)^2}\right)$ GA 225(100)

3. $F\left(\alpha, \alpha+\dfrac{1}{2}-\beta; \beta+\dfrac{1}{2}; z^2\right) = (1+z)^{-2\alpha}F\left(\alpha, \beta; 2\beta; \dfrac{4z}{(1+z)^2}\right)$ GA 225(101)

9.135 $F\left(\alpha, \beta; \alpha+\beta+\dfrac{1}{2}; \sin^2\varphi\right) = F\left(2\alpha, 2\beta; \alpha+\beta+\dfrac{1}{2}; \sin^2\dfrac{\varphi}{2}\right)$

$\qquad\qquad\qquad\qquad\qquad\left[x = \sin^2\dfrac{\varphi}{2}\ \text{real}; \quad \dfrac{1-\sqrt{2}}{2} < x < \dfrac{1}{2}\right]$

$\qquad\qquad\qquad\qquad\qquad\qquad\qquad\qquad\qquad\qquad\qquad\qquad\qquad\qquad\qquad$ MO 13

9.136[8]   We set

$$A = \frac{\Gamma\left(\alpha+\beta+\frac{1}{2}\right)\sqrt{\pi}}{\Gamma\left(\alpha+\frac{1}{2}\right)\Gamma\left(\beta+\frac{1}{2}\right)}, \qquad B = \frac{-\Gamma\left(\alpha+\beta+\frac{1}{2}\right)2\sqrt{\pi}}{\Gamma(\alpha)\,\Gamma(\beta)};$$

then

1. $F\left(2\alpha, 2\beta; \alpha+\beta+\dfrac{1}{2}; \dfrac{1-\sqrt{z}}{2}\right) = AF\left(\alpha, \beta; \dfrac{1}{2}; z\right) + B\sqrt{z}\,F\left(\alpha+\dfrac{1}{2}, \beta+\dfrac{1}{2}; \dfrac{3}{2}; z\right)$

$\qquad\qquad\qquad\qquad\qquad\qquad\qquad\qquad\qquad\qquad\qquad\qquad\qquad\qquad\qquad$ GA 227(106)

2. $F\left(2\alpha, 2\beta; \alpha+\beta+\dfrac{1}{2}; \dfrac{1+\sqrt{z}}{2}\right) = AF\left(\alpha, \beta; \dfrac{1}{2}; z\right) - B\sqrt{z}\,F\left(\alpha+\dfrac{1}{2}, \beta+\dfrac{1}{2}; \dfrac{3}{2}; z\right)$

$\qquad\qquad\qquad\qquad\qquad\qquad\qquad\qquad\qquad\qquad\qquad\qquad\qquad\qquad\qquad$ GA 227(107)

3. $\dfrac{\left(\alpha-\frac{1}{2}\right)\left(\beta-\frac{1}{2}\right)}{\alpha+\beta-\frac{1}{2}}A\sqrt{z}\,F\left(\alpha, \beta; \dfrac{3}{2}; z\right) = F\left(2\alpha-1, 2\beta-1; \alpha+\beta-\dfrac{1}{2}; \dfrac{1+\sqrt{z}}{2}\right)$

$\qquad\qquad\qquad\qquad\qquad\qquad - F\left(2\alpha-1, 2\beta-1; \alpha+\beta-\dfrac{1}{2}; \dfrac{1-\sqrt{z}}{2}\right)$

$\qquad\qquad\qquad\qquad\qquad\qquad\qquad\qquad\qquad\qquad\qquad\qquad\qquad\qquad\qquad$ GA 229(110)

9.137[7] Gauss' recursion functions:

1. $\gamma[\gamma - 1 - (2\gamma - \alpha - \beta - 1)z] F(\alpha, \beta; \gamma; z) + (\gamma - \alpha)(\gamma - \beta)z F(\alpha, \beta; \gamma + 1; z) + \gamma(\gamma - 1)(z - 1) F(\alpha, \beta; \gamma - 1; z) = 0$

2. $(2\alpha - \gamma - \alpha z + \beta z) F(\alpha, \beta; \gamma; z) + (\gamma - \alpha) F(\alpha - 1, \beta; \gamma; z) + \alpha(z - 1) F(\alpha + 1, \beta; \gamma; z) = 0$

3. $(2\beta - \gamma - \beta z + \alpha z) F(\alpha, \beta; \gamma; z) + (\gamma - \beta) F(\alpha, \beta - 1; \gamma; z) + \beta(z - 1) F(\alpha, \beta + 1; \gamma; z) = 0$

4. $\gamma F(\alpha, \beta - 1; \gamma; z) - \gamma F(\alpha - 1, \beta; \gamma; z) + (\alpha - \beta)z F(\alpha, \beta; \gamma + 1; z) = 0$

5.[8] $\gamma(\alpha - \beta) F(\alpha, \beta; \gamma; z) - \alpha(\gamma - \beta) F(\alpha + 1, \beta; \gamma + 1; z) + \beta(\gamma - \alpha) F(\alpha, \beta + 1; \gamma + 1; z) = 0$

6. $\gamma(\gamma + 1) F(\alpha, \beta; \gamma; z) - \gamma(\gamma + 1) F(\alpha, \beta; \gamma + 1; z) - \alpha\beta z F(\alpha + 1, \beta + 1; \gamma + 2; z) = 0$

7. $\gamma F(\alpha, \beta; \gamma; z) - (\gamma - \alpha) F(\alpha, \beta + 1; \gamma + 1; z) - \alpha(1 - z) F(\alpha + 1, \beta + 1; \gamma + 1; z) = 0$

8. $\gamma F(\alpha, \beta; \gamma; z) + (\beta - \gamma) F(\alpha + 1, \beta; \gamma + 1; z) - \beta(1 - z) F(\alpha + 1, \beta + 1; \gamma + 1; z) = 0$

9. $\gamma(\gamma - \beta z - \alpha) F(\alpha, \beta; \gamma; z) - \gamma(\gamma - \alpha) F(\alpha - 1, \beta; \gamma; z) + \alpha\beta z(1 - z) F(\alpha + 1, \beta + 1; \gamma + 1; z) = 0$

10. $\gamma(\gamma - \alpha z - \beta) F(\alpha, \beta; \gamma; z) - \gamma(\gamma - \beta) F(\alpha, \beta - 1; \gamma; z) + \alpha\beta z(1 - z) F(\alpha + 1, \beta + 1; \gamma + 1; z) = 0$

11. $\gamma F(\alpha, \beta; \gamma; z) - \gamma F(\alpha, \beta + 1; \gamma; z) + \alpha z F(\alpha + 1, \beta + 1; \gamma + 1; z) = 0$

12.[8] $\gamma F(\alpha, \beta; \gamma; z) - \gamma F(\alpha + 1, \beta; \gamma; z) + \beta z F(\alpha + 1, \beta + 1; \gamma + 1; z) = 0$

13. $\gamma[\alpha - (\gamma - \beta)z] F(\alpha, \beta; \gamma; z) - \alpha\gamma(1 - z) F(\alpha + 1, \beta; \gamma; z) + (\gamma - \alpha)(\gamma - \beta)z F(\alpha, \beta; \gamma + 1; z) = 0$

14. $\gamma[\beta - (\gamma - \alpha)z] F(\alpha, \beta; \gamma; z) - \beta\gamma(1 - z) F(\alpha, \beta + 1; \gamma; z) + (\gamma - \alpha)(\gamma - \beta)z F(\alpha, \beta; \gamma + 1; z) = 0$

15.[8] $\gamma(\gamma + 1) F(\alpha, \beta; \gamma; z) - \gamma(\gamma + 1) F(\alpha, \beta + 1; \gamma + 1; z) + \alpha(\gamma - \beta)z F(\alpha + 1, \beta + 1; \gamma + 2; z) = 0$

16. $\gamma(\gamma + 1) F(\alpha, \beta; \gamma; z) - \gamma(\gamma + 1) F(\alpha + 1, \beta; \gamma + 1; z) + \beta(\gamma - \alpha)z F(\alpha + 1, \beta + 1; \gamma + 2; z) = 0$

17. $\gamma F(\alpha, \beta; \gamma; z) - (\gamma - \beta) F(\alpha, \beta; \gamma + 1; z) - \beta F(\alpha, \beta + 1; \gamma + 1; z) = 0$

18.[8] $\gamma F(\alpha, \beta; \gamma; z) - (\gamma - \alpha) F(\alpha, \beta; \gamma + 1; z) - \alpha F(\alpha + 1, \beta; \gamma + 1; z) = 0$ MO 13–14

9.14 A generalized hypergeometric series

The series

1. $$_pF_q(\alpha_1, \alpha_2, \ldots, \alpha_p; \beta_1, \beta_2, \ldots, \beta_q; z) = \sum_{k=0}^{\infty} \frac{(\alpha_1)_k (\alpha_2)_k \ldots (\alpha_p)_k}{(\beta_1)_k (\beta_2)_k \ldots (\beta_q)_k} \frac{z^k}{k!}$$ MO 14

is called a *generalized hypergeometric series* (see also 9.210).

2. $_2F_1(\alpha, \beta; \gamma; z) \equiv F(\alpha, \beta; \gamma; z)$ MO 15

For integral representations, see **3.254** 2, **3.259** 2, and **3.478** 3.

9.15 The hypergeometric differential equation

9.151 A hypergeometric series is one of the solutions of the differential equation

$$z(1 - z)\frac{d^2u}{dz^2} + [\gamma - (\alpha + \beta + 1)z]\frac{du}{dz} - \alpha\beta u = 0,$$ WH

which is called the *hypergeometric equation*.

The solution of the hypergeometric differential equation

9.152   The hypergeometric differential equation **9.151** possesses *two linearly independent solutions.* These solutions have analytic continuations to the entire z-plane except possibly for the three points 0, 1, and ∞. Generally speaking, the points $z = 0, 1, \infty$ are branch points of at least one of the branches of each solution of the hypergeometric differential equation. The ratio $w(z)$ of two linearly independent solutions satisfies the differential equation

$$2\frac{w'''}{w'} - 3\left(\frac{w''}{w'}\right)^2 = \frac{1 - a_1^2}{z^2} + \frac{1 - a_2^2}{(z-1)^2} + \frac{a_1^2 + a_2^2 - a_3^2 - 1}{z(z-1)},$$

where

$$a_1^2 = (1 - \gamma)^2, \quad a_2^2 = (\gamma - \alpha - \beta)^2, \quad a_3^2 = (\alpha - \beta)^2$$

If α, β, γ are real, the function $w(z)$ maps the upper (Im $z > 0$) or the lower (Im $z < 0$) half-plane onto a curvilinear triangle whose angles are $\pi a_1, \pi a_2, \pi a_3$. The vertices of this triangle are the images of the points $z = 0, z = 1$, and $z = \infty$.

9.153   Within the unit circle $|z| < 1$, the linearly independent solutions $u_1(z)$ and $u_2(z)$ of the hypergeometric differential equation are given by the following formulas:

1.   If γ is not an integer,

$$u_1 = F(\alpha, \beta; \gamma; z),$$
$$u_2 = z^{1-\gamma} e\, F(\alpha - \gamma + 1, \beta - \gamma + 1; 2 - \gamma; z)$$

2.   If $\gamma = 1$, then

$$u_1 = F(\alpha, \beta; 1; z),$$
$$u_2 = F(\alpha, \beta; 1; z) \ln z + \sum_{k=1}^{\infty} z^k \frac{(\alpha)_k (\beta)_k}{(k!)^2}$$
$$\times \{\psi(\alpha + k) - \psi(\alpha) + \psi(\beta + k) - \psi(\beta) - 2\psi(k+1) + 2\psi(1)\}$$

(see 9.14 2)

3.   If $\gamma = m + 1$ (where m is a natural number), and if neither α nor β is a positive number not exceeding m, then

$$u_1 = F(\alpha, \beta; m + 1; z),$$
$$u_2 = F(\alpha, \beta; m + 1; z) \ln z + \sum_{k=1}^{\infty} z^k \frac{(\alpha)_k (\beta)_k}{(1 + m)_k} \{h(k) - h(0)\} - \sum_{k=1}^{m} \frac{(k-1)!(-m)_k}{(1-\alpha)_k (1-\beta)_k} z^{-k}$$

(see 9.14 2)

where

$$h(n) = \psi(\alpha + n) + \psi(\beta + n) - \psi(m + 1 + n) - \psi(n + 1) \qquad [n + 1 \text{ is a natural number}]$$

4.[7]   Suppose that $\gamma = m + 1$ (where m is a natural number) and that α or β is equal to $m' + 1$, where $0 \le m' < m$. Then, for example, for $\alpha = m' + 1$, we obtain

$$u_1 = F(1 + m', \beta; 1 + m; z),$$
$$u_2 = z^{-m} F(1 + m', -m, \beta - m; 1 - m; z)$$

In this case, u_2 is a polynomial in z^{-1}.

5.   If $\gamma = 1 - m$ (where m is a natural number) and if α and β are both different from the numbers $0, -1, -2, \ldots, 1 - m$, then

$$u_1 = z^m \, F(\alpha + m, \beta + m; 1 + m; z),$$

$$u_2 = z^m \, F(\alpha + m, \beta + m; 1 + m; z) \ln z + \sum_{k=1}^{\infty} z^k \frac{(\alpha + m)_k (\beta + m)_k}{(1 + m)_k k!} \{h^*(k) - h^*(0)\}$$

$$- \sum_{k=1}^{\infty} \frac{(k-1)!(-m)_k}{(1 - \alpha - m)_k (1 - \beta - m)_k} z^{m-n}$$

(see **9.14** 2)

where

$$h^*(n) = \psi(\alpha + m + n) + \psi(\beta + m + n) - \psi(1 + m + n) - \psi(1 + n)$$

We note that

$$\psi(\alpha + n) - \psi(\alpha) = \frac{1}{\alpha} + \frac{1}{\alpha + 1} + \cdots + \frac{1}{\alpha + n - 1}$$ (cf. **8.365** 3)

and that, for $\alpha = -\lambda$, where λ is a natural number or zero and $n = \lambda + 1, \lambda + 2, \ldots$ the expression

$$(\alpha)_k \left[\psi(\alpha + n) - \psi(\alpha) \right]$$

in formulas **9.153** 2–5 should be replaced with the expression

$$(-1)^{\lambda} \lambda!(n - \lambda - 1)!$$

6.   Suppose that $\gamma = 1 - m$ (where m is a natural number) and that α or β is an integer $(-m')$, where m' is one of the following numbers: $0, 1, \ldots, m - 1$. Suppose, for example, that $\alpha = -m'$. Then,

$$u_1 = F\left(-m', \beta; 1 - m; z\right),$$
$$u_2 = F\left(-m' + m, \beta + m; 1 + m; z\right)$$

MO 18

7.   For $\gamma = \frac{1}{2}(\alpha + \beta + 1)$

$$u_1 = F\left(\alpha, \beta; \frac{1}{2}(\alpha + \beta + 1); z\right),$$
$$u_2 = F\left(\alpha, \beta; \frac{1}{2}(\alpha + \beta + 1); 1 - z\right)$$

are two linearly independent solutions of the hypergeometric differential equation provided α, β and γ are not zero or negative integers.

MO 17–19

The analytic continuation of a solution that is regular at the point $z = 0$

9.154   Formulas **9.153** make possible the analytic continuation, by means of the hypergeometric series, of the function $F(\alpha, \beta; \gamma; z)$ defined inside the circle $|z| < 1$ to the region $|z| > 1$, and $|\arg(-z)| < \pi$. Here, it is assumed that $\alpha - \beta$ is not an integer. In the event that $\alpha - \beta$ is an integer (for example, if $\beta = \alpha + m$, where m is a natural number), then, for $|z| > 1$, and $|\arg(-z)| < \pi$ we have:

1. $$\frac{\Gamma(\alpha)\,\Gamma(\alpha + m)}{\Gamma(\gamma)} F(\alpha, \alpha + m; \gamma; z)$$

$$= \frac{\sin \pi(\gamma - \alpha)}{\pi} \left\{ \sum_{k=0}^{m-1} \frac{\Gamma(\alpha + k)\,\Gamma(1 - \gamma + \alpha + k)\,\Gamma(m - k)}{k!} (-z)^{-\alpha - k} \right.$$

$$\left. + (-z)^{-\alpha - m} \sum_{k=0}^{\infty} \frac{\Gamma(\alpha + m + k)\,\Gamma(1 - \gamma + \alpha + m + k)}{k!\,(k + m)!} g(k) z^{-k} \right\}$$

       where

2. $$g(n) = \ln(-z) + \pi \cot \pi(\gamma - \alpha) + \psi(n + 1) + \psi(n + m + 1)$$

$$- \psi(\alpha + m + n) - \psi(1 - \gamma + \alpha + m + n)$$

       For $m = 0$, we should set $\displaystyle\sum_{k=0}^{m-1} = 0$.

9.155   This formula loses its meaning when α, γ, or $\alpha - \gamma + 1$ is equal to one of the numbers $0, -1, -2, \ldots$. In this last case, we have

1.   If α is a non-positive integer and γ is not an integer, $F(\alpha, \alpha + m; \gamma; z)$ is a polynomial in z.

2.   Suppose that γ is a non-positive integer and that α is not an integer. We then set $\gamma = -\lambda$, where $\lambda = 0, 1, 2, \ldots$. Then,

$$\frac{\Gamma(\alpha + \lambda + 1)\,\Gamma(\alpha + \lambda + m + 1)}{\Gamma(\lambda + 2)} z^{\lambda + 1} F(\alpha + \lambda + 1, \alpha + \lambda + m + 1; \lambda + 2; z)$$

       is a solution of the hypergeometric equation that is regular at the point $z = 0$. This solution is equal to the right hand member of formula **9.154** 1 if we replace γ with λ in this equation and in formula **9.154** 2.

3.   If $\alpha - \gamma + 1$ is a non-positive integer and if α and γ are not themselves integers, we may use the formula

$$F(\alpha, \alpha + m; \gamma; z) = (1 - z)^{\gamma - 2\alpha - m} F(\gamma - \alpha - m, \gamma - \alpha; \gamma; z)$$

       and apply formula **9.154** 1 to its right hand member provided $\gamma - \alpha - m > 0$. However, if $\alpha - \gamma - m \leq 0$, the right member of this expression is a polynomial taken to the $(1 - z)^{\text{th}}$ power.

4.   If α, β, and γ are integers, the hypergeometric differential equation always has a solution that is regular for $z = 0$ and that is of the form

$$R_1(z) + \ln(1-z)R_2(z),$$

where $R_1(z)$ and $R_2(z)$ are rational functions of z. To get a solution of this form, we need to apply formulas **9.137** 1–**9.137** 3 to the function $F(\alpha, \beta; \gamma; z)$. However, if $\gamma = -\lambda$, where $\lambda + 1$ is a natural number, formulas **9.137** 1 and **9.137** 2 should be applied not to $F(\alpha, \beta; \gamma; z)$ but to the function $z^{\lambda+1} F(\alpha + \lambda + 1, \beta + \lambda + 1; \lambda + 2, z)$.

By successive applications of these formulas, we can reduce the positive values of the parameters to the pair, unity and zero. Furthermore, we can obtain the desired form of the solution from the formulas

$$F(1, 1; 2; z) = -z^{-1}\ln(1-z),$$
$$F(0, \beta; \gamma; z) = F(\alpha, 0; \gamma; z) = 1$$

MO 19–20

9.16   Riemann's differential equation

9.160   The hypergeometric differential equation is a particular case of Riemann's differential equation

1.
$$\frac{d^2u}{dz^2} + \left[\frac{1-\alpha-\alpha'}{z-a} + \frac{1-\beta-\beta'}{z-b} + \frac{1-\gamma-\gamma'}{z-c}\right]\frac{du}{dz}$$
$$+ \left[\frac{\alpha\alpha'(a-b)(a-c)}{z-a} + \frac{\beta\beta'(b-c)(b-a)}{z-b} + \frac{\gamma\gamma'(c-a)(c-b)}{z-c}\right]\frac{u}{(z-a)(z-b)(z-c)} = 0$$

WH

The coefficients of this equation have poles at the points a, b, and c, and the numbers α, α'; β, β'; γ, γ' are called the indices corresponding to these poles. The indices α, α'; β, β'; γ, γ' are related by the following equation:

$$\alpha + \alpha' + \beta - \beta' + \gamma + \gamma' - 1 = 0 \qquad\qquad \text{WH}$$

2.   The differential equations **9.160** 1 are written diagramatically as follows:

3.
$$u = P\left\{\begin{matrix} a & b & c \\ \alpha & \beta & \gamma & z \\ \alpha' & \beta' & \gamma' \end{matrix}\right\}$$

The singular points of the equation appear in the first row in this scheme, the indices corresponding to them appear beneath them, and the independent variable appears in the fourth column.   WH

9.161   The two following transformation formulas are valid for Riemann's P-equation:

1.
$$\left(\frac{z-a}{z-b}\right)^k \left(\frac{z-c}{z-b}\right)^l P\left\{\begin{matrix} a & b & c \\ \alpha & \beta & \gamma & z \\ \alpha' & \beta' & \gamma' \end{matrix}\right\} = P\left\{\begin{matrix} a & b & c \\ \alpha+k & \beta-k-1 & \gamma+l & z \\ \alpha'+k & \beta'-k-1 & \gamma'+l \end{matrix}\right\} \qquad \text{WH}$$

2.
$$P\left\{\begin{matrix} a & b & c \\ \alpha & \beta & \gamma & z \\ \alpha' & \beta' & \gamma' \end{matrix}\right\} = P\left\{\begin{matrix} a_1 & b_1 & c_1 \\ \alpha & \beta & \gamma & z_1 \\ \alpha' & \beta' & \gamma' \end{matrix}\right\} \qquad \text{WH}$$

The first of these formulas means that if

$$u = P \left\{ \begin{matrix} a & b & c & \\ \alpha & \beta & \gamma & z \\ \alpha' & \beta' & \gamma' & \end{matrix} \right\},$$

then the function

$$u_1 = \left(\frac{z-a}{z-b} \right)^k \left(\frac{z-c}{z-b} \right)^l u$$

satisfies a second-order differential equation having the same singular points as equation **9.161** 2 and indices equal to $\alpha + k, \alpha' + k; \beta - k - l, \beta' - k - l; \gamma + l, \gamma' + l$. The second transformation formula converts a differential equation with singularities at the points a,b, and c, indices $\alpha, \alpha'; \beta, \beta'; \gamma, \gamma'$, and an independent variable z into a differential equation with the same indices, singular points a_1, b_1, and c_1, and independent variable z_1. The variable z_1 is connected with the variable z by the fractional transformation

$$z = \frac{Az_1 + B}{Cz_1 + D} \qquad [AD - BC \neq 0]$$

The same transformation connects the points a_1, b_1, and c_1 with the points a, b, and c.

WH, MO 20

9.162   By the successive application of the two transformation formulas **9.161** 1 and **9.161** 2, we can convert Riemann's differential equation into the hypergeometric differential equation. Thus, the solution of Riemann's differential equation can be expressed in terms of a hypergeometric function.

For $k = -\alpha, l = -\gamma$, and $z_1 = \frac{(z-a)(c-b)}{(z-b)(c-a)}$, we have

1.
$$u = P \left\{ \begin{matrix} a & b & c & \\ \alpha & \beta & \gamma & z \\ \alpha' & \beta' & \gamma' & \end{matrix} \right\} = \left(\frac{z-a}{z-b} \right)^\alpha \left(\frac{z-c}{z-b} \right)^\gamma P \left\{ \begin{matrix} a & b & c & \\ 0 & \beta + \alpha + \gamma & 0 & z \\ \alpha' - \alpha & \beta' + \alpha + \gamma & \gamma' - \gamma & \end{matrix} \right\}$$

$$= \left(\frac{z-a}{z-b} \right)^\alpha \left(\frac{z-c}{z-b} \right)^\gamma P \left\{ \begin{matrix} 0 & \infty & 1 & \\ 0 & \beta + \alpha + \gamma & 0 & \frac{(z-a)(c-b)}{(z-b)(c-a)} \\ \alpha' - \alpha & \beta' + \alpha + \gamma & \gamma' - \gamma & \end{matrix} \right\}$$

MO 23

Thus, this solution can be expressed as a hypergeometric series as follows:

2.
$$u = \left(\frac{z-a}{z-b} \right)^\alpha \left(\frac{z-c}{z-b} \right)^\gamma F \left(\alpha + \beta + \gamma, \alpha + \beta' + \gamma; 1 + \alpha - \alpha'; \frac{(z-a)(c-b)}{(z-b)(c-a)} \right)$$

If the constants $a, b, c; \alpha, \alpha'; \beta, \beta'; \gamma, \gamma'$ are permuted in a suitable manner, Riemann's equation remains unchanged. Thus, we obtain a set of 24 solutions of differential equations having the following form (provided none of the differences $\alpha - \alpha', \beta - \beta', \gamma - \gamma'$ are integers): WH, MO 23

9.163

1.
$$u_1 = \left(\frac{z-a}{z-b} \right)^\alpha \left(\frac{z-c}{z-b} \right)^\gamma F \left\{ \alpha + \beta + \gamma, \alpha + \beta' + \gamma; 1 + \alpha - \alpha'; \frac{(c-b)(z-a)}{(c-a)(z-b)} \right\}$$

2.
$$u_2 = \left(\frac{z-a}{z-b} \right)^{\alpha'} \left(\frac{z-c}{z-b} \right)^\gamma F \left\{ \alpha' + \beta + \gamma, \alpha' + \beta' + \gamma; 1 + \alpha' - \alpha; \frac{(c-b)(z-a)}{(c-a)(z-b)} \right\}$$

3.
$$u_3 = \left(\frac{z-a}{z-b} \right)^\alpha \left(\frac{z-c}{z-b} \right)^{\gamma'} F \left\{ \alpha + \beta + \gamma', \alpha + \beta' + \gamma'; 1 + \alpha - \alpha'; \frac{(c-b)(z-a)}{(c-a)(z-b)} \right\}$$

4.
$$u_4 = \left(\frac{z-a}{z-b} \right)^{\alpha'} \left(\frac{z-c}{z-b} \right)^{\gamma'} F \left\{ \alpha' + \beta + \gamma', \alpha' + \beta' + \gamma; 1 + \alpha' - \alpha; \frac{(c-b)(z-a)}{(c-a)(z-b)} \right\}$$

9.164

1.[10] $u_5 = \left(\dfrac{z-b}{z-c}\right)^{\beta} \left(\dfrac{z-a}{z-c}\right)^{\alpha} F\left\{\beta + \gamma + \alpha, \beta + \gamma' + \alpha; 1 + \beta - \beta'; \dfrac{(a-c)(z-b)}{(a-b)(z-c)}\right\}$

2. $u_6 = \left(\dfrac{z-b}{z-c}\right)^{\beta'} \left(\dfrac{z-a}{z-c}\right)^{\alpha} F\left\{\beta' + \gamma + \alpha, \beta' + \gamma' + \alpha; 1 + \beta' - \beta; \dfrac{(a-c)(z-b)}{(a-b)(z-c)}\right\}$

3. $u_7 = \left(\dfrac{z-b}{z-c}\right)^{\beta} \left(\dfrac{z-a}{z-c}\right)^{\alpha'} F\left\{\beta + \gamma + \alpha', \beta + \gamma' + \alpha'; 1 + \beta - \beta'; \dfrac{(a-c)(z-b)}{(a-b)(z-c)}\right\}$

4. $u_8 = \left(\dfrac{z-b}{z-c}\right)^{\beta'} \left(\dfrac{z-a}{z-c}\right)^{\alpha'} F\left\{\beta' + \gamma + \alpha', \beta' + \alpha' + \gamma'; 1 + \beta' - \beta; \dfrac{(a-c)(z-b)}{(a-b)(z-c)}\right\}$

9.165

1. $u_9 = \left(\dfrac{z-c}{z-a}\right)^{\gamma} \left(\dfrac{z-b}{z-a}\right)^{\beta} F\left\{\gamma + \alpha + \beta, \gamma + \alpha' + \beta; 1 + \gamma - \gamma'; \dfrac{(b-a)(z-c)}{(b-c)(z-a)}\right\}$

2. $u_{10} = \left(\dfrac{z-c}{z-a}\right)^{\gamma'} \left(\dfrac{z-b}{z-a}\right)^{\beta} F\left\{\gamma' + \alpha + \beta, \gamma + \alpha' + \beta; 1 + \gamma' - \gamma; \dfrac{(b-a)(z-c)}{(b-c)(z-a)}\right\}$

3. $u_{11} = \left(\dfrac{z-c}{z-a}\right)^{\gamma} \left(\dfrac{z-b}{z-a}\right)^{\beta'} F\left\{\gamma + \alpha + \beta', \gamma + \alpha' + \beta'; 1 + \gamma - \gamma'; \dfrac{(b-a)(z-c)}{(b-c)(z-a)}\right\}$

4. $u_{12} = \left(\dfrac{z-c}{z-a}\right)^{\gamma'} \left(\dfrac{z-b}{z-a}\right)^{\beta'} F\left\{\gamma' + \alpha + \beta', \gamma' + \alpha' + \beta'; 1 + \gamma' - \gamma; \dfrac{(b-a)(z-c)}{(b-c)(z-a)}\right\}$

9.166

1. $u_{13} = \left(\dfrac{z-a}{z-c}\right)^{\alpha} \left(\dfrac{z-b}{z-c}\right)^{\beta} F\left\{\alpha + \gamma + \beta, \alpha + \gamma' + \beta; 1 + \alpha - \alpha'; \dfrac{(b-c)(z-a)}{(b-a)(z-c)}\right\}$

2. $u_{14} = \left(\dfrac{z-a}{z-c}\right)^{\alpha'} \left(\dfrac{z-b}{z-c}\right)^{\beta} F\left\{\alpha' + \gamma + \beta, \alpha' + \gamma' + \beta; 1 + \alpha' - \alpha; \dfrac{(b-c)(z-a)}{(b-a)(z-c)}\right\}$

3. $u_{15} = \left(\dfrac{z-a}{z-c}\right)^{\alpha} \left(\dfrac{z-b}{z-c}\right)^{\beta'} F\left\{\alpha + \gamma + \beta', \alpha + \gamma' + \beta'; 1 + \alpha - \alpha'; \dfrac{(b-c)(z-a)}{(b-a)(z-c)}\right\}$

4. $u_{16} = \left(\dfrac{z-a}{z-c}\right)^{\alpha'} \left(\dfrac{z-b}{z-c}\right)^{\beta'} F\left\{\alpha' + \gamma + \beta', \alpha' + \gamma' + \beta'; 1 + \alpha' - \alpha; \dfrac{(b-c)(z-a)}{(b-a)(z-c)}\right\}$

9.167

1. $u_{17} = \left(\dfrac{z-c}{z-b}\right)^{\gamma} \left(\dfrac{z-a}{z-b}\right)^{\alpha} F\left\{\gamma + \beta + \alpha, \gamma + \beta' + \alpha; 1 + \gamma - \gamma'; \dfrac{(a-b)(z-c)}{(a-c)(z-b)}\right\}$

2. $u_{18} = \left(\dfrac{z-c}{z-b}\right)^{\gamma'} \left(\dfrac{z-a}{z-b}\right)^{\alpha} F\left\{\gamma' + \beta + \alpha, \gamma' + \beta' + \alpha; 1 + \gamma' - \gamma; \dfrac{(a-b)(z-c)}{(a-c)(z-b)}\right\}$

3. $u_{19} = \left(\dfrac{z-c}{z-b}\right)^{\gamma} \left(\dfrac{z-a}{z-b}\right)^{\alpha'} F\left\{\gamma + \beta + \alpha', \gamma + \beta' + \alpha'; 1 + \gamma - \gamma'; \dfrac{(a-b)(z-c)}{(a-c)(z-b)}\right\}$

4. $u_{20} = \left(\dfrac{z-c}{z-b}\right)^{\gamma'} \left(\dfrac{z-a}{z-b}\right)^{\alpha'} F\left\{\gamma' + \beta + \alpha', \gamma' + \beta' + \alpha'; 1 + \gamma' - \gamma; \dfrac{(a-b)(z-c)}{(a-c)(z-b)}\right\}$

9.168

1. $u_{21} = \left(\dfrac{z-b}{z-a}\right)^{\beta} \left(\dfrac{z-c}{z-a}\right)^{\gamma} F\left\{\beta + \alpha + \gamma, \beta + \alpha' + \gamma; 1 + \beta - \beta'; \dfrac{(c-a)(z-b)}{(c-b)(z-a)}\right\}$

2. $u_{22} = \left(\dfrac{z-b}{z-a}\right)^{\beta'} \left(\dfrac{z-c}{z-a}\right)^{\gamma} F\left\{\beta' + \alpha + \gamma, \beta' + \alpha' + \gamma; 1 + \beta' - \beta; \dfrac{(c-a)(z-b)}{(c-b)(z-a)}\right\}$

3. $u_{23} = \left(\dfrac{z-b}{z-a}\right)^{\beta} \left(\dfrac{z-c}{z-a}\right)^{\gamma'} F\left\{\beta + \alpha + \gamma', \beta + \alpha' + \gamma'; 1 + \beta - \beta'; \dfrac{(c-a)(z-b)}{(c-b)(z-a)}\right\}$

4. $u_{24} = \left(\dfrac{z-b}{z-a}\right)^{\beta'} \left(\dfrac{z-c}{z-a}\right)^{\gamma'} F\left\{\beta' + \alpha + \gamma', \beta' + \alpha' + \gamma'; 1 + \beta' - \beta; \dfrac{(c-a)(z-b)}{(c-b)(z-a)}\right\}$ WH

9.17 Representing the solutions to certain second-order differential equations using a Riemann scheme

9.171 The hypergeometric equation (see **9.151**):

$$u = P\left\{\begin{matrix} 0 & \infty & 1 & \\ 0 & \alpha & 0 & z \\ 1-\gamma & \beta & \gamma - \alpha - \beta & \end{matrix}\right\}$$ WH

9.172 The associated Legendre's equation defining the functions $P_n^m(z)$ for n and m integers (see **8.700** 1):

1. $$u = P\left\{\begin{matrix} 0 & \infty & 1 & \\ \tfrac{1}{2}m & n+1 & \tfrac{1}{2}m & \dfrac{1-z}{2} \\ -\tfrac{1}{2}m & -n & -\tfrac{1}{2}m & \end{matrix}\right\}$$ WH

2. $$u = P\left\{\begin{matrix} 0 & \infty & 1 & \\ -\tfrac{1}{2}n & \tfrac{1}{2}m & 0 & \dfrac{1}{1-z^2} \\ \dfrac{n+1}{2} & -\tfrac{1}{2}m & \tfrac{1}{2} & \end{matrix}\right\}$$ WH

9.173 The function $P_n^m\left(1 - \dfrac{z^2}{2n^2}\right)$ satisfies the equation

$$u = P\left\{\begin{matrix} 4n^2 & \infty & 0 & \\ \tfrac{1}{2}m & n+1 & \tfrac{1}{2}m & z^2 \\ -\tfrac{1}{2}m & -n & -\tfrac{1}{2}m & \end{matrix}\right\}$$ WH

The function $J_m(z)$ satisfies the limiting form of this equation obtained as $n \to \infty$.

9.174   The equation defining the Gegenbauer polynomials $C_n^\lambda(z)$, (see **8.938**):

$$u = P \left\{ \begin{matrix} -1 & \infty & 1 \\ \frac{1}{2} - \lambda & n + 2\lambda & \frac{1}{2} - \lambda & z \\ 0 & -n & 0 \end{matrix} \right\}$$
WH

9.175   Bessel's equation (see **8.401**) is the limiting form of the equations:

1. $$u = P \left\{ \begin{matrix} 0 & \infty & c \\ n & ic & \frac{1}{2} + ic & z \\ -n & -ic & \frac{1}{2} - ic \end{matrix} \right\}$$
WH

2. $$u = e^{iz} P \left\{ \begin{matrix} 0 & \infty & c \\ n & \frac{1}{2} & 0 & z \\ -n & \frac{3}{2} - 2ic & 2ic - 1 \end{matrix} \right\}$$
WH

3. $$u = P \left\{ \begin{matrix} 0 & \infty & c^2 \\ \frac{1}{2}n & \frac{1}{2}(c - n) & 0 & z^2 \\ -\frac{1}{2}n & -\frac{1}{2}(c + n) & n + 1 \end{matrix} \right\}$$
WH

as $c \to \infty$.

9.18 Hypergeometric functions of two variables

9.180

1. $$F_1(\alpha, \beta, \beta', \gamma; x, y) = \sum_{m=0}^{\infty} \sum_{n=0}^{\infty} \frac{(\alpha)_{m+n} (\beta)_m (\beta')_n}{(\gamma)_{m+n} m! n!} x^m y^n$$

$[|x| < 1, \quad |y| < 1]$
EH I 224(6), AK 14(11)

2. $$F_2(\alpha, \beta, \beta', \gamma, \gamma'; x, y) = \sum_{m=0}^{\infty} \sum_{n=0}^{\infty} \frac{(\alpha)_{m+n} (\beta)_m (\beta')_n}{(\gamma)_m (\gamma')_n m! n!} x^m y^n$$

$[|x| + |y| < 1]$ EH I 224(7), AK 14(12)

3. $$F_3(\alpha, \alpha', \beta, \beta', \gamma; x, y) = \sum_{m=0}^{\infty} \sum_{n=0}^{\infty} \frac{(\alpha)_m (\alpha')_n (\beta)_m (\beta')_n}{(\gamma)_{m+n} m! n!} x^m y^n$$

$[|x| < 1, \quad |y| < 1]$
EH I 224(8), AK 14(13)

4. $$F_4(\alpha, \beta, \gamma, \gamma'; x, y) = \sum_{m=0}^{\infty} \sum_{n=0}^{\infty} \frac{(\alpha)_{m+n} (\beta)_{m+n}}{(\gamma)_m (\gamma')_n m! n!} x^m y^n \qquad [|\sqrt{x}| + |\sqrt{y}| < 1]$$

EH I 224(9), AK 14(14)

9.181   The functions F_1, F_2, F_3, and F_4 satisfy the following systems of partial differential equations for z:

1.   System of equations for $z = F_1$:

$$x(1-x)\frac{\partial^2 z}{\partial x^2} + y(1-x)\frac{\partial^2 z}{\partial x \, \partial y} + [\gamma - (\alpha + \beta + 1)x]\frac{\partial z}{\partial x} - \beta y\frac{\partial z}{\partial y} - \alpha\beta z = 0,$$
EH I 233(9)

$$y(1-y)\frac{\partial^2 z}{\partial y^2} + x(1-y)\frac{\partial^2 z}{\partial x \, \partial y} + [\gamma - (\alpha + \beta' + 1)\, y]\frac{\partial z}{\partial x} - \beta' x\frac{\partial z}{\partial x} - \alpha\beta' z = 0$$

2. System of equations for $z = F_2$:

$$x(1-x)\frac{\partial^2 z}{\partial x^2} - xy\frac{\partial^2 z}{\partial x\,\partial y} + [\gamma - (\alpha + \beta + 1)x]\frac{\partial z}{\partial x} - \beta y\frac{\partial z}{\partial y} - \alpha\beta z = 0, \qquad \text{EH I 234(10)}$$

$$y(1-y)\frac{\partial^2 z}{\partial y^2} - xy\frac{\partial^2 z}{\partial x\,\partial y} + [\gamma' - (\alpha + \beta' + 1)\,y]\frac{\partial z}{\partial y} - \beta' x\frac{\partial z}{\partial x} - \alpha\beta' z = 0$$

3. System of equations for $z = F_3$:

$$x(1-x)\frac{\partial^2 z}{\partial x^2} + y\frac{\partial^2 z}{\partial x\,\partial y} + [\gamma - (\alpha + \beta + 1)x]\frac{\partial z}{\partial x} - \alpha\beta z = 0,$$

$$y(1-y)\frac{\partial^2 z}{\partial y^2} + x\frac{\partial^2 z}{\partial x\,\partial y} + [\gamma - (\alpha' + \beta' + 1)\,y]\frac{\partial z}{\partial y} - \alpha'\beta' z = 0$$

EH I 234(11)

4. System of equations for $z = F_4$:

$$x(1-x)\frac{\partial^2 z}{\partial x^2} - y^2\frac{\partial^2 z}{\partial y^2} - 2xy\frac{\partial^2 z}{\partial x\,\partial y} + [\gamma - (\alpha + \beta + 1)x]\frac{\partial z}{\partial x} - (\alpha + \beta + 1)y\frac{\partial z}{\partial y} - \alpha\beta z = 0,$$

EH I 234(12)

$$(1-y)\frac{\partial^2 z}{\partial y^2} - x^2\frac{\partial^2 z}{\partial x^2} - 2xy\frac{\partial^2 z}{\partial x\,\partial y} + [\gamma' - (\alpha + \beta + 1)y]\frac{\partial z}{\partial y} - (\alpha + \beta + 1)x\frac{\partial z}{\partial x} - \alpha\beta z = 0$$

AK 44

9.182 For certain relationships between the parameters and the argument, hypergeometric functions of two variables can be expressed in terms of hypergeometric functions of a single variable or in terms of elementary functions:

1. $\qquad F_1\left(\alpha, \beta, \beta', \beta + \beta'; x, y\right) = (1-y)^{-\alpha}\, F\left(\alpha, \beta; \beta + \beta'; \dfrac{x-y}{1-y}\right)$ $\qquad$ EH I 238(1), AK 24(28)

2. $\qquad F_2\left(\alpha, \beta, \beta', \beta, \gamma'; x, y\right) = (1-x)^{-\alpha}\, F\left(\alpha, \beta'; \gamma'; \dfrac{y}{1-x}\right)$ $\qquad$ EH I 238(2), AK 23

3. $\qquad F_2\left(\alpha, \beta, \beta', \alpha, \alpha; x, y\right) = (1-x)^{-\beta}(1-y)^{-\beta'}\, F\left(\beta, \beta'; \alpha; \dfrac{xy}{(1-x)(1-y)}\right)$ $\qquad$ EH I 238(3)

4. $\qquad F_3\left(\alpha, \gamma - \alpha, \beta, \gamma - \beta, \gamma; x, y\right) = (1-y)^{\alpha + \beta - \gamma}\, F(\alpha, \beta; \gamma; x + y - xy)$ $\qquad$ EH I 238(4), AK 25(35)

5. $\qquad F_4\left(\alpha, \gamma + \gamma' - \alpha - 1, \gamma, \gamma'; x(1-y), y(1-x)\right)$

$$= F\left(\alpha, \gamma + \gamma' - \alpha - 1; \gamma; x\right) F\left(\alpha, \gamma + \gamma' - \alpha - 1; \gamma'; y\right)$$

EH I 238(5)

6. $\qquad F_4\left(\alpha, \beta, \alpha, \beta; -\dfrac{x}{(1-x)(1-y)}, \dfrac{-y}{(1-x)(1-y)}\right) = \dfrac{(1-x)^\beta(1-y)^\alpha}{(1-xy)}$ $\qquad$ EH I 238(6)

7. $\qquad F_4\left(\alpha, \beta, \beta, \beta; -\dfrac{x}{(1-x)(1-y)}, -\dfrac{y}{(1-x)(1-y)}\right) = (1-x)^\alpha(1-y)^\alpha\, F\left(\alpha, 1 + \alpha - \beta; \beta; xy\right)$

EH I 238(7)

8. $F_4 \left(\alpha, \beta, 1 + \alpha - \beta, \beta; -\dfrac{x}{(1-x)(1-y)}, -\dfrac{y}{(1-x)(1-y)} \right)$

$$= (1-y)^\alpha F\left[\alpha, \beta; 1 + \alpha - \beta; -\frac{x(1-y)}{1-x} \right]$$

EH I 238(8)

9. $F_4 \left(\alpha, \alpha + \dfrac{1}{2}, \gamma, \dfrac{1}{2}; x, y \right) = \dfrac{1}{2}(1 + \sqrt{y})^{-2\alpha} F\left(\alpha, \alpha + \dfrac{1}{2}; \gamma; \dfrac{x}{(1 + \sqrt{y})^2} \right)$

$$+ \frac{1}{2}(1 - \sqrt{y})^{-2\alpha} F\left(\alpha, \alpha + \frac{1}{2}; \gamma; \frac{x}{(1 - \sqrt{y})^2} \right)$$

AK 23

10. $F_1(\alpha, \beta, \beta', \gamma; x, 1) = \dfrac{\Gamma(\gamma)\, \Gamma(\gamma - \alpha - \beta')}{\Gamma(\gamma - \alpha)\, \Gamma(\gamma - \beta')} F(\alpha, \beta : \gamma - \beta'; x)$ EH I 239(10), AK 22(23)

11. $F_1(\alpha, \beta, \beta', \gamma; x, x) = F(\alpha, \beta + \beta'; \gamma; x)$ EH I 239(11), AK 23(25)

9.183   Functional relations between hypergeometric functions of two variables:

1. $F_1(\alpha, \beta, \beta', \gamma; x, y) = (1-x)^{-\beta}(1-y)^{-\beta} F_1 \left(\gamma - \alpha, \beta, \beta', \gamma; \dfrac{x}{x-1}, \dfrac{y}{y-1} \right)$

EH I 239(1)

$$= (1-x)^{-\alpha} F_1 \left(\alpha, \gamma - \beta - \beta', \beta', \gamma; \frac{x}{x-1}, \frac{y-x}{1-x} \right)$$

EH I 239(2)

$$= (1-y)^{-\alpha} F_1 \left(\alpha, \beta, \gamma - \beta - \beta', \gamma; \frac{y-x}{y-1}, \frac{y}{y-1} \right)$$

EH I 239(3)

$$= (1-x)^{\gamma-\alpha-\beta}(1-y)^{-\beta'} F_1 \left(\gamma - \alpha, \gamma - \beta - \beta', \beta', \gamma; x, \frac{x-y}{1-y} \right)$$

EH I 240(4)

$$= (1-x)^{-\beta}(1-y)^{\gamma-\alpha-\beta'} F_1 \left(\gamma - \alpha, \beta, \gamma - \beta - \beta', \gamma; \frac{x-y}{x-1}, y \right)$$

EH I 240(5), AK 30(5)

2.[8] $F_2(\alpha, \beta, \beta', \gamma, \gamma'; x, y) = (1-x)^{-\alpha} F_2 \left(\alpha, \gamma - \beta, \beta', \gamma, \gamma'; \dfrac{x}{x-1}, \dfrac{y}{1-x} \right)$

EH I 240(6)

$$= (1-y)^{-\alpha} F_2 \left(\alpha, \beta, \gamma' - \beta', \gamma, \gamma'; \frac{x}{1-y}, \frac{y}{y-1} \right)$$

EH I 240(7)

$$= (1-x-y)^{-\alpha} F_2 \left(\alpha, \gamma - \beta, \gamma' - \beta', \gamma, \gamma'; \frac{x}{x+y-1}, \frac{y}{x+y-1} \right)$$

EH I 240(8), AK 32(6)

3.7 $F_4\left(\alpha,\beta,\gamma,\gamma';x,y\right)=\dfrac{\Gamma\left(\gamma'\right)\Gamma\left(\beta-\alpha\right)}{\Gamma\left(\gamma'-\alpha\right)\Gamma\left(\beta\right)}\left(-y\right)^{-\alpha}F_4\left(\alpha,\alpha+1-\gamma',\gamma,\alpha+1-\beta;\dfrac{x}{y},\dfrac{1}{y}\right)$

$$+\dfrac{\Gamma(\gamma'(\Gamma(\alpha-\beta)}{\Gamma\left(\gamma'-\beta\right)\Gamma\left(\alpha\right)}\left(-y\right)^{\beta}F_4\left(\beta+1-\gamma',\beta,\gamma,\beta+1-\alpha;\dfrac{x}{y},\dfrac{1}{y}\right)$$

EH I 240(9), AK 26(37)

9.184   Integral representations:

Double integrals of the Euler type

1. $F_1\left(\alpha,\beta,\beta',\gamma;x,y\right)=\dfrac{\Gamma(\gamma)}{\Gamma(\beta)\,\Gamma\left(\beta'\right)\Gamma\left(\gamma-\beta-\beta'\right)}$

$$\times\iint\limits_{\substack{u\geq0,v\geq0\\u+v\leq1}}u^{\beta-1}v^{\beta'-1}(1-u-v)^{\gamma-\beta-\beta'-1}(1-ux-vy)^{-\alpha}\,du\,dv$$

$[\operatorname{Re}\beta>0,\quad\operatorname{Re}\beta'>0,\quad\operatorname{Re}\left(\gamma-\beta-\beta'\right)>0]$ EH I 230(1), AK 28(1)

2. $F_2\left(\alpha,\beta,\beta',\gamma,\gamma';x,y\right)=\dfrac{\Gamma(\gamma)\,\Gamma\left(\gamma'\right)}{\Gamma(\beta)\,\Gamma\left(\beta'\right)\Gamma(\gamma-\beta)\,\Gamma\left(\gamma'-\beta'\right)}$

$$\times\int_0^1\int_0^1 u^{\beta-1}v^{\beta'-1}(1-u)^{\gamma-\beta-1}(1-v)^{\gamma'-\beta'-1}(1-ux-vy)^{-\alpha}\,du\,dv$$

$[\operatorname{Re}\beta>0,\quad\operatorname{Re}\beta'>0,\quad\operatorname{Re}\left(\gamma-\beta\right)>0,\quad\operatorname{Re}\left(\gamma'-\beta'\right)>0]$ EH I 230(2), AK 28(2)

3. $F_3\left(\alpha,\alpha',\beta,\beta',\gamma;x,y\right)$

$$=\dfrac{\Gamma(\gamma)}{\Gamma(\beta)\,\Gamma\left(\beta'\right)\Gamma\left(\gamma-\beta-\beta'\right)}$$

$$\times\iint\limits_{\substack{u\geq0,v\geq0\\u+v\leq1}}u^{\beta-1}v^{\beta'-1}(1-u-v)^{-\gamma-\beta-\beta'-1}(1-ux)^{-\alpha}(1-vy)^{-\alpha'}\,du\,dv$$

$[\operatorname{Re}\beta>0,\quad\operatorname{Re}\beta'>0,\quad\operatorname{Re}\left(\gamma-\beta-\beta'\right)>0]$ EH I 230(3), AK 28(3)

4. $F_4\left(\alpha,\beta,\gamma,\gamma';x(1-y),y(1-x)\right)$

$$=\dfrac{\Gamma(\gamma)\,\Gamma\left(\gamma'\right)}{\Gamma(\alpha)\,\Gamma(\beta)\,\Gamma(\gamma-\alpha)\,\Gamma\left(\gamma'-\beta\right)}\int_0^1\int_0^1 u^{\alpha-1}v^{\beta-1}(1-u)^{\gamma-\alpha-1}(1-v)^{\gamma'-\beta-1}$$

$$\times(1-ux)^{\alpha-\gamma-\gamma'+1}(1-vy)^{\beta-\gamma-\gamma'+1}(1-ux-vy)^{\gamma+\gamma'-\alpha-\beta-1}\,du\,dv$$

$[\operatorname{Re}\alpha>0,\quad\operatorname{Re}\beta>0,\quad\operatorname{Re}\left(\gamma-\alpha\right)>0,\quad\operatorname{Re}\left(\gamma'-\beta\right)>0]$ EH I 230(4)

Integrals of the Mellin–Barnes type

9.185   The functions F_1, F_2, F_3 and F_4 can be represented by means of double integrals of the following form:

$$F(x,y)=\dfrac{\Gamma(\gamma)}{\Gamma(\alpha)\,\Gamma(\beta)(2\pi i)^2}\int_{-i\infty}^{i\infty}\int_{-i\infty}^{i\infty}\Psi(s,t)\,\Gamma(-s)\,\Gamma(-t)(-x)^s(-y)^t\,ds\,dt$$

$\Psi(s,t)$	$F(x,y)$
$\dfrac{\Gamma(\alpha+s+t)\,\Gamma(\beta+s)\,\Gamma(\beta'+t)}{\Gamma(\beta')\,\Gamma(\gamma+s+t)}$	$F_1\left(\alpha,\beta,\beta',\gamma;x,y\right)$
$\dfrac{\Gamma(\alpha+s+t)\,\Gamma(\beta+s)\,\Gamma(\beta'+t)\,\Gamma(\gamma')}{\Gamma(\beta')\,\Gamma(\gamma+s)\,\Gamma(\gamma'+t)}$	$F_2\left(\alpha,\beta,\beta',\gamma,\gamma';x,y\right)$
$\dfrac{\Gamma(\alpha+s)\,\Gamma(\alpha'+t)\,\Gamma(\beta+s)\,\Gamma(\beta'+t)}{\Gamma(\alpha')\,\Gamma(\beta')\,\Gamma(\gamma+s+t)}$	$F_3\left(\alpha,\alpha',\beta,\beta',\gamma;x,y\right)$
$\dfrac{\Gamma(\alpha+s+t)\,\Gamma(\beta+s+t)\,\Gamma(\gamma')}{\Gamma(\gamma+s)\,\Gamma(\gamma'+t)}$	$F_4\left(\alpha,\beta,\gamma,\gamma';x,y\right)$
$[\alpha,\alpha',\beta,\beta'$ may not be negative integers]	EH I 232(9–13), AK 41(33)

9.19 A hypergeometric function of several variables

$F_A\left(\alpha;\beta_1,\ldots,\beta_n;\gamma_1,\ldots,\gamma_n;z_1,\ldots,z_n\right)$

$$= \sum_{m_1=0}^{\infty}\sum_{m_2=0}^{\infty}\cdots\sum_{m_n=0}^{\infty}\frac{(\alpha)_{m_1+\cdots+m_n}\,(\beta_1)_{m_1+\ldots}\,(\beta_n)_{m_n}}{(\gamma_1)_{m_1\ldots}(\gamma_n)_{m_n}\,m_1!\ldots m_n!}z_1^{m_1}z_2^{m_2}\ldots z_n^{m_n}$$

ET I 385

9.2 Confluent Hypergeometric Functions

9.20 Introduction

9.201[10]   A *confluent hypergeometric function* is obtained by taking the limit as $c \to \infty$ in the solution of Riemann's differential equation

$$u = P\left\{\begin{matrix} 0 & \infty & c & \\ \frac{1}{2}+\mu & -c & c-\lambda & z \\ \frac{1}{2}-\mu & 0 & \lambda & \end{matrix}\right\}$$

WH

9.202   The equation obtained by means of this limiting process is of the form

1. $\dfrac{d^2u}{dz^2} + \dfrac{du}{dz} + \left(\dfrac{\lambda}{z} + \dfrac{\frac{1}{4}-\mu^2}{z^2}\right)u = 0$

WH

Equation **9.202** 1 has the following two linearly independent solutions:

2. $z^{\frac{1}{2}+\mu}e^{-z}\,\Phi\left(\frac{1}{2}+\mu-\lambda, 2\mu+1; z\right)$

3. $z^{\frac{1}{2}-\mu}e^{-z}\,\Phi\left(\frac{1}{2}-\mu-\lambda, -2\mu+1; z\right)$

which are defined for all values of $\mu \neq \pm\frac{1}{2}, \pm\frac{2}{2}, \pm\frac{3}{2}, \ldots$

MO 111

9.21   The functions $\Phi(\alpha, \gamma; z)$ and $\Psi(\alpha, \gamma; z)$

9.210[10]   The series

1. $$\Phi(\alpha, \gamma; z) = 1 + \frac{\alpha}{\gamma} \frac{z}{1!} + \frac{\alpha(\alpha+1)}{\gamma(\gamma+1)} \frac{z^2}{2!} + \frac{\alpha(\alpha+1)(\alpha+2)}{\gamma(\gamma+1)(\gamma+2)} \frac{z^3}{3!} + \ldots$$

     is also called a *confluent hypergeometric function.*

     A second notation: $\Phi(\alpha, \gamma; z) = {}_1F_1(\alpha; \gamma; z)$.

2. $$\Psi(\alpha, \gamma; z) = \frac{\Gamma(1-\gamma)}{\Gamma(\alpha-\gamma+1)} \Phi(\alpha, \gamma; z) + \frac{\Gamma(\gamma-1)}{\Gamma(\alpha)} z^{1-\gamma} \Phi(\alpha-\gamma+1, 2-\gamma; z)$$ EH I 257(7)

3.   Bateman's function $k_\nu(x)$ is defined by

$$k_\nu(x) = \frac{2}{\pi} \int_0^{\pi/2} \cos(x \tan\theta - \nu\theta)\, d\theta \qquad [x, \nu \text{ real}]$$ EH I 267

9.211   Integral representation:

1. $$\Phi(\alpha, \gamma; z) = \frac{2^{1-\gamma} e^{\frac{1}{2}z}}{B(\alpha, \gamma-\alpha)} \int_{-1}^1 (1-t)^{\gamma-\alpha-1} (1+t)^{\alpha-1} e^{\frac{1}{2}zt}\, dt$$

$$[0 < \operatorname{Re}\alpha < \operatorname{Re}\gamma] \qquad \text{MO 114}$$

2. $$\Phi(\alpha, \gamma; z) = \frac{1}{B(\alpha, \gamma-\alpha)} z^{1-\gamma} \int_0^z e^t t^{\alpha-1} (z-t)^{\gamma-\alpha-1\, dt}$$

$$[0 < \operatorname{Re}\alpha < \operatorname{Re}\gamma] \qquad \text{MO 114}$$

3. $$\Phi(-\nu, \alpha+1; z) = \frac{\Gamma(\alpha+1)}{\Gamma(\alpha+\nu+1)} e^z z^{-\frac{\alpha}{2}} \int_0^\infty e^{-t} t^{\nu+\frac{\alpha}{2}} J_\alpha\left(2\sqrt{zt}\right) dt$$

$$\left[\operatorname{Re}(\alpha+\nu+1) > 0, \quad |\arg z| < \frac{\pi}{2}\right]$$
$$\text{MO 115}$$

4.[8] $$\Psi(\alpha, \gamma; z) = \frac{1}{\Gamma(\alpha)} \int_0^\infty e^{-zt} t^{\alpha-1} (1+t)^{\gamma-\alpha-1}\, dt \qquad [\operatorname{Re}\alpha > 0, \quad \operatorname{Re} z > 0] \qquad \text{EH I 255(2)}$$

Functional relations

9.212

1. $\Phi(\alpha, \gamma; z) = e^z \Phi(\gamma-\alpha, \gamma; -z)$ MO 112

2. $\dfrac{z}{\gamma} \Phi(\alpha+1, \gamma+1; z) = \Phi(\alpha+1, \gamma; z) - \Phi(\alpha, \gamma; z)$ MO 112

3. $\alpha \Phi(\alpha+1, \gamma+1; z) = (\alpha-\gamma) \Phi(\alpha, \gamma+1; z) + \gamma \Phi(\alpha, \gamma; z)$ MO 112

4. $\alpha \Phi(\alpha+1, \gamma; z) = (z+2a-\gamma) \Phi(\alpha, \gamma; z) + (\gamma-\alpha) \Phi(\alpha-1, \gamma; z)$ MO 112

9.213 $\dfrac{d\,\Phi}{dz} = \dfrac{\alpha}{\gamma} \Phi(\alpha+1, \gamma+1; z)$ MO 112

9.214 $\displaystyle \lim_{\gamma \to -n} \frac{1}{\Gamma(\gamma)} \Phi(\alpha, \gamma; z) = z^{n+1} \binom{\alpha+n}{n+1} \Phi(\alpha+n+1, n+2; z) \qquad [n = 0, 1, 2, \ldots]$ MO 112

9.215[10]

1. $\Phi(\alpha, \alpha; z) = e^z$ MO 15

2. $\Phi(\alpha, 2\alpha; 2z) = 2^{\alpha - \frac{1}{2}} \exp\left[\frac{1}{4}(1 - 2\alpha)\pi i\right] \Gamma\left(\alpha + \frac{1}{2}\right) e^z z^{\frac{1}{2} - \alpha} J_{\alpha - \frac{1}{2}}\left(ze^{\frac{\pi}{2}i}\right)$ MO 112

3. $\Phi\left(p + \frac{1}{2}, 2p + 1; 2iz\right) = \Gamma(p + 1)\left(\dfrac{z}{2}\right)^{-p} e^{iz} J_p(z)$ MO 15

For a representation of special functions in terms of a confluent hypergeometric function $\Phi(\alpha, \gamma; z)$, see:

- for the probability integral, **9.236**;
- for integrals of Bessel functions, **6.631** 1;
- for Hermite polynomials, **8.953** and **8.959**;
- for Laguerre polynomials, **8.972** 1;
- for parabolic cylinder functions, **9.240**;
- for the Whittaker functions $M_{\lambda,\mu}(z)$, **9.220** 2 and **9.220** 3.

9.216   The function $\Phi(\alpha, \gamma; z)$ is a solution of the differential equation

1. $z\dfrac{d^2F}{dz^2} + (\gamma - z)\dfrac{dF}{dz} - \alpha F = 0$ MO 111

This equation has two linearly independent solutions:

2. $\Phi(\alpha, \gamma; z)$

3. $z^{1-\gamma}\,\Phi(\alpha - \gamma + 1, 2 - \gamma; z)$ MO 112

9.22–9.23 The Whittaker functions $M_{\lambda,\mu}(z)$ and $W_{\lambda,\mu}(z)$

9.220   If we make the change of variable $u = e^{-\frac{z}{2}}W$ in equation **9.202** 1, we obtain the equation

1. $\dfrac{d^2W}{dz^2} + \left(-\dfrac{1}{4} + \dfrac{\lambda}{z} + \dfrac{\frac{1}{4} - \mu^2}{z^2}\right)W = 0$ MO 115

Equation **9.220** 1 has the following two linearly independent solutions:

2. $M_{\lambda,\mu}(z) = z^{\mu + \frac{1}{2}} e^{-z/2}\,\Phi\left(\mu - \lambda + \frac{1}{2}, 2\mu + 1; z\right)$

3. $M_{\lambda,-\mu}(z) = z^{-\mu + \frac{1}{2}} e^{-z/2}\,\Phi\left(-\mu - \lambda + \frac{1}{2}, 2\mu + 1; z\right)$ MO 115

To obtain solutions that are also suitable for $2\mu = \pm 1, \pm 2, \ldots$, we introduce Whittaker's function

4. $W_{\lambda,\mu}(z) = \dfrac{\Gamma(-2\mu)}{\Gamma\left(\frac{1}{2} - \mu - \lambda\right)} M_{\lambda,\mu}(z) + \dfrac{\Gamma(2\mu)}{\Gamma\left(\frac{1}{2} + \mu - \lambda\right)} M_{\lambda,-\mu}(z)$ WH

which, for 2μ approaching an integer, is also a solution of equation **9.220** 1.

For the functions $M_{\lambda,\mu}(z)$ and $W_{\lambda,\mu}(z)$, $z = 0$ is a branch point and $z = \infty$ is an essential singular point. Therefore, we shall examine these functions only for $|\arg z| < \pi$.

These functions $W_{\lambda,\mu}(z)$ and $W_{-\lambda,\mu}(-z)$ are linearly independent solutions of equation **9.220** 1.

Integral representations

9.221 $M_{\lambda,\mu}(z) = \dfrac{z^{\mu+\frac{1}{2}}}{2^{2\mu}\,\mathrm{B}\left(\mu+\lambda+\frac{1}{2},\mu-\lambda+\frac{1}{2}\right)} \displaystyle\int_{-1}^{1}(1+t)^{\mu-\lambda-\frac{1}{2}}(1-t)^{\mu+\lambda-\frac{1}{2}}e^{\frac{1}{2}zt}\,dt,$ \hfill WH

if the integral converges. See also **6.631** 1 and **7.623** 3.

9.222

1. $\quad W_{\lambda,\mu}(z) = \dfrac{z^{\mu+\frac{1}{2}}e^{-z/2}}{\Gamma\left(\mu-\lambda+\frac{1}{2}\right)}\displaystyle\int_0^\infty e^{-zt}t^{\mu-\lambda-\frac{1}{2}}(1+t)^{\mu-\lambda-\frac{1}{2}}\,dt$

$$\left[\operatorname{Re}(\mu-\lambda)>-\tfrac{1}{2},\quad |\arg z|<\frac{\pi}{2}\right]$$

\hfill MO 118

2. $\quad W_{\lambda,\mu}(z) = \dfrac{z^{\lambda}e^{-z/2}}{\Gamma\left(\mu-\lambda+\frac{1}{2}\right)}\displaystyle\int_0^\infty t^{\mu-\lambda-\frac{1}{2}}e^{-t}\left(1+\frac{t}{z}\right)^{\mu+\lambda-\frac{1}{2}}\,dt$

$$\left[\operatorname{Re}(\mu-\lambda)>-\tfrac{1}{2},\quad |\arg z|<\pi\right]$$ \hfill WH

9.223 $W_{\lambda,\mu}(z) = \dfrac{e^{-\frac{z}{2}}}{2\pi i}\displaystyle\int_{-i\infty}^{i\infty}\dfrac{\Gamma(u-\lambda)\,\Gamma\left(-u-\mu+\frac{1}{2}\right)\Gamma\left(-u+\mu+\frac{1}{2}\right)}{\Gamma\left(-\lambda+\mu+\frac{1}{2}\right)\Gamma\left(-\lambda-\mu+\frac{1}{2}\right)}z^u\,du$

[the path of integration is chosen in such a way that the poles of the function $\Gamma(u-\lambda)$ are separated from the poles of the functions $\Gamma\left(-u-\mu+\frac{1}{2}\right)$ and $\Gamma\left(-u+\mu+\frac{1}{2}\right)$]. See also **7.142**. \hfill MO 118

9.224 $W_{\mu,\frac{1}{2}+\mu}(z) = z^{\mu+1}e^{-\frac{1}{2}z}\displaystyle\int_0^\infty(1+t)^{2\mu}e^{-zt}\,dt = z^{-\mu}e^{\frac{1}{2}z}\displaystyle\int_z^\infty t^{2\mu}e^{-t}\,dt \qquad [\operatorname{Re}z>0]$ \hfill WH

9.225

1. $\quad W_{\lambda,\mu}(x)\,W_{-\lambda,\mu}(x) = -x\displaystyle\int_0^\infty\tanh^{2\lambda}\frac{t}{2}\left\{J_{2\mu}(x\sinh t)\sin(\mu-\lambda)\pi\right.$

$$\left.+\,Y_{2\mu}(x\sinh t)\cos(\mu-\lambda)\pi\right\}\,dt$$

$$\left[|\operatorname{Re}\mu|-\operatorname{Re}\lambda<\tfrac{1}{2};\quad x>0\right]$$ \hfill MO 119

2. $\quad W_{\kappa,\mu}(z_1)\,W_{\lambda,\mu}(z_2) = \dfrac{(z_1z_2)^{\mu+\frac{1}{2}}\exp\left[-\frac{1}{2}(z_1+z_2)\right]}{\Gamma(1-\kappa-\lambda)}$

$$\times\displaystyle\int_0^\infty e^{-t}t^{-\kappa-\lambda}(z_1+t)^{-\frac{1}{2}+\kappa-\mu}(z_2+t)^{-\frac{1}{2}+\lambda-\mu}$$

$$\times\,F\left(\tfrac{1}{2}-\kappa+\mu,\tfrac{1}{2}-\lambda+\mu;1-\kappa-\lambda;\Theta\right)\,dt$$

$$\Theta = \dfrac{t(z_1+z_2+t)}{(z_1+t)(z_2+t)},\qquad [z_1\neq0,\quad z_2\neq0,\quad |\arg z_1|<\pi,\quad |\arg z_2|<\pi,\quad \operatorname{Re}(\kappa+\lambda)<1]$$

\hfill MO 119

See also **3.334**, **3.381** 6, **3.382** 3, **3.383** 4, 8, **3.384** 3, **3.471** 2.

9.226 Series representations

$$M_{0,\mu}(z) = z^{\frac{1}{2}+\mu}\left\{1+\sum_{k=1}^\infty\frac{z^{2k}}{2^{4k}k!(\mu+1)(\mu+2)\ldots(\mu+k)}\right\}$$ \hfill WH

Asymptotic representations

9.227[7]   For large values of $|z|$

$$W_{\lambda,\mu}(z) \sim e^{-z/2} z^\lambda \left(1 + \sum_{k=1}^\infty \frac{\left[\mu^2 - \left(\lambda - \frac{1}{2}\right)^2\right]\left[\mu^2 - \left(\lambda - \frac{3}{2}\right)^2\right]\cdots\left[\mu^2 - \left(\lambda - k + \frac{1}{2}\right)^2\right]}{k! z^k} \right)$$

$$[|\arg z| \le \pi - \alpha < \pi]$$ WH

9.228   For large values of $|\lambda|$

$$M_{\lambda,\mu}(z) \sim \frac{1}{\sqrt{\pi}} \Gamma(2\mu + 1) \lambda^{-\mu - \frac{1}{4}} z^{1/4} \cos\left(2\sqrt{\lambda z} - \mu\pi - \frac{1}{4}\pi\right)$$ MO 118

9.229

1. $W_{\lambda,\mu} \sim -\left(\frac{4z}{\lambda}\right)^{\frac{1}{4}} e^{-\lambda + \lambda \ln \lambda} \sin\left(2\sqrt{\lambda z} - \lambda\pi - \frac{\pi}{4}\right)$ MO 118

2. $W_{-\lambda,\mu} \sim \left(\frac{z}{4\lambda}\right)^{\frac{1}{4}} e^{\lambda - \lambda \ln \lambda - 2\sqrt{\lambda z}}$ MO 118

Formulas **9.228** and **9.229** are applicable for
$|\lambda| \gg 1, \quad |\lambda| \gg |z|, \quad |\lambda| \gg |\mu|, \quad z \ne 0, \quad |\arg\sqrt{z}| < \frac{3\pi}{4}$ and $|\arg\lambda| < \frac{\pi}{2}$.   MO 118

Functional relations

9.231

1. $M_{n+\mu+\frac{1}{2},\mu}(z) = \dfrac{z^{\frac{1}{2}-\mu} e^{\frac{1}{2}z}}{(2\mu+1)(2\mu+2)\dots(2\mu+n)} \dfrac{d^n}{dz^n}\left(z^{n+2\mu} e^{-z}\right)$

$$[n = 0, 1, 2, \dots; \quad 2\mu \ne -1, -2, -3, \dots]$$
MO 117

2. $z^{-\frac{1}{2}-\mu} M_{\lambda,\mu}(z) = (-z)^{-\frac{1}{2}-\mu} M_{-\lambda,\mu}(-z)$ $[2\mu \ne -1, -2, -3, \dots]$ WH

9.232

1. $W_{\lambda,\mu}(z) = W_{\lambda,-\mu}(z)$ MO 116

2. $W_{-\lambda,\mu}(-z) = \dfrac{\Gamma(-2\mu)}{\Gamma\left(\frac{1}{2} - \mu + \lambda\right)} M_{-\lambda,\mu}(-z) + \dfrac{\Gamma(2\mu)}{\Gamma\left(\frac{1}{2} + \mu + \lambda\right)} M_{-\lambda,-\mu}(-z)$

$$\left[|\arg(-z)| < \frac{3}{2}\pi\right]$$ WH

9.233

1. $M_{\lambda,\mu}(z) = \dfrac{\Gamma(2\mu+1)}{\Gamma\left(\mu - \lambda + \frac{1}{2}\right)} e^{i\pi\lambda} W_{-\lambda,\mu}\left(e^{i\pi} z\right) + \dfrac{\Gamma(2\mu+1)}{\Gamma\left(\mu + \lambda + \frac{1}{2}\right)} \exp\left[i\pi\left(\lambda - \mu - \frac{1}{2}\right)\right] W_{\lambda,\mu}(z)$

$$\left[-\frac{3}{2}\pi < \arg z < \frac{1}{2}\pi; \quad 2\mu \ne -1, -2, \dots\right]$$
MO 117

2. $M_{\lambda,\mu}(z) = \dfrac{\Gamma(2\mu+1)}{\Gamma\left(\mu - \lambda + \frac{1}{2}\right)} e^{-i\pi\lambda} W_{-\lambda,\mu}\left(e^{-i\pi} z\right) + \dfrac{\Gamma(2\mu+1)}{\Gamma\left(\mu + \lambda + \frac{1}{2}\right)} \exp\left[-i\pi\left(\lambda - \mu - \frac{1}{2}\right)\right] W_{\lambda,\mu}(z)$

$$\left[-\frac{1}{2}\pi < \arg z < \frac{3}{2}\pi; \quad 2\mu \ne -1, -2, \dots\right]$$
MO 117

9.234   Recursion formulas

1. $W_{\mu,\lambda}(z) = \sqrt{z}\, W_{\mu-\frac{1}{2},\lambda-\frac{1}{2}}(z) + \left(\frac{1}{2} + \lambda - \mu\right) W_{\mu-1,\lambda}(z)$ WH

2. $W_{\mu,\lambda}(z) = \sqrt{z}\, W_{\mu-\frac{1}{2},\lambda-\frac{1}{2}}(z) + \left(\frac{1}{2} - \lambda - \mu\right) W_{\mu-1,\lambda}(z)$ WH

3. $z\dfrac{d}{dz}\, W_{\lambda,\mu}(z) = \left(\lambda - \frac{1}{2}z\right) W_{\lambda,\mu}(z) - \left[\mu^2 - \left(\lambda - \frac{1}{2}\right)^2\right] W_{\lambda-1,\mu}(z)$ WH

4. $\left[\left(\mu + \dfrac{1-z}{2}\right) W_{\lambda,\mu}(z) - z\dfrac{d}{dz}\, W_{\lambda,\mu}(z)\right]\left(\mu + \tfrac{1}{2} + \lambda\right)$

$$= \left[\left(\mu + \dfrac{1+z}{2}\right) W_{\lambda,\mu+1}(z) + z\dfrac{d}{dz}\, W_{\lambda,\mu+1}(z)\right]\left(\mu + \tfrac{1}{2} - \lambda\right)$$

MO 117

5. $\left(\frac{3}{2} + \lambda + \mu\right)\left(\frac{1}{2} + \lambda + \mu\right) z\, W_{\lambda,\mu}(z) = z(z + 2\mu + 1)\dfrac{d}{dz}\, W_{\lambda+1,\mu+1}(z)$

$$+ \left[\tfrac{1}{2}z^2 + \left(\mu - \lambda - \tfrac{1}{2}\right)z + 2\mu^2 + 2\mu + \tfrac{1}{2}\right] W_{\lambda+1,\mu+1}(z)$$

MO 117

Connections with other functions

9.235
1. $M_{0,\mu}(z) = 2^{2\mu}\, \Gamma(\mu+1)\sqrt{z}\, I_\mu\left(\dfrac{z}{2}\right)$ MO 125a

2. $W_{0,\mu}(z) = \sqrt{\dfrac{z}{\pi}}\, K_\mu\left(\dfrac{z}{2}\right)$ MO 125

9.236

1. $\Phi(x) = 1 - \dfrac{e^{\frac{x^2}{2}}}{\sqrt{\pi x}}\, W_{-\frac{1}{4},\frac{1}{4}}\left(x^2\right) = \dfrac{2x}{\sqrt{\pi}}\,\Phi\left(\tfrac{1}{2}, \tfrac{3}{2}; -x^2\right)$ WH, MO 126

2. $\mathrm{li}(z) = -\dfrac{\sqrt{z}}{\sqrt{\ln\frac{1}{z}}}\, W_{-\frac{1}{2},0}\left(-\ln z\right)$ WH

3. $\Gamma(\alpha, x) = e^{-x}\,\Psi(1 - \alpha, 1 - \alpha; x)$ EH I 266(21)

4. $\gamma(\alpha, x) = \dfrac{x^\alpha}{\alpha}\,\Phi(\alpha, \alpha + 1; -x)$ EH I 266(22)

9.237
1. $W_{\lambda,\mu}(z) = \dfrac{(-1)^{2\mu}\, z^{\mu+\frac{1}{2}}\, e^{-\frac{1}{2}z}}{\Gamma\left(\frac{1}{2} - \mu - \lambda\right)\Gamma\left(\frac{1}{2} + \mu - \lambda\right)}$

$$\times \left\{\sum_{k=0}^{\infty} \dfrac{\Gamma\left(\mu + k - \lambda + \frac{1}{2}\right)}{k!(2\mu + k)!}\, z^k \left[\Psi(k+1) + \Psi(2\mu + k + 1) - \Psi\left(\mu + k - \lambda + \tfrac{1}{2}\right) - \ln z\right]\right.$$

$$\left. + (-z)^{-2\mu} \sum_{k=0}^{2\mu-1} \dfrac{\Gamma(2\mu - k)\,\Gamma\left(k - \mu - \lambda + \frac{1}{2}\right)}{k!}\, (-z)^k \right\}$$

$\left[|\arg z| < \tfrac{3}{2}\pi; \quad 2\mu + 1 \text{ is a natural number}\right]$ MO 116

2.   Set $\lambda - \mu - \frac{1}{2} = l$, where $l + 1$ is a natural number. Then

3. $W_{l+\mu+\frac{1}{2},\mu}(z) = (-1)^l z^{\mu+\frac{1}{2}} e^{-\frac{1}{2}z} (2\mu+1)(2\mu+2) \cdots (2\mu+l) \, \Phi(-l, 2\mu+1; z)$

$= (-1)^l z^{\mu+\frac{1}{2}} e^{-\frac{1}{2}z} L_l^{2\mu}(z)$

MO 116

9.238

1. $J_\nu(x) = \dfrac{2^{-\nu}}{\Gamma(\nu+1)} x^\nu e^{-ix} \, \Phi\left(\frac{1}{2}+\nu, 1+2\nu; 2ix\right)$ EH I 265(9)

2. $I_\nu(x) = \dfrac{2^{-\nu}}{\Gamma(\nu+1)} x^\nu e^{-x} \, \Phi\left(\frac{1}{2}+\nu, 1+2\nu; 2x\right)$ EH I 265(10)

3. $K_\nu(x) = \sqrt{\pi} e^{-x} (2x)^\nu \, \Psi\left(\frac{1}{2}+\nu, 1+2\nu; 2x\right)$ EH I 265(13)

9.24–9.25 Parabolic cylinder functions $D_p(z)$

9.240 $D_p(z) = 2^{\frac{1}{4}+\frac{p}{2}} \, W_{\frac{1}{4}+\frac{p}{2}, -\frac{1}{4}}\left(\dfrac{z^2}{2}\right) z^{-1/2}$

$$= 2^{\frac{p}{2}} e^{-\frac{z^2}{4}} \left\{ \frac{\sqrt{\pi}}{\Gamma\left(\dfrac{1-p}{2}\right)} \Phi\left(-\frac{p}{2}, \frac{1}{2}; \frac{z^2}{2}\right) - \frac{\sqrt{2\pi}\,z}{\Gamma\left(-\dfrac{p}{2}\right)} \Phi\left(\frac{1-p}{2}, \frac{3}{2}; \frac{z^2}{2}\right) \right\}$$

MO 120a

are called *parabolic cylinder functions*.

Integral representations

9.241

1. $D_p(z) = \dfrac{1}{\sqrt{\pi}} 2^{p+\frac{1}{2}} e^{-\frac{\pi}{2}pi} e^{\frac{z^2}{4}} \displaystyle\int_{-\infty}^{\infty} x^p e^{-2x^2+2ixz}\,dx$ $[\operatorname{Re} p > -1; \quad \text{for } x < 0, \quad \arg x^p = p\pi i]$

MO 122

2. $D_p(z) = \dfrac{e^{-\frac{z^2}{4}}}{\Gamma(-p)} \displaystyle\int_0^{\infty} e^{-xz-\frac{x^2}{2}} x^{-p-1}\,dx$ $[\operatorname{Re} p < 0]$ (cf. **3.462** 1)   MO 122

9.242

1.[10] $D_p(z) = -\dfrac{\Gamma(p+1)}{2\pi i} e^{-\frac{1}{4}z^2} \displaystyle\int_{\infty}^{(0+)} e^{-zt-\frac{1}{2}t^2} (-t)^{-p-1}\,dt$ $[|\arg(-t)| \le \pi]$ WH

2. $D_p(z) = 2^{\frac{1}{2}(p-1)} \dfrac{\Gamma\left(\frac{p}{2}+1\right)}{i\pi} \displaystyle\int_{-\infty}^{(-1+)} e^{\frac{1}{4}z^2 t} (1+t)^{-\frac{1}{2}p-1} (1-t)^{\frac{1}{2}(p-1)}\,dt$

$$\left[|\arg z| < \frac{\pi}{4}; \quad |\arg(1+t)| \le \pi\right]$$ WH

3. $D_p(z) = \dfrac{1}{2\pi i} e^{-\frac{1}{4}z^2} \displaystyle\int_{-\infty i}^{\infty i} \dfrac{\Gamma\left(\frac{1}{2}t - \frac{1}{2}p\right) \Gamma(-t)}{\Gamma(-p)} \left(\sqrt{2}\right)^{t-p-2} z^t\,dt$

$$\left[|\arg z| < \tfrac{3}{4}\pi; \quad p \text{ is not a positive integer}\right]$$ WH

4. $D_p(z) = \dfrac{1}{2\pi i} e^{-\frac{1}{4}z^2} \displaystyle\int_\infty^{(0-)} \dfrac{\Gamma\left(\frac{1}{2}t - \frac{1}{2}p\right)\Gamma(-t)}{\Gamma(-p)} \left(\sqrt{2}\right)^{t-p-2} z^t\, dt$

[for all values of arg z; also, the contours encircle the poles of the function $\Gamma(-t)$ but they do not encircle the poles of the function $\Gamma\left(\frac{1}{2}t - \frac{1}{2}p\right)$].   WH

9.243

1. $D_n(z) = (-1)^\mu \left(\dfrac{\pi}{2}\right)^{-1/2} \left(\sqrt{n}\right)^{n+1} e^{\frac{1}{4}z^2 - \frac{1}{2}n} \left\{ \displaystyle\int_{-\infty}^\infty e^{-n(t-1)^2} \dfrac{\cos}{\sin}\left(zt\sqrt{n}\right) dt \right.$

           $\left. + \displaystyle\int_0^\infty \left[e^{\frac{1}{2}n(1-t^2)} t^n - e^{-n(t-1)^2}\right] \dfrac{\cos}{\sin}\left(zt\sqrt{n}\right) dt - \displaystyle\int_{-\infty}^0 e^{-n(t-1)^2} \dfrac{\cos}{\sin}\left(zt\sqrt{n}\right) dt \right\}$

                                   [n is a natural number]   WH

2. $D_n(z) = (-1)^\mu 2^{n+2}(2\pi)^{-1/2} e^{\frac{1}{4}z^2} \displaystyle\int_0^\infty t^n e^{-2t^2} \dfrac{\cos}{\sin}(2zt)\, dt$

[n is a natural number, $\mu = \left\lfloor \dfrac{n}{2} \right\rfloor$, and the cosine or sine is chosen according as n is even or odd]

                                                       WH

9.244

1. $D_{-p-1}[(1+i)z] = \dfrac{e^{-\frac{iz^2}{2}}}{2^{\frac{p-1}{2}}\,\Gamma\left(\frac{p+1}{2}\right)} \displaystyle\int_0^\infty \dfrac{e^{-ix^2 z^2} x^p}{(1+x^2)^{1+\frac{p}{2}}}\, dx$ $\left[\operatorname{Re} p > -1, \quad \operatorname{Re} iz^2 \geq 0\right]$ MO 122

2. $D_p[(1+i)z] = \dfrac{2^{\frac{p+1}{2}}}{\Gamma\left(-\frac{p}{2}\right)} \displaystyle\int_1^\infty e^{-\frac{1}{2}z^2 x} \dfrac{(x+1)^{\frac{p-1}{2}}}{(x-1)^{1+\frac{p}{2}}}\, dx$ $\left[\operatorname{Re} p < 0; \quad \operatorname{Re} iz^2 \geq 0\right]$ MO 122

See also **3.383** 6, 7, **3.384** 2, 6, **3.966** 5, 6.

9.245

1.[10] $D_p(x)\, D_{-p-1}(x) = -\dfrac{1}{\sqrt{\pi}} \displaystyle\int_0^\infty \coth^{p+\frac{1}{2}} \dfrac{t}{2} \dfrac{1}{\sqrt{\sinh t}} \sin\left(\dfrac{x^2 \sinh t + p\pi}{2}\right) dt$

                                   [x is real, $\operatorname{Re} p < 0$]   MO 122

2. $D_p\left(ze^{\frac{\pi}{4}i}\right) D_p\left(ze^{-\frac{\pi}{4}i}\right) = \dfrac{1}{\Gamma(-p)} \displaystyle\int_0^\infty \coth^p t \exp\left(-\dfrac{z^2}{2}\sinh 2t\right) \dfrac{dt}{\sinh t}$

                                   $\left[|\arg z| < \dfrac{\pi}{4}; \quad \operatorname{Re} p < 0\right]$ MO 122

See also **6.613**.

9.246   Asymptotic expansions. If $|z| \gg 1$ and $|z| \gg |p|$, then

1. $D_p(z) \sim e^{-\frac{z^2}{4}} z^p \left(1 - \dfrac{p(p-1)}{2z^2} + \dfrac{p(p-1)(p-2)(p-3)}{2\cdot 4 z^4} - \cdots\right)$

                                     $\left[|\arg z| < \dfrac{3}{4}\pi\right]$ MO 121

2. $D_p(z) \sim e^{-z^2/4} z^p \left(1 - \dfrac{p(p-1)}{2z^2} + \dfrac{p(p-1)(p-2)(p-3)}{2\cdot 4 z^4} - \cdots\right)$

          $- \dfrac{\sqrt{2\pi}}{\Gamma(-p)} e^{p\pi i} e^{z^2/4} z^{-p-1} \left(1 + \dfrac{(p+1)(p+2)}{2z^2} + \dfrac{(p+1)(p+2)(p+3)(p+4)}{2\cdot 4 z^4} + \cdots\right)$

                                   $\left[\dfrac{1}{4}\pi < \arg z < \dfrac{5}{4}\pi\right]$ MO 121

3. $D_p(z) \sim e^{-z2/4} z^p \left(1 - \dfrac{p(p-1)}{2z^2} + \dfrac{p(p-1)(p-2)(p-3)}{2 \cdot 4z^4} - \cdots \right)$

$$- \frac{\sqrt{2\pi}}{\Gamma(-p)} e^{-p\pi i} e^{z2/4} z^{-p-1} \left(1 + \frac{(p+1)(p+2)}{2z^2} + \frac{(p+1)(p+2)(p+3)(p+4)}{2 \cdot 4z^4} + \cdots \right)$$

$$\left[-\tfrac{1}{4}\pi > \arg z > \tfrac{5}{4}\pi \right] \qquad \text{MO 121}$$

Functional relations

9.247   Recursion formulas:

1. $D_{p+1}(z) - z\, D_p(z) + p\, D_{p-1}(z) = 0$ WH

2. $\dfrac{d}{dz} D_p(z) + \dfrac{1}{2} z\, D_p(z) - p\, D_{p-1}(z) = 0$ WH

3. $\dfrac{d}{dz} D_p(z) - \dfrac{1}{2} z\, D_p(z) + D_{p+1}(z) = 0$ MO 121

9.248   Linear relations:

1. $D_p(z) = \dfrac{\Gamma(p+1)}{\sqrt{2\pi}} \left[e^{\pi p i/2} D_{-p-1}(iz) + e^{-\pi p i/2} D_{-p-1}(-iz) \right]$

$$= e^{-p\pi i} D_p(-z) + \frac{\sqrt{2\pi}}{\Gamma(-p)} e^{-\pi(p+1)i/2} D_{-p-1}(iz)$$

$$= e^{p\pi i} D_p(-z) + \frac{\sqrt{2\pi}}{\Gamma(-p)} e^{\pi(p+1)i/2} D_{-p-1}(-iz)$$

$$\text{MO 121}$$

9.249[10] $D_p[(1+i)x] + D_p[-(1+i)x] = \dfrac{2^{1+p/2}}{\Gamma(-p)} \exp\left[-\dfrac{i}{2}\left(x^2 + p\dfrac{\pi}{2} \right) \right] \displaystyle\int_0^\infty \frac{\cos xt}{t^{p+1}} e^{-it^2/4}\, dt$

$$[x \text{ real}; \quad -1 < \operatorname{Re} p < 0] \qquad \text{MO 122}$$

9.251[10] $D_n(z) = (-1)^n e^{z^2/4} \dfrac{d^n}{dz^n}\left(e^{-z^2/2} \right)$ $[n = 0, 1, 2, \ldots]$ WH

9.252 $D_p(ax+by) = \exp \dfrac{(bx-ay)^2}{4} \left(\dfrac{a}{\sqrt{a^2+b^2}} \right)^p \displaystyle\sum_{k=0}^\infty \binom{p}{k} D_{p-k}\left(\sqrt{a^2+b^2}\, x \right) D_k\left(\sqrt{a^2+b^2}\, y \right) \left(\dfrac{b}{a} \right)^k$

$$[a > b > 0, \quad x > 0, \quad y > 0, \quad \operatorname{Re} p \geq 0] \qquad \text{``summation theorem''} \quad \text{MO 124}$$

Connections with other functions

9.253[10] $D_n(z) = -2^{-\frac{n}{2}} e^{-\frac{z^2}{4}} H_n\left(\dfrac{z}{\sqrt{2}} \right)$ MO 123a

9.254

1. $D_{-1}(z) = e^{\frac{z^2}{4}} \sqrt{\dfrac{\pi}{2}} \left[1 - \Phi\left(\dfrac{z}{\sqrt{2}} \right) \right]$ MO 123

2.[3] $D_{-2}(z) = e^{\frac{z^2}{4}} \sqrt{\dfrac{\pi}{2}} \left\{ \sqrt{\dfrac{\pi}{2}} e^{-\frac{z^2}{2}} - z\left[1 - \Phi\left(\dfrac{z}{\sqrt{2}} \right) \right] \right\}$ MO 123

9.255   Differential equations leading to parabolic cylinder functions:

1. $\dfrac{d^2u}{dz^2} + \left(p + \dfrac{1}{2} - \dfrac{z^2}{4}\right)u = 0$

      The solutions are $u = D_p(z)$, $D_p(-z)$, $D_{-p-1}(iz)$, and $D_{-p-1}(-iz)$

      (These four solutions are linearly dependent. See **9.248**)

2. $\dfrac{d^2u}{dz^2} + \left(z^2 + \lambda\right)u = 0,$ $u = D_{-\frac{1+i\lambda}{2}}\left[\pm(1+i)z\right]$

                                                                    EH II 118(12,13)a, MO 123

3.[7] $\dfrac{d^2u}{dz^2} + z\dfrac{du}{dz} + (p+1)u = 0,$ $u = e^{-\frac{z^2}{4}}D_p(z)$ MO 123

9.26   Confluent hypergeometric series of two variables

9.261

1.[6] $\Phi_1(\alpha, \beta, \gamma, x, y) = \displaystyle\sum_{m,n=0}^{\infty} \dfrac{(\alpha)_{m+n}(\beta)_m}{(\gamma)_{m+n}m!n!}x^m y^n$ $[|x| < 1]$ EH I 225(20)

2. $\Phi_2(\beta, \beta', \gamma, x, y) = \displaystyle\sum_{m,n=0}^{\infty} \dfrac{(\beta)_m(\beta')_m}{(\gamma)_{m+n}m!n!}x^m y^n$ EH I 225(21)a, ET I 385

3. $\Phi_3(\beta, \gamma, x, y) = \displaystyle\sum_{m,n=0}^{\infty} \dfrac{(\beta)_m}{(\gamma)_{m+n}m!n!}x^m y^n$ EH I 225(22)

The functions Φ_1, Φ_2, Φ_3 satisfy the following systems of partial differential equations:

9.262

1. $z = \Phi_1(\alpha, \beta, \gamma, x, y)$ EH I 235(23)

$$x(1-x)\frac{\partial^2 z}{\partial x^2} + y(1-x)\frac{\partial^2 z}{\partial x\,\partial y} + [\gamma - (\alpha+\beta+1)x]\frac{\partial z}{\partial x} - \beta y\frac{\partial z}{\partial y} - \alpha\beta z = 0,$$

$$y\frac{\partial^2 z}{\partial y^2} + x\frac{\partial^2 z}{\partial x\,\partial y} + (\gamma - y)\frac{\partial z}{\partial y} - x\frac{\partial z}{\partial x} - \alpha z = 0$$

2. $z = \Phi_2(\beta, \beta', \gamma, x, y)$ EH I 235(24)

$$x\frac{\partial^2 z}{\partial x^2} + y\frac{\partial^2 z}{\partial x\,\partial y} + (\gamma - x)\frac{\partial z}{\partial x} - \beta z = 0,$$

$$y\frac{\partial^2 z}{\partial y^2} + x\frac{\partial^2 z}{\partial x\,\partial y} + (\gamma - y)\frac{\partial z}{\partial y} - \beta' z = 0$$

3. $z = \Phi_3(\beta, \gamma, x, y)$ EH I 235(25)

$$x\frac{\partial^2 z}{\partial x^2} + y\frac{\partial^2 z}{\partial x\, \partial y} + (\gamma - x)\frac{\partial z}{\partial x} - \beta z = 0,$$

$$y\frac{\partial^2 z}{\partial y^2} + x\frac{\partial^2 z}{\partial x\, \partial y} + \gamma\frac{\partial z}{\partial y} - z = 0$$

9.3 Meijer's G-Function

9.30 Definition

9.301 $G_{p,q}^{m,n}\left(x \left| \begin{matrix} a_1, \ldots, a_p \\ b_1, \ldots, b_q \end{matrix} \right. \right) = \dfrac{1}{2\pi i} \displaystyle\int \dfrac{\displaystyle\prod_{j=1}^{m} \Gamma(b_j - s) \prod_{j=1}^{n} \Gamma(1 - a_j + s)}{\displaystyle\prod_{j=m+1}^{q} \Gamma(1 - b_j + s) \prod_{j=n+1}^{p} \Gamma(a_j - s)} x^s\, ds$

$[0 \leq m \leq q, \quad 0 \leq n \leq p$, and the poles of $\Gamma(b_j - s)$ must not coincide with the poles of $\Gamma(1 - a_k + s)$ for any j and k (where $j = 1, \ldots, m; \quad k = 1, \ldots, n]$). Besides **9.301**, the following notations are also used:

$$G_{pq}^{mn}\left(x \left| \begin{matrix} a_r \\ b_s \end{matrix} \right. \right), \qquad G_{pq}^{mn}(x), \qquad G(x) \qquad\qquad \text{EH I 207(1)}$$

9.302   Three types of integration paths L in the right member of **9.301** can be exhibited:

1.   The path L runs from $-\infty$ to $+\infty$ in such a way that the poles of the functions $\Gamma(1 - a_k + s)$ lie to the left, and the poles of the functions $\Gamma(b_j - s)$ lie to the right of L (for $j = 1, 2, \ldots, m$ and $k = 1, 2, \ldots, n$). In this case, the conditions under which the integral **9.301** converges are of the form

$$p + q < 2(m + n), \quad |\arg x| < \left(m + n - \tfrac{1}{2}p - \tfrac{1}{2}q \right)\pi \qquad \text{EH I 207(2)}$$

2. L is a loop, beginning and ending at $+\infty$, that encircles the poles of the functions $\Gamma(b_j - s)$ (for $j = 1, 2, \ldots, m$) once in the negative direction. All the poles of the functions $\Gamma(1 - a_k + s)$ must remain outside this loop. Then, the conditions under which the integral **9.301** converges are:

$$q \geq 1 \text{ and either } p < q \text{ or } p = q \text{ and } |x| < 1 \qquad \text{EH I 207(3)}$$

3. L is a loop, beginning and ending at $-\infty$, that encircles the poles of the functions $\Gamma(1 - a_k + s)$ (for $k = 1, 2, \ldots, n$) once in the positive direction. All the poles of the functions $\Gamma(b_j - s)$ (for $j = 1, 2, \ldots, m$) must remain outside this loop.

The conditions under which the integral in **9.301** converges are

$$p \geq 1 \text{ and either } p > q \text{ or } p = q \text{ and } |x| > 1 \qquad \text{EH I 207(4)}$$

The function $G_{pq}^{mn}\left(x \left| \begin{matrix} a_r \\ b_s \end{matrix} \right. \right)$ is analytic with respect to x; it is symmetric with respect to the parameters $a_1, \ldots, a_n$ and also with respect to $a_{n+1}, \ldots, a_p$; $b_1, \ldots, b_m$; $b_{m+1}, \ldots, b_q$.

9.303[7]   If no two b_j (for $j = 1, 2, \ldots, n$) differ by an integer, then, under the conditions that either $p < q$ or $p = q$ and $|x| < 1$,

$$
G_{pq}^{mn} \left(x \left| \begin{matrix} a_r \\ b_s \end{matrix} \right. \right) = \sum_{h=1}^{m} \frac{\prod\limits_{j=1}^{m}{}' \Gamma\left(b_j - b_h\right) \prod\limits_{j=1}^{n}{}' \Gamma\left(1 + b_h - a_j\right)}{\prod\limits_{j=m+1}^{q}{}' \Gamma\left(1 + b_h - b_j\right) \prod\limits_{j=n+1}^{p}{}' \Gamma\left(a_j - b_h\right)} x^{b_h}
$$

$$
\times \; {}_pF_{q-1}\left[1 + b_h - a_1, \ldots, 1 + b_h - a_p; \quad 1 + b_h - b_1, \ldots \right.
$$

$$
\left. \ldots, *, \ldots, 1 + b_h - b_q; \quad (-1)^{p-m-n} x \right]
$$

EH I 208(5)

       The prime by the product symbol denotes the omission of the product when $j = h$. The asterisk in the function $\,{}_pF_{q-1}$ denotes the omission of the hth parameter.

9.304[7]   If no two a_k (for $k = 1, 2, \ldots, n$) differ by an integer, then, under the conditions that $q < p$ or $q = p$ and $|x| > 1$,

$$
G_{pq}^{mn} \left(x \left| \begin{matrix} a_r \\ b_s \end{matrix} \right. \right) = \sum_{h=1}^{n} \frac{\prod\limits_{j=1}^{n}{}' \Gamma\left(a_h - a_j\right) \prod\limits_{j=1}^{m} \Gamma\left(b_j - a_h + 1\right)}{\prod\limits_{j=n+1}^{p} \Gamma\left(a_j - a_h + 1\right) \prod\limits_{j=m+1}^{q} \Gamma\left(a_h - b_j\right)} x^{a_h - 1}
$$

$$
\times \; {}_qF_{p-1}\left[1 + b_1 - a_h, \ldots, 1 + b_q - a_h; \quad 1 + a_1 - a_h, \ldots \right.
$$

$$
\left. \ldots, *, \ldots, 1 + a_p - a_h; \quad (-1)^{q-m-n} x^{-1} \right]
$$

EH I 208(6)

9.31   Functional relations

If one of the parameters a_j (for $j = 1, 2, \ldots, n$) coincides with one of the parameters b_j (for $j = m + 1, \; m + 2, \ldots, q$), the order of the G-function decreases. For example,

1. $$G_{pq}^{mn} \left(x \left| \begin{matrix} a_1, \ldots, a_p \\ b_1, \ldots, b_{q-1}, a_1 \end{matrix} \right. \right) = G_{p-1, q-1}^{m, n-1} \left(x \left| \begin{matrix} a_2, \ldots, a_p \\ b_1, \ldots, b_{q-1} \end{matrix} \right. \right)$$

$$[n, p, q \geq 1]$$

       An analogous relationship occurs when one of the parameters b_j (for $j = 1, 2, \ldots, m$) coincides with one of the a_j (for $j = n + 1, \ldots, p$). In this case, it is m and not n that decreases by one unit.

       The G-function with $p > q$ can be transformed into the G-function with $p < q$ by means of the relationships:

2. $$G_{pq}^{mn} \left(x^{-1} \left| \begin{matrix} a_r \\ b_s \end{matrix} \right. \right) = G_{qp}^{nm} \left(x \left| \begin{matrix} 1 - b_s \\ 1 - a_r \end{matrix} \right. \right)$$ EH I 209(9)

3. $x \dfrac{d}{dx} G_{pq}^{mn} \left(x \left| \begin{matrix} a_r \\ b_s \end{matrix} \right. \right) = G_{pq}^{mn} \left(x \left| \begin{matrix} a_1 - 1, a_2, \ldots, a_p \\ b_1, \ldots, b_q \end{matrix} \right. \right) + (a_1 - 1) G_{pq}^{mn} \left(x \left| \begin{matrix} a_r \\ b_s \end{matrix} \right. \right)$

$$[n \geq 1] \qquad \text{EH I 210(13)}$$

4.* $G_{p+1,q+1}^{m+1,n} \left(z \left| \begin{matrix} \mathbf{a}_p, 1 - r \\ 0, \mathbf{b}_q \end{matrix} \right. \right) = (-1)^r G_{p+1,q+1}^{m,n+1} \left(z \left| \begin{matrix} 1 - r, \mathbf{a}_p \\ \mathbf{b}_q, 1 \end{matrix} \right. \right)$

$$[r = 0, 1, 2, \ldots] \qquad \text{MS2 6 (1.2.2)}$$

5.* $z^k G_{pq}^{mn} \left(z \left| \begin{matrix} \mathbf{a}_p \\ \mathbf{b}_q \end{matrix} \right. \right) = G_{pq}^{mn} \left(z \left| \begin{matrix} \mathbf{a}_p + k \\ \mathbf{b}_q + k \end{matrix} \right. \right)$

$$\text{MS2 7 (1.2.7)}$$

9.32 A differential equation for the G-function

$G_{pq}^{mn} \left(x \left| \begin{matrix} a_r \\ b_s \end{matrix} \right. \right)$ satisfies the following linear q^{th}-order differential equation

$$\left[(-1)^{p-m-n} x \prod_{j=1}^{p} \left(x \frac{d}{dx} - a_j + 1 \right) - \prod_{j=1}^{q} \left(x \frac{d}{dx} - b_j \right) \right] y = 0 \qquad [p \leq q] \qquad \text{EH I 210(1)}$$

9.33 Series of G-functions

$$G_{pq}^{mn} \left(\lambda x \left| \begin{matrix} a_1, \ldots, a_p \\ b_1, \ldots, b_q \end{matrix} \right. \right) = \lambda^{b_1} \sum_{r=0}^{\infty} \frac{1}{r!} (1 - \lambda)^r G_{pq}^{mn} \left(x \left| \begin{matrix} a_1, \ldots, a_p \\ b_1 + r, b_2, \ldots, b_q \end{matrix} \right. \right)$$

$$[|\lambda - 1| < 1, \quad m \geq 1, \quad \text{if } m = 1 \text{ and } p < q, \lambda \text{ may be arbitrary}]$$
$$\text{EH I 213(1)}$$

$$= \lambda^{b_q} \sum_{r=0}^{\infty} \frac{1}{r!} (\lambda - 1)^r G_{pq}^{mn} \left(x \left| \begin{matrix} a_1, \ldots, a_p \\ b_1, \ldots, b_{q-1}, b_q + r \end{matrix} \right. \right)$$

$$[m < q, \quad |\lambda - 1| < 1]$$
$$\text{EH I 213(2)}$$

$$= \lambda^{a_1 - 1} \sum_{r=0}^{\infty} \frac{1}{r!} \left(\lambda - \frac{1}{\lambda} \right)^r G_{pq}^{mn} \left(x \left| \begin{matrix} a_1 - r, a_2, \ldots, a_p \\ b_1, \ldots, b_q \end{matrix} \right. \right)$$

$$\left[n \geq 1, \quad \text{Re} \, \lambda > \tfrac{1}{2}, \quad (\text{if } n = 1 \text{ and } p > q, \text{ then } \lambda \text{ may be arbitrary}) \right]$$
$$\text{EH I 213(3)}$$

$$= \lambda^{a_p - 1} \sum_{r=0}^{\infty} \frac{1}{r!} \left(\frac{1}{\lambda} - 1 \right)^r G_{pq}^{mn} \left(x \left| \begin{matrix} a_1, \ldots, a_{p-1}, a_p - r \\ b_1, \ldots, b_q \end{matrix} \right. \right)$$

$$\left[n < p, \quad \text{Re} \, \gamma > \tfrac{1}{2} \right]$$
$$\text{EH I 213(4)}$$

For integrals of the G-function, see **7.8**.

9.34 Connections with other special functions

1. $J_\nu(x) x^\mu = 2^\mu G_{02}^{10} \left(\frac{1}{4} x^2 \left| \begin{matrix} \\ \frac{1}{2}\nu + \frac{1}{2}\mu, \frac{1}{2}\mu - \frac{1}{2}\nu \end{matrix} \right. \right)$ EH I 219(44)

2. $Y_\nu(x) x^\mu = 2^\mu G_{13}^{20} \left(\frac{1}{4} x^2 \left| \begin{matrix} \frac{1}{2}\mu - \frac{1}{2}\nu - \frac{1}{2} \\ \frac{1}{2}\mu - \frac{1}{2}\nu, \frac{1}{2}\mu + \frac{1}{2}\nu, \frac{1}{2}\mu - \frac{1}{2}\nu - \frac{1}{2} \end{matrix} \right. \right)$ EH I 219(46)

3. $\quad K_\nu(x) x^\mu = 2^{\mu-1} \, G_{02}^{20} \left(\frac{1}{4} x^2 \, \middle| \, {\textstyle\frac{1}{2}\mu + \frac{1}{2}\nu, \frac{1}{2}\mu - \frac{1}{2}\nu} \right)$ $\qquad$ EH I 219(47)

4. $\quad K_\nu(x) = e^x \sqrt{\pi} \, G_{12}^{20} \left(2x \, \middle| \, {\textstyle \frac{1}{2} \atop \nu, -\nu} \right)$ $\qquad$ EH I 219(49)

5. $\quad \mathbf{H}_\nu(x) x^\mu = 2^\mu \, G_{13}^{11} \left(\frac{1}{4} x^2 \, \middle| \, {\textstyle \frac{1}{2} + \frac{1}{2}\nu + \frac{1}{2}\mu \atop \frac{1}{2} + \frac{1}{2}\nu + \frac{1}{2}\mu, \frac{1}{2}\mu - \frac{1}{2}\nu, \frac{1}{2}\mu + \frac{1}{2}\nu} \right)$ $\qquad$ EH I 220(51)

6. $\quad S_{\mu,\nu}(x) = 2^{\mu-1} \dfrac{1}{\Gamma\left(\frac{1-\mu-\nu}{2}\right) \Gamma\left(\frac{1-\mu+\nu}{2}\right)} \, G_{13}^{31} \left(\frac{1}{4} x^2 \, \middle| \, {\textstyle \frac{1}{2} + \frac{1}{2}\mu \atop \frac{1}{2} + \frac{1}{2}\mu, \frac{1}{2}\nu, -\frac{1}{2}\nu} \right)$ $\qquad$ EH I 220(55)

7.[7] $\quad {}_2F_1(a,b;c;-x) = \dfrac{\Gamma(c) x}{\Gamma(a)\Gamma(b)} \, G_{22}^{12} \left(x \, \middle| \, {\textstyle -a, -b \atop -1, -c} \right)$ $\qquad$ EH I 222(74)a

8. $\quad {}_pF_q\left(a_1,\ldots,a_p; b_1,\ldots,b_q; x\right) = \dfrac{\prod_{j=1}^q \Gamma(b_j)}{\prod_{j=1}^p \Gamma(a_j)} \, G_{p,q+1}^{1,p} \left(-x \, \middle| \, {\textstyle 1-a_1,\ldots,1-a_p \atop 0, 1-b_1,\ldots,1-b_q} \right)$

$\qquad\qquad\qquad = \dfrac{\prod_{j=1}^q \Gamma(b_j)}{\prod_{j=1}^p \Gamma(a_j)} \, G_{q+1,p}^{p,1} \left(-\frac{1}{x} \, \middle| \, {\textstyle 1, b_1,\ldots,b_q \atop a_1,\ldots,a_p} \right)$

$\qquad$ EH I 215(1)

9. $\quad W_{k,m}(x) = \dfrac{2^k \sqrt{x}\, e^{\frac{1}{2}x}}{\sqrt{2\pi}} \, G_{24}^{40} \left(\frac{x^2}{4} \, \middle| \, {\textstyle \frac{1}{4} - \frac{1}{2}k, \frac{3}{4} - \frac{1}{2}k \atop \frac{1}{2} + \frac{1}{2}m, \frac{1}{2} - \frac{1}{2}m, \frac{1}{2}m, -\frac{1}{2}m} \right)$ $\qquad$ EH I 221(70)

9.4 MacRobert's E-Function

9.41 Representation by means of multiple integrals

$E\left(p; \alpha_r : q; \varrho_s : x\right) = \dfrac{\Gamma(\alpha_{q+1})}{\Gamma(\varrho_1 - \alpha_1)\Gamma(\varrho_2 - \alpha_2)\cdots\Gamma(\varrho_q - \alpha_q)}$

$\qquad \times \displaystyle\prod_{\mu=1}^{q} \int_0^\infty \lambda_\mu^{\varrho_\mu - \alpha_\mu - 1} (1 - \lambda_\mu)^{-\varrho_\mu} \, d\lambda_\mu \prod_{\nu=2}^{p-q-1} \int_0^\infty e^{-\lambda_{q+\nu}} \lambda_{q+\nu}^{\alpha_{q+\nu} - 1} \, d\lambda_{q+\nu}$

$\qquad \times \displaystyle\int_0^\infty e^{-\lambda_p} \lambda_p^{\alpha_p - 1} \left[1 + \dfrac{\lambda_{q+2}\lambda_{q+3}\cdots\lambda_p}{(1+\lambda_1)\cdots(1+\lambda_q) x} \right]^{-\alpha_{q+1}} d\lambda_p$

$[|\arg x| < \pi$, $p \geq q+1$, α_r and ϱ_s are bounded by the condition that the integrals on the right be convergent.] $\qquad$ EH I 204(3)

9.42 Functional relations

1. $\quad \alpha_1 x \, E\left(\alpha_1,\ldots,\alpha_p : \varrho_1,\ldots,\varrho_q : x\right) = x \, E\left(\alpha_1 + 1, \alpha_2,\ldots,\alpha_p : \varrho_1,\ldots,\varrho_q : x\right)$

$\qquad\qquad + E\left(\alpha_1 + 1, \alpha_2 + 1,\ldots,\alpha_p + 1 : \varrho_1 + 1,\ldots,\varrho_q + 1 : x\right)$

$\qquad$ EH I 205(7)

2. $(\varrho_1 - 1)\, x\, E\,(\alpha_1, \ldots, \alpha_p : \varrho_1, \ldots, \varrho_q : x) = x\, E\,(\alpha_1, \ldots, \alpha_p : \varrho_1 - 1, \varrho_2, \ldots, \varrho_q : x)$

$$+ E\,(\alpha_1 + 1, \ldots, \alpha_p + 1 : \varrho_1 + 1, \ldots, \varrho_q + 1 : x)$$

EH I 205(9)

3. $\dfrac{d}{dx}\, E\,(\alpha_1, \ldots, \alpha_p : \varrho_1, \ldots, \varrho_q : x) = x^{-2}\, E\,(\alpha_1 + 1, \ldots, \alpha_p + 1 : \varrho_1 + 1, \ldots, \varrho_q + 1 : x)$

EH I 205(8)

9.5 Riemann's Zeta Functions $\zeta(z,q)$, and $\zeta(z)$, and the Functions $\Phi(z,s,v)$ and $\xi(s)$

9.51 Definition and integral representations

9.511 $\zeta(z,q)$

$$= \frac{1}{\Gamma(z)} \int_0^\infty \frac{t^{z-1} e^{-qt}}{1 - e^{-t}}\, dt; \qquad\qquad\qquad\qquad\qquad \text{WH}$$

$$= \frac{1}{2} q^{-z} + \frac{q^{1-z}}{z-1} + 2 \int_0^\infty \left(q^2 + t^2\right)^{-\frac{z}{2}} \left[\sin\left(z \arctan \frac{t}{q}\right)\right] \frac{dt}{e^{2\pi t} - 1} \qquad [0 < q < 1, \quad \operatorname{Re} z > 1] \quad \text{WH}$$

9.512 $\zeta(z,q) = -\dfrac{\Gamma(1-z)}{2\pi i} \displaystyle\int_\infty^{(0+)} \dfrac{(-\theta)^{z-1} e^{-q\theta}}{1 - e^{-\theta}}\, d\theta$

    This equation is valid for all values of z except for $z = 1, 2, 3, \ldots$. It is assumed that the path of integration (see drawing below) does not pass through the points $2n\pi i$ (where n is a natural number).

    See also **4.251** 4, **4.271** 1, 4, 8, **4.272** 9, 12, **4.294** 11.

9.513

1. $\zeta(z) = \dfrac{1}{(1 - 2^{1-z})\, \Gamma(z)} \displaystyle\int_0^\infty \dfrac{t^{z-1}}{e^t + 1}\, dt$ $[\operatorname{Re} z > 0]$ WH

2. $\zeta(z) = \dfrac{2^z}{(2^z - 1)\, \Gamma(z)} \displaystyle\int_0^\infty \dfrac{t^{z-1} e^t}{e^{2t} - 1}\, dt$ $[\operatorname{Re} z > 1]$ WH

3. $\zeta(z) = \dfrac{\pi^{\frac{z}{2}}}{\Gamma\left(\frac{z}{2}\right)} \left[\dfrac{1}{z(z-1)} + \displaystyle\int_1^\infty \left(t^{\frac{1-z}{2}} + t^{\frac{z}{2}}\right) t^{-1} \sum_{k=1}^\infty e^{-k^2 \pi i}\, dt\right]$ WH

4. $\zeta(z) = \dfrac{2^{z-1}}{z-1} - 2^z \displaystyle\int_0^\infty \left(1 + t^2\right)^{-\frac{z}{2}} \sin(z \arctan t)\, \dfrac{dt}{e^{\pi t} + 1}$ WH

5. $\zeta(z) = \dfrac{2^{z-1}}{2^z - 1} \dfrac{z}{z-1} + \dfrac{2}{2^z - 1} \displaystyle\int_0^\infty \left(\dfrac{1}{4} + t^2\right)^{-z/2} \sin(z \arctan 2t)\, \dfrac{dt}{e^{2\pi t} - 1}$ WH

See also **3.411** 1, **3.523** 1, **3.527** 1, 3, **4.271** 8.

9.52 Representation as a series or as an infinite product

9.521

1. $\zeta(z, q) = \sum_{n=0}^{\infty} \dfrac{1}{(q+n)^z}$ [$\operatorname{Re} z > 1, \quad q \neq 0, -1, -2, \ldots$]   WH

2. $\zeta(z, q) = \dfrac{2\,\Gamma(1-z)}{(2\pi)^{1-z}} \left[\sin \dfrac{z\pi}{2} \sum_{n=1}^{\infty} \dfrac{\cos 2\pi q n}{n^{1-z}} + \cos \dfrac{z\pi}{2} \sum_{n=1}^{\infty} \dfrac{\sin 2\pi q n}{n^{1-z}} \right]$

                                             [$\operatorname{Re} z < 0, \quad 0 < q \leq 1$]   WH

3.[8] $\zeta(z, q) = \sum_{n=0}^{N} \dfrac{1}{(q+n)^z} - \dfrac{1}{(1-z)(N+q)^{z-1}} - \sum_{n=N}^{\infty} F_n(z),$

where

$$F_n(z) = \dfrac{1}{1-z} \left(\dfrac{1}{(n+1+q)^{z-1}} - \dfrac{1}{(n+q)^{z-1}} \right) - \dfrac{1}{(n+1+q)^z}$$

$$= z \int_n^{n+1} \dfrac{(t-n)\,dt}{(t+q)^{z+1}}$$

     WH

9.522

1. $\zeta(z) = \sum_{n=1}^{\infty} \dfrac{1}{n^z}$ [$\operatorname{Re} z > 1$]   WH

2. $\zeta(z) = \dfrac{1}{1 - 2^{1-z}} \sum_{n=1}^{\infty} (-1)^{n+1} \dfrac{1}{n^z}$ [$\operatorname{Re} z > 0$]   WH

9.523 The following product and summation are taken over all primes p:

1.[7] $\zeta(z) = \prod_p \dfrac{1}{1 - p^{-z}}$ [$\operatorname{Re} z > 1$]   WH

2. $\ln \zeta(z) = \sum_p \sum_{k=1}^{\infty} \dfrac{1}{k p^{kz}}$ [$\operatorname{Re} z > 1$]   WH

9.524[7] $\dfrac{\zeta'(z)}{\zeta(z)} = -\sum_{k=1}^{\infty} \dfrac{\Lambda(k)}{k^z},$ [$\operatorname{Re} z > 1$]

where $\Lambda(k) = 0$ when k is not a power of a prime and $\Lambda(k) = \ln p$ when k is a power of a prime p.   WH

9.53 Functional relations

9.531 $\zeta(-n, q) = -\dfrac{B'_{n+2}(q)}{(n+1)(n+2)} = \dfrac{-B_{n+1}(q)}{n+1}$

                         [n is a nonnegative integer]   see EH I 27 (11)   WH

9.532 $\sum_{k=2}^{\infty} \dfrac{(-1)^{k-1}}{k} z^k\, \zeta(k, q) = \ln \dfrac{e^{-Cz}\,\Gamma(q)}{\Gamma(z+q)} - \dfrac{z}{q} + \sum_{k=1}^{\infty} \dfrac{qz}{k(q+k)}$ [$|z| < q$]   WH

9.533

1. $\displaystyle\lim_{z \to 1} \frac{\zeta(z, q)}{\Gamma(1 - z)} = -1$ WH

2. $\displaystyle\lim_{z \to 1} \left\{ \zeta(z, q) - \frac{1}{z - 1} \right\} = -\Psi(q)$ WH

3. $\displaystyle\left\{ \frac{d}{dz} \zeta(z, q) \right\}_{z=0} = \ln \Gamma(q) - \frac{1}{2} \ln 2\pi$ WH

9.534 $\zeta(z, 1) = \zeta(z)$

9.535

1. $\displaystyle\zeta(z) = \frac{1}{2^z - 1} \zeta\left(z, \tfrac{1}{2}\right)$ [$\operatorname{Re} z > 1$]   WH

2. $\displaystyle 2^z \, \Gamma(1 - z) \, \zeta(1 - z) \sin \frac{z\pi}{2} \pi^{1-z} \zeta(z)$ WH

3. $\displaystyle 2^{1-z} \, \Gamma(z) \, \zeta(z) \cos \frac{z\pi}{2} = \pi^z \, \zeta(1 - z)$ WH

4. $\displaystyle \Gamma\left(\frac{z}{2}\right) \pi^{-\frac{z}{2}} \zeta(z) = \Gamma\left(\frac{1 - z}{2}\right) \pi^{\frac{z-1}{2}} \zeta(1 - z)$ WH

9.536 $\displaystyle\lim_{z \to 1} \left\{ \zeta(z) - \frac{1}{z - 1} \right\} = C$

9.537   Set $z = \tfrac{1}{2} + it$. Then, $\Xi(t) = \dfrac{(z - 1)\, \Gamma\left(\frac{z}{2} + 1\right)}{\sqrt{\pi^z}} \zeta(z) = \Xi(-t)$ is an even function of t with real coefficients in its expansion in powers of t^2.   JA

9.54 Singular points and zeros

9.541[7]

1. $z = 1$ is the only singular point of the function $\zeta(z)$ WH

2.   The function $\zeta(z)$ has simple zeros at the points $-2n$, where n is a natural number. All other zeros of the function $\zeta(z)$ lie in the strip $0 \le \operatorname{Re} z < 1$.

3.[8]   Riemann's hypothesis: All zeros of the function $\zeta(z)$ lie on the straight line $\operatorname{Re} z = \tfrac{1}{2}$. It has been shown that a countably infinite set of zeros of the zeta function lie on this line. The first $1{,}500{,}000{,}001$ zeros lying in $0 < \operatorname{Im} z < 545{,}439{,}823.215$ are known to have $\operatorname{Re} z = \tfrac{1}{2}$.   WH

9.542   Particular values:

1. $\displaystyle\zeta(2m) = \frac{2^{2m-1}\pi^{2m} |B_{2m}|}{(2m)!}$ [m is a natural number]   WH

2. $\displaystyle\zeta(1 - 2m) = -\frac{B_{2m}}{2m}$ [m is a natural number]   WH

3. $\zeta(-2m) = 0$ [m is a natural number]   WH

4. $\zeta'(0) = -\tfrac{1}{2} \ln 2\pi$ WH

9.55   The Lerch function $\Phi(z, s, v)$

9.550   Definition:

$$\Phi(z, s, v) = \sum_{n=0}^{\infty} (v+n)^{-s} z^n \qquad [|z| < 1, \quad v \neq 0, -1, \ldots] \qquad \text{EH I 27(1)}$$

Functional relations

9.551 $\Phi(z, s, v) = z^m \, \Phi(z, s, m+v) + \sum_{n=0}^{m-1} (v+n)^{-s} z^n \qquad [m = 1, 2, 3, \ldots, \quad v \neq 0, -1, -2, \ldots]$

EH I 27(1)

9.552 $\Phi(z, s, v)$

$$= iz^{-v}(2\pi)^{s-1} \Gamma(1-s) \left[e^{-i\pi\frac{s}{2}} \, \Phi\left(e^{-2\pi i v}, 1-s, \frac{\ln z}{2\pi i} \right) - e^{i\pi\left(\frac{s}{2}-2v\right)} \, \Phi\left(e^{2\pi i v}, 1-s, 1-\frac{\ln z}{2\pi i} \right) \right]$$

EH I 29(7)

Series representation

9.553 $\Phi(z, s, v) = z^{-v} \, \Gamma(1-s) \sum_{n=-\infty}^{\infty} (-\ln z + 2\pi n i)^{s-1} e^{2\pi n v i}$

$$[0 < v \leq 1, \quad \operatorname{Re} s < 0, \quad |\arg(-\ln z + 2\pi n i)| \leq \pi] \quad \text{EH I 28(6)}$$

9.554 $\Phi(z, m, v) = z^{-v} \left\{ \sum_{n=0}^{\infty}{}' \zeta(m-n, v) \frac{(\ln z)^n}{n!} + \frac{(\ln z)^{m-1}}{(m-1)!} \left[\Psi(m) - \Psi(v) - \ln\left(\ln \frac{1}{z} \right) \right] \right\}^{*}$

$$[m = 2, 3, 4, \ldots, \quad |\ln z| < 2\pi, \quad v \neq 0, -1, -2, \ldots] \quad \text{EH I 30(9)}$$

9.555 $\Phi(z, -m, v) = \frac{m!}{z^v} \left(\ln \frac{1}{z} \right)^{-m-1} - \frac{1}{z^v} \sum_{r=0}^{\infty} \frac{B_{m+r+1}(v)(\ln z)^r}{r!(m+r+1)} \qquad [|\ln z| < 2\pi]$

EH I 30(11)

Integral representation

9.556 $\Phi(z, s, v) = \dfrac{1}{\Gamma(s)} \displaystyle\int_0^{\infty} \dfrac{t^{s-1} e^{-vt}}{1 - z e^{-t}} \, dt = \dfrac{1}{\Gamma(s)} \displaystyle\int_0^{\infty} \dfrac{t^{s-1} e^{-(v-1)t} \, dt}{e^t - z}$

$$[\operatorname{Re} v > 0, \text{ or } |z| \leq 1, \quad z \neq 1, \quad \operatorname{Re} s > 0, \text{ or } z = 1, \quad \operatorname{Re} s > 1] \quad \text{EH I 27(3)}$$

Limit relationships

9.557 $\lim_{z \to 1} (1-z)^{1-s} \, \Phi(z, s, v) = \Gamma(1-s) \qquad\qquad\qquad [\operatorname{Re} s < 1]$ EH I 30(12)

9.558 $\lim_{z \to 1} \dfrac{\Phi(z, 1, v)}{-\ln(1-z)} = 1$ EH I 30(13)

A connection with a hypergeometric function

9.559 $\Phi(z, 1, v) = v^{-1} \, {}_2F_1(1, v; 1+v; z) \qquad\qquad\qquad [|z| < 1]$ EH I 30(10)

---

*In 9.554 the prime on the symbol $\sum$ means that the term corresponding to $n = m - 1$ is omitted.

9.56 The function $\xi(s)$

9.561 $\xi(s) = \dfrac{1}{2}s(s-1)\dfrac{\Gamma\left(\frac{1}{2}s\right)}{\pi^{\frac{1}{2}s}}\zeta(s)$ EH III 190(10)

9.562 $\xi(1-s) = \xi(s)$ EH III 190(11)

9.6 Bernoulli Numbers and Polynomials, Euler Numbers, the Functions $\nu(x)$, $\nu(x,\alpha)$, $\mu(x,\beta)$, $\mu(x,\beta,\alpha)$, $\lambda(x,y)$ and Euler Polynomials

9.61 Bernoulli numbers

9.610   The numbers B_n, representing the coefficients of $\dfrac{t^n}{n!}$ in the expansion of the function

$$\frac{t}{e^t - 1} = \sum_{n=0}^{\infty} B_n \frac{t^n}{n!} \qquad [0 < |t| < 2\pi],$$

are called *Bernoulli* numbers. Thus, the function $\dfrac{t}{e^t - 1}$ is a generating function for the Bernoulli numbers.

                                                         GE 48(57), FI II 520

9.611   Integral representations

1. $B_{2n} = (-1)^{n-1} 4n \displaystyle\int_0^{\infty} \frac{x^{2n-1}}{e^{2\pi x} - 1}\, dx$ $[n = 1, 2, \ldots]$ (cf. **3.411** 2, 4)

                                                            FI II 721a

2. $B_{2n} = (-1)^{n-1} \pi^{-2n} \displaystyle\int_0^{\infty} \frac{x^{2n}}{\sinh^2 x}\, dx$ $[n = 1, 2, \ldots]$

3. $B_{2n} = (-1)^{n-1} \dfrac{2n(1-2n)}{\pi} \displaystyle\int_0^{\infty} x^{2n-2} \ln\left(1 - e^{-2\pi x}\right) dx$

                                                      $[n = 1, 2, \ldots]$

See also **3.523** 2, **4.271** 3.

Properties and functional relations

9.612[8]   A symbolic notation:

$$(B + \alpha)^{[n]} = \sum_{k=0}^{n} \binom{n}{k} B_k \alpha^{n-k} \qquad [n \geq 2]$$

in particular

$$B_n = (B + 1)^{[n]} = \sum_{k=0}^{n} \binom{n}{k} B_k \qquad [n \geq 2]$$

hence by recursion

$$B_n = -n \sum_{k=0}^{n-1} \frac{B_k}{k!(n+1-k)!} \qquad [n \geq 2]$$

9.613   All the Bernoulli numbers are rational numbers.

9.614   Every number B_n can be represented in the form

$$B_n = C_n - \sum \frac{1}{k+1},$$

where C_n is an integer and the sum is taken over all $k > 0$ such that $k+1$ is a prime and k is a divisor of n.

GE 64

9.615[8] All the Bernoulli numbers with odd index are equal to zero except that $B_1 = -\frac{1}{2}$; that is, $B_{2n+1} = 0$ for n a natural number.

GE 52, FI II 521

$$B_{2n} = -\frac{1}{2n+1} + \frac{1}{2} - \sum_{k=1}^{n-1} \frac{2n(2n-1)\ldots(2n-2k+2)}{(2k)!} B_{2/2} \qquad [n \geq 1]$$

9.616 $B_{2n} = \frac{(-1)^{n-1}(2n)!}{2^{2n-1}\pi^{2n}} \zeta(2n) \qquad [n \geq 0]$ (cf. **9.542**)   GE 56(79), FI II 721a

9.617[7] $B_{2n} = (-1)^{n-1} \frac{2(2n)!}{(2\pi)^{2n}} \dfrac{1}{\prod\limits_{p=2}^{\infty} \left(1 - \dfrac{1}{p^{2n}}\right)} \qquad [n \geq 1] \qquad$ (cf. **9.523**)

(where the product is taken over all primes p).

- For a connection with Riemann's zeta function, see **9.542**.
- For a connection with the Euler numbers, see **9.635**.
- For a table of values of the Bernoulli numbers, see **9.71**

9.618[6] Symbolic notation: $(B + \alpha)^{[n]} \equiv \sum_{k=0}^{n} \binom{n}{k} B_k \alpha^{n-k} \qquad [n = 0, 1, \ldots]$ CE 337

9.619   An inequality $\left|(B - \theta)^{[n]}\right| \leq |B_n| \qquad [0 < \theta < 1]$

9.62 Bernoulli polynomials

9.620   The Bernoulli polynomials $B_n(x)$ are defined by

$$B_n(x) = \sum_{k=0}^{n} \binom{n}{k} B_k x^{n-k}$$ GE 51(62)

or symbolically, $B_n(x) = (B + x)^{[n]}$. GE 52(68)

9.621   The generating function

$$\frac{e^{xt}}{e^t - 1} = \sum_{n=0}^{\infty} B_n(x) \frac{t^{n-1}}{n!} \qquad [0 < |t| < 2\pi] \qquad \text{(cf. 1.213)}$$ GE 65(89)a

9.622   Series representation

1.[7] $B_n(x) = -2\frac{n!}{(2\pi)^n} \sum_{k=1}^{\infty} \frac{\cos\left(2\pi kx - \frac{1}{2}\pi n\right)}{k^n}$

$$[n > 1, \quad 1 \geq x \geq 0; \quad n = 1, \quad 1 > x > 0] \quad \text{AS 805(23.1.16)}$$

2.[7] $B_{2n-1}(x) = 2\frac{(-1)^n 2(2n-1)!}{(2\pi)^{2n-1}} \sum_{k=1}^{\infty} \frac{\sin 2k\pi x}{k^{2n-1}}$

$$[n > 1, \quad 1 \geq x \geq 0; \quad n = 1, \quad 1 > x > 0] \quad \text{AS 805(23.1.17)}$$

$3.^{10}$ $B_{2n}(x) = \dfrac{(-1)^{n-1}2(2n)!}{(2\pi)^{2n}} \displaystyle\sum_{k=1}^{\infty} \dfrac{\cos 2k\pi x}{k^{2n}}$ $[0 \leq x \leq 1, \quad n = 1, 2, \ldots]$ GE 71

9.623 Functional relations and properties:

1. $B_{m+1}(n) = B_{m+1} + (m+1) \displaystyle\sum_{k=1}^{n-1} k^m$

[n and m are natural numbers]   (see also **0.121**)   GE 51(65)

2. $B_n(x+1) - B_n(x) = nx^{n-1}$ GE 65(90)

3. $B'_n(x) = n\,B_{n-1}(x)$ $[n = 1, 2, \ldots]$ GE 66

4. $B_n(1-x) = (-1)^n\,B_n(x)$ GE 66

$5.^{10}$ $(-1)^n\,B_n(-x) = B_n(x) + nx^{n-1}$ $[n = 0, 1, \ldots]$ AS 804(23.1.9)

9.624[7] $B_n(mx) = m^{n-1} \displaystyle\sum_{k=0}^{m-1} B_n\left(x + \dfrac{k}{m}\right)$

$[m = 1, 2, \ldots n = 0, 1, \ldots]$;   "summation theorem"   GE 67

9.625 For n odd, the differences
$$B_n(x) - B_n$$
vanish on the interval $[0, 1]$ only at the points $0, \frac{1}{2}$, and 1. They change sign at the point $x = \frac{1}{2}$. For n even, these differences vanish at the end points of the interval $[0, 1]$. Within this interval, they do not change sign and their greatest absolute value occurs at the point $x = \frac{1}{2}$.

9.626 The polynomials
$$B_{2n}(x) - B_{2n} \text{ and } B_{2n+2}(x) - B_{2n+2}$$
have opposite signs in the interval $(0, 1)$.   GE 87

9.627 Special cases:

1. $B_1(x) = x - \frac{1}{2}$ GE 70

2. $B_2(x) = x^2 - x + \frac{1}{6}$ GE 70

3. $B_3(x) = x^3 - \frac{3}{2}x^2 + \frac{1}{2}x$ GE 70

4. $B_4(x) = x^4 - 2x^3 + x^2 - \frac{1}{30}$ GE 70

5. $B_5(x) = x^5 - \frac{5}{2}x^4 + \frac{5}{3}x^3 - \frac{1}{6}x$ GE 70

9.628 Particular values:

1. $B_n(0) = B_n$

2. $B_1(1) = -B_1 = \frac{1}{2}, \quad B_n(1) = B_n$ $[n \neq 1]$ GE 76

9.63 Euler numbers

9.630   The numbers E_n, representing the coefficients of $\frac{t^n}{n!}$ in the expansion of the function

$$\frac{1}{\cosh t} = \sum_{n=0}^{\infty} E_n \frac{t^n}{n!} \qquad \left[|t| < \frac{\pi}{2}\right]$$

are known as the *Euler numbers*. Thus, the function $\frac{1}{\cosh t}$ is a generating function for the Euler numbers.

CE 330

9.631   A recursion formula

$$(E+1)^{[n]} + (E-1)^{[n]} = 0 \qquad [n \geq 1], \qquad E_0 = 1$$

CE 329

Properties of the Euler numbers

9.632   The Euler numbers are integers.

9.633   The Euler numbers of odd index are equal to zero; the signs of two adjacent numbers of even indices are opposite; that is,

$$E_{2n+1} = 0, \quad E_{4n} > 0, \quad E_{4n+2} < 0$$

CE 329

9.634   If $\alpha, \beta\gamma, \ldots$ are the divisors of the number $n - m$, the difference $E_{2n} - E_{2m}$ is divisible by those of the numbers $2\alpha + 1, 2\beta + 1, 2\gamma + 1, \ldots,$ that are primes.

9.635   A connection with the Bernoulli numbers (symbolic notation):

1.[6] $\displaystyle E_{n-1} + 4(-1)^n \left(3^{n-1} - 1\right) B_1 = \frac{(4B-1)^{[n]}(4B-3)^{[n]}}{2n} + 4(-1)^{n+1} \left(3^{n-1} - 1\right) B_1$

CE 330

2. $\displaystyle B_n = \frac{n(E+1)^{[n-1]}}{2^n \left(2^n - 1\right)}$ $[n \geq 2]$

CE 330

3.[6] $\displaystyle \left(B + \tfrac{1}{4}\right)^{[2n+1]} = -4^{-2n-1}(2n+1)E_{2n}$ $[n \geq 0]$

CE 341

For a table of values of the Euler numbers, see **9.72**.

9.64 The functions $\nu(x)$, $\nu(x, \alpha)$, $\mu(x, \beta)$, $\mu(x, \beta, \alpha)$, $\lambda(x, y)$

9.640

1. $\displaystyle \nu(x) = \int_0^{\infty} \frac{x^t \, dt}{\Gamma(t+1)}$

EH III 217(1)

2. $\displaystyle \nu(x, \alpha) = \int_0^{\infty} \frac{x^{\alpha+t} \, dt}{\Gamma(\alpha+t+1)}$

EH III 217(1)

3. $\displaystyle \mu(x, \beta) = \int_0^{\infty} \frac{x^t t^\beta \, dt}{\Gamma(\beta+1)\,\Gamma(t+1)}$

EH III 217(2)

4. $\displaystyle \mu(x, \beta, \alpha) = \int_0^{\infty} \frac{x^{\alpha+t} t^\beta \, dt}{\Gamma(\beta+1)\,\Gamma(\alpha+t+1)}$

EH III 217(2)

5. $\displaystyle \lambda(x, y) = \int_0^{y} \frac{\Gamma(u+1)\, du}{x^u}$

MI 9

9.65^{10} Euler polynomials

9.650   The Euler polynomials are defined by

$$E_n(x) = \sum_{k=0}^{n} \binom{n}{k} \frac{E_k}{2^k} \left(x - \frac{1}{2} \right)^{n-k}$$

     AS 804 (23.1.7)

9.651   The generating function:

$$\frac{2e^{xt}}{e^t + 1} = \sum_{n=0}^{\infty} E_n(x) \frac{t^n}{n!}$$

     AS 804 (23.1.1)

9.652   Series representation:

1. $$E_n(x) = 4 \frac{n!}{\pi^{n+1}} \sum_{k=0}^{\infty} \frac{\sin\left((2k+1)\pi x - \frac{1}{2}\pi n \right)}{(2k+1)^{n+1}}$$

$$[n > 0, \quad 1 \geq x \geq 0, \quad n = 1, \quad 1 > x > 0] \quad \text{AS 804 (23.1.16)}$$

2.^{10} $$E_{2n-1}(x) = \frac{(-1)^n 4(2n-1)!}{\pi^{2n}} \sum_{k=0}^{\infty} \frac{\cos(2k+1)\pi x}{(2k+1)^{2n}} \qquad [n = 1, 2, \ldots, \quad 1 \geq x \geq 0]$$

$$\text{AS 804 (23.1.17)}$$

3. $$E_{2n}(x) = \frac{(-1)^n 4(2n)!}{\pi^{2n+1}} \sum_{k=0}^{\infty} \frac{\sin(2k+1)\pi x}{(2k+1)^{2n+1}}$$

$$[n > 0, \quad 1 \geq x \geq 0, \quad n = 0, \quad 1 > x > 0] \quad \text{AS 804 (23.1.18)}$$

9.653   Functional relations and properties:

1. $$E_m(n+1) = 2 \sum_{k=1}^{n} (-1)^{n-k} k^m + (-1)^{n+1} E_m(0), \qquad [m \text{ and } n \text{ are natural numbers}]$$

$$\text{AS 804 (23.1.4)}$$

2. $E_n'(x) = nE_{n-1}(x).$ $[n = 1, 2, \ldots]$ AS 804 (23.1.5)

3. $E_n(x+1) + E_n(x) = 2x^n$ $[n = 0, 1, \ldots]$ AS 804 (23.1.6)

4.^{8} $$E_n(mx) = m^n \sum_{k=0}^{m-1} (-1)^k E_n \left(x - \frac{k}{m} \right) \qquad [n = 0, 1, \ldots, m = 1, 3, \ldots]$$

$$\text{AS 804 (23.1.10)}$$

5. $$E_n(mx) = \frac{-2}{n+1} m^n \sum_{k=0}^{m-1} (-1)^k B_{n+1} \left(x + \frac{k}{m} \right) \qquad [n = 0, 1, \ldots, m = 2, 4, \ldots]$$

$$\text{AS 804 (23.1.10)}$$

9.654   Special cases:

1. $E_1(x) = x - \frac{1}{2}$

2. $E_2(x) = x^2 - x$

3. $E_3(x) = x^3 - \frac{3}{2}x^2 + \frac{1}{4}$

4. $E_4(x) = x^4 - 2x^3 + x$

5. $E_5(x) = x^5 - \frac{5}{2}x^4 + \frac{5}{2}x^2 - \frac{1}{2}$

9.655   Particular values:

1. $E_{2n+1} = 0.$ $[n = 0, 1, \ldots]$ AS 805 (23.1.19)

2. $E_n(0) = -E_n(1) = -2(n+1)^{-1}\left(2^{n+1} - 1\right)B_{n+1}$ $[n = 1, 2, \ldots]$ AS 805 (23.1.20)

3. $E_n\left(\frac{1}{2}\right) = 2^{-n}E_n$ $[n = 0, 1, \ldots]$ AS 805 (23.1.21)

4. $E_{2n-1}\left(\frac{1}{3}\right) = -E_{2n-1}\left(\frac{2}{3}\right) = -(2n)^{-1}\left(1 - 3^{1-2n}\right)\left(2^{2n} - 1\right)B_{2n}$

                                                     $[[n = 1, 2, \ldots]]$ AS 806 (23.1.22)

9.7 Constants

9.71 Bernoulli numbers

- $B_0 = 1$
- $B_1 = -\frac{1}{2}$
- $B_2 = \frac{1}{6}$
- $B_4 = \frac{1}{30}$
- $B_6 = -\frac{1}{42}$
- $B_8 = \frac{1}{30}$
- $B_{10} = -\frac{5}{66}$
- $B_{12} = -\frac{691}{2730}$
- $B_{14} = -\frac{7}{6}$
- $B_{16} = -\frac{3617}{510}$

- $B_{18} = \frac{43\,867}{798}$
- $B_{20} = -\frac{174\,611}{330}$
- $B_{22} = \frac{854\,513}{138}$
- $B_{24} = \frac{236\,364\,091}{2730}$
- $B_{26} = \frac{8\,553\,103}{6}$
- $B_{28} = -\frac{23\,749\,461\,029}{870}$
- $B_{30} = \frac{8\,615\,841\,276\,005}{14\,322}$
- $B_{32} = -\frac{7\,709\,321\,041\,217}{510}$
- $B_{34} = \frac{2\,577\,687\,858\,367}{6}$

9.72 Euler numbers

- $E_0 = 1$
- $E_2 = -1$
- $E_4 = 5$
- $E_6 = -61$
- $E_8 = 1385$
- $E_{10} = -50\,521$

- $E_{12} = 2\,702\,765$
- $E_{14} = -199\,360\,981$
- $E_{16} = 19\,391\,512\,145$
- $E_{18} = -2\,404\,879\,675\,441$
- $E_{20} = 370\,371\,188\,237\,525$

The Bernoulli and Euler numbers of odd index (with the exception of B_1) are equal to zero.

9.73 Euler's and Catalan's constants

Euler's constant

$$C = 0.577\,215\,664\,901\,532\,860\,606\,512\ldots \qquad\qquad\qquad \text{(cf. \textbf{8.367})}$$

Catalan's constant

$$G = \sum_{k=0}^{\infty} \frac{(-1)^k}{(2k+1)^2} = 0.915\,965\,594\ldots$$

9.74^{10} Stirling numbers

9.740   The **Stirling number of the first kind** $S_n^{(m)}$ is defined by the requirement that $(-1)^{n-m} S_n^{(m)}$ is the number of permutations of n symbols which have exactly m cycles.   AS 824 (23.1.3)

9.741   Generating functions:

1. $$x(x-1)\cdots(x-n+1) = \sum_{m=0}^{n} S_n^{(m)} x^m \qquad\qquad\qquad\qquad \text{AS 824 (24.1.3)}$$

2. $$\{\ln(1+x)\}^m = m! \sum_{n=m}^{\infty} S_n^{(m)} \frac{x^n}{n!} \qquad\qquad [|x| < 1] \qquad\qquad \text{AS 824 (24.1.3)}$$

9.742   Recurrence relations:

1.^{8} $$S_{n+1}^{(m)} = S_n^{(m-1)} - n S_n^{(m)}; \quad S_n^{(0)} = \delta_{0n}; \quad S_n^{(1)} = (-1)^{n-1}(n-1)!; \quad S_n^{(n)} = 1$$
$$[n \geq m \geq 1] \qquad\qquad \text{AS 824 (24.1.3)}$$

2. $$\binom{m}{r} S_n^{(m)} = \sum_{k=m-r}^{n-r} \binom{n}{k} S_{n-k}^{(r)} S_k^{(m+r)} \qquad\qquad [n \geq m \geq r] \qquad\qquad \text{AS 824 (24.1.3)}$$

9.743   Functional relations and properties

1. $$x(x-h)(x-2h)\cdots(x-mh+h) = \frac{h^m \Gamma\left(\frac{x}{h}+1\right)}{\Gamma\left(\frac{x}{h}-m+1\right)} = h^m \sum_{k=1}^{m} \left(\frac{x}{h}\right)^k S_k^{(m)}$$

2. $$[(x+1)(x+2)\cdots(x+m)]^{-1} = \left[\binom{x+m}{m} m!\right]^{-1} = \left[\sum_{k=1}^{p} (x+m)^k S_k^{(m)}\right]^{-1}$$

3. $$[(x+h)(x+2h)\cdots(x+mh)]^{-1} = \frac{\Gamma\left(\frac{x}{h}+1\right)}{h^m \Gamma\left(\frac{x}{h}+m+1\right)} = \left[h^m \sum_{k=1}^{m} \left(\frac{x}{h}+m\right)^k S_k^{(m)}\right]^{-1}$$

9.744   The Stirling number of the second kind $\mathfrak{S}_n^{(m)}$ is the number of ways of partitioning a set of n elements into m non-empty subsets.

9.745   Generating functions:

1. $x^n = \sum_{m=0}^{n} \mathfrak{S}_n^{(m)} x(x-1)\cdots(x-m+1)$ AS 824 (24.1.4)

2. $(e^x - 1)^m = m! \sum_{n=m}^{\infty} \mathfrak{S}_n^{(m)} \dfrac{x^n}{n!}$ AS 824 (24.1.4)

3. $[(1-x)(1-2x)\cdots(1-mx)]^{-1} = \sum_{n=m}^{\infty} \mathfrak{S}_n^{(m)} x^{n-m}$ $\left[|x| < m^{-1}\right]$ AS 824 (24.1.4)

9.746   Closed form expression:

1. $\mathfrak{S}_n^{(m)} = \dfrac{1}{m!} \sum_{k=0}^{m} (-1)^{m-k} \dbinom{m}{k} k^n$ AS 824 (24.1.4)

9.747   Recurrence relations:

1.[8] $\mathfrak{S}_{n+1}^{(m)} = m\mathfrak{S}_n^{(m)} + \mathfrak{S}_n^{(m-1)}, \quad \mathfrak{S}_n^{(0)} = \delta_{0n}, \quad \mathfrak{S}_n^{(1)} = \mathfrak{S}_n^{(n)} = 1$

                                           $[n \geq m \geq 1]$ AS 825(24.1.4)

2. $\dbinom{m}{r} \mathfrak{S}_n^{(m)} = \sum_{k=m-r}^{n-r} \dbinom{n}{k} \mathfrak{S}_{n-k}^{(r)} \mathfrak{S}_k^{(m-r)}$ $[n \geq m \geq r]$ AS 825 (24.1.4)

3. $S_n^{(m)} = \sum_{k=0}^{n-m} (-1)^k \dbinom{n-1+k}{n-m+k} \dbinom{2n-m}{n-m-k} \mathfrak{S}_{n-m+k}^{(k)}$ AS 824 (24.1.3)

9.748[7]   Particular values:

Stirling numbers of the first kind $S_n^{(m)}$

m	$S_1^{(m)}$	$S_2^{(m)}$	$S_3^{(m)}$	$S_4^{(m)}$	$S_5^{(m)}$	$S_6^{(m)}$	$S_7^{(m)}$	$S_8^{(m)}$	$S_9^{(m)}$
1	1	-1	2	-6	24	-120	720	-5040	40320
2		1	-3	11	-50	274	-1764	13068	-109584
3			1	-6	35	-225	1624	-13132	118121
4				1	-10	85	-735	6769	-67284
5					1	-15	175	-1960	22449
6						1	-21	332	-4536
7							1	-28	546
8								1	-36
9									1

Stirling numbers of the second kind $\mathfrak{S}_n^{(m)}$

m	$\mathfrak{S}_1^{(m)}$	$\mathfrak{S}_2^{(m)}$	$\mathfrak{S}_3^{(m)}$	$\mathfrak{S}_4^{(m)}$	$\mathfrak{S}_5^{(m)}$	$\mathfrak{S}_6^{(m)}$	$\mathfrak{S}_7^{(m)}$	$\mathfrak{S}_8^{(m)}$	$\mathfrak{S}_9^{(m)}$
1	1	1	1	1	1	1	1	1	1
2		1	3	7	15	31	63	127	255
3			1	6	25	90	301	966	3025
4				1	10	65	350	1701	7770
5					1	15	140	1050	6951
6						1	21	266	2646
7							1	28	462
8								1	36
9									1

9.749[8] Relationship between Stirling numbers of the first kind and derivatives of $(\ln x)^{-m}$:

1. $$\frac{d^n}{dx^n}\left(\frac{1}{\ln^m x}\right) = \frac{1}{\ln^m x}\sum_{k=1}^{n}\frac{(-1)^k (m)_k S_n^{(k)}}{x^n \ln^k x}$$

$$\text{where } (m)_k = \Gamma(m+k)/\Gamma(m), \qquad [m, n \text{ are positive integers}]$$

10 Vector Field Theory

10.1–10.8 Vectors, Vector Operators, and Integral Theorems

10.11 Products of vectors

Let $\mathbf{a} = (a_1, a_2, a_3)$, $\mathbf{b} = (b_1, b_2, b_3)$, and $\mathbf{c} = (c_1, c_2, c_2)$ be arbitrary vectors, and $\mathbf{i}, \mathbf{j}, \mathbf{k}$ be the set of orthogonal unit vectors in terms of which the components of $\mathbf{a}$, $\mathbf{b}$, and $\mathbf{c}$ are expressed. Two different products involving pairs of vectors are defined, namely, the scalar product, written $\mathbf{a} \cdot \mathbf{b}$, and the vector product, written either $\mathbf{a} \times \mathbf{b}$ or $\mathbf{a} \wedge \mathbf{b}$. Their properties are as follows:

1. $\mathbf{a} \cdot \mathbf{b} = a_1 b_1 + a_2 b_2 + a_3 b_3$ (scalar product)

2. $\mathbf{a} \times \mathbf{b} = \begin{vmatrix} \mathbf{i} & \mathbf{j} & \mathbf{k} \\ a_1 & a_2 & a_3 \\ b_1 & b_2 & b_3 \end{vmatrix}$ (vector product)

3. $\mathbf{a} \times \mathbf{b} \cdot \mathbf{c} = \begin{vmatrix} a_1 & a_2 & a_3 \\ b_1 & b_2 & b_3 \\ c_1 & c_2 & c_3 \end{vmatrix}$ (triple scalar product)

4. $\mathbf{a} \times (\mathbf{b} \times \mathbf{c}) = (\mathbf{a} \cdot \mathbf{c}) \mathbf{b} - (\mathbf{a} \cdot \mathbf{b}) \mathbf{c}$ (triple vector product)

10.12 Properties of scalar product

1. $\mathbf{a} \cdot \mathbf{b} = \mathbf{b} \cdot \mathbf{a}$ (commutative)

2. $\mathbf{a} \times \mathbf{b} \cdot \mathbf{c} = \mathbf{b} \times \mathbf{c} \cdot \mathbf{a} = \mathbf{c} \times \mathbf{a} \cdot \mathbf{b} = -\mathbf{a} \times \mathbf{c} \cdot \mathbf{b} = -\mathbf{b} \times \mathbf{a} \cdot \mathbf{c} = -\mathbf{c} \times \mathbf{b} \cdot \mathbf{a}$.

    Note: $\mathbf{a} \times \mathbf{b} \cdot \mathbf{c}$ is also written $[\mathbf{a}, \mathbf{b}, \mathbf{c}]$; thus (2) may also be written

3. $[\mathbf{a}, \mathbf{b}, \mathbf{c}] = [\mathbf{b}, \mathbf{c}, \mathbf{a}] = [\mathbf{c}, \mathbf{a}, \mathbf{b}] = -[\mathbf{a}, \mathbf{c}, \mathbf{b}] = -[\mathbf{b}, \mathbf{a}, \mathbf{c}] = -[\mathbf{c}, \mathbf{b}, \mathbf{a}]$

10.13 Properties of vector product

1. $\mathbf{a} \times \mathbf{b} = -\mathbf{b} \times \mathbf{a}$ (anticommutative)

2. $\mathbf{a} \times (\mathbf{b} \times \mathbf{c}) = -\mathbf{a} \times (\mathbf{c} \times \mathbf{b}) = -(\mathbf{b} \times \mathbf{c}) \times \mathbf{a}$

3. $\mathbf{a} \times (\mathbf{b} \times \mathbf{c}) + \mathbf{b} \times (\mathbf{c} \times \mathbf{a}) + \mathbf{c} \times (\mathbf{a} \times \mathbf{b}) = 0$

10.14 Differentiation of vectors

If $\mathbf{a}(t) = (a_1(t), a_2(t), a_3(t))$, $\mathbf{b}(t) = (b_1(t), b_2(t), b_3(t))$, $\mathbf{c}(t) = (c_1(t), c_2(t), c_3(t))$, $\phi(t)$ is a scalar and all functions of t are differentiable, then

1. $$\frac{d\mathbf{a}}{dt} = \frac{da_1}{dt}\mathbf{i} + \frac{da_2}{dt}\mathbf{j} + \frac{da_3}{dt}\mathbf{k}$$

2. $$\frac{d}{dt}(\mathbf{a} + \mathbf{b}) = \frac{d\mathbf{a}}{dt} + \frac{d\mathbf{b}}{dt}$$

3. $$\frac{d}{dt}(\phi\mathbf{a}) = \frac{d\phi}{dt}\mathbf{a} + \phi\frac{d\mathbf{a}}{dt}$$

4. $$\frac{d}{dt}(\mathbf{a} \cdot \mathbf{b}) = \frac{d\mathbf{a}}{dt} \cdot \mathbf{b} + \mathbf{a} \cdot \frac{d\mathbf{b}}{dt}$$

5. $$\frac{d}{dt}(\mathbf{a} \times \mathbf{b}) = \frac{d\mathbf{a}}{dt} \times \mathbf{b} + \mathbf{a} \times \frac{d\mathbf{b}}{dt}$$

6. $$\frac{d}{dt}(\mathbf{a} \times \mathbf{b} \cdot \mathbf{c}) = \frac{d\mathbf{a}}{dt} \times \mathbf{b} \cdot \mathbf{c} + \mathbf{a} \times \frac{d\mathbf{b}}{dt} \cdot \mathbf{c} + \mathbf{a} \times \mathbf{b} \cdot \frac{d\mathbf{c}}{dt}$$

7. $$\frac{d}{dt}\{\mathbf{a} \times (\mathbf{b} \times \mathbf{c})\} = \frac{d\mathbf{a}}{dt} \times (\mathbf{b} \times \mathbf{c}) + \mathbf{a} \times \left(\frac{d\mathbf{b}}{dt} \times \mathbf{c}\right) + \mathbf{a} \times \left(\mathbf{b} \times \frac{d\mathbf{c}}{dt}\right)$$

10.21 Operators grad, div, and curl

In cartesian coordinates $O\{x_1, x_2, x_3\}$, in which system it is convenient to denote the triad of unit vectors by $\mathbf{e}_1, \mathbf{e}_2, \mathbf{e}_3$, the vector operator ∇, called either "del" or "nabla", has the form

1. $$\nabla \equiv \mathbf{e}_1\frac{\partial}{\partial x_1} + \mathbf{e}_2\frac{\partial}{\partial x_2} + \mathbf{e}_3\frac{\partial}{\partial x_3}$$

    If $\Phi(x, y, z)$ is any differentiable scalar function, the gradient of Φ, written grad Φ, is

2. $$\text{grad } \Phi \equiv \nabla\Phi = \frac{\partial\Phi}{\partial x_1}\mathbf{e}_1 + \frac{\partial\Phi}{\partial x_2}\mathbf{e}_2 + \frac{\partial\Phi}{\partial x_3}\mathbf{e}_3$$

    The divergence of the differentiable vector function $\mathbf{f} = (f_1, f_2, f_3)$, written div $\mathbf{f}$, is

3. $$\text{div } f \equiv \nabla \cdot \mathbf{f} = \frac{\partial f_1}{\partial x_1} + \frac{\partial f_2}{\partial x_2} + \frac{\partial f_3}{\partial x_3}$$

    The curl, or rotation, of the differentiable vector function $\mathbf{f} = (f_1, f_2, f_3)$, written either curl $\mathbf{f}$ or rot $\mathbf{f}$, is

4. $$\text{curl } \mathbf{f} \equiv \text{rot } \mathbf{f} \equiv \nabla \times \mathbf{f} = \left(\frac{\partial f_3}{\partial x_2} - \frac{\partial f_2}{\partial x_3}\right)\mathbf{e}_1 + \left(\frac{\partial f_1}{\partial x_3} - \frac{\partial f_3}{\partial x_1}\right)\mathbf{e}_2 + \left(\frac{\partial f_2}{\partial x_1} - \frac{\partial f_1}{\partial x_2}\right)\mathbf{e}_3,$$

    or equivalently,

    $$\text{curl } \mathbf{f} = \begin{vmatrix} \mathbf{e}_1 & \mathbf{e}_2 & \mathbf{e}_2 \\ \frac{\partial}{\partial x_1} & \frac{\partial}{\partial x_2} & \frac{\partial}{\partial x_3} \\ f_1 & f_2 & f_3 \end{vmatrix}$$

10.31 Properties of the operator ∇

Let $\Phi\,(x_1, x_2, x_3)$, $\Psi\,(x_1, x_2, x_3)$ be any two differentiable scalar functions, $\mathbf{f}\,(x_1, x_2, x_3)$, $\mathbf{g}\,(x_1, x_2, x_3)$ any two differentiable vector functions, and $\mathbf{a}$ an arbitrary vector. Define the scalar operator ∇^2, called the Laplacian, by

$$\nabla^2 \equiv \frac{\partial^2}{\partial x_1^2} + \frac{\partial^2}{\partial x_2^2} + \frac{\partial^2}{\partial x_3^2}$$

Then, in terms of the operator ∇, we have the following:

MF I 114

1. $\nabla(\Phi + \Psi) = \nabla\,\Phi + \nabla\,\Psi$

2. $\nabla(\Phi\Psi) = \Phi\,\nabla\,\Psi + \Psi\,\nabla\,\Phi$

3. $\nabla\,(\mathbf{f}\cdot\mathbf{g}) = (\mathbf{f}\cdot\nabla)\,\mathbf{g} + (\mathbf{g}\cdot\nabla)\,\mathbf{f} + \mathbf{f}\times(\nabla\times\mathbf{g}) + \mathbf{g}\times(\nabla\times\mathbf{f})$

4. $\nabla\cdot(\Phi\mathbf{f}) = \Phi\,(\nabla\cdot\mathbf{f}) + \mathbf{f}\cdot\nabla\,\Phi$

5. $\nabla\cdot(\mathbf{f}\times\mathbf{g}) = \mathbf{g}\cdot(\nabla\times\mathbf{f}) - \mathbf{f}\cdot(\nabla\times\mathbf{g})$

6. $\nabla\times(\Phi\mathbf{f}) = \Phi\,(\nabla\times\mathbf{f}) + (\nabla\,\Phi)\times\mathbf{f}$

7. $\nabla\times(\mathbf{f}\times\mathbf{g}) = \mathbf{f}\,(\nabla\cdot\mathbf{g}) - \mathbf{g}\,(\nabla\cdot\mathbf{f}) + (\mathbf{g}\cdot\nabla)\,\mathbf{f} - (\mathbf{f}\cdot\nabla)\,\mathbf{g}$

8. $\nabla\times(\nabla\times\mathbf{f}) = \nabla\,(\nabla\cdot\mathbf{f}) - \nabla^2\mathbf{f}$

9. $\nabla\times(\nabla\,\Phi) \equiv \mathbf{0}$

10. $\nabla\cdot(\nabla\times\mathbf{f}) \equiv 0$

11.[10] $\nabla^2(\Phi\Psi) = \Phi\nabla^2\Psi + 2\,(\nabla\,\Phi)\cdot(\nabla\,\Psi) + \Psi\nabla^2\Phi$

The equivalent results in terms of grad, div, and curl are as follows:

1.$'$ $\operatorname{grad}(\Phi + \Psi) = \operatorname{grad}\Phi + \operatorname{grad}\Psi$

2.$'$ $\operatorname{grad}(\Phi\Psi) = \Phi\operatorname{grad}\Psi + \Psi\operatorname{grad}\Phi$

3.$'$ $\operatorname{grad}\,(\mathbf{f}\cdot\mathbf{g}) = (\mathbf{f}\cdot\operatorname{grad})\,\mathbf{g} + (\mathbf{g}\cdot\operatorname{grad})\,\mathbf{f} + \mathbf{f}\times\operatorname{curl}\mathbf{g} + \mathbf{g}\times\operatorname{curl}\mathbf{f}$

4.$'$ $\operatorname{div}\,(\Phi\mathbf{f}) = \Phi\operatorname{div}\mathbf{f} + \mathbf{f}\cdot\operatorname{grad}\Phi$

5.$'$ $\operatorname{div}\,(\mathbf{f}\times\mathbf{g}) = \mathbf{g}\cdot\operatorname{curl}\mathbf{f} - \mathbf{f}\cdot\operatorname{curl}\mathbf{g}$

6.$'$ $\operatorname{curl}\,(\Phi\mathbf{f}) = \Phi\operatorname{curl}\mathbf{f} + \operatorname{grad}\Phi\times\mathbf{f}$

7.$'$ $\operatorname{curl}\,(\mathbf{f}\times\mathbf{g}) = \mathbf{f}\operatorname{div}\mathbf{g} - \mathbf{g}\operatorname{div}\mathbf{f} + (\mathbf{g}\cdot\operatorname{grad})\,\mathbf{f} - (\mathbf{f}\cdot\operatorname{grad})\,\mathbf{g}$

8.$'$ $\operatorname{curl}\,(\operatorname{curl}\mathbf{f}) = \operatorname{grad}\,(\operatorname{div}\mathbf{f}) - \nabla^2\mathbf{f}$

9.$'$ $\operatorname{curl}\,(\operatorname{grad}\Phi) \equiv \mathbf{0}$

10.$'$ $\operatorname{div}\,(\operatorname{curl}\mathbf{f}) \equiv 0$

11.$'$ $\nabla^2(\Phi\Psi) = \Phi\nabla^2\Psi + 2\operatorname{grad}\Phi\cdot\operatorname{grad}\Psi + \Psi\nabla^2\Phi$

The expression $(\mathbf{a}\cdot\nabla)$ or, equivalently $(\mathbf{a}\cdot\operatorname{grad})$, defined by

$$(\mathbf{a}\cdot\nabla) \equiv a_1\frac{\partial}{\partial x_1} + a_2\frac{\partial}{\partial x_2} + a_3\frac{\partial}{\partial x_3},$$

is the directional derivative operator in the direction of vector $\mathbf{a}$.

10.41 Solenoidal fields

A vector field $\mathbf{f}$ is said to be solenoidal if div $\mathbf{f} \equiv 0$. We have the following representation.

10.411 *Representation theorem for vector Helmholtz equation.* If u is a solution of the scalar Helmholtz equation

$$\nabla^2 u + \lambda^2 u = 0,$$

and $\mathbf{m}$ is a constant unit vector, then the vectors

$$\mathbf{X} = \operatorname{curl}(\mathbf{m}u), \qquad \mathbf{Y} = \frac{1}{\lambda}\operatorname{curl}\mathbf{X}$$

are independent solutions of the vector Helmholtz equation

$$\nabla^2\mathbf{H} + \lambda^2\mathbf{H} = \mathbf{0}$$

involving a solenoidal vector $\mathbf{H}$. The general solution of the equation is

$$\mathbf{H} = \operatorname{curl}(\mathbf{m}u) + \frac{1}{\lambda}\operatorname{curl}\operatorname{curl}(\mathbf{m}u)$$

10.51–10.61 Orthogonal curvilinear coordinates

Consider a transformation from the cartesian coordinates $O\{x_1, x_2, x_3\}$ to the general orthogonal curvilinear coordinates $O\{u_1, u_2, u_3\}$:

$$x_1 = x_1(u_1, u_2, u_3), \qquad x_2 = x_2(u_1, u_2, u_3), \qquad x_3 = x_3(u_1, u_2, u_3)$$

Then,

1. $\quad dx_i = \dfrac{\partial x_i}{\partial u_1}\,du_1 + \dfrac{\partial x_i}{\partial u_2}\,du_2 + \dfrac{\partial x_i}{\partial u_3}\,du_3 \qquad\qquad (i = 1, 2, 3),$

   and the length element dl may be determined from

2. $\quad dl^2 = g_{11}\,du_1^2 + g_{22}\,du_2^2 + g_{33}\,du_3^2 + 2g_{23}\,du_2\,du_3 + 2g_{31}\,du_3\,du_1 + 2g_{12}\,du_1\,du_2,$

   where

3.3 $\quad g_{ij} = \dfrac{\partial x_1}{\partial u_i}\dfrac{\partial x_1}{\partial u_j} + \dfrac{\partial x_2}{\partial u_i}\dfrac{\partial x_2}{\partial u_j} + \dfrac{\partial x_3}{\partial u_i}\dfrac{\partial x_3}{\partial u_j} = g_{ji}, \qquad g_{ij} = 0, \quad i \neq j.$

   provided the Jacobian of the transformation

4. $\quad J = \begin{vmatrix} \dfrac{\partial x_1}{\partial u_1} & \dfrac{\partial x_2}{\partial u_1} & \dfrac{\partial x_3}{\partial u_1} \\[6pt] \dfrac{\partial x_1}{\partial u_2} & \dfrac{\partial x_2}{\partial u_2} & \dfrac{\partial x_3}{\partial u_2} \\[6pt] \dfrac{\partial x_1}{\partial u_3} & \dfrac{\partial x_2}{\partial u_3} & \dfrac{\partial x_3}{\partial u_3} \end{vmatrix}$

   does not vanish (see **14.313**).

   Define the metrical coefficients

5. $\quad h_1 = \sqrt{g_{11}}, \quad h_2 = \sqrt{g_{22}}, \quad h_3 = \sqrt{g_{33}};$

   then the volume element dV in orthogonal curvilinear coordinates is

6. $\quad dV = h_1 h_2 h_3\,du_1\,du_2\,du_3,$

   and the surface elements of area ds_i on the surfaces $u_i = $ constant, for $i = 1, 2, 3$, are

7. $\quad ds_1 = h_2 h_3\,du_2\,du_3, \quad ds_2 = h_1 h_3\,du_1\,du_3, \quad ds_3 = h_1 h_2\,du_1\,du_2$

   Denote by $\mathbf{e}_1, \mathbf{e}_2$, and $\mathbf{e}_3$ the triad of orthogonal unit vectors that are tangent to the u_1, u_2, and u_3 coordinate lines through any given point P, and choose their sense so that they form a right-handed set in this order. Then in terms of this triad of vectors and the components f_{u_1}, f_{u_2}, and f_{u_3} of $\mathbf{f}$ along the coordinate line,

8. $\mathbf{f} = f_{u_1}\mathbf{e}_1 + f_{u_2}\mathbf{e}_2 + f_{u_3}\mathbf{e}_3$ MF I 115

10.611 $\nabla\Phi$, div $\mathbf{f}$, curl $\mathbf{f}$, and ∇^2 *in general orthogonal curvilinear coordinates.*

1. $\operatorname{grad}\Phi = \dfrac{\mathbf{e}_1}{h_1}\dfrac{\partial\Phi}{\partial u_1} + \dfrac{\mathbf{e}_2}{h_2}\dfrac{\partial\Phi}{\partial u_2} + \dfrac{\mathbf{e}_3}{h_3}\dfrac{\partial\Phi}{\partial u_3}$

2.[3] $\operatorname{div}\mathbf{f} = \dfrac{1}{h_1 h_2 h_3}\left(\dfrac{\partial}{\partial u_1}(h_2 h_3 f_{u_1}) + \dfrac{\partial}{\partial u_2}(h_3 h_1 f_{u_2}) + \dfrac{\partial}{\partial u_3}(h_1 h_2 f_{u_3})\right)$

3. $\operatorname{curl}\mathbf{f} = \dfrac{1}{h_1 h_2 h_3}\begin{vmatrix} h_1\mathbf{e}_1 & h_2\mathbf{e}_2 & h_3\mathbf{e}_3 \\ \frac{\partial}{\partial u_1} & \frac{\partial}{\partial u_2} & \frac{\partial}{\partial u_3} \\ h_1 f_{u_1} & h_2 f_{u_2} & h_3 f_{u_3} \end{vmatrix}$

4. $\nabla^2 \equiv \dfrac{1}{h_1 h_2 h_3}\left(\dfrac{\partial}{\partial u_1}\left(\dfrac{h_2 h_3}{h_1}\dfrac{\partial}{\partial u_1}\right) + \dfrac{\partial}{\partial u_2}\left(\dfrac{h_3 h_1}{h_2}\dfrac{\partial}{\partial u_2}\right) + \dfrac{\partial}{\partial u_3}\left(\dfrac{h_1 h_2}{h_3}\dfrac{\partial}{\partial u_3}\right)\right)$ MF I 21-31

10.612   *Cylindrical polar coordinates.* In terms of the coordinates $O\{r,\phi,z\}$, that is, $u_1 = r$, $u_2 = \phi$, $u_3 = z$, where $x_1 = r\cos\phi$, $x_2 = r\sin\phi$, $x_3 = z$ for $-\pi < \phi \le \pi$, it follows that

1. $h_1 = 1, \quad h_2 = r, \quad h_3 = 1,$

       and

2. $\operatorname{grad}\Phi = \dfrac{\partial\Phi}{\partial r}\mathbf{e}_r + \dfrac{1}{r}\dfrac{\partial\Phi}{\partial\phi}\mathbf{e}_\phi + \dfrac{\partial\Phi}{\partial z}\mathbf{e}_z,$

3. $\operatorname{div}\mathbf{f} = \dfrac{1}{r}\dfrac{\partial}{\partial r}(rf_r) + \dfrac{1}{r}\dfrac{\partial f_\phi}{\partial\phi} + \dfrac{\partial f_z}{\partial z},$

4. $\operatorname{curl}\mathbf{f} = \dfrac{1}{r}\begin{vmatrix} \mathbf{e}_r & r\mathbf{e}_\phi & \mathbf{e}_z \\ \frac{\partial}{\partial r} & \frac{\partial}{\partial\phi} & \frac{\partial}{\partial z} \\ f_r & rf_\phi & f_z \end{vmatrix},$

5. $\nabla^2 \equiv \dfrac{1}{r}\dfrac{\partial}{\partial r}\left(r\dfrac{\partial}{\partial r}\right) + \dfrac{1}{r^2}\dfrac{\partial^2}{\partial\phi^2} + \dfrac{\partial^2}{\partial z^2}$ MF I 116

10.613   *Spherical polar coordinates.* In terms of the coordinates $O\{r,\theta,\phi\}$, that is, $u_1 = r$, $u_2 = \theta$, $u_3 = \phi$, where $x_1 = r$, $\sin\theta\cos\phi$, $x_2 = r\sin\theta$, $\sin\phi$, $x_3 = r$, $\cos\theta$, for $0 \le \theta \le \pi$, $-\pi < \phi \le \pi$, we have

1. $h_1 = 1, \quad h_2 = r, \quad h_3 = r\sin\theta d,$

       and

2.[10] $\operatorname{grad}\Phi = \dfrac{\partial\Phi}{\partial r}\mathbf{e}_r + \dfrac{1}{r}\dfrac{\partial\Phi}{\partial\theta}\mathbf{e}_\theta + \dfrac{1}{r\sin\theta}\dfrac{\partial\Phi}{\partial\phi}\mathbf{e}_\phi,$

3. $\operatorname{div}\mathbf{f} = \dfrac{1}{r^2}\dfrac{\partial}{\partial r}(r^2 f_r) + \dfrac{1}{r\sin\theta}\dfrac{\partial}{\partial\theta}(f_\theta\sin\theta) + \dfrac{1}{r\sin\theta}\dfrac{\partial f_\phi}{\partial\phi},$

4. $\operatorname{curl}\mathbf{f} = \dfrac{1}{r^2\sin\theta}\begin{vmatrix} \mathbf{e}_r & r\mathbf{e}_\theta & r\sin\theta\mathbf{e}_\phi \\ \frac{\partial}{\partial r} & \frac{\partial}{\partial\theta} & \frac{\partial}{\partial\phi} \\ f_r & rf_\theta & r\sin\theta f_\phi \end{vmatrix},$

5. $\nabla^2 \equiv \dfrac{1}{r^2}\dfrac{\partial}{\partial r}\left(r^2\dfrac{\partial}{\partial r}\right) + \dfrac{1}{r^2\sin\theta}\dfrac{\partial}{\partial\theta}\left(\sin\theta\dfrac{\partial}{\partial\theta}\right) + \dfrac{1}{r^2\sin^2\theta}\dfrac{\partial^2}{\partial\phi^2}$ MF I 116

Special Orthogonal Curvilinear Coordinates and their Metrical Coefficients h_1, h_2, h_3

10.614   *Elliptic cylinder coordinates* $O\{u_1, u_2, u_3\}$.

1. $x_1 = u_1 u_2, \qquad x_2 = \sqrt{\left(u_1^2 - c^2\right)\left(1 - u_2^2\right)}, \qquad x_3 = u_3$

2. $h_1 = \sqrt{\dfrac{u_1^2 - c^2 u_2^2}{u_1^2 - c^2}}, \qquad h_2 = \sqrt{\dfrac{u_1^2 - c^2 u_2^2}{1 - u_2^2}}, \qquad h_3 = 1$ MF I 657

10.615   *Parabolic cylinder coordinates* $O\{u_1, u_2, u_3\}$.

1. $x_1 = \dfrac{1}{2}\left(u_1^2 - u_2^2\right), \qquad x_2 = u_1 u_2, \qquad x_3 = u_3$

2. $h_1 = \sqrt{u_1^2 + u_2^2}, \qquad h_2 = \sqrt{u_1^2 + u_2^2}, \qquad h_3 = 1$ MF I 658

10.616   *Conical coordinates* $O\{u_1, u_2, u_3\}$.

1. $x_1 = \dfrac{u_1}{a}\sqrt{\left(a^2 - u_2^2\right)\left(a^2 + u_3^2\right)}, \qquad x_2 = \dfrac{u_1}{b}\sqrt{\left(b^2 + u_2^2\right)\left(b^2 - u_3^2\right)}, \qquad x_3 = \dfrac{u_1 u_2 u_3}{ab}$

$$\text{with } a^2 + b^2 = 1$$

2. $h_1 = 1, \qquad h_2 = u_1\sqrt{\dfrac{u_2^2 + u_3^2}{\left(a^2 - u_2^2\right)\left(b^2 + u_2^2\right)}}, \qquad h_3 = u_1\sqrt{\dfrac{u_2^2 + u_3^2}{\left(a^2 + u_3^2\right)\left(b^2 - u_3^2\right)}}$ MF I 659

10.617   *Rotational parabolic coordinates* $O\{u_1, u_2, u_3\}$.

1. $x_1 = u_1 u_2 u_3, \qquad x_2 = u_1 u_2\sqrt{1 - u_3^2}, \qquad x_3 = \dfrac{1}{2}\left(u_1^2 - u_2^2\right)$

2. $h_1 = \sqrt{u_1^2 + u_2^2}, \qquad h_2 = \sqrt{u_1^2 + u_2^2}, \qquad h_3 = \dfrac{u_1 u_2}{\sqrt{1 - u_3^2}}$ MF I 660

10.618   *Rotational prolate spheroidal coordinates* $O\{u_1, u_2, u_3\}$.

1. $x_1 = \sqrt{\left(u_1^2 - a^2\right)\left(1 - u_2^2\right)}, \qquad x_2 = \sqrt{\left(u_1^2 - a^2\right)\left(1 - u_2^2\right)\left(1 - u_3^2\right)}, \qquad x_3 = u_1 u_2$

2. $h_1 = \sqrt{\dfrac{u_1^2 - a^2 u_2^2}{u_1^2 - a^2}}, \qquad h_2 = \sqrt{\dfrac{u_1^2 - a^2 u_2^2}{1 - u_2^2}}, \qquad h_3 = \sqrt{\dfrac{\left(u_1^2 - a^2\right)\left(1 - u_2^2\right)}{1 - u_3^2}}$ MF I 661

10.619   *Rotational oblate spheroidal coordinates* $O\{u_1, u_2, u_3\}$.

1. $x_1 = u_3\sqrt{\left(u_1^2 + a^2\right)\left(1 - u_2^2\right)}, \qquad x_2 = \sqrt{\left(u_1^2 + a^2\right)\left(1 - u_2^2\right)\left(1 - u_3^2\right)}, \qquad x_3 = u_1 u_2$

2. $h_1 = \sqrt{\dfrac{u_1^2 + a^2 u_2^2}{u_1^2 + a^2}}, \qquad h_2 = \sqrt{\dfrac{u_1^2 + a^2 u_2^2}{1 - u_2^2}}, \qquad h_3 = \sqrt{\dfrac{\left(u_1^2 + a^2\right)\left(1 - u_2^2\right)}{1 - u_3^2}}$ MF I 662

10.620   *Ellipsoidal coordinates* $O\{u_1, u_2, u_3\}$.

1. $$x_1 = \sqrt{\frac{(u_1^2 - a^2)(u_2^2 - a^2)(u_3^2 - a^2)}{a^2(a^2 - b^2)}}, \quad x_2 = \sqrt{\frac{(u_1^2 - b^2)(u_2^2 - b^2)(u_3^2 - b^2)}{b^2(b^2 - a^2)}}, \quad x_3 = \frac{u_1 u_2 u_3}{ab}$$

2. $$h_1 = \sqrt{\frac{(u_1^2 - u_2^2)(u_1^2 - u_3^2)}{(u_1^2 - a^2)(u_1^2 - b^2)}}, \quad h_2 = \sqrt{\frac{(u_2^2 - u_1^2)(u_2^2 - u_3^2)}{(u_2^2 - a^2)(u_2^2 - b^2)}}, \quad h_3 = \sqrt{\frac{(u_3^2 - u_1^2)(u_3^2 - u_2^2)}{(u_3^2 - a^2)(u_3^2 - b^2)}}$$

MF I 663

10.621   *Paraboloidal coordinates $O\{u_1, u_2, u_3\}$.*

1. $$x_1 = \sqrt{\frac{(u_1^2 - a^2)(u_2^2 - a^2)(u_3^2 - a^2)}{a^2 - b^2}}, \quad x_2 = \sqrt{\frac{(u_1^2 - b^2)(u_2^2 - b^2)(u_3^2 - b^2)}{b^2 - a^2}},$$
$$x_3 = \tfrac{1}{2}\left(u_1^2 + u_2^2 + u_3^2 - a^2 - b^2\right)$$

2. $$h_1 = \sqrt{\frac{(u_1^2 - u_2^2)(u_1^2 - u_3^2)}{(u_1^2 - a^2)(u_1^2 - b^2)}}, \quad h_2 = u_2\sqrt{\frac{(u_3^2 - u_1^2)(u_3^2 - u_2^2)}{(u_2^2 - a^2)(u_2^2 - b^2)}}, \quad h_3 = u_3\sqrt{\frac{(u_3^2 - u_1^2)(u_3^2 - u_2^2)}{(u_3^2 - a^2)(u_3^2 - b^2)}}$$

MF I 664

10.622   *Bispherical coordinates $O\{u_1, u_2, u_3\}$.*

1. $$x_1 = au_3\frac{\sqrt{1 - u_2^2}}{u_1 - u_2}, \quad x_2 = a\frac{\sqrt{(1 - u_2^2)(1 - u_3^2)}}{u_1 - u_2}, \quad x_3 = \frac{\sqrt{u_1^2 - 1}}{u_1 - u_2}$$

2. $$h_1 = \frac{a}{(u_1 - u_2)\sqrt{u_1^2 - 1}},$$
$$h_2 = \frac{a}{(u_1 - u_2)\sqrt{1 - u_2^2}}, \quad h_3 = \left(\frac{a}{u_1 - u_2}\right)\sqrt{\frac{1 - u_2^2}{1 - u_3^2}} \quad \text{MF I 665}$$

10.71–10.72   Vector integral theorems

10.711   *Gauss's divergence theorem.* Let V be a volume bounded by a simple closed surface S and let $\mathbf{f}$ be a continuously differentiable vector field defined in V and on S. Then, if $d\mathbf{S}$ is the outward drawn vector element of area,

$$\int_S \mathbf{f} \cdot d\mathbf{S} = \int_V \operatorname{div} \mathbf{f} \, dV \qquad \text{KE 39}$$

10.712   *Green's theorems.* Let Φ and Ψ be scalar fields which, together with $\nabla^2\Phi$ and $\nabla^2\Psi$, are defined both in a volume V and on its surface S, which we assume to be simple and closed. Then, if $\partial/\partial n$ denotes differentiation along the outward drawn normal to S, we have

10.713   *Green's first theorem*

$$\int_S \Phi \frac{\partial \Psi}{\partial n} \, dS = \int_V \left(\Phi \nabla^2 \Psi + \operatorname{grad} \Phi \cdot \operatorname{grad} \Psi \right) dV$$

KE 212

10.714   *Green's second theorem*

$$\int_S \left(\Phi \frac{\partial \Psi}{\partial n} - \Psi \frac{\partial \Phi}{\partial n} \right) dS = \int_V \left(\Phi \nabla^2 \Psi - \Psi \nabla^2 \Phi \right) dV$$

KE 215

10.715   *Special cases*

1. $$\int_S (\Phi \operatorname{grad} \Phi) \cdot d\mathbf{S} = \int_V \left(\Phi \nabla^2 \Phi + (\operatorname{grad} \Phi)^2 \right) dV$$

2. $$\int_S \frac{\partial \Phi}{\partial n} \, dS = \int_V \nabla^2 \Phi \, dV$$

MV 81

10.716   *Green's reciprocal theorem.* If Φ and Ψ are harmonic, so that $\nabla^2 \Phi = \nabla^2 \Psi = 0$, then

3. $$\int_S \Phi \frac{\partial \Psi}{\partial n} \, dS = \int_S \Psi \frac{\partial \Phi}{\partial n} \, dS$$

MM 105

10.717   *Green's representation theorem.* If Φ and $\nabla^2 \Phi$ are defined within a volume V bounded by a simple closed surface S, and P is an interior point of V, then in three dimensions

4. $$\Phi(P) = -\frac{1}{4\pi} \int_V \frac{1}{r} \nabla^2 \Phi \, dV + \frac{1}{4\pi} \int_S \frac{1}{r} \frac{\partial \Phi}{\partial n} \, dS - \frac{1}{4\pi} \int_S \Phi \frac{\partial}{\partial n} \left(\frac{1}{r} \right) dS$$

KE 219

If Φ is harmonic within V, so that $\nabla^2 \Phi = 0$, then the previous result becomes

5. $$\Phi(P) = \frac{1}{4\pi} \int_S \frac{1}{r} \frac{\partial \Phi}{\partial n} \, dS - \frac{1}{4\pi} \int_S \Phi \frac{\partial}{\partial n} \left(\frac{1}{r} \right) dS$$

In the case of two dimensions, result (4) takes the form

6. $$\Phi(p) = \frac{1}{2\pi} \int_S \nabla^2 \Phi(q) \ln |p - q| \, dS$$
$$+ \frac{1}{2\pi} \int_C \Phi(q) \frac{\partial}{\partial n_q} \ln |p - q| \, dq - \frac{1}{2\pi} \int \ln |p - q| \frac{\partial}{\partial n_q} \Phi(q) \, dq$$

MM 116

where C is the boundary of the planar region S, and result (5) takes the form

7. $$\Phi(p) = \frac{1}{2\pi} \int_C \Phi(q) \frac{\partial}{\partial n_q} \ln |p - q| \, dq - \frac{1}{2\pi} \int_C \ln |p - q| \frac{\partial}{\partial n_q} \Phi(q) \, dq$$

VL 280

10.718   *Green's representation theorem in R^n.* If Φ is twice differentiable within a region Ω in R^n bounded by the surface Σ with outward drawn unit normal $\mathbf{n}$, then for $p \notin \Sigma$ and $n > 3$

$$\Phi(p) = \frac{-1}{(n-2)\sigma_n} \int_\Omega \frac{\nabla^2 \Phi(q)}{|p - q|^{n-2}} \, d\Omega_q + \frac{1}{(n-2)\sigma_n} \int_\Sigma \left(\frac{1}{|p - q|^{n-2}} \frac{\partial \Phi(q)}{\partial n_q} - \Phi(q) \frac{\partial}{\partial n_q} \frac{1}{|p - q|^{n-2}} \right) d\Sigma_q,$$

where

$$\sigma_n = \frac{2\pi^{n/2}}{\Gamma(n/2)} \qquad \text{VL 279}$$

is the area of the unit sphere in R^n.

10.719 *Green's theorem of the arithmetic mean.* If Φ is harmonic in a sphere, then the value of Φ at the center of the sphere is the arithmetic mean of its value on the surface.

KE 223

10.720 *Poisson's integral in three dimensions.* If Φ is harmonic in the interior of a spherical volume V of radius R and is continuous on the surface of the sphere on which, in terms of the spherical polar coordinates (r, θ, ϕ), it satisfies the boundary condition $\Phi(R, \theta, \phi) = f(\theta, \phi)$, then

$$\Phi(r, \theta, \phi) = \frac{R(R^2 - r^2)}{4\pi} \int_0^\pi \int_{-\pi}^\pi \frac{f(\theta', \phi') \sin\theta' \, d\theta' \, d\phi'}{(r^2 + R^2 - 2rR\cos\gamma)^{3/2}},$$

where

$$\cos\gamma = \cos\theta\cos\theta' + \sin\theta\sin\theta'\cos(\phi - \phi') \qquad \text{KE 241}$$

10.721 *Poisson's integral in two dimensions.* If Φ is harmonic in the interior of a circular disk S of radius R and is continuous on the boundary of the disk on which, in terms of the polar coordinates (r, θ), it satisfies the boundary condition $\Phi(R, \theta) = f(\theta)$, then

$$\Phi(r, \theta) = \frac{(R^2 - r^2)}{2\pi} \int_{-\pi}^\pi \frac{f(\phi) \, d\phi}{r^2 + R^2 - 2rR\cos(\theta - \phi)}$$

10.722 *Stokes' theorem.* Let a simple closed curve C be spanned by a surface S. Define the positive normal $\mathbf{n}$ to S, and the positive sense of description of the curve C with line element $d\mathbf{r}$, such that the positive sense of the contour C is clockwise when we look through the surface S in the direction of the normal. Then, if $\mathbf{f}$ is continuously differentiable vector field defined on S and C with vector element $\mathbf{S} = \mathbf{n} \, dS$,

$$\oint_C \mathbf{f} \cdot d\mathbf{r} = \int_S \operatorname{curl} \mathbf{f} \cdot d\mathbf{S}, \qquad \text{MM 143}$$

where the line integral around C is taken in the positive sense.

10.723 *Planar case of Stokes' theorem.* If a region R in the (x, y)-plane is bounded by a simple closed curve C, and $f_1(x, y), f_2(x, y)$ are any two functions having continuous first derivatives in R and on C, then

$$\oint_C (f_1 \, dx + f_2 \, dy) = \int\int_R \left(\frac{\partial f_2}{\partial x} - \frac{\partial f_1}{\partial y} \right) dx \, dy, \qquad \text{MM 143}$$

where the line integral is taken in the anticlockwise sense.

10.81 Integral rate of change theorems

10.811 *Rate of change of volume integral bounded by a moving closed surface.* Let f be a continuous scalar function of position and time t defined throughout the volume $V(t)$, which is itself bounded by a simple closed surface $S(t)$ moving with velocity $\mathbf{v}$. Then the rate of change of the volume integral of f is given by

$$\frac{D}{Dt} \int_{V(t)} f \, dV = \int_{V(t)} \frac{\partial f}{\partial t} \, dV + \int_{S(t)} f\mathbf{v} \cdot d\mathbf{S},$$

where $d\mathbf{S}$ is the outward drawn vector element of area, and

$$\frac{D}{Dt} \equiv \frac{\partial}{\partial t} + \mathbf{v} \cdot \nabla.$$

By virtue of Gauss's theorem this also takes the form

$$\frac{D}{Dt} \int_{V(t)} f \, dV = \int_{V(t)} \left(\frac{Df}{Dt} + f \operatorname{div} \mathbf{v} \right) dV \qquad \text{MV 88}$$

10.812 *Rate of change of flux through a surface.* Let $\mathbf{q}$ be a vector function that may also depend on the time t, and $\mathbf{n}$ be the unit outward drawn normal to the surface S that moves with velocity $\mathbf{v}$. Defining the flux of $\mathbf{q}$ through S as

$$m = \int_S \mathbf{q} \cdot \mathbf{n} \, dS,$$

then

$$\frac{Dm}{Dt} = \int_S \left(\frac{\partial \mathbf{q}}{\partial t} + \mathbf{v} \operatorname{div} \mathbf{q} + \operatorname{curl} (\mathbf{q} \times \mathbf{v}) \right) \cdot \mathbf{n} \, dS \qquad \text{MV 90}$$

10.813 *Rate of change of the circulation around a given moving curve.* Let C be a closed curve, moving with velocity $\mathbf{v}$, on which is defined a vector field $\mathbf{q}$. Defining the circulation ζ of $\mathbf{q}$ around C by

$$\zeta = \int_C \mathbf{q} \cdot d\mathbf{r},$$

then

$$\frac{D\zeta}{Dt} = \int_C \left(\frac{\partial \mathbf{q}}{\partial t} + (\operatorname{curl} \mathbf{q}) \times \mathbf{v} \right) \cdot d\mathbf{r} \qquad \text{MV 94}$$

11 Algebraic Inequalities

11.1–11.3 General Algebraic Inequalities

11.11 Algebraic inequalities involving real numbers

11.111 *Lagrange's identity.* Let $a_1, a_2, \ldots, a_n$ and $b_1, b_2, \ldots, b_n$ be any two sets of real numbers; then

$$\left(\sum_{k=1}^{n} a_k b_k \right)^2 = \left(\sum_{k=1}^{n} a_k^2 \right) \left(\sum_{k=1}^{n} b_k^2 \right) - \sum (a_k b_j - a_j b_k)^2 \qquad \text{BB 3}$$

11.112 *Cauchy–Schwarz–Buniakowsky inequality.* Let $a_1, a_2, \ldots, a_n$ and $b_1, b_2, \ldots, b_n$ be any two arbitrary sets of real numbers; then

$$\left(\sum_{k=1}^{n} a_k b_k \right)^2 \leq \left(\sum_{k=1}^{n} a_k^2 \right) \left(\sum_{k=1}^{n} b_k^2 \right)$$

The equality holds if, and only if, the sequences $a_1, a_2, \ldots, a_n$ and $b_1, b_2, \ldots, b_n$ are proportional.

MT 30

11.113 *Minkowski's inequality.* Let $a_1, a_2, \ldots, a_n$ and $b_1, b_2, \ldots, b_n$ be any two sets of nonnegative real numbers and let $p > 1$; then

$$\left(\sum_{k=1}^{n} (a_k + b_k)^p \right)^{1/p} \leq \left(\sum_{k=1}^{n} a_k^p \right)^{1/p} + \left(\sum_{k=1}^{n} b_k^p \right)^{1/p}$$

The equality holds if, and only if, the sequences $a_1, a_2, \ldots, a_n$ and $b_1, b_2, \ldots, b_n$ are proportional.

MT 55

11.114 *Hölder's inequality.* Let $a_1, a_2, \ldots, a_n$ and $b_1, b_2, \ldots, b_n$ be any two sets of nonnegative real numbers, and let $\dfrac{1}{p} + \dfrac{1}{q} = 1$, with $p > 1$; then

$$\left(\sum_{k=1}^{n} a_k^p \right)^{1/p} \left(\sum_{k=1}^{n} b_k^q \right)^{1/q} \geq \sum_{k=1}^{n} a_k b_k$$

The equality holds if, and only if, the sequences $a_1^p, a_2^p, \ldots, a_n^p$ and $b_1^q, b_2^q, \ldots, b_n^q$ are proportional.

MT 50

11.115 *Chebyshev's inequality.* Let $a_1, a_2, \ldots, a_n$ and $b_1, b_2, \ldots, b_n$ be two arbitrary sets of real numbers such that either $a_1 \geq a_2 \geq \cdots \geq a_n$ and $b_1 \geq b_2 \geq \cdots \geq b_n$, or $a_1 \leq a_2 \leq \cdots \leq a_n$ and $b_1 \leq b_2 \leq \cdots \leq b_n$; then

$$\left(\frac{a_1 + a_2 + \cdots + a_n}{n}\right)\left(\frac{b_1 + b_2 + \cdots + b_n}{n}\right) \leq \frac{1}{n}\sum_{k=1}^{n} a_k b_k$$

The equality holds if, and only if, either $a_1 = a_2 = \cdots = a_n$ or $b_1 = b_2 = \cdots = b_n$.

11.116 *Arithmetic-geometric inequality.* Let $a_1, a_2, \ldots, a_n$ be any set of positive numbers, with arithmetic mean

$$A_n = \left(\frac{a_1 + a_2 + \cdots + a_n}{n}\right)$$

and geometric mean

$$G_n = (a_1 a_2 \ldots a_n)^{1/n};$$

then $A_n \geq G_n$ or, equivalently,

$$\left(\frac{a_1 + a_2 + \cdots + a_n}{n}\right) \geq (a_1 a_2 \ldots a_n)^{1/n}$$

The equality holds only in the event that all of the numbers a_i are equal.   BB 4

11.117 *Carleman's inequality.* If $a_1, a_2, \ldots, a_n$ is any set of positive numbers, then the geometric and arithmetic means satisfy the inequality

$$\sum_{r=1}^{n} G_r \leq A_n$$

or, equivalently,

$$\sum_{r=1}^{n} (a_1 a_2 \ldots a_r)^{1/r} \leq e\left(\frac{a_1 + a_2 + \cdots + a_n}{n}\right),$$

where e is the best possible constant in this inequality.   MT 131

11.118 *An inequality involving absolute values.* Let $a_1, a_2, \ldots, a_n$ and $b_1, b_2, \ldots, b_n$ be two arbitrary sets of real numbers; then

$$\sum_{i,j=1}^{n} \left\{|a_i - b_j|^p + |b_i - a_j|^p - |a_i - a_j|^p - |b_i - b_j|^p\right\} \geq 0, \qquad 0 < p \leq 2.$$

11.21 Algebraic inequalities involving complex numbers

If α, β are any two real numbers, the complex number $z = \alpha + i\beta$ with real part α and imaginary part β has for its modulus $|z|$ the nonnegative number

$$|z| = \sqrt{\alpha^2 + \beta^2},$$

and for its argument (amplitude)$\arg z$ the angle $\arg z = \theta$ such that

$$\cos\theta = \frac{\alpha}{|z|} \text{ and } \sin\theta = \frac{\beta}{|z|},$$

where $-\pi < \theta \leq \pi$. The complex number $\overline{z} = \alpha - i\beta$ is said to be the **complex conjugate** of $z = \alpha + i\beta$.

$$\text{If } z = re^{i\theta} = r\left(\cos\theta + i\sin\theta\right),$$

then

$$z^n = r^n e^{in\theta} = r^n \left(\cos n\theta + i\sin n\theta\right),$$

and setting $r = 1$ we have **de Moivre's theorem**

$$(\cos\theta + i\sin\theta)^n = \cos n\theta + i\sin n\theta$$

It follows directly that, if $z = e^{i\theta}$, then

$$\cos\theta = \frac{1}{2}\left(z + \frac{1}{z}\right), \qquad \sin\alpha = -\frac{i}{2}\left(z - \frac{1}{z}\right),$$

and

$$\cos r\theta = \frac{1}{2}\left(z^r + \frac{1}{z^r}\right), \qquad \sin r\theta = -\frac{i}{2}\left(z^r - \frac{1}{z^r}\right)$$

If $w = z^{p/q}$ with p, q integral, and $z = re^{i\theta}$, then the q roots of $w_0, w_1, \ldots, w_{q-1}$ of z are

$$w_k = r^{p/q}\left[\cos\left(\frac{p\theta + 2k\pi}{q}\right) + i\sin\left(\frac{p\theta + 2k\pi}{q}\right)\right],$$

with $k = 0, 1, 2, \ldots, q - 1$.

11.211[7]   *Simple properties and inequalities involving the modulus and the complex conjugate.* If the real part of z is denoted by $\operatorname{Re} z$ and the imaginary part by $\operatorname{Im} z$, then

$$z + \overline{z} = 2\operatorname{Re} z = 2\alpha,$$
$$z - \overline{z} = 2\operatorname{Im} z = 2i\beta,$$
$$z = \overline{(\overline{z})},$$
$$\frac{1}{\overline{z}} = \overline{\left(\frac{1}{z}\right)},$$
$$\overline{(z^n)} = (\overline{z})^n,$$
$$\left|\frac{\overline{z_1}}{\overline{z_2}}\right| = \frac{|\overline{z_1}|}{|\overline{z_2}|},$$
$$\overline{(z_1 + z_2 + \cdots + z_n)} = \overline{z_1} + \overline{z_2} + \cdots + \overline{z_n},$$
$$\overline{z_1 z_2 \cdots z_n} = \overline{z_1}\,\overline{z_2}\cdots\overline{z_n}.$$

11.212   *Inequalities for pairs of complex numbers.*   If a,b are any two complex numbers, then

(i) $|a + b| \leq |a| + |b|$ (triangle inequality),

(ii) $|a - b| \geq ||a| - |b||$.

11.31 Inequalities for sets of complex numbers

11.311   *Complex Cauchy–Schwarz–Buniakowsky inequality.*   Let $a_1, a_2, \ldots, a_n$ and $b_1, b_2, \ldots, b_n$ be any two arbitrary sets of complex numbers; then

$$\left|\sum_{k=1}^{n} a_k b_k\right|^2 \leq \left(\sum_{k=1}^{n} |a_k|^2\right)\left(\sum_{k=1}^{n} |b_k|^2\right)$$

The equality holds if, and only if, the sequences $\overline{a_1}, \overline{a_2}, \ldots, \overline{a_n}$ and $b_1, b_2, \ldots, b_n$ are proportional.

MT 42

11.312   *Complex Minkowski inequality.*   Let $a_1, a_2, \ldots, a_n$ and $b_1, b_2, \ldots, b_n$ be any two arbitrary sets of complex numbers, and let the real number p be such that $p > 1$; then

$$\left(\sum_{k=1}^{n} |a_k + b_k|^p\right)^{1/p} \leq \left(\sum_{k=1}^{n} |a_k|^p\right)^{1/p} + \left(\sum_{k=1}^{n} |b_k|^p\right)^{1/p}$$

MT 56

11.313   *Complex Hölder inequality.*   Let $a_1, a_2, \ldots, a_n$ and $b_1, b_2, \ldots, b_n$ be any two arbitrary sets of complex numbers, and let the real numbers p,q be such that $p > 1$ and $\dfrac{1}{p} + \dfrac{1}{q} = 1$; then

$$\left(\sum_{k=1}^{n}|a_k|^p\right)^{1/p}\left(\sum_{k=1}^{n}|b_k|^q\right)^{1/p} \geq \left|\sum_{k=1}^{n}a_k b_k\right|$$

The equality holds if, and only if, the sequences

$$|a_1|^p, \quad |a_2|^p, \ldots, |a_n|^p \text{ and } |b_1|^p, |b_2|^p, \ldots |b_n|^p,$$

are proportional and $\arg a_k b_k$ is independent of k for $k = 1, 2, \ldots, n$.   MT 53

12 Integral Inequalities

12.11 Mean value theorems

12.111 First mean value theorem.

Let $f(x)$ and $g(x)$ be two bounded functions integrable in $[a, b]$ and let $g(x)$ be of one sign in this interval. Then

$$\int_a^b f(x)g(x)\,dx = f(\xi)\int_a^b g(x)\,dx,$$

CA 105

with $a \leq \xi \leq b$.

12.112 Second mean value theorem.

(i)   Let $f(x)$ be a bounded, monotonic decreasing, and nonnegative function in $[a, b]$, and let $g(x)$ be a bounded integrable function. Then,

$$\int_a^b f(x)g(x)\,dx = f(a)\int_a^\xi g(x)\,dx,$$

with $a \leq \xi \leq b$.

(ii)   Let $f(x)$ be a bounded, monotonic increasing, and nonnegative function in $[a, b]$, and let $g(x)$ be a bounded integrable function. Then,

$$\int_a^b f(x)g(x)\,dx = f(b)\int_\eta^b g(x)\,dx,$$

with $a \leq \eta \leq b$.

(iii)   Let $f(x)$ be bounded and monotonic in $[a, b]$, and let $g(x)$ be a bounded integrable function which experiences only a finite number of sign changes in $[a, b]$. Then,

$$\int_a^b f(x)g(x)\,dx = f(a+0)\int_a^\xi g(x)\,dx + f(b-0)\int_\xi^b g(x)\,dx,$$

CA 107

with $a \leq \xi \leq b$.

12.113 First mean value theorem for infinite integrals.

Let $f(x)$ be bounded for $x \geq a$, and integrable in the arbitrary interval $[a, b]$, and let $g(x)$ be of one sign in $x \geq a$ and such that $\int_a^\infty g(x)\,dx$ is finite. Then,

$$\int_a^\infty f(x)g(x)\,dx = \mu \int_a^\infty g(x)\,dx, \qquad\qquad \text{CA 123}$$

where $m \le \mu \le M$ and m, M are, respectively, the lower and upper bounds of $f(x)$ for $x \ge a$.

12.114 Second mean value theorem for infinite integrals.

Let $f(x)$ be bounded and monotonic when $x \ge a$, and $g(x)$ be bounded and integrable in the arbitrary interval $[a, b]$ in which it experiences only a finite number of changes of sign. Then, provided $\int_a^\infty g(x)\,dx$ is finite,

$$\int_a^\infty f(x)g(x)\,dx = f(a+0)\int_a^\xi g(x)\,dx + f(\infty)\int_\xi^\infty g(x)\,dx, \qquad\qquad \text{CA 123}$$

with $a \le \xi \le \infty$.

12.21 Differentiation of definite integral containing a parameter

12.211 Differentiation when limits are finite.

Let $\phi(\alpha)$ and $\psi(\alpha)$ be twice differentiable functions in some interval $c \le \alpha \le d$, and let $f(x, \alpha)$ be both integrable with respect to x over the interval $\phi(\alpha) \le x \le \psi(\alpha)$ and differentiable with respect to α. Then,

$$\frac{d}{d\alpha}\int_{\phi(\alpha)}^{\psi(\alpha)} f(x,\alpha)\,dx = \left(\frac{d\psi}{d\alpha}\right) f\left(\psi(\alpha),\alpha\right) - \left(\frac{d\phi}{d\alpha}\right) f\left(\phi(\alpha),\alpha\right) + \int_{\phi(\alpha)}^{\psi(\alpha)} \frac{\partial f}{\partial \alpha}\,dx \qquad \text{FI II 680}$$

12.212 Differentiation when a limit is infinite.

Let $f(x, \alpha)$ and $\partial f/\partial \alpha$ both be integrable with respect to x over the semi-infinite region $x \ge a$, $b \le \alpha < c$. Then, if the integral

$$f(\alpha) = \int_a^\infty f(x, \alpha)\,dx$$

exists for all $b \le \alpha \le c$, and if $\int_a^\infty \frac{\partial f}{\partial \alpha}\,dx$ is uniformly convergent for α in $[b, c]$, it follows that

$$\frac{d}{d\alpha}\int_a^\infty f(x, \alpha)\,dx = \int_a^\infty \frac{\partial f}{\partial \alpha}\,dx$$

12.31 Integral inequalities

12.311 Cauchy–Schwarz–Buniakowsky inequality for integrals.

Let $f(x)$ and $g(x)$ be any two real integrable functions on $[a, b]$. Then,

$$\left(\int_a^b f(x)g(x)\,dx\right)^2 \le \left(\int_a^b f^2(x)\,dx\right)\left(\int_a^b g^2(x)\,dx\right),$$

and the equality will hold if, and only if, $f(x) = kg(x)$, with k real.   BB 21

12.312 Hölder's inequality for integrals.

Let $f(x)$ and $g(x)$ be any two real functions for which $|f(x)|^p$ and $|g(x)|^q$ are integrable on $[a, b]$ with $p > 1$ and $\frac{1}{p} + \frac{1}{q} = 1$; then

$$\int_a^b f(x)g(x)\, dx \le \left(\int_a^b |f(x)|^p\, dx \right)^{1/p} \left(\int_a^b |g(x)|^q\, dx \right)^{1/q}.$$

The equality holds if, and only if, $\alpha |f(x)|^p = \beta |g(x)|^q$, where α and β are positive constants. **BB 21**

12.313 Minkowski's inequality for integrals.

Let $f(x)$ and $g(x)$ be any two real functions for which $|f(x)|^p$ and $|g(x)|^p$ are integrable on $[a, b]$ for $p > 0$; then

$$\left(\int_a^b |f(x) + g(x)|^p\, dx \right)^{1/p} \le \left(\int_a^b |f(x)|^p\, dx \right)^{1/p} + \left(\int_a^b |g(x)|^p\, dx \right)^{1/p}.$$

The equality holds if, and only if, $f(x) = kg(x)$ for some real $k \ge 0$. n **BB 21**

12.314 Chebyshev's inequality for integrals.

Let $f_1, f_2, \ldots, f_n$ be nonnegative integrable functions on $[a, b]$ which are all either monotonic increasing or monotonic decreasing; then

$$\int_a^b f_1(x)\, dx \int_a^b f_2(x)\, dx \ldots \int_a^b f_n(x)\, dx \le (b - a)^{n-1} \int_a^b f_1(x) f_2(x) \ldots f_n(x)\, dx$$ **MT 39**

12.315 Young's inequality for integrals.

Let $f(x)$ be a real-valued continuous strictly monotonic increasing function on the interval $[0, a]$, with $f(0) = 0$ and $b \le f(a)$. Then

$$ab \le \int_0^a f(x)\, dx + \int_0^b f^{-1}(y)\, dy,$$

where $f^{-1}(y)$ denotes the function inverse to $f(x)$. The equality holds if, and only if, $b = f(a)$. **BB 15**

12.316 Steffensen's inequality for integrals.

Let $f(x)$ be nonnegative and monotonic decreasing in $[a, b]$ and $g(x)$ be such that $0 \le g(x) \le 1$ in $[a, b]$. Then

$$\int_{b-k}^b f(x)\, dx \le \int_a^b f(x)g(x)\, dx \le \int_a^{a+k} f(x)\, dx,$$

where $k = \int_a^b g(x)\, dx$. **MT 107**

12.317 Gram's inequality for integrals.

Let $f_1(x), f_2(x), \ldots, f_n(x)$ be real square integrable functions on $[a, b]$; then

$$\begin{vmatrix} \int_a^b f_1^2(x)\, dx & \int_a^b f_1(x)f_2(x)\, dx & \cdots & \int_a^b f_1(x)f_n(x)\, dx \\ \int_a^b f_2(x)f_1(x)\, dx & \int_a^b f_2^2(x)\, dx & \cdots & \int_a^b f_2(x)f_n(x)\, dx \\ \vdots & \vdots & \ddots & \vdots \\ \int_a^b f_n(x)f_1(x)\, dx & \int_a^b f_n(x)f_2(x)\, dx & \cdots & \int_a^b f_n^2(x)\, dx \end{vmatrix} \ge 0.$$ **MT 47**

12.318   Ostrowski's inequality for integrals.

Let $f(x)$ be a monotonic function integrable on $[a, b]$, and let $f(a)f(b) \geq 0, |f(a)| \geq |f(b)|$. Then, if g is a real function integrable on $[a, b]$,

$$\left| \int_a^b f(x)g(x)\,dx \right| \leq |f(a)| \max_{a \leq \xi \leq b} \left| \int_a^\xi g(x)\,dx \right|.$$

12.41 Convexity and Jensen's inequality

A function $f(x)$ is said to be **convex** on an interval $[a, b]$ if for any two points x_1, x_2 in $[a, b]$

$$f\left(\frac{x_1 + x_2}{2} \right) \leq \frac{f(x_1) + f(x_2)}{2}.$$

A function $f(x)$ is said to be **concave** on an interval $[a, b]$ if for any two points x_1, x_2 in $[a, b]$ the function $-f(x)$ is convex in that interval.

    If the function $f(x)$ possesses a second derivative in the interval $[a, b]$, then a necessary and sufficient condition for it to be convex on that interval is that $f''(x) \geq 0$ for all x in $[a, b]$.

    A function $f(x)$ is said to be **logarithmically convex** on the interval $[a, b]$ if $f > 0$ and $\log f(x)$ is concave on $[a, b]$.

    If $f(x)$ and $g(x)$ are logarithmically convex on the interval $[a, b]$, then the functions $f(x) + g(x)$ and $f(x)g(x)$ are also logarithmically convex on $[a, b]$.   MT 17

12.411   Jensen's inequality.

Let $f(x), p(x)$ be two functions defined for $a \leq x \leq b$ such that $\alpha \leq f(x) \leq \beta$ and $p(x) \geq 0$, with $p(x) \not\equiv 0$. Let $\phi(u)$ be a convex function defined on the interval $\alpha \leq u \leq \beta$; then

$$\phi\left(\frac{\int_a^b f(x)p(x)\,dx}{\int_a^b p(x)\,dx} \right) \leq \frac{\int_a^b \phi(f)\,p(x)\,dx}{\int_a^b p(x)\,dx}. \qquad\qquad \text{HL 151}$$

12.51 Fourier series and related inequalities

The trigonometric **Fourier series** representation of the function $f(x)$ integrable on $[-\pi, \pi]$ is

$$f(x) \sim \frac{a_0}{2} + \sum_{n=1}^{\infty} (a_n \cos nx + b_n \sin nx),$$

where the **Fourier coefficients** a_n and b_n of $f(x)$ are given by

$$a_n = \frac{1}{2\pi} \int_{-\pi}^{\pi} f(x) \cos nx\,dx, \qquad b_n = \frac{1}{2\pi} \int_{-\pi}^{\pi} f(x) \sin nx\,dx.$$

(See **0.320–0.328** for convergence of Fourier series on $(-l, l)$.)   TF 1

12.511   Riemann–Lebesgue lemma

If $f(x)$ is integrable on $[-\pi, \pi]$, then

$$\lim_{t \to \infty} \int_{-\pi}^{\pi} f(x) \sin tx\,dx \to 0$$

and

$$\lim_{t \to \infty} \int_{-\pi}^{\pi} f(x) \cos tx \, dx \to 0.$$

TF 11

12.512 Dirichlet lemma

$$\int_{0}^{\pi} \frac{\sin\left(n + \frac{1}{2}\right)x}{2\sin\frac{1}{2}x} \, dx = \frac{\pi}{2},$$

in which $\sin\left(n + \frac{1}{2}\right)x \left/ 2\sin\frac{1}{2}x \right.$ is called the **Dirichlet kernel**.

ZY 21

12.513 Parseval's theorem for trigonometric Fourier series

If $f(x)$ is square integrable on $[-\pi, \pi]$, then

$$\frac{a_0^2}{2} + \sum_{r=1}^{\infty}\left(a_r^2 + b_r^2\right) = \frac{1}{\pi}\int_{-\pi}^{\pi} f^2(x) \, dx.$$

Y 10

12.514 Integral representation of the nth partial sum.

If $f(x)$ is integrable on $[-\pi, \pi]$, then the nth partial sum

$$s_n(x) = \frac{a_0}{2} + \sum_{r=1}^{n}\left(a_r \cos rx + b_r \sin rx\right)$$

has the following integral representation in terms of the Dirichlet kernel,

$$s_n(x) = \frac{1}{\pi}\int_{-\pi}^{\pi} f(x - t)\frac{\sin\left(n + \frac{1}{2}\right)t}{2\sin\frac{1}{2}t} \, dt.$$

Y 20

12.515 Generalized Fourier series

Let the set of functions $\{\phi_n\}_{n=0}^{\infty}$ form an **orthonormal set** over $[a, b]$, so that

$$\int_{a}^{b} \phi_m(x)\phi_n(x) \, dx = \begin{cases} 1 & \text{for} \quad m = n, \\ 0 & \text{for} \quad m \neq n. \end{cases}$$

Then the **generalized Fourier series** representation of an integrable function $f(x)$ on $[a, b]$ is

$$f(x) \sim \sum_{n=0}^{\infty} c_n\phi_n(x),$$

where the generalized Fourier coefficients of $f(x)$ are given by

$$c_n = \int_{a}^{b} f(x)\phi_n(x) \, dx.$$

12.516 Bessel's inequality for generalized Fourier series

For any square integrable function defined on $[a, b]$,

$$\sum_{n=0}^{\infty} c_n^2 \leq \int_{a}^{b} f^2(x) \, dx,$$

where the c_n are the generalized Fourier coefficients of $f(x)$.

12.517 Parseval's theorem for generalized Fourier series.

If $f(x)$ is a square integrable function defined on $[a, b]$ and $\{\phi_n(x)\}_{n=0}^{\infty}$ is a **complete orthonormal** set of continuous functions defined on $[a, b]$, then

$$\sum_{n=0}^{\infty} c_n^2 = \int_a^b f^2(x)\, dx,$$

where the c_n are generalized Fourier coefficients of $f(x)$.

13 Matrices and related results

13.11-13.12 Special matrices

13.111 Diagonal matrix

A square matrix $\mathbf{A}$ of the form

$$\mathbf{A} = \begin{bmatrix} \lambda_1 & 0 & 0 & \ldots & 0 \\ 0 & \lambda_2 & 0 & \ldots & 0 \\ 0 & 0 & \lambda_3 & & 0 \\ \vdots & \vdots & & \ddots & \\ 0 & 0 & 0 & & \lambda_n \end{bmatrix}$$

in which all entries away from the **leading diagonal** are zero.

13.112 Identity matrix and null matrix

The **identity matrix** is a diagonal matrix $\mathbf{I}$ in which all entries in the leading diagonal are unity. The **null matrix** is all zeros.

13.113 Reducible and irreducible matrices

The $n \times n$ matrix $\mathbf{A} = [a_{ij}]$ is said to be **reducible**, if the indices $1, 2, \ldots, n$ can be divided into two disjoint non-empty sets $i_1, i_2, \ldots, i_\mu; j_1, j_2, \ldots, j_\nu$ with $(\mu + \nu = n)$, such that

$$a_{i_\alpha j_\beta} = 0 \qquad (\alpha = 1, 2, \ldots, \mu; \quad \beta = 1, 2, \ldots, \nu).$$

Otherwise $\mathbf{A}$ will be said to be irreducible. GA 61

13.114 Equivalent matrices

An $m \times n$ matrix $\mathbf{A}$ is **equivalent** to an $m \times n$ matrix $\mathbf{B}$ if, and only if, $\mathbf{B} = \mathbf{PAQ}$ for suitable non-singular $m \times m$ and $n \times n$ matrices $\mathbf{P}$ and $\mathbf{Q}$, respectively.

13.115 Transpose of a matrix

If $\mathbf{A} = [a_{ij}]$ is an $m \times n$ matrix with element a_{ij} in the ith row and the jth column, then the transpose $\mathbf{A}^{\text{T}}$ of $\mathbf{A}$ is the $n \times m$ matrix

$$\mathbf{A}^{\text{T}} = [b_{ij}] \quad \text{with} \quad b_{ij} = a_{ji},$$

that is, the matrix derived from $\mathbf{A}$ by interchanging rows and columns.

13.116 Adjoint matrix

If $\mathbf{A}$ is an $n \times n$ matrix, then its **adjoint**, denoted by adj $\mathbf{A}$, is the transpose of the matrix of cofactors A_{ij} of $\mathbf{A}$, so that

$$\text{adj}\,\mathbf{A} = [A_{ij}]^{\mathrm{T}} \qquad (\text{see } \mathbf{14.13})$$

13.117 Inverse matrix

If $\mathbf{A} = [a_{ij}]$ is an $n \times n$ matrix with a nonsingular determinant $|\mathbf{A}|$, then its **inverse** $\mathbf{A}^{-1}$ is given by

$$\mathbf{A}^{-1} = \frac{\text{adj}\,\mathbf{A}}{|\mathbf{A}|}.$$

13.118 Trace of a matrix

The trace of an $n \times n$ matrix $\mathbf{A} = [a_{ij}]$, written tr $\mathbf{A}$, is defined to be the sum of the terms on the leading diagonal, so that

$$\text{tr}\,\mathbf{A} = a_{11} + a_{22} + \ldots + a_{nn}.$$

13.119 Symmetric matrix

The $n \times n$ matrix $\mathbf{A} = [a_{ij}]$ is **symmetric** if $a_{ij} = a_{ji}$ for $i, j = 1, 2, \ldots, n$.

13.120 Skew-symmetric matrix

The $n \times n$ matrix $\mathbf{A} = [a_{ij}]$ is **skew-symmetric** if $a_{ij} = -a_{ji}$ for $i, j = 1, 2, \ldots, n$.

13.121 Triangular matrices

An $n \times n$ matrix $\mathbf{A} = [a_{ij}]$ is of **upper triangular type** if $a_{ij} = 0$ for $i > j$ and of **lower triangular type** if $a_{ij} = 0$ for $j > i$.

13.122 Orthogonal matrices

A real $n \times n$ matrix $\mathbf{A}$ is **orthogonal** if, and only if, $\mathbf{A}\mathbf{A}^{\mathrm{T}} = \mathbf{I}$.

13.123 Hermitian transpose of a matrix

If $\mathbf{A} = [a_{ij}]$ is an $n \times n$ matrix with complex elements, then its **hermitian transpose** $\mathbf{A}^{\mathrm{H}}$ is defined to be

$$\mathbf{A}^{\mathrm{H}} = [\bar{a}_{ji}],$$

with the bar denoting the complex conjugate operation.

13.124 Hermitian matrix

An $n \times n$ matrix $\mathbf{A}$ is **hermitian** if $\mathbf{A} = \mathbf{A}^{\mathrm{H}}$, or equivalently, if $\mathbf{A} = \overline{\mathbf{A}}^{\mathrm{T}}$, with the bar denoting the complex conjugate operation.

13.125 Unitary matrix

An $n \times n$ matrix $\mathbf{A}$ is **unitary** if $\mathbf{A}\mathbf{A}^{\mathrm{H}} = \mathbf{A}\mathbf{A} = \mathbf{I}$.

13.126 Eigenvalues and eigenvectors

If $\mathbf{A}$ is an $n \times n$ matrix, each eigenvector $\mathbf{x}$ corresponding to λ satisfies the equation
$$\mathbf{A}\mathbf{X} = \lambda \mathbf{x},$$
while the **eigenvalues** λ satisfy the **characteristic equation**
$$|\mathbf{A} - \lambda \mathbf{I}| = 0 \qquad \text{(see } \mathbf{15.61})$$

13.127 Nilpotent matrix

An $n \times n$ matrix $\mathbf{A}$ is **nilpotent** if $\mathbf{A}^{k} = \mathbf{0}$ for some k.

13.128 Idempotent matrix

An $n \times n$ matrix $\mathbf{A}$ is **idempotent** if $\mathbf{A}^{2} = \mathbf{A}$.

13.129 Positive definite

An $n \times n$ matrix $\mathbf{A}$ is **positive definite** if $\mathbf{x}^{\mathrm{T}}\mathbf{A}\mathbf{x} > 0$, for $\mathbf{x} \neq \mathbf{0}$ an n element column vector.

13.130 Non-negative definite

An $n \times n$ matrix $\mathbf{A}$ is **non-negative definite** if $\mathbf{x}^{\mathrm{T}}\mathbf{A}\mathbf{x} \geq 0$, for $\mathbf{x} \neq \mathbf{0}$ an n element column vector.

13.131 Diagonally dominant

An $n \times n$ matrix $\mathbf{A}$ is **diagonally dominant** if $|a_{ii}| > \sum_{j \neq i} |a_{ij}|$ for all i.

13.21 Quadratic forms

A **quadratic form** involving the n real variables $x_1, x_2, \ldots, x_n$ that are associated with the real $n \times n$ matrix $\mathbf{A} = [a_{ij}]$ is the scalar expression
$$Q(x_1, x_2, \ldots, x_n) = \sum_{i=1}^{n} \sum_{j=1}^{n} a_{ij} x_i x_j.$$
In terms of matrix notation, if $\mathbf{x}$ is the $n \times 1$ column vector with real elements $x_1, x_2, \ldots, x_n$, and $\mathbf{x}^{\mathrm{T}}$ is the transpose of $\mathbf{x}$, then
$$Q(\mathbf{x}) = \mathbf{x}^{\mathrm{T}}\mathbf{A}\mathbf{x}.$$
Employing the inner product notation, this same quadratic form may also be written
$$Q(\mathbf{x}) \equiv (\mathbf{x}, \mathbf{A}\mathbf{x}).$$
If the $n \times n$ matrix $\mathbf{A}$ is hermitian, so that $\overline{\mathbf{A}}^{\mathrm{T}} = \mathbf{A}$, where the bar denotes the complex conjugate operation, then the quadratic form associated with the hermitian matrix $\mathbf{A}$ and the vector $\mathbf{x}$ which may have complex elements is the real quadratic form

$$Q(\mathbf{x}) = (\mathbf{x}, \mathbf{A}\mathbf{x}).$$

It is always possible to express an arbitrary quadratic form

$$Q(\mathbf{x}) = \sum_{i=1}^{n} \sum_{j=1}^{n} \alpha_{ij} x_i x_j$$

in the form

$$Q(\mathbf{x}) = (\mathbf{x}, \mathbf{A}\mathbf{x}),$$

where $\mathbf{A} = [a_{ij}]$ is a symmetric matrix, by defining

$$a_{ii} = \alpha_{ii} \qquad \text{for } i = 1, 2, \ldots, n$$

and

$$a_{ij} = \tfrac{1}{2}(\alpha_{ij} + \alpha_{ji}) \qquad \text{for } i, j = 1, 2, \ldots, n \quad \text{and} \quad i \neq j.$$

13.211 Sylvester's law of inertia

When a quadratic form Q in n variables is reduced by a nonsingular linear transformation to the form

$$Q = y_1^2 + y_2^2 + \ldots + y_p^2 - y_{p+1}^2 - y_{p+2}^2 - \ldots - y_r^2,$$

the number p of positive squares appearing in the reduction is an invariant of the quadratic form Q, and does not depend on the method of reduction itself.   ML 377

13.212 Rank

The **rank** of the quadratic form Q in the above canonical form is the total number r of squared terms (both positive and negative) appearing in its reduced form.   ML 360

13.213 Signature

The **signature** of the quadratic form Q above is the number s of positive squared terms appearing in its reduced form. It is sometimes also defined to be $2s - r$.   ML 378

13.214 Positive definite and semidefinite quadratic form

The quadratic form $Q(\mathbf{x}) = (\mathbf{x}, \mathbf{A}\mathbf{x})$ is said to be **positive definite** when $Q(\mathbf{x}) > 0$ for $\mathbf{x} \neq \mathbf{0}$. It is said to be **positive semidefinite** if $Q(x) \geq 0$ for $x \neq 0$.   ML 394

13.215 Basic theorems on quadratic forms

1.   Two real quadratic forms are **equivalent** under the group of linear transformations if, and only if, they have the same rank and the same signature.

2.   A real quadratic form in n variables is positive definite if, and only if, its canonical form is

$$Q = z_1^2 + z^2 + \ldots + z_n^2.$$

3.   A real symmetric matrix $\mathbf{A}$ is positive definite if, and only if, there exists a real nonsingular matrix bfM such that $\mathbf{A} = \mathbf{M}\mathbf{M}^{\mathrm{T}}$.

4.   Any real quadratic form in n variables may be reduced to the diagonal form

$$Q = \lambda_1 z_1^2 + \lambda_2 z_2^2 + \ldots + \lambda_n z_n^2, \lambda_1 \geq \lambda_2 \geq \ldots \geq \lambda_n$$

by a suitable orthogonal point-transformation.

5. The quadratic form $Q = (\mathbf{x}, \mathbf{A}\mathbf{x})$ is positive definite if, and only if, every eigenvalue of $\mathbf{A}$ is positive; it is positive semidefinite if, and only if, all the eigenvalues of $\mathbf{A}$ are nonnegative, and it is indefinite if the eigenvalues of $\mathbf{A}$ are of both signs.

6. The necessary conditions for an hermitian matrix $\mathbf{A}$ to be positive definite are

    (i) $a_{ii} > 0$ for all i,

    (ii) $a_{ii}a_{ij} > |a_{ij}|^2$ for $i \neq j$,

    (iii)   the element of largest modulus must lie on the leading diagonal,

    (iv) $|\mathbf{A}| > 0$.

7. The quadratic form $Q = (\mathbf{x}, \mathbf{A}\mathbf{x})$ with $\mathbf{A}$ hermitian will be positive definite if all the principal minors in the top left-hand corner of $\mathbf{A}$ are positive, so that

$$a_{11} > 0, \quad \begin{vmatrix} a_{11} & a_{12} \\ a_{21} & a_{22} \end{vmatrix} > 0, \quad \begin{vmatrix} a_{11} & a_{12} & a_{13} \\ a_{21} & a_{22} & a_{23} \\ a_{31} & a_{32} & a_{33} \end{vmatrix} > 0, \ldots . \qquad \text{ML 353-379}$$

13.31 Differentiation of matrices

If the $n \times m$ matrices $\mathbf{A}(t)$ and $\mathbf{B}(t)$ have elements that are differentiable functions of t, so that
$$\mathbf{A}(t) = [a_{ij}(t)], \qquad \mathbf{B}(t) = [b_{ij}(t)]$$
then

1. $$\frac{d}{dt}\mathbf{A}(t) = \left[\frac{d}{dt}a_{ij}(t) \right]$$

2. $$\frac{d}{dt}[\mathbf{A}(t) \pm \mathbf{B}(t)] = \left[\frac{d}{dt}a_{ij}(t) \pm \frac{d}{dt}b_{ij}(t) \right]$$
    $$= \frac{d}{dt}\mathbf{A}(t) \pm \frac{d}{dt}\mathbf{B}(t)$$

3. If the matrix product $\mathbf{A}(t)\mathbf{B}(t)$ is defined, then
    $$\frac{d}{dt}[\mathbf{A}(t)\mathbf{B}(t)] = \left(\frac{d}{dt}\mathbf{A}(t) \right)\mathbf{B}(t) + \mathbf{A}(t)\left(\frac{d}{dt}\mathbf{B}(t) \right)$$

4. If the matrix product $\mathbf{A}(t)\mathbf{B}(t)$ is defined, then
    $$\frac{d}{dt}[\mathbf{A}(t)\mathbf{B}(t)]^{\text{T}} = \left(\frac{d}{dt}\mathbf{B}(t) \right)^{\text{T}}\mathbf{A}^{\text{T}}(t) + \mathbf{B}^{\text{T}}(t)\left(\frac{d}{dt}\mathbf{A}(t) \right)^{\text{T}}$$

5. If the square matrix $\mathbf{A}$ is nonsingular, so that $|\mathbf{A}| \neq 0$, then

$$\frac{d}{dt}\left[\mathbf{A}^{-1}\right] = -\mathbf{A}^{-1}(t)\left(\frac{d}{dt}\mathbf{A}(t)\right)\mathbf{A}^{-1}(t)$$

6. $$\int_{t_0}^{T}\mathbf{A}(\tau)\,d\tau = \left[\int_{t_0}^{T}a_{ij}(\tau)\,d\tau\right]$$

13.41 The matrix exponential

If $\mathbf{A}$ is a square matrix, and z is any complex number, then the matrix exponential e^{Az} is defined to be

$$e^{Az} = \mathbf{I} + \mathbf{A}z + \ldots + \frac{\mathbf{A}^n z^n}{n!} + \ldots = \sum_{r=0}^{\infty}\frac{1}{r!}\mathbf{A}^r z^r$$

3.411 Basic properties

1. $e^0 = \mathbf{I}, \quad e^{Iz} = \mathbf{I}e^z, \quad e^{\mathbf{A}(z_1+z_2)} = e^{\mathbf{A}z_1}\cdot e^{\mathbf{A}z_2},$ [when $\mathbf{A}+\mathbf{B}$ is defined and $\mathbf{AB} = \mathbf{BA}$]

$$e^{-\mathbf{A}z} = \left(e^{\mathbf{A}z}\right)^{-1}, \quad e^{\mathbf{A}z}\cdot e^{\mathbf{B}z} = e^{(\mathbf{A}+\mathbf{B})z}$$

2. $$\frac{d^r}{dz^r}\left(e^{\mathbf{A}z}\right) = \mathbf{A}^r e^{\mathbf{A}z} = e^{\mathbf{A}z}\mathbf{A}^r.$$

ML 340

3.   If the square matrix $\mathbf{A}$ can be expressed in the form $\mathbf{A} = \begin{bmatrix}\mathbf{B} & \mathbf{0} \\ \mathbf{0} & \mathbf{C}\end{bmatrix}$, with $\mathbf{B}$ and $\mathbf{C}$ square matrices, then

$$e^{\mathbf{A}z} = \begin{bmatrix}e^{\mathbf{B}z} & \mathbf{0} \\ \mathbf{0} & e^{\mathbf{C}z}\end{bmatrix}$$

14 Determinants

14.11 Expansion of second- and third-order determinants

1. $\begin{vmatrix} a_{11} & a_{12} \\ a_{21} & a_{22} \end{vmatrix} = a_{11}a_{22} - a_{12}a_{21}.$

2. $\begin{vmatrix} a_{11} & a_{12} & a_{13} \\ a_{21} & a_{22} & a_{23} \\ a_{31} & a_{32} & a_{33} \end{vmatrix} = a_{11}a_{22}a_{33} - a_{11}a_{23}a_{32} + a_{12}a_{23}a_{31} - a_{12}a_{21}a_{33} + a_{13}a_{21}a_{32} - a_{13}a_{22}a_{31}.$

14.12 Basic properties

Let $\mathbf{A} = [a_{ij}]$ and $\mathbf{B} = [b_{ij}]$ be $n \times n$ matrices. Then the following results are true:

1. If any two adjacent rows (or columns) of a square matrix are interchanged, then the sign of the associated determinant is changed.

2. If any two rows (or columns) of a determinant are identical, the determinant is zero.

3. A determinant is not changed in value if any multiple of a row (or column) is added to any other row (or column).

4. $|k\mathbf{A}| = k^n|\mathbf{A}|$ for any scalar k.

5. $|\mathbf{A}^{\mathrm{T}}| = |\mathbf{A}|$ where $\mathbf{A}^{\mathrm{T}}$ is the transpose of $\mathbf{A}$.

6. $|\mathbf{AB}| = |\mathbf{A}||\mathbf{B}|$.

7. $|\mathbf{A}^{-1}| = \frac{1}{|\mathbf{A}|}$ when the inverse exists.

8. If the elements a_{ij} of $\mathbf{A}$ are functions of x, then

$$\frac{d|\mathbf{A}|}{dx} = \sum_{i,j=1}^{n} \frac{da_{ij}}{dx} A_{ij} \qquad \text{(see 14.13)}$$

14.13 Minors and cofactors of a determinant

The **minor** M_{ij} of the element a_{ij} in the nth-order determinant $|\mathbf{A}|$ associated with the square $n \times n$ matrix $\mathbf{A}$ is the $(n-1)^{\text{th}}$-order determinant derived from $\mathbf{A}$ by deletion of the ith row and j^{th} column. The cofactor A_{ij} of the element a_{ij} is defined to be

$$A_{ij} = (-1)^{i+j} M_{ij}.$$ ML 20

14.14 Principal minors

A **principal minor** is one whose elements are situated symmetrically with respect to the leading diagonal of $\mathbf{A}$.
                                                                   ML 197

14.15* Laplace expansion of a determinant

The nth-order determinant denoted by $|\mathbf{A}|$, or det $\mathbf{A}$, associated with the $n \times n$ matrix $\mathbf{A} = [a_{ij}]$ may be expanded either by elements of the ith row as

$$|\mathbf{A}| = \sum_{j=1}^{n} a_{ij} A_{ij},$$

or by elements of the jth column as

$$|\mathbf{A}| = \sum_{i=1}^{n} a_{ij} A_{ij},$$

where A_{ij} is the cofactor of element a_{ij}. The cofactors A_{ij} satisfy the following n linear equations:

$$\sum_{j=1}^{n} a_{ij} A_{kj} = \delta_{ik} |\mathbf{A}|, \qquad \sum_{i=1}^{n} a_{ij} A_{ik} = \delta_{jk} |\mathbf{A}|,$$
                                                                   ML 21
$$\text{for } i, j, k = 1, 2, \ldots, n \text{ and } \delta_{ij} = \begin{cases} 1 & \text{for } i = j \\ 0 & \text{for } i \neq j \end{cases}$$

14.16 Jacobi's theorem

Let M_r be an r-rowed minor of the nth-order determinant $|\mathbf{A}|$, associated with the $n \times n$ matrix $\mathbf{A} = [a_{ij}]$, in which the rows $i_1, i_2, \ldots, i_r$ are represented together with the columns $k_1, k_2, \ldots, k_r$.

Define the **complementary minor** to M_r to be the $(n-k)$-rowed minor obtained from $|\mathbf{A}|$ by deleting all the rows and columns associated with M_r, and the **signed complementary minor** $M^{(r)}$ to M_r to be

$$M^{(r)} = (-1)^{i_1+i_2+\cdots+i_r+k_1+k_2+\cdots+k_r} \times (\text{complementary minor to } M_r).$$

Then, if Δ is the matrix of cofactors given by

$$\Delta = \begin{vmatrix} A_{11} & A_{12} & \cdots & A_{1n} \\ A_{21} & A_{22} & \cdots & A_{2n} \\ \vdots & \vdots & \ddots & \vdots \\ A_{n1} & A_{n2} & \cdots & A_{nn} \end{vmatrix},$$

and M_r and M_r' are corresponding r-rowed minors of $|\mathbf{A}|$ and Δ, it follows that

$$M_r' = |\mathbf{A}|^{r-1} M^{(r)}.$$ ML 25

Corollary. If $|\mathbf{A}| = 0$, then

$$A_{pk} A_{nq} = A_{nk} A_{pq}.$$

14.17 Hadamard's theorem

If $|\mathbf{A}|$ is an $n \times n$ determinant with elements a_{ij} that may be complex, then $|\mathbf{A}| \neq 0$ if

$$|a_{ii}| > \sum_{j=1, j \neq i}^{n} |a_{ij}|.$$

14.18 Hadamard's inequality

Let $\mathbf{A} = [a_{ij}]$ be an arbitrary $n \times n$ nonsingular matrix with real elements and determinant $|\mathbf{A}|$. Then

$$|\mathbf{A}|^2 \leq \prod_{i=1}^{n} \left(\sum_{k=1}^{n} a_{ik}^{2} \right).$$

This result is also true when $\mathbf{A}$ is hermitian.   ML 418

Deductions.

1.   If $M = \max |a_{ij}|$, then

$$|\mathbf{A}| \leq M^n n^{n/2}.$$ ML 419

2.   If the $n \times n$ matrix $\mathbf{A} = [a_{ij}]$ is positive definite, then

$$|\mathbf{A}| \leq a_{11} a_{22} \ldots a_{nn}.$$ BL 126

3.   If the real $n \times n$ matrix $\mathbf{A}$ is diagonally dominant, so that $\sum_{j \neq 1}^{n} |a_{ij}| < |a_{ii}|$ for $i = 1, 2, \ldots, n$, then $|\mathbf{A}| \neq 0$.

14.21 Cramer's rule

If the n linear equations

$$
\begin{array}{ccccccc}
a_{11}x_1 & + & a_{12}x_2 & + & \cdots & + & a_{1n}x_n & = & b_1, \\
a_{21}x_1 & + & a_{22}x_2 & + & \cdots & + & a_{2n}x_n & = & b_2, \\
\vdots & & \vdots & & \ddots & & \vdots & & \vdots \\
a_{n1}x_1 & + & a_{n2}x_2 & + & \cdots & + & a_{nn}x_n & = & b_n,
\end{array}
$$

have a nonsingular coefficient matrix $\mathbf{A} = [a_{ij}]$, so that $|\mathbf{A}| \neq 0$, then there is a unique solution

$$x_j = \frac{A_{1j}b_1 + A_{2j}b_2 + \cdots + A_{nj}b_j}{|\mathbf{A}|}$$

for $j = 1, 2, \ldots, n$, where A_{ij} is the cofactor of element a_{ij} in the coefficient matrix $\mathbf{A}$.   ML 134

14.31 Some special determinants

14.311 Vandermonde's determinant (alternant).

Third order.

$$\begin{vmatrix} 1 & 1 & 1 \\ x_1 & x_2 & x_3 \\ x_1^2 & x_2^2 & x_3^2 \end{vmatrix} = (x_3 - x_2)(x_3 - x_1)(x_2 - x_1),$$

and, in general, the nth-order Vandermonde's determinant is

$$\begin{vmatrix} 1 & 1 & \cdots & 1 \\ x_1 & x_2 & \cdots & x_n \\ x_1^2 & x_2^2 & \cdots & x_n^2 \\ \vdots & \vdots & \ddots & \vdots \\ x_1^{n-1} & x_2^{n-1} & \cdots & x_n^{n-1} \end{vmatrix} = \prod_{1 \le i < j \le n} (x_j - x_i),$$

where the right-hand side is the continued product of all the differences that can be formed from the $\frac{1}{2}n(n-1)$ pairs of numbers taken from $x_1, x_2, \ldots, x_n$, with the order of the differences taken in the reverse order of the suffixes that are involved.   ML 17

14.312 Circulants.

Second order.

$$\begin{vmatrix} x_1 & x_2 \\ x_2 & x_1 \end{vmatrix} = (x_1 + x_2)(x_1 - x_2).$$

Third order.

$$\begin{vmatrix} x_1 & x_2 & x_3 \\ x_3 & x_1 & x_2 \\ x_2 & x_3 & x_1 \end{vmatrix} = (x_1 + x_2 + x_3)\left(x_1 + \omega x_2 + \omega^2 x_3\right)\left(x_1 + \omega^2 x_2 + \omega x_3\right),$$

where ω and ω^2 are the complex cube roots of 1. In general, the nth-order circulant determinant is

$$\begin{vmatrix} x_1 & x_2 & x_3 & \cdots & x_n \\ x_n & x_1 & x_2 & \cdots & x_{n-1} \\ x_{n-1} & x_n & x_1 & \cdots & x_{n-2} \\ \vdots & \vdots & \vdots & \ddots & \vdots \\ x_2 & x_3 & x_4 & \cdots & x_1 \end{vmatrix} = \prod_{j=1}^{n} \left(x_1 + x_2\omega_j + x_3\omega_j{}^2 + \cdots + x_n\omega_j{}^{n-1}\right),$$

where ω_j is an nth root of 1. The eigenvalues λ (see **15.61**) of an $n \times n$ circulant matrix are

$$\lambda_j = x_1 + x_2\omega_j + x_3\omega_j{}^2 + \cdots + x_n\omega_j{}^{n-1},$$

where ω_j is again an nth root of 1.   ML 36

14.313 Jacobian determinant.

If $f_1, f_2, \ldots, f_n$ are n real-valued functions which are differentiable with respect to $x_1, x_2, \ldots, x_n$, then the Jacobian $J_f(x)$ of the f_i with respect to the x_j is the determinant

$$J_f(x) = \begin{vmatrix} \frac{\partial f_1}{\partial x_1} & \frac{\partial f_1}{\partial x_2} & \cdots & \frac{\partial f_1}{\partial x_n} \\ \frac{\partial f_2}{\partial x_1} & \frac{\partial f_2}{\partial x_2} & \cdots & \frac{\partial f_2}{\partial x_n} \\ \vdots & \vdots & \ddots & \vdots \\ \frac{\partial f_n}{\partial x_1} & \frac{\partial f_n}{\partial x_2} & \cdots & \frac{\partial f_n}{\partial x_n} \end{vmatrix}$$

The notation

$$\frac{\partial(f_1, f_2, \ldots, f_n)}{\partial(x_1, x_2, \ldots, x_n)}$$

is also used to denote the Jacobian $J_f(x)$.

14.314 Hessian determinants.

The Jacobian of the derivatives $\dfrac{\partial \phi}{\partial x_1}, \dfrac{\partial \phi}{\partial x_2}, \ldots, \dfrac{\partial \phi}{\partial x_n}$ of a function $\phi(x_1, x_2, \ldots, x_n)$ with respect to x_1, x_2, $\ldots$, x_n is called the Hessian H of ϕ, so that

$$H = \begin{vmatrix} \frac{\partial^2 \phi}{\partial x_1^2} & \frac{\partial^2 \phi}{\partial x_1 \partial x_2} & \frac{\partial^2 \phi}{\partial x_1 \partial x_3} & \cdots & \frac{\partial^2 \phi}{\partial x_1 \partial x_n} \\ \frac{\partial^2 \phi}{\partial x_2 \partial x_1} & \frac{\partial^2 \phi}{\partial x_2^2} & \frac{\partial^2 \phi}{\partial x_2 \partial x_3} & \cdots & \frac{\partial^2 d2\phi}{\partial x_2 \partial x_n} \\ \vdots & \vdots & \vdots & \ddots & \vdots \\ \frac{\partial^2 \phi}{\partial x_n \partial x_1} & \frac{\partial^2 \phi}{\partial x_n \partial x_2} & \frac{\partial^2 \phi}{\partial x_n \partial x_3} & \cdots & \frac{\partial^2 \phi}{\partial x_n^2} \end{vmatrix}.$$

14.315 Wronskian determinants.

Let f_1, f_2, $\ldots$, f_n be n functions each n times differentiable with respect to x in some open interval (a, b). Then the Wronskian $W(x)$ of $f_1, f_2, \ldots, f_n$ is defined by

$$W(x) = \begin{vmatrix} f_1 & f_2 & \cdots & f_n \\ f_1^{(1)} & f_2^{(1)} & \cdots & f_n^{(1)} \\ f_1^{(2)} & f_2^{(2)} & \cdots & f_n^{(2)} \\ \vdots & \vdots & \ddots & \vdots \\ f_1^{(n-1)} & f_2^{(n-1)} & \cdots & f_n^{(n-1)} \end{vmatrix},$$

where $f_i^{(r)} = \dfrac{d^r f_i}{dx^r}$.

14.316 Properties.

1. $\dfrac{dW}{dx}$ follows from $W(x)$ by replacing the last row of the determinant defining $W(x)$ by the nth derivatives $f_1^{(n)}, f_2^{(n)}, \ldots, f_n^{(n)}$.

2. If constants $k_1, k_2, \ldots, k_n$ exist, not all zero, such that

$$k_1 f_1 + k_2 f_2 + \cdots + k_n f_n = 0$$

for all x in (a, b), then $W(x) = 0$ for all x in (a, b).

3. The vanishing of the Jacobian throughout (a, b) is necessary, but not sufficient, for the linear dependence of $f_1, f_2, \ldots, f_n$.

14.317   Gram–Kowalewski theorem on linear dependence.

A necessary and sufficient condition for n functions $f_1, f_2, \ldots, f_n$ square integrable over $a \leq n \leq b$ to be linearly dependent in this interval is the vanishing of the Gram determinant

$$G(f_1, f_2, \ldots, f_n) = \begin{vmatrix} \int_a^b f_1^2(x)\,dx & \int_a^b f_1(x)f_2(x)\,dx & \cdots & \int_a^b f_1(x)f_n(x)\,dx \\ \int_a^b f_2(x)f_1(x)\,dx & \int_a^b f_2^2(x)\,dx & \cdots & \int_a^b f_2(x)f_n(x)\,dx \\ \vdots & \vdots & \ddots & \vdots \\ \int_a^b f_n(x)f_1(x)\,dx & \int_a^b f_n(x)f_2(x)\,dx & \cdots & \int_a^b f_n^2(x)\,dx \end{vmatrix}. \qquad \text{SA 2 (Theorem 3)}$$

14.318   If the n functions $f_1, f_2, \ldots, f_n$ are square integrable over $a \leq n \leq b$, then the Gram determinant

$$G(f_1, f_2, \ldots, f_n) \geq 0,$$

and the equality sign holds only when the functions are linearly dependent in $a \leq n \leq b$.

SA 4 (Corollary 1)

14.319   The rank of the matrix corresponding to the Gram determinant $G(f_1, f_2, \ldots, f_n)$ gives the maximum number of linearly independent functions $f_1, f_2, \ldots, f_n$ in $a \leq x \leq b$. If the rank is r, then r of the functions are linearly independent, and the other $n - r$ functions are linearly dependent on these.

SA 3 (Theorem 4)

15 Norms

15.1–15.9 Vector Norms

15.11 General properties

The **vector norm** $||\mathbf{x}||$ of an $n \times 1$ column vector $\mathbf{x}$ is a nonnegative number having the property that

1. $||\mathbf{x}|| > 0$ when $\mathbf{x} \neq \mathbf{0}$ and $||\mathbf{x}|| = 0$ if, and only if, $\mathbf{x} = \mathbf{0}$;
2. $||k\mathbf{x}|| = |k|||\mathbf{x}||$ for any scalar k;
3. $||\mathbf{x} + \mathbf{y}|| \leq ||\mathbf{x}|| + ||\mathbf{y}||$.

15.21 Principal vector norms

15.211 The norm $||\mathbf{x}||_1$

If $\mathbf{x}$ is a vector with complex components $x_1, x_2, \ldots, x_n$, then

$$||\mathbf{x}||_1 = \sum_{r=1}^{n} |x_r|. \qquad \text{VA 15}$$

15.212 The norm $||\mathbf{x}||_2$ (Euclidean or L_2 norm)

If $\mathbf{x}$ is a vector with complex components $x_1, x_2, \ldots, x_n$, then

$$||\mathbf{x}||_2 = \left(\sum_{r=1}^{n} |x_r|^2 \right)^{1/2}. \qquad \text{VA 8}$$

15.213 The norm $||\mathbf{x}||_\infty$

If $\mathbf{x}$ is a vector with complex components $x_1, x_2, \ldots, x_n$, then
$$||\mathbf{x}||_\infty = \max_i |x_i|. \qquad \text{VA 15}$$

15.31 Matrix norms

15.311 General properties

The **matrix norm** $||\mathbf{A}||$ of a square matrix $\mathbf{A}$ is a nonnegative number associated with $\mathbf{A}$ having the property that

1. $||\mathbf{A}|| > 0$ when $\mathbf{A} \neq \mathbf{0}$ and $||\mathbf{A}|| = 0$ if, and only if, $\mathbf{A} = \mathbf{0}$;
2. $||k\mathbf{A}|| = |k|||\mathbf{A}||$ for any scalar k;
3. $||\mathbf{A} + \mathbf{b}|| \leq ||\mathbf{A}|| + ||\mathbf{b}||$;
4. $||\mathbf{Ab}|| \leq ||\mathbf{A}||||\mathbf{b}||$.   VA 9

The matrix norm $||\mathbf{A}||$ associated with $\mathbf{A} = [a_{ij}]$, and the vector norm $||\mathbf{x}||$ associated with the column vector $\mathbf{x}$ for which the matrix product $\mathbf{Ax}$ is defined, are said to be **compatible** if

$$||\mathbf{Ax}|| \leq ||\mathbf{A}||||\mathbf{x}||$$

15.312 Induced norms

When a vector $\mathbf{z}$ with norm $||\mathbf{z}||$ exists such that the maximum is attained in the expression

$$||\mathbf{A}|| = \max_{||\mathbf{z}||=1} ||\mathbf{Az}||,$$

then $||\mathbf{A}||$ is a matrix norm and is said to be the **natural norm induced** by, or **subordinate** to, the vector norm $||\mathbf{z}||$.   NO 428

15.313 Natural norm of unit matrix

If $\mathbf{I}$ is the unit matrix, then for any natural norm

$$||\mathbf{I}|| = 1$$ NO 429

15.41 Principal natural norms

The natural matrix norms induced on matrix $\mathbf{A} = [a_{ij}]$ by the 1, 2, and ∞ vector norms are as follows:

15.411 Maximum absolute column sum norm

$$||\mathbf{A}||_1 = \max_j \sum_{i=1}^{n} |a_{ij}|.$$ NO 429

15.412 Spectral norm

If $\mathbf{A}^{\mathrm{H}}$ denotes the Hermitian transpose of the square matrix $\mathbf{A} = [a_{ij}]$, so that $\mathbf{A}^{\mathrm{H}} = [\overline{a_{ji}}]$ with a bar denoting the complex conjugate operation, then

$$||\mathbf{A}||_2 = \sqrt{\text{maximum eigenvalue of } \mathbf{A}^{\mathrm{H}}\mathbf{A}},$$

or, equivalently,

$$||\mathbf{A}||_2 = \max_{||x||_2 \neq 0} \frac{||\mathbf{Ax}||_2}{||\mathbf{x}||_2}.$$ NO 429

15.413 Maximum absolute row sum norm

$$||\mathbf{A}||_{\infty} = \max_{i} \sum_{j=1}^{n} |a_{ij}|.$$
NO 429

15.51 Spectral radius of a square matrix

Let $\mathbf{A} = [a_{ij}]$ be an $n \times n$ matrix with elements that may be complex, and with eigenvalues $\lambda_1, \lambda_2, \ldots, \lambda_n$. Then the **spectral radius** $\rho(\mathbf{A})$ of $\mathbf{A}$ is the number

$$\rho(\mathbf{A}) = \max_{1 \leq i \leq n} |\lambda_i|.$$
VA 9

15.511 Inequalities concerning matrix norms and the spectral radius

1. $||\mathbf{A}||_2^2 \leq ||\mathbf{A}||_1 ||\mathbf{A}||_{\infty}.$
NO 431

2.   If $\mathbf{A}$ is any arbitrary $n \times n$ matrix with elements that may be complex, and the $n \times n$ matrix $\mathbf{U}$ is unitary, so that $\mathbf{U}^{\mathrm{H}} = \mathbf{U}^{-1}$, with ^{H} denoting the Hermitian transpose of $\mathbf{A}$ (see **13.123**), then

$$||\mathbf{A}\mathbf{U}|| = ||\mathbf{U}\mathbf{A}|| = ||\mathbf{A}||.$$
VA 15

3.   If $\mathbf{A}$ is any nonsingular $n \times n$ matrix with elements that may be complex with eigenvalues λ_1, λ_2, λ_n, then

$$\frac{1}{||\mathbf{A}^{-1}||} \leq |\lambda| \leq ||\mathbf{A}||.$$
VA 16

4.   For any square matrix $\mathbf{A}$ with spectral radius $\rho(\mathbf{A})$ and any natural norm $||\mathbf{A}||$,

$$\rho(\mathbf{A}) \leq ||\mathbf{A}||.$$
NO 430

5.   If the square matrix $\mathbf{A}$ is Hermitian, then

$$\rho(\mathbf{A}) = ||\mathbf{A}||.$$

6.   If the square matrix $\mathbf{A}$ is Hermitian and $P_m(x)$ is any polynomial of degree m with real coefficients, then

$$||P_m(\mathbf{A})|| = \rho(P_m(\mathbf{A})).$$

7.   If $\mathbf{A}$ is any arbitrary $n \times n$ matrix with elements that may be complex, then the sequence of matrices $\mathbf{A}, \mathbf{A}^2, \mathbf{A}^3, \ldots$ converges to the null matrix as $n \to \infty$ if, and only if, $\rho(\mathbf{A}) < 1$.
NO 303

15.512 Deductions from Gerschgorin's theorem (see **15.814**)

1.   Let $\mathbf{A}$ be any arbitrary $n \times n$ matrix with elements that may be complex; then $\rho(\mathbf{A}) \leq$

$$\min\left(\max_{1 \leq i \leq n} \sum_{j=1}^{n} |a_{ij}|, \max_{1 \leq j \leq n} \sum_{i=1}^{n} |a_{ij}| \right).$$
VA 17

2. Let $\mathbf{A}$ be any arbitrary $n \times n$ matrix with elements that may be complex, and $x_1, x_2, \ldots, x_n$ be any set of n positive numbers; then $\quad \rho(\mathbf{A}) \le \min \left(\max_{1 \le i \le n} \left(\frac{\sum_{j=1}^{n} |a_{ij}| x_j}{x_i} \right), \max_{1 \le j \le n} \left(x_j \sum_{i=1}^{n} \frac{|a_{ij}|}{x_i} \right) \right)$.

VA 18

15.61 Inequalities involving eigenvalues of matrices

The **eigenvalues** (**characteristic values** or **latent roots**) λ of an $n \times n$ matrix $\mathbf{A} = [a_{ij}]$ are the solutions to the characteristic equation

$$|\mathbf{A} - \lambda\mathbf{I}| = 0.$$

When expanded, the determinant $|\mathbf{A} - \lambda\mathbf{I}|$ is called the **characteristic polynomial** and it has the form

$$|\mathbf{A} - \lambda\mathbf{I}| = (-1)^n \lambda^n + c_{n-1}\lambda^{n-1} + c_{n-2}\lambda^{n-2} + \cdots + c_1\lambda + c_0.$$

The zeros of this polynomial satisfy the characteristic equation and so are the eigenvalues of $\mathbf{A}$. In the characteristic polynomial the coefficients have the form

$$c_{n-r} = (-1)^{n-r} \quad \text{(sum of all principal minors of } |\mathbf{A}| \text{ of order } r)$$

It then follows that

$$b_{n-1} = (-1)^n (a_{11} + a_{22} + \cdots + a_{nn}),$$

$$b_{n-2} = (-1)^n \sum_{i<j} (a_{ii}a_{jj} - a_{ij}a_{ji}),$$

$$b_0 = |\mathbf{A}|.$$

Since the sum of the elements of the leading diagonal of $\mathbf{A}$ is called the **trace** of $\mathbf{A}$, written $\operatorname{tr}\mathbf{A}$, it follows that $b_{n-1} = (-1)^n \operatorname{tr}\mathbf{A}$.

ML 198

15.611 Cayley–Hamilton theorem

Every square matrix $\mathbf{A}$ satisfies its characteristic equation, so that

$$(-1)^n \mathbf{A}^n + c_{n-1}\mathbf{A}^{n-1} + c_{n-2}\mathbf{A}^{n-2} + \cdots + c_1\mathbf{A} + c_0\mathbf{I} = \mathbf{0}.$$

ML 206

15.612 Corollaries

1. If $\mathbf{A}$ is nonsingular then its adjoint, denoted by $\operatorname{adj}\mathbf{A}$, is

$$\operatorname{adj}\mathbf{A} = -\left[(-1)^n \mathbf{A}^{n-1} + c_{n-1}\mathbf{A}^{n-2} + c_{n-2}\mathbf{A}^{n-3} + \cdots + c_2\mathbf{A} + c_1\mathbf{I}\right].$$

2. If $\mathbf{A}$ is nonsingular, then the characteristic polynomial of $\mathbf{A}^{-1}$ is

$$(-1)^n \left(\lambda^n + \frac{c_1}{|\mathbf{A}|}\lambda^{n-1} + \frac{c_2}{|\mathbf{A}|}\lambda^{n-2} + \cdots + \frac{(-1)^n}{|\mathbf{A}|} \right).$$

15.71 Inequalities for the characteristic polynomial

The first group of inequalities that follow, which relate to the characteristic polynomial of an $n \times n$ matrix $\mathbf{A}$ whose elements may be complex, refer directly to the coefficients of the polynomial when written in the form

$$P(\lambda) \equiv |\lambda\mathbf{I} - \mathbf{A}| = \lambda^n + b_1\lambda^{n-1} + b_2\lambda^{n-2} + \cdots + b_{n-1}\lambda + b_n,$$

and only implicitly to the coefficients a_{ij} of $\mathbf{A}$ that give rise to the b_i.

15.711 Named and unnamed inequalities

The first group of inequalities relating to the eigenvalues λ satisfying $P(\lambda) = 0$ are unnamed and are as follows:

1. All the eigenvalues λ lie within or on the circle $||z|| \leq r$, where r is the positive root of

    $$|b_n| + |b_{n-1}|z + |b_{n-2}|z^2 + \cdots + |b_1|z^{n-1} - z^n = 0 \qquad \text{MG 122}$$

2. All the eigenvalues λ lie within the circle

    $$|z| < 1 + \max_i |b_i|; \qquad \text{MG 123}$$

3. When $b_n \neq 0$ the eigenvalue λ of smallest modulus lies in the annulus $R \leq |z| \leq \dfrac{R}{2^{1/n} - 1}$, where R is the positive root of

    $$|b_n| - |b_{n-1}|z - |b_{n-2}|z^2 - \cdots - z^n = 0 \qquad \text{MG 126}$$

4. All the eigenvalues λ lie on or outside the circle

    $$|z| = \min_k \left[\frac{|b_n|}{(|b_n| + |b_k|)} \right] \qquad \text{MG 126}$$

5. If the eigenvalues λ are ordered so that

    $$|\lambda_1| \geq |\lambda_2| \geq \cdots \geq |\lambda_p| > 1 \geq |\lambda_{p+1}| \geq \cdots \geq |\lambda_n|,$$

    then

    $$|z_1 z_2 \ldots z_p| \leq N, \qquad |z_p| \leq N^{\frac{1}{p}},$$

    where

    $$N^2 = 1 + |b_1|^2 + |b_2|^2 + \cdots + |b_n|^2 \qquad \text{MG 129}$$

6. All the eigenvalues λ lie in or on the circle

    $$|z| \leq \sum_{j=1}^{n} |b_j|^{1/j} \qquad \text{MG 126}$$

7. All the eigenvalues λ lie on the disk

    $$\left| z + \frac{b_1}{2} \right| \leq \left| \frac{b_1}{2} \right| + |b_2|^{1/2} + |b_3|^{1/3} + \cdots + |b_n|^{1/n} \qquad \text{MG 145}$$

8. All the eigenvalues λ lie in the annulus $m \leq ||z|| \leq M$, where

    $$m^2 = \max \left\{ 0, \min_{1 \leq j \leq n-1} \left[1 - |b_j|, |b_n|^2 \right] \right\}$$

    and

$$M^2 = \max \left\{ 1 + |b_j|, |b_n|^2 + 2 \sum_{j=1}^{n-1} |b_j|^2 \right\}.$$

The next group of inequalities are named theorems that apply to the explicit form of the characteristic polynomial $P(\lambda)$.   MG 145

15.712 Parodi's theorem

The eigenvalues λ satisfying $P(\lambda) = 0$ lie in the union of the disks

$$|z| \leq 1, \qquad |z + b_1| \leq \sum_{j=1}^{n} |b_j|.$$ MG 143

15.713 Corollary of Brauer's theorem

If

$$|b_1| > 1 + \sum_{j=2}^{n} |b_j|,$$

then one and only one eigenvalue satisfying $P(\lambda) = 0$ lies on the disk

$$|z + b_1| \leq \sum_{j=2}^{n} |b_j|.$$ MG 141

15.714 Ballieu's theorem

For any set $\mu = (\mu_1, \mu_2, \ldots, \mu_n)$ of positive numbers, let $\mu_0 = 0$ and

$$M_\mu = \max_{0 \leq k \leq n-1} \left[\frac{\mu_k + \mu_n |b_{n-k}|}{\mu_{k+1}} \right].$$

Then all the eigenvalues satisfying $P(\lambda) = 0$ lie on the disk $||z|| \leq M_\mu$.   MG 144

15.715 Routh–Hurwitz theorem

Consider the characteristic equation

$$|\lambda \mathbf{I} - \mathbf{A}| = \lambda^n + b_1 \lambda^{n-1} + \cdots + b_{n-1}\lambda + b_n = 0$$

determining the n eigenvalues λ of the real $n \times n$ matrix $\mathbf{A}$. Then the eigenvalues λ all have negative real parts if

$$\Delta_1 > 0, \qquad \Delta_2 > 0, \qquad \ldots, \qquad \Delta_n > 0,$$

where

$$\Delta_k = \begin{vmatrix} b_1 & 1 & 0 & 0 & 0 & 0 & \cdots & 0 \\ b_3 & b_2 & b_1 & 1 & 0 & 0 & \cdots & 0 \\ b_5 & b_4 & b_3 & b_2 & b_1 & 0 & \cdots & 0 \\ \vdots & \vdots & \vdots & \vdots & \vdots & & & \\ b_{2k-1} & b_{2k-2} & b_{2k-3} & b_{2k-4} & b_{2k-5} & b_{2k-6} & \cdots & b_k \end{vmatrix}$$ GM 230

15.81–15.82 Named theorems on eigenvalues

In the following theorems involving eigenvalue inequalities the elements a_{ij} of matrix $\mathbf{A}$ enter directly, and not in the form of the coefficients of the characteristic polynomial.

15.811 Schur's inequalities

If $\mathbf{A} = [a_{ij}]$ is an $n \times n$ matrix with elements that may be complex, and eigenvalues $\lambda_1, \lambda_2, \ldots, \lambda_n$, then

1. $$\sum_{i=1}^{n} |\lambda_i|^2 \leq \sum_{i,j=1}^{n} |a_{ij}|^2$$

2. $$\sum_{i=1}^{n} |\text{Re}\,\lambda_i|^2 \leq \sum_{i,j=1}^{n} \left| \frac{a_{ij} + \overline{a_{ji}}}{2} \right|^2$$

3. $$\sum_{i=1}^{n} |\text{Im}\,\lambda_i|^2 \leq \sum_{i,j=1}^{n} \left| \frac{a_{ij} - \overline{a_{ji}}}{2} \right|^2$$

ML 309

15.812 Sturmian separation theorem

Let $\mathbf{A}_r = [a_{ij}]$ with $i,\, j = 1, 2, \ldots, r$ and $r = 1, 2, \ldots, N$ be a sequence of N symmetric matrices of increasing order. Then if $\lambda_k(\mathbf{A}_r)$ for $k = 1, 2, \ldots, r$ denotes the kth eigenvalue of A_r, where the ordering is such that

$$\lambda_1(A_r) \geq \lambda_2(A_r) \geq \cdots \geq \lambda_r(A_r),$$

it follows that

$$\lambda_{k+1}(A_{i+1}) \leq \lambda_k(A_i) \leq \lambda_k(A_{i+1})$$

BL 115

15.813 Poincare's separation theorem

Let $\{\mathbf{y}^k\}$, with $k = 1, 2, \ldots, K$, be a set of orthonormal vectors so that the inner product $(\mathbf{y}^k, \mathbf{y}^k) = 1$. Set

$$\mathbf{x} = \sum_{k=1}^{K} u_k \mathbf{y}^k,$$

so that for any square matrix $\mathbf{A}$ for which the product $\mathbf{Ax}$ is defined, the quadratic form

$$(\mathbf{x}, \mathbf{Ax}) = \sum_{k,l=1}^{K} u_k u_l \left(\mathbf{y}^k, \mathbf{Ay}^l \right).$$

Then if

$$\mathbf{b}_K = \left(\mathbf{y}^k, \mathbf{Ay}^l \right) \text{ for } k, l = 1, 2, \ldots, K,$$

it follows that

$$\lambda_i(\mathbf{b}_K) \leq \lambda_i(\mathbf{A}) \qquad \text{for } i = 1, 2, \ldots, K,$$
$$\lambda_{K-j}(\mathbf{b}_K) \geq \lambda_{N-j}(\mathbf{A}) \qquad \text{for } j = 0, 1, 2, \ldots, K - 1,$$

BL 117

15.814 Gerschgorin's theorem

Let $\mathbf{A} = [a_{ij}]$ be any arbitrary $n \times n$ matrix with elements that may be complex, and let

$$\Lambda_i \equiv \sum_{j=1, i \neq j}^{n} |a_{ij}| \text{ for } i = 1, 2, \dots, n.$$

Then all of the eigenvalues λ_i of $\mathbf{A}$ lie in the union of the n disks Γ_i, where

$$\Gamma_i : \quad |z - a_{ii}| \leq \Lambda_i \text{ for } i = 1, 2, \dots, n. \qquad \text{VA 16}$$

15.815 Brauer's theorem

If in Gerschgorin's theorem for a given m

$$|a_{jj} - a_{mm}| \geq \Lambda_j + \Lambda_m$$

for all $j \neq m$, then one and only one eigenvalue of $\mathbf{A}$ lies in the disk Γ_m. $\qquad \text{MG 141}$

15.816 Perron's theorem

If $\boldsymbol{\mu} = (\mu_1, \mu_2, \dots, \mu_n)$ is an arbitrary set of positive numbers, then all the eigenvalues λ of the $n \times n$ matrix $\mathbf{A} = [a_{ij}]$ lie on the disk $|z| \leq \mathbf{M}_\mu$, where

$$\mathbf{M}_\mu = \max_{1 \leq i \leq n} \sum_{j=1}^{n} \frac{\mu_j}{\mu_i} |a_{ij}|. \qquad \text{MG 141}$$

15.817 Frobenius theorem

If $\mathbf{A} = [a_{ij}]$ is a matrix with positive coefficients, so that $a_{ij} > 0$ for all $i, j = 1, 2, \dots, n$, then $\mathbf{A}$ has a positive eigenvalue λ_0 and all its eigenvalues lie on the disk

$$|z| \leq \lambda_0 \qquad \text{MG 142}$$

15.818 Perron–Frobenius theorem

If all elements a_{ij} of an irreducible matrix $\mathbf{A}$ are nonnegative, then $R = \min M_\lambda$ is a simple eigenvalue of $\mathbf{A}$ and all the eigenvalues of $\mathbf{A}$ lie on the disk

$$|z| \leq R,$$

where, if $\boldsymbol{\lambda} = (\lambda_1, \lambda_2, \dots, \lambda_n)$ is a set of nonnegative numbers, not all zero,

$$M_\lambda = \inf \left\{ \mu : \mu \lambda_i > \sum_{j=1}^{n} |a_{ij}| \lambda_j, 1 \leq i \leq n \right\}$$

and $R = \min M_\lambda$.

Furthermore, if $\mathbf{A}$ has exactly p eigenvalues $(p \leq n)$ on the circle $|z| = R$, then the set of all its eigenvalues is invariant under rotations $2\pi/p$ about the origin. $\qquad \text{GM 69}$

15.819 Wielandt's theorem

If the $n \times n$ matrix $\mathbf{A}$ satisfies the conditions of the Perron–Frobenius theorem and if in the $n \times n$ matrix $\mathbf{C} = [c_{ij}]$

$$|c_{ij}| \leq a_{ij}, \qquad i,j = 1, 2, \ldots, n,$$

then any eigenvalue λ_0 of $\mathbf{C}$ satisfies the inequality $|\lambda_0| \leq R$. The equality sign holds only when there exists an $n \times n$ matrix $\mathbf{D} = [\pm \delta_{ij}]$ such that $\delta_{ii} = 1$ for all i, $\delta_{ij} = 0$ for all $i \neq j$, and

$$\mathbf{C} = (\lambda_0 / R)\, \mathbf{D} \mathbf{A} \mathbf{D}^{-1}.$$

GM 69

15.820 Ostrowski's theorem

If $\mathbf{A} = [a_{ij}]$ is a matrix with positive coefficients and λ_0 is the positive eigenvalue in Frobenius' theorem, then the $n-1$ eigenvalues $\lambda_j \neq \lambda_0$ satisfy the inequality

$$|\lambda_j| \leq \lambda_0 \frac{M^2 - m^2}{M^2 + m^2},$$

where

$$M = \max a_{ij}, \qquad m = \min a_{ij} \qquad \text{for} \quad i,j = 1, 2, \ldots, n$$

MG 145

15.821 First theorem due to Lyapunov

In order that all the eigenvalues of the real $n \times n$ matrix $\mathbf{A}$ have negative real parts, it is necessary and sufficient that if V is an $n \times n$ matrix, the equation

$$\mathbf{A}^\mathrm{T}\mathbf{V} + \mathbf{V}\mathbf{A} = -\mathbf{I}$$

has as a solution the matrix of coefficients $\mathbf{V}$ of some positive-definite quadratic form $(\mathbf{x}, \mathbf{V}\mathbf{x})$ (see **13.21**).

GM 224

15.822 Second theorem due to Lyapunov

If all the eigenvalues of the real matrix $\mathbf{A}$ have negative real parts, then to an arbitrary negative-definite quadratic form $(\mathbf{x}, \mathbf{W}\mathbf{x})$ with $\mathbf{x} = \mathbf{x}(t)$ there corresponds a positive-definite quadratic form $(\mathbf{x}, \mathbf{V}\mathbf{x})$ such that if one takes

$$\frac{d\mathbf{x}}{dt} = \mathbf{A}\mathbf{x}$$

then $(\mathbf{x}, \mathbf{V}\mathbf{x})$ and $(\mathbf{x}, \mathbf{W}\mathbf{x})$ satisfy

$$\frac{d}{dt}(\mathbf{x}, \mathbf{V}\mathbf{x}) = (\mathbf{x}, \mathbf{W}\mathbf{x}).$$

Conversely, if for some negative-definite form $(\mathbf{x}, \mathbf{W}\mathbf{x})$ there exists a positive-definite form $(\mathbf{x}, \mathbf{V}\mathbf{x})$ connected to $(\mathbf{x}, \mathbf{W}\mathbf{x})$ by the preceding two equations, then all the eigenvalues of $\mathbf{A}$ have negative real parts (see **13.21**, **13.31**).

GM 222

15.823 Hermitian matrices and diophantine relations involving circular functions of rational angles due to Calogero and Perelomov

1.　The off-diagonal Hermitian matrix $\mathbf{A}$ of rank n whose elements are given by

$$a_{jk} = (1 - \delta_{jk})\left\{1 + i\cot\left[\frac{(j-k)\pi}{n}\right]\right\},$$

has the integer eigenvalues

$$\lambda_a^{(a)} = 2s - n - 1 \text{ for } s = 1, 2, \ldots, n,$$

and the corresponding eigenvectors $v^{(s)}$ have the components

$$v_j{}^{(s)} = \exp\left(-\frac{2\pi i s j}{n}\right) \qquad \text{for} \quad j = 1, 2, \ldots, n.$$

2.   The two off-diagonal Hermitian matrices $\mathbf{b}$ and $\mathbf{C}$ whose elements are defined by the formulas

$$b_{jk} = (1 - \delta_{jk}) \sin^{-2}\left[\frac{(j-k)\pi}{n}\right],$$

$$c_{jk} = (1 - \delta_{jk}) \sin^{-4}\left[\frac{(j-k)\pi}{n}\right],$$

are related to the matrix $\mathbf{A}$ in (1) by the equations

$$\mathbf{b} = \frac{1}{2}\left(\mathbf{A}^2 + 2\mathbf{A} - \sigma_n{}^{(1)}\mathbf{I}\right),$$

$$\mathbf{C} = -\frac{1}{6}\left(\mathbf{B}^2 - 2\left(2 + \sigma_n{}^{(1)}\right)\mathbf{B} - \sigma_n{}^{(2)}\mathbf{I}\right),$$

where $\mathbf{I}$ is the unit matrix and

$$\sigma_n{}^{(1)} = \frac{1}{3}\left(n^2 - 1\right), \qquad \sigma_n{}^{(2)} = \frac{1}{45}\left(n^2 - 1\right)\left(n^2 + 11\right).$$

The eigenvalues of $\mathbf{b}$ and $\mathbf{C}$ corresponding to the eigenvector $v_j{}^{(s)}$ in (1) have the form

$$\lambda_s{}^{(b)} = \sigma_n{}^{(1)} - 2s(n - s) \qquad\qquad \text{for } s = 1, 2, \ldots, n,$$

$$\lambda_s{}^{(c)} = \sigma_n{}^{(2)} - 2s(n - s)\frac{s(n - s) + 2}{3} \qquad\qquad \text{for } s = 1, 2, \ldots, n.$$

3.   Together, the above two results imply the following diophantine summation rules:

(a) $\displaystyle\sum_{k=1}^{n-1} \cot\left(\frac{k\pi}{n}\right)\sin\left(\frac{2sk\pi}{n}\right) = n - 2s \qquad \text{for } s = 1, 2, \ldots, n - 1$

(b) $\displaystyle\sum_{k=1}^{n-1} \sin^{-2}\left(\frac{k\pi}{n}\right)\cos\left(\frac{2sk\pi}{n}\right) = b_s \qquad \text{for } s = 1, 2, \ldots, n - 1,$

(c) $\displaystyle\sum_{k=1}^{n-1} \sin^{-4}\left(\frac{k\pi}{n}\right)\cos\left(\frac{2sk\pi}{n}\right) = c_s \qquad \text{for } s = 1, 2, \ldots, n - 1,$

(d) $\displaystyle\sum_{k=1}^{n-1} \sin^{-2p}\left(\frac{k\pi}{n}\right) = \sigma_n{}^{(p)},$

with $b_s, c_s, \sigma_n{}^{(1)}$ and $\sigma_n{}^{(2)}$ as defined in (2), and

$$\sigma_n{}^{(3)} = \sigma_n{}^{(1)}\frac{2n^4 + 23n^2 + 191}{315},$$

$$\sigma_n{}^{(4)} = \sigma_n{}^{(2)}\frac{3n^4 + 10n^2 + 227}{315}.$$

15.91 Variational principles

15.911 Rayleigh quotient

If $\mathbf{A}$ is an Hermitian matrix, the Rayleigh quotient $\rho(\mathbf{x})$ is the expression

$$\rho(\mathbf{x}) = \frac{(\mathbf{x}, \mathbf{Ax})}{(\mathbf{x}, \mathbf{x})}.$$

NO 407

15.912 Basic theorems

1. If the $n \times n$ matrix A is Hermitian and has eigenvalues $\lambda_1 \le \lambda_2 \le \cdots \le \lambda_n$, then

    $$\lambda_1 \le \rho \le \lambda_n,$$

    where ρ is the Rayleigh quotient for any $\mathbf{x} \ne \mathbf{0}$, and

    $$\lambda_1 = \min_{x \ne 0} \frac{(\mathbf{x}, \mathbf{Ax})}{(\mathbf{x}, \mathbf{x})} \qquad \text{and} \qquad \lambda_n = \max_{x \ne 0} \frac{(\mathbf{x}, \mathbf{Ax})}{(\mathbf{x}, \mathbf{x})}.$$

    NO 407

2. If the $n \times n$ matrix $\mathbf{A}$ is Hermitian and has eigenvalues $\lambda_1 \le \lambda_2 \le \cdots \le \lambda_n$ corresponding to the eigenvectors $\mathbf{x}_1, \mathbf{x}_2, \dots, \mathbf{x}_n$, respectively, and $\mathbf{x} \ne \mathbf{0}$ is such that

    $$(\mathbf{x}, \mathbf{x}_1) = (\mathbf{x}, \mathbf{x}_2) = \cdots = (\mathbf{x}, \mathbf{x}_n) = 0,$$

    then

    $$\lambda_j = \min_{x} \frac{(\mathbf{x}, \mathbf{Ax})}{(\mathbf{x}, \mathbf{x})},$$

    and

    $$\lambda_j \le \frac{(\mathbf{x}, \mathbf{Ax})}{(\mathbf{x}, \mathbf{x})} \le \lambda_n.$$

    NO 410

3. If the $n \times n$ matrix $\mathbf{A}$ is Hermitian, then the eigenvalue

    $$\lambda_r = \max \left(\min \frac{(\mathbf{x}, \mathbf{Ax})}{(\mathbf{x}, \mathbf{x})} \right),$$

    where first the minimum over $\mathbf{x}$ is taken subject to $(\mathbf{b}_i, \mathbf{x}) = 0, i = 1, 2, \dots, r-1$, with the $\mathbf{b}_i$ regarded as fixed vectors, and then the maximum over all possible $\mathbf{b}_i$. Also, the eigenvalue

    $$\lambda_r = \min \left(\max \frac{(\mathbf{x}, \mathbf{Ax})}{(\mathbf{x}, \mathbf{x})} \right),$$

    where now the maximum over $\mathbf{x}$ is taken first subject to $(\mathbf{b}_i, \mathbf{x}) = 0, i = r+1, r+2, \dots, n$ for fixed $\mathbf{b}_i$, and then the minimum over all possible $\mathbf{b}_i$.

    NO 414

4. The $(n-1)$ eigenvalues $\lambda_1', \lambda_2', \dots, \lambda_{n-1}'$ obtained from the $(n-1) \times (n-1)$ matrix derived from an Hermitian matrix $\mathbf{A}$ from which the last row and column have been omitted separate the n eigenvalues of $\mathbf{A}$, so that

    $$\lambda_1 < \lambda_1' < \lambda_2 < \lambda_2' < \cdots < \lambda_{n-1}' < \lambda_n \qquad \text{(see 15.812)}$$

16 Ordinary differential equations

16.1–16.9 Results relating to the solution of ordinary differential equations

16.11 First-order equations

16.111 Solution of a first-order equation

Consider the real function $f(t,x)$ that is defined and continuous in an open set $D \subset R^2$. Then a **solution** to the first-order differential equation

$$\frac{dx}{dt} = f(t,x)$$

in the open interval $I \subset R$ is a real function $u(t)$ that is defined and is both continuous and differentiable in I, with the property that

(i) $\quad (t, u(t)) \in D$ for $t \in I$,

(ii) $\quad \dfrac{du}{dt} = f(t, u(t))$ for $t \in I$.

16.112 Cauchy problem

The **Cauchy problem** for the differential equation

$$\frac{dx}{dt} = f(t,x)$$

is the problem of existence and uniqueness of the solution to this equation satisfying the initial condition

$$u(t_0) = x_0,$$

where $(t_0, u(t_0)) \in D$, the open set defined above. The solution to the initial value problem may be expressed in the form of the integral equation

$$u(t) = x_0 + \int_{t_0}^{t} f\left(\tau, u(\tau)\right) \, d\tau \qquad \text{(see \textbf{16.316})}$$

16.113 Approximate solution to an equation

The real function $\phi(t)$ is said to be an **approximate solution**, to within the error ϵ, of the differential equation

$$\frac{dx}{dt} = f(t, x)$$

if ϕ' is piecewise continuous, and for a given $\epsilon > 0$ and an open interval $I \subset R$,

$$|\phi'(t) - f(t, \phi(t))| \leq \epsilon,$$

except at points of discontinuity of the derivative.   HU 3

16.114 Lipschitz continuity of a function

The real function $f(t, x)$ defined and continuous in some open set $D \subset R^2$ is said to be **Lipschitz continuous** with respect to x for some constant $k > 0$ if, for all points (t, x_1) and (t, x_2) belonging to D

$$|f(t, x_1) - f(t, x_2)| \leq k|x_1 - x_2| \qquad \text{HU 5}$$

16.21 Fundamental inequalities and related results

16.211 Gronwall's lemma

Let the three piecewise continuous, non-negative functions u, v, and w be defined in the interval $[0, a]$ and satisfy the inequality

$$w(t) \leq u(t) + \int_0^t v(\tau)w(\tau)\, d\tau,$$

except at points of discontinuity of the functions. Then, except at these same points,

$$w(t) \leq u(t) + \int_0^t u(\tau)v(\tau) \exp\left(\int_\tau^t v(\sigma)\, d\sigma\right) d\tau \qquad \text{BB 135}$$

16.212 Comparison of approximate solutions of a differential equation

Let f be a real function that is defined in an open set $D \subset R^2$, in which it is both continuous and Lipschitz continuous. In addition, let u_1 and u_2 be two approximate solutions of

$$\frac{dx}{dt} = f(t, x)$$

in an open set $I \subset R$ in the sense already defined, with

$$|u_1'(t) - f(t, u_1(t))| \leq \epsilon_1, \quad |u_2'(t) - f(t, u_2(t))| \leq \epsilon_2,$$

except where the derivatives are discontinuous. Then, if for all $t_0 \in I$

$$|u_1(t_0) - u_2(t_0)| \leq \delta,$$

it follows that

$$|u_1(t) - u_2(t)| \leq \delta \exp\{|t - t_0|\} + \left(\frac{\epsilon_1 + \epsilon_2}{k}\right)[\exp\{k|t - t_0|\} - 1] \qquad \text{HU 6}$$

16.31 First-order systems

16.311 Solution of a system of equations

The **system** of n first-order differential equations

$$\frac{dx_1}{dt} = f_1\left(t, x_1, x_2, \ldots, x_n\right),$$

$$\frac{dx_2}{dt} = f_2\left(t, x_1, x_2, \ldots, x_n\right),$$

$$\vdots$$

$$\frac{dx_n}{dt} = f_n\left(t, x_1, x_2, \ldots, x_n\right),$$

in which the functions $f_1, f_2, \ldots, f_n$ are real and continuous in an open set $D \subset R^{n+1}$ may be written in the concise matrix form

$$\frac{d\mathbf{x}}{dt} = \mathbf{f}\left(t, \mathbf{x}\right),$$

where $\mathbf{x}$ and $\mathbf{f}$ are $n \times 1$ column vectors. Its solution in the open interval $I \subset R$ is the vector $\mathbf{u}(t)$ with elements $u_1(t), u_2(t), \ldots, u_n(t)$ with the property that

(i) $(t, \mathbf{u}(t)) \in D$ for $t \in I$,

(ii) $\dfrac{d\mathbf{u}}{dt} = \mathbf{f}(t, \mathbf{u}(t))$ for $t \in I$. HU 24

16.312 Cauchy problem for a system

The **Cauchy problem** for the system

$$\frac{d\mathbf{x}}{dt} = \mathbf{f}\left(t, \mathbf{x}\right)$$

is the problem of existence and uniqueness of the solution to this system satisfying the **initial vector** condition

$$\mathbf{u}\left(t_0\right) = \mathbf{x}_0,$$

where $(t_0, \mathbf{u}(t_0)) \in D$, the open set defined above in connection with the system. The solution to the initial value problem may be expressed in the form of the **vector integral equation**

$$\mathbf{u}(t) = \mathbf{x}_0 + \int_{t_0}^{t} \mathbf{f}(\tau, \mathbf{u}(\tau))\, d\tau$$

16.313 Approximate solution to a system

The real vector $\phi(t)$ is said to be an **approximate vector solution**, to within the order ϵ, of the system

$$\frac{d\mathbf{x}}{dt} = \mathbf{f}\left(t, \mathbf{x}\right),$$

if the elements of ϕ' are piecewise continuous, and for a given $\epsilon > 0$ and open interval $I \subset R$,

$$\|\phi'(t) - \mathbf{f}\left(t, \phi(t)\right)\| \leq \epsilon,$$

except at points of discontinuity of the derivative, where $\|\mathbf{w}\|$ denotes the supremum norm

$$\|\mathbf{w}\| = \sup\left(|w_1|, |w_2|, \ldots, |w_n|\right)$$ HU 25

16.314 Lipschitz continuity of a vector

The real vector $\mathbf{f}(t, x)$ defined and continuous in some open set $D \subset R^n$ is said to be **Lipschitz contin-**
uous with respect to x for some constant $k > 0$ if, for all points $(t, \mathbf{x}_1)$, $(t, \mathbf{x}_2)$ belonging to D,

$$||\mathbf{f}(t, \mathbf{x}_1) - \mathbf{f}(t, \mathbf{x}_2)|| \le k||\mathbf{x}_1 - \mathbf{x}_2||$$

HU 26

16.315 Comparison of approximate solutions of a system

Let $\mathbf{f}$ be a real vector defined in an open set $D \subset R \times R^n$ in which it is both continuous and Lipschitz
continuous. In addition, let $\mathbf{u}_1$ and $\mathbf{u}_2$ be two approximate solutions of the system

$$\frac{d\mathbf{x}}{dt} = \mathbf{f}(t, \mathbf{x})$$

in an open set $I \subset R$ in the sense already defined, with

$$|\mathbf{u}_1'(t) - \mathbf{f}(t, \mathbf{u}_1(t))| \le \epsilon_1, \qquad |\mathbf{u}_2'(t) - \mathbf{f}(t, \mathbf{u}_2(t))| \le \epsilon_2,$$

except where the derivatives are discontinuous. Then, if for all $t_0 \in I$

$$||\mathbf{u}_1(t_0) - \mathbf{u}_2(t_0)|| \le \delta,$$

it follows that

$$||\mathbf{u}_1(t) - \mathbf{u}_2(t)|| \le \delta \exp\{k|t - t_0|\} + \left(\frac{\epsilon_1 + \epsilon_2}{k}\right)[\exp\{k|t - t_0|\} - 1]$$

HU 27

16.316 First-order linear differential equation

The **first-order linear differential equation** when expressed in the canonical form

$$\frac{dy}{dt} + P(t)y = Q(t)$$

has an integrating factor

$$\mu(t) = \exp\left(\int P(t)\, dt\right),$$

and a general solution

$$y(t) = \frac{1}{\mu(t)}\left(\mu(t_0)\, y_0 + \int_{t_0}^t \mu(\xi)Q(\xi)\, d\xi\right),$$

where $y_0 = y(t_0)$.

16.317 Linear systems of differential equations

Consider the **homogeneous system** of linear differential equations

$$\frac{d\mathbf{x}}{dt} = \mathbf{A}(t)\mathbf{x},$$

where $\mathbf{x}$ is an $n \times 1$ column vector and $\mathbf{A}(t)$ an $n \times n$ matrix. Then a **fundamental system** of solutions of
this system is a set of n linearly independent solution vectors $\phi_1(t), \phi_2(t), \dots, \phi_n(t)$, The square matrix
$\mathbf{K}(t)$ whose columns comprise the vectors $\phi_1(t), \phi_2(t), \dots, \phi_n(t)$ is called the **fundamental matrix** of
the differential equation, and we have the representation

$$|\mathbf{K}(t)| = |\mathbf{K}(t_0)| \exp\left(\int_{t_0}^t \operatorname{tr} \mathbf{A}(\tau)\, d\tau\right).$$

Using the fundamental matrix $\mathbf{K}(t)$ defined in terms of the homogeneous system, the unique solution to
the inhomogeneous system

$$\frac{d\mathbf{x}}{dt} = \mathbf{A}(t)\mathbf{x} + \mathbf{b}(t),$$

assuming the initial value $\mathbf{x}(t_0) = \mathbf{x}_0$, is

$$\phi(t) = \mathbf{K}(t)[\mathbf{K}(t_0)]^{-1}\mathbf{x}_0 + \mathbf{K}(t)\int_{t_0}^{t} [\mathbf{K}(\tau)]^{-1}\mathbf{b}(\tau)\,d\tau, \qquad \text{HU 43}$$

where $\mathbf{b}(t)$ is an $n \times 1$ column vector.

CL 69

16.41 Some special types of elementary differential equations

16.411 Variables separable

A first-order differential equation is said to be **variables separable** if it is of the form

$$\frac{dy}{dx} = M(x)N(y),$$

or

$$P(x)Q(y)\,dx + R(x)S(y)\,dy = 0.$$

It may then be written in the form

$$M(x)\,dx - \frac{1}{N(y)}\,dy = 0,$$

or

$$\frac{P(x)}{R(x)}\,dx + \frac{S(y)}{Q(y)}\,dy = 0,$$

provided $R(x)Q(y) \neq 0$.

16.412 Exact differential equations

A differential equation

$$M(x, y)\,dx + N(x, y)\,dy = 0$$

is said to be **exact** if there exists a function $h(x, y)$ such that

$$d\,[h(x, y)] = M(x, y)\,dx + N(x, y)\,dy \qquad \text{IN 16}$$

16.413 Conditions for an exact equation

A necessary and sufficient condition that an equation of this form is exact is that the functions $M(x, y)$ and $N(x, y)$ together with their partial derivatives $\partial M/\partial y$ and $\partial N/\partial x$ exist and are continuous in a region in which

$$\frac{\partial M}{\partial y} = \frac{\partial N}{\partial x} \qquad \text{IN 16}$$

16.414 Homogeneous differential equations

A differential equation

$$M(x, y)\,dx + N(x, y)\,dy = 0$$

is said to be **algebraically homogeneous** if, for arbitrary k,

$$\frac{M(kx, ky)}{N(kx, ky)} = \frac{M(x, y)}{N(x, y)}.$$

Setting $y = sx$, it may then be expressed in the form

$$[M(1, s) + sN(1, s)]\, dx + xN(1)\, dx = 0,$$

in which the variables s and x are separable.

IN 18

16.51 Second-order equations

16.511 Adjoint and self-adjoint equations

The linear second-order differential equation

$$L(u) \equiv a(x)\frac{d^2 u}{dx^2} + b(x)\frac{du}{dx} + c(x)u = 0,$$

has associated with it the adjoint equation

$$M(v) \equiv \frac{d^2}{dx^2}\left[a(x)v\right] - \frac{d}{dx}\left[b(x)v\right] + c(x)v = 0.$$

The equation $L(u) = 0$ is said to be **self-adjoint** if $L(u) \equiv M(u)$.

A linear self-adjoint second-order differential equation defined on $[\alpha, \beta]$ can always be expressed in the form

$$\frac{d}{dx}\left(p(x)\frac{du}{dx}\right) + q(x)u = 0,$$

where $p(x)$ and $q(x)$ are continuous on $[\alpha, \beta]$ and $p(x) > 0$. The general equation $L(u) = 0$ can always be made self-adjoint and written in this form by multiplication by the factor

$$\frac{1}{a(x)}\left[\exp \int \frac{b(x)}{a(x)}\, dx\right],$$

when

$$p(x) = \exp \int \frac{b(x)}{a(x)}\, dx \quad \text{and} \quad q(x) = \frac{c(x)}{a(x)}\left[\exp \int \frac{b(x)}{a(x)}\, dx\right].$$

In general, if

$$L(u) = p_0 \frac{d^n u}{dx^n} + p_1 \frac{d^{n-1} u}{dx^{n-1}} \ldots + p_{n-1}\frac{du}{dx} + p_n u,$$

then its adjoint is

$$M(v) = (-1)^n \frac{d^n}{dx^n}\left[p_0 v\right] + (-1)^{n-1}\frac{d^{n-1}}{dx^{n-1}}\left[p_1 v\right] + \ldots - \frac{d}{dx}\left[p_{n-1}v\right] + p_n v.$$

HI 391

16.512 Abel's identity

If $p(x)$ and $q(x)$ are continuous in $[\alpha, \beta]$ in which $p(x) > 0$, and $u(x)$ and $v(x)$ are suitably differentiable with

$$\frac{d}{dx}\left(p(x)\frac{du}{dx}\right) + q(x)u = 0,$$

then the result

$$p(x)\left(u\frac{dv}{dx} - v\frac{du}{dx}\right) \equiv \text{const.}$$

is known as **Abel's identity**.

More generally, if we consider the linear n^{th}-order equation

$$p_0 \frac{d^n u}{dx^n} + p_1 \frac{d^{n-1} u}{dx^{n-1}} + \ldots + p_{n-1}\frac{du}{dx} + p_n = 0,$$

and Δ is the Wronskian of a (fundamental) set of linearly independent solutions $u_1, u_2, \ldots, u_n$, the Abel identity takes the form

$$\Delta = \Delta_0 \exp\left(-\int_{x_0}^x \frac{p_1(x)}{p_0(x)}\,dx\right),$$

where Δ_0 is the value of Δ at $x = x_0$.

IN 119

16.513 Lagrange identity

If the linear n^{th}-order equation $L(u) = 0$ is defined by

$$L(u) \equiv p_0 \frac{d^n u}{dx^n} + p_1 \frac{d^{n-1}u}{dx^{n-1}} + \ldots + p_{n-1}\frac{du}{dx} + p_n u,$$

then the expression

$$vL(u) - uM(v) = \frac{d}{dx}\{P(u,v)\},$$

where $M(v)$ is the adjoint of $L(u)$, is called the **Lagrange identity**. The expression $P(u,v)$, which is linear and homogeneous in

$$u, \frac{du}{dx}, \ldots, \frac{d^{n-1}u}{dx^{n-1}} \quad \text{and} \quad v, \frac{dv}{dx}, \ldots, \frac{d^{n-1}v}{dx^{n-1}},$$

is then known as the **bilinear concomitant**. In the case of the second-order equation

$$L(u) = a(x)\frac{d^2 u}{dx^2} + b(x)\frac{du}{dx} + c(x)u = 0,$$

with adjoint $M(v)$, the Lagrange identity becomes

$$vL(u) - uM(v) = \frac{d}{dx}\left(a(x)v\frac{du}{dx} - \frac{d}{dx}(a(x)v)u + b(x)uv\right)$$

IN 124

16.514 The Riccati equation

The general **Riccati equation** has the form

$$\frac{dz}{dx} + a(x)z + b(x)z^2 + c(x) = 0,$$

and an equation of this form results from the substitution

$$z = \frac{\left(p(x)\frac{du}{dx}\right)}{u}$$

in the general self-adjoint equation

$$\frac{d}{dx}\left(p(x)\frac{du}{dx}\right) + q(x)u = 0.$$

The further substitution $v = u\left(\exp\int_\alpha^x a(x)\,dx\right)$ in the Riccati equation then gives the more convenient form

$$\frac{dv}{dx} + r(x)v^2 + s(x) = 0,$$

with

$$r(x) = b(x)\exp\left(-\int_\alpha^x a(x)\,dx\right) \quad \text{and} \quad s(x) = c(x)\exp\left(\int_\alpha^x a(x)\,dx\right)$$

HI 273

16.515 Solutions of the Riccati equation

If in the Riccati equation

$$\frac{dv}{dx} + r(x)v^2 + s(x) = 0$$

$r(x) \neq 0$, while $r(x)$ and $s(x)$ are continuous on the interval $[\alpha, \beta]$, then every solution $v(x)$ may be expressed in the form

$$\frac{1}{r(x)} \frac{Au'(x) + Bv'(x)}{Au(x) + Bv(x)},$$

with A, B arbitrary constants, not both zero, and the prime denoting differentiation, while u and v are linearly independent solutions of

$$\frac{d}{dx}\left(\frac{1}{r(x)}\frac{dz}{dx}\right) + s(x)z = 0.$$

Conversely, if $u(x)$ and $v(x)$ are linearly independent solutions of this last equation and A and B are arbitrary constants, not both zero, the function

$$\frac{1}{r(x)} \frac{Au'(x) + Bv'(x)}{Au(x) + Bv(x)}$$

is a solution of the Riccati equation wherever $Au(x) = Bv(x) \neq 0$.   IN 24

16.516 Solution of a second-order linear differential equation

A **fundamental system** of solutions of a homogeneous second-order linear differential equation in the canonical form

$$\frac{d^2x}{dt^2} + a(t)\frac{dx}{dt} + b(t)x = 0$$

is a system of two linearly independent solutions $\phi_1(t)$ and $\phi_2(t)$. The Wronskian of these solutions is

$$W(t) = \begin{vmatrix} \phi_1(t) & \phi_2(t) \\ \phi_1'(t) & \phi_2'(t) \end{vmatrix} = \phi_1(t)\phi_2'(t) - \phi_2(t)\phi_1'(t),$$

and the solution to the inhomogeneous equation

$$\frac{d^2x}{dt^2} + a(t)\frac{dx}{dt} + b(t)x = f(t),$$

with $x(t_0) = x_0$ may be written

$$x(t) = x_0 + \int_{t_0}^{t} \frac{\phi_1(\xi)\phi_2(t) - \phi_2(\xi)\phi_1(t)}{W(\xi)} f(\xi)\,d\xi.$$

The linear combination $c_1\phi_1(t) + c_2\phi_2(t)$ is known as the **complementary function** where c_1 and c_2 are arbitrary constants.

16.61–16.62 Oscillation and non-oscillation theorems for second-order equations

Equations whose solutions possess an infinite number of zeros in the interval $(0, \infty)$ are said to have **oscillatory** solutions. The following theorems relate to such properties.

16.611 First basic comparison theorem

If all solutions of the equation

$$\frac{d^2u}{dx^2} + \phi(x)u = 0$$

are oscillatory, and if

$$\psi(x) \geq \phi(x),$$

then all the solutions of

$$\frac{d^2v}{dx^2} + \psi(x)v = 0$$

are oscillatory, and conversely. That is, if $\psi(x) \geq \phi(x)$ and some solutions v are non-oscillatory, then so also must some solutions u be non-oscillatory.

BS 119

16.622 Second basic comparison theorem

If all the solutions of the self-adjoint equation

$$\frac{d}{dx}\left(p_1(x)\frac{du}{dx}\right) + q_1(x)u = 0$$

are oscillatory as $x \to \infty$, and if

$$q_2(x) \geq q_1(x),$$
$$p_2(x) \geq p_1(x) > 0,$$

then all the solutions of the self-adjoint equation

$$\frac{d}{dx}\left(p_2(x)\frac{dv}{dx}\right) + q_2(x)v = 0$$

are oscillatory.

BS 120

16.623 Interlacing of zeros

Let $y_1(x)$ and $y_2(x)$ be two linearly independent solutions of

$$\frac{d^2y}{dx^2} + F(x)y = 0,$$

and suppose that $y_1(x)$ has at least two zeros in the interval (a, b). Then if x_1 and x_2 are two consecutive zeros of $y_1(x)$, the function $y_2(x)$ has one, and only one, zero in the interval (x_1, x_2).

HI 374

16.624 Sturm separation theorem

Let $u(x)$ and $v(x)$ be two linearly independent solutions of the self-adjoint equation

$$\frac{d}{dx}\left(p(x)\frac{dy}{dx}\right) + q(x) = 0,$$

in which $p(x) > 0$ and $p(x), q(x)$ are continuous on $[a, b]$. Then, between any two consecutive zeros of $u(x)$ there will be one, and only one, zero of $v(x)$.

IN 224

16.625 Sturm comparison theorem

Let $p_1(x) \geq p_2(x) > 0$ and $q_1(x) \geq q_2(x)$ be continuous functions in the differential equations

$$\frac{d}{dx}\left(p_1(x)\frac{du}{dx}\right) + q_1(x)u = 0,$$

$$\frac{d}{dx}\left(p_2(x)\frac{dv}{dx}\right) + q_2(x)v = 0.$$

Then between any two zeros of a non-trivial solution $u(x)$ of the first equation there will be at least one zero of every non-trivial solution $v(x)$ of the second equation.

IN 228

16.626 Szegö's comparison theorem

Suppose, under the conditions of the Sturm comparison theorem, that $p_1(x) \equiv p_2(x), q_1(x) \not\equiv q_2(x)$, and $u(x) > 0, v(x) > 0$ for $a < x < b$, together with

$$\lim_{x \to a} p_1(x) \left(\frac{du}{dx} v - \frac{dv}{dx} u \right) = 0.$$

Then, if $u(b) = 0$, there is a point ξ in (a, b) such that $v(\xi) = 0$.   HI 379

16.627 Picone's identity

Consider the equations

$$\frac{d}{dx} \left(p_1(x) \frac{du}{dx} \right) + q_1(x)u = 0,$$

$$\frac{d}{dx} \left(p_2(x) \frac{dv}{dx} \right) + q_2(x)v = 0,$$

with p_1, p_2, q_1, and q_2 positive and continuous for $a < x < b$, where $q_2(x) > q_1(x)$ and $p_1(x) > p_2(x)$. Then with $a < \alpha < \beta < b$, Picone's identity is

$$\left(\frac{u}{v} \left(p_1 \frac{du}{dx} v - p_2 \frac{dv}{dx} u \right) \right)_\alpha^\beta = \int_\alpha^\beta (q_2 - q_1) \, u^2 \, ds + \int_\alpha^\beta (p_1 - p_2) \left(\frac{du}{ds} \right)^2 ds + \int_\alpha^\beta \frac{p_2}{v^2} \left(v \frac{du}{ds} - u \frac{dv}{ds} \right)^2 ds$$

IN 226

16.628 Sturm–Picone theorem

Consider the self-adjoint equations

$$\frac{d}{dx} \left(p_1(x) \frac{du}{dx} \right) + q_1(x)u = 0$$

and

$$\frac{d}{dx} \left(p_2(x) \frac{dv}{dx} \right) + q_2(x)v = 0.$$

Let p_1, p_2, q_1, and q_2 be positive and continuous for $a < x < b$, where $q_2(x) > q_1(x)$ and $p_1(x) > p_2(x)$. Then, if x_1 and x_2 is a pair of consecutive zeros of $u(x)$ in $(a, b), v(x)$ has at least one zero in the open interval (a, b).   IN 225

16.629 Oscillation on the half line

Consider the self-adjoint equation

$$\frac{d}{dx} \left(p(x) \frac{du}{dx} \right) + q(x)u = 0.$$

We then have the following results.

(i)   Let $p(x) > 0$ and p, q be continuous on $[0, \infty)$. If the two improper integrals

$$\int_1^\infty \frac{dx}{p(x)} \quad \text{and} \quad \int_1^\infty q(x) \, dx$$

diverge, then every solution $u(x)$ has infinitely many zeros on the interval $[1, \infty)$. Also, if the two integrals

$$\int_0^1 \frac{dx}{p(x)} = +\infty \quad \text{and} \quad \int_0^1 q(x)\,dx = +\infty,$$

then every solution $u(x)$ has infinitely many zeros on the interval $(0,1)$.

(ii)   (Moore's theorem). Every non-trivial solution $u(x)$ has at most a finite number of zeros on the interval $[a, \infty)$ if the improper integral

$$\int_a^\infty \frac{dx}{p(x)}$$

converges, and if

$$\left| \int_a^x q(s)\,ds \right| < M \quad \text{for} \quad a \le x < \infty$$

with $M > 0$ a finite constant.

16.71 Two related comparison theorems

16.711 Theorem 1

Consider the equations in the Sturm comparison theorem with the same assumptions on $p(x)$ and $q(x)$, and let $u(x), v(x)$ be solutions such that

$$u(x_1) = v(x_1) = 0, \quad u'(x) = v'(x_1) > 0.$$

Then if $u(x)$ is increasing in $[x_1, x_2]$ and reaches a maximum at x_2, the function $v(x)$ reaches a maximum at some point x_3 such that $x_1 < x_3 < x_2$.

HI 376

16.712 Theorem 2

Consider the equation

$$\frac{d^2 y}{dx^2} + F(x)y = 0,$$

in which $F(x)$ is continuous in (a, b) and such that

$$0 < m \le F(x) \le M.$$

Then, if the solution $y(x)$ has two successive zeros x_1, x_2, it follows that

$$\pi M^{-1/2} \le x_2 - x_1 \le \pi m^{-1/2}$$

16.81–16.82 Non-oscillatory solutions

The real solution $y(x)$ of

$$\frac{\partial^2 y}{\partial x^2} + F(x)y = 0$$

is said to be **non-oscillatory** in the wide sense in $(0, \infty)$ if there exists a finite number c such that the solution has no zeros in $[c, \infty)$.

HI 376

16.811 Kneser's non-oscillation theorem

Consider the equation

$$\frac{d^2y}{dx^2} + F(x)y = 0,$$

and let

$$\limsup \left[x^2 F(x)\right] = \gamma^*,$$
$$\liminf \left[x^2 F(x)\right] = \gamma_*.$$

Then the solution $y(x)$ is non-oscillatory if $\gamma^* < \frac{1}{4}$, oscillatory if $\frac{1}{4} < \gamma_*$ and no conclusion can be drawn if either γ^* or γ_* equals $\frac{1}{4}$.   HI 461

16.822 Comparison theorem for non-oscillation

Consider the differential equations

$$\frac{d^2y}{dx^2} + F(x)y = 0, \quad f(x) = x\int_x^\infty F(s)\,ds,$$

$$\frac{d^2y}{dx^2} + G(x)y = 0, \quad g(x) = x\int_x^\infty G(s)\,ds,$$

where $0 < g(x) < f(x)$. Then if the first equation is non-oscillatory in the wide sense, so also is the second.   HI 460

16.823 Necessary and sufficient conditions for non-oscillation

Consider the equation

$$\frac{d^2y}{dx^2} + F(x)y = 0.$$

Then, if

$$\lim_{x\to\infty} \sup \left(x\int_x^\infty F(s)\,ds \right) = F^*,$$

$$\lim_{x\to\infty} \inf \left(x\int_x^\infty F(s)\,ds \right) = F_*,$$

it follows that:

(i)   a necessary condition that the solution $y(x)$ be non-oscillatory is that $F_* \leq \frac{1}{4}$ and $F^* \leq 1$;

(ii)   a sufficient condition that the solution $y(x)$ be non-oscillatory is that $F^* < \frac{1}{4}$.

16.91 Some growth estimates for solutions of second-order equations

16.911 Strictly increasing and decreasing solutions

Suppose that $G(x) > 0$ be continuous in $(-\infty, \infty)$ and such that $xG(x) \notin L(0, \infty)$. Then the equation

$$\frac{d^2y}{dx^2} - G(x)y = 0$$

has one, and only one, solution $y_+(x)$ passing through the point $(0, 1)$ which is positive and strictly monotonic decreasing for all x, and one and only one solution $y_-(x)$ through the point $(0, 1)$ which is positive and strictly increasing for all x. The solution $y_+(x)$ has the property that

$$[G(x)]^{1/2}y_+(x) \in L_2(0,\infty) \quad \text{and} \quad \frac{dy_+(x)}{dx} \in L_2(0,\infty).$$

If, in addition, $0 < \alpha^2 \le G(x) \le \beta^2 < \infty$, then

$$e^{-\beta x} \le y_+(x) \le e^{-\alpha x} \quad \text{for} \quad x > 0.$$

HI 359

16.912 General result on dominant and subdominant solutions

Consider the equations

$$\frac{d^2y}{dx^2} - g(x)y = 0, \quad \frac{d^2Y}{dx^2} - G(x)Y = 0,$$

where g and G are continuous on $(0,\infty)$ with $0 < g(x) < G(x)$, and $xg(x) \notin L(0,\infty)$. In addition, let y_α and Y_α be the solutions of these respective equations corresponding to

$$y_\alpha(0) = Y_\alpha(0) = 1, \quad y_\alpha'(0) = Y_\alpha'(0) = \alpha \text{ for } -\infty < \alpha < \infty.$$

Let y_ω and Y_ω be determined, respectively, by

$$y_\omega(0) = Y_\omega(0) = 0, \quad y_\omega'(0) = Y_\omega'(0) = 1,$$

and let y_+ and Y_+ be the **subdominant solutions** for which

$$y_+(0) = Y_+(0) = 1$$

while $[y_+'(x)]^2$, $g(x)[y_+(x)]^2$, $[Y_+'(x)]^2$, and $G(x)[Y_+'(x)]^2$ belong to $L(0,\infty)$. Then, if β and γ are such that $y_{-\beta} = y_+$ and $Y_{-\gamma} = Y_+$, it follows that $\beta < \gamma$ and

$$y_\alpha(x) < Y_\alpha(x), \quad 0 < x < \infty, \quad -\gamma \le \alpha,$$
$$y_\omega(x) < Y_\omega(x),$$
$$y_+(x) > Y_+(x).$$

HI 440

16.913 Estimate of dominant solution

Let $G(x)$ be positive and continuous with continuous first- and second-order derivatives satisfying

$$G(x)G'(x) < \tfrac{5}{4}[G'(x)]^2.$$

Then there exists a **dominant solution** $y(x)$ of the fundamental solutions $Y_0(x)$ and $Y_1(x)$ of

$$\frac{d^2y}{dx^2} - G(x)y = 0,$$

determined by the initial conditions

$$2Y_0(0) = 0, \quad Y_1(0) = 1,$$
$$Y_0'(0) = 1, \quad Y_1'(0) = 0,$$

such that

$$y(x) < [G(x)]^{-1/4}\exp\left(\int_0^x [G(\xi)]^{1/2}\,d\xi\right),$$

and a positive constant C such that the normalized subdominant solution $y_+(x)$, for which $y_+(0) = 1$ and $[y_+'(x)]^2 \in L(0,\infty)$, $G(x)[y_+(x)]^2 \in L(0,\infty)$, satisfies

$$y_+(x) > CG(x)^{-1/4}\exp\left(-\int_0^x [G(\xi)]^{1/2}\,d\xi\right)$$

HI 443

16.914   A theorem due to Lyapunov

Let $y(x)$ be any solution of

$$\frac{d^2y}{dx^2} - G(x)y = 0$$

with $G(x)$ positive and continuous in $(0, \infty)$ with $xG(x) \in L(0, \infty)$. Then

$$\exp\left(-\int_0^x [G(\xi) + 1]\, d\xi\right) < [y(x)]^2 + [y'(x)]^2$$

$$< C \exp\left(\int_0^x [G(\xi) + 1]\, d\xi\right),$$

HI 446

where $C = [y(0)]^2 + [y'(0)]^2$.

16.92   Boundedness theorems

16.921[6]   All solutions of the equation

$$\frac{d^2u}{dx^2} + (1 + \phi(x) + \psi(x))\, u = 0$$

are bounded, provided that

(i) $\int^\infty |\phi(x)|\, dx < \infty$,

(ii) $\displaystyle\int^\infty |\psi(x)|\, dx < \infty$ and $\psi(x) \to 0$ as $x \to \infty$.

BS 112

16.922   If all solutions of the equation

$$\frac{d^2u}{dx^2} + a(x)u = 0$$

are bounded, then all solutions of

$$\frac{d^2u}{dx^2} + (a(x) + b(x))u = 0$$

are also bounded if

$$\int^\infty |b(x)|\, dx < \infty.$$

BS 112

16.923   If $a(x) \to \infty$ monotonically as $x \to \infty$, then all solutions of

$$\frac{d^2u}{dx^2} + a(x)u = 0$$

are bounded as $x \to \infty$.

BS 113

16.924   Consider the equation

$$\frac{d^2u}{dx^2} + a(x)u = 0$$

in which

$$\int^{\infty} x|a(x)|\,dx < \infty.$$

Then $\displaystyle\lim_{x\to\infty}\left(\dfrac{du}{dx}\right)$ exists, and the general solution is asymptotic to $d_0 + d_1 x$ as $x \to \infty$, where d_0 and d_1 may be zero, but not simultaneously.

BS 114

16.93^{10} Growth of maxima of $|y|$

Sonin's theorem generalized by Pólya may be stated as follows: *Let $y(x)$ satisfy the differential equation*

$$\{k(x)y'\}' + \phi(x)y = 0,$$

where $k(x) > 0, \phi(x) > 0$, and both functions $k(x), \phi(x)$ have a continuous derivative. Then the relative maxima of $|y|$ form an increasing or decreasing sequence according as $k(x)\phi(x)$ is decreasing or increasing.

SZ 164

17 Fourier, Laplace, and Mellin Transforms

17.1– 17.4 Integral Transforms

17.11 Laplace transform

The **Laplace transform** of the function $f(x)$, denoted by $F(s)$, is defined by the integral

$$F(s) = \int_0^\infty f(x)e^{-sx}\,dx, \qquad \operatorname{Re} s > 0.$$

The functions $f(x)$ and $F(s)$ are called a **Laplace transform pair**, and knowledge of either one enables the other to be recovered.

If f is summable over all finite intervals, and there is a constant c for which

$$\int_0^\infty |f(x)|e^{-c|x|}\,dx,$$

is finite, then the Laplace transform exists when $s = \sigma + i\tau$ is such that $\sigma \geq c$.

Setting

$$F(s) = \mathcal{L}\left[f(x); s\right],$$

to emphasize the nature of the transform, we have the symbolic inverse result

$$f(x) = \mathcal{L}^{-1}\left[F(s); x\right].$$

The inversion of the Laplace transform is accomplished for analytic functions $F(s)$ of order $O\left(s^{-k}\right)$ with $k > 1$ by means of the **inversion integral**

$$f(x) = \frac{1}{2\pi i} \int_{\gamma - i\infty}^{\gamma + i\infty} F(s)e^{sx}\,ds,$$

where γ is a real constant that exceeds the real part of all the singularities of $F(s)$. SN 30

17.12 Basic properties of the Laplace transform

$1.^8$ For a and b arbitrary constants,

$$\mathcal{L}\left[af(x) + bg(x)\right] = aF(s) + bG(s) \qquad \text{(linearity)}$$

2.   If $n > 0$ is an integer and $\lim_{x \to \infty} f(x)e^{-sx} = 0$, then for $x > 0$,

$$\mathcal{L}\left[f^{(n)}(x); s\right] = s^n F(s) - s^{n-1}f(0) - s^{n-2}f^{(1)}(0) - \cdots - f^{(n-1)}(0) \qquad \text{(transform of a derivative)}$$

SN 32

3.   If $\lim_{x \to \infty} \left(e^{-sx} \int_0^x f(\xi) \, d\zeta \right) = 0$, then

$$\mathcal{L}\left[\int_0^x f(\xi) \, d\xi; s \right] = \frac{1}{s} F(s) \qquad \text{(transform of an integral)} \qquad \text{SN 37}$$

4. $\mathcal{L}\left[e^{-ax} f(x); s \right] = F(s + a)$ (shift theorem)   SU 143

5.   The **Laplace convolution** $f * g$ of two functions $f(x)$ and $g(x)$ is defined by the integral

$$f * g(x) = \int_0^x f(x - \xi) g(\xi) \, d\xi,$$

and it has the property that $f * g = g * f$ and $f * (g * h) = (f * g) * h$. In terms of the convolution operation

$$\mathcal{L}\left[f * g(x); s \right] = F(s) G(s) \qquad \text{(convolution (Faltung) theorem)} \qquad \text{SN 30}$$

17.13^{10} Table of Laplace transform pairs

	$f(x)$		$F(s)$		
1	1		$1/s$		
2	x^n,	$n = 0, 1, 2, \ldots$	$\dfrac{n!}{s^{n+1}}$,	$\operatorname{Re} s > 0$	ET I 133(3)
3	x^ν,	$\nu > -1$	$\dfrac{\Gamma(\nu + 1)}{s^{\nu+1}}$,	$\operatorname{Re} s > 0$	ET I 137(1)
4	$x^{n-\frac{1}{2}}$		$\dfrac{\Gamma\left(n + \frac{1}{2}\right)}{s^{n+\frac{1}{2}}}$,	$\operatorname{Re} s > 0$	ET I 135(17)
5	$x^{-1/2}(x + a)^{-1}$,	$|\arg a| < \pi$	$\pi a^{-1/2} e^{as} \operatorname{erfc}\left(a^{1/2} s^{1/2}\right)$, $\operatorname{Re} s \geq 0$		ET I 136(25)
6	$\begin{cases} x & 0 < x < 1 \\ 1 & x > 1 \end{cases}$		$\dfrac{1 - e^{-s}}{s^2}$,	$\operatorname{Re} s > 0$	ET I 142(14)
7	e^{-ax}		$\dfrac{1}{s + a}$,	$\operatorname{Re} s > -\operatorname{Re} a$	ET I 143(1)
8	xe^{-ax}		$\dfrac{1}{(s + a)^2}$,	$\operatorname{Re} s > -\operatorname{Re} a$	ET I 144(2)
9	$\dfrac{e^{-ax} - e^{-bx}}{b - a}$		$(s + a)^{-1}(s + b)^{-1}$, $\operatorname{Re} s > \{-\operatorname{Re} a, -\operatorname{Re} b\}$		AS 1022(29.3.12)

continued on next page

	$f(x)$	$F(s)$
	continued from previous page	
10	$\dfrac{ae^{-ax} - be^{-bx}}{b - a}$	$s(s+a)^{-1}(s+b)^{-1}$, $\operatorname{Re}_s > \{-\operatorname{Re} a, -\operatorname{Re} b\}$ $\qquad$ AS 1022(29.3.13)
11	$\dfrac{e^{ax} - 1}{a}$	$s^{-1}(s-a)^{-1}$, $\qquad\qquad$ $\operatorname{Re} s > \operatorname{Re} a$
12	$\dfrac{e^{ax} - ax - 1}{a^2}$	$s^{-2}(s-a)^{-1}$, $\qquad\qquad$ $\operatorname{Re} s > \operatorname{Re} a$
13	$\dfrac{\left(e^{ax} - \frac{1}{2}a^2x^2 - ax - 1\right)}{a^3}$	$s^{-3}(s-a)^{-1}$, $\qquad\qquad$ $\operatorname{Re} s > \operatorname{Re} a$
14	$(1 + ax)e^{ax}$	$\dfrac{s}{(s-a)^2}$, $\qquad\qquad$ $\operatorname{Re} s > \operatorname{Re} a$
15	$\dfrac{1 + (ax - 1)e^{ax}}{a^2}$	$s^{-1}(s-a)^{-2}$, $\qquad\qquad$ $\operatorname{Re} s > \operatorname{Re} a$
16	$\dfrac{2 + ax + (ax - 2)e^{ax}}{a^3}$	$s^{-2}(s-a)^{-2}$, $\qquad\qquad$ $\operatorname{Re} s > \operatorname{Re} a$
17	$x^n e^{ax}$, $\qquad\qquad n = 0, 1, 2, \ldots$	$n!(s-a)^{-(n+1)}$, $\qquad$ $\operatorname{Re} s > \operatorname{Re} a$
18	$\left(x + \frac{1}{2}ax^2\right)e^{ax}$	$\dfrac{s}{(s-a)^3}$, $\qquad\qquad$ $\operatorname{Re} s > \operatorname{Re} a$
19	$\left(1 + 2ax + \frac{1}{2}a^2x^2\right)e^{ax}$	$\dfrac{s^2}{(s-a)^3}$, $\qquad\qquad$ $\operatorname{Re} s > \operatorname{Re} a$
20	$\frac{1}{6}x^3 e^{ax}$	$(s-a)^{-4}$, $\qquad\qquad$ $\operatorname{Re} s > \operatorname{Re} a$
21	$\left(\frac{1}{2}x^2 + \frac{1}{6}ax^3\right)e^{ax}$	$\dfrac{s}{(s-a)^4}$, $\qquad\qquad$ $\operatorname{Re} s > \operatorname{Re} a$
22	$\left(x + ax^2 + \frac{1}{6}a^2x^3\right)e^{ax}$	$s^2(s-a)^{-4}$, $\qquad\qquad$ $\operatorname{Re} s > \operatorname{Re} a$
23	$\left(1 + 3ax + \frac{3}{2}a^2x^2 + \frac{1}{6}a^3x^3\right)e^{ax}$	$s^3(s-a)^{-4}$, $\qquad\qquad$ $\operatorname{Re} s > \operatorname{Re} a$
24	$\dfrac{ae^{ax} - be^{bx}}{a - b}$	$s(s-a)^{-1}(s-b)^{-1}$, $\qquad$ $\operatorname{Re} s > \{\operatorname{Re} a, \operatorname{Re} b\}$

continued on next page

	$f(x)$		$F(s)$	
	continued from previous page			
25	$\dfrac{\left(\frac{1}{a}e^{ax} - \frac{1}{b}e^{bx} + \frac{1}{b} - \frac{1}{a}\right)}{a-b}$		$s^{-1}(s-a)^{-1}(s-b)^{-1},\qquad \operatorname{Re}s > \{\operatorname{Re}a, \operatorname{Re}b\}$	
26	$x^{\nu-1}e^{-ax},$	$\operatorname{Re}\nu>0$	$\Gamma(\nu)(s+a)^{-\nu},\quad \operatorname{Re}s > -\operatorname{Re}a$	ET I 144(3)
27	$xe^{-x^2/(4a)},$	$\operatorname{Re}a>0$	$2a - 2\pi^{1/2}a^{3/2}se^{as^2}\operatorname{erfc}\left(sa^{1/2}\right)$ ET I 146(22)	
28	$\exp\left(-ae^{x}\right),$	$\operatorname{Re}a>0$	$a^{s}\,\Gamma\left(-s,a\right)$ ET I 147(37)	
29[8]	$x^{1/2}e^{-a/(4x)},$	$\operatorname{Re}a\ge 0$	$\frac{1}{2}\pi^{1/2}s^{-3/2}\left(1 + a^{1/2}s^{1/2}\right)\exp\left[(-as)^{1/2}\right],$ $\operatorname{Re}s>0$ ET I 146(26)	
30[8]	$x^{-1/2}e^{-a/(4x)},$	$\operatorname{Re}a\ge 0$	$\pi^{1/2}s^{-1/2}\exp\left[(-as)^{1/2}\right],$ $\operatorname{Re}s>0$ ET I 146(27)	
31[8]	$x^{-3/2}e^{-a/(4x)},$	$\operatorname{Re}a>0$	$2\pi^{1/2}a^{-1/2}\exp\left[(-as)^{1/2}\right],$ $\operatorname{Re}s\ge 0$ ET I 146(28)	
32	$\sin(ax)$		$a\left(s^2+a^2\right)^{-1},\qquad \operatorname{Re}s>|\operatorname{Im}a|$	ET I 150(1)
33	$\cos(ax)$		$s\left(s^2+a^2\right)^{-1},\qquad \operatorname{Re}s>|\operatorname{Im}a|$	ET I 154(3)
34	$|\sin(ax)|,$	$a>0$	$a\left(s^2+a^2\right)^{-1}\coth\left(\dfrac{\pi s}{2a}\right),$ $\operatorname{Re}s>0$ ET I 150(2)	
35	$|\cos(ax)|,$	$a>0$	$\left(s^2+a^2\right)^{-1}\left[s + a\operatorname{cosech}\left(\dfrac{\pi}{2a}\right)\right],$ $\operatorname{Re}s>0$ ET I 155(44)	
36	$\dfrac{1-\cos(ax)}{a^2}$		$s^{-1}\left(s^2+a^2\right)^{-1},$ $\operatorname{Re}s>|\operatorname{Im}a|$ AS 1022(29.3.19)	

continued on next page

continued from previous page

	$f(x)$	$F(s)$
37	$\dfrac{ax - \sin(ax)}{a^3}$	$s^{-2}\left(s^2 + a^2\right)^{-1},$ $\qquad \operatorname{Re} s > \lvert \operatorname{Im} a \rvert$ $\qquad$ AS 1022(29.3.20)
38	$\dfrac{\sin(ax) - ax\cos(ax)}{2a^3}$	$\left(s^2 + a^2\right)^{-2}, \quad \operatorname{Re} s > \lvert \operatorname{Im} a \rvert$ $\qquad$ AS 1022(29.3.21)
39	$\dfrac{x\sin(ax)}{2a}$	$s\left(s^2 + a^2\right)^{-2}, \qquad \operatorname{Re} s > \lvert \operatorname{Im} a \rvert$ $\qquad$ ET I 152(14)
40	$\dfrac{\sin(ax) + ax\cos(ax)}{2a}$	$s^2\left(s^2 + a^2\right)^{-2},$ $\qquad \operatorname{Re} s > \lvert \operatorname{Im} a \rvert$ $\qquad$ AS 1023(29.3.23)
41	$x\cos(ax)$	$\left(s^2 - a^2\right)\left(s^2 + a^2\right)^{-2},$ $\qquad \operatorname{Re} s > \lvert \operatorname{Im} a \rvert$ $\qquad$ ET I 157(57)
42	$\dfrac{\cos(ax) - \cos(bx)}{b^2 - a^2}$	$s\left(s^2 + a^2\right)^{-1}\left(s^2 + b^2\right)^{-1},$ $\qquad \operatorname{Re} s > \{\lvert \operatorname{Im} a \rvert, \lvert \operatorname{Im} b \rvert\}$ $\qquad$ AS 1023(29.3.25)
43	$\dfrac{\left[\frac{1}{2}a^2x^2 - 1 + \cos(ax)\right]}{a^4}$	$s^{-3}\left(s^2 + a^2\right)^{-1}, \qquad \operatorname{Re} s > \lvert \operatorname{Im} a \rvert$
44	$\dfrac{\left[1 - \cos(ax) - \frac{1}{2}ax\sin(ax)\right]}{a^4}$	$s^{-1}\left(s^2 + a^2\right)^{-2}, \qquad \operatorname{Re} s > \lvert \operatorname{Im} a \rvert$
45	$\dfrac{\left[\frac{1}{b}\sin(bx) - \frac{1}{a}\sin(ax)\right]}{a^2 - b^2}$	$\left(s^2 + a^2\right)^{-1}\left(s^2 + b^2\right)^{-1}, \quad \operatorname{Re} s > \{\lvert \operatorname{Im} a \rvert, \lvert \operatorname{Im} b \rvert\}$
46	$\dfrac{\left[1 - \cos(ax) + \frac{1}{2}ax\sin(ax)\right]}{a^2}$	$s^{-1}\left(s^2 + a^2\right)^{-2}\left(2s^2 + a^2\right), \qquad \operatorname{Re} s > \lvert \operatorname{Im} a \rvert$
47	$\dfrac{a\sin(ax) - b\sin(bx)}{a^2 - b^2}$	$s^2\left(s^2 + a^2\right)^{-1}\left(s^2 + b^2\right)^{-1},$ $\qquad \operatorname{Re} s > \{\lvert \operatorname{Im} a \rvert, \lvert \operatorname{Im} b \rvert\}$
48	$\sin(a + bx)$	$\left(s\sin a + b\cos a\right)\left(s^2 + b^2\right)^{-1}, \qquad \operatorname{Re} s > \lvert \operatorname{Im} b \rvert$
49	$\cos(a + bx)$	$\left(s\cos a - b\sin a\right)\left(s^2 + b^2\right)^{-1}, \qquad \operatorname{Re} s > \lvert \operatorname{Im} b \rvert$

continued on next page

	$f(x)$	$F(s)$
	continued from previous page	
50	$\dfrac{\left[\frac{1}{a}\sinh(ax) - \frac{1}{b}\sin(bx)\right]}{a^2 + b^2}$	$\left(s^2 - a^2\right)^{-1}\left(s^2 + b^2\right)^{-1}, \quad \operatorname{Re} s > \{|\operatorname{Re} a|, |\operatorname{Im} b|\}$
51	$\dfrac{\cosh(ax) - \cos(bx)}{a^2 + b^2}$	$s\left(s^2 - a^2\right)^{-1}\left(s^2 + b^2\right)^{-1},$ $\operatorname{Re} s > \{|\operatorname{Re} a|, |\operatorname{Im} b|\}$
52	$\dfrac{a\sinh(ax) + b\sin(bx)}{a^2 + b^2}$	$s^2\left(s^2 - a^2\right)^{-1}\left(s^2 + b^2\right)^{-1},$ $\operatorname{Re} s > \{|\operatorname{Re} a|, |\operatorname{Im} b|\}$
53	$\sin(ax)\sin(bx)$	$2abs\left[s^2 + (a-b)^2\right]^{-1}\left[s^2 + (a+b)^2\right]^{-1},$ $\operatorname{Re} s > \{|\operatorname{Im} a|, |\operatorname{Im} b|\}$
54	$\cos(ax)\cos(bx)$	$s\left(s^2 + a^2 + b^2\right)\left[s^2 + (a-b)^2\right]^{-1}\left[s^2 + (a+b)^2\right]^{-1},$ $\operatorname{Re} s > \{|\operatorname{Im} a|, |\operatorname{Im} b|\}$
55	$\sin(ax)\cos(bx)$	$a\left(s^2 + a^2 - b^2\right)\left[s^2 + (a-b)^2\right]^{-1}\left[s^2 + (a+b)^2\right]^{-1},$ $\operatorname{Re} s > \{|\operatorname{Im} a|, |\operatorname{Im} b|\}$
56	$\sin^2(ax)$	$2a^2 s^{-1}\left(s^2 + 4a^2\right)^{-1}, \qquad \operatorname{Re} s > |\operatorname{Im} a|$
57	$\cos^2(ax)$	$\left(s^2 + 2a^2\right)s^{-1}\left(s^2 + 4a^2\right)^{-1}, \qquad \operatorname{Re} s > |\operatorname{Im} a|$
58	$\sin(ax)\cos(ax)$	$a\left(s^2 + 4a^2\right)^{-1}, \qquad \operatorname{Re} s > |\operatorname{Im} a|$
59	$e^{-ax}\sin(bx)$	$b\left[(s+a)^2 + b^2\right]^{-1}, \qquad \operatorname{Re} s > \{-\operatorname{Re} a, \quad |\operatorname{Im} b|\}$
60	$e^{-ax}\cos(bx)$	$(s+a)\left[(s+a)^2 + b^2\right]^{-1},$ $\operatorname{Re} s > \{-\operatorname{Re} a, \quad |\operatorname{Im} b|\}$
61	$x^{-1}\sin(ax)$	$\arctan(a/s), \qquad \operatorname{Re} s > |\operatorname{Im} a| \qquad$ ET I 152(16)

continued on next page

	$f(x)$	$F(s)$
	continued from previous page	
62	$x^{-1}\left[1 - \cos(ax)\right]$	$\frac{1}{2}\ln\left(1 + a^2/s^2\right),$ $\operatorname{Re} s > \lvert\operatorname{Im} a\rvert$ ET I 157(59)
63	$\sinh(ax)$	$a\left(s^2 - a^2\right)^{-1}, \qquad \operatorname{Re} s > \lvert\operatorname{Re} a\rvert$ ET I 162(1)
64	$\cosh(ax)$	$s\left(s^2 - a^2\right)^{-1}, \qquad \operatorname{Re} s > \lvert\operatorname{Re} a\rvert$ ET I 162(2)
65	$x^{\nu-1}\sinh(ax), \qquad \operatorname{Re}\nu > -1$	$\frac{1}{2}\Gamma(\nu)\left[(s-a)^{-\nu} - (s+a)^{-\nu}\right],$ $\operatorname{Re} s > \lvert\operatorname{Re} a\rvert$ ET I 164(18)
66	$x^{\nu-1}\cosh(ax), \qquad \operatorname{Re}\nu > 0$	$\frac{1}{2}\Gamma(\nu)\left[(s-a)^{-\nu} + (s+a)^{-\nu}\right],$ $\operatorname{Re} s > \lvert\operatorname{Re} a\rvert$ ET I 164(19)
67	$x\sinh(ax)$	$2as\left(s^2 - a^2\right)^{-2}, \qquad\qquad \operatorname{Re} s > \lvert\operatorname{Re} a\rvert$
68	$x\cosh(ax)$	$\left(s^2 + a^2\right)\left(s^2 - a^2\right)^{-2}, \qquad\qquad \operatorname{Re} s > \lvert\operatorname{Re} a\rvert$
69	$\sinh(ax) - \sin(ax)$	$2a^3\left(s^4 - a^4\right)^{-1},$ $\operatorname{Re} s > \{\lvert\operatorname{Re} a\rvert, \lvert\operatorname{Im} a\rvert\}$ AS 1023(29.3.31)
70	$\cosh(ax) - \cos(ax)$	$2a^2 s\left(s^4 - a^4\right)^{-1},$ $\operatorname{Re} s > \{\lvert\operatorname{Re} a\rvert, \lvert\operatorname{Im} a\rvert\}$ AS 1023(29.3.32)
71	$\sinh(ax) + ax\cosh(ax)$	$2as^2\left(a^2 - s^2\right)^{-2}, \qquad\qquad \operatorname{Re} s > \lvert\operatorname{Re} a\rvert$
72	$ax\cosh(ax) - \sinh(ax)$	$2a^3\left(a^2 - s^2\right)^{-2}, \qquad\qquad \operatorname{Re} s > \lvert\operatorname{Re} a\rvert$
73	$x\sinh(ax) - \cosh(ax)$	$s\left(a^2 + 2a - s^2\right)\left(a^2 - s^2\right)^{-2}, \qquad \operatorname{Re} s > \lvert\operatorname{Re} a\rvert$
74	$\dfrac{\left[\frac{1}{a}\sinh(ax) - \frac{1}{b}\sinh(bx)\right]}{a^2 - b^2}$	$\left(a^2 - s^2\right)^{-1}\left(b^2 - s^2\right)^{-1}, \quad \operatorname{Re} s > \{\lvert\operatorname{Re} a\rvert, \lvert\operatorname{Re} b\rvert\}$

continued on next page

	$f(x)$	$F(s)$

continued from previous page

	$f(x)$	$F(s)$
75	$\dfrac{\cosh(ax) - \cosh(bx)}{a^2 - b^2}$	$s\left(s^2 - a^2\right)^{-1}\left(s^2 - b^2\right)^{-1},$ $\operatorname{Re} s > \{|\operatorname{Re} a|, |\operatorname{Re} b|\}$
76	$\dfrac{a\sinh(ax) - b\sinh(bx)}{a^2 - b^2}$	$s^2\left(s^2 - a^2\right)^{-1}\left(s^2 - b^2\right)^{-1},$ $\operatorname{Re} s > \{|\operatorname{Re} a|, |\operatorname{Re} b|\}$
77	$\sinh(a + bx)$	$(b\cosh a + s\sinh a)\left(s^2 - b^2\right)^{-1}, \qquad \operatorname{Re} s > |\operatorname{Re} b|$
78	$\cosh(a + bx)$	$(s\cosh a + b\sinh a)\left(s^2 - b^2\right)^{-1}, \qquad \operatorname{Re} s > |\operatorname{Re} b|$
79	$\sinh(ax)\sinh(bx)$	$2abs\left[s^2 - (a+b)^2\right]^{-1}\left[s^2 - (a-b)^2\right]^{-1},$ $\operatorname{Re} s > \{|\operatorname{Re} a|, |\operatorname{Re} b|\}$
80[8]	$\cosh(ax)\cosh(bx)$	$s\left(s^2 - a^2 - b^2\right)\left[s^2 - (a+b)^2\right]^{-1}\left[s^2 - (a-b)^2\right]^{-1},$ $\operatorname{Re} s > \{|\operatorname{Re} a|, |\operatorname{Re} b|\}$
81	$\sinh(ax)\cosh(bx)$	$a\left(s^2 - a^2 + b^2\right)\left[s^2 - (a+b)^2\right]^{-1}\left[s^2 - (a-b)^2\right]^{-1},$ $\operatorname{Re} s > \{|\operatorname{Re} a|, |\operatorname{Re} b|\}$
82	$\sinh^2(ax)$	$2a^2 s^{-1}\left(s^2 - 4a^2\right)^{-1}, \qquad \operatorname{Re} s > |\operatorname{Re} a|$
83	$\cosh^2(ax)$	$\left(s^2 - 2a^2\right) s^{-1}\left(s^2 - 4a^2\right)^{-1}, \qquad \operatorname{Re} s > |\operatorname{Re} a|$
84	$\sinh(ax)\cosh(ax)$	$a\left(s^2 - 4a^2\right)^{-1}, \qquad \operatorname{Re} s > |\operatorname{Re} a|$
85	$\dfrac{\cosh(ax) - 1}{a^2}$	$s^{-1}\left(s^2 - a^2\right)^{-1}, \qquad \operatorname{Re} s > |\operatorname{Re} a|$
86	$\dfrac{\sinh(ax) - ax}{a^3}$	$s^{-2}\left(s^2 - a^2\right)^{-1}, \qquad \operatorname{Re} s > |\operatorname{Re} a|$
87	$\dfrac{\left[\cosh(ax) - \frac{1}{2}a^2x^2 - 1\right]}{a^4}$	$s^{-3}\left(s^2 - a^2\right)^{-1}, \qquad \operatorname{Re} s > |\operatorname{Re} a|$

continued on next page

continued from previous page

	$f(x)$	$F(s)$
88	$\dfrac{\left[1 - \cosh(ax) + \frac{1}{2}ax\sinh(ax)\right]}{a^4}$	$s^{-1}\left(s^2 - a^2\right)^{-2},$ $\qquad\qquad \operatorname{Re} s > |\operatorname{Re} a|$
89	$x^{1/2}\sinh(ax)$	$\left(\pi^{1/2}/4\right)\left[(s-a)^{3/2} - (s+a)^{3/2}\right],$ $\qquad\qquad\qquad\qquad \operatorname{Re} s > |\operatorname{Re} a|$
90	$\ln x$	$-s^{-1}\ln\left(\mathbf{C}s\right),\qquad\qquad \operatorname{Re} s > 0 \qquad$ ET I 148(1)
91	$\ln(1 + ax),\qquad\qquad |\arg a| < \pi$	$s^{-1}e^{s/a}\operatorname{Ei}(-s/a),\qquad \operatorname{Re} s > 0 \qquad$ ET I 148(4)
92	$x^{-1/2}\ln x$	$-(\pi/s)^{1/2}\ln\left(4\mathbf{C}s\right),\qquad \operatorname{Re} s > 0 \qquad$ ET I 148(9)
93	$\mathrm{H}(x - a) = \begin{cases} 0 & x < a \\ 1 & x > a \end{cases}$ (Heaviside step function)	$s^{-1}e^{-as},\qquad\qquad\qquad\qquad a \geq 0$
94	$\delta(x)\qquad$ (Dirac delta function)	1
95	$\delta(x - a)$	$e^{-as},\qquad\qquad\qquad\qquad\qquad a \geq 0$
96	$\delta'(x - a)$	$se^{-as},\qquad\qquad\qquad\qquad\qquad a \geq 0$
97	$\operatorname{Si}(x) \equiv \displaystyle\int_0^x \frac{\sin\xi}{\xi}\,d\xi \equiv \frac{1}{2}\pi + \operatorname{si}(x)$	$s^{-1}\operatorname{arccot} s,\qquad\qquad \operatorname{Re} s > 0 \qquad$ ET I 177(17)
98	$\operatorname{Ci}(x) \equiv \operatorname{ci}(x) \equiv -\displaystyle\int_x^\infty \frac{\cos\xi}{\xi}\,d\xi$	$-\dfrac{1}{2}s^{-1}\ln\left(1 + s^2\right),\qquad \operatorname{Re} s > 0 \qquad$ ET I 178(19)
99[8]	$\operatorname{erf}\left(\dfrac{x}{2a}\right)$	$s^{-1}e^{a^2 s^2}\operatorname{erfc}(as),$ $\qquad\qquad \operatorname{Re} s > 0, |\arg a| < \pi/4 \qquad$ ET I 176(2)
100	$\operatorname{erf}\left(a\sqrt{x}\right)$	$as^{-1}\left(s + a^2\right)^{-1/2},$ $\qquad\qquad\qquad \operatorname{Re} s > \left\{0, -\operatorname{Re} a^2\right\} \qquad$ ET I 176(4)

continued on next page

continued from previous page

	$f(x)$		$F(s)$
101	erfc $\left(a\sqrt{x}\right)$		$s^{-1}\left(s+a^2\right)^{-\frac{1}{2}}\left[\left(s+a^2\right)^{1/2}-a\right]$, $\operatorname{Re}s>0$ ET I 177(9)
102[8]	erfc $\left(\dfrac{a}{\sqrt{x}}\right)$		$s^{-1}e^{-2a\sqrt{s}}$, $\operatorname{Re}s>0,\quad \operatorname{Re}a>0$ ET I 177(11)
103[8]	$J_\nu(ax)$,	$\operatorname{Re}\nu>-1$	$a^{-\nu}\left(\sqrt{s^2+a^2}-s\right)^\nu\left(s^2+a^2\right)^{-1/2}$, $\operatorname{Re}s>\lvert\operatorname{Im}a\rvert$, ET I 182(1)
104	$x\,J_\nu(ax)$,	$\operatorname{Re}\nu>-2$	$a^\nu\left[s+\nu\left(s^2+a^2\right)^{1/2}\right]\left[s+\left(s^2+a^2\right)^{1/2}\right]^{-\nu}$ $\times\left(s^2+a^2\right)^{-3/2}$, $\operatorname{Re}s>\lvert\operatorname{Im}a\rvert$, ET I 182(2)
105	$\dfrac{J_\nu(ax)}{x}$		$a^\nu\nu^{-1}\left[s+\left(s^2+a^2\right)^{1/2}\right]^{-\nu}$, $\operatorname{Re}s\geq\lvert\operatorname{Im}a\rvert$ ET I 182(5)
106	$x^n\,J_n(ax)$		$1\cdot3\cdot5\cdots(2n-1)a^n\left(s^2+a^2\right)^{-\left(n+\frac{1}{2}\right)}$, $\operatorname{Re}s>\lvert\operatorname{Im}a\rvert$ ET I 182(4)
107	$x^\nu\,J_\nu(ax)$,	$\operatorname{Re}\nu>-\tfrac{1}{2}$	$2^\nu\pi^{-1/2}\,\Gamma\left(\nu+\tfrac{1}{2}\right)a^\nu\left(s^2+a^2\right)^{-\left(\nu+\frac{1}{2}\right)}$, $\operatorname{Re}s>\lvert\operatorname{Im}a\rvert$, ET I 182(7)
108	$x^{\nu+1}\,J_\nu(ax)$,	$\operatorname{Re}\nu>-1$	$2^{\nu+1}\pi^{-1/2}\,\Gamma\left(\nu+\tfrac{3}{2}\right)a^\nu s\left(s^2+a^2\right)^{-\left(\nu+\frac{3}{2}\right)}$, $\operatorname{Re}s>\lvert\operatorname{Im}a\rvert$ ET I 182(8)
109[8]	$I_\nu(ax)$,	$\operatorname{Re}\nu>-1$	$a^{-\nu}\left[s-\sqrt{s^2-a^2}\right]^\nu\left(s^2-a^2\right)^{-1/2}$, $\operatorname{Re}s>\lvert\operatorname{Re}a\rvert$ ET I 195(1)
110	$x^\nu\,I_\nu(ax)$,	$\operatorname{Re}\nu>-\tfrac{1}{2}$	$2^\nu\pi^{-1/2}\,\Gamma\left(\nu+\tfrac{1}{2}\right)a^\nu\left(s^2-a^2\right)^{-\left(\nu+\frac{1}{2}\right)}$, $\operatorname{Re}s>\lvert\operatorname{Re}a\rvert$ ET I 195(6)

continued on next page

continued from previous page	
$f(x)$	$F(s)$
111 $\quad x^{\nu+1} I_\nu(ax), \qquad \operatorname{Re}\nu > -1$	$2^{\nu+1}\pi^{-1/2}\Gamma\left(\nu+\tfrac{3}{2}\right)a^\nu s\left(s^2-a^2\right)^{-\left(\nu+\frac{3}{2}\right)},$ $\operatorname{Re} s > \lvert\operatorname{Re} a\rvert \qquad$ ET I 196(7)
112 $\quad x^{-1} I_\nu(ax), \qquad \operatorname{Re}\nu > 0$	$\nu^{-1}a^\nu\left[s+\left(s^2-a^2\right)^{1/2}\right]^{-\nu},$ $\operatorname{Re} s > \lvert\operatorname{Re} a\rvert \qquad$ ET I 195(4)
113 $\quad \sin\left(2a^{1/2}x^{1/2}\right)$	$(\pi a)^{1/2}s^{-3/2}e^{-a/s}, \qquad \operatorname{Re} s > 0 \qquad$ ET I 153(32)
114 $\quad x^{-1/2}\cos\left(2a^{1/2}x^{1/2}\right)$	$\pi^{1/2}s^{-1/2}e^{-a/s}, \qquad \operatorname{Re} s > 0 \qquad$ ET I 158(67)
115 $\quad x^{-1}e^{-ax} I_1(ax)$	$\left[(s+2a)^{1/2}-s^{1/2}\right]\left[(s+2a)^{1/2}+s^{1/2}\right]^{-1},$ $\operatorname{Re} s > \lvert\operatorname{Re} a\rvert \qquad$ AS 1024(29.3.52)
116 $\quad \dfrac{J_k(ax)}{x}$	$k^{-1}a^{-k}\left[\left(s^2+a^2\right)^{1/2}-s\right]^k,$ $\operatorname{Re} s > \lvert\operatorname{Im} a\rvert, k > -1 \qquad$ AS 1025(29.3.58)
117 $\quad \left(\dfrac{x}{2a}\right)^{k-\frac{1}{2}} J_{k-\frac{1}{2}}(ax)$	$\Gamma(k)\pi^{-1/2}\left(s^2+a^2\right)^k,$ $\operatorname{Re} s > \lvert\operatorname{Im} a\rvert, \quad k > 0 \qquad$ AS 1024(29.3.57)
118 $\quad J_0(ax) - ax\,J_1(ax)$	$s^2\left(s^2+a^2\right)^{-3/2}, \qquad \operatorname{Re} s > \lvert\operatorname{Im} a\rvert$
119 $\quad I_0(ax) + ax\,I_1(ax)$	$s^2\left(s^2-a^2\right)^{-3/2}, \qquad \operatorname{Re} s > \lvert\operatorname{Im} a\rvert$

17.21 Fourier transform

The **Fourier transform**, also called the **exponential** or **complex Fourier transform**, of the function $f(x)$, denoted by $F(\xi)$, is defined by the integral

$$F(\xi) = \frac{1}{\sqrt{2\pi}}\int_{-\infty}^\infty f(x)e^{i\xi x}\,dx.$$

The functions $f(x)$ and $F(\xi)$ are called a **Fourier transform pair**, and knowledge of either one enables the other to be recovered. Setting

$$F(\xi) = \mathcal{F}\left[f(x);\xi\right],$$

to emphasize the nature of the transform, we have the symbolic inverse result

$$f(x) = \mathcal{F}^{-1}\left[F(\xi);x\right].$$

The inversion of the Fourier transform is accomplished by means of the **inversion integral**

$$f(x) = \frac{1}{\sqrt{2\pi}} \int_{-\infty}^{\infty} F(\xi) e^{-i\xi x} \, d\xi.$$

17.22 Basic properties of the Fourier transform

1. For a and b arbitrary constants,

$$\mathcal{F}\left[af(x) + bg(x)\right] = aF(\xi) + bG(\xi) \qquad \text{(linearity)}$$

2. If $n > 0$ is an integer, and $\lim_{|x| \to \infty} f^{(r)}(x) = 0$ for $r = 0, 1, \ldots, n-1$ with $f^{(0)}(x) \equiv f(x)$, then

$$\mathcal{F}\left[f^{(n)}(x); \xi\right] = (-i\xi)^n F(\xi) \qquad \text{(transform of a derivative)} \qquad \text{SN 27}$$

3. The **Fourier convolution** $f * g$ of two functions $f(x)$ and $g(x)$ is defined by the integral

$$f * g(x) = \frac{1}{\sqrt{2\pi}} \int_{-\infty}^{\infty} f(x - \xi) g(\xi) \, d\xi,$$

and it has the property $f * g = g * f$, and $f * (g * h) = (f * g) * h$. In terms of the convolution operation.

$$\mathcal{F}\left[f * g(x); \xi\right] = F(\xi) G(\xi) \qquad \text{(convolution (Faltung) theorem)} \qquad \text{SN 24}$$

17.23[10] Table of Fourier transform pairs

	$f(x)$		$F(\xi)$	
1	1		$(2\pi)^{1/2} \delta(\xi)$	SU 496
2[7]	$\dfrac{1}{x}$		$(\pi/2)^{1/2} i \operatorname{sgn} \xi$	SU 50
3	$\delta(x)$		$(2\pi)^{-1/2}$	SU 496
4[8]	$\delta(ax + b),$	$a, b \in \mathbb{R}, \quad a \neq 0$	$(2\pi)^{-1/2} e^{ib\xi/a}$	SU 517
5	$\begin{cases} 1 & |x| < a \\ 0 & |x| > a \end{cases},$	$a > 0$	$(2/\pi)^{1/2} \xi^{-1} \sin(a\xi)$	
6[8]	$\mathrm{H}(x) = \begin{cases} 0 & x < 0 \\ 1 & x > 0 \end{cases}$		$-\dfrac{1}{i\xi\sqrt{2\pi}} + \sqrt{\dfrac{\pi}{2}}\,\delta(\xi)$	SN 523
7	$\dfrac{1}{|x|^a},$	$0 < \operatorname{Re} a < 1$	$\dfrac{(2/\pi)^{1/2}\, \Gamma(1 - a) \sin\left(\frac{1}{2} a\pi\right)}{|\xi|^{1-a}}$	SN 523

continued on next page

	$f(x)$		$F(\xi)$		

continued from previous page

8	e^{iax},	$a \in \mathbb{R}$	$(2\pi)^{1/2}\,\delta(\xi + a)$		SU 50
9	$e^{-a|x|}$,	$a > 0$	$\dfrac{a(2/\pi)^{1/2}}{a^2 + \xi^2}$		SU 50
10[7]	$xe^{-a|x|}$,	$a > 0$	$\dfrac{2ai\xi(2/\pi)^{1/2}}{\left(a^2 + \xi^2\right)^2}$,	$\xi > 0$	SU 50
11	$|x|e^{-a|x|}$,	$a > 0$	$\dfrac{(2/\pi)^{1/2}\left(a^2 - \xi^2\right)}{\left(a^2 + \xi^2\right)^2}$		SU 50
12	$\dfrac{e^{-a|x|}}{|x|^{1/2}}$,	$a > 0$	$\dfrac{\left[a + \left(a^2 + \xi^2\right)^{1/2}\right]^{1/2}}{x(a^2 + \xi^2)^{1/2}}$		SN 523
13	$e^{-a^2 x^2}$,	$a > 0$	$\left(a\sqrt{2}\right)^{-1} e^{-\xi^2/4a^2}$		SU 51
14	$\dfrac{1}{a^2 + x^2}$,	$\operatorname{Re} a > 0$	$\dfrac{(\pi/2)^{1/2} e^{-a|\xi|}}{a}$		SU 51
15[7]	$\dfrac{x}{a^2 + x^2}$,	$\operatorname{Re} a > 0$	$i\,\operatorname{sgn}\xi(\pi/2)^{1/2} e^{-a|\xi|}$		
16[9]	$\sin\left(ax^2\right)$		$\dfrac{1}{(2a)^{1/2}} \cos\left(\dfrac{\xi^2}{4a} + \dfrac{\pi}{4}\right)$		SN 523
17	$\cos\left(ax^2\right)$		$\dfrac{1}{(2a)^{1/2}} \cos\left(\dfrac{\xi^2}{4a} - \dfrac{\pi}{4}\right)$		SN 523
18	$e^{-a|x|}\cos(bx)$,	$a > 0, \quad b > 0$	$a(2\pi)^{-1/2}\left[\dfrac{1}{a^2 + (b+\xi)^2} + \dfrac{1}{a^2 + (b-\xi)^2}\right]$		
19	$e^{-\frac{1}{2}ax^2}\sin(bx)$,	$a > 0, \quad b > 0$	$\dfrac{1}{2}ia^{-1/2}\left\{\exp\left[-\dfrac{1}{2}\dfrac{(\xi-b)^2}{a}\right]\right.$ $\left. - \exp\left[-\dfrac{1}{2}\dfrac{(\xi+b)^2}{a}\right]\right\}$		
20[9]	$\dfrac{\sinh(ax)}{\sinh(bx)}$,	$|a| < |b|$	$\dfrac{(\pi/2)^{1/2}\sin(\pi a/b)}{b\left[\cosh\left(\pi\xi/b\right) + \cos(\pi a/b)\right]}$		SU 123
21[9]	$\dfrac{\cosh(ax)}{\sinh(bx)}$,	$|a| < |b|$	$\dfrac{i(\pi/2)^{1/2}\sinh\left(\pi\xi/b\right)}{b\left[\cosh\left(\pi\xi/b\right) + \cos(\pi a/b)\right]}$		SU 123

continued on next page

continued from previous page

	$f(x)$	$F(\xi)$	
22	$\dfrac{\sin(ax)}{x}$	$\begin{cases} (\pi/2)^{1/2} & \|\xi\| < a, \\ 0 & \|\xi\| > a \end{cases}$	SN 523
23[7]	$\dfrac{x}{\sinh x}$	$\dfrac{\left(2/\pi^3\right)^{1/2} e^{\pi\xi}}{\left(1 + e^{\pi\xi}\right)^2}$	SU 123
24[7]	$x^n \operatorname{sgn} x, \qquad n = 1, 2, \ldots$	$(2/\pi)^{1/2} (-i\xi)^{-(1+n)} n!$	SU 506
25[7]	$\|x\|^\nu,$ $-1 < \nu < 0,$ but not integral	$(2/\pi)^{1/2} \Gamma(\nu + 1) \|\xi\|^{-\nu-1} \cos\left[\pi(\nu + 1)/2\right]$ SU506
26[7]	$\|x\|^\nu \operatorname{sgn} x,$ $-1 < \nu < 0,$ but not integral	$\dfrac{i \operatorname{sgn}\xi (2/\pi)^{1/2} \sin\left[(\pi/2)\,(\nu + 1)\right] \Gamma(\nu + 1)}{\|\xi\|^{\nu+1}}$ SU 506
27	$e^{-ax} \log\left\|1 - e^{-x}\right\|,$ $-1 < \operatorname{Re} a < 0$	$\left(\dfrac{\pi}{2}\right)^{1/2} \dfrac{\cot\left(\pi a - i\xi\pi\right)}{a - i\xi}$	ET I 121(26)
28	$e^{-ax} \log\left(1 + e^{-x}\right),$ $-1 < \operatorname{Re} a < 0$	$\left(\dfrac{\pi}{2}\right)^{1/2} \dfrac{\csc\left(\pi a - i\xi\pi\right)}{a - i\xi}$	ET I 121 (27)

In deriving results for the preceding table from ET I, account has been taken of the fact that the normalization factor $1/(2\pi)^{1/2}$ employed in our definition of F has not been used in those tables, and that there is a difference of sign between the exponents used in the definitions of the exponential Fourier transform.

17.24* Table of Fourier transform pairs for spherically symmetric functions

$f(\|\mathbf{r}\|) = \dfrac{1}{(2\pi)^{3/2}} \iiint E(\|\mathbf{k}\|) e^{i\mathbf{k}\cdot\mathbf{r}}\, d\mathbf{k}$	$E(\|\mathbf{k}\|) = \dfrac{1}{(2\pi)^{3/2}} \iiint f(\|\mathbf{r}\|) e^{-i\mathbf{k}\cdot\mathbf{r}}\, d\mathbf{r}$
$f(r) = \sqrt{\dfrac{2}{\pi}}\dfrac{1}{r}\displaystyle\int_0^\infty E(k) \sin(kr) k\, dk$	$E(k) = \sqrt{\dfrac{2}{\pi}}\dfrac{1}{k}\displaystyle\int_0^\infty f(r) \sin(kr) r\, dr$
e^{-ar}	$\dfrac{2}{2\pi}\dfrac{2a}{\left(a^2 + k^2\right)^2}$
$\dfrac{e^{-ar}}{r}$	$\dfrac{2}{2\pi}\dfrac{1}{\left(a^2 + k^2\right)^2}$
1	$(2\pi)^{3/2}\, \delta(\mathbf{k})$

17.31 Fourier sine and cosine transforms

The **Fourier sine** and **cosine transforms** of the function $f(x)$, denoted by $F_s(\xi)$ and $F_c(\xi)$, respectively, are defined by the integrals

$$F_s(\xi) = \sqrt{\frac{2}{\pi}} \int_0^\infty f(x) \sin(\xi x)\, dx \quad \text{and} \quad F_c(\xi) = \sqrt{\frac{2}{\pi}} \int_0^\infty f(x) \cos(\xi x)\, dx$$

The functions $f(x)$ and $F_s(\xi)$ are called a **Fourier sine transform pair**, and the functions $f(x)$ and $F_c(\xi)$ a **Fourier cosine transform pair**, and knowledge of either $F_s(\xi)$ or $F_c(\xi)$ enables $f(x)$ to be recovered.

Setting

$$F_s(\xi) = \mathcal{F}_s\left[f(x); \xi\right] \quad \text{and} \quad F_c(\xi) = \mathcal{F}_c\left[f(x); \xi\right],$$

to emphasize the nature of the transforms, we have the symbolic inverses

$$f(x) = \mathcal{F}_s^{-1}\left[F_s(\xi); x\right] \quad \text{and} \quad f(x) = \mathcal{F}_c^{-1}\left[F_c(\xi); x\right]$$

The inversion of the Fourier sine transform is accomplished by means of the **inversion integral**

$$f(x) = \sqrt{\frac{2}{\pi}} \int_0^\infty F_s(\xi) \sin(\xi x)\, d\xi \qquad (x \geq 0)$$

and the inversion of the Fourier cosine transform is accomplished by means of the **inversion integral**

$$f(x) = \sqrt{\frac{2}{\pi}} \int_0^\infty F_c(\xi) \cos(\xi x)\, d\xi \qquad (x \geq 0) \qquad \text{SN 17}$$

17.32 Basic properties of the Fourier sine and cosine transforms

1.   For a and b arbitrary constants,

$$\mathcal{F}_s\left[af(x) + bg(x)\right] = aF_s(\xi) + bG_s(\xi)$$

and

$$\mathcal{F}_c\left[af(x) + bg(x)\right] = aF_c(\xi) + bG_c(\xi) \qquad \text{(linearity)}$$

2.   If $\lim_{x \to \infty} f^{(r-1)}(x) = 0$ and $\lim_{x \to \infty} \sqrt{\frac{2}{\pi}} f^{(r-1)}(x) = a_{r-1}$, then denoting the Fourier sine and cosine transforms of $f^{(r)}(x)$ by $F_s^{(r)}$ and $F_c^{(r)}$, respectively,

(i) $F_c^{(r)}(\xi) = -a_{r-1} + \xi F_s^{(r-1)}$.

(ii) $F_s^{(r)}(\xi) = -\xi F_c^{(r-1)}(\xi)$,

(iii) $F_c^{(2r)}(\xi) = -\sum_{n=0}^{r-1} (-1)^n a_{2r-2n-1} \xi^{2n} + (-1)^r \xi^{2n} F_c(\xi)$,

(iv) $F_c^{(2r+1)}(\xi) = -\sum_{n=0}^{r-1}{}' (-1)^n a_{2r-2n} \xi^{2n} + (-1)^r \xi^{2r+1} F_s(\xi)$,

(v) $F_s^{(r)}(\xi) = \xi a_{r-2} - \xi^2 F_s^{(r-2)}(\xi)$,

(vi)[6] $F_s^{(2r)}(\xi) = -\sum_{n=1}^{r} (-1)^n \xi^{2n-1} a_{2r-2n} + (-1)^r \xi^{2r} F_s(\xi)$,

(vii) $\quad F_s{}^{(2r+1)}(\xi) = -\sum_{n=1}^{r}{}'(-1)^n \xi^{2n-1} a_{2r-2n+1} + (-1)^{r+1} \xi^{2r+1} F_c(\xi).$ SN 28

3. (i) $\quad \displaystyle\int_0^\infty F_s(\xi) G_s(\xi) \cos(\xi x)\, d\xi = \frac{1}{2} \int_0^\infty g(s)\,[f(s+x) + f(s-x)]\, ds,$

   (ii) $\quad \displaystyle\int_0^\infty F_c(\xi) G_c(\xi) \cos(\xi x)\, d\xi = \frac{1}{2} \int_0^\infty g(s)\,[f(s+x) + f(|x-s|)]\, ds$

   (convolution (Faltung) theorem) SN 24

4. (i) If $F_s(\xi)$ is the Fourier sine transform of $f(x)$, then the Fourier sine transform of $F_s(x)$ is $f(\xi)$.

   (ii) If $F_c(\xi)$ is the Fourier cosine transform of $f(x)$, then the Fourier cosine transform of $F_c(x)$ is $f(\xi)$.

   (iii) If $f(x)$ is an odd function in $(-\infty, \infty)$, then the Fourier sine transform of $f(x)$ in $(0, \infty)$ is $-iF(\xi)$.

   (iv) If $f(x)$ is an even function in $(-\infty, \infty)$, then the Fourier cosine transform of $f(x)$ in $(0, \infty)$ is $F(\xi)$.

   (v) The Fourier sine transform of $f(x/a)$ is $aF_s(a\xi)$.

   (vi) The Fourier cosine transform of $f(x/a)$ is $aF_c(a\xi)$.

   (vii) $\mathcal{F}_s\,[f(x); \xi] = F_s\,(|\xi|)\,\mathrm{sgn}\,\xi$ SU 45

17.33^{10} Table of Fourier sine transforms

	$f(x)$		$F_s(\xi)$ $\quad (\xi > 0)$		
1	x^{-1}		$(\pi/2)^{1/2},$	$\xi > 0$	ET I 64(3)
2	$x^{-\nu},$	$0 < \mathrm{Re}\,\nu < 2$	$(2/\pi)^{1/2} \xi^{\nu-1}\, \Gamma(1-\nu)\cos(\nu\pi/2),$		
				$\xi > 0$	ET I 68(1)
3	$x^{-1/2}$		$\xi^{-1/2},$	$\xi > 0$	ET I 64(6)
4	$x^{-3/2}$		$2\xi^{1/2},$	$\xi > 0$	ET I 64(9)
5	$\begin{cases}1 & 0 < x < a \\ 0 & x > a\end{cases}$		$(2/\pi)^{1/2}\xi^{-1}\,[1 - \cos(a\xi)],$	$\xi > 0$	ET I 63(1)
6	$\begin{cases}x^{-1} & 0 < x < a \\ 0 & x > a\end{cases}$		$(2/\pi)^{1/2}\,\mathrm{Si}(a\xi),$	$\xi > 0$	ET I 64(4)

continued on next page

continued from previous page		
	$f(x)$	$F_s(\xi)$ $\quad(\xi > 0)$
7	$\dfrac{1}{a-x},$ $\qquad a > 0$	$(2/\pi)^{1/2}\left\{\sin(a\xi)\operatorname{Ci}(a\xi) - \cos(a\xi)\left[\tfrac{1}{2}\pi + \operatorname{Si}(a\xi)\right]\right\},$ $\qquad\qquad\qquad \xi > 0 \qquad$ ET I 64(11)
8[7]	$\dfrac{1}{x^2+a^2},$ $\qquad a > 0$	$(2\pi)^{-1/2}a^{-1}\left[e^{-a\xi}\operatorname{Ei}(a\xi) - e^{a\xi}\operatorname{Ei}(-a\xi)\right],$ $\qquad\qquad\qquad \xi > 0 \qquad$ ET I 65(14)
9	$x\left(x^2+a^2\right)^{-3/2},$ $\qquad \operatorname{Re} a > 0$	$(2/\pi)^{1/2}\xi\,K_0(a\xi), \qquad\qquad \xi > 0 \qquad$ ET I 66(27)
10	$x^{-1/2}\left(x^2+a^2\right)^{-1/2},$ $\qquad \operatorname{Re} a > 0$	$\xi^{1/2}I_{\frac{1}{4}}\left(\tfrac{1}{2}a\xi\right)K_{\frac{1}{4}}\left(\tfrac{1}{2}a\xi\right), \quad \xi > 0 \qquad$ ET I 66(28)
11[7]	$x\left(x^2+a^2\right)^{-\nu-\frac{3}{2}},$ $\qquad \operatorname{Re}\nu > -1, \quad \operatorname{Re} a > 0$	$\dfrac{\xi^{\nu+1}}{\sqrt{2}(2a)^\nu\,\Gamma\left(\nu+\frac{3}{2}\right)}\,K_\nu(a\xi),$
12	$\dfrac{x}{a^2+x^2},$ $\qquad \operatorname{Re} a > 0$	$\left(\dfrac{\pi}{2}\right)^{1/2}e^{-a\xi}, \qquad\qquad \xi > 0 \qquad$ ET I 65(15)
13	$\dfrac{x}{\left(a^2+x^2\right)^2}$	$\sqrt{\pi/8}\,a^{-1}\xi e^{-a\xi}, \qquad\qquad \xi > 0 \qquad$ ET I 67(35)
14	$x^{-1}\left(x^2+a^2\right)^{-1},$ $\qquad \operatorname{Re} a > 0$	$\dfrac{\sqrt{\pi/2}}{a^2}\left(1 - e^{-a\xi}\right), \qquad\qquad \xi > 0 \qquad$ ET I 65(20)
15	$x^{-1}e^{-ax},$ $\qquad \operatorname{Re} a > 0$	$(2/\pi)^{1/2}\tan^{-1}\left(\dfrac{\xi}{a}\right), \qquad\qquad \xi > 0 \qquad$ ET I 72(2)
16	$x^{\nu-1}e^{-ax},$ $\qquad \operatorname{Re}\nu > -1, \quad \operatorname{Re} a > 0$	$(2/\pi)^{1/2}\Gamma(\nu)\left(a^2+\xi^2\right)^{-\nu/2}\sin\left[\nu\tan^{-1}\left(\dfrac{\xi}{a}\right)\right],$ $\qquad\qquad\qquad \xi > 0 \qquad$ ET I 72(7)
17	$e^{-ax},$ $\qquad \operatorname{Re} a > 0$	$\dfrac{\sqrt{2/\pi}\,\xi}{a^2+\xi^2}, \qquad\qquad \xi > 0 \qquad$ ET I 72(1)
18	$xe^{-ax},$ $\qquad \operatorname{Re} a > 0$	$\dfrac{(2/\pi)^{1/2}2a\xi}{\left(a^2+\xi^2\right)^2}, \qquad\qquad \xi > 0 \qquad$ ET I 72(3)
19	$xe^{-ax^2},$ $\qquad |\arg a| < \pi/2$	$(2a)^{-3/2}\xi\exp\left(\dfrac{-\xi^2}{4a}\right), \qquad \xi > 0 \qquad$ ET I 73(19)
20	$\dfrac{\sin ax}{x},$ $\qquad a > 0$	$\dfrac{1}{(2\pi)^{1/2}}\log\left|\dfrac{\xi+a}{\xi-a}\right|, \qquad\qquad \xi > 0 \qquad$ ET I 78(1)
		continued on next page

	continued from previous page		
	$f(x)$		$F_s(\xi) \qquad (\xi > 0)$
21	$\dfrac{\sin ax}{x^2},$	$a > 0$	$\begin{cases} \xi\left(\frac{\pi}{2}\right)^{1/2} & 0 < \xi < a \\ a\left(\frac{\pi}{2}\right)^{1/2} & a < \xi < \infty \end{cases}, \quad \xi > 0 \qquad$ ET I 78(2)
22	$\sin\left(\dfrac{a^2}{x}\right),$	$a > 0$	$a\left(\dfrac{\pi}{2}\right)^{1/2} \xi^{-1/2} J_1\left(2a\xi^{\frac{1}{2}}\right), \\ \qquad\qquad\qquad\qquad \xi > 0 \qquad$ ET I 83(6)
23	$x^{-1}\sin\left(\dfrac{a^2}{x}\right),$	$a > 0$	$\left(\dfrac{\pi}{2}\right)^{1/2} Y_0\left(2a\xi^{1/2}\right) + \left(\dfrac{2}{\pi}\right)^{1/2} K_0\left(2a\xi^{1/2}\right) \\ \qquad\qquad\qquad\qquad\qquad$ ET I 83(7)
24	$x^{-2}\sin\left(\dfrac{a^2}{x}\right),$	$a > 0$	$\left(\dfrac{\pi}{2}\right)^{1/2} a^{-1}\xi^{1/2} J_1\left(2a\xi^{1/2}\right), \\ \qquad\qquad\qquad\qquad \xi > 0 \qquad$ ET I 83(8)
25[10]	$\operatorname{cosech}(ax),$	$\operatorname{Re} a > 0$	$(\pi/2)^{1/2} a^{-1} \tanh\left(\frac{1}{2}\pi a^{-1}\xi\right), \\ \qquad\qquad\qquad\qquad \xi > 0 \qquad$ ET I 88(2)
26	$\coth\left(\dfrac{1}{2}ax\right) - 1,$	$\operatorname{Re} a > 0$	$(2\pi)^{1/2} a^{-1} \coth\left(\pi a^{-1}\xi\right) - \xi, \\ \qquad\qquad\qquad\qquad \xi > 0 \qquad$ ET I 88(3)
27	$\left(1-x^2\right)^{-1}\sin(\pi x)$		$\begin{cases} (2/\pi)^{1/2}\sin\xi & 0 \le \xi \le \pi \\ 0 & \pi < \xi \end{cases} \qquad$ ET I 78(4)
28	$e^{-ax^2}\sin(bx),$	$\operatorname{Re} a > 0$	$(2a)^{-1/2}\exp\left[-\left(\xi^2+b^2\right)/(4a)\right]\sinh\left(b\xi/2a\right), \\ \qquad\qquad\qquad\qquad \xi > 0 \qquad$ ET I 78(7)
29	$\dfrac{\sin^2(ax)}{x},$	$a > 0$	$\begin{cases} \pi^{1/2}2^{-3/2} & 0 < \xi < 2a \\ \pi^{1/2}2^{-5/2} & \xi = 2a \\ 0 & 2a < \xi \end{cases} \qquad$ ET I 78(8)
30	$\sin\left(ax^2\right),$	$a > 0$	$a^{-1/2}\left\{\cos\left(\xi^2/4a\right) C\left[(2\pi a)^{-1/2}\xi\right]\right\} \\ + \sin\left(\xi^2/4a\right) S\left[(2\pi a)^{-1/2}\xi\right], \\ \qquad\qquad\qquad\qquad \xi > 0 \qquad$ ET I 82(1)
			continued on next page

continued on next page

	continued from previous page	
	$f(x)$	$F_s(\xi)$ $\qquad (\xi > 0)$
31	$\cos\left(ax^2\right),\qquad\qquad a > 0$	$a^{-1/2}\left\{\sin\left(\xi^2/4a\right) C\left[(2\pi a)^{-1/2}\xi\right]\right\}$ $\quad -\cos\left(\xi^2/4a\right) S\left[(2\pi a)^{-1/2}\xi\right],$ $\qquad\qquad\qquad\qquad\qquad\qquad \xi > 0$
32	$\arctan\left(\dfrac{x}{a}\right),\qquad\qquad a > 0$	$(\pi/2)^{1/2}\xi^{-1}e^{-a\xi},\qquad\qquad \xi > 0\qquad$ ET I 87(3)
33[7]	$\arctan\left(\dfrac{2a}{x}\right),\qquad\qquad \operatorname{Re}a > 0$	$(2\pi)^{-1/2}e^{-a\xi}\sinh(a\xi),\qquad \xi > 0\qquad$ ET I 87(8)
34	$\dfrac{\ln x}{x}$	$-(\pi/2)^{1/2}\left(\boldsymbol{C}+\ln\xi\right),\qquad \xi > 0\qquad$ ET I 76(2)
35	$\ln\left\|\dfrac{x+a}{x-a}\right\|,\qquad\qquad a > 0$	$(2\pi)^{1/2}\xi^{-1}\sin(a\xi),\qquad\qquad \xi > 0\qquad$ ET I 77(11)
36[7]	$\dfrac{\ln\left(1+a^2x^2\right)}{x},\qquad\qquad a > 0$	$-(2\pi)^{1/2}\operatorname{Ei}\left(-\xi/a\right),\qquad \xi > 0\qquad$ ET I 77(14)
37	$J_0(ax),\qquad\qquad a > 0$	$\begin{cases} 0 & 0 < \xi < a \\ (2/\pi)^{1/2}\left(\xi^2-a^2\right)^{-1/2} & a < \xi < \infty \end{cases}$ $\qquad\qquad\qquad\qquad\qquad\qquad\qquad$ ET I 99(1)
38	$J_\nu(ax),\qquad \operatorname{Re}\nu > -2,\quad a > 0$	$(2/\pi)^{1/2}\left(a^2-\xi^2\right)^{-1/2}\sin\left[\nu\sin^{-1}\left(\dfrac{\xi}{a}\right)\right]$ $\qquad\qquad\qquad\qquad\qquad\qquad\text{for } 0 < \xi < a$ $\dfrac{a^\nu\cos\left(\frac{1}{2}\nu\pi\right)}{\left(\xi^2-a^2\right)^{1/2}\left[\xi+\left(\xi^2-a^2\right)^{1/2}\right]^\nu}\quad\text{for } a < \xi < \infty$ $\qquad\qquad\qquad\qquad\qquad\qquad\qquad$ ET I 99(3)
39	$\dfrac{J_0(ax)}{x},\qquad\qquad a > 0$	$\begin{cases} (2/\pi)^{1/2}\sin^{-1}\left(\frac{\xi}{a}\right) & 0 < \xi < a \\ (\pi/2)^{1/2} & a < \xi < \infty \end{cases}$ $\qquad\qquad\qquad\qquad\qquad\qquad\qquad$ ET I 99(4)
40[7]	$\left(x^2+b^2\right)^{-1}J_0(ax),$ $\qquad\qquad a > 0,\quad \operatorname{Re}b > 0$	$(2/\pi)^{1/2}\sinh(b\xi)\,K_0(ab)/b,$ $\qquad\qquad\qquad 0 < \xi < a\qquad$ ET I 100(12)
		continued on next page

	continued from previous page	
	$f(x)$	$F_s(\xi) \qquad (\xi > 0)$
41	$x\left(x^2 + b^2\right)^{-1} J_0(ax),$	$(\pi/2)^{1/2} e^{-b\xi} I_0(ab),$
	$\qquad a > 0, \quad \operatorname{Re} b > 0$	$\qquad a < \xi < \infty \qquad$ ET I 100(13)

In deriving results for the preceding table from ET I, account has been taken of the fact that the normalization factor $\sqrt{2/\pi}$ employed in our definition of F_s has not been used in those tables.

17.34[10] Table of Fourier cosine transforms

	$f(x)$	$F_c(\xi)$
1	$x^{-\nu}, \qquad\qquad 0 < \operatorname{Re}\nu < 1$	$(\pi/2)^{1/2}[\Gamma(\nu)]^{-1} \sec\left(\tfrac{1}{2}\nu\pi\right)\xi^{\nu-1},$
		$\xi > 0 \qquad$ ET I 10(1)
2	$\begin{cases} 1 & 0 < x < a \\ 0 & x > a \end{cases}$	$(2/\pi)^{1/2}\dfrac{\sin(a\xi)}{\xi}, \qquad\qquad \xi > 0 \qquad$ ET I 7(1)
3	$\begin{cases} 0 & 0 < x < a \\ 1/x & x > a \end{cases}$	$-(2/\pi)^{1/2}\operatorname{Ci}(a\xi), \qquad\qquad \xi > 0 \qquad$ ET I 8(3)
4	$\begin{cases} x^{-1/2} & 0 < x < a \\ 0 & x > a \end{cases}$	$2\xi^{-1/2}\,C(a\xi), \qquad\qquad \xi > 0 \qquad$ ET I 8(5)
5	$\begin{cases} 0 & 0 < x < a \\ x^{-1/2} & x > a \end{cases}$	$2\xi^{-1/2}\left[\tfrac{1}{2} - C(a\xi)\right], \qquad\qquad \xi > 0 \qquad$ ET I 8(6)
6[9]	$x^{\nu-1}, \qquad\qquad 0 < \nu < 1$	$(2/\pi)^{1/2}\,\Gamma(\nu)\xi^{-\nu}\cos\left(\tfrac{1}{2}\nu\pi\right),$
		$0 < \nu < 1 \qquad$ ET I 10(1)
7	$\dfrac{1}{x^2 + a^2}, \qquad \operatorname{Re} a > 0$	$\dfrac{(\pi/2)^{1/2}e^{-a\xi}}{a}, \qquad\qquad \xi > 0 \qquad$ ET I 11(7)
8	$\dfrac{1}{x^2 + a^2}, \qquad \operatorname{Re} a > 0$	$\dfrac{(\pi/2)^{\frac{1}{2}}(1 + a\xi)e^{-a\xi}}{2a^3}, \qquad\qquad \xi > 0 \qquad$ ET I 11(7)
9	$\left(x^2 + a^2\right)^{-\nu-\frac{1}{2}},$ $\qquad\qquad \operatorname{Re} a > 0, \operatorname{Re}\nu > -\dfrac{1}{2}$	$\sqrt{2}\left(\dfrac{\xi}{2a}\right)^\nu \dfrac{K_\nu(a\xi)}{\Gamma\left(\nu+\frac{1}{2}\right)}, \qquad \xi > 0 \qquad$ ET I 11(7)
10	$\begin{cases} \left(a^2 - x^2\right)^\nu & 0 < x < a \\ 0 & x > a \end{cases},$ $\qquad\qquad\qquad \operatorname{Re}\nu > -1$	$2^\nu\,\Gamma(\nu+1)(a/\xi)^{\nu+\frac{1}{2}}\,J_{\nu+\frac{1}{2}}(a\xi),$ $\qquad\qquad \xi > 0 \qquad$ ET I 11(8)
		continued on next page

	$f(x)$	$F_c(\xi)$
	continued from previous page	
11	$\begin{cases} 0 & 0 < x < a \\ \left(x^2 - a^2\right)^{-\nu-\frac{1}{2}} & x > a \end{cases}$, $\qquad -\dfrac{1}{2} < \operatorname{Re}\nu < \dfrac{1}{2}$	$-2^{-\left(\nu+\frac{1}{2}\right)} \Gamma\left(\tfrac{1}{2} - \nu\right) (\xi/a)^\nu\, Y_\nu(a\xi),$ $\xi > 0 \qquad$ ET I 11(9)
12	$e^{-ax}, \qquad\qquad \operatorname{Re} a > 0$	$(2/\pi)^{1/2} a\left(a^2 + \xi^2\right)^{-1}, \qquad \xi > 0 \qquad$ ET I 14(1)
13	$xe^{-ax}, \qquad\qquad \operatorname{Re} a > 0$	$(2/\pi)^{1/2}\left(a^2 - \xi^2\right)\left(a^2 + \xi^2\right)^{-2},$ $\xi > 0 \qquad$ ET I 15(7)
14[7]	$x^{\nu-1} e^{-ax},$ $\operatorname{Re} a > 0, \quad \operatorname{Re}\nu > a$	$(2/\pi)^{1/2} \Gamma(\nu)\left(a^2 + \xi^2\right)^{-\nu/2} \cos\left[\nu \tan^{-1}\left(\dfrac{\xi}{a}\right)\right],$ $\xi > 0 \qquad$ ET I 15(7)
15	$x^{-1/2} e^{-ax}, \qquad\qquad \operatorname{Re} a > 0$	$\left(a^2 + \xi^2\right)^{-1/2}\left[\left(a^2 + \xi^2\right)^{1/2} + a\right]^{1/2},$ $\xi > 0 \qquad$ ET I 14(4)
16[7]	$e^{-a^2 x^2}, \qquad\qquad \operatorname{Re} a > 0$	$2^{-1/2}|a|^{-1} e^{-\xi^2/4a^2}, \qquad \xi > 0 \qquad$ ET I 15(11)
17	$x^{-1} e^{-x} \sin x$	$(2\pi)^{-1/2} \tan^{-1}\left(\dfrac{2}{\xi^2}\right), \qquad \xi > 0 \qquad$ ET I 19(7)
18	$\sin\left(ax^2\right), \qquad\qquad a > 0$	$\dfrac{1}{2\sqrt{a}}\left[\cos\left(\dfrac{\xi^2}{4a}\right) - \sin\left(\dfrac{\xi^2}{4a}\right)\right],$ $\xi > 0 \qquad$ ET I 23(1)
19	$\cos\left(ax^2\right), \qquad\qquad a > 0$	$\dfrac{1}{2\sqrt{a}}\left[\cos\left(\dfrac{\xi^2}{4a}\right) + \sin\left(\dfrac{\xi^2}{4a}\right)\right],$ $\xi > 0 \qquad$ ET I 24(7)
20	$\dfrac{\sin(ax)}{x}, \qquad\qquad a > 0$	$\begin{cases} (\pi/2)^{1/2} & \xi < a \\ \frac{1}{2}(\pi/2)^{1/2} & \xi = a \\ 0 & \xi > a \end{cases}$ ET I 18(1)
21[7]	$\dfrac{\sin^2(ax)}{x^2}, \qquad\qquad a > 0$	$\begin{cases} (\pi/2)^{1/2}\left(a - \frac{1}{2}\xi\right) & \xi < 2a \\ 0 & 2a < \xi \end{cases}$ ET I 19(8)
22[7]	$e^{-bx}\sin(ax), \quad a > 0, \quad \operatorname{Re} b > 0$	$(2\pi)^{-1/2}\left[\dfrac{a+\xi}{b^2 + (a+\xi)^2} + \dfrac{a-\xi}{b^2 + (a-\xi)^2}\right],$ $\xi > 0 \qquad$ ET I 19(6)

continued on next page

	continued from previous page	
	$f(x)$	$F_c(\xi)$
23	$\dfrac{\sin\left[b\left(x^2+a^2\right)^{1/2}\right]}{\left(x^2+a^2\right)^2}\qquad a>0$	$(b/a)(\pi/2)^{1/2}e^{-a\xi},\qquad \xi>0\qquad$ ET I 26(29)
24	$\left(x^2+a^2\right)^{-1/2}\sin\left[b\left(x^2+a^2\right)^{1/2}\right],$ $\qquad a>0$	$\begin{cases}(\pi/2)^{1/2}J_0\left[a\left(b^2-\xi^2\right)^{1/2}\right] & 0<\xi<b\\ 0 & b<\xi\end{cases}$ ET I 26(30)
25	$\dfrac{1-\cos(ax)}{x^2},\qquad a>0$	$\begin{cases}(\pi/2)^{1/2}(a-\xi) & \xi<a\\ 0 & a<\xi\end{cases}$ ET I 20(16)
26	$e^{-ax^2}\sin\left(bx^2\right),\qquad \operatorname{Re}a>\lvert\operatorname{Im}b\rvert$	$2^{-1/2}\left(a^2+b^2\right)^{-1/4}\exp\left\{-a\xi^2/\left[4\left(a^2+b^2\right)\right]\right\}$ $\times\sin\left[\tfrac{1}{2}\arctan(b/a)-\tfrac{1}{4}b\xi^2\left(a^2+b^2\right)^{-1}\right],$ $\qquad\qquad \xi>0\qquad$ ET I 23(5)
27	$e^{-ax^2}\cos\left(bx^2\right),\qquad \operatorname{Re}a>\lvert\operatorname{Im}b\rvert$	$2^{-1/2}\left(a^2+b^2\right)^{-1/4}\exp\left\{-a\xi^2/\left[4\left(a^2+b^2\right)\right]\right\}$ $\times\cos\left[\tfrac{1}{4}b\xi^2\left(a^2+b^2\right)^{-1}-\tfrac{1}{2}\arctan(b/a)\right],$ $\qquad\qquad \xi>0\qquad$ ET I 24(6)
28	$\dfrac{\sinh(ax)}{\sinh(bx)}\qquad \lvert\operatorname{Re}a\rvert<\operatorname{Re}b$	$\left(\dfrac{\pi}{2}\right)^{1/2}\dfrac{\sin(\pi a/b)}{b\left[\cosh\left(\pi\xi/b\right)+\cos(\pi a/b)\right]},$ $\qquad\qquad \xi>0\qquad$ ET I 31(14)
29	$\dfrac{\cosh(ax)}{\cosh(bx)},\qquad \lvert\operatorname{Re}a\rvert<\operatorname{Re}b$	$\dfrac{(2\pi)^{1/2}\cos(\pi a/2b)\cosh\left(\pi\xi/2b\right)}{b\left[\cosh\left(\pi\xi/b\right)+\cos(\pi a/b)\right]},$ $\qquad\qquad \xi>0\qquad$ ET I 31(12)
30	$\operatorname{sech}(ax),\qquad \operatorname{Re}a>0$	$a^{-1}(\pi/2)^{1/2}\operatorname{sech}\left(\pi\xi/2a\right),\quad \xi>0\qquad$ ET I 30(1)
31	$\left(x^2+a^2\right)\operatorname{sech}\left(\dfrac{\pi x}{2a}\right),\quad \operatorname{Re}a>0$	$2(2/\pi)^{1/2}a^3\operatorname{sech}^3(a\xi),\qquad \xi>0\qquad$ ET I 32(19)
32	$\ln\left(1+\dfrac{a^2}{x^2}\right),\qquad \operatorname{Re}a>0$	$(2\pi)^{1/2}\xi^{-1}\left(1-e^{-a\xi}\right),\qquad \xi>0\qquad$ ET I 18(10)
33[7]	$\ln\left(\dfrac{a^2+x^2}{b^2+x^2}\right),$ $\qquad \operatorname{Re}a>0,\operatorname{Re}b>0$	$(2\pi)^{1/2}\left(e^{-b\xi}-e^{-a\xi}\right),\qquad \xi>0\qquad$ ET I 18(12)
		continued on next page

continued from previous page	
$f(x)$	$F_c(\xi)$
34 $\quad \left(x^2 + b^2\right)^{-1} J_0(ax),$ $\qquad a > 0, \quad \operatorname{Re} b > 0$	$(\pi/2)^{1/2} b^{-1} e^{-b\xi} I_0(ab),$ $\qquad a < \xi < \infty \qquad$ ET I 45(14)
35 $\quad x\left(x^2 + b^2\right)^{-1} J_0(ax),$ $\qquad a > 0, \quad \operatorname{Re} b > 0$	$(2/\pi)^{1/2} \cosh(b\xi) K_0(ab),$ $\qquad 0 < \xi < a \qquad$ ET I 45(15)

In deriving results for the preceding table from ET I, account has been taken of the fact that the normalization factor $\sqrt{2/\pi}$ employed in our definition of F_c has not been used in those tables.

17.35 Relationships between transforms

The following relationships exist between transforms and they may be used to derive further transform pairs from among the results given in Sections 17.13–17.34. The appropriate sections of the main body of the tables may also be used to extend the list of transform pairs.

17.351

Fourier cosine transform and Laplace transform relationship

$$\mathcal{F}_c\left[f(x); \xi\right] = \frac{1}{\sqrt{2\pi}} \mathcal{L}\left[f(x); i\xi\right] + \frac{1}{\sqrt{2\pi}} \mathcal{L}\left[f(x); -i\xi\right].$$

17.352

Fourier sine transform and Laplace transform relationship.

$$\mathcal{F}_s\left[f(x); \xi\right] = \frac{i}{\sqrt{2\pi}} \mathcal{L}\left[f(x); i\xi\right] - \frac{i}{\sqrt{2\pi}} \mathcal{L}\left[f(x); -i\xi\right].$$

17.353

Exponential Fourier transform and Laplace transform relationship

$$\mathcal{F}\left[f(x); \xi\right] = \sqrt{2\pi}\mathcal{L}\left[f(x); -i\xi\right] + \sqrt{2\pi}\mathcal{L}\left[f(-x); i\xi\right].$$

17.41[10] Mellin transform

The **Mellin transform** of the function $f(x)$, denoted by $f^*(s)$, is defined by the integral

$$f^*(s) = \int_0^\infty f(x) x^{s-1}\, dx.$$

The functions $f(x)$ and $f^*(s)$ are called a **Mellin transform pair**, and knowledge of either one enables the other to be recovered.

The transform exists provided the integral

$$\int_0^\infty |f(x)| x^{k-1}\, dx$$

is bounded for some $k > 0$, and then the inversion of the Mellin transform is accomplished by means of the **inversion integral**

$$f(x) = \frac{1}{2\pi i} \int_{c-i\infty}^{c+i\infty} f^*(s) x^{-s}\, ds,$$

where $c > k$.

Setting

$$f^*(s) = \mathcal{M}[f(x); s]$$

to denote the Mellin transform, we have the symbolic expression for the inverse result

$$f(x) = \mathcal{M}^{-1}[f^*(s); x] \qquad\qquad \text{MS 397(6)}$$

17.42 Basic properties of the Mellin transform

1. For a and b arbitrary constants,

$$\mathcal{M}[af(x) + bg(x)] = af^*(s) + bg^*(s) \qquad \text{(linearity)}$$

2. If $\lim_{x \to 0} x^{s-r-1} f^{(r)}(x) = 0, \quad r = 0, 1, \dots, n-1,$

    (i) $\mathcal{M}\left[f^{(n)}(x); s\right] = (-1)^n \dfrac{\Gamma(s)}{\Gamma(s-n)} f^*(s-n)$

    $$\text{(transform of a derivative)} \quad \text{SU 267 (4.2.3)}$$

    (ii) $\mathcal{M}\left[x^n f^{(n)}(x); s\right] = (-1)^n \dfrac{\Gamma(s+n)}{\Gamma(s)} f^*(s)$

    $$\text{(transform of a derivative)} \quad \text{SU 267 (4.2.5)}$$

3. Denoting the nth repeated integral of $f(x)$ by $I_n[f(x)]$, where

    $$I_n[f(x)] = \int_0^x I_{n-1}[f(u)]\, du,$$

    (i) $\mathcal{M}[I_n[f(x)]; s] = (-1)^n \dfrac{\Gamma(s)}{\Gamma(n+s)} f^*(s+n)$

    $$\text{(transform of an integral)} \quad \text{SU 269 (4.2.15)}$$

    (ii) $\mathcal{M}[I_n^\infty[f(x)]; s] = \dfrac{\Gamma(s)}{\Gamma(s+n)} f^*(s+n),$

    where

    $$I_n^\infty[f(x)] = \int_x^\infty I_{n-1}^\infty f(u)\, du \qquad \text{(transform of an integral)} \qquad \text{SU 269 (4.2.18)}$$

4. $\mathcal{M}[f(x)g(x); s] = \dfrac{1}{2\pi i} \int_{c-i\infty}^{c+i\infty} f^*(u) g^*(s-u)\, du$

    $$\text{(Mellin convolution theorem)} \quad \text{SU 275(4.4.1)}$$

17.43^{10} Table of Mellin cosine transforms

$f(x)$	$f^*(s)$	
1 e^{-x}	$\Gamma(s),$ $\qquad\qquad \operatorname{Re}s>0$	SU 521(M13)
2 e^{-x^2}	$\frac{1}{2}\,\Gamma\left(\frac{1}{2}s\right),$ $\qquad\qquad \operatorname{Re}s>0$	SU 521(M14)
3 $\cos x$	$\Gamma(s)\cos\left(\frac{1}{2}\pi s\right),\quad 0<\operatorname{Re}s<1$	SU 521(M15)
4 $\sin x$	$\Gamma(s)\sin\left(\frac{1}{2}\pi s\right),\quad 0<\operatorname{Re}s<1$	SU 521(M16)
5 $\dfrac{1}{1-x}$	$\pi\cot(\pi s),\qquad 0<\operatorname{Re}s<1$	SU 521(M1)
6 $\dfrac{1}{1+x}$	$\pi\operatorname{cosec}(\pi s),\qquad 0<\operatorname{Re}s<1$	SU 521(M2)
7 $(1+x^a)^{-b}$	$\dfrac{\Gamma(s/a)\,\Gamma(b-s/a)}{a\,\Gamma(b)},\quad 0<\operatorname{Re}s<ab$	SU 521(M3)
8 $\dfrac{T_n(x)\,\mathrm{H}(1-x)}{\sqrt{(1-x^2)}}$	$\dfrac{2^{-s}\pi\,\Gamma(s)}{\Gamma\left(\frac{1}{2}+\frac{1}{2}s+\frac{1}{2}n\right)\Gamma\left(\frac{1}{2}+\frac{1}{2}s-\frac{1}{2}n\right)},$ $\operatorname{Re}s>0$	SU 521(M4)
9 $\dfrac{T_n\left(x^{-1}\right)\mathrm{H}(1-x)}{\sqrt{(1-x^2)}}$	$\dfrac{2^{s-2}\,\Gamma\left(\frac{1}{2}n+\frac{1}{2}s\right)\Gamma\left(\frac{1}{2}s-\frac{1}{2}n\right)}{\Gamma(s)},$ $\operatorname{Re}s>n$	SU 521(M5)
10 $P_n(x)\,\mathrm{H}(1-x)$	$\dfrac{\Gamma\left(\frac{1}{2}s\right)\Gamma\left(\frac{1}{2}s+\frac{1}{2}\right)}{2\,\Gamma\left(\frac{1}{2}s-\frac{1}{2}n+\frac{1}{2}\right)\Gamma\left(\frac{1}{2}s+\frac{1}{2}n+1\right)},$ $\operatorname{Re}s>0$	SU 521(M6)
11 $P_n\left(x^{-1}\right)\mathrm{H}(1-x)$	$\dfrac{2^{s-1}\,\Gamma\left(\frac{1}{2}s+\frac{1}{2}n+\frac{1}{2}\right)\Gamma\left(\frac{1}{2}s-\frac{1}{2}n\right)}{\sqrt{\pi}\,\Gamma(s+1)},$ $\operatorname{Re}s>n$	SU 521(M7)
12 $\dfrac{1+x\cos\phi}{1-2x\cos\phi+x^2}$	$\dfrac{\pi\cos(s\phi)}{\sin(s\pi)},\qquad 0<\operatorname{Re}s<1$	SU 521(M11)
13 $\dfrac{x\sin\phi}{1-2x\cos\phi+x^2},\quad -\pi<\phi<\pi$	$\dfrac{\pi\sin(s\phi)}{\sin(s\pi)},\qquad 0<\operatorname{Re}s<1$	SU 521(M12)

continued on next page

continued from previous page

$f(x)$	$f^*(s)$
14 $\quad e^{-x\cos\phi}\cos\left(x\sin\phi\right),$ $\qquad\qquad \frac{1}{2}\pi < \phi < \frac{1}{2}\pi$	$\Gamma(s)\cos(s\phi), \qquad\quad \operatorname{Re} s > 0 \qquad$ SU 522(M17)
15 $\quad e^{-x\sin\phi}\sin\left(x\sin p\phi\right),$ $\qquad\qquad -\frac{1}{2}\pi < \phi < \frac{1}{2}\pi$	$\Gamma(s)\sin(s\phi), \qquad\quad \operatorname{Re} s > -1 \qquad$ SU 522(M18)
16 $\quad x^{-\nu}J_\nu(x), \qquad\qquad \nu > -\frac{1}{2}$	$\dfrac{2^{s-\nu-1}\,\Gamma\left(\frac{1}{2}s\right)}{\Gamma\left(\nu-\frac{1}{2}s+1\right)}, \quad 0 < \operatorname{Re} s < 1 \qquad$ SU 522(M19)
17 $\quad Y_\nu(x), \qquad\qquad\quad \nu \in \mathbb{R}$	$-2^{s-1}\pi^{-1}\,\Gamma\left(\frac{1}{2}s+\frac{1}{2}\nu\right)\Gamma\left(\frac{1}{2}s-\frac{1}{2}\nu\right)$ $\times \cos\left(\frac{1}{2}s-\frac{1}{2}\nu\right)\pi,$ $\qquad\qquad |\nu| < \operatorname{Re} s < \frac{3}{2} \qquad$ SU 522(M20)
18 $\quad K_\nu(x), \qquad\qquad\quad \nu \in \mathbb{R}$	$2^{s-2}\,\Gamma\left(\frac{1}{2}s+\frac{1}{2}\nu\right)\Gamma\left(\frac{1}{2}s-\frac{1}{2}\nu\right),$ $\qquad\qquad \operatorname{Re} s > \nu > 0 \qquad$ SU 522(M21)
19 $\quad \mathbf{H}_\nu(x), \qquad\qquad\quad \nu \in \mathbb{R}$	$\dfrac{2^{s-1}\tan\left(\frac{1}{2}\pi s+\frac{1}{2}\pi\nu\right)\Gamma\left(\frac{1}{2}s+\frac{1}{2}\nu\right)}{\Gamma\left(\frac{1}{2}\nu-\frac{1}{2}s+1\right)},$ $-1-\nu < \operatorname{Re} s < \min\left(\frac{3}{2}, 1-\nu\right) \qquad$ SU 522(M22)
20 $\quad \dfrac{1}{a+x^n},$ $\qquad |\arg a| < \pi, \quad n = 1,2,3,\dots,$	$\pi n^{-1}\operatorname{cosec}\left(\dfrac{\pi s}{n}\right)a^{(s/n)-1},$ $\qquad\qquad 0 < \operatorname{Re} s < n \qquad$ MS 453
21 $\quad \left(1+ax^h\right)^{-\nu},$ $\qquad\qquad h > 0, \quad |\arg a| < \pi$	$h^{-1}a^{-s/h}\,\mathrm{B}\left(s/h, \nu-(s/h)\right)$ $\qquad\qquad 0 < \operatorname{Re} s < h\operatorname{Re}\nu \qquad$ MS 454
22 $\quad \begin{cases}\left(1-x^h\right)^{\nu-1} & \text{for } 0 < x < 1 \\ 0 & \text{for } x > 1\end{cases},$ $\qquad\qquad h > 0, \quad \operatorname{Re}\nu > 0$	$h^{-1}\,\mathrm{B}\left(\nu, s/h\right) \qquad\qquad$ MS 454
23 $\quad \ln(1+ax), \qquad\qquad |\arg a| < \pi$	$\pi s^{-1}a^{-s}\operatorname{cosec}(\pi s), \quad -1 < \operatorname{Re} s < 0 \qquad$ MS 454
24 $\quad \arctan x$	$-\frac{1}{2}\pi s^{-1}\sec(\pi s/2), \quad -1 < \operatorname{Re} s < 0 \qquad$ MS 454

continued on next page

	$f(x)$	$f^*(s)$	
	continued from previous page		
25	$\operatorname{arccot} x$	$\frac{1}{2}\pi s^{-1}\sec(\pi s/2),\qquad 0 < \operatorname{Re} s < 1$	MS 454
26	$\operatorname{cosech}(ax)\qquad\operatorname{Re} a > 0$	$a^{-s}2\left(1 - 2^{-s}\right)\Gamma(s)\zeta(s),\qquad \operatorname{Re} s > 1$	MS 454
27	$\operatorname{sech}^2(ax),\qquad \operatorname{Re} a > 0$	$4a^{-s}(1 - 2^{2-s})\,\Gamma(s)2^{-s}\,\zeta(s-1),$ $\operatorname{Re} s > 2$	MS 454
28	$\operatorname{cosech}^2(ax),\qquad \operatorname{Re} a > 0$	$4a^{-s}\,\Gamma(s)2^{-s}\,\zeta(s-1),\qquad \operatorname{Re} s > 2$	MS 454
29	$\left(x^2 + b^2\right)^{-\frac{1}{2}\nu}J_\nu\left[a\left(x^2 + b^2\right)^{1/2}\right]$	$2^{\frac{1}{2}s-1}a^{-\frac{1}{2}s}b^{\frac{1}{2}s-\nu}\,\Gamma\left(\frac{1}{2}s\right)J_{\nu-\frac{1}{2}}(ab),$ $0 < \operatorname{Re} s < \frac{3}{2} + \operatorname{Re}\nu$	MS 454
30	$\begin{cases}\left(a^2 - x^2\right)^{\frac{1}{2}\nu}J_\nu\left[a\left(b^2 - x^2\right)^{1/2}\right] \\ \qquad\qquad \text{for } 0 < x < a \\ 0 \qquad\qquad \text{for } x > a \end{cases}$ $\operatorname{Re}\nu > -1$	$2^{\frac{1}{2}s-1}\Gamma\left(\frac{1}{2}s\right)b^{-\frac{1}{2}s}a^{\nu+\frac{1}{2}s}\,J_{\nu+\frac{1}{2}s}(ab),$ $\operatorname{Re} s > 0$	MS 455
31	$\begin{cases}\left(a^2 - x^2\right)^{-\frac{1}{2}\nu}J_\nu\left[b\left(a^2 - x^2\right)^{1/2}\right] \\ \qquad\qquad \text{for } 0 < x < a \\ 0 \qquad\qquad \text{for } x > a \end{cases}$	$2^{1-\nu}[\Gamma(\nu)]^{-1}a^{\frac{1}{2}s-\nu}b^{-\frac{1}{2}\nu}\,s_{\nu-1+\frac{1}{2}s,\frac{1}{2}s-\nu}(ab),$ $\operatorname{Re} s > 0$	MS 455
32	$K_\nu(\alpha x)$	$\alpha^{-s}2^{s-2}\,\Gamma\left(\frac{1}{2}s - \frac{1}{2}\nu\right)\Gamma\left(\frac{1}{2}s + \frac{1}{2}\nu\right),$ $\operatorname{Re} s > |\operatorname{Re}\nu|$	MS 455
33	$\left(\beta a^2 + x^2\right)^{-\frac{1}{2}\nu}$ $\times K_\nu\left[\alpha\left(\beta a^2 + x^2\right)^{1/2}\right]$ $\operatorname{Re}(\alpha, \beta) > 0$	$\alpha^{-\frac{1}{2}s}2^{\frac{1}{2}s-1}\beta^{\frac{1}{2}s-\nu}\,\Gamma,\left(\frac{1}{2}s\right)K_{\nu-\frac{1}{2}s}(\alpha\beta),$ $\operatorname{Re} s > 0$	MS 455

18* The z-transform

18.1–18.3 Definition, Bilateral, and Unilateral z-Transforms

18.1 Definitions

The z-**transform** converts a numerical sequence $x[n]$ into a function of the complex variable z and it takes two different forms. The **bilateral** or **two-sided** z-**transform**, denoted here by $Z_b\{x[n]\}$, is used mainly in signal and image processing, while the **unilateral** or **one-sided** z-**transform**, denoted here by $Z_u\{x[n]\}$, is used mainly in the analysis of discrete time systems and the solution of linear difference equations.

The **bilateral** z-**transform**, $X_b(z)$ of the sequence $x[n] = \{x_n\}_{n=-\infty}^{\infty}$ is defined as

$$Z_b\{x[n]\} = \sum_{n=-\infty}^{\infty} x_n z^{-n} = X_b(z)$$

and the **unilateral** z-**transform** $X_u(z)$ of the sequence $x[n] = \{x_n\}_{n=0}^{\infty}$ is defined as

$$Z_b\{x[n]\} = \sum_{n=0}^{\infty} x_n z^{-n} = X_u(z),$$

where each has its own domain of convergence (DOC). The series $X_b(z)$ is a Laurent series, and $X_u(z)$ is the principal part of the Laurent series for $X_b(z)$. When $x_n = 0$ for $n < 0$, the two z-transforms $X_b(z)$ and $X_u(z)$ are identical. In each case the sequence $x[n]$ and its associated z-transform is a called a z-**transform pair**.

The inverse z-transformation $x[n] = Z^{-1}\{X(z)\}$ is given by

$$x[n] = \frac{1}{2\pi i} \int_{\Gamma} X(z) z^{n-1} \, dz$$

where $X(z)$ is either $X_b(z)$ or $X_u(z)$, and Γ is a simple closed contour containing the origin and lying entirely within the domain of convergence of $X(z)$. In many practical situations the z-transform is either found by using a series expansion of $X(z)$ in the inversion integral or, if $X(z) = N(z)/D(z)$ where $N(z)$ and $D(z)$ are polynomials in z, by means of partial fractions and the use of an appropriate table of z-transform pairs. In order for the inverse z-transform to be unique it is necessary to specify the domain of convergence, as can be seen by comparison of entries 3 and 4 of Table 18.2. Table 18.1 lists general properties of the bilateral z-transform, and Table 18.2 lists some bilateral z-transform pairs. In what follows, use is made of the **unit integer function** $h(n) = \begin{cases} 0 & \text{for } n < 0 \\ 1 & \text{for } n \geq 0 \end{cases}$, that is a generalization of the Heaviside step function, and the **unit integer pulse function** $\Delta(n - k) = \begin{cases} 1 & \text{for } n = k \\ 0 & \text{for } n \neq k \end{cases}$, that is a generalization of the delta function.

18.2　Bilateral z-transform

Table 18.1 General properties of the bilateral z-transform $X_b(n) = \displaystyle\sum_{n=-\infty}^{\infty} x_n z^{-n}$.

	Term in sequence	z-Transform $X_b(z)$	Domain of Convergence
1	$\alpha x_n + \beta y_n$	$\alpha X_b(z) + \beta Y_b(z)$	Intersection of DOC's of $X_b(z)$ and $Y_b(z)$ with α, β constants
2	x_{n-N}	$z^{-n} X_b(z)$	DOC of $X_b(z)$ to which it may be necessary to add or delete the origin or the point at infinity
3	$n x_n$	$-z \dfrac{dX_b(z)}{dz}$	DOC of $X_b(z)$ to which it may be necessary to add or delete the origin and the point at infinity
4	$z_0^n x_n$	$X_b\left(\dfrac{z}{z_0}\right)$	DOC of $X_b(z)$ scaled by $|z_0|$
5	$n z_0^n x_n$	$-z \dfrac{dX_b(z/z_0)}{dz}$	DOC of $X_b(z)$ scaled by $|z_0|$ to which it may be necessary to add or delete the origin and the point at infinity
6	x_{-n}	$X_b(1/z)$	DOC of radius $1/R$, where R is the radius of convergence of DOC of $X_b(z)$
7	$n x_{-n}$	$-z \dfrac{dX_b(1/z)}{dz}$	DOC of radius $1/R$, where R is the radius of convergence of DOC of $X_b(z)$
8	$\bar{x}_n$	$\overline{X_b(\bar{z})}$	The same DOC as x_n
9	$\operatorname{Re} x_n$	$\frac{1}{2}\left[X_b(z) + \overline{X_b(\bar{z})}\right]$	DOC contains the DOC of x_n
10	$\operatorname{Im} x_n$	$\frac{1}{2i}\left[X_b(z) - \overline{X_b(\bar{z})}\right]$	DOC contains the DOC of x_n
11	$\displaystyle\sum_{k=-\infty}^{\infty} x_k y_{n-k}$	$X_b(z) Y_b(z)$	DOC contains the intersection of the DOCs of $X_b(z)$ and $Y_b(z)$ (convolution theorem)
12	$x_n y_n$	$\dfrac{1}{2\pi i}\displaystyle\int_\Gamma X_b(\xi) Y_b\left(\dfrac{z}{\xi}\right)\xi^{-1}\,d\xi$	DOC contains the DOCs of $X_b(z)$ and $Y_b(z)$, with Γ inside the DOC and containing the origin (convolution theorem)
13	Parseval formula	$\displaystyle\sum_{n=-\infty}^{\infty} x_n \bar{y}_n = \dfrac{1}{2\pi i}\int_\Gamma X_b(\xi)\overline{Y_b}\left(\dfrac{z}{\bar{\xi}}\right)\xi^{-1}\,d\xi$	DOC contains the intersection of DOCs of $X_b(z)$ and $Y_b(z)$, with Γ inside the DOC and containing the origin
14	Initial value theorem for $x_n h(n)$	$x_0 = \displaystyle\lim_{z\to\infty} X_b(z)$	

Table 18.2 Basic bilateral z-transforms

	Term in sequence	z-Transform $X_b(z)$	Domain of Convergence
1	$\Delta(n)$	1	Converges for all z
2	$\Delta(n - N)$	z^{-n}	When $N > 0$ convergence is for all z except at the origin. When $N < 0$ convergence is for all z except at ∞
3	$a^n h(n)$	$\dfrac{z}{z - a}$	$|z| > |a|$
4	$a^n h(-n - 1)$	$\dfrac{z}{z - a}$	$|z| < |a|$
5	$n a^n h(n)$	$\dfrac{az}{(z - a)^2}$	$|z| > a > 0$
6	$n a^n h(-n - 1)$	$\dfrac{az}{(z - a)^2}$	$|z| < a, \quad a > 0$
7	$n^2 a^n h(n)$	$\dfrac{az(z + a)}{(z - a)^3}$	$|z| > a > 0$
8	$\left(\dfrac{1}{a^n} + \dfrac{1}{b^n}\right) h(n)$	$\dfrac{az}{az - 1} + \dfrac{bz}{bz - 1}$	$|z| > \max\left(\frac{1}{|a|}, \frac{1}{|b|}\right)$
9	$a^n h(n - N)$	$\dfrac{z\left(1 - (a/z)^N\right)}{z - a}$	$|z| > 0$
10	$a^n h(n) \sin \Omega n$	$\dfrac{az \sin \Omega}{z^2 - 2az \cos \Omega + a^2}$	$|z| > a > 0$
11	$a^n h(n) \cos \Omega n$	$\dfrac{z(z - a \cos \Omega)}{z^2 - 2az \cos \Omega + a^2}$	$|z| > a > 0$
12	$e^{an} h(n)$	$\dfrac{z}{z - e^a}$	$|z| > e^{-a}$
13	$e^{-an} h(n) \sin \Omega n$	$\dfrac{ze^a \sin \Omega}{z^2 e^{2a} - 2ze^a \cos \Omega + 1}$	$|z| > e^{-a}$
14	$e^{-an} h(n) \cos \Omega n$	$\dfrac{ze^a(ze^a - \cos \Omega)}{z^2 e^{2a} - 2ze^a \cos \Omega + 1}$	$|z| > e^{-a}$

18.3   Unilateral z-transform

The relationship between the Laplace transform of a continuous function $x(t)$ sampled at $t = 0$, T, $2T$, ... and the unilateral z-transform of the function $\hat{x}(t) = \sum_{n=0}^{\infty} x(nT)\delta(t - nT)$ follows from the result

$$\mathcal{L}\{\hat{x}(t)\} = \int_0^{\infty} \left[\sum_{k=0}^{\infty} x(kT)\delta(t - kT) \right] e^{-st}\, dt$$

$$= \sum_{k=0}^{\infty} x(kT)e^{-ksT}$$

Setting $z = e^{sT}$ this becomes:

$$\mathcal{L}\{\hat{x}(t)\} = \sum_{k=0}^{\infty} x(kT)z^{-k} = X(z),$$

showing that the unilateral z-transform $X_u(z)$ can be considered to be the Laplace transform of a continuous function $x(t)$ for $t \geq 0$ sampled at $t = 0$, T, $2T$, ...

Table 18.3 lists some general properties of the unilateral z-transform, and Table 18.4 lists some unilateral z-transform pairs.

Table 18.3 General properties of the unilateral z-transform

	Term in sequence	z-Transform $X_u(z)$	Domain of Convergence
1	$\alpha x_n + \beta y_n$	$\alpha X_u(z) + \beta Y_u(z)$	Intersection of DOC's of $X_u(z)$ and $Y_u(z)$ with α, β constants
2	x_{n+k}	$z^k X_u(z) - z^k x_0 - z^{k-1}x_1$ $-z^{k-2}x_2 - \cdots - zx_{k-1}$	
3	nx_n	$-z\dfrac{dX_u(z)}{dz}$	DOC of $X_u(z)$ to which it may be necessary to add or delete the origin and the point at infinity
4	$z_0^n x_n$	$X_u\left(\dfrac{z}{z_0}\right)$	DOC of $X_b(z)$ scaled by $|z_0|$ to which it may be necessary to add or delete the origin and the point at infinity
5	$nz_0^n x_n$	$-z\dfrac{dX_u(z/z_0)}{dz}$	DOC of $X_u(z)$ scaled by $|z_0|$ to which it may be necessary to add or delete the origin and the point at infinity
6	$\bar{x}_n$	$\overline{X_u(\bar{z})}$	The same DOC as x_n
7	$\operatorname{Re} x_n$	$\frac{1}{2}\left[X_u(z) + \overline{X_u(\bar{z})}\right]$	DOC contains the DOC of x_n
8	$\dfrac{\partial}{\partial \alpha}x_n(\alpha)$	$\dfrac{\partial}{\partial \alpha}X_u(z,\alpha)$	Same DOC as $x_n(\alpha)$
9	Initial value theorem	$x_0 = \lim\limits_{z\to\infty} X_u(z)$	
10	Final value theorem	$\lim\limits_{n\to\infty} x_n = \lim\limits_{z\to 1}\left[\left(\dfrac{z-1}{z}\right)X_u(z)\right]$	when $X_u(z) = N(z)/D(z)$ with $N(z)$, $D(z)$ polynomials in z and the zeros of $D(z)$ inside the unit circle $|z| = 1$ or at $z = 1$

Table 18.4 Basic unilateral z-transforms

	Term in sequence	z-Transform $X_u(z)$	Domain of Convergence
1	$\Delta(n)$	1	Converges for all z
2	$\Delta(n-k)$	z^{-k}	Convergence for all $z \neq 0$
3	$a^n h(n)$	$\dfrac{z}{z-a}$	$|z| > |a|$
4	$na^n h(n)$	$\dfrac{az}{(z-az)^2}$	$|z| > a > 0$
5	$n^2 a^n h(n)$	$\dfrac{az(z+a)}{(z-a)^3}$	$|z| > a > 0$
6	$na^{n-1} h(n)$	$\dfrac{z}{(z-a)^2}$	$|z| > a > 0$
7	$(n-1)a^n h(n)$	$\dfrac{z(2a-z)}{(z-a)^2}$	$|z| > a > 0$
8	$e^{-an} h(n)$	$\dfrac{ze^a}{ze^a - 1}$	$|z| > e^{-a}$
9	$ne^{-an} h(n)$	$\dfrac{ze^a}{(ze^a - 1)^2}$	$|z| > e^{-a}$
10	$n^2 e^{-an} h(n)$	$\dfrac{ze^a(1 + ze^a)}{(ze^a - 1)^3}$	$|z| > e^{-a}$
11	$e^{-an} h(n) \sin \Omega n$	$\dfrac{ze^a \sin \Omega}{z^2 e^{2a} - 2ze^a \cos \Omega + 1}$	$|z| > e^{-a}$
12	$e^{-an} h(n) \cos \Omega n$	$\dfrac{ze^a(ze^a - \cos \Omega)}{z^2 e^{2a} - 2ze^a \cos \Omega + 1}$	$|z| > e^{-a}$
13	$h(n) \sinh an$	$\dfrac{z \sinh a}{z^2 - 2z \cosh a + 1}$	$|z| > e^{-a}$
14	$h(n) \cosh an$	$\dfrac{z(z - \cosh a)}{z^2 - 2z \cosh a + 1}$	$|z| > e^{-a}$
15	$h(n) a^{n-1} e^{-an} \sin \Omega n$	$\dfrac{ze^a \sin \Omega}{z^2 e^{2a} - 2zae^a \cos \Omega + a^2}$	$|z| > e^{-a}$
16	$h(n) a^n e^{-an} \cos \Omega n$	$\dfrac{ze^a(ze^a - a\cos \Omega)}{z^2 - 2zae^a \cos \Omega + a^2}$	$|z| > e^{-a}$

Bibliographic references used in preparation of text

(See the introduction for an explanation of the letters preceding each bibliographic reference.)

AS Abramowitz, M. and Stegun, I. A., *Handbook of Mathematical Functions*, Dover Publications, New York, 1972.

AD Adams, E. P. and Hippisley, R. L., *Smithsonian Mathematical Formulae and Tables of Elliptic Functions*, Smithsonian Institute, Washington, D.C., 1922.

AK Appell, P. and Kampé de Fériet, *Fonctions hypergéometriques et hypersphériques, polynomes d'Hermite*, Paris, 1926.

BB Beckenbach, E. F. and Bellman, R., *Inequalities*, 3rd printing. Springer Verlag, Berlin, 1971.

BE Bertrand, J., *Traite de calcul différentiel et de calcul intégral*, vol. 2, *Calcul intégral, intégrales définies et indéfinies.* Gauthier-Villars, Paris, 1870.

BI Bierens de Haan, D., *Nouvelles tables d'intégrales définies.* Amsterdam, 1867.

BL Bellman, R., *Introduction to Matrix Analysis.* McGraw Hill, New York, 1960.

BR * Bromwich, T. I'A., *An Introduction to the Theory of Infinite Series.* Macmillan, London, 1908, 2nd edition, 1926.

BS Bellman, R., *Stability Theory of Differential Equations.* McGraw-Hill, New York, 1953.

BU Buchholz, H., *Die konfluente hypergeometrische Funktion mit besonderer Berücksichtigung ihrer Anwendungen.* Springer Verlag, Berlin, 1953. Also an English edition: *The confluent Hypergeometric Function*, Springer-Verlag, Berlin, 1969.

BY Byrd, P. F. and Friedman, M. D., *Handbook of Elliptic Integrals for Engineers and Physicists.* Springer Verlag, Berlin, 1954.

CA Carslaw, H. S., *Introduction to the Theory of Fourier's Series and Integrals.* Macmillan, London, 1930.

CE Cesàro, Z., *Elementary Class Book of Algebraic Analysis and the Calculation of Infinite Limits*, 1st ed. ONTI, Moscow and Leningrad, 1936.

CL Coddington, E. A. and Levinson, N., *Theory of Ordinary Differential Equations.* McGraw Hill, New York, 1955.

CO Courant, R. and Hilbert, D., *Methods of Mathematical Physics*, vol. I. Wiley (Interscience), New York, 1953.

DW Dwight, H. B., *Tables of Integrals and Other Mathematical Data.* Macmillan, New York, 1934.

DW61 Dwight, H. B., *Tables of Integrals and Other Mathematical Data.* Macmillan, New York, 1961.

The Bibliographic Reference BR refers to the 1908 edition of Bromwich T. I.'A., *An Introduction to the Theory of Infinite Series.* BR refers to the 1926 edition.

EF   Efros, A. M. and Danilevskiy, A. M., *Operatsionnoye ischisleniye i konturnyye integraly* (Operational calculus and contour integrals). GNTIU, Khar'kov, 1937.

EH   Erdélyi, A., et al., *Higher Transcendental Functions*, vols. I, II, and III. McGraw Hill, New York, 1953-1955.

ET   Erdélyi, A. et al., *Tables of Integral Transforms*, vols. I and II. McGraw Hill, New York, 1954.

EU   Euler, L., *Introductio in Analysin Infinitorum*. Lausanne, 1748.

FI   Fikhtengol'ts, G. M., *Kurs differentsial'nogo i integral'nogo ischisleniya* (Course in differential and integral calculus), vols. I, II, and III. Gostekhizdat, Moscow and Leningrad, 1947-1949. Also a German edition: *Differential-und Integralrechnung I–III*, VEB Deutscher Verlag der Wissenschaften, Berlin, 1986–1987.

GA   Gauss, K. F., *Werke*, Bd. III. Göttingen, 1876.

GE   Gel'fond, A. O., *Ischisleniye konechnykh raznostey* (Calculus of finite differences), part I. ONTI, Moscow and Leningrad, 1936.

GH2   Gröbner, W. and Hofreiter, N., *Integraltafel*, vol. 2, *Bestimmte Integrale*. Springer, Wien, 1961.

GI   Giunter, N. M. and Kuz'min, R. O. (eds.), *Sbornik zadach po vysshey matematike* (Collection of problems in higher mathematics), vols. I, II, and III. Gostekhizdat, Moscow and Leningrad, 1947.

GM   Gantmacher, F. R., *Applications of the Theory of Matrices*, translation by J. L. Brenner. Wiley (Interscience), New York, 1959.

GO   Goursat, E. J. B., *Cours d'Analyse*, vol. I. Gauthier-Villars, Paris, 1923.

GU   Gröbner, W. et al., *Integraltafel*, Teil I, *Unbestimmte Integrale*. Braunschweig, 1944.

GW   Gröbner, W. and Hofreiter, N., *Integraltafel*, Teil II, *Bestimmte Integrale*. Springer Verlag, Wien and Innsbruck, 1958.

HI   Hille, E., *Lectrues on Ordinary Differential Equations*. Addison- Wesley, Reading, Massachusetts, 1969.

HL   Hardy, G. H., Littlewood, J. E., and Polya, G., *Inequalities*. Cambridge University Press, London, 2nd ed., 1952.

HO   Hobson, E. W., *The Theory of Spherical and Ellipsoidal Harmonics*. Cambridge University Press, 1931.

HU   Hurewicz, W., *Lectures on Ordinary Differential Equations*. MIT Press, Cambridge, Massachusetts, 1958.

IN   Ince, E. L., *Ordinary Differential Equations*. Dover, New York, 1944.

JA   Jahnke, E. and Emde, F., *Tables of Functions with Formulas and Curves*. Dover, New York, 1943.

JAC   Jackson, J. D., *Classical Electrodynamics*, Wiley, New York, 1975.

JE   James, H. M. et al. (eds.), *Theory of Servomechanisms*. McGraw Hill, New York, 1947.

JO   Jolley, L., *Summation of Series*. Chapman and Hall, London, 1925.

KE   Kellogg, O. D., *Foundations of Potential Theory*. Dover, New York, 1958.

KR   Krechmar, V. A., *Zadachnik po algebre* (Problem book in algebra), 2nd ed. Gostekhizdat, Moscow and Leningrad, 1950.

KU   Kuzmin, R. O., *Besselevy funktsii* (Bessel functions). ONTI, Moscow and Leningrad, 1935.

LA   Laska, W., *Sammlung von Formeln der reinen und angewandten Mathematik*. Friedrich Viewig und Sohn, Braunschweig, 1888.

LE   Legendre, A. M., *Exercises calcul intégral*. Paris, 1811.

LI   Lindeman, C. E., *Examen des nouvelles tables d'intégrales définies de M. Bierens de Haan*, Amsterdam, 1867, Norstedt, Stockholm, 1891.

LO   Lobachevskiy, N. I., *Poloye sobraniye sochineniy* (Complete works), vols. I, III, and V. Gostekhizdat, Moscow and Leningrad, 1946-1951.

LUKE Luke, Y. L., *Mathematical Functions and their Approximations*, Academic press, New York, 1975.

LW   Lawden, D. F., *Elliptic Functions and Applications*. Springer Verlag, Berlin, 1989.

MA   McLachlan, N. W., *Theory and Application of Mathieu Functions*. Oxford University Press, 1947.

MC   Computation by Mathematica.

ME   McLachlan, N. W. and Humbert, P., *Formulaire pour le calcul symbolique*. L'Acad. des Sciences de Paris et al., Fasc. 100, 1950.

MF   Morse, M. P. and Feshbach, H., *Methods of Theoretical Physics*, vol. I. McGraw Hill, New York, 1953.

MG   Marden, M., *Geometry of Polynomials*. American Mathematical Society, Mathematical Survey 3, Providence, Rhode Island, 1966.

MI   McLachlan, N. W. et al., *Supplément au formulaire pour le calcul symbolique*. L'Acad. des Sciences de Paris et al., Fasc. 113, 1950.

ML   Mirsky L., *An Introduction to Linear Algebra*. Oxford University Press, 1963.

MM   MacMillan, W. D., *The Theory of the Potential*. Dover, New York, 1958.

MO   Magnus, W. and Oberhettinger, F., *Formeln und Sätze für die speziellen Funktionen der mathematischen Physik*. Springer Verlag, Berlin, 1948.

MS   Magnus, W., Oberhettinger, F. and Soni, R. P., *Formulas and Theorems for the Special Functions of Mathematical Physics*, 3rd ed. Springer Verlag, Berlin, 1966.

MS2 Mathai, A. M. and Saxens, R. K. , *Generalized Hypergeometrics Functions With Applications in Statistics and Physical Science*, Springer-Verlag, 1973.

MT   Mitrinović, D. S., *Analytic Inequalities*. Springer Verlag, Berlin, 1970.

MV   Milne, E. A., *Vectorial Mechanics*. Methuen, London, 1948.

MZ   Meyer Zur Capellen, W., *Integraltafeln, Sammlung unbestimmer Integrale elementarer Funktionen*. Springer Verlag, Berlin, 1950.

NA   Natanson, I. P., *Konstruktivnaya teoriya funktsiy* (Constructive theory of functions). Gostekhizdat, Moscow and Leningrad, 1949.

NH   Nielsen, N., *Handbuch der Theorie der Gammafunktion*. Teubner, Leipzig, 1906.

NO   Noble, B., *Applied Linear Algebra*. Prentice Hall, Englewood Cliffs, New Jersey, 1969.

NT   Nielsen, N., *Theorie des Integrallogarithmus und verwandter Transcendenten*. Teubner, Leipzig, 1906.

NV   Novoselov, S. I., *Obratnyye trigonometricheskiye funktsii, posobive dlya uchiteley* (Inverse trigonometric functions, textbook for students), 3rd ed. Uchpedgiz, Moscow and Leningrad, 1950.

OB   Oberhettinger, F., *Tables of Bessel Transforms*, Springer-Verlag, New York: 1972.

PBM Prudnikov, A. P., Brychkov, Yu. A., and Marichev, O. I., *Integrals and Series*, Gordan and Breach, New York, vols. I (1986), II (1986), III (1990).

PE   Peirce, B. O., *A Short Table of Integrals*, 3rd ed. Ginn, Boston, 1929.

SA   Sansone, G., *Orthogonal Functions* (Revised English Edition), Interscience, New York, 1959.

SI Sikorskiy, Yu. S., *Elementy teorii ellipticheskikh funktsiy s prilozheniyama k mekhanike* (Elements of theory of elliptic functions with applications to mechanics). ONTI, Moscow and Leningrad, 1936.

SN Sneddon, I. N., *Fourier Transforms*, 1st ed. McGraw Hill, New York, 1951.

SM Smirnov, V. I., *Kurs vysshey matematiki* (A course of higher mathematics), vol. III, Part 2, 4th ed. Gostekhizdat, Moscow and Leningrad, 1949.

ST Strutt, M. J. O., *Lamésche, Mathieusche und Verwandte Funktionen in Physik and Technik.* Springer Verlag, Berlin, 1932.

STR Straton, J. C., *Phys. Rev A*, **43**(3), pages 1381–1388, 1991.

SU Sneddon, I. N., *The Use of Integral Transforms.* McGraw Hill, New York, 1972.

SZ Szegö, G., *Orthogonal Polynomials*, Revised Edition, Colloquium Publications XXIII, American Mathematical Society, New York, 1959.

TF Titchmarsh, E. C., *Introduction to the Theory of Fourier Integrals*, 2nd ed. Oxford University Press, London, 1948.

TI Timofeyev, A. F. *Integrirovaniye funktsiy* (Integration of functions), part I. GTTI, Moscos and Leningrad, 1933.

VA Varga, R. S., *Matrix Iterative Analysis.* Prentice Hall, Englewood Cliffs, New Jersey, 1963.

VL Vladimirov, V. S., *Equations of Mathematical Physics.* Dekker, New York, 1971.

WA Watson, G. N., *A Treatise on the Theory of Bessel Functions*, 2nd ed. Cambridge University Press, 1966.

WH Whittaker, E. T. and Watson, G. N., *Modern Analysis*, 4th ed. Cambridge University Press, 1927, part II, 1934.

ZH Zhuravskiy, A. M., *Spravochnik po ellipticheskim funktsiyam* (Reference book on elliptic functions). Izd. Akad. Nauk. U.S.S.R., Moscow and Leningrad, 1941.

ZY Zygmund, A., *Trigonometric Series*, 2nd ed. Chelsea, New York, 1952.

Classified supplementary references

(Prepared by Alan Jeffrey for the English language edition.)

General reference books

1. Bromwich, T. I'A., *An Introduction to the Theory of Infinite Series*. Macmillan, London. 2nd ed. 1926 (Reprinted 1942).
2. Carlitz, L., *Generating Functions*, Duke University, Durham, 1969.
3. Copson, E. T., *An Introduction to the Theory of Functions of a Complex Variable*. Oxford University Press, 1935.
4. Courant, R. and Hilbert, D., *Methods of Mathematical Physics*, vol. I. Interscience, New York, 1953.
5. Davies, H. T., *Summation of Series*, San Antonio, Texas, 1962.
6. Erdélyi, A. et al. *Higher Transcendental Functions*, vols. I to III, McGraw Hill, New York 1953 to 1955.
7. Erdélyi, A. et al., *Tables of Integral Transforms*, vols. I and II. McGraw Hill, New York, 1954.
8. Fletcher, A., Miller, J. C. P., and Rosenhead, L., *An Index of Mathematical Tables*. Scientific Computing Service, London, 2nd ed., 1962.
9. Gröbner, W. and Hofreiter, N., *Integraltafel*. I, II. Springer Verlag, Wien and Innsbruck, 1949.
10. Hardy, G. H., Littlewood, J. E., and Pólya, G., *Inequalities*. Cambridge University Press, 2nd ed., 1952.
11. Hartley, H. O. and Greenwood, J. A., *Guide to Tables in Mathematical Statistics*. Princeton University Press, 1962.
12. Jeffreys, H. and Jeffreys, B. S., *Methods of Mathematical Physics*. Cambridge University Press, 1956.
13. Jolley, L. B. W., *Summation of Series*, Dover, New York, 1962.
14. Knopp, K., *Theory and Application of Infinite Series*. Blackie, London, 1946, Hafner, New York, 1948.
15. Lebedev, N. N., *Special Functions and their Applications*, Prentice Hall, 1965.
16. Magnus, W. and Oberhettinger, F., *Formulas and Theorems for the Special Functions of Mathematical Physics*. Chelsea, New York, 1949.
17. McBride, E. B., *Obtaining Generating Functions*, Springer-Verlag, Berlin, 1971.
18. National Bureau of Standards, *Handbook of Mathematical Functions*. U.S. Government Printing Office, Washington, D.C., 1964.
19. Prudnikov, A. P., Brychkov, Yu. A., and Marichev, O. I., *Integrals and Series*, Vols. 1–4, Gordon and Breach, New York, 1986–1992.
20. Truesdell, C. *A Unified Theory of Special Functions*. Princeton University Press, New Jersey, 1948.
21. Vein, R. and Dale, P., *Determinants and Their Applications in Mathematical Physics*, Springer Verlag, New York, 1999.
22. Whittaker, E. T. and Watson, G. N., *A Course of Modern Analysis*. Cambridge University Press, 4th ed. 1940.

Exponential integrals, gamma function and related functions

1. Artin, E., *The Gamma Function*. Holt, Rinehart, and Winston, New York, 1964.
2. Busbridge, I. W., *The Mathematics of Radiative Transfer*. Cambridge University Press, 1960.
3. Erdélyi, A. et al., *Higher Transcendental Functions*, vol. II. McGraw Hill, New York, 1953.
4. Erdélyi, A. et al., *Tables of Integral Transforms*, vols. I and II. McGraw Hill, New York, 1954.
5. Hastings, Jr., C., *Approximations for Digital Computers*. Princeton University Press. New Jersey, 1955.
6. Kourganoff, V., *Basic Methods in Transfer Problems*. Oxford University Press, 1952.
7. Losch, F. and Schoblik, F., *Die Fakultät (Gammafunktion) und Verwandte Funktionen*, Teubner, Leipzig, 1951.
8. Neilsen, N., *Handbuch der Theorie der Gammafunktion*, Teubner, Leipzig, 1906.
9. Oberhettinger, F., *Tabellen zur Fourier Transformation*, Springer Verlag, Berlin, 1957.

Error function and Fresnel integrals

1. Erdélyi, A. et al., *Higher Transcendental Functions*, vol. II. McGraw Hill, New York, 1953.
2. Erdélyi, A. et al., *Tables of Integral Transforms*, vol. I. McGraw Hill, New York, 1954.
3. Slater, L. J., *Confluent Hypergeometric Functions*. Cambridge University Press, 1960.
4. Tricomi, F. G., *Funzioni ipergeometriche confluenti*. Edizioni Cremonese, Italy, 1954.
5. Watson, G. N., *A Treatise on the Theory of Bessel Functions*. Cambridge University Press, 2nd ed., 1958.

Legendre and related functions

1. Erdélyi, A. et al., *Higher Transcendental Functions*, vol. I. McGraw Hill, New York, 1953.
2. Helfenstein, H., *Ueber eine Spezielle Lamésche Differentialgleichung*. Brunner and Bodmer, Zurich, 1950 (Bibliography).
3. Hobson, E. W., *The Theory of Spherical and Ellipsoidal Harmonics*. Cambridge University Press, 1931. Reprinted by Chelsea, New York, 1955.
4. Lense, J., *Kugelfunktionen*. Geest and Portig, Leipzig, 1950.
5. MacRobert, T. M., *Spherical Harmonics: An Elementary Treatise on Harmonic Functions with Applications*, Methuen, England, 1927. (Revised ed. 1947; reprinted Dover, New York, 1948).
6. Snow, C., *The Hypergeometric and Legendre Functions with Applications to Integral Equations of Potential Theory*. National Bureau of Standards, Washington, D.C., 2nd ed. 1952.
7. Stratton, J. A., Morse, P. M., Chu, L. J. and Hunter, R. A., *Elliptic Cylinder and Spheroidal Wave Functions Including Tables of Separation Constants and Coefficients*. Wiley, New York, 1941.

Bessel functions

1. Bickley, W. G., *Bessel Functions and Formulae*. Cambridge University Press, 1953.
2. Erdélyi, A. et al., *Higher Transcendental Functions*, vols. I and II. McGraw Hill, New York, 1954.
3. Erdélyi, A. et al., *Tables of Integral Transforms*, vols. I and II. McGraw Hill, New York, 1954.
4. Gray, A., Mathews, G. B. and MacRobert, T. M., *A Treatise on Bessel Functions and Their Applications to Physics*. Macmillan, 2nd ed. 1922.
5. McLachlan, N. W., *Bessel Functions for Engineers*. Oxford University Press, 2nd ed. 1955.
6. Luke, Y. L., *Integrals of Bessel Functions*, McGraw-Hill, New York, 1962.
7. Petiau, G., *La théorie des fonctions de Bessel*. Centre National de la Recherche Scientifique, Paris, 1955.

8. Relton, F. E., *Applied Bessel Functions*. Blackie, London, 1946.
9. Watson, G. N., *A Treatise on the Theory of Bessel Functions*. Cambridge University Press, 2nd ed. 1958.
10. Wheelon, A. D., *Tables of Summable Series and Integrals Involving Bessel Functions*, Holden-Day, San Francisco, 1968.

Struve functions

1. Erdélyi, A. et al., *Higher Transcendental Functions*, vol. II. McGraw Hill, New York, 1954.
2. Gray, A., Mathews, G. B. and MacRobert, T. M., *A Treatise on Bessel Functions and Their Applications to Physics*. Macmillan, London, 2nd ed., 1922.
3. Watson, G. N., *A Treatise on the Theory of Bessel Functions*. Cambridge University Press, 2nd ed. 1958.

Hypergeometric and confluent hypergeometric functions

1. Appell, P., *Sur les Fonctions Hypergéometriques de Plusieures Variables*, Gauthier-Villars, Paris, 1926.
2. Bailey, W. N., *Generalized Hypergeometric Functions*. Cambridge University Press, 1935.
3. Buchholz, H., *Die konfluente hypergeometrische Funktion*. Springer Verlag, Berlin, 1953.
4. Erdélyi, A. et al., *Higher Transcendental Functions*, vol. I. McGraw Hill, New York, 1953.
5. Jeffreys, H. and Jeffreys, B. S., *Methods of Mathematical Physics*. Cambridge University Press, 1956.
6. Klein, F., *Vorlesungenüber die hypergeometrische Funktion*. Springer Verlag, Berlin, 1933.
7. Nörlund, N. E., *Sur les Fonctions Hypergéometriques d'Ordre Superior*, Copenhagen, 1956.
8. Slater, L. J., *Confluent Hypergeometric Functions*. Cambridge University Press, 1960.
9. Slater, L. J. *Generalized Hypergeometric Functions*, Cambridge University Press, London, 1966.
10. Snow, C., *The Hypergeometric and Legendre Functions with Applications to Integral Equations of Potential Theory*. National Bureau of Standards, Washington, D.C., 2nd ed. 1952.
11. Swanson, C. A. and Erdélyi, A., *Asymptotic Forms of Confluent Hypergeometric Functions*. Memoir 25, American Mathematical Society, 1957.
12. Tricomi, F. G., *Lezioni sulla funzioni ipergeometriche confluenti*. Gheroni, Torino, 1952.

Jacobian and Weierstrass elliptic functions and related functions

1. Erdélyi, A. et al., *Higher Transcendental Functions*, vol. II. McGraw Hill, New York, 1953.
2. Byrd, P. F. and Friedman, M. D., *Handbook of Elliptic Integrals for Engineers and Physicists*, Springer Verlag, Berlin, 1954.
3. Graeser, E., *Einführung in die Theorie der Elliptischen Funktionen und deren Anwendungen*. Oldenbourg, Munich, 1950.
4. Hancock, H., *Lectures on the Theory of Elliptic Functions*, vol. I. Dover, New York, 1958.
5. Neville, E. H., *Jacobian Elliptic Functions*. Oxford University Press, 1944 (2nd ed. 1951).
6. Oberhettinger, F. and Magnus, W., *Anwendungen der Elliptischen Funktionen in Physik und Technik*. Springer Verlag, Berlin, 1949.
7. Roberts, W. R. W., *Elliptic and Hyperelliptic Integrals and Allied Theory*. Cambridge University Press, 1938.
8. Tannery, J. and Molk, J., *Eléments de la Théorie des Fonctions Elliptiques*, 4 volumes. Gauthier-Villars, Paris, 1893–1902.
9. Tricomi, F. G., *Elliptische Funktionen*. Akad. Verlag, Leipzig, 1948.

Parabolic cylinder functions

1. Buchholz, H., *Die konfluente hypergeometrische Funktion.* Springer Verlag, Berlin, 1953.
2. Erdélyi, A. et al., *Higher Transcendental Functions*, vol. II. McGraw Hill, New York, 1954.

Orthogonal polynomials and functions

1. *Bibliography on Orthogonal Polynomials.* Bulletin of National Research Council No. 103, Washington, D.C., 1940.
2. Courant, R. and Hilbert, D., *Methods of Mathematical Physics*, vol. I. Interscience, New York, 1953.
3. Erdélyi, A. et al., *Higher Transcendental Functions*, vol. II. McGraw Hill, New York, 1954.
4. Kaczmarz, St. and Steinhaus, H., *Theorie der Orthogonalreihen.* Chelsea, New York, 1951.
5. Lorentz, G. G., *Bernstein Polynomials.* University of Toronto Press, Toronto, 1953.
6. Sansone, G., *Orthogonal Functions.* Interscience, New York, 1959.
7. Shohat, J. A. and Tamarkin, J. D., *The Problem of Moments.* American Mathematical Society, 1943.
8. Szegö, G., *Orthogonal Polynomials*, American Mathematical Society Colloquim Pub. No. 23, 1959.
9. Titchmarsh, E. C., *Eigenfunction Expansions Associated with Second Order Differential Equations.* Oxford University Press, part I (1946), part II (1958).
10. Tricomi, F. G., *Vorlesungenüber Orthogonalreihen.* Springer Verlag, Berlin, 1955.

Riemann zeta function

1. Titchmarsh, E. C., *The Zeta Function of Riemann.* Cambridge University Press, 1930.
2. Titchmarsh, E. C., *The Theory of the Riemann Zeta Function.* Oxford University Press, 1951.

Probability function

1. Cramer, H., *Mathematical Methods of Statistics.* Princeton University Press, 1951.
2. Erdélyi, A. et al., *Higher Transcendental Functions*, vols. I, II, and III. McGraw Hill, New York, 1953 to 1955.
3. Kendall, M. G. and Stuart, A., *The Advanced Theory of Statistics*, vol. I: *Distribution Theory*, Griffin, London, 1958.

Mathieu functions

1. Erdélyi, A., *Higher Transcendental Functions*, vol. III. McGraw Hill, New York, 1955.
2. McLachlan, N. W., *Theory and Application of Mathieu Functions.* Oxford University Press, 1947.
3. Meixner, J. and Schäfke, F. W., *Mathieusche Funktionen und Spheroid- funktionen mit Anwendungen auf Physikalische und Technische Probleme.* Springer Verlag, Heidelberg, 1954.
4. Strutt, M. J. O., *Lamésche, Mathieusche und verwandte Funktionen in Physik und Technik.* Ergeb. Math. Grenzgeb. *1*, 199-323 (1932). Reprint Edwards Bros., Ann Arbor, Michigan, 1944.

Integral transforms

1. Bochner, S., *Vorlesungenüber Fouriersche Integrale.* Akad. Verlag, Leipzig, 1932. Reprint Chelsea, New York, 1948.

2. Bochner, S. and Chandrasekharan, K., *Fourier Transforms.* Princeton University Press, 1949.

3. Campbell, G. and Foster, R., *Fourier Integrals for Practical Applications*, Van Nostrand, New York, 1948.

4. Carslaw, H. S. and Jaeger, J. C., *Conduction of Heat in Solids.* Oxford University Press, 1948.

5. Doetsch, G., *Theorie und Anwendung der Laplace-Transformation.* Springer Verlag, Berlin, 1937. (Reprinted by Dover, New York, 1943)

6. Doetsch, G., *Theory and Application of the Laplace-Transform.* Chelsea, New York, 1965.

7. Doetsch, G., *Handbuch der Physik, Mathematische Methoden II*, 1st ed., Berlin, 1955.

8. Doetsch, G., *Guide to the Applications of the Laplace and Z-Transforms*, 2nd Ed., Van Nostrand-Reinhold, London, 1971.

9. Doetsch, G., Handbuch der Laplace-Transformation, Vols. I–IV, Birkhäuser Verlag, Basel, 1950–56.

10. Doetsch, G., Kniess, H., and Voelker, D., Tabellen zur Laplace- Transformation, Springer-Verlag, Berlin, 1947.

11. Erdélyi, A., *Operational Calculus and Generalized Functions*, Holt, Rinehart and Winston, New York, 1962.

12. Exton, H., *Multiple Hypergeometric Functions and Applications,* Horwood, Chichester, 1976.

13. Exton, H., *Handbook of Hypergeometric Integrals: Theory, Applications, Tables, Computer Programs,* Horwood, Chichester, 1978.

14. Hirschmann, J. J. and Widder, D. V., *The Convolution Transformation.* Princeton University Press, New Jersey, 1955.

15. Marichev, O. I., *Handbook of Integral Transforms of Higher Transcendental Functions, Theory and Algorithmic Tables*, Ellis Horwood Ltd., Chichester (1982).

16. Oberhettinger, F., *Tabellen zur Fourier Transformation,* Springer-Verlag, Berlin (1957).

17. Oberhettinger, F., *Tables of Bessel Transforms*, Springer-Verlag, New York (1972).

18. Oberhettinger, F., *Fourier Expansions: A Collection of Formulas,* Academic Press, New York, 1973.

19. Oberhettinger, F., *Fourier Transforms of Distributions and Their Inverses*, Academic Press, New York, 1973.

20. Oberhettinger, F., *Tables of Mellin Transforms*, Springer-Verlag, Berlin, 1974.

21. Oberhettinger, F. and Badii, L., *Tables of Laplace Transforms,* Springer-Verlag, Berlin, 1973.

22. Oberhettinger, F. and Higgins, T. P., *Tables of Lebedev, Mehler and Generalized Mehler Transforms*, Math. Note No. 246, Boeing Scientific Research Laboratories, Seattle, Wash., 1961.

23. Roberts, G. E. and Kaufman, H., *Table of Laplace Transforms,* McAinsh, Toronto, 1966.

24. Sneddon, I. N., *Fourier Transforms.* McGraw Hill, New York, 1951.

25. Titchmarsh, E. C., *Introduction to the Theory of Fourier Integrals.* Oxford University Press, 1937.

26. Van der Pol, B. and Bremmer, H., *Operational Calculus Based on the Two Sided Laplace Transformation.* Cambridge University Press, 1950.

27. Widder, D. V., *The Laplace Transform.* Princeton University Press, New Jersey, 1941.

28. Wiener, N., *The Fourier Integral and Certain of its Applications.* Dover, New York, 1951.

Asymptotic expansions

1. De Bruijn, N. G., *Asymptotic Methods in Analysis.* North-Holland Publishing Co., Amsterdam, 1958.

2. Cesari, L., *Asymptotic Behavior and Stability Problems in Ordinary Differential Equations.* 3rd ed. New York: Springer, 1971.

3. Copson, E. T., *Asymptotic Expansions.* Cambridge University Press, 1965.

4.   Erdélyi, A., *Asymptotic Expansions*. Dover, New York, 1956.

5.   Ford, W. B., *Studies on Divergent Series and Summability*. Macmillan, New York, 1916.

6.   Hardy, G. H., *Divergent Series*. Clarendon Press, Oxford, 1949.

7.   Watson, G. N., *A Treatise on the Theory of Bessel Functions*. Cambridge University Press, 2nd ed., 1958.

Complex analysis

1.   Ahlfors, L. V., *Complex Analysis*. 3rd ed. New York: McGraw-Hill, 1979.

2.   Ahlfors, L. V. and Sario, L., *Riemann Surfaces*. Princeton, NJ: Princeton University Press, 1971.

3.   Bieberbach, L., *Conformal Mapping*. New York: Chelsea, 1964.

4.   Henrici, P., *Applied and Computational Complex Analysis*. 3 vols. New York: Wiley, 1988, 1991, 1977.

5.   Hille, E., *Analytic Function Theory*. 2 vols. 2nd ed. New York: Chelsea, 1990, 1987.

6.   Kober, H., *Dictionary of Conformal Representations*, Dover, 1952.

7.   Titchmarsh, E. C., *The Theory of Functions*. 2nd ed. London: Oxford University Press, 1939. (Reprinted 1975).

Index of Functions and Constants

This index shows the occurance of functions and constants used in the expressions within the text. The numbers refer to pages on which the function or constant appears.

Symbols and Greek Letters

C

Index of Concepts

This index refers to concepts appearing in the text.

N

O

P

Q

R